Symbols, Weights, and Measures

Units of Length

m	meter
km	kilometer (10^3 m)
cm	centimeter (10^{-2} m)
mm	millimeter (10^{-3} m)
μm	micrometer (10^{-6} m)
nm	nanometer (10^{-9} m)

Units of Mass and Weight

amu	atomic mass unit
MW	molecular weight
mole	MW in grams
g	gram
kg	kilograms (10^3 g)
mg	milligrams (10^{-3} g)

Units of Pressure

atm	atmospheres (1 atm = 760 mmHg)
mmHg	millimeters of mercury
P_x	partial pressure of gas x (as in P_{O_2})

Conversion Factors

1 in. = 2.54 cm		1 cm = 0.394 in.
1 fl oz = 29.6 mL		1 mL = 0.034 fl oz
1 qt = 0.946 L		1 L = 1.057 qt
1 g = 0.0035 oz		1 oz = 28.38 g
1 lb = 0.45 kg		1 kg = 2.2 lb
$°C = (5/9)(°F - 32)$		$°F = (9/5)(°C) + 32$

Units of Volume

L	liter
dL	deciliter (= 100 mL) (10^{-1} L)
mL	milliliter (10^{-3} L)
μL	microliter (= 1 mm^3) (10^{-6} L)

Units of Concentration

mEq/L	milliequivalents per liter
Osm/L	osmoles per liter
mOsm/L	milliosmoles per liter
M	molar
mM	millimolar
pH	negative log of H^+ molarity

Units of Heat

cal	"small" calories
kcal	kilocalories (Calories; 1 kcal = 1,000 cal)
Cal	"large" (dietary) calories (1 Cal = 1,000 cal)

Greek Letters

α	alpha
β	beta
γ	gamma
Δ	delta (uppercase)
δ	delta (lowercase)
η	eta
θ	theta
μ	mu

Anatomy & Physiology

THE UNITY OF FORM AND FUNCTION

Kenneth S. Saladin

Georgia College and State University

with

Carol Mattson Porth, R.N., M.S.N., Ph.D.
Clinical Consultant

Boston, Massachusetts   Burr Ridge, Illinois   Dubuque, Iowa   Madison, Wisconsin
New York, New York   San Francisco, California   St. Louis, Missouri

WCB/McGraw-Hill

A Division of The McGraw·Hill Companies

ANATOMY & PHYSIOLOGY: THE UNITY OF FORM AND FUNCTION

This book is printed on acid-free paper.

1 2 3 4 5 6 7 8 9 0 VNH/VNH 9 0 9 8 7

ISBN 0–697–23087–2

Vice president, editorial director: Kevin T. Kane
Publisher: Michael D. Lange
Sponsoring editor: Kristine Noel Tibbetts
Developmental editor: Kelly A. Drapeau
Marketing manager: Keri L. Witman
Project manager: Sue Dillon
Production supervisor: Deb Donner
Designer: K. Wayne Harms
Interior and cover design: Jeff Storm
Photo research coordinator: John C. Leland
Art editor: Kathleen Timp
Compositor: Shepherd Inc.
Typeface: 10/12 Melior
Printer: Von Hoffman Press
Cover image: The Observatory Group

The credits section for this book begins on page 1088 and is considered an extension of the copyright page.

Library of Congress Cataloging-in Publication Data

Saladin, Kenneth S.
    Anatomy and physiology: the unity of form and function / Kenneth
S. Saladin. — 1st ed.
      p.   cm.
    Includes index.
    ISBN 0–697–23087–2
    1. Human physiology.   2. Human anatomy.
    [DNLM: 1. Anatomy.   2. Physiology.   QS4 S159a   1998]
QP34.5.S23   1998
612—dc21
DNLM/DLC
for Library of Congress   97–34190
                                                              CIP

When ordering this title, use ISBN 0–07–115547–3

http://www.mhhe.com

the unity of form and function

about the author

Kenneth S. Saladin is professor of biology at Georgia College and State University. He obtained his B.S. degree in zoology from Michigan State University and his Ph.D. in biological sciences from Florida State University. He has taught human anatomy and physiology, histology, and other courses for over twenty years, and is a three-time recipient of the Phi Kappa Phi award for outstanding mentor to his undergraduate students. Ken is a member of the Human Anatomy and Physiology Society and serves on the Competency Examination Committee. He served as a developmental reviewer and wrote test items for several other WCB/McGraw-Hill anatomy and physiology textbooks for a number of years before beginning his own book. Ken is married to C. Diane Saladin, a registered nurse in obstetrics and gynecology; they have two children in high school.

This book is dedicated to two outstanding teachers who had a transforming effect on my career in science:

Donald R. Sly,
the first to encourage my early interest in zoology and
set me on the road that led to this book;

and

Professor Robert B. Short,
whose example and instruction in the art of teaching,
writing, and scientific illustration I have tried,
imperfectly, to reflect in these pages.

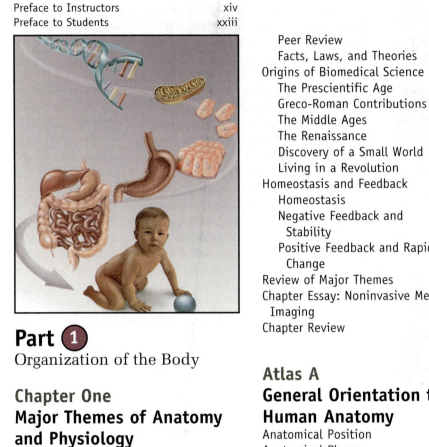

Part 1
Organization of the Body

Chapter One
Major Themes of Anatomy and Physiology

Atlas A
General Orientation to Human Anatomy

Chapter Two
Matter and Energy

Chapter Three
The Molecules of Life

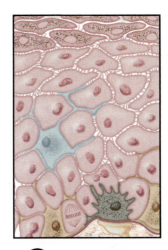

Part ② Support and Movement

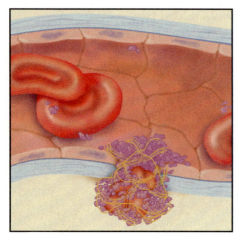

Part ④
Regulation and Maintenance

Chapter Eighteen
The Circulatory System: Blood

Chapter Nineteen
The Circulatory System: The Heart

Chapter Twenty
The Circulatory System: Blood Vessels and Circulation

Part 5
Reproduction and
Development

A New Approach to the Study of Anatomy and Physiology

It may seem surprising that anyone would launch another textbook of human anatomy and physiology when there are already so many to choose from. I set out to express a unique vision of how to present the subject. During the development of the manuscript, the comments of more than 300 reviewers assured the editorial team and me that we were, indeed, on to something that would be received as a novel and welcome entry into this field. We pulled out all the stops to make this a textbook without equal. I explain here how it differs from others, and in the Preface to Students that follows, I discuss pedagogic features of special relevance to them. I encourage instructors to read that as well, since it deals with issues important to the effective use of the book and with some of the most fundamental differences between this A&P textbook and others.

Intended Audience

I designed *Anatomy and Physiology: The Unity of Form and Function* for a two-semester college course, primarily for students who hope to enter the health professions. I assume little or no prior college chemistry or cell biology. Chapters 2 to 5 are a comprehensive introduction to these subjects. I also assume that many students, like mine, are still forming the study and thinking habits essential for success in the technical disciplines. Thus, the book aims not only to convey information, but also to promote the development of these skills through a consistent framework of pedagogic devices in each chapter. A "Brushing Up" list at the start of each chapter reminds students that each chapter builds on knowledge gained from those before it and gives instructors more flexibility to cover the chapters in a different order than presented.

Novel and Unifying Themes

Certain themes are indispensable to the teaching of A&P and are well represented in these pages. Students are repeatedly reminded of the central importance of homeostasis, the complementarity of form and function, and the cellular basis of all anatomy and physiology. Clinical applications and dysfunctions are used generously, not only to lend deeper insight into the normal function of the body, but also to demonstrate why this subject is relevant to the student's career aspirations.

But in addition to these time-honored themes, the book introduces some that are more unusual in the A&P marketplace of ideas—scientific method, biomedical history, and human evolution. These are among the elements that reviewers found most noteworthy and welcome. They enrich and humanize the subject and promote analytical thinking. The textbook controversies that have raged in American newspapers, school board meetings, and federal courts have made it glaringly apparent how few people understand what science is, how to weigh competing claims of truth, and how to distinguish science from pseudoscience. Many biology textbooks have redressed these misunderstandings by paying increasing attention to scientific thought and method and to the personalities and history of science. Anatomy and physiology textbooks, however, have lagged behind. A&P is a field jam-packed with information. Students and instructors want clinical relevance, and it may be hard to see how there could be room to include any history, biography, or the philosophy of science. But without compromising information on the basic biology of the human body, and without glossing over clinical applications of this information, I have woven these other threads into the story and produced a textbook no longer than others written for the same market niche.

Scientific Method   Most A&P textbooks do little more than present facts, creating the impression that science is a massive body of data rather than a method of discovery. If students have no appreciation of how such information was obtained, then they have no particular reason to believe that it's more trustworthy than any other claims to truth. If any science is vulnerable to boisterous fads and costly frauds, it is surely the health sciences. In this field, we have a special duty to cultivate scientific skepticism and sound judgment. I introduce some basic ideas of scientific method in chapter 1 and reinforce them with asides and thought questions throughout the book.

A Historical and Humanistic Perspective A&P has a colorful, engrossing history. I often make time in my lectures for little stories of biomedical history, and these are among the things students remember most favorably in their course evaluations, spontaneous feedback, and return visits. Thus, I have instilled that approach into this book. Chapter 1 describes the growth of rationalism in medical thought from the ancient Greeks to the dawn of modern science. Later chapters carry on the story with vignettes of the discoverers, their tribulations, and their sacrifices in engrossing, sometimes poignant and tragic stories—for example, Rosalind Franklin's under-celebrated role in the discovery of DNA structure; William Beaumont's studies of gastric function on the reluctant "man with the hole in his stomach"; the bitter feud that followed Crawford Long's discovery of ether anesthesia; and Charles Drew's pioneering work in blood banking before he, himself, bled to death. We can hope that such stories inspire our students to enter their professions with more historical and humane vision and make their reading more satisfying.

Human Evolution The human body can never be fully appreciated without a sense of its evolutionary past. Much of what we know about it has been won through comparative anatomy and physiology—twin sciences grounded in evolutionary theory. Evolution provides a deep and inspiring vision of human form and function, as recognized by several books of "evolutionary (Darwinian) medicine" that appeared in the 1990s for general readers and by a growing number of research and review papers in the medical journals. Yet the recent textbook controversies have left legions of teachers wary of covering this subject, emboldened some to denounce it, and flooded our colleges with students who have never been exposed to the most revolutionary idea and unifying principle in all of biology. This is perhaps the first A&P textbook that gives more than a brief nod to the subject. Chapter 1 introduces the logic of natural selection and a few high points of human evolution. Later chapters reinforce the point with evolutionary insights ranging from muscle anatomy to menopause. I use comparative A&P to explain such features as skeletal adaptations to bipedalism, the four-chambered heart, and the nephron loop—not merely to state that we differ from other animals, but to show *why* our differences are adaptive.

Developmental Biology and Aging Some instructors will want to know how much embryology the book covers. I describe the embryonic development of the bones, central nervous system, pituitary, heart, and reproductive systems, but not of the integumentary, muscular, lymphatic, respiratory, urinary, or digestive systems. Some omissions are necessary in any book, and I had two reasons for these particular ones: a tepid response from re-

viewers on their necessity and my feeling that the space would be better dedicated to the *aging* of the organ systems. Demographers and journalists frequently remind us of the health-care implications of a population that is markedly increasing in age. Our students today will be caring for this aging population tomorrow. In this context, I find it more important that they understand, for example, how the skin ages and how this affects other organ systems than how it develops in the embryo. In addition, the quality of life in old age depends on how we treat our bodies in our youth. Most of our students are at an age when they can best benefit from this foresight. Chapter 29 therefore has a system-by-system overview of age-related changes in the body.

Clinical Concepts and Consultants

The best understanding of human form and function often comes from an appreciation of what happens when things go wrong. It could be rather dull for a student to study ion pumps, for example, just because scientists have decreed that they are important. But if the student can empathize with a child with cystic fibrosis, then understand how the symptoms of CF result from a defect in chloride pumps, cell membranes and ion pumps take on striking relevance to a future nurse or respiratory therapist. Thus, I have repeatedly used clinical examples and thought questions to lead the student to deeper insights into normal human form and function.

In developing these applications, I was aided not only by my local colleagues and the instructors who reviewed the manuscript, but also by Dr. Carol Mattson Porth, author of a leading textbook, *Pathophysiology: Concepts of Altered Health States* (J. B. Lippincott Company), who served as clinical consultant to this project. I was also significantly aided in this by my in-house clinical consultant, a registered nurse to whom I have been married for 18 years and who played an active role in the development of this book. I am grateful to both for their valuable input.

Chapter Content and Order

The order and content of the chapters is clear enough from the Table of Contents, but let me call your attention to a few places where these differ from other A&P textbooks.

Major Themes of A&P Chapter 1, Major Themes of Anatomy and Physiology, has two purposes: to set a historical and philosophical stage for the study of A&P and to identify five major themes that run through the rest of the book—the unity of form and function, the hierarchy of human structure, cell theory, evolution, and homeostasis. This chapter addresses the meaning of human life through two questions: What is life? and What is a human? It places humans in taxonomic context and describes the characteristics that define our species. It introduces the logic of Darwinian thought and describes how the habitats of our ancestors shaped several aspects of human form and function we now take for granted. The section on scientific method discusses inductive and hypothetico-deductive reasoning, experimental design, and peer review, and clarifies the widely misunderstood concepts of fact, law, and theory in science. This is followed by a short history of medical thought, extending roughly from Hippocrates to Harvey, with emphasis on people who challenged and overthrew entrenched superstitions and dogmas. Some reviewers felt that this section ended prematurely and wished it had continued into the twentieth century, but those stories are told in the sidebars and vignettes of later chapters. The twentieth-century ideas of homeostasis and feedback, however, have such central importance that they are presented in chapter 1.

We have grown accustomed to finding directional terminology, body planes, membranes, cavities, and so forth treated in chapter 1 of most A&P textbooks. I found it more effective, however, to consolidate this information in a 15-page atlas following that chapter. The beautiful illustrations subsequently prepared for the atlas further enhance its value as a stand-alone reference. Since the terminology introduced here is fundamental to the rest of the book, I conclude the atlas with a set of self-testing questions similar to those in the chapters, and I include it in the test items in the *Instructor's Manual* and *Student Study Guide.*

Chemistry

Chemistry   Well-prepared students can probably skip chapter 2, Matter and Energy, and focus on chapter 3, The Molecules of Life. Chapter 2 deals with general and inorganic chemistry and elementary principles of thermodynamics. Chapter 3 is concerned with biochemistry, including enzymology but not the nucleic acids (see chapter 5). It gives a clear but concise introduction to the synthesis and uses of ATP, sufficient for understanding such later topics as muscle physiology but without going prematurely into such details as glycolysis and the citric acid cycle. Those are treated in chapter 26.

Cytology and Histology   Chapter 4 focuses on cell structure and membrane transport processes, while chapter 5 deals with the nucleic acids, protein synthesis and secretion, mitosis and the cell cycle, and fundamentals of heredity. Most textbooks bury heredity in the last chapter, which I feel is a mistake. It would be difficult to explain blood types, sickle-cell anemia, or color blindness, for example, if a student had no concept of dominant and recessive alleles, codominance, sex linkage, and pleiotropy. I reserve anomalies such as nondis-

junction and trisomy to the final chapter, but I feel it is important to introduce the basic ideas of normal heredity early. Chapter 6, Histology, concludes part 1. The student is then prepared to embark on a study of the organ systems.

Systems of Support and Movement   The next six chapters and atlas B concern the integumentary, skeletal, and muscular systems. Beginning in chapter 7, each organ system has a feature we call Connective Issues—a chart that shows how that system affects others and is affected by them. The order of the muscular system chapters is opposite that found in most books. I treat gross anatomy of the muscular system (chapter 11) before its physiology (chapter 12). There are two reasons for this: (1) It provides a continuous story line from the morphology of the bones, through the joints and their actions, to the muscles and tendons that produce those actions and whose origins and insertions are explained with reference to bone morphology. (2) It links muscle as an excitable tissue to the ensuing chapter on neurons. Resting membrane potentials, action potentials, and synaptic function are introduced for the first time in the muscle physiology chapter (12) and segue into the more detailed investigation of these processes in the neuron chapter (13). Atlas B, Surface Anatomy, is a set of photographs of living subjects showing muscular and skeletal anatomy. It ends with a photographic quiz that can engage your students in relating surface structure to the skeletomuscular anatomy described in chapters 9 to 11.

Systems of Integration and Control   The next five chapters deal with the nervous and endocrine systems. Today's major two-semester textbooks have at least five chapters on the nervous system, and some have six or seven. Most of us are struggling to finish the first semester at this point, and many instructors and students are finding this much material on the nervous system unmanageable at term's end—all the more so for students because this is one of the hardest systems for them to understand. Among instructors, there seem to be increasing calls for a more concise treatment of the nervous system. I have therefore limited it to four chapters: chapter 13, Nervous Tissue; chapter 14, The Central Nervous System; chapter 15, The Peripheral Nervous System and Reflexes; and chapter 16, Sense Organs. Under reflexes in chapter 15, I include both somatic reflexes and visceral autonomic reflexes. Under sense organs in chapter 16, I include not only the special senses but also the general (somesthetic) senses, which some authors treat in separate chapters.

In the endocrine system (chapter 17), I depart from convention by treating the glands and their hormones before the cellular and molecular mechanisms of hormone synthesis, transport, and action. This enables students to start with the perspective they usually find easiest to grasp and then move on to the finer cellular and molecular details, which tend to give them more trouble. It also provides a "cast of characters"—an inventory of glands and hormones—that we can call upon to explain how hormones work. This has worked better in my teaching than the molecules-to-organs approach, which discourages some chemophobic students from hope of understanding the subject. This chapter also covers topics a little beyond the scope of the endocrine system—the eicosanoids and stress.

Systems of Regulation and Maintenance

The next nine chapters concern the circulatory, immune, respiratory, urinary, and digestive systems. Chapter 18, on the blood, introduces antigens and antibodies in the context of blood typing, transfusions, and Rh compatibility, but the details of immunology and leukocyte functions are treated in chapter 21. Chapters 19 and 20 discuss the heart and the blood vessels and circulation, respectively. Chapter 21 embraces the lymphatic system, nonspecific resistance, and specific immunity.

Except for the detour into immunology, we return quickly to systems that regulate the blood composition, pH, and blood pressure, namely the respiratory and urinary systems. Most textbooks treat the urinary system in association with the reproductive system. If this were only an anatomy book, I would do so too, since they have several points of embryological and anatomical relationship. But from a physiological standpoint, the urinary system has a much closer working relationship with the circulatory and respiratory systems. The kidneys regulate the hematocrit, they have more influence on blood pressure than any other organ, they collaborate with the respiratory system in controlling the acid-base balance of the blood, and they regulate the concentrations of other blood solutes. In turn, the kidneys are regulated by hormones that are secreted in response to blood pressure variations, and the capillary fluid exchange mechanisms studied in chapter 20 are needed for students to understand glomerular filtration and tubular reabsorption. In light of this intimate, bidirectional relationship between the organ systems, it seems a mistake to separate the circulatory and urinary systems any more than necessary. Chapter 24, Water, Electrolyte, and Acid-Base Balance, ties together many of the concepts from the preceding circulatory, respiratory, and urinary chapters.

This section of the book ends with chapters 25 and 26 on digestion, nutrition, and metabolism. Chapter 26 also covers the issues of body heat and thermoregulation. In chapter 26, I kept the tables of nutrient requirements and functions relatively concise. Some books go on for page after page listing nutrient functions that require a knowledge of enzymes and metabolic pathways beyond the scope of an introductory textbook. Health science students who need that knowledge usually receive it in a later nutrition course. In discussing intermediary metabolism, I have again tried to convey the clearest possible "big picture" unbelabored by the structural formulae of metabolic intermediates. This chapter discusses mechanisms of appetite and satiety more than most textbooks do, and covers blood lipoproteins in a way that reviewers found especially clear and original.

Reproduction and Development

Chapter 27 covers not only the male reproductive system but also some general issues of reproductive biology: What is sex? What defines male and female? What prenatal factors govern sexual differentiation? Meiosis is also introduced here. Chapter 28 runs the normal gamut of female reproductive biology through prenatal differentiation, puberty and adolescence, adult reproductive cycles, pregnancy, and lactation. Chapter 29 covers prenatal development, neonatal adaptations, and congenital anomalies, then jumps to the other end of the life span and discusses the senescence of each organ system in some depth. The last topic can serve as an excellent capstone to a two-semester course, not only sensitizing students to some issues of gerontology, but also refreshing their memory of concepts treated in earlier chapters. It integrates the organ systems by showing how each is affected by senescence of the others, thus reinforcing the concept of system interactions conveyed through the Connective Issues pages. Chapter 29 concludes with molecular to evolutionary theories of senescence; issues of longevity and death; and in the final chapter essay, new beginnings through reproductive technology.

End Matter

At the back of the book are a periodic table of the elements, color-coded with reference to human physiology and with a short discussion of the history and logic of the table; answers to the objective chapter review questions; a brief discussion, Understanding Biomedical Vocabulary, which gives insight into understanding word derivations; and a 1,000-word glossary. A well-crafted glossary is an important working tool for the student, and I did not want this to be a last-minute rush job done as the book was being readied for press. I began early to gather terms from the manuscript that I felt the student would refer back to most often and to write clear, complete, unambiguous definitions.

Answers to the more analytical Testing Your Comprehension questions at the end of each chapter exceeded our space limits for this book, but are printed in the *Student Study Guide* and *Instructor's Manual*. Answers to the Think About It questions dispersed through each chapter are also in the *Instructor's Manual* but not in the *Study Guide.* You can make these available to your students or withhold them and assign those questions to your class. Answers are not provided for the Key Point Review questions that come at intervals through each chapter, mostly for lack of space but also because these usually are simple recall questions that can be answered from the material within the preceding five or six pages.

Inside the back cover are 400 of the word roots and affixes most often footnoted throughout the text, and inside the front cover are reference tables of symbols, weights, measures, and biomedical abbreviations.

Key Features

Art Program

Today's A&P students are the heirs to a rich tradition of medical illustration extending from the historic art of Vesalius through *Gray's Anatomy* and its twentieth-century successors. That tradition just got richer, as the medical illustrators of The Observatory Group, Inc. (OGI) of Cincinnati, Ohio, combined their anatomical training and state-of-the-art technology to produce the illustrations in this textbook and its ancillaries. Many of these began as my own black-ink and colored-pencil drawings of new concepts that I felt would be helpful or necessary. But it was the artists of OGI who brought these concepts to life with a wonderful palette, translucency, and three-dimensionality that exceeded my expectations.

Whatever success I may enjoy with this writing endeavor, I will owe a great deal to accredited medical illustrators Lisa Petkun-Klancher, Quade Paul, and

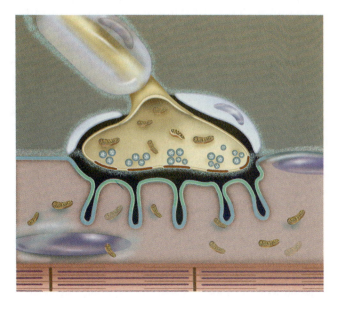

Emiko Koike; graphic illustrators Jenny Robinson, Troy Hitch, and Susan Young; and contributing medical illustrators Stephanie Orr and David Baker. We worked together very closely to ensure the best combination of scientific accuracy and artistic skill. When you look at such figures as the neuromuscular junction shown on this page, the muscle spindle on page 526, or the middle ear on page 565 and compare them to the same subjects illustrated in other textbooks, perhaps you will agree that these illustrators have created an entirely new echelon of artistic quality in A&P textbooks. These figures are available to adopters in formats that can be integrated into your lectures, ranging from color transparencies to a CD-ROM for Powerpoint presentations (see Supplemental Materials).

Supplemental Materials

In addition to this book itself, the production team has commissioned and assembled a rich package of supplemental materials, some of them obtainable by students, and others provided to qualified instructors who adopt the book.

1. *Anatomy and Physiology Laboratory Manual* by Eric Wise (0-697-20554-1) is a new manual written to support this textbook, but is suitable for independent use. This manual uses the same four-color art program as this book and follows the same order of presentation. Dissection exercises are based on the cat but are not indispensable to the manual. It is accompanied by a separate *Instructor's Manual* (0-697-20555-X), which contains solutions and keys for grading laboratory reports.
2. *Instructor's Manual and Test Item File* by David Evans (0-697-23097-X) includes suggested lecture outlines, chapter overviews, key concepts, discussion topics, transparency list, suggested readings, learning strategies, lists of related media and suppliers, answers to textbook questions, 75 new test questions for each chapter, and a visual testbank of black-and-white transparency masters.
3. *Student Study Guide* by Kenneth S. Saladin (0-697-23096-1) discusses study habits, time management, and A&P "survival tips"; for each chapter, a set of vocabulary-building and content-testing exercises and a practice exam; and for each of the five main sections of the book, a comprehensive practice exam calling for more integration and comparison of information from related chapters. Answer keys are provided for all practice exams and for the Testing Your Comprehension questions in the textbook.
4. *MicroTest III* (Windows 0-697-23091-0, Macintosh 0-697-23092-9) is a computerized test generator,

available free to qualified adopters, which enables instructors to generate tests from questions in the Instructor's Manual.

5. *Transparencies* (0-697-23095-3) include 500 color illustrations and photographs from this book reprinted as overhead lecture transparencies, packaged in a 3-ring binder.

6. *QuickStudy* by Kenneth S. Saladin (IBM 0-697-29787-X, Macintosh 0-697-29788-8) is a computerized study guide for the student with a wide selection of questions for each chapter in true/false, multiple choice, and short answer formats.

7. *The Dynamic Human Powerpoint Presentation and Visual Resource Library* (0-697-39933-8) is a CD-ROM with all of the color art in this textbook and related animations from *The Dyamic Human;* allows for easy incorporation of all the textbook art into computer-assisted lecture presentations.

8. *The Dynamic Human CD-ROM,* Version 2.0 (Windows 0-697-38935-9; Macintosh 0-697-38936-7) consists of three-dimensional and other visualizations of relationships between human structure and function. A *Dynamic Human* icon (✗)appears in relevant figure legends in this book. A list of these correlations by Jeffrey and Karianne Prince is on page xxvi.

9. *The Dynamic Human Videodisc* (0-697-38937-5) contains all of the CD-ROM animations, with a barcode directory.

10. *Explorations in Human Biology CD-ROM* by George Johnson (Windows, 0-697-37906-X, and Macintosh 0-697-37907-8) consists of 16 interactive animations on human biology.

11. *Explorations in Cell Biology, Metabolism, and Genetics CD-ROM* by George Johnson (Windows and Macintosh 0-697-29214-2) provides 17 colorful animations that afford an engrossing way for students to delve into these often-challenging topics.

12. *Life Sciences Living Lexicon CD-ROM* by William Marchuk (0-697-29266-5) provides interactive vocabulary-building exercises closely related to appendix C, "Understanding Biomedical Vocabulary." It includes the meanings of word roots, prefixes, and suffixes, with illustrations and audio pronunciations.

13. *Virtual Physiology Lab* (0-697-37994-9) has 10 simulations of animal-based experiments common in the physiology component of a laboratory course; allows students to repeat experiments for improved mastery.

14. *WCB Anatomy and Physiology Videodisc* (0-697-27716-X) has more than 30 physiological animations, line art, and photomicrographs, with a barcode directory.

15. *WCB's Life Science Animations* (LSA) contains 53 animations on five VHS videocassettes: Chemistry, the Cell, and Energetics (0-697-25068-7); Cell Division, Heredity, Genetics, Reproduction, and Development (0-697-25069-5); Animal Biology No. 1 (0-697-25070-9); Animal Biology No. 2 (0-697-25071-7); and Plant Biology, Evolution, and Ecology (0-697-26600-1). A videocassette icon (▣) appears in figure legends in this book to alert the reader to related animations. A list of these correlations by Jeffrey and Karianne Prince is on page xxix. Another available videotape is Physiological Concepts of Life Science (0-697-21512-1).

16. *WCB Anatomy and Physiology Videotape Series* consists of four videotapes, free to qualified adopters, including Blood Cell Counting, Identification, and Grouping (0-697-11629-8); Introduction to the Human Cadaver and Prosection (0-697-11177-6); Introduction to Cat Dissection: Cat Musculature (0-697-11630-1); and Internal Organs and Circulatory System of the Cat (0-697-13922-0).

17. *Human Anatomy and Physiology Study Cards* by Kent Van De Graaff, Ward Rhees, and Christopher Creek (0-697-26447-5) is a boxed set of 300 illustrated cards (3 × 5 in), each of which concisely summarizes a concept of structure or function, defines a term, and provides a concise table of related information.

18. *Coloring Guide to Anatomy and Physiology* by Robert and Judith Stone (0-697-17109-4) consists of outline drawings and text that emphasize learning through color association. Students retain information through a meditative exercise in color-coding structures and correlated labels. This can be an especially effective aid for students who remember visual concepts more easily than verbal ones.

19. *Atlas of the Skeletal Muscles* by Robert and Judith Stone (0-697-13790-2) illustrates each skeletal muscle in a diagram that the student can color, and provides a concise table of the origin, insertion, action, and innervation of each muscle.

20. *Laboratory Atlas of Anatomy and Physiology,* 2/e, by Douglas Eder et al. (0-697-39480-8) is a full-color atlas including histology, skeletal and muscular anatomy, dissections, and reference tables.

21. *Case Histories in Human Physiology,* 2/e, by Donna Van Wynsberghe and Gregory Cooley (0-697-13791-0) stimulates analytical thinking through case studies and problem solving; includes an instructor's answer key.

22. *Survey of Infectious and Parasitic Diseases* by Kent Van De Graaff (0-697-27535-3) is a booklet of essential information on 100 of the most significant infectious diseases.

Acknowledgments

Any project of this magnitude involves an enormous team effort, and I am gratefully indebted to the people who made it happen. Thank you, first of all, to Colin Wheatley for sensing that this book was in me waiting to come out, and for securing the opportunity to produce it. Thanks to Michael Lange for making me feel so much at home with WCB/McGraw-Hill and for sparing no effort to put the company's confidence and resources into making this book second to none. The team of Kris Noel Tibbetts, Kelly Drapeau, Sue Dillon, Carrie Burger, John Leland, Kathleen Timp, and Darlene Schueller were wonderful in their ability to make all the trains run on time and keep me reasonably sane and optimistic, even through the periods of daunting details and relentless deadlines. Copyeditors Ann Mirels and Laura Beaudoin provided many helpful content and style suggestions and contributed greatly to the congenial spirit of this collaborative effort.

I am deeply grateful to the artists of OGI, listed earlier, for making this such a visually stimulating book. For original photographic contributions, I am also indebted to Dennis Strete, who prepared histological and cadaver photographs to my specifications, and photographer Joe DeGrandis, who worked with me at Milledgeville to photograph the living models. Many thanks to the students, friends, and family members who modeled for these photographs: Christopher Allen, Laura Ammons, Elizabeth Brown, Amy Burmeister, Mae Carpenter, Valeria Champion, Adam Fraley, Yashica Marshall, Wang Xiaodan, Nathan Williams, Danielle Wychoff; Diane, Emory, and Nicole Saladin; and some who prefer to remain anonymous but to whom I am no less grateful.

For factual content and accuracy, I am indebted to clinical consultant Carol Porth; to Kelly Griffith of Creighton University and the cybervillage of Olde Wundee; and to Harold Reed and Thomas Toney of Georgia College and State University, whose boundless intellects I often picked when technical questions arose. The WCB/McGraw-Hill focus group members and hundreds of reviewers gave tremendously important information, thoughtful critiques, and heartening encouragement throughout the development of this book. Thanks especially to David Evans, Richard Roller, and Al Baccari, who reviewed the entire manuscript and significantly contributed to the development of this book. Many thanks again to David Evans for providing an extensive set of World Wide Web links for each chapter, accessible to students through WCB/McGraw-Hill's home page for *Anatomy and Physiology: The Unity of Form and Function* at: http://www.mhhe.com/sciencemath/biology/saladin/

My students were a special source of encouragement as they watched this project develop from raw manuscript to page proofs with keen interest. Many of them read portions of the manuscript and compared it to other books in this field. Sherylyn Bond and Jane Talisman were especially enthusiastic and thorough; their input is reflected at several places in the textbook and study guide.

Thanks to the many colleagues, students, and friends who, in the flesh and on the Internet, provided continual support, encouragement, and suggestions. It was great to have such a community of friends with whom I could vent my frustrations, share ideas, and celebrate the triumphs.

Special thanks to Emory and Nicole for standing by your dad even as he wandered absent-mindedly around the house muttering about editors, page proofs, and deadlines. Last but best of all, thank you, Diane, for sharing the adventure; for modeling your facial expressions for the muscle anatomy chapter; for all your work in reading the manuscript, correcting errors, and checking the glossary and other details—but especially for your love and patience. I married well.

Kenneth S. Saladin

Reviewers

I am very grateful to the many reviewers, most of them anonymous to me until now, who provided substantive information, thoughtful critiques, and heartening encouragement as the manuscript of this book was in development:

M. Abbott
Nottingham University School of Nursing

Harold S. Adams
Dabney S. Lancaster Community College

Jerri Adler
Lane Community College

Irene Allan
University of Dundee, School of Nursing and Midwifery

Nabil Amer
Jordan University

Dennis I. Anderson
Oklahoma City Community College

Edith Applegate
Kettering College of Medical Arts

R. K. Atkinson
University of Southern Queensland

Bert Atsma
Union County College

Albert A. Baccari, Jr.
Montgomery County Community College

Darwish Hasan Badran
University of Jordan

Frank Baker
Golden West College

Steve Baker
Oxford College of Emory University

Linda Banta
Cosumnes River College–El Dorado Center

A. D. Barber
Plymouth School of Podiatry

Debra Barnes
Contra Costa College

Robert Bauman
Amarillo College

Dean Beckwith
Illinois Central College

Irwin Beitch
Quinnipiac College

Edwin Bessler
Franciscan University of Steubenville

Peter Biesemeyer
North Country Community College

David J. Bird
University of the West of England

Richard O. Blackburn
University of Greenwich

P. K. Bourne
School of Biomedical Sciences, Curtin University of Technology

D. L. Bovell
Glasgow Caledonian University

Julie Harrill Bowers
East Tennessee State University

Bradford Boyer
Suffolk County Community College

James H. Boyson
San Antonio College

J. D. Brammer
North Dakota State University

A. Glenn Brice
University of Wisconsin-LaCrosse

James Bridger
Prince George's Community College

R. F. Brightwell
Edith Cowan University

Carol Bryan
School of Health and Community Studies, King Alfreds College

Alvin M. Burt
Vanderbilt University, School of Nursing and School of Medicine

Mark A. Burton
Charles Sturt University

Claranne M. Bush
Florence Darlington Technical College

Colin Butler
Coventry Technical College

John S. Cameron
Wellesley College

John R. Capeheart
University of Houston-Downtown

Christine Lorraine Carline
School of Health Staffordshire University

Pamela J. Carlton
College of Staten Island

William M. Chamberlain
Indiana State University

Carol Fordham Clarke
The Nightingale Institute, King's College

Conrad A. Claytor
Community Hospital/College of Health Sciences

John Patrick Click
Indiana University Southeast

William Cliff
Niagara University

Craig W. Clifford
Northeastern State University

Joe Coelho
Western Illinois University

D. Colborn
East London and the City Health Authority

B. Theodore Cole
University of South Carolina

Wade L. Collier
Manatee Community College, South Campus

David T. Corey
Midlands Technical College

Marion E. Cornelius
Central Arizona College

Desmond Cornes
Glasgow College of Nursing and Midwifery

Laurie A. Cree
University of Technology, Sydney Kuring-gai Campus

James A. Crowder, Jr.
Muhlenberg College

Clementine A. de Angelis
Tarrant County Junior College

Harold F. Delisle
Moorpark College

Brent G. DeMars
Lakeland Community College

Fiona Ann Dick
Forth Valley College

Elissa N. Ditto
Red Rocks Community College

Michele Don
Griffith University

Maureen S. Donaldson
Bishop State Community College

Gerald R. Dotson
Front Range Community College

Douglas Duff
Indiana University–South Bend

William E. Dunscombe
Union County College

John Dziak
Community College of Allegheny County–Allegheny Campus

Phillip Eichman
University of Rio Grande

Victor P. Eroschenko
University of Idaho

David L. Evans
Pennsylvania College of Technology/Penn State University

Robert Farrell
Penn State University

Stanley W. Ferguson
Community College of Aurora

Sharon Flanagan
Nunez Community College

Kathleen Flickinger
Iowa State University

Margaret D. Folsom
Methodist College

Pamela B. Fouché
Walters State Community College

Sarah Fowler
Holyoke Community College

Alice M. Fox
Harold Washington City College

Wayne Frair
The King's College

Marc Franco
South Seattle Community College

Frederick R. Frank, Jr.
Volunteer State Community College

Eugenia M. Fulcher
Swainsboro Technical Institute

Robert T. Galbraith
Crafton Hills College

Frances H. Gallacher
Pensacola Junior College

Elizabeth A. Gardner
Alvernia College

Elizabeth Gargus
Jefferson State Community College

Greg Garman
Centralia College

Elizabeth C. Gayton
Liverpool Institute of Higher Education

Dalia Giedrimiene
Saint Joseph College

Iain F. Gilbert
University of Paisley

Mac F. Given
Neumann College

Sister Terence Glum
University of Mary

Helen Godfrey
University of the West of England

David L. Goldstein
Wright State University

John Gole
Central Missouri State University

Nancy L. Goodyear
Bainbridge College

Glenn A. Gorelick
Citrus College

Harold J. Grau
Christopher Newport University

Donald W. Green
New Mexico Junior College

William Hairston
Harrisburg Area Community College/ Penn State University

James E. Hall
Central Piedmont Community College

Gary B. Hanson
Concordia University–Portland

Ruth L. Hays
Clemson University

Mary D. Healey
Springfield College

Kathryn Hedges
Indiana University Northwest

Robert E. Herrington
Georgia Southwestern State University

William F. Hibschman
Harford Community College

Nelda W. Hinckley
John A. Logan College

Dawn Holtzmeier
Hocking College

Jacqueline A. Homan
South Plains College

Robin H. Hooper
Loughborough University of Technology

Julie G. Horsch
Casper College

James Horwitz
Palm Beach Community College

Herbert W. House
Elon College

J. M. Hulbert
Birmingham and Solihull College of Nursing and Midwifery

Reinhold Hutz
University of Wisconsin–Milwaukee

Fred L. Jackson
Community College of Southern Nevada

Patrick Jackson
Canadian Memorial Chiropractic College

James M. Janik
Miami University

Rosemary C. Jernigan
Jefferson Davis Community College

Eldridge F. Johnson
University of Tennessee, Memphis College of Medicine

Ronald L. Johnson
Arkansas State University

Russell Johnson
Ricks College

Drusilla Beal Jolly
Forsyth Technical Community College

Geoffrey Jowett
Brunswick College

Valerie G. Kalter
Wilkes University

George Karleskint, Jr.
St. Louis Community College-Meramec

M. E. Kelly
School of Pharmacy, University of Bradford

David B. Kieffer
Montgomery College

Glenn E. Kietzmann
Wayne State College

Paul G. Kimball
Northeast Iowa Community College

Jamie King
Craven Community College

Helmut Koch
Husson College

Nirmala Kotagal
Maryville University

Jeanne Kowalczyk
University of South Carolina

Marilyn Lacy

A. J. Ladman
Allegheny University of the Health Sciences MCP ◆ Hahnemann School of Medicine

Paul D. Langer
Gwynedd Mercy College

William Langley
Butler County Community College

Stephen G. Lebsack
Linn-Benton Community College

Donald C. Leynaud
Wabash Valley College

Earl F. Lindberg
Davidson Community College

Jerri K. Lindsey
Tarrant County Junior College

Norman V. Martin
University of Southampton School of Nursing and Midwifery

Grace D. Matzen
Molloy College

Craighton S. Mauk
University of Kentucky–Prestonsburg Community College

Marlene McCall
Community College of Allegheny County

David McCaugherty
Bath and Swindon College of Health

Vikki L. McCleary
University of North Dakota School of Medicine and Health Sciences

Carolyn McCracken
Northeast State Technical Community College

Daphne L. McCulloch
Glasgow Caledonian University

E. McDonald
North West Lancashire College of Nursing and Health Studies

Paul McGrath
University of Newcastle

Pamela S. McLaughlin
University of Kentucky, Madisonville Community College

Christopher McNair
Hardin-Simmons University

Margaret S. Merkley
Delaware Technical and Community College

Donna Z. Merrill
Great Lakes Junior College

Wayne E. Meyer
Austin College

Ann Miele
Broome Community College

Lewis M. Milner
North Central Technical College

James B. Mitchell
Moravian College

Rose M. Morgan
Minot State University

M. Morris
University of the West of England

Peter J. Murray
Chester College of Higher Education, School of Nursing and Midwifery

John J. Natalini
Quincy University

Ava Nickerson
North Central Texas College

Lynne Nicoll
Buckinghamshire College of Nursing and Midwifery

Alan Nowicki
Highland Community College

Catherine H. O'Brien
San Jacinto College South

Omokere E. Odje
Central State University

Jonas E. Okeagu
Fayetteville State University

Thomas E. Oldfield
Ferris State University

Daniel R. Olson
Northern Illinois University

Kurt Olson
College of St. Catherine

Betsy Ott
Tyler Junior College

Michael A. Palladino
Brookdale Community College

David L. Parker
Northern Virginia Community College

John D. Pasto
Middle Georgia College

Mark Paternostro
Pennsylvania College of Technology

Michael J. Patrick
Penn State Altoona

Brian K. Paulson
California University of Pennsylvania

D. F. Peach
Centre for Radiographic Studies Cranfield University

Peter Pearson
North Park College

Andrew C. Petersen
Cork Regional Technical College

William J. Pietraface
SUNY-Oneonta

David J. Porta
Bellarmine College/University of Louisville School of Medicine

Dan Porter
Amarillo College

Michael Postula
Parkland College

David R. Pratt
Texas A&M University-Kingsville

Nikki Privacky
Palm Beach Community College

Donald W. Puder
College of Southern Idaho

David M. Quincey
Bournemouth University

Abdool Farook Rajbally
School of Health, Biological and Environmental Science Middlesex University

Dell P. Redding
Evergreen Valley College

Ralph E. Reiner
College of the Redwoods

Dennis Rich
Naugatuck Valley Community and Technical College

Dan Rogers
Somerset Community College

Richard A. Roller
Baylor College of Medicine

Terence B. Rooney
Cowal Training Institute/University of Paisley

Roscoe B. Root
Lansing Community College

Geraldine Ross
Highline Community College

Evelyn Rutty
School of Nursing and Midwifery, University of Wolverhampton

Patrick Saintas
Institute of Nursing and Midwifery University of Brighton

Jane J. Salisbury
Edison State Community College

Veena Sallan
Owensboro Community College, University of Kentucky

David Saltzman
Santa Fe Community College

Stephen N. Sarikas
Lasell College

Kathryn J. Saunders
Glasgow Caledonian University

Annette Schaefer
New York City Technical College/CUNY

Fred H. Schindler
Indian Hills Community College

Linda A. Serrianne
Trocaire College

Marilyn L. Shaver
Ashland Community College-University of Kentucky

Judith T. Shea
Kutztown University

Jacqueline R. Shepperson
Winston-Salem State University

Janet Anne Sherman
Pennsylvania College of Technology

Brian R. Shmaefsky
Kingwood College

Casey A. Shonis
Bloomsburg University

John R. Sibbald
University of Wollongong

George Simpson
Liverpool John Moores University

Robert A. Sinclair
Emeritus San Antonio College

Paul Keith Small
Eureka College

Rosalyn Snellen
Southwest Baptist University

Gordon Snyder
Schoolcraft College

Fitzgerald Spencer
Southern University

George F. Spiegel, Jr.
Mid-Plains Community College

Wilma M. Steedman
Queen Margaret College

Ralph W. Stevens III
Old Dominion University

Kerstin Stoedefalke
Colby-Sawyer College

Ken Strudwick
West Sussex College of Nursing and Midwifery

Ruth D. Stutts
Bishop State Community College

Eric L. Sun
Macon College

R. Bruce Sundrud
Harrisburg Area Community College

Cynthia A. Surmacz
Bloomsburg University

R. S. Taylor
Postgraduate Medical School/University of Exeter

Kenneth Wm. Thomulka
Phila College of Pharmacy and Science

Karen Thrasher
Lifeline Home Health

Colin Torrance
University of Wales Swansea

Brian Tsukimura
California State University, Fresno

Robin Vance
Union College

Frank V. Veselovsky
South Puget Sound Community College

Raghunath A. Virkar
Kean College of New Jersey

Eugene Volz
Sacramento City College

Samuel E. Wages
South Plains College

Roger Watson
The University of Edinburgh

M. Anne Waugh
Lothian College of Health Studies

S. G. West
University of Huddersfield

Katherine Whelchel
Anoka-Ramsey Community College

John Whetton
Nottingham Trent University

Billy J. Wilbanks
Jacksonville College

Janet Wilcoxon
Shropshire and Staffordshire College of Nursing and Midwifery

Peter J. Wilkin
Purdue University North Central

Betty M. Williams
Allied School of Remedial Massage

Ron Williamson
Copiah-Lincoln Community College

Clarence C. Wolfe
Northern Virginia Community College

David M. Wolfrom
Paducah Community College

Bob Wyatt
Seminole State College

Xiaobo Yu
Kean University of New Jersey

Samuel J. Zeakes
Radford University

1996 Focus Group Participants (Art Program)

Craig W. Clifford
Northeastern State University

Darla H. Culmer
Broward Community College

Gregory R. Garman
Centralia College

Gary B. Hanson
Concordia University-Portland

Margaret S. Merkley
Delaware Technical and Community College

Shirley Mulcahy
San Diego Mesa College

Geraldine Y. Ross
Highline Community College

George F. Spiegel, Jr.
Mid-Plains Community College

1997 Focus Group Participants (Ancillaries, Internet, and Technology)

David M. Bastedo
San Bernardino Valley College

Philip J. Costa
Queensborough Community College

Jacqueline A. Homan
South Plains College

Fred L. Jackson
Community College of Southern Nevada

Theodore Markus
Kingsborough Community College

Michael A. Palladino
Brookdale Community College

Nikki Privacky
Palm Beach Community College

Norman A. Scherzer
Essex County College

Henry L. Scurry
Miami-Dade Community College

s you embark on your study of human A&P, I count you among my wider sphere of students and consider it my job not only to present you with information you will need in your career, but also to make it enjoyable and provide you with aids to understanding it. Your success may depend partly on how effectively you use these features, so I would like to familiarize you with them.

Chapter Outline   Before you begin a chapter, it is important to have a broad overview of what it covers. This is provided by an outline on the first page of each chapter, page-referenced to facilitate your later review and study.

Brushing Up   Knowledge doesn't come in little compartments that you can forget as soon as an exam is over. In the health professions, you must have a comprehensive understanding of the entire body, and as you begin each new book chapter, you will need to remember concepts covered in earlier ones. Chapters 3 to 29 have a "Brushing Up" box that lists major concepts, with page references, that you should understand before continuing. If your memory is rusty, review those pages before you start the new material. "Brushing Up" will also help in the event that your instructor covers the chapters in a different order than the book does and you haven't yet studied information that is prerequisite to understanding the new chapter.

Objectives and Key Point Review   There's no escaping the fact that human A&P encompasses a lot of information, but you don't have to swallow it all in one bite. I divide each chapter into short, digestible sections averaging five to six pages each. Each section begins with a short list of objectives and ends with a few self-testing questions titled "Key Point Review." This makes it easier for you to break down the subject matter, plan your goals for a study session, and test your progress at frequent intervals. Use the Key Point Review as a test of your memory and ability to explain an idea, not your ability to look it up. To get the most out of them, answer the questions in writing and without looking back at the text. It's doubtful that your instructor will allow you to look up answers during an exam, so you don't cheat yourself by doing so on practice questions. I regret that there isn't enough space to print the answers to these, but most of them are based on simple recall and you can check your answers by reading back just a few pages and discussing the questions with your study partners. They don't cover every important concept in a section. Getting them right doesn't guarantee that you're completely prepared for your instructor's exam, but getting several of them wrong does indicate that you need to study a section more carefully.

Think About It   Success in the health professions requires far more than memorization. More important is your insight and ability to *apply* what you remember to new cases and problems. Scattered throughout each chapter are four or five "Think About It" questions identified by a brain icon. Pause for a moment at these points, reflect on what you have just read, and see how well you can apply that knowledge to these questions. Ideally, you'll find yourself figuring out more about the body than I directly tell you, and the most exciting kind of knowledge is that which you discover for yourself. These questions may lead to new insights into the relationship between concepts presented in different chapters—for example, how an egg's way of blocking excess sperm resembles the way a nerve cell releases its chemical signals. The answers to these questions are in the *Instructor's Manual.* Your instructor might provide them for you, or might prefer that you think them through for yourself before he or she discloses the answer.

Special Topics and Chapter Essays   Each chapter has three to five Special Topics intended as diversionary reading for enjoyment, not as part of the core information essential to a chapter. Most of these fall into three categories: the clinical relevance of a concept; historical sketches of the personalities and events behind the facts of A&P; and evolutionary insights into the body's structure and function. The essay at the end of each chapter has a similar purpose but allows a topic to be explored in a little more depth. They range from technical subjects such as medical imaging and genetic engineering to engrossing historical accounts of how a frontier military doctor studied digestion in a man with a shotgun wound, and how ether went from being a party drug to a surgical anesthetic.

Vocabulary Aids: Word Origins and Pronunciation Guides
At first, the field of A&P seems to present a bewildering array of new terms that many find difficult to spell, pronounce, and remember. This is a field with a large, rich vocabulary that can be either a friend or foe to

your progress. I find that my students use terms with more confidence and accuracy if they recognize the roots that compose them. I've taught a course on this since 1983, and many of my students have reported back from schools of medicine, nursing, and physical therapy to say how much it helped them in their professional coursework.

Therefore, when I introduce a new term, I frequently include a footnote that gives you the derivation of the word. A list of the 400 most common word derivations is included inside the back cover, and appendix C, Understanding Biomedical Vocabulary, gives you some pointers on how to become more comfortable with new terms. You will soon recognize that such elements as *hypo-, natri-, -cyte,* and *-itis* occur over and over in the book. As their meanings become familiar, you will approach pronunciation, spelling, and comprehension with more confidence and proficiency.

Pronunciation is a difficult issue for many students; it's hard to remember words that we can't pronounce in the first place. You will find pronunciation guides in parentheses following new terms. I worked with many of my students to develop a style for these, like *pro-NUN-see-AY-shun,* that they found simple and easy to understand. I encourage you to sound out words as you read and review. Good pronunciation will greatly improve your memory and understanding and will create a more professional impression on your future instructors, supervisors, and peers.

Study Outline
Each chapter ends with a chapter review. Part of this is a Study Outline that can be used as a basis for organizing your work. It organizes the key points you should know about and gives page references so that you can easily check your knowledge, organize your facts, and review the relevant pages.

Vocabulary Checklist
The boldfaced terms in a chapter are listed in a page-referenced vocabulary checklist at the end of each chapter. These terms are listed in the order presented in the chapter, which has two advantages over an alphabetical list: (1) it enables you to review the terms one section at a time and to correlate the vocabulary list with the study outline that precedes it, and (2) it keeps related terms together, which makes it easier to remember their meanings. To save space, I do not repeat terms that are listed and defined in tables within the body of the chapter, but I include a note that advises you which tables to study. Terms in the Special Topics and Chapter Essays are not listed. Add these to your list if your instructor tests on that material.

Testing Your Recall
Each chapter has 10 multiple choice and 10 short answer questions you can use to check your knowledge. Answer the questions without looking up the answers, then check your answers in appendix B and restudy any concepts you missed.

Connective Issues
The human organ systems do not, of course, exist in isolation from each other. Diseases of the circulatory system can lead to failure of the urinary system and aging of the skin can lead to weakening of the skeleton, for example. For each organ system, I include a page called "Connective Issues" to show how it affects other systems of the body and is affected by them. These beautifully illustrated pages will help you get the big picture and appreciate the body as an integrated whole.

Connective Issues

Interactions Between the NERVOUS SYSTEM and Other Organ Systems

Integumentary System
- Provides sensations of heat, cold, pressure, pain, and vibration; protects peripheral nerves
- Nervous system regulates piloerection and sweating; controls cutaneous blood flow to regulate heat loss

Skeletal System
- Serves as reservoir of Ca^{2+} needed for neural function; protects CNS and some peripheral nerves
- Nervous stimulation generates muscle tension essential for bone development and remodeling

Muscular System
- Gives expression to thoughts, emotions, and motor commands that arise in the CNS
- Somatic nervous system activates skeletal muscles and maintains muscle tone

Endocrine System
- Many hormones affect neuronal growth and metabolism; hormones control electrolyte balance essential for neural function
- Hypothalamus controls pituitary gland; sympathetic nervous system stimulates adrenal medulla

Circulatory System
- Delivers O_2 and carries away wastes; transports hormones to and from CNS; CSF produced from and returned to blood
- Nervous system regulates heartbeat, blood vessel diameters, blood pressure, and routing of blood; influences blood clotting

Lymphatic/Immune Systems
- Immune cells provide protection and promote tissue repair
- Nerves innervate lymphoid organs and influence development and activity of immune cells; nervous system plays a role in regulating immune response; emotional states influence susceptibility to infection

Respiratory System
- Provides O_2, removes CO_2, and helps to maintain proper pH for neural function
- Nervous system regulates rate and depth of respiration

Urinary System
- Disposes of wastes and maintains electrolyte and pH balance
- Nervous system regulates renal blood flow, thus affecting rate of urine formation; controls emptying of bladder

Digestive System
- Provides nutrients; liver provides stable level of blood glucose for neural function during periods of fasting
- Nervous system regulates appetite, feeding behavior, digestive secretion and motility, and defecation

Reproductive System
- Sex hormones influence CNS development and sexual behavior; hormones of the menstrual cycle stimulate or inhibit hypothalamus
- Nervous system regulates sex drive, arousal, and orgasm; secretes or stimulates pituitary release of many hormones involved in menstrual cycle, sperm production, pregnancy, and lactation

chapter 15 The Peripheral Nervous System and Reflexes 547

Testing Your Comprehension
Each chapter has five discussion questions that go beyond memorization to require a deeper level of analysis. These questions, like the Think About It items, call for you to apply your knowledge to a new situation or to think through some implications of what you have learned. There was no space to print the answers to these in the textbook, but they are in the *Study Guide,* which you can obtain separately.

In summary, you can see that each chapter has an abundance of opportunities to evaluate your knowledge of the subject. Each has a total of 50 to 55 questions within the book itself, another 150 or more in the *Student Study Guide,* and still more on the *QuickStudy* computerized study guide (see Preface to Instructors). You can use the Objectives, Study Outlines, and Selected Vocabulary lists to design additional review and testing drills or questions of your own (vocabulary flashcards, for example). Following are some additional aids that go beyond the scope of individual chapters.

Inside Covers   Take a moment to look inside the front and back covers of the book. I've placed features here to provide you with convenient reference to the symbols, weights, measures, abbreviations, and word origins most commonly used in the book.

Glossary   The glossary was carefully planned to provide concise but comprehensive definitions of the 1,000 terms I thought you would most likely need to look up. Many glossary entries include pronunciation guides and cite illustrations that help to clarify the concept.

Supplemental Materials   A variety of materials can be purchased separately to supplement this book, including flashcards, coloring atlases, self-testing computer software, and a *Student Study Guide*. Please see the Preface to Instructors for a list and description of these items. To order, call the WCB/McGraw-Hill Customer Service Department at 1-800-338-3987.

World Wide Web Links   For your research needs and personal interest, we have created the Saladin Home Page on the World Wide Web. Point your web browser to

http://www.mhhe.com/sciencemath/biology/saladin/

For each chapter of the book, you will find links that take you to an international variety of web sites com-

piled by Dr. David Evans of the Pennsylvania College of Technology. These provide a wealth of supplemental information and images as diverse as updates on AIDS or osteoporosis, medical art and history, histology and cadaver photographs, MRI images, and supplemental readings and research references.

E-mail the Author

Do you ever wish you could ask the author of your textbook a question the way you can ask your own instructor? Now you can. Electronic mail has made long-distance communication so quick and easy that I can extend this opportunity to you. You're invited to write to me at the e-mail address below. There are certain things I *can't* do by e-mail: I am a biologist, not a medical doctor, and I can't give medical advice. I can't answer your homework questions, research your term paper for you, or tutor by e-mail. But I would like to clarify anything in the book you find difficult to understand, and I am interested in knowing what you think of the book—how it could be better, features that should *not* be dropped or changed, and so forth. Many of my past students had a hand in shaping this edition, and you're invited to have a hand in shaping those to come.

Kenneth S. Saladin
Georgia College and State University
Milledgeville, Georgia 31061
ksaladin@mail.gac.peachnet.edu

continued next page

Life Science Animations Correlation Guide

Chapter 1			
1.4	Tape 2	Concept 21	Human Embryonic Development
Chapter 2			
2.5	Tape 1	Concept 1	Formation of an Ionic Bond
Chapter 3			
3.14	Tape 2	Concept 17	Protein Synthesis
3.18, 3.20	Tape 6	Concept 1	Lock and Key Model of Enzyme Action
3.24, 3.25	Tape 1	Concept 11	ATP as an Energy Carrier
3.26	Tape 1	Concept 5	Glycolysis
	Tape 1	Concept 6	Oxidative Respiration (including Krebs Cycle)
	Tape 1	Concept 7	Electron Transport Chain and the Production of ATP
	Tape 6	Concept 5	Electron Transport Chain and Oxidative Phosphorylation
Chapter 4			
4.4, 4.26	Tape 6	Concept 3	Active Transport Across a Cell Membrane
4.7	Tape 1	Concept 2	Journey into a Cell
4.22	Tape 6	Concept 2	Osmosis
4.30	Tape 1	Concept 3	Endocytosis
4.32	Tape 1	Concept 4	Cellular Secretion
Chapter 5			
5.6	Tape 2	Concept 16	Transcription of a Gene
5.8	Tape 2	Concept 17	Protein Synthesis
5.12, 5.13	Tape 2	Concept 14	DNA Replication
5.15	Tape 2	Concept 12	Mitosis
Chapter 12			
12.1–12.3, 12.5	Tape 3	Concept 29	Levels of Muscle Structure
12.12	Tape 3	Concept 31	Regulation of Muscle Contraction
12.13	Tape 3	Concept 30	Sliding Filament Model of Muscle Contraction
Chapter 13			
13.5	Tape 3	Concept 22	Formation of Myelin Sheath
13.13	Tape 3	Concept 23	Saltatory Nerve Conduction
	Tape 6	Concept 6	Conduction of Nerve Impulses
13.14	Tape 3	Concept 23	Saltatory Nerve Conduction
	Tape 6	Concept 6	Conduction of Nerve Impulses
13.16, 13.18	Tape 6	Concept 8	Synaptic Transmission
13.19	Tape 6	Concept 8	Synaptic Transmission
	Tape 3	Concept 28	Peptide Hormone Action (cAMP)
13.21	Tape 6	Concept 7	Temporal and Spatial Summation
13.25	Tape 3	Concept 24	Signal Integration
Chapter 15			
15.12	Tape 3	Concept 25	Reflex Arcs
Chapter 16			
16.14	Tape 3	Concept 27	Organ of Corti
16.20	Tape 3	Concept 26	Organ of Static Equilibrium
16.35	Tape 6	Concept 9	Visual Accommodation

Chapter 17			
17.22	Tape 6	Concept 10	Action of Steroid Hormone on Target Cells
17.23	Tape 6	Concept 11	Action of Thyroid Hormone on Target Cells
17.24	Tape 6	Concept 12	Cyclic AMP Action
	Tape 3	Concept 28	Peptide Hormone Action (cAMP)
Chapter 18			
18.13, 18.14	Tape 4	Concept 40	A, B, O Blood Types
Chapter 19			
19.1, 19.8	Tape 4	Concept 37	Blood Circulation
19.15, 19.16	Tape 4	Concept 38	Production of Electrocardiogram
19.19	Tape 4	Concept 32	The Cardiac Cycle and Production of Sounds
Chapter 21			
21.17, 21.18	Tape 4	Concept 42	Structure and Function of Antibodies
21.19, 21.20, 21.22	Tape 4	Concept 41	B-Cell Immune Response
21.26	Tape 4	Concept 43	Types of T Cells
	Tape 4	Concept 44	Relationship of Helper T Cells and Killer T Cells
21.27, 21.28	Tape 4	Concept 43	Types of T Cells
	Tape 4	Concept 44	Relationship of Helper T Cells and Killer T Cells
Figure E-1	Tape 6	Concept 13	Life Cycle of HIV
Chapter 25			
25.10	Tape 4	Concept 33	Peristalsis
25.28	Tape 4	Concept 34	Digestion of Carbohydrates
25.30	Tape 4	Concept 35	Digestion of Proteins
25.31	Tape 4	Concept 36	Digestion of Lipids
Chapter 26			
26.3	Tape 1	Concept 5	Glycolysis
26.4	Tape 1	Concept 6	Oxidative Respiration (including Krebs Cycle)
26.5	Tape 1	Concept 7	Electron Transport Chain and the Production of ATP
	Tape 6	Concept 5	Electron Transport and Oxidative Phosphorylation
26.6	Tape 1	Concept 7	Electron Transport Chain and the Production of ATP
	Tape 1	Concept 11	ATP as an Energy Carrier
26.7	Tape 1	Concept 11	ATP as an Energy Carrier
Chapter 27			
27.14	Tape 2	Concept 12	Mitosis
	Tape 2	Concept 13	Meiosis
27.16, 27.17	Tape 2	Concept 19	Spermatogenesis
Chapter 28			
28.11	Tape 2	Concept 20	Oogenesis
Chapter 29			
29.15	Tape 2	Concept 21	Human Embryonic Development

[Major Themes of Anatomy [and Physiology

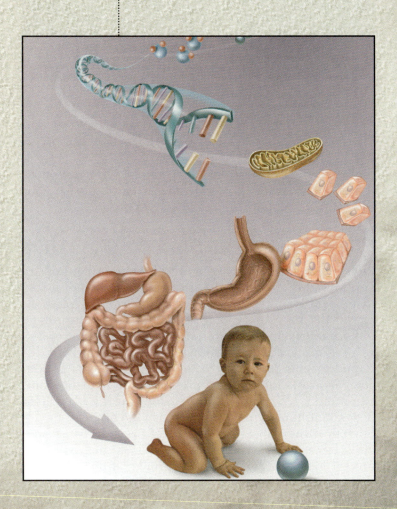

Special Topics

This book is about human anatomy and physiology—the structure and function of the body. It is meant to provide you with a foundation for later study in pathology, health care, exercise physiology, and other fields related to health and fitness. Even an introductory course in anatomy and physiology requires the assimilation of a large amount of information, but the effort is very rewarding. You will find your knowledge of the body not only fascinating but of highly practical value. In addition to preparing you for a possible career in health science, the knowledge you gain can serve you well in managing your own health. If you have not already done so, please read the Preface to Students, which includes tips for the most effective use of this book and its companion materials in the enjoyment and mastery of this subject.

The Scope of Anatomy and Physiology

This chapter discusses some unifying principles that underlie the study of anatomy and physiology. It gives some insight into the history of this science and into the scientific way of thinking that underlies our present knowledge of human form and function. These topics will put the rest of the book in perspective and, ideally, enrich your appreciation of the design of the body and its intricate workings.

Anatomy—The Study of Form

Anatomy[1] is the study of structure. The most productive method of studying anatomy, dating to the era of ancient Greece, has been **dissection**[2]—carefully cutting and separating tissues to reveal their relationships. (The words *anatomy* and *dissection* both mean "cutting apart.") Many students of health science begin their training by dissecting a **cadaver,**[3] or dead body (fig. 1.1). Until modern times, cadaver dissection was called *anatomizing.*

   Gross anatomy refers to structure that can be seen with the naked eye. **Microscopic anatomy** is the structure of cells and tissues as revealed by the microscope, and **ultrastructure** refers to even finer details revealed by the electron microscope. **Comparative anatomy** is the study of more than one species in order to learn generalizations and evolutionary trends in structure. Students of human anatomy often begin by dissecting other mammals, with which we share a common ancestry and many structural similarities.

   Dissection, of course, is not the method of choice when studying the anatomy of a living person! One alternative, **palpation,** is the process of feeling organs with the hands—for example, palpating a swollen liver, examining the prostate gland, or taking a pulse. **Auscultation** (AWS-cul-TAY-shun) is listening to the natural sounds of the body, such as heart and lung sounds. In **percussion,** the examiner taps on the body and listens to the echo for indications of abnormal pockets of air or fluid.

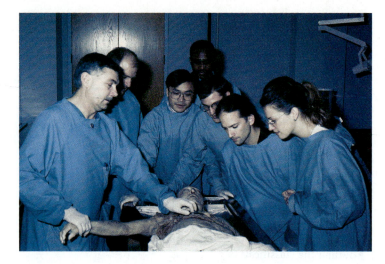

Figure 1.1 Students of the health sciences often begin their professional training by dissecting a cadaver.

Physiology—The Study of Function

Physiology is the study of bodily function. It primarily employs the methods of experimental science. It has many subdisciplines such as *neurophysiology* (physiology of the nervous system), *endocrinology* (physiology of hormones), and *pathophysiology* (mechanisms of disease). **Comparative physiology** is the study of how different species have solved the same problems of life, such as regulation of blood pressure, water balance, and respiration. Like comparative anatomy, it reveals a great deal about human function. It is also the basis for development of new drugs and medical procedures. A vaccine, for example, cannot be used on human subjects until it has been demonstrated through animal research that it confers significant benefits with minimal risks.

Unity of Form and Function

Anatomy and physiology cannot be separated from each other. The functions of the kidney, for example, would be incomprehensible without knowing its gross and microscopic anatomy, while at the same time the study of anatomy without function would be a sterile exercise in description without meaning. Indeed, *all anatomy is the result of physiology,* of cells building tissues and organs. Conversely, *all physiology is made possible by anatomy,* by specific structural arrangements of molecules, cells, and tissues. The subtitle of this book, "The Unity of Form and Function," is a reminder that these two aspects of the body are complementary and inseparable.

---

1. *ana* = apart + *tomy* = cutting
2. *dis* = apart + *sect* = to cut
3. Latin, *cadere* = to fall or die

Architect Louis Henry Sullivan coined the phrase, "Form follows function." What do you think he meant by this? Discuss how this idea could be applied to the human body. When you have completed the chapter, cite a specific example of human anatomy to support this argument.

The Nature of Human Life

▼Objectives

When you have completed this section, you should be able to

- list some characteristics that distinguish living organisms from nonliving objects;
- outline the classification of humans within the animal kingdom;
- describe the anatomical features that define "human" and distinguish humans from other animals;
- list the levels of human structure from simplest to most complex; and
- discuss the value of both reductionistic and holistic views of human function.

This book is a study of human life, so it is fitting that we begin by considering what we mean by this expression, "human life." This is really two questions: What is life? and What is a human?

What Is Life?

Why is a growing child alive and a growing crystal not alive? Is abortion the taking of a human life? If so, what about a contraceptive foam that kills only sperm cells? At what point is it ethical to disconnect life-support equipment from a patient and remove organs for donation? As long as the organs are "alive," as they must be to serve someone else, then why shouldn't the patient be considered alive? Questions such as these have no easy answers, but they demand a concept of what constitutes life—a concept that may differ with one's biological, medical, or legal perspective.

From a biological viewpoint, life is not a single property or entity—it is a collection of properties that helps to distinguish a living from a nonliving thing.

- **Cellular organization.** All living things are composed of one or more cells. Cells are the microscopic units of structure and function that carry out all the processes of the body.
- **Biochemical unity.** All living organisms have a similar chemical composition: proteins, lipids, carbohydrates, DNA, vitamins, and so forth. Such compounds are universal to life on earth but are not found in anything of nonbiological origin.

- **Metabolism.**[4] Some nonliving things can grow, such as crystals of sugar when a sugar solution evaporates. When a crystal grows, however, there is no chemical change in its composition; it merely adds more molecules of the same sugar to its surface. The growth of an organism is different. It occurs by **assimilation**—a process in which the organism takes in environmental chemicals, converts them to the chemicals that constitute a living body, and adds these to itself. Assimilation and all other chemical reactions in the body are collectively called *metabolism.* There is a constant turnover of molecules in the body. All of your nerve and muscle cells have been present since your early childhood, but there are few atoms in your body that have been there more than a year. Although you sense a continuity of personality and experience from childhood to the present, it is food for thought that nearly all of your body has been replaced within the past year. Most metabolism serves to maintain a state of internal stability, or *homeostasis,* discussed later in this chapter.
- **Responsiveness (excitability).** All organisms sense and respond to changes in their environment. Changes detected by a cell are called **stimuli;** they may come from outside the body (such as the odor of food) or inside (such as a full bladder). The cell or organ that detects a stimulus is called a **receptor**—such as a taste cell of the tongue or the eye. The cell or organ (usually a muscle or gland) that carries out a response is called an **effector.** All organisms exhibit responsiveness, but in most animals it is especially obvious because of highly sensitive nerve cells, rapid transmission of information, and quick reactions of muscles.
- **Development.** Development is any change in form or function over the lifetime of the organism. The two major aspects of development are growth and differentiation. **Growth** is simply an increase in size, whether the organism spends its entire life as one cell or grows from a single-celled egg to a complex multicellular individual. **Differentiation** is a change from cells with no specialized function to cells that are committed to a particular physiological or structural task.
- **Reproduction.** All living organisms can produce copies of themselves. Any organism eventually wears out and dies, but its DNA may live on in its offspring. With our complex brains and long lives extending well beyond our reproductive years, humans find more meaning in life than to reproduce; yet our ultimate function is to stay alive and well long enough to produce new and younger containers for our DNA.

---

4. *metabol* = change + *ism* = process

Clinical and legal criteria of life differ from these biological criteria. If a person has shown no brain waves for 30 minutes or longer, and has no reflexes, respiration, or heartbeat other than what is provided by artificial life support, in most cases the person can be declared legally dead. At such time, however, most of the body is still biologically alive, and organs may be removed for transplant.

What Is a Human?

Our second question was, What is a human? To answer this we must consider our place in the animal kingdom. Modern humans are classified according to the following hierarchy, where each level is a smaller subdivision of the one above it:

Kingdom Animalia
   Phylum Chordata
      Subphylum Vertebrata
         Class Mammalia
            Order Primates
               Family Hominidae
                  Genus *Homo*
                     Species *Homo sapiens*

Our Animal Characteristics

Biologists classify all living things into approximately five large categories called kingdoms (fig. 1.2). (Some say more than five, but this debate concerns mainly the bacteria and is beyond our immediate interests here.) These five kingdoms comprise the bacteria (Monera); other one-celled and simple colonial organisms (Protista); mushrooms, molds, and yeasts (Fungi); plants (Plantae); and animals (Animalia).

    Humans belong to the kingdom **Animalia,** which is characterized by the following features:

- Animals are **eukaryotes**[5] (you-CARE-ee-oats), meaning the genetic material (DNA) of each cell is contained in a nucleus surrounded by a double-walled membrane called the nuclear envelope. Protists, fungi, and plants are also eukaryotic, so this feature sets these kingdoms apart only from the bacteria (prokaryotes), which have no membranes enclosing their DNA.
- Animals are **multicellular.** The human body consists of an estimated 50 trillion (5×10^{13}) cells, which are very diverse in form and function. Monera, Protista,

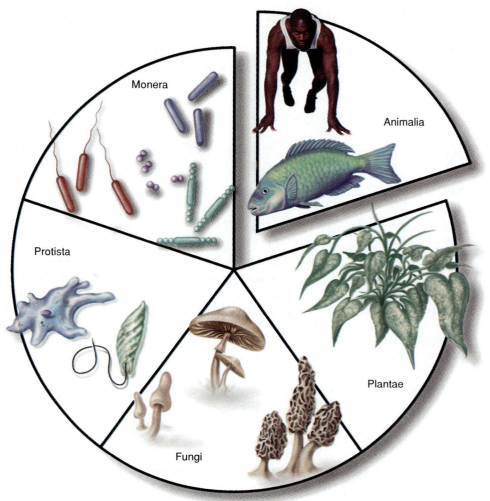

Figure 1.2 The five kingdoms of living organisms. Some authorities subdivide the Monera or Protista and thus recognize more than five kingdoms.

and some of the Fungi, by contrast, consist of single cells or simple colonies of similar cells.
- Animals are **heterotrophic,**[6] meaning they cannot synthesize their own nutrients as do most plants and some bacteria and protists. Unlike these *autotrophic* ("self-feeding") organisms, animals must consume other organisms as food. Fungi are also heterotrophic, but they acquire nourishment by secreting enzymes onto the organic matter around them (living tissues or dead bodies of other organisms) and absorbing the predigested nutrients. Most animals ingest (swallow) their food first and break it down within a digestive tract.
- Most animals have **nerve** and **muscle cells,** which enable them to respond quickly to stimuli and to move freely. Because of this responsiveness and mobility, animals constitute the most diverse, "successful" kingdom, with more known species than the other four kingdoms combined.

---

5. *eu* = true + *karyo* = nucleus

6. *hetero* = other

Thus, we can define animals as eukaryotic, multicellular, heterotrophic organisms that usually possess nerve and muscle cells and correspondingly quick responses. Humans clearly fit this definition.

Our Chordate Characteristics

Figure 1.3 summarizes our place within the Animalia. The animal kingdom is divided into 30 to 35 major categories called **phyla** (singular, *phylum*). Humans belong to the phylum **Chordata** (cor-DAH-tuh), which is far from the largest phylum but contains the most complex species. The following characteristics distinguish the Chordata:

- A **notochord.** This is a flexible dorsal (upper) rod, present in some stage of the animal's development. It is usually replaced by the spinal column as the embryo develops.
- A **dorsal hollow nerve cord.** This is a column of nervous tissue that passes along the animal's upper side and has a central canal filled with fluid. Nonchordates either lack nerve cords or have solid threadlike cords.
- **Pharyngeal gill pouches.** These develop as a series of bulges in the pharyngeal (throat) region of all chordate embryos. They develop into gill slits in fish and amphibians, but human gill pouches (fig. 1.4) seldom break through to form slits.
- A **postanal tail.** Chordates have a tail that extends beyond the anus. In humans it is present only in the embryo (fig. 1.4), but the small bones of the coccyx ("tailbone") are a remnant of it. Many nonchordates have tails, but their digestive tracts generally extend to the end of the tail, and the anus is located there.

Our Vertebrate Characteristics

All Chordata except for a few marine species belong to the subphylum **Vertebrata.**[7] Vertebrates, including humans, are characterized by the following:

- An internal skeleton.
- A jointed vertebral (spinal) column.
- A protective enclosure for the brain called the *cranium.*
- A well-developed brain and sense organs.

Our Mammalian Characteristics

The Vertebrata are divided into fish, amphibians, reptiles, birds, and mammals. We belong to the class **Mammalia,**[8] most members of which exhibit the following characteristics:

- Mammary glands for nourishment of the young.
- Hair, which serves most mammals in the retention of body heat.
- **Endothermy.**[9] This is the ability to generate most of their body heat by metabolic means, as opposed to the ectothermy of fish, amphibians, and reptiles, which must obtain their heat from the environment.
- **Heterodonty.**[10] This is the possession of varied types of teeth (incisors, canines, premolars, and molars) specialized to puncture, cut, and grind food. The food can thus be digested quickly, which is essential if mammals are to maintain the high metabolic rate needed for endothermy.
- A single jawbone. In reptiles, the jaw joint has three bones and is not very strong. As the early mammals evolved from reptiles, the jaw retained only one of these bones and became stronger and better adapted for chewing food.
- Three middle-ear bones (commonly called the *hammer, anvil,* and *stirrup*). Reptiles have only one. The other two bones in mammals arise from the two bones lost from the jaw joint.

Humans As Primates

Mammals are divided into 19 orders. We belong to the order **Primates,** which also contains the apes, monkeys, lemurs, and a few other species. The major human characteristics that place us in this order are as follows:

- Four upper and four lower incisors, the cutting teeth at the front of the jaw.
- A pair of clavicles (collarbones).
- Only two mammary glands.
- A pendulous penis, attached only at the base.
- Forward-facing eyes with stereoscopic vision.
- Flat nails in place of claws.
- Usually opposable thumbs, which allow the tip of the thumb to be touched to the fingertips and enable primates to hold small objects and manipulate them more precisely.

Humans As a Family

Humans belong to the family **Hominidae** (ho-MIN-ih-dee), which may be defined as the **bipedal**[11] primates. That is, in addition to the previously listed characteristics, hominids habitually walk on two legs. The first primates known to do this, between 3 and 4 million

---

7. *vertebr* = backbone + *ata* = possessing
8. *mamma* = breast + *alia* = possessing
9. *endo* = within + *therm* = heat
10. *hetero* = different, varied + *odont* = teeth
11. *bi* = two + *ped* = foot

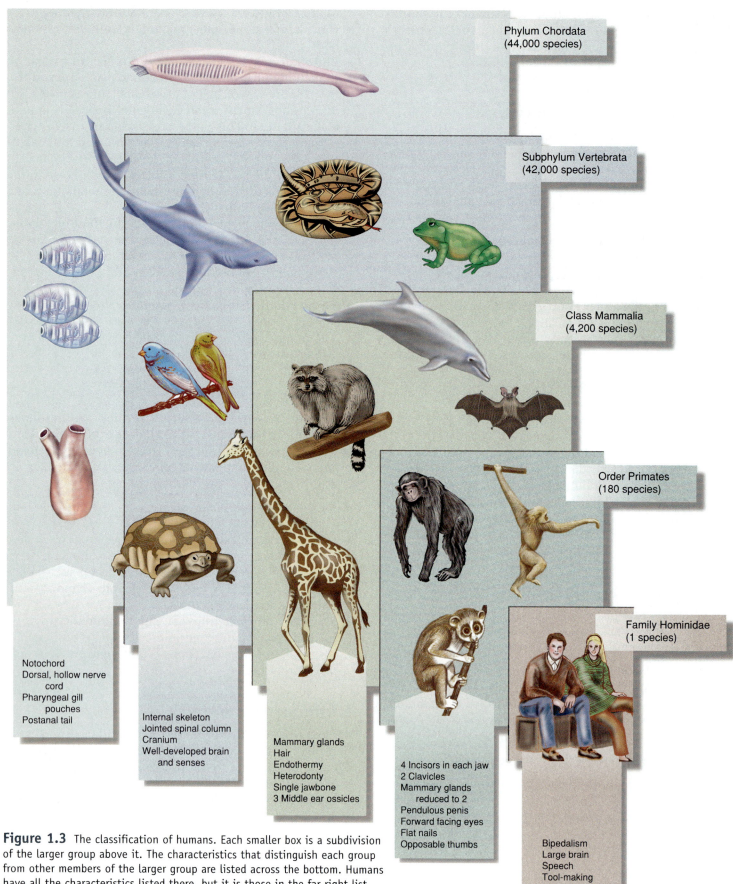

Phylum Chordata
(44,000 species)

Subphylum Vertebrata
(42,000 species)

Class Mammalia
(4,200 species)

Order Primates
(180 species)

Family Hominidae
(1 species)

Notochord
Dorsal, hollow nerve
  cord
Pharyngeal gill
  pouches
Postanal tail

Internal skeleton
Jointed spinal column
Cranium
Well-developed brain
  and senses

Mammary glands
Hair
Endothermy
Heterodonty
Single jawbone
3 Middle ear ossicles

4 Incisors in each jaw
2 Clavicles
Mammary glands
  reduced to 2
Pendulous penis
Forward facing eyes
Flat nails
Opposable thumbs

Bipedalism
Large brain
Speech
Tool-making

Figure 1.3 The classification of humans. Each smaller box is a subdivision of the larger group above it. The characteristics that distinguish each group from other members of the larger group are listed across the bottom. Humans have all the characteristics listed there, but it is those in the far right list that distinguish them from all other chordates.

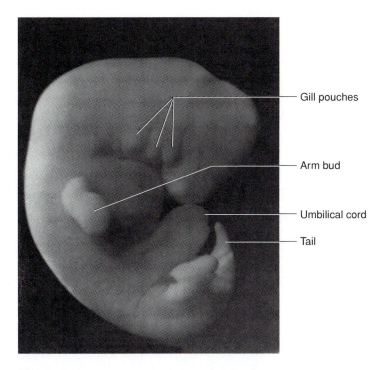

Figure 1.4 A human embryo at 38 days, showing the tail and gill pouches that characterize all members of the Chordata at some stage of their development.

- Gill pouches
- Arm bud
- Umbilical cord
- Tail

Table 1.1	Brain Volumes of the Hominidae	
Genus or Species	Time of Origin (millions of years ago)	Brain Volume (cubic centimeters)
Australopithecus	3.4	400
Homo habilis	2.5	650
Homo erectus	1.1	1,100
Homo sapiens	0.3	1,350

years ago, are placed in the genus **Australopithecus** (aus-TRAL-oh-PITH-eh-cus) (fig. 1.5). *Homo habilis,* the earliest member of our genus, is believed to have evolved from *Australopithecus* about 2.5 million years ago. Some authorities restrict the term *human* to the genus **Homo.** *Homo habilis* was probably the earliest primate able to speak, and it also differed from *Australopithecus* in brain volume (table 1.1), height, tool-making ability, and some details of skull anatomy. *Homo habilis* gave rise to *Homo erectus* about 1.1 million years ago, which in turn gave rise to our own species, *Homo sapiens,* about 300,000 years ago (fig. 1.6). *Homo sapiens* includes the extinct Neanderthal and Cro-Magnon people as well as modern humans.

Human Organization—A Hierarchy of Structural Complexity

Just as all words in the English language are composed of various combinations of 26 letters, the complex structures of living organisms are made up of various combinations of simpler structures. The hierarchy of structural complexity shown in figure 1.7 pertains to humans and most other multicellular organisms.

The organism is composed of organ systems,
organ systems are composed of organs,
organs are composed of tissues,
tissues are composed of cells,
cells are composed partially of organelles,
organelles are composed of molecules,
molecules are composed of atoms, and
atoms are composed of subatomic particles.

The **organism** is *a single complete, individual.* An **organ system** is *a group of organs with a unique collective function such as circulation, respiration, or digestion.* The human body consists of the *integumentary, skeletal,*

Figure 1.5 *Australopithecus afarensis* lived about 3.4 million years ago. It was one of the earliest bipedal primates and was ancestral to the genus *Homo.*

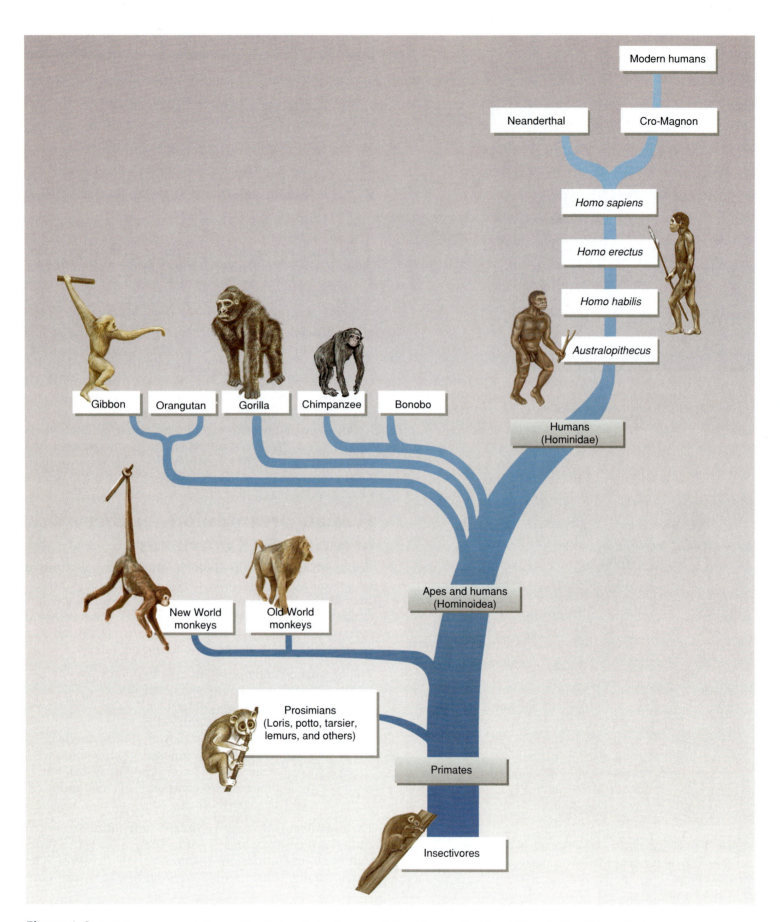

Figure 1.6 A phylogeny, or evolutionary "family tree," showing our relationship to extinct hominids and other living primates.

muscular, nervous, endocrine, cardio-vascular, lymphatic, respiratory, digestive, urinary, and reproductive systems. These 11 organ systems are described and illustrated in Atlas A, immediately following this chapter.

An **organ** is *a structure composed of two or more tissue types that work together to carry out a particular function.* Organs have definite anatomical boundaries and are visibly distinguishable from adjacent structures. The urinary system has six major organs, for example: two kidneys, two ureters, a bladder, and a urethra. It is important to note, however, that the large organs of gross anatomy typically contain other organs much smaller in size. The skin, for example, is the body's largest organ, but included within it are thousands of smaller organs. Each hair, nail, gland, and blood vessel of the skin is an organ in itself, as are some of the skin's microscopic sensory structures.

A **tissue** is *a group of similar cells and nonliving cell products that forms a discrete region of an organ and performs a specific function.* The entire body is composed of only four primary classes of tissue—muscular, nervous, connective, and epithelial. The study of tissues is called **histology.**[12]

A **cell** is *the smallest unit of an organism capable of carrying out all the basic functions of life.* Cells are surrounded by a membrane made of lipid and protein. All of them (except bacteria) contain a nucleus at some time in their lives and are filled with a gelatinous fluid called *cytoplasm.* Some human cells have more than one nucleus (some liver and bone marrow cells, for example), and some lose their nuclei before they mature (red blood cells). Cells can be further analyzed at simpler levels of complexity, but nothing less than a cell is considered to be alive. The study of cells is called **cytology.**[13]

Organelles[14] are *microscopic structures in the cytoplasm that are specialized to carry out individual functions*

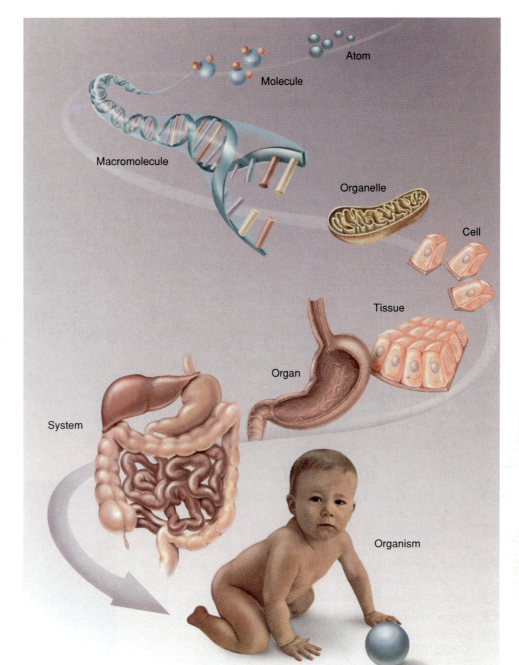

Figure 1.7 The hierarchy of human structure from atom to organism.

of the cell. Examples include mitochondria, centrioles, lysosomes, and endoplasmic reticulum (see chapter 4).

The components of cells are composed of **molecules,** and cellular structure and function can be adequately understood only at a molecular level. Molecules in turn are composed of **atoms,** and atoms are composed of **subatomic particles** (protons, neutrons, and electrons) that determine their physical properties and chemical behavior.

The next five chapters of this book are organized around this hierarchy from atoms to tissues. The remainder

---

12. *histo* = web, fabric, tissue
13. *cyto* = cell + *logy* = study of
14. *elle* = little

of the book is organized around the organ systems. To make sense of organ function, however, we must refer continually to the cellular and molecular levels of organization.

The theory that a large, complex system such as the human body can be understood by studying its simpler components is called **reductionism.** Historically, this has proven to be a highly productive approach; indeed, it is essential to scientific thinking. Yet the reductionistic view is not the last word in understanding human life. Just as it would be very difficult to predict the workings of an automobile transmission merely by looking at a disassembled pile of the gears and levers, one could never predict the human personality from a complete knowledge of the circuitry of the brain or the genetic sequence of DNA. **Holism**[15] is the theory that there are properties of the whole that cannot be known from an examination of the separate parts—human beings are more than the sum of their parts. To be most effective, a health-care provider does not merely treat a disease or an organ system but rather a whole person. The patient's perceptions, emotional responses to life, and confidence in the nurse, therapist, or physician profoundly affect the outcome of the treatment. In fact, these psychological factors may play a greater role in the patient's recovery than the physical treatments administered.

.... **Key Point Review** ....

1. Explain why dissection of other animals is relevant to students of human anatomy.
2. List four biological criteria of life and one clinical criterion. Explain why a person could be clinically dead but biologically alive.
3. Explain why humans are classified as animals, as mammals, and as primates.
4. Do you consider *Australopithecus* to be human? Why or why not?
5. List the levels of human organization from atom to organism.
6. Explain why both reductionism and holism are important to human biology and therapy.

Human Evolution

▼**Objectives**
When you have completed this section, you should be able to
• define *evolution* and *natural selection;*
• describe some human characteristics that can be attributed to the tree-dwelling habits of the first primates;
• describe other human characteristics that can be attributed to the bipedal locomotion of early ground-dwelling hominids; and
• explain why evolution is relevant to understanding human anatomy and physiology.

Our evolutionary history helps to explain why our bodies look and function as they do. In the following account, we briefly consider the relevance of evolution to human anatomy and physiology. Later chapters reiterate this theme and show how an evolutionary perspective sheds light on why we are subject to some of the frailties and diseases that require medical care. The science of *darwinian medicine* traces some of our diseases and anatomical imperfections to our species' evolutionary past.

Darwin and Natural Selection

Charles Darwin (fig. 1.8) was undoubtedly the most influential biologist who ever lived. His book *On the Origin of Species by Means of Natural Selection* (1859) has been called "the book that shook the world." In presenting the first well-supported theory of evolution, this monumental work not only gave impetus to a complete restructuring of biology, it also profoundly changed the prevailing view of our origin, nature, and place in the universe.

Since at least the time of Aristotle, speculation about evolution had been central to biological thought. But it was left to Darwin to amass a convincing body of evidence and develop a credible theory, called **natural selection,** to explain how evolution worked. While the *Origin of Species* scarcely touched upon human biology, the unmistakable implications for humans gave rise to a

Figure 1.8 Charles Darwin (1809–82) at the age of 60.

15. *holo* = whole, entire

storm of controversy. In 1871, Darwin published *The Descent of Man,* which directly addressed the issue of human evolution and emphasized features of anatomy and behavior that showed evidence of our relationship to other forms of life.

Evolution simply means change in the genetic composition of a population of organisms. The development of bacterial resistance to antibiotics, the appearance of new strains of HIV (the AIDS virus), and the emergence of new species of organisms are examples. Natural selection is a process that favors the reproduction of particular individuals because they possess hereditary advantages over their competitors; for example, better camouflage, disease resistance, or ability to attract mates. In passing on their genetic advantages to the next generation, these individuals bring about the genetic change that defines evolution.

Natural forces that cause some individuals to be more reproductively successful than others are called **selection pressures.** They include such things as climate, predators, disease, competition, and availability of food. **Adaptations** are features of an organism's anatomy, physiology, and behavior that have evolved in response to these selection pressures and enable it to cope with the challenges of its environment. We will consider some of the selection pressures and adaptations that were important to human evolution shortly.

Darwin could scarcely have predicted the overwhelming mass of genetic, molecular, fossil, and other evidence of human evolution that would be discovered in the twentieth century and further substantiate his theory. A technique called DNA hybridization, for example, shows a difference of only 1.6% in DNA structure between humans and chimpanzees. Chimpanzees and gorillas differ from each other by 2.3%. This suggests that the chimpanzee's closest living relative is not the gorilla or any other ape—it is us.

Several aspects of our anatomy make little sense without an awareness that the human body has a history. Certain structures of our bodies reflect this history but no longer play significant roles (see special topic 1.1). Our evolutionary relationship with other species is also important in choosing animals for biomedical research. If there were no issues of cost, availability, or ethics, we might test drugs on our nearest living relatives, the chimpanzees, before approving them for human use. The genetics, anatomy, and physiology of these animals are most similar to ours, and their reactions to drugs therefore afford the best prediction of how the human body would react. On the other hand, if we had no genetic relationship to other species, it would scarcely matter which animals were used to test drug safety; we might as well use frogs or goldfish. In reality, we compromise. Rats and mice are used extensively for biomedical research because they are fellow mammals with a physiology similar to ours, and they present fewer problems of cost, availability, maintenance, and ethics than do other mammals. An animal species or strain that is selected for research on a particular human disease is called a **model;** for example, a mouse model for leukemia.

Life in the Trees

The story of primate origins began 55 to 60 million years ago, when certain squirrel-sized, insect-eating mammals (insectivores) took up life in the trees. This **arboreal**[16] habitat probably afforded greater safety from predators, less competition, and a rich food supply of insects, lizards, leaves, and fruit. Despite these advantages, the forest canopy is a challenging world, with dim and dappled sunlight, swaying branches, and prey darting about in the dense foliage. Any new feature that enabled arboreal animals to forage more efficiently for food would have been strongly favored by natural selection. The hands became **prehensile**—adapted for grasping branches by encircling them with the thumb and fingers (fig. 1.9). The thumb is so important that it

16. *arbor* = tree + *eal* = pertaining to

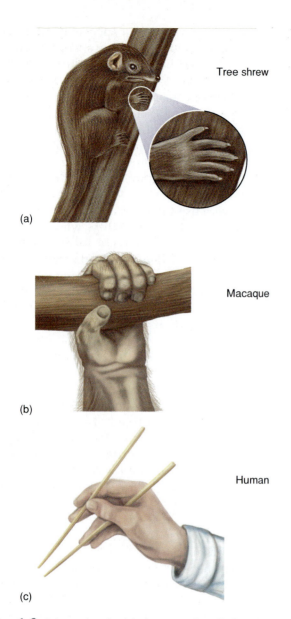

(a)

Tree shrew

Macaque

(b)

Human

(c)

Figure 1.9 Primate hands. (*a*) The nonprehensile hand of the tree shrew. (*b*) The prehensile hand of a monkey. (*c*) The prehensile hand of the human.

receives highest priority in the repair of hand injuries. If the thumb can be saved, the hand can be reasonably functional; if the thumb is lost, hand functions are severely diminished.

The eyes of primates moved to a more forward-facing position (fig. 1.10), which enabled better **stereoscopic vision** (depth perception). This adaptation provided better hand-eye coordination in catching and manipulating prey, with the added advantage of making it easier to judge distances accurately in leaping from tree to tree. Color vision, rare among mammals, is also a primate hallmark. Primates eat mainly fruit and leaves. The ability to distinguish subtle shades of orange and red enables them to distinguish ripe, sugary fruits from unripe

Figure 1.10 The forward-facing eyes of most primates enables them to have stereoscopic vision and good depth perception. Opposable thumbs, another primate characteristic, enable some species to use tools. This ability became supremely important in human evolution.

ones. Distinguishing subtle shades of green helps them to differentiate between tender young leaves and tough, more toxic older foliage.

Various fruits ripen at different times and at widely separated places in the tropical forest. This requires a good memory of what will be available when and where and of how to get there. Larger brains may have evolved in response to the challenge of efficient food finding and in turn laid the foundation for more sophisticated and complex social organization.

None of this is meant to imply that humans evolved from monkeys or apes. This is a common misconception about evolution and is something no biologist believes. Observations of monkeys and apes, however, provide insight into how primates adapt to the arboreal habitat and how certain human adaptations originated.

The Impact of Bipedalism

Approximately 4 to 5 million years ago, Africa became hotter and drier. Much of the forest was replaced by grassland, and some primates adapted to living there.

This was a dangerous place, with more predators and less protection. Just as squirrels and monkeys stand briefly on their hind legs to look around for danger, so would these early ground-dwellers. Being able to stand up not only helps an animal to stay on the alert, it also frees the forelimbs for purposes other than locomotion. Chimpanzees carry sticks and rocks as weapons, and it is reasonable to suppose that early hominids did so too. They could also carry infants and food.

These advantages are so great that they favored skeletal modifications that made it easier to stand and more efficient to walk on two legs. The anatomy of the human pelvis, femur, knee, great toe, foot arches, spinal column, skull, and arms are all related to the origin of this bipedal locomotion, as are many aspects of human social behavior.

As the pelvis was remodeled for bipedalism, brain volume increased dramatically. It must have become increasingly difficult for a fully developed, large-brained infant to pass through the pelvic outlet. This may explain why human infants are born in a relatively immature, helpless state compared to other mammals, before their nervous systems have matured and the bones of the skull have fused.

.... **Key Point Review** ....

7 Define *adaptation* and *selection pressure*. Why are these concepts important in understanding human anatomy and physiology?

8 Select any two human characteristics and explain how they may have originated in primate adaptations to an arboreal habitat.

9 Select two other human characteristics and explain how they may have resulted from adaptation to a grassland habitat.

Scientific Method

Our knowledge of the human body has been obtained by centuries of scientific study. When we say "scientific," we mean that such study was based on certain assumptions and methods that yielded reliable, objective, and testable information about nature. The assumptions of science are ideas that have proven most fruitful in the past; for example, the ideas that natural phenomena have natural causes and that nature is predictable and understandable. The methods of science are highly variable. There is no manual on how to conduct research, but scientists cultivate the habits of disciplined creativity, careful observation, logical thinking, and honest analysis of their observations

and conclusions. Some aspects of the scientific method are explored here because it is important to appreciate how scientists obtained the information in this book. Space does not permit detailed descriptions of individual experiments, but we can at least explore the general intellectual habits that underlie them.

The Inductive Method

The **inductive method** is a process in which numerous observations are made until the investigator feels confident enough to make generalizations and predictions from them. What we know of human anatomy is a product of the inductive method. For example, nineteenth-century biologists studied many species of animals and plants with the microscope until they were confident in the generalization that all organisms are made of cells. This became a cornerstone of the cell theory, which is now so vital to biomedical science.

This raises the issue of what is considered proof in science. We can never prove a claim beyond all possible refutation. We cannot be sure that no one will ever discover an organism that is not made of cells, for example. We can, however, consider a statement as proved *beyond reasonable doubt* if it was arrived at by reliable methods of observation, tested and confirmed repeatedly, and not falsified by any credible observation. In science, all truth is tentative; there is no room for dogma. We must always be prepared to abandon yesterday's truth if tomorrow's facts disprove it.

The Hypothetico-Deductive Method

Most of what we know about physiology comes from the **hypothetico-deductive method.** The investigator begins by asking a question and formulating a **hypothesis**—an educated speculation or reasonable answer to the question. A good hypothesis must be (1) consistent with what is already known and (2) capable of being tested and possibly falsified by evidence. For example, the hypothesis that epilepsy is caused by bursts of abnormal activity in nerve cells of the brain is falsifiable. We could record the brain waves of people with epilepsy, and if no abnormal activity was seen during seizures, it would falsify that hypothesis.

 Think About It
Why would it be unscientific to hypothesize that epilepsy is caused by invisible demons in the body?

The purpose of a hypothesis is to suggest a method for answering a question. From the hypothesis, the investigator makes a **deduction,** typically in the form of an "if-then" prediction: *If* my hypothesis is correct and I record the brain waves of patients during epileptic seizures, *then* I should observe abnormal bursts of activity. A properly conducted experiment will yield observations that may (1) support a hypothesis, (2) require it to be modified, or (3) require it to be abandoned, a new hypothesis formulated, and that one tested. Hypothesis testing may therefore operate in cycles of conjecture and disproof until a hypothesis is found that is supported by the evidence.

Experimental Design

Doing an experiment properly involves several important considerations. What shall I measure, and how can I measure it? What effects should I watch for, and which ones should I ignore? How can I be sure that my results are due to the factors (*variables*) I manipulate and not due to something else? When working with human subjects, how can I eliminate psychosomatic effects (effects of a subject's state of mind on his or her physiology)? Most importantly, how can I eliminate my own biases and be sure that even the most skeptical critics will have as much confidence in my conclusions as I do? Several elements of experimental design address such issues.

- **Sample size.** The number of subjects (animals or people) used in a study is the sample size. An adequate sample size controls for chance events and individual variations in response—therefore, it enables one to place more confidence in the outcome. A drug-effectiveness study involving 500 patients, for example, would be considered more reliable than a study in which only 5 patients were used.
- **Controls.** A **control group** is a group of subjects that are as much like the experimental subjects (**treatment group**) as possible, except with respect to the variable or procedure being tested. For example, there is some evidence that garlic lowers blood cholesterol levels. In one study, a treatment group of patients with high cholesterol levels was given 800 mg of garlic powder daily for 4 months and exhibited an average 12% reduction in cholesterol concentration. Is this a significant reduction, and was it due to something in the garlic? It is impossible to say without comparison to a control group of similar patients who did not receive the treatment. In this study, the control group averaged only a 3% reduction in cholesterol, so garlic *seems* to have made a difference.
- **Psychosomatic effects.** These are physiological effects, such as changes in blood pressure or immune responses, that result from a person's state of mind. Such effects can have an undesirable impact on experimental results if one does not control for them. In drug research, it is therefore common to give the control group a **placebo** (plah-SEE-bo)—an inert substance used to distinguish between

pharmacological and psychosomatic effects. To illustrate, a treatment group may be given daily tablets of a drug, and a control group may be given identical tablets of a cornstarch placebo, with no subject aware of which substance he or she received. If any physiological difference appeared between the two groups, we could then be confident that it was not due to knowledge of what they were taking.
- **Experimenter bias.** In the competitive, high-stakes world of biomedical research, experimenters may want certain results so much that their interpretation of the data may be tainted. They may be more inclined to reject data that disagree with the outcome they hope for than data that support their expectations. A researcher may even be unaware that he or she is interpreting data in a biased manner. The **double-blind method** is one way of controlling for bias. With this method, the subject does not know whether he or she receives an experimental drug or a placebo and neither does the person giving the treatment and recording the results. A researcher, for example, might prepare identical-looking tablets, some containing a drug and some only placebo, and then label these with code numbers and distribute them to participating physicians. The physicians themselves would not know whether they were administering drug or placebo but would give the tablets to volunteer patients and record the results. This way, neither the patients nor the physicians would be influenced by knowing the composition of the tablets. When the data were collected, the researcher would correlate them with the composition of the tablets and determine whether the drug had any more effect than the placebo.
- **Statistical testing.** If you tossed a coin 100 times, you would expect it to come up 50 heads and 50 tails. If it actually came up 48:52, you would probably attribute this to random error rather than bias in the coin. But what if it came up 40:60? At what point would you begin to suspect a biased coin? This type of problem is faced routinely in research—how great a difference must there be between a control and experimental group before we feel confident that the treatment really had an effect? What if a treatment group exhibited a 12% reduction in cholesterol level and the placebo group a 10% reduction? Would this be enough to conclude that the treatment was effective? Scientists are well grounded in statistical tests that can be applied to the data (perhaps you know of the chi square test, the *t* test, or analysis of variance, for example). A typical outcome of a statistical test might be expressed, "We can be 99.5% sure that the difference between group A and group B was due to the experimental treatment and not to random variation."

Peer Review

When a scientist applies for funds to support a research project or submits results for publication, the application or manuscript is submitted to **peer review**—a critical evaluation by other scientists knowledgeable in that field. Even after a report is published, if the results are important or unconventional, other scientists may attempt to reproduce the experiment in their own laboratories to see if the original author was correct. At every stage from planning to postpublication, scientists are therefore subject to the intense scrutiny of their colleagues. This is one mechanism for ensuring honesty, objectivity, and quality in science.

Facts, Laws, and Theories

The most important product of scientific research is an understanding of how nature works—whether it be the nature of a pond to an ecologist or the nature of a liver cell to a physiologist. Science conducted primarily for the sake of understanding is often called **basic research.** Most medical research is **applied science,** conducted because of its potential application to human welfare, such as new treatments for disease.

Numerous facts, laws, and theories of human form and function are discussed throughout this book. In scientific usage, the terms *fact, law,* and *theory,* have different meanings, and it is significant to appreciate the differences. A **scientific fact** is information that can be independently observed and verified by any trained person. A **law of nature** is a generalization about the predictable ways in which matter and energy behave. It is the result of inductive reasoning based on repeated, confirmed observations. Some laws are expressed as concise verbal statements, such as the first law of thermodynamics: Energy can be converted from one form to another but cannot be created or destroyed. Others are expressed as mathematical formulae. In your later study of the respiratory system, for example, you will encounter the law of Laplace: $F = 2T/r,$ where F is the force that tends to cause one of the microscopic air sacs of the lung to collapse, T is the surface tension of the fluid lining the sac, and r is the sac's radius.

A **theory** is an explanatory statement, or set of statements, derived from facts, laws, and confirmed hypotheses. Some theories have names; for example, the cell theory, the fluid-mosaic theory of cell membranes, and the sliding filament theory of muscle contraction. Most, however, remain unnamed, and numerous theories are stated in this book without making an issue of the fact that they are theories. The purpose of a theory is not only to concisely summarize what we already know but also to suggest directions for further study and to help predict what the findings should be if the theory is correct.

Law and theory mean something different in science than they do in lay usage. In common usage, *law* refers to a rule created and enforced by people—one must obey it or risk a penalty. A law of nature, however, is merely a description of the way matter or energy behaves. *Theory,* in lay usage, often means the same thing a scientist means by a hypothesis—for example, "I have a theory why my car won't start." This causes significant confusion when it leads some people to think that a scientific theory (such as the theory of evolution) is merely a guess or conjecture, instead of recognizing it as a summary of conclusions drawn from a large body of observed facts. The concepts of gravity and electrons are theories, too, but this does not mean that they are merely speculations.

Key Point Review

10 Describe the general process involved in the inductive method.

11 Discuss some sources of potential bias in biomedical research. What are some of the ways in which it can be minimized?

12 Is there more information in a scientific fact or in a theory? Explain.

Origins of Biomedical Science

▼Objectives
When you have completed this section, you should be able to
- give examples of how modern biomedical science emerged from an era of superstition and authoritarianism; and
- describe the contributions of some key people who helped to bring about this transformation.

There has been more progress in health science in the last 25 years than in the 2,500 years before that, but the field did not spring up overnight. It rests on a foundation built from centuries of thought and controversy. We can more deeply appreciate the health professions of today if we at least know a little bit about the dramatic history shaped by people who had the curiosity to try new things, the vision to look at human form and function in new ways, and the courage to question prevailing opinions.

The Prescientific Age

There is written evidence that more than 3,000 years ago physicians were treating diseases with herbal drugs, salts, physical therapy, and faith healing. Ancient people thought disease was caused by forces beyond human understanding, such as displeasure of the gods or evil spirits (the demonic theory of disease). They knew of no natural way to account for times when the spirit seemed to leave the body—in sleep and dreams, illness, and death, for example. In some cultures, the tribal medicine man, or *shaman* ("he who knows"), was believed to cure illness by pursuing and capturing the soul of the afflicted person and returning it to the body. Failing that, he would supposedly accompany the soul to the next world; then he would emerge from his trance and tell a vivid story of the journey.

We cannot smugly dismiss supernatural concepts of disease as the naive beliefs of primitive people. Even in modern scientific cultures, surprising numbers of people adhere to these beliefs. For example, some believe that a prescription drug is supposed to be left on a shelf in the house so it can "work its spirit" on the people who live there or that when someone has died of diabetes, the disease will "come out of" that person's body and try to enter the body of someone attending the funeral. Such beliefs can frustrate the well-meaning efforts of health-care providers. Often parents refuse medical treatment for their children on the basis of beliefs such as these.

Greco-Roman Contributions

The Greek physician **Hippocrates** (c. 460–c. 377 B.C.E.) was perhaps the first to advocate the ideas that natural events have natural causes (**naturalism**) and that there is an underlying order to the world that we can discover by rational inquiry (**rationalism**). He discredited the divine and demonic notions of disease that prevailed in his time and displayed an early awareness of the need to isolate variables from each other in any meaningful experimental or therapeutic design. His teaching is reflected in a manuscript on epilepsy, entitled "On the Sacred Disease":

> It seems to me that the disease called sacred . . . has a natural cause, just as other diseases have. Men think it divine merely because they do not understand it. But if they called everything divine that they did not understand, there would be no end of divine things! . . . If you watch those fellows treating the disease, you see them use all kinds of incantations and magic—but they are also very careful in regulating diet. Now if food makes the disease better or worse, how can they say it is the gods who do this? . . . It does not really matter whether you call such things divine or not. In Nature, all things are alike in this, in that they can be traced to preceding causes.

In Hippocrates' view, an essential aspect of the physician's job was to record the symptoms and course of a disease. Then, if the same symptoms were encountered in another patient, the physician could diagnose the disease and give a prognosis (predict its outcome). Hippocrates based his treatment plan on the assumptions that nature has a healing power of its own and that the physician's main role was to assist this process through nutrition, exercise, and rest rather than try to direct it through aggressive medication or surgery. The oldest code of medical ethics, the Hippocratic Oath, memorialized his creed. Variations of the oath are still recited by many graduating medical students today.

Unlike Hippocrates, whose interest was in applied science, **Aristotle** (384–322 B.C.E.) pursued scientific knowledge for its own sake. He distinguished supernatural causes, which he called the *theologi,* from natural ones, which he called the *physici* or *physiologi.* This is the origin of the words *physician* and *physiology.*[17] Until the nineteenth century, physicians were called "doctors of physic"; the word *medicine* is relatively recent.

In his anatomy text, *Of the Parts of Animals,* Aristotle tried to identify unifying themes in nature. He was an early proponent of reductionism; he argued that complex structures are built from a smaller variety of simpler components and are best understood in terms of those components. Some of his observations were fairly accurate. He recognized, for example, that blood "percolates" through the kidneys, which separate wastes from it and send these by way of two "sturdy channels" (ureters) to the bladder. No one contributed any further significant knowledge on the kidney for more than 2,000 years. On the other hand, Aristotle wrongly identified the heart as the seat of human emotions and viewed the brain as merely a radiator to cool the blood.

Some of the early Greeks had the opportunity to dissect cadavers, and some carried this practice to the horrendous extreme of dissecting living people for study or public demonstrations. This was eventually outlawed—but so was cadaver dissection and the opportunity to learn anatomy from the human body itself. **Claudius Galen** (129–c. 199) longed for the days when it had been allowed. As physician to the Roman gladiators, he was able to learn some human anatomy while attending to their combat wounds. In greater part, however, he had to be satisfied with dissecting pigs and monkeys and guessing how much of their anatomy was similar to ours. As a result, medical textbooks of the next 1,500 years contained grossly inaccurate descriptions and illustrations of human anatomy.

Galen was a careful observer and was the first to describe the circulation of blood through the chambers of the heart. He believed, however, that blood flowed back and forth in the veins, rather than traveling in a complete circuit around the body, and that the brain made muscles contract by pumping "psychic pneuma" (a vapor) into them by way of hollow, tubelike nerves.

More important than his contribution to anatomy, however, was Galen's view of science. For him, it was a way of knowing—not a body of fact to be accepted on faith. He was aware that many physicians of his time

---

17. *physio* = nature, natural cause

were wrong about anatomy, and he realized that he himself might be wrong. He urged readers not to put too much faith in books, including his own, but to believe above all in the evidence of their own senses.

The Middle Ages

The Middle Ages (from about 450 to about 1500) were dominated by theology. Saint Augustine, for example, believed that all natural phenomena had supernatural causes, and he discouraged the effort to understand them scientifically. The demonic theory replaced naturalism, and magic replaced therapy. It was thought that everything worth knowing could be found in the writings of "ancient masters" such as Aristotle and Galen; therefore, there was no need for further research.

A number of medical schools were established in Europe during this period, but the professors taught medicine primarily as a commentary on Galen and Aristotle, not as a field of original research. When anatomy was found to disagree with Galen's descriptions, it was assumed that either the new observations were incorrect or that the body had changed since Galen's time. Authorities refused to consider the possibility that Galen could have been wrong. Whereas Galen had encouraged skepticism, his writings were twisted into a dogma that dominated medicine until the Renaissance.

Free inquiry was less inhibited in the Muslim world, where Ibn Sina (980–1037), known in the West as **Avicenna,** studied Galen and Aristotle, combined their findings with original discoveries, and questioned authority when necessary. Muslim medicine soon became superior to Western medicine, and even in Western society Avicenna came to be revered as "the Galen of Islam." His textbook *The Canon of Medicine* was the leading authority in European medical schools until the sixteenth century.

The Renaissance

Unlike scholars of the Middle Ages, the Swiss physician **Paracelsus** (1493–1541) was not intimidated by the weight of ancient authorities or the prevailing opinions of his own time. He is reputed to have thrown the books of Galen and Avicenna into a student bonfire, denouncing them as obstacles to learning. While most medical professors taught in Latin to protect their secrets from the general public, Paracelsus lectured in his everyday language (German) to make knowledge more accessible to others. He disputed Hippocrates' theories of disease and argued correctly that many diseases arise from factors outside the body. While other physicians of his time relied on little more than herbal cures, Paracelsus used mercury, sulfur, arsenic, and lead as drugs for syphilis and other diseases.

Think About It

How might Hippocrates and Paracelsus differ in the way they would have treated syphilis?

By this time cadaver dissection had come back into use for training medical students, but it was not done by the professors themselves, who considered it beneath their dignity. In these days before refrigeration or embalming, the odor from the decaying cadaver was unbearable, and medical students had to fight the urge to vomit lest they incur the wrath of an overbearing professor. Dissections were conducted outdoors in a nonstop 4-day race against decay. The professor sat in an elevated chair—the cathedra—reading dryly from Galen. Meanwhile, a lower-ranking *barber-surgeon* removed the putrefying organs from the cadaver and held them up for the students to see. Barbering and surgery were considered to be "kindred arts of the knife"; today's barber poles date from this era—their red and white stripes symbolizing blood and bandages.

Andreas Vesalius (1514–64), a Flemish physician teaching in Italy, broke with tradition by coming down from the cathedra and doing the dissections himself. He urged that every medical student learn anatomy by dissecting a cadaver, not merely watching someone else do it. Opportunities to do so were rare, however. A typical medical college had access to only one cadaver every year or two, and most of these were either mutilated bodies of executed prisoners or bodies stolen from fresh graves by "resurrection men." American medical schools were supplied by professional grave robbers as recently as the nineteenth century.

Vesalius was quick to point out that much of the anatomy in Galen's books was wrong, and he was among the first to develop accurate illustrations for teaching anatomy. Before Vesalius, a typical medical illustration was an astrological chart that showed which sign of the zodiac was thought to influence each organ of the body. From this pseudoscience originated the word *influenza,* Italian for *influence.* When Vesalius's illustrations began to be plagiarized, he published the first comprehensive atlas of anatomy, *De Humani Corporis Fabrica (On the Structure of the Human Body),* in 1543 (fig. 1.11). A rich tradition of medical illustration, started by Vesalius, has been handed down to you through *Gray's Anatomy* (1856) to the richly illustrated books of today.

While Vesalius was a pioneer of modern anatomy, his counterpart in physiology was the Englishman **William Harvey** (fig. 1.12). Harvey is particularly remembered for a small book he published in 1628, *On the Movement of the Heart and Blood in Animals.* Authorities before Harvey believed that digested food was turned into blood by the liver and then distributed to the rest of the body by veins. Harvey measured the rate

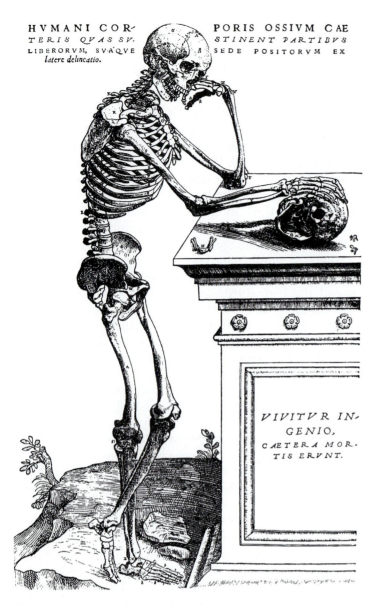

Figure 1.11 An illustration from Andreas Vesalius's *De Humani Corporis Fabrica* (1543).

Figure 1.12 William Harvey (1578–1657) *(fourth from left)* was the first modern experimental physiologist. In 1628, he demonstrated that blood flows in a complete circuit from the heart through the blood vessels and back to the heart, rather than flowing back and forth in the veins.

of cardiac output, however, and reasoned that the amount of food eaten could not possibly account for such a large volume of blood. Thus, he concluded that the blood must be recycled—pumped out of the heart by way of arteries, which were somehow connected to the veins, and then returned to the heart. This route was hypothetical because capillaries, the connections between arteries and veins, could not be seen with the naked eye, and the microscope was yet to be invented.

Discovery of a Small World

Although crude microscopes existed long before his time, it was the Dutch textile merchant **Antony van Leeuwenhoek** (an-TOE-nee vahn LAY-wen-hook) who

invented the first microscopes capable of visualizing single cells. In order to examine the weave of a fabric more closely, he designed a metal plate with a hole in it to pass light, a movable pin for mounting specimens over the hole, and a beadlike lens. This **simple** (single-lens) **microscope** (fig. 1.13) magnified objects 200 to 300 times. Out of curiosity, Leeuwenhoek examined a drop of lake water and was astonished to find a variety of microorganisms—"little animalcules," he called them, "very prettily a-swimming." He went on to observe practically anything he could get his hands on—red blood cells, blood capillaries, sperm cells, and muscular tissue. Few people in history have looked at nature in such a revolutionary way. Leeuwenhoek opened the door to an entirely new understanding of human structure and the causes of disease. He was praised at first, and reports of his observations were eagerly received by scientific societies, but public enthusiasm did not last. By the end of the seventeenth century, the microscope had declined to a mere toy for the upper classes, as amusing and meaningless as a kaleidoscope. Leeuwenhoek had become a target of satire.

Think About It

Would you consider Leeuwenhoek more like Hippocrates or Aristotle in the motives underlying his biological work? Why?

(a)

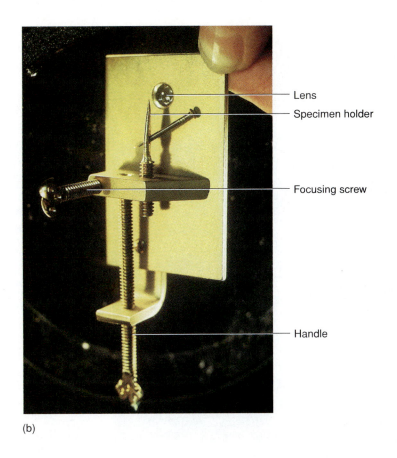

Lens

Specimen holder

Focusing screw

Handle

(b)

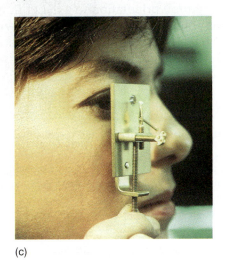

(c)

Figure 1.13 (*a*) Antony van Leeuwenhoek (1632–1723). (*b*) A brass replica of a Leeuwenhoek microscope made by Dr. William Walter of Montana State University. (*c*) Demonstration of the use of a Leeuwenhoek microscope.

Leeuwenhoek's most faithful admirer was the Englishman **Robert Hooke** (1635–1703), who developed the first practical **compound microscope**—a tube with a lens at each end, in which the second lens further magnified the image produced by the first one (fig. 1.14). Galileo had already made several compound microscopes by 1612, but Hooke invented many of the features common to the microscopes used by students today: a stage to hold the specimen, an illuminator, and coarse and fine focus controls. His microscopes produced poor images with blurry edges (*sphyerical aberration*) and rainbow-colored distortions (*chromatic aberration*), but poor images were better than none.

Although Leeuwenhoek was the first to see cells, Hooke named them. In 1663, when Hooke observed a thin shaving of cork under his compound microscope, he noted that it consisted of a great many little boxes, which he named cells after the barren cubicles of a monastery (fig. 1.14). He published these observations in his book *Micrographia* in 1665.

In 1830, the compound microscope was considerably improved to reduce distortions of the image, and biologists began eagerly examining a wider variety of specimens. Botanist **Matthias Schleiden** (1804–81) concluded by 1838 that all plant tissues were cellular. Animal cells were more difficult to study because they did not have thick, supportive cell walls, but methods were developed to better preserve animal tissues, cut thinner slices, and stain them to bring out detail. This enabled **Theodor Schwann** (1810–82) to study a wide range of animal tissues, and he concluded in 1839 that all animals were likewise made of cells. The inference that all organisms are made of cells was the founding tenet of the **cell theory,** to which other tenets were later added (see chapter 4). The cell theory was perhaps the most important breakthrough in the history of biology and medicine because all functions of the body were eventually interpreted as the effects of cellular function.

Although much of the philosophical foundation of modern practice was established in the time of Leeuwenhoek, Hooke, and Harvey, clinical medicine was still in a dismal state. Only a few seventeenth-century doctors had any formal education in basic science or human anatomy. Physicians tended to be ignorant, ineffective, and pompous. They relied heavily on expelling imaginary toxins from the body by bleeding, vomiting, sweating,

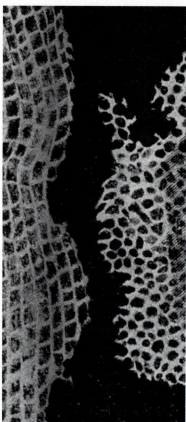

Figure 1.14 The compound microscope developed by Robert Hooke, along with his drawings of thin shavings of cork, showing the cell walls characteristic of plant tissues.

chemotherapy, immunization, anesthesia, surgery, tissue and organ transplants, and human genetics. We are on the threshold of having a library of the molecular structure of every human gene, and gene substitution therapy is being used to treat children born with genetic diseases recently considered incurable. As medical historians of the next century look back on the end of this one, they may exult about the Genetic Revolution in which we are now living.

Several discoveries of the nineteenth and twentieth centuries, and the men and women behind them, are covered in short historical sketches in later chapters. Yet, the stories told in this chapter are different in a significant way. The people discussed here were pioneers in establishing the scientific way of thinking. They helped to transform a supernatural orientation to a belief in natural law. Without this intellectual revolution, those who followed could not have conceived of the right questions to ask, much less a method for answering them.

Key Point Review

13 In what important ways did Paracelsus and Vesalius break from tradition?

14 How does medical science today depend strongly on what Leeuwenhoek and Hooke did in the seventeenth century?

15 Define *naturalism* and explain why biomedical science today could not function without this point of view.

Homeostasis and Feedback

▼Objectives
When you have completed this section, you should be able to
- define *homeostasis* and explain why this concept is central to physiology;
- define *negative feedback,* give an example of this process, and explain its importance to homeostasis; and
- define *positive feedback* and give examples of the beneficial and harmful effects of this process.

The human body has a remarkable capacity for self-restoration. Hippocrates commented that it usually returns to a state of equilibrium by itself, and people recover from most illnesses even without the help of a physician. This tendency results from the body's ability to detect change and activate mechanisms that oppose it.

Homeostasis

French physiologist **Claude Bernard** (1813–78) observed that the internal conditions of the body remain fairly stable even when external conditions are highly variable. For example, whether it is freezing cold or

or purging their patients with useless herbal drugs. Operations were performed with dirty hands and instruments, and there was no anesthesia to lessen the pain. Fractured limbs often became gangrenous and had to be amputated. Disease was still widely attributed to the supernatural powers of demons and witches, and many considered it to be against God's will to interfere by trying to cure it.

Living in a Revolution

This short history brings us only to the threshold of modern biomedical science; it stops short of such discoveries as the germ theory of disease, the mechanisms of heredity, and the structure of DNA. In the twentieth century, basic biology and biochemistry have given us a much deeper understanding of how the body works, and technological advances such as noninvasive medical imaging (see chapter essay, p. 24) have enhanced our diagnostic ability and life-support strategies. We have witnessed monumental developments in swelteringly hot outdoors, the internal temperature of your body stays within a range of about 36° to 37°C (97° to 99°F). The American physiologist **Walter Cannon** (1871–45) coined the term **homeostasis**[18] (ho-me-oh-STAY-sis) for this tendency of the body to maintain relatively stable internal conditions. Homeostasis has been one of the most enlightening concepts in physiology. Physiology is largely a group of mechanisms for maintaining homeostasis, and the loss of homeostatic control tends to cause illness or death. Pathophysiology is essentially the study of unstable conditions that result when our homeostatic control mechanisms go awry.

We should not overestimate the degree of internal stability, however. Internal conditions are not absolutely constant; they may fluctuate within a limited range, such as the range of body temperatures noted. The internal state of the body is best described as one of **dynamic equilibrium** (balanced change), in which there is a certain **set point** or average for a given variable (such as 37°C for body temperature), and conditions fluctuate slightly around this point.

Negative Feedback and Stability

The fundamental mechanism that keeps a variable close to its set point is **negative feedback**—a process in which the body senses a change and activates mechanisms that negate or reverse that change. By maintaining stability, negative feedback is a means of maintaining health.

These principles can be understood by analogy to a home heating system (fig. 1.15). Suppose it is a cold winter day and the thermostat has been set to 20°C (68°F)—the set point. If the room becomes too cold, a temperature-sensitive switch in the thermostat turns on the furnace. The temperature rises until it is slightly above the set point, and then the switch breaks the circuit and turns the furnace off. This is a negative feedback process that reverses the falling temperature and restores it to something close to the set point. When the furnace turns off, the temperature slowly drops again until the switch is reactivated—thus, the furnace cycles on and off all day. The room temperature does not stay at exactly 20°C; rather, it *fluctuates* a few degrees either way—the system maintains a state of dynamic equilibrium in which the temperature averages 20°C and deviates from this set point by only a few degrees. Because feedback mechanisms alter the original changes (temperature, for example) that triggered them, they are often called **feedback loops.**

Body temperature is also regulated by a "thermostat"—a group of nerve cells in the base of the brain that monitors

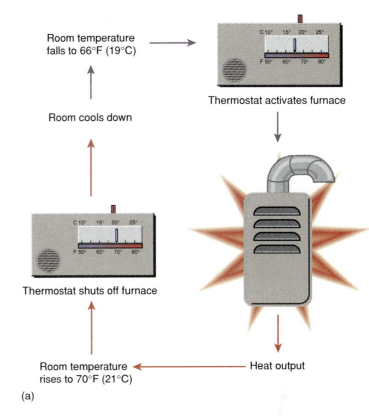

Thermostat activates furnace

Room temperature falls to 66°F (19°C)

Room cools down

Thermostat shuts off furnace

Room temperature rises to 70°F (21°C)

Heat output

(a)

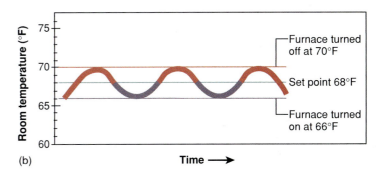

(b)

Furnace turned off at 70°F

Set point 68°F

Furnace turned on at 66°F

Time →

Figure 1.15 (*a*) Home heating systems are based on a negative feedback loop between a thermostat and furnace. (*b*) Negative feedback keeps the room temperature within a few degrees of a set point.

the temperature of the blood. If you become overheated, for example, the thermostat triggers heat-losing mechanisms (fig. 1.16). Blood vessels of the skin widen (**vasodilation** [VAY-zo-dy-LAY-shun]), which allows warm blood to flow closer to the body surface and lose more heat. If this is not enough to return the temperature to normal, sweating occurs, and evaporation of water from the skin has a powerful cooling effect (see special topic 1.2). Conversely, if it is cold outside and your body temperature drops much below the 37°C set point, these nerve cells activate heat-conserving mechanisms. Blood vessels of the skin become narrower (**vasoconstriction** [VAY-zo-con-STRIC-shun]), which

18. *homeo* = the same + *stas* = to place, stand, stay

keeps the warm blood deeper in the body and reduces heat loss through the skin. If this is not enough, the brain activates shivering—muscle tremors that generate heat. Thus, we can see that homeostasis is maintained by self-correcting negative feedback loops. Many more examples will be found throughout this book.

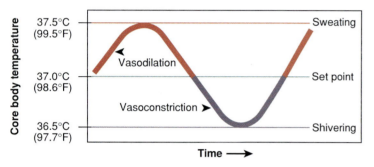

Figure 1.16 Negative feedback mechanisms keep human body temperature within about 0.5°C of a set point of 37.0°C. When body temperature rises, vasodilation in the skin helps rid the body of excess heat. If this is insufficient, sweating helps lower the body temperature. When body temperature falls, vasoconstriction keeps warm blood deeper in the body, and if the body temperature continues to fall, shivering helps raise it.

Positive Feedback and Rapid Change

Positive feedback is a self-amplifying cycle in which a physiological change leads to even greater changes in the same direction, rather than producing the corrective effects of negative feedback. Positive feedback is often a normal way of producing rapid change. When a woman is giving birth, for example, the head of the baby pushes against her cervix and stimulates nerve endings there (fig. 1.17). Nerve signals are sent to the brain, which, in turn, stimulates the pituitary gland to secrete the hormone oxytocin. Oxytocin travels in the blood and stimulates the uterus to contract. This pushes the baby downward, stimulating the cervix still more and causing the positive feedback loop to be repeated. Labor contractions therefore become more and more intense until the baby is expelled.

It should be noted, however, that the overall process of childbirth is a negative feedback loop—it is a response to pregnancy that terminates the pregnancy. But within this negative feedback loop, there is the smaller positive feedback loop we have just described. Beneficial positive feedback loops are often part of larger negative feedback loops. We will encounter other cases of beneficial positive feedback later in the

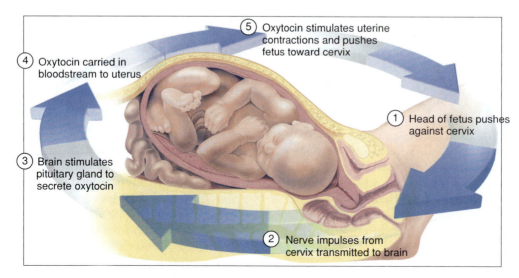

Figure 1.17 Positive feedback is beneficial in childbirth and several other physiological situations in which rapid change is important.

book; for example, in connection with the generation of nerve signals, blood clotting, and the stomach's digestion of protein.

Frequently, however, positive feedback is a harmful and even life-threatening process. This is because its self-amplifying nature can quickly change the internal state of the body to something far from its homeostatic set point. Consider a high fever, for example. A fever triggered by infection is beneficial up to a point, but if the body temperature rises much above 42°C (108°F), it may create a dangerous positive feedback loop (fig. 1.18). This high temperature raises the metabolic rate, which makes the body produce heat faster than it can get rid of it. Thus, temperature rises still further, increasing the metabolic rate and heat production still more. This "vicious circle" becomes fatal at approximately 45°C (113°F). Such temperatures are so high that they destroy the proteins that cells need to function. Convulsions and coma are some outward signs of this damage. Thus, positive feedback loops often create dangerously out-of-control situations that require emergency medical treatment.

.... **Key Point Review** ....

16 What is meant by *dynamic equilibrium?* Why would it be wrong to say homeostasis prevents internal change?

17 Explain why stabilizing mechanisms are called *negative* feedback.

18 Explain why positive feedback is more likely than negative feedback to disturb homeostasis.

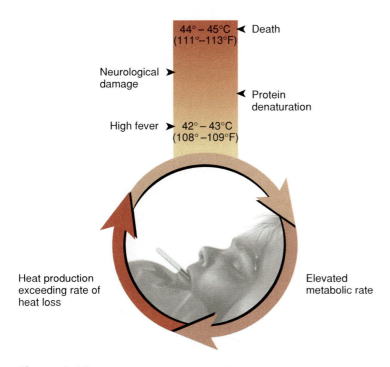

44°–45°C (111°–113°F) ◄ Death

Neurological damage ►

◄ Protein denaturation

High fever ► 42°–43°C (108°–109°F)

Heat production exceeding rate of heat loss

Elevated metabolic rate

Figure 1.18 Positive feedback can be life-threatening in such cases as a high fever.

Review of Major Themes

To close this chapter, let's distill a few major points from it. The following themes can provide you with a sense of perspective that will make the rest of the book more meaningful—not just a collection of disconnected facts. Some later discussions will explicitly remind you of where these themes are relevant, but at many other points you will probably recognize them without being told. In either event, these are important unifying principles behind all study of human anatomy and physiology.

Unity of form and function. *Form and function complement each other; physiology cannot be divorced from anatomy.* This point was made on page 2, with the kidney as an example, but the unity holds true even down to the molecular level. Our very molecules, such as DNA and proteins, are structured in ways that enable them to carry out their functions. Slight changes in molecular structure can destroy their activity and threaten life.

Hierarchy of structure. *Human structure can be viewed as a series of levels of complexity.* Each level is composed of a smaller number of simpler subunits than the level above it. These subunits are arranged in different ways to form diverse structures of higher complexity. For example, all the body's organs are composed of just four primary classes of tissue, and the thousands of proteins are made of various combinations of just 20 amino acids. Understanding these simpler components is the key to understanding higher levels of structure.

Cell theory. *All structure and function result from the activity of cells.* Every physiological concept in this book ultimately must be understood from the standpoint of how cells function. Even anatomy is a result of cellular function. Physiology is a complex web of mechanisms for providing cells with the conditions necessary for their survival and function. Any serious departure from these conditions can be harmful or fatal to cells—and if cells are damaged or destroyed, we see the results in disease symptoms of the whole person.

Evolution. *The human body is a product of evolution.* Like every other living organism, we have been molded by millions of years of natural selection to function in a changing environment. Many aspects of human anatomy and physiology reflect our ancestors' adaptations to their environment. Unless they are considered in the light of evolutionary history, these aspects cannot be completely understood.

Homeostasis. *The purpose of most normal physiology is to maintain stable conditions within the body.* Cells depend on a favorable internal environment to carry out their functions. Human physiology is essentially a group of mechanisms that, having survived the test of natural selection, produce favorable, stable conditions within the body.

Noninvasive Medical Imaging

The development of techniques for looking into the body for signs of disease without having to do exploratory surgery has greatly accelerated progress in twentieth-century medicine. A few of these techniques are described and illustrated here.

Radiography

X rays, a form of high-energy radiation, were discovered by William Roentgen in 1885. Because of their penetration ability, X rays can pass through soft tissues of the body and darken photographic film on the other side. They are absorbed, however, by dense tissues such as bone, teeth, tumors, and tuberculosis nodules, which leave the film lighter in these areas (fig. E.1*a*). The process of examining the body with X rays is called **radiography.** The term *X ray* also applies to the photograph (**radiograph**) made by this technique. Radiography is commonly used in dentistry, mammography, diagnosis of fractures, and studying structures of the chest. Hollow organs can be visualized by filling them with a *radiopaque substance* that absorbs X rays. Barium sulfate is given orally for examination of the esophagus, stomach, and small intestine or by enema for examination of the large intestine. Other substances are given by injection for *angiography,* the examination of blood vessels (fig. E.1*b*). Some disadvantages of radiography are that images of overlapping organs can be confusing, and slight differences in tissue density are not easily detected. Nevertheless, radiography still accounts for over 50% of all clinical imaging. Until about 30 years ago, it was the only method available.

Sonography

The second oldest and second most widely used method of imaging is **sonography,**[19] an outgrowth of sonar technology developed in World War II. A handheld device held firmly to the skin produces high-frequency ultrasound waves and receives the signals that echo back from internal organs. Although sonography was first used medically in the 1950s, images of significant clinical value had to wait until computer technology had developed enough to analyze differences in the way tissues reflect ultrasound. Sonography is not very useful for examining bones or lungs, but it is the method of choice in obstetrics, where the ultrasound image (**sonogram**) can be used to locate

19. *sono* = sound + *graphy* = recording process

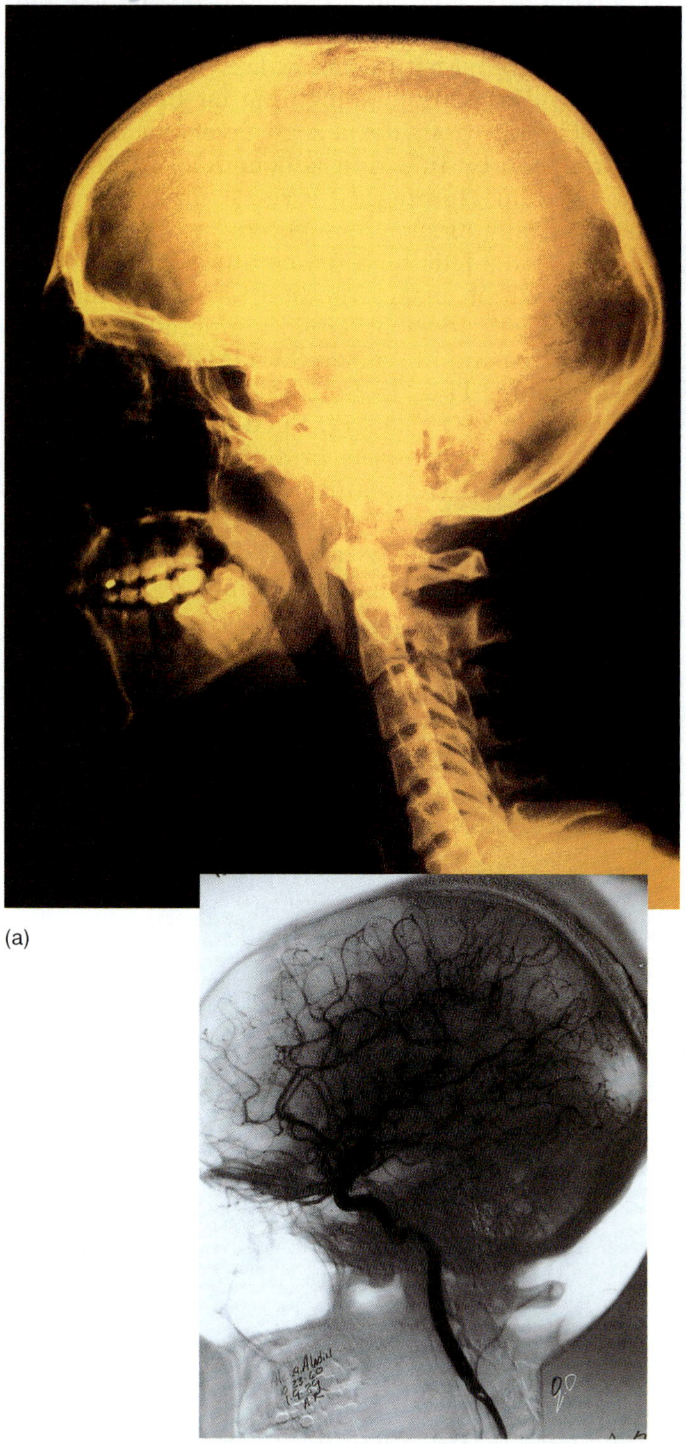

(a)

(b)

Figure E.1 Radiography. (*a*) An X ray (radiograph) of the head and neck. (*b*) A cerebral angiogram, made by injecting a substance opaque to X rays into the circulation and then taking an X ray of the head to visualize the blood vessels.

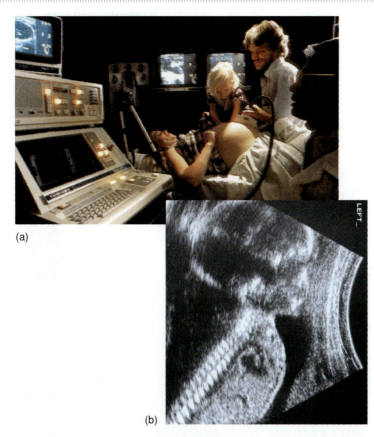

(a)

(b)

Figure E.2 Sonography. (*a*) A pregnant woman undergoing sonography. (*b*) A sonogram of the fetus.

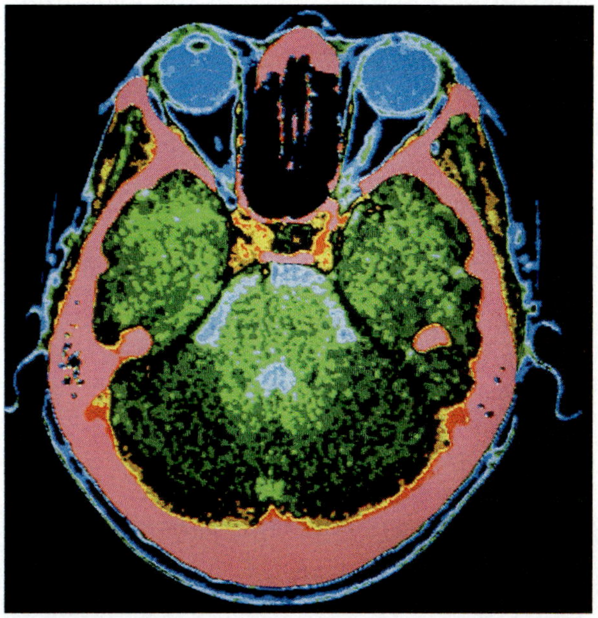

Figure E.3 Computed tomographic (CT) scan of the head at the level of the eyes. ✗

the placenta and evaluate fetal age, position, and development. Sonography avoids the harmful effects of X rays, and the equipment is inexpensive and portable. The primary disadvantage is that it does not produce a very sharp image (fig. E.2).

Computed Tomography (CT)

The **CT scan,** formerly called a *computerized axial tomographic[20] (CAT) scan,* is a more sophisticated application of X rays developed in 1972. As the patient is moved through a ring-shaped machine, low-intensity X rays are emitted on one side, and a detector directly across from this receives the X rays that emerge from the body. Signals from the detector are analyzed by a computer, which produces an image of a "slice" of the body about as thin as a coin (fig. E.3). The CT machine can then be moved a short distance to X-ray another slice,

and these slices can be "stacked" by computer, if desired, to reconstruct a three-dimensional image of the body. CT scanning has the advantage of imaging thin sections of the body, so there is little overlap of organs and the image is much sharper than a conventional X ray. CT scanning is useful for identifying tumors, aneurysms, kidney stones, cerebral hemorrhages, and other abnormalities. It has virtually eliminated exploratory surgery.

Magnetic Resonance Imaging (MRI)

MRI, until recently known as *nuclear magnetic resonance (NMR) imaging,* was developed in the 1970s as a technique superior to CT scanning for visualizing soft tissues. The patient lies within a cylindrical chamber surrounded by a large electromagnet that creates a magnetic field 3,000 to 60,000 times as strong as the earth's. This magnetic field causes hydrogen atoms in the tissues to align. The patient is then bombarded with radio waves, which cause the hydrogen atoms to absorb additional energy and align in a different direction. When the radio waves are turned off, the hydrogen atoms abruptly realign themselves to the magnetic field and give off their excess energy at different rates, depending

20. *tomo* = section, cut, slice

—continued

on the type of tissue. The emitted energy is analyzed by computer to produce an image of the body. MRI can "see" clearly through bone of the skull and spinal column to produce images of the nervous tissue; moreover, it is better than CT for distinguishing between soft tissues, such as the white and gray matter of the nervous system (fig. E.4).

Positron Emission Tomography (PET)

The **PET scan,** developed in the 1970s, is used to assess the metabolic state of a tissue at a given moment in time (fig. E.5) and to determine which tissues are most active. The procedure begins with an injection of radioactively labeled glucose. The labeled glucose emits particles called positrons (like electrons, but with a positive charge). When a positron and electron meet, they annihilate each other and give off a pair of gamma rays that can be detected by sensors and analyzed by computer. The result is a color image that shows which tissues were using the most glucose at the moment. In cardiology, the image can reveal the extent of damaged heart tissue. Since it consumes little or no glucose, the damaged tissue will appear dark. The PET scan is an example of *nuclear medicine*—the use of radioactive isotopes to form anatomical images of the body (see chapter 2 essay, p. 72). Some of the applications of the PET scan to neuroscience are described in chapter 14.▲

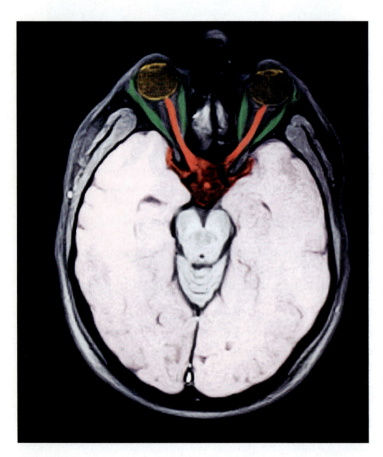

Figure E.4 Magnetic resonance imaging (MRI) scan of the head at the level of the eyes. The optic nerves appear in red and the muscles that move the eyes appear in green. ✗

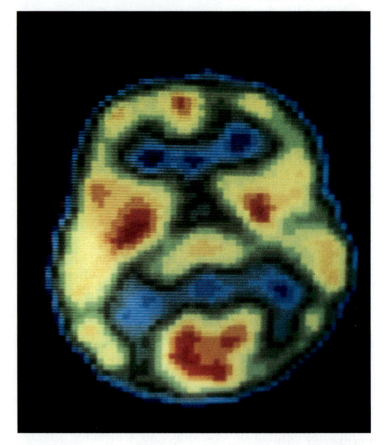

Figure E.5 Positron emission tomographic (PET) scan of the brain of an unmedicated schizophrenic patient. Red areas indicate high glucose use (high metabolism). In this patient, the visual center at the posterior region of the brain was especially active when the scan was made.

The Scope of Anatomy and Physiology (pp. 2–3)
1. Anatomy and its subdisciplines
2. Physiology and its subdisciplines
3. Unity of form and function

The Nature of Human Life (pp. 3–10)
1. What is life?
   a. Biological criteria
      - Cellular organization
      - Biochemical unity
      - Metabolism
      - Responsiveness
      - Development
      - Reproduction
   b. Clinical and legal criteria
2. What is a human?
   a. Animal characteristics
   b. Chordate characteristics
   c. Vertebrate characteristics
   d. Mammalian characteristics
   e. Primate characteristics
   f. Uniquely human characteristics
3. Hierarchy of structural complexity
4. Reductionist and holistic perspectives

Human Evolution (pp. 10–13)
1. Darwin and natural selection
   a. Impact of Darwinism
   b. Natural selection
   c. Meaning of evolution
   d. Selection pressures and adaptations
   e. Biomedical relevance of evolution
2. Life in the trees
   a. Selection pressures of arboreal life
   b. Adaptations to arboreal life

3. Impact of bipedalism
   a. Habitat changes and bipedalism
   b. Advantages of bipedalism
   c. Adaptations to bipedalism

Scientific Method (pp. 13–15)
1. Inductive method
2. Scientific proof and truth
3. Hypothetico-deductive method
   a. Standards for a valid hypothesis
   b. Deductions and predictions
4. Experimental design
   a. Sample size
   b. Controls
   c. Psychosomatic effects and placebos
   d. Experimenter bias
   e. Statistical testing
5. Peer review
6. Facts, laws, and theories
   a. Basic and applied science
   b. Scientific facts
   c. Laws of nature
   d. Theories

Origins of Biomedical Science (pp. 15–20)
1. Prescientific age
2. Greco-Roman contributions
   a. Hippocrates
      - Naturalism and rationalism
      - Role of the physician
   b. Aristotle
      - Natural versus supernatural causes
      - Hierarchy of structure
      - Observations on physiology
   c. Galen
      - Studies of anatomy
      - Views on circulation
      - Encouragement of skepticism

3. Middle Ages
   a. Suppression of science
   b. Establishment of medical schools
   c. Avicenna and Muslim medicine
4. Renaissance
   a. Paracelsus and skepticism
   b. Revival of cadaver dissection
   c. Vesalius and anatomy
   d. Harvey and physiology
5. Discovery of a small world
   a. Leeuwenhoek and the microscope
   b. Hooke and cytology
   c. Schleiden, Schwann, and the cell theory
6. Seventeenth-century medicine
7. Living in a revolution

Homeostasis and Feedback (pp. 20–23)
1. Bernard, Cannon, and homeostasis
2. Set point and dynamic equilibrium
3. Negative feedback and stability
4. Positive feedback and rapid change
   a. Beneficial effects
   b. Pathological effects

Review of Major Themes (p. 23)
1. Unity of form and function
2. Hierarchy of structure
3. Cell theory
4. Evolution
5. Homeostasis

Selected Vocabulary

anatomy 2
dissection 2
cadaver 2
gross anatomy 2
microscopic anatomy 2
ultrastructure 2
comparative anatomy 2
palpation 2
auscultation 2
percussion 2
physiology 2
comparative physiology 2
cellular organization 3
biochemical unity 3
metabolism 3

assimilation 3
responsiveness 3
stimulus 3
receptor 3
effector 3
development 3
growth 3
differentiation 3
reproduction 3
Animalia 4
eukaryote 4
multicellular 4
heterotrophic 4
nerve cells 4
muscle cells 4

phyla 5
Chordata 5
notochord 5
dorsal hollow nerve cord 5
pharyngeal gill pouches 5
postanal tail 5
Vertebrata 5
Mammalia 5
endothermy 5
heterodonty 5
Primates 5
Hominidae 5
bipedal 5
Australopithecus 7
Homo 7

organism 7
organ system 7
organ 9
tissue 9
histology 9
cell 9
cytology 9
organelle 9
molecule 9
atom 9
subatomic particle 9
reductionism 9
holism 10
natural selection 10
evolution 11

Testing Your Recall   Answers in Appendix C

1. Structure that can be observed with the naked eye is called
   a. gross anatomy.
   b. ultrastructure.
   c. microscopic anatomy.
   d. macroscopic anatomy.
   e. cytology.

2. Any organism with _____ is said to be eukaryotic.
   a. homeostatic controls
   b. nuclei in its cells
   c. true tissues
   d. heterotrophic nutrition
   e. true cells

3. The simplest structures considered to be alive are
   a. organisms.
   b. organs.
   c. tissues.
   d. cells.
   e. organelles.

4. Which of the following individuals revolutionized the teaching of gross anatomy?
   a. Vesalius
   b. Aristotle
   c. Galen
   d. Leeuwenhoek
   e. Cannon

5. Which of the following embodies the greatest amount of scientific information?
   a. a fact
   b. a law of nature
   c. a theory

   d. a deduction
   e. a hypothesis

6. An informed, uncertain, but testable conjecture is
   a. a natural law.
   b. a scientific theory.
   c. a hypothesis.
   d. a deduction.
   e. a scientific fact.

7. A self-amplifying chain of physiological events is called
   a. positive feedback.
   b. negative feedback.
   c. dynamic constancy.
   d. homeostasis.
   e. physiological instability.

8. Which of the following is *not* a human organ system?
   a. integumentary
   b. muscular
   c. epithelial
   d. nervous
   e. endocrine

9. Which of the following characteristics do humans *not* share with all other chordates?
   a. pharyngeal gill pouches
   b. heterotrophic nutrition
   c. multicellular organization
   d. eukaryotic cells
   e. a vertebral column

10. The word root *hetero* means
    a. same.
    b. different.
    c. both.

    d. solid.
    e. below.

11. Cutting and separating tissues to reveal structural relationships is called _____.

12. The inventer of the compound microscope and founder of cytology was _____.

13. By the process of _____, a scientist predicts what the result of a certain experiment will be if his or her hypothesis is correct.

14. Physiological effects of a person's mental state are called _____ effects.

15. The tendency of the body to maintain stable internal conditions is called _____.

16. Blood pH averages 7.4 but fluctuates between 7.35 and 7.45. A pH of 7.4 can therefore be considered the _____ for this system.

17. Self-correcting mechanisms in physiology are called _____ loops.

18. A/an _____ is the simplest body structure to be composed of two or more types of tissue.

19. Depth perception, or the ability to form three-dimensional images, is called _____ vision.

20. Our thumbs are said to be _____ because the tip of the thumb can touch the tips of the fingers, allowing us to grasp objects.

1. What aspect of William Harvey's view of blood circulation could be considered a scientific hypothesis? What would you predict from that hypothesis? What observation could you carry out today to test this hypothesis?

2. Which of the characteristics of living things are possessed by an automobile? What bearing does this have on the definition of life?

3. Suppose our primate ancestors had never gone through an arboreal phase but had originated in the grasslands and fed mainly on plant tubers. How might humans be different today?

4. Why is a horse not classified as a primate? (Hint: What characteristics must an animal have to be a primate, and which of these are horses lacking?) Why is a monkey not classified as human?

5. Suppose you are doing heavy yard work on a hot day and sweating profusely. You become very thirsty, so you drink a tall glass of lemonade. Explain how your thirst relates to the concept of homeostasis. Which type of feedback—positive or negative—does this illustrate?

Web Site Link

For a listing of the most current web sites related to this chapter, please visit the Saladin homepage at:

http://www.mhhe.com/sciencemath/biology/saladin/

atlas a

A

[General Orientation to Human Anatomy

the unity of form and functio

Anatomical Position

Anatomical position is a stance in which the subject is standing erect, with eyes directed forward, feet flat on the floor and slightly apart, arms down to the sides, and palms facing forward. This position, shown in figure A.1, provides a precise and standardized frame of reference for anatomical description and dissection. Without such a frame of reference, to say that a structure such as the sternum, thymus, or aorta is "above the heart" would be vague, since it would depend on whether the subject was standing, lying face down, or lying face up. From the perspective of anatomical position, however, we can describe the thymus as *superior* to the heart, the sternum as *anterior* or *ventral* to it, and the aorta as *posterior* or *dorsal* to it, and these descriptions remain valid regardless of the subject's position. Unless stated otherwise, it can be assumed that all terms of direction that describe the relationship of one body part to another are made with reference to anatomical position.

Bear in mind that if a subject is facing the viewer in anatomical position, the subject's left will be on the viewer's right and vice versa. In most anatomical illustrations, for example, the left atrium of the heart appears toward the right side of the page, and while the appendix is located in the right lower quadrant of the abdomen, it appears on the left side of most illustrations.

The forearm is said to be **supine** when the palms face forward and **prone** when they face to the rear (fig. A.2). The difference is particularly important to descriptions of anatomy of this region. In the supine position, the two forearm bones (radius and ulna) are parallel, and the radius is lateral to the ulna. In the prone position, the radius and ulna cross, and the radius is lateral to the ulna at the elbow but medial to it at the wrist. Descriptions of nerves, muscles, blood vessels, and other structures of the arm assume that the arm is supine.

Figure A.1 In anatomical position, the feet are flat on the floor and slightly separated, the arms are down and supinated, and the face and eyes are directed forward.

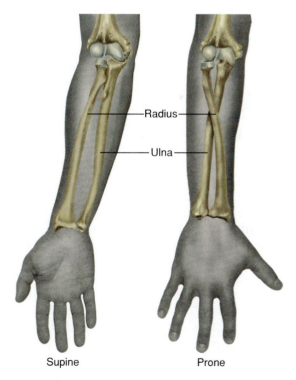

Radius

Ulna

Supine   Prone

Figure A.2 When the forearm is supine, the palm faces forward and the radius and ulna are parallel. When the forearm is prone, the palm faces to the rear and the radius crosses the ulna. The relative positions of many muscles, nerves, and blood vessels also depend on whether the forearm is supine or prone.

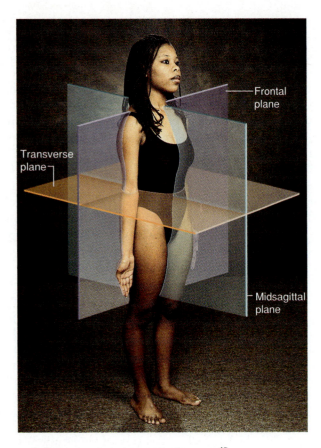

Figure A.3 Anatomical planes of reference. ✗

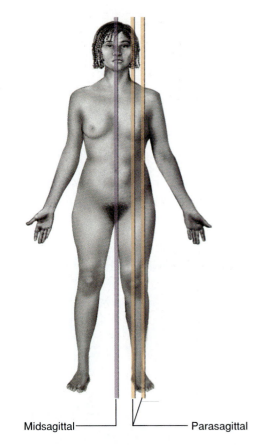

Figure A.4 The body has only one true midsagittal plane but an indefinite number of parasagittal planes. ✗

Midsagittal ——— ⌐ ⌐ ——— Parasagittal

Anatomical Planes

Many views of the body are based on real or imaginary "slices" called *sections* or *planes.* "Section" implies an actual cut or slice to reveal internal anatomy, whereas "plane" implies an imaginary flat surface passing through the body. The three major anatomical planes are *sagittal, frontal,* and *transverse* (fig. A.3).

A **sagittal**[1] (SADJ-ih-tul) **plane** extends vertically and divides the body (or an organ of the body) into right and left portions. A **midsagittal plane** passes through the midline of the body and divides it into equal right and left halves. A **parasagittal**[2] **plane** lies parallel to the midsagittal plane but to the right or left of the midline. While there is only one true midsagittal plane, there are an infinite number of parasagittal planes (fig. A.4). The head and pelvic organs are commonly illustrated in midsagittal views (fig. A.5; see also figs. A.33, A.38, and A.39).

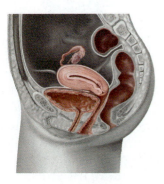

Figure A.5 A midsagittal section of the pelvic region.

A **frontal plane,** or **coronal plane,** also extends vertically, but it is perpendicular to the sagittal plane; therefore, it divides the body into anterior (front) and posterior (back) portions. In a frontal section of the head, for example, one portion contains the face and the other the back of the head. Contents of the thoracic and abdominal cavities are most commonly shown in frontal section (fig. A.6; see also figs. A.34 and A.36).

....

1. *sagitta* = arrow
2. *para* = next to

Figure A.6 A frontal (coronal) section of the thoracic region. ✗

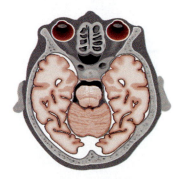

Figure A.7 A transverse (cross) section of the head at the level of the eyes. ✗

A **transverse plane**—also called a **cross-sectional,** or **horizontal,** plane—passes across the body or an organ perpendicular to its long axis (fig. A.7; see also figs. A.35 and A.37); therefore, it divides the body or organ into superior (upper) and inferior (lower) portions. CT scans are typically transverse sections (see chapter 1, fig. E.3).

Directional Terms

Table A.1 summarizes frequently used terms that describe the position of one structure relative to another. Because of the bipedal, upright stance of humans, some directional terms have different meanings for humans than they do for other animals. *Anterior,* for example, denotes the region of the body that leads the way in normal locomotion. For a four-legged animal such as a cat, this is the head end of the body; for a human, however, it is the area of the chest and abdomen. Thus, *anterior* has the same meaning as *ventral* for a human but not for a cat. *Posterior* denotes the region of the body that comes last in normal locomotion—the tail end of a cat but the dorsal side of a human. These differences must be kept in mind when dissecting other animals for comparison to human anatomy. Intermediate directions are often indicated

Table A.1	Directional Terms	
Term	**Meaning**	**Examples of Usage**
Ventral	Toward the front or belly	The aorta is *ventral* to the spinal column.
Dorsal	Toward the back or spine	The spinal column is *dorsal* to the aorta.
Anterior	Toward the ventral side	The sternum is *anterior* to the heart.
Posterior	Toward the dorsal side	The esophagus is *posterior* to the trachea.
Superior	Above	The heart is *superior* to the diaphragm.
Inferior	Below	The liver is *inferior* to the diaphragm.
Medial	Toward the midsagittal plane	The heart is *medial* to the lungs.
Lateral	Away from the midsagittal plane	The clavicles are *lateral* to the sternum.
Proximal	Closer to the point of attachment or origin	The elbow is *proximal* to the wrist.
Distal	Farther from the point of attachment or origin	The fingernails are at the *distal* ends of the fingers.
Central	Near or toward the midline of the body	The brain and spinal cord make up the *central* nervous system.
Peripheral	Away from the midline or center of the body	*Peripheral* nerves lead from the spinal cord to the skeletal muscles. Blood drawn from a fingerstick is *peripheral* blood.
Superficial	Closer to the body surface	The skin is *superficial* to the muscles.
Deep	Farther from the body surface	The bones are *deep* to the muscles.

by combinations of the terms in table A.1. For example, one structure may be described as *dorsolateral* to another (toward the back and side).

Surface Anatomy

Knowledge of the external regional anatomy and landmarks of the body is important in performing a physical examination and in physical therapy and other clinical procedures. For purposes of study, the body is divided into two major regions called the *axial* and *appendicular regions.* Smaller areas within the major regions are described in the following paragraphs and illustrated in figures A.8 and A.9.

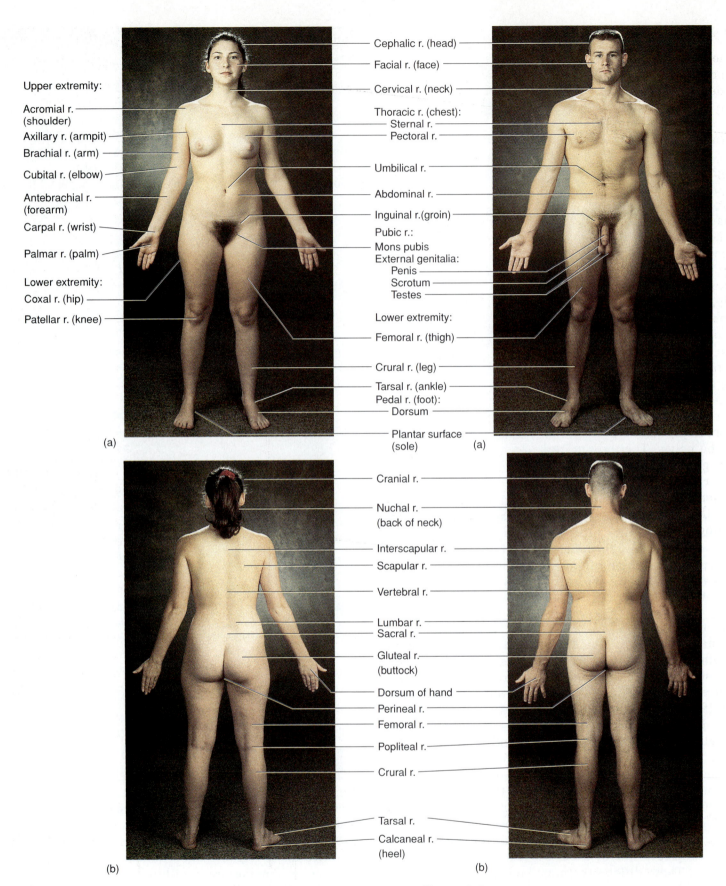

Upper extremity:

Acromial r.
(shoulder)

Axillary r. (armpit)

Brachial r. (arm)

Cubital r. (elbow)

Antebrachial r.
(forearm)

Carpal r. (wrist)

Palmar r. (palm)

Lower extremity:

Coxal r. (hip)

Patellar r. (knee)

Cephalic r. (head)

Facial r. (face)

Cervical r. (neck)

Thoracic r. (chest):
Sternal r.
Pectoral r.

Umbilical r.

Abdominal r.

Inguinal r.(groin)

Pubic r.:
Mons pubis
External genitalia:
Penis
Scrotum
Testes

Lower extremity:

Femoral r. (thigh)

Crural r. (leg)

Tarsal r. (ankle)
Pedal r. (foot):
Dorsum

Plantar surface
(sole)

(a)

(a)

Cranial r.

Nuchal r.
(back of neck)

Interscapular r.

Scapular r.

Vertebral r.

Lumbar r.
Sacral r.

Gluteal r.
(buttock)

Dorsum of hand

Perineal r.

Femoral r.

Popliteal r.

Crural r.

Tarsal r.

Calcaneal r.
(heel)

(b)

(b)

Figure A.8 The adult female body. (*a*) Ventral aspect. (*b*) Dorsal aspect. (*r* = region.)

Figure A.9 The adult male body. (*a*) Ventral aspect. (*b*) Dorsal aspect. (*r* = region.)

Axial Region

The **axial region** of the body consists of the **head, neck** (*cervical*[3] *region*), and **trunk.** The trunk is further divided into the **thoracic region** above the diaphragm and the **abdominal region** below the diaphragm.

One way of referring to the locations of abdominal structures is to divide the region into **quadrants.** As shown in figure A.10, the **right upper quadrant (RUQ), right lower quadrant (RLQ), left upper quadrant (LUQ),** and **left lower quadrant (LLQ)** are defined by two perpendicular lines that intersect at the umbilicus (navel). Medical personnel typically use the quadrant scheme to locate the site of an abdominopelvic pain or abnormality.

The abdomen also can be divided into **regions** defined by four lines that intersect like a tic-tac-toe grid (fig. A.11). Each vertical line is called a **midclavicular line** because it passes through the midpoint of the clavicle (collarbone). The superior transverse line is called the **subcostal**[4] **line** because it connects the inferior borders of the right and left tenth costal cartilages (cartilage connecting the tenth rib on each side to the inferior end of the sternum). The inferior transverse line is called the **intertubercular**[5] **line** because it passes from left to right between the tubercles of the iliac crest of the pelvis. The three lateral regions on each side of this grid, from upper to lower, are the **hypochondriac,**[6] **lateral abdominal,** and **inguinal**[7] (IN-gwih-nul) **regions.** The three medial regions from upper to lower are the **epigastric,**[8] **umbilical,** (um-BIL-ih-cul) and **hypogastric** regions.

Appendicular Region

The **appendicular** (AP-en-DIC-you-lur) **region** of the body consists of the appendages: the **upper extremities** and the **lower extremities.** The upper extremity includes the **brachium**[9] (BRAY-kee-um) (arm), **antebrachium**[10] (AN-teh-BRAY-kee-um) (forearm), **carpus** (wrist), **manus** (hand), and **digits** (fingers). The lower extremity includes the **thigh, crus**[11] (leg), **tarsus** (ankle), **pes** (foot), and **digits** (toes).

It should be noted that anatomically, "arm" refers only to that part of the upper extremity between the

---

3. *cervix* = neck
4. *sub* = below + *costa* = rib
5. *inter* = between + *tubercle* = little swelling
6. *hypo* = below + *chondri* = cartilage
7. *inguin* = groin
8. *epi* = above, over + *gastr* = stomach
9. *brachi* = arm
10. *ante* = fore, before
11. *crus* = leg

Table A.2	Body Cavities and Membranes	
Name of Cavity	**Principal Contents**	**Membranous Lining**
Dorsal Body Cavity		
Cranial cavity	Brain	Meninges
Vertebral canal	Spinal cord	Meninges
Ventral Body Cavity		
Thoracic Cavity		
Pleural cavities	Lungs	Pleurae
Pericardial cavity	Heart	Pericardium
Abdominopelvic Cavity		
Abdominal cavity	Digestive organs, spleen, kidneys	Peritoneum
Pelvic cavity	Bladder, reproductive organs	Peritoneum

shoulder and elbow. "Leg" refers only to that part of the lower extremity between the knee and ankle.

Body Cavities and Membranes

The body is internally divided into three major **body cavities** (fig. A.12). The organs within them are called the **viscera** (VISS-er-uh) (singular, *viscus*[12]). Various membranes line the body cavities, cover the viscera, and hold the viscera in place (table A.2).

Dorsal Body Cavity

The **dorsal body cavity** has two subdivisions: (1) the **cranial** (CRAY-nee-ul) **cavity,** which is enclosed by the cranium (braincase) and contains the brain, and (2) the **vertebral canal,** which is enclosed by the spinal column (backbone) and contains the spinal cord. The dorsal body cavity is lined by three membrane layers called the **meninges** (meh-NIN-jeez) (singular, *meninx*). Among other functions, the meninges protect the delicate nervous tissue from the hard protective bone that encloses it.

Ventral Body Cavity

During embryonic development, a space called the **coelom** (SEE-lum) forms within the trunk and eventually gives rise to the **ventral body cavity.** This cavity later

---

12. *viscus* = body organ

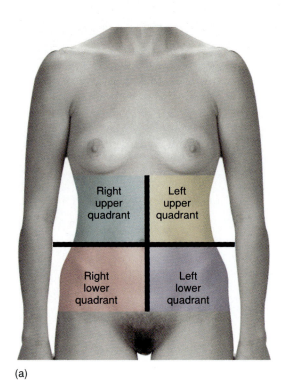

(a)

Right upper quadrant

Left upper quadrant

Right lower quadrant

Left lower quadrant

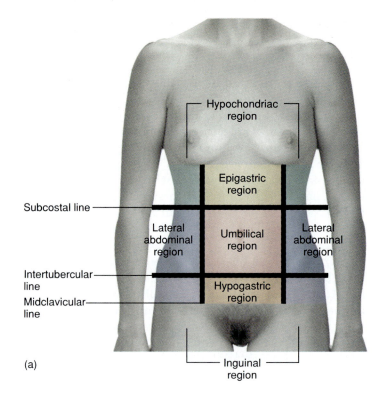

(a)

Hypochondriac region

Epigastric region

Subcostal line

Lateral abdominal region

Umbilical region

Lateral abdominal region

Intertubercular line

Midclavicular line

Hypogastric region

Inguinal region

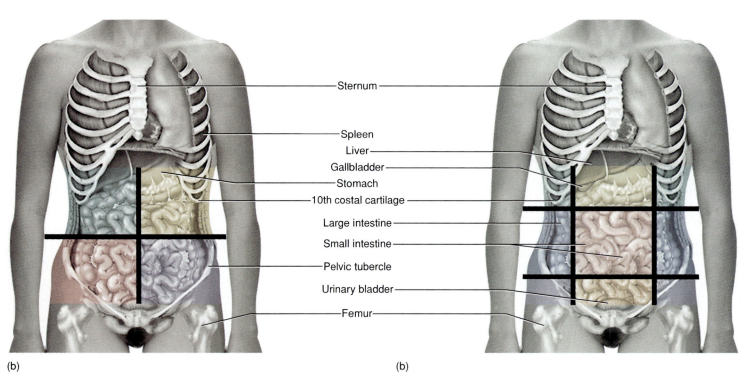

(b)

Sternum

Spleen

Liver

Gallbladder

Stomach

10th costal cartilage

Large intestine

Small intestine

Pelvic tubercle

Urinary bladder

Femur

(b)

Figure A.10 The four abdominal quadrants, defined by perpendicular lines crossing at the navel. (*a*) External landmarks. (*b*) Internal landmarks.

Figure A.11 The nine abdomen regions. (*a*) External landmarks. (*b*) Internal landmarks. (*r* = region.)

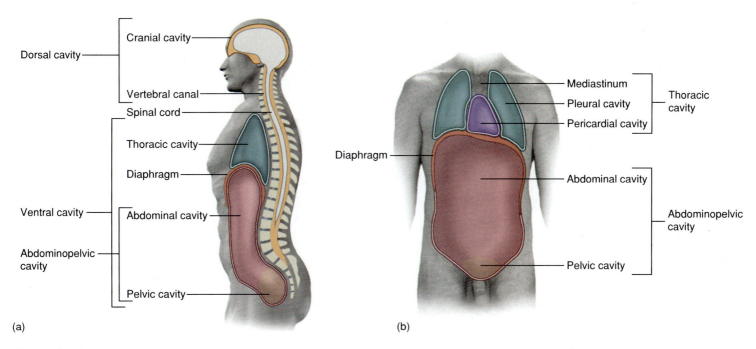

(a)

(b)

Figure A.12 The major body cavities. (*a*) Left lateral view. (*b*) Anterior view.

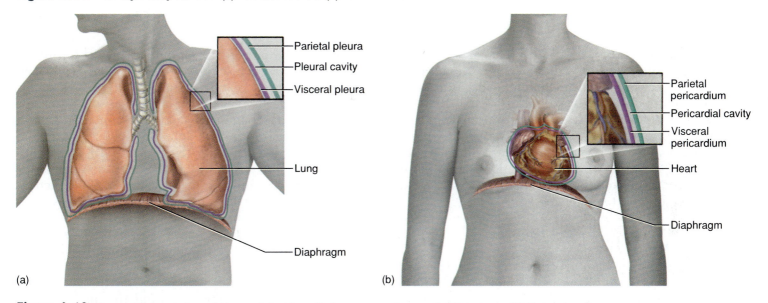

(a)

(b)

Figure A.13 The parietal and visceral layers of double-walled serous membranes. (*a*) The pleura. (*b*) The pericardium.

becomes partitioned by a muscular sheet, the **diaphragm,** into a superior **thoracic cavity** and an inferior **abdominopelvic cavity.** The thoracic and abdominopelvic cavities are lined with thin **serous membranes.** These membranes secrete a lubricating film of moisture similar to blood serum (hence the name *serous*).

Thoracic Cavity

The thoracic cavity is divided into right, left, and median portions by a partition called the **mediastinum**[13]

(ME-dee-ass-TY-num) (fig. A.12). The right and left sides contain the lungs and are lined by a two-layered membrane called the **pleura**[14] (PLOOR-uh) (fig. A.13*a;* see also fig. A.34). The outer layer, or **parietal**[15] (pa-RY-eh-tul) **pleura,** lies against the inside of the rib cage; the inner layer, or **visceral** (VISS-er-ul) **pleura,** forms the external surface of the lung. The narrow, moist space between the visceral and parietal pleurae is called the **pleural cavity.**

13. *mediastinum* = in the middle

14. *pleur* = rib, side
15. *pariet* = wall

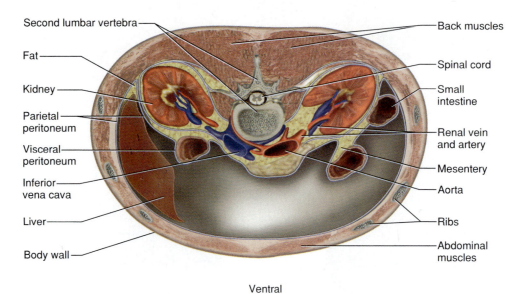

Dorsal

Second lumbar vertebra

Fat

Kidney

Parietal peritoneum

Visceral peritoneum

Inferior vena cava

Liver

Body wall

Back muscles

Spinal cord

Small intestine

Renal vein and artery

Mesentery

Aorta

Ribs

Abdominal muscles

Ventral

Figure A.14 Transverse section of the body at the level of the kidneys and second lumbar vertebra, showing the peritoneum, peritoneal cavity, and some retroperitoneal organs. 🏃

The median portion of the thoracic cavity, within the mediastinum, is occupied by the heart. The heart is enclosed by a two-layered membrane called the **pericardium**. The **parietal pericardium** is farthest from the heart, and the **visceral pericardium** forms the heart surface (fig. A.13b; see also fig. A.34). The **pericardial**[16] **cavity** is the space between the visceral and parietal pericardia. The mediastinum also contains the major blood vessels connected to the heart and a gland called the **thymus**, superior to the heart.

Abdominopelvic Cavity

The abdominopelvic cavity consists of an **abdominal cavity** above the brim of the pelvis and a **pelvic cavity** below the brim (see fig. A.32). The abdominal cavity contains most of the digestive organs, as well as the kidneys and ureters. The pelvic cavity contains the distal part of the large intestine, the urinary bladder and urethra, and the reproductive organs. There is no membrane that physically separates the abdominal and pelvic cavities, but the pelvic cavity is markedly narrower and its inferior end tilts posteriorly (see fig. A.12a).

The abdominopelvic cavity contains a moist serous membrane called the **peritoneum**[17] (PERR-ih-toe-

NEE-um). The **parietal peritoneum** lines the walls of the abdominopelvic cavity, while the **visceral peritoneum** covers the external surfaces of most digestive organs. The **peritoneal cavity** is the space between the parietal and visceral portions of the peritoneum.

Some organs of the abdominopelvic cavity lie outside the peritoneum, so they are said to have a **retroperitoneal**[18] position (fig. A.14). These include the kidneys, ureters, adrenal glands, most of the pancreas, and abdominal portions of two major blood vessels—the aorta and inferior vena cava (see fig. A.31).

The intestines are suspended from the dorsal abdominal wall by a translucent membrane called the **mesentery**[19] (MESS-en-tare-ee). This is a continuation of the peritoneum that wraps around the intestines, forming a moist membrane called the **serosa** (seer-OH-sa) on their outer surfaces (fig. A.15a). The mesentery of the large intestine is also called the **mesocolon.**

A fatty membrane called the **greater omentum**[20] hangs like an apron from the inferolateral margin of the stomach and overlies the intestines (fig. A.15b). It is unattached at its inferior border and can be lifted to reveal the intestines. A smaller **lesser omentum** extends from the superomedian border of the stomach to the liver.

16. *peri* = around + *cardi* = heart
17. *peri* = around + *toneum* = stretched

18. *retro* = behind
19. *mes* = in the middle + *enter* = intestine
20. *omentum* = covering

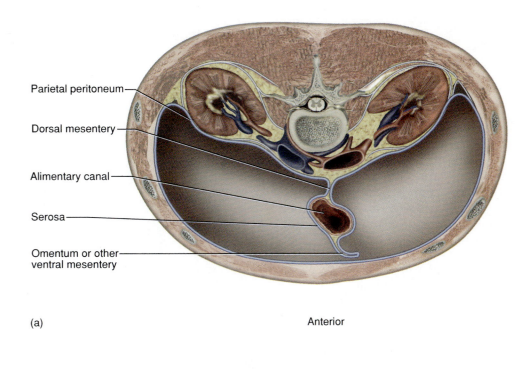

Parietal peritoneum

Dorsal mesentery

Alimentary canal

Serosa

Omentum or other
ventral mesentery

(a)   Anterior

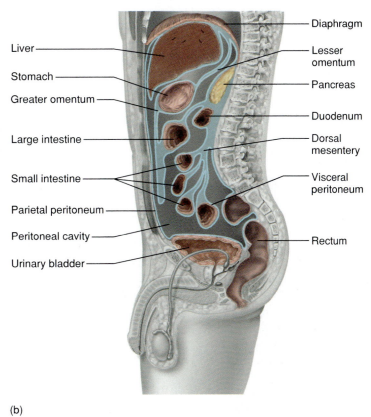

Liver

Stomach

Greater omentum

Large intestine

Small intestine

Parietal peritoneum

Peritoneal cavity

Urinary bladder

Diaphragm

Lesser
omentum

Pancreas

Duodenum

Dorsal
mesentery

Visceral
peritoneum

Rectum

(b)

Figure A.15 Serous membranes of the abdominal cavity. (*a*) Transverse section. (*b*) Midsagittal section, left lateral view.

Organ Systems

The human body has 11 **organ systems** and an immune system, which is better described as a population of cells than as an organ system. These systems are pictured and described in figures A.16 through A.27. They are classified in the following list by their principal functions, but this is an unavoidably flawed classification. Some organs belong to two or more of these systems—for example, the male urethra is part of both the urinary and reproductive systems; the pharynx is part of the respiratory and digestive systems; and the mammary glands can be considered part of the integumentary and female reproductive systems.

Protection, Support, and Movement
Integumentary system
Skeletal system
Muscular system

Internal Communication and Integration
Nervous system
Endocrine system

Fluid Transport
Circulatory system
Lymphatic system

Defense
Immune system

Input and Output
Respiratory system
Urinary system
Digestive system

Reproduction
Reproductive system

Figure A.16 The integumentary system.

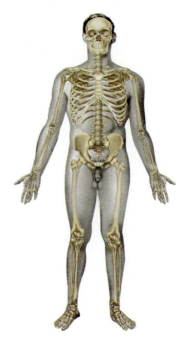

Figure A.17 The skeletal system. ⚼

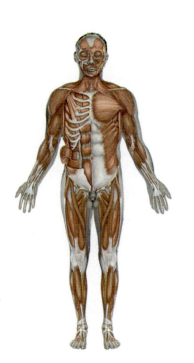

Figure A.18 The muscular system.

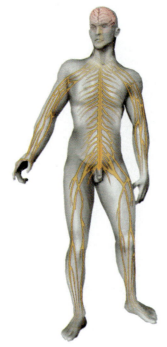

Figure A.19 The nervous system.

Figure A.20 The endocrine system. ⚼

Figure A.21 The circulatory system.

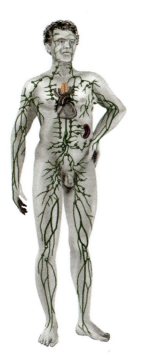

Figure A.22 The lymphatic system. ⚹

Figure A.23 The immune system, which is not an organ system but a collection of disease-fighting cells that populate the lymphatic and other organ systems. ⚹

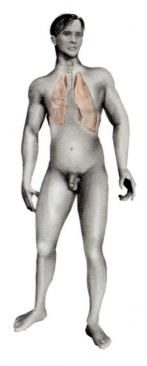

Figure A.24 The respiratory system. ⚹

Figure A.25 The urinary system. ⚹

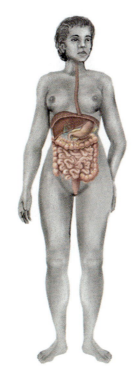

Figure A.26 The digestive system. ⚹

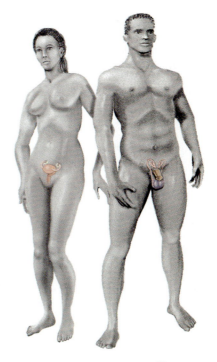

Figure A.27 The reproductive system. ⚹

The illustrations in figures A.28 through A.32 provide an overview of the anatomy of the trunk and internal organs of the thoracic and abdominopelvic cavities. Fig- ures A.33 through A.39 show the major organs of the dorsal and ventral body cavities in frontal, transverse, and midsagittal sections of the cadaver.

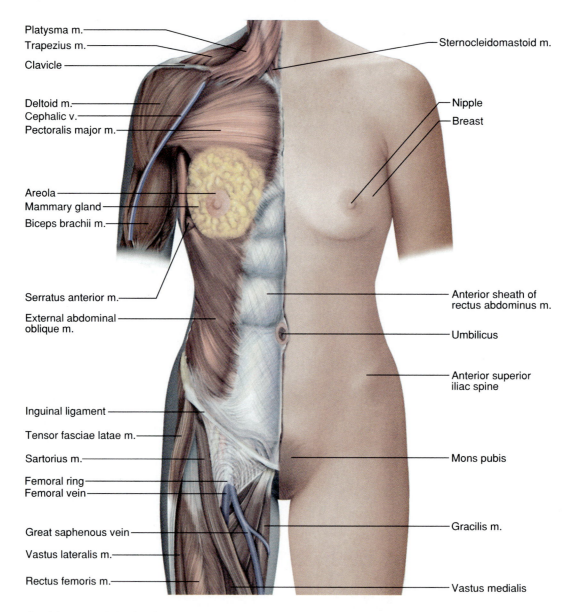

Platysma m.

Trapezius m.

Clavicle

Deltoid m.

Cephalic v.

Pectoralis major m.

Areola

Mammary gland

Biceps brachii m.

Serratus anterior m.

External abdominal oblique m.

Inguinal ligament

Tensor fasciae latae m.

Sartorius m.

Femoral ring

Femoral vein

Great saphenous vein

Vastus lateralis m.

Rectus femoris m.

Sternocleidomastoid m.

Nipple

Breast

Anterior sheath of rectus abdominus m.

Umbilicus

Anterior superior iliac spine

Mons pubis

Gracilis m.

Vastus medialis

Figure A.28 Superficial features of the female body. The left side is a surface view and the right side shows structures immediately deep to the skin (*m.* = muscle).

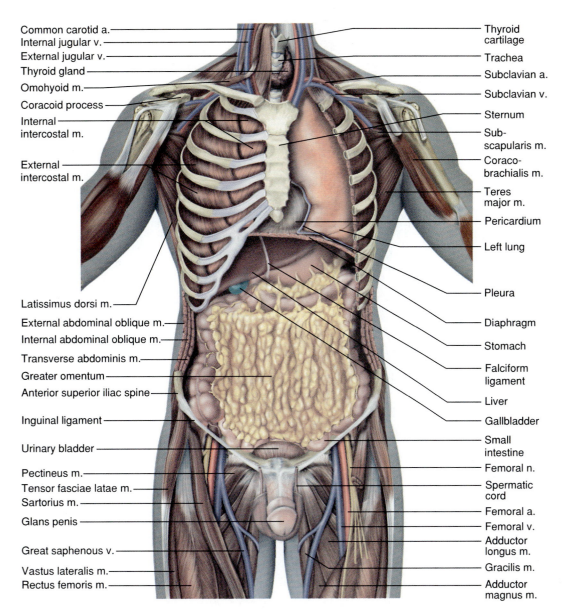

Common carotid a.
Internal jugular v.
External jugular v.
Thyroid gland
Omohyoid m.
Coracoid process
Internal intercostal m.
External intercostal m.
Latissimus dorsi m.
External abdominal oblique m.
Internal abdominal oblique m.
Transverse abdominis m.
Greater omentum
Anterior superior iliac spine
Inguinal ligament
Urinary bladder
Pectineus m.
Tensor fasciae latae m.
Sartorius m.
Glans penis
Great saphenous v.
Vastus lateralis m.
Rectus femoris m.

Thyroid cartilage
Trachea
Subclavian a.
Subclavian v.
Sternum
Sub-scapularis m.
Coraco-brachialis m.
Teres major m.
Pericardium
Left lung
Pleura
Diaphragm
Stomach
Falciform ligament
Liver
Gallbladder
Small intestine
Femoral n.
Spermatic cord
Femoral a.
Femoral v.
Adductor longus m.
Gracilis m.
Adductor magnus m.

Figure A.29 The male body, dissected to the level of the rib cage and greater omentum (*a.* = artery; *v.* = vein; *m.* = muscle; *n.* = nerve).

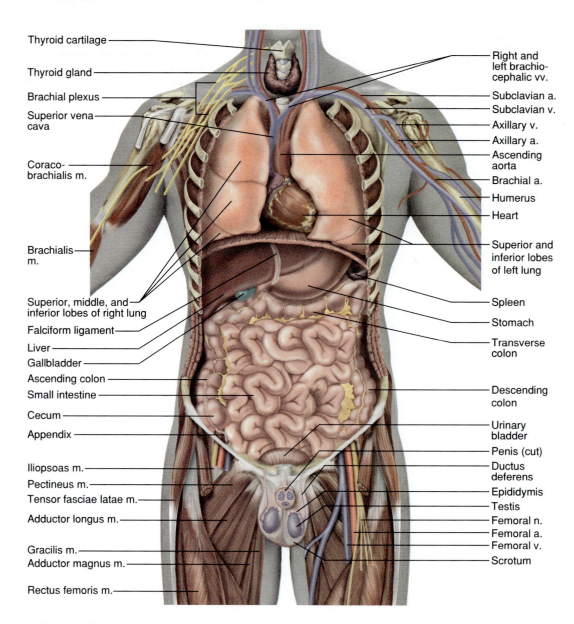

Thyroid cartilage

Thyroid gland

Brachial plexus

Superior vena cava

Coraco-brachialis m.

Brachialis m.

Superior, middle, and inferior lobes of right lung

Falciform ligament

Liver

Gallbladder

Ascending colon

Small intestine

Cecum

Appendix

Iliopsoas m.

Pectineus m.

Tensor fasciae latae m.

Adductor longus m.

Gracilis m.

Adductor magnus m.

Rectus femoris m.

Right and left brachio-cephalic vv.

Subclavian a.

Subclavian v.

Axillary v.

Axillary a.

Ascending aorta

Brachial a.

Humerus

Heart

Superior and inferior lobes of left lung

Spleen

Stomach

Transverse colon

Descending colon

Urinary bladder

Penis (cut)

Ductus deferens

Epididymis

Testis

Femoral n.

Femoral a.

Femoral v.

Scrotum

Figure A.30 The male body, dissected to the level of the lungs and intestines.

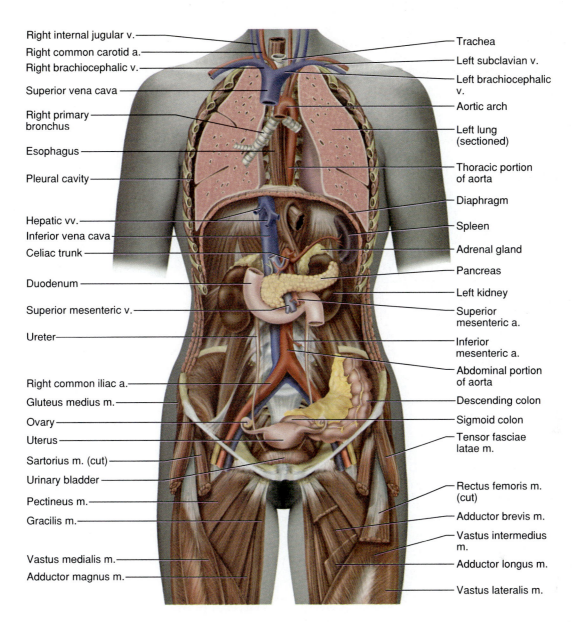

Right internal jugular v.
Right common carotid a.
Right brachiocephalic v.
Superior vena cava
Right primary bronchus
Esophagus
Pleural cavity
Hepatic vv.
Inferior vena cava
Celiac trunk
Duodenum
Superior mesenteric v.
Ureter
Right common iliac a.
Gluteus medius m.
Ovary
Uterus
Sartorius m. (cut)
Urinary bladder
Pectineus m.
Gracilis m.
Vastus medialis m.
Adductor magnus m.

Trachea
Left subclavian v.
Left brachiocephalic v.
Aortic arch
Left lung (sectioned)
Thoracic portion of aorta
Diaphragm
Spleen
Adrenal gland
Pancreas
Left kidney
Superior mesenteric a.
Inferior mesenteric a.
Abdominal portion of aorta
Descending colon
Sigmoid colon
Tensor fasciae latae m.
Rectus femoris m. (cut)
Adductor brevis m.
Vastus intermedius m.
Adductor longus m.
Vastus lateralis m.

Figure A.31 The female body, dissected to the level of the retroperitoneal viscera.

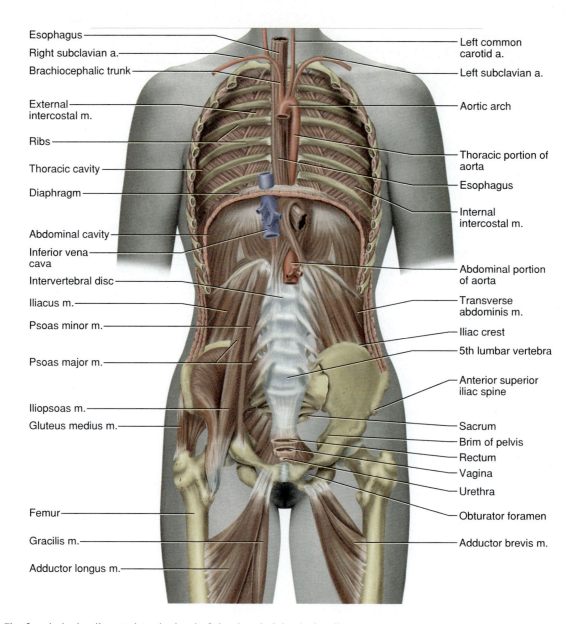

Esophagus

Right subclavian a.

Brachiocephalic trunk

External intercostal m.

Ribs

Thoracic cavity

Diaphragm

Abdominal cavity

Inferior vena cava

Intervertebral disc

Iliacus m.

Psoas minor m.

Psoas major m.

Iliopsoas m.

Gluteus medius m.

Femur

Gracilis m.

Adductor longus m.

Left common carotid a.

Left subclavian a.

Aortic arch

Thoracic portion of aorta

Esophagus

Internal intercostal m.

Abdominal portion of aorta

Transverse abdominis m.

Iliac crest

5th lumbar vertebra

Anterior superior iliac spine

Sacrum

Brim of pelvis

Rectum

Vagina

Urethra

Obturator foramen

Adductor brevis m.

Figure A.32 The female body, dissected to the level of the dorsal abdominal wall.

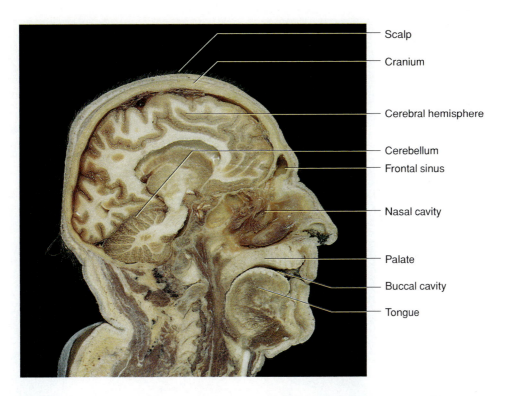

Scalp

Cranium

Cerebral hemisphere

Cerebellum
Frontal sinus

Nasal cavity

Palate

Buccal cavity

Tongue

sagittal section of the head, showing contents of the cranial, nasal, and buccal cavities. ⚥

Figure A.34 Frontal view of the thoracic cavity. ✗

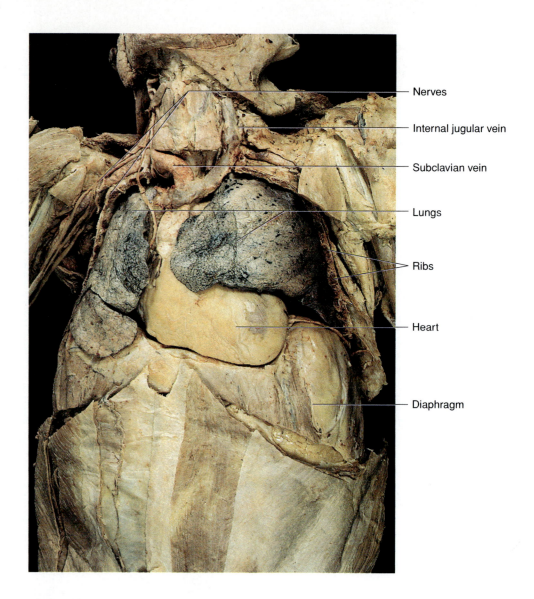

Nerves

Internal jugular vein

Subclavian vein

Lungs

Ribs

Heart

Diaphragm

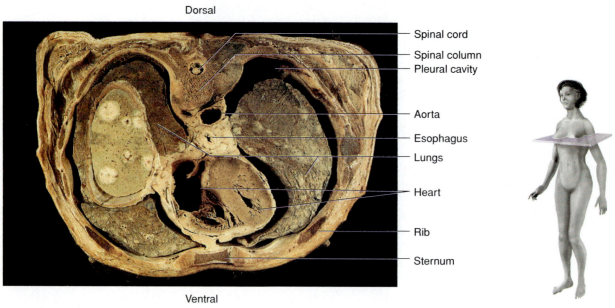

Dorsal

Spinal cord

Spinal column

Pleural cavity

Aorta

Esophagus

Lungs

Heart

Rib

Sternum

Ventral

Figure A.35 Transverse section of the thoracic cavity at the level shown on the inset. ✗

General Orientation to Human Anatomy

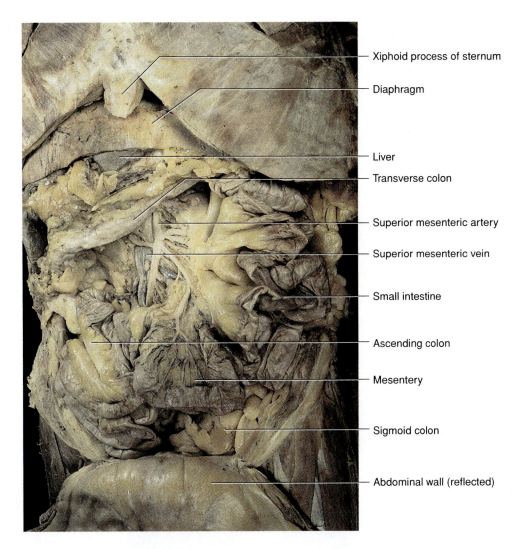

Xiphoid process of sternum

Diaphragm

Liver

Transverse colon

Superior mesenteric artery

Superior mesenteric vein

Small intestine

Ascending colon

Mesentery

Sigmoid colon

Abdominal wall (reflected)

Figure A.36 Frontal view of the abdominal cavity. ⃕

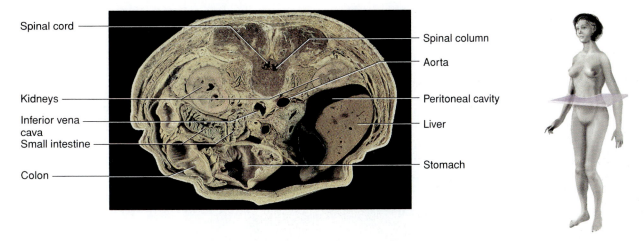

Spinal cord

Kidneys

Inferior vena cava

Small intestine

Colon

Spinal column

Aorta

Peritoneal cavity

Liver

Stomach

Figure A.37 Transverse section of the abdominal cavity at the level shown on the inset. ⃕

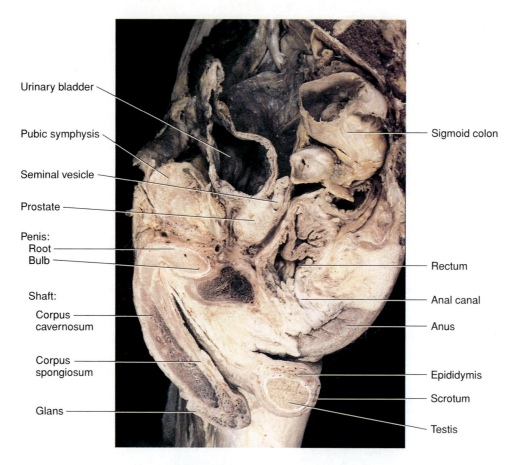

Urinary bladder

Pubic symphysis

Seminal vesicle

Prostate

Penis:
  Root
  Bulb

Shaft:
  Corpus
  cavernosum

  Corpus
  spongiosum

Glans

Sigmoid colon

Rectum

Anal canal

Anus

Epididymis

Scrotum

Testis

Figure A.38 Midsagittal section of the male pelvic cavity, viewed from the left. ✗

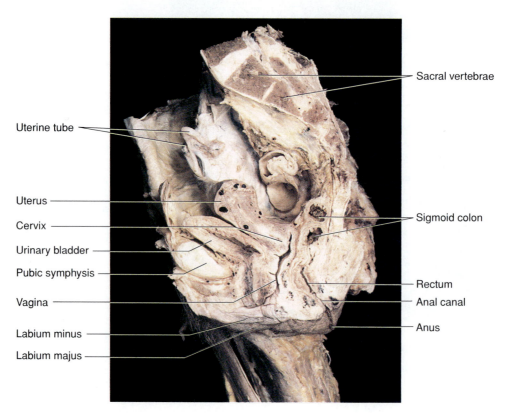

Uterine tube

Uterus

Cervix

Urinary bladder

Pubic symphysis

Vagina

Labium minus

Labium majus

Sacral vertebrae

Sigmoid colon

Rectum

Anal canal

Anus

Figure A.39 Midsagittal section of the female pelvic cavity, viewed from the left. ✗

1. Which of the following is *not* an essential part of anatomical position?
   a. eyes facing forward
   b. feet flat on the floor
   c. forearms supine
   d. mouth closed
   e. arms down to the sides

2. A ring-shaped section of the small intestine would be a _____ section.
   a. parasagittal
   b. midsagittal
   c. transverse
   d. frontal
   e. medial

3. The tarsal region is _____ to the popliteal region.
   a. medial
   b. superficial
   c. superior
   d. dorsal
   e. distal

4. The greater omentum is _____ to the small intestine.
   a. posterior
   b. parietal
   c. deep
   d. superficial
   e. proximal

5. A _____ line passes through the sternum, umbilicus, and mons pubis.
   a. central
   b. proximal
   c. midclavicular
   d. midsagittal
   e. intertubercular

6. The _____ region is immediately medial to the coxal region.
   a. inguinal
   b. hypochondriac
   c. umbilical
   d. popliteal
   e. antecubital

7. Which of the following regions is *not* part of the upper extremity?
   a. plantar
   b. carpal
   c. antecubital
   d. brachial
   e. palmar

8. Which of these organs is within the peritoneal cavity?
   a. urinary bladder
   b. kidneys
   c. heart
   d. small intestine
   e. brain

9. In which area do you think pain from the gallbladder would be felt?
   a. umbilical region
   b. right upper quadrant
   c. hypogastric region
   d. left hypochondriac region
   e. left lower quadrant

10. Which of the following is *not* an organ system?
   a. muscular system
   b. integumentary system
   c. endocrine system
   d. lymphatic system
   e. immune system

11. The hand is said to be _____ when the palms are facing forward.

12. The more superficial layer of the pleura is called the _____ pleura.

13. The right and left pleural cavities are separated by a thick wall called the _____.

14. The back of the head is called the _____ region, and the neck is the _____ region.

15. Of the three major anatomical planes, which one could not pass through both eyes at once?

16. The dorsal body cavity is lined by membranes called the _____.

17. Organs that lie within the abdominal cavity but not within the peritoneal cavity are said to have a _____ position.

18. Nerves and blood vessels far from the medial axis of the body are said to be _____.

19. The pelvic cavity can be described as _____ to the abdominal cavity in position.

20. The anterior "pit" of the elbow is the _____ region, and the corresponding (but posterior) "pit" of the knee is the _____ fossa.

1. A college freshman in a physical education course is accidentally shot on the archery range. The arrow enters the pectoral region and emerges from the lumbar region, narrowly missing the heart and spleen but causing the collapse of a lung. Which serous membranes did the arrow pass through?

2. Lay people often misunderstand anatomical terminology. What do you think people really mean when they say they have "planter's warts"?

3. Name one structure or anatomical feature that could be found in each of the following locations relative to the ribs: medial, lateral, superior, inferior, deep, superficial, posterior, and anterior. Try not to use the same example twice.

4. Based on the illustrations in this atlas, identify an internal organ that is (a) in the upper left quadrant and retroperitoneal, (b) in the lower right quadrant of the peritoneal cavity, (c) in the hypogastric region, (d) in the right hypochondriac region, and (e) in the pectoral region.

5. Why do you think people with imaginary illnesses came to be called hypochondriacs?

chapter two

2

[Matter and Energy

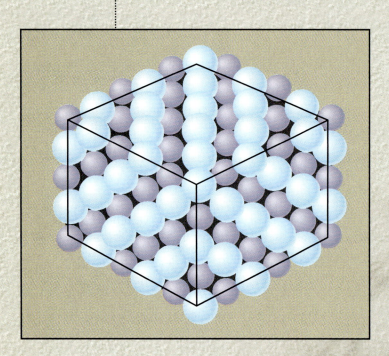

Why is too much sodium or cholesterol harmful? Why does iron deficiency cause anemia and iodine deficiency cause a goiter? Why does a pH imbalance make some drugs less effective? Why might a pregnant woman go into convulsions if she has suffered several days of vomiting? Why can radiation cause cancer as well as cure it?

None of these questions can be answered, nor would the rest of this book be intelligible, without understanding the chemistry of life. A little knowledge of chemistry can help you choose a healthy diet, use medications more wisely, avoid worthless health fads and frauds, and explain treatments and procedures to clients. Thus, we begin our study of the human body with basic chemistry—the simplest grade of the body's structural organization.

Chemical Elements and Atomic Structure

▼Objectives

When you have completed this section, you should be able to
- list the major chemical elements involved in human structure and function;
- describe the structure of atoms;
- distinguish atomic mass from atomic weight; and
- define *ion, electrolyte,* and *free radical.*

Matter and the Elements

Matter is anything that has mass and occupies space—the substance of which all things are composed. The simplest form of matter to have unique chemical properties is an **element.** Water, for example, has unique properties, but it can be broken down into two elements, hydrogen and oxygen, which have unique chemical properties of their own. If we carry this process further, however, we find that hydrogen and oxygen are made of protons, neutrons, and electrons—and none of these are unique. A proton of gold is identical to a proton of oxygen. Hydrogen and oxygen are therefore the simplest chemically unique components of water.

Each element can be identified by an **atomic number,** which is the number of protons in its nucleus. The atomic number of carbon is 6, and that of oxygen is 8, for example. The periodic table of the elements (see appendix A) arranges the elements in order by their atomic numbers. As the periodic table also shows, the elements are represented by one- or two-letter symbols, usually based on their English names: C for carbon, Mg for magnesium, Cl for chlorine, and so forth. A few symbols are based on Latin names, such as K for potassium (*kalium*), Na for sodium (*natrium*), and Fe for iron (*ferrum*).

There are 92 naturally occurring elements, 24 of which play normal physiological roles in humans. Table 2.1 groups these according to their total weight in the body. Six of them account for 98.5% of the body's weight: oxygen, carbon, hydrogen, nitrogen, calcium, and phosphorus. Hydrogen, the lightest of the elements,

Table 2.1	Elements of the Human Body		
Name	**Symbol**	**Percentage of Body Weight**	
Major Elements (total 98.5% of body weight)			
Oxygen	O	65.0	
Carbon	C	18.0	
Hydrogen	H	10.0	
Nitrogen	N	3.0	
Calcium	Ca	1.5	
Phosphorus	P	1.0	
Lesser Elements (total 0.8% of body weight)			
Potassium	K	0.20	
Sulfur	S	0.25	
Sodium	Na	0.15	
Chlorine	Cl	0.15	
Magnesium	Mg	0.05	
Iron	Fe	0.006	
Name	**Symbol**	**Name**	**Symbol**
Trace Elements (total 0.7% of body weight)			
Chromium	Cr	Molybdenum	Mo
Cobalt	Co	Selenium	Se
Copper	Cu	Silicon	Si
Fluorine	F	Tin	Sn
Iodine	I	Vanadium	V
Manganese	Mn	Zinc	Zn

is the most abundant in terms of number of atoms. It constitutes only 10% of the body's weight but 63% of its atoms (fig. 2.1).

Another 6 elements make up 0.8% of body weight: potassium, sulfur, sodium, chlorine, magnesium, and iron. The remaining 12 account for 0.7% of body weight, and no one of them accounts for more than 0.02%. These 12 elements, known as **trace elements,** play vital roles in physiology despite their minute quantities in the body. Several other elements without natural physiological roles can contaminate the body and severely disrupt its functions (see special topic 2.1).

As figure 2.1 indicates, the body is not simply composed of whatever is most readily available in the environment. Except in the case of oxygen, what is most abundant on earth bears little resemblance to what is most abundant in the body. Silicon and aluminum are among the most abundant elements on earth, for example, but the body contains no aluminum and only a tiny amount of silicon. Carbon constitutes about 18% of the body weight but only 0.09% of the earth's crust. Such differences show that the body is selective in its use of elements to form its own molecules.

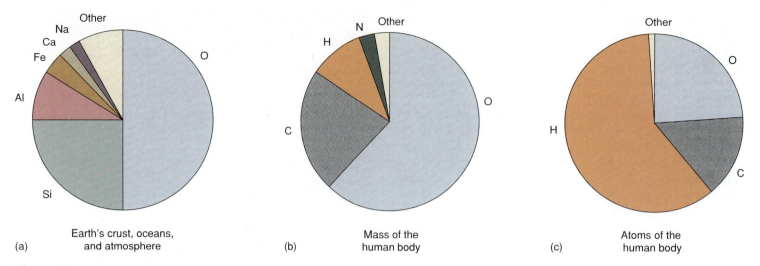

Figure 2.1 The relative abundance of elements (*a*) at the surface of the earth by mass, (*b*) in the human body by mass, and (*c*) in the human body by number of atoms.

Atomic Structure

In the fifth century B.C.E., the Greek philosopher Democritus reasoned that gold nuggets could be cut into tiny pieces, but eventually the pieces would be so small that not even the finest instruments imaginable would be able to divide them. He called these imaginary particles **atoms**[1] ("uncut, indivisible"). However, the concept of atoms remained a philosophical abstraction until the period from 1803 to 1807, when English chemist John Dalton developed an atomic theory based on experimental evidence. In 1913, Danish physicist Niels Bohr proposed a model of atomic structure similar to planets orbiting the sun (fig. 2.2). Although this *planetary model* is too simple to account for many of the properties of atoms, it remains useful for elementary purposes.

At the center of an atom is the **nucleus** (NEW-clee-us), composed of protons and neutrons. **Protons (p⁺)** have a single positive charge and **neutrons (n⁰)** have no charge. Each proton or neutron weighs approximately 1 **atomic mass unit (amu),** defined as one-twelfth the mass of an atom of carbon-12 (table 2.2). The **atomic**

1. *a* = not + *tom* = cut

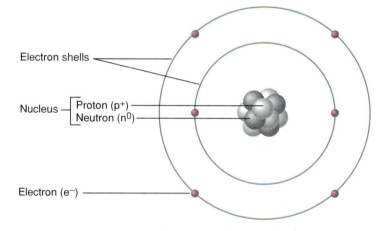

Figure 2.2 The Bohr planetary model of a carbon atom.

Table 2.2	The Primary Subatomic Particles			
Particle	**Symbol**	**Charge**	**Mass**	**Location**
Proton	p⁺	+1	1.0073 amu	Nucleus
Neutron	n⁰	0	1.0087 amu	Nucleus
Electron	e⁻	−1	0.0005 amu	Energy shells around nucleus

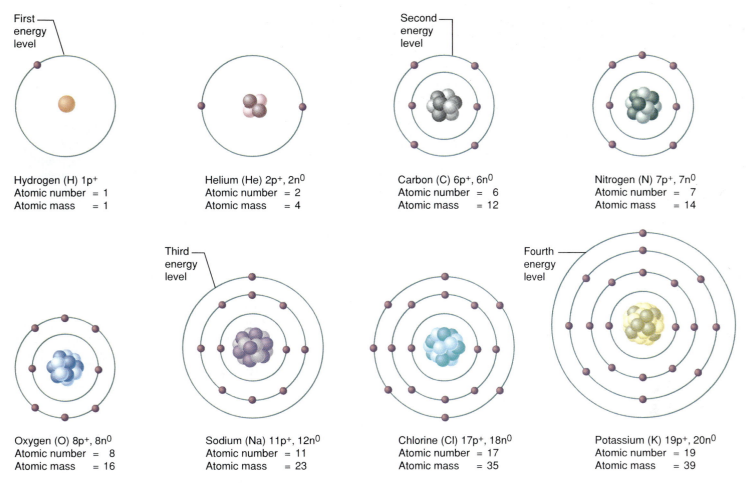

Hydrogen (H) 1p+
Atomic number = 1
Atomic mass = 1

Helium (He) 2p+, 2n0
Atomic number = 2
Atomic mass = 4

Carbon (C) 6p+, 6n0
Atomic number = 6
Atomic mass = 12

Nitrogen (N) 7p+, 7n0
Atomic number = 7
Atomic mass = 14

Oxygen (O) 8p+, 8n0
Atomic number = 8
Atomic mass = 16

Sodium (Na) 11p+, 12n0
Atomic number = 11
Atomic mass = 23

Chlorine (Cl) 17p+, 18n0
Atomic number = 17
Atomic mass = 35

Potassium (K) 19p+, 20n0
Atomic number = 19
Atomic mass = 39

Figure 2.3 Planetary models of eight selected elements, showing the filling of successive electron shells. Protons are represented p+ and neutrons are represented n^0.

mass of an element is approximately equal to its total number of protons and neutrons.

Around the nucleus are one or more concentric clouds of **electrons (e⁻),** which are tiny particles with a single negative charge and very low mass. It takes 1,836 electrons to equal 1 amu, so for most purposes we can disregard their mass. In a person who weighs 64 kg (140 lb), electrons account for less than 24 g (1 oz). This hardly means that we can ignore electrons, however. They determine the chemical properties of an atom, thereby governing what molecules can exist and what chemical reactions can occur. The number of electrons equals the number of protons, so their charges cancel each other, and an atom is electrically neutral.

Electrons swarm about the nucleus in concentric regions called **electron shells,** or **energy levels.** The more energy an electron has, the farther away from the nucleus its orbit lies. Each electron shell holds a limited number of electrons (fig. 2.3). The one closest to the nucleus holds a maximum of 2 electrons, the second one holds a maximum of 8, and the third holds a maximum of 18. The elements known to date range up to seven

electron shells, but those ordinarily involved in human physiology do not exceed four.

The electrons of the outermost shell, called the **valence electrons,** determine the chemical binding properties of an atom. At atom tends to bond with other atoms that will fill its outer shell and produce a stable number of valence electrons. A hydrogen atom, with only a single electron shell and one electron, tends to react with other atoms that provide another electron and fill this shell with a stable number of two electrons. All other atoms react in ways that produce eight electrons in the valence shell. This tendency is called the **octet rule,** or **rule of eights.**

Atoms of biological interest range from about 0.07 to 0.35 nm in diameter (1 nm = 10^{-9} m). It would take about 500,000 atoms laid side by side to be barely visible to the naked eye. (The limit of our vision is near 0.1 mm.)

Isotopes and Atomic Weight

Dalton believed that every atom of an element was identical, but we now know this to be untrue. All elements have varieties called **isotopes**[2] that differ from each

2. *iso* = same + *top* = place (i.e., same position in the periodic table)

other only in number of neutrons and therefore in atomic mass. Most hydrogen atoms have only one proton, for example, but an isotope called *deuterium* has one proton and one neutron, and another called *tritium* has one proton and two neutrons (fig. 2.4). These isotopes are symbolized ^1H, ^2H, and ^3H to indicate their atomic masses. Over 99% of carbon atoms have an atomic mass of 12 ($6p^+$, $6n^0$) and are called carbon-12 (^{12}C), but a small percentage of carbon atoms are carbon-13 (^{13}C), with seven neutrons, and carbon-14 (^{14}C) with eight. All isotopes of a given element behave the same chemically. Deuterium (^2H), for example, reacts with oxygen in the same way ^1H does to produce water.

Think About It

Why is there no difference in the chemical behavior of different isotopes of the same element?

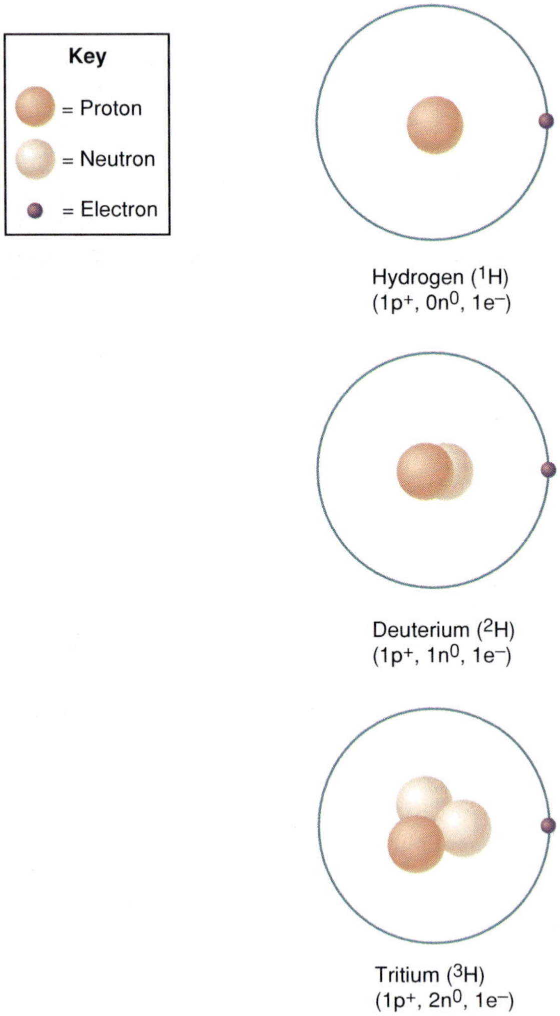

Table 2.3	Terms Related to Atomic Weights
Atomic Number	Number of protons in an element, equal to its number of electrons.
Atomic Mass	Total number of protons and neutrons in an element (approximately).
Atomic Mass Unit (amu)	One-twelfth the mass of a ^{12}C atom, or approximately the mass of one proton or neutron.
Atomic Weight	Average atomic mass of the atoms in a sample of an element. Considers the atomic masses of all isotopes in the sample and the relative abundance of each isotope.

Atomic weight considers the fact that an element is a mixture of isotopes. If all carbon were ^{12}C, the atomic weight of carbon would be the same as its atomic mass, 12.000. But since a sample of carbon also contains small amounts of the heavier isotopes ^{13}C and ^{14}C, the atomic weight is slightly higher, 12.011. These concepts are summarized in table 2.3.

Ions

Ions are charged particles in which there is no longer an equal number of protons and electrons. Elements with one to three valence electrons have a tendency to give them up, and those with four to seven electrons have a tendency to gain more. If an atom of the first kind is exposed to an atom of the second, electrons may transfer from one to the other and turn both of them into ions. Formation of ions is called **ionization.** The particle that gains electrons acquires a negative charge and is called an **anion** (AN-eye-on). The one that loses electrons acquires a positive charge (because it then has a relative surplus of protons) and is called a **cation** (CAT-eye-on).

Sodium, for example, has three electron shells with a total of 11 electrons: 2 in the first shell, 8 in the second, and 1 in the third (fig. 2.5). If it can give up the electron in the third shell, its second shell becomes the valence shell and has the stable configuration of 8 electrons described by the octet rule. Chlorine, by contrast, has 17 electrons: 2 in the first shell, 8 in the second, and 7 in the third. If it can gain one more electron, it can fill the third shell with 8 electrons and become stable. Sodium and chlorine seem "made for each other"—one needs to lose an electron, and the other needs to gain one. This is just what they do. When exposed to each other, 1 electron transfers from Na to Cl. Now, Na has 11 protons in its nucleus but only 10 electrons. This gives Na one surplus positive charge, and the sodium ion is symbolized Na^+. Chlorine has been changed to a chloride ion with a surplus negative charge, symbolized Cl^-.

Key
= Proton
= Neutron
= Electron

Hydrogen (^1H)
($1p^+$, $0n^0$, $1e^-$)

Deuterium (^2H)
($1p^+$, $1n^0$, $1e^-$)

Tritium (^3H)
($1p^+$, $2n^0$, $1e^-$)

Figure 2.4 The three isotopes of hydrogen differ only in the number of neutrons present.

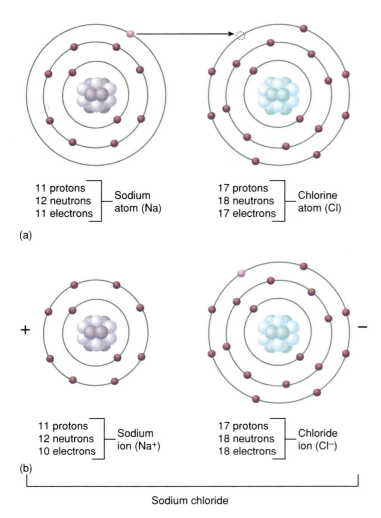

11 protons
12 neutrons — Sodium
11 electrons   atom (Na)

17 protons
18 neutrons — Chlorine
17 electrons   atom (Cl)

(a)

+   −

11 protons
12 neutrons — Sodium
10 electrons   ion (Na⁺)

17 protons
18 neutrons — Chloride
18 electrons   ion (Cl⁻)

(b)

Sodium chloride

Figure 2.5 (*a*) A sodium atom readily donates an electron to a chlorine atom. (*b*) This electron transfer converts these atoms to a positive sodium ion and a negative chloride ion, which are mutually attracted by their opposite charges and form an ionic bond. [icon]

Some elements can exist in two or more ionic forms, such as iron, which has ferrous (Fe^{2+}) and ferric (Fe^{3+}) ions. Note that some ions have a single positive or negative charge, while others have charges of ±2 or ±3 because they gain or lose more than one electron. The charge of an ion is called its **valence**. Ions are not always single atoms that have become charged; some are groups of atoms—phosphate (PO_4^{3-}) and bicarbonate (HCO_3^-) ions, for example.

Electrolytes

Electrolytes are molecules that ionize in water and form a solution capable of conducting electricity. The most abundant electrolytes in the body are salts of sodium (Na^+), potassium (K^+), calcium (Ca^{2+}), magnesium (Mg^{2+}), chloride (Cl^-), phosphate (PO_4^{3-}), and bicarbonate (HCO_3^-). We can detect electrical activity of the muscles, heart, and brain with electrodes on the skin because electrolytes in the body fluids conduct currents from these organs to the body surface. Electrolytes are essential to nerve and muscle function, and electrolyte balance is one of the most important considerations in patient care. Electrolyte imbalances cause symptoms ranging from muscle cramps and brittle bones to coma and cardiac arrest.

Free Radicals

A **free radical** is a particle with an odd number of electrons. A common example is the *superoxide anion.* Oxygen is composed of two oxygen atoms, O_2; if an additional electron is added, it becomes superoxide, $O_2^-\bullet$. The dot in the formula represents the odd electron.

Superoxide and other free radicals are extremely reactive and are destructive to other molecules. They are produced by certain metabolic reactions of the body (for example, in a mechanism used by some white blood cells to kill bacteria), by radiation (sunlight, ultraviolet radiation, and X rays), and by chemicals (carbon tetrachloride, a cleaning solvent, and nitrites, used as preservatives in wine, meat, and other foods). Free radicals are short-lived and combine quickly with fats, proteins, DNA, and lipids, converting these into free radicals and triggering chain reactions that lead to cellular damage and death. Free radicals have been implicated in heart disease, some forms of cancer, and other diseases. One theory of aging is that it results in part from lifelong cellular damage caused by free radicals.

There are a number of mechanisms for neutralizing free radicals. An **antioxidant** is any substance that reacts with and neutralizes free radicals of oxygen. *Superoxide dismutase (SOD)* is an enzyme that converts superoxide into oxygen and hydrogen peroxide. *Selenium* and *vitamin E* (alpha-tocopherol) are two antioxidants obtained from the diet. In some parts of the country, the soil is deficient in selenium, and people who grow much of their own food have a higher incidence of heart attacks as a result. In the United States, food on the public market is pooled from a variety of locations to ensure that the food sold in a given locality is not consistently deficient in selenium or other nutrients. Animals deprived of vitamin E exhibit sterility, muscular dystrophy, and other symptoms of free radical damage. Fortunately, few people lack vitamin E because it is so abundant in vegetables, eggs, and meat.

Key Point Review

1 What is an element? Which six elements are most abundant in the human body? What is a trace element?

2 What is an electron shell?

3 Define *atomic number, atomic mass,* and *atomic weight.*

4 How do isotopes of the same element differ from each other?

5 How is an ion different from an atom?

▼Objectives

When you have completed this section, you should be able to
- describe the different types of chemical bonds and explain how they form;
- show how compounds are represented by chemical formulae;
- define *isomer* and discuss the significance of isomers; and
- calculate the molecular weight of compounds.

Molecules and Compounds

Atoms are most biologically interesting when they chemically bond with each other to form **molecules.** Some molecules are composed of two or more atoms of the same element, such as oxygen (O_2), nitrogen (N_2), and hydrogen (H_2). Molecules composed of two or more different elements are called **compounds,** such as carbon dioxide (CO_2) and glucose ($C_6H_{12}O_6$).

Chemical Formulae

Molecules are represented by formulae that indicate the types and numbers of atoms present. The simplest is a **molecular formula,** which itemizes each element only once, with subscripts indicating how many atoms of each are present.

Molecular formulae conveniently fit on a single line of print and are the shortest way to represent compounds, but they are sometimes not informative enough. For example, both ethanol (grain alcohol) and ethyl ether have the molecular formula C_2H_6O, yet there is little doubt which one would better complement a steak dinner! They have the same numbers of the same atoms, but their chemical properties are different because their atoms are arranged differently (fig. 2.6). Molecules with identical atomic compositions but different arrangements of those atoms are called **structural isomers**[3] of each other. To show the difference between them, we use a **structural formula,** like those in figure 2.6, which shows the location of each atom. It is also possible to indicate their differences with a **condensed structural formula,** which can be written on one line. Ethanol is CH_3CH_2OH, and ether is CH_3OCH_3. Note how these condensed structural formulae relate to the structural formulae in figure 2.6. With a little practice, you will be able to look at a formula of either type and visualize the other one.

3. *mer* = parts

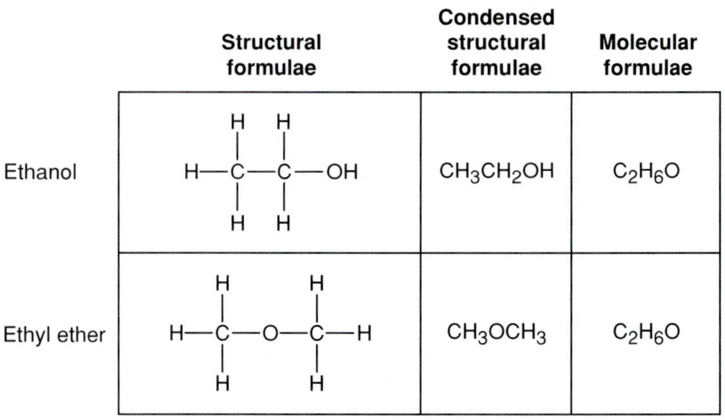

	Structural formulae	Condensed structural formulae	Molecular formulae
Ethanol		CH_3CH_2OH	C_2H_6O
Ethyl ether		CH_3OCH_3	C_2H_6O

Figure 2.6 Ethanol and ethyl ether are isomers of each other. Their molecular formulae are identical, but their structures and chemical properties are different.

Molecular Weight

The **molecular weight (MW)** of a compound is obtained by adding the atomic weights of its atoms. Rounding the atomic mass units (amu) to whole numbers, we can calculate the approximate MW of glucose ($C_6H_{12}O_6$), for example, as

$$
\begin{aligned}
6 \text{ C atoms} \times 12 \text{ amu each} &= 72 \text{ amu} \\
12 \text{ H atoms} \times 1 \text{ amu each} &= 12 \text{ amu} \\
6 \text{ O atoms} \times 16 \text{ amu each} &= \underline{96 \text{ amu}} \\
\text{Molecular weight (MW)} &= 180 \text{ amu}
\end{aligned}
$$

Molecular weight is needed to compute some measures of concentration, as we will see later in the chapter.

Chemical Bonds

A **chemical bond** is a force that attracts one atom to another. The three types of chemical bonds of greatest physiological interest are *ionic bonds, covalent bonds,* and *hydrogen bonds.*

Ionic Bonds

In the earlier discussion of ionization, it was shown that an element such as sodium easily gives up an electron to an element such as chlorine. The transfer of an electron creates two stable ions, the sodium ion (Na^+) and the chloride ion (Cl^-). Because these ions have opposite charges, and because oppositely charged particles are attracted to each other, the sodium and chloride ions tend to cling together and form the compound sodium chloride (NaCl), common table salt. This attraction of oppositely charged ions to each other is called an **ionic bond.** Ionic compounds generally form geometrically regular arrays called *crystals* (fig. 2.7).

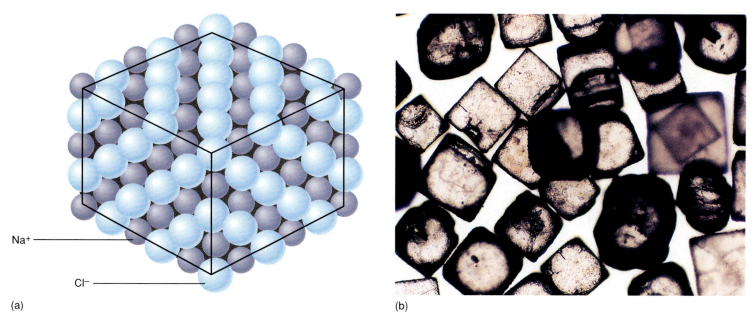

(a)   (b)

Figure 2.7 (*a*) Ionic bonds hold sodium and chloride ions together in a regular geometric lattice that produces a crystal of sodium chloride. (*b*) Sodium chloride (table salt) crystals (×400).

Some ions bond with two or more other ions at once. For example, calcium has two valence electrons. It can become stable by donating one electron to one chlorine atom, and the other electron to another chlorine, producing a calcium ion, Ca^{2+}, and two chloride ions, Cl^-. The resulting ionic compound is calcium chloride, $CaCl_2$.

Ionic bonds are weak because ions tend to separate (dissociate) from each other in the presence of something more attractive, such as water. The ionic bonds of NaCl are easily broken as the salt dissolves in water, because both the sodium and chloride ions are more attracted to water molecules than they are to each other. As we will see in the next chapter, water molecules cluster around each ion in a way that keeps them separated.

 Think About It

Would it be logical to assume that ionic bonds are common in the human body? Why or why not?

Covalent Bonds

Another way for atoms to form stable valence shells is to share electrons. For example, two hydrogen atoms can share their valence electrons to form a hydrogen molecule, H_2 (fig. 2.8). The two electrons, one donated by each atom, swarm around both nuclei in a dumbbell-shaped cloud. Bonds formed by the sharing of electrons

are called **covalent bonds.** If a single pair of electrons are shared, the bond is called a **single covalent bond.** This type of bond is symbolized with a single line between atomic symbols, for example, H—H. It is also common for atoms to share two pairs of electrons. In carbon dioxide, for example, the central carbon atom shares two electron pairs with each oxygen atom. Such bonds are called **double covalent bonds** and are symbolized by two lines—for example O—C—O.

When electrons spend approximately equal time around each nucleus, the bond is called a **nonpolar covalent bond** (fig. 2.9). The bond between two carbon atoms is nonpolar, for example. Nonpolar covalent bonds are the strongest of all chemical bonds. If two nuclei are not equally attractive to electrons, their atoms may form a **polar covalent bond,** in which the electrons spend more time orbiting the more attractive nucleus. When hydrogen bonds with oxygen, for example, the electrons are more attracted to the oxygen nucleus and orbit that nucleus more than they do the hydrogen nucleus. Since electrons carry a negative charge, this makes the oxygen region of the molecule slightly negative and the hydrogen region slightly positive. The Greek letter delta (δ) is used to symbolize a charge less than that of one electron or proton. A slightly negative region of a molecule is represented δ−, and a slightly positive region is represented δ+. Such a molecule is said to be a **dipole** because it has two charged poles. Water molecules provide a noteworthy example, for the central oxygen carries a δ− charge and each hydrogen carries a δ+ charge (fig. 2.10).

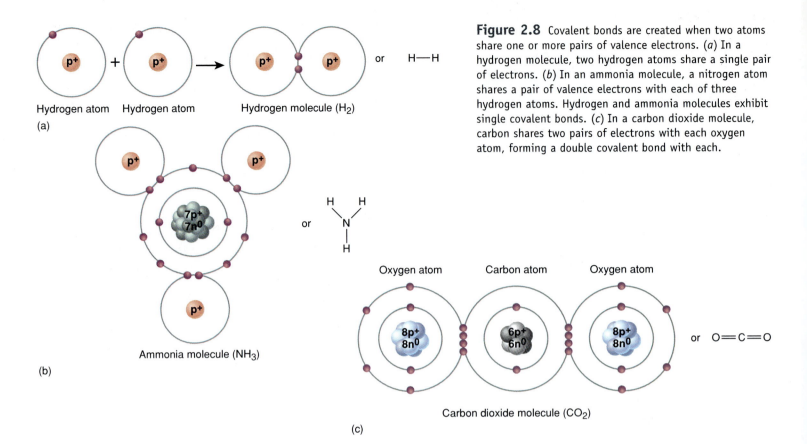

(a)

Hydrogen atom   Hydrogen atom   Hydrogen molecule (H_2)   or   H—H

(b)

Ammonia molecule (NH_3)

Oxygen atom   Carbon atom   Oxygen atom

or   O=C=O

Carbon dioxide molecule (CO_2)

(c)

Figure 2.8 Covalent bonds are created when two atoms share one or more pairs of valence electrons. (a) In a hydrogen molecule, two hydrogen atoms share a single pair of electrons. (b) In an ammonia molecule, a nitrogen atom shares a pair of valence electrons with each of three hydrogen atoms. Hydrogen and ammonia molecules exhibit single covalent bonds. (c) In a carbon dioxide molecule, carbon shares two pairs of electrons with each oxygen atom, forming a double covalent bond with each.

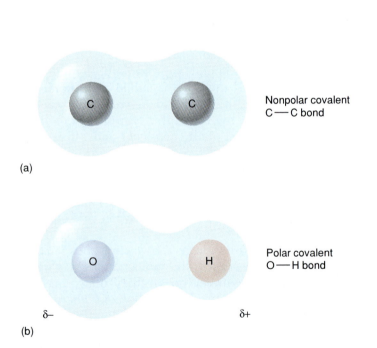

(a)

Nonpolar covalent
C—C bond

Polar covalent
O—H bond

(b)

Figure 2.9 (a) A nonpolar covalent bond, in which electrons spend equal time around both nuclei. (b) A polar covalent bond, in which two atoms share valence electrons unequally. In hydroxyl (—OH) groups, the electrons spend more time orbiting the oxygen nucleus than they do the hydrogen nucleus, giving the oxygen a slightly negative ($\delta-$) charge and the hydrogen a slightly positive ($\delta+$) charge.

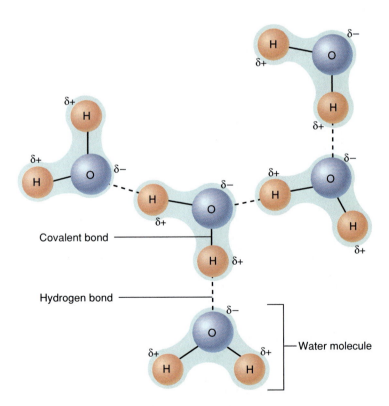

Covalent bond

Hydrogen bond

Water molecule

Figure 2.10 The polar covalent bonds of water molecules enable each oxygen atom to form a hydrogen bond with a hydrogen atom of a neighboring molecule. Thus the water molecules are weakly attracted to each other.

Hydrogen Bonds

A weak attraction called a **hydrogen bond** often forms between the slightly positive hydrogen of one molecule and the slightly negative oxygen or nitrogen of another. Water molecules, for example, are weakly attracted to each other by hydrogen bonds (fig. 2.10); this attraction is responsible for many of the biologically important properties of water to be discussed in the next chapter. Hydrogen bonds can also form between different regions of the same molecule. This is especially true of very large molecules such as proteins and DNA. These molecules fold or coil into precise three-dimensional shapes. Hydrogen bonds are symbolized by dotted or broken lines between atoms: —C—O···H—N—. Hydrogen bonds are the weakest of all the bond types discussed here, but they are enormously important to physiology.

Chemical bonds are summarized in table 2.4.

Key Point Review

6. How do isomers differ from each other? What is the difference between an isomer and an isotope?

7. What does chemical bonding have to do with full electron shells?

8. How is an ionic bond formed?

9. What is a covalent bond? How do single and double covalent bonds differ? How do nonpolar and polar covalent bonds differ?

10. What is a hydrogen bond? Why do hydrogen bonds depend on the existence of polar covalent bonds?

Table 2.4	Types of Chemical Bonds
Bond Type	**Definition and Remarks**
Ionic Bond	Relatively weak attraction between an anion and a cation. Easily disrupted in water, as when salt dissolves.
Covalent Bond	Sharing of one or more pairs of electrons between nuclei. Strongest type of chemical bond.
Single Covalent	Sharing of one electron pair.
Double Covalent	Sharing of two electron pairs. Often occurs between carbon atoms, between carbon and oxygen, and between carbon and nitrogen.
Nonpolar Covalent	Covalent bond in which electrons are equally attracted to both nuclei.
Polar Covalent	Covalent bond in which electrons are more attracted to one nucleus than to the other, resulting in slightly positive and negative regions on one molecule.
Hydrogen Bond	Weak attraction between polarized molecules or between polarized regions of the same molecule. Important in the three-dimensional folding and coiling of large molecules. Weakest of all bonds; easily disrupted by temperature and pH changes.

Mixtures

▼Objectives

When you have completed this section, you should be able to

• define *mixture* and distinguish between mixtures and compounds;

• show how three kinds of mixtures differ from each other; and

• describe some of the ways in which the concentration of a mixture can be expressed, and explain why different expressions are used for different purposes.

A **mixture** consists of substances that are physically blended together but not chemically combined. Each substance retains its own chemical properties. To contrast a mixture with a compound, consider sodium chloride again. Sodium is a lightweight metal that bursts into flame if exposed to water, and chlorine is a yellow-green poisonous gas that was used for chemical warfare in World War I. When these elements chemically react, they form ordinary table salt—sodium chloride. Clearly, the properties of the compound are very different from those of the elements of which it is composed. But if you were to put a little salt on your watermelon, the watermelon would taste salty and sweet because the sugar of the melon and the salt you added merely form a mixture in which each compound retained its individual properties.

Solutions, Colloids, and Suspensions

In human physiology, we are mostly concerned with mixtures of various substances in water. These mixtures can be classified as *solutions, colloids,* and *suspensions.*

Solutions

A **solution** consists of particles of matter called the **solute** mixed with a more abundant substance (usually water) called the **solvent.** The solute can be a gas, solid, or liquid—as in a solution of oxygen, sodium chloride, or alcohol in water, respectively. Solutions are defined by the following properties:

• The solute particles are under 1 nanometer (nm) in size. This is too small to scatter light appreciably, so solutions are usually transparent (fig. 2.11*a*).

• The solute and solvent cannot be visually distinguished from each other, even with a microscope.

• The solute does not separate from the solvent when the solution is allowed to stand.

• The solute particles will pass through most artificial selectively permeable membranes (as discussed in chapter 4).

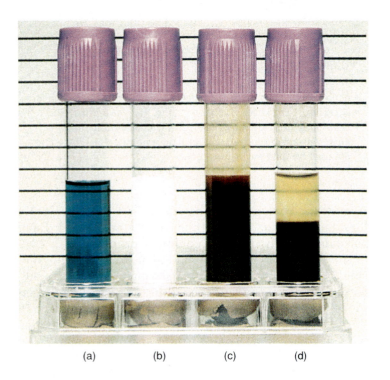

Figure 2.11 Solution, colloid, and suspension. (*a*) In solutions such as Benedict reagent (a copper sulfate solution), the particles are so small that they remain permanently mixed and the solution is transparent. (*b*) In colloids such as this milk, the particles are larger and scatter light, making the mixture cloudy or opaque; yet they are still small enough to remain permanently mixed. (*c*) In suspensions such as this freshly mixed blood, the particles also scatter light and make the mixture opaque. (*d*) The particles of a suspension, however, are too large to remain permanently mixed. Thus they settle out of the mixture, as seen here in a sample of blood that was left standing overnight, allowing the red blood cells to settle out of the plasma.

Colloids

A **colloid**[4] is defined by the following properties:

- The colloidal particles range from 1 to 100 nm in size. This is large enough to scatter light, so colloidal mixtures are usually cloudy (fig. 2.11*b*).
- The particles are still small enough to remain permanently mixed with the solvent when the mixture stands.
- The colloidal particles are too large to pass through most selectively permeable membranes.

Many colloids can change from liquid to gel states—gelatin desserts, agar culture media, the material that fills the space between cells in many of our tissues, and the fluid within our cells, for example. Proteins are the most common colloidal particles in the body.

Suspensions

A **suspension** is defined by the following properties:

- The suspended particles exceed 100 nm in size; therefore, suspensions are cloudy or opaque.
- The particles are too heavy to remain permanently suspended, so suspensions separate on standing.

....

4. *collo* = glue + *oid* = like, resembling

Blood is a suspension of cells in plasma. If allowed to stand, blood cells settle to the bottom of a tube (fig. 2.11*c, d*). An **emulsion** is a suspension of one liquid in another, such as oil and vinegar salad dressing. Many medications, including Kaopectate and milk of magnesia, are emulsions.

A single mixture can fit into more than one of these categories. Blood is a perfect example—a solution of sodium chloride, a colloid of protein, and a suspension of cells. Another example, milk, is a solution of calcium, a colloid of protein, and an emulsion of butterfat. Table 2.5 summarizes these types of mixtures and provides additional examples.

Think About It

Air is a mixture of nitrogen, oxygen, carbon dioxide, and other gases. Would you classify it as a solution, colloid, or suspension? Explain.

Measures of Concentration

Solutions are often described in terms of their **concentration**—the amount of solute in a given volume of solution. Concentration is expressed in different ways for different purposes, some of which are explained here. You may find the table of symbols, weights, and measures on the inside the front cover (and in appendix B) a helpful reference as you study this section.

Table 2.5	Types of Mixtures		
	Solution	**Colloid**	**Suspension**
Particle Size	<1 nm	1–100 nm	>100 nm
Appearance	Clear	Often cloudy	Cloudy-opaque
Will particles settle out?	No	No	Yes
Will particles pass through a selectively permeable membrane?	Yes	No	No
Examples	Glucose in blood	Proteins in blood	Blood cells
	Dissolved O_2 in water	Intracellular fluid	Cornstarch in water
	Saline solutions	Milk protein	Muddy water
	Sugar in coffee	Gelatin	Kaopectate

Weight per Volume

A simple expression of concentration is the weight of solute in a given volume of solution. For example, intravenous (I.V.) saline typically contains 8.5 g of NaCl per liter of solution (8.5 g/L). For many biological purposes, however, we must deal with smaller quantities such as milligrams per deciliter (mg/dL; 1 dL = 100 mL). For example, a person's serum cholesterol concentration may be 200 mg/dL, also expressed 200 mg/100 mL or 200 milligram-percent (mg-%).

Percentages

Percent concentrations are also simple to compute, but it is necessary to specify whether the percentage refers to the weight or the volume of solute in a given volume of solution. For example, if we begin with 5 g of dextrose (an isomer of glucose) and add enough water to make 100 mL of solution, the result will be 5% weight per volume (w/v) dextrose. A common intravenous fluid is D5W, which stands for 5% w/v dextrose in distilled water. By contrast, if we begin with 70 mL of ethanol and dilute it with water to make 100 mL of solution, the result will be 70% volume per volume (v/v) ethanol. Volume per volume percentages are commonly used when the solute is a liquid such as ethanol.

Molarity

Percent concentrations are easy to prepare, but they can be misleading. For example, if you were asked to determine whether the human tongue is more sensitive to glucose or sucrose, you might choose to prepare solutions of 1% glucose and 1% sucrose, place the same number of drops of each on the tongues of blindfolded subjects, and ask the subjects to rate their sweetness. You would not have a valid comparison, however, because even if you had measured each sugar precisely, in a more important sense you would not have had equal amounts of each.

If you had prepared a 1% w/v solution of each sugar, you would have used 1 g of each per 100 mL of solution. But glucose has a molecular weight of 180 and sucrose a molecular weight of 342. Therefore, 1 g of glucose would have many more molecules than 1 g of sucrose—just as a ton of bricks contains about 365 bricks of clay but only 65 bricks of lead. Each lead brick, like each sucrose molecule, weighs more. It is the number of molecules, not their weight, that determines a physiological effect. If one solution were judged sweeter than the other, it might have been because there were more molecules per drop in that solution, not because that sugar was a more effective stimulus.

When comparing the effects of different solutes, it is therefore necessary to use a unit of concentration that ensures a known *number of molecules* of a solute in a given volume, not merely a known *weight* of solute. This requires a knowledge of the molecular weight (MW) of the solute. Glucose has a MW of 180. One hundred eighty grams of it is known as its *gram molecular weight,* or 1 **mole.** Sucrose has a MW of 342, and thus 1 mole of sucrose is 342 g.

The Italian chemist Amedeo Avogadro (1776–1856) determined that 1 g molecular weight of any compound will contain the same number of molecules, regardless of the identity of the compound. Other chemists subsequently determined that this number—now called *Avogadro's number*—is 6.023×10^{23} (table 2.6). In other words, if you weigh out 1 mole of any substance, you will always have 6.023×10^{23} molecules of it. Such a large number is hard to imagine. If each molecule were the size of a pea, 6.023×10^{23} molecules would cover 60 earth-sized planets 10 feet deep!

Molarity is a measure of the number of moles of solute per liter of solution. A **one-molar** (1.0 M) **solution** contains 1 mole of solute per liter of solution. A 1.0 M solution of glucose would be 180 g/L, and 1.0 M sucrose would be 342 g/L. Both would have the same number of solute molecules in a given volume. A 1.0 M solution is

Table 2.6	Terms Related to Molecular Weight
Molecular Weight	Sum of the atomic weights of all atoms in a molecule.
Mole	The molecular weight of a substance in grams; for example, methane has a molecular weight of 16.043, so 1 mole of methane is 16.043 g.
Avogadro's Number	6.023×10^{23}, the number of molecules, atoms, or ions in 1 mole of any substance.

too concentrated for most biological applications, however. More often, biologists work with **millimolar solutions.** A millimole (mmol) is one-thousandth of a mole. For example, a 1.0 mM solution of glucose would be 0.18 g/L. Figure 2.12 shows the contrast between solutions of equivalent percent concentration and solutions of equivalent molarity.

Electrolyte Concentrations

The physiological effect of an electrolyte depends on two factors: the concentration of ions in solution and the charge on the ions. A calcium ion (Ca^{2+}) has twice the effect of a sodium ion (Na^+), for example, because it carries twice the charge. One **equivalent (Eq)** of an electrolyte is the amount that would neutralize the charges in 1 mole of hydrogen ions (H^+) or hydroxyl ions (OH^-). For example, 1 mole (58.4 g) of NaCl yields 1 mole, or 1 Eq, of Na^+. A solution of 58.4 g of NaCl per liter would be 1 Eq of Na^+ per liter. One mole of sulfuric acid (H_2SO_4, MW 98) yields 2 moles of positive charges (H^+), so 98 g of sulfuric acid per liter would be a solution of 2 Eq/L.

These, too, are very high concentrations for living systems, so more often we speak of **milliequivalents per liter (mEq/L).** A milliequivalent is one-thousandth of an equivalent. If you know the millimolar concentration of an electrolyte, you can easily convert this to mEq/L by multiplying it by the valence of the ion.

$$1 \text{ mM Na}^+ = 1 \text{ mEq/L}$$
$$1 \text{ mM Ca}^{2+} = 2 \text{ mEq/L}$$
$$1 \text{ mM Fe}^{3+} = 3 \text{ mEq/L}$$

.... Key Point Review ....

11 Distinguish between a mixture, solution, colloid, and suspension. Give examples of the last three.

12 Explain why it may be inappropriate for some purposes to express chemical concentrations as percentages.

13 What is a mole? How many particles are in 1 mole?

14 Explain how molarity is calculated.

15 Explain what is meant by milliequivalents per liter. What substances are measured in this way?

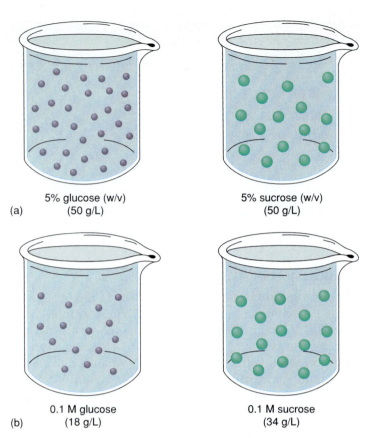

(a) 5% glucose (w/v) (50 g/L)   5% sucrose (w/v) (50 g/L)

(b) 0.1 M glucose (18 g/L)   0.1 M sucrose (34 g/L)

Figure 2.12 (a) Five percent glucose and sucrose solutions have the same weight of sugar per volume of solution, but not the same number of molecules, because each sucrose molecule is nearly twice as heavy as a glucose molecule. (b) A 0.1 molar (M) glucose solution has the same number of molecules per volume as 0.1 M sucrose. Molarity is a unit of concentration that takes differences in molecular weight into account.

Acids, Bases, and pH

▼**Objectives**
When you have completed this section, you should be able to
• define *acid, base,* and *pH;*
• describe the pH scale; and
• discuss the role of buffers in the body's acid-base balance.

Acids and Bases

Most people have some sense of what acids and bases are. Advertisements are full of references to excess stomach acid and pH-balanced shampoos. We know that drain cleaners (which are strong bases) or battery acid can cause serious chemical burns. But what exactly do "acidic" and "basic" mean, and how can they be quantified?

An **acid** is any molecule that releases a proton (H^+) in water—it is a *proton donor.* A **base,** or *alkali,* is a proton acceptor. Since hydroxyl (hi-DROCK-sil) ions (OH^-) accept H^+ and produce water, many bases are sub-

Table 2.7   The pH Scale

Molarity of OH⁻		Molarity of H⁺		Log [H⁺]	pH = −log [H⁺]
0.00000000000001	$= 10^{-14}$	1.0	$= 10^{0}$	0	0
0.0000000000001	$= 10^{-13}$	0.1	$= 10^{-1}$	−1	1
0.000000000001	$= 10^{-12}$	0.01	$= 10^{-2}$	−2	2
0.00000000001	$= 10^{-11}$	0.001	$= 10^{-3}$	−3	3
0.0000000001	$= 10^{-10}$	0.0001	$= 10^{-4}$	−4	4
0.000000001	$= 10^{-9}$	0.00001	$= 10^{-5}$	−5	5
0.00000001	$= 10^{-8}$	0.000001	$= 10^{-6}$	−6	6
0.0000001	$= 10^{-7}$	0.0000001	$= 10^{-7}$	−7	7
0.000001	$= 10^{-6}$	0.00000001	$= 10^{-8}$	−8	8
0.00001	$= 10^{-5}$	0.000000001	$= 10^{-9}$	−9	9
0.0001	$= 10^{-4}$	0.0000000001	$= 10^{-10}$	−10	10
0.001	$= 10^{-3}$	0.00000000001	$= 10^{-11}$	−11	11
0.01	$= 10^{-2}$	0.000000000001	$= 10^{-12}$	−12	12
0.1	$= 10^{-1}$	0.0000000000001	$= 10^{-13}$	−13	13
1.0	$= 10^{0}$	0.00000000000001	$= 10^{-14}$	−14	14

stances that release hydroxyl ions—sodium hydroxide (NaOH), for example. A base does not have to be a hydroxyl donor, however. Ammonia, NH_3, is also a base. It does not release hydroxyl ions, but it readily accepts hydrogen ions to become the ammonium ion, NH_4^+.

In pure water, one out of every 10 million molecules spontaneously ionizes to form hydrogen and hydroxyl ions: $H_2O \leftrightarrow H^+ + OH^-$. Strictly speaking, hydrogen ions do not exist in this free state; rather they combine with other water molecules to form hydronium ions, H_3O^+. Thus, the preceding equation can be rewritten: $2\ H_2O \leftrightarrow H_3O^+ + OH^-$. For convenience, however, we will use H^+ in chemical equations throughout this book.

pH

Acidity is expressed in terms of **pH,** a measure of the molarity of H^+. Molarity is represented by square brackets, so the molarity of H^+ is symbolized $[H^+]$. pH is defined as the negative logarithm of hydrogen ion molarity—that is, $pH = -\log [H^+]$.

Pure water has a neutral pH because it contains equal amounts of hydrogen (or hydronium) and hydroxyl ions. The molarity of hydrogen ion and the pH of water are therefore

$$[H^+] = 0.0000001 \text{ molar} = 10^{-7}\text{ M}$$
$$\log [H^+] = -7$$
$$pH = -\log [H^+] = 7$$

The pH scale (table 2.7) was invented in 1909 by Danish biochemist and brewer Sören Sörensen as a way of measuring the acidity of beer. The scale extends from 0.0 to 14.0. A solution with a pH of 7 is **neutral;** solu-

tions with a pH below 7 are **acidic;** and solutions with a pH greater than 7 are **basic,** or **alkaline.** Note that the lower the pH value, the more hydrogen ions a solution has and the more acidic it is. Since the pH scale is logarithmic, a change of one whole number on the scale represents a 10-fold change in the H^+ concentration. In other words, a solution with a pH of 4 is 10 times as acidic as one with a pH of 5 and 100 times as acidic as one with a pH of 6. The pH values of some body fluids and common household substances are listed in table 2.8.

Slight disturbances of pH can seriously disrupt normal physiological functions, so it is important that the body be able to carefully control pH (see special topic 2.2). Blood, for example, normally has a pH ranging from 7.35 to 7.45. Deviations from this range can cause dizziness, fainting, coma, tremors, paralysis, or even death. Chemical solutions that resist changes in pH are called **buffers.** Buffers and pH regulation are considered in detail in chapter 24.

Think About It
A pH of 7.20 is slightly alkaline, yet a blood pH of 7.20 is called acidosis. Why do you think it is called this?

.... **Key Point Review** ....

16 Define *acid* and *base,* and explain how acids and bases affect the pH of a solution.

17 Explain why each digit of the pH scale represents a 10-fold difference in hydrogen ion concentration.

18 What is a buffer? Why are buffers important?

pirin is in the acidic environment of the stomach, for example, it is uncharged and passes easily through the stomach lining into the bloodstream. Here it encounters an alkaline pH, whereupon it becomes ionized. In this charged state, it is unable to pass back through the membrane; thus it accumulates in the blood. This series of events, called *ion trapping*, or *pH partitioning*, can be controlled to help clear poisons from the body. The pH of the urine, for example, can be manipulated so that poisons become trapped there and thus excreted from the body rapidly.

Table 2.8 pH Values of Selected Body Fluids and Other Substances at 25°C

Substance	pH
1 M hydrochloric acid (HCl)	0.0
Gastric juice	1.2–3.0
Carbonated soft drinks, citrus juices	3.0–3.5
Coffee (black)	5.0
Urine	4.6–8.0
Cow's milk	6.5–6.6
Saliva	6.3–7.3
Pure water	7.0
Blood	7.35–7.45
Egg white	7.6–8.0
Pancreatic juice	7.1–8.2
Bile	7.6–8.6
Milk of magnesia	10.0–11.0
Household ammonia	10.5–11.8
Oven cleaner, drain cleaner, lye	13.5
1 M sodium hydroxide (NaOH)	14.0

Chemical Reactions

Objectives
When you have completed this section, you should be able to
- define *chemical reaction*;
- recognize the reactants and products in a chemical equation and determine whether the equation is balanced;
- list and define the fundamental types of chemical reactions; and
- describe the factors that govern the speed and direction of a reaction.

Chemical Equations

A **chemical reaction** is a process in which a chemical bond is formed or broken. **Reactants** are substances that enter into a reaction, and **products** are the new sub-

stances produced by it. A **chemical equation** is a symbolic representation of the course of a reaction. The reactants are typically indicated on the left and the products on the right, with an arrow pointing from the reactants to the products (the arrow represents the word *yields*). The left-to-right rule is not always followed, however. In many biochemical equations, reaction chains are written vertically or even in circles.

As an example of a chemical reaction, consider this common occurrence: If you open a bottle of wine and let it sit for several days, you may be disappointed to find it tastes sour. Wine "turns to vinegar" because oxygen gets into the bottle and reacts with ethanol to produce acetic acid and water. Acetic acid gives the tart flavor to vinegar and spoiled wine. The equation for this reaction is

$$CH_3CH_2OH \ + \ O_2 \ \rightarrow \ CH_3COOH \ + \ H_2O$$

Ethanol	Oxygen	Acetic acid	Water

Ethanol and oxygen are the reactants, and acetic acid and water are the products.

According to the **law of conservation of mass,** there is no change in mass in a chemical reaction—matter cannot appear out of nothing nor can it disappear. If the equation is **balanced,** every atom of the reactants is accounted for in the products. In the case of the preceding reaction, our tally could be shown as follows:

Reactants		Products	
Ethanol, C_2H_6O	MW 46.07	Acetic acid, $C_2H_4O_2$	MW 60.05
Oxygen, O_2	MW 32.00	Water, H_2O	MW 18.02
Total: 2 C, 6 H, 3 O	78.07	Total: 2 C, 6 H, 3 O	78.07

We can see that our equation is balanced with respect to number of atoms and total molecular weight.

This was a simple equation in which there was a 1:1 ratio of reactants to products. Some chemical equations, however, are more complex. When we completely "burn" glucose to carbon dioxide and water, there is a 6:1 ratio of oxygen, carbon dioxide, and water to glucose:

$$C_6H_{12}O_6 + 6\ O_2 \rightarrow 6\ CO_2 + 6\ H_2O$$

Classes of Reactions

Chemical reactions can be classified as *decomposition, synthesis,* or *exchange reactions.* In **decomposition reactions,** a large molecule is broken down into two or more smaller ones (fig. 2.13a); symbolically, AB → A + B. When you eat a baked potato, for example, digestive enzymes decompose starch into thousands of molecules of glucose, and most cells further decompose glucose to water and carbon dioxide. Starch, a very large molecule, ultimately yields about 36,000 molecules of H_2O and CO_2.

Synthesis reactions are just the opposite of decomposition reactions. Two or more small molecules are combined to form a larger one; symbolically, A + B → AB (fig. 2.13b). When the body makes proteins, for example, it may combine several hundred molecules called amino acids into one protein molecule.

In **exchange reactions,** two molecules exchange atoms or groups of atoms (AB + CD → AC + BD) (fig. 2.14). For example, when stomach acid, HCl, enters the small intestine, the pancreas secretes sodium bicarbonate, $NaHCO_3$, to neutralize it. The reaction between the two is $NaHCO_3 + HCl \rightarrow NaCl + H_2CO_3$. We could say that the sodium atom has exchanged its bicarbonate (HCO_3) group for a chlorine atom.

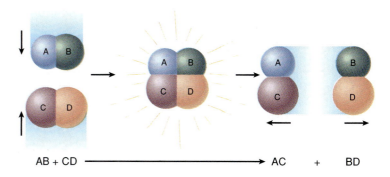

AB + CD ⟶ AC + BD

Figure 2.14 An exchange reaction occurs when two molecules collide and exchange groups of atoms.

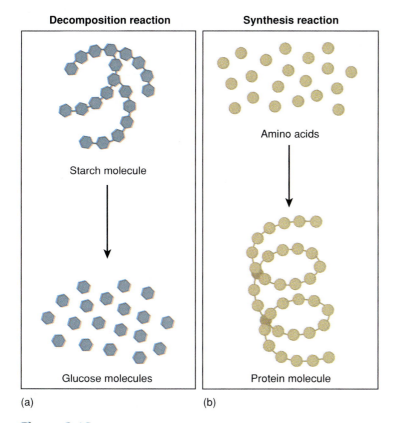

Decomposition reaction	Synthesis reaction

Starch molecule

Amino acids

Glucose molecules

Protein molecule

(a)   (b)

Figure 2.13 (*a*) In a decomposition reaction, a large, complex molecule is broken down into its simpler subunits. (*b*) In a synthesis reaction, molecular subunits are joined to form a larger, more complex molecule.

Reversible Reactions

Reversible reactions are represented by a double-headed arrow. For example, carbon dioxide combines with water to produce carbonic acid, which in turn decomposes into bicarbonate ions and hydrogen ions:

$$CO_2 + H_2O \leftrightarrow H_2CO_3 \leftrightarrow HCO_3^- + H^+$$

Carbon dioxide	Water	Carbonic acid	Bicarbonate ion	Hydrogen ion

The direction in which the reaction goes is determined by which substances are most abundant. If there is a surplus of CO_2, the reaction proceeds toward the right and produces bicarbonate and hydrogen ions. If bicarbonate and hydrogen ions are present in excess, the reaction proceeds to the left and produces carbon dioxide and water. Reversible reactions follow the **law of mass action:** The direction of a reversible reaction is from the side with the greater quantity of reactants to the side with the lesser quantity. This law will help to clarify concepts discussed in later chapters. It explains, for example, why hemoglobin picks up oxygen in the lungs and then releases it to muscle cells (chapter 22).

In the absence of any upsetting influence, reversible reactions exist in an **equilibrium state,** in which the ratio of products and reactants is stable. The carbonic acid reaction, for example, normally maintains a 20:1 ratio of bicarbonate ions to carbonic acid molecules. This equilibrium can be upset, however, by such influences as an excess of hydrogen ions. This would drive the reaction to the left and produce a ratio lower than 20:1. Another equilibrium reaction we have already touched upon is the ionization of water—$H_2O \leftrightarrow H^+ + OH^-$.

Reaction Rates

The basis for chemical reactions is molecular motion and collisions. All molecules are in constant motion, and chemical reactions occur when mutually reactive molecules collide with each other (fig. 2.14). The rate of a chemical reaction depends on the frequency and force of these collisions and the nature of the reactants. Some of the factors that affect reaction rates are as follows:

- **Concentration.** Reaction rates increase when the reactants are more concentrated. This is because the molecules are more crowded and collide with each other more frequently.
- **Temperature.** Reaction rates increase as the temperature rises. This is because heat causes molecules to move more rapidly and collide with greater force and frequency.
- **Catalysts** (CAT-uh-lists). These are substances that temporarily bind to reactants and hold them in a favorable position to react with each other. By reducing the element of chance in molecular collisions, a catalyst speeds up a reaction. The catalyst then releases the products and is available to repeat the process with more reactants. The catalyst itself is not permanently consumed or changed by the reaction. The most important biological catalysts are *enzymes,* discussed in detail in the next chapter.

Key Point Review

19 What are the substances that enter into a chemical reaction called? What is the term for new substances that result from the reaction?

20 What are the differences between decomposition, synthesis, and exchange reactions? Write a generalized equation for each type using the letters *A* through *D*.

21 List three factors that affect the rates of chemical reactions. Explain how each factor works.

Energy

▼Objectives
When you have completed this section, you should be able to
- define *work* and *energy;*
- list some biologically important forms of energy; and
- discuss ionizing radiation and its biological effects.

Work and Energy

To do **work** means to move something, whether it is a muscle or a molecule. Some examples of physiological work are breaking chemical bonds, building molecules, pumping blood, and contracting skeletal muscles. All of the body's activities are forms of work.

Energy is the capacity to do work and may be broadly classified as *kinetic* or *potential.* **Kinetic energy** is the energy of motion. Skeletomuscular movements, the flow of ions into a cell, and vibration of the eardrum are examples. **Potential energy** is the capacity for two stationary objects to interact with each other and release kinetic energy. For example, there is potential energy between the earth and the water behind a dam. It can be converted to kinetic energy if a spillway is opened and water is allowed to flow over the dam, and this kinetic energy can be tapped for such purposes as generating electricity. Like water behind a dam, ions concentrated on one side of a cell membrane have potential energy that can be released by opening gates in the membrane and allowing them to flow through. Their kinetic energy can be tapped to contract a muscle, send a nerve signal, or perform other bodily functions.

Within the two broad categories of potential and kinetic energy, there are several forms of energy relevant to human physiology. **Chemical energy** is potential energy stored in the bonds of chemical compounds. Chemical reactions can release this energy and make it available for physiological work. **Heat** is the kinetic energy of molecular motion. The temperature of a substance is a measure of the rate of this motion, and adding heat to a substance increases this rate. **Electromagnetic energy** is the kinetic energy of moving "packets" of radiation called *photons.* The most familiar form of electromagnetic energy is light, but other, invisible forms—infrared and ultraviolet rays, X rays, and gamma rays—also have important physiological effects. **Electrical energy** exists in both potential and kinetic forms. It is potential energy when there is a concentration of charged particles that have accumulated at a certain point (for example, on a battery terminal or on one side of a cell membrane). It is kinetic energy when these particles begin to move and create an electrical current (for example, when electrons move through your household wiring or sodium ions move through a nerve cell membrane). The rest of this chapter is concerned mainly with chemical energy, but first we will consider radioactivity and its biomedical relevance.

Radioisotopes and Ionizing Radiation

We saw earlier that different isotopes of an element behave the same chemically. This is because chemical behavior is determined by the number of electrons in an atom, and isotopes differ only in their number of neutrons. Isotopes do, however, exhibit differences in physical behavior because many of them are unstable. They **decay** (break down) to more stable isotopes by giving off nuclear particles or radiation. Unstable isotopes are called **radioisotopes,** and the process of decay is called **radioactivity** (see special topic 2.3). Every element has at least one radioisotope. Oxygen, for example, has five radioactive isotopes and three stable ones. All of us contain radioisotopes such as ^{14}C and ^{40}K—that is, we are all mildly radioactive!

French scientist Henri Becquerel discovered radioactivity in 1896 when he found that uranium darkened photographic plates even through several thick layers of paper. Marie and Pierre Curie discovered that polonium and radium can do the same, and Marie Curie coined the term *radioactivity*. The Curies and Becquerel shared the 1903 Nobel Prize for Physics. **Marie Curie** (1867–1934) was the first woman in France to receive a Ph.D. and the first woman anywhere to receive a Nobel Prize. She invented radiation therapy for breast and uterine cancer, for which she received a second Nobel Prize for Chemistry in 1911. Curie crusaded to train women for careers in science, and in World War I she and her daughter Irene trained physicians in the use of X-ray machines.

In the wake of such discoveries, radium was regarded as a wonder drug. Slow to realize its danger, people drank radium tonics and flocked to health spas to bathe in radium-enriched waters. Also oblivious to the danger, Marie Curie herself suffered extensive damage to her hands from handling radioactive material and died at age 67 from radiation poisoning. The year after her mother died, Irene and her husband Frederick Joliot were awarded a Nobel Prize for work in artificial radioactivity and synthetic radioisotopes. Also a martyr to her science, **Irene Joliot-Curie** (1897–1956) died of leukemia, possibly induced by radiation exposure.

When a nucleus decays, it emits **ionizing radiation**—particles or electromagnetic rays with enough energy to eject electrons from atoms, which produces highly reactive ions and free radicals. Radioactivity is a two-edged sword that can cause and cure disease. It causes immediate death at high radiation doses, and at lesser doses it can result in leukemia, lung cancer, and birth defects (see special topic 2.4). Yet the field of **nuclear medicine** is based on the controlled use of ionizing radiation for medical diagnosis and cancer treatment (see chapter essay, p. 72).

There are three kinds of ionizing radiation: *alpha (α) particles, beta (β) particles,* and *gamma (γ) rays.* An **α particle** is composed of two protons and two neutrons (equivalent to a helium nucleus), and a **β particle** is a free electron. Alpha particles are too large to penetrate the skin, and β particles can penetrate only a few millimeters. Thus, these particles are relatively harmless when emitted by sources outside the body. They are very dangerous, however, when emitted by radioisotopes that have gotten into the body. Strontium-90 (^{90}Sr), for example, has been released by atmospheric testing of nuclear weapons and by accidents such as the 1986 explosion and fire at the Chernobyl nuclear power plant in the Ukraine. Strontium-90 settles onto pastures, gets into cow's milk, and behaves chemically like calcium. It becomes incorporated into the bones and emits β particles for many years. Electromagnetic rays emitted by such elements as uranium and plutonium are **γ rays.** They have high energy and penetrating power and are very dangerous even when emitted by sources outside the body.

Each radioisotope has a characteristic **physical half-life,** the time required for 50% of its atoms to decay to a more stable state. One milligram of ^{90}Sr, for example, would be half gone in 28 years. In 56 years, there would still be 0.25 mg left; in 84 years, 0.125 mg; and so forth. Many radioisotopes are much longer-lived. The half-life of ^{40}K, for example, is 1.3 billion years. Nuclear power plants produce hundreds of radioisotopes that will be intensely radioactive for at least 10,000 years—longer than the life of any disposal container yet conceived. The **biological half-life** of a radioisotope is the time required for half of it to disappear from the body. This is a function of both physical decay and physiological clearance from the body. Cesium-137, for example, has a physical half-life of 30 years but a biological half-life of only 17 days. Chemically, it behaves like potassium, and it is quite mobile and rapidly excreted by the kidneys.

Radiation Exposure

There are several ways to measure the intensity of ionizing radiation, the amount absorbed by the body, and the relative biological effects of different forms of radiation. To understand these measurements requires a grounding in physics beyond the scope of this book, but we can appreciate some relative values in millirems[5] (mrem). The average American absorbs about 295 mrem of naturally occurring **background radiation** annually (table 2.9). Much of this is from *radon,* a gas produced by the radioactive decay of uranium in granite. Radon enters buildings through cracks in their foundations and through well water. It emits α particles and can cause lung cancer if inhaled in sufficient quantity. Radon exposure varies with the amount of granite in different geographic localities but averages about 200 mrem/year. In recent years, there has been a move to make buildings more airtight to better conserve energy, but this has also resulted in trapping more radon. The best protection against radon is well-sealed foundations and adequate ventilation.

5. rem = **R**oentgen **e**quivalent in **m**an (1 rem = 1,000 mrem)

Table 2.9 Annual Doses of Radiation Received by the Average U.S. Resident	
Source	Millirem/year
Background Radiation (natural sources)	
Cosmic rays	28
Radioactive earth and air	28
Radioisotopes in body tissues	39
Inhaled radon	≈200
Human Devices and Uses	
Consumer products	10
Medical and dental diagnostics	39
Radiation therapy	14
Total	**358**

Source: "Ionizing radiation exposure of the population of the United States," Report No. 93 of the National Council on Radiation Protection and Measurement, 1987.

We are also exposed to radiation from medical X rays, radiation therapy, and consumer products such as color televisions, smoke detectors, and luminous watch dials. Such voluntary exposure must be considered from the standpoint of its risk-to-benefit ratio. The benefits of a smoke detector or mammogram far outweigh the risk from exposure to this amount of radiation. Radiation therapists and radiologists face a greater risk than their patients, however. U.S. federal standards set a limit of 5 rem/year as acceptable occupational exposure to ionizing radiation.

Key Point Review

22 Define *energy* and distinguish kinetic from potential energy.

23 Discuss the three forms of ionizing radiation, noting the distinctions between them.

24 What is the greatest source of background radiation for the average U.S. resident? How can it be avoided?

25 How much radiation does the average U.S. resident absorb in a year? What is considered the safe limit of occupational exposure for a radiation therapist?

Thermodynamics and Metabolism

▼Objectives
When you have completed this section, you should be able to
• state the first two laws of thermodynamics and explain why these laws are relevant to biological function;
• describe the nature of endergonic and exergonic reactions;
• describe the nature of oxidation and reduction reactions; and
• define *metabolism* and explain how catabolism and anabolism differ.

Human physiology depends on the transfer of chemical energy from one molecule to another and the conversion of one form of energy to another—as when muscles contract and chemical energy is changed to heat and motion. **Thermodynamics** is the science of heat and energy conversions.

The Laws of Thermodynamics

The **first law of thermodynamics** states that *energy can be converted from one form to another, but it can neither be created nor destroyed.* This is not strictly true of nuclear reactions, where small amounts of matter are converted into enormous amounts of energy, but it is valid for the purposes of chemistry and physiology. It means, for example, that as you consume a certain amount of glucose jogging a few laps around the track, the energy that was once in the chemical bonds of glucose is converted to other forms, such as the movement of your body and the heat you feel; however, none of it simply disappears.

The **second law of thermodynamics** states that *in every energy transfer, some energy becomes heat and can no longer do useful work.* For example, when we oxidize glucose we transfer much of its energy to a molecule called ATP, and then ATP can be used to do other work. It might be nice if we could transfer *all* the energy of our dietary glucose to ATP, but the second law says this is impossible—some will always be lost. Indeed, we only conserve about 40% of it and lose the other 60% as body heat, so we say this process has a 40% efficiency. This may seem poor, but it is much better than the efficiency of most machines. Your automobile, for example, probably has about a 17% efficiency. For every $100 spent on gasoline, $17 moves your car down the road and $83 results merely in air pollution and engine heat. This heat may provide winter comfort, however, and we also make use of the heat generated by our physiology. The body requires a certain amount of heat for its metabolic processes to work normally, and it sometimes burns fat or other fuel just for the heat that is generated.

Free energy is the energy available in a system to do work. The second law says that unless a system exchanges energy with its surroundings, it loses free energy with time. A system's degree of disorder is called its **entropy.** The less free energy a system has, the greater its entropy. The human body represents a highly ordered state of matter, and it takes a lot of work to keep it that way. We see this principle at work all around us. If we do not maintain our house, it falls apart. If we do not maintain our body, it too falls apart. We can forestall this by putting energy into the system (food and health maintenance), but entropy inevitably

Acute radiation sickness is caused by high doses of ionizing radiation, usually as the result of accidental exposure. The nervous system, digestive tract, and hemopoietic[6] (HE-mo-poy-ET-ic) tissues suffer the most damage. Whole-body doses up to 100 rem produce nausea and vomiting but are not usually fatal. Doses of 100 to 200 rem are associated with an increased incidence of cancer. Doses of 200 to 1,000 rem produce *hemopoietic* *syndrome*—death of lymphatic tissue and bone marrow, resulting in a deficiency of infection-fighting white blood cells. Overwhelming infection or hemorrhage, especially around the body orifices, typically kills the individual within 8 to 50 days. Doses of 600 to 3,000 rem produce *GI syndrome*. The intestinal lining is destroyed, massive amounts of blood plasma seep into the intestines, the body becomes swollen, and there is copious vomiting and diarrhea. So much fluid leaves the bloodstream for the intestines and tissues that the circulatory system fails, and death usually occurs within 3 to 10 days. A whole-body dose of 3,000 rem or more causes *cerebral syndrome,* characterized by vomiting and drowsiness. Convulsions follow, and death occurs within a few hours. The median lethal dose (LD_{50}) of radiation is the dose that is fatal to 50% of exposed individuals—about 400 to 500 rem.

wins in the end, and we are forced to reconcile ourselves to the inevitability of death.

Exergonic and Endergonic Reactions

Metabolism refers to all the chemical reactions of the body necessary to maintain life. There are two divisions of metabolism called *catabolism* and *anabolism.* **Catabolism**[7] (ca-TAB-oh-lizm) consists of energy-releasing decomposition reactions. Such reactions break covalent bonds, produce smaller molecules from larger ones, and release energy that can be used for other physiological work. Energy-releasing reactions are called **exergonic**[8] **reactions.** If you hold a beaker of water in your hand and pour sulfuric acid into it, the beaker will get so hot you may have to put it down. If you break down energy-storage molecules to run a race, you too will get hot. In both cases, the heat signifies that exergonic reactions are occurring.

Anabolism[9] (ah-NAB-oh-lizm) consists of synthesis reactions, such as the production of protein or fat molecules. These reactions require an energy input and are therefore called **endergonic**[10] **reactions.** Anabolism is driven by energy released from other molecules by catabolism, so endergonic and exergonic processes—anabolism and catabolism—are inseparably linked in human metabolism.

---

6. *hemo* = blood + *poietic* = forming
7. *cata* = down, to break down
8. *ex, exo* = out + *erg* = work
9. *ana* = up, to build up
10. *end* = in

Oxidation and Reduction

Oxidation refers to any chemical reaction in which a molecule gives up electrons and releases free energy. A molecule that has undergone this process is said to be **oxidized,** and whatever took the electrons from it is said to be an **oxidizing agent,** or **electron acceptor.** The term *oxidation* stems from the fact that oxygen has a strong tendency to accept electrons from other atoms; it is a strong oxidizing agent. Therefore, we can sometimes conclude that an oxidation reaction has occurred from the fact that oxygen has been added to a molecule. The rusting of iron, for example, is a slow oxidation process in which oxygen is added to iron (Fe) to form iron oxide (Fe_2O_3). It should be noted, however, that many oxidation reactions do not involve oxygen at all. For example, when yeast ferments glucose to alcohol, no oxygen is required; indeed, the alcohol *contains less oxygen* than the sugar originally did, but it is *more oxidized* than the sugar:

$$C_6H_{12}O_6 \rightarrow 2\ CH_3CH_2OH + 2\ CO_2$$
$$\text{Glucose} \qquad\qquad \text{Ethanol}$$

When electrons are removed from one molecule, they must be accepted by another, the oxidizing agent. When a molecule accepts electrons, it is said to be **reduced,** and it has a higher free energy content than it did before the reaction. A molecule that donates electrons to another is called a **reducing agent,** or **electron donor.** The oxidation of one molecule is always accompanied by the reduction of another, so these electron transfers are known as **oxidation-reduction reactions,** or simply **redox reactions.**

It is not necessary that *only* electrons be transferred in a redox reaction. Often, the electrons are transferred in the form of a hydrogen atom or pair of

Table 2.10 Energy Transfer Reactions in the Human Body

Exergonic Reactions	Reactions in which there is a net release of energy. The products have less total free energy than the reactants did.
Oxidation	An exergonic reaction in which electrons are removed from a reactant. Electrons may be removed one or two at a time and may be removed in the form of hydrogen atoms (H or H_2). The product is then said to be oxidized.
Decomposition	A reaction in which larger molecules are broken down into smaller ones, as in digestion and cell respiration.
Catabolism	The sum of all decomposition reactions in the body.
Endergonic Reactions	Reactions in which there is a net input of energy. The products have more total free energy than the reactants did.
Reduction	An endergonic reaction in which electrons are donated to a reactant. The product is then said to be reduced.
Synthesis	A reaction in which two or more smaller molecules are combined into a larger one, as in the synthesis of protein and glycogen.
Anabolism	The sum of all synthesis reactions in the body.

hydrogen atoms. The fact that a proton (the hydrogen nucleus) is also transferred is immaterial to whether we consider a reaction oxidation or reduction.

We can summarize oxidation and reduction as follows, letting H_2 represent a pair of electrons (hydrogen atoms):

$$AH_2 + B \rightarrow A + BH_2$$

High-energy reduced state   Low-energy oxidized state   Low-energy oxidized state   High-energy reduced state

AH_2 is a reducing agent because it reduces B, and B is an oxidizing reagent because it oxidizes AH_2. Table 2.10 summarizes these energy transfer reactions.

Key Point Review

26 Define *metabolism, catabolism,* and *anabolism.*

27 What does *oxidation* mean? What does *reduction* mean? Which of them is endergonic and which is exergonic?

28 When sodium chloride is formed, which element—sodium or chlorine—is oxidized? Which one is reduced?

CHAPTER ESSAY

Radioisotopes in Research, Diagnosis, and Therapy

Although radioisotopes can be dangerous, they are commonly used by researchers and clinicians for beneficial purposes. In research, isotopes are used as *tracers.* Suppose, for example, we want to know which cells of the uterus are stimulated by estrogen and where in the cell the estrogen becomes concentrated. Estrogen molecules can be synthesized with a radioisotope such as tritium (^3H) in place of some of the hydrogen (^1H). This "tagged" hormone is injected into laboratory animals, and tissue sections (thin slices) of the uterus are made later. These sections are mounted on slides coated with photographic emulsion. Over a time, radiation emitted by the tritium will cause tiny black dots to appear in the emulsion wherever the hormone molecules are concentrated. When the slides are developed like a photograph and examined under the microscope, these dots indicate which cells took up the hormone and where the hormone became localized in the cell (fig. E.1). This procedure, called *autoradiography,* has yielded a great deal of knowledge of hormone action and other physiology.

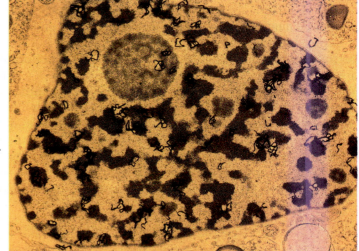

Figure E.1 An autoradiogram of the nucleus of a protist. Radioactive thymidine, labeled with tritium (^3H), has accumulated in the nucleus and exposed the photographic emulsion there, as indicated by the dark blotches. This identifies the nucleus as the site where thymidine is incorporated into newly synthesized DNA.

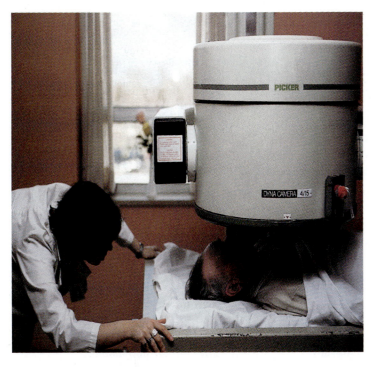

Figure E.2 Scanning a patient's radioactively labeled thyroid gland with a scintillation counter.

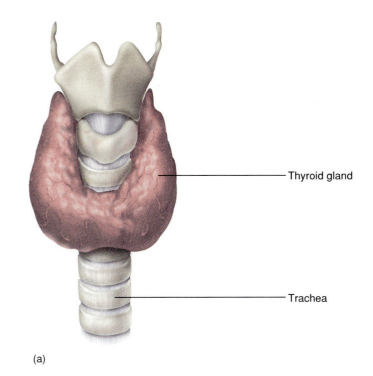

(a)

Radioactive tracers are also used in clinical diagnosis. Thallium-201 is used as a tracer to determine the location and extent of cardiac damage in heart attack patients, and cobalt-58 is used to evaluate intestinal absorption of vitamin B_{12}. A procedure called a *thyroid scan* uses radioactive iodine, ^{131}I, to determine the size and activity of the thyroid gland and detect thyroid cancer. The patient swallows a solution containing ^{131}I, which becomes concentrated in the thyroid—the only organ that absorbs and uses iodine. Twenty-four hours later, the patient is scanned with a machine called a *scintillation counter* (fig. E.2). A computer integrates information from the counter and produces an image for interpretation by the radiologist (fig. E.3). Since ^{131}I has a half-life of only 8.1 days, it presents little long-term hazard to the patient; and the slight risk from radiation exposure is overridden by the value of the diagnosis.

Larger doses of ^{131}I are used to treat thyroid cancer. Radioactive iodine is absorbed by the gland and by any cancerous thyroid cells that have spread to other parts of the body. The radiation kills those cells with little risk to tissues that do not take up iodine.

Radium can also be introduced into or near a tumor to destroy it. The α particles it emits are highly destructive to cells in the immediate vicinity but penetrate so poorly through human tissues that they present little risk of damage to adjacent normal tissues. Cobalt-60, which emits both β particles and γ rays, is also used to destroy tumor cells. Treatment of cancer with radioisotopes, invented by Marie Curie, is called **radiotherapy.**▲

(b)

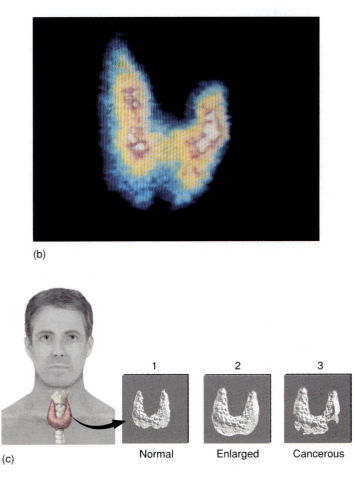

(c)

Figure E.3 (*a*) The thyroid gland wraps around the trachea (windpipe) immediately inferior to the larynx (voicebox). (*b*) A normal thyroid scan. (*c*) Scans of normal and diseased thyroid glands. ✗

Chemical Elements and Atomic Structure (pp. 53–57)

1. Matter and the elements
   a. Nature of elements
   b. Atomic number
   c. Symbols
   d. Relative abundance
2. Atomic structure
   a. Protons, neutrons, and atomic mass
   b. Electrons and energy levels
   c. Filling of the valence shell
3. Isotopes and atomic weight
   a. Meaning of isotope
   b. Atomic weight
4. Ions
   a. Process of ionization
   b. Anions and cations
   c. Valence of ions
5. Electrolytes
6. Free radicals
   a. How they are produced
   b. Biological effects
   c. Antioxidants

Molecules, Compounds, and Chemical Bonds (pp. 58–61)

1. Molecules and compounds
2. Chemical formulae
   a. Molecular formulae
   b. Structural isomers
   c. Structural formulae
   d. Condensed structural formulae
3. Molecular weight
4. Chemical bonds
   a. Ionic bonds
   b. Covalent bonds
      • Single and double
      • Polar and nonpolar
   c. Hydrogen bonds

Mixtures (pp. 61–64)

1. Mixtures versus compounds
2. Solutions, colloids, and suspensions
   a. Properties of a solution
   b. Properties of a colloid
   c. Properties of a suspension
3. Measures of concentration
   a. Weight per volume and percent concentration
      • How these are computed
      • When these can be misleading
   b. Molarity
      • Meaning of a mole
      • Avogadro's number
      • Molar and millimolar solutions
   c. Electrolyte concentrations
      • Meaning of equivalents
      • Computing mEq/L
      • Converting mM to mEq/L

Acids, Bases, and pH (pp. 64–66)

1. Acids and bases
2. pH
   a. The pH scale
   b. Buffers

Chemical Reactions (pp. 66–68)

1. Chemical equations
   a. How they are written
   b. Balanced equations
2. Classes of reactions
   a. Decomposition
   b. Synthesis
   c. Exchange
3. Reversible reactions
   a. Law of mass action
   b. Chemical equilibrium
4. Factors that affect reaction rates
   a. Concentration
   b. Temperature
   c. Catalysts

Energy (pp. 68–70)

1. Work and energy
   a. Kinetic versus potential energy
   b. Forms of energy
      • Chemical
      • Heat
      • Electromagnetic
      • Electrical
2. Radioisotopes and ionizing radiation
   a. Radioisotopes and radioactivity
   b. Types of ionizing radiation
      • α particle: two protons and two neutrons
      • β particle: electron
      • γ ray: high-energy photon
   c. Half-life: physical and biological
   d. Radiation exposure

Thermodynamics and Metabolism (pp. 70–72)

1. Laws of thermodynamics
   a. First law of thermodynamics
   b. Second law of thermodynamics
      • Free energy
      • Entropy
2. Exergonic and endergonic reactions
   a. Exergonic and catabolic reactions
   b. Endergonic and anabolic reactions
3. Oxidation and reduction
   a. Meaning of oxidation
   b. Meaning of reduction
   c. Oxidizing and reducing agents

S e l e c t e d   V o c a b u l a r y

matter 53	energy level 55	free radical 57	covalent bond 59
element 53	valence electron 55	antioxidant 57	single covalent bond 59
atomic number 53	octet rule 55	molecule 58	double covalent bond 59
trace element 53	rule of eights 55	compound 58	nonpolar covalent bond 59
atom 54	isotope 55	molecular formula 58	polar covalent bond 59
nucleus 54	atomic weight 56	structural isomer 58	dipole 59
proton (p^+) 54	ion 56	structural formula 58	hydrogen bond 61
neutron (n^0) 54	ionization 56	condensed structural	mixture 61
atomic mass unit (amu) 54	anion 56	formula 58	solution 61
atomic mass 54	cation 56	molecular weight (MW) 58	solute 61
electron (e^-) 55	valence 57	chemical bond 58	solvent 61
electron shell 55	electrolyte 57	ionic bond 58	colloid 62

Testing Your Recall   Answers in Appendix C

1. The simplest substance with unique
   chemical properties is
   a. an isotope.
   b. an element.
   c. a subatomic particle.
   d. a molecule.
   e. a compound.

2. Which of the following substances
   does *not* play a role in normal human
   physiology?
   a. fluorine
   b. sulfur
   c. silver
   d. phosphorus
   e. iron

3. Which of these particles determine
   the chemical properties of an atom?
   a. electrons
   b. protons
   c. neutrons
   d. *a* and *b* only
   e. *b* and *c* only

4. How many electrons must carbon
   gain from other atoms to have a
   stable valence shell?
   a. one
   b. two
   c. four
   d. six
   e. eight

5. Which of the following is the least
   penetrating type of nuclear radiation?
   a. α particles
   b. β particles
   c. γ rays
   d. π particles
   e. λ rays

6. A substance that _____ is
   considered a chemical compound.
   a. contains at least two different
      elements
   b. contains at least two
      atoms
   c. has a chemical bond
   d. has a stable valence shell
   e. has covalent bonds

7. An ionic bond is formed when
   a. two anions meet.
   b. two cations meet.
   c. electrons are shared equally
      between two nuclei.
   d. electrons are unequally shared
      between nuclei.
   e. electrons transfer completely from
      one atom to another.

8. The molecular weight of ethanol
   (CH_3CH_2OH) is
   a. 6.
   b. 9.
   c. 29.
   d. 46.
   e. 58.

9. In keeping with the law of
   conservation of mass,
   a. all chemical equations must be
      balanced.
   b. entropy increases in all chemical
      reactions.
   c. all decomposition reactions are
      exergonic.
   d. energy can change form but cannot
      be created or destroyed.
   e. all synthesis reactions are anabolic
      reactions.

10. A solution that resists changes in pH is
    a. a catalyst.
    b. an oxidizing agent.
    c. a reducing agent.
    d. an enzyme.
    e. a buffer.

11. The electrons of an atom's outermost
    shell are called its _____ electrons.

12. When an atom gives up an electron
    and acquires a positive charge, it is
    called a/an _____.

13. If a pair of shared electrons are more
    attracted to one nucleus than the
    other, they form a/an _____
    bond.

14. Two molecules with the same atoms
    arranged in different order are called
    _____.

15. What law of nature says that the total
    mass of the products of a reaction must
    equal the total mass of the reactants?

16. Any substance that increases the rate
    of a reaction without being consumed
    by it is a/an _____.

17. A/an _____ is a mixture of two
    liquids that will separate from each
    other on standing.

18. Avogadro's number is the number of
    molecules in 1 _____ of a
    substance.

19. A solution with a pH of 5 has a
    hydroxyl ion molarity of _____.

20. All the synthesis reactions in the
    body form a division of metabolism
    called _____.

1. A pregnant woman with severe morning sickness has been vomiting steadily for several days. How will loss of stomach acid affect the pH in her body? Explain.

2. A person with a severe anxiety attack hyperventilates and exhales carbon dioxide faster than his body produces it. Consider the carbonic acid reaction on page 67 and explain how this hyperventilation will affect his blood pH. (Hint: Remember the law of mass action.)

3. The complete oxidation of glucose has the equation $C_6H_{12}O_6 + 6\ O_2 \rightarrow 6\ CO_2 + 6\ H_2O$. Set up a table of reactants and products like the one on page 66 to demonstrate that this is a balanced equation.

4. Glycine has a molecular weight of 75. How much glycine would be needed to make 500 mL of a 100 mM solution?

5. How would the body's metabolic rate be affected it there were no such thing as enzymes? Explain.

Web Site Link

For a listing of the most current web sites related to this chapter, please visit the Saladin homepage at:

http://www.mhhe.com/sciencemath/biology/saladin/

3

[The Molecules of Life

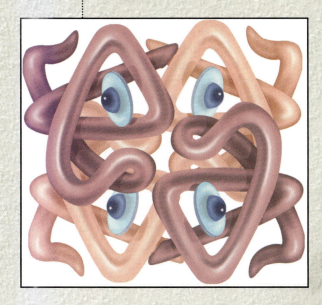

Special Topics

Brushing up

To understand this chapter, it is essential that you understand or brush up on the following concepts:

▶ Polar and nonpolar covalent bonds (p. 59)
▶ Hydrogen bonds (p. 61)
▶ Decomposition and synthesis reactions (p. 67)
▶ Catalysts (p. 68)
▶ Oxidation and reduction (p. 71)

I t is common knowledge that we need proteins, fats, carbohydrates, vitamins, and minerals in our diet. We are also aware that too much saturated fat and cholesterol puts us at risk for heart disease. Yet many of us have only a vague concept of what these molecules are, much less how they function in the human body. Study of the molecules that compose living organisms is called **biochemistry.** It overlaps with **organic chemistry,** the study of carbon-containing (organic) compounds, because the most distinctive biomolecules are organic compounds. In this chapter, we begin our discussion with some inorganic components of the body and progress to the organic ones. This will lay a foundation for understanding human form and function at many points later in the book.

Inorganic Matter

▼Objectives

When you have completed this section, you should be able to
• discuss the properties of water, how they result from its molecular structure, and why they are important to human life; and
• explain the roles of minerals and gases in human form and function.

Water

Water is a deceptively simple substance with many deep biological implications in its structure and behavior. It makes up about 50% to 75% of total body mass, depending on a person's age and sex. Fluid intake and output, along with related variables such as blood pressure and volume, are critically important to health.

The unique properties of water stem from two aspects of its structure. First, its atoms are joined by polar covalent bonds, and second, the molecule is V-shaped, with a 105° bond angle (fig. 3.1). This makes the molecule as a whole polar, with a slight negative ($\delta-$) charge on the oxygen atom and a slight positive ($\delta+$) charge on each hydrogen atom. Like little magnets, water molecules are attracted to each other by hydrogen bonds (see fig. 2.10, p. 60). This gives water a set of properties that account for its ability to support life: namely, *solvency, cohesion, thermal stability,* and *chemical reactivity.*

Solvency

Solvency is the ability to dissolve matter, and water is sometimes called the *universal solvent* because it can dissolve a broader range of substances than any other liquid. Molecules that dissolve in water, such as sugar, are said to be **hydrophilic**[1] (HY-dro-FILL-ic); the relatively few molecules that do not, such as fats, are called **hydrophobic**[2] (HY-dro-FOE-bic).

To be soluble in water, a substance must be charged, or polarized, so that its charges can interact with those of the water molecule. When NaCl is dropped into water, for example, the ionic bonds between Na^+ and Cl^- are overpowered by the attraction of each ion to water molecules. Water molecules form a cluster, or **hydration sphere,** around each sodium ion, with the $O^{\delta-}$ pole of each water molecule facing the Na^+. They also form a hydration sphere around each chloride ion, with the $H^{\delta+}$ poles facing it. This isolates the sodium ions from the chloride ions and keeps them dissolved (fig. 3.1).

Virtually all metabolic reactions that occur in the body depend on the solvency of water. Biological molecules must be dissolved in fluid to move around freely, come together, and react. This solvency also makes water the body's primary means of transporting substances from place to place.

Cohesion

Cohesion is the tendency of molecules of the same substance to cling together. The hydrogen bonds of water make it a very cohesive liquid in comparison to nonpolar substances. For example, if you spill liquid nitrogen on a countertop, it will dance about and evaporate in seconds, like a drop of water in a hot frying pan. This is because nitrogen molecules have no hydrogen bonds and no attraction for each other, so the little bit of heat provided by the countertop is enough to disperse them into the air. Spilled water, by contrast, forms a puddle and evaporates slowly because its molecules are hydrogen bonded, and it takes more energy to separate them.

The cohesion of water is especially evident at its surface, where it forms an elastic layer called the *surface film.* This layer is held together by a force called **surface tension**—essentially the pull of water molecules below the surface on those at the surface. This is what causes water to hang in drops from a leaky faucet and travel in rivulets down a window. Surface tension also causes water to form films on the surfaces of cells and body membranes, such as the pericardium and pleurae.

---

1. *hydro* = water + *philic* = loving, attracted to
2. *phobic* = fearing, avoiding

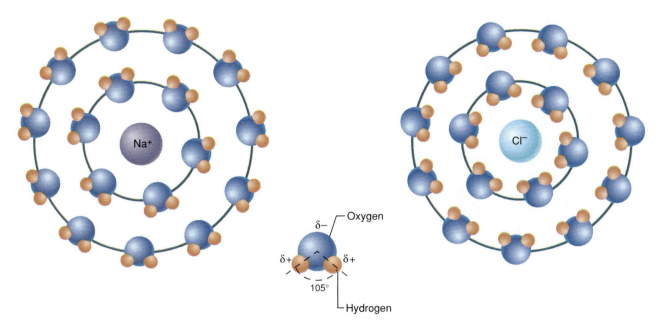

Figure 3.1 Water molecules form a hydration sphere around a sodium ion with their negatively charged oxygen poles facing the positively charged sodium. The positively charged hydrogen poles face a negatively charged chloride ion to form a hydration sphere around chloride as well. These hydration spheres keep the ions separated and account for the solubility of sodium chloride in water. The polarity and bond angle of a water molecule are shown in the inset.

Thermal Stability

Water helps to make the internal temperature of our bodies more stable than that of the air around us. This is because water has a high **heat capacity** compared to other liquids. The heat capacity of a substance is the amount of heat required to raise the temperature of 1 g of it by 1°C. Heat is measured in **calories**[3] (cal)—1 cal is the amount of heat that will raise the temperature of 1 g of water 1°C. It takes an unusually large amount of absorbed heat to raise the temperature of water—and, again, hydrogen bonding is the reason. To cause an increase in temperature, the molecules of a substance must move around more actively. The hydrogen bonding of water molecules inhibits their movement, so it takes more energy input to raise the temperature of water than it does to raise the temperature of liquid nitrogen and other nonpolar liquids.

Also, when it finally does change from a liquid to a vapor, water carries a large amount of heat away with it—thus, it serves as an effective coolant. Just 1 mL of perspiration evaporating from the skin removes about 500 cal of heat from the body. When you are hot and sweaty and stand in front of a fan, this effect is apparent.

Think About It

Why are heat and temperature not the same thing?

---

3. *calor* = heat

Chemical Reactivity

Reactivity is the ability to participate in chemical reactions, and water is a reactant or product in many metabolic reactions. This ability results from the fact that each polar covalent O—H bond of water is a weak, ionizable bond, and the attraction of water molecules to each other pulls some of them apart into hydroxyl and hydrogen ions. These ions are then available to react with other molecules. Later in this chapter, we examine several reactions involving water as a reactant or product.

In summary, then, we can see that the polarity and hydrogen bonding of water make it an effective solvent, transport medium, coolant, lubricant, and participant in chemical reactions (table 3.1). No other liquid could serve in its place.

Minerals

Minerals are inorganic elements that are extracted from the soil by plants and passed up the food chain to humans and other organisms. They constitute about 4% of the mass of the human body. Nearly three-quarters of this is calcium and phosphorus, and the rest is mainly potassium, sulfur, sodium, chlorine, and magnesium. Other minerals present in lesser quantities (trace elements) were itemized in the previous chapter in table 2.1.

Minerals contribute to the structure of various body components. Bone, for example, consists partly of crystals of *hydroxyapatite,* a complex arrangement of calcium carbonate, calcium phosphate, magnesium,

Table 3.1	Summary of the Properties of Water
Polarity	Polar covalent bonds and a 105° bond angle make the water molecule polar; this is the basis for all of its other properties.
Solvency	Water is the "universal solvent," dissolving all but the relatively few hydrophobic biological molecules. It forms hydration spheres around ions and polar molecules.
Medium of Transport	Solvency also makes water an effective medium of transport.
Medium for Chemical Reactions	Since so many substances dissolve in water, it is the medium for a great variety of chemical reactions.
Cohesion	The attraction of water molecules to each other accounts for the surface tension and thermal stability of water.
Surface Tension	Water forms films on body membranes, lubricates them, and aids the transport of solutes through membranes.
Thermal Stability	The high heat capacity of water stabilizes body temperature and makes water an effective coolant.
Chemical Reactivity	Water causes many other molecules to ionize, and water itself ionizes. It can thus participate as a reactant in hydrolysis reactions and form as a product of condensation reactions.

fluoride, sulfate, and hydroxyl ions. Tooth enamel is about 96% minerals by weight—mainly, calcium phosphate. Sulfur is a component of many proteins, and phosphorus is a major component of nucleic acids, ATP, and cell membranes.

Minerals also serve as cofactors that enable enzymes and other organic molecules to function. Iodine is a component of thyroid hormone, and iron is a component of hemoglobin. Some enzymes can function only when manganese, zinc, copper, or other minerals are bound to them.

Electrolytes, which are mainly ionized minerals, are responsible for the electrical events in cells of the nervous system, heart, muscular system, and elsewhere. Certain electrolytes also help to maintain normal osmolarity and water balance of body fluids and the acid-base balance required for normal cellular activities.

Gases

The gases of greatest physiological importance are oxygen and carbon dioxide. Oxygen is carried mainly by the red blood cells, but there is some dissolved oxygen in all body fluids. It serves as the final electron acceptor in aerobic respiration (see chapter 4) and some other oxidation reactions. Carbon dioxide is a product of aerobic respiration and must be constantly expelled from the body to prevent it from upsetting the body's pH balance.

Biochemists have long studied organic chemicals that are released by nerve and gland cells to communicate with other cells. More recently, however, it has been discovered that certain gases are also used as chemical messengers. Many cells contain an enzyme called *nitric oxide synthetase,* which catalyzes the formation of nitric oxide, NO, from the amino acid arginine. (Nitric oxide is not to be confused with *nitrous oxide,* N_2O, or "laughing gas.") NO is a highly reactive free radical that diffuses quickly to neighboring cells, where it stimulates a variety of reactions (see special topic 3.1). Some brain cells also use carbon monoxide (CO) to communicate with each other.

.... **Key Point Review** ....

1. Explain how the bond angle and polar covalent bonds of the water molecule account for its biologically important properties.
2. Define the term *calorie.*
3. Discuss the major functions of minerals in the body.

Special Topic — Communicating with Gas — 3.1

Nitric oxide was formerly regarded only as an air pollutant, a destroyer of ozone, and possibly a cause of cancer. The discovery that it functions as an intercellular messenger has greatly expanded our view of cellular communication.

One of the functions of nitric oxide is to control blood pressure. For example, high blood pressure stimulates the linings of the blood vessels to produce NO, which then causes the vessels to relax, thus lowering the blood pressure. In some bacterial infections, too much NO is produced and blood pressure drops to a dangerously low level. In such cases, NO inhibitors have been used to bring the blood pressure back to a safe level. As another example, NO released by pelvic nerves during sexual arousal dilates the arteries of the penis and causes the increased blood flow needed for erection; NO inhibitors prevent erection.

The functions of nitric oxide are not limited to blood vessels. Although the precise role of NO in the brain is not fully understood, some forms of learning and memory seem to involve nerve cells communicating with each other by means of this gas.

Carbon and Organic Molecules

▼**Objectives**

When you have completed this section, you should be able to
- explain why carbon is better suited than any other element to serve as the structural foundation of most biomolecules;
- name and diagram some common functional groups of organic molecules; and
- explain what monomers and polymers are, and how they are related to each other through dehydration synthesis and hydrolysis.

Carbon

Compounds of carbon are called *organic* because it was once thought that they were made only by *organisms.* With some exceptions, this is true. Carbon is an especially versatile atom that can serve as the base of a wide variety of structures. It has four valence electrons, and so it must covalently bond with other atoms that can provide it with four more to complete its valence shell. It can do this partly by bonding with one or more other carbon atoms, which gives carbon the ability to form long chains, branched molecules, and rings—an enormous variety of "backbones" for organic molecules. It also commonly forms covalent bonds with other elements; for example, hydrogen, oxygen, nitrogen, and sulfur. The only other element with such versatility is silicon (Si), but Si–Si bonds are unstable in the presence of oxygen, so silicon would be a poor substitute for carbon.

Functional Groups

Carbon backbones are also versatile in the number of **functional groups** that may be attached to them. A functional group is a small cluster of atoms that determines many of the properties of an organic molecule. For example, organic acids bear a **carboxyl** (car-BOC-sil) **group,** and ATP is named for its three **phosphate groups.** These and other functional groups are illustrated in figure 3.2.

Monomers and Polymers

Since carbon can form long chains, some organic molecules attain molecular weights that range from the thousands (as in starch and proteins) to the millions (as in DNA). These are called **macromolecules.** Most macromolecules are **polymers**[4]—molecules made of a repetitive series of identical or similar subunits called **monomers** (MON-oh-murs). Starch, for example, consists of a chain

Name and Symbol	Structure	Occurs in
Hydroxyl (—OH)	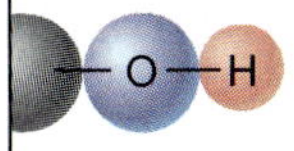	Sugars, alcohols
Methyl (—CH$_3$)		Fats, oils, steroids, amino acids
Carboxyl (—COOH)	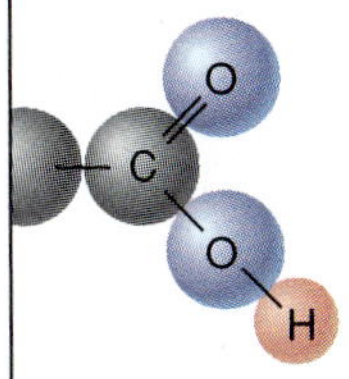	Amino acids, sugars, proteins
Amino (—NH$_2$)		Amino acids, proteins
Phosphate	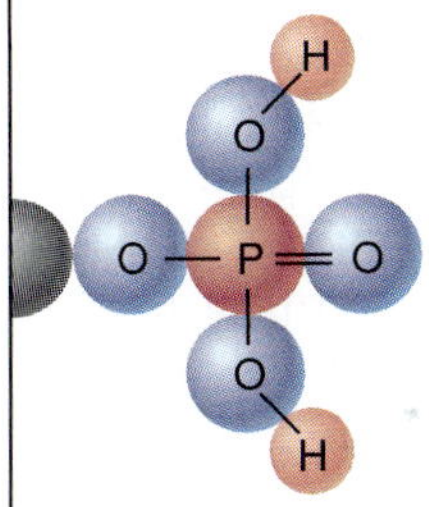	Nucleic acids, ATP

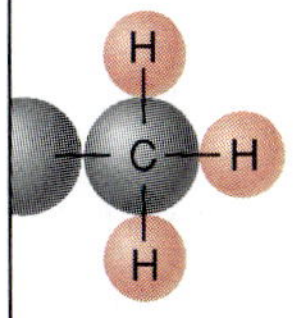

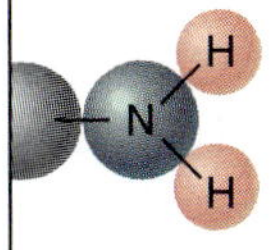

Figure 3.2 Structures of some important functional groups of organic molecules.

of about 3,000 glucose monomers. In some polymers the monomers are identical, such as the glucose monomers of starch, while in other cases they have a basic structural similarity but differ in detail. Four kinds of monomers make up the structure of DNA, for example, and 20 monomers are used to make proteins.

The joining of monomers to form a polymer is called **polymerization,** and in cells this is achieved by **dehydration synthesis** reactions (fig. 3.3a). A hydroxyl (—OH) group is removed from one monomer and a hydrogen nucleus (—H) from another, leaving each monomer in an unstable, reactive state. The two monomers become joined by a covalent bond, and the —OH and —H unite to form water as a by-product. Repetitive dehydration synthesis reactions add one monomer at a time to the growing end of a macromolecule.

---

4. *poly* = many + *mer* = part

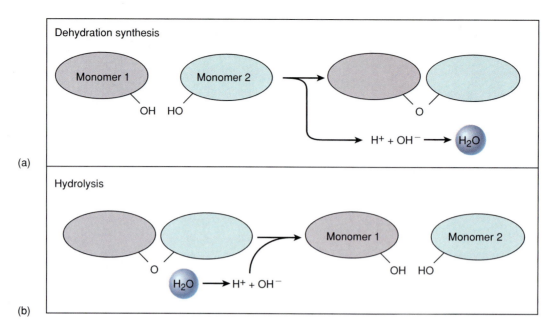

Dehydration synthesis

Monomer 1   Monomer 2
OH   HO

$H^+ + OH^- \longrightarrow H_2O$

(a)

Hydrolysis

O

$H_2O \longrightarrow H^+ + OH^-$

Monomer 1   Monomer 2
OH   HO

(b)

Figure 3.3 (*a*) In a dehydration synthesis reaction, a hydrogen atom is removed from one monomer and a hydroxyl group is removed from another. The monomers become joined through the remaining oxygen atom, and water is produced as a by-product. (*b*) In a hydrolysis reaction, water dissociates into hydrogen and hydroxyl ions. A covalent bond between two monomers is broken, the hydrogen ion is added to one monomer, and the hydroxyl ion is added to the other.

The opposite of a dehydration synthesis reaction is **hydrolysis**[5] (hy-DROL-ih-sis), which literally means a "splitting apart" by means of water (fig. 3.3*b*). In a hydrolysis reaction, a water molecule must first ionize to OH^- and H^+. A covalent bond linking one monomer to another is then broken, and the OH^- is added to one monomer and the H^+ to the other one. All digestion consists of hydrolysis reactions.

.... **Key Point Review** ....

4 Explain why no other element could serve as well as carbon as a backbone for biomolecules.

5 Define *monomer* and *polymer*.

6 To what branch of metabolism do you think polymerization belongs—oxidation or reduction? Synthesis or decomposition? Explain each answer.

Carbohydrates

▼**Objectives**
When you have completed this section, you should be able to
• explain how carbohydrates differ from other biomolecules;
• list the major monosaccharides, disaccharides, and polysaccharides; and
• list the physiological functions of carbohydrates.

By 1900, biochemists had classified the biological macromolecules into four primary categories: *carbohy-*

drates, lipids, proteins, and *nucleic acids.* In the following discussion, we examine the characteristics and subclasses of the first three of these. Nucleic acids, which are concerned with heredity and genetic regulation of cellular function, are described in chapter 5.

A **carbohydrate**[6] is a hydrophilic organic molecule with a 2:1 ratio of hydrogen to oxygen and a generalized formula $(CH_2O)_n$, where *n* represents the number of carbon atoms. In the sugar glucose, for example, $n = 6$ and the formula is therefore $C_6H_{12}O_6$. The names of individual carbohydrates are often built on the word root *sacchar-* or the suffix *-ose,* which both mean "sugar" or "sweet." The most familiar examples of carbohydrates are the sugars and starches.

Think About It
Why is *carbohydrate* an appropriate name for this class of compounds? Relate this name to the generalized formula of carbohydrates.

Monosaccharides

The simplest carbohydrates are called simple sugars or **monosaccharides**[7] (MON-oh-SAC-uh-rides). The three of primary importance are **glucose, fructose,** and **galactose,** all of which have the molecular formula $C_6H_{12}O_6$. They differ only in the arrangement of these 24 atoms and are therefore structural isomers of each other (fig. 3.4).

....
6. *carbo* = carbon + *hydr* = water
7. *mono* = one + *sacchar* = sugar

....
5. *lysis* = splitting apart

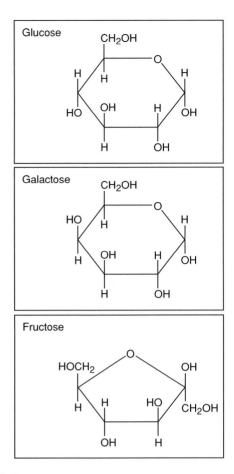

Figure 3.4 Structures of the three major monosaccharides.

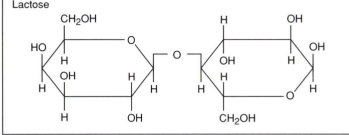

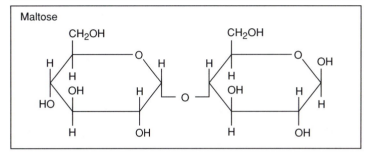

Figure 3.5 Structures of the three major disaccharides.

Disaccharides

Monosaccharides can be combined in pairs to form sugars called **disaccharides** (dy-SAC-uh-rides). The three of greatest importance are **sucrose** (made of glucose + fructose), **lactose** (glucose + galactose), and **maltose** (glucose + glucose) (fig. 3.5). Sucrose is produced by sugarcane and sugar beets and used as common table sugar. Lactose is milk sugar. Maltose is a product of starch digestion that is present in germinating wheat and other foods.

The C—O—C bonds that hold disaccharides together are called **glycosidic bonds.** Figure 3.6a shows how a glycosidic bond is formed by dehydration synthesis; figure 3.6b shows how these bonds are broken by hydrolysis.

Polysaccharides

Glucose can also form long chains called **polysaccharides** (POL-ee-SAC-uh-rides), some of which have molecular weights of 500,000 amu or more. Three polysaccharides of interest to human physiology are *cellulose, starch,* and *glycogen.* The first two are made by plants, and glycogen is made by animals, including ourselves.

Cellulose is a structural polysaccharide that gives strength to the cell walls of plants. It is the principal component of wood, cotton, and paper. A cellulose molecule consists of a few thousand glucose monomers joined together, with every other one "upside down" relative to the next (fig. 3.7). Although cellulose is the most abundant organic compound on earth and a common component of the diets of humans and other animals, very few animals have the enzyme needed to digest it. In fact, humans derive no energy or other nutrition from cellulose. In spite of being indigestible, however, cellulose is important to the diet as "fiber," "bulk," or "roughage." It swells with water in the digestive tract and helps move other materials through the intestine that might otherwise stagnate, ferment, and generate toxins. Colon cancer may be triggered by a diet poor in cellulose because there is not enough bulk to keep potentially harmful substances moving through and out of the colon.

Starch differs from cellulose mainly in that every glucose molecule faces the same way. It is a storage

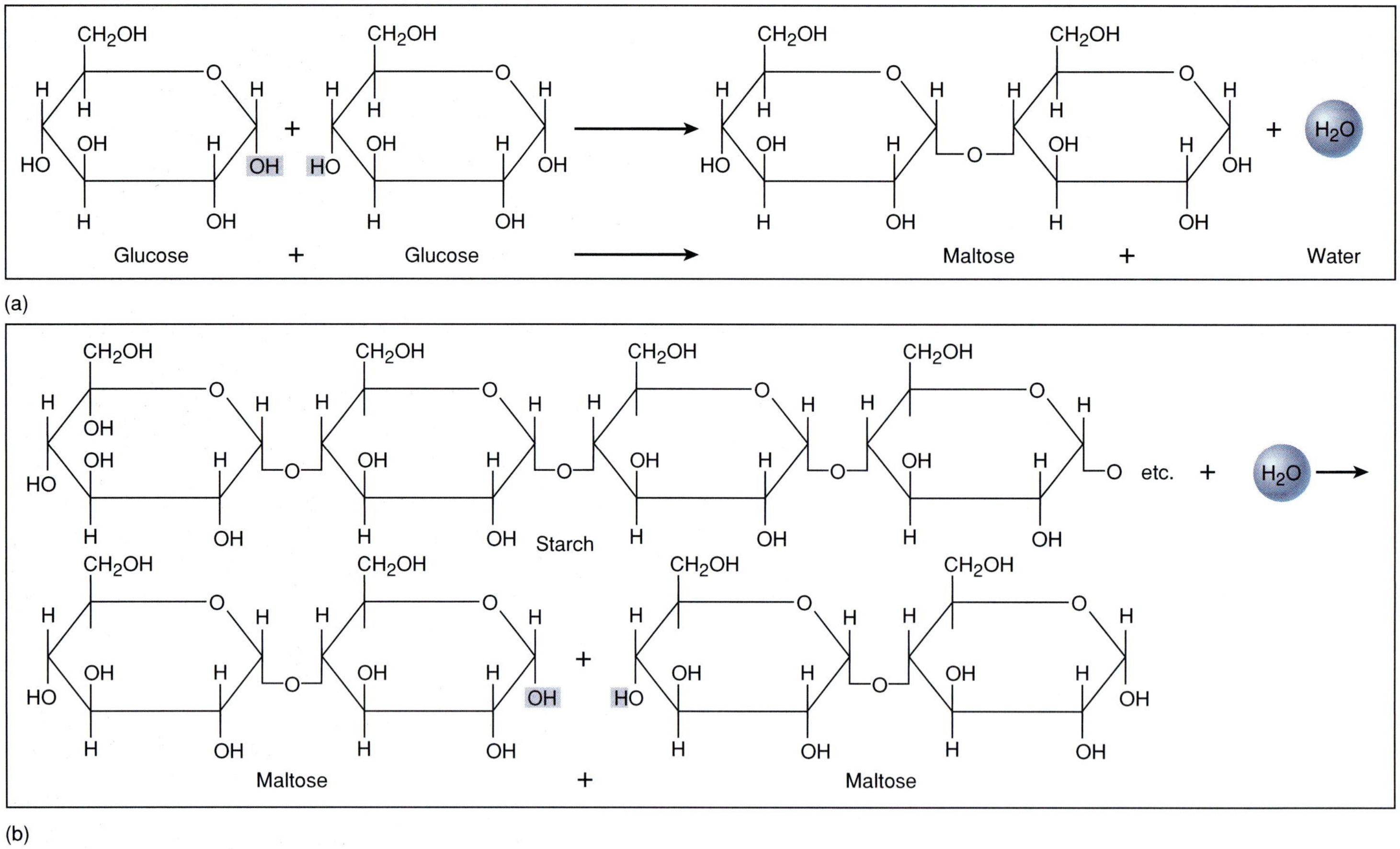

Figure 3.6 (*a*) Dehydration synthesis of maltose from glucose. (*b*) Production of maltose by the hydrolysis of starch.

polysaccharide made by plants to store reserve energy, and it is drawn upon when photosynthesis is not possible (at night, for example, or in winter when a plant has shed its leaves). There are two forms of starch: a straight-chain molecule called *amylose* and a branched form called *amylopectin.* Both forms are digested by an enzyme called *amylase,* present in our saliva and pancreatic juice.

Glycogen[8] (GLY-co-jen) is a storage polysaccharide made by animals. In humans, it is synthesized by cells of the liver, muscles, uterus, and vagina. Glycogen is similar to amylopectin but more highly branched. Liver cells produce glycogen after a meal, when blood glucose levels are high; they break it down again between meals to maintain blood glucose levels when there is no food intake. Muscle cells store glycogen for their own energy needs, and uterine cells use it in pregnancy to provide nutrition to the embryo.

Carbohydrate Functions

Carbohydrates are, above all, a source of energy that can be quickly mobilized. All digested carbohydrate is ultimately converted to glucose, and glucose is oxidized to make ATP. But carbohydrates have other functions as well (table 3.2). They are often **conjugated**[9] with (covalently bound to) proteins and lipids. Many of the lipid and protein molecules at the external surface of the cell membrane have chains of up to 12 carbohydrate monomers attached to them, forming **glycolipids** (GLY-co-LIP-ids) and **glycoproteins** (GLY-co-PRO-teens), respectively. The carbohydrate component forms a fuzzy, somewhat sticky layer called the *glycocalyx* (GLY-co-CAY-licks) on the surface of every cell—in a sense, every cell is "sugar-coated." The glycocalyx absorbs and retains water at the cell surface and serves as an identification tag that enables the immune system to distinguish its own cells from potentially dangerous foreign cells (see table 4.4). Glycoproteins are also a major component of mucus, which traps particles in the respiratory system, resists infection, and protects the digestive tract from its own acid and enzymes.

When discussing conjugated macromolecules it is convenient to refer to each chemically different component as a **moiety**[10] (MOY-eh-tee). Glycoproteins have a protein moiety and a carbohydrate moiety, for example.

---

8. *glyco* = sugar + *gen* = producing

9. *con* = together + *jug* = join
10. *moiet* = half

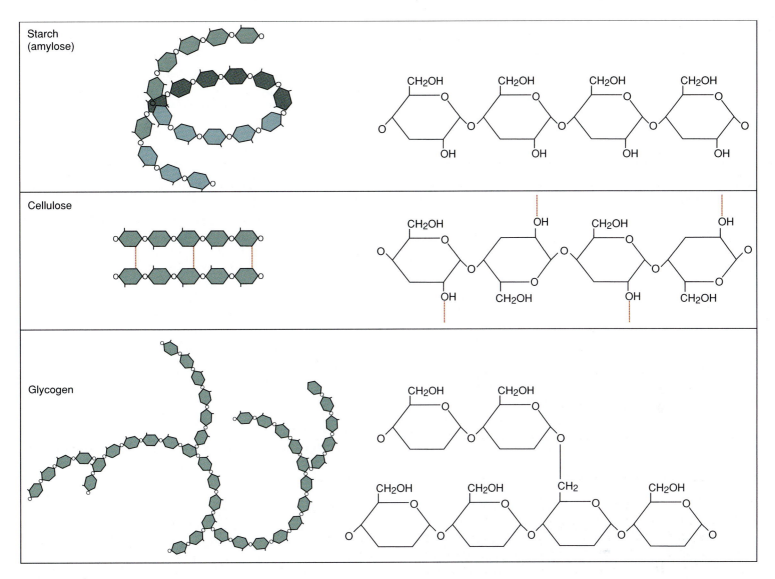

Figure 3.7 Structures of the three major polysaccharides.

Proteoglycans (once called mucopolysaccharides) are macromolecules in which the carbohydrate component is dominant and a peptide or protein forms a smaller moiety. They form gels that help hold cells and tissues together, lubricate the joints of the skeletal system, make up much of the rubbery extracellular material of cartilage, and form a gelatinous filler in the umbilical cord and eye. The structure and functions of proteoglycans are further considered in chapter 6.

Lipids

▼**Objectives**
When you have completed this section, you should be able to
• explain how lipids differ from other biomolecules;
• list six major classes of lipids and describe their differences; and
• list some of the functions of lipids.

A **lipid** is a hydrophobic organic molecule, usually composed only of carbon, hydrogen, and oxygen, with a high ratio of hydrogen to oxygen. A fat called *tristearin* (tri-STEE-uh-rin), for example, has the molecular formula $C_{57}H_{110}O_6$—more than 18 hydrogens for every oxygen. Thus, lipids are *less oxidized* than carbohydrates, and they have *more calories* per gram. Beyond these criteria, however, it is difficult to generalize about lipids; they are much more variable in structure than the other macromolecules we are considering. We will consider the five

Table 3.2 Carbohydrates and Their Functions

Type	Function
Monosaccharides	
Glucose	Blood sugar—energy source for most cells
Galactose	Converted to glucose and catabolized
Fructose	Fruit sugar—converted to glucose and catabolized
Disaccharides	
Sucrose	Cane sugar—digested to glucose and fructose
Lactose	Milk sugar—digested to glucose and galactose
Maltose	Malt sugar—product of starch digestion; further digested to glucose
Polysaccharides	
Cellulose	Structural polysaccharide of plants; dietary fiber
Starch	Energy storage in plant cells
Glycogen	Energy storage in animal cells (liver, muscle, uterus, vagina)
Conjugated Carbohydrates	
Glycoproteins	Cellular recognition; water retention at the cell surface; component of mucus
Glycolipids	Same as glycoproteins—both form the glycocalyx
Proteoglycans	Cell adhesion; lubrication; supportive filler of some tissues and organs

primary types of lipids in humans—*fatty acids, triglycerides, phospholipids, prostaglandins,* and *steroids.*

Fatty Acids

A **fatty acid** is a chain of 4 to 24 carbon atoms, with a carboxyl group at one end and a methyl group at the other. It is a polymer of two-carbon *acetyl groups;* thus, a fatty acid as a whole has an even number of carbon atoms. Fatty acids, and the fats made from them, are classified as **saturated** or **unsaturated.** Palmitic acid, for example, is a saturated fatty acid; every carbon atom has as much hydrogen as it can possibly carry, except in the carboxyl group (fig. 3.8). No more hydrogen could be added without exceeding four covalent bonds per carbon atom—thus, it is "saturated" with hydrogen. In linoleic (LIN-oh-LEE-ic) acid, however, some carbon

atoms are joined by double covalent bonds. Each of these could potentially share one pair of electrons with another hydrogen atom instead of the adjacent carbon. Hydrogen therefore could be added to this molecule, and we say it is unsaturated. **Polyunsaturated** fatty acids are those with multiple C=C bonds.

Triglycerides

A **triglyceride** (try-GLISS-ur-ide) is a molecule consisting of three fatty acids covalently bound to a three-carbon alcohol called **glycerol** (GLISS-er-ol). As shown in figure 3.8, each bond between a fatty acid and glycerol is formed by dehydration synthesis. Once joined to glycerol, a fatty acid can no longer donate a proton to solution and is therefore no longer an acid. For this reason, triglycerides are also called **neutral fats.** Triglycerides are broken down by hydrolysis reactions, which split each of these bonds apart by the addition of water. Enzymes that do this are called **lipases** (LY-pace-es).

Triglycerides that are liquid at room temperature are also called oils, but the difference between a fat and oil is fairly arbitrary. Coconut oil, for example, is solid at room temperature. Animal fats are usually made of saturated fatty acids, so they are called *saturated fats.* They are solid at room or body temperature. Most plant triglycerides are *polyunsaturated fats,* which generally remain liquid at room temperature. Examples are peanut, olive, corn, and linseed oils. Saturated fats contribute more to cardiovascular disease than do unsaturated fats, which is why it is healthier to cook with vegetable oils than with lard or bacon fat.

The primary function of fat is energy storage. When concentrated in *adipose tissue,* however, fat also provides thermal insulation and acts as a shock-absorbing cushion for various organs (see chapter 6).

Phospholipids

Phospholipids are similar to neutral fats but contain only two fatty acids. In place of a third fatty acid, they possess a phosphate group (PO_4), which in turn may be linked to other functional groups. Lecithin is a common phospholipid in which the phosphate is bound to a nitrogenous group called *choline* (CO-leen) (fig. 3.9). Phospholipids have a dual nature. The two fatty acid "tails" of the molecule are hydrophobic, but the phosphate "head" is hydrophilic. Molecules with both hydrophobic and hydrophilic regions are called **amphiphilic**[11] (AM-fih-FIL-ic). The most important function of phospholipids is to serve as the structural foundation of cell membranes; in chapter 4 you will see how this role depends on their amphiphilic property.

---

11. *amphi* = both

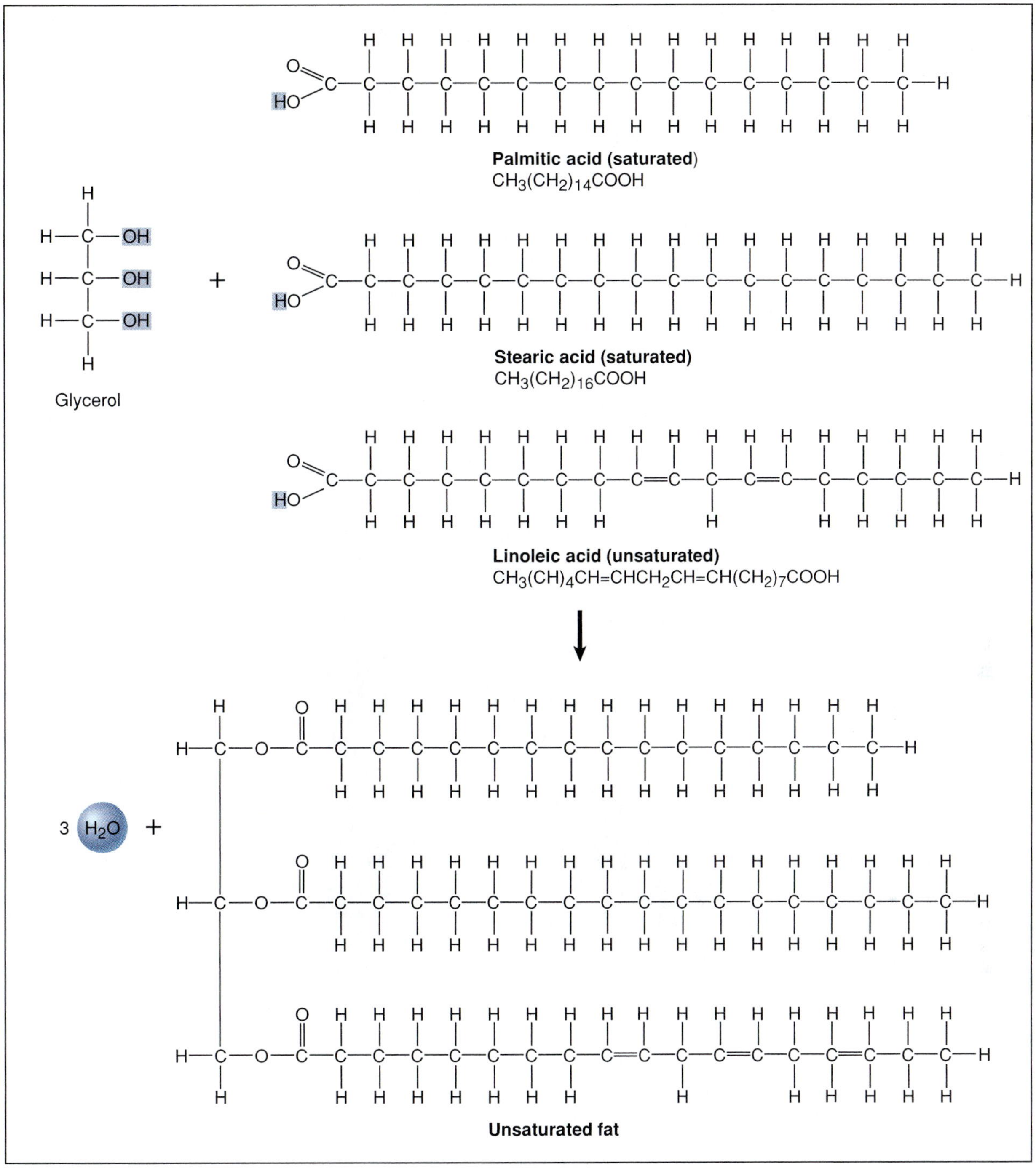

Figure 3.8 Synthesis of a triglyceride from glycerol and three fatty acids. Note the difference between saturated and unsaturated fatty acids, and the production of 3 H_2O as a by-product of this dehydration synthesis reaction.

Prostaglandins

Prostaglandins (PROSS-ta-GLAN-dins) are modified fatty acids in which five of the carbon atoms are arranged in a ring (fig. 3.10). They were originally found in the secretions of bovine prostate glands, hence their name, but they are now known to be produced in almost all tissues. They are intercellular messengers that play a variety of roles in inflammation, blood clotting, hormone action, labor contractions, control of blood vessel diameter, and other processes.

Steroids

A **steroid** (STERR-oyd, STEER-oyd) is a lipid with 17 of its carbon atoms arranged in four rings (fig. 3.11). Steroids differ from each other in the location of C=C bonds within the rings and in the functional groups

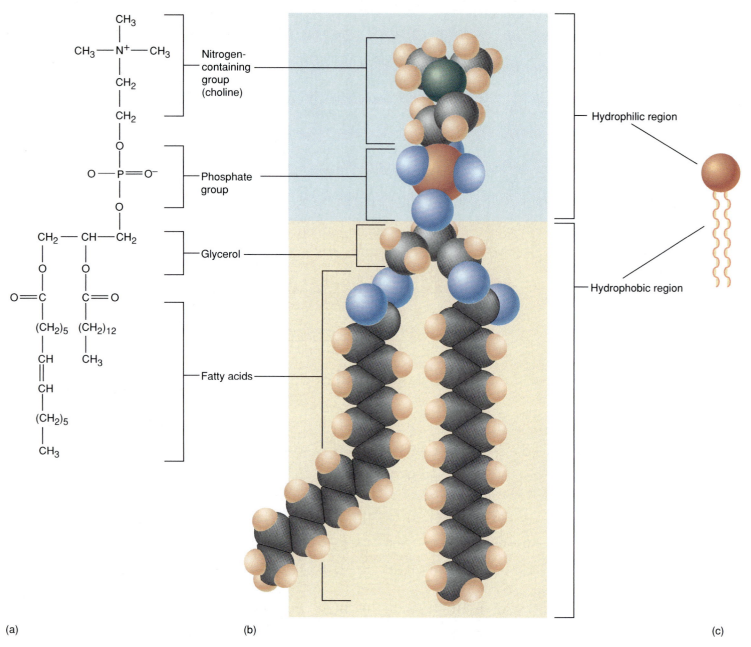

(a)   (b)   (c)

Figure 3.9 Structure of lecithin, a phospholipid. (*a*) Structural formula, (*b*) a space-filling model that gives some idea of the actual shape of the molecule, and (*c*) a simplified representation of the phospholipid molecule used in diagrams of cell membranes.

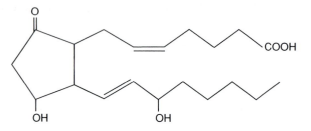

Figure 3.10 Structure of a prostaglandin, a modified fatty acid with some of its carbon atoms arranged in a five-sided ring.

attached to the rings. Examples of steroids include cholesterol, cortisol, bile salts, progesterone, estrogens, and testosterone (see chapter essay, p. 102). The average adult body contains over 200 g (half a pound) of cholesterol. Cholesterol is synthesized only by animal cells (especially liver cells) and is not present in vegetable oils or other plant products. It is the precursor from which the other steroids are synthesized.

Cholesterol has a bad reputation as a factor in cardiovascular disease (see special topic 3.2), and it is true

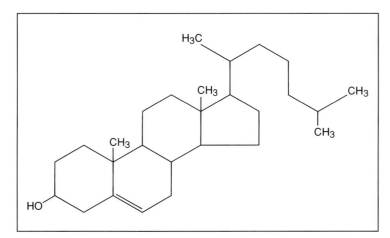

Figure 3.11 Structure of cholesterol. All steroids have this basic four-ringed structure, with variations in the functional groups and locations of double bonds within the rings.

that hereditary and dietary factors can elevate serum cholesterol to dangerously high levels. Nevertheless, cholesterol is a natural product of the body. Only about 15% of the body's cholesterol comes from the diet; the other 85% is internally synthesized. In addition to being the precursor of the other steroids, cholesterol is an important component of cell membranes. It is also required for proper nervous system function.

The primary lipids and their functions are summarized in table 3.3.

Key Point Review

11. What are lipids? What do they have in common with carbohydrates and how do they differ from them?
12. How do prostaglandins differ from other fatty acids?
13. What is the difference between saturated and unsaturated fats?
14. How do phospholipids differ from triglycerides?
15. State one function of each of the following molecules: triglycerides, phospholipids, prostaglandins, and steroids.

Table 3.3	Lipids and Their Functions
Type	**Function**
Fatty acids	Precursor of triglycerides; source of energy
Triglycerides	Energy storage; thermal insulation; space filling; binding organs together; cushioning organs
Phospholipids	Major component of cell membranes
Prostaglandins	Chemical messengers between cells
Fat-soluble vitamins	Involved in blood clotting, wound healing, vision, and calcium absorption; perform a wide variety of other specialized functions
Cholesterol	Component of cell membranes; precursor of other steroids
Steroid hormones	Chemical messengers between cells
Bile salts	Steroids that aid in fat digestion and nutrient absorption

Proteins

▼Objectives
When you have completed this section, you should be able to
- diagram the structure of a generalized amino acid and show how peptide bonds are formed between amino acids;
- describe the four levels of protein structure;
- describe some nonprotein moieties that are often bound to proteins; and
- list the physiological functions of proteins.

Amino Acids

The term *protein* is derived from the Greek word *proteios*, meaning "of first importance." Proteins are the most versatile molecules in the body, and it is especially important to understand their structure and diversity.

Special Topic "Good" and "Bad" Cholesterol   3.2

There is only one kind of cholesterol, and it does far more good than harm. When the popular press refers to "good" and "bad" cholesterol, it is actually referring to droplets in the blood called *lipoproteins,* which are complexes of cholesterol, fat, phospholipids, and protein. So-called bad cholesterol actually refers to low-density lipoprotein (LDL), which has a high ratio of lipid to protein and contributes to cardiovascular disease. So-called good cholesterol refers to high-density lipoprotein (HDL), which has a lower ratio of lipid to protein and may help to prevent cardiovascular disease. Even when food products are advertised as cholesterol-free, they may be high in saturated fat, which stimulates the body to produce more cholesterol. Palmitic acid seems to be the greatest culprit in stimulating elevated cholesterol levels, while linoleic acid has a cholesterol-lowering effect. Both are shown in figure 3.8. LDLs, HDLs, and heart disease are discussed further in later chapters.

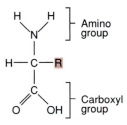

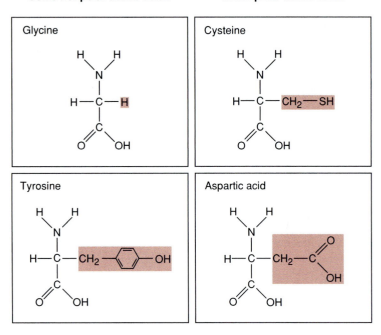

Figure 3.12 Generalized structure of an amino acid, and structures of four representative amino acids. Amino acids differ only in the structures of their functional R groups.

Table 3.4	The 20 Amino Acids and Their Abbreviations		
Alanine	Ala	Leucine	Leu
Arginine	Arg	Lysine	Lys
Asparagine	Asn	Methionine	Met
Aspartic acid	Asp	Phenylalanine	Phe
Cysteine	Cys	Proline	Pro
Glutamine	Gln	Serine	Ser
Glutamic acid	Glu	Threonine	Thr
Glycine	Gly	Tryptophan	Trp
Histidine	His	Tyrosine	Tyr
Isoleucine	Ile	Valine	Val

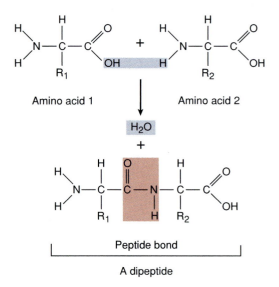

Figure 3.13 Dehydration synthesis of a dipeptide, which produces a peptide bond between two amino acids.

A protein is a polymer of **amino acids.** An amino acid has a central carbon atom with an amino (—NH$_2$) group and a carboxyl (—COOH) group bound to it (fig. 3.12). The 20 kinds of amino acids involved in protein structure are listed in table 3.4, along with their abbreviations. All of them are identical except with respect to a third functional group called the *radical,* or *R group.* In the simplest amino acid, glycine, R is merely a hydrogen atom, while in the largest amino acids it includes rings of carbon.

Peptides

A **peptide** is any molecule that consists of two or more amino acids. A **peptide bond,** formed by dehydration synthesis, joins the amino group of one monomer to the carboxyl group of the next (fig. 3.13). Peptides are named for the number of amino acids they contain—for example, dipeptide (2), tripeptide (3), oligopeptide ("a few"), polypeptide ("many"). An example of an oligopeptide is oxytocin, a childbirth hormone that is

8 amino acids long. Adrenocorticotropic hormone (ACTH) is a representative polypeptide that contains 39 amino acids. Polypeptides with more than 50 to 100 amino acids (the number depending on personal opinion) are considered to be proteins. A typical amino acid has a molecular weight of about 80 amu, and the molecular weights of the smallest proteins range from about 4,000 to 8,000 amu. The average protein weighs in at about 30,000 amu, and some of them have molecular weights in the hundreds of thousands.

Levels of Protein Structure

Proteins typically have complex coiled and folded structures that are critically important to the roles they play. Even slight changes in three-dimensional shape may render a protein unable to function. Protein molecules have three to four levels of complexity, from primary through quaternary structure (fig. 3.14).

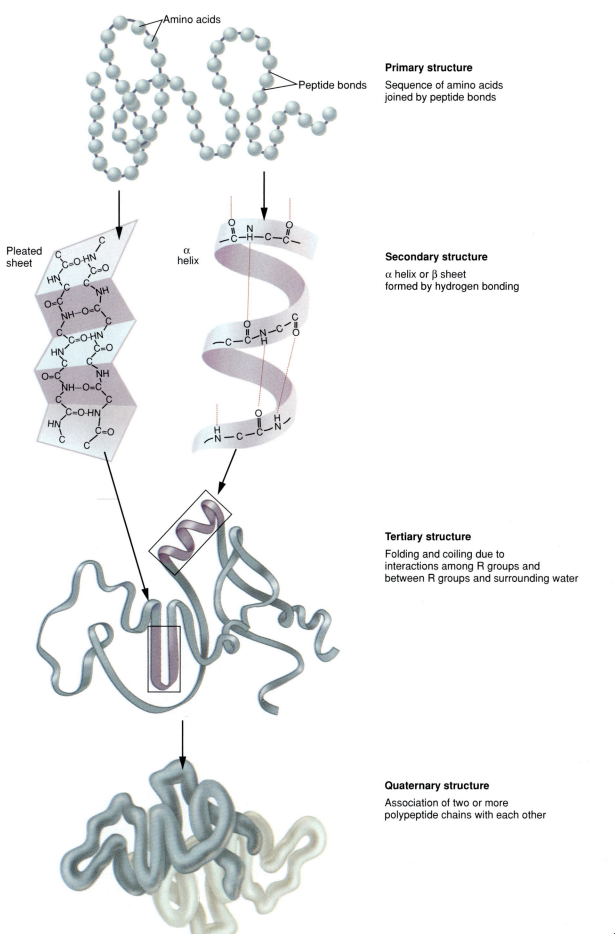

Primary structure

Sequence of amino acids joined by peptide bonds

Secondary structure

α helix or β sheet formed by hydrogen bonding

Tertiary structure

Folding and coiling due to interactions among R groups and between R groups and surrounding water

Quaternary structure

Association of two or more polypeptide chains with each other

Figure 3.14 Four levels of protein structure.

Primary structure is the order in which the amino acids occur—a sequence that is encoded in the genes. In chapter 5 we examine how the genetic code directs the assembly of a protein.

Secondary structure is a coiled or folded shape held together by hydrogen bonds between the slightly negative C=O group of one peptide bond and the slightly positive N—H group of another peptide bond some distance away. The most common secondary structures are a springlike shape called the **α helix** and a pleated, ribbonlike shape, the **β sheet** (or **β pleated sheet**). Many proteins have multiple α-helical and β-pleated regions joined together by short segments with a less regular geometry. A single protein molecule may fold back on itself and have two or more β-pleated regions linked to each other by hydrogen bonds. Separate, parallel protein molecules also may be linked to each other through their β-pleated regions.

Tertiary[12] (TUR-she-air-ee) **structure** is formed by the further bending and folding of proteins into various globular and fibrous shapes. It results from interaction of the R groups with each other and with the surrounding water. Hydrophobic R groups avoid water and tend to face the inside of the protein, while hydrophilic R groups are attracted to water and tend to face the outside. **Globular proteins,** somewhat resembling a wadded ball of yarn, have a compact tertiary structure well suited for proteins embedded in cell membranes and proteins that must move around freely in the body fluids, such as enzymes and antibodies. **Fibrous proteins** such as myosin, keratin, and collagen are slender filaments better suited for such roles as muscle contraction and providing strength to skin, hair, and tendons.

The amino acid cysteine (Cys), whose R group is —CH₂—SH (see fig. 3.12), often stabilizes a protein's tertiary structure by forming covalent *disulfide bridges.* When two cysteines align with each other, each can release a hydrogen atom, leaving the sulfur atoms to form a disulfide (—S—S—) bridge:

$$Cys—CH_2—SH + HS—CH_2—Cys \rightarrow$$
$$Cys—CH_2—S—S—CH_2—Cys + 2\ H$$

Disulfide bridges hold separate polypeptide chains together in antibody molecules and in the hormone insulin, for example (fig. 3.15).

Quaternary[13] (QUA-tur-nare-ee) **structure** is the association of two or more polypeptide chains with each other by noncovalent forces such as ionic bonds and hydrophilic-hydrophobic interactions. It occurs in only some proteins. Hemoglobin, for example, consists of four polypeptides—two identical α chains and two identical, slightly longer β chains (fig. 3.16).

....

12. *tert* = third
13. *quater* = fourth

Figure 3.15 Insulin, a protein in which two polypeptide chains are joined by disulfide bridges.

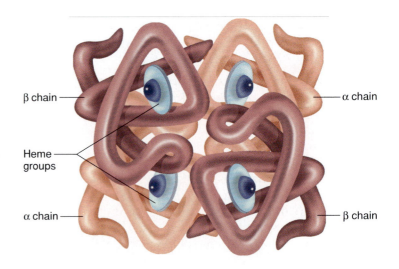

Figure 3.16 Hemoglobin, a conjugated protein (having a nonprotein heme moiety in each polypeptide chain) with a quaternary structure (multiple polypeptide chains joined together).

Protein Conformation and Denaturation

The **conformation** of a macromolecule, its overall three-dimensional shape, is highly important to its ability to function. In proteins, conformation refers especially to tertiary structure. One of the most important properties of proteins is their ability to change conformation. This can be triggered by circumstances such as voltage changes on a cell membrane during the action of nerve cells, the binding of a hormone to a protein, or the dissociation (detachment) of a molecule from a protein. Subtle conformational changes are important in processes such as enzyme function, muscle contraction, and the opening and closing of channels through cell membranes. **Denaturation** is a more drastic conformational change in response to conditions such as extreme heat or pH. It is seen, for example, when you cook an egg and the egg white protein (albumin) turns from clear to opaque. Denaturation is sometimes reversible when normal conditions are restored, but often, it permanently destroys the function of a protein.

Conjugated Proteins

Conjugated proteins have a non-amino acid moiety called a **prosthetic**[14] **group** covalently bound to them. Hemoglobin, for example, not only has a quaternary structure of four polypeptide chains, but each chain also has a complex iron-containing ring called a *heme* moiety attached to it (fig. 3.16). Hemoglobin cannot transport oxygen unless this group is present. In glycoproteins, as described earlier, the prosthetic group is a small carbohydrate.

Protein Functions

Proteins are more diverse in their functions than other macromolecules. Among their many roles are the following:

- **Structure.** Keratin, a tough structural protein, adds strength to hair, nails, and skin. The deeper layers of the skin, as well as bones, cartilage, and teeth, contain an abundance of the durable protein collagen.
- **Catalysis.** Most metabolic pathways of the body are controlled by enzymes, which are globular proteins that function as catalysts.
- **Communication.** Some hormones and other cell-to-cell signals are proteins, and so are the receptors to which the signal molecules bind in the receiving cell. A hormone or other molecule that reversibly binds to a protein is called a **ligand**[15] (LIG-and, LY-gand).
- **Membrane transport.** Some proteins form channels in cell membranes that govern which types of solutes are allowed to pass through the membranes, and when. Other proteins act as carriers that briefly bind to solute particles and transport them through cell membranes. Among their other roles, such proteins help to turn nerve and muscle activity on and off.
- **Recognition and protection.** The role of glycoproteins in immune recognition was mentioned earlier. When invading organisms are detected, they are often attacked and neutralized by antibodies and other proteins. Several clotting proteins help protect the body against blood loss.
- **Movement.** Muscle contraction, the beating of hairlike processes (cilia) in the respiratory tract, and other kinds of movement are due to special contractile proteins.

Key Point Review

16 Draw two amino acids side by side and show how they would be joined by a peptide bond.

17 Compare and contrast the four levels of protein structure.

18 What level of protein structure could be disrupted by a chemical that breaks disulfide bonds?

19 Discuss how the tertiary structure of proteins exemplifies the unity of form and function.

Enzymes and Metabolism

▼Objectives
When you have completed this section, you should be able to
- define *enzyme* and explain how enzymes are named;
- explain how enzymes accelerate metabolic reactions;
- describe the structure of an enzyme and relate this structure to its function; and
- explain how metabolic pathways can be regulated by enzymes.

What Enzymes Do

Enzymes are proteins that function as biological catalysts. Nearly every metabolic reaction in the body is controlled by an enzyme. Even the simple bacterium *Escherichia coli* contains as many as 3,000 different enzymes. Enzymes were first discovered in single-celled yeasts (see special topic 3.3).

---

14. *prosthe* = appendage, addition

15. *lig* = to bind

Yeast has been used since ancient times to ferment sugar to alcohol, but even as late as the nineteenth century no one had figured out how to duplicate this or other biochemical reactions in the absence of living cells. Many scientists therefore fell back on a *vitalist* philosophy—the belief that organisms or cells contain a mysterious "vital force" beyond the realm of science and the physical world. The German biochemist Eduard Buchner (1860–1917), however, was less content with easy answers. He eventually extracted a juice from yeast that fermented glucose. Not knowing what chemicals in the extract did this, he named the hypothetical substances *enzymes,* from the Greek *en* = within + *zym* = yeast.

Reaction occurring without a catalyst   **Reaction occurring with a catalyst**

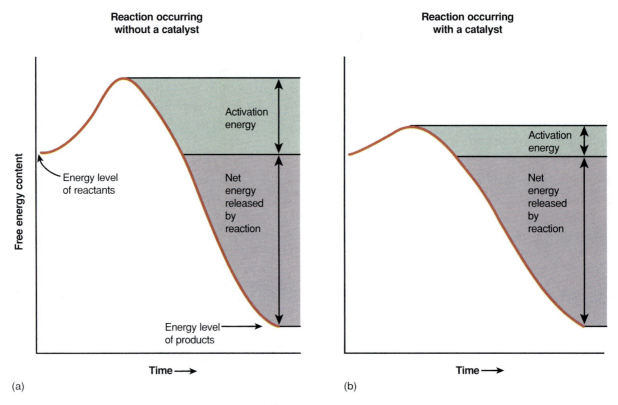

Figure 3.17 (*a*) Without catalysts, some chemical reactions proceed slowly because of the high activation energy needed to get molecules to react. (*b*) A catalyst facilitates molecular interaction, thus lowering the activation energy and making the reaction proceed more rapidly.

To appreciate what enzymes do, consider what would happen if you ignited a pile of glucose in the presence of oxygen. It would burn, producing water and carbon dioxide:

$$C_6H_{12}O_6 + 6\ O_2 \rightarrow 6\ CO_2 + 6\ H_2O$$

This book is composed mainly of glucose (in the form of its polymer cellulose), and yet it does not spontaneously burst into flame while you are reading it. Why? Because few of its molecules have enough kinetic energy to react. Holding a match to the paper, however, would raise the kinetic energy of its molecules enough to initiate rapid oxidation (combustion). This energy input needed to get the reaction started is called the **activation energy** (fig. 3.17*a*).

In the body, we "burn" glucose to yield the same end products, CO_2 and H_2O, and we extract energy from it to meet our metabolic needs. We could not tolerate the heat of a flame in our bodies, however, so we must oxidize glucose in a more controlled way at a biologically feasible and safe temperature. The way to do this is to require less energy input to get the reaction started—that is, lower the activation energy—and release the energy in small steps rather than in a single burst of heat. Enzymes are the key to doing this. They reduce the activation energy (fig. 3.17*b*) by holding two ligands in a way that enables them to react more easily or by binding to a single ligand and applying stresses to it that break covalent bonds (fig. 3.18). Any ligand acted upon and changed by an enzyme is called its **substrate.**

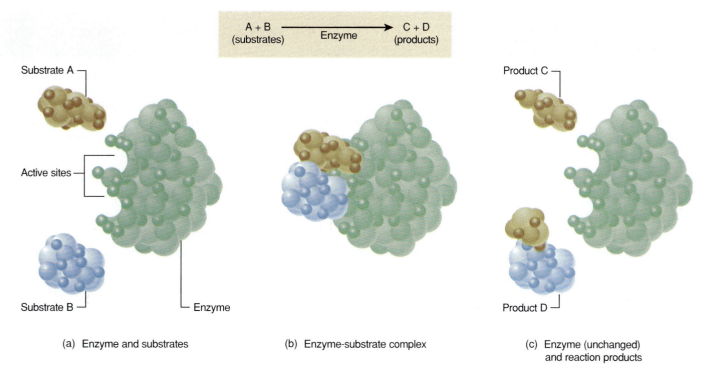

Figure 3.18 (*a*) An enzyme molecule has one or more active sites that fit specific substrate molecules. (*b*) When substrates bind temporarily to an enzyme, they are held in positions that favor a chemical reaction between them. (*c*) The reaction products dissociate from the active sites, leaving the enzyme unchanged and available to catalyze another such reaction.

After a reaction has occurred, the enzyme releases the products and is free to bind to more substrate molecules and repeat the reaction. Since the enzyme is not consumed, a small amount of it can catalyze the conversion of a large amount of substrate. For example, consider the breakdown of carbonic acid (H_2CO_3) by the enzyme carbonic anhydrase (CAH):

$$H_2CO_3 \rightarrow CO_2 + H_2O$$

Because each CAH molecule is used over and over, 1 molecule can process 36 million molecules of carbonic acid each minute. One ounce of pepsin, a digestive enzyme, could digest 2 tons of egg white in a few hours at room temperature. The same reaction without pepsin would require boiling the egg white in about 15 tons of strong acid for up to 48 hours.

Enzymes cannot bring about reactions that would not occur in their absence; they only speed up chemical reactions or enable them to occur at lower temperatures. In many reversible reactions, the same enzyme causes the reaction to proceed in either direction. Which way the reaction goes is dictated by the law of mass action.

How Enzymes Are Named

At one time, enzymes were given somewhat arbitrary names such as *pepsin* and *trypsin*. The modern system of naming enzymes, however, is more systematic and informative. It identifies the substrate and sometimes the function of the enzyme and adds the suffix *-ase*. For example, lipase is an enzyme that digests a lipid, amylase digests starch (*amyl-* = starch), and carbonic anhydrase removes water (*anhydr-*) from carbonic acid.

Enzymes can also be categorized according to their general function. *Phosphatases* remove phosphate groups from molecules and *kinases* add them. *Dehydrogenases* remove hydrogen atoms. *Synthetases* catalyze dehydration synthesis reactions, and *hydrolases* catalyze hydrolysis. *Polymerases* assemble polymers, such as proteins and RNA. *Isomerases* rearrange atoms to convert one isomer to another—for example, galactose to glucose. Enzyme names may be further modified to distinguish different forms of the same enzyme found in different tissues (see special topic 3.4).

Enzyme Structure

To help you visualize some of the working properties of enzymes, you can make a model by wadding a sheet of paper into a ball. Choose one of the pockets or grooves in the paper wad and make three marks on facing surfaces of the groove with a pen (fig. 3.19).

Like this paper wad, an enzyme has pockets, ridges, and other irregularities in its surface, some of

Different cells may possess slightly different forms of the same enzyme. These different forms, called **isoenzymes,** catalyze the same chemical reactions but differ enough structurally to allow them to be distinguished by standard laboratory techniques. This is useful in the diagnosis of disease. When organs are diseased, some of their cells die and break down, releasing specific isoenzymes that can be detected in the blood serum. Normally these enzymes would not be present in the serum or would have very low concentrations. If their serum levels are elevated, it can help pinpoint which cells in the body have been damaged.

For example, the enzyme *creatine phosphokinase (CPK)* occurs in different forms in different cells. An elevated serum level of CPK-1 indicates that skeletal muscle is breaking down, and this is one of the signs of muscular dystrophy. An elevated level of CPK-2 indicates heart disease, because this isoenzyme can come only from heart muscle. There are five isoenzymes of *lactate dehydrogenase (LDH).* High serum levels of LDH-1 can indicate a tumor of the ovaries or testes, while LDH-5 can indicate liver disease or muscular dystrophy. *Phosphatase* occurs in the forms acid phosphatase, named for the fact that it functions best at pH 5.5, and alkaline phosphatase, which has an optimal pH of 9.0. Acid phosphatase is found in the prostate gland and alkaline phosphatase in bone, so elevated levels of one or the other may indicate prostate or bone disease.

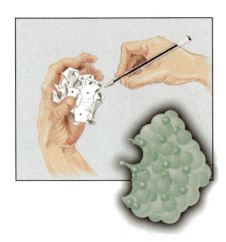

Figure 3.19 A marked paper ball can serve as a simple model for enzyme structure and the effects of denaturation.

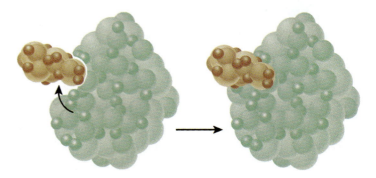

Figure 3.20 According to the induced-fit theory, substrate molecules may cause slight changes in the conformation of an enzyme so the active site fits the substrate better.

which serve as **active sites** where substrate molecules can attach. An active site typically has three amino acid R groups that bind to the substrate, analogous to the three marks you made on your paper model. For example, a substrate might bind to the —COOH, —SH, and —OH groups of three amino acids of the enzyme. It is important that the enzyme have just the right conformation; otherwise, its three R groups may not be positioned properly to bind to the substrate. Generally, an enzyme can bind to only one specific substrate or at most to a few that are very similar—somewhat like the way only one key fits a given lock. This selectivity of enzymes for substrates is called **enzyme-substrate specificity.**

The lock-and-key analogy, however, is now known to be oversimplified. A substrate may not fit perfectly into the active site at first, but as it begins to bind to the enzyme, it changes the enzyme's shape a little so it can better fit the active site. This is called the **induced-fit model** of enzyme-substrate interaction (fig. 3.20). When enzyme and substrate are bonded, they form a temporary **enzyme-substrate complex.** Enzymes often have two or more active sites that allow different kinds of substrates to bind to them simultaneously.

Once a substrate is bound to an enzyme, the enzyme may change shape slightly, in a way that puts tension on the substrate and weakens or breaks its covalent bonds. Thus, one substrate molecule may be converted into two product molecules. Or, an enzyme may bind two substrates and hold them close together, with the necessary orientation for them to react and combine into one product molecule. One way or the other, the enzyme reduces the element of chance in a chemical reaction and makes the reaction proceed more rapidly.

Think About It

Why do polysaccharides lack the diversity of tertiary structures that proteins have? Why are proteins better suited to serve as enzymes?

Effects of Temperature and pH on Enzymes

Changes in temperature or pH can easily disrupt the weak hydrogen bonds and hydrophobic-hydrophilic interactions that hold enzymes in their normal conformations. This can denature enzymes, just as it does other proteins such as egg albumin, and destroy an enzyme's catalytic ability.

Enzymes function best at the body temperature of the species in which they occur. Fishes that live under the arctic ice cap, for example, have enzymes that function best at freezing temperatures, while the enzymes of certain fishes and algae from hot springs work best at near-boiling temperatures that would destroy any human enzyme. Human enzymes function best at the internal temperature of the human body (fig. 3.21a). They work more slowly at temperatures below 37°C

(99°F) not because low temperatures denature them, but because the cold slows down molecular motion and the frequency of contact between reactants. As temperature rises, molecular motion increases, reactants contact each other and the enzyme more often, and reaction rates increase. If the temperature gets too high, however, the enzyme begins to denature and eventually may not function at all. Human enzymes begin to denature at about 45°C (113°F); they are usually destroyed by temperatures over 60° C (140°F).

Think About It
Why is a high fever dangerous? What does this have to do with homeostasis?

The body's pH varies more from place to place than temperature does, and enzymes show more variation in pH optimum. The pH of most body fluids ranges from 5 to 8, but the stomach contents have a pH that ranges from 1.5 to 2.5 due to hydrochloric acid, and the small intestine may have a pH of 9 or higher due to bicarbonate ions. The optimal pH for pepsin, a protein-digesting enzyme of the stomach, is about 2; the optimum for salivary amylase, a starch-digesting enzyme that functions in the mouth, is 7; and the optimum for trypsin, a protein-digesting enzyme secreted by the pancreas, is 9.5 (fig. 3.21b). Unlike temperature, extremes of pH denature enzymes both above and below the optimum.

You can use your paper model to see why denaturation destroys an enzyme's catalytic property. Begin to slowly unfold it, especially around the groove or pocket that you marked. This represents the conformational change in an enzyme under the influence of heat or abnormal pH. Note that the marks on the model grow farther apart, just as the three functional groups of an enzyme molecule do—making the enzyme's active site incapable of binding to the substrate.

Cofactors
Many enzymes also require nonprotein components to function. Metal ions required for enzyme function are called **cofactors.** They include the ions of iron, copper, magnesium, manganese, and calcium. Some cofactors attach to the active site of an enzyme, and some attach to sites remote from this, causing the enzyme to fold into a conformation that activates it (fig. 3.22).

Enzymes and Metabolic Pathways
In chapter 2, we considered some simple one- and two-step metabolic reactions. In living cells, chemical reactions are often linked together in long sequences called **metabolic pathways.** Each step of a metabolic pathway

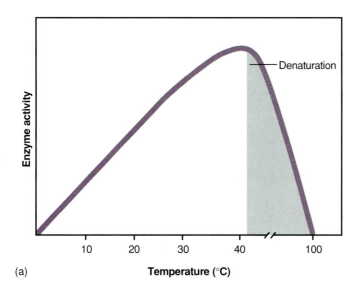

(a)

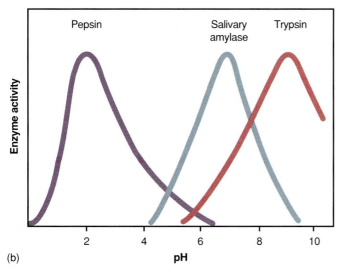

(b)

Figure 3.21 Effects of temperature and pH on enzyme activity. (a) Enzymatic reaction rates increase as the temperature rises up to a point and then rapidly decline as higher temperatures denature the enzyme and destroy its catalytic ability. (b) Human enzymes vary in optimum pH because they function in environments with very different pH levels. Above and below the optimum pH for each enzyme, the enzyme is denatured and reaction rate declines.

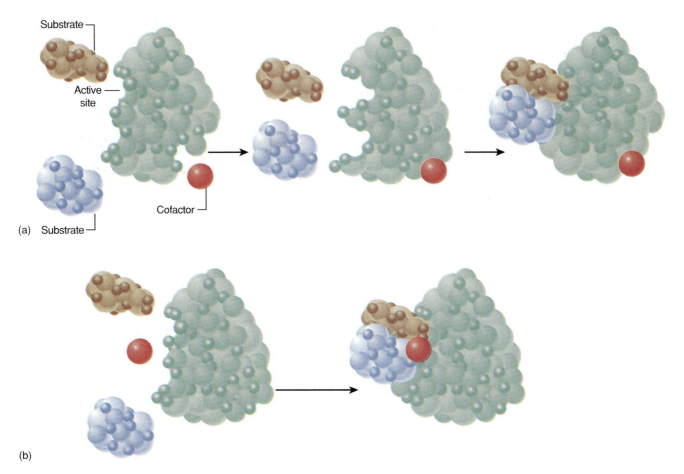

Figure 3.22 (*a*) In some cases, a cofactor binds to a site on an enzyme remote from the active site and causes a conformational change that enables the active site to bind its substrates. (*b*) In other cases, the cofactor binds to the active site along with the substrate molecules.

is usually catalyzed by a specific enzyme. A simple metabolic pathway can be symbolized

$$\begin{array}{ccc} \alpha & \beta & \gamma \\ A \rightarrow B \rightarrow C \rightarrow D \end{array}$$

where *A* is the *initial reactant, B* and *C* are called *intermediates,* and *D* is the *end product.* The Greek letters above the reaction arrows represent specific enzymes that catalyze each step of the reaction. Because these steps are controlled by enzymes, and because enzymes are so capable of conformational changes when something binds to them, it is easy to turn a reaction pathway on and off when its end product is needed or not needed.

One mechanism for turning a pathway off is **allosteric[16] inhibition.** The end product of the pathway binds to an enzyme earlier in the pathway and causes it to fold into a new shape in which it can no longer bind to its substrate (fig. 3.23). For example, end product D could bind to a site on enzyme α, causing a conformational change in the enzyme that renders it temporarily

inactive. The pathway would therefore shut down when there was plenty of end product D available and no need to synthesize more.

Coenzymes

Many enzymes can work only in conjunction with organic compounds called **coenzymes** (co-EN-zimes), which are usually derived from niacin, riboflavin, and other water-soluble vitamins. Coenzymes accept electrons from an enzyme in one metabolic pathway and transfer them to an enzyme in another pathway—thus they serve to transfer energy from one pathway to another. For example, cells partially oxidize glucose through an enzymatic pathway called glycolysis. A coenzyme called NAD^+,[17] derived from niacin, shuttles electrons and energy from this pathway to another metabolic pathway called *aerobic respiration,* which uses the energy to make ATP.

Some of the functional properties of enzymes are summarized in table 3.5.

16. *allo* = different + *ster* = three-dimensional shape

17. Nicotinamide adenine dinucleotide

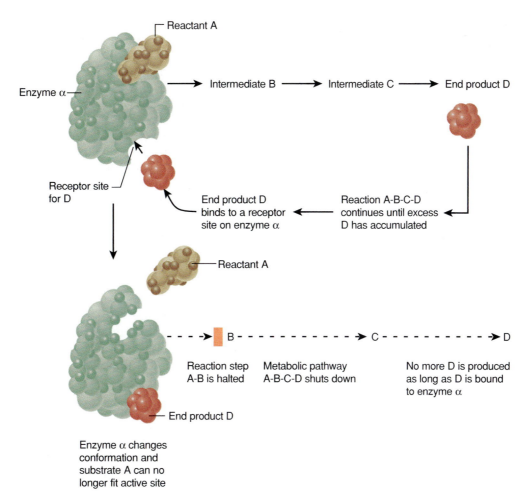

Figure 3.23 Allosteric inhibition, a mechanism that shuts down a metabolic pathway when a sufficient amount of the end product has accumulated.

Table 3.5 Functional Properties of Enzymes

Enzymes lower activation energy.

Enzymes enable molecules with relatively low kinetic energy to react more easily, speeding up the rate of a reaction.

Enzymes are not consumed by the reactions they catalyze.

An enzyme molecule is temporarily changed during a reaction but is restored to its original conformation and can be reused. Thus, very small quantities of enzymes can process large amounts of substrate.

Enzymes are substrate-specific.

The binding site of an enzyme will fit only a particular substrate or a few similar substrates.

Enzymes are temperature- and pH-sensitive.

Abnormal temperature or pH can denature an enzyme and make it incapable of carrying out its catalytic function.

Enzymes can be reversibly inhibited.

Some molecules and ions can bind to an enzyme, or dissociate from it, causing a reversible change in conformation that temporarily activates or deactivates its enzymatic activity. Metabolic pathways can be turned on or off by means of reversible inhibition and activation.

.... **Key Point Review** ....

20 How is enzyme function related to activation energy?

21 To say that a substrate fits its enzyme like a key fits a specific lock is an oversimplification. Explain why.

22 Explain how tertiary structure is related to enzyme-substrate specificity.

23 Explain why denaturation destroys enzyme activity.

24 How is a coenzyme different from a cofactor?

Adenosine Triphosphate (ATP)

▼Objectives

When you have completed this section, you should be able to

- describe or diagram the structure of ATP;
- discuss the metabolic importance of ATP;
- explain how ATP releases energy; and
- distinguish between aerobic and anaerobic pathways of ATP synthesis.

Enzymes are essentially involved in controlling energy transfers between one molecule and another. Many of these transfers involve a universal energy-carrying molecule called **adenosine** (ah-DEN-oh-seen) **triphosphate (ATP).** ATP briefly stores energy gained from exergonic reactions, such as glucose oxidation, and releases it for physiological work—polymerization reactions, muscle contraction, and pumping ions through cell membranes, for example. The production or consumption of ATP is often drawn as a separate reaction linked to the arrow of the primary reaction:

$$C_6H_{12}O_6 + 6\,O_2 \longrightarrow 6\,CO_2 + 6\,H_2O$$

$$ADP + P_i \qquad ATP$$

ATP is a short-lived molecule, generally consumed within 60 seconds of its formation. The entire amount present in the body would support life for less than a minute if it were not being continually replenished. The reason cyanide is so lethal is that it halts ATP synthesis. At a moderate rate of physical activity, a full day's supply of ATP would weigh about twice as much as you do. Even if you never got out of bed, you would need about 45 kg (99 lb) of ATP to stay alive for a day.

ATP Structure, Hydrolysis, and Energy Release

ATP consists of a double carbon-nitrogen ring called adenine; the five-carbon sugar, ribose; and a chain of three phosphate groups (fig. 3.24). Most energy transfers involving ATP involve breaking or forming a bond between the third phosphate group and the rest of the molecule. The body's cells have various enzymes called **ATPases** (adenosine triphosphatases) specialized to hydrolyze this bond. Hydrolysis produces adenosine diphosphate (ADP) and an inorganic phosphate group (P_i). Breaking the ADP~P_i bond is an energy-consuming reaction, but it is linked to other reactions that release more energy than this one consumes, such as the binding of ATP to ATPase just before hydrolysis, or **phosphorylation** (foss-FOR-uh-LAY-shun) reactions— binding the free P_i group to another molecule. The net effect is the release of 7.3 kcal of energy for every mole (505 g) of ATP hydrolyzed. (One kilocalorie, or kcal, equals 1,000 calories.)

Phosphorylation reactions are generally carried out by enzymes called *kinases* or *phosphokinases.* Some enzymes are inactive until they are phosphorylated, and this event is sometimes the "switch" that turns on a metabolic pathway.

Most of the energy released from ATP escapes as heat, but we live on that portion of it that does metabolic work. We can summarize all of this as follows:

$$ATP + H_2O \rightarrow ADP + P_i + Energy \begin{array}{l} \nearrow \text{Heat} \\ \searrow \text{Work} \end{array}$$

ATP Synthesis

Glucose oxidation is the body's primary source of energy for ATP synthesis (fig. 3.25). Although ATP synthesis is explained in detail in chapter 26, a general idea of this process will be useful to you before you reach that chapter—in understanding muscle physiology (see chapter 12), for example.

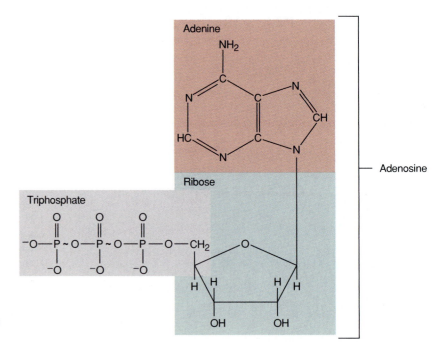

Figure 3.24 The structure of adenosine triphosphate (ATP). Adenosine is composed of the nitrogenous base adenine and the sugar ribose. The last two P~O bonds of the triphosphate group are high-energy bonds. High-energy phosphate bonds are indicated by wavy lines.

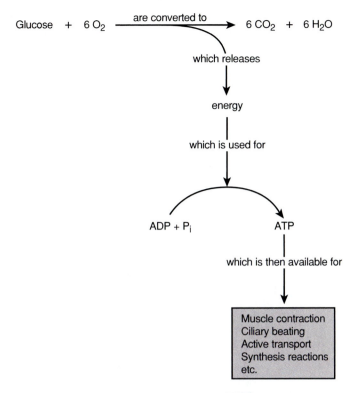

$$\text{Glucose} + 6\,O_2 \xrightarrow{\text{are converted to}} 6\,CO_2 + 6\,H_2O$$

which releases

↓

energy

↓

which is used for

ADP + P$_i$ ATP

which is then available for

↓

Muscle contraction
Ciliary beating
Active transport
Synthesis reactions
etc.

Figure 3.25 Production and uses of ATP.

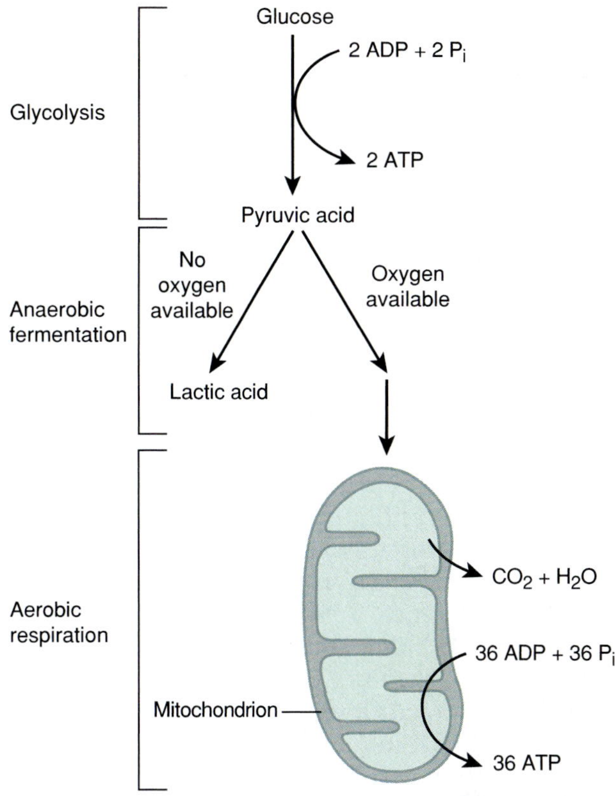

Glycolysis

Glucose
2 ADP + 2 P$_i$
2 ATP
Pyruvic acid

Anaerobic fermentation

No oxygen available
Oxygen available
Lactic acid

Aerobic respiration

$CO_2 + H_2O$
36 ADP + 36 P$_i$
Mitochondrion
36 ATP

Figure 3.26 The production of ATP in glycolysis and aerobic respiration. In the absence of oxygen, anaerobic fermentation is necessary to keep glycolysis running and producing a small amount of ATP. In aerobic respiration, mitochondria oxidize pyruvic acid and produce a much greater amount of ATP than does fermentation.

The first stage in the oxidation of glucose (fig. 3.26) is a reaction pathway called **glycolysis** (gly-COLL-ih-sis). This literally means "sugar splitting," and indeed its major effect is to split the six-carbon glucose molecule into two three-carbon molecules of *pyruvic acid.* A little ATP is produced in this stage (a net yield of two ATP per glucose), but most of the chemical energy of the glucose is still in the pyruvic acid.

What happens to pyruvic acid depends on whether or not oxygen is available. If oxygen is not available, it is converted to lactic acid by a pathway called **anaerobic**[18] (AN-err-OH-bic) **fermentation.** This pathway has two noteworthy disadvantages: first, it does not extract any more energy from pyruvic acid; second, the lactic acid it produces is toxic, thus most cells can use anaerobic fermentation only as a temporary measure. The only advantage to this pathway is that it enables glycolysis to continue (for reasons explained in chapter 26), and thus a cell can continue to produce a small amount of ATP.

If oxygen is available, a much more efficient pathway called **aerobic respiration** occurs. This pathway breaks pyruvic acid down to carbon dioxide and water and generates up to 36 more molecules of ATP for each of the original glucose molecules. The reactions of aerobic respiration are carried out in cellular compartments called *mitochondria* (described in the next chapter), so mitochondria are regarded as a cell's principal "ATP factories."

18. *an* = without + *aer* = air + *obic* = pertaining to life

Molecules Related to ATP

Another energy-transfer molecule similar to ATP is **guanosine** (GWAH-no-seen) **triphosphate (GTP).** We will encounter reactions from time to time where GTP phosphorylates other molecules. In some pathways, it donates its third phosphate group to ADP to regenerate ATP.

ATP is also closely related to the nucleic acids **DNA** (deoxyribonucleic [dee-OCK-see-RY-bo-new-CLEE-ic] acid) and **RNA** (ribonucleic [RY-bo-NEW-clee-ic] acid). Nucleic acids are polymers of nucleotides, which have a structure very similar to ATP.

Key Point Review

25. What are the three main structural components of an ATP molecule?

26. Explain how the structure of ATP is changed in the process of releasing energy.

27. Name two metabolic pathways in which ATP is produced. Compare the amount of ATP produced by each pathway.

28. Discuss the similarities and differences between ATP and the nucleotide shown in figure 5.3.

Synthetic Steroids and Artificial Athletes

The sex hormone testosterone stimulates muscular growth and aggressive behavior, especially in males. In Nazi Germany, testosterone was given to SS troops in an effort to make them more aggressive, but with no proven success. In the 1950s, pharmaceutical companies developed compounds similar to testosterone, called **anabolic steroids,** to treat anemia, breast cancer, osteoporosis, and some muscle diseases and to prevent the shrinkage of muscles in immobilized patients. By the early 1960s, anabolic steroids were being used illegitimately by athletes to stimulate muscle growth, accelerate the repair of tissues damaged in training or competition, and stimulate the aggressive behavior needed to excel at contact sports such as football and hockey. Such usage is officially condemned by the American Medical Association and American College of Sports Medicine and banned by the National Football League, National Collegiate Athletic Association, and International Olympic Committee.

Aside from the fact that muscular mass or strength gained from steroid use is "ill-gotten gain" compared to the benefits of honest exercise and training, there are compelling medical reasons for the widespread condemnation of steroid use by athletes. The doses used to stimulate muscle development are 10 to 1,000 times higher than the doses used for medical purposes. They cause a rise in cholesterol levels, which promotes fatty degeneration of the arteries (atherosclerosis). This often leads to coronary disease, heart and kidney failure, and stroke. Deteriorating blood circulation may result in gangrene and require amputation of the extremities. As the liver attempts to dispose of the high concentration of steroids, liver cancer and other liver diseases may ensue. In addition, these steroids suppress the immune system, so the user is more susceptible to infection and cancer. They also cause a premature curtailment of bone growth, so adolescents who abuse steroids may be permanently stunted.

Anabolic steroids have the same effect on the brain as natural testosterone. Thus, when steroid levels are high, the brain and pituitary stop producing the hormones that promote sperm production and testosterone secretion. In men, this leads to atrophy of the testes, impotence (inability to achieve or maintain an erection), a low sperm count, and infertility. Ironically, anabolic steroids have feminizing effects on men and masculinizing effects on women. Men may develop enlarged breasts (gynecomastia), while in women the breasts and uterus atrophy and the clitoris may become enlarged. Ovulation is inhibited and menstrual periods become irregular. Women develop excessive facial and body hair and a deeper voice, and both sexes show increased tendency toward baldness.

Especially in men, anabolic steroid use is often linked to severe emotional disorders. The steroids themselves stimulate heightened aggressiveness and unpredictable mood swings, so the abuser may vacillate between depression and violence. It surely doesn't help matters that impotence, shrinkage of the testes, infertility, and enlargement of the breasts are so incongruous with the "macho" self-image of a male athlete who abuses steroids (fig. E.1).

One would think that the dangers of steroids would be enough to deter abuse, yet some athletes still swear by them, and steroids remain available through unscrupulous coaches, physicians, and other sources. By some estimates, up to 80% of weightlifters, 30% of college and professional athletes, and 20% of male high school athletes now use anabolic steroids. ▲

Figure E.1 Dying to be macho.

Inorganic Matter (pp. 78–80)
1. Water
   a. Solvency
      • Hydrophobic and hydrophilic substances
      • Hydration spheres
   b. Cohesion and surface tension
   c. Thermal stability
      • Heat capacity
      • Meaning of calories
      • Cooling effect
   d. Chemical reactivity
2. Minerals
   a. As structural components
   b. As activators of organic compounds
   c. As electrolytes
3. Gases
   a. Oxygen and carbon dioxide
   b. Nitric oxide and carbon monoxide

Carbon and Organic Molecules (pp. 81–82)
1. Special properties of carbon
2. Functional groups of organic compounds
3. Monomers, polymers, and macromolecules
   a. Dehydration synthesis
   b. Hydrolysis

Carbohydrates (pp. 82–85)
1. Definition and general formula
2. Monosaccharides: glucose, fructose, and galactose
3. Disaccharides
   a. Sucrose, lactose, and maltose
   b. Synthesis and hydrolysis
4. Polysaccharides: cellulose, starch, and glycogen

5. Carbohydrate functions
   a. Energy sources
   b. Conjugated carbohydrates
      • Glycolipids and glycoproteins
      • The glycocalyx
      • Proteoglycans

Lipids (pp. 85–89)
1. Definition and general properties
2. Fatty acids
   a. General properties
   b. Saturated and unsaturated forms
3. Triglycerides (neutral fats)
   a. Synthesis and hydrolysis
   b. Fats and oils
   c. Functions
4. Phospholipids
   a. Structure
   b. Amphiphilic character
5. Prostaglandins
6. Steroids

Proteins (pp. 89–93)
1. Amino acids
   a. General structure
   b. Variety of functional R groups
2. Peptides and peptide bonds
3. Levels of protein structure
   a. Primary structure: amino acid sequence
   b. Secondary structure
      • α helix
      • β sheet
   c. Tertiary structure (conformation)
      • Globular and fibrous proteins
      • Disulfide bridges
      • Reversible changes
      • Denaturation
   d. Quaternary structure
4. Conjugated proteins and prosthetic groups

5. Protein functions
   a. Structure
   b. Catalysis
   c. Communication
   d. Membrane transport
   e. Recognition and protection
   f. Movement

Enzymes and Metabolism (pp. 93-99)
1. What enzymes do
   a. Definition of an enzyme
   b. Energy of activation
   c. Substrates
   d. Reusability
2. How enzymes are named
3. Enzyme structure
   a. Active sites
   b. Enzyme-substrate specificity
   c. Induced-fit model
   d. Enzyme-substrate complex
4. Effects of temperature and pH on enzymes
5. Role of cofactors
6. Enzymes and metabolic pathways
   a. Reactants, intermediates, and end products
   b. Allosteric inhibition
7. Role of coenzymes

Adenosine Triphosphate (ATP) (pp. 99–101)
1. ATP structure, hydrolysis, and energy release
   a. ATPase
   b. Phosphorylation reactions
2. ATP synthesis
   a. Glycolysis
   b. Anaerobic fermentation
   c. Aerobic respiration
3. Molecules related to ATP
   a. Guanosine triphosphate (GTP)
   b. Relationship of ATP to nucleic acids

Selected Vocabulary

biochemistry 78
organic chemistry 78
solvency 78
hydrophilic 78
hydrophobic 78
hydration sphere 78
cohesion 78
surface tension 78
heat capacity 79
calorie 79
reactivity 79

mineral 79
functional group 81
carboxyl group 81
phosphate group 81
macromolecule 81
polymer 81
monomer 81
polymerization 81
dehydration synthesis 81
hydrolysis 82
carbohydrate 82

monosaccharide 82
glucose 82
fructose 82
galactose 82
disaccharide 83
sucrose 83
lactose 83
maltose 83
glycosidic bond 83
polysaccharide 83
cellulose 83

starch 83
glycogen 84
conjugated 84
glycolipid 84
glycoprotein 84
moiety 84
proteoglycan 85
lipid 85
fatty acid 86
saturated 86
unsaturated 86

polyunsaturated 86
triglyceride 86
glycerol 86
neutral fat 86
lipase 86
phospholipid 86
amphiphilic 86
prostaglandin 87
steroid 87
amino acid 90
peptide 90
peptide bond 90
primary structure 92

secondary structure 92
α helix 92
β sheet 92
tertiary structure 92
globular protein 92
fibrous protein 92
quaternary structure 92
conformation 93
denaturation 93
prosthetic group 93
ligand 93
enzyme 93

activation energy 94
substrate 94
active site 96
enzyme-substrate
    specificity 96
induced-fit model 96
enzyme-substrate complex 96
cofactor 97
metabolic pathway 97
allosteric inhibition 98
coenzyme 98

adenosine triphosphate
    (ATP) 100
ATPase 100
phosphorylation 100
glycolysis 101
anaerobic fermentation 101
aerobic respiration 101
guanosine triphosphate
    (GTP) 101
DNA 101
RNA 101

T e s t i n g   Y o u r   R e c a l l   Answers in Appendix C

1. A substance that dissolves freely in water is said to be
   a. hydrophilic.
   b. hydrophobic.
   c. hydrolyzed.
   d. hydrated.
   e. amphiphilic.

2. The major effect of glycolysis is
   a. glycogen hydrolysis.
   b. glycogen synthesis.
   c. glucose synthesis.
   d. glucose metabolism in the absence of oxygen.
   e. splitting of glucose into pyruvic acid molecules.

3. A carboxyl group is symbolized
   a. —OH.
   b. —NH$_2$.
   c. —CH$_3$.
   d. —CH$_2$OH.
   e. —COOH.

4. Which of the following macromolecules is/are produced only by plants?
   a. starch
   b. glycogen
   c. cholesterol
   d. triglycerides
   e. polypeptides

5. The arrangement of a polypeptide into a fibrous or globular shape is called its
   a. primary structure.
   b. secondary structure.
   c. tertiary structure.
   d. quaternary structure.
   e. conjugated structure.

6. Which of the following functions is more characteristic of carbohydrates than of proteins?
   a. contraction
   b. energy storage
   c. catalyzing reactions
   d. immune defense
   e. intercellular communication

7. A substance acted upon and changed by an enzyme is called the enzyme's
   a. coenzyme.
   b. cofactor.
   c. substrate.
   d. reactant.
   e. intermediate.

8. The reason liquid nitrogen evaporates so easily is that it lacks
   a. polarity.
   b. cohesion.
   c. hydrogen bonds.
   d. surface tension.
   e. all of the above.

9. Glycogen is most similar structurally to
   a. glucose.
   b. starch.
   c. phospholipids.
   d. cellulose.
   e. maltose.

10. The feature that most distinguishes a lipid from a carbohydrate is that the lipid
    a. has more phosphate.
    b. has more sulfur.
    c. has a lower ratio of carbon to oxygen.

    d. has a lower ratio of oxygen to hydrogen.
    e. is larger in size and molecular weight.

11. Molecules dissolve in water if the water can form zones called _____ around them.

12. A chemical reaction that produces water as a by-product is called _____.

13. Organic compounds obtain their unique chemical characteristics less from their carbon backbones than from their _____.

14. The suffix _____ denotes a sugar, while the suffix _____ denotes an enzyme.

15. When oxygen is unavailable, cells employ a metabolic pathway called _____ to produce ATP.

16. A saturated fat is one to which no more _____ atoms can be added.

17. A unique type of bond formed between two amino acids is called a/an _____ bond.

18. The amphiphilic lipids of cell membranes are called _____.

19. _____ occurs when the end product of a metabolic pathway deactivates the enzyme that controls an earlier step in the pathway.

20. The monomers of a polysaccharide are called _____ and those of a protein are called _____.

1. Explain why water would have none of its biologically important properties if its three atoms formed a straight line rather than a 105° angle.

2. Carbon monoxide is now known to be a chemical messenger between cells, but it is highly unlikely that carbon dioxide will be found to play a similar role. Why?

3. Examine the structures of cellulose and amylose. Based on this observation, why do you think amylase cannot digest cellulose? To which functional group(s) on its substrate might you suspect amylase to bind?

4. Explain how allosteric inhibition exemplifies the concept of negative feedback.

5. What are the similarities and differences between a glycoprotein and a proteoglycan?

Web Site Link

For a listing of the most current web sites related to this chapter, please visit the Saladin homepage at:

http://www.mhhe.com/sciencemath/biology/saladin/

4

[Cellular Form and Function

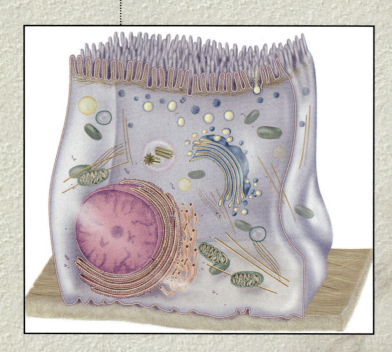

Special Topics

Brushing up

To understand this chapter, it is essential that you understand or brush up on the following concepts:

▶ Glycolipids and glycoproteins (p. 84)
▶ Phospholipids and their amphiphilic character (p. 86)
▶ Tertiary structure in relation to protein and enzyme functions (pp. 92–93, 94–97)
▶ Functions of proteins (p. 93)

All organisms, from the simplest to the most complex, are composed of cells. A bacterium or an amoeba consists of just one cell, whereas the human body is made up of trillions of cells. All of the properties of any living organism derive from the structure and functions of the cells of which it is made. Therefore, to understand life in all of its forms, a study of the cell is essential.

Concepts of Cellular Structure

▼**Objectives**
When you have completed this section, you should be able to
• discuss the development and modern tenets of the cell theory;
• describe the shapes and sizes of cells and their components; and
• explain how early concepts of cellular structure were revised with the advent of new technology.

Development of the Cell Theory

As you may recall from chapter 1, Robert Hooke had observed only the empty cell walls of cork when he first named the cell in 1663. Later, he observed fresh wood under his microscope and saw cells "filled with juices"—a fluid later named **protoplasm**.[1] In the nineteenth century, Theodor Schwann studied a wide range of animal tissues and concluded that all animals are made of cells. Schwann and other biologists originally believed that cells came from nonliving body fluid that somehow congealed and acquired a membrane and nucleus. The idea of **spontaneous generation**—that living things arise from nonliving matter—was rooted in the scientific thought of the times. For centuries, it had been considered simple common sense that decaying meat turns into maggots, stored grain into rodents, and mud into frogs. Schwann and his colleagues merely extended this idea to cells. But common sense often proves wrong—for example, it would have us believe that the sun orbits around the earth—and only careful scientific observation was able to prove otherwise.

In 1858, the German pathologist Rudolph Virchow (VEER-co) (1832–1902) declared that cells can arise only from preexisting cells, a conclusion encapsulated in the now famous Latin phrase *omnis cellula e cellula* ("every cell from a cell"). He carried the thought further and argued that nothing can arise in a cell except by the action of the cell itself; this led him to conclude that all disease results from disorders in cell structure and function. Unfortunately, he was so dogmatic about his pet theory that he refused to believe disease could be caused by such external factors as bacteria.

Virchow's pronouncement did not settle the issue of spontaneous generation, however. In 1859, the French Academy of Sciences sponsored a competition to "try by well-performed experiment to throw new light on the question of spontaneous generation." French microbiologist Louis Pasteur (1822–95) accepted the challenge and conducted experiments proving that sterile meat broth gave rise to bacteria only if their airborne spores were able to get into it. By the end of the nineteenth century, it had been established beyond all reasonable doubt that cells do indeed arise only from other cells. The development of biochemistry from the late nineteenth to the twentieth century made it further apparent that all physiological processes of the body are based on cellular activity and that the cells of all species exhibit remarkable biochemical unity. Thus emerged the modern cell theory (table 4.1).

Cell Shapes

The shapes of cells vary widely (fig. 4.1). **Squamous**[2] (SQUAY-mus) cells are thin and flat, and they often have polygonal (angular) contours when viewed from above. Such cells line the esophagus and cover the skin. **Cuboidal**[3] (cue-BOY-dul) cells are approximately as tall

Table 4.1 Tenets of the Modern Cell Theory

1. All organisms are composed of cells and cell products.
2. The cell is the simplest structural and functional unit of life. There are no smaller subdivisions of a cell or organism that, in themselves, are alive.
3. An organism's structure and all of its functions are ultimately due to the activities of its cells.
4. All cells come only from preexisting cells, not from nonliving matter. All life, therefore, traces its ancestry to the same original cells.
5. Because of this common ancestry, the cells of all species have many fundamental similarities in their chemical composition and metabolic mechanisms.

1. *proto* = first + *plasm* = formed
2. *squam* = scale
3. *cub* = cube + *oid* = like, resembling

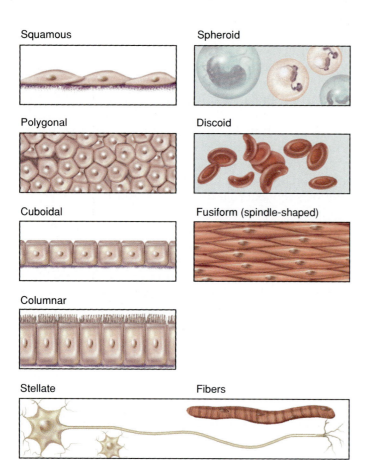

Squamous	Spheroid
Polygonal	Discoid
Cuboidal	Fusiform (spindle-shaped)
Columnar	
Stellate	Fibers

Figure 4.1 Some common cell shapes.

as they are wide; liver cells are a good example. **Columnar** cells, such as those lining the intestines, are markedly taller than wide. Egg cells and fat cells are **spheroid** to **ovoid** (round to egg-shaped). Red blood cells are **discoid** (disc-shaped). **Fusiform**[4] (FEW-zih-form) cells are thick in the middle and tapered toward the ends—somewhat cigar-shaped; the cells of smooth muscle are fusiform. Skeletal muscle cells are called muscle **fibers** because of their long, slender shape. Some nerve cells have multiple extensions, which give them a starlike or **stellate**[5] shape.

Epithelial cells, which form the linings of organs, have three surfaces that are often quite different from each other in membrane composition and function. The "lower" surface on which an epithelial cell rests is the **basal surface,** the sides of the cell are its **lateral surfaces,** and its exposed "upper" surface is the **apical surface.**

Cell Size

The size of a cell is usually expressed in micrometers (μm), and sizes of the smaller structures within a cell

4. *fusi* = spindle
5. *stell* = star

are expressed in nanometers (nm). (Units of measure are presented in appendix B and on the inside front cover.) Most human cells range from 10 to 15 μm in diameter. The human egg cell, an exceptionally large 100 μm in diameter, is barely visible to the naked eye. The longest human cells are nerve cells (sometimes over a meter long) and muscle cells (up to 30 cm long), but both are so slender that they cannot be seen with the naked eye.

There is a limit as to how large a cell can be, largely due to the relationship between its volume and its surface area. The surface area of a cell is proportional to the square of its diameter, and volume is proportional to the cube of its diameter. Picture a cuboidal cell 10 μm on each side (fig. 4.2). It would have a surface area of 600 μm^2 (10 μm $\times$ 10 μm $\times$ 6 sides) and a volume of 1,000 μm^3 (10 $\times$ 10 $\times$ 10 μm). Now suppose it grew by another 10 μm on each side. Its new surface area would be 20 μm $\times$ 20 μm $\times$ 6 = 2,400 μm^2, and its volume would be 20 $\times$ 20 $\times$ 20 μm = 8,000 μm^3. For a given increase in diameter, cell volume therefore increases much faster than surface area. The 20 μm cell has eight times as much protoplasm needing nourishment and waste removal, but only four times as much membrane surface through which wastes and nutrients can be exchanged. A cell with too much volume in proportion to its surface area—that is, a cell that grows too large—cannot support itself.

Think About It

Can you conceive of some other reasons for an organ to consist of many small cells rather than fewer, larger ones?

An Evolving Perspective on Cells

In Schwann's time, little was known about cell structure except that cells were enclosed in a membrane and contained a nucleus. The fluid between the nucleus and surface membrane, called **cytoplasm,** was considered to be little more than a gelatinous mixture of chemicals. This concept changed very little until the mid-twentieth century, when the **transmission electron microscope (TEM)** was invented. Using a beam of electrons in place of light, the TEM enabled biologists to see far more detail in cells (fig. 4.3). Since that time, the cytoplasm has been conceived as an intricate maze of passages and compartments, embedded in a clear fluid called the **cytosol,** or **intracellular fluid (ICF)** (fig. 4.4).

Schwann and his contemporaries were unaware of the details of cell structure simply because the structures are so small. Most of them are beyond the limits of the ordinary **light microscope (LM).** The most important thing about a good microscope is not magnification but

resolution—the ability to reveal detail. Any image can be photographed and enlarged as much as we wish, but if this fails to reveal any more useful detail, it is *empty magnification*. A big fuzzy image is not nearly as informative as one that is smaller but sharp. The TEM reveals far more detail than the LM, even at the same magnification (fig. 4.5). Because of its superior resolution, the TEM also allows for much greater magnifications and is now used to observe structure even at the molecular level.

The minute structures within a cell are called **organelles** (literally "little organs") because they are to the cell what organs are to the body. Each plays a role necessary to the survival of the whole. A cell may have 10 billion protein molecules, some of which are potent enzymes with the potential to destroy the cell if they are not contained and isolated from other cellular components. You can imagine the enormous problem of keeping track of all this material, directing molecules to the correct destinations and maintaining order against the incessant tug of disorder. Cells maintain order partly by compartmentalizing their contents in organelles.

Table 4.2 gives the sizes of some cells and subcellular objects relative to the resolution of the naked eye, light microscope, and electron microscope. You can see why the very existence of cells

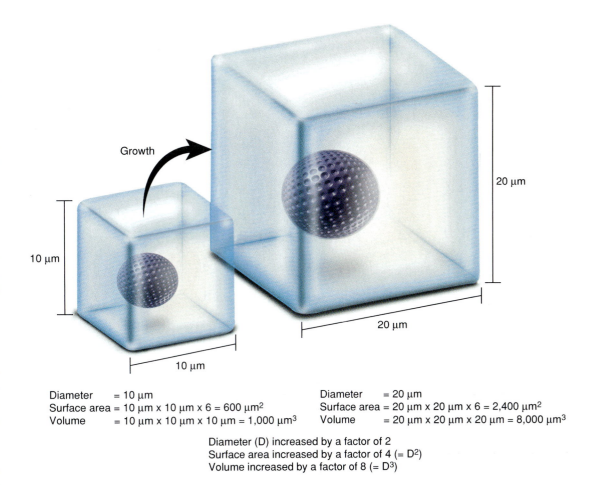

Diameter = 10 µm	Diameter = 20 µm
Surface area = 10 µm x 10 µm x 6 = 600 µm²	Surface area = 20 µm x 20 µm x 6 = 2,400 µm²
Volume = 10 µm x 10 µm x 10 µm = 1,000 µm³	Volume = 20 µm x 20 µm x 20 µm = 8,000 µm³

Diameter (D) increased by a factor of 2
Surface area increased by a factor of 4 ($= D^2$)
Volume increased by a factor of 8 ($= D^3$)

Figure 4.2 The relationship between cell surface area and volume. As a cell doubles in diameter, its volume increases eightfold, but its surface area increases only fourfold. A cell that is too large may have too little plasma membrane area to support the metabolic needs of its increased volume of cytoplasm.

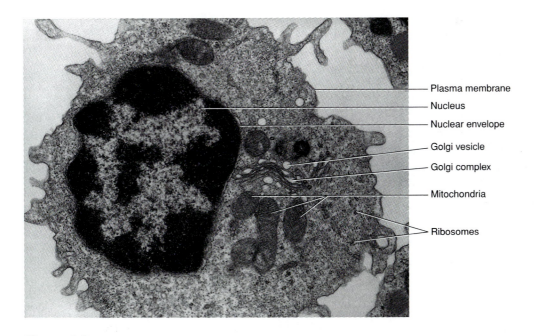

Figure 4.3 Structure of a white blood cell as seen by TEM.

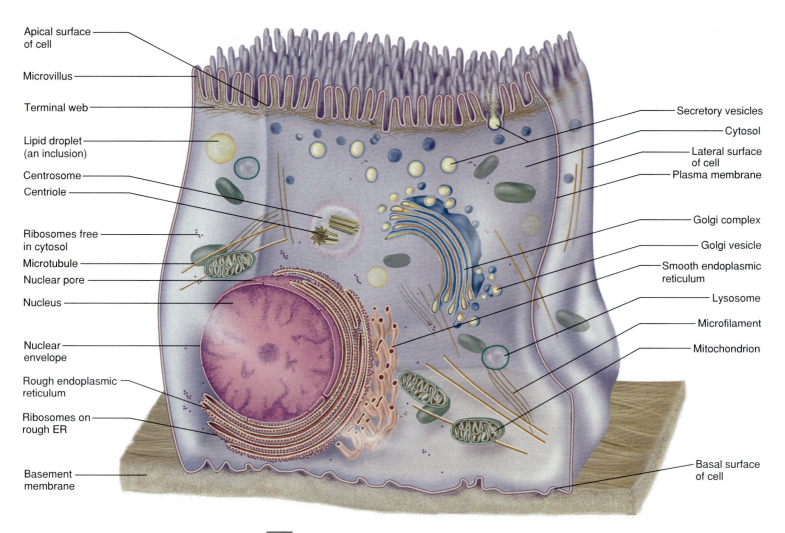

Apical surface of cell
Microvillus
Terminal web
Lipid droplet (an inclusion)
Centrosome
Centriole
Ribosomes free in cytosol
Microtubule
Nuclear pore
Nucleus
Nuclear envelope
Rough endoplasmic reticulum
Ribosomes on rough ER
Basement membrane

Secretory vesicles
Cytosol
Lateral surface of cell
Plasma membrane
Golgi complex
Golgi vesicle
Smooth endoplasmic reticulum
Lysosome
Microfilament
Mitochondrion
Basal surface of cell

Figure 4.4 Structure of a representative cell.

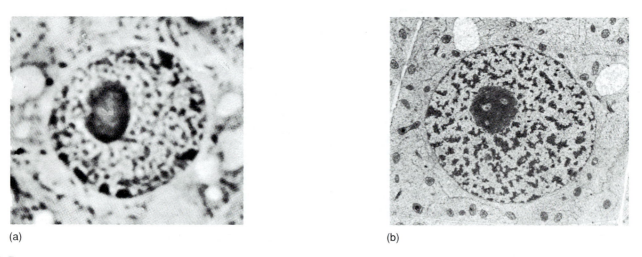

(a)   (b)

Figure 4.5 The value of an electron microscope is superior resolution, not merely magnification. These cells were photographed through (a) a light microscope at ×740 magnification and (b) a transmission electron microscope (TEM) at ×770 magnification. Much more detail is visible in the TEM micrograph because of its superior resolving power.

Table 4.2 Sizes of Some Biological Structures in Relation to the Resolving Power of the Human Eye, Light Microscope (LM), and Transmission Electron Microscope (TEM)

Object	Size	Eye	LM	TEM
Human egg, diameter	100 μm			
Most human cells, diameter	10–15 μm			
Cilia, length	7–10 μm			
Mitochondria, width × length	0.2 × 4 μm			
Bacteria (E. coli), length	1–3 μm			
Microvilli, length	1–2 μm			
Lysosomes, diameter	0.5 μm = 500 nm			
Nuclear pores, diameter	30–100 nm			
Centriole, diameter × length	20 × 50 nm			
Polio virus, diameter	30 nm			
Ribosomes, diameter	15 nm			
Globular proteins, diameter	5–10 nm			
Plasma membrane, thickness	7.5 nm			
DNA molecule, diameter	2.0 nm			
Plasma membrane channels, diameter	0.8 nm			
Carbon atom, diameter	0.15 nm			
Hydrogen atom, diameter	0.07 nm			

was unsuspected until the light microscope was invented and why little was known about the organelles until the TEM became available. Another instrument, the **scanning electron microscope (SEM),** produces dramatic three-dimensional images at high magnification and resolution (see chapter essay, on p. 135).

(see chapter essay, on p. 135)

Key Point Review

1. When was the idea of spontaneous generation of cells laid to rest? Who were some people who questioned the idea?
2. What are the tenets of the cell theory?
3. Define *cytoplasm, cytosol,* and *organelle.*
4. What is the main advantage of an electron microscope over a light microscope?

The Cell Surface

▼**Objectives**

When you have completed this section, you should be able to

- describe the structure of unit membranes, especially the plasma membrane;
- explain the functions of the protein, lipid, and carbohydrate components of the cell surface; and
- describe the structure and function of microvilli, cilia, and flagella.

The Plasma Membrane

The electron microscope reveals that the cell and many of the organelles within it are bordered by **unit membranes,** each of which appears as a pair of dark parallel lines with a total thickness of about 7.5 nm (fig. 4.6*a*). The unit membrane around the cell as a whole is called the **cell membrane,** or **plasma membrane.** It defines the boundaries of the cell, controls the passage of materials into and out of it, and governs its interactions with other cells. The side that faces the cytoplasm is the *intracellular face* of the membrane, and the side that faces outward is the *extracellular face.*

By the 1930s, it was known that the plasma membrane consists mainly of phospholipids and proteins, and multiple theories were proposed as to how these molecules are arranged. The **fluid-mosaic model,** put forth in 1972, best accounts for the evidence. It describes the plasma membrane (and other unit membranes) as an array of mobile globular proteins embedded in an oily film of phospholipids. Further refinements of this model have given rise to the concept illustrated in figure 4.6*b.*

Membrane Lipids

About 75% of the lipid molecules of the plasma membrane are phospholipids. As we saw in the previous chapter, these are amphiphilic molecules, each with a hydrophilic phosphate-containing head and two hydrophobic fatty acid tails. Phospholipids organize themselves into a bilayer, with their heads facing the water on each side of the membrane and their tails directed toward the center of the membrane, away from the water within and around the cell. The phospholipids drift laterally from place to place, they spin on their axes, and their tails flex. These movements keep the membrane fluid.

Think About It

What would happen if a plasma membrane were made primarily of a hydrophilic substance such as carbohydrate? Which of the major themes at the end of chapter 1 does this exemplify?

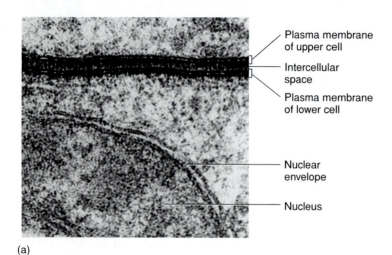

Plasma membrane of upper cell

Intercellular space

Plasma membrane of lower cell

Nuclear envelope

Nucleus

(a)

About 20% of the membrane lipid molecules are cholesterol, which is found in the middle of the membrane amid the fatty acid tails. By interacting with the phospholipids and "holding them still," cholesterol can stiffen the membrane (make it less fluid) in spots. Higher concentrations of cholesterol, however, can increase membrane fluidity by preventing the phospholipids from becoming packed too closely together.

The remaining 5% of the membrane lipid molecules are glycolipids—phospholipids with short oligosaccharide chains covalently bound to them. These are found only on the extracellular face of the membrane.

An important quality of the plasma membrane is its capacity for self-repair. When a physiologist inserts a probe into a cell, it does not pop the cell as it would a balloon. The probe slips through the oily film, and the membrane seals itself around it. When cells take in matter by endocytosis (described later in this chapter), they pinch off bits of their own membrane, which form bubblelike vesicles in the cytoplasm. As these vesicles pull away from the membrane, they do not leave gaping holes; the lipids immediately flow together to seal the break.

Membrane Proteins

Only 1% to 10% of the molecules in the plasma membrane are proteins, but proteins are larger than lipids and constitute about half the membrane weight. Some of them, called **integral (transmembrane) proteins,** pass through the membrane and are exposed to both the cytoplasm and extracellular fluid. Most of these are glycoproteins, which are conjugated with oligosaccharides on the extracellular side of the membrane. Some of the integral proteins are anchored to the *cytoskeleton*—an intracellular system of tubules and filaments (discussed later). **Peripheral proteins** are those that do not protrude into the phospholipid layer but adhere to the intracellular face of the membrane. A peripheral protein is typically associated with an integral protein and tethered to the cytoskeleton.

The functions of membrane proteins fall within the general scope of protein functions described in the previous chapter:

- **Receptors** (fig. 4.7a). Cells communicate with each other by means of chemical messengers that include hormones (released by endocrine gland cells) and

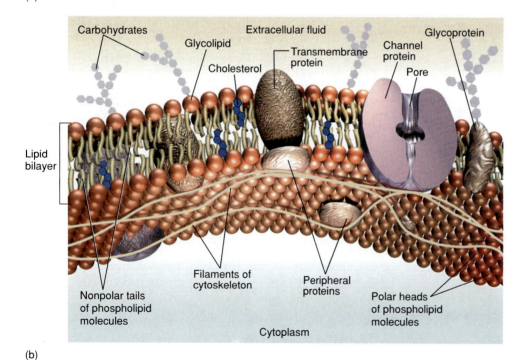

Carbohydrates

Extracellular fluid

Glycolipid

Glycoprotein

Cholesterol

Transmembrane protein

Channel protein

Pore

Lipid bilayer

Nonpolar tails of phospholipid molecules

Filaments of cytoskeleton

Peripheral proteins

Polar heads of phospholipid molecules

Cytoplasm

(b)

Figure 4.6 (a) The plasma membrane of two adjacent cells (electron micrograph) (b) Molecular structure of the plasma membrane.

neurotransmitters (released by nerve cells). Some of these messengers (prolactin, for example) cannot enter target cells but must bind to a membrane protein called a **receptor** on the target cell surface. Receptors are usually specific for one particular messenger, much like an enzyme that is specific for one substrate.

- **Second messenger systems** (fig. 4.7b). When a messenger binds to a receptor, it changes the shape of the receptor molecule. Since the receptor spans the plasma membrane and emerges on its intracellular face, this change can activate metabolic changes within the cell. In many cases, it activates a peripheral **G protein.** G proteins are named for the ATP-like chemical, guanosine triphosphate (GTP), to which they bind. Upon binding to GTP, a G protein moves along the intracellular face of the membrane and binds to an "effector" protein. In some cases, the effector is an enzyme (fig. 4.7c) that catalyzes the formation of a **second messenger** within the target cell. The second messenger relays the message, in effect, and tells the cell how to respond to the original signal. G proteins have an enormous range of roles in physiology and disease. Their discoverers, Martin Rodbell and Alfred Gilman, were awarded the 1994 Nobel Prize in Medicine or Physiology.

- **Messenger deactivators** (fig. 4.7d). A chemical messenger binds to a receptor for a millisecond or less and then detaches. It can bind again and restimulate the target cell, but this must not go on indefinitely. Eventually, the messenger must either be removed or broken down so that the signal will stop. Some membrane proteins are enzymes that degrade hormones and neurotransmitters into inactive, nonstimulatory fragments.

- **Channel proteins** (fig. 4.7e, f). These are integral proteins with water-filled pores that allow passage of hydrophilic solutes that cannot pass through the lipid part of the membrane. Some channels are always open (fig. 4.7e), and others are **gates** that open and close (fig. 4.7f), thus determining when solutes may be admitted (see special topic 4.1). A cell controls when these gates open and close in one of two ways. **Ligand-gated channels** open or close when chemical messengers bind to them. **Voltage-gated channels** open or close when the electrical potential (voltage) across the plasma membrane changes. Gated channels play an important role in the timing of nerve signals and muscle contraction, which depend on the movements of electrolytes through the membrane at the appropriate times.

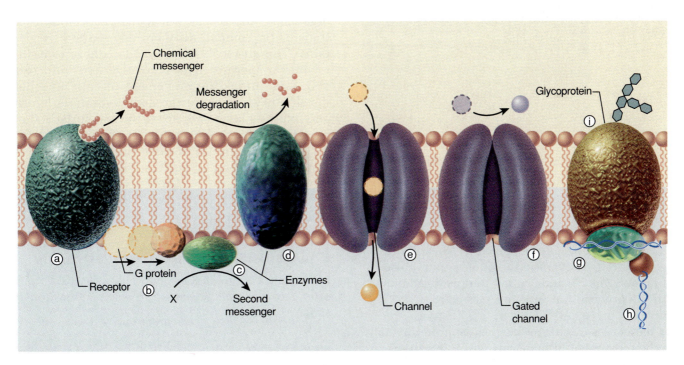

Figure 4.7 Some functions of the proteins of a plasma membrane. (a) A receptor protein that binds to intercellular chemical messengers such as hormones and neurotransmitters. (b) A G protein, which is activated by receptor protein a and which, in turn, activates enzyme c. (c) An enzyme that catalyzes the production of a second messenger within the cell. (d) An enzyme that degrades the chemical messenger so it will not go on stimulating the cell indefinitely. (e) A protein channel that allows particles to diffuse through the membrane. (f) A gated channel that can open or close in response to stimuli. (g) Fibrous proteins of the membrane skeleton, parallel to the inner surface of the membrane. (h) Proteins approaching the plasma membrane from the cytoskeleton deeper in the cell. (i) A glycoprotein, with a short carbohydrate chain conjugated to its extracellular side.

The diameter of an artery and the amount of blood flowing through it depend on the contraction or relaxation of smooth muscle in its wall. Excessive vasoconstriction can cause hypertension (high blood pressure) or heart pain (angina) due to inadequate blood flow to the cardiac muscle. In order to contract, a smooth muscle cell must open calcium channels in its plasma membrane and allow calcium to enter from the extracellular fluid. Drugs called *calcium channel blockers* prevent calcium channels from opening and therefore help to relax the arteries, relieve angina, and lower blood pressure.

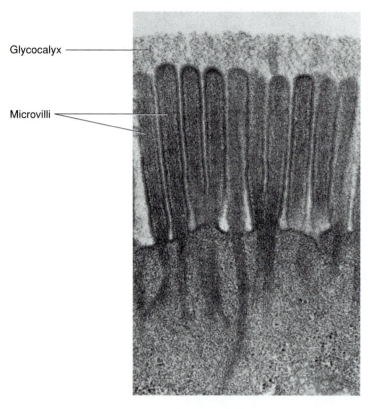

Glycocalyx

Microvilli

Figure 4.8 Microvilli and glycocalyx on the apical surface of an absorptive cell of the small intestine.

approach the membrane from deeper within the cell and attach to its peripheral proteins (fig. 4.7*h*).

- **Cell identity markers.** Glycoproteins (fig. 4.7*i*) contribute to the glycocalyx, a carbohydrate surface coating whose functions we will consider shortly.

The Glycocalyx

External to the plasma membrane, all animal cells have a fuzzy coat called the **glycocalyx**[6] (GLY-co-CAY-licks). As mentioned in chapter 3, the glycocalyx consists of the carbohydrate moieties of the membrane glycolipids and glycoproteins (fig. 4.8). It is chemically unique in everyone but identical twins, and it is like a personal identification tag that enables the body to distinguish its own healthy cells from transplanted tissues, invading organisms, and diseased cells. Human blood types and transfusion compatibility are determined by glycoproteins. The glycocalyx helps cells adhere to each other, enables a sperm cell to recognize and bind to an egg cell, and guides the movement of cells in embryonic development. Because the carbohydrate is hydrophilic and retains a layer of water at the cell surface, the glycocalyx assists in cellular uptake of substances from the extracellular fluid. Most molecules must dissolve in the water film at the cell surface before they can pass into the cell.

Some of the functions of the glycocalyx are summarized in table 4.3.

Extensions of the Cell Surface

Many cells have surface extensions called *microvilli, cilia,* and *flagella.* These aid in absorption, motility, and sensory events.

Microvilli

Microvilli[7] (MY-cro-VIL-eye; singular, *microvillus*) are extensions of the plasma membrane that serve primarily to increase a cell's surface area (figs. 4.8 and 4.9*a, b*).

- **Carriers** (see figs. 4.24 and 4.26). Carriers are integral proteins that bind to glucose, electrolytes, and other solutes and transfer them to the other side of the membrane. Some carriers, called **pumps,** consume ATP in the process. Carriers and pumps are explained in more detail later in this chapter.
- **Motor molecules.** Motor molecules are proteins that cause the movement of materials within a cell or cause the cell as a whole to move or change shape— for example, to pinch in two (cell division), surround and engulf foreign particles (phagocytosis), or cave in at points to imbibe droplets of extracellular fluid (pinocytosis). Such processes depend on the action of fibrous proteins, especially actin and myosin, that lie parallel to the plasma membrane (fig. 4.7*g*) or

---

6. *glyco* = sugar + *calyx* = cup, vessel
7. *micro* = small + *villi* = hairs

Table 4.3 Functions of the Glycocalyx

Immunity to Infection	Enables the immune system to recognize and selectively attack foreign organisms
Transplant Compatibility	Forms the basis for compatibility of blood transfusions, tissue grafts, and organ transplants
Recognition of Cancer	Changes in the glycocalyx when cells turn cancerous enable the immune system to recognize and destroy them
Cell Adhesion	Binds cells to each other so that tissues hold together
Fertilization	Enables sperm cells and egg cells to recognize and bind to each other
Embryonic Development	Guides cell migration during development
Cellular Uptake	Retains a film of water at the cell surface, which enables a cell to take up dissolved matter

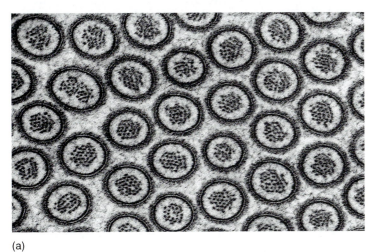

(a)

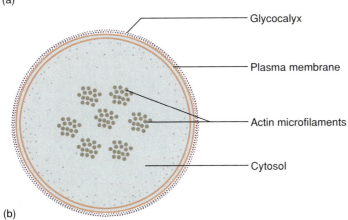

(b)

Glycocalyx

Plasma membrane

Actin microfilaments

Cytosol

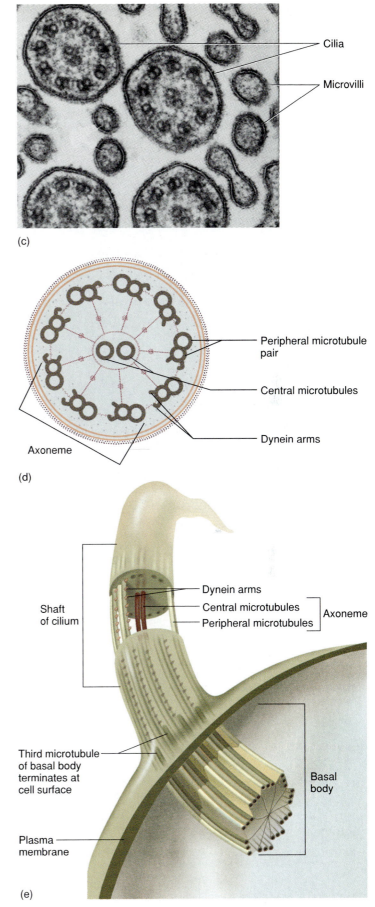

Cilia

Microvilli

(c)

Peripheral microtubule pair

Central microtubules

Dynein arms

Axoneme

(d)

Dynein arms

Central microtubules

Peripheral microtubules

Axoneme

Shaft of cilium

Third microtubule of basal body terminates at cell surface

Basal body

Plasma membrane

(e)

Figure 4.9 (*a–b*) Electron micrograph and diagram of microvilli in cross section. (*c–d*) Electron micrograph and diagram of cilia in cross section. The smaller objects among the four cilia in the photograph are microvilli. (*e*) Three-dimensional structure of a cilium at the point where it meets the cell surface.

Microvilli are best developed in cells specialized for absorption, such as the epithelial cells of the intestines and kidney tubules. They give such cells 15 to 40 times as much absorptive surface area as they would have if their apical surfaces were flat. Individual microvilli cannot be distinguished very well with the light microscope because they are only about 1 to 2 μm long. They appear as a fringe called the **brush border** at the apical cell surface. With the scanning electron microscope, they resemble a shag carpet.

With the transmission electron microscope, microvilli typically looks like finger-shaped projections of the cell surface. They show little internal structure, but often a bundle of stiff filaments of a protein called *actin* can be discerned. Actin filaments attach to the plasma membrane at the tip of the microvillus, and at its base they extend a little way into the cell and anchor the microvillus to a protein mesh called the **terminal web.** When tugged by another protein in the cytoplasm, actin can shorten a microvillus to "milk" its absorbed contents downward into the cell.

On many cells, microvilli are little more than tiny bumps on the plasma membrane. On cells of the taste buds and inner ear, they are well developed but serve sensory rather than absorptive functions.

Cilia

Cilia (SIL-ee-uh; singular, *cilium*[8]) (figs. 4.9c–e and 4.10) are hairlike processes about 7 to 10 μm long. There may be 50 to 200 cilia on the surface of one cell. Cilia are usually motile and serve primarily to move materials such as mucus in the respiratory tract and egg cells in the uterine tubes.

Cilia beat in waves that sweep across the surface of an epithelium, always in the same direction (fig. 4.11). They bend stiffly forward, producing a **power stroke** that pushes along the mucus or other matter riding on their tips. Shortly after a cilium begins its power stroke, the one just ahead of it begins, and the next and the next—collectively they exhibit a wavelike motion. After a cilium completes its power stroke, it is pulled limply back by a **recovery stroke** that restores it to the upright position, ready to flex again.

Think About It

How would it affect mucus of the respiratory tract if cilia were equally stiff on both the power and recovery strokes?

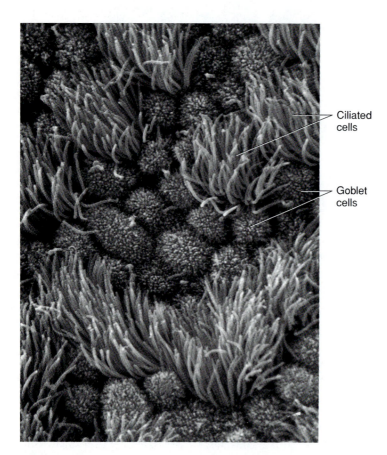

Figure 4.10 Cilia of the trachea. Several nonciliated, mucus-secreting goblet cells are visible among the ciliated cells. The small surface projects on the goblet cells are microvilli.

Ciliated cells

Goblet cells

8. *cilium* = eyelash

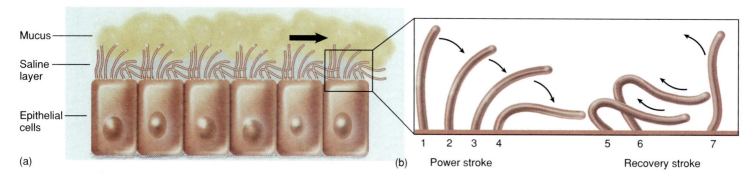

Figure 4.11 (a) Cilia of an epithelium moving mucus along a surface layer of saline. (b) Power and recovery strokes of a cilium.

Mucus

Saline layer

Epithelial cells

(a)

(b)   Power stroke   Recovery stroke

1 2   3   4   5   6   7

Cystic fibrosis (CF) is a hereditary disease occurring in about 1 out of 10,000 white children. It is caused by a defect in which cells make chloride pumps but fail to install them in the plasma membrane. Consequently, there is no saline layer, mucus rests on the cell surface, and the cilia cannot beat freely. The respiratory tract becomes congested with thick mucus, and chronic respiratory infection and pulmonary collapse are common. The pancreas is also affected; its ducts become blocked, digestive enzymes cannot be secreted, and digestion and nutrition are impaired. Individuals with CF seldom survive past early adulthood.

Cilia could not beat freely if they were embedded in sticky mucus. Instead, they beat within a saline (saltwater) layer at the cell surface. This layer is produced by *chloride pumps* in the apical plasma membrane, which pump chloride ions into the extracellular fluid. Sodium ions follow because of the *electrostatic attraction* of oppositely charged particles to each other, and water follows the Na⁺ by osmosis. Mucus essentially floats on the surface of this layer and is pushed along by the tips of the cilia (see special topic 4.2).

A cross section of a cilium shows the structural basis for its contraction (see fig. 4.9*c–e*). It contains a core called an **axoneme**[9] (ACK-so-neem), which consists of an array of thin protein cylinders called **microtubules**. There are two microtubules at the center, surrounded by a pinwheel-like ring of nine microtubule pairs—an arrangement called the **9 + 2 structure**. The two central microtubules stop at the cell surface, but the peripheral microtubules continue a short distance into the cell as part of a **basal body** that anchors the cilium to the cell. In each pair of peripheral microtubules, one member has two little **dynein** (DINE-een) **arms** on it. Dynein, a motor protein, is an ATPase that uses energy from ATP to "crawl" up the adjacent pair of microtubules. When microtubules on the front of the cilium crawl up the microtubules behind them, the cilium bends toward the front.

In some organs, cilia have lost their motility and assumed sensory roles. The outer half of each sensory cell in the retina of the eye, for example, is a modified cilium specialized for absorbing light. Some cilia in the nasal cavity bind to odor molecules and initiate the sense of smell. In the inner ear, cilia function along with microvilli in the senses of hearing and balance.

Flagella

A **flagellum**[10] (fla-JEL-um) is a long whiplike structure with an axoneme identical to that of a cilium. The only functional flagellum in humans is the tail of a sperm cell, but vestigial (stunted, nonfunctional) flagella are found on the epithelial cells of several organs, including the uterus, testis, kidney, pancreas, and pituitary gland.

5. How does the structure of a plasma membrane depend on the amphiphilic nature of phospholipids?
6. Distinguish between integral and peripheral proteins.
7. Why is the glycocalyx important to survival?
8. How do microvilli and cilia differ in structure and function? What is one role that both of them sometimes play?

The Cell Interior

▼**Objectives**
When you have completed this section, you should be able to
• list the main organelles of a cell and state their functions;
• describe the cytoskeleton and its functions; and
• explain how cell inclusions differ from organelles and give some examples of inclusions.

We will now probe more deeply into the cell to study the structures in the cytoplasm. These are classified into three groups—*organelles, cytoskeleton,* and *inclusions*—all embedded in the clear, gelatinous cytosol.

Organelles

Organelles are internal structures of a cell that carry out specialized metabolic tasks. Some are surrounded by one or two layers of unit membrane and are therefore referred to as *membranous organelles*. These are the nucleus, mitochondria, lysosomes, peroxisomes, rough endoplasmic reticulum, smooth endoplasmic reticulum, and Golgi complex. Organelles that are not surrounded by membranes include the ribosomes, centrosome, centrioles, and basal bodies.

The Nucleus

The **nucleus** is the largest organelle and usually the only one visible with the light microscope. With the TEM, it can be distinguished by the *two* unit membranes surrounding it, which together form the **nuclear envelope** (see fig. 4.6*a*). The nucleus houses the DNA,

---

9. *axo* = axis + *nem* = thread
10. *flagellum* = whip

which is a genetic program for the cell's structure and function. Nuclear structure and function are described in the next chapter, which deals in detail with the nucleus and its genetic role.

Rough Endoplasmic Reticulum

Endoplasmic reticulum (ER) literally means "little network within the cytoplasm." The **rough endoplasmic reticulum** (fig. 4.12) is a system of branching channels that may occupy much of the cell. It is continuous with the outer unit membrane of the nuclear envelope. Each channel is a flattened sac with a unit membrane on each side and a space called the **cisterna**[11] (sis-TUR-nuh) between them. The rough ER gets its name from the fact that its outer surface is studded with granular ribosomes, described next. The rough ER serves as a site of

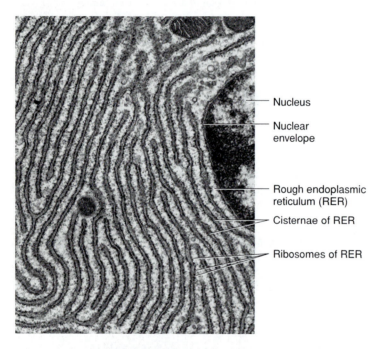

Nucleus

Nuclear envelope

Rough endoplasmic reticulum (RER)

Cisternae of RER

Ribosomes of RER

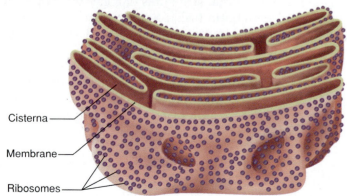

Cisterna

Membrane

Ribosomes

Figure 4.12 The rough endoplasmic reticulum.

protein synthesis but has other functions as well. It synthesizes phospholipids and is the cell's "membrane factory," producing most of its plasma membrane and internal unit membranes. Its role in protein synthesis is discussed in the next chapter.

Ribosomes

Ribosomes are small granules of protein and RNA found in the cytosol and on the outer surfaces of the rough ER and nuclear envelope. They "read" the coded genetic messages (messenger RNA) from the nucleus and assemble amino acids into proteins specified by the code. This process is described in detail in the next chapter.

Golgi Complex and Vesicles

The **Golgi**[12] (GOAL-jee) **complex** resembles a stack of pancakes. With the TEM it looks like a series of parallel sacs (cisternae) with swollen edges (fig. 4.13). The Golgi complex receives polypeptide chains from the rough ER, modifies their structure, and packages them into spheroid **Golgi vesicles,** which are abundant in the area of the complex. Some of these vesicles remain in the cell as lysosomes (described next), and some become **secretory vesicles** that migrate to the surface and release their products to the outside of the cell. The Golgi complex is also a "carbohydrate factory"; it makes and attaches the carbohydrate components of glycoproteins and proteoglycans. It receives the polypeptide portion from the rough ER and adds the carbohydrate with enzymes in the Golgi cisternae. The Golgi complex also synthesizes some hormones and lipids. Its involvement in protein synthesis and secretion is further discussed in the next chapter.

Lysosomes

A **lysosome** (LY-so-some) (fig. 4.14*a*) is a package of enzymes contained in a single unit membrane. Although often round or oval, lysosomes are extremely variable in shape. When viewed with the TEM, they often exhibit dark gray contents devoid of structure, but sometimes crystals or parallel layers of protein can be seen. At least 50 lysosomal enzymes have been identified; they hydrolyze proteins, nucleic acids, complex carbohydrates, phospholipids, and other substrates.

The unit membrane normally prevents lysosomal enzymes from escaping into the cytosol and killing the cell. Many cells, however, are made to do a certain job and then die. Their membranes become bubbled, the nuclei condense, and the cells shrink and disintegrate. This pattern is called **programmed cell death,** or

---

11. *cistern* = reservoir

12. Camillo Golgi (1843–1926), Italian histologist

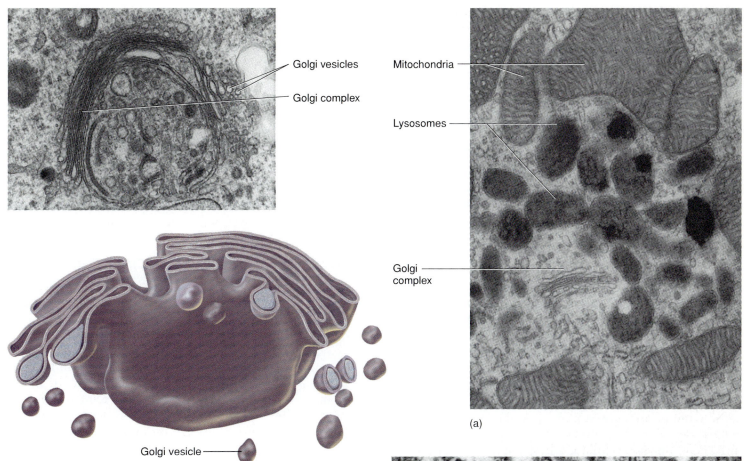

Golgi vesicles

Golgi complex

Golgi vesicle

Mitochondria

Lysosomes

Golgi complex

(a)

Figure 4.13 The Golgi apparatus.

apoptosis[13] (AP-oh-TOE-sis), and is sometimes executed by the lysosomes (which have been nicknamed "suicide bags"). The uterus, for example, weighs about 900 g at full-term pregnancy and shrinks to 60 g within 5 or 6 weeks after birth. This shrinkage is due to digestion of surplus cells by lysosomal enzymes. Lysosomes are very abundant in white blood cells, where they digest bacteria and other foreign matter engulfed by the cell. Lysosomes also digest worn-out organelles in a process called **autophagy**[14] (aw-TOFF-uh-jee). In the liver, they break down stored glycogen to mobilize glucose.

Peroxisomes

Peroxisomes (fig. 4.14b) resemble lysosomes but have different enzymes. They are especially abundant in liver and kidney cells, where they help to detoxify alcohol and other drugs and neutralize free radicals. One peroxisomal enzyme is *catalase,* which degrades the toxic hydrogen peroxide (H_2O_2) to water and oxygen.

13. *apo* = from, away + *ptosis* = dropping, falling
14. *auto* = self + *phagy* = eating

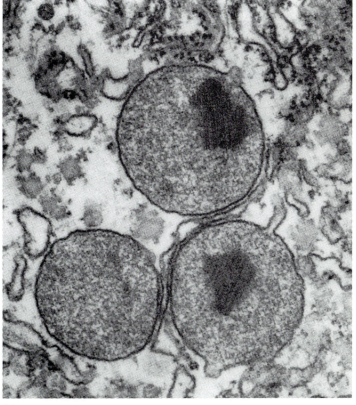

(b)

Figure 4.14 (*a*) Lysosomes. (*b*) Peroxisomes.

There is evidence that mitochondria evolved from *endosymbiotic bacteria* that invaded the ancestors of modern eukaryotic cells and evolved a mutually beneficial metabolic relationship with them. In size and physiology, mitochondria resemble bacteria that now live within other cells in a state of symbiosis. Mitochondria use a host cell's pyruvic acid and supply the cell with ATP. Mitochondrial DNA resembles bacterial DNA more than it does nuclear DNA and is replicated independently of nuclear DNA.

While nuclear DNA is reshuffled in each generation by the process of sexual reproduction, mtDNA remains unchanged from generation to generation except by the slow pace of random mutation. Biologists and anthropologists have used mtDNA as a "molecular clock" to trace evolutionary lineages in humans and other species. They have gained some evidence, although still controversial, that all humans today are the descendants of a single female, the "mitochondrial Eve," who lived about 200,000 years ago.

Mitochondria

Mitochondria[15] (MY-toe-CON-dree-uh) are spheroid, rod-shaped, or threadlike structures (fig. 4.15). Like the nucleus, a mitochondrion is surrounded by a double unit membrane. The inner mitochondrial membrane, however, is usually thrown into folds called **cristae**[16] (CRIS-tee), which project like shelves across the organelle. Mitochondria are the "powerhouses" of the cell. Energy is not *made* here, but it is extracted from organic compounds and stored in ATP, primarily by enzymes located on the cristae. The role of mitochondria in ATP synthesis is explained in detail in chapter 26.

The space between the cristae is called the **mitochondrial matrix**. It contains a small, circular DNA molecule called *mitochondrial DNA (mtDNA)*, which is genetically different from the DNA in the cell's nucleus (see special topic 4.3). Mutations in mtDNA are responsible for some muscle, heart, and eye diseases. Mitochondria also have their own ribosomes in the matrix.

Smooth Endoplasmic Reticulum

The **smooth endoplasmic reticulum** (fig. 4.16) is a network of branching tubules with a smooth surface, free of ribosomes. Smooth ER has a wider variety of functions than rough ER. In liver cells, it converts alcohol, phenobarbital, and other drugs to less toxic, water-soluble forms that can be excreted by the kidneys. Long-term use of some legal and illicit drugs leads to *tolerance* partly because the smooth ER becomes highly developed and detoxifies the drug more rapidly. Smooth ER also synthesizes triglycerides, cholesterol, and steroid hormones. It is very abundant, for example, in cells that produce testosterone. In skeletal muscle cells, the smooth ER is a storage reservoir for calcium ions and releases them as part of the process that triggers muscle contraction.

15. *mito* = thread; *chondr* = grain
16. *crista* = crest

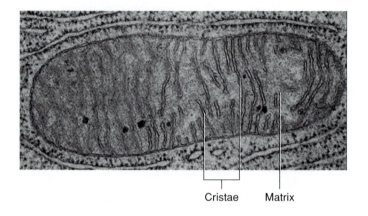

Cristae   Matrix

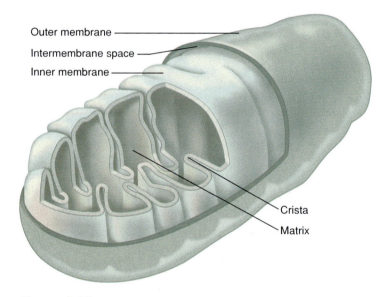

Outer membrane
Intermembrane space
Inner membrane

Crista

Matrix

Figure 4.15 A mitochondrion.

Centrioles

A **centriole** (SEN-tree-ole) is a short cylindrical assembly of microtubules, arranged in nine groups of three microtubules each (fig. 4.17). Two centrioles lie perpendicular to each other within a small clear area of cytoplasm

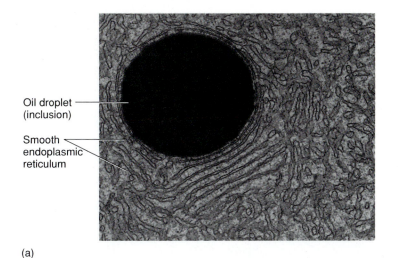

Oil droplet
(inclusion)

Smooth
endoplasmic
reticulum

(a)

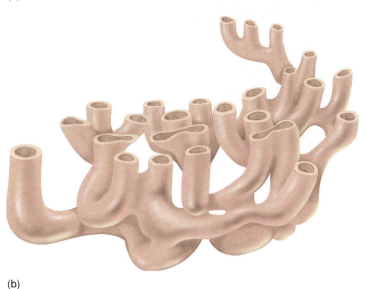

(b)

Figure 4.16 (*a*) The smooth endoplasmic reticulum and a lipid droplet. The latter is an inclusion. (*b*) Diagram of the smooth ER.

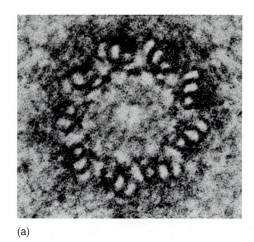

(a)

(b)

Figure 4.17 (*a*) Cross section of a centriole. (*b*) A pair of centrioles in their typical mutually perpendicular orientation.

called the **centrosome**[17] and play a role in cell division, described in the next chapter. Flagella and cilia each have a basal body composed of a single centriole oriented perpendicular to the plasma membrane. Basal bodies form by a vaguely understood process at a *centriolar organizing center,* from which they migrate to the plasma membrane. Two microtubules of each triplet then elongate to form the nine pairs of peripheral microtubules of the axoneme. A cilium can grow to its full length in less than an hour.

Cytoskeleton

The **cytoskeleton** (fig. 4.18) is a collection of protein filaments and cylinders that lends structural support to a cell, helps organize its contents, moves substances

---

17. *centro* = central + *some* = body

through the cell, and contributes to movements of the cell as a whole. Its components are called *microfilaments, intermediate filaments,* and *microtubules.*

Microfilaments are about 6 nm in diameter and are made of the protein actin. They form a network on the cytoplasmic side of the plasma membrane called the **membrane skeleton.** It is thought that the phospholipid bilayer would break up into little droplets were it not for the fact that it is layered on top of this membrane skeleton. The roles of actin in supporting microvilli and producing cell movements were discussed earlier. In conjunction with another protein, *myosin,* microfilaments are also responsible for muscle contraction.

Intermediate filaments (8–10 nm in diameter) are thicker and stiffer than microfilaments. They participate in junctions that hold epithelial cells together and resist stresses placed on a cell. In epidermal cells, they are made of the tough protein keratin and occupy most of the cytoplasm.

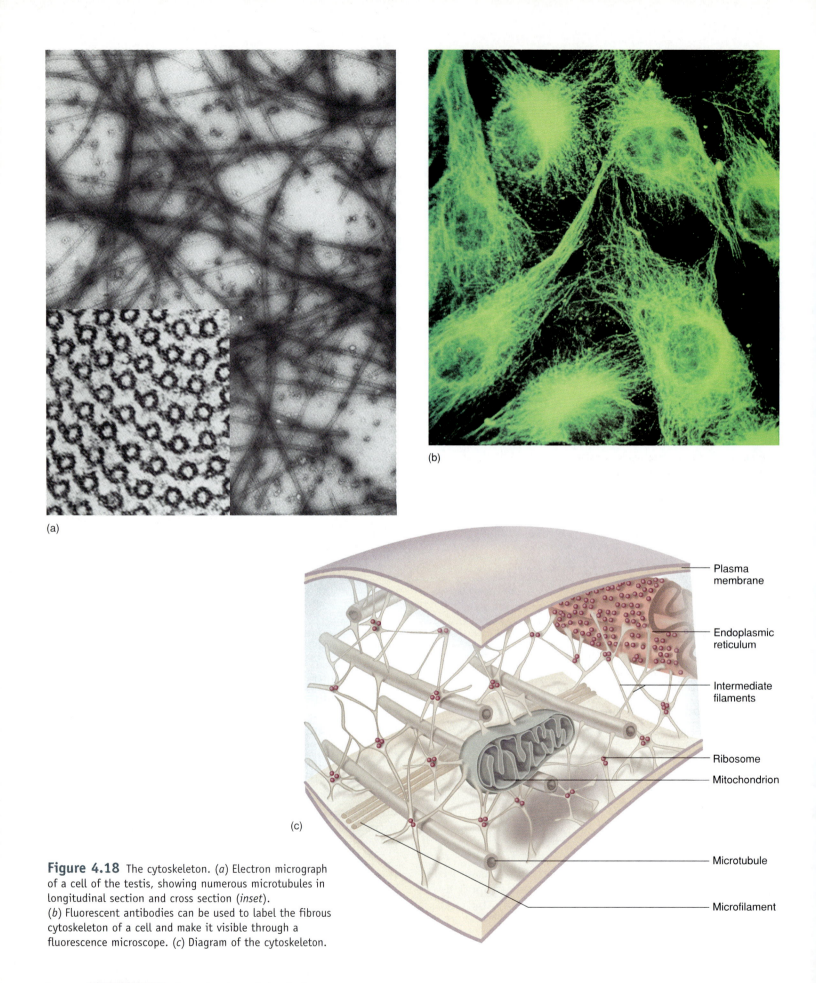

(a)

(b)

(c)

Plasma membrane

Endoplasmic reticulum

Intermediate filaments

Ribosome

Mitochondrion

Microtubule

Microfilament

Figure 4.18 The cytoskeleton. (*a*) Electron micrograph of a cell of the testis, showing numerous microtubules in longitudinal section and cross section (*inset*). (*b*) Fluorescent antibodies can be used to label the fibrous cytoskeleton of a cell and make it visible through a fluorescence microscope. (*c*) Diagram of the cytoskeleton.

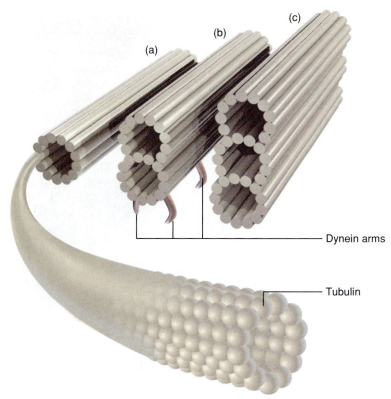

Dynein arms

Tubulin

Figure 4.19 (*a*) A microtubule is composed of 13 protofilaments. Each protofilament is a spiral chain of globular proteins called tubulin. (*b*) One of the nine microtubule pairs that form the axonemes of cilia and flagella. (*c*) One of the nine microtubule triplets that form a centriole.

A **microtubule** (25 nm in diameter) is a cylinder made of 13 parallel strands called *protofilaments*. Each protofilament is a long chain of globular proteins called **tubulin** (fig. 4.19). Microtubules radiate from the centrosome and hold organelles in place, form bundles that maintain cell shape and rigidity, and act somewhat like railroad tracks to guide organelles and molecules to specific destinations in a cell. They form the axonemes of cilia and flagella and are responsible for their beating movements. They also form the mitotic spindle that pulls the chromosomes apart during cell division. Microtubules are not permanent structures. They come and go moment by moment as tubulin molecules assemble into a tubule and then suddenly break apart again to be used somewhere else in the cell. The double and triple sets of microtubules in cilia, flagella, basal bodies, and centrioles, however, are more stable.

Inclusions

Inclusions have traditionally been regarded as temporary cellular contents that are not essential to cell sur-

vival, having no metabolic roles of their own. They have never been very precisely defined, however, and some are proving to have unexpected metabolic activity. Some inclusions are accumulated cell products, such as glycogen granules, oil droplets (see fig. 4.16), mucus, or pigments. Others include the indigestible residue of intracellular digestion, and foreign bodies such as dust particles and viruses that the cell has taken in.

The major features of a cell are summarized in table 4.4.

Key Point Review

9 Distinguish between organelles and inclusions. Give two examples of each.

10 Briefly state how each of the following cell components can be microscopically recognized: nucleus, mitochondrion, lysosome, and centriole. What is the primary function of each?

11 Which three organelles are involved in protein synthesis?

12 In what ways do rough and smooth endoplasmic reticulum differ?

13 Define *centriole, microtubule, cytoskeleton,* and *axoneme.* How do these structures relate to each other?

Membrane Transport—Passive Mechanisms

▼Objectives

When you have completed this section, you should be able to
- distinguish between passive and active mechanisms of membrane transport;
- describe the passive mechanisms for moving substances through plasma membranes;
- explain how an equilibrium can arise that balances osmosis and filtration;
- explain the concepts of osmotic pressure and tonicity; and
- compare the characteristics of carrier-mediated transport with enzyme function.

The plasma membrane is both a barrier and gateway between the cytoplasm and extracellular fluid (ECF). While it must prevent vital substances from escaping from the cell and prevent other substances from entering, it also must allow nutrients, wastes, and other matter to pass. Thus, the membrane is **selectively permeable**—able to allow some things through and exclude others.

The ways of moving substances into or out of a cell can be classified as *passive* or *active mechanisms.* Passive mechanisms do not require any energy expenditure by the cell. In most cases, the energy is provided by the random molecular motion (kinetic energy) of the particles themselves. Active mechanisms, however, require the cell to consume ATP.

Table 4.4   Summary of Organelles and Other Cellular Features

Structure	Description*	Function
Plasma membrane (fig. 4.6)	Two dark lines at surface of cell, separated by lighter space	Prevents escape of cell contents; regulates exchange of materials between cytoplasm and extracellular fluid; involved in intercellular communication and recognition
Glycocalyx (fig. 4.8)	Fuzzy coat on surface of all cells, external to plasma membrane	Involved in cell recognition, adhesion, sperm-egg binding, etc. (see table 4.3)
Microvilli (figs. 4.8 and 4.9)	Short, densely spaced hairlike projections or bumps on plasma membrane; interior featureless or with bundle of microfilaments	Increase absorptive surface area; some sensory roles (hearing, equilibrium, taste)
Cilia (figs. 4.9 and 4.10)	Long hairlike projections of apical plasma membrane; axoneme with 9 + 2 array of microtubules	Aid movement of substances along cell surface; some sensory roles (smell, hearing, equilibrium, vision)
Flagella	Very long single whiplike process with axoneme	Enable sperm motility
Nucleus (figs. 4.3 and 4.4)	Usually the largest structure within a cell, surrounded by double unit membrane with nuclear pores	Genetic control center of cell; directs protein synthesis
Rough ER (fig. 4.12)	Extensive parallel sheets with unit membrane on each side and cisterna (clear space) between membranes; granular surface studded with ribosomes	Involved in synthesis of proteins, especially those destined for exocytosis or deployment in lysosomes
Ribosomes (fig. 4.12)	Small dark granules lying singly or clustered in cytosol or on surface of rough ER	Interpret the genetic code and assemble polypeptides
Golgi complex (fig. 4.13)	Several closely spaced parallel cisternae with thick edges, often with many Golgi vesicles nearby	Involved in modification of newly synthesized proteins; lipid and carbohydrate synthesis; formation of lysosomes and secretory vesicles
Golgi vesicles (fig. 4.13)	Round to irregular sacs clustered near Golgi complex, usually with light, featureless contents	Become secretory vesicles and carry cell products to cell surface for exocytosis, or become lysosomes
Lysosomes (fig. 4.14)	Usually round or oval sacs with single unit membrane, uniform dark gray interior, sometimes with crystals or layers of protein	Contain enzymes for intracellular digestion, autophagy, programmed cell death, and glucose mobilization
Peroxisomes (fig. 4.14)	Round sacs with single unit membrane; light gray interior	Contain enzymes for detoxification of alcohol, free radicals, etc.
Mitochondria (fig. 4.15)	Round, rod-shaped, or threadlike structures surrounded by double unit membrane; inner membrane with numerous folds (cristae)	Site of most ATP synthesis
Smooth ER (fig. 4.16)	Branching network of tubules with smooth surface (no ribosomes); usually appears as numerous small pieces in TEM photos	Involved in lipid synthesis, detoxification, and calcium storage and release
Centrioles (fig. 4.17)	Short cylindrical bodies, each composed of a circle of nine triplets of microtubules; occur in pairs lying perpendicular to each other	Form the mitotic spindle during cell division; form the basal bodies of cilia and flagella
Centrosome	Small clear area near nucleus; contains centrioles	Serves as organizing center for microtubules of cytoskeleton and mitotic spindle
Basal bodies (fig. 4.9)	Structurally identical to unpaired centrioles; oriented perpendicular to the plasma membrane at the base of each cilium or flagellum	Point of origin and growth of a cilium or flagellum; produces axoneme of each and anchors cilium or flagellum to cell
Microfilaments (fig. 4.18)	Microscopic filaments (6 nm diameter); often occur in dense networks or parallel bundles in the cytoplasm	Support microvilli; involved in cellular contraction and motility, endocytosis, cell division, and muscle contraction
Intermediate filaments (fig. 4.18)	Thicker filaments (8–10 nm diameter), ramifying through cytosol or concentrated at cell-to-cell junctions	Provide physical support; resist stresses; hold cells together
Microtubules (figs. 4.18 and 4.19)	Hollow cylinders (25 nm diameter)	Form axonemes of cilia and flagella, centrioles, basal bodies, and mitotic spindles; enable motility of cell parts or direct the movement of organelles and cell products to their destinations
Inclusions (fig. 4.16)	Highly variable—oil droplets, glycogen granules, protein crystals, dust, viruses, and other matter; never enclosed in unit membranes	Storage products or other products of cellular metabolism, accumulated foreign matter, infectious microorganisms, etc.

*Features are described as they usually appear in transmission electron micrographs.

Filtration

Filtration is a process in which particles are driven through a selectively permeable membrane by **hydrostatic pressure,** the force exerted by water on a membrane. A coffee filter provides an everyday example. The weight of the water on the filter is a force that drives water and dissolved matter through, while the filter holds back larger particles (the coffee grounds). In physiology, the most important instance of filtration is in the blood capillaries, where blood pressure forces fluid through the capillary wall. This is how water, salts, nutrients, and other solutes are transferred from the bloodstream to the tissue fluid and how the kidneys filter wastes from the blood. Capillaries hold back larger particles such as blood cells and proteins.

Simple Diffusion

Simple diffusion is the movement of particles from place to place as a result of their own kinetic energy and constant spontaneous motion. Imagine, for example, that several people are in a small room where the air is still. Someone opens a bottle of concentrated ammonia, and before long people closest to the bottle begin to smell the fumes. As time passes, the fumes reach people farther and farther from the bottle. This is because the ammonia molecules exhibit *net* diffusion (more molecules diffusing in one direction than in another) from the point of origin, where their concentration is high, to more remote parts of the room, where their concentration is low. When the concentration of a substance differs from one point to another, we say there is a **concentration gradient.** Particle movement from a region of high concentration toward a region of lower concentration is said to go *down,* or *with, the gradient,* and movement in the other direction is said go *up,* or *against, the gradient* (fig. 4.20*a*).

The reason net diffusion proceeds *down* a concentration gradient, and not up, can be explained in terms of probability. Each particle is in a state of constant random motion, changing directions every time it collides with another particle. Of all the possible directions in which a particle may move, there is only one that would take it toward the place where it originated—to the mouth of the ammonia bottle, for example. There are innumerable directions, however, that would take it farther away from its point of origin (fig. 4.20*b*). Thus, there is a much greater likelihood that a particle will move away from its origin than back to it.

Diffusion occurs in air, water, and even (slowly) in solids; it has no need of a membrane. However, if there is a selectively permeable membrane in the path of the diffusing molecules, and its pores allow the molecules to pass, then they will pass from one side of the membrane to the other. This is how oxygen leaves the blood-

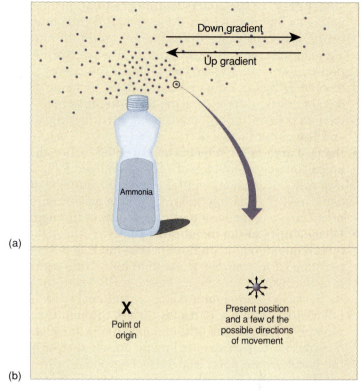

(a)

(b)

Figure 4.20 Diffusion and concentration gradients. (*a*) Diffusion of ammonia in air. Molecules are less and less concentrated the farther away from the source they have traveled. The direction from a high concentration to a low concentration is considered to be "down" the concentration gradient. The opposite direction is "up" the gradient. (*b*) A molecule at any point in space is more likely to diffuse away from the source (down the gradient) than up. Of the innumerable directions in which in may randomly move, only one would take it back toward the point of origin.

stream and enters cells and how carbon dioxide leaves the cells that produce it. Dialysis treatment for kidney disease patients is based on diffusion of solutes through artificial *dialysis membranes.*

Diffusion rates are very important to cell survival because they often determine how quickly a cell can acquire nutrients or rid itself of wastes. Some factors that affect the rate of diffusion through a membrane are as follows:

- **Temperature.** Diffusion is driven by the kinetic energy of the particles, and temperature is a measure of that kinetic energy. The warmer a substance is, the more rapidly its particles diffuse. This is why sugar diffuses more quickly through a cup of hot tea than through a glass of iced tea.
- **Molecular weight.** Heavy molecules such as proteins move more sluggishly and diffuse more slowly than lighter particles such as electrolytes and gases; smaller molecules pass through membrane pores more easily than larger ones.

- **"Steepness" of the concentration gradient.** Just as a ball rolls down a steep hill faster than it rolls down a gentle slope, particles diffuse more rapidly if there is a greater concentration difference between two points. For example, we can increase the rate of oxygen diffusion into a patient's blood by using an oxygen mask, thus increasing the difference in oxygen concentration between the surrounding air and the patient's blood.
- **Surface area of the membrane.** As noted earlier, the apical surface membrane of cells specialized for absorption (for example, in the small intestine) is often extensively folded into microvilli. This makes more membrane available for particles to diffuse through.
- **Permeability of the membrane.** Diffusion through a membrane depends on how permeable it is to the particles. For example, potassium ions diffuse more rapidly through plasma membranes than sodium ions do because the membrane is more permeable to potassium. Whether particles diffuse through the phospholipid regions of the membrane or through its channel proteins depends largely on whether or not the particles are polar (fig. 4.21). Nonpolar, lipid-soluble substances such as oxygen, alcohol, and steroids diffuse through the phospholipid. Charged or hydrophilic particles such as electrolytes and water itself will not mix with lipids, but they can diffuse through channel proteins in the membrane. Cells can adjust their permeability to a substance by adding channel proteins to the plasma membrane or taking them away. Kidney tubules, for example, do

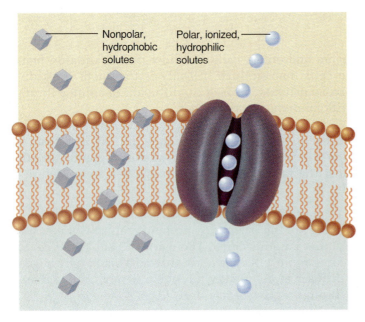

Figure 4.21 Hydrophobic solutes diffuse through the phospholipid regions of a plasma membrane. Hydrophilic solutes cannot pass through the phospholipid but can pass through the water-filled pores of a channel protein.

Nonpolar, hydrophobic solutes

Polar, ionized, hydrophilic solutes

this as a way of controlling the amount of water eliminated from the body.

Osmosis

Osmosis[18] (oz-MO-sis) is the net diffusion of water through a selectively permeable membrane, from the side where it is more concentrated to the side where it is less so. Cells exchange a tremendous amount of water by osmosis. The amount of water passing into and out of a red blood cell *each second,* for example, is 100 times the total cell volume.

It is important to note that the higher the concentration of solutes is, the lower the concentration of water will be, since the solutes take up some of the space that would otherwise be occupied by water molecules. Therefore, the direction of osmosis will be from a more dilute solution to a more concentrated one. In figure 4.22a, for example, we see a chamber divided by a semipermeable membrane. Side A contains a solution of large particles that cannot pass through the membrane pores—a *nonpermeating* solute such as albumin (egg white protein). Side B contains distilled water. Since albumin takes up some of the space on side A, water is less concentrated there than on side B, and it diffuses down its concentration gradient from B to A (fig. 4.22b). This is essentially because more water molecules strike the membrane per second on side B than they do on side A, where water concentration is lower.

Under these conditions, the water level on side B would fall and the level on side A would rise. It might seem as if this would go on indefinitely until side B dried up. This would not happen, however, because as water accumulated on side A, it would become heavier and heavier, exerting more hydrostatic pressure on that side of the membrane. This would cause some filtration of water from side A back to B. At some point, the rate of filtration would equal the rate of osmosis, water would pass through the membrane equally in both directions, and net osmosis would slow down and stop. At this point, an equilibrium (balance between opposing forces) would exist. The hydrostatic pressure on side A that would stop osmosis is called **osmotic pressure.** The more solute there is on side A, the greater its osmotic pressure will be.

Think About It

Would the fluid level on side A rise, fall, or remain the same if the concentration of albumin on side A were reduced? Why?

The equilibrium between osmosis and filtration will be an important consideration as we study fluid exchange

---

18. *osmo* = push, thrust

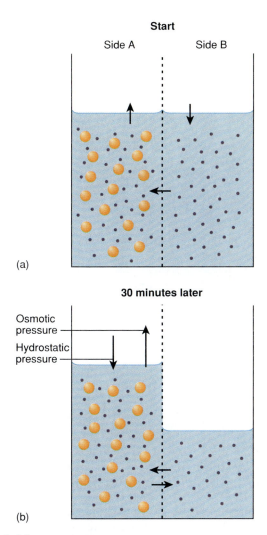

Start

Side A   Side B

(a)

30 minutes later

Osmotic
pressure

Hydrostatic
pressure

(b)

Figure 4.22 Osmosis. The dashed line represents a selectively permeable membrane dividing the chamber in half. The large particles on side A represent albumin or any other solute that cannot pass through the pores. The small particles are water molecules. (*a*) Water diffuses through the membrane from side B, where it is relatively concentrated, to side A, where it is less concentrated. (*b*) Net diffusion stops when the weight (hydrostatic pressure) of the liquid in side A balances the osmotic pressure, so water passes at equal rates from A to B by filtration and from B to A by osmosis. At this point, filtration and osmosis are in equilibrium.

through blood capillaries later on in chapter 20. Blood plasma also contains albumin. In the preceding discussion, side A is analogous to the bloodstream and side B is analogous to the tissue fluid surrounding the capillaries (although tissue fluid is not distilled water). Water leaves the capillaries by filtration, but this is approximately balanced by water moving back into the capillaries by osmosis.

Tonicity

It is essential for cells to be in a state of osmotic equilibrium with the fluid around them, and this requires that the extracellular fluids have the same concentration of nonpermeating solutes as the intracellular fluid. If cells lose more water than they gain, they *crenate* (shrivel), their membranes may tear, and they may die. If they gain excess water, they swell and the delicate plasma membrane may *lyse* (rupture). The ability of a solution to affect the fluid volume and pressure within a cell is called its **tonicity.**[19] **Hypotonic**[20] solutions are more dilute than intracellular fluid, so cells in a hypotonic solution absorb water by osmosis and are at risk of lysis and death. Distilled water is the extreme example; given to a person intravenously, it would cause lysis of the blood cells. A **hypertonic**[21] solution is more concentrated than intracellular fluid, so cells in hypertonic solutions lose water and become crenated. In **isotonic**[22] solutions, the total concentration of nonpermeating solutes is the same as that of the intracellular fluid—hence, these solutions cause no change in cell volume or shape (fig. 4.23). Intravenous fluids given to patients are generally isotonic solutions, but hypertonic or hypotonic fluids are sometimes given for special purposes. A 0.9% solution of NaCl, called **normal saline,** is isotonic to human blood cells.

19. *ton* = tension
20. *hypo* = below
21. *hyper* = above
22. *iso* = same

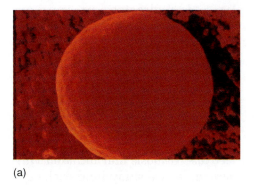

(a)

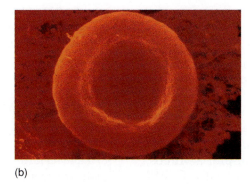

(b)

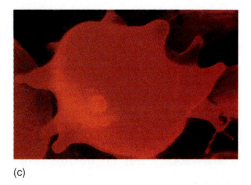

(c)

Figure 4.23 (*a*) In a hypotonic medium such as distilled water, red blood cells (RBCs) absorb water, swell, and may rupture. (*b*) In an isotonic medium such as 0.9% NaCl, RBCs gain and lose water at equal rates and maintain their normal, concave disc shapes. (*c*) In a hypertonic medium such as 2% NaCl, RBCs lose more water than they gain and become shrunken and spiky (crenated).

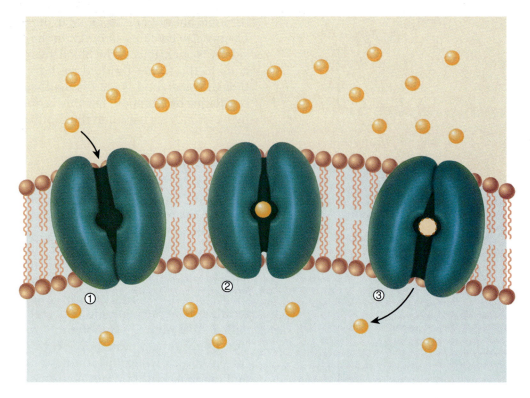

Figure 4.24 Facilitated diffusion. (*1*) A solute particle enters the channel of a membrane protein. (*2*) The solute binds to a receptor site on the protein and the protein changes conformation. (*3*) The protein ejects the solute on the other side of the membrane. Note that the solute particles move down the concentration gradient.

Facilitated Diffusion

Facilitated[23] **diffusion** is the movement of a solute through a cellular membrane *down its concentration gradient* with the aid of carrier proteins in the membrane. It is used to transport solutes that cannot pass through the membrane unaided. Some cells use facilitated diffusion to absorb glucose, for example. Glucose attaches to a binding site on the carrier, causing the carrier to change shape and expel it on the intracellular side of the membrane (fig. 4.24).

Carrier-Mediated Transport

Facilitated diffusion and active transport are called **carrier-mediated transport** because of their use of carrier proteins. Carriers do not chemically change the substances they transport, but in other ways they behave like enzymes: the solute is a *ligand* for the carrier, it binds to a specific *receptor site,* and the carrier *changes conformation* when the ligand binds to it. Like the active site of an enzyme, the binding site of a carrier exhibits **specificity**—the ability to bind only one ligand or a few closely related ligands. For example, a glucose carrier will not transport fructose. Carriers also exhibit

saturation. As solute concentration rises, its rate of transport increases—up to a point. When every carrier is transporting solute, the addition of more solute cannot speed up the process any further—there are no more carriers available to handle the increased demand. At this point, we say the carriers are saturated and transport has reached a rate called the **transport maximum (T_m)** (fig. 4.25). As we will see later in the book, the transport maximum explains why glucose appears in the urine of people with diabetes mellitus.

Key Point Review

14 How is filtration relevant to human physiology?

15 How does osmosis help maintain blood volume?

16 Why is it important that saline solutions given intravenously to a patient be isotonic?

17 How is net diffusion different from simple diffusion? In what ways are they similar?

Membrane Transport—Active Mechanisms

▼Objectives
When you have completed this section, you should be able to
• define active transport and explain how this process works;
• distinguish between symport and antiport systems;
• describe the functions of the sodium-potassium pump; and
• describe the different types of bulk transport.

23. *facil* = easy

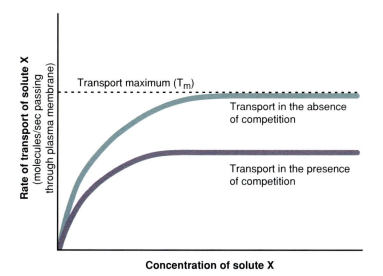

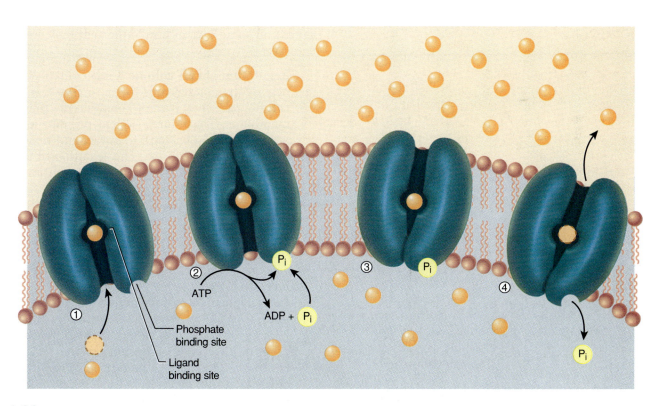

Figure 4.25 The transport maximum and competition characteristic of carrier-mediated transport processes. *Top curve:* Up to a point, the rate of solute transport through a membrane increases with solute concentration. At the transport maximum (T_m), however, all transport proteins are "busy" and the solute cannot be transported any faster even if more of it is present. *Bottom curve:* Some proteins can transport two or more solutes. If a second, competing solute is present, transport of the first solute is slower.

In this section, we consider the mechanisms of membrane transport that require an energy (ATP) expenditure by the cell.

Active Transport

Active transport is a carrier-mediated process that moves solutes *up their concentration gradients* instead of down. Such movement requires energy provided by ATP. All cells continually pump out sodium and calcium ions by active transport, even though these ions are already more concentrated in the extracellular fluid than within the cell. Active transport also brings solutes into a cell that are already highly concentrated there. All cells have a high concentration of amino acids and potassium, for example, but continue to pump in more by active transport.

As shown in figure 4.26, active transport is a four-step process:

1. A ligand from one side of the plasma membrane attaches to a binding site on the carrier.
2. The carrier hydrolyzes an ATP molecule and becomes phosphorylated.

Figure 4.26 Active transport. (*1*) A ligand from one side of the plasma membrane attaches to a binding site on the carrier. (*2*) The carrier hydrolyzes ATP and becomes phosphorylated. (*3*) Phosphorylation causes the carrier to undergo a conformational change. (*4*) The carrier releases the ligand on the other side of the plasma membrane, releases the phosphate group, and returns to its original conformation, ready to repeat the process. Note that the solute particles are moving up the concentration gradient. ▭

3. Phosphorylation causes the carrier to undergo a conformational change.
4. The carrier releases the ligand on the other side of the plasma membrane, releases the phosphate group, and returns to its original conformation, ready to repeat the process.

Symport and Antiport Systems

Carrier-mediated transport proteins sometimes carry more than one solute at a time through the plasma membrane. For example, absorptive cells of the small intestine have apical membrane proteins called *sodium-dependent glucose transporters (SGLTs)*. SGLTs bind to glucose in the digested food but cannot transport it into the cell unless they simultaneously bind and transport sodium ions. The SGLT is called a **symport**[24] system because it moves both solutes in the same direction (fig. 4.27)

In this particular case and some others like it, the transport mechanism is also called **secondary active transport.** The SGLT works by facilitated diffusion. Sodium and glucose move from a region of high concentration in the intestine to a region of lower concentration within the absorptive cell. However, the low intracellular Na+ concentration is maintained by an active transport pump on the basal side of the cell, which continually pumps Na+ out of the cell on that side. So

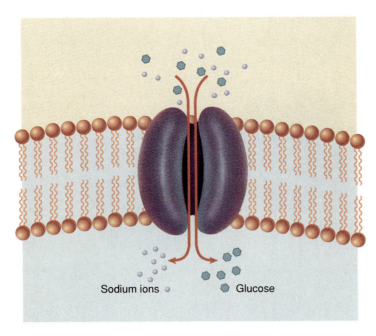

Sodium ions ∘   Glucose ⬡

Figure 4.27 The sodium-dependent glucose transporter (SGLT) of the small intestine is a symport. It moves glucose and sodium ions simultaneously in the same direction.

while the SGLT itself does not use ATP, it does depend on active transport occurring elsewhere in the plasma membrane.

An **antiport**[25] system transports two solutes in opposite directions—one into a cell and one out of it. The best known example is the **sodium-potassium (Na+-K+) pump.** Like all active transport systems, this pump uses ATP and is therefore also known as **Na+-K+ ATPase.** Figure 4.28 shows how the pump works. It first binds to three sodium ions on the cytoplasmic side of the membrane and then hydrolyzes a molecule of ATP. The phosphate group taken from ATP phosphorylates the pump, which changes conformation and releases the Na+ to the extracellular fluid (ECF). While in this conformation, the pump binds two potassium ions from the ECF. The phosphate group then separates from the pump, which returns to its original conformation and releases the K+ into the cytoplasm. Thus one cycle of the pump uses one molecule of ATP and exchanges three Na+ for two K+.

This pump keeps the intracellular K+ concentration higher and the Na+ concentration lower than the extracellular concentrations. These ions continually leak through the membrane channels, however, and flow back down their concentration gradients. The Na+-K+ pump compensates for this, like bailing out a leaky boat. About half the calories your body consumes each day are used just to operate your Na+-K+ pumps, which gives you some idea of how important they are.

Na+-K+ pumps have at least four functions:

1. **Regulation of cell volume.** Cells contain many negatively charged ions that are "fixed" within the cell because they cannot penetrate the plasma membrane. These ions, such as proteins and other organic molecules, tend to attract and retain positively charged ions. The retention of these ions would cause the cell to swell and eventually rupture if it were not corrected. Cellular swelling, however, stimulates the Na+-K+ pumps, and by pumping out three sodium ions for every two potassium ions they bring in, the pumps help to reduce ion concentration and swelling. Thus, cellular volume is controlled by a negative feedback loop involving the Na+-K+ pumps.

2. **Cotransport (secondary active transport).** The pump maintains a great concentration difference in Na+ and K+ from one side of the membrane to the other (a "steep concentration gradient"). Like water behind a dam that can be tapped to generate electricity, this ion gradient has a high potential energy that can drive other processes. Since Na+ has a high concentration outside the cell, it has a

---

24. *sym* = together + *port* = carry

25. *anti* = opposite, against

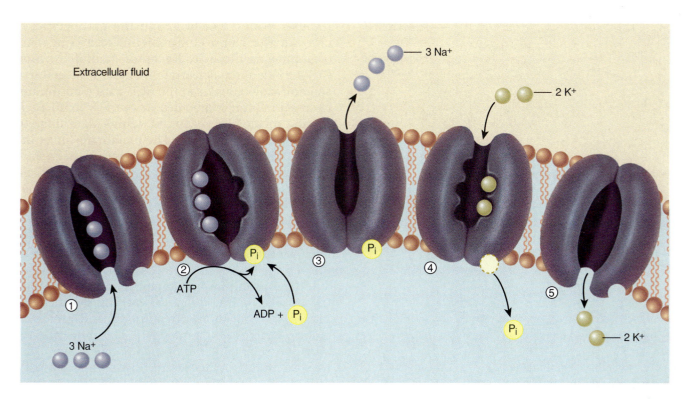

Figure 4.28 The sodium-potassium pump, an antiport system that transports Na⁺ and K⁺ in opposite directions through a plasma membrane. (*1*) Three sodium ions bind to receptor sites on the carrier. (*2*) The carrier hydrolyzes ATP and becomes phosphorylated. (*3*) The carrier protein changes conformation and releases the Na⁺ to the extracellular fluid. (*4*) Two potassium ions bind to receptor sites, and the carrier releases the phosphate group. (*5*) The carrier returns to its original conformation and releases K⁺ into the cytoplasm.

tendency to diffuse back in. In kidney tubule cells, its diffusion is linked to a glucose carrier that recovers glucose from the urinary filtrate so that it is not wastefully excreted in the urine (a symport system). Other cells pump out Ca^{2+} (calcium ions) by means of an antiport system linked to the inward diffusion of Na⁺. Thus, cotransport embraces both symport and antiport mechanisms.

3. **Heat production.** When the weather turns chilly, we not only turn up the furnace in our home but also the "furnace" in our body. Thyroid hormone stimulates cells throughout the body to produce more Na⁺-K⁺ pumps. Each cycle of the pump hydrolyzes a molecule of ATP, and each ATP hydrolysis releases heat—therefore the greater the number of active transport pumps that are functioning, the more heat we release from chemical storage.

4. **Maintenance of a membrane potential.** All living cells have an electrical charge called the *resting membrane potential* across the plasma membrane. It stems from the unequal distribution of ions on the two sides of the membrane, maintained by the Na⁺-K⁺ pump. The resting membrane potential is essential to the function of nerve and muscle cells, as we will study in later chapters.

Bulk Transport

The processes we have considered so far move particles through the plasma membrane one or a few at a time. **Bulk transport** includes a variety of processes by which cells take up or release large particles and droplets of fluid. Those that release material from a cell are called **exocytosis**[26] (EC-so-sy-TOE-sis) and those that bring matter into a cell are called **endocytosis**[27] (EN-doe-sy-TOE-sis). There are two forms of endocytosis: *phagocytosis* and *pinocytosis.*

    Phagocytosis[28] (FAG-oh-sy-TOE-sis), or "cell eating," is the process of engulfing particles such as bacteria, dust particles, and cellular debris. White blood cells called *neutrophils,* for example, phagocytize and kill bacteria, protecting the body from infection. A neutrophil spends most of its life crawling about in the connective tissues, using blunt extensions of its cytoplasm called **pseudopods**[29] (SOO-doe-pods). When a neutrophil encounters a bacterium, it reaches around it with the pseudopods. Once the bacterium is

26. *exo* = out of + *cyt* = cell + *osis* = process
27. *endo* = within, into
28. *phago* = eat
29. *pseudo* = false + *pod* = foot

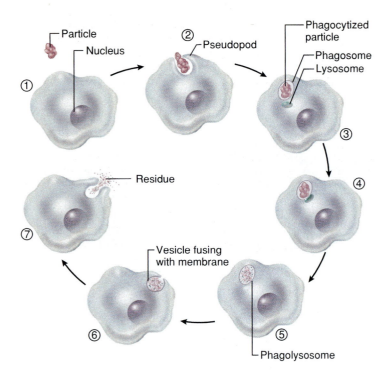

Figure 4.29 Phagocytosis, intracellular digestion, and exocytosis. (*1*) A phagocytic cell encounters a particle of foreign matter. (*2*) The cell surrounds the particle with its pseudopods. (*3*) The particle is phagocytized, becoming incorporated into a phagosome within the cell. (*4*) A lysosome fuses with the phagosome, forming a phagolysosome. (*5*) Enzymes from the lysosome digest the foreign matter. (*6*) The phagolysosome fuses with the plasma membrane. (*7*) The indigestible residue is voided by exocytosis.

surrounded, the tips of the pseudopods flow together, so that the bacterium becomes trapped in a **vacuole** or **phagosome**—a "bubble" in the cytoplasm surrounded by a unit membrane (fig. 4.29). A lysosome subsequently merges with the vacuole, converting it to a **phagolysosome,** and contributes enzymes that destroy the bacterium. Several other kinds of phagocytic cells are described in chapter 21. In general, phagocytosis is a way of keeping the tissues free of debris and infectious microorganisms. Some cells called *macrophages* (literally, "big eaters") phagocytize the equivalent of 25% of their own volume per hour.

Pinocytosis[30] (PIN-oh-sy-TOE-sis), or "cell drinking," is the process of taking in droplets of extracellular fluid (ECF) containing molecules of some use to the cell. The plasma membrane caves in at points, forming small membrane-bounded **pinocytotic vesicles** in the cytoplasm. While phagocytosis occurs in only a few specialized cells, pinocytosis occurs in all human cells.

There are two kinds of pinocytosis. **Fluid-phase pinocytosis** is a nonselective process in which the cell

imbibes droplets of ECF, and the pinocytotic vesicles contain the same composition and concentrations of solutes as the ECF. **Receptor-mediated pinocytosis** (fig. 4.30) is more selective. It enables the cell to take in specific molecules from the ECF with a minimum of unnecessary fluid. Molecules in the ECF bind to specific plasma membrane receptors, which then migrate laterally and cluster together. The membrane sinks in at this point, creating a pit coated with a protein called **clathrin.**[31] This pit soon pinches off to form a **clathrin-coated vesicle** in the cytoplasm. Clathrin is one of the peripheral proteins of the plasma membrane. It may serve as an "address label" on the coated vesicle that directs it to an appropriate destination in the cell, or it may inform other structures in the cell what to do with the vesicle.

One example of receptor-mediated pinocytosis is the uptake of low-density lipoproteins (LDLs)—complexes of cholesterol, other lipids, and proteins in the blood plasma (see chapter 26). The thin cells that line blood vessels, called *endothelial cells*, have LDL receptors on their surfaces and absorb LDLs in clathrin-coated vesicles. Inside the cell, the LDL is freed from the vesicle and metabolized, and the membrane with its LDL receptors is recycled back to the cell surface. Another case of receptor-mediated pinocytosis is insulin transport. Insulin is too large a molecule to pass through a plasma membrane, yet it must somehow get out of the blood and reach the surrounding cells if it is to have any effect. Endothelial cells take up insulin by receptor-mediated pinocytosis, and the vesicles cross over to the other side of the cell and release the insulin into the tissue fluid by exocytosis, described next. Such transport of a substance across a cell is called **transcytosis**[32] (fig. 4.31). Receptor-mediated pinocytosis is not always beneficial; hepatitis, polio, and AIDS viruses exploit this process to invade cells.

Exocytosis (fig. 4.32) is the process of discharging material from a cell. It occurs, for example, when endothelial cells release insulin to the tissue fluid, breast cells secrete milk, nerve cells release neurotransmitters, or sperm cells release enzymes for penetrating an egg. It bears a superficial resemblance to endocytosis in reverse. A secretory vesicle in the cell migrates to the surface and becomes linked to peripheral proteins of the plasma membrane. These proteins pull on the membrane, creating a dimple that eventually fuses with the vesicle and allows for release of its contents.

The question might occur to you, If pinocytosis and phagocytosis continually take away bits of plasma

30. *pino* = drink

31. *clathr* = lattice
32. *trans* = across

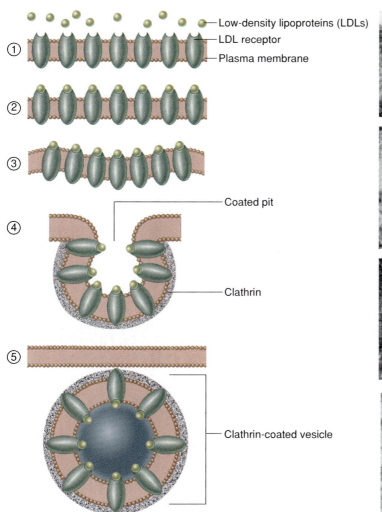

- Low-density lipoproteins (LDLs)
- LDL receptor
- Plasma membrane

①

②

③

Coated pit

④

Clathrin

⑤

Clathrin-coated vesicle

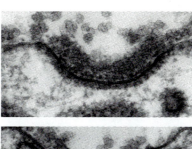
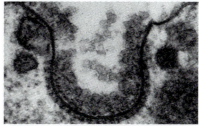
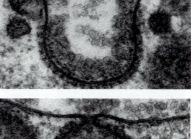
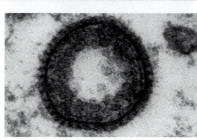

Figure 4.30 Receptor-mediated endocytosis of low-density lipoproteins (LDLs). (*1*) Endothelial cells of the blood capillaries have LDL receptors in their plasma membranes. (*2*) LDLs bind to their receptors. (*3*) Receptors migrate laterally and become clustered in the membrane. (*4*) The membrane caves in at that point, forming a clathrin-coated pit. (*5*) The pit separates from the plasma membrane and becomes a clathrin-coated vesicle.

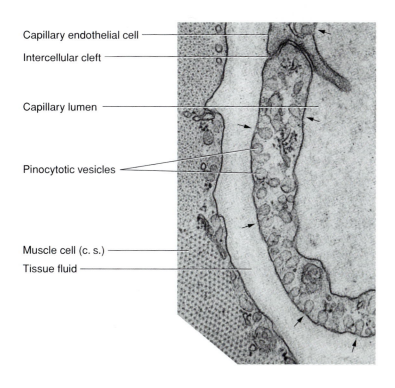

- Capillary endothelial cell
- Intercellular cleft
- Capillary lumen
- Pinocytotic vesicles
- Muscle cell (c. s.)
- Tissue fluid

Figure 4.31 Transcytosis across an endothelial cell of a blood capillary. Fluid droplets are imbibed by endocytosis on one side of the cell (*arrows*), pinocytotic vesicles are transported across the cell, and the vesicle contents are released by exocytosis on the other side. This is especially common in muscle capillaries and transfers a significant amount of blood albumin to the tissue fluid.

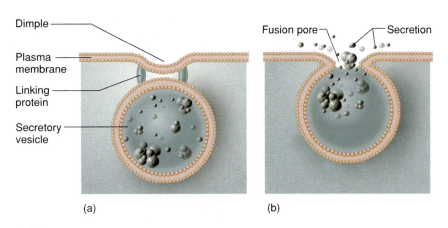

Dimple
Plasma membrane
Linking protein
Secretory vesicle

Fusion pore
Secretion

(a)

(b)

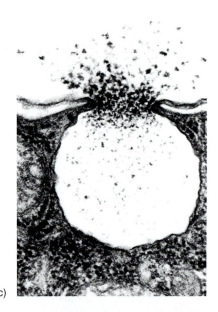

(c)

Figure 4.32 Exocytosis. (*a*) A secretory vesicle approaches the plasma membrane and becomes attached to it by linking proteins. The plasma membrane caves in at that point to meet the vesicle. (*b*) The plasma membrane and vesicle membrane unite to form a fusion pore through which the cell's secretion is released. (*c*) Electron micrograph of exocytosis.

Table 4.5	Methods of Membrane Transport
Filtration	Movement of particles through a selectively permeable membrane as a result of hydrostatic pressure
Passive Transport (Diffusion)	Movement of particles down a concentration gradient as a result of their kinetic energy; involves no expenditure of ATP
Simple Diffusion	Diffusion of particles through a medium such as water or air or through a selectively permeable living or artificial membrane without the aid of membrane carriers
Osmosis	Simple diffusion of water through a selectively permeable membrane
Facilitated Diffusion	Diffusion of particles through a selectively permeable cellular membrane with the aid of membrane carriers
Active Transport	Transport processes by which particles move up their concentration gradients with the expenditure of ATP
Cotransport	Linked active transport systems that transport two or more solutes simultaneously
Symport System	Cotransport system that moves two or more solutes through a membrane in the same direction
Antiport System	Cotransport system that moves two or more solutes through a membrane in opposite directions
Bulk Transport	Movement of particles or fluid through a plasma membrane by means of vacuoles or vesicles of unit membrane; involves expenditure of ATP
Endocytosis	Bulk transport processes that move matter into a cell
Phagocytosis	"Cell eating"; process by which a cell engulfs large particles with its pseudopods and encloses them in cytoplasmic vacuoles
Pinocytosis	"Cell drinking"; process by which the plasma membrane caves in and pinches off small vesicles containing droplets of extracellular fluid
Fluid-Phase Pinocytosis	Nonspecific pinocytosis in which the fluid imbibed has the same composition and concentration as the ECF
Receptor-Mediated Pinocytosis	Specific pinocytosis in which certain molecules in the ECF bind to receptors on the plasma membrane and are taken into clathrin-coated vesicles with a minimum of extracellular fluid
Exocytosis	Process by which a cytoplasmic vesicle fuses with the plasma membrane and discharges its contents from the cell; used by some cells to release secretions and by all cells to replace plasma membrane removed by endocytosis

membrane to form intracellular vesicles and vacuoles, why doesn't the plasma membrane grow smaller and smaller? Another purpose of exocytosis, however, is to replace plasma membrane that has been removed by endocytosis, as well as to replace old or damaged membrane. Plasma membrane is continually being recycled, passing from the cell surface into the cytoplasm and back to the surface.

Table 4.5 summarizes the mechanisms of transport we have discussed.

Key Point Review

18. How is active transport similar to facilitated diffusion? How is it different?

19. How does the Na⁺-K⁺ pump exchange sodium ions for potassium ions across the plasma membrane? What is the pump used for?

20. How does phagocytosis differ from pinocytosis?

21. Distinguish between the two kinds of pinocytosis.

22. Describe the process of exocytosis. What are some of the purposes it serves?

CHAPTER ESSAY

Advances in Microscopy

The simple microscope of Antony van Leeuwenhoek magnified specimens about 200 times, and modern electron microscopes magnify up to 600,000 times. Yet the most significant aspect of the history of microscopy has not been magnification but *resolution,* the ability to see detail.

Light Microscope

The modern light microscope (LM) was developed in the early nineteenth century, chiefly by Carl Zeiss. By using many lenses in combination, he corrected for the deficiencies of each, including the spherical and chromatic aberrations of Robert Hooke's seventeenth-century microscopes (see chapter 1). The light microscope is still the most routinely used type. It illuminates the specimen with visible light, which has wavelengths from about 400 nm at the violet end of the spectrum to 750 nm at the red end. This range of wavelengths limits the resolution of the light microscope to about 200 nm (0.2 μm) and its useful magnification to about 1,200 times. There are many varieties of light microscope, including the fluorescence microscope used to photograph figure 4.18*b.*

Transmission Electron Microscope

One of the laws of optics is that resolution improves when objects are viewed with radiation of shorter wavelength. This is the reason electron microscopes use an electron beam instead of visible light. In the transmission electron microscope (TEM), a beam of electrons, focused with electromagnets, passes through a specimen and strikes a fluorescent screen or photographic film. The electrons have a wavelength of 0.005 nm (compare visible light), and

the TEM can resolve biological objects as small as 0.5 nm. This makes it possible to obtain meaningful detail at magnifications up to 600,000 times, which is sufficient to see proteins, nucleic acids, and other large molecules.

In order to absorb electrons, specimens for the TEM must be sliced ultrathin with diamond knives and stained with heavy metals such as osmium. All images seen by the TEM are really images of where these metal ions have collected in the specimen, not images of the specimen itself. One disadvantage of the TEM is that living cells usually cannot be studied. Since the electrons would be scattered by air, the specimen must be placed in a vacuum chamber. Recent advances have somewhat overcome this limitation, however, with TEMs that operate at ultrahigh voltages (1–2 million volts) and use wet specimen chambers. The TEM produces two-dimensional black-and-white images, but TEM photomicrographs are often colorized for instructional purposes.

Scanning Electron Microscope

The scanning electron microscope (SEM) uses a specimen coated with vaporized metal ions (usually gold). An electron beam strikes the specimen and discharges secondary electrons from the metal coating. These electrons then produce an image on a fluorescent screen or photographic film. The SEM yields less resolution than the TEM and is used at lower magnification. It can be used only to view the surface of an object; it does not see through an object like the LM and TEM. Cell interiors can be viewed, however, by a

—continued

freeze-fracture method in which a cell is frozen, cracked open, and then coated with gold vapor. An advantage of the SEM is that it produces dramatic three-dimensional images that are sometimes more informative than the flat images of the LM and TEM (see, for example, the contrasting views of nuclear pores in figure 5.1).

Micrographs produced by the LM, TEM, and SEM are used throughout this book. Figure E.1 shows red blood cells photographed by all three methods. See if you can identify at least two other TEM photomicrographs and two other SEM photomicrographs in this chapter.▲

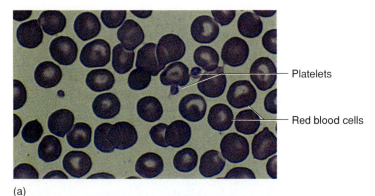

(a)

Platelets

Red blood cells

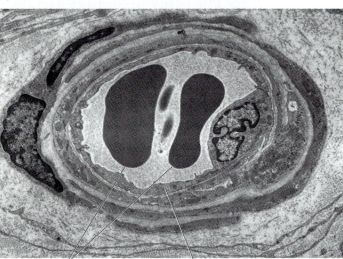

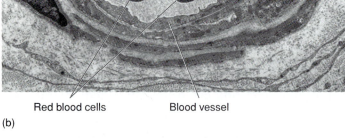

Red blood cells   Blood vessel

(b)

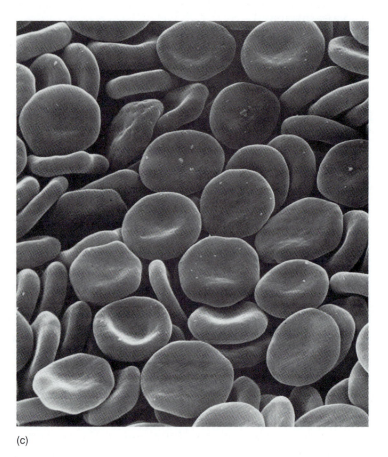

(c)

Figure E.1 Red blood cells photographed with (*a*) the light microscope (LM), (*b*) the transmission electron microscope (TEM), and (*c*) the scanning electron microscope (SEM).

Chapter Review   Study Outline

Concepts of Cellular Structure (pp. 107–111)
1. Development of the cell theory
   a. Early studies of Hooke and Schwann
   b. The spontaneous generation issue
   c. Tenets of the cell theory
2. Cell shapes
   a. Terms for cell shapes
   b. Surfaces of epithelial cells
3. Cell size
   a. Sizes of human cells
   b. Limitations on size

4. An evolving perspective on cells
   a. Microscopes and resolution
   b. Internal organization of a cell

The Cell Surface (pp. 111–117)
1. The plasma membrane
   a. Membrane lipids
      • Phospholipid bilayer
      • Cholesterol
      • Glycolipids
   b. Membrane proteins
      • Integral proteins
      • Peripheral proteins

   c. Functions of membrane proteins
      • Receptors for chemical messengers
      • Second messenger systems
      • Messenger deactivators
      • Channels and gates
        •• Ligand-gated channels
        •• Voltage-gated channels
      • Carriers and pumps
      • Motor molecules
      • Cell identity markers
2. The glycocalyx
   a. Composition
   b. Functions

3. Extensions of the cell surface
   a. Microvilli (brush border)
      • Core of microfilaments
      • Absorptive and sensory roles
   b. Cilia
      • Power and recovery strokes
      • Importance of saline layer
      • 9 + 2 structure of axoneme
      • Basal body
      • Sensory roles
   c. Flagella

The Cell Interior (pp. 117–123)
1. Organelles
   a. The nucleus
   b. Rough endoplasmic reticulum
   c. Ribosomes
   d. Golgi complex and vesicles
   e. Lysosomes
   f. Peroxisomes
   g. Mitochondria
   h. Smooth endoplasmic reticulum
   i. Centrioles

2. Cytoskeleton
   a. Microfilaments
   b. Intermediate filaments
   c. Microtubules

3. Inclusions

Membrane Transport—Passive Mechanisms (pp. 123–128)
1. Selective permeability
2. Filtration
3. Simple diffusion
   a. Concentration gradients
   b. Factors affecting diffusion rate
4. Osmosis
   a. Osmotic pressure
5. Tonicity
6. Facilitated diffusion
7. Carrier-mediated transport
   a. Specificity
   b. Saturation

Membrane Transport—Active Mechanisms (pp. 128–135)
1. Active transport
   a. Movement up a gradient
   b. Use of ATP
   c. Steps in active transport
2. Symport and antiport systems
   a. The Na⁺-K⁺ pump as an antiport system
      • Mechanism
      • Functions
        •• Regulation of cell volume
        •• Cotransport
        •• Heat production
        •• Maintenance of membrane potential
3. Bulk transport
   a. Endocytosis
      • Phagocytosis
      • Pinocytosis
        •• Fluid-phase pinocytosis
        •• Receptor-mediated pinocytosis
   b. Exocytosis

Selected Vocabulary

Also review the terms in tables 4.4 and 4.5, which are not repeated here.

protoplasm 107
spontaneous generation 107
squamous 107
cuboidal 107
columnar 108
spheroid 108
ovoid 108
discoid 108
fusiform 108
fibers 108
stellate 108
basal surface 108
lateral surface 108
apical surface 108
cytoplasm 108
transmission electron microscope (TEM) 108
cytosol 108

intracellular fluid (ICF) 108
light microscope (LM) 108
resolution 109
organelles 109
scanning electron microscope (SEM) 111
unit membrane 111
cell membrane 111
fluid-mosaic model 111
integral (transmembrane) protein 112
peripheral protein 112
receptor 113
G protein 113
second messenger 113
channel protein 113
gates 113
ligand-gated channel 113
voltage-gated channel 113
carrier 114
pump 114
brush border 116

terminal web 116
power stroke 116
recovery stroke 116
axoneme 117
9 + 2 structure 117
dynein arms 117
nuclear envelope 117
cisterna 118
secretory vesicle 118
programmed cell death 118
apoptosis 119
autophagy 119
crista 120
mitochondrial matrix 120
cytoskeleton 121
membrane skeleton 121
tubulin 123
selectively permeable membrane 123
hydrostatic pressure 125
concentration gradient 125
osmotic pressure 126

tonicity 127
hypotonic 127
hypertonic 127
isotonic 127
normal saline 127
carrier-mediated transport 128
specificity 128
saturation 128
transport maximum (T_m) 128
secondary active transport 130
sodium-potassium (Na⁺-K⁺) pump 130
Na⁺-K⁺ ATPase 130
pseudopod 131
vacuole 132
phagosome 132
phagolysome 132
pinocytotic vesicle 132
clathrin 132
clathrin-coated vesicle 132
transcytosis 132

Testing Your Recall   Answers in Appendix C

1. The clear, structureless colloid in a cell is its
   a. nucleoplasm.
   b. protoplasm.
   c. cytoplasm.
   d. neoplasm.
   e. cytosol.

2. The Na⁺-K⁺ pump is a/an
   a. peripheral protein.
   b. integral protein.
   c. G protein.
   d. glycolipid.
   e. phospholipid.

3. Which of the following processes could only occur in a living plasma membrane?
   a. facilitated diffusion
   b. simple diffusion
   c. filtration
   d. active transport
   e. osmosis

4. Cells specialized for absorption of matter from the extracellular fluid are likely to show an abundance of
   a. lysosomes.
   b. microvilli.
   c. mitochondria.
   d. secretory vesicles.
   e. smooth endoplasmic reticulum.

5. Osmosis is a special case of
   a. pinocytosis.
   b. carrier-mediated transport.
   c. active transport.
   d. facilitated diffusion.
   e. simple diffusion.

6. Membrane carriers resemble enzymes except for the fact that carriers
   a. are not proteins.
   b. do not have binding sites.
   c. are not selective for the ligands they bind.
   d. change conformation when they bind a ligand.
   e. do not chemically change their ligands.

7. The cotransport of glucose derives energy from
   a. a sodium concentration gradient.
   b. the glucose it transports.

c. a calcium gradient.
d. the membrane voltage.
e. body heat.

8. Programmed cell death often involves
   a. release of lysosomal enzymes.
   b. exocytosis of the organelles.
   c. anaerobic fermentation of the cell.
   d. nuclear disintegration.
   e. nuclear holocaust.

9. Most cellular membranes are made by
   a. the nucleus.
   b. the cytoskeleton.
   c. enzymes in the peroxisomes.
   d. the rough endoplasmic reticulum.
   e. replication of existing membranes.

10. Matter can leave a cell by any of the following means except
    a. active transport.
    b. fluid-phase pinocytosis.
    c. an antiport system.
    d. simple diffusion.
    e. exocytosis.

11. Most human cells are 10 to 15 _____ in diameter.

12. When a hormone cannot get into a cell, it activates the formation of a/an _____ in the cell.

13. _____ channels in the plasma membrane open or close in response to changes in the electrical charge on a membrane.

14. The force exerted on a membrane by water is called _____.

15. A concentrated solution that causes a cell to shrink is said to be _____ to that cell.

16. Fusion of a secretory vesicle with the plasma membrane, and release of the vesicle's contents, is called _____.

17. Two organelles that are surrounded by a double unit membrane are the _____ and _____.

18. Liver cells can detoxify alcohol with two organelles, the _____ and _____.

19. An ion gate in the plasma membrane that opens or closes when a chemical binds to it is called a/an _____.

20. The space enclosed by unit membranes of the Golgi complex, smooth ER, and rough ER is called the _____.

Testing Your Comprehension — Answers in *Study Guide*

1. If someone bought a saltwater fish in a pet shop and put it in a freshwater aquarium at home, what would happen to the fish's cells? What would happen if you put a freshwater fish in a saltwater aquarium? Explain.

2. A worker's hand and forearm are badly crushed in a piece of machinery, and there is extensive cellular damage. Given what you know about electrolyte concentrations in the cytoplasm and

extracellular fluid, how would you expect this injury to affect the plasma K+ concentration? Explain.

3. Many children worldwide suffer severe dietary protein deficiency, which greatly lowers the amount of albumin in the blood plasma. How do you think this affects the water content and volume of their blood?

4. It is often said that mitochondria make energy for a cell. Why is this statement false?

5. Kartagener's syndrome is a hereditary disease in which dynein arms are lacking from the microtubules of cilia and flagella. Predict what effect this would have on a man's ability to father a child. What respiratory problem would you expect him to have? Explain your answers

Web Site Link

For a listing of the most current web sites related to this chapter, please visit the Saladin homepage at:

http://www.mhhe.com/sciencemath/biology/saladin/

[Genetics and Cellular Function

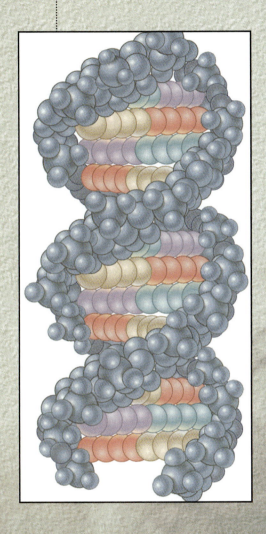

Special Topics

Brushing up

To understand this chapter, it is essential that you understand or brush up on the following concepts:

▶ Levels of protein structure (pp. 90–92)
▶ Ribosomes, rough endoplasmic reticulum, and Golgi complex (p. 118)
▶ Centrioles and microtubules (pp. 120–123)
▶ Exocytosis (pp. 132–134)

An awareness of some of the basic principles of heredity dates to antiquity, but a scientific understanding of how traits are passed from parent to offspring began with the Austrian monk Gregor Mendel (1822–84) and his famous experiments on garden peas. In the early twentieth century, the importance of Mendel's work was realized and chromosomes were first seen with the microscope.

Cytogenetics uses techniques of cytology and microscopy to study chromosome structure and number and their relationship to hereditary traits. **Molecular genetics** uses the techniques of biochemistry to study DNA's structure, replication, and control of cellular function. In this chapter, we bring together some of the findings of molecular genetics, cytogenetics, and Mendelian heredity to explore what the genes are, how they regulate cellular function, and how they are passed on when cells divide and individuals reproduce. Some basic concepts of heredity are also introduced to provide a foundation for understanding concepts in the chapters that follow.

The Nucleus and Nucleic Acids

▼Objectives
When you have completed this section, you should be able to
- describe the structure of the cell nucleus;
- describe how DNA is organized in the nucleus;
- compare the molecular structures of DNA and RNA; and
- compare the functions of DNA and RNA.

With improvements in the microscope, nineteenth-century cytologists saw that the nucleus divides in preparation for cell division, and they came to regard the nucleus as the most likely center of heredity. In the late 1800s, biochemists set out to identify what was in the nucleus, and they discovered the nucleic acids (see special topic 5.1). The nucleus is the location of deoxyribonucleic acid (DNA), which directly or indirectly regulates all cellular form and function, and it is the site where ribonucleic acid (RNA) is synthesized.

Nuclear Structure

Most cells have a single nucleus, but there are exceptions. Mature red blood cells have none, and a few cells have two or more nuclei, including some liver cells, skeletal muscle cells, certain cells that break down bone tissue, and cells that produce blood platelets. The nucleus is usually spheroid to elliptical in shape and averages about 5 μm in diameter. It is surrounded by a **nuclear envelope** consisting of two parallel unit membranes (fig. 5.1). Passages called **nuclear pores,** about 30 to 100 nm in diameter, allow two-way traffic of molecules through the envelope. Several substances must be allowed into the nucleus, including raw materials for the synthesis of DNA and RNA, enzymes that are made in the cytoplasm but function in the nucleus, and hormones that activate certain genes. RNA must be able to leave through these pores to do its job in the cytoplasm. A nuclear pore is made of a ring-shaped complex of proteins that not only regulates passage through the envelope but also acts like a rivet to hold the two unit membranes together at a uniform distance.

The material within the nucleus is called **nucleoplasm.** Its fine-grained appearance is due to coagulation of proteins by the chemicals used to preserve tissue. It usually contains one or more dark-staining masses of RNA called **nucleoli,** which are sites of ribosome pro-

---

Special Topic — Miescher and the Discovery of DNA   5.1

Swiss biochemist Johann Friedrich Miescher (1844–95) was one of the first scientists intent on identifying the hereditary material contained within nuclei. In order to isolate nuclei with minimal contamination, Miescher chose to work with cells that have large nuclei and very little cytoplasm. At first he chose white blood cells extracted from the pus in used surgical bandages; later, he used the sperm of salmon (probably more agreeable to work with than used bandages!). Miescher isolated a substance that was acidic and rich in phosphorus. He named it *nuclein* and correctly guessed that it was the hereditary matter of the cell. He was unable to provide strong evidence for this supposition, however, and his work was harshly criticized. Miescher died of tuberculosis at the age of 51. One of his students, Richard Altmann, named the substance *nucleic acid* in 1889, and this term has since been extended to include RNA.

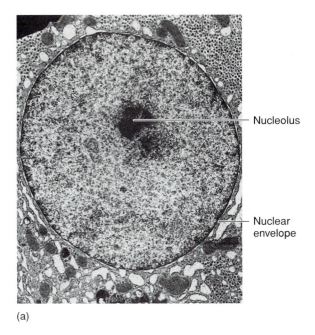

(a)

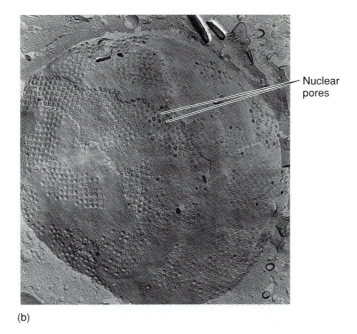

(b)

Figure 5.1 The nucleus. (*a*) Transmission electron micrograph showing the nuclear envelope and internal structure. (*b*) Scanning electron micrograph of the nuclear envelope's surface, showing the high density of nuclear pores.

duction. The DNA and its associated proteins constitute the **chromatin**[1] (CRO-muh-tin).

Organization of the Chromatin

A human cell usually has 46 molecules of DNA with a total length of slightly over 2 m. DNA molecules are 2 nm in diameter and average 44 mm long. To put this in perspective, if an average DNA molecule were the thickness of a telephone pole (20 cm, or 8 in.), it would reach about 4,400 km (2,700 mi) into space. Imagine trying to build a pole that was 20 cm thick and 4,400 km long without breaking it! The problem for a cell is even greater. It has 46 DNA molecules packed together in a single nucleus, and when the cell divides, it has to make an exact copy of every one of them and distribute these equally to its two daughter cells. Keeping the DNA molecules organized and intact is a tremendous task.

Molecular biology and high-resolution electron microscopy have provided some insight into how this task is accomplished. Chromatin looks like a granular thread (fig. 5.2*a*). Each granule, called a **nucleosome**, consists of a cluster of eight proteins called **histones,** with the DNA molecule wound around each cluster. Histones serve as spools that protect and organize the DNA. Other nuclear proteins called **nonhistones** seem to pro-

vide structural support for the chromatin and regulate gene activity.

Winding DNA around the histones makes the chromatin shorter and more compact, but chromatin also has higher orders of structure. The "beads on a string" fiber, about 10 nm wide, coils and thickens to about 30 nm; then, as a cell prepares to undergo division, the chromatin supercoils into a fiber about 200 nm in diameter (fig. 5.2*b*). Thus, the 2 m of DNA in each cell becomes shortened and compacted in an orderly way that prevents tangling and breakage without interfering with genetic function.

Nucleotides

Nucleic acids are polymers of **nucleotides** (NEW-clee-oh-tides), each of which consists of a monosaccharide, a phosphate group, and a single- or double-ringed **nitrogenous** (ny-TRODJ-eh-nus) **base**. Three bases—**cytosine, thymine,** and **uracil**—have a single carbon-nitrogen ring and are thus classified as *pyrimidines* (py-RIM-ih-deens). The other two bases—**adenine** and **guanine**—have double rings and are classified as *purines* (fig. 5.3).

DNA Structure

The structure of DNA resembles a ladder (fig. 5.4*a*). Each sidepiece is a backbone of alternating sugar (deoxyribose) and phosphate groups, and the steplike

1. *chrom* = color

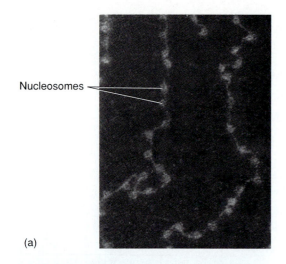

(a)

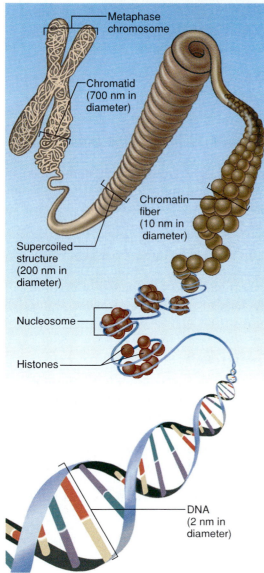

(b)

Metaphase
chromosome

Chromatid
(700 nm in
diameter)

Chromatin
fiber
(10 nm in
diameter)

Supercoiled
structure
(200 nm in
diameter)

Nucleosome

Histones

DNA
(2 nm in
diameter)

Figure 5.2 The structure of chromatin. (*a*) Transmission electron micrograph of the chromatin. Each beadlike swelling is a nucleosome, and the thin strands connecting them are DNA. (*b*) The coiling of chromatin and its relationship to the histones.

Nucleosomes

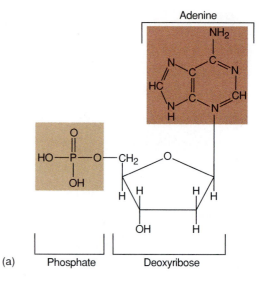

Adenine

(a)   Phosphate   Deoxyribose

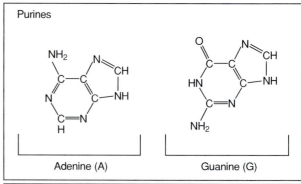

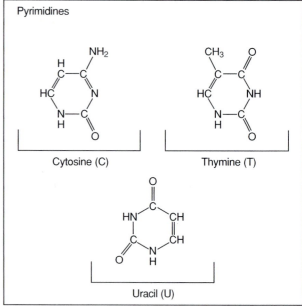

Purines

Adenine (A)   Guanine (G)

Pyrimidines

Cytosine (C)   Thymine (T)

Uracil (U)

(b)

Figure 5.3 (*a*) The structure of a nucleotide, one of the monomers of DNA and RNA. (*b*) The five nitrogenous bases found in the nucleotides of DNA and RNA.

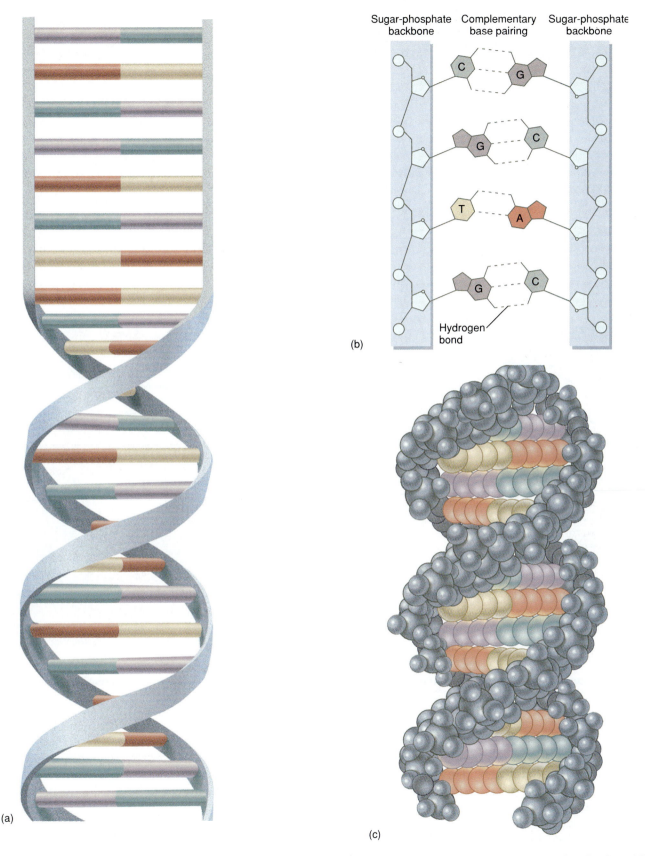

(a)

(b)

Sugar-phosphate backbone Complementary base pairing Sugar-phosphate backbone

C - - - - - G

G - - - - - C

T - - - - - A

G - - - - - C

Hydrogen bond

(c)

Figure 5.4 The structure of DNA. (*a*) The "twisted ladder" structure of DNA. The two sugar-phosphate backbones twine around each other while complementary bases (colored bars) face each other on the inside of the double helix. (*b*) A small segment of DNA showing the composition of the backbone and complementary pairing of the nitrogenous bases. (*c*) A molecular space-filling model of DNA giving some impression of its actual geometry.

The components of DNA were known by 1900—the sugar, phosphate, and four bases—but the technology did not exist to determine how they were put together. The credit for that discovery went mainly to James Watson and Francis Crick in 1953 (fig. 1). The events surrounding their discovery of the double helix represent one of the most dramatic stories of modern science—the subject of many books and a movie.

When Watson and Crick came to share a laboratory at Cambridge University in 1951, both had barely begun their careers. Watson, age 23, had just completed his Ph.D. in the United States, and Crick, 11 years older, was a doctoral candidate. Yet the two were about to become the most famous molecular biologists of the twentieth century, and the discovery that won them such acclaim came without a single laboratory experiment of their own.

Others were fervently at work on DNA, and the one closest to discovering its structure was Rosalind Franklin, working under Maurice Wilkins at King's College in London. Using a technique called X-ray diffraction, she had determined that DNA had a repetitious helical structure with sugar and phosphate on the outside of the helix. Without her permission, Wilkins showed one of Franklin's best X-ray photographs to Watson. It provided a flash of insight that allowed the Watson and Crick team to beat Franklin to the finish line. They were quickly able to piece together a scale model from cardboard and sheet metal that fully accounted for the known geometry of DNA. They rushed a paper into print in 1953 describing the double helix, barely mentioning that it was Franklin's painstaking X-ray diffraction work that had unlocked the mystery of life's most important molecule.

Watson, Crick, and Wilkins shared the Nobel Prize in 1962 for this discovery. Nobel Prizes are awarded only to the living, and in the final irony of her career, Rosalind Franklin had died of cancer in 1958, at the age of 37.

(a)

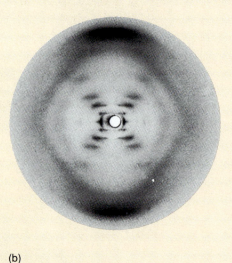
(b)

(c)

Figure 1 Discovery of the double helical structure of DNA.
(a) Rosalind Franklin, whose painstaking X-ray diffraction photographs revealed important information about the basic geometry of DNA.
(b) One of Franklin's X-ray diffraction photographs. (c) James Watson (*left*) and Francis Crick (*right*), with their model of the double helix.

connections between the backbones are pairs of nitrogenous bases. Imagine this as a soft rubber ladder that you can twist, so that the two backbones become entwined to resemble a spiral staircase. This is analogous to the shape of the DNA molecule, described as a *double helix* (see special topic 5.2).

The nitrogenous bases face the inside of the helix and hold the two backbones together with hydrogen bonds. Across from a purine on one backbone, there is normally a pyrimidine on the other. This is because two purines would be too large to fit within the 2 nm width of the molecule, and two pyrimidines would be too small to reach and bond with each other across this distance. A purine across from a pyrimidine, however, fit just right, and the consistency of this pairing gives DNA a uniform diameter all along the length of the molecule.

Furthermore, a given purine cannot arbitrarily bind to just any pyrimidine. Adenine (A) and thymine (T) form two hydrogen bonds with each other, and guanine (G) and cytosine (C) form three, as shown in figure 5.4*b*. Therefore, wherever there is an A on one backbone, there is a T across from it, and every C is paired with a G. A–T and C–G are called the **base pairs.** The fact that one strand governs the base sequence of the other is

Table 5.1	Comparison of DNA and RNA	
Feature	**DNA**	**RNA**
Sugar	Deoxyribose	Ribose
Nitrogenous bases	A, T, C, G	A, U, C, G
Number of nucleotide chains	Two (double helix)	One
Number of nitrogenous bases	10^8 to 10^9 base pairs	70 to 10^4 unpaired bases
Site of action	Functions in nucleus; cannot leave	Leaves nucleus; functions in cytoplasm
Function	Codes for synthesis of RNA and protein	Carries out the instructions in DNA; assembles proteins

called the **law of complementary base pairing**. It enables us to predict the base sequence of one strand if we know the sequence of the complementary strand.

Think About It

What would be the base sequence of the DNA strand across from ATTGACTCG? If a DNA molecule were known to be 20% adenine, predict its percentage of cytosine and explain your answer.

DNA Function

The essential function of DNA is to serve as a code for the structure of proteins synthesized by a cell. A **gene** is a sequence of DNA nucleotides that codes for a protein. The next section of this chapter explains in some detail how the genes direct protein synthesis. A human being has about 100,000 genes, collectively called the **genome** (JEE-nome). Surprisingly, however, only 2% to 4% of the DNA codes for proteins. The other 96% to 98% does not code for anything. It has sometimes been labeled "junk DNA," but it may play more important organizing roles than this dismissive name would suggest.

RNA Structure and Function

DNA directs the synthesis of proteins by means of its smaller cousins, the ribonucleic acids (RNAs). There are three types of RNA, to be defined and distinguished shortly. For now we consider what they have in common and how all three differ from DNA (table 5.1).

The most significant difference is that while DNA is a double helix, RNA consists of only one nucleotide chain, not held together by complementary base pairs except in certain regions of tRNA where the molecule folds back on itself. RNA molecules are also much smaller, ranging from about 70 to 90 bases in tRNA to slightly over 10,000 bases in the largest mRNA. DNA, by contrast, may be over a billion base pairs long. The

sugar in RNA is ribose instead of deoxyribose, and one of the pyrimidines of DNA, thymine, is replaced by one called uracil (U) in RNA (see fig. 5.3).

The essential function of RNA is to interpret the code in DNA and direct the synthesis of proteins. RNA works in the cytoplasm, while DNA remains safely behind in the nucleus, "giving orders" from there. This process is described in the next section of this chapter.

Protein Synthesis and Secretion

▼Objectives

When you have completed this section, you should be able to
- explain what is meant by the genetic code and describe the way in which DNA codes for protein structure;
- describe the events of genetic transcription and translation, and explain the importance of these processes;
- explain what happens to a protein after its amino acid sequence has been synthesized; and
- explain how DNA, which codes only for proteins, indirectly regulates the synthesis of other classes of biomolecules.

Everything a cell synthesizes ultimately results from the action of enzymes, and DNA directs the synthesis of all enzymes as well as other proteins. Cells, of course, synthesize many substances other than proteins—glycogen, fat, phospholipids, steroids, pigments, and so on. There are no genes for these cell products, but their synthesis depends on enzymes that are coded for by the genes. Even though a cell of the testis has no genes for testosterone, for example, testosterone synthesis is indirectly under genetic control (fig. 5.5). Since testosterone strongly influences such behaviors as sexual drive (in both sexes), we can see that genes also make a significant contribution to behavior. In this section, we examine how protein synthesis results from the instructions given in the genes.

Preview

Before studying the details of protein synthesis, it is useful to consider the big picture. In brief, DNA contains a genetic code that specifies which proteins a cell can make. All the body's cells except the sex cells contain identical genes, but different genes are activated in

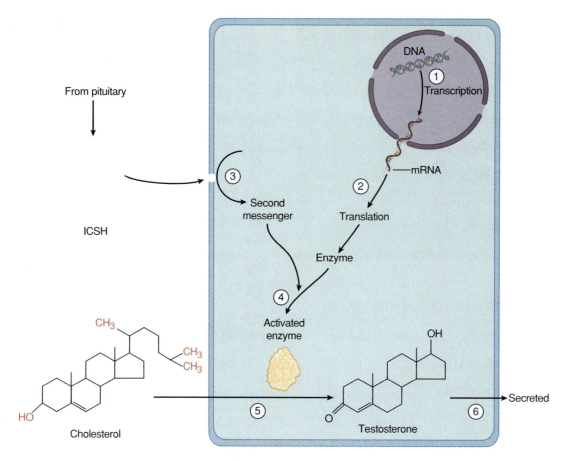

Figure 5.5 DNA codes only for protein synthesis, but through the enzymes encoded by it, DNA also indirectly controls the synthesis of nonprotein molecules. Here it is shown indirectly controlling the synthesis of testosterone by an interstitial cell of the testis. (*1*) DNA codes for an mRNA molecule, which leaves the nucleus. (*2*) In the cytoplasm, mRNA is translated and an enzyme is synthesized. (*3*) When testosterone is needed, interstitial cell-stimulating hormone (ICSH) acts through a second messenger system to (*4*) activate this enzyme. (*5*) The enzyme converts cholesterol to testosterone. (*6*) Testosterone is secreted from the cell and exerts various anatomical, physiological, and behavioral effects.

different cells; for example, the genes for blood-clotting proteins are not active in muscle cells. When a gene is activated, a copy of it called **messenger RNA (mRNA)** is made. Messenger RNA migrates from the nucleus to the cytoplasm, where its code is "read" by a ribosome composed of **ribosomal RNA (rRNA)** and enzymes. **Transfer RNA (tRNA)** molecules deliver amino acids to the ribosome, and the ribosome chooses from among these to assemble amino acids in the order directed by the mRNA. In summary, you can think of the process of protein synthesis as DNA → mRNA → protein, reading each arrow as "codes for the production of." After the protein is made, it may be modified in various ways by the rough ER and Golgi complex. Then it may continue on to any of several destinations to be considered later.

The Genetic Code

The human genome codes for about 100,000 proteins. All of these are made of the same 20 amino acids, and the genes are all made of the same four nucleotides (A, T, C, G)—a striking illustration of how a great variety of complex structures can be made from a smaller variety of simpler components arranged in different ways. The **genetic code** is a system in which four nucleotides code for the amino acid sequences of all proteins.

It is not unusual for simple codes to represent complex information. The complete text of an encyclopedia could be transmitted in Morse code, if we wished, with only the two symbols • and −. Computers store and transmit information, including pictures and sounds, in a binary code with only the symbols 1 and 0. It is not surprising, then, that a mere 20 amino acids can be represented by a code of four nucleotides. All that is required is various combinations of these symbols. A combination of two nucleotides for each amino acid would not suffice, because A, U, C, and G can only be combined in 16 different pairs (AA, AU, AC, AG, UA, UU, etc.). The minimum code to symbolize 20 amino acids would require at least three symbols, and indeed

Table 5.2	Examples of the Genetic Code		
Base Triplet of DNA	Codon of mRNA	Name of Amino Acid	Abbreviation for Amino Acid
CCT	GGA	Glycine	GLY
CCA	GGU	Glycine	GLY
CCC	GGG	Glycine	GLY
CTC	GAG	Glutamic acid	GLU
CGC	GCG	Alanine	ALA
CGT	GCA	Alanine	ALA
TGG	ACC	Threonine	THR
TGC	ACG	Threonine	THR
GTA	CAU	Valine	VAL
TAC	AUG	Methionine	MET

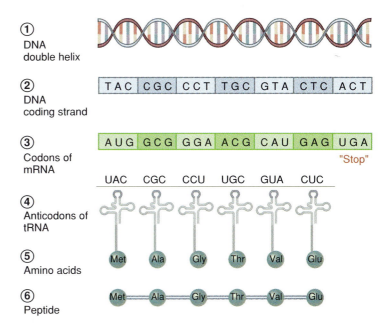

Figure 5.6 Relationship of a DNA base sequence to a protein's amino acid sequence. (*1*) The DNA double helix. (*2*) A series of base triplets in the coding strand of DNA. (*3*) The corresponding codons that would be in an mRNA molecule transcribed from this DNA sequence. (*4*) The binding of tRNA to mRNA by the complementary anticodons of the tRNA. (*5*) The amino acids bound to these tRNA molecules. (*6*) Linkage of the amino acids into a peptide that was ultimately encoded in the DNA molecule.

this is the case. A sequence of three DNA nucleotides that stands for one amino acid is called a **base triplet.** The "mirror image" sequence in mRNA is called a **codon.** The genetic code is expressed in terms of mRNA codons.

A few representative triplets and codons are listed in table 5.2, along with the amino acids they represent. You can see from this listing that two or more codons sometimes represent the same amino acid. The reason for this is easy to explain mathematically. Four symbols (*N*) taken three at a time (*x*) can be combined in N^x different ways; that is, there are $4^3 = 64$ possible codons available to represent only 20 amino acids. Of the 64 codons, however, only 61 code for amino acids. The other three—UAG, UGA, and UAA—are called **stop codons** because they signal "end of message," like the period at the end of a sentence. This is how the cell's protein-synthesizing machinery senses that it has reached the end of the code for a particular protein. The codon AUG plays two roles—it serves as a code for methionine and as a **start codon.** This dual function is explained shortly.

Figure 5.6 shows how a nucleotide sequence translates to a hypothetical peptide of 6 amino acids. A protein 500 amino acids long would have to be represented, at a minimum, by a sequence of 1,503 nucleotides (3 for each amino acid, plus a stop codon). The average gene is probably around 1,200 nucleotides long; a few may be as much as ten times this long.

Transcription

Protein synthesis occurs in the cytoplasm, but DNA is too large a molecule to leave the nucleus. It is necessary, then, to make a small RNA copy of the information in DNA that can migrate through a nuclear pore into the cytoplasm. Just as we might transcribe (copy) a document, **transcription** in genetics means the process of copying genetic instructions from DNA to RNA. It is triggered by chemical messengers from the cytoplasm that enter the nucleus and bind to the chromatin at the site of the relevant gene. An enzyme called **RNA polymerase** (po-LIM-ur-ase) then binds to the DNA at this point and begins making RNA. Certain base sequences (often TATATA or TATAAA) inform the polymerase where to begin transcription.

RNA polymerase opens up the DNA helix about 17 base pairs at a time. It transcribes the bases from one strand of the DNA, making a corresponding RNA. Where it finds a C on the DNA, it adds a G to the RNA; where it finds an A, it adds a U; and so forth. The enzyme then rewinds the DNA helix behind it. Another RNA polymerase may follow closely behind the first one; thus, a gene may be transcribed by several polymerase molecules at once, and numerous copies of the same RNA are made. At the end of the gene is a base sequence that serves as a terminator. It signals the polymerase to release the RNA molecule and separate from the DNA.

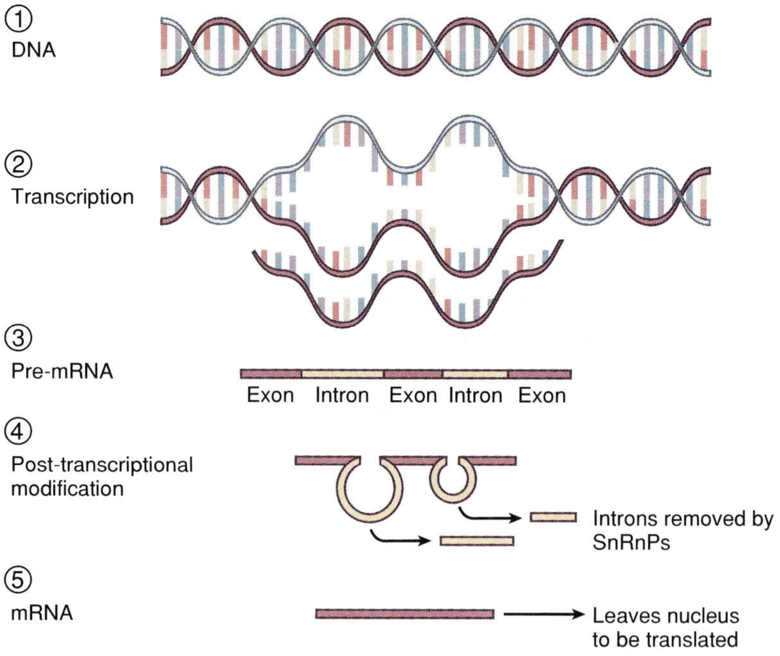

Figure 5.7 Transcription of DNA and posttranscriptional modification of RNA. (*1*) The DNA double helix has a coding strand (green) and a noncoding strand. (*2*) The coding strand is the template from which RNA is made during transcription. (*3*) This RNA molecule, called pre-mRNA, has segments called exons that code for protein synthesis, and noncoding segments called introns that must be removed. (*4*) In posttranscriptional modification, enzymes called SnRNPs remove the introns and splice together the exons. (*5*) The result is mRNA, which leaves the nucleus to be translated in the cytoplasm.

Posttranscriptional Modification

The RNA produced by transcription is not quite ready to direct protein synthesis. It is in a form called **pre-mRNA,** which requires a cutting and splicing process called **posttranscriptional modification** to produce the functional mRNA (fig. 5.7). This is because pre-mRNA contains both "sense" portions, or **exons,** which will be translated into an amino acid sequence, and "nonsense" portions, or **introns,** which are never translated to anything. Introns range from 100 to over 1,000 bases long, and there may be 2 to 50 of them in a gene. Introns are snipped out of the pre-mRNA molecule, and the ends of the exons are joined together by macromolecules called SnRNPs[2] (pronounced "snurps"). The result is a functional mRNA molecule that leaves the nucleus and carries its genetic message into the cytoplasm.

Translation

Just as we might translate a work from the French language into English, **genetic translation** translates the language of nucleotides into the language of amino acids (fig. 5.8). This job is done by ribosomes, which are found on the rough ER, on the outer membrane of the nuclear envelope, and free in the cytosol. A ribosome consists of two granular subunits, large and small, each made of several rRNA and enzyme molecules.

The mRNA molecule begins with a **leader sequence** of bases that are not translated to protein but serve as a binding site for the ribosome. The small ribosomal subunit binds to it, the large subunit joins the complex, and the ribosome begins pulling the mRNA through it like a tickertape, reading bases as it goes. When it reaches the start codon, AUG, it begins making protein. Since AUG codes for methionine, all proteins begin with methionine when first synthesized, although this may be removed later.

Translation requires the participation of 61 types of transfer RNA (tRNA), one for each codon used in protein synthesis. Transfer RNA is a small RNA molecule that turns back and coils on itself to form a cloverleaf shape, which is then twisted into an angular L-shape (fig. 5.9). One end of the L includes three nucleotides called an **anticodon,** and the other end has a binding site specific for one amino acid. One ATP molecule is used to bind the amino acid to this site. This bond energy is used later to join that amino acid to the growing protein. Thus, protein synthesis consumes one ATP for each peptide bond formed.

When the small ribosomal subunit reads a codon such as CGC, it must find an activated tRNA with the corresponding anticodon; in this case, GCG. This particular tRNA would have alanine at its other end. The ribosome binds and holds this tRNA and then reads the next codon—say GGU. Here, it would bind a tRNA with anticodon CCA, which carries glycine.

The large ribosomal subunit contains an enzyme that forms peptide bonds, and now that alanine and

2. Small nuclear ribonucleoproteins

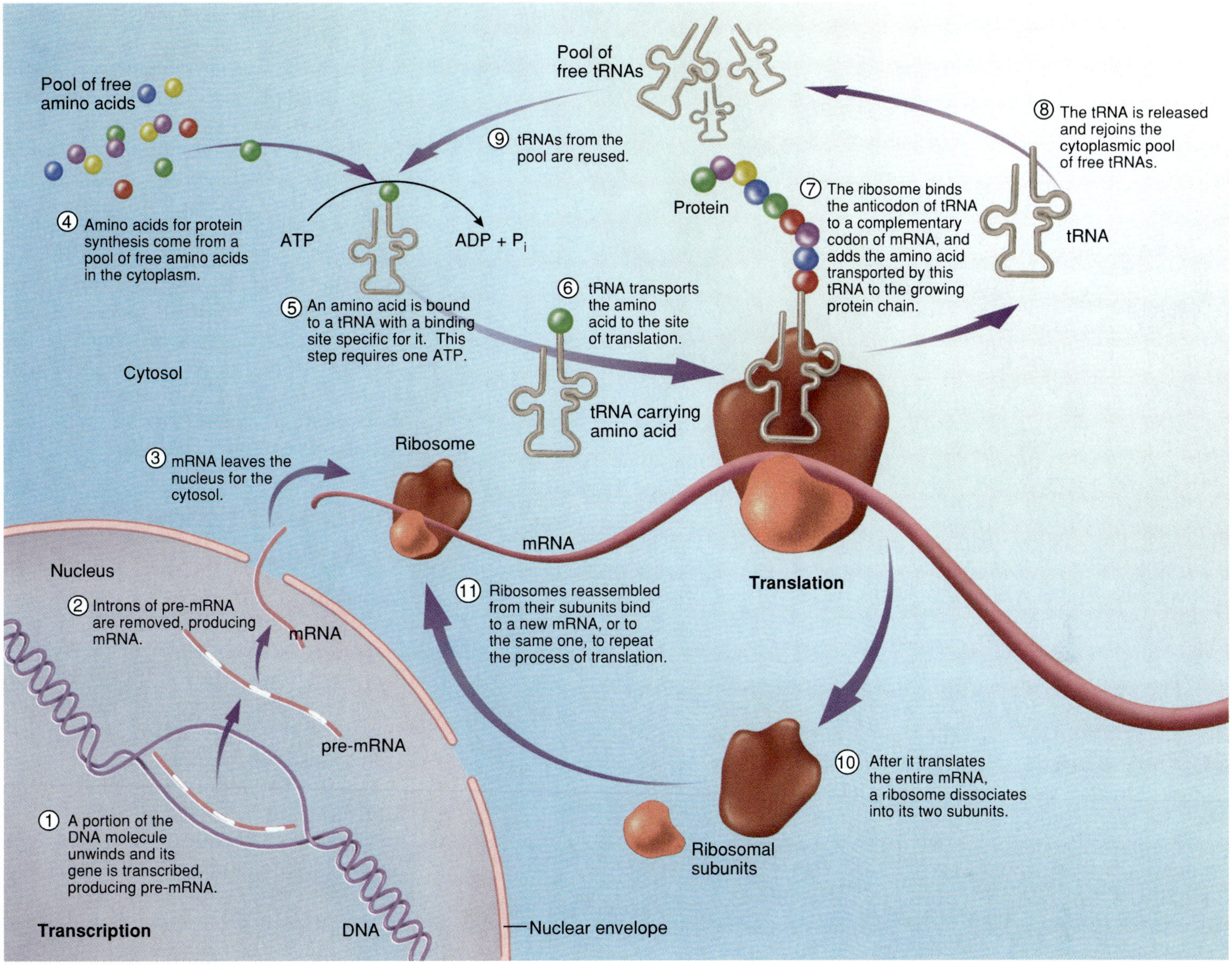

Pool of free amino acids
④ Amino acids for protein synthesis come from a pool of free amino acids in the cytoplasm.
Pool of free tRNAs
⑨ tRNAs from the pool are reused.
ATP
ADP + P_i
⑤ An amino acid is bound to a tRNA with a binding site specific for it. This step requires one ATP.
⑥ tRNA transports the amino acid to the site of translation.
Protein
⑦ The ribosome binds the anticodon of tRNA to a complementary codon of mRNA, and adds the amino acid transported by this tRNA to the growing protein chain.
⑧ The tRNA is released and rejoins the cytoplasmic pool of free tRNAs.
tRNA
Cytosol
tRNA carrying amino acid
Ribosome
③ mRNA leaves the nucleus for the cytosol.
mRNA
Nucleus
② Introns of pre-mRNA are removed, producing mRNA.
mRNA
Translation
⑪ Ribosomes reassembled from their subunits bind to a new mRNA, or to the same one, to repeat the process of translation.
pre-mRNA
⑩ After it translates the entire mRNA, a ribosome dissociates into its two subunits.
① A portion of the DNA molecule unwinds and its gene is transcribed, producing pre-mRNA.
Ribosomal subunits
Transcription
DNA
Nuclear envelope

Figure 5.8 Overview of the processes of genetic transcription and translation. Posttranscriptional modification of RNA is omitted to simplify the diagram.

glycine are side by side, it links them together. The first tRNA is no longer needed, so it is released from the ribosome. The second tRNA is used, temporarily, to anchor the growing peptide to the ribosome. Now, the ribosome reads the third codon—say GUA. It finds the tRNA with the anticodon CAU, which carries the amino acid valine. The large subunit adds valine to the growing chain, now three amino acids long. By repetition of this process, the entire protein is assembled. Eventually, the ribosome reaches a stop codon and is finished translating this mRNA. The polypeptide is turned loose, and the ribosome dissociates into its two subunits.

A ribosome can assemble about 15 amino acids per second; for example, it takes about 10 seconds to complete a molecule of β hemoglobin, 146 amino acids long. But an mRNA molecule is usually translated by several ribosomes at once. After its leader sequence passes through one ribosome, it may enter another and begin to be translated by that one before the first ribosome has finished. One mRNA often holds several ribosomes together in a cluster called a **polyribosome** (fig. 5.10). Not only is each mRNA translated by several ribosomes at once, but a cell may have 300,000 identical mRNA molecules, all undergoing simultaneous translation. Thus, a cell may produce 100,000 protein molecules per second—a remarkably

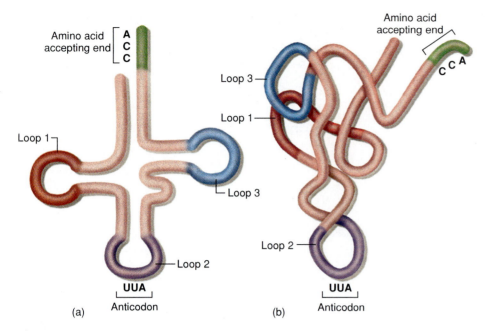

Figure 5.9 Transfer RNA (tRNA). (*a*) tRNA has an amino acid accepting end that binds to one specific amino acid, and an anticodon that binds to a complementary codon of mRNA. (*b*) The three-dimensional shape of a tRNA molecule.

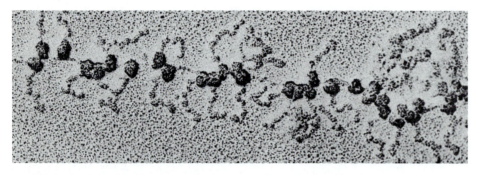

Figure 5.10 Several ribosomes attached to a single mRNA molecule, forming a polyribosome. The fine horizontal filament is mRNA; the large granules attached to it are ribosomes; and the beadlike chains projecting from each ribosome are newly formed proteins.

productive protein factory! Ribosomes make up as much as 25% of the dry weight of liver cells and some others that are highly active in protein synthesis.

Chaperones and Protein Structure

The amino acid sequence of a protein (primary structure) is only the beginning; the end of translation is not the end of protein synthesis. The protein now coils or folds into its secondary and tertiary structures and, in some cases, associates with other polypeptide chains (quaternary structure) or conjugates with a nonprotein moiety, such as a vitamin or carbohydrate. It is essential that these processes not begin prematurely as the amino acid sequence is being assembled, since the correct final shape may depend on

amino acids that have not been added yet. Therefore, as proteins are assembled by ribosomes, they are sometimes picked up by other proteins called **chaperones** (also known as *heat-shock proteins,* or *Hsps,* because they are often produced in response to thermal stress). A chaperone prevents a new protein from folding prematurely and assists in its proper folding once the amino acid sequence has been completed. The chaperone may also escort a newly synthesized protein to the correct destination in a cell, such as the plasma membrane. As in the colloquial sense of the word, a chaperone is an "older protein" that escorts and regulates the behavior of the "youngsters."

Posttranslational Modification

If a protein is going to be used in the cytosol (for example, the enzymes of glycolysis), it is likely to be made by free ribosomes in the cytosol. If it is going to be packaged into a lysosome or secreted from the cell, however, it is more likely to be made by ribosomes on the rough ER and its production completed by the Golgi complex. Thus, we turn to the function of the rough ER and Golgi complex in the modification, packaging, and secretion of a protein (fig. 5.11).

A protein produced on the rough ER begins with a leader sequence of about 30 hydrophobic amino acids. This sequence is repelled by water in the cytosol and drawn into the phospholipid membrane of the ER. When the protein emerges into the cisterna, the leader sequence is enzymatically removed. The new protein is then trapped in the cisterna, because without the leader sequence it cannot pass backward through the phospholipid.

Membranes of the rough ER contain enzymes that modify the structure of newly synthesized proteins. A protein may be folded, have some amino acid segments removed, have disulfide bridges added to stabilize the teritary structure, and so forth. Such changes are called **posttranslational modification.** Insulin, for example, is first synthesized as a polypeptide of 86 amino acids. In posttranslational modification, the chain folds back on itself, three disulfide bridges are formed, and 35 amino acids are removed. The final insulin molecule is therefore made of two chains of 21 and 30 amino acids held together by disulfide bridges (see fig. 17.22).

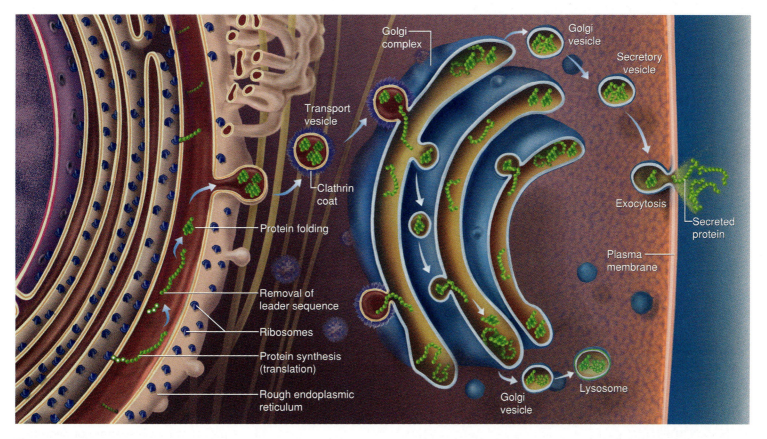

Figure 5.11 Functional relationship of the rough endoplasmic reticulum to the Golgi complex. Proteins are synthesized by ribosomes on the rough ER and carried in transport vesicles to the nearest cisterna of the Golgi complex. The Golgi complex modifies the structure of the protein, transferring it from one cisterna to the next, and finally packages it in Golgi vesicles. Some Golgi vesicles may remain within the cell and function as lysosomes, while others may migrate to the plasma membrane and release the cell product by exocytosis.

When the rough ER is finished modifying and transporting a protein, it pinches off clathrin-coated **transport vesicles.** Like the address on a letter, clathrin may direct the vesicle to its destination, the Golgi complex. The Golgi complex removes the clathrin, fuses with the vesicle, and takes the protein into its cisterna. Here, it may further modify the protein, for example, by adding carbohydrate to it. Such modifications begin in the cisterna closest to the RER. Each cisterna can form transport vesicles that carry the protein to the next cisterna, where different enzymes may further modify the new protein.

Packaging and Secretion

When the protein finally enters the last Golgi cisterna, farthest from the rough ER, that cisterna pinches off spherical, membrane-bounded **Golgi vesicles** containing the finished product. Some Golgi vesicles become **secretory vesicles,** which migrate to the plasma membrane and release the product by exocytosis. This is how a cell of the salivary gland, for example, secretes mucus and digestive enzymes. The destinations of these and some other newly synthesized proteins are summarized in table 5.3.

Table 5.3	Some Destinations and Functions of Newly Synthesized Proteins
Destination or Function	**Proteins**
Deposited as a structural protein within cells	Components of the cytoskeleton Keratin of the epidermis
Used in the cytosol as an enzyme that controls metabolic processes	ATPase Enzymes of glycolysis
Returned to the nucleoplasm for use in nuclear metabolism	Histones of the chromatin RNA polymerase
Packaged in lysosomes for such purposes as autophagy and programmed cell death	Numerous lysosomal enzymes
Delivered to other cytoplasmic organelles for use in their metabolism	Catalase of peroxisomes Mitochondrial enzymes of aerobic respiration
Delivered to the plasma membrane to serve transport and other functions	Hormone and neurotransmitter receptors Ion pumps and channel proteins
Secreted by exocytosis for functions external to the cell	Digestive enzymes Mucus (mucoprotein)

Key Point Review

5. Define *genetic code* and *codon*.

6. Describe the roles of RNA polymerase and SnRNPs in mRNA synthesis.

7. Describe the roles of ribosomes and tRNA in translation.

8. Why are chaperones important in determining tertiary protein structure?

9. What roles do the rough ER and Golgi complex play in the production and secretion of proteins?

DNA Replication and the Cell Cycle

▼Objectives

When you have completed this section, you should be able to

- describe how DNA is replicated;
- discuss the consequences of replication errors;
- describe the life history of a cell, including the events of mitosis;
- explain how the timing of cell division is regulated;
- describe some of the genetic and environmental causes of cancer; and
- explain why untreated cancer is fatal.

When a cell divides, each daughter cell receives half of its DNA. At some point in the cell's life history, therefore, it must replicate its DNA to restore the original amount and be able to divide again. Since DNA controls all cellular function, this replication process must be very exact. We now examine how it is accomplished and consider the consequences of mistakes.

DNA Replication

The law of complementary base pairing shows that we can predict the base sequence of one DNA strand if we know the sequence of the other. More importantly, it enables a cell to reproduce one strand based on information in the other. This immediately occurred to Watson and Crick when they discovered the structure of DNA (see special topic 5.2). Watson was hesitant to make such a grandiose claim in their first publication, but Crick implored, "Well, we've got to say *something!* Otherwise people will think these two unknown chaps are so dumb they don't even realize the implications of their own work!" Thus, the last sentence of their first paper modestly states, "It has not escaped our notice that the specific pairing we have postulated . . . immediately suggests a possible copying mechanism for the genetic material." Five weeks later they published a second paper pressing this point more vigorously.

The basic idea of DNA replication is evident from its base pairing, but the complex way in which DNA is organized in the chromatin introduces some complications that were not apparent when Watson and Crick first wrote. As shown in figure 5.12, the replication process occurs in several steps.

1. The double helix unwinds from the histones in each nucleosome.
2. Like a zipper, an enzyme called **DNA helicase** opens up a short segment of the helix, exposing its nitrogenous bases.
3. An enzyme called **DNA polymerase** moves along the opened strands, reads the exposed bases, and assembles complementary bases across from them. If the polymerase finds the sequence TCG, for example, it assembles bases AGC across from it. A polymerase molecule replicates a long stretch of one DNA strand and then folds back on itself and replicates several short stretches of the other strand, alternating between the two strands in this fashion until replication is complete.
4. Another enzyme, **DNA ligase** (LY-gase), splices the short segments of new DNA together. The final result is two DNA double helices.

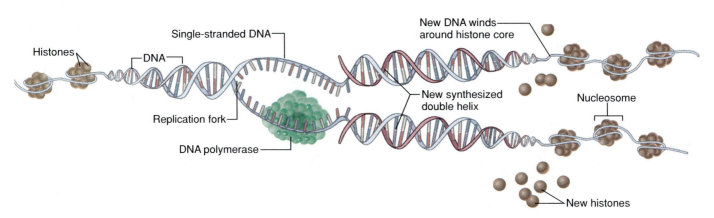

Figure 5.12 Replication of DNA. At the far left, DNA unwinds from the histones. The double helix unwinds to open up a replication fork where its nitrogenous bases are exposed. DNA polymerase reads the exposed bases on each side and assembles a new nucleotide strand complementary to each of the pre-existing strands. The result is two DNA double helices, which are rewound around histones at the right side of the diagram. ▭

5. Each helix needs histones to wind around, so as DNA is being replicated in the nucleus, new histones are synthesized in the cytoplasm and transported into the nucleus. Each new helix is wrapped around the histones to make new nucleosomes.

Despite the complexity of this process, each DNA polymerase works at an impressive rate of about 100 base pairs per second. Even at this rate, however, it would take weeks for one polymerase molecule to replicate one chromosome. The reason the process is completed in a mere 6 to 8 hours is that thousands of polymerase molecules work simultaneously on each DNA molecule.

When replication is complete, each new DNA molecule consists of one new helix that has been assembled from newly synthesized nucleotides and one helix conserved from the parent molecule (fig. 5.13). The process is therefore called **semiconservative replication.**

Errors and Mutations

DNA polymerase is fast and accurate, but it makes mistakes. For example, it might read A and place a C across from it where it should have placed a T. In bacteria (*Escherichia coli*), where DNA replication has been most thoroughly studied, about three errors are made for every 100,000 bases copied. At this rate of error, every generation of cells would have about 1,000 faulty proteins—coded for by DNA that had been miscopied. To help prevent catastrophic damage to the organism as a result of miscopied nucleotides, the DNA is continuously scanned for errors. After DNA polymerase has replicated a strand, a smaller polymerase comes along, "proofreads" it, and makes corrections where needed—for example, removing C and replacing it with T. This improves the accuracy of replication to one error for every billion bases copied—only one faulty protein for every 10 cell divisions (in *E. coli*).

Changes in DNA structure, called **mutations,**[3] can result from replication errors or from environmental factors. Uncorrected mutations can be passed on to the descendents of that cell, but some of these have no adverse effect. One reason is that a new base sequence sometimes codes for the same thing as the old one. For example, ACC and ACG both code for an amino acid called threonine (see table 5.2), so a mutation from C to G in the third

place would cause no change in protein structure. Another reason is that a change in protein structure is not always critical to its function. For example, out of the 146 amino acids that make up the β hemoglobin of humans and horses, 25 differ from one species to the other—yet the hemoglobin is fully functional in both.

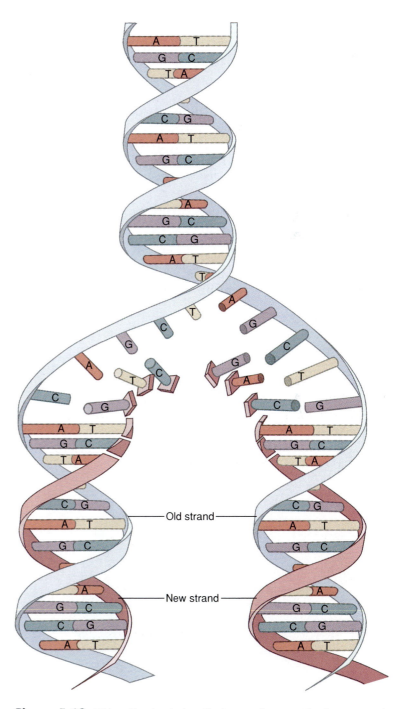

Figure 5.13 DNA replication is described as semiconservative because each new DNA molecule consists of one strand from the original parent molecule (blue) and one that has been newly synthesized (violet).

3. *muta* = change

Some mutations, however, may kill a cell, turn it cancerous, or cause genetic defects in future generations. When a mutation changes the sixth amino acid of β hemoglobin from glutamic acid to valine, for example, the result is a crippling disorder called sickle-cell disease. Clearly some amino acid substitutions are more critical than others, and this affects the severity of a mutation.

The Cell Cycle

Most cells periodically divide into two daughter cells, so a cell has a life cycle extending from one division to the next. This **cell cycle,** shown in figure 5.14, is divided into four main phases: G_1, S, G_2, and M.

In **G_1,** the **first growth phase,** a cell synthesizes proteins, grows, and carries out its preordained tasks for the body. Almost all of the discussion in this book relates to what cells do in the G_1 phase. Even then, however, cells are preparing to divide again. For example, a cell normally has a pair of centrioles and gives one to each daughter cell when it divides. Each centriole begins to replicate in the G_1 phase, so that a pair will be available again at the time of the next division.

In **S,** the **synthesis phase,** a cell undergoes the semiconservative replication of DNA previously de-scribed. This produces two identical sets of DNA molecules, which are then available to be divided up between daughter cells at the next cell division.

G_2, the **second growth phase,** is a relatively brief interval in which a cell synthesizes enzymes that control cell division. Centriole replication is completed during G_2.

In **M,** the **mitotic phase,** a cell replicates its nucleus and then pinches in two to form two new daughter cells. The details of this phase are considered in the next section. Phases G_1, S, and G_2 are collectively called **interphase**—the time between M phases.

Some idea of a time frame for these phases can be gained from *in vitro*[4] (IN VEE-tro) cultures of fetal cells called fibroblasts, which divide every 18 to 24 hours. The G_1 phase lasts 8 to 10 hours, the S phase (DNA replication) 6 to 8 hours, the G_2 phase 4 to 6 hours, and the M phase 1 to 2 hours. The length of the cell cycle varies greatly from one cell type to another, however. Stomach and skin cells divide rapidly, bone and cartilage cells slowly, and nerve and skeletal muscle cells not at all. Cells that never divide are said to have left the cell cycle and become permanently arrested in a **G_0 phase** (G-zero phase).

Mitosis

Mitosis (my-TOE-sis), in the sense used here, is the process by which a cell divides into two daughter cells with identical copies of its DNA. (Some define it only as division of the nucleus and do not include the subsequent cytoplasmic division.) Mitosis has four main functions:

1. formation of a multicellular embryo from a fertilized egg cell,
2. tissue growth,
3. replacement of old and dead cells, and
4. repair of tissues injured by disease or trauma.

All cells of the body are produced by mitosis, except for some cells that eventually give rise to mature egg and sperm cells. These are produced through a combination of mitosis and another form of cell division, *meiosis,* described in chapter 27. As shown in figure 5.15, four phases of mitosis are recognizable—*prophase, metaphase, anaphase,* and *telophase.*

In **prophase**[5] (fig. 5.15a), the chromatin becomes supercoiled into short, dense **chromosomes** (fig. 5.16).

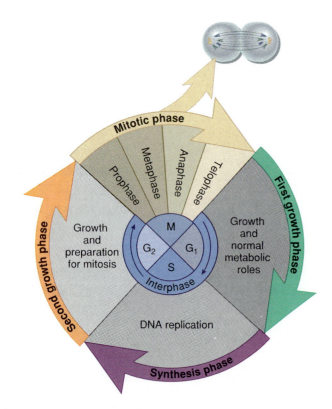

Figure 5.14 The cell cycle.

4. *vitre* = glass (*in vitro* = in glassware vessels)
5. *pro* = first

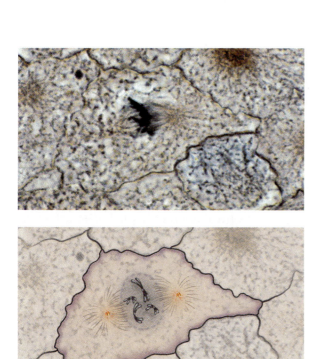

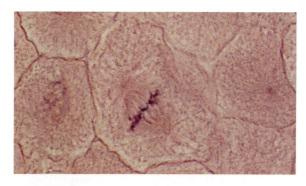

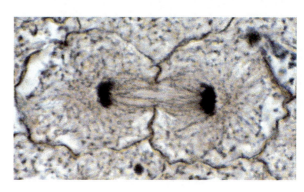

Prophase
Chromatin condenses into chromosomes.
Nucleoli and nuclear envelope break down.
Spindle fibers grow from centrioles.
Centrioles migrate to opposite poles of cell.

Metaphase
Chromosomes lie along equator of cell.
Some spindle fibers attach to kinetochores.
Fibers of aster attach to plasma membrane.

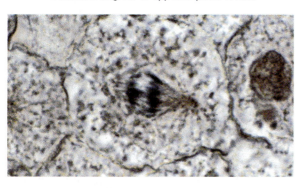

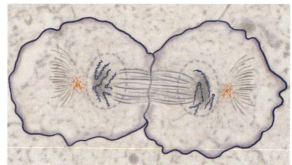

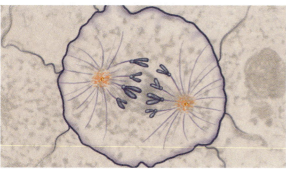

Anaphase
Centromeres divide in two.
Spindle fibers pull sister chromatids
    to opposite poles of cell.
Each pole (future daughter cell) now
    has an identical set of genes.

Telophase
Chromosomes gather at each pole of cell.
Chromatin decondenses.
New nuclear envelope appears at each pole.
New nucleoli appear in each nucleus.
Mitotic spindle vanishes.
(Above photo also shows cytokinesis.)

Figure 5.15 Mitosis. The drawings show a hypothetical cell with only two chromosome pairs; in humans, there are 23 pairs. The photographs show mitosis in whitefish eggs, where the chromosomes are relatively easy to observe (×1,000). ▭

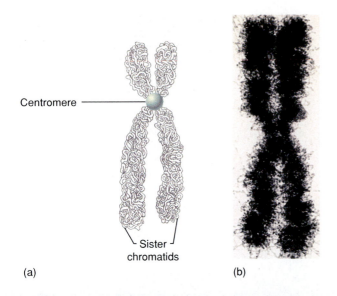

Centromere

Sister chromatids

(a)

(b)

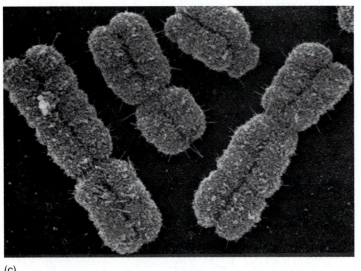

(c)

Figure 5.16 (*a*) The structure of a metaphase chromosome. (*b*) Transmission electron micrograph. (*c*) Scanning electron micrograph.

These are easier to distribute to daughter cells than the long, delicate chromatin. Like string, DNA is useful only when uncoiled, but it is much easier to distribute when it is packaged in a ball. A chromosome consists of two *genetically identical* bodies called **sister chromatids,** joined together by proteins at a pinched spot called the **centromere.** There are 46 chromosomes, two chromatids per chromosome, and one molecule of DNA in each chromatid.

The nuclear envelope disintegrates during prophase and releases the chromosomes into the cytosol. Each centriole begins to sprout elongated microtubules, which push the two centrioles apart as they grow. Eventually, the centrioles come to lie at opposite poles of the cell.

In **metaphase**[6] (fig. 5.15*b*), the chromosomes line up along the midline (equator) of the cell. Microtubules grow from each centriole toward the chromosomes, and some of them attach to the centromere. This forms a football-shaped array called the **mitotic spindle.** Shorter microtubules also radiate from each centriole to form a star-shaped array called an *aster.*[7] These microtubules anchor the centriole to the nearby plasma membrane.

In **anaphase**[8] (fig. 5.15*c*), each centromere divides in two, and its chromosome is pulled apart into two chromatids. One chromatid migrates to each pole of the cell, with its centromere leading the way and the arms trailing behind. There is some evidence that the spindle fiber acts a little like a railroad track, and a protein complex in the centromere called the **kinetochore**[9] (kih-NEE-toe-core) acts as a molecular motor that propels the chromatid along the track. One of the kinetochore proteins is dynein, the same motor molecule that causes movement of cilia and flagella (see chapter 4). Since sister chromatids are genetically identical, and since each daughter cell receives one sister chromatid from each chromosome, you can see why the daughter cells of mitosis are genetically identical.

In **telophase**[10] (fig. 5.15*d*), the chromosomes cluster on each side of the cell. The rough ER produces a new nuclear envelope around each cluster, and the chromosomes begin to uncoil and return to the thinly dispersed chromatin form. The mitotic spindle breaks up and vanishes. Each new nucleus forms nucleoli, indicating it has already begun making RNA and preparing for protein synthesis.

Telophase is the end of nuclear division but overlaps with **cytokinesis**[11] (SY-toe-kih-NEE-sis), division of the cytoplasm. Cytokinesis is achieved by microfilaments of actin and myosin pulling on the membrane skeleton around the equator of the cell. A crease called the *cleavage furrow* is created, and the cell eventually pinches in two. Interphase has now begun for these new cells.

Timing of Cell Division

One of the most important questions in biology is what signals cells when to divide and when to stop. Although the trigger for cell division is not yet well understood, it depends on two kinds of protein—a **CDC** (cell division cycle) **protein,** present throughout the

---

6. *meta* = next in a series
7. *aster* = star
8. *ana* = apart
9. *kineto* = motion + *chore* = place
10. *telo* = last, end
11. *kinesis* = action, motion

Ionizing radiation can damage tissues indirectly, by generating ions and free radicals that react with biomolecules, or by a direct hit on DNA and other critical biomolecules (see chapter 2). It is more likely to damage DNA that is condensed into mitotic chromosomes than DNA that is thinly dispersed in the chromatin of interphase. Therefore, dividing cells and mitotically active tissues are the most vulnerable to radiation damage—especially bone marrow and the intestinal lining, where there is a rapid rate of mitosis. Tumors also exhibit rapid rates of cell division, so radiation can be used to kill cancer cells with minimal risk to adjacent healthy cells.

cell cycle, and at least eight **cyclins,** which are produced as a cell enters the M phase and decline to very low levels by the start of the next interphase. When cyclins and CDC protein combine, they activate enzymes called **cyclin-dependent kinases (Cdks).** A Cdk phosphorylates and activates other proteins in the cell, including those that mediate the activities of mitosis and cytokinesis.

In tissue culture, cell division stops when cells are closely surrounded by others and touch each other—a phenomenon called **contact inhibition.** The activation and inhibition of cell division are subjects of intense research for obvious reasons such as management of cancer and tissue repair.

Cancer

Proper tissue development depends on a balance between cell division and cell death. When cells multiply faster than they die, abnormal growths called tumors or **neoplasms**[12] may occur. The study of tumors is called **oncology.**[13]

Benign[14] (beh-NINE) tumors are surrounded by a connective tissue capsule, grow slowly, and do not spread to other organs. They are still potentially lethal—even slow-growing tumors can kill by compressing brain tissue, nerves, blood vessels, or airways. The term **cancer** refers to **malignant**[15] (muh-LIG-nent) tumors, which are fast-growing and unencapsulated and which spread easily to other organs by way of the blood or lymph. This spreading is called **metastasis**[16] (meh-TASS-tuh-sis). Malignant cells exhibit no contact inhibition, so they grow into other tissues and send out branches that are somewhat reminiscent of a crab's legs, hence the name *cancer* ("crab"). The cells involved in some common types of cancer are listed in table 5.4.

Table 5.4   Types of Cancer

Type of Cancer	Cells Involved
Carcinoma	Epithelial cells
Melanoma	Pigment-producing skin cells
Sarcoma	Bone, other connective tissue, or muscle
Leukemia	Blood-forming tissues
Lymphoma	Lymph nodes

Causes of Cancer

The World Health Organization estimates that 60% to 70% of all human cancers are caused by environmental agents called **carcinogens**[17] (car-SIN-oh-jens). These agents are of three kinds:

1. **Chemical carcinogens.** The most notorious chemical carcinogen is cigarette tar, responsible for 30% of cancer deaths in the United States. Others include food preservatives, such as nitrites, and asbestos fibers from old insulation, roofing tile, and other sources.
2. **Radiation.** Cancer can be caused by ultraviolet light, γ rays, α particles, and β particles (see special topic 5.3).
3. **Viruses.** The type 2 herpes simplex virus is implicated in uterine cancer, and hepatitis B virus is associated with liver cancer. Human immunodeficiency virus (HIV), the AIDS virus, indirectly leads to cancer in many patients by destroying immune cells that protect against cancer.

Carcinogens cause cancer by producing *mutations*—permanent structural changes in genes. Therefore, carcinogens are **mutagens**[18] (MEW-tuh-jens). The body has several defenses against mutagens: (1) scavenger cells remove them before they can cause genetic damage;

---

12. *neo* = new + *plasm* = growth, formation
13. *onco* = tumor
14. *benign* = mild, gentle
15. *mal* = bad
16. *meta* = beyond + *stasis* = being stationary

17. *carcino* = cancer + *gen* = producing
18. *muta* = change + *gen* = causing

(2) peroxisomes neutralize nitrites, free radicals, superoxides, and other carcinogenic oxidizing agents; and (3) nuclear enzymes detect and repair damaged DNA. If these mechanisms fail, or if they are overworked by heavy exposure to mutagens, a cell may die of genetic damage, it may be recognized and destroyed by the immune system before it can multiply, or it may multiply and produce a tumor. Even then, tumors may be destroyed by a substance called *tumor necrosis factor (TNF),* secreted by macrophages and certain white blood cells.

Recent research has revealed three particularly important factors in the development of cancer: growth factors, oncogenes, and tumor suppressor genes. **Growth factors** are polypeptides released by blood platelets, kidney cells, and other cells. They stimulate mitosis and differentiation in target cells that have membrane receptors for them. In fetal development, growth factors promote rapid tissue growth, but they continue to function at lower levels throughout life. When fetal development is complete, it is necessary to reduce the secretion of growth factors or the sensitivity of target cells to them in order for cell division to slow down.

Oncogenes are named for the fact that they cause tumor growth. When functioning normally, they are called **proto-oncogenes** and apparently code for growth factors or their receptors. Proto-oncogenes may stimulate normal fetal tissue growth and then be deactivated. If a carcinogen should later reactivate one of these, transforming a proto-oncogene to an oncogene, it may trigger abnormal tissue growth. About 15% to 20% of the various types of human cancer have been traced to oncogenes. Breast and ovarian cancer, for example, are associated with multiple copies of an oncogene called *erbB2.*

Tumor suppressor (TS) genes inhibit the development of cancer. There are several hypotheses about their mechanism of action. They may, for example, oppose the action of oncogenes, promote DNA repair, or control the normal histological organization of tissues, which is notably lacking in malignancies. A TS gene called *p16* acts by inhibiting one of the cyclin-dependent kinases of the cell cycle. Thus, if oncogenes are thought of as the "accelerator" of the cell cycle, then TS genes are like the "brake." Like other genes, TS genes occur in *homologous pairs* (explained in the following section), and even one normal TS gene in a pair is enough to suppress cancer. Damage to both members of a pair, however, removes normal controls over cell division and may trigger cancer. A cancer of the eyes of infants, called *retinoblastoma,* is caused by damage to a TS gene called the *RB gene* on chromosome 13. Colon cancer is caused by abnormal function in five or six genes, including damage to at least three TS genes on chromosomes 5, 17, and 18 and activation of an oncogene on chromosome 12. Thus, it may take more than one mutation for a cell to turn cancerous, and this may be why cancer is slow to de-velop and is more common in old age. Furthermore, the longer we live, the more carcinogens we are exposed to, the less efficient our DNA and tissue repair mechanisms become, and the less effective our immune system is at recognizing and destroying malignant cells.

Effects of Cancer

Cancer is almost always fatal if not treated. The primary reasons are as follows:

- Cancer cells displace normal cells, so the function of the affected organ deteriorates. Lung cancer, for example, can destroy so much tissue that oxygenation of the blood becomes inadequate to support life. You might think that an increased number of cells in an organ would enable it to function better. In cancer, however, accelerated cell multiplication has the opposite effect. Cancer cells generally exhibit **anaplasia**[19] (AN-uh-PLAY-zee-uh)—they revert back to an undifferentiated and nonfunctional form and do not resemble the normal cells of that organ. Also, a tumor may consist of cells that have metastasized from elsewhere and are unable to function in the host organ. A colon cancer whose malignant cells have metastasized to the liver, for example, destroys liver tissue but cannot perform any of the liver's essential functions.
- Tumors can block vital passageways. The growth of a tumor can put pressure on a bronchus of the lung, obstructing air flow and causing pulmonary collapse, or it can compress a major blood vessel, reducing the delivery of blood to a vital organ or its return to the heart.
- Tumors have a high metabolic rate and compete with healthy tissues for nutrition. Other organs of the body may even break down their own proteins to nourish the tumor. This leads to general weakness, fatigue, emaciation, and susceptibility to infections.

.... Key Point Review ....

10. Describe the roles of DNA helicase, DNA polymerase, and DNA ligase.

11. Why is DNA replication called *semiconservative?*

12. Define *mutation.* Why are some mutations harmless and others harmful?

13. List the four stages of the cell cycle and describe the main events of each.

14. Describe the structure of a chromosome at metaphase.

15. Explain how an oncogene may cause cancer and how a tumor suppressor gene may prevent cancer.

19. *ana* = back + *plasia* = formation, growth

▼Objectives

When you have completed this section, you should be able to
- describe the structure and paired arrangement of human chromosomes;
- define *allele* and discuss the various ways in which alleles interact to affect the traits of the individual; and
- discuss the interaction of heredity and environment in producing individual traits.

Heredity is the transmission of genetic characteristics from parent to offspring. Several traits and diseases discussed in the forthcoming chapters are hereditary: baldness, blood types, color blindness, and hemophilia, for example. Thus it is appropriate at this point to lay the groundwork for these discussions by introducing a few basic principles of normal heredity. Hereditary birth defects are described in chapter 29.

The Karyotype

A **karyotype** (fig. 5.17) is a chart of the chromosomes isolated from a cell at metaphase and arranged in order by size and structure. It shows that most human cells, with the exception of germ cells (described below), contain 46 chromosomes, consisting of 23 similar-looking **homologous** (ho-MOLL-uh-gus) **pairs.** Two of the chromosomes, designated X and Y, are called **sex chromosomes** and the other 22 pairs are called **autosomes** (AW-toe-somes). A female normally has a pair of X chromosomes, whereas a male has one X chromosome and a much smaller Y chromosome.

Think About It

When a karyotype is needed, why does a cytogeneticist use cells at metaphase instead of interphase?

The paired state of the homologous chromosomes results from the fact that a sperm cell bearing 23 chromosomes fertilizes an egg, which also has 23. Sperm and egg cells, and the cells set aside to become sperm and eggs, are called **germ cells.** All other cells of the body are called **somatic cells.** Somatic cells are described as **diploid**[20] because their chromosomes are in pairs, whereas most germ cells are **haploid,**[21] which means they contain half as many chromosomes as the somatic cells. In meiosis (described in detail in chapter 27), homologous chromosomes become **segregated** from each other into separate daughter cells leading to the haploid sex cells. At fertilization, one set of *paternal* (sperm) chromosomes unites with one set of *maternal* (egg) chromosomes to restore the diploid number to the fertilized egg and the somatic cells that arise from it. Although the two chromosomes of a homologous pair appear to be identical, they come from different parents and therefore are not *genetically* identical.

Genes and Alleles

Each gene is located at a predictable site called a **locus** on a particular chromosome. An oncogene for colon cancer, for example, is found at the same locus of chromosome 12 in every human being. The Human Genome Project (see special topic 5.4), aimed at identifying the locus and nucleotide sequence of every human gene, may open the door to increasingly effective *gene substitution therapy* for hereditary diseases (see chapter essay, p. 164).

For each gene at a particular locus on a chromosome, there is a similar gene at the same point on the homologous chromosome. When the two genes of a pair differ somewhat, the alternative forms are called **alleles**[22] (ah-LEELS). Alleles tend to produce alternative forms of a particular trait. One allele, for example, codes for dimples in the

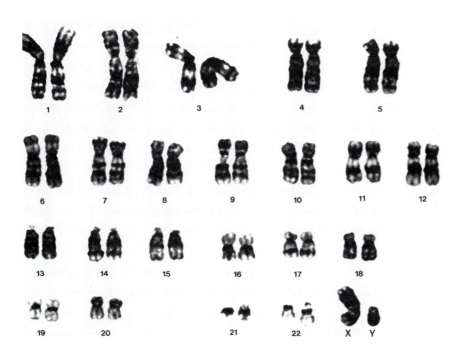

Figure 5.17 Karyotype of a normal human male. The only difference in a female karyotype would be the presence of a pair of X chromosomes in place of the XY pair of the male.

20. *diplo* = double
21. *haplo* = half
22. *allo* = other, different

The Human Genome Project is a massive effort to map the locations and determine the nucleotide sequences of all 100,000 genes of the genome. The project was proposed in 1985 and began operation on October 1, 1990. Now involving numerous laboratories in the United States, Europe, and Japan, it is slated for completion by 2005. In essence, the outcome will be an "instruction book" for human development and maintenance. Understanding the instructions, however, is likely to take much longer than obtaining them. Among other difficulties, 96% to 98% of the nucleotide sequences will consist of DNA that does not code for anything. At a cost of $3 billion, this is the first biological research project of such magnitude, and questions have been raised about spending such a large sum of money to obtain information that may be 98% meaningless. One thing is certain, however—we cannot advance very far in the treatment of genetic diseases until we know where the responsible genes lie. The Human Genome Project will provide much of the information we need to locate these genes, as well as genes that underlie many normal aspects of human form and function.

cheeks, and its alternative allele codes for an absence of dimples. These alleles can be represented as *D* and *d*, respectively. If both chromosomes of a homologous pair have identical alleles, such as *DD* or *dd*, the individual is said to be **homozygous**[23] (HO-mo-ZY-gus) for that trait. If the two homologous chromosomes have different alleles for that gene (*Dd*), the individual is **heterozygous**[24] (HET-er-oh-ZY-gus). The alleles an individual possesses for a trait constitute the **genotype** (JEE-no-type). A detectable trait resulting from the genotype, such as the presence or absence of dimples, is called the **phenotype**[25] (FEE-no-type).

We say that an allele is *expressed* if it shows up in the phenotype of an individual. Allele *D,* if present on one or both homologous chromosomes, is normally expressed as cheek dimples, whereas allele *d* is not expressed unless it occurs on both chromosomes. That is, *D* masks the effect of *d,* and *D* is therefore said to be **dominant.** A **recessive** allele (in this case, *d*) is not expressed if it occurs in a heterozygous state (*Dd*) but is expressed when it occurs in a homozygous recessive state (*dd*). It is customary to represent a dominant allele with a capital letter and a recessive allele with its lowercase equivalent. The only way most recessive alleles can be expressed is for an individual to inherit them from both parents.

Recessive traits can "skip" one or more generations. For example, albinism is a condition in which the brown pigment melanin is lacking in the body. People who are albino have white hair and very pale skin, and their eyes may appear pink as the blood shows through where melanin is lacking. Albinism can result from any of several recessive alleles that lead to the lack of an enzyme in the melanin synthesis pathway. In figure 5.18, a simple diagram called a *Punnett square* shows how

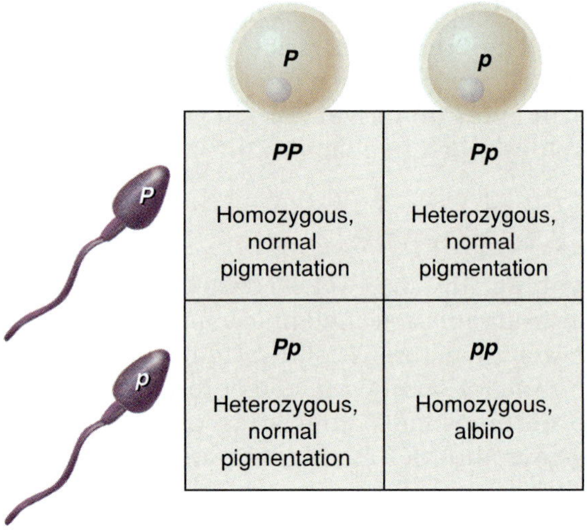

Figure 5.18 The genetics of albinism shows why recessive traits may "skip a generation" and may appear in the offspring of parents who do not have that trait but carry the gene for it.

two heterozygous, normally pigmented parents can produce an albino child. The allele for normal pigmentation is designated *P,* and the recessive allele is designated *p.* Across the top are the two genetically possible types of eggs the mother could produce, and on the left side, the possible types of sperm cells from the heterozygous father. The four cells of the square are filled in with the genotypes that would result from each possible combination of sperm cell and egg. You can see that three of the possible combinations would produce a normally pigmented child (genotypes *PP* and *Pp*), but one combination (*pp*) would produce an albino child. Therefore, albinism "skipped" the parental generation in this case but could be expressed in their child.

Think About It

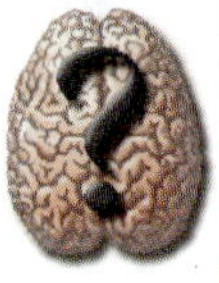

Would it be possible for an albino woman to have children with normal pigmentation? Use a Punnett square and one or more hypothetical genotypes for the father to demonstrate your point.

---

23. *homo* = same + *zygo* = union
24. *hetero* = different, unlike
25. *pheno* = evident, showing

An individual who carries a recessive allele but does not phenotypically express it is called a **carrier.** For some hereditary diseases, tests are available to detect carriers and allow couples to weigh their risk for having children with genetic disorders. *Genetic counselors* perform or refer clients for genetic testing, advise couples on the probability of transmitting genetic diseases, and assist people in coping with genetic disease.

Multiple Alleles, Codominance, and Incomplete Dominance

Many genes have more than two alleles in a population. The ABO blood group affords one example of **multiple alleles.** Phenotypically, a person may have blood type A, B, AB, or O, owing to the presence of three alleles in the population. Two of the alleles are dominant and symbolized with a capital I (for immunoglobulin) and a superscript: I^A and I^B. There is one recessive allele, symbolized with a lowercase i. An individual can inherit any two of these three alleles, resulting in one of the following:

Genotype	Phenotype
$I^A I^A$	Type A
$I^A i$	Type A
$I^B I^B$	Type B
$I^B i$	Type B
$I^A I^B$	Type AB
ii	Type O

Think About It

Why couldn't one person have all three of these alleles?

In some cases, there are far more alleles of one gene in the population. Over 100 alleles for cystic fibrosis have been discovered, for example.

Some alleles are equally dominant, or **codominant** (co-DOM-ih-nent). When both of them are present, both are phenotypically expressed. A person who inherits allele I^A from one parent and allele I^B from the other has blood type AB. These alleles code for enzymes that produce the surface glycoproteins and glycolipids of red blood cells. Type AB means that both A and B glycoproteins are present, and type O means that neither of them is present.

Other alleles exhibit **incomplete dominance.** When two different alleles are present, the phenotype is intermediate between that which either allele would produce alone. For example, *familial hypercholesterolemia*[26]

---

26. *familial* = running in the family + *hyper* = excessive + *cholesterol* + *emia* = blood condition

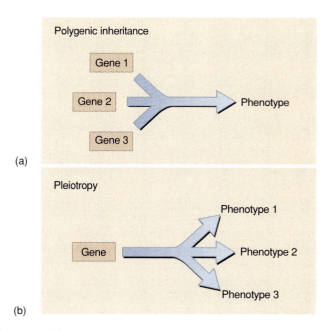

Figure 5.19 (*a*) In polygenic inheritance, two or more separate genes combine their effects to produce a single phenotypic trait, such as skin color. (*b*) In pleiotropy, a single gene causes multiple phenotypic effects, as in sickle-cell disease.

is a hereditary disease in which the blood cholesterol concentration is abnormally high because liver cells lack the receptors needed to remove it from the blood. Individuals with two abnormal alleles die of heart attacks in childhood, those with only one abnormal allele typically die as young adults, and those with two normal alleles have normal life expectancies.

Polygenic Inheritance and Pleiotropy

Polygenic (multiple-gene) inheritance (fig. 5.19*a*) is a phenomenon in which genes at two or more loci, or even on different chromosomes, are required to produce a single phenotypic trait. Human eye and skin colors are normal polygenic traits, for example. Several diseases are also thought to stem from polygenic inheritance, including some forms of alcoholism, mental illness, cancer, and heart disease.

Pleiotropy (ply-OT-roe-pee) (fig. 5.19*b*) is a phenomenon in which one gene produces multiple phenotypic effects. Sickle-cell disease, for example, is caused by a recessive allele that changes just one amino acid in the structure of hemoglobin. It causes red blood cells (RBCs) to assume an abnormally elongated, pointed shape when oxygen levels are low, and it makes them sticky and fragile. As RBCs rupture, a person becomes anemic and the spleen becomes enlarged as it attempts to dispose of the ruptured cells.

Because of the deficiency of RBCs the blood carries insufficient oxygen to the tissues, resulting in far-reaching effects on different parts of the body.

Sex Linkage

Sex-linked traits are carried on the X or Y chromosome and therefore tend to be inherited by one sex more than the other. Certain forms of hemophilia, baldness, and color blindness, for example, are more common in men than women because the causative alleles are recessive and located on the X chromosome (*X-linked*). Men have only one X chromosome and normally express any recessive allele found there. Women, on the other hand, have two X chromosomes; if a recessive allele occurs on one of them, there is a chance that its effects will be masked by a dominant allele on the other. The X chromosome is thought to carry about 260 genes, most of which have nothing to do with determining an individual's sex. There are so few functional genes on the Y chromosome—concerned mainly with development of the testes—that virtually all sex-linked traits are associated with the X chromosome.

Consider red-green color blindness, for example (fig. 5.20), which is caused by an X-linked recessive allele (*c*). The dominant allele (*C*) codes for normal color vision. Red-green color blindness occurs in about 8% of American white males but in only 0.5% of American white females (the percentages are considerably lower among blacks). Even though it is far more common among men, a man can inherit red-green color blindness only from his mother. Why? Because if he inherits *c* on his mother's X chromosome, he will be color-blind. There is no "second chance" to inherit a normal color vision allele on a second X chromosome. A woman, however, gets an X chromosome from each parent. Even if one parent transmits the recessive allele to her, the chances are high that she will inherit an allele for normal color vision from her other parent. She would have to have the extraordinarily bad luck to inherit it from both parents in order for her to be color-blind.

Penetrance and Environmental Effects

People do not inevitably exhibit the phenotypes that would be predicted from their genotypes. For example, there is a dominant allele that causes **polydactyly**,[27] the presence of extra fingers or toes. We might predict that since it is dominant, anyone who inherited the allele would exhibit this trait. Some do, but there are others known to have the allele who have the normal number of digits. If 80% of people with the allele actually exhibit polydactyly, the allele is said to have 80% **penetrance.** Penetrance is the percentage of the population with a given genotype who actually exhibit the predicted phenotype.

Another reason the connection between genotype and phenotype is not inevitable is that environmental factors play an important role in the expression of all genes. At the very least, all gene expression depends on nutrition (fig. 5.21), which provides the building blocks for all bodily structures including DNA itself. Children born with the hereditary disease *phenylketonuria* (FEN-il-KEE-toe-NEW-ree-uh) (PKU), for example, become retarded if they eat a normal diet. However, if PKU is detected early, a child can be placed on a diet low in phenylalanine (an amino acid) and enjoy normal mental development.

No gene can produce any phenotypic effect without nutritional and other environmental input, and no nutrients can produce a body or specific phenotype without genetic instructions that tell cells what to do with them. In short, heredity + environment → phenotype.

Dominant and Recessive Alleles at the Population Level

It is a common misconception that dominant alleles must be more common in the population than recessive alleles. The truth is that dominance and recessiveness

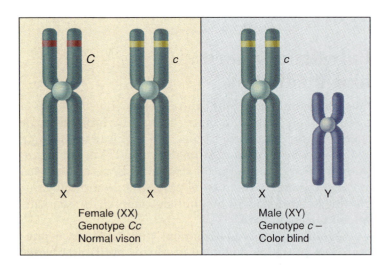

Figure 5.20 Sex-linked inheritance of color blindness. *Left:* A female who inherits a single recessive allele (*c*) for color blindness from one parent may not exhibit the trait, because she is likely to inherit the allele (*c*) for normal color vision from her other parent. *Right:* A male who inherits the recessive allele from his mother will exhibit color blindness, because the Y chromosome inherited from his father does not have a gene locus for color vision and therefore has no potential to mask the effect of the recessive allele on the X chromosome.

27. *poly* = many + *dactyl* = finger, toe

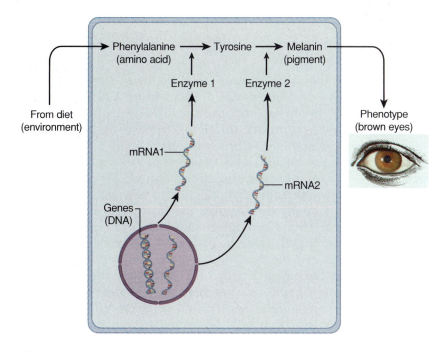

From diet
(environment)

Phenylalanine
(amino acid) → Tyrosine → Melanin
(pigment)

Enzyme 1   Enzyme 2

mRNA1

mRNA2

Genes
(DNA)

Phenotype
(brown eyes)

Figure 5.21 All phenotypes result from a combination of environmental and hereditary causes. Brown eye color, for example, requires dietary phenylalanine and two genetically coded enzymes to convert phenylalanine to melanin, the eye pigment.

have nothing to do with how common an allele is. For example, type O is the most common ABO blood type in North America, but it is caused by the recessive allele *i*. Polydactyly, by contrast, is caused by a dominant allele, yet it is rare in the population.

Definitions of some basic genetic terms are summarized in table 5.5.

Key Point Review

16 Define these terms: *homologous chromosomes, locus, gene,* and *allele.*

17 Why must a carrier of a genetic disease be heterozygous?

18 Distinguish between multiple gene inheritance and pleiotropy.

19 Give at least three reasons why a person's genotype cannot necessarily be inferred from the phenotype.

20 Explain why some traits occur mainly in males, even though they can be inherited only from their mothers.

Table 5.5	The Basic Terminology of Heredity
Term	**Definition**
Homologous pair	A pair of identical-looking chromosomes with the same gene loci but not necessarily the same alleles; one is of maternal origin and the other paternal
Sex chromosomes	A pair of chromosomes (X and Y) that determine an individual's sex
Autosomes	All chromosomes except the sex chromosomes
Gene	A segment of DNA that codes for a protein
Locus	The place on a chromosome where a particular gene is located
Allele	Any of the varieties a given gene may take
Genotype	The pair of alleles an individual possesses at a particular locus or for a particular trait
Phenotype	The detectable traits of an individual correlated with a particular genotype, such as blood type or eye color
Recessive	Pertaining to an allele that is not phenotypically expressed in the presence of a dominant allele; usually represented by a lowercase letter
Dominant	Pertaining to an allele that is phenotypically expressed, even in the presence of any other allele; usually represented by a capital letter
Codominance	A condition in which two alleles are both fully expressed when present in the same individual
Incomplete dominance	A condition in which two alleles are both expressed when present in the same individual, and the phenotype is intermediate between that which either allele would produce alone
Homozygous	Having identical alleles for a given gene
Heterozygous	Having different alleles for a given gene
Carrier	An individual who carries a recessive allele but does not phenotypically express it
Polygenic inheritance	A type of inheritance in which a single phenotype results from the action of genes at two or more loci, as in eye color
Pleiotropy	A situation in which a single gene produces multiple phenotypic effects, as in sickle-cell disease
Sex linkage	Inheritance of a gene on the X or Y chromosome, so the associated phenotype tends to be expressed in one sex more than the other
Penetrance	The percentage of individuals with a given genotype who actually exhibit the phenotype predicted from it

Genetic Engineering

Hemophilia, cystic fibrosis, muscular dystrophy, and sickle-cell disease are a few of the genetic diseases discussed at various places in this book. Historically, genetic diseases have been regarded as incurable, and treatment has centered mainly on controlling symptoms. This is still largely the case, but in the past decade **genetic engineering** has opened the door to new therapeutic approaches and captured the public imagination.

Genetic engineering depends largely on *recombinant DNA technology.* **Recombinant DNA (rDNA)** is any DNA molecule produced by splicing together the DNA of two different species. One benefit that has resulted from this is the commercial production of a blood-clotting enzyme called *factor VIII,* or *antihemophiliac factor A,* which is normally produced by liver cells but is deficient or absent in the most common form of hemophilia. Genes for human factor VIII have been engineered into bacterial cells, and huge cultures of bacteria now synthesize human factor VIII in abundance.

The procedure by which this is accomplished is shown in figure E.1. First, factor VIII mRNA is extracted from human liver cells. The mRNA is used to make DNA by a process that is essentially the opposite of genetic transcription. It relies on an enzyme appropriately named *reverse transcriptase* and results in short segments called **complementary DNA (cDNA).** These segments are copies of the gene for factor VIII, except that they lack the introns, or "junk DNA," of the original gene.

The next thing needed is a vehicle, or **gene vector,** that can get this cDNA into bacterial cells. Bacteria contain small circular DNA molecules called **plasmids.** These are isolated from bacteria and snipped open with **restriction enzymes.** DNA ligase (discussed earlier in connection with its role in DNA replication) is used to join human cDNA to the bacterial plasmid and close the loop. This recombinant plasmid—now composed partly of bacterial DNA and partly of human DNA—is the rDNA molecule. When bacteria are exposed to recombinant plasmids, a few of them take up the rDNA. These can be identified, isolated, and cultured in industrial vats. The bacterium *E. coli* reproduces every 20 to 30 minutes, and when it does, it replicates not only its own DNA but also the human factor VIII gene. Thus, in a short time, millions of copies of the factor VIII gene are produced; hence this procedure is known as **gene amplification,** or **gene cloning.** These cultured bacteria produce human factor VIII that can be isolated from the culture and purified for pharmaceutical use.

The procedure is simple to understand, but in practice it can be tedious and frustrating. It often fails, but the many failures are more than compensated for by the success stories. Besides factor VIII, some valuable products produced in this way include insulin, growth hormone, interferon (a virus-fighting agent), erythropoietin (a hormone that stimulates red blood cell production), and a number of vaccines.

Another exciting dimension of DNA technology is **gene therapy**—introduction of new genes into a patient's body to take over the functions of a defective gene. This, too, sounds simple in principle: Remove specific cells from the body such as bone marrow stem cells, replace the defective gene, reintroduce the cells into the patient's body, and let them multiply and produce a gene product that was formerly lacking—perhaps normal hemoglobin or factor VIII. In reality, there are formidable obstacles at each of these steps, and gene therapy has produced only modest results to date.

First of all, isolating target cells from the patient is difficult. Bone marrow stem cells, for example, are scarce and hard to obtain in quantity. Then, even if the target cells can be isolated and the desired gene cloned, getting the gene into them presents a problem. Unlike bacteria, human cells do not take up plasmids. Something else is needed to serve as a vector. Since viruses insert their genes into other cells as part of their normal life cycle, they have been the most promising candidates: get a functional human gene into a virus, and let the virus bind to the target cells and inject the gene into them. However, it is hard to find viruses that are not only capable of this but that also can be purified and concentrated and still retain their activity. Viruses are too small to carry many of the genes that require replacement therapy—the gene for Duchenne muscular dystrophy, for example, is gigantic. Furthermore, viruses bind only to certain target cells that have receptors for them, and too little is known yet about viral receptors on plasma membranes to design viruses that will deliver DNA to the right cells. It is also hard to control what happens to the DNA when it enters a host cell; for every DNA that inserts at the right place on the right chromosome, a thousand times as many trials result in DNA insertion at the wrong place.

There are still other problems to overcome. For example, some of the viruses with the greatest vector potential may cause cancer if they multiply in the host body, so it is necessary to engineer viruses that will deliver the DNA and then perish without reproducing. In addition, incorporation of foreign DNA into the host cell's DNA requires that the host cells divide, so gene replacement therapy may not be feasible for cells that divide slowly or not at all, such as

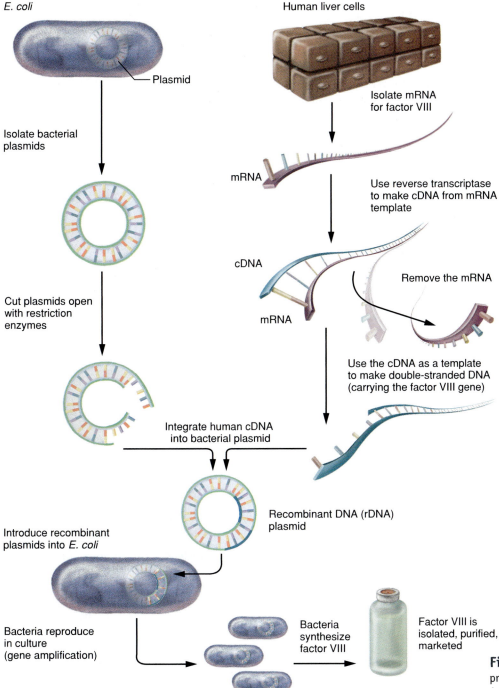

E. coli

Plasmid

Isolate bacterial
plasmids

Cut plasmids open
with restriction
enzymes

Introduce recombinant
plasmids into *E. coli*

Bacteria reproduce
in culture
(gene amplification)

Human liver cells

Isolate mRNA
for factor VIII

mRNA

Use reverse transcriptase
to make cDNA from mRNA
template

cDNA

Remove the mRNA

mRNA

Use the cDNA as a template
to make double-stranded DNA
(carrying the factor VIII gene)

Integrate human cDNA
into bacterial plasmid

Recombinant DNA (rDNA)
plasmid

Bacteria
synthesize
factor VIII

Factor VIII is
isolated, purified,
marketed

Figure E.1 The recombinant DNA technique for
producing gene products such as factor VIII
(a treatment for hemophilia).

nerve and muscle cells. These are just a few of the technical problems confronting molecular biologists in this field. The social and ethical implications of manipulating human genes are yet another consideration.

Despite all the obstacles, gene substitution therapy has scored an apparent success in the treatment of an inherited disease called *severe combined immunodeficiency disease (SCID)*. A child with SCID is born without the ability to ward off infection (see chapter 21). About 25% of cases of SCID are caused by a defective gene for the enzyme adenosine deaminase (ADA). For-

tunately, this gene is small and relatively manageable. The first federally approved clinical trials of gene substitution therapy for ADA deficiency began in September 1990. White blood cells called *T cells* were isolated from children with SCID and viruses were used to introduce normal ADA genes into the cells. The engineered T cells were then reintroduced into the children, where they have produced ADA and allowed them to live normal lives. This is not a lifetime cure, however. These children will require periodic treatment with more T cells as the engineered cells die out.▲

The Nucleus and Nucleic Acids
(pp. 140–145)
1. Nuclear structure
   a. Nuclear envelope and pores
   b. Nucleoplasm and nucleoli
   c. Chromatin

2. Organization of chromatin
   a. Nucleosome structure
   b. Supercoiling of DNA
   c. Nonhistone proteins

3. Nucleotides—monomers of DNA and RNA
   a. Sugar-phosphate backbone
   b. Nitrogenous bases
      • Pyrimidines—single ring
         •• Cytosine (C)
         •• Thymine (T)
         •• Uracil (U)
      • Purines—double ring
         •• Adenine (A)
         •• Guanine (G)

4. DNA structure
   a. Sugar-phosphate backbone
   b. Nitrogenous base
   c. Double helix
   d. Complementary base pairing
      • Purine-pyrimidine pairs
         •• A across from T
         •• G across from C
      • Reasons for these pairs
         •• Size of purine and pyrimidine
         •• Number of hydrogen bonds

5. DNA function
   a. Codes for structure of proteins
   b. Genes and the genome

6. RNA structure and function
   a. Smaller and single-stranded
   b. Ribose replaces deoxyribose
   c. Uracil replaces thymine
   d. Interprets code in DNA
   e. Directs protein synthesis

Protein Synthesis and Secretion
(pp. 145–152)
1. The genetic code
   a. Four nucleotides representing 20 amino acids
   b. DNA base triplets and RNA codons
   c. Stop and start codons

2. Transcription

3. Posttranscriptional modification

4. Movement of mRNA into cytoplasm

5. Translation
   a. Attachment of ribosome to mRNA
   b. Binding to tRNA
   c. Growth of peptide chain
   d. Polyribosomes

6. Chaperones and protein structure

7. Posttranslational modification
   a. Role of leader sequence
   b. Changes occurring in ER cisterna
   c. Formation of transport vesicles
   d. Changes occurring in Golgi complex

8. Packaging and secretion
   a. Golgi vesicles
   b. Secretory vesicles and exocytosis

DNA Replication and the Cell Cycle
(pp. 152–158)
1. DNA replication
   a. Unwinding from histones
   b. Uncoiling by DNA helicase
   c. Replication by DNA polymerase
   d. Splicing by DNA ligase
   e. Rewinding around new histones
   f. Semiconservative nature of DNA replication

2. Errors and mutations
   a. Error rates in bacteria
   b. Proofreading and error correction
   c. Effects of mutations
      • None if new base sequence codes for same amino acid sequence as old
      • None if amino acid substitution is in a noncritical part of a protein
      • Genetic defects if substitution is in a critical part of the protein

3. The cell cycle
   a. G_1 phase: normal cellular functions
   b. S phase: DNA replication
   c. G_2 phase: preparation for mitosis
   d. M phase: nuclear and cytoplasmic division
   e. G_0 phase: cells that have left the cycle

4. Functions of mitosis
   a. Embryonic development
   b. Tissue growth
   c. Replacement of old and dead cells
   d. Repair of injury

5. Stages of mitosis
   a. Prophase
      • Nuclear envelope disintegrates
      • Chromatin condenses into chromosomes
      • Mitotic spindle begins to form
   b. Metaphase
      • Chromosomes line up on equator
      • Spindle fibers attach to centromeres
   c. Anaphase
      • Centromeres divide
      • Sister chromatids move apart

   d. Telophase
      • Chromosomes decondense
      • Nuclear envelopes form
      • Mitotic spindle breaks down

6. Cytokinesis: cytoplasmic division

7. Timing of cell division
   a. CDC proteins present throughout cell cycle
   b. Cyclins appear in M phase
   c. Cyclin-dependent kinases (Cdks)
   d. Mitosis suppressed by contact inhibition

8. Cancer
   a. Tumors and oncology
   b. Benign versus malignant tumors
   c. Causes of cancer
      • Carcinogens—environmental agents
      • Mutations
   d. Growth factors
   e. Oncogenes
      • Involved in embryonic development
      • Reactivated later by carcinogens
      • Code for growth factors or receptors
   f. Tumor suppressor (TS) genes
      • Possible modes of action
      • TS gene mutations and cancer
   g. Effects of cancer
      • Loss of organ function
      • Compression of tissues and organs
      • Energy diversion

Chromosomes and Heredity (pp. 159–163)
1. The karyotype
   a. Twenty-three pairs of homologous chromosomes
      • Autosomes (22 pairs)
      • Sex chromosomes (X and Y)
   b. Germ cells and somatic cells
   c. Haploid and diploid cells
   d. Maternal and paternal chromosomes

2. Genes and alleles
   a. Gene loci
   b. Alleles
   c. Genotype and phenotype
   d. Homozygous and heterozygous genotypes
   e. Dominant and recessive alleles
   f. Heterozygous carriers

3. Multiple alleles, codominance, and incomplete dominance

4. Polygenic inheritance and pleiotropy

5. Sex linkage

6. Penetrance and environmental effects

7. Alleles at the population level

Also review the terms in table 5.5, which are not repeated here.

cytogenetics 140
molecular genetics 140
nuclear envelope 140
nuclear pore 140
nucleoplasm 140
nucleolus 140
chromatin 141
nucleosome 141
histone 141
nonhistone 141
nucleotide 141
nitrogenous base 141
cytosine 141
thymine 141
uracil 141
adenine 141
guanine 141
base pair 144
law of complementary base
  pairing 145
gene 145
genome 145
messenger RNA (mRNA) 146

ribosomal RNA (rRNA) 146
transfer RNA (tRNA) 146
genetic code 146
base triplet 147
codon 147
stop codon 147
start codon 147
transcription 147
RNA polymerase 147
pre-mRNA 148
posttranscriptional
  modification 148
exon 148
intron 148
genetic translation 148
leader sequence 148
anticodon 148
polyribosome 149
chaperone 150
posttranslational modification
  150
transport vesicle 151
Golgi vesicle 151
secretory vesicle 151
DNA helicase 152
DNA polymerase 152

DNA ligase 152
semiconservative replication
  153
mutation 153
cell cycle 154
G_1 phase 154
S phase 154
G_2 phase 154
M phase 154
interphase 154
G_0 phase 154
mitosis 154
prophase 154
chromosome 154
sister chromatids 156
centromere 156
metaphase 156
mitotic spindle 156
anaphase 156
kinetochore 156
telophase 156
cytokinesis 156
CDC protein 156
cyclin 157
cyclin-dependent kinase
  (Cdk) 157

contact inhibition 157
neoplasm 157
oncology 157
benign 157
cancer 157
malignant 157
metastasis 157
carcinogen 157
mutagen 157
growth factor 158
oncogene 158
proto-oncogene 158
tumor suppressor (TS) gene
  158
anaplasia 158
heredity 159
karyotype 159
germ cell 159
somatic cell 159
diploid 159
haploid 159
segregation 159
multiple alleles 161
polydactyly 162

Testing Your Recall — Answers in Appendix C

1. Production of more than one phenotypic trait by a single gene is called
   a. pleiotropy.
   b. genetic determinism.
   c. codominance.
   d. penetrance.
   e. genetic recombination.

2. When a ribosome reads a codon on mRNA, it must bind to the _____ of a corresponding tRNA.
   a. start codon
   b. stop codon
   c. intron
   d. exon
   e. anticodon

3. The normal functions of a liver cell—synthesizing proteins, detoxifying wastes, storing glycogen, and so forth—are done during its
   a. anaphase.
   b. telophase.
   c. G_1 phase.
   d. G_2 phase.
   e. synthesis phase.

4. Two genetically identical strands of a chromosome, joined at the centromere, are its
   a. kinetochores.
   b. centrioles.
   c. sister chromatids.
   d. homologous chromatids.
   e. nucleosomes.

5. Which of the following is *not* found in DNA?
   a. thymine
   b. phosphate
   c. cytosine
   d. deoxyribose
   e. uracil

6. Genetic transcription is performed by
   a. SnRNPs.
   b. RNA polymerase.
   c. DNA polymerase.
   d. DNA ligase.
   e. chaperones.

7. After DNA is replicated, the short segments of new DNA are joined end-to-end by
   a. histones.
   b. DNA ligase.

   c. DNA polymerase.
   d. DNA kinase.
   e. SnRNPs.

8. An allele that is not phenotypically expressed in the presence of an alternative allele of the same gene is said to be
   a. codominant.
   b. lacking penetrance.
   c. heterozygous.
   d. recessive.
   e. subordinate.

9. Semiconservative replication occurs during
   a. transcription.
   b. translation.
   c. posttranslational modification.
   d. the S phase of the cell cycle.
   e. mitosis.

10. A cell is sometimes unharmed by a mutagen for all of the following reasons *except*
   a. Some mutagens are natural harmless products of the cell itself.
   b. Some mutagens are detoxified by peroxisomes before they can do any harm.

c. The body has DNA repair mechanisms that can detect and correct genetic damage.

d. Change in a codon does not always change the amino acid encoded by it.

e. Some mutations change protein structure in ways that are not critical to normal function.

11. The cytoplasmic division at the end of mitosis is called _____.

12. The alternative forms in which a single gene can occur are called _____.

13. The pattern of nitrogenous bases that represents the 20 amino acids of a protein is called the _____.

14. Several ribosomes attached to one mRNA, which they are all transcribing, form a cluster called a/an _____.

15. Environmental agents that can cause cells to become malignant are called _____.

16. Newly synthesized proteins may be escorted to their destination in a cell by other proteins called _____.

17. At prophase, a cell has _____ chromosomes, _____ chromatids, and _____ molecules of DNA.

18. Tumors that can metastasize to other organs are said to be _____.

19. Cells are stimulated to divide when proteins called _____ appear and bind to their CDC proteins.

20. All chromosomes except the sex chromosomes are called _____.

Testing Your Comprehension
Answers in *Study Guide*

1. Reexamine table 4.2 in the previous chapter. In light of what you have learned in the present chapter, explain why nuclear pores are so much larger than pores in the plasma membrane. What is the functional significance of this difference?

2. Suppose the DNA double helix had a backbone of alternating nitrogenous bases and phosphates, with the deoxyribose components facing each other across the middle of the helix. Why couldn't such a molecule function as a genetic code?

3. Given the information in this chapter, present an argument that evolution is not merely possible but inevitable. (Hint: Review the definition of evolution in chapter 1.)

4. What would be the minimum length (approximate number of bases) of an mRNA that coded for a protein 300 amino acids long?

5. Give three examples from this chapter of the complementarity of structure and function.

Web Site Link

For a listing of the most current web sites related to this chapter, please visit the Saladin homepage at:

http://www.mhhe.com/sciencemath/biology/saladin/

chapter six

6

[Histology

Special Topics

The chapters following this one focus on the organ systems of the body. These systems derive their functions not from their cells alone but from how the cells are organized into tissues. The focus of this chapter, then, is **histology**[1]—the study of the organization of cells and extracellular material into tissues and organs. Histology bridges the gap between cytology and gross anatomy and provides a basis for understanding system physiology.

The Study of Tissues

▼Objectives

When you have completed this section, you should be able to
- visualize the three-dimensional shape of a structure from a two-dimensional tissue section;
- name the three embryonic primary germ layers and the adult tissues derived from each; and
- list the four primary categories into which all adult tissues can be classified.

Interpretation of Tissue Sections

The histological specimens you may study have an artificial appearance in some respects, and you must know how to interpret them in order to derive true biological meaning from them. Histologists use numerous "cookbooks" of techniques for preserving, sectioning (slicing), and staining tissues to show their structural details as clearly as possible. Tissue specimens are preserved in a **fixative**—a solution such as formalin that prevents decay but also tends to shrink tissues and sometimes distorts them. After fixation, most tissues are cut into very thin slices called **histological sections,** one or two cells thick, to allow light to pass through and reduce confusion of the image that would result from many layers of overlapping cells. Finally, tissue specimens are artificially colored with histological **stains** to bring out detail. Unless they are stained, most tissues are seen as a very pale gray. With stains that bind to different components of the cells, however, you may see pink cytoplasm, violet nuclei, and blue or green connective tissue fibers, depending on the stain.

Sectioning a tissue reduces a three-dimensional structure to a two-dimensional image. You must keep this in mind and be able to translate the image in the microscope into a three-dimensional mental image of the structure. Furthermore, the two-dimensional appearance of an object can be different depending on the plane in which the section was cut. Figure 6.1*a* and *b* show how different two familiar objects—a boiled egg and a piece of elbow macaroni—would look in different planes of section. A coiled tube, such as a gland of the uterus (fig. 6.1*c*), may appear to be broken into several small segments in a tissue section, since it meanders in and out of the plane of section. Nevertheless, it can be inferred that the visible duct segments are part of a tube leading to the surface.

Many anatomical structures are significantly longer in one direction than another—the humerus and esophagus, for example. A tissue (or whole body) section cut in the long direction is called a **longitudinal section** (abbreviated l.s.). A section at a right angle to the longitudinal section—in the shortest distance across the organ—is called a **cross section** (c.s. or x.s.), or **transverse section** (t.s.). A section cut at an angle between a longitudinal and cross section (on a slant) is called an **oblique section.** Figure 6.2 defines these three sections and shows how a bone and a hollow tube would look in each case.

Think About It

Elastic tissue can occur as stretchy fibers or as sheets similar to layers of lasagna. In a tissue section stained for elastic tissue, how could you tell whether you were looking at a fiber or sheet?

The Primary Germ Layers

All mature tissues of the body arise from three layers of cells in the embryo called the **primary germ layers,** which form in the first 2 weeks after fertilization. One of them, the **ectoderm,**[2] is an outer layer of cells that eventually gives rise to the epidermis and nervous system. An inner layer, the **endoderm,**[3] gives rise to the lining of the digestive tract, associated glands, and the lining of the respiratory tract. After the outer and inner layers form, a layer of more loosely organized cells, the **mesoderm**[4] (MEZ-oh-durm), develops between them and eventually gives rise to muscle, bone, and blood, as well as other tissues. Most organs are composed of tissues arising from two or more primary germ layers.

---

1. *histo* = web

2. *ecto* = outer + *derm* = skin
3. *endo* = inner
4. *meso* = middle

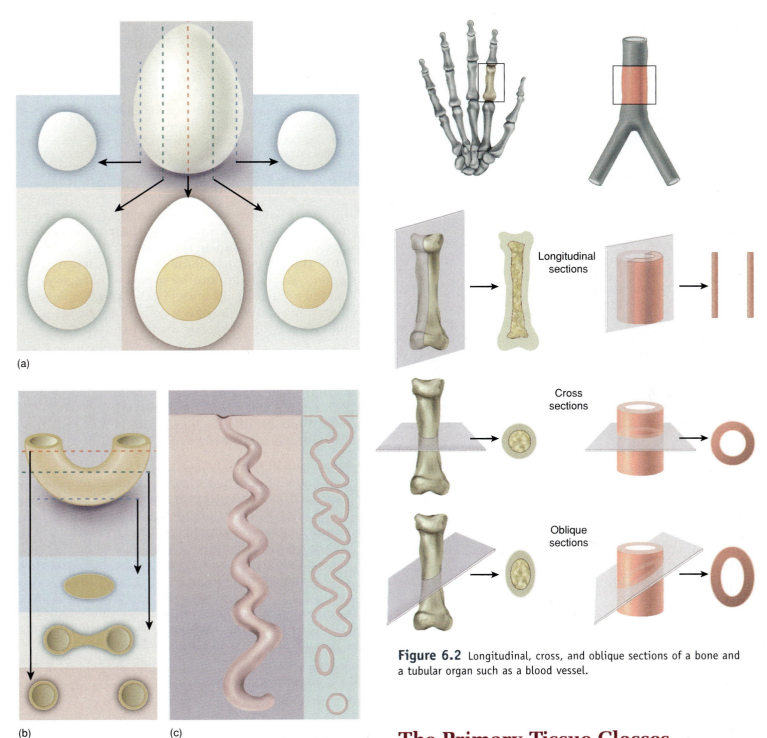

(a)

(b)

(c)

Figure 6.1 Some three-dimensional objects and their appearance in two-dimensional sections. (*a*) A boiled egg. Note that grazing sections (*far left* and *right*) would miss the yolk. (*b*) Elbow macaroni, which resembles many curved anatomical ducts and tubes. A section far from the bend would give the appearance of two separate tubes; a section near the bend shows two lumina (cavities) and the connecting tissue; and a section low in the bend may miss the lumen completely. (*c*) A coiled gland in three-dimensions (*left*) and as it would appear in a vertical tissue section (*right*).

Longitudinal sections

Cross sections

Oblique sections

Figure 6.2 Longitudinal, cross, and oblique sections of a bone and a tubular organ such as a blood vessel.

The Primary Tissue Classes

The human body has approximately 50 trillion (5×10^{13}) cells and thousands of organs—an array of structures that would be utterly bewildering if we could not classify them. This diversity is more comprehensible when we realize that all human structure is composed of just four **primary tissues**—muscular, nervous, connective, and epithelial (table 6.1). In this chapter, we define and characterize these four tissue classes and survey the subclasses of each, so you will be better able to understand the structure and function of organs later.

Table 6.1 The Four Primary Tissue Classes

Type	Definition	Representative Locations
Muscular	Tissue composed of elongated, electrically excitable cells specialized for contraction	Skeletal muscles Cardiac (heart) muscle Muscular walls of viscera
Nervous	Tissue containing electrically excitable cells specialized for rapid transmission of coded information	Brain Spinal cord Nerves
Connective	Tissue with usually more matrix than cell volume, often specialized to support, bind together, and protect organs	Tendons and ligaments Cartilage and bone Blood and lymph
Epithelial	Tissue composed of layers of closely spaced cells that cover organ surfaces and serve for protection, secretion, and absorption	Surface of the skin (epidermis) Inner lining of digestive tract Outer surface of stomach

All tissues consist of two components: cells and the nonliving **extracellular material,** or **matrix,** in which they are embedded. The matrix consists of water; small solutes such as gases, nutrients, and metabolic wastes; and macromolecules such as glycoproteins, polysaccharides, and fibrous proteins. The space between the cells and fibers of a tissue is usually occupied by **tissue fluid,** also called **interstitial** (IN-tur-STISH-ul) **fluid.**

The classes and subclasses of tissue differ from each other in the types and functions of their cells, the composition of their matrix, and the relative volume occupied by cells and matrix. In muscle and epithelium, the cells are so closely packed that the matrix is scarcely visible with the light microscope. In cartilage and tendons, by contrast, the cells are much farther apart, and most of the tissue is matrix.

Key Point Review

1 Define *histology* and explain its place in the hierarchy of structure discussed in chapter 1.

2 Discuss three ways in which the microscopic appearance of a prepared tissue specimen is unnatural.

3 Name the four primary tissue types and concisely state what distinguishes each one from the other three.

Excitable Tissues

▼Objectives

When you have completed this section, you should be able to
- list the functions of muscular tissue;
- explain how the three types of muscular tissue differ from each other;
- describe the basic cell types of nervous tissue; and
- define the major parts of a nerve cell.

All living cells have an electrical charge (voltage) called the *membrane potential* across their plasma membranes. Muscular and nervous tissue are called **excitable tissues** because their cells are specialized to respond quickly to stimuli by means of changes in the membrane potential. In muscle cells, these changes result in contraction; in nerve cells, they result in the rapid transmission of signals to other cells.

Muscular Tissue

Muscular tissue consists of cells specialized for contraction. Most of it arises from the embryonic mesoderm. The general function of muscular tissue is movement, but movement achieves a variety of more specific purposes:

- movement of the body as a whole (locomotion);
- movement of body parts such as a finger or eyelid;
- movement of substances through body passages, as in swallowing food and pumping blood;
- sitting or standing upright (posture) by resisting unwanted movement;
- speech, which involves muscular control of the vocal cords, tongue, and lips;
- respiration, which employs skeletal muscles of the chest and diaphragm to ventilate the lungs; and
- heat production by releasing energy from ATP in the course of shivering and other muscle contractions.

There are three types of muscle—*skeletal, cardiac,* and *smooth.* They differ in microscopic appearance, function, and physiology.

Skeletal Muscle

Skeletal muscle (fig. 6.3*a*) consists of long, cylindrical, parallel cells called **muscle fibers.** Most of it is attached to bones, but there are exceptions such as the diaphragm, wall of the upper esophagus, some facial muscles, and some **sphincter**[5] (SFINK-tur) muscles (ringlike or cufflike muscles that open and close body

---

5. *sphinc* = squeeze, bind tightly

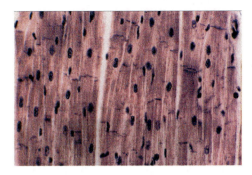
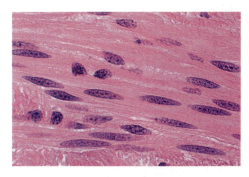

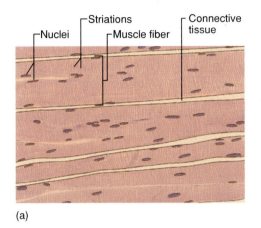

Nuclei — Striations — Muscle fiber — Connective tissue

(a)

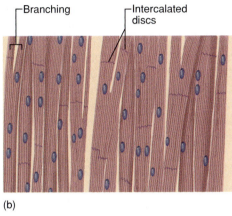

Branching — Intercalated discs

(b)

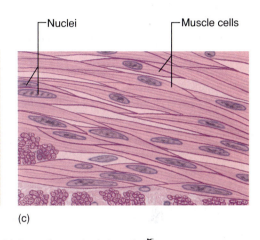

Nuclei — Muscle cells

(c)

Figure 6.3 The three types of muscle. (*a*) Skeletal muscle (×400). (*b*) Cardiac muscle (×1,000). (*c*) Smooth muscle (×1,000). ✗

passages). Each skeletal muscle fiber contains multiple nuclei adjacent to the plasma membrane. Most of the fiber is filled with parallel protein filaments that overlap in such a regular way that the fiber exhibits alternating light and dark bands in stained specimens. These bands are called **striations** (stry-AY-shuns), and skeletal muscle is therefore a type of *striated muscle*. It is also called *voluntary muscle* because almost any skeletal muscle can be contracted voluntarily, and it normally does not contract at all without signals from the nervous system. The term "skeletal muscle" refers not only to this type of *tissue* but also to the muscles—the *organs* composed of this tissue along with connective tissue, blood vessels, and nervous tissue. The word *muscle* literally means "little mouse," apparently referring to the appearance of rippling muscles under the skin.

Think About It

How is the shape of a skeletal muscle fiber adapted to its function? What cell shapes would not serve this function as well?

Cardiac Muscle

Cardiac muscle (fig. 6.3*b*) makes up most of the heart and extends a short distance into some of the major blood vessels connected to it. Like skeletal muscle, it is striated, but it differs in other features. Its cells are much shorter, so they are commonly called **myocytes**[6] rather than muscle fibers. The myocytes are branched and contain only one nucleus, which is located near the center. A light-staining region of glycogen often surrounds the nucleus. Cardiac myocytes are joined end-to-end by junctions called **intercalated**[7] (in-TUR-ku-LAY-tid) **discs.** The electrical connections at these junctions enable a wave of excitation to travel rapidly from cell to cell. Thus, all the myocytes of a heart chamber can be stimulated at once, and they contract almost simultaneously. Intercalated discs appear as dark transverse lines separating each myocyte from the next. They may be only faintly visible, however, unless the tissue has been specially stained for

6. *myo* = muscle + *cyte* = cell
7. *inter* = between + *cala* = inserted

them. Cardiac muscle is considered *involuntary* because it is not usually under conscious control; it contracts even if all nerve connections to it are severed.

Smooth Muscle

Smooth muscle (fig. 6.3c) gets its name from the fact that it lacks striations. Like cardiac muscle, it is *involuntary*—it is not usually subject to conscious control and is able to contract without nervous stimulation. Smooth muscle cells are relatively short and fusiform—thick in the middle and tapered at the ends. Each cell contains a single centrally located nucleus. Smooth muscle forms layers in the digestive, respiratory, and urinary tracts; in the uterus; and in other viscera. The smooth muscle in these organs is also called *visceral muscle.* In locations such as the esophagus and small intestine, smooth muscle forms adjacent layers, with the cells of one layer encircling the organ and the cells of the other layer running longitudinally. When the circular smooth muscle contracts, it may propel contents such as food through an organ, and it tends to make the organ narrower and longer—like squeezing bread dough between your fingers. When the longitudinal layer contracts, it makes the organ shorter and thicker. In the blood vessels, smooth muscle controls vessel diameter, blood pressure, and blood flow. Both smooth and skeletal muscle form sphincters that control the emptying of the bladder and rectum. Small bundles of smooth muscle attach to the hair follicles and make the hair "stand on end" in response to chills or fear.

The three muscle types are compared in table 6.2.

Nervous Tissue

Nervous tissue (fig. 6.4) consists of **neurons** (NOOR-ons) (nerve cells) and a greater number of small cells called **neuroglia**[8] (noo-ROG-lee-uh), or **glial** (GLY-ul) **cells,** that protect and assist the neurons. Neurons are specialized to detect stimuli, respond quickly, and transmit coded information to other cells. Each neuron has a prominent **soma** (cell body) with several fibrous processes extending from it. The soma, which contains the nucleus and other organelles, is the cell's primary site of biochemical synthesis. There are usually multiple processes called **dendrites**[9] that receive information from other sources and transmit it toward the cell body. Dendrites are usually short and highly branched. The soma also has one **axon,** or **nerve fiber,** extending from it. Axons are usually much longer than dendrites and have relatively few branches, except at their distal ends. They conduct electrical signals away from the soma toward another cell and may extend from a few millimeters to more than a meter.

8. *glia* = glue
9. *dendr* = tree, branch + *ite* = little

Table 6.2	Comparison of the Three Types of Muscle Tissue		
	Skeletal	**Cardiac**	**Smooth**
Striated	Yes	Yes	No
Intercalated Discs	No	Yes	No
Cell Shape	Long, cylindrical	Shorter, branched	Shorter, spindle-shaped
Nuclei per Cell	Many	One	One
Position of Nuclei	Near plasma membrane	Central	Central
Control	Voluntary	Involuntary	Involuntary

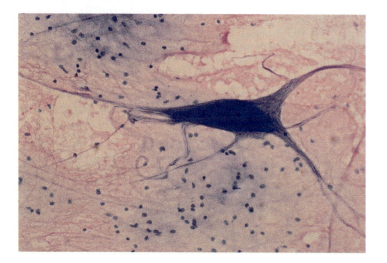

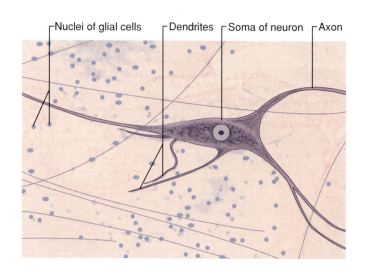

Figure 6.4 Nervous tissue (×200).

4 List at least five functions of muscle tissue.

5 Explain how the three types of muscle tissue can be distinguished from each other. Where can each type be found?

6 Which two types of muscle tissue are striated? Which two types are involuntary? What is meant by the terms *striated* and *involuntary*?

7 Describe the three principal parts of a neuron and state their functions.

Embryonic and Fibrous Connective Tissues

▼**Objectives**
When you have completed this section, you should be able to
• list the properties most connective tissues have in common;
• list the four major types of connective tissue;
• list the types of cells and fibers in connective tissue;
• discuss the composition of the ground substance; and
• describe the various types of embryonic and fibrous connective tissues and state their functions.

Overview of Connective Tissue

Of the primary tissue types, **connective tissue** is the most variable, abundant, and widely distributed. One of the general characteristics of connective tissue is that its cells occupy less space than the extracellular matrix. Many types of connective tissue serve to support other tissues or bind organs together, but there are exceptions. For example, blood does not bind other organs together, but it is produced by another connective tissue (bone marrow) and it consists mostly of extracellular material (blood plasma). Adipose tissue (fat) has more cellular volume than matrix, although most of this volume consists of large lipid droplets.

There are four major subclasses of connective tissue: (1) embryonic connective tissue, (2) fibroconnective tissue, (3) supportive connective tissue, and (4) fluid connective tissue. Examples of their functional diversity are as follows:

• **Binding of organs.** Tendons bind muscle to bone, ligaments bind one bone to another, and other fibrous tissue binds skin to underlying muscle.
• **Support.** Bones support the body as a whole, and cartilage supports the ears, nose, trachea, and bronchi.
• **Movement.** Bones provide the lever system for body movement, and cartilages are involved in movement of the vocal cords.
• **Physical protection.** Such bones as the cranium, ribs, and sternum protect delicate organs such as the brain, lungs, and heart; fatty cushions around the kidneys and eyes protect these organs; mucous connective tissue protects the umbilical blood vessels.

• **Immune defense.** Areolar tissue forms a battlefield in which leukocytes can be easily mobilized against microorganisms that penetrate through the epithelia; immune responses are carried out by leukocytes and other connective tissue cells.
• **Energy storage.** Adipose tissue is the body's major energy reserve.
• **Mineral storage.** Bone provides a reservoir of calcium and phosphorus that can be drawn upon when needed.
• **Heat production.** Brown fat generates heat in infants and children.
• **Transport.** Blood and lymph transport gases, nutrients, wastes, hormones, and blood cells.

Embryonic Connective Tissue

The embryonic mesoderm mentioned earlier gives rise to an embryonic connective tissue called **mesenchyme**[10] (MEZ-en-kime), which in turn produces all permanent connective tissues, as well as most muscle. It exhibits fine, wispy collagen fibers and branching cells called fibroblasts (discussed shortly), embedded in a gelatinous ground substance (fig. 6.5).

The other type of embryonic connective tissue is called **mucous connective tissue.** It looks very much like mesenchyme but is almost entirely limited to a substance called *Wharton's jelly* that fills the umbilical cord and cushions its blood vessels. In contrast to mesenchyme, mucous connective tissue is a temporary tissue. It exists only in the fetus and never gives rise to permanent tissue.

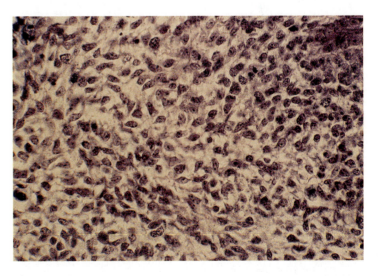

Figure 6.5 Mesenchyme, showing loosely organized cells with delicate processes (×400).

10. *mes* = middle + *enchym* = poured in

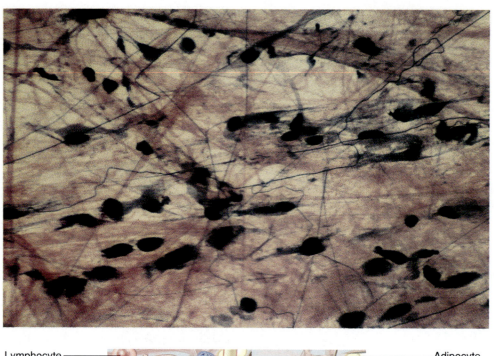

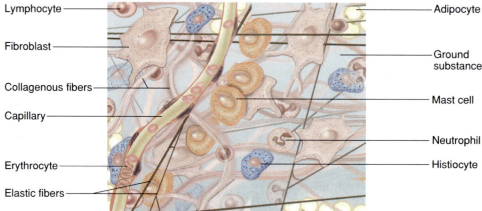

Lymphocyte

Fibroblast

Collagenous fibers

Capillary

Erythrocyte

Elastic fibers

Adipocyte

Ground substance

Mast cell

Neutrophil

Histiocyte

Figure 6.6 Areolar tissue in a spread of the mesentery (×200).

Components of Fibroconnective Tissue

The most widespread and varied type of permanent connective tissue is called **fibroconnective tissue,** or **fibrous connective tissue.** Although nearly all connective tissues contain fibers, they are especially conspicuous in this type.

Cells

As shown in figure 6.6, fibroconnective tissue contains several cell types:

• **Fibroblasts.**[11] These are the most common and the only cells present in tendons and ligaments. They

also occur in the embryonic connective tissues mentioned earlier. Fibroblasts are large flat cells with slender branches; in tissue sections, they often appear tapered at the ends. They produce the fibers and ground substance that together form the matrix of fibroconnective tissue.

• **Histiocytes**[12] (HISS-tee-oh-sites). The body has many types of large phagocytic cells called *macrophages,*[13] discussed most extensively in chapter 21. These cells destroy bacteria and other foreign particles, rid the tissues of dead white blood cells and debris, and help activate other cells of the immune system. The macrophages of fibroconnective tissues are called *histiocytes.* They arise from white blood cells called *monocytes*

11. *fibro* = fiber + *blast* = produce, give rise to

12. *histio* = tissue + *cyte* = cell
13. *macro* = large + *phag* = eat

and are the second most abundant cells in fibroconnective tissue.

- **Leukocytes**[14] (LOO-co-sites), or **white blood cells.** Leukocytes travel briefly in the bloodstream after they are formed, then migrate through the walls of the capillaries into the connective tissues and spend most of their lives there. *Neutrophils* are generally the most abundant leukocytes in fibroconnective tissue. They wander about and attack any bacteria that have invaded the tissue. In the mucous membranes of the respiratory and digestive tracts, among other places, there are often dense masses of *lymphocytes*—tiny leukocytes that react against bacteria, toxins, and other foreign matter.
- **Plasma cells.** These antibody-producing cells develop from certain lymphocytes in response to infections or toxins. They are rarely seen except in the wall of the digestive tract and in inflamed tissue.
- **Mast cells.** These are found especially in the vicinity of blood vessels. They produce **heparin** (HEP-uh-rin), a substance that inhibits blood clotting, and **histamine,** a substance that increases blood flow by dilating the vessels.
- **Adipocytes** (AD-ih-po-sites), or **fat cells.** These appear in small clusters in some types of fibroconnective tissues. Each adipocyte contains a large triglyceride droplet that pushes the cytoplasm and nucleus to the edge of the cell, adjacent to the plasma membrane. The triglyceride is not enclosed in a unit membrane; it is an inclusion, not an organelle. Histological preparation usually dissolves the fat out of these cells, so in routine tissue sections adipocytes tend to appear shriveled and empty.

Fibers

Three types of protein fibers are found in connective tissue: collagenous, reticular, and elastic. **Collagenous** (col-LADJ-eh-nus) **fibers** are tough and flexible and resist stretching. In fresh tissue they often have a glistening white appearance, as in tendons (fig. 6.7) and some meat; thus, they are also called *white fibers.* In tissue sections they tend to form coarse wavy bundles, often colored pink, blue, or green, depending on the stain used. These fibers are made of **collagen** (COLL-uh-jen), which constitutes about 25% of the body's total protein. A molecule of collagen consists of three polypeptide chains, each about a thousand amino acids long, twined around each other in a triple helix called *tropocollagen.*[15] It takes many tropocollagen molecules bundled together to make

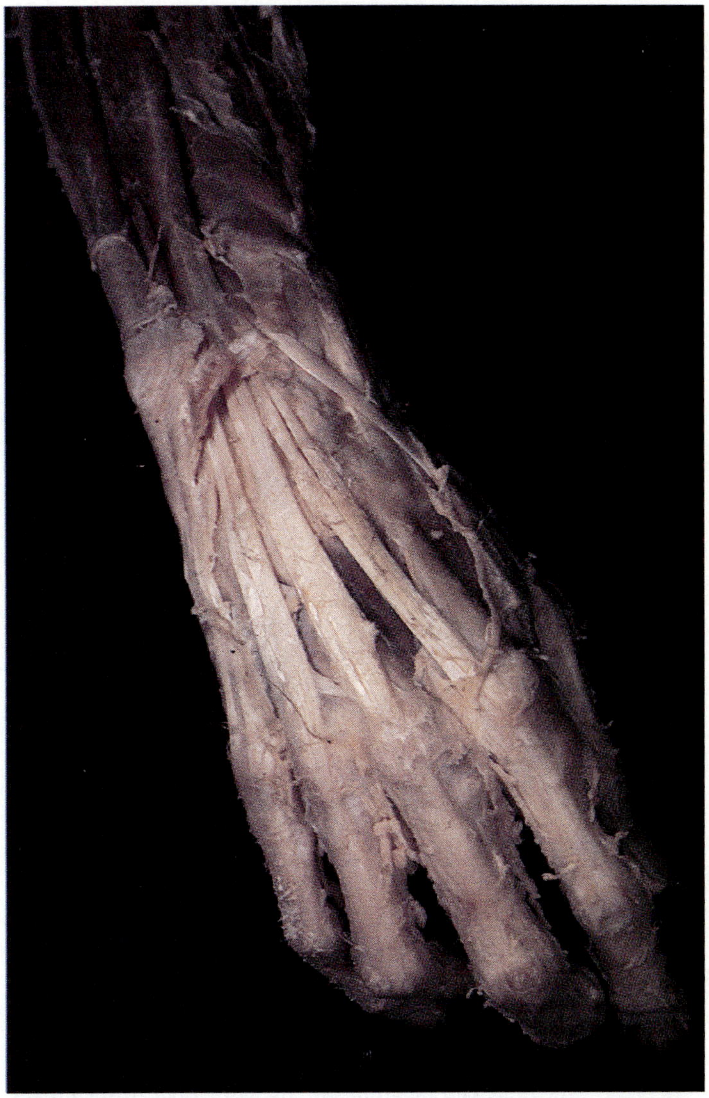

Figure 6.7 Tendons of the hand are composed of collagenous fibroconnective tissue.

even the finest collagen fibers visible with the light microscope. Collagen is the source of such animal products as gelatin, leather, and glue; it is named for the last of these.[16]

Reticular[17] **fibers** are made of thin collagen fibers coated with glycoprotein. They are found especially in reticular connective tissue, where they branch extensively to form delicate networks.

Elastic fibers are thinner than collagenous fibers, and they branch and rejoin each other frequently along their course. They are made of a protein called **elastin,** whose coiled structure allows it to stretch and recoil like a rubber band. Elastic fibers contribute to the ability

14. *leuko* = white
15. *tropo* = turn, turn into

16. *colla* = glue + *gen* = producing
17. *ret* = network + *icul* = little

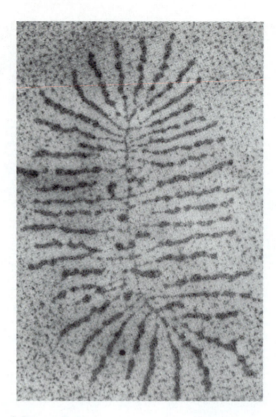

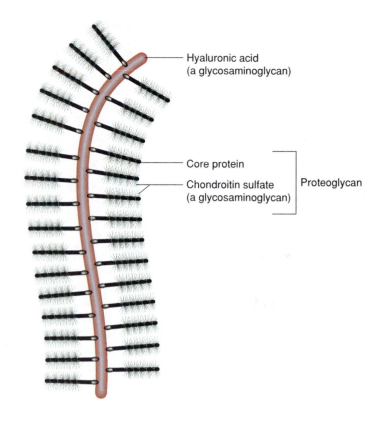

Hyaluronic acid
(a glycosaminoglycan)

Core protein

Chondroitin sulfate
(a glycosaminoglycan)

Proteoglycan

Figure 6.8 The "bottle brush" structure of a proteoglycan.

of the skin, lungs, and arteries to recoil after they have been stretched (the term *elastic* means "able to recoil"). Since fresh elastic fibers appear yellow, they are sometimes called *yellow fibers.*

Think About It

How does the meaning of the word *fiber* differ in the following usages: muscle fiber, nerve fiber, and connective tissue fiber?

Ground Substance

The space amid the cells and fibers of connective tissue looks empty on prepared slides, but it is occupied in life by a featureless mixture of chemicals called the **ground substance.** Components of the ground substance include tissue fluid, minerals, and proteoglycans. Tissue fluid is composed of water and small solutes (nutrients, gases, and wastes), which are transported to and from the cells of the tissue. The minerals are mostly calcium salts, found in bones and teeth. Proteoglycans were briefly introduced in chapter 3 and now warrant a closer look. Like glycoproteins, they are composed of carbohydrate and protein. While a glycoprotein is composed mostly of protein, a proteoglycan is composed mostly of polysaccharides called **glycosaminoglycans** (GLY-cose-am-ih-no-GLY-cans), or **GAGs.** Unlike the polysaccharides surveyed in chap-

ter 3, a GAG is made of modified sugars with amino groups. A proteoglycan consists of numerous GAGs covalently bound to a core protein in a manner reminiscent of a test tube brush (fig. 6.8). The entire proteoglycan may in turn be bound to the collagen and elastin fibers of connective tissue.

GAGs have numerous negatively charged functional groups that bind sodium and potassium ions. These in turn attract water, so GAGs play an important role in the water and electrolyte balance of connective tissues. Such large hydrophilic molecules absorb water and form thick colloids, such as those we see in gravy, pudding, gelatin, and glue. Proteoglycans are especially large colloidal particles. They form a viscous **tissue gel** that stabilizes the arrangement of cells and fibers and functions as a sieve to regulate which substances can diffuse freely through the tissue matrix.

The most common GAG is **chondroitin** (con-DRO-ih-tin) **sulfate.** It is abundant in blood vessels and bones and is responsible for the relative stiffness of the matrix of cartilage. **Hyaluronic** (HY-uh-loo-RON-ic) **acid** is a viscous and slippery GAG that forms much of the matrix of the fibroconnective tissues (see special topic 6.1). It is a very effective lubricant in joints of the skeletal system and constitutes much of the jellylike vitreous humor of the eyeball and the Wharton's jelly of the umbilical cord. Heparin, the previously mentioned anticoagulant from mast cells, is also a GAG.

Hyaluronic acid is a widespread and effective barrier to the spread of toxins and micro-organisms through the connective tissues. Thus, it is not surprising that many infectious organisms and toxins contain an enzyme called *spreading factor,* or *hyaluronidase,* which breaks the connective tissue matrix down to a more watery consistency. Hyaluronidase in bee and snake venoms enables these toxins to spread more quickly through the tissues. It is produced by some bacteria, such as *Staphylococcus aureus* (the cause of "staph" infections), as well as by sperm and white blood cells. Hyaluronidase is included in some subcutaneous drug injections to promote faster drug absorption and reduce pain.

Types of Fibroconnective Tissue

There are two broad categories of fibroconnective tissue, distinguished by their relative abundance of fiber: (1) In **loose connective tissue,** much of the space is occupied by ground substance. This is dissolved by routine histological preparation, so on prepared slides it appears as empty space. The three subclasses of loose fibroconnective tissue are *areolar, adipose,* and *reticular tissue.* (2) In **dense connective tissue,** the fibers are closely packed, and the cells and ground substance occupy relatively little space. The two subclasses of dense connective tissue are *dense regular* and *dense irregular connective tissue,* distinguished by the way in which the fibers are arranged.

Areolar Tissue

Areolar[18] (AIR-ee-OH-lur) **connective tissue** shows a lot of seemingly empty space and loosely organized fibers (see fig. 6.6). All six of the aforementioned cell types are found in areolar tissue, with fibroblasts being the most abundant. The fibers run in random directions and are mostly collagenous, but elastic and reticular fibers are also present. Because of the abundance of open, fluid-filled space, white blood cells can move about freely in areolar tissue and easily find and destroy microorganisms and foreign substances. Areolar tissue is highly vascular (has an abundance of blood vessels), which promotes the arrival and action of white blood cells in times of infection and inflammation.

At least some areolar tissue is found in tissue sections from almost every part of the body. It surrounds blood vessels and nerves and penetrates with them even into the small spaces of muscles, tendons, and other tissues. Nearly every epithelium rests on a layer of areolar tissue and depends on its blood vessels for nutrition and waste removal, since epithelia lack any blood vessels of their own. Most foreign agents enter the body through breaks in its epithelia, but the underlying areolar tissue is a formidable battleground where armies of defensive cells wait to intercept and destroy them. Areolar tissue can be seen very easily beneath the epidermis and the epithelia of the respiratory and digestive tracts. Beneath the epidermis, it forms the superficial portion of the dermis. Beneath the respiratory, digestive, and other mucous membranes, it forms a layer called the *lamina propria* (see fig. 6.23).

Areolar tissue is highly variable in appearance. A stained spread of the *mesenteries*[19] (MESS-en-tare-eez)—membranes that suspend the intestines from the body wall—looks like figure 6.6. Areolar tissue of the skin and mucous membranes, by contrast, is more compact, as shown in figures 6.16 and 6.17. In the skin, especially, it may grade into dense irregular connective tissue, with little perceptible difference between the two tissues. Some advice on how to tell them apart is given after the discussion of dense irregular connective tissue.

Adipose Tissue

Adipocytes occur singly or in small clusters in areolar tissue, but when they become the dominant cell type, the tissue is called **adipose tissue,** or **fat** (fig. 6.9*a*). Adipocytes usually range from 70 to 120 μm in diameter, but they may be five times as large in obese people. The space between adipocytes is occupied by areolar tissue, reticular tissue, and blood capillaries.

Fat is the body's primary energy reservoir. The quantity of stored triglyceride and the number of adipocytes are quite stable in a person, but this doesn't mean stored fat is stagnant. New triglycerides are constantly being synthesized and stored, as others are hydrolyzed and released into circulation. Thus, there is a constant turnover of stored triglyceride, with an equilibrium between synthesis and hydrolysis, energy storage and energy use.

Most body fat is located between the skin and muscle in a layer called *subcutaneous fat.* There are also significant amounts in the bone marrow and mesenteries and around the eyes, heart, kidneys, and spinal cord. Adipose tissue also provides thermal

---

18. *areola* = little space

19. *enter* = gut, intestine

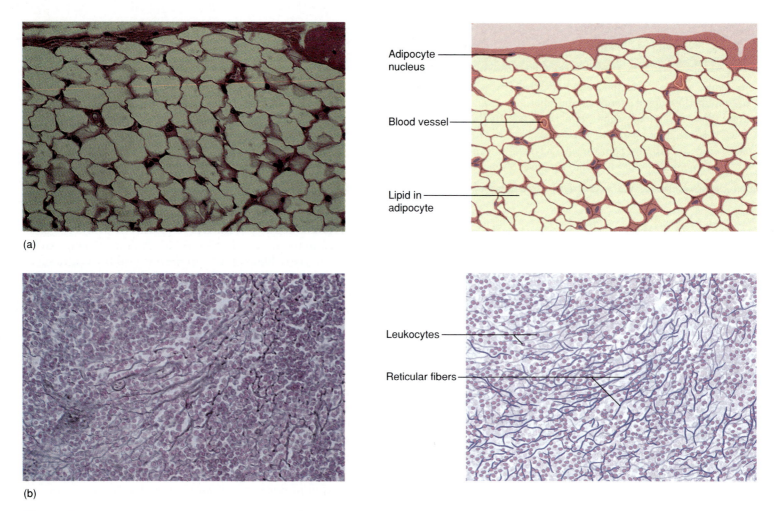

(a)

Adipocyte nucleus

Blood vessel

Lipid in adipocyte

(b)

Leukocytes

Reticular fibers

Figure 6.9 (a) Adipose tissue (×400). (b) Reticular tissue (×400).

Special Topic (Obesity) 6.2

As so many dieters learn, it is very difficult for an adult to lose weight significantly. Obesity is usually untreatable, and most diets are unsuccessful over the long run as dieters repeatedly lose and regain the same weight. A predisposition to obesity seems to be established by overfeeding in infancy and childhood. During childhood, excess calories cause the adipocytes to increase in size and number. Beyond childhood, adipocytes cannot divide. Except in some extreme weight gains, their number seems to be constant, and weight gains and losses are due to changes in cell size. There is some evidence that, in adulthood, fat cells secrete chemicals that stimulate hunger. In rare cases, obesity is caused by hereditary defects in fat metabolism.

Excess weight shortens life expectancy, puts a person at higher risk for cardiovascular disease, complicates surgery, and increases the risk for kidney stones, gallbladder disorders, diabetes mellitus, and uterine and breast cancer. It interferes with breathing and causes carbon dioxide accumulation, which in turn may cause sleepiness and lack of vitality.

insulation, and it contributes to body contours such as the female breasts and hips. It is considered the ideal for a man to have 15% to 18% body fat and a woman to have 20% to 22%. **Obesity** (see special topic 6.2) is regarded as a condition in which a person is more than 20% above the ideal weight for his or her sex, age, and height as a result of excessive fat.

Most adipose tissue is of a type called *white fat,* but fetuses, infants, and children also have a heat-generating tissue called *brown fat.* Brown fat may account for up to 6% of an infant's weight. It gets its color from an unusual abundance of blood vessels and lysosomes. It stores lipid in the form of multiple droplets rather than one large one. Brown fat has numerous mitochondria,

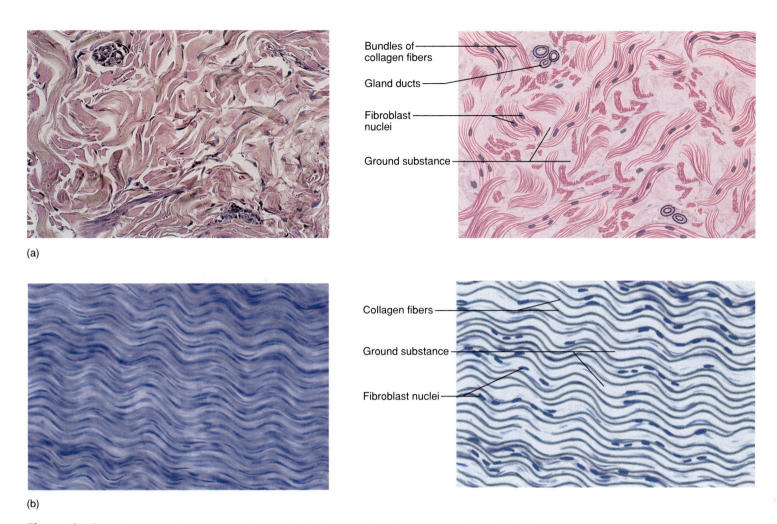

Figure labels:
- Bundles of collagen fibers
- Gland ducts
- Fibroblast nuclei
- Ground substance
- Collagen fibers
- Ground substance
- Fibroblast nuclei

(a)

(b)

Figure 6.10 (a) Dense regular connective tissue (×400). (b) Dense irregular connective tissue (×400).

but their electron-transport system is not linked to ATP synthesis. Therefore, when triglycerides are oxidized in these cells, all the energy is released as heat. Hibernating animals accumulate brown fat in preparation for winter.

 Think About It

Why do infants and children need more brown fat than adults do?

Reticular Tissue

Reticular tissue (fig. 6.9b) is a mesh of reticular fibers and fibroblasts that forms the structural framework (stroma) of such organs and tissues as the lymph nodes, spleen, thymus, and bone marrow. The space amid the fibers is filled with blood cells. If you can imagine a kitchen sponge soaked with blood, the sponge fibers would be analogous to the reticular tissue stroma. Reticular fibers are hard to see with routine stains because they are obscured by the cells. Silver stains turn them brown to black, however, and make them more conspicuous.

Dense Regular Connective Tissue

Dense regular connective tissue (fig. 6.10a) is found especially in tendons and ligaments. It is called "regular" because all the fibers run in the same direction, parallel to each other. Thus, it is adapted to the fact that tendons pull in predictable directions; the direction of stress on the ligaments is also relatively predictable. The only cells in a tendon or ligament are fibroblasts, visible by their slender, violet-staining nuclei, which are squeezed between bundles of collagen. This type of tissue has few blood vessels, and injured tendons and ligaments are therefore slow to heal.

The vocal cords, suspensory ligament of the penis, and some ligaments of the spinal column are made of a type of dense regular connective tissue called **yellow elastic tissue.** In addition to branching elastic fibers, it exhibits densely packed collagen fibers and a greater number of fibroblasts. The fibroblasts have larger, more conspicuous nuclei than are usually found in dense regular connective tissue.

The walls of large and medium arteries consist partly of wavy elastic sheets. When the heart pumps blood into the arteries, these sheets enable them to expand and relieve some of the pressure on smaller vessels downstream. When the heart relaxes, the arterial wall springs back and keeps the blood pressure from dropping too low between heartbeats. The importance of this elastic tissue becomes especially clear when there is not enough of it—for example, in Marfan syndrome (see chapter essay, p. 197)—or when it is compromised by arteriosclerosis (see chapter 19 essay).

Dense Irregular Connective Tissue

Like dense regular connective tissue, **dense irregular connective tissue** (fig. 6.10b) has thick bundles of collagen and relatively little room for cells and ground substance. In the latter, however, the collagen bundles run in random directions, which enables the tissue to resist stresses that may be applied in unpredictable directions. Dense irregular connective tissue constitutes most of the dermis, where it resists stress and binds the skin to the underlying muscle and connective tissue. It forms a protective capsule around organs such as the kidneys, testes, and spleen and a tough fibrous sheath around the bones, nerves, and most cartilages.

It is sometimes difficult to judge whether a tissue is areolar or dense irregular. In the dermis, for example, these tissues occur side by side, and the transition from one to the other is not at all obvious. A relatively large amount of clear space suggests areolar tissue, and thicker bundles of collagen with relatively little clear space suggests dense irregular tissue.

Key Point Review

(8) What are the two principal components of the matrix of connective tissue?

(9) What is a GAG and what are some of its functions?

(10) What three kinds of fibers are found in connective tissue?

(11) Describe four types of cells found in connective tissue and state the functions of each.

(12) List four types of fibrous connective tissue and state where each could be found.

Supportive and Fluid Connective Tissues

▼**Objectives**
When you have completed this section, you should be able to
• explain how the three kinds of cartilage differ in structure and function;
• compare the two types of bone;
• list the connective tissue components of teeth; and
• compare and contrast blood and lymph.

Three types of connective tissue—cartilage, bone, and some of the dental tissues—have a stiff or hard matrix and function mainly to provide physical support for the body or its parts. Cartilage and bone are closely related because cartilage is the precursor of most bones during embryonic development and constitutes a growth zone in the long bones of children and adolescents.

Cartilage

Cartilage (fig. 6.11) is a connective tissue with a flexible rubbery matrix, rich in chondroitin sulfate. It gives shape to the external ear, the tip of the nose, and the larynx (voicebox)—the most easily palpated cartilages in the body. Cells called **chondroblasts**[20] (CON-dro-blasts) produce the matrix and then become trapped in little cavities called **lacunae**[21] (la-CUE-nee). From then on, the cells are called **chondrocytes** (CON-dro-sites). Cartilage is free of blood vessels except when transforming into bone, thus nutrition and waste removal depend on solute diffusion through the stiff matrix. Because this is a slow process, chondrocytes have low rates of metabolism and cell division, and injured cartilage heals slowly. The matrix contains collagen fibers that range in thickness from invisibly fine to conspicuously coarse. Differences in the fibers provide a basis for classifying cartilage into three types: *hyaline cartilage, elastic cartilage,* and *fibrocartilage.*

Hyaline Cartilage

Hyaline[22] (HY-uh-lin) **cartilage** (fig. 6.11a) has a clear, glassy appearance. The fibers are thinly dispersed and usually invisible. The clear ground substance is white or faintly bluish in life and is usually stained light blue in prepared sections. The chondrocytes are often found in little clusters called *cell nests,* consisting of two or more cells in one lacuna. Hyaline cartilage is easy to find in various grocery items—it is the "gristle" at the ends of pork ribs, on chicken leg and breast bones, and at the joints of pigs' feet, for example. In dissections, cartilage can be found at almost any movable joint, where it forms a layer called the *articular cartilage* that enables adjacent bones to glide easily over each other. It also attaches the ends of most ribs to the sternum (breastbone), forms supportive plates and rings around the trachea and bronchi, and forms the "Adam's apple" of the larynx. Most of the skeleton of a fetus begins as hyaline cartilage, and most of the joints (elbows, knees, knuckles, hips, etc.) are still made of cartilage, not bone,

---

20. *chondro* = cartilage, gristle
21. *lacuna* = lake, basin, cavity
22. *hyal* = glass

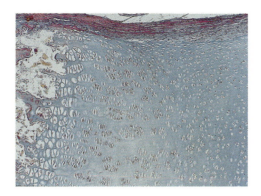 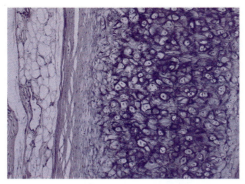 

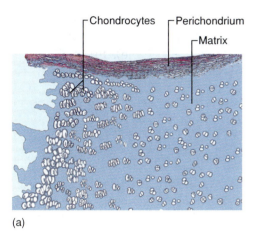 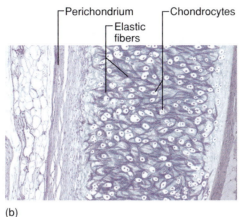 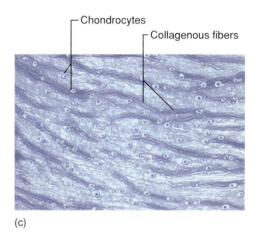

(a)　　　　　　　　　　　(b)　　　　　　　　　　　(c)

Figure 6.11 Cartilage. (*a*) Hyaline cartilage at the head of a developing bone (×100). (*b*) Elastic cartilage from the epiglottis (×100). (*c*) Fibrocartilage from an intervertebral disc (×100). ✗

when a baby is born. Hyaline cartilage is usually covered by a fibrous tissue sheath called **perichondrium**[23] (PERR-ih-CON-dree-um), except at the articular cartilages. A reserve population of chondroblasts between the perichondrium and cartilage contributes to cartilage growth throughout life.

Elastic Cartilage

Elastic cartilage (fig. 6.11*b*) is yellowish in life and always covered by a perichondrium. Elastic fibers arise from the perichondrium and pass through the cartilage matrix, branching and rejoining each other to form a weblike mesh amid the lacunae. Elastic cartilage gives support and shape to the outer ear, epiglottis, and auditory (eustachian) tube.

23. *peri* = around

Fibrocartilage

Fibrocartilage (fig. 6.11*c*) is a transitional tissue between dense fibroconnective tissue and hyaline cartilage. Where a tendon attaches to a bone near a hyaline articular cartilage, for example, there is a zone where the collagenous fibers of the tendon overlap with the hyaline cartilage and produce fibrocartilage. Fibrocartilage exhibits parallel collagen fibers similar to those of a tendon, with chondrocytes often arranged in rows between them. Fibrocartilage never has a perichondrium. The vertebrae (bones of the spinal column) are separated by *intervertebral discs,* each of which consists of a ring of fibrocartilage surrounding a jellylike core. The two halves of the pelvis are joined together in the pubic region by a pad of fibrocartilage called the *pubic symphysis* (SIM-fih-sis) (see special topic 6.3). Between the bones at the knee and some other joints, a pad of fibrocartilage aids in joint motion and absorbs some of the shock of walking and running.

In late pregnancy, many women develop a characteristic waddle in their walk. This is because progesterone softens the pubic symphysis and other cartilaginous joints of the pelvis to allow some movement of the pelvic bones during birth of the baby. Softening of the cartilage also contributes to the backache of late pregnancy.

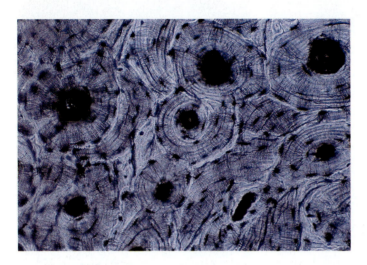

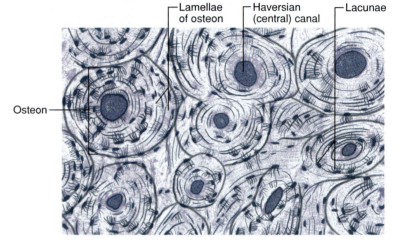

┌Lamellae ┌Haversian ┌Lacunae
 of osteon   (central) canal

Osteon─

Figure 6.12 Compact bone, (osseous tissue), c.s. (×200).

Bone (Osseous Tissue)

There are two kinds of bone, or **osseous tissue** (fig. 6.12): (1) **Spongy bone** fills the heads of the long bones and lines the marrow spaces. Although calcified and hard, its delicate slivers and plates give it a spongelike appearance. (2) **Compact bone** is dense calcified tissue, with no spaces visible to the naked eye. It forms the external surface of all bones, so that spongy bone, if present, is always surrounded by compact bone. The formation of both types of bone and the differences between them are discussed more fully in chapter 8.

The most typical study specimens are cross sections of compact bone. Bone is too hard to slice with a blade, so it is often prepared by drying it, cutting a thin slice with a saw, and then grinding it to microscopic thinness with abrasives. When the specimens are examined with the microscope, they show round to oval regions called **osteons.** At the center of each osteon is a **central canal,** or **haversian**[24] (ha-VUR-zhun) **canal,** surrounded by layers of calcified tissue called **lamellae** (la-MELL-ee) that resemble the layers of an onion slice. Visible between the lamellae are small dark spaces called lacunae that were formerly occupied by mature bone cells called **osteocytes.** All of the osteocytes of one osteon are nourished by blood vessels that travel through its haversian canal. A bone, as a whole, is surrounded by a tough fibrous **periosteum,**[25] (PERR-ee-OSS-tee-um), similar to the perichondrium of cartilage.

About a third of the dry weight of bone is composed of collagen fibers and chondroitin sulfate; two-thirds consists of minerals (mainly calcium salts) deposited around the collagen fibers. Details of bone histology and physiology are discussed in chapter 8.

Dental Tissues

Teeth are not bones, but they are composed mostly of bonelike calcified tissues. Two tissues, the **dentin** and **cementum,** are connective tissues composed of cells, collagen fibers, other proteins, glycosaminoglycans, and calcium salts. Dentin forms most of the bulk of a tooth. It underlies and supports a thinner layer of harder enamel, which is not a connective tissue but a noncellular secretion. Cementum and other tissues bind the root of a tooth to its socket in the jaw. The histology of teeth is discussed in more detail in chapter 25.

24. Clopton Havers (1650–1702), English anatomist

25. *peri* = around + *oste* = bone

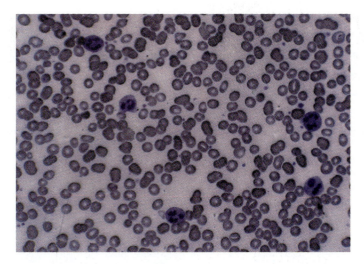

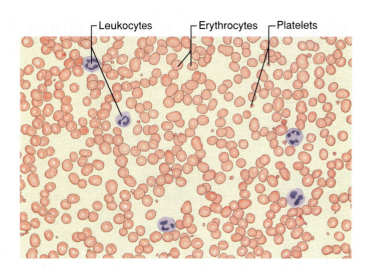

Leukocytes ⎤ Erythrocytes ⎤ Platelets ⎤

Figure 6.13 A blood film (×400).

Blood and Lymph

Blood and lymph are fluid connective tissues that travel through tubular vessels. Their primary function is to transport cells and dissolved matter from place to place, so they are also referred to as *transport tissues.*

Blood (fig. 6.13) consists of a ground substance called **plasma** and of cells and cell fragments collectively called **formed elements. Erythrocytes**[26] (eh-RITH-ro-sites), or red blood cells, are the most abundant formed elements. They look like pale pink doughnuts, but the "doughnut hole" is not truly a hole through the cell; it is a thin spot where the cell once had a nucleus. Erythrocytes transport oxygen and carbon dioxide. Leukocytes, or white blood cells, serve various roles in defense and immunity. They travel from one organ to another in the bloodstream and lymph but spend most of their lives in the fibroconnective tissues. Leukocytes are somewhat larger than erythrocytes, and their conspicuous nuclei usually appear violet in stained preparations. There are five kinds, distinguished partly by variations in nuclear shape. Their individual characteristics are considered in detail in chapter 18. **Platelets** are small cell fragments scattered singly or in clusters amid the blood cells. They are involved in clotting and other mechanisms for minimizing blood loss.

Lymph is a clear fluid that flows through lymphatic vessels. It is formed as these vessels collect tissue fluid from other tissues. Lymphocytes are added as lymph flows through the lymph nodes, but other leukocytes are scarce, and there are no erythrocytes or platelets in the lymph. The lymphatic vessels ultimately lead to two large veins located just beneath the clavicles (collarbones), where the lymph flows into the bloodstream.

The connective tissues are summarized in table 6.3.

Key Point Review

⓭ What is a perichondrium? What type of connective tissue is it made of? What type of connective tissue does it surround?

⓮ Where in the body can each type of cartilage be found?

⓯ What are lacunae? What two types of cells occur in lacunae?

⓰ What are the formed elements of blood? Name the three major types. What is the matrix of blood called?

⓱ List five functions of connective tissue, including a connective tissue responsible for each one.

Epithelial Tissue

▼Objectives

When you have completed this section, you should be able to
- describe the defining characteristics and functions of epithelium;
- describe the basement membrane on which epithelium rests; and
- explain how epithelia are classified and describe the different types.

Epithelial[27] **tissue** covers the surface of the body, lines the body cavities, and forms the external coverings or linings of many organs. Most glands are also composed primarily of epithelium. Epithelia have no blood vessels, but they almost always lie on a layer of areolar connective tissue and depend on its blood vessels for nourishment and waste removal.

An epithelium adheres to the underlying connective tissue by means of a layer called the **basement membrane,** usually too thin to be visible with the light microscope. It contains collagen, a GAG called *heparan sulfate,* and glycoproteins called *laminin* and *fibronectin.* It gradually blends with collagenous and

---

26. *erythro* = red

27. *epi* = upon + *theli* = nipple, tender, female

Table 6.3   Classification and Summary of the Connective Tissues

Type	Appearance in Tissue Sections	Representative Locations
Fibroconnective Tissue	Relatively abundant conspicuous fibers	—
Loose Fibroconnective Tissue	Loosely organized fibers, with ample ground substance	—
Areolar tissue (fig. 6.6)	Loose array of fibers running in random directions; various kinds of widely scattered cells; ample ground substance seen as white space in tissue preparations	Beneath almost every epithelium; dermis of skin; lamina propria of mucous membranes
Adipose tissue (fig. 6.9*a*)	Globose cells with large fat droplets and cytoplasm confined to a thin peripheral rim	Deep to skin; around kidneys, heart, and eyeballs; in mesenteries
Reticular tissue (fig. 6.9*b*)	Branching fibers and cells forming a spongelike matrix	Framework of spleen, lymph nodes, and thymus
Dense Fibroconnective Tissue	Abundant fibers, leaving relatively little space for ground substance	—
Dense regular connective tissue (fig. 6.10*a*)	Closely packed parallel fibers, often wavy; fibroblasts visible as slender nuclei squeezed between collagen fibers	Tendons and ligaments
Yellow elastic tissue	Similar to dense regular, but fibers are elastic, fibroblasts are much more numerous, and fibroblast nuclei are thicker and sausage-shaped	Ligaments of penis, spinal column, and vocal cords
Dense irregular connective tissue (fig. 6.10*b*)	Coarse bundles of collagen running in random directions	Capsules of some viscera; sheaths around bones, nerves, and cartilages
Supportive Connective Tissues	Relatively stiff or hard (calcified) matrix	—
Cartilage	Individual cells, or cells clustered in small cell nests within lacunae, embedded in a stiff rubbery matrix devoid of blood vessels	—
Hyaline cartilage (fig. 6.11*a*)	Clear, glassy matrix; fibers usually too fine to be noticeable	Covering bones at joints, ends of ribs, and trachea; fetal skeleton
Elastic cartilage (fig. 6.11*b*)	Conspicuous elastic fibers between lacunae	Ear, epiglottis, and auditory tube
Fibrocartilage (fig. 6.11*c*)	Bundles of collagen fibers; essentially a zone of overlap between dense regular fibroconnective tissue and hyaline cartilage	Intervertebral discs; pubic symphysis; some tendon-bone junctions and joint cavities
Bone (osseous tissue)	Calcified matrix; cells in lacunae	Skeleton
Spongy bone	Delicate plates and slivers of bone, giving a spongelike appearance	Within heads of long bones, bodies of vertebrae, and lining marrow cavities
Compact bone (fig. 6.12)	Dense white tissue; cells and matrix arranged in concentric cylinders around a haversian canal	Outer surfaces and shafts of bones
Dental Connective Tissue (dentin and cementum)	Dense calcified tissue without osteons or lacunae	Teeth, deep to the enamel
Fluid Connective Tissue	Liquid matrix; suspended cells; no fibers except when clotted	—
Blood (fig. 6.13)	Red and white blood cells scattered through featureless plasma	Within blood vessels, bone marrow, and blood sinuses
Lymph	Clear featureless liquid; contains lymphocytes but few other cells	Within lymphatic vessels and lymph nodes

reticular fibers on the connective tissue side. Epithelia are classified as *simple* or *stratified,* depending on how many cell layers there are and whether or not every cell touches the basement membrane (fig. 6.14).

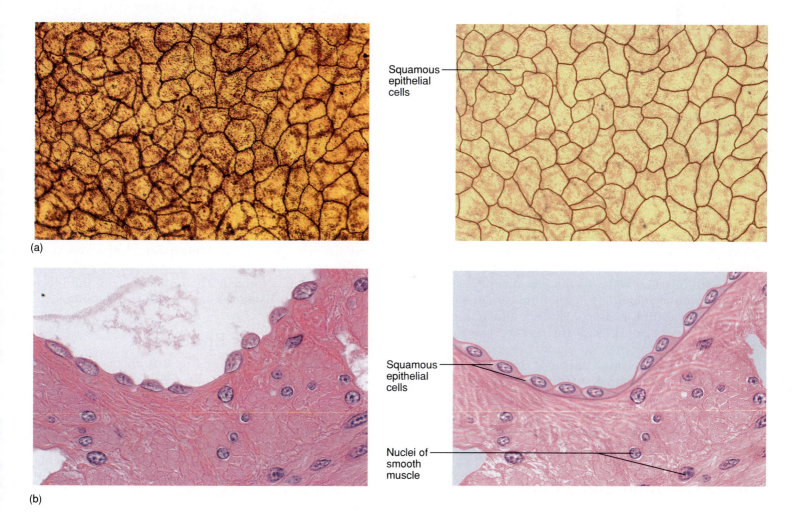

Simple epithelium

Basement membrane

Stratified epithelium

Figure 6.14 Comparison of simple and stratified epithelia.

Simple Epithelia

A **simple epithelium** is composed of one layer of cells, all of which rest on the basement membrane. In most cases, every cell also reaches the free surface of the epithelium. There are four categories of simple epithelium: *simple squamous, simple cuboidal, simple columnar,* and *pseudostratified.*

Simple Squamous Epithelium

Simple squamous epithelium (fig. 6.15) is composed of a single layer of thin cells well adapted to the rapid transport of substances through a membrane. Each cell is shaped like a fried egg, with a broad thin mass of cytoplasm and a bulge where the nucleus is located. The nucleus is usually somewhat flattened in the plane of the cell. Seen from the surface, the cells form irregular polygons and the nucleus looks round or ovoid. In tissue sections where the cells are seen on edge, the epithelium may be too thin to see clearly, except at the nuclear bulge of each cell.

(a)

Squamous epithelial cells

(b)

Squamous epithelial cells

Nuclei of smooth muscle

Figure 6.15 Simple squamous epithelium. (*a*) Surface view of mesothelium (×400). (*b*) Vertical section of intestinal wall (×400). ✗

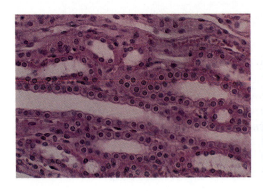
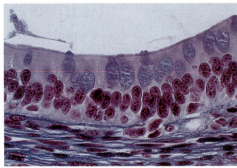
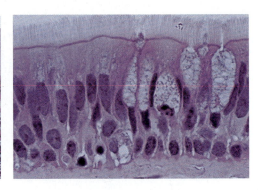

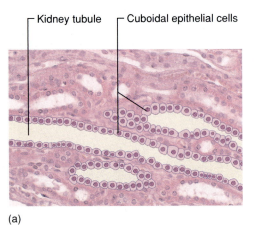
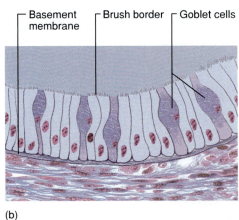
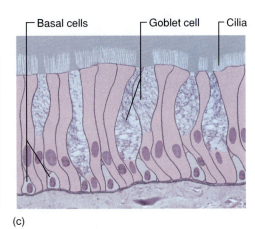

(a) — Kidney tubule — Cuboidal epithelial cells

(b) — Basement membrane — Brush border — Goblet cells

(c) — Basal cells — Goblet cell — Cilia

Figure 6.16 Three types of simple epithelia. (*a*) Simple cuboidal epithelium (×400). (*b*) Simple columnar epithelium (×1,000). (*c*) Pseudostratified epithelium (×1,000).

Simple squamous epithelium occurs where dissolved matter needs to pass through a membrane quickly. For example, it lines the air sacs (alveoli) of the lungs, where rapid diffusion of oxygen and carbon dioxide is essential. The walls of the blood capillaries are made of a simple squamous epithelium called **endothelium** (EN-doe-THEEL-ee-um), which allows the bloodstream to quickly unload nutrients, gases, and hormones, as well as pick up wastes. The endothelium forms a continuous lining through the heart and all blood vessels. The external surfaces of the stomach, intestines, heart, lungs, and some other viscera, as well as the linings of the body cavities, are covered with a moist simple squamous epithelium called the **mesothelium** (MEZ-oh-THEE-lee-um).

Simple Cuboidal Epithelium

Simple cuboidal epithelium (fig. 6.16*a*) consists of a single layer of cells, usually with squarish profiles and round, centrally located nuclei. It is common in glands such as the liver, thyroid, mammary, and salivary glands, where the cells perform a secretory role. It lines many of the kidney tubules, where it secretes and reabsorbs solutes as urine is formed. The bronchioles of the lungs are lined with a ciliated simple cuboidal epithelium.

Simple Columnar Epithelium

Simple columnar epithelium (fig. 6.16*b*) is composed of a single layer of tall, thin cells. The nuclei are usually elongated vertically and located in the basal one-third of the cell. These cells often have a secretory function, with the apical cytoplasm (the cytoplasm above the nucleus) filled with secretory vesicles that are visible with the electron microscope. In the small intestine, they absorb nutrients, and in the uterine tubes they are ciliated and propel an egg or embryo toward the uterus. Other locations include the internal linings of the uterus, stomach and large intestine, and some kidney tubules.

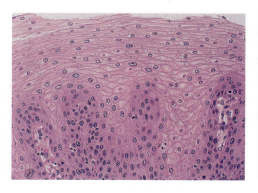

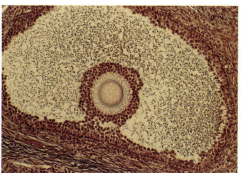

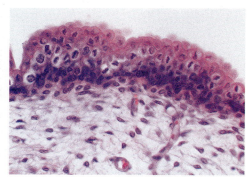

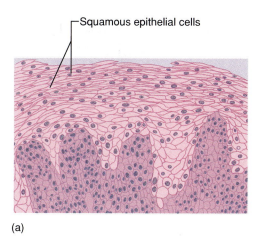

Squamous epithelial cells

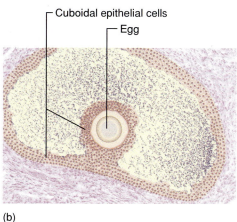

Cuboidal epithelial cells

Egg

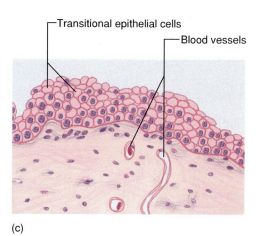

Transitional epithelial cells

Blood vessels

(a)   (b)   (c)

Figure 6.17 Three types of stratified epithelia. (*a*) Stratified squamous epithelium (×400). (*b*) Stratified cuboidal epithelium (×200). (*c*) Transitional epithelium (×400).

Pseudostratified Epithelium

At a glance, pseudostratified[28] epithelium (fig. 6.16*c*) appears to be stratified. Nuclei are seen at many levels of the epithelium, and some cells seem to rest on top of others. It is actually simple, however, not every cell reaches the free surface, but with an electron microscope, it can be seen that every cell does contact the basement membrane, if only by sending fine tendrils between other cells. This epithelium is most abundant in the respiratory tract from the nasal cavity through the trachea and bronchi, where it is conspicuously ciliated and both secretes and propels the respiratory mucus. It also occurs in a nonciliated form in some portions of the male reproductive tract. Pseudostratified epithelium often contains mucus-secreting *goblet cells,* described later in this chapter.

28. *pseudo* = false

Stratified Epithelia

Stratified epithelia (see fig. 6.14*b*) range from 2 to 20 or more layers of cells, with some cells resting directly on others and only the deepest layer resting on the basement membrane. They are classified as *stratified squamous, stratified cuboidal, stratified columnar,* and *transitional epithelium,* mainly by the shape of the surface cells. The deeper cells may be of a different shape than the surface cells.

Stratified Squamous Epithelium

The surface cells of stratified squamous epithelium (fig. 6.17*a*) are flattened, whereas the deeper cells may have cuboidal to low columnar shapes. The deep cells are the only ones close to a blood supply, and as they divide, their daughter cells are pushed farther and farther toward the surface, away from this source of nutrition. The daughter cells therefore begin to die and flatten, until they flake off the epithelial surface. Their

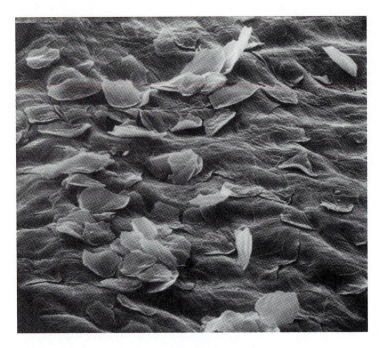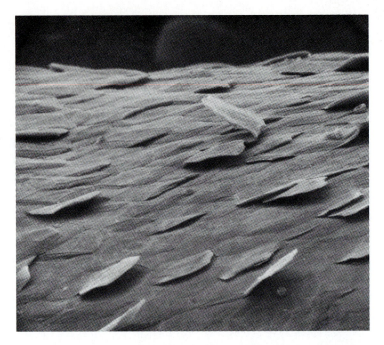

Figure 6.18 The flaking of cells from the surface of a stratified squamous epithelium, called exfoliation or desquamation, is seen here on the epithelium of the (a) vagina, and (b) hard palate. (SEM).

From R. G. Kessel and R. H. Kardon, *Scanning Electron Microscopy of Tissues and Organs,* 1979.

separation from the surface is called *desquamation* or *exfoliation* (fig. 6.18). The study of exfoliated cells, called *exfoliate cytology,* is the basis of the Pap smear, a test for cancer of the cervix. You may have studied some exfoliate cytology yourself if you have done the classic exercise in which you stain and examine cells scraped from the inside of your cheek.

Some stratified squamous epithelia are **keratinized,** meaning that the surface cells have died after having secreted a large amount of the tough protein keratin. The surface thus consists of scaly layers of dead cells, especially that of the epidermis. The tongue, esophagus, vagina, and anal canal, by contrast, are lined by **nonkeratinized** stratified squamous epithelia. Stratified squamous epithelial cells are held tightly together by intercellular junctions (discussed later), enabling this type of epithelium to resist abrasion. It is therefore well suited to the skin surface (where it is especially thick on the palms and soles) and to the passages subject to abrasion by sexual intercourse, the swallowing of food, and the passage of feces.

Stratified Cuboidal Epithelium

Stratified cuboidal epithelium (fig. 6.17*b*), has cuboidal or rounded surface cells. It lines fluid-filled bubbles called *follicles* in the ovaries and sperm-producing ducts called *seminiferous tubules* in the testes. Sweat gland ducts are lined by a stratified cuboidal epithelium

two cells thick—a surface layer of cuboidal cells resting on a basal layer of squamous cells.

Stratified Columnar Epithelium

Stratified columnar epithelium is a scarce type in which columnar surface cells rest on cuboidal basal cells. It is generally found in short transitional zones where a stratified squamous epithelium grades into a columnar or pseudostratified type, as in limited regions of the pharynx, larynx, anal canal, and male urethra.

Transitional Epithelium

Transitional epithelium (fig. 6.17*c*) is limited to the urinary system, where it occurs in part of the kidney, the ureters, the urinary bladder, part of the urethra, and the allantoic duct of the umbilical cord, which is part of the fetal urinary system. It is sometimes confused with stratified squamous epithelium, but the differences can be easily discerned. The surface cells of stratified squamous epithelium are conspicuously flat and are often dead and devoid of nuclei. The surface cells of transitional epithelium are living, have conspicuous nuclei (sometimes two per cell), and are usually cuboidal to globose in shape. They often show markedly convex apical surfaces, like a loaf of home-baked bread bulging from the pan. This type of epithelium is adapted to stretching. When the urinary

Table 6.4 Types of Epithelium

Simple Epithelia	Each cell contacts basement membrane and usually reaches to free surface (except in pseudostratified epithelium)
Simple Squamous (fig. 6.15)	*Description:* Cells flat (scaly or fried egg-shaped)
	Function: Diffusion, secretion, absorption
	Locations: Endothelium, mesothelium, alveoli
Simple Cuboidal (fig. 6.16a)	*Description:* Cells squarish in profile, about equal in height and width, occasionally ciliated
	Function: Secretion, absorption, ciliary movement of mucus
	Locations: Bronchioles, kidney tubules, liver, thyroid, many other glands
Simple Columnar (fig. 6.16b)	*Description:* Cells taller than wide
	Function: Secretion, absorption, ciliary movement
	Locations: Stomach, intestines, uterus, uterine tubes
Pseudostratified (fig. 6.16c)	*Description:* Some cells fail to reach free surface, giving false appearance of multiple layers; all touch basement membrane, however; often ciliated
	Function: Secretion, ciliary movement
	Locations: Nasal cavity, trachea, bronchi
Stratified Epithelia	Some cells rest entirely on top of other cells and do not reach basement membrane
Stratified Squamous (figs. 6.17a and 6.18)	*Description:* Several layers of cells ranging from cuboidal at base to squamous at surface
	Function: Abrasion resistance
	Locations: Epidermis, oral cavity, esophagus, anal canal, vagina
Stratified Cuboidal (fig. 6.17b)	*Description:* Two or more layers of cells, with rounded to cuboidal surface cells
	Function: Sex cell production, lining of ducts
	Locations: Ducts of sweat glands and testes, follicles of ovaries
Stratified Columnar	*Description:* Two or more layers of cells, typically with columnar surface cells resting on cuboidal basal cells
	Function: Transitional zone between stratified squamous and simple columnar or pseudostratified
	Locations: Scarce; some regions of larynx and rectum
Transitional (fig. 6.17c)	*Description:* Two to six or more cell layers, with basal cells more or less cuboidal, and surface cells varying from globose to flattened
	Function: Allows for distension as an organ fills with fluid
	Locations: Parts of urinary system only—renal pelvis, ureters, urinary bladder, parts of urethra

bladder is empty, for example, the epithelium is up to six cells thick. However, as the bladder becomes distended with urine, the epithelial cells slide over each other, the epithelium becomes thinner (only two to three cells thick), and the surface cells flatten somewhat.

The epithelial tissues are summarized in table 6.4.

························· **Key Point Review** ·························

18 Give two reasons why an epithelium almost always rests on a layer of areolar tissue.

19 What is the criterion for dividing all eight kinds of epithelia into two large groups?

 List the eight types of epithelia and state one of the most common functions performed by each.

Intercellular Junctions, Glands, and Membranes

▼Objectives
When you have completed this section, you should be able to
- describe three types of junctions that hold cells together;
- distinguish between endocrine and exocrine glands;
- describe the typical anatomy of an exocrine gland;
- compare the different types of exocrine secretions and the methods by which they are produced;
- describe how tissues are organized into membranes;
- list the body's major membrane types.

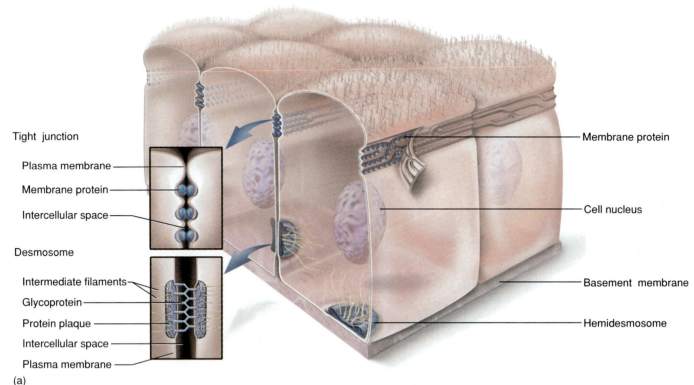

Tight junction
Plasma membrane
Membrane protein
Intercellular space

Desmosome
Intermediate filaments
Glycoprotein
Protein plaque
Intercellular space
Plasma membrane

(a)

Membrane protein
Cell nucleus
Basement membrane
Hemidesmosome

Figure 6.19 (*a*) Intercellular junctions commonly occurring between epithelial cells. (*b*) A gap junction between two cardiac muscle cells. 𝓧

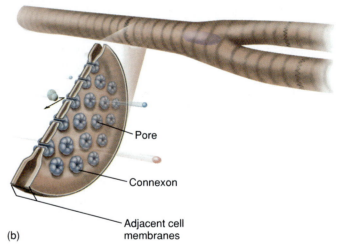

Pore
Connexon
Adjacent cell membranes
(b)

Intercellular Junctions

In some tissues, the cells are firmly attached to each other by various kinds of **intercellular junctions.** These attachments enable the cells to resist stresses or communicate with each other. Without them, cardiac muscle cells would pull apart when they contracted, and every swallow of food would scrape away the lining of your esophagus. The principal types of intercellular junctions are shown in figure 6.19.

Tight Junctions

A **tight junction** completely encircles an epithelial cell near its apex and joins it tightly to the neighboring cells. Proteins in the membranes of two adjacent cells form a zipperlike pattern of complementary grooves and ridges. This seals off the intercellular space and makes it difficult for some substances to pass between the cells. In the stomach and intestines, tight junctions prevent digestive juices from seeping between epithelial cells and digesting the underlying connective tissue. They also help to prevent intestinal bacteria from invading the tissues, and they ensure that most digested nutrients pass *through* the epithelial cells and not *between* them.

Desmosomes

A **desmosome**[29] (DEZ-mo-some) is like a "spot weld" between two cells, a patch that holds cells together and enables a tissue to resist mechanical stress. Desmosomes are common in the epidermis, cardiac muscle, and cervix of the uterus. The neighboring cells are separated by a small gap, which is spanned by a fine mesh of glycoprotein filaments. These filaments terminate in a thickened protein plaque at the surface of each cell. On the cytoplasmic side of each plaque, intermediate filaments from the cytoskeleton approach and penetrate the plaque, turn like a J, and reach a short distance back

---

29. *desmo* = band, bond, ligament + *som* = body

into the cytoplasm. Each cell contributes half of the complete junction. The basal cells of epithelial tissue have *hemidesmosomes*—half-desmosomes that anchor them to the underlying basement membrane.

Think About It

Why would a desmosome not be a suitable intercellular junction for the epithelium of the stomach?

Gap (Communicating) Junctions

A **gap junction** is formed by a ringlike *connexon*, which consists of six transmembrane proteins surrounding a water-filled channel. Ions, glucose, amino acids, and other small solutes can pass directly from the cytoplasm of one cell into the next through these channels. In the embryo, nutrients pass from cell to cell through gap junctions until the circulatory system forms and takes over the role of nutrient distribution. Gap junctions are found in the intercalated discs of cardiac muscle and between the cells of some smooth muscle. The flow of ions through these junctions allows electrical excitation to pass directly from cell to cell, so that the cells contract in near-unison. Gap junctions are absent from skeletal muscle.

Glands

A **gland** is a cell or organ that synthesizes substances or that removes and modifies substances from the tissues, secretes them for use elsewhere in the body, or releases them for elimination from the body. Multicellular glands are composed predominantly of epithelial tissue.

Endocrine and Exocrine Glands

Glands are broadly classified as endocrine or exocrine. **Endocrine**[30] (EN-doe-crin) **glands** have a high density of blood capillaries and secrete their products directly into the blood. They have no tubular passageways (ducts) to carry the products to their destinations. The secretions of endocrine glands, called *hormones,* function as chemical messengers to stimulate cells elsewhere in the body. Endocrine glands are the subject of chapter 17 and are not further considered here. **Exocrine**[31] (EC-so-crin) **glands,** by contrast, usually have ducts that convey secretions to their destinations, and these secretions do not function as hormones. The destination of a secretion may be the body surface, as in the case of sweat, mammary, and tear glands, but more often it is the cavity (lumen) of another organ, such as the mouth or intestine.

---

30. *endo* = in, into + *crin* = to separate, secrete
31. *exo* = out

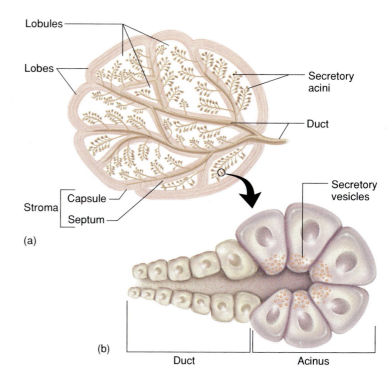

Figure 6.20 (*a*) General structure of a typical exocrine gland. The gland duct branches into finer and finer divisions, at first following the connective tissue septa between lobes and lobules and finally terminating in saccular acini. (*b*) An acinus, where the gland product is synthesized and secreted into the duct system. ✗

These distinctions are not always clear. The liver is an exocrine gland that secretes one of its products, bile, through a system of ducts but secretes other products directly into the bloodstream. Several glands have both exocrine and endocrine cells, such as the pancreas, testis, ovary, and kidney; and nearly all of the viscera have at least some cells that secrete hormones, even though most of these organs are not usually thought of as glands (for example, the heart).

Unicellular glands are exocrine cells found in an epithelium that is predominantly nonsecretory. For example, the respiratory tract, which is lined mainly by ciliated cells, also has a liberal scattering of nonciliated **goblet cells.** These are mucus-secreting gland cells that get their name from their resemblance to a wine goblet (see fig. 6.23).

Exocrine Gland Structure

Figure 6.20 shows a generalized multicellular exocrine gland—a structural arrangement found in such organs as the mammary gland, pancreas, and salivary glands. Most glands are enclosed in a fibrous **capsule.** The capsule often gives off extensions called **septa,** or **trabeculae** (trah-BEC-you-lee), that divide the interior of the gland into compartments called **lobes,** which are visible to the naked eye. Finer connective tissue septa may further

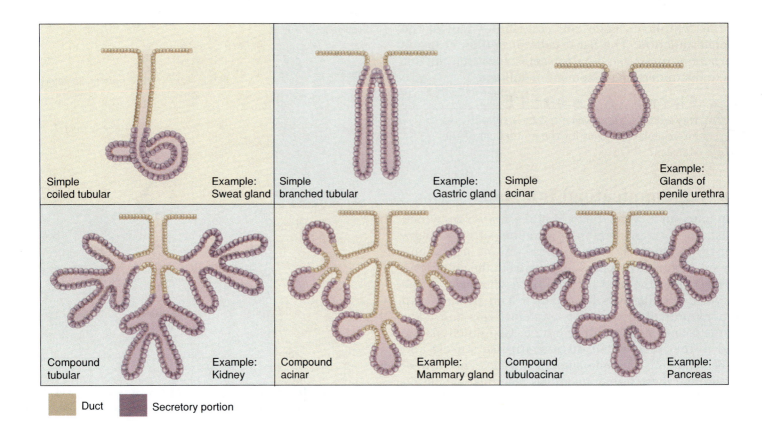

Figure 6.21 Some exocrine gland types classified according to whether the duct system branches (simple versus compound glands) and according to the distribution of secretory versus ductile epithelium (acinar, tubular, and tubuloacinar glands).

subdivide each lobe into microscopic **lobules** (LOB-yools). Blood vessels, nerves, and the gland's own ducts generally travel through these septa. The connective tissue framework of the gland, called its **stroma,** supports and organizes the glandular tissue. The cells that perform the tasks of synthesis and secretion are collectively called the **parenchyma** (pa-REN-kih-muh). This is typically simple cuboidal or simple columnar epithelium.

Ducts are epithelium-lined tubes that convey a gland product away from the secretory cells. Exocrine glands are classified as **simple** if they have a single unbranched duct and **compound** if they have a branched duct. If the duct and secretory portion are of uniform diameter, the gland is called **tubular.** If the secretory cells form a dilated sac at the end of the duct, the gland is called **acinar** and the sac is an **acinus**[32] (ASS-ih-nus) or **alveolus**[33] (AL-vee-OH-lus). Figure 6.21 illustrates some of the ways in which these terms are applied.

Types of Secretions

Glands are classified as serous, mucous, or mixed depending on the nature of their secretions. **Serous** (SEER-us) **glands** produce relatively thin, watery fluids such as per-

spiration, milk, tears, and digestive juices. **Mucous glands** secrete a glycoprotein called *mucin* (MEW-sin) that mixes with water after it is secreted and forms the sticky product *mucus.* (Note that *mucus,* the secretion, is spelled differently from *mucous,* the adjective form of the word.) **Mixed glands** contain both serous and mucous cells and produce a mixture of the two types of secretions. The testes and ovaries are called **cytogenic**[34] **glands** because their secretions are the sperm and egg cells.

Methods of Secretion

Glands are classified as merocrine or holocrine depending on how they produce their secretions. **Merocrine**[35] (MERR-oh-crin) **glands,** also called **eccrine**[36] (EC-rin) **glands,** have secretory vesicles that release their secretion by exocytosis, as described in chapter 4 (fig. 6.22*a*). These include the tear glands, pancreas, gastric glands, and many others. In **holocrine**[37] **glands,** cells accumulate a product and then the entire cell disintegrates, so the secretion is a mixture of cell fragments and the substance the cell had synthe-

32. *acinus* = berry
33. *alveol* = cavity, pit

34. *cyto* = cell + *genic* = producing
35. *mero* = part + *crin* = to separate
36. *ec* = *ex* = out
37. *holo* = whole, entire

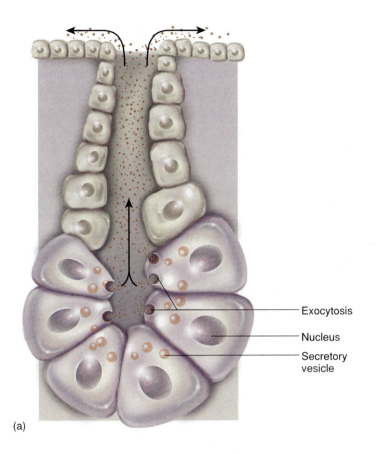

(a)

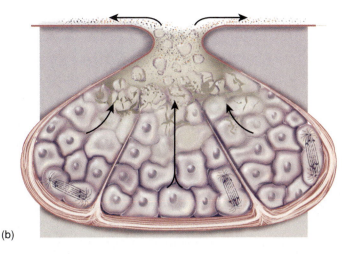

(b)

Figure 6.22 Methods of exocrine secretion. (*a*) An eccrine gland, which secretes its products by exocytosis at the apical cell surface. (*b*) A holocrine gland, whose secretion is formed by the breakdown of whole cells in the center of the gland. These cells are replaced by mitosis of the peripheral cells.

Labels in figure (a): Exocytosis, Nucleus, Secretory vesicle

sized prior to its disintegration (fig. 6.22*b*). The oil-producing glands of the scalp are an example. Holocrine secretions tend to be thicker than merocrine secretions.

Some glands, such as the axillary (armpit) sweat glands and mammary glands, are named **apocrine**[38] **glands** from a former belief that the secretion was produced by bits of apical cytoplasm breaking away from the cell. Closer study showed that these glands employ the merocrine method of secretion and that the appearance of the glands was merely an artifact (an artificial appearance caused by the method of tissue preparation, not a natural feature of the tissue). These glands are nevertheless different from eccrine glands in function and histological appearance, and they are still referred to as apocrine glands even though this is not a unique mode of secretion.

Membranes

In Atlas A the major cavities of the body were described, as well as some of the membranes that line them and cover their viscera. We now consider some histological aspects of the major body membranes.

The largest membrane of the body is the **cutaneous** (cue-TAY-nee-us) **membrane**—or more simply, the skin. It

consists of a stratified squamous epithelium (epidermis) resting on a layer of fibroconnective tissue (dermis). Unlike the other membranes to be considered, it is relatively dry. It retards dehydration of the body and provides an inhospitable environment for the growth of infectious organisms.

The two principal kinds of internal membranes are mucous and serous membranes. **Mucous membranes** (fig. 6.23), also called **mucosae** (mew-CO-see), line passageways that open to the exterior environment: the digestive, respiratory, urinary, and reproductive tracts. A mucosa consists of two to three layers: (1) an epithelium, (2) an areolar connective tissue layer called the **lamina propria**[39] (LAM-ih-nuh PRO-pree-uh), and sometimes (3) a layer of smooth muscle called the **muscularis** (MUSK-you-LAIR-iss) **mucosae.** Mucous membranes have absorptive, secretory, and protective functions. They are covered with mucus secreted by goblet cells, multicellular mucous glands, or both. The mucus traps bacteria and foreign particles, keeping them from invading the tissues and aiding in their removal from the body. The epithelium of a mucous membrane may also include absorptive, ciliated, and other types of cells.

A **serous membrane (serosa)** is composed of a simple squamous mesothelium resting on a thin layer of areolar connective tissue. Serous membranes produce watery **serous** (SEER-us) **fluid,** which arises from the blood. It derives its name from the fact that it is similar to blood serum in composition. Serous membranes line the insides of some body cavities and form a smooth surface on the outer surfaces of some of the viscera, such as the digestive tract. The pleurae, pericardium, and peritoneum described in Atlas A are serous membranes.

---

38. *apo* = from, off, away

39. *lamina* = layer + *propri* = one's own

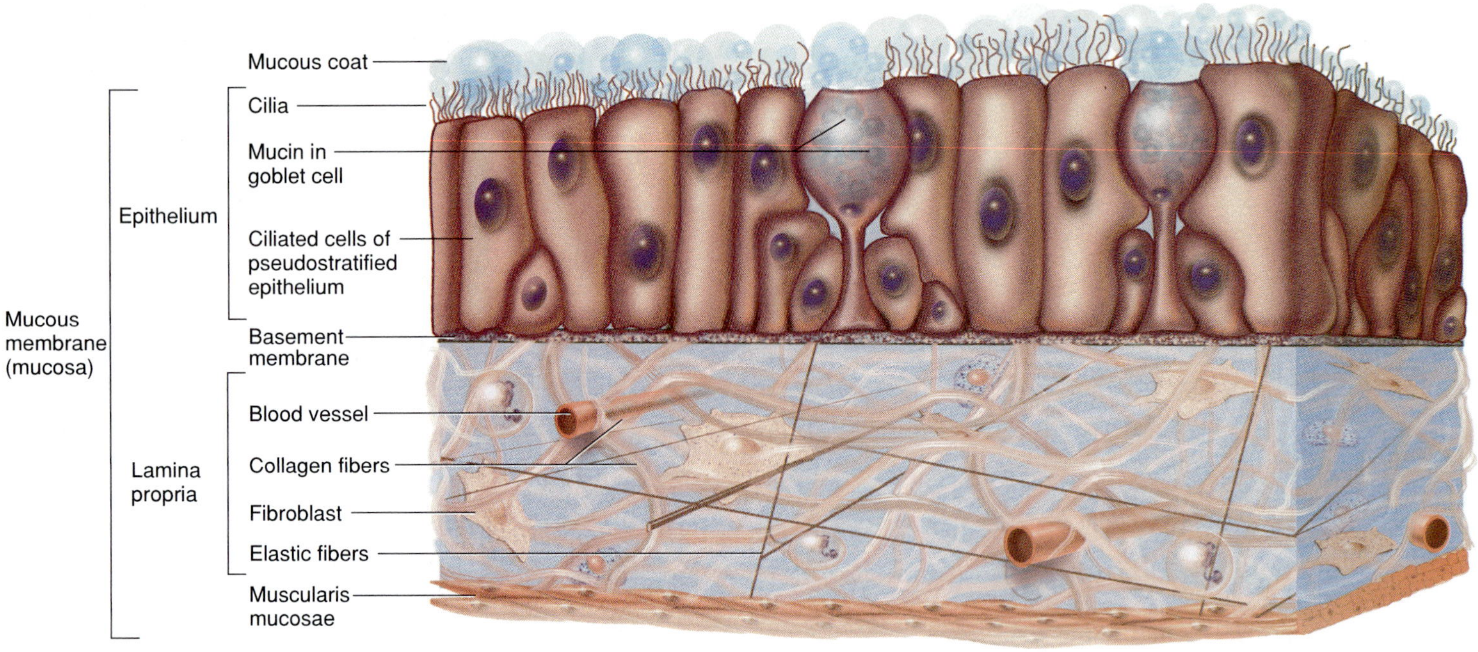

Mucous coat
Cilia
Mucin in goblet cell
Epithelium
Ciliated cells of pseudostratified epithelium
Mucous membrane (mucosa)
Basement membrane
Blood vessel
Lamina propria
Collagen fibers
Fibroblast
Elastic fibers
Muscularis mucosae

Figure 6.23 Histology of a mucous membrane such as the lining of the respiratory tract.

The circulatory system is lined with a simple squamous epithelium called endothelium, derived from endoderm. The endothelium rests on a thin layer of areolar tissue, which often rests in turn on an elastic sheet. Collectively, these tissues make up a membrane called the *tunica interna* of the blood vessels and *endocardium* of the heart.

Some joints of the skeletal system are lined by fibrous **synovial** (sih-NO-vee-ul) **membranes.** These membranes span the gap from one bone to the next and secrete slippery *synovial fluid* (rich in hyaluronic acid) into the joint.

Key Point Review

21 Describe the structure of tight junctions and gap junctions. In each case, relate structure to function.

22 Distinguish a simple gland from a compound gland and a tubular gland from an acinar gland.

23 Contrast the merocrine and holocrine methods of secretion.

24 What are the differences between mucous and serous membranes?

Tissue Transformations

▼Objectives

When you have completed this section, you should be able to

- describe the growth processes, degenerative changes, and other transformations that take place in tissues;
- explain how damaged tissue is repaired.

Tissues exhibit many changes over the course of their lifetime. Unspecialized embryonic tissues undergo **differentiation,** giving rise to tissues specialized to perform specific functions. As mentioned in chapter 5, cancer cells and some other types reverse this process, undergoing *anaplasia,* or *dedifferentiation,* reverting to unspecialized cells resembling those of the embryo.

Tissue growth due to the enlargement of already-existing cells is called **hypertrophy**[40] (hy-PUR-truh-fee), whereas growth due to cellular multiplication is called **hyperplasia** (HY-pur-PLAY-zhuh). In athletic conditioning, for example, muscle growth is not due primarily to multiplication of muscle cells (which are incapable of mitosis); rather, it is due to hypertrophy of cells that already exist. The development of a tumor (neoplasm) composed of abnormal, nonfunctional tissue is called **neoplasia**[41] (NEE-oh-PLAY-zee-uh).

Metaplasia[42] is the transformation of one tissue type to another. For example, in a young girl the vagina is lined with simple cuboidal epithelium, but in puberty this changes to stratified squamous epithelium, which is better adapted for intercourse. The nasal cavity is lined mainly with ciliated pseudostratified epithelium, but if one nostril is blocked so that a person is forced to breathe through the other one for several days, patches of epithelium in the unblocked nasal cavity change to stratified squamous.

The shrinkage of a tissue is called **atrophy**[43] (AT-ruh-fee). *Senescence atrophy* is due to normal aging; *disuse atrophy* is due to lack of use. Muscles exhibit

40. *hyper* = excessive + *trophy* = nourishment
41. *neo* = new
42. *meta* = change + *plasia* = formation, growth
43. *a* = without + *trophy* = nourishment

disuse atrophy, for example, when a broken limb is set in a cast. As another example, unless astronauts maintain a program of resistance exercises, their muscles exhibit disuse atrophy in prolonged space flights. The first astronauts to remain in space for several days were too weak to walk across the deck of a recovery ship when they returned to normal gravity.

The pathological death of tissue is called **necrosis**[44] (neh-CRO-sis). It may result from infection, lack of nourishment, or inadequate waste removal. If a patient remains motionless in a bed or wheelchair for several days, for example, pressure on the skin cuts off blood flow, and the tissue in that region dies—sometimes all the way to the bone. This produces a *decubitus* (de-CUE-bih-tus) *ulcer,* or bedsore. Local tissue death that is due to an inadequate blood supply is called **infarction** (in-FARK-shun). When it occurs in heart muscle (*myocardial infarction*), it can lead to the functional irregularities colloquially known as a heart attack. More massive necrosis due to infection, injury, or lack of

blood supply is called **gangrene** (GANG-green). A common cause of gangrene is poor circulation stemming from diabetes mellitus. Whatever the cause, gangrene often involves bacterial invasion of the decaying tissue and may require amputation to halt its spread. Massive crushing injuries and contamination with dirt and anaerobic bacteria may cause *gas gangrene,* characterized by decay, foul-smelling discharge, and gas bubbles under the skin.

Damaged tissues can be repaired by regeneration, fibrosis, or a combination of these. **Regeneration** is replacement of the damaged tissue with the same kind of cells that previously existed, and **fibrosis** is replacement with scar tissue, composed mainly of collagen produced by fibroblasts.

Key Point Review

(25) Define *hypertrophy, neoplasia, hyperplasia,* and *metaplasia.*

(26) How does atrophy differ from necrosis?

CHAPTER ESSAY

Connective Tissue Diseases

Collagen and elastin contribute greatly to the physical support and resilience of tissues and organs. Organs with defective or scanty fibers may be weakened, with consequences ranging from mild deformity to death.

Marfan[45] **syndrome** occurs in about 1 out of 20,000 live births. This disease is usually due to the mutation of a gene on chromosome 15 for fibrillin—a glycoprotein that forms the scaffold for elastin fibers. Marfan syndrome involves mainly the skeleton, eyes, and cardiovascular system. Clinical symptoms include unusually tall structure, long limbs and spidery fingers, abnormal spinal curvature, and a protruding "pigeon breast." Some authorities postulate that Abraham Lincoln's tall, gangly physique and spindly fingers were signs of Marfan syndrome, which perhaps would have killed him prematurely had he not been assassinated. Hyperextensible joints and hernias of the groin are other common symptoms. The eyeballs are weakened and become abnormally long, and the lenses may be deformed or displaced, causing visual impairments. Most seriously, however, the heart valves and elastic walls of the blood vessels are weakened. The aorta, where blood pressure is highest, is sometimes greatly dilated close to the heart and can suddenly tear or rupture. Most people with Marfan syndrome die by their mid-30s.

Ehlers-Danlos[46] **syndrome** is a hereditary defect in collagen synthesis. It results in abnormally long, loosely structured collagen fibers, which is reflected in abnormally stretchy skin, loose joints, impaired wound healing, and abnormalities in the blood vessels, intestine, and urinary bladder. Hip dislocations are often present at birth.

Osteogenesis imperfecta is due to insufficient collagen deposition in the bones. With a collagen deficiency, the bones are abnormally brittle. They commonly fracture before a child is born, and spontaneous fractures in childhood are common. Deformities of the teeth (which also contain collagen) and middle-ear bones, with corresponding hearing loss, are also common.

Several other collagen diseases are known, and since collagen is such a widespread protein in the body, the effects are very diverse. Not all collagen diseases are hereditary. **Scurvy,** for example, is characterized by bleeding gums, subcutaneous and intramuscular hemorrhages, and poor wound healing, all due to a dietary deficiency of ascorbic acid (vitamin C). Collagen is rich in the amino acids proline and lysine, and ascorbic acid is a necessary cofactor in the metabolism of these two amino acids.▲

---

44. *necr* = death + *osis* = condition, process
45. Antoine Bernard-Jean Marfan (1858–1942), French pediatrician

46. Edward L. Ehlers (1863–1937), Danish dermatologist; Henri A. Danlos (1844–1912), French dermatologist

The Study of Tissues (pp. 170–172)
1. Interpretation of tissue sections
2. The primary germ layers
3. The primary tissue classes (table 6.1)
4. Composition of tissues
   a. Cells
   b. Matrix (extracellular material)

Excitable Tissues (pp. 172–175)
1. Muscular tissue (table 6.2)
   a. General properties and functions
   b. Skeletal muscle
   c. Cardiac muscle
   d. Smooth muscle
2. Nervous tissue
   a. Neurons
   b. Neuroglia

Embryonic and Fibrous Connective Tissues (pp. 175–182)
1. Overview of connective tissue (table 6.3)
   a. General characteristics
      • Most variable tissue type
      • Usually more matrix than cell volume
      • Often supports other tissues or binds organs
   b. Functions
2. Embryonic connective tissue
   a. Mesenchyme
   b. Mucous connective tissue
3. Components of fibroconnective tissue
   a. Cells
      • Fibroblasts
      • Histiocytes (tissue macrophages)
      • Leukocytes (white blood cells)
      • Plasma cells
      • Mast cells
      • Adipocytes (fat cells)
   b. Fibers
      • Collagenous (white) fibers
      • Reticular fibers
      • Elastic (yellow) fibers
   c. Ground substance
      • Tissue (interstitial) fluid
      • Minerals (mainly calcium salts)
      • Proteoglycans
         •• Protein (minor component)
         •• Glycosaminoglycans (GAGs)
4. Types of fibroconnective tissue
   a. Loose connective tissue
      • Areolar tissue
      • Adipose tissue: white and brown fat
      • Reticular tissue

   b. Dense connective tissues
      • Dense regular connective tissue
      • Yellow elastic tissue
      • Dense irregular connective tissue

Supportive and Fluid Connective Tissues (pp. 182–185)
1. Cartilage
   a. Stiff matrix due to chondroitin sulfate
   b. Chondrocytes and lacunae
   c. Hyaline cartilage
   d. Elastic cartilage
   e. Fibrocartilage
2. Bone (osseous tissue)
   a. Types: Spongy and compact
      • Osteocytes in lacunae
      • Lamellae surrounding haversian canal
3. Dental tissues
4. Blood and lymph
   a. Plasma (matrix)
   b. Formed elements
      • Erythrocytes
      • Leukocytes
      • Platelets
   c. Lymph

Epithelial Tissue (pp. 185–191)
1. Definition and general properties
   a. Layers of cells covering or lining surfaces
   b. Scanty extracellular material
   c. No blood vessels
   d. Supported on basement membrane
   e. Classification (table 6.4)
2. Simple epithelia
   a. Simple squamous
   b. Simple cuboidal
   c. Simple columnar
   d. Pseudostratified
3. Stratified epithelia
   a. Stratified squamous
   b. Stratified cuboidal
   c. Stratified columnar
   d. Transitional

Intercellular Junctions, Glands, and Membranes (pp. 191–196)
1. Intercellular junctions
   a. Tight junctions
   b. Desmosomes and hemidesmosomes
   c. Gap (communicating) junctions
2. Glands
   a. Endocrine glands
      • Ductless
      • Secrete hormones into blood

   b. Exocrine glands
      • Usually have ducts
      • Produce secretions other than hormones
      • Some unicellular
   c. Exocrine gland structure
      • Stroma (connective tissue)
         •• Fibrous capsule
         •• Septa (trabeculae)
      • Parenchyma (secretory portion)
         •• Epithelial in nature
         •• Divided into lobes and lobules
         •• May form acini (alveoli)
      • Ducts
         •• Simple glands: unbranched ducts
         •• Compound glands: branched ducts
   d. Classification by type of secretion
      • Serous glands
      • Mucous glands
      • Mixed glands
      • Cytogenic glands
   e. Classification by method of secretion
      • Merocrine glands
      • Holocrine glands
      • Apocrine glands (modified merocrine glands)
3. Membranes
   a. Cutaneous membrane
   b. Mucous membranes (mucosae)
      • Epithelium
      • Lamina propria
      • Muscularis mucosae
   c. Serous membranes (serosae)
      • Mesothelium
      • Areolar tissue
   d. Endothelium of circulatory system
   e. Synovial membranes of joints

Tissue Transformations (pp. 196–197)
1. Differentiation
2. Anaplasia (dedifferentiation)
3. Hyperplasia (cell multiplication)
4. Hypertrophy (cell growth)
5. Neoplasia (tumor development)
6. Metaplasia (change in tissue type)
7. Atrophy
8. Necrosis (tissue death)
9. Repair

histology 170
fixative 170
histological section 170
stain 170
longitudinal section 170
cross section 170
transverse section 170
oblique section 170
primary germ layers 170
ectoderm 170
endoderm 170
mesoderm 170
primary tissues 171
extracellular material 172
matrix 172
tissue (interstitial) fluid 172
excitable tissue 172
muscular tissue 172
skeletal muscle 172
muscle fiber 172
sphincter 172
striation 173
cardiac muscle 173
myocyte 173
intercalated disc 173
smooth muscle 174
nervous tissue 174
neuron 174
neuroglia 174
glial cell 174
soma 174
dendrite 174
axon 174
nerve fiber 174
connective tissue 175
mesenchyme 175
mucous connective tissue 175
fibroconnective tissue 176
fibroblast 176
histiocyte 176

leukocyte 177
plasma cell 177
mast cell 177
heparin 177
histamine 177
adipocyte 177
collagenous fiber 177
collagen 177
reticular fiber 177
elastic fiber 177
elastin 177
ground substance 178
glycosaminoglycan (GAG) 178
chondroitin sulfate 178
hyaluronic acid 178
loose connective tissue 179
dense connective tissue 179
areolar connective tissue 179
adipose tissue 179
fat 179
obesity 180
reticular tissue 181
dense regular connective
    tissue 181
yellow elastic tissue 181
dense irregular connective
    tissue 182
cartilage 182
chondroblast 182
lacuna 182
chondrocyte 182
hyaline cartilage 182
perichondrium 183
elastic cartilage 183
fibrocartilage 183
osseous tissue 184
spongy bone 184
compact bone 184
osteon 184
haversian canal 184

lamella 184
osteocyte 184
periosteum 184
dentin 184
cementum 184
blood 185
plasma 185
formed element 185
erythrocyte 185
platelet 185
lymph 185
epithelial tissue 185
basement membrane 185
simple epithelium 187
simple squamous
    epithelium 187
endothelium 188
mesothelium 188
simple cuboidal epithelium 188
simple columnar
    epithelium 188
pseudostratified
    epithelium 189
stratified epithelium 189
stratified squamous
    epithelium 189
keratinized 190
nonkeratinized 190
stratified cuboidal
    epithelium 190
stratified columnar
    epithelium 190
transitional epithelium 190
intercellular junction 192
tight junction 192
desmosome 192
gap junction 193
gland 193
endocrine gland 193
exocrine gland 193

goblet cell 193
capsule 193
septum 193
trabecula 193
lobe 193
lobule 194
stroma 194
parenchyma 194
simple gland 194
compound gland 194
tubular gland 194
acinar gland 194
acinus 194
alveolus 194
serous gland 194
mucous gland 194
mixed gland 194
cytogenic gland 194
merocrine (eccrine) gland 194
holocrine gland 194
apocrine gland 195
cutaneous membrane 195
mucous membrane
    (mucosa) 195
lamina propria 195
muscularis mucosae 195
serous membrane (serosa) 195
serous fluid 195
synovial membrane 196
differentiation 196
hypertrophy 196
hyperplasia 196
neoplasia 196
metaplasia 196
atrophy 196
necrosis 197
infarction 197
gangrene 197
regeneration 197
fibrosis 197

Answers in Appendix C

1. Transitional epithelium can be found in
   a. the urinary system.
   b. the respiratory system.
   c. the digestive system.
   d. the reproductive system.
   e. any of the above.

2. The surface of the stomach facing the peritoneal cavity is covered by its
   a. mucosa.
   b. serosa.
   c. peritoneum.
   d. lamina propria.
   e. basement membrane.

3. Which of the following is a primary germ layer?
   a. epidermis
   b. mucosa
   c. ectoderm
   d. endothelium
   e. mesothelium

4. A seminiferous tubule of the testis is lined with
   a. simple cuboidal epithelium.
   b. pseudostratified ciliated epithelium.
   c. stratified squamous epithelium.
   d. transitional epithelium.
   e. stratified cuboidal epithelium.

5. Fluids can be kept from seeping between epithelial cells by
   a. glycosaminoglycans.
   b. hemidesmosomes.
   c. tight junctions.
   d. communicating junctions.
   e. a basement membrane.

6. A fixative serves to
   a. stop tissue decay.
   b. improve contrast.
   c. repair a damaged tissue.
   d. bind cells together in an epithelium.
   e. attach one cardiac myocyte to another.

7. The macrophages of areolar tissue are called
   a. histiocytes.
   b. fibroblasts.
   c. mast cells.
   d. basophils.
   e. lymphocytes.

8. Tendons are composed of _____ connective tissue.
   a. skeletal
   b. areolar
   c. dense irregular
   d. yellow elastic
   e. dense regular

9. The shape of the external ear is due to
   a. skeletal muscle.
   b. elastic cartilage.
   c. fibrocartilage.
   d. articular cartilage.
   e. hyaline cartilage.

10. The most abundant formed element(s) of blood is/are
    a. plasma.
    b. erythrocytes.
    c. platelets.
    d. leukocytes.
    e. protein.

11. Any form of tissue death is called _____.

12. The simple squamous epithelium that lines the peritoneal cavity is called the _____.

13. Osteocytes and chondrocytes occupy little cavities called _____.

14. Muscle cells and axons are often called _____ because of their shape.

15. Tendons and ligaments are made predominantly of the protein _____.

16. The only type of muscle that does not have gap junctions is _____ muscle.

17. An epithelium rests on a layer called the _____ between its deepest cells and the underlying connective tissue.

18. Fibers and ground substance collectively make up the _____ of a connective tissue.

19. Polysaccharide chains bound to a core protein form macromolecules called _____, an important part of the connective tissue matrix.

20. Any epithelium in which every cell contacts the basement membrane is called a/an _____ epithelium.

Testing Your Comprehension
Answers in *Study Guide*

1. A woman in the final stages of labor is often told to push. By pushing, is she consciously contracting her uterus to expel the baby? Justify your answer based on the muscular composition of the uterus.

2. A major premise of the cell theory is that all bodily structure and function is based on cells. The structural characteristics of bone, cartilage, and tendons, however, are due more to their intercellular material than to their cells. Is this an exception to the cell theory? Why or why not?

3. When cartilage is compressed, water is squeezed out of the proteoglycans, and when pressure is taken off, water flows back in. This being the case, why do you think cartilage at weight-bearing joints of the body (such as the knees) can degenerate from lack of exercise?

4. Explain how it is possible for an adult to develop a brain tumor in spite of the fact that neurons are incapable of mitosis.

5. Which do you think would heal faster, cartilage or bone? Stratified squamous or simple columnar epithelium? Why?

Web Site Link

For a listing of the most current web sites related to this chapter, please visit the Saladin homepage at:

http://www.mhhe.com/sciencemath/biology/saladin/

[The Integumentary System

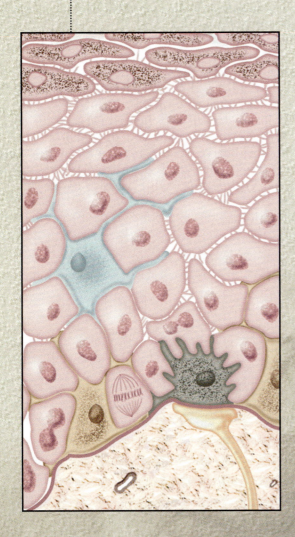

Special **Topics**

the unity of form and function

Brushing up

T he **integumentary system** consists of the skin, hair, nails, and cutaneous glands. We are perhaps more conscious of this system than of any other in the body. Few of us would venture out of the house each day without first looking in a mirror to see if our skin and hair are presentable. We are well aware of how important this system is to individual recognition, social acceptance, and self-image—and social considerations aside, self-image is important to a person's overall health.

Dermatology is the scientific study and medical treatment of the integumentary system. This is the most easily examined of the organ systems, and its appearance provides important clues not only to its own health but also to deeper disorders such as liver cancer, anemia, and heart failure. It is also the most vulnerable organ system, exposed to radiation, trauma, infection, and injurious chemicals. Consequently, it needs medical attention more often than any other.

Structure of the Skin

The **skin,** or **integument,** is the body's largest organ (fig. 7.1). In adults, it covers an area of about 1.5 to 2.0 square meters and accounts for about 15% of the body weight. It consists of two layers: (1) a stratified squamous epithelium called the *epidermis*[1] and (2) a deeper connective tissue layer called the *dermis.* Below the skin is another connective tissue layer called the *hypodermis,*[2] which is also discussed in this chapter.

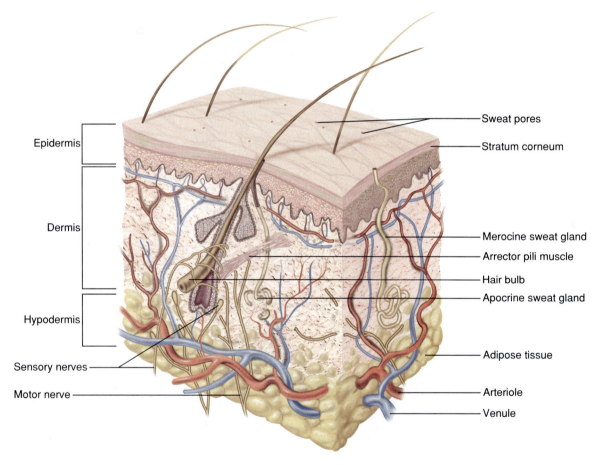

Figure 7.1 Structure of the skin, hypodermis, and accessory organs of the skin (hair and glands).

1. *epi* = upon, above + *derm* = skin
2. *hypo* = below

Most of the skin is 1 to 2 mm thick—about half as thick as the cover of this book. It ranges, however, from about 0.5 mm thick on the eyelids to 6 mm thick between the shoulders. The difference in total thickness is due mainly to variation in the thickness of the dermis. The epidermis accounts for only 0.007 to 0.12 mm of the total.

A few special areas of the skin are classified as *thick* or *thin skin* based on the relative thickness of the epidermis alone, especially the surface layer of dead cells called the *stratum corneum.* **Thick skin** has a very thick, tough stratum corneum (see fig. 7.2b). It is limited to the palms, the volar (anterior) surfaces of the fingers, and the plantar surfaces (soles) of the feet. **Thin skin** has a very thin stratum corneum (see fig. 7.7b). It occurs in especially sensitive areas, such as the lips, eyelids, eardrums, and some parts of the genitals. Most of the skin, however, is "average" and does not fall into either one of these special categories.

The Epidermis

The **epidermis** is a stratified squamous epithelium. Like other epithelia, it lacks blood vessels and depends on diffusion of nutrients from the underlying connective tissue. It has sparse nerve endings for the senses of touch and pain, but most sensations of the skin are due to nerve endings in the dermis. Unlike the stratified squamous epithelia of the esophagus and vagina, the epidermis is **cornified (keratinized)**[3]—covered with several layers of dead cells with heavy deposits of the durable protein **keratin,** the same substance that forms the horns and hooves of mammals such as cows.

The epidermis usually consists of four layers of cells (five in thick skin). They are described here in order from deep to superficial and are shown in figure 7.2.

Stratum Basale

The **stratum basale** (bah-SAY-lee) consists of a single layer of cuboidal to low columnar cells resting on the basement membrane of the epithelium. There are three types of cells in this layer (fig. 7.3). Most of them are **keratinocytes** (keh-RAT-ih-no-sites), named for their role in synthesizing keratin. They are the source of most new epidermal cells. Keratinocytes are pushed toward the surface as others divide below them, and eventually they become the dead cells of the keratinized surface.

Melanocytes synthesize the pigment melanin. They have long branching processes that spread among the basal keratinocytes and continually shed melanin-containing fragments from their tips. The keratinocytes phagocytize these fragments and accumulate melanin

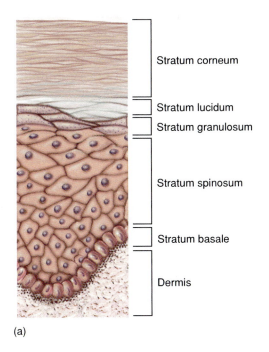

(a)

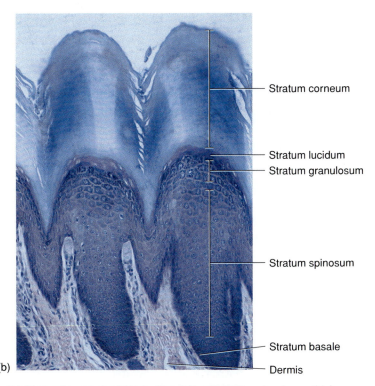

(b)

Figure 7.2 Layers of the epidermis. (*a*) Drawing of the epidermal layers. (*b*) Photomicrograph of thick skin of the fingertip, showing a thick stratum corneum and two friction ridges (×100). Compare this heavily keratinized skin to figure 7.7.

3. *corn* = *kera* = horn

granules on the "sunny side" of the nucleus. Like a little parasol, the pigment shields the DNA from ultraviolet radiation. People of all races have about equal numbers of melanocytes, but in blacks there is a higher rate of melanocyte activity and melanin is well dispersed throughout each keratinocyte. In whites, the melanin is less abundant and tends to remain clumped near the nucleus, so it imparts less color to the skin.

A few cells of the stratum basale are called **Merkel**[4] **cells.** The Merkel cell and a dermal nerve fiber form a tactile (touch) receptor called a *Merkel disc.*

Stratum Spinosum

The **stratum spinosum** (spy-NO-sum) consists of several layers of keratinocytes. Some of these cells are able to divide, but as they are pushed farther and farther from the dermal blood vessels (their source of nutrition and route of waste removal), they lose this ability. The nuclei and organelles begin to degenerate and the cells become flatter. The keratinocytes are tightly joined by desmosomes. When skin is histologically fixed, these cells shrink and pull away from each other, but remain attached at their desmosomes. This creates bridgelike extensions where one keratinocyte reaches out to another across the gap—a little like two people holding hands while standing farther apart. These extensions give the cells the spiny appearance for which this epidermal layer is named.

The stratum spinosum and stratum granulosum also contain macrophages called **Langerhans**[5] **cells** that migrate to the skin from their origin in the bone marrow. These cells help to protect the body against foreign substances and microorganisms by "capturing and presenting" foreign matter to the immune system for a response.

Stratum Granulosum

The **stratum granulosum** consists of two to five layers of flat cells containing coarse, dark-staining grains of **keratohyalin.** There is no broad agreement yet as to

Stratum
corneum

Stratum
lucidum

Stratum
granulosum

Stratum
spinosum

Stratum
basale

Keratinocytes

Langerhans cell

Merkel cell

Melanocyte

Sensory
nerve ending

Dermis

Figure 7.3 Cell types of the epidermis.

4. F. S. Merkel (1845–1919), German anatomist
5. Paul Langerhans (1847–88), German anatomist
6. *dermat* = skin + *phag* = eat

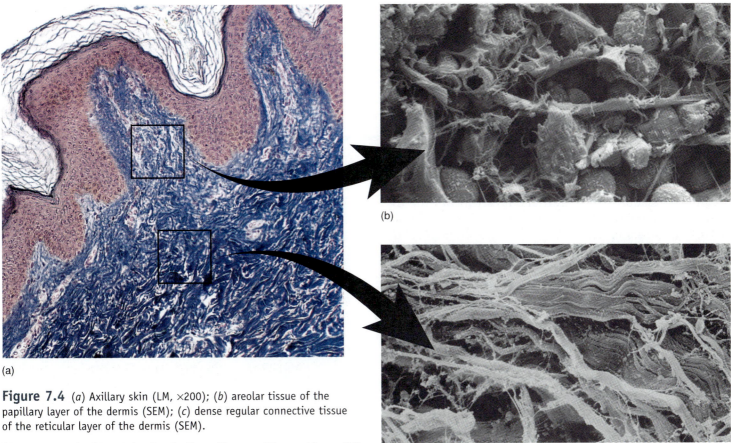

Figure 7.4 (*a*) Axillary skin (LM, ×200); (*b*) areolar tissue of the papillary layer of the dermis (SEM); (*c*) dense regular connective tissue of the reticular layer of the dermis (SEM).

(b) From R. G. Kessel and R. H. Kardon, *Scanning Electron Microscopy of Tissues and Organs*, 1979.

(a)

(b)

(c)

what becomes of keratohyalin, or whether or not it is a precursor of the keratin at the skin surface.

Stratum Lucidum

The **stratum lucidum**[7] (LOO-sih-dum) is a thin translucent zone superficial to the stratum granulosum, seen only in thick skin. Here, the keratinocytes have lost their nuclei and produced more extracellular material. Cellular boundaries are very indistinct, giving this zone its clear, featureless appearance.

Stratum Corneum

The **stratum corneum** consists of 25 to 30 layers of dead, scaly, keratinized cells. Those at the surface flake off (*desquamate*) as tiny scales called *dander* (see special topic 7.1). Dandruff is composed of clumps of desquamated cells stuck together by oil from the scalp. It takes about 40 days from the time a cell is produced in the stratum basale to the time it flakes off the surface. The process is slower in old age but faster in skin that has been injured or stressed. Physical stress stimulates multiplication of cells in the stratum basale, and this eventually results in a thicker stratum corneum, seen, for

example, in the calluses on the fingertips of a violinist or palms of a carpenter. Injury also accelerates the multiplication of keratinocytes, and injured epidermis regenerates more rapidly than any other tissue in the body.

The Dermis

The **dermis** is composed mainly of collagen but also contains elastic and reticular fibers, the usual cells of fibroconnective tissue (described in chapter 6), and an abundance of blood vessels, sweat glands, hair follicles, nail roots, and several kinds of sensory nerve endings (see fig. 7.1). Smooth muscle fibers of the dermis make the hairs stand on end in response to cold or fear, and they wrinkle the skin in areas such as the scrotum and areola, particularly in response to cold or touch. Skeletal muscles of the face act on dermal collagen fibers and produce such facial expressions as a smile, a wrinkle of the forehead, or a wink of the eye.

There are two layers of dermis called the papillary and reticular layers (fig. 7.4). The **papillary**[8] (PAP-ih-lerr-ee) **layer** is a zone of areolar tissue immediately adjacent to the epidermis, typically constituting about one-fifth of the thickness of the dermis. The loosely

7. *lucid* = light, clear

8. *papill* = nipple

Collagen fibers of the reticular layer are organized in bundles with gaps between them called *tension lines*, or *Langer's lines*. A surgical incision along these lines closes easily and heals with a minimum of scarring. By contrast, an incision across the tension lines tends to gape open because of the pull of the collagen fibers, leaving a more conspicuous scar. Surgeons therefore cut along the tension lines when possible. Tension lines run in a circular to oblique direction around the neck and trunk and longitudinally over most of the extremities.

organized tissue of the papillary layer allows for free mobility of white blood cells and other defenses against organisms introduced through breaks in the epidermis.

If you closely examine the skin on the back of your hand, you will see delicate furrows that divide it into tiny rectangular to rhomboidal areas. In tissue sections, the furrows can be seen as a wavy boundary between the epidermis and dermis, a little reminiscent of corrugated cardboard. This interlocking arrangement resists slippage of the epidermis across the dermis when stress is applied to the skin. The upward projections of dermis are called **dermal papillae.** They form the ridges of the fingerprints, and in such highly sensitive areas as the lips and genitals, they are especially tall and allow dermal nerve fibers to extend close to the skin surface. The margins of the lips get their red color from blood capillaries that approach the surface in unusually high dermal papillae.

Think About It

Dermal papillae are relatively high and numerous in palmar and plantar skin but low and few in number in skin of the face and abdomen. What do you think is the functional significance of this difference?

The **reticular layer** of the dermis consists of dense irregular connective tissue, with dense bundles of collagen oriented in multiple directions (see special topic 7.2). The boundary between the papillary and reticular layers is often indistinct, but in the reticular layer the collagen forms thicker bundles and leaves less room for ground substance. This layer often contains small clusters of adipocytes. Stretching of the skin, as occurs in obesity and pregnancy, can tear the collagen fibers and produce **linea albicantes**[9] (LIN-ee-uh AL-bih-CON-teez), commonly called "stretch marks" (fig. 7.5). These are especially common on the thighs, buttocks, abdomen, and breasts—the areas most subject to stretching as a result of weight gain.

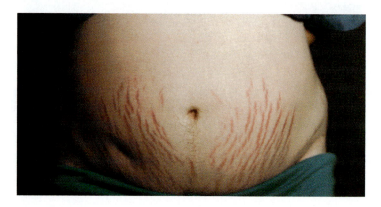

Figure 7.5 Stretch marks (linea albicantes) of the abdomen.

The Hypodermis

Beneath the skin is a layer called the **hypodermis, subcutaneous tissue,** or **superficial fascia**[10] (FASH-ee-uh). It is composed of areolar and adipose tissue; when adipose tissue predominates, the hypodermis is called the **panniculus**[11] **adiposus,** or **subcutaneous fat.** The hypodermis binds the skin to the underlying tissues, pads the body, serves as an energy reservoir, and provides thermal insulation. Infants and elderly people are more sensitive to cold than others because they have less fat in this layer. Obesity is due mainly to the accumulation of neutral fat (triglyceride) in the adipocytes of this tissue. The hypodermis is about 8% thicker in women than in men and differs in distribution between the sexes (fig. 7.6). Drugs are introduced here by hypodermic injection because the subcutaneous tissue is highly vascular and absorbs them quickly.

The skin and hypodermis are summarized in table 7.1.

Color

Normal skin colors are due to the pigments hemoglobin, melanin, and carotene. **Hemoglobin** in the dermal capillaries shows through the white collagen fibers to impart

9. *linea* = lines + *alb* = white

10. *fascia* = a band
11. *panni* = rag, cloth + *culus* = little

pink and flesh tones to the skin. The skin is redder where capillaries come closer to the surface—for example, on the lips.

Melanin in the stratum basale and stratum spinosum produces a variety of brown, black, tan, yellowish, and reddish hues (fig. 7.7). A combination of heredity and exposure to light determines the abundance of melanin and the resulting skin color. Ultraviolet light stimulates melanin synthesis and darkens the skin. A suntan fades as melanized keratinocytes migrate to the surface and flake off. The amount of melanin varies substantially from place to place on the body. In whites, it is concentrated in freckles and moles, in the nipple and surrounding area (areola) of the breast, around the anus, and in the scrotum and penis. In blacks, the dorsal skin of the hands and feet is darker than the palmar and plantar skin, the areola is darker than the rest of the breast, and the lateral

Table 7.1	Stratification of the Skin and Hypodermis, from Superficial to Deep
Layer	**Description**
Epidermis	Stratified squamous epithelium
Stratum Corneum	Dead, keratinized cells of the skin surface
Stratum Lucidum	Clear, featureless, thin layer in thick skin only
Stratum Granulosum	Two to five layers of cells with dark-staining keratohyalin granules
Stratum Spinosum	Many layers of keratinocytes, typically shrunken in fixed tissues but attached to each other by desmosomes, which give them a spiny look; progressively flattened the farther they are from the dermis. Langerhans cells occur here but are not visible in routinely stained preparations.
Stratum Basale	Single layer of cuboidal to columnar cells resting on basal lamina; site of most mitosis; consists of keratinocytes, melanocytes, and Merkel cells, but these are not distinguishable with routine stains. Melanin is conspicuous in keratinocytes of this layer in black to brown skin.
Dermis	Fibroconnective tissue, richly endowed with blood vessels and nerve endings. Sweat glands and hair follicles originate here and in hypodermis.
Papillary Layer	Superficial one-fifth of dermis; composed of areolar tissue; often extends upward as dermal papillae
Reticular Layer	Deeper four-fifths of dermis; dense irregular connective tissue
Hypodermis	Areolar or adipose tissue between skin and muscle

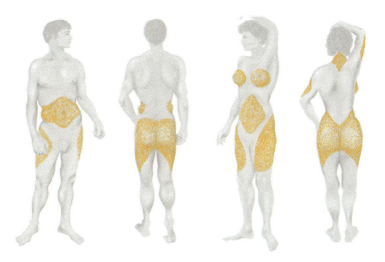

Figure 7.6 The major deposits of subcutaneous fat in males and females.

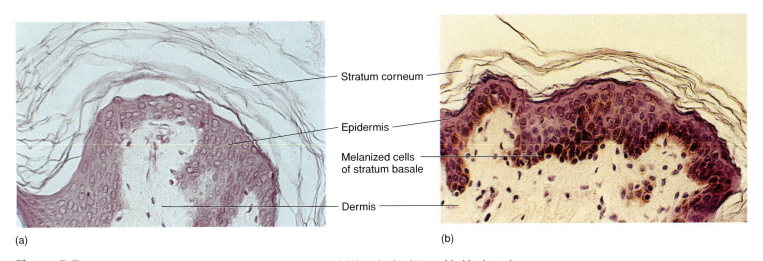

Stratum corneum

Epidermis

Melanized cells of stratum basale

Dermis

(a)

(b)

Figure 7.7 Comparison of (*a*) nonmelanized Caucasian skin and (*b*) melanized Negroid skin (×400).

Table 7.2	Abnormal Coloration of the Skin	
Condition	**Appearance of Skin**	**Causes**
Cyanosis	Blue	Oxygen-poor blood resulting from respiratory obstruction, respiratory arrest, or lung disease, or from poor cutaneous blood circulation as may occur in cold weather or in cardiac arrest
Pallor	Pale	Anemia or severe reduction in cutaneous circulation resulting from circulatory shock or cold temperatures
Albinism	White	Hereditary inability to synthesize melanin
Erythema	Red	Increased circulation resulting from exercise, heat, sunburn, radiation exposure, emotional states, and other conditions
Jaundice	Yellow	Elevated bilirubin concentration in blood resulting from rapid red blood cell breakdown or inability of diseased or immature liver to dispose of bilirubin
Hematoma	Bruised	Mass of clotted blood visible through the skin; may indicate hemophilia, accidental trauma, or physical abuse

surfaces of the female genitals are melanized while the medial region is not.

Carotene[12] is a yellow pigment acquired from egg yolks and yellow and orange vegetables. Depending on the diet, it becomes concentrated to various degrees in the stratum corneum and subcutaneous fat. It is often most conspicuous in skin of the heel and in "corns" or calluses of the feet because this is where the stratum corneum is thickest. The yellowish tint of some Asiatic complexions, once attributed to carotene, is now regarded as being due to melanin.

The skin may also exhibit abnormal colors of diagnostic value (table 7.2). **Cyanosis,** a blueness of the skin, is due to a deficiency of oxygen in the circulating blood, which turns the hemoglobin a reddish violet color. It can result from conditions that prevent the blood from picking up a normal load of oxygen in the lungs, such as airway obstructions in drowning and choking, lung diseases such as emphysema, or respiratory arrest. Cyanosis can also result from reduced circulation through the skin as oxygen is removed from the blood and fresh blood does not arrive quickly. This can occur in cardiac arrest or in cold weather, when the dermal blood vessels constrict to retain heat.

Pallor is a pale or "ashen" color that occurs when there is so little blood flow through the skin that the predominant color is that of the dermal collagen. It can result from emotional stress, circulatory shock, cold temperatures, severe anemia, or low blood pressure. **Albinism** is a genetic lack of melanin, resulting in white hair, pale skin, and pink eyes. Melanin is synthesized from the amino acid tyrosine by the enzyme tyrosinase. People with albinism have inherited a recessive, nonfunctional tyrosinase allele from both parents.

Erythema[13] (ERR-ih-THEE-muh) is abnormal redness of the skin as a result of increased blood flow. It can be caused by exercise, heat, sunburn, radiation exposure, and such emotional states as anger or embarrassment. **Jaundice** is a yellow color of the skin and the whites of the eyes resulting from high levels of bilirubin in the blood. It occurs when there is an excessive rate of red blood cell breakdown, since bilirubin is derived from free hemoglobin; when liver diseases (cancer, hepatitis, or cirrhosis, for example) interfere with the ability of this organ to dispose of bilirubin; or when the liver is insufficiently developed to dispose of bilirubin adequately (as in premature infants). A bruise, or **hematoma** ("black-and-blue mark"), is a mass of clotted blood showing through the skin. It is usually due to accidental trauma (blows to the skin), but it also can indicate hemophilia, other metabolic or nutritional disorders, or physical abuse.

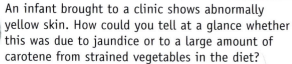

Think About It

An infant brought to a clinic shows abnormally yellow skin. How could you tell at a glance whether this was due to jaundice or to a large amount of carotene from strained vegetables in the diet?

Markings

Hemangiomas[14] (he-MAN-jee-OH-mas), or birthmarks, are patches of discolored skin caused by benign tumors of the dermal blood capillaries. *Capillary hemangiomas* (strawberry birthmarks) are bright red to deep purple and are slightly swollen; they usually disappear in childhood. *Cavernous hemangiomas* (port wine stains) are flat, duller in color, and last for life.

Moles and freckles are tan to black aggregations of melanocytes. A **nevus,** or **mole,** is an elevated patch of melanized skin, often with hair. Moles are harmless and sometimes even regarded as "beauty marks," but they

12. *carot* = carrot

13. *eryth* = red
14. *hem* = blood + *angi* = vessel + *oma* = mass, tumor

should be watched for possible changes that suggest malignancy. **Freckles** are flat patches that vary with heredity and exposure to the sun.

The skin is also marked by many lines, creases, and ridges. **Friction ridges** (see fig. 7.2*b*) are characteristic of most primates. They help prevent monkeys, for example, from slipping off a branch as they walk across it, and they enhance the manual dexterity typical of primates. Friction ridges form during fetal development. Their pattern is unique to every individual and remains essentially unchanged throughout life. Because sweat pores open along the tops of the friction ridges, the fingertips leave identifying fingerprints on almost any object they touch. Not even identical twins have identical fingerprints.

Flexion creases are the lines of the palms formed by repeated closing of the hand. They also occur at the wrist (see fig. B.7) and other joints. Flexion creases develop after birth.

········· **Key Point Review** ·········

1 Name four types of epidermal cells and state the function of each.

2 List the five layers of epidermis in order from deep to superficial. What are the distinctive features of each?

3 What are the two layers of the dermis?

4 Name the pigments responsible for normal skin colors and explain how certain conditions produce symptomatic discoloration of the skin.

Functions of the Skin

▼Objectives
When you have completed this section, you should be able to
• describe the barrier functions of the skin; and
• describe the roles of the skin in synthesis, absorption, sensation, temperature control, and social interaction.

The skin is much more than a container for the body. It has a wide variety of important functions that go well beyond appearance. It protects the body from infections, physical injury, solar radiation, and chemicals; it prevents excessive loss or gain of water; it excretes wastes; it carries out a step in vitamin D synthesis; it helps control body temperature; it is the body's most extensive sense organ; and it is an important means of nonverbal communication. Although its structural and physiological complexity is commonly underestimated, it is indeed a marvel of biological engineering. It is hard to conceive of any other organ that could serve functions so varied yet display such strength, flexibility, and remarkable capacity for growth and self-repair.

The Skin as a Barrier

The skin bears the brunt of most physical injuries to the body, but it is better able to resist and recover from trauma than other organs. The toughness of keratin and strength of the epidermal desmosomes make the skin a barrier that is not easily breached. Few infectious organisms can penetrate the skin on their own. Those that do either rely on accidental breaks in the skin or have life cycles that involve such animals as mosquitoes, fleas, lice, and ticks, whose mouthparts are strong enough to puncture the skin. Animals that carry a parasite and introduce it into its next host are called *vectors.*

The epidermal surface is populated by great numbers of bacteria, fungi, and other pathogens poised for any opportunity to get inside. Even vigorous scrubbing in a hot shower does not rid the skin of bacteria. They are, however, discouraged from multiplying on the skin by its relatively dry, unfavorable habitat. The sebum (oil of the skin) contains bactericidal substances, and sweat forms a film called the **acid mantle** (pH 4–6) that is unfavorable to microbial growth.

Even when a pathogen breaches the epidermis, it faces an army of immune cells that can quickly migrate to the site of infection and mount a defense. These include the Langerhans cells of the epidermis, and macrophages and leukocytes in the papillary layer of the dermis.

The skin is also important as a barrier to water. When it is wet, it prevents the body from absorbing excess water. Even more importantly, it prevents the body from losing excess water. This is especially evident in burn patients; when a large area of skin has been destroyed, fluid replacement therapy becomes one of the most critical needs for survival.

The skin is also a barrier to solar radiation, especially to its ultraviolet (UV) component. Most UV radiation is filtered out by atmospheric ozone, but even the small fraction that reaches us is enough to cause sunburns and skin cancer. As noted earlier, melanocyte activity increases in response to UV radiation. The ozone layer is destroyed by chemicals called CFCs (chlorofluorocarbons) emitted from air conditioners, refrigerators, spray cans, and other sources. One reason scientists are so apprehensive about this problem is that ozone depletion could lead to a catastrophic increase in skin cancer. Many countries have agreed to phase out CFC production by the year 2000, but efforts to accelerate this process are now underway.

Vitamin D Synthesis

Although UV radiation cannot penetrate far into the skin, some of it does reach the capillaries of the dermis, where it plays a vital role in the synthesis of vitamin D.

The ability of the skin to absorb chemicals makes it possible to administer several medicines as ointments or lotions, or by means of adhesive patches that release the medicine steadily through a semipermeable membrane. For example, inflammation can be treated with a hydrocortisone ointment, nitroglycerine patches are used to relieve heart pain, nicotine patches are used to help overcome tobacco addiction, and other medicated patches are used to control high blood pressure and motion sickness.

Unfortunately, the skin can also be a route for absorption of poisons. These include toxic alkaloids from poison ivy and other plants; metals such as mercury, arsenic, and lead; and solvents such as carbon tetrachloride (dry cleaning fluid), acetone (nail polish remover), paint thinner, and insecticides. Some of these can cause brain damage, liver failure, or kidney failure, which is good reason for using protective gloves when handling such substances.

The blood contains a steroid called *dehydrocholesterol,* which is converted by UV radiation into *cholecalciferol* (CO-lee-cal-SIF-ur-ol). The liver and kidneys then convert this to an active form of vitamin D called *calcitriol,* which is needed for normal bone and calcium metabolism. Vitamin D is discussed more extensively in chapter 8, along with the diseases (rickets and osteomalacia) that result from inadequate exposure of the skin to sunlight.

Cutaneous Absorption

Although the skin is impermeable to most chemicals, there are exceptions. The blood receives 1% to 2% of its oxygen by diffusion through the skin, and it gives off some carbon dioxide and volatile organic chemicals. Amino acids and steroids diffusing through the skin are one factor that attracts mosquitoes to people. The fat-soluble vitamins A, D, E, and K can be readily absorbed through the skin, as can many drugs and poisons (see special topic 7.3).

Sensory Roles

A variety of nerve endings in the skin react to heat, cold, touch, texture, pressure, vibration, and tissue injury. These cutaneous sensory receptors are especially abundant on the face, palms, fingers, soles, nipples, and genitals. There are relatively few on the back and in skin overlying joints such as the knees and elbows. Some of them are naked dendrites that penetrate into the epidermis. Most, however, are limited to the dermis and hypodermis, where specialized connective tissues give the nerve cells more selective sensitivity to particular types of stimuli. Cutaneous receptors are discussed with other sense organs in chapter 16.

Thermoregulation

The nervous, endocrine, muscular, and integumentary systems are all involved in regulating body temperature. The details of thermoregulation are discussed in chapter 26, but here we briefly consider the role of the skin, which functions as an adjustable radiator. If we want to heat a room with a steam radiator, we adjust a valve that allows more steam to flow through it. Similarly, the skin can be made to radiate more heat by means of *cutaneous vasodilation*—expanding the blood vessels (arterioles) of the dermis so that more blood flows through the skin and heat is lost to the surrounding air. If this does not cool the body adequately, additional heat can be lost by perspiration; evaporating water carries a lot of heat with it. Conversely, when it is necessary to retain body heat, *cutaneous vasoconstriction* reduces blood flow close to the body surface, which keeps the blood and heat deeper within the body.

Social Interaction

In all groups of vertebrate animals, the integumentary system plays an important role in the social relations among members of a species, including their ability to identify members of their own species and distinguish the sexes. Among mammals, recognition is often based on the color and distribution of hair. Animals may also accept or reject one another's company and choose mates based on the appearance of the integument, which may indicate an animal's state of health, including the presence of contagious diseases. Thus, it is not surprising that animals allocate a lot of time to grooming.

Humans are no exception. A little personal reflection will emphasize just how much impact an integumentary condition can have on someone's self-image and emotional state—whether the ravages of adolescent acne, the presence of a birthmark or a disfiguring scar, or just a "bad hair day." The skin is also our most significant means of nonverbal communication. The faces of primates are far more expressive than those of other mammals because complex skeletal muscles insert on dermal collagen

Figure 7.8 Nonverbal communication is an important function of the facial skin. The highly variable and subtle expressions of humans and other primates are made possible by a complex array of facial muscles that insert on collagen fibers of the dermis.

fibers and create subtle and varied facial expressions (fig. 7.8). Thus the skin has very important psychological and social functions.

.... **Key Point Review** ....

5 How does the skin inhibit the survival of pathogens on its surface?

6 To what besides pathogenic organisms does the skin present a barrier?

7 Describe the step in the synthesis of vitamin D that occurs in the skin.

8 Describe two ways in which the skin helps to control body temperature.

Hair and Nails

▼**Objectives**

When you have completed this section, you should be able to
- describe the three types of hair;
- describe the histology of a hair and hair follicle;
- explain how hair grows and discuss the causes of baldness;
- discuss some theories of the purpose served by human hair; and
- describe the structure and functions of nails.

The hair, nails, and cutaneous glands are the **accessory organs,** or **appendages,** of the skin. Hair and nails are composed mostly of dead, keratinized cells. While the

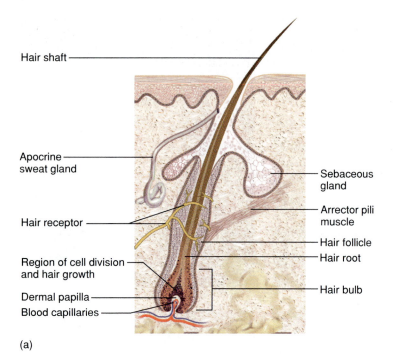

Hair shaft

Apocrine sweat gland

Hair receptor

Region of cell division and hair growth

Dermal papilla

Blood capillaries

Sebaceous gland

Arrector pili muscle

Hair follicle

Hair root

Hair bulb

(a)

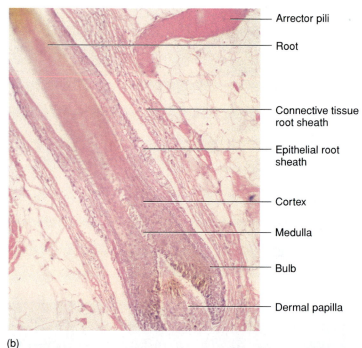

Arrector pili

Root

Connective tissue root sheath

Epithelial root sheath

Cortex

Medulla

Bulb

Dermal papilla

(b)

Figure 7.9 (a) Anatomy of a hair follicle and its associated structures; (b) light micrograph of the deep end of a hair follicle (×100); (c) two hairs emerging from their follicles (SEM). Notice the scaly cells of the stratum corneum encircling each follicle and the overlapping scales of the hair cuticle.

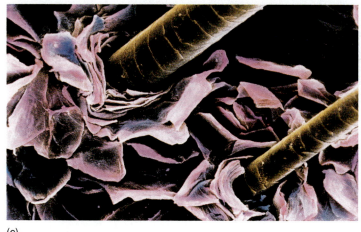

(c)

stratum corneum of the skin is made of pliable **soft keratin,** the hair and nails are composed mostly of **hard keratin.** Hard keratin is more compact than soft keratin and is toughened by a greater number of disulfide bridges between the protein molecules.

Types and Distribution of Hair

A hair is also known as a **pilus** (PY-lus); in the singular, *pili* (PY-lye). It is a slender filament of keratin that grows from an oblique tube in the skin called a **follicle** (FOLL-ih-cul) (fig. 7.9). Hair is found almost everywhere on the body except the lips, nipples, parts of the genitals, palmar and plantar skin, lateral surfaces of the fingers, toes, and feet, and distal segment of the fingers. The extremities and trunk have about 55 to 70 hairs per square centimeter, and the face has about 10 times as many. There are an estimated 100,000 hair follicles in the scalp and about 30,000 in a man's beard. The number of hairs in a given area does not differ greatly from one person to another or even between the sexes. Differences in apparent hairiness are due mainly to differences in the texture and pigmentation of the hair.

Hair may be classified as lanugo, vellus, or terminal hair. **Lanugo** is fine, downy, unpigmented hair that appears on the fetus in the last 3 months of development. By the time of birth, most of it is replaced by sim-

ilarly fine, unpigmented hair called **vellus.** Except for the eyebrows, eyelashes, and hair of the scalp, all of the hair of children, about two-thirds of the hair of women, and one-tenth of the hair of men is vellus. **Terminal hair** is longer, coarser, and pigmented. It occurs on the scalp, eyebrows, and eyelashes; at puberty it replaces the vellus in the axillary and pubic regions, on the face of males (to form the beard), and to varying degrees on the trunk and extremities.

Structure of the Hair and Follicle

A hair is divisible into three zones along its length: (1) the **bulb,** a swelling at the base, where the hair originates in the dermis; (2) the **root,** which is the remainder of the hair within the follicle; and (3) the **shaft,** which is the portion above the skin surface. Except near the bulb,

Laboratory analysis of hair can provide important clues to metabolic diseases, specific nutritional deficiencies, and poisoning. A deficiency of zinc in the hair may indicate malnutrition. A deficiency of both magnesium and calcium is symptomatic of phenylketonuria. A deficiency of calcium and an excess of sodium may indicate cystic fibrosis.

Several poisons accumulate in the hair. Arsenic, cadmium, lead, and mercury, for example, are 10 times as concentrated in hair as in blood or urine, and hair analysis can therefore aid in establishing a cause of death. There is some evidence that President Zachary Taylor's death in 1850 was due to foul play, and in 1991 a historian convinced authorities to exhume his body so his hair could be analyzed for arsenic. None was found, however.

While hair analysis has valuable medical and legal applications, it is also exploited by health-food charlatans who charge high fees for hair analysis and special diets based on supposed deficiencies revealed by the analysis. Hair analysis has not proven very useful for assessing a person's overall nutritional state.

(a)   (b)   (c)   (d)

Figure 7.10 Structural basis for the color and texture of hair. Straight hair (a–b) is round in cross section, whereas curly hair (c–d) is flattened. (a) Blonde hair has scanty melanin granules in the hair cortex. (b) Brown and black hair have higher concentrations of melanin. (c) Red hair is colored by the iron-containing pigment trichosiderin. (d) Gray and white hair lack cortical pigment and have air in the medulla.

all the tissue is dead. A bit of vascular connective tissue called the **papilla** grows into the bulb and provides the hair with its sole source of nutrition. Although ingested substances can become incorporated into hair (see special topic 7.4), nothing you apply externally to your hair can nourish it, notwithstanding advertising pitches to the gullible.

In cross section, a hair reveals three layers: (1) the **medulla,** a core of loosely arranged cells and air spaces; (2) the **cortex,** composed of densely packed keratinized cells; and (3) the **cuticle,** a single layer of scaly cells that overlap each other like roof shingles, with their free edges directed upward (fig. 7.9c). Cells lining the follicle are like shingles facing in the opposite direction. They interlock with the scales of the hair cuticle and

resist pulling on the hair. When a hair is pulled out, this layer of follicle cells comes with it.

The texture of hair is related to differences in cross-sectional shape (fig. 7.10)—straight hair is round, wavy hair is oval, and kinky hair is relatively flat. The color of black and brown hair is due to melanin produced by melanocytes in the bulb and transferred to cells of the cortex. A blond color is simply due to a scanty amount of melanin. Red hair is colored by an iron-containing pigment called **trichosiderin.**[15] White hair is due to air in the medulla and lack of pigment in the cortex, and gray hair is a mixture of white and pigmented hairs.

15. *tricho* = hair + *sider* = iron

The follicle is a diagonal tube that dips deeply into the dermis and sometimes extends as far as the hypodermis. It has two principal layers: an **epithelial root sheath** and a **connective tissue root sheath.** The epithelial root sheath, which is an extension of the epidermis, lies immediately adjacent to the hair root. The connective tissue root sheath, derived from the dermis, surrounds the epithelial sheath. Associated with the follicle are nerve and muscle fibers. Nerve fibers called **hair receptors** entwine each follicle and are sensitive to hair movements. You can feel their effect by lightly running your finger over the hairs of your arm without touching the skin or by carefully moving a single hair with a pin. Also associated with each hair is an **arrector pili**[16] (ah-REC-tur PY-lye; plural, *arrectores pilorum*), a bundle of smooth muscle fibers extending from dermal collagen fibers to the connective tissue root sheath of the follicle (see figs. 7.1 and 7.9). In response to cold, fear, or other stimuli, the sympathetic nervous system stimulates these muscles and makes the hair stand on end. In mammals other than humans, this traps an insulating layer of warm air next to the skin or makes the animal appear larger and less vulnerable to a potential enemy. In humans, it pulls the follicles into a vertical position and causes "goose bumps" but serves no useful purpose.

Growth of Hair

The growth of a hair is due to mitosis of cells in the stratum basale of the epithelial root sheath. Both hair cells and follicle cells are pushed toward the skin surface as the cells below them multiply, and the hair cells, like epidermal keratinocytes, become progressively keratinized and die as they are pushed upward, away from the blood supply.

A hair of the scalp grows about 1 mm in 3 days (10–18 cm per year) and may attain a length of a meter or more. It normally grows for 2 to 4 years and then enters a dormant phase lasting 3 to 4 months. A new hair begins to grow beneath it, pushing the older hair ahead of it until the older hair falls from the follicle. The scalp normally loses 10 to 100 hairs a day. Hairs of the eyebrows and eyelashes grow for only 3 to 4 months; therefore, they never attain the length that scalp hairs do.

Hair grows fastest from adolescence until the 40s. After that, terminal hairs of the scalp are increasingly replaced by wispy vellus, and the hair grows thinner. Thinning of the hair, or baldness, is called **alopecia**[17] (AL-oh-PEE-she-uh). It occurs to some degree in both sexes and may be caused by disease, poor nutrition, fever, emotional trauma, radiation, or chemotherapy. In the great majority of cases, however, it is simply a matter of aging.

Pattern baldness is the condition in which only some regions of the scalp become devoid of hair, rather than a uniform thinning of the hair across the entire scalp. It results from a combination of heredity and hormones. First, a man must inherit a recessive allele on his X chromosome, or a woman must inherit the recessive allele on both of her X chromosomes. Even one dominant allele will prevent pattern baldness from developing. Given the right genotype, pattern baldness then depends on having a high (typically male) level of testosterone. The genotype makes the follicles so sensitive to testosterone that they have an abnormally short growth cycle, hairs fail to emerge from the follicles, and baldness occurs. Even with a homozygous recessive genotype, women develop pattern baldness only if their testosterone levels are higher than usual for a female.

Here again, advertisers are ever ready to dupe the desperate with ineffective miracle cures. The only drug that has been found to stimulate hair growth is minoxidil (Rogaine), a potent vasodilator available as a scalp ointment. Minoxidil sometimes stimulates hair growth when the hair is beginning to thin, but has no effect on scalp that is already bald. Another treatment for baldness is to remove patches of hairy skin from other places on the body and graft them to the scalp, a little like sodding a lawn.

Think About It

A young man says his grandfather and father were both bald by the age of 35, so he expects to go bald at an early age as well. What is the flaw in his reasoning?

Excessive or undesirable hairiness in areas that are not usually hairy, especially in women and children, is called **hirsutism** or **hypertrichosis.**[18] It tends to run in families and is usually a result of either masculinizing ovarian tumors or hypersecretion of androgens by the pituitary gland or adrenal cortex. It is often associated with menopause.

Functions of Hair

The relative hairlessness of humans is so unusual among mammals it raises the question, Why do we have any hair at all? What purpose does it serve? There are probably different answers for the different types of hair; furthermore, some of the answers would make little sense if we limited our frame of reference to industrialized societies, where barbers and hairdressers are engaged to alter the natural state of the hair. It is more useful to take a comparative approach

---

16. *arrect* = to erect + *pil* = hair
17. *alopecia* = fox mange

18. *hyper* = excessive + *trich* = hair + *osis* = condition

to this question and consider the purposes hair serves in other species of mammals.

Most hair of the human trunk and extremities is probably best interpreted as vestigial, with little present function. Its function in our ancestors was undoubtedly for warmth, but in modern humans it is too scanty for this purpose. Stimulation of the hair receptors, however, alerts us to parasites crawling on the skin, such as lice and fleas.

The scalp is the only place where the hair is thick enough to retain heat. Heat loss from a bald scalp can be extensive and quite uncomfortable for several reasons. First, the brain receives a rich supply of warm blood, and heat from it is easily conducted through the bone of the skull. In addition, most of the scalp lacks an insulating fat layer, and without hair there is nothing to break the wind and stop it from carrying away heat. Hair of the scalp may also protect it from sunburn, since the scalp is most directly exposed to the sun's rays. These may be the reasons humans have retained hair on their heads while losing most of it from the rest of the body.

Tufts and patches of hair, sometimes with contrasting colors, are important among mammals in advertising species, age, sex, and individual identity. For the less groomed members of the human species, scalp hair may play a similar role. The indefinitely growing hair of a man's scalp and beard, for example, could provide a striking contrast to an otherwise almost hairless face and create a badge of recognition instantly visible at a distance.

Pubic, axillary, and male facial hair signify sexual maturity and aid in the transmission of sexual scents. We will further reflect on this in a later discussion of apocrine sweat glands, whose distribution and function add significant evidence to support this theory.

Stout protective **guard hairs,** or **vibrissae** (vy-BRISS-ee), guard the nostrils and ear canals and prevent foreign particles from entering easily. The eyelashes can shield the eye from windblown debris with a quick blink. In windy or rainy conditions, we can squint so that the eyelashes protect the eye without completely obstructing our vision.

The eyebrows are often presumed to keep debris out of the eyes, but this seems doubtful. It is more plausible that they function mainly to enhance facial expression. Movements of the eyebrows are an important means of nonverbal communication in humans of all cultures. In addition, many species of monkeys and apes greet each other, assert their dominance, and break up quarrels with a quick flash of the eyebrows.

Nails

Fingernails and toenails are clear, hard derivatives of the stratum corneum. They are composed of very thin, scale-like dead cells, densely packed together and filled with

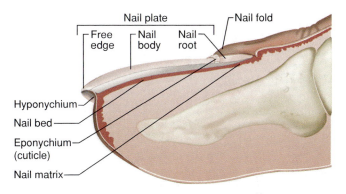

Figure 7.11 Anatomy of a fingernail.

parallel fibers of hard keratin. Most mammals have claws, whereas flat nails are one of the distinguishing characteristics of primates (see chapter 1). Flat nails allow for more fleshy and sensitive fingertips, while they also serve as strong keratinized "tools" that can be used for digging, grooming, picking apart food, and other manipulations. Fingernails grow at a rate of about 1 mm per week and toenails somewhat more slowly. New cells are added to the nail plate by mitosis in the *nail matrix* at its proximal end. Contrary to some advertising claims, adding gelatin to the diet has no effect on the growth or hardness of the nails.

The anatomical features of a nail are shown in figure 7.11 and defined in table 7.3. The most important of these are the aforementioned **nail matrix,** a growth zone concealed beneath the skin at the proximal edge of the nail, and the **nail plate,** which is the visible portion covering the dorsum of the fingertip. The nail groove and *hyponychium* (HY-po-NICK-ee-um) tend to accumulate dirt and bacteria and require special attention when scrubbing for duty in an operating room or nursery. The appearance of the nails can be valuable to medical diagnosis. An iron deficiency, for example, may cause the nails to become flat or concave (spoonlike) rather than convex. The nails and fingertips become clubbed in conditions of long-term hypoxemia—deficiency of oxygen in the blood—resulting from congenital heart defects and other causes.

Table 7.3 Anatomical Features of Nails

Nail Bed	The skin upon which the nail plate rests
Nail Plate	The clear, keratinized portion of a nail
Root	The proximal end of a nail, underlying the nail fold
Body	The major portion of the nail plate, overlying the nail bed
Free Edge	The portion of the nail plate that extends beyond the end of the digit
Hyponychium[19]	The space beneath the free edge of the nail plate, where dirt accumulates
Nail Fold	The fold of skin around the margins of the nail plate
Nail Groove	The groove where the nail fold meets the nail plate
Eponychium[20]	The dead epidermis that covers the proximal end of the nail plate; commonly called the cuticle
Nail Matrix	The growth zone (mitotic tissue) at the proximal end of the nail, corresponding to the stratum basale of the epidermis
Lunule[21]	The region at the base of the nail that appears as a small white crescent because it overlies a thick stratum basale that obscures dermal blood vessels from view

Key Point Review

9. Distinguish between vellus and terminal hair.

10. Describe the three regions of a hair along its length and the three layers of a hair in cross section.

11. State the function of the hair papilla, hair receptors, and arrector pili.

12. Give a reasonable explanation for the presence of scalp, nostril, and axillary hair and for eyebrows and eyelashes.

13. Define nail body, nail fold, eponychium, hyponychium, and nail matrix.

Cutaneous Glands

▼Objectives

When you have completed this section, you should be able to
- name two types of sweat glands and describe the structure and function of each;
- describe sebaceous and ceruminous glands; and
- distinguish between breasts and mammary glands and discuss their respective functions.

The skin has five types of glands: *merocrine sweat glands, apocrine sweat glands, sebaceous glands, ceruminous glands,* and *mammary glands.*

19. *onycho* = nail
20. *ep,* from *epi* = upon, above
21. *lun* = moon + *ule* = little

Sweat Glands

Sudoriferous[22] (soo-dor-IF-er-us) **glands,** or **sweat glands,** are the most numerous cutaneous glands. There are two kinds, merocrine and apocrine, which were defined in the previous chapter. **Merocrine sweat glands** (fig. 7.12c) produce watery perspiration that primarily serves to cool the body. There are about 3 to 4 million merocrine sweat glands in the adult skin, with a total weight about equal to that of a kidney. They are especially abundant on the palms, soles, and forehead, but they are widely distributed over the rest of the body as well. Each is a simple tubular gland with a twisted coil in the dermis or hypodermis and an undulating or coiled duct leading to a sweat pore on the skin surface. This duct is lined by a stratified cuboidal epithelium in the dermis and by keratinocytes in the epidermis. Amid the secretory cells at the deep end of the gland, there are specialized **myoepithelial**[23] **cells** with properties similar to smooth muscle. They contract in response to stimuli from the sympathetic nervous system and squeeze perspiration up the duct.

Sweat begins as a protein-free filtrate of the blood plasma produced by the deep secretory portion of the gland. Most sodium chloride is reabsorbed from this filtrate as the secretion passes through the duct. Potassium ions, urea, lactic acid, ammonia, and some sodium chloride remain in the sweat. Some drugs are also excreted in the perspiration. On average, sweat is 99% water and has a pH ranging from 4 to 6. Although the sweat glands play a minor excretory role in ridding the body of wastes and foreign substances, they function primarily to cool the body. Each day they secrete about 500 mL of **insensible perspiration,** which does not produce noticeable wetness of the skin. Under conditions of exercise or heat, however, a person may lose as much as a liter of perspiration each hour. In fact, so much fluid can be lost from the bloodstream by sweating as to cause circulatory shock. Sweating with visible wetness of the skin is called **diaphoresis**[24] (DY-uh-foe-REE-sis).

Apocrine sweat glands occur in the groin, anal region, axilla, and areola and on the faces of mature males. They are absent from the axillary region of Koreans, however, and are very sparse in the Japanese. Their ducts lead into nearby hair follicles rather than opening directly onto the skin surface (fig. 7.12b). They produce their secretion in the same way that merocrine glands do—that is, by exocytosis. The distal, secretory part of an apocrine gland, however, has a much larger lumen than that of a merocrine gland, and so these glands have continued to be referred to as apocrine glands to distinguish them functionally and histologically from the merocrine

22. *sudor* = sweat + *fer* = to carry
23. *myo* = muscle
24. *dia* = through + *phoresis* = carrying

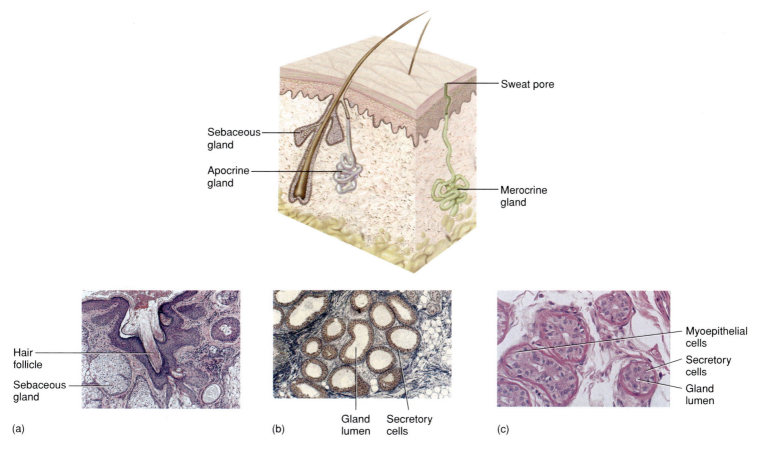

Figure 7.12 Principal types of cutaneous glands. (*a*) Sebaceous (holocrine) glands associated with a hair follicle (×100). (*b*) Apocrine sweat gland, showing several sections through a single coiled tubule (×100). (*c*) Merocrine (eccrine) sweat gland (×200).

type. Apocrine sweat is thicker and more milky than merocrine sweat because it has more fatty acids in it.

Apocrine sweat glands are scent glands that respond especially to stress and sexual stimulation. They do not develop until puberty, and they apparently correspond to the scent glands that develop in other mammals on attainment of sexual maturity. In women, they enlarge and shrink in synchrony with the menstrual cycle. Apocrine sweat does not have a disagreeable odor, and indeed it is considered attractive or arousing in some cultures, where it is as much a part of courtship as artificial perfume is to other people. Clothing, however, traps stale sweat long enough for bacteria to degrade the secretion and release free fatty acids with a rancid odor. Disagreeable body odor is called *bromhidrosis.* It occasionally indicates a metabolic disorder, but more often it reflects poor hygiene.

Many mammals have apocrine scent glands associated with specialized tufts of hair; in humans their bodily distribution matches that of the pubic, axillary, and male facial hair, suggesting that these glands are similar to other mammalian scent glands in function. The hair serves to retain the aromatic secretion and regulate its rate of evaporation from the skin. Thus, it seems no mere coincidence that women's faces lack both apocrine scent glands and a beard. It has also been demonstrated that men's beards grow faster when they live in the presence of women than when they live apart from them.

Sebaceous Glands

Sebaceous (see-BAY-shus) **glands** produce an oily secretion called **sebum** (SEE-bum). They are flask-shaped, with short ducts that usually open into a hair follicle (fig. 7.12*a*), although some of them open directly onto the skin surface. These are holocrine glands with little visible lumen. Their secretion consists of broken-down cells that are replenished by mitosis around the gland periphery. Sebum keeps the skin and hair from becoming dry, brittle, and cracked. The sheen of well-brushed hair is due to sebum distributed by the hairbrush. Ironically, we go to great lengths to wash sebum from the skin, only to replace it with various skin creams and hand lotions made of little more than sheep sebum (*lanolin*).

Ceruminous Glands

Ceruminous (seh-ROO-mih-nus) **glands** are simple, coiled, tubular glands with ducts leading to the skin surface. They are found only in the external ear canal, where their secretion combines with sebum and dead epidermal cells to form earwax, or **cerumen.** Cerumen is waterproof and bitter to the taste. It keeps the eardrum pliable and its bitterness may repel mites, insects, and other pests that would otherwise invade the ear canal.

Breasts and Mammary Glands

The breasts (mammae) and **mammary glands** are often mistakenly regarded as one and the same. Breasts, however, are present in both sexes, and even in females they rarely contain more than small traces of mammary glands. The female breast is one of the secondary sexual characteristics—anatomical features whose function lies primarily in their appeal to the opposite sex. The **mammary glands,** by contrast, are the milk-producing glands that develop within the female breast only during pregnancy and lactation. Mammary glands are modified apocrine sweat glands that produce a richer secretion and channel it through ducts to a nipple for more efficient conveyance to the offspring. The anatomy and physiology of the mammary gland are discussed in more detail in chapter 28.

In most mammals, two rows of mammary glands form along lines called the *mammary ridges,* or *milk lines.* The most anterior glands on the milk line produce the most milk, so in mammals with litters of young there is an advantage to nursing from those. Primates have dispensed with all but these two most productive glands. A few women, however, develop additional nipples or mammae along the milk line inferior to the primary mammae. In Colonial America this condition, called *polythelia,*[25] was used to incriminate women as supposed "witches," a practice dating from the Middle Ages.

The glands of the skin are summarized in table 7.4.

Key Point Review

14 How do merocrine and apocrine sweat glands differ in appearance and function?

15 What types of hair are associated with apocrine glands? Why?

16 What other type of gland is associated with a hair follicle? How does its mode of secretion differ from that of sweat glands?

17 Distinguish between the breasts and the mammary glands.

Integumentary Aging and Diseases

▼**Objectives**
When you have completed this section, you should be able to
• discuss the changes associated with puberty;
• discuss the changes associated with senescence;
• describe the three most common skin cancers; and
• describe the three classes of burns and discuss the most urgent concerns in burn treatment.

25. *poly* = many + *theli* = nipple

Table 7.4	Cutaneous Glands
Gland Type	**Definition**
Sudoriferous glands	Sweat glands
Merocrine	Sweat glands that function for evaporative cooling; widely distributed over the body surface; open by ducts onto the skin surface
Apocrine	Sweat glands that function as scent glands; found in the regions covered by the pubic, axillary, and male facial hair; open by ducts into hair follicles
Sebaceous glands	Holocrine oil-producing glands associated with hair follicles
Ceruminous glands	Glands of the ear canal that produce cerumen (earwax)
Mammary glands	Milk-producing glands located in the breasts

Aging

The integumentary system undergoes its most noteworthy changes at puberty and after middle age. At puberty, a girl's ovaries begin producing estrogen, which makes her skin thicken somewhat, although it remains thinner than a boy's. The skin of females also contains more blood vessels than that of males; consequently, females tend to blush more easily, bleed more freely, and have warmer skin than males. In adolescent boys, testosterone stimulates the growth of facial hair and causes the skin to become darker and thicker than in girls. In both sexes, testosterone stimulates the growth of the pubic and axillary hair and increases the secretion of sebum, which in turn causes various degrees of acne in most adolescents.

With age, all organ systems undergo degenerative changes called **senescence** (seh-NESS-ense). Integumentary senescence becomes most noticeable around the late 40s. Senescene of the integumentary and other organ systems is discussed in chapter 29.

Skin Cancer

With age and UV exposure, the probability of developing **skin cancer** increases. Skin cancer is most common on the face and neck, where exposure to the sun is greatest, and most common among fair-skinned people and people over the age of 50. There is simply no such thing as a healthy suntan.

Think About It
Skin cancer is relatively rare among blacks. Why do you think this is so?

Skin cancer is the most common of all cancers, but it is also one of the easiest to treat and has one of the highest survival rates when it is detected and treated early. There are three types of skin cancer named for the epidermal cells in which they originate and distinguished from each other by the types of lesions they produce. (A **lesion**[26] is any circumscribed zone of tissue injury in the skin or any other organ.)

Basal cell carcinoma is the most common type but also the least dangerous because it seldom metastasizes. It arises from cells of the stratum basale and eventually invades the dermis. On the surface, the lesion first appears as a small, shiny bump. As the bump enlarges, it often develops a central depression and a beaded "pearly" edge (fig. 7.13a). These lesions usually occur on the face and are treated by surgical removal and radiation therapy.

Squamous cell carcinoma arises from keratinocytes of the stratum spinosum, usually on the scalp, ears, lower lip, or back of the hand. The lesion has a raised, reddened, scaly appearance, later forming a concave ulcer with raised edges (fig. 7.13b). The chance of recovery is good with early detection and surgical removal, but if it goes unnoticed or is neglected, this cancer can metastasize to the lymph nodes and can be lethal.

Malignant melanoma is the most deadly skin cancer but accounts for only 5% of all cases. It usually arises from melanocytes of a preexisting mole. It metastasizes quickly and is often fatal if not treated immediately. The risk for malignant melanoma is greatest in people who experienced severe sunburns as children, especially redheads.

It is important to distinguish a mole from malignant melanoma. A mole usually has a uniform color and even contour, and it is no larger in diameter than the end of a pencil eraser (about 6 mm). If it becomes malignant, however, it forms a large, flat, spreading lesion with a scalloped border (fig. 7.13c). The American Cancer Society suggests an "ABCD rule" for recognizing malignant melanoma: *A* for asymmetry (one side of the lesion looks different from the other); *B* for border irregularity (the contour is not uniform but wavy or scalloped); *C* for color (often a mixture of brown, black, tan, and sometimes red and blue); and *D* for diameter (greater than 6 mm).

Burns

Burns are the leading cause of accidental death. They are usually caused by fires, kitchen spills, or excessively hot bath water, but they also can be caused by sunlight, ionizing radiation, strong acids and bases, or electrical shock. Burn deaths are primarily due to circulatory shock from fluid loss, infection from loss of the protective barrier of skin, and the toxic effects of **eschar** (ESS-car), the burned, dead tissue.

---

26. *lesio* = to injure

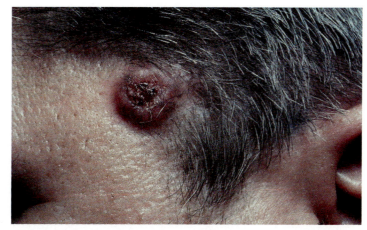

(a)

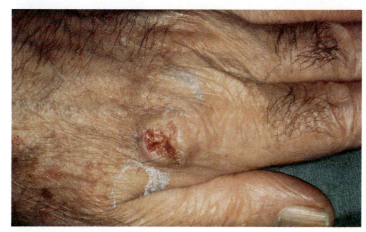

(b)

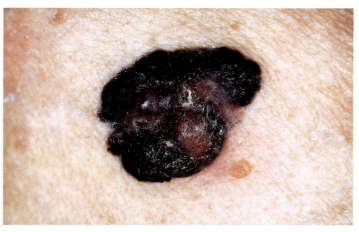

(c)

Figure 7.13 Typical skin cancer lesions. (*a*) Basal cell carcinoma. (*b*) Squamous cell carcinoma. (c) Malignant melanoma.

Burns are classified according to the depth of tissue involvement (fig. 7.14). **First-degree burns** involve only the epidermis and are marked by redness, slight edema, and pain. They heal in a few days and seldom leave scars. Most sunburns are first-degree burns.

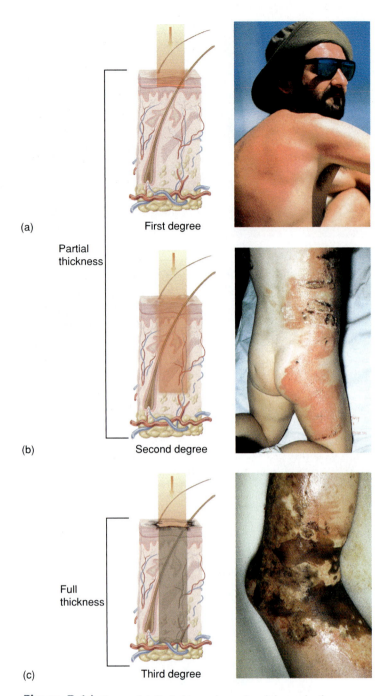

(a) Partial thickness — First degree

(b) — Second degree

(c) Full thickness — Third degree

Figure 7.14 Burns. (*a*) First-degree burn, involving only the epidermis. (*b*) Second-degree burn, involving the epidermis and part of the dermis. First- and second-degree burns are also called partial thickness burns. (*c*) Third-degree burn, which destroys all of the epidermis, dermis and often part or all of the hypodermis.

Second-degree burns involve the epidermis and part of the dermis but leave at least some of the dermis intact. First- and second-degree burns are therefore also known as **partial-thickness burns.** A second-degree burn may be red, tan, or white and is blistered and very painful. It may take from 2 weeks to several

months to heal and may leave scars. The epidermis regenerates by division of epithelial cells in the hair follicles and sweat glands and those around the edges of the lesion. Some sunburns and many scalds are second-degree burns.

Third-degree burns are also called **full-thickness burns** because the epidermis, dermis, and often some deeper tissue are completely destroyed. Since no dermis remains, the skin can regenerate only from the edges of the wound. Third-degree burns often require skin grafts (see chapter essay, p. 221). If a third-degree burn is left to itself to heal, contracture (abnormal connective tissue fibrosis) and severe disfigurement may result.

Think About It

A third-degree burn may be surrounded by painful areas of first- and second-degree burns, but the region of the third-degree burn is painless. Explain the basis for this lack of pain.

The two most urgent considerations in treating a burn patient are fluid replacement and infection control. A patient can lose several liters of water, electrolytes, and protein each day from the burned area. As fluid is lost from the tissues, more is transferred from the bloodstream to replace it, and the volume of circulating blood declines. A patient may lose up to 75% of the blood plasma within a few hours, potentially leading to circulatory shock and cardiac arrest—the principal cause of death in burn patients. Intravenous fluid must be given to make up for this loss. A burn patient also requires thousands of extra calories daily to compensate for protein loss and the demands of tissue repair. Supplementary nutrients are given intravenously or through a gastric tube.

Infection is controlled by keeping the patient in an aseptic (germ-free) environment and administering antibiotics. The eschar is sterile for the first 24 hours, but then it quickly becomes infected and may have toxic effects on the digestive, respiratory, and other systems. Its removal, called **debridement** (deh-BREED-ment), is essential to infection control.

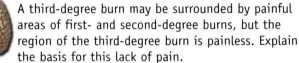

Key Point Review

18 What two hormones especially affect the integument at puberty? What effects do they have?

19 What types of cells are involved in the three types of skin cancer? Which of the three is most dangerous and what are its early warning signs?

20 Describe first-, second-, and third-degree burns.

21 What are the two most urgent concerns in the treatment of a burn patient? How do health-care workers deal with these concerns?

Skin Grafts and Artificial Skin

Third-degree burns leave no dermal tissue as a foundation for regenerating what was lost, and therefore they generally require skin grafts. The ideal graft is an **autograft**—tissue taken from another location on the same person's body—since this is not rejected by the patient's immune system. An autograft is generally performed by taking epidermis and part of the dermis from an undamaged area such as the thigh or buttock and grafting it to a burned area. This method is called a *split-skin graft* because part of the dermis is left behind to proliferate and replace the epidermis that was removed—the same way a second-degree burn heals.

An autograft may not be possible, however, if the burns are too extensive. The best treatment option in this case is an **isograft,** using skin from an identical twin. Because the donor and recipient are genetically identical, the recipient's immune system is unlikely to reject the graft. Since identical twins are rare, however, the best that can be hoped for in most cases is donor skin from another close relative.

A **homograft,** or **allograft,** is a graft from an unrelated person. Cadaver skin from a skin bank is often used. The immune system attempts to reject homografts, but they may suffice as temporary coverings for the burned area, and they can be replaced by autografts when the patient is well enough for healthy skin to be removed from an undamaged area of the body.

Pig skin is sometimes used on burn patients but presents the same problem of immune rejection. A graft of tissue from a different species is called a **heterograft,** or **xenograft.** This is a special case of a *heterotransplant,* which also includes transplantations of organs such as baboon hearts or livers into humans. Heterografts and heterotransplants are short-term methods of maintaining a patient until a better, long-term solution is possible. The immune reaction can be suppressed by drugs called *immunosuppressants,* but this procedure is risky because it also lowers the patient's resistance to infection, which is already compromised in a burn patient.

Some alternatives to skin grafts are also being used. Burns are sometimes temporarily covered with amniotic membrane (the membrane that surrounds a developing fetus). In addition, tiny keratinocyte patches cultured with growth stimulants have produced sheets of epidermal tissue as large as the entire body surface. These can replace large areas of burned tissue. Dermal fibroblasts also have been successfully cultured and used for skin grafts. A drawback of these approaches is that the culture process requires 3 or 4 weeks, which is too long a wait for some patients with severe burns. Efforts are underway to develop techniques of cryopreservation (cold storage) of cultured tissue so that it will be more readily available for homografts.

Various kinds of artificial skin have also been developed as a temporary burn covering. Some of these are sheets with an upper layer of silicone and a lower layer of collagen and chondroitin sulfate. This stimulates growth of connective tissue and blood vessels from the patient's underlying tissue. The artificial skin can be removed after about 3 weeks and replaced with a thin layer of cultured or grafted epidermis. Alternatively, epidermal cells can be seeded onto the lower layer of the artificial skin before it is placed on the patient.▲

Connective Issues

c o n n e c t i v e   i s s u e s

Interactions Between the INTEGUMENTARY SYSTEM and Other Organ Systems

- Role of skin in vitamin D synthesis promotes calcium absorption needed for muscle contraction; skin dissipates excess heat generated by muscular activity

Nervous System
- Regulates diameter of cutaneous blood vessels; stimulates perspiration and contraction of arrector pili muscles
- Sensory impulses from skin transmitted to nervous system

Endocrine System
- Sex hormones cause changes in integumentary features at puberty (oily skin and facial, pubic, and axillary hair); some hormone imbalances have pathological effects on skin
- Skin converts some steroid hormones to their active forms

Circulatory System
- Delivers O_2, nutrients, and hormones to skin and carries away wastes; hemoglobin colors skin
- Dermal vasoconstriction diverts blood to other organs; skin prevents loss of fluid from cardiovascular system; dermal mast cells cause vasodilation and increased blood flow

Lymphatic/Immune Systems
- Lymphatic system controls fluid balance and prevents edema; immune cells protect skin from infection and promote tissue repair
- Langerhans cells detect foreign substances

Respiratory System
- Provides O_2 and removes CO_2
- Nasal hairs filter macroparticles that might otherwise be inhaled

Urinary System
- Disposes of wastes and maintains electrolyte and pH balance
- Skin excretes salts and some nitrogenous wastes in sweat; keratinized epidermis limits fluid loss from skin

Digestive System
- Provides nutrients needed for integumentary development and function; nutritional disorders often reflected in appearance of skin, hair, and nails
- Role of skin in vitamin D synthesis enables intestinal absorption of calcium

Reproductive System
- Gonadal sex hormones promote growth, maturation, and maintenance of skin
- Cutaneous receptors respond to erotic stimuli; mammary glands produce milk to nourish infants

All Systems
The integumentary system serves all other systems by providing a physical barrier to environmental hazards.

Skeletal System
- Supports skin at scalp and other places where bone lies close to surface
- Role of skin in vitamin D synthesis promotes calcium absorption needed for bone growth and maintenance

Muscular System
- Active muscles generate heat to warm skin; contractions of skeletal muscles pull on skin, producing facial expressions

Structure of the Skin (pp. 202–209)

1. General structure
   a. Covers area of 1.5 to 2.0 square meters
   b. Accounts for about 15% of body weight
   c. Usually 1 to 2 mm thick
   d. Variations: thick and thin skin
2. Layers (table 7.1)
   a. Epidermis
   b. Dermis
3. Hypodermis (subcutaneous tissue)
4. Color
   a. Sources of normal color
      • Hemoglobin
      • Melanin
      • Carotene
   b. Abnormal coloration (table 7.2)
5. Markings
   a. Birthmarks (hemangiomas)
   b. Moles and freckles
   c. Friction ridges
   d. Flexion creases

Functions of the Skin (pp. 209–211)

1. Barrier functions
   a. Barrier to pathogens
      • Difficulty of physical penetration
      • Antibacterial effect of sebum
      • Acid mantle
      • Inflammation
   b. Barrier to water
   c. Barrier to UV radiation
2. Vitamin D synthesis
3. Cutaneous absorption
   a. Gas exchange
   b. Permeability to organic compounds
   c. Absorption of vitamins, drugs, and poisons
4. Sensory roles
5. Thermoregulation
   a. Heat-losing mechanisms
      • Vasodilation
      • Perspiration
   b. Heat-retaining mechanism
      • Vasoconstriction
6. Social interaction

Hair and Nails (pp. 211–216)

1. Hard and soft keratin
2. Hair (pili)
   a. Lanugo, vellus, and terminal hair
   b. Bulb, root, and shaft
   c. Medulla, cortex, and cuticle
   d. Pigments: melanin, trichosiderin
3. Hair follicle
   a. Epithelial root sheath
   b. Connective tissue root sheath
   c. Hair receptors
   d. Arrector pili muscle
4. Growth of hair
   a. Growth and dormant phases
   b. Alopecia (baldness)
   c. Pattern baldness
   d. Hypertrichosis (hirsutism)
5. Functions of hair
   a. Vestigial (most body hair)
   b. Thermoregulation (scalp hair)
   c. Protection from sun (scalp hair)
   d. Recognition of species, sex, etc.
   e. Association with scent glands (pubic and axillary hair, beard)
   f. Guard hairs (vibrissae)
   g. Facial expression (eyebrows)
6. Nails
   a. Structure (table 7.3)
   b. Diagnostic value

Cutaneous Glands (pp. 216–218)

1. Sudoriferous (sweat) glands
   a. Merocrine glands and cooling
      • Location and structure
      • Formation of perspiration
      • Insensible perspiration
      • Diaphoresis
   b. Apocrine glands and scent
      • Location and structure
      • Development and function
2. Sebaceous (oil) glands and sebum
3. Ceruminous glands and cerumen (earwax)
4. Breasts and mammary glands

Integumentary Aging and Diseases (pp. 218–220)

1. Puberty
   a. Estrogen and female skin
   b. Testosterone and male skin
   c. Effects of testosterone on both sexes
2. Senescence
3. Skin cancer
   a. Basal cell carcinoma
   b. Squamous cell carcinoma
   c. Malignant melanoma
4. Burns
   a. First-, second-, and third-degree
   b. Partial- versus full-thickness
   c. Treatment concerns
      • Fluid replacement
      • Infection and asepsis

Selected Vocabulary

Also review the terms in tables 7.1 through 7.4, which are not repeated here.

integumentary system 202
dermatology 202
skin (integument) 202
thick skin 203
thin skin 203
cornified 203
keratinized 203
keratin 203
keratinocyte 203
melanocyte 203
Merkel cell 204
Langerhans cell 204

keratohyalin 204
dermal papilla 206
linea albicantes 206
subcutaneous tissue 206
superficial fascia 206
panniculus adiposus 206
subcutaneous fat 206
hemoglobin 206
melanin 207
carotene 208
hemangioma 208
nevus 208
mole 208
freckle 209
friction ridge 209
flexion crease 209

acid mantle 209
accessory organ 211
appendage 211
soft keratin 212
hard keratin 212
pilus 212
follicle 212
lanugo 212
terminal hair 212
hair bulb 212
hair root 212
hair shaft 212
papilla 213
hair medulla 213
hair cortex 213
hair cuticle 213

trichosiderin 213
epithelial root sheath 214
connective tissue root sheath 214
arrector pili 214
alopecia 214
pattern baldness 214
hirsutism 214
hypertrichosis 214
guard hair 215
vibrissa 215
sweat gland 216
myoepithelial cell 216
insensible perspiration 216
diaphoresis 216
sebum 217
cerumen 218

Testing Your Recall   Answers in Appendix C

1. Cells of the _____ are keratinized and dead.
   a. papillary layer
   b. stratum spinosum
   c. stratum basale
   d. stratum corneum
   e. stratum granulosum

2. Which of the following terms is *least* closely related to the rest?
   a. panniculus adiposus
   b. superficial fascia
   c. reticular layer
   d. hypodermis
   e. subcutaneous tissue

3. Which of the following skin conditions would most likely result from liver failure?
   a. pallor
   b. erythema
   c. jaundice
   d. albinism
   e. melanization

4. All of the following mechanisms interfere with microbial invasion of the skin *except*
   a. the acid mantle.
   b. melanization.
   c. inflammation.
   d. keratinization.
   e. sebum.

5. The hair on a six-year-old's arms is
   a. vellus.
   b. lanugo.
   c. trichosiderin.
   d. definitive hair.
   e. eponychium.

6. Which of the following terms is *least* closely related to the rest?
   a. lunula
   b. nail plate
   c. hyponychium
   d. free edge
   e. cortex

7. Which of the following is a scent gland?
   a. eccrine gland
   b. sebaceous gland
   c. apocrine gland
   d. merocrine gland
   e. ceruminous gland

8. _____ are skin cells with a sensory function.
   a. Merkel cells
   b. Langerhans cells
   c. Prickle cells
   d. Melanocytes
   e. Keratinocytes

9. Which of the following glands produce the skin's acid mantle?
   a. merocrine sweat glands
   b. apocrine sweat glands
   c. mammary glands
   d. ceruminous glands
   e. sebaceous glands

10. Which of the following cells alert the immune system to pathogens invading the epidermis?
   a. fibroblasts
   b. melanocytes

   c. keratinocytes
   d. Langerhans cells
   e. Merkel cells

11. _____ perspiration refers to sweating without noticeable wetness of the skin.

12. A muscle that causes a hair to stand on end is called a/an _____.

13. The process of removing burned skin from a patient is called _____.

14. Pattern baldness results from not only the right genotype but also an adequate concentration of _____ in the blood.

15. Blueness of the skin due to low oxygen concentration in the blood is called _____.

16. Projections of the dermis toward the epidermis are called _____.

17. Cerumen is more commonly known as _____.

18. The holocrine glands attached to a hair follicle are called _____ glands.

19. The space underneath the fingernails that accumulates dirt is known as the _____.

20. A hair is nourished by blood vessels in a connective tissue projection called the _____.

Testing Your Comprehension   Answers in *Study Guide*

1. Many organs of the body contain numerous smaller organs, perhaps even thousands. Use the skin as an example to illustrate this point.

2. Certain aspects of human form and function are easier to understand when viewed from the standpoint of

comparative anatomy and evolution. Apply this concept to the integumentary system. What features of the human integument would be difficult to interpret without this perspective?

3. Explain how unity of form and function is reflected in the fact that

the dermis has two histologically distinct layers, instead of just one.

4. Cold weather does not interfere with oxygen uptake by the blood, but it can cause cyanosis anyway. Why?

5. Why is it important that the epidermis be effective, but not *too* effective, in screening out UV radiation?

Web Site Link

For a listing of the most current web sites related to this chapter, please visit the Saladin homepage at:

http://www.mhhe.com/sciencemath/biology/saladin/

[Bone Tissue

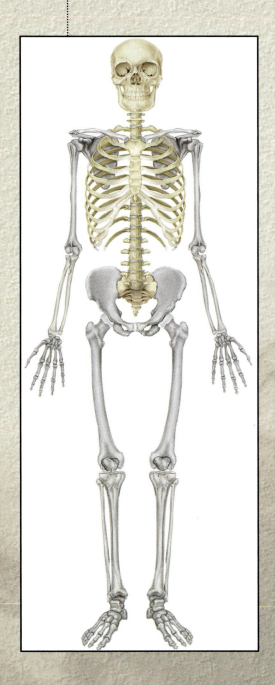

Special/**Topics**

Nothing symbolizes death as much as a skull or skeleton.[1] The dry bones presented for laboratory study suggest that the skeleton is an inert scaffold for the body, like the steel girders of a building. Seeing it in such a sanitized form makes it easy to forget that the living skeleton is made of dynamic tissues, full of cells—that it continually remodels itself and interacts physiologically with all of the other organ systems of the body. The skeleton is permeated with nerves and blood vessels, and it receives a blood flow of about 200 to 400 mL/min in adults, attesting to its sensitivity and metabolic activity.

The skeletal system is composed of three types of organs—bones, cartilages, and ligaments—tightly joined together to form a strong, flexible organ system. *Cartilage,* the forerunner of most bones in embryonic and childhood development, covers many joint surfaces in the mature skeleton. *Ligaments* hold bones together at the joints and are discussed in chapter 10. *Tendons* are structurally similar to ligaments but attach muscle to bone; they are discussed with the muscular system in chapter 11.

Tissues and Organs of the Skeletal System

▼Objectives
When you have completed this section, you should be able to
- summarize the organs and tissues that constitute the skeletal system;
- defend the argument that osseous tissue is not inert but a living, dynamic tissue with numerous functions;
- distinguish between bone as a tissue and as an organ; and
- state the number of bones in a typical adult skeleton and explain the reasons for deviations from this number.

Functions of the Skeleton

Physical support is the most obvious of the skeleton's functions, but it is by no means its only function. As summarized in table 8.1, the skeleton plays a variety of roles in the body, many of which go unnoticed by most people.

Bones and Osseous Tissue

Each bone is a discrete organ composed of multiple tissues. There is the hard, calcified **osseous[2] tissue** that is preserved so well after death, but also present in the bones are blood,

Table 8.1	Functions of the Skeleton
Function	**Examples or Explanation**
Support	Bones of the legs, pelvis, and vertebral column hold up the body; the mandible supports the teeth; nearly all bones provide support for muscles; many other soft organs are directly or indirectly supported by nearby bones.
Protection	The cranium encloses the brain; the spinal column encloses the spinal cord; the rib cage encloses the heart and lungs; the pelvis encloses the organs of the reproductive system and lower urinary and digestive tracts; most bones enclose bone marrow.
Movement	Skeletal muscles would serve no useful purpose if not for the rigid attachment and leverage provided by bones. Leg and arm movements are the most obvious examples; a less obvious one is that ventilation of the lungs depends on movement of the ribs by skeletal muscles.
Blood formation	Red bone marrow is the major producer of blood cells, including most cells of the immune system.
Electrolyte balance	The skeleton is the body's main mineral reservoir. It stores calcium and phosphate and releases them according to the body's physiological needs.
Acid-base balance	Bone buffers the blood against excessive pH changes by absorbing or releasing alkaline mineral salts.
Detoxification	Bone tissue removes heavy metals and other foreign elements from the blood and helps reduce their effects on nervous and other tissues. It can later release these more slowly for excretion.

bone marrow, cartilage, and adipose, nervous, and fibroconnective tissues. The word *bone* has different meanings in different contexts. At the level of gross anatomy, it denotes an organ composed of all the aforementioned tissue types. At the histological level, it denotes osseous tissue, which is only one component of a bone.

The scientific study of bones is called **osteology.** The student typically begins by examining dried bones that are either held together by wires and rods to show their spatial relationships to each other (an articulated[3] skeleton) or taken apart (disarticulated) so that the anatomy of individual bones can be studied in more detail.

In this chapter, we examine the diverse roles and the general form and function of osseous tissue. In chapter 9, we study the gross anatomy of the skeletal system, and in chapter 10, the form and function of the joints.

---

1. *skelet* = dried up
2. *os, osse, oste* = bone

3. *artic* = joint

Table 8.2   The 206 Bones of a Typical Adult Skeleton

The number of bones of each type is indicated in parentheses.

Axial Skeleton

Skull	Total 22

Cranial bones

Frontal bone (1)

Parietal bones (2)

Occipital bone (1)

Temporal bones (2)

Sphenoid bone (1)

Ethmoid bone (1)

Facial bones

Maxillae (2)

Palatine bones (2)

Zygomatic bones (2)

Lacrimal bones (2)

Nasal bones (2)

Vomer (1)

Inferior nasal conchae (2)

Mandible (1)

Auditory (middle-ear) Ossicles	Total 6

Malleus (2)

Incus (2)

Stapes (2)

Hyoid Bone (1)	Total 1
Vertebral Column	Total 26

Cervical vertebrae (7)

Thoracic vertebrae (12)

Lumbar vertebrae (5)

Sacrum (1)

Coccyx (1)

Thoracic Cage	Total 25

Ribs (24)

Sternum (1)

Appendicular Skeleton

Pectoral Girdle	Total 4

Scapulae (2)

Clavicles (2)

Upper Extremity	Total 60

Humerus (2)

Radius (2)

Ulna (2)

Carpals (16)

Metacarpals (10)

Phalanges (28)

Pelvic Girdle	Total 2

Ossa coxae (2)

Lower Extremity	Total 60

Femur (2)

Patella (2)

Tibia (2)

Fibula (2)

Tarsals (14)

Metatarsals (10)

Phalanges (28)

Grand Total: 206

Bones of the Skeletal System

It is often stated that there are 206 bones in the skeleton, but this is only a typical adult count. At birth there are about 270, and during childhood even more are formed. With age, however, the number of bones decreases as separate bones gradually fuse. For example, each half of the adult pelvis is a single bone called the *os coxae* (oss COC-see), which results from the fusion of three child-hood bones—the ilium, ischium, and pubis. The fusion of several bones, completed by late adolescence or the early 20s, brings about the average adult number of 206.

This number varies even among adults. One reason is that new bones called **sesamoid**[4] **bones** form within some tendons. Most of these are small, rounded bones in such locations as the knuckles. The *patella* (kneecap) is the largest sesamoid bone. Another reason the number varies is that some adults have extra bones in the skull called **sutural** (SOO-chure-ul), or **wormian,**[5] **bones.**

The skeleton is divisible into two main portions: (1) the **axial skeleton** forms the median axis of the body and includes the skull, middle-ear bones, hyoid bone, vertebral column, rib cage, and sternum; (2) the **appendicular skeleton** includes the upper and lower extremities and the pectoral and pelvic girdles, which join the appendages to the axial skeleton. An overview of the skeleton is shown in figure 8.1, and the typical adult bones are listed in table 8.2.

4. *sesam* = sesame seed + *oid* = resembling

5. Ole Worm (1588–1654), Danish physician

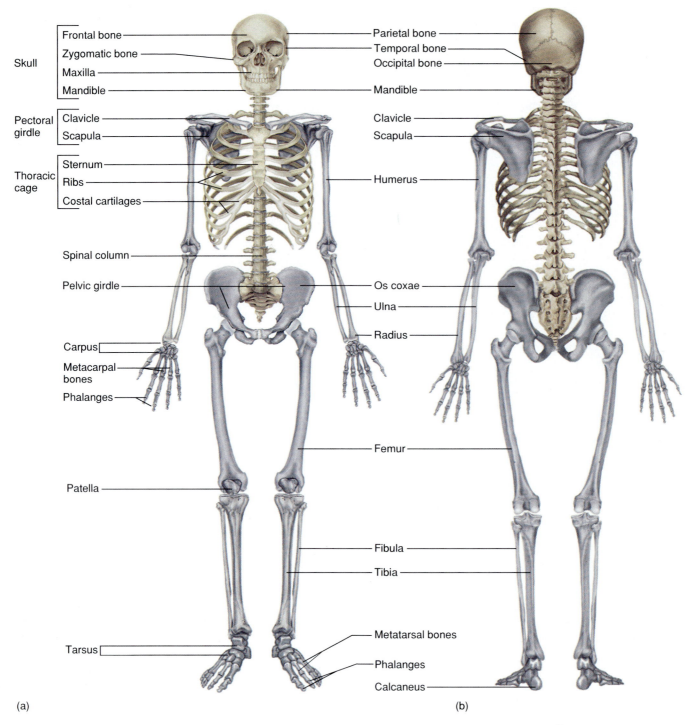

Skull
- Frontal bone
- Zygomatic bone
- Maxilla
- Mandible

Pectoral girdle
- Clavicle
- Scapula

Thoracic cage
- Sternum
- Ribs
- Costal cartilages

Spinal column

Pelvic girdle

Carpus
Metacarpal bones
Phalanges

Patella

Tarsus

Parietal bone
Temporal bone
Occipital bone
Mandible

Clavicle
Scapula

Humerus

Os coxae
Ulna
Radius

Femur

Fibula
Tibia

Metatarsal bones
Phalanges
Calcaneus

(a)   (b)

Figure 8.1 The human skeleton. (*a*) Anterior view. (*b*) Posterior view. The appendicular skeleton is colored blue. ⚔

Gross Anatomy of Bones

▼Objectives

When you have completed this section, you should be able to
- describe the four major groups of bones as classified by shape and function; and
- define terms that describe the structural features of bones.

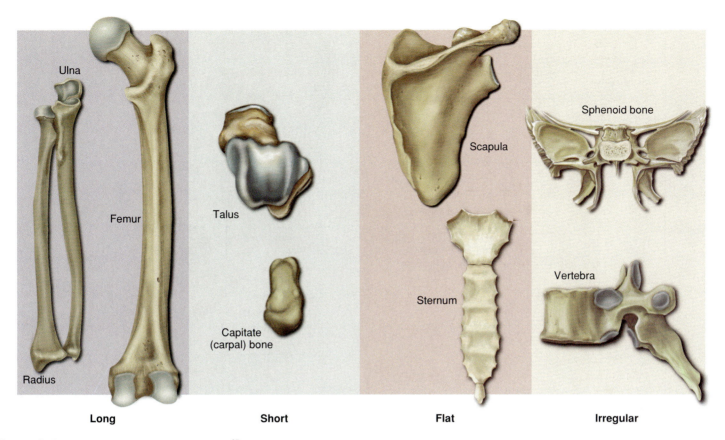

Figure 8.2 Classification of bones by shape.

Long   Short   Flat   Irregular

Long bones: Ulna, Radius, Femur. Short bones: Talus, Capitate (carpal) bone. Flat bones: Scapula, Sternum. Irregular bones: Sphenoid bone, Vertebra.

The Shapes of Bones

As shown in figure 8.2, the bones can be classified into four groups by shape and function:

1. **Long bones.** These bones are conspicuously longer than wide. Like crowbars, they serve as rigid levers that are acted upon by the skeletal muscles to produce body movements. Long bones include the humerus, radius, and ulna of the arm; the metacarpals and phalanges of the hand; the femur of the thigh; the tibia and fibula of the leg; and the metatarsals and phalanges of the feet.

2. **Short bones.** These are more nearly equal in length and width. They include the carpal (wrist) and tarsal (ankle) bones. They have limited motion and merely glide across one another, enabling the ankles and wrists to bend in multiple directions.

3. **Flat bones.** These enclose and protect soft organs and provide broad surfaces for muscle attachment. They include the cranium, mandible, ribs, sternum, scapulae, and pelvis.

4. **Irregular bones.** These are bones with elaborate shapes that do not fit into any of the preceding

categories. They include the vertebrae and some of the skull bones, such as the sphenoid.

General Features of Bones

Knowing the terms used to describe a long bone will help you to understand the anatomy of other bones. A longitudinal section through a long bone is shown in figure 8.3a. It is immediately apparent that much of the bone is composed of a cylinder of thick, dense, white osseous tissue; this is called **compact,** or **dense, bone.** The cylinder encloses a space called the **medullary** (MED-you-lerr-ee) **cavity,** which contains bone marrow. At the ends of the bone, the central space is occupied by a form of osseous tissue called **spongy,** or **cancellous, bone,** which looks like a sponge although it is not soft. The skeleton is about three-quarters compact bone and one-quarter spongy bone by weight. Spongy bone is found at the ends of the long bones and in the middle of nearly all others. It is always enclosed by more durable compact bone. A flat bone is like a sandwich, with a layer of spongy bone called **diploe** (DIP-lo-ee) lying between two layers of compact bone (fig. 8.3b). A moderate blow to the skull can fracture the outer layer of

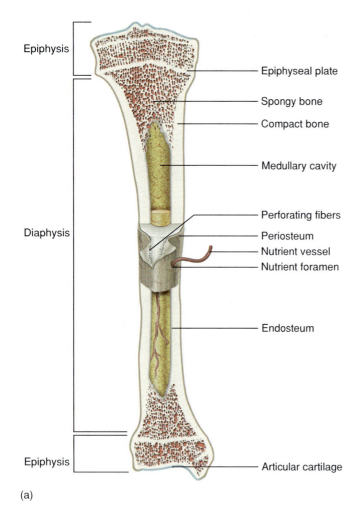

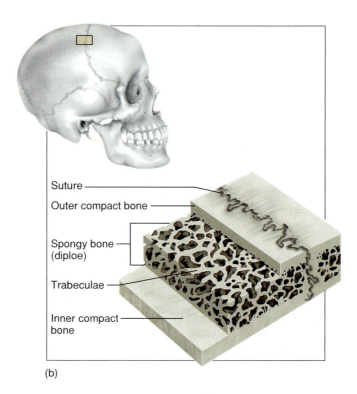

(b)

(a)

Figure 8.3 Anatomy of (a) a long bone and (b) a flat bone. 🖢

compact bone, but the diploe may absorb the damage and leave the inner layer of compact bone unharmed.

Another obvious feature of a long bone is its shaft, called the **diaphysis**[6] (dy-AF-ih-sis), with an expanded head at each end called the **epiphysis**[7] (eh-PIF-ih-sis). The diaphysis provides leverage, while the epiphysis is enlarged to strengthen the joint and provide added surface area for the attachment of tendons and ligaments. The joint surface where one bone meets another is covered with a layer of hyaline cartilage called the **articular cartilage.** Together with a lubricating fluid secreted between the bones, this cartilage enables a joint to move far more freely than it would if one bone rubbed directly against the other. Blood vessels penetrate into the diaphysis through minute holes called **nutrient foramina** (for-AM-ih-nuh); we will trace where they go when we consider the histology of bone.

Externally, a bone is covered with a dense irregular fibrous sheath called the **periosteum**[8] (fig. 8.3a). Some collagen fibers of the periosteum are continuous with the tendons that attach a muscle to the bone. Others continue into the calcified matrix of the bone, where they are called **perforating (Sharpey**[9]**) fibers.** The periosteum thus provides strong attachment and continuity from tendon to bone. (The continuity from tendon to muscle is described in chapter 11.) The periosteum is also important to the growth of bone and healing of fractures in ways explained later in this chapter. There is no periosteum over the articular cartilage. The internal surface of the bone is covered with a layer of sparse reticular connective tissue called **endosteum,**[10] which lines the medullary cavity.

In children and adolescents, an **epiphyseal** (EP-ih-FIZZ-ee-ul) **plate** of hyaline cartilage separates the marrow spaces of the epiphysis and diaphysis (fig 8.3a). On X rays, it appears as a transparent line at the end of a long bone (see fig. 8.13). The plate functions as a growth zone where the bones can elongate. Between the ages of 18 and 20, the epiphyseal plates undergo *closure*—their cartilage is exhausted and the epiphysis and diaphysis become joined by bone. When closure of the plates is complete, the bones cannot grow any longer and a person cannot grow any taller. In adults, the site of the former epiphyseal plate is marked by an *epiphyseal line* on the bone surface.

---

6. *dia* = across, separate + *physis* = growth; originally named for a ridge of the tibial shaft
7. *epi* = upon, above

8. *peri* = around + *oste* = bone
9. William Sharpey (1802–80), Scottish histologist
10. *end* = within

Table 8.3	Anatomical Features of Bones
Term	**Description and Example**
Articulations	
Condyle	A rounded knob (occipital condyles of the skull)
Facet	A smooth, flat, slightly concave or slightly convex articular surface (articular facets of the vertebrae)
Head	The prominent expanded end of a bone, sometimes rounded (head of the femur)
Extensions and Projections	
Crest	A narrow ridge (iliac crest of the pelvis)
Line	A slightly raised, elongated ridge (nuchal lines of the skull)
Epicondyle	A projection superior to a condyle (medial epicondyle of the femur)
Process	Any bony prominence (mastoid process of the skull)
Trochanter	A massive process unique to the femur
Spine	A sharp, slender, or narrow process (spine of the scapula)
Tubercle	A small, rounded process (greater tubercle of the humerus)
Tuberosity	A rough surface (tibial tuberosity)
Depressions	
Alveolus	A pit or socket (tooth socket)
Fossa	A shallow, broad, or elongated basin (mandibular fossa)
Fovea	A small pit (fovea capitis of the femur)
Sulcus	A groove for a tendon, nerve, or blood vessel (intertubercular sulcus of the humerus)
Passages	
Foramen	A hole through a bone, usually round (foramen magnum of the skull)
Canal or Meatus	A tubular passage or tunnel through a bone (auditory meatus of the ear)
Fissure	A slit through a bone (orbital fissures behind the eye)
Sinuses	Spaces or cavities within a bone (frontal sinus of the skull)

Elaborations of Bone Structure

The surface of a bone is marked by elevations, depressions, canals, pores, slits, and articular surfaces. The terms used for these features are listed in table 8.3, and some of the features are illustrated in figure 8.4. You must be familiar with the basic vocabulary of osteology in order to understand the gross anatomy in chapter 9; some of the terms will become clearer as you study that chapter and encounter additional examples.

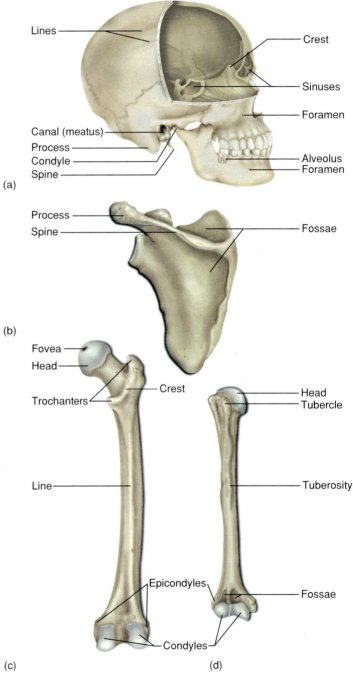

Figure 8.4 Common surface elaborations of bones, exemplified on (a) the skull, (b) the scapula, (c) the femur, and (d) the humerus.

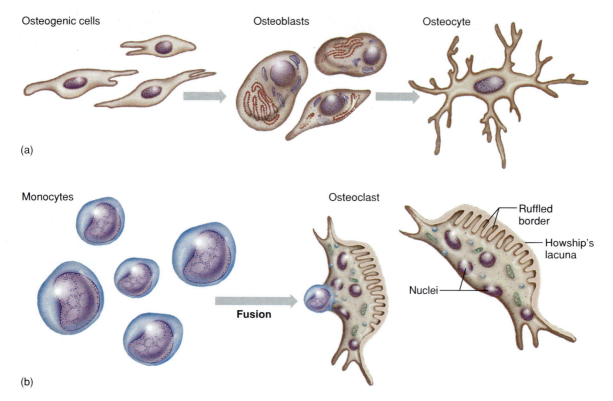

Figure 8.5 Types and origins of bone cells. (*a*) Osteogenic cells give rise to osteoblasts, which become osteocytes when trapped in lacunae of the matrix. (*b*) Osteoclasts arise by the fusion of blood monocytes.

Key Point Review

⑤ Name the four shapes of bones and give one example of each.

⑥ What is the difference between compact and spongy bone, and what is their spatial relationship to each other?

⑦ State the anatomical terms for the shaft, head, growth zone, and fibrous covering of a long bone.

⑧ Concisely define *condyle, process, tubercle, fossa,* and *foramen.*

Histology of Osseous Tissue

▼**Objectives**
When you have completed this section, you should be able to
• list and describe the cells, fibers, and ground substance of osseous tissue and state the importance of each constituent;
• compare the two types of osseous tissue with respect to their histology; and
• distinguish between the three types of bone marrow.

The Cells of Osseous Tissue

Like any other connective tissue, bone consists of cells, fibers, and ground substance. As shown in figure 8.5, there are four principal types of bone cells:

1. **Osteogenic (osteoprogenitor) cells.** These develop from embryonic mesenchyme but are a little more specialized and have a narrower range of developmental potential. They occur in the endosteum, inner surface of the periosteum, and within the haversian canals. Unlike the other bone cells, they remain capable of mitosis and are therefore the only source of new cells of types 2 and 3 following.

2. **Osteoblasts.**[11] These are bone-forming cells that differentiate from osteogenic cells. They synthesize the collagen and glycosaminoglycans (GAGs) of the bone matrix and play a role in the mineralization of bone (discussed later). Osteoblasts are incapable of mitosis, but osteogenic cells multiply rapidly under the influence of stress or fractures and differentiate into large numbers of osteoblasts. Osteoblasts often line up in rows on the surface of a bone, somewhat resembling a cuboidal epithelium, and here they build new bone matrix.

3. **Osteocytes.** As osteoblasts deposit matrix, they become trapped within tiny spaces (*lacunae*). From that time they are known as osteocytes. Osteocytes stop producing matrix but remain active in its maintenance. They also assist in maintaining the proper calcium and phosphate balance between the blood and bone.

---

11. *blast* = form, produce, build up

Bone can play a detoxifying role by absorbing lead and other heavy metals. It protects the body from short-term high blood concentrations of the contaminant, and later releases it slowly for excretion by the kidneys. Less fortunately, bone also absorbs uranium, plutonium, radium, strontium-90, and other radioactive contaminants. These cause dangerous long-term irradiation of the bone and may cause a form of bone cancer called *osteosarcoma*. A well-publicized case involved the millionaire playboy and championship golfer Eben Byers, whose death in 1932 ended the radium tonic health fad that followed the discovery of radioactivity (see special topic 2.3, p. 69). Byers drank several bottles of radium tonic a day and touted its virtues as a wonder drug and aphrodisiac. His bones and teeth became so radioactive they exposed photographic film in the dark. By the time of his death, the once-handsome socialite had his entire upper jaw and most of his mandible removed in an effort to arrest the osteosarcoma. Holes had begun to form in his skull and brain damage had left him unable to speak, yet he was mentally alert to the bitter end. Osteosarcoma also afflicted many women who worked in watch factories painting luminous numbers on the dials. They used radium paint and licked their brushes to maintain the points.

4. **Osteoclasts.**[12] These are bone-dissolving cells that form by fusion of monocytes, one of the five kinds of white blood cells. They secrete acids and enzymes that break down the mineral salts and organic matter of the matrix, and they release minerals into the blood plasma. Osteoclasts are very large cells. They contain as many as 50 nuclei (one contributed by each monocyte) and often rest in little depressions on the bone surface called *Howship's*[13] *lacunae,* which they create by dissolving the matrix. On the side facing the matrix, an osteoclast has a *ruffled border* as a result of many deep infoldings of the cell membrane. These infoldings increase the surface area available for secretion of enzymes or for absorption of bone components. Tiny crystals of matrix are found between the infoldings when bone is being actively resorbed. Osteoclasts also contain many lysosomes near the ruffled border; they produce a bone-dissolving enzyme to be discussed later.

Think About It

What two organelles do you think are especially prominent in osteoblasts? (Hint: Consider the major substances osteoblasts synthesize.)

The Matrix of Osseous Tissue

The matrix of bone consists of both organic and inorganic components. Organic matter, which accounts for about one-third of the dry weight, includes collagen, GAGs (especially chondroitin sulfate), proteoglycans, and glycoproteins. The mineral component accounts for the other two-thirds. A mineral salt called **hydroxyapatite,** $Ca_{10}(PO_4)_6(OH)_2$, constitutes about 85% of the mineral component. Calcium carbonate, $CaCO_3$, constitutes another 10%. The remainder includes magnesium, sodium, potassium, fluoride, sulfate, carbonate, and hydroxyl ions. These do not form crystals but are apparently adsorbed by the surfaces of hydroxyapatite crystals. (To be *ad*sorbed means to cling in a thin layer to a surface; to be *ab*sorbed means to become incorporated into another substance.) Several foreign elements behave chemically like bone minerals and become incorporated into osseous tissue as contaminants (see special topic 8.1).

The combination of collagen and minerals gives bones a combination of strength and resilience similar to that of reinforced concrete. Like concrete, the minerals resist compression (crumbling or sagging when weight is applied); like steel reinforcement bars, the collagen resists tension (bending). If the minerals are dissolved by immersing a bone in a weak acid such as vinegar, the bone becomes rubbery (fig. 8.6). If only the collagen is removed, it becomes very brittle. Without collagen, a jogger's bones would shatter under the impact of running. (Recall from the chapter 6 essay [p. 197] that defective collagen is responsible for "brittle bone disease," or osteogenesis imperfecta.)

Histology of Compact Bone

Compact bone (fig. 8.7a) is usually histologically studied in dried cross sections ground to translucent thinness. This destroys the cells and most other organic content but reveals fine details of the inorganic matrix (fig. 8.7b, c). Such sections show onionlike layers of matrix, called **lamellae,** concentrically arranged around a **haversian** (ha-VUR-zhun) **(central) canal.** A haversian canal and its concentric lamellae constitute an **osteon** (*haversian system*)—the basic structural unit of compact bone. Between the lamellae are small almond-shaped cavities, the **lacunae,** which are occupied by osteocytes when bone is alive.

---

12. *clast* = destroy, break down
13. J. Howship (1781–1891), English surgeon

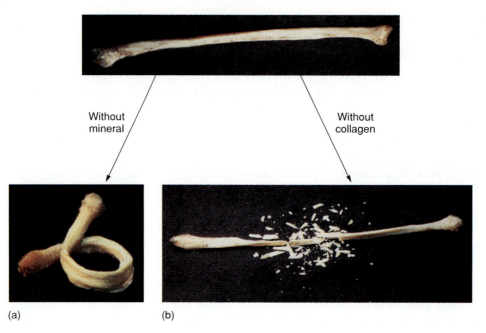

Figure 8.6 The properties of bone depend on the presence of both minerals and collagen. (*a*) When the fibula is demineralized, it becomes rubbery and can easily be twisted. (*b*) When its collagen is removed, the bone becomes very brittle and is easily shattered.

Osteocytes receive nutrients and eliminate wastes by way of blood vessels in the haversian canal. These vessels do not reach the osteocytes themselves; instead, the osteocytes must reach out toward the canal, or toward neighboring osteocytes directly or indirectly in touch with the canal. On close inspection of the matrix, delicate channels called **canaliculi**[14] (CAN-uh-LIC-you-lye) can be seen extending from each lacuna to its neighbors. Within the canaliculi are delicate cytoplasmic processes (see fig. 8.5) that join one osteocyte to the next by gap junctions. Canaliculi from the innermost lamella open into the haversian canal (fig. 8.7*d*). Nutrients pass from the haversian blood vessels to the nearest osteocytes by way of the osteocyte processes in these canaliculi, and the inner osteocytes pass them on to more remote ones. Wastes travel in the opposite direction.

In longitudinal views and three-dimensional reconstructions, we can see that an osteon is actually a cylinder of tissue surrounding a haversian canal (fig. 8.7*b*) and that the haversian canals branch and rejoin each other by transverse or oblique connections. Narrower **perforating (Volkmann**[15]**) canals** also enter the bone from the periosteal and endosteal surfaces and feed into the haversian systems, carrying nerves and blood vessels deep into the compact bone. On the periosteal surface, these are continuations of the nutrient foramina mentioned earlier.

Not all of the matrix is organized into osteons. The inner and outer boundaries of dense bone are arranged in lamellae that run parallel to the bone surface. Between osteons, irregular regions called *interstitial lamellae* also can be seen. These are fragments of old osteons that are broken down as bone grows and remodels itself.

Histology of Spongy Bone

Spongy bone (fig. 8.7*a*) consists of a lattice of slender rods, plates, and spines called **trabeculae.**[16] In living bone, the spaces between the trabeculae are occupied by bone marrow. Like compact bone, the matrix is arranged in lamellae. There are few osteons, however; they are not needed because no osteocyte is very far from the blood in the marrow. Spongy bone is well designed to impart strength to a bone while adding a minimum of weight. Its trabeculae are not randomly arranged as they might seem at a glance; rather, they develop along lines of stress applied to the bone (fig. 8.8).

Bone Marrow

Bone marrow is a general term for soft tissue that occupies the medullary cavity of a long bone, the spaces amid the trabeculae of spongy bone, and the larger haversian canals. There are three kinds of bone marrow, and we can best appreciate their differences by considering a long bone over the life span of an individual.

In a child, the medullary cavity is filled with **red marrow,** or **myeloid tissue.** This is a **hemopoietic**[17] (HE-mo-poy-ET-ic) tissue—that is, it produces blood

---

14. *canal* = canal + *icul* = little
15. Alfred Volkmann (1800–1877), German physiologist
16. *trab* = plate
17. *hemo* = blood + *poietic* = forming

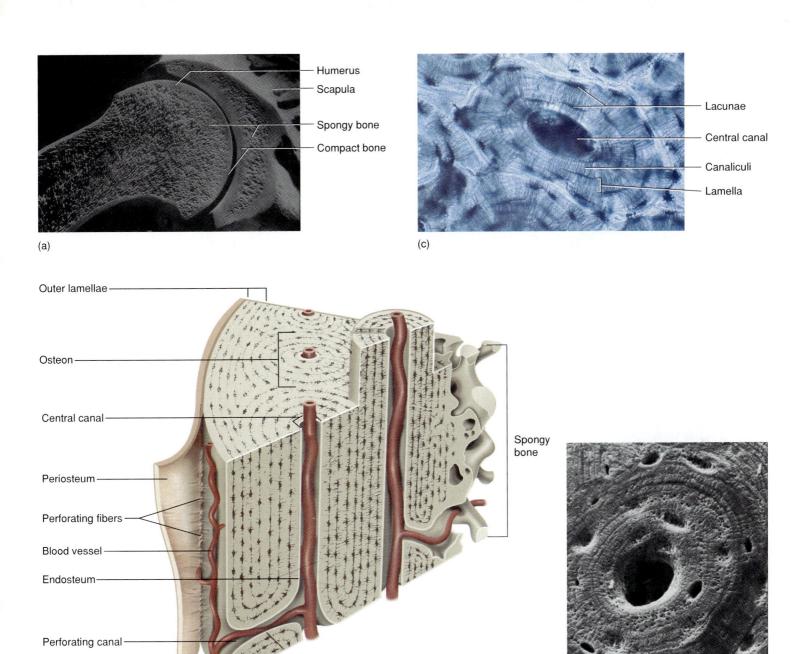

(a) Humerus
Scapula
Spongy bone
Compact bone

(c) Lacunae
Central canal
Canaliculi
Lamella

Outer lamellae
Osteon
Central canal
Periosteum
Perforating fibers
Blood vessel
Endosteum
Perforating canal
Spongy bone

(b)

(d)

Figure 8.7 (a) Frontal section of the shoulder joint, showing spongy and compact bone. (b) Three-dimensional structure of compact bone. (c) Cross section of dried compact bone (×400). (d) Scanning electron micrograph of one osteon. ✗

(d) From R. G. Kessel and R. H. Kardon, Tissues and Organs: *A Text-Atlas of Scanning Electron Microscopy*, 1979, W. H. Freeman and Company.

cells. When drawn from a bone by suction (aspiration), myeloid tissue looks like blood but with a thicker consistency. It consists of a delicate mesh of reticular tissue, with developing blood cells amid the reticular fibers. Myeloid tissue also contains scattered adipocytes.

If we were to examine the same bone when the child had grown to an adult of 30, we would find most of it full of fat (like the fat at the center of a ham bone). This **yellow marrow** no longer performs a hemopoietic role, but it can temporarily revert to red bone marrow and produce blood in the event of severe or chronic anemia.

Now, suppose the person had reached the age of 70. Upon examining the same bone, we would find that most of the fat had been replaced by a reddish, jellylike tissue called **gelatinous marrow.**

Fetuses and children have only red bone marrow. With few exceptions, such as the middle-ear bones and hyoid bone, it is found in every bone of their skeleton. In an adult, yellow bone marrow occupies the shafts of the long bones and red bone marrow is limited to the axial skeleton and proximal heads of the humerus and femur (fig. 8.9).

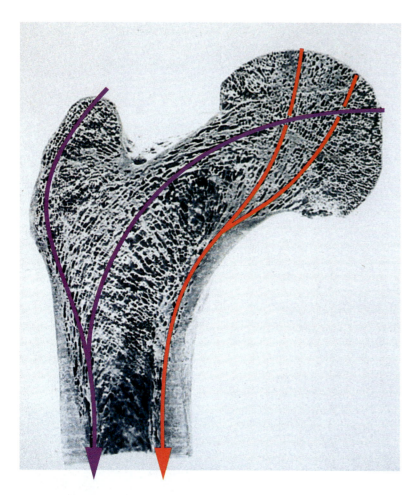

Figure 8.8 The trabeculae of spongy bone develop along the lines of stress applied to the femur.

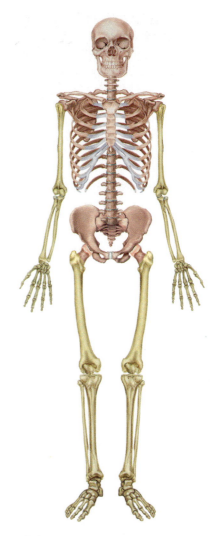

Figure 8.9 The distribution of red and yellow bone marrow in the adult skeleton. The red bone marrow is hemopoietic (blood-forming) tissue. The yellow bone marrow is adipose tissue.

Key Point Review

9 If you had electron micrographs of the four kinds of bone cells and their neighboring tissue, how would you ascertain which was which? Name the four cell types in your answer.

10 What are three organic constituents of the bone matrix?

11 What are the mineral crystals of osseous tissue called, and what are they made of?

12 Sketch a cross section of an osteon and label its major parts.

13 List the names of the three kinds of bone marrow and state where each is found. Which one is a hemopoietic tissue? What does this mean?

Bone Development

▼Objectives

When you have completed this section, you should be able to
- describe two methods of bone formation (ossification); and
- explain how the growth and remodeling of mature bones occurs.

The formation of bone is called **ossification,** or **osteogenesis.** The cranial bones of the skull develop by a method called **intramembranous ossification,** (IN-tra-MEM-bruh-nus OSS-ih-fih-CAY-shun) whereas most other bones (the vertebrae, pelvis, and bones of the extremities) form by **endochondral** (EN-doe-CON-drul) **ossification.** The clavicles, mandible, and facial bones form by combination of the two methods. Bone tissue sometimes forms unnaturally in sites other than the skeleton (see special topic 8.2).

Intramembranous Ossification

Intramembranous means "within a membrane." This type of ossification begins as some of the embryonic connective tissue (mesenchyme) transforms into a

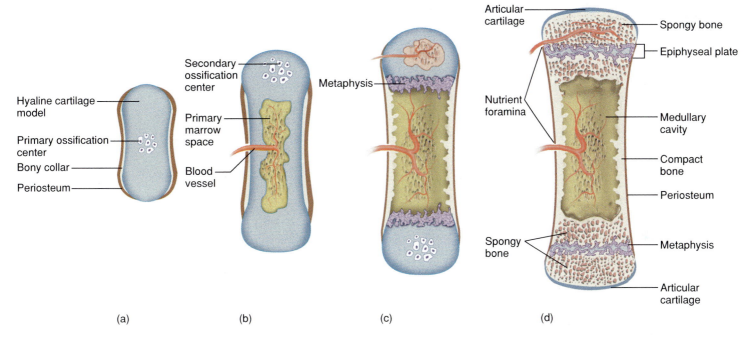

Figure 8.10 Endochondral ossification. (*a*) Hypertrophy of chondrocytes at the center of the cartilage model, and formation of a supportive bony collar. (*b*) Invasion of the model by blood vessels and creation of a primary marrow space. (*c*) Typical state of the bone at the time of birth, with vascular invasion of the secondary marrow space and well-defined metaphyses at each end of the primary marrow space. (*d*) Appearance of the bone in childhood and adolescence. Exhaustion of the epiphyseal plates will unite the primary and secondary marrow spaces by early adulthood.

highly vascular sheet or membrane of soft tissue that represents the future location of a flat bone. Its cells enlarge and differentiate into osteogenic cells and osteoblasts, and some of the mesenchyme condenses into a network of soft trabeculae. Osteogenic cells gather on the trabeculae, become osteoblasts, and deposit an organic matrix called **osteoid**[20] **tissue**—soft collagenous tissue similar to bone except for a lack of calcium salts. The trabeculae grow thicker, calcium phosphate is deposited in the matrix, and some osteoblasts become trapped in lacunae. Once trapped, they differentiate into osteocytes. Some of the now-calcified trabeculae form spongy bone. At the surface of a flat bone, other trabeculae undergo further calcification, which closes up much of the space in the spongy bone and converts it to compact bone.

Endochondral Ossification

Endochondral means "within cartilage"; it denotes a method in which a bone arises from a hyaline cartilage model with roughly the shape of the bone to come. The cartilage model arises from embryonic mesenchyme and is surrounded by a perichondrium. This cartilage must be broken down, reorganized, and hardened (**mineralized**) to form a bone—a process that is only partially underway even by the time of birth. The process is shown in figure 8.10.

The Primary Ossification Center

The first sign that a cartilage model is transforming to bone is that some of its cells (chondrocytes) near the center of the model begin to multiply and swell. The region

18. *ec*, from *ex* = out of + *top* = place
19. *calc* = calcium, stone + *ulus* = little
20. *oste* = bone + *oid* = like, resembling

where this occurs is called the **primary ossification center.** As the lacunae enlarge, the matrix between them is reduced to thin walls in which calcium carbonate crystals begin to accumulate. Some cells of the perichondrium become osteogenic cells and osteoblasts, which produce a bony collar around the cartilage model. This collar acts like a splint to provide temporary support as the middle of the model is weakened by the hypertrophy of its cells and loss of matrix. Once the collar has formed, the fibrous sheath around it is considered periosteum rather than perichondrium.

Buds of connective tissue grow from this periosteum into the cartilage model and take blood vessels with them. These buds penetrate the thin walls between the enlarged lacunae, break them down, and form a space in the center of the model called the **primary marrow space.** Osteogenic cells brought in by the connective tissue buds transform into osteoblasts and line this space, while the hypertrophied cartilage cells die. The osteoblasts deposit osteoid tissue and then mineralize it to form a temporary framework of bony trabeculae. By this process, the primary marrow space grows larger and progresses toward the ends of the bone.

The Metaphysis

While the primary marrow space is enlarging, each head of the bone still consists of hyaline cartilage. The transitional zone between the head and primary marrow space is called the **metaphysis** (meh-TAF-ih-sis). As shown in figure 8.11, it exhibits five histological zones representing stages of ossification:

1. **The zone of reserve cartilage.** In this zone, farthest from the marrow space, the "resting" hyaline cartilage as yet shows no sign of transforming into bone.
2. **The zone of cell proliferation.** A little closer to the marrow space, chondrocytes are dividing and the daughter cells are arranged in longitudinal columns of flattened lacunae.
3. **The zone of cell hypertrophy.** Next, the chondrocytes cease to divide and begin to hypertrophy, just as they did in the primary ossification center. The cartilage walls between columns become very thin. Cell multiplication in zone 2 and hypertrophy in zone 3 make the bone grow longer.
4. **The zone of calcification.** Minerals are deposited in the matrix between columns.
5. **The zone of bone deposition.** Within each column, the walls between lacunae break down and the chondrocytes die. This converts each column into a longitudinal channel, which is quickly invaded by marrow and blood vessels from the primary marrow space. Osteoblasts line up along the walls of these channels and begin depositing bone matrix in concentric layers. The channel therefore grows

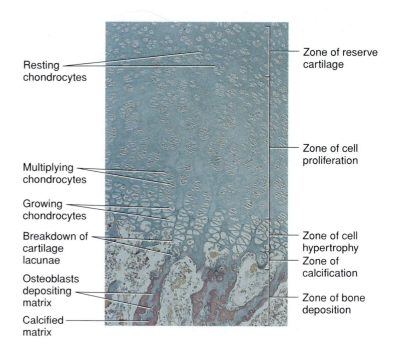

Figure 8.11 Zones of endochondral ossification (×100).

narrower and narrower as one layer after another is laid down, until only a narrow channel remains in the middle—now a haversian canal. Osteoblasts trapped in their own matrix become osteocytes and stop producing matrix.

 Think About It

In a given osteon, which lamellae are the oldest—those immediately adjacent to the haversian canal or those around the perimeter of the osteon?

The primary ossification centers of a 12-week-old fetus are shown in figure 8.12. The joints are translucent because they are still composed of cartilage that has not yet ossified. This is the state of the skeleton even at the time of birth, which is one reason human newborns cannot walk.

The Secondary Ossification Center

Around the time of birth, **secondary ossification centers** begin to form in the epiphyses (see fig. 8.10c). Here, too, chondrocytes enlarge, the walls of matrix between them dissolve, and the chondrocytes die. Vascular buds arise from the periosteum and grow into the cartilage, bringing osteogenic cells with them. Erosion of the cartilage progresses in all directions from the secondary ossification center, so the epiphysis is hollowed out from the center outward. Thin trabeculae of cartilage become the calcified trabeculae of spongy bone. Hyaline cartilage persists in two places—on the epiphyseal surfaces, as the articular cartilages, and at the junction of the diaphysis and epiphysis, where it forms the epiphyseal plate (fig. 8.13).

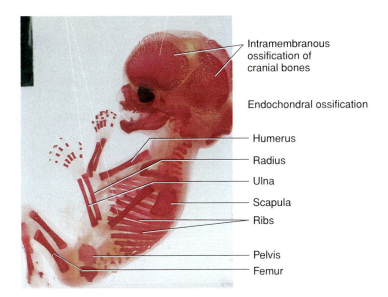

Figure 8.12 A 12-week-old human fetus stained to show where the bones have become calcified. The skull sutures and joints such as the elbow, wrist, knee, and ankle joints appear translucent because they are still cartilagenous.

Intramembranous ossification of cranial bones

Endochondral ossification

Humerus
Radius
Ulna
Scapula
Ribs
Pelvis
Femur

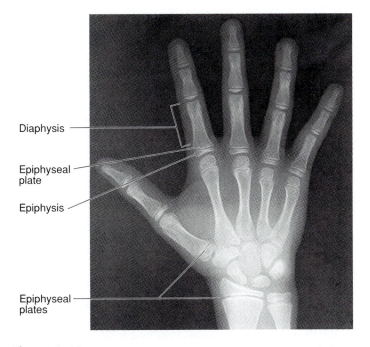

Diaphysis
Epiphyseal plate
Epiphysis
Epiphyseal plates

Figure 8.13 X ray of a child's hand, in which the epiphyseal plates are still evident. By early adulthood, these plates close and the epiphyses and diaphyses fuse.

Bone Growth and Remodeling

Ossification is by no means the end of bone development. Each year, bone exchanges about 18% of its calcium with the blood in an adult, and at any one time about 5% of the adult skeleton is undergoing **remodeling**—structural changes at the microscopic to gross level. One type of remodeling is a change in size

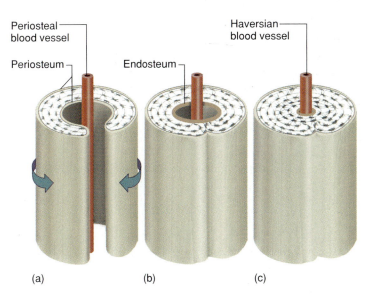

Periosteal blood vessel
Periosteum
Endosteum
Haversian blood vessel

(a)   (b)   (c)

Figure 8.14 Appositional growth at the surface of a bone. (*a*) Bone deposition along the margins of surface grooves progressively encloses blood vessels that lie in them. (*b*) The groove eventually closes, becoming a channel containing the vessel. The periosteum of the groove is now the endosteum of the channel. (*c*) New lamellae of compact bone are deposited within the channel, progressively constricting its lumen until all that remains of it is the haversian canal. A new osteon has now been formed.

and contour to accommodate the growing body. The femurs become longer, the curvature of the cranium changes to accommodate a larger brain, and so forth. Remodeling also adapts bones to the amount of gravitational force and muscular tension exerted on them. Many of the bony processes listed in table 8.3 develop as a child begins to walk and the muscles exert tension on the bones. The prominence of these processes and the density of bone depend on the amount of stress to which the bone has been subjected. On average, the bones have more density and mass in athletic people than in sedentary ones. Anthropologists who study ancient skeletal remains can distinguish between members of different social classes by the degree of bone development—a reflection of the individual's history of manual labor and nutrition. The skeleton also yields information about sex, race, height, weight, and medical history that can be useful to forensic pathology. A heavily developed greater trochanter of the femur, for example, suggests a life of heavy lifting.

Cartilage can grow by adding more matrix internally (**interstitial growth**) or by adding more to the surface (**appositional growth**). In bone, however, only the latter occurs. Osteocytes have little room as it is and none to spare for the deposition of more matrix. The only way bone can grow, therefore, is by increasing in length and adding more osseous tissue to the surface. In figure 8.14, we see how a new osteon forms by appositional growth, adding to the width of a bone. As a diaphysis grows in diameter by addition of new osteons at

Figure 8.15 The student at the right, pictured with her roommate of normal height, is an achondroplastic dwarf with a height of about 122 cm (48 in.). Her parents were of normal height. Note that while the limbs are shortened, her head and trunk are of normal proportions since achondroplasia has little effect outside the metaphyses of the long bones.

the surface, osteoclasts on the medullary side erode matrix and make the medullary cavity larger.

Growth in length results from the multiplication and hypertrophy of chondrocytes in zones 2 and 3 of the metaphysis. If these chondrocytes fail to multiply, the long bones grow slowly and stop growing early, while other bones of the body are unaffected. This occurs in **achondroplastic**[21] (ah-con-dro-PLAS-tic) **dwarfism** (fig. 8.15). Two people of normal height with no family history of dwarfism can have a child with achondroplastic dwarfism, since this condition results from a spontaneous mutation, or error in DNA replication. The resulting allele is dominant, however, so the children of a heterozygous achondroplastic dwarf have at least a 50% chance of exhibiting dwarfism, or higher depending on the genotype of their other parent.

There is a critical balance between bone deposition and removal. If one process outpaces the other, or if both of them occur too rapidly, various bone deformities and other developmental abnormalities occur (see table 8.5, p. 245, especially osteitis deformans).

21. *a* = without + *chondro* = cartilage + *plas* = growth

(14) Describe the sequence of events that occurs in intramembranous ossification.

(15) List the five zones of a metaphysis and describe the major event(s) that occur in each one.

(16) Describe the major gross features of a long bone at the time of birth.

(17) Use a sketch to describe the process of appositional bone growth.

Physiology of Osseous Tissue

▼Objectives

When you have completed this section, you should be able to

- explain how bone becomes mineralized and how the minerals are removed;
- discuss the role of bones in regulating blood levels of calcium and phosphorus;
- explain how vitamin D is synthesized and describe its effects on the skeleton; and
- list the hormones that influence the formation and resorption of bone and briefly describe the effects of each.

Even after a bone is fully formed, it remains a metabolically active organ with many roles to play. Not only is it involved in its own maintenance, growth, and remodeling, it also exerts a profound influence on the rest of the body by exchanging minerals with the tissue fluid. Disturbances of bone physiology can disrupt neuromuscular function, for example, and even cause a person to die of suffocation. At this point, we turn our attention to the physiology of mature osseous tissue.

Mineral Deposition

Deposition, or **mineralization,** is the process by which calcium and phosphate ions from the blood plasma are deposited in bone tissue. It is a crystallization process that can be understood best, perhaps, by way of a simple demonstration. If you dissolve as much table salt as possible in hot water you will have a saturated solution. As the water cools, it becomes an unstable *supersaturated* solution because cold water cannot hold as much salt as hot water can. If you hang a string in this solution, it will act as a focal point for crystallization. The first grains of salt to crystallize on the string act as **seed crystals,** quickly inducing more ions from the solution to crystallize around them. The more salt crystal that forms, the faster additional salt crystallizes out.

The mineralization of bone is similarly based on the action of seed crystals on an unstable solution of Ca^{2+} and PO_4^{3-} (phosphate) salts. Collagen fibers take

The urinary, respiratory, and skeletal systems cooperate to maintain the body's acid-base balance. If the pH of the blood drops below 7.35, a state of *acidosis* exists and triggers corrective mechanisms in these three organ systems. The role of the skeleton is to release calcium phosphate. As a base, calcium phosphate helps to prevent the blood pH from dropping any lower. Patients with chronic kidney disease may have impaired hydrogen ion excretion in the urine. Their pH stabilizes at a level below normal but is kept from dropping indefinitely by the buffering action of the skeleton. This can have adverse effects on the skeleton, however, leading to rickets or osteomalacia. Treatment of the acidosis with intravenous bicarbonate restores the osseous tissue to normal.

the place of the string in our demonstration, so the mineralization of bone requires that the collagen be deposited first. Osteoblasts deposit collagen in a helical pattern along the length of an osteon.

Tissue fluid has a high concentration of calcium ions, but calcium phosphate does not precipitate out unless the product of calcium and phosphate concentration, represented $[Ca^{2+}] \bullet [PO_4^{3-}]$, reaches a critical value called the **solubility product.** Most tissues have inhibitors to prevent this occurrence, so they do not become calcified. It is thought that osteoblasts, however, neutralize these inhibitors and thus allow the salts to precipitate on the collagen. Minerals first accumulate as hydroxyapatite seed crystals in the spaces between collagen fibers. These crystals promote the accelerated precipitation of more hydroxyapatite until the matrix is thoroughly calcified.

Mineral Resorption

Resorption is the process of dissolving bone so its minerals can be released into the blood and used for other purposes in the body. It is done by osteocytes and osteoclasts under the influence of hormones. Osteoclasts dissolve bone with a combination of acid and an enzyme, **acid phosphatase,** which digests the collagen of the matrix. Hydrogen ion pumps in the ruffled border of an osteoclast pump out H^+, producing an extracellular pH of 4 that dissolves the mineral component of bone. When orthodontic appliances (braces) are used to reposition teeth, movement of the tooth is achieved by osteoclasts dissolving the jawbone in front of the tooth (where the appliance creates greater pressure of the tooth against the bone) and osteoblasts depositing bone in the low-pressure zone behind the tooth.

Calcium and Phosphorus Homeostasis

The adult body contains approximately 1,100 g of calcium and 500 to 800 g of phosphorus. About 99% of the former and between 85% and 90% of the latter are present in the skeleton, but calcium and phosphorus are used for much more than skeletal structure. Phosphorus is essential for the synthesis of DNA, RNA, ATP, phospholipids, and many other compounds. Calcium is needed for nerve cell action, muscle contraction, blood clotting, exocytosis, and other membrane transport processes. It also serves as a second messenger in many cells and as a cofactor for many enzymes. Calcium phosphate from the skeleton also helps to correct acid-base imbalances in the blood (see special topic 8.3).

The skeleton is a reservoir for these minerals. They are deposited there when the supply is ample and withdrawn later when they are needed for other physiological processes. Bone has two calcium reserves—a stable pool of calcium that is incorporated into hydroxyapatite and is not readily exchanged with the blood, and exchangeable calcium, which is 1% or less of the total but easily released to the tissue fluid.

Blood plasma normally has a calcium concentration ranging from 9.2 to 10.4 mg/dL. This is a rather narrow margin of safety, as we shall soon see. About 45% of this is in the ionized form (Ca^{2+}), which can diffuse through capillary walls and exert physiological effects. The rest of the blood calcium is bound to plasma proteins and other solutes. It is not used physiologically, but it serves as a reserve from which Ca^{2+} can be withdrawn as needed. The plasma phosphorus concentration ranges between 3.5 and 4.0 mg/dL. Phosphorus occurs in the form of HPO_4^{2-} and $H_2PO_4^{-}$ ions (monohydrogen and dihydrogen phosphate, respectively).

Changes in phosphorus concentration have little immediate effect on the body, but changes in calcium can be serious. Excessively low calcium concentration is called **hypocalcemia**[22] (HY-po-cal-SEE-me-uh). In this condition, the nervous system becomes hyperexcitable and sends spontaneous impulses to the skeletal muscles, causing tremors, spasms, or **tetany**—a sustained contraction in which the muscle cannot relax. Tetany begins to appear at a plasma Ca^{2+} concentration of 6 mg/dL. One of the warning signs of hypocalcemia is

---

22. *hypo* = below normal + *calc* = calcium + *emia* = blood condition

Figure 8.16 Hypocalcemia is marked by carpopedal spasm, a characteristic tetany of the hands and feet. The wrist and fingers are spasmodically flexed.

carpopedal spasm of the hands and feet (fig. 8.16). A drop to 4 mg/dL can cause fatal *laryngospasm*—tight constriction of the muscles of the larynx (voicebox), causing suffocation. An excessive Ca^{2+} concentration is called **hypercalcemia.**[23] Symptoms begin to appear at about 12 mg/dL; they include depression of the nervous system, emotional disturbances, muscle weakness, sluggish reflexes, and sometimes cardiac arrest.

You can see how critical blood calcium levels are, but what causes them to deviate from the norm, and how does the body correct this? Hypercalcemia is rare, but hypocalcemia can result from a wide variety of causes including vitamin D deficiency, diarrhea, thyroid tumors, or underactive parathyroid glands. Pregnancy and lactation also put women at risk for hypocalcemia because of the calcium demanded by ossification of the fetal skeleton and synthesis of milk.

Calcium homeostasis depends on a balance between dietary intake, urinary and fecal losses, and exchanges with the osseous tissue. The balance between osseous deposition and resorption is regulated by two hormones—calcitonin and parathyroid hormone.

Calcitonin is released by certain thyroid gland cells (C cells) when plasma calcium concentration rises too high. Calcitonin lowers the concentration by two principal mechanisms:

1. **Osteoclast inhibition.** Within 15 minutes, calcitonin reduces osteoclast activity by as much as 70% (in children), so osteoclasts release less calcium from the skeleton.
2. **Osteoblast stimulation.** Within an hour, calcitonin increases the number and activity of osteoblasts, which deposit calcium into the skeleton.

Calcitonin secretion is an effective homeostatic mechanism in children. Their osteoclasts are very active and release 5 g or more of calcium into the circulation

each day. This gives calcitonin considerable latitude to significantly influence blood calcium levels. Calcitonin is not very effective in adults, however, because their osteoclasts release only about 0.8 g of calcium into the circulation each day. Suppressing such a minor contribution makes little impact, and calcitonin deficiency is not known to cause any adult disease. Calcitonin may, however, prevent bone loss in pregnant and lactating women.

Calcium homeostasis in children is regulated by both calcitonin and **parathyroid hormone (PTH).** PTH is secreted by four tiny parathyroid glands on the posterior surface of the thyroid gland. Adult calcium homeostasis is almost entirely under the control of PTH. A mere 1% drop in the level of blood calcium doubles the secretion of PTH. PTH raises blood calcium levels by three mechanisms:

1. **Osteocyte and osteoclast stimulation.** PTH stimulates osteocytes to dissolve bone matrix in the immediate vicinity of their lacunae and release minerals into the blood. It also stimulates a rapid increase in the number of osteoclasts, which dissolve osseous tissue at the bone surfaces.
2. **Urinary excretion.** PTH reduces excretion of calcium while increasing excretion of phosphate. This lowers the plasma concentration of phosphate to a level that prevents the formation of hydroxyapatite. Thus, calcium remains dissolved in the plasma and is accessible to other organs.
3. **Vitamin D synthesis.** PTH stimulates the production of an enzyme in the kidneys that activates vitamin D. Since vitamin D is needed for the absorption of calcium by the small intestine, PTH indirectly enhances intestinal calcium absorption.

Think About It

Parathyroid hormone does *not* promote intestinal absorption of phosphate. Explain why it would defeat the purpose of PTH to do so.

Since the parathyroid glands are so small and adhere to the surface of the thyroid, they are sometimes accidentally removed in thyroid surgery. Without hormone substitution therapy, this can lead to fatal tetany within 4 days. Accidental removal of the parathyroids is the leading cause of hypocalcemic tetany.

Vitamin D

The hormone **vitamin D** is produced by the coordinated action of the skin, liver, and kidneys and then carried in the blood to other organs, where it exerts its effect. It is called a vitamin only because it is also added to the diet, mainly in fortified milk, as a safeguard for people who do not get enough sunlight to meet their vitamin D needs.

23. *hyper* = above normal

The most active form of vitamin D is **calcitriol** (CAL-sih-TRY-ol). The placenta, some macrophages, and keratinocytes of the epidermis synthesize calcitriol, but most of it is made by the following process (fig. 8.17):

1. A cholesterol derivative called 7-dehydrocholesterol circulates through dermal capillaries close to the skin surface, where the UV radiation in sunlight converts it to cholecalciferol (CO-lee-cal-SIF-ur-ol), or vitamin D_3.
2. The liver adds a hydroxyl group to it, forming calcidiol.
3. The kidneys add another hydroxyl group, forming calcitriol (1,25-dihydroxycholecalciferol). As previously described, this is the step promoted by PTH.

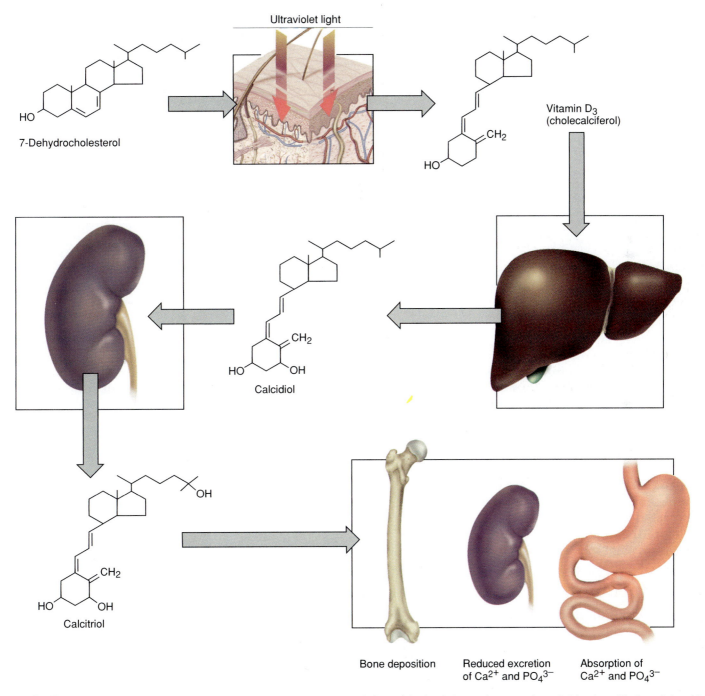

Figure 8.17 The synthesis and action of calcitriol. Starting at the upper left, 7-dehydrocholesterol passes through blood capillaries of the skin, where ultraviolet light converts it to vitamin D_3. The liver adds a hydroxyl group to this molecule, converting it to calcidiol. The kidneys add another hydroxyl group, converting it to calcitriol. Calcitriol exerts its effects on the bones, kidneys, and small intestine, raising blood calcium and phosphate concentrations and promoting bone deposition.

Table 8.4 Agents Affecting Calcium and Bone Metabolism

Name	Effect
Hormones	
Parathyroid Hormone	Raises blood Ca^{2+} concentration by stimulating osteoclasts and osteocytes to resorb bone, inhibiting urinary Ca^{2+} excretion, and promoting urinary PO_4^{3-} excretion, intestinal Ca^{2+} absorption, and calcitriol synthesis
Calcitonin	Almost no effects in adults; promotes mineralization and lowers blood Ca^{2+} concentration in children; may prevent bone loss in pregnant and lactating women
Calcitriol (vitamin D)	Promotes intestinal absorption of Ca^{2+} and PO_4^{3-}; reduces urinary excretion of both; promotes mineralization; stimulates osteoclast activity
Growth Hormone	Stimulates bone elongation and cartilage proliferation at epiphyseal plate; increases urinary excretion of Ca^{2+} but also increases intestinal Ca^{2+} absorption, which compensates for the loss
Glucocorticoids	Inhibit osteoclast activity, but if secreted in excess (Cushing disease) can cause osteoporosis by reducing bone deposition through inhibition of cellular division and protein synthesis
Thyroid Hormones	Essential to bone growth; enhance effects of growth hormone, but excesses can cause hypercalcemia, increased Ca^{2+} excretion in urine, and osteoporosis
Estrogens	Stimulate osteoblasts and prevent osteoporosis
Testosterone	Stimulates osteoblasts and promotes protein synthesis, thus promoting epiphyseal growth and closure
Insulin	Stimulates bone formation; significant bone loss occurs in untreated diabetes mellitus
Growth Factors	At least 12 hormonelike substances produced in bone itself that stimulate neighboring bone cells, promote collagen synthesis, stimulate epiphyseal growth, and produce many other effects
Vitamins	
Vitamin A	Promotes synthesis of chondroitin sulfate
Vitamin C (ascorbic acid)	Promotes collagen cross-linking, bone growth, and fracture repair
Vitamin D	Normally functions as a hormone (see calcitriol)

Calcitriol promotes intestinal absorption of *both* calcium and phosphate (unlike PTH) and reduces urinary excretion of both. This raises the $[Ca^{2+}] \cdot [PO_4^{3-}]$ level to the solubility product at which calcium phosphate can precipitate and mineralize the bone. An insufficient amount of sunlight or supplemental vitamin D can therefore result in abnormal softness of the bones, a condition called **rickets** in children and **osteomalacia** in adults. Calcitriol also promotes osteoclast activity. In the absence of other factors, this would erode bone, but the increased absorption of dietary calcium and phosphate more than compensates for any erosion, so calcitriol causes a net increase in bone mass.

Other Hormones

At least 20 more hormones, growth factors, and vitamins affect osseous tissue in complex ways that are still not well understood (table 8.4). Osteoblast activity and growth at the epiphyseal plates are stimulated by growth hormone, thyroid hormone, insulin, and sex steroids. Estrogens and testosterone are responsible for the growth spurt of adolescent girls and boys and eventually cause the epiphyseal plates to close and growth to cease. Excessive or deficient secretion of these steroids can therefore cause abnormalities ranging from stunted growth to very tall stature (see chapter 17). The use of anabolic steroids by adolescent athletes can cause premature closure of the epiphyseal plates and abnormally short adult stature.

Key Point Review

18 How are collagen and seed crystals relevant to bone mineralization?

19 Why must blood calcium levels be kept within such a narrow range?

20 What effect does calcitonin have on blood calcium, and how does it achieve this effect?

21 What effect does parathyroid hormone have on blood calcium, and how does it achieve this effect?

22 Explain how vitamin D is synthesized and how it exerts an influence on blood calcium concentration.

Bone Disorders

▼Objectives

When you have completed this section, you should be able to
- describe the various types of fractures;
- explain how a fracture is repaired; and
- discuss some clinical treatments for fractures and other skeletal disorders.

Table 8.5　Bone Diseases

Rickets (in children) and *Osteomalacia* (in adults)	Defective mineralization of bone, usually as a result of insufficient sunlight or vitamin D. Sometimes it is due to a dietary deficiency of calcium or phosphate or to liver or kidney diseases that interfere with calcitriol synthesis. Osteomalacia is most common in poorly nourished women who have had multiple pregnancies. Bones become softened, deformed, and more susceptible to fractures.
Osteoporosis	Loss of bone mass, especially spongy bone, usually as a result of lack of exercise or a deficiency of estrogen after menopause. It results in increasing brittleness and susceptibility to fractures (see chapter essay for details).
Osteitis Deformans (Paget[24] disease)	Excessive proliferation of osteoclasts and resorption of excess bone, with osteoblasts attempting to compensate by depositing extra bone. This results in rapid, disorderly bone remodeling with weak, deformed bones. Paget disease usually passes unnoticed, but in some cases it causes pain, disfiguration, and fractures. It is most common in males over the age of 50.
Osteomyelitis	Inflammation of bone and marrow as a result of bacterial infection. This disease was often fatal before the discovery of antibiotics.
Osteoma	A benign bone tumor, especially in the flat bones of the skull. It may grow into the orbits or sinuses.
Osteochondroma	A benign tumor involving both bone and cartilage. It often forms bone spurs at the ends of long bones.
Osteosarcoma (osteogenic sarcoma)	The most common and deadly form of bone cancer. It occurs most often in the tibia, femur, and humerus of males between the ages of 10 and 25. In 10% of the cases, it metastasizes to the lungs or other organs; if untreated, death occurs within 1 year.
Chondrosarcoma	A slow-growing cancer of hyaline cartilage, most common in middle age. It requires surgical removal; chemotherapy is ineffective.

Earlier in the chapter, we mentioned rickets and osteomalacia. These and several other bone disorders are summarized in table 8.5. Here we take a more in-depth look at fractures, their healing, and the treatment of some skeletal disorders. The most common bone disease, however, is **osteoporosis,** which receives special consideration in the chapter essay (p. 248).

Fractures and Their Repair

We seldom give much thought to our skeletons unless we break a bone. Fractures may be classified according to the direction of the fracture line, whether or not the skin is broken, and whether a bone is broken into separate pieces or merely cracked. The different types are summarized in table 8.6 and illustrated in figure 8.18.

Some fracture terms refer to specific bones and locations. A **Colles**[25] (COL-eez) **fracture,** common in osteoporosis, occurs at the distal end of the radius. A **Pott**[26] **fracture,** a common sports injury, occurs at the distal end of the tibia, fibula, or both. Juveniles are subject to **epiphyseal fractures,** in which the epiphysis separates from the diaphysis along the epiphyseal plate.

24. Sir James Paget (1814–99), English surgeon
25. Abraham Colles (1773–1843), Irish surgeon
26. Sir Percivall Pott (1713–88), British surgeon

The Healing of Fractures

A fracture heals in about 8 to 12 weeks in uncomplicated cases. Complex fractures take longer, and all fractures heal more slowly in older people. As shown in figure 8.19, the healing process occurs in the following stages:

1. **Hematoma formation.** A bone fracture also breaks blood vessels of the bone and periosteum, causing bleeding and the formation of a clot (*fracture hematoma*).
2. **Formation of granulation tissue.** This soft tissue forms as blood vessels grow into the hematoma. Macrophages arrive by way of these vessels and clean up tissue debris. Osteoclasts, osteogenic cells, and fibroblasts also migrate into the tissue from the periosteal and medullary sides of the fracture.
3. **Callus formation.** Fibroblasts deposit collagen in the granulation tissue, while some osteogenic cells become chondroblasts and produce patches of fibrocartilage. The result is a **soft callus,** which subsequently becomes calcified into a **hard (bony) callus.** The hard callus forms a collar around the periosteal and endosteal surfaces of the fracture, acting like a splint to join the broken ends of the bone to each other. It takes about 4 to 6 weeks for a bony callus to form. During this period, it is

Table 8.6	Classification of Bone Fractures
Type	**Description**
Closed	Skin is not broken (formerly called *simple*)
Open	Skin is broken; bone protrudes through skin or wound extends to fractured bone (formerly called *compound*)
Complete	Bone is broken into two or more pieces
Incomplete	Partial fracture that extends only partway across bone; pieces remain joined
Greenstick	Bone is bent on one side and has incomplete fracture on opposite side
Hairline	Fine crack in which sections of bone remain aligned; common in skull
Comminuted	Bone is broken into three or more pieces
Displaced	The portions of a fractured bone are out of anatomical alignment
Nondisplaced	The portions of bone are still in correct anatomical alignment
Impacted	One bone fragment is driven into the medullary space or spongy bone of the other
Depressed	Broken portion of bone forms a concavity, as in skull fractures
Linear	Fracture parallel to long axis of bone
Transverse	Fracture perpendicular to long axis of bone
Oblique	Diagonal fracture, between linear and transverse
Spiral	Fracture spirals around axis of long bone, the result of a twisting stress
Epiphyseal	Epiphysis is separated from the diaphysis along the epiphyseal plate
Avulsion	Body part (such as a finger) is completely severed

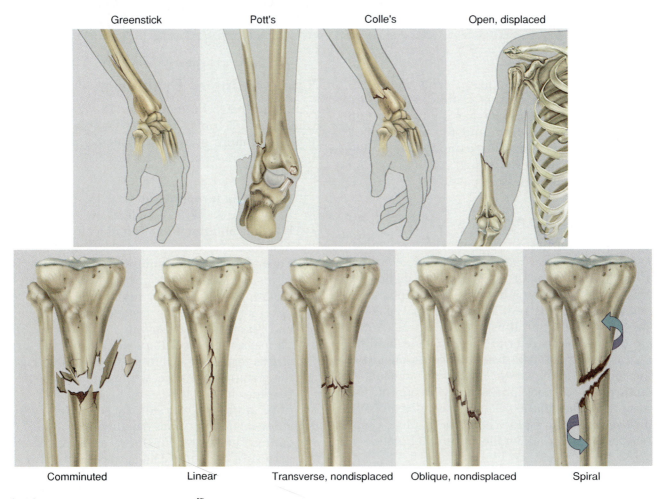

Figure 8.18 Some types of bone fractures.

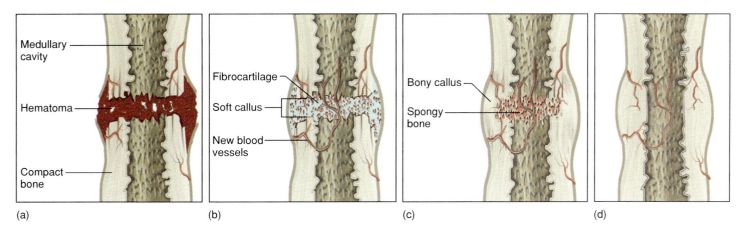

Figure 8.19 The healing of a fracture. (*a*) Blood vessels are broken at the fracture line; the blood clots and forms a fracture hematoma. (*b*) Blood vessels grow into the fracture and a fibrocartilage soft callus forms. (*c*) The fibrocartilage becomes ossified and forms a bony callus made of spongy bone. (*d*) Osteoclasts remove excess tissue from the bony callus and the bone eventually resembles its original appearance.

important that a broken bone be immobilized by traction or a cast to prevent refracture.

4. **Remodeling.** The bony callus persists for about 3 to 4 months as osteoclasts dissolve small fragments of broken bone and osteoblasts bridge the gap between the broken ends with spongy bone. This spongy bone is subsequently remodeled into compact bone. Usually the fracture leaves a slight thickening of the bone visible by X ray, but in some cases healing is so complete that no trace of the fracture can be found.

The Treatment of Fractures

Most fractures can be set by **closed reduction**, a procedure in which the bone fragments are manipulated into their normal positions without surgery. **Open reduction** involves the surgical exposure of the bone and the use of various plates, screws, or pins to realign the fragments (fig. 8.20*b*). To stabilize the bone during healing, fractures have traditionally been set in plaster casts. Fiberglass casts have become more common, however, because they are lighter, stronger, washable, porous, and cause less itching. *Traction* is used to treat fractures of the femur in children. It aids in the alignment of the bone fragments by overriding the force of the strong thigh muscles. Traction is rarely used for elderly patients, however, because the risks from long-term confinement to bed outweigh the benefits. Hip fractures are usually pinned and early ambulation (walking) is encouraged because it promotes blood circulation and healing. Fractures that take longer than 2 months to heal may be treated with electrical stimulation, which accelerates repair by suppressing the action of parathyroid hormone on osteoclasts.

Orthopedics[27] is the branch of medicine that deals with the prevention and correction of injuries and dis-

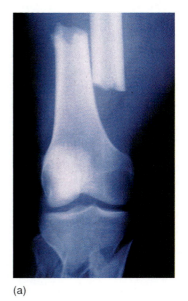
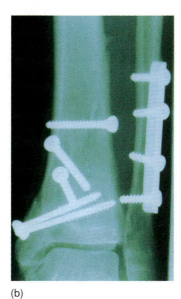

(a)   (b)

Figure 8.20 (*a*) X ray of a displaced fracture of the femur and comminuted fracture of the tibia and fibula. (*b*) A fracture of the tibia and fibula at the ankle that has been set by open reduction, a process of surgically exposing the bone and realigning the fragments with the aid of plates, screws, or pins. ✗

orders of the bones, joints, and muscles. As the word suggests, this field was originally concerned with the correction of skeletal deformities in children. The chapter 10 essay (p. 321) describes a newer branch of orthopedics especially relevant to elderly people—the design of artificial joints.

Key Point Review

23 Name and describe at least five types of fractures.

24 Why would osteomyelitis be more likely to occur in an open fracture than a closed fracture?

25 What is a callus? How does it contribute to fracture repair?

27. *ortho* = straight + *ped* = child, foot

Osteoporosis

The most common bone disease is osteoporosis (OSS-tee-oh-pore-OH-sis), literally, "porous bones." This is a disease in which the bones lose mass and become increasingly brittle and subject to fractures. It involves loss of both organic matrix and minerals, and it affects spongy bone in particular, since this is the most metabolically active type (fig. E.1). The highest incidence of osteoporosis is among elderly white women, where it is closely linked to age and menopause. White and black men develop osteoporosis about equally but less frequently than white women do. Black women are rarely afflicted by it.

Fractures are the most serious consequence of osteoporosis. They occur especially in the hip, wrist, and vertebral column and under stresses as slight as sitting down too quickly. Hip fractures usually occur at the neck of the femur and wrist fractures at the distal end of the radius and ulna (Colles' fracture). Among the elderly, hip fractures lead to fatal complications in 12% to 20% of cases; they involve a long, costly recovery for half of those who survive. As the weight-bearing bodies of the vertebrae lose spongy bone, they become compressed like marshmallows. Consequently, many people lose stature after middle age, and in some women, the spine becomes deformed into a "widow's hump," a condition called **kyphosis** (fig. E.2).

Postmenopausal white women are at greatest risk for osteoporosis because women have less bone mass than men to begin with and then tend to lose mass after age 35. By age 70, the average white woman has lost 30% of her bone mass, and some have lost as much as 50%. In men, the loss begins at around age 60 and seldom exceeds 25%. Osteoblasts are stimulated by estrogens, but the ovaries stop producing estrogens after menopause and osteoblast activity is correspondingly low. Ironically, osteoporosis also occurs among some female runners and dancers in spite of their vigorous exercise. Their percentage of body fat is so low that they stop ovulating and the ovaries secrete unusually low levels of estrogen.

Estrogen replacement therapy cannot reverse osteoporosis, but it can slow its progress. Alternatives to estrogen therapy are becoming available, but each has its own undesirable side effects. Milk and other calcium sources and moderate exercise can also slow the progress of osteoporosis, but only slightly. As is so often true, an ounce of prevention is worth a pound of cure. The time to minimize the risk for osteoporosis is between the ages of 25 and 40, when the skeleton is building to its maximum mass. The more bone mass a person has going into middle age, the less he or she will be affected by osteoporosis later. Ample exercise and calcium intake (850–1,000 mg/day) are the best preventative measures.

Although osteoporosis has become a major public health problem because of the increasing age of the population, it is not limited to the elderly. **Disuse osteoporosis** can occur at any age as a result

(a)

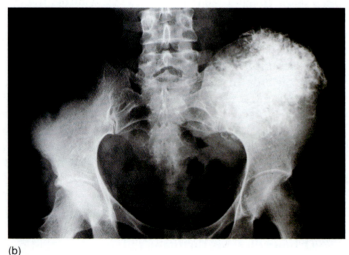
(b)

Figure E.1 (*a*) Portion of a lumbar vertebra, comparing normal spongy bone on the left and spongy bone with osteoporosis on the right. (*b*) X ray of the pelvis of a person with osteoporosis.

of immobilization or inadequate weight-bearing exercise. In the absence of preventive exercise, astronauts on prolonged microgravity missions have experienced disuse osteoporosis. Other risk factors include smoking, low calcium and protein intake, vitamin C deficiency, and diabetes mellitus.▲

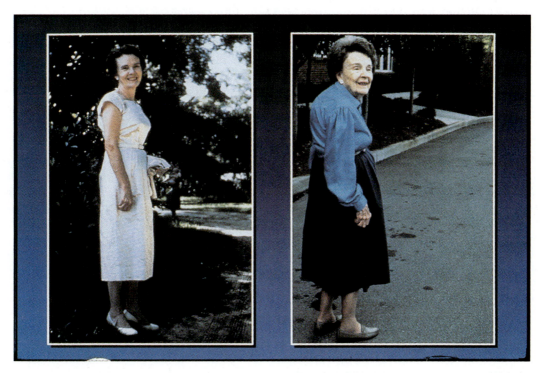

Figure E.2 The same woman before and after the visible onset of osteoporosis. Notice the kyphosis, or exaggerated thoracic curvature of the spine, that has resulted from erosion and compression of the vertebrae.

Chapter Review   Study Outline

Tissues and Organs of the Skeletal System (p. 226–228)
1. Dynamic nature of bone
2. Organs of the skeletal system
3. Functions of the skeleton (table 8.1)
4. Bones and osseous tissue
5. Bones of the skeletal system
   a. Number of bones
   b. Changes and variations in number
   c. Subdivisions of the skeleton
      • Axial
      • Appendicular
   d. Catalog of bones (table 8.2)

Gross Anatomy of Bones (p. 228–232)
1. Shapes and examples
   a. Long (femur, humerus)
   b. Short (carpals, tarsals)
   c. Flat (cranium, sternum)
   d. Irregular (vertebrae, facial bones)
2. General features of bones
   a. Dense and spongy bone
   b. Medullary cavity
   c. Diaphysis and epiphysis
   d. Articular cartilage
   e. Nutrient foramina
   f. Periosteum and endosteum
   g. Epiphyseal plate and line
3. Elaborations of bone structure (table 8.3)

Histology of Osseous Tissue (p. 232–236)
1. The cells of osseous tissue
   a. Osteogenic cells
   b. Osteoblasts
   c. Osteocytes
   d. Osteoclasts

2. The matrix of osseous tissue
   a. Organic components
      • Collagen
      • GAGs, proteoglycans, and glycoproteins
   b. Mineral component
      • Hydroxyapatite (85%)
      • Calcium carbonate (10%)
      • Cations: Mg^{2+}, Na^+, K^+
      • Anions: F^-, SO_4^{2-}, CO_3^{2-}, OH^-
   c. Combination of strength and resilience
3. Histology of compact bone
   a. Haversian canal and lamellae
   b. Lacunae and canaliculi
   c. Nutrient canals
4. Histology of spongy bone
   a. Trabeculae
   b. Strength with minimal weight

5. Bone marrow
   a. Red marrow (myeloid tissue)
   b. Yellow marrow
   c. Gelatinous marrow
   d. Changes in distribution with age

Bone Development (p. 236–240)
1. Intramembranous ossification
   a. Condensation of mesenchyme
   b. Differentiation of osteoblasts
   c. Deposition of osteoid tissue
   d. Cell entrapment in lacunae
   e. Formation of spongy bone trabeculae
2. Endochondral ossification
   a. Primary ossification center
      • Chondrocyte hypertrophy
      • Formation of bony collar
      • Vascular buds
      • Creation of primary marrow space
      • Deposition of osteoid tissue
      • Mineralization
   b. Metaphysis
      • Zone of reserve cartilage
      • Zone of cell proliferation
      • Zone of cell hypertrophy
      • Zone of calcification
      • Zone of bone deposition
   c. Secondary ossification center
3. Bone growth and remodeling
   a. Changes in size and contour
   b. Development of surface processes
   c. Increase in width
      • Appositional growth on outside
      • Erosion on medullary side
   d. Closure of epiphyseal plates

Physiology of Osseous Tissue (p. 240–244)
1. Mineral deposition
   a. Collagen as a prerequisite
   b. Formation of seed crystals
   c. Importance of solubility product
2. Mineral resorption
   a. Hydrogen ion pumps of osteoclasts
   b. Role of acid phosphatase
3. Calcium and phosphorus homeostasis
   a. Uses of calcium and phosphate ions
   b. Importance of blood calcium levels
      • Effects of hypocalcemia
      • Effects of hypercalcemia
   c. Calcitonin (from thyroid)
      • Inhibits osteoclasts
      • Stimulates osteoblasts
      • Lowers plasma Ca^{2+} concentration in children
      • Negligible effect in adults
   d. Parathyroid hormone (PTH)
      • Stimulates osteocytes and osteoclasts to resorb bone
      • Promotes renal phosphate excretion
      • Promotes renal calcium retention
      • Promotes vitamin D synthesis
      • Promotes intestinal absorption of calcium
      • Raises blood calcium levels
   e. Vitamin D
      • Synthesis by skin, liver, and kidneys
      • Promotes intestinal absorption of calcium and phosphate ions
      • Reduces urinary excretion of both
      • Causes bone deposition
   f. Other hormones and growth factors affecting bone
      • Sex steroids
      • Insulin, thyroid hormone, growth hormone
   g. Vitamins A and C

Bone Disorders (p. 244–247)
1. Bone diseases (table 8.6)
2. Fractures and their repair
   a. Classification (table 8.6)
   b. The healing of fractures
      • Fracture hematoma
      • Granulation tissue
      • Soft and bony callus
      • Remodeling
   c. The treatment of fractures
      • Closed and open reduction
      • Orthopedics

Selected Vocabulary

Also review the terms in tables 8.3 and 8.5, which are not repeated here.

osseous tissue 226
osteology 226
sesamoid bone 227
sutural (wormian) bone 227
axial skeleton 227
appendicular skeleton 227
long bone 229
short bone 229
flat bone 229
irregular bone 229
compact (dense) bone 229
medullary cavity 229
spongy (cancellous) bone 229
diploe 229
diaphysis 230
epiphysis 230
articular cartilage 230
nutrient foramina 230
periosteum 230

perforating (Sharpey) fibers 230
endosteum 230
epiphyseal plate 230
osteogenic cells 232
osteoblasts 232
osteocytes 232
osteoclasts 233
hydroxyapatite 233
bone lamella 233
haversian (central) canal 233
osteon 233
lacuna 233
canaliculus 234
perforating (Volkmann) canal 234
trabecula 234
bone marrow 234
red marrow 234
myeloid tissue 234
hemopoietic 234
yellow marrow 235
gelatinous marrow 235

ossification 236
osteogenesis 236
intramembranous ossification 236
endochondral ossification 236
osteoid tissue 237
mineralized 237
primary ossification center 238
primary marrow space 238
metaphysis 238
zone of reserve cartilage 238
zone of cell proliferation 238
zone of cell hypertrophy 238
zone of calcification 238
zone of bone deposition 238
secondary ossification center 238
remodeling 239
interstitial growth 239
appositional growth 239
achondroplastic dwarfism 240
deposition 240

mineralization 240
seed crystal 240
solubility product 241
resorption 241
acid phosphatase 241
hypocalcemia 241
tetany 241
hypercalcemia 242
calcitonin 242
parathyroid hormone (PTH) 242
vitamin D (calcitriol) 242
rickets 244
osteomalacia 244
osteoporosis 245
Colles fracture 245
Pott fracture 245
epiphyseal fracture 245
callus 245
closed reduction 247
open reduction 247
orthopedics 247

Answers in Appendix C

1. Which cells have a ruffled border and secrete acid?
   a. C cells
   b. osteocytes
   c. osteogenic cells
   d. osteoblasts
   e. osteoclasts

2. A skeleton that is taken apart for study of the individual bones is said to be
   a. appendicular.
   b. disarticulated.
   c. cancellous.
   d. inarticulate.
   e. articulated.

3. The medullary cavity of an adult bone may contain
   a. myeloid tissue.
   b. hyaline cartilage.
   c. periosteum.
   d. osteocytes.
   e. articular cartilage.

4. The spurt of growth in puberty is due to cell proliferation and hypertrophy in the
   a. epiphysis.
   b. epiphyseal line.
   c. dense bone.
   d. epiphyseal plate.
   e. spongy bone.

5. Osteoclasts develop from
   a. osteoprogenitor cells.
   b. osteogenic cells.
   c. monocytes.
   d. fibroblasts.
   e. osteoblasts.

6. Which of the following has the most bones?
   a. a fetus
   b. a newborn infant
   c. a 10-year-old child
   d. a 40-year-old woman
   e. none of the above (all have same number)

7. The walls between individual cartilage lacunae break down in the zone of
   a. cell proliferation.
   b. calcification.
   c. reserve cartilage.
   d. bone deposition.
   e. cell hypertrophy.

8. Which of these is *not* an effect of PTH?
   a. a rise in blood phosphate levels
   b. a reduction of calcium excretion
   c. an increase in intestinal absorption of calcium
   d. an increase in the number of osteoclasts
   e. an increase in the rate of vitamin D synthesis

9. A hole through a bone is called
   a. a foramen.
   b. a facet.
   c. a condyle.
   d. a fossa.
   e. a sulcus.

10. A child jumps from the top of a playground "jungle gym" to the ground. His leg bones do not shatter mainly because they contain
    a. plenty of glycosaminoglycans.
    b. young, resilient osteocytes.
    c. plenty of calcium phosphate.
    d. collagen fibers.
    e. hydroxyapatite crystals.

11. Calcium phosphate crystallizes in bone as a mineral called _____.

12. Osteocytes contact each other through channels in the bone matrix called _____.

13. A bone can increase in diameter only by _____ growth, the addition of new osteons to the surface.

14. Seed crystals of hydroxyapatite first form in the spaces between _____.

15. A calcium deficiency, called _____, may lead to suffocation by laryngospasm.

16. The cells responsible for secreting collagen and GAGs and then stimulating calcium phosphate precipitation are _____.

17. The most active form of vitamin D, made mainly in the kidneys, is _____.

18. The most common bone disease is _____.

19. The transitional region between epiphyseal cartilage and the primary marrow cavity of a young bone is called the _____.

20. A pregnant, poorly nourished woman may suffer a softening of the bones called _____.

Answers in *Study Guide*

1. Most osteocytes of an osteon are far removed from blood vessels, yet they are still able to respond to blood-borne hormones. Explain how this is possible.

2. What positive feedback loop can you recognize in the process of bone mineralization?

3. How does the regulation of the blood calcium concentration exemplify negative feedback and homeostasis?

4. Suppose the microscope had never been invented and nothing was known about cells. Given these circumstances, write a textbook paragraph explaining how a bone grows in length during the teenage years. After doing so, comment briefly on the importance of cell theory to osteology.

5. Describe how the arrangement of trabeculae in spongy bone demonstrates the unity of form and function.

Web Site Links

For a listing of the most current web sites related to this chapter, please visit the Saladin homepage at:

http://www.mhhe.com/sciencemath/biology/saladin/

The Skeletal System

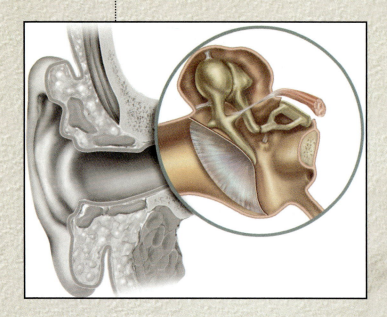

Special **Topics**

A knowledge of skeletal anatomy will be useful to you as you study later chapters. It provides a foundation for studying the gross anatomy of other organ systems because many organs are named for their relationships to nearby bones. The subclavian artery and vein, for example, are located beneath the clavicles; the temporalis muscle is attached to the temporal bone; the ulnar nerve and radial artery travel beside the ulna and radius of the forearm; and the frontal, parietal, temporal, and occipital lobes of the brain are named for bones of the cranium. An understanding of how the muscles produce body movements also depends on a knowledge of skeletal anatomy. Additionally, the positions, shapes, and processes of bones can serve as landmarks for a clinician in determining where to give an injection or record a pulse, what to look for in an X ray, or how to perform physical therapy and other medical procedures.

This chapter is divided into four sections: (1) the skull, (2) the vertebral column and thoracic cage, (3) the pectoral girdle and upper extremity, and (4) the pelvic girdle and lower extremity. The bones of the first two sections constitute the axial skeleton, and those of the last two constitute the appendicular skeleton. Following each major section is a checklist of skeletal features you should know (tables 9.2, 9.4, 9.5, and 9.7). The terms in each checklist are defined in the pages immediately preceding it.

As you study this chapter, use yourself as a model. You can easily palpate (feel) many of the bones and some of their details through the skin. Rotate your forearm, cross your legs, palpate your skull, and think about what is happening beneath the surface or what you are feeling through the skin. You will gain the most from this chapter (and indeed, the entire book) if you are conscious of your own body at every step.

The Skull

▼Objectives
When you have completed this section, you should be able to
• list the bones of the skull and their anatomical features; and
• describe the development of the skull from infancy through childhood.

The skull is the most complex part of the skeleton. Although it may seem to consist only of the mandible (lower jaw) and "the rest," in reality the skull is composed of 22 bones. All but the mandible are locked together by immovable joints called **sutures** (SOO-chures). Figures 9.1 through 9.4 present an overview of the major features of the skull from several perspectives. We examine its individual bones in more detail in the pages to come.

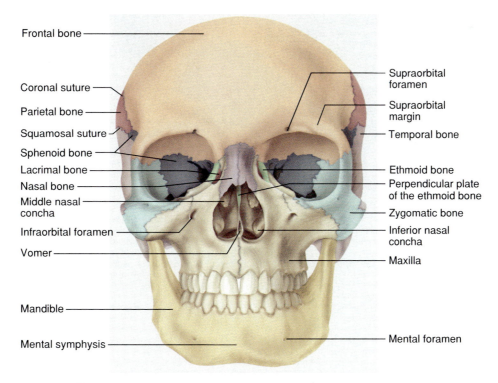

Figure 9.1 The skull, anterior view.

Labels (left): Frontal bone, Coronal suture, Parietal bone, Squamosal suture, Sphenoid bone, Lacrimal bone, Nasal bone, Middle nasal concha, Infraorbital foramen, Vomer, Mandible, Mental symphysis

Labels (right): Supraorbital foramen, Supraorbital margin, Temporal bone, Ethmoid bone, Perpendicular plate of the ethmoid bone, Zygomatic bone, Inferior nasal concha, Maxilla, Mental foramen

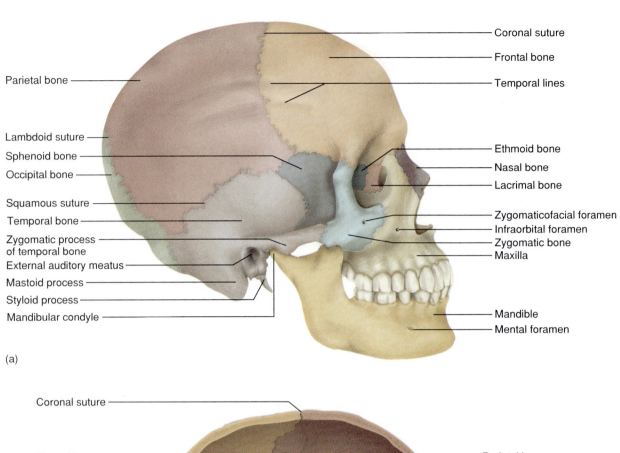

Coronal suture
Frontal bone
Temporal lines
Parietal bone
Lambdoid suture
Sphenoid bone
Occipital bone
Ethmoid bone
Nasal bone
Lacrimal bone
Squamous suture
Temporal bone
Zygomaticofacial foramen
Infraorbital foramen
Zygomatic process of temporal bone
Zygomatic bone
Maxilla
External auditory meatus
Mastoid process
Styloid process
Mandibular condyle
Mandible
Mental foramen

(a)

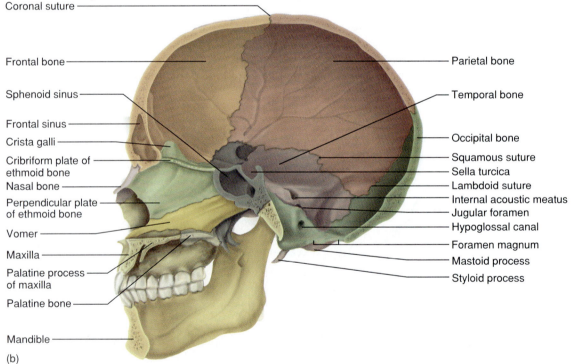

Coronal suture
Frontal bone
Parietal bone
Sphenoid sinus
Temporal bone
Frontal sinus
Crista galli
Occipital bone
Cribriform plate of ethmoid bone
Squamous suture
Sella turcica
Nasal bone
Lambdoid suture
Perpendicular plate of ethmoid bone
Internal acoustic meatus
Jugular foramen
Vomer
Hypoglossal canal
Maxilla
Foramen magnum
Palatine process of maxilla
Mastoid process
Styloid process
Palatine bone
Mandible

(b)

Figure 9.2 The skull. (*a*) Right lateral view; (*b*) interior of the right half, viewed in midsagittal section.

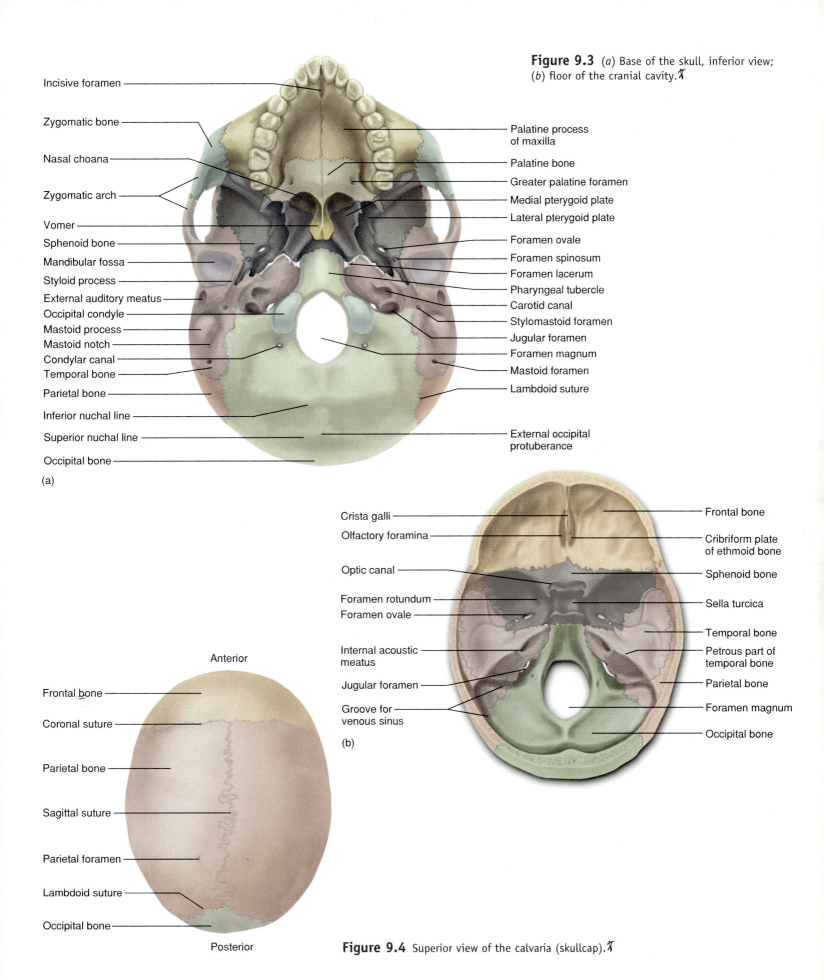

Figure 9.3 (a) Base of the skull, inferior view; (b) floor of the cranial cavity.

Incisive foramen

Zygomatic bone

Nasal choana

Zygomatic arch

Vomer

Sphenoid bone

Mandibular fossa

Styloid process

External auditory meatus

Occipital condyle

Mastoid process

Mastoid notch

Condylar canal

Temporal bone

Parietal bone

Inferior nuchal line

Superior nuchal line

Occipital bone

(a)

Palatine process of maxilla

Palatine bone

Greater palatine foramen

Medial pterygoid plate

Lateral pterygoid plate

Foramen ovale

Foramen spinosum

Foramen lacerum

Pharyngeal tubercle

Carotid canal

Stylomastoid foramen

Jugular foramen

Foramen magnum

Mastoid foramen

Lambdoid suture

External occipital protuberance

Crista galli

Olfactory foramina

Optic canal

Foramen rotundum

Foramen ovale

Internal acoustic meatus

Jugular foramen

Groove for venous sinus

(b)

Frontal bone

Cribriform plate of ethmoid bone

Sphenoid bone

Sella turcica

Temporal bone

Petrous part of temporal bone

Parietal bone

Foramen magnum

Occipital bone

Anterior

Frontal bone

Coronal suture

Parietal bone

Sagittal suture

Parietal foramen

Lambdoid suture

Occipital bone

Posterior

Figure 9.4 Superior view of the calvaria (skullcap).

The skull contains several prominent cavities, as shown in figure 9.5. The largest, with an adult volume of 1,300 to 1,350 mL, is the **cranial cavity,** which encloses the brain. Other cavities include the **orbits** (eye sockets), **nasal cavity, buccal** (BUCK-ul) **cavity** (mouth), **middle-** and **inner-ear cavities,** and **paranasal sinuses** (air-filled spaces connected with the nasal cavity). The paranasal sinuses are named for the bones in which they occur (fig. 9.6)—the **frontal, sphenoid, ethmoid,** and **maxillary sinuses.** They lighten the anterior portion of the skull and act as echo chambers that add resonance to the voice. The latter effect can be sensed in the way your voice changes when you have a cold

and mucus obstructs the travel of sound into the sinuses and back.

The skull has especially conspicuous **foramina**—singular, *foramen* (fo-RAY-men). These holes allow nerves and blood vessels to pass between the cranial cavity and the more superficial tissues of the head and neck. The major foramina are summarized in table 9.1. The details of this table will mean more to you when you study cranial nerves and blood vessels in later chapters.

Cranial Bones

The cranial cavity is enclosed by the **cranium**[1] (braincase), which protects the brain and associated sensory organs. The cranium is composed of eight **cranial bones,** as follows:

1 frontal bone	1 occipital bone
2 parietal bones	1 sphenoid bone
2 temporal bones	1 ethmoid bone

The cranium is a rigid structure with an opening, the **foramen magnum** (literally "large hole"), where the spinal cord enters. An important consideration in treatment of head injuries is swelling of the brain. Since the cranium cannot enlarge, swelling puts pressure on the brain and results in even more tissue damage. Severe swelling may force the brainstem out through the foramen magnum, usually with fatal consequences.

The delicate brain tissue does not come directly into contact with the cranial bones but is separated from them by three membranes called the *meninges* (meh-NIN-jeez) (see chapter 14). The thickest and toughest of these, the *dura mater*[2] (DUE-rah MAH-tur), lies loosely against the inside of the cranium in most places but is firmly attached to it at a few points.

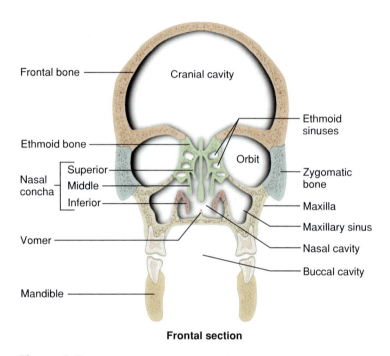

Frontal section

Figure 9.5 Major cavities of the skull, frontal section.

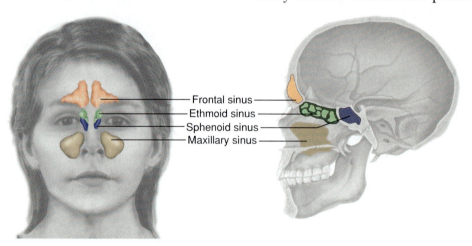

Figure 9.6 The paranasal sinuses.

Frontal sinus
Ethmoid sinus
Sphenoid sinus
Maxillary sinus

---

1. *cran* = helmet
2. *dura* = tough, strong + *mater* = mother

Table 9.1 Foramina of the Skull and the Nerves and Blood Vessels Transmitted through Them

Bones and Their Foramina*	Structures Transmitted
Frontal Bone	
Supraorbital Foramen or Notch	Supraorbital nerve, artery, and vein; ophthalmic nerve
Parietal Bone	
Parietal Foramen	Emissary vein of superior longitudinal sinus
Temporal Bone	
Carotid Canal	Internal carotid artery
External Auditory Meatus	Sound waves to eardrum
Internal Auditory Meatus	Vestibulocochlear nerve; internal auditory vessels
Stylomastoid Foramen	Facial nerve
Mastoid Foramen	Meningeal artery
Temporal–Occipital Region	
Jugular Foramen	Internal jugular vein; glossopharyngeal, vagus, and accessory nerves
Temporal–Occipital–Sphenoid Region	
Foramen Lacerum	No major nerves or vessels; closed by cartilage
Occipital Bone	
Foramen Magnum	Spinal cord; accessory nerve; vertebral arteries
Hypoglossal Canal	Hypoglossal nerve to muscles of tongue
Condylar Canal	Vein from transverse sinus
Sphenoid Bone	
Foramen Ovale	Mandibular branch of trigeminal nerve; accessory meningeal artery
Foramen Rotundum	Maxillary branch of trigeminal nerve
Foramen Spinosum	Middle meningeal artery; spinosal nerve
Optic Foramen	Optic nerve; ophthalmic artery
Superior Orbital Fissure	Oculomotor, trochlear, and abducens nerves; ophthalmic branch of trigeminal nerve; ophthalmic veins
Ethmoid Bone	
Olfactory Foramina	Olfactory nerves
Maxilla	
Infraorbital Foramen	Infraorbital nerve and vessels
Incisive Foramen	Nasopalatine nerves
Maxilla–Sphenoid Region	
Inferior Orbital Fissure	Infraorbital nerve; zygomatic nerve; infraorbital vessels
Lacrimal Bone	
Lacrimal Foramen	Tear duct leading to nasal cavity
Palatine Bone	
Greater Palatine Foramen	Palatine nerves
Zygomatic Bone	
Zygomaticofacial Foramen	Zygomaticofacial nerve
Zygomaticotemporal Foramen	Zygomaticotemporal nerve
Mandible	
Mental Foramen	Mental nerve and vessels
Mandibular Foramen	Inferior dental nerves and vessels

*When two or more bones listed together (for example, temporal–occipital), it indicates that the foramen passes between them.

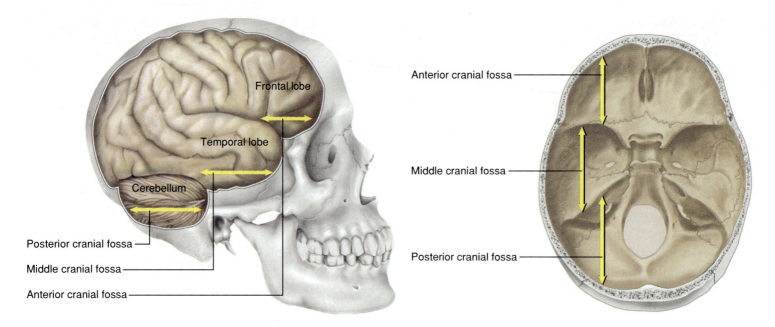

Figure 9.7 The three cranial fossae conform to the contours of the base of the brain.

The cranium consists of two major parts—the calvaria and the base. The **calvaria**[3] (see fig. 9.4), also known as the *skullcap* or *cranial vault,* forms the roof and walls. In study skulls it is often sawed so that part of it can be lifted off for examination of the interior. This reveals the **base** (floor) of the cranial cavity (see fig. 9.3*b*), which is divided into three basins corresponding to the contour of the inferior surface of the brain (fig. 9.7). The relatively shallow **anterior cranial fossa** is shaped like a crescent moon and accommodates the frontal lobes of the brain. The **middle cranial fossa,** which drops abruptly deeper, is shaped like a pair of outstretched bird's wings and accommodates the temporal lobes. The **posterior cranial fossa** is deepest and houses a large posterior division of the brain called the cerebellum.

We now consider the eight cranial bones and their major features. These are illustrated in figures 9.1 through 9.10.

Frontal Bone

The **frontal bone** extends from the forehead back to a prominent **coronal**[4] **suture,** which crosses the crown of the head from right to left (see figs. 9.2*a* and 9.4). It forms the anterior wall and about one-third of the roof of the cranial cavity, and it turns inward to form nearly all of the anterior cranial fossa and the superior walls of the orbits. Deep to the eyebrows it thickens to form a ridge called the **supraorbital margin.** The center of each margin is perforated by a single **supraorbital foramen**

(see figs. 9.1 and 9.12), which provides passage for a nerve, artery, and veins. In some people, the edge of this foramen breaks through the margin of the orbit. The foramen is then called the *supraorbital notch.* The frontal bone also contains the frontal sinus. You may not see this sinus on some study skulls if the calvaria is cut too high, and it is absent in some individuals. Along the cut calvaria, you can see the diploe—the layer of spongy bone in the middle of the cranial bones.

Parietal Bones

The two **parietal** (pa-RY-eh-tul) **bones** (see figs. 9.2*a* and 9.4) extend posteriorly from the coronal suture to the **lambdoid** (LAM-doyd) **suture** at the rear of the head (named for its resemblance to the Greek letter lambda, λ). The **sagittal suture** separates the right and left parietal bones from each other. Small sutural (wormian) bones are often seen along the course of the sagittal and lambdoid sutures, like little islands of bone with the suture lines passing around them. The parietal bones form most of the roof of the cranial cavity and part of its walls. Internally, the parietal and frontal bones have markings that look a bit like aerial photographs of river tributaries (see fig. 9.2*b*). These represent places where the bone has been molded around blood vessels of the dura mater.

Externally, the parietal bones are almost featureless. A **parietal foramen** sometimes occurs just anterior to the lambdoid suture and lateral to the sagittal suture (see fig. 9.4). A pair of slight thickenings, the superior and inferior **temporal lines,** form an arc across the parietal and frontal bones (see fig. 9.2*a*). They mark the attachment of the large, fan-shaped *temporalis* muscle,

---

3. *calv, calvar* = bald, skull
4. *corona* = crown

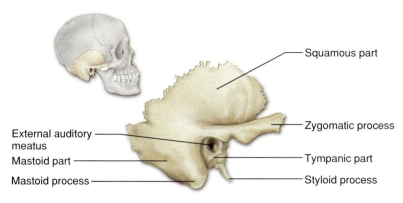

Figure 9.8 The right temporal bone, lateral external view.

which passes between the zygomatic arch and temporal bone and inserts on the mandible.

Temporal Bones

If you palpate your skull just anterior to the ear—that is, the temporal region—you can feel the **temporal bone,** which forms the lower lateral wall and part of the floor of the cranial cavity (fig. 9.8). The temporal bone derives its name from the fact that people often develop their first gray hairs on the temples with the passage of time.[5] The relatively complex shape of the temporal bone is best understood by dividing it into four parts.

The **squamous**[6] **part** (which you just palpated) is relatively flat and vertical. It is encircled by the **squamous suture,** which separates it from the occipital bone to the rear, the parietal bones above, and the sphenoid bone to the front. The squamous part bears two prominent features: (1) the **zygomatic process,** which extends anteriorly to form part of the cheekbone, or **zygomatic arch,** and (2) the **mandibular fossa,** which is a depression beneath the proximal end of the zygomatic process, where the mandible articulates with the cranium.

The **tympanic**[7] **part** is a small ring that borders the **external auditory meatus** (me-AY-tus), the ear canal. The **styloid process,** a pointed spine on its inferior surface, is named for its resemblance to the stylus used by ancient Greeks and Romans to write on wax tablets. It provides attachment for muscles of the tongue, pharynx, and hyoid bone.

The **mastoid**[8] **part** lies posterior to the tympanic part. It bears a heavy **mastoid process,** which you can palpate as a prominent lump behind the earlobe. The mastoid process is filled with small air sinuses that communicate with the middle-ear cavity. The sinuses are subject to infection and inflammation (mastoiditis), which can erode the bone and spread to the brain. Ven-

trally, a groove called the **mastoid notch** is positioned medial to the mastoid process (see fig. 9.3a). The mastoid notch is perforated by the **stylomastoid foramen** at its anterior end and by the **mastoid foramen** at its posterior end.

The **petrous**[9] **part** can be seen in the cranial floor, where it resembles a small mountain range separating the middle cranial fossa from the posterior fossa (see fig. 9.3b). It houses the middle- and inner-ear cavities. The **internal auditory meatus** (see fig. 9.2b), an opening on its posteromedial surface, allows passage of the vestibulocochlear (vess-TIB-you-lo-COC-lee-ur) nerve, which carries sensations of hearing and balance from the inner ear to the brain. On the ventral surface of the petrous part are two prominent foramina named for the major blood vessels that pass through them (see fig. 9.3a). The **carotid canal** is a passage for the internal carotid artery, a major blood supply to the brain. This artery is so close to the inner ear that you can sometimes hear the pulsing of its blood when your ear is resting on a pillow or your heart is beating hard. The **jugular foramen** is a large, irregular opening just medial to the styloid process, between the temporal and occipital bones. This is the passage by which the internal jugular vein drains blood from the brain. Three cranial nerves also pass through this foramen.

Occipital Bone

The **occipital** (oc-SIP-ih-tul) **bone** forms the rear of the skull (occiput) and much of its base (see fig. 9.3). Its most conspicuous feature, the foramen magnum, admits the spinal cord to the cranial cavity and provides a point of attachment for the dura mater. Anterior to the foramen magnum, the bone continues as a thick medial plate called the **pharyngeal tubercle** (fah-RIN-jee-ul TOO-ber-cul). On either side of the foramen magnum is a smooth knob called the **occipital condyle** (CON-dile), where the skull rests on the vertebral column. At the anterolateral edge of each condyle is a **hypoglossal**[10] **canal,** named for the hypoglossal nerve that passes through it to supply the muscles of the tongue. A **condylar** (CON-dih-lur) **canal** sometimes occurs posterior to each occipital condyle.

Internally, the occipital bone displays impressions left by large venous sinuses that drain blood from the brain (see fig. 9.3b). One of these grooves travels along the midsagittal line. Just before reaching the foramen magnum, it branches into right and left grooves that wrap around the occipital bone like outstretched arms before terminating at the jugular foramina. Here, the venous sinuses drain their blood into the internal jugular veins, which exit the cranial cavity and pass down the neck.

5. *tempor* = time + *al* = pertaining to
6. *squam* = flat + *ous* = characterized by
7. *tympan* = drum (eardrum) + *ic* = pertaining to
8. *mast* = breast + *oid* = shaped, resembling

9. *petr* = stone, rock
10. *hypo* = below + *gloss* = the tongue + *al* = pertaining to

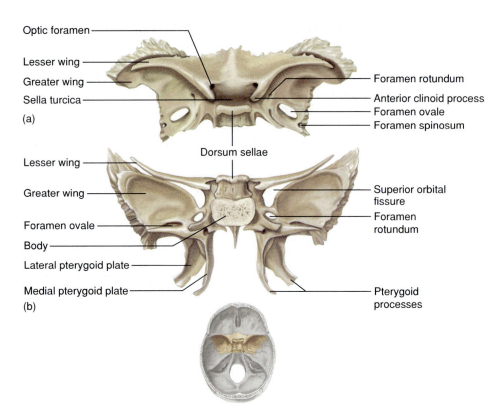

Optic foramen
Lesser wing
Greater wing
Sella turcica
(a)
Foramen rotundum
Anterior clinoid process
Foramen ovale
Foramen spinosum
Dorsum sellae

Lesser wing
Greater wing
Foramen ovale
Body
Lateral pterygoid plate
Medial pterygoid plate
(b)
Superior orbital fissure
Foramen rotundum
Pterygoid processes

Figure 9.9 The sphenoid bone. (*a*) Superior view; (*b*) posterior view.

Other features of the occipital bone can be palpated on the back of your head. One is a prominent medial bump called the **external occipital protuberance**—the attachment for the **nuchal** (NEW-kul) **ligament,** which binds the skull to the vertebral column. A ridge, the **superior nuchal line,** can be traced horizontally from the external occipital protuberance toward the mastoid process (see fig. 9.3*a*). It defines the superior limit of the neck and provides attachment for several neck and back muscles to the skull. By pulling down on the occipital bone, some of these muscles help to keep the head erect. Another attachment for these muscles is the deeper **inferior nuchal line.** This inconspicuous ridge cannot be palpated but is visible on an isolated skull.

Sphenoid Bone

The **sphenoid**[11] (SFEE-noyd) (fig. 9.9) is a complex bone with a thick medial **body** and outstretched **greater** and **lesser wings,** which give the bone as a whole a somewhat ragged, mothlike shape. Most of it is best seen from the superior perspective. In this view, the lesser wings form the posterior margin of the anterior cranial fossa and end at a sharp bony crest, where the sphenoid drops abruptly to the greater wings. These

form about half of the middle cranial fossa (the temporal bone forming the rest) and are perforated by several foramina, to be discussed shortly.

The greater wing also forms part of the lateral surface of the cranium just anterior to the temporal bone (see fig. 9.2*a*). The lesser wing forms the posterior wall of the orbit and contains the **optic foramen,** which permits passage of the optic nerve and ophthalmic artery (see fig. 9.12). Inside the brainpan, a pair of bony spines of the lesser wing called the **anterior clinoid processes** appear to guard the optic foramina. A gash in the posterior wall of the orbit, the **superior orbital fissure,** angles upward lateral to the optic foramen. It serves as a passage for nerves that supply some of the muscles that move the eyes.

The pituitary gland lies in a deep medial depression in the sphenoid, named the **sella turcica** (SEL-la TUR-sih-ca) after its resemblance to a saddle.[12] A membrane is stretched over this depression in life, penetrated by a stalk that attaches the pituitary to the floor of the brain. The posterior edge of the sella turcica is a bony transverse process called the **dorsum sellae** (DOR-sum SEL-lee).

Lateral to the sella turcica, the sphenoid is perforated by several foramina (see fig. 9.3*a*). The **foramen ovale** (oh-VAY-lee) and **foramen rotundum** are passages for two branches of the trigeminal nerve. The **foramen spinosum,** about the diameter of a pencil lead, provides passage for an artery of the meninges. An irregular gash called the **foramen lacerum**[13] (LASS-eh-rum) occurs at the junction of the sphenoid, temporal, and occipital bones. It is filled with cartilage in life, and no major vessels or nerves pass through it.

In an inferior view, the sphenoid can be located just anterior to the pharyngeal tubercle. The internal openings of the nasal cavity seen here are called the **nasal choanae**[14] (co-AH-nee), or **internal nares.** Lateral to each choana, the sphenoid bone exhibits a pair of parallel plates—the **medial pterygoid**[15] (TERR-ih-goyd) **plate** and the **lateral pterygoid plate** (see fig. 9.3*a*). These provide attachment for some of the jaw muscles. The **sphenoid sinus** occurs within the body of the sphenoid bone.

11. *sphen* = wedge

12. *sella* = saddle + *turcica* = Turkish
13. *lacerum* = torn, lacerated
14. *choana* = funnel
15. *pterygo* = wing

Ethmoid Bone

The **ethmoid**[16] (ETH-moyd) **bone** is located between the orbital cavities and forms the roof of the nasal cavity (fig. 9.10). An inferior projection of the ethmoid, called the

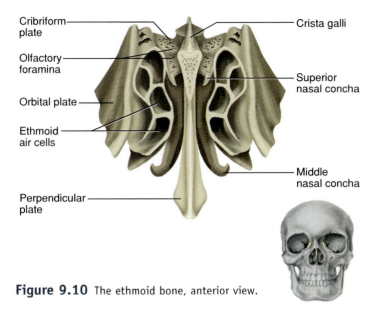

Cribriform plate

Olfactory foramina

Orbital plate

Ethmoid air cells

Perpendicular plate

Crista galli

Superior nasal concha

Middle nasal concha

Figure 9.10 The ethmoid bone, anterior view.

perpendicular plate, forms the superior part of the **nasal septum,** which divides the nasal cavity into right and left **nasal fossae** (FOSS-ee). Three curled, scroll-like **nasal conchae**[17] (CON-kee), or **turbinate bones,** project into each fossa from the lateral wall (see figs. 9.5 and 9.11). The superior and middle conchae are extensions of the ethmoid bone. The inferior concha—a separate bone—is included in the discussion of facial bones in the next section. The conchae are covered with a ciliated mucous membrane that serves to cleanse, warm, and humidify the inhaled air as it swirls around them. The superior concha and the adjacent region of the nasal septum also bear the receptor cells for the sense of smell (olfactory sense). Flanking the perpendicular plate on each side is a large but delicate mass of bone permeated with passages called **ethmoid air cells,** collectively constituting the ethmoid sinus.

The superior part of the ethmoid, viewed from the interior of the skull, exhibits a midsagittal crest called the **crista galli**[18] (GAL-eye) (see fig. 9.3b), a point of attachment for the meninges. On either side of the crista is a horizontal **cribriform**[19] (CRIB-rih-form) **plate** marked by numerous perforations through which olfactory nerve fibers pass from the nasal cavity to the brain (see special topic 9.1).

Facial Bones

The remaining 14 bones of the skull, called **facial bones,** have no contact with the brain. These bones, all of which are paired except for the vomer and mandible, are as follows:

2 maxillae	2 nasal bones
2 palatine bones	2 inferior nasal conchae
2 zygomatic bones	1 vomer
2 lacrimal bones	1 mandible

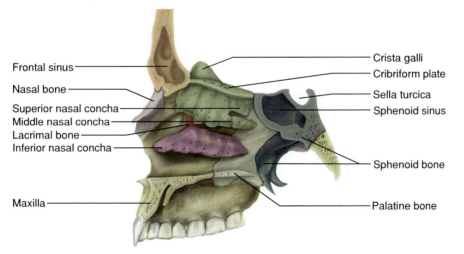

Frontal sinus

Nasal bone

Superior nasal concha

Middle nasal concha

Lacrimal bone

Inferior nasal concha

Maxilla

Crista galli

Cribriform plate

Sella turcica

Sphenoid sinus

Sphenoid bone

Palatine bone

Figure 9.11 The right nasal cavity, midsagittal section.

Special Topic — Injury to the Ethmoid Bone — 9.1

The ethmoid bone is very delicate and easily injured by a sharp upward blow to the nose, such as a person might suffer by striking an automobile dashboard in a collision. The force of the blow can drive bone fragments through the cribriform plate into the meninges or brain tissue. Such an injury, indicated by leakage of cerebrospinal fluid into the nasal cavity, can allow infection to spread from the nasal cavity to the brain.

16. *ethmo* = sieve, strainer

17. *concha* = conch, a large marine snail
18. *crista* = crest + *galli* = of a rooster
19. *cribri* = sieve

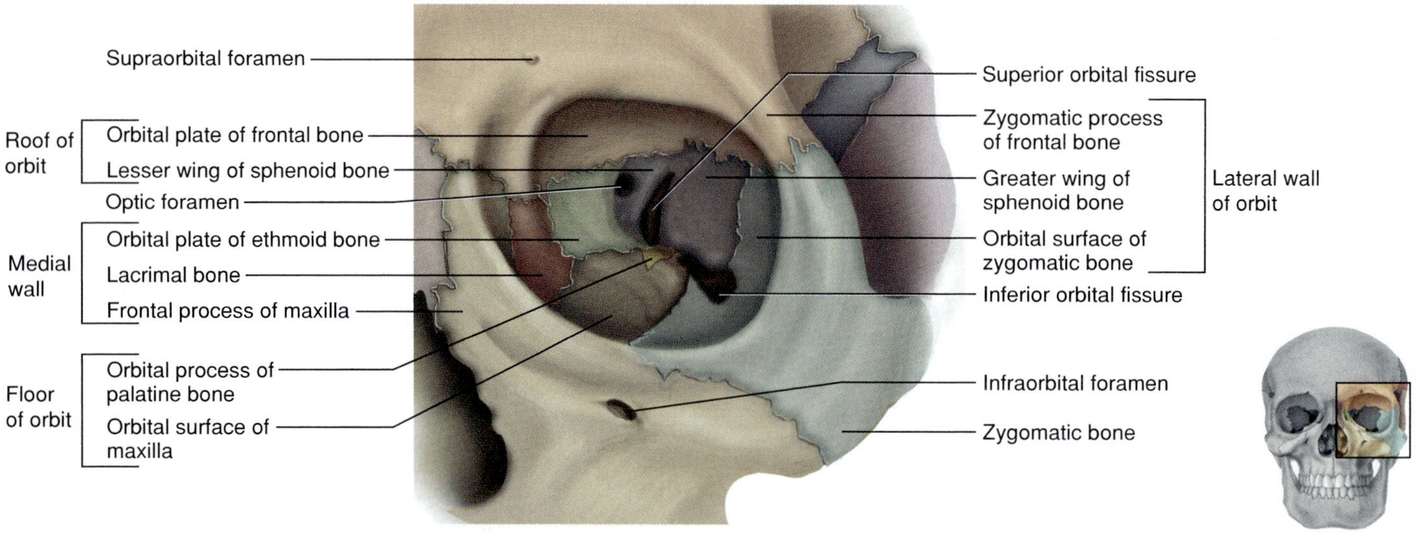
Supraorbital foramen
Roof of orbit
Orbital plate of frontal bone
Lesser wing of sphenoid bone
Optic foramen
Medial wall
Orbital plate of ethmoid bone
Lacrimal bone
Frontal process of maxilla
Floor of orbit
Orbital process of palatine bone
Orbital surface of maxilla
Superior orbital fissure
Zygomatic process of frontal bone
Greater wing of sphenoid bone
Orbital surface of zygomatic bone
Lateral wall of orbit
Inferior orbital fissure
Infraorbital foramen
Zygomatic bone

Figure 9.12 The left orbit, anterior view.

Special Topic: Evolutionary Significance of the Palate — 9.2

The palate is a mammalian characteristic, not found in reptiles or other vertebrates. By separating the buccal cavity from the nasal cavity, it allows breathing to continue while food is being chewed. This is necessary to support the relatively high metabolic rate of mammals, which requires that food be thoroughly broken up by the teeth before being swallowed and subjected to chemical digestion.

The facial bones support the teeth, give shape and individuality to the face, form part of the orbital and nasal cavities, and provide attachment for the muscles of facial expression and mastication.

Maxillae

The **maxillae** (mac-SILL-ee) are the largest facial bones. They form the upper jaw and meet each other at a midsagittal suture (see figs. 9.1 and 9.2*a*). Small points of maxillary bone called **alveolar processes** grow into the spaces between the bases of the teeth. The root of each tooth is inserted into a deep socket, or **alveolus.** If a tooth is lost or extracted so that chewing no longer puts stress on the maxilla, the alveolar processes are resorbed, leaving a smooth area on the maxilla. The teeth are discussed in detail in chapter 25.

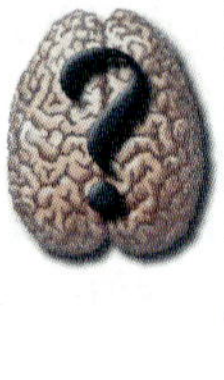

Think About It

Suppose you are studying a skull with some teeth missing from the mandible. How could you tell whether the teeth had been lost from the skull after the person's death or years before he or she had died?

Each maxilla extends from the teeth to the inferomedial wall of the orbit. Just below the orbit, it exhibits an **infraorbital foramen,** which provides passage for a blood vessel to the face and a nerve that receives sensations from the nasal region. This nerve emerges through the foramen rotundum into the cranial cavity. The maxilla forms part of the floor of the orbit, where it exhibits a gash called the **inferior orbital fissure** that angles downward and medially (fig. 9.12). The inferior and superior orbital fissures form a sideways V whose apex lies near the optic foramen. The inferior orbital fissure is a passage for blood vessels and a nerve that supply more of the muscles that control eye movements.

The **palate** forms the roof of the mouth and separates the oral and nasal cavities (see special topic 9.2). It consists of a bony **hard palate** in front and a fleshy **soft palate** in the rear. Most of the hard palate is formed by horizontal maxillary extensions called the **palatine** (PAL-uh-tine) **processes** (see fig. 9.3*a*). Near the anterior margin of each palatine process, just behind the incisors, is an **incisive foramen.** The palatine processes normally fuse at about 12 weeks of gestation. Failure to fuse results in a *cleft palate,* often accompanied by a *cleft lip* lateral to the midline. Cleft palate and lip can be surgically corrected with good cosmetic results, but a cleft palate makes it difficult for an infant to generate the suction needed for nursing.

Palatine Bones

The **palatine bones** form the rest of the hard palate and part of the wall of the nasal cavity (see fig. 9.3a). At the posterolateral corners of the hard palate are two large **greater palatine foramina**.

Zygomatic Bones

The **zygomatic**[20] **bones** (cheekbones) form the angles of the cheeks at the inferolateral margins of the orbits and part of the lateral wall of each orbit; they extend about halfway to the ear (see figs. 9.2a and 9.3a). Each zygomatic bone has a ⊥-shape and usually a small **zygomaticofacial** (ZY-go-MAT-ih-co-FAY-shul) **foramen** near the intersection of the stem and crossbar of the ⊥. The prominent zygomatic arch that flares from each side of the skull is formed by the union of the zygomatic and temporal bones.

Lacrimal Bones

The **lacrimal**[21] (LACK-rih-mul) **bones** form part of the medial wall of each orbit (fig. 9.12). A groove called the **lacrimal sulcus** houses a membranous *lacrimal canal* in life. Tears from the eye collect in this canal and drain into the nasal cavity.

Nasal Bones

Two small rectangular **nasal bones** form the bridge of the nose (see fig. 9.1) and support cartilages that give shape to the lower portion of the nose. If you palpate the bridge, you can easily feel where the nasal bones end and the cartilages begin. The nasal bones are often fractured by blows to the nose.

Inferior Nasal Concha

The nasal conchae were mentioned in connection with the ethmoid bone, which contributes the superior and middle conchae. The **inferior nasal concha (inferior turbinate bone)**—a separate bone—is the largest of the three conchae (see fig. 9.11).

Vomer

The **vomer** forms the inferior half of the nasal septum (see figs. 9.1 and 9.2b). Its name literally means "plowshare," which refers to its resemblance to the blade of a

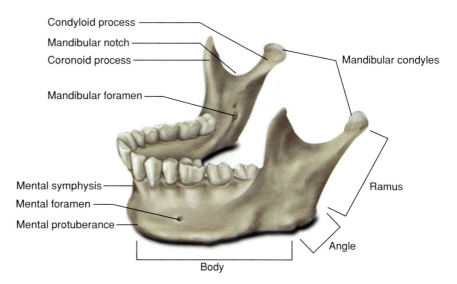

Figure 9.13 The mandible.

plow. The superior half of the nasal septum is formed by the perpendicular plate of the ethmoid bone, as mentioned earlier. The vomer and ethmoid bones support a layer of *septal cartilage* that forms most of the anterior part of the nasal septum.

Mandible

The **mandible** (fig. 9.13) is the strongest bone of the skull and the only one that can move. It supports the lower teeth and provides attachment for muscles of mastication and facial expression. The horizontal portion is called the **body;** the posterior corner, where it turns upward, is the **angle;** and the vertical-to-oblique posterior portion is the **ramus** (RAY-mus)—plural, *rami* (RAY-my). The mandible develops as separate right and left bones in the fetus. In early childhood, they fuse at a midsagittal line called the **mental symphysis** (SIM-fih-sis). The part of the mandible that juts forward beneath the symphysis is the **mental protuberance.** On the anterolateral surface of the body, the **mental foramen** permits the passage of nerves and blood vessels of the chin. The inner surface of the body has a number of shallow depressions and ridges to accommodate muscles and the salivary glands. The angle of the mandible has a rough surface for insertion of the *masseter,* a muscle of mastication.

The ramus is somewhat Y-shaped. Its posterior branch, called the **condyloid** (CON-dih-loyd) **process,** bears the **mandibular condyle.** This condyle inserts into the mandibular fossa of the temporal bone and forms the hinge of the mandible—the **temporomandibular joint (TMJ).** The anterior branch, called the **coronoid process,** is the point of insertion for the temporalis muscle. The U-shaped arch between the two processes is called the **mandibular notch.** Just below the notch,

---

20. *zygo* = union, to join
21. *lacrim* = tear, to cry

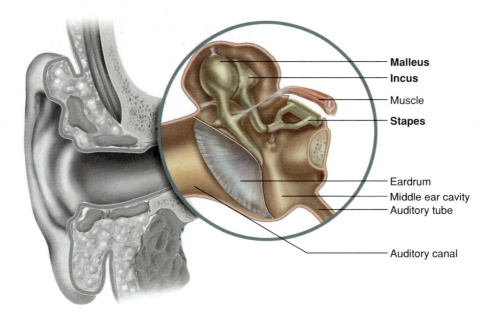

Figure 9.14 The auditory ossicles of the middle ear.

on the medial surface of the ramus, is the **mandibular foramen.** From this foramen emerge the blood vessels and nerve that supply the lower teeth (see special topic 9.3).

Bones Associated with the Skull

Seven bones are closely associated with the skull but not considered part of it. These are the three **auditory ossicles**[22] in each middle-ear cavity and the **hyoid bone** beneath the chin. The auditory ossicles are the **malleus,**[23] attached to the medial side of the tympanic membrane (eardrum); the **incus,**[24] attached to the malleus; and the **stapes**[25] (STAY-peez), which connects the incus to the inner ear (fig. 9.14). These bones transfer vibrations from the eardrum to the inner ear (see chapter 16).

22. *os* = bone + *icle* = little
23. *malleus* = hammer
24. *incus* = anvil
25. *stapes* = stirrup

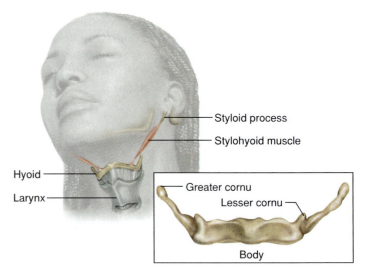

Figure 9.15 The hyoid bone.

The word *hyoid,* literally meaning "U-shaped," describes this slender bone between the chin and larynx (fig. 9.15). It is one of the few bones that does not articulate with any other. The hyoid is suspended from

Obstetric nurses routinely assess the fontanels of newborns by palpation. In a difficult delivery, one cranial bone may override another along a suture line, which calls for close monitoring of the infant. Abnormally wide sutures may indicate hydrocephalus, the accumulation of excessive amounts of cerebrospinal fluid, which causes the cranium to swell. Bulging fontanels suggest abnormally high intracranial pressure, while depressed fontanels indicate dehydration.

the styloid processes of the skull, somewhat like a hammock, by the small *stylohyoid* muscles. The medial **body** of the hyoid is flanked on either side by hornlike projections—a **greater** and **lesser cornu**[26] (COR-new)—plural, *cornua* (COR-new-uh). Forensic pathologists look for a fractured hyoid as evidence of strangulation.

The Skull in Infancy and Childhood

The head of an infant could not fit through the mother's pelvic outlet at birth were it not for the fact that the bones of its skull are not yet fused. The shifting of the skull bones during birth may cause the infant to appear deformed, but the head soon assumes a more normal shape. Spaces between the unfused cranial bones are called **fontanels,**[27] after the fact that pulsation of the infant's blood can be felt there. Palpation of the fontanels is part of the routine physical examination of the newborn (see special topic 9.4). The bones are joined at these points only by fibrous membranes, in which intramembranous ossification will be completed later. Four of these sites are especially prominent and regular in location: the **anterior, posterior, sphenoid,** and **mastoid fontanels** (fig. 9.16). Most fontanels ossify by the time the infant is a year old, but the largest one—the anterior fontanel—can still be palpated 18 to 24 months after birth.

The frontal bone and mandible, represented by separate right and left bones at birth, fuse medially in early childhood. The frontal bones usually fuse by age five or six, but in some children a *metopic*[28] *suture* persists between them. Traces of this suture are evident in some adult skulls.

The face of a newborn is flat and the cranium relatively large. To accommodate the growing brain, the skull grows more rapidly than the rest of the skeleton during childhood. It reaches about half its adult size by 9 months

26. *cornu* = horn
27. *fontan* = fountain + *el* = little
28. *met* = beyond + *op* = the eyes

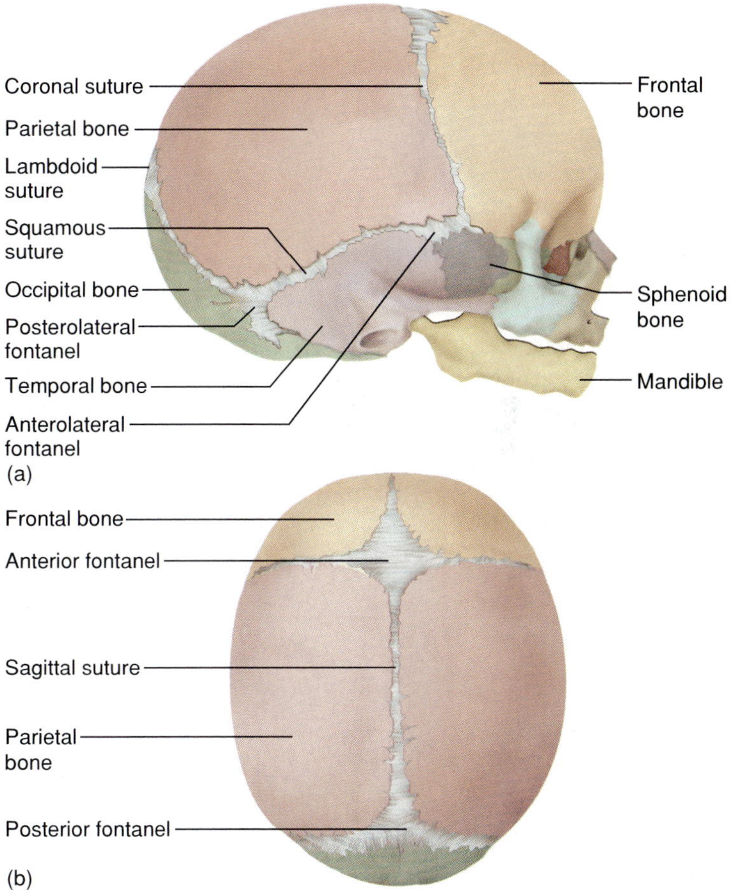

Figure 9.16 The fetal skull near the time of birth. (*a*) Right lateral view; (*b*) superior view.

Labels (a): Coronal suture, Parietal bone, Lambdoid suture, Squamous suture, Occipital bone, Posterolateral fontanel, Temporal bone, Anterolateral fontanel, Frontal bone, Sphenoid bone, Mandible

Labels (b): Frontal bone, Anterior fontanel, Sagittal suture, Parietal bone, Posterior fontanel

of age, three-quarters adult size by age two, and nearly final size by age eight or nine. The heads of babies and children are therefore much larger in proportion to the trunk than the heads of adults—an attribute thoroughly exploited by cartoonists and advertisers who draw big-headed characters to give them a more endearing or immature appearance. In humans and other animals, the large rounded heads of the young are thought to promote survival by stimulating parental caregiving instincts.

The bones of the skull are summerized in table 9.2.

Table 9.2 **Anatomical Checklist for the Skull and Associated Bones**

Cranial Bones

Frontal Bone (figs. 9.2 and 9.4)
    Supraorbital margin
    Supraorbital foramen or notch
    Frontal sinus

Parietal Bones (figs. 9.2a and 9.4)
    Temporal lines
    Parietal foramen

Temporal Bones (figs. 9.2, 9.3b, and 9.8)
    Squamous part
        Zygomatic process
        Mandibular fossa
    Tympanic part
        External auditory meatus
        Styloid process
    Mastoid part
        Mastoid process
        Mastoid notch
        Mastoid foramen
        Stylomastoid foramen
    Petrous part
        Internal auditory meatus
        Carotid canal
        Jugular foramen

Occipital Bone (figs. 9.2, 9.3, and 9.4)
    Foramen magnum
    Hypoglossal canal
    Condylar canal
    Pharyngeal tubercle
    Occipital condyles
    External occipital protuberance
    Superior nuchal line
    Inferior nuchal line

Sphenoid Bone (figs. 9.2a, 9.5, and 9.9)
    Body
    Lesser wing
        Optic foramen
        Superior orbital fissure
        Anterior clinoid process
    Greater wing
        Foramen ovale
        Foramen rotundum
        Foramen spinosum
        Foramen lacerum
        Medial and lateral pterygoid plates
    Nasal choanae
    Sphenoid sinus
    Sella turcica
    Dorsum sellae

Ethmoid Bone (figs. 9.2, 9.5, and 9.10)
    Perpendicular plate
    Superior nasal concha (superior turbinate bone)
    Middle nasal concha (middle turbinate bone)
    Ethmoid sinus (air cells)
    Crista galli
    Cribriform plate

Facial Bones

Maxilla (figs. 9.1, 9.2, and 9.3a)
    Alveoli
    Alveolar processes
    Infraorbital foramen
    Inferior orbital fissure
    Palatine processes
    Incisive foramen
    Maxillary sinus

Palatine Bones (figs. 9.3a and 9.12)
    Greater palatine foramen

Zygomatic Bones (figs. 9.2a and 9.3a)
    Zygomaticofacial foramen

Lacrimal Bones (figs. 9.1 and 9.12)
    Lacrimal sulcus

Nasal Bones (figs. 9.1 and 9.11)

Inferior Nasal Concha (fig. 9.11)

Vomer (figs. 9.1 and 9.2b)

Mandible (figs. 9.1 and 9.13)
    Body
        Mental symphysis
        Mental protuberance
        Mental foramen
    Angle
    Ramus
        Condyloid process
        Coronoid process
        Mandibular condyle
        Mandibular notch
        Mandibular foramen

Bones Associated with the Skull

Auditory Ossicles (fig. 9.14)
    Malleus (hammer)
    Incus (anvil)
    Stapes (stirrup)

Hyoid Bone (fig. 9.15)
    Body
    Greater Cornu
    Lesser Cornu

1. Name the four paranasal sinuses and four other cavities found in the skull.

2. What is the difference between a cranial bone and a facial bone? Name four of each.

3. Draw an oval representing a superior view of the calvaria. Draw in lines representing the coronal, sagittal, and lambdoid sutures, and then label each of the bones bordered by these sutures.

4. Match each of the following features with the appropriate bone: squamous part, hypoglossal foramen, greater cornu, condyloid process, greater wing, and cribriform plate.

5. List all of the other bones with which each of the following articulates: parietal, temporal, zygomatic, and ethmoid.

6. Palpate any of the following structures you can on your own head and identify those that cannot be palpated on a living subject: mastoid process, crista galli, superior orbital fissure, palatine processes, zygomatic bone, mental symphysis, and stapes.

The Spinal Column and Thoracic Cage

▼Objectives

When you have completed this section, you should be able to

- describe the general features of the spinal column and those of a typical vertebra;
- describe the special features of vertebrae in different regions of the spinal column and discuss the functional significance of the regional differences; and
- describe the anatomy of the sternum and ribs and explain how the ribs articulate with the thoracic vertebrae.

General Features of the Spinal Column

The **spinal (vertebral) column** physically supports the skull and trunk, allows for their movement, protects the spinal cord, and absorbs stresses produced by walking, running, and lifting. It also provides attachment for the extremities, thoracic cage, and postural muscles. Although commonly called the backbone, it does not consist of a single bone but a chain of 33 **vertebrae** with **intervertebral discs** of fibrocartilage between most of them. The adult spinal column averages about 71 cm (28 in.) long, with the 23 intervertebral discs accounting for about one-quarter of the length.

As shown in figure 9.17, the vertebrae are divided into five groups: 7 *cervical (SUR-vih-cul) vertebrae* in the neck, 12 *thoracic vertebrae* in the chest, 5 *lumbar vertebrae* in the lower back, 5 *sacral vertebrae* at the base of the spine, and 4 tiny *coccygeal* (coc-SIDJ-ee-ul) *vertebrae.*

Variations in the typical organization of vertebrae occur in about 1 person in 20. For example, the last lumbar vertebra is sometimes incorporated into the sacrum, producing four lumbar and six sacral vertebrae. In other cases, the first sacral vertebra fails to fuse with

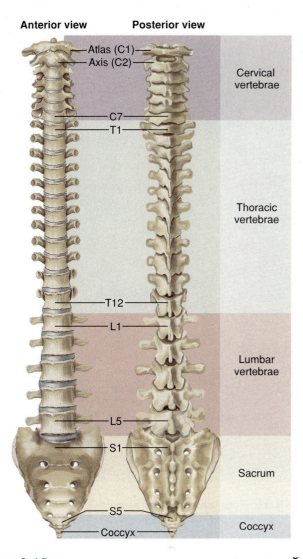

Anterior view **Posterior view**

Atlas (C1)
Axis (C2)
Cervical vertebrae
C7
T1
Thoracic vertebrae
T12
L1
Lumbar vertebrae
L5
S1
Sacrum
S5
Coccyx
Coccyx

Figure 9.17 The spinal column, anterior and posterior views.

the second, producing six lumbar and four sacral vertebrae. The cervical and thoracic vertebrae are much more constant in number. To help remember the typical numbers of cervical, thoracic, and lumbar vertebrae—7, 12, and 5—you might think of a typical work day: go to work at 7, have lunch at 12, and go home at 5.

In an infant, the spine exhibits one continuous C-shaped curve (fig. 9.18), as it does in monkeys, apes, and most other four-legged animals. As an infant begins to crawl and lift its head, however, the cervical region becomes curved toward the dorsal side, which enables an infant on its belly to look forward. As a toddler begins walking, another curve develops in the same direction in the lumbar region. These two curvatures result in the characteristic S-shape of the mature spine (fig. 9.19), which makes bipedal locomotion possible (see chapter essay, p. 289). The four bends of the adult vertebral column are called the **cervical, thoracic, lumbar,** and **pelvic curvatures.** The thoracic and pelvic

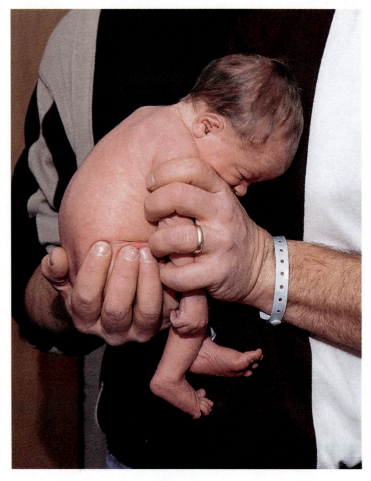

Figure 9.18 A newborn infant has a single C-shaped spinal curvature. The secondary curvatures of the cervical and lumbar regions develop as the child begins to crawl and walk.

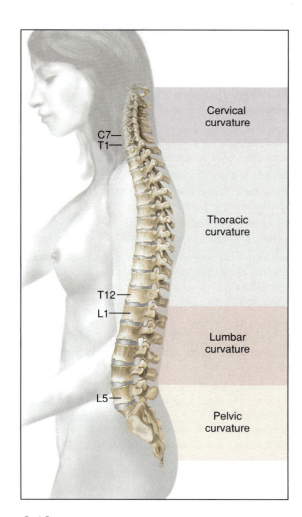

Figure 9.19 Curvatures of the adult spinal column.

2nd lumbar vertebra: superior view

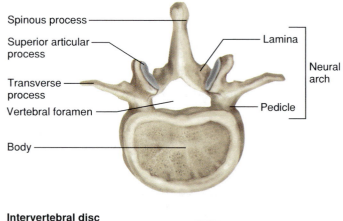

Intervertebral disc

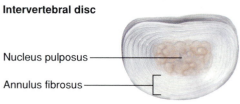

Figure 9.20 A representative vertebra and its intervertebral disc; superior views.

curvatures are called *primary curvatures* because they are remnants of the original infantile curvature. The cervical and lumbar curvatures are called *secondary curvatures* because they are not well developed until approximately age three, having resulted from the child's crawling and walking.

General Structure of a Vertebra

A representative vertebra and intervertebral disc are shown in figure 9.20. The most obvious feature of a vertebra is the **body,** or **centrum**—a mass of spongy bone and red bone marrow covered with a thin layer of compact bone. This is the weight-bearing portion of the vertebra. Its rough superior and inferior surfaces provide firm attachment to the intervertebral discs.

Think About It

The vertebral bodies and intervertebral discs low on the spinal column are larger than those higher up. What is the functional significance of this?

Abnormal spinal curvatures (fig. 1) can result from disease, weakness, or paralysis of the trunk muscles; poor posture; or congenital defects in vertebral anatomy. The most common deformity is an abnormal lateral curvature called **scoliosis.** It occurs most often in the thoracic region, particularly among adolescent girls. It sometimes results from a developmental abnormality in which the body and arch (described later) fail to develop on one side of a vertebra. If the person's skeletal growth is not yet complete, scoliosis can be corrected with a back brace.

An exaggerated thoracic curvature is called kyphosis (hunchback, in lay language). It is usually a result of osteoporosis, but it also occurs in athletic adolescent boys and can be a consequence of osteomalacia or spinal tuberculosis. An exaggerated lumbar curvature is called lordosis (swayback, in lay language). It may result from the same disorders that cause kyphosis, or it may be associated with pregnancy or obesity.

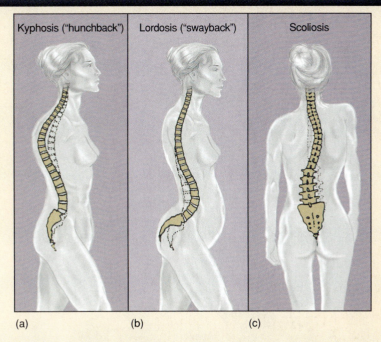

Figure 1 Abnormal spinal curvatures. (*a*) Kyphosis, an exaggerated thoracic curvature. (*b*) Lordosis, an exaggerated lumbar curvature. (*c*) Scoliosis, an abnormal lateral deviation.

Dorsal to the body of each vertebra is a triangular canal called the **vertebral foramen.** The vertebral foramina collectively form the **vertebral canal,** a passage for the spinal cord. The foramen is bordered by a bony **neural arch** composed of two parts on each side: a pillarlike **pedicle** and platelike **lamina.** Extending from the apex of the arch, a projection called the **spinous process** is directed toward the rear and downward (fig. 9.20). You can see and feel the spinous processes as a row of bumps along the spine. A **transverse process** extends laterally from the point where the pedicle and lamina meet. The spinous and transverse processes provide points of attachment for the spinal muscles.

A pair of **superior articular processes** that project upward from one vertebra and meet a similar pair of **inferior articular processes** that project downward from the vertebra just above (fig. 9.21*a*). Each process has a flat articular surface (facet) facing that of the adjacent vertebra. These processes restrict twisting of the vertebral column, which could otherwise severely damage the spinal cord.

When two vertebrae are joined, they exhibit an opening between their pedicles called the **intervertebral foramen.** This allows passage for spinal nerves that connect with the spinal cord at regular intervals. Each foramen is formed by an **inferior vertebral notch** in the pedicle of the superior vertebra and a **superior vertebral notch** in the pedicle of the one just below it (fig. 9.21*b*).

Intervertebral Discs

An **intervertebral disc** is a pad consisting of an inner gelatinous **nucleus pulposus** surrounded by a ring of fibrocartilage, the **annulus fibrosus** (see fig. 9.20). The discs help to bind adjacent vertebrae together, support the weight of the body, and absorb shock. Under stress—for example, when you lift a heavy weight—the discs bulge laterally. Excessive stress can crack the annulus and cause the nucleus to ooze out. This is called a *herniated disc* ("ruptured" or "slipped" disc in lay terms) and may put painful pressure on the spinal cord or a spinal nerve. To relieve the pressure, a procedure called a *laminectomy* may be performed—each lamina is cut and the laminae and spinous processes are removed. This procedure is also used to expose the spinal cord for anatomical study or surgery.

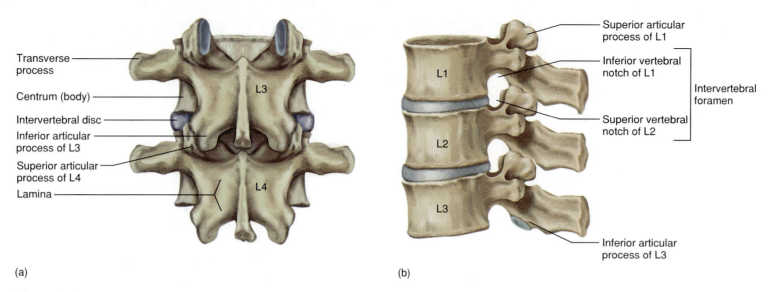

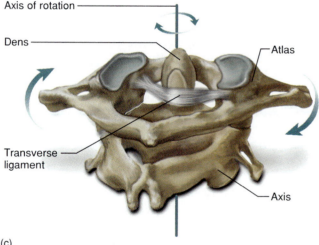

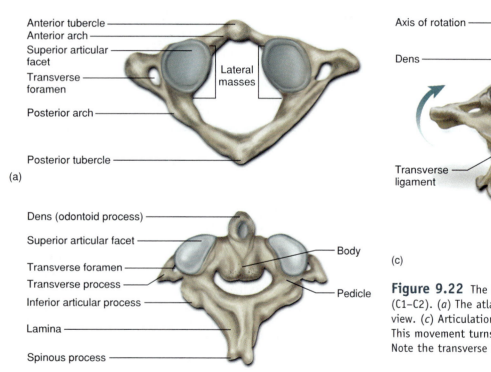

Figure 9.21 Articulated lumbar vertebrae. (*a*) Dorsal view of vertebrae L3 to L4; (*b*) left lateral view of vertebrae L1 to L3.

Figure 9.22 The atlas and axis, the first two cervical vertebrae (C1–C2). (*a*) The atlas, superior view. (*b*) The axis, posterosuperior view. (*c*) Articulation of the atlas and axis and rotation of the atlas. This movement turns the head from side to side, as in gesturing "no." Note the transverse ligament holding the dens of the axis in place.

Regional Characteristics of Vertebrae

We are now prepared to consider how vertebrae differ from one region of the spinal column to another and from the generalized anatomy just described. Knowing these variations will enable you to identify the region of the spine from which an isolated vertebra was taken. More importantly, these modifications in form reflect functional differences among the vertebrae.

Cervical Vertebrae

The **cervical vertebrae** (C1–C7) are the smallest and lightest. The first two (C1 and C2) have unique structures that allow for head movements (fig. 9.22). Vertebra C1 is called the **atlas** because it supports the head in a manner reminiscent of the Titan of Greek mythology who was condemned by Zeus to carry the world on his shoulders. It scarcely resembles the typical vertebra; it is little more than a delicate ring surrounding a large

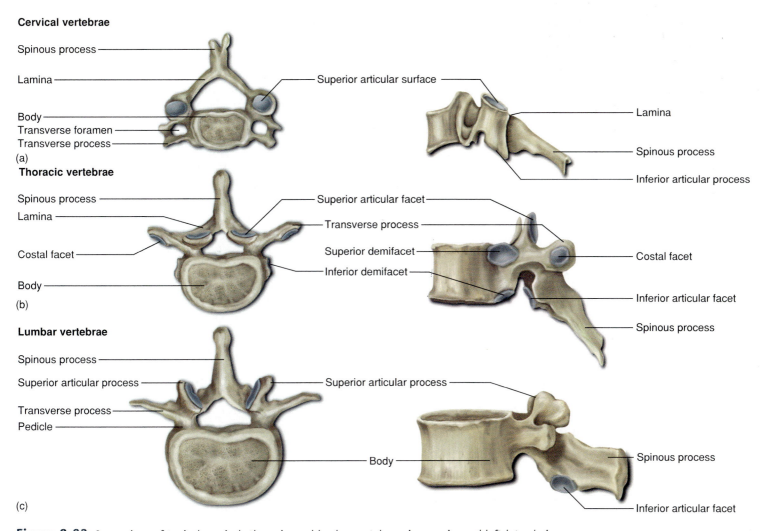

Cervical vertebrae

Spinous process

Lamina

Body
Transverse foramen
Transverse process
(a)

Superior articular surface

Lamina

Spinous process

Inferior articular process

Thoracic vertebrae

Spinous process

Lamina

Costal facet

Body
(b)

Superior articular facet

Transverse process

Superior demifacet

Inferior demifacet

Costal facet

Inferior articular facet

Spinous process

Lumbar vertebrae

Spinous process

Superior articular process

Transverse process

Pedicle

Superior articular process

Body

Spinous process

Inferior articular facet

(c)

Figure 9.23 Comparison of typical cervical, thoracic, and lumbar vertebrae, in superior and left lateral views.

vertebral foramen. On each side is a **lateral mass** with a deeply concave **superior articular facet** that articulates with the occipital condyle of the skull. A nodding motion of the skull, as in gesturing "yes," causes the occipital condyles to rock back and forth on these facets. The **inferior articular facets,** which are comparatively flat or only slightly concave, articulate with C2. The lateral masses are connected by an **anterior arch** and a **posterior arch,** which bear slight protuberances called the **anterior** and **posterior tubercle,** respectively.

Vertebra C2, the **axis,** allows rotation of the head as in gesturing "no." Its most distinctive feature is a prominent knob called the **dens** (denz), or **odontoid**[29] **process,** on its anterosuperior side. No other vertebra has a dens. It is thought to be the "missing body" of the atlas because it separates from the atlas during embryonic development and fuses with the axis. It projects into the vertebral foramen of the atlas, where it is nestled in

a facet and held in place by a **transverse ligament** (see fig. 9.22*b*). A heavy blow to the top of the head can cause a fatal injury in which the dens is driven through the foramen magnum into the brain stem. The articulation between the atlas and the cranium is called the **atlanto-occipital joint;** the one between the atlas and axis is called the **atlantoaxial joint.**

The axis is the first vertebra that exhibits a spinous process. In vertebrae C2 to C6, the process is forked, or *bifid,*[30] at its tip (fig. 9.23*a*). This fork provides attachment for the nuchal ligament of the back of the neck. All seven cervical vertebrae, have a prominent round **transverse foramen** in each transverse process. These foramina provide passage and protection for the *vertebral arteries,* which supply blood to the brain. Transverse foramina occur in no other vertebrae and thus provide an easy means of recognizing a cervical vertebra.

---

29. *dens = odont* = tooth

30. *bifid* = cleft in two parts

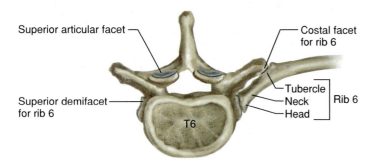

Figure 9.24 Articulation of thoracic vertebra T6 with rib 6. Note that the rib articulates with the vertebra at two points: the vertebral body and transverse process.

Think About It

How would head movements be affected if vertebrae C1 and C2 had the same structure as C3? Why does C1 lack a spinous process?

Cervical vertebrae C3 to C6 are similar to the typical vertebra described earlier, with the addition of the transverse foramina and bifid spinous processes. Vertebra C7 is a little different—its spinous process is not bifid, but it is especially long and forms a prominent bump on the lower back of the neck. This feature is a convenient landmark for counting vertebrae.

Thoracic Vertebrae

There are 12 **thoracic vertebrae** (T1–T12), corresponding to the 12 pairs of ribs attached to them. They lack the transverse foramina and bifid processes that distinguish the cervicals, but possess the following distinctive features of their own (fig. 9.23b):

- The spinous processes are relatively pointed and angle sharply downward.
- The body is somewhat heart-shaped, more massive than in the cervical vertebrae but less than in the lumbar vertebrae.
- The body has articular surfaces (to be described shortly) for attachment of the ribs.
- On vertebrae T1 to T10, a shallow, cuplike **costal**[31] **facet** at the end of each transverse process provides a second point of articulation for each rib. There are no costal facets on T11 and T12 because ribs 11 and 12 attach only to the bodies of the vertebrae.

Thoracic vertebrae vary among themselves mainly because of variations in the way the ribs articulate with them, so we must consider the anatomy of the ribs and the thoracic vertebrae jointly. The most common type of articulation between a rib and the spinal column is

shown in figure 9.24. This figure should be examined along with figure 9.27, in which the anatomy of the ribs is shown. Most ribs have a *head* that inserts *between* two adjacent vertebrae. The vertebra above has a notch in its body called the **inferior demifacet**[32] (DEM-ee-FASS-it) that articulates with the **superior articular facet** of the rib. The vertebra below has a **superior demifacet** on its body that articulates with the **inferior articular facet** of the same rib.

Distal to the head of the rib is a *tubercle* with another articular facet that articulates with the costal facet on the end of the transverse process of a thoracic vertebra. Thus, most ribs have two points of attachment to the spinal column—the head and tubercle. The exceptions to this pattern are ribs 11 and 12.

Rib 1 attaches to a complete facet on the body of vertebra T1. This vertebra also has an inferior demifacet that articulates with rib 2. Vertebrae T10 to T12 each have a complete facet on the body where they articulate with ribs 10 to 12; they do not have demifacets. Vertebrae T11 and T12 also lack costal facets because ribs 11 and 12 articulate only with the vertebral bodies. These variations in articulation are summarized in table 9.3.

Lumbar Vertebrae

There are five **lumbar vertebrae** (L1–L5). Their most distinctive features are a thick, stout body and a blunt, squarish spinous process (see fig. 9.23c). In addition, their articular processes are oriented differently than on other vertebrae. In thoracic vertebrae, the superior processes face forward and the inferior processes face to the rear. In lumbar vertebrae, the superior processes face medially (like the palms of your hands about to clap), and the inferior processes of the vertebrae above meet them by facing laterally. This arrangement makes the lumbar region of the spine especially resistant to twisting. These differences are best observed on an articulated skeleton.

Sacrum

The **sacral vertebrae** (S1–S5) begin to fuse between the ages of 16 and 18. By age 26, they are fused into a single triangular bone called the **sacrum** (fig. 9.25). This bone is named for the fact that it was once considered the seat of the soul.[33] The anterior surface of the sacrum forms the wall of the pelvic cavity. It is relatively smooth and concave and has four transverse lines that indicate where the five vertebrae have fused. Four pairs

---

31. *costa* = rib

32. *demi* = half
33. *sacr* = sacred

Rib	Type	Costal Cartilage	Articulating Vertebral Bodies	Articulating with a Costal Facet?	Rib Tubercle
1	True	Individual	T1	Yes	Present
2	True	Individual	T1 and T2	Yes	Present
3	True	Individual	T2 and T3	Yes	Present
4	True	Individual	T3 and T4	Yes	Present
5	True	Individual	T4 and T5	Yes	Present
6	True	Individual	T5 and T6	Yes	Present
7	True	Individual	T6 and T7	Yes	Present
8	False	Shared with rib 7	T7 and T8	Yes	Present
9	False	Shared with rib 7	T8 and T9	Yes	Present
10	False	Shared with rib 7	T10	Yes	Present
11	False, floating	None	T11	No	Absent
12	False, floating	None	T12	No	Absent

*Ribs 2 to 9 articulate with demifacets on the bodies of two adjacent vertebrae. Ribs 1 and 10 to 12 each articulate with a complete facet on the body of a single vertebra. Note that only those ribs that articulate with the costal facets have tubercles.

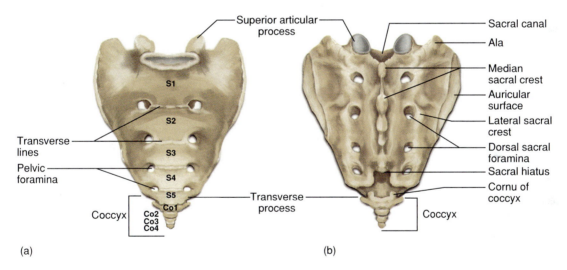

Figure 9.25 The sacrum and coccyx. (*a*) Anterior surface, which faces the viscera of the pelvic cavity. (*b*) Posterior surface. The processes of this surface can be palpated in the sacral region.

of large **pelvic** (anterior sacral) **foramina** occur on this surface. These allow for passage of nerves and arteries to the pelvic organs.

The dorsal surface of the sacrum is very rough. The spinous processes of the vertebrae are fused into a dorsal ridge called the **median sacral crest**. Again, there are four pairs of openings for spinal nerves, the **dorsal sacral foramina**. A **sacral canal** runs through the sacrum and ends in an inferior opening called the **sacral hiatus** (hy-AY-tus). On each side of the sacrum is an ear-shaped region called

the **auricular**[34] (aw-RIC-you-lur) **surface.** This articulates with a similarly shaped surface on the os coxae, forming the strong, immovable **sacroiliac** (SAY-cro-ILL-ee-ac) **joint.** At the superior end of the sacrum, lateral to the median crest, are a pair of **superior articular processes** that articulate with vertebra L5. Lateral to these are a pair of large, rough, winglike extensions called the **alae**[35] (AIL-ee).

34. *auri* = ear + *cul* = little
35. *alae* = wings

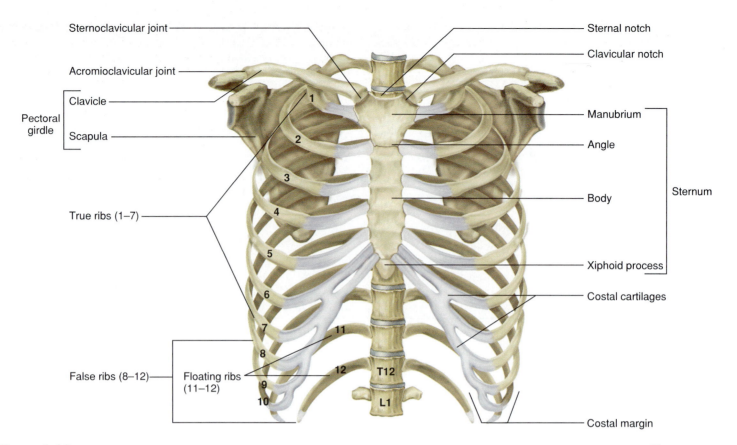

Sternoclavicular joint

Acromioclavicular joint

Pectoral girdle

Clavicle

Scapula

True ribs (1–7)

False ribs (8–12)

Floating ribs (11–12)

Sternal notch

Clavicular notch

Manubrium

Angle

Body

Sternum

Xiphoid process

Costal cartilages

Costal margin

Figure 9.26 Anterior view of the thoracic cage (ribs, sternum, and thoracic vertebrae) and pectoral girdle (clavicle and scapula).

Coccyx

The **coccyx**[36] (fig. 9.25) usually consists of four (sometimes five) small vertebrae, Co1 to Co4, fused by the age of 20 to 30 into a single triangular bone. Vertebra Co1 has a pair of hornlike projections, the **cornua,** which serve as attachment points for ligaments that bind the coccyx to the sacrum. The coccyx can be fractured by a difficult childbirth or a hard fall to the buttocks. Although it is the vestige of a tail, it is not entirely useless; it provides attachment for muscles of the pelvic floor.

The Thoracic Cage

The **thoracic cage** (fig. 9.26) consists of the sternum and ribs. It forms a more or less conical enclosure for the lungs and heart and provides attachment for the pectoral girdle and upper extremity. It has a narrow superior apex and broad base, and it is rhythmically expanded by the respiratory muscles to create a vacuum that draws air into the lungs. The inferior border of the thoracic cage is formed by a downward arc of the ribs called the **costal margin.** The ribs protect not only the thoracic organs but also the spleen, most of the liver, and to some extent the kidneys.

Sternum

The **sternum** (breastbone) is a bony plate anterior to the heart. It is subdivided into three regions: the manubrium, body, and xiphoid process. The **manubrium**[37] (ma-NOO-bree-um) is the broad superior portion. It has a superomedial **sternal notch** (jugular notch), which you can easily palpate between your clavicles (collarbones), and right and left **clavicular notches,** where the clavicles articulate with it. The **body,** or **gladiolus,**[38] is the longest part of the sternum. It joins the manubrium at the **sternal angle,** which can be palpated as a transverse ridge at the point where the sternum projects farthest forward. In some people, however, it is rounded or concave. The second rib attaches here, making the sternal angle a useful landmark for counting ribs in a physical examination. The manubrium and body have scalloped lateral margins where cartilages of the ribs are attached. At the inferior end of the sternum is a small, pointed **xiphoid**[39] (ZIF-oyd) **process** that provides attachment for some of the abdominal muscles.

36. *coccyx* = cuckoo; named for its resemblance to a cuckoo's beak

37. *manubri* = handle
38. *gladiolus* = sword
39. *xipho* = sword

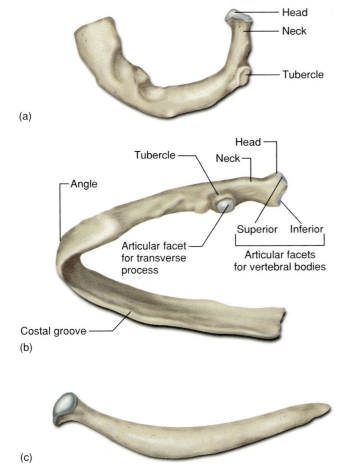

(a)

(b)

(c)

Figure 9.27 Anatomy of the ribs. (*a*) Rib 1 is an atypical flat plate. (*b*) Typical features of ribs 2 to 10. (*c*) Appearance of the floating ribs, 11 and 12. ✗

Ribs

There are 12 pairs of **ribs.** Each is attached at its posterior (proximal) end to the spinal column. A strip of hyaline cartilage called the **costal cartilage** extends from the anterior (distal) ends of ribs 1 to 7 to the sternum. Ribs 1 to 7 are thus called **true ribs.** Ribs 8 to 10 attach to the costal cartilage of rib 7, and ribs 11 and 12 do not attach to anything at the distal end but are embedded in thoracic muscle. Ribs 8 to 12 are therefore called **false ribs,** and ribs 11 and 12 are also called **floating ribs** for lack of a connection to the sternum.

As described earlier, ribs 1 to 10 each have a **head** and **tubercle,** connected by a narrow **neck;** ribs 11 and 12 have a head only, in keeping with their mode of attachment to the spinal column (fig. 9.27 and table 9.3). Ribs 2 to 9 have angled heads. The upper portion of the head, the **superior articular facet,** articulates with the demifacet of the vertebra above that rib; the lower portion, the **inferior articular facet,** articulates with the demifacet of the vertebra below. Ribs 2 to 10 have a sharp turn called the **angle,** just distal to the tubercle, and the remainder consists of a flat blade called the

shaft. Along the inferior margin of the shaft is a **costal groove** that marks the path of the intercostal blood vessels and nerve. Unlike the others, the first rib is a flat horizontal plate.

Features of the spinal column and thoracic cage are summarized in table 9.4.

The Pectoral Girdle and Upper Extremity

▼Objectives
When you have completed this section, you should be able to
• identify the clavicle and scapula, and describe their features;
• identify the humerus, radius, and ulna, and describe their features; and
• name the bones of the wrist and hand, and describe their features.

Pectoral Girdle

The **pectoral girdle** (shoulder girdle) supports the arm. It consists of two bones on each side of the body: the *clavicle* (collarbone) and the *scapula* (shoulder blade). The clavicle articulates with the sternum at the **sternoclavicular joint** (see fig. 9.26). The scapula articulates with the clavicle and the humerus. These are loose attachments that result in a shoulder far more flexible than that of most other mammals, but they also make the shoulder joint easy to dislocate.

Think About It
How do you think the unusual flexibility of the human shoulder is related to the habitat of our primate ancestors?

Clavicle

The **clavicle**[40] (fig. 9.28) is a slightly S-shaped bone, somewhat flattened dorsoventrally and easily seen and palpated on the upper thorax. The superior surface is relatively smooth, whereas the inferior surface is marked by grooves and ridges for muscle attachment. The medial **sternal end** has a rounded, hammerlike head, and the lateral **acromial end** is markedly flattened. Near the

40. *clav* = hammer, club + *icle* = little

Table 9.4 Anatomical Checklist for the Spinal Column and Thoracic Cage

Spinal Column

Spinal Curvatures (fig. 9.19)
- Cervical curvature
- Thoracic curvature
- Lumbar curvature
- Pelvic curvature

General Vertebral Structure (figs. 9.20 and 9.21)
- Body (centrum)
- Vertebral foramen
- Vertebral canal
- Neural arch
  - Pedicle
  - Lamina
- Spinous process
- Transverse process
- Superior articular process
- Inferior articular process
- Intervertebral foramen
  - Inferior vertebral notch
  - Superior vertebral notch

Intervertebral Discs (fig. 9.20)
- Annulus fibrosus
- Nucleus pulposus

Cervical Vertebrae (figs. 9.22 and 9.23)
- Transverse foramina
- Bifid spinous processes
- Atlas
  - Anterior arch
  - Anterior tubercle
  - Posterior arch
  - Posterior tubercle
  - Lateral mass
  - Superior articular facet
  - Inferior articular facet
  - Transverse ligament
- Axis
  - Dens (odontoid process)

Thoracic Vertebrae (figs. 9.23 and 9.24)
- Superior demifacet
- Inferior demifacet
- Costal facet

Lumbar Vertebrae (figs. 9.21 and 9.23)

Sacral Vertebrae (fig. 9.25)
- Sacrum
- Pelvic foramina
- Dorsal sacral foramina
- Median sacral crest
- Sacral canal
- Sacral hiatus
- Auricular surface
- Superior articular processes
- Alae

Coccygeal Vertebrae (fig. 9.25)
- Coccyx
- Cornua

Thoracic Cage

Sternum (fig. 9.26)
- Manubrium
  - Sternal notch
  - Clavicular notches
  - Sternal angle
- Body (gladiolus)
- Xiphoid process

Ribs (figs. 9.24, 9.26, and 9.27)
- True ribs (ribs 1–7)
- False ribs (ribs 8–12)
- Floating ribs (ribs 11 and 12)
- Head
- Superior articular facet
- Inferior articular facet
- Neck
- Tubercle
- Angle
- Shaft
- Costal groove
- Costal cartilage

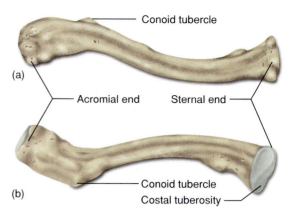

Figure 9.28 The right clavicle (collarbone). (*a*) Superior view; (*b*) inferior view. 𝒳

acromial end is a rough tuberosity called the **conoid tubercle**—a ligament attachment that faces toward the rear and slightly downward. The clavicle braces the shoulder and is thickened in people who do heavy manual labor. Without it, the pectoralis major muscles would pull the shoulders forward and medially, as occurs when a clavicle is fractured. Indeed, the clavicle is the most commonly fractured bone in the body because it is so close to the surface and because people often reach out with their arms to break a fall.

Scapula

The **scapula** (fig. 9.29) is a triangular plate that dorsally overlies ribs 2 to 7. The three sides of the triangle are called the **superior, medial** (vertebral), and **lateral** (axillary) **borders,** and its three angles are the **superior, inferior,** and **lateral angles.** A conspicuous **scapular notch** in the superior border provides passage for a nerve. The broad anterior surface of the scapula, called the **subscapular fossa,** is slightly concave and relatively featureless. The posterior surface has a transverse ridge called the **spine,** a deep transverse indentation superior to the spine called the **supraspinous fossa,** and a broad surface inferior to it called the **infraspinous fossa.**[41] The scapula is held in place by numerous muscles attached to these three fossae.

The most complex region of the scapula is its lateral angle, which has three main features:

1. The **acromion**[42] (ah-CRO-me-on) is a platelike extension of the scapular spine that forms the apex of the shoulder. The **acromioclavicular joint** between the acromion and clavicle (see fig. 9.26) is the only point at which the scapula articulates with the axial skeleton.

---

41. *supra* = above; *infra* = below
42. *acr* = extremity, point + *omi* = shoulder

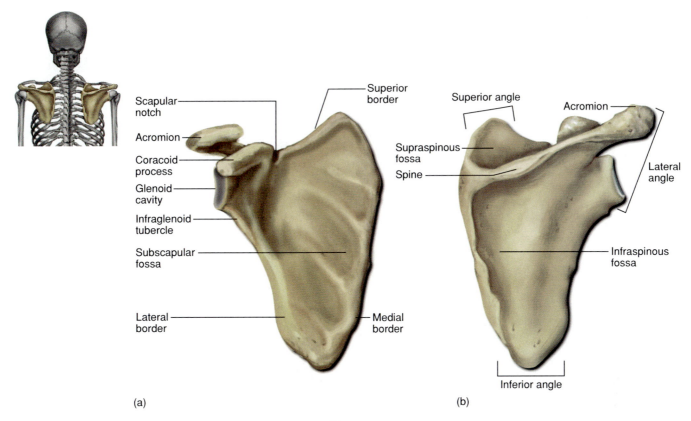

(a)   (b)

Figure 9.29 The left scapula. (*a*) Anterior view; (*b*) posterior view.

2. The **coracoid**[43] (COR-uh-coyd) **process** is shaped like a finger but named for a vague resemblance to a crow's beak; it provides attachment for the biceps muscle.
3. The **glenoid** (GLEN-oyd) **cavity** is a shallow socket that articulates with the head of the humerus.

Upper Extremity

The upper extremity is divided into four regions containing a total of 30 bones.

1. The **brachium**[44] (BRAY-kee-um), or arm proper, extends from shoulder to elbow. It contains only one bone, the *humerus.*
2. The **antebrachium,**[45] or forearm, extends from elbow to wrist and contains two bones—the *radius* and *ulna.* In anatomical position, these bones are parallel and the radius is lateral to the ulna.
3. The **carpus,**[46] or wrist, contains eight small bones arranged in two rows.
4. The **manus,**[47] or hand, contains 19 bones in two groups—5 *metacarpals* in the palm and 14 *phalanges* in the fingers.

----

43. *corac* = crow
44. *brachi* = arm
45. *ante* = before
46. *carp* = wrist; to seize
47. *man* = hand

Humerus

The **humerus** has a hemispherical **head** at its proximal end that articulates with the glenoid cavity of the scapula (fig. 9.30). The smooth surface of the head (covered with articular cartilage in life) is bordered by a groove called the **anatomical neck,** which is the epiphyseal line of this bone. The **surgical neck,** a common fracture site, is where the head tapers to the shaft. Other prominent features of the proximal end are muscle attachments called the **greater** and **lesser tubercles** and an **intertubercular groove** between them that accommodates a tendon of the biceps muscle.

The shaft has a rough area called the **deltoid tuberosity** on its lateral surface. This is an insertion for the deltoid muscle—the thick shoulder muscle where injections are given. The distal end of the humerus has two smooth condyles. The lateral one, called the **capitulum**[48] (ca-PIT-you-lum), is shaped somewhat like a flat tire and articulates with the radius. The medial one, called the **trochlea**[49] (TROCK-lee-uh), is pulleylike and articulates with the ulna. Immediately proximal to these condyles, the humerus flares out to form two bony processes, the **lateral** and **medial epicondyles.** The

----

48. *capit* = head + *ul* = little
49. *troch* = wheel, pulley

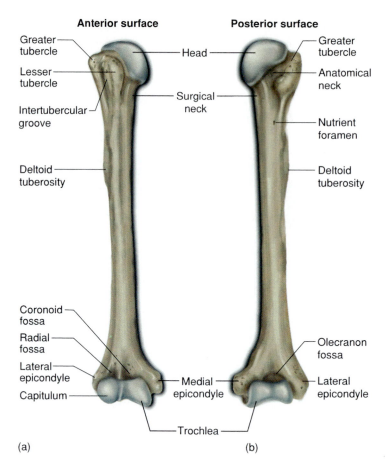

Anterior surface

Greater tubercle
Head
Lesser tubercle
Intertubercular groove
Surgical neck
Deltoid tuberosity

Coronoid fossa
Radial fossa
Lateral epicondyle
Capitulum
Medial epicondyle
Trochlea

(a)

Posterior surface

Greater tubercle
Anatomical neck
Nutrient foramen
Deltoid tuberosity

Olecranon fossa
Lateral epicondyle

(b)

Figure 9.30 The right humerus. (*a*) Anterior view; (*b*) posterior view. ✗

medial epicondyle protects the ulnar nerve, which passes close to the surface across the back of the elbow. This epicondyle is popularly known as the "funny bone" because striking the elbow on the edge of a table stimulates the ulnar nerve and produces a sharp tingling sensation.

The distal end of the humerus also shows three deep pits—two anterior and one posterior. The posterior pit, called the **olecranon** (oh-LEC-ruh-non) **fossa,** accommodates the olecranon process of the ulna when the arm is extended. On the anterior surface, a medial pit called the **coronoid fossa** accommodates the coronoid process of the ulna when the arm is flexed. The lateral pit is the **radial fossa,** named for the nearby head of the radius.

Radius

The proximal head of the **radius** (fig. 9.31) is a distinctive disc that rotates freely on the humerus when the palm is turned forward and back. It articulates with the capitulum of the humerus and radial notch of the ulna. On the shaft, immediately distal to the head, is a medial rough elevation, the **tuberosity,** which is the insertion of the biceps muscle.

The distal end of the radius has four prominent features. From lateral to medial, these are

1. a bony point, the **styloid process,** which can be palpated proximal to the thumb;
2. a shallow depression, the **navicular facet,** which articulates with the navicular bone of the wrist;
3. another depression, the **lunate facet,** which articulates with the lunate bone of the wrist; and
4. the **ulnar notch,** which articulates with the end of the ulna.

Ulna

At the proximal end of the **ulna** (fig. 9.31) is a deep, C-shaped **trochlear notch** that wraps around the trochlea of the humerus. The posterior side of this notch is formed by a prominent **olecranon**—the bony point where you rest your elbow on a table. The anterior side is formed by a less prominent **coronoid process.** Medially, the head of the ulna has a less conspicuous **radial notch,** which accommodates the head of the radius.

At the distal end of the ulna is a medial **styloid process.** The bony lumps you can palpate on each side of your wrist are the styloid processes of the radius and ulna. The radius and ulna are attached along their shafts

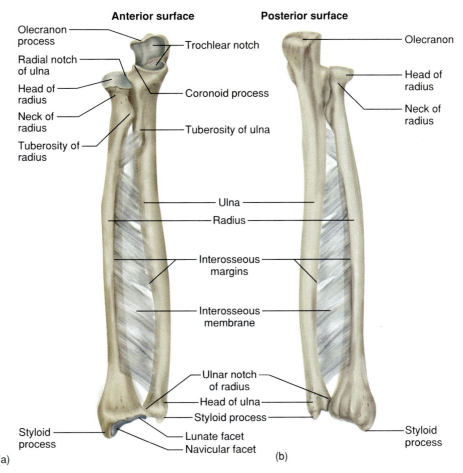

Olecranon process

Trochlear notch

Radial notch of ulna

Head of radius

Neck of radius

Tuberosity of radius

Coronoid process

Tuberosity of ulna

Olecranon

Head of radius

Neck of radius

Ulna

Radius

Interosseous margins

Interosseous membrane

Ulnar notch of radius

Head of ulna

Styloid process

Styloid process

Lunate facet

Navicular facet

Styloid process

(a)

(b)

Figure 9.31 The radius and ulna of the right antebrachium (forearm). (*a*) Anterior view; (*b*) posterior view. ⅄

by a ligament called the **interosseous** (IN-tur-OSS-ee-us) **membrane,** which is attached to an angular ridge called the **interosseous margin** on the medial side of each bone.

Carpal Bones

The **carpal bones,** which form the wrist, are arranged in two rows of four bones each (fig. 9.32). These short bones allow movements of the wrist from side to side and up and down. The carpal bones of the proximal row, starting at the lateral (thumb) side, are the **navicular** (or scaphoid), **lunate, triquetral** (tri-QUEE-trul), and **pisiform.** Translating from the Latin, these words mean boat-, moon-, triangle-, and pea-shaped, respectively. Unlike the other carpal bones, the pisiform is a sesamoid bone; it develops within a tendon and modifies the direction in which the associated muscle pulls.

The bones of the distal row, again starting on the lateral side, are the **trapezium,**[50] **trapezoid, capitate,**[51] and **hamate.**[52] The hamate can be recognized by a prominent hook, or **hamulus,** on the palmar side.

Metacarpal Bones

Bones of the palm are called **metacarpals.**[53] Metacarpal I is located at the base of the thumb and metacarpal V at the base of the little finger. On a skeleton, the metacarpals look like extensions of the fingers, so that the fingers seem much longer than they really are. The proximal end of a metacarpal bone is called the **base,** the shaft is called the **body,** and the distal end is called the **head.** The heads of the metacarpals form knuckles when you clench your fist.

Phalanges

The bones of the fingers are called **phalanges** (fah-LAN-jeez); in the singular, *phalanx* (FAY-lanks). There are two phalanges in the **pollex** (thumb) and three in each of the other digits. Phalanges are identified by Roman numerals preceded by *proximal, middle,* and *distal.* For example, proximal phalanx I is in the basal segment of the thumb (the first segment beyond the web between the thumb and palm); the left proximal phalanx IV is where people usually wear wedding rings; and distal

50. *trapez* = table, grinding surface
51. *capit* = head
52. *ham* = hook

53. *meta* = beyond

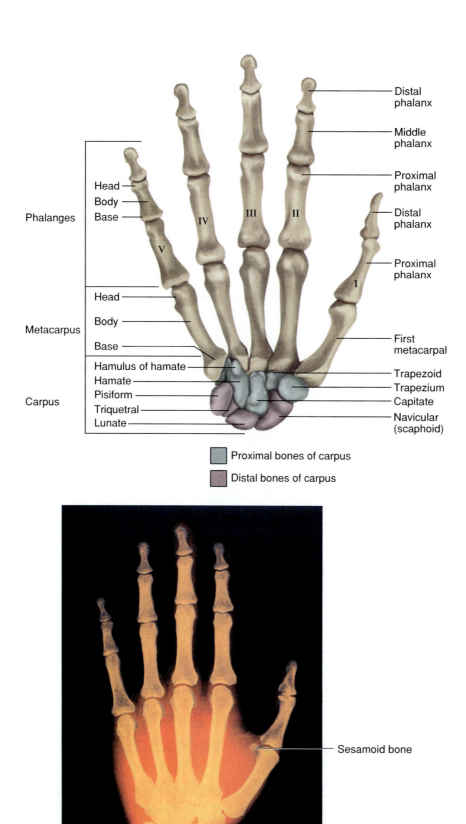

Phalanges
- Head
- Body
- Base

Distal phalanx

Middle phalanx

Proximal phalanx

Distal phalanx

Proximal phalanx

IV III II I

V

Metacarpus
- Head
- Body
- Base

First metacarpal

Carpus
- Hamulus of hamate
- Hamate
- Pisiform
- Triquetral
- Lunate

Trapezoid
Trapezium
Capitate
Navicular (scaphoid)

Proximal bones of carpus

Distal bones of carpus

Sesamoid bone

Figure 9.32 Bones of the right wrist and hand, anterior (palmar) view; drawing and X ray. Identify the unlabeled bones in the X ray by comparing it to the drawing. ⚡

phalanx V forms the tip of the little finger. The three parts of a phalanx are the same as in a metacarpal: base, body, and head. The ventral surface of a phalanx is slightly concave from end to end and flattened from side to side; the dorsal surface is rounder and slightly convex.

The bones of the pectoral girdle and upper extremity are summarized in table 9.5.

Key Point Review

11 How you can distinguish the medial end of the clavicle from the lateral end and the superior surface from the inferior?

12 Name the three fossae of the scapula and describe the location of each.

13 Which three bones meet at the elbow? Describe the elbow joint in terms of the fossae, articular surfaces, and processes that meet there.

14 List each bone of the medial side of the hand, in order from the palm to the tip of the little finger.

The Pelvic Girdle and Lower Extremity

▼**Objectives**
When you have completed this section, you should be able to
• describe the features of the pelvic girdle;
• contrast the anatomy of the male and female pelves;
• identify the femur, patella, tibia, and fibula, and describe their features; and
• name the bones of the ankle and foot, and describe their features.

Pelvic Girdle

The **pelvic girdle** (fig. 9.33) supports the trunk on the legs and encloses and protects viscera of the pelvic cavity—mainly the lower colon, urinary bladder, and reproductive organs. Each half of the adult pelvis is a single bone called the **os coxae**—plural, **ossa coxae.** Another term for the os coxae—arguably the most self-contradictory term in anatomy—is the *innominate*[54] (ih-NOM-ih-nate) *bone* ("unnamed bone").

The os coxae is joined to the spinal column at one point, the sacroiliac joint, where its **auricular surface** matches the one on the sacrum. On the anterior side of the pelvis, the left and right ossa coxae are joined by a fibrocartilage disc called the **pubic symphysis,**[55] which can be palpated immediately above the genitalia.

The **pelvis**[56] has a bowl-like shape with the broad **greater (false) pelvis** between the flare of the hips and the narrower **lesser (true) pelvis** below. The two are separated by a somewhat round margin called the **pelvic brim.** The opening circumscribed by the brim is called

54. *in* = without + *nom* = name
55. *sym* = together + *physis* = growth
56. pelvis = basin

Table 9.5 Anatomical Checklist for the Pectoral Girdle and Upper Extremity

Pectoral Girdle

Clavicle (fig. 9.28)
  Sternal end
  Acromial end
  Conoid tubercle
Scapula (figs. 9.29)
  Borders
    Superior border
    Medial (vertebral) border
    Lateral (axillary) border
  Angles
    Superior angle
    Inferior angle
    Lateral angle
  Scapular notch
  Spine
  Fossae
    Subscapular fossa
    Supraspinous fossa
    Infraspinous fossa
  Acromion
  Coracoid process
  Glenoid cavity

Upper Extremity

Humerus (fig. 9.30)
  Proximal end
    Head
    Anatomical neck
    Surgical neck
    Greater tubercle
    Lesser tubercle
    Intertubercular groove
  Shaft
    Deltoid tuberosity
  Distal end
    Capitulum
    Trochlea
    Lateral epicondyle
    Medial epicondyle
    Olecranon fossa
    Coronoid fossa
    Radial fossa

Radius (fig. 9.31)
  Head
  Tuberosity
  Styloid process
  Navicular facet
  Lunate facet
  Ulnar notch
Ulna (fig. 9.31)
  Trochlear notch
  Olecranon
  Coronoid process
  Radial notch
  Styloid process
  Interosseous margin
  Interosseous membrane
Carpal Bones (fig. 9.32)
  Proximal group
    Navicular
    Lunate
    Triquetral
    Pisiform
  Distal group
    Trapezium
    Trapezoid
    Capitate
    Hamate
      Hook (hamulus)
Bones of the Hand (fig. 9.32)
  Metacarpal bones I–V
    Base
    Body
    Head
  Phalanges I–V
    Proximal phalanx
    Middle phalanx
    Distal phalanx

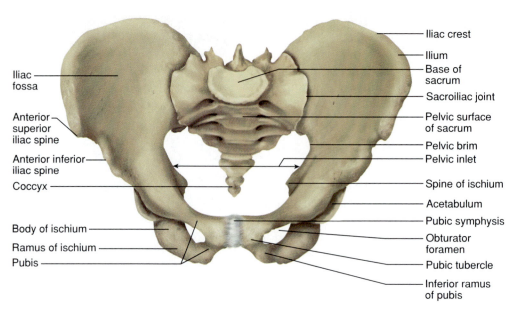

Iliac
fossa

Anterior
superior
iliac spine

Anterior inferior
iliac spine

Coccyx

Body of ischium

Ramus of ischium

Pubis

Iliac crest

Ilium

Base of
sacrum

Sacroiliac joint

Pelvic surface
of sacrum

Pelvic brim

Pelvic inlet

Spine of ischium

Acetabulum

Pubic symphysis

Obturator
foramen

Pubic tubercle

Inferior ramus
of pubis

Figure 9.33 The pelvic girdle and sacrum, anterosuperior view.

the **pelvic inlet**—an entry into the lesser pelvis through which an infant's head passes during birth. The lower margin of the lesser pelvis is called the **pelvic outlet.**

The os coxae has three distinctive features that will serve as landmarks for further description. These are the **iliac**[57] **crest** (superior crest of the hip); **acetabulum**[58] (hip socket—named for its resemblance to vinegar cups used in ancient Rome); and **obturator**[59] **foramen** (a large round-to-triangular hole below the acetabulum, closed by a ligament called the *obturator membrane* in life).

The adult os coxae forms by the fusion of three childhood bones called the *ilium* (ILL-ee-um), *ischium* (ISS-kee-um), and *pubis* (PEW-biss), identified by color in figure 9.34. The largest of these is the **ilium,** which extends from the iliac crest to the superior wall of the acetabulum. The iliac crest extends from a point or angle on the anterior side, called the **anterior superior spine,** to a sharp posterior angle, called the **posterior superior spine.** In a lean person, the anterior superior spines form visible anterior protrusions, and the posterior superior spines are sometimes marked by dimples above the buttocks where connective tissue attached to the spines pulls inward on the skin (fig. 9.35).

Below the superior spines are the **anterior** and **posterior inferior spines.**

Below the posterior inferior spine is a deep **greater sciatic** (sy-AT-ic) **notch,** named for the sciatic nerve that passes through it and continues down the posterior side of the thigh. The ilium is thickened along a line called the **iliac pillar,** extending from the iliac crest to the superior margin of the acetabulum. Both the notch and pillar result from evolution of the pelvis for bipedal locomotion, as explained in the chapter essay (p. 289).

The posterolateral surface of the ilium is relatively rough-textured because it serves for attachment of several muscles of the buttocks and thighs. The anteromedial surface of the ilium is the smooth, slightly concave **iliac fossa,** covered in life by the broad *iliacus* muscle. Medially, the ilium exhibits an auricular surface that matches the one on the sacrum, so that the two bones form the sacroiliac joint.

The **ischium** forms the inferoposterior portion of the os coxae. Its heavy **body** is marked with a prominent **ischial spine.** Inferior to the spine is a thick, rough-surfaced **ischial tuberosity,** which supports your body when you are sitting. It can be easily palpated by sitting on your fingers. The **ramus** of the ischium joins the inferior ramus of the pubis anteriorly.

The **pubis** (pubic bone) is the most anterior portion of the os coxae. It has a **superior** and **inferior ramus** and a triangular **body.** The body of one pubis meets the body of the other at the pubic symphysis. The pubis and ischium enclose the obturator foramen.

57. *ilia* = flank, loin
58. *acetabulum* = vinegar cup
59. *obtur* = to close, stop up

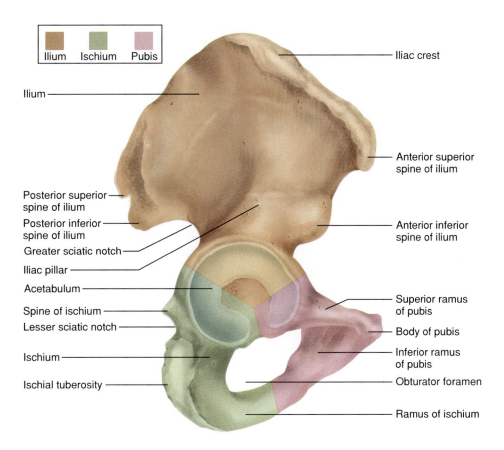

Figure 9.34 The right os coxae of an adult, showing the boundaries of its three constituent childhood bones—the ilium, ischium, and pubis. ⚜

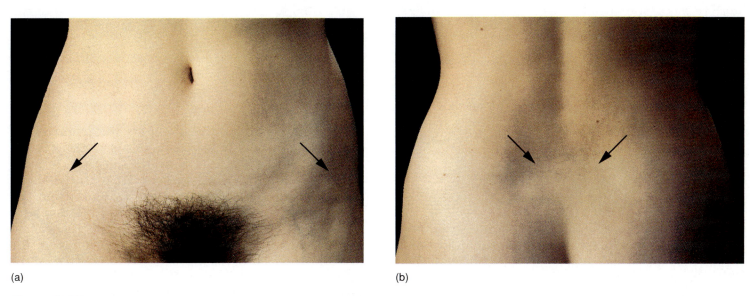

(a)   (b)

Figure 9.35 External landmarks of the pelvic girdle. (*a*) Anterior aspect, showing the location of the anterior superior spine of the ilium (*arrows*). (*b*) Posterior aspect, where dimples of the sacral region (*arrows*) mark the location of the posterior superior spines of the ilium. ⚜

Table 9.6	Sexual Dimorphism of the Pelvis	
	Female	**Male**
General Appearance	Less massive; processes more delicate; smoother	More massive; heavier processes; rougher
Tilt	Upper end of pelvis tilted forward	More vertical
Ilium	Shallower; does not project as far above sacroiliac joint	Deeper; projects farther above sacroiliac joint
Sacrum	Shorter and wider	Narrower and longer
Coccyx	More movable; tilted backward	Less movable; more vertical
Width of Greater Pelvis	Anterior superior spines farther apart; hips more flared	Anterior superior spines closer together; hips less flared
Pelvic Inlet	Round or oval	Heart-shaped
Pubic Symphysis	Shorter	Taller
Greater Sciatic Notch	Narrower	Wider
Obturator Foramen	Triangular	Oval
Acetabulum	Faces slightly anteriorly	Faces more laterally
Pubic arch	Usually greater than 100°	Usually 90° or less

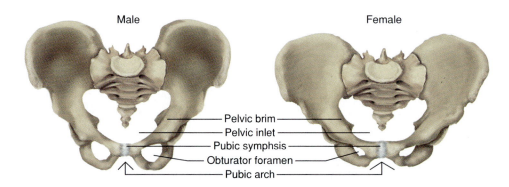

Figure 9.36 Comparison of the male and female pelvic girdles.

The female pelvis is adapted to the needs of pregnancy and childbirth. Some of the differences between the male and female pelves are described in table 9.6 and illustrated in figure 9.36.

Lower Extremity

The lower extremity includes the **femoral region** from hip to knee, the **patella** (kneecap), the **crural** (CROO-rul) **region** from knee to ankle, the **tarsus** (ankle), and the **pes** (foot). In anatomical terminology, *leg* refers only to the part of the lower extremity extending from the knee to the ankle. The number and arrangement of bones in the lower extremity are similar to those of the upper extremity. In the lower extremity, however, they are adapted for weight-bearing and locomotion and are therefore shaped and articulated differently.

Femur

The thigh contains only one bone, the **femur** (FEE-mur), which is the longest and strongest bone of the body (fig. 9.37). The head is nearly spherical, forming a quintessential *ball-and-socket joint* with the acetabulum of the pelvis. A ligament extends from the acetabulum to a pit, the **fovea capitis** (FOE-vee-uh CAP-ih-tiss), in the head of the femur. The **greater** and **lesser trochanters** (tro-CAN-turs) are insertions for the powerful muscles of the hip. They are connected on the posterior side by a thick oblique ridge of bone, the **intertrochanteric crest,** and on the anterior side by a more delicate **intertrochanteric line.**

The primary feature of the shaft is a posterior ridge called the **linea aspera** (LIN-ee-uh ASS-peh-ruh) at its midpoint. It branches into less conspicuous lateral and medial ridges at its inferior and superior ends.

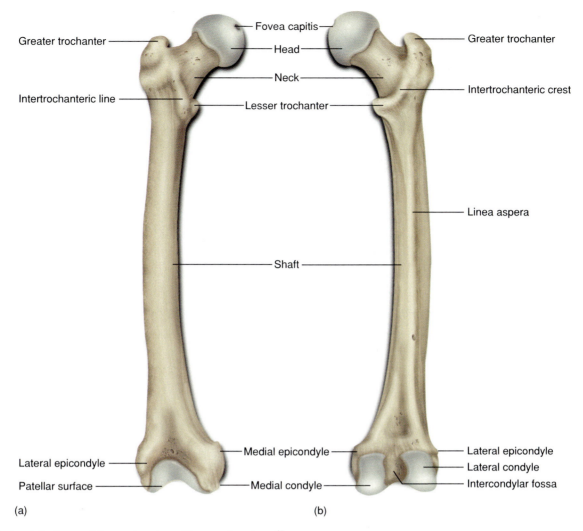

Greater trochanter
Fovea capitis
Head
Neck
Intertrochanteric line
Lesser trochanter
Greater trochanter
Intertrochanteric crest
Linea aspera
Shaft
Lateral epicondyle
Patellar surface
Medial epicondyle
Medial condyle
Lateral epicondyle
Lateral condyle
Intercondylar fossa

(a)
(b)

Figure 9.37 The right femur. (*a*) Anterior view; (*b*) posterior view. 🏃

The distal end of the femur exhibits prominent **medial** and **lateral condyles,** which form a strong knee joint. They are separated by a groove called the **intercondylar** (IN-tur-CON-dih-lur) **fossa.** The **medial** and **lateral epicondyles** flank the condyles superiorly and serve as sites of muscle and ligament attachment. On the anterior side of the femur, a smooth medial depression called the **patellar surface** articulates with the patella.

Patella

The patella, or kneecap (fig. 9.38), is a roughly triangular sesamoid bone that forms within the tendon of the knee as a child begins to walk. It has a broad superior **base,** a pointed inferior **apex,** and a pair of shallow **articular facets** on its posterior surface. The *patellar tendon* extends from the anterior muscle of the thigh (the quadriceps) to the patella, and it continues as the *patellar ligament* from the patella to the tibia.

Tibia

The leg has two bones—a thick, strong tibia (TIB-ee-uh) and a slender, lateral fibula (FIB-you-luh) (fig. 9.38). The **tibia** is the only weight-bearing bone of the crural region. Its broad superior head has two fairly flat articular surfaces, the **medial** and **lateral condyles,** separated by a ridge called the **intercondylar eminence.** The tibial condyles articulate with the condyles of the femur. The rough anterior surface of the tibia, the **tibial tuberosity,** can be palpated just below the patella. This is where the patellar ligament inserts and the thigh muscles exert their pull when they extend the leg. Distal to this, the shaft has a sharply angular **anterior crest,** which can be palpated in the shin region. At the ankle, just above the

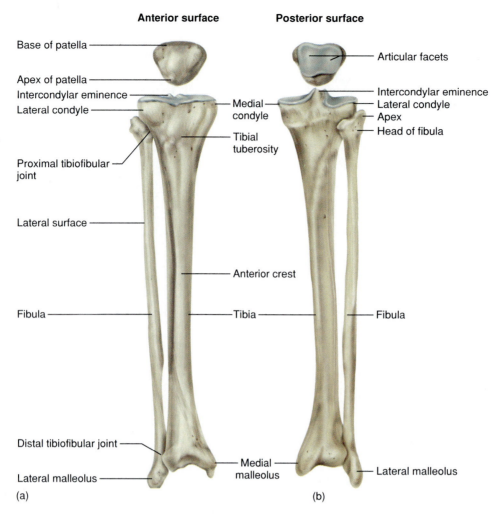

Anterior surface

Base of patella
Apex of patella
Intercondylar eminence
Lateral condyle
Proximal tibiofibular joint
Lateral surface
Fibula
Distal tibiofibular joint
Lateral malleolus
(a)

Medial condyle
Tibial tuberosity
Anterior crest
Tibia
Medial malleolus

Posterior surface

Articular facets
Intercondylar eminence
Lateral condyle
Apex
Head of fibula
Fibula
Lateral malleolus
(b)

Figure 9.38 Bones of the right leg and knee—the patella, tibia, and fibula. (*a*) Anterior view; (*b*) posterior view. ✗

rim of a standard dress shoe, you can palpate a prominent bony knob on each side. These are the **medial** and **lateral malleoli.**[60] The medial malleolus is part of the tibia, and the lateral malleolus is the distal head of the fibula.

Fibula

The **fibula** is somewhat thicker and broader at its proximal end than at the distal. The point of the proximal head is called the **apex,** or **styloid process.** The distal head is the lateral malleolus. The fibula is a lateral strut that helps to stabilize the ankle, but it does not bear any of the body's weight. Orthopedic surgeons sometimes remove the fibula and use it to replace damaged or missing bone elsewhere in the body.

The Ankle and Foot

The **tarsal bones** of the ankle are arranged in proximal and distal groups somewhat like the carpal bones of the

wrist (fig. 9.39). Because of the load-bearing role of the ankle, however, their shapes and arrangement are conspicuously different from those of the carpal bones. The largest tarsal bone is the **calcaneus**[61] (cal-CAY-nee-us), which forms the heel. Its posterior end is the point of attachment for the **calcaneal (Achilles) tendon** from the calf muscles. The second-largest tarsal bone, and the most superior, is the *talus.* It has three articular surfaces: an inferoposterior one that articulates with the calcaneus, a superior **trochlear surface** that articulates with the tibia, and an anterior surface that articulates with a short, wide tarsal bone called the **navicular.** The talus, calcaneus, and navicular are considered the proximal row of tarsal bones. (Remember that there is also a carpal bone called the navicular; be careful not to confuse the two.)

The distal group forms a row of four bones. Proceeding from the great toe to the little one, these are the **first, second,** and **third cuneiforms**[62] (cue-NEE-ih-forms) and the **cuboid.** The cuboid is the largest.

---

60. *malle* = hammer + *olus* = little

61. *calc* = stone, chalk
62. *cunei* = wedge

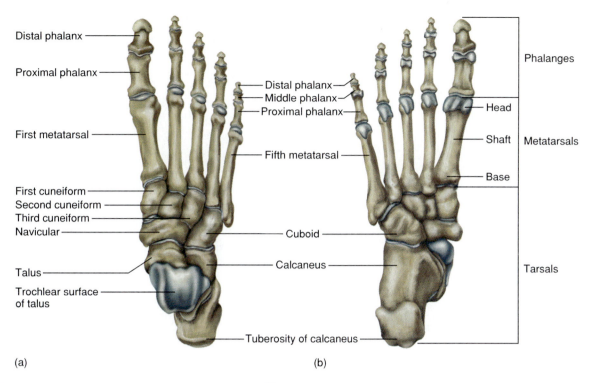

Distal phalanx
Proximal phalanx
First metatarsal
First cuneiform
Second cuneiform
Third cuneiform
Navicular
Talus
Trochlear surface
of talus

Distal phalanx
Middle phalanx
Proximal phalanx
Fifth metatarsal
Cuboid
Calcaneus
Tuberosity of calcaneus

Phalanges
Head
Shaft   Metatarsals
Base
Tarsals

(a)

(b)

Figure 9.39 The right foot. (*a*) Superior view; (*b*) inferior view. 𝑋

Bones of the foot are similar in arrangement and name to those of the hand. The proximal **metatarsals** are similar to the metacarpals. They are **metatarsals I to V** from medial to lateral, metatarsal I being proximal to the great toe. (Note that Roman numeral I represents the *medial* group of bones in the foot but the *lateral* group in the hand. In both cases, however, Roman numeral I refers to the largest digit of the extremity.)

Metatarsals I to III articulate with the first through third cuneiforms; metatarsals IV and V both articulate with the cuboid. Bones of the toes, like those of the fingers, are called phalanges. The great toe is the **hallux** and contains only two bones, the proximal and distal phalanx I. The other toes each contain a proximal, middle, and distal phalanx. The metatarsal and phalangeal bones each have a base, body, and head, like the bones of the hand. All of them, especially the phalanges, are slightly concave on the ventral side.

The sole of the foot normally does not rest flat on the ground; rather, it has three springy arches that absorb the stress of walking (fig. 9.40). The **medial longitudinal arch,** which essentially extends from heel to hallux, is formed from the calcaneus, talus, navicular, cuneiforms, and metatarsals I to III. The **lateral longitudinal arch** extends from heel to little toe and includes the calcaneus,

cuboid, and metatarsals IV and V. The **transverse arch** includes the cuboid, cuneiforms, and proximal heads of the metatarsals. These arches are held together by short, strong ligaments. Excessive weight, repetitious stress, or congenital weakness of these ligaments can stretch them, resulting in *pes planis* (commonly called flat feet or fallen arches). This condition makes a person less tolerant of prolonged standing and walking. A comparison of the flat-footed apes with humans underscores the significance of the human foot arches (see chapter essay, p. 289).

The pelvic girdle and lower extremity are summarized in table 9.7.

> ············· **Key Point Review** ·············
>
> ⑮ Which two bones make up the adult pelvic girdle? Which six bones make up a child's pelvic girdle?
>
> ⑯ Name four structures of the pelvis that you can palpate. Describe where to palpate them.
>
> ⑰ What parts of the femur are involved in the hip joint? What parts are involved in the knee joint?
>
> ⑱ What are the prominent knobs on each side of your ankle called? Which bones contribute to these structures?
>
> ⑲ Name all of the bones that articulate with the talus and describe the location of each.

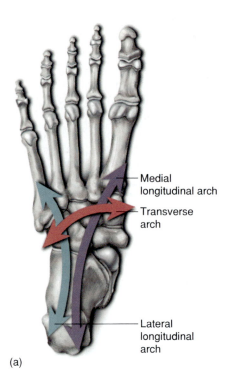

(a)

- Medial longitudinal arch
- Transverse arch
- Lateral longitudinal arch

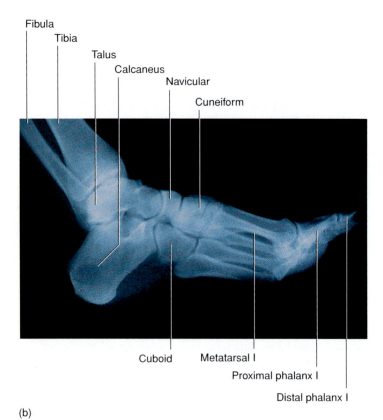

Fibula
Tibia
Talus
Calcaneus
Navicular
Cuneiform

Cuboid   Metatarsal I
Proximal phalanx I
Distal phalanx I

(b)

Figure 9.40 Arches of the foot. (a) Inferior view of the right foot. (b) X ray of the right foot, lateral view, showing the lateral longitudinal arch.

Table 9.7	Anatomical Checklist for the Pelvic Girdle and Lower Extremity

Pelvic Girdle

Os Coxae (figs. 9.33 and 9.34)
- Pubic symphysis
- Greater (false) pelvis
- Lesser (true) pelvis
- Pelvic brim
- Pelvic inlet
- Pelvic outlet
- Acetabulum
- Obturator foramen
- Ilium
  - Iliac crest
  - Anterior superior spine
  - Anterior inferior spine
  - Posterior superior spine
  - Posterior inferior spine
  - Greater sciatic notch
  - Iliac pillar
  - Iliac fossa
  - Auricular surface
- Ischium
  - Body
  - Ischial spine
  - Ischial tuberosity
  - Ramus
- Pubis
  - Superior ramus
  - Inferior ramus
  - Body
  - Pubic arch

Lower Extremity

Femur (fig. 9.37)
- Proximal end
- Head
- Fovea capitis
- Neck
- Greater trochanter
- Lesser trochanter
- Intertrochanteric crest
- Intertrochanteric line
- Shaft
  - Linea aspera

Distal end
- Medial condyle
- Lateral condyle
- Intercondylar fossa
- Medial epicondyle
- Lateral epicondyle
- Patellar surface

Patella (fig. 9.38)
- Base
- Apex
- Articular facets

Tibia (fig. 9.38)
- Medial condyle
- Lateral condyle
- Intercondylar eminence
- Tibial tuberosity
- Anterior crest
- Medial malleolus

Fibula (fig. 9.38)
- Apex (styloid process)
- Lateral malleolus

Tarsal Bones (fig. 9.39)
- Proximal group
  - Calcaneus
  - Talus
  - Navicular
- Distal group
  - First cuneiform
  - Second cuneiform
  - Third cuneiform
  - Cuboid

Bones of the Foot (figs. 9.39 and 9.40)
- Metatarsal bones I–V
- Phalanges
  - Proximal phalanx
  - Middle phalanx
  - Distal phalanx
- Arches of the foot
  - Medial longitudinal arch
  - Lateral longitudinal arch
  - Transverse arch

Skeletal Adaptations to Bipedalism

Some mammals can stand, hop, or walk briefly on their hind legs, but humans are the only mammals that are habitually bipedal. Footprints preserved in a layer of volcanic ash in Tanzania indicate that hominids walked upright as early as 3.6 million years ago. Several adaptations of the human feet, legs, spine, and skull make bipedalism possible (fig. E.1). These features are so distinctive that paleoanthropologists (those who study human fossil remains) can tell with considerable certainty whether a fossil species was able to walk upright.

As important as the hand has been to human evolution, the foot may be an even more significant adaptation. Unlike other mammals, humans support their entire body weight on two feet. While apes are flat-footed, humans have strong, springy foot arches that absorb shock as the body jostles up and down during walking and running. The tarsal bones are tightly articulated with each other, and the calcaneus is strongly developed. The hallux (great toe) is not opposable as it is in most Old World monkeys and apes, but it is highly developed so that it provides the "toe-off" that pushes the body forward in the last phase of the stride. For this reason, loss of the hallux has a more crippling effect than the loss of any other toe.

While the femurs of an ape are nearly vertical, in a human they angle medially from the hip to the knee. This places our knees closer together, beneath the body's center of gravity. We lock our knees when standing, allowing us to maintain an erect posture with little muscular effort. Apes cannot do this, and they cannot stand on two legs for very long without tiring—much as you would if you tried to maintain an erect posture with your knees slightly bent.

In an ape or other quadrupedal (four-legged) mammal, the abdominal viscera are supported by the muscular wall of the abdomen. In a human, the viscera bear down on the floor of the pelvic cavity, and a bowl-shaped pelvis is necessary to support their weight. This has resulted in a narrower pelvic outlet—a condition quite incompatible with the fact that we, including our infants, are such a large-brained species. The pain of childbirth is unique to humans and, one might say, a price we must pay for having both a large brain and a bipedal stance.

The largest muscle of the buttock, the *gluteus maximus,* serves in apes primarily as an abductor of the thigh—that is, it draws the leg laterally. In humans, however, the ilium has expanded posteriorly, so the gluteus maximus originates behind the hip joint. This changes the function of the muscle—

instead of abducting the thigh, it pulls the thigh back in the second half of a stride (pulling back on your right thigh, for example, when your left foot is off the ground and swinging forward). This action accounts for the smooth, efficient stride of a human as compared to the awkward, shuffling gait of a chimpanzee or gorilla when it is walking upright. The posterior growth of the ilium is the reason the greater sciatic notch is so deeply concave.

The iliac pillar is a line of attachment for some of the gluteal muscles. When the left foot is lifted off the ground, these muscles shift the trunk of the body slightly toward the right, and vice versa, which enables us to keep our balance. The presence of an iliac pillar on a fossil primate's pelvis is taken as good evidence that the species was bipedal.

The lumbar curvature of the human spine allows for efficient bipedalism by shifting the body's center of gravity to the rear, above and slightly behind the hip joint. Because of their C-shaped spines, chimpanzees cannot stand as easily. Their center of gravity is anterior to the hip joint when they stand; they must exert a continual muscular effort to keep from falling forward, and fatigue sets in relatively quickly. Humans, by contrast, require little muscular effort to keep their balance. Our australopithecine ancestors probably could travel all day with relatively little fatigue.

The human head is balanced on the spinal column with the gaze directed forward. The cervical curvature of the spine and remodeling of the skull have made this possible. The foramen magnum has moved to a more inferior location, and the face is much flatter than in an ape, so there is less weight anterior to the occipital condyles. Being balanced on the spine, the head does not require strong muscular attachments to hold it erect. Apes have prominent supraorbital ridges for the attachment of muscles that pull back on the skull. In humans these ridges are much lighter and the muscles of the forehead serve only for facial expression, not to hold the head up.

The forelimbs of apes are longer than the hindlimbs; indeed, some species such as the orangutan and gibbons hold their long forelimbs over their heads when they walk on their hind legs. By contrast, our arms are shorter than our legs and far less muscular than the forelimbs of apes. No longer needed for locomotion, our forelimbs have become better adapted for precise manipulation of objects.▲

—continued

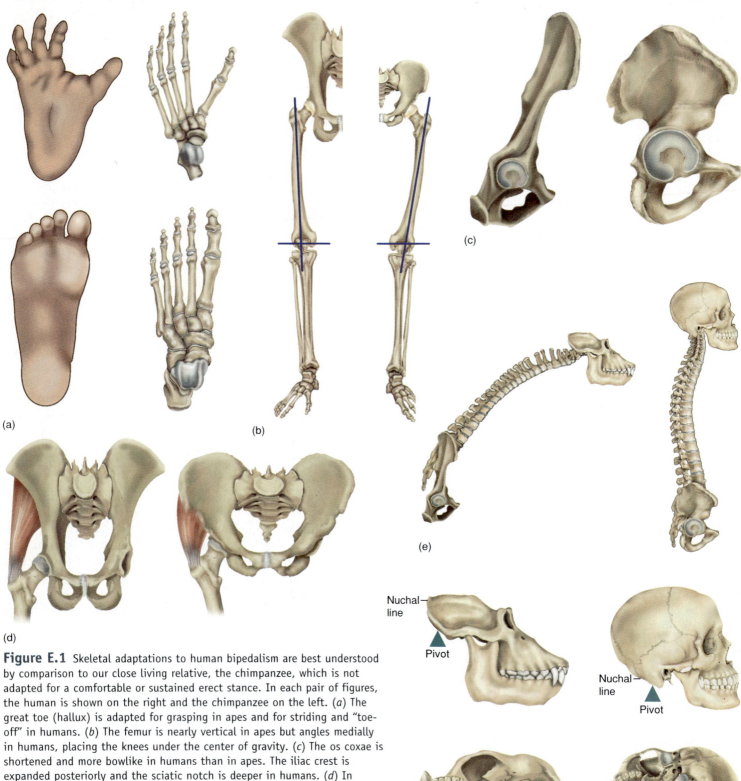

Figure E.1 Skeletal adaptations to human bipedalism are best understood by comparison to our close living relative, the chimpanzee, which is not adapted for a comfortable or sustained erect stance. In each pair of figures, the human is shown on the right and the chimpanzee on the left. (*a*) The great toe (hallux) is adapted for grasping in apes and for striding and "toe-off" in humans. (*b*) The femur is nearly vertical in apes but angles medially in humans, placing the knees under the center of gravity. (*c*) The os coxae is shortened and more bowlike in humans than in apes. The iliac crest is expanded posteriorly and the sciatic notch is deeper in humans. (*d*) In humans, the gluteus medius and minimus help to balance the body weight over one leg when the other leg is lifted from the ground. (*e*) The curvature of the human spine centers the body's weight over the pelvis, so humans can stand more effortlessly than apes. (*f*) The foramen magnum is shifted ventrally and the face is flatter in humans; thus the skull is balanced on the spinal column and the gaze is directed forward when a person is standing.

Interactions between the Skeletal System and Other Organ Systems

Integumentary System
- Initiates synthesis of vitamin D needed for bone deposition
- Bones lying close to body surfaces shape the skin

Muscular System
- Muscles move bones; stress produced by muscles affects patterns of ossification and remodeling, as well as shape of mature bones
- Bones provide leverage and sites of attachment for muscles; provide calcium needed for muscle contraction

Nervous System
- Sensory receptors provide sensations of body position and pain from bones and joints
- Cranium and vertebral column protect brain and spinal cord; bones provide calcium needed for neural function

Endocrine System
- Hormones regulate mineral deposition and resorption, bone growth, and skeletal mass and density
- Bones protect endocrine organs in head, chest, and pelvis

Circulatory System
- Delivers O_2, nutrients, and hormones to bone tissue and carries away wastes; delivers blood cells to marrow
- Myeloid tissue forms blood cells; bone matrix stores calcium needed for cardiac muscle activity

Lymphatic/Immune Systems
- Maintains balance of interstitial fluid within bone tissue; lymphocytes assist in defense and repair of bone tissue
- Most types of blood cells produced in myeloid tissue function as part of immune system

Respiratory System
- Provides O_2 and removes CO_2
- Bones form respiratory passageway through nasal cavity; protect lungs and aid in ventilation

Urinary System
- Kidneys activate vitamin D and regulate calcium and phosphate excretion
- Skeleton physically supports and protects organs of urinary system

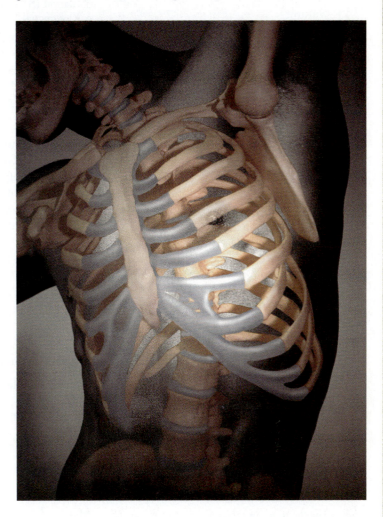

Digestive System
- Provides nutrients needed for bone growth and maintenance
- Skeleton provides bony protection for digestive organs

Reproductive System
- Gonads produce hormones that affect bone growth and closure of epiphyseal plates
- Skeleton protects some reproductive organs

Study Outline

Features of specific bones are outlined in tables throughout the chapter and are not repeated here.

The Skull (pp. 253–267)
1. General features
   a. Sutures
   b. Major cavities
   c. Paranasal sinuses
   d. Foramina
2. Cranium (general features)
3. Cranial bones
   a. Frontal bone (1)
   b. Parietal bones (2)
   c. Temporal bones (2)
      - Squamous part
      - Tympanic part
      - Mastoid part
      - Petrous part
   d. Occipital bone (1)
   e. Sphenoid bone (1)
   f. Ethmoid bone (1)
4. Facial bones
   a. Maxillae (2)
   b. Palatine bones (2)
   c. Zygomatic bones (2)
   d. Lacrimal bones (2)
   e. Nasal bones (2)
   f. Inferior nasal conchae (2)
   g. Vomer (1)
   h. Mandible (1)
5. Bones associated with skull
   a. Auditory ossicles
   b. Hyoid bone
6. The skull in infancy and childhood

The Spinal Column and Thoracic Cage (pp. 267–275)
1. Classes of vertebrae
   a. Cervical (C1–C7)
   b. Thoracic (T1–T12)
   c. Lumbar (L1–L5)
   d. Sacral (S1–S5)
   e. Coccygeal (Co1–Co4)
2. Spinal curvatures
   a. Primary curvatures
      - Thoracic
      - Pelvic
   b. Secondary curvatures
      - Cervical
      - Lumbar
3. General structure of a vertebra
4. Intervertebral discs
   a. Nucleus pulposus
   b. Annulus fibrosus
5. Regional characteristics of vertebrae
   a. Cervical vertebrae
   b. Thoracic vertebrae
   c. Lumbar vertebrae
   d. Sacrum
   e. Coccyx
6. The thoracic cage
   a. Sternum
   b. Ribs
      - True, false, and floating ribs
      - General anatomy

The Pectoral Girdle and Upper Extremity (pp. 275–281)
1. Pectoral girdle
   a. Clavicle
   b. Scapula
2. Upper Extremity
   a. Humerus
   b. Radius
   c. Ulna
   d. Carpal bones
   e. Metacarpal bones
   f. Phalanges

The Pelvic Girdle and Lower Extremity (p. 281–288)
1. Pelvic girdle (os coxae)
   a. General features
   b. Ilium
   c. Ischium
   d. Pubis
2. Lower extremity
   a. Femur
   b. Patella
   c. Tibia
   d. Fibula
   e. The ankle and foot
      - Tarsal bones
      - Metatarsal bones
      - Phalanges
      - Arches of the foot

Selected Vocabulary

Also review the terms in tables 9.2, 9.4, 9.5, and 9.7, which are not repeated here.

suture 253
cranial cavity 256
orbit 256
nasal cavity 256
buccal cavity 256
middle-ear cavity 256

inner-ear cavity 256
paranasal sinuses 256
cranium 256
calvaria 258
anterior cranial fossa 258
middle cranial fossa 258
posterior cranial fossa 258
coronal suture 258
lambdoid suture 258
sagittal suture 258

squamous suture 259
zygomatic arch 259
nuchal ligament 260
nasal septum 261
hard palate 262
soft palate 262
anterior fontanel 265
posterior fontanel 265
sphenoid fontanel 265
mastoid fontanel 265

atlanto-occipital joint 271
atlantoaxial joint 271
sacroiliac joint 273
costal margin 274
sternoclavicular joint 275
acromioclavicular joint 276
pollex 279
trochlear surface 286
hallux 287

1. Which of these is *not* a paranasal sinus?
   a. frontal
   b. temporal
   c. sphenoid
   d. ethmoid
   e. maxillary

2. Which of these is a facial bone?
   a. frontal
   b. ethmoid
   c. occipital
   d. temporal
   e. lacrimal

3. Which of these cannot be palpated on a living person?
   a. the crista galli
   b. the mastoid process
   c. the zygomatic arch
   d. the superior nuchal line
   e. the hyoid

4. All of the following are groups of vertebrae except for _____, which is one of the spinal curvatures.
   a. thoracic
   b. cervical
   c. lumbar
   d. pelvic
   e. sacral

5. Thoracic vertebrae do *not* have
   a. transverse foramina.
   b. costal facets.
   c. demifacets.
   d. transverse processes.
   e. pedicles.

6. The tubercle of a rib articulates with
   a. the sternal notch.
   b. the margin of the gladiolus.
   c. the demifacets of two vertebrae.
   d. the body of a vertebra.
   e. the transverse process of a vertebra.

7. The disc-shaped head of the radius articulates with the _____ of the humerus.
   a. radial tuberosity
   b. trochlea
   c. capitulum
   d. olecranon process
   e. glenoid cavity

8. All of the following are carpal bones except the _____, which is a tarsal bone.
   a. trapezium
   b. cuboid
   c. trapezoid
   d. triquetral
   e. pisiform

9. The bone that supports your body weight when you are sitting down is
   a. the acetabulum.
   b. the pubis.
   c. the ilium.
   d. the coccyx.
   e. the ischium.

10. Which of these is the bone of the heel?
    a. cuboid
    b. calcaneus
    c. navicular

    d. trochlear
    e. talus

11. Gaps between the cranial bones of an infant are called _____.

12. The external auditory meatus is a canal in the _____ bone.

13. The most medial of the three auditory ossicles is the _____.

14. The _____ bone has greater and lesser wings and protects the pituitary gland.

15. A herniated disc occurs when a ring called the _____ is cracked.

16. The "missing body" of the atlas is thought to be represented by the _____ of the axis.

17. The sacroiliac joint is formed where the _____ surface of the sacrum articulates with that of the ilium.

18. The _____ processes of the radius and ulna form bony protuberances on each side of the wrist.

19. The thumb is also known as the _____ and the great toe is also known as the _____.

20. The _____ arch of the foot extends from the heel to the great toe.

1. A child was involved in an automobile collision. She was not wearing a safety restraint, and her chin struck the dashboard hard. When the attending physician looked into her auditory canal, he could see into her throat. What do you think is the nature of her injury?

2. By palpating the hind leg of a cat or dog or by examining laboratory skeletons, you can see that cats and dogs stand on the heads of the metatarsals; the calcaneus does not touch the ground. How is their stance similar to that of a woman wearing high-heeled shoes? How is it different?

3. Contrast the tarsal bones with the carpal bones. Which ones are similar in name, location, or both? Which ones are different?

4. Name some bones that you think would be well suited for withdrawing red bone marrow for biopsy or transfusion. Why do you suggest these particular bones over others?

5. A teenage boy is babysitting a younger child who is shocked by a frayed extension cord. The child is not breathing and there is no pulse. The teenager has not been trained in cardiopulmonary resuscitation, but he has watched numerous rescue programs on television and feels he can do it. He performs rhythmic chest compressions on the sternum, but the child dies. Autopsy shows that the liver has been punctured and has hemorrhaged fatally. From the standpoint of skeletal anatomy, explain how this could have happened.

Web Site Links

For a listing of the most current web sites related to this chapter, please visit the Saladin homepage at:

http://www.mhhe.com/sciencemath/biology/saladin/

chapter ten

10

[Joints

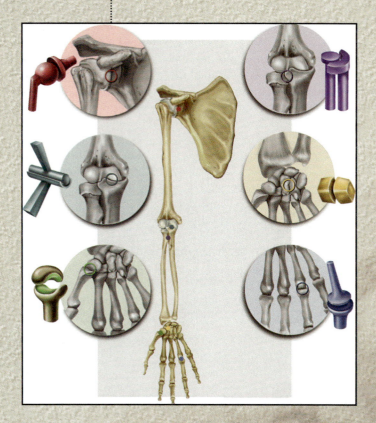

Special Topics

the unity of form and functio

Brushing up

To understand this chapter, it is essential that you understand or brush up on the following concepts:

▸ Surface features of bones, especially of their articular surfaces (table 8.3, p. 231)
▸ Names of the bones (all of chapter 9)

I n order for the skeleton to serve the purposes of protection and movement, the bones must be joined together. This chapter describes the **joints,** or **articulations,** of the skeleton and discusses some basic principles of biomechanics relevant to athletic performance and patient care. Chapter 11, in which the actions of skeletal muscles are described, will build on the discussion of joint anatomy and function presented here.

Joints and Their Classification

▼Objectives

When you have completed this section, you should be able to

• explain what joints are, how they are named, and what functions they serve;
• define *arthrology, kinesiology,* and *biomechanics;* and
• name and describe the three major structural classes and three major functional classes of joints.

Arthrology is the science concerned with the anatomy, function, dysfunction, and treatment of joints. The study of musculoskeletal movements is called **kinesiology** (kih-NEE-see-OL-oh-jee). This is a subdiscipline of **biomechanics,** which deals with a broad range of motions and mechanical processes, including the physics of blood circulation, respiration, and hearing.

Joints such as the shoulder, elbow, and knee are remarkable specimens of biological design—self-lubricating, almost frictionless, and able to bear heavy loads and withstand compression while executing smooth and precise movements (fig. 10.1). Yet, it is equally important that other joints be less movable or even immovable. The spinal column, for example, must provide a combination of support and flexibility; thus its joints are only moderately movable. The immovable joints between the cranial bones afford the best possible protection for the brain and sense organs.

The name of a joint is typically derived from the names of the bones involved. For example, the *atlanto-occipital joint* is where the occipital condyles meet the atlas, the *humeroscapular joint* is where the humerus meets the scapula, and the *coxal joint* is where the femur meets the os coxae.

Figure 10.1 These acrobats demonstrate the flexibility, precision, and weight-bearing capacity of many of the body's joints.

Joints can be classified according to their relative freedom of movement. A **diarthrosis**[1] (DY-ar-THRO-sis) is a freely movable joint such as the elbow. An **amphiarthrosis**[2] (AM-fee-ar-THRO-sis) is a joint that is slightly movable, such as an intervertebral or intercarpal joint. A **synarthrosis**[3] (SIN-ar-THRO-sis) is a joint that is capable of little or no movement, such as a suture of the skull.

Joints are also classified according to how the adjacent bones are joined. In *fibrous joints,* collagen fibers from the bone matrix cross the gap between adjacent bones and bind them together. These include sutures of the skull, the joint between the radius and ulna, and the joint between the tibia and fibula. *Cartilaginous joints* are those in which two bones are held together by cartilage. These include the epiphyseal plates of the long bones, the joints between vertebral bodies, and the pubic symphysis. In *synovial* (sih-NO-vee-ul) *joints,* the bones are separated by a space that contains a slippery lubricating fluid. Most synovial joints, including the jaw, elbow, hip, and knee joints, are freely movable. A *synostosis*[4] (SIN-oss-TOE-sis) is a joint in which two bones, once separate, have become fused by osseous tissue and in most cases are then regarded as a single bone.

The preceding structural and functional schemes of classification overlap. For example, synovial joints may be either diarthroses or amphiarthroses, and there are amphiarthroses of all three structural types—synovial, fibrous, and cartilaginous (fig. 10.2).

---

1. *dia* = separate, apart + *arthr* = joint + *osis* = condition
2. *amphi* = on all sides
3. *syn* = together
4. *ost* = bone

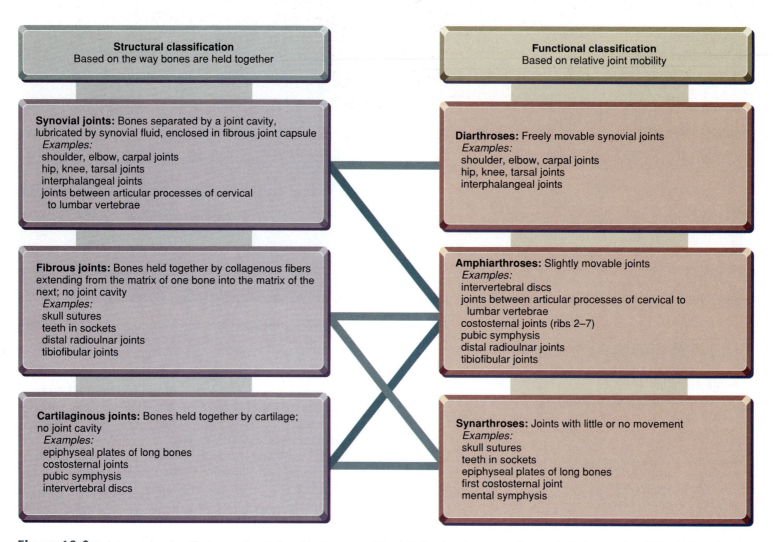

Structural classification Based on the way bones are held together	Functional classification Based on relative joint mobility

Synovial joints: Bones separated by a joint cavity, lubricated by synovial fluid, enclosed in fibrous joint capsule
Examples:
shoulder, elbow, carpal joints
hip, knee, tarsal joints
interphalangeal joints
joints between articular processes of cervical
  to lumbar vertebrae

Fibrous joints: Bones held together by collagenous fibers extending from the matrix of one bone into the matrix of the next; no joint cavity
Examples:
skull sutures
teeth in sockets
distal radioulnar joints
tibiofibular joints

Cartilaginous joints: Bones held together by cartilage; no joint cavity
Examples:
epiphyseal plates of long bones
costosternal joints
pubic symphysis
intervertebral discs

Diarthroses: Freely movable synovial joints
Examples:
shoulder, elbow, carpal joints
hip, knee, tarsal joints
interphalangeal joints

Amphiarthroses: Slightly movable joints
Examples:
intervertebral discs
joints between articular processes of cervical to
  lumbar vertebrae
costosternal joints (ribs 2–7)
pubic symphysis
distal radioulnar joints
tibiofibular joints

Synarthroses: Joints with little or no movement
Examples:
skull sutures
teeth in sockets
epiphyseal plates of long bones
first costosternal joint
mental symphysis

Figure 10.2 Joints can be classified according to how the bones are joined (*left column*) or according to their degree of mobility (*right column*). Connecting lines indicate overlap between the classification systems. For example, synovial joints can be either diarthroses or amphiarthroses, all diarthroses are synovial joints, and amphiarthroses include joints of the synovial, fibrous, and cartilaginous types.

1. What joins two bones together at synovial, fibrous, and cartilaginous joints? Give one example of each.
2. Which type of joint is a diarthrosis—synovial, fibrous, or cartilaginous?

Fibrous, Cartilaginous, and Bony Joints

▼Objectives
When you have completed this section, you should be able to
• describe the three types of fibrous joints and give an example of each;
• distinguish between the three types of sutures;
• describe the two types of cartilaginous joints and give an example of each; and
• name some joints that become synostoses as they age.

Fibrous Joints

As discussed in chapter 8, the matrix of a bone contains fibers of collagen. At a **fibrous joint,** these fibers emerge from one bone, cross the space between bones, and penetrate into the matrix of the next (fig. 10.3). There are three types of fibrous joints: *sutures, gomphoses,* and *syndesmoses.* In sutures and gomphoses, the collagen fibers are very short and allow for little movement. In syndesmoses, the fibers are longer and the attached bones are more movable.

Sutures

Sutures, described in chapter 9, are limited to the skull. In chapter 9 we did not take much notice of the differences between one suture and another, but some of these differences may have caught your attention as

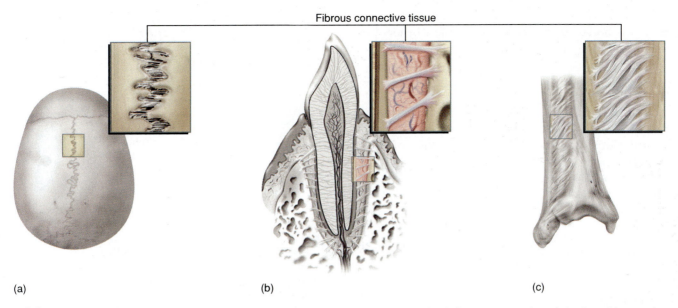

Fibrous connective tissue

(a)   (b)   (c)

Figure 10.3 Types of fibrous joints. (*a*) A suture between the parietal bones. (*b*) A gomphosis between a tooth and the jaw. (*c*) A syndesmosis between the tibia and fibula. ⚕

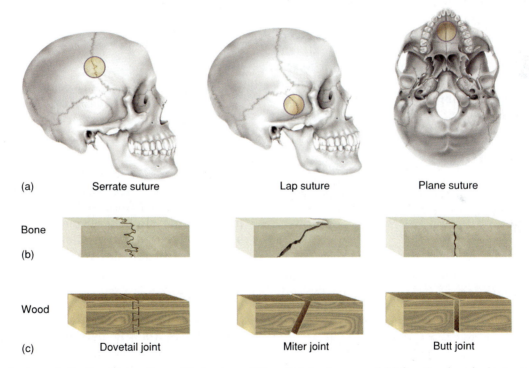

(a)   Serrate suture   Lap suture   Plane suture

Bone
(b)

Wood

(c)   Dovetail joint   Miter joint   Butt joint

Figure 10.4 Types of sutures indicating (*a*) locations, (*b*) structure of the adjoining bones, and (*c*) functional analogies to some common wood joints. ⚕

you studied the diagrams in that chapter or examined laboratory specimens. Sutures can be classified as *serrate, lap,* and *plane sutures.* Readers with some background in woodworking may recognize that the structures and functional properties of these sutures have something in common with basic types of carpentry joints (fig. 10.4).

Serrate sutures form distinctly wavy lines, as seen at the coronal, sagittal, and lambdoid sutures of the calvaria.

The adjoining bones firmly interlock with each other along their serrated margins, like a dovetail wood joint.

Lap (squamous) sutures occur where two bones have overlapping beveled edges, like a miter joint in carpentry. On the surface, a lap suture appears as a relatively smooth (nonserrated) line. An example is the squamous suture between the temporal and parietal bones.

Plane (butt) sutures occur where two bones have straight, nonoverlapping edges, for example in the roof of the mouth, between the palatine processes of the maxillae. The two bones merely border on each other, like two boards glued together in a butt joint.

Gomphoses

Even though the teeth are not regarded as bones, the attachment of a tooth to its socket is classified as a joint called a **gomphosis** (gom-FOE-sis). The term refers to its similarity to a nail hammered into wood.[5] The tooth is held firmly in place by a fibrous **periodontal membrane,** which consists of collagen fibers that extend from the bone matrix of the jaw into the dental tissue (see fig. 10.3b). The periodontal membrane allows the tooth to move or "give" a little under the stress of chewing.

Syndesmoses

Syndesmoses[6] (SIN-dez-MO-seez) are the most movable fibrous joints. The bones of a syndesmosis are joined by ligaments, which are bands of relatively long collagen fibers that do not bind the bones as tightly as the fibers of a suture or gomphosis. In the side-by-side attachments of the radius to the ulna and the tibia to the fibula, there is a relatively broad, sheetlike ligament called an **interosseous membrane** along the shafts of the two bones (see fig. 10.3c).

Cartilaginous Joints

In **cartilaginous joints,** the two bones are bound to each other by cartilage. The two types of cartilaginous joints are *synchondroses* and *symphyses.*

Synchondroses

In a **synchondrosis**[7] (SIN-con-DRO-sis), the bones are joined by *hyaline cartilage.* Examples include the epiphyseal plate between the epiphysis and diaphysis of a growing long bone and the attachment of a rib to the sternum by a hyaline costal cartilage (fig. 10.5a).

Symphyses

In a **symphysis,**[8] two bones are joined by *fibrocartilage* (fig. 10.5b, c). One example is the pubic symphysis, a pad of fibrocartilage between the pubic bones. Another is the joint between the bodies of two vertebrae, formed by an intervertebral disc. The surface of each vertebral

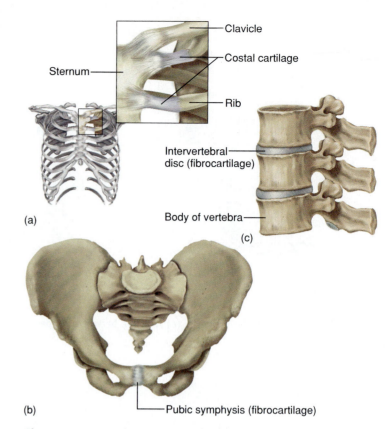

(a)

(b)

(c)

Figure 10.5 Cartilaginous joints. (a) Costal cartilages. (b) The pubic symphysis. (c) Intervertebral discs. ✗

body is covered with hyaline cartilage. Between the vertebrae, this cartilage becomes infiltrated with collagen bundles to form fibrocartilage. Each intervertebral disc permits only slight movement between adjacent vertebrae, but the collective effect of all 23 discs gives the spine considerable flexibility.

Synostoses (Bony Joints)

Some fibrous and cartilaginous joints ossify with age—that is, the gap between adjacent bones becomes filled with osseous tissue until the two bones appear as one. As mentioned earlier, the joint is then called a **synostosis.** In the skull, for example, both the frontal bone and mandible are represented at birth by separate right and left bones; in early childhood, these bones become fused. (In spite of its name, the mental symphysis of the mandible is a synostosis.) In old age, some sutures become obliterated by ossification and adjacent cranial bones fuse seamlessly together. The shafts and heads of the long bones are joined by cartilaginous joints in childhood and adolescence, and these become synostoses in early adulthood. The attachment of the first rib to the sternum also becomes a synostosis with age.

---

5. *gompho* = nail, bolt
6. *syn* = together + *desm* = band + *osis* = condition
7. *syn* = together + *chondr* = cartilage + *osis* = condition
8. *sym* = together + *physis* = growth

Think About It

The intervertebral joints are symphyses only in the thoracic through the lumbar region. How would you classify the intervertebral joints of the sacrum and coccyx in a middle-aged adult?

Key Point Review

3 Define *suture, gomphosis,* and *syndesmosis,* and explain what these three joints have in common.

4 Name the three types of sutures and describe how they differ.

5 Name two synchondroses and two symphyses.

6 Give some examples of joints that become synostoses with age.

Form and Function of Synovial Joints

▼Objectives

When you have completed this section, you should be able to
- describe the anatomy of a synovial joint and its associated structures;
- describe the six types of synovial joints;
- list and demonstrate the types of movements that occur at diarthroses;
- discuss the factors that affect the range of motion of a joint;
- give an anatomical example of a first-, second-, and third-class lever and explain why each is classified as it is; and
- relate the concept of mechanical advantage to the power and speed of joint action.

The rest of this chapter is concerned with synovial joints. These are not only the most common and familiar joints in the body, but they are also the most structurally complex and the most likely to develop uncomfortable and crippling dysfunctions.

General Anatomy

The study of dry bones and models in the laboratory can easily give the impression that the knee, elbow, or hip is a point at which one bone rubs against another. In life, however, the bones do not touch each other; rather, fluid and soft tissues separate them and hold them in proper alignment (fig. 10.6).

The bones of a **synovial joint** are separated by a **joint cavity** containing a lubricant called **synovial fluid.** Synovial fluid is rich in albumin and hyaluronic acid, which give it a viscous, slippery texture similar to that of egg white.[9] It also contains phagocytes that continually clean up tissue debris resulting from wear of the joint cartilages. The adjoining surfaces of the bones are covered with a layer of hyaline **articular cartilage** about 2 mm thick in young, healthy joints (see special topic 10.1). The cartilages and synovial fluid make movements at synovial joints almost friction-free. A fibrous **joint capsule** encloses the cavity and retains the

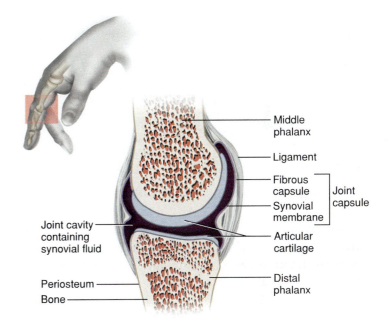

- Middle phalanx
- Ligament
- Fibrous capsule
- Synovial membrane
- Joint capsule
- Joint cavity containing synovial fluid
- Articular cartilage
- Periosteum
- Bone
- Distal phalanx

Figure 10.6 Structure of a simple synovial joint. 𝓍

Special Topic — Exercise and Articular Cartilage
10.1

When synovial fluid is warmed by exercise, it becomes thinner (less viscous) and more easily absorbed by the articular cartilage. Thus, the cartilage swells and provides a more effective cushion against compression. For this reason, a warm-up period before vigorous exercise helps protect the articular cartilage from undue wear and tear.

Since cartilage is nonvascular, it depends on repetitive compression for nutrition and waste removal. Each time a cartilage is compressed, fluid and metabolic wastes are squeezed out of it. When weight is taken off the joint, the cartilage absorbs synovial fluid like a sponge, and the fluid carries oxygen and nutrients to the chondrocytes. Lack of exercise causes the articular cartilages to deteriorate more rapidly from lack of nutrition, oxygenation, and waste removal.

Weight-bearing exercise builds bone mass and strengthens the muscles that stabilize many of the joints, thus reducing the risk of joint dislocations. Excessive joint stress, however, can hasten the progression of osteoarthritis (see chapter essay, p. 321) by damaging the articular cartilage. Swimming is a good way of exercising the joints with minimal damage.

9. *ovi* = egg

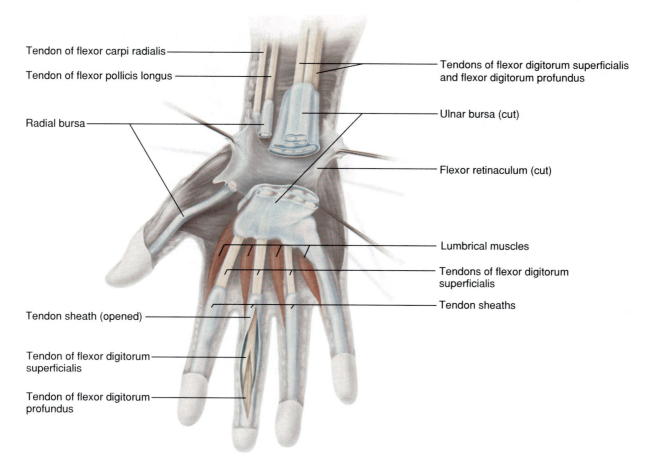

Figure 10.7 Tendon sheaths and other bursae in the hand and wrist.

Labels (from top left, clockwise):
- Tendon of flexor carpi radialis
- Tendon of flexor pollicis longus
- Radial bursa
- Tendons of flexor digitorum superficialis and flexor digitorum profundus
- Ulnar bursa (cut)
- Flexor retinaculum (cut)
- Lumbrical muscles
- Tendons of flexor digitorum superficialis
- Tendon sheaths
- Tendon sheath (opened)
- Tendon of flexor digitorum superficialis
- Tendon of flexor digitorum profundus

fluid. It has an outer **fibrous capsule** continuous with the periosteum of the adjoining bones and an inner **synovial membrane** of areolar tissue, which secretes the fluid.

In the jaw, sternoclavicular, and knee joints, cartilage grows inward from the joint capsule and forms a pad called a **meniscus**[10] between the articulating bones. The meniscus absorbs shock and pressure, guides the bones across each other, and reduces the chance of dislocation.

The most important accessory structures of a synovial joint are tendons, ligaments, and bursae. A **tendon** is a strip or sheet of tough, collagenous connective tissue that attaches a muscle to a bone. Tendons are often the most important structures in stabilizing a joint. A **ligament** is a similar tissue that attaches one bone to another. Several ligaments are named and illustrated in our later discussion of individual joints, and tendons are more fully considered in the next chapter along with gross anatomy of the muscles.

A **bursa**[11] is a fibrous sac filled with synovial fluid, located between adjacent muscles or where a tendon passes over a bone. Bursae cushion muscles, help tendons slide more easily over the joints, and sometimes enhance the mechanical effect of a muscle by modifying the direction in which its tendon pulls. Bursae called **tendon sheaths** are elongated and wrapped around a tendon like a hotdog bun. These are especially numerous in the hand and foot (fig. 10.7).

Types of Synovial Joints

There are six types of synovial joints with distinctive patterns of motion determined by the shapes of the articular surfaces of the bones. These joints are illustrated in figure 10.8 and summarized in table 10.1.

1. **Ball-and-socket joints.** These occur at the shoulder and hip, where one bone has a smooth hemispherical head that fits within a cuplike depression on the other. The head of the humerus

---

10. *meniscus* = crescent

11. *bursa* = purse

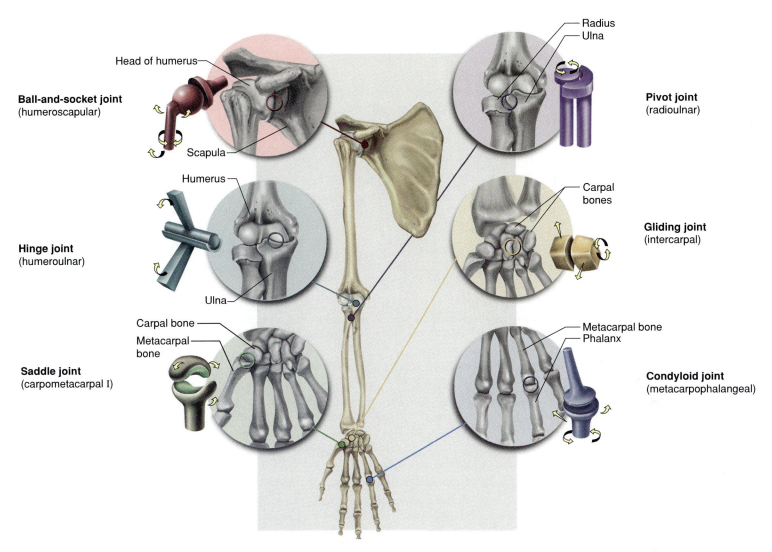

Figure 10.8 The six types of diarthroses can all be found in the forelimb. Mechanical models show the types of motion possible at each joint. ✗

Ball-and-socket joint
(humeroscapular)

Head of humerus

Scapula

Hinge joint
(humeroulnar)

Humerus

Ulna

Saddle joint
(carpometacarpal I)

Carpal bone
Metacarpal bone

Radius
Ulna

Pivot joint
(radioulnar)

Carpal bones

Gliding joint
(intercarpal)

Metacarpal bone
Phalanx

Condyloid joint
(metacarpophalangeal)

fits into the glenoid cavity of the scapula, and the head of the femur fits into the acetabulum of the os coxae. These highly movable joints are described as *multiaxial* because the humerus or femur can be moved in more than two directions or planes—for example, in the motion called *circumduction* to be described shortly.

2. **Hinge joints.** At a hinge joint, one bone has a convex surface that fits into a concave depression of the other one. Hinge joints are *monaxial*—like a door hinge, they can only move in one, usually parasagittal, plane. Examples include the elbow, knee, and interphalangeal (finger and toe) joints.

3. **Condyloid (ellipsoid) joints.** These joints exhibit an oval convex surface on one bone that fits into a similarly shaped depression on the next. The radiocarpal joint of the wrist and the metacarpophalangeal (MET-uh-CAR-po-fuh-LAN-jee-ul) joints at the bases of the fingers are examples.

These are considered *biaxial* joints because they can move in two directions, for example up-and-down and side-to-side. To demonstrate, hold your hand with your palm facing you. Flex your index finger back and forth as if gesturing to someone, "come here," and then move the finger from side to side toward the thumb and away. This shows the biaxial motion of the condyloid joint.

4. **Saddle joint.** The only saddle joint is the trapeziometacarpal (tra-PEE-zee-oh-MET-uh-CAR-pul) joint at the base of the thumb. Each bone—metacarpal I and the trapezium of the wrist—is shaped like a saddle, concave in one direction and convex in the other. If you compare the range of motion of your thumb with that of your fingers, you can see that a saddle joint is more movable than a condyloid or hinge joint. This is the joint responsible for that hallmark of primate anatomy, the opposable thumb.

Table 10.1	Anatomical Classification of the Joints
Joint	**Characteristics and Examples**
Fibrous Joints	Adjacent bones bound by collagen fibers extending from the matrix of one into the matrix of the other
Sutures (fig. 10.4)	Immovable fibrous joints between cranial and facial bones
Serrate suture	Bones joined by a wavy line formed by interlocking teeth along the margins. Examples: coronal, sagittal, and lambdoid sutures
Lap suture	Bones beveled to overlap each other; superficial appearance is a smooth line. Example: squamous suture around temporal bone
Plane suture	Bones butted against each other without overlapping or interlocking. Example: palatine suture
Gomphosis (fig. 10.3b)	Insertion of a tooth into a socket, held in place by collagen fibers of periodontal membrane
Syndesmosis (fig. 10.3c)	Slightly movable joint held together by ligaments or interosseous membranes. Examples: tibiofibular joint and radioulnar joint
Cartilaginous Joints	Adjacent bones bound by cartilage
Synchondrosis (fig. 10.5a)	Bones held together by hyaline cartilage. Examples: articulation of ribs and sternum and epiphyseal plate uniting the head and shaft of a long bone
Symphysis (fig. 10.5b, c)	Slightly movable joint held together by fibrocartilage. Examples: intervertebral discs and pubic symphysis
Synostoses	Former fibrous or cartilaginous joint in which adjacent bones have become fused by ossification. Examples: midsagittal lines of frontal bone and mandible, joint between first rib and sternum, and fusion of epiphysis and diaphysis of a long bone
Synovial Joints (fig. 10.6)	Adjacent bones covered with hyaline articular cartilage, separated by lubricating synovial fluid and enclosed in a fibrous joint capsule
Ball-and-socket	Multiaxial diarthrosis in which a smooth hemispherical head of one bone fits into a cuplike depression of another. Examples: shoulder (humeroscapular) joint and hip (coxal) joint
Hinge	Monaxial diarthrosis, able to flex and extend in only one plane. Examples: elbow, metacarpophalangeal, interphalangeal, knee, and metatarsophalangeal joints
Condyloid	Biaxial diarthrosis in which an oval convex surface of one bone articulates with an elliptical depression of another. Examples: radiocarpal and metacarpophalangeal joints
Saddle	Joint in which each bone is saddle-shaped (concave on one axis and convex on the axis perpendicular to this). Unique to the thumb (trapeziometacarpal joint), where it allows opposition (touching of the thumb to the fingertips)
Pivot	Joint in which a projection of one bone fits into a ringlike ligament of another, allowing one bone to rotate on its longitudinal axis. Examples: atlantoaxial joint and proximal radioulnar joint.
Gliding	Synovial amphiarthrosis with slightly concave or convex bone surfaces that slide across each other. Examples: intercarpal, intertarsal, and sternoclavicular joints; joints between the articular processes of the vertebrae

5. **Pivot joints.** In these joints, one bone has a projection that fits into a ringlike ligament of another, and the first bone rotates on its longitudinal axis relative to the other. One example is the atlantoaxial joint between the first two vertebrae—the dens of the axis projects into the spinal foramen of the atlas, where it is held against the arch of the axis by a ligament (see fig. 9.23c). This joint pivots when you rotate your head as in gesturing "no." Another example is the proximal radioulnar joint, where the *annular ligament* on the ulna encircles the head of the radius (see fig. 10.22a) and permits the radius to rotate during pronation and supination of the forearm (motions to be described shortly).

6. **Gliding (plane) joints.** Here, the articular surfaces are flat or only slightly concave and convex. The adjacent bones slide over each other and have rather limited movement; they are amphiarthroses in contrast to the other five types listed, which are diarthroses. Gliding joints occur between the carpal and tarsal bones, between the articular processes of the vertebrae, and at the sternoclavicular joint. To feel a gliding joint in motion, palpate your sternoclavicular joint as you raise your arm above your head.

In table 10.1 the joints are classified by structural criteria. Some of joints are difficult to classify, however, because they have elements of more than one type. The jaw joint, for example, has some aspects of condyloid, hinge, and gliding joints for reasons that will be apparent later.

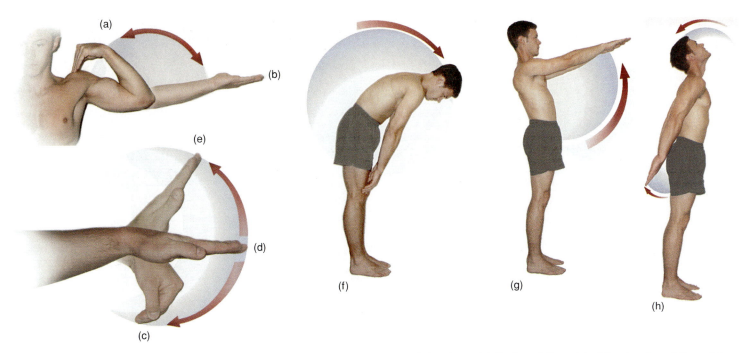

Figure 10.9 Types of joint movement. (*a*) Flexion of the elbow; (*b*) extension of the elbow; (*c*) flexion of the wrist; (*d*) extension of the wrist; (*e*) hyperextension of the wrist; (*f*) flexion of the spine; (*g*) extension of the spine and flexion of the shoulder; (*h*) hyperextension of the neck and shoulder.

Movements of Diarthroses

In physical therapy, kinesiology, and other medical and scientific fields, specific terms are used to describe the movements of diarthroses. You will need a command of these terms to understand the muscle actions in the next chapter. In the following discussion, many of them are grouped to describe opposite or contrasting movements.

Flexion, Extension, and Hyperextension

Flexion (figs. 10.9 and 10.10*c*) is movement that decreases the angle of a joint, usually in a sagittal plane. Examples are bending the elbow or knee and bending the neck to look down at the floor. Bending at the waist, as if taking a bow, is flexion of the spine. Flexion of the shoulder consists of raising the arm from anatomical position in a sagittal plane, as if to point in front of you or toward the ceiling.

Extension is movement that straightens a joint, usually to about 180°—for example, straightening the elbow or knee, raising the head to look directly forward, straightening at the waist, or moving the arm back to anatomical position.

Hyperextension is movement that increases the angle of a joint beyond 180°. For example, if you extend your arm and hand with the palm down, and then raise the back of your hand as if admiring a new ring, you are hyperextending the wrist. If you look up toward the ceiling, you are hyperextending your neck. If you move

your arm to a position posterior to the shoulder, you are hyperextending your shoulder.

Think About It

Some synovial joints have articular surfaces or ligaments that prevent them from being hyperextended. Try hyperextending some of your synovial joints and list a few for which this is impossible.

Abduction and Adduction

Abduction[12] (ab-DUC-shun) (fig 10.10) is movement of a body part away from the midsagittal line—for example, raising the arm or leg to one side of the body. To abduct the fingers is to spread them apart. **Adduction**[13] (ah-DUC-shun) is movement toward the midsagittal line or median axis of the middle digit—that is, returning the body part to anatomical position. Some movements are open to alternative interpretations. Bending the head to one side or bending sideways at the waist may be regarded as abduction or lateral flexion.

Elevation and Depression

Elevation (fig. 10.11*a*) is movement that raises a bone vertically. The mandible is elevated when biting off a piece of food, and the clavicles are elevated when shrugging the

12. *ab* = away + *duc* = to carry
13. *ad* = toward

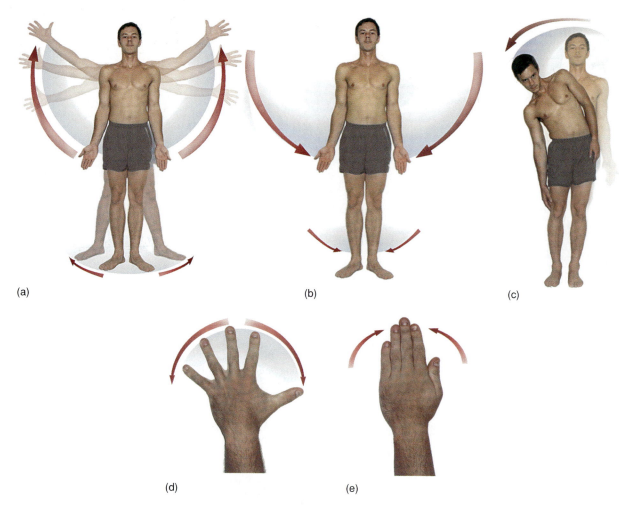

Figure 10.10 Types of joint movement, continued. (*a*) Abduction of the arms and legs; (*b*) adduction of the arms and legs; (*c*) abduction (lateral flexion) of the spine; (*d*) abduction of the fingers; (*e*) adduction of the fingers.

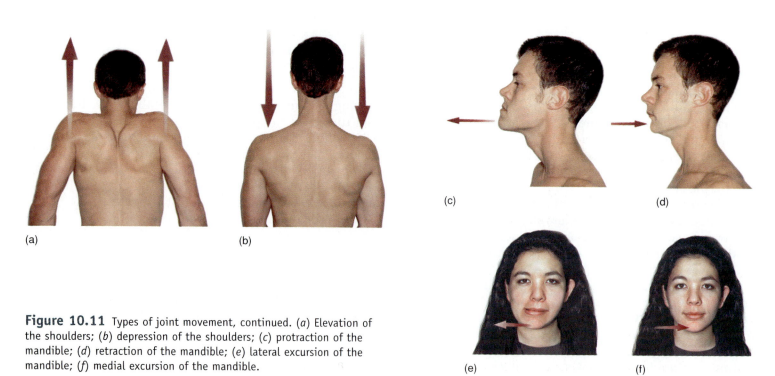

Figure 10.11 Types of joint movement, continued. (*a*) Elevation of the shoulders; (*b*) depression of the shoulders; (*c*) protraction of the mandible; (*d*) retraction of the mandible; (*e*) lateral excursion of the mandible; (*f*) medial excursion of the mandible.

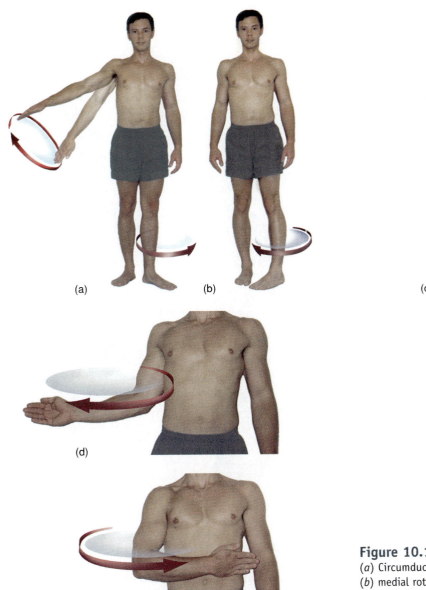

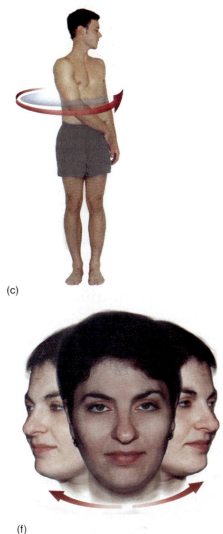

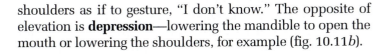

Figure 10.12 Types of joint movement, continued.
(*a*) Circumduction of the shoulder and lateral rotation of the femur; (*b*) medial rotation of the femur; (*c*) rotation of the spine; (*d*) lateral rotation of the humerus; (*e*) medial rotation of the humerus; (*f*) rotation of the neck.

shoulders as if to gesture, "I don't know." The opposite of elevation is **depression**—lowering the mandible to open the mouth or lowering the shoulders, for example (fig. 10.11*b*).

Protraction and Retraction

Protraction is movement of a bone anteriorly (forward) on a horizontal plane, and **retraction** is movement posteriorly (fig. 10.11*c, d*). Jutting the jaw outward, hunching the shoulders forward, or thrusting the pelvis forward are examples of protraction. The clavicles are retracted when standing at military attention. Most people have some degree of overbite and so must protract the mandible to make the incisors meet when taking a bite of fruit, for example. The mandible is then retracted to make the molars meet and grind food between them.

Lateral and Medial Excursion

Biting and chewing food require several movements of the jaw: up and down (elevation-depression), forward and back (protraction-retraction), and side-to-side grinding movements. The last of these are called **lateral excursion** (sideways movement to the right or left) and **medial excursion** (movement back to the midline) (fig. 10.11*e, f*).

Circumduction

Circumduction (fig. 10.12*a*) is movement in which one end of an appendage remains relatively stationary while the other end makes a circular motion. The appendage as a whole thus describes a conical space. For example, if an artist standing at an easel reaches out and draws a circle on the canvas, the shoulder remains stationary while the hand makes a circle. The extremity as a whole thus exhibits circumduction. A baseball player winding up for the pitch circumducts the arm in a more extreme "windmill" fashion. Circumduction is actually a sequence of flexion, abduction, extension, and adduction.

Rotation

Rotation is a movement in which a bone turns on its longitudinal axis. Figure 10.12 shows the limb movements that occur in **lateral** and **medial rotation** of the femur and humerus. Twisting at the waist is rotation of the trunk. When the head is turned from side to side, the atlas rotates on the axis.

Supination and Pronation

These movements are limited to the forearm. **Supination** (SOO-pih-NAY-shun) (fig. 10.13a) is rotation of the forearm so that the palm faces forward or upward; in anatomical position, the forearm is already supine. **Pronation** (fig. 10.13b) is rotation of the forearm so that the palm faces toward the rear or downward. As an aid to memory, think of it this way: you are *prone* to stand in the most comfortable position, which is with the palm *pronated*. If you were holding a bowl of *soup* in your hand, your forearm would have to be *supinated*. These movements are achieved with muscles discussed in the next chapter. The *supinator* is the most powerful, and supination is the sort of movement you would usually make to turn a doorknob or to drive a screw into a piece of wood.

Opposition and Reposition

Opposition is movement of the thumb to approach or touch the fingertips, and **reposition** is its movement back to anatomical position, parallel to the index finger (fig. 10.13c, d). This is the movement that enables the hand to grasp objects and is the single most important hand function.

Dorsiflexion and Plantar Flexion

These movements are limited to the foot. In anatomical position, the foot is extended. **Dorsiflexion** (DOR-sih-FLEC-shun) is a movement in which the toes are raised (as one might do to apply toenail polish) (fig. 10.14a). The foot is dorsiflected in each step you take as your foot comes forward. Dorsiflexion prevents your toes from scraping on the ground and results in a "heel strike" when that foot touches down in front of you. **Plantar** (PLAN-tur) **flexion** is hyperextension of the foot

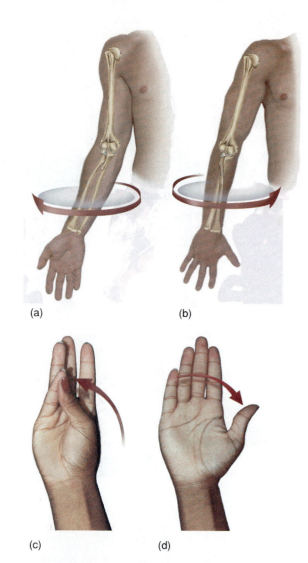

Figure 10.13 Types of joint movement unique to the upper extremity. (*a*) Supination of the forearm; (*b*) pronation of the forearm; (*c*) opposition of the thumb; (*d*) reposition of the thumb.

so that the toes point downward, as in standing on tiptoe or pressing the gas pedal of a car (fig. 10.14c). This motion also produces the "toe-off" in each step you take, as the heel of the foot behind you lifts off the ground.

Inversion and Eversion

These movements are also unique to the feet (fig. 10.14d, e). **Inversion** is a movement in which the soles are turned medially; **eversion** is a turning of the soles to face laterally. Inversion and eversion are common in fast sports such as tennis and football and often result in ankle sprains. These terms also refer to congenital deformities of the feet, which are often corrected by orthopedic shoes or braces.

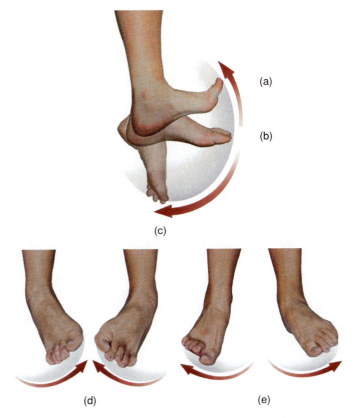

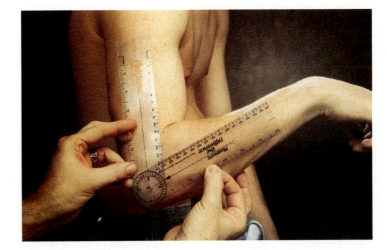

Figure 10.15 The use of a goniometer to measure the range of motion (ROM) of the elbow.

Figure 10.14 Types of joint movement unique to the foot. (*a*) Dorsiflexion; (*b*) extension; (*c*) plantar flexion; (*d*) inversion; and (*e*) eversion.

Think About It

A chimpanzee sitting on the ground reaches out and grasps an object between its fingertips. Then it raises its hand to its face and turns the object to examine it. List the diarthrotic movements that would occur and identify the joint at which each one would occur.

Range of Motion

We can see from the movements just described that the **range of motion (ROM)** of a joint varies greatly from one type to another. ROM obviously affects a person's functional independence and quality of life. It is also an important consideration in training for athletics or dance, in clinical diagnosis, and in monitoring the progress of rehabilitation. ROM can be measured with a simple device called a *goniometer*[14] (fig. 10.15).

Several factors affect the ROM and stability of a joint:

- **Structure and action of the muscles.** The two most important factors in stabilizing a joint are tendons and muscle tone (a state of partial contraction of a

14. *gonio* = angle + *meter* = measuring device

"resting" muscle). Tendons, ligaments, and muscles have sensory nerve endings called **proprioceptors** (PRO-pree-oh-SEP-turs) that continually monitor joint angle and tension. Upon receiving this information, the spinal cord sends nerve signals back to the muscles to increase or decrease their state of contraction and adjust the position of the joint and tautness of the tendons.

- **Structure of the articular surfaces of the bones.** You cannot hyperextend your elbow, for example, because the olecranon of the ulna fits into the olecranon fossa of the humerus and prevents further movement in that direction.
- **Strength and tautness of ligaments, tendons, and the joint capsule.** You cannot hyperextend your knee, for example, because of a *cruciate ligament* that is pulled tight when the knee is extended, preventing further motion. Gymnasts and acrobats increase the ROM of their joints by gradually stretching their ligaments during training. "Double jointed" people have unusually large ROMs at some joints, not because the joint is actually double or fundamentally different from normal in its anatomy, but because the ligaments are unusually long or slack.

Levers and Biomechanics of the Joints

A **lever** is any elongated, rigid object that rotates around a fixed point called the **fulcrum** (fig. 10.16). Familiar examples include a seesaw and a crowbar. Rotation occurs when an **effort** applied to one point on the lever overcomes a **resistance** located at some other point. The part of a lever from the fulcrum to the point of effort is called the **effort arm.** The part from the fulcrum to the point of resistance is called the **resistance arm**. In the body, a long bone acts as a lever, a joint serves as the fulcrum, and the effort is generated by a muscle attached to the bone by its tendon.

The function of a lever is to confer an advantage—either to exert more force against a resisting object than the force applied to the lever (for example, in moving a heavy boulder with a crowbar) or to move the resisting object farther or faster than the effort arm is moved (as in swinging a baseball bat). The same lever cannot confer both advantages. There is a trade-off between force on one hand and speed or distance on the other—as one increases, the other decreases.

The **mechanical advantage (MA)** of a lever is the ratio of its output force to its input force. It can be predicted from the length of the effort arm, L_E, divided by the length of the resistance arm, L_R:

$$MA = \frac{L_E}{L_R}$$

If MA is greater than 1.0, the lever will produce more force, but less speed or distance, than the force exerted on it. If MA is less than 1.0, the lever will produce more speed or distance, but less force, than the input. Consider the elbow joint, for example (fig. 10.17a). Its resistance arm is longer than its effort arm, so we know from the preceding formula that the mechanical advantage is going to be less than 1.0. The figure shows some representative values for L_E and L_R that yield MA = 0.15. The biceps brachii muscle puts more power into the lever than we get out of it, but the hand moves farther and faster than the point where the biceps tendon inserts on the ulna. The ankle, depending on how it is interpreted (see later discussion), may have an MA greater than 1.0 (fig. 10.17b).

In the next chapter, you will see that several joints have two or more muscles acting on them, seemingly producing the same effect. At first, you might consider this arrangement redundant, but it makes sense if the tendinous insertions of the muscles are at slightly different

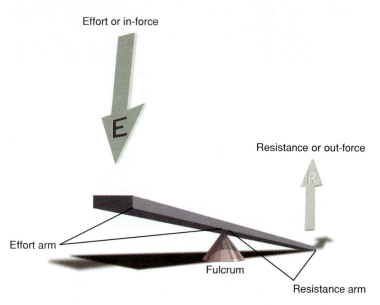

Figure 10.16 The basic components of a lever. This example is a first-class lever.

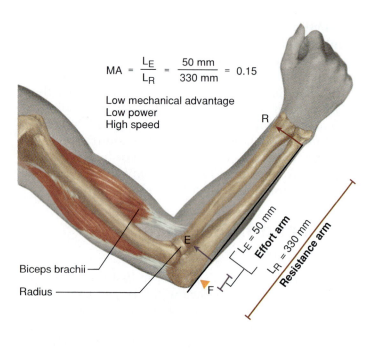

$$MA = \frac{L_E}{L_R} = \frac{50\ mm}{330\ mm} = 0.15$$

Low mechanical advantage
Low power
High speed

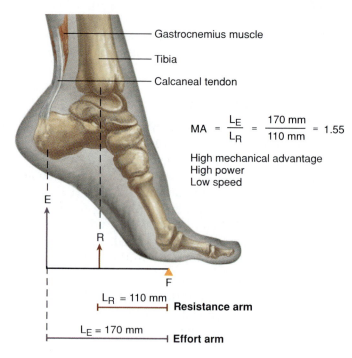

$$MA = \frac{L_E}{L_R} = \frac{170\ mm}{110\ mm} = 1.55$$

High mechanical advantage
High power
Low speed

Figure 10.17 Mechanical advantage calculated from the lengths of the effort arm and resistance arm. (a) The elbow, a third-class lever. (b) The ankle, if interpreted as a second-class lever.

Figure 10.18 Some of the body's joints and muscles are structured to form a "low gear" with high mechanical advantage, important for acceleration at the beginning of a race. Others form a "high gear" with a lower mechanical advantage, important for speed.

places and produce different mechanical advantages. A runner taking off from the starting line, for example (fig. 10.18), uses "low-gear" muscles that do not generate much speed but have the power to overcome the inertia of the body. A runner then "shifts into high gear" by using muscles with different insertions that have a lower mechanical advantage but produce more speed at the feet. This is analogous to the way an automobile transmission works to get a car to move and then cruise at high speed.

Physicists recognize three classes of levers that differ with respect to which component—the fulcrum (F), effort (E), or resistance (R)—is in the middle. A **first-class lever** (EFR) is one with the fulcrum in the middle (fig. 10.19*a*), such as a pair of scissors. An anatomical example is the atlanto-occipital joint of the neck, where the muscles of the back of the neck pull down on the nuchal lines of the skull and oppose the tendency of the head to tip forward. Loss of muscle tone here can be embarrassing if you nod off in class.

A **second-class lever** (FRE) is one in which the resistance is in the middle (fig. 10.19*b*). Lifting the handles of a wheelbarrow, for example, causes it to pivot on its wheel at the opposite end, lifting a load in the middle. It is questionable whether there are any second-class

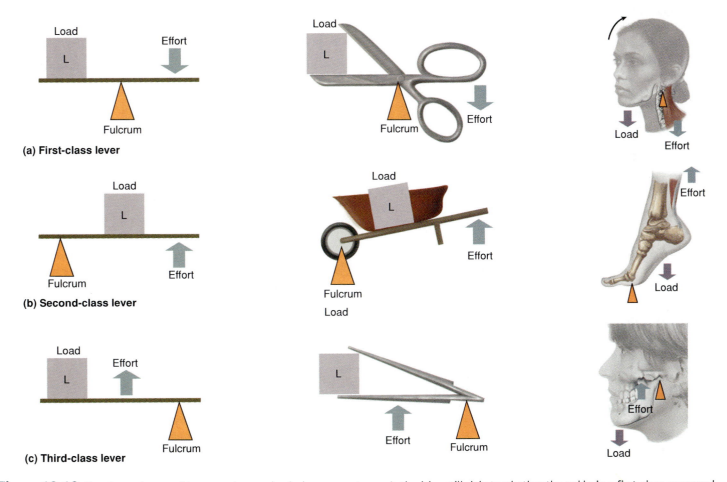

(a) First-class lever

(b) Second-class lever

(c) Third-class lever

Figure 10.19 The three classes of levers and examples in human anatomy. Authorities still debate whether the ankle is a first-class or second-class lever.

While the talocrural (ankle) joint seems analogous to a wheelbarrow, the analogy has an important flaw. If you lift a load in a wheelbarrow, you are standing on the ground and your upward pull on the handles is countered by a downward force of your feet on the ground. The calf muscles that pull up on the heel, however, are part of the load bearing down on the ankle joint. To compare this arrangement to a wheelbarrow is like picturing someone standing in a wheelbarrow trying to lift himself by pulling up on the handles (fig. 1). In the alternative view, the talocrural joint is the fulcrum and the resistance is the earth, making this a first-class lever whose action, in effect, is to push the earth away from the body. Since all motion is relative, this is equivalent to lifting the body away from the earth.

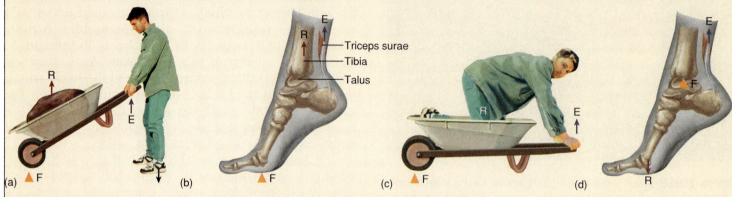

Figure 1 Two interpretations of the ankle joint. *E* = effort, *R* = resistance or load, and *F* = fulcrum. (*a*) In lifting a wheelbarrow (a second-class lever), the effort produces a force exerted through the person's feet to the ground. (*b*) Those who classify the ankle as a second-class lever interpret the weight of the body as the resistance and the ball of the foot (tarsometatarsal joint) as the fulcrum. (*c*) Those who argue that the ankle is not a second-class lever point out that the force generated by the gastrocnemius muscle adds to the resistance, which is analogous to trying to lift a wheelbarrow while standing in it. (*d*) In the latter interpretation, the talocrural joint is the fulcrum and resistance is the earth, making this a first-class lever whose action is to push the earth farther away from the body.

levers in the human body. The talocrural (ankle) joint is often presented as an example. It would seem that the fulcrum is the ball of the foot (distal heads of the metatarsal bones), the resistance is the weight of the body on the talus, and the effort is generated by the calf muscles (the *triceps surae* described in the next chapter) lifting the heel. This may be a misinterpretation, however. Some authorities interpret the ankle as a first-class lever (see special topic 10.2).

In a **third-class lever** (FER), the effort is applied between the fulcrum and resistance. A pair of forceps, for example, consists of two third-class levers joined together. Most levers in the human body are third-class levers. In the *temporomandibular joint*, for example, the mandibular condyle is the fulcrum; the effort is applied by the *temporalis* muscle, which inserts on the coronoid process anterior to the joint; and the resistance is supplied by the item of food being bitten (fig. 10.19c). At the elbow, the fulcrum is the joint between the ulna and humerus; the effort is applied by the tendon of the *biceps brachii* muscle, which inserts on the ulna just distal to the joint; and the resistance can be provided by any weight in the hand or the weight of the forearm itself.

Key Point Review

7   What are the two components of a joint capsule? What is the function of each?

8   What types of joints are described as monaxial, biaxial, and multiaxial? Give an example of each and explain the reason for its classification.

9   Name the joints that would be involved if you reached directly overhead and screwed a lightbulb into a ceiling fixture. Describe the joint actions that would occur.

10   Where are the effort, fulcrum, and resistance in the act of dorsiflexion? What class of lever is this? Would you expect it to have a mechanical advantage greater or less than 1.0? Why?

Anatomy of Selected Diarthroses

▼**Objectives**

When you have completed this section, you should be able to
- identify the major anatomical features of the jaw, shoulder, elbow, hip, knee, and ankle joints; and
- explain how the anatomical differences between these joints are related to differences in function.

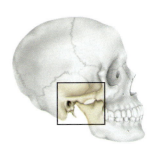

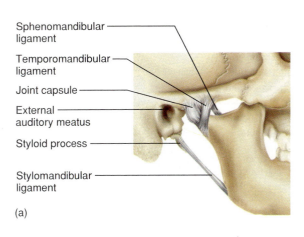

Sphenomandibular ligament

Temporomandibular ligament

Joint capsule

External auditory meatus

Styloid process

Stylomandibular ligament

(a)

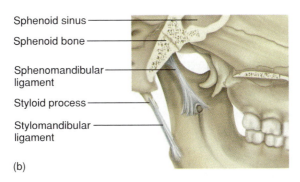

Sphenoid sinus

Sphenoid bone

Sphenomandibular ligament

Styloid process

Stylomandibular ligament

(b)

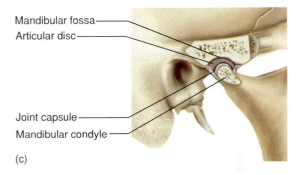

Mandibular fossa

Articular disc

Joint capsule

Mandibular condyle

(c)

Figure 10.20 The temporomandibular joint (TMJ). (*a*) Lateral view; (*b*) medial view; (*c*) sagittal section through the joint cavity.

We now examine the gross anatomy of certain diarthroses. It is beyond the scope of this book to discuss all of them, but the ones selected here most often require medical attention and many of them have a strong bearing on athletic performance.

The Temporomandibular Joint

The **temporomandibular joint (TMJ)** is the insertion of the mandibular condyle into the mandibular fossa of the temporal bone (fig. 10.20). You can feel its action by pressing your fingertips against the jaw immediately anterior to the ear while opening and closing your mouth. This joint combines elements of condyloid, hinge, and gliding joints. It functions in a hingelike fashion when the mandible is elevated and depressed, it glides slightly forward when the jaw is protracted to take a bite, and it glides from side to side to grind food between the molars.

The synovial cavity of the TMJ is divided into superior and inferior chambers by a meniscus called the **articular disc,** which permits lateral and medial excursion of the mandible. Two ligaments support the joint. The **temporomandibular ligament** on the lateral side, deep to the parotid salivary gland, prevents posterior displacement of the mandible. If the jaw receives a hard blow, this ligament normally prevents the condyloid process from being driven upward and fracturing the base of the skull. The other supporting ligament, the **sphenomandibular ligament** on the medial side of the joint, extends from the spine of the sphenoid bone to the ramus of the mandible. A *stylomandibular ligament* extends from the styloid process to the angle of the mandible but is not part of the TMJ proper.

A deep yawn or other strenuous depression of the mandible can dislocate the TMJ by making the condyle pop out of the fossa and slip forward. The joint can be relocated by pressing down on the molars while pushing the jaw backward. *TMJ syndrome* is another common disorder of this joint (see special topic 10.3).

The Humeroscapular Joint

The shoulder joint is called the **humeroscapular,** or **glenohumeral, joint** (fig. 10.21). It is the most freely movable joint of the body but also one of the most commonly injured—indeed, this is a case in which stability is sacrificed for freedom of movement.

The joint is enclosed in a loose capsule that does not interfere with its movement. The glenoid cavity is a rather shallow socket, deepened slightly by a ring of

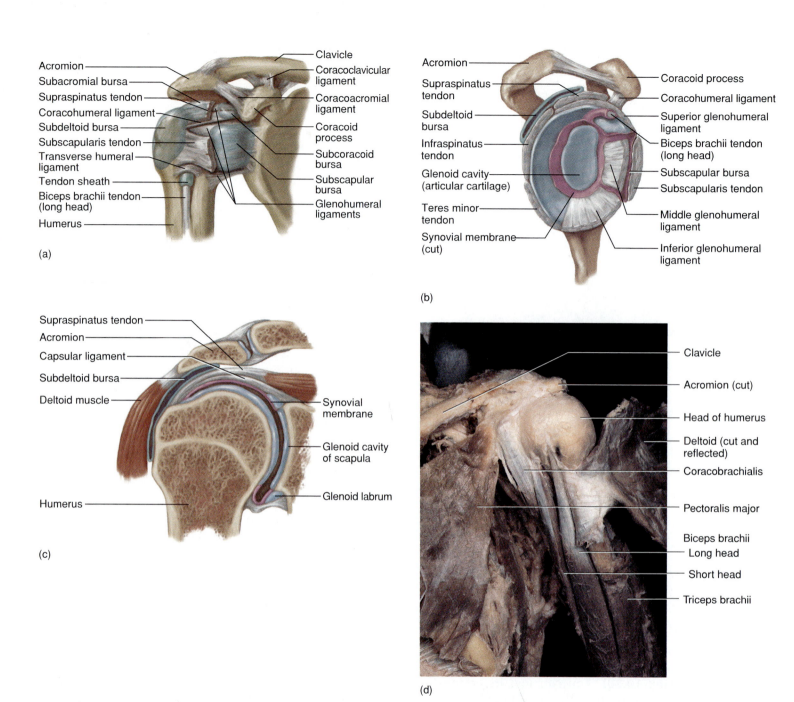

Figure 10.21 The shoulder (humeroscapular) joint. (*a*) Anterior view; (*b*) lateral view of the glenoid cavity and labrum with the humerus removed; (*c*) frontal section of the right shoulder joint, anterior view; (*d*) photograph of the joint. ⚡

Special Topic (TMJ Syndrome)   10.3

TMJ syndrome has received medical recognition only recently, although it may affect as many as 75 million Americans. It can cause moderate intermittent facial pain, clicking sounds in the jaw, limitation of jaw movement, and in some people, more serious symptoms—severe headaches, vertigo (dizziness), tinnitus (ringing in the ears), and pain radiating from the jaw down the neck, shoulders, and back. It seems to be caused by a combination of psychological tension and malocclusion (misalignment of the teeth). Treatment may involve psychological management, physical therapy, analgesic and anti-inflammatory drugs, and sometimes corrective dental appliances to align the teeth properly.

fibrocartilage called the **glenoid labrum.**[15] Five principal ligaments support the joint. Three of them, called **glenohumeral ligaments,** are relatively weak and sometimes absent. The other two are the **coracohumeral ligament,** which extends from the coracoid process of the scapula to the greater tubercle of the humerus, and the **transverse humeral ligament,** which extends from the greater to the lesser tubercle of the humerus, creating a tunnel through which a tendon of the biceps brachii passes.

This tendon is the most important stabilizer of the shoulder. It originates on the margin of the glenoid cavity, passes through the joint capsule, and emerges into the intertubercular groove of the humerus, where it is held by the transverse humeral ligament. Inferior to this groove, it merges into the biceps brachii. Thus, the tendon functions as a taut, adjustable strap that holds the humerus against the glenoid cavity.

In addition to the biceps brachii, four muscles important in stabilizing the humeroscapular joint are the *subscapularis, supraspinatus, infraspinatus,* and *teres minor,* all of which are shown and described more fully in the next chapter. The tendons of these four muscles form the **rotator cuff,** which is fused to the joint capsule on all sides except ventrally. Rotator cuff injuries are common in baseball pitchers because of vigorous circumduction of the arm.

Shoulder dislocations are very painful and can result in permanent damage. For two reasons, the most common dislocation is downward displacement of the humerus: (1) because the rotator cuff protects the joint in all directions except ventrally, and (2) because the joint is protected from above by the coracoid process, acromion process, and clavicle. Dislocations most often occur when the arm is abducted and then receives a blow from above—for example, when the outstretched arm is struck by heavy objects falling off a shelf. They also occur in children who are jerked off the ground by one arm or forced to follow by a hard tug on the arm. The injury is due not only to the inherent stress caused by such abuse but also to the fact that a child's shoulder is not fully ossified and the rotator cuff is not strong enough to withstand such stress. Because this joint is so easily dislocated, you should never attempt to move an immobilized person by pulling on his or her arm.

Four bursae are associated with the shoulder joint. Their names clearly describe their locations—the **subdeltoid, subacromial, subcoracoid,** and **subscapular bursae** (fig. 10.21).

---

15. *labrum* = lip

Think About It

Comment on this proposition from the standpoint of human evolution and functional anatomy: There would be no professional baseball players if *Homo sapiens* had not descended from arboreal primates.

The Elbow Joint

The elbow is a hinge joint composed of two articulations—the **humeroulnar joint,** where the trochlea of the humerus joins the trochlear notch of the ulna, and the **humeroradial joint,** where the capitulum of the humerus meets the head of the radius (fig. 10.22). Both are enclosed in a single joint capsule. On the posterior side of the elbow, there is a prominent **olecranon bursa** to ease the movement of tendons over the elbow. Side-to-side motions of the elbow joint are restricted by a pair of ligaments, the **radial (lateral) collateral ligament** and **ulnar (medial) collateral ligament.**

Another joint occurs in the elbow region, the **proximal radioulnar joint,** but it is not involved in the hinge. At this joint, the disclike head of the radius fits into the radial notch of the ulna and is held in place by the **annular ligament,** which encircles the head of the radius and attaches at each end to the ulna.

The elbow joint is a common site of inflammation and pain because so many muscles originate or insert near it (see the discussion of tennis elbow in the chapter 11 essay, p. 374).

The Coxal Joint

The **coxal** (hip) **joint** is the point where the head of the femur inserts into the acetabulum of the os coxae (fig. 10.23). Because the coxal joints bear much of the body's weight, they have deep sockets and are much more stable than the shoulder joint. The depth of the socket is somewhat greater than you see on dried bones because a horseshoe-shaped ring of fibrocartilage called the **acetabular labrum** is attached to its rim. Dislocations of the hip are rare, but some infants suffer congenital dislocations because the acetabulum is not deep enough to hold the head of the femur in place. This condition can be treated by placing the infant in traction until the acetabulum develops enough strength to support the body's weight (fig. 10.24).

Ligaments that support the coxal joint include the **iliofemoral** (ILL-ee-oh-FEM-oh-rul) and **pubofemoral** (PYU-bo-FEM-or-ul) **ligaments** on the anterior side and the **ischiofemoral** (ISS-kee-oh-FEM-or-ul) **ligament** on the posterior side. When you stand up, these ligaments become twisted and pull the head of the femur tightly

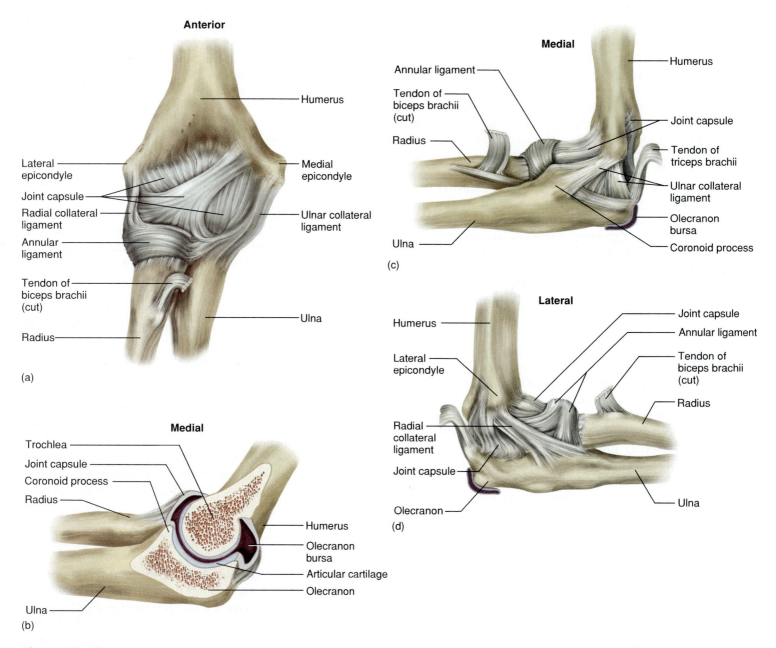

Figure 10.22 The elbow joint. (*a*) Anterior view; (*b*) midsagittal section, medial view; (*c*) medial view; (*d*) lateral view.

into the acetabulum. The head of the femur has a conspicuous pit called the *fovea capitis.* The **ligamentum teres**[16] (TERR-eez) arises here and attaches to the lower margin of the acetabulum. This is a relatively slack ligament, so it is uncertain whether it plays a significant role in holding the femur in its socket. It does, however, contain an artery that supplies blood to the head of the femur. A **transverse acetabular ligament** bridges a gap in the inferior margin of the acetabular labrum.

16. *teres* = round

The Knee Joint

The knee joint, or **tibiofemoral joint,** is the largest and most complex diarthrosis of the body (fig. 10.25). It is a primarily a hinge joint, but when the knee is flexed it is also capable of slight rotation and lateral gliding. The patella and patellar ligament also form a gliding **patellofemoral joint** with the femur.

The joint capsule encloses only the lateral and posterior aspects of the knee joint, not the anterior. The anterior aspect is covered by the patellar ligament and the **lateral** and **medial patellar retinacula** (not illustrated). These are extensions of the tendon of the *quadriceps femoris* muscle, the large anterior muscle of

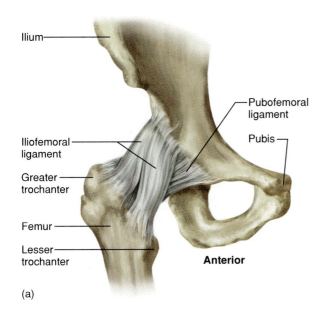

Ilium

Pubofemoral ligament

Iliofemoral ligament

Pubis

Greater trochanter

Femur

Lesser trochanter

Anterior

(a)

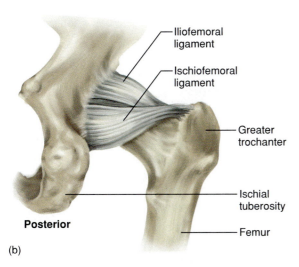

Iliofemoral ligament

Ischiofemoral ligament

Greater trochanter

Ischial tuberosity

Posterior

Femur

(b)

Figure 10.23 The coxal (hip) joint. (*a*) Anterior view; (*b*) posterior view; (*c*) the acetabulum with the femoral head retracted; (*d*) photograph of the right hip with the femoral head retracted, anterior view. ⚕

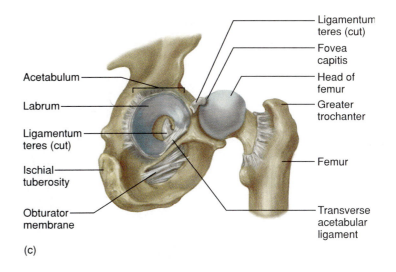

Ligamentum teres (cut)

Fovea capitis

Acetabulum

Head of femur

Labrum

Greater trochanter

Ligamentum teres (cut)

Ischial tuberosity

Femur

Obturator membrane

Transverse acetabular ligament

(c)

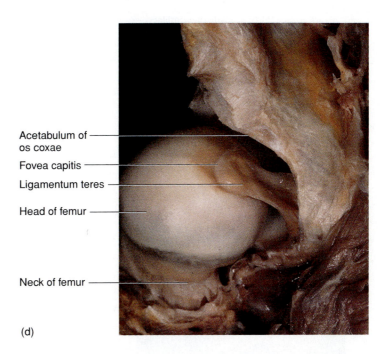

Acetabulum of os coxae

Fovea capitis

Ligamentum teres

Head of femur

Neck of femur

(d)

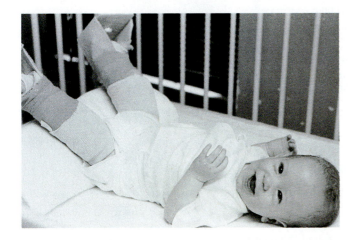

Figure 10.24 Infants are sometimes placed in traction to treat congenital hip dislocations.

the thigh. The knee is stabilized mainly by the quadriceps tendons in front and the tendon of the *semimembranosus* muscle on the rear of the thigh. Developing strength in these muscles therefore reduces the risk of knee injury.

The joint cavity contains two cartilages called the **lateral meniscus** and **medial meniscus,** joined by a **transverse ligament.** These menisci absorb the shock of the body weight jostling up and down on the knee joint and prevent the femur from rocking from side to side on the tibia.

The posterior "pit" of the knee, the **popliteal** (po-LIT-ee-ul) **region,** is supported by a complex array of **intracapsular ligaments** within the joint capsule and **extracapsular ligaments** external to it. The extracapsular ligaments are the **oblique popliteal ligament** (an extension

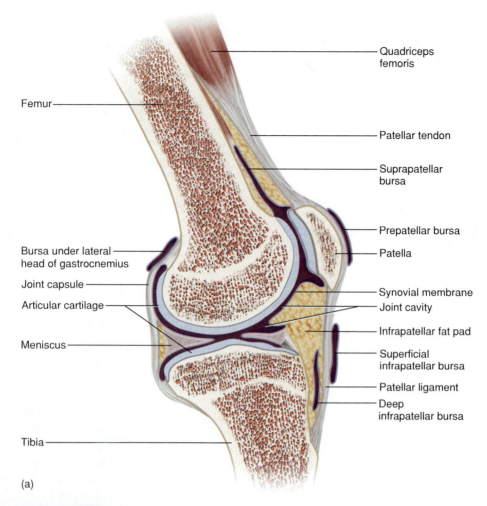

Femur

Quadriceps
femoris

Patellar tendon

Suprapatellar
bursa

Prepatellar bursa

Patella

Bursa under lateral
head of gastrocnemius

Joint capsule

Articular cartilage

Synovial membrane

Joint cavity

Meniscus

Infrapatellar fat pad

Superficial
infrapatellar bursa

Patellar ligament

Deep
infrapatellar bursa

Tibia

(a)

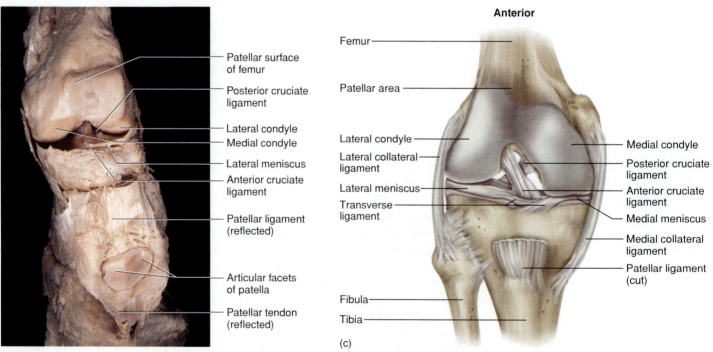

Patellar surface
of femur

Posterior cruciate
ligament

Lateral condyle

Medial condyle

Lateral meniscus

Anterior cruciate
ligament

Patellar ligament
(reflected)

Articular facets
of patella

Patellar tendon
(reflected)

(b)

Anterior

Femur

Patellar area

Lateral condyle

Lateral collateral
ligament

Lateral meniscus

Transverse
ligament

Fibula

Tibia

Medial condyle

Posterior cruciate
ligament

Anterior cruciate
ligament

Medial meniscus

Medial collateral
ligament

Patellar ligament
(cut)

(c)

Figure 10.25 The knee joint. (a) Diagram of a midsagittal section; (b) photograph of the right knee in which the patellar tendon has been cut and folded (reflected) downward, exposing the joint cavity and the posterior surface of the patella; (c) anterior view of structures in the joint cavity of the right knee;—*Cont'd* ↗

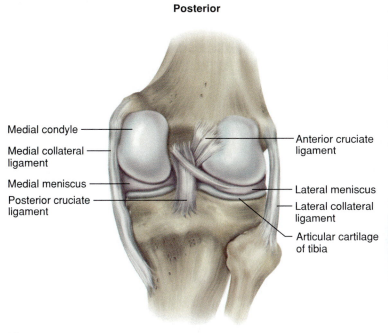

Posterior

Medial condyle

Medial collateral ligament

Medial meniscus

Posterior cruciate ligament

Anterior cruciate ligament

Lateral meniscus

Lateral collateral ligament

Articular cartilage of tibia

(d)

Figure 10.25—Cont'd (*d*) posterior view of the right knee. ✗

Figure 10.26 As many as half of professional football players suffer serious knee injuries as a result of the stresses placed on the joint and the vulnerability of the knee to horizontal and rotational stress.

of the semimembranosus tendon), **arcuate** (AR-cue-et) **popliteal ligament, lateral (fibular) collateral ligament,** and **medial (tibial) collateral ligament.** The two collateral ligaments prevent the knee from rotating when the joint is extended.

There are two intracapsular ligaments deep within the joint cavity. The synovial membrane folds around them, however, so that they are excluded from the fluid-filled synovial cavity. These ligaments cross each other in the form of an X; hence, they are called the **anterior** and **posterior cruciate**[17] (CROO-she-ate) **ligaments.** They are named according to whether they attach to the anterior or posterior side of the tibia, not for their attachments to the femur. When the knee is extended, the anterior cruciate ligament is pulled tight and prevents hyperextension. The posterior cruciate ligament prevents the femur from sliding off the front of the tibia or the tibia from being displaced backward.

An important aspect of human bipedalism is the ability to "lock" the knees and stand erect without tiring the extensor muscles of the leg (see chapter 9 essay, page 289). When the knee is extended to the fullest degree allowed by the anterior cruciate ligament, the femur rotates medially on the tibia. This action locks the knee, and in this state all the major knee ligaments are twisted and taut. To unlock the knee, the *popliteus* muscle rotates the femur laterally and untwists the ligaments.

The knee joint has at least 13 bursae. Four of these are anterior—the **superficial infrapatellar, suprapatellar, prepatellar,** and **deep infrapatellar.** Located in the popliteal region are the **popliteal bursa** and **semimembranosus bursa** (not illustrated). At least seven more bursae are found on the lateral and medial sides of the knee joint. From figure 10.25*a*, your knowledge of the relevant word elements (*infra-, supra-, pre-*), and the terms *superficial* and *deep,* you should be able to work out the reasoning behind most of these names and develop a system for remembering the locations of these bursae.

Although the knee is capable of withstanding a great deal of vertical compression, it is highly vulnerable to rotational and horizontal stress, especially blows from the lateral side. As many as half of all professional football players receive serious knee injuries (fig. 10.26). Knee injuries heal slowly because ligaments and tendons have a very scanty blood supply and cartilage is completely avascular. *Arthroscopic surgery,* however, has made it possible to treat many knee injuries and joint disorders with a minimum of surgical invasion (see special topic 10.4).

The Ankle Joint

The ankle, or **talocrural joint,** includes two articulations—the medial joint between the tibia and talus and the lateral joint between the fibula and talus, both enclosed in one joint capsule. The malleoli of the tibia and fibula overhang the talus on each side like a cap, preventing

17. *cruci* = cross

Arthroscopy is a procedure in which the interior of a joint is viewed with a pencil-thin instrument called an *arthroscope* inserted through an incision as small as 5 mm long (fig. 1). The arthroscope has a light source, a lens, and fiber optics and can be connected to a video camera. It enables a physician to examine the synovial membranes, articular cartilages, and cruciate ligaments; to take photographs or videotapes of the knee joint; and to withdraw samples of synovial fluid. If surgery is required, additional small incisions can be made for the surgical instruments and the procedures can be observed through the arthroscope. Saline is often introduced through one incision to expand the joint and provide a clearer view of its structures. Arthroscopic surgery produces much less tissue damage than conventional surgery, enabling patients to recover more quickly.

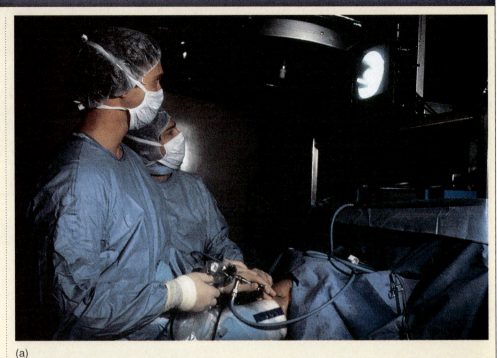

(a)

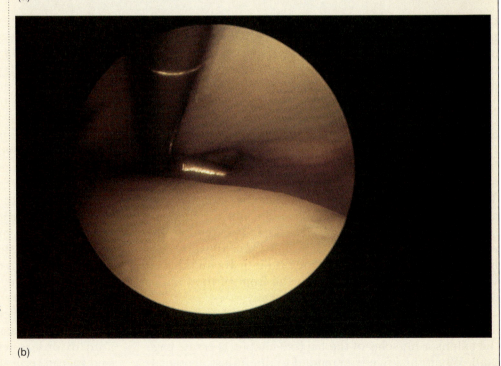

(b)

Figure 1 (*a*) Performing a knee arthroscopy. The arthroscope and, if necessary, surgical instruments are inserted through separate incisions. (*b*) A view of the joint cavity through the arthroscope. Saline is often introduced into the joint cavity to expand it and make the internal structures more visible. 𝒳

most side-to-side motion (fig. 10.27). The ankle therefore has a more restricted range of motion than the wrist.

The ligaments of the ankle (fig. 10.28) include (1) **anterior** and **posterior tibiofibular ligaments,** binding the tibia to the fibula, (2) a multipart **deltoid liga-**ment binding the tibia to the foot on the medial side, and (3) a multipart **lateral collateral ligament** binding the fibula to the foot on the lateral side. The **calcaneal (Achilles) tendon** extends from the calf muscles to the calcaneus. It enables plantar flexion and limits dorsi-

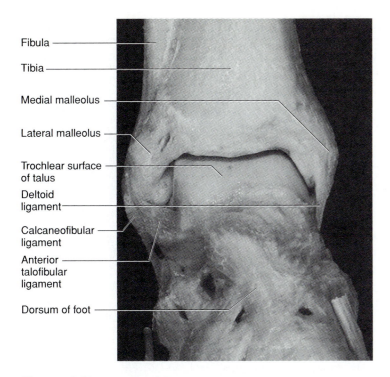

Fibula

Tibia

Medial malleolus

Lateral malleolus

Trochlear surface
of talus

Deltoid
ligament

Calcaneofibular
ligament

Anterior
talofibular
ligament

Dorsum of foot

Figure 10.27 Photograph of the talocrural (ankle) joint, anterior view.

flexion of the joint. Plantar flexion is limited by extensor tendons on the anterior side of the ankle and by the anterior part of the joint capsule.

Sprains are especially common at the ankle. Excessive eversion can tear the deltoid ligament, and excessive inversion often tears the anterior talofibular ligament and calcaneofibular ligament. These sprains are painful and usually accompanied by immediate swelling. They are best treated by immobilizing the joint and reducing swelling with an ice pack, but in extreme cases they may require a cast or surgery.

The synovial joints described in this section are summarized in table 10.2. A summary of some common joint disorders, including sprains, is presented in table 10.3.

Key Point Review

11 What keeps the mandibular condyle from slipping out of its fossa in a posterior direction?

12 Explain how the biceps tendon braces the shoulder joint.

13 What structure elsewhere in the skeletal system has a structure and function similar to the acetabular labrum of the os coxae?

14 What keeps the femur from slipping backward off the tibia?

15 What keeps the tibia from slipping sideways off the talus?

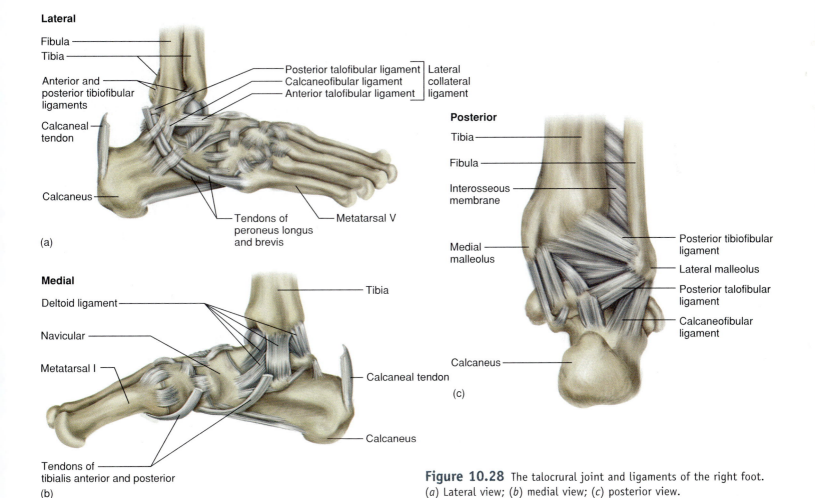

Lateral

Fibula

Tibia

Anterior and
posterior tibiofibular
ligaments

Calcaneal
tendon

Calcaneus

Posterior talofibular ligament ⎤ Lateral
Calcaneofibular ligament   ⎥ collateral
Anterior talofibular ligament ⎦ ligament

Tendons of
peroneus longus
and brevis

Metatarsal V

(a)

Medial

Deltoid ligament

Navicular

Metatarsal I

Tendons of
tibialis anterior and posterior

Tibia

Calcaneal tendon

Calcaneus

(b)

Posterior

Tibia

Fibula

Interosseous
membrane

Medial
malleolus

Posterior tibiofibular
ligament

Lateral malleolus

Posterior talofibular
ligament

Calcaneofibular
ligament

Calcaneus

(c)

Figure 10.28 The talocrural joint and ligaments of the right foot. (a) Lateral view; (b) medial view; (c) posterior view.

Table 10.2   Review of the Principal Diarthroses

Joint	Principal Anatomical Features
Temporomandibular (fig. 10.20)	*Type:* condyloid, hinge, and gliding joint *Movements:* elevation-depression, protraction-retraction, lateral-medial excursion *Articulation:* condyle of mandible, mandibular fossa of temporal bone *Ligaments:* temporomandibular, sphenomandibular *Cartilage:* articular disc
Humeroscapular (fig. 10.21)	*Type:* ball-and-socket *Movements:* adduction-abduction, flexion-extension, circumduction *Articulation:* head of humerus, glenoid fossa of scapula *Ligaments:* coracohumeral, transverse humeral, three glenohumerals *Tendons:* rotator cuff (tendons of subscapularis, supraspinatus, infraspinatus, teres minor), tendon of biceps brachii *Bursae:* subdeltoid, subacromial, subcoracoid, subscapular *Cartilage:* glenoid labrum
Elbow (fig. 10.22)	*Type:* hinge *Movements:* flexion-extension, pronation-supination, rotation *Articulations:* humeroulnar—trochlea of humerus, trochlear notch of ulna; humeroradial—capitulum of humerus, head of radius; radioulnar—head of radius, radial notch of ulna *Ligaments:* radial collateral, ulnar collateral, annular *Bursa:* olecranon
Coxal (fig. 10.23)	*Type:* ball-and-socket *Movements:* adduction-abduction, flexion-extension, circumduction *Articulations:* head of femur, acetabulum of os coxae *Ligaments:* iliofemoral, pubofemoral, ischiofemoral, ligamentum teres, transverse acetabular *Cartilage:* acetabular labrum
Knee (fig. 10.25)	*Type:* primarily hinge *Movements:* flexion-extension, abduction-adduction, slight rotation *Articulations:* tibiofemoral, patellofemoral *Ligaments:* anterior—lateral patellar retinaculum, medial patellar retinaculum; popliteal intracapsular—anterior cruciate, posterior cruciate; popliteal extracapsular—oblique popliteal, arcuate popliteal, lateral collateral, medial collateral *Bursae:* anterior—superficial infrapatellar, suprapatellar, prepatellar, deep infrapatellar; popliteal—popliteal, semimembranosus; medial and lateral—seven other bursae not named in this chapter *Cartilages:* lateral meniscus, medial meniscus (connected by transverse ligament)
Ankle (figs. 10.27 and 10.28)	*Type:* hinge *Articulations:* tibia-talus, fibula-talus *Movements:* dorsiflexion, plantar flexion, extension, inversion, eversion *Ligaments:* anterior and posterior tibiofibular, deltoid, lateral collateral *Tendon:* calcaneal (Achilles)

Table 10.3	Some Common Joint Disorders
Strain	Painful overstretching of a tendon or muscle without serious tissue damage. Often results from inadequate warmup before exercise.
Sprain	Torn ligament or tendon, sometimes with damage to a meniscus or other cartilage.
Synovitis	Inflammation of a joint capsule, often as a complication of a sprain.
Dislocation	Displacement of a bone from its normal position at a joint, usually accompanied by a sprain of the adjoining connective tissues. Most common at the fingers, thumb, shoulder, and knee.
Rheumatism	Broad term for any pain in the supportive and locomotory organs of the body, including bones, ligaments, tendons, and muscles. See chapter essay.
Arthritis	Broad term embracing more than 100 types of joint rheumatism. See chapter essay.
Osteoarthritis (OA)	The most common form of arthritis, also known as "wear-and-tear arthritis" because it is a normal consequence of aging. Associated with softening and degeneration of the articular cartilage, exposure of the epiphyseal bone, and development of bony spurs in the joint cavity, causing pain and restricting movement. See chapter essay.
Rheumatoid Arthritis (RA)	A more severe form of arthritis resulting from an autoimmune attack against the joint tissues (failure of the immune system to recognize the tissues as one's own). See chapter essay.
Gout	A hereditary disease, most common in men, in which uric acid crystals accumulate in the joints, irritating the articular cartilage and synovial membrane. Causes gouty arthritis, with swelling, pain, tissue degeneration, and sometimes fusion of the joint. Most commonly affects the great toe.
Bursitis	Inflammation of a bursa, usually due to overexertion of a joint.
Tendinitis	A form of bursitis in which a tendon sheath is inflamed.

CHAPTER ESSAY

Chapter Essay

Arthritis and Artificial Joints

Arthritis[18] is a broad term for pain and inflammation of a joint and embraces more than a hundred different diseases of largely obscure or unknown causes. In all of its forms, it is the most common crippling disease in the United States; nearly everyone past middle age develops arthritis to some degree.

The most common form of arthritis is **osteoarthritis (OA),** also called "wear-and-tear arthritis" because it is apparently a normal consequence of years of wear on the joints. As joints age, the articular cartilage softens and degenerates. As it becomes roughened by wear, joint movement may be accompanied by crunching or crackling sounds called **crepitus.** OA affects especially the fingers, intervertebral joints, hips, and knees. As the articular cartilage wears away, exposed bone tissue often develops spurs that grow into the joint cavity, restrict movement, and cause pain. OA rarely occurs before age 40, but it affects about 85% of people older than 70. It usually does not cripple, but in severe cases it can immobilize the hip.

Rheumatoid arthritis (RA), which is far more severe than osteoarthritis, results from an autoimmune attack against the joint tissues. It begins when the body produces defensive proteins (antibodies) to fight an infection. Failing to recognize the body's own tissues, a misguided antibody known as **rheumatoid factor** also attacks the synovial membranes. Inflammatory cells accumulate in the synovial fluid and produce enzymes that degrade the articular cartilage. The synovial membrane thickens and adheres to the articular cartilage, fluid accumulates in the joint capsule, and the capsule is invaded by fibroconnective tissue. As articular cartilage degenerates, the joint begins to ossify, and sometimes the bones become solidly fused and immobilized, a condition called **ankylosis** (fig. E.1). The disease tends to develop symmetrically—if the right wrist or hip develops RA, so does the left.

—continued

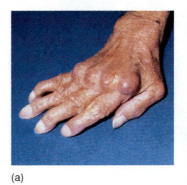 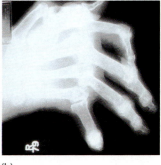

(a)   (b)

Figure E.1 (*a*) Rheumatoid arthritis (RA) of the hand. (*b*) X ray of a hand with RA.

18. *arthr* = joint + *itis* = inflammation

Rheumatoid[19] arthritis is named for the fact that symptoms tend to flare up and subside (go into remission) periodically. It affects women far more often than men, and because RA typically begins between the ages of 30 and 40, it can cause decades of pain and disability. There is no cure, but joint damage can be slowed with hydrocortisone or other steroids. Since long-term use of steroids weakens the bone, however, aspirin is the treatment of first choice to control the inflammation. Physical therapy is also used to preserve the joint's range of motion and the patient's functional ability.

Arthroplasty,[20] a treatment of last resort, is the replacement of a diseased joint with an artificial device called a **prosthesis.** Joint prostheses were first developed to treat war injuries in World War II and the Korean War. Total hip replacement (THR), first performed in 1963 by English orthopedic surgeon Sir John Charnley, is now the most common orthopedic procedure for the elderly. The first knee replacements were performed in the 1970s.

Joint prostheses are now available for finger, shoulder, and elbow joints, as well as for hip and knee joints. Arthroplasty is now performed on over 250,000 patients per year in the United States, primarily to relieve pain and restore function in elderly people with OA or RA.

Arthroplasty presents ongoing challenges for biomedical engineering. An effective prosthesis must be strong, nontoxic, and corrosion-resistant. In addition it must bond firmly to the patient's bones and enable a normal range of motion with a minimum of friction. The heads of long bones are usually replaced with prostheses made of a metal alloy such as cobalt-chrome, titanium alloy, or stainless steel. Joint sockets are made of polyethylene (fig. E.2). Prostheses are bonded to the patient's bone with screws or bone cement.

About 80% to 90% of hip replacements and at least 60% of ankle replacements are still functional 2 to 10 years later. The most common form of failure is detachment of the prosthesis from the bone. This problem has been reduced by using *porous coated prostheses,* which become infiltrated by the patient's own bone and create a firmer bond. A prosthesis is not as strong as a natural joint, however, and is not an option for many young, active patients.

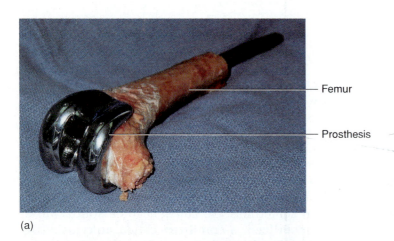

(a)

    — Femur

    — Prosthesis

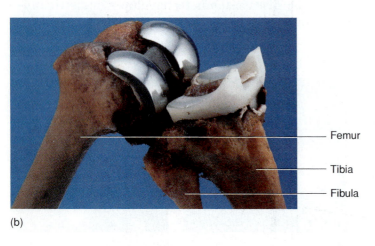

(b)

    — Femur

    — Tibia

    — Fibula

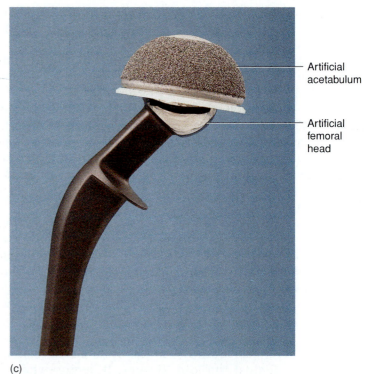

(c)

    — Artificial acetabulum

    — Artificial femoral head

Figure E.2 Joint prostheses. (*a*) An artificial femoral head inserted into the femur. (*b*) An artificial knee joint bonded to a natural femur and tibia. (*c*) A porous-coated hip prosthesis. The caplike portion replaces the acetabulum of the os coxae, and the ball and shaft shown below are bonded to the proximal end of the femur—*Cont'd.*

19. *rheumat* = tending to change
20. *plasty* = surgical repair

Perhaps the most promising recent advance in arthroplasty is *computer-assisted design and manufacture (CAD/CAM).* A computer can scan radiographs from the patient and present several design possibilities for review. Once a design is selected, the computer can generate a program to operate the machinery that produces the prosthesis. CAD/CAM has reduced the waiting period for a prosthesis from 12 weeks to 2 and has lowered the cost dramatically.▲

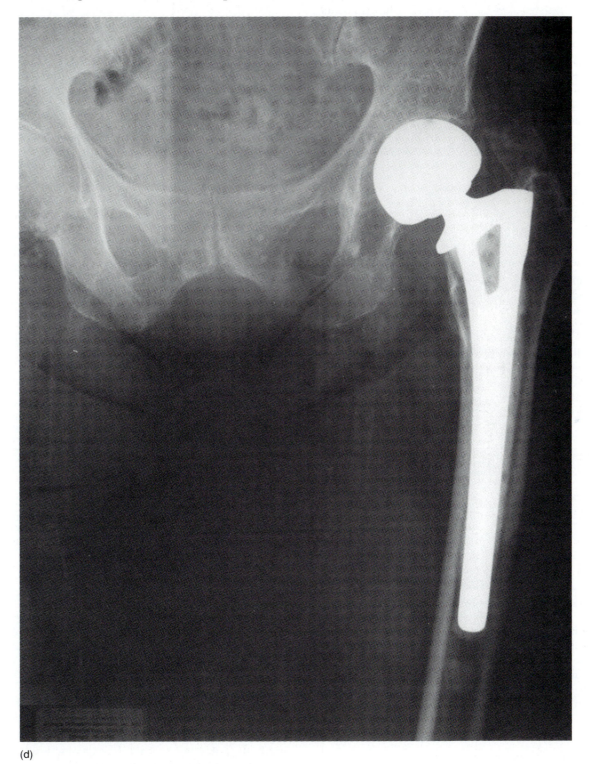

(d)

Figure E.2—Cont'd (*d*) X ray of a patient with a total hip replacement.

Joints and Their Classification (pp. 295–296)

1. Naming of joints
2. Classification by freedom of movement
   a. Diarthroses
   b. Amphiarthroses
   c. Synarthroses
3. Classification by mode of attachment
   a. Fibrous joints
   b. Cartilaginous joints
   c. Synovial joints
   d. Synostoses (bony joints)

Fibrous, Cartilaginous, and Bony Joints (pp. 296–299)

1. Fibrous joints
   a. Sutures
      • Serrate sutures
      • Lap sutures
      • Plane sutures
   b. Gomphoses
   c. Syndesmoses
2. Cartilaginous joints
   a. Synchondroses
   b. Symphyses
3. Synostoses (bony joints)

Form and Function of Synovial Joints (pp. 299–310)

1. General anatomy
   a. Joint cavity
   b. Synovial fluid
   c. Articular cartilages
   d. Joint capsule
   e. Synovial membrane
   f. Menisci
   g. Tendons
   h. Ligaments
   i. Bursae
2. Types of synovial joints
   a. Ball-and-socket
   b. Hinge
   c. Condyloid
   d. Saddle
   e. Pivot
   f. Gliding
3. Movements of diarthroses
   a. Flexion, extension, and hyperextension
   b. Abduction and adduction
   c. Elevation and depression
   d. Protraction and retraction
   e. Lateral and medial excursion
   f. Circumduction
   g. Rotation
   h. Supination and pronation
   i. Opposition and reposition
   j. Dorsiflexion and plantar flexion
   k. Inversion and eversion
4. Range of motion (ROM)
   a. Importance and measurement
   b. Determining factors
5. Levers and biomechanics of the joints
   a. General features of a lever
   b. Advantages of a lever
      • Increased force
      • Increased speed or distance
   c. Mechanical advantage
      • $MA = L_E/L_R$
      • Power-speed tradeoff
   d. Classes of levers
      • First-class (fulcrum in middle)
      • Second-class (resistance in middle)
      • Third-class (effort in middle)

Anatomy of Selected Diarthroses (pp. 310–321)

See table 10.2 for summaries.

1. Temporomandibular (jaw) joint
2. Humeroscapular (shoulder) joint
3. Elbow joint—humeroulnar, humeroradial, and radioulnar
4. Coxal (hip) joint
5. Knee joint—tibiofemoral and patellofemoral
6. Talocrural (ankle) joint

Selected Vocabulary

Also review the terms in tables 10.1 and 10.2, which are not repeated here.

articulation 295
arthrology 295
kinesiology 295
biomechanics 295
diarthrosis 295
amphiarthrosis 295
synarthrosis 295
periodontal membrane 298
interosseous membrane 298
joint cavity 299
synovial fluid 299

articular cartilage 299
joint capsule 299
fibrous capsule 300
synovial membrane 300
meniscus 300
tendon 300
ligament 300
bursa 300
tendon sheath 300
flexion 303
extension 303
hyperextension 303
abduction 303
adduction 303
elevation 303

depression 305
protraction 305
retraction 305
lateral excursion 305
medial excursion 305
circumduction 305
rotation 306
lateral rotation 306
medial rotation 306
supination 306
pronation 306
opposition 306
reposition 306
dorsiflexion 306
plantar flexion 306

inversion 306
eversion 306
range of motion (ROM) 307
proprioceptor 307
lever 307
fulcrum 307
effort 307
resistance 307
effort arm 307
resistance arm 308
mechanical advantage (MA) 308
first-class lever 309
second-class lever 309
third-class lever 310

1. Which of the following joint types are unique to the thumb?
   a. gliding
   b. hinge
   c. saddle
   d. condyloid
   e. pivot

2. Which of the following joint types is the least movable?
   a. diarthrosis
   b. synarthrosis
   c. symphysis
   d. syndesmosis
   e. condyloid

3. Which of the following pair of movements is (are) unique to the foot?
   a. dorsiflexion and inversion
   b. elevation and depression
   c. circumduction and rotation
   d. abduction and adduction
   e. opposition and reposition

4. Which of the following joints cannot be circumducted?
   a. carpometacarpal
   b. metacarpophalangeal
   c. humeroscapular
   d. coxal
   e. interphalangeal

5. Which of the following terms denotes a general condition that includes the other four?
   a. gout
   b. arthritis
   c. rheumatism
   d. osteoarthritis
   e. rheumatoid arthritis

6. In an adult, the ischium and pubis are united by
   a. a synchondrosis.
   b. a diarthrosis.
   c. a synostosis.
   d. an amphiarthrosis.
   e. a symphysis.

7. In a second-class lever, the effort
   a. is applied at the end opposite the fulcrum.
   b. is applied to the fulcrum itself.
   c. is applied between the fulcrum and resistance.
   d. always produces MA less than 1.0.
   e. is applied at one side of the fulcrum to move a resistance on the other side.

8. Which of the following joints has anterior and posterior cruciate ligaments?
   a. shoulder
   b. elbow
   c. hip
   d. knee
   e. ankle

9. To bend backward at the waist involves _____ of the spinal column.
   a. rotation
   b. hyperextension
   c. dorsiflexion
   d. abduction
   e. flexion

10. The rotator cuff includes the tendons of all of the following muscles *except*
    a. the subscapularis.
    b. the supraspinatus.
    c. the infraspinatus.
    d. the biceps brachii.
    e. the teres minor.

11. The lubricant of a diarthrosis is called _____.

12. A fluid-filled sac that eases the movement of a tendon over a bone is called a/an _____.

13. A _____ joint allows one bone to swivel on another.

14. _____ is the science of movement.

15. The joint between a tooth and the mandible is called a/an _____.

16. In a _____ suture, the articulating bones have interlocking wavy margins, somewhat like a dovetail joint in carpentry.

17. In kicking a football, what type of action does the knee joint exhibit?

18. The angle through which a joint can be moved is called the _____.

19. Both the shoulder and hip joints are somewhat deepened and supported by a ring of fibrocartilage called a/an _____.

20. The femur is prevented from slipping sideways off the tibia in part by a pair of cartilages called the lateral and medial _____.

1. All second-class levers produce a mechanical advantage greater than 1.0, and all third-class levers produce a mechanical advantage less than 1.0. Explain why.

2. Suppose a lever measures 17 cm from effort to fulcrum and 11 cm from resistance to fulcrum. (a) Calculate its mechanical advantage. (b) Would this lever produce more force or less force than that exerted on it? Explain. (c) Which of the three classes of levers could *not* have these measurements? Explain.

3. In order of occurrence, list the joint actions (flexion, pronation, etc.) and the joints where they would occur as you (a) sat down at a table, (b) reached out and picked up an apple, (c) took a bite, and (d) chewed it. Assume that you started in anatomical position.

4. There is a continuous strip of fibroconnective tissue from the quadriceps femoris muscle of the anterior aspect of the thigh to the tibial tuberosity of the tibia. Why is it called the patellar *tendon* superior to the patella and the patellar *ligament* inferior to it?

5. Suppose you are dissecting a cat or fetal pig with the task of finding examples of each type of synovial joint. Which type of human synovial joint would not be found in either of those animals? For lack of that joint, what human joint action would the animals be unable to perform?

Web Site Link

For a listing of the most current web sites related to this chapter, please visit the Saladin homepage at:

http://www.mhhe.com/sciencemath/biology/saladin/

The Muscular System

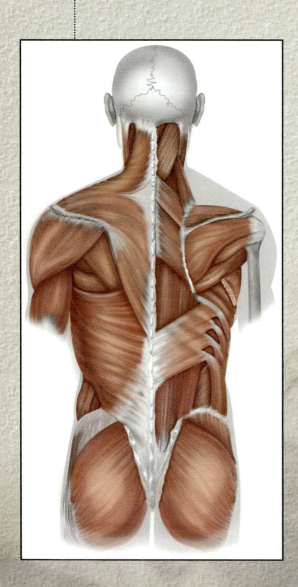

Special Topics

Brushing up

T he **muscular system** consists of about 600 skeletal muscles and accounts for about 40% of the body's weight. **Myology,**[1] the study of muscle, adds meaning to what you already know about bones and joints. It relates muscle attachments to the bone structures described in chapter 9 and muscle function to the joint movements described in chapter 10. It is therefore appropriate to study the muscular system in close connection with those chapters. In the next chapter, muscle contraction is considered in finer detail, from the cellular to molecular level.

Muscular anatomy and function occupy a place of central importance in several fields of health care and athletics. Physical and occupational therapists must be well acquainted with the muscular system to design and carry out rehabilitation programs. Many health professionals must move patients who are physically incapacitated, and to do this safely and effectively requires understanding of joints and muscles. Even to give intramuscular injections safely requires knowledge of the muscles and their associated nerves and blood vessels. Coaching, movement science, and sports medicine focus much of their attention on skeletomuscular anatomy and mechanics.

The Study of Skeletal Muscles

▼**Objectives**
When you have completed this section, you should be able to
• explain why a knowledge of muscular anatomy is important in a variety of professions; and
• interpret Latin words used in the naming of muscles.

How Muscles are Named

Some of the principal superficial muscles are shown in figure 11.1. Most of this chapter is a descriptive inventory

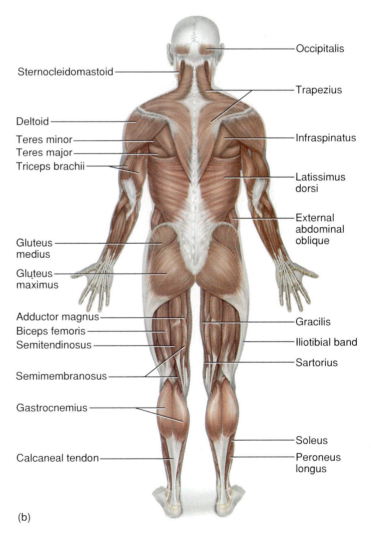

Figure 11.1 The major superficial muscles. (*a*) Anterior and (*b*) posterior views.

1. *my, myo, mys* = muscle

of muscles, including their location, function (action), origin, insertion, and innervation. Learning about these muscles and remembering their names is much easier when you have some appreciation of the meaning behind the names.

Early anatomists coined names for the muscles and other structures in their native languages. This quickly led to a state of confusion and a recognized need for standard international terminology. In 1895, an international committee of anatomists developed a system of Latin names called the *Nomina Anatomica* (N.A.), which has been periodically updated. Although in this book, most terms are given in English rather than Latin, the customary English names for skeletal muscles are, at most, only slight modifications of the Latin names—for example, *anterior scalene* is a derivative of the N.A. term, *musculus scalenus anterior.*

Some muscle names are several words long—for example, the *flexor digiti minimi brevis,* a "short muscle that flexes the little finger." These names may seem intimidating at first, but they are really more of a help than an obstacle to understanding if you gain a little insight into the most commonly used Latin words. Several of these words are interpreted in table 11.1, and others are explained in footnotes throughout the chapter. Familiarity with these terms can help you translate muscle names and remember the location, appearance, and action of the muscles.

A Learning Strategy

In the remainder of this chapter, we consider approximately 160 muscles. The following suggestions may help you develop a rational learning strategy for the muscular system:

- Examine models, cadavers, dissected animals, or a photographic atlas as you read about these muscles. Visual images are often easier to remember than words, and direct observation of a muscle may stick in your memory better than descriptive text or two-dimensional drawings.
- When studying a particular muscle, palpate it on yourself if possible. Contract the muscle to feel it bulge and see its action. This will make muscle locations and actions less abstract. Atlas B, "Surface Anatomy," following this chapter, shows where you can see and palpate several of these muscles on the living body.
- Locate the origins and insertions of muscles on an articulated skeleton. Some study skeletons are painted and labeled to show these. This will help you visualize the locations of muscles and understand how they produce particular joint actions.

- Study the derivation of each muscle name; the name of a muscle usually describes its location, appearance, origin, insertion, or action.
- Say the names aloud to yourself or a study partner. It is harder to remember and spell terms you cannot pronounce, and silent pronunciation is not nearly as effective as hearing the name spoken. Pronunciation guides are provided in the muscle tables for all but the most obvious pronunciations.

1. Name at least five criteria by which muscles are named.
2. In muscle names, what do the words *brevis, teres, digitorum, pectoralis, triceps,* and *profundus* mean?

Functions and Structural Organization of Muscles

▼Objectives
When you have completed this section, you should be able to
- list the functions of the muscular system;
- identify the connective tissue components of a skeletal muscle and explain how they contribute to the organization of the muscle;
- describe some anatomical features that most skeletal muscles have in common;
- classify muscles according to the orientation of their fascicles; and
- explain how skeletal muscles can work together to produce coordinated movements at a joint.

Functions of Muscles

Generally speaking, a muscle converts the chemical energy of ATP into the kinetic energy of motion and exerts a useful pull on another tissue. More specifically, muscle contraction has the following effects:

- **Movement.** Most obviously, the muscles enable us to move from place to place and to move and position individual body parts. Muscular contractions also move body contents in the course of breathing, circulation, digestion, urination, defecation, and childbirth.
- **Communication.** Muscle contractions enable us to communicate by means of writing, speech, and body language such as facial expressions.
- **Stability.** Muscles maintain posture by resisting the pull of gravity. They hold articulating bones in place by maintaining tension on the tendons of joints.
- **Control of body openings.** Circular muscles called **sphincters** close the eyes, mouth, and other orifices; regulate the passage of stomach contents, bile, urine, and feces; and regulate the diameter of the pupil.
- **Heat production.** As much as 85% of our body heat is produced by contraction of the skeletal muscles. This heat is vital to the proper functioning of enzymes and therefore to all of our metabolism.

Table 11.1 Words Commonly Used to Name Muscles

Criterion	Term and Meaning	Examples of Usage
Size	Major (large)	Pectoralis major
	Maximus (largest)	Gluteus maximus
	Minor (small)	Pectoralis minor
	Minimus (smallest)	Gluteus minimus
	Longus (long)	Abductor pollicis longus
	Brevis (short)	Extensor pollicis brevis
Shape	Rhomboideus (rhomboidal)	Rhomboideus major
	Trapezius (trapezoidal)	Trapezius
	Teres (round, cylindrical)	Pronator teres
	Deltoid (triangular)	Deltoid
Location	Capitis (of the head)	Splenius capitis
	Pectoralis (of the chest)	Pectoralis major
	Abdominis (of the abdomen)	Rectus abdominis
	Femoris (of the femur)	Quadriceps femoris
	Brachii (of the arm)	Biceps brachii
	Intercostal (between the ribs)	External intercostals
	Digiti (of a finger or toe)	Extensor digiti minimi
	Digitorum (of fingers or toes, plural)	Flexor digitorum profundus
	Pollicis (of the thumb)	Opponens pollicis
	Hallucis (of the great toe)	Abductor hallucis
Relative Position	Lateral	Lateral pterygoid
	Medial	Medial pterygoid
	Internal	Internal intercostals
	External	External intercostals
	Superficialis (superficial)	Flexor digitorum superficialis
	Profundus (deep)	Flexor digitorum profundus
Number of Heads	Biceps (two heads)	Biceps brachii
	Triceps (three heads)	Triceps surae
	Quadriceps (four heads)	Quadriceps femoris
Orientation	Rectus (straight)	Rectus abdominis
	Transversus (transverse)	Transversus abdominis
	Oblique (slanted)	External abdominal oblique
Action	Adductor	Adductor pollicis
	Abductor	Abductor digiti minimi
	Flexor	Flexor carpi radialis
	Extensor	Extensor carpi radialis
	Pronator	Pronator teres
	Supinator	Supinator
	Levator	Levator scapulae
	Depressor	Depressor anguli oris

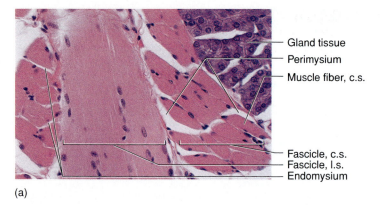

(a)

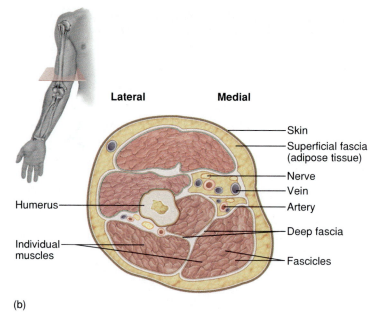

Lateral   Medial

- Skin
- Superficial fascia (adipose tissue)
- Nerve
- Vein
- Artery
- Deep fascia
- Fascicles

Humerus

Individual muscles

(b)

Figure 11.2 Connective tissues associated with skeletal muscle. (*a*) Muscle fascicles in the tongue (×400). Vertical fascicles passing between the dorsal and ventral surfaces of the tongue are seen alternating with cross-sectioned horizontal fascicles that pass from the tip to the rear of the tongue. A fibrous perimysium separates each fascicle from the others, while a fibrous endomysium separates each muscle cell within a fascicle. (*b*) Cross-section of the arm, showing muscle fascicles on a smaller scale, deep fascia separating neighboring muscles from each other, and superficial fascia separating the muscles from the skin. ⚚

These functions are collectively performed by skeletal, cardiac, and smooth muscle. The remainder of this chapter, however, concerns only the skeletal muscles.

Connective Tissues of a Skeletal Muscle

Skeletal muscle cells, or muscle *fibers,* are about 10 to 100 μm in diameter and up to 30 cm long. Each fiber is surrounded by a layer of areolar tissue called the **endomysium**[2] (EN-doe-MIZ-ee-um) that allows passage for blood capillaries and nerve fibers (fig. 11.2*a*).

Muscle fibers are grouped in bundles called **fascicles**[3] (FASS-ih-culs), which are visible to the naked eye as parallel strands. You can see these as the "grain" in a cut of meat; tender meat is most easily pulled apart along its fascicles. Each fascicle is separated from neighboring ones by a connective tissue sheath called the **perimysium.**[4] The muscle as a whole is surrounded by still another connective tissue layer, the **epimysium.**[5]

The epimysium grades imperceptibly into connective tissue sheets called the *fascia* (FASH-ee-uh)—**deep fascia** between neighboring muscle and **superficial fascia** (hypodermis) between muscle and skin. The superficial fascia is largely adipose in areas such as the buttocks and abdomen, but the deep fascia is devoid of fat (fig. 11.2*b*).

There are two ways a muscle can attach to a bone. In a **direct** (fleshy) **attachment,** collagen fibers of the epimysium are continuous with the periosteum of the bone; on gross examination, the red muscle tissue appears to emerge directly from the bone. The masseter muscle of the jaw shows this type of attachment to the zygomatic arch. In an **indirect attachment,** the collagen of the deep fascia continues as a strong fibrous tendon that merges into the periosteum of a nearby bone. The attachment of the biceps brachii to the scapula is one of many examples of this type. Since some collagen fibers of the periosteum continue into the bone matrix as *perforating (Sharpey's) fibers* (see chapter 8), there is a strong structural continuity of collagen fibers: endomysium → perimysium → epimysium → fascia → tendon → periosteum → bone matrix. Excessive stress is more likely to tear a tendon than pull it loose from the muscle or bone. The physiological relevance of this series of tissues is discussed in the next chapter.

In some cases, the fascia of one muscle attaches to the fascia or tendon of another or to collagen fibers of the dermis. The ability of a skeletal muscle to produce facial expressions depends on the latter type of attachment. Some muscles are connected to a broad, flat, sheetlike tendon called an **aponeurosis** (AP-oh-new-RO-sis).[6] This term originally referred to the tendon located beneath the scalp, but now it also refers to tendons associated with certain abdominal, lumbar, hand, and foot muscles (see fig. 11.15). In some places, groups of tendons from separate muscles pass under a band of connective tissue

---

2. *endo* = within
3. *fasc* = bundle + *icle* = little
4. *peri* = around
5. *epi* = above, upon

6. *apo* = above, upon + *neuro* = nerve

called a **retinaculum**[7] (RET-ih-NAC-you-lum). One of these covers each surface of the wrist like a bracelet, for example; the tendons of several forearm muscles pass under them on the way to the hand (see fig. 11.30).

General Anatomy of Skeletal Muscles

Most skeletal muscles are attached to a different bone at each end, so either the muscle or its tendon spans at least one joint. When the muscle contracts, it moves one bone relative to the other. A muscle attachment at the relatively stationary end is called the **origin,** or **head,** of the muscle, and an attachment at the relatively mobile end is its **insertion.** Many muscles are narrow at the origin and insertion and have a thicker middle region called the **belly** (figs. 11.3 and 11.4).

The strength of a muscle and the direction in which it pulls are determined partly by the orientation of its fascicles. Differences in orientation are the basis for classifying muscles into five types (fig. 11.4). These are good examples of the way form and function complement each other.

1. **Fusiform**[8] **muscles** are thick in the middle and tapered at each end. Their contractions are moderately strong. The *biceps brachii* of the arm and the *gastrocnemius* of the calf are examples of this type.
2. **Parallel muscles** are long, straplike muscles of uniform width and parallel fascicles. They can span a great distance and shorten more than other muscle types, but they are weaker than fusiform muscles. Examples include the *rectus abdominis* of the abdomen, *sartorius* of the thigh, and *zygomaticus major* and *minor* of the face.

3. **Convergent muscles** are fan-shaped—broad at the origin, with fascicles converging toward a narrower insertion. These are relatively strong muscles because the tension exerted by numerous fascicles is concentrated at this small insertion. The *pectoralis major* of the chest is an example of this type.
4. **Pennate**[9] **muscles** are feather-shaped. Their fascicles insert obliquely on a tendon that runs the length of the muscle, like the shaft of a feather. In a

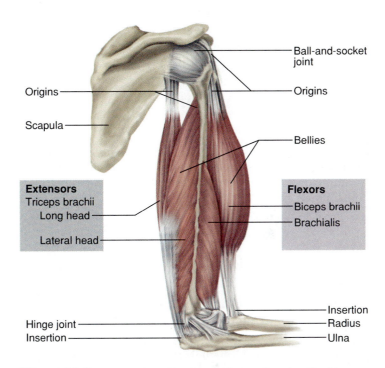

Figure 11.3 Synergistic and antagonistic muscle pairs. The biceps brachii and brachialis muscles act as synergists to flex the elbow. The triceps brachii is an antagonist of flexion and the prime mover in extension of the elbow. ⚡

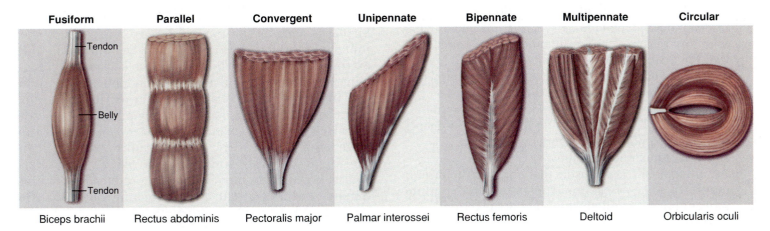

Figure 11.4 Classification of muscles according to the arrangement of their fascicles.

---
7. *retinac* = retainer, bracelet
8. "Spindle-shaped"
9. *penna* = feather

unipennate muscle, all fascicles approach the tendon from one side. The *palmar interosseous* muscles between the metacarpal bones and the *semimembranosus* muscle of the thigh are examples. In a **bipennate** muscle, fascicles approach the tendon from both sides. The *rectus femoris* of the thigh is of this type. **Multipennate** muscles, such as the deltoid of the shoulder, are shaped like a bunch of feathers whose quills converge toward a single point.

5. **Circular muscles (sphincters)** surround body openings. They are the weakest of all muscle types. Examples include the *orbicularis oris* of the lips and *orbicularis oculi* of the eyelids.

Coordinated Action of Muscle Groups

The movement produced by a muscle is called its **action.** Skeletal muscles seldom act independently; instead, they function in groups whose combined actions produce the coordinated motion of a joint. Muscles can be classified into at least four categories according to their actions, but it must be stressed that a particular muscle can act in a certain way during one joint action and in a different way during other actions of the same joint.

1. The **prime mover (agonist)** is the muscle that produces most of the force during a particular joint action. In flexing the elbow, for example, the prime mover is the biceps brachii.

2. A **synergist**[10] (SIN-ur-jist) is a muscle that works with the prime mover. Several synergists acting on a joint can produce more power than a single larger muscle could. The actions of a prime mover and its synergist are not necessarily identical and redundant, however. If the prime mover worked alone at a joint, it might cause rotation or other undesirable movements. A synergist may stabilize a joint and restrict these unintended movements, so that the action of the prime mover is more coordinated and specific. The actions of synergistic muscles also may modify the direction of movement at the joint and produce more accurate movements. The *brachialis,* for example, lies deep to the biceps brachii and works with it as a synergist to extend the elbow.

3. An **antagonist**[11] is a muscle that opposes the prime mover. In some cases, it may relax to give the prime mover almost complete control over an action. More often, however, both muscles work simultaneously and the antagonist moderates the

speed or range of the agonist, thus preventing excessive movement and joint injury. If you extend your arm to reach out and pick up a cup of tea, your triceps brachii is the prime mover and your biceps brachii acts as an antagonist to slow the extension and stop it at the appropriate point. If you extend your arm rapidly to throw a dart, the biceps must be quite relaxed at first, but still it contracts at the end of the motion to help prevent overextension of the arm.

The biceps and triceps brachii represent an **antagonistic pair** of muscles that act on opposite sides of a joint (see fig. 11.3). Antagonistic pairs are needed to produce opposite movements at a joint because a given muscle can only pull, not push— one muscle cannot flex and extend the elbow, for example. Which member of the pair acts as the agonist depends on the motion under consideration. In flexion of the elbow, for example, the biceps is the agonist and the triceps is the antagonist; when the elbow is extended, their roles are reversed.

4. A **fixator** is a muscle that prevents the movement of a bone. To *fix* a structure means to hold it steady, allowing another muscle attached to it to pull on something else. For example, consider again the flexion of the elbow by the biceps brachii. The biceps originates on the scapula and inserts on the radius. The scapula is very loosely attached to the axial skeleton, so when the biceps contracts, it seems likely that it would pull the scapula laterally. There are fixator muscles attached to the scapula, however, that contract at the same time. By holding the scapula firmly in place, they ensure that the force generated by the biceps moves the radius rather than the scapula.

Intrinsic and Extrinsic Muscles

In places such as the tongue, larynx, back, hand, and foot, anatomists distinguish between **intrinsic** and **extrinsic muscles.** An intrinsic muscle is entirely contained within a particular region, having both its origin and insertion within that region; an extrinsic muscle, however, acts upon a designated region but has its origin elsewhere. For example, some movements of the fingers are produced by extrinsic muscles located in the forearm, whose long tendons reach to the phalanges; other finger movements are produced by the intrinsic muscles of the hand, located between the metacarpal bones.

Innervation

Innervation refers to the nerve supply to an organ. Knowing the innervation to each skeletal muscle

---

10. *syn* = together + *erg* = work
11. *ant, anti* = against + *agonist* = competitor

enables clinicians to diagnose nerve and spinal cord injuries from their effects on muscle function and to set realistic goals for rehabilitation.

In the muscle tables of this chapter, the nerve supply to each muscle is identified. This information will be more meaningful after you have studied the peripheral nervous system (see chapter 15), but a brief orientation will be helpful here. Muscles of the head and neck are supplied by *cranial nerves* that arise from the base of the brain and emerge through the skull foramina. *Spinal nerves* originate in the spinal cord, emerge through the intervertebral foramina, and then branch into a *dorsal (posterior) ramus*[12] and a *ventral (anterior) ramus.* They are identified by letters and numbers that refer to the vertebrae—for example, T6 for the sixth thoracic nerve and S2 for the second sacral nerve. You will note references to nerve numbers and rami in many of the muscle tables. The term *plexus* in some of the tables refers to certain weblike networks of spinal nerves adjacent to the vertebral column.

<hr>

Key Point Review

3. List some of the functions of muscles other than moving the body or its individual parts.
4. Describe four layers of fibroconnective tissue associated with a skeletal muscle.
5. Explain why a muscle does not easily pull loose from a bone, even under great stress.
6. Define the terms *origin, insertion, belly, action,* and *innervation.*
7. Describe the five basic shapes of muscles.
8. Explain the function of prime movers, synergists, antagonists, and fixators.

Muscles of the Head and Neck

▼Objectives

When you have completed this section, you should be able to
- name and locate the muscles that produce facial expressions;
- name and locate the muscles used for chewing and swallowing;
- name and locate the neck muscles that move the head; and
- state the origin, insertion, and action of any head or neck muscles.

Muscles of Facial Expression

One of the most striking contrasts between a human face and that of a rat, horse, or dog, for example, is the variety and subtlety of human facial expression. This is made possible by a complex array of small muscles that insert in the dermis. Contraction of these muscles tenses the skin and produces effects as diverse as a pleasant smile, a threatening scowl, a puzzled frown,

and a flirtatious wink (fig. 11.5). The muscles of facial expression are enormously important in nonverbal communication and add subtle shades of meaning even to our spoken words.

We will briefly "tour" the scalp and face to get a general idea of the locations and actions of these muscles (figs. 11.6 and 11.7). Their origins, insertions, and innervations are listed in table 11.2, and the pronunciation of their names is indicated. All but one are innervated by branches of the facial nerve (cranial nerve VII). Because of its location, lacerations and skull fractures can easily damage this nerve, paralyze the innervated muscles, and cause parts of the face to sag.

The **frontalis** lies under the skin of the forehead, and the **occipitalis** is at the rear of the head. They are connected by a broad aponeurosis, the **galea aponeurotica**[13] (GAY-lee-uh AP-oh-new-ROT-ih-cuh) underlying the scalp of the parietal region (fig. 11.6). Collectively, these scalp muscles are also known as the **epicranius**[14] or **occipitofrontalis.** They move the scalp, forehead skin, and eyebrows.

Each orbit is encircled by the **orbicularis oculi,**[15] a sphincter of the eyelid that closes the eye (figs. 11.6 and 11.7). The eye is opened mainly by the **levator palpebrae superioris,**[16] which lies deep to the orbicularis oculi in the eyelid and roof of the orbit. Other muscles of the orbital and nasal regions—the **corrugator supercilii,**[17] **procerus,**[18] and **nasalis**—are adequately described in table 11.2. Movements of the eye itself are produced by muscles in the orbit, described in chapter 16.

The mouth is the most expressive part of the face, so it is not surprising that the muscles here are especially diverse. It is surrounded by a sphincter, the **orbicularis oris,**[19] which is approached from all directions by several other muscles. Beginning near the nose and progressing laterally, the first is a little muscle with a big name—the **levator labii superioris alaeque nasi** ("levator of the upper lip and wings of the nose"). It is a narrow slip of muscle that originates near the medial corner of the eye and passes vertically alongside the nose. Lateral to this muscle is the **levator labii**[20] **superioris,** a triangular muscle that originates at the middle of the orbicularis oculi. Next is the **zygomaticus**[21] **minor,** which originates near the lateral corner of the eye. All three of these converge, fanlike, on the orbicularis oris. The **zygomaticus major** originates in front of the ear

<hr>

12. *ramus* = branch

13. *galea* = helmet + *apo* = above + *neuro* = nerves, the brain
14. *epi* = above + *crani* = skull, cranium
15. *orb* = circle + *ocul* = eye
16. *levator* = that which raises + *palpebr* = eyelid + *superior* = upper
17. *corrug* = wrinkle + *supercili* = eyebrow
18. *procer* = tall, long, slender
19. *oris* = mouth
20. *labi* = lip
21. Refers to the zygomatic arch

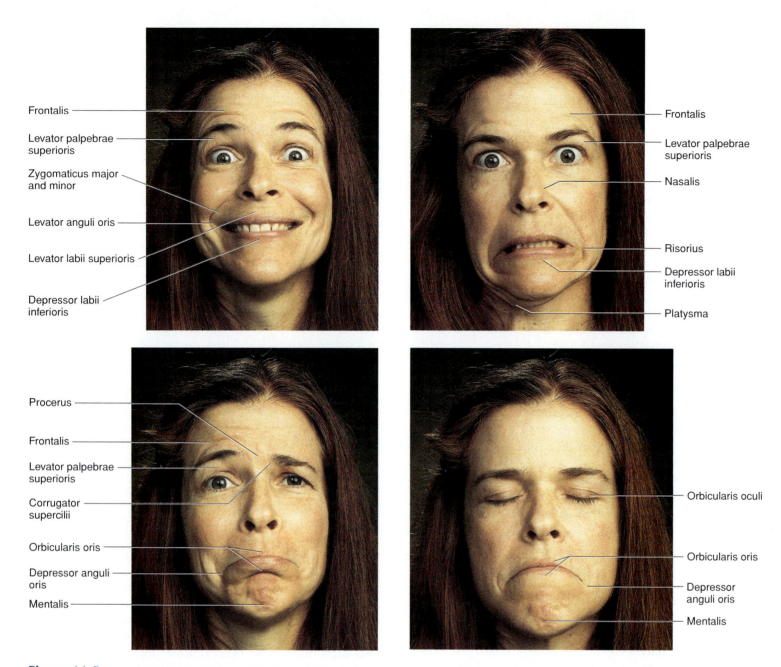

Figure 11.5 Expressions produced by several of the facial muscles.

and inserts on the superolateral corner of the mouth. The **risorius,** a horizontal muscle whose name means "laughter," inserts at the junction of the upper and lower lips. From their insertions, you can probably guess that these five muscles draw the upper lip upward and laterally in such expressions as smiling and laughing.

Along the lower lip are muscles that draw it downward. The most lateral is the **depressor anguli oris,**[22] also known as the *triangularis* because of its

shape. Lying deep to it and a little more medially is the **depressor labii inferioris.** Most medially, near the mental symphysis, is a pair of tiny **mentalis**[23] muscles. Unlike the other two, these do not depress the lip. They originate on the mandible and insert in the dermis of the chin. When they contract, they pull the soft tissues of the chin upward, which pushes the lower lip outward and creates a pouting expression. People with especially thick mentalis muscles have a groove between them, the *mental cleft,* externally visible as a dimple of the chin.

22. *depressor* = that which lowers + *angul* = corner, angle + *oris* = mouth

23. *mental* = chin

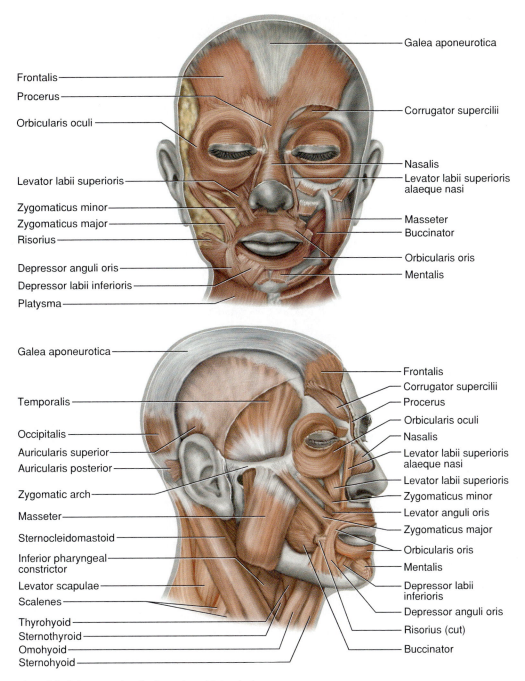

Frontalis
Procerus
Orbicularis oculi

Levator labii superioris

Zygomaticus minor
Zygomaticus major
Risorius

Depressor anguli oris
Depressor labii inferioris
Platysma

Galea aponeurotica

Corrugator supercilii

Nasalis
Levator labii superioris
alaeque nasi

Masseter
Buccinator

Orbicularis oris
Mentalis

Galea aponeurotica

Temporalis

Occipitalis
Auricularis superior
Auricularis posterior

Zygomatic arch

Masseter

Sternocleidomastoid

Inferior pharyngeal
constrictor
Levator scapulae

Scalenes

Thyrohyoid
Sternothyroid
Omohyoid
Sternohyoid

Frontalis
Corrugator supercilii
Procerus

Orbicularis oculi
Nasalis
Levator labii superioris
alaeque nasi

Levator labii superioris
Zygomaticus minor
Levator anguli oris
Zygomaticus major

Orbicularis oris
Mentalis

Depressor labii
inferioris
Depressor anguli oris

Risorius (cut)

Buccinator

Figure 11.6 The muscles of facial expression in frontal and lateral views.

The **buccinator**[24] muscles in the cheeks push food between the teeth to be chewed. The name literally means "trumpeter"—if the cheeks are inflated with air, compression of the buccinator muscles blows it out. If the buccinators are fully contracted to draw in the cheeks and then relaxed, the cheeks expand and produce suction—an especially important muscle action for nursing infants. To appreciate this action, hold your fingertips lightly on your cheeks as you make a kissing noise. You will feel the relaxation of the buccinators at the moment air is sharply drawn in through the pursed lips.

A broad superficial muscle called the **platysma**[25] arises from the shoulder and upper chest and inserts broadly along the mandible and the overlying skin. It depresses the mandible, widens the mouth, and tenses the skin of the neck (for example, when men shave).

---

24. *bucc* = cheek; *buccinator* = trumpeter

25. *platy* = flat

Table 11.2 Muscles of Facial Expression (see figs. 11.6 and 11.7)

O = origin, I = insertion, and N = innervation; n. and nn. = nerve and nerves

Muscles of the Scalp

Epicranius (EP-ih-CRAY-nee-us), or Occipitofrontalis (oc-SIP-ih-toe-frun-TAY-liss)

Frontalis

When occipitalis fixes galea aponeurotica, frontalis raises eyebrows and creates horizontal wrinkles in forehead; when occipitalis is relaxed, frontalis draws scalp forward

O: galea aponeurotica   I: skin of forehead   N: facial n.

Occipitalis

Retracts scalp; fixes galea aponeurotica

O: superior nuchal line   I: galea aponeurotica   N: facial n.

Muscles of the Face

Orbicularis Oculi (or-BIC-you-LERR-iss OC-you-lye)

Closes eye, as in blinking, winking, squinting, and during sleep; compresses lacrimal gland and promotes flow of tears to moisten eye

O: medial wall of orbit   I: eyelids   N: facial n.

Levator Palpebrae (leh-VAY-tur pal-PEE-bree) Superioris

Raises upper eyelid; opens eye

O: roof of orbit   I: skin of upper   N: oculomotor n.
                              eyelid

Corrugator Supercilii (COR-oo-GAY-tur SOO-per-SIL-ee-eye)

Medially depresses eyebrows and draws them closer together; wrinkles skin between eyebrows

O: superciliary ridge   I: skin of eyebrow   N: facial n.

Procerus (pro-SEE-rus)

Wrinkles skin between eyebrows

O: skin over bridge of nose I: skin of forehead   N: facial n.

Nasalis (nay-SAY-liss)

One part widens nostrils; another depresses nasal cartilages and compresses nostrils

O: maxilla and nasal   I: bridge and alae   N: facial n.
   cartilages   of nose

Orbicularis Oris

Closes lips; protrudes lips as in kissing; aids in speech

O: muscle fibers   I: mucous membranes   N: facial n.
   around mouth   of lips

Levator Labii Superioris Alaeque Nasi (leh-VAY-tur LAY-bee-eye soo-PEER-ee-or-is ay-LEE-quee NAY-zye)

Elevates upper lip, flares nostrils

O: maxilla   I: skin of upper lip   N: facial n.
                              and alae of nose

Levator Labii Superioris

Elevates upper lip

O: zygomatic bone   I: skin of upper lip   N: facial n.
   and maxilla

Levator Anguli (ANG-you-lye) Oris

Elevates corners of the mouth, as in smiling and laughing

O: maxilla   I: superior corner   N: facial n.
                              of mouth

Zygomaticus (ZY-go-MAT-ih-cus) Minor and Major

Draw corners of mouth laterally and upward, as in smiling and laughing

O: zygomatic bone   I: superolateral corner N: facial n.
                              of mouth

Risorius (rih-SOR-ee-us)

Draws angle of mouth laterally, as in grimacing

O: fascia anterior to ear I: skin at angle   N: facial n.
                              of mouth

Depressor Anguli Oris (= triangularis)

Depresses corners of mouth, as in frowning

O: mandible   I: inferolateral corner N: facial n.
                              of mouth

Depressor Labii Inferioris

Depresses lower lip

O: mandible near symphysis I: skin of lower lip   N: facial n.

Mentalis (men-TAY-lis)

Pulls skin of chin upward; elevates and protrudes lower lip, as in pouting

O: mandible near symphysis I: skin of chin   N: facial n.

Buccinator (BUCK-sin-AY-tur)

Compresses cheek; pushes food between teeth; expels air or liquid from mouth; creates suction

O: lateral aspects of maxilla I: orbicularis oris N: facial n.
   and mandible

Platysma (plah-TIZ-muh)

Depresses mandible; widens mouth; tenses skin of neck

O: fascia of deltoid and   I: mandible, skin of   N: facial n.
   pectoralis major muscles   lower face, and muscles
                              at angle of mouth

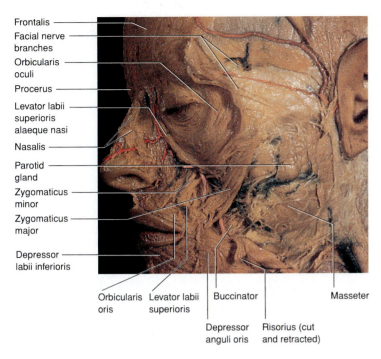

Frontalis
Facial nerve branches
Orbicularis oculi
Procerus
Levator labii superioris alaeque nasi
Nasalis
Parotid gland
Zygomaticus minor
Zygomaticus major
Depressor labii inferioris

Orbicularis oris   Levator labii superioris   Buccinator   Masseter

Depressor anguli oris   Risorius (cut and retracted)

Figure 11.7 Some muscles of facial expression seen in a dissection of the cadaver.

Muscles of Chewing and Swallowing

The following muscles contribute to facial expression and speech but are concerned primarily with manipulation of food, including tongue movements, chewing (mastication), and swallowing (table 11.3).

The tongue is a very agile organ. Both intrinsic and extrinsic muscle groups (fig. 11.8) are responsible for its complex movements. The intrinsic muscles consist of variable numbers of vertical muscles that extend from the superior to inferior side of the tongue, transverse muscles that extend from left to right, and longitudinal muscles that extend from root to tip. The extrinsic muscles connect the tongue to other structures in the head and neck. These include the **genioglossus,**[26] **hyoglossus, styloglossus,** and **palatoglossus.** The last three names refer to the bony origins and insertions listed in table 11.3. The tongue and buccinator muscle shift food into position between the molars for grinding. The tongue presses chewed food against the palate to form a

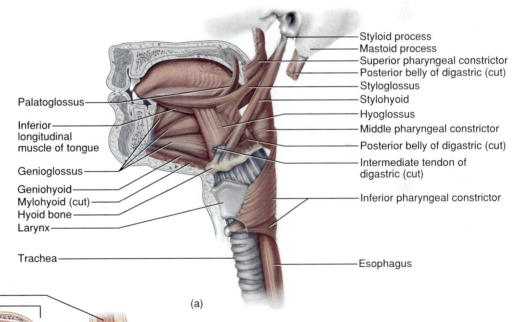

Palatoglossus
Inferior longitudinal muscle of tongue
Genioglossus
Geniohyoid
Mylohyoid (cut)
Hyoid bone
Larynx
Trachea

Styloid process
Mastoid process
Superior pharyngeal constrictor
Posterior belly of digastric (cut)
Styloglossus
Stylohyoid
Hyoglossus
Middle pharyngeal constrictor
Posterior belly of digastric (cut)
Intermediate tendon of digastric (cut)
Inferior pharyngeal constrictor
Esophagus

(a)

Buccinator
Tongue
Superior longitudinal muscle
Styloglossus
Vertical and transverse muscles
Inferior longitudinal muscle
Hyoglossus
Genioglossus
Mylohyoid
Submandibular salivary gland
Mandible
Hyoid bone

(b)

Figure 11.8 Muscles of the tongue and pharynx. (*a*) Lateral view of the extrinsic muscles of the tongue and constrictors of the pharynx. (*b*) Anterior view of a frontal section of the tongue and mandible behind the first molar tooth, showing the intrinsic and some extrinsic tongue muscles.

26. *genio* = chin + *gloss* = tongue

Table 11.3 Muscles of Chewing and Swallowing (see figs. 11.8–11.10)

O = origin, I = insertion, and N = innervation; n. and nn. = nerve and nerves

Extrinsic Muscles of the Tongue

Genioglossus (JEE-nee-oh-GLOSS-us)

Depresses and protrudes tongue; creates groove in tongue that enables infant to grasp nipple and channel milk to pharynx

| O: mental spines of mandible | I: hyoid bone and lateral aspect of tongue | N: hypoglossal n. |

Hyoglossus

Depresses sides of tongue

| O: body and greater cornu of hyoid | I: lateral aspect of tongue | N: hypoglossal n. |

Styloglossus

Elevates and retracts tongue

| O: styloid process | I: lateral aspect of tongue | N: hypoglossal n. |

Palatoglossus

Elevates posterior part of tongue; constricts fauces (opening from oral cavity to pharynx)

| O: soft palate | I: lateral aspect of tongue | N: accessory n. |

Muscles of Mastication

Temporalis (TEM-poe-RAY-liss)

Elevates mandible for biting and chewing; retracts mandible

| O: temporal lines | I: coronoid process | N: trigeminal n., mandibular branch |

Masseter (ma-SEE-tur)

Elevates mandible; produces forceful bite and some lateral excursion

| O: zygomatic arch | I: lateral surface of mandibular ramus and angle | N: trigeminal n., mandibular branch |

Medial Pterygoid (TERR-ih-goyd)

Elevates mandible; produces lateral excursion

| O: pterygoid process of sphenoid | I: medial aspect of mandibular angle | N: trigeminal n., mandibular branch |

Lateral Pterygoid

Protracts mandible; produces lateral excursion

| O: pterygoid process of sphenoid | I: slightly anterior to mandibular condyle | N: trigeminal n., mandibular branch |

Pharyngeal (fah-RIN-jee-ul) Constrictors

Constrict pharynx to force food bolus into esophagus

| O: *superior*—mandible and medial pterygoid plate; *middle*—hyoid; *inferior*—cartilages of larynx | I: posterior median raphe (a fibrous seam) of pharynx | N: vagus and glossopharyngeal nn. |

Hyoid Muscles, Suprahyoid Group

Digastric

Retracts mandible; elevates and fixes hyoid; depresses mandible when hyoid is fixed

| O: mastoid notch and inner aspect of mandible near symphysis | I: lesser cornu of hyoid via fascial sling | N: facial and mylohyoid nn. |

Geniohyoid (JEE-nee-oh-HY-oyd)

Elevates and protracts hyoid; dilates pharynx to receive food; depresses mandible when hyoid is fixed

| O: posterior (inner) aspect of mental symphysis | I: body of hyoid | N: ansa cervicalis |

Mylohyoid

Forms floor of mouth; elevates hyoid; depresses mandible when hyoid is fixed

| O: inferior margin of mandible | I: body of hyoid | N: trigeminal n., mandibular branch |

Stylohyoid

Elevates hyoid

| O: styloid process | I: body of hyoid | N: facial n. |

Hyoid Muscles, Infrahyoid Group

Omohyoid

Depresses hyoid; fixes hyoid during depression of mandible

| O: superior border of scapula | I: hyoid | N: ansa cervicalis |

Sternohyoid

Depresses hyoid; fixes hyoid during depression of mandible

| O: manubrium and costal cartilage 1 | I: hyoid | N: ansa cervicalis |

Sternothyroid

Depresses larynx; fixes hyoid during depression of mandible

| O: manubrium and costal cartilage 1 or 2 | I: thyroid cartilage of larynx | N: ansa cervicalis |

Thyrohyoid

Depresses hyoid; elevates larynx; fixes hyoid during depression of mandible

| O: thyroid cartilage of larynx | I: greater cornu of hyoid | N: ansa cervicalis |

soft, sticky mass called a *bolus* and then pushes it into the pharynx (throat) for swallowing.

There are four paired muscles of mastication: the temporalis, masseter, and lateral and medial pterygoids (fig. 11.9). The **temporalis,** named for the temporal bone, is a broad, fan-shaped muscle that arises from the temporal lines of the skull, passes behind the zygomatic arch, and inserts on the coronoid process of the mandible (fig. 11.9*a*). The **masseter**[27] is shorter and su-

perficial to the temporalis, arising from the zygomatic arch and inserting on the outer surface of the angle of the mandible (see fig. 11.6). It is a thick muscle easily palpated on the side of your jaw. The temporalis and masseter elevate the mandible to bite and chew food; they are two of the most powerful muscles in the body. Similar action is provided by the **medial** and **lateral pterygoid muscles.** They arise from the pterygoid processes of the sphenoid bone and insert on the inner surface of the mandible (fig. 11.9*b*). The pterygoids elevate and protract the mandible and produce the lateral excursions used to grind food between the molars. (See also special topic 11.1.)

Several of the actions of chewing and swallowing are aided by a group of eight **hyoid muscles** attached to the hyoid bone. Four of them, superior to the hyoid, are called the **suprahyoid group**—the *digastric, geniohyoid, mylohyoid,* and *stylohyoid.* (See fig. 11.8*a* for the geniohyoid and fig. 11.10 for the others.) The four that are inferior to the hyoid constitute the **infrahyoid group**—the *omohyoid, sternohyoid, sternothyroid,* and *thyrohyoid.* Most of the hyoid muscles receive their innervation from the *ansa cervicalis,* a loop of nerve at the side of the neck formed by certain fibers of the first through third cervical nerves.

The **digastric**[28] muscle arises from the mastoid process and thickens into a *posterior belly* beneath the margin of the mandible. It then narrows, passes through a connective tissue loop (*fascial sling*) attached to the hyoid bone, widens into an *anterior belly,* and attaches to the mandible near the symphysis. When it contracts, it pulls on the sling like a basket handle and elevates the hyoid bone. If the hyoid bone is fixed by the infrahyoid muscles, however, the digastric muscle opens the mouth. The mouth normally opens by itself when the temporalis and masseter muscles are relaxed,

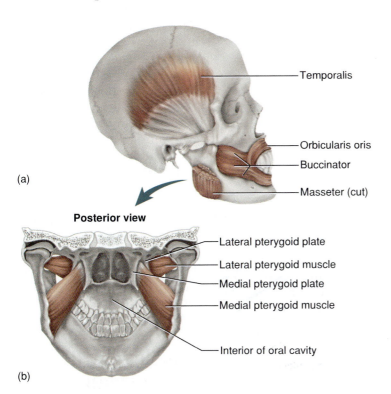

Temporalis

Orbicularis oris

Buccinator

Masseter (cut)

(a)

Posterior view

Lateral pterygoid plate

Lateral pterygoid muscle

Medial pterygoid plate

Medial pterygoid muscle

Interior of oral cavity

(b)

Figure 11.9 Muscles of chewing. (*a*) The masseter and temporalis muscles, lateral view. (*b*) Posterior view of the oral cavity showing the pterygoid muscles.

Special Topic Discovery of a New Muscle 11.1

New discoveries in physiology are an everyday occurrence, but you would think all the skeletal muscles of the human body had been discovered long ago. Some have even assumed that human gross anatomy is a completed science, a "dead discipline." Thus a great deal of excitement attended the 1996 announcement of a new muscle of mastication discovered by U.S. dentists Gary Hack and Gwendolyn Dunn.

Hack and Dunn were studying the muscles of mastication using an unorthodox dissection technique in which they entered the head from the front rather than from the side. "There it was," Hack said, "just staring at us"—a muscle, extending from the greater wing of the sphenoid to the medial side of the mandible, that everyone else had either overlooked or dismissed as part of the temporalis or medial pterygoid. Hack and Dunn named it the **sphenomandibularis.** Unlike the temporalis or pterygoids, it is innervated by the maxillary branch of the trigeminal nerve.

In chapter 1 we saw that some of history's greatest advances in scientific thinking came from people with the imagination to view things from a different angle than everyone else had done. In the discovery of the sphenomandibularis, we see that even little steps are made this way, and even the "finished" sciences hold surprises for people with imaginative approaches.

27. *masset* = to chew

28. *di* = two + *gastr* = belly

but the digastric can open it more widely. The **geniohyoid** muscle protracts the hyoid bone to widen the pharynx when food is swallowed. The **mylohyoid**[29] muscles fuse at the midline, form the floor of the mouth, and work synergistically with the digastric to forcibly open the mouth. The **stylohyoid,** named for its origin and insertion, elevates the hyoid bone.

When food enters the pharynx, the **superior, middle,** and **inferior pharyngeal constrictors** contract in that order and force the bolus downward, into the esophagus. The **thyrohyoid** muscle, named for the hyoid bone and large *thyroid cartilage* of the larynx, helps to prevent choking. It elevates the thyroid cartilage so that the larynx becomes sealed by a flap of tissue, the epiglottis. You can feel this effect by placing your fingers on your "Adam's apple" (a prominence of the thyroid cartilage) and feeling it bob up as you swallow. The **sternothyroid** muscle then pulls the larynx down again. These infrahyoid muscles that act on the larynx are called the extrinsic muscles of the larynx. The larynx also has intrinsic muscles, which are concerned with control of the vocal cords and laryngeal opening (see chapter 22).

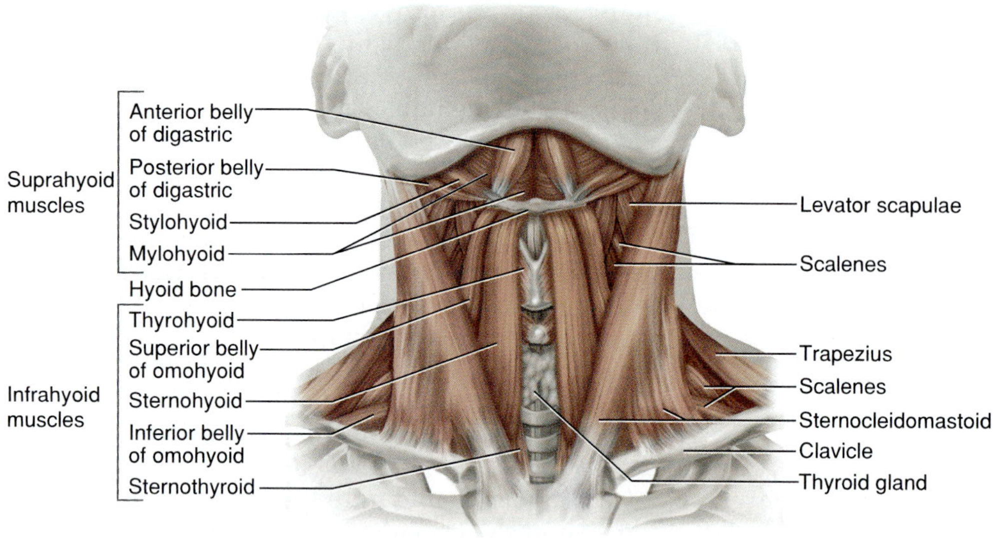

Figure 11.10 The hyoid muscles, anterior view. The geniohyoid is deep to the mylohyoid and can be seen in figure 11.8a.

Table 11.4 Muscles Acting on the Head (see figs. 11.11–11.13)

O = origin, I = insertion, and N = innervation; n. and nn. = nerve and nerves

Flexors

Sternocleidomastoid (STIR-no-CLY-doe-MASS-toyd)

Contraction of both draws head forward and down, as in looking between the feet; contraction of only one draws head down and to the side opposite the contracting muscle

O: clavicle and manubrium	I: mastoid process	N: accessory n.

Scalenes (SCAY-leens)

Elevate ribs 1 and 2 in inspiration; flex neck laterally

O: transverse processes of vertebrae C2–C6	I: ribs 1 and 2, anterolateral aspect	N: C3 and C4

Extensors

Trapezius (tra-PEE-zee-us)

Abducts and extends head (see other functions in table 11.9)

O: external occipital protuberance, ligamentum nuchae, and spinous processes of vertebrae C7–T12	I: clavicle, acromion and scapular spine	N: accessory n.

Splenius Capitis (SPLEE-nee-us CAP-ih-tis)

Rotates head; prime mover of neck extension

O: spinous processes of vertebrae C4–T6	I: mastoid process and superior nuchal line	N: middle and lower cervical nn.

Semispinalis (SEM-ee-spy-NAY-liss) Capitis

Rotates and extends head (see other parts of semispinalis in table 11.7)

O: transverse processes of vertebrae T1–T6; articular processes of C4–C7	I: occipital bone	N: dorsal rami of cervical nn.

Muscles Acting on the Head

Muscles that move the head originate on the spinal column, thoracic cage, and pectoral girdle and insert on the cranial bones (table 11.4). The principal flexors of the head are the **sternocleidomastoid**[30] and the **anterior, middle,** and **posterior scalenes** (fig. 11.11a). The prime mover is the sternocleidomastoid, a thick cordlike muscle that extends from the sternum and clavicle to the mastoid process behind the ear. It is most easily seen and palpated when the head is turned to one side and

29. *mylo* = mill, molar teeth

30. *sterno* = sternum + *cleido* = clavicle + *mastoid* = mastoid process

slightly extended. As it passes obliquely across the neck, the sternocleidomastoid divides it into **anterior** and **posterior triangles.** Other muscles and landmarks subdivide each of these into smaller triangles that are important in neck surgery (fig. 11.11*b*).

When both sternocleidomastoids contract, the neck flexes forward; for example, when you look down at something between your feet. When only the left one contracts, the head tilts down and to the right, and when the right one acts alone, it draws the head down and to the left. This may be a little hard to understand at first, but you can visualize it this way: Hold the index finger of your left hand on your left mastoid process and the index finger of your right hand on your sternal notch. Now contract the left sternocleidomastoid in a way that brings the two fingertips as close together as possible. You will note that this action causes you to look downward and to the right.

The extensors of the head are located in the back of the neck. Their actions include extension (holding the head erect), hyperextension (as in looking upward toward the sky), abduction (tilting the head to one side), and rotation (as in looking to the left and right). Extension and hyperextension involve equal action of the right and left members of a pair; the other actions require the muscle on one side to contract more strongly than the opposite muscle. Many head movements result from a combination of these actions—for example, looking up over the shoulder involves a combination of rotation and extension.

We will consider only three primary extensors, the trapezius, splenius capitis, and semispinalis capitis (figs. 11.12 and 11.13). The **trapezius** is a vast triangular muscle of the back and neck; together, the right and left trapezius muscles form a trapezoid. The origin of the trapezius extends from the occipital protuberance of the skull to the twelfth thoracic vertebra. It converges to an insertion on the shoulder. The **splenius**[31] **capitis,** which lies just deep to the trapezius on the neck, has oblique fascicles that diverge from the spinal column toward the ears. It is nicknamed the "bandage muscle" because of the way it tightly binds deeper neck muscles. The **semispinalis capitis** muscle is slightly deeper, and its fascicles travel vertically up

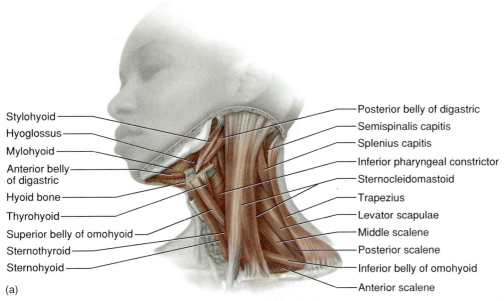

Stylohyoid
Hyoglossus
Mylohyoid
Anterior belly of digastric
Hyoid bone
Thyrohyoid
Superior belly of omohyoid
Sternothyroid
Sternohyoid

Posterior belly of digastric
Semispinalis capitis
Splenius capitis
Inferior pharyngeal constrictor
Sternocleidomastoid
Trapezius
Levator scapulae
Middle scalene
Posterior scalene
Inferior belly of omohyoid
Anterior scalene

(a)

Anterior triangles
A1. Muscular
A2. Carotid
A3. Submandibular
A4. Suprahyoid

Posterior triangles
P1. Occipital
P2. Subclavian

Sternocleidomastoid
Scalenes

(b)

Figure 11.11 Muscles of the neck. (*a*) Superficial muscles, lateral view. (*b*) Surface view of the triangles of the neck.

the back of the neck to insert on the occipital bone. A complex array of smaller, deeper extensors are synergists of the primary extensors; they extend the head, rotate it, or both.

Think About It

Of the muscles you have studied so far, name three that you would consider intrinsic muscles of the head and three that you would classify as extrinsic. Give the reasons for your classification.

---

31. *spleni* = bandage

Muscles of Respiration

The lungs are ventilated primarily by muscles that enclose the thoracic cavity—the *diaphragm*, which forms its floor; 11 pairs of *external intercostal muscles* between the ribs; and 11 pairs of *internal intercostal muscles* between the ribs deep to the external intercostals (fig. 11.14 and table 11.5). The lungs themselves contain very little muscle tissue; they do not play an active part in their own ventilation.

The **diaphragm**[32] is a muscular partition between the abdominal and thoracic cavities. It is domed upward and has openings that allow the esophagus and major blood vessels to pass through it. Its fascicles converge from the margins toward a fibrous **central tendon** in the middle. When the diaphragm contracts, it flattens slightly, increasing the volume of the thoracic cage and creating a partial vacuum that draws air into the lungs. Contraction of the diaphragm also raises pressure in the abdominal cavity below, thus helping to expel the contents of the bladder and rectum and facilitating childbirth—which is why an individual may take a deep breath and hold it during these functions.

The **external intercostal**[33] muscles extend obliquely downward and anteriorly from each rib to the rib below it. When the scalene muscles fix the first rib, the external intercostals lift the others. Each rib is pulled up somewhat like the handle of a bucket, which pulls the ribs closer together and draws the entire rib cage upward and outward. Thus, the thoracic cage enlarges, promoting inhalation.

It takes no muscular effort to exhale—when the diaphragm and external intercostals relax, the thoracic cage springs back to its prior size and expels the air. However, forced expiration—exhaling more than the usual amount of air or exhaling quickly as in blowing out a candle—is achieved mainly by the **internal intercostal muscles.** These also extend from one rib to the next, but they lie deep to the external intercostals and have fascicles at right angles to them. The abdominal muscles also aid in forced expiration by pushing the viscera up against the diaphragm.

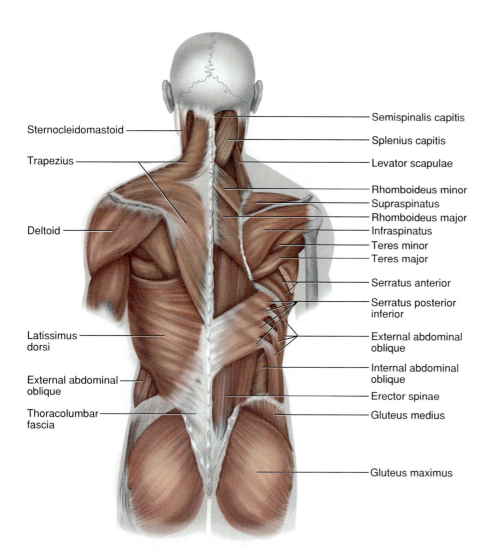

Figure 11.12 Muscles of the neck, back, and gluteal region. The most superficial muscles are shown on the left, and the next deeper layer on the right.

Sternocleidomastoid

Trapezius

Deltoid

Latissimus dorsi

External abdominal oblique

Thoracolumbar fascia

Semispinalis capitis

Splenius capitis

Levator scapulae

Rhomboideus minor

Supraspinatus

Rhomboideus major

Infraspinatus

Teres minor

Teres major

Serratus anterior

Serratus posterior inferior

External abdominal oblique

Internal abdominal oblique

Erector spinae

Gluteus medius

Gluteus maximus

Key Point Review

9 What are the two divisions of the epicranius and what do they do?

10 Name any two muscles that elevate the upper lip and two that depress the lower lip.

11 Name the four muscles of mastication and state where they insert on the mandible.

12 Distinguish between the functions of suprahyoid and infrahyoid muscles.

13 Identify the prime movers of head flexion and extension.

Muscles of the Trunk

▼Objectives

When you have completed this section, you should be able to
- name and locate the muscles that ventilate the lungs and explain how they affect abdominal pressure; and
- name and locate the muscles of the abdominal wall, spinal column, and pelvic floor.

---

32. *dia* = across + *phragm* = partition
33. *inter* = between + *costa* = rib

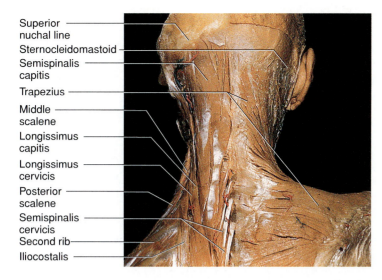

- Superior nuchal line
- Sternocleidomastoid
- Semispinalis capitis
- Trapezius
- Middle scalene
- Longissimus capitis
- Longissimus cervicis
- Posterior scalene
- Semispinalis cervicis
- Second rib
- Iliocostalis

Figure 11.13 Muscles of the dorsal shoulder and nuchal regions.

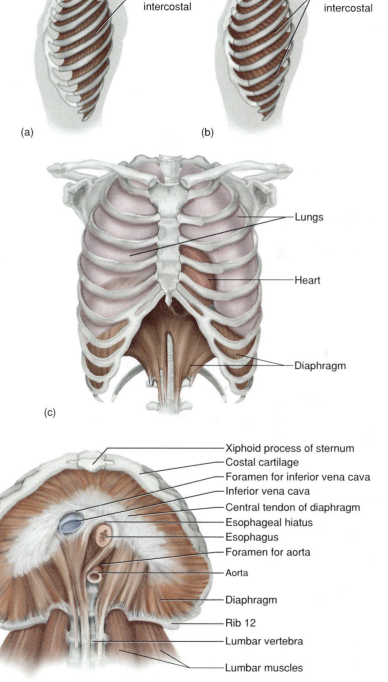

(a) External intercostal

(b) Internal intercostal

(c)
- Lungs
- Heart
- Diaphragm

(d)
- Xiphoid process of sternum
- Costal cartilage
- Foramen for inferior vena cava
- Inferior vena cava
- Central tendon of diaphragm
- Esophageal hiatus
- Esophagus
- Foramen for aorta
- Aorta
- Diaphragm
- Rib 12
- Lumbar vertebra
- Lumbar muscles

Figure 11.14 Muscles of respiration. (*a*) Lateral view of the external intercostal muscles between the ribs. (*b*) Lateral view of the internal intercostal muscles, between the ribs and deep to the external intercostals. (*c*) Frontal view of the diaphragm in relation to the rib cage and spinal column. (*d*) Inferior view of the diaphragm.

Think About It

If you eat spare ribs, what muscles are you eating?

Abdominal Muscles

The anterior and lateral walls of the abdomen exhibit four pairs of sheetlike muscles (figs. 11.15 and 11.16) that support the viscera, stabilize the spinal column during heavy lifting, and aid in respiration, urination, defecation, vomiting, and childbirth. They are the *rectus abdominis, external oblique, internal oblique,* and *transversus abdominis (table 11.6).*

The **rectus**[34] **abdominis** is a medial straplike muscle extending vertically from the pubis to the sternum. It is separated into four segments by transverse **tendinous insertions** that give the abdomen a segmented appearance in well-muscled individuals. The rectus abdominis is enclosed in a sleeve of fibroconnective tissue called the **rectus sheath,** and the right and left rectus muscles are separated by a vertical fibrous strip called the **linea alba.**[35]

34. *rect* = straight
35. *linea* = line + *alb* = white

Table 11.5 Muscles of Respiration (see fig. 11.14)

O = origin, I = insertion, and N = innervation; n. and nn. = nerve and nerves

Diaphragm (DY-uh-fram)

Prime mover of inspiration; compresses abdominal viscera to help expel contents, as in urination, defecation, and childbirth

O: xiphoid process, costal cartilages 5–9, ribs 10–12, and lumbar vertebrae	I: central tendon	N: phrenic n.

External Intercostals (IN-tur-COSS-tulz)

When rib 1 is fixed by scalenes, external intercostals draw ribs 2–12 upward and outward to expand thoracic cavity and inflate lungs

O: inferior margins of ribs 1–11	I: superior margins of ribs 2–12	N: intercostal nn.

Internal Intercostals

When rib 12 is fixed by quadratus lumborum and other muscles, internal intercostals draw the ribs downward and inward to compress thoracic cavity and cause forced expiration; not needed for relaxed expiration

O: inferior margins of ribs 1–11	I: superior margins of ribs 2–12	N: intercostal nn.

Table 11.6 Muscles of the Abdominal Wall (see figs. 11.15 and 11.16)

O = origin, I = insertion, and N = innervation; n. and nn. = nerve and nerves

Rectus Abdominis (ab-DOM-ih-niss)

Supports abdominal viscera; flexes abdomen as in sit-ups; depresses ribs; stabilizes pelvis in walking; increases intra-abdominal pressure for expulsion of urine, feces, or fetus

O: pubis	I: xiphoid process and costal cartilages 5–7	N: ventral rami of T7–T12

External Oblique (oh-BLEEK)

Flexes abdomen as in sit-ups; rotates trunk and flexes spine laterally when contracting on one side

O: ribs 5–12	I: xiphoid process and linea alba	N: T7–T12

Internal Oblique

Similar to external oblique

O: inguinal ligament, iliac crest, and lumbodorsal fascia	I: xiphoid process, linea alba, pubis, and ribs 10–12	N: T7–T12 and iliohypogastric n.

Transversus Abdominis

Compresses abdomen; increases intra-abdominal pressure; flexes vertebral column

O: inguinal ligament, iliac crest, lumbodorsal fascia, and costal cartilages 7–12	I: linea alba, xiphoid process, pubis, and inguinal ligament	N: T7–T12 and iliohypogastric n.

The **external oblique** is the most superficial muscle of the lateral abdominal wall. Its fascicles run anteriorly and downward. Deep to the external oblique is the **internal oblique,** whose fascicles run anteriorly and upward, perpendicular to those of former muscle. Deepest of all is the **transversus abdominis,** whose fascicles run horizontally across the abdomen.

Unlike the thoracic cavity, the abdominal cavity lacks a protective bony enclosure. However, its three muscle layers form a very strong wall, enhanced by the way in which their fascicles run in different directions like layers of plywood (fig. 11.15).

The tendons of the abdominal muscles are aponeuroses. They continue medially as the rectus sheath and terminate at the linea alba. At the inferior margin, each aponeurosis forms a strong, cordlike **inguinal ligament** that extends from the pubis to the anterior superior spine of the pelvis. Some abdominal muscles originate on the *lumbodorsal fascia,* explained in the next section.

Muscles Acting on the Spine

Here we consider muscles of the back that extend, rotate, and abduct the spine (figs. 11.17 and 11.18). These powerful muscles moderate your motion when you bend forward and contract strongly to return the trunk to the erect position (see special topic 11.2). They are classified into two groups—a *superficial group* of muscles, which extend from the vertebrae to the ribs, and a *deep group* of muscles, which connect adjacent vertebrae (table 11.7).

Most muscles of the superficial group belong to a complex called the **erector spinae,** the prime mover of spinal extension and the only one in this group that we will consider. It is divided into three "columns"—the **iliocostalis thoracis, longissimus thoracis,** and **spinalis thoracis.** Most of the lower back (lumbar) muscles are in the longissimus group.

The major deep thoracic muscle is the **semispinalis.** We have already studied one part of it, the semispinalis capitis (see table 11.4). The two lower portions are the **semispinalis cervicis**[36] and **semispinalis thoracis.**[37] The major deep lumbar muscle is the **quadratus**[38] **lumborum.** The erector spinae and quadratus lumborum are enclosed in a fibroconnective sheath called the **lumbodorsal fascia,** which is the origin of some of the abdominal and lumbar muscles.

36. *cervicis* = of the neck
37. *thoracis* = of the thorax
38. *quadrat* = four-sided

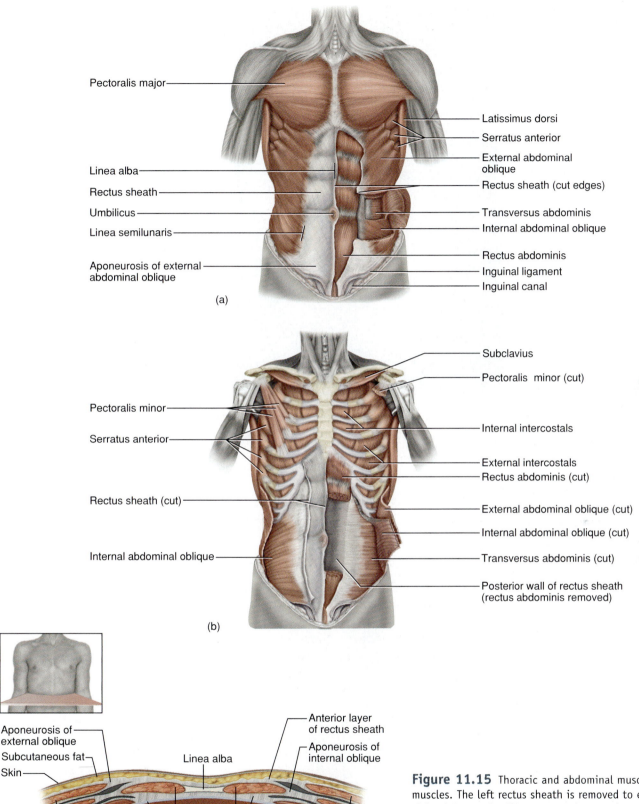

Pectoralis major

Latissimus dorsi

Serratus anterior

External abdominal
oblique

Linea alba

Rectus sheath

Umbilicus

Linea semilunaris

Rectus sheath (cut edges)

Transversus abdominis

Internal abdominal oblique

Rectus abdominis

Inguinal ligament

Inguinal canal

Aponeurosis of external
abdominal oblique

(a)

Subclavius

Pectoralis minor (cut)

Pectoralis minor

Serratus anterior

Internal intercostals

External intercostals

Rectus abdominis (cut)

Rectus sheath (cut)

External abdominal oblique (cut)

Internal abdominal oblique (cut)

Internal abdominal oblique

Transversus abdominis (cut)

Posterior wall of rectus sheath
(rectus abdominis removed)

(b)

Aponeurosis of
external oblique

Subcutaneous fat

Skin

Linea alba

Anterior layer
of rectus sheath

Aponeurosis of
internal oblique

Rectus
abdominis

Posterior layer
of rectus sheath

Aponeurosis of
transversus abdominis

Transverse abdominis

Internal oblique

External oblique

(c)

Figure 11.15 Thoracic and abdominal muscles. (*a*) Superficial
muscles. The left rectus sheath is removed to expose the rectus
abdominis muscle. (*b*) Deep muscles. On the anatomical right,
the external oblique has been removed to expose the internal
oblique. On the anatomical left, the pectoralis major is removed
to expose the pectoralis minor, the internal oblique has been cut
to expose the transversus abdominis, and the rectus abdominis
has been cut to expose the posterior rectus sheath. (*c*) Cross
section of the anterior abdominal wall.

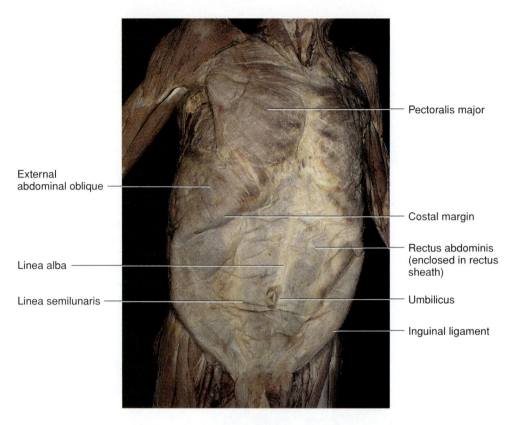

External
abdominal oblique

Linea alba

Linea semilunaris

Pectoralis major

Costal margin

Rectus abdominis
(enclosed in rectus
sheath)

Umbilicus

Inguinal ligament

Figure 11.16 Some abdominal muscles seen in a dissection of the cadaver.

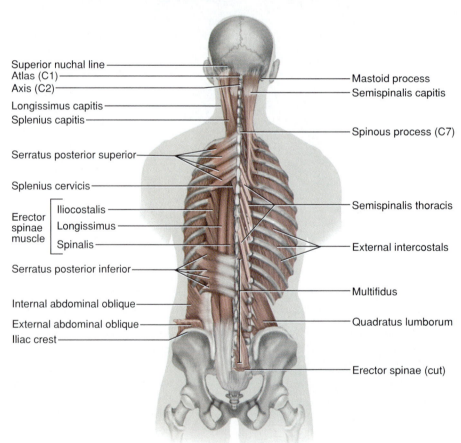

Superior nuchal line
Atlas (C1)
Axis (C2)

Longissimus capitis
Splenius capitis

Serratus posterior superior

Splenius cervicis

Erector
spinae
muscle
Iliocostalis
Longissimus
Spinalis

Serratus posterior inferior

Internal abdominal oblique

External abdominal oblique
Iliac crest

Mastoid process
Semispinalis capitis

Spinous process (C7)

Semispinalis thoracis

External intercostals

Multifidus

Quadratus lumborum

Erector spinae (cut)

Figure 11.17 Muscles acting on the spine. The muscles shown on the left are relatively superficial, and those on the right are deep.

Muscles of the Pelvic Floor

The floor of the pelvic cavity is formed by three layers of muscles and fasciae that span the pelvic outlet and support the viscera (table 11.8). The floor is penetrated by the anal canal, urethra, and vagina, which open into a region between the thighs called the **perineum** (PERR-ih-NEE-um). This is a diamond-shaped area defined by four bony landmarks—the pubic symphysis anteriorly, the coccyx posteriorly, and the ischial tuberosities laterally (fig. 11.19). The anterior half of the perineum is called the **urogenital triangle,** and the posterior half is

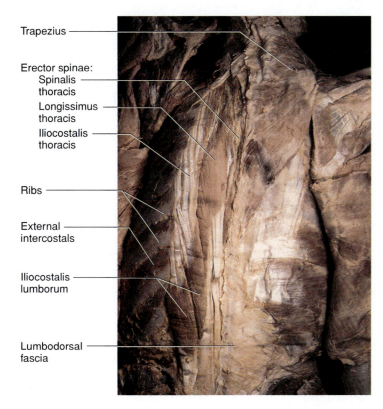

Trapezius

Erector spinae:
 Spinalis thoracis
 Longissimus thoracis
 Iliocostalis thoracis

Ribs

External intercostals

Iliocostalis lumborum

Lumbodorsal fascia

Figure 11.18 Some deep muscles of the back seen in a dissection of the cadaver.

Table 11.7 Muscles Acting on the Spine (see figs. 11.17 and 11.18)

O = origin, I = insertion, and N = innervation; n. and nn. = nerve and nerves

Superficial Group—Erector Spinae (ee-RECK-tur SPY-nee)

Iliocostalis Thoracis (ILL-ee-oh-coss-TAH-liss tho-RA-sis)

Extends and laterally flexes vertebral column

O: iliac crests, ribs 3–12, lumbar vertebrae, sacrum, and lumbodorsal fascia	I: angles of all 12 ribs and transverse processes of C4–C6	N: dorsal rami of spinal nn.

Longissimus (lawn-JISS-ih-muss) Thoracis

Extends and laterally flexes vertebral column

O: transverse processes of C4–T6, spinous processes of L1–S5, and iliac crest	I: mastoid process, transverse processes of cervical to upper lumbar vertebrae, and ribs	N: dorsal rami of spinal nn.

Spinalis (spy-NAY-liss) Thoracis

Extends head and vertebral column; rotates trunk; flexes trunk laterally

O: spinous processes of lower cervical to upper lumbar vertebrae	I: occipital bone and spinous processes of C2–C4, T4–T8	N: dorsal rami of spinal nn.

Deep Group

Semispinalis Cervicis (SEM-ee-spy-NAY-liss SUR-vih-sis) and Thoracis (tho-RA-sis)

Extend neck; extend and rotate vertebral column

O: transverse processes of C4–T12	I: spinous processes of C2–T4	N: dorsal rami of spinal nn.

Quadratus Lumborum (quad-RAY-tus lum-BORE-um)

Laterally flexes vertebral column, depresses rib 12

O: iliac crest, lower lumbar vertebrae and lumbodorsal fascia	I: upper lumbar vertebrae and rib 12	N: ventral rami of L1–L3

Special Topic Heavy Lifting and Back Injuries 11.2

When you are fully bent over forward, as in touching your toes, the erector spinae muscles are fully stretched. Because of the length-tension relationship explained in the next chapter, these muscles cannot contract when stretched to such extremes. Standing up is therefore initiated by the hamstring muscles on the back of the thigh and the gluteus maximus of the buttocks. The erector spinae muscles join the action when they are partially contracted. Standing too suddenly or improperly lifting a heavy weight, however, can strain the erector spinae muscles, cause painful muscular spasms, tear tendons and ligaments of the lower back, and rupture intervertebral discs. These muscles are adapted to maintaining posture, not to lifting. This is why it is important, in heavy lifting, to kneel down and use the powerful extensor muscles of the thighs and buttocks to lift the load.

Table 11.8 Muscles of the Pelvic Floor (see fig. 11.20).

O = origin, I = insertion, and N = innervation; n. and nn. = nerve and nerves

Superficial Muscles of the Perineum

Ischiocavernosus (ISS-kee-oh-CAV-er-NO-sus)

Aids in erection of penis and clitoris

O: ischial and pubic rami and ischial tuberosity   I: penis or clitoris   N: perineal n.

Bulbospongiosus (BUL-bo-SPUN-jee-OH-sus)

Compresses male urethra to expel semen or urine; compresses vaginal orifice in female

O: central tendon of both sexes and bulb of penis in male   I: fascia of perineum and penis or clitoris   N: perineal n.

Superficial Transverse Perineus (PERR-ih-NEE-us)

Fixes central tendon; supports perineum

O: ischial ramus   I: central tendon   N: perineal n.

Muscles of the Urogenital Diaphragm

Deep Transverse Perineus

Supports pelvic floor; fixes central tendon; expels last drops of urine in both sexes and semen in males

O: ischial ramus   I: central tendon   N: perineal n.

External Urethral Sphincter

Compresses urethra; enables conscious control of urination

O: rami of ischium and pubis   I: medial raphe of male or vaginal wall of female   N: perineal n.

Muscle of the Anal Triangle

External Anal Sphincter

Closes anal orifice; enables conscious control of defecation

O: anococcygeal raphe   I: central tendon   N: S4 and inferior rectal n.

Muscles of the Pelvic Diaphragm

Levator Ani (leh-VAY-tur AY-nye)

Supports viscera; resists pressure surges in abdominal cavity; elevates anus during defecation; forms vaginal and anorectal sphincters

O: os coxae from pubis to ischial spine   I: coccyx, anal canal, and anococcygeal raphe   N: S3, S4, and perineal n.

Coccygeus (coc-SIDJ-ee-us)

Draws coccyx anteriorly after defecation or childbirth; supports and elevates pelvic floor; resists abdominal pressure surges

O: ischial spine   I: lower sacrum to upper coccyx   N: S3 or S4

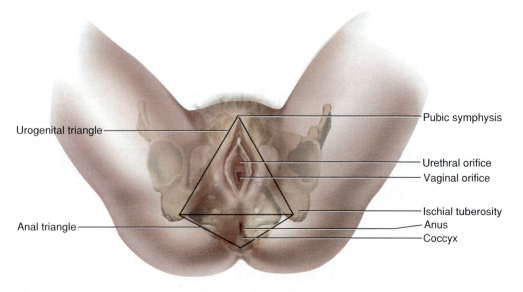

Figure 11.19 Triangles of the perineum. The urogenital triangle is defined by the pubic symphysis and ischial tuberosities. It contains the urinary orifice and, in females, the vaginal orifice. The anal triangle is defined by the ischial tuberosities and coccyx. It contains the anus.

the **anal triangle.** These are especially important landmarks in obstetrics.

The "compartment" of the pelvic floor just beneath the skin is called the **superficial perineal space** (fig. 11.20a). It contains two muscles that function primarily during sexual intercourse. The **ischiocavernosus** muscles converge like a V from the ischial tuberosities toward the penis or clitoris and assist in erection. In males, the **bulbospongiosus (bulbocavernosus)** muscle forms a sheath around the base (bulb) of the penis; it expels semen during ejaculation. In females, it encloses the vagina like a pair of parentheses and tightens on the penis

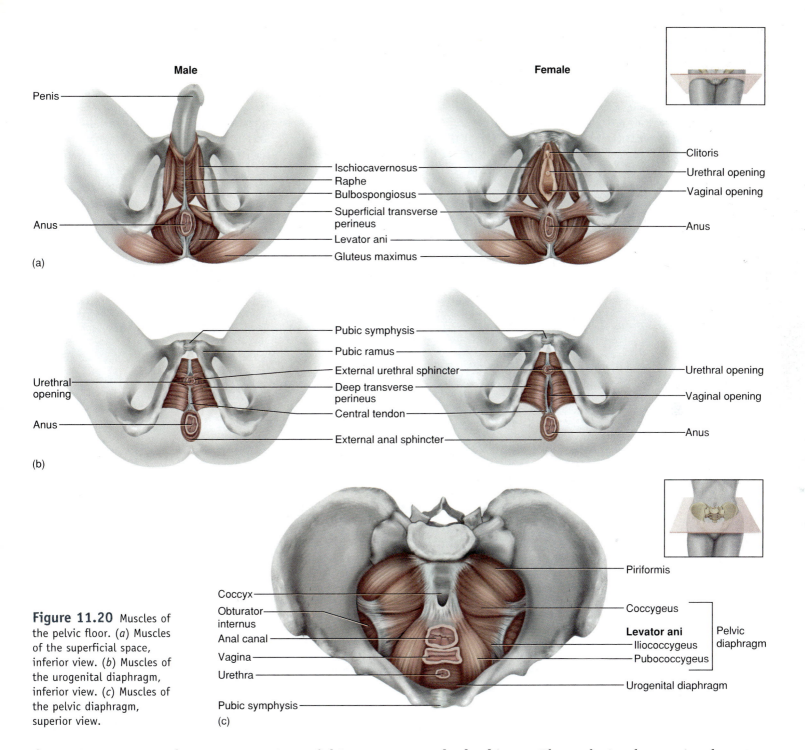

Male

Penis

Anus

(a)

Female

Clitoris
Urethral opening
Vaginal opening

Anus

Ischiocavernosus
Raphe
Bulbospongiosus
Superficial transverse perineus
Levator ani
Gluteus maximus

Urethral opening

Anus

(b)

Pubic symphysis
Pubic ramus
External urethral sphincter
Deep transverse perineus
Central tendon
External anal sphincter

Urethral opening

Vaginal opening

Anus

Coccyx
Obturator internus
Anal canal
Vagina
Urethra

Pubic symphysis

(c)

Piriformis

Coccygeus
Levator ani
Iliococcygeus
Pubococcygeus

Urogenital diaphragm

Pelvic diaphragm

Figure 11.20 Muscles of the pelvic floor. (*a*) Muscles of the superficial space, inferior view. (*b*) Muscles of the urogenital diaphragm, inferior view. (*c*) Muscles of the pelvic diaphragm, superior view.

during intercourse. Voluntary contractions of this muscle in both sexes also help void the last few milliliters of urine. A third muscle at this level is the **superficial transverse perineus,** extending from the ischial tuberosities to a strong **central tendon** of the perineum (fig. 11.20*b*).

In the middle layer of the pelvic floor, the urogenital triangle is spanned by a thin triangular sheet called the **urogenital diaphragm** (fig. 11.20*c*). This is composed of a fibrous membrane and two muscles— the **deep transverse perineus** and the **external**

urethral sphincter. The anal triangle contains the **external anal sphincter.**

The deepest layer of the pelvic floor, called the **pelvic diaphragm,** is similar in both sexes. It consists of two muscle pairs shown in figure 11.20*c*—the **levator ani** and **coccygeus.**

This concludes our discussion of the muscles that enclose the abdominopelvic cavity. Special topic 11.3, however, concerns occasional weaknesses in this muscular enclosure that allow abdominal viscera to protrude through the wall—a condition called *hernia.*

A *hernia* is any condition in which the abdominopelvic viscera protrude through a weak point in the muscular body wall. The most common type to require treatment is an **inguinal hernia** (fig. 1*a*). In male fetuses, each testis descends from the pelvic cavity to the scrotum by way of an *inguinal canal* through the muscles of the groin. This canal is a weak point, especially in male infants and children. Pressure surges in the abdominal cavity can force part of the intestine or bladder into the canal or even into the scrotum. This can also occur in men who hold their breath while lifting heavy weights. As the diaphragm and abdominal muscles contract, pressure in the abdominal cavity can soar to 1,500 pounds per square inch (psi)—more than 100 times normal pressure. Inguinal hernias are rare in women.

In a **hiatal hernia**, the superior curvature of the stomach protrudes through the diaphragm into the thoracic cavity (fig. 1*b*). This is most common in overweight people over 40. It may cause pain and heartburn due to regurgitated stomach acid, but most cases go unnoticed. In an **umbilical hernia**, abdominal viscera protrude through the navel.

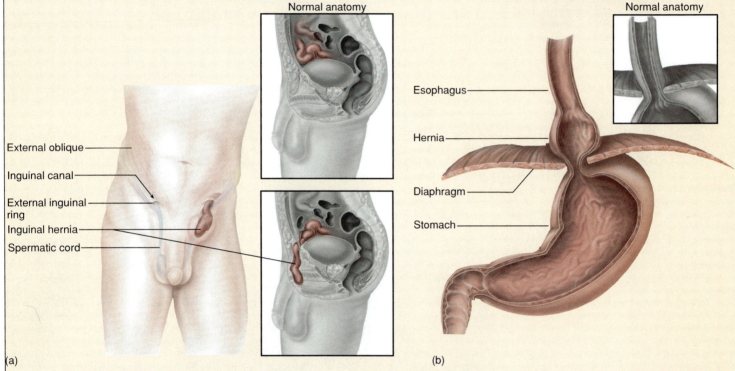

Figure 1 Hernias occur when abdominal pressure forces viscera through weak points in the abdominal or pelvic musculature. (*a*) An inguinal hernia with a loop of small intestine protruding through the inguinal canal; (*b*) a hiatal hernia with a portion of the stomach protruding through the diaphragm into the thoracic cavity.

Key Point Review

14. Which muscles are used more often—the external intercostals or internal intercostals? Why?
15. Explain how pulmonary ventilation affects abdominal pressure and vice versa.
16. Describe the relationship of the abdominal aponeurosis to the rectus sheath, linea alba, and inguinal ligament.
17. Name a major superficial muscle and two major deep muscles of the back.
18. Define *perineum, urogenital triangle,* and *anal triangle.*
19. Name one muscle in each of the following regions: pelvic diaphragm, urogenital diaphragm, and superficial perineal space.

Muscles Acting on the Upper Extremity

▼ Objectives
When you have completed this section, you should be able to
- name and locate the muscles that move the pectoral girdle, arm, forearm, and hand; and
- describe the origins, insertions, and actions of these muscles.

Muscles Acting on the Pectoral Girdle

The scapula is loosely attached to the thorax and is capable of considerable movement—rotation (as in raising and lowering the apex of the shoulder), elevation and

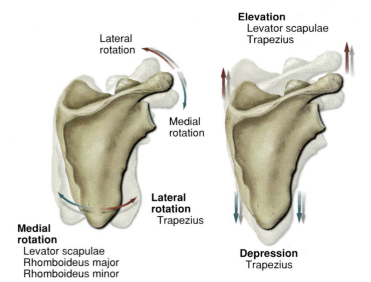

Lateral
rotation

Elevation
Levator scapulae
Trapezius

Medial
rotation

**Medial
rotation**
Levator scapulae
Rhomboideus major
Rhomboideus minor

**Lateral
rotation**
Trapezius

Depression
Trapezius

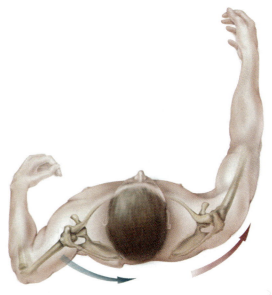

Retraction
Rhomboideus major
Rhomboideus minor
Trapezius

Protraction
Pectoralis minor
Serratus anterior

Figure 11.21 Actions of some thoracic muscles on the scapula. Note that an individual muscle can contribute to multiple actions, depending on which of its fibers are active and what synergists act with it.

Table 11.9	Muscles Acting on the Pectoral Girdle (see figs. 11.12, 11.13, and 11.15)	
O = origin, I = insertion, and N = innervation; n. and nn. = nerve and nerves		
Anterior Group		
Pectoralis (PEC-toe-RAY-liss) Minor		
Protracts and depresses scapula when ribs are fixed; elevates ribs when scapula is fixed		
O: ribs 3–5	I: coracoid process	N: medial and lateral anterior thoracic nn.
Serratus (serr-AY-tus) Anterior		
Holds scapula against rib cage; elevates ribs; abducts and rotates scapula to tilt glenoid cavity upward; abducts and raises arm; prime mover in forward thrusting, throwing, and pushing ("boxer's muscle")		
O: ribs 1–9	I: medial border of scapula	N: long thoracic n.
Posterior Group		
Trapezius (tra-PEE-zee-us)		
Superior fibers elevate scapula or rotate it to tilt glenoid cavity upward; middle fibers retract scapula; inferior fibers depress scapula; when scapula is fixed, one trapezius acting alone flexes neck laterally, and both trapezius muscles working together extend neck		
O: external occipital protuberance, ligamentum nuchae, and spinous processes of C7–T12	I: clavicle, acromion, and scapular spine	N: accessory n.
Levator Scapulae (leh-VAY-tur SCAP-you-lee)		
Rotates scapula to tilt glenoid cavity downward; flexes neck when scapula is fixed; elevates scapula when acting with superior fibers of trapezius		
O: transverse processes of C1–C4	I: superior angle to medial margin of scapula	N: C3–C5
Rhomboideus (rom-BOY-dee-us) Major		
Retracts, elevates, and fixes scapula; rotates scapula to tilt glenoid cavity downward		
O: spinous processes of T2–T5	I: medial border of scapula	N: dorsal scapular n.
Rhomboideus Minor		
Retracts and elevates scapula		
O: spinous processes of C7–T1	I: medial border of scapula	N: dorsal scapular n.

depression (as in shrugging and lowering the shoulders), and protraction and retraction (pulling both shoulders forward or back) (fig. 11.21). The clavicles brace the shoulders and moderate these movements.

The muscles that act on the pectoral girdle originate on the axial skeleton and insert on the clavicle and scapula. They are divided into *anterior* and *posterior groups* (table 11.9). The important muscles of the anterior group are the **pectoralis minor** and **serratus anterior** (see fig. 11.15*b*). In the posterior group, there is one large superficial muscle, the trapezius (see fig. 11.12), and three deep muscles, the **levator scapu-**

lae and **rhomboideus major** and **minor.** The action of the trapezius depends on whether its superior, middle, or inferior fibers contract and whether it acts alone or with other muscles. The levator scapulae and superior fibers of the trapezius rotate the scapula in opposite directions if either of them acts alone. If both act

Table 11.10 Muscles Acting on the Humerus (see figs. 11.1, 11.22–11.24)

O = origin, I = insertion, and N = innervation; n. and nn. = nerve and nerves

Pectoralis (PEC-toe-RAY-liss) Major

Prime mover of shoulder flexion; adducts and medially rotates humerus; depresses pectoral girdle; elevates ribs; aids in climbing, pushing, and throwing

O: clavicle, sternum, and costal cartilages 1–6	I: intertubercular groove of humerus	N: medial and lateral anterior thoracic nn.

Latissimus Dorsi (la-TISS-ih-muss DOR-sye)

Adducts and medially rotates humerus; extends shoulder joint; produces strong downward strokes of arm, as in hammering or swimming ("swimmer's muscle"); pulls body upward in climbing

O: vertebrae T7–L5, lower three or four ribs, lumbodorsal fascia, iliac crest, and inferior angle of scapula	I: intertubercular groove of humerus	N: thoracodorsal n.

Deltoid

Lateral fibers abduct humerus; anterior fibers flex and medially rotate it; posterior fibers extend and laterally rotate it

O: clavicle, scapular spine, and acromion process	I: deltoid tuberosity of humerus	N: axillary n.

Teres (TERR-eez) Major

Adducts and medially rotates humerus; extends shoulder joint

O: inferior angle to lateral margin of scapula	I: medial aspect of proximal shaft of humerus	N: subscapular n.

Coracobrachialis (COR-uh-co-BRAY-kee-AL-iss)

Adducts and flexes arm

O: coracoid process	I: medial aspect of shaft of humerus	N: musculocutaneous n.

Rotator Cuff

All rotator cuff muscles hold head of humerus in glenoid cavity and stabilize shoulder joint in addition to performing these actions:

Infraspinatus (IN-fra-spy-NAY-tus)

Extends and laterally rotates humerus

O: infraspinous fossa of scapula	I: greater tubercle of humerus	N: suprascapular n.

Supraspinatus

Abducts humerus; resists downward displacement when carrying heavy weight

O: supraspinous fossa of scapula	I: greater tubercle of humerus	N: suprascapular n.

Subscapularis (SUB-SCAP-you-LERR-iss)

Medially rotates humerus

O: subscapular fossa of scapula	I: lesser tubercle of humerus	N: subscapular n.

Teres Minor

Adducts and laterally rotates humerus

O: lateral border of scapula	I: greater tubercle of humerus	N: axillary n.

---

together, their opposite rotational effects balance each other and they elevate the scapula and shoulder, as when one carries a heavy weight on the shoulder. Depression of the scapula is mainly by gravitational pull, but the trapezius and serratus anterior cause forcible depression against resistance, as in swimming, hammering, or rowing.

Muscles Acting on the Humerus

Nine muscles cross the shoulder joint and insert on the humerus (table 11.10). Two are called *axial muscles* because they originate primarily on the axial skeleton—the **pectoralis major** and **latissimus dorsi**[39] (figs. 11.1, 11.15, and 11.22). The pectoralis major is the thick, fleshy muscle of the mammary region, and the latissimus dorsi is a broad muscle of the back that extends from the waist to the axilla. These muscles bear the primary responsibility for attachment of the arm to the

trunk, and they are prime movers of the shoulder joint. The pectoralis major flexes the shoulder and the latissimus dorsi extends it—thus, they are antagonists.

The other seven muscles of the shoulder are called *scapular muscles* because they originate on the scapula. Among these, the prime mover is the **deltoid**—the thick muscle that caps the shoulder. It acts like three different muscles. Its anterior fibers flex the shoulder, its posterior fibers extend it, and its lateral fibers abduct it. Abduction by the deltoid is antagonized by the combined action of the pectoralis major and latissimus dorsi. Extension is assisted by the **teres major** (see fig. 11.12), and flexion and adduction are assisted by the **coracobrachialis** (fig. 11.22d).

Think About It

Earlier it was remarked that since a muscle can only pull (not push), antagonistic pairs of muscles are necessary to produce opposite actions at a joint. Reconcile this fact with the observation that the deltoid muscle both flexes and extends the shoulder.

---

39. *latissimus* = broadest + *dorsi* = of the back

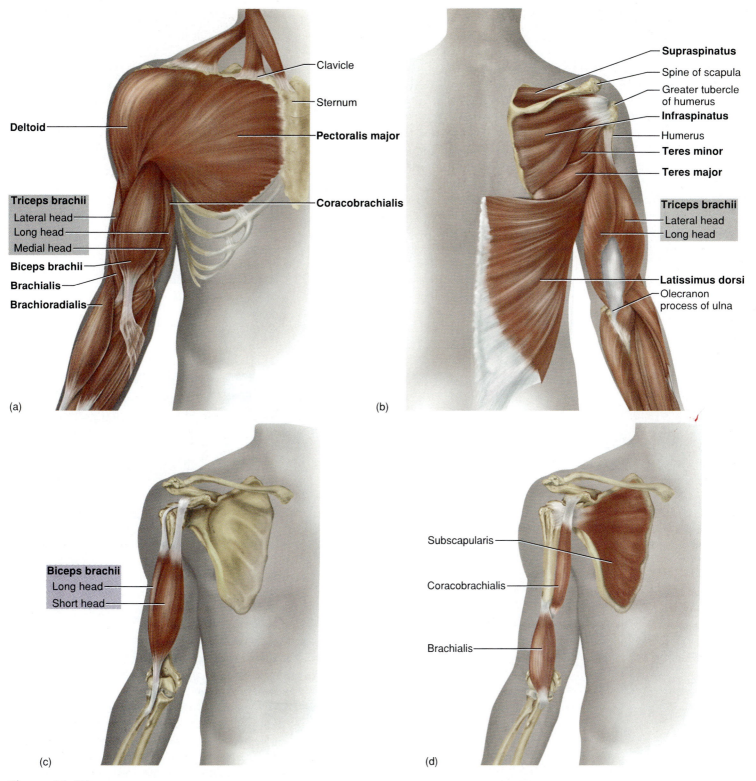

Figure 11.22 Muscles of the pectoral and brachial regions. (*a*) Anterior view; (*b*) posterior view; (*c*) the biceps brachii, the superficial flexor of the elbow; (*d*) the brachialis, the deep flexor of the elbow; and the coracobrachialis and subscapularis, which act on the humerus.

Tendons of the remaining four scapular muscles form the **rotator cuff**—the **infraspinatus, supraspinatus, subscapularis,** and **teres minor** (figs. 11.12, 11.23, and 11.24). The subscapularis fills most of the subscapular fossa on the anterior surface of the scapula. The other three originate on the posterior surface. The supraspinatus and infraspinatus occupy the corresponding fossae above and below the scapular spine, with the teres minor lying inferior to the infraspinatus. The tendons of these muscles merge with the joint capsule as they pass

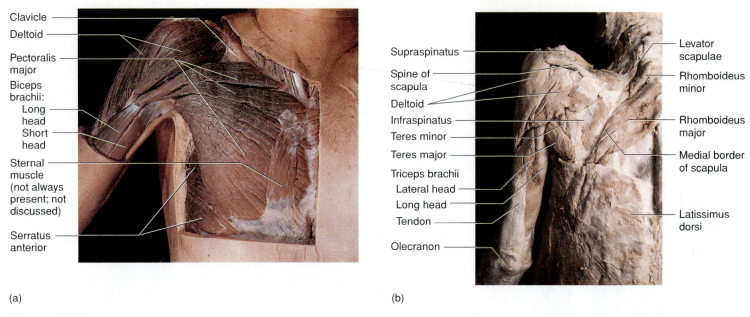

Clavicle
Deltoid
Pectoralis major
Biceps brachii:
  Long head
  Short head
Sternal muscle (not always present; not discussed)
Serratus anterior

(a)

Supraspinatus
Spine of scapula
Deltoid
Infraspinatus
Teres minor
Teres major
Triceps brachii
  Lateral head
  Long head
  Tendon
Olecranon

Levator scapulae
Rhomboideus minor
Rhomboideus major
Medial border of scapula
Latissimus dorsi

(b)

Figure 11.23 Some muscles of the pectoral and brachial regions seen in a dissection of the cadaver. (a) Anterior view and (b) posterior view, with the trapezius removed to expose the deeper muscles of the upper back.

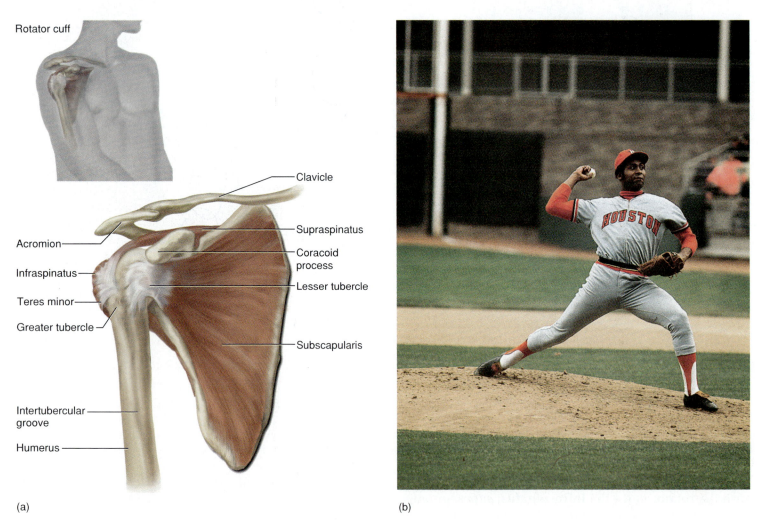

Rotator cuff

Acromion
Infraspinatus
Teres minor
Greater tubercle
Intertubercular groove
Humerus

Clavicle
Supraspinatus
Coracoid process
Lesser tubercle
Subscapularis

(a)

(b)

Figure 11.24 The rotator cuff. (a) Anterolateral view of right shoulder. (b) Vigorous circumduction of the shoulder joint often causes rotator cuff injuries in baseball pitchers.

Table 11.11 Actions of the Shoulder (humeroscapular) Joint

Prime movers are in **boldface;** the muscle in parentheses has only a slight effect

Flexion	Extension
Anterior deltoid	**Posterior deltoid**
Pectoralis major	**Latissimus dorsi**
Coracobrachialis	Teres major
Biceps brachii	

Abduction	Adduction
Lateral deltoid	**Pectoralis major**
Supraspinatus	**Latissimus dorsi**
	Coracobrachialis
	Triceps brachii
	Teres major
	(Teres minor)

Medial Rotation	Lateral Rotation
Subscapularis	**Infraspinatus**
Teres major	**Teres minor**
Latissimus dorsi	Deltoid
Deltoid	
Pectoralis major	

Table 11.12 Muscles Acting on the Forearm (see figs. 11.22, 11.25, and 11.26)

O = origin, I = insertion, and N = innervation; n. and nn. = nerve and nerves

With Bellies in the Arm

Biceps Brachii (BY-seps BRAY-kee-eye)

Flexes elbow; supinates forearm; abducts arm; holds head of humerus in glenoid cavity

O: long head–supraglenoid tubercle; short head–coracoid process	I: radial tuberosity	N: musculocutaneous n.

Brachialis (BRAY-kee-AL-iss)

Flexes elbow

O: anterior distal shaft of humerus	I: coronoid process of ulna and capsule of elbow joint	N: musculocutaneous and radial nn.

Triceps Brachii

Extends elbow; long head adducts humerus

O: long head–infraglenoid tubercle of scapula; lateral head–proximal posterior shaft of humerus; medial head–posterior aspect of humeral shaft	I: olecranon process of ulna	N: radial n.

With Bellies in the Forearm

Brachioradialis (BRAY-kee-oh-RAY-dee-AH-liss)

Flexes elbow

O: lateral supracondylar ridge of humerus	I: styloid process of radius	N: radial n.

Anconeus (an-CO-nee-us)

Extends elbow

O: lateral epicondyle of humerus	I: olecranon process and posterior aspect of ulna	N: radial n.

Pronator Teres (PRO-nay-tur TERR-eez)

Pronates forearm

O: medial epicondyle of humerus and coronoid process of ulna	I: lateral midshaft of radius	N: median n.

Pronator Quadratus (quad-RAY-tus)

Pronates forearm

O: anterior distal shaft of ulna	I: anterior distal shaft of radius	N: palmar interosseous n.

Supinator (SOO-pih-NAY-tur)

Supinates forearm

O: lateral epicondyle of humerus and proximal shaft of ulna	I: proximal shaft of radius	N: dorsal interosseos nn.

it en route to the humerus. They insert on the proximal end of the humerus, forming a partial sleeve around it. The rotator cuff muscles reinforce the joint capsule and hold the head of the humerus in the glenoid cavity. They are synergists in effecting shoulder movements. As mentioned previously (see chapter 10), the rotator cuff is easily damaged by strenuous circumduction, and this damage is common among baseball pitchers. Since the humeroscapular joint is capable of such a wide range of movements and is acted upon by so many muscles, its actions are summarized in table 11.11.

Muscles Acting on the Forearm

The forearm is capable of four motions: flexion, extension, pronation, and supination. The prime movers of flexion are on the anterior side of the humerus and include the superficial **biceps brachii**[40] and deeper **brachialis** (see fig. 11.22; table 11.12). During flexion, the biceps elevates the radius while the brachialis elevates the ulna. The biceps is named for its two heads, which have separate proximal tendons. The tendon of the long head is important in holding the humerus to the glenoid cavity and stabilizing the shoulder joint. The

40. *bi* = two + *ceps* = head + *brachi* = arm. Note that *biceps* is singular; there is no such word as *bicep*.

two heads converge close to the elbow on a single distal tendon leading to the radial tuberosity.

The **brachioradialis** is a synergist in elbow flexion. Its belly lies in the forearm beside the radius, rather than beside the humerus with those of the other two flexors (fig. 11.22a). This muscle forms the thick, fleshy mass on the lateral side of the forearm just distal to the elbow. Its origin is on the distal end of the humerus, and its insertion is on the distal end of the radius. Since its insertion is so far from the fulcrum, the brachioradialis does not generate as much power as the prime movers; it is effective mainly when the prime movers have partially flexed the elbow.

The prime mover of extension is the **triceps brachii** on the posterior side of the humerus (see figs. 11.3 and 11.22). The **anconeus**[41] is a weaker synergist of extension that crosses the posterior side of the elbow (see fig. 11.29).

Pronation is achieved by two muscles in the anterior forearm—the **pronator teres** near the elbow and **pronator quadratus** near the wrist. Supination is achieved by the biceps brachii and the **supinator** of the posterior forearm (fig. 11.26). Although the pronators and supinators are located in the forearm, they are included in this section because two of them span the elbow joint and they all produce movements of the forearm. Muscles acting on the elbow joint are classified by their actions in table 11.13.

Muscles Acting on the Wrist, Hand, and Fingers

The hand is acted upon by extrinsic muscles located in the forearm and intrinsic muscles located in the hand itself (table 11.14). The fleshy roundness of the proximal forearm consists mainly of the bellies of extrinsic

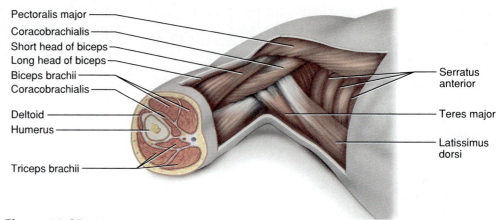

Figure 11.25 Muscles of the axillary region and arm.

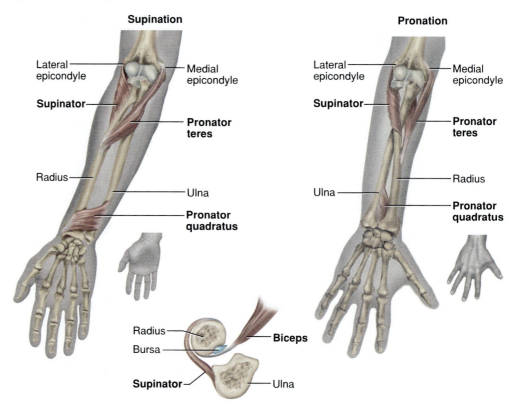

Figure 11.26 Actions of the rotator muscles of the forearm. (a) Supination; (b) pronation; (c) cross section of the proximal forearm in pronated position, showing how contraction of the biceps brachii aids in supination.

Table 11.13	Actions of the Forearm		
Prime movers are in **boldface;** the muscle in parentheses has only a slight effect			
Flexion	**Extension**	**Pronation**	**Supination**
Biceps brachii	**Triceps brachii**	**Pronator teres**	**Supinator**
Brachialis	Anconeus	Pronator quadratus	Biceps brachii
Brachioradialis			
Flexor carpi radialis			
(Pronator teres)			

41. *ancon* = elbow

O = origin, I = insertion, and N = innervation; n. and nn. = nerve and nerves

Anterior Side of Forearm—Superficial Group

Flexor Carpi Radialis (CAR-py ray-dee-AY-liss)

Powerful wrist flexor; abducts hand; synergist in elbow flexion

O: medial epicondyle of humerus	I: base of metacarpals II and III	N: median n.

Flexor Carpi Ulnaris (ul-NAY-riss)

Flexes and adducts wrist; stabilizes wrist during extension of fingers

O: medial epicondyle of humerus	I: pisiform, hamate, and metacarpal V	N: ulnar n.

Flexor Digitorum Superficialis (DIDJ-ih-TOE-rum SOO-per-FISH-ee-AY-liss)

Flexes fingers II–V at proximal interphalangeal joints; aids in flexion of wrist and metacarpophalangeal joints

O: medial epicondyle of humerus, radius, coronoid process of ulna	I: four tendons leading to middle phalanges II–V	N: median n.

Palmaris Longus (pall-MERR-iss LONG-us)

Weakly flexes wrist; often absent

O: medial epicondyle of humerus	I: palmar aponeurosis and flexor retinaculum	N: median n.

Anterior Side of Forearm—Deep Group

Flexor Digitorum Profundus

Flexes wrist and distal interphalangeal joints

O: shaft of ulna and interosseous membrane	I: four tendons to distal phalanges II–V	N: median and ulnar nn.

Flexor Pollicis (PAHL-ih-sis) Longus

Flexes interphalangeal joint of thumb; weakly flexes wrist

O: radius and interosseous membrane	I: distal phalanx I	N: median n.

Posterior Side of Forearm—Superficial Group

Extensor Carpi Radialis Brevis

Extends and abducts wrist; fixes wrist during finger flexion

O: lateral epicondyle of humerus	I: base of metacarpal III	N: radial n.

Extensor Carpi Radialis Longus

Extends and abducts wrist

O: lateral epicondyle of humerus	I: base of metacarpal II	N: radial n.

Extensor Carpi Ulnaris

Extends and adducts wrist

O: lateral epicondyle of humerus and posterior shaft of ulna	I: base of metacarpal V	N: radial n.

Extensor Digiti Minimi (DIDJ-ih-ty MIN-ih-my)

Extends metacarpophalangeal joint of little finger; may be considered a detached portion of extensor digitorum

O: lateral epicondyle of humerus	I: distal and middle phalanges V	N: radial n.

Extensor Digitorum (DIDJ-ih-TOE-rum)

Extends fingers at metacarpophalangeal joints

O: lateral epicondyle of humerus	I: dorsal aspect of phalanges II–V	N: radial n.

Posterior Side of Forearm—Deep Group

Abductor Pollicis Longus

Abducts and extends thumb; abducts wrist

O: posterior ulna and radius and interosseous membrane	I: trapezium, base of metacarpal I	N: radial n.

Extensor Indicis (IN-dih-sis)

Extends index finger at metacarpophalangeal joint

O: shaft of ulna and interosseous membrane	I: middle and distal phalanges II	N: dorsal interosseous n.

Extensor Pollicis Longus

Extends thumb at metacarpophalangeal joint

O: shaft of ulna and interosseous membrane	I: distal phalanx I	N: dorsal interosseous n.

Extensor Pollicis Brevis

Extends thumb at metacarpophalangeal joint

O: shaft of radius and interosseous membrane	I: proximal phalanx I	N: dorsal interosseous n.

muscles, whose tendons extend into the wrist and hand. Their actions are mainly flexion and extension, but the wrist and fingers can be abducted and adducted, and the thumb and fingers can be opposed. It may seem as if the tendons would stand up like taut bowstrings when these muscles were contracted, but this is prevented by the fact that most of them pass under a **flexor retinaculum (transverse carpal ligament)** on the anterior side of the wrist and an **extensor retinaculum (dorsal carpal ligament)** on the posterior side (see fig. 11.30). The

The **carpal tunnel** is a tight space between the carpal bones and flexor retinaculum (fig. 1). The flexor tendons passing through here are enveloped in tendon sheaths that enable the tendons to slide back and forth more easily. Prolonged repetitive wrist motions, however, can cause tissues in the carpal tunnel to become inflamed, swollen, or fibrotic. Since the carpal tunnel cannot expand, swelling puts pressure on the median nerve of the wrist. This causes tingling and muscular weakness in the palm and medial side of the hand and pain that may radiate to the arm and shoulder. This condition, called **carpal tunnel syndrome**, is common among pianists, keyboard operators, meat cutters, and others who spend long hours making repetitive wrist motions. Carpal tunnel syndrome is treated with aspirin and other anti-inflammatory drugs, immobilization of the wrist, and sometimes surgical removal of all or part of the flexor retinaculum to relieve pressure on the nerve.

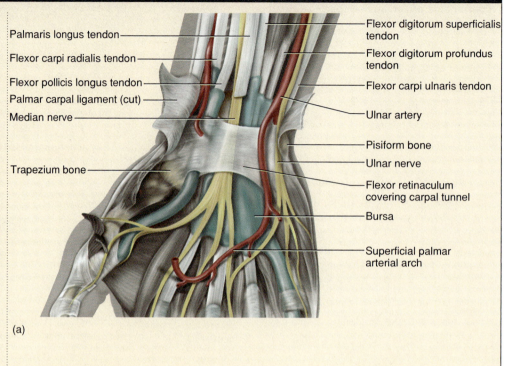

Palmaris longus tendon
Flexor carpi radialis tendon
Flexor pollicis longus tendon
Palmar carpal ligament (cut)
Median nerve
Trapezium bone

Flexor digitorum superficialis tendon
Flexor digitorum profundus tendon
Flexor carpi ulnaris tendon
Ulnar artery
Pisiform bone
Ulnar nerve
Flexor retinaculum covering carpal tunnel
Bursa
Superficial palmar arterial arch

(a)

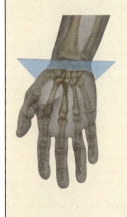

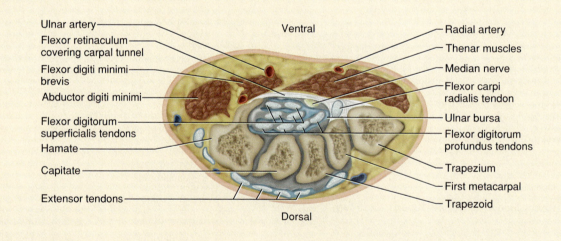

Ulnar artery
Flexor retinaculum covering carpal tunnel
Flexor digiti minimi brevis
Abductor digiti minimi
Flexor digitorum superficialis tendons
Hamate
Capitate
Extensor tendons

Ventral

Radial artery
Thenar muscles
Median nerve
Flexor carpi radialis tendon
Ulnar bursa
Flexor digitorum profundus tendons
Trapezium
First metacarpal
Trapezoid

Dorsal

(b)

Figure 1 The carpal tunnel. (*a*) Dissection of the wrist, anterior view, showing the tendons, bursae, and nerves that pass under the flexor retinaculum. (*b*) Cross section through the wrist. Note how the flexor tendons and median nerve are confined in the tight space between the carpal bones and the flexor retinaculum.

space under the flexor retinaculum is called the *carpal tunnel* (see special topic 11.4).

Several of these muscles originate on the humerus; therefore, they cross the elbow joint and weakly contribute to flexion and extension of the elbow. This action is relatively negligible, however, and we will focus on their action at the wrist and fingers. Although these muscles are numerous and complex, most of their names suggest their

actions, and from their actions their approximate locations in the forearm can generally be deduced.

The deep fascia divide the muscles of the forearm into anterior and posterior compartments and each compartment into superficial and deep layers (fig. 11.27). The muscles are listed and classified this way in table 11.14. Most muscles of the anterior compartment are flexors of the wrist and fingers that arise from a common

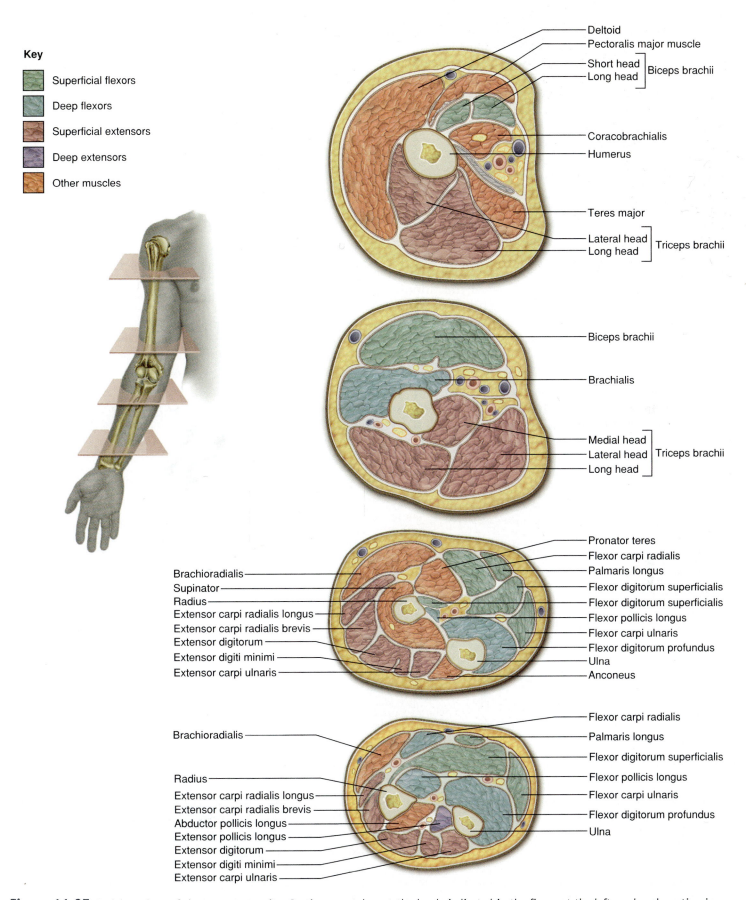

Key

- Superficial flexors
- Deep flexors
- Superficial extensors
- Deep extensors
- Other muscles

Deltoid
Pectoralis major muscle
Short head ⎤ Biceps brachii
Long head ⎦
Coracobrachialis
Humerus
Teres major
Lateral head ⎤ Triceps brachii
Long head ⎦

Biceps brachii
Brachialis
Medial head ⎤
Lateral head ⎥ Triceps brachii
Long head ⎦

Pronator teres
Flexor carpi radialis
Palmaris longus
Flexor digitorum superficialis
Flexor digitorum superficialis
Flexor pollicis longus
Flexor carpi ulnaris
Flexor digitorum profundus
Ulna
Anconeus

Brachioradialis
Supinator
Radius
Extensor carpi radialis longus
Extensor carpi radialis brevis
Extensor digitorum
Extensor digiti minimi
Extensor carpi ulnaris

Flexor carpi radialis
Palmaris longus
Flexor digitorum superficialis
Flexor pollicis longus
Flexor carpi ulnaris
Flexor digitorum profundus
Ulna

Brachioradialis
Radius
Extensor carpi radialis longus
Extensor carpi radialis brevis
Abductor pollicis longus
Extensor pollicis longus
Extensor digitorum
Extensor digiti minimi
Extensor carpi ulnaris

Figure 11.27 Serial sections of the upper extremity. Sections are taken at the levels indicated in the figure at the left, and each section is pictured with the posterior compartment facing down.

tendon on the humerus (fig. 11.28). At the distal end, the tendon of the **palmaris longus** passes over the flexor retinaculum while the other tendons pass beneath it. The two prominent tendons that you can palpate at the wrist belong to the palmaris longus on the medial side and the **flexor carpi radialis** on the lateral side. The latter is an important landmark for finding the radial artery, where the pulse is usually taken.

Muscles of the posterior compartment are mostly wrist and finger extensors that arise from a common tendon on the humerus (fig. 11.29). One of the superficial muscles on this side, the **extensor digitorum,** has four tendons that can easily be seen and palpated on the back of the hand when the fingers are strongly hyperextended. By strongly abducting the thumb into a "hitchhiker's" position, you should also be able to see a deep dorsolateral pit at the base of the thumb, with a taut tendon on each side of it. This depression is called the *anatomical snuffbox* because it was once fashionable to place a pinch of snuff here and inhale it (see fig. B.7*b* in the atlas following this chapter). It is bordered laterally by the tendon of the **abductor pollicis longus** and medially by tendons of the **extensor pollicis longus** and **brevis.**

Other muscles of the forearm were considered earlier because they act on the radius and ulna rather than on the hand. These are the pronator quadratus, pronator teres, supinator, anconeus, and brachioradialis.

Think About It

Why are the prime movers of finger extension and flexion located in the forearm rather than in the hand, closer to the fingers?

Intrinsic Muscles of the Hand

The intrinsic muscles of the hand assist the flexors and extensors of the forearm and make finger movements more precise (table 11.15). They are contained entirely within the hand, and can be divided into three groups:

1. Four *thenar muscles* make up the **thenar eminence,** the thick part of the hand at the base of the thumb. One of them, the *adductor pollicis,* forms the fleshy web between the thumb and palm.

Table 11.15 Intrinsic Muscles of the Hand (see fig. 11.30)

O = origin, I = insertion, and N = innervation; n. and nn. = nerve and nerves

Thenar Group

Abductor Pollicis (PAHL-ih-sis) Brevis

Abducts thumb

O: scaphoid, trapezium, and flexor retinaculum	I: lateral aspect of proximal phalanx I	N: median n.

Adductor Pollicis

Adducts and opposes thumb

O: trapezium, trapezoid, capitate, and metacarpals II–IV	I: medial aspect of proximal phalanx I	N: ulnar n.

Flexor Pollicis Brevis

Flexes thumb at metacarpophalangeal joint

O: trapezium and flexor retinaculum	I: proximal phalanx I	N: median n.

Opponens (op-OH-nenz) Pollicis

Opposes thumb

O: trapezium and flexor retinaculum	I: metacarpal I	N: median n.

Hypothenar Group

Abductor Digiti Minimi (DIDJ-ih-tye MIN-ih-my)

Abducts little finger

O: pisiform and tendon of flexor carpi ulnaris	I: medial aspect of proximal phalanx V	N: ulnar n.

Flexor Digiti Minimi Brevis

Flexes little finger at metacarpophalangeal joint

O: hamulus of hamate and flexor retinaculum	I: median aspect of proximal phalanx V	N: ulnar n.

Opponens Digiti Minimi

Opposes little finger; deepens pit of palm

O: hamulus of hamate and flexor retinaculum	I: median aspect of metacarpal V	N: ulnar n.

Midpalmar Group

Dorsal Interossei (IN-tur-OSS-ee-eye) (four muscles)

Abduct digits II–IV

O: two heads on facing sides of adjacent metacarpals	I: proximal phalanges II–IV	N: ulnar n.

Palmar Interossei (three muscles)

Adduct digits II, IV, and V

O: metacarpals II, IV, and V	I: proximal phalanges II, IV, and V	N: ulnar n.

Lumbricals

Flexes metacarpophalangeal joints; extends interphalangeal joints

O: tendons of flexor digitorum profundus	I: proximal phalanges II–V	N: median and ulnar nn.

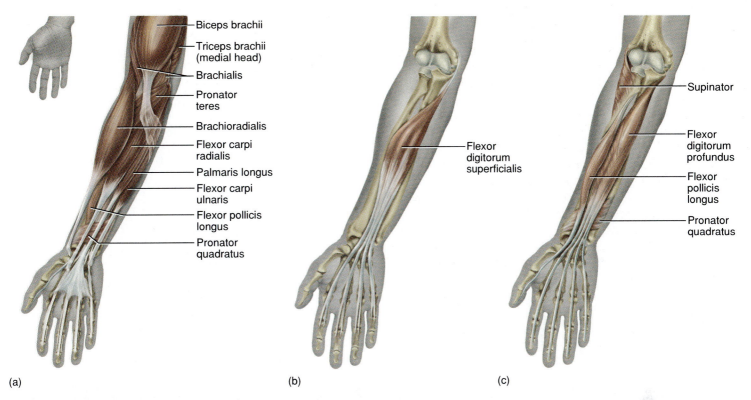

(a)   (b)   (c)

Figure 11.28 Muscles of the forearm, anterior compartment. (*a*) Muscles of the superficial layer; (*b*) the flexor digitorum superficialis, which is deep to the muscles in *a* but still considered a member of the superficial layer; (*c*) muscles of the deep layer.

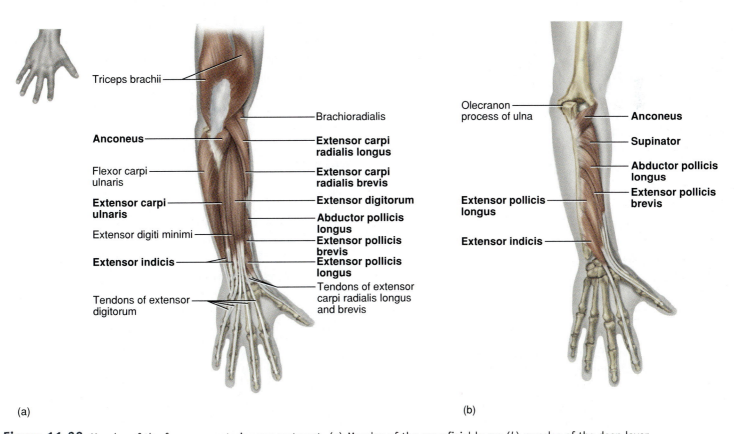

(a)   (b)

Figure 11.29 Muscles of the forearm, posterior compartment. (*a*) Muscles of the superficial layer; (*b*) muscles of the deep layer.

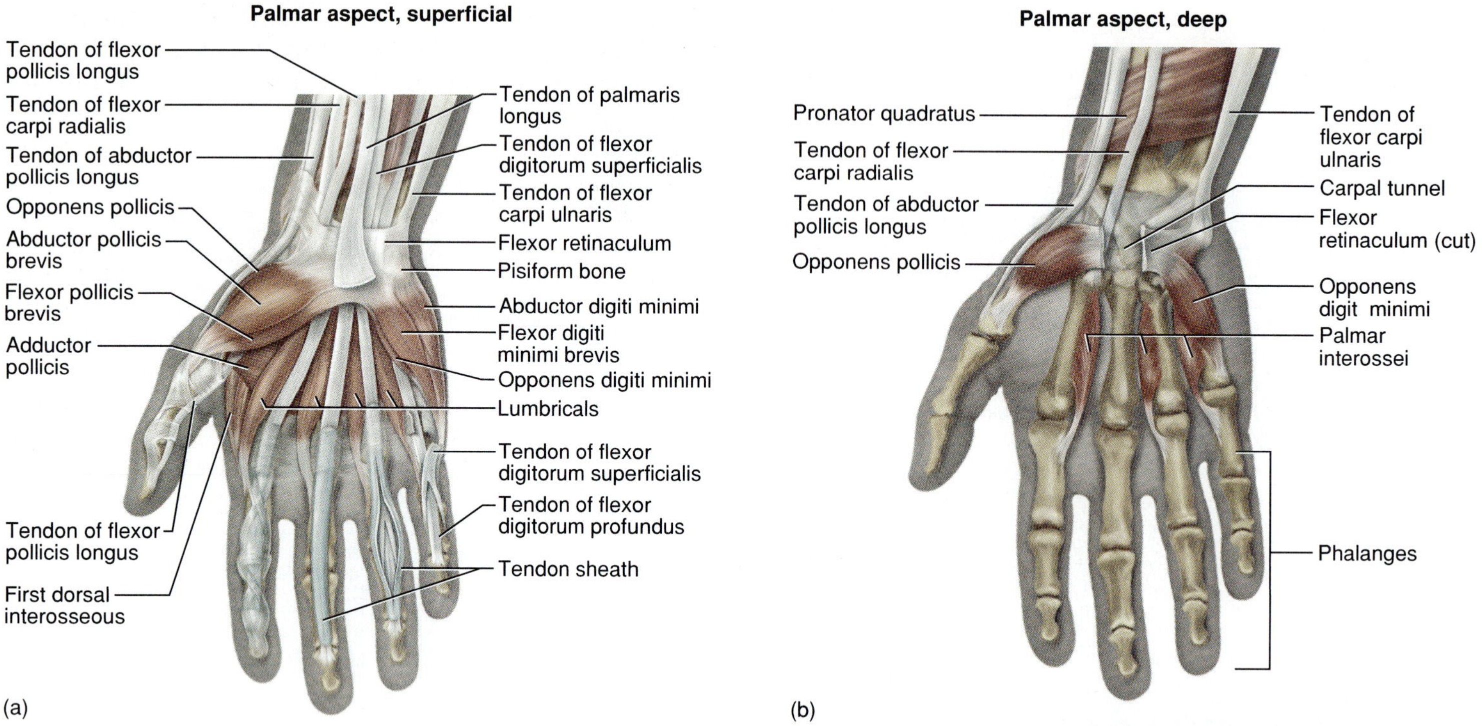

Palmar aspect, superficial

Tendon of flexor pollicis longus
Tendon of flexor carpi radialis
Tendon of abductor pollicis longus
Opponens pollicis
Abductor pollicis brevis
Flexor pollicis brevis
Adductor pollicis
Tendon of flexor pollicis longus
First dorsal interosseous

Tendon of palmaris longus
Tendon of flexor digitorum superficialis
Tendon of flexor carpi ulnaris
Flexor retinaculum
Pisiform bone
Abductor digiti minimi
Flexor digiti minimi brevis
Opponens digiti minimi
Lumbricals
Tendon of flexor digitorum superficialis
Tendon of flexor digitorum profundus
Tendon sheath

(a)

Palmar aspect, deep

Pronator quadratus
Tendon of flexor carpi radialis
Tendon of abductor pollicis longus
Opponens pollicis

Tendon of flexor carpi ulnaris
Carpal tunnel
Flexor retinaculum (cut)
Opponens digit minimi
Palmar interossei
Phalanges

(b)

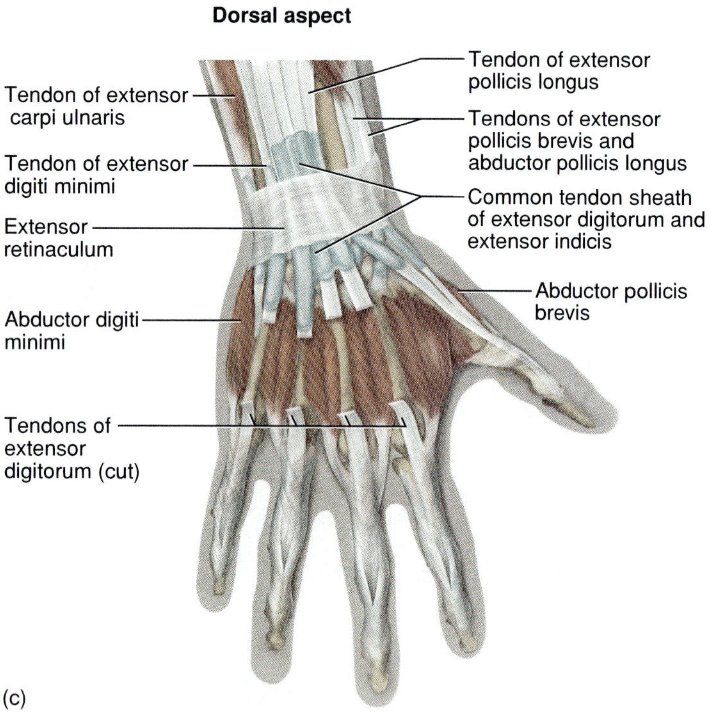

Dorsal aspect

Tendon of extensor carpi ulnaris
Tendon of extensor digiti minimi
Extensor retinaculum
Abductor digiti minimi
Tendons of extensor digitorum (cut)

Tendon of extensor pollicis longus
Tendons of extensor pollicis brevis and abductor pollicis longus
Common tendon sheath of extensor digitorum and extensor indicis
Abductor pollicis brevis

(c)

Figure 11.30 Intrinsic muscles of the hand. (*a*) Superficial muscles, anterior (palmar) view. (*b*) Deep muscles, anterior view. (*c*) Superficial muscles, posterior (dorsal) view.

2. Three *hypothenar muscles* make up the thick **hypothenar eminence** on the opposite side of the palm, at the base of the little finger.
3. Between the thenar and hypothenar muscles are 11 muscles of the **midpalmar (intermediate) group** (fig. 11.30). They can be divided into three

Table 11.16	Actions of Extrinsic and Intrinsic Muscles on the Hand		
Prime movers are in **boldface;** minor contributors to the joint action are in parentheses			
Wrist Flexion	**Wrist Extension**		
Flexor carpi radialis	**Extensor digitorum**		
Flexor carpi ulnaris	Extensor carpi radialis longus		
Flexor digitorum superficialis	Extensor carpi radialis brevis		
(Palmaris longus)	Extensor carpi ulnaris		
(Flexor pollicis longus)			
Wrist Abduction	**Wrist Adduction**		
Flexor carpi radialis	Flexor carpi ulnaris		
Extensor carpi radialis longus	Extensor carpi ulnaris		
Extensor carpi radialis brevis			
Abductor pollicis longus			
Finger Flexion	**Finger Extension**	**Thumb Opposition**	
Flexor digitorum superficialis	Extensor digitorum	Opponens pollicis	
Flexor digitorum profundus	Extensor pollicis longus	Opponens digiti minimi	
Flexor pollicis longus	Extensor pollicis brevis		
	Extensor indicis		

subgroups: (1) Four **dorsal interosseous**[42] **muscles (interossei)** are bipennate muscles attached to adjacent metacarpal bones on both sides; they fan (abduct) the fingers. (2) Three **palmar interosseous muscles** are unipennate muscles arising from metacarpals II, IV, and V; they draw these three fingers toward the middle finger (adduct them). (3) Four wormlike **lumbrical**[43] **muscles** flex the proximal (metacarpophalangeal) knuckles but extend the others.

Table 11.16 summarizes the muscles responsible for the major movements of the wrist and hand.

Key Point Review

20 Name a muscle that inserts on the scapula and plays a significant part in (a) pushing a stalled car, (b) paddling a canoe, (c) squaring the shoulders in military attention, (d) lifting the shoulder to carry a heavy box on it, and (e) lowering the shoulder to pick up a suitcase.

21 Describe three contrasting actions of the deltoid muscle.

22 Name the four rotator cuff muscles and describe the scapular surfaces against which their bellies lie.

23 Name the prime movers of elbow flexion and extension.

24 State three different functions of the biceps brachii.

25 Name three extrinsic muscles and two intrinsic muscles that flex the phalanges.

Muscles Acting on the Lower Extremity

▼Objectives

When you have completed this section, you should be able to
- identify the muscles that act on the thigh, leg, and foot; and
- state the action, origin, insertion, and innervation of these muscles.

The largest and strongest muscles in the body are found in the lower extremity. Unlike those of the upper extremity, they are adapted less for precision than for the strength needed to stand, walk, run, and maintain balance. Several of them cross two or more joints, such as the hip and knee, and act on both or all of them.

To avoid confusion in this discussion, remember that in the anatomical sense the word *leg* refers only to that part of the limb between the knee and ankle. The term *foot* includes the tarsus (ankle), metatarsus, and toes.

Muscles Acting on the Femur

Most muscles that act on the femur (table 11.17) originate on the os coxae. The two principal anterior muscles both flex the hip joint. These are the **iliacus**, which fills most of the broad anteromedial iliac fossa of the pelvis, and the **psoas major**—a thick, rounded muscle that originates mainly on the lumbar vertebrae. These muscles converge on a single tendon that inserts on the femur. Therefore, they produce the same action and are collectively called the **iliopsoas** muscle (fig. 11.31).

On the posterolateral aspect of the hip are three gluteal muscles—the **gluteus maximus, medius,** and **minimus**—and the **tensor fasciae latae** (figs. 11.32 and 11.33). The gluteus medius is often used as a site for intramuscular drug injections (see special topic 11.5). The gluteus maximus is the largest of these and forms most of the mass of the buttocks. It is an extensor of the hip joint that produces its most powerful contraction when the thigh is flexed at a 45° angle to the trunk. This is the advantage in starting a foot race from a crouched position. The **fascia lata**[44] is a deep fascia that encircles

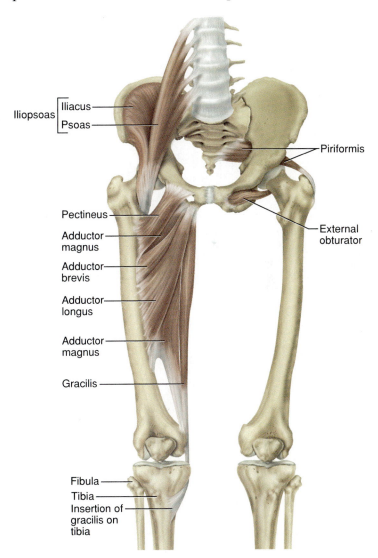

Figure 11.31 Muscles that act on the waist and femur, anterior view.

---

42. *inter* = between + *osse* = bones
43. *lumbric* = earthworm
44. *lata* = broad

Table 11.17 Muscles Acting on the Femur (see figs. 11.31 and 11.32)

O = origin, I = insertion, and N = innervation; n. and nn. = nerve and nerves

Anterior Group

Iliopsoas (ILL-ee-oh-SO-us)

Iliacus (ih-LY-uh-cuss)

Flexes hip joint; medially rotates femur

O: iliac fossa   I: lesser trochanter of   N: femoral n.
                      femur and coxal
                      joint capsule

Psoas major (SO-ass)

Flexes hip joint; medially rotates femur

O: vertebral bodies   I: lesser trochanter
  N: ventral rami of
  T12–L4   L2–L3

Posterolateral Group

Gluteus Maximus (GLOO-tee-us MAX-ih-mus)

Extends hip joint; abducts and laterally rotates femur; important in the backswing of the stride

O: posterior ilium   I: gluteal tuberosity of   N: inferior gluteal n.
  and sacrum   femur and fascia lata

Gluteus Medius (ME-dee-us) and Minimus (MIN-ih-mus)

Abduct and medially rotate femur; maintain balance by shifting body weight during walking

O: posterior ilium   I: greater trochanter   N: superior gluteal n.
                         of femur

Tensor Fasciae Latae (TEN-sor FASH-ee-ee LAY-tee)

Flexes hip joint; abducts and medially rotates femur; tenses fascia lata

O: iliac crest, near   I: lateral condyle   N: superior gluteal n.
  anterior superior   of tibia
  spine

Deep Lateral Rotators

Gemellus (jeh-MEL-us) Inferior and Superior

Laterally rotate femur

O: body of ischium   I: obturator internus tendon   N: sacral plexus

Obturator (OB-too-RAY-tur) Externus

Laterally rotates femur

O: margin of   I: greater trochanter of femur   N: obturator n.
  obturator foramen,
  anterior aspect

Obturator Internus

Abducts and laterally rotates femur

O: margin of   I: greater trochanter   N: sacral plexus
  obturator foramen,   of femur
  posterior aspect

Piriformis (PIR-ih-FOR-miss)

Abducts and laterally rotates femur

O: anterolateral   I: greater trochanter   N: ventral rami
  aspect of   of femur   of S1–S2
  sacroiliac region

Quadratus Femoris (quad-RAY-tus FEM-oh-riss)

Adducts and laterally rotates femur

O: ischial tuberosity I: intertrochanteric   N: sacral plexus
                         ridge of femur

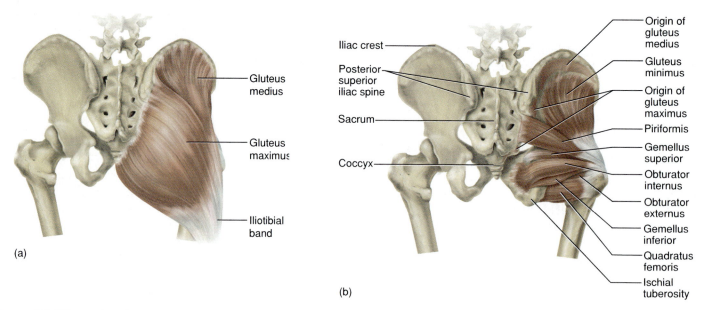

Figure 11.32 Muscles of the gluteal region, posterior view. (*a*) Superficial muscles; (*b*) deep muscles.

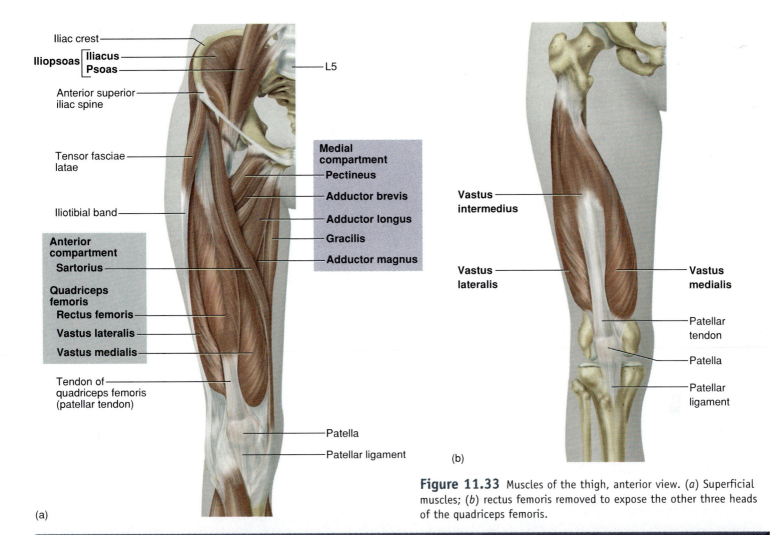

Iliac crest

Iliopsoas — **Iliacus** / **Psoas**

L5

Anterior superior iliac spine

Tensor fasciae latae

Iliotibial band

Medial compartment
- **Pectineus**
- **Adductor brevis**
- **Adductor longus**
- **Gracilis**
- **Adductor magnus**

Anterior compartment
Sartorius

Quadriceps femoris
Rectus femoris
Vastus lateralis
Vastus medialis

Tendon of quadriceps femoris (patellar tendon)

Patella

Patellar ligament

(a)

Vastus intermedius

Vastus lateralis

Vastus medialis

Patellar tendon

Patella

Patellar ligament

(b)

Figure 11.33 Muscles of the thigh, anterior view. (*a*) Superficial muscles; (*b*) rectus femoris removed to expose the other three heads of the quadriceps femoris.

Special Topic — Intramuscular Injections

11.5

Muscles with thick bellies are commonly used for intramuscular (I.M.) drug injections. Since drugs injected into these muscles are absorbed into the bloodstream gradually, it is possible to administer relatively large doses (up to 5 mL) that could be dangerous or even fatal if injected directly into the bloodstream. I.M. injections also cause less tissue irritation than subcutaneous injections.

Knowledge of subsurface anatomy is essential to avoid damaging nearby nerves or accidentally injecting a drug into a blood vessel. Such knowledge also enables a clinician to position a patient so that the muscle is relaxed, making the injection less uncomfortable.

Amounts up to 2 mL are commonly injected into the *deltoid* muscle, about two finger widths below the acromion. Amounts over 2 mL are often injected into the *gluteus medius,* in the superolateral quadrant of the gluteal area, away from the sciatic nerve and major gluteal blood vessels. The *vastus lateralis* of the thigh is often used for infants and young children, whose deltoid and gluteal muscles are not well developed.

the thigh and tightly binds its muscles. On the lateral surface, it combines with the tendons of the gluteus maximus and tensor fasciae latae to form the **iliotibial band,** which extends from the iliac crest to the lateral condyle of the tibia.

 ### Think About It

What muscles are used for the downstroke in pedaling a bicycle? Why are racing bicycles designed to make the rider lean forward?

The deep *lateral rotators* of the pelvic region (table 11.17; fig. 11.32*b*) rotate the femur laterally and therefore oppose medial rotation by the gluteus medius and minimus. Most of them also abduct or adduct the femur, which is important when we walk because it shifts the body weight to the leg that is on the ground and prevents us from falling over when the other leg is lifted.

The thigh itself contains a number of muscles that act on the hip joint, knee, or both. Any attempt to group

Table 11.18 Muscles Located in the Thigh (see figs. 11.31, 11.33, and 11.34)

O = origin, I = insertion, and N = innervation; n. and nn. = nerve and nerves

Anterior Compartment

Quadriceps Femoris (QUAD-rih-seps FEM-oh-riss)

Rectus femoris

Extends knee joint; flexes hip joint

O: anterior inferior iliac spine	I: tibial tuberosity via patella	N: femoral n.

Vastus lateralis

Extends knee joint

O: posterolateral femoral shaft	I: tibial tuberosity via patella	N: femoral n.

Vastus medialis

Extends knee joint

O: linea aspera of femur	I: tibial tuberosity via patella	N: femoral n.

Vastus intermedius

Extends knee joint

O: anterior femoral shaft	I: tibial tuberosity via patella	N: femoral n.

Sartorius (sar-TORE-ee-us)

Flexes hip and knee joints; rotates femur medially; rotates tibia laterally; used to cross legs

O: anterior superior iliac spine	I: medial side of tibial tuberosity	N: femoral n.

Medial Compartment

Adductor (ah-DUCK-tur) Longus and Brevis

Adduct and laterally rotate femur; flex hip joint

O: pubis	I: posterior shaft of femur	N: obturator n.

Adductor Magnus

Anterior part adducts and laterally rotates femur and flexes hip joint; posterior part extends hip joint

O: ischium	I: posterior shaft of femur	N: obturator and tibial nn.

Gracilis (GRASS-ih-liss)

Adducts femur; flexes knee joint; medially rotates tibia

O: pubis	I: proximal medial tibia	N: obturator n.

Pectineus (pec-TIN-ee-us)

Adducts and laterally rotates femur; flexes hip joint

O: pubis	I: proximal posterior shaft of femur	N: femoral n.

Posterior Compartment (Hamstring Muscles)

Biceps (BY-seps) Femoris

Long head

Flexes knee joint; extends hip joint; laterally rotates leg

O: ischial tuberosity	I: head of fibula	N: tibial n.

Short head

Same as long head

O: posterior midshaft of femur	I: head of fibula	N: peroneal n.

Semimembranosus (SEM-ee-MEM-bran-OH-sus)

Flexes knee joint; tenses capsule of joint; extends hip joint; medially rotates tibia

O: ischial tuberosity	I: medial condyle of tibia and collateral ligament of knee	N: tibial n.

Semitendinosus (SEM-ee-TEN-din-OH-sus)

Flexes knee joint; extends hip joint; medially rotates leg

O: ischial tuberosity	I: near tibial tuberosity	N: tibial n.

muscles of the lower extremity by function results in some illogical groupings by location, and vice versa. It is probably easiest to learn them by location, however, which is why they are discussed in the next section in spite of their actions on the hip joint.

Muscles Located in the Thigh

The muscles that make up the bulk of the thigh act across the hip and knee joints to move the femur, tibia, and fibula (table 11.18). Flexion of the hip and extension of the knee are important to the forward swing of the leg during walking, and extension of the hip and flexion of the knee are important to the backswing that thrusts the body forward.

The fascia lata of the thigh divides the thigh muscles into three compartments, each with its own nerve and blood supply: the *anterior (extensor) compartment*, *medial (adductor) compartment*, and *posterior (flexor) compartment*. Muscles of the anterior compartment function mainly as extensors of the knee, those of the medial compartment as adductors of the femur, and those of the posterior compartment as extensors of the hip and flexors of the knee.

The anterior compartment contains the large **quadriceps femoris** muscle, the prime mover of knee extension and the most powerful muscle of the body (figs. 11.33 and 11.34). As the name implies, it has four heads—the **rectus femoris, vastus lateralis, vastus medialis,** and **vastus intermedius.** All four converge on a single **patellar tendon,**

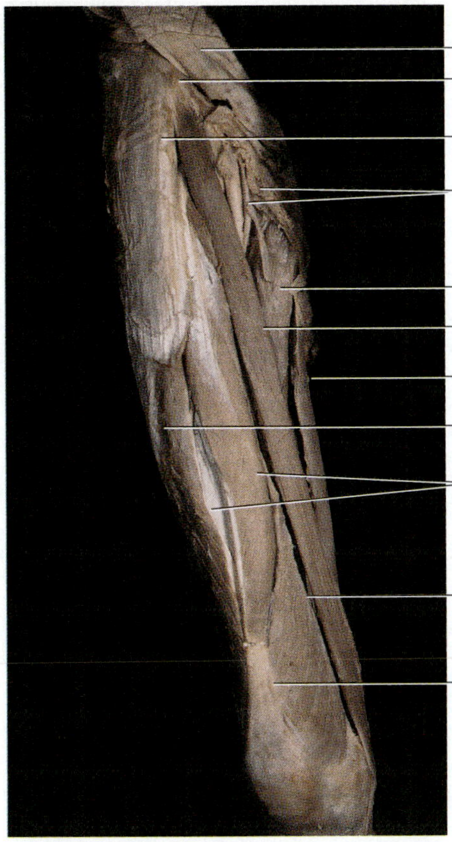
Iliacus
Anterior superior iliac spine
Tensor fasciae latae
Femoral artery and vein (in femoral triangle)
Adductor longus
Sartorius
Gracilis
Iliotibial band
Rectus femoris
Vastus medialis
Patellar tendon

Figure 11.34 Some superficial anterior muscles of the right thigh seen in a dissection of the cadaver.

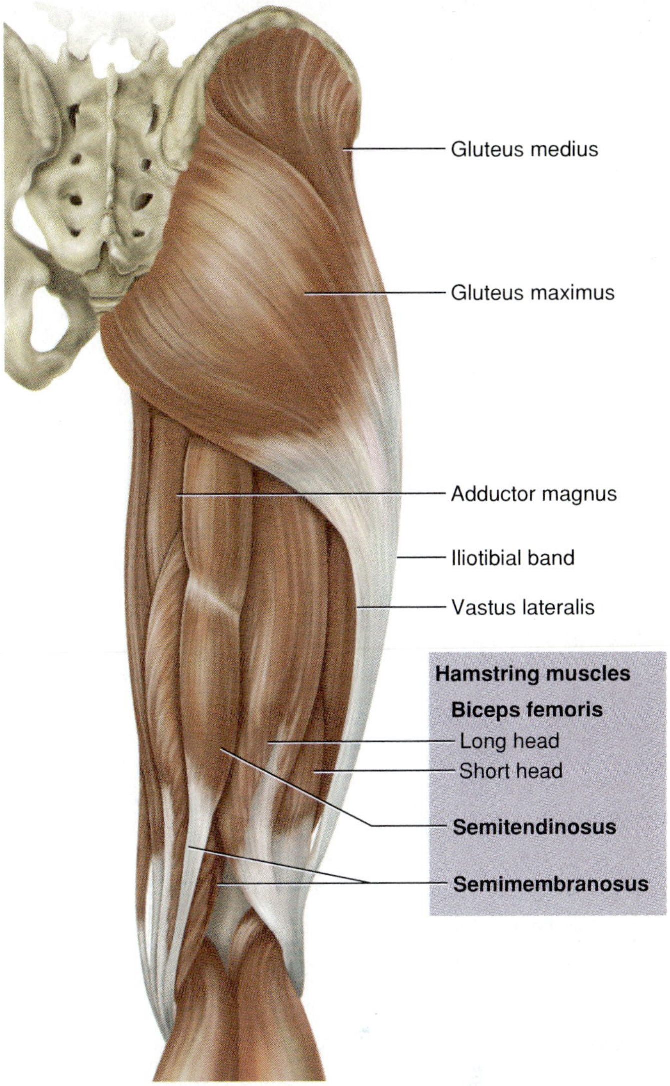
Gluteus medius
Gluteus maximus
Adductor magnus
Iliotibial band
Vastus lateralis
Hamstring muscles
Biceps femoris
Long head
Short head
Semitendinosus
Semimembranosus

Figure 11.35 Muscles of the thigh, posterior view.

which extends to the patella, then continues as the **patellar ligament,** and inserts on the tibial tuberosity. (Remember that a tendon usually extends from muscle to bone, and a ligament from bone to bone.) The patellar ligament is struck with a rubber mallet to test the knee-jerk reflex. The quadriceps extends the knee when you stand up, take a step, or kick a ball. It is very important in running because, together with the iliopsoas, it acts to flex the hip in each airborne phase of the leg's cycle of motion. The rectus femoris also flexes the hip in such actions as high kicks or simply in drawing the leg forward during a stride.

Crossing the quadriceps from the lateral side of the hip to the medial side of the knee is the narrow, strap-like **sartorius,**[45] which is the longest muscle of the body. It flexes the hip and knee joints and laterally rotates the thigh, as in crossing the legs. It is colloquially called the "tailor's muscle" because tailors are traditionally depicted sitting cross-legged to support their work on the raised knee.

In the medial compartment are five muscles that act on the hip joint—the **adductor longus, adductor brevis, adductor magnus, gracilis,**[46] and **pectineus**[47] (see fig. 11.31). All of them adduct the thigh, but some cross both the hip and knee joints and have additional actions noted in table 11.18.

Muscles of the posterior compartment are the **biceps femoris, semimembranosus,** and **semitendinosus** (figs. 11.34 and 11.35). These three muscles are colloquially known as the "hamstrings" because their tendons at the knee of a hog are commonly used to hang a ham for curing. They flex the knee and extend the hip during walking and running, aided by the gluteus maximus. The pit at the rear of the knee, called the **popliteal fossa,** is bordered by the biceps tendon on the lateral side and the tendons of the semimembranosus and semitendinosus on the medial side. When wolves attack large prey, they often attempt to sever the hamstring tendons, because this renders the prey helpless. Hamstring injuries are common among sprinters, soccer players, and other athletes who rely on quick acceleration.

45. *sartor* = tailor
46. *gracil* = slender
47. *pectin* = comb

Table 11.19 The Crural Muscles (see figs. 11.37–11.40)

O = origin, I = insertion, and N = innervation; n. and nn. = nerve and nerves

Anterior Compartment

Extensor Digitorum (DIDJ-ih-TOE-rum) Longus

Extension of toes II–V; dorsiflexion and eversion of foot

O: lateral condyle of tibia, shaft of fibula and interosseous membrane   I: middle and distal phalanges II–V   N: deep peroneal n.

Extensor Hallucis (hal-OO-sis) Longus

Extension of great toe (hallux); dorsiflexion and inversion of foot

O: medial aspect of fibula and interosseous membrane   I: distal phalanx I   N: deep peroneal n.

Peroneus Tertius (perr-OH-nee-us TUR-she-us)

Dorsiflexion and eversion of foot

O: distal shaft of fibula   I: metatarsal V   N: deep peroneal n.

Tibialis (TIB-ee-AY-liss) Anterior

Dorsiflexion and inversion of foot

O: medial tibia and interosseous membrane   I: medial cuneiform and metatarsal I   N: deep peroneal n.

Posterior Compartment—Superficial Group

Triceps Surae (SOOR-ee)

   Gastrocnemius (GAS-trock-NEE-me-us)

   Flexion of knee joint; plantar flexion of foot

   O: medial and lateral epicondyles of femur   I: calcaneus   N: tibial n.

   Soleus (SO-lee-us)

   Plantar flexion of foot

   O: proximal third of tibia and fibula   I: calcaneus   N: tibial n.

Plantaris (plan-TERR-is) (sometimes absent)

Flexion of knee joint; plantar flexion of foot

O: distal femur   I: calcaneus   N: tibial n.

Posterior Compartment—Deep Group

Flexor Digitorum Longus

Flexion of toes II–V; plantar flexion and inversion

O: midshaft of tibia   I: distal phalanges II–V   N: tibial n.

Flexor Hallucis Longus

Flexion of great toe; plantar flexion and inversion

O: shaft of fibula   I: distal phalanx I   N: tibial n.

Tibialis Posterior

Plantar flexion and inversion

O: proximal half of tibia, fibula, and interosseous membrane   I: navicular, cuneiforms, and metatarsals II–IV   N: tibial n.

Popliteus (pop-LIT-ee-us)

Unlocks knee so it can flex; flexes knee; medially rotates tibia

O: lateral condyle of femur   I: posterior proximal tibia   N: tibial n.

Lateral (Peroneal) Compartment

Peroneus Brevis

Eversion and plantar flexion of foot

O: shaft of fibula   I: base of metatarsal V   N: superficial peroneal n.

Peroneus Longus

Eversion and plantar flexion of foot

O: proximal half of fibula and lateral condyle of tibia   I: medial cuneiform and metatarsal I   N: superficial peroneal n.

Muscles Acting on the Foot

The fleshy mass of the leg proper (below the knee) is formed by the **crural muscles,** which act on the foot (table 11.19). These muscles are tightly bound together by deep fasciae, which compress the muscles and aid in the return of blood from the legs. They also separate the crural muscles into anterior, lateral, and posterior compartments. All the muscles in one compartment share a common nerve and blood supply (fig. 11.36).

Muscles of the *anterior compartment* dorsiflex the ankle and prevent the toes from scuffing against the ground during walking. These are the **extensor dig-** itorum longus (extensor of toes II–V), **extensor hallucis**[48] **longus** (extensor of the great toe), **peroneus**[49] **tertius,** and **tibialis anterior.** Their tendons are held tightly against the ankle and kept from bowing by retinacula similar to those of the wrist—the **superior extensor retinaculum (transverse ligament)** and **inferior extensor retinaculum (cruciate ligament)** of the ankle (fig. 11.37).

48. *hallucis* = of the great toe (hallux)
49. *peroneo* = fibular

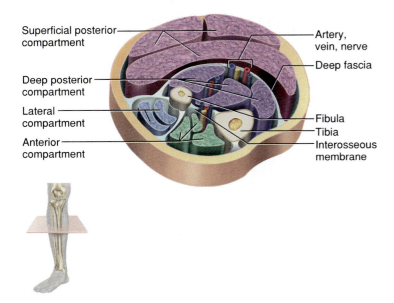

Figure 11.36 Musculoskeletal compartments of the leg, showing fascial sheaths around the muscle groups and associated blood vessels and nerves.

The *posterior compartment* has superficial and deep muscle groups. The three muscles of the superficial group are plantar flexors—the **gastrocnemius**,[50] **soleus**,[51] and **plantaris**[52] (fig. 11.38). The first two of these, collectively known as the **triceps surae**,[53] insert on the calcaneus by way of the **calcaneal (Achilles) tendon.** This is the strongest tendon of the body but is nevertheless a common site of sports injuries resulting from sudden stress (see chapter essay, p. 374, on other sports-related injuries to the skeletomuscular system). The plantaris inserts medially on the calcaneus by a tendon of its own.

There are four muscles in the deep group. The **flexor digitorum longus, flexor hallucis longus,** and **tibialis posterior** are plantar flexors. The **popliteus** unlocks the knee joint so that it can be flexed, and it functions in flexion and medial rotation at the knee.

50. *gastro* = belly + *cnem* = lower leg
51. *solea* = sandal, sole of foot
52. *planta* = sole of foot
53. *sura* = calf of leg

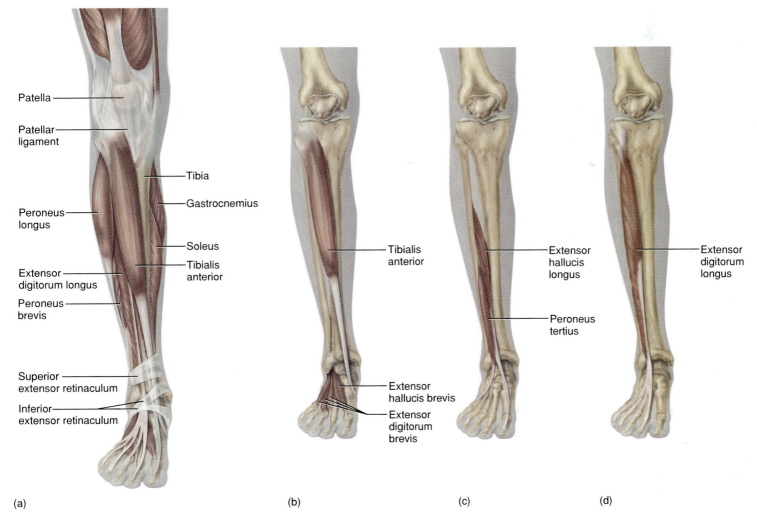

Figure 11.37 Muscles of the leg, anterior view. (*a*) Anterior view with minimal dissection, showing some muscles of the anterior, lateral, and posterior compartments; (*b–d*) individual muscles of the anterior compartment of the leg and dorsum of the foot.

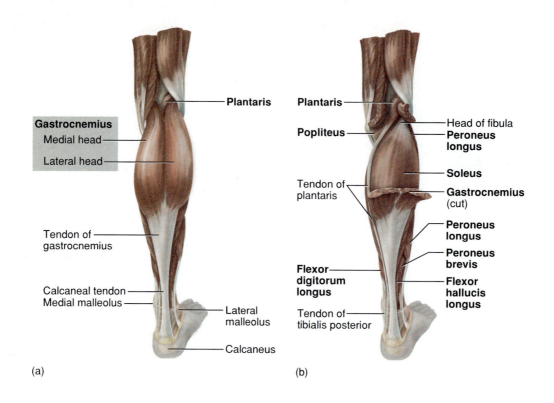

(a)

(b)

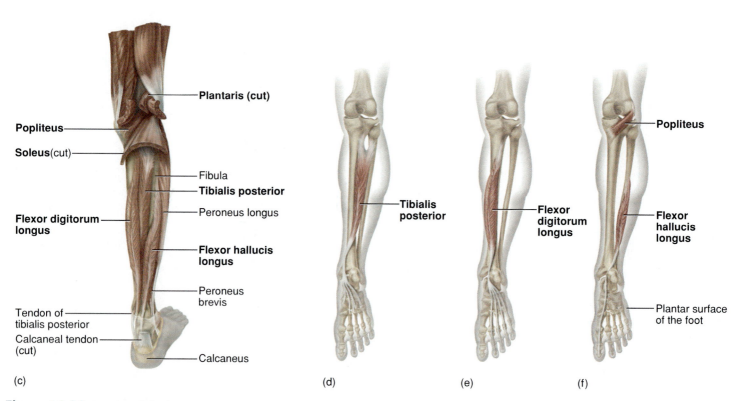

(c)

(d)

(e)

(f)

Figure 11.38 Muscles of the leg, posterior compartment. (*a*) Superficial muscles; (*b* and *c*) progressively deeper dissections; (*d–f*) some individual deep muscles.

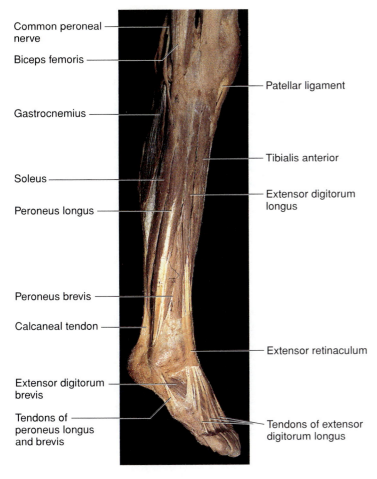

Figure 11.39 Superficial muscles of the leg, right lateral view, as seen in a dissection of the cadaver.

The muscles of the *lateral (peroneal) compartment* are the **peroneus brevis** and **peroneus longus** (figs. 11.37*a*, 11.38*b, c*, and 11.39). They plantar flex and evert the foot. Plantar flexion is important not only in standing on tiptoes but in providing lift and forward thrust each time a step is taken.

Intrinsic Muscles of the Foot

The intrinsic muscles of the foot support the arches and act on the toes in ways that aid locomotion (table 11.20). Several of them are similar in name and location to the intrinsic muscles of the hand. One of these muscles is dorsal to the tarsal and metatarsal bones—the **extensor digitorum brevis**.[54] The others are ventral or lie between the metatarsals. They can be grouped in four layers:

1. The most superficial layer includes the stout **flexor digitorum brevis** medially, with four tendons that supply all the digits except the hallux. It is flanked

....

54. "Short extensor of the digits"

Table 11.20   Intrinsic Muscles of the Foot (see fig. 11.41)

O = origin, I = insertion, and N = innervation; n. and nn. = nerve and nerves

Dorsal Muscle

Extensor Digitorum (DIDJ-ih-TOE-rum) brevis

Extends toes

O: dorsal aspect of calcaneus	I: tendons of extensor digitorum longus	N: deep peroneal n.

Ventral Layer 1 (Superficial)

Flexor Digitorum Brevis

Flexes toes II–V

O: calcaneus and plantar aponeurosis	I: middle phalanges II–V	N: medial plantar n.

Abductor Digiti Minimi (DIDJ-ih-ty MIN-ih-my)

Abducts and flexes little toe; supports lateral longitudinal arch

O: calcaneus and plantar aponeurosis	I: proximal phalanx I	N: lateral plantar n.

Abductor Hallucis (hal-OO-sis)

Flexes great toe; supports medial longitudinal arch

O: calcaneus and plantar aponeurosis	I: proximal phalanx I	N: medial plantar n.

Ventral Layer 2

Quadratus Plantae (quad-RAY-tus PLAN-tee)

Flexes toes

O: calcaneus and plantar aponeurosis	I: tendons of flexor digitorum longus	N: lateral plantar n.

Lumbricals

Flex metatarsophalangeal joints; extend interphalangeal joints

O: tendons of flexor digitorum longus	I: extensor tendons to digits II–V	N: lateral and medial plantar nn.

Ventral layer 3

Adductor Hallucis

Adducts great toe

O: metatarsals II–IV	I: proximal phalanx I	N: lateral plantar n.

Flexor Digiti Minimi Brevis

Flexes little toe

O: metatarsal V and plantar aponeurosis	I: proximal phalanx V	N: lateral plantar n.

Flexor Hallucis Brevis

Flexes great toe

O: cuboid and plantar aponeurosis	I: proximal phalanx I	N: medial plantar n.

Ventral Layer 4 (Deepest)

Dorsal Interossei (IN-tur-OSS-ee-eye)

Abduct toes II–IV

O: each with two heads on facing sides of adjacent metatarsals	I: proximal phalanges II–IV	N: lateral plantar n.

Plantar Interossei

Adduct toes III–V

O: medial aspect of metatarsals III–V	I: proximal phalanges III–V	N: lateral plantar n.

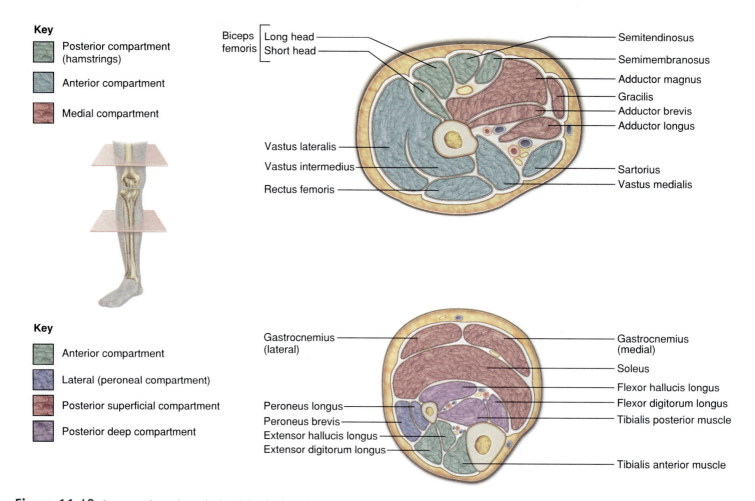

Biceps femoris
- Long head
- Short head

Semitendinosus

Semimembranosus

Adductor magnus

Gracilis

Adductor brevis

Adductor longus

Vastus lateralis

Vastus intermedius

Rectus femoris

Sartorius

Vastus medialis

Gastrocnemius (lateral)

Gastrocnemius (medial)

Soleus

Flexor hallucis longus

Peroneus longus

Peroneus brevis

Extensor hallucis longus

Extensor digitorum longus

Flexor digitorum longus

Tibialis posterior muscle

Tibialis anterior muscle

Figure 11.40 Cross sections through the right thigh and leg.

by the **abductor digiti minimi**[55] laterally and **abductor hallucis**[56] medially; the tendons of these two muscles serve the little toe and great toe, respectively (fig. 11.41a).

2. Deep to the first layer are the thick medial **quadratus plantae,** which joins the tendons of the flexor digitorum longus, and the four **lumbrical** muscles located between the metatarsals (fig. 11.41b).

3. The third layer includes the **adductor hallucis, flexor digiti minimi brevis,** and **flexor hallucis brevis** (fig. 11.41c). The adductor hallucis has an oblique head crossing the foot and inserting at the base of the great toe and a transverse head that passes across the bases of digits II–V to meet the long head at the base of the hallux.

4. The deepest layer consists of four **dorsal** and three **plantar interosseous muscles (interossei)** located

between the metatarsals. Each dorsal interosseous muscle is bipennate and originates on the metatarsals on both sides of it. The plantar interossei are unipennate and originate on only one metatarsal each (fig. 11.41d, e).

Several of the muscles in the first three layers originate on a broad **plantar aponeurosis,** which lies between the plantar skin and muscles; it diverges like a fan from the calcaneus to the bases of all five toes.

Key Point Review

26 In the middle of a stride, you have your left foot on the ground and you are about to swing your right leg forward. What muscles produce the motion of the right leg?

27 Name some muscles whose tendons cross both the hip and knee joints and act on both.

28 List the major actions performed by muscles of the anterior, medial, and posterior compartments of the thigh.

29 What is the role of plantar flexion and dorsiflexion in walking? Which muscles produce these actions?

55. "Abductor of the little toe"
56. "Abductor of the hallux"

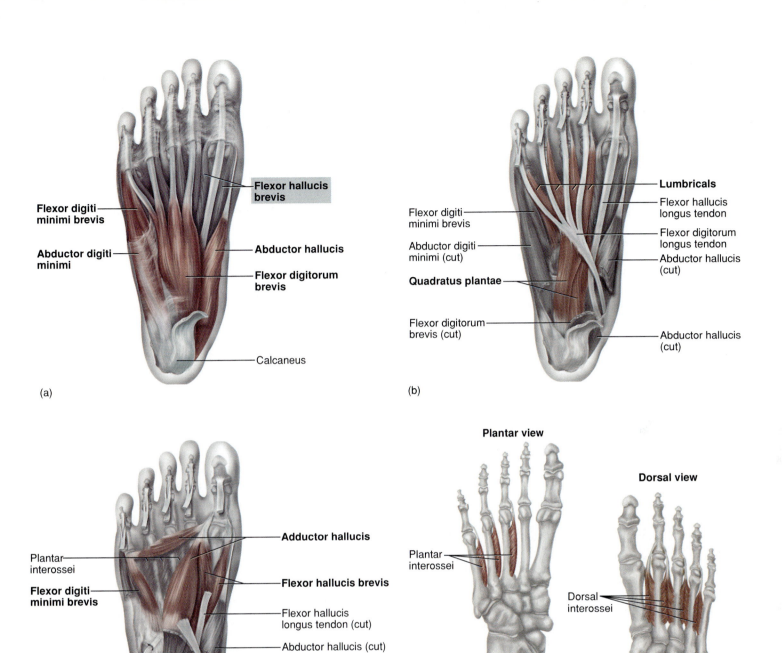

Flexor digiti minimi brevis

Abductor digiti minimi

Flexor hallucis brevis

Abductor hallucis

Flexor digitorum brevis

Calcaneus

(a)

Lumbricals

Flexor digiti minimi brevis

Abductor digiti minimi (cut)

Quadratus plantae

Flexor digitorum brevis (cut)

Flexor hallucis longus tendon

Flexor digitorum longus tendon

Abductor hallucis (cut)

Abductor hallucis (cut)

(b)

Adductor hallucis

Plantar interossei

Flexor digiti minimi brevis

Flexor hallucis brevis

Flexor hallucis longus tendon (cut)

Abductor hallucis (cut)

Quadratus plantae (cut)

Flexor digitorum longus tendon (cut)

(c)

Plantar view

Dorsal view

Plantar interossei

Dorsal interossei

(d)

(e)

Figure 11.41 Intrinsic muscles of the foot. All are ventral (plantar) views except figure *e*. (*a*) First layer (superficial muscles); (*b*) second layer; (*c*) third layer; (*d*) fourth layer (interosseous muscles), plantar view; (*e*) fourth layer, dorsal view.

Sports-Related Injuries

Although the muscular system is subject to fewer diseases than most organ systems, it is particularly vulnerable to sports-related injuries resulting from sudden and intense stress placed on muscles and tendons. Each year, thousands of athletes from the high school to professional level sustain some type of injury to their muscles. These injuries are also common among the increasing numbers of people who have taken up running and other forms of physical conditioning. Overzealous exertion without proper conditioning and warm-up is frequently the cause. Sports-related injuries include the following:

Shinsplints—a general term embracing several kinds of injury with pain in the crural region: tendinitis of the tibialis posterior muscles, inflammation of the tibial periosteum, and anterior compartment syndrome. Shinsplints may result from unaccustomed jogging or walking on snowshoes or from any vigorous activity of the legs after a period of relative inactivity.

Compartment syndrome—a condition in which overuse, contusion, or muscle strain damages blood vessels in a compartment of the arm or leg. Since the fascia enclosing a compartment are tight and cannot stretch, blood or tissue fluid accumulating in the compartment can put pressure on the muscles, nerves, and blood vessels. The lack of blood flow in the compartment can cause destruction of nerves if untreated within 2 to 4 hours and death of muscle tissue if it goes untreated for 6 hours or more. Nerves can regenerate if blood flow is restored, but muscle damage is irreversible. Depending on its severity, compartment syndrome may be treated with immobilization and rest, drainage of fluid from the compartment, or an incision to relieve pressure.

Pulled hamstrings—strained hamstring muscles or a partial tear in the tendinous origin, often with a hematoma (blood clot) in the fascia lata. This condition is frequently caused by repetitive kicking (as in football and soccer) or long, hard running.

Tennis elbow—inflammation at the origin of the extensor carpi muscles at the lateral epicondyle of the humerus. It occurs when these muscles are repeatedly tensed during backhand strokes and then strained by sudden impact with the tennis ball. Any activity that requires rotary movements of the forearm and a firm grip of the hand (for example, using a screwdriver) can cause the symptoms of tennis elbow.

Tennis leg—a partial tear in the lateral origin of the gastrocnemius muscle. It results from repeated strains put on the muscle while supporting the body weight on the toes.

Rider's bones—abnormal ossification in the tendons of the adductor muscles of the medial thigh. It results from prolonged abduction of the thighs when riding horses.

Pulled groin—strain in the adductor muscles of the thigh. It is common in gymnasts and dancers who perform splits and high kicks.

Charley horse—any painful tear, stiffness, and blood clotting in a muscle. A charley horse of the quadriceps femoris is often caused by football tackles.

Blocker's arm—ectopic ossification in the lateral margin of the forearm as a result of repeated impact with opposing players.

Pitcher's arm—inflammation at the origin of the flexor carpi resulting from hard wrist flexion in releasing a baseball.

Baseball finger—tears in the extensor tendons of the fingers resulting from the impact of a baseball with the extended fingertip.

Most athletic injuries can be prevented by proper conditioning. A person who suddenly takes up vigorous exercise may not have sufficient muscle and bone mass to withstand the stresses such exercise entails. These must be developed gradually. Stretching exercises keep ligaments and joint capsules supple and therefore reduce injuries. Warm-up exercises promote more efficient and less injurious musculoskeletal function in several ways, discussed in the next chapter. Most of all, moderation is important, as most injuries simply result from overuse of the muscles. "No pain, no gain" is a dangerous misconception.

Muscular injuries can be treated initially with **r**est, **i**ce, **c**ompression, and **e**levation **("RICE")**. Rest prevents further injury and allows repair processes to occur; ice reduces swelling; compression with an elastic bandage helps to prevent fluid accumulation and swelling; and elevation of an injured limb promotes drainage of blood from the affected area and limits further swelling. If these measures are not enough, anti-inflammatory drugs may be employed, including corticosteroids as well as aspirin and other nonsteroidal agents.▲

Connective Issues

c o n n e c t i v e   i s s u e s

Interactions Between the MUSCULAR SYSTEM and Other Organ Systems

Integumentary System

- Covers and protects superficial muscles; initiates synthesis of vitamin D needed for muscle contraction; dissipates heat generated by muscles
- Facial muscles pull on skin to provide facial expressions

Skeletal System

- Provides levers that enable muscles to act; stores calcium needed for muscle contraction
- Muscles produce movements at joints and stabilize joints; produce stress that affects ossification, bone remodeling, and shapes of mature bones

Nervous System

- Stimulates muscle contraction; monitors and adjusts muscle tension; adjusts cardiopulmonary functions to meet needs of muscles during exercise
- Muscles give expression to thoughts, emotions, and motor commands that arise in the central nervous system

Endocrine System

- Hormones stimulate growth and development of muscles; regulate levels of glucose and electrolytes important for muscle contraction
- Exercise stimulates secretion of stress hormones; skeletal muscles protect some endocrine organs

Circulatory System

- Delivers O_2 and nutrients; carries away wastes and heat generated by muscles; cardiovascular efficiency and RBC count greatly affect muscular endurance
- Skeletal muscle contractions help to move blood through veins; exercise stimulates growth of new blood vessels and promotes cardiac strength

Lymphatic/Immune Systems

- Lymphatic system drains fluid from muscles; immune system protects muscles from pathogens and promotes tissue repair
- Muscle contractions promote lymph flow; exercise elevates levels of immune cells and antibodies; excess exercise inhibits immune responses

Respiratory System

- Provides O_2 and removes CO_2; respiratory efficiency greatly affects muscular endurance
- Muscle contractions ventilate lungs; muscles of larynx and pharynx regulate air flow; CO_2 generated by exercise stimulates respiratory rate and depth

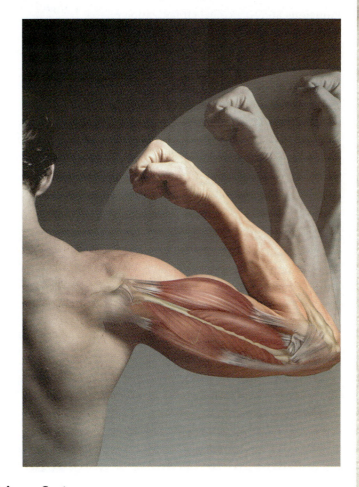

Urinary System

- Eliminates wastes generated by muscles; regulates levels of electrolytes important for muscle contraction
- Muscles control voluntary urination; muscles of pelvic floor support bladder

Digestive System

- Absorbs nutrients needed by muscles; liver regulates blood glucose levels and metabolizes lactic acid generated by anaerobic muscle metabolism
- Muscles enable chewing and swallowing; control voluntary defecation; abdominal and lumbar muscles protect lower digestive organs

Reproductive System

- Gonadal steroids affect muscular growth and development
- Muscles contribute to erection and ejaculation; abdominal and pelvic muscles aid childbirth

The Study of Skeletal Muscles (pp. 327–328)
1. How muscles are named
   a. *Nomina Anatomica*
   b. Common terms in muscle names
2. A learning strategy

Functions and Structural Organization of Muscles (pp. 328–333)
1. Functions of muscles
   a. Movement
   b. Communication
   c. Stability
   d. Control of body openings
   e. Heat production
2. Connective tissues of a skeletal muscle
   a. Endomysium around fibers
   b. Perimysium around fascicles
   c. Epimysium around muscle
   d. Deep fascia between muscles
   e. Superficial fascia between muscle and skin
   f. Direct and indirect attachments to bone
   g. Aponeuroses
   h. Retinacula
3. General anatomy of skeletal muscles
   a. Origin (head)
   b. Insertion
   c. Belly
4. Classification by orientation of fascicles
   a. Fusiform muscles
   b. Parallel muscles

c. Convergent muscles
d. Pennate muscles
   • Unipennate
   • Bipennate
   • Multipennate
e. Circular muscles
5. Coordinated action of muscle groups
   a. Prime movers (agonists)
   b. Synergists
   c. Antagonists
   d. Fixators
6. Intrinsic and extrinsic muscles
7. Innervation of muscles

Muscles of the Head and Neck (pp. 333–342)
1. Muscles of facial expression (table 11.2)
2. Muscles of chewing and swallowing (table 11.3)
   a. Intrinsic tongue muscles
   b. Extrinsic tongue muscles
   c. Muscles of mastication
   d. Hyoid muscles
   e. Pharyngeal constrictors
3. Muscles acting on the head (table 11.4)

Muscles of the Trunk (pp. 342–350)
1. Muscles of respiration (table 11.5)
2. Abdominal muscles (table 11.6)
3. Muscles acting on the spine (table 11.7)

4. Muscles of the pelvic floor (table 11.8)

Muscles Acting on the Upper Extremity (pp. 350–363)
1. Muscles acting on the pectoral girdle (table 11.9)
2. Muscles acting on the humerus (tables 11.10 and 11.11)
   a. Axial muscles
   b. Scapular muscles
   c. Rotator cuff
3. Muscles acting on the forearm (tables 11.12 and 11.13)
4. Muscles acting on the wrist, hand, and fingers (table 11.14)
5. Intrinsic muscles of the hand (tables 11.15 and 11.16)

Muscles Acting on the Lower Extremity (pp. 363–373)
1. Muscles acting on the femur (table 11.17)
2. Muscles located in the thigh (table 11.18)
3. Muscles acting on the foot (table 11.19)
4. Intrinsic muscles of the foot (table 11.20)

S e l e c t e d   V o c a b u l a r y

Also review the terms in tables 11.2 to 11.10, 11.12, 11.14, 11.15, and 11.17 to 11.20, which are not repeated here.

muscular system 327
myology 327
sphincter 328
endomysium 330
fascicle 330
perimysium 330
epimysium 330
deep fascia 330
superficial fascia 330
direct attachment 330
indirect attachment 330
aponeurosis 330
retinaculum 331
origin (head) 331

insertion 331
belly 331
fusiform muscle 331
parallel muscle 331
convergent muscle 331
pennate muscle 331
unipennate 332
bipennate 332
multipennate 332
circular muscle 332
action 332
prime mover 332
synergist 332
antagonist 332
antagonistic pair 332
fixator 332
intrinsic muscle 332
extrinsic muscle 332
innervation 332

galea aponeurotica 333
anterior triangle 341
posterior triangle 341
central tendon of diaphragm 342
tendinous insertions 343
rectus sheath 343
linea alba 343
inguinal ligament 344
lumbodorsal fascia 344
perineum 347
urogenital triangle 347
anal triangle 348
superficial perineal space 348
central tendon of perineum 349
urogenital diaphragm 349
pelvic diaphragm 349
flexor retinaculum 357

extensor retinaculum 357
thenar eminence 360
hypothenar eminence 362
fascia lata 363
iliotibial band 365
patellar tendon 366
patellar ligament 367
popliteal fossa 367
crural muscles 368
superior extensor retinaculum 368
inferior extensor retinaculum 368
calcaneal (Achilles) tendon 369
plantar aponeurosis 372

1. Which of the following muscles is the prime mover in spitting out a mouthful of liquid?
   a. platysma
   b. buccinator
   c. risorius
   d. masseter
   e. palatoglossus

2. Each muscle fiber has a sleeve of areolar connective tissue around it called
   a. the deep fascia.
   b. the superficial fascia.
   c. the perimysium.
   d. the epimysium.
   e. the endomysium.

3. Which of these is *not* a suprahyoid muscle?
   a. genioglossus
   b. geniohyoid
   c. stylohyoid
   d. mylohyoid
   e. digastric

4. Which of these muscles is an extensor of the head?
   a. external oblique
   b. sternocleidomastoid
   c. splenius capitis
   d. iliocostalis
   e. latissimus dorsi

5. Which of these muscles of the pelvic floor is the deepest?
   a. superficial transverse perineus
   b. bulbospongiosus
   c. ischiocavernosus
   d. deep transverse perineus
   e. levator ani

6. Which of these actions is *not* performed by the trapezius?
   a. extension of the neck
   b. depression of the scapula
   c. elevation of the scapula
   d. rotation of the scapula
   e. adduction of the humerus

7. Both the hands and feet are acted upon by a muscle or muscles called
   a. the extensor digitorum.
   b. the abductor digiti minimi.
   c. the flexor digitorum profundus.
   d. the abductor hallucis.
   e. the flexor digitorum longus.

8. Which of the following muscles does *not* extend the hip joint?
   a. quadriceps femoris
   b. gluteus maximus
   c. biceps femoris
   d. semitendinosus
   e. semimembranosus

9. The calcaneal tendon leads from the triceps surae, which consists of the gastrocnemius and _____ muscles.
   a. semimembranosus
   b. tibialis posterior
   c. tibialis anterior
   d. soleus
   e. plantaris

10. Which of the following muscles raises the upper lip?
    a. levator palpebrae superioris
    b. orbicularis oris
    c. zygomaticus minor
    d. masseter
    e. mentalis

11. The _____ of a muscle is the point at which it attaches to a relatively stationary bone.

12. A bundle of muscle fibers surrounded by perimysium is called a/an _____.

13. The _____ is the muscle primarily responsible for a given movement at a joint.

14. The three large muscles on the posterior side of the thigh are commonly known as the _____ muscles.

15. Connective tissue bands called _____ prevent flexor tendons from rising up like bowstrings.

16. The anterior half of the perineum is a region called the _____.

17. The abdominal aponeuroses converge on a midsagittal fibrous band on the abdomen called the _____.

18. A muscle that works with another to produce the same or similar movement is called a/an _____.

19. A muscle somewhat like a feather, with fibers obliquely approaching its tendon from both sides, is called a/an _____.

20. A circular muscle that closes an opening in the body is called a/an _____.

1. Radical mastectomy, once a common treatment for breast cancer, involves removal of the pectoralis major along with the breast. What functional impairments would result from this? What synergists could a physical therapist train a patient to use to recover some lost function?

2. Removal of cancerous lymph nodes from the neck sometimes requires removal of the sternocleidomastoid on that side. How would this affect a patient's range of head movement?

3. In a disease called tick paralysis, the saliva from a tick bite paralyzes skeletal muscles beginning with the lower extremities and progressing superiorly. What would be the most urgent threat to the life of a tick paralysis patient?

4. Women who habitually wear high heels may suffer painful "high heel syndrome" when they go barefoot or wear flat shoes. What muscle(s) and tendon(s) do you think are involved? Explain.

5. A student moving out of the dormitory kneels down, in correct fashion, to lift a heavy box of books. What prime movers are involved as he straightens his legs to lift the box?

Web Site Link

For a listing of the most current web sites related to this chapter, please visit the Saladin homepage at:

http://www.mhhe.com/sciencemath/biology/saladin/

atlas b

B

[Surface Anatomy

The Importance of External Anatomy

In the study of human anatomy, it is easy to become so preoccupied with internal structure that we forget the importance of what we can see and feel externally. Yet external anatomy and appearances are major concerns in giving a physical examination and in many aspects of patient care. A knowledge of the body's surface landmarks is essential to one's competence in such practices as physical therapy, cardiopulmonary resuscitation, surgery, making X rays and electrocardiograms, giving injections, drawing blood, listening to heart and respiratory sounds, measuring the pulse and blood pressure, and finding pressure points to stop arterial bleeding. A misguided attempt to perform some of these procedures while disregarding or misunderstanding external anatomy can be very harmful and even fatal to a person.

Having just studied skeletal and muscular anatomy in the preceding chapters, this is an opportune time to study the body surface. Much of what we see there reflects the underlying structure of the superficial bones and muscles. A broad photographic overview of surface anatomy is given in atlas A (figs. A.8 and A.9). In the following pages, we examine the body literally from head (fig. B.1) to toe (fig. B.10), studying its regions in more detail. To make the most profitable use of this atlas, refer back to the skeletal and muscular anatomy in chapters 9 to 11. Relate drawings of the clavicles in chapter 9 to the photograph in figure B.1, for example. Study the shape of the scapula in chapter 9 and see how much of it you can trace on the photographs in figure

B.4. See if you can relate the tendons visible on the hand (fig. B.6) to the muscles of the forearm illustrated in chapter 11.

For learning surface anatomy, there is a resource available to you that is far more valuable than any laboratory model or textbook illustration—your own body. For the best understanding of human structure, compare the art and photographs in this book with your body or with structures visible on a study partner. In addition to bones and muscles, you can palpate a number of superficial arteries, veins, tendons, ligaments, and cartilages, among other structures. By palpating regions such as the shoulder, elbow, or ankle, you can develop a mental image of the subsurface structures better than you can obtain by looking at two-dimensional textbook images. And the more you can study with other people, the more you will appreciate the variations in human structure and be able to apply your knowledge to your future patients or clients, who will not look quite like any textbook diagram or photograph you have ever seen. Through comparisons of art, photography, and the living body, you will get a much deeper understanding of the body than if you were to study this atlas in isolation from the earlier chapters.

At the end of this atlas, you can test your knowledge of externally visible muscle anatomy. The two photographs in figure B.11 have 30 numbered muscles and a list of 26 names, some of which are shown more than once in the photographs and some of which are not shown at all. Identify the muscles to your best ability without looking back at the previous illustrations, and then check your answers in appendix B.

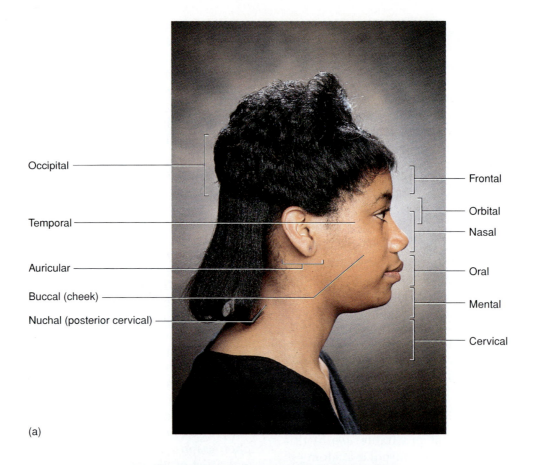

Occipital

Temporal

Auricular

Buccal (cheek)

Nuchal (posterior cervical)

Frontal

Orbital

Nasal

Oral

Mental

Cervical

(a)

Superciliary ridge

Superior palpebral sulcus

Inferior palpebral sulcus

Auricle (pinna) of ear

Philtrum

Labia (lips)

Trapezius muscle

Supraclavicular fossa

Frons (forehead)

Root of nose

Bridge of nose

Lateral commissure

Medial commissure

Dorsum nasi

Apex of nose

Ala nasi

Mentolabial sulcus

Mentum (chin)

Sternoclavicular joints

Clavicle

Sternal notch

Sternum

(b)

Figure B.1 The head and neck. (*a*) Anatomical regions of the head, lateral aspect. (*b*) Features of the facial region and upper thorax.

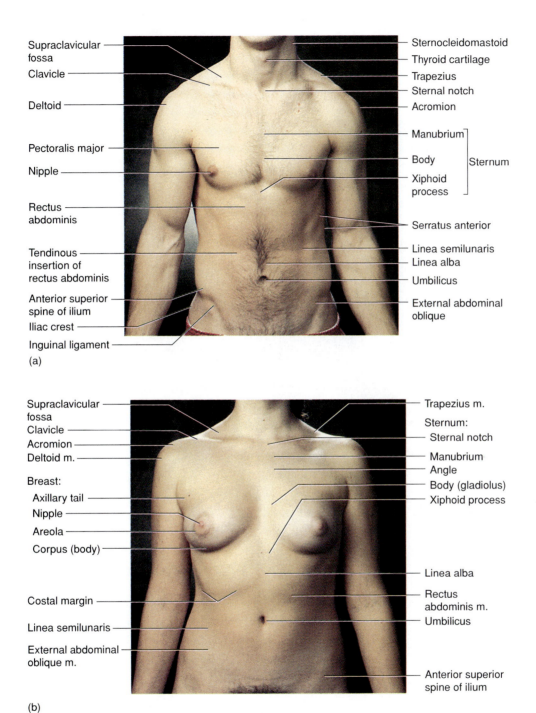

Supraclavicular fossa

Clavicle

Deltoid

Pectoralis major

Nipple

Rectus abdominis

Tendinous insertion of rectus abdominis

Anterior superior spine of ilium

Iliac crest

Inguinal ligament

Sternocleidomastoid

Thyroid cartilage

Trapezius

Sternal notch

Acromion

Manubrium

Body   Sternum

Xiphoid process

Serratus anterior

Linea semilunaris

Linea alba

Umbilicus

External abdominal oblique

(a)

Supraclavicular fossa

Clavicle

Acromion

Deltoid m.

Breast:

  Axillary tail

  Nipple

  Areola

  Corpus (body)

Costal margin

Linea semilunaris

External abdominal oblique m.

Trapezius m.

Sternum:

  Sternal notch

Manubrium

Angle

Body (gladiolus)

Xiphoid process

Linea alba

Rectus abdominis m.

Umbilicus

Anterior superior spine of ilium

(b)

Figure B.2 The thorax and abdomen, anterior (ventral) aspect. (*a*) Male; (*b*) female. Except for the breast, all of the features labeled are common to both sexes, though some are labeled only on the photograph that shows them best.

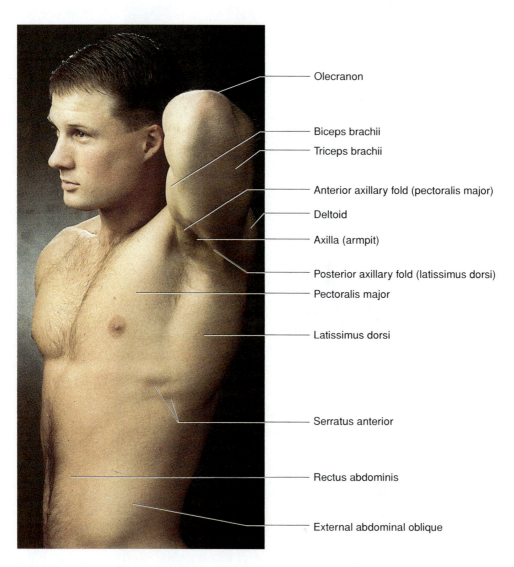

Olecranon

Biceps brachii
Triceps brachii

Anterior axillary fold (pectoralis major)

Deltoid

Axilla (armpit)

Posterior axillary fold (latissimus dorsi)
Pectoralis major

Latissimus dorsi

Serratus anterior

Rectus abdominis

External abdominal oblique

Figure B.3 The axillary region.

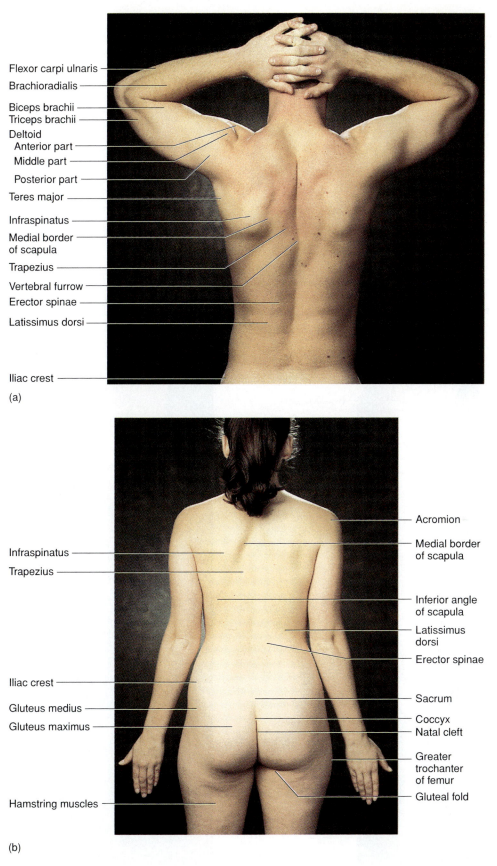

Flexor carpi ulnaris
Brachioradialis
Biceps brachii
Triceps brachii
Deltoid
  Anterior part
  Middle part
  Posterior part
Teres major
Infraspinatus
Medial border
of scapula
Trapezius
Vertebral furrow
Erector spinae
Latissimus dorsi

Iliac crest

(a)

Infraspinatus
Trapezius

Acromion
Medial border
of scapula

Inferior angle
of scapula
Latissimus
dorsi
Erector spinae

Iliac crest
Gluteus medius
Gluteus maximus

Sacrum
Coccyx
Natal cleft
Greater
trochanter
of femur
Gluteal fold

Hamstring muscles

(b)

Figure B.4 The back and gluteal region. (*a*) Male; (*b*) female. All of the features labeled are common to both sexes, though some are labeled only on the photograph that shows them best.

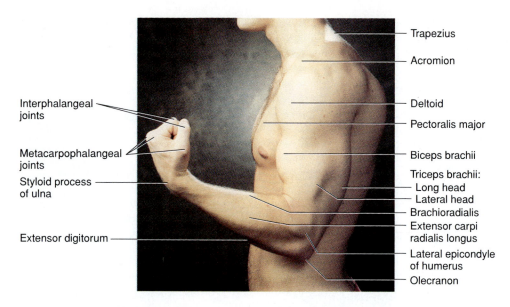

Interphalangeal joints

Metacarpophalangeal joints

Styloid process of ulna

Extensor digitorum

Trapezius

Acromion

Deltoid

Pectoralis major

Biceps brachii

Triceps brachii:
Long head
Lateral head

Brachioradialis

Extensor carpi radialis longus

Lateral epicondyle of humerus

Olecranon

Figure B.5 The upper extremity, lateral aspect.

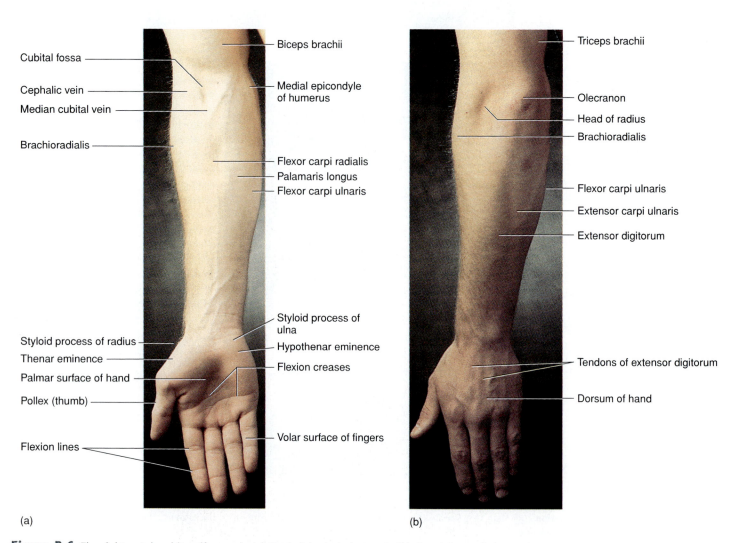

Cubital fossa

Cephalic vein

Median cubital vein

Brachioradialis

Biceps brachii

Medial epicondyle of humerus

Flexor carpi radialis
Palamaris longus
Flexor carpi ulnaris

Styloid process of radius
Thenar eminence
Palmar surface of hand
Pollex (thumb)

Flexion lines

Styloid process of ulna

Hypothenar eminence

Flexion creases

Volar surface of fingers

(a)

Triceps brachii

Olecranon

Head of radius

Brachioradialis

Flexor carpi ulnaris

Extensor carpi ulnaris

Extensor digitorum

Tendons of extensor digitorum

Dorsum of hand

(b)

Figure B.6 The right antebrachium (forearm). (*a*) Ventral (anterior) aspect; (*b*) dorsal (posterior) aspect.

Figure B.7 The left wrist and hand. (*a*) Ventral (anterior) aspect; (*b*) dorsal (posterior) aspect.

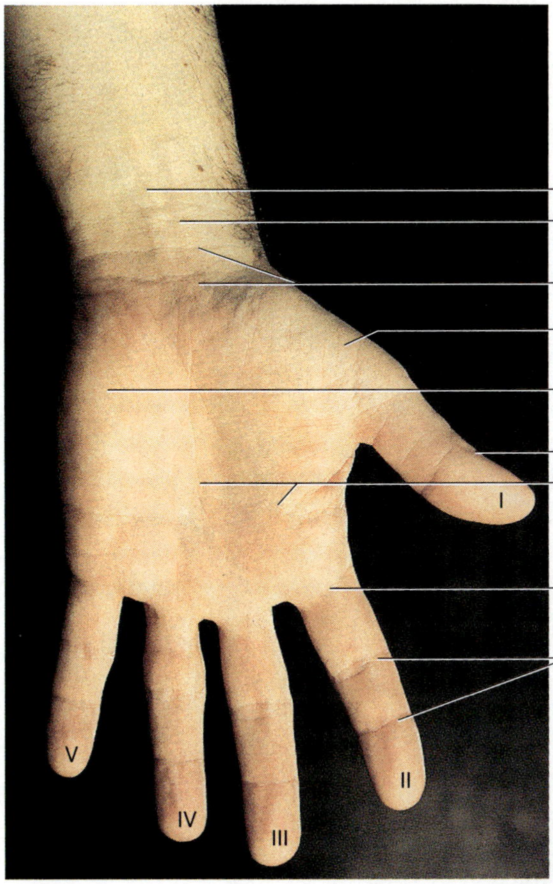

(a)

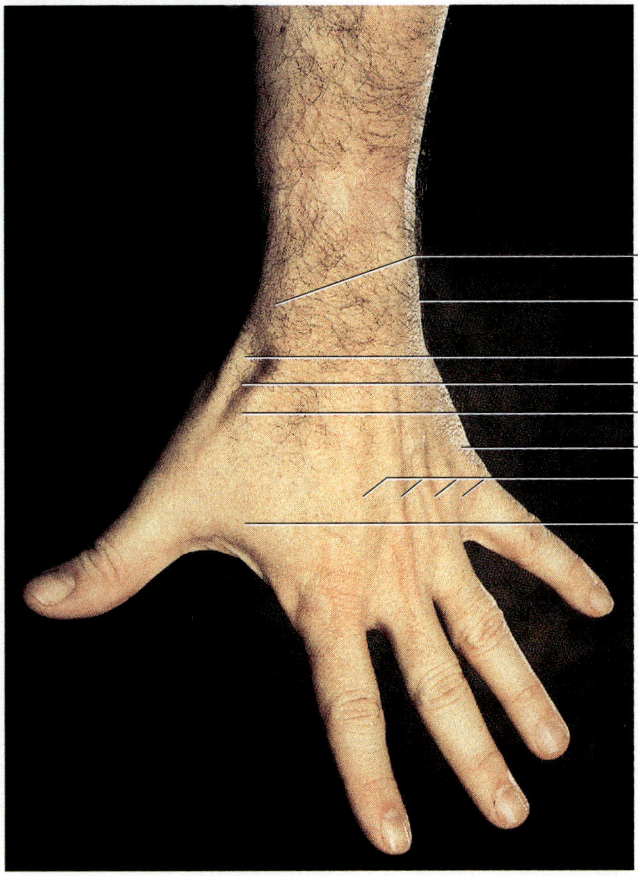

- Palmaris longus tendon
- Flexor carpi radialis tendon
- Flexion creases
- Thenar eminence
- Hypothenar eminence
- Pollex (thumb)
- Flexion creases
- Metacarpophalangeal joint
- Interphalangeal joints

I
II
III
IV
V

(b)

- Styloid process of radius
- Styloid process of ulna
- Extensor pollicis brevis tendon
- Anatomical snuffbox
- Extensor pollicis longus tendon
- Extensor digiti minimi tendon
- Extensor digitorum tendons
- Adductor hallucis

Lateral ←————|————→ Medial

Tensor fasciae latae —

Rectus femoris —

Gracilis —

Vastus lateralis —

Vastus medialis —

Patellar tendon —

Iliotibial band —

Patella —

Patellar ligament —

Tibial tuberosity —

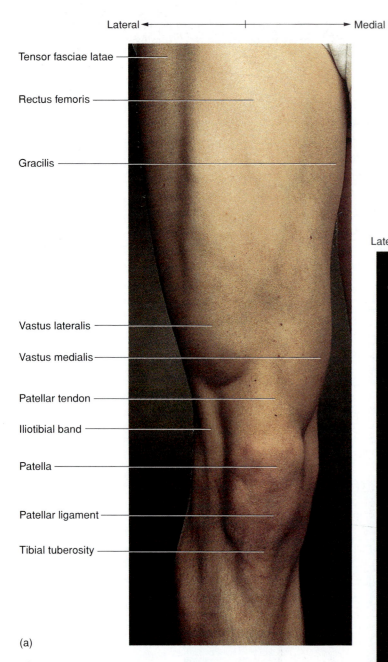

(a)

Figure B.8 The right thigh and knee. (*a*) Ventral (anterior) aspect; (*b*) dorsal (posterior) aspect.

Lateral ←————|————→ Medial

— Vastus lateralis

— Biceps femoris (long head)

— Semitendinosus

— Semimembranosus

— Gracilis

— Popliteal fossa

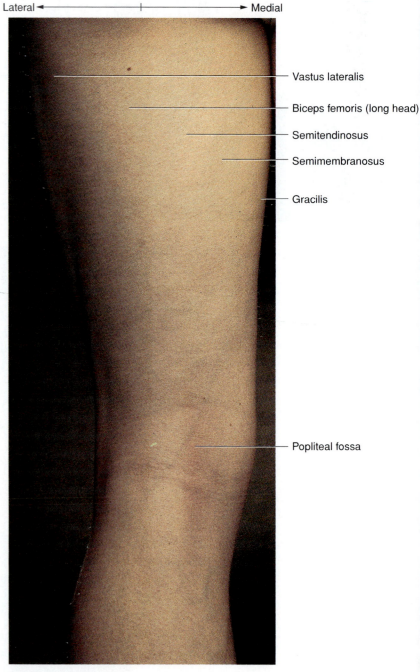

(b)

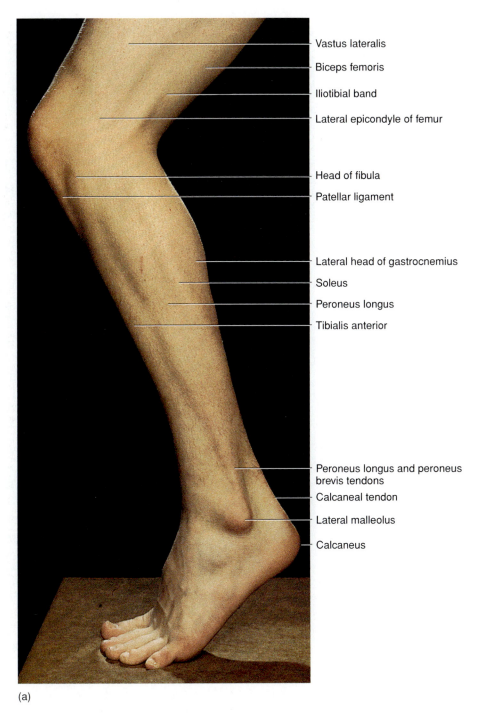

Vastus lateralis

Biceps femoris

Iliotibial band

Lateral epicondyle of femur

Head of fibula

Patellar ligament

Lateral head of gastrocnemius

Soleus

Peroneus longus

Tibialis anterior

Peroneus longus and peroneus brevis tendons

Calcaneal tendon

Lateral malleolus

Calcaneus

(a)

Figure B.9 The left leg and foot. (*a*) Lateral aspect—*Cont'd next page.*

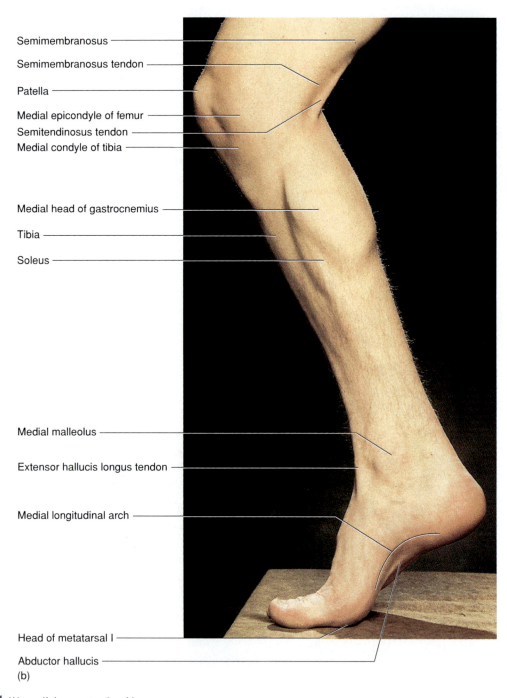

Semimembranosus

Semimembranosus tendon

Patella

Medial epicondyle of femur

Semitendinosus tendon

Medial condyle of tibia

Medial head of gastrocnemius

Tibia

Soleus

Medial malleolus

Extensor hallucis longus tendon

Medial longitudinal arch

Head of metatarsal I

Abductor hallucis

(b)

Figure B.9—Cont'd (*b*) medial aspect—*Cont'd.*

Medial ← | → Lateral

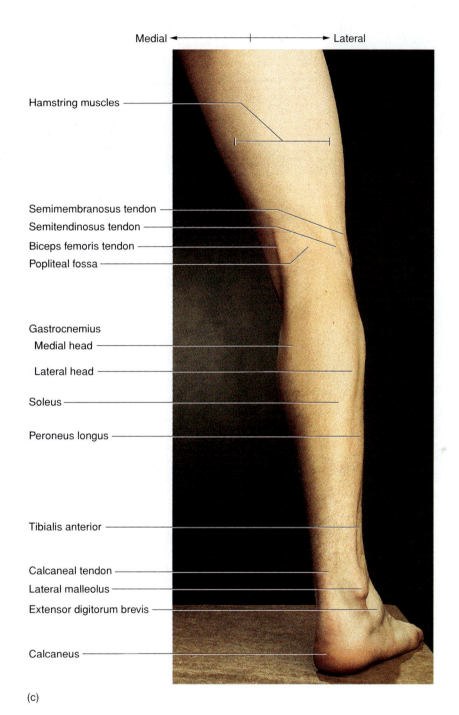

Hamstring muscles ──────

Semimembranosus tendon ──
Semitendinosus tendon ──
Biceps femoris tendon ──
Popliteal fossa ──

Gastrocnemius
  Medial head ──

  Lateral head ──

Soleus ──

Peroneus longus ──

Tibialis anterior ──

Calcaneal tendon ──
Lateral malleolus ──
Extensor digitorum brevis ──

Calcaneus ──

(c)

Figure B.9—Cont'd (c) dorsal (posterior) aspect.

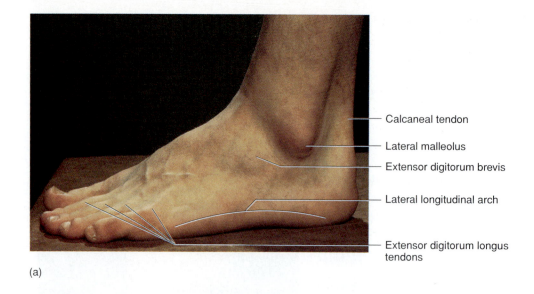

Calcaneal tendon

Lateral malleolus

Extensor digitorum brevis

Lateral longitudinal arch

Extensor digitorum longus tendons

(a)

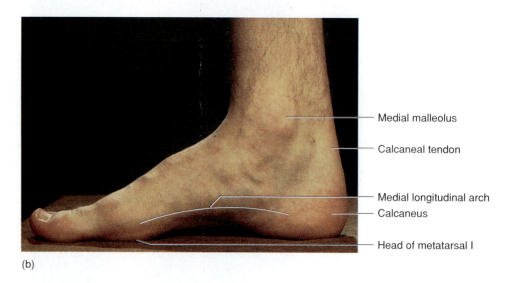

Medial malleolus

Calcaneal tendon

Medial longitudinal arch

Calcaneus

Head of metatarsal I

(b)

Figure B.10 The left foot. (*a*) Lateral aspect; (*b*) medial aspect—*Cont'd.*

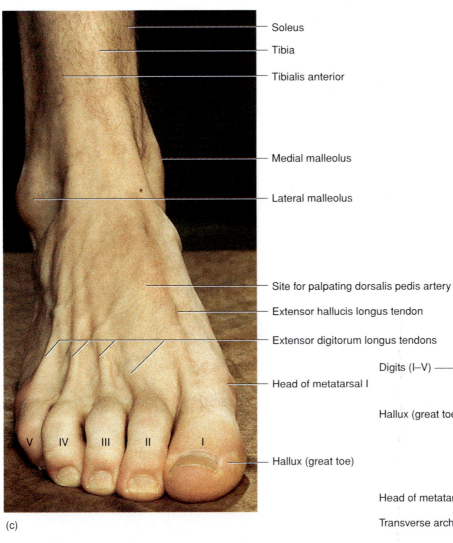

- Soleus
- Tibia
- Tibialis anterior
- Medial malleolus
- Lateral malleolus
- Site for palpating dorsalis pedis artery
- Extensor hallucis longus tendon
- Extensor digitorum longus tendons
- Head of metatarsal I
- Hallux (great toe)

V IV III II I

(c)

Figure B.10—Cont'd (c) dorsal aspect; (d) plantar (ventral) aspect.

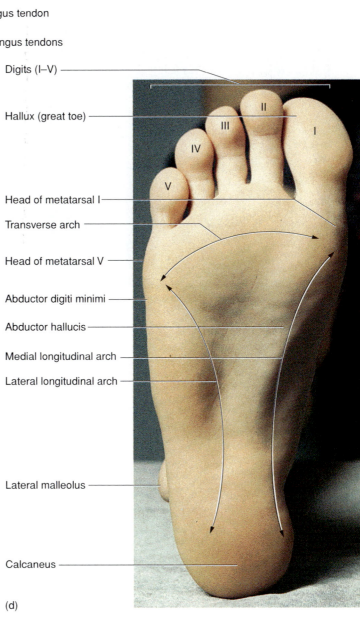

- Digits (I–V)
- Hallux (great toe)
- Head of metatarsal I
- Transverse arch
- Head of metatarsal V
- Abductor digiti minimi
- Abductor hallucis
- Medial longitudinal arch
- Lateral longitudinal arch
- Lateral malleolus
- Calcaneus

II
III
I
IV
V

(d)

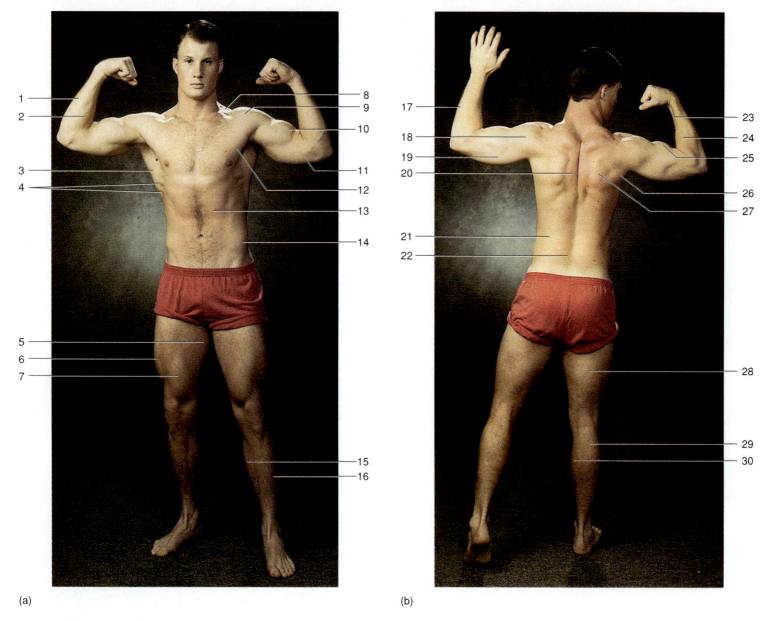

(a)   (b)

Figure B.11 To test your knowledge of muscle anatomy, match the 30 labeled muscles on these photographs to the alphabetical list of muscles below. Answer as many as possible without referring back to the previous illustrations. Some of these names will be used more than once, since the same muscle may be shown from different perspectives, and some of these names will not be used at all. The answers are in appendix B.

a. biceps brachii
b. brachioradialis
c. deltoid
d. erector spinae
e. external oblique
f. flexor carpi ulnaris
g. gastrocnemius
h. gracilis
i. hamstrings

j. infraspinatus
k. latissimus dorsi
l. pectineus
m. pectoralis major
n. rectus abdominis
o. rectus femoris
p. serratus anterior
q. soleus
r. splenius capitis

s. sternocleidomastoid
t. subscapularis
u. teres major
v. tibialis anterior
w. transversus abdominis
x. trapezius
y. triceps brachii
z. vastus lateralis

chapter twelve

12

Muscular Tissue

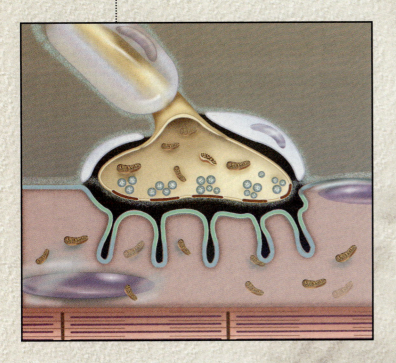

Special Topics

the unity of form and function

Brushing up

To understand this chapter, it is essential that you understand or brush up on the following concepts:

▶ Aerobic and anaerobic metabolism (p. 101)
▶ Plasma membranes and the functions of membrane proteins (pp. 112–114)
▶ The three types of muscle tissue (pp. 172–174)
▶ Structure of a neuron (p. 174)
▶ Gap junctions (p. 193)

Muscle tissue is composed of elongated cells that are capable of shortening when stimulated. A muscle cell is essentially a device for converting the chemical energy of ATP into the mechanical energy of contraction. This chapter discusses contraction at the cellular and molecular level and explains the basis of such aspects of performance as warm-up, strength, endurance, and fatigue.

The three types of muscle tissue—*skeletal, cardiac,* and *smooth*—were described and compared in chapter 6. This chapter is concerned primarily with the microscopic anatomy and physiology of skeletal muscle. Smooth muscle is discussed in depth in the last section of this chapter and cardiac muscle in chapter 19. The functions of muscle were listed in the preceding chapter.

Types and Characteristics of Muscle Tissue

▼Objectives
When you have completed this section, you should be able to
• describe the physiological properties that all muscle types have in common;
• list the defining characteristics of skeletal muscle; and
• state the functions of the series-elastic components of a muscle.

Universal Characteristics of Muscle

Regardless of type, all muscle tissue has the following characteristics essential to its functions:

• **Excitability (irritability).** Excitability is a property of all living cells, but muscle cells respond to stimuli in a unique way: they produce an electrical current that travels along the plasma membrane and triggers contraction of the cell.
• **Contractility.** Muscle fibers are unique in their ability to shorten substantially when stimulated. This enables them to exert tension on other tissues to which they are mechanically coupled.
• **Extensibility.** In order to contract, a muscle cell must also be extensible—able to stretch again between contractions. Most cells rupture if they are stretched even a little, but skeletal muscle fibers can safely stretch to as much as three times their contracted length.

• **Elasticity.** When a muscle cell is stretched and the tension is then released, it recoils to its original resting length. Elasticity, commonly misunderstood as the ability to stretch, refers to this tendency of a muscle cell to return to its original length when tension is released.

Skeletal Muscle

Skeletal muscle may be defined as voluntary striated muscle that is usually attached to one or more bones. A typical skeletal muscle cell is about 10 μm in diameter and 100 μm long; some are as thick as 100 μm and as long as 30 cm. Because of their extraordinary length, skeletal muscle cells are usually called muscle *fibers.* A skeletal muscle fiber exhibits alternating light and dark transverse bands, or **striations,** that reflect the overlapping arrangement of the internal contractile proteins. Cardiac muscle cells are also striated but differ structurally from skeletal muscle fibers (see chapter 19). Skeletal muscle is called **voluntary** because it is usually subject to conscious control. The other types of muscle are **involuntary** (not usually under conscious control), and they are never attached to the skeleton.

Series-Elastic Components

As described in the preceding chapter (pp. 330–331), skeletal muscles are attached to the bones by a series of connective tissue elements—endomysium, perimysium, epimysium, fascia, and tendon—called the **series-elastic components** of the muscular system (fig. 12.1). Unlike muscle tissue, they are not excitable or contractile, but they are extensible and elastic—they stretch under tension and recoil to their original length when released.

When a muscle begins to contract, it generates a force called **internal tension** that stretches the series-elastic components. When these are taut, they deliver a force, the **external tension,** to the load to be moved—usually a bone at the muscle's insertion. By analogy, imagine lifting a weight off a table with a rubber band. At first, the internal tension stretches the rubber band; then the external tension lifts the weight. Because no energy transfer is 100% efficient, the external tension that a muscle exerts on a bone is always somewhat less than the internal tension. The series-elastic components have the additional function of helping to stretch a muscle to its resting length when contraction ceases.

········· **Key Point Review** ·········

1 How is skeletal muscle different from any other type of muscle?
2 Define *excitability, contractility, extensibility,* and *elasticity.*
3 Why would the skeletal muscles perform poorly if not for their series-elastic components?

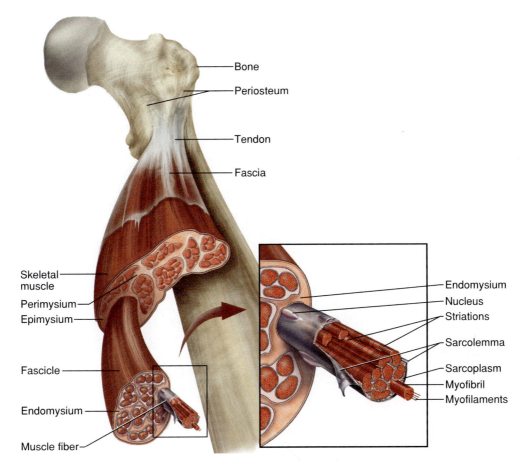

Figure 12.1 The series-elastic components of a muscle. The endomysium, perimysium, epimysium, fascia, tendon, and periosteum provide a tough, continuous series of collagen fibers attaching a muscle to a bone. *Inset:* Detail of a single muscle fiber. ✗ 🔳

Microscopic Anatomy of Skeletal Muscle

▼**Objectives**
When you have completed this section, you should be able to
• describe the structural components of a muscle fiber;
• relate the striations of a muscle fiber to the overlapping arrangement of its protein myofilaments; and
• name the proteins that compose the myofilaments and state the function of each.

The Muscle Fiber

In order to understand muscle contraction, you must know how the organelles and macromolecules of a muscle fiber are arranged. The plasma membrane of the fiber, called the **sarcolemma,**[1] has tunnel-like infoldings called **transverse tubules (T tubules)** that penetrate the inside of the fiber. Several flattened or sausage-shaped nuclei lie immediately beneath the sarcolemma. The reason for the cell's multinucleate condition is that it forms by the fusion of embryonic cells

called *myoblasts*, each of which contributes one nucleus to the mature fiber.

The cytoplasm of the fiber, called **sarcoplasm,** is occupied mainly by long, threadlike **myofibrils** about 1 μm in diameter (fig. 12.2). Each myofibril consists of a bundle of parallel protein microfilaments called **myofilaments.** Most other organelles of the cell, such as mitochondria and smooth endoplasmic reticulum, are located between adjacent myofibrils. The smooth ER of a muscle fiber is called **sarcoplasmic reticulum (SR).** It forms a network around each myofibril, and alongside the T tubules it exhibits dilated sacs called **terminal cisternae.** The SR is a reservoir for calcium ions and has gated channels in its membrane that can release a flood of calcium into the cytosol. A T tubule and the terminal cisternae on each side of it form a *triad.* The function of a T tubule is to carry an electrical current from the surface to the interior of the cell, where it stimulates the terminal cisternae to release calcium. The sarcoplasm also contains an abundance of **glycogen** and a red pigment called **myoglobin.** Glycogen provides stored energy for muscle contraction, and myoglobin binds oxygen until it is needed for muscular activity. The

---

1. *sarco* = flesh, muscle + *lemma* = husk

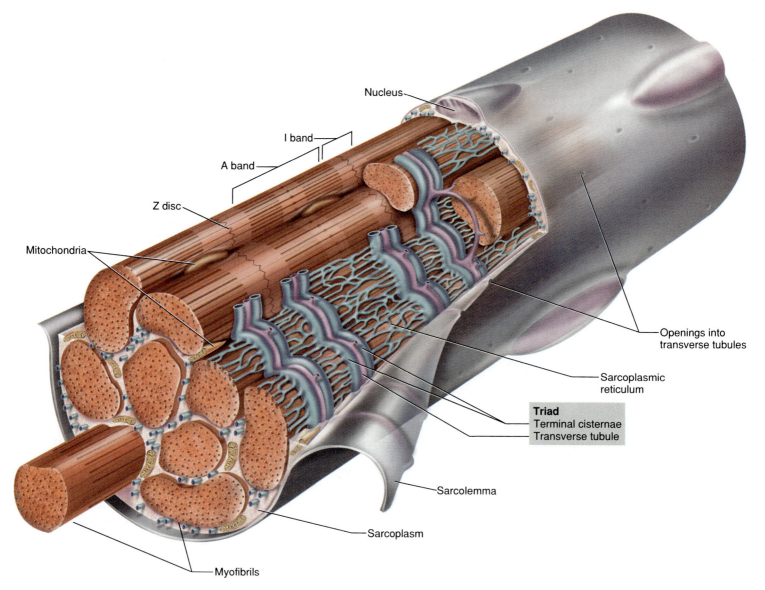

Figure 12.2 The structure of a skeletal muscle fiber. This is a single cell containing 11 myofibrils (9 shown at the left end and 2 cut off at midfiber). ✗ ▭

roles of glycogen and myoglobin are further discussed later in this chapter.

Myofilaments

The key to muscle contraction is the composition and arrangement of the myofilaments. There are two kinds of myofilaments, called thick and thin filaments. A **thick filament** (fig. 12.3*a, b*) is about 11 nm in diameter and is made of several hundred molecules of a protein called **myosin.** Each myosin molecule is shaped somewhat like a golf club, with a shaftlike *tail* and a globular *head* projecting from the tail at an angle. A thick filament may be likened to a bundle of 200 to 500 such "golf clubs," with their heads directed outward in a spiral array around the bundle. The heads on one half of the thick filament angle to the left, and the heads on the

other half angle to the right; in the middle is a *bare zone* with no heads.

A **thin filament,** approximately 5 to 6 nm in diameter, is composed primarily of 300 to 400 molecules of a protein called **fibrous (F) actin.** Each F actin molecule is a string of beadlike subunits called **globular (G) actin.** Each G actin has an **active site** that can bind to the head of a myosin molecule. Some authorities hold that a thin filament consists of two F actin molecules twined around each other (fig. 12.3*c*), and others believe it consists of only one twisted strand of F actin.

F actin has a spiral groove along its length that is occupied by 40 to 60 molecules of yet another protein called **tropomyosin.** When a muscle fiber is relaxed, tropomyosin blocks the active sites of actin and prevents myosin from binding to it. Each tropomyosin molecule, in turn, has smaller proteins called **troponin**

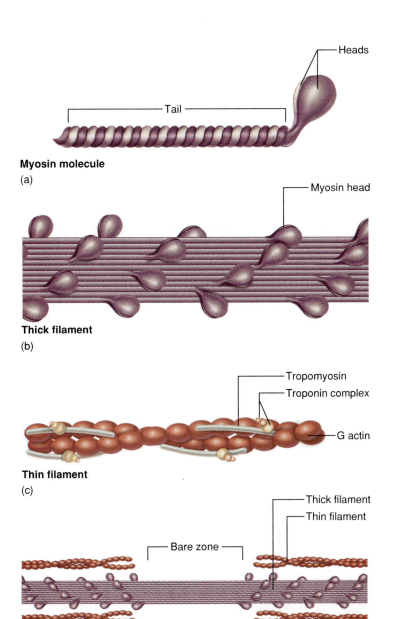

Myosin molecule

(a)

Thick filament

(b)

Tropomyosin

Troponin complex

G actin

Thin filament

(c)

Thick filament

Thin filament

Bare zone

Portion of a sarcomere showing the overlap of thick and thin filaments

(d)

Figure 12.3 Molecular structure of the myofilaments. (*a*) A single myosin molecule consists of two intertwined polypeptides forming a filamentous tail and a double globular head. (*b*) A thick myofilament consists of 200 to 500 myosin molecules bundled together with the heads projecting outward in a spiral array. (*c*) A thin myofilament consists of two intertwined chains of G actin molecules, smaller filamentous tropomyosin molecules, and a three-part protein called troponin associated with the tropomyosin. (*d*) A region of overlap between the thick and thin myofilaments.

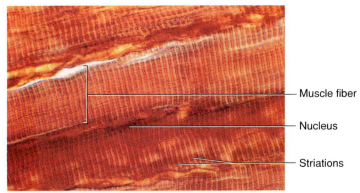

Figure 12.4 Photomicrograph of a skeletal muscle (×1,000).

Muscle fiber

Nucleus

Striations

bound to it at regular intervals. Troponin is an aggregate of three polypeptides: *troponin I*, which binds to actin; *troponin T*, which binds to tropomyosin; and *troponin C*, which binds to calcium ions. This is a good example of the quaternary protein structure described in chapter 3.

Tropomyosin and troponin are called the **regulatory proteins** of a muscle because they act like a switch to determine when it can contract and when it cannot. Several clues as to how they do this may be apparent from what has already been said—calcium ions are released into the sarcoplasm to activate contraction; troponin C is a calcium receptor; troponin C is bound to tropomyosin; and tropomyosin blocks the active sites of actin, so that myosin cannot bind to it when the muscle is not stimulated. Perhaps you already have some idea of the contraction mechanism to be explained shortly.

Striations

Myosin and actin are not unique to muscle; these proteins occur in all cells, where they function in cellular motility, mitosis, and transport of intracellular materials. In skeletal and cardiac muscle they are especially abundant, however, and are organized into a geometrically precise array that accounts for the striations of these two muscle types (figs. 12.4 and 12.5).

Striated muscle has dark **A bands** alternating with lighter **I bands**. (*A* stands for *anisotropic* and *I* for *isotropic*, which refers to the way these bands affect polarized light.) Each A band consists of thick filaments lying side by side. Part of the A band, where thick and thin filaments overlap, is especially dark. Each thick filament is surrounded by a hexagonal array of thin filaments. In the middle of the A band, there is a lighter region called the **H band**,[2] into which the thin filaments do not reach.

2. H = *helle* = bright

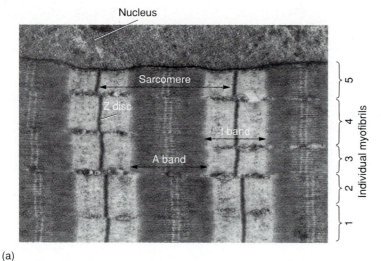

Nucleus

Sarcomere

Z disc

I band

A band

Individual myofibrils

5 4 3 2 1

(a)

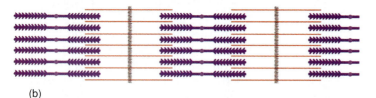

(b)

(c)

Figure 12.5 Molecular basis of the striations of skeletal muscle. (*a*) Five myofibrils of a single muscle fiber, showing the striations in the relaxed state (TEM ×20,000). (*b*) The overlapping pattern of thick and thin myofilaments that accounts for the striations seen in figure *a*. (*c*) The pattern of myofilaments in a fully contracted muscle fiber. Note that all myofilaments are the same length as before, but they overlap to a greater extent. Which band narrows or disappears when muscle contracts? 🏃 ▭

Each I band is bisected by a dark narrow line, traditionally called the *Z line*. The Z line is actually a disc of protein to which the thin filaments are anchored. We will thus refer to it as a **Z disc,**[3] which is closer to the original German origin of the term. Each segment of a myofibril from one Z disc to the next, called a **sarcomere**[4] (SAR-co-meer), is a functional contractile unit of the muscle fiber. A muscle shortens because its individual sarcomeres shorten and pull the Z discs closer to each other. The sarcolemma appears to be attached by way of the cytoskeleton to the Z discs of the sarcomeres. As the Z discs are pulled closer together during contraction, they may therefore pull on the sarcolemma to achieve overall shortening of the cell.

3. Z = *Zwischenscheibe* (ZVISH-en-shy-beh) = "between disc"
4. *mer* = part, segment

The terminology of muscle fiber structure is summarized in table 12.1; this table may be a useful reference as you study the mechanism of contraction.

Muscle Innervation and Membrane Potentials

▼**Objectives**

When you have completed this section, you should be able to
• identify the structures of a neuromuscular junction and explain their functions;
• explain what a motor unit is and how motor units relate to muscle contraction; and
• explain how a cell develops an electrical charge across its plasma membrane, and why this charge is essential to the function of nerve and muscle cells.

Unlike cardiac or smooth muscle, skeletal muscle does not contract unless it is stimulated by a nerve; if its nervous connections are severed or poisoned, the muscle is paralyzed. Thus, muscle contraction cannot be understood without first understanding the relationship between nerve and muscle cells.

Motor Neurons

Skeletal muscles are innervated by *somatic motor neurons* located in the brain stem and spinal cord; they issue axons called **somatic motor fibers** to the skeletal muscles. (Later these are contrasted with *autonomic motor fibers*, which innervate cardiac and smooth muscle.) A somatic motor fiber commonly divides into 200 or so branches at its distal end. Each branch travels through the endomysium and terminates on the middle portion of a single muscle fiber (fig. 12.6). Each muscle fiber is supplied by only one motor nerve fiber.

The Motor Unit

When a nerve impulse arrives at the end of an axon, it stimulates all the muscle fibers supplied by that neuron to contract in unison. Since they behave as a single functional unit, one nerve fiber and all the muscle fibers innervated by it are called a **motor unit.** The muscle fibers of a single motor unit are not all clustered together—rather, they are dispersed throughout a muscle

Table 12.1 Structures of a Muscle Fiber

Term	Definition
General Structure and Contents of the Muscle Fiber	
Sarcolemma	The plasma membrane of a muscle fiber
Sarcoplasm	The cytoplasm of a muscle fiber
Glycogen	An energy-storage polysaccharide abundant in muscle
Myoglobin	A red muscle pigment that binds and stores oxygen
Sarcoplasmic reticulum	The smooth ER of a muscle fiber; a Ca^{2+} reservoir
Terminal cisternae	The dilated ends of sarcoplasmic reticulum adjacent to a T tubule
T tubule	A tunnel-like extension of the sarcolemma extending from one side of the muscle fiber to the other; conveys electrical signals from the cell surface to its interior
Triad	A T tubule and the terminal cisternae on each side of it
Myofibril	A bundle of myofilaments
Myofilaments and Constituent Proteins	
Myofilament	A threadlike complex of several hundred contractile protein molecules
Thick filament	A myofilament about 11 nm diameter composed of bundled myosin molecules
Thin filament	A myofilament about 5 to 6 nm diameter composed of actin, troponin, and tropomyosin
Myosin	A protein with a long shaftlike tail and a globular head
F actin	A fibrous protein made of a long chain of G actin molecules twisted into a helical pattern
G actin	A globular or bean-shaped subunit of F actin with an active site for binding myosin
Regulatory proteins	Troponin and tropomyosin, proteins that do not directly engage in the sliding filament process of muscle contraction but regulate the process of myosin-actin binding
Tropomyosin	A regulatory protein that lies in the groove of F actin and, in relaxed muscle, blocks the myosin-binding active sites
Troponin	A regulatory protein associated with tropomyosin that acts as a calcium receptor; composed of three subunits that bind to actin (troponin I), tropomyosin (troponin T), and calcium (troponin C)
Striations and Sarcomeres	
Striations	Alternating light and dark bands that cross a myofibril
A band	Dark band formed by parallel thick filaments that partly overlap the thin filaments
H band	A lighter region in the middle of an A band that contains thick filaments only; thin filaments do not reach this far into the A band in relaxed muscle
I band	A light band composed of thin filaments only
Z disc	A disc to which thin filaments are anchored at each end of a sarcomere; appears as a narrow dark line in the middle of the I band
Sarcomere	The distance from one Z disc to the next; the contractile unit of a muscle fiber

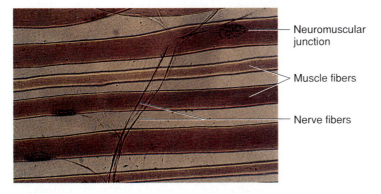

Figure 12.6 Motor nerve fibers innervating a group of skeletal muscle fibers. ✗

- Neuromuscular junction
- Muscle fibers
- Nerve fibers

(fig. 12.7). Thus, when they are stimulated, they cause a weak contraction over a wide area—not just a localized twitch in one small region.

Where fine control is needed, as in the muscles of speech and eye movements, motor units are relatively small—that is, each nerve fiber innervates relatively few muscle fibers. The muscles of eye movement, for example, average about 23 muscle fibers per motor neuron. The neurons of small motor units have relatively small somas and tend to be easily excited. By contrast, the motor units tend to be larger in muscles where strength is more important than precision. Each nerve fiber innervates about 500 muscle fibers in the quadriceps femoris and about 1,000 muscle fibers in the

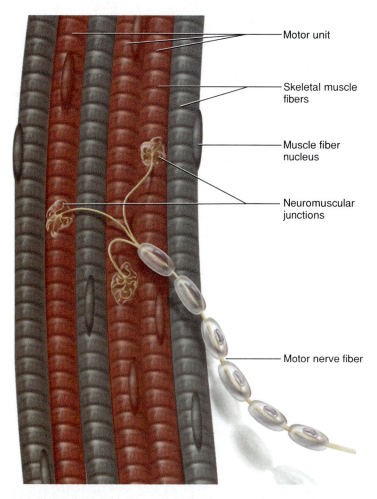

Motor unit

Skeletal muscle fibers

Muscle fiber nucleus

Neuromuscular junctions

Motor nerve fiber

Figure 12.7 A motor nerve unit consists of one somatic motor nerve fiber and all the muscle fibers innervated by it. Note that the muscle fibers of this motor unit (color) are not clustered together but distributed among muscle fibers belonging to other motor units (gray). ⚡

gastrocnemius, for example. The somas of large motor units are relatively large and less easily excited—it takes stronger stimuli to activate them.

One advantage of having multiple motor units in a muscle is that they are able to "work in shifts." Muscle fibers fatigue when subjected to continual stimulation. If all of the fibers in one of your postural muscles fatigued at once, for example, you might collapse. To prevent this, other motor units take over while the fatigued ones rest, and the muscle as a whole can sustain long-term contraction. The role of motor units in muscular strength is discussed later in the chapter.

The Neuromuscular Junction

The functional connection or point of close contact between a nerve fiber and its target cell is called a **synapse** (SIN-aps). When the second cell is a muscle fiber, the synapse is called a **neuromuscular junction** (fig. 12.8).

Each branch of a motor nerve fiber ends in a bulbous swelling called a **synaptic** (sih-NAP-tic) **knob,** which is nestled in a depression on the sarcolemma called the **motor end plate.** The two cells do not actually touch each other but are separated by a tiny gap, the **synaptic cleft,** about 10 nm wide (only a little wider than the 7 nm thickness of a typical plasma membrane). A third cell, called a *Schwann cell,* envelops the entire neuromuscular junction and isolates it from the surrounding tissue fluid.

The electrical signal (nerve impulse) traveling down a nerve fiber cannot cross the synaptic cleft like a spark jumping between two electrodes—rather, it causes the nerve fiber to release a neurotransmitter from secretory organelles called **synaptic vesicles.** Although many chemicals function as neurotransmitters, the one released at the neuromuscular junction is **acetylcholine** (ASS-eh-till-CO-leen) **(ACh).** Synaptic vesicles are especially concentrated above areas of the plasma membrane called **active zones,** the ACh release sites.

Directly across from the active zones, the sarcolemma exhibits infoldings called **junctional folds,** about 1 μm deep. There are about 50 million **ACh receptors**—integral proteins of the sarcolemma that bind acetylcholine—concentrated across from the active zones and extending about halfway into the junctional folds. Very few ACh receptors are found anywhere else on a muscle fiber. Junctional folds increase the surface area for receptor sites and ensure a more effective response to ACh. Several muscle cell nuclei clustered beneath the junctional folds are specifically dedicated to the synthesis of ACh receptors and other proteins of the motor end plate. A deficiency of ACh receptors leads to muscle paralysis in the disease myasthenia gravis (see chapter essay, p. 422).

The entire muscle fiber is surrounded by a basement membrane that passes through the synaptic cleft and virtually fills this space. In the cleft, it contains **acetylcholinesterase** (ASS-eh-till-CO-lin-ESS-ter-ase) **AChE),** an enzyme that breaks down ACh, shuts down the stimulation of muscle fibers, and allows a muscle fiber to relax (see special topic 12.1).

You will need to be very familiar with the foregoing terms to understand how a nerve stimulates a muscle fiber and how the fiber contracts. These terms are summarized in table 12.2 for your later reference.

Membrane Potentials

Muscle fibers and neurons are regarded as *electrically excitable cells* because their plasma membranes exhibit voltage changes in response to stimulation. The study of the electrical activity of cells, called **electrophysiology,** is a key to understanding nervous activity, muscle contraction, the heartbeat, and other physiological phenomena.

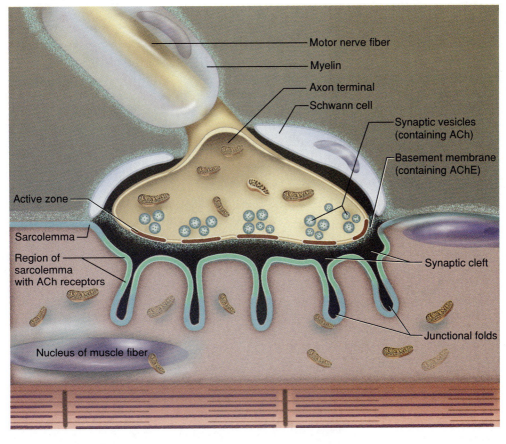

Figure 12.8 A neuromuscular junction.

Motor nerve fiber
Myelin
Axon terminal
Schwann cell
Synaptic vesicles (containing ACh)
Basement membrane (containing AChE)
Active zone
Sarcolemma
Region of sarcolemma with ACh receptors
Synaptic cleft
Junctional folds
Nucleus of muscle fiber

Electrical potential, which is measured in volts (V), is a form of potential energy that results from a greater concentration of charged particles at one point than at another. If a car battery has a *voltage,* for example, it has the *potential* to power the car and the battery is said to be charged, or **polarized.** To measure voltage, we connect an electrode to each pole and obtain a reading such as 12 V. If the electrode is attached to only one pole, there is no reading because voltage is a measure of the *difference* in number of charges (electrons) at one pole compared to the other.

A cell is essentially a little battery with a charge across its plasma membrane. More delicate instruments are needed to record its electrical potential, however (fig. 12.9). One electrode, called a *reference electrode,* can be a fine wire located anywhere in the extracellular fluid. The other one, called a *recording electrode,* or *microelectrode,* is inserted into the cell. It is usually a glass tube that has been heated and stretched to a thinness much finer than a hair so that it can pierce the plasma membrane without damaging the cell. It is filled with a concentrated electrolyte solution, such as

Table 12.2 Structures of the Neuromuscular Junction

Term	Definition
Neuromuscular junction	A functional connection between the distal end of a nerve fiber and the middle portion of a muscle fiber; consists of a synaptic knob and motor end plate
Synaptic knob	The dilated tip of a nerve fiber that contains synaptic vesicles
Motor end plate	A depression in the sarcolemma, near the middle of the muscle fiber, that receives the synaptic knob; contains acetylcholine receptors
Synaptic cleft	A gap of about 10 nm between the synaptic knob and motor end plate
Synaptic vesicle	A secretory vesicle in the synaptic knob that contains acetylcholine
Active zones	Regions of the plasma membrane of the synaptic knob where synaptic vesicles are clustered; acetylcholine release sites
Acetylcholine (ACh)	The neurotransmitter released by a somatic motor fiber that stimulates a skeletal muscle fiber (also used elsewhere in the nervous system)
ACh receptor	An integral protein in the sarcolemma of the motor end plate that binds to ACh
Acetylcholinesterase	An enzyme in the basement membrane of the muscle fiber where it crosses the synaptic cleft; responsible for degrading ACh and stopping the stimulation of the muscle fiber
Junctional folds	Invaginations of the membrane of the motor end plate where ACh receptors are especially concentrated; located across from the active zones

Figure 12.9 The method of recording electrical potentials across a plasma membrane. A hollow glass microelectrode filled with an electrolyte solution pierces the cell at one end and is connected to an amplifier by a wire at the other end. The reference electrode is a wire in contact with the nearby extracellular fluid. The fluorescent screen of the oscilloscope displays tiny voltage changes (millivolts) occurring across the plasma membrane over very brief periods of time (milliseconds).

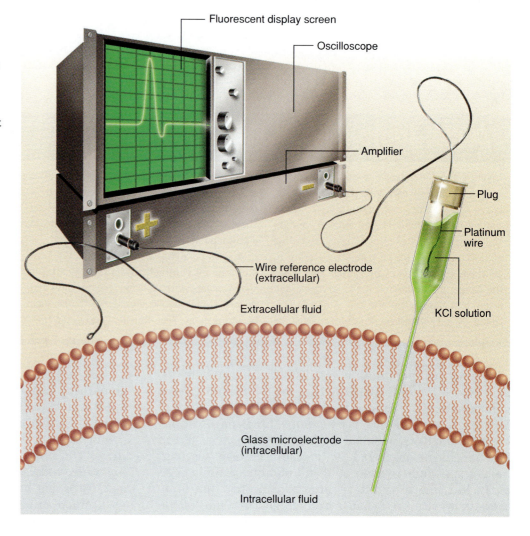

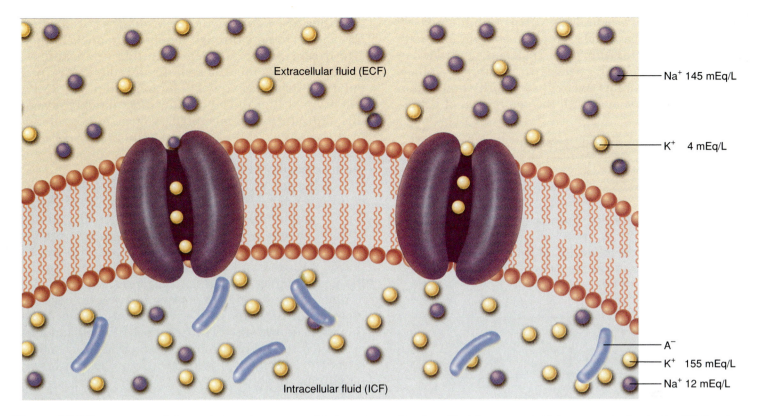

Extracellular fluid (ECF)

Na⁺ 145 mEq/L

K⁺ 4 mEq/L

A⁻
K⁺ 155 mEq/L
Na⁺ 12 mEq/L

Intracellular fluid (ICF)

Figure 12.10 Ionic basis of the resting membrane potential. Note that sodium ions (Na^+) have a much higher concentration in the ECF than in the ICF. The symbol A^- represents all cytoplasmic anions (proteins, phosphates, and so forth) that cannot pass through the plasma membrane but that contribute to the negative intracellular charge.

5 M potassium chloride, that conducts electricity to a wire at its larger end. The electrodes are connected to an amplifier and a device called an *oscilloscope.* An oscilloscope looks a little like a small television; it is sensitive enough to record membrane voltages in the millivolt (mV) range and events that occur in a matter of milliseconds.

The voltage across the plasma membrane of a resting cell is called the **resting membrane potential (RMP).** RMPs are much lower than the voltages recorded from storage batteries—typically about −75 mV for a neuron and −90 mV for a skeletal muscle fiber. These are stated as negative numbers to denote that the inside of the cell is negative compared to the outside. The RMP is determined by a combination of three factors: (1) *diffusion* of ions down their concentration gradients, (2) *selective permeability* of the plasma membrane, and (3) *electrostatic attraction,* the fact that cations (positive ions) are attracted to anions.

The RMP is largely due to an unequal concentration of potassium ions (K^+) across the plasma membrane. Potassium is about 40 times as concentrated in the intracellular fluid (ICF) as in the extracellular fluid (ECF) (fig. 12.10), and so it tends to diffuse down its concentration gradient, out of the cell. This leaves be-

hind a number of cytoplasmic anions that cannot diffuse through the plasma membrane because of their size or charge. These include ATP, other phosphates, sulfates, small organic acids, proteins, and nucleic acids.

As potassium leaves the cell and these anions stay behind, the ICF grows increasingly negative and more attractive to the positively charged potassium ions. The opposing forces of diffusion and electrostatic attraction soon reach an equilibrium at which K^+ is moving through the membrane in both directions at equal rates—therefore, the *net* diffusion of K^+ stops.

If K^+ diffusion were the only factor affecting the RMP, the plasma membrane of a resting neuron would have a potential of about −90 mV. Sodium ions, however, are about 12 times as concentrated in the ECF as in the ICF and have a tendency to diffuse into the cell. A resting plasma membrane is not very permeable to Na^+, so this diffusion is only a trickle; nevertheless, it is sufficient to cancel some of the negative internal charge and reduce the voltage across the membrane to about −75 mV—the resting membrane potential. The RMP is maintained by the Na^+-K^+ pump, which removes three Na^+ from the cell for every two K^+ it brings in, and therefore has the net effect of contributing to the negative intracellular charge.

When a nerve or muscle cell is stimulated, ion gates in the plasma membrane open, sodium ions rush into the cell, potassium ions rush out, and we can record quick, dramatic changes in membrane voltage called **action potentials.** We will see shortly how action potentials are associated with the contraction of a muscle fiber.

Key Point Review

8 Distinguish between acetylcholine, an acetylcholine receptor, and acetylcholinesterase. State where each is found and describe the function it serves.

9 Distinguish between a resting membrane potential and an action potential.

10 What two forces tend to drive potassium ions in opposite directions through a plasma membrane? Why is there a negative charge across the membrane when these forces are in equilibrium?

Contraction and Relaxation of Skeletal Muscle Fibers

▼Objectives

When you have completed this section, you should be able to
- explain how a nerve fiber stimulates a skeletal muscle fiber;
- explain how stimulation of a muscle fiber activates its contractile mechanism;
- explain the sliding filament mechanism of skeletal muscle contraction;
- explain how a muscle fiber relaxes; and
- explain why the force of a muscular contraction depends on the length of the muscle prior to stimulation.

The process of muscle contraction can be viewed as occurring in four major phases, each with several smaller steps:

1. **Excitation.** Action potentials in the nerve fiber give rise to action potentials in the muscle fiber.
2. **Excitation-contraction coupling.** Action potentials in the muscle fiber lead to activation of the myofilaments.
3. **Contraction.** Sliding of the thin myofilaments past the thick ones causes the muscle fiber to shorten.
4. **Relaxation.** When nervous stimulation ceases, the muscle relaxes.

Excitation

The events of excitation are listed and illustrated in figure 12.11. In brief, action potentials in the synaptic knob trigger the exocytosis of synaptic vesicles, which release acetylcholine (ACh) into the synaptic cleft. One action potential causes exocytosis of about 60 synaptic vesicles, and each vesicle releases about 10,000 molecules of ACh. The acetylcholine receptors on the mus-

cle fiber are *ligand-gated ion channels* that allow Na^+ and K^+ to pass through the same channels in opposite directions. As these channels open and Na^+ quickly diffuses into the cell, the muscle fiber reverses polarity—its voltage quickly jumping from the RMP of -90 mV to $+70$ mV. The outflow of K^+ quickly reverses this and returns the membrane voltage to a level close to the original RMP. This rapid fluctuation in membrane voltage at the motor end plate is called the **end-plate potential (EPP).**

Adjacent to the end plate, the sarcolemma has *voltage-gated ion channels,* some of which are specific for Na^+ and others for K^+. The EPP triggers the opening of these channels, and Na^+ and K^+ movements produce action potentials. The details of how an action potential is produced and propagates itself are explained in the next chapter. For present purposes, it is sufficient to note that action potentials spread away from the end plate in all directions, like ripples on a pond. When this wave of excitation reaches the T tubules, it continues down the tubules into the sarcoplasm and stimulates the terminal cisternae of the sarcoplasmic reticulum.

Think About It

An impulse begins at the middle of a 100-μm-long muscle fiber and travels 5 m/sec. How long would it take to reach the ends of the muscle fiber?

Excitation-Contraction Coupling

Excitation-contraction coupling refers to the physiological events that link the action potentials on the SR to the initial events in muscle contraction (fig. 12.12). Action potentials in the T tubules excite the adjacent terminal cisternae of the SR. Ion channels in the SR open and release a flood of calcium ions into the cytosol. The calcium ions bind to troponin C of the thin filaments, causing the troponin-tropomyosin complex to change shape and shift to a new position. This exposes the active sites on the actin filaments, allowing the heads of the myosin molecules to bind to the active sites and initiate contraction.

Contraction

Until sophisticated techniques in electron microscopy enabled cytologists to see the molecular organization of muscle fibers, the way in which the fibers became so much shorter remained a mystery. One hypothesis was that the plasma membrane folded like an accordion. In 1954, however, two researchers at the Massachusetts Institute of Technology, Jean Hanson and Hugh Huxley, proposed a model now called the **sliding filament theory.** This theory holds that the thin filaments slide over

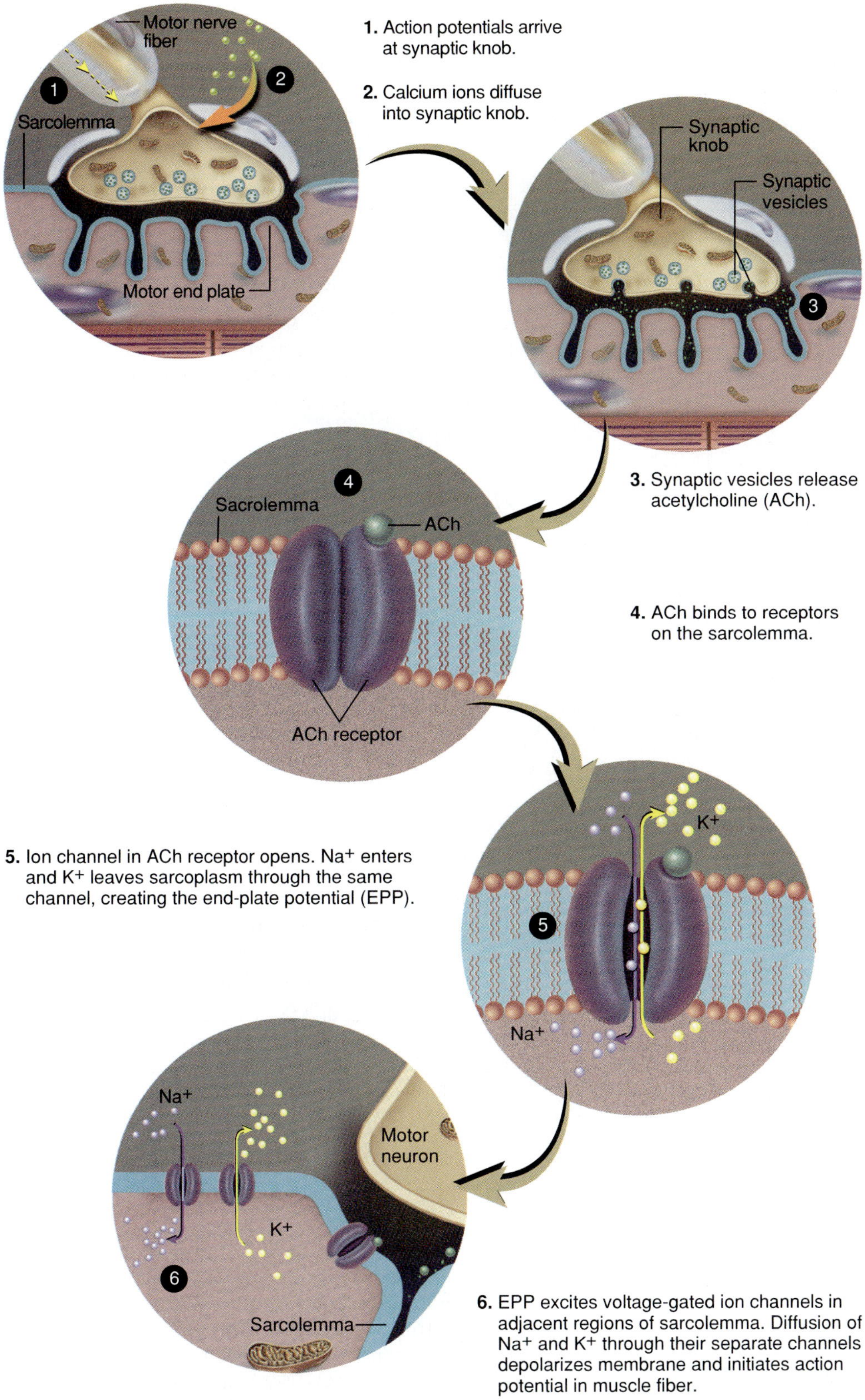
Motor nerve fiber
1
2
Sarcolemma
Motor end plate
1. Action potentials arrive at synaptic knob.
2. Calcium ions diffuse into synaptic knob.
Synaptic knob
Synaptic vesicles
3
3. Synaptic vesicles release acetylcholine (ACh).
4
Sacrolemma
ACh
ACh receptor
4. ACh binds to receptors on the sarcolemma.
5. Ion channel in ACh receptor opens. Na+ enters and K+ leaves sarcoplasm through the same channel, creating the end-plate potential (EPP).
K+
5
Na+
Na+
Motor neuron
K+
6
Sarcolemma
6. EPP excites voltage-gated ion channels in adjacent regions of sarcolemma. Diffusion of Na+ and K+ through their separate channels depolarizes membrane and initiates action potential in muscle fiber.

Figure 12.11 Excitation of the muscle fiber; events leading from action potentials in the nerve fiber to action potentials in the muscle fiber.

the thick ones and pull the Z discs behind them, causing the cell as a whole to shorten. The individual steps in this mechanism are shown in figure 12.13.

Some aspects of the sliding filament process can be compared to the way you might pull in the rope on a boat anchor. The clublike head of each myosin molecule contains an enzyme called **myosin ATPase** that can release energy from ATP by hydrolyzing it to ADP and phosphate (P_i). In preparation for action, a myosin head binds an ATP molecule and hydrolyzes it. ADP and P_i

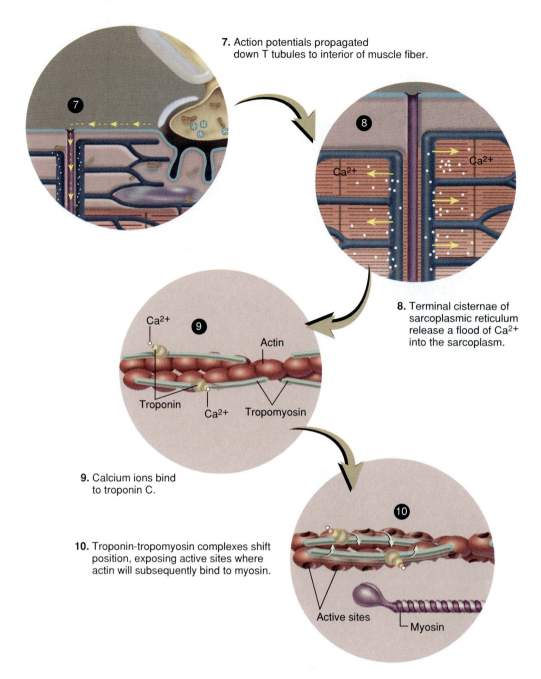

7. Action potentials propagated down T tubules to interior of muscle fiber.

8. Terminal cisternae of sarcoplasmic reticulum release a flood of Ca^{2+} into the sarcoplasm.

Ca²⁺

Ca²⁺

Actin

Ca²⁺

Troponin   Ca²⁺   Tropomyosin

9. Calcium ions bind to troponin C.

10. Troponin-tropomyosin complexes shift position, exposing active sites where actin will subsequently bind to myosin.

Active sites   Myosin

Figure 12.12 Excitation-contraction coupling; events leading from excitation of the muscle fiber to enabling myosin and actin to form cross bridges. The numbered steps in this figure begin where the previous figure left off. ⚡

remain temporarily bound to the head, and the head extends into a high-energy "cocked" position, like your hand reaching out to grasp an anchor rope. When an active site on the actin filament is exposed, the myosin head can bind to it, like your hand grasping the rope. Upon doing so, it releases the ADP and P_i, flexes into a lower energy position, and tugs on the actin filament, like your elbow flexing to pull on the rope. This is called the **power stroke** of the myosin head.

At the end of the power stroke, myosin binds to a new ATP molecule, releases the actin, and extends to its original position. This is the **recovery stroke.** You might expect that the actin filament would slip back to its original position, just as the boat anchor would sink again if you let go of the rope. To pull in an anchor, however, you would hold onto the rope with one hand as you released it with the other, using a hand-over-hand motion to repeatedly grasp the rope farther down its length. Similarly, there are many myosin heads pulling on a thin filament at once. At a given moment during contraction, about half of the heads are bound to the thin filament and the other half are extending forward to grasp the filament farther down. That is, the myosin heads of a thick filament do not all "stroke" at

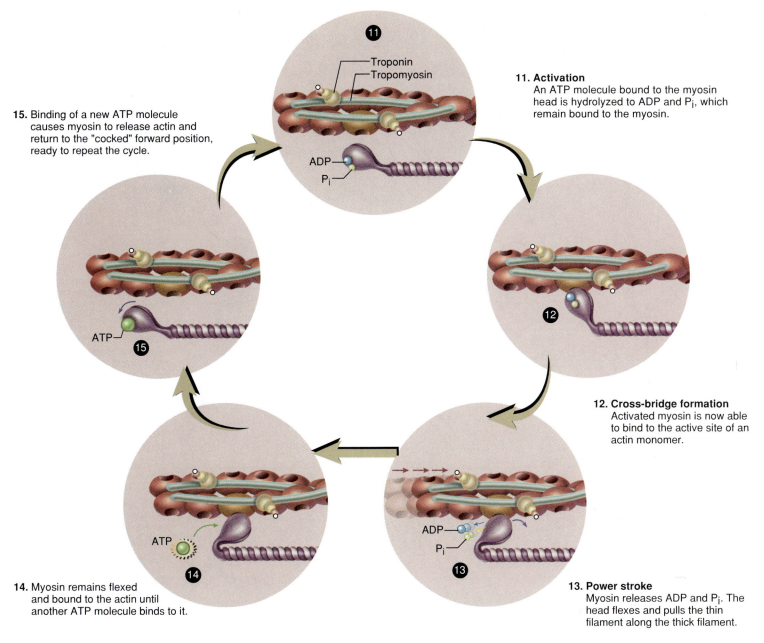

15. Binding of a new ATP molecule causes myosin to release actin and return to the "cocked" forward position, ready to repeat the cycle.

11. Activation
An ATP molecule bound to the myosin head is hydrolyzed to ADP and P_i, which remain bound to the myosin.

Troponin
Tropomyosin

ADP
P_i

ATP
15

11

12

ATP
14

ADP
P_i
13

12. Cross-bridge formation
Activated myosin is now able to bind to the active site of an actin monomer.

14. Myosin remains flexed and bound to the actin until another ATP molecule binds to it.

13. Power stroke
Myosin releases ADP and P_i. The head flexes and pulls the thin filament along the thick filament.

Figure 12.13 The sliding filament mechanism of muscle contraction, a cycle of repetitive events that generates tension in the muscle fiber and typically cause it to shorten. The numbered steps in this figure begin where the previous figure left off.

once but contract sequentially. By analogy, consider a millipede—a little wormlike animal with a few hundred tiny legs. Each leg takes individual jerky steps, but all the legs working together produce smooth, steady locomotion—just as all the heads of a thick filament collectively produce a smooth, steady pull on the thin filament. Note that even though the muscle fiber contracts, the *myofilaments do not become shorter,* any more than a rope becomes shorter as you pull in an anchor. The thin filaments slide over the thick ones, as the name of the theory implies.

A single cycle of power and recovery strokes by all the myosin heads in a muscle fiber would shorten the fiber by about 1%. A fiber, however, may shorten by as much as 40% of its resting length, so obviously the cycle of power and recovery must be repeated many times by each myosin head. Each head can carry out about five strokes per second, and each cycle of the head consumes one molecule of ATP.

Relaxation

When its work is done, a muscle fiber must relax and return to its resting length (fig. 12.14). The first step in achieving this is for the motor nerve fiber to stop generating action potentials and releasing ACh.

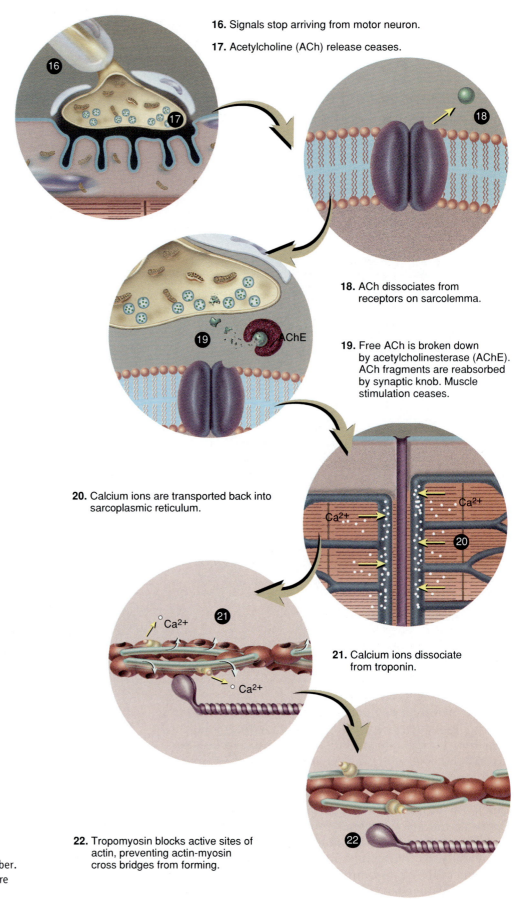

16. Signals stop arriving from motor neuron.

17. Acetylcholine (ACh) release ceases.

18. ACh dissociates from receptors on sarcolemma.

19. Free ACh is broken down by acetylcholinesterase (AChE). ACh fragments are reabsorbed by synaptic knob. Muscle stimulation ceases.

20. Calcium ions are transported back into sarcoplasmic reticulum.

21. Calcium ions dissociate from troponin.

22. Tropomyosin blocks active sites of actin, preventing actin-myosin cross bridges from forming.

Figure 12.14 Relaxation of the muscle fiber. The numbered steps in this figure begin where the previous figure left off.

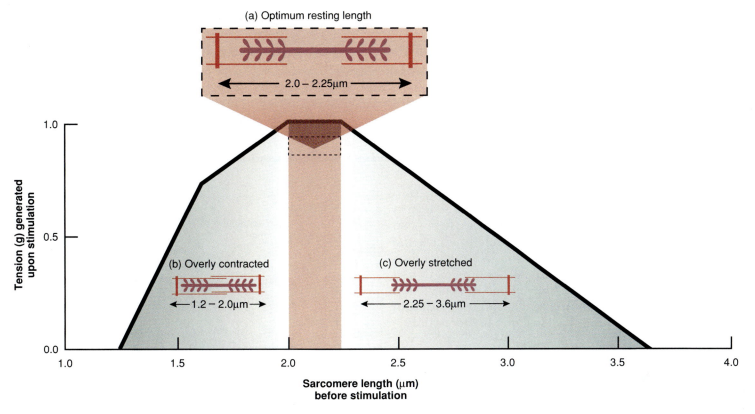

Figure 12.15 The length-tension relationship of skeletal muscle. (*a*) In a resting muscle fiber, the sarcomeres are usually 2.0–2.25 mm long, the optimum length for producing maximum tension when the muscle is stimulated to contract. Note how this relates to the degree of overlap between the thick and thin myofilaments. (*b*) If the muscle is overly contracted, the thick filaments butt against the Z discs and the fiber cannot contract very much more when it is stimulated. (*c*) If the muscle is overly stretched, there is so little overlap between the thick and thin myofilaments that few cross bridges can form between myosin and actin.

Acetylcholinesterase in the synaptic cleft breaks down the ACh that is already present into nonstimulatory fragments that are reabsorbed and recycled by the neuron. Then, in the absence of ACh, the muscle fiber stops generating its own action potentials. Active transport pumps in the sarcoplasmic reticulum pump calcium ions from the cytosol back into the cisternae. Here, the calcium binds to a protein called **calsequestrin** (CAL-see-QUES-trin) and is stored until the fiber is stimulated again. Since active transport requires ATP, you can see that *ATP is needed for muscle relaxation as well as for muscle contraction.* This is a factor in rigor mortis (see special topic 12.2). As calcium ions dissociate from troponin C and are pumped back into the sarcoplasmic reticulum, tropomyosin moves back into the position where it blocks the active sites of the actin filament. Myosin therefore can no longer bind to actin and exert a pull on it. The muscle returns to its resting length with the aid of two forces: (1) like a recoiling rubber band, the series-elastic components stretch the muscle; and (2) since muscles often occur in antagonistic pairs, the contraction of an antagonist helps lengthen the re-

laxed muscle. Contraction of the triceps brachii, for example, extends the elbow and lengthens the biceps brachii.

The Length-Tension Relationship and Tonus

Muscle fibers exhibit a **length-tension relationship**—the tension generated by contraction depends on how stretched or contracted the fiber was before it was stimulated. If a fiber is overly contracted at rest, its thick filaments are rather close to the Z discs (fig. 12.15). The stimulated muscle may contract a little, but then the thick filaments butt up against the Z discs and can go no farther. The contraction is therefore a weak one. On the other hand, if a muscle fiber is overly stretched before it is stimulated, there is relatively little overlap between its thick and thin filaments. When the muscle is stimulated, its myosin heads cannot "get a good grip" on the thin filaments, and again the contraction is weak. (As mentioned in chapter 11, this is one reason you should not bend at the waist to pick up a heavy object.

Rigor mortis[5] is the hardening of the muscles and stiffening of the body that begins 3 to 4 hours after death. The deteriorating sarcoplasmic reticulum releases calcium ions into the cytosol, and deterioration of the sarcolemma admits more calcium ions from the extracellular fluid. The calcium ions activate myosin-actin cross bridging and muscle contraction. Since the muscle cannot relax without ATP, and ATP is no longer produced after death, the fibers remain contracted until the myofilaments begin to decay. Rigor mortis peaks about 12 hours after death and then diminishes over the next 48 to 60 hours.

Muscles of the back become overly stretched and cannot contract effectively to straighten your spine against a heavy resistance.) Between these extremes, there is an optimum resting length at which a muscle produces the greatest force when it contracts. The central nervous system continually monitors and adjusts the length of a resting muscle, maintaining a state of partial contraction called **tonus (muscle tone).** This maintains optimum length and makes the muscles ideally ready for action.

Key Point Review

11 What change does ACh cause in an ACh receptor? How does this electrically affect the muscle fiber?

12 How does an end-plate potential lead to an increase in Ca^{2+} concentration in the sarcoplasm?

13 How do troponin and tropomyosin regulate the interaction between myosin and actin?

14 Describe the roles of ATP in the power and recovery strokes of a myosin head.

15 What steps are necessary for a contracted muscle fiber to return to its resting length?

Behavior of Whole Muscles

▼Objectives

When you have completed this section, you should be able to
- describe a muscle twitch and explain how it relates to threshold and latency;
- reconcile the fact that muscles contract with variable strength with the fact that individual muscle fibers contract in an all-or-none manner;
- explain how temporal summation produces tetanus and why tetanus is necessary to muscle function;
- distinguish between isometric, isotonic, concentric, and eccentric contractions; and
- describe treppe and explain how it relates to muscle warm-up.

Now that we have examined muscle contraction at the molecular and cellular level, we are prepared to move up to the organ grade of construction and consider the behavior of a muscle as a whole. The gastrocnemius (calf) muscle of a frog, which can be taken from the leg along with its connected sciatic nerve, is widely used to demonstrate a muscle's contractile properties (see spe-

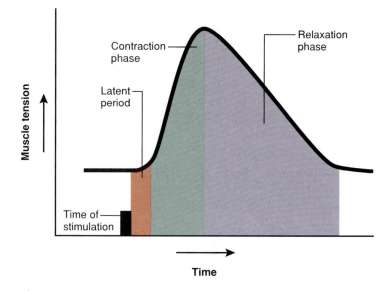

Figure 12.16 A muscle twitch.

cial topic 12.3). This nerve-muscle preparation can be attached to stimulating electrodes and to a recording device that produces a *myogram,* a chart of the timing and strength of the muscle's contraction.

Threshold, Latent Period, and Twitch

A weak electrical stimulus to a muscle or its nerve causes no contraction. By gradually increasing the voltage and stimulating the muscle again, we can determine the **threshold**—the minimum voltage necessary to produce muscle contraction. If the muscle is given a single brief stimulus at or above threshold, it shows a quick cycle of contraction and relaxation called a **twitch** (fig. 12.16). There is a delay of typically 2 milliseconds (msec) or so between the stimulus and the onset of the twitch. This very brief **latent period** represents the time required for excitation, for excitation-contraction coupling, and for tautening of the series-elastic components by the contracting muscle. The muscle then becomes

5. *rigor* = rigidity + *mortis* = of death

visibly shorter and moves a resisting object (load) at its distal end. The load could be the sensor of a recording apparatus or, in the body, the bone on which the muscle inserts. The interval during the twitch in which the muscle shortens is called the **contraction phase,** and the interval during which it returns to its resting length is the **relaxation phase.** The total twitch lasts from 7 to 100 msec depending on the type of muscle fiber.

Graded and All-or-None Responses

We have seen that below threshold, a muscle fiber does not contract at all. At threshold and above, increasing the stimulus voltage does not increase the force of contraction in the muscle fiber—the fiber contracts with maximum force regardless of stimulus strength. The muscle fiber behaves according to the **all-or-none law**—either it exhibits a maximum response or it exhibits none at all.

This may seem at odds with the behavior of an intact muscle, which must be able to contract with varying degrees of speed and strength—differently in lifting a glass of champagne than in lifting a heavy barbell, for example. If the frog sciatic nerve is stimulated with increasing voltages, the gastrocnemius muscle produces stronger twitches (fig. 12.17)—that is, the strength of a twitch is **graded.** How can the contraction strength vary if every muscle fiber behaves in an all-or-none fashion? The answer to this paradox is that higher voltages activate more axons in the sciatic nerve, which in turn stimulate a larger number of motor units in the muscle. The contraction strength of the muscle does not depend on *how strongly* individual fibers contract; it depends, rather, on *how many* motor units contract. This is discussed further in a later section on factors that determine strength.

Treppe

Consider a muscle in which a twitch lasts 30 msec. If it is stimulated every 50 msec, it has ample time to recover between stimuli, and each twitch is identical (fig. 12.18*a*). If it is stimulated more frequently—say every 35 msec—it still fully relaxes between stimuli, but successive twitches develop more tension (fig. 12.18*b*).

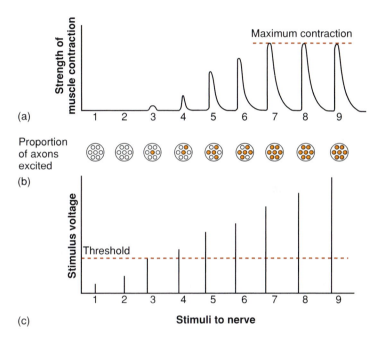

(a)

Proportion of axons excited

(b)

(c)   **Stimuli to nerve**

Figure 12.17 The relationship between stimulus intensity (*a*), the excitation of motor nerve fibers (*b*), and the strength of muscle contraction (*c*). Each vertical line in *a* represents one stimulus. At voltages below threshold (stimuli *1–2*), there is no response from the motor nerve fibers or muscle. As stimulus voltage reaches threshold (stimulus *3*), the most sensitive nerve fibers fire and weak muscle twitches are observed, representing the contractions of a small number of motor units. Stimuli *3–7*, of increasing voltage, activate more and more motor units until all motor units are active and the muscle contracts at maximum strength. Beyond that point, further increases in stimulus intensity (stimuli *8–9*) do not produce stronger contractions.

This increase in the strength of the twitches in response to stimuli of the same strength is called **treppe**[6] (TREP-eh), or the *staircase phenomenon,* after the appearance of the myogram. According to one hypothesis, treppe occurs because stimuli arrive so rapidly that the sarcoplasmic reticulum does not have time between stimuli to completely reabsorb all the calcium it has released. Thus, the calcium concentration in the cytosol

6. *treppe* = staircase

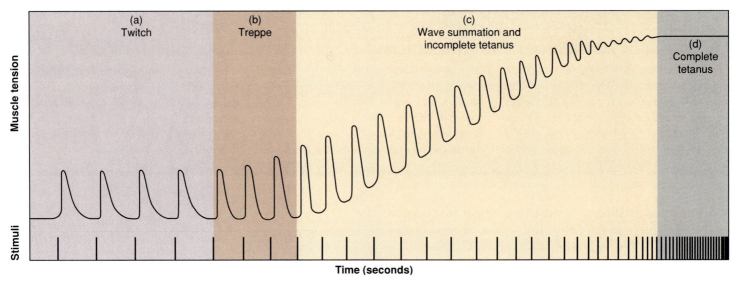

Figure 12.18 The relationship between stimulus frequency and contraction strength. *Top:* each peak represents one muscle twitch; the height of the twitch represents the amount of tension generated by the muscle. *Bottom:* each vertical line represents one stimulus; the stimuli are all of the same voltage but occur at closer time intervals from left to right. (*a*) At low frequency, the muscle relaxes completely between stimuli and shows twitches of uniform length. (*b*) Treppe; at a moderate frequency of stimulation, the muscle relaxes fully between contractions, but successive twitches are stronger. (*c*) Wave summation and incomplete tetanus; at still higher stimulus frequency, the muscle does not have time to relax completely between twitches, and the force of each twitch builds on the previous one. (*d*) Complete tetanus; at high stimulus frequency, the muscle does not have time to relax at all between stimuli and exhibits a state of continual contraction with about four times as much tension as a single twitch.

rises higher and higher with each stimulus, causing subsequent twitches to be stronger. Another hypothesis is that the heat released by each muscle twitch causes enzymes in the muscle to work more efficiently and produce stronger twitches as the muscle warms up. One purpose of warm-up exercises before athletic competition is to induce treppe, so that the muscle contracts more effectively when the competition begins.

Refractory Period, Temporal Summation, and Tetanus

Just after a twitch has begun, there is a short **refractory period** of about 5 msec in which the muscle will not respond to another stimulus. If a second stimulus arrives after the refractory period but before the twitch has ended, it may initiate a new contraction that "rides piggyback" on the first and achieves a higher level of tension (fig. 12.18*c*). This phenomenon of increasing tension is called **temporal**[7] **summation** because it results from two stimuli arriving close together; it is also known as **wave summation** because it results from one wave of contraction added on top of another.

To say that muscle twitches become stronger when the stimuli are spaced close together may seem to contradict the all-or-none law, which says that a fiber either

exhibits a maximum response or none at all. The distinction is that the all-or-none law means the strength of a twitch does not vary with the *strength* (voltage) of the stimulus, whereas treppe shows that the strength of the twitches can vary with the *frequency* of stimulation.

If stimuli are so frequent that the muscle cannot completely relax between them (20 msec apart, or 50 stimuli/sec, for example), wave is added upon wave. Each twitch reaches a higher level of tension than the one before, and the muscle relaxes only partially between stimuli. This is a state called **incomplete tetanus.** At a still higher frequency (such as 80–100 stimuli/sec), the muscle has no time to relax at all between stimuli, and the twitches fuse into a smooth, prolonged contraction called **complete tetanus.** This produces about four times as much force as a single twitch (fig. 12.18*d*).

Tetanus should not be confused with the disease of the same name caused by a bacterial neurotoxin. The tetanus caused by repetitive stimulation is a normal part of muscle physiology and results in smooth, sustained contractions rather than jerky, spasmodic ones. Imagine trying to lift a cup of scalding hot tea to your lips if your muscles exhibited short spasmodic contractions separated by partial relaxations!

Isometric and Isotonic Contraction

Suppose you are kneeling to lift a heavy box of books. When you first contract the muscles of your arms and

7. *tempor* = time

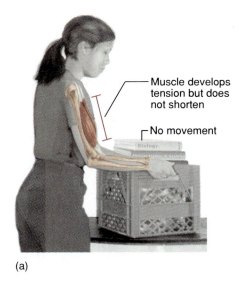

Muscle develops tension but does not shorten

No movement

(a)

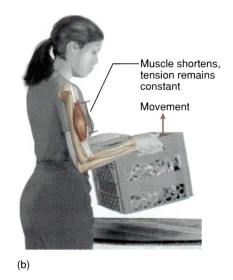

Muscle shortens, tension remains constant

Movement

(b)

Figure 12.19 (*a*) Isometric contraction, in which a muscle develops tension but does not shorten. This occurs at the beginning of any muscle contraction but is prolonged in actions such as lifting heavy weights. (*b*) Isotonic contraction, in which the muscle changes length while maintaining a constant degree of tension. In this phase, the muscle moves a load. 𝒳

legs, you can feel the tension in them even though the box is not yet moving. At this point, your muscles are contracting at a cellular level, but their tension is being absorbed by the series-elastic components; the muscle as a whole is not producing any external movement. This phase is called **isometric**[8] **contraction**—contraction without a change in length. When tension is great enough to overcome the resistance (weight of the box), the muscle shortens, moves the load, and maintains essentially the same tension from then on (fig. 12.19). This phase is called **isotonic**[9] **contraction**—contraction with a change in length. Isometric and isotonic contraction are both phases of normal muscular action.

Isotonic contractions are of two types: concentric and eccentric. In a **concentric contraction,** a muscle shortens as it contracts—for example, when the biceps brachii contracts and flexes the elbow. In an **eccentric contraction,** a muscle lengthens as it contracts. This may seem contradictory, but when the elbow is extended by the triceps brachii, the biceps contracts (produces tension) even as it is being stretched and thus acts as a brake on the action of the triceps. Because of the eccentric contraction of the biceps, you can set a coffee cup down gently on a table rather than slamming it down, as would occur if the triceps acted alone. A weight lifter uses concentric contraction when lifting a barbell and eccentric contraction when lowering it to the floor.

In summary, during isometric contraction, a muscle develops tension without changing length; in concentric contraction, a muscle develops tension and shortens; and in eccentric contraction, a muscle develops tension while it is lengthening.

8. *iso* = same, uniform + *metr* = length
9. *ton* = tension

Muscle Metabolism

▼**Objectives**
When you have completed this section, you should be able to
• explain how skeletal muscle meets its ATP demands during rest and exercise;
• explain the physiological basis of muscle fatigue and soreness;
• define *oxygen debt* and list the purposes for which extra oxygen is needed after exercise;
• distinguish between slow- and fast-twitch muscle fibers and explain the functional roles of these two types;
• discuss the factors that affect muscular strength; and
• discuss the effects of resistance and endurance exercises on muscle.

Energy Transfer

The key to muscle function and endurance is ATP, the immediate energy source required for all contractile activity. With an adequate supply of ATP, contraction can

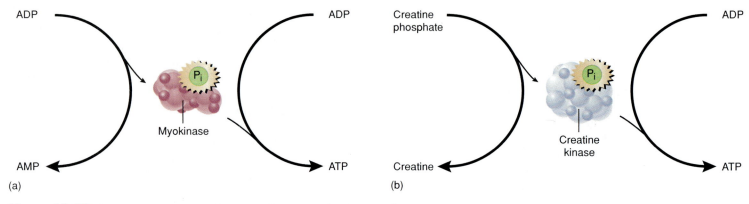

(a)
(b)

Figure 12.20 The phosphagen system for generating ATP in the absence of oxygen. (*a*) Myokinase transfers an inorganic phosphate group (P_i) from one ADP to another, turning the second one into ATP. (*b*) Creatine kinase transfers a P_i from creatine phosphate to ADP, converting the latter to ATP.

continue for long periods of time. The supply of ATP in turn depends on the availability of oxygen and organic energy sources. To understand how muscle manages its ATP budget, you must be familiar with the two main pathways of ATP synthesis—aerobic respiration and anaerobic fermentation (see fig. 3.26 p. 101). Anaerobic fermentation enables a cell to produce a limited amount of ATP in the absence of oxygen, but it produces a toxic end product, lactic acid, which is a major factor in muscle fatigue. By contrast, aerobic respiration produces far more ATP and end products that are less toxic, but it requires a continual supply of oxygen. Although aerobic respiration is best known as a pathway for glucose oxidation, it is also used to extract energy from other organic compounds. In a resting muscle, almost all ATP is generated by the aerobic respiration of fatty acids.

During the course of exercise, different mechanisms of ATP synthesis are used, depending on the intensity and duration of the exercise. We will view these mechanisms from the standpoint of immediate, short-term, and long-term energy, but it must be stressed that muscle does not make sudden shifts from one mechanism to another like an automobile transmission shifting gears. Rather, these mechanisms blend and overlap as exercise continues.

Immediate Energy

In a short, intense exercise such as a 100 m dash, the respiratory and cardiovascular systems cannot deliver oxygen to the muscles quickly enough to meet the increased ATP demand aerobically. The myoglobin in a muscle fiber supplies oxygen for a limited amount of aerobic respiration, but in brief exercises a muscle meets most of its ATP demand by borrowing phosphate groups from other molecules and transferring them to ADP. Two enzyme systems control these phosphate transfers: (1) **myokinase** (MY-oh-KY-nase) transfers P_i

groups from one ADP to another, converting the latter to ATP, and (2) **creatine kinase** (CREE-uh-tin KY-nase) obtains P_i groups from an energy storage molecule, **creatine phosphate (CP)** (fig. 12.20).

ATP and CP, collectively called the **phosphagen system,** provide nearly all the energy used for short bursts of intense activity. Muscle contains about 5 millimoles of ATP and 15 millimoles of CP per kilogram of tissue, which is enough to power about 1 minute of brisk walking or 6 seconds of sprinting or fast swimming. The phosphagen system is especially important in activities requiring brief but maximal effort, such as football, baseball, and weight lifting.

Short-Term Energy

After the phosphagen system is exhausted, the muscles rely on anaerobic fermentation to "buy time" until cardiopulmonary function can supply enough oxygen to meet the muscle's ATP needs aerobically. During this period, the muscles use glucose obtained from the blood and from their own stored glycogen. The pathway from glycogen to lactic acid, called the **glycogen–lactic acid system,** produces enough ATP for 30 to 40 seconds of maximum activity. To play basketball or to run a home run in baseball, for example, depends heavily on this energy-transfer system.

Long-Term Energy

After 40 seconds or so, the respiratory and cardiovascular systems "catch-up" and deliver enough oxygen to the muscles for aerobic respiration to meet most of the ATP demand. A person's rate of oxygen consumption rises for 3 to 4 minutes and then levels off at a *steady state* in which aerobic ATP production keeps pace with the demand. In exercises lasting more than 10 minutes, more than 90% of the ATP is produced aerobically.

Little lactic acid accumulates under steady state conditions, but this does not mean that the exercise could continue indefinitely or that it is limited only by a person's willpower. The depletion of glycogen and blood glucose, together with the loss of fluid and electrolytes through sweating, sets limits to endurance and performance even when lactic acid does not.

Fatigue and Endurance

Muscle fatigue is the progressive weakness and loss of contractility that results from prolonged use of the muscles. For example, if you hold this book at arm's length for a minute, you will feel your muscles growing weaker, and eventually you will be unable to hold it up. Repeatedly squeezing a rubber ball, pushing a video game button, or trying to take lecture notes from a fast-talking professor produces fatigue in the hand and finger muscles. Fatigue has multiple causes:

- ATP synthesis declines as glycogen is consumed.
- The ATP shortage slows down the sodium-potassium pumps, which are needed to maintain the resting membrane potential and excitability of the muscle fibers.
- Lactic acid lowers the pH of the sarcoplasm and impairs the action of enzymes needed for contraction, ATP synthesis, and other aspects of muscle function.
- Each action potential releases potassium ions from the sarcoplasm to the extracellular fluid. The accumulation of extracellular K^+ reduces the membrane potential and excitability of the muscle fiber.
- Motor nerve fibers use up their acetylcholine and become unable to stimulate muscle fibers. This is called *junctional fatigue.*
- The central nervous system, where all motor commands originate, fatigues by processes not yet understood.

Think About It

Suppose you repeatedly stimulated the sciatic nerve in a frog nerve-muscle preparation until the muscle stopped contracting. What simple test could you do to determine whether this was due to junctional fatigue or to one of the other fatigue mechanisms?

A person's ability to maintain high-intensity exercise for more than 4 to 5 minutes is determined in large part by his or her **maximum oxygen uptake,** or $\dot{V}_{O_2}$ **max**—the point at which the rate of oxygen consumption reaches a plateau and does not increase further with an added workload. $\dot{V}_{O_2}$ max is proportional to body size; it peaks at around age 20; it is usually greater in males than in females; and it can be twice as great in a trained endurance athlete as in an untrained person (see the later discussion on effects of conditioning).

Physical endurance also depends on the supply of organic nutrients—fatty acids, amino acids, and especially glucose. Many endurance athletes use a dietary strategy called *carbohydrate loading* to "pack" as much as 4–5 g of glycogen into every 100 g of muscle. This can significantly increase endurance, but an extra 2.7 g of water is also stored with each added gram of glycogen. The resulting "heaviness" and other side effects sometimes outweigh the benefits of carbohydrate loading.

Oxygen Debt

You have probably noticed that you breathe heavily not only during a strenuous exercise, but also for several minutes afterwards. This is because your body accrues an oxygen debt that must be "repaid." **Oxygen debt** is the difference between the resting rate of oxygen consumption and the elevated rate following an exercise. The total amount of extra oxygen consumed after a strenuous exercise is typically about 11 L. It is used for the following purposes:

- *Replacing the body's oxygen reserves* that were depleted in the first minute of exercise. These reserves include 0.3 L of oxygen bound to muscle myoglobin, 1.0 L bound to blood hemoglobin, 0.25 L dissolved in the blood plasma and other extracellular fluids, and 0.5 L in the air in the lungs.
- *Replenishing the phosphagen system.* This involves synthesizing ATP and using some of it to donate phosphate groups back to creatine until the resting levels of ATP and CP are restored.
- *Oxidizing lactic acid.* About 80% of the lactic acid produced by muscle enters the bloodstream and is reconverted to pyruvic acid in the kidneys, the cardiac muscle, and especially the liver. Some of this pyruvic acid enters the aerobic (mitochondrial) pathway to make ATP, but the liver converts most of it back to glucose. Glucose is then available to replenish the glycogen stores of the muscle.
- *Serving the elevated metabolic rate.* As long as the body temperature remains elevated by exercise, the total metabolic rate remains high, so that extra oxygen is required.

Slow- and Fast-Twitch Fibers

Muscle fibers fall into two primary categories: *slow-twitch fibers* and *fast-twitch fibers* (table 12.3). All fibers of a single motor unit are of the same type, but all muscles have a mixture of slow- and fast-twitch motor units in different proportions according to each muscle's function.

Table 12.3 Classification of Skeletal Muscle Fibers

Structural and Functional Features	Fiber Type	
	Slow-Twitch (also called slow oxidative [SO] fibers, red fibers, or type I fibers)	Fast-Twitch (also called fast glycolytic [FG] fibers, white fibers, or type II [or IIB] fibers)
Relative diameter	Smaller	Larger
Primary mode of ATP synthesis	Aerobic	Anaerobic
Fatigue resistance	Good	Poor
ATP hydrolysis	Slow	Fast
Glycolysis	Moderate	Fast
Myoglobin content	Abundant	Low
Glycogen content	Low	Abundant
Mitochondria	Abundant and large	Fewer and smaller
Capillaries	Abundant	Fewer
Color	Red	White, pale
Representative Muscles in Which Fiber Type is Predominant		
	Soleus	Gastrocnemius
	Erector spinae	Biceps brachii
	Quadratus lumborum	Muscles of eye movement

Table 12.4 Proportion of Slow- and Fast-Twitch Fibers in the Quadriceps Femoris Muscles of Male Athletes

Sample Population	Slow-Twitch	Fast-Twitch
Marathon runners	82%	18%
Swimmers	74	26
Average males	45	55
Sprinters and jumpers	37	63

Slow-twitch fibers are relatively small and produce twitches up to 100 msec long. They have more mitochondria, myoglobin, and blood capillaries than fast-twitch fibers do. These features adapt slow-twitch fibers for aerobic respiration—and because aerobic respiration does not generate lactic acid, these fibers are relatively resistant to fatigue. The soleus muscle of the calf and the postural muscles of the spinal column, for example, are composed mainly of slow-twitch, high-endurance fibers.

Fast-twitch fibers are larger than the slow-twitch type and produce twitches as short as 7.5 msec. These fibers are specialized to use anaerobic "quick energy" systems—they are rich in enzymes for the phosphagen and glycogen–lactic acid systems. Their sarcoplasmic reticulum releases and reabsorbs calcium ions quickly, which partially accounts for their quick, forceful contractions. Fast-twitch fibers are especially important in sports such as basketball, with stop-and-go activity and frequent changes of pace.

In the preceding chapter, we noted that the gastrocnemius and soleus muscles of the calf insert on the heel (calcaneus) through a common tendon and have the same action. The muscles may seem redundant until we realize that the soleus is composed mainly of slow-twitch fibers and the gastrocnemius is composed mainly

of fast-twitch fibers. The gastrocnemius is well adapted for quick, powerful actions such as jumping, whereas the soleus does most of the work in endurance exercises such as jogging and cross-country skiing.

Some authorities recognize two subcategories of fast-twitch fibers, IIA and IIB. Type IIB is the common type previously described, whereas type IIA (intermediate) fibers are relatively rare except in some endurance athletes. They combine fast-twitch responses with aerobic fatigue-resistant metabolism.

Muscles composed mainly of slow-twitch fibers are relatively dark in color because of their high concentrations of blood capillaries, mitochondria, and myoglobin; they are called *red muscles*. Muscles composed mainly of fast-twitch fibers are called *white muscles* because they are paler. The "dark meat" and "white meat" of poultry are red and white muscles, respectively.

The proportion of one fiber type to another differs among people with different types and levels of physical activity (table 12.4). This does not necessarily mean, however, that athletic training can change one fiber type to another. Whether fibers can change type remains an unsettled issue. The preponderance of evidence suggests that people are born with genetically determined differences in the ratios of fiber types and training produces little if any change. Athletes discover those sports at which they excel and gravitate toward those for which heredity has best equipped them. In effect, some people are "born marathoners" and others are "born sprinters."

Muscular Strength

Muscular strength depends on a variety of anatomical and physiological factors:

- **Muscle size.** The strength of a muscle depends primarily on its size; this is why weight lifting increases the size and strength of a muscle simultaneously. A muscle can exert a tension of about 3 to 4 kg/cm² (50 lb/in.²) of cross-sectional area. The triceps surae can easily support the entire body weight, as when a person stands on tiptoe. The gluteus maximus can generate 1,200 kg of tension, and all the muscles of the body can produce a total tension of 22,000 kg (nearly 25 tons). Indeed, the muscles can generate more tension than the bones and tendons can withstand—a fact that accounts for many injuries to the patellar and calcaneal tendons.

- **Fascicle arrangement.** Pennate muscles such as the quadriceps femoris are stronger than parallel muscles such as the sartorius, which in turn are stronger than circular muscles such as the orbicularis oculi.
- **Size of active motor units.** Large motor units produce stronger contractions than small ones.
- **Multiple motor unit summation.** When a stronger muscle contraction is desired, the nervous system activates more motor units in a process called **multiple motor unit (MMU) summation,** or **recruitment.** This phenomenon has been known to produce extraordinary feats of strength under desperate conditions—rescuing a loved one pinned under an automobile, for example. Getting "psyched up" for athletic competition is also partly a matter of recruitment.
- **Temporal summation.** Nerve impulses usually arrive at a muscle in a series of closely spaced action potentials. The greater the frequency of stimulation, the more strongly a muscle contracts. This is due to *temporal summation,* explained earlier in the chapter.
- **The length-tension relationship.** As noted earlier, a muscle resting at optimum length is prepared to contract more forcefully than a muscle that is excessively contracted or stretched.
- **Fatigue.** Muscles contract more weakly when they are fatigued.

Muscular Conditioning and Atrophy

Resistance exercise, such as weight lifting, is the contraction of muscles against a load that resists movement. A few minutes of resistance exercise at a time, a few times each week, is enough to stimulate muscle growth. Growth results primarily from cellular enlargement, not cellular division. The muscle fibers synthesize more myofilaments and the myofibrils grow thicker. Myofibrils split longitudinally when they reach a certain size, so a well-conditioned muscle has more myofibrils than a poorly conditioned one. Muscle fibers themselves are incapable of mitosis, but there is some evidence that as they enlarge, they too may split longitudinally. A small part of muscle growth may therefore result from an increase in the number of fibers, but most results from the enlargement of fibers that have existed since childhood.

Think About It

Is muscle growth mainly the result of hypertrophy or hyperplasia?

Endurance (aerobic) exercise, such as jogging and swimming, improves the fatigue-resistance of the muscles. Slow-twitch fibers, especially, produce more mitochondria and glycogen with conditioning and acquire a greater density of blood capillaries. Endurance exercise also improves skeletal strength, increases the red blood cell count and the oxygen transport capacity of the blood, and enhances the function of the cardiovascular, respiratory, and nervous systems. Endurance training does not significantly increase muscular strength, and resistance training does not improve endurance. Optimal performance and skeletomuscular health requires **cross-training,** which incorporates elements of both types.

If muscles are not kept sufficiently active, they become *deconditioned*—weaker and more easily fatigued. *Disuse atrophy*—muscular shrinkage from lack of use, is discussed in chapter 6. *Senescence atrophy,* muscle shrinkage in old age, is discussed in chapter 29.

Delayed Onset Muscle Soreness

Delayed onset muscle soreness (DOMS) is unusual pain, stiffness, or tenderness that is felt several hours to a day after vigorous exercise. It is accompanied by elevated blood levels of myoglobin, lactate dehydrogenase, and creatine kinase released by damaged muscle fibers. Electron microscopic examination confirms that muscle fibers exhibit various forms of *microtrauma* after a heavy workout, including disrupted Z discs and damage to the sarcolemma and myofibrils.

Cramps

Earlier we saw that the central nervous system (CNS) monitors and adjusts the tension in a muscle, maintaining tonus. At times, the CNS triggers painful, spasmodic contractions known as **cramps.** Cramps are initiated by irritating factors such as extreme cold, heavy exercise, lack of blood flow, electrolyte depletion, dehydration, and low blood glucose. They begin when such factors trigger sensory impulses from the muscle to the spinal cord. The spinal cord then triggers reflex contraction of that muscle, which further stimulates the sensory nerve endings that started the process. Thus a positive feedback cycle is initiated that builds to extreme, painful contraction.

Key Point Review

21 From which two molecules can ADP borrow a phosphate group to become ATP? What is the enzyme that catalyzes each transfer?

22 In a long period of intense exercise, why does muscle generate ATP anaerobically at first, and then switch to aerobic respiration?

23 List four causes of muscle fatigue.

24 List three causes of oxygen debt.

25 What properties of fast- and slow-twitch fibers adapt them for different physiological purposes?

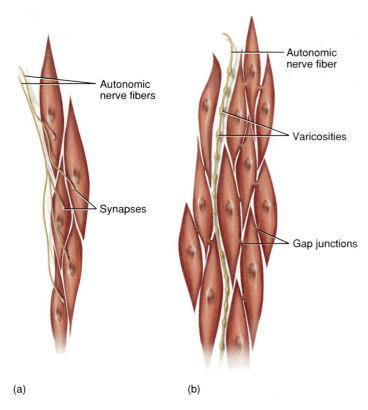

(a)   (b)

Figure 12.21 Innervation of (*a*) multiunit and (*b*) single-unit smooth muscle. ✗

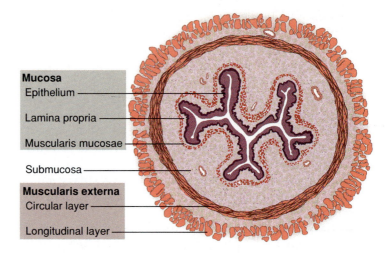

Figure 12.22 Layers of visceral (single-unit) smooth muscle in the wall of the esophagus. Many hollow organs have alternating circular and longitudinal layers of smooth muscle.

In the figure 12.22, the labeled structures are:

Mucosa
Epithelium
Lamina propria
Muscularis mucosae

Submucosa

Muscularis externa
Circular layer
Longitudinal layer

Smooth Muscle

▼**Objectives**

When you have completed this section, you should be able to
- describe the ways in which smooth muscle differs structurally and physiologically from skeletal muscle; and
- relate the unique physiological properties of smooth muscle to its locations and functions.

Smooth muscle plays important roles in all organ systems. Here we examine how it differs from skeletal muscle and how these differences relate to the roles it must play.

Types and Functions

There are two functional categories of smooth muscle (fig. 12.21). **Multiunit smooth muscle** occurs in some of the largest arteries and pulmonary air passages, in the arrector pili muscles that attach to hair follicles, and in the iris of the eye. Its innervation is similar to that of skeletal muscle—the terminal branches of an axon synapse with individual muscle cells and form a motor unit. Each motor unit contracts independently of the others, hence the name of this muscle type. In **single-unit smooth muscle,** a nerve fiber does not synapse with any specific muscle cell—rather, it passes through the tissue and releases neurotransmitter at several points along the way, stimulating many cells at once. Furthermore, the muscle cells are electrically coupled to each other by gap junctions; they directly stimulate each other, and a large number of cells contract as a single unit.

Single-unit smooth muscle is much more common than the multiunit type. It occurs in the walls of the blood vessels and in the digestive, respiratory, urinary, and reproductive tracts—thus, it is also called **visceral muscle.** In the digestive tract and some other locations, it forms two layers—an inner *circular layer,* in which the fibers are arranged around the circumference of the organ, and an outer *longitudinal layer,* in which the fibers run parallel to the long axis of the organ (fig. 12.22). Together they produce a traveling wave of contraction called **peristalsis** (PERR-ih-STAL-sis) that propels the contents of an organ such as the esophagus or intestines. Smooth muscle cells encircle the walls of the blood vessels, where their contraction and relaxation play an important role in regulating blood pressure and redirecting blood flow from one organ to another.

Microscopic Anatomy

Smooth muscle differs from skeletal muscle in several aspects of structure and function (table 12.5). A smooth muscle cell is fusiform, much shorter than a skeletal muscle fiber, and has only one nucleus. Although thick and thin filaments are both present, they are not in register with each other and produce no visible striations or sarcomeres. Z discs are absent; instead, the thin filaments are indirectly attached to **dense bodies** scattered throughout the sarcoplasm and on the inner

Table 12.5	Comparison of Skeletal and Smooth Muscle	
Feature	Skeletal Muscle	Smooth Muscle
Location	Associated with skeletal system	Walls of viscera and blood vessels, iris of eye, arrector pili muscles
Cell shape	Long cylindrical fibers	Fusiform cells
Length	100 μm to 30 cm	50–200 μm
Diameter	10–100 μm	2–10 μm
Striations	Present because of regular arrangement of myofilaments	Absent because of irregular arrangement of myofilaments
Nuclei	Multiple nuclei, adjacent to sarcolemma	One nucleus, in center
Connective tissues	Endomysium, perimysium, epimysium	Sparse endomysium only
Regulatory proteins	Tropomyosin and troponin	Tropomyosin and calmodulin
Ca^{2+} receptor	Troponin of thin filament	Calmodulin of thick filament
Source of Ca^{2+}	Sarcoplasmic reticulum	Mainly extracellular fluid
Myosin heads	Absent from middle of thick filament	Occur throughout thick filament
Thin-filament attachment	Attached to Z discs at ends of each sarcomere	Attached to dense bodies scattered throughout sarcoplasm and on inner face of sarcolemma
Sarcoplasmic reticulum	Abundant	Scanty
T tubules	Present	Absent
Innervation	Somatic motor fibers synapse with a motor end plate on each fiber	Innervation of multiunit smooth muscle similar to that of skeletal muscle; in single-unit smooth muscle, autonomic fibers form synapses en passant, not with any specific muscle cell
Need for nervous stimulation	Requires nervous stimulation	Can contract without nervous stimulation
Effect of neural stimulation	Excitatory only	Excitatory or inhibitory
Neurotransmitter receptors	ACh receptors only, concentrated at motor end plate	ACh and norepinephrine receptors throughout sarcolemma
Gap junctions	Absent	Present in single-unit smooth muscle

face of the sarcolemma. The sarcoplasmic reticulum is very scanty, and there are no T tubules. Although smooth muscle contraction is activated by calcium ions, most of the calcium comes from the extracellular fluid (ECF) by way of Ca^{2+} channels in the sarcolemma. During relaxation, calcium is pumped back out of the cell.

Innervation and Stimulation

Smooth muscle is innervated by nerve fibers that can trigger or modify its contractions, yet it is also capable of contracting without nervous stimulation—in response to such stimuli as stretch and chemicals. In childbirth, for example, oxytocin stimulates smooth muscle of the uterus to contract.

Smooth muscle is innervated by fibers of the *autonomic nervous system* (ANS; see chapter 15), not by the somatic motor fibers that stimulate skeletal muscle. The ANS is divided into *sympathetic* and *parasympathetic divisions.* In many cases, called *dual innervation,* both divisions innervate the same smooth muscle and exert opposite effects on it. For example, the parasympathetic division causes contraction of smooth muscle of the bronchioles (small air passages in the lungs), whereas sympathetic fibers relax the muscle and dilate the bronchioles.

In single-unit smooth muscle, each branch of the autonomic nerve fiber has up to 20,000 beadlike swellings called **varicosities** along its length (fig. 12.23). Each varicosity contains synaptic vesicles and a few mitochondria. Instead of approaching any one smooth muscle cell particularly closely, the nerve fiber passes amid several cells and stimulates all of them at once when it releases its neurotransmitter (see fig. 12.21b). The muscle cells do not have motor end plates or any other specialized area of sarcolemma to bind the neurotransmitter; rather, they have receptor sites scattered throughout the surface. These autonomic nerve endings are known as **diffuse junctions** because there is no one-to-one relationship between nerve fiber and muscle cell. They are also called **synapses en passant**[10] because the nerve fiber associates with some smooth muscle cells at one varicosity and then passes on and associates with other muscle cells farther along.

10. *en passant* = in passing

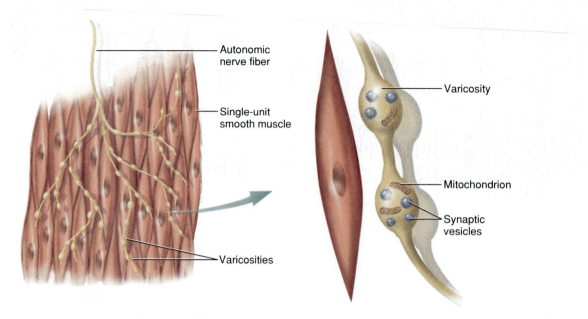

Figure 12.23 Varicosities of an autonomic nerve fiber innervating single-unit smooth muscle.

Another difference between smooth and skeletal muscle lies in the neurotransmitters involved and their effects on the muscle cells. As mentioned previously, the only neurotransmitter that stimulates skeletal muscle is acetylcholine, and its effect is always excitatory. Action potentials in the nerve fiber lead to action potentials in the muscle fiber, followed by contraction of the muscle. None of this is necessarily true of smooth muscle. The parasympathetic and a few sympathetic fibers release acetylcholine but most sympathetic endings release **norepinephrine** (nor-EP-ih-NEF-rin). Either of these neurotransmitters can have an excitatory effect called an **excitatory junction potential (EJP)** or an inhibitory effect called an **inhibitory junction potential (IJP).** If enough EJPs are produced, they add up (summate) to reach a threshold membrane voltage, and the muscle cell contracts. This type of summation does not occur in skeletal muscle.

Whether an EJP or IJP is produced is not determined by the neurotransmitter but by the kind of membrane receptor present on the muscle cells. Norepinephrine relaxes smooth muscle cells of the bronchioles but excites those of the arteries because the norepinephrine receptors of the smooth muscle differ in these two locations (discussed more fully in chapter 15). Furthermore, smooth muscle cells are capable of graded responses rather than all-or-none contractions— a weak stimulus produces a weak contraction.

Gap Junctions and Pacemakers

Single-unit muscle cells can electrically stimulate each other—something skeletal muscle fibers and multiunit smooth muscle cells cannot do. When a nerve fiber stimulates one single-unit muscle cell to contract, this cell directly stimulates its neighbors to do likewise, producing synchronous or wavelike contractions of the tissue as a whole. This is possible because the cells have gap junctions between them that allow ions to flow from the sarcoplasm of one cell directly into the next (see chapter 6).

Skeletal muscle fibers maintain a stable resting potential and wait for a nerve to stimulate them. In some single-unit smooth muscle, there is more leakage of Na$^+$ and K$^+$ across the membranes, and the membrane potential continually drifts toward threshold. When it reaches threshold, an action potential is set off and the muscle contracts. Thus, some single-unit smooth muscle contracts spontaneously and sometimes rhythmically without any external stimulation. The first cells to reach threshold stimulate neighboring cells through the gap junctions. The first cells may thus act as **pacemakers** that set off rhythmic contractions of an entire layer of smooth muscle tissue. Such contractions mix and propel semidigested food through the digestive tract.

Contraction and Relaxation

In both smooth and skeletal muscle, contraction is triggered by a rise in intracellular Ca^{2+} concentration, energized by ATP, and achieved by the sliding of thin filaments over and between the thick filaments. The mechanism of excitation-contraction coupling, however, is very different in smooth muscle. Only minor amounts of Ca^{2+} come from the sarcoplasmic reticulum. Most of it comes from the extracellular fluid, entering the cell through calcium channels in the sarcolemma. Some of these channels are voltage-gated, some are ligand-gated, and some are opened by stretching of the fiber.

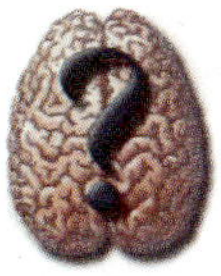

Think About It

How is smooth muscle contraction affected by the drugs called calcium channel blockers? (See p. 114)

There is no troponin in smooth muscle. Calcium ions bind instead to a similar protein called **calmodulin**[11] (cal-MOD-you-lin), associated with the thick filaments. Calmodulin then activates an enzyme called **myosin light-chain kinase.** This enzyme transfers a phosphate group from ATP to the head of the myosin molecule, thus activating the myosin head. The myosin can now form cross bridges with actin and engage in power and recovery strokes just as it does in skeletal muscle. As thick filaments pull on the thin ones, the thin filaments pull on intermediate filaments, which in turn pull on the dense bodies of the plasma membrane. This shortens the entire cell. When a smooth muscle cell contracts, it twists in a spiral fashion, somewhat like wringing out a wet towel except that the "towel" wrings itself (fig. 12.24).

The ATP that phosphorylates myosin only activates the myosin ATPase. The myosin head must still bind and hydrolyze an additional ATP for each stroke. In skeletal muscle, there is typically a 2 msec latent period between stimulation and the onset of contraction. In smooth muscle, by contrast, the latent period is considerably longer—from 50 to 100 msec. Tension peaks about 500 msec (0.5 sec) after the stimulus and then declines over a period of 1 to 2 seconds. Compared to skeletal muscle, then, smooth muscle is very slow to contract and relax. The slowness of contraction is largely because the myosin ATPase of smooth muscle is a slow enzyme. The slowness of relaxation is due in part to slowness of the pumps that remove Ca^{2+} from the sarcoplasm. As the Ca^{2+} level falls, myosin is dephosphorylated and is no longer able to hydrolyze ATP and execute power strokes. However, it does not necessarily detach from actin immediately. It has a **latch-bridge mechanism** that enables it to remain attached to actin for some time without expending more energy.

Smooth muscle is very resistant to fatigue. It has few mitochondria and low energy requirements and makes most of its ATP anaerobically. Its fatigue-resistance and latch-bridge mechanism are important in enabling it to maintain a state of continual **smooth muscle tone (tonic contraction).** The intestines are much longer in a cadaver than they are in a living person because of the loss of smooth muscle tone at death. A loss of smooth muscle tone in the arteries can cause a dangerous drop in blood pressure.

---

11. Acronym for *calcium modulating protein*

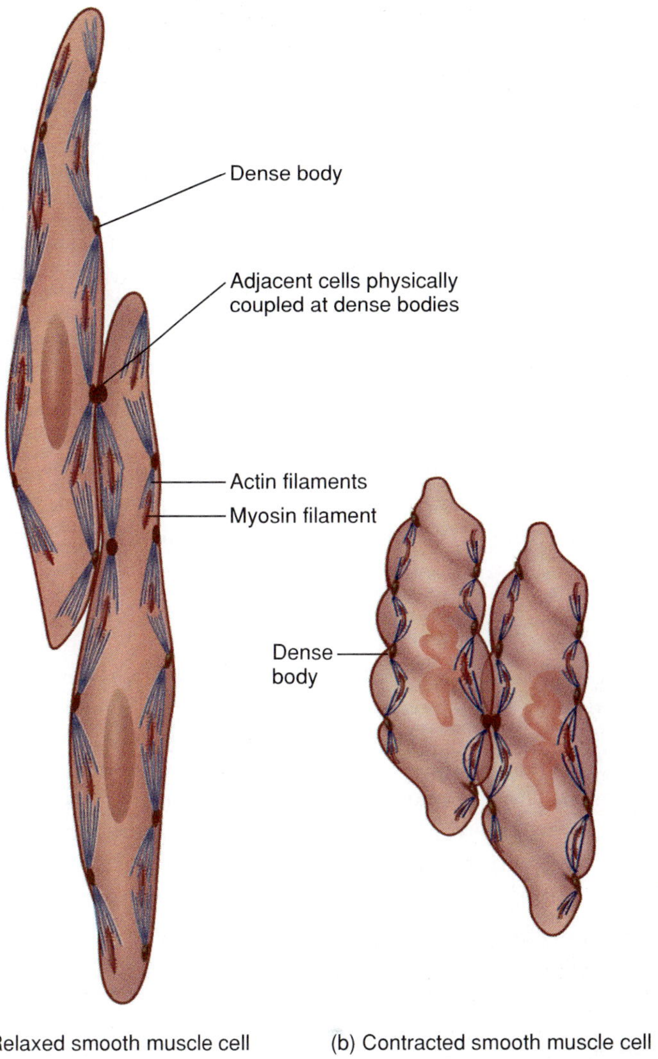

(a) Relaxed smooth muscle cell   (b) Contracted smooth muscle cell

Figure 12.24 Smooth muscle cells in the (*a*) relaxed and (*b*) contracted states. Actin myofilaments are tethered to dense bodies on the plasma membrane by way of intermediate filaments of the cytoskeleton.

Response to Stretch

Smooth muscle sometimes contracts simply in response to stretch, which physically opens calcium channels in the sarcolemma. Distension of the esophagus with food or the colon with feces, for example, evokes a contraction that propels the contents along the organ. Smooth muscle exhibits a reaction called the **stress-relaxation response.** Upon being stretched, it reacts momentarily by contracting and resisting, but then it relaxes. The significance of this response is apparent in the urinary bladder, whose wall consists of three layers of smooth muscle. If the muscle did not have a stress-relaxation response, the bladder would expel urine almost as soon as it began to fill, thus failing to stretch enough to store the urine until an opportune time.

Remember that skeletal muscle cannot contract very forcefully if it is overstretched. Smooth muscle is not subject to the limitations of this length-tension relationship. It must be able to contract forcefully even when greatly stretched, so that hollow organs such as the stomach and bladder can fill and then expel their contents efficiently. Skeletal muscle must be within 30% of optimum length in order to contract strongly when stimulated. Smooth muscle, by contrast, can be anywhere from half to twice the average resting length and still contract powerfully. There are three reasons for this: (1) there are no Z discs, so thick filaments cannot butt against these and stop the contraction; (2) since the thick and thin filaments are not arranged in orderly sarcomeres, stretching of the muscle does not cause a situation where there is too little myofilament overlap for cross bridges to form; and (3) since the thick filaments have myosin heads along their entire length, cross bridges can form anywhere, not just at the ends of the filaments. Smooth muscle also exhibits **plasticity**—the ability to adjust its tension to the degree of stretch. Thus, a hollow organ such as the bladder can be greatly stretched yet not become flabby when it is empty.

26 How do single-unit and multiunit smooth muscle differ in innervation and contractile behavior?

27 How does smooth muscle differ from skeletal muscle with respect to its source of calcium and its calcium receptor?

28 Explain why the same neurotransmitter can cause some smooth muscle cells to contract and others to relax.

29 Explain why the stress-relaxation response is an important factor in smooth muscle function.

CHAPTER ESSAY

Muscular Dystrophy and Myasthenia Gravis

The muscular system suffers fewer diseases than most organ systems. However, two serious disorders of the muscles—muscular dystrophy and myasthenia gravis—warrant more than just a mention.

Muscular dystrophy[12] is a collective term for several hereditary diseases in which the skeletal muscles degenerate, lose strength, and are gradually replaced by adipose and fibrous tissue. This new connective tissue impedes blood circulation, which in turn accelerates muscle degeneration in a fatal spiral of positive feedback. The most common form of the disease is **Duchenne**[13] **muscular dystrophy (DMD)**, caused by a sex-linked recessive allele. Like other sex-linked traits (see chapter 5), DMD is mainly a disease of males (fig. E.1), occurring in about 1 in 3,500 male live births. DMD is not usually diagnosed, however, until sometime between the ages of 2 and 10. Difficulties begin to appear early on, as a child begins to walk. Falls are frequent, and the child has difficulty standing up again. The disease affects the hips first, then the legs, and progresses to the abdominal and spinal muscles. The muscles shorten as they atrophy, causing postural abnormalities such as scoliosis. DMD is incurable but is treated with exercise to slow the atrophy and with braces to reinforce the weakened hips and correct the posture. Usually, patients are confined to a wheelchair by early adolescence and rarely live beyond the age of 20.

The DMD gene was identified in 1987, and genetic screening is now available to inform prospective parents of whether or not they are carriers. The normal allele of this gene makes *dystrophin,* a large protein present in the sarcolemma of healthy muscle but absent in people with DMD. Dystrophin seems to

Figure E.1 A boy with muscular dystrophy in physical therapy.

12. *dys* = bad, abnormal + *trophy* = growth
13. Guillaume B. A. Duchenne (1806–75), French physician

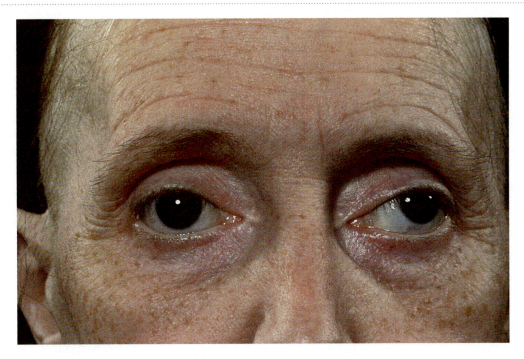

Figure E.2 Myasthenia gravis tends especially to affect the muscles of the head. It is characterized by drooping of the eyelids, weakness of the muscles of eye movement, and double vision resulting from the divergence (*strabismus*) of the eyes.

be involved in controlling the T tubule–sarcoplasmic reticulum mechanism or the calcium channels of the sarcoplasmic reticulum (SR). In DMD, calcium ions leak uncontrollably out of the SR and apparently activate *phospholipase A,* an enzyme that attacks phospholipids, destroys cellular membranes, and kills the muscle fiber. Some success has been achieved in supplying dystrophin to defective muscle fibers in mice, but not yet in humans.

A less severe form of muscular dystrophy is **facioscapulohumeral (Landouzy–Dejerine[14]) muscular dystrophy,** an autosomal dominant trait that begins in adolescence and affects both sexes. It involves the facial and shoulder muscles more than the pelvic muscles and disables some individuals while it barely affects others. A third form, **limb-girdle dystrophy,** is a combination of several diseases of intermediate severity that affect the shoulder, arm, and pelvic muscles.

Myasthenia gravis[15] (MY-ass-THEE-nee-uh GRAV-is) **(MG)** usually occurs in women between the ages of 20 and 40. It is an autoimmune disease in which antibodies attack the neuromuscular junctions and bind ACh

receptors together in clusters. The muscle fiber then removes the clusters from the sarcolemma by endocytosis. As a result, the muscle fibers become less and less sensitive to ACh. The effects often appear first in the facial muscles (fig. E.2) and commonly include drooping eyelids and double vision (due to weakness of the eye muscles). The initial symptoms are often followed by difficulty in swallowing, weakness of the limbs, and poor physical endurance. Some people with MG die quickly as a result of respiratory failure, but others have normal life spans. One method of assessing the progress of the disease is to use *bungarotoxin,* a protein from cobra venom that binds to ACh receptors. The amount that binds is proportional to the number of receptors that are still functional. The muscle of an MG patient sometimes binds less than one-third as much bungarotoxin as normal muscle does.

Myasthenia gravis is often treated with cholinesterase inhibitors. These drugs retard the breakdown of ACh in the neuromuscular junction and enable it to stimulate the muscle longer. Prednisone, Imuram, or other drugs may be used to suppress the immune system and reduce the production of the antibodies that destroy ACh receptors. Removal of the thymus, or *thymectomy,* is sometimes helpful. Also, a technique called *plasmapheresis* may be used to remove the harmful antibodies from the blood plasma.▲

14. Louis T. J. Landouzy (1845–1917), and Joseph J. Dejerine (1849–1917), French neurologists

15. *my* = muscle + *asthen* = weakness + *grav* = severe, heavy

Types and Characteristics of Muscle Tissue (p. 394)

1. Universal characteristics of muscle
   a. Excitability
   b. Contractility
   c. Extensibility
   d. Elasticity
2. Unique properties of skeletal muscle
3. Series-elastic components
   a. Connective tissue components
   b. Functions

Microscopic Anatomy of Skeletal Muscle (pp. 395–398)

1. The muscle fiber
   a. Sarcolemma and T tubules
   b. Nuclei
   c. Myofibrils and myofilaments
   d. Sarcoplasmic reticulum (SR)
   e. Glycogen and myoglobin
2. Myofilaments
   a. Thick filaments and myosin
   b. Thin filaments
      • F actin and G actin
      • Tropomyosin
      • Troponin I, T, and C
   c. Striations
      • Due to arrangement of myofilaments
      • Sarcomeres

Muscle Innervation and Membrane Potentials (pp. 398–404)

1. Somatic motor neurons
2. The motor unit
   a. One motor neuron and all muscle fibers innervated by it
   b. Relationship to strength and control
   c. Compensation for fatigue
3. The neuromuscular junction
   a. Structures of the nerve fiber
      • Synaptic knob
      • Synaptic vesicles containing acetylcholine (ACh)
      • Active zones
   b. Structures of the muscle fiber
      • Motor end plate
      • Junctional folds containing ACh receptors
   c. Other structures
      • Synaptic cleft
      • Schwann cell
      • Basement membrane containing acetylcholinesterase
4. Membrane potentials
   a. Concept of electrical potential
   b. Resting membrane potential (RMP)
      • Diffusion of K^+ out of cell
      • Retention of organic anions in cell

   • Electrostatic attraction of K^+ into cell
   • Diffusion of Na^+ into cell
   • Role of Na^+-K^+ pump
   c. Action potentials

Contraction and Relaxation of Skeletal Muscle Fibers (pp. 404–410)

1. Excitation
   a. ACh release by motor nerve ending
   b. Binding of ACh to receptor sites
   c. End-plate potential (EPP) of muscle fiber
   d. Action potential of muscle fiber
2. Excitation-contraction coupling
   a. Stimulation of terminal cisternae
   b. Release of calcium ions
   c. Binding of Ca^{2+} by troponin C
   d. Movement of tropomyosin away from active sites of actin
3. Contraction (sliding filament theory)
   a. Activation and "cocking" of myosin head
   b. Formation of myosin-actin cross bridges
   c. Power stroke
   d. Binding and hydrolysis of ATP by myosin head
   e. Recovery stroke
4. Relaxation
   a. Cessation of nerve signal
   b. Degradation of ACh
   c. Reabsorption of Ca^{2+} by sarcoplasmic reticulum
   d. Blocking of active sites by tropomyosin
   e. Stretching of muscle by series-elastic components and antagonists
5. Length-tension relationship and tonus
   a. Resting length and contraction strength
   b. Muscle tone

Behavior of Whole Muscles (pp. 410–413)

1. Threshold
2. Latent period
3. Twitch
   a. Contraction phase
   b. Relaxation phase
4. Graded and all-or-none responses
   a. All-or-none reaction of muscle fibers
   b. Graded contractions of whole muscle
5. Treppe
   a. Theories of mechanism
   b. Relevance to warm-up exercises
6. Refractory period

7. Temporal (wave) summation
8. Tetanus
9. Isometric and isotonic contraction
10. Concentric and eccentric contraction

Muscle Metabolism (pp. 413–417)

1. Energy transfer
   a. Review of aerobic respiration and anaerobic fermentation
   b. Immediate energy from the phosphagen system
      • Myokinase catalyzes ADP + ADP → AMP + ATP
      • Creatine kinase catalyzes CP + ADP → creatine + ATP
   c. Short-term energy from the glycogen–lactic acid system
   d. Long-term energy from aerobic respiration
2. Fatigue and endurance
   a. Causes of fatigue
   b. Maximal oxygen uptake ($\dot{V}_{O_2}$ max)
3. Oxygen debt
   a. Replacing oxygen reserves
   b. Replenishing phosphagen system
   c. Oxidizing lactic acid
   d. Serving elevated metabolic rate
4. Metabolic classes of muscle fibers
   a. Slow-twitch fibers
      • Twitches up to 100 msec
      • Adaptations for aerobic respiration
      • Resistance to fatigue
      • Important in posture and endurance exercises
   b. Fast-twitch fibers
      • Twitches as brief as 7.5 msec
      • Adaptations for anaerobic energy transfer
      • Quick release and reuptake of Ca^{2+}
      • Susceptibility to fatigue
      • Important in sports with quick stop-and-go action
   c. Relationship to synergistic muscle groups
   d. Intermediate fibers
   e. Red and white muscles
   f. Heredity and muscle composition
5. Muscular strength
   a. Muscle size (cross-sectional area)
   b. Fascicle arrangement
   c. Size of active motor units
   d. Multiple motor unit summation
   e. Temporal summation
   f. Length-tension relationship
   g. Fatigue

6. Muscular conditioning and atrophy
   a. Effects of resistance exercises
   b. Effects of endurance exercises
   c. Muscle atrophy
7. Delayed onset muscle soreness
8. Cramps

Smooth Muscle (pp. 418–422)

1. Types and functions
   a. Multiunit smooth muscle
      • Locations (relatively limited)
      • Innervation similar to skeletal muscle
      • Independent contraction of fibers
   b. Single-unit smooth muscle
      • Locations (more widespread, visceral)
      • Innervation
      • Cells coupled by gap junctions

2. Microscopic anatomy
   a. Fusiform uninucleate cells
   b. Dense bodies instead of Z discs
   c. Lack of T tubules and sarcomeres
   d. Scanty SR; calcium obtained primarily from ECF

3. Innervation and stimulation
   a. Autonomic nervous system
      • Sympathetic and parasympathetic divisions
      • Dual innervation
   b. Varicosities and diffuse junctions (synapses en passant)
   c. Neurotransmitters and receptors
      • Acetylcholine and norepinephrine
      • Excitatory and inhibitory junction potentials
      • Role of receptor types in excitation and inhibition
   d. Graded responses

4. Gap junctions and pacemakers
5. Contraction and relaxation
   a. Influx of Ca^{2+} from the ECF
   b. Binding of Ca^{2+} to calmodulin
   c. Action of myosin light-chain kinase
   d. Power and recovery strokes similar to skeletal muscle
   e. Long latent period and slow contraction
   f. Latch-bridge mechanism and smooth muscle tone
6. Response to stretch
   a. Stress-relaxation response
   b. Lack of length-tension relationship
   c. Plasticity

Selected Vocabulary

Also review the terms in tables 12.1 and 12.2, which are not repeated here.

excitability 394
contractility 394
extensibility 394
elasticity 394
striations 394
voluntary 394
involuntary 394
series-elastic components 394
internal tension 394
external tension 394
active site 396
somatic motor fiber 398
motor unit 398
synapse 400
electrophysiology 400
electrical potential 401
polarized 401
resting membrane potential (RMP) 403

action potential 404
end-plate potential (EPP) 404
excitation-contraction coupling 404
sliding filament theory 404
myosin ATPase 405
power stroke 406
recovery stroke 406
calsequestrin 409
length-tension relationship 409
tonus 410
threshold 410
twitch 410
latent period 410
contraction phase 411
relaxation phase 411
all-or-none law 411
graded twitch 411
treppe 411
refractory period 412
temporal (wave) summation 412
incomplete tetanus 412

complete tetanus 412
isometric contraction 413
isotonic contraction 413
concentric contraction 413
eccentric contraction 413
myokinase 414
creatine kinase 414
creatine phosphate (CP) 414
phosphagen system 414
glycogen–lactic acid system 414
muscle fatigue 415
maximum oxygen uptake ($\dot{V}_{O_2}$ max) 415
oxygen debt 415
slow-twitch fiber 416
fast-twitch fiber 416
multiple motor unit (MMU) summation 417
resistance exercise 417
endurance exercise 417
cross-training 417
delayed onset muscle soreness (DOMS) 417

cramp 417
multiunit smooth muscle 418
single-unit smooth muscle 418
visceral muscle 418
peristalsis 418
dense bodies 418
varicosities 419
diffuse junctions 419
synapses en passant 419
norepinephrine 420
excitatory junction potential (EJP) 420
inhibitory junction potential (IJP) 420
pacemakers 420
calmodulin 421
myosin light-chain kinase 421
latch-bridge mechanism 421
smooth muscle tone 421
stress-relaxation response 421
plasticity 422

Testing Your Recall .... Answers in Appendix C

1. To make a muscle contract more strongly, the nervous system can activate more of its motor units. This process is called
   a. recruitment.
   b. summation.
   c. incomplete tetanus.
   d. twitch.
   e. treppe.

2. The _____ is a depression in the sarcolemma that receives a motor nerve ending.
   a. T tubule
   b. terminal cisterna
   c. sarcomere
   d. motor end plate
   e. synapse

3. Before a muscle fiber can contract, ATP must bind to
   a. a Z disc.
   b. the myosin head.
   c. tropomyosin.
   d. troponin.
   e. actin.

4. Skeletal muscle fibers have _____, whereas smooth muscle fibers do not.
   a. T tubules
   b. a sarcolemma
   c. thick myofilaments
   d. thin myofilaments
   e. ACh receptors

5. Smooth muscle fibers have _____, whereas skeletal muscle fibers do not.
   a. an extensive sarcoplasmic reticulum
   b. tropomyosin
   c. calmodulin
   d. Z discs
   e. myosin ATPase

6. ACh receptors are found mainly in
   a. synaptic vesicles.
   b. terminal cisternae.
   c. thick filaments.
   d. thin filaments.
   e. junctional folds.

7. The calcium receptor of skeletal muscle is
   a. in the junctional folds.
   b. tropomyosin.
   c. troponin.

   d. calmodulin.
   e. on the thick filament.

8. Single-unit smooth muscle cells can stimulate each other because they have
   a. a latch-bridge mechanism.
   b. diffuse junctions.
   c. gap junctions.
   d. tight junctions.
   e. calcium pumps.

9. Warm-up exercises take advantage of _____ to enable muscles to perform at peak strength.
   a. the stress-relaxation response
   b. the length-tension relationship
   c. excitatory junction potentials
   d. oxygen debt
   e. treppe

10. Slow-twitch muscle fibers have all of the following features *except for*
    a. a high capacity to synthesize ATP aerobically.
    b. an abundance of myoglobin.
    c. high resistance to fatigue.
    d. an abundance of glycogen.
    e. a red color.

11. The minimum stimulus intensity that will make a muscle contract is called _____.

12. A state of prolonged maximum contraction of a muscle is called _____.

13. Parts of the sarcoplasmic reticulum called _____ lie on each side of a T tubule.

14. Thick myofilaments consist mainly of the protein _____.

15. The neurotransmitter that stimulates skeletal muscle is _____.

16. Muscle contains an oxygen-binding pigment called _____.

17. The _____ of skeletal muscle play the same role as the dense bodies of smooth muscle.

18. In autonomic fibers that innervate single-unit smooth muscle, the neurotransmitter is contained in swellings called _____.

19. A state of continual partial muscle contraction is called _____.

20. _____ is an end product of anaerobic fermentation that causes muscle fatigue.

Testing Your Comprehension
Answers in *Study Guide*

1. Without ATP, relaxed muscle cannot contract and muscle that is already contracted cannot relax. Explain why.

2. It was noted in chapter 3 that enzymes are very sensitive to pH changes and that pH variations may cause enzymes to change conformation and become less active. Explain how these facts relate to muscle fatigue.

3. Why would skeletal muscle be unsuitable for the wall of the urinary bladder? Discuss this in terms of unity of form and function at the cellular and molecular level.

4. As skeletal muscle contracts, one or more bands of the sarcomere become narrower and disappear, and one or more of them remain the same width. Which bands will change—A, H, and/or I—and why?

5. Botulism occurs when a bacterium, *Clostridium botulinum*, releases a neurotoxin that prevents motor neurons from releasing ACh. In view of this, what early symptoms of botulism would you predict? Explain how a person with botulism could die of suffocation as the disease progressed.

Web Site Link

For a listing of the most current web sites related to this chapter, please visit the Saladin homepage at:

http://www.mhhe.com/sciencemath/biology/saladin/

Nervous Tissue

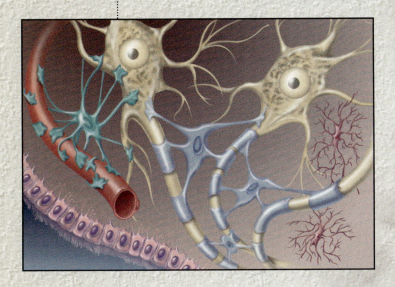

Special Topics

To understand this chapter, it is essential that you understand or brush up on the following concepts:

▶ Ligand- and voltage-gated ion channels (p. 113)
▶ Simple diffusion and active transport (pp. 125, 129)
▶ The sodium-potassium pump (pp. 130–131)
▶ Synapses and neurotransmitters (p. 400)
▶ The ionic basis of a resting membrane potential (RMP) (pp. 400–403)

An Overview of the Nervous System

▼Objectives
When you have completed this section, you should be able to
• describe the subdivisions and organization of the nervous system;
• explain the general functions of the nervous system;
• distinguish between afferent neurons, interneurons, and efferent neurons;
• discuss the general physiological properties of neurons; and
• compare the nervous and endocrine systems.

I f the body is to maintain homeostasis and function effectively, its trillions of cells must work together in a coordinated fashion. Should each cell behave without regard to what others were doing, the result would be physiological chaos and death. Two communication systems prevent this—the nervous and endocrine systems. A particularly important aspect of these systems is that they can detect changes in an organ system and modify its physiology as well as the physiology of other systems. Thus, they functionally unite the systems of the body and play a central coordinating role in homeostasis. The nervous system is the focus of this chapter and the three that follow; the endocrine system is covered in chapter 17.

Subdivisions of the Nervous System

The two major anatomical subdivisions of the nervous system are the **central nervous system (CNS)** and **peripheral nervous system (PNS)** (fig. 13.1). The CNS consists of the brain and spinal cord, which are enclosed and protected by the cranium and spinal column. The PNS originates with 43 pairs of nerves that arise from the CNS and divide into finer and finer branches serving all of the other organs. Twelve pairs, called *cranial nerves,* arise from the base of the brain and emerge through foramina of the skull. The other 31 pairs, called *spinal nerves,* arise from the spinal cord and emerge through the intervertebral foramina. Along the course of these nerves, the PNS has knotlike *ganglia* where the cell bodies of many neurons are located.

From a functional standpoint, the PNS can be divided into the *somatic* and *autonomic nervous systems.* These are distinguished from each other primarily by the types of effectors they innervate, but both systems also have sensory nerve fibers coming from different types of receptors. Further subdivision of the autonomic nervous system and functional contrasts between the somatic and autonomic systems are discussed in chapter 15.

The foregoing terminology may give the impression that the body has several nervous systems—central, peripheral, somatic, and autonomic. These are just terms of convenience, however. There is only one nervous system, and these subsystems are interconnected parts of the whole.

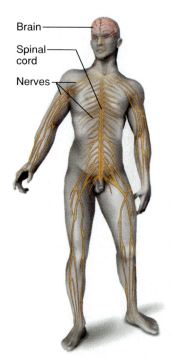

Brain
Spinal cord
Nerves

Major structural divisions of the nervous system

Central nervous system (CNS)	Brain	Gray matter
		White matter
	Spinal cord	Gray matter
		White matter
Peripheral nervous system (PNS)	Nerves	Cranial nerves (12 pairs)
		Spinal nerves (31 pairs)
	Ganglia	

Functional classes of peripheral nerve fibers

Somatic	Sensory
	Motor
Autonomic	Sensory
	Motor

Figure 13.1 The nervous system and its major subdivisions.

Functions of the Nervous System

There are three broad categories of nervous system function, corresponding to three categories of neurons (fig. 13.2):

1. **Sensory.** The nervous system detects stimuli that originate outside the body, such as sights and sounds, and others that originate within organs, such as joint movements and blood pressure. Cells or organs called **receptors** detect stimuli, and **sensory,** or **afferent**[1] (AFF-uh-rent), **neurons** conduct stimulus information to the CNS. A single cell may be both a receptor and afferent neuron, such as the receptors for pain and smell.

2. **Integrative.** About 90% of neurons are **interneurons**[2] **(association neurons).** They lie entirely within the CNS and bridge the gap from sensory to efferent neurons (discussed next). Interneurons receive information from multiple sources, interpret it, "decide" whether to transmit a message about it to other cells, and sometimes store information in memory.

3. **Responsive.** Responses are carried out by muscle and gland cells, which are called **effectors.** Neurons leading from the CNS to effectors are called **motor neurons** (because most of them innervate muscles) or **efferent**[3] (EFF-ur-unt) **neurons.**

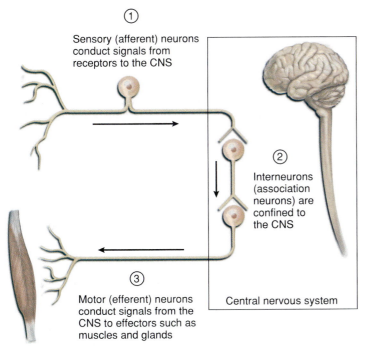

Figure 13.2 The three functional categories of neurons—sensory neurons, interneurons, and motor neurons.

① Sensory (afferent) neurons conduct signals from receptors to the CNS

② Interneurons (association neurons) are confined to the CNS

③ Motor (efferent) neurons conduct signals from the CNS to effectors such as muscles and glands

Central nervous system

1. *af* = toward + *fer* = to carry
2. *inter* = between
3. *ef* = out of

Properties of Nerve Cells

The functions of the nervous system depend on the following physiological properties of neurons:

- **Excitability.** Neurons have a low threshold of excitability (irritability); that is, they are highly responsive to changes around them.
- **Conductivity.** Neurons respond to stimuli by producing traveling signals (nerve impulses) that quickly alert other cells.
- **Secretion.** Neurons communicate with other cells by secreting neurotransmitters and other chemical messengers.

Think About It

What basic physiological properties do a nerve cell and a muscle cell have in common? What physiological properties does each one have that the other one lacks?

The Nervous and Endocrine Systems Compared

Although the nervous and endocrine systems are both means of internal communication, they are not redundant. When we compare the two, we see that they complement rather than duplicate each other's function (table 13.1).

One important difference between the two systems is the speed with which they start and stop responding to a stimulus. The nervous system typically begins responding in just a few milliseconds, whereas it takes from a few seconds to days for a hormone to produce an effect. When a stimulus ceases, the nervous system stops responding almost immediately. Some endocrine glands, however, go on responding for several days after a stimulus has ceased. On the other hand,

Table 13.1 Comparison of the Nervous and Endocrine Systems	
Nervous System	**Endocrine System**
Based on neurons and neurotransmitters	Based on endocrine glands and hormones
Communication by electrical impulses and neurotransmitters	Communication by hormones released into the bloodstream
Reacts quickly to stimuli, usually within 1 to 10 msec	Slower to react; often takes seconds to days
Stops reacting quickly when stimulus ceases	May continue responding long after stimulus ceases
Always adapts to prolonged stimulation	Slower to adapt; with continual stimulation, may respond for months or years
Neurons have relatively local, specific effects	Some hormones have more widespread, general effects

under long-term stimulation, receptors of the nervous system invariably *adapt* and exhibit a declining response. The endocrine system is more capable of responding to continual stimulation. For example, thyroid hormone secretion rises in response to cold weather and remains elevated as long as it remains cold.

Finally, an efferent nerve fiber innervates only one organ and a limited number of cells within that organ; its effects, therefore, are precisely targeted and relatively specific. A hormone, by contrast, circulates throughout the body, and while some hormones are very specific in their effects, some of them have widespread effects on many organs.

The two systems of communication are not independent of each other, and indeed they are sometimes difficult to distinguish. Some neurons release hormones, and some endocrine cells are modified neurons. A single chemical, such as epinephrine, may serve as both a neurotransmitter and hormone. Neurons often trigger hormone secretion, and hormones often stimulate or inhibit neurons. The two systems continually regulate and communicate with each other as they coordinate the activities of other organ systems.

Key Point Review

1 List the three basic functions of the nervous system and name the three basic classes of neurons that correspond to these functions.

2 What three physiological properties of neurons enable them to perform their communicative function?

3 Contrast the physiological properties of the nervous and endocrine systems.

Cells of the Nervous System

▼Objectives

When you have completed this section, you should be able to

• list the types of neuroglia and describe their locations and functions;

• define terms pertaining to the general structure of a neuron and classify neurons on the basis of structure;

• explain how a myelin sheath is formed;

• describe the various modes of axonal transport; and

• explain how a nerve fiber regenerates.

Neuroglia

There are about a trillion (10^{12}) neurons in the nervous system and 100 billion (10^{11}) in the brain alone—about as many neurons in your brain as there are stars in our galaxy! Still, they are outnumbered as much as 50 to 1 by attendant cells called **neuroglia** (noo-ROG-lee-uh), or **glial cells.** Neuroglia protect neurons and aid their function. The word *glia*, which means "glue," implies one of their roles—they bind neurons together and provide a supportive framework for the nervous tissue. In the fetus, glial cells form a scaffold that guides young migrating neurons to their destinations. Wherever a mature neuron is not in synaptic contact with another cell, it is covered with glial cells. This prevents neurons from contacting each other except at points specialized for signal transmission and thus gives precision to their conduction pathways.

There are six kinds of neuroglia, each of which has a unique function. *Schwann cells* and *satellite cells* are found in the peripheral nervous system, and *oligodendrocytes, astrocytes, ependymal cells,* and *microglia* (fig. 13.3) are found in the central nervous system.

1. **Schwann**[4] (shwon) **cells** form a sheath called the **neurilemma**[5] (noor-ih-LEM-ah) around each nerve fiber of the PNS. An average Schwann cell only covers 1 mm of nerve fiber, and some nerve fibers are more than a meter long; thus each nerve fiber is ensheathed in many Schwann cells, like a mile-long hot dog in a series of 6-inch buns. In most cases, a Schwann cell spirals around the nerve fiber like a jelly roll, forming a thick **myelin** (MY-eh-lin) **sheath** beneath the neurilemma. Neuromuscular junctions are also enclosed and insulated by Schwann cells (see chapter 12). Schwann cells are also necessary for regeneration of damaged PNS fibers. Later in this chapter, we will explore the structure and function of the myelin sheath and the role of Schwann cells in regeneration.

2. **Satellite cells** surround the nerve cell bodies in the PNS. Little is known of their function.

3. **Oligodendrocytes**[6] (OL-ih-go-DEN-dro-sites) form myelin sheaths around nerve fibers of the CNS but unlike Schwann cells, they do not form a neurilemma. Each oligodendrocyte has as many as 15 processes that extend from the bulbous cell body like the arms of an octopus. Each process forms the myelin sheath around a portion of one nerve fiber. Again, the nerve fiber is much longer than the reach of a single glial cell, so while each oligodendrocyte myelinates several nerve fibers, each nerve fiber requires many oligodendrocytes to cover its entire length.

4. **Astrocytes**[7] are many-branched star-shaped glial cells. They are the most abundant type in the CNS, constituting over 90% of the tissue in some areas of the brain. One type, **protoplasmic astrocytes,** cover the entire brain surface and most nonsynaptic regions of the neurons. Their numerous extensions, called **vascular processes** or **perivascular feet,**

----

4. Theodore Schwann (1810–1882), German histologist
5. *lemma* = peel, husk, sheath
6. *oligo* = few + *dendro* = branches + *cyte* = cell
7. *astro* = star

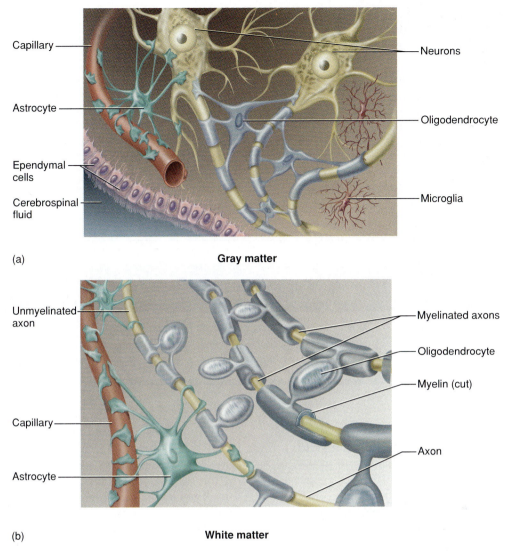

Capillary

Astrocyte

Ependymal cells

Cerebrospinal fluid

Neurons

Oligodendrocyte

Microglia

(a) **Gray matter**

Unmyelinated axon

Capillary

Astrocyte

Myelinated axons

Oligodendrocyte

Myelin (cut)

Axon

(b) **White matter**

Figure 13.3 Neuroglia of the central nervous system. (*a*) Glial cells of the gray matter. (*b*) Glial cells of the white matter.

wrap around blood vessels. Astrocytes help to form a *blood-brain barrier* that strictly controls which substances are able to get from the bloodstream into the tissue fluid of the brain (see chapter 14). They also help to regulate the chemical composition of the tissue fluid in the brain. As neurons transmit signals, they release potassium ions and neurotransmitters. Protoplasmic astrocytes take up these substances and prevent them from reaching excessive levels in the tissue fluid. A second type, **fibrous astrocytes,** form a physically supportive framework for the CNS. When neurons are damaged, fibrous astrocytes form hardened masses of scar tissue and fill space formerly occupied by neurons. This process is called *astrocytosis* or *sclerosis.*[8]

5. **Ependymal**[9] (ep-EN-dih-mul) **cells** produce and circulate cerebrospinal fluid (CSF), a liquid similar to blood plasma that fills cavities in the spinal cord and brain. Superficially, the ependyma looks like a cuboidal epithelium lining these cavities, with patches of cilia that circulate the fluid. Unlike epithelial cells, however, they have long rootlike processes at the basal surface that extend into the underlying nervous tissue. Ependymal cells and CSF are considered in more detail in the next chapter.

6. **Microglia** are small mobile macrophages that develop from white blood cells called monocytes. Like all macrophages, microglia thus arise from embryonic mesoderm, whereas all other glial cells

8. *sclero* = hard, hardening + *osis* = process

9. Greek, "upper garment"

Table 13.2 Neuroglia

Type	Location	Functions
Schwann cells	PNS	Form neurilemma around all PNS nerve fibers and myelin around most of them; aid in regeneration of damaged nerve fibers; encapsulate neuromuscular junctions
Satellite cells	PNS	Surround neurosomas; function uncertain
Oligodendrocytes	CNS	Form myelin in brain and spinal cord
Astrocytes	CNS	
Protoplasmic		Cover nonsynaptic regions of neurons; contribute to blood-brain barrier; remove neurotransmitters and K^+ from ECF of brain and spinal cord; help to regulate composition of ECF
Fibrous		Form supportive framework in CNS; form scar tissue to replace damaged neurons
Ependymal cells	CNS	Line cavities of brain and spinal cord; secrete and circulate cerebrospinal fluid
Microglia	CNS	Phagocytize and destroy microorganisms, foreign matter, and dead nervous tissue

Special Topic — Glial Cells and Brain Tumors   13.1

A tumor consists of a mass of rapidly dividing cells. Mature neurons, however, are incapable of mitosis and seldom form tumors. Brain tumors sometimes arise from the meninges (protective membranes of the CNS) or pituitary gland, but adult brain tumors are more often composed of glial cells, which are mitotic. Such tumors, called **gliomas**,[12] grow rapidly and are highly malignant.

are of ectodermal origin, like the neurons. Microglia wander freely through the CNS and phagocytize dead nervous tissue, microorganisms, and other foreign matter. They become concentrated in areas damaged by infection, trauma, or stroke. Pathologists look for clusters of microglia in histological sections of the brain as a clue to sites of injury.

Neuroglia are summarized in table 13.2, and their role in brain cancer is discussed in special topic 13.1.

Structure of a Representative Neuron

There are several varieties of neurons. A generalized neuron combining features of different types is shown in figure 13.4. The control center of the neuron is its **cell body,** also called the **soma**[10] or **perikaryon**[11] (PERR-ih-CAR-ee-on). It has a single centrally located nucleus with a large nucleolus. The cytoplasm contains mitochondria, lysosomes, a Golgi complex, numerous inclusions, extensive rough endoplasmic reticulum, and a cytoskeleton. The cytoskeleton consists of a dense mesh of microtubules and **neurofibrils** (intermediate filaments) that compartmentalize the rough ER into dark-staining regions called **Nissl**[13] **bodies** (fig. 13.4b,c).

Nissl bodies are unique to neurons and a helpful clue to identifying them in tissue sections with mixed cell types. Mature neurons lack centrioles and undergo no further mitosis after adolescence, but they are unusually long-lived cells, capable of functioning for over a hundred years.

The major cytoplasmic inclusions are glycogen granules, lipid droplets, melanin, and a golden brown pigment called **lipofuscin** (LIP-oh-FEW-sin)—an end-product of lysosomal digestion of worn-out organelles and other products. Lipofuscin collects with age and pushes the nucleus to one side of the cell. Lipofuscin granules are also called "wear-and-tear granules" because they are most abundant in old neurons, but they are apparently harmless to the neuron.

The soma of a neuron usually gives rise to a few thick processes that branch into a vast number of **dendrites**[14]—named for their striking resemblance to the bare branches of a tree in winter. The dendrites have receptors for neurotransmitters and are the primary site for receiving signals from other neurons. Dendrites do not produce action potentials but conduct information toward the soma in the form of local potentials, to be discussed later. Some neurons have only one dendrite, but 200 or more are not unusual. The more dendrites a

10. *soma* = body
11. *peri* = around + *karyo* = nucleus

12. *oma* = mass, tumor
13. Franz Nissl (1860–1919), German neuropathologist
14. *dendr* = tree + *ite* = little

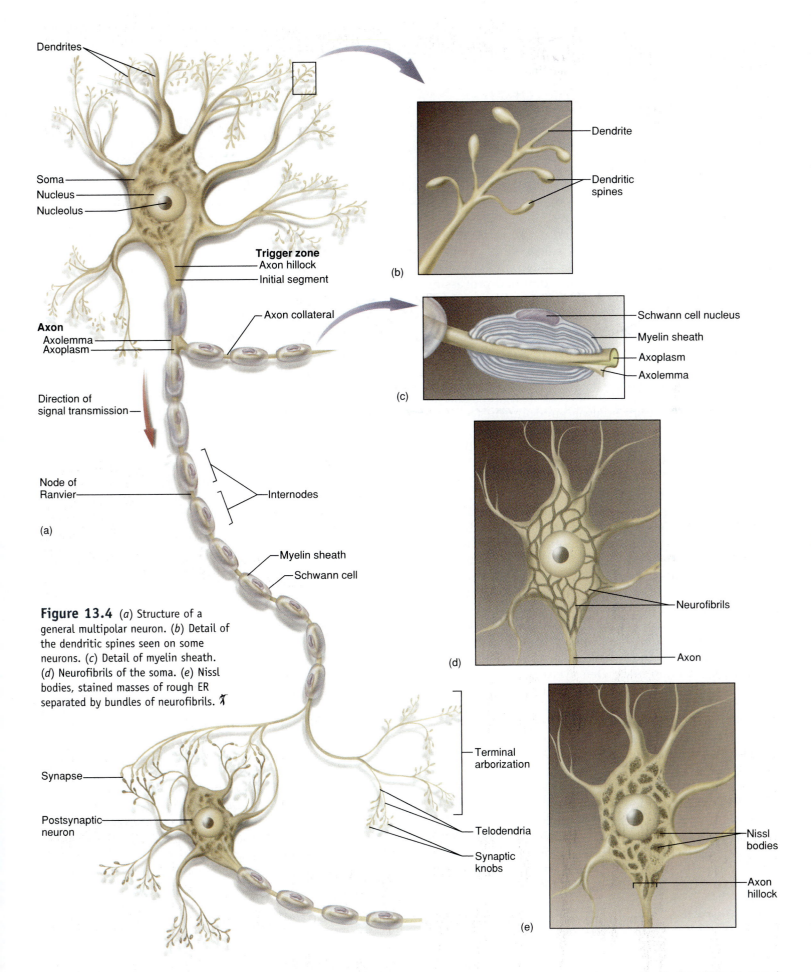

Dendrites

Soma
Nucleus
Nucleolus

Trigger zone
Axon hillock
Initial segment

Axon collateral

Axon
Axolemma
Axoplasm

Direction of
signal transmission

Node of
Ranvier

Internodes

(a)

Myelin sheath
Schwann cell

Figure 13.4 (*a*) Structure of a
general multipolar neuron. (*b*) Detail of
the dendritic spines seen on some
neurons. (*c*) Detail of myelin sheath.
(*d*) Neurofibrils of the soma. (*e*) Nissl
bodies, stained masses of rough ER
separated by bundles of neurofibrils.

Synapse

Postsynaptic
neuron

Terminal
arborization

Telodendria

Synaptic
knobs

(b)

Dendrite

Dendritic
spines

(c)

Schwann cell nucleus
Myelin sheath
Axoplasm
Axolemma

(d)

Neurofibrils

Axon

(e)

Nissl
bodies

Axon
hillock

Table 13.3 Classes of Nerve Fibers (Axons)

Fiber Type	Diameter	Myelinated	Conduction Velocity	Functions
A	2–20 µm	Yes	12–120 m/sec	Somatic motor fibers; sensory fibers for balance, position, touch, pressure, and pain
B	<3 µm	Yes	3–15 m/sec	Fibers from CNS to autonomic ganglia; some sensory fibers for temperature and pain
C	2–4 µm	No	0.5–2.0 m/sec	Fibers from autonomic ganglia to smooth muscle and gland cells; sensory fibers for smell, pain, temperature, and some mechanical stimuli

neuron has, the more information it can receive from other cells and incorporate into its decision making. As tangled as the dendrites may seem, they provide exquisitely precise pathways for the reception and processing of neural information. On some neurons, the dendrites are studded with innumerable **dendritic spines** (fig. 13.4b) that increase the area of synaptic contact. Studies on fish and other experimental animals have shown that social and sensory deprivation causes these spines to decline in number, while a richly stimulatory environment causes them to proliferate—an intriguing clue to the importance of a stimulating environment to infant and child development.

On one side of the soma is a thick region called the **axon hillock.** It is from this thickened area that the **axon,**[15] or **nerve fiber,** originates. The axon is cylindrical and relatively unbranched for most of its length. It is specialized for rapid transmission of nerve impulses to points remote from the soma. Its cytoplasm is called the **axoplasm** and its membrane the **axolemma.** A neuron never has more than one axon, and there are some neurons in the retina and brain that have no axons at all; these neurons communicate with each other by local potentials transmitted through their dendrites.

Somas range from 5 to 135 µm in diameter. By contrast, axons range from 1 to 20 µm in diameter and may be a few millimeters to more than a meter long. Such dimensions are more impressive when we scale them up to the size of familiar objects. If the soma of a spinal motor neuron were the size of a tennis ball, its dendrites would form a huge bushy mass that could fill a 30-seat classroom from floor to ceiling. Its axon would be about a mile long but a little narrower than a garden hose. This is a worthy point to ponder. The neuron must assemble molecules and organelles in its "tennis ball" soma and deliver them through its "mile-long garden hose" to the end of the axon. How it does this will be explained shortly.

Along its length, an axon sometimes gives rise to a few branches called **collaterals.** At the distal end, axons usually have a **terminal arborization**—an extensive complex of fine branches called **telodendria.**[16] Each

branch ends in a little swelling called a **synaptic knob (terminal button)** that forms a junction with a muscle cell, gland cell, or another neuron. These junctions are called **synapses,**[17] and the synaptic knob contains **synaptic vesicles** full of neurotransmitter.

Nerve fibers may be classified into three primary groups on the basis of their diameters and the presence or absence of myelin (table 13.3). Large fibers conduct impulses more rapidly than small ones because conduction occurs only along the surface of a fiber, not deep within its axoplasm, and larger fibers have more surface area. Large fibers therefore serve to conduct information of special urgency, such as motor commands to skeletal muscles and sensory information for vision and balance. If all nerve fibers were this large, however, the nervous system would either be impossibly bulky or limited to far fewer fibers. Therefore smaller fibers are used where speed is less important—for motor commands to glands and smooth muscle, for example. The fastest fibers are large myelinated ones (type A), and the slowest are small unmyelinated ones (type C).

Schwann Cells and Myelin

All axons of the PNS have a sheath of Schwann cells around them. In most cases (type A and B fibers), each Schwann cell spirals repeatedly around the axon, putting down as many as a hundred layers of its own membrane with almost no cytoplasm between layers. These layers form the myelin sheath (fig. 13.5). Myelin is about 20% protein and 80% lipid, including cholesterol, phospholipid, and glycolipid.

The outermost coil of the Schwann cell, the neurilemma, is surrounded in turn by a basement membrane and then a thin connective tissue sleeve called the *endoneurium.* The nucleus and cytoplasm of the Schwann cell are squeezed into the neurilemma, forming a bulge external to the myelin sheath. By analogy, suppose you wrapped an almost-empty tube of toothpaste tightly around a pencil. The pencil represents the axon, and the spiral layers of toothpaste tube (with

15. *axo* = axle, axis
16. *telo* = end + *dendr* = tree, branch

17. *syn* = together + *aps* = to touch, join

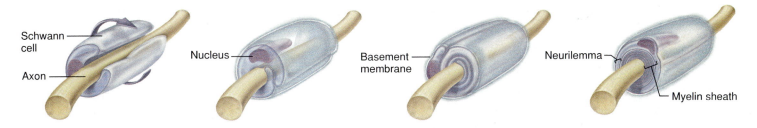

Schwann cell

Axon

Nucleus

Basement membrane

Neurilemma

Myelin sheath

Figure 13.5 Formation of the myelin sheath.

---

the toothpaste squeezed out) represent the myelin. The toothpaste would be forced to one end of the tube, which would form a bulge on the external surface of the wrapping, like the cell body of the Schwann cell.

Between adjacent Schwann cells are gaps called **nodes of Ranvier**[18] (RON-vee-AY), and the segments from one gap to the next are called **internodes** (see fig. 13.4). The short segment of axon between the axon hillock and the first Schwann cell is called the **initial segment.** Since the axon hillock and initial segment play an important role in initiating a nerve impulse, they are collectively called the **trigger zone.**

In the CNS, nerve fibers are myelinated by oligodendrocytes rather than Schwann cells and have no neurilemma, basement membrane, or endoneurium. Each oligodendrocyte myelinates a portion of several axons.

Myelin begins to form in the fourteenth week of gestation, but the process of myelination is far from finished at birth. Thus, the nervous system of an infant does not function as effectively as that of an older person, and an infant's responses to stimuli are less precise and coordinated. Most myelin sheaths have begun to develop by the time a child starts to walk, but myelination continues through adolescence. Abnormal development or later degeneration of the myelin sheath can be fatal (see special topic 13.2).

Unmyelinated Nerve Fibers

Small type C nerve fibers are described as **unmyelinated** because they have no myelin sheaths (fig. 13.6). From 1 to 12 type C fibers lie within grooves in the surface of a single Schwann cell. The Schwann cell wraps only once around each nerve fiber and somewhat overlaps itself. A basement membrane surrounds the entire Schwann cell along with its nerve fibers.

Structural Diversity in Neurons

Neurons are classified structurally according to the number of processes extending from the soma (fig. 13.7). Most neurons are like the generalized neuron described earlier—they have one axon and several dendrites, and are thus called **multipolar** neurons. Others are called **bipolar** because they have only one dendrite and one axon—for example, the olfactory cells of the nasal cavity and some neurons of the retina. The somatic sensory neurons that lead from the muscles and skin to the spinal cord are **unipolar.** During their development, the axon and dendrite fuse and thus there is only one process extending from the soma of the mature neuron. A short distance away, this process branches like a T, with a *peripheral fiber* leading to the skin or other source of

---

18. L. A. Ranvier (1835–1922), French histologist and pathologist

19. Warren Tay (1843–1927), English physician; Bernard Sachs (1858–1944), American neurologist

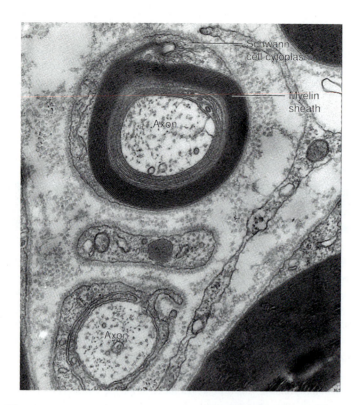

Figure 13.6 Transmission electron micrograph of a myelinated axon (*above*) and an unmyelinated axon (*below*). Note that Schwann cells cover both types of fibers but do not wrap repeatedly around the unmyelinated fibers.

sensation and a *central fiber* continuing into the spinal cord. Usually, a dendrite carries signals toward a soma and an axon carries them away. In this case, however, there is one long myelinated fiber that bypasses the soma and carries nerve impulses directly to the spinal cord. The entire fiber can be considered the axon, and the dendrites are the branching receptive endings in the skin.

Axonal Transport

Axonal transport is the two-way transport of materials in an axon—some molecules and organelles are moved from the soma to the distal end of the axon, and others are moved from the axon terminal back to the soma. If you think again of the tennis ball–garden hose analogy, you can visualize the problem of transporting materials such long distances.

Microtubules in the axoplasm act like railroad tracks to direct materials to their destinations. Movement from the soma to the telodendria, called **anterograde transport,** employs a motor protein called *kinesin* to move substances along the microtubules. Movement in the opposite direction, called **retrograde[20] transport,** uses a different motor protein called *dynein.*

....

20. *antero* = forward; *retro* = backward; *grad* = to walk or step

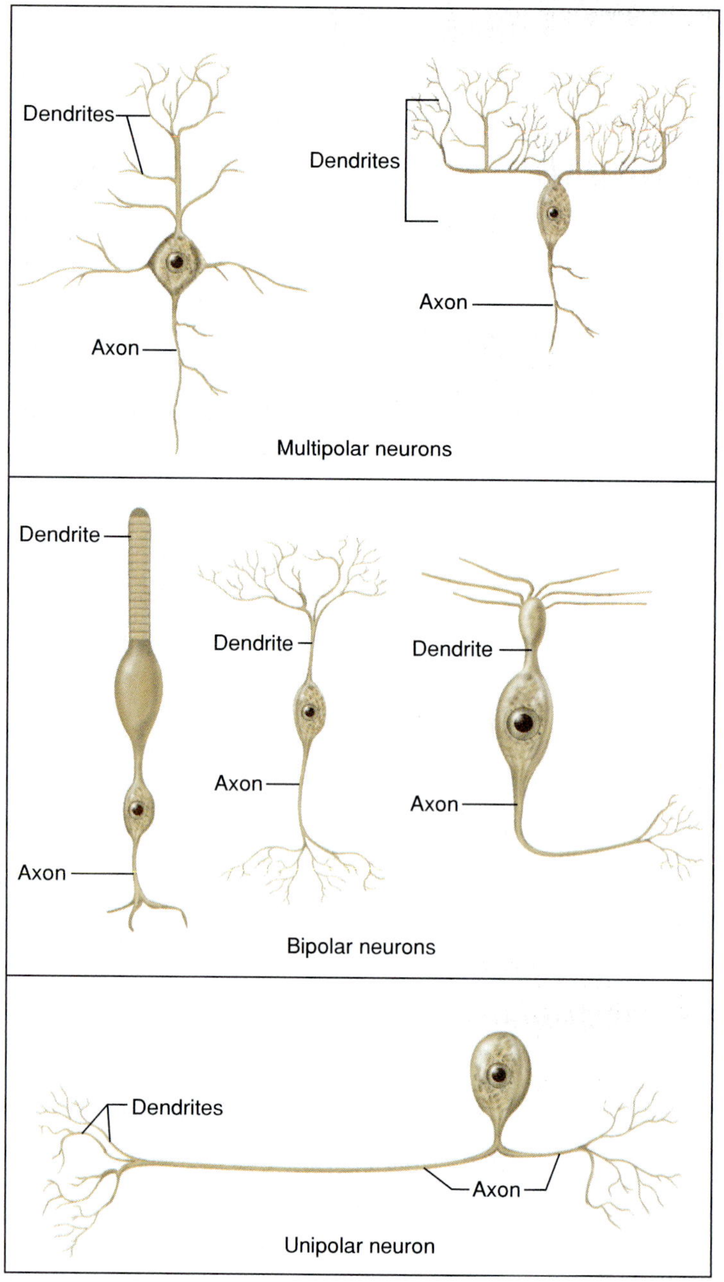

Figure 13.7 Variation in neuronal structure. *Top row, left to right:* two multipolar neurons of the brain—a pyramidal cell and Purkinje cell. *Middle row, left to right:* three bipolar neurons—a rod cell of the retina, a bipolar cell of the retina, and an olfactory neuron. *Bottom:* a unipolar neuron of the type involved in the senses of touch and pain.

Fast axonal transport occurs at a rate of 20 to 400 mm/day and may be either anterograde or retrograde (table 13.4). *Fast anterograde transport* moves mitochondria, synaptic vesicles, other organelles, and low-molecular-weight substances such as sugars, amino acids, nucleotides, and calcium ions away from the soma. *Fast retrograde transport* returns some synaptic vesicles and other materials to the soma to be degraded or recycled. It may also serve as a means of informing

Nerve growth factor (NGF) belongs to a family of proteins called *neurotrophins* that regulate the development of neurons. NGF is essential to the development of sympathetic nerve fibers and some sensory fibers. It is secreted by effector (gland and muscle) cells, picked up by axon terminals, and transported back to the soma. NGF guides developing nerve fibers to the appropriate target cells and thus plays a key role in the proper "wiring" of the nervous system to its effectors. The study of NGF and other growth factors is one of the most rapidly developing areas of biomedical research today, especially since recombinant DNA technology has made it possible to produce large quantities of these chemicals for study and clinical use. The use of NGF to treat Alzheimer disease is discussed in the chapter essay on page 458.

Table 13.4	Modes of Axonal Transport	
Method	**Speed**	**Functions**
Fast	20–400 mm/day	
Anterograde		Transport of mitochondria, synaptic vesicles, acetylcholinesterase, components of axolemma, glucose, amino acids, nucleotides, and calcium from soma to telodendria
Retrograde		Return of used synaptic vesicles and other axoplasmic substances to soma to be recycled or degraded; exploited by some viruses and toxins to reach neurosomas
Slow	0.5–10.0 mm/day	Anterograde transport of enzymes and cytoskeletal components; delivery of axoplasmic components to regenerating tips of damaged axons

the soma of conditions at the telodendria. Some pathogens exploit retrograde transport to invade neurons, including tetanus toxin and the herpes simplex, rabies, and polio viruses. In such infections, the delay between infection and the onset of symptoms corresponds to the time needed for the pathogens to reach the somas.

Slow axonal transport, also called *axoplasmic flow,* occurs at a rate of 0.5 to 10.0 mm/day and is always anterograde. It moves enzymes and cytoskeletal elements down the axon, renews worn-out axoplasmic components in mature neurons, and supplies new axoplasm for developing or regenerating neurons. Damaged nerve fibers regenerate at a speed governed by slow axonal transport.

Regeneration of Nerve Fibers

Because they are not enclosed in bone like the CNS, nerves of the PNS are subject to cuts, crushing injuries, and other trauma. A damaged peripheral nerve fiber can regenerate, however, if the soma is intact and at least some neurilemma remains. The process is shown in figure 13.8. The severed distal end of the axon and its myelin sheath degenerate within a few weeks and macrophages remove the debris. A **regeneration tube** forms from the neurilemma and the endoneurium surrounding it. The regeneration tube can guide a growing axon back to its original destination. The axon stump puts out several sprouts until one of them finds its way into the regeneration tube. This sprout begins to grow more rapidly (about 3–5 mm/day), possibly under the influence of chemicals secreted by the tube (see special topic 13.3), while the other sprouts are reabsorbed. If the nerve originally led to a skeletal muscle, the muscle atrophies in the absence of innervation but regrows after the nerve connection is reestablished.

Think About It
Damaged nerve fibers in the CNS cannot regenerate. Why?

Key Point Review

4 How is a glial cell different from a neuron? Name the six types of glial cells and state the functions of each.

5 Sketch a multipolar neuron and label its soma, dendrites, axon, telodendria, synaptic knobs, myelin sheath, and nodes of Ranvier.

6 How is myelin produced? How does myelin production in the CNS differ from that in the PNS?

7 How are materials synthesized in the soma delivered to the axon terminals?

8 How can a severed peripheral nerve fiber regenerate and find its way back to the original target cells?

Electrophysiology of Neurons

▼**Objectives**
When you have completed this section, you should be able to
• describe the properties of the resting membrane potential;
• explain how a local potential results from stimulation of a neuron;
• explain how a neuron generates an action potential; and
• explain how a nerve impulse is propagated along an axon.

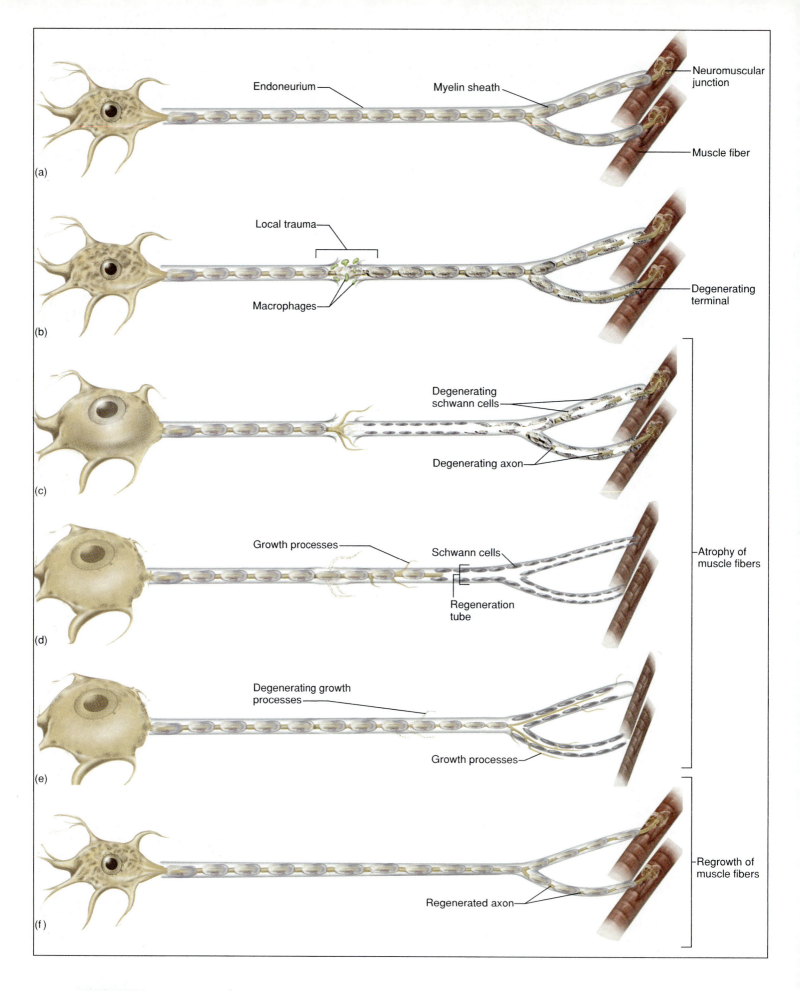

(a)

Endoneurium

Myelin sheath

Neuromuscular junction

Muscle fiber

(b)

Local trauma

Macrophages

Degenerating terminal

(c)

Degenerating schwann cells

Degenerating axon

Atrophy of muscle fibers

(d)

Growth processes

Schwann cells

Regeneration tube

(e)

Degenerating growth processes

Growth processes

(f)

Regenerated axon

Regrowth of muscle fibers

◀ **Figure 13.8** Regeneration of a damaged nerve fiber. (*a*) Normal nerve fiber. (*b*) Injured fiber with macrophages cleaning up tissue debris at the point of injury and early degeneration of the nerve fiber, myelin sheath, and axon terminals distal to that. (*c*) Early regeneration; the soma is swollen, the Nissl bodies have dispersed, and the axon stump has begun producing multiple growth processes, while the severed distal part of the nerve fiber shows further degeneration and (in the case of somatic motor neurons) muscle fibers atrophy for lack of innervation. (*d*) Continued regeneration; the Schwann cells, basement membrane, and neurilemma form a regeneration tube, one growth process grows into the tube, and the other growth processes are reabsorbed. (*e*) Continued regeneration; the growth processes have almost reestablished contact with the muscle fibers. (*f*) Regeneration complete; the muscle fibers are reinnervated and have regrown, and the soma of the neuron has returned to its original appearance.

The nervous system has intrigued scientists and philosophers since the ancient era. Erasistratus and Galen thought that the brain produced a fluid called *psychic pneuma* that was pumped through hollow nerves and squirted into the muscles to make them contract. René Descartes (1596–1650), the French philosopher-scientist, still argued for this theory 2,000 years later, but it fell out of favor in the eighteenth century when Luigi Galvani discovered the role of electricity in muscle contraction. When microscopes and histological staining methods were better developed, the Spanish histologist Santiago Ramón y Cajal (1852–1934) was able, with great effort, to trace the course of neurons and demonstrate that the nervous pathway was not a continuous "wire" or tube but a series of separate cells, functionally joined at the synapses. His theory, now called the *neuron doctrine,* suggested another question and direction for research—how do neurons communicate? We now know they employ an electrical and chemical code. Our objective in this section is to study how this code is produced and transmitted.

Concepts in Review

Several concepts introduced in earlier chapters are basic to the study of neurophysiology. We will review these here and enhance them a little, but you should reread the earlier material if necessary to comprehend the following discussion.

Neurons typically have a **resting membrane potential (RMP)** of about −70 mV, with the cytoplasm negative to the ECF. As long as a plasma membrane has a charge difference across it, it is said to be **polarized. Depolarization** is any shift in this potential toward 0 mV, and **repolarization** is a shift back toward the RMP. Most of the RMP is due to the balance between the diffusion and electrostatic attraction of potassium and sodium ions through the selectively permeable plasma membrane, as explained in the previous chapter. About −3 mV of the RMP is contributed by the sodium-potassium pump, which removes more positive charges (three Na^+) from the cell than it takes in (two K^+).

The Na^+-K^+ pump accounts for about 70% of the energy requirements of the nervous system. Later you will see that every action potential generated by a neuron slightly upsets the distribution of Na^+ and K^+ just described, and the Na^+-K^+ pump must work continually to restore equilibrium. This is why nervous tissue has one of the highest rates of ATP consumption of any tissue in the body and why it is so demanding of glucose and oxygen. Although a neuron is said to be resting when it is not producing action potentials, it is highly active during this time, maintaining its RMP and "waiting," as it were, for something to happen.

An electrical charge is called a *potential* because it has the potential to make charged particles move. Any flow of charged particles is called a *current;* by convention a current is considered to flow in the direction in which positive charges are moving. With these general principles in mind, we can begin to examine how they apply specifically to neurons.

Local Potentials

Neurons can be stimulated by chemicals, light, heat, or mechanical distortion of the plasma membrane. The effect of a stimulus is to open some of the gated ion channels in the plasma membrane, especially in the dendrites and soma. This allows ions in and out of the cell and produces a small deviation from the RMP called a **local potential.**

Suppose the neuron in figure 13.9 is chemically stimulated on its dendrite. Ligand-gated sodium channels open and there is a sodium influx (inward flow) at this point. This creates a region of low Na^+ concentration outside the cell. Sodium ions diffuse toward this point from neighboring regions of the ECF and enter the cell through the open gate—like water in a bathtub flowing toward an open drain. The area of membrane with open gates is appropriately called a **current sink.** As it draws positive charges away from adjacent areas of membrane, it makes the outside of the membrane less positive than it was at rest.

The Na^+ that enters the cytosol diffuses along the inside of the membrane to areas of lower concentration. Everywhere it goes, it cancels some of the negative charge of the cytosol. Therefore, the external surface of the membrane becomes less positive while the internal surface becomes less negative—that is, the membrane depolarizes in these regions of sodium flow. The more intense or prolonged the stimulus, the greater the number of gated channels that open, the greater the amount of Na^+ that enters the cell, and the greater the change in the membrane voltage. Thus, the local potential is *graded* (variable in magnitude) and proportional to stimulus intensity.

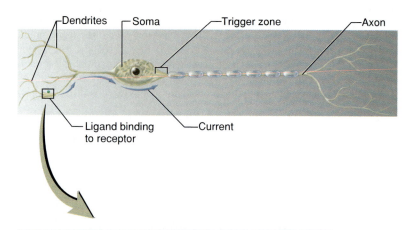

Figure 13.9 Excitation of a neuron by a chemical stimulus. When the chemical (ligand) binds to a receptor on the neuron, the receptor acts as a ligand-gated ion channel and current sink, through which Na⁺ diffuses into the cell. This depolarizes the plasma membrane.

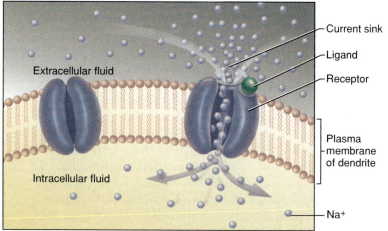

Like any diffusing solute, the Na⁺ diffusing along the inside of the membrane becomes more diluted the farther it goes. Furthermore, the Na⁺-K⁺ pumps along the membrane remove some of it from the cytoplasm. Therefore, Na⁺ exerts less and less of an effect on the RMP the farther it travels—the signal is *decremental* (gets weaker with distance). This explains why it is called a *local potential*—its effects are not felt at any substantial distance from the point of origin.

If the local potential travels far enough and is still strong enough, it can trigger an action potential by the mechanism described in the next section. If the stimulus is weak or brief, however, the local potential fades out (decays), returning to the RMP as the Na⁺-K⁺ pump returns Na⁺ ions to the ECF. That is, a local potential is *reversible*.

Finally, local potentials can be either *excitatory* or *inhibitory*. The first type makes a neuron more sensitive and ready to fire an action potential, and the second makes it less sensitive and less ready to fire than it was at the RMP. This will be further explored when we come to synaptic transmission.

In summary, the following five characteristics define a local potential and distinguish it from the action potentials to be studied shortly: a local potential is (1) graded, (2) decremental, (3) ineffective beyond a short distance, (4) reversible, and (5) either excitatory or inhibitory.

Action Potentials

An **action potential** can be generated only in places where the plasma membrane has an adequate density of voltage-gated ion channels. On the soma there are only about 50 to 75 of these channels per square micrometer (μm^2) of membrane—not enough to support an action potential. On the axon hillock, however, they range in density from 350 to 500 per μm^2. If an excitatory local potential travels all the way to the axon hillock and is still strong enough when it arrives, it can excite these ion gates and generate an action potential. An action potential can also occur if local potentials arrive from multiple points of origin on the cell and their combined effect at the axon hillock is great enough to excite these gates.

The action potential consists of rapid dramatic changes in membrane voltage as shown in figure 13.10a. The numbers in this figure correspond to the description that follows. Figure 13.11 correlates these voltage changes with events in the plasma membrane.

1. When current sinks open, sodium ions of the ECF diffuse away from the trigger zone and into the sink, while in the ICF they diffuse toward the trigger zone. The trigger zone thus becomes less positive externally and more positive internally. This can be detected as a steadily rising voltage at the axon hillock; it is called the **generator potential** because of its ability to generate an action potential.

2. For anything else to happen, the generator potential must rise to a critical voltage (typically about −55 mV) called the **threshold.** This is the minimum voltage needed to open voltage-regulated K⁺ and Na⁺ gates. When these gates open, the neuron *fires* (produces an action potential).

3. At threshold, K⁺ gates slowly begin to open, whereas Na⁺ gates open quickly. Only a few Na⁺ gates open at first, but as sodium ions rush in from the ECF, they depolarize the membrane still more, and this stimulates still more voltage-regulated Na⁺

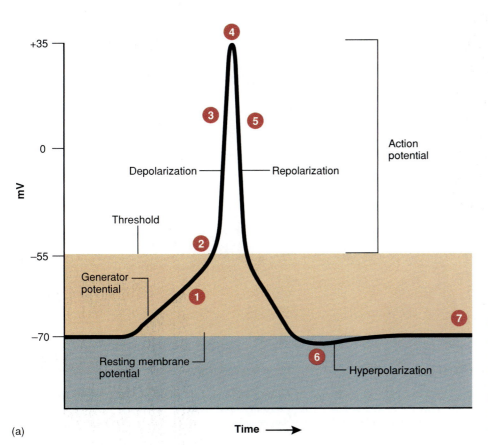

(a)

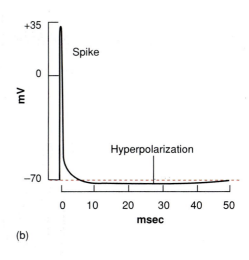

(b)

Figure 13.10 An action potential.
(a) Diagrammed with a distorted timescale to make details of the action potential visible; numbers correspond to stages discussed in the text. (b) On an accurate timescale, the generator potential is so brief it is imperceptible, the action potential appears as a spike, and the hyperpolarization is very prolonged.

gates to open. Thus, a positive feedback cycle is created that makes the membrane voltage rise rapidly. Indeed, the cell not only depolarizes but overshoots the 0 mV mark, peaking at approximately +35 mV. At this point, the membrane is positive on the inside and negative on the outside—its polarity is reversed in comparison to the RMP.

4. Membrane depolarization stimulates Na^+ gates to close. As a result, Na^+ influx ceases (a process called *sodium inactivation*), and the voltage stops rising.

5. By this time, the slow K^+ gates are fully open. Potassium ions rush out, repelled by the now-positive cytosol just inside the membrane. Their efflux causes the membrane voltage to drop rapidly. The action potential consists of the voltage changes that occur from the time the threshold is first reached to the time the voltage returns to threshold.

6. K^+ gates stay open longer than the Na^+ gates do, so the number of potassium ions that leave the cell is greater than the number of sodium ions that entered. Therefore the membrane voltage becomes 1 to 2 mV more negative than the original RMP, producing a negative overshoot called **hyperpolarization.**

7. As you can see, Na^+ and K^+ switch places across the membrane during an action potential. During hyperpolarization, ion diffusion through the membrane gradually restores the original distribution of Na^+ and K^+ until equilibrium and RMP are reestablished.

At the risk of being misleading, figure 13.11 is drawn as if the majority of sodium and potassium ions had traded places. In reality, only about one ion in a million crosses the membrane to produce an action potential, and an action potential only affects ion distributions in a thin layer close to the membrane. If the illustration tried to represent these points accurately, the difference would be so slight you could not see it; indeed the changes in ion concentrations inside and outside the cell are so slight they cannot be measured in the laboratory unless a neuron has been stimulated for a long time. Even after many action potentials, the cytosol still has a higher concentration of K^+ and lower concentration of Na^+ than the ECF does.

Figure 13.10a also has been deliberately distorted. In order to demonstrate the different phases of the generator potential and action potential, the magnitudes of the generator potential and hyperpolarization

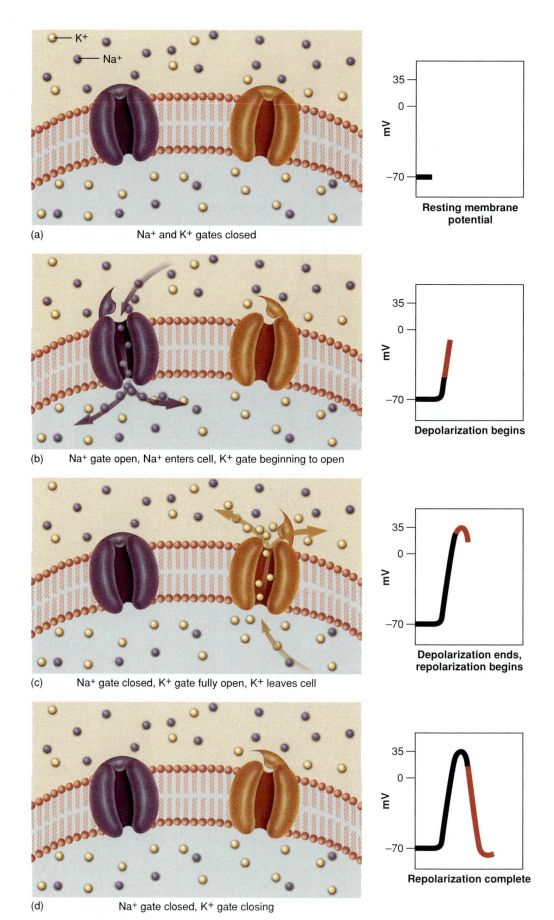

K+
Na+

(a)   Na+ and K+ gates closed

mV
35
0
−70

Resting membrane potential

(b)   Na+ gate open, Na+ enters cell, K+ gate beginning to open

mV
35
0
−70

Depolarization begins

(c)   Na+ gate closed, K+ gate fully open, K+ leaves cell

mV
35
0
−70

Depolarization ends, repolarization begins

(d)   Na+ gate closed, K+ gate closing

mV
35
0
−70

Repolarization complete

Figure 13.11 Actions of the sodium and potassium gates associated with the phases of an action potential.

Table 13.5 Comparison of a Local Potential and Action Potential	
Local Potential	**Action Potential**
Graded; membrane voltage proportional to stimulus intensity	All-or-none; either does not occur at all or exhibits same peak voltage regardless of stimulus intensity
Decremental; signal grows weaker as it travels farther from origin	Nondecremental; signal maintains same strength regardless of distance from origin
Local; effects occur no more than a short distance from where signal originated	Self-propagating; effects occur a great distance from where signal originated
Reversible; returns to RMP if stimulation ceases before threshold is reached	Irreversible; goes to completion if threshold is reached
Produced by ligand-gated channels on the dendrites and soma	Produced by voltage-gated channels on the axon hillock and axon
May be positive (depolarizing) or negative (hyperpolarizing) voltage charges	Always involves depolarization

are exaggerated, the generator potential is stretched out to make it seem longer, and the duration of hyperpolarization is shrunken so that the graph does not run off the page. When these events are plotted on a more realistic timescale, they look like figure 13.10*b*. The generator potential is so brief it is unnoticeable, and hyperpolarization is very long but only slightly more negative than the RMP. An action potential is often called a **spike;** it is easy to see why from this figure.

Although a local potential is proportional to stimulus strength, an action potential is not. No matter how strong a stimulus is, if it depolarizes the neuron to its threshold, the action potential always peaks at the same voltage (approximately +35 mV). The action potential obeys an **all-or-none law**—either the neuron fires at maximum voltage, or it does not fire at all. Also, whereas a local potential is reversible, an action potential is not. If a neuron reaches threshold, the action potential goes to completion—it cannot be stopped.

In these respects, you could compare the firing of a neuron to the firing of a gun. As the trigger is squeezed, a gun either fires with maximum force or does not fire at all (analogous to the all-or-none law). You cannot fire a fast bullet by squeezing the trigger hard or a slow bullet by squeezing it gently—once the trigger is pulled to its "threshold," the bullet always leaves the muzzle at the same velocity. And, like an action potential, the firing of a gun is irreversible once the threshold is reached. Table 13.5 contrasts a local potential with an action potential, including some characteristics of action potentials explained in the next section.

Refractory Period

A new stimulus generally does not affect a neuron while it is generating an action potential. From the time the membrane reaches threshold until repolarization is

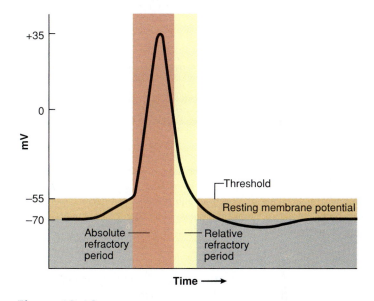

Figure 13.12 The absolute and relative refractory periods in relation to the action potential.

about one-third complete, it is completely insensitive to additional stimulation. This period, called the **absolute refractory period** (fig. 13.12), corresponds to the period in which the Na⁺ gates are open and continues briefly after they close. It is followed by a **relative refractory period,** which lasts until the membrane voltage returns to threshold and the K⁺ gates close. During this time, a new stimulus can reopen the Na⁺ gates and trigger a new action potential, but only if it is stronger than a threshold stimulus. After the K⁺ gates close, the membrane is essentially as sensitive as it was before the action potential, and it is capable of being reexcited by any stimulus of threshold strength or greater.

Be careful to note that the refractory period refers only to a small patch of membrane where an action potential has just occurred, not to the entire neuron. Other parts of the neuron can still be stimulated while a small

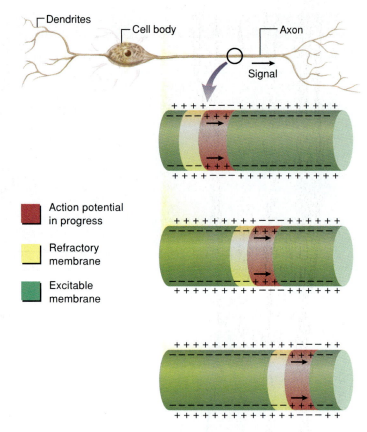

Figure 13.13 Conduction of a nerve impulse in an unmyelinated fiber. Note that the membrane polarity is reversed in the region of the action potential (red). A region of membrane in its refractory period (yellow) trails the action potential and prevents the nerve impulses from going backward toward the soma. The other membrane areas (green) are fully polarized and ready to respond.

Legend:
- **Action potential in progress** (red)
- **Refractory membrane** (yellow)
- **Excitable membrane** (green)

area of it is refractory, and even this area quickly recovers once the nerve impulse has passed on.

Signal Conduction in Unmyelinated Fibers

If a neuron is to communicate with another cell, an impulse has to travel to the end of the axon. We will now see how this is achieved, beginning with the relatively simple case of unmyelinated fibers (fig. 13.13).

The process is not hard to understand based on what we have already covered. An unmyelinated fiber has voltage-gated Na^+ channels along its entire length. An action potential is generated when events at one point on the membrane activate these gates at another point just distal to that. When an action potential occurs at the trigger zone, Na^+ enters the axoplasm and diffuses to adjacent regions just beneath the plasma membrane; outside the membrane, Na^+ flows toward the current sink. The resulting depolarization excites voltage-regulated ion gates immediately distal to the action

potential. Na^+ and K^+ gates open and close just as they did at the trigger zone; thus, a new action potential is produced here. By repetition, this excites the membrane immediately in front of that, and so forth, until the traveling impulse reaches the end of the axon. In unmyelinated fibers, the impulse travels at a rate of 0.5 to 2.0 m/sec.

Note that an action potential itself does not travel along an axon; rather, it stimulates the production of a new action potential in the membrane just ahead of it. Thus, we distinguish between *action potential* and *nerve impulse.* The nerve impulse is a traveling signal or wave of excitation produced by self-propagating action potentials. This is a little like a line of falling dominoes. No one domino travels to the end of the line, but there is a transmission of energy from the first domino to the last. No one action potential travels to the end of an axon, but a nerve impulse is transmitted by means of a chain reaction of action potentials.

If one action potential can stimulate the production of a new one in an adjacent region of membrane, you might think that the impulse could also start traveling backward and return to the soma. This does not occur, however, because the membrane behind the nerve impulse is still in its refractory period and cannot be restimulated. Only the membrane ahead of the nerve impulse is sensitive to stimulation. The refractory period thus ensures that nerve impulses are conducted in the proper direction.

The traveling impulse is an electrical current, but it is not the same as a current traveling through a wire. A current in a wire has a speed of 300 million m/sec, and it is decremental—it gets weaker with distance. A nerve impulse is much slower, but it is *nondecremental.* Even in the longest axons, the last action potential generated in the telodendria has the same voltage as the first one generated in the trigger zone. To clarify this concept we can compare the nerve impulse to a burning fuse. When a fuse is lit, the heat ignites powder immediately in front of this point, and this event repeats itself in self-propagating fashion until the end of the fuse is reached. At the end, the fuse burns just as hot as it did at the beginning. In an axon, the electrochemical gradient across the axolemma is a source of potential energy that the nerve impulse taps at every point along the way. Thus, the impulse does not grow weaker with distance; it is self-propagating, like the burning of a fuse.

Signal Conduction in Myelinated Fibers

Matters are somewhat different in myelinated fibers, because voltage-gated Na^+ channels are very scarce in the myelin-covered internodes—fewer than $25/\mu m^2$ in these regions compared to 2,000 to 12,000/μm^2 at the nodes of Ranvier. There would be little point in having ion gates in the internodes—myelin insulates the fiber from the ECF at these points, and sodium ions from the ECF could not flow into the axoplasm even if more gates were present. Thus the only way a nerve impulse can travel along an internode is for Na^+ that had entered at the previous node to diffuse down the fiber under the axolemma (fig. 13.14a).

This is a very fast process, but these diffusing ions become diluted with distance, and the signal becomes weaker the farther it goes. Therefore, this aspect of transmission is decremental. The signal cannot travel much farther than 1 mm before it becomes too weak to open any voltage-gated channels. This is precisely the distance to the next node of Ranvier, where the axolemma is exposed to ECF and there is an abundance of voltage-gated channels. When the diffusing ions reach this point, the signal is just strong enough to open these gates and create a new action potential. This action potential has the same strength as the one at the previous node, so each node of Ranvier essentially boosts the signal back to its original strength (35 mV). This is called **saltatory**[21] **conduction**—the propagation of a nerve impulse that seems to jump from node to node (fig. 13.14b).

In the internodes, saltatory conduction is therefore based on a process that is very fast (diffusion of ions along the fiber) but decremental. At each node, transmission is slower but nondecremental. Since most of the axon is covered with myelin, transmission along a myelinated fiber occurs mainly

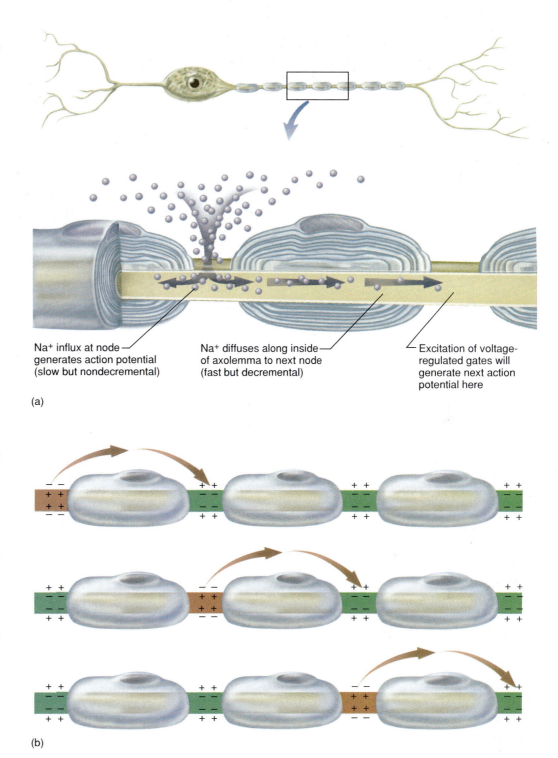

(a)

Na⁺ influx at node generates action potential (slow but nondecremental)

Na⁺ diffuses along inside of axolemma to next node (fast but decremental)

Excitation of voltage-regulated gates will generate next action potential here

(b)

Figure 13.14 Saltatory conduction of a nerve impulse in a myelinated fiber. (a) Ions can be exchanged with the ECF only at the nodes of Ranvier. In the internodes, the nerve signal travels by the rapid diffusion of ions along the inside of the plasma membrane. (b) Action potentials (red) occur only at the nodes, and the nerve impulse therefore appears to jump from node to node.

by the fast longitudinal diffusion process, with occasional pauses (every millimeter) to reamplify the signal. This is why myelinated fibers transmit signals much faster than unmyelinated ones—up to 130 m/sec (almost 300 mi/hr).

21. *saltare* = to leap or dance

9 What is a current sink? Define *current* in relation to the movement of cations.

10 How does the plasma membrane at the trigger zone differ from that on the soma? How does it resemble the membrane at a node of Ranvier?

11 What makes an action potential stop rising at +35 mV and start falling back toward the RMP?

12 Explain why myelinated fibers can transmit signals much faster than unmyelinated fibers.

Synapses

▼**Objectives**

When you have completed this section, you should be able to
- explain how messages are transmitted from one neuron to another;
- explain how stimulation of a postsynaptic cell is stopped; and
- give examples of different types of neurotransmitters and describe their actions.

A nerve impulse soon reaches the telodendria and can go no farther. In most cases, it must trigger the release of a neurotransmitter that can stimulate a new wave of electrical activity in a cell on the other side of a synapse. The most well-studied type of synapse is the neuromuscular junction described in chapter 12. Here we consider synapses between two neurons—a **presynaptic neuron,** which releases the neurotransmitter, and a **postsynaptic neuron,** which responds to it.

The synaptic knob of a presynaptic neuron may synapse with a dendrite, the soma, or the axon of a postsynaptic neuron, forming an *axodendritic, axosomatic,* or *axoaxonic synapse,* respectively. A neuron can have an enormous number of synapses (fig. 13.15). For example, a spinal motor neuron is covered with the synaptic knobs of about 10,000 presynaptic neurons—8,000 ending on its dendrites and another 2,000 on the soma. Some cells in the cerebellum of the brain have as many as 100,000 synapses.

The Discovery of Neurotransmitters

As the neuron doctrine became generally accepted, it raised the question of how neurons communicate with each other. In the early twentieth century, it was assumed that synaptic communication was electrical—a logical hypothesis given that neurons seemed to touch each other and signals were transmitted so quickly from one to the next. Closer histological examination, however, revealed a 20 to 30 nm gap between neurons—the **synaptic cleft**—casting doubt on the hypothesis of electrical conduction. In 1921, a study by German pharmacologist Otto Loewi (1873–1961) conclusively demonstrated that neurons communicate through the release of chemicals.

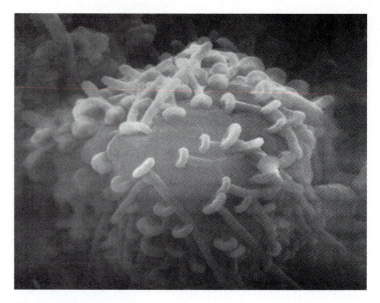

Figure 13.15 Synaptic knobs on the soma of a neuron in a marine slug (TEM).

The *vagus nerves* supply the heart, among other organs, and cause it to slow down. Loewi stimulated a frog's vagus nerve, and the heart slowed down as expected. He then removed some of the saline in which that heart was bathed and used it to bathe the heart of a second frog. The solution alone caused the second frog's heart to slow down, because it contained some substance released by the vagus nerve of the first frog. Loewi called the unidentified substance *Vagusstoffe*;[22] it was later found to be acetylcholine—the first known neurotransmitter.

Think About It

As described, does the previous experiment conclusively prove that the second frog's heart slowed as a result of something released by the vagus nerves? If you were Loewi, what control experiment would you do to rule out alternative explanations?

Following Loewi's work, the idea of electrical communication between cells fell into disrepute. Now, however, we realize that some neurons, neuroglia, and muscle cells do indeed have **electrical synapses,** where adjacent cells are only 2 nm apart and are joined by gap junctions (p. 193). These junctions have the advantage of quick transmission because they do not entail a delay for the release and binding of neurotransmitter. Their disadvantage, however, is that they cannot integrate information and make decisions. This is a property of **chemical synapses,** in which neurons communicate by neurotransmitters. Chemical synapses form the basis of the sophisticated information-processing capability of the nervous system.

---

22. *stoffe* = stuff, substance

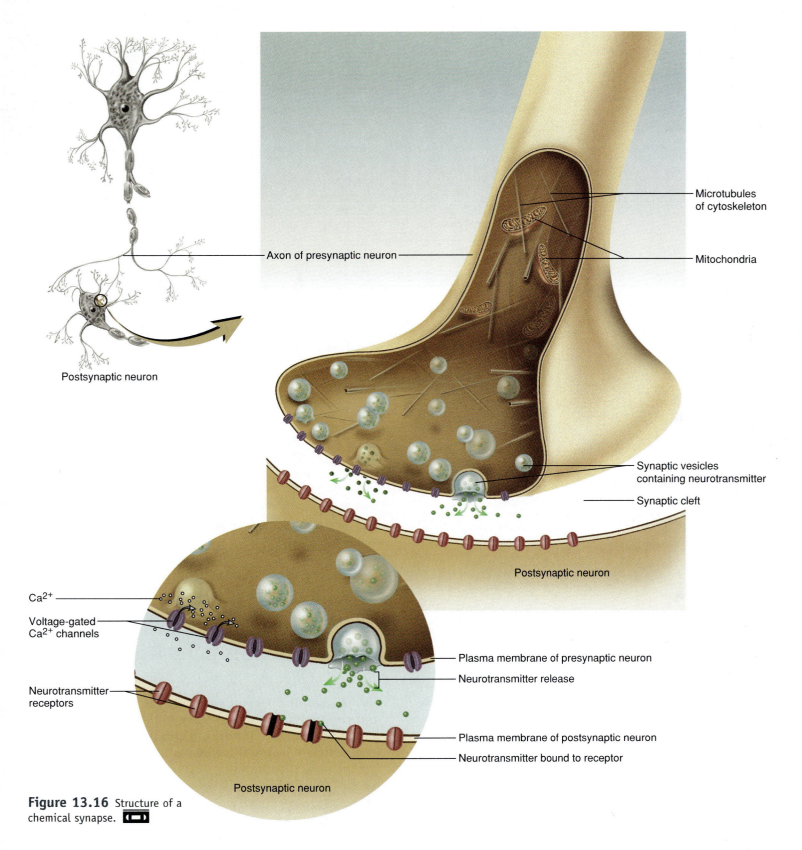

Figure 13.16 Structure of a chemical synapse.

Labels in figure:
- Axon of presynaptic neuron
- Microtubules of cytoskeleton
- Mitochondria
- Synaptic vesicles containing neurotransmitter
- Synaptic cleft
- Postsynaptic neuron
- Postsynaptic neuron
- Ca^{2+}
- Voltage-gated Ca^{2+} channels
- Neurotransmitter receptors
- Plasma membrane of presynaptic neuron
- Neurotransmitter release
- Plasma membrane of postsynaptic neuron
- Neurotransmitter bound to receptor
- Postsynaptic neuron

Structure of a Chemical Synapse

The synaptic knob (fig. 13.16) was described in the previous chapter. It contains synaptic vesicles, many of which are "docked" at release sites on the plasma membrane, ready to release their neurotransmitter on demand. A reserve pool of synaptic vesicles are located a little farther away from the membrane; they are clustered together near the release sites and tethered to the cytoskeleton and each other by protein microfilaments.

Table 13.6 Some Neurotransmitters and Neuropeptides

Neurotransmitter	Locations and Actions
Acetylcholine (ACh)	Neuromuscular junctions, all presynaptic autonomic fibers, postsynaptic parasympathetic fibers, some postsynaptic sympathetic fibers, retina, and many parts of the brain
Amino Acids with Excitatory Effects	
Glutamic Acid (Glutamate)	Cerebral cortex and brainstem; accounts for about 75% of all excitatory synaptic transmission in the brain; involved in learning and memory
Aspartic Acid (Aspartate)	Spinal cord; effects similar to those of glutamic acid
Amino Acids with Inhibitory Effects	
Glycine	Inhibitory neurons of the brain, spinal cord, and retina; most common inhibitory neurotransmitter in the spinal cord
GABA (γ-Aminobutyric Acid)	Thalamus, hypothalamus, cerebellum, occipital lobes of cerebrum, and retina; most common inhibitory neurotransmitter in the brain
Biogenic Amines (Monoamines)	
Catecholamines	
Norepinephrine	Most sympathetic postsynaptic fibers, cerebral cortex, hypothalamus, brainstem, cerebellum, and spinal cord; involved in dreaming, waking, and mood
Epinephrine	Hypothalamus, thalamus, spinal cord, and adrenal medulla; effects similar to those of norepinephrine
Dopamine	Hypothalamus, limbic system, cerebral cortex, and retina; highly concentrated in substantia nigra of midbrain; involved in elevation of mood and control of skeletal muscles
Indolamines	
Serotonin	Hypothalamus, limbic system, cerebellum, retina, and spinal cord; also secreted by blood platelets and intestinal cells; involved in sleepiness, alertness, thermoregulation, and mood
Histamine	Hypothalamus; also a potent vasodilator released by mast cells of connective tissue
Neuropeptides	
Substance P	Basal nuclei, midbrain, hypothalamus, cerebral cortex, and afferent pain-receptor neurons; also secreted by intestine; mediates pain transmission
Enkephalins	Hypothalamus, limbic system, pituitary, pain pathways of spinal cord, and nerve endings of digestive tract; act as analgesics (pain-relievers) by inhibiting substance P; inhibit intestinal motility; secretion increases sharply in women in labor
β-Endorphin	Digestive tract and many parts of the brain; also secreted as a hormone by the pituitary; suppresses pain; reduces perception of fatigue and produces "runner's high" in athletes
Cholecystokinin (CCK)	Cerebral cortex and small intestine; suppresses appetite

The postsynaptic neuron does not show such conspicuous specializations. Its plasma membrane runs closely parallel to the presynaptic membrane. It contains no synaptic vesicles and cannot release neurotransmitters. Its membrane does, however, contain proteins that function as receptors and ion gates.

Neurotransmitters and Related Messengers

More than 100 confirmed or suspected neurotransmitters have been identified since Loewi discovered acetylcholine. Some of the best-known ones are listed in table 13.6. Parts of the brain referred to in this table will become familiar to you as you study the next chapter, and you may wish to refer back to this table then to enhance your understanding of brain function. Neurotransmitters fall into three major categories according to chemical composition (fig. 13.17):

1. **Acetylcholine** is in a class by itself. It is formed by the condensation of acetic acid (acetate) and choline.
2. **Amino acid** neurotransmitters include glycine, glutamic acid, aspartic acid, and gamma-(γ-) aminobutyric acid (GABA).
3. **Biogenic amines** are synthesized from amino acids by replacing the —COOH group with another functional group. They retain the —NH$_2$ (amino

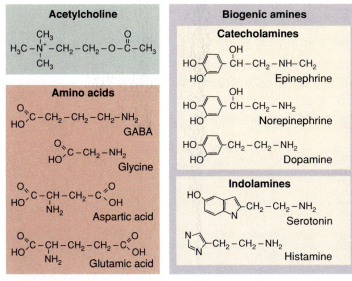

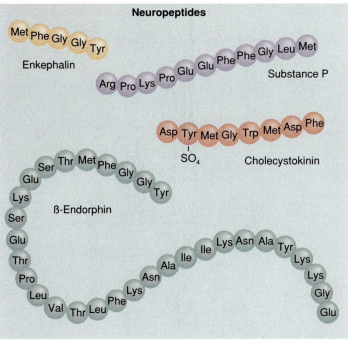

Figure 13.17 Structure and classification of some neurotransmitters and neuropeptides.

group) and are therefore also known as **monoamines.** The major neurotransmitters in this class fall into two subclasses defined by the group that replaces —COOH: **catecholamines** (CAT-eh-COAL-uh-meen) (epinephrine, norepinephrine, and dopamine) and **indolamines** (serotonin and histamine).

Neuropeptides are chains of 2 to 40 amino acids that sometimes modify the action of neurotransmitters. It is common, for example, for serotonin to be released together with a neuropeptide that influences its effect

on the postsynaptic cell. Neuropeptides are classified as neurotransmitters by some authorities and, for reasons discussed later, are classified separately by others.

Synaptic Transmission

Different neurotransmitters can be excitatory or inhibitory, and some exert both effects on a postsynaptic cell. Moreover, they can use either ligand-gated ion channels or second-messenger systems as a mechanism of action. Bearing this diversity in mind, let us begin by considering a chemical synapse at which acetylcholine opens ligand-gated channels in the postsynaptic cell, thus depolarizing and exciting it. A synapse where transmission is mediated by acetylcholine (ACh) is called a **cholinergic** (CO-lin-UR-jic) **synapse.**

The mechanism of transmission is detailed in figure 13.18. In brief, the arrival of a nerve impulse at the synaptic knob triggers ACh release from synaptic vesicles docked at the release sites. Empty vesicles drop back into the cytoplasm to be refilled with ACh, while synaptic vesicles in the reserve pool move forward to the release sites and release their ACh—a bit like a line of Revolutionary War soldiers firing their flintlock rifles and falling back to reload as another line moves to the front.

At the postsynaptic neuron, ACh binds to ligand-gated channels. This causes the channels to open, allowing Na^+ and K^+ to enter and leave the cell, respectively. It is the Na^+ influx, especially, that depolarizes the postsynaptic cell and produces a local **postsynaptic potential (PSP).** If it is strong and persistent enough, the PSP opens voltage-gated ion channels in the trigger zone and causes the postsynaptic neuron to fire. As complex as this may seem, it all happens in an interval of only 0.5 msec called **synaptic delay**—the time from the arrival of an impulse at the axon terminal of a presynaptic cell to the beginning of an action potential in the postsynaptic cell.

The mechanism just described is called an **ionotropic**[23] (eye-ON-oh-TRO-pic) **effect** because the neurotransmitter opens ion gates in the postsynaptic membrane and causes immediate changes in the membrane potential. It is the effect typically achieved by acetylcholine and amino acid neurotransmitters. Biogenic amines and neuropeptides have **metabotropic** (meh-TAB-oh-TRO-pic) **effects** mediated by a second messenger such as cyclic AMP (fig. 13.19). The receptor is not an ion channel but is associated on its intracellular side with a G protein that activates the enzyme **adenylate** (ad-DEN-ih-late) **cyclase.** When a

---

23. *iono* = ions + *trop* = to turn, change, affect

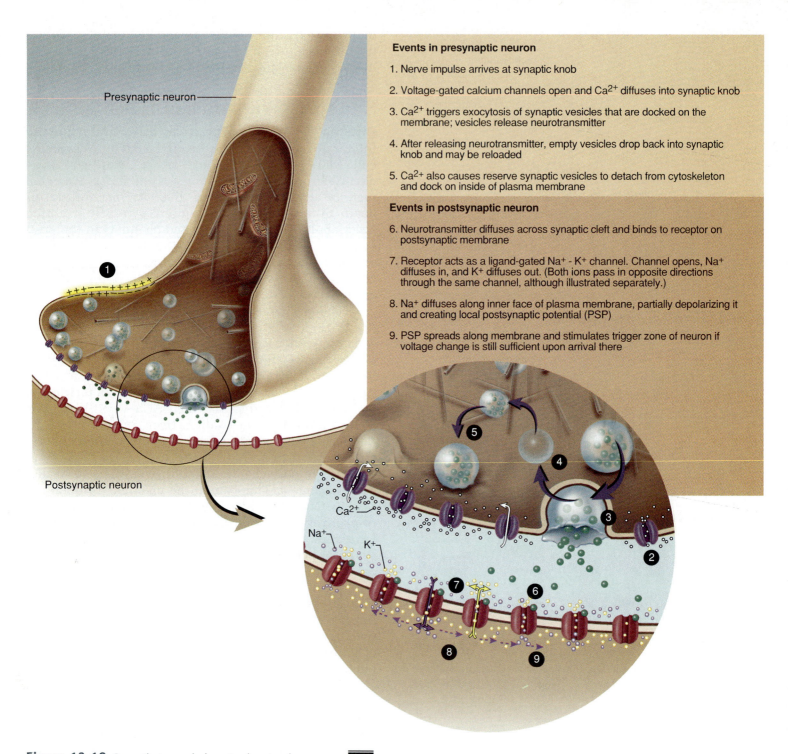

Events in presynaptic neuron

1. Nerve impulse arrives at synaptic knob

2. Voltage-gated calcium channels open and Ca²⁺ diffuses into synaptic knob

3. Ca²⁺ triggers exocytosis of synaptic vesicles that are docked on the membrane; vesicles release neurotransmitter

4. After releasing neurotransmitter, empty vesicles drop back into synaptic knob and may be reloaded

5. Ca²⁺ also causes reserve synaptic vesicles to detach from cytoskeleton and dock on inside of plasma membrane

Events in postsynaptic neuron

6. Neurotransmitter diffuses across synaptic cleft and binds to receptor on postsynaptic membrane

7. Receptor acts as a ligand-gated Na⁺ - K⁺ channel. Channel opens, Na⁺ diffuses in, and K⁺ diffuses out. (Both ions pass in opposite directions through the same channel, although illustrated separately.)

8. Na⁺ diffuses along inner face of plasma membrane, partially depolarizing it and creating local postsynaptic potential (PSP)

9. PSP spreads along membrane and stimulates trigger zone of neuron if voltage change is still sufficient upon arrival there

Figure 13.18 Synaptic transmission at a ionotropic synapse.

neurotransmitter such as epinephrine binds to the receptor, adenylate cyclase converts ATP to cAMP by splitting off two phosphate groups (PP_i). Cyclic AMP activates an enzyme called **protein kinase** that phosphorylates other cytoplasmic enzymes.

In some cases, this process activates an enzyme and in some cases it deactivates one; but either way, it changes cellular metabolism. When an enzyme is activated by this process, it may (1) bind to the intracellular side of a channel protein and open ion gates in the membrane, (2) turn on a metabolic pathway so that the postsynaptic cell begins to synthesize something new, or (3) enter the nucleus and activate genetic transcription, in which the cell synthesizes new proteins. A number of hormones also act this way, underscoring the similarities between the endocrine and nervous systems. Indeed, epinephrine is both a metabotropic neurotransmitter and a hormone.

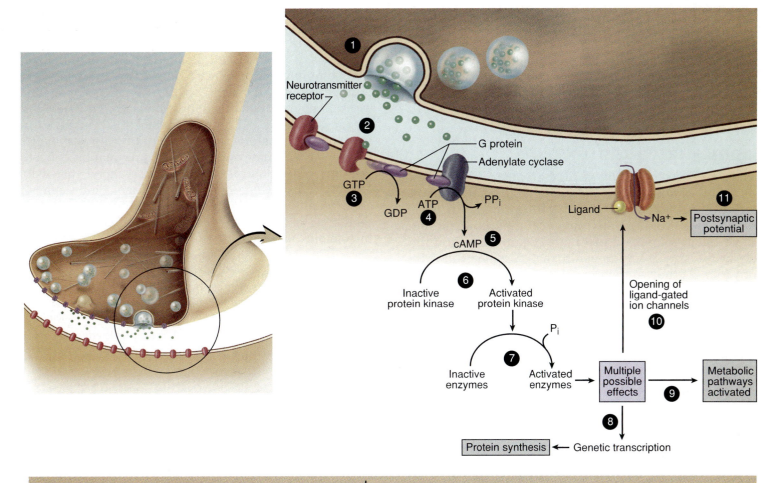

Figure 13.19 Synaptic transmission at a metabotropic synapse. In contrast to the events in the preceding figure, the neurotransmitter receptor is not an ion channel. It activates a second messenger system with a variety of possible effects within the postsynaptic cell.

1. Presynaptic neuron releases neurotransmitter.

2. Neurotransmitter binds to receptor on postsynaptic neuron.

3. Receptor activates a G protein, which hydrolyzes GTP and migrates along the plasma membrane.

4. G protein binds to adenylate cyclase.

5. Adenylate cyclase removes two phosphate groups (PP_i) from ATP, producing cAMP.

6. cAMP activates a protein kinase.

7. Protein kinase phosphorylates other cytoplasmic enzymes. This can lead to any of the following effects.

8. Activated enzymes trigger genetic transcription and synthesis of new proteins.

9. Activated enzymes activate other metabolic pathways.

10. Activated enzymes open ligand-gated ion channels in the plasma membrane, allowing Na^+ to enter cell and create a postsynaptic potential.

Cessation of the Signal

It is not only important to stimulate the postsynaptic cell but equally important to be able to turn off the stimulus. A single neurotransmitter does not bind to its receptor for long. ACh binds for only 1 msec or so, during which time the ion channel opens and thousands of sodium and potassium ions rush through the membrane. ACh then dissociates from the receptor site and the channel closes. If the presynaptic cell continues to secrete ACh into the cleft, one ACh molecule is quickly replaced by another, and the ion channel opens again and lets another burst of ions through. This immediately suggests a way of stopping synaptic transmission—stop adding new transmitter to the cleft and get rid of that which is already there. The first step is achieved simply by the cessation of impulses in the presynaptic nerve fiber. The second can be achieved in the following ways:

- **Diffusion.** Some neurotransmitter diffuses out of the synaptic cleft into the nearby ECF, where it may be taken up by astrocytes.
- **Reuptake.** Neurotransmitter is actively transported back into the synaptic knob and recycled. This is a major mechanism for removing all neurotransmitters except ACh and neuropeptides.
- **Degradation.** As we saw at the neuromuscular junction (see chapter 12), ACh is degraded by

acetylcholinesterase to nonstimulatory fragments—acetate and choline. Choline is reabsorbed by the presynaptic cell and used to synthesize more ACh. Norepinephrine, epinephrine, and serotonin are degraded by an enzyme called **monoamine oxidase (MAO)** after reuptake by the presynaptic neuron.

Other Modes of Chemical Communication

The rapid pace of discovery in neurophysiology and endocrinology has made it increasingly difficult to distinguish neurotransmitters from hormones and other messengers. Traditionally, however, neurotransmitters have been conceived as small organic compounds, about the size of amino acids, that meet the following criteria: (1) they are synthesized by presynaptic neurons, (2) they are released in response to stimulation, (3) they bind to specific receptors on adjacent postsynaptic cells, and (4) they alter the physiology of the postsynaptic cell. Neurons have other means of communication, however, that fall outside the scope of this traditional concept of neurotransmission.

Neuropeptides are included among neurotransmitters by some authorities and classified separately by others. While neurotransmitters are stored in synaptic vesicles about 50 nm in diameter, neuropeptides are stored in larger (100 nm) *secretory granules,* also called *dense-core vesicles* because they exhibit a dense mass of peptide within them. Some neurons contain both neuropeptides and neurotransmitters. They may release them simultaneously or release one substance or the other under different conditions—for example, a neurotransmitter under mild stimulation and a neuropeptide under stronger stimulation.

Some neuropeptides are produced not only by neurons but also by the digestive tract; thus they are known as *gut-brain peptides.* Some of these cause cravings for specific nutrients such as fat or sugar and may be associated with certain eating disorders (see chapter 26).

Think About It

Small neurotransmitters such as ACh and monoamines can be synthesized in the synaptic knobs, but neuropeptides can be synthesized only in the soma and must be transported to the synaptic knobs from there. Why is the synthesis of neuropeptides limited to the soma?

Neuromodulators are messengers other than neurotransmitters that indirectly influence synaptic communication. They include hormones and neuropeptides released by endocrine and nerve cells, respectively. Neuromodulators alter the rate of synthesis, release, reuptake, or enzymatic degradation of neurotransmitters or alter the sensitivity of the postsynaptic neuron to a neu-

rotransmitter. The effectiveness of synaptic transmission therefore can be influenced by chemicals released at distant points in the nervous or endocrine system.

It has recently been discovered that neurons can also communicate backward across the synapse (from postsynaptic to presynaptic cell) by means of inorganic gases. Some postsynaptic neurons release nitric oxide or carbon monoxide. Nitric oxide seems to be involved in learning and memory. When released by a postsynaptic neuron, it stimulates not only the presynaptic cell but also any other neurons in the neighborhood that were actively transmitting signals at the time. This seems to contribute to a learning process called long-term potentiation, to be discussed later. The role of carbon monoxide remains less clear.

Key Point Review

13 Concisely describe four steps that occur in the interval between the arrival of an action potential at the telodendria and the beginning of a new action potential in the postsynaptic neuron.

14 Distinguish between ionotropic and metabotropic effects.

15 Describe three mechanisms that cause synaptic transmission to stop.

16 What is the function of neuromodulators? Name two of them.

Neural Integration

▼Objectives

When you have completed this section, you should be able to
- explain how postsynaptic potentials determine whether or not a neuron fires;
- explain how complex information about the environment is converted into a meaningful code by the nervous system;
- describe how learning and memory are related to changes in synaptic physiology; and
- explain how neurons work together in groups to process information and produce effective output.

Synaptic delay slows down the transmission of nervous information; the more synapses there are in a neural pathway, the longer it takes information to get from its origin to its destination. You might wonder, therefore, why we have synapses at all—why a nervous pathway is not, indeed, a continuous tube or "wire" as biologists believed before the neuron doctrine. The presence of synapses is not due to limitations in the length of a neuron—after all, one nerve fiber can reach from your toes to your brainstem, and it is fascinating to consider how long some nerve fibers may be in a giraffe or a whale. We have also seen that cells can communicate through gap junctions much more quickly than they can through chemical synapses. So why have chemical synapses at all?

What we value most about our nervous system is its ability to process and store information and make decisions—and chemical synapses are the decision-making devices of the system. The more synapses a neuron has, the greater its information-processing capability is. The cells of the cerebral cortex that you are using to comprehend this passage, called *pyramidal cells* (see fig. 13.7), each have about 40,000 synaptic contacts with other neurons (about 98% on the dendrites and 2% on the soma). It is estimated that there are about 100 trillion (10^{14}) synapses in the cerebral cortex alone. To get some impression of this number, imagine trying to count them. Even if you could count two synapses per second day and night without stopping, it would take you 1.6 million years.

Postsynaptic Potentials

The information-processing, decision-making, and memory mechanisms of neurons are collectively called **neural integration.** Such processes are based on the postsynaptic potentials (PSPs) produced by a neurotransmitter. A voltage shift toward the threshold is a depolarization, and since it makes the postsynaptic cell more likely to fire an action potential, it is called an **excitatory postsynaptic potential (EPSP)** (fig. 13.20*a*). This is usually the result of Na^+ flowing into the cell and canceling some of the negative charge on the membrane.

In other cases, a neurotransmitter makes the voltage on the postsynaptic cell even more negative than it was at rest. This is hyperpolarization, and since it makes the postsynaptic cell less likely to fire, it is called an **inhibitory postsynaptic potential (IPSP)** (fig. 13.20*b*). In many cases, a neurotransmitter produces this effect by opening a ligand-gated chloride channel, causing Cl^- to flow into the cell and make the cytosol more negative. A less common way of producing IPSPs is to open selective K^+ channels, accelerating the diffusion of K^+ out of the cell without letting Na^+ in.

Glutamic and aspartic acid tend to produce EPSPs and are excitatory neurotransmitters. Glycine and GABA produce IPSPs and are therefore inhibitory. Acetylcholine (ACh) and norepinephrine are excitatory to some cells and inhibitory to others, depending on the type of receptors present on the target cells. For example, ACh is excitatory to skeletal muscle and inhibitory to cardiac muscle because the two types of muscle have different types of ACh receptors. This is discussed more fully in the context of the autonomic nervous system in chapter 15.

Summation and Facilitation

A brain neuron may receive input from thousands of presynaptic cells simultaneously. Some of these incoming nerve fibers may produce EPSPs while others produce IPSPs. Whether or not the neuron fires depends on whether the *net* input is excitatory or inhibitory. If the EPSPs override the IPSPs, threshold may be reached; but if the IPSPs prevail, the neuron will not fire.

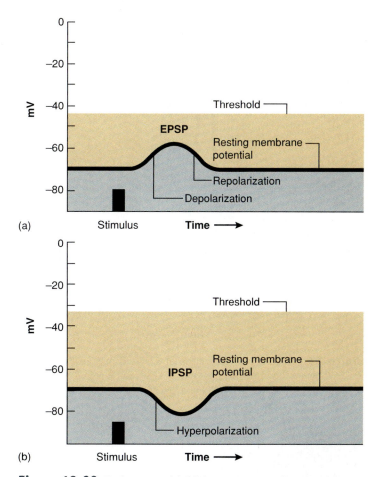

(a)

(b)

Figure 13.20 Excitatory and inhibitory postsynaptic potentials. Note that a single EPSP is not enough to reach threshold and make a neuron fire. The size of the EPSP is greatly exaggerated here for clarity; compare the following figure.

Suppose, for example, you were working in the kitchen and had turned off a stove just a minute earlier. Reaching for something on a high shelf above the stove, you accidentally put your other hand on the still-hot burner. EPSPs in your motor neurons cause you to jerk your hand back quickly and avoid being burned. Yet a moment later, you pick up a cup of tea hotter than the burner was and put it to your lips. Since you were expecting the cup to be hot, you do not jerk your hand away. You have learned that it will not injure you, and so IPSPs prevail.

Thus, it is fundamentally a balance between EPSPs and IPSPs that enables the nervous system to make decisions. A postsynaptic neuron is like a little cellular democracy acting on the "majority vote" of hundreds of presynaptic cells. Some may send messages that signify "Hot! Danger!" in the form of EPSPs, and at the same time others may be producing IPSPs that signify "Safe." Whether the postsynaptic neurons cause a hand-withdrawal reflex depends on whether the EPSPs or IPSPs prevail. The process of adding up incoming information and responding to the net effect, called **summation,** occurs in the trigger zone of the neuron.

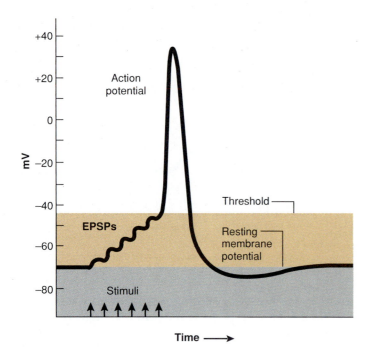

Figure 13.21 Temporal summation of EPSPs. If new EPSPs are produced before the preceding ones decay, EPSPs can build on each other to bring the neuron to threshold and trigger an action potential. ◁▭▷

One action potential in a synaptic knob generally does not produce enough activity to cause a postsynaptic cell to fire. An EPSP may be produced, but it decays before reaching threshold (fig. 13.20a). A typical EPSP lasts 15 to 20 msec and has a voltage of about 0.5 mV. A depolarization of 10 to 15 mV, however, is needed to bring the postsynaptic cell to its threshold. Thus it takes 20 to 30 EPSPs to reach threshold, and these must arrive quickly, before the effects of the earliest ones decay.

If EPSPs arrive in quick succession, each one can add its effects to the preceding ones, adding up to a voltage that reaches threshold (fig. 13.21). This is called **temporal summation** because it depends on short time intervals between successive EPSPs. Threshold can also be reached by **spatial summation**. This occurs when local potentials arrive at the trigger zone from several different synapses and produce EPSPs that add up to the threshold. Since each synapse makes it easier for the others to cause firing, presynaptic cells converging on a single neuron are said to **facilitate** each other.

Neural Coding

The nervous system must convey and interpret both quantitative and qualitative information about its environment—whether a light is dim or bright, red or green; whether a taste is mild or intense, salty or sour; whether a sound is loud or soft, bass or treble. Considering the complexity of information to be communicated about conditions within and outside of the body, it is a marvel that it can be done in the form of something as

simple as action potentials—particularly since a neuron follows the all-or-none law, and all of its action potentials are identical.

Yet when we considered the genetic code in chapter 5, we saw that complex messages can indeed be expressed in simple codes. The way in which the nervous system converts information to a meaningful pattern of action potentials is called **neural coding** (or *sensory coding* when it occurs in the sense organs).

Qualitative information is coded in terms of which neurons are firing. Red light and green light, for example, excite different fibers in the optic nerve; a high-pitched sound and a low-pitched sound excite different fibers in the auditory nerve; a sweet substance and a sour one excite different taste cells. The brain is programmed to interpret input from different fibers in terms of these stimulus qualities.

There are two ways of encoding quantitative information—that is, information about the intensity of a stimulus. One way depends on the fact that different neurons have different thresholds of excitation. A weak stimulus excites those with the lowest thresholds, while a strong stimulus excites less sensitive high-threshold neurons. This is called **recruitment;** it enables the nervous system to judge stimulus strength by which neurons, and how many of them, are firing.

Think About It
How does neuronal recruitment resemble the process of multiple motor unit summation described in chapter 12?

Another way of encoding stimulus strength depends on the fact that the more strongly a neuron is stimulated, the more frequently it fires. A weak stimulus may cause a nerve fiber to transmit 6 action potentials per second and a strong stimulus, 600 per second. Thus, the central nervous system can judge stimulus strength from the firing frequency of afferent neurons (fig. 13.22).

The absolute refractory period sets an upper limit to how often a neuron can fire. Think of an electronic camera flash by analogy. If you take a photograph and your flash unit takes 15 seconds to recharge, then obviously you cannot take more than four photographs per minute. Similarly, if a nerve fiber takes 1 msec to repolarize after it has fired, then it cannot fire more than 1,000 times per second. Refractory periods may be as short as 0.5 msec, setting a theoretical limit to firing frequency of 2,000 nerve impulses per second. The highest frequencies actually observed, however, are between 500 and 1,000 per second.

Synaptic Potentiation and Inhibition

All of your thoughts, feelings, and actions, and many of your unconscious physiological responses, are based on

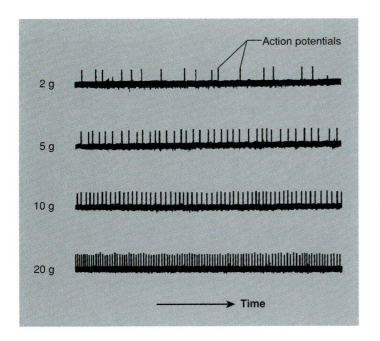

Figure 13.22 An example of neural coding. This figure is based on recordings made from a sensory fiber of the frog sciatic nerve as the gastrocnemius muscle was stretched by suspending increasing weights from it. As the stimulus strength (weight) and stretch increases, the firing frequency of the neuron increases. Thus firing frequency is a coded message that informs the CNS of stimulus intensity.

nerve impulses traveling from cell to cell along frequently used pathways. Just as the frequent practice of a task makes it easier to perform, the frequent use of a nerve pathway makes it easier for impulses to travel that route. At the cellular level, repeated use of a particular synapse enhances the ability of one neuron to stimulate another and results in larger postsynaptic potentials. Since this makes it easier for nerve signals to travel across that synapse in the future, the process is called **synaptic potentiation** or **presynaptic facilitation.**

Prolonged high-frequency (tetanic) stimulation of a synapse causes more calcium ions to accumulate in the synaptic knob and thus enhances neurotransmitter release by the presynaptic cell. This is called **tetanic potentiation.** In some cases, called **posttetanic potentiation,** the effect persists for hours to weeks after stimulation has ceased. Either type of potentiation increases the efficiency of future transmission along certain pathways—that is, potentiation is a learning process. This phenomenon is especially pronounced in a region of the brain called the hippocampus, known to be an important center of learning and memory.

Long-term potentiation (LTP) resembles posttetanic potentiation but lasts much longer. In the hippocampus, glutamic acid stimulates postsynaptic neurons and opens voltage-gated calcium channels called *NMDA*[24] *receptors.* Ca^{2+} enters the postsynaptic

24. N-methyl-D-aspartate

cell and alters the phosphorylation of enzymes in the cytoplasm. These enzymes affect the membrane structure of the postsynaptic cell in a way that makes it more sensitive to the neurotransmitter. Future releases of neurotransmitter cause larger EPSPs in the postsynaptic cell, and thus nerve signals are communicated more easily across this synapse. In addition, the postsynaptic cell releases *nitric oxide* (*NO*). NO diffuses across the synapse to the presynaptic cell and causes it to release more glutamic acid in response to future stimuli. Thus, each neuron at the synapse alters the other, and future communication across the synapse is facilitated. Whether these synapses are involved in riding a bicycle, playing the French horn, or solving calculus problems, the relevance of LTP to learned skills is obvious.

In **presynaptic inhibition,** one neuron suppresses the release of neurotransmitter by another. In figure 13.23*a*, neuron S (stimulator) releases a neurotransmitter that excites neuron R (responder). Neuron I (inhibitor) forms an axoaxonic synapse with S but is presently quiet. Under some circumstances, it may be desirable to reduce transmission across this synapse. In figure 13.23*b*, neuron I is active. It releases a neurotransmitter that inhibits neuron S. As a result, S no longer stimulates R.

Neuronal Pools

So far, we have dealt with interactions involving only two or three neurons at a time. Actually, neurons function in much larger groups called **neuronal pools,** each of which consists of thousands to millions of interneurons concerned with a particular body function. The brain is thought to have from a few hundred to a few thousand neuronal pools. At this point, we will explore a few ways in which neuronal pools collectively process information.

As an input fiber enters a neuronal pool, it branches repeatedly and synapses with numerous neurons. In some cases, it forms many synapses with the same postsynaptic cell. It can produce EPSPs at all points of contact with that cell, and these can summate to make that cell fire. Such postsynaptic neurons are said to be in the **discharge zone** of the input fiber (fig. 13.24). Neurons in its **facilitated zone,** by contrast, receive only one or a few contacts from the input fiber. That input fiber alone cannot make them fire, but it partially depolarizes them and makes them more responsive to input from other presynaptic fibers.

Neuronal Circuits

The functioning of a radio can be understood from a circuit diagram showing its electronic components and their connections. Similarly, the functions of a neuronal pool are determined by its **neuronal circuit**—the connection

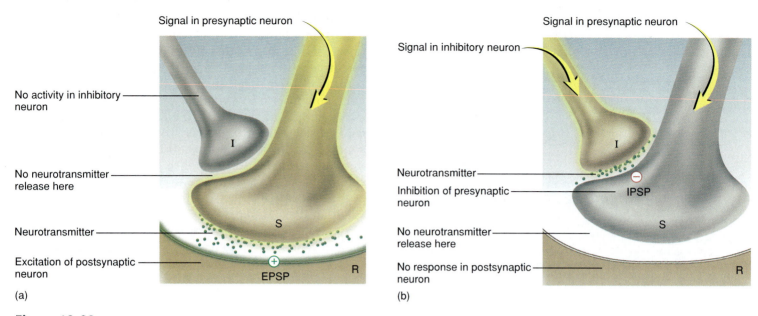

Signal in presynaptic neuron

No activity in inhibitory neuron

No neurotransmitter release here

Neurotransmitter

Excitation of postsynaptic neuron

I

S

R

EPSP

(a)

Signal in presynaptic neuron

Signal in inhibitory neuron

Neurotransmitter

Inhibition of presynaptic neuron

No neurotransmitter release here

No response in postsynaptic neuron

I

IPSP

S

R

(b)

Figure 13.23 Presynaptic inhibition. (a) Neuron R is active because it is stimulated by neuron S. (b) In presynaptic inhibition, neuron I halts the release of neurotransmitter by S.

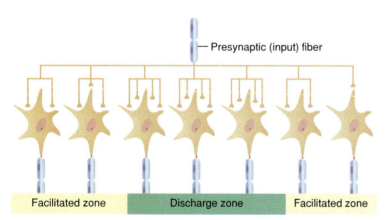

Presynaptic (input) fiber

Facilitated zone | Discharge zone | Facilitated zone

Figure 13.24 Facilitated and discharge zones in a neuronal pool. In a facilitated zone, the input has few synaptic contacts with the output neurons. It makes it easier for those neurons to respond to input from other sources, but it cannot, by itself, make those neurons fire. In the discharge zone, the input neuron has extensive connections with each output neuron and is capable, by itself, of making the output neurons fire.

pathways among its neurons. Just as a wide variety of electronic devices are constructed from a relatively limited number of circuit types, a wide variety of neuronal functions result from the operation of four principal kinds of neuronal circuits: *diverging, converging, reverberating,* and *parallel after-discharge circuits* (fig. 13.25).

In a **diverging circuit,** one nerve fiber synapses with several postsynaptic cells, and each of them synapses with several more, so a larger and larger number of neurons are affected farther away from the initial one. This produces a chain-reaction amplifying effect. In this way, one neuron of the cerebral cortex can ulti-

mately cause the contraction of thousands of muscle fibers, and sensory input from one neuron in the skin can be directed toward separate parts of the brain concerned with motor response, pleasure, and memory.

In **converging circuits,** input from several different sources is funneled to one neuron or neuronal pool. For example, your sense of balance depends on input from the eyes, inner ears, and stretch receptors in the neck. Your rhythm of breathing is set by a neuronal pool in the brainstem and receives input from other parts of the brain, receptors for blood chemistry and blood pressure in the arteries, and stretch receptors in the lungs.

In a **reverberating (oscillating) circuit,** neurons stimulate each other in a linear sequence: A → B → C → D. However, neuron C sends an axon collateral back to A and starts the process over. As a result, every time C fires it restimulates A. Neuron D is an *output neuron;* it produces a prolonged or repetitious effect that lasts until one or more neurons in the circuit fail to fire. This may be due to adaptation or to inhibitory input from another circuit. In adaptation, a neuron responds to long-term stimulation by decreasing its output. Such a circuit can continue to reverberate for seconds, hours, or a lifetime. A reverberating circuit sends repetitious impulses to your diaphragm and intercostal muscles, for example, to make you inhale. When the circuit stops firing, you exhale; the next time it fires, you inhale again. Reverberating circuits are also thought to be involved in short-term memory. For example, if someone tells you a telephone number, a reverberating circuit may enable you to remember it just long enough to get to a telephone and dial it. A short time later, when this circuit has stopped firing, you can no longer remember the

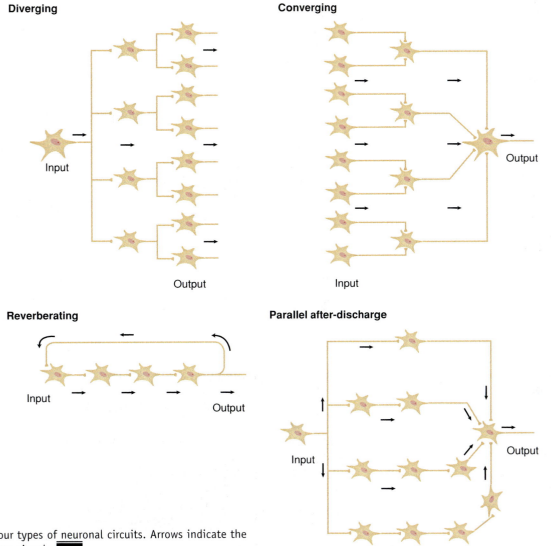

Diverging

Input

Converging

Input

Reverberating

Input

Parallel after-discharge

Input

Output

Figure 13.25 Four types of neuronal circuits. Arrows indicate the direction of the nerve signal.

number. Reverberating circuits may also be involved in the uncontrolled "storms" of neuronal activity that occur in grand mal epilepsy.

In a **parallel after-discharge circuit,** an input neuron diverges to stimulate several parallel neuron chains. Each of these chains has a different number of neurons and synapses and therefore varying degrees of total synaptic delay. Eventually, however, they all reconverge on a single output neuron. Since each circuit differs in number of synapses and total synaptic delay, their signals arrive at the output neuron at different times, and the output neuron may go on firing for some time after input has ceased. This continued firing after the input stops is called **after-discharge.** Unlike a reverberating circuit, this type has no feedback. When all the neurons in the circuit have fired, the output ceases.

One use for a parallel after-discharge circuit is the withdrawal reflex. If you step on something sharp, you experience a painful but brief stimulus. Withdrawing your foot, however, requires a longer-lasting output to the leg muscles. You can see how such a circuit may produce a prolonged output from a brief input. Parallel after-discharge circuits are also thought to be involved in carrying out complex mental tasks such as working mathematics problems.

Key Point Review

17 Explain how EPSPs and IPSPs differ. How do they affect the probability that a postsynaptic neuron will fire?

18 Contrast the two types of summation at a synapse.

19 Describe how the nervous system communicates quantitative and qualitative information about stimuli.

20 How does long-term potentiation enhance the transmission of nerve signals along certain pathways?

21 List the four types of neuronal circuits and describe their similarities and differences. Discuss the unity of form and function in these four types—that is, explain why each type would not perform as it does if its neurons were connected differently.

Alzheimer and Parkinson Diseases

Alzheimer and Parkinson diseases are degenerative disorders of the brain associated with neurotransmitter deficiencies.

Alzheimer[25] **disease (AD)** may begin before the age of 50 with symptoms so slight and ambiguous that early diagnosis is difficult. One of the first symptoms is memory loss, especially for recent events. A person with AD may ask the same questions repeatedly, show a reduced attention span, and become disoriented and lost in previously familiar places. Family members often feel helpless and confused as they watch their loved one's personality gradually deteriorate beyond recognition. The AD patient may become moody, confused, paranoid, combative, or hallucinatory—he or she may ask irrational questions such as, Why is this room full of snakes? The patient may eventually lose even the ability to read, write, talk, walk, and eat. Death ensues from pneumonia or other complications of confinement and immobility. AD affects about 11% of the U.S. population over the age of 65; the incidence rises to 47% by age 85. It accounts for nearly half of all nursing home admissions and is a leading cause of death among the elderly. AD claims about 100,000 lives per year in the U.S.

Diagnosis of AD is confirmed on autopsy. There is atrophy of some of the gyri of the cerebral cortex and hippocampus, which are important centers of memory. Nerve cells exhibit *neurofibrillary tangles*—dense masses of broken and twisted cytoskeleton (fig. E.1). These were first observed by Alzheimer in 1907 in the brain of a patient who had died of senile dementia. The more severe the disease symptoms, the more neurofibrillary tangles are seen at autopsy. In the intercellular spaces, there are *senile plaques* consisting of aggregations of cells, altered nerve fibers, and a core of *β-amyloid protein*—a breakdown product of a glycoprotein of plasma membranes rarely seen in elderly people without AD.

The cause of AD remains uncertain and controversial. There is disagreement, for example, about whether β-amyloid protein is a product or the cause of neuronal degeneration. Some researchers attribute AD to a protein called *apolipoprotein-E* that is involved in cholesterol transport. Whatever the ultimate cause, AD is marked by degeneration of cholinergic neurons and a deficiency of ACh. Treatment with ACh precursors has proven ineffective, but therapy with cholinesterase inhibitors (to prolong the life of the ACh that *is* produced) has been of some value. Nerve growth factor (see special topic 13.3) stimulates ACh synthesis in the brain, and AD patients show a deficiency of NGF in some brain regions. NGF has been shown to retard degeneration of brain cells in animals and humans and to improve memory in some AD patients. Intense biomedical research efforts are geared toward identifying the cause of AD and developing treatment strategies.

Parkinson[26] **disease (PD),** also called *paralysis agitans* or *parkinsonism,* is a progressive loss of motor function beginning in a person's 50s or 60s. It is due to degeneration of dopamine-releasing neurons in a portion of the brain called the *substantia nigra* (see chapter 14). There is little or no evidence of a hereditary cause; some suspect environmental neurotoxins.

Dopamine (DA) is an inhibitory neurotransmitter that normally prevents excessive activity in motor centers of the brain called the basal nuclei. Degeneration of the dopamine-releasing neurons leads to an excessive ratio of ACh to DA, which causes the basal nuclei to become hyperactive. As a result, a person with PD suffers involuntary muscle contractions. These take such forms as shaking of the hands (tremor) and compulsive "pill-rolling" motions of the thumb and fingers. In addition, the facial muscles may become rigid, producing a staring, expressionless face with a slightly open mouth. The patient's range of motion diminishes. He or she takes smaller steps and develops a slow, shuffling gait with a forward-bent posture and often a tendency to fall forward. Speech becomes slurred and handwriting becomes cramped and eventually illegible. Tasks

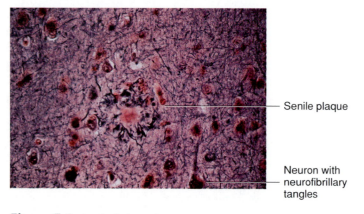

— Senile plaque

— Neuron with neurofibrillary tangles

Figure E.1 Cerebral tissue from a person with Alzheimer disease (×250). Neurofibrillary tangles appear within the neurons, and senile plaques are evident in the extracellular matrix.

---

25. Alois Alzheimer (1864–1915), German neurologist

26. James Parkinson (1755–1824), British physician

such as buttoning clothes and preparing food become increasingly laborious.

Patients cannot be expected to recover from PD, but the symptoms can be alleviated with drugs and physical therapy. Treatment with dopamine is ineffective because it cannot cross the blood-brain barrier, but its precursor, levodopa (L-DOPA), does cross the barrier and has been used to treat PD since the 1960s. L-DOPA affords some relief from symptoms, but it does not slow progression of the disease and it has undesirable side effects on the liver and heart. A newer drug, Deprenyl, is a monoamine oxidase (MAO) inhibitor that retards neuronal degeneration and delays the development of symptoms. Encouraging results have also been obtained by implanting other dopamine-producing tissues into the brains of PD patients—namely, adrenal medulla and fetal brain tissue. Even though the latter tissue has not come from elective abortions, it has raised ethical questions concerning this line of therapy. A surgical technique called *pallidotomy* has been used since the 1940s to alleviate severe tremors. It involves the destruction of a small portion of cerebral tissue called the *globus pallidus*. Pallidotomy fell out of favor when L-DOPA came into common usage, but it has made a comeback now that computer-guided equipment has improved the precision and reduced the risks of surgery.▲

Chapter Review · Study Outline

An Overview of the Nervous System (pp. 428–430)

1. Subdivisions of the nervous system
   a. Central nervous system (brain and spinal cord)
   b. Peripheral nervous system
      • Cranial and spinal nerves
      • Somatic and autonomic divisions
2. Functions of the nervous system
   a. Sensory function and afferent neurons
   b. Integrative function and interneurons
   c. Responsive function and efferent neurons
3. Properties of nerve cells
   a. Excitability
   b. Conductivity
   c. Secretion of chemical messengers
4. The nervous and endocrine systems compared (table 13.1)

Cells of the Nervous System (pp. 430–437)

1. Neuroglia (table 13.2)
   a. Schwann cells
   b. Satellite cells
   c. Oligodendrocytes
   d. Astrocytes
   e. Ependymal cells
   f. Microglia
2. Structure of a representative neuron
   a. Soma (cell body)
   b. Dendrites
   c. Axon hillock
   d. Axon (nerve fiber)
      • Axoplasm and axolemma
      • Collaterals
      • Terminal arborization
      • Synaptic knobs
      • Fiber types A, B, and C (table 13.3)
3. Schwann cells and myelin
   a. Neurilemma
   b. Formation of myelin sheath
   c. Nodes of Ranvier and internodes
   d. Initial segment and trigger zone
4. Unmyelinated nerve fibers
5. Structural diversity in neurons
   a. Multipolar neurons
   b. Bipolar neurons
   c. Unipolar neurons
6. Axonal transport (table 13.4)
   a. General functions
   b. Fast axonal transport
      • Anterograde
      • Retrograde
   c. Slow axonal transport
7. Regeneration of nerve fibers

Electrophysiology of Neurons (pp. 437–446)

1. Concepts in review
   a. Resting membrane potential
   b. Depolarization and repolarization
   c. Role of the Na⁺-K⁺ pump
   d. Electrical potentials and currents
2. Local potentials
   a. Role of the current sink
   b. Characteristics of local potentials
      • Graded
      • Decremental
      • Local
      • Reversible
      • Excitatory or inhibitory
3. Action potentials
   a. Necessity of voltage-gated ion channels
   b. Current sink and generator potential
   c. Threshold
   d. Action of Na⁺ and K⁺ gates
   e. Depolarization and repolarization
   f. Hyperpolarization
   g. Spikes
   h. All-or-none law
4. Refractory period
   a. Absolute
   b. Relative
5. Signal conduction in unmyelinated fibers
6. Signal conduction in myelinated fibers
   a. Conduction in internodes
   b. Nodes of Ranvier
   c. Saltatory conduction

Synapses (pp. 446–452)

1. Presynaptic and postsynaptic neurons
2. Axodendritic, axosomatic, and axoaxonic synapses
3. Discovery of neurotransmitters
4. Electrical and chemical synapses
5. Structure of a chemical synapse
   a. Presynaptic neuron
      • Synaptic knob
      • Synaptic vesicles
   b. Postsynaptic neuron
      • Neurotransmitter receptors
      • Absence of synaptic vesicles
6. Neurotransmitters and related messengers (table 13.6)
   a. Acetylcholine
   b. Amino acids
   c. Biogenic amines (monoamines)
   d. Neuropeptides

7. Synaptic transmission
   a. Arrival of nerve impulse
   b. Opening of Ca^{2+} gates
   c. Exocytosis of synaptic vesicles
   d. Reloading of synaptic vesicles
   e. Binding of transmitter to receptor
   f. Na^+ influx, K^+ efflux
   g. Local postsynaptic potential (PSP)
   h. Synaptic delay
   i. Ionotropic and metabotropic effects
8. Cessation of the signal
   a. Diffusion from synaptic cleft
   b. Reuptake of neurotransmitter
   c. Degradation of neurotransmitter

9. Other modes of chemical communication
   a. Neuropeptides
   b. Neuromodulators
   c. Gases

Neural Integration (pp. 452–457)
1. Why synapses exist
2. Postsynaptic potentials
   a. Excitatory (EPSPs)
   b. Inhibitory (IPSPs)
3. Summation and facilitation
   a. Temporal summation
   b. Spatial summation
   c. Facilitation
4. Neural coding
   a. Qualitative information

   b. Quantitative information
      • Recruitment
      • Firing frequency
      • Relation to refractory period
5. Synaptic potentiation and inhibition
   a. Tetanic potentiation
   b. Posttetanic potentiation
   c. Long-term potentiation
   d. Presynaptic inhibition
6. Neuronal pools
   a. Discharge zone
   b. Facilitated zone
7. Neuronal circuits
   a. Diverging
   b. Converging
   c. Reverberating
   d. Parallel after-discharge

Selected Vocabulary

central nervous system (CNS) 428
peripheral nervous system (PNS) 428
receptor 429
afferent neuron 429
interneuron 429
effector 429
efferent neuron 429
neuroglia 430
Schwann cell 430
neurilemma 430
myelin sheath 430
satellite cell 430
oligodendrocyte 430
protoplasmic astrocyte 430
perivascular feet 430
fibrous astrocyte 431
ependymal cell 431
microglia 431
soma 432
neurofibril 432
Nissl bodies 432
lipofuscin 432
dendrites 432
dendritic spines 434
axon hillock 434
axon (nerve fiber) 434

axoplasm 434
axolemma 434
collateral 434
terminal arborization 434
telodendria 434
synaptic knob 434
synapse 434
synaptic vesicle 434
node of Ranvier 435
internode 435
initial segment 435
trigger zone 435
unmyelinated fiber 435
multipolar neuron 435
bipolar neuron 435
unipolar neuron 435
anterograde transport 436
retrograde transport 436
fast axonal transport 436
slow axonal transport 437
regeneration tube 437
resting membrane potential (RMP) 439
polarized 439
depolarization 439
repolarization 439
local potential 439

current sink 439
action potential 440
generator potential 440
threshold 440
hyperpolarization 441
spike 443
all-or-none law 443
absolute refractory period 443
relative refractory period 443
saltatory conduction 445
presynaptic neuron 446
postsynaptic neuron 446
synaptic cleft 446
electrical synapse 446
chemical synapse 446
acetylcholine 448
biogenic amine 448
monoamine 449
catecholamine 449
indolamine 449
neuropeptide 449
cholinergic synapse 449
postsynaptic potential (PSP) 449
synaptic delay 449
ionotropic effect 449
metabotropic effect 449

adenylate cyclase 449
protein kinase 450
monoamine oxidase (MAO) 452
neuromodulator 452
neural integration 453
EPSP 453
IPSP 453
temporal summation 454
spatial summation 454
facilitation 454
neural coding 454
recruitment 454
synaptic potentiation 455
tetanic potentiation 455
posttetanic potentiation 455
long-term potentiation (LTP) 455
presynaptic inhibition 455
neuronal pool 455
discharge zone 455
facilitated zone 455
neuronal circuit 455
diverging circuit 456
converging circuit 456
reverberating circuit 456
parallel after-discharge circuit 457

Testing Your Recall   Answers in Appendix C

1. The integrative functions of the nervous system are performed mainly by
   a. afferent neurons.
   b. neuroglia.
   c. efferent neurons.
   d. special nerve fibers.
   e. interneurons.

2. The highest density of voltage-regulated ion channels is found on the _____ of a neuron.
   a. dendrites
   b. soma
   c. nodes of Ranvier
   d. internodes
   e. synaptic knobs

3. The soma of a mature neuron lacks
   a. rough ER.
   b. lysosomes.
   c. ribosomes.
   d. centrioles.
   e. synapses.

4. The glial cells responsible for destroying microorganisms in the CNS are
   a. microglia.
   b. satellite cells.
   c. ependymal cells.
   d. oligodendrocytes.
   e. astrocytes.

5. In pyramidal cells of the cerebral cortex, most synapses are found on
   a. the axon hillock.
   b. the axon collaterals.
   c. the dendritic spines.
   d. the telodendria.
   e. the soma.

6. An IPSP is a _____ of the postsynaptic neuron.
   a. generator potential
   b. traveling action potential
   c. depolarization
   d. repolarization
   e. hyperpolarization

7. Saltatory conduction occurs only in or at
   a. chemical synapses.
   b. the initial segment of a neuron.
   c. the axon and initial segment.
   d. myelinated nerve fibers.
   e. type C nerve fibers.

8. Some neurotransmitters can have either excitatory or inhibitory effects depending on the type of
   a. receptors the postsynaptic neuron has.
   b. synaptic vesicles the axon terminal has.
   c. postsynaptic potentials produced on the synaptic knob.
   d. neuronal circuits involved.
   e. synaptic facilitation involved.

9. Differences in the volume of a sound are most likely to be encoded in terms of different _____ produced by neurons of the inner ear.
   a. neurotransmitters
   b. forms of potentiation
   c. types of postsynaptic potentials
   d. firing frequencies
   e. degrees of potentiation

10. Motor effects that depend on repetitive output from a neuronal pool are most likely to employ _____ circuits.
    a. parallel after-discharge
    b. reverberating
    c. facilitated
    d. diverging
    e. converging

11. Neurons that convey information to the CNS are called sensory, or _____, neurons.

12. To perform their role, neurons must have the properties of excitability, _____, and secretion.

13. The _____ is a period of time in which a neuron is producing an action potential and cannot be reexcited by a stimulus of any strength.

14. The processes of a neuron specialized to receive incoming signals are called _____.

15. Myelin is produced by _____ cells.

16. In a myelinated nerve fiber, action potentials are produced only at regions called _____.

17. The trigger zone of a neuron consists of its _____ and _____.

18. Neurotransmitters that work through second-messenger systems are said to have _____ effects.

19. A presynaptic nerve fiber cannot cause other neurons in its _____ to fire, but it can cause them to be more sensitive to input from other presynaptic neurons.

20. _____ are substances that may be released along with a neurotransmitter and modify the effects of the neurotransmitter on a postsynaptic cell.

Testing Your Comprehension   Answers in *Study Guide*

1. Schizophrenia is sometimes treated with drugs such as chlorpromazine that inhibit dopamine receptors. A side effect is that patients begin to develop muscle tremors, speech impairment, and other disorders characteristic of Parkinson disease. Explain.

2. Hyperkalemia is a condition in which there is excess K$^+$ in the extracellular fluids. What effect would this have on the resting membrane potentials of the nervous system and on neuronal excitability?

3. Explain how saltatory conduction combines a slow, nondecremental form of signal transmission and a fast, decremental form.

4. The unity of form and function is an important concept in understanding synapses. Give two structural reasons why nerve signals cannot travel backward across a synapse. What might the consequences be if they did travel freely in both directions?

5. The local anesthetics procaine (Novocain) and tetracaine prevent voltage-regulated Na$^+$ gates from opening. Explain why this would block pain transmission in a sensory nerve.

Web Site Link

For a listing of the most current web sites related to this chapter, please visit the Saladin homepage at:

http://www.mhhe.com/sciencemath/biology/saladin/

The Central Nervous System

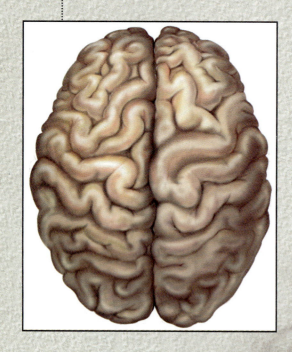

Special Topics

Brushing up

T he mystique of the brain continues to intrigue modern biologists and psychologists even as it did the philosophers of antiquity. Aristotle dismissed the brain as merely a radiator for cooling the blood. Generations earlier, however, Hippocrates had expressed a more accurate view of its functions. "Men ought to know," he argued, "that from the brain, and from the brain only, arise our pleasures, joy, laughter and jests, as well as our sorrows, pains, griefs and tears. Through it, in particular, we think, see, hear, and distinguish the ugly from the beautiful, the bad from the good, the pleasant from the unpleasant." Brain function is so strongly associated with what it means to be alive and human that the cessation of brain activity is considered a clinical criterion of death even when other organs of the body are still functioning.

With its hundreds of neuronal pools, 35 billion neurons, and trillions of synapses, the brain performs sophisticated tasks beyond our present understanding. Still, all of our mental functions, no matter how complex, are ultimately based on the cellular activities described in the previous chapter. The relationship of the mind or personality to the cellular function of the brain is a question that will provide fertile ground for scientific and philosophical debate long into the future.

The previous chapter discussed the nervous system at the cellular level. We now move up the structural hierarchy to consider the central nervous system at the organ and system levels. This chapter will lay a foundation for understanding some relatively simple reflex actions of the spinal cord and brainstem considered in the next chapter, and it will plumb some of the mysteries of motor control, sensation, emotional drives, analytical thought, language, personality, memory, dreams, and plans. Your study of this chapter is essentially one brain's attempt to understand itself.

Characteristics of the Central Nervous System

▼**Objectives**
When you have completed this section, you should be able to
• describe the major subdivisions of the brain; and
• describe the development of the embryonic neural tube into the fully formed central nervous system.

Major Landmarks

Before we consider the form and function of specific regions of the central nervous system (CNS), it will be helpful to get a general impression of its major landmarks (fig. 14.1). These will provide important points of reference as we progress through a more detailed study of the system.

The average adult brain weighs about 1,600 g (3.5 lb) in men and 1,450 g in women. Its size is proportional to body size, not intelligence—the Neanderthal people had larger brains than modern humans.

The brain can be divided into three major portions—the *cerebrum, cerebellum,* and *brainstem.* The **cerebrum** (SERR-eh-brum; seh-REE-brum) constitutes about 83% of its volume and consists of a pair of **cerebral hemispheres** (fig. 14.1a). Each hemisphere is marked by thick folds called **gyri**[1] (JY-rye; singular, *gyrus*) separated by shallow grooves called **sulci**[2] (SUL-sigh; singular, *sulcus*) and a few deeper grooves called **fissures.** The **longitudinal fissure** separates the right and left hemispheres from each other. The hemispheres are connected by a thick bundle of nerve fibers called the **corpus callosum**[3]—a prominent and useful landmark for anatomical description (fig. 14.2b).

The **cerebellum**[4] (SER-eh-BEL-um) lies inferior to the cerebrum and occupies the posterior cranial fossa. It is also marked by gyri, sulci, and fissures, but its gyri are more delicate and more nearly parallel than those of the cerebrum.

The **brainstem** is what remains if the cerebrum and cerebellum are removed. In the living subject, it is oriented like a vertical stalk with the cerebrum perched on top of it. Postmortem changes give it a more oblique angle in the cadaver, and consequently in many medical illustrations. Some major components of the brainstem, from superior to inferior, are the *thalamus, hypothalamus, midbrain, pons,* and *medulla oblongata* (see fig. 14.1). Some authorities, however, regard only the midbrain, pons, and medulla as the brainstem.

Two major tissue types of the CNS are the **gray matter** and **white matter.** It is within the gray matter that neurosomas, dendrites, and synapses occur; this is therefore the site of neural integration. White matter consists of nerve fibers that constitute an "information highway" for signal transmission from one part of the CNS to another. Myelin gives it a bright, glistening white appearance in fresh preparations, while the gray matter is somewhat duller. In preserved material and histological sections, both are often stained tan, brown, or orange, but they are still distinguishable from each other. In the cerebrum and cerebellum, the gray matter forms a surface layer of tissue called the *cortex,* (see fig. 14.6c). In the spinal cord, by contrast, the gray matter is deep to the white matter.

Much of the CNS is hollow. At its core is a system of spaces called *ventricles* in the brain and the *central canal* in the spinal cord. These spaces are filled with a liquid called *cerebrospinal fluid (CSF).*

1. *gyr* = turn, twist
2. *sulc* = furrow, groove
3. *corpus* = body + *call* = thick
4. "little brain"

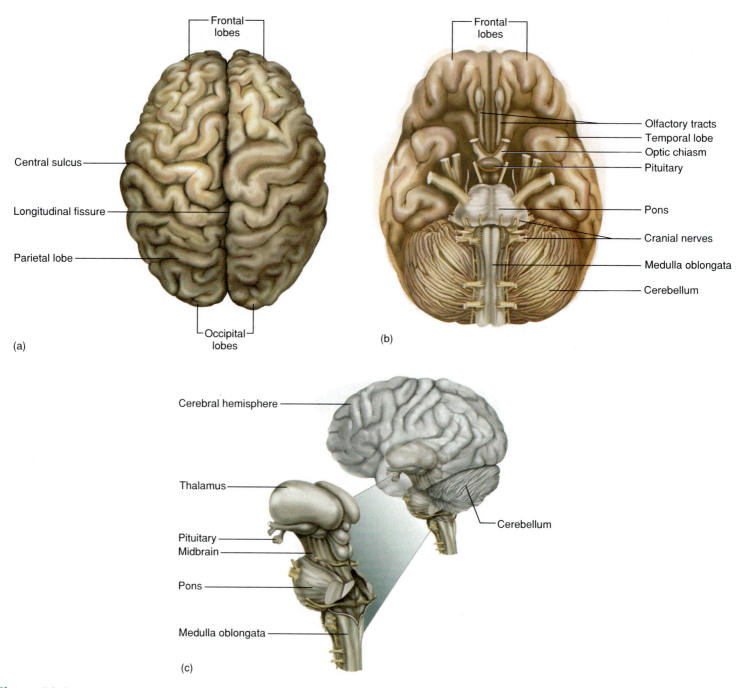

Figure 14.1 Overview of brain anatomy. (*a*) Superior view of the cerebral hemispheres. (*b*) Inferior view (the base of the brain). (*c*) The brainstem, which is what remains when the cerebral hemispheres and cerebellum are removed. ✗

Embryonic Development

To understand why the CNS is hollow, and to understand the classification of its anatomical and functional regions, it is necessary to be aware of its embryonic development.

The nervous system develops from the outermost of the three primary germ layers of an embryo, the ectoderm. By the third week of embryonic development, a dorsal streak called the *neuroectoderm* appears along the entire length of the embryo and thickens to form a **neural plate** (fig. 14.3), destined to give rise to all neurons and glia except microglia. As development progresses, the neural plate sinks and forms a **neural groove,** while the cells along its margins continue to proliferate and form a **neural fold** on each side. The neural folds eventually meet and fuse along the midline, somewhat like a closing zipper. By 4 weeks, this

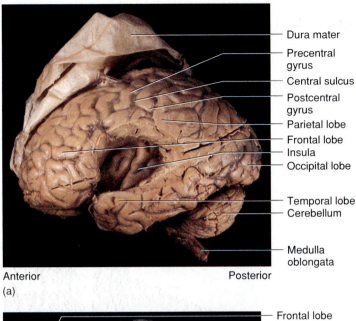

Dura mater
Precentral gyrus
Central sulcus
Postcentral gyrus
Parietal lobe
Frontal lobe
Insula
Occipital lobe
Temporal lobe
Cerebellum
Medulla oblongata

Anterior   Posterior
(a)

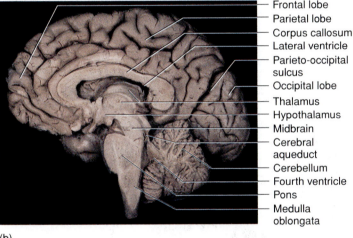

Frontal lobe
Parietal lobe
Corpus callosum
Lateral ventricle
Parieto-occipital sulcus
Occipital lobe
Thalamus
Hypothalamus
Midbrain
Cerebral aqueduct
Cerebellum
Fourth ventricle
Pons
Medulla oblongata

(b)

Figure 14.2 Major features of the brain. (*a*) Left lateral view with the dura mater reflected and part of the left cerebral hemisphere cut away to expose the insula. The pia mater is visible as a delicate translucent membrane over many of the sulci of the frontal lobe. (*b*) Midsagittal section, left lateral view. ✗

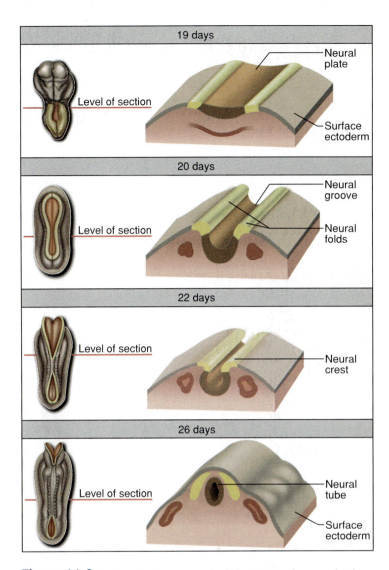

19 days
Neural plate
Level of section
Surface ectoderm

20 days
Neural groove
Level of section
Neural folds

22 days
Level of section
Neural crest

26 days
Level of section
Neural tube
Surface ectoderm

Figure 14.3 Embryonic development of the CNS to the neural tube stage (26 days). The left column represents dorsal views of the embryo and the right column shows three-dimensional reconstructions at the indicated levels.

creates a hollow channel called the **neural tube** (see special topic 14.1). The neural tube separates from the overlying ectoderm, sinks a little deeper, and grows lateral processes that later form motor nerve fibers. The lumen of the neural tube develops into the central canal and ventricles of the CNS.

Some ectodermal cells separate from the neural tube and form a longitudinal column on each side called the **neural crest.** Some neural crest cells develop processes and become sensory neurons, while some others migrate to other locations and become sympathetic neurons, Schwann cells, and other cell types.

By the fourth week, the neural tube exhibits three anterior dilations, or *primary vesicles,* called the **forebrain**

(*prosencephalon*[5]) (PROSS-en-SEF-uh-lon), **midbrain** (*mesencephalon*[6]) (MEZ-en-SEF-uh-lon), and **hindbrain** (*rhombencephalon*[7]) (ROM-ben-SEF-uh-lon) (fig. 14.4).

By the fifth week, the neural tube undergoes further flexion and subdivision and exhibits five secondary vesicles. The forebrain divides into two secondary vesicles—the **telencephalon**[8] (tel-en-SEFF-uh-lon) and **diencephalon**[9] (DY-en-SEF-uh-lon); the midbrain remains undivided and retains the same name; and the hindbrain divides into two secondary vesicles—the

5. *pros* = before, in front + *encephal* = brain
6. *mes* = middle
7. *rhomb* = rhombus
8. *tele* = end, remote
9. *di* = through, between

Neural tube defects are abnormalities present at birth that result from abnormal development of the neural tube and spinal column. **Spina bifida** (SPY-nuh BIF-ih-duh) is a defect that occurs when one or more vertebrae fail to form a complete neural arch (laminae and spinous process) for enclosure of the spinal cord. It is especially common in the lumbosacral region. One form, *spina bifida occulta*,[10] involves only one to a few vertebrae. It produces no functional problems and the only external sign of the condition is a dimple or hairy pigmented spot. *Spina bifida cystica* is more serious. A sac (cyst) protrudes from the spine and may contain meninges, CSF, and parts of the spinal cord and nerve roots (fig. 1). In extreme cases, inferior spinal cord function is absent, causing lack of bowel control and paralysis of the lower limbs and urinary bladder. The last of these conditions can lead to chronic urinary infections and renal failure.

10. *occult* = hidden

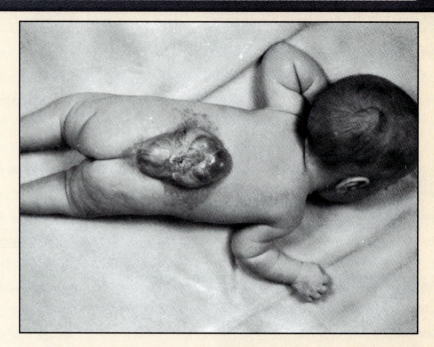

Figure 1 Spina bifida cystica.

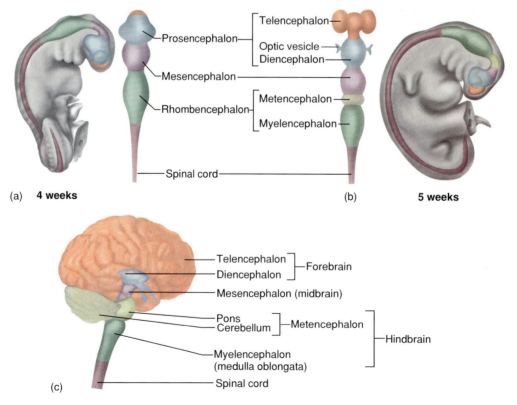

(a) **4 weeks**

Prosencephalon
Mesencephalon
Rhombencephalon

Telencephalon
Optic vesicle
Diencephalon
Metencephalon
Myelencephalon

Spinal cord

(b) **5 weeks**

Telencephalon ⎤
Diencephalon ⎦ Forebrain
Mesencephalon (midbrain)
Pons ⎤
Cerebellum ⎦ Metencephalon ⎤
⎦ Hindbrain
Myelencephalon (medulla oblongata)
Spinal cord

(c)

Figure 14.4 (*a*) Development of the three primary vesicles of the embryonic brain at 4 weeks. (*b*) By 5 weeks, the prosencephalon and rhombencephalon have subdivided, so there are now five secondary vesicles. (*c*) The mature brain structures derived from each secondary vesicle, colored to match *b*.

Figure 14.5 The meninges associated with the brain and spinal cord. (*a*) Frontal section of the head. (*b*) Cross section of a vertebra and the spinal cord. 𝒳

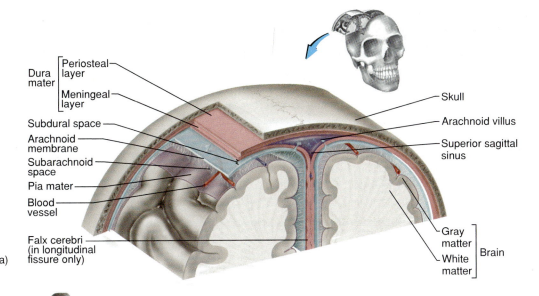

(a)

- Dura mater
  - Periosteal layer
  - Meningeal layer
- Subdural space
- Arachnoid membrane
- Subarachnoid space
- Pia mater
- Blood vessel
- Falx cerebri (in longitudinal fissure only)
- Skull
- Arachnoid villus
- Superior sagittal sinus
- Gray matter
- White matter
- Brain

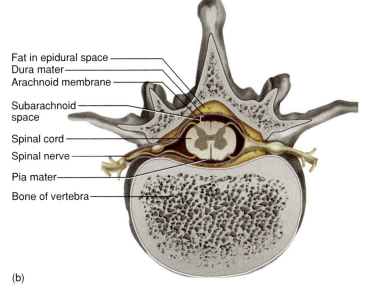

(b)

- Fat in epidural space
- Dura mater
- Arachnoid membrane
- Subarachnoid space
- Spinal cord
- Spinal nerve
- Pia mater
- Bone of vertebra

metencephalon[11] (MET-en-SEF-uh-lon) and **myelencephalon**[12] (MY-el-en-SEF-uh-lon). The telencephalon is marked by a pair of lateral outgrowths that later become the cerebral hemispheres, and the diencephalon exhibits a pair of small cuplike *optic vesicles* that become the retinas of the eyes. Figure 14.4*c* shows structures of the fully developed brain that arise from each of the secondary vesicles.

11. *met* = behind, beyond, distal to
12. *myel* = spinal cord

Meninges

The **meninges** (meh-NIN-jeez)—in the singular, *meninx*[13] (MEN-inks)—are three protective fibrous membranes that separate the soft tissue of the brain and spinal cord from the bones of the skull and vertebrae (fig. 14.5) (see special topic 14.2). The outermost meninx, the **dura mater**[14] (DUE-rah MAH-tur), is a tough membrane similar in thickness and texture to a household rubber glove. In the cranial cavity, it consists of two layers—an outer *periosteal layer,* equivalent to the periosteum of the cranial bone, and an inner *meningeal layer.* In some places, the two layers are separated by **dural sinuses,** which collect blood that has circulated through the brain and empty into the internal jugular veins of the neck.

In certain places, the meningeal layer of the dura mater folds inward to separate major parts of the brain: the **falx**[15] **cerebri** (falks SER-eh-bry) extends into the

13. *menin* = membrane
14. *dura* = tough
15. *falx* = sickle

Meningitis—inflammation of the meninges—is one of the most serious diseases of infancy and childhood, occurring especially between 3 months and 2 years of age. It is caused by a variety of bacteria and viruses that invade the CNS by way of the nose and throat, often following respiratory, throat, or ear infections. The pia mater and arachnoid are most likely to be affected, and from here the inflammation can spread to the adjacent nervous tissue. In bacterial meningitis, the brain is swollen, the ventricles are enlarged, and the brainstem may have hemorrhages. Symptoms include a high fever, stiff neck, drowsiness, and intense headache and may progress to vomiting, loss of sensory and motor functions, coma, and death.

Meningitis is diagnosed partly by examining the CSF for bacteria and white blood cells. The CSF is obtained by making a *lumbar puncture* (*spinal tap*) between two lumbar vertebrae and drawing fluid from the subarachnoid space. This site is chosen because it has an abundance of CSF and there is no risk of injury to the spinal cord, which does not extend into the lower lumbar vertebrae.

longitudinal fissure between the right and left cerebral hemispheres and the **tentorium**[16] (ten-TOE-ree-um) **cerebelli** stretches like a roof over the posterior cranial fossa and separates the cerebellum from the overlying cerebrum. The dura also forms a roof called the **diaphragma sellae** (DY-uh-FRAG-muh SEL-lee) over the sella turcica, which contains the pituitary gland.

In the vertebral canal, the periosteal layer of the dura is absent, and the meningeal layer forms a **dural sheath** around the spinal cord. The space between the sheath and vertebral bone, called the **epidural space,** is occupied by a protective pad of adipose and loose fibroconnective tissue (fig. 14.5*b*). During childbirth, some women are given *epidural anesthesia*—introduction of an anesthetic into this space to block pain signals from the pelvic region.

The other two meninges are deep to the dura mater: the *arachnoid membrane* in the middle and the *pia mater* on the brain surface. The dura and arachnoid are separated by a *subdural space,* and the arachnoid and pia are separated by the *subarachnoid space.* The **arachnoid**[17] (ah-RACK-noyd) **membrane** adheres to the inner surface of the dura and issues fibrous extensions, like a spider web (hence the name), that span the subarachnoid space and attach to the pia. The **pia mater**[18] (PEE-uh MAH-tur) is a thin, translucent, richly vascular layer of loose connective tissue. It resembles a sheet of cellophane closely following the contours of the brain, extending into its sulci and fissures.

Cerebrospinal Fluid

Cerebrospinal fluid (CSF) is a clear, colorless liquid that fills and surrounds the CNS. Its functions are fourfold:

1. **Buoyancy.** Because the brain and CSF are very similar in density, the brain neither sinks nor floats in the CSF but remains suspended in it—that is, the brain has *neutral buoyancy*. A human brain removed from the body weighs about 1,500 g, but when suspended in CSF its effective weight is only about 50 g. By analogy, consider how much harder it is to lift another adult on land than it is to lift the same person in a lake. Neutral buoyancy allows the brain to attain considerable size without being impaired by its own weight. If the brain rested heavily on the floor of the cranium, the pressure would kill the nervous tissue.

2. **Protection.** CSF also protects the brain from striking the cranium when the head is jolted. If the jolt is severe, however, the brain still may strike the inside of the cranium or suffer shearing injury from contact with the angular surfaces of the cranial floor. This is one of the common findings in child abuse (shaken child syndrome) and head injuries (concussion) from auto accidents, boxing, and the like.

3. **Waste removal.** CSF is secreted by the epithelium of each choroid plexus and is ultimately absorbed into the bloodstream. It provides a route for the removal of metabolic wastes from the CNS.

4. **Stability of chemical environment.** CSF homeostatically regulates the chemical environment of the CNS neurons. Slight changes in its composition can cause malfunctions of the nervous system. For example, a high glycine concentration disrupts temperature and blood pressure control, and a high pH causes dizziness and fainting.

The CNS produces about 500 mL of CSF per day, but only 100 to 160 mL of CSF is present at one time because the fluid is constantly reabsorbed at a rate matching its secretion. It is produced by a structure on the wall of each ventricle called the **choroid** (CO-royd) **plexus**—named for its histological resemblance to the chorion, one of the membranes surrounding a fetus. A choroid plexus consists of a network of blood capillaries covered by a simple cuboidal epithelium. Tight junctions between the epithelial cells form a **blood-CSF barrier** that protects the brain from potentially harmful substances in the blood.

16. *tentorium* = tent
17. *arachn* = spider, spider web + *oid* = resembling
18. *pia* = tender + *mater* = mother

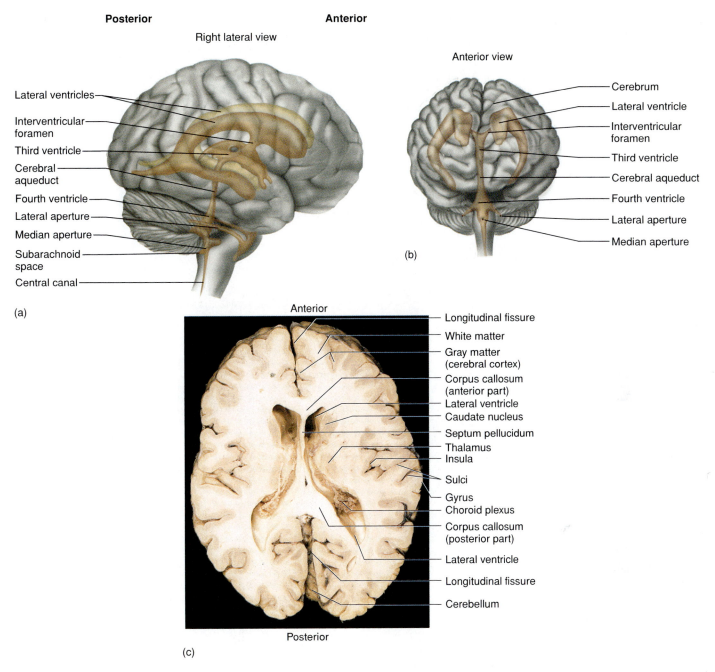

Posterior   **Anterior**

Right lateral view

Lateral ventricles

Interventricular foramen

Third ventricle

Cerebral aqueduct

Fourth ventricle

Lateral aperture

Median aperture

Subarachnoid space

Central canal

(a)

Anterior view

Cerebrum

Lateral ventricle

Interventricular foramen

Third ventricle

Cerebral aqueduct

Fourth ventricle

Lateral aperture

Median aperture

(b)

Anterior

Longitudinal fissure

White matter

Gray matter (cerebral cortex)

Corpus callosum (anterior part)

Lateral ventricle

Caudate nucleus

Septum pellucidum

Thalamus

Insula

Sulci

Gyrus

Choroid plexus

Corpus callosum (posterior part)

Lateral ventricle

Longitudinal fissure

Cerebellum

Posterior

(c)

Figure 14.6 Ventricles of the brain. (*a*) Right lateral view. (*b*) Anterior view. (*c*) Superior view of a horizontal section of the brain, showing the lateral ventricles and other features of the cerebrum.

Ventricles and CSF Circulation

The brain has four CSF-filled chambers called **ventricles** (fig. 14.6). The first two are the large **lateral ventricles,** one in each cerebral hemisphere, forming a butterfly shape in cross sections of the brain. On the midsagittal plane inferior to the corpus callosum is the **third ventricle,** connected to each lateral ventricle by a tiny passage called the **interventricular foramen.** A canal called the **cerebral aqueduct** passes through the midbrain and connects the third ventricle with the

fourth ventricle, a small triangular chamber between the pons and cerebellum.

The CSF begins its journey at the choroid plexus in each lateral ventricle (fig. 14.7). It flows through the interventricular foramen into the third ventricle and then down the cerebral aqueduct to the fourth ventricle. Choroid plexuses in the third and fourth ventricles add more CSF along the way. Some CSF fills the central canal of the spinal cord, but ultimately, all of it escapes through three pores in the walls of the fourth ventricle—a *median aperture* and two *lateral apertures.*

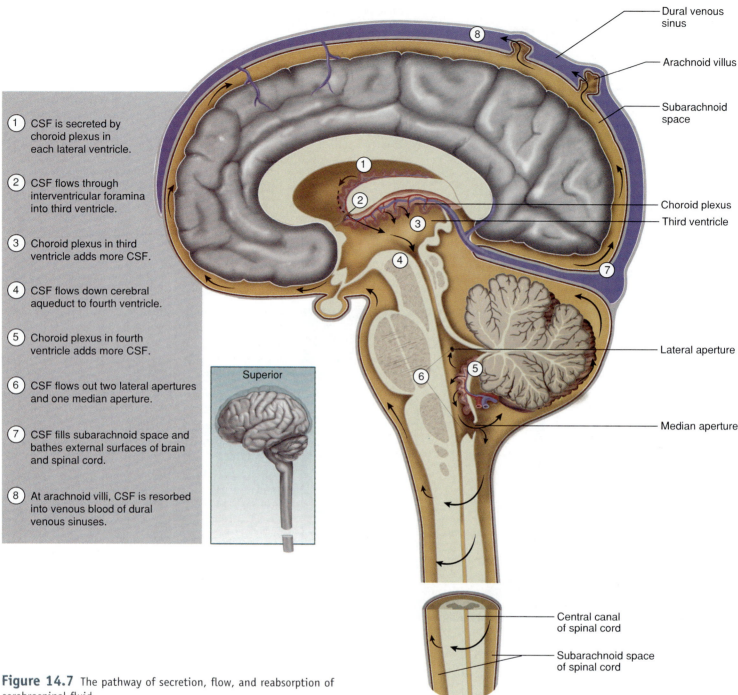

1. CSF is secreted by choroid plexus in each lateral ventricle.

2. CSF flows through interventricular foramina into third ventricle.

3. Choroid plexus in third ventricle adds more CSF.

4. CSF flows down cerebral aqueduct to fourth ventricle.

5. Choroid plexus in fourth ventricle adds more CSF.

6. CSF flows out two lateral apertures and one median aperture.

7. CSF fills subarachnoid space and bathes external surfaces of brain and spinal cord.

8. At arachnoid villi, CSF is resorbed into venous blood of dural venous sinuses.

Superior

Dural venous sinus

Arachnoid villus

Subarachnoid space

Choroid plexus

Third ventricle

Lateral aperture

Median aperture

Central canal of spinal cord

Subarachnoid space of spinal cord

Figure 14.7 The pathway of secretion, flow, and reabsorption of cerebrospinal fluid.

These apertures lead into the subarachnoid space. Here, the CSF flows around the posterior side of the brain and spinal cord, then up the anterior side of both, and finally reaches the **arachnoid villi** on the superior surface of the brain. These are cauliflower-like extensions of the arachnoid membrane that protrude through the dura mater into a large, blood filled *superior sagittal sinus* along the top of the head. CSF permeates the walls of the arachnoid villi and mixes with the blood in this sinus.

Hydrocephalus[19] is the abnormal accumulation of CSF in the brain, usually resulting from a blockage in its route of flow and reabsorption. It causes expansion of the ventricles and potentially fatal compression and destruction of the nervous tissue. In a fetus or infant, it can cause the entire head to become enlarged because the cranial bones are not yet fused. Good recovery can be achieved with prompt surgical intervention to drain excess CSF from the ventricles.

19. *hydro* = water + *cephal* = head

Blood Supply and the Blood-Brain Barrier

Although the brain constitutes only 2% of the adult body weight, it receives 20% of the blood and consumes 20% of the body's oxygen and glucose. The brain receives about 750 mL of blood per minute. Blood flow can be shifted from one region of the brain to another as different regions are called into play to perform specific tasks (see chapter essay, p. 499).

Glucose is the only energy substrate that neurons can metabolize. Because neurons have such a high demand for ATP, and because they cannot fall back on any alternative source of energy, the constancy of the blood supply is especially critical to the nervous system. An interruption in blood flow of only 10 seconds can cause loss of consciousness; an interruption of 1 to 2 minutes can significantly impair neural function; and 4 minutes without blood causes irreversible brain damage because lysosomes begin to release enzymes that digest neurons from within.

The CNS is well protected by a **blood-brain barrier (BBB),** which, like the blood-CSF barrier, carefully regulates which substances leave the bloodstream and get to the neurons. The perivascular feet of the astrocytes contribute to the BBB, but it is largely formed by tight junctions between the endothelial cells of the capillaries and by a continuous basement membrane around them. The BBB is highly permeable to water; to lipid-soluble substances such as oxygen, carbon dioxide, alcohol, caffeine, nicotine, and anesthetics; and to carrier-transported solutes such as glucose. It is slightly permeable to sodium, potassium, chloride, and the waste products urea and creatinine. While the BBB is an important protective device, it poses a problem in administering drugs for treatment of infections or neurological disorders of the brain. Some drugs are administered as nasal sprays because they can travel up the olfactory nerve fibers into the brain and bypass the BBB. Trauma and inflammation sometimes damage the BBB and allow toxins and pathogenic organisms to enter the brain tissue.

It is essential that the brain be able to monitor and respond to fluctuations in the composition of the blood. For this reason the BBB is entirely absent in limited regions called **circumventricular organs (CVOs).** CVOs occur in the walls of the third and fourth ventricles, where they afford a means for neurons to monitor blood glucose, pH, salinity, osmolarity, and pressure. Unfortunately, the CVOs also afford a route for the human immunodeficiency virus (HIV) to enter the brain, causing dementia in advanced cases of AIDS.

Key Point Review

4 Name the three meninges from superficial to deep.

5 Describe four functions of the cerebrospinal fluid.

6 Distinguish between the blood-CSF barrier and the blood-brain barrier.

7 Where does the CSF originate and what route does it take through and around the CNS?

The Spinal Cord

▼**Objectives**

When you have completed this section, you should be able to

- describe the gross and microscopic anatomy of the spinal cord; and
- name the major ascending and descending columns and tracts of the spinal cord and state their functions.

Functions

The spinal cord (fig. 14.8) serves three principal functions:

1. **Conduction.** The white matter provides conduction pathways that carry information up and down the cord. This allows sensory information to reach the brain, motor commands to reach the effectors, and input received at one level of the cord to affect output from another level.
2. **Locomotion.** Movement of the limbs is initiated by neurons in the cerebral cortex. Walking, however, also involves reverberating circuits called **central pattern generators (CPGs)** in the gray matter of the spinal cord. These circuits produce the necessary sequence of outputs to the extensor and flexor muscles to cause alternating movements of the legs. CPGs put out rhythmic bursts of motor impulses even in slices of spinal cord isolated from the rest of the nervous system, but in the body they are turned on and off by centers of voluntary motor control in the brain.
3. **Control of reflex activity.** Reflexes are involuntary stereotyped responses to stimuli that are mediated through both the brain and spinal cord. They involve the concerted action of the CNS and PNS and are discussed in detail in the next chapter.

Gross Anatomy

The spinal cord begins at the foramen magnum and passes through the vertebral canal of the spinal column. It averages about 1.8 cm in diameter and 45 cm in length in the adult, ending within the first lumbar vertebra (L1). Thus, it occupies only the upper two-thirds of the vertebral canal; the lower one-third is described shortly.

The spinal cord is divided into **cervical, thoracic, lumbar,** and **sacral regions.** It may seem peculiar that it has a sacral region when the cord itself ends well above the sacrum, at L1. These regions, however, are named for the part of the spinal column from which the nerves emerge, not for the vertebrae that contain the cord itself.

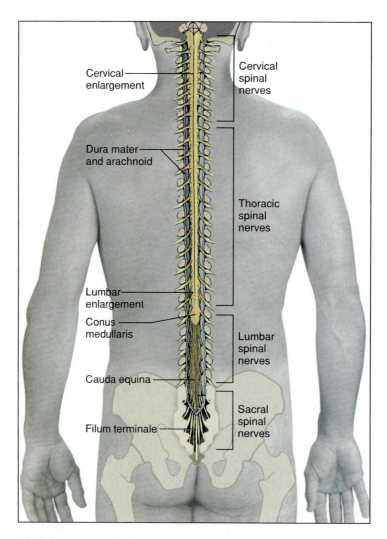

Figure 14.8 The spinal cord, posterior view.

The spinal cord gives rise to 31 pairs of *spinal nerves* that pass through the intervertebral foramina. The portion of the spinal cord connected to each pair of nerves is called a *segment* of the cord.

In the inferior cervical region, a **cervical enlargement** of the cord gives rise to nerves of the upper extremities. In the lumbosacral region, there is a similar **lumbar enlargement** where nerves to the pelvic region and lower extremities arise. Inferior to the lumbar enlargement, the cord tapers to a point called the **conus medullaris.** The pia mater continues beyond this point as a fibrous strand, the **filum terminale** (FY-lum TUR-mih-NAY-lee), which extends to the coccyx and anchors the lower end of the cord. The canal of vertebrae L2 to S5 contains a thick bundle of nerve roots that arise from the lumbar enlargement and conus medullaris. This bundle, named the **cauda equina**[20]

(CAW-duh ee-KWY-nah) for its resemblance to a horse's tail, innervates the pelvic organs and lower extremities.

Cross-Sectional Anatomy

Figure 14.5*b* shows the relationship of the spinal cord to a vertebra and spinal nerve in cross section, and figure 14.9 shows the spinal cord in more detail. The spinal cord exhibits longitudinal grooves on its anterior and posterior sides—the **anterior median fissure** and **posterior median sulcus,** respectively. The pia mater follows the contour of the cord into each groove.

20. *cauda* = tail + *equin* = horse

Gray Matter

The cord consists of a central region of gray matter surrounded by bundles of white matter. The butterfly- or H-shaped gray matter consists of two **dorsal (posterior)** horns, which extend toward the dorsolateral surfaces of the cord, and two thicker **ventral (anterior) horns,** which extend toward the ventrolateral surfaces. The right and left regions of gray matter are connected by a **gray commissure,** which crosses the middle of the cord. In the middle of the gray commissure is the **central canal,** which is filled with CSF and lined with ependymal cells.

Sensory nerve fibers approach the cord by way of a *dorsal root* of the spinal nerve, enter the dorsal horn, and sometimes synapse with an interneuron there. Such interneurons are especially numerous in the cervical and lumbar enlargements and are quite evident in histological sections at these levels. The large somas of the somatic motor neurons are found in the ventral horns. Their fibers exit by way of the *ventral root* of the spinal nerve and lead to the skeletal muscles. The spinal nerve roots are described more fully in the next chapter. In the thoracic and lumbar regions, an additional **lateral horn** is visible on each side of the gray matter. Motor fibers of the sympathetic nervous system arise from the lateral horn and send their axons out of the cord by way of the ventral root, along with the somatic efferent fibers. Special topic 14.3 concerns two well-known diseases of the spinal motor neurons.

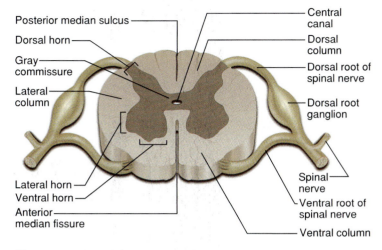

Posterior median sulcus
Dorsal horn
Gray commissure
Lateral column
Lateral horn
Ventral horn
Anterior median fissure

Central canal
Dorsal column
Dorsal root of spinal nerve
Dorsal root ganglion
Spinal nerve
Ventral root of spinal nerve
Ventral column

Figure 14.9 Cross section of the thoracic spinal cord showing its relationship to the spinal nerves.

Table 14.1　Major Spinal Tracts

Tract	Column	Decussation	Functions
Ascending Tracts			
Fasciculus gracilis	Dorsal	Yes, in medulla	Limb position, movement, touch, pressure
Fasciculus cuneatus	Dorsal	Yes, in medulla	Same as fasciculus gracilis
Dorsal spinocerebellar	Lateral	No	Feedback from muscles (proprioception)
Ventral spinocerebellar	Lateral	Yes, in cord	Same as dorsal spinocerebellar
Spinothalamic	Lateral and ventral	Yes, in cord	Touch, tickle, itch, temperature, pain, pressure
Descending Tracts			
Tectospinal	Lateral and ventral	Yes, in midbrain	Reflex turning of head and neck in response to visual and auditory stimuli
Lateral corticospinal	Lateral	Yes, in medulla	Fine control of limb movements
Ventral corticospinal	Ventral	No	Fine control of limb movements
Lateral reticulospinal	Lateral	No	Balance and posture
Medial reticulospinal	Ventral	No	Balance and posture
Vestibulospinal	Ventral	No	Balance

The male and female spinal cords differ in the lumbar region. In men there is a large cluster of motor neurons that controls the ischiocavernosus and bulbocavernosus muscles concerned with erection and ejaculation. These muscles and spinal neurons have testosterone receptors that contribute to the prenatal development and adult sexual function of males. The neuron cluster and aforementioned pelvic muscles of females are smaller.

White Matter

The white matter of the spinal cord consists of bundles of myelinated axons that course up and down the cord, providing an avenue of communication between different levels of the CNS. The white matter is arranged in three pairs of bundles known as **columns** or **funiculi**[23] (few-NIC-you-lie)—a **dorsal (posterior), lateral,** and **ventral (anterior) column** on each side. Each column of white matter consists of subdivisions called **tracts** or **fasciculi**[24] (fah-SIC-you-lye).

Spinal Tracts

Knowledge of the locations and functions of the spinal tracts is essential in managing spinal cord injuries, functional losses, and rehabilitation. **Ascending tracts** carry sensory information up the spinal cord, and **descending tracts** conduct motor impulses down the cord. All nerve fibers in a given tract have a similar origin,

destination, and function. Many of the fibers exhibit **decussation**[25] (DEE-cuh-SAY-shun)—they cross from one side of the body to the other within the spinal cord or medulla oblongata. As a result, each side of the brain receives sensory information from the opposite (*contralateral*[26]) side of the body and issues motor impulses to that side—that is, the left cerebral hemisphere senses and controls the right side of the body, and the right hemisphere senses and controls the left side. A stroke that damages motor regions of the right cerebral cortex, therefore, may cause paralysis on the left side of the body and vice versa. Not all spinal nerve fibers decussate, however; those that do not are said to remain on the *ipsilateral*[27] side of the body.

Ascending Tracts

The major spinal cord tracts are summarized in table 14.1 and figure 14.10. The names of most ascending tracts consist of the prefix *spino-* followed by a root denoting the destination of its fibers. The major ascending tracts are described here, and some of the ascending nervous pathways are shown in figure 14.11.

- The **fasciculus gracilis**[28] (fah-SIC-you-lus GRASS-ih-liss) travels up the medial part of the dorsal column. It carries sensory signals concerning body movements, positions of the limbs, fine touch discrimination, and pressure, originating on the ipsilateral side of the body below the midthoracic

---

23. *funicul* = little rope, cord
24. *fascicul* = little bundle

25. *decuss* = to cross, form an X
26. *contra* = opposite
27. *ipsi* = same
28. *gracilis* = thin, slender

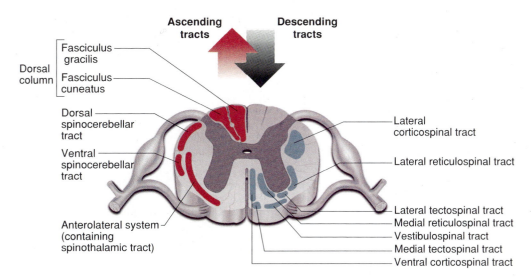

Figure 14.10 Major sensory and motor tracts of the spinal cord. The left side shows the ascending sensory tracts (red), and the right side shows the descending motor tracts (green). All of the illustrated tracts occur on both sides of the cord, however. ✗

(T6) level. From the T6 level down, the fasciculus gracilis makes up the entire dorsal column. Its fibers end superiorly in the medulla oblongata.

- The **fasciculus cuneatus**[29] (CUE-nee-ay-tus) joins the fasciculus gracilis at the T6 level. It carries the same type of sensory signals, originating from this level of the body and up. It, too, ends in the ipsilateral side of the medulla oblongata. In the brainstem, the fasciculus gracilis and fasciculus cuneatus form the **medial lemniscal**[30] (lem-NIS-cul) **system.**
- The **dorsal** and **ventral spinocerebellar** (SPY-no-SERR-eh-BEL-ur) **tracts** travel through the lateral column and carry sensory signals from the muscles and tendons of the legs and trunk to the cerebellum. This provides the cerebellum with feedback that it uses to coordinate muscle action, as discussed later in this chapter.
- The **spinothalamic** (SPY-no-tha-LAM-ic) **tract** and some smaller tracts form the *anterolateral system,* which passes up the anterior and lateral columns of the spinal cord. The spinothalamic tract carries signals for light touch, tickle, itch, temperature, pain, and pressure; it ends in the thalamus, which relays these signals to the cerebrum.

Descending Tracts

The names of most descending tracts consist of a word root denoting the point of origin in the brain, followed by the suffix *-spinal.* The major descending tracts are

described here, and some of their nervous pathways are shown in figure 14.12.

- The **tectospinal** (TEC-toe-SPY-nul) **tract** begins in the midbrain, crosses to the opposite side of the brainstem, and in the lower medulla it branches into *lateral* and *medial tectospinal tracts* of the upper spinal cord. It is involved in reflex movements of the head and neck, especially in response to visual and auditory stimuli.
- The **corticospinal** (COR-tih-co-SPY-nul) **tracts** convey motor impulses from the cerebral cortex for precise, finely coordinated limb movements. The fibers of this system form pyramid-shaped ridges on the anterior surface of the medulla oblongata; they were once called *pyramidal tracts,* although this term is becoming obsolete. Most corticospinal fibers cross to the contralateral side of the lower medulla and then form the **lateral corticospinal tract** of the cord. A few fibers remain uncrossed and form the **ventral corticospinal tract.**
- The **lateral** and **medial reticulospinal** (reh-TIC-you-lo-SPY-nul) **tracts** originate in the *reticular formation* of the brainstem. They control flexor and extensor muscles of the upper and lower extremities, especially to maintain posture and balance.
- The **vestibulospinal** (vess-TIB-you-lo-SPY-nul) **tract** begins with a *vestibular nucleus* that receives impulses for balance from the inner ear. The tract passes down the ventral column of the spinal cord and, like the reticulospinal tracts, it controls limb muscles that maintain balance.

Rubrospinal tracts are prominent in other mammals, where they aid in muscle coordination. Although

---

29. *cune* = wedge
30. *lemnisc* = ribbon

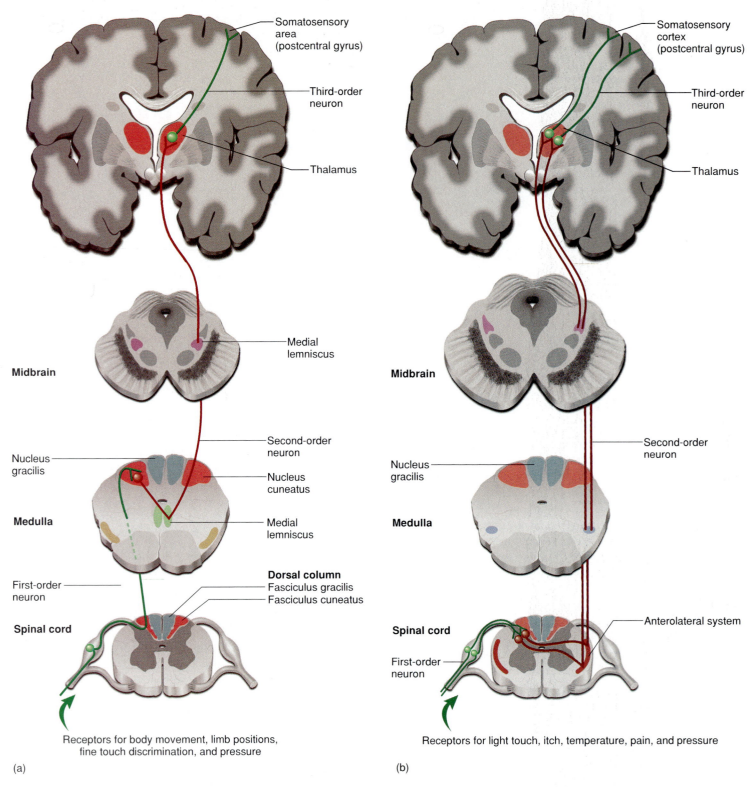

Somatosensory area (postcentral gyrus)

Third-order neuron

Thalamus

Midbrain

Medial lemniscus

Second-order neuron

Nucleus gracilis

Nucleus cuneatus

Medulla

Medial lemniscus

First-order neuron

Dorsal column
Fasciculus gracilis
Fasciculus cuneatus

Spinal cord

Receptors for body movement, limb positions, fine touch discrimination, and pressure

(a)

Somatosensory cortex (postcentral gyrus)

Third-order neuron

Thalamus

Midbrain

Second-order neuron

Nucleus gracilis

Medulla

Anterolateral system

Spinal cord

First-order neuron

Receptors for light touch, itch, temperature, pain, and pressure

(b)

Figure 14.11 Some ascending (sensory) pathways of the CNS. Nerve signals enter the spinal cord at the bottom of the figure and carry somatosensory information up to the cerebral cortex. (a) The fasciculus cuneatus and medial lemniscal system. (b) The spinothalamic tract. ⚡

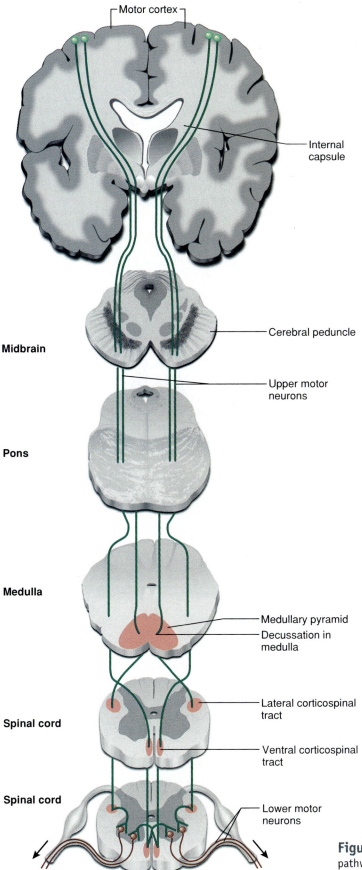

Motor cortex

Internal capsule

Midbrain

Cerebral peduncle

Upper motor neurons

Pons

Medulla

Medullary pyramid
Decussation in medulla

Spinal cord

Lateral corticospinal tract

Ventral corticospinal tract

Spinal cord

Lower motor neurons

To skeletal muscles   To skeletal muscles

often pictured in illustrations of human anatomy, they are almost nonexistent in humans and have little functional importance.

Think About It

You are blindfolded and either a hamster or an ice cube is placed in your right hand. What spinal tracts would carry the signals that enable you to discriminate between these two objects?

Key Point Review

8 Name the four major regions and two enlargements of the spinal cord.

9 Describe the distal (inferior) end of the spinal cord and the contents of the vertebral canal from level L2 to S5.

10 Sketch a cross section of the spinal cord showing the dorsal and ventral horns. Where are the gray and white matter? Where are the columns and tracts?

11 Give an anatomical explanation as to why a stroke in the right cerebral hemisphere can paralyze the limbs on the left side of the body.

The Hindbrain and Midbrain

▼Objectives

When you have completed this section, you should be able to
• list the anatomical components of the hindbrain and midbrain and describe the functions of each;
• discuss the role of the cerebellum in movement and equilibrium;
• define the term *brainstem* and describe the anatomical relationship of this part of the brain to the cerebellum and forebrain; and
• describe the location and functions of the reticular formation.

Our study of the form and function of the brain is organized around the five secondary vesicles of the embryonic brain and their mature derivatives. We begin with the hindbrain and its relatively simple functions and work our way upward, finishing with the forebrain. The regional anatomy and corresponding functions of the brain are summarized at the end of this discussion in table 14.2.

The Myelencephalon

As noted earlier, the embryonic hindbrain differentiates into two subdivisions, the myelencephalon and metencephalon (see fig. 14.4). The myelencephalon develops into just one structure, the **medulla oblongata** (meh-DULL-uh OB-long-GAH-ta). The medulla is about

Figure 14.12 The corticospinal tracts, which are descending (motor) pathways of the CNS. Nerve signals originate in the cerebral cortex at the top of the figure and carry motor commands down the spinal cord. ⚹

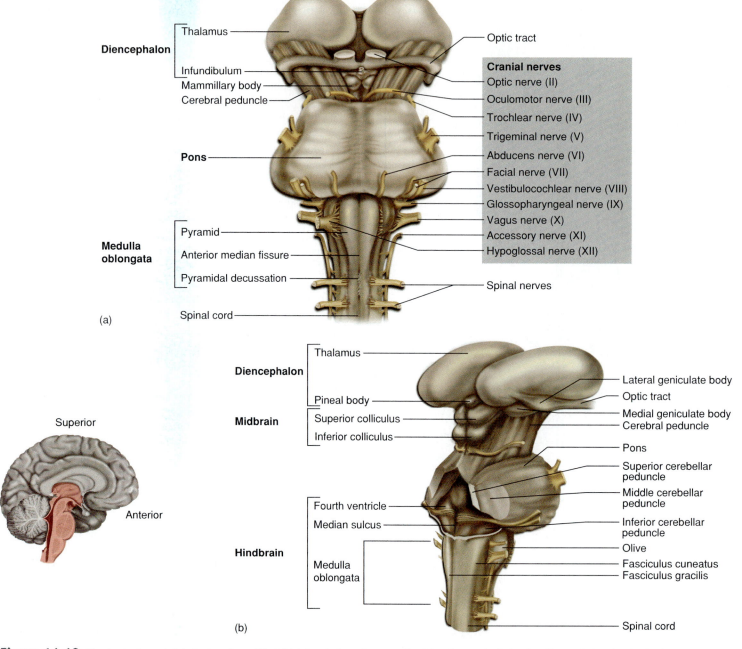

Figure 14.13 The brainstem. (*a*) Anterior view; (*b*) right lateral view. Some authorities do not include the diencephalon in the brainstem.

3 cm long and superficially looks like an extension of the spinal cord, only slightly wider. Significant differences are seen, however, on closer inspection of its gross and microscopic anatomy.

On the anterior surface are a pair of ridges, the **pyramids,** that contain the nerve fibers of the corticospinal (pyramidal) tracts. The pyramids are wider at the superior end, taper inferiorly, and are separated by an *anterior median fissure* continuous with that of the spinal cord (fig. 14.13). Most pyramidal nerve fibers decussate at a visible point near the inferior end of the medulla. Lateral to each pyramid is an elevated area

called the **olive.** It contains a wavy layer of gray matter, the **inferior olivary nucleus,** which is a relay center for signals going to the cerebellum. The medulla contains ascending and descending nerve tracts and is the origin of the last four cranial nerves: IX (glossopharyngeal), X (vagus), XI (accessory), and XII (hypoglossal).

The medulla includes nuclei that control coughing, sneezing, hiccuping, swallowing, vomiting, and sweating, as well as the following major neuronal pools:

- the **cardiac center,** which adjusts the rate and force of the heartbeats;

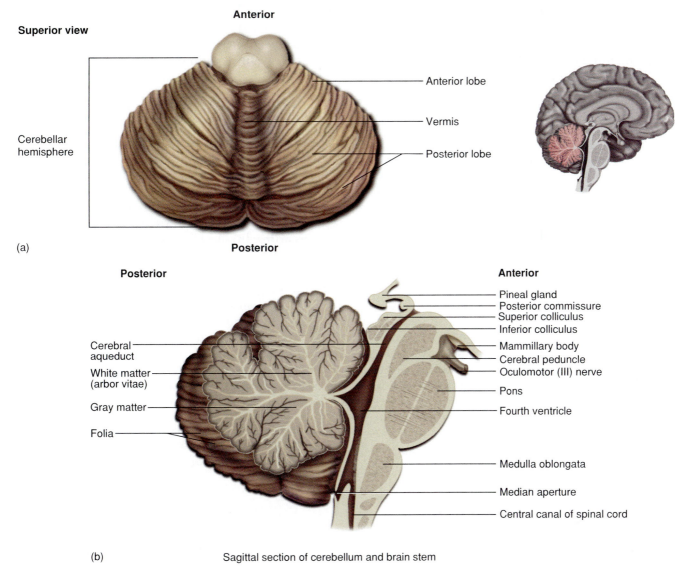

Superior view

Anterior

Cerebellar hemisphere

Anterior lobe

Vermis

Posterior lobe

(a)

Posterior

Posterior

Anterior

Cerebral aqueduct

White matter (arbor vitae)

Gray matter

Folia

Pineal gland
Posterior commissure
Superior colliculus
Inferior colliculus

Mammillary body
Cerebral peduncle
Oculomotor (III) nerve

Pons

Fourth ventricle

Medulla oblongata

Median aperture

Central canal of spinal cord

(b)   Sagittal section of cerebellum and brain stem

Figure 14.14 The cerebellum. (*a*) Superior view and (*b*) right lateral view, midsagittal section, showing its relationship to the brainstem.

- the **vasomotor center,** which adjusts blood vessel diameter to regulate blood pressure and reroute blood from one part of the body to another; and
- **respiratory centers,** which control the rate and depth of breathing.

The Metencephalon

Two structures arise from the metencephalon: the pons and cerebellum. The **pons**[31] shows as an anterior bulge in the brainstem superior to the medulla oblongata (fig. 14.13). Its white matter consists of nerve fibers passing up and down between the cerebrum and medulla and many fibers that conduct motor information from the cerebrum to the cerebellum. Its gray matter includes relay nuclei for signals from the cerebrum to the cere-

bellum and nuclei concerned with sleep, posture, respiration, swallowing, and bladder control, discussed later in this and other chapters. Cranial nerve V (trigeminal) arises from the pons, and cranial nerves VI (abducens), VII (facial), and VIII (vestibulocochlear) arise from the junction of the pons and medulla. Their functions are discussed in the next chapter.

The cerebellum is the largest part of the hindbrain (fig. 14.14). It lies in the posterior cranial fossa, posterior to the fourth ventricle, brainstem, and foramen magnum. It is separated from the cerebrum by a deep transverse fissure and the tentorium cerebelli of the dura mater. It consists of right and left **cerebellar hemispheres** connected by a narrow **vermis.**[32] Each hemisphere exhibits numerous folds called **folia**[33] (gyri),

31. *pons* = bridge

32. *verm* = worm
33. *foli* = leaf

which are separated by shallow sulci. Three paired fiber tracts called **cerebellar peduncles**[34] (peh-DUN-culs) connect the cerebellum to the brainstem: the *inferior peduncles* to the medulla oblongata, the *middle peduncles* to the pons, and the *superior peduncles* to the midbrain.

The cerebellum has a surface cortex of gray matter and a deeper layer of white matter. In a sagittal section, the white matter, called the **arbor vitae,**[35] exhibits a branching, fernlike pattern. Embedded in the white matter are masses of gray matter called the *deep nuclei.* Although the cerebellum constitutes only 10% of the volume of the brain, it contains over 50% of the brain's neurons. The most distinctive of these are the **Purkinje**[36] (pur-KIN-jee) **cells**—unusually large, globose neurons with a tremendous profusion of dendrites (see fig. 13.7), arranged in a single row in the cerebellar cortex. Their axons synapse in the deep nuclei, and postsynaptic cells from these nuclei send output to sites in the cerebral cortex and brainstem.

The cerebellum modulates and coordinates voluntary movement of the limbs, maintains muscle tone and posture, coordinates eye movements with body movements, and aids in the learning of motor skills. Commands to the skeletal muscles originate in **upper motor neurons** of the cerebral cortex and travel by way of the corticospinal tracts to **lower motor neurons** in the spinal cord. These neurons then send their axons to the muscles. The cerebellum receives information from the cerebral cortex about the intended movements and receives feedback about the actual performance from sense organs called *proprioceptors*[37] in the muscles and joints (fig. 14.15). The Purkinje cells compare what the muscles are doing with what they were told to do and with stored information on learned skills. When there is a discrepancy between the intent and performance, the cells send signals to the deep cerebellar nuclei. Neurons of the nuclei transmit signals to the cerebral cortex and brainstem that correct the performance to match the intent. Lesions in the cerebellum can result in a clumsy, awkward gait and make some tasks such as climbing a flight of stairs virtually impossible.

The Mesencephalon (Midbrain)

The **midbrain** (figs. 14.1, 14.2*b*, and 14.13) is a short segment of the brainstem that connects the hindbrain and forebrain. It contains the cerebral aqueduct, which connects the third and fourth ventricles, and it is the origin of cranial nerves III (oculomotor) and IV

(trochlear). The four principal regions of the midbrain are the *cerebral peduncles, substantia nigra, tegmentum,* and *tectum* (fig. 14.16).

The **cerebral peduncles** contain the corticospinal tracts described earlier. The **substantia nigra**[38] (sub-STAN-she-uh NY-gruh) is a dark gray to black nucleus pigmented with melanin, located between the peduncles and tegmentum. It is a motor center that relays information to the thalamus (part of the diencephalon to be discussed shortly) and to the basal nuclei (motor centers of the cerebrum, discussed later). The **tegmentum**[39] contains the **red nucleus,** which has a pink color in life because of its high density of blood vessels. This is the origin of fibers of the rubrospinal tract, but as mentioned earlier, this tract is vestigial and insignificant in humans.

The **tectum**[40] consists of four nuclei called the **corpora quadrigemina,**[41] which bulge from the midbrain roof. The two superior nuclei, called the **superior colliculi**[42] (col-LIC-you-lye), function in visual attention, visually tracking moving objects, and such reflexes as turning the eyes and head in response to a sound, sight, or touch. The two **inferior colliculi** receive all afferent signals from the inner ear and relay them to other parts of the brain, especially the thalamus.

Think About It

Why are the inferior colliculi shown in figure 14.14 but not in 14.16? How are these two figures related?

The Reticular Formation

The midbrain, pons, and medulla have two kinds of nuclei—those in which cranial nerves III to XII begin or end, and the **reticular formation,** a group of more than 100 nuclei scattered throughout the core of this region (fig. 14.17). The neurons of the reticular formation have unusual branched axons, with one branch extending down into the spinal cord and the other extending up to the thalamus, hypothalamus, or cerebral cortex. The functions of these nuclei fall into four categories:

1. **Somatic motor control.** Some upper motor neurons of the cerebral cortex project (send their axons) to reticular formation nuclei, which then give rise to the reticulospinal tracts of the spinal cord. While the upper motor neurons generate the primary signals for muscle contraction, reticular formation nuclei modulate the actions of the skeletal muscles. They

---

34. *ped* = foot + *uncle* = little
35. "tree of life"
36. Johannes E. von Purkinje (1787–1869), Bohemian anatomist
37. *proprio* = one's own + *ceptor* = from receptor

38. *substantia* = substance + *nigra* = black
39. *tegmen* = cover
40. *tectum* = roof, cover
41. *corpora* = bodies + *quadrigemina* = quadruplets
42. *colli* = hill + *cul* = little

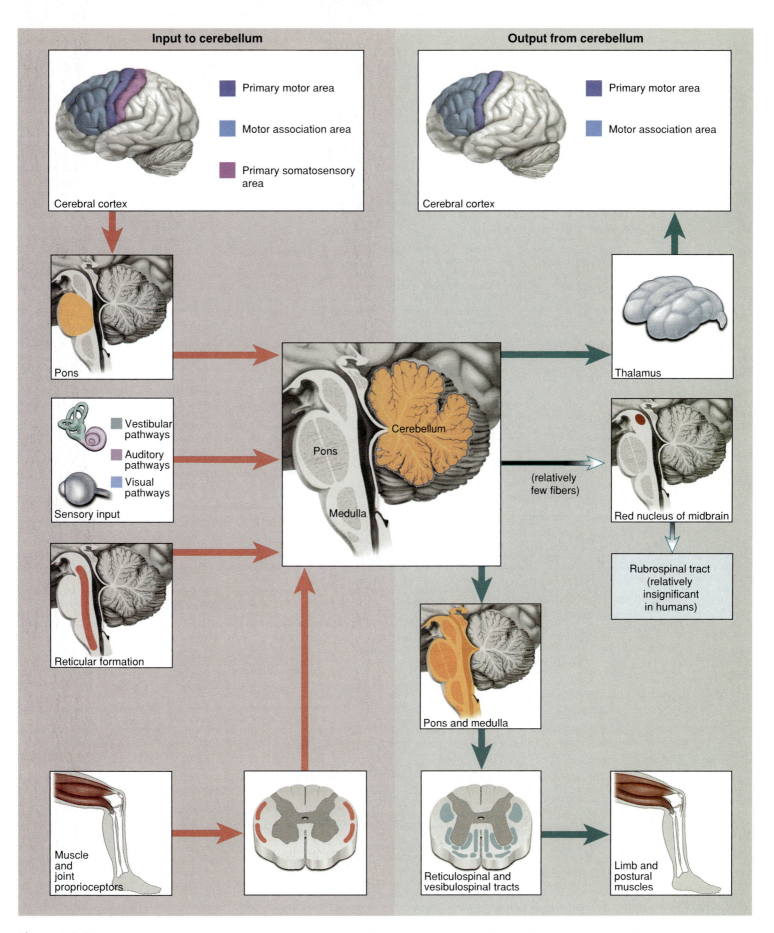

Figure 14.15 Pathways of motor control involving the cerebellum. The cerebellum receives its input from the afferent pathways (red) on the left and sends its output through the efferent pathways (green) on the right.

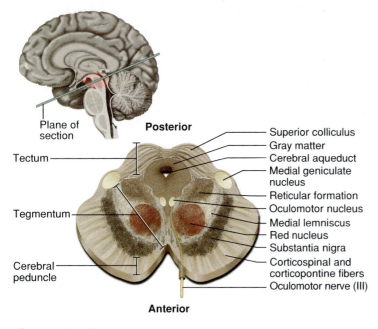

Figure 14.16 Cross section of the midbrain.

Labels for Figure 14.16:
Plane of section
Posterior
Tectum
Tegmentum
Cerebral peduncle
Anterior
Superior colliculus
Gray matter
Cerebral aqueduct
Medial geniculate nucleus
Reticular formation
Oculomotor nucleus
Medial lemniscus
Red nucleus
Substantia nigra
Corticospinal and corticopontine fibers
Oculomotor nerve (III)

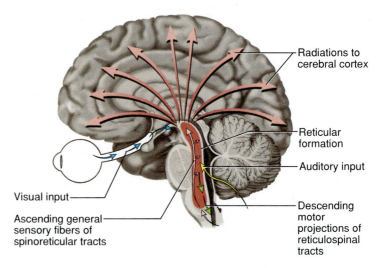

Figure 14.17 General location of the reticular formation and its projections to the cerebral cortex. Shown as a single red bar in this figure, the reticular formation consists of over 100 brainstem nuclei.

Labels for Figure 14.17:
Radiations to cerebral cortex
Reticular formation
Auditory input
Descending motor projections of reticulospinal tracts
Visual input
Ascending general sensory fibers of spinoreticular tracts

may stimulate antagonists and fixators (see chapter 11) or suppress stretch reflexes (see chapter 15), resulting in smoother, more finely coordinated muscular actions. The reticular formation is a relay for sensory signals going into the cerebellum and for motor signals coming out of it, enabling motor responses to visual, auditory, and equilibrium stimuli. The reticular formation also stimulates muscle tone, aids in posture, helps us keep our balance when one foot is raised from the ground, and has central pattern generators that produce the sequential or cyclic muscle contractions of breathing and swallowing.

2. **Autonomic control.** Some reticular formation nuclei are cardiovascular and respiratory centers that alter heart rate and contraction strength, blood pressure, respiratory rate, and the depth of breathing.

3. **Arousal.** Some reticular formation neurons send fibers to synapse in the thalamus, which relays their signals to the cerebral cortex. A greater number, however, bypass the thalamus and project directly to the cerebral cortex. These fibers modulate (increase or decrease) activity of the cortex in various ways, such as enhancing or suppressing its response to sensory input. This system is involved in **habituation**—a process in which the brain learns to ignore repetitive, inconsequential stimuli while remaining sensitive to others. In a noisy city, for example, a person can learn to sleep through traffic sounds but wakes promptly to the sound of an alarm clock or crying baby. The reticular formation also functions in alertness, attentiveness, and the sleep-wake cycle. Reticular nuclei and fibers that modulate cortical

activity used to be called the *reticular activating system,* but this term has fallen out of use with the realization that they do more than activate the cortex. More recently, this network has been called the *extrathalamic cortical modulatory system,* which calls attention to the facts that its fibers bypass the thalamus and that they modulate (not merely activate) the cerebral cortex.

4. **Pain modulation.** Fibers that descend from the reticular formation through the spinal cord can block pain messages from reaching the brain. The mechanism for this is described in chapter 16.

Key Point Review

12 Name any six visceral functions controlled by nuclei of the medulla.

13 Explain how the cerebellum improves the performance of intentional movements.

14 What are some functions of the midbrain nuclei?

15 Describe the reticular formation and list several of its functions.

The Diencephalon

▼Objectives
When you have completed this section, you should be able to
• name the three major components of the diencephalon and describe their locations; and
• describe the functions of the thalamus, epithalamus, and hypothalamus.

The forebrain consists of the diencephalon and telencephalon. The major derivatives of the diencephalon are the *thalamus, hypothalamus,* and *epithalamus.* The diencephalon encloses the third ventricle.

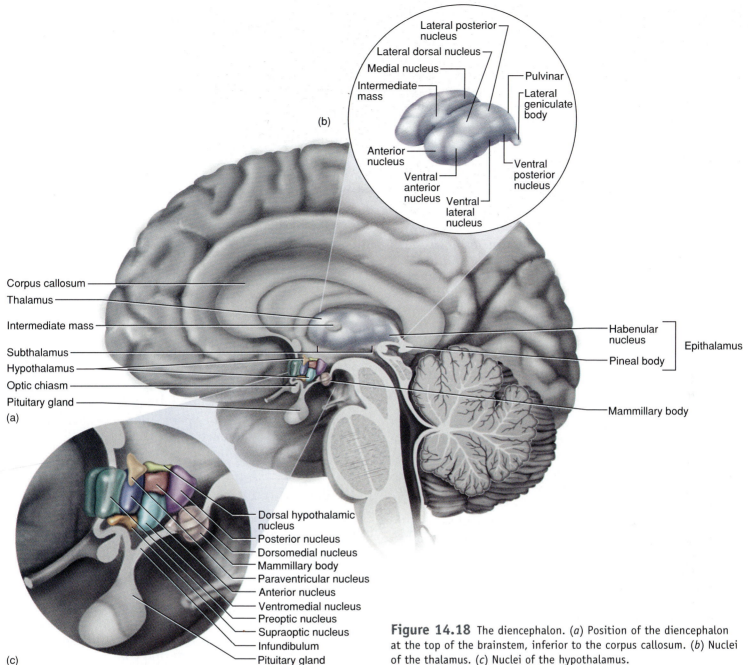

(b)

Lateral posterior nucleus
Lateral dorsal nucleus
Medial nucleus
Intermediate mass
Pulvinar
Lateral geniculate body
Anterior nucleus
Ventral anterior nucleus
Ventral lateral nucleus
Ventral posterior nucleus

Corpus callosum
Thalamus
Intermediate mass
Subthalamus
Hypothalamus
Optic chiasm
Pituitary gland
(a)

Habenular nucleus
Pineal body
Epithalamus

Mammillary body

Dorsal hypothalamic nucleus
Posterior nucleus
Dorsomedial nucleus
Mammillary body
Paraventricular nucleus
Anterior nucleus
Ventromedial nucleus
Preoptic nucleus
Supraoptic nucleus
Infundibulum
Pituitary gland

(c)

Figure 14.18 The diencephalon. (*a*) Position of the diencephalon at the top of the brainstem, inferior to the corpus callosum. (*b*) Nuclei of the thalamus. (*c*) Nuclei of the hypothalamus.

The Thalamus

The **thalamus**[43] (fib. 14.18*b*) constitutes about four-fifths of the diencephalon. It consists of two oval masses of gray matter, each of which underlies the cerebral cortex and lateral ventricle on one side of the brain. The two masses bulge medially into the third ventricle, where they join each other by a narrow *intermediate mass.*

The thalamus may be regarded as a "gateway to the cerebral cortex." Nearly all information going to the cerebrum passes by way of synapses in the thalamus, including all sensory input except for smell. Specialized thalamic nuclei (fig. 14.18*b*) integrate sensory information originating throughout the body and direct it to appropriate processing centers of the cerebrum.

The Hypothalamus

The **hypothalamus** (fig. 4.18*c*) forms part of the walls and floor of the third ventricle. It extends anteriorly to the *optic chiasma* (ky-AZ-muh) where the optic nerves meet, and posteriorly to a pair of humps called the **mammillary**[44]

---

43. *thalamus* = chamber, inner room

44. *mamm* = breast + *illa* = little

bodies. The mammillary bodies relay signals from the limbic system, discussed later, to the thalamus. The pituitary gland is attached to the hypothalamus by a stalk located between the optic chiasma and mammillary bodies.

The hypothalamus is the major control center of the autonomic nervous system and endocrine system and plays an essential role in the homeostatic regulation of nearly all organs of the body. Its nuclei include centers concerned with a wide variety of visceral functions, including the following:

- **Food and water intake.** Neurons of the *hunger* and *satiety centers* produce sensations of hunger and fullness in response to the levels of glucose and amino acids in the blood. **Osmoreceptors** are hypothalamic neurons that monitor the osmolarity of the blood. They trigger the sense of thirst and the mechanisms that cause the kidneys to retain water.
- **Thermoregulation.** The *hypothalamic thermostat* is a nucleus that monitors blood temperature. When the temperature becomes too high or too low, the thermostat signals other hypothalamic nuclei—the *heat-losing center* or *heat-producing center,* respectively, which control the heat-losing, heat-retaining, and heat-producing mechanisms discussed in chapter 1.
- **Cardiovascular regulation.** Hypothalamic nuclei send nerve fibers to the cardiac center of the medulla oblongata, which in turn adjusts the heart rate.
- **Hormone secretion.** The hypothalamus is not only part of the CNS but also part of the endocrine system. It produces and secretes hormones that regulate the functions of the pituitary gland, kidneys, uterus, and mammary glands. Through the pituitary, it indirectly influences many other endocrine glands. These relationships are explored in detail in later chapters.
- **Sleep and waking.** The hypothalamus regulates daily cycles of sleep and waking.
- **Emotional behavior.** Hypothalamic centers are involved in a variety of emotional responses including anger, fear, pleasure, and contentment; and in sexual drive, copulation, and orgasm.

The Epithalamus

The **epithalamus** consists mainly of the **pineal gland** (an endocrine gland discussed in chapter 17), the **habenula** (a relay from the limbic system to the midbrain, discussed later), and a thin roof over the third ventricle.

Think About It

General anesthesia prevents sensory information from reaching the cerebrum during surgery, while smelling salts are used to arouse an unconscious person. How do you think these relate to the function of the reticular formation?

The Telencephalon (Cerebrum)

▼Objectives
When you have completed this section, you should be able to
- name and locate the three major types of tracts in the cerebral white matter;
- name and define the five lobes of the cerebral cortex;
- describe the distinctive cell types and histological arrangement of the cerebral cortex;
- describe the locations and functions of the basal nuclei; and
- describe the location and functions of the limbic system.

The telencephalon (see fig. 14.4) develops chiefly into the *cerebrum,* the largest and most conspicuous part of the human brain. It enables you to turn these pages, read and comprehend the words, remember ideas, talk about them with your peers, and take an examination. It is the seat of your sensory perception, memory, thought, judgment, and voluntary motor actions. It is the most challenging frontier of neurobiology.

The Cerebral White Matter

The white matter of the cerebrum is not a decision-making (integrative) center, but it constitutes most of the cerebral volume. It consists of three types of fiber tracts (fig. 14.19):

1. **Projection tracts** extend vertically from higher to lower brain or spinal cord centers and carry information between the cerebrum and the rest of the body. Superior to the brainstem, they form a dense band called the **internal capsule** that lies between the thalamus and basal nuclei. They then radiate in a diverging, fanlike array (the **corona radiata**[45]) to specific areas of the cortex.
2. **Commissural tracts** cross from one cerebral hemisphere to the other through bridges called **commissures** (COM-ih-shurs). The great majority of commissural tracts pass through the large C-shaped corpus callosum (see fig. 14.2*b*), which forms the floor of the longitudinal fissure. A few tracts pass through the much smaller **anterior** and **posterior commissures.** Commissural tracts enable the two sides of the cerebrum to "talk to each other."

45. *corona* = crown + *radiata* = radiating

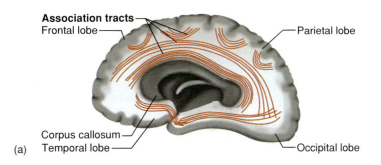

(a)

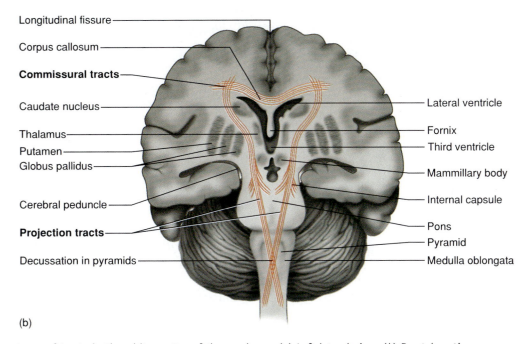

(b)

Figure 14.19 Three types of tracts in the white matter of the cerebrum. (*a*) Left lateral view. (*b*) Frontal section.

3. **Association tracts** connect different regions within the same hemisphere; they do not cross from one side of the brain to the other. The *lobes* of a cerebral hemisphere, described later, are interconnected by *long association fibers,* whereas *short association fibers* connect adjacent gyri within a single lobe. Among their other roles, association tracts link perceptual and memory centers of the brain, which enables you to smell something, for example, and picture what it looks like.

Gross Anatomy of the Cerebral Cortex

The telencephalon has three major integrative divisions—the cerebral cortex, basal nuclei, and limbic system. We begin here with the **cerebral cortex**,[46] which is a layer of gray matter covering the surfaces of the cere-

bral hemispheres. Even though it is only 2 to 3 mm thick, the cerebral cortex constitutes about 40% of the mass of the brain and has a surface area of about 2,500 cm², comparable to 4.5 pages of this textbook. If the cortex were smooth-surfaced, it would have only one-third as much area. As noted earlier, however, it is folded into gyri that provide expanded gray matter. Some cerebral gyri have consistent and predictable anatomy; others vary from brain to brain and from the right hemisphere to the left.

Certain fissures and unusually deep sulci divide each hemisphere into five anatomically and functionally distinct lobes. Four of these are visible superficially and are named for the cranial bones overlying them (fig. 14.20); the fifth lobe is not visible from the surface.

1. The **frontal lobe** lies immediately behind the frontal bone, superior to the orbits. Its posterior boundary is the **central sulcus.** The frontal lobe is chiefly concerned with voluntary motor functions, motivation, foresight, planning, mood, social judgment, and aggression.

---

46. *cortex* = bark, rind

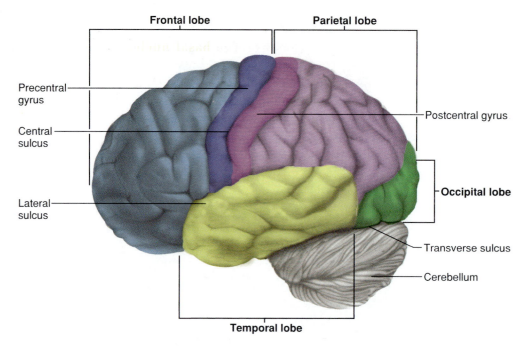

Figure 14.20 The four superficial lobes of the cerebrum. The fifth lobe, the insula, is not visible from the surface (see fig. 14.2a). ✗

Frontal lobe
Parietal lobe
Precentral gyrus
Postcentral gyrus
Central sulcus
Occipital lobe
Lateral sulcus
Transverse sulcus
Cerebellum
Temporal lobe

2. The **parietal lobe,** which underlies the parietal bone, extends from the central sulcus posteriorly to a relatively indistinct **parieto-occipital sulcus,** visible on the medial surface of each hemisphere. It is largely concerned with integration of sensory information except hearing, smell, and vision.
3. The **occipital lobe** is located behind the parieto-occipital sulcus and underlies the occipital bone. It receives and interprets visual signals.
4. The **temporal lobe** is a long horizontal lobe deep to the temporal bone, separated from the parietal lobe above it by a deep **lateral sulcus.** It is concerned with hearing, learning, memory, visual recognition, and emotional behavior.
5. The **insula**[47] is a relatively small mass of cortex deep to the lateral sulcus (see figs. 14.2a and 14.22), visible only by retracting or cutting away some of the overlying cerebrum. Little is known about its function.

Histology of the Cerebral Cortex

Two principal types of neurons can be identified in the cortex (fig. 14.21). **Stellate cells** have spheroidal somas with dendrites projecting for short distances in all directions. They are concerned largely with receiving sensory input and processing information on a local level. **Pyramidal cells** are tall and conical (triangular in tissue sections). The apex points toward the brain surface and has a

thick dendrite with many branches and dendritic spines. The base gives rise to horizontally oriented dendrites and an axon that passes into the white matter. Pyramidal cells are the output neurons of the cerebrum—they transmit signals to other parts of the CNS. Their axons have collaterals that synapse with neurons in other layers of the cortex or in deeper regions of the brain. Collectively, the cortical neurons have a staggering 100 trillion synapses—powerful testimony to the remarkable capacity of the cerebrum to store and process information.

About 90% of the human cerebral cortex is a six-layered tissue called **neocortex**[48] because of its relatively recent evolutionary origin. Although vertebrates have existed for about 600 million years, the neocortex did not develop significantly until the rise of mammals about 60 million years ago, and it attained its highest development by far in the primates. The six layers of neocortex vary in cellular composition, synaptic connections, size of their neurons, destination of their axons, and relative thickness from one part of the cerebrum to another. Layer IV is thickest in sensory regions and layer V in motor regions, for example. All axons that leave the cortex and enter the white matter arise from layers III, IV, and VI. Some olfactory regions of the cerebral cortex are composed of a one- to five-layered tissue called *paleocortex* (PALE-ee-oh-cor-tex), and the limbic system (discussed shortly) has a three-layered *archicortex* (AR-kee-cor-tex). These are of much greater evolutionary age than the neocortex.

---

47. *insula* = island

48. *neo* = new

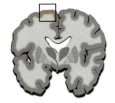

Cortical surface

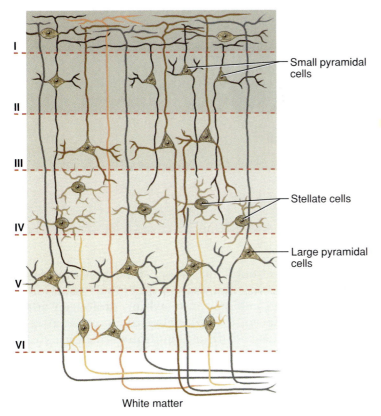

I

II

III

Small pyramidal cells

IV

Stellate cells

V

Large pyramidal cells

VI

White matter

Figure 14.21 Neurons of the cerebral neocortex are arranged in six histologically recognizable layers.

The Basal Nuclei

The **basal nuclei** are masses of cerebral gray matter buried deep in the white matter, lateral to the thalamus (fig. 14.22). They are often called basal ganglia, but the word *ganglion* is best restricted to clusters of neurons outside the CNS. There are five basal nuclei in each cerebral hemisphere: the **caudate**[49] **nucleus, putamen,**[50] **globus pallidus,**[51] **amygdala,**[52] and **claustrum.**[53] The putamen and globus pallidus are also collectively called the *lentiform*[54] *nucleus*, while the caudate nucleus and putamen are collectively called the *corpus striatum* after their striped appearance in nonhuman animals.

The basal nuclei are involved in motor control and cognition (thought). They receive input from the entire cerebral cortex and then issue output fibers back to the cerebrum by way of the thalamus. Output signals go to the prefrontal cortex (just behind the forehead) and to motor control centers of the frontal lobe. Thus the basal nuclei seem to be part of a feedback circuit involved in the planning and execution of movement. Their exact mode of action is not yet known, but they may facilitate some muscle contractions while inhibiting others, or they may be part of the system that compares ongoing muscle actions with the original conscious intent. People with damage to the basal nuclei tend to move slowly

49. *caudate* = tailed, taillike
50. *putam* = pod, husk
51. *glob* = globe, ball + *pall* = pale
52. *amygdal* = almond
53. *claustr* = lock, bar, door
54. *lenti* = lens + *form* = shape

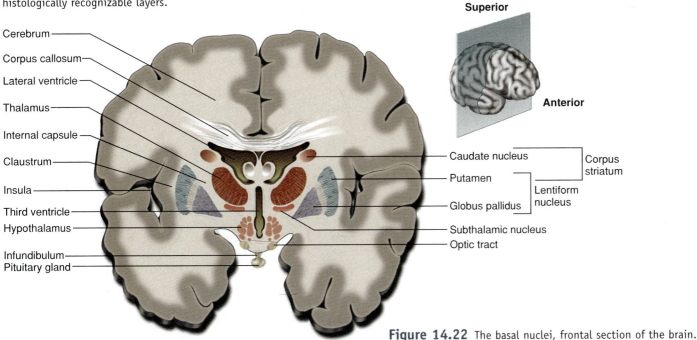

Cerebrum
Corpus callosum
Lateral ventricle
Thalamus
Internal capsule
Claustrum
Insula
Third ventricle
Hypothalamus
Infundibulum
Pituitary gland

Superior

Anterior

Caudate nucleus ⎤ Corpus
Putamen ⎦ striatum
Globus pallidus ⎦ Lentiform nucleus
Subthalamic nucleus
Optic tract

Figure 14.22 The basal nuclei, frontal section of the brain.

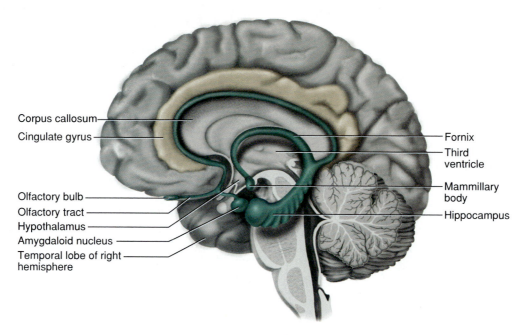

Figure 14.23 The limbic system.

Labels in figure:
- Corpus callosum
- Cingulate gyrus
- Olfactory bulb
- Olfactory tract
- Hypothalamus
- Amygdaloid nucleus
- Temporal lobe of right hemisphere
- Fornix
- Third ventricle
- Mammillary body
- Hippocampus

and have difficulty even initiating movement. They also exhibit changes in muscle tone and posture and involuntary muscle contractions such as flailing of the limbs or the tremors of Parkinson disease (see chapter 13 essay, p. 458).

The Limbic System

The **limbic**[55] **system,** named for the medial border of the temporal lobe, is a loop of cortical structures surrounding the corpus callosum and thalamus (fig. 14.23). It is the oldest component of the cerebral cortex, present even in fish and amphibians, but some parts of it are most highly developed in humans. It is closely associated with the hypothalamus in its function.

Functions of the limbic system have been investigated by such techniques as surgical removal, ablation (destruction) of small regions with electrodes, and stimulation with electrodes and chemical implants. Changes in an animal's behavior following removal or ablation give clues to the functions that the region performs. However, interconnections between parts of the limbic system and their connections to other parts of the brain make interpretation of the results difficult and controversial.

When specific regions of the limbic system are destroyed, neurobiologists observe impaired or exaggerated expressions of anger, fear, aggression, self-defense, pleasure, pain, love, sexuality, and parental affection, as well as abnormalities in learning, mem-

ory, and motivation. Thus, many important aspects of human personality depend on an intact, functional limbic system. Limbic circuits exhibit prolonged *after-discharge*—the continued firing of neurons after a stimulus has ceased; thus, emotional responses can last much longer than the stimuli that provoked them. This system also affects such basic responses as feeding, copulation, and orgasm.

Much of our behavior is shaped by learned associations between stimuli, our responses to them, and the rewards or punishments that result. Reward and punishment have been localized to certain neuronal pools in the limbic systems of cats, rats, monkeys, and other animals. A test animal can be placed in an apparatus with a pedal or bar wired to a stimulating electrode implanted in the limbic system. Sooner or later, the animal accidentally presses the bar and stimulates itself. If the sensation is pleasurable, the animal soon learns to press the bar over and over, and may spend most of its time doing so—even to the point of neglecting food and water. Rats have been known to bar-press 5,000 to 12,000 times an hour, and monkeys up to 17,000 times an hour, to stimulate their pleasure centers.

These animals cannot tell us what they are feeling, but electrode implants have also been used for people who suffer otherwise incurable schizophrenia, pain, or epilepsy. These people also bar-press repeatedly, although they report feelings short of joy or ecstasy. Some are unable to explain why they press the bar; others report "relief from tension" or "a quiet, relaxed feeling." With electrodes in other areas of the limbic system, subjects report sensations of fear or terror when stimulated.

55. *limbus* = border

19 Distinguish commissural, association, and projection tracts from each other.

20 Name the five lobes of the cerebral cortex and describe their locations relative to each other.

21 What are the basal nuclei and where are they located? What is their function and what is the result of degenerative changes in these nuclei?

22 Describe the location and functions of the limbic system.

Functions of the Neocortex

▼Objectives

When you have completed this section, you should be able to

- list the four types of brain waves and discuss their relationship to sleep and other mental states;
- explain how the brain controls the skeletal muscles;
- identify the parts of the cerebrum that receive and interpret somatic sensory signals;
- identify the parts of the cerebrum that receive and interpret signals from the special senses;
- discuss the functional relationship between the right and left cerebral hemispheres;
- describe the locations and functions of the language centers; and
- discuss the different types of memory and state which regions of the cerebrum are associated with memory.

Brain Waves

Brain waves are rhythmic voltage changes resulting predominantly from synchronized postsynaptic potentials in the superficial layers of the cerebral cortex. They can be recorded by placing electrodes on the scalp and connecting them to an amplifier and recording apparatus. The recording, called an **electroencephalogram (EEG)**, is useful in studying normal brain functions such as sleep and consciousness and in diagnosing degenerative brain diseases, metabolic abnormalities, brain tumors, sites of trauma, and so forth. The complete and persistent absence of brain waves in the EEG is often used as a clinical and legal criterion for brain death.

Brain waves are classified as *alpha, beta, theta,* and *delta waves* (fig. 14.24). These are distinguished by differences in amplitude (millivolts) and frequency (cycles per second, cps; or hertz, Hz).

- **Alpha (α) waves** have a frequency of 8 to 13 Hz and are recorded especially in the parieto-occipital area. They are seen when the subject is awake and resting, with eyes closed and mind wandering. They disappear when the subject opens the eyes, receives sensory stimulation, or engages in a mental task such as performing mathematical calculations. They are also absent during deep sleep.

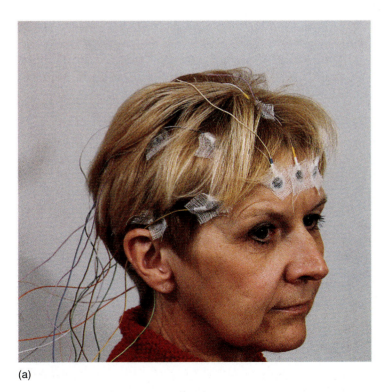

(a)

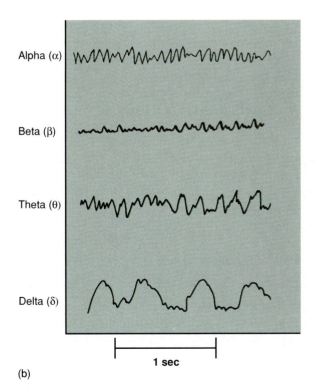

(b)

Figure 14.24 The electroencephalogram (EEG), a record of the electrical activity of the cerebral cortex. (*a*) An electroencephalogram (EEG) is recorded from an array of electrodes on the forehead and scalp. (*b*) Four classes of brain waves are seen in EEGs.

chapter 14 The Central Nervous System **489**

- **Beta (β) waves** have a frequency of 14 to 30 Hz and occur in the frontal to parietal region. They are seen during mental activity and sensory perception.
- **Theta (θ) waves** have a frequency of 4 to 7 Hz. They are normal in children and sleeping adults, but in awake adults they accompany emotional stress and brain disorders.
- **Delta (δ) waves** are high-amplitude "slow waves" with a frequency of less than 3.5 Hz. Infants exhibit delta waves when awake and adults exhibit them in deep sleep. When present in awake adults, they indicate serious brain damage.

Sleep

Sleep is a temporary state of unconsciousness from which a person can be aroused by stimulation. **Coma,** on the other hand, is a state of unconsciousness from which no amount of stimulation arouses a person. We spend about one-third of our lives asleep, yet no one knows exactly what the purpose of sleep is. It seems to be essential for most people because volunteers who undergo prolonged sleep deprivation (over 200 hours in record cases) usually experience hallucinations, paranoia, difficulty in focusing the eyes, and loss of concentration and motivation; interestingly, appetite, pain sensitivity, and sexual drive often increase. Conventional wisdom says we need to sleep for our bodies to rest and recover. This is consistent with the observation that people sleep more deeply after heavy exercise, which would be expected to create a greater need for tissue repair. Also, the onset of sleep stimulates the secretion of growth hormone, which promotes protein synthesis and tissue repair. However, this hypothesis would suggest that animals with comparable metabolic rates need comparable amounts of sleep, and this is not the case. Bats and shrews have similar high metabolic rates, yet bats sleep about 19 hours each day and shrews rarely sleep at all. Moreover, the human brain certainly does not rest during sleep—EEGs show that the brain is highly active at this time; in fact, its metabolic rate is often higher during sleep than in waking hours. Finally, the rest and recovery hypothesis does not explain why we should need to lose consciousness for 8 hours a day. People who are permitted to rest in bed for 8 hours, but not allowed to fall asleep, still feel fatigued the next day.

Comparative studies on animal sleep suggest that its primary function may be to force animals to find a safe place and remain immobile at times when the risks of venturing out override the benefits. This may not seem very relevant to an understanding of human sleep today, but certainly it could have applied to our ancestors, who were at great risk from such predators as hyenas and large cats.

The neural control centers for sleep are in the hypothalamus and brainstem. The upper brainstem has neurons whose activity stimulates the cortex and keeps us awake, and the lower brainstem has neurons whose activity puts us to sleep. The biological clock that regulates the daily rhythm of sleep and waking is apparently in the **suprachiasmatic** (SOO-pra-KY-az-MAT-ic) **nucleus,** located in the anterior hypothalamus above the optic chiasma. Some nerve fibers from the eyes lead to this nucleus and serve to synchronize sleep and other physiological rhythms to the external rhythm of night and day.

We pass through four stages of brain activity in the first 30 to 45 minutes of sleep, during which time the brain waves decline in frequency but increase in amplitude (fig. 14.25a). In **stage 1** the eyes are closed, we begin to relax, thoughts come and go, and a drifting sensation occurs. The EEG is dominated by alpha waves. We awaken easily if stimulated. In **stage 2** the EEG is more irregular, with short bursts of 12 to 14 Hz brain waves called **sleep spindles.** We are less easily aroused from this stage. In **stage 3,** about 20 minutes after stage 1, sleep deepens, the muscles relax, and the vital signs (body temperature, blood pressure, pulse, and respiratory rate) decline. Theta and delta waves appear in the EEG. **Stage 4** is also called **slow-wave sleep (SWS)** because the EEG is dominated by delta waves. The vital signs are at their lowest levels, the muscles are very relaxed, and it is difficult to arouse the sleeper.

About five times a night, a sleeper "backtracks" from stage 4 to stage 1 and enters episodes of **rapid eye movement (REM) sleep.** The eyes flicker about under the eyelids as if watching a movie. Vital signs increase and the brain consumes more oxygen than when the sleeper is awake. In males, REM sleep is nearly always accompanied by penile erection. REM sleep is also called **paradoxical sleep** because the EEG resembles that of the waking state, yet the sleeper is harder to arouse than in any other stage. Most dreams occur during REM sleep, but nightmares are associated more with stages 3 and 4. Except for the muscles of eye movement, the skeletal muscles are inhibited during REM sleep, which prevents us from acting out our dreams. The first bout of REM sleep occurs about 90 minutes after falling asleep. In the second half of the night, bouts of REM sleep become longer and more frequent (fig. 14.25b).

Motor Control

Voluntary muscle contractions are initiated by activity in the **motor association (premotor) area** of the frontal lobes (see fig. 14.28). Neurons here determine the degree and sequence of muscle contraction required for an intended action such as dancing, typing, speaking,

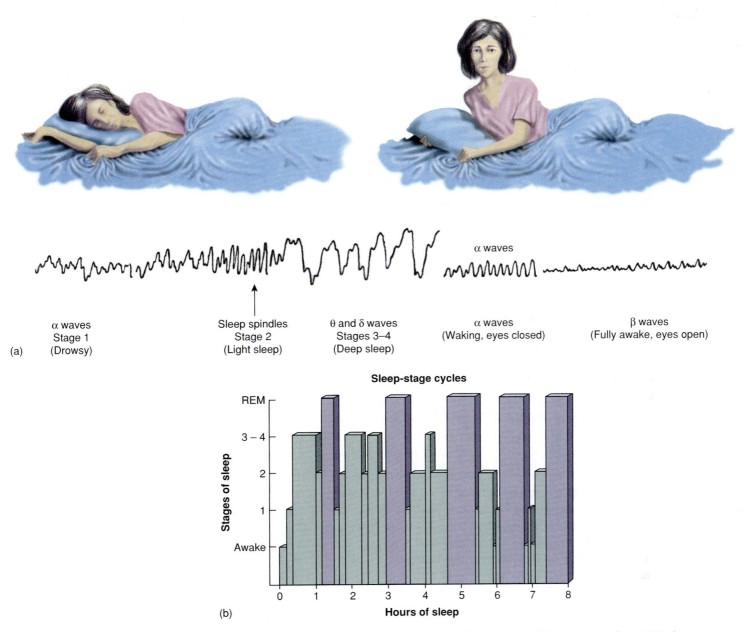

(a)

α waves
Stage 1
(Drowsy)

Sleep spindles
Stage 2
(Light sleep)

θ and δ waves
Stages 3–4
(Deep sleep)

α waves
(Waking, eyes closed)

β waves
(Fully awake, eyes open)

α waves

Sleep-stage cycles

Stages of sleep

REM

3 – 4

2

1

Awake

0   1   2   3   4   5   6   7   8

Hours of sleep

(b)

Figure 14.25 Sleep stages and their relationship to brain activity. (*a*) Correlation of the brain waves with the stages of non-REM sleep. (*b*) Stages of sleep over an 8-hour night in an average young adult. Stages 3 to 4 dominate the first half of the night and REM sleep dominates the second half. Most dreaming occurs during REM sleep.

or playing a musical instrument. The plan of muscle contraction is then transmitted to neurons of the **precentral gyrus (primary motor area)** (fig. 14.26*a*). This is the most posterior gyrus of the frontal lobe; it lies immediately anterior to the central sulcus and extends into the longitudinal fissure. Fibers from this gyrus descend through the pyramids of the medulla oblongata and decussate there. Thus, the right precentral gyrus controls muscles on the left side of the body and vice versa.

The precentral gyrus exhibits **somatotopy**—a point-for-point correspondence between groups of

neurons in the gyrus and muscles on the opposite side of the body. The neurons for toe movements, for example, are deep in the longitudinal fissure on the medial side of the gyrus. Neurons near the summit of the gyrus control the trunk, shoulder, and arm; those in the inferolateral region control the facial muscles. The precentral gyrus is like an upside-down map of the muscles on the opposite side of the body. This map is traditionally diagrammed as a *motor homunculus*[56] (fig. 14.26*b*)

....

56. *hom* = man, person + *unculus* = little

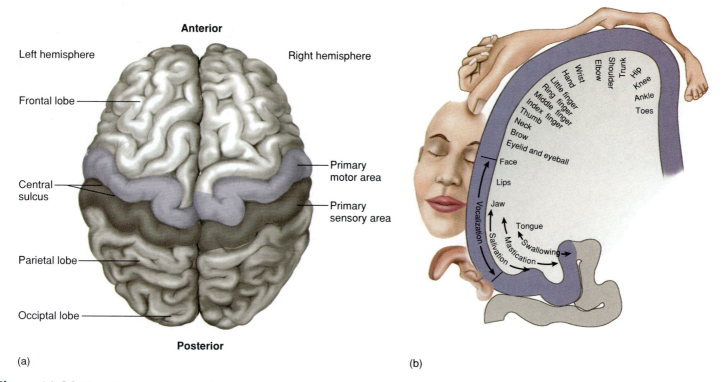

Figure 14.26 The primary motor cortex (precentral gyrus) of the cerebrum. (a) Location, superior view. (b) Motor homunculus, drawn so that body parts are in proportion to the amount of primary motor cortex dedicated to their control. ✗

relating regions of the gyrus to regions of the body. The reason for its bizarre, distorted appearance is that the amount of cerebral tissue devoted to a given body region is proportional to the number of muscles and motor units in that region, not to the size of the region. Areas of fine control, such as the hands, have more muscles, more motor units per muscle, and larger areas of motor cortex to control them than areas such as the trunk and thigh.

Somatic Sensation

Immediately posterior to the central sulcus is the **postcentral gyrus,** which functions as the **primary sensory area** or **somesthetic**[57] (SO-mess-THET-ic) **cortex** (fig. 14.27). Neurons here receive general sensory information pertaining to touch, temperature, pain, pressure, and proprioception. Sensory pathways to the brain decussate in the spinal cord or brainstem and then synapse in the thalamus with neurons whose fibers project to the postcentral gyrus.

Pathways from the lower parts of the body project to the upper parts of the gyrus, and vice versa. Thus, sensations from the foot project to a medial part of the

gyrus deep in the longitudinal fissure, and sensations from the face project to the inferolateral part of this gyrus. Consequently, the postcentral gyrus also exhibits somatotopy, as shown by the *sensory homunculus* in figure 14.27b. Here again, we see that the largest areas of cortex are associated with the most richly innervated parts of the body, especially the hands and face.

Special Senses

The senses of smell, vision, hearing, and equilibrium do not issue fibers to the postcentral gyrus but to other specialized regions of the cerebral cortex. These sensory pathways are described in chapter 16. For now, we will simply state where the sensory signals are ultimately received in the cortex (fig. 14.28):

- visual—posterior region of the occipital lobe;
- auditory—superior margin of the temporal lobe;
- olfactory—medial surface of the temporal lobe and inferior surface of the frontal lobe;
- gustatory—near the lower lateral end of the postcentral gyrus and part of the insula.

Equilibrium receptors of the inner ear project mainly to the cerebellum, but some follow a pathway via the thalamus to a still-uncertain destination in the cerebral cortex.

57. *som* = body + *esthet* = feeling

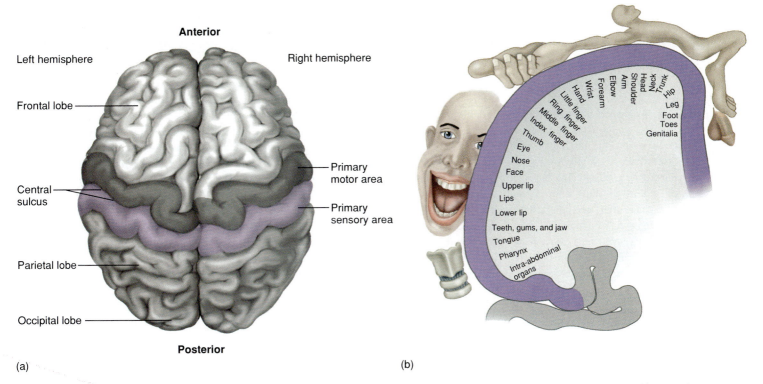

Anterior

Left hemisphere

Right hemisphere

Frontal lobe

Central sulcus

Parietal lobe

Occipital lobe

Posterior

Primary motor area

Primary sensory area

(a)

Neck
Head
Shoulder
Arm
Elbow
Forearm
Wrist
Hand
Little finger
Ring finger
Middle finger
Index finger
Thumb
Eye
Nose
Face
Upper lip
Lips
Lower lip
Teeth, gums, and jaw
Tongue
Pharynx
Intra-abdominal organs

Trunk
Hip
Leg
Foot
Toes
Genitalia

(b)

Figure 14.27 The primary sensory cortex (postcentral gyrus) of the cerebrum. (*a*) Location, superior view. (*b*) Sensory homunculus, drawn so that body parts are in proportion to the amount of cortex dedicated to their sensation. 🔭

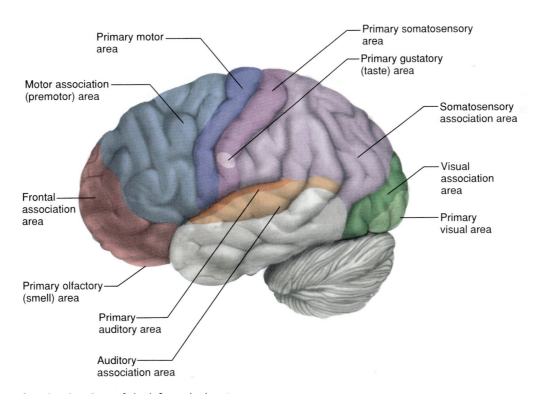

Primary motor area

Motor association (premotor) area

Frontal association area

Primary olfactory (smell) area

Primary auditory area

Auditory association area

Primary somatosensory area

Primary gustatory (taste) area

Somatosensory association area

Visual association area

Primary visual area

Figure 14.28 Some functional regions of the left cerebral cortex.

Accidental but nonfatal destruction of parts of the brain has afforded many clues to the function of various regions. One of the most famous incidents occurred in 1848 to Phineas Gage, a laborer on a railroad construction project in Vermont. Gage was packing blasting powder into a hole with a 3½-foot tamping iron when the powder prematurely exploded. The tamping rod was blown out of the hole and passed through Gage's maxilla, orbit, and frontal lobe of his brain before emerging from his skull near the hairline and landing 50 feet away. Gage went into convulsions but later sat up and conversed with his crewmates as they drove him to a physician in an oxcart. On arrival, he stepped out on his own and told the physician, "Doctor, here is business enough for you." The physician, John Harlow, reported that he could insert his index finger all the way into Gage's wound. Yet 2 months later, Gage was walking around town, carrying on his normal business.

He was not, however, the Phineas Gage people had known. Before the accident, Gage was a competent, responsible, financially prudent man, well liked by his associates. In his 1868 publication on the incident, Harlow said that following the accident, Gage was "fitful, irreverent, indulging at times in the grossest profanity." He became irresponsible, lost his job, and died a vagrant 12 years later.

A 1994 computer analysis of Gage's skull indicated that the brain injury was primarily to the ventromedial region of both frontal lobes. In Gage's time, scientists were reluctant to attribute social behavior and moral judgment to any region of the brain. These functions were strongly tied to issues of religion and ethics and were considered inaccessible to scientific analysis. Based partly on Phineas Gage and other brain injury patients like him, neuroscientists today recognize that planning, moral judgment, and emotional control are among the functions of the prefrontal cortex.

Association Areas

For sensory information to be useful, it must be identified and interpreted with respect to memory and foresight—for example, in associating a name with a face, foreseeing the consequences of our reactions to a stimulus, or planning a motor response. These functions belong to the **association areas**—essentially all areas of cortex except those designated *primary* (fig. 14.28). The following are some of the better-known association areas:

- The **somatosensory (somesthetic) association area** is located in the parietal lobe, immediately posterior to the primary sensory area. It enables you to interpret information about the positions of parts of your body and to determine the shape or texture of an object without having to look at it, for example.
- The **visual association area** is located in the occipital lobe, between the somesthetic association area and the primary visual cortex. It enables you to identify the objects that you see. Recognition of faces resides in the inferior region of the temporal lobes. Lesions here may make it impossible to recognize even family members on sight yet leave intact the ability to name a person upon hearing his or her voice.
- The **auditory association area** is located in the temporal lobe, deep within the lateral sulcus. It enables you to remember the name of a piece of music, to identify a person by his or her voice, and so forth.
- The **frontal association area** is located in the **prefrontal area** anterior to the premotor area discussed earlier. This area is well developed only in primates, especially humans. Many of our cognitive (thought) processes reside here. It is the area concerned with foresight, planning, judgment, temperament, and deciding on an appropriate behavior for a given set of circumstances (see special topic 14.4).

Cerebral Lateralization

The two cerebral hemispheres appear to be identical at a glance, but close examination reveals a number of differences. In left-handed people, for example, the frontal, parietal, and occipital lobes of the right hemisphere are usually narrower than those of the left. In women, the left temporal lobe is longer than the right. The two hemispheres also differ in some of their functions (fig. 14.29). Neither hemisphere is "dominant," but each is specialized for certain tasks.

One hemisphere, usually the left, is called the *categorical hemisphere.* It is specialized for spoken and written language and for the sequential and analytical reasoning employed in such fields as science and mathematics. This hemisphere seems to break information into fragments and analyze it in a linear way. The other hemisphere, usually the right, is called the *representational hemisphere.* It perceives information in a more integrated, unfragmented way. It is the seat of imagination and insight, musical and artistic skills, perception of patterns and spatial relationships, and comparison of sights, sounds, smells, and tastes. Lesions to the inferior parietal lobe of this hemisphere lead to *unilateral inattention*—a tendency to ignore stimuli coming from the contralateral side of the body and failure to care for

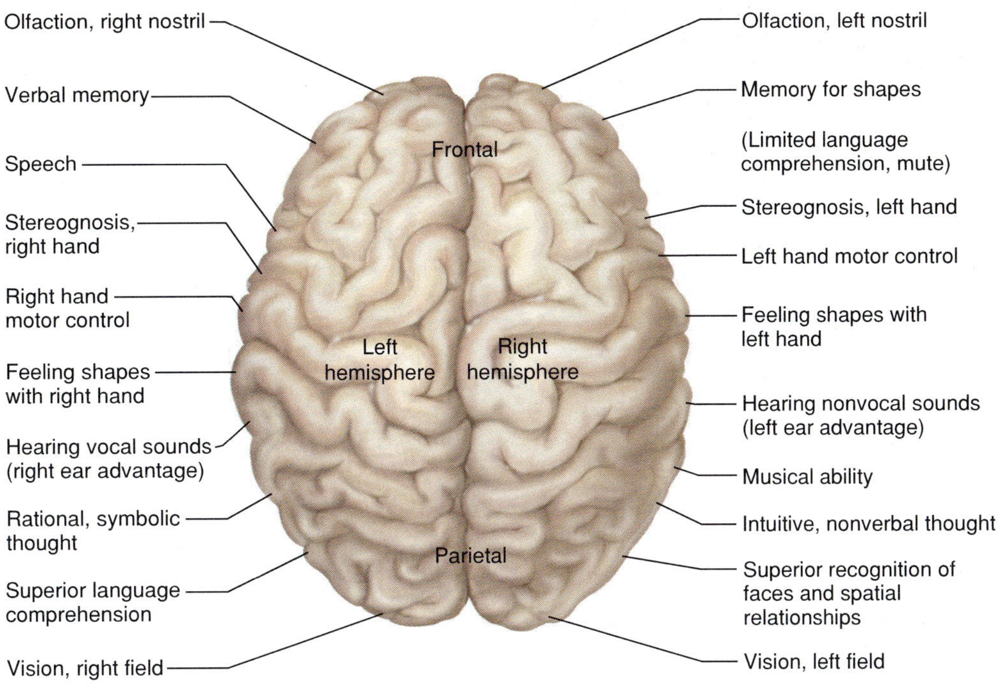
Olfaction, right nostril
Verbal memory
Speech
Stereognosis, right hand
Right hand motor control
Feeling shapes with right hand
Hearing vocal sounds (right ear advantage)
Rational, symbolic thought
Superior language comprehension
Vision, right field
Frontal
Left hemisphere
Right hemisphere
Parietal
Olfaction, left nostril
Memory for shapes
(Limited language comprehension, mute)
Stereognosis, left hand
Left hand motor control
Feeling shapes with left hand
Hearing nonvocal sounds (left ear advantage)
Musical ability
Intuitive, nonverbal thought
Superior recognition of faces and spatial relationships
Vision, left field

Figure 14.29 Lateralization of cerebral functions.

that side. A person with this disorder may shave only one side of the face, dress only half of the body, or read only half of each page.

Cerebral lateralization, the assignment of different tasks to different hemispheres, is correlated with handedness. Ninety-one percent of Americans are right-handed. The left hemisphere is the categorical one in 96% of these people and the right hemisphere in 4%. Among left-handed people, the right hemisphere is categorical in 15%, the left in 70%, and in the remaining 15% neither hemisphere is distinctly specialized. Lateralization develops with age. Brain cancer sometimes requires removal of one cerebral hemisphere and if this is done in young children, the functions of the missing hemisphere may be taken over by the remaining one. Males exhibit more lateralization than females and show more functional loss when one hemisphere is damaged. Damage to the left hemisphere, for example, causes loss of speech three times as often in men as in women. The reason for this difference between the sexes is not yet clear, but it may be related to the corpus callosum. In men, the corpus callosum has a fairly uniform thickness, but in women its posterior portion is thickened by additional commissural fibers, suggesting that women have a more extensive communication between hemispheres.

Language

Language includes several abilities—reading, writing, speaking, and understanding words—assigned to different regions of cerebral cortex (fig. 14.30). The recognition of spoken and written language resides in a region called **Wernicke's**[58] (VERR-nick-ez) **area,** just posterior to the lateral sulcus, usually in the left hemisphere. This is a sensory association area that receives visual and auditory input from the respective regions of primary sensory cortex. The **angular gyrus** just posterior to Wernicke's area processes the words we read into a form that can be converted to speech.

Wernicke's area formulates phrases according to learned rules of grammar and transmits this plan to **Broca's**[59] **area,** located in the inferior part of the prefrontal cortex of the same hemisphere. Broca's area generates a motor program to cause the muscles of the larynx, tongue, cheeks, and lips to produce speech. This program is then transmitted to the primary motor cortex, which initiates the appropriate muscular contractions. PET scans (see chapter essay, p. 499) show that Broca's area becomes active as we prepare to speak.

The emotional aspect of language is controlled by regions in the opposite hemisphere that mirror Wernicke's and Broca's areas. Opposite Broca's area is the **affective language area.** Lesions to this area result in **aprosodia,** a condition in which speech is flat and emotionless, lacking the intonations that modify the meaning of our spoken words. The cortex opposite

58. Karl Wernicke (1848–1905), German neurologist
59. Pierre Paul Broca (1824–80), French surgeon and anthropologist

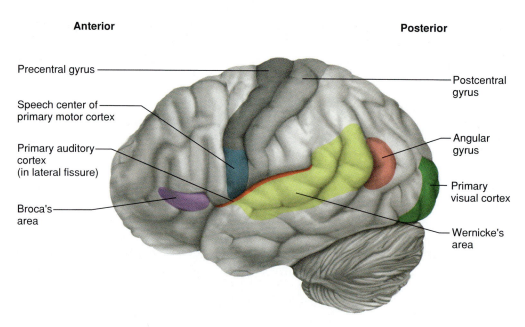

Anterior

Posterior

Precentral gyrus

Speech center of primary motor cortex

Primary auditory cortex (in lateral fissure)

Broca's area

Postcentral gyrus

Angular gyrus

Primary visual cortex

Wernicke's area

Figure 14.30 Cortical centers involved in speech.

Wernicke's area is concerned with recognizing the emotional content of another person's speech.

Aphasia[60] (ah-FAY-zee-uh) is any language defect resulting from lesions of the categorical hemisphere. The many forms of aphasia are difficult to classify. *Nonfluent aphasia,* which is due to a lesion to Broca's area, results in slow speech, difficulty in choosing words, or use of words that only approximate the correct word. For example, a person may say "tssair" when asked to identify a picture of a chair. In extreme cases, the person's entire vocabulary consists of two or three words, sometimes those that were being spoken when a stroke occurred. A lesion to Wernicke's area may result in *fluent aphasia,* in which a person speaks normally, and sometimes excessively, but uses jargon and invented words that make little sense (for example, "choss" for chair). The person also fails to comprehend written and spoken words. In *anomic aphasia,* a person can speak normally and understand speech but cannot identify written words or pictures. Shown a picture of a chair, the person may say, "I know what it is. . . . I have a lot of them," but be unable to name the object.

This represents only a small sample of the complex and puzzling linguistic effects of brain lesions. Other lesions to small areas of cortex can cause impaired mathematical ability, a tendency to write only consonants, or difficulty understanding the second half of each word a person reads. Lesions in the language centers of the representational hemisphere interfere with perception of the emotional content of speech and with a person's ability to understand or tell a joke.

Memory

Memory may be defined as the storage and retrieval of acquired information or skills. At the cellular level, it is a process that creates new pathways of signal transmission or makes transmission across existing synapses easier (see previous chapter). New or facilitated pathways are called *memory traces.*

Although memory has been studied for many years, it is still not well understood. Psychologists and physiologists have multiple theories about memory and different ways of classifying the forms of memory. The classification system and theories presented here are by no means the last word on the subject, but they will give you some idea of how memory is thought to work.

At least two forms of memory can be distinguished by how long they last—short-term memory and long-term memory. **Short-term memory (STM)** lasts from a few seconds to a few hours and is limited to about 7 to 12 bits of information. It lasts long enough for you to dial the telephone number you just looked up, for example, and perhaps to dial it again from memory several minutes later. Information stored in STM is quickly forgotten if we stop reciting it mentally or if we are called upon to remember something new. One possible mechanism of STM is the reverberating circuit described in the preceding chapter. A circuit's output,

60. *a* = without + *phas* = speech

such as one mental repetition of a telephone number, could reactivate neurons earlier in the circuit, causing a repetition of the chain of events that leads to your recollection of the number. Another mechanism of STM could be synaptic potentiation, also discussed in the preceding chapter. Signals traveling a particular pathway can cause the accumulation of calcium ions in the synaptic knobs, for example, resulting in facilitated neurotransmitter release and easier transmission of signals along that pathway.

Long-term memory (LTM) not only lasts longer than STM (sometimes for a lifetime) but is far less limited in the amount of information that can be stored. It allows you to memorize the lines of a play or song or to amass information for an examination. Several circumstances can result in the transfer of short-term memories to LTM. Whether memorizing textbook information or learning to play the guitar, repetition promotes LTM. Another factor is **memory consolidation**—the process of classifying information and associating it with what you already know. Experiences associated with strong emotional states are also more likely to be registered in LTM. So are facts learned when we are alert rather than fatigued.

Not all LTM is permanent, of course; you may no longer remember a telephone number you knew many years ago, and your instructor may not remember your name if you go back to visit years later. Yet distant memories beyond voluntary recall may still reside within the brain. Canadian neurosurgeon Wilder Penfield stimulated areas of the cerebral cortex in patients who were conscious and could describe the resulting sensations to him. This sometimes evoked vivid memories of seemingly trivial events from long ago in the patient's life.

There are two forms of LTM—declarative and procedural memory—that are processed by different mechanisms and parts of the brain. **Declarative memory** is the retention of events and facts that you can put into words, or "declare"—numbers, names, and dates, for example. This type of memory is associated with conscious thought and symbolic language. The **hippocampus,**[61] a part of the temporal lobe and limbic system, is especially important in the retention of facts. Another limbic system component, the **amygdaloid**[62] (ah-MIG-duh-loyd) **nucleus,** is involved in associating pleasant or unpleasant feelings with memories—for example, in remembering a person with affection or a roller coaster ride with fear. The thalamus, hypothalamus, and prefrontal cortex also play a part in declarative memory. Memory is not nearly as localized a function as the sensory and motor processes described earlier. The visual, olfactory, auditory, and other sensory aspects of our memories of past events apparently reside in various areas of association cortex located near the primary sensory cortex for these functions.

The other form of LTM, **procedural memory,** includes skills and habits that are often unconscious, yet not easily forgotten. These include simple habits, skilled motor tasks such as typing or riding a bicycle, and **conditioned reflexes,** which are patterns of behavior we have learned to avoid or repeat depending on past associations with reward and punishment. Such skills and habits seem almost "instinctive," even though they were in fact learned and are not inborn. Procedural memory is stored, at least in part, by the cerebellum and premotor cortex.

The physiological mechanisms of long-term memory are believed to involve structural changes in neurons. Electron microscopic studies of nervous tissue following repetitive stimulation, as would occur in learning a motor skill or memorizing a fact, reveal the following changes: an increased number of presynaptic endings (axon terminals), larger synaptic knobs, increased branching and conductivity of dendrites, and more neuroglia. Chapter 13 describes the roles of nitric oxide, NMDA receptors, and long-term potentiation in synaptic changes that seem to provide a cellular basis for long-term memory.

The complexity of the brain vastly exceeds what we have been able to consider within the confines of one chapter. A brief summary of the major structural components we have studied is presented in table 14.2.

Key Point Review

(23) Suppose you are reading a novel and gradually fall asleep and begin to dream. How would your brain waves change during this sequence of events?

(24) Describe the somatotopy of the primary motor area and primary sensory area.

(25) Describe the locations and functions of the somesthetic, visual, auditory, and frontal association areas.

(26) List the brain centers concerned with understanding and producing speech and describe their locations and functions.

(27) Distinguish between declarative and procedural memory and identify the parts of brain responsible for each.

---

61. *hippocampus* = sea horse, named for its shape
62. *amygdal* = almond + *oid* = resembling

Table 14.2 Summary of the Structures of the Brain and Their Functions

Regions/Components	Functions	Regions/Components	Functions
Rhombencephalon (hindbrain)		**Prosencephalon (forebrain)**	
Myelencephalon		*Diencephalon*	
Medulla oblongata (figs. 14.1 and 14.13)	Transmission pathways between spinal cord and higher brain centers	Thalamus (figs. 14.1, 14.13, and 14.18)	Relay of most signals to cerebral cortex from the rest of the CNS
	Relay of signals between cerebellum and brainstem	Hypothalamus (fig. 14.18)	Relay of signals from limbic system to thalamus
	Coughing, sneezing, hiccuping, swallowing, vomiting		Control of anterior pituitary secretion
	Autonomic regulation of many functions including breathing, blood circulation, digestion, and sweating		Synthesis of hormones released by posterior pituitary
			Hunger and satiety
Metencephalon			Thirst and satiety, urine output, water balance
Pons (fig. 14.13)	Transmission pathways for hearing and equilibrium		Thermoregulation
	Transmission pathways for hearing and equilibrium		Cardiovascular regulation
	Relay of signals between cerebellum and brainstem		Sleep-waking cycle and other physiological rhythms
	Relay of motor signals from cerebrum to cerebellum		Emotional responses and sexual behavior
	Modulation of respiratory rhythm	Epithalamus	Endocrine (pineal) function
	Swallowing		Relay from limbic system to midbrain
	Bladder control	*Telencephalon (Cerebrum)* (figs. 14.20, 14.28, and 14.30)	
Cerebellum (figs. 14.14 and 14.15)	Modulation of muscle contraction	Frontal lobe	Thought
	Muscle tone and posture		Memory
	Coordination of eye and body movements		Foresight, planning, judgment
	Equilibrium		Temperament, mood
	Creating procedural memory (motor skills)		Speech
			Control of skeletal muscles
Mesencephalon (midbrain) (figs. 14.1, 14.13, and 14.16)			Consciousness of smell
	Transmission pathways between cerebrum and lower brainstem	Parietal lobe	Somesthetic senses, taste
	Relay of signals between cerebellum and brainstem		Language comprehension
	Relay of signals to thalamus from other parts of CNS	Occipital lobe	Vision
	Visual attention and tracking	Temporal lobe	Emotional behavior
	Visual and auditory reflexes		Visual recognition
			Hearing
Reticular Formation (in medulla, pons, and midbrain) (fig. 14.17)			Smell
	Includes the nuclei for many of the functions previously listed	Insula	Taste
			Other functions largely unknown
	Modulation of cerebral responses to stimuli; arousal and habituation	Basal nuclei (fig. 14.22)	Modulation of voluntary muscular action
	Sleep and waking		Muscle tone and posture
	Modulation of pain signals in the spinal cord		Cognition
		Limbic system (fig. 14.23)	Emotions
			Sexual behavior
			Motivation
			Senses of reward, punishment
			Learning and memory

Images of the Mind

Enclosed as it is in the cranium, there is no easy way to observe a living brain directly. This has long frustrated neurobiologists, who once had to content themselves with glimpses of brain function afforded by electroencephalograms, patients with brain lesions, and patients who remain awake and conversant during brain surgery and who have consented to experimentation while the brain is exposed. New methods of biomedical imaging, however, are beginning to yield dramatic images of brain function. Two of these—*positron emission tomography* (*PET*) and *magnetic resonance imaging* (*MRI*)—were explained in the essay at the end of chapter 1. In brain imaging, both techniques rely on transient increases in blood flow to parts of the brain called into action to perform specific tasks. If these blood flow changes can be detected, neuroscientists can identify which parts of the brain are involved in specific tasks.

A **PET scan** uses water that has been synthesized with a radioisotope of oxygen, ^{15}O, and injected into an arm vein. There is little or no risk to the subject because ^{15}O is used in very low doses and decays rapidly to a nonradioactive isotope; it is virtually gone in 10 minutes. During its decay, however, ^{15}O emits positrons (positively charged particles with the mass of an electron) that immediately collide with electrons. A colliding positron and electron annihilate each other and emit a pair of γ rays traveling in opposite directions. A detector senses the γ rays and a computer pinpoints their site of origin in the tissue within a few millimeters. The regions of greatest brain activity exhibit increased blood circulation, increased delivery of labeled water, and thus the highest rate of positron emission.

Shortly after injection, the subject is given a specific task to perform. For example, the examiner may say the word *car* and ask the subject to repeat this word and then give a verb related to it, such as *drive.* A PET image of the brain is made in a *control state* before the task is begun, and another image is made in the *task state* while the subject performs the task. Neither image is very revealing by itself, but the computer subtracts the control-state data from the task-state data and presents a color-coded image of the difference. To compensate for chance events and individual variation, the computer also produces an image that is either averaged from several trials with one person or from trials with several different people (fig. E.1).

—continued

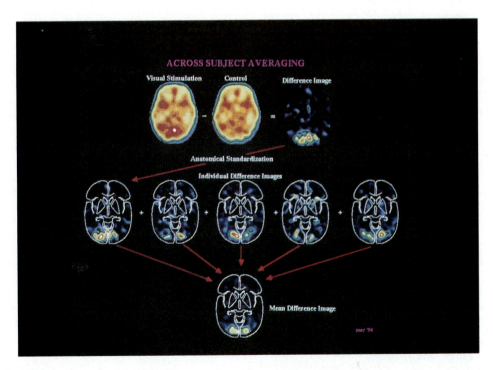

Figure E.1 Positron emission tomography (PET) scans of the brain made during the performance of a visual task. In image subtraction (*top row*), a computer subtracts control-state data from task-state data to produce an image of the difference, showing the region of the brain most active during performance of the designated task. To reduce the effects of random fluctuations and statistical variation among subjects, the computer can average images from several subjects or from multiple scans of the same subject and produce a composite mean difference image (*bottom*).

In such averaged images, the busiest areas of the brain seem to "light up" from moment to moment as the task is performed. This identifies the regions used for various stages of the task, such as seeing the word *car,* interpreting it, saying it, thinking of a verb to go with it, planning to speak the word *drive,* and actually saying it. Among other things, such experiments demonstrate that Broca's and Wernicke's areas are not involved in simply repeating words; they are active, however, when a subject must evaluate a word and choose an appropriate response—that is, they function in formulating the new word the subject is going to say. They also show that different neuronal pools take over a function as we practice and become more proficient at a task.

Magnetic resonance imaging (MRI) depends on a surprising new finding about neuronal metabolism. Until recently, neurons were thought to employ only aerobic respiration to metabolize glucose. Now it is known that when they are stimulated to heightened activity, they employ anaerobic fermentation. They receive an increased blood flow but do not increase their oxygen consumption. The *supply* of oxygen increases during brain activity, but the *demand* for it does not.

Thus, the blood leaving a region of highly active brain tissue contains more oxygen than the blood leaving less active regions. Since the magnetic properties of hemoglobin depend on how much oxygen is bound to it, changes in brain circulation can be detected by MRI. MRI is more precise than PET, pinpointing regions of brain activity with a precision of 1 to 2 mm. MRI also has the advantage of requiring no injected substances and no exposure to radioisotopes.

PET and MRI scanning have enhanced our knowledge of neurobiology by identifying shifting patterns of brain activity associated with attention and consciousness, sensory perception, memory, emotion, motor control, reading, speaking, musical judgment, planning a chess strategy, and so forth. In addition to their contribution to basic neuroscience, these techniques show promise as an aid to neurosurgery and psychopharmacology. They are also enhancing our understanding of brain dysfunctions such as depression, schizophrenia, and attention deficit disorders. We are on the threshold of an exciting era in the safe visualization of normal brain function, virtually producing pictures of the mind at work.▲

Chapter Review

Study Outline

Characteristics of the Central Nervous System (pp. 463–467)
1. Major landmarks
   a. Cerebrum
      - Cerebral hemispheres
      - Gyri, sulci, and fissures
      - Corpus callosum
   b. Cerebellum
   c. Brainstem
   d. Gray and white matter
   e. Ventricles and central canal
2. Embryonic development
   a. Neural plate, groove, and folds
   b. Neural tube
   c. Neural crest
   d. Prosencephalon (forebrain)
      - Telencephalon
      - Diencephalon
   e. Mesencephalon (midbrain)
   f. Rhombencephalon (hindbrain)
      - Metencephalon
      - Myelencephalon

Meninges, Cerebrospinal Fluid, Ventricles, and Blood Supply (pp. 467–471)
1. Meninges
   a. Dura mater
      - Periosteal layer
      - Meningeal layer

- Dural sinuses
   b. Meningeal specializations
      - Falx cerebri
      - Tentorium cerebelli
      - Diaphragma sellae
   c. Variations in spinal meninges
      - Dural sheath
      - Epidural space
   d. Arachnoid membrane
   e. Pia mater
2. Cerebrospinal fluid
   a. Functions
   b. Choroid plexuses
   c. Blood-CSF barrier
3. Ventricles and CSF circulation
4. Blood supply and the blood-brain barrier
   a. Rate of flow
   b. Blood-brain barrier (BBB)
   c. Circumventricular organs

The Spinal Cord (pp. 471–477)
1. Functions
   a. Conduction
   b. Locomotion
   c. Reflex control
2. Gross anatomy
   a. Diameter and length

   b. Four regions
   c. Segmental arrangement
   d. Cervical and lumbar enlargements
   e. Conus medullaris and filum terminale
   f. Cauda equina
3. Cross-sectional anatomy
   a. Relationship to vertebrae and meninges
   b. Anterior median fissure and posterior median sulcus
   c. Gray matter
      - Dorsal and ventral horns
      - Gray commissure and central canal
      - Lateral horns in thoracic region
   d. White matter
      - Columns (funiculi)
      - Tracts (fasciculi)
      - Decussation of nerve fibers
4. Spinal tracts
   a. Ascending tracts
      - Fasciculus gracilus
      - Fasciculus cuneatus
      - Dorsal and ventral spinocerebellar
      - Spinothalamic

Testing Your Recall   Answers in Appendix C

1. Which of these structures is *not* part of the brainstem, as defined in this chapter?
   a. thalamus
   b. hypothalamus
   c. pons
   d. cerebellum
   e. medulla oblongata

2. Locomotion is controlled partly by central pattern generators located in
   a. the basal nuclei of the cerebrum.
   b. the primary motor cortex.
   c. the cerebellum.
   d. the gray matter of the spinal cord.
   e. the white matter of the spinal cord.

3. The ventricles of the brain are most comparable to the _____ of the spinal cord in structure and function.
   a. central canal
   b. epidural space
   c. dorsal horns
   d. funiculi
   e. fascicles

4. Which secondary vesicle of the embryonic brain gives rise to the medulla oblongata?
   a. prosencephalon
   b. myelencephalon
   c. mesencephalon
   d. telencephalon
   e. rhombencephalon

5. The reticular formation is involved in all of the following functions *except*
   a. stretch reflexes.
   b. regulation of respiratory rate.
   c. short-term memory.
   d. alertness.
   e. stimulus filtering.

6. The vertebral canal below level L2 is occupied by a bundle of spinal nerve roots called
   a. the filum terminale.
   b. the descending tracts.
   c. the fasciculus gracilis.
   d. the conus medullaris.
   e. the cauda equina.

7. Sneezing, vomiting, and sweating are functions controlled by
   a. the primary motor cortex.
   b. the hypothalamus.
   c. the medulla oblongata.
   d. the pons.
   e. the thalamus.

8. The _____ uses input from proprioceptors and some of the special senses to improve motor coordination.
   a. cerebellum
   b. primary motor area
   c. premotor area
   d. vasomotor center
   e. postcentral gyrus

9. All sensory information passes through the thalamus except the sense of
   a. vision.
   b. proprioception.
   c. smell.
   d. hearing.
   e. touch.

10. The sense of hearing is associated with
    a. the insula.
    b. the frontal lobe.
    c. the parietal lobe.
    d. the occipital lobe.
    e. the temporal lobe.

11. The right and left cerebral hemispheres are connected by a thick C-shaped bundle of fibers called the _____.

12. The brain has four major chambers called _____, which are filled with cerebrospinal fluid.

13. The white matter of the cerebellum is arranged in a branching pattern called the _____.

14. The crossing of CNS pathways from the right side of the body to the left, or vice versa, is called _____.

15. Each ventricle contains a _____ that secretes cerebrospinal fluid.

16. The primary motor area of the cerebrum is the _____ gyrus of the frontal lobe.

17. The sense of reward and punishment resides in a ring of structures called the _____ system.

18. Areas of cerebral cortex that identify or interpret sensory input are called _____.

19. Linear, verbal, and analytical thinking occur in the _____ hemisphere of the cerebrum, which is on the left in most people.

20. The motor pattern for speech is generated in an area of cortex called _____ and then transmitted to the primary motor cortex to be executed.

Testing Your Comprehension

Answers in *Study Guide*

1. Explain what the blood-brain barrier is and why it presents a problem for clinical medicine.

2. How would a lesion in the cerebellum affect skeletal muscular function differently than a lesion in the basal nuclei?

3. You are driving at night and suddenly hear a car horn and see headlights in your right peripheral vision. What is your immediate response? What structures of the brain mediate this response?

4. A person can survive destruction of an entire cerebral hemisphere but cannot survive destruction of the hypothalamus, which is a much smaller mass of tissue. Explain this difference, and describe some ways in which cerebral hemispheric destruction could affect the quality of life.

5. Contrast the motor effects of three types of CNS injuries: (a) a stroke that destroys tissue of the left precentral gyrus, (b) a diving accident that severs the cervical spinal cord, and (c) a shrapnel wound that shatters the lumbar vertebrae and severs most nerves of the cauda equina.

Web Site Link

For a listing of the most current web sites related to this chapter, please visit the Saladin homepage at:

http://www.mhhe.com/sciencemath/biology/saladin/

[The Peripheral Nervous System and Reflexes

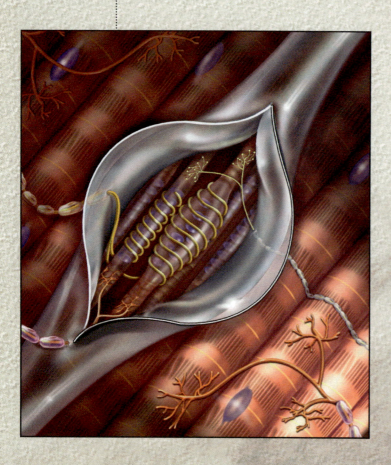

Special Topics

Brushing up

To understand this chapter, it is essential that you understand or brush up on the following concepts:

- Foramina of the skull (p. 257)
- Function of antagonistic muscles (p. 332)
- Names and locations of skeletal muscles (pp. 333–373)
- Innervation of smooth muscle (pp. 419–420)
- Functional divisions of the peripheral nervous system (p. 428)
- Myelinated and unmyelinated nerve fibers (pp. 434–435)
- Neurotransmitters and synaptic transmission (pp. 446–451)
- Parallel after-discharge circuits (p. 457)

The homeostatic controls exerted by the central nervous system require fast and efficient input and output routes to maintain a continuous flow of information. These are provided by the peripheral nervous system (PNS), composed of nerves and ganglia. If the entire PNS were suddenly disabled, the most immediate effects would be complete paralysis and loss of sensation from the outside world. But this state of affairs would not last for long, for homeostatic controls throughout the body would break down and death would soon ensue.

Nerves, Nerve Fibers, and Ganglia

▼Objectives

When you have completed this section, you should be able to

- describe the structure of a nerve;
- explain how nerves and nerve fibers are classified; and
- describe the structure of a ganglion and the relationship of ganglia to peripheral nerves.

Anatomy of a Nerve

It is important to make the distinction between a nerve and a nerve fiber. A nerve fiber is the axon of a single neuron. A **nerve** is an organ composed of multiple nerve fibers bound together by sheaths of connective tissue (fig. 15.1). If a nerve fiber is analogous to a wire carrying an electrical current in one direction only, a nerve is analogous to an electrical cable composed of thousands of wires carrying currents in both directions. A nerve contains anywhere from a few nerve fibers to more than a million. Nerves usually have a pearly white color; they often look like frayed string as they divide into smaller and smaller branches.

All nerve fibers of the peripheral nervous system are ensheathed in Schwann cells, which form a *neurilemma* around the axon (see chapter 13). In myelinated fibers, the Schwann cell wraps around repeatedly to deposit layers of its membrane (*myelin*) between the neurilemma and axon. External to the

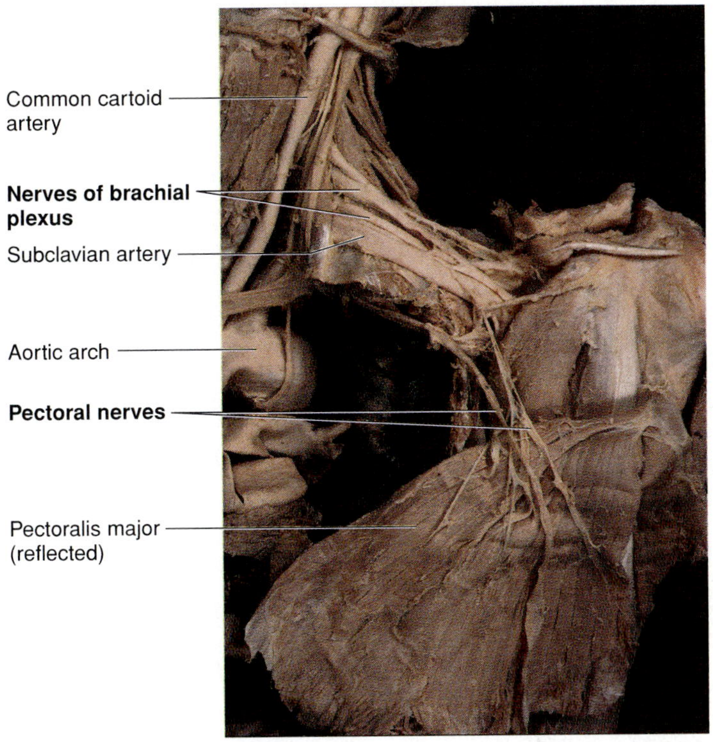

Figure 15.1 Nerves of the brachial and pectoral regions, anterior view. The pectoralis major has been cut and reflected (folded) downward.

neurilemma, each fiber is surrounded by a basement membrane and a sparse connective tissue **endoneurium.** Nerve fibers have a high metabolic rate and need a good supply of blood-borne glucose and oxygen. The endoneurium provides a way for blood capillaries to reach each nerve fiber.

In large nerves, there is usually a mixture of myelinated and unmyelinated fibers gathered together in bundles called **nerve fascicles** (fig. 15.2). Each fascicle is wrapped in a fibrous sheath called the **perineurium.** Finally, the entire nerve is enclosed in a fibrous **epineurium.** In nerves large enough to be subdivided into fascicles, extensions of the epineurium penetrate between them. Blood vessels and adipocytes are common in these ingrowths of the epineurium.

Think About It

How does the structure of a nerve compare to that of a skeletal muscle (see chapter 11)? Which of the descriptive terms for nerves have similar counterparts in muscle histology?

Functional Classes of Nerve Fibers and Nerves

As shown in table 15.1, there are several criteria for classifying nerve fibers: the direction in which they

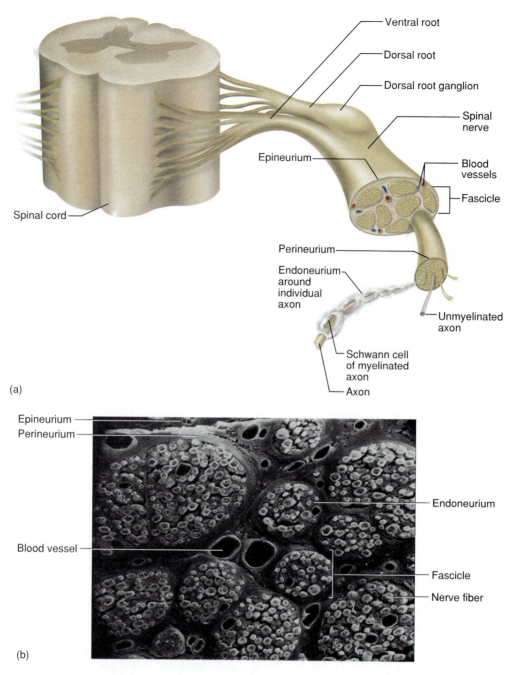

(a)

(b)

Figure 15.2 (*a*) The structure of a nerve. (*b*) Cross section of a nerve (SEM).

From Richard E. Kessel and Randy H. Kardon, *Tissues and Organs: A Text-Atlas of Scanning Electron Microscopy*, 1979, W. H. Freeman and Company

Table 15.1	Classification of Nerve Fibers in the PNS
Class	**Description**
Afferent fibers	Carry sensory signals from receptors to the CNS
Efferent fibers	Carry motor signals from the CNS to effectors
Somatic fibers	Innervate skin, skeletal muscles, bones, and joints
Visceral fibers	Innervate blood vessels, glands, and viscera
General fibers	Innervate widespread organs such as muscles, skin, glands, viscera, and blood vessels
Special fibers	Innervate more localized organs in the head, including the eyes, ears, olfactory and taste receptors, and muscles of chewing, swallowing, and facial expression

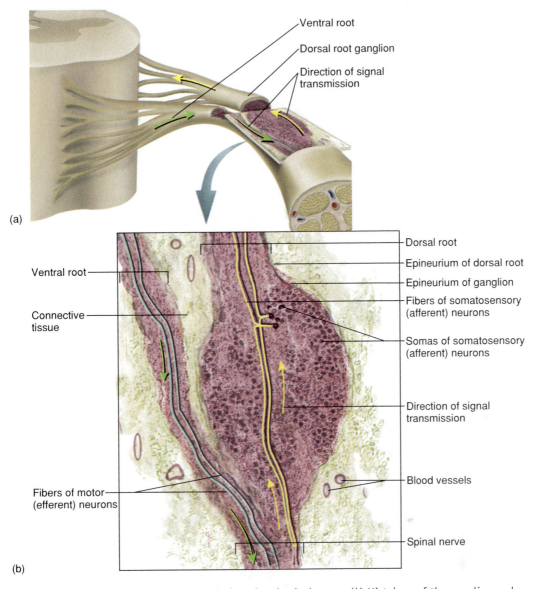

(a)

(b)

Figure 15.3 (*a*) A dorsal root ganglion in relation to the spinal cord and spinal nerve. (*b*) Histology of the ganglion and conduction pathways of spinal nerve fibers. ✗

transmit signals (**afferent** and **efferent fibers**), the types of organs they innervate (**somatic** and **visceral fibers**), and the widespread or localized distribution of the innervated organs (**general** and **special fibers**). General sensory neurons are also called **somatosensory neurons,** as opposed to the special senses localized in sense organs of the head. **Mixed nerves** contain a mixture of sensory (afferent) fibers and motor (efferent) fibers and therefore transmit signals in two directions. Most nerves are mixed. Purely **sensory nerves** are less common; they include the optic and olfactory nerves. Nerves with only efferent fibers are called **motor nerves.** Many nerves often listed as motor are actually mixed because they carry sensory feedback (proprioception) from the muscles they innervate.

Ganglia

A *ganglion*[1] (GANG-lee-un; plural, *ganglia*) is a cluster of neuron cell bodies outside the CNS (fig. 15.3). It forms a knotlike swelling in a nerve and is covered by an epineurium continuous with that of the nerve.

···· **Key Point Review** ····

1 Distinguish between a nerve and a nerve fiber.

2 Describe the relationship of the endoneurium, perineurium, and epineurium to the nerve fibers of the PNS.

3 Distinguish between somatic and visceral nerve fibers and between general and special nerve fibers.

4 Describe the structure of a ganglion.

1. *gangli* = knot in a string

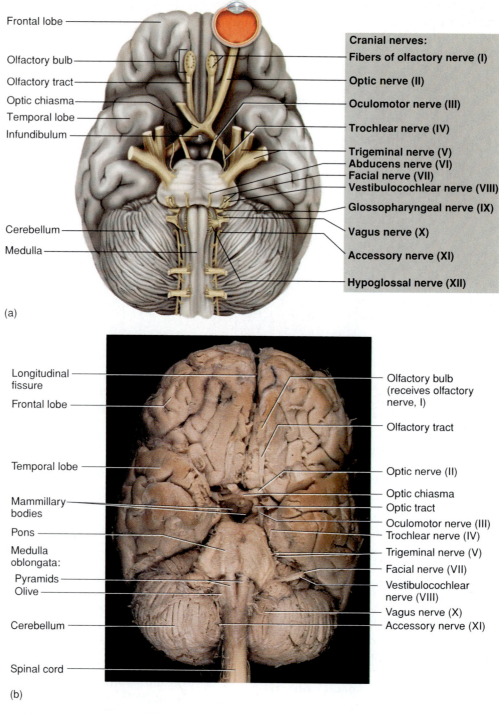

Frontal lobe

Olfactory bulb

Olfactory tract

Optic chiasma

Temporal lobe

Infundibulum

Cerebellum

Medulla

(a)

Cranial nerves:

Fibers of olfactory nerve (I)

Optic nerve (II)

Oculomotor nerve (III)

Trochlear nerve (IV)

Trigeminal nerve (V)
Abducens nerve (VI)
Facial nerve (VII)
Vestibulocochlear nerve (VIII)

Glossopharyngeal nerve (IX)

Vagus nerve (X)

Accessory nerve (XI)

Hypoglossal nerve (XII)

Longitudinal fissure

Frontal lobe

Temporal lobe

Mammillary bodies

Pons

Medulla oblongata:

Pyramids
Olive

Cerebellum

Spinal cord

(b)

Olfactory bulb (receives olfactory nerve, I)

Olfactory tract

Optic nerve (II)

Optic chiasma
Optic tract

Oculomotor nerve (III)
Trochlear nerve (IV)

Trigeminal nerve (V)

Facial nerve (VII)

Vestibulocochlear nerve (VIII)

Vagus nerve (X)
Accessory nerve (XI)

Figure 15.4 (*a*) The cranial nerves. (*b*) Base of the brain showing most of the cranial nerves.

The Cranial Nerves

▼Objectives

When you have completed this section, you should be able to
• list the cranial nerves by name and number;
• identify where each cranial nerve originates and terminates; and
• state the functions of each cranial nerve.

Cranial nerves emerge primarily from the base of the brain, exit the cranium through its foramina, and lead to muscles and sense organs primarily in the head and neck (fig. 15.4). There are 12 pairs of cranial nerves, numbered I to XII starting with the most anterior. Each nerve also has a descriptive name (see special topic 15.1). Cranial nerves I and II are purely sensory, whereas the others are mixed nerves. Some of the mixed nerves are predominantly motor, however, with sensory functions limited to proprioception from the muscles they innervate.

The motor fibers of the cranial nerves arise from brainstem nuclei and lead to glands and muscles. The sensory fibers arise from receptors located mainly in the head and neck and carry impulses to the brain. All of these except the olfactory fibers terminate in the thalamus, where they synapse with other neurons that relay the signal to the cerebral cortex. For now we will not trace these sensory fibers any farther than the thalamus; their pathways are discussed more fully in chapter 16.

Each of the cranial nerves is illustrated and described in table 15.2.

Special Topic 15.1 — A Mnemonic for the Cranial Nerves

Generations of biology and medical students have relied on numerous mnemonic (memory-aiding) phrases and ditties—ranging from the sublimely silly to the unprintably ribald—to help them remember the cranial nerves and other anatomy. An old classic began, "On Old Olympus' towering tops. . . . ," with the first letter of each word matching the first letter of each cranial nerve (olfactory, optic, oculomotor, etc.). Some cranial nerves have changed names, however, since that passage was devised. One of the author's students[2] devised a new and better phrase that can remind you of the first two to four letters of most cranial nerves:

Old	**ol**factory (I)
Opie	**op**tic (II)
occasionally	**oc**ulomotor (III)
tries	**tr**ochlear (IV)
trigonometry	**trig**eminal (V)
and	**ab**ducens (VI)
feels	**f**acial (VII)
very	**ve**stibulocochlear (VIII)
gloomy,	**glo**ssopharyngeal (IX)
vague,	**vagu**s (X)
and	**a**ccessory (XI)
hypoactive.	**hypo**glossal (XII)

2. Courtesy of Ms. Marti Haykin

Table 15.2 The Cranial Nerves

Origins of proprioceptive fibers are not tabulated; they are the muscles innervated by the motor fibers. Under the listing "composition," *sensory* refers to sensory functions other than proprioception.

I. Olfactory Nerve

Composition: Sensory

Function: Smell

Origin: Olfactory mucosa in nasal cavity

Course: Fascicles pass through cribriform plate of ethmoid bone; terminate in olfactory bulbs beneath frontal lobes of brain

Effects of damage: Impaired sense of smell

Clinical test: Ask subject to identify aromatic substances such as vanilla, coffee, tobacco, clove oil, and soap; test each nostril separately

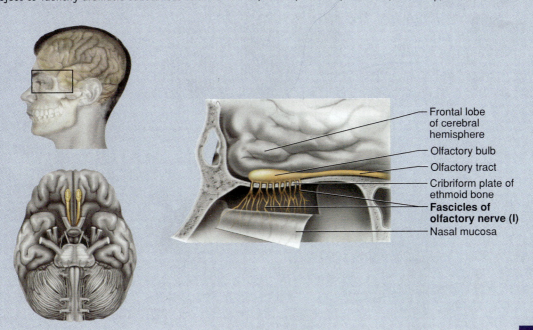

Frontal lobe of cerebral hemisphere
Olfactory bulb
Olfactory tract
Cribriform plate of ethmoid bone
Fascicles of olfactory nerve (I)
Nasal mucosa

continued next page

Table 15.2 **Continued**

II. Optic Nerve

Composition: Sensory

Function: Vision

Origin: Retina

Course: Fibers pass through optic foramen; form X-shaped optic chiasma below diencephalon and continue via optic tracts, terminating mainly in thalamus

Effects of damage: Blindness in part or all of the visual field

Clinical test: Inspect retina with ophthalmoscope; test peripheral vision and visual acuity

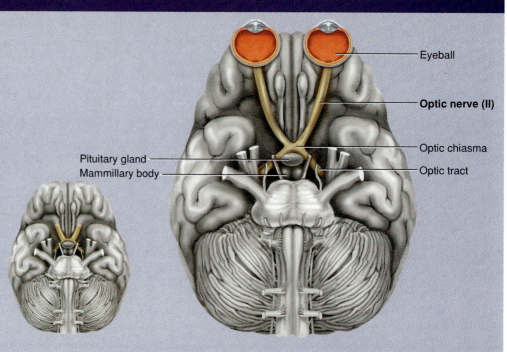

Eyeball

Optic nerve (II)

Optic chiasma

Optic tract

Pituitary gland

Mammillary body

III. Oculomotor (OC-you-lo-MO-tur) Nerve

Composition: Motor and proprioceptive

Function: Eye movements, opening of eyelid, constriction of pupil, focusing

Origin: Ventral midbrain

Course: Passes through superior orbital fissure; divides there into two branches.

Superior branch leads to superior rectus muscle, which turns eye upward, and to levator palpebrae superioris muscle, which raises upper eyelid. *Inferior branch* leads to medial rectus, inferior rectus, and inferior oblique muscles, which turn eye medially, turn eye downward, and rotate eyeball, respectively; also contains parasympathetic fibers that enter rear of eyeball and lead to constrictor muscle of iris and to ciliary muscle, which focuses lens.

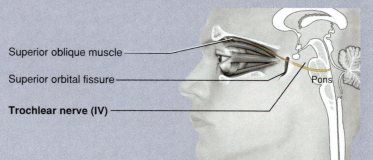

Superior branch

Inferior branch

Levator palpebrae muscle (cut)

Superior rectus muscle

Medial rectus muscle

Ciliary ganglion

Inferior oblique muscle

Inferior rectus muscle

Parasympathetic motor fibers

Superior orbital fissure

Pons

Oculomotor nerve (III)

Effects of damage: Drooping eyelid, dilated pupil, inability to move eye in certain directions, tendency of eye to rotate laterally at rest, double vision, and difficulty focusing

Clinical test: Examine size and shape of pupils; look for differences between right and left pupil; test constriction of pupils in response to light and ability to follow moving objects up, down, side to side, and diagonally and to look cross-eyed at an object as it approaches bridge of nose

IV. Trochlear (TROCK-lee-ur) Nerve

Composition: Motor and proprioceptive

Function: Inferior and lateral movement of eyeball

Origin: Dorsal midbrain

Course: Travels around to ventral midbrain; exits via superior orbital fissure and leads to superior oblique muscle

Effects of damage: Double vision and inability to rotate eye inferolaterally

Clinical test: Same as for cranial nerve III

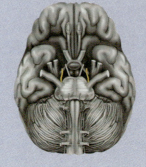

Superior oblique muscle

Superior orbital fissure

Pons

Trochlear nerve (IV)

Table 15.2   Continued

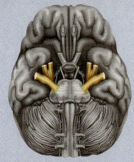

Infraorbital nerve
Superior alveolar nerves
Lingual nerve
Inferior alveolar nerve

Superior orbital fissure
Ophthalmic division (V₁)
Trigeminal ganglion
Trigeminal nerve (V)
Mandibular division (V₃)
Foramen ovale
Maxillary division (V₂)
Foramen rotundum
Anterior trunk to chewing muscles

V. Trigeminal (tri-JEM-ih-nul) Nerve ⅄

Composition: Predominantly sensory, some motor and proprioceptive fibers. Largest of the cranial nerves; consists of three divisions designated V_1 to V_3. V_1 and V_2 are purely sensory nerves, whereas V_3 is mixed.

V_1, Ophthalmic Division

Function: Main sensory nerve of upper face (touch, temperature, pain)

Origin: Skin of anterior scalp, forehead, upper eyelid, surface of eyeball, lacrimal (tear) gland, side of nose, superior nasal mucosa, frontal and ethmoid sinuses

Course: Enters cranium via superior orbital fissure; terminates in pons

Effects of damage: Loss of sensation (but see special topic 15.2)

Clinical test: Test corneal reflex—blinking in response to light touch to eyeball

V_2, Maxillary Division

Function: Same sensations as V_1 from different regions of face

Origin: Lateral forehead, temporal region, palate, upper teeth and gums, upper lip, skin and mucosa of nose and cheek, and lower eyelid

Course: Enters cranium via infraorbital foramen; terminates in pons

Effects of damage: Loss of sensation

Clinical test: Test sensations of touch, pain, and temperature over face with light touch, pinpricks, and hot and cold objects

V_3, Mandibular Division

Function: Same sensations as V_1 from different regions of face; mastication

Origin: Sensory fibers arise from anterior two-thirds of tongue (but not taste buds), lower teeth and gums, floor of mouth, chin, temporal area of scalp, external ear, and dura mater

Course: Passes through foramen ovale. Motor fibers lead to anterior belly of digastric, masseter, temporalis, mylohyoid, medial and lateral pterygoids, and tensor tympani of middle ear. Sensory fibers terminate in pons.

Effects of damage: Loss of sensation; impaired chewing

Clinical test: Assess motor functions by palpating masseter and temporalis muscles while subject clenches teeth; test ability of subject to move mandible from side to side and to open mouth against resistance

Note: Dentists commonly anesthetize this division in order to work on teeth of the mandible; anesthetic is injected posterior to the molars

Distribution of sensory fibers of each division

V_1
V_2
V_3

Inset shows motor branches of the mandibular division (V_3)

Temporalis muscle
Lateral pterygoid muscle
Medial pterygoid muscle
Masseter muscle
Anterior belly of digastric muscle

VI. Abducens (ab-DOO-senz) Nerve

Composition: Motor and proprioceptive

Function: Lateral movement (abduction) of eye

Origin: Inferior pons

Course: Passes through superior orbital fissure to lateral rectus muscle

Effects of damage: Inability to rotate eye laterally; at rest, eye rotates medially because of action of antagonistic muscles

Clinical test: Test with cranial nerve III, paying special attention to lateral eye movements

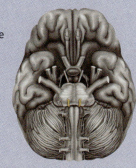

Lateral rectus muscle
Superior orbital fissure
Abducens nerve (VI)
Pons

continued next page

Table 15.2   Continued

VII. Facial Nerve

Composition: Motor, proprioceptive, and sensory

Function: Major motor nerve of facial expression; autonomic control of tear glands, nasal and palatine glands, submandibular and sublingual salivary glands; sense of taste

Origin: Motor fibers arise from lower pons. Sensory fibers arise from taste buds on anterior two-thirds of tongue.

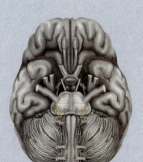

Course: Gustatory (taste) fibers lead to geniculate ganglion and then pons. *Motor fibers* enter temporal bone via internal acoustic meatus, run through middle-ear cavity, and emerge by way of stylomastoid foramen near parotid salivary gland anterior to ear; motor division then gives rise to five major branches called the *temporal, zygomatic, buccal, mandibular,* and *cervical branches;* motor fibers lead to posterior belly of digastric, stapedius muscle of middle ear, and muscles of facial expression; autonomic fibers lead to submandibular and sublingual salivary glands.

Effects of damage: Inability to control muscles on affected side of face; sagging resulting from loss of muscle tonus (see special topic 15.2); distorted sense of taste, especially for sweets

Clinical test: Test right and left sides of anterior two-thirds of tongue using substances such as sugar, salt, vinegar (sour), and quinine (bitter); test response of tear glands using ammonia fumes; test motor functions by asking subject to close eyes, smile, whistle, frown, raise eyebrows, etc.

Parasympathetic efferent and sensory afferent fibers

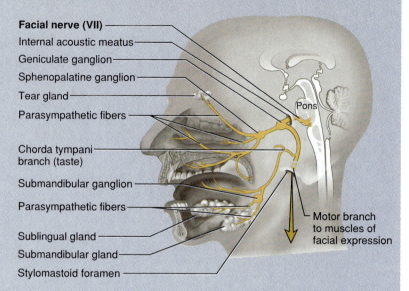

Somatic motor fibers to facial and scalp muscles

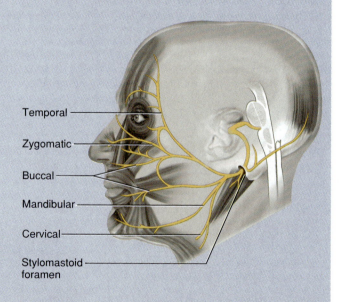

A way to remember the distribution of the five major branches of the facial nerve

Table 15.2   Continued

VIII. Vestibulocochlear (vess-TIB-you-lo-COC-lee-ur) Nerve (formerly called *auditory*)

Composition: Predominantly sensory; some motor fibers

Function: Hearing and equilibrium

Origin: Sensory fibers originate in vestibular and spiral ganglia of inner ear. Motor fibers originate in superior olivary nucleus of medulla.

Course: Sensory fibers from vestibule and cochlea merge and exit inner ear via internal acoustic meatus, terminating near junction of pons and medulla oblongata. Motor fibers travel through cochlear nerve and terminate on or near bases of sensory cells in cochlea.

Effects of damage: Damage to cochlear branch causes nerve deafness; damage to vestibular branch causes dizziness, nausea, vomiting, loss of balance, and nystagmus (involuntary oscillation of the eyes from side to side).

Clinical test: Test subject's balance and ability to walk a straight line (similar to tests for intoxication); test hearing

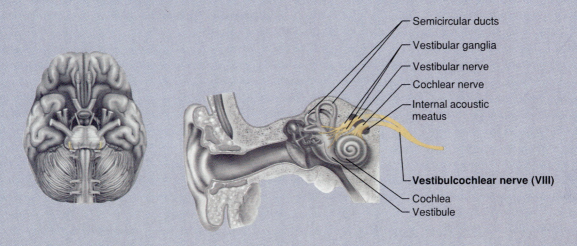

IX. Glossopharyngeal (GLOSS-oh-fah-RIN-jee-ul) Nerve

Composition: Motor, proprioceptive, and sensory

Function: Swallowing, salivation, gag reflex; regulation of blood pressure and respiration; sensations of touch, pressure, taste, and pain from tongue and pharynx; sensations of touch, pain, and temperature from external ear

Origin: Motor fibers arise from medulla oblongata. Sensory fibers arise from pharynx; middle-ear cavity, pinna, and external auditory canal of ear; posterior one-third of tongue (including taste buds); and receptors for blood chemistry and pressure in carotid arteries.

Course: Passes through jugular foramen; parasympathetic fibers travel to tympanic cavity, to auditory tube, and via otic ganglion to parotid salivary gland; some sensory fibers enter cranium with vagus nerve

Effects of damage: Loss of bitter and sour taste; impaired swallowing

Clinical test: Test gag reflex, swallowing, and coughing; note any speech impediments; test taste on posterior one-third of tongue using bitter and sour stimuli (quinine and vinegar)

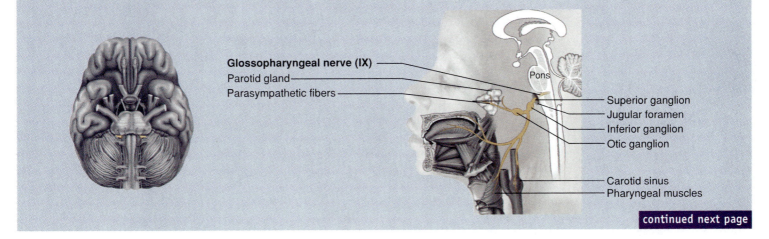

continued next page

Table 15.2 Continued

X. Vagus[3] (VAY-gus) Nerve

Composition: Includes most motor and sensory fibers of parasympathetic nervous system, as well as proprioceptive, other somatosensory, and motor fibers

Function: Swallowing; speech; regulation of pulmonary, cardiovascular, and gastrointestinal function; sensations of hunger, fullness, and intestinal discomfort

Origin: Motor fibers originate in medulla oblongata; sensory fibers originate in widespread thoracic and abdominal viscera, pharynx, larynx, external ear, and dura mater

Course: The only cranial nerve to extend to organs beyond the head-neck region; exits via jugular foramen and descends through neck and thoracic region, sends branches to tongue, carotid arteries, pharynx, larynx, and heart; penetrates diaphragm and sends branches to abdominal viscera

Effects of damage: Hoarseness or loss of voice; impaired swallowing and gastrointestinal motility; fatal if both vagus nerves are damaged

Clinical test: Test with cranial nerve IX

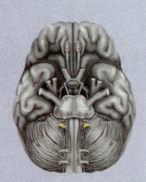

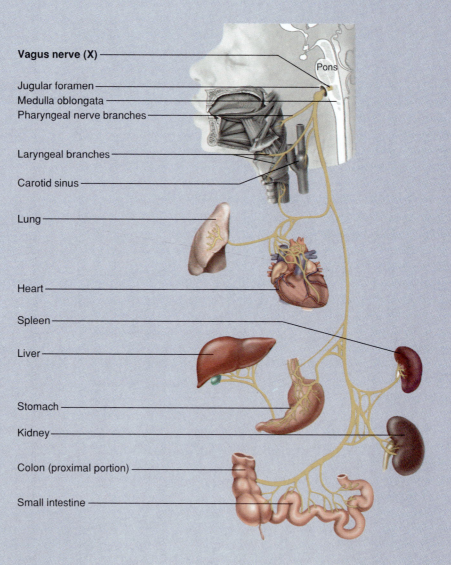

Vagus nerve (X)
Pons
Jugular foramen
Medulla oblongata
Pharyngeal nerve branches
Laryngeal branches
Carotid sinus
Lung
Heart
Spleen
Liver
Stomach
Kidney
Colon (proximal portion)
Small intestine

3. *vagus* = wandering

Table 15.2 Continued

XI. Accessory Nerve (sometimes called *spinal accessory*)

Composition: Motor and proprioceptive

Function: Swallowing; movements of head, neck, and shoulders

Origin: Upper cervical segments of spinal cord

Course: Passes upward alongside spinal cord, enters foramen magnum, then exits via jugular foramen; fibers lead to sternocleidomastoid and trapezius muscles

Effects of damage: Impaired movement of head, neck, and shoulders; difficulty in shrugging shoulders on damaged side; paralysis of sternocleidomastoid, causing head to turn toward injured side

Clinical test: Have subject rotate head or shrug shoulders against resistance

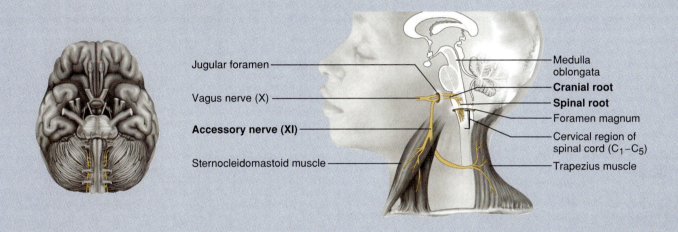

XII. Hypoglossal (HY-po-GLOSS-ul) Nerve

Composition: Motor and scanty proprioceptive fibers

Function: Tongue movements of speech, food manipulation, and swallowing

Origin: Medulla oblongata

Course: Passes via hypoglossal canal to extrinsic and intrinsic muscles of tongue

Effects of damage: Difficulty in speech and swallowing; when both right and left hypoglossal nerves are injured, tongue cannot be protruded; when only one is injured, tongue is deflected toward and atrophies on that side, becoming shrunken and furrowed

Clinical test: Note deviations of tongue as subject protrudes and retracts it; test ability to protrude tongue against resistance from a tongue depressor

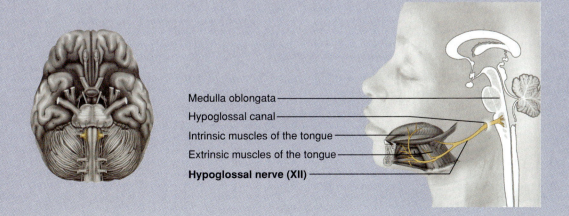

Trigeminal neuralgia,[4] also called *tic doloreux*[5] (DOO-loo-ROO), is a syndrome characterized by brief recurring episodes of intense stabbing pain in the trigeminal nerve. The cause is unknown; there is no visible change in the nerve. It occurs most commonly over the age of 50 and more in women than in men. The pain lasts from a few seconds to a minute or two, but it strikes at unpredictable intervals and sometimes up to a hundred times a day. The pain usually occurs in a specific zone of the face, such as around the mouth and nose. It may be triggered by brushing the teeth, washing the face, drinking, or touch. Analgesics (pain relievers) give only limited relief. In severe cases the nerve is surgically cut proximal to the trigeminal ganglion, but this also deadens most other sensation in that side of the face.

Bell[6] **palsy** is a degenerative disorder of the facial nerve, probably due to a virus. It is characterized by paralysis of the facial muscles on one side with resulting distortion of the facial features, such as sagging of the mouth or lower eyelid. The paralysis may interfere with speech, prevent closure of the eye, and cause excessive secretion of tears. There may also be a partial loss of the sense of taste. Bell palsy may appear abruptly, sometimes overnight, and often disappears spontaneously within 3 to 5 weeks.

Key Point Review

5. List the purely sensory cranial nerves and state the function of each.

6. What is the only cranial nerve to extend beyond the head-neck region?

7. If the oculomotor, trochlear, or abducens nerve were damaged, the effect would be similar in all three cases. What would that effect be?

8. Which cranial nerve supplies more of the head than any other?

9. Name two cranial nerves involved in the sense of taste.

The Spinal Nerves

▼Objectives

When you have completed this section, you should be able to
- describe the attachment of a spinal nerve to the spinal cord;
- trace the branches of a spinal nerve distal to its attachment;
- name the five plexuses of spinal nerves and describe their general anatomy;
- name some major nerves that arise from each plexus;
- explain the relationship of dermatomes to the spinal nerves.

There are 31 pairs of **spinal nerves:** 8 cervical (C1–C8), 12 thoracic (T1–T12), 5 lumbar (L1–L5), 5 sacral (S1–S5), and 1 coccygeal (Cx) (fig. 15.5). The first cervical nerve emerges between the skull and atlas, and the others emerge through intervertebral foramina, including the pelvic and dorsal foramina of the sacrum.

Proximal Branches

Within the intervertebral foramen, each spinal nerve branches into a **dorsal root** and **ventral root** (fig. 15.6).

A swelling of the dorsal root, the **dorsal root ganglion,** is occupied by cell bodies of the afferent neurons. The dorsal root then breaks up into six to eight *rootlets* that enter the dorsal horn of one segment of the cord (fig. 15.7). Efferent fibers leave the ventral horn of the gray matter by way of another six to eight rootlets. They converge and form the ventral root, which passes around the ventral (anterior) side of the spinal cord and merges with the dorsal root. The convergence of the dorsal and ventral roots forms the spinal nerve. The spinal nerve is a mixed nerve, but all of its sensory fibers travel through the dorsal root and all of its motor fibers travel through the ventral root. Some viruses invade the nervous system by way of these roots (see special topic 15.3).

The dorsal and ventral roots are shortest in the cervical region and become longer inferiorly. The roots that arise from segments L2 to Cx of the cord form the *cauda equina,* which occupies the vertebral canal from level L2 to S5. After they merge and penetrate the dura mater, they become spiral nerves.

Distal Branches

Distal to the vertebrae, the branches of a spinal nerve are more complex (fig. 15.8). Immediately after emerging from the intervertebral foramen, the nerve divides into a **dorsal ramus,**[7] a **ventral ramus,** and a small **meningeal branch.** The meningeal branch (see fig. 15.6) reenters the

---

4. *algia* = pain
5. *douloureux* = painful
6. Sir Charles Bell (1774–1842), Scottish physician

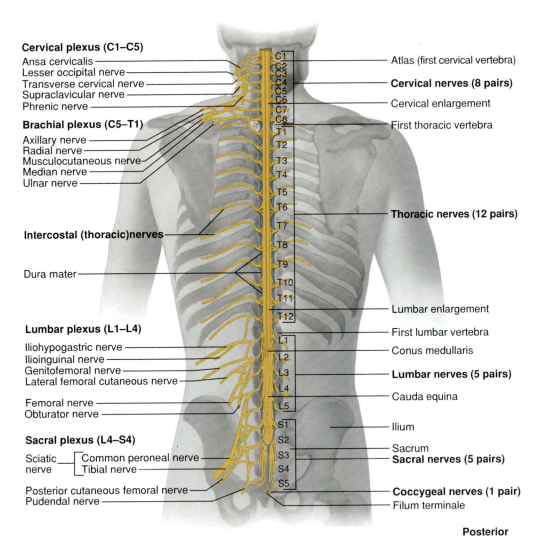

Figure 15.5 The spinal nerve roots and plexuses, dorsal view.

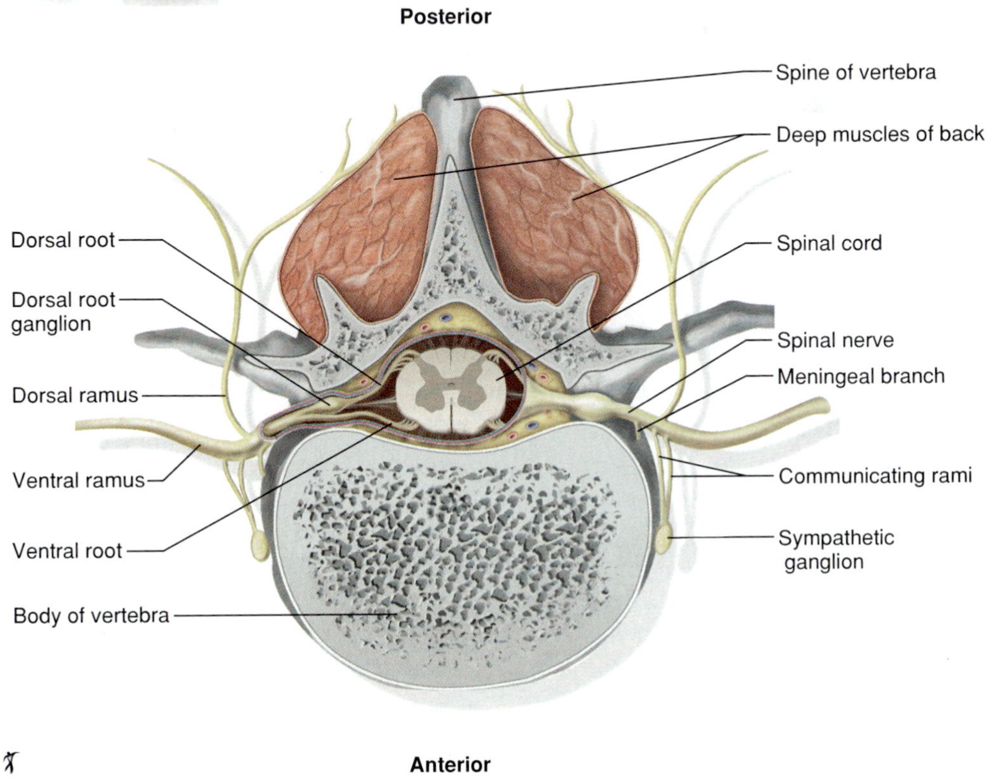

Cervical plexus (C1–C5)
Ansa cervicalis
Lesser occipital nerve
Transverse cervical nerve
Supraclavicular nerve
Phrenic nerve

Brachial plexus (C5–T1)
Axillary nerve
Radial nerve
Musculocutaneous nerve
Median nerve
Ulnar nerve

Intercostal (thoracic) nerves

Dura mater

Lumbar plexus (L1–L4)
Iliohypogastric nerve
Ilioinguinal nerve
Genitofemoral nerve
Lateral femoral cutaneous nerve
Femoral nerve
Obturator nerve

Sacral plexus (L4–S4)
Sciatic nerve — Common peroneal nerve / Tibial nerve
Posterior cutaneous femoral nerve
Pudendal nerve

Atlas (first cervical vertebra)
Cervical nerves (8 pairs)
Cervical enlargement
First thoracic vertebra

Thoracic nerves (12 pairs)

Lumbar enlargement
First lumbar vertebra
Conus medullaris
Lumbar nerves (5 pairs)
Cauda equina
Ilium
Sacrum
Sacral nerves (5 pairs)
Coccygeal nerves (1 pair)
Filum terminale

C1
C2
C3
C4
C5
C6
C7
C8
T1
T2
T3
T4
T5
T6
T7
T8
T9
T10
T11
T12
L1
L2
L3
L4
L5
S1
S2
S3
S4
S5

Posterior

Spine of vertebra
Deep muscles of back
Spinal cord
Spinal nerve
Meningeal branch
Communicating rami
Sympathetic ganglion

Dorsal root
Dorsal root ganglion
Dorsal ramus
Ventral ramus
Ventral root
Body of vertebra

Figure 15.6 Relationship of a spinal nerve to the spinal cord and vertebra (cross section).

Anterior

chapter 15 The Peripheral Nervous System and Reflexes **517**

Posterior median sulcus

Fasciculus gracilis

Fasciculus cuneatus

Lateral column

Segment C5

Cross section

Arachnoid

Dura mater

Neural arch of vertebra C3 (cut)

Vertebral artery

Spinal nerve C5
Rootlets
Dorsal root
Dorsal root ganglion
Ventral root

Figure 15.7 The point of entry of several spinal nerve roots into the spinal cord, dorsal view. Each root is divided into several rootlets that enter one segment of the spinal cord.

vertebral canal and innervates the meninges, vertebrae, and spinal ligaments. The dorsal ramus innervates the muscles and joints in that region of the spine and the skin of the back. The ventral ramus innervates the ventral and lateral skin and muscles of the trunk and gives rise to nerves of the extremities.

 Think About It

Do you think the meningeal branch is composed of afferent fibers, efferent fibers, or both? Explain your reasoning.

The ventral ramus differs from one region of the trunk to another. In the thoracic region, each ventral

ramus forms an **intercostal nerve** that travels along the inferior margin of a rib and innervates the skin and intercostal muscles (thus contributing to breathing). All other ventral rami form the *nerve plexuses* described next.

Nerve Plexuses

Except in the thoracic region, the ventral rami branch and anastomose (merge) repeatedly to form five web-like nerve plexuses: the small **cervical plexus** deep in the neck, the **brachial plexus** near the shoulder, the **lumbar plexus** of the lower back, the **sacral plexus** immediately inferior to this, and finally the tiny **coccygeal plexus** adjacent to the lower sacrum and coccyx. A general view of these plexuses is shown in figure 15.5; they are illustrated and described in tables 15.3 through 15.6.

7. *ramus* = branch

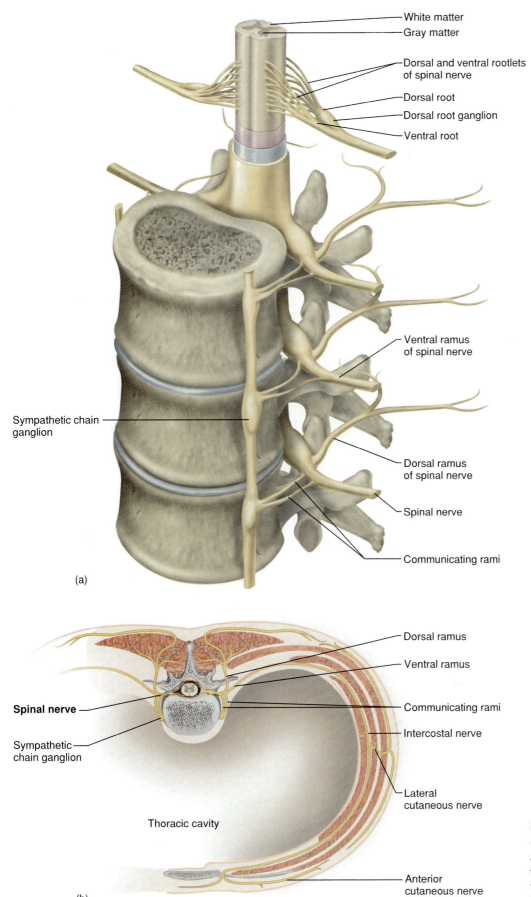

White matter
Gray matter
Dorsal and ventral rootlets of spinal nerve
Dorsal root
Dorsal root ganglion
Ventral root

Ventral ramus of spinal nerve

Sympathetic chain ganglion

Dorsal ramus of spinal nerve

Spinal nerve

Communicating rami

(a)

Dorsal ramus
Ventral ramus

Spinal nerve

Communicating rami

Sympathetic chain ganglion

Intercostal nerve

Lateral cutaneous nerve

Thoracic cavity

Anterior cutaneous nerve

(b)

Figure 15.8 Rami of the spinal nerves. (*a*) Anterolateral view of the spinal nerves and their subdivisions. (*b*) Cross section of the body at the thoracic level, showing innervation of muscles of the chest and back. ⚡

Table 15.3   The Cervical Plexus

The cervical plexus receives fibers from the ventral rami of nerves C1 to C5 and gives rise to the nerves listed, in order from superior to inferior. The most important of these are the *phrenic*[8] *nerves*, which travel down each side of the mediastinum, innervate the diaphragm, and play an essential role in ventilation of the lungs. In addition to the major nerves listed, there are several motor branches that innervate the geniohyoid, thyrohyoid, scalene, levator scapulae, trapezius, and sternocleidomastoid muscles.

Lesser Occipital Nerve

Composition: Somatosensory

Innervation: Skin of lateral scalp and posterior part of external ear

Great Auricular Nerve

Composition: Somatosensory

Innervation: Skin of and around external ear

Transverse Cervical Nerve

Composition: Somatosensory

Innervation: Skin of anterior and lateral aspect of neck

Ansa Cervicalis

Composition: Motor

Innervation: Omohyoid, sternohyoid, and sternothyroid muscles

Supraclavicular Nerves

Composition: Somatosensory

Innervation: Skin of lower anterior and lateral neck, shoulder, and anterior chest

Phrenic (FREN-ic) Nerve

Composition: Motor

Innervation: Diaphragm

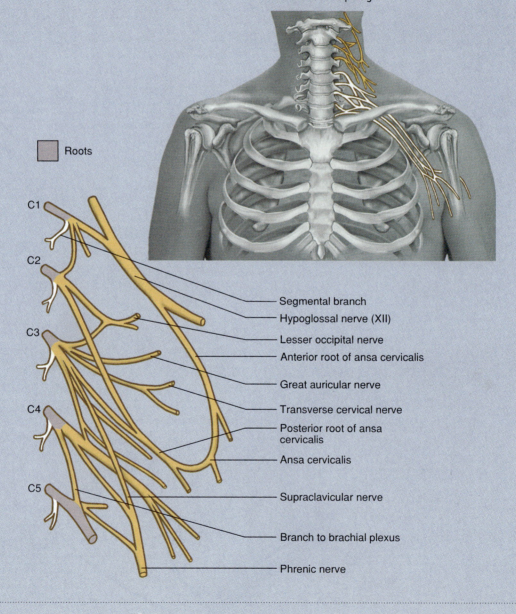

Roots

C1

C2

C3

C4

C5

Segmental branch
Hypoglossal nerve (XII)
Lesser occipital nerve
Anterior root of ansa cervicalis
Great auricular nerve
Transverse cervical nerve
Posterior root of ansa cervicalis
Ansa cervicalis
Supraclavicular nerve
Branch to brachial plexus
Phrenic nerve

8. *phren* = diaphragm

Table 15.4 The Brachial Plexus

The brachial plexus (fig. 15.9) receives fibers from the ventral rami of nerves C4 to T2. The plexus passes over the first rib into the axilla and innervates the upper extremity and some muscles of the neck and shoulder. It gives rise to nerves for cutaneous sensation, muscle contraction, and proprioception from the joints and muscles. The muscle actions stimulated by these nerves are described in the tables in chapter 11.

The subdivisions of this plexus are called *roots, trunks, divisions,* and *cords* (color-coded in the accompanying figure). The five **roots** are the ventral rami of nerves C5 to T1, which provide most of the fibers to this plexus (C4 and T2 contribute partially). The five roots unite to form three segments called the **upper, middle,** and **lower trunks.** Each trunk divides into an **anterior** and **posterior division,** and finally the six divisions merge to form three large fiber bundles—the **posterior, medial,** and **lateral cords.**

Axillary Nerve

Composition: Motor and somatosensory

Origin: Posterior cord of brachial plexus

Sensory innervation: Skin of lateral shoulder and arm; shoulder joint

Motor innervation: Deltoid and teres minor

Radial Nerve

Composition: Motor and somatosensory

Origin: Posterior cord of brachial plexus

Sensory innervation: Skin of posterior aspect of arm, forearm, and wrist; joints of elbow, wrist, and hand

Motor innervation: Muscles of posterior arm and forearm: triceps brachii, supinator, anconeus, brachioradialis, extensor carpi radialis brevis, extensor carpi radialis longus, and extensor carpi ulnaris

Musculocutaneous Nerve

Composition: Motor and somatosensory

Origin: Lateral cord of brachial plexus

Sensory innervation: Skin of lateral aspect of forearm

Motor innervation: Muscles of anterior arm: coracobrachialis, biceps brachii, and brachialis

Median Nerve

Composition: Motor and somatosensory

Origin: Medial cord of brachial plexus

Sensory innervation: Skin of lateral two-thirds of hand, joints of hand

Motor innervation: Flexors of anterior forearm (flexor carpi ulnaris and flexor digitorum); lateral palm; first and second lumbricals

Ulnar Nerve

Composition: Motor and somatosensory

Origin: Medial cord of brachial plexus

Sensory innervation: Skin of medial part of hand; joints of hand

Motor innervation: Flexor carpi ulnaris, flexor digitorum, adductor pollicis, hypothenar muscles, interosseous muscles, and third and fourth lumbricals

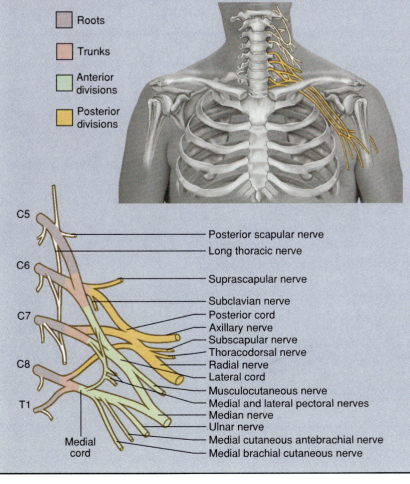

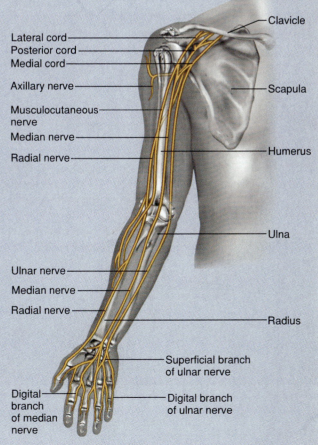

Table 15.5 The Lumbar Plexus

The lumbar plexus is formed from the ventral rami of nerves L1 to L4 and some fibers from T12. With only five roots and two divisions, it is less complex than the brachial plexus.

Iliohypogastric Nerve

Composition: Motor and somatosensory

Sensory innervation: Skin of anterior abdominal wall

Motor innervation: Internal and external obliques and transversus abdominis

Ilioinguinal Nerve

Composition: Motor and somatosensory

Sensory innervation: Skin of upper medial thigh; male scrotum and root of penis; female labia majora

Motor innervation: Joins iliohypogastric nerve and innervates the same muscles

Genitofemoral Nerve

Composition: Somatosensory

Sensory innervation: Skin of middle anterior thigh; male scrotum and cremaster muscle; female labia majora

Lateral Femoral Cutaneous Nerve

Composition: Somatosensory

Sensory innervation: Skin of lateral aspect of thigh

Femoral Nerve

Composition: Motor and somatosensory

Sensory innervation: Skin of anterior and lateral thigh; medial leg and foot

Motor innervation: Anterior muscles of thigh and extensors of leg; iliacus, psoas major, pectineus, quadriceps femoris, and sartorius

Saphenous (sah-FEE-nus) Nerve

Composition: Somatosensory

Sensory innervation: Skin of medial aspect of leg and foot; knee joint

Obturator Nerve

Composition: Motor and somatosensory

Sensory innervation: Skin of superior medial thigh; hip and knee joints

Motor innervation: Adductor muscles of leg: external obturator, pectineus, adductor longus, adductor brevis, adductor magnus, and gracilis

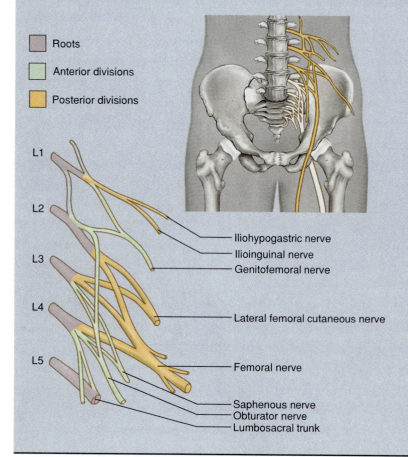

Roots

Anterior divisions

Posterior divisions

L1
L2
L3
L4
L5

Iliohypogastric nerve
Ilioinguinal nerve
Genitofemoral nerve

Lateral femoral cutaneous nerve

Femoral nerve

Saphenous nerve
Obturator nerve
Lumbosacral trunk

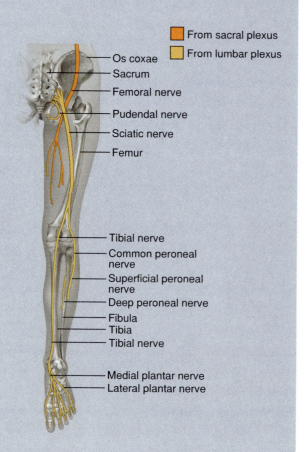

From sacral plexus
From lumbar plexus

Os coxae
Sacrum
Femoral nerve
Pudendal nerve
Sciatic nerve
Femur

Tibial nerve
Common peroneal nerve
Superficial peroneal nerve
Deep peroneal nerve
Fibula
Tibia
Tibial nerve

Medial plantar nerve
Lateral plantar nerve

Table 15.6 The Sacral and Coccygeal Plexuses

The sacral plexus is formed from the ventral rami of nerves L4, L5, and S1 to S4. It has six roots and anterior and posterior divisions. Since it is connected to the lumbar plexus by fibers that run through the *lumbosacral trunk,* the two plexuses are sometimes referred to collectively as the *lumbosacral plexus.* The coccygeal plexus is a tiny plexus formed from the ventral rami of S4, S5, and Cx (the coccygeal nerve).

The *tibial* and *common peroneal nerves* listed in this table travel together through a connective tissue sheath; they are referred to collectively as the **sciatic** (sy-AT-ic) **nerve.** The sciatic nerve passes through the greater sciatic notch of the pelvis, extends for the length of the thigh, and ends at the popliteal fossa. Here, the nerves diverge and follow their separate paths into the leg. The sciatic nerve is a common focus of injury and pain (see special topic 15.4).

Superior Gluteal Nerve

Composition: Motor

Motor innervation: Gluteus minimus, gluteus medius, and tensor fasciae latae

Inferior Gluteal Nerve

Composition: Motor

Motor innervation: Gluteus maximus

Nerve to Piriformis

Composition: Motor

Motor innervation: Piriformis

Nerve to Quadratus Femoris

Composition: Motor and somatosensory

Sensory innervation: Hip joint

Motor innervation: Quadratus femoris and gemellus inferior

Nerve to Internal Obturator

Composition: Motor

Motor innervation: Internal obturator and gemellus superior

Perforating Cutaneous Nerve

Composition: Somatosensory

Sensory innervation: Skin of posterior aspect of buttock

Posterior Cutaneous Nerve

Composition: Somatosensory

Sensory innervation: Skin of lower lateral buttock, anal region, upper posterior thigh, upper calf, scrotum, and labia majora

Tibial Nerve

Composition: Motor and somatosensory

Sensory innervation: Skin of posterior leg and sole of foot; knee and foot joints

Motor innervation: Gastrocnemius, soleus, flexor digitorum longus, flexor hallucis longus, tibialis posterior, popliteus, and intrinsic muscles of foot

Common Peroneal Nerve

Composition: Motor and somatosensory

Sensory innervation: Skin of anterior distal third of leg, dorsum of foot, and toes I and II; knee joint

Motor innervation: Short head of biceps femoris, peroneus tertius, peroneus brevis, peroneus longus, tibialis anterior, extensor hallucis longus, extensor digitorum longus, and extensor digitorum brevis

Pudendal Nerve

Composition: Motor and somatosensory

Sensory innervation: Skin of penis and scrotum of male; clitoris, labia majora and minora, and lower vagina of female

Motor innervation: Muscles of perineum

Coccygeal Nerve

Composition: Motor and somatosensory

Sensory innervation: Skin over coccyx

Motor innervation: Muscles of pelvic floor

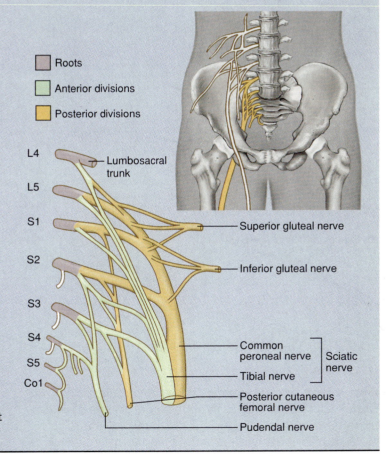

- Roots
- Anterior divisions
- Posterior divisions

L4 — Lumbosacral trunk
L5
S1 — Superior gluteal nerve
S2 — Inferior gluteal nerve
S3
S4
S5 — Common peroneal nerve — Sciatic nerve
Co1 — Tibial nerve
— Posterior cutaneous femoral nerve
— Pudendal nerve

Shingles is a disease caused by the *herpes zoster* virus. This virus causes varicella (chicken pox) in childhood and then lies dormant in the dorsal root ganglia. The immune system keeps it in check if it attempts to spread. If the immune system is compromised, however, the virus can travel down the sensory nerves by fast axonal transport, leaving a painful trail of skin discoloration and fluid-filled vesicles along the path of the nerve. The symptoms usually appear in the chest and waist, often on just one side of the body. Shingles is most likely to occur after the age of 50. While it is quite uncomfortable and symptoms may last 6 months or longer, it eventually heals spontaneously and requires no special treatment other than aspirin and steroidal ointment to relieve pain and inflammation.

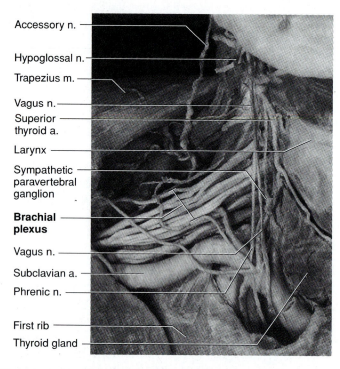

Figure 15.9 The brachial plexus, anterior view of the right shoulder. This photograph also shows three of the cranial nerves, the sympathetic trunk, and the phrenic nerve (a branch of the cervical plexus). Most of the other structures resembling nerves in this photograph are blood vessels.

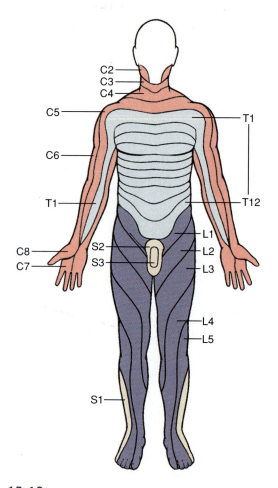

Figure 15.10 A dermatome map, anterior view, showing areas of skin with sensory innervation by individual spinal nerves. Nerve C1 does not innervate the skin.

Cutaneous Innervation and Dermatomes

Each spinal nerve except C1 receives sensory input from a specific area of skin called a **dermatome**.[9] A *dermatome map* (fig. 15.10) is a diagram of cutaneous regions innervated by each spinal nerve. Such a map is oversimplified, however, because the dermatomes overlap at their edges by as much as 50%. Therefore severance of one sensory nerve root does not entirely deaden sensation from a dermatome. It is necessary to sever or anesthetize three successive spinal nerves to produce a total loss of sensation from a dermatome. Spinal nerve damage is assessed by testing the dermatomes with pinpricks and noting the areas in which the patient has no sensation.

9. *derma* = skin + *tom* = segment

The radial and sciatic nerves are especially vulnerable to injury. The radial nerve may be compressed against the humerus by improperly adjusted crutches, causing **crutch paralysis.** A similar injury often resulted from the now-discredited practice of trying to correct a dislocated shoulder by putting a foot in the armpit and pulling on the arm. One consequence of radial nerve injury is **wrist drop**—the fingers, hand, and wrist are chronically flexed because the extensor muscles supplied by the radial nerve are paralyzed.

Because of its position and length, the sciatic nerve is the most vulnerable nerve in the body. Trauma to this nerve produces **sciatica,** a sharp pain that travels from the gluteal region along the posterior side of the thigh and leg as far as the ankle. Ninety percent of cases are due to a herniated intervertebral disc or osteoarthritis of the lower spine, but sciatica can also be caused by pressure from a pregnant uterus, dislocation of the hip, injections in the wrong area of the buttock, or sitting for a long time on the edge of a hard chair. Men sometimes suffer sciatica from the habit of sitting on a wallet carried in the hip pocket.

Key Point Review

10 What is meant by the dorsal and ventral roots of a spinal nerve? Which of these is sensory and which is motor?

11 Where are the somas of the dorsal root located? Where are the somas that supply the ventral root?

12 List the five plexuses of spinal nerves and state where each one is located.

13 State which plexus gives rise to each of the following nerves: axillary, ilioinguinal, obturator, phrenic, pudendal, radial, and sciatic.

Somatic Reflexes

▼**Objectives**

When you have completed this section, you should be able to
• define *reflex* and explain how reflexes differ from other motor actions;
• describe the general components of a typical reflex arc; and
• explain how the basic types of spinal reflexes function.

Most of us have had our reflexes tested with a little rubber hammer; a tap near the knee or elbow produces an uncontrollable jerk of the leg or forearm. In this section, we discuss what reflexes are and how they are produced by an assembly of receptors, neurons, and effectors. We also survey the different types of neuromuscular reflexes.

The Nature of Reflexes

Reflexes are quick, involuntary, stereotyped reactions of peripheral effectors to stimulation. This definition sums up four important properties of a reflex:

1. Reflexes *require stimulation*—they are not spontaneous actions but responses to sensory input.
2. Reflexes are *quick*—they generally involve a minimum of interneurons and synaptic delay.
3. Reflexes are *involuntary*—they occur without intent, often without our awareness, and they are difficult to suppress. Given an adequate stimulus, the response is essentially automatic. You may become conscious of the stimulus that evoked a reflex—and this awareness may enable you to correct or avoid a potentially dangerous situation—but awareness is not a part of the reflex itself. It may come after the reflex action has been completed, and spinal reflexes occur even if the spinal cord has been severed at the neck so that no stimuli can reach the brain.
4. Reflexes are *stereotyped*—they occur in essentially the same way every time; the response is very predictable.

Reflexes include glandular secretion and contractions of all three types of muscle. They also include some learned responses, such as the salivation of dogs in response to a sound they have come to associate with feeding time, first studied by Ivan Pavlov and named *conditioned reflexes.* In this section, however, we are concerned with unlearned reflexes that are mediated by the brainstem and spinal cord. They result in the involuntary contraction of skeletal muscles—for example, the quick withdrawal of your hand from a hot stove or the lifting of your foot when you step on something sharp. We may think of these as **somatic reflexes,** since the receptors are part of the somatic nervous system. Later in this chapter, we consider **visceral reflexes.** The somatic reflexes have traditionally been called **spinal reflexes,** although some visceral reflexes also involve the spinal cord.

You are prepared to understand somatic reflexes now that you have studied spinal cord anatomy in the previous chapter and the spinal nerves in this one. A spinal reflex functions by means of a **reflex arc** that includes the following components:

• **somatic receptors** in the skin, a muscle, or a tendon;
• **afferent nerve fibers,** which carry information from these receptors into the dorsal horn of the spinal cord;

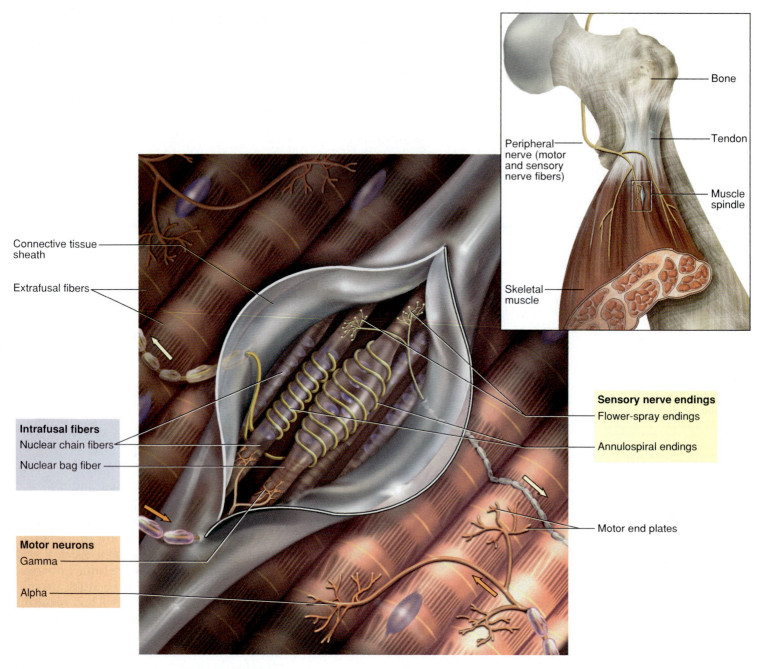

Connective tissue sheath

Extrafusal fibers

Intrafusal fibers
Nuclear chain fibers
Nuclear bag fiber

Motor neurons
Gamma
Alpha

Bone

Peripheral nerve (motor and sensory nerve fibers)

Tendon

Muscle spindle

Skeletal muscle

Sensory nerve endings
Flower-spray endings
Annulospiral endings

Motor end plates

Figure 15.11 A muscle spindle and its innervation.

- **interneurons,** which integrate information and generate multiple outputs, although these may be lacking from some reflex arcs;
- **efferent nerve fibers,** which carry motor impulses back to the skeletal muscles; and
- **skeletal muscles,** the somatic effectors that carry out the response.

Many spinal reflexes involve stretch receptors in the muscles called *muscle spindles.* These are among the body's most important proprioceptors—sense organs that monitor the position and movements of body parts. Before we can understand spinal reflexes, we must understand the structure of a muscle spindle.

The Muscle Spindle

Muscle spindles are cigar-shaped organs about 4 to 10 mm long scattered throughout the fleshy part of a muscle (fig. 15.11). A spindle contains 3 to 12 modified muscle fibers and a few nerve fibers, all wrapped in a fibrous capsule. The muscle fibers within a spindle are called **intrafusal fibers**[10] to distinguish them from the **extrafusal fibers** that make up most of the bulk of a muscle. The middle of an intrafusal fiber lacks sarcomeres and cannot contract; this portion acts as the stretch receptor. The ends are contractile. There are two classes of intrafusal fibers: *nuclear chain fibers,* which have a row of nuclei extending in single file through the noncontractile region, and *nuclear bag fibers,* which are about twice as long and have nuclei clustered together in a thickened midregion.

Three types of nerve fibers are associated with a muscle spindle. **Primary afferent fibers** terminate in an *annulospiral ending* that coils around the middle of both nuclear chain and nuclear bag fibers. **Secondary afferent fibers** have *flower-spray endings,* somewhat resembling the dried head of a wildflower, wrapped primarily around nuclear chain fibers on each side of the midregion. Primary fibers respond mainly to the onset of muscle stretch, whereas secondary fibers respond mainly to sustained stretch. **Gamma (γ) motor neurons,** the third type, originate in the ventral horn of the spinal cord and lead to the contractile ends of the intrafusal fibers. They are distinguished from the **alpha (α) motor neurons,** which innervate the extrafusal fibers of the muscle. Gamma motor neurons function continually to adjust the tension in a muscle spindle, causing the ends of the intrafusal fibers to contract slightly and keep the fibers taut and responsive, regardless of whether the surrounding muscle is contracted or relaxed. This is evidently a very important function, because γ motor neurons constitute about one-third of all the motor fibers in a spinal nerve.

The Stretch Reflex

When a muscle is stretched, it tends to "fight back"—it contracts, maintains an elevated level of tonus, and thus feels stiffer than an unstretched muscle. This response is called the **stretch (myotatic[11]) reflex.** For example, if you stand with your feet flat on the floor and then start to tip forward (without bending at the waist), the dorsiflexion of your feet stretches the triceps surae muscles of the calf. This stimulates muscle spindles and triggers afferent signals to the spinal cord. Through the cerebellum and the motor cortex of the cerebrum, this information is integrated and sent back to the muscles by way of the α motor neurons of the spinal cord. The triceps surae contracts, causing plantar flexion and righting your body.

Stretch reflexes often feed back not to a single muscle but to a set of synergists and antagonists. Since the contraction of a muscle on one side of a joint stretches the antagonistic muscle on the other side, strong flexion of a joint triggers a stretch reflex in the extensors, and strong extension likewise stimulates a stretch reflex in the flexors. Consequently, stretch reflexes are valuable in stabilizing joints by balancing the tension of the extensors and flexors. They also dampen, or smoothe, muscle action. Without stretch reflexes, a person's movements tend to be jerky. Stretch reflexes are especially important in coordinating vigorous and precise movements such as dance.

Although a stretch reflex is mediated primarily by the brain and is not, therefore, strictly a spinal reflex, a weak component of it is spinal and occurs even if the spinal cord is severed below the brain. Even this component can be more pronounced if a muscle is stretched very suddenly. This occurs in a **tendon reflex**—the reflexive contraction of a muscle when its tendon is tapped, as in the familiar knee jerk (patellar) reflex. Tapping the patellar ligament with a reflex hammer creates a sudden stretch in the quadriceps femoris muscle of the thigh (fig. 15.12). This stimulates numerous muscle spindles in the quadriceps and sends an intense volley of signals, mainly by way of the primary afferent fibers, to the spinal cord. These fibers form **monosynaptic reflex arcs** with the α motor neurons that return to the muscle. That is, there is only one synapse between the afferent and efferent neurons, therefore little synaptic delay and a very prompt response. The α motor neurons send excitatory signals back to the quadriceps muscle, making it contract and creating the knee jerk.

There are many other tendon reflexes (fig. 15.13). A tap on the calcaneal tendon causes plantar flexion of the foot, a tap on the tendon of the triceps brachii causes extension of the elbow, and a tap on the side of the face causes clenching of the jaw. The abdominal reflex, included in figure 15.13 for convenience, is not a tendon reflex but involves a spinal reflex arc between the abdominal skin and muscles. Testing the spinal reflexes is valuable in diagnosing many diseases that cause exaggeration, inhibition, or absence of reflexes: neurosyphilis, diabetes mellitus, multiple sclerosis, alcoholism, electrolyte imbalances, and lesions of the nervous system, among others.

Reciprocal Inhibition

The knee jerk response, and others like it, would not be very pronounced if the antagonistic muscles of the same joint contracted and opposed the agonist. For the quadriceps femoris to produce a knee jerk, it is necessary for its antagonists, the hamstring muscles,

10. *intra* = within + *fus* = spindle
11. *myo* = muscle + *tatic* (from *tasis*) = stretch

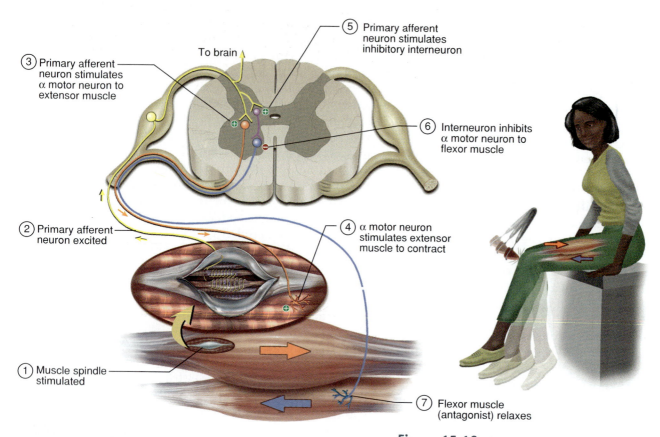

③ Primary afferent neuron stimulates α motor neuron to extensor muscle

To brain

⑤ Primary afferent neuron stimulates inhibitory interneuron

⑥ Interneuron inhibits α motor neuron to flexor muscle

② Primary afferent neuron excited

④ α motor neuron stimulates extensor muscle to contract

① Muscle spindle stimulated

⑦ Flexor muscle (antagonist) relaxes

Figure 15.12 The patellar tendon reflex arc and reciprocal inhibition of the antagonistic muscle. *Plus* signs indicate excitation of a postsynaptic cell and *minus* signs indicate inhibition. 𝑋 ▭

Abdominal reflex

Supinator reflex

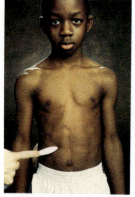

Patellar reflex

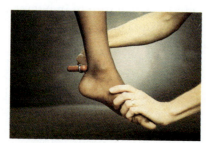

Achilles reflex

Figure 15.13 Some common tests for spinal reflexes.

to relax. Concurrently with the stretch reflex, relaxation of the antagonists is ensured by the phenomenon of **reciprocal inhibition.** Some branches from the afferent fibers of the muscle spindle stimulate spinal interneurons, which, in turn, inhibit α motor neurons to the antagonistic muscles.

The Flexor (Withdrawal) Reflex

Flexor reflexes are important when a limb must be quickly pulled away from an injurious situation, such as jerking your arm back when you burn yourself on a hot cooking pot or quickly lifting your foot when you step on something sharp. The protective function of a flexor reflex requires more than the quick jerk of a tendon reflex and therefore involves more complex neural pathways.

Suppose you are wading in a lake and step on a broken bottle with your right foot (fig. 15.14). Even before you are consciously aware of the pain, you contract the flexors and relax the extensors in your right leg; thus you pull your foot away quickly before the glass penetrates any deeper. To produce a sustained contraction of the flexors involves a parallel after-

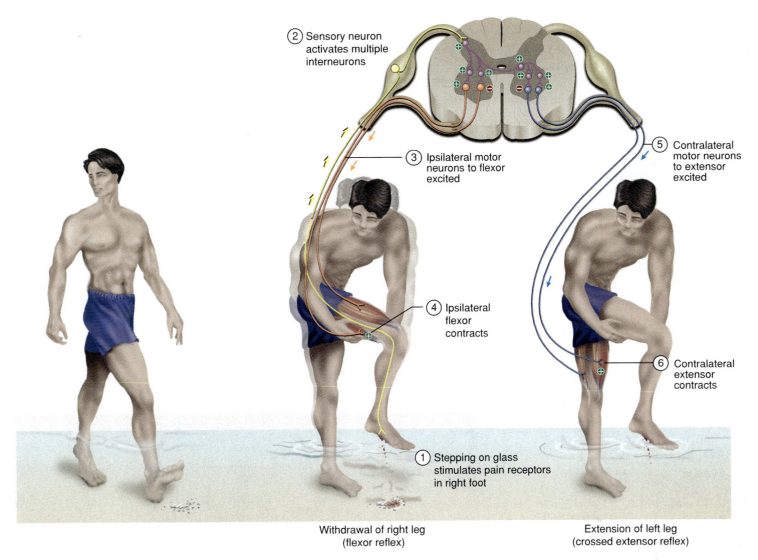

② Sensory neuron
activates multiple
interneurons

③ Ipsilateral motor
neurons to flexor
excited

⑤ Contralateral
motor neurons
to extensor
excited

④ Ipsilateral
flexor
contracts

⑥ Contralateral
extensor
contracts

① Stepping on glass
stimulates pain receptors
in right foot

Withdrawal of right leg
(flexor reflex)

Extension of left leg
(crossed extensor reflex)

Figure 15.14 The withdrawal (flexor) reflex and crossed extensor reflex. The pain stimulus triggers a withdrawal reflex, which results in contraction of flexor muscles of the injured limb, and a crossed extensor reflex, which results in contraction of extensor muscles of the opposite limb. The latter aids in balance when the injured limb is raised. Note that for each limb, while the agonist contracts, the α motor neuron to its antagonist is inhibited, as indicated by the red *minus* signs in the spinal cord.

discharge circuit in the spinal cord (see fig. 13.25, p. 457). This is a **polysynaptic reflex arc**—signals travel over many synapses on their way back to the leg. Some signals follow pathways with only a few synapses and return to the flexor muscles quickly. Other signals follow pathways with more synapses, and therefore more delay, so they reach the flexor muscles a little later. Consequently, the flexor muscles receive prolonged output from the spinal cord and not just one sudden stimulus as in a tendon reflex. By the time these efferent signals begin to die out, you will probably be consciously aware of the pain and begin taking voluntary, not merely reflexive, action to prevent further harm.

The Crossed Extensor Reflex

In the preceding situation, if *all* you did was to quickly lift the injured leg from the lake bottom, you would fall over. To prevent this and maintain your balance, other reflexes shift your center of gravity over the leg that is still on the ground. These responses are partly cerebellar and partly spinal. In the **crossed extensor reflex** (fig. 15.14), some branches of the afferent nerve fibers cross to the contralateral (opposite) side of the spinal cord. Here they synapse with interneurons, which, in turn, excite or inhibit α motor neurons to the muscles of the contralateral leg. In the ipsilateral leg (the one on the side that was hurt), you would contract your flexors and relax your extensors to lift the leg from the ground. On the contralateral side, you would relax your flexors and contract the extensors to stiffen the leg, since you must suddenly support your entire body on that one limb. At the same time, signals may travel up the spinal cord and cause contraction of contralateral muscles of the hip and abdomen to help shift your center of gravity over the extended leg. To a large extent, the coordination of all these muscles and maintenance of equilibrium is mediated by the cerebellum and cerebral cortex.

The flexor reflex is an **ipsilateral**[12] **reflex arc**—one in which the sensory input and motor output are on the same sides of the spinal cord. The crossed extensor reflex is a **contralateral**[13] **reflex arc**, in which the input and output are on opposite sides of the cord. When pain to the foot causes contractions of abdominal and hip muscles higher up the body, it is because nerve signals travel up the spinal cord and produce an output from a different segment of the cord than the one that received the input. This is called an **intersegmental reflex arc.**

The Golgi Tendon Reflex

Golgi tendon organs are proprioceptors located at the junction between a muscle and tendon (fig. 15.15). A tendon organ is about 1 mm long and consists of an encapsulated tangle of knobby nerve endings intertwined with collagen fibers of the tendon. The tendon organ is attached to a few fibers of the skeletal muscle at one end and to the tendon at the other end. As long as the tendon is slack, its collagen fibers are slightly spread and do not put pressure on the nerve endings woven among them. When muscle contraction pulls on the tendon, the collagen fibers come together like the two sides of a stretched rubber band and squeeze the Golgi nerve endings between them. The nerve fiber sends signals to the spinal cord that provide the CNS with feedback on the degree of muscle tension at the joint. If the tension is excessive, α motor neurons to the muscle fibers are inhibited and the muscle fibers relax.

This inhibitory response is the **Golgi tendon reflex.** One of its functions is to stop strong muscle contraction before it tears a tendon or pulls it loose from the muscle or bone. Nevertheless, strong muscles and quick movements sometimes damage a tendon before the reflex can occur, causing such athletic injuries as a ruptured calcaneal tendon. The reflex also functions when some extrafusal fibers in a muscle are contracting more than others. It inhibits the fibers

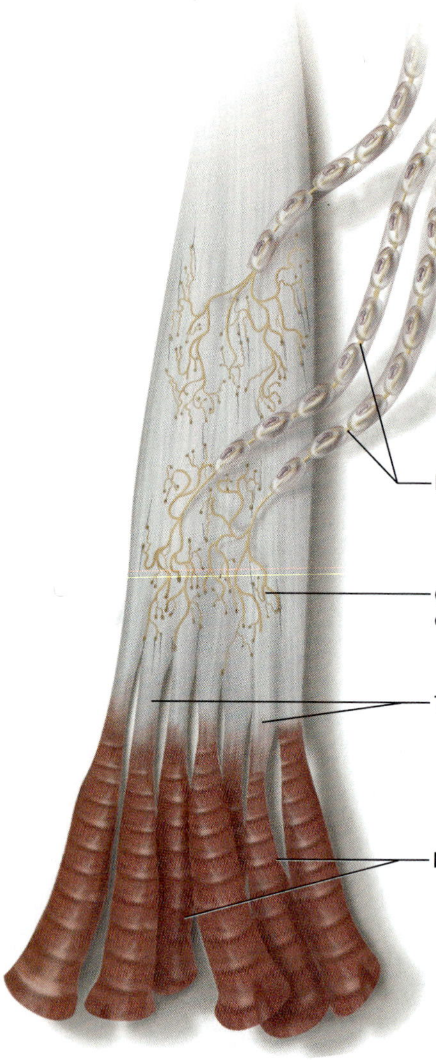

Figure 15.15 A Golgi tendon organ.

connected with overstimulated tendon organs and stimulates stronger contractions in fibers connected to the less stimulated tendon organs. This spreads the workload more evenly over the entire muscle. This is beneficial in such actions as maintaining a steady grip on a tool.

Key Point Review

14. Name five structural components of a typical spinal reflex arc. Which one of these is absent from a monosynaptic reflex arc?

15. State the function of each of the following in a muscle spindle: intrafusal fiber, annulospiral ending, and motor neuron.

16. Explain how nerve fibers in a tendon sense the degree of tension in a muscle.

17. Why must the withdrawal reflex, but not the stretch reflex, involve a polysynaptic reflex arc?

18. Explain why the crossed extensor reflex must accompany a withdrawal reflex of the leg.

12. *ipsi* = same + *later* = side
13. *contra* = opposite

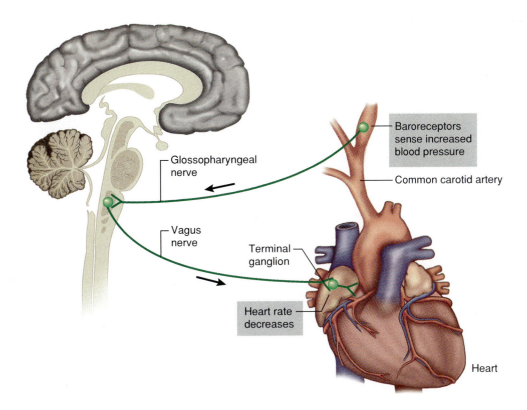

Figure 15.16 An autonomic reflex arc in which a rise in blood pressure is detected by baroreceptors in the carotid artery and the vagus nerve decreases the heart rate, resulting in a corrective drop in blood pressure.

The Autonomic Nervous System: Introduction and Anatomy

▼**Objectives**

When you have completed this section, you should be able to

• explain how the autonomic nervous system and the somatic nervous system differ in form and function;
• explain how the two divisions of the autonomic nervous system differ in general function;
• identify the anatomical components of the sympathetic and parasympathetic divisions; and
• discuss the relationship of the adrenal glands to the sympathetic nervous system.

General Properties

We now turn to visceral reflexes, which are mediated through the **autonomic nervous system (ANS).** The ANS has two divisions, the *sympathetic* and *parasympathetic,* whose differences will be discussed shortly. The ANS acts on all effectors other than skeletal muscle—its target organs are glands, cardiac muscle, and smooth muscle. The homeostasis of all organ systems depends on the ANS. Among other roles, it helps to regulate heart rate, blood pressure, body temperature, digestion, and waste elimination.

Autonomic literally means "self-governed." The ANS usually carries out its actions without our conscious intent or awareness. We are usually unaware, for example, of changes in the diameters of our pupils or blood vessels. Indeed, it is difficult to consciously alter or suppress autonomic responses, and for this reason some of them form a basis for polygraph ("lie detector") tests. Responses of the somatic nervous system have traditionally been called *voluntary* responses, and those of the autonomic nervous system have been called *involuntary.* This is not as clear-cut a distinction as it once seemed, however. Experiments in meditation and biofeedback, discussed later in this chapter, have shown that people can learn to control some autonomic functions. Furthermore, some skeletal muscle responses are quite involuntary, such as the spinal reflexes, and there are some skeletal muscles that cannot be voluntarily controlled at all, such as the middle-ear muscles.

The receptors for autonomic reflex arcs are nerve endings that detect stretch, pain, blood chemistry, and other internal stimuli. For example, high blood pressure stimulates nerve endings called *baroreceptors*[14] in the carotid arteries and aorta (fig. 15.16). They transmit signals via the glossopharyngeal nerves to the medulla oblongata. The medulla integrates this input with other

14. *baro* = pressure

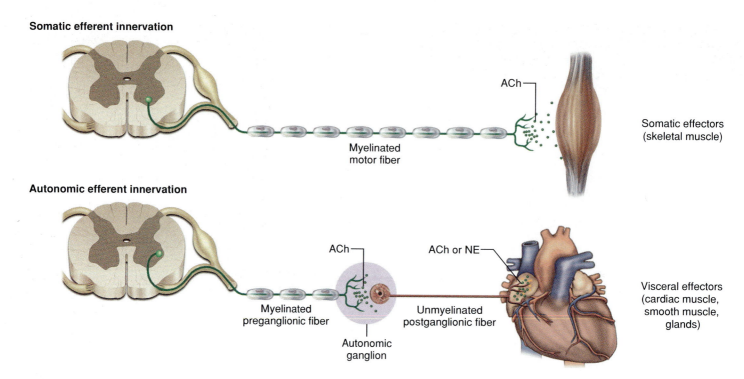

Somatic efferent innervation

ACh

Somatic effectors
(skeletal muscle)

Myelinated
motor fiber

Autonomic efferent innervation

ACh

ACh or NE

Visceral effectors
(cardiac muscle,
smooth muscle,
glands)

Myelinated
preganglionic fiber

Unmyelinated
postganglionic fiber

Autonomic
ganglion

Figure 15.17 Comparison of efferent pathways of the somatic and autonomic nervous systems.

information and transmits efferent signals back to the heart by way of parasympathetic fibers in the vagus nerves. These fibers slow down the heart, reducing blood pressure. A similar reflex arc acts through sympathetic fibers to accelerate the heart when blood pressure drops below normal.

Think About It

Draw a negative feedback loop to show how autonomic effects on the heart contribute to homeostasis.

Fibers of the ANS travel through the cranial and spinal nerves you have already studied and arise from somas located in both the central and peripheral nervous systems. The ANS, however, differs from the somatic motor division in several respects. The most conspicuous anatomical difference between them is their efferent pathways. A somatic motor neuron has its soma in the gray matter of the CNS and its axon extends all the way to the skeletal muscle fibers; there are no ganglia along the way. In the autonomic nervous system, by contrast, *two* neurons span the distance from CNS to effector (fig. 15.17). The first neuron has its soma in the CNS and its axon terminates in a ganglion of the PNS. This neuron is therefore called the **preganglionic (presynaptic) neuron.** In the ganglion, it synapses with a **postganglionic (post-**

synaptic) neuron whose axon extends the rest of the way to the target cells.

A skeletal muscle cannot contract unless stimulated by motor neurons. In the absence of innervation, it exhibits flaccid paralysis. Effectors of the ANS, however, function even in the absence of nervous stimulation—for example, the heart goes on beating even if it is *denervated* (all nerve connections to it are severed). The ANS merely *modifies* its activity, for example by accelerating or decelerating the heart. When autonomic nerves are severed from smooth or cardiac muscle, the muscle exhibits not flaccid paralysis but exaggerated sensitivity, called *denervation hypersensitivity.* Differences between the somatic and autonomic nervous systems are summarized in table 15.7.

Divisions

The two divisions of the autonomic nervous system differ in anatomy and function and often have contrasting effects on the same target organs. The **sympathetic division** prepares the body in many ways for physical activity—it makes the heart beat faster and harder, enhances pulmonary air flow, elevates blood glucose concentration, increases blood flow to cardiac muscle and the skeletal muscles, and so forth. Walter Cannon, father of the homeostasis concept (see chapter 1), referred to these sympathetic responses as the "fight

Table 15.7	Comparison of the Motor Components of the Somatic and Autonomic Nervous Systems	
Feature	**Somatic**	**Autonomic**
Effectors	Skeletal muscle	Smooth muscle, cardiac muscle, and glands
Efferent pathways	One nerve fiber from CNS to effector; no motor ganglia	Two nerve fibers from CNS to effector; synapse at a motor ganglion
Neurotransmitters	Acetylcholine (ACh)	ACh or norepinephrine (NE)
Effect on target cells	Always excitatory	Excitatory or inhibitory
Effect of denervation	Flaccid paralysis	Denervation hypersensitivity
Degree of voluntary control	Usually voluntary	Usually involuntary

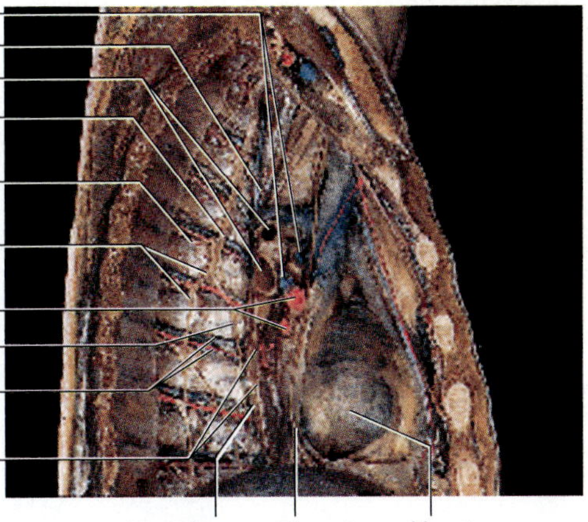

Pulmonary a.
Cardiac n.
Bronchi
Thoracic ganglion
Communicating ramus
Sympathetic chain
Pulmonary v.
Splanchnic n.
Intercostal a. and v.
Vagus n.

Esophagus   Phrenic n.   Heart

Figure 15.18 The sympathetic chain (paravertebral) ganglia, right lateral view of the thoracic cavity.

or flight" reaction. In animals it comes into play when the individual must attack, defend itself, or flee from danger. In our own lives it occurs in many situations involving arousal, passion, competition, stress, danger, anger, or fear.

The **parasympathetic division,** by comparison, has a calming effect on many body functions. It is associated with reduced energy expenditure and normal bodily maintenance, including such functions as digestion, defecation, and urination. This is sometimes referred to as the "resting and digesting" state to contrast it with fight or flight.

Anatomy of the Sympathetic Division

The sympathetic division is also called the **thoracolumbar division** because its preganglionic fibers arise from the thoracic and lumbar regions of the spinal cord. Its somas are in the lateral horns of the gray matter at these levels. Their fibers exit by way of spinal nerves T1 to L2 and lead to the **sympathetic chain** of ganglia (**paravertebral**[15] **ganglia**) along each side of the spinal column (figs. 15.18 and 15.19). Although this chain receives input only from the thoracolumbar region of the cord, it extends into the cervical and sacral regions as well. Sympathetic ganglia in these regions receive their input

from nerve fibers traveling up and down the chain. The number of ganglia varies from person to person, but usually there are 3 cervical (*superior, middle,* and *inferior*), 11 thoracic, 4 lumbar, 4 sacral, and 1 coccygeal ganglia.

In the thoracolumbar region, each paravertebral ganglion is connected to a spinal nerve by two branches called **communicating rami** (fig. 15.20). The preganglionic fibers from the spinal cord to the ganglion are small myelinated (type B) fibers that travel from the spinal nerve to the ganglion by way of the **white communicating ramus,** which gets its color from the myelin. The postganglionic fibers are unmyelinated (type C) fibers that leave the ganglion by way of the **gray communicating ramus** and other routes and extend the rest of the way to the target cells.

Preganglionic fibers may follow any of these courses after they enter the sympathetic chain:

- Some synapse with a postganglionic neuron in the ganglion that they enter.
- Some travel up or down the chain and synapse with postganglionic neurons at other levels. It is these fibers that link the paravertebral ganglia into a chain, and they are the only route by which ganglia at the cervical and sacral-to-coccygeal levels receive input.
- Some pass through the ganglion without synapsing and continue as *splanchnic* (SPLANK-nic) *nerves,* to be considered shortly.

Most preganglionic neurons synapse with postganglionic neurons in the paravertebral ganglia, but by no

15. *para* = next to

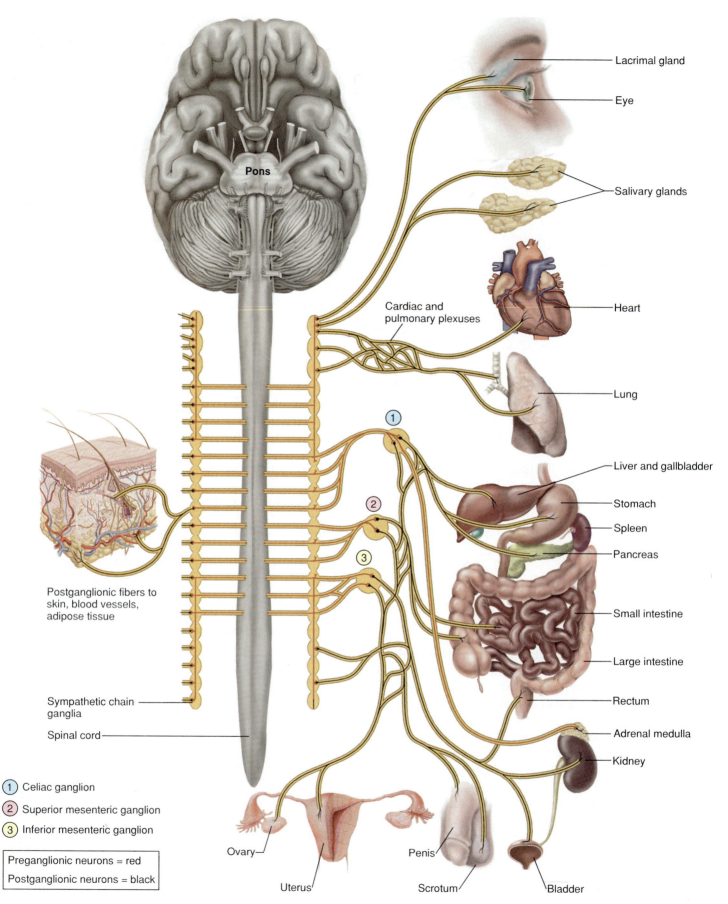

Lacrimal gland

Eye

Salivary glands

Cardiac and
pulmonary plexuses

Heart

Lung

① (Celiac ganglion)

Liver and gallbladder

Stomach

② (Superior mesenteric ganglion)

Spleen

Pancreas

③ (Inferior mesenteric ganglion)

Small intestine

Postganglionic fibers to
skin, blood vessels,
adipose tissue

Large intestine

Rectum

Sympathetic chain
ganglia

Adrenal medulla

Spinal cord

Kidney

① Celiac ganglion

② Superior mesenteric ganglion

③ Inferior mesenteric ganglion

Ovary

Penis

Uterus

Scrotum

Bladder

Pons

Preganglionic neurons = red
Postganglionic neurons = black

Figure 15.19 Efferent pathways of the sympathetic nervous system.

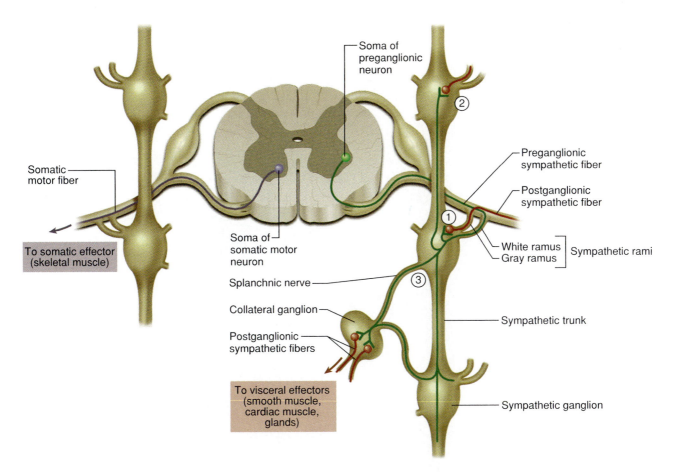

Figure 15.20 Efferent pathways of the sympathetic nervous system (*right*) compared to the somatic nervous system (*left*). Sympathetic efferent fibers can follow any of the three numbered routes. (*1*) The spinal nerve route. (*2*) The sympathetic nerve route. (*3*) The splanchnic nerve route.

means do they have a simple one-to-one relationship. For one thing, each postganglionic cell may receive synapses from multiple preganglionic cells, exhibiting the principle of *neuronal convergence* discussed in chapter 13. Furthermore, each preganglionic fiber branches and synapses with multiple postganglionic fibers, thus showing *neuronal divergence.* Indeed, there are about 17 postganglionic neurons for every preganglionic neuron in the sympathetic division. This means that when one preganglionic neuron fires, it can excite multiple postganglionic fibers that lead to different target organs. Thus when the sympathetic division is excited, it can affect several organ systems at once—as suggested by the name sympathetic.[16] This effect, called **mass activation,** would not be possible without the divergent neural pathways established in the paravertebral ganglia. It is a fine example of the unity of form and function.

Nerve fibers leave the paravertebral ganglia by three routes: spinal, sympathetic, and splanchnic

nerves. These are numbered in figure 15.20 to correspond to the following descriptions:

1. **The spinal nerve route.** Some postganglionic fibers exit by way of the gray communicating ramus, return to the spinal nerve or its subdivisions, and travel the rest of the way to the target organ by way of spinal nerve branches. This is the route taken by sympathetic fibers to targets such as most sweat glands and arrector pili muscles, and blood vessels of the skin and skeletal muscles.

2. **The sympathetic nerve route.** Other postganglionic fibers leave by way of **sympathetic nerves** that extend to the heart, lungs, esophagus, and thoracic blood vessels. Some of these nerves form a plexus around each carotid artery and issue fibers from there to effectors in the head—including sweat, salivary, and nasal glands, arrector pili muscles, blood vessels, and dilators of the iris. Some fibers from the superior cervical ganglion form the **cardiac nerves** to the heart.

3. **The splanchnic nerve route.** This is the route formed by fibers that pass through the paravertebral

16. *sym* = together + *path* = feeling

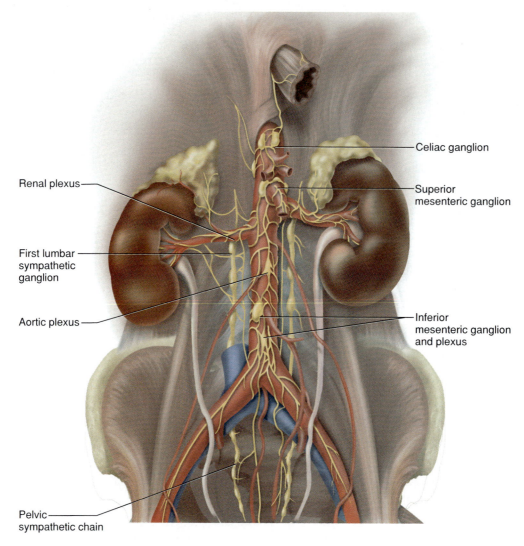

Figure 15.21 Sympathetic collateral ganglia and the abdominal aortic plexus.

Labels on figure:
- Celiac ganglion
- Renal plexus
- Superior mesenteric ganglion
- First lumbar sympathetic ganglion
- Aortic plexus
- Inferior mesenteric ganglion and plexus
- Pelvic sympathetic chain

ganglia without synapsing. These fibers originate predominantly from spinal nerves T5 to T12. After passing through the paravertebral ganglia, they form **splanchnic**[17] **nerves,** which lead to synapses in another set of ganglia, the **collateral (prevertebral) ganglia.** These contribute to a network, known as the **abdominal aortic plexus,** which is wrapped around the aorta (fig. 15.21). There are three major collateral ganglia in this plexus, which occur at points where certain major arteries branch off the aorta—the **celiac (SEE-lee-ac) ganglion, superior mesenteric ganglion,** and **inferior mesenteric ganglion.** Their postganglionic fibers accompany these arteries and their branches to the target organs. (The *solar plexus* is regarded as a collective

designation for the celiac and superior mesenteric ganglia by some authorities and as a synonym for only the celiac ganglion by others.) Innervation to and from the three major collateral ganglia is summarized in table 15.8.

The Adrenal Glands

The paired **adrenal**[18] **glands** rest like hats, one on the superior pole of each kidney (fig. 15.21). Each adrenal is actually two separate glands with different functions and embryonic origins. The outer rind, the **adrenal cortex,** secretes several steroid hormones discussed in chapter 17. The inner core, the **adrenal medulla,** consists of closely packed rounded cells that are modified

---

17. *splanchn* = viscera

18. *ad* = toward, near + *ren* = kidney

Table 15.8

Table 15.8   Innervation to and from the Collateral Ganglia

Sympathetic Ganglia and Splanchnic Nerve →	Collateral Ganglion →	Postganglionic Target Organs
From thoracic ganglion 5 to 9 or 10 via greater splanchnic nerve	Celiac ganglion	Stomach, spleen, liver, small intestine, and kidneys
From thoracic ganglia 9 and 10 via lesser splanchnic nerve	Celiac and superior mesenteric ganglia	Small intestine and colon
From lumbar ganglia via lumbar splanchnic nerve	Celiac and inferior mesenteric ganglia	Distal colon, rectum, urinary bladder, and reproductive organs

neurons of the sympathetic nervous system—that is, the adrenal medulla is actually a modified sympathetic ganglion. Preganglionic sympathetic fibers from the thoracic spinal cord penetrate through the cortex and terminate on cells of the medulla. When stimulated, the medulla secretes a mixture of hormones into the bloodstream—about 85% epinephrine (adrenalin), 15% norepinephrine (noradrenalin), and a trace of dopamine. These hormones, the *catecholamines,* were briefly considered in chapter 13 because they also function as neurotransmitters. The adrenal catecholamines complement the action of other sympathetic postganglionic fibers, and so the adrenal medulla is involved in the mass activation typical of this division. The sympathetic nervous system and adrenal medulla are so closely related in development and function that they are referred to collectively as the *sympathoadrenal system.*

Anatomy of the Parasympathetic Division

The parasympathetic nervous system is also called the **craniosacral division** because its fibers travel in certain cranial and sacral nerves. The preganglionic neurons have their somas in the pons, medulla oblongata, and segments S2 to S4 of the spinal cord (fig. 15.22). The parasympathetic ganglia, called **terminal ganglia** (see fig. 15.16), lie in or near the target organs. When they are embedded in the wall of a target organ, they are also known as **intramural**[19] **ganglia.** Thus, the parasympathetic division has long preganglionic fibers, reaching almost all the way to the target cells, and short postganglionic fibers that cover the rest of the distance.

There is a slight degree of neuronal divergence in the parasympathetic division, with a ratio of about two postganglionic fibers to each preganglionic. This divergence is much less than in the sympathetic division, however, and parasympathetic effects do not exhibit the mass activation characteristic of sympathetic effects. The parasympathetic division is relatively selective in its stimulation of target organs since the preganglionic fibers reach the organs before any postganglionic divergence occurs.

Parasympathetic fibers leave the brainstem by way of four cranial nerves: the oculomotor (III), facial (VII), glossopharyngeal (IX), and vagus (X). The first three of these nerves supply all parasympathetic innervation to the head, and the vagus supplies parasympathetic innervation to viscera of the thoracic cavity and most of the abdominal cavity. The following discussion traces each of these in more detail.

Oculomotor nerve. Oculomotor fibers begin in a nucleus of the midbrain, enter the orbit, and end there in the *ciliary ganglion.* Postganglionic fibers enter the eyeball and innervate the *ciliary muscle,* which controls the lens, and the *pupillary constrictor,* which narrows the pupil.

Facial nerve. The facial nerve arises from the superior salivatory nucleus of the lower pons. Parasympathetic fibers soon split away from the rest of the facial nerve to form two smaller branches. The upper branch ends at a ganglion near the junction of the maxilla and palatine bones. Postganglionic fibers then continue to the lacrimal (tear) glands and scattered glands of the nasal cavity, palate, and other areas of the oral cavity. The lower branch crosses the middle-ear cavity and ends at a ganglion near the angle of the mandible. Postganglionic fibers from here supply salivary glands in the floor of the mouth. Thus the facial nerve parasympathetics primarily regulate the secretion of tears, saliva, and nasal secretions.

Glossopharyngeal nerve. This nerve begins in the inferior salivatory nucleus of the medulla oblongata. Its parasympathetic fibers soon leave the nerve and form the *tympanic nerve,* which crosses the eardrum and ends in the *otic*[20] *ganglion* near the foramen ovale of the skull. Postganglionic fibers from here follow the trigeminal nerve and innervate a salivary gland just in front of the earlobe. Thus they complement the fibers of the facial nerve in regulating salivation.

---

19. *intra* = within + *mur* = wall

20. *ot* = ear + *ic* = pertaining to

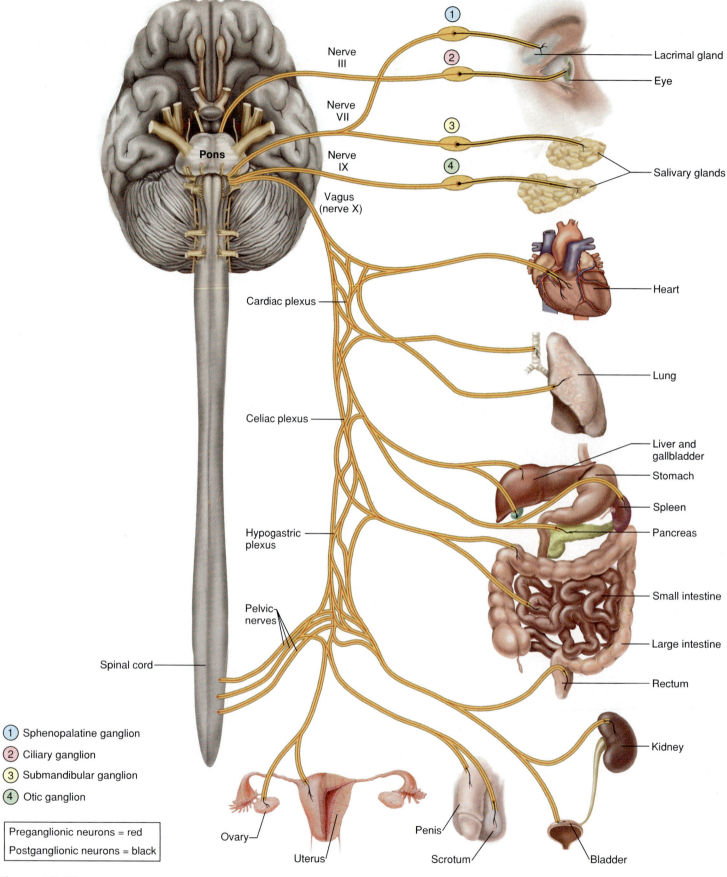

Nerve III

Nerve VII

Nerve IX

Vagus (nerve X)

Pons

Cardiac plexus

Celiac plexus

Hypogastric plexus

Pelvic nerves

Spinal cord

1 Sphenopalatine ganglion
2 Ciliary ganglion
3 Submandibular ganglion
4 Otic ganglion

Preganglionic neurons = red
Postganglionic neurons = black

Lacrimal gland

Eye

Salivary glands

Heart

Lung

Liver and gallbladder

Stomach

Spleen

Pancreas

Small intestine

Large intestine

Rectum

Kidney

Bladder

Ovary

Uterus

Penis

Scrotum

Figure 15.22 Efferent pathways of the parasympathetic nervous system.

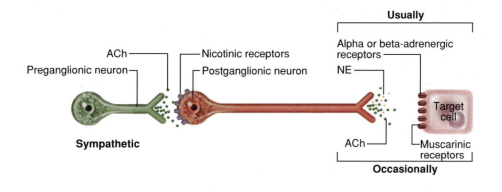

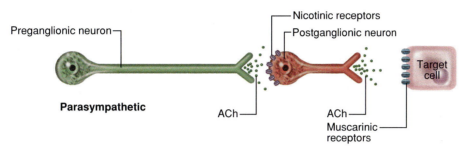

Figure 15.23 Neurotransmitters and receptors of the autonomic nervous system.

Table 15.9	Comparison of the Sympathetic and Parasympathetic Divisions	
Feature	**Sympathetic**	**Parasympathetic**
Origin in CNS	Thoracolumbar	Craniosacral
Location of ganglia	Paravertebral ganglia adjacent to spinal column and prevertebral ganglia anterior to it	Terminal ganglia near or within target organs
Fiber lengths	Short preganglionic Long postganglionic	Long preganglionic Short postganglionic
Neuronal divergence	Extensive (about 1:17)	Minimal (about 1:2)
Effects of system	Mass activation common	More specific and local effects

Vagus nerve. About 90% of all parasympathetic preganglionic fibers travel in the vagus nerve. Vagal preganglionic fibers arise from the *dorsal motor nuclei of the vagus* in the medulla oblongata. The vagus nerves travel down the neck in the *carotid sheath,* a fibrous sleeve that also encloses the carotid artery and internal jugular vein. Vagal fibers then form three nerve plexuses in the mediastinum. From superior to inferior, these are the **cardiac plexus,** which supplies fibers to the heart; the **pulmonary plexus,** whose fibers accompany the bronchi and blood vessels into the lungs; and the **esophageal plexus,** whose fibers regulate esophageal peristalsis.

At the lower end of the esophagus, these plexuses give off the **anterior** and **posterior vagal trunks,** each of

which contains fibers from the right and left vagus. These travel with the esophagus through the diaphragm, enter the abdominal cavity, and contribute to the extensive abdominal aortic plexus. This plexus, consisting of intermingled sympathetic and parasympathetic fibers, has already been considered in the discussion of the sympathetic division. As we have seen, the sympathetic fibers synapse here. The parasympathetic fibers, however, pass through without synapsing and lead to the liver, pancreas, stomach, small intestine, and proximal half of the large intestine.

The remaining parasympathetic efferent fibers arise from levels S2 to S4 of the spinal cord. They travel a short distance in the ventral rami of the spinal nerves and then form **pelvic splanchnic nerves** that lead to the **inferior hypogastric (pelvic) plexus.** Some parasympathetic fibers synapse here, but most pass through the pelvic plexus and travel by way of **pelvic nerves** to the terminal ganglia in their target organs: the distal half of the large intestine, the rectum, ureters, urinary bladder, and reproductive organs.

The sympathetic and parasympathetic divisions of the ANS are compared in figure 15.23 and table 15.9.

19 In what ways is an autonomic reflex arc like a somatic motor reflex arc? In what ways does it differ?

20 Explain why the sympathetic division is also called the thoracolumbar division in spite of the fact that its chain of paravertebral ganglia extends all the way from the cervical to the sacral region.

21 Describe or diagram the structural relationships among the following: preganglionic fiber, postganglionic fiber, ventral ramus, gray ramus, white ramus, and paravertebral ganglion.

22 Explain in anatomical terms why the parasympathetic division affects target organs more selectively than the sympathetic division does.

23 Trace the pathway of a parasympathetic fiber of the vagus nerve from the medulla oblongata to the small intestine.

The Autonomic Nervous System: Physiology

▼Objectives
When you have completed this section, you should be able to
• name the neurotransmitters that are employed at different synapses of the ANS;
• name the kinds of receptors that exist for these neurotransmitters and explain how these receptor types relate to autonomic effects on target organs;
• explain how the ANS controls many target organs through dual innervation;
• explain how control is exerted in the absence of dual innervation; and
• discuss the ways in which the ANS is governed by various levels of the CNS.

Neurotransmitters and Receptors

The key to understanding the effects of autonomic activity lies in knowing which neurotransmitters are released by the autonomic fibers and what kind of receptors occur on the target cells. The ANS has **cholinergic fibers,** which secrete acetylcholine (ACh), and **adrenergic fibers,** which secrete norepinephrine (NE). Cholinergic fibers include all preganglionic fibers of both divisions, all postganglionic fibers of the parasympathetic division, and a few sympathetic postganglionic fibers (those that innervate sweat glands and some blood vessels in the skeletal muscles). The great majority of sympathetic postganglionic fibers are adrenergic (table 15.10).

The receptors for acetylcholine and norepinephrine are called **cholinergic** and **adrenergic receptors,** respectively. They are located on the dendrites and somas of the postganglionic neurons and on the effector (muscle and gland) cells.

Table 15.10 Locations of Cholinergic and Adrenergic Fibers in the ANS

Division	Preganglionic Fibers	Postganglionic Fibers
Sympathetic	Always cholinergic	Usually adrenergic; a few cholinergic
Parasympathetic	Always cholinergic	Always cholinergic

Both the sympathetic and parasympathetic divisions have excitatory effects on some effectors and inhibitory effects on others. For example, the parasympathetic division stimulates the wall of the urinary bladder to contract but stimulates the internal urinary sphincter to relax—both of which are necessary for the expulsion of urine. In both cases, the neurotransmitter employed is acetylcholine. Similarly, some adrenergic fibers of the sympathetic division cause contraction of the iris and most blood vessels but relaxation of the bronchioles and coronary arteries. Clearly, the difference is not due to the neurotransmitter. Rather, it is due to the fact that different effector cells have *different kinds of receptors* for the neurotransmitter. Knowledge of these different receptor types is essential in the field of neuropharmacology, as you will see.

Cholinergic Receptors

Acetylcholine binds to two classes of receptors—**nicotinic** (NIC-oh-TIN-ic) and **muscarinic** (MUSS-cuh-RIN-ic) **receptors**—named for the drugs that were used to identify and distinguish them. Nicotine binds only to the former type, while muscarine, a mushroom poison, binds only to the latter. Other drugs also selectively bind to one type or the other—atropine binds only to muscarinic receptors and curare only to nicotinic receptors, for example (see chapter essay, p. 545).

Nicotinic receptors occur on all postganglionic somas of the ANS, on the adrenal medulla cells, and at the neuromuscular junctions discussed in chapter 12. Muscarinic receptors occur on all cholinergic target cells of the ANS. When ACh binds to a nicotinic receptor, it always produces an excitatory postsynaptic potential (EPSP). When it binds to a muscarinic receptor, however, the effect can be excitatory (as in intestinal smooth muscle) or inhibitory (as in cardiac muscle).

Adrenergic Receptors

There are likewise different classes of adrenergic receptors that account for the different effects of norepinephrine (NE) on different target cells. NE receptors fall into two broad classes called **alpha-(α-)adrenergic** and **beta-(β-)adrenergic receptors.** The binding of NE to α receptors is usually excitatory, and its binding to β receptors is usually inhibitory, but there are exceptions to both. For example, NE binds to β receptors in cardiac muscle but has an excitatory effect.

The exceptions result from the existence of subclasses of each receptor type, called α_1 and α_2, β_1 and β_2 receptors. All four types function by means of second-messenger systems. Both types of β receptors activate the production of cyclic AMP (cAMP) when they bind norepinephrine. The α_2 receptors suppress cAMP production, and the α_1 receptors act by mobilizing calcium ions, which serve as the second messenger. Some target cells have both α and β receptors.

The binding of NE to α-adrenergic receptors on the smooth muscle fibers of blood vessels causes vasoconstriction. The arteries that supply the heart and skeletal muscles, however, have β-adrenergic receptors. When NE binds to these receptors it relaxes the arterial walls, causing vasodilation and increased blood flow to these organs. NE also relaxes the smooth muscle of the bronchioles when it binds to β-adrenergic receptors.

The locations and effects of many cholinergic and adrenergic receptor types are summarized in table 15.11. In a number of cases, it is not known yet which subclass of α or β receptor is present in a given organ, which is why several entries in the table have no subscripts. The autonomic effects on glandular secretion are often achieved through the adjustment of blood flow to the gland rather than direct stimulation of the gland cells.

Think About It

It is noted in table 15.11 that the sympathetic nervous system has an α-adrenergic effect on blood platelets and promotes clotting. How can this be, considering that the platelets are drifting cell fragments in the bloodstream and have no nerve fibers leading to them?

Dual Innervation

Most of the viscera receive nerve fibers from both the sympathetic and parasympathetic divisions, and thus they are said to have **dual innervation.** In such cases, the two divisions often have contrasting effects on the same organ, as seen in table 15.11.

In such cases, the two divisions may have *antagonistic* or *cooperative* effects. **Antagonistic effects** oppose each other. For example, the sympathetic division increases heart rate, and the parasympathetic division decreases it; the sympathetic division inhibits gastrointestinal movement and secretion, and the parasympathetic division stimulates them; the sympathetic division dilates the pupil, and the parasympathetic division constricts it. In some cases, these effects are exerted through dual innervation of the same effector cells, as in the heart, where nerve fibers of both divisions terminate on the same muscle cells. In other cases, antagonistic effects arise because each division innervates different effector cells that have opposite effects on organ function. In the iris of the eye, for example, contractile fibers that dilate the pupil receive sympathetic innervation, whereas a different set of fibers that constrict the pupil receive parasympathetic innervation (fig. 15.24).

Cooperative effects are seen when the two divisions act on different effectors to produce an overall effect. Salivation is a good example of this. The parasympathetic division stimulates serous cells of the salivary glands to secrete a watery, enzyme-rich secretion, while the sympathetic division stimulates mucous cells of the same glands to secrete mucus. The enzymes and mucus are both necessary components of the saliva.

Even when both divisions innervate a single organ, they do not always innervate it equally or exert equal influence. For example, the parasympathetic division forms an extensive plexus in the wall of the digestive tract and exerts much more influence over digestive function than does the sympathetic division. In the ventricles of the heart, by contrast, parasympathetic innervation is substantially less than that of the sympathetic division.

Control Without Dual Innervation

Dual innervation is not always necessary for the ANS to produce opposite effects on an organ. The adrenal medulla, arrector pili muscles, sweat glands, and many blood vessels receive only sympathetic fibers. The most significant example of control without dual innervation is regulation of blood pressure and routes of blood flow. The sympathetic fibers to a blood vessel have a baseline firing frequency called **sympathetic tone.** This keeps the vessels in a state of partial constriction called **vasomotor tone.** An increase in firing frequency causes vasoconstriction by stimulating a greater smooth muscle contraction. A drop in firing frequency causes vasodilation as the smooth muscle relaxes. Thus, the sympathetic division alone can exert opposite effects on the vessels.

Sympathetic control of vasomotor tone can shift blood flow from one organ to another according to the changing needs of the body. In times of stress or danger,

Table 15.11 Effects of the Sympathetic and Parasympathetic Nervous Systems on Various Targets

Target	Sympathetic Effect and Receptor Type*	Parasympathetic Effect and Receptor Type*
Adipose Tissue	Decreased fat breakdown (α_2)	No effect
	Increased fat breakdown (β_1)	
Eye		
Dilator of Pupil	Pupillary dilation (α_1)	No effect
Constrictor of Pupil	No effect	Pupillary constriction (M)
Ciliary Muscle and Lens	Relaxation for far vision (β_1)	Contraction for near vision (M)
Lacrimal (Tear) Gland	Less secretion (α)	More secretion (M)
Integumentary System		
Merocrine Sweat Glands (Cooling)	Secretion (M)	No effect
Apocrine Sweat Glands (Scent)	Secretion (α)	No effect
Arrector Pili Muscles	Hair erection (α)	No effect
Adrenal Medulla	Hormone secretion (N)	No effect
Circulatory System		
Heart Rate and Force	Increased (α)	Decreased (M)
Deep Coronary Arteries	Vasodilation (β_2)	Slight vasodilation (M)
Blood Vessels of Most Viscera	Vasoconstriction (α)	Vasodilation (M)
Blood Vessels of Skeletal Muscles	Vasodilation (β_2)	No effect
Blood Vessels of Skin	Vasoconstriction (α)	Vasodilation, blushing (M)
Platelets (Blood Clotting)	Increased clotting (α_2)	No effect
Respiratory System		
Bronchi and Bronchioles	Bronchodilation (β_2)	Bronchoconstriction (M)
Mucous Glands	Decreased secretion (β_1)	No effect
	Increased secretion (β_2)	
Urinary System		
Kidneys	Reduced urine output (α)	No effect
Bladder Contraction	No effect	Stimulation (M)
Internal Urinary Sphincter	Contraction (α)	Relaxation (M)
Digestive System		
Salivary Glands	Thick mucous secretion (α_1)	Thin serous secretion (M)
Gastrointestinal Peristalsis	Inhibition (α_1)	Stimulation (M)
Gastrointestinal Secretion	No effect	Stimulation (M)
Liver	Glycogen breakdown (β_2)	Glycogen synthesis (M)
Pancreatic Secretion (Exocrine)	Decreased secretion (α)	Increased secretion (M)
Pancreatic Insulin Secretion	Decreased secretion (α_2)	No effect
	Increased secretion (β_2)	
Reproductive System		
Orgasm, Smooth Muscle Roles	Stimulation (α)	No effect
Penile or Clitoral Erection	No effect	Stimulation (M)
Glandular Secretion	No effect	Stimulation (M)
Uterus in Orgasm	Dilation of cervical canal	No effect
Uterus in Labor	Contraction (α)	No effect

*Adrenergic receptors are indicated α and β, with subclass if known. Cholinergic receptors are indicated by M for muscarinic and N for nicotinic.

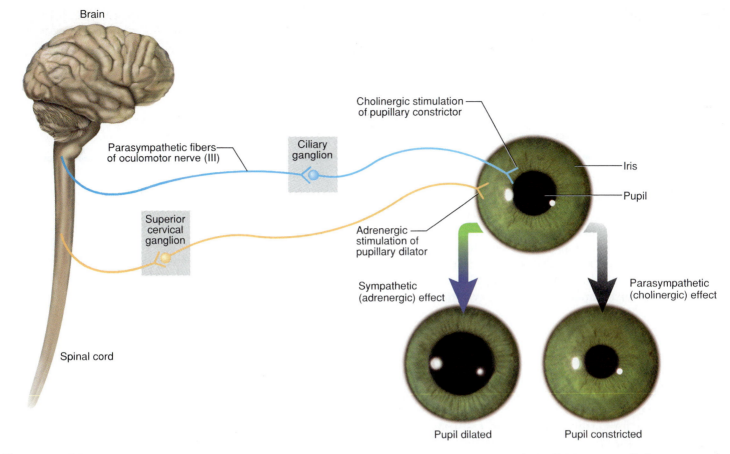

Figure 15.24 Dual innervation of the iris and antagonistic effects of the sympathetic and parasympathetic divisions on pupil diameter.

the skeletal muscles and heart receive a high circulatory priority, and the sympathetic division dilates the arteries that supply them. Such processes as digestion, nutrient absorption, and urine formation can wait; thus the sympathetic division constricts arteries to the gastrointestinal tract and kidneys. It also reduces blood flow through the skin, which may help to minimize bleeding in the event that the stress-producing situation leads to injury. Furthermore, since there is not enough blood in the body to abundantly supply all the organ systems at once, it is necessary to temporarily divert blood away from other organs in order to supply the muscular system adequately during times of stress.

Central Control of Autonomic Function

In spite of its name, the ANS is not an independent nervous system but is regulated by several levels of the CNS. In this section we consider how the cerebrum, hypothalamus, brainstem, and spinal cord influence autonomic function.

Cerebral Control

Conscious processes in the cerebrum can cause autonomic effects—for example, when anger makes the blood pressure rise or anxiety inhibits sexual function. Recent studies and therapeutic uses of meditation and biofeedback show that the ANS is not entirely free of conscious control (see special topic 15.5). Hindu yogis and other people adept at deep meditation techniques can voluntarily slow their heart rates and lower their metabolism and oxygen use. Some can make one hand several degrees warmer than the other by selectively dilating or constricting the arteries in the skin. The deep cerebral components of the limbic system also initiate many emotional responses that are carried out by way of the hypothalamus and ANS.

Hypothalamic Control

The hypothalamus is the most important integrating center for autonomic function. As we saw in the previous chapter, it contains nuclei for numerous visceral functions including salivation, sweating, vasodilation and vasoconstriction, cardiac and pulmonary function, sexual response, and many emotions and drives that act through the ANS. Signals from the limbic system act through the hypothalamus to produce the fight or flight response characteristic of the sympathetic nervous system.

Brainstem Control

Output from the hypothalamus travels largely to the brainstem and from there to the cranial nerves and spinal cord. The most direct influence over autonomic function is exerted by the reticular formation, which contains centers for cardiac, vasomotor, respiratory, and gastrointestinal function. The reticular formation also receives extensive sensory input from the viscera by way of the vagus nerve, and it can integrate and respond directly to this even without cerebral, cerebellar, or hypothalamic input. Most autonomic functions can be produced experimentally by direct electrical stimulation of the brainstem.

Spinal Control

Some autonomic responses, such as the defecation and micturition (urination) reflexes, are integrated in the spinal cord without the involvement of the brain. Fortunately, however, the brain is able to inhibit these responses consciously.

Key Point Review

24 Summarize the locations of nicotinic and muscarinic receptors in the ANS.

25 What second messenger is involved in the action of α_2, β_1, and β_2 adrenergic receptors but not α_1 receptors? What second messenger is involved in α_1 receptors?

26 Explain what is meant by dual innervation and give some examples.

27 Describe how the sympathetic division can control blood flow in the absence of dual innervation.

28 Cite evidence indicating that the ANS is subject to cerebral influence.

CHAPTER ESSAY

Drugs and the Nervous System

Neuropharmacology is the branch of medicine that deals with the effects of drugs on the nervous system, especially those that mimic, enhance, or inhibit the action of neurotransmitters. A few examples will help illustrate the clinical relevance of understanding neurotransmitter and receptor functions.

A number of drug actions are due to their effects on adrenergic and cholinergic nerve fibers and receptors. **Sympathomimetics**[21] are drugs that enhance sympathetic action by stimulating adrenergic receptors or promoting norepinephrine release. For example phenylephrine, found in such cold medicines as Chlor-Trimeton and Dimetapp, aids breathing by stimulating α_1 receptors and dilating the bronchioles. **Sympatholytics**[22] are drugs that suppress sympathetic action by binding to adrenergic receptors without stimulating them or by inhibiting norepinephrine release. Propanolol, for example, is a **beta-blocker.** It reduces hypertension (high blood pressure) by blocking β-adrenergic receptors and interfering with the effects of epinephrine and norepinephrine on the heart and blood vessels.

Parasympathomimetics enhance parasympathetic effects. Pilocarpine, for example, is used to treat glaucoma (excessive pressure within the eyeball) by dilating a vessel that drains fluid (aqueous humor) from the eye. **Parasympatholytics** inhibit ACh release or block its receptors. Atropine, for example, blocks muscarinic receptors and is sometimes used to dilate the pupils for eye examinations and to dry the mucous membranes of the respiratory tract before inhalation anesthesia. It is an extract of the deadly nightshade plant, *Atropa belladonna.* Women of the Middle Ages used nightshade to dilate their pupils, which was thought to enhance their beauty.[23]

The drugs we have mentioned so far act on the peripheral nervous system and its effectors. Numerous other drugs act on the central nervous system. Strychnine, for example, blocks the inhibitory action of glycine on spinal motor neurons, causing spastic paralysis of the skeletal muscles and sometimes death by suffocation.

Sigmund Freud predicted that psychiatry would eventually draw upon biology and chemistry to deal with emotional problems originally treated only by counseling and psychoanalysis. A branch of neuropharmacology called **psychopharmacology** is beginning to fulfill his prediction. Psychopharmacology dates to the 1950s, when chlorpromazine, an antihistamine, was accidentally found to alleviate the symptoms of schizophrenia. The field is still in its infancy and is still based more on trial-and-error experimentation and chance findings than on mechanistic predictions of how psychoactive drugs will work.

The management of clinical depression is one example of how contemporary psychopharmacology has supplemented counseling approaches. Some cases of depression are due to deficiencies of the monoamine neurotransmitters and yield to drugs that prolong the effects of the monoamines already present at the synapses. One of the earliest discovered antidepressants was imipramine, which blocks the synaptic reuptake of serotonin and norepinephrine. However, it produces undesirable side effects such as dry mouth and irregular cardiac rhythms; it has been largely replaced by Prozac (fluoxetine), which prolongs the mood-elevating effect of serotonin by blocking its reuptake. Prozac is also used to treat fear of rejection, excess sensitivity to criticism, lack of self-esteem, and inability to experience pleasure, all of which were long handled only through counseling, group therapy, and psychoanalysis. After monoamines are taken up from the synapse, they are degraded by monoamine oxidase (MAO). Another approach to the treatment of depression and other effects of monoamine deficiency is afforded by MAO inhibitors.

Our growing understanding of neurochemistry also gives deeper insight into the action of dangerously addictive drugs such as amphetamines and cocaine. Amphetamines ("speed") chemically resemble norepinephrine and dopamine, two neurotransmitters associated with elevated mood. Dopamine is especially important in sensations of pleasure. Cocaine blocks dopamine reuptake and thus produces a brief rush of good feelings. But when dopamine is not reabsorbed by the neurons, it diffuses out of the synaptic cleft and is degraded elsewhere. Cocaine therefore depletes the neurons of dopamine faster than they can synthesize it, so that finally there is no longer an adequate supply to maintain normal mood. The postsynaptic neurons make new dopamine receptors as if "searching" for the neurotransmitter—all of which

21. *mimet* = imitate, mimic
22. *lyt* = break down, destroy
23. *bella* = beautiful, fine + *donna* = woman

—continued

leads ultimately to anxiety, depression, and the inability to experience pleasure without cocaine.

Caffeine exerts its stimulatory effect by competing with adenosine. Adenosine, which you know as a component of DNA, RNA, and ATP, also functions as an inhibitory neurotransmitter in the brain. Caffeine has enough structural similarity to adenosine (fig. E.1) to bind to its receptors, but it does not inhibit the post-synaptic cell as adenosine would. In the absence of this inhibition, a person experiences elevated alertness and mood (although, in excessive amounts, caffeine can be harmful in many ways).▲

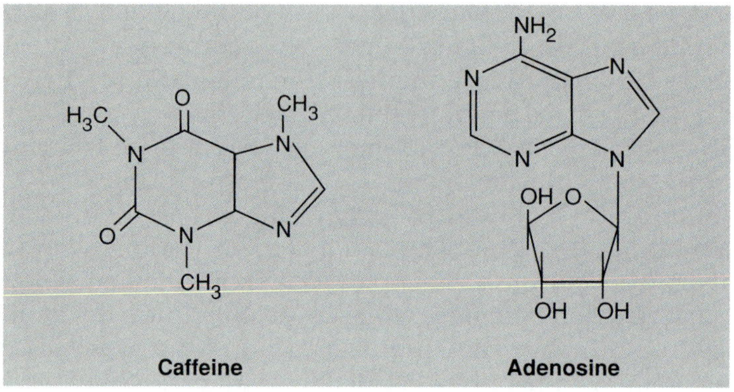

Figure E15.1 The structural similarity of caffeine to the neurotransmitter adenosine enables it to bind to adenosine receptors, blocking the inhibitory action of adenosine.

Connective Issues

Connective Issues

Interactions Between the NERVOUS SYSTEM and Other Organ Systems

Integumentary System
- Provides sensations of heat, cold, pressure, pain, and vibration; protects peripheral nerves
- Nervous system regulates piloerection and sweating; controls cutaneous blood flow to regulate heat loss

Skeletal System
- Serves as reservoir of Ca^{2+} needed for neural function; protects CNS and some peripheral nerves
- Nervous stimulation generates muscle tension essential for bone development and remodeling

Muscular System
- Gives expression to thoughts, emotions, and motor commands that arise in the CNS
- Somatic nervous system activates skeletal muscles and maintains muscle tone

Endocrine System
- Many hormones affect neuronal growth and metabolism; hormones control electrolyte balance essential for neural function
- Hypothalamus controls pituitary gland; sympathetic nervous system stimulates adrenal medulla

Circulatory System
- Delivers O_2 and carries away wastes; transports hormones to and from CNS; CSF produced from and returned to blood
- Nervous system regulates heartbeat, blood vessel diameters, blood pressure, and routing of blood; influences blood clotting

Lymphatic/Immune Systems
- Immune cells provide protection and promote tissue repair
- Nerves innervate lymphoid organs and influence development and activity of immune cells; nervous system plays a role in regulating immune response; emotional states influence susceptibility to infection

Respiratory System
- Provides O_2, removes CO_2, and helps to maintain proper pH for neural function
- Nervous system regulates rate and depth of respiration

Urinary System
- Disposes of wastes and maintains electrolyte and pH balance
- Nervous system regulates renal blood flow, thus affecting rate of urine formation; controls emptying of bladder

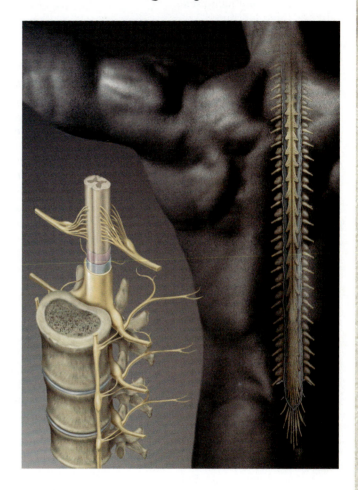

Digestive System
- Provides nutrients; liver provides stable level of blood glucose for neural function during periods of fasting
- Nervous system regulates appetite, feeding behavior, digestive secretion and motility, and defecation

Reproductive System
- Sex hormones influence CNS development and sexual behavior; hormones of the menstrual cycle stimulate or inhibit hypothalamus
- Nervous system regulates sex drive, arousal, and orgasm; secretes or stimulates pituitary release of many hormones involved in menstrual cycle, sperm production, pregnancy, and lactation

Chapter Review — Study Outline

Nerves, Nerve Fibers, and Ganglia (pp. 505–507)
1. Anatomy of a nerve
   a. Sheaths of a nerve fiber
      - Myelin sheath
      - Neurilemma
      - Basement membrane
      - Endoneurium
   b. Nerve fascicles
   c. Perineurium
   d. Epineurium
2. Sensory, motor, and mixed nerves
3. Types of nerve fibers
4. Ganglia

The Cranial Nerves (pp. 508–516)
Review table 15.2 for details.
   I. Olfactory
   II. Optic
   III. Oculomotor
   IV. Trochlear
   V. Trigeminal
      V_1 Ophthalmic
      V_2 Maxillary
      V_3 Mandibular
   VI. Abducens
   VII. Facial
   VIII. Vestibulocochlear
   IX. Glossopharyngeal
   X. Vagus
   XI. Accessory
   XII. Hypoglossal

The Spinal Nerves (pp. 516–525)
1. Classes
   a. Cervical nerves (C1–C8)
   b. Thoracic nerves (T1–T12)
   c. Lumbar nerves (L1–L5)
   d. Sacral nerves (S1–S5)
   e. Coccygeal nerve (Cx)
2. Proximal branches
   a. Dorsal root and rootlets
   b. Dorsal root ganglion
   c. Ventral root and rootlets
3. Distal branches
   a. Meningeal branch
   b. Dorsal ramus
   c. Ventral ramus
      - Intercostal nerves
      - Nerve plexuses
4. Cervical plexus (table 15.3)
5. Brachial plexus (table 15.4)
   a. Roots (ventral rami of C5–T1)
   b. Trunks: upper, lower, and middle
   c. Divisions: anterior and posterior
   d. Cords: posterior, medial, and lateral

6. Lumbar plexus (table 15.5)
7. Sacral and coccygeal plexuses (table 15.6)
8. Cutaneous innervation and dermatomes

Somatic Reflexes (pp. 525–530)
1. Nature of reflexes
   a. Require stimulation
   b. Quick
   c. Involuntary
   d. Stereotyped
2. Somatic and visceral reflexes
3. Components of a spinal reflex arc
   a. Somatic receptors in skin, muscle, or tendon
   b. Afferent nerve fibers
   c. Interneurons (sometimes)
   d. Efferent nerve fibers
   e. Skeletal muscles
3. Stretch reflex
   a. Structure of a muscle spindle
      - Intrafusal and extrafusal muscle fibers
      - Annulospiral and flower-spray nerve endings
      - α and γ motor neurons
   b. Mechanism
      - Muscle stretch stretches intrafusal fibers
      - Sensory nerve endings excited
      - Spinal integration
      - Alpha motor neurons stimulate extrafusal fibers
      - Muscle contracts
      - Reciprocal innervation inhibits antagonistic muscles
      - γ motor neurons maintain optimal tension in muscle spindle
4. Golgi tendon reflex
   a. Structure of Golgi tendon organ
   b. Excitation of nerve endings in tendon
   c. Inhibition of muscle contraction
5. Flexor (withdrawal) reflex
   a. Role of parallel after-discharge circuits
   b. Role of intersegmental reflex arcs
6. Crossed extensor reflex
   a. Normally accompanies withdrawal reflex
   b. Example of contralateral reflex arc

The Autonomic Nervous System: Introduction and Anatomy (pp. 531–540)
1. General properties
   a. Effectors other than skeletal muscle
   b. Two-neuron efferent pathways

   c. Multiple neurotransmitters
   d. Excitatory and inhibitory effects
   e. Responses usually involuntary
2. Sympathetic (thoracolumbar) division
   a. Chain of paravertebral ganglia
   b. Communicating rami
   c. Various destinations of fibers entering paravertebral ganglia
   d. Neuronal convergence in ganglia
   e. Neuronal divergence and mass activation
   f. Routes out of paravertebral ganglia
      - Via spinal nerves
      - Via sympathetic nerves
      - Via splanchnic nerves
3. Adrenal glands
4. Parasympathetic (craniosacral) division
   a. Terminal ganglia near or in target organ
   b. Only slight neuronal divergence
   c. Cranial nerve routes out of CNS
      - Oculomotor nerve
      - Facial nerve
      - Glossopharyngeal nerve
      - Vagus nerve
   d. Sacral nerve routes out of CNS
      - Pelvic splanchnic nerves
      - Inferior hypogastric plexus
      - Pelvic nerves

The Autonomic Nervous System: Physiology (pp. 540–544)
1. Neurotransmitters
   a. Cholinergic fibers
   b. Adrenergic fibers
2. Receptors for neurotransmitters
   a. Cholinergic receptors
      - Nicotinic
      - Muscarinic
   b. Adrenergic receptors
      - α_1 and α_2
      - β_1 and β_2
      - Second messengers
3. Dual innervation
   a. Antagonistic effects
   b. Cooperative effects
4. Control without dual innervation
   a. Effectors with only sympathetic fibers
   b. Sympathetic and vasomotor tone
5. Central control of autonomic function
   a. Cerebral
   b. Hypothalamic
   c. Brainstem
   d. Spinal

Also review the terms in tables, 15.1 through 15.6, which are not repeated here.

nerve 505
endoneurium 505
nerve fascicle 505
perineurium 505
epineurium 505
somatosensory neuron 507
mixed nerve 507
sensory nerve 507
motor nerve 507
cranial nerve 508
spinal nerve 516
dorsal root 516
ventral root 516
dorsal root ganglion 516
dorsal ramus 518
ventral ramus 518
meningeal branch 518
intercostal nerve 518
cervical plexus 518
brachial plexus 518
lumbar plexus 518
sacral plexus 518

coccygeal plexus 518
dermatome 524
reflex 525
somatic reflex 525
visceral reflex 525
spinal reflex 525
reflex arc 525
muscle spindle 526
intrafusal fiber 527
extrafusal fiber 527
primary afferent fiber 527
secondary afferent fiber 527
γ motor neuron 527
α motor neuron 527
stretch reflex 527
tendon reflex 527
monosynaptic reflex arc 527
reciprocal inhibition 528
flexor reflex 528
polysynaptic reflex arc 529
crossed extensor reflex 529
ipsilateral reflex arc 530
contralateral reflex arc 530
intersegmental reflex arc 530
Golgi tendon organ 530

Golgi tendon reflex 530
autonomic nervous system
   (ANS) 531
preganglionic neuron 532
postganglionic neuron 532
sympathetic division 532
parasympathetic division 533
thoracolumbar division 533
sympathetic chain 533
white communicating
   ramus 533
gray communicating
   ramus 533
mass activation 535
sympathetic nerves 535
cardiac nerves 535
splanchnic nerve 536
collateral ganglion 536
abdominal aortic plexus 536
celiac ganglion 536
superior mesenteric
   ganglion 536
inferior mesenteric
   ganglion 536
adrenal glands 536

adrenal cortex 537
adrenal medulla 537
craniosacral division 537
terminal ganglion 537
intramural ganglion 537
cardiac plexus 539
pulmonary plexus 539
esophageal plexus 539
vagal trunks 539
pelvic splanchnic nerves 539
inferior hypogastric plexus 539
pelvic nerves 539
cholinergic fiber 540
adrenergic fiber 540
cholinergic receptor 540
adrenergic receptor 540
nicotinic receptor 540
muscarinic receptor 540
α-adrenergic receptors 541
β-adrenergic receptors 541
dual innervation 541
antagonistic effects 541
cooperative effects 541
sympathetic tone 541
vasomotor tone 541

Answers in Appendix C

1. The body of a Schwann cell forms
   a. the axolemma.
   b. the neurilemma.
   c. the basal lamina.
   d. the endoneurium.
   e. the perineurium.

2. The sense of taste is a function of
   a. the glossopharyngeal nerve.
   b. the hypoglossal nerve.
   c. the maxillary nerve.
   d. the abducens nerve.
   e. the trochlear nerve.

3. All of the following *except* the
   _____ nerve have origins or
   terminations in the orbit.
   a. optic
   b. oculomotor
   c. trochlear
   d. abducens
   e. accessory

4. The brachial plexus gives rise to all
   of the following nerves *except*
   a. the axillary nerve.
   b. the radial nerve.
   c. the saphenous nerve.
   d. the median nerve.
   e. the ulnar nerve.

5. Nerve fibers that adjust the tension in
   a muscle spindle are called
   a. intrafusal fibers.
   b. extrafusal fibers.
   c. α motor neurons.
   d. γ motor neurons.
   e. annulospiral fibers.

6. A stretch reflex requires the action of
   _____ to keep the antagonistic
   muscle from contracting and
   interfering with the agonist.
   a. γ motor neurons
   b. a withdrawal reflex
   c. a crossed-extensor reflex
   d. reciprocal innervation
   e. a contralateral reflex

7. The autonomic nervous system
   innervates all of the following
   effectors *except*
   a. cardiac muscle cells.
   b. skeletal muscle fibers.
   c. smooth muscle cells.
   d. salivary glands.
   e. blood vessels.

8. A gray ramus communicans contains
   a. visceral sensory fibers.
   b. parasympathetic motor fibers.

   c. sympathetic preganglionic fibers.
   d. sympathetic postganglionic fibers.
   e. somatic motor fibers.

9. Throughout the autonomic nervous
   system, the neurotransmitter released
   by the preganglionic neuron binds to
   _____ receptors on the
   postganglionic neuron.
   a. nicotinic
   b. muscarinic
   c. adrenergic
   d. α_1
   e. β_2

10. Which of these is *not* an effect of
    sympathetic stimulation?
    a. dilation of the pupil
    b. acceleration of the heart
    c. increased digestive secretion
    d. enhanced blood clotting
    e. piloerection

11. Outside the CNS, somas are clustered
    in swellings called _____.

12. Distal to the intervertebral foramen, a
    spinal nerve branches into a dorsal
    and ventral _____.

13. The _____ nerves arise from the cervical plexus and innervate the diaphragm.

14. Most parasympathetic preganglionic fibers are found in the _____ nerve.

15. Modified muscle fibers serving primarily to detect stretch are called _____.

16. The sciatic nerve is a composite of two nerves, the _____ and _____.

17. The adrenal medulla consists of modified postganglionic neurons of the _____ nervous system.

18. Nicotinic and muscarinic receptors both bind the neurotransmitter _____.

19. Adrenergic receptors classified as α_2, β_1, and β_2 act by changing the level of _____ in the target cell.

20. Sympathetic fibers to blood vessels maintain a background rate of firing called _____, and this causes a state of partial vasoconstriction called _____.

Testing Your Comprehension
Answers in *Study Guide*

1. Stand with your right shoulder, hip, and foot firmly against a wall. Now raise your left foot from the floor. What happens? Why? What principle of this chapter does this illustrate?

2. What is the main difference between a somatic reflex and a visceral reflex? In what ways are these reflexes similar?

3. You are dicing raw onions while preparing dinner, and your eyes are watering profusely. Describe the afferent and efferent neural pathways involved in this reflex.

4. Which cranial nerves convey pain signals to the brain when (a) sand blows into your eye, (b) you bite the rear of your tongue, and (c) your stomach hurts from eating too much?

5. A surgical resident who is learning thoracic surgery by operating on a pig accidentally severs the right phrenic nerve. How do you think this will affect pulmonary ventilation? Will both lungs be affected equally?

Web Site Link

For a listing of the most current web sites related to this chapter, please visit the Saladin homepage at:

http://www.mhhe.com/sciencemath/biology/saladin/

Sense Organs

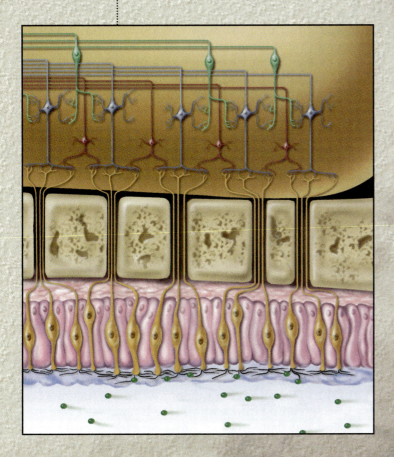

Special **Topics**

To understand this chapter, it is essential that you understand or brush up on the following concepts:

► Type A and C nerve fibers (p. 434)
► Generator potentials, threshold, and action potentials (pp. 440–443)
► Excitatory and inhibitory postsynaptic potentials (EPSPs and IPSPs) (p. 453)
► Spatial summation (pp. 453–454)
► Neural coding (p. 454)
► Converging circuits of neurons (p. 456)
► Spinal tracts (pp. 474–477)
► Anatomy of the brain (pp. 477–489)

Anyone who enjoys music, art, fine food, or a good conversation appreciates the human senses. Yet their importance extends beyond deriving pleasure from the environment. In the 1950s, behavioral scientists at Princeton University studied the methods used by Soviet Communists to extract confessions from political prisoners, including solitary confinement and sensory deprivation. Student volunteers were immobilized in dark soundproof rooms or suspended in dark chambers of water. In a short time, they began to exhibit visual, auditory, and tactile hallucinations; incoherent thought patterns; deterioration of intellectual performance; and sometimes morbid fear or panic. Similar effects have been observed in burn patients who are immobilized and extensively bandaged (including the eyes) and who thus suffer prolonged lack of sensory input. Patients connected to life-support equipment and confined under oxygen tents sometimes become delirious. Sensory input is vital to the integrity of personality and intellectual function. Furthermore, much of the information conveyed by the sense organs never comes to our conscious attention—blood pressure, body temperature, and muscle tension, for example. By monitoring such conditions, however, the sense organs initiate somatic and visceral reflexes that are indispensable to homeostasis and to our very survival in a ceaselessly changing and challenging environment.

Properties and Types of Sensory Receptors

▼**Objectives**
When you have completed this section, you should be able to
• define *receptor* and *sense organ;*
• list the four kinds of information obtained from sensory receptors and describe how the nervous system codes for each type; and
• outline three ways of classifying receptors.

A **receptor** is any structure specialized to detect a stimulus. Some receptors are simple nerve endings (sensory dendrites), whereas others are **sense organs**—nerve endings combined with connective, epithelial, or muscular tissues that enhance or moderate the response to a stimulus. Our eyes and ears are obvious examples of sense organs, but there are also innumerable microscopic sense organs in our skin, muscles, joints, and viscera.

General Properties of Receptors

All sensory receptors are transducers. A *transducer* is any device that converts one form of energy to another—a microphone, light bulb, heating coil, or gasoline engine, for example. Sensory transducers convert stimulus energy into nerve energy—a meaningful pattern of action potentials.

The effect of a stimulus on any receptor cell is to produce a **receptor potential**—a voltage change on its plasma membrane. The receptor potential may cause the cell to release a neurotransmitter that stimulates an afferent neuron. If the receptor itself is a neuron and the receptor potential reaches threshold, the stimulus triggers the firing of nerve impulses to the central nervous system (CNS). These impulses may result in a **sensation**—a conscious awareness of the stimulus—but much of the sensory information reaching the CNS produces no sensation. We are seldom aware of sensory information concerning muscle tension and blood pH, for example, but this information triggers somatic and autonomic reflexes important to homeostasis.

Sensory receptors transmit four kinds of information—stimulus *modality*, *location*, *intensity* and *duration*.

1. **Modality** refers to the type of stimulus or the sensation it produces. Vision, hearing, and taste are examples of sensory modalities. The nervous system uses a *labeled line code* to distinguish stimulus modalities from each other. We can think of the brain as having numerous "lines" (nerve fibers) feeding information into it, and each line as being "labeled" to represent a certain modality. All the nerve impulses that arrive at the brain are essentially identical, but impulses coming in on one line have a different meaning than impulses arriving on another. Any impulses from the optic nerve, for example, are interpreted as light. Thus, a blow to the eye can be perceived as a flash of light even if no light enters the eye.
2. **Location** is also encoded by which nerve fibers are firing. **Sensory projection** is the ability of the brain to identify the site of stimulation, including very small and specific areas within a receptor such as the retina. The pathways followed by sensory signals to their ultimate destinations in the CNS are called *projection pathways.*
3. **Intensity** can be encoded in three ways: (a) as stimulus intensity rises, the firing frequencies of sensory nerve fibers rise; (b) intense stimuli recruit larger numbers of nerve fibers to fire; and (c) weak

stimuli activate only the most sensitive nerve fibers, whereas strong stimuli may activate a different group of fibers with higher thresholds. Thus, the brain can distinguish intensities based on the number and kind of fibers that are firing and the time intervals between action potentials. These concepts were discussed under *neural coding* in chapter 13.

4. **Duration** is encoded in the way nerve fibers change their firing frequencies over time. Some receptors, such as the lamellated corpuscles to be discussed shortly, fire briefly when a stimulus begins, then become "silent," and fire briefly again when stimulation ends. Others fire more constantly, but all receptors exhibit sensory **adaptation**—if the stimulus is prolonged, firing frequency and conscious sensation decline. Adapting to hot bath water is an example. **Phasic receptors** generate a burst of action potentials when first stimulated; then they quickly adapt and stop transmitting impulses even if the stimulus continues. Lamellated corpuscles, touch receptors, and smell receptors are rapidly adapting phasic receptors. **Tonic receptors** adapt slowly and generate nerve impulses continually. Proprioceptors are among the most slowly adapting tonic receptors because the brain always must be aware of body position, muscle tension, and joint motions.

Think About It

Although you may find it difficult to immerse yourself in a hot tub of water or a cold lake, you soon adapt and become more comfortable. In light of this, do you think cold and warm receptors are phasic or tonic?

Classification of Receptors

Receptors can be classified by several overlapping systems. One way is by stimulus modality:

- **Chemoreceptors** respond to chemicals, including odors, tastes, and chemicals in the body fluids.
- **Thermoreceptors** respond to heat and cold.
- **Nociceptors**[1] (NO-sih-SEP-turs) are pain receptors; they respond to tissue damage resulting from trauma (blows, cuts, crushing injuries), ischemia (poor blood flow), or excessive stimulation by agents such as heat and chemicals.
- **Mechanoreceptors** respond to physical deformation of the plasma membrane caused by touch, pressure, stretch, tension, or vibration. They include the organs of hearing and balance and many receptors of the skin, viscera, and joints.
- **Photoreceptors,** the eyes, respond to light.

The senses can also be classified into two categories called the general and special senses. The **general (somatic, somatosensory,** or **somesthetic) senses** have receptors that are widely distributed in the body rather than limited to specific localities. These receptors occur in the skin, muscles, tendons, joint capsules, and viscera. They detect touch, pressure, stretch, heat, cold, and pain, as well as many stimuli that we do not perceive consciously, such as blood pressure and chemistry. The **special senses** are vision, hearing, equilibrium, taste, and smell. Their receptors are limited to the head and are innervated by the cranial nerves.

Finally, receptors can be classified according to the origins of their stimuli. **Interoceptors** detect stimuli that originate in the internal organs; they are responsible for feelings of visceral pain, nausea, stretch, and pressure. **Proprioceptors** sense the position and movements of the body or its parts. They occur in muscles, tendons, joint capsules, and the inner ear. **Exteroceptors** sense changes external to the body; they include the receptors for vision, hearing, taste, smell, touch, and cutaneous pain.

Key Point Review

1 What is the difference between a receptor and a nerve ending?

2 Distinguish between general and special senses.

3 Three schemes of receptor classification were presented in this section. In each scheme, how would you classify the receptors for a full bladder? How would you classify taste receptors?

4 What does it mean to say sense organs are transducers? What form of energy do all receptors have as their output?

5 Nociceptors are tonic rather than phasic receptors. Speculate on why this is beneficial to homeostasis.

The General Senses

▼**Objectives**
When you have completed this section, you should be able to
- list several types of receptors for the general senses;
- describe the somesthetic projection pathways in the CNS; and
- explain the mechanisms of pain.

Receptors for the general senses are relatively simple in structure and physiology. They consist of one or a few sensory nerve fibers and, usually, a sparse amount of connective tissue that enhances the sensitivity or specificity of their responses.

Unencapsulated Nerve Endings

Unencapsulated nerve endings are sensory dendrites that lack a connective tissue wrapping. They include the following:

- **Free nerve endings.** These are bare dendrites that have no special association with specific accessory

1. *noci* = pain

cells or tissues. They are most abundant in epithelial and connective tissues. They include *warm receptors,* which respond to rising temperatures; *cold receptors,* which respond to falling temperatures; and *nociceptors,* or pain receptors.

- **Merkel[2] discs.** These are flattened nerve endings associated with specialized *Merkel cells* at the base of the epidermis (see fig. 7.3, p. 204). They are tonic receptors for light touch and pressure.
- **Hair receptors (peritrichial[3] endings).** These consist of a few dendrites entwined around the base of a hair follicle. They are stimulated by light touch, which bends the hairs. Because they adapt quickly, we are not constantly annoyed by our clothing bending the body hairs. However, when an ant crawls across our skin, bending one hair after another, we are very aware of it.

Encapsulated Nerve Endings

Encapsulated nerve endings are dendrites wrapped in glial cells or fibroconnective tissue. Most of them are mechanoreceptors for touch, pressure, and stretch. The connective tissue "investments" around the dendrites enhance the sensitivity or specificity of the receptor. We have already considered some encapsulated nerve endings in chapter 15—muscle spindles and Golgi tendon organs. Others include the following:

- **Tactile (Meissner[4]) corpuscles.** These receptors occur in the dermal papillae of the skin, especially in sensitive hairless areas such as the fingertips, palms, eyelids, lips, nipples, and genitals. They are tall, ovoid- to pear-shaped, and consist of two or three nerve fibers meandering upward through a mass of connective tissue. Tactile corpuscles are phasic receptors for light touch, texture, and low-frequency vibration. They enable you to tell the difference between silk and sandpaper, for example, through light strokes of your fingertips.
- **Krause[5] end bulbs.** These are similar to tactile corpuscles but occur in mucous membranes rather than in the skin.
- **Lamellated (pacinian[6]) corpuscles.** These receptors occur in the pancreas, some other viscera, and deep in the dermis—especially on the hands, feet, breasts, and genitals. They consist of numerous concentric lamellae of Schwann cells surrounding a core of one to several sensory nerve fibers. They are phasic receptors for deep pressure, stretch, and high-frequency vibration.

2. Friedrich S. Merkel (1845–1911), German anatomist and physiologist
3. *peri* = around + *trich* = hair
4. George Meissner (1829–1905), German histologist
5. William J. F. Krause (1833–1910), German anatomist
6. Filippo Pacini (1812–83), Italian anatomist

- **Ruffini[7] corpuscles.** These are located in the dermis, subcutaneous tissues, and joint capsules. Each is a flattened, elongated capsule containing a few nerve fibers. They respond tonically to constant heavy pressure and to joint movements.

Nerve endings involved in general sensation are summarized in table 16.1.

Somesthetic Projection Pathways

The afferent neuron of a sensory pathway is called the **first-order neuron.** Fibers for touch, pressure, and proprioception are large myelinated type A fibers. They enter the spinal cord by way of the dorsal root, travel through various ascending columns (see table 14.1, p. 474) on the same side of the body, and terminate in the medulla oblongata. In the medulla, they synapse with **second-order neurons** that cross (decussate) to the opposite side of the brainstem and continue by way of the medial lemniscus to the thalamus. Here, they synapse with **third-order neurons** that project to the somesthetic cortex of the postcentral gyrus.

Impulses for heat and cold are carried to the spinal cord by small unmyelinated type C fibers. Near the point of entry, they synapse with second-order neurons that decussate at that level and travel up the opposite side of the cord in the spinothalamic tract. In the thalamus, they synapse with third-order neurons that continue to the somesthetic cortex.

Pain

Nociceptors are found in virtually all organs, especially in the skin and mucous membranes, but they are lacking from the brain. Brain surgery in which a patient must be awake and able to communicate with the surgeon requires only a local scalp anesthetic. (See the chapter essay, p. 595–596, for a discussion of anesthesia.)

There are two types of nociceptors corresponding to different pain sensations. Myelinated type A pain fibers conduct at speeds of 12 to 30 m/sec and produce the sensation known as **fast pain**—the initial feeling of sharp, localized, stabbing pain perceived at the time of injury. Unmyelinated type C pain fibers conduct at speeds of 0.5 to 2.0 m/sec and produce the **slow pain** that follows—a longer-lasting, dull, diffuse feeling.

Types and Sources of Pain

Somatic pain arises from the skin, muscles, and joints. *Superficial somatic pain* has a sharp, stabbing, or prickling quality. *Deep somatic pain* is less localized, longer

7. Angelo Ruffini (1864–1929), Italian anatomist

Table 16.1 Receptors for the General Senses

	Receptor Type	Locations	Modality
Unencapsulated Endings			
	Free Nerve Endings	Widespread, especially in epithelia and connective tissues	Pain, heat
	Merkel Discs	Stratum basale of epidermis	Touch
	Hair Receptors	Around hair follicle	Light touch, movement of hairs
Encapsulated Nerve Endings			
	Tactile Corpuscles	Dermal papillae of fingertips, palms, eyelids, lips, tongue, nipples, and genitals	Light touch, texture, low-frequency vibration
	Krause End Bulbs	Mucous membranes	Similar to Meissner corpuscles
	Lamellated Corpuscles	Dermis, joint capsules, breasts, genitals, and some viscera	Deep pressure, stretch, high-frequency vibration
	Ruffini Corpuscles	Dermis, subcutaneous tissue, and joint capsules	Heavy continuous touch or pressure
	Muscle Spindles	Skeletal muscles near tendon	Muscle stretch (proprioception)
	Golgi Tendon Organs	Tendons	Tension on tendons (proprioception)

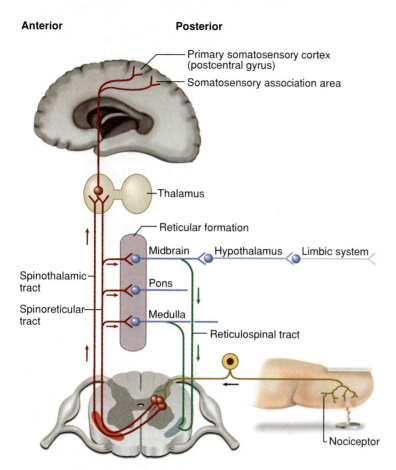

Figure 16.1 The projection pathway for pain signals. A first-order neuron conducts a pain signal to the dorsal horn of the spinal cord, a second-order neuron conducts it to the thalamus, and a third-order neuron conducts it to the cerebral cortex.

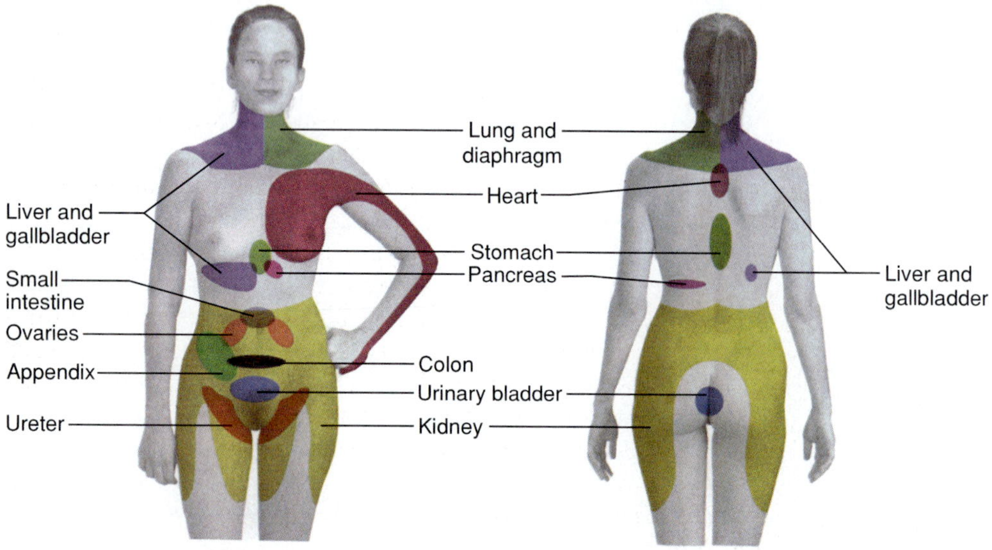

Anterior   *Posterior*

- Primary somatosensory cortex (postcentral gyrus)
- Somatosensory association area
- Thalamus
- Reticular formation
- Midbrain   Hypothalamus   Limbic system
- Spinothalamic tract
- Pons
- Spinoreticular tract
- Medulla
- Reticulospinal tract
- Nociceptor

Figure 16.2 Referred pain, showing areas of the skin in which pain from the viscera is commonly felt.

- Lung and diaphragm
- Heart
- Liver and gallbladder
- Stomach
- Pancreas
- Small intestine
- Ovaries
- Appendix
- Colon
- Urinary bladder
- Ureter
- Kidney
- Liver and gallbladder

lasting, and has a more aching, burning, or nauseating quality. **Visceral pain** arises from organs of the thoracic and abdominal cavities. It commonly results from stretch, chemical irritants, or ischemia. It is often associated with nausea and may be very intense, but it is poorly localized because the viscera have relatively few nociceptors.

Projection Pathways for Pain

The first-order neurons for pain follow the usual somatosensory route to the spinal cord by way of the dorsal root. In the dorsal horn, they synapse with interneurons whose fibers travel primarily up the spinothalamic tract to the thalamus. The thalamus relays the signals to the cerebral cortex (fig. 16.1), where the type and site of pain are identified. Pain signals also travel up the spinoreticular tract to the reticular formation and other destinations in the brain and may thus affect arousal—for example, when we are awakened by indigestion or alarmed by a bee sting. From the reticular formation, some third-order nerve fibers travel to the hypothalamus and limbic system, triggering visceral, emotional, and behavioral reactions to pain.

Referred Pain

It is sometimes difficult to precisely locate the source of pain, especially visceral pain. **Referred pain** is a phenomenon in which pain from the viscera is misinterpreted as coming from the skin or other superficial structures. People suffering heart attacks, for example, often perceive cardiac pain as if it were "radiating" along the left shoulder and medial side of the left arm. Spinal cord segments T1 to T5 innervate both the heart and the chest and arm region. Pain fibers from the heart and skin in this region are thought to converge on the same spinal interneurons and to follow the same pathway from there to the thalamus and cerebral cortex. The skin has many more pain receptors than the heart and suffers injury far more often. Thus the brain "assumes" that signals arriving by this labeled line are most likely coming from the skin, even when the heart is the actual source. Knowledge of the origins of referred pain is useful in the diagnosis of organ dysfunctions (fig. 16.2).

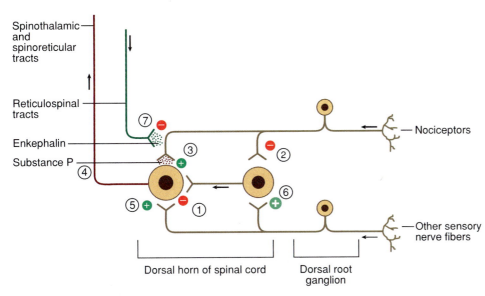

1. In the absence of a pain stimulus, an inhibitory interneuron of the spinal cord prevents transmission of pain signals.
2. When tissue damage stimulates a nociceptor, the nociceptor inhibits the inhibitory interneuron.
3. The nociceptor also releases substance P, which stimulates the projection neuron.
4. The projection neuron sends a pain signal to the brain.
5. Some sensory neurons other than nociceptors also stimulate the projection neuron.
6. These sensory neurons, however, have an even stronger effect on the inhibitory interneuron, thus blocking the transmission of pain signals.
7. Neurons of the reticular formation release enkephalin, which, by presynaptic inhibition, blocks the release of substance P. Thus the brain can reduce the transmission of the pain signal to itself.

Figure 16.3 Mechanisms for the spinal gating of pain signals.

Chemical Agents of Pain

Injured tissues release a number of chemicals that stimulate nociceptors or sensitize them to other stimuli—bradykinin, serotonin, prostaglandins, histamine, and potassium ions. **Bradykinin** (BRAD-ee-KY-nin), derived from a protein in the blood plasma, is one of the most potent stimuli; it is intensely painful when experimentally injected under the skin. Bradykinin not only makes us aware of injuries but also triggers a cascade of reactions that promote healing (see chapter 19).

In the dorsal horn of the spinal cord, the first-order nerve fiber releases multiple neurotransmitters, including glutamtic acid and a variety of neuropeptides. The most studied neuropeptide involved in pain is **substance P.**

CNS Modulation of Pain

The intensity of pain is not simply a function of the extent of tissue injury and firing of the afferent nociceptors; it can be greatly affected by states of mind, such as the joy of childbirth or fear of dental work. The CNS has **analgesic**[8] (pain-relieving) mechanisms that are just beginning to be understood. The analgesic effects of opium, morphine, and heroin have been known and exploited for many centuries, but in 1974, neurophysiologists discovered receptor sites for them in the brain. Since these opiates do not occur naturally in the body, this discovery touched off a search to find the natural ligands for the receptors. Two oligopeptides with 200 times the analgesic potency of morphine were identified and named **enkephalins.**[9] Larger neuropeptides, the **endorphins,**[10] were discovered later.

Enkephalins and endorphins, known collectively as *endogenous opiates,* are secreted by the brain, pituitary gland, digestive tract, and other organs in states of stress or exercise. They act as *neuromodulators* (see p. 452) to block the transmission of pain signals and produce feelings of pleasure and euphoria. Enkephalin secretion rises sharply in women giving birth. Endorphins may be responsible for the "second wind" or "runner's high" experienced by athletes and perhaps for the reports of many mortally wounded soldiers that they feel no pain. Unfortunately, efforts to employ endorphins and enkephalins for pain therapy have been disappointing. Exercise, however, is an effective part of therapy for chronic pain, perhaps because of endorphin secretion.

The reticular formation may modulate sensitivity to pain by means of endorphins. A proposed mechanism is shown in figure 16.3. According to this hypothesis, fibers from the reticular formation travel down the reticulospinal tracts and form axoaxonic synapses with the first-order pain neurons. The axonal endings of the first-order fibers are known to have endorphin receptors. By secreting endorphins here, the reticulospinal fibers may suppress the secretion of substance P (a case of presynaptic inhibition—see p. 455) and thus reduce the transmission of pain signals.

Some interneurons of the dorsal horn inhibit second-order neurons of the pain pathway and receive input from touch fibers. When the touch fibers

8. *an* = without + *alges* = pain

9. *en* = within + *kephal* = head
10. Acronym, from *endogenous morphine*like substance

are stimulated, they stimulate the inhibitory interneurons, and those neurons block the transmission of signals by the second-order pain fibers. This may be why rubbing a sore area makes it feel less painful. Such mechanisms are referred to as the *gating* of pain in the dorsal horn.

The Chemical Senses

▼**Objectives**

When you have completed this section, you should be able to
- explain how taste and smell receptors are stimulated; and
- describe the structure and projection pathways of olfactory and gustatory receptors.

Taste and smell are the chemical senses. In both cases, receptor potentials are created by the action of environmental chemicals on sensory cells.

Taste (Gustation)

Taste **(gustation)** is a sensation that results from the action of chemicals on the 10,000 **taste buds** located on the tongue, cheeks, soft palate, pharynx, and epiglottis.

Anatomy

The tongue, where the sense of taste is best developed, is marked by numerous bumps called **lingual papillae** (fig. 16.4*a*). There are four types:

1. **Filiform**[11] **papillae** are tiny spikes without taste buds. They are responsible for the rough feel of a cat's tongue and are important to many mammals for grooming the fur. They are the most numerous papillae on the human tongue, but they are poorly developed and play no sensory role.
2. **Foliate**[12] **papillae** are also weakly developed in humans and have few taste buds. They form parallel ridges on the sides of the tongue about two-thirds of the way back from the tip.

3. **Fungiform**[13] (FUN-jih-form) **papillae** are shaped somewhat like mushrooms. Each one has about five taste buds, located mainly on the apex. These papillae are widely distributed but especially concentrated at the tip and sides of the tongue.
4. **Vallate**[14] **(circumvallate) papillae** are large papillae surrounded by deep circular trenches. There are only 7 to 12 of them, arranged in a V at the rear of the tongue. Each vallate papilla has about 100 taste buds located on the walls of the papilla, facing the trench (fig. 16.4*b*).

Regardless of location and sensory specialization, all taste buds look alike (fig. 16.4*c,d*). They are lemon-shaped groups of 40 to 60 cells of three kinds—**taste cells, supporting cells,** and **basal cells.** Taste and supporting cells are more or less banana-shaped. Each taste cell has a tuft of apical microvilli called **taste hairs** that serve as receptor surfaces for taste molecules. The hairs project into a pit called a **taste pore** on the epithelial surface of the tongue. A taste cell lives 7 to 10 days and is then replaced by mitosis and differentiation of basal cells. Taste cells are not neurons, but they synapse with sensory nerve fibers at their base.

Physiology

To be tasted, molecules must dissolve in saliva and flood the taste pore. On a dry tongue, sugar or salt has as little taste as a sprinkle of sand. The traditional view is that all tastes are due to mixtures of four primary taste sensations:

1. **Salty**—produced by metal ions such as sodium and potassium. Since these are vital electrolytes, there is obvious value in the ability to taste them. Electrolyte deficiencies can cause a craving for salty foods; many animals such as deer, elephants, and parrots thus seek out salt deposits when necessary. Pregnancy can lower a woman's electrolyte concentrations and create a craving for salty food. The taste buds most sensitive to salt are located on the tip and sides of the tongue.
2. **Sweet**—produced by many organic compounds, especially sugars. Sweetness is associated with carbohydrates and foods of high caloric value. Many flowering plants have evolved sweet nectar and fruits that entice animals to eat them and disperse their pollen and seeds. Thus, our fondness for fruit has coevolved with plant reproductive strategies. The tip of the tongue is the region most sensitive to sweet compounds.

---

11. *fili* = thread, filament + *form* = shape
12. *foliate* = leaflike

13. *fungi* = mushroom, fungus
14. *vall* = wall

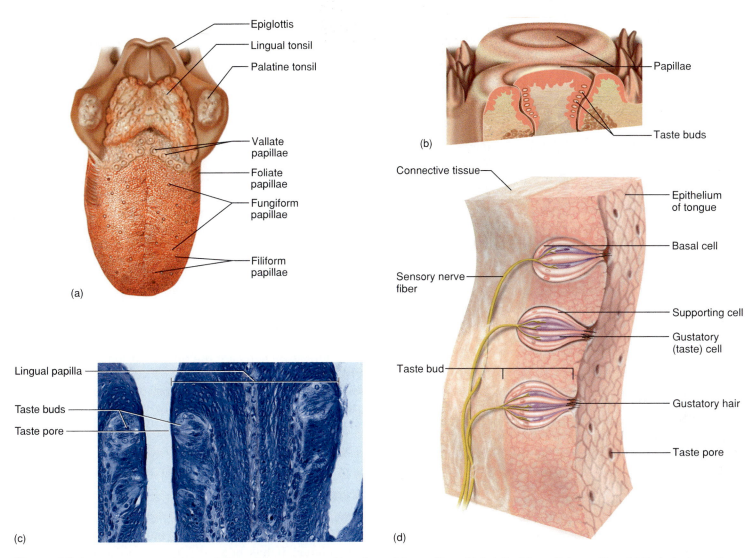

Figure 16.4 Taste receptors. (*a*) Dorsal view of the tongue and locations of its papillae. (*b*) Detail of the vallate papillae. (*c*) Taste buds on the walls of two adjacent foliate papillae (×200). (*d*) Structure of the taste buds.

3. **Sour**—usually associated with acids in such foods as citrus fruits. Taste buds along the sides of the tongue are especially sensitive to acids.
4. **Bitter**—often associated with spoiled foods or with alkaloids such as nicotine, caffeine, quinine, and morphine. Bitter alkaloids are often poisonous, and this sensation usually induces a human or animal to reject a food. While flowering plants make their fruits temptingly sweet, they often load their leaves with bitter, toxic alkaloids—it is of no advantage to a plant to have its leaves eaten! The bitter sensation resides in the taste buds of the vallate papillae.

The threshold for the bitter taste is the lowest of all—that is, it takes less of an alkaloid to be tasted than it does an acid. Our senses of sweet and salty are the least sensitive. Any taste bud is sensitive to all of the primary tastes, but is much more sensitive to one category than to others. Figure 16.5 shows the regions of the tongue most sensitive to each primary taste.

Bitter and sweet chemicals bind to proteins in the plasma membranes of the taste cells and trigger the formation of intracellular second messengers. Acids and salts have no specific receptor sites; they penetrate the plasma membrane and affect the taste cell without the aid of a second messenger.

The flavor of food involves much more than taste. It is affected by texture, temperature, aroma, and appearance, among other things. Some flavors such as pepper are due to stimulation of free endings of the trigeminal nerve. Many flavors depend on smell; without the sense of smell, cinnamon merely has a faintly sweet taste, and coffee and peppermint are bitter, for example.

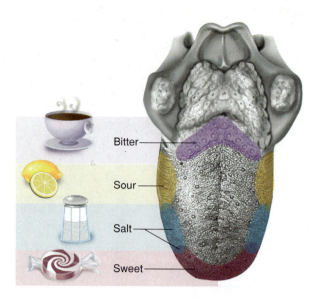

Figure 16.5 Regional specialization of the tongue for the four primary taste sensations. All of the red and blue areas are responsive to salts, but only the red area is specialized for sweets. ✗

Projection Pathways

Taste buds on the anterior two-thirds of the tongue are innervated by the facial nerve (VII); those on the posterior one-third of the tongue, by the glossopharyngeal nerve (IX); and those in the palate and pharynx, by the vagus nerve (X). All taste fibers project to the *gustatory nucleus* of the medulla oblongata. Second-order neurons from this nucleus relay signals to two destinations: (1) other brainstem nuclei that activate autonomic reflexes such as salivation, gagging, and vomiting and (2) the thalamus. The thalamus, in turn, relays signals by way of a third-order neuron to the gustatory cortex in the inferior area of the postcentral gyrus.

Smell (Olfaction)

The sense of smell (**olfaction**) is mediated by a patch of epithelium called the **olfactory mucosa** in the roof of the nasal cavity (fig. 16.6a). This location places the olfactory cells close to the brain, but it is poorly ventilated; forcible sniffing is often needed to identify an odor or locate its source. Nevertheless, the sense of smell is highly sensitive. We can detect odor concentrations as low as a few parts per trillion. Most people can distinguish 2,000 to 4,000 different odors, and some can distinguish up to 10,000. On average, women are more sensitive to odors than men are, and they are measurably more sensitive to some odors near the time of ovulation than during other phases of the menstrual cycle. Olfaction is highly important in the social interactions of other animals and, in more subtle ways, to humans (see special topic 16.1).

Anatomy

The olfactory mucosa covers about 5 cm² of the superior concha and nasal septum. It consists of 10 to 20 million **olfactory neurons,** as well as epithelial supporting cells and small basal cells (fig. 16.6b). The mucosa has a yellowish tinge due to lipofuscin in the supporting cells. The rest of the nasal cavity is lined by an unpigmented nonsensory *respiratory mucosa.*

An olfactory neuron is shaped a little like a bowling pin. Its widest part, the soma, contains the nucleus. The neck and head of the cell are a modified dendrite with a swollen tip bearing 10 to 20 cilia called **olfactory hairs.** They have binding sites for odor molecules and are immobile; they lie in a tangled mass embedded in a thin layer of mucus. The basal end of each cell tapers to an axon. These axons collect into small fascicles and leave the nasal cavity by way of the pores in the cribriform plate of the ethmoid bone. Collectively, these fascicles are regarded as cranial nerve I (the olfactory nerve).

Olfactory cells are the only neurons in the body directly exposed to the external environment. Because of this exposure, they have a short life span of about 60 days. Unlike most neurons, however, they are replaceable. The basal cells continually divide and differentiate into new olfactory cells.

Physiology

Efforts to identify a few primary odors comparable to the four primary tastes have been controversial and indecisive. It has been difficult even to specify what properties are needed to give a molecule an odor. At a minimum, the molecule must be volatile—that is, able to evaporate and be carried by the inhaled air stream. The intensity of an odor is not simply proportional to volatility, however. Water is highly volatile and has no odor, while musk has a pronounced odor but is poorly volatile. A second requirement, if a molecule is to have an odor, is that it must be water soluble so that it can dissolve in the mucous layer and reach the olfactory hairs.

Sensory transduction begins when a molecule binds to a receptor of an olfactory hair. The receptor triggers the production of a second messenger, which in turn opens ion channels in the membrane. Na⁺ enters the cell, depolarizes it, and creates a receptor potential. In many cases, the second messenger is cAMP, but some odors activate other second-messenger systems. Olfactory receptors adapt quickly—we may therefore be unaware of our own body odors or have difficulty locating a gas leak in a room. Adaptation does not occur in the receptor cells but is due to inhibition in the olfactory bulbs of the brain.

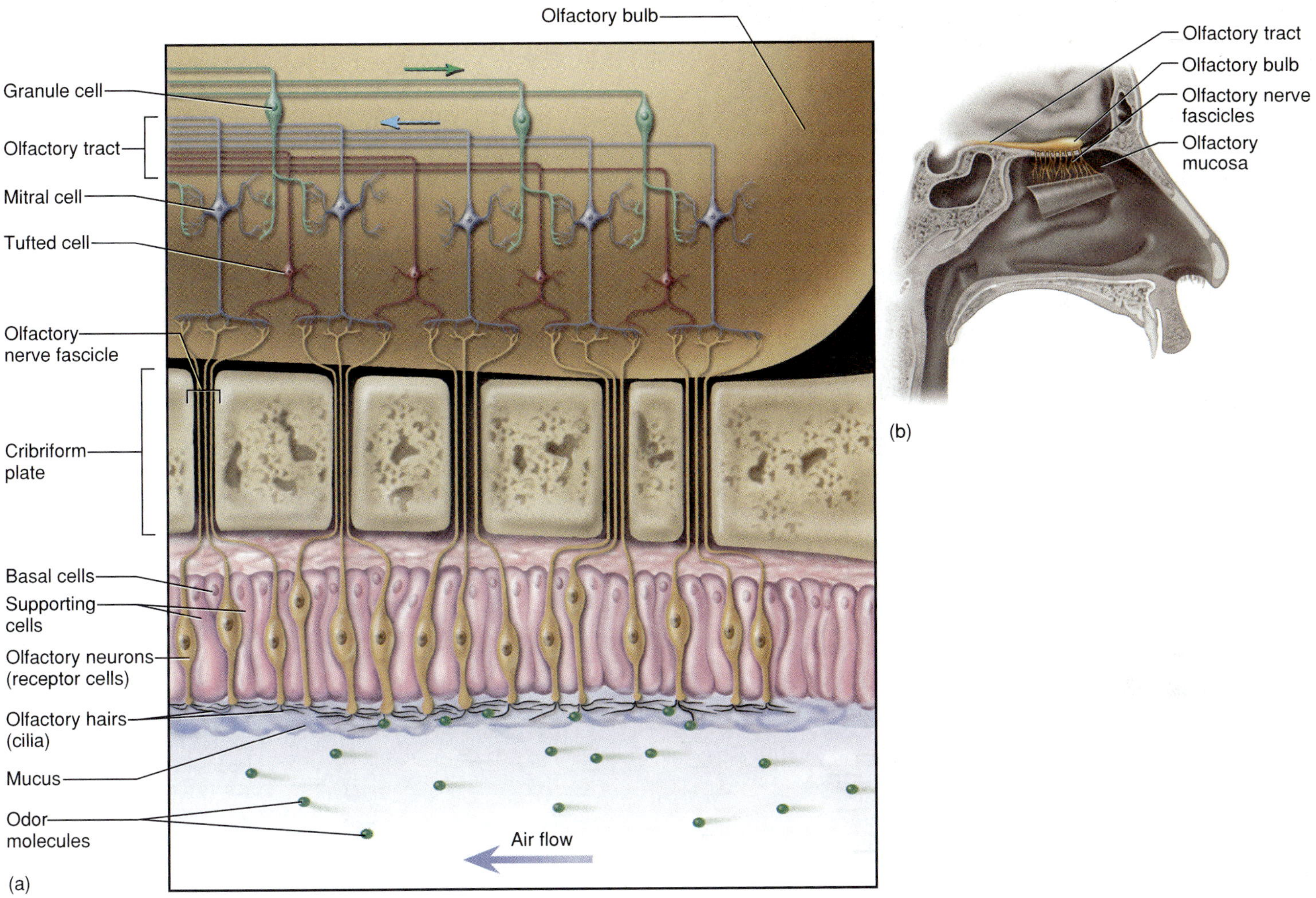
Olfactory bulb
Granule cell
Olfactory tract
Mitral cell
Tufted cell
Olfactory nerve fascicle
Cribriform plate
Basal cells
Supporting cells
Olfactory neurons (receptor cells)
Olfactory hairs (cilia)
Mucus
Odor molecules
Air flow
(a)
Olfactory tract
Olfactory bulb
Olfactory nerve fascicles
Olfactory mucosa
(b)

Figure 16.6 Olfactory receptors. (*a*) A small area of olfactory mucosa and projection pathways in the olfactory bulb. (*b*) Location of the major structures in relation to the nasal and cranial cavities.

Special Topic: Human Pheromones 16.1

Many mammals employ odors called **pheromones** to attract mates or communicate territory, status, and group membership. Often these are secreted by apocrine scent glands, sometimes associated with specialized tufts of hair. Some female sex pheromones, which communicate the time of fertility, are produced by bacteria acting on vaginal secretions.

There is an abundance of anecdote, but no clear experimental evidence, that human body odors affect sexual behavior. There is more adequate evidence, however, that a person's apocrine and vaginal secretions affect other people's sexual physiology, even when the odors cannot be consciously smelled. Experimental evidence shows that a woman's apocrine sweat can influence the timing of other women's menstrual cycles. The *dormitory effect* refers to the tendency for the menstrual cycles of female roommates to synchronize. Men's beards grow faster in the presence of women than in all-male settings, and the presence of men seems to influence female ovulation.

Some odors produce sensations by stimulating nociceptors of the trigeminal nerve; these include ammonia, menthol, chlorine, and hot peppers. "Smelling salts" are used to revive unconscious persons by strongly stimulating the trigeminal nerve with ammonia fumes.

Projection Pathways

When olfactory axons pass through the cribriform plate, they enter a pair of **olfactory bulbs** beneath the frontal lobes of the brain. Here, they synapse with neurons called *mitral cells* and *tufted cells* (fig. 16.6*b*), whose axons form bundles called the **olfactory tracts.** These tracts follow a complex pathway leading to the medial side of the temporal lobes (fig. 16.7). Input to the limbic system and hypothalamus can trigger emotional and autonomic reactions even without our conscious awareness of an odor. For example, we may react emotionally to the odor of certain foods, someone's perfume, a hospital, or decaying flesh and may exhibit such visceral reflexes as sneezing, coughing, salivating, vomiting, or secreting gastric juice. Unlike other sensory modalities, therefore, olfactory signals reach the cerebral cortex before the thalamus. Signals concerned with the conscious awareness of smell, however, ultimately pass through the thalamus and are relayed to an area of neocortex in the frontal lobes called the *orbitofrontal gyri.*

The sense of smell is also subject to feedback from the cerebral cortex, which sends signals back to *granule cells* in the olfactory bulbs. The granule cells, in turn, inhibit the mitral cells. One effect of this is that odors may change in quality and significance under different conditions. Food may smell more appetizing when you are hungry, for example, than it does after you have just eaten.

Think About It

Which taste sensations could be lost after damage to (1) the facial nerve or (2) the glossopharyngeal nerve? Fractures of which cranial bone would most likely disrupt the sense of smell?

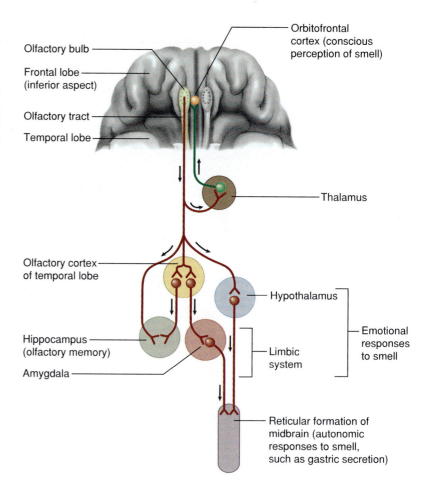

Figure 16.7 Olfactory projection pathways in the brain.

Key Point Review

---

11 What is the difference between a lingual papilla and a taste bud?

12 List the four primary taste sensations and discuss the adaptive significance (survival value) of each.

13 Which cranial nerves carry gustatory impulses to the brain?

14 What part of an olfactory cell bears the binding sites for odor molecules?

15 Which regions of the brain receive and process input from the olfactory cells?

Hearing and Equilibrium

▼Objectives
When you have completed this section, you should be able to
- identify the physical properties of sound waves that account for pitch and loudness;
- describe the gross and microscopic structures of the ear;
- explain how the ear converts vibrations to nerve impulses and discriminates between sounds of different intensity and pitch;
- explain how the vestibular apparatus enables the brain to interpret the body's position and movements; and
- describe the pathways taken by auditory and vestibular signals to the brain.

Hearing is a response to vibrating air molecules, and *equilibrium* is the sense of motion and balance. These senses reside in the inner ear, a maze of fluid-filled passages and sensory cells. This section explains how the fluid is set in motion and how the sensory cells convert this motion into an informative pattern of action potentials.

The Nature of Sound

To understand the physiology of hearing, it is necessary to appreciate some basic properties of sound. **Sound** may be defined as any audible vibration of molecules. It can be transmitted through water, solids, or air, but not through a vacuum. Our discussion is limited to airborne sound.

Sound is produced by a vibrating object such as a tuning fork, a loudspeaker, or the vocal cords. Consider a loudspeaker producing a pure tone (fig. 16.8). When the speaker cone moves forward, it pushes air molecules closer together, producing a high-density *zone of compression.* When it vibrates back, it leaves a low-density *zone of rarefaction* where air molecules are less closely spaced than at rest. Individual molecules do not travel from the vibrating object to the eardrum. Rather, a molecule collides with one immediately in front of it, that one collides with another, and so on, like a series of billiard balls; finally, some molecules collide with the eardrum and make it vibrate. The sensations we perceive as the pitch and loudness of the sound are related to the physical properties of these vibrations.

Pitch

Pitch is determined by the frequency with which the sound source, the eardrum, and other parts of the ear

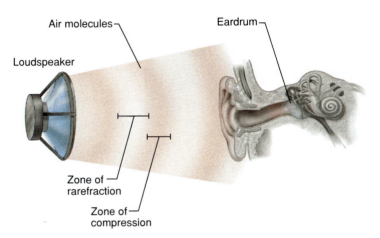

Figure 16.8 The production of sound and its transmission to the ear. Air molecules are pushed close together in the zones of compression, created by each forward movement of the speaker. In the zones of rarefaction, molecules are more widely spaced. 𝒯

vibrate. One movement of a vibrating object back and forth is called a cycle, and the number of cycles per second (cps or hertz, Hz) is called **frequency.** The lowest note on a piano, for example, is 30 Hz, middle C is 261 Hz, and the highest note is 5,000 Hz. The most sensitive human ears can hear frequencies from 20 to 20,000 Hz. The *infrasonic* frequencies below 20 Hz are not detected by the ear, but we sense them through vibrations of the skull and skin, and they play a significant role in our appreciation of music. The inaudible sounds above 20,000 Hz are called *ultrasonic.* Human ears are most sensitive to frequencies ranging from 1,500 to 4,000 Hz. In this range, we can hear sounds of relatively low energy (volume), whereas sounds above or below this range must be louder to be audible (fig. 16.9). Normal speech falls within this frequency range. Most of the hearing loss suffered with age is in the range of 250 to 2,050 Hz.

Loudness

Loudness is the perception of sound energy, intensity, or **amplitude** of vibration. In the speaker example, amplitude is a measure of how far forward and back the cone vibrates on each cycle and how much it compresses the air molecules in front of it. Loudness is expressed in decibels (dB), with 0 dB being the threshold of hearing. Every 10 dB step up the scale represents a sound with 10 times greater intensity. Thus 10 dB is 10 times threshold, 20 dB is 100 times threshold, 30 dB is 1,000 times threshold, and so forth. At most frequencies, the threshold of pain is 120 to 140 dB. The decibel

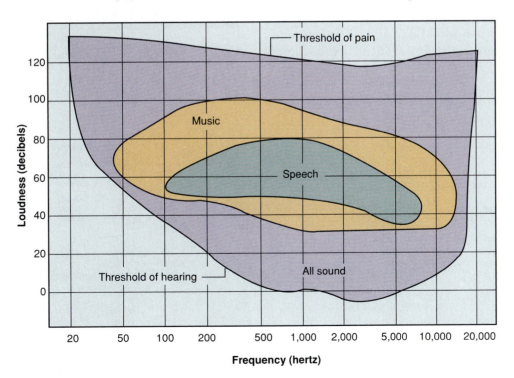

Figure 16.9 The range of human hearing.

levels of some common sounds are listed in table 16.2. Prolonged exposure to sounds greater than 90 dB can cause permanent loss of hearing.

Anatomy of the Ear

The ear has three parts called the *external, middle,* and *inner ear.* The first two are concerned only with transmitting sound to the inner ear, which houses the transducer that converts fluid motion to action potentials.

External Ear

The **external ear** is essentially a funnel and tube for conducting air vibrations to the eardrum. It begins with the fleshy **auricle,** or **pinna,** on the side of the head, shaped and supported by elastic cartilage except for the earlobe. It has a predictable arrangement of whorls and recesses that direct sound into the auditory canal (fig. 16.10).

The **auditory canal (external auditory meatus)** is slightly S-shaped and about 3 cm long in adults (fig. 16.11). It is lined with skin and supported by fibrocartilage at its opening and by the temporal bone for the rest of its length. Ceruminous and sebaceous glands in the canal produce secretions that mix with dead skin cells and form cerumen (earwax). Cerumen normally dries up and falls from the canal, but sometimes it becomes impacted and interferes with hearing.

The eardrum, or **tympanic[15] membrane,** closes the inner end of the auditory canal and separates it from the middle ear. It is about 1 cm in diameter and slightly concave on its outer surface. It is suspended in a ring-shaped groove in the temporal bone and vibrates freely in response to sound. It is innervated by sensory

Table 16.2	Some Familiar Sounds on the Decibel Scale
Threshold of Pain	120–140 dB
Loud Thunder	120 dB
Subway Train	100 dB
Conversation	60 dB
Quiet Office	40 dB
Whisper	30 dB
Threshold of Hearing	0 dB

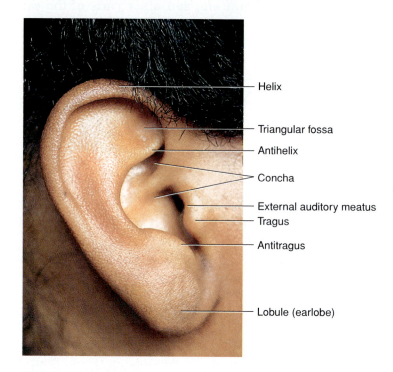

Figure 16.10 Anatomy of the auricle (pinna) of the ear.

- Helix
- Triangular fossa
- Antihelix
- Concha
- External auditory meatus
- Tragus
- Antitragus
- Lobule (earlobe)

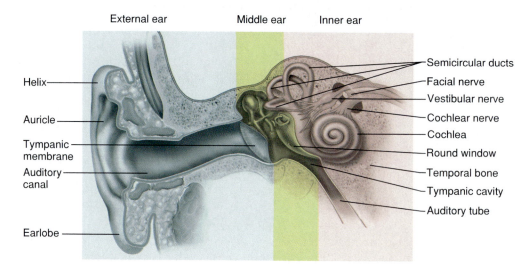

External ear   Middle ear   Inner ear

- Helix
- Auricle
- Tympanic membrane
- Auditory canal
- Earlobe
- Semicircular ducts
- Facial nerve
- Vestibular nerve
- Cochlear nerve
- Cochlea
- Round window
- Temporal bone
- Tympanic cavity
- Auditory tube

Figure 16.11 Internal anatomy of the right ear.

15. *tympan* = drum

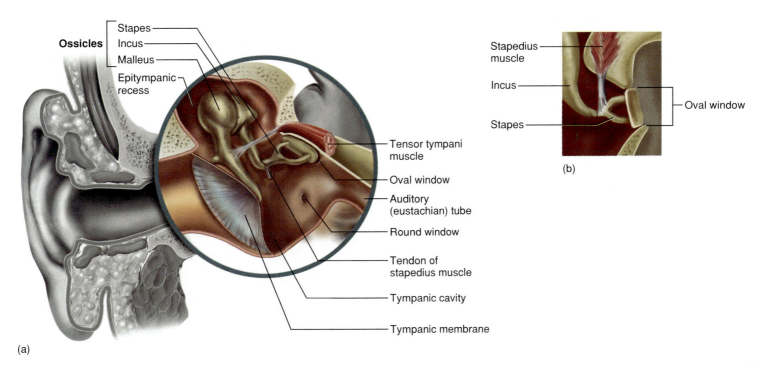

Figure labels (left diagram a):
Ossicles — Stapes, Incus, Malleus
Epitympanic recess
Tensor tympani muscle
Oval window
Auditory (eustachian) tube
Round window
Tendon of stapedius muscle
Tympanic cavity
Tympanic membrane
(a)

Figure labels (right diagram b):
Stapedius muscle
Incus
Stapes
Oval window
(b)

Figure 16.12 (*a*) Anatomy of the middle ear. (*b*) Detail of the relationship of the stapes to the stapedius muscle, its tendon, and the oval window. 𝑋

Special Topic ⟨ Otitis Media ⟩   16.2

Otitis[16] media (middle-ear infection) is especially common in children because their auditory tubes are relatively short and horizontal. Upper respiratory infections can easily spread from the throat to the tympanic cavity and mastoidal air cells. Fluid accumulates in the cavity and produces pressure, pain, and impaired hearing. If otitis media goes untreated, it may spread from the mastoidal air cells and cause meningitis, a potentially deadly infection (see chapter 14). Otitis media can also cause fusion of the middle-ear bones, resulting in hearing loss. It is sometimes necessary to drain fluid from the tympanic cavity by lancing the eardrum and inserting a tiny drainage tube—a procedure called *myringotomy*.[18] The tube, which is eventually sloughed out of the ear, relieves the pressure and permits the infection to heal.

branches of the vagus and trigeminal nerves and is highly sensitive to pain.

Middle Ear

The **middle ear** is located in the **tympanic cavity** of the temporal bone (fig. 16.12). Posteriorly, the tympanic cavity is continuous with the mastoidal air cells in the mastoid process. It is filled with air that enters by way of the **auditory (eustachian[17]) tube,** a passageway to the nasopharynx. (Be careful not to confuse *auditory tube* with *auditory canal.*) The auditory tube is normally flattened and closed, but swallowing or yawning opens it and allows air to enter or leave the tympanic cavity.

This equalizes air pressure on both sides of the eardrum, allowing it to vibrate freely. Excessive pressure on one side or the other dampens the sense of hearing. The auditory tube also allows infections to spread to the middle ear (see special topic 16.2).

The tympanic cavity contains the three smallest bones and two smallest skeletal muscles of the body. The bones, called the **auditory ossicles,**[19] span the 2 to 3 mm distance from the tympanic membrane to the inner ear. Progressing inward, the first is the **malleus,**[20] which has an elongated *handle* attached to the inner surface of the eardrum; a *head,* which is suspended from the wall of the tympanic cavity; and a *short process,* which articulates with the next ossicle. That

16. *ot* = ear + *itis* = inflammation
17. Bartholomeo Eustachio (1520–74), Italian anatomist

18. *myringo* = eardrum + *tomy* = cutting into
19. *oss* = bone + *icle* = little
20. *malleus* = hammer

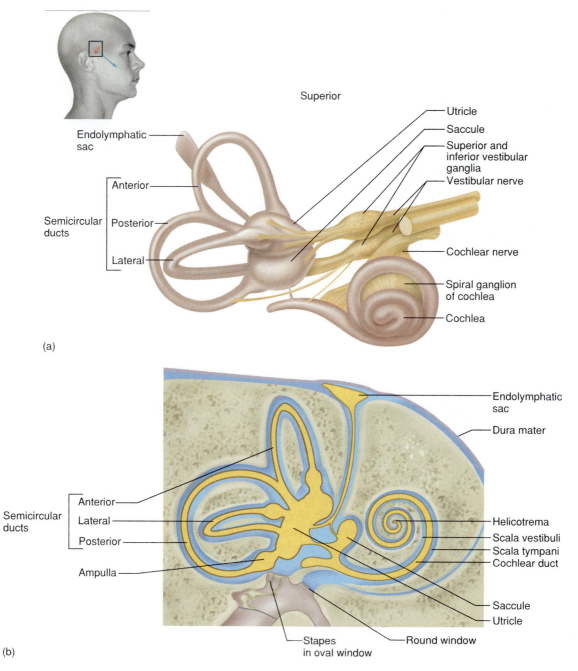

(a)

Superior

Endolymphatic sac

Semicircular ducts
- Anterior
- Posterior
- Lateral

Utricle

Saccule

Superior and inferior vestibular ganglia

Vestibular nerve

Cochlear nerve

Spiral ganglion of cochlea

Cochlea

(b)

Endolymphatic sac

Dura mater

Semicircular ducts
- Anterior
- Lateral
- Posterior

Ampulla

Helicotrema

Scala vestibuli

Scala tympani

Cochlear duct

Saccule

Utricle

Stapes in oval window

Round window

Figure 16.13 (a) Anatomy of the inner ear. (b) Relationship of the perilymph (blue) and endolymph (yellow) to the labyrinth of the inner ear. ✗

bone, called the **incus,**[21] articulates in turn with the **stapes**[22] (STAY-peez), named for the shape created by its arch and footplate. The *footplate,* shaped like the sole of a steam iron, is held by a ringlike ligament in an opening called the **oval window,** where the inner ear begins.

The muscles of the middle ear are the stapedius and tensor tympani. The **stapedius** (stay-PEE-dee-us) arises from the posterior wall of the cavity and inserts on the stapes. The **tensor tympani** (TEN-sor TIM-pan-

eye) arises from the wall of the auditory tube, travels alongside it, and inserts on the malleus. The protective function of these muscles is discussed later.

Inner Ear

The **inner ear** (fig. 16.13a) is housed in a maze of passageways in the temporal bone called the **bony labyrinth.** These passageways are lined by a system of fleshy tubes called the **membranous labyrinth.** Between the bony and membranous labyrinths is a cushion of fluid, similar to cerebrospinal fluid, called **perilymph** (PER-ih-limf). Within the membranous labyrinth is a

21. *incus* = anvil
22. *stapes* = stirrup

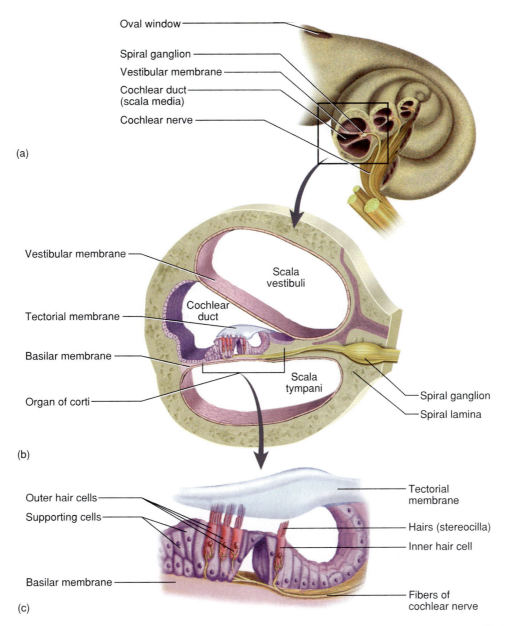

(a)

Oval window
Spiral ganglion
Vestibular membrane
Cochlear duct (scala media)
Cochlear nerve

(b)

Vestibular membrane
Scala vestibuli
Cochlear duct
Tectorial membrane
Basilar membrane
Scala tympani
Organ of corti
Spiral ganglion
Spiral lamina

(c)

Outer hair cells
Supporting cells
Tectorial membrane
Hairs (stereocilia)
Inner hair cell
Basilar membrane
Fibers of cochlear nerve

Figure 16.14 Anatomy of the cochlea. (*a*) Vertical section. The apex of the cochlea faces downward and anterolaterally in anatomical position. (*b*) Detail of one section through the membranous labyrinth of the cochlea. (*c*) Detail of the organ of Corti. 𝒳 ▭

fluid, similar to intracellular fluid, called **endolymph** (fig. 16.13*b*). (Despite their names, neither of these fluids is related to the lymph in the lymphatic system.) The labyrinths begin at an area called the **vestibule,** which contains organs of equilibrium to be discussed later.

The organ of hearing is the **cochlea**[23] (COC-lee-uh), a coiled tube that arises from the anterior side of the vestibule. In other vertebrates, it is straight or slightly curved. In most mammals, however, it has assumed the form of a snail-like spiral, allowing a longer cochlea to

fit in a compact space. In humans, the spiral is about 9 mm wide at the base and 5 mm high. Its apex points anterolaterally. The cochlea winds for about 2.5 coils around an axis of spongy bone called the **modiolus**[24] (mo-DY-oh-lus). The modiolus is shaped like a screw; its threads form a spiral platform that supports the fleshy tube of the cochlea. A vertical section cuts through the cochlea about five times (fig. 16.14*a*). A single cross section through it looks like figure 16.14*b*. It is important to realize that the structures seen in cross section actually

23. *cochlea* = snail

24. *modiolus* = hub

have the form of spiral strips winding around the modiolus from base to apex.

The cochlea has three fluid-filled chambers. The superior one is called the **scala**[25] **vestibuli** (SCAY-la vess-TIB-you-lye) and the inferior one is the **scala tympani.** These are filled with perilymph and communicate with each other through a narrow channel called the *helicotrema* (HEL-ih-co-tree-muh) at the tip of the cochlea. The scala vestibuli begins near the oval window and spirals to the tip; from there, the scala tympani spirals back down to the base and terminates at the **round window.** The round window is covered by a membrane called the *secondary tympanic membrane* (see fig. 16.13*b*).

The middle chamber is a triangular space, the **cochlear duct (scala media),** where sensory transduction occurs. It is separated from the scala vestibuli above by a thin **vestibular membrane** and from the scala tympani below by a much thicker **basilar membrane.** Unlike those chambers, it is filled with endolymph rather than perilymph.

The basilar membrane supports the **organ of Corti**[26] (COR-tee), a thick epithelium with associated structures (fig. 16.14*c*). It is the transducer that converts vibrations into nerve impulses, so we must pay particular attention to its structural details. The epithelium consists of **hair cells** and **supporting cells.** Hair cells are named for the long, stiff microvilli called **stereocilia**[27] on their apical surfaces. (Stereocilia should not be confused with true cilia. They do not have an axoneme of microtubules as seen in cilia, and they do not move by themselves.) Resting on top of the stereocilia is a gelatinous **tectorial**[28] **membrane.**

The organ of Corti has four rows of hair cells spiraling along its length (fig. 16.15). About 3,500 of these, called **inner hair cells (IHCs),** are arranged in a row by themselves on the medial side of the basilar membrane (facing the modiolus). Each of these has a cluster of 50 to 60 stereocilia, graded from short to tall. Another 20,000 **outer hair cells (OHCs)** are neatly arranged in three rows across from the inner hair cells. Each outer hair cell has about 100 stereocilia arranged in the form of a V, with their tips embedded in the tectorial membrane. All that we hear comes from the IHCs, which supply 90% to 95% of the sensory fibers of the cochlear nerve. The function of the OHCs is to adjust the response of the cochlea to different frequencies and enable the IHCs to work with greater precision. We will see shortly how this is done. At their basal ends, IHCs synapse with sensory dendrites and OHCs with both sensory and motor nerve fibers.

25. *scala* = staircase
26. Alfonso Corti (1822–88), Italian anatomist
27. *stereo* = solid
28. *tect* = roof

Figure 16.15 Apical surfaces of the cochlear hair cells (SEM). On the left are the three rows of outer hair cells, which serve only to tune the cochlea. Each cell has a V-shaped row of stereocilia. On the right is the single row of inner hair cells, which generate the signals we hear.

From R.G. Kessel and R.H. Kardon, *Scanning Electon Microscopy of Tissues and Organs,* 1979.

The Physiology of Hearing

We can now examine the way in which sound affects the ear and results in a meaningful pattern of action potentials.

The Middle Ear

The auditory ossicles provide no mechanical advantage—vibrations of the stapes against the inner ear normally have the same amplitude as vibrations of the eardrum against the malleus. Why have auditory ossicles, then? There are two answers to this. One is that the eardrum, which moves in air, vibrates quite easily, whereas the stapes footplate must vibrate against the fluid of the inner ear. This fluid puts up a much greater resistance (impedance) to motion. If airborne sound waves struck the footplate directly, they would not have enough energy to overcome this impedance and move the stapes. The area of the eardrum, however, is 18 times that of the oval window. By concentrating the energy of the vibrating eardrum on an area 1/18 that size, the ossicles create a greater force per unit area at the oval window and overcome the impedance of the endolymph. This function is called **impedance matching.**

The ossicles and their muscles also have a protective function. In response to a loud noise, the tensor tympani pulls the eardrum inward and tenses it, while the stapedius reduces mobility of the stapes. This **tympanic reflex** reduces the transfer of vibrations from the eardrum to the oval window. The reflex probably evolved in part for protection from loud but slowly building noises such as thunder. It has a latency of about 40 msec, which is not quick enough to protect the inner ear from sudden noises such as gunshots. Such noises can irreversibly damage the hair cells of the inner ear by fracturing their stereocilia. It is therefore imperative to wear ear protection when using firearms or working in noisy environments.

The middle-ear muscles also help to coordinate speech with hearing. Without them, the sound of your own speech would be so loud it could damage your inner ear, and it would drown out soft or high-pitched sounds from other sources. Just as you are about to speak, however, the brain signals these muscles to contract. This dampens the sense of hearing in phase with the inflections of your own voice and makes it possible to hear other people while you are speaking.

Think About It

What type of muscle fibers—Type I or II (see chapter 12)—do you think constitute the stapedius and tensor tympani? That is, which type would best suit the purpose of these muscles?

Stimulation of Cochlear Hair Cells

To produce a sensation of sound, vibration of the auditory ossicles must vibrate the basilar membrane on which the hair cells rest. A simplified mechanical model of the ear (fig. 16.16) makes it easy to see how this happens. The stapes pushes on the perilymph of the scala vestibuli; the perilymph pushes the vestibular membrane down; the vestibular membrane pushes on the endolymph of the cochlear duct; and the endolymph pushes the basilar membrane down. (The vestibular membrane is omitted from the diagram for simplicity; it has no significant effect on the mechanics of the cochlea.) The basilar membrane puts pressure on the perilymph of the scala tympani below it, and the secondary tympanic membrane bulges outward to relieve this pressure. In short, as the stapes goes in-out-in, the secondary tympanic membrane goes out-in-out, and the basilar membrane goes down-up-down. It is not difficult to see how this happens—the only thing hard to imagine is that it can happen as often as 20,000 times per second!

The vestibular membrane separates the perilymph of the scala vestibuli from the endolymph of the cochlear duct. In order for the hair cells to function properly, the tips of their stereocilia must be bathed in endolymph. Endolymph has a much higher K^+ concentration and an electrical potential about 80 mV higher than the perilymph. This creates a strong electrochemical gradient from the tip to the base of a hair cell, which is the source of potential energy that ultimately enables a hair cell to work.

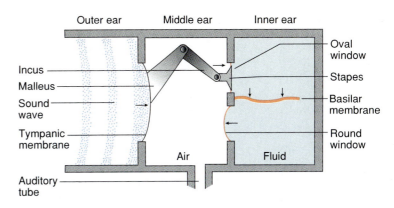

Figure 16.16 Mechanical model of the ear. Each inward movement of the tympanic membrane pushes inward on the auditory ossicles of the middle ear and fluid of the inner ear. This pushes down on the basilar membrane, and pressure is relieved by an outward bulge of the round window. Thus the basilar membrane vibrates up and down in synchrony with the vibrations of the tympanic membrane.

The tectorial membrane is especially important in cochlear mechanics. Remember that the stereocilia of the outer hair cells have their tips embedded in it, and those of the inner hair cells come very close to it. The tectorial membrane is anchored to the modiolus (see fig. 16.14c), which holds it relatively still as the basilar membrane and hair cells vibrate up and down. Movement of the basilar membrane thus bends the hair cell stereocilia back and forth.

At the tip of each stereocilium of the inner hair cells is a single transmembrane protein that functions as a mechanically gated ion channel. A fine, stretchy protein filament called a **tip link** extends like a spring from the ion channel of one stereocilium to the side of the stereocilium next to it (fig. 16.17). The stereocilia increase in height progressively, so that all but the tallest ones have tip links leading to taller stereocilia beside them. When a taller stereocilium bends away from a shorter one, it pulls on the tip link and opens the ion channel of the shorter stereocilium. The channel is nonselective, but since the predominant ion of the endolymph is K^+, the primary effect of this gating is to allow a brief K^+ influx into each hair cell. This depolarizes the hair cell while the channel is open, and when the stereocilium bends the other way its channel closes and the cell becomes briefly hyperpolarized. During the moments of depolarization, a hair cell releases a neurotransmitter that stimulates the sensory dendrites synapsing with its base. Each depolarization thus generates action potentials in the cochlear nerve.

Sensory Coding

The cochlea responds differently to vibrations of different amplitude and frequency. Loud sounds produce more

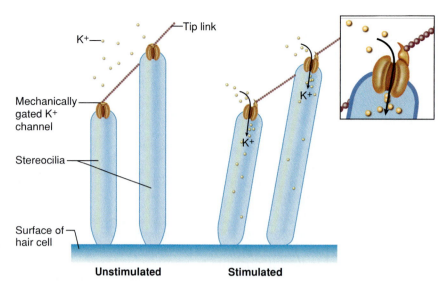

Figure 16.17 Two stereocilia of a cochlear hair cell, showing how the tip link from a taller stereocilium to a shorter one mechanically opens the ion gate of the shorter stereocilium.

vigorous vibrations of the organ of Corti. This excites a greater number of hair cells over a broader area of basilar membrane and triggers a higher frequency of action potentials in the cochlear nerve fibers. If the brain detects intense activity in nerve fibers from a broad region of the organ of Corti, it interprets this as a loud sound.

Frequency discrimination requires a more sophisticated mechanism. The basilar membrane is spanned by short, stiff collagen fibers of varying lengths, somewhat like the reeds of a harmonica. At its proximal (basal) end, the basilar membrane is attached, relatively narrow, and stiff. At its distal (apical) end, it is unattached, five times wider than at the base, and more flexible. Think of the basilar membrane as analogous to a rope stretched tightly between two posts. If you pluck the rope at one end, a wave of vibration travels down its length and back. This produces a standing wave, with some regions of the rope vibrating more than others. Similarly, a sound causes a standing wave in the basilar membrane. The peak amplitude of this wave is near the distal end in the case of low-frequency sounds. With sounds of higher frequencies, it is closer to the proximal end. The brain interprets auditory input as low- and high-pitched sounds, respectively, when it detects that inner hair cells are responding most strongly at the distal or proximal end of the basilar membrane (fig. 16.18). Speech, music, and other everyday sounds, of course, are not pure tones—they create complex patterns of vibration in the basilar membrane that must be sorted out by interneurons of the brain.

Cochlear Tuning

Just as we tune a radio to receive a certain frequency, the cochlea can be tuned to receive some frequencies better than others. The outer hair cells are supplied with a few sensory fibers (5%–10% of those in the cochlear nerve), but more importantly, they receive motor fibers from the brain. Therefore, while the vestibulocochlear nerve is typically listed as a sensory nerve, it must actually be regarded as mixed.

Sound activates a feedback loop from the outer hair cells (OHCs) to the brainstem and back. Motor feedback to an OHC causes it to shorten by 10% to 15% of its resting length. Because an OHC is anchored to the basilar membrane below and its stereocilia are anchored to the tectorial membrane above, contractions of the OHCs tense the basilar membrane and modify its vibrations. If the OHCs are experimentally incapacitated, the inner hair cells (IHCs) respond much less precisely to differences in pitch. Another mechanism of cochlear tuning involving the inner hair cells is explained in the next section.

The Auditory Projection Pathway

Sensory dendrites leaving the cochlear hair cells lead to bipolar neurons in the **spiral ganglion** that winds around the modiolus. The axons of these cells lead away from the base of the modiolus as the **cochlear nerve.** The cochlear nerve joins the vestibular nerve, to be discussed later, and the two together become the vestibulocochlear nerve (cranial nerve VIII).

Cochlear nerve fibers project to the *cochlear nucleus* on each side of the medulla oblongata and synapse with second-order neurons leading to the nearby *superior olivary nucleus* (fig. 16.19). This nucleus issues efferent fibers that return to the cochlea and synapse with the sensory fibers near the bases of the IHCs. By transmitting efferent signals back to these

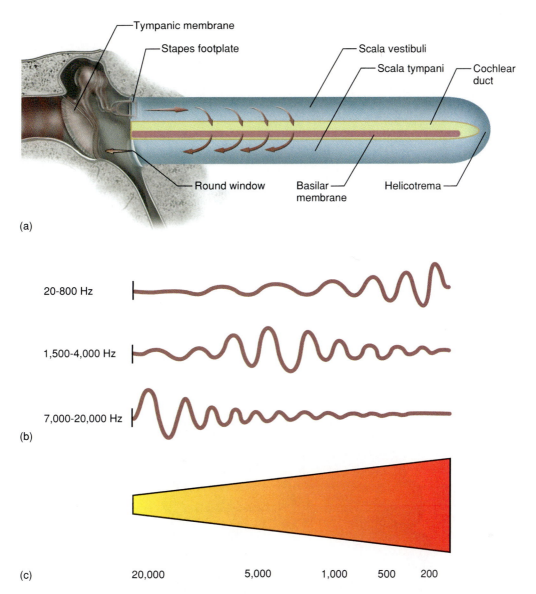

Figure 16.18 Frequency response of the basilar membrane of the cochlea. (*a*) The cochlea, uncoiled and laid out straight. (*b*) Sounds produce a standing wave of vibration along the basilar membrane. The peak amplitude of the wave varies with the frequency of the sound, as shown here. The amount of vibration is greatly exaggerated in this diagram to clarify the standing wave. (*c*) The taper of the basilar membrane and its correlation with sound frequencies. High frequencies (7,000–20,000 Hz) are best detected by hair cells near the basal (proximal) end at the left and low frequencies (20–800 Hz) by hair cells near the apical (distal) end at the right. ⚡

sensory fibers, the nucleus inhibits sensory output from some hair cells and enhances the contrast between signals from the more responsive and less responsive regions of the organ of Corti. Combined with the previously described role of the OHCs, this sharpens the tuning of the cochlea and the ability to discriminate sounds of different pitch. Other efferent fibers from the superior olivary nucleus innervate the stapedius and tensor tympani muscles. The superior olivary nucleus also plays a role in **binaural**[29]

hearing—comparing signals from the right and left ears to identify the direction from which a sound is coming.

Other neurons from the superior olivary nucleus continue to higher centers of the brain. They ascend to the inferior colliculi, where they synapse with neurons that lead to the thalamus. In the thalamus, those neurons synapse with others that continue to the primary auditory cortex in each temporal lobe. The temporal lobe is the site of conscious perception of sound; it completes the information processing essential to binaural hearing.

29. *bin* = two + *aur* = ear

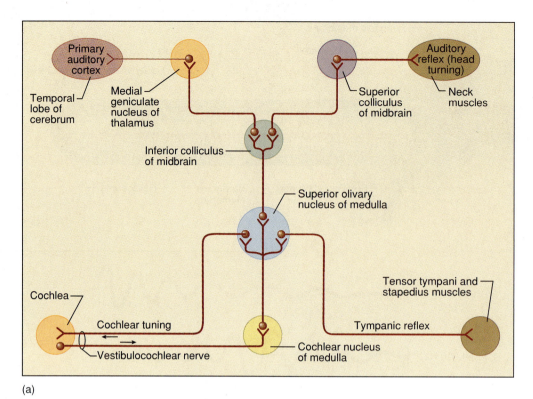

(a)

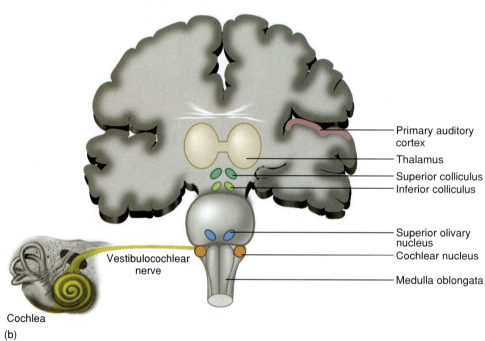

(b)

Figure 16.19 Auditory pathways in the brain. (*a*) Schematic. Each red circle represents the soma of a neuron, and the fork represents the distal end of the axon. (*b*) Frontal section of the cerebrum and brainstem, showing the locations of auditory processing centers.

Other nerve fibers pass from the inferior to the superior colliculus. Both pairs of colliculi seem to be involved in startle responses and rapid head movements in response to loud or sudden noises. Because of extensive decussation in the auditory pathway, damage to the right or left auditory cortex does not cause a unilateral loss of hearing.

Deafness

Deafness means any hearing loss, from mild and temporary to complete and irreversible. **Conduction deafness** is a type that results from the impaired transmission of vibrations to the inner ear. It may result from a damaged eardrum, otitis media, or blockage of the auditory canal with cerumen or foreign objects. In *otosclerosis*,[30] another cause of conduction deafness, the auditory ossicles become fused to one another or the stapes becomes fused to the oval window, and the ossicles cannot vibrate freely. **Sensorineural deafness** is due to the death of hair cells or any nervous elements concerned with hearing, from the cochlear nerve to the auditory cortex. It usually results from the death of hair cells, a common effect of aging and often a consequence of exposure to loud noises. It is a common occupational disease among musicians. Deafness leads some people to develop delusions of being talked about, disparaged, or cheated. Beethoven said his deafness drove him nearly to suicide.

Equilibrium

The original function of the ear in vertebrate history was **equilibrium**—providing sensory input for coordination and balance. Only later did vertebrates evolve the cochlea, middle-ear structures, and auditory function of the ear. In humans, the receptors for equilibrium constitute the **vestibular apparatus,** which consists of three **semicircular ducts** and two chambers—an anterior **saccule** (SAC-yule) and a posterior **utricle**[31] (YOU-trih-cul). The sense of equilibrium is divided into **static equilibrium,** the perception of the orientation of the head when the body is stationary, and **dynamic equilibrium,** the perception of motion or acceleration. Acceleration is divided into *linear acceleration,* a change in velocity in a straight line, as when riding in a car or elevator, and *angular acceleration,* a change in the rate of rotation. The saccule and utricle are responsible for static equilibrium and the sense of linear acceleration; the semicircular ducts detect only angular acceleration.

The Saccule and Utricle

Each of these chambers has a 2 by 3 mm patch of hair cells and supporting cells called a **macula.**[32] The **macula sacculi** has a nearly vertical orientation on the wall of the saccule. The **macula utriculi** lies on a near-horizontal plane on the floor of the utricle (fig. 16.20*a*).

Each hair cell of the maculae has 40 to 70 stereocilia and one motile true cilium, called a **kinocilium.**[33] The tips of the stereocilia and kinocilium are embedded in a gelatinous coat called the **otolithic membrane;** this membrane is weighted with granules of calcium carbonate and protein called **otoliths**[34]

(fig. 16.20*b*). Otoliths add to the inertia of the membrane and enhance the sense of gravity and motion.

Figure 16.20*c* shows how the macula utriculi detects tilt of the head. With the head erect, the otolithic membrane bears directly down on the hair cells and stimulation is minimal. When the head is tilted, however, the weight of the membrane bends the stereocilia and stimulates the hair cells. Any orientation of the head causes a combination of stimulation to the utricles and saccules of the two ears, which enables the brain to sense head orientation by comparing these inputs to each other and to other types of input from the eyes and stretch receptors in the neck.

The inertia of the otolithic membranes is especially important in detecting linear acceleration. Suppose you are sitting in a car at a stoplight and then begin to move. The heavy otolithic membrane of the macula utriculi briefly lags behind the rest of the tissues, bends the stereocilia backward, and stimulates the cells. When you stop at the next intersection, the macula stops but the otolithic membrane keeps on going for a moment, bending the stereocilia forward. The hair cells convert this pattern of stimulation to nerve signals, and the brain is thus advised of changes in your linear velocity.

If you are standing in an elevator and it begins to move up, the otolithic membrane of the vertical macula sacculi lags behind briefly and pulls down on the hair cells. When the elevator stops, the otolithic membrane keeps on going for a moment and bends the hairs upward. The macula sacculi thus detects vertical acceleration.

The Semicircular Ducts

Rotational acceleration is detected by the three *semicircular ducts* (fig. 16.21), each housed in an osseous *semicircular canal* of the temporal bone. The **anterior** and **posterior semicircular ducts** are positioned vertically, at right angles to each other. The **lateral semicircular duct** is about 30° from the horizontal plane. The orientation of the ducts causes a different duct to be stimulated by rotation of the head in different planes—turning it from side to side as in gesturing "no," nodding up and down as in gesturing "yes," or tilting it from side to side as in touching your ears to your shoulders.

The semicircular ducts are filled with endolymph. Each duct opens into the utricle and has a dilated sac at one end called an **ampulla.**[35] Within the ampulla is a mound of hair cells and supporting cells called the **crista**[36] **ampullaris.** The hair cells have stereocilia and a kinocilium embedded in the **cupula,**[37] a gelatinous

---

30. *oto* = ear + *sclerosis* = hardening
31. *saccule* = little sac; *utricle* = little bag
32. *macula* = spot
33. *kino* = movement
34. *oto* = ear + *lith* = stone

35. *ampulla* = little jar
36. *crista* = ridge, crest
37. *cupula* = little tub

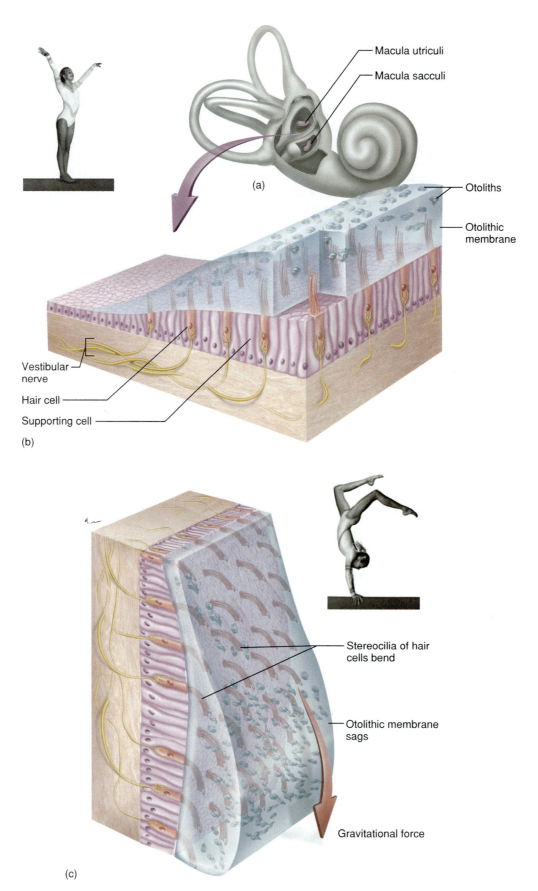

Macula utriculi

Macula sacculi

(a)

Otoliths

Otolithic
membrane

Vestibular
nerve

Hair cell

Supporting cell

(b)

Stereocilia of hair
cells bend

Otolithic membrane
sags

Gravitational force

(c)

Figure 16.20 (*a*) Location of the macula utriculi and macula sacculi. (*b*) Structure of a macula. (*c*) Action of the otolithic membrane on the hair cells when the head is tilted.

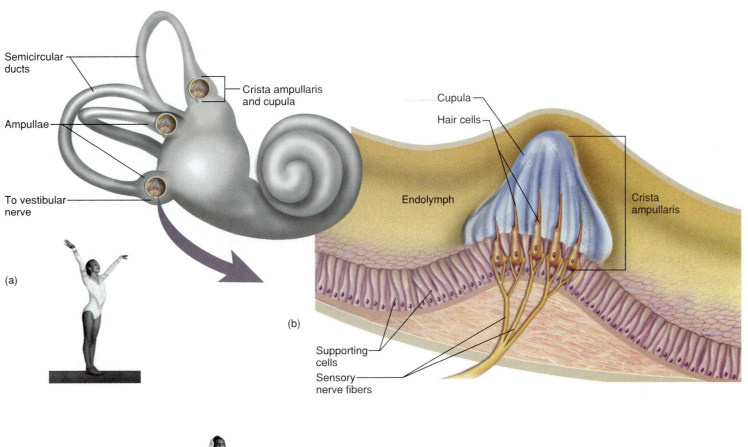

(a)

Semicircular
ducts

Crista ampullaris
and cupula

Ampullae

To vestibular
nerve

Cupula

Hair cells

Endolymph

Crista
ampullaris

(b)

Supporting
cells

Sensory
nerve fibers

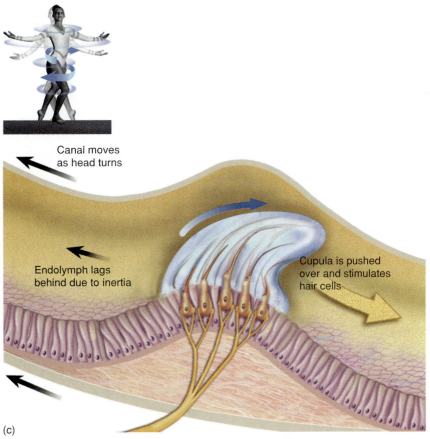

(c)

Canal moves
as head turns

Endolymph lags
behind due to inertia

Cupula is pushed
over and stimulates
hair cells

Figure 16.21 (*a*) Structure of the semicircular ducts, with each ampulla opened to show the crista ampullaris and cupula. (*b*) Detail of the crista ampullaris and cupula. (*c*) Action of the endolymph on the cupula and hair cells when the head is rotated. ⅄

membrane that extends from the crista to the roof of the ampulla. When the head turns the duct rotates, but because of its inertia the endolymph lags behind. This pushes the cupula, bends the stereocilia, and stimulates the hair cells. After 25 to 30 seconds of continual rotation, however, the endolymph catches up with the movement of the duct, and stimulation of the hair cells ceases.

Think About It

The semicircular ducts do not detect motion itself, but only acceleration—a *change in the rate* of motion. Explain.

Equilibrium Projection Pathways

Hair cells of the macula sacculi, macula utriculi, and semicircular ducts synapse at their bases with sensory fibers of the **vestibular nerve.** This nerve joins the cochlear nerve, forming the vestibulocochlear nerve (VIII). Most fibers from the vestibular apparatus terminate in the **vestibular nucleus** of the pons, while others project to the cerebellum. Projection fibers from the vestibular nucleus lead to the cervical spinal cord, and from there fibers lead to the brainstem nuclei for the cranial nerves that control eye movements—the oculomotor (III), trochlear (IV), and abducens (VI) nerves. Other fibers from the cervical spinal cord lead by way of the accessory nerve (XI) to muscles that move the head and neck.

Reflex pathways to the extrinsic eye muscles enable us to fixate visually on a point in space while the head is moving. To observe this effect, hold this book in front of you at a comfortable reading distance and fixate on the middle of the page. Move the book left and right about once per second, and you will be unable to read it. Now hold the book still and shake your head from side to side at the same rate. This time you will be able to read it because the reflex pathway compensates for your head movements and keeps your eyes fixated on the target.

Before we move on to the sense of vision, you may wish to review the anatomy of the ear (table 16.3) and think back on the function of each component.

Table 16.3	Anatomical Checklist for the Ear
External Ear (fig. 16.10)	Organ of Corti
Auricle (pinna)	Supporting cells
Auditory canal (external auditory meatus)	Inner hair cells
	Outer hair cells
Ceruminous glands	Stereocilia
Tympanic membrane (eardrum)	Tectorial membrane
Helix	Basilar membrane
Antihelix	Neural components
Triangular fossa	Spiral ganglion
Concha	Cochlear nerve
Tragus	
Antitragus	**Vestibular Apparatus (figs. 16.20 and 16.21)**
Lobe	Oval window
Middle Ear (Tympanic Cavity) (fig. 16.12)	Vestibule
Auditory Ossicles	Saccule
Malleus	Macula sacculi
Incus	Utricle
Stapes	Macula utriculi
Footplate	Otolithic membranes
Muscles	Semicircular ducts
Stapedius	Ampulla
Tensor tympani	Crista ampullaris
Auditory (Eustachian) Tube	Cupula
	Vestibular nerve
Inner Ear (fig. 16.13)	Round window
Labyrinths and Fluids	Secondary tympanic membrane
Bony labyrinth	**Projection Pathways (fig. 16.19)**
Membranous labyrinth	
Perilymph	Cochlear nerve
Endolymph	Vestibular nerve
	Vestibulocochlear nerve
Cochlea (fig. 16.14)	Cochlear nucleus
Modiolus	Vestibular nucleus
Scala vestibuli	Superior olivary nucleus
Scala tympani	Inferior and superior colliculi
Helicotrema	Thalamus
Cochlear duct	Primary auditory cortex
Vestibular membrane	

Key Point Review

(16) What physical properties of sound waves correspond to the sensations of loudness and pitch?

(17) What is the benefit of having three auditory ossicles and two muscles in the middle ear?

(18) Explain how vibration of the tympanic membrane ultimately results in fluctuations of membrane voltage in a cochlear hair cell.

(19) How does the brain recognize the difference between high C and middle C of a piano? Between a loud sound and a soft one?

(20) How does the function of the semicircular ducts differ from the function of the saccule and utricle?

(21) How is the mode of sensory transduction in the semicircular ducts similar to that in the saccule and utricle?

Vision and Light

Vision is the perception of light. *Light* is electromagnetic radiation with wavelengths ranging from about 400 to 750 nm. Most solar radiation that reaches the surface of the earth falls within this range; radiation of shorter and longer wavelengths is generally filtered out by ozone, carbon dioxide, and water vapor in the atmosphere. Vision is thus adapted to take advantage of the radiation that is most available to us.

Yet there is further reason for vision to be limited to this range of wavelengths. If it is to produce a physiological response, light must cause a *photochemical reaction*—a change in chemical structure caused by the energy of a photon. When an electron absorbs a photon, it is boosted to a higher energy level (orbit) around its nucleus and can transfer to another atom. The transfer of an electron from one atom to another is the essence of a chemical reaction. Radiation below a wavelength of 400 nm has so much energy that it ionizes organic molecules and kills cells. It is useful for sterilizing food and medical instruments, but it has too much energy for the biochemical processes of vision. Radiation above 750 nm has too little energy to break chemical bonds. It warms the tissues (heat lamps are based on this principle) but does not cause chemical reactions. These other forms of electromagnetic radiation include ultraviolet (UV) rays, X rays, and gamma rays at the high-energy (short-wavelength) end of the spectrum; at the low-energy (long-wavelength) end, they include infrared (IR) rays, microwaves, and radio waves. Some other organisms can see radiation extending in both directions slightly beyond the range visible to humans.

Accessory Structures of the Orbit

Before considering the eye itself, let's survey the accessory structures located in and around the orbit (fig. 16.22). These include the *eyebrows, eyelids, conjunctiva, lacrimal apparatus,* and *extrinsic eye muscles.*

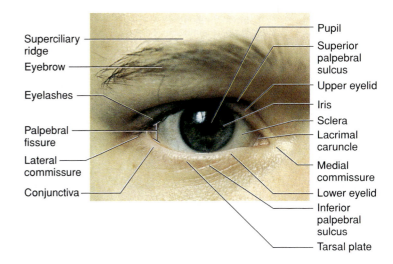

Figure 16.22 External anatomy of the orbital region.

- The **eyebrows** probably serve mainly to enhance facial expressions and nonverbal communication (see chapter 7), but they may also protect the eyes from glare and help to keep perspiration from running into the eye.
- The **eyelids,** or **palpebrae** (pal-PEE-bree), close periodically to moisten the eye with tears, sweep debris from the surface, block foreign objects from the eye, and prevent visual stimuli from disturbing our sleep. The two eyelids are separated by the **palpebral fissure,** and the corners where they meet are called the **medial** and **lateral commissures (canthi).** The eyelid consists largely of the orbicularis oculi muscle covered with skin (fig. 16.23). It also has a thick connective tissue ridge called the **tarsal plate** along its free edge. Within this plate are 20 to 25 **tarsal glands** that open along the edge of the eyelid. They secrete an oil that coats the eye and reduces tear evaporation. The **eyelashes** help to keep debris from the eye. Touching the eyelashes stimulates hair receptors and triggers the blink reflex.
- The **conjunctiva** (CON-junk-TY-vuh) is a transparent mucous membrane that covers the inner surface of the eyelid and anterior surface of the eyeball. Its primary purpose is to secrete a thin mucous film that prevents the eyeball from drying. It is richly innervated and highly sensitive to pain. It is also very vascular, which is especially evident when the vessels are dilated and the eyes are "bloodshot." Because it is vascular and the cornea is not, the conjunctiva heals more readily than the cornea when injured.
- The **lacrimal**[38] **apparatus** (fig. 16.24) consists of the lacrimal (tear) gland and a series of ducts that drain

38. *lacrima* = tear

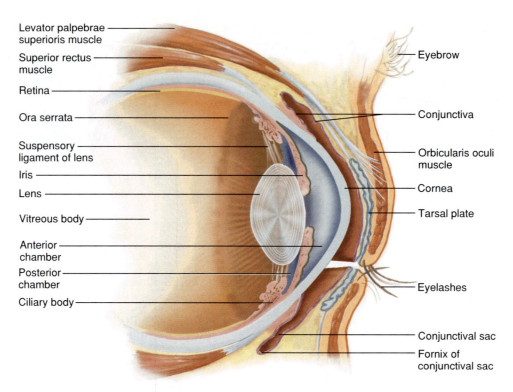

Levator palpebrae
superioris muscle

Superior rectus
muscle

Retina

Ora serrata

Suspensory
ligament of lens

Iris

Lens

Vitreous body

Anterior
chamber

Posterior
chamber

Ciliary body

Eyebrow

Conjunctiva

Orbicularis oculi
muscle

Cornea

Tarsal plate

Eyelashes

Conjunctival sac

Fornix of
conjunctival sac

Figure 16.23 Anterior portion of the eye and accessory structures of the orbit, in sagittal section.

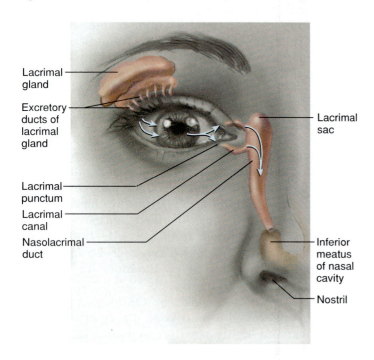

Lacrimal
gland

Excretory
ducts of
lacrimal
gland

Lacrimal
punctum

Lacrimal
canal

Nasolacrimal
duct

Lacrimal
sac

Inferior
meatus
of nasal
cavity

Nostril

Figure 16.24 The lacrimal apparatus.

the tears into the nasal cavity. The **lacrimal gland,** about the size and shape of an almond, is nestled in a shallow fossa of the frontal bone in the superolateral corner of the orbit. About 12 short ducts lead from the lacrimal gland to the surface of the conjunctiva. Tears wash across the conjunctiva

and collect at the **lacrimal caruncle**[39] (CAR-un-cul), the pink fleshy mass at the medial commissure of the eye. The caruncle has two tiny pores, the **puncta,**[40] that lead via short **lacrimal canals** into a **lacrimal sac.** From here, a **nasolacrimal duct** carries the tears to the inferior meatus of the nasal cavity—thus an abundance of tears from crying or "watery eyes" can result in a runny nose. Normally, the tears drain from the nasal cavity to the pharynx and are swallowed. When you have a cold, the nasolacrimal ducts become swollen and obstructed, the tears cannot drain, and tears may overflow from the brim of your eye. By moistening the conjunctiva, the tears not only wash foreign particles from the eye but also allow for the diffusion of oxygen and carbon dioxide between the air and tissues; such gases cannot diffuse adequately through a dry membrane. The tears also contain a bactericidal substance called lysozyme that keeps the eye surface fairly sterile.

• The **extrinsic eye muscles** are the six muscles attached to the orbit and to the external surface of each eyeball. *Extrinsic* means arising from without and distinguishes these from the *intrinsic* muscles inside the eyeball, to be considered later. The extrinsic muscles are responsible for movements of

39. *car* = fleshy mass + *uncle* = small
40. *punct* = point

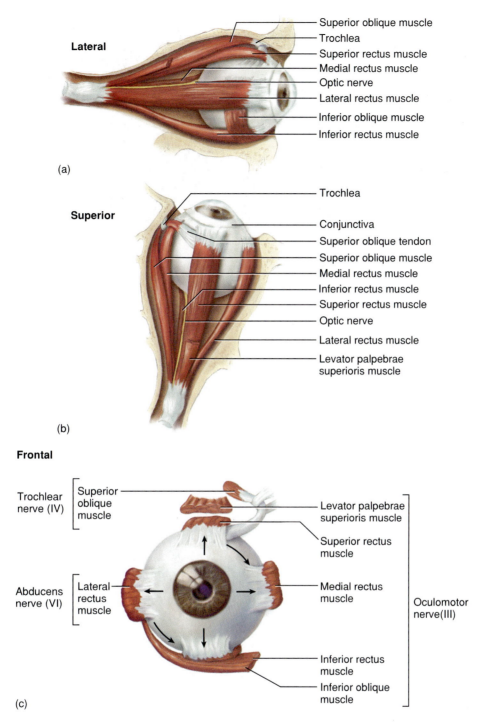

Lateral
- Superior oblique muscle
- Trochlea
- Superior rectus muscle
- Medial rectus muscle
- Optic nerve
- Lateral rectus muscle
- Inferior oblique muscle
- Inferior rectus muscle

(a)

Superior
- Trochlea
- Conjunctiva
- Superior oblique tendon
- Superior oblique muscle
- Medial rectus muscle
- Inferior rectus muscle
- Superior rectus muscle
- Optic nerve
- Lateral rectus muscle
- Levator palpebrae superioris muscle

(b)

Frontal

Trochlear nerve (IV) — Superior oblique muscle

Levator palpebrae superioris muscle

Superior rectus muscle

Abducens nerve (VI) — Lateral rectus muscle

Medial rectus muscle

Oculomotor nerve(III)

Inferior rectus muscle

Inferior oblique muscle

(c)

Figure 16.25 Extrinsic muscles of the eye. (*a*) Lateral view of the right eye. (*b*) Superior view of the right eye. (*c*) Innervation of the extrinsic muscles and movements produced by them.

the eye (fig. 16.25). They include four *rectus* ("straight") muscles and two *oblique* muscles. The **superior, inferior, medial,** and **lateral rectus** originate on the posterior wall of the orbit and insert on the anterior region of the eyeball, just beyond the visible "white of the eye." They move the eye up, down, medially, and laterally. The **superior oblique** travels along the medial wall of the orbit. Its tendon

passes through a fibrocartilage ring, the **trochlea**[41] (TROCK-lee-uh), and inserts on the superolateral aspect of the eyeball. It rotates the eye downward and medially. The **inferior oblique** extends from the medial wall of the orbit to the inferolateral aspect of

41. *trochlea* = pulley

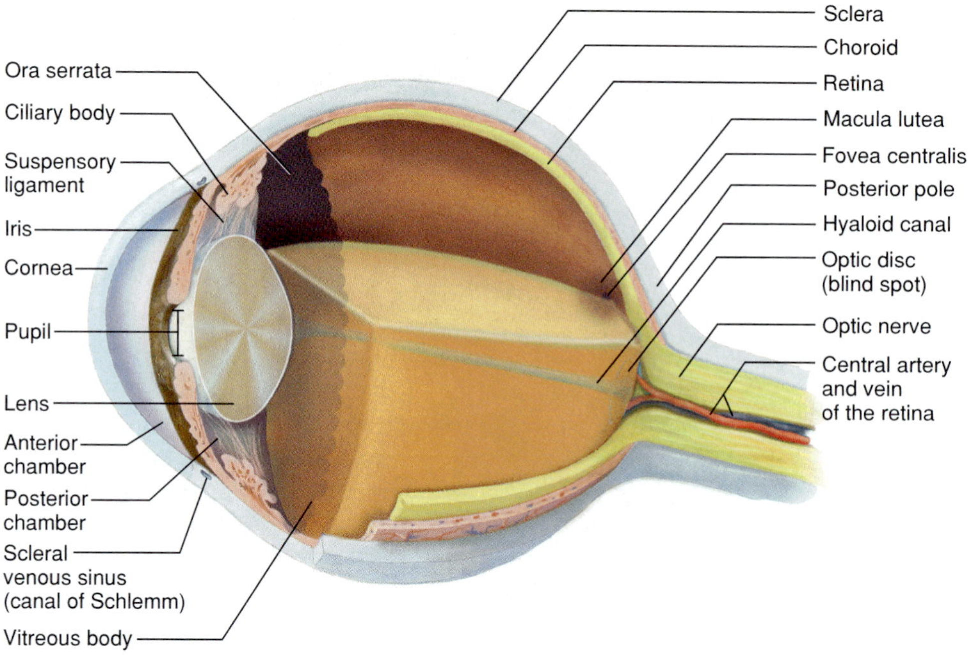
Ora serrata
Ciliary body
Suspensory ligament
Iris
Cornea
Pupil
Lens
Anterior chamber
Posterior chamber
Scleral venous sinus (canal of Schlemm)
Vitreous body
Sclera
Choroid
Retina
Macula lutea
Fovea centralis
Posterior pole
Hyaloid canal
Optic disc (blind spot)
Optic nerve
Central artery and vein of the retina

Figure 16.26 Anatomy of the eye in sagittal section. The vitreous body has been omitted from the upper half to reveal structures behind it.

the eye and rotates the eye upward and laterally. Most of these muscles are supplied by the oculomotor nerve (III), but the lateral rectus is innervated by the abducens nerve (VI) and the superior oblique by the trochlear nerve (IV).

The eye is surrounded on the sides and back by **orbital fat.** It cushions the eye, gives it freedom of motion, and protects blood vessels and nerves as they pass through the rear of the orbit.

Anatomy of the Eye

The eyeball itself is a sphere about 24 mm in diameter (fig. 16.26) with three principal components: (1) the *tunics,* three layers that form the wall of the eyeball; (2) the *optical apparatus,* which admits and focuses light; and (3) the *neural apparatus,* which consists of the retina and optic nerve.

The Tunics

The three tunics of the eyeball are as follows:

1. The **tunica fibrosa,** the outermost layer, is divided into two regions: the sclera and cornea. The **sclera**[42] (white of the eye) covers most of its surface and consists of dense collagenous connective tissue perforated by blood vessels and nerves. The **cornea** is the transparent region of modified sclera that admits light into the eye.
2. The **tunica vasculosa** is the second layer from the surface. It is also called the **uvea**[43] (YOU-vee-uh) because it resembles a peeled grape in fresh dissection. It consists of three regions—the choroid, ciliary body, and iris. The **choroid** (CO-royd) is a highly vascular, deeply pigmented layer of tissue behind the retina. It gets its name from a histological resemblance to the chorion of the pregnant uterus. The **ciliary body,** a thickened extension of the choroid, forms a muscular ring around the lens. It supports the iris and lens and secretes a fluid called the aqueous humor. The **iris** is an adjustable diaphragm that controls the diameter of the **pupil** (its central opening). At the rear of the iris is a *pigment epithelium* that prevents light from reaching the retina except through the pupil and lens. It also has an *anterior border layer* containing pigment cells called **chromatophores.**[44] If there is enough melanin in the chromatophores, the eyes are black, brown, or hazel. If there is a paucity of melanin, light reflected from the posterior pigment epithelium gives the eyes a blue, green, or gray color.
3. The **tunica interna** consists of the *retina,* which faces the interior of the eyeball. It is the focus of most of the remainder of this chapter.

42. *scler* = hard, tough

43. *uvea* = grape

44. *chromato* = color, pigment + *phor* = to bear, carry

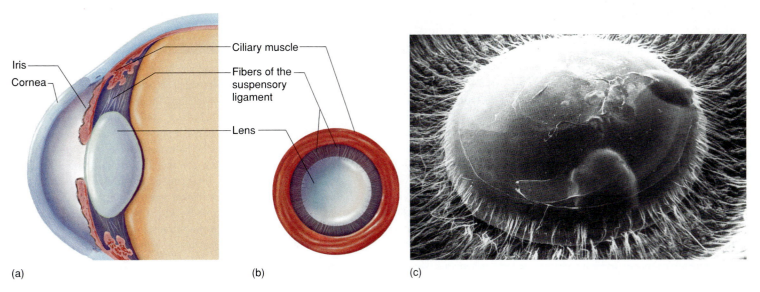

(a)   (b)   (c)

Figure 16.27 The lens of the eye, (a) Lateral view showing its relationship to the suspensory ligament and ciliary muscle. (b) Posterior view. (c) Posterior view of the lens and suspensory ligament as seen by SEM. ⚡

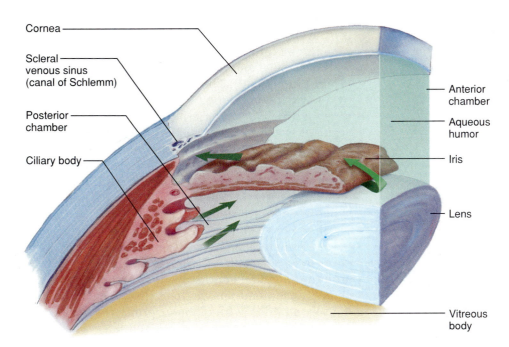

Figure 16.28 Production and reabsorption of aqueous humor.

The Optical Apparatus

The **optical apparatus** of the eye consists of a series of transparent media that admit light rays, bend (refract) them, and focus images on the retina. They include the *cornea,* the *lens,* the watery *aqueous humor,* and the gelatinous *vitreous body.* The cornea has been described already.

The **lens** is connected to the ciliary body by a ring of fibers called the **suspensory ligament** (fig. 16.27). Tension on the ligament somewhat flattens the lens, so it is about 9.0 mm in diameter and 3.6 mm thick at the middle. When the lens is removed from the eye and not under tension, it relaxes into a spherical shape and resembles a plastic bead.

The space between the cornea and iris is the **anterior chamber,** and the space between the iris and lens is the **posterior chamber** (fig. 16.28); they are continuous with each other via the pupil. Both are filled with **aqueous humor,** a serous fluid secreted by the ciliary body. Aqueous humor enters the posterior chamber and nourishes the lens. It then flows through the pupil into the anterior chamber, where it is reabsorbed by a ringlike vessel called the **scleral venous sinus** *(canal of Schlemm).*[45] Normally the rate of reabsorption balances the rate of secretion, but not in glaucoma (see special topic 16.3).

....
45. Friedrich S. Schlemm (1795–1858), German anatomist

The two most common causes of blindness are cataracts and glaucoma. A **cataract** is a clouding of the lens. It occurs as the lens thickens with age, and it is a common complication of diabetes mellitus. It causes the vision to appear milky or as if one were looking from behind a waterfall.[47] Cataracts may also stem from heavy smoking and exposure to the UV radiation of the sun. They are usually treated by replacing the natural lens with a plastic one. The implanted lens improves vision almost immediately, but glasses still may be needed for near vision.

Glaucoma[48] is a state of elevated pressure within the eye that occurs when the scleral venous sinus is obstructed and the reabsorption of aqueous humor cannot keep pace with its secretion. Pressure in the anterior and posterior chambers drives the lens back and puts pressure on the vitreous body. This presses the retina against the choroid and compresses the blood vessels that nourish the retina. Without a good blood supply, retinal cells die and the optic nerve may atrophy, producing blindness. Symptoms often go unnoticed until the damage is irreversible. In late stages, they include dimness of vision, reduced visual field, and colored halos around artificial lights. Glaucoma can be halted with drugs or surgery, but lost vision cannot be restored. This disease can be detected at an early stage in the course of regular eye examinations. The field of vision is checked, the optic nerve is examined, and the intraocular pressure is measured with an instrument called a *tonometer*.

The large space behind the lens is filled with a transparent jelly called the **vitreous**[46] **body** (*vitreous humor*). An oblique channel through this body, called the *hyaloid canal,* is the remnant of a *hyaloid artery* present in the embryo (see fig. 16.26).

The Neural Apparatus

The **neural apparatus** includes the retina and optic nerve. The retina forms from a cup-shaped outgrowth of the telencephalon (see chapter 14); it is actually a part of the brain—the only part that can be viewed without dissection. It is a thin transparent membrane attached at only two points—the **optic disc,** where the optic nerve leaves the eye, and its scalloped anterior margin, the **ora serrata.** Otherwise, the retina is held smoothly against the rear of the eyeball by the pressure of the vitreous body. Insufficient pressure can cause it to buckle, however, and separate from the wall of the eye. Such a *detached retina* may cause blurry areas in the field of vision. It can lead to blindness if allowed to remain isolated from the chorion on which it depends for oxygen, nutrition, and waste removal.

The most posterior layer of the retina is the **pigment epithelium,** a layer of darkly pigmented cuboidal cells whose basal processes interdigitate with receptor cells of the retina. This epithelium acts like the blackened inside of a camera to prevent the visual image from being degraded by light reflection within the eye. Indeed, the vertebrate eye is often called a *camera eye* for its many resemblances to the mechanisms of a camera. (How many can you think of?)

The neural apparatus of the retina consists of three principal cell layers (fig. 16.29). Progressing from the rear of the eye forward, these are composed of *photoreceptor (rod* and *cone), bipolar,* and *ganglion cells.*

1. **Photoreceptor cells.** These elaborately modified bipolar neurons are called **rods** and **cones.** Each type has an **outer segment** that points toward the wall of the eye and an **inner segment** facing the interior (fig. 16.30). The two segments are separated by a narrow constriction containing a cylindrical array of nine pairs of microtubules; the outer segment is actually a highly modified cilium specialized to absorb light. The inner segment contains the nucleus, mitochondria, and other organelles. At its base, it gives rise to a cell body, which contains the nucleus, and to processes that synapse with other retinal neurons.

   In a rod, the outer segment is more or less cylindrical and somewhat resembles a stack of coins in a paper roll—there is a cell membrane around the outside and a neatly arrayed stack of about 1,000 membranous discs inside. Each disc is densely studded with globular proteins—the visual pigment *rhodopsin,* to be discussed later. The membranes hold these pigment molecules in the most efficient orientation for light absorption. Rod cells are responsible for vision in dim light (**night, or scotopic,**[49] **vision**); they cannot distinguish colors from each other.

   A cone cell is similar except that the outer segment tapers to a point and the discs are not detached from the cell membrane but are parallel infoldings of it. Cones function in bright light; they are responsible for **day (photopic) vision,** as well as for color vision.

---

46. *vitre* = glass

47. *cataract* = waterfall
48. *glauc* = grayness
49. *scoto* = dark + *op* = vision

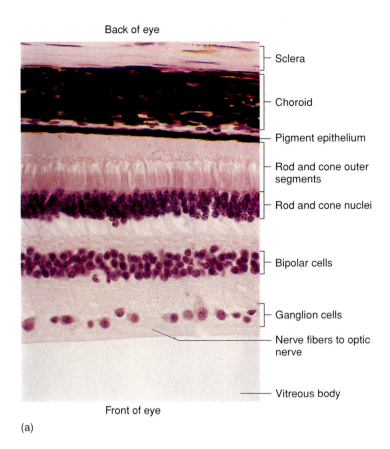

Back of eye

- Sclera
- Choroid
- Pigment epithelium
- Rod and cone outer segments
- Rod and cone nuclei
- Bipolar cells
- Ganglion cells
- Nerve fibers to optic nerve
- Vitreous body

Front of eye

(a)

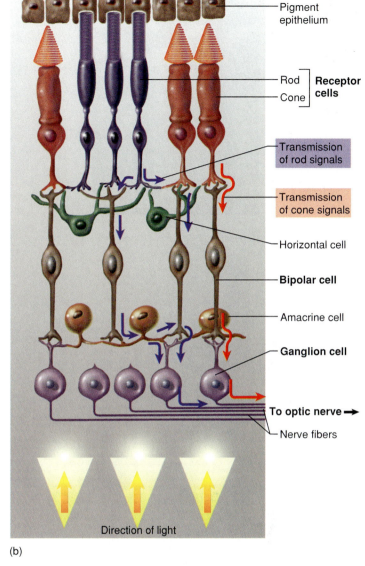

Back of eye

- Pigment epithelium
- Rod } Receptor
- Cone } cells
- Transmission of rod signals
- Transmission of cone signals
- Horizontal cell
- **Bipolar cell**
- Amacrine cell
- **Ganglion cell**
- **To optic nerve** →
- Nerve fibers

Direction of light

(b)

Figure 16.29 Histology of the retina. (a) Photomicrograph (×400). (b) Schematic of the layers and synaptic relationships of the retinal neurons. ✗

2. **Bipolar cells.** Rods and cones synapse with the dendrites of bipolar neurons, which in turn synapse with the ganglion cells described next (see fig. 16.29b). There are approximately 130 million rods and 6.5 million cones in one retina but only 1.2 million nerve fibers in the optic nerve. With a ratio of 114 receptor cells to 1 optic nerve fiber, it is obvious that there must be substantial *neuronal convergence* and information processing in the retina itself before signals are transmitted to the brain proper. It begins with the bipolar cells.

3. **Ganglion cells.** These are the largest neurons of the retina, arranged in a single layer close to the vitreous body. Most ganglion cells receive input from multiple bipolar cells. The axons of the ganglion cells form the fibers of the optic nerve.

There are other retinal cells, but they do not form layers of their own. **Horizontal cells** and **amacrine**[50] **cells** form horizontal connections among rod, cone, and bipolar cells. They play diverse roles in enhancing the perception of contrast, the edges of objects, and changes in light intensity. In addition, much of the mass of the retina is composed of astrocytes and other types of glial cells.

Correlation of Retinal Histology with Gross Anatomy

The interior rear of the eye (*fundus*) can be examined with an illuminating and magnifying instrument called

50. *a* = without + *macr* = long + *in* = fiber

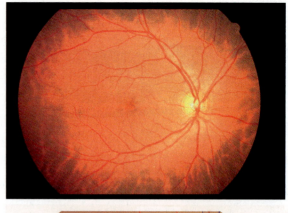

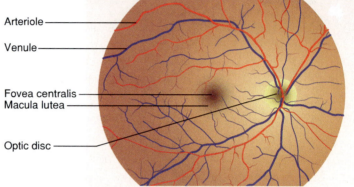

Arteriole

Venule

Fovea centralis
Macula lutea

Optic disc

Figure 16.31 View of the back of the right eye as seen with an ophthalmoscope. Note the blood vessels diverging from the optic disc, where they enter the eye with the optic nerve.

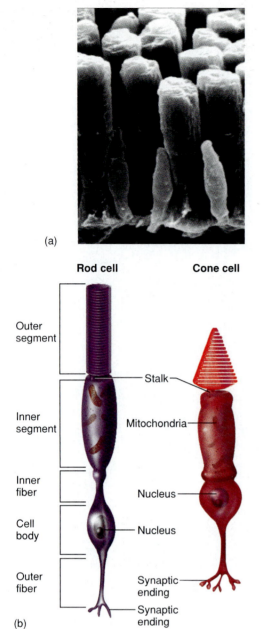

(a)

Rod cell   **Cone cell**

Outer
segment

Stalk

Inner
segment

Mitochondria

Inner
fiber

Nucleus

Cell
body

Nucleus

Outer
fiber

Synaptic
ending

Synaptic
ending

(b)

Figure 16.30 (a) Photoreceptor cells in a salamander retina (SEM). The tall cylindrical cells are rods and the short-tapered cells in the foreground are cones. (b) Structure of rod and cone cells.

an *ophthalmoscope.* Directly posterior to the center of the lens, on the visual axis of the eye, is a patch of cells called the **macula lutea**[51] about 3 mm in diameter (fig. 16.31). In the center of the macula is a tiny pit, the **fovea**[52] **centralis,** that produces the most finely detailed images in photopic vision. The reason for this will be apparent later. About 3 mm medial to the macula lutea is the *optic disc,* where the optic nerve originates and

blood vessels enter the eye (see special topic 16.4). Nerve fibers from all the ganglion cells of the retina converge on this point. Because there are no receptor cells in the optic disc, it forms a **blind spot** in the visual field of each eye. To visualize this, close your right eye and gaze straight ahead with your left. Hold your finger about 30 cm (1 ft) from your face, with your fingertip at eye level. Begin moving your finger toward the left, but be sure you keep your gaze directed straight ahead. When your fingertip is about 15° away from your line of vision, it will become invisible because its image falls on the blind spot of your left eye. The reason you do not normally notice a blind patch in your visual field is that the brain uses the image surrounding the blind spot to fill in the spot with similar, essentially "imaginary" information (fig. 16.32).

Before we begin to study visual physiology, it would be advisable to review and be sure you thoroughly understand the anatomy of the eye (table 16.4).

Formation of an Image

The visual process begins when light rays enter the eye, become focused on the retina, and produce a tiny inverted image. Each photoreceptor cell absorbs light

51. *macula* = spot + *lutea* = yellow
52. *fovea* = pit, depression

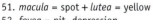

Blood vessels travel through the optic nerve and enter the eye at the optic disc. This is the only place in the body that affords opportunity for the noninvasive visual examination of blood vessels. Eye examinations are therefore valuable not only for evaluating the visual system but also for diagnosing hypertension, diabetes mellitus, atherosclerosis, and other vascular diseases.

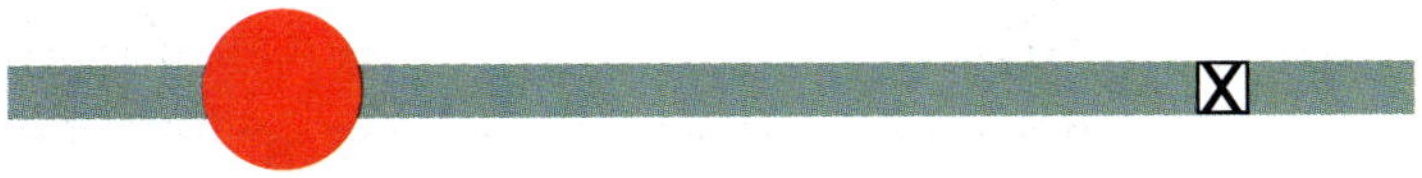

Figure 16.32 Demonstration of the blind spot and visual filling. Close your right eye, hold this page about 30 cm (1 ft) from your face, and fixate on the X. Without moving your eyes, move the page slightly back and forth, or sideways, until the red dot disappears. This occurs because its image is falling on your blind spot. Note that the green bar fills in the space where the red dot used to be. This occurs because the brain is "making it up"—filling in the visual field with what it sees adjacent to the blind spot.

from one small area of the image and initiates a nerve signal that codes for the amount of light being absorbed.

Admittance of Light to the Eye

When fully dilated, the pupil admits five times as much light into the eye as it does when fully constricted. Its diameter is controlled by two sets of contractile elements in the iris. The **pupillary constrictor** consists of concentric circles of smooth muscle cells around the pupil. When stimulated by the parasympathetic nervous system, it narrows the pupil and admits less light to the eye. The **pupillary dilator** consists of a spokelike arrangement of modified contractile epithelial cells called *myoepithelial cells.* When stimulated by the sympathetic nervous system, these cells contract, the pupil widens, and more light enters the eye. Pupillary constriction and dilation are sometimes called *miosis* and *mydriasis,* respectively. Dilation and constriction occur in response to changes in light intensity and in conjunction with focusing on distant or nearby objects. Pupillary constriction in response to light is called the **photopupillary reflex.** Both pupils constrict even if only one eye is illuminated; this is called the **consensual light reflex.**

When light intensity increases, impulses are transmitted from the retina to the pretectal region near the junction of the midbrain and thalamus. Parasympathetic preganglionic fibers return to the orbit and synapse near the eye with postganglionic fibers that innervate the pupillary constrictor. Sympathetic innervation to the pupil originates, with all other sympathetic efferent

Table 16.4	Anatomical Checklist for the Eye
Accessory Organs (figs. 16.22 and 16.23)	*Tunica Interna*
	Pigment epithelium
Eyebrows	Retina
Eyelids (Palpebrae)	Optic disc
Orbicularis oculi	Ora serrata
Tarsal plate	**Optical Apparatus (figs. 16.27 and 16.28)**
Tarsal glands	
Medial and lateral commissures	Cornea
Eyelashes	Pupil
Palpebral fissure	Anterior chamber
Conjunctiva	Posterior chamber
Lacrimal apparatus (fig. 16.24)	Aqueous humor
	Scleral venous sinus
Lacrimal gland	Lens
Lacrimal caruncle	Suspensory ligament
Lacrimal puncta	Vitreous body
Lacrimal canal	Hyaloid canal
Lacrimal sac	**Neural Apparatus (figs. 16.29–16.31)**
Nasolacrimal duct	
Extrinsic Muscles (fig. 16.25)	Optic nerve
Superior oblique	Optic disc
Trochlea	Macula lutea
Inferior oblique	Fovea centralis
Superior rectus	Retinal cells
Inferior rectus	Rod cells
Lateral rectus	Cone cells
Medial rectus	Bipolar cells
Orbital Fat	Ganglion cells
Tunics of the Eye (fig. 16.26)	Horizontal cells
	Amacrine cells
Tunica Fibrosa	
Sclera	
Cornea	
Tunica Vasculosa (Uvea)	
Choroid	
Ciliary body	
Iris	

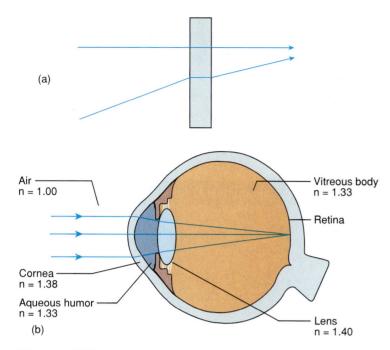

(a)

Air
n = 1.00

Vitreous body
n = 1.33

Retina

Cornea
n = 1.38

Aqueous humor
n = 1.33

(b)

Lens
n = 1.40

Figure 16.33 Principles of refraction. (*a*) A refractive medium does not bend light rays that strike it at a 90° angle but does bend light rays that enter or leave it at any other angle. (*b*) Refractive indices of the optic media from air to retina. The greater the difference between the refractive indices of two media, the more strongly light rays are refracted when passing from one to the other. In vision, most refraction occurs as light passes from air to cornea. ✗

fibers, in the spinal cord. Preganglionic fibers lead from the thoracic cord to the superior cervical ganglion. From there, postganglionic fibers follow the carotid arteries into the head and lead ultimately to the pupillary dilator.

Refraction

Image formation depends on **refraction,** the bending of light rays. Light travels at a speed of 300,000 km/sec (186,000 mi/sec) in a vacuum, but it slows down slightly in air, water, glass, and other media. The *refractive index* of a medium (symbolized *n*) is a measure of how much it retards light rays relative to air, which has a refractive index arbitrarily set at $n = 1.00$. If light traveling through air strikes a medium of higher refractive index at a 90° *angle of incidence,* it slows down but does not change course—the light rays are not bent. If it strikes at any other angle, however, the light is refracted (fig. 16.33*a*). The greater the difference in refractive index between the two media, and the greater the angle of incidence, the stronger the refraction is.

As light enters the eye, it passes from a medium with $n = 1.00$ (air) to one with $n = 1.38$ (the cornea) (fig. 16.33*b*). Light rays striking the very center of the cornea pass straight through, but because of the curvature of the cornea, light rays striking off-center are bent toward

the center. The aqueous humor has a refractive index of 1.33 and does not greatly alter the path of the light. The lens has a refractive index of 1.40. As light passes from air to cornea, the refractive index changes by 0.38; but as it passes from aqueous humor to lens, the refractive index changes by only 0.02. Thus, the cornea plays a greater role in focusing light than the lens does. The lens merely fine-tunes the image, especially as you shift your focus between near and distant objects.

The Near Response

When you focus on an object more than 6 m (20 ft) away, your eye is in a relaxed state called **emmetropia**[53] (EM-eh-TRO-pee-uh). At this distance, light rays entering the eye are essentially parallel; the relaxed eye focuses them on the retina without effort. If you shift your focus to something closer, light rays from the source are too divergent to be focused without additional effort. In other words, the eye is automatically focused on things in the distance unless you make an effort to focus elsewhere. For a wild animal, or our ancestors, this would be adaptive because it allows for a quick response to predators at a distance. The adjustments to close-range vision, collectively called the **near response** (fig. 16.34), involve three events that bring the image into focus on the retina:

1. **Convergence of the eyes.** Move your finger gradually closer to a baby's nose and the baby will go cross-eyed. This **convergence** of the eyes orients the visual axis of each eye toward the object in order to focus its image on the fovea centralis of each. If the eyes cannot converge accurately—for example, when the extrinsic muscles are weaker in one eye than in the other—double vision or *diplopia*[54] results. The images fall on different parts of the two retinas and the brain sees two images. If you press very gently near the lateral commissure of one eye as you look at this page, the image of the print falls on noncorresponding regions of the two eyes and causes you to see double.

Think About It

Which extrinsic muscles of the eyes are the prime movers in convergence?

2. **Constriction of the pupil.** Lenses cannot refract light rays at their edges as well as they can refract rays closer to the center. The image produced by any lens is therefore somewhat blurry around the edges; this *spherical aberration* is quite evident in

53. *em* = in + *metr* = measure + *opia* = vision
54. *diplo* = double

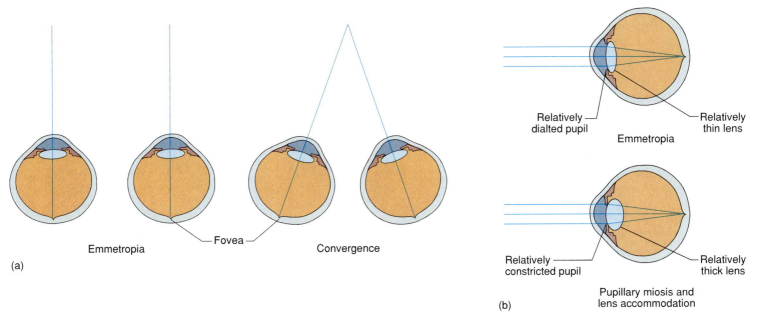

(a)

Emmetropia — Fovea — Convergence

Relatively dialted pupil — Relatively thin lens

Emmetropia

Relatively constricted pupil — Relatively thick lens

Pupillary miosis and lens accommodation

(b)

Figure 16.34 Emmetropia and the near response. (*a*) Superior view of both eyes fixated on an object more than 6 m away (*left*), and both eyes fixated on an object closer than 6 m (*right*). (*b*) Lateral view of the eye fixated on a distant object (*top*) and nearby object (*bottom*).

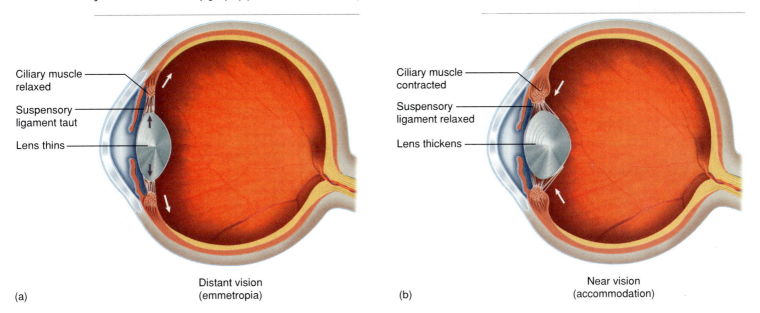

Ciliary muscle relaxed

Suspensory ligament taut

Lens thins

Distant vision (emmetropia)

(a)

Ciliary muscle contracted

Suspensory ligament relaxed

Lens thickens

Near vision (accommodation)

(b)

Figure 16.35 Accommodation of the lens. (*a*) In the relaxed (emmetropic) eye, the ciliary muscle is relaxed and dilated. It puts tension on the suspensory ligament and flattens the lens. (*b*) In accommodation, the ciliary muscle contracts and narrows in diameter. This reduces tension on the suspensory ligament and allows the lens to relax into a more convex shape.

an inexpensive microscope. It can be minimized by screening out these peripheral light rays, and for this reason, the pupil constricts as you focus on nearby objects. Like the diaphragm of a camera, the pupil thus has a dual purpose—to adjust the eye to variations in brightness and to reduce spherical aberration.

3. **Accommodation of the lens. Accommodation** is a change in the curvature of the lens that enables you to focus on a nearby object. When you look at

something close, the ciliary muscle surrounding the lens contracts. This narrows the diameter of the ciliary body, relaxes the fibers of the suspensory ligament, and allows the lens to relax into a more convex shape (fig. 16.35). In emmetropia, the lens is about 3.6 mm thick at the center; in accommodation, it thickens to about 4.5 mm. A more convex lens refracts light more strongly, bringing the divergent light rays into focus on the retina. The closest an object can be and still come

Table 16.5   Common Disorders in Visual Image Formation

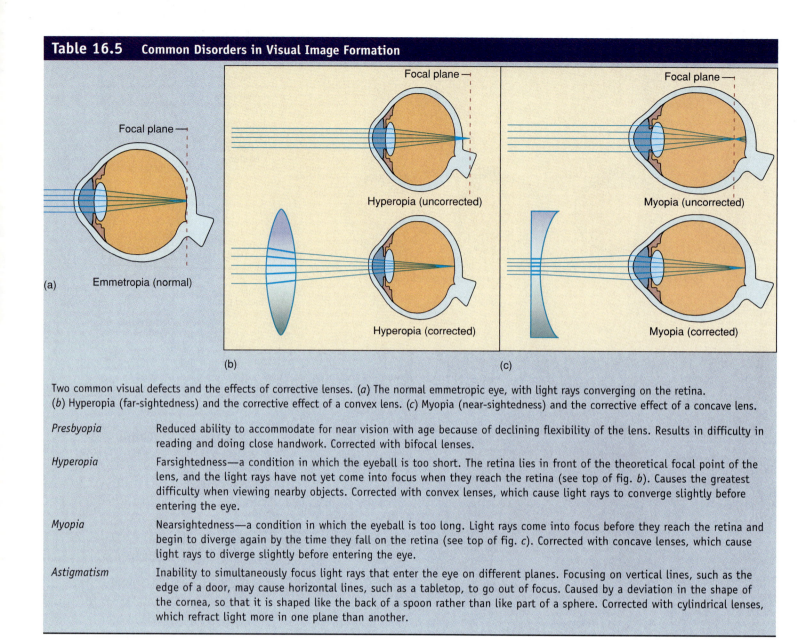

Two common visual defects and the effects of corrective lenses. (*a*) The normal emmetropic eye, with light rays converging on the retina. (*b*) Hyperopia (far-sightedness) and the corrective effect of a convex lens. (*c*) Myopia (near-sightedness) and the corrective effect of a concave lens.

Presbyopia	Reduced ability to accommodate for near vision with age because of declining flexibility of the lens. Results in difficulty in reading and doing close handwork. Corrected with bifocal lenses.
Hyperopia	Farsightedness—a condition in which the eyeball is too short. The retina lies in front of the theoretical focal point of the lens, and the light rays have not yet come into focus when they reach the retina (see top of fig. *b*). Causes the greatest difficulty when viewing nearby objects. Corrected with convex lenses, which cause light rays to converge slightly before entering the eye.
Myopia	Nearsightedness—a condition in which the eyeball is too long. Light rays come into focus before they reach the retina and begin to diverge again by the time they fall on the retina (see top of fig. *c*). Corrected with concave lenses, which cause light rays to diverge slightly before entering the eye.
Astigmatism	Inability to simultaneously focus light rays that enter the eye on different planes. Focusing on vertical lines, such as the edge of a door, may cause horizontal lines, such as a tabletop, to go out of focus. Caused by a deviation in the shape of the cornea, so that it is shaped like the back of a spoon rather than like part of a sphere. Corrected with cylindrical lenses, which refract light more in one plane than another.

into focus is called the **near point of vision** and depends on the flexibility of the lens. The lens stiffens with age—thus, the near point averages about 9 cm at the age of 10 and 83 cm by the age of 60.

Some common defects in image formation are listed in table 16.5.

Sensory Transduction in the Retina

We now consider the events that convert light energy into action potentials—what happens when a rod or cone cell absorbs a photon of light. We begin with the visual pigments responsible for light absorption.

Visual Pigments

The visual pigment of rod cells is called **rhodopsin** (ro-DOP-sin), or *visual purple.* Each molecule consists of two major parts (moieties)—a vitamin A derivative called **retinal (retinene)** and a protein called **opsin** (fig. 16.36). Opsin is embedded in the disc membranes of the rod's outer segment. All rod cells contain a single kind of rhodopsin with an absorption peak at a wavelength of 500 nm. The rods are less sensitive to light of other wavelengths.

In cones, the pigment is called **photopsin,** or **iodopsin.** Its retinal moiety is the same as that of rhodopsin, but the opsin moieties have different amino acid sequences that modify the wavelengths of light absorbed. There are three kinds of cones, structurally

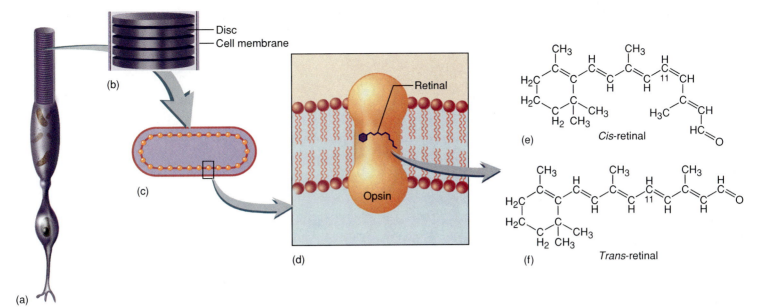

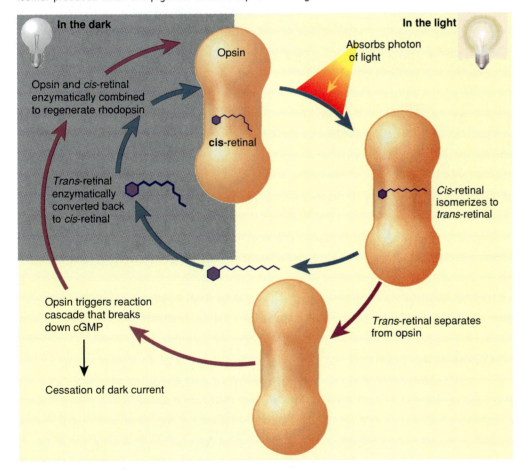

Figure 16.36 Structure and location of the visual pigments. (*a*) A rod cell. (*b*) Detail of the rod outer segment. (*c*) One disc of the outer segment, showing the membrane studded with pigment molecules. (*d*) A pigment molecule, embedded in the unit membrane of the disc, showing the protein moiety, opsin, and the vitamin A derivative, retinal. (*e*) *Cis*-retinal, the isomer present in the absence of light. (*f*) *Trans*-retinal, the isomer produced when the pigment absorbs a photon of light.

Figure 16.37 The bleaching and regeneration of rhodopsin. The yellow background indicates the bleaching events that occur in the light; the gray background indicates the regenerative events that are independent of light. The latter occur in light and dark but are able to outpace bleaching only in the dark.

identical in appearance but different with respect to the wavelengths of light they absorb most effectively. These differences, as you will see shortly, enable us to perceive different colors.

The Photochemical Reaction

The events of sensory transduction are probably the same in rods and cones. We describe them here from the standpoint of rhodopsin, the visual pigment of the rods. In the dark, retinal has a bent shape called the 11-*cis* isomer, or **cis-retinal**. When it absorbs a photon of light, it is converted to a straight form called the all-trans isomer, or **trans-retinal**. This isomer dissociates from the opsin (fig. 16.37), leading to the production of a nerve signal by a mechanism explained next. Rhodopsin can be chemically isolated from animal eyes in a darkroom. It has a violet color when first examined in the light, but it fades very quickly to a colorless solution as the opsin

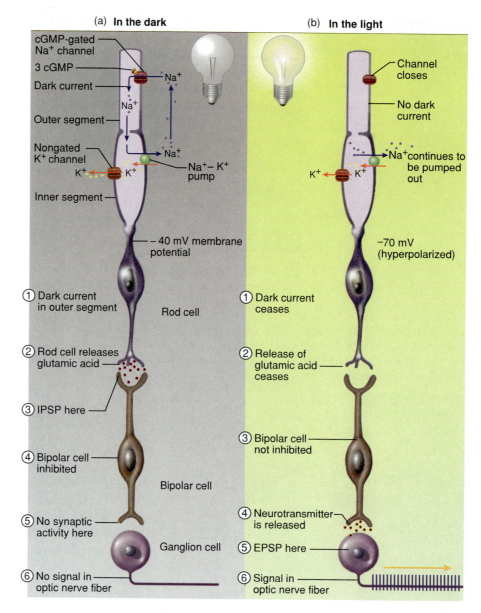

(a) In the dark

- cGMP-gated Na⁺ channel
- 3 cGMP
- Dark current
- Outer segment
- Nongated K⁺ channel
- Inner segment
- Na⁺
- Na⁺
- Na⁺
- Na⁺ – K⁺ pump
- K⁺ K⁺
- −40 mV membrane potential
- Rod cell

① Dark current in outer segment

② Rod cell releases glutamic acid

③ IPSP here

④ Bipolar cell inhibited

Bipolar cell

⑤ No synaptic activity here

Ganglion cell

⑥ No signal in optic nerve fiber

(b) In the light

- Channel closes
- No dark current
- Na⁺ continues to be pumped out
- K⁺ K⁺
- −70 mV (hyperpolarized)

① Dark current ceases

② Release of glutamic acid ceases

③ Bipolar cell not inhibited

④ Neurotransmitter is released

⑤ EPSP here

⑥ Signal in optic nerve fiber

Figure 16.38 Mechanism of generating impulses in the optic nerve in response to light. (*a*) In the dark, a dark current in the rod cells inhibits the production of nerve impulses. (*b*) In the light, cGMP breaks down, shuts off the dark current, and allows the generation of a signal in the optic nerve.

and retinal dissociate. This process, called **bleaching,** occurs in the eye as well as in the test tube.

Generating the Optic Nerve Signal

In the absence of light, rod cells do not sit quietly doing nothing. They are very active, producing a steady ion flow called the **dark current.** The production of a nerve signal stems from the fact that light *shuts off* this dark current. Here we will see how the current is produced and how light deactivates it.

The outer segment of a rod cell has ligand-gated Na⁺ channels. When cyclic guanosine monophosphate (cGMP) binds to the intracellular side of the channel proteins, it opens the gates. Thus, sodium continually

diffuses into the cell and the cell is partially depolarized. The expression "dark current" refers specifically to this sodium influx. The inner segment of the rod cell has nongated K⁺ channels, which allow K⁺ to diffuse out continually. It also has a high density of Na⁺ –K⁺ pumps, which compensate for both the Na⁺ and K⁺ flows by continually pumping these ions, respectively, back out of and into the cell (fig. 16.38). As long as the dark current continues and depolarizes the cell, the cell releases glutamic acid from its synaptic end. Its effect on the bipolar cell is considered shortly.

When a rod absorbs light, the dark current drops or ceases. The intact rhodopsin molecule is essentially a dormant enzyme. When it bleaches, it becomes enzymatically active and triggers a cascade of chemical reactions that ultimately break down several hundred thousand molecules of cGMP. This amplifying effect—the ability of one rhodopsin molecule to affect so many cGMP molecules—makes rod cells extremely sensitive to light. As cGMP is degraded, the gated Na^+ channels in the outer segment close, the dark current declines, and glutamic acid secretion is reduced or halted.

There are two kinds of bipolar cells. One type is inhibited (hyperpolarized) by glutamic acid and excited (depolarized) when its secretion drops. The other type is excited by glutamic acid and inhibited when its secretion drops. Thus, the former bipolar cells are excited by rising light intensity and the latter are excited by falling light intensity. As your eye scans a scene, you see objects of greater and lesser brightness. Their images on the retina cause a rapidly changing pattern of bipolar cell responses as the light intensity on a given patch of retina rises and falls.

When bipolar cells detect fluctuations in light intensity, they communicate this to the ganglion cells. Ganglion cells are the only retinal cells that produce all-or-none propagated action potentials; all other retinal neurons produce only graded local potentials. The ganglion cells respond with rising and falling firing frequencies which, via the optic nerve, provide the brain with a basis for interpreting the image on the retina.

Each ganglion cell receives input from a circular patch of retina called its **receptive field.** The principal function of the ganglion cells is to code for contrast between the center and the edge of its receptive field—that is, between an object and its surroundings.

Regeneration of Rhodopsin

For a rod to continue to function, it must regenerate rhodopsin at a rate that keeps pace with bleaching. When *trans*-retinal dissociates from rhodopsin, it is transported to the pigment epithelium at the rear of the retina, converted again to *cis*-retinal, transported back to the rod outer segment, and reunited with a bleached opsin moiety (see fig. 16.37). It takes about 5 minutes to regenerate half of the bleached rhodopsin. Cone cells are less dependent on the pigment epithelium and regenerate half of their pigment in about 90 seconds.

Light and Dark Adaptation

Light adaptation is a process that occurs when we go from the dark into bright light. If you wake up in the middle of the night and turn on a lamp, at first you see a harsh glare; you may even experience pain from the overstimulated retinas. The pupils quickly constrict to reduce the intensity of stimulation, but color vision and visual acuity (the ability to see fine detail) do not become optimal for 5 to 10 minutes—the time needed for pigment bleaching to adjust retinal sensitivity to this light intensity. The rods bleach quickly in bright light, and cones take over.

On the other hand, suppose you are sitting in a bright room watching television at night, and suddenly there is a power failure. Your eyes will need to undergo **dark adaptation** before you can see well enough to find your way around in the dark. Your rod pigment was bleached by the lights in the room while the power was on, but now in the relative absence of light, the rate of rhodopsin regeneration begins to exceed the rate of bleaching. In 20 to 30 minutes, the amount of rhodopsin is sufficient for your eyes to have reached essentially the maximum possible sensitivity in the dark. Dilation of the pupils also helps by admitting more light to the eye.

The Duplicity Theory

You may wonder why we need two types of photoreceptor cells, the rods and cones. Why can't we simply have one type that would produce detailed color vision, both day and night? The **duplicity theory** of vision is that a single type of receptor cell cannot produce both high sensitivity and high resolution. It takes one type of cell and neuronal circuit, working at its maximum capacity, to provide sensitive night (scotopic) vision. A different type of receptor and circuit are required to provide high-resolution daytime (photopic) vision.

The high sensitivity of rods in dim light stems partly from the cascade of reactions leading to cGMP breakdown described earlier, but it is also due to the extensive neuronal convergence that occurs between the rods and ganglion cells. Up to 600 rods converge on each bipolar cell, and several bipolar cells converge on each ganglion cell. This allows for a high degree of *spatial summation* in the scotopic system. Weak stimulation of many rod cells can produce an additive effect on one bipolar cell, and several bipolar cells may be enough to excite one ganglion cell. Thus, a ganglion cell can respond in dim light that only weakly stimulates individual rods. A shortcoming of this system is that it cannot resolve finely detailed images. One ganglion cell receives input from all the rods in about 1 mm² of retina—its receptive field. What the brain perceives is therefore a coarse, grainy image similar to an overenlarged newspaper photograph.

Around the edges of the retina, receptor cells are especially large and widely spaced. If you fixate on the middle of this page, you will notice that you cannot read the words near the margins. Visual acuity decreases rapidly as the image falls away from the fovea centralis. Our peripheral vision is a low-resolution system that serves mainly to alert us to motion in the periphery and to stimulate us to look that way to identify what is there.

When you look directly at something, its image falls on the fovea, which is occupied by about 4,000 tiny cone cells and no rods. The other neurons of the fovea are displaced to one side, so they do not interfere with light falling on the cones. The smallness of these cones is like the smallness of the dots in a high-quality photograph; it is partially responsible for the high-resolution images formed at the fovea. In addition, the cones here show no neuronal convergence. Each cone synapses with only one bipolar cell and each bipolar cell with only one ganglion cell. This gives each foveal cone a "private line" to the brain, and each ganglion cell of the fovea reports to the brain on a receptive field of just 2 μm^2 of retinal area. Cones distant from the fovea exhibit some neuronal convergence but not nearly as much as rods do. The price of this lack of convergence at the fovea, however, is that cone cells have little spatial summation, and the cone system therefore has less sensitivity to light. Moonlight, for example, is not bright enough to stimulate the photopic system, and the cones provide us with no night vision.

 Think About It

If you look directly at a dim star in the night sky, it disappears, and if you look slightly away from it, it reappears. Why?

Color Vision

Most nocturnal vertebrates have only rod cells, but many diurnal animals are endowed with cones and color vision. Color vision is especially well developed in primates for evolutionary reasons discussed in chapter 1. It is based on the existence of three populations of cones named for the absorption peaks of their photopsins: **blue cones,** with peak sensitivity at 420 nm; **green cones,** which peak at 531 nm; and **red cones,** which peak at 558 nm (fig. 16.39). Red cones do not peak in the red part of the spectrum (558 nm light is perceived as orange-yellow), but they are the only cones that respond at all to red light. Our perception of different colors is based on a mixture of nerve signals representing cones with different absorption peaks. In figure 16.39, note that light at 400 nm excites only the blue cones. At 500 nm, however, all three types of cones are stimulated. The red cones respond at 60% of their maximum capability, green cones respond at 82% of their maximum, and blue cones respond at 20% of their maximum. The brain interprets this mixture of signals as blue-green. The table in figure 16.39 shows how other color sensations are generated by other response ratios.

Some individuals have a hereditary lack of one photopsin or another and consequently exhibit **color blindness.** The most common form is *red-green color blindness,* resulting from a lack of either red or green cones, which renders a person incapable of distinguish-

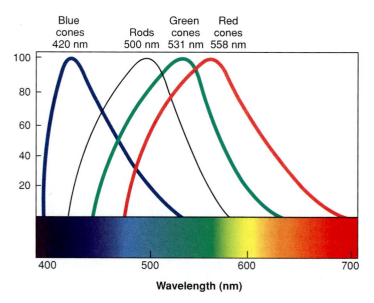

Wavelength (nm)	Percent of maximum cone response (red:green:blue)	Perceived hue
400	0:0:50	Violet
450	0:30:72	Blue
500	60:82:20	Blue-green
550	97:85:0	Yellow
625	35:3:0	Orange
675	5:0:0	Red

Figure 16.39 Absorption spectra of the three iodopsins of the cone cells. In the middle column of the table, each number indicates how strongly the respective cone cells are responding as a percentage of their maximum response. At 450 nm, for example, red cones do not respond at all, green cones respond at 30% of their maximum, and blue cones respond at 72% of their maximum.

ing these and related shades from each other. For example, a person with normal *trichromatic* color vision sees figure 16.40 as the number 74, whereas a person with red-green color blindness sees the number 21.

Stereoscopic Vision

Stereoscopic vision (stereopsis) is depth perception—the ability to judge relative distances to different points in the visual field. It depends on having two eyes with overlapping visual fields, which allows each eye to look at the same object from a different angle. Stereoscopic vision contrasts with the *panoramic vision* of mammals such as rodents and horses, where the eyes are on opposite sides of the head and virtually 100% of the optic nerve fibers decussate to the opposite cerebral hemisphere. Although stereoscopic vision covers a smaller visual field than panoramic vision and provides less alertness to sneaky predators, it has the advantage of depth perception. The evolutionary basis of depth perception in primates was considered in chapter 1.

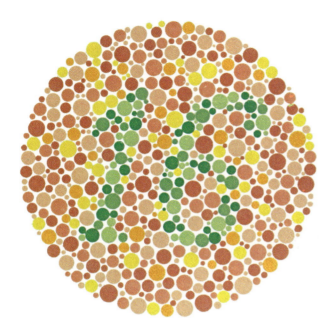

Figure 16.40 A test for red-green color blindness. Persons with normal vision see the number 16. Persons with red-green color blindness see no discernible number.

When you fixate on something within a distance of 30 m (100 ft), each eye views it from a slightly different angle and focuses its image on the fovea centralis. The point on which the eyes are focused is called the *fixation point.* Objects farther away than the fixation point cast an image somewhat medial to the foveas, and closer objects cast their images more laterally (fig. 16.41). The distance of an image from the two foveas provides the brain with information used to judge the position of other points relative to the fixation point.

The Visual Projection Pathway

The optic nerves converge on each other and form an X, the **optic chiasma**[55] (ky-AZ-muh), immediately inferior to the hypothalamus and anterior to the pituitary. Beyond this, the fibers continue as a pair of **optic tracts** (see p. 510). Within the chiasma, half the fibers of each optic nerve cross over to the opposite side of the brain (fig. 16.42). This is called **hemidecussation,** since only half of the fibers decussate. As a result, objects in the left visual field, whose images fall on the right half of each retina (the medial half of the left eye and lateral half of the right eye), are perceived by the right cerebral hemisphere. Objects in the right visual field are perceived by the left hemisphere. Since the right brain controls motor responses on the left side of the body and vice versa, each side of the brain needs to see what is on the side of the body where it exerts motor control.

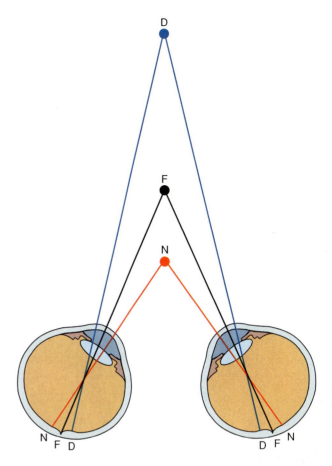

Figure 16.41 The basis of stereoscopic vision (depth perception). When the eyes are fixated on the fixation point (*F*), more distant objects (*D*) are focused on the retinas medial to the fovea and the brain interprets them as being farther away than the fixation point. Nearby objects (*N*) are focused lateral to the fovea and interpreted as being closer.

The optic tracts pass laterally around the hypothalamus, and most of their axons end at synapses in the **lateral geniculate**[56] (jeh-NIC-you-late) **body** of the thalamus. Second-order neurons arise here and form the **optic radiation** of fibers in the white matter of the cerebrum. These project to the primary visual cortex of the occipital lobe, where the conscious perception of an image occurs. A stroke that destroys occipital lobe tissue can cause blindness even if the eyes are fully functional. Association tracts connect the primary visual cortex to the visual association area just anterior to it. The association area stores visual memories and enables the brain to interpret what we are seeing—for example, to recognize printed words, name the objects we see, or identify another person.

A few optic nerve fibers take a different route in which they project to the midbrain and terminate in the superior colliculi and pretectal nuclei. The superior

---

55. *chiasm* = cross, X

56. *geniculate* = bent like a knee

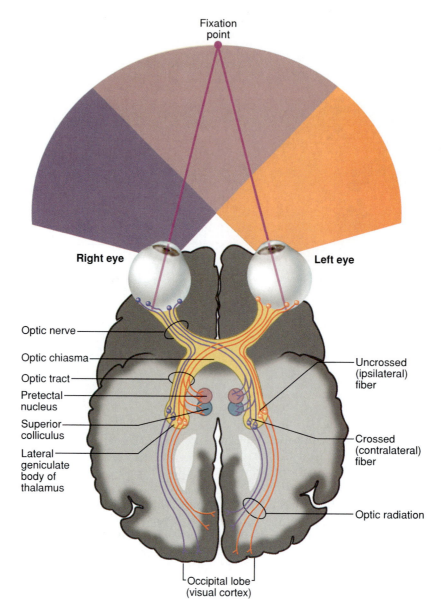

Fixation
point

Right eye

Left eye

Optic nerve

Optic chiasma

Optic tract

Pretectal
nucleus

Superior
colliculus

Lateral
geniculate
body of
thalamus

Uncrossed
(ipsilateral)
fiber

Crossed
(contralateral)
fiber

Optic radiation

Occipital lobe
(visual cortex)

Figure 16.42 The visual projection pathway. Diagram of hemidecussation and projection to the primary visual cortex.

colliculi control the visual reflexes of the extrinsic muscles, and the pretectal nuclei are involved in the photopupillary and accommodation reflexes.

Space does not allow us to consider much about the very complex processes of visual information processing in the brain. We have seen that some information processing, such as contrast, motion, and stereopsis, begins in the retina. The occipital lobe, posterior part of the parietal lobe, and inferior part of the temporal lobe process retinal data in ways beyond our present consideration to extract information about color, motion, contours, depth, and the boundaries between one object and another. What is yet to be learned about visual processing will no doubt have important implications for biology, medicine, psychology, and even philosophy.

Key Point Review

22 Why can't we see wavelengths below 350 nm or above 750 nm?

23 Why are light rays bent (refracted) more by the cornea than by the lens of the eye?

24 List as many structural and functional differences between rods and cones as you can.

25 Explain how the absorption of a photon of light leads to depolarization of a bipolar retinal cell.

26 Discuss the duplicity theory of vision, summarizing the advantage of having two types of retinal photoreceptor cells.

Anesthesia—From Ether Frolics to Modern Surgery

Surgery is as old as civilization. Even Stone Age people practiced *trephination*—cutting a hole in the skull to let out "evil spirits" that were thought to cause headaches. The ancient Hindus were expert surgeons for their time, and the Greeks and Romans pioneered military surgery. But until the nineteenth century, surgery was a miserable and dangerous business, done only as a last resort and with little hope of the patient's survival. Surgeons rarely attempted anything more complex than amputations or kidney stone removal. A surgeon had to be somewhat indifferent to the struggles and screams of his patient. Most operations had to be completed in 3 minutes or less, and a strong arm and stomach were more important qualifications for a surgeon than extensive anatomical knowledge.

At least three things were needed before surgery could be effective: better knowledge of anatomy, *asepsis*[57] for the control of infection, and *anesthesia*[58] for the control of pain. Early efforts to control pain during surgery were crude and usually ineffective, such as choking a patient into unconsciousness and trying to complete the surgery before he or she awoke. Alcohol and opium were often used as anesthetics, but the dosage was poorly controlled; some patients were underanesthetized and suffered great pain anyway, and others died of overdoses. Often there was no resort but for a few strong men to hold the struggling patient down as the surgeon worked. Charles Darwin originally intended to become a physician, but in medical school he was sickened by observing "two very bad operations, one on a child," in the days before anesthesia.

In 1799, Sir Humphrey Davy suggested using nitrous oxide for anesthesia. His student, Michael Faraday, suggested ether. Neither of these ideas caught on for several decades, however. Nitrous oxide ("laughing gas") was a popular amusement in the 1800s, as traveling showmen went from town to town demonstrating its effects on volunteers from the audience. In 1841, at a medicine show in Georgia, some students were impressed with the volunteers' euphoric giggles and antics and asked a young physician, Crawford W. Long, if he could make some nitrous oxide for them. Long lacked the equipment to synthesize it, but he recommended they try ether. Ether was commonly used in small oral doses for toothaches and "nervous ailments," but its main claim to popularity was its use as a party drug for so-called ether frolics. Long himself was a bit of a bon vivant; he enjoyed ether frolics and put on demonstrations for some of the young ladies, with the disclaimer that he could not be held responsible for whatever he might do under its influence (such as stealing a kiss).

At some of these parties, Long noted that people sometimes suffered considerable injuries without feeling pain. In 1842, he was confronted with the case of a young man, terrified of pain, who needed a tumor removed from his neck. Long excised the tumor without difficulty as his patient sniffed ether from a towel. The operation created a sensation in town, but other physicians ridiculed Long and pronounced anesthesia dangerous. Long's medical practice declined as people became afraid to go to him, but over the next 4 years he performed eight minor surgeries on patients under ether. He even compared surgeries done on the same person with and without ether, struggling to overcome criticisms that its effects were merely due to hypnotic suggestion or individual variation in sensitivity to pain.

Long failed to publish his results quickly enough, and in 1844 he was scooped by a Connecticut dentist, Horace Wells, who had tried nitrous oxide as a dental anesthetic. Another dentist, William T. G. Morton of Boston, had tried everything from champagne to opium to kill pain in his patients. He too became interested in ether and gave a public demonstration at Massachusetts General Hospital, where a tumor was removed from a patient under ether. Morton patented a "secret formula" he called Morton's Letheon,[59] which smelled suspiciously of ether, but eventually he went broke trying to monopolize ether anesthesia and died a pauper. Wells, who had engaged in a bitter feud to establish himself as the inventor of anesthesia, committed suicide at the age of 34. Crawford Long went on to a successful career as an Atlanta pharmacist, but until his death he remained disappointed that he had not received due credit as the first to perform surgery with ether anesthesia.

Ether and chloroform became obsolete with the development of safer anesthetics such as cyclopropane, ethylene, and nitrous oxide. These are **general anesthetics** that render a patient unconscious by crossing the blood-brain barrier and blocking nervous transmission through the brainstem. They have little structural similarity, but what they do have in common is a tendency to dissolve in the phospholipids of plasma membranes

—continued

---

57. *a* = without + *sepsis* = infection
58. *an* = without + *esthesia* = feeling, sensation

59. *lethe* = oblivion, forgetfulness

and make them more fluid. This increases their K$^+$ permeability, hyperpolarizes neurons, and makes them less likely to fire. Diazepam (Valium) deadens pain by activating gamma-(γ-)aminobutyric acid (GABA) receptors and causing an influx of Cl$^-$, which hyperpolarizes neurons.

Local anesthetics such as procaine (Novocain) and tetracaine selectively deaden specific nerves. They decrease the permeability of membranes to Na$^+$, reducing their ability to produce action potentials. Spinal anesthesia (spinal block) uses local anesthetics injected into the subarachnoid space for surgery of the ab-

dominopelvic region and lower extremity. It is also an effective and safe method to use in childbirth, since little of the anesthetic enters the mother's bloodstream and the effect on the fetus is minimal.

A sound knowledge of anatomy, control of infection and pain, and development of better tools converged to allow surgeons time to operate more carefully. As a result, surgery became more intellectually challenging and interesting. It attracted a more educated class of practitioner, which put it on the road to becoming the remarkable lifesaving approach that it is today.▲

Chapter Review Study Outline

Properties and Types of Sensory Receptors (pp. 552–553)
1. General properties of receptors
   a. Receptors as transducers
   b. Information transmitted
      • Modality
      • Location
      • Intensity
      • Duration
         •• Sensory adaptation
         •• Phasic receptors
         •• Tonic receptors
2. Classification of receptors
   a. Stimulus modality
   b. General and special senses
   c. Sources of stimuli

The General Senses (pp. 553–558) (table 16.1)
1. Unencapsulated nerve endings
2. Encapsulated nerve endings
3. Somesthetic projection pathways
4. Pain
   a. Causes
   b. Fast and slow pain fibers
   c. Somatic and visceral pain
   d. Projection pathways
   e. Referred pain
   f. Chemical agents of pain
      • Bradykinin
      • Substance P
   g. CNS modulation of pain
      • Enkephalins and endorphins
      • Pain gating

The Chemical Senses (pp. 558–562)
1. Taste (gustation)
   a. Anatomy and histology
      • Lingual papillae
      • Taste buds
      • Taste cells

   b. Physiology
      • Four primary taste sensations
      • Taste and flavor
   c. Projection pathways
      • Cranial nerves VII, IX, and X
      • Role of nuclei in medulla oblongata
      • Projection via thalamus to parietal lobe
2. Smell (olfaction)
   a. Anatomy and histology
      • Location of olfactory mucosa
      • Olfactory neurons
   b. Physiology
      • Properties of odorant molecules
      • Olfactory transduction
      • Olfactory adaptation
   c. Projection pathways
      • Olfactory nerves
      • Olfactory bulbs
      • Olfactory tracts
      • Projections to cerebral cortex

Hearing and Equilibrium (pp. 562–576)
1. Nature of sound
   a. Production of sound
   b. Frequency and pitch
   c. Amplitude and loudness
2. Anatomy of the ear (table 16.3)
   a. External ear
      • Auricle (pinna)
      • Auditory canal
      • Tympanic membrane
   b. Middle ear (tympanic cavity)
      • Auditory (eustachian) tube
      • Malleus, incus, and stapes
      • Stapedius and tensor tympani
   c. Inner ear
      • Bony and membranous labyrinths
      • Perilymph and endolymph
      • Vestibular system and cochlea
   d. Structure of the cochlea (table 16.3)

3. Physiology of hearing
   a. Middle-ear functions
      • Impedance matching
      • Tympanic reflex
   b. Hair cell stimulation
      • Vibration of stapes
      • Movement of perilymph
      • Vibration of basilar membrane
      • Role of tectorial membrane
      • Hair cell tip links and ion channels
   c. Sensory coding
      • Coding for amplitude
      • Coding for frequency
   d. Cochlear tuning
      • By contraction of outer hair cells
      • By feedback to inner hair cells
4. Auditory projection pathway
   a. Projections to medulla oblongata
      • Cochlear nucleus
      • Superior olivary nucleus
   b. Projection to primary auditory cortex
   c. Projections to superior and inferior colliculi
5. Deafness
6. Equilibrium
   a. Vestibular apparatus
   b. Static and dynamic equilibrium
   c. Saccule and utricle
      • Saccule and macula sacculi
      • Utricle and macula utriculi
      • Otolithic membrane
      • Static equilibrium
      • Detection of linear acceleration
   d. Semicircular ducts
      • Orientation and structure
      • Crista ampullaris and cupula
      • Detection of angular acceleration
   e. Projection pathways
      • Vestibular nerve
      • Vestibular nucleus of medulla

- • Nuclei of cranial nerves III, IV, and VI
- • Terminations in cerebellum, spinal cord, and cerebral cortex
- • Visual reflexes

Vision (pp. 577–594)
1. Light energy and photochemical reactions
2. Accessory organs of the orbit (table 16.4)
   a. Eyebrows
   b. Eyelids
   c. Conjunctiva
   d. Lacrimal apparatus
   e. Extrinsic muscles
   f. Orbital fat
3. Anatomy of the eye (table 16.4)
   a. Tunics
   b. Optical apparatus
   c. Neural apparatus
   d. Correlation of retinal histology with gross anatomy
4. Formation of an image
   a. Admittance of light
      • Pupillary dilation and constriction
      • Photopupillary and consensual light reflexes
   b. Refraction

c. Near response
   • Convergence of eyes
   • Pupillary constriction
   • Accommodation of lens
5. Sensory transduction
   a. Visual pigments
      • Rhodopsin (visual purple)
      • Photopsin
   b. The photochemical reaction
      • Isomerization of retinal
      • Dissociation from opsin
      • Bleaching
   c. Generating an optic nerve signal
      • The dark current
      • Glutamic acid secretion
      • Events occurring in light
         •• Bleaching of rhodopsin
         •• Breakdown of cGMP
         •• Cessation of dark current
         •• Hyperpolarization of receptor cells
         •• Effects on bipolar and ganglion cells
         •• Signals in optic nerve
   d. Regeneration of rhodopsin
      • Conversion of *trans*- to *cis*-retinal
      • Linkage of *cis*-retinal to opsin

6. Sensory adaptation
   a. Light adaptation
      • Photopupillary reflex (constriction)
      • Bleaching of photopigment
   b. Dark adaptation
      • Photopupillary reflex (dilation)
      • Regeneration of rhodopsin
7. Duplicity theory
   a. Rod function (scotopic vision)
      • Extensive neuronal convergence
      • High sensitivity
      • Low-resolution image
      • Role of peripheral vision
   b. Cone function (photopic vision)
      • Minimal neuronal convergence
      • Low light sensitivity
      • High-resolution image
8. Color vision
   a. Differential stimulation of blue, green, and red cones
   b. Color blindness
9. Stereoscopic vision
10. Visual projection pathway
   a. Optic nerves and chiasma
   b. Optic tracts
   c. Lateral geniculate body
   d. Optic radiation
   e. Primary visual cortex
   f. Superior colliculi and pretectal nuclei

Selected Vocabulary

Also review the terms in tables 16.1, 16.3, and 16.4, which are not repeated here.

receptor 552
sense organ 552
receptor potential 552
sensation 552
modality 552
sensory projection 552
adaptation 553
phasic receptor 553
tonic receptor 553
chemoreceptor 553
thermoreceptor 553
nociceptor 553
mechanoreceptor 553
photoreceptor 553
general senses 553
special senses 553
interoceptor 553
proprioceptor 553
exteroceptor 553
first-order neuron 554
second-order neuron 554
third-order neuron 554
fast pain 554

slow pain 554
somatic pain 554
visceral pain 556
referred pain 556
bradykinin 557
substance P 557
analgesic 557
enkephalins 557
endorphins 557
gustation 558
taste bud 558
lingual papilla 558
filiform papilla 558
foliate papilla 558
fungiform papilla 558
vallate papilla 558
taste cell 558
supporting cell 558
basal cell 558
taste hairs 558
taste pore 558
olfaction 560
olfactory mucosa 560
olfactory neuron 560
olfactory hairs 560
olfactory bulb 561
olfactory tract 562

sound 562
pitch 563
frequency 563
loudness 563
amplitude 563
impedance matching 568
tympanic reflex 568
tip link 569
binaural hearing 571
conduction deafness 572
sensorineural deafness 573
static equilibrium 573
dynamic equilibrium 573
kinocilium 573
otoliths 573
chromatophores 580
outer segment 582
inner segment 582
scotopic vision 582
photopic vision 582
blind spot 584
pupillary constrictor 585
pupillary dilator 585
photopupillary reflex 585
consensual light reflex 585
refraction 586
emmetropia 586

near response 586
convergence 586
accommodation 587
near point 588
rhodopsin 588
retinal 588
opsin 588
photopsin 588
cis-retinal 589
trans-retinal 589
bleaching 590
dark current 590
receptive field 591
light adaptation 591
dark adaptation 591
duplicity theory 591
blue cones 592
green cones 592
red cones 592
color blindness 592
stereoscopic vision 592
optic chiasma 593
optic tracts 593
hemidecussation 593
lateral geniculate body 593
optic radiation 593

1. Hot and cold stimuli are detected by
   a. free nerve endings.
   b. proprioceptors.
   c. bulbs of Krause.
   d. pacinian corpuscles.
   e. Meissner corpuscles.

2. _____ is a neurotransmitter that transmits pain sensations to second-order spinal neurons.
   a. Endorphin
   b. Enkephalin
   c. Substance P
   d. Acetylcholine
   e. Norepinephrine

3. _____ is a neuromodulator that blocks the transmission of pain sensations to second-order spinal neurons.
   a. An endorphin
   b. An enkephalin
   c. Substance P
   d. Acetylcholine
   e. Norepinephrine

4. Taste buds of the circumvallate papillae are most sensitive to _____ tastes.
   a. bitter
   b. sour
   c. sweet
   d. peppery
   e. salty

5. The higher the frequency of a sound,
   a. the louder it sounds.
   b. the harder it is to hear.
   c. the more it stimulates hair cells at the distal end of the organ of Corti.
   d. the faster it travels through air.
   e. the higher its pitch.

6. Cochlear hair cells rest on
   a. the tympanic membrane.
   b. the secondary tympanic membrane.
   c. the tectorial membrane.
   d. the vestibular membrane.
   e. the basilar membrane.

7. The acceleration you feel when an elevator begins to rise is sensed by
   a. the anterior semicircular duct.
   b. the organ of Corti.
   c. the crista ampullaris.
   d. the macula sacculi.
   e. the macula utriculi.

8. The color of a light is determined by
   a. its velocity.
   b. its amplitude.
   c. its wavelength.
   d. its refraction.
   e. the degree to which it stimulates the rod cells.

9. The retina receives its oxygen supply from the _____ and will deteriorate if it becomes detached from this structure.
   a. hyaloid artery
   b. vitreous body
   c. choroid
   d. pigment epithelium
   e. scleral venous sinus

10. Which of the following statements about photopic vision is *not* true?
    a. It is mediated by the cones.
    b. It has a low stimulus threshold.
    c. It produces fine resolution.
    d. It does not function in dim light.
    e. It does not employ rhodopsin.

11. The most finely detailed vision occurs when an image falls on a pit in the retina called the _____.

12. The only cells of the retina that generate action potentials are the _____ cells.

13. The retinal dark current is due to the flow of _____ in and out of the receptor cells.

14. The gelatinous membranes of the macula sacculi and macula utriculi are weighted by calcium carbonate and protein granules called _____.

15. Three rows of _____ in the cochlea have V-shaped arrays of stereocilia and tune the frequency sensitivity of the cochlea.

16. The _____ is a tiny bone that vibrates in the oval window and thereby transfers sound vibrations to the inner ear.

17. The _____ of the midbrain receives auditory input and triggers the head-turning auditory reflex.

18. The apical stereocilia of a gustatory cell are known as _____.

19. Olfactory neurons synapse with mitral cells and tufted cells in the _____, which lie inferior to the frontal lobes.

20. In the phenomenon of _____, pain from the viscera is perceived as coming from an area of the skin.

1. The principle of neuronal convergence was explained in chapter 13. Discuss its relevance to referred pain, and scotopic vision.

2. A child feels a head louse crawling through her hair. What type of cutaneous receptor is involved? What type of receptor is most important to the ability of a blind person to read braille? What type of cutaneous receptors would enable you to palpate a patient's pulse?

3. Contraction of a muscle usually puts *more* tension on a structure, but contraction of the ciliary muscle puts *less* tension on the lens. Explain how.

4. Discuss the similarities and differences between presbyopia and hyperopia.

5. Color blindness affects about 10% of males but only 0.7% of females. What does this suggest about its mode of inheritance (see chapter 5)? How does this compare to pattern baldness (see chapter 7)?

Web Site Link

For a listing of the most current web sites related to this chapter, please visit the Saladin homepage at:

http://www.mhhe.com/sciencemath/biology/saladin/

chapter *seventeen*

17

[The Endocrine System

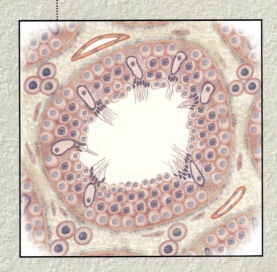

the unity of form and function

Brushing up

We noted in chapter 13 that if a multicellular organism is to function as an integrated homeostatic whole, it needs mechanisms of intercellular communication—mainly the nervous and endocrine systems. **Endocrinology** is the study of the endocrine system and the diagnosis and treatment of its dysfunctions. To begin this chapter, we survey the endocrine glands, their hormones, their target organs, and the effects the hormones produce. Having established this "cast of characters," we then delve into the details of how hormones act on their target cells.

An Overview of the Endocrine System

▼**Objectives**
When you have completed this section, you should be able to
• define *hormone* and *endocrine system;*
• compare and contrast the nervous and endocrine systems; and
• list the major organs of the endocrine system.

The term **hormone**[1] is defined in various ways, but for our purposes it means a chemical messenger that is secreted into the bloodstream and affects the metabolism of target cells usually located in another organ (fig. 17.1). An **endocrine gland** or **endocrine cell** is one that secretes a hormone. Whereas exocrine glands secrete their products through ducts onto the body surface or into the lumen of the digestive tract (which is technically external to the body), endocrine glands lack ducts and secrete their products into the surrounding tissue fluid.[2] The blood then takes up the hormones and carries them throughout the body. The **endocrine system** is a collective term for all such glands and the hormone-secreting cells distributed throughout many other organs.

The classic inventory of endocrine glands (fig. 17.2) leaves a lot to be desired. For one thing, some organs such as the pancreas have both endocrine and exocrine functions. For another, hormones are also produced by endocrine cells in many organs that most people do not think of as glands—the brain, heart, stomach, small intestine, and placenta, for example.

As discussed in chapter 13, the nervous and endocrine systems use complementary modes of communication. Their differences, listed in table 13.1, must not obscure their numerous similarities. Both systems communicate by chemical means, and several chemicals function as both neurotransmitters and

. . . .
1. *hormone* = to excite, set in motion
2. *endo* = into; *exo* = out of; *crin* = to separate or secrete

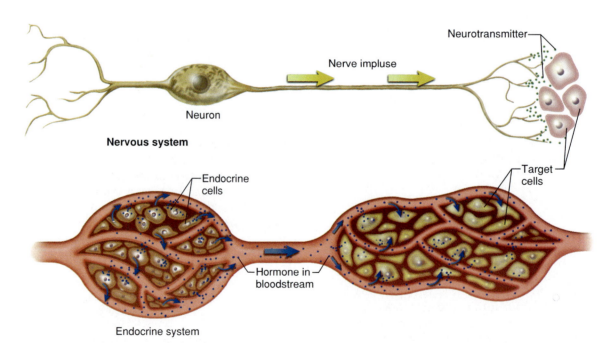

Figure 17.1 A comparison of communication by the nervous and endocrine systems. (*a*) A neuron has a long fiber that delivers its neurotransmitter to the immediate vicinity of its target cells. (*b*) Endocrine cells secrete a hormone into the bloodstream. The hormone leaves the blood capillaries and binds to target cells at places often remote from the gland cells. 𝕏

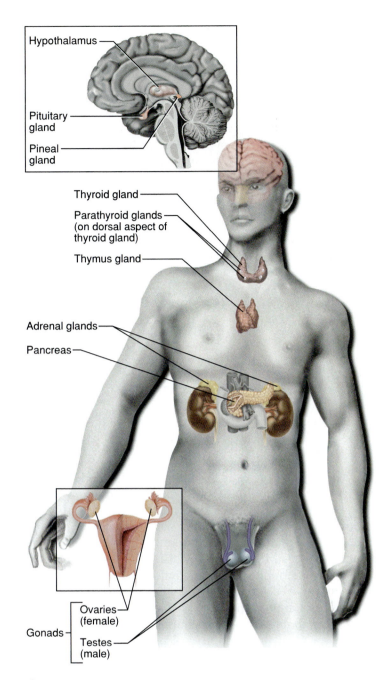

Hypothalamus

Pituitary gland

Pineal gland

Thyroid gland

Parathyroid glands (on dorsal aspect of thyroid gland)

Thymus gland

Adrenal glands

Pancreas

Ovaries (female)

Gonads

Testes (male)

Figure 17.2 Major organs of the endocrine system. This system also includes gland cells in many other organs not shown here. 𝒳

Table 17.1	Hormone Terminology	
Abbreviation	**Name**	**Source**
ACTH	Adrenocorticotropic hormone (corticotropin)	Anterior pituitary
ADH	Antidiuretic hormone (vasopressin)	Posterior pituitary
ANF	Atrial natriuretic factor	Heart
CRH	Corticotropin-releasing hormone	Hypothalamus
FSH	Follicle-stimulating hormone	Anterior pituitary
GH	Growth hormone (somatotropin)	Anterior pituitary
GHIH	Growth hormone–inhibiting hormone (somatostatin)	Hypothalamus
GHRH	Growth hormone–releasing hormone	Hypothalamus
GnRH	Gonadotropin-releasing hormone	Hypothalamus
LH	Luteinizing hormone	Anterior pituitary
OT	Oxytocin	Posterior pituitary
PIF	Prolactin-inhibiting factor	Hypothalamus
PRF	Prolactin-releasing factor	Hypothalamus
PRL	Prolactin (luteotropin)	Anterior pituitary
PTH	Parathyroid hormone (parathormone)	Parathyroids
T_3	Triiodothyronine	Thyroid
T_4	Thyroxine (tetraiodothyronine)	Thyroid
TRH	Thyrotropin-releasing hormone	Hypothalamus
TSH	Thyroid-stimulating hormone	Anterior pituitary

tially reaches every cell in the body. Yet it produces a response only in certain **target cells** that possess receptors for it.

Hormones are commonly referred to by abbreviations or acronyms that vary slightly in different sources. It is easy to forget or confuse these when first learning the endocrine system, so the ones used repeatedly in this chapter are listed alphabetically in table 17.1 for easy reference. The first name given for each hormone is the one used in this chapter; synonyms commonly used in other sources are indicated in parentheses.

Key Point Review

❶ Define the word *hormone* and distinguish a hormone from a neurotransmitter. Why is this an imperfect distinction?

❷ Name some sources of hormones other than purely endocrine glands.

hormones—norepinephrine and dopamine, for example. Some hormones, such as oxytocin and the catecholamines, are secreted by **neuroendocrine cells**—neurons that release their secretions into the blood. Some neurotransmitters and hormones produce identical effects on the same target cells. For example, norepinephrine and glucagon cause hydrolysis of glycogen by the liver. Unlike a neurotransmitter, which is delivered to a specific target cell by a nerve fiber, a hormone travels everywhere the blood goes and poten-

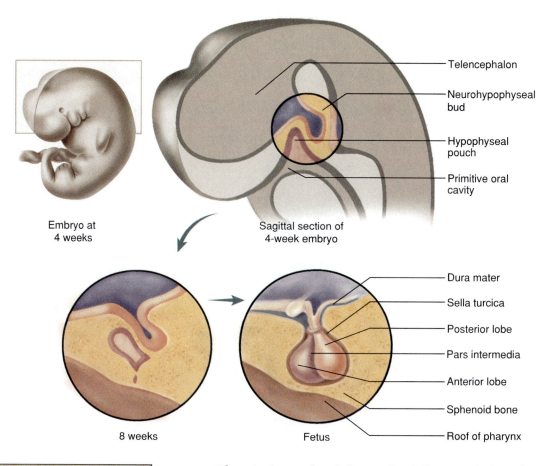

Figure 17.3 Embryonic development of the pituitary gland. (*a*) Anterior end of the embryo at about 4 weeks. (*b*) Sagittal section of the embryo showing the early beginnings of the adenohypophysis and neurohypophysis. (*c*) Separation of the hypophyseal pouch from the pharynx at about 8 weeks. (*d*) Development nearly completed. The pars intermedia largely disappears by birth.

Embryo at
4 weeks

Sagittal section of
4-week embryo

Telencephalon

Neurohypophyseal
bud

Hypophyseal
pouch

Primitive oral
cavity

Dura mater

Sella turcica

Posterior lobe

Pars intermedia

Anterior lobe

Sphenoid bone

Roof of pharynx

8 weeks

Fetus

The Hypothalamus and Pituitary Gland

▼**Objectives**

When you have completed this section, you should be able to

- list the hormones produced by the hypothalamus and pituitary gland; and
- explain how the hypothalamus and pituitary are controlled and coordinated with each other.
- describe the functions of growth hormone; and
- describe the effects of pituitary hypo- and hypersecretion.

There is no "master control center" that regulates the entire endocrine system, but the hypothalamus and pituitary gland secrete more hormones and have broader effects than any other endocrine glands. It is therefore appropriate that we begin our inventory with them.

Anatomy

The hypothalamus forms the floor and walls of the third ventricle of the brain, immediately posterior to the optic chiasma (see fig. 14.18, p. 483). It regulates primitive functions of the body ranging from water balance to sex drive. Many of these are carried out by way of the pituitary gland, which is closely associated with the hypothalamus.

The **pituitary gland (hypophysis[3])** is nestled in the sella turcica of the sphenoid bone and attached to the hypothalamus by a *stalk* (*infundibulum*). It is usually about 1.3 cm in diameter but can grow to twice that large during pregnancy. It is actually composed of two structures—the *adenohypophysis* and *neurohypophysis.* These arise independently in the embryo and have entirely separate functions (fig. 17.3).

The **adenohypophysis[4]** (AD-eh-no-hy-POFF-ih-sis) constitutes about three quarters of the pituitary (fig. 17.4*a*). It develops from the hypophyseal pouch, an outgrowth of the roof of the embryonic pharynx. In adults it has two parts: a large **anterior lobe,** also called the *pars distalis* ("distal part") because it is most distal to the pituitary stalk, and the **pars tuberalis,** a small mass of cells adhering to the anterior side of the stalk. In the fetus there is also a **pars intermedia,** a strip of tissue between the anterior lobe and the neurohypophysis. During subsequent fetal development, its cells mingle with those of the anterior lobe; in adults, there is no longer a separate pars intermedia.

The **neurohypophysis** (fig. 17.4*b*) constitutes the posterior one quarter of the pituitary. It has three parts: the **median eminence,** which is a funnel-like inferior extension of the hypothalamus; the **stalk** mentioned earlier; and the **posterior lobe** (*pars nervosa*), which constitutes

3. *hypo* = below + *physis* = growth
4. *adeno* = gland

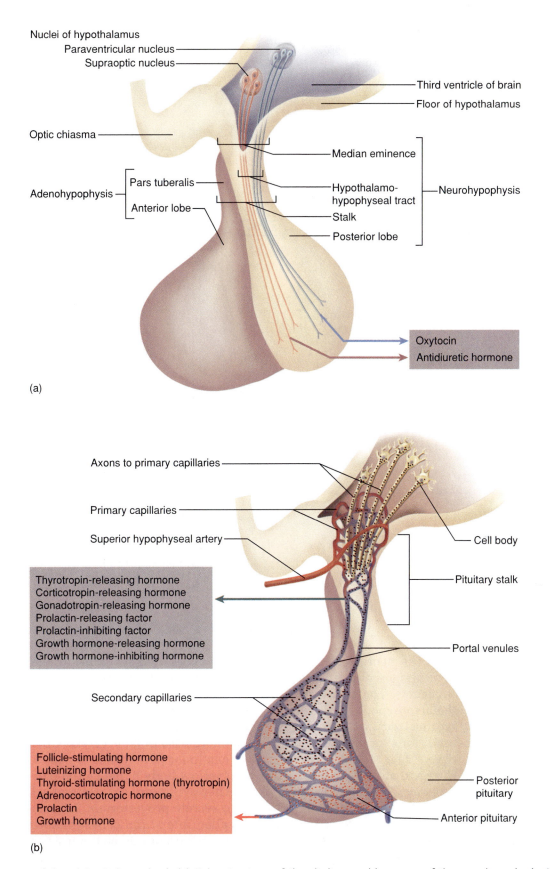

Nuclei of hypothalamus
Paraventricular nucleus
Supraoptic nucleus
Third ventricle of brain
Floor of hypothalamus
Optic chiasma
Median eminence
Adenohypophysis
Pars tuberalis
Hypothalamo-
hypophyseal tract
Neurohypophysis
Anterior lobe
Stalk
Posterior lobe

Oxytocin
Antidiuretic hormone

(a)

Axons to primary capillaries

Primary capillaries

Superior hypophyseal artery

Cell body

Pituitary stalk

Thyrotropin-releasing hormone
Corticotropin-releasing hormone
Gonadotropin-releasing hormone
Prolactin-releasing factor
Prolactin-inhibiting factor
Growth hormone-releasing hormone
Growth hormone-inhibiting hormone

Portal venules

Secondary capillaries

Follicle-stimulating hormone
Luteinizing hormone
Thyroid-stimulating hormone (thyrotropin)
Adrenocorticotropic hormone
Prolactin
Growth hormone

Posterior
pituitary

Anterior pituitary

(b)

Figure 17.4 Anatomy of the adult pituitary gland. (*a*) Major structures of the pituitary and hormones of the neurohypophysis. Note that these hormones are produced by two nuclei in the hypothalamus and later released from the posterior lobe of the pituitary. (*b*) The hypothalamo-hypophyseal portal system. The hormones in the violet box are secreted by the hypothalamus and travel in the portal system to the anterior pituitary. The hormones in the red box are secreted by the anterior pituitary under the control of the hypothalamic releasers and inhibitors. ⍟

Figure 17.5 Hormones of the anterior pituitary gland and their target organs. The three axes physiologically link pituitary function to the function of other endocrine glands. ⚡

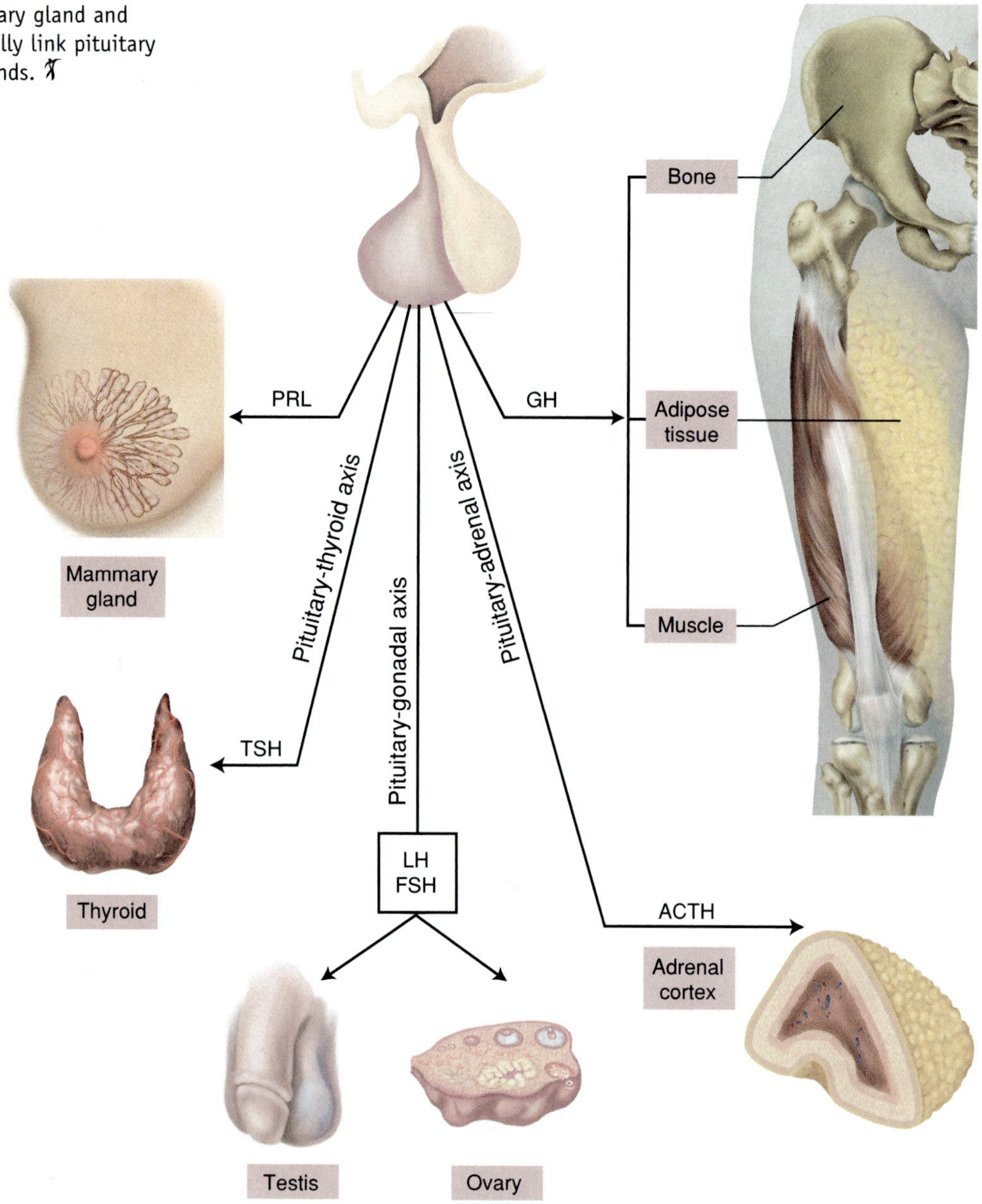

Bone

Adipose tissue

GH

Muscle

PRL

Mammary gland

Pituitary–thyroid axis

Pituitary–gonadal axis

Pituitary–adrenal axis

TSH

Thyroid

LH
FSH

ACTH

Adrenal cortex

Testis

Ovary

the bulk of the neurohypophysis. The neurohypophysis is not a true gland but a mass of nervous tissue whose somas are located in the hypothalamus. The axons from these neurons form the **hypothalamo-hypophyseal tract** in the stalk. Hormones are synthesized in the somas, transported down their axons, and stored in the posterior pituitary until a nerve signal triggers their release.

Hereafter, we can largely disregard all parts of the pituitary except the anterior and posterior lobes; these secrete or release all of the pituitary hormones we will consider.

Anterior Lobe Hormones

The anterior lobe synthesizes and secretes six hormones (fig. 17.5):

1. **Follicle-stimulating hormone (FSH)** is secreted by cells called *gonadotropes*. In the ovaries, it stimulates the development of eggs and the follicles (vesicles) that contain them. In the testes, it stimulates the production of sperm.

2. **Luteinizing hormone (LH)** is also secreted by the gonadotropes. In females, it stimulates *ovulation* (the release of an egg). After ovulation, the remainder of a follicle is called the *corpus luteum* ("yellow body"), hence the name of this hormone. After ovulation, LH stimulates the corpus luteum to secrete progesterone. In males, LH is sometimes called *interstitial cell–stimulating hormone* (ICSH); it stimulates the testes to secrete testosterone.

3. **Thyroid-stimulating hormone (TSH),** or **thyrotropin,** is secreted by cells called *thyrotropes.* It stimulates growth of the thyroid gland and the secretion of thyroid hormones.

4. **Adrenocorticotropic hormone (ACTH)** is secreted by *corticotropes*. It stimulates growth of the adrenal cortex and the secretion of its hormones; fat catabolism in adipose tissue; and insulin secretion by the pancreas.

5. **Prolactin[5] (PRL)** is secreted by *lactotropes* (*mammotropes*), which increase greatly in size and number during pregnancy. They begin to secrete PRL immediately after a woman gives birth and continue secreting it for as long as she nurses. PRL acts on the mammary gland to promote milk synthesis. In males, it has a gonadotropic effect and makes the testes more sensitive to LH. Thus, it indirectly enhances their secretion of testosterone. Excessive levels of PRL in men can cause breast enlargement (gynecomastia) and impotence.

6. **Growth hormone (GH), or somatotropin,** is secreted by *somatotropes,* the most numerous cells in the anterior pituitary. The pituitary produces at least a thousand times as much GH as any other hormone. The general effect of GH is to stimulate cellular growth, mitosis, and differentiation. Therefore, it promotes overall tissue and organ growth.

The first four hormones listed are called **tropic,** or **trophic,**[6] hormones—pituitary hormones whose target organs are other endocrine glands. More specifically, FSH and LH are called **gonadotropins** because their target organs are the gonads (ovaries and testes). GH and PRL have nonendocrine target organs.

The relationship between the pituitary, its tropic hormones, and their target endocrine glands is called an *axis*—a convenient term for referring to the complex of hormones released in response to a stimulus and to the way these endocrine glands influence each other. There are three such axes: the **pituitary-thyroid axis, pituitary-adrenal axis,** and **pituitary-gonadal axis.**

The Pars Intermedia

As mentioned earlier, the pars intermedia is absent from the adult human pituitary, but it is present in other animals and in the human fetus. Until recently, it was thought to secrete *melanocyte-stimulating hormone (MSH),* which was believed to influence pigmentation of the skin. This is true for other species, but evidence now indicates that there is no circulating MSH in humans. Some cells of the anterior lobe, derived from the fetal pars intermedia, produce a large polypeptide called **pro-opiomelanocortin (POMC).** POMC is not secreted, but is processed within the pituitary to yield smaller fragments such as ACTH and endorphins.

Posterior Lobe Hormones

The posterior lobe stores and releases two hormones synthesized by the hypothalamus:

1. **Antidiuretic[7] hormone (ADH)** acts on the kidneys to increase water retention, lower urine volume, and help prevent dehydration. It is also called *vasopressin* because it causes vasoconstriction at high concentrations. These concentrations are so unnaturally high for the human body, however, that this effect is of doubtful significance except in pathological states. ADH is synthesized in the supraoptic nucleus of the hypothalamus.

2. **Oxytocin[8] (OT)** stimulates labor contractions of the uterus and milk release by the mammary glands. Based on the evidence to date, however, it remains uncertain whether it is essential to either of these functions in humans. OT is synthesized in the paraventricular nucleus of the hypothalamus.

Hormones of the pituitary gland are summarized in table 17.2.

Control of Pituitary Secretion

Pituitary hormones are not secreted at a steady rate. Growth hormone is secreted mainly at night, luteinizing hormone peaks at the middle of the menstrual cycle, and oxytocin secretion surges during labor and nursing, for example. The timing and amount of pituitary secretion is regulated by the hypothalamus, by higher brain centers, and by feedback from the target organs.

Hypothalamic Control of the Anterior Lobe

Although the anterior pituitary has no nervous connection to the hypothalamus, it is connected to it by a complex of blood vessels called the **hypothalamo-hypophyseal portal system** (see fig. 17.4a). A *portal system* is a circulatory route in which the blood flows through two consecutive capillary networks before returning to the heart. In most routes, by contrast, blood leaves the heart, passes through one capillary bed, and then returns to the heart. The median eminence of the hypothalamus secretes hormones into its capillaries,

---

5. *pro* = favoring + *lact* = milk
6. *trop* = to turn, change; *troph* = to feed, nourish

7. *anti* = against + *diuret* = to pass through, urinate
8. *oxy* = sharp, quick + *toc* = childbirth

Table 17.2 Pituitary Hormones

Hormone	Target Organ or Cells	Principal Effects
Anterior Pituitary		
FSH: Follicle-stimulating hormone	Ovaries and testes	Growth of ovarian follicles and secretion of estrogens; development of eggs; production of sperm
LH: Luteinizing hormone, also called interstitial cell–stimulating hormone (ICSH)	Ovaries and testes	In females, ovulation and maintenance of corpus luteum of ovary; in males, secretion of testosterone
TSH: Thyroid-stimulating hormone	Thyroid gland	Growth of thyroid gland and secretion of thyroid hormone
ACTH: Adrenocorticotropic hormone, also called corticotropin	Adrenal cortex	Growth of adrenal cortex and secretion of glucocorticoids
PRL: Prolactin	Mammary glands and testes	In females, milk synthesis; in males, enhanced sensitivity of testis to LH and increased testosterone secretion
GH: Growth hormone, also called somatotropin	Most tissues of the body	Amino acid uptake; Ca^{2+} absorption by small intestine; synthesis of collagen and other proteins; cell division; tissue growth, especially of cartilage and bone
Posterior Pituitary		
ADH: Antidiuretic hormone, also called vasopressin	Kidneys	Water retention; reduced urine output; reduced dehydration
OT: Oxytocin	Uterus, mammary glands	Labor contractions; milk ejection

Special Topic The Many Roles of TRH   17.1

The actions of thyrotropin-releasing hormone (TRH) underscore the relationship between the nervous and endocrine systems, the impossibility of rigidly classifying chemical messengers, and the limited extent of our understanding about hormone function. TRH promotes secretion of TSH and PRL from the pituitary, but in some psychiatric disorders it also stimulates GH secretion. TRH also acts as a neurotransmitter in many parts of the brain. Elevated levels of TRH occur in schizophrenia, Huntington[9] chorea, and some cases of depression and Alzheimer disease. TRH is also produced by the digestive tract, placenta, and prostate, where its functions are not yet known.

which lead to veins that travel down the pituitary stalk. In the anterior lobe, these veins give rise to a second capillary bed, where the hypothalamic hormones leave the bloodstream.

Some of these hormones are **releasing hormones,** which stimulate secretion by the pituitary cells. Others are **inhibiting hormones,** which suppress pituitary secretion (table 17.3). For example, **thyrotropin-releasing hormone (TRH)** stimulates thyrotropes of the anterior lobe to secrete TSH, among other functions (see special topic 17.1). **Prolactin-inhibiting factor (PIF)** suppresses PRL secretion by the lactotropes. It is the same as the neurotransmitter dopamine.

Some hypothalamic hormones affect the output of two pituitary hormones: gonadotropin-releasing hormone (GnRH), for example, stimulates the release of both FSH and LH. Some pituitary cells are under dual control: lactotropes, for example, are under the control of prolactin-releasing factor and prolactin-inhibiting factor, which oppose each other's action.

Hypothalamic Control of the Posterior Lobe

The posterior pituitary is controlled by **neuroendocrine reflexes**—the release of hormones in response to signals from the nervous system. For example, the suckling of an infant stimulates nerve endings in the nipple. Signals are transmitted to the hypothalamus and from there to the posterior pituitary. This triggers the release of oxytocin into the blood, and oxytocin causes a milk-ejection reflex.

Antidiuretic hormone (ADH) is also controlled by a neuroendocrine reflex. Dehydration raises the osmolarity of the blood, which is detected by hypothalamic neurons called osmoreceptors. Once stimulated, osmoreceptors trigger the release of ADH from the

9. George Huntington (1850–1916), American physician

Hormone	Principal Effects
TRH: Thyrotropin-releasing hormone	Promotes TSH and PRL secretion
CRH: Corticotropin-releasing hormone	Promotes ACTH secretion
GnRH: Gonadotropin-releasing hormone	Promotes FSH and LH secretion
PRF: Prolactin-releasing factor	Promotes PRL secretion
PIF: Prolactin-inhibiting factor (dopamine)	Inhibits PRL secretion
GHRH: Growth hormone-releasing hormone	Promotes GH secretion
GHIH: Growth hormone-inhibiting hormone (somatostatin)	Inhibits GH and TSH secretion

posterior pituitary. ADH slows the rate of urine output and stimulates the sense of thirst; thus, it opposes dehydration. Excessive blood pressure, by contrast, stimulates stretch receptors in the heart and certain arteries. By another neuroendocrine reflex, this inhibits ADH release, increases urine output, and brings blood volume and pressure back to normal.

Think About It

Which of the unifying themes at the end of chapter 1 is best exemplified by the neuroendocrine reflexes that govern ADH secretion?

Influence of Higher Brain Centers

Neuroendocrine reflexes can involve higher brain centers and both lobes of the pituitary. For example, the milk-ejection reflex can be triggered when a lactating mother simply hears a baby cry. Emotional stress can affect secretion of gonadotropins and disrupt ovulation, the menstrual rhythm, and fertility. Stress also increases the secretion of CRH, which in turn stimulates ACTH output and the secretion of hormones by the adrenal cortex.

Feedback from Target Organs

The pituitary not only controls several other organs, but those organs also send controlling signals back to the pituitary. Their influence is usually in the form of **negative feedback inhibition**—the pituitary stimulates another endocrine gland to secrete its own hormone, and that hormone inhibits further secretion by the pituitary. All the pituitary axes are controlled by negative feedback inhibition. Consider the pituitary-thyroid axis, for example (fig. 17.6). The hypothalamus secretes TRH, TRH stimulates the anterior pituitary to secrete TSH, and TSH stimulates the thyroid gland to secrete thyroid hormones. Thyroid hormones stimulate most cells throughout the body and exert negative feedback inhibition on the pituitary, rendering the pituitary unrespon-

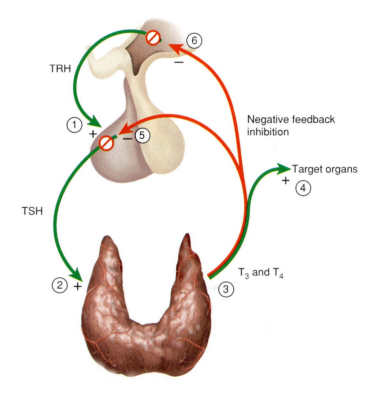

Figure 17.6 Negative feedback inhibition in the pituitary-thyroid axis. (*1*) The hypothalamus secretes TRH. (*2*) TRH stimulates the anterior pituitary to secrete TSH. (*3*) TSH stimulates the thyroid gland to secrete T_3 and T_4. (*4*) T_3 and T_4 stimulate a wide variety of target organs. (*5*) T_3 and T_4 also inhibit TSH release by the anterior pituitary, so that when T_3 and T_4 levels are high, TSH secretion is reduced. (*6*) To a lesser extent, T_3 and T_4 also inhibit the release of TRH. Pathways 5 and 6, in red, constitute negative feedback inhibition of the pituitary and hypothalamus. ✗

sive to TRH. Thus, as long as thyroxine levels are high, TSH secretion remains low. If thyroxine secretion drops, however, TSH secretion rises, and this stimulates the thyroid to secrete more hormone. This keeps thyroxine levels oscillating around a set point in typical homeostatic fashion.

Think About It

How would the level of pituitary tropic hormones be affected by the surgical removal of their target organs from an experimental animal? Explain.

Feedback from a target organ is not always inhibitory, however. During labor, oxytocin triggers a positive feedback cycle from the posterior pituitary to the uterus to the hypothalamus and back to the posterior pituitary. The effect is increased OT release (see fig. 1.17, p. 22).

We see then that there is no strict chain of command "from the top down," with the hypothalamus dictating the actions of the pituitary and the pituitary dictating the actions of its target organs. Rather, there is a reciprocal interaction in which the hypothalamus, pituitary, and target organs moderate each other's activities.

Growth Hormone

Most of the pituitary hormones are considered in later sections of this chapter or in subsequent chapters that deal with their target organs: ADH in chapter 24 on water balance; the gonadotropins, oxytocin, and prolactin in chapters 27 and 28 on reproduction; and TSH and ACTH later in this chapter. Now is the most opportune time, however, for a further consideration of growth hormone (GH). GH is not targeted to a specific organ. Its general effect is to stimulate tissue and organ growth by promoting cellular growth, mitosis, and cellular differentiation. It achieves this by a variety of mechanisms:

- **Protein synthesis.** Within minutes of its secretion, GH stimulates the translation of preexisting mRNA. Within hours, it stimulates genetic transcription to make more mRNA. To provide the raw material for protein synthesis, GH also enhances amino acid transport into cells. To ensure that protein synthesis outpaces breakdown, it also suppresses protein catabolism.
- **Lipid metabolism.** GH stimulates adipocytes to catabolize fat and release free fatty acids (FFAs) and glycerol into the blood. With this energy source available, cells catabolize less protein and carbohydrate. During the long stretch between the evening and morning meals, GH secretion rises and mobilizes FFAs for energy.
- **Carbohydrate metabolism.** By mobilizing FFAs and reducing the body's dependence on glucose for energy, GH promotes glycogen synthesis and storage.
- **Electrolyte balance.** GH promotes Na^+, K^+, and Cl^- retention by the kidneys, enhances Ca^{2+} absorption by the small intestine, and makes these electrolytes available to the growing tissues.

The most obvious effect of GH is on cartilage, bone, and muscle growth. It promotes the multiplication of chondrocytes and osteogenic cells and stimulates protein deposition in the cartilage or bone matrix. In childhood and adolescence, it causes the growth of long bones at the epiphyseal plates. In adulthood, it stimulates osteoblast activity and the appositional growth of bone; thus, it continues to influence bone thickening and remodeling. Although a person ceases to grow in height when the epiphysial plates close, there is no abrupt drop in GH secretion at the end of adolescence. GH levels decline gradually with age—averaging about 6 nanograms (ng) per mL of blood plasma in adolescence and about one-quarter of that in very old age. The resulting decline in protein synthesis may contribute to aging of the tissues, including wrinkling of the skin and decreasing muscular mass and strength. At age 30, the average adult body is 10% bone, 30% muscle, and 20% fat by weight; at age 75, the body averages 8% bone, 15% muscle, and 40% fat.

GH concentration fluctuates greatly over the course of a day. It rises to 20 ng/mL or higher during the first 2 hours of deep sleep and may reach 30 ng/mL in response to vigorous exercise. Smaller peaks occur after high-protein meals, but high-carbohydrate meals tend to suppress GH secretion. Trauma, hypoglycemia (low blood sugar), and other conditions also stimulate GH secretion.

At least some effects of GH seem to be produced indirectly. GH stimulates various organs, especially the liver, to produce small polypeptides called **somatomedins.**[10] These are growth factors with effects similar to those of insulin—thus, they are also called **insulin-like growth factors (IGFs).** At least four IGFs have been identified, but one of them, called IGF-I, seems most important in the action of GH on bone and cartilage. Its exact role is not yet certain. One hypothesis is that GH stimulates chondrocytes to synthesize IGF-I receptors, and then IGF-I acts as the stimulus for cartilage growth. Another hypothesis is that IGF-I may prolong the effect of GH. GH is cleared from the blood plasma (excreted) relatively quickly, whereas IGF-I is cleared very slowly. By stimulating the production of IGF-I, GH may therefore indirectly exert its effects long after it has been removed from the blood.

Pituitary Disorders

Many diseases result from underactivity (*hyposecretion*) or overactivity (*hypersecretion*) of the endocrine glands.

---

10. Acronym for *somato*tropin *medi*ating *prote*in

(a)

(b)

Figure 17.7 (*a*) Pituitary dwarfism, the effect of childhood hyposecretion of growth hormone. (*b*) Pituitary gigantism, the effect of childhood hypersecretion of growth hormone.

Hypopituitarism can result from prolonged steroid treatment or from hemorrhages, blood clots, or tumors. Childhood hypopituitarism results in **pituitary dwarfism** (fig. 17.7*a*) because of the deficiency of growth hormone. The complete loss of anterior pituitary function (*panhypopituitarism*[11]) causes atrophy of the thyroid, adrenal cortex, and gonads. Diminished secretion by those glands leads, in turn, to a broad range of disorders: loss of gonadal function; loss of pubic and axillary hair; disruption of carbohydrate, fat, and protein metabolism; and fluid and electrolyte imbalances. To prevent these effects following removal of a cancerous pituitary, a patient requires lifelong *hormone replacement therapy* with thyroxine, cortisone, growth hormone, and sex steroids.

Hyperpituitarism in childhood causes excessive growth, or **gigantism** (fig. 17.7*b*). In adulthood, it causes **acromegaly**—thickening of the bones and soft tissues with especially noticeable effects on the hands, feet, and face (fig. 17.8).

The chief consequence of posterior lobe hyposecretion is **diabetes insipidus,** resulting from lack of ADH. The term *diabetes* simply means any chronic excessive urine output; not all forms of diabetes are related to insulin and glucose metabolism. Diabetes insipidus usually stems from destruction of the hypothalamus or hypothalamo-hypophyseal tract by a tumor, infection, or basal skull fracture. It causes weakness, intense thirst, and a urine output that may exceed 15 L per day—10 times the normal output. It also creates electrolyte imbalances that place a person at risk for neurological and cardiac disorders. The disease can be treated with ADH administered by injection or as a nasal spray.

11. *pan* = all

(a)   (b)   (c)   (d)

Figure 17.8 The progression of acromegaly is dramatically shown by these four photographs of the same person taken at the ages of 9, 16, 33, and 52. Note the characteristic thickening of the face and hands.

Key Point Review

3 What are two good reasons for considering the pituitary to be two separate glands?

4 Name three anterior lobe hormones that have reproductive functions and three that have nonreproductive roles. What target organs are stimulated by each of these hormones?

5 Briefly contrast hypothalamic control of the anterior pituitary with its control of the posterior pituitary.

6 In what sense does the pituitary "take orders" from the target organs under its command?

Other Endocrine Glands

▼Objectives

When you have completed this section, you should be able to
- describe the structure and location of the remaining organs of the endocrine system;
- name the hormones these endocrine organs produce and state their functions; and
- discuss some common effects of endocrine dysfunctions and hormone imbalances.

The Pineal Gland

The **pineal**[12] **gland (epiphysis cerebri)** is a pine cone–shaped growth attached to the roof of the third ventricle, beneath the posterior end of the corpus callosum (see fig. 17.2). The philosopher René Descartes (1596–1650) thought it was the seat of the human soul. If so, children must have more soul than adults—a child's pineal gland is about 8 mm long and 5 mm wide, but after age seven it regresses rapidly and is no more than a tiny shrunken mass of fibrous tissue in the adult. Pineal

secretion peaks between the ages of 1 and 5 years; by the end of puberty it is reduced by three-quarters.

We no longer look for the human soul in the pineal, but this little gland remains an intriguing mystery. In animals with seasonal breeding, it regulates the gonads and the annual breeding cycle. The pineal produces **serotonin** by day and converts it to **melatonin** at night. Melatonin may suppress gonadotropin secretion; removal of the pineal from animals causes premature sexual maturation. Some physiologists think that the pineal gland may regulate the timing of puberty in humans, but a clear demonstration of its role has remained elusive. Pineal tumors can cause premature onset of puberty in young boys, but this may be because they also involve damage to the hypothalamus. There seems to be a relationship between melatonin and mood, including depression and sleep disorders (see chapter essay, p. 632).

The Thymus

The **thymus** is located in the mediastinum, superior to the heart (fig. 17.9). Like the pineal, it is large in infants and children, but it undergoes **involution** (shrinkage) after puberty. In adults, it is a shriveled vestige of its former self, with most of its parenchyma replaced by fibrous and adipose tissue. The thymus secretes *thymopoietin* and *thymosins,* hormones that regulate the development and later activation of disease-fighting blood cells called T lymphocytes (*T* for *thymus*). This is discussed in detail in chapter 21.

The Thyroid

The **thyroid** is the largest endocrine gland; it weighs between 20 and 25 g and receives one of the body's highest rates of blood flow per gram of tissue. It is wrapped around the anterior and lateral sides of the trachea,

---

12. *pineal* = pine cone

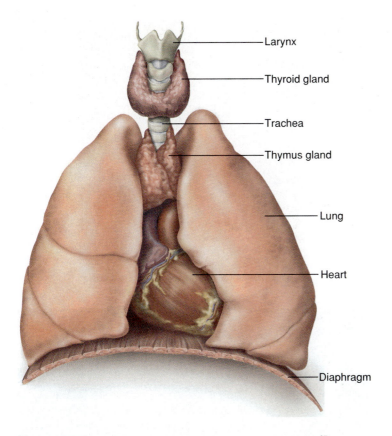

Figure 17.9 Locations of the thyroid and thymus glands. ⚚

immediately below the larynx. It consists of two large lobes, one on each side of the trachea, connected by a narrow anterior *isthmus* (fig. 17.10*a*).

The main histological feature of the thyroid is a large number of sacs called **thyroid follicles** (fig. 17.10*b*). Each is filled with a protein-rich colloid and lined by a simple cuboidal epithelium of **follicular cells.** These cells secrete two main thyroid hormones—**thyroxine,** also known as **T_4,** or **tetraiodothyronine** (TET-ra-EYE-oh-doe-THY-ro-neen), and **T_3,** or **triiodothyronine** (try-EYE-oh-doe-THY-ro-neen).

Target cells convert T_4 to T_3 (as described later), and only T_3 has a significant metabolic effect. It increases the concentration and activity of mitochondrial enzymes that make ATP and it stimulates the activity of Na^+–K^+ pumps. These effects increase the oxygen consumption and heat production of a cell, accounting for the **calorigenic**[13] **effect** of thyroxine. Cold weather stimulates thyroxine secretion and warm weather suppresses it. Consequently, the body consumes more calories and generates more heat in cold weather. In pregnancy, the placenta produces a TSH-like hormone that promotes fetal growth and development. Thyroid hormones are especially important in the development of the nervous system in the fetus and child.

13. *calor* = heat + *genic* = producing

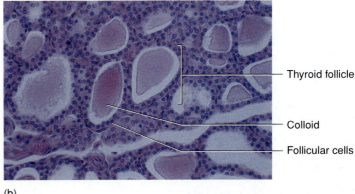

(a)

(b)

Figure 17.10 The thyroid gland. (*a*) Gross anatomy. (*b*) Histology (×400). ⚚

Between the thyroid follicles are clusters of less numerous **C cells (parafollicular cells).** As discussed more fully in chapter 8, hypercalcemia stimulates C cells to secrete a hormone called **calcitonin.** Calcitonin lowers blood calcium concentration by antagonizing the action of parathyroid hormone (described shortly) and promotes calcium deposition and bone formation by stimulating osteoblast activity. The effects of calcitonin are significant in children but relatively unimportant in adults.

Thyroid Disorders

Thyroid hyposecretion present from birth is called **congenital hypothyroidism** (formerly *cretinism*) (fig. 17.11*a*). Hypothyroidal infants have a normal appearance at birth because they have received an adequate supply of thyroxine from their mothers. Removed from this source, however, the infants begin to show stunting and other abnormalities of bone development, thickened facial features, low body temperature, lethargy, and mental retardation. Adult hypothyroidism causes **myxedema** (MIX-eh-DEE-muh), a syndrome characterized by low metabolic rate, sluggishness and sleepiness, weight gain, constipation, dry skin and hair, abnormal sensitivity to cold, elevated blood pressure, and swelling of the tissues. Both infant and adult hypothyroidism are treatable with hormone replacement therapy.

A **goiter** is any pathological enlargement of the thyroid gland. **Endemic goiter** (fig. 17.11b) is due to dietary iodine deficiency, which results in insufficient thyroxine secretion. Lacking feedback from the thyroid, the pituitary secretes extra TSH, which stimulates hypertrophy of the thyroid gland. There is little iodine in soil or most foods, but seafood and iodized salt are good sources.

Thyroid hypertrophy also occurs in **toxic goiter (Graves[14] disease).** This is an autoimmune disease in which abnormal antibodies mimic the stimulatory effect of TSH on the thyroid. This causes an elevated secretion of thyroid hormones (in contrast to endemic goiter), and all of the body's tissues suffer from excessive thyroxine stimulation. A person with toxic goiter exhibits elevated metabolic rate and heart rate, nervousness, sleeplessness, weight loss, abnormal sweating and heat sensitivity, and *exophthalmos* (EX-off-THAL-mos)—bulging of the eyes because of edema in the orbit (fig. 17.11c).

The Parathyroids

The tiny **parathyroid glands** are partially embedded in the posterior surface of the thyroid (fig. 17.12). There are usually four—a superior pair and an inferior pair. Each gland is about 3 to 8 mm long and 2 to 5 mm wide. Hypocalcemia stimulates the **chief cells** of the parathyroids to secrete **parathyroid hormone (PTH).** PTH raises blood calcium levels by promoting intestinal calcium absorption, inhibiting urinary calcium excretion, and stimulating osteoclasts to resorb bone. PTH and calcium metabolism are discussed in detail in chapter 8.

Because of their location and small size, the parathyroids are sometimes accidentally removed during thyroid surgery. Without hormone replacement therapy, this can cause a rapid decline in blood calcium levels and fatal tetany within 3 or 4 days. **Hyperparathyroidism**—excess PTH secretion—is usually caused by a parathyroid tumor. It causes the bones to become soft, deformed, and fragile; it elevates plasma levels of calcium and phosphate ions; and it promotes

14. Robert James Graves (1796–1853), Irish physician

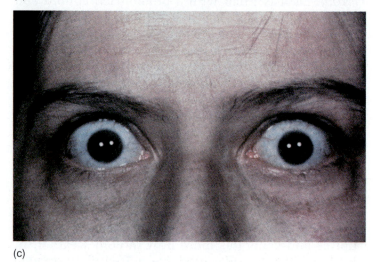

(a)

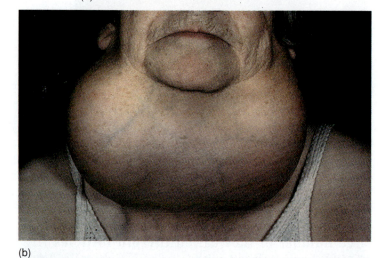

(b)

(c)

Figure 17.11 Effects of thyroid hormone imbalances. (*a*) Congenital hypothyroidism, characterized by lethargy, thickened facial features, a protruding tongue, and stunted growth. (*b*) Endemic goiter, a hypertrophy of the thyroid gland resulting from iodine deficiency. (*c*) Exophthalmia, the protuberant eyes and startled-seeming expression resulting from thyroid hypersecretion.

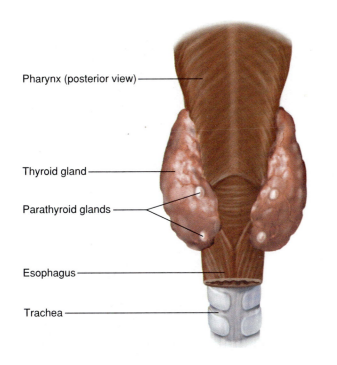

Pharynx (posterior view)

Thyroid gland

Parathyroid glands

Esophagus

Trachea

Figure 17.12 Posterior view of the pharynx and thyroid showing the location of the four parathyroid glands. ✗

the formation of renal calculi (kidney stones), composed of calcium phosphate.

The Adrenals

The **adrenal (suprarenal) glands** sit like caps on the superior poles of the kidneys (fig. 17.13). Like the kidneys, they are retroperitoneal—located between the peritoneum and the posterior body wall. In adults, the adrenal is about 5 cm (2 in.) long, 3 cm (1.2 in.) wide, and weighs about 4 g; it weighs about twice this much at birth. Like the pituitary, the adrenal gland is formed by the merger of two fetal glands with different origins and functions. Its inner core, the *adrenal medulla,* is a small portion of the total gland. Surrounding it is a much thicker *adrenal cortex.*

The Adrenal Medulla

The **adrenal medulla** was discussed as part of the sympathoadrenal system in chapter 15 and is reviewed only briefly here. It arises from the neural crest and is not fully formed until the age of three. It is actually a sympathetic ganglion consisting of modified neurons devoid of dendrites and axons. These cells are richly innervated by preganglionic fibers and respond to sympathetic stimulation by secreting catecholamines, especially epinephrine and norepinephrine. These hormones mimic the effects of the sympathetic nervous system, but their effects last much longer because they circulate in the blood.

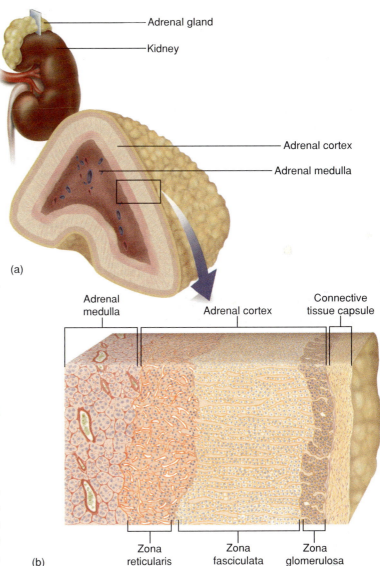

Adrenal gland

Kidney

Adrenal cortex

Adrenal medulla

(a)

Adrenal medulla

Adrenal cortex

Connective tissue capsule

(b)

Zona reticularis

Zona fasciculata

Zona glomerulosa

Figure 17.13 The adrenal gland. (*a*) Gross anatomy. (*b*) Histology (×400). ✗

The Adrenal Cortex

The **adrenal cortex** consists of three tissue layers (fig. 17.13*b*)—an outer **zona glomerulosa** (glo-MER-you-LO-suh), a thick middle **zona fasciculata** (fah-SIC-you-LAH-ta), and an inner **zona reticularis.** The adrenal cortex synthesizes more than 25 steroid hormones known collectively as the **corticosteroids,** or **corticoids.** They fall into three categories:

1. **Sex steroids.** These are weak androgens and smaller amounts of estrogens secreted by the zona reticularis. **Androgens** are hormones that control many aspects of male development and reproductive physiology. The most potent androgen is **testosterone,** secreted by the testes, but the main androgen secreted by the adrenal cortex is

dehydroepiandrosterone (DHEA). Although DHEA is much weaker than testosterone, tremendous amounts of it are produced by the large fetal adrenal glands and enter the mother's circulation. Androgens stimulate the libido (sex drive) and the growth of pubic and axillary hair in both sexes. In males, these effects result mainly from testosterone from the testes, but in females, they are due solely to DHEA from the adrenal cortex.

2. **Mineralocorticoids.** These steroids, secreted by the zona glomerulosa, control electrolyte balance by acting on the kidneys. The principal mineralocorticoid is **aldosterone,** secreted in response to low Na^+ and high K^+ concentrations in the blood and in response to another hormone called **angiotensin II.** Aldosterone acts on the kidneys to promote Na^+ retention and K^+ excretion. It is discussed more fully in chapter 24.

3. **Glucocorticoids.** These steroids are secreted mainly by the zona fasciculata in response to ACTH. They stimulate fat and protein catabolism, *gluconeogenesis* (the synthesis of glucose from amino acids), and the release of fatty acids and glucose into the blood. They help the body adapt to stress and to repair damaged tissues. Nearly all glucocorticoid effects are caused by one member of the family, **cortisol (hydrocortisone); corticosterone** is a less potent glucocorticoid. Recently it has been discovered that some cells of the adrenal medulla extend into the cortex. When stress activates the sympathoadrenal system, catecholamines from the medulla stimulate cells of the cortex to secrete corticosterone. Thus, it is no mere coincidence that the cortex and medulla are combined into a single gland; they have an intimate physiological interaction.

Adrenal Disorders

Hypersecretion by the adrenal medulla may result from a tumor called a **pheochromocytoma** (FEE-oh-CRO-mo-sy-TOE-muh). The catecholamines excess mimics overactivity of the sympathetic nervous system, causing hypertension, elevated metabolic rate, hyperglycemia (excess glucose in the blood), glycosuria (glucose in the urine), nervousness, indigestion, and sweating. Ultimately it can lead to fatigue, exhaustion, and high susceptibility to other diseases.

Hypersecretion by the adrenal cortex can result from a cortical tumor or ACTH hypersecretion. It causes **Cushing**[15] **syndrome,** characterized by disturbances of carbohydrate and protein metabolism with hyper-

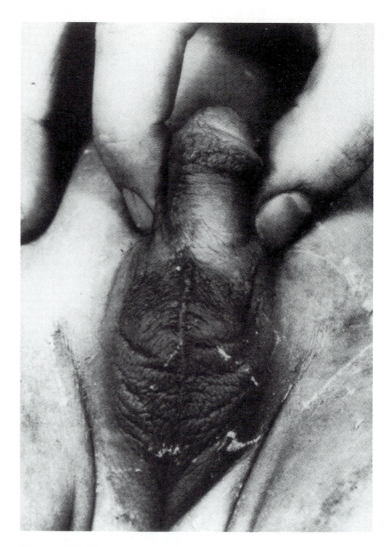

Figure 17.14 Genitals of a female (XX) baby with adrenogenital syndrome (AGS), masculinized by prenatal hypersecretion of adrenal androgens. Such infants are easily mistaken for boys and raised as such. Note the fusion of the labia majora to resemble a scrotum and enlargement of the clitoris to resemble a penis.

glycemia, hypertension, muscular weakness, and edema. Muscle and bone mass are lost as protein is rapidly catabolized. Some people with Cushing syndrome exhibit "moon face" or "buffalo hump" because of fat deposition in the face or between the shoulders, respectively. Long-term corticosteroid therapy has similar effects.

Adrenogenital syndrome (AGS), caused by hypersecretion of adrenal androgens, commonly accompanies Cushing syndrome. In children, it often involves enlargement of the penis or clitoris and the premature onset of puberty. Newborn girls with AGS can easily be misidentified as boys (see chapter 28). In women, AGS produces such masculinizing effects as increased body hair, deepening of the voice, and beard growth (fig. 17.14).

15. Harvey Cushing (1869–1939), American physician

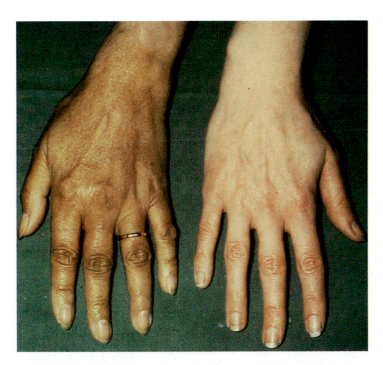

Figure 17.15 Darkening or "bronzing" of the skin in Addison disease (*left*) compared to a normal hand.

Pancreatic Hormones

The pancreatic islets secrete at least five hormones or hormonelike messengers, the two most important of which are *insulin* and *glucagon.*

1. **Insulin** is secreted by **beta (β) cells** in response to rising blood concentrations of critical nutrients, especially glucose and amino acids. It stimulates many cells to absorb these nutrients from the blood and to either store or metabolize them. Glucose uptake by liver and brain cells, however, is independent of insulin. Insulin promotes the synthesis of glycogen, fat, and protein and antagonizes the effects of glucagon discussed next.
2. **Glucagon** is secreted by **alpha (α) cells** when blood glucose levels fall. In the liver, it stimulates glycogenolysis (glycogen hydrolysis), gluconeogenesis, and the release of glucose into circulation. In adipose tissue, it stimulates fat catabolism and the release of free fatty acids and ketone bodies. Glucagon is also secreted in response to rising amino acid levels in the blood after a high-protein meal. Amino acids are the raw material for gluconeogenesis.

Somatostatin is another important pancreatic secretion, chemically identical to hypothalamic GHIH. Pancreatic somatostatin is not secreted into the blood; it does not function here as a hormone but as a *paracrine* secretion—a chemical messenger that diffuses to neighboring target cells rather than traveling in the blood. Paracrines are discussed later in this chapter in the section on eicosanoids. Somatostatin is produced by **delta (δ) cells** of the islets when blood glucose and amino acids rise after a meal. It diffuses to α and β cells and modulates their secretion of glucagon and insulin. Somatostatin is also secreted by the brain, spinal cord, anterior pituitary, digestive tract, gonads, and other organs. It plays a variety of neurotransmitter, hormonal, and paracrine roles.

You may have noticed that glucagon is not the only hormone that raises blood glucose levels; so do growth hormone, epinephrine, norepinephrine, cortisol, and corticosterone. Any hormone that does this is called a *hyperglycemic hormone.* Insulin is called a *hypoglycemic hormone* because it lowers blood glucose levels.

Hyposecretion of glucocorticoids and mineralocorticoids is called **Addison[16] disease.** Its symptoms include hypoglycemia, Na+ and K+ imbalance, dehydration, hypotension (low blood pressure), weight loss, weakness, and a loss of stress resistance. The lack of negative feedback inhibition from the adrenal cortex causes the pituitary to secrete excessive amounts of ACTH. This stimulates melanin synthesis and causes the skin to darken (fig. 17.15). Unless treated with corticosteroids, Addison disease can cause fatal dehydration and electrolyte imbalances.

Think About It

Which could a person more easily live without—the adrenal medulla or adrenal cortex? Why?

The Pancreas

The elongated spongy **pancreas** is located retroperitoneally, below and behind the stomach (fig. 17.16). It is approximately 15 cm long and 2.5 cm thick. Most of it is an exocrine digestive gland, but scattered among the exocrine acini are endocrine cell clusters called the **pancreatic islets (islets of Langerhans[17]).**

Diabetes Mellitus

Diabetes mellitus[18] (DM) is any disease syndrome resulting from the hyposecretion or inaction of insulin. Its three classic signs are **polyuria[19]** (excessive urine

---

16. Thomas Addison (1793–1860), English physician
17. Paul Langerhans (1847–88), German anatomist

18. *diabet* = to flow through + *melli* = honey
19. *poly* = much, excessive + *uri* = urine

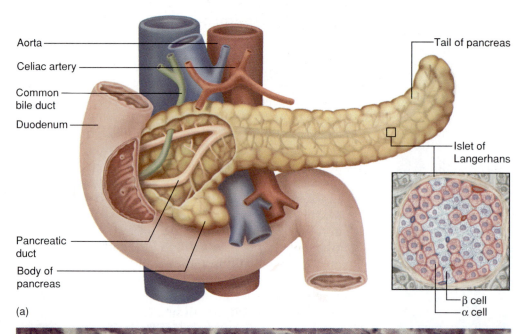

- Aorta
- Celiac artery
- Common bile duct
- Duodenum
- Pancreatic duct
- Body of pancreas

(a)

- Tail of pancreas
- Islet of Langerhans
- β cell
- α cell

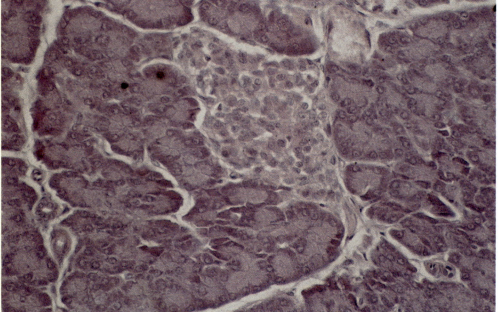

(b)

Figure 17.16 The pancreas. (*a*) Gross anatomy; (*b*) a pancreatic islet amid the exocrine acini of the pancreas (×400).

To understand why glycosuria and polyuria occur, a little knowledge of kidney physiology is required. The kidneys filter blood plasma and process the filtrate to produce urine. Normally, the kidney tubules remove all glucose from the filtrate by active transport and return it to the blood. Thus there is little or no glucose in the urine of a healthy person. By osmosis, the tubules also reclaim most of the water in the filtrate.

Like any other active transport systems, the glucose transporters of the kidney tubules have a transport maximum (T_m) (see chapter 4). In diabetes mellitus, there is so much glucose in the filtrate that it exceeds the T_m. Excess glucose passes through the tubules into the urine before it can be reabsorbed. Glucose in the kidney tubules also raises the osmolarity of the filtrate and causes **osmotic diuresis**—glucose opposes the osmotic reabsorption of water, and large amounts of water are therefore passed in the urine, hence the polyuria, dehydration, and thirst characteristic of diabetes.

There are two forms of diabetes mellitus—*type I* and *type II*. Type I, also called **insulin-dependent diabetes mellitus (IDDM),** results from the destruction of β cells by an autoimmune response. In individuals genetically susceptible to IDDM, the immune system fails to distinguish the β cells from foreign matter and destroys them. Up to a point, the body can tolerate this without symptoms, but when 90% of the β cells are destroyed, insulin secretion falls to critically low levels. The signs of IDDM then appear suddenly, typically before the age of 15; IDDM used to be called *juvenile-onset DM.*

When cells cannot absorb glucose, they fall back on protein and fat for energy. Protein catabolism results in a loss of muscle mass, emaciation, and weakness. Rapid fat catabolism elevates blood concentrations of free fatty acids and their breakdown products, the

output), **polydipsia**[20] (intense thirst); and **polyphagia**[21] (ravenous hunger); but DM is confirmed by clinical tests. Urinalysis typically reveals **glycosuria**[22] and **ketonuria**— the abnormal presence of glucose and ketones, respectively, in the urine. A blood test reveals **hyperglycemia.** Blood glucose is elevated because, without insulin activity, most cells do not absorb it from the blood.

20. *dipsia* = drinking
21. *phagia* = eating
22. *glyco* = glucose, sugar + *uria* = urine condition

ketones (ketone bodies)—acetoacetic acid, acetone, and β-hydroxybutyric acid. Ketones in the urine (ketonuria) contribute to osmotic diuresis. Ketones also carry Na+ and K+ out of the body, creating electrolyte deficiencies that can lead to abdominal pain, vomiting, irregular heartbeat, and nervous system disorders. As acids, ketones also lower the pH of the blood—a condition called **ketoacidosis.** This depresses the nervous system and can result in coma and death.

IDDM leads to long-term degenerative cardiovascular and neurological diseases. The high cholesterol and fat levels in the blood lead to atherosclerosis—fatty obstruction of the blood vessels that can lead, in turn, to heart attack, stroke, blindness, and the leading cause of death from IDDM, kidney failure. Because of poor circulation, especially in the lower extremities, the body begins to lose its ability to fight bacterial infections and becomes increasingly vulnerable to gangrene. This often requires the amputation of toes, feet, or legs. The neurological complications include impotence, incontinence, and loss of sensation.

IDDM is treated with periodic insulin injections or with the continual delivery of insulin by a pump worn externally. One area of active diabetes research aims to transplant pancreatic islets into diabetic patients.

Type II, also called **non-insulin-dependent diabetes mellitus (NIDDM),** accounts for 90% of cases and usually exhibits normal or even elevated insulin levels. The problem in NIDDM is *insulin resistance*—a failure of target cells to respond to insulin because they lack the receptors for it. The three major risk factors for NIDDM are heredity, age, and obesity. It tends to run in families; if an identical twin contracts NIDDM it is virtually certain that the other twin will also contract it. NIDDM has a gradual onset, with symptoms usually not appearing until age 40 or beyond; it used to be called *mature-onset DM.* Type II diabetics are usually overweight. In obesity, the adipocytes produce a hormone-like secretion that indirectly interferes with glucose transport into most kinds of cells. For reasons not yet known, ketoacidosis is usually absent from NIDDM, and the cardiovascular and neurological effects are less severe. NIDDM can often be managed through a weight-loss program of diet and exercise.

Hyperinsulinism

Hyperinsulinism usually occurs when a diabetic injects too much insulin, but it occasionally results from a tumor of the pancreatic islets. Hyperinsulinism causes such rapid glucose uptake that a person becomes hypoglycemic, weak, and hungry. Hypoglycemia, in turn, triggers the secretion of corrective hyperglycemic hormones—epinephrine, glucagon, and growth hormone—which have side effects including anxiety, sweating,

and elevated heart rate. Uncorrected hyperinsulinism can produce **insulin shock**—a condition in which the brain is so deprived of glucose that the individual exhibits disorientation, convulsions, or unconsciousness. It can be fatal if the blood glucose level is not quickly elevated. If the person is conscious, a spoonful of sugar mixed into a glass of fruit juice can be given to restore blood sugar levels.

The Gonads

Like the pancreas, the **gonads** are both endocrine and exocrine glands (fig. 17.17). Their exocrine product is eggs and sperm, and their endocrine product is the **gonadal hormones,** most of which are steroids.

Each follicle of the ovary contains an egg cell surrounded by a wall of **granulosa cells** (fig. 17.17a). The granulosa cells produce an estrogen called **estradiol** in the first half of the menstrual cycle. After ovulation, the corpus luteum secretes **progesterone** for the next 12 days or so, or for several weeks in the event of pregnancy. The functions of estradiol and progesterone are discussed in chapter 28. In brief, they contribute to the development of the reproductive system and feminine physique, they regulate the menstrual cycle, they sustain pregnancy, and they prepare the mammary glands for lactation. Another ovarian hormone, **inhibin,** secreted by the follicle and corpus luteum, suppresses FSH secretion by means of negative feedback inhibition of the anterior pituitary.

The testis consists mainly of microscopic tubules that produce sperm. Nestled between these are clusters of **interstitial cells (cells of Leydig[23])** (fig. 17.17b), which produce testosterone and lesser amounts of weaker androgens and estradiol. Testosterone stimulates development of the male reproductive system in the fetus and adolescent, the development of the masculine physique in adolescence, and the sex drive. It sustains sperm production and the sexual instinct throughout adult life. **Sustentacular (Sertoli[24]) cells** of the testis secrete inhibin. In the male, its control over FSH secretion serves to homeostatically stabilize the rate of sperm production.

Endocrine Cells in Other Organs

Several other organs have hormone-secreting cells:

- The *heart* is stretched by high blood pressure, and this stimulates muscle cells in the atria to secrete **atrial natriuretic[25] factor (ANF).** ANF increases

---

23. Franz von Leydig (1821–1908), German histologist
24. Enrico Sertoli (1824–1910), Italian histologist
25. *natri* = sodium + *uretic* = pertaining to urine

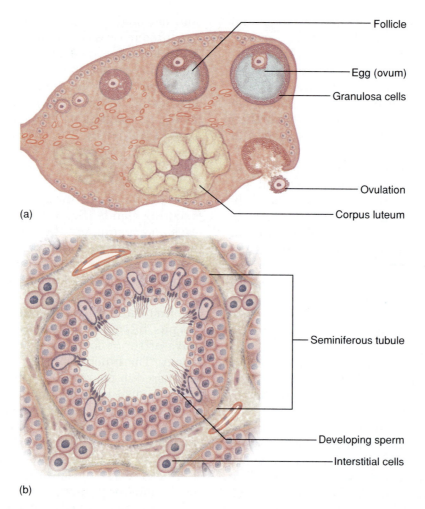

Follicle

Egg (ovum)

Granulosa cells

Ovulation

Corpus luteum

(a)

Seminiferous tubule

Developing sperm

Interstitial cells

(b)

Figure 17.17 (*a*) Histology of the ovary; (*b*) histology of the testis. The granulosa cells of the ovary secrete estrogens and progesterone. The interstitial cells of the testis secrete testosterone. ✗

urine output and sodium excretion and interferes with the actions of angiotensin II, described shortly. Together, these effects lower the blood pressure.

- The *kidneys* produce two hormones—calcitriol and erythropoietin. **Calcitriol** affects the handling of calcium by the kidneys, small intestines, and bones (see chapter 8). **Erythropoietin** (eh-RITH-ro-POY-eh-tin) stimulates the bone marrow to produce red blood cells (see chapter 18).

- The *liver* secretes somatomedins, which mediate the action of somatotropin; about 15% of the body's erythropoietin; and a protein called *angiotensinogen,* a prohormone (hormone precursor) that is converted in the blood to angiotensin II. Angiotensin II stimulates vasoconstriction and aldosterone secretion by the adrenal cortex. Together, these effects raise blood pressure (see chapter 20).

- The *stomach* and *small intestine* have various *enteroendocrine cells,* which secrete at least 10 different **enteric hormones.** These are discussed in chapter 25. In general, they coordinate the different regions and glands of the digestive system with each other.

- The *placenta* performs many functions in pregnancy, including the secretion of estrogen, progesterone, and other hormones that regulate pregnancy and stimulate development of the fetus and the mother's mammary glands. These are discussed in chapter 28.

You can see that the endocrine system is extensive. It includes numerous discrete glands as well as individual cells in the tissues of other organs. The endocrine organs and tissues other than the hypothalamus and pituitary are reviewed in table 17.4.

Key Point Review

7 Name two endocrine glands that are larger in children than in adults.

8 What hormone increases the body's heat production in cold weather?

9 Name a glucocorticoid, a mineralocorticoid, and a catecholamine secreted by the adrenal gland.

10 Does the action of glucocorticoids more closely resemble that of glucagon or insulin? Explain.

11 What is the difference between a gonadal hormone and a gonadotropin?

Pineal Gland

Hormones:	Melatonin and serotonin
Target organ or cells:	Brain
Effect:	May regulate timing of puberty; influences mood

Thymus

Hormones:	Thymopoietin and thymosins
Target organ or cells:	T lymphocytes
Effect:	Promote T lymphocyte development and activation

Thyroid

Hormones:	Triiodothyronine (T_3) and thyroxine (T_4)
Target organ or cells:	Most tissues
Effect:	Stimulate Na^+-K^+ pumps, metabolic rate, heat production, alertness, protein synthesis, and fetal and childhood growth and CNS development
Hormone:	Calcitonin
Target organ or cells:	Osteoblasts of bone
Effect:	Stimulates calcium deposition and ossification; reduces blood calcium concentration

Parathyroids

Hormone:	Parathyroid hormone (PTH)
Target organ or cells:	Small intestine, kidneys, and osteoclasts of bone
Effect:	Stimulates calcium absorption and retention; promotes bone resorption; elevates blood calcium concentration

Adrenal Medulla

Hormones:	Epinephrine, norepinephrine, and dopamine
Target organ or cells:	Most tissues
Effect:	Complement action of sympathetic nervous system

Adrenal Cortex

Hormone:	Aldosterone (mineralocorticoid)
Target organ or cells:	Kidney
Effect:	Promotes Na^+ retention and K^+ excretion; maintains blood pressure and volume
Hormones:	Cortisol and corticosterone (glucocorticoids)
Target organ or cells:	Most tissues
Effect:	Promote fat and protein catabolism, gluconeogenesis, stress resistance, and tissue repair; inhibit inflammation
Hormone:	Androgen (DHEA) and negligible amounts of other sex steroids
Target organ or cells:	Bone, muscle, integument, many other organs
Effect:	Stimulate growth of pubic and axillary hair in both sexes; stimulate libido; negligible effects in males compared to testosterone from testes

Pancreatic Islets

Hormone:	Glucagon
Target organ or cells:	Primarily liver
Effect:	Stimulates glycogen and fat hydrolysis, mobilization of glucose and fatty acids, and gluconeogenesis
Hormone:	Insulin
Target organ or cells:	Most tissues except brain and liver
Effect:	Promotes glucose and amino acid uptake and synthesis of glycogen, fat, and protein

Ovaries

Hormone:	Estradiol
Target organ or cells:	Ovaries, uterus, mammary glands, brain, many other tissues
Effect:	Regulates egg production; stimulates adolescent growth; promotes development of female secondary sex characteristics; prepares mammary glands for lactation; prepares uterus for pregnancy
Hormone:	Progesterone
Target organ or cells:	Ovaries, uterus, mammary glands, many other tissues
Effect:	Stimulates mammary development in puberty and pregnancy; prepares uterus for pregnancy
Hormone:	Inhibin
Target organ or cells:	Anterior pituitary
Effect:	Suppresses FSH secretion

Testes

Hormone:	Testosterone
Target organ or cells:	Most tissues
Effect:	Regulates sperm production; promotes development of male reproductive system and physique; stimulates adolescent growth, libido, and sexual behavior
Hormone:	Inhibin
Target organ or cells:	Anterior pituitary
Effect:	Suppresses FSH secretion

Heart

Hormone:	Atrial natriuretic factor (ANF)
Target organ or cells:	Kidneys
Effect:	Increases Na^+ excretion and urine output; lowers blood pressure

Kidneys

Hormone:	Calcitriol (vitamin D)
Target organ or cells:	Small intestine, kidneys, and bone
Effect:	Aids in absorption of dietary calcium and phosphorus; promotes ossification
Hormone:	Erythropoietin
Target organ or cells:	Red bone marrow
Effect:	Increases production of red blood cells

Liver

Hormone:	Angiotensinogen (a prohormone)
Target organ or cells:	Blood vessels and adrenal cortex (after conversion to angiotensin II)
Effect:	Angiotensin II acts as a vasoconstrictor and stimulates aldosterone secretion
Hormone:	Erythropoietin
Target organ or cells:	Red bone marrow
Effect:	Increases production of red blood cells

Stomach and Small Intestine

Hormones:	Enteric hormones
Target organ or cells:	Digestive tract and its accessory glands
Effect:	Coordinate secretion and motility in digestion

Placenta

Hormones:	Estrogen, progesterone, and others
Target organ or cells:	Maternal and fetal tissues
Effect:	Promote fetal growth; regulate pregnancy; prepare mammary glands for lactation

In this section, we go from the organ level to the cellular and molecular levels of endocrinology.

The Chemical Identity of Hormones

Hormones fall into three major chemical classes—*steroids, biogenic amines,* and *peptides* (table 17.5):

- **Steroid hormones** are derived from cholesterol. They include the sex steroids produced by the gonads (estrogens, progesterone, and androgens) and the corticosteroids (mainly cortisol, corticosterone, and aldosterone) produced by the adrenal cortex.
- **Biogenic amines** are synthesized from amino acids and retain an amino group. They include epinephrine, norepinephrine, prolactin-inhibiting factor (= dopamine), and the thyroid hormones. All of the hormones in this category are synthesized from the amino acid tyrosine.
- **Peptide hormones** are chains of 3 to over 200 amino acids (fig. 17.18). Some of them are glycoproteins—polypeptides conjugated with short carbohydrate chains. All glycoprotein hormones have an identical α chain of 92 amino acids and a variable β chain, which distinguishes the hormones from each other. All releasing and inhibiting hormones produced by the hypothalamus are polypeptides, and most hormones of the anterior pituitary are either unconjugated polypeptides or glycoproteins.

Think About It

Some hormones can be taken orally (such as birth-control pills and thyroxine), but insulin must be injected. Why would insulin be ineffective as an oral medication?

Synthesis and Transport

The classes of hormones differ substantially in how they are synthesized and transported in the blood.

Table 17.5	Chemical Classification of Hormones
Steroids	
Aldosterone	
Cortisol	
Cortisone	
Estrogens	
Progesterone	
Testosterone	
Biogenic Amines	
Dopamine	
Epinephrine	
Melatonin	
Norepinephrine	
Serotonin	
Thyroxine (T_4)	
Triiodothyronine (T_3)	
Oligopeptides (3–10 Amino Acids)	
Angiotensin II	
Antidiuretic hormone	
Gonadotropin-releasing hormone	
Oxytocin	
Thyrotropin-releasing hormone	
Polypeptides (14–199 Amino Acids)	
Adrenocorticotropic hormone	
Atrial natriuretic factor	
Corticotropin-releasing hormone	
Glucagon	
Growth hormone	
Growth hormone–inhibiting hormone	
Growth hormone–releasing hormone	
Insulin	
Parathyroid hormone	
Prolactin	
Glycoproteins (92 Amino Acids in the α Chain, 112–118 Amino Acids in the β Chain)	
Follicle-stimulating hormone	
Human chorionic gonadotropin	
Inhibin	
Luteinizing hormone	
Thyroid-stimulating hormone	

Steroids

Steroid hormones are derived from cholesterol and differ only in the functional groups attached to the four-ring steroid backbone (fig. 17.19). Steroids are hydrophobic and thus dissolve poorly in blood plasma.

To be transported, they must bind to hydrophilic **transport proteins**—albumin and globulins synthesized in the liver. A hormone attached to a transport protein is called **bound** hormone, and one that is not attached is called **unbound (free)** hormone. Only the unbound form can be taken up by target cells and have a physiological effect. In the absence of target cell uptake, there is an equilibrium between bound and unbound forms:

$$\text{Bound hormone} \leftrightharpoons \text{Unbound hormone}$$

Hormone uptake shifts the equilibrium toward the right; thus, more hormone dissociates from the transport protein to become available to the target cells:

$$\begin{array}{ccccc} \text{Bound} & \leftrightharpoons & \text{Unbound} & \rightarrow & \text{Target cell} \\ \text{hormone} & & \text{hormone} & & \text{uptake} \end{array}$$

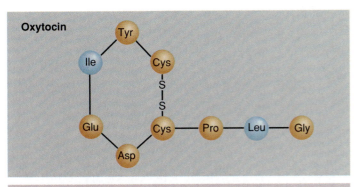

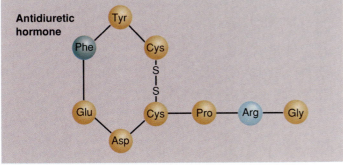

Figure 17.18 The two hormones released by the posterior pituitary, oxytocin and antidiuretic hormone, differ by only two amino acids.

Not only do transport proteins enable steroids to travel in the bloodstream, they also prolong their **half-life**—the time it takes for half the amount of a circulating hormone to be deactivated or excreted from the body. Small molecules are quickly filtered from the blood by the kidneys and may then be excreted in the urine. A hormone attached to a large protein, however, resists filtration and stays in the blood longer. This provides a hormone reservoir from which the target cells can draw as needed.

Thyroid Hormones

Thyroid hormone has a unique mode of synthesis that involves iodine (fig. 17.20). Ingested iodine (I) is reduced to iodide ion (I^-) in the stomach and enters the circulation as I^-. The follicular cells of the thyroid gland have **iodide pumps** that capture I^- from the tissue fluid, oxidize it back to iodine, and concentrate it in the colloid (see special topic 17.2). The colloid consists largely of a protein called **thyroglobulin.** One of the largest known proteins, thyroglobulin contains 115 tyrosine residues to which iodine can be added. When one iodine atom binds to tyrosine, it forms **monoiodotyrosine** (MON-oh-eye-OH-doe-TIE-ro-seen) **(MIT).** When a second iodine atom is added, MIT becomes **diiodotyrosine (DIT).** Next, the modified tyrosine residues bond with each other to form the two thyroid hormones. The union of MIT and DIT forms triiodothyronine (T_3), and two DITs form tetraiodothyronine (T_4), also known as thyroxine. At this point, T_3 and T_4 are still part of thyroglobulin, a storage form that normally provides several weeks' supply of thyroid hormone.

Upon stimulation by TSH, the follicular cells grow microvilli and take up colloid droplets by pinocytosis. A pinocytotic vesicle fuses with a lysosome, thyroglobulin is hydrolyzed, and T_4 and a much smaller amount of T_3 are released into the blood. These hormones are hydrophobic and must be transported by carrier proteins, the most important of which is **thyroxine-binding globulin (TBG).** About 99.98% of the thyroxine is bound to carriers, and the other 0.02% is unbound.

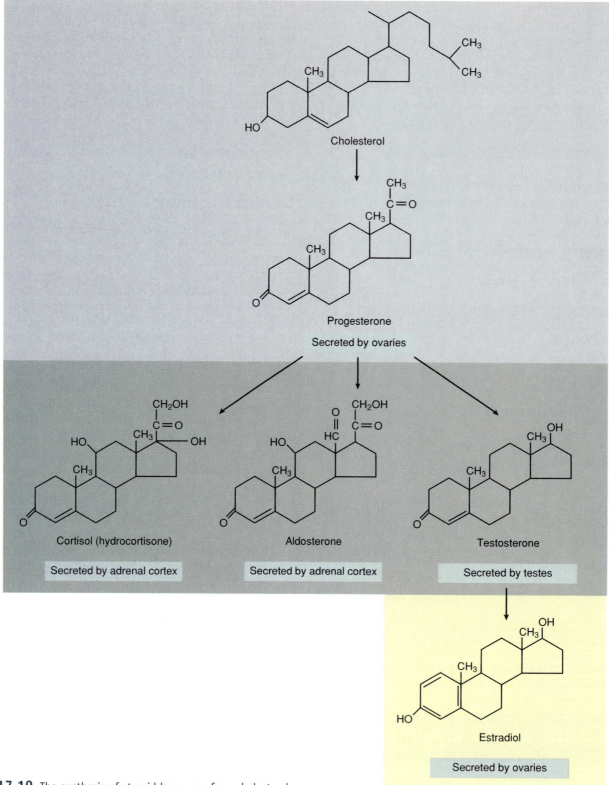

Figure 17.19 The synthesis of steroid hormones from cholesterol.

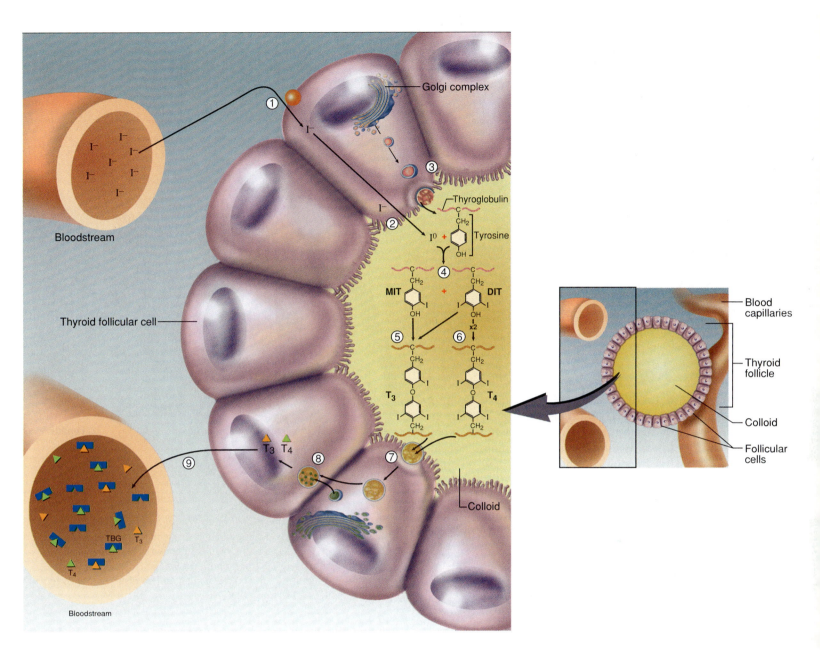

Figure 17.20 Synthesis of the thyroid hormones. (*1*) A follicular cell transports I^- into its cytoplasm. (*2*) An enzyme on the apical surface of the cell converts I^- to neutral iodine atoms (I^0) as they pass into the lumen of the thyroid follicle. (*3*) The Golgi complex packages thyroglobulin into secretory vesicles, which release it into the lumen. (*4*) Iodine binds to the tyrosine residues of thyroglobulin to form MIT and DIT. (*5*) MIT and DIT become linked to form T_3. (*6*) Two DITs become linked to form T_4. At this point the thyroid hormones are still bound to the rest of the thyroglobulin molecule. (*7*) Droplets of colloid are taken into the follicular cells by pinocytosis. (*8*) Pinocytotic vesicles fuse with lysosomes, which contribute an enzyme that hydrolyzes thyroglobulin. This releases T_3 and T_4 from the parent protein. (*9*) T_3 and T_4 are secreted into the bloodstream, where most of the T_4 and some of the T_3 bind to thyroxine-binding globulin (TBG). Inset shows relationship to entire thyroid follicle.

Since bound thyroxine serves as a blood reservoir of the hormone, there is little sign of thyroxine deficiency for about 2 weeks even if the thyroid is surgically removed.

Only unbound T_3 and T_4 can enter target cells. Oddly enough, the thyroid secretes mainly T_4, which has little direct metabolic effect on the target cells. The target cells convert most of it to T_3, which stimulates them by a mechanism to be described shortly.

Peptides

Peptide hormones are often synthesized from longer hormonally inactive polypeptides. These **prohormones** are cut and spliced to make the active hormones. Prohormones, in turn, come from even larger precursors called **preprohormones.** The synthesis of insulin, for example, is through the sequence preproinsulin → proinsulin → insulin. The ribosomes of a pancreatic β cell assemble

preproinsulin, a large polypeptide with a leader sequence to guide it into the cisterna of the rough endoplasmic reticulum. The rough ER cleaves off the leader sequence, and the remaining polypeptide folds back on itself and forms three disulfide bridges. The product of this is **proinsulin,** which is stored in the Golgi complex. Immediately before insulin is secreted, enzymes in the Golgi complex remove a large middle segment of the proinsulin molecule called the **connecting peptide (C peptide).** The remaining 51 amino acid residues constitute the active hormone, insulin, composed of two peptide chains joined by disulfide bonds (fig. 17.21). Most peptides are transported in the plasma as free hormones, but some, such as GH, are bound to transport proteins.

Think About It

During the synthesis of glycoprotein hormones, where in the cell would the carbohydrate be added? (See chapter 5.)

Hormone Receptors

Hormones stimulate only those cells that have receptors for them. Nearly all cells have receptors for T_3, GH, and some other hormones, but a much narrower range of cells have receptors for aldosterone, the hypothalamic releasing and inhibiting hormones, and others. The receptors are protein or glycoprotein molecules and the hormones are their ligands. Receptor-hormone interactions are similar to enzyme-substrate interactions and show comparable specificity and saturation. Unlike enzymes, however, receptors do not chemically change their ligands. A target cell usually has a few thousand receptor proteins for a given hormone. They act like switches to turn certain metabolic pathways on or off as they bind hormone molecules. Hormone receptor defects lie at the heart of several endocrine diseases (see special topic 17.3).

When a hormone reaches a target cell, it binds to a receptor either on the plasma membrane or in its nucleus. The difference is determined by whether or not the hormone can pass through the plasma membrane. Peptides and catecholamines cannot do so and therefore bind to receptors on the cell surface. Steroids and thyroid hormones readily diffuse into the cell and bind to nuclear receptors, which are proteins of the chromatin (fig. 17.22).

Modes of Action

Hormones can act on a target cell in several ways: (1) they can directly activate genetic transcription and synthesis of new enzymes that affect cell physiology; (2) they can activate or deactivate enzymes that are already present; and (3) they can alter plasma membrane permeability and membrane potential. Steroids and

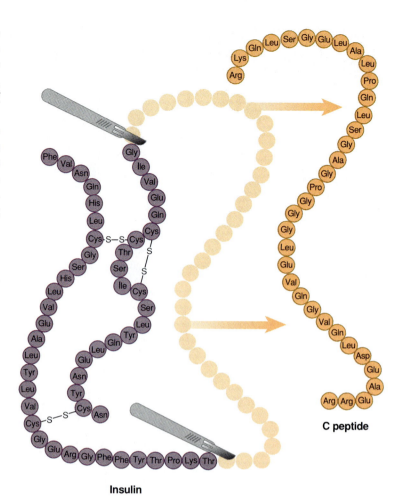

Insulin

Figure 17.21 Synthesis of insulin. Proinsulin is a polypeptide of 86 amino acids with three disulfide bonds. Immediately before secretion, the Golgi complex removes the middle 35 amino acids, called the connecting peptide (C peptide), shown in orange. Insulin (shown in violet) is a chain of 30 amino acids joined by two disulfide bonds to another chain of 21 amino acids.

thyroid hormones work by the first mechanism, whereas peptides and biogenic amines work primarily by the second and third mechanisms.

Think About It

Which hormone would produce the quickest effect on target cell metabolism—estradiol or insulin? Why?

The Action of Thyroid Hormones

The target cells for thyroid hormones absorb free T_3 and T_4 and convert most T_4 into T_3 in the cytoplasm. T_3 then diffuses into the nucleus and binds to receptors in specific regions of the chromatin (fig. 17.23). These receptors then activate the transcription of DNA (fig. 17.23). One of the gene products produced as a result of this is Na^+-K^+ ATPase—the sodium-potassium pump. This leads to an increased metabolic rate and greater heat production

In treating endocrine disorders, it is essential to understand the role of hormone receptors. For example, 90% of diabetes mellitus cases are caused by a defect or deficiency of insulin receptors. No amount of insulin replacement can correct this. And while genetically engineered bacteria can now produce human growth hormone for treating pituitary dwarfism, GH is of no use to children with *Laron dwarfism*. These children have normal amounts of GH but have a hereditary defect in the receptors. This is also the case in African pygmies. *Androgen insensitivity syndrome* is due to defective or deficient androgen receptors; it causes genetic males to develop feminine genitalia and other features (see special topic 27.1, p. 966). Some malignant tumors have estrogen receptors and depend on estrogen for their growth. For this reason, hormone replacement therapy should not be given to patients with estrogen-dependent cancer.

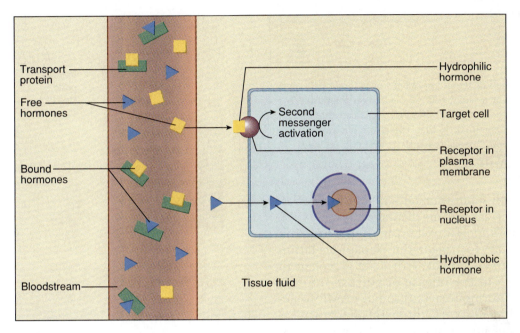

Figure 17.22 Upon arrival at a target cell, hydrophilic hormones (*yellow squares*) such as the peptides and catecholamines bind to receptors on the plasma membrane and activate second-messenger systems. Hydrophobic hormones (*blue triangles*) such as steroids and thyroid hormones diffuse into the cell and bind to receptors in the nucleus. Note that most hormone in the blood is bound to transport proteins, but only the unbound hormone can leave the blood and stimulate the target cell.

(see p. 131). Somatotropes of the anterior pituitary also respond to T_3 by synthesizing more growth hormone. The pituitary thus stimulates the thyroid with TSH, and the thyroid in turn stimulates the pituitary to synthesize GH.

The Action of Steroids

Steroid hormones and calcitriol (a steroid derivative) dissociate from their carrier proteins, diffuse through the phospholipid regions of the plasma membrane, and pass into the nucleus of a target cell. A steroid receptor has three functional regions that explain its action on the DNA: (1) a region that binds the hormone, (2) a region that binds to an **acceptor site** on the chromatin, and (3) a region that activates DNA transcription at that site. Transcription leads to the synthesis of proteins, including enzymes that alter the metabolism of the target cell. Progesterone, for example, acts on cells of the uterine lining to induce the synthesis of enzymes, which, in turn, synthesize glycogen. This prepares the uterus for the possibility of pregnancy.

The Action of Peptides and Biogenic Amines

Peptide hormones and biogenic amines other than thyroid hormones generally do not enter their target cells. They bind to receptor sites on the cell surface and exert their effects through second messengers. The best known second messenger is cyclic adenosine monophosphate (cAMP). Its role in neurotransmitter action was explained in chapter 13 (see fig. 13.19, p. 451), and it works in precisely the same way in hormone action. This is not surprising, since some of the same

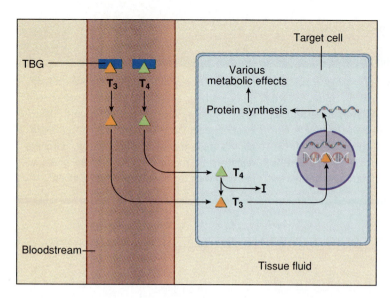

Figure 17.23 The action of thyroid hormones on a target cell. Most thyroid hormone arriving at a target cell is T_4; a smaller fraction is T_3. Within the cytoplasm of the target cell, however, T_4 is converted to T_3 by the removal of one iodine atom. T_3 then enters the nucleus and binds to DNA or to a receptor associated with it, where it activates genetic transcription.

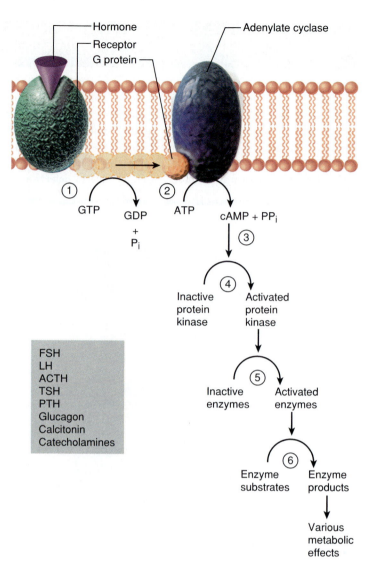

Figure 17.24 Cyclic AMP as a second messenger in hormone action. (*1*) The binding of a hormone to a membrane receptor activates a G protein. (*2*) The G protein activates adenylate cyclase. (*3*) Adenylate cyclase produces cAMP. (*4*) cAMP activates protein kinases. (*5*) Protein kinases phosphorylate enzymes and other proteins in the cytoplasm. Some enzymes are activated and others are deactivated by this phosphorylation. (*6*) Activated enzymes catalyze metabolic reactions with a wide range of possible effects on the cell, such as synthesis, secretion, and changes in plasma membrane potential. Some hormones acting through cAMP are listed in the box.

chemicals act as both neurotransmitters and hormones. Figure 17.24 summarizes this mechanism.

To take some specific examples, epinephrine works through cAMP to trigger the phosphorylation and activation of enzymes that hydrolyze glycogen in the liver. This leads to the release of glucose into the blood. Luteinizing hormone acts through cAMP to activate enzymes that convert cholesterol to progesterone and testosterone. Some other hormones that work through cAMP are listed in figure 17.24. GHIH works by *inhibiting* cAMP synthesis. ANF works through a similar second messenger, cyclic guanosine monophosphate (cGMP).

Calcium ions are involved in the action of several hormones. Most cells pump Ca^{2+} into the extracellular fluid (ECF) or smooth endoplasmic reticulum to maintain low concentrations in the cytosol. Calcium is 5,000 to 10,000 times more concentrated in the ECF than it is in the cytosol. Such a steep concentration gradient can be exploited for various purposes. Oxytocin (OT) mobilizes Ca^{2+} by the mechanism shown in figure 17.25. OT binds to its receptor, which activates a G protein in the plasma membrane. The G protein, in turn, activates a membrane enzyme, *phospholipase*. Phospholipase breaks down certain phospholipids of the plasma membrane into two molecules with second-messenger roles—a **diacylglycerol (diglyceride)** and **inositol triphosphate (IP₃).** Diacylglycerol activates protein kinases just as cAMP does, with effects as diverse as those noted for cAMP. IP₃ triggers the release of Ca^{2+} from the smooth ER or the entry of Ca^{2+} from the ECF. Calcium

ions can alter the membrane permeability and membrane potential of the cell; directly activate certain enzymes; or bind to a cytoplasmic receptor called **calmodulin,** which in turn activates protein kinases.

Some hormones employ different second-messenger systems in different target cells. ADH, for example, employs the IP₃-calcium system in smooth muscle but employs the cAMP system in kidney tubules. Insulin is unusual in comparison to other peptide hormones.

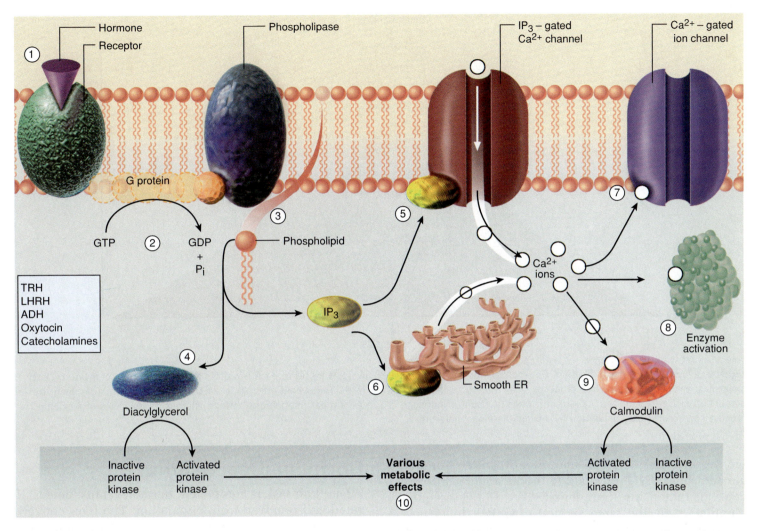

Figure 17.25 Hormone action through three second messengers: diacylglycerol, inositol triphosphate (IP_3), and calcium ions (Ca^{2+}). (*1*) The binding of a hormone to a membrane receptor activates a G protein. (*2*) The G protein activates a membrane enzyme, phospholipase. (*3*) Phospholipase breaks down membrane phospholipids into diacylglycerol and inositol triphosphate (IP_3). (*4*) Diacylglycerol activates protein kinases. (*5*) Inositol triphosphate can bind to gated calcium channels in the plasma membrane, admitting calcium to the cell from the ECF or (*6*) it can trigger calcium release from the smooth ER. Calcium ions can have a variety of effects: (*7*) Ca^{2+} can open other ion channels in the plasma membrane, (*8*) Ca^{2+} can function as cofactors that activate enzymes, or (*9*) Ca^{2+} can bind to calmodulin, a protein which then activates protein kinases. (*10*) Protein kinases exert the same variety of metabolic effects here as they do in the cAMP system.

Rather than using a second-messenger system, it binds to a plasma membrane enzyme, *tyrosine kinase,* that directly phosphorylates cytoplasmic proteins.

Enzyme Amplification

One hormone molecule does not trigger the synthesis or activation of just one enzyme molecule. It activates thousands of enzyme molecules through a cascade effect called **enzyme amplification.** The cAMP system is an example of this (fig. 17.26). One hormone molecule binds to a membrane receptor. Through the G proteins, this activates many molecules of adenylate cyclase. Each adenylate cyclase makes many molecules of cAMP, and each cAMP activates many molecules of pro-

tein kinase. Each protein kinase phosphorylates numerous enzyme molecules, and each enzyme molecule can act on many molecules of its substrate. A single molecule of epinephrine, for example, may thus stimulate the release of millions of molecules of glucose from a liver cell. Because of enzyme amplification, hormones can work in small quantities. Their circulating concentrations are very low compared to other blood solutes (on the order of ng/dL), and target cells do not need a great number of receptors.

Effects of Hormone Concentration

Hormones influence the number of target cell receptor sites for themselves or for other hormones to be secreted

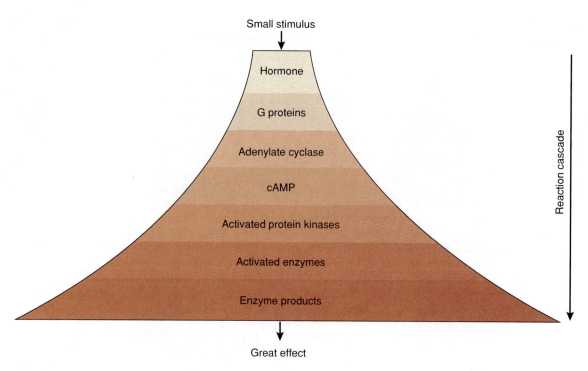

Small stimulus

Hormone

G proteins

Adenylate cyclase

cAMP

Activated protein kinases

Activated enzymes

Enzyme products

Reaction cascade

Great effect

Figure 17.26 Enzyme amplification. One molecule of hormone activates several G proteins; each G protein activates several adenylate cyclase molecules; each adenylate cyclase produces many molecules of cAMP; and so forth. The overall effect is a reaction cascade, with larger quantities of product at each reaction step. As a result, a small amount of hormone can cause a great physiological effect. Millions of molecules of a cellular product may result from stimulation by a single hormone molecule.

later. For example, small amounts of GnRH stimulate gonadotropes of the pituitary to produce more GnRH receptors, increasing their sensitivity to it. Subsequent stimulation by the same amount of GnRH thus causes a stronger response. This is called **up-regulation** (fig. 17.27*a*).

Long-term exposure to high hormone concentration can reduce the sensitivity of target cells—an effect called **down-regulation** (fig. 17.27*b*). This apparently occurs because target cells reduce the number of receptors they have, thus escaping overstimulation. Subsequent exposure to normal hormone levels then produces a subnormal response. Adipocytes down-regulate when exposed to high concentrations of insulin, and cells of the testis down-regulate in response to high concentrations of LH.

Hormone therapy often requires long-term use of abnormally high *pharmacological doses* of hormone, which may have undesirable side effects. Long-term treatment of inflammatory diseases with hydrocortisone, for example, undesirably alters bone metabolism. Such abnormal effects are produced by several mechanisms, including the following: (1) excess hormone may bind to receptor sites for other, related hormones and activate effects that mimic those of the other hormones, and (2) a target cell may convert one hormone into another, such as testosterone to estrogen. Thus, long-term high doses of testosterone can, paradoxically, have feminizing effects.

Hormone Deactivation

When their task is completed, hormones must be deactivated and removed from circulation. The liver converts most hormones to physiologically inactive compounds that are excreted by way of the urine and bile. Many hormones are so rapidly cleared from the blood they have a half-life of less than 2 minutes. Hormonal effects mediated by cAMP are quickly halted by an enzyme called **phosphodiesterase** that converts cAMP into an inactive form.

Hormone Interactions

No hormone travels in the bloodstream alone, and no cell is bombarded by only one hormone. Rather, there are many hormones circulating in the blood and tissue fluid at once. All cells ignore the majority of them because they have no receptors for them, but most cells are sensitive to more than one. In these cases, the hormones may have three kinds of interactive effects:

1. **Synergistic effects,** in which two or more hormones act together to produce an effect that is greater than the sum of their separate effects. Neither FSH nor testosterone alone, for example, can stimulate significant sperm production. When they act together, however, the testes produce some 300,000 spermatozoa per minute.

Up-regulation

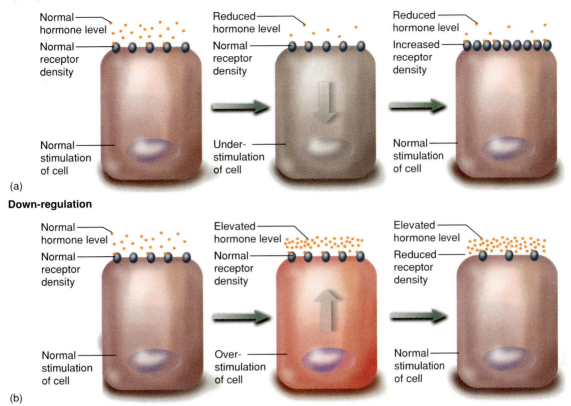

Down-regulation

Figure 17.27 Adaptation of target cells to hormone concentration. (*a*) In up-regulation, the target cell responds to a drop in hormone concentration by producing more receptors for it, thus heightening its sensitivity to the small amount of hormone that is present. (*b*) In down-regulation, an elevated hormone level causes a target cell to produce fewer receptors, or actively remove some receptors from the plasma membrane, so that it is not overstimulated by the hormone.

2. **Permissive effects,** in which one hormone enhances the target organ's response to a second hormone that is secreted later. FSH, for example, stimulates the gonads to develop LH receptors, and estradiol stimulates uterine cells to develop progesterone receptors. FSH and estradiol therefore prepare their respective target organs for the LH and progesterone to follow. A target organ would respond poorly to the second hormone, if at all, had it not been primed by the first hormone.

3. **Antagonistic effects,** in which one hormone opposes the action of another. Insulin and glucagon, for example, have opposite effects on blood sugar and on the synthesis or hydrolysis of glycogen. During pregnancy, placental estrogen suppresses the secretion of prolactin; thus milk is not synthesized until the placenta is shed in the afterbirth.

> **Key Point Review**

12 What are the three chemical classes of hormones? Name at least one hormone in each class.

13 Why do corticosteroids and thyroid hormones require transport proteins to travel in the bloodstream?

14 Explain how MIT, DIT, T_3, and T_4 relate to each other structurally.

15 What are the two locations of hormone receptors in target cells? Name one hormone that employs each receptor location.

16 How can one hormone molecule lead to the synthesis of thousands of enzyme molecules?

Eicosanoids and Other Chemical Messengers

▼**Objectives**

When you have completed this section, you should be able to
- define *eicosanoid*, name some of the major classes of eicosanoids, and explain their chemical origin; and
- describe several physiological roles of the prostaglandins.

Neurotransmitters and hormones are not the only chemical messengers in the body. Here we briefly consider some others. Unlike neurotransmitters, they are not produced by neurons. Unlike hormones, they are not carried in the blood to distant target organs.

Although we do not use the term *hormone* in such a broad sense in this chapter, some authorities call these messengers **local hormones** because they diffuse only short distances and stimulate cells close to the ones that

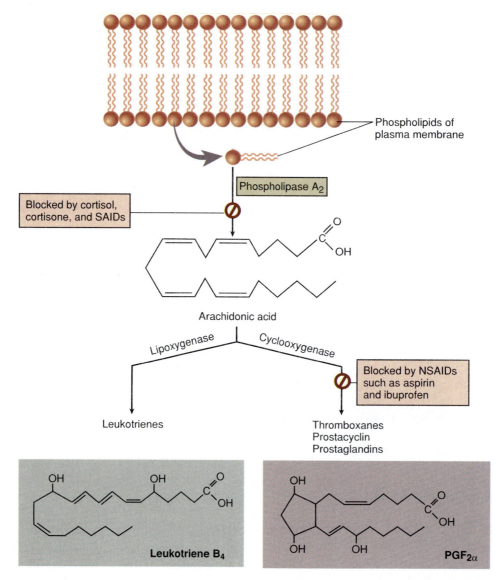

Figure 17.28 Pathways of eicosanoid synthesis, showing points of drug action. SAIDs are steroidal anti-inflammatory drugs such as hydrocortisone; NSAIDs are nonsteroidal anti-inflammatory drugs such as aspirin and ibuprofen.

produced them. They are also known as **paracrine**[26] **secretions.** Somatostatin, for example, is produced by pancreatic δ cells and diffuses to α and β cells in the same pancreatic islet, where it inhibits their secretion of glucagon and insulin. Catecholamines can also function as paracrine secretions; the adrenal medulla secretes catecholamines that diffuse to the adrenal cortex and stimulate it to secrete corticosterone. Histamine is a paracrine secretion of mast cells, which often line up along the blood vessels of areolar connective tissue. It stimulates the smooth muscle of a blood vessel to relax, thus causing vasodilation.

A family of paracrine secretions called the **eicosanoids**[27] (eye-CO-sah-noyds) are commanding a great deal of attention lately. Eicosanoids are 20-carbon derivatives of a polyunsaturated fatty acid called **arachidonic** (ah-RACK-ih-DON-ic) **acid,** part of a phospholipid of the plasma membrane. It is liberated from the membrane under the influence of many peptide hormones and other chemicals and then oxidized by either of two enzymes that convert it to various eicosanoids (fig. 17.28).

Lipoxygenase catalyzes the first step of the conversion of arachidonic acid to **leukotrienes,**

---

26. *para* = next to + *crin* = secrete

27. *eicosa* (variation of *icosa*) = 20

eicosanoids that mediate allergic and inflammatory reactions (see chapter 21). **Cyclooxygenase** converts arachidonic acid to three other types of eicosanoids: prostacyclin, thromboxanes, and prostaglandins. **Prostacyclin** is produced in the walls of the blood vessels, where it inhibits blood clotting and vasoconstriction. **Thromboxanes** are produced by blood platelets. In the event of injury, they can override prostacyclin and stimulate vasoconstriction and clotting. Prostacyclin and thromboxanes are further discussed in chapter 18. **Prostaglandins (PGs)** are the most diverse group. These eicosanoids have a five-sided carbon ring in their backbone. They are named PG for *prostaglandin,* plus a third letter that indicates the type of ring structure (PGE, PGF, etc.) and a subscript that indicates the number of C=C double bonds in the side chain. Named for their discovery in the bovine prostate gland and semen, they are now thought to be produced in most or all organs of the body. The PGEs are usually antagonized by PGFs. For example, the PGE family relaxes smooth muscle in the bladder, intestines, bronchioles, and uterus and stimulates contraction of the smooth muscle of blood vessels. $PGF_{2\alpha}$ has precisely the opposite effects. Some other roles of prostaglandins are described in table 17.6.

Understanding the pathways of eicosanoid synthesis makes it possible to understand the action of several familiar drugs (see special topic 17.4). The roles of prostaglandins and other eicosanoids are further explored in later chapters on blood, immunity, and reproduction.

17 What are eicosanoids and how do they differ from neurotransmitters and hormones?

18 Distinguish between a paracrine and endocrine effect.

19 State any four functions of prostaglandins.

Table 17.6 A Few of the Known Roles of Prostaglandins (PGs)

Inflammatory	Promote fever and pain, two cardinal signs of inflammation
Endocrine	Mimic effects of TSH, ACTH, and other hormones; alter sensitivity of anterior pituitary to hypothalamic hormones; work with glucagon, catecholamines, and other hormones in regulation of fat mobilization
Nervous	Function as neuromodulators, altering the release or effects of neurotransmitters in the brain
Reproductive	Promote ovulation and formation of corpus luteum; induce labor contractions
Gastrointestinal	Inhibit gastric secretion
Vascular	Act as vasodilators and vasoconstrictors
Respiratory	Constrict or dilate bronchioles
Renal	Promote blood circulation through the kidney, increase water and electrolyte excretion

Stress and Adaptation

▼Objectives

When you have completed this section, you should be able to
• Give a physiological definition of stress; and
• discuss how the body adapts to stress through its endocrine system.

We all must adapt to stress at some time—whether fighting an illness, dealing with a family crisis, or worrying about an upcoming examination. Even such pleasant experiences as graduating from school, getting married, or buying a new car produce similar physiological reactions in the body. In this section, we consider the meaning of stress and the physiological adaptations to it.

In the 1930s, Canadian biochemist Hans Selye (1907–82) injected extracts of cattle ovary into laboratory rats. This caused growth of the adrenal cortex, atrophy of the lymphatic organs, and bleeding ulcers. At first, Selye thought a hormone in the extracts produced these effects, but then he discovered that the same effects were produced by a wide range of injected chemicals, and even by exposing laboratory

Special Topic — Anti-Inflammatory Drugs 17.4

Cortisol and corticosterone inhibit inflammation by blocking the release of arachidonic acid from plasma membranes, thus inhibiting the synthesis of all eicosanoids. A disadvantage of these steroidal anti-inflammatory drugs (SAIDs) is that prolonged use causes side effects that mimic Cushing syndrome.

Aspirin and ibuprofen (Motrin) are nonsteroidal anti-inflammatory drugs (NSAIDs) with more selective effects. They change the structure of cyclooxygenase and destroy its activity; thus, they block prostaglandin synthesis without affecting lipoxygenase and the leukotrienes. For similar reasons, aspirin inhibits blood clotting (see chapter 18). One still-controversial theory of fever is that it results from the action of prostaglandins on the hypothalamus. If true, this would explain the antipyretic (antifever) effect of NSAIDs.

animals to temperature extremes or making them swim until exhausted. He found that any form of stress stimulated the pituitary-adrenal axis and caused elevated levels of ACTH and cortisol. Some now define **stress** as any situation that stimulates the pituitary-adrenal axis. Even pleasant changes in our lives come within this definition. In a very broad sense, others define stress as any deviation from a homeostatic set point or any immediate threat to physical or emotional well-being. Stress may be physical (infection, injury, or intense stimulation) or psychological (anger, grief, depression, or guilt).

Selye observed that there is "a nonspecific response of the body to readjust itself following any demand made upon it." This response involves the secretion of cortisol by the adrenal cortex and catecholamines by the adrenal medulla. These responses supply cells with increased amounts of fuel (glucose and fatty acids), amino acids for protein synthesis, and oxygen (due to increased pulmonary ventilation). Blood circulation speeds up in order to meet these demands of the cells.

Selye called this pattern of response the **general adaptation syndrome (GAS),** and he broke it down into three stages: an alarm reaction, a stage of resistance, and if this failed, a stage of exhaustion. The **alarm reaction** is the sequence of secretions from CRH to ACTH to glucocorticoids, followed by the usual glucocorticoid effects such as glucose and fatty acid mobilization. The

alarm reaction also involves the sympathoadrenal system and fight-or-flight state described in chapter 15.

The **stage of resistance** entails several adjustments of bodily function, including an accelerated heartbeat and elevated blood concentrations of glucose, glycerol, amino acids, fatty acids, and epinephrine. Glucagon and somatotropin levels rise and mobilize glucose and amino acids for tissue repair. ADH level rises and promotes sodium and water retention, which would be beneficial if the stressful situation led to sweating or bleeding.

If these reactions fail to restore homeostasis, the **stage of exhaustion** follows and is commonly accompanied by illness. In the short run, high cortisol levels are lifesaving, but in the long run they can be very harmful. Continued high levels of ACTH and cortisol cause ulcers, headaches, irregular heartbeat, and personality disorders. Cortisol also inhibits the immune system, which makes us more vulnerable to infectious diseases and cancer. Collectively, these disturbances of homeostasis can lead to death.

<hr>

Key Point Review

20 Define *stress* from the standpoint of endocrinology.

21 Describe the stages of the general adaptation syndrome.

22 List seven hormones that show increased secretion in the general adaptation syndrome. Describe how each one contributes to recovery from stress.

CHAPTER ESSAY

Hormones and Circadian Rhythms

If you are accustomed to waking up at 0600 every workday, you may have noticed that you wake up about the same time on weekends even without an alarm clock. This is because the body has its own internal **biological clock** that regulates daily cycles, or **circadian**[28] **rhythms**, of physiology and behavior. These cycles are approximately 25 hours long in the absence of rhythmic external stimuli, but in the presence of rhythmic stimuli such as daylight, our body becomes *entrained* by them and follows a rhythm of more nearly 24 hours. The study of these rhythms is called *chronobiology*.[29]

The biological clock consists of a pair of *suprachiasmatic* (SOO-pra-KY-az-MAT-ic) *nuclei (SCN)* located in the hypothalamus, above the optic chiasma. Some fibers of the optic nerve follow a tortuous *retinohypothalamic pathway* rather than going to the visual cortex. These fibers leave the optic

nerve and terminate in the SCN. From here, signals are relayed to the reticular formation and travel through the thoracic spinal cord to the superior cervical ganglion. From there, sympathetic postganglionic fibers travel back to the brain and end on the pineal gland. These nerve fibers carry information on *photoperiod*—the daily ratio of light to dark. In daylight hours, the pineal gland synthesizes serotonin. In the absence of light, it converts serotonin to melatonin—the serotonin level falls and the melatonin level rises at night. The balance between serotonin and melatonin seems to affect mood and other physiological functions, as discussed shortly.

Other hormones also exhibit circadian rhythms of secretion. Most GH and TSH are secreted shortly after we fall asleep, while ACTH levels begin to rise shortly before we wake up. In fact, 75% of the day's total ACTH secretion occurs between 0400 and 1000. The diagnosis of endocrine diseases such as Addison disease and Cushing syndrome must take into account the time of day at which blood samples are

<hr>

28. *circa* = approximately + *dia* = a day
29. *chrono* = time

taken relative to cycles of hormone secretion. Body temperature also shows a circadian rhythm, oscillating between 36.4°C at 0400 to 37.0°C at 0900 (fig. E.1). These cycles have many health implications, of which we consider three: shift work, seasonal affective disorder, and drug chronotherapy.

Shift work. Many night-shift workers sleep poorly during the day even if undisturbed by ringing telephones and salespeople. They exhibit high levels of melatonin during the day instead of at night; their cycle of body temperature remains out of synchrony with the sleep-waking cycle for several days after a shift change; and fatigue results both from the lack of sleep and from this out-of-phase temperature cycle. Digestive complaints and irritability are common. The greatest disturbances occur in people who change shift every 3 to 4 days—for example, those who work a night shift during the week and then try to adapt to the daytime activities of family and friends on weekends. The least disruptive effects on the temperature cycle occur among people who change shift every 2 days. Shift workers experience some relief if they are exposed to 3 to 6 hours of bright light each day. Normal indoor light intensities are about 300 lux, whereas it takes at least 4 hours of exposure to 2,500 lux to produce a beneficial effect (or exposure for as little as 30 minutes to 10,000 lux).

Seasonal affective disorder (SAD). This is a sense of unaccountable depression, sleepiness, irritability, carbohydrate craving, and weight gain that many people experience in winter when the nights are longer and they get less exposure to sunlight. It has an incidence ranging from 2.5% of the population in Florida to as high as 20% in Alaska. Symptoms are relieved by 2 to 3 hours of exposure to bright light each day. This suggests

—continued

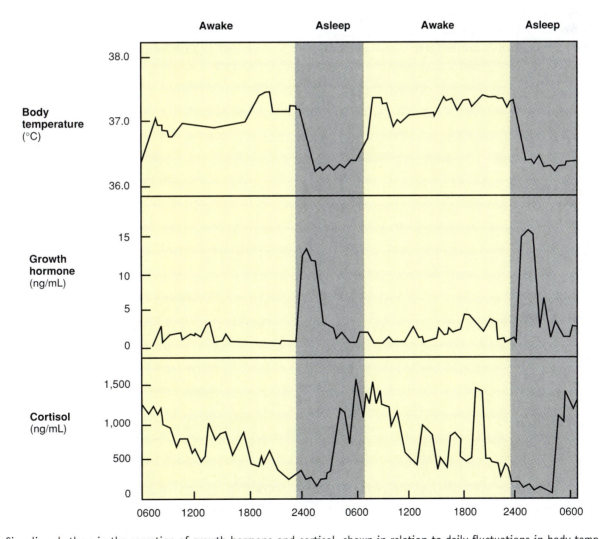

Figure E.1 Circadian rhythms in the secretion of growth hormone and cortisol, shown in relation to daily fluctuations in body temperature.

that high melatonin levels, which accumulate during dark hours, are responsible for SAD. **Premenstrual syndrome (PMS)** is similar to SAD, including high melatonin levels. PMS also can be treated with phototherapy. Although phototherapy lowers melatonin levels, there is also evidence that casts doubt on the hypothesis that high melatonin levels cause SAD and PMS. The exact neuroendocrine basis for phototherapy remains uncertain and controversial.

Chronotherapy. This is a treatment strategy that takes a patient's circadian rhythms into account when planning the timing and dosage of drugs. With most over-the-counter drugs, we use a method called *homeostatic therapy*—taking a constant dose at regular time intervals to maintain a fairly stable concentration of drug in the body fluids. Chronotherapy, by contrast, uses a complex schedule of drugs and doses determined by cycles of the patient's biological rhythms. The effects of drugs on animals can vary greatly depending on the time of day they are administered. For example, a given dose of amphetamine (26 mg per kg of body weight) was found lethal to 80% of rats if administered at 0300 but to only 5% if administered 3 hours later. Similar varia-

tions have been found in the effects of alcohol, insulin, nicotine, morphine, X rays, and many other agents. In one study, mice were inoculated with leukemia cells and divided into three groups. In the control group, which received no treatment, all mice died within 10 days. The other two groups were treated with an anticancer drug called *arabinosylcytosine*—one group on a homeostatic therapy schedule and the other on a chronotherapy schedule. As indicated in the table, chronotherapy resulted in a substantial improvement in survival rate.

	Mice Surviving 10 Days or Longer	Mice Surviving 60 Days or Longer
Control (no therapy)	0%	—
Homeostatic Therapy	50%	18%
Chronotherapy	90%	37%

The striking implications for chemotherapy are striking. Chronotherapy has demonstrated superior effectiveness against some human cancers.▲

Interactions between the Endocrine System and Other Organ Systems

Most Systems

The development and metabolism of most organ systems are affected by somatotropin, insulin, thyroid hormone, adrenal catecholamines, and glucocorticoids.

Integumentary System

- Protects superficial endocrine organs; plays role in synthesis of vitamin D, which functions as a hormone
- Sex hormones affect skin pigmentation, development of body hair and apocrine sweat glands, thickening of skin in males, and deposition of subcutaneous fat

Skeletal System

- Provides protection for some endocrine organs; stores calcium needed for endocrine function
- Many hormones affect bone growth and development; erythropoietin stimulates bone marrow to make RBCs; sex hormones help to maintain bone mass in adults

Muscular System

- Skeletal muscles protect some endocrine organs
- Somatotropin and testosterone stimulate muscular growth and development; other hormones affect electrolyte balance needed for muscle contraction

Nervous System

- Much of endocrine system controlled by hypothalamo-pituitary axis; sympathetic nervous system stimulates adrenal medulla
- Nervous system is subject to negative feedback inhibition from endocrine target organs; several hormones affect CNS development, mood, and behavior; hormones help to regulate electrolyte balance needed for neural function

Circulatory System

- Distributes hormones throughout body; blood pressure variations stimulate corrective secretions of hormones affecting cardiovascular physiology and blood osmolarity
- Blood volume and pressure regulated by angiotensin, aldosterone, and atrial natriuretic factor; heart rate affected by epinephrine and norepinephrine; vasoconstriction affected by angiotensin

Lymphatic/Immune Systems

- Lymphatic system maintains balance of interstitial fluid within endocrine organs; lymphocytes provide defense against infection
- Thymosin and other hormones activate immune cells; glucocorticoids have anti-inflammatory effects and suppress immune responses

Respiratory System

- Provides O_2 and removes CO_2; converts angiotensin I to angiotensin II
- Epinephrine and norepinephrine affect rate and depth of respiration and airflow through bronchioles

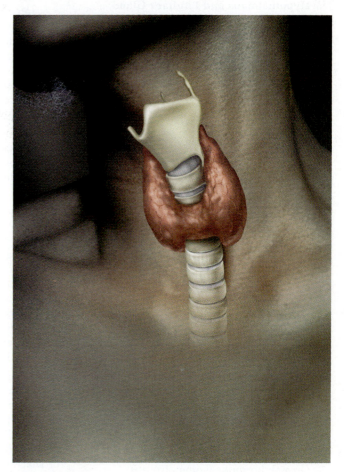

Urinary System

- Eliminates deactivated hormones and other metabolic wastes; kidneys produce calcitrol and erythropoietin
- PTH and aldosterone regulate electrolyte excretion; ADH regulates water excretion

Digestive System

- Provides nutrients; glucose absorption regulates insulin-glucagon balance; gut-brain peptides stimulate specific hungers via hypothalamus
- PTH affects nutrient absorption; insulin and glucagon regulate glucose storage and metabolism; several hormones coordinate GI secretion and motility

Reproductive System

- Sex hormones and inhibin suppress secretory activity in hypothalamus and pituitary gland
- Hormones play major role in regulating sexual development, gamete formation, sex drive, pregnancy, the menstrual cycle, fetal development, and lactation

An Overview of the Endocrine System (p. 600–601)
1. Meaning of *hormone*
2. Meaning of *endocrine*
3. Comparisons with nervous system

The Hypothalamus and Pituitary Gland (p. 602–610)
1. Hypothalamus
2. Pituitary Gland
   a. Adenohypophysis
      • Anterior lobe
      • Pars tuberalis
      • Pars intermedia
   b. Neurohypophysis
      • Median eminence
      • Stalk
      • Posterior lobe
      • Hypothalamo-hypophyseal tract
3. Anterior lobe hormones
   a. Six hormones (table 17.2)
   b. Tropic hormones
   c. Pituitary axes
4. Pars intermedia
5. Posterior lobe hormones (table 17.2)
6. Control of pituitary secretion
   a. Hypothalamic control
      • Hypothalamo-hypophyseal portal system
      • Releasing and inhibiting hormones (table 17.3)
      • Neuroendocrine reflexes
   b. Control by higher brain centers
   c. Feedback from target organs
      • Negative feedback inhibition
      • Positive feedback
7. Growth hormone (somatotropin)
   a. Functions
   b. Relationship to growth and aging
   c. Stimuli and cycles of secretion
   d. Role of somatomedins
8. Pituitary disorders
   a. Hypopituitarism
   b. Hyperpituitarism
   c. Diabetes insipidus

Other Endocrine Glands (p. 610–619)
1. Pineal gland (epiphysis cerebri)
   a. Changes with age
   b. Melatonin and serotonin
   c. Relation to sexual development
2. Thymus
   a. Changes with age
   b. Thymopoietin and thymosins

3. Thyroid
   a. Anatomy
      • Two lobes connected by isthmus
      • Follicles and follicular cells
      • C cells (parafollicular cells)
   b. Hormones
      • Triiodothyronine (T_3) and tetraiodothyronine (T_4)
      • Calcitonin
   c. Disorders
      • Congenital hypothyroidism
      • Myxedema
      • Endemic goiter and iodine
      • Toxic goiter (Graves disease)
4. Parathyroids
   a. Size and location
   b. Chief cells and parathyroid hormone (PTH)
   c. PTH deficiency and excess
5. Adrenals
   a. Adrenal medulla
   b. Zones of adrenal cortex
   c. Sex steroids
   d. Mineralocorticoid: aldosterone
   e. Glucocorticoids: cortisol and corticosterone
   f. Disorders
      • Pheochromocytoma
      • Cushing syndrome
      • Adrenogenital syndrome
      • Addison disease
6. Pancreas
   a. Anatomy
   b. Cells and secretions of the islets
      • α cells: glucagon
      • β cells: insulin
      • δ cells: somatostatin
   c. Diabetes mellitus (DM)
      • Signs and symptoms
      • Relationship to kidney physiology
      • Types, causes, and pathology
         •• Insulin-dependent DM (IDDM)
         •• Non-insulin-dependent DM (NIDDM)
   d. Hyperinsulinism
7. Gonads
   a. Ovarian anatomy
   b. Ovarian hormones (table 17.4)
   c. Testicular anatomy
   d. Testicular hormones (table 17.4)
8. Endocrine cells in other organs (table 17.4)
   a. Heart
   b. Kidneys
   c. Liver

   d. Stomach and small intestine
   e. Placenta

Hormones and Their Actions (p. 620–629)
1. Chemical identity of hormones (table 17.5)
   a. Steroids
   b. Biogenic amines
   c. Peptides
2. Synthesis and transport
   a. Steroids
      • Synthesized from cholesterol
      • Carried by transport proteins
      • Bound and unbound hormone
   b. Thyroid hormones
      • Synthesized from tyrosine and iodine
      • Formation of T_3 and T_4
      • Storage as thyroglobulin
      • Transport by thyroxine-binding globulin (TBG)
      • $T_4 \rightarrow T_3$ conversion in target cell
   c. Peptides
      • Preprohormones and prohormones
      • Insulin synthesis as example
3. Hormone receptors
   a. Analogy to enzyme-substrate binding
   b. Locations in target cells
      • Plasma membrane
      • Nucleus
4. Thyroid hormone action
   a. Binding to nuclear receptor
   b. Representative effects
      • Production of Na^+-K^+ pumps
      • Stimulation of GH secretion
5. Steroid hormone action
   a. Binding to nuclear receptor
   b. Induction of protein synthesis
6. Peptide and biogenic amine action
   a. Binding to membrane receptor
   b. Activation of second messengers
      • cAMP and cGMP
      • Diacylglycerol
      • Inositol triphosphate and calcium
7. Enzyme amplification
8. Effects of hormone concentration
   a. Up-regulation
   b. Down-regulation
   c. Effects of pharmacological doses
9. Hormone deactivation

10. Hormone interactions
    a. Synergistic
    b. Permissive
    c. Antagonistic

Eicosanoids and Other Chemical Messengers (p. 629–631)
1. Paracrine secretions (local hormones)

2. Eicosanoid synthesis
    a. Release of arachidonic acid
    b. Lipoxygenase pathway
    c. Cyclooxygenase pathway
3. Prostaglandin actions

Stress and Adaptation (p. 631–632)
1. Stress and the pituitary-adrenal axis

2. General adaptation syndrome (GAS)
    a. Alarm reaction
    b. Stage of resistance
    c. Stage of exhaustion

Selected Vocabulary

Testing Your Recall   Answers in Appendix C

1. CRH secretion would *not* raise the plasma concentration of
    a. ACTH.
    b. thyroxine.
    c. cortisol.
    d. corticosterone.
    e. glucose.

2. Which of the following hormones has the *least* in common with the others?
    a. adrenocorticotropic hormone
    b. follicle-stimulating hormone
    c. thyrotropin
    d. thyroxine
    e. prolactin

3. Which hormone would no longer be secreted if the hypothalamo-hypophyseal tract were destroyed?
    a. oxytocin
    b. follicle-stimulating hormone
    c. growth hormone
    d. adrenocorticotropic hormone
    e. corticosterone

4. Which of the following is *not* a hormone?
   a. prolactin
   b. prolactin-inhibiting factor
   c. thyroxine-binding globulin
   d. cortisol
   e. atrial natriuretic factor

5. Where are the receptors for thyroxine located?
   a. in the follicular cells
   b. in the blood plasma
   c. on the target cell plasma membrane
   d. in the target cell cytoplasm
   e. in the target cell nucleus

6. What would be the consequence of defective ADH receptors?
   a. diabetes mellitus
   b. adrenogenital syndrome
   c. dehydration and electrolyte depletion
   d. seasonal affective disorder
   e. none of the above

7. Which gland has more exocrine than endocrine tissue?
   a. the pineal gland
   b. the adenohypophysis
   c. the thyroid
   d. the pancreas
   e. the adrenal gland

8. Which of these cells stimulate bone deposition?
   a. α cells
   b. β cells
   c. C cells
   d. G cells
   e. T cells

9. Which of these hormones employs cAMP as a second messenger?
   a. ACTH
   b. progesterone
   c. thyroxine
   d. testosterone
   e. epinephrine

10. Prostaglandins are derived from
    a. pro-opiomelanocortin.
    b. cyclooxygenase.
    c. leukotriene.
    d. lipoxygenase.
    e. arachidonic acid.

11. The _____ develops from the hypophyseal pouch of the embryo.

12. Thyroxine (T_4) is synthesized by combining two residues called _____.

13. Growth hormone hypersecretion in adulthood results in a disease called _____.

14. Kidney stones may be caused by hypersecretion of the _____ gland(s).

15. Adrenal steroids that regulate glucose metabolism are called _____.

16. Sex steroids are secreted by the _____ cells of the ovary and _____ cells of the testis.

17. Target cells can reduce pituitary secretion by the process of _____.

18. Hypothalamic releasing factors are delivered to the anterior pituitary by way of a network of blood vessels called the _____.

19. A hormone is said to have a/an _____ effect when it stimulates target cells to develop receptors for another hormone to follow.

20. The 20-carbon paracrine derivatives of arachidonic acid are all called _____.

Testing Your Comprehension
Answers in Study Guide

1. Propose a model of enzyme amplification for a steroid hormone and construct a diagram similar to figure 17.26 for your model. A review of protein synthesis in chapter 5 may be helpful.

2. Suppose you are browsing in a health food store and see a product advertised: "Put an end to heart disease! This herbal medicine will totally rid your body of cholesterol!" Would you buy the product? Why or why not? If the claim were true, what other effects would the drug produce?

3. A person with a toxic goiter tends to sweat profusely. Explain this in terms of homeostasis.

4. How is the action of a peptide hormone similar to the action of a metabotropic neurotransmitter?

5. Review the effects of long-term abuse of anabolic steroids (chapter 3 essay, p. 000) and explain how some of these effects may relate to the concept of down-regulation discussed in this chapter.

Web Site Link

For a listing of the most current web sites related to this chapter, please visit the Saladin homepage at:

http://www.mhhe.com/sciencemath/biology/saladin/

chapter *eighteen*

18

[The Circulatory System: Blood

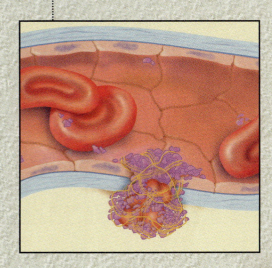

Special Topics

the unity of form and function

Brushing up

Blood has always had a special mystique. From time immemorial, people have seen blood flow from the body and with it, the life of the individual. The ancients thus presumed that blood carries a mysterious "vital force," and Roman gladiators drank it to fortify themselves for battle. Even today, we become especially alarmed when we find ourselves bleeding, and the emotional impact of blood is enough to make many people faint at the sight of it. From ancient Egypt to nineteenth-century America, physicians drained "bad blood" from their patients to treat everything from gout to headaches, from menstrual cramps to mental illness. It was long thought that hereditary traits were transmitted through the blood; people still use such unfounded expressions as "I have one-quarter Cherokee blood."

Scarcely anything meaningful was known about blood until the first microscopes revealed the blood cells. Blood is a uniquely accessible tissue for study, yet most of what we know about it dates only to the last 50 years. Recent developments in **hematology**[1]—the study of blood—have empowered us to save and improve the lives of countless people who would otherwise suffer or die.

Functions and Properties of Blood

▼Objectives

When you have completed this section, you should be able to
• list several functions of blood;
• list the blood's fluid and cellular components; and
• explain the causes and significance of the blood's viscosity and osmolarity.

The blood plays a central role in the body's respiration, nutrition, waste elimination, thermoregulation, immune defense, acid-base balance, water balance, and internal communication (table 18.1). It is a connective tissue with two main components—a clear extracellular fluid matrix called **plasma** and cells and cell fragments called **formed elements** (fig. 18.1).

Table 18.1	Functions of the Blood
Transport	
Carries oxygen from the lungs to other organs	
Carries nutrients from the digestive system and storage depots to other organs	
Carries carbon dioxide from all organs to the lungs for removal	
Carries other wastes to the liver and kidneys for detoxification or removal	
Carries hormones from the endocrine glands to their target cells	
Carries metabolic heat to the skin for removal; helps to stabilize body temperature	
Protection	
Plays several roles in inflammation	
Leukocytes destroy some microorganisms and cancerous cells	
Antibodies and complement proteins neutralize toxins or help to destroy microorganisms	
Platelet factors initiate clotting and minimize blood loss	
Regulation	
Transfers water to and from the tissues; helps to stabilize the water content of cells	
Buffers acids and bases; helps to stabilize body pH	

The formed elements of blood are classified as follows. All of them are cells except for platelets, which are fragments of certain bone marrow cells.

Erythrocytes[2] (red blood cells, RBCs)

Platelets

Leukocytes[3] (white blood cells, WBCs)

  Granulocytes

    Neutrophils

    Eosinophils

    Basophils

  Agranulocytes

    Lymphocytes

    Monocytes

The formed elements can be separated from the plasma by placing a sample of blood in a tube and spinning it for a few minutes in a centrifuge (fig. 18.2). Erythrocytes become packed into the bottom of the tube and typically constitute about 45% of the total volume (a value

---

1. *hem, hemato* = blood + *logy* = the study of

2. *erythro* = red
3. *leuko* = white

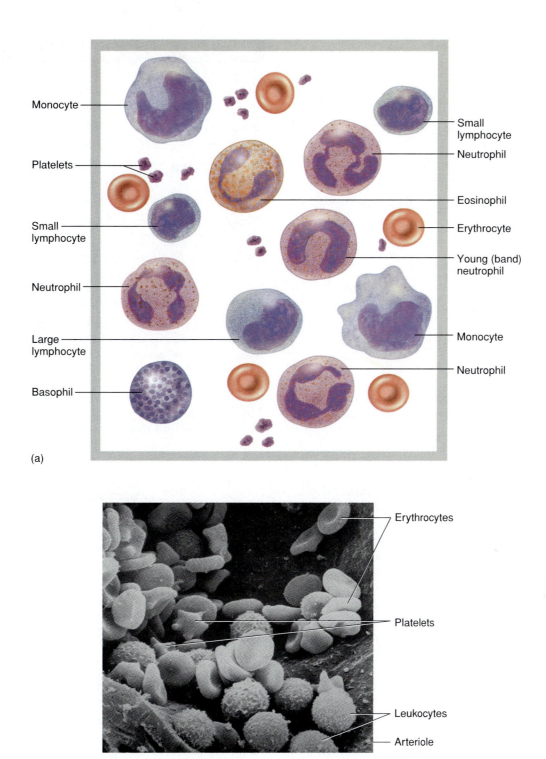

(a)

(b)

Monocyte

Platelets

Small lymphocyte

Neutrophil

Large lymphocyte

Basophil

Small lymphocyte

Neutrophil

Eosinophil

Erythrocyte

Young (band) neutrophil

Monocyte

Neutrophil

Erythrocytes

Platelets

Leukocytes

Arteriole

Figure 18.1 (*a*) The formed elements of blood. (*b*) Blood cells in an arteriole (SEM).

called the *hematocrit*). Leukocytes and platelets make up a narrow whitish zone called the *buffy coat,* just above the erythrocytes. At the top of the tube is the plasma, which has a pale yellow color and accounts for nearly 55% of the total volume.

Several properties of the blood are given in table 18.2. Its viscosity and osmolarity warrant special attention. **Viscosity** is the resistance of a fluid to flow because of cohesion between its particles. At a given temperature, mineral oil is more viscous than water, for

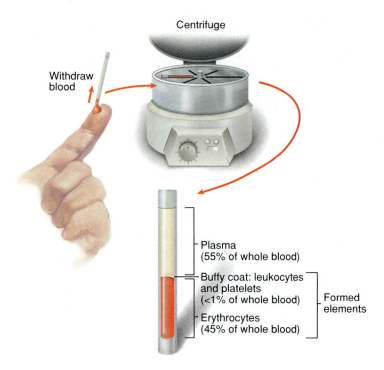

Centrifuge

Withdraw
blood

Plasma
(55% of whole blood)

Buffy coat: leukocytes
and platelets
(<1% of whole blood)

Erythrocytes
(45% of whole blood)

Formed
elements

Figure 18.2 Procedure for separating the formed elements of blood from the plasma. The hematocrit is the percentage of the whole blood composed of erythrocytes (45% in this figure).

Table 18.2	General Properties of Whole Blood*
Fraction of body weight	8%
Volume in the adult body	Female: 4–5 L; male: 5–6 L
Volume/body weight	80–85 mL/kg
Mean temperature	38°C (100.4°F)
pH	7.35–7.45
Viscosity (relative to water)	Whole blood: 4.5–5.5; plasma: 2.0
Osmolarity	280–296 mOsm/L
Mean salinity (mainly NaCl)	0.9%
Hematocrit (percent RBCs by volume)	Female: 37%–48%; male: 45%–52%
Hemoglobin	Female: 12–16 g/dL; male: 13–18 g/dL
Mean RBC count	Female: 4.8 million/μL; male: 5.4 million/μL
Platelet count	130,000–360,000/μL
Total WBC count	4,000–11,000/μL

*Values may vary slightly depending on the testing methods used.

example, and honey is more viscous than mineral oil. Whole blood is 4.5 to 5.5 times as viscous as water. This is due mainly to the RBCs; plasma alone is 2.0 times as viscous as water, mainly because of its protein. Viscosity is important in circulatory function because it partially governs the flow of blood through the vessels. An RBC or protein deficiency reduces viscosity and causes blood to flow too easily, whereas an excess causes blood to flow too sluggishly. Either of these conditions puts a strain on the heart that may lead to cardiac arrest or other cardiovascular problems if not corrected.

The **osmolarity** of blood is another important factor in cardiovascular function. In order to nourish surrounding cells and remove their wastes, substances must pass between the bloodstream and tissue fluid through the capillary walls. This transfer of fluids depends on a balance between the filtration of fluid from the capillary and its reabsorption by osmosis (see fig. 4.22, p. 127). If the osmolarity of the blood is too high, the bloodstream absorbs too much fluid, resulting in high blood pressure and a potentially dangerous strain on the heart and arteries. If osmolarity drops too low, the tissues become edematous (swollen with fluid) and the blood pressure can drop to dangerously low levels. It is therefore important that the blood maintain an optimal osmolarity. The osmolarity of the blood results mainly from its sodium ions, protein, and erythrocytes.

The contribution of protein to blood osmotic pressure—called the **colloid osmotic pressure (COP)**—is especially important (see special topic 18.1).

········ **Key Point Review** ········

1 About how much blood is in your body? Give answers in both units of weight and volume.

2 What are the principal components of the blood?

3 What percentage of the blood is plasma?

4 Why is blood viscosity important? What are the main factors that contribute to blood viscosity?

5 Why is osmolarity important? What are the main factors that contribute to blood osmolarity?

Plasma

▼Objectives
When you have completed this section, you should be able to
• distinguish between plasma and serum;
• list the proteins of blood plasma and state their functions;
• name the nonprotein nitrogenous compounds of blood plasma and explain their significance; and
• list the major nutrients, gases, and electrolytes found in plasma.

Blood plasma is a complex mixture of proteins, enzymes, nutrients, wastes, hormones, and gases (table 18.3). Blood **serum** is the fluid that remains after blood clots and the solids are removed. Except for the absence of clotting proteins, serum is identical to plasma. Serum is used as a vehicle for vaccines.

Several conditions can lead to **hypoproteine-mia,** a deficiency of plasma protein: extreme starvation or a diet severely deficient in protein; liver diseases that interfere with protein synthesis; kidney diseases that allow protein to be excreted in the urine; and severe burns, which can cause a patient to lose several liters of plasma per day through the body surface. As plasma protein concentration drops, so does the osmolarity of the blood. Consequently, the bloodstream loses more fluid by filtration than it absorbs by osmosis. The tissues become edematous and a pool of fluid may accumulate in the abdominal cavity—a condition called *ascites* (ah-SY-teez).

Severely malnourished children often exhibit a condition called *kwashiorkor* (kwah-she-OR-cor) in which the arms and legs are emaciated for lack of muscle, the skin is tight and shiny because of edema, and the abdomen is swollen by ascites (fig. 1). *Kwashiorkor* is an African word for a "deposed" child who is no longer breast-fed. Symptoms appear when a child is weaned and placed on a diet consisting mainly of rice or other cereals. Dehydration and diarrhea develop for lack of protein, often leading to death.

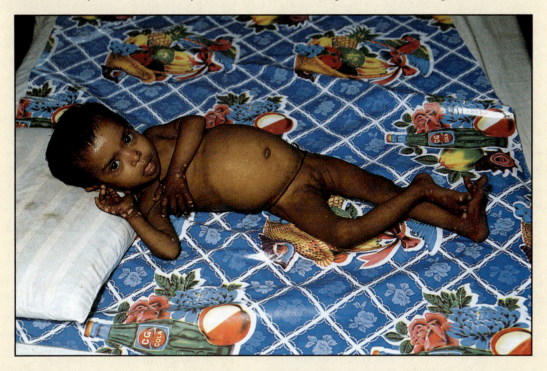

Figure 1 A child with kwashiorkor.

Proteins

Protein is the most abundant plasma solute by weight, totaling 6 to 9 g/dL. Plasma proteins play a variety of roles including clotting, defense, and transport. They can be separated from each other by *electrophoresis*[4]—a technique in which a layer of blood plasma is placed over a layer of gel and exposed to an electrical field. Proteins are negatively charged and therefore migrate toward the positive pole (anode) of the field at different rates determined by their charge and molecular size. This procedure separates the proteins into three main categories—*albumins, globulins,* and *fibrinogen* (table 18.4). Many other plasma proteins are indispensable to survival, but they account for less than 1% of the total.

Albumins are the smallest and most abundant plasma proteins. Because of their major contributions to viscosity and osmolarity, pathological changes in albumin concentration strongly influence blood pressure, flow, and fluid balance. **Globulins** are divided into three subclasses; from smallest to largest in molecular weight, they are the alpha (α), beta (β), and gamma (γ) globulins. **Fibrinogen** is a soluble precursor of *fibrin,* a sticky protein that forms the framework of a blood clot. Some of the other plasma proteins are enzymes involved in the clotting process.

---

4. *electro* = electricity + *phoresis* = transport

Table 18.3   Composition of Blood Plasma*

Water	92% by weight
Proteins	Total 6–9 g/dL
Albumins	60% of total protein, 3.2–5.5 g/dL
Globulins	36% of total protein, 2.3–3.5 g/dL
Fibrinogen	4% of total protein, 0.2–0.3 g/dL
Nutrients	
Glucose (dextrose)	70–110 mg/dL
Amino acids	33–51 mg/dL
Lactic acid	6–16 mg/dL
Total lipid	450–850 mg/dL
Cholesterol	120–220 mg/dL
Fatty acids	190–420 mg/dL
High-density lipoprotein (HDL)	30–80 mg/dL
Low-density lipoprotein (LDL)	62–185 mg/dL
Neutral fats (triglycerides)	40–150 mg/dL
Phospholipids	6–12 mg/dL
Iron	50–150 μg/dL
Trace elements	Traces
Vitamins	Traces
Electrolytes	
Sodium (Na^+)	135–145 mEq/L
Potassium (K^+)	3.5–5.0 mEq/L
Magnesium (Mg^{2+})	1.3–2.1 mEq/L
Calcium (Ca^{2+})	9.2–10.4 mEq/L
Chloride (Cl^-)	100–106 mEq/L
Bicarbonate (HCO_3^-)	23.1–26.7 mEq/L
Phosphate (HPO_4^{2-})	1.4–2.7 mEq/L
Sulfate (SO_4^{2-})	0.6–1.2 mEq/L
Nitrogenous Wastes	
Ammonia	0.02–0.09 mg/dL
Urea	8–25 mg/dL
Creatine	0.2–0.8 mg/dL
Creatinine	0.6–1.5 mg/dL
Uric acid	1.5–8.0 mg/dL
Bilirubin	0–1.0 mg/dL
Other Components	
Respiratory gases (O_2, CO_2, N_2)	—
Enzymes of diagnostic value	—

*This table is limited to substances of greatest relevance to this and later chapters. Concentrations refer to plasma only, not to whole blood.

Table 18.4   The Major Plasma Proteins

Proteins	Functions
Albumins (60%)*	Responsible for colloid osmotic pressure Major contributor to blood viscosity Transport lipids, hormones, Ca^{2+}, and other solutes Buffer the pH of plasma
Globulins (36%)*	
Alpha (α) Globulins	
Haptoglobulin	Transports hemoglobin released by dead erythrocytes
Ceruloplasmin	Transports copper
Prothrombin	Promotes blood clotting
Others	Transport lipids, fat-soluble vitamins, and hormones
Beta (β) Globulins	
Transferrin	Transports iron
Complement proteins	Aid in destruction of toxins and pathogenic microorganisms
Others	Transport lipids
Gamma (γ) Globulins	Antibodies that combat pathogens
Fibrinogen (4%)*	Forms fibrin, the major component of blood clots

*Percentage of the total plasma protein by weight.

Think About It

What would be the rationale for giving intravenous albumin to a patient who has experienced fluid loss and low blood volume?

The liver produces all of the major plasma proteins except γ globulins, secreting as much as 4 g of protein per hour. The γ globulins come from *plasma cells*—connective tissue cells that are descended from white blood cells called *B lymphocytes.*

Nonprotein Nitrogenous Substances

Blood plasma contains several important nitrogenous compounds in addition to protein—notably, amino acids and nitrogenous wastes. The amino acids come from the digestion of dietary protein or the catabolism of tissue proteins. **Nitrogenous wastes** are toxic end products of catabolism (see table 18.3). The most abundant is *urea*, a product of amino acid catabolism. Nitrogenous wastes are normally cleared from the blood and excreted by the kidneys at a rate that balances their rate of production.

Nutrients

Nutrients absorbed by the digestive tract are transported in the blood plasma. They include glucose, amino acids, fats, cholesterol, phospholipids, vitamins, and minerals.

Gases

Plasma transports some of the oxygen and carbon dioxide carried by the blood. It also contains a substantial amount of dissolved nitrogen, which normally has no physiological role in the body, but becomes important under circumstances such as diving and aviation.

Electrolytes

Electrolytes of the blood plasma are listed in table 18.3. Sodium ions constitute about 90% of the plasma cations and account for more of the blood's osmolarity than any other solute. Sodium therefore has a major influence on blood volume and pressure; people with high blood pressure are thus advised to limit their sodium intake. Electrolyte concentrations are carefully regulated by the body and have rather stable concentrations in the plasma.

Key Point Review

6 List the three major classes of plasma proteins. Which one is missing from blood serum?

7 What are the functions of blood albumin?

8 List some organic and inorganic components of plasma other than protein.

Blood Cell Formation

▼Objectives

When you have completed this section, you should be able to
- explain where the formed elements of blood are produced in fetuses, children, and adults;
- describe the early stages of blood cell production and state the factors that influence its rate; and
- explain how uncommitted stem cells become committed to forming specific types of blood cells.

A knowledge of **hemopoiesis**[5] (HE-mo-poy-EE-sis), production of the formed elements of blood, provides a foundation for understanding leukemia, anemia, and other blood disorders. Hemopoiesis occurs in the *hemopoietic tissues.* The earliest traces of it are seen in the *yolk sac,* a membrane associated with all vertebrate embryos. In most vertebrates, it encloses the yolk of the egg and functions in both hemopoiesis and the transfer of

yolk nutrients to the embryo. Even animals that do not lay eggs, however, have a yolk sac that retains its hemopoietic function. (It is also the source of cells that later colonize the gonads and produce eggs and sperm.) Cell clusters called *blood islands* form in the yolk sac by the third week of human development. They produce primitive *stem cells* that colonize the fetal bone marrow, liver, spleen, and thymus, where they subsequently produce blood cells.

The liver stops producing blood cells around the time of birth. The spleen stops producing RBCs soon after birth, but it continues to produce lymphocytes for life. From infancy onward, all types of formed elements are produced by **myeloid**[6] **hemopoiesis** in the red bone marrow (see chapter 8). Lymphocytes are also produced by **lymphoid hemopoiesis** in widely distributed lymphoid tissues and organs—especially the thymus, tonsils, lymph nodes, spleen, and patches of lymphoid tissue in the intestines.

The stages of myeloid hemopoiesis are shown in figure 18.3. All hemopoiesis begins with hemopoietic stem cells, or **hemocytoblasts.**[7] These cells multiply continually to keep up their own numbers, and they are *pluripotent*—capable of differentiating into multiple cell lines that give rise to all of the formed elements. Differentiation begins when hemocytoblasts develop surface receptors for specific hormones—*erythropoietin, interleukins,* and *colony-stimulating factors (CSFs).* At this point, they can no longer produce any more hemocytoblasts; they are now called *committed cells* because each type is destined to continue down one specific developmental pathway. Let us now examine these different types of hemopoiesis—*erythropoiesis, leukopoiesis,* and *thrombopoiesis.*

Erythrocyte Production

Erythropoiesis (eh-RITH-ro-poy-EE-sis), the production of erythrocytes, produces about 20 mL of RBCs per day, or 2.5 million cells per second, which matches the rate at which old RBCs die and maintains a fairly stable RBC count. The committed cell, called a *proerythroblast,* has receptors for **erythropoietin.** This hormone, secreted by the kidneys and to a lesser extent by the liver, stimulates proerythroblasts to transform into *erythroblasts.* Erythroblasts multiply, synthesize hemoglobin, and then their nuclei degenerate and pop out of the cell. Once the nucleus is lost, the cell is called a *reticulocyte*—named for the fact that it retains a fine network of endoplasmic reticulum. The transformation from hemocytoblast to reticulocyte takes 3 to 5 days and involves four major developments—a reduction in cell

---

5. *poiesis* = formation, production

6. *myelo* = bone marrow
7. *cyto* = cell + *blast* = precursor, forming

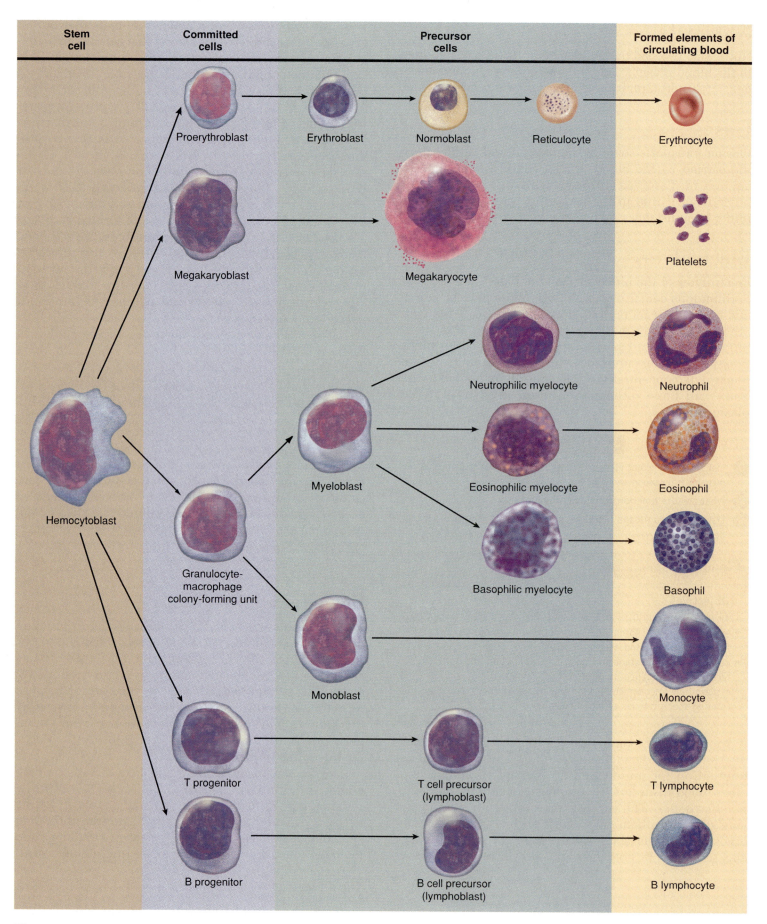

Figure 18.3 The stages of hemopoiesis, or production of the formed elements of blood.

Stem cell

Committed cells

Precursor cells

Formed elements of circulating blood

Proerythroblast
Erythroblast
Normoblast
Reticulocyte
Erythrocyte

Megakaryoblast
Megakaryocyte
Platelets

Hemocytoblast

Granulocyte-macrophage colony-forming unit

Myeloblast
Neutrophilic myelocyte
Neutrophil
Eosinophilic myelocyte
Eosinophil
Basophilic myelocyte
Basophil

Monoblast
Monocyte

T progenitor
T cell precursor (lymphoblast)
T lymphocyte

B progenitor
B cell precursor (lymphoblast)
B lymphocyte

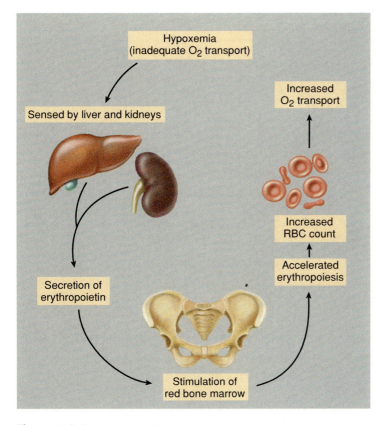

Figure 18.4 A negative feedback loop for correcting hypoxemia.

size, an increase in cell numbers, the synthesis of hemoglobin, and the loss of the nucleus.

Reticulocytes leave the bone marrow and enter the bloodstream. In another day or two, the remains of the endoplasmic reticulum disappear and the cell is considered a mature erythrocyte. About 0.5% to 1.5% of the circulating RBCs are reticulocytes, but this percentage increases under some circumstances. Blood loss, for example, stimulates accelerated erythropoiesis and leads to an increasing number of reticulocytes in circulation—as if the bone marrow were in such a hurry to replenish the lost RBCs that it let many developing RBCs into circulation a little early.

The major factor affecting the rate of erythropoiesis is oxygenation of the tissues. **Hypoxia** is an oxygen deficiency in any tissue; **hypoxemia**[8] refers specifically to an abnormally low level of oxygen in the blood. When the kidneys are hypoxic, they secrete more erythropoietin. Three or 4 days later, the RBC count begins to rise and reverses the hypoxemia that started the process—a good example of negative feedback and homeostasis (fig. 18.4).

8. *hyp* = below normal + *ox* = oxygen + *emia* = blood condition

One cause of hypoxemia is reduced availability of oxygen. If a person moves from Miami to Denver, for example, the lower level of oxygen at the high altitude of Denver produces temporary hypoxemia and stimulates increased erythropoiesis. The blood of an average adult has about 5 million RBCs/μL, but people who live at high altitudes may have counts of 7 to 8 million RBCs/μL. Another cause of hypoxemia is an increase in the body's oxygen demand. If a lethargic person suddenly takes up tennis or aerobics, for example, the muscles extract oxygen from the blood more rapidly than before, creating a state of hypoxemia that stimulates erythropoiesis. Endurance-trained athletes commonly have RBC counts as high as 6.5 million RBCs/μL.

Some causes of hypoxemia cannot be corrected by increasing erythropoiesis. In emphysema, for example, there is less lung tissue capable of oxygenating the blood. Raising the RBC count cannot correct this, but the kidneys and bone marrow have no way of knowing it and act as if the hypoxemia were due to anemia. The bone marrow thus produces more red cells, resulting in a dangerous excess called *polycythemia,* discussed later.

The nutritional requirements for erythropoiesis include iron, a critical part of the hemoglobin molecule. Men lose about 0.9 mg of iron per day through the urine, feces, and bleeding; women of reproductive age lose an average of 1.7 mg/day because of the added factor of menstruation. Since we absorb only a fraction of the iron present in our food, we must consume 5 to 20 mg/day to replace our losses. Pregnant women need 20 to 48 mg/day, especially in the last 3 months, to meet not only their own need but also that of the fetus.

Pathways of iron utilization are shown in figure 18.5. Dietary iron exists in two forms, the ferric (Fe^{3+}) and ferrous (Fe^{2+}) ions. Stomach acid converts most Fe^{3+} to Fe^{2+}, the only form that can be absorbed by the small intestine. Ferrous ions bind to **gastroferritin,** a transport protein produced by the stomach, and travel to the small intestine. Here they are absorbed into the blood, bind to a plasma protein called **transferrin,** and travel to the bone marrow, liver, and other tissues. Bone marrow uses the iron for hemoglobin synthesis; muscle cells use it to make their oxygen-storage pigment, myoglobin; and nearly all cells use iron to make the cytochromes of the mitochondrial electron-transport system. The liver stores surplus iron in the form of **ferritin,** an iron-protein complex, and releases the iron into circulation when needed.

Erythropoiesis involves rapid cell division and DNA synthesis, which in turn requires vitamin B_{12} and folic acid. Other nutritional requirements for hemopoiesis include copper and vitamin C, both of which serve as cofactors for some of the enzymes that synthesize hemoglobin. Blood plasma contains an α globulin, *ceruloplasmin,* for transporting copper (see table 18.4).

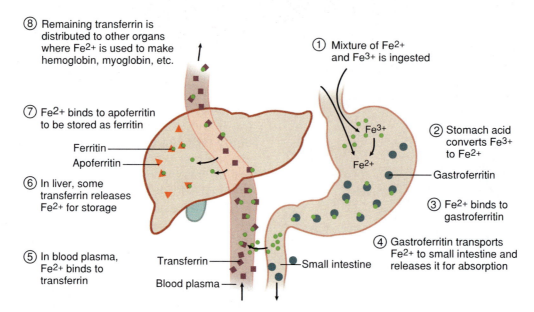

Figure 18.5 The pathway of iron absorption, transport, and storage.

⑧ Remaining transferrin is distributed to other organs where Fe^{2+} is used to make hemoglobin, myoglobin, etc.

① Mixture of Fe^{2+} and Fe^{3+} is ingested

⑦ Fe^{2+} binds to apoferritin to be stored as ferritin

Ferritin
Apoferritin

② Stomach acid converts Fe^{3+} to Fe^{2+}

Fe^{3+}
Fe^{2+}

Gastroferritin

⑥ In liver, some transferrin releases Fe^{2+} for storage

③ Fe^{2+} binds to gastroferritin

④ Gastroferritin transports Fe^{2+} to small intestine and releases it for absorption

⑤ In blood plasma, Fe^{2+} binds to transferrin

Transferrin

Small intestine

Blood plasma

Leukocyte Production

The production of white blood cells, called **leukopoiesis** (LOO-co-poy-EE-sis), also begins with hemocytoblasts (see fig. 18.3). As hemocytoblasts develop interleukin and CSF receptors, they become one of three types of committed cells—*B progenitors,* which give rise to B lymphocytes, *T progenitors,* which give rise to T lymphocytes, or *granulocyte-macrophage colony-forming units,* which give rise to the granulocytes and monocytes. Interleukins and CSFs are secreted mainly by mature T lymphocytes and macrophages in response to infections and other challenges to the immune system. There are a variety of these hormones that stimulate the production of specific types of WBCs in response to specific needs, such as fighting a bacterial or viral infection.

Granulocytes and monocytes are stored in the red bone marrow and released into circulation when needed. The red bone marrow contains 10 to 20 times more of these cells than the circulating blood does. Lymphocytes begin their development in the bone marrow, migrate to the lymphoid tissues to mature, and many of them recolonize the bone marrow. Chapter 21 discusses leukocyte development in more detail.

Leukocytes do not stay in the bloodstream very long. Granulocytes circulate for 4 to 8 hours and then migrate into the tissues, where they live another 4 or 5 days. Monocytes travel in the blood for 10 to 20 hours and then migrate into the tissues and transform into a variety of **macrophages** (MAC-ro-fay-jes). Macrophages, discussed in detail in chapter 21, can live for as long as a few years. Lymphocytes survive from a few weeks to decades. They leave the bloodstream for the tissues and eventually enter the lymphatic system, which empties them back into the bloodstream. Thus, they are continually recycled from blood to tissue fluid to lymph and finally back to the blood.

Platelet Production

The production of platelets, called **thrombopoiesis,** begins when a hemocytoblast becomes a committed cell called a *megakaryoblast.* In response to the hormone *thrombopoietin,* a megakaryoblast replicates its DNA repeatedly without undergoing nuclear or cytoplasmic division. The result is a gigantic cell (up to 100 μm in diameter) called a **megakaryocyte**[9] (meg-ah-CAR-ee-site), with a huge multilobed nucleus and multiple sets of chromosomes (fig. 18.6). Most megakaryocytes live in the bone marrow, but some of them colonize the lungs. A megakaryocyte exhibits infoldings of the plasma membrane that divide its marginal cytoplasm into little compartments. The cytoplasm breaks up along these lines of weakness into tiny fragments that enter the bloodstream. Some of these are functional platelets, while others are larger platelet precursors that break up

9. *mega* = huge + *karyo* = nucleus

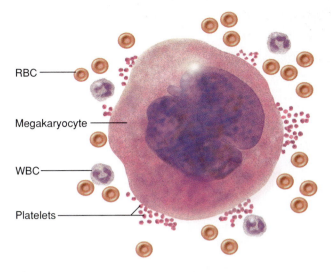

Figure 18.6 A megakaryocyte producing platelets. Several red and white blood cells are shown for size comparison.

RBC

Megakaryocyte

WBC

Platelets

into platelets as they pass through the lungs. About 25% to 40% of the platelets are stored in the spleen and released as needed. The remainder circulate freely in the blood and live for about 4 days.

Key Point Review

9 List the fetal tissues and organs that produce blood.

10 How do the sites of hemopoiesis differ between children and adults?

11 Distinguish between lymphoid and myeloid hemopoiesis.

12 How is a hemocytoblast different from a committed hemopoietic cell?

Erythrocytes

▼Objectives
When you have completed this section, you should be able to
- describe the structure of erythrocytes;
- describe the structure and function of hemoglobin;
- describe the ways in which the erythrocyte and hemoglobin content of the blood are quantified and explain why these values differ between men and women;
- describe the life cycle of erythrocytes; and
- describe the types of anemia and polycythemia and their causes and effects.

Form and Function

Erythrocytes have two principal functions: (1) to pick up oxygen from the lungs and deliver it to tissues elsewhere and (2) to pick up carbon dioxide from other tissues and unload it in the lungs. An erythrocyte is a

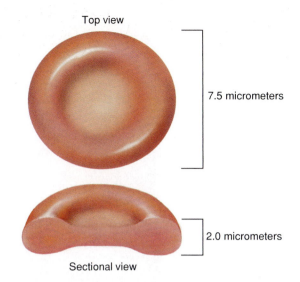

Top view

7.5 micrometers

2.0 micrometers

Sectional view

(a)

(b)

Figure 18.7 (*a*) Structure of an erythrocyte. (*b*) Erythrocytes on the tip of a hypodermic needle (SEM).

disc-shaped cell with a thick rim and a thin sunken center where the nucleus used to be. It is about 7.5 μm in diameter and 2.0 μm thick at the rim (fig. 18.7).

The plasma membrane of a mature RBC has glycoproteins and glycolipids that determine a person's blood type. On its inner surface are two peripheral proteins, *spectrin* and *actin,* that give the membrane resilience and durability. This is especially important when RBCs pass through small blood capillaries and sinusoids (fig. 18.8), since many of these passages are narrower than the diameter of an RBC. RBCs often must stretch, bend, or fold to squeeze through them. When they enter larger vessels, they spring back to their discoid shape.

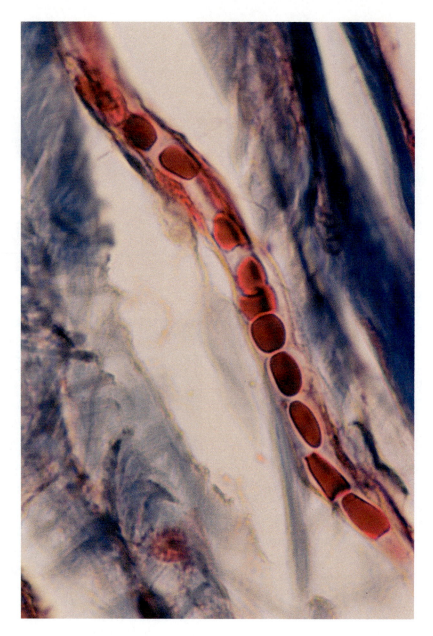

Figure 18.8 Erythrocytes in a blood capillary (×400).

Most cells, including white blood cells, have an abundance of organelles. RBCs, however, lose nearly all of their organelles during maturation and are almost devoid of internal structure (fig. 18.9). Because they lack mitochondria, RBCs are incapable of aerobic respiration. This prevents them from consuming the oxygen they are meant to transport. Erythrocytes are the only cells in the body that can carry on anaerobic fermentation indefinitely.

The cytoplasm of an RBC consists mainly of a 33% solution of **hemoglobin (Hb),** the red pigment that gives the RBC its color and name. Hemoglobin carries most of the oxygen and some of the carbon dioxide transported by the blood. The cytoplasm also contains an enzyme, *carbonic anhydrase (CAH),* that catalyzes a reaction between CO_2 and water to produce carbonic acid (H_2CO_3). The role of CAH in gas transport and pH balance is discussed in chapters 22 and 24. The lack of a nucleus makes an RBC unable to repair itself but it has an overriding advantage: the biconcave shape gives the cell a much greater ratio of surface area to volume, which enables O_2 and CO_2 to diffuse quickly to and from the hemoglobin and CAH.

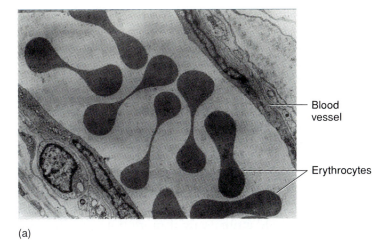

(a)

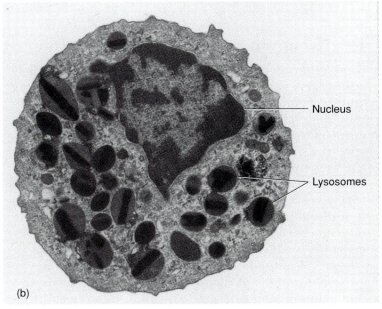

(b)

Blood vessel

Erythrocytes

Nucleus

Lysosomes

Figure 18.9 Comparison of RBCs and WBCs as seen by TEM. (*a*) RBCs have a strongly biconcave shape, lack organelles, and exhibit no internal microscopic structure. (*b*) Leukocytes, such as this eosinophil, exhibit a nucleus, many lysosomes, and other organelles.

Hemoglobin

Each erythrocyte contains about 280 million molecules of hemoglobin (see special topic 18.2). Hemoglobin consists of four protein chains called **globins** (fig. 18.10*a*). Two of these, the *alpha (α) chains,* are 141 amino acids long, and the other two, the *beta (β) chains,* are 146 amino acids long. Each chain is conjugated with a nonprotein moiety called the **heme** group (fig. 18.10*b*), which binds oxygen to a ferrous ion (Fe^{2+}) at its center. Each heme can carry one molecule of O_2; thus, the hemoglobin molecule as a whole can transport up to 4 O_2. About 20% of the carbon dioxide in the bloodstream is also transported by hemoglobin, but is bound to the globin moiety rather than to the heme. Gas transport by hemoglobin is discussed in detail in chapter 22.

T h i n k   A b o u t   I t

Hemoglobin is sometimes considered an enzyme. Based on what you know of hemoglobin and what you have learned about enzymes in chapter 3, do you agree? Why or why not?

Hemoglobin exists in several forms that display slight differences in the globin chains. The form we have just described is called *adult hemoglobin* (HbA). About 2.5% of an adult's hemoglobin, however, is of a form called HbA$_2$, which has two *delta (δ) chains* in place of the β chains. The fetus produces a form called *fetal hemoglobin* (HbF), which has two *gamma (γ) chains* in place of the adult β chains. HbF has a higher oxygen-binding capacity than adult hemoglobin and enables the fetus to extract oxygen from the mother's bloodstream. The δ and γ chains are the same length as the β chains but differ in amino acid sequence.

Special Topic (The Packaging of Hemoglobin)   18.2

The gas-transport pigments of earthworms, snails, and many other animals are dissolved in the plasma rather than contained in blood cells. You might wonder why human hemoglobin must be contained in RBCs. The main reason is osmotic. Remember that the osmolarity of blood depends on the number of particles in solution. A "particle," for this purpose, can be a sodium ion, an albumin molecule, or a whole RBC. If all the hemoglobin contained in the RBCs were free in the plasma, it would increase osmolarity 280 million times (since each RBC "particle" would yield 280 million "particles" of hemoglobin). The circulatory system would become enormously congested with fluid, and circulation would be severely impaired. The blood simply could not contain that much free hemoglobin and support life. On the other hand, if it contained a safe level of free hemoglobin, it could not transport oxygen fast enough to support the high metabolic demand of the human body. By having our hemoglobin packaged in RBCs, we are able to have much more of it and hence more efficient gas transport and more active metabolism.

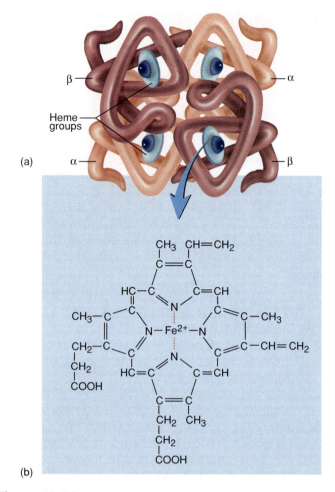

(a)

Heme groups

(b)

Figure 18.10 The structure of hemoglobin. (*a*) The hemoglobin molecule consists of two α proteins and two β proteins, each conjugated to a nonprotein heme group. (*b*) Structure of the heme group. Oxygen binds to Fe^{2+} at the center of the heme.

Quantities of Erythrocytes and Hemoglobin

The RBC count and hemoglobin concentration are important clinical data because they determine the amount of oxygen the blood can carry. Three of the most common measurements are hematocrit, hemoglobin concentration, and RBC count. The **hematocrit**[10] is the percentage of whole blood volume composed of RBCs (see fig. 18.2). In men, it normally ranges between 42% and 52%; in women, between 37% and 48%. The **hemoglobin concentration** of whole blood is normally 13 to 18 g/dL in men and 12 to 16 g/dL in women. The **RBC count** is normally 4.6 to 6.2 million RBCs/μL in men and 4.2 to 5.4 million/μL in women. This is

often expressed as cells per cubic millimeter (mm^3); 1 μL = 1 mm^3.

Notice that these values tend to be lower in women than in men. There are three physiological reasons for this: (1) androgens stimulate RBC production, and men have higher androgen levels than women; (2) women of reproductive age have periodic menstrual losses; and (3) hematocrit is inversely proportional to percent body fat, which is higher in women than in men. The blood also clots faster and the skin has fewer blood vessels in men than in women. Such differences are not limited to humans. From the evolutionary standpoint, the adaptive value of these differences may lie in the fact that male animals fight more than females and suffer more injuries. The hematologic traits described here may serve to minimize or compensate for their blood loss.

Erythrocyte Destruction

Circulating erythrocytes live for about 120 days. Their life cycle is summarized in figure 18.11. As an RBC ages and its membrane proteins (especially spectrin) deteriorate, the membrane grows increasingly fragile. Without a nucleus or ribosomes, an RBC cannot synthesize new spectrin. Eventually, it ruptures as it tries unsuccessfully to flex its way through narrow capillaries and blood sinusoids. The spleen has been called an "erythrocyte graveyard" because RBCs have an especially difficult time passing through its small channels. Here the old cells become trapped, broken up, and destroyed. An enlarged and tender spleen may indicate diseases in which RBCs are rapidly breaking down.

The process of disposing of old erythrocytes and hemoglobin is outlined in table 18.5. **Hemolysis**[11] (he-MOLL-ih-sis), the rupture of RBCs, releases hemoglobin and cell fragments. The cell fragments are easily destroyed by macrophages in the liver and spleen, but hemoglobin disposal is a bit more complicated. It must be disposed of efficiently, however, or it can block kidney tubules and filtration slits and cause renal failure.

The first step is to separate the heme from the globin. The globin polypeptide is hydrolyzed to free amino acids, which are incorporated into the body's general pool of amino acids available for protein synthesis or energy-releasing catabolism. Disposing of the heme is another matter. First, the iron is removed; it combines with transferrin in the blood and is used or stored in the same way as dietary iron. The rest of the heme becomes a greenish pigment called **biliverdin**[12] (BIL-ih-VUR-din). Most of this is further modified to a yellow-green

10. *crit* = to separate

11. *lysis* = breakdown, splitting
12. *bili* = bile + *verd* = green

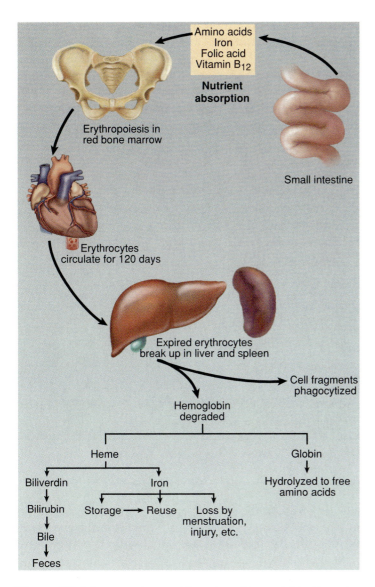

Figure 18.11 The erythrocyte life cycle and hemoglobin disposal.

Table 18.5	The Fate of Expired Erythrocytes and Hemoglobin

1. RBCs lose elasticity with age
2. RBCs break down while squeezing through blood capillaries and sinusoids
3. Cell fragments are phagocytized by macrophages in the spleen and liver
4. Hemoglobin decomposes into

   Globin portion—hydrolyzed to amino acids, which can be reused

   Heme portion—further decomposed into

     Iron

       Transported by albumin to liver and bone marrow

       Some used in bone marrow to make new hemoglobin

       Excess stored in liver as ferritin

     Biliverdin

       Converted to bilirubin and bound to albumin

       Removed by liver and secreted in bile

       Stored and concentrated in gallbladder

       Discharged into small intestine

       Converted by intestinal bacteria to urobilinogen

       Excreted in feces

pigment called **bilirubin**[13], which binds to blood albumin for transport. The liver removes bilirubin from albumin and secretes it into the bile, to which it imparts a dark green color as the bile becomes concentrated in the gallbladder. Biliverdin and bilirubin are collectively known as **bile pigments.** The gallbladder subsequently discharges the bile into the small intestine, where bacteria further modify bilirubin to *urobilinogen,* responsible for the brown color of the feces. Another hemoglobin breakdown pigment, *urochrome,* produces the yellow color of urine. A high level of bilirubin in the blood causes *jaundice,* a yellowish cast in light-colored skin and the whites of eyes. Jaundice may be a sign of rapid hemolysis or a liver disease that interferes with bilirubin disposal.

Erythrocyte Disorders

Any imbalance between the rates of erythropoiesis and RBC destruction may produce an excess or deficiency of red cells. An RBC excess is called *polycythemia*[14] (POL-ee-sy-THEE-me-uh), and a deficiency of either RBCs or hemoglobin is called *anemia.*[15]

Polycythemia

Primary polycythemia (*polycythemia vera*) is due to cancer of the erythropoietic line of the myeloid tissue. It can result in an RBC count as high as 11 million RBCs/μL and a hematocrit as high as 80%. Polycythemia from all other causes, called **secondary polycythemia,** is characterized by RBC counts as high as 6 to 8 million RBCs/μL. Secondary polycythemia can result from dehydration because water is lost from the bloodstream while erythrocytes remain and become concentrated. More often, it is caused by smoking, air pollution, emphysema, high altitude, strenuous physical conditioning, or other factors that create a state of hypoxemia and stimulate erythropoietin secretion.

---

13. *rub* = red

14. *poly* = many + *cyt* = cell + *hemia* = blood condition
15. *an* = without

The principal dangers of polycythemia are the increased blood volume, pressure, and viscosity that result from it. Blood volume can double in primary polycythemia, causing the circulatory system to become tremendously engorged. Blood viscosity may rise to three times normal. Circulation is poor, the capillaries are clogged with viscous blood, and the heart is dangerously strained. Chronic (long-term) polycythemia can lead to embolism, stroke, or heart failure. The deadly consequences of emphysema and some other lung diseases are due in part to polycythemia.

Anemia

Anemia compromises the oxygen-carrying capacity of the blood. It can have any of three consequences:

1. *The tissues become hypoxic.* The individual is lethargic and becomes short of breath upon physical exertion. The skin is pallid because of the deficiency of hemoglobin. Severe anemic hypoxia can lead to life-threatening necrosis of brain, heart, and kidney tissues.
2. *The osmolarity of the blood is reduced.* More fluid is thus transferred from the bloodstream to the intercellular spaces, resulting in edema.
3. *The blood pressure drops* because of the reduced volume and viscosity. Because the blood puts up so little resistance to flow, the heart beats faster than normal and cardiac failure may ensue.

The general causes of anemia fall into three categories: (1) bleeding, which produces **hemorrhagic anemia;** (2) rapid hemolysis, which produces **hemolytic anemia;** and (3) inadequate erythropoiesis or hemoglobin synthesis. For each category, specific examples are given in table 18.6; they are largely self-explanatory, but we will give special attention to the deficiencies of erythropoiesis and to some forms of hemolytic anemia.

Inadequate erythropoiesis usually results from lack of exercise. Erythropoiesis declines with age, however, and for this reason, along with poor nutritional habits and a more sedentary lifestyle, anemia is common among the elderly. *Nutritional anemia* results from a dietary deficiency of any of the requirements for erythropoiesis discussed earlier. Its most common form is *iron-deficiency anemia. Pernicious anemia* can result from a deficiency of vitamin B_{12}, but this vitamin is so abundant in meat that a B_{12} deficiency is rare except in strict vegetarians. More often, it occurs when gland cells of the stomach fail to produce a substance called *intrinsic factor* that the small intestine needs to absorb vitamin B_{12}. This becomes more common in old age because of atrophy of the stomach. Pernicious anemia

Table 18.6	Types and Causes of Anemia

Anemia Due to Inadequate Erythropoiesis

Inadequate exercise

Inadequate nutrition

    Iron-deficiency anemia

    Folic acid, vitamin B_{12}, or vitamin C deficiency

    Pernicious anemia (deficiency of intrinsic factor)

Old age

Destruction of myeloid tissue (hypoplastic and aplastic anemia)

    Radiation exposure

    Viral infection

    Autoimmune disease

    Some drugs and poisons (arsenic, mustard gas, benzene, etc.)

Hemorrhagic Anemia, Due to Excessive Bleeding

Trauma, hemophilia, menstruation, ulcer, ruptured aneurysm, etc.

Hemolytic Anemia, Due to Erythrocyte Destruction

Mushroom toxins, snake and spider venoms

Some drug reactions (such as penicillin allergy)

Malaria (invasion and destruction of RBCs by certain parasites)

Sickle-cell anemia and thalassemia (defective hemoglobin)

Hemolytic disease of the newborn (mother-fetus Rh mismatch)

can also be hereditary. An oral B_{12} supplement would be useless because the digestive tract cannot absorb it, but vitamin B_{12} injections are effective in treating pernicious anemia.

Hypoplastic[16] *anemia* is caused by a decline in erythropoiesis; in *aplastic anemia,* erythropoiesis ceases entirely as a result of destruction or failure of the myeloid tissue. Aplastic anemia leads to grotesque tissue necrosis and blackening of the skin. Most victims die within a year. About half of all cases are of unknown or hereditary cause, especially in adolescents and young adults. Other causes are given in table 18.6.

Sickle-Cell Anemia and Thalassemia

Sickle-cell anemia and thalassemia are hereditary hemoglobin defects that occur mostly among people of African and Mediterranean descent, respectively. About 1.3% of African Americans have **sickle-cell anemia.**

16. *plast* = form

Figure 18.12 Blood of a person with sickle cell anemia (SEM). Note the deformed, pointed erythrocytes.

Sickle-cell hemoglobin (HbS) differs from normal HbA only in the sixth amino acid of the β chain, where glutamic acid is replaced by valine. Sickle-cell anemia is caused by a recessive allele for HbS. People who are heterozygous for it are resistant to malaria—a significant survival advantage in Africa, where sickle-cell anemia originated. They are said to have *sickle-cell trait* but rarely have severe symptoms. People who are homozygous recessive exhibit sickle-cell anemia. Approximately 1 in every 12 African Americans is a heterozygous carrier of the HbS allele. If two carriers marry, their children each have a one-in-four chance of being homozygous and having the disease. Without treatment, a child with sickle-cell anemia has little chance of living to age 2, but even with the best available treatment, few victims live to the age of 50.

At low oxygen concentrations, HbS turns to a gel and causes the erythrocytes to become elongated and pointed at the ends (fig. 18.12), hence the name of the

disease. Sickled erythrocytes are sticky; they *agglutinate*[17] (clump together) and block small blood vessels, causing intense pain in oxygen-starved tissues. Blockage of the circulation can also lead to kidney or heart failure, rheumatism, or paralysis. Hemolysis of the fragile cells causes anemia and hypoxemia, which triggers further sickling in a deadly positive feedback loop. Chronic hypoxemia also causes fatigue, weakness, mental deficiency, and deterioration of the heart and other organs. In a futile effort to counteract the hypoxemia, the hemopoietic tissues become so active that bones of the cranium and elsewhere become enlarged and misshapen. As the spleen attempts to process the debris from hemolyzed erythrocytes, it becomes enlarged and fibrous. Sickle-cell anemia is a prime example of *pleiotropy*—the occurrence of multiple phenotypic effects from a change in a single gene (see chapter 5).

Thalassemia[18] (thal-ah-SEE-me-ah) refers to a family of hereditary anemias that are typically seen among Greeks, Italians, and other people of the Mediterranean area and their descendants. Thalassemia is characterized by a deficiency or absence of the α or β polypeptide and by RBC counts that may be less than 2 million cells/μL.

Key Point Review

13 Describe the shape, size, and contents of an erythrocyte and explain how it acquires its unusual shape.

14 What is the function of hemoglobin? What are its protein and nonprotein moieties called?

15 What happens to each of these moieties when old erythrocytes break up?

16 What is the body's primary mechanism for correcting hypoxemia? How does this illustrate homeostasis?

17 What are the three primary causes or categories of anemia? What are its three primary consequences?

Blood Types

▼Objectives
When you have completed this section, you should be able to
• explain what determines a person's ABO blood type and how this relates to transfusion compatibility;
• explain what determines a person's Rh blood type and how this relates to transfusions and mother-fetus compatibility; and
• list some blood groups other than ABO and Rh and explain how they may be useful.

17. *ag* (from *ad*) = together + *glutin* = glue

18. *thalass* = the sea; once thought to be caused by "sea air"

Charles Drew (fig. 1) understood blood banking in a deep and ultimately tragic way. After receiving his M.D. from McGill University of Montreal in 1933, Drew became the first black person to pursue the advanced degree of Doctor of Science in Medicine, for which he studied transfusion and blood-banking procedures at Columbia University. He became the director of a new blood bank at Columbia Presbyterian Hospital in 1939 and organized numerous blood banks during World War II.

It was Drew who convinced physicians to use plasma rather than whole blood for battlefield and other emergency transfusions. Whole blood could be stored for only a week and given to recipients with compatible blood types. Plasma could be stored longer and was less likely to cause transfusion reactions. The plasma transfusions he pioneered saved millions of lives.

When the U.S. War Department issued a directive forbidding the mixing of Caucasian and Negro blood in military blood banks, Drew denounced the order and moved on to Howard University in Washington, D.C. There, he became a professor of surgery and later chief of staff at Freedmen's Hospital. He mentored numerous young black physicians and campaigned to get them accepted into the medical community. However, he never suc-

ceeded in getting the American Medical Association to admit black members, even himself.

Late one night in 1950, Drew and three colleagues set out to volunteer their services at an annual free clinic in Tuskegee, Alabama. Drew fell asleep at the wheel and was critically injured in the resulting accident. Doctors at the nearest hospital administered blood and attempted to revive him, but he bled to death at the age of 45.

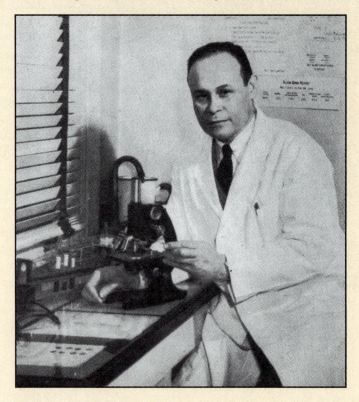

Figure 1 Charles Drew (1904–50).

Blood types and transfusion compatibility are a matter of interaction between plasma proteins and erythrocytes. Ancient Greek physicians attempted transfusions using a pig's bladder and porcupine quill to transfer blood from one person to another. While some recipients benefited from the procedure, others died from it. The reason some people's blood is compatible and some is not remained obscure until 1900, when Karl Landsteiner discovered blood types A, B, and O—a discovery that won him the 1930 Nobel Prize for Physiology or Medicine. World War II stimulated great improvements in transfusions, blood banking, and blood substitutes (see special topic 18.3).

We now know that all cells have an inherited combination of proteins, glycoproteins, and glycolipids on their surfaces that enable the immune system to distinguish our own cells from foreign invaders. These *anti-gen* molecules can stimulate a defensive immune reaction, including production of γ globulins called *antibodies* to combat the invader. The antigens of RBCs that govern blood type are called **agglutinogens** (ah-glue-TIN-oh-jens) because they are partially responsible for RBC *agglutination* in mismatched transfusions. The plasma antibodies that react against them are called **agglutinins** (ah-GLUE-tih-nins).

The ABO Group

The **ABO blood group** consists of blood types A, B, AB, and O. The hereditary presence or absence of RBC agglutinogens A and B determines a person's ABO blood type (table 18.7). The genetic determination of blood types was explained in chapter 5, p. 161. The

Table 18.7	The ABO Blood Group			
	ABO Blood Type			
	Type O	**Type A**	**Type B**	**Type AB**
Possible Genotypes	ii	$I^A I^A$, $I^A i$	$I^B I^B$, $I^B i$	$I^A I^B$
RBC Agglutinogen (Antigen)	None	A	B	A, B
Plasma Agglutinin (Antibody)	Anti-A, anti-B	Anti-B	Anti-A	None
Compatible Donor Blood	O	O, A	O, B	O, A, B, AB
Incompatible Donor Blood	A, B, AB	B, AB	A, AB	None
Frequency in U.S. Population				
White	45%	40%	11%	4%
Black	49%	27%	20%	4%
Hispanic	63%	14%	20%	3%
Japanese	31%	38%	22%	9%
Native American	79%	16%	4%	<1%

agglutinins of the ABO group begin to appear in the plasma within 2 to 8 months after birth. They reach their maximum concentrations between 8 and 10 years of age and then slowly decline throughout a person's life. They are produced mainly in response to the bacteria that inhabit our intestines, but they happen to cross-react with RBC antigens and are therefore best known for their significance in transfusions. Agglutinins react against any AB agglutinogen except those present on a person's own RBCs. The agglutinin that reacts against antigen A is called *α agglutinin,* or *anti-A;* it is present in the plasma of people with type O or type B blood—that is, anyone who does *not* possess agglutinogen A. The agglutinin that reacts against antigen B is *β agglutinin,* or *anti-B,* and is present in type O and type A individuals—those who do not possess agglutinogen B. Each agglutinin molecule has 10 binding sites where it can attach to an A or B agglutinogen. An agglutinin can therefore attach to several RBCs at once and bind them together (fig. 18.13). **Agglutination** is the process in which RBCs adhere to each other in masses that are bound by these agglutinins.

A person's ABO blood type can be determined by placing one drop of blood in a pool of anti-A serum and another drop in a pool of anti-B serum. Blood type AB will exhibit conspicuous agglutination in both antisera; type A or B will agglutinate only in the corresponding antiserum; and type O will not agglutinate in either one (fig. 18.14). The anti-A and anti-B reagents most widely used now, however, are not blood sera but proteins called *lectins* that are extracted from lentils and other plant seeds. These are simpler to obtain and have a longer shelf life than blood serum.

The ABO blood types are summarized in table 18.7. Note that type O is the most common and AB is

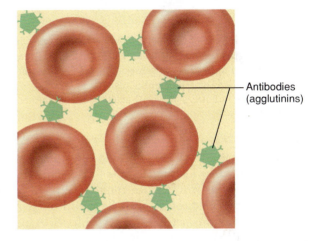

Figure 18.13 Agglutination of erythrocytes by an agglutinin such as anti-A or anti-B. Each agglutinin molecule has 10 binding sites, located at the tips of the Ys, and can therefore bind multiple RBCs to each other.

the rarest blood type in the United States. Percentages differ from one region of the world to another and among ethnic groups because people tend to marry within their locality and ethnic group and perpetuate statistical variations particular to that group.

In giving transfusions, it is imperative that the donor's blood not agglutinate as it enters the recipient's bloodstream. For example, if type B blood were transfused into a type A recipient, the recipient's anti-B agglutinins would immediately agglutinate the donor's RBCs (fig. 18.15). A mismatched transfusion causes a **transfusion reaction**—the agglutinated RBCs block small blood vessels, hemolyze, and release their

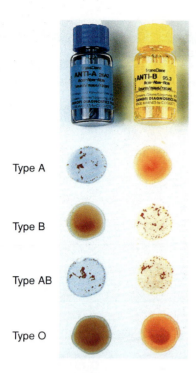

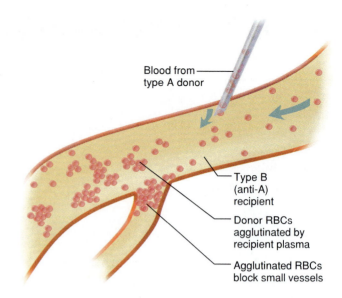

Figure 18.15 Agglutination of mismatched donor RBCs by agglutinins in the recipient's blood plasma. The clumped RBCs may lodge in smaller blood vessels downstream from this point and cut off the blood flow to vital tissues.

Figure 18.14 Reactions of the four human ABO blood types in antisera. Blood cells remain uniformly suspended if they lack the agglutinogens for the respective antiserum. Thus type A agglutinates only in anti-A antiserum; type B agglutinates only in anti-B; type AB agglutinates in both; and type O agglutinates in neither of them.

hemoglobin over the next few hours to days. Free hemoglobin can block the kidney tubules and cause death within a week or so from acute renal failure. For this reason, a person with type A (anti-B) blood must never be given a transfusion of type B or AB blood. A person with type B (anti-A) must never receive type A or AB blood. Type O (anti-A and anti-B) individuals cannot safely receive type A, B, or AB blood.

Type AB was once called the *universal recipient* because this blood type lacks both anti-A and anti-B agglutinins; thus, it will not agglutinate donor RBCs of any ABO type. However, this overlooks the fact that the *donor's* plasma can agglutinate the *recipient's* RBCs if it contains anti-A, anti-B, or both. For similar reasons, type O was once called the *universal donor*. The plasma of a type O donor, however, can agglutinate the RBCs of a type A, B, or AB recipient. There are procedures for reducing the risk of a transfusion reaction in certain mismatches, such as giving packed RBCs with a minimum of plasma; nevertheless, the concept of universal donors and recipients is obsolete and dangerously misleading.

Contrary to some people's belief, blood type is not changed by transfusion. It is fixed at conception and remains the same for life.

The Rh Group

The **Rh blood group** is named for the rhesus monkey, in which the Rh agglutinogens were first discovered in 1940. Agglutinogen D is the most important of the several known Rh agglutinogens. If any type of Rh agglutinogen is present on the RBCs, however, a person is considered to be *Rh positive (Rh+)*. If the RBCs lack Rh agglutinogens, the person is *Rh negative (Rh−)*. The Rh type is usually combined with the ABO type in a single expression such as O+ for type O, Rh positive, or AB− for type AB, Rh negative. About 85% of white Americans are Rh+ and 15% are Rh−. ABO blood type has no influence on Rh type, or vice versa. If the frequency of type O whites in the United States is 45%, and 85% of these are also Rh+, then the frequency of O+ individuals is the product of these separate frequencies: 0.45 • 0.85 ≈0.38, or 38%. Rh frequencies vary among ethnic groups just as ABO frequencies do. About 99% of Asians are Rh+, for example.

Think About It

What percentage of Japanese Americans would be expected to have type B− blood?

The presence or absence of Rh agglutinogens is an inherited trait, like that of the ABO agglutinogens. In contrast to the ABO group, however, *anti-Rh* agglutinins are not normally present. They form only in Rh− individuals who are exposed to Rh+ blood. If an

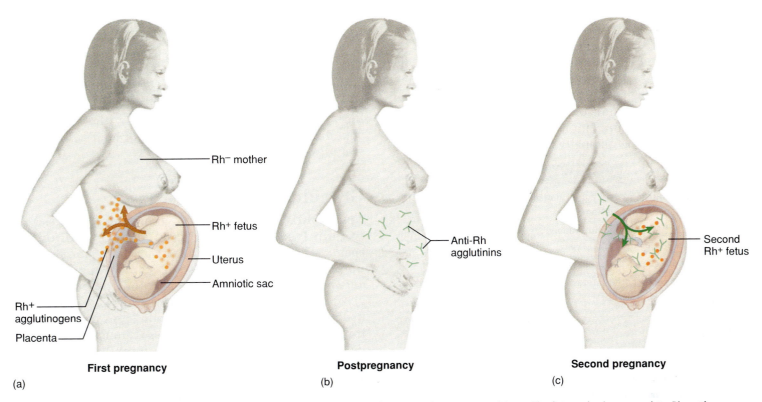

First pregnancy	Postpregnancy	Second pregnancy

(a) (b) (c)

Figure 18.16 Hemolytic disease of the newborn (HDN). (*a*) When an Rh⁻ woman is pregnant with an Rh⁺ fetus, she is exposed to Rh antigens, especially during childbirth. (*b*) Following that pregnancy, her immune system produces anti-Rh agglutinins. (*c*) If she later becomes pregnant with another Rh⁺ fetus, these agglutinins can cross the placenta and agglutinate the blood of that fetus, causing that child to be born with HDN.

Rh⁻ person receives an Rh⁺ transfusion, the recipient's antibody-producing cells are stimulated to produce anti-Rh. Since anti-Rh does not appear instantaneously, this presents little danger in the first mismatched transfusion. But if a sensitized Rh⁻ individual should later receive another Rh⁺ transfusion, the recipient's plasma agglutinins would agglutinate the donor's RBCs.

A related condition may occur when an Rh⁻ woman carries an Rh⁺ fetus. The first pregnancy is likely to be uneventful because the placenta normally prevents maternal and fetal blood from mixing. However, at the time of birth, or if a miscarriage occurs, placental tearing may expose the mother to Rh⁺ fetal blood. She then begins to produce anti-Rh agglutinins (fig. 18.16). If she becomes pregnant again with an Rh⁺ fetus, her anti-Rh agglutinins can pass through the placenta and agglutinate the fetal erythrocytes. Agglutinated RBCs hemolyze, and the baby is born with a severe anemia called **hemolytic disease of the newborn (HDN),** or *erythroblastosis fetalis.* HDN, like so many other disorders, is easier to prevent than to treat. If an Rh⁻ woman gives birth to an Rh⁺ child, or has a miscarriage or induced abortion, she can be given an anti-Rh γ globulin such as RhoGAM. The γ globulin binds fetal agglutinogens so they cannot stimulate her immune system to produce anti-Rh agglutinin. Some physicians now give RhoGAM throughout the Rh⁺ pregnancy of any Rh⁻ woman.

If an Rh⁻ woman has had one or more previous Rh⁺ pregnancies, her subsequent Rh⁺ infants have about a 17% probability of being born with HDN. Infants with HDN are usually severely anemic. As the fetal hemopoietic tissues respond to the need for more RBCs, erythroblasts (immature RBCs) are prematurely released into circulation—hence the name *erythroblastosis fetalis.* Hemolyzed RBCs release hemoglobin, which is converted to bilirubin. High bilirubin levels can cause *kernicterus,* a syndrome of toxic brain damage that may kill the infant or leave it with motor, sensory, and mental deficiencies. HDN can be treated with *phototherapy*—exposing the infant to ultraviolet light, which degrades bilirubin as blood passes through the capillaries of the skin. In more severe cases, the infant's Rh⁺ blood may be replaced with Rh⁻ blood by *exchange transfusion.* In time, the infant's hemopoietic tissues will replace the donor's RBCs with Rh⁺ cells, and by then the maternal agglutinin will have disappeared from the infant's blood.

Think About It

A baby with HDN typically has jaundice and an enlarged spleen. Explain these symptoms.

Other Blood Groups

There are at least 300 other detectable blood groups in addition to ABO and Rh, including M, N, Duffy, Kell, Kidd, and Lewis. These rarely cause transfusion reactions, but they are useful for such legal purposes as paternity and criminal cases and for research in anthropology and population genetics. The Kell, Kidd, and Duffy groups occasionally cause HDN.

· · · · · · · · · · · · · · · · · · **Key Point Review** · · · · · · · · · · · · · · · · · ·

18 What are agglutinins and agglutinogens? How do they interact to cause a transfusion reaction?

19 What agglutinins and agglutinogens are present in people with each of the four ABO blood types?

20 Describe the cause, prevention, and treatment of HDN.

21 Why might someone be interested in determining a person's blood type other than ABO/Rh?

Leukocytes

▼Objectives

When you have completed this section, you should be able to

- state the general function of leukocytes;
- name and describe the five types of leukocytes; and
- name the different types of abnormal leukocyte counts and discuss their causes and effects.

Leukocytes (LOO-co-sites), or white blood cells (WBCs), play a number of roles in the body's defense against pathogens. Their individual functions are summarized in figure 18.17, but they are discussed more extensively in chapter 21. Although they travel in the bloodstream, leukocytes spend most of their lives in the connective tissues and are much less numerous than RBCs in blood films (fig. 18.18). Most of them live only from a few hours to a few days. Some lymphocytes, however, provide immune memory and live for many years or even decades.

There are five kinds of WBCs. They are easily distinguished from erythrocytes in stained blood films because they contain conspicuous nuclei that stain from light violet to dark purple with the most common blood stains. Three WBC types—the *neutrophils, eosinophils,* and *basophils*—are called **granulocytes** because their cytoplasm contains organelles that appear as colored granules through the microscope. These are missing or relatively scanty in the two types known as **agranulocytes**—the *lymphocytes* and *monocytes.*

Types of Leukocytes

The five leukocyte types are compared in figure 18.17. From the photographs and data, take note of their sizes relative to each other and to the size of erythrocytes (which are about 7.5 μm in diameter). Also note how the leukocytes differ from each other in relative abundance—from neutrophils, which constitute about two-thirds of the WBC count, to basophils, which usually account for less than 1%. Nuclear shape is an important key to identifying leukocytes. The granulocytes are further distinguished from each other by the coarseness, abundance, and staining properties of their cytoplasmic granules.

Granulocytes

Neutrophils have very fine cytoplasmic granules that contain lysozyme, peroxidase, and other antibiotic agents. They are named for the way these granules take up blood stains at pH 7—some stain with acidic dyes and others with basic dyes, and the combined effect gives the cytoplasm a pale lilac color. The nucleus is usually divided into three or four lobes connected by thin strands of nucleoplasm. These strands are often so delicate and hard to see that a neutrophil may appear to have several small nuclei, although there is only one. Young neutrophils often exhibit an undivided nucleus shaped like a band or a knife puncture; they are thus called *band,* or *stab, cells.* Neutrophils are also called *polymorphonuclear* leukocytes (PMNs) because of their variety of nuclear shapes.

Eosinophils (EE-oh-SIN-oh-fills) have coarse cytoplasmic granules that stain from pinkish orange to rosy with the acidic dye *eosin.* The nucleus typically exhibits two globose lobes connected by a thin strand, like two balloons tied together.

Basophils have a U- or S-shaped nucleus, but it is obscured from view by coarse cytoplasmic granules that stain dark purple with basic dyes. It is sometimes difficult to distinguish a basophil from a lymphocyte, but basophils are conspicuously grainy while the purple lymphocyte nucleus is more homogeneous.

Agranulocytes

Lymphocytes are usually comparable to erythrocytes in size, or only slightly larger. They are sometimes classified by size as small (5–8 μm), medium (10–12 μm), and large (14–17 μm) lymphocytes, but there are gradations between these categories. Medium and large lymphocytes are usually seen in connective tissues and only occasionally in the circulating blood. The lymphocyte nucleus is rounded, often with a slight dimple on one side. It fills almost the entire cell of a small lymphocyte, leaving only a narrow rim of clear, light blue cytoplasm. Large lymphocytes, however, have ample cytoplasm around the nucleus and are sometimes difficult to distinguish from monocytes. There are several subclasses of lymphocytes

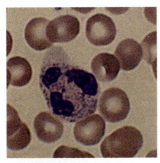

Neutrophils

Percent of WBCs	60%–70%
Mean count	4,150 cells/μL
Diameter	9–12 μm

Characteristic appearance in stained blood films
- Nucleus usually with 3–5 lobes in S- or C-shaped array
- Fine reddish to violet granules in cytoplasm

Differential count
- Increases in bacterial infections

Functions
- Phagocytosis of bacteria
- Release of antimicrobial chemicals

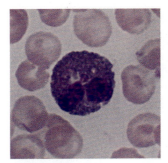

Eosinophils

Percent of WBCs	2%–4%
Mean count	165 cells/μL
Diameter	10–14 μm

Characteristic appearance in stained blood films
- Nucleus usually has two large lobes connected by thin strand
- Conspicuous orange-pink granules in cytoplasm

Differential count
- Increases in parasitic infections, allergies, collagen diseases, and diseases of the spleen and central nervous system
- Fluctuates greatly from day to night, seasonally, and with phase of menstrual cycle

Functions
- Phagocytosis of antigen-antibody complexes, allergens, and inflammatory chemicals
- Aggregate near parasites such as worms and release enzymes that weaken or destroy them

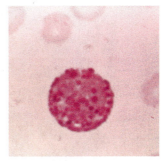

Basophils

Percent of WBCs	<0.5%–1%
Mean count	44 cells/μL
Diameter	8–10 μm

Characteristic appearance in stained blood films
- Nucleus large and irregularly shaped, but typically obscured from view
- Coarse, abundant, dark violet granules in cytoplasm

Differential count
- Relatively stable
- Increases in chicken pox, sinusitis, diabetes mellitus, myxedema, and polycythemia

Functions
- Secrete histamine (a vasodilator), which increases blood flow to a tissue
- Secrete heparin (an anticoagulant), which promotes mobility of other WBCs by preventing clotting

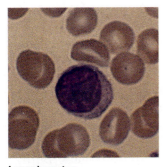

Lymphocytes

Percent of WBCs	25%–33%
Mean count	2,185 cells/μL

Diameter
- Small class — 5–8 μm
- Medium class — 10–12μm
- Large class — 14–17 μm

Characteristic appearance in stained blood films
- Nucleus round or ovoid to slightly dimpled on one side, of uniform dark violet color
- In small lymphocytes, nucleus fills nearly all of the cell and leaves only a scanty rim of clear, light blue cytoplasm
- In large lymphocytes, cytoplasm is more abundant; large lymphocytes may be hard to differentiate from monocytes

Differential count
- Increases in diverse infections and immune responses

Functions
- Several functional classes indistinguishable by light microscopy
- Natural killer (NK) lymphocytes attack cells of the body that are infected with viruses or that have turned cancerous
- B lymphocytes "present" antigens and activate other cells of immune system; differentiate into plasma cells that secrete antibodies; and serve as memory cells in humoral immunity
- T lymphocytes destroy foreign cells, coordinate action of other immune system cells, limit immune response, and serve as memory cells in cellular immunity

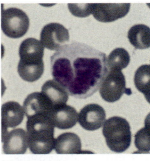

Monocytes

Percent of WBCs	3%–8%
Mean count	456 cells/μL
Diameter	12–15 μm

Characteristic appearance in stained blood films
- Nucleus ovoid, kidney-shaped, or horseshoe-shaped; light violet
- Abundant cytoplasm with sparse, fine granules
- Sometimes very large with stellate or polygonal shapes

Differential count
- Increases in viral infections and inflammation

Functions
- Differentiate into numerous types of macrophages (large phagocytic cells of the tissues
- Macrophages phagocytize pathogens, dead neutrophils, and debris of dead cells; "present" antigens and activate other cells of the immune system

Figure 18.17 White blood cells.

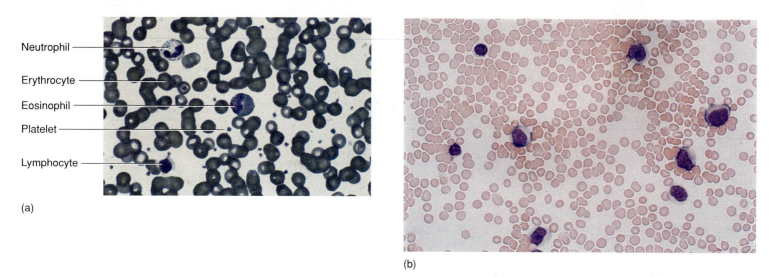

Neutrophil

Erythrocyte

Eosinophil

Platelet

Lymphocyte

(a)

(b)

Figure 18.18 (*a*) A normal blood smear compared to (*b*) blood from a person with acute monocytic leukemia (×400). Note the abnormally high number of white blood cells, especially monocytes, in *b*.

with different immune functions (see chapter 21), but they look alike through the light microscope.

Monocytes are the largest of the formed elements, typically about twice the size of an erythrocyte but sometimes approaching three times as large. Their nuclei are quite variable in shape—round, oval, lobed, kidney-shaped, or C-shaped. The nucleus tends to stain a lighter blue than most leukocyte nuclei. The cytoplasm is abundant and relatively clear. In stained blood films monocytes sometimes appear as very large cells with bizarre stellate (star-shaped) or polygonal contours.

Abnormalities of Leukocyte Count

The total WBC count is normally 5,000 to 10,000 WBCs/μL. A count below this range, called **leukopenia**[19] (LOO-co-PEE-nee-uh), is seen in lead, arsenic, and mercury poisoning; radiation sickness; and such infectious diseases as measles, mumps, chicken pox, poliomyelitis, influenza, typhoid fever, and AIDS. It can also be produced by glucocorticoids, anticancer drugs, and immunosuppressant drugs given to organ transplant patients. A count above 10,000 WBCs/μL, called **leukocytosis**,[20] usually indicates infection, allergy, or other diseases, but can also occur in response to dehydration or emotional disturbances. More useful than a total WBC count is a *differential WBC count*, which identifies what percentage of the total WBC count consists of each type of leukocyte. A high neutrophil count is a sign of bacterial infection; neutrophils become

sharply elevated in appendicitis, for example. A high eosinophil count usually indicates an allergy or a parasitic infection such as hookworms or tapeworms.

Leukemia is a cancer of the hemopoietic tissues that usually produces an extraordinarily high number of circulating leukocytes and their precursors (fig. 18.18). **Myelocytic leukemia** is characterized by uncontrolled granulocyte production, whereas **lymphocytic leukemia** involves uncontrolled lymphocyte production. **Acute leukemia** appears suddenly and progresses rapidly, causing death within a few months if it is not treated. **Chronic leukemia** develops more slowly and may go undetected for many months. Untreated chronic leukemia causes death in about 3 years. Both myelocytic and lymphocytic leukemia occur in acute and chronic forms. The greatest success in treatment and cure has been with acute lymphocytic leukemia, the most common type of childhood cancer. Treatment employs chemotherapy and marrow transplants along with the control of side effects such as anemia, hemorrhaging, and infection.

As leukemic cells proliferate, they replace normal bone marrow and a person suffers from a deficiency of normal granulocytes, erythrocytes, and platelets. Although enormous numbers of leukocytes are produced and spill over into the bloodstream, they are immature cells incapable of performing their normal defensive roles. The deficiency of competent WBCs leaves the patient vulnerable to *opportunistic infection*—the establishment of pathogenic organisms that usually cannot get a foothold in people with healthy immune systems. The RBC deficiency renders the patient anemic and fatigued, and the platelet deficiency results in impaired blood clotting. Death from leukemia is usually due to bleeding and opportunistic infection. Cancerous

19. *penia* = deficiency
20. *osis* = increase

hemopoietic tissue tends to metastasize from the bone marrow or lymph nodes to other organs of the body, where the cells displace or compete with normal cells. Metastasis to the bone tissue itself is common and leads to bone and joint pain.

Key Point Review

22. What is the overall function of leukocytes?
23. What can cause abnormally high or low leukocyte counts?
24. Define *leukemia*. Distinguish between myelocytic and lymphocytic leukemia.

Hemostasis

▼**Objectives**

When you have completed this section, you should be able to
- describe the mechanisms that stop bleeding in the event of injury;
- list the functions of platelets;
- describe the two pathways of blood clotting;
- explain what happens to blood clots when they are no longer needed;
- explain what keeps blood from clotting in the absence of injury; and
- describe some disorders of blood clotting.

Circulatory systems developed very early in animal evolution, and with them evolved mechanisms for stopping leaks that could otherwise be fatal. **Hemostasis**[21] is the stoppage of bleeding. Although hemostatic mechanisms may not stop a hemorrhage from a large blood vessel, they are quite effective at closing breaks in small ones. Platelets play multiple roles in hemostasis, so we begin with a consideration of their form and function.

Platelets

Platelets (see fig. 18.1) are not cells but small fragments of megakaryocyte cytoplasm. Although they were once called *thrombocytes,* this term is now usually reserved for nucleated true cells in other species of animals such as birds and reptiles. Platelets are about 2 to 4 μm in diameter and possess lysosomes, endoplasmic reticulum, a Golgi complex, and Golgi vesicles, or "granules," that contain a variety of factors involved in platelet function. Platelets have pseudopods and are capable of ameboid movement and phagocytosis. In normal blood from a fingerstick, the platelet count ranges from 130,000 to 400,000 platelets/μL (average about 250,000/μL). The

count can vary greatly, however, under different physiological conditions and in blood from different places in the body. When a blood specimen dries on a slide, platelets clump together; therefore in stained blood films, they often appear in clusters.

The importance of platelets in blood clotting has been known for several decades, but a much broader range of functions has come to light in recent years:

- They secrete growth factors that stimulate mitosis in fibroblasts and smooth muscle and help to maintain the linings of blood vessels.
- They secrete vasoconstrictors that cause *vascular spasms* in broken vessels.
- They form temporary *platelet plugs* to stop bleeding.
- They phagocytize and destroy bacteria.
- They secrete chemicals that attract neutrophils and monocytes to sites of inflammation.
- They dissolve blood clots that have outlasted their usefulness.

There are three hemostatic mechanisms—*vascular spasm, platelet plug formation,* and *blood clotting (coagulation)* (fig. 18.19). Platelets play an important role in all three.

Vascular Spasm

The most immediate protection against blood loss is **vascular spasm,** a prompt constriction of the broken vessel. Several things trigger this reaction. An injury stimulates pain receptors, some of which directly innervate nearby blood vessels and cause them to constrict. This effect lasts only a few minutes, but other mechanisms take over by the time it subsides. Injury to the smooth muscle of the blood vessel itself causes a longer-lasting vasoconstriction, and platelets release serotonin, a chemical vasoconstrictor. Thus, the vascular spasm is maintained long enough for the other two hemostatic mechanisms to come into play.

Platelet Plug Formation

Platelets will not adhere to the endothelium (inner lining) of undamaged blood vessels. The endothelium is normally very smooth and coated with **prostacyclin,** a platelet repellent. When a vessel is broken, however, collagen fibers of its wall are exposed to the bloodstream. Upon contact with collagen or other rough surfaces, platelets put out long spiny pseudopods that adhere to the vessel and to other platelets; the pseudopods then contract and draw the walls of the vessel together. The mass of platelets thus formed, called a **platelet plug,** may reduce or stop minor bleeding.

21. *stasis* = stability, to stay the same

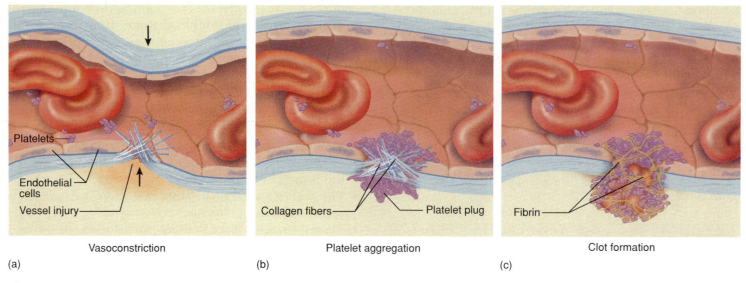

(a) Vasoconstriction   (b) Platelet aggregation   (c) Clot formation

Figure 18.19 Hemostasis at a break in a blood vessel. (*a*) Vasoconstriction reduces blood loss. (*b*) A platelet plug forms as platelets adhere to exposed collagen fibers of the vessel wall. The platelet plug temporarily seals the break. (*c*) A blood clot forms as platelets and erythrocytes become enmeshed in fibrin threads. This forms a longer-lasting seal and gives the vessel a chance to repair itself.

As platelets aggregate, they undergo **degranulation**—the exocytosis of their cytoplasmic granules and release of factors that promote hemostasis. Among these are serotonin, a vasoconstrictor; adenosine diphosphate (ADP), which attracts more platelets to the area and stimulates their degranulation; and **thromboxane A$_2$**, an eicosanoid that promotes platelet aggregation, degranulation, and vasoconstriction. Thus, a positive feedback cycle is activated that can quickly seal a small break in a blood vessel.

Coagulation

Coagulation (clotting) of the blood is the last but most effective defense against bleeding. It is important for the blood to clot quickly when a vessel has been broken, but equally important for it not to clot in the absence of vessel damage. Because of this delicate balance, coagulation is one of the most complex processes in the body, involving over 30 chemical reactions. It is presented here in a very simplified form.

Perhaps clotting is best understood if we first consider its goal. The objective is to convert the plasma protein fibrinogen into **fibrin,** a sticky protein that adheres to the walls of a vessel. As blood cells and platelets arrive, they become stuck to the fibrin like insects sticking to a spider web (fig. 18.20). The resulting mass of fibrin, blood cells, and platelets ideally seals the break in the blood vessel. The complexity of clotting lies in how the fibrin is formed.

There are two reaction pathways to coagulation (fig. 18.21). One of them, the **extrinsic mechanism,** is initiated by clotting factors released by the damaged

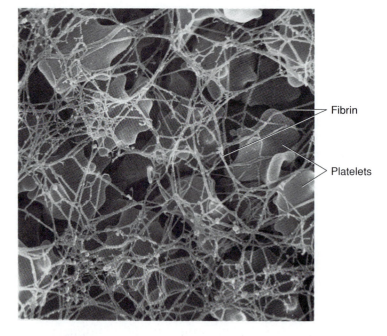

Figure 18.20 A blood clot seen by scanning electron microscopy, showing platelets caught in a mesh of fibrin.

blood vessel and perivascular[22] tissues. The word *extrinsic* refers to the fact that these factors come from sources other than the blood itself. Blood may also clot, however, without these tissue factors—for example, when platelets adhere to a fatty plaque of atherosclerosis or to a test tube. The reaction pathway in this case is

---

22. *peri* = around + *vas* = vessel

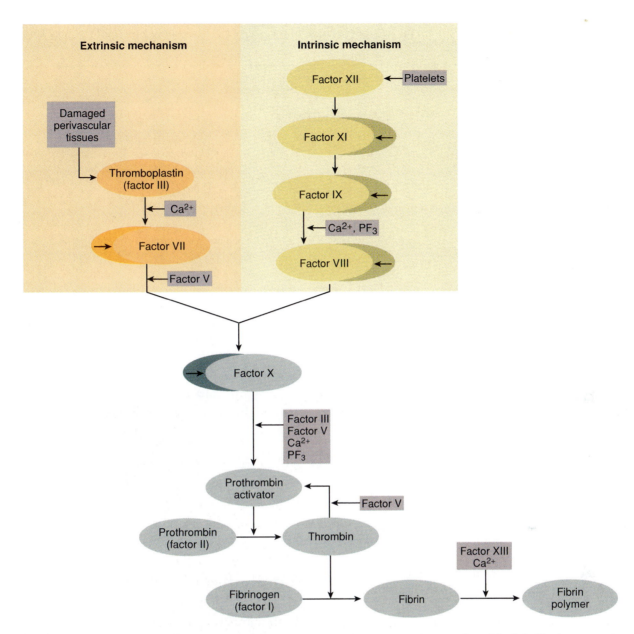

Figure 18.21 The pathways of coagulation. Several clotting factors are converted from an inactive form (*shaded ellipse*) to an active enzyme (*lighter ellipse*) by the preceding clotting factor.

called the **intrinsic mechanism** because it uses only clotting factors found in the blood itself.

Clotting factors (table 18.8) are called **procoagulants,** in contrast to the **anticoagulants** discussed later (see chapter essay, p. 669). Most procoagulants are proteins produced by the liver. They are always present in the plasma in inactive form, but when one factor is activated, it functions as an enzyme that activates the next one in the pathway. That factor activates the next, and so on, in a sequence called a **reaction cascade**—a series of reactions, each of which depends on the product of the preceding one. Many of the clotting factors are identified by Roman numerals, which indicate the order in

which they were discovered, not the order of the reactions. Factors IV and VI are not included in table 18.8. These terms were abandoned when it was found that factor IV was calcium and factor VI was activated factor V. The last four procoagulants in the table are called *platelet factors* (PF$_1$ through PF$_4$) because they are produced by the platelets.

Initiation of Coagulation

The extrinsic mechanism is diagrammed on the left side of figure 18.21. The damaged blood vessel and perivascular tissues release a lipoprotein mixture called **tissue**

Table 18.8 Clotting Factors (Procoagulants)

Number	Name	Origin	Function
I	Fibrinogen	Liver	Precursor of fibrin
II	Prothrombin	Liver	Precursor of thrombin, which converts fibrinogen to fibrin
III	Tissue thromboplastin	Perivascular tissues	Activates factor VII
V	Proaccelerin (labile factor)	Liver	Activates factor VII; combines with factor X to form prothrombin activator
VII	Proconvertin (stable factor)	Liver	Activates factor X in extrinsic pathway
VIII	Antihemophilic factor A	Liver	Activates factor X in intrinsic pathway
IX	Antihemophilic factor B (plasma thromboplastin; Christmas factor)	Liver	Activates factor VIII
X	Thrombokinase (Stuart-Prower factor)	Liver	Combines with factor V to form prothrombin activator
XI	Plasma thromboplastin antecedent (PTC) (antihemophilic factor C)	Liver	Activates factors IX
XII	Hageman factor	Liver and platelets	Activates factor XI and plasmin; converts prekallikrein to kallikrein, the first step in dissolving clots that are no longer needed
XIII	Fibrin-stabilizing factor (fibrinase)	Platelets and plasma	Cross-links fibrin filaments to make fibrin polymer and stabilize clot
PF_1	Platelet accelerator	Platelets	Same role as factor V; also accelerates platelet activation
PF_2	Thrombin accelerator	Platelets	Accelerates thrombin formation
PF_3	Platelet thromboplastic factor	Platelets	Aids in activation of factor VIII and prothrombin activator
PF_4	Platelet factor 4	Platelets	Binds heparin during clotting to inhibit its anticoagulant effect

thromboplastin (factor III). In the presence of Ca^{2+}, tissue thromboplastin activates factor VII, which then activates factor X. The extrinsic and intrinsic pathways differ only in how they arrive at active factor X. Therefore, before examining their common pathway from factor X to the end, let's consider how the intrinsic pathway reaches this step.

The intrinsic mechanism is diagrammed on the right side of figure 18.21. Everything needed to initiate it is present in the plasma or platelets. When platelets degranulate, they release factor XII (Hageman factor, named for the patient in whom it was discovered). Through a cascade of reactions, this leads to activated factors XI, IX, and VIII, in that order—each serving as an enzyme that catalyzes the next step—and finally to factor X. This pathway also requires calcium ions and platelet thromboplastic factor (PF_3).

The cascade of enzymatic reactions acts as an amplifying mechanism to ensure the rapid clotting of blood (fig. 18.22). Each activated enzyme in the pathway produces a larger number of enzyme molecules at the following step. One activated molecule of factor XII at the start of the intrinsic pathway, for example, causes thousands of fibrin molecules to be produced very quickly. Note the similarity of this process to the *enzyme amplification* that occurs in hormone action (see chapter 17, fig. 17.26).

Completion of Coagulation

Once factor X is activated, the remaining events are identical in the intrinsic and extrinsic mechanisms. Factor X combines with factors III and V in the presence of Ca^{2+} and PF_3 to produce an enzyme, *prothrombin activator.* This enzyme acts on a globulin called **prothrombin** (factor II), converting it to the enzyme **thrombin.** Thrombin then converts fibrinogen to fibrin. Fibrin forms a loose mesh at first, but factor XIII causes the formation of covalent cross-links that convert this to *fibrin polymer*—a dense aggregation of fibers that forms the structural basis of the clot.

Once a clot begins to form, it launches a self-accelerating positive feedback process that seals off the damaged vessel more quickly. Thrombin works with factor V to accelerate the production of prothrombin activator, which in turn produces more thrombin. This is a positive feedback cycle, but it must be noted that it serves as part of a larger self-limiting negative feedback loop: trauma results in bleeding, which activates coagulation, which stops the bleeding.

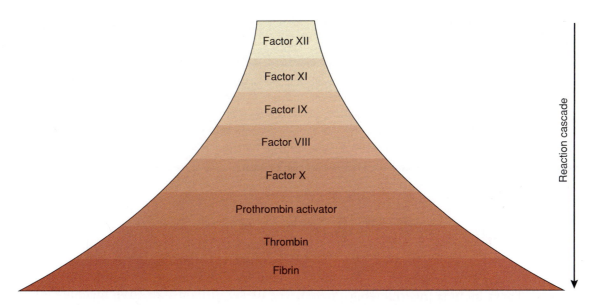

Figure 18.22 Enzyme amplification in blood clotting. Each clotting factor produces many molecules of the next one, so the number of active clotting factors increases rapidly and a large amount of fibrin is quickly formed. The example shown here is for the intrinsic mechanism.

Notice that the extrinsic mechanism requires fewer steps to activate factor X than the intrinsic mechanism does; it is a "shortcut" to coagulation. It takes 3 to 6 minutes for a clot to form by the intrinsic pathway but only 15 seconds or so by the extrinsic pathway. For this reason, when a small wound bleeds, you can stop the bleeding sooner by massaging the site. This releases thromboplastin from the perivascular tissues and activates or speeds up the extrinsic pathway.

A number of laboratory tests are used to evaluate the efficiency of coagulation. Normally, the bleeding of a fingerstick should stop within 2 to 3 minutes, and a sample of blood in a clean test tube should clot within 15 minutes. Other techniques are available that can separately assess the effectiveness of the intrinsic and extrinsic mechanisms.

The Fate of Blood Clots

After a clot has formed, spinous pseudopods of the platelets adhere to strands of fibrin and contract. This pulls on the fibrin threads and draws the edges of the broken vessel together, like a drawstring closing a purse. Through this process of **clot retraction,** the clot becomes more compact within about 30 minutes.

Platelets and endothelial cells secrete a mitotic stimulant named *platelet-derived growth factor (PDGF).* PDGF stimulates fibroblasts and smooth muscle cells to multiply and repair the damaged blood vessel. Fibroblasts also invade the clot and produce fibrous connective tissue, which helps to strengthen and seal the vessel while the repairs take place.

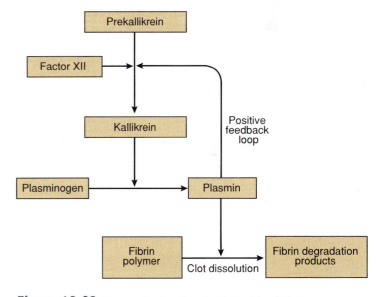

Figure 18.23 The mechanism for dissolving blood clots.

Eventually, tissue repair is completed and the clot must be disposed of. **Fibrinolysis,** the dissolution of a clot, is achieved by a small cascade of reactions with a positive feedback component. In addition to promoting clotting, factor XII catalyzes the formation of a plasma enzyme called **kallikrein** (KAL-ih-KREE-in). Kallikrein, in turn, converts the inactive protein *plasminogen* into **plasmin,** a fibrin-dissolving enzyme that breaks up the clot. Thrombin also activates plasmin, and plasmin indirectly promotes the formation of more kallikrein, thus completing a positive feedback loop (fig. 18.23).

The liver synthesizes most clotting factors. Liver disease such as hepatitis and cirrhosis can therefore lead to deficiencies of blood clotting. Vitamin K is needed as a cofactor for synthesizing factors II, VII, IX, and X, and it is not absorbed well from the small intestine unless bile is present. Gallstones can therefore lead to a clotting deficiency by blocking the bile duct and interfering with bile secretion and vitamin K absorption. Pregnant women should take supplemental vitamin K to ensure that their blood and the baby's will clot efficiently at birth. A vitamin K injection is frequently given to newborn infants.

Prevention of Inappropriate Coagulation

Precise controls are required to prevent coagulation when it is not needed. These include the following:

- **Platelet repulsion.** As noted earlier, platelets do not adhere to the smooth prostacyclin-coated endothelium of undamaged blood vessels.
- **Dilution.** Small amounts of thrombin form spontaneously in the plasma, but at normal rates of blood flow the thrombin is diluted so quickly that a clot has little chance to form. If flow decreases, however, enough thrombin can accumulate to cause clotting. This can happen in circulatory shock, for example, when output from the heart is diminished and circulation slows down.
- **Anticoagulants.** Thrombin formation is suppressed by anticoagulants that are present in the plasma. **Antithrombin,** secreted by the liver, deactivates thrombin before it can act on fibrinogen. **Heparin,** secreted by basophils and mast cells, interferes with the formation of prothrombin activator, blocks the action of thrombin on fibrinogen, and promotes the action of antithrombin. Heparin is given by injection to patients with abnormal clotting tendencies.

Coagulation Disorders

In a process as complex as coagulation, it is not surprising that things can go wrong. Clotting deficiencies can result from causes as diverse as malnutrition, leukemia, and gallstones (see special topic 18.4). Destruction of the bone marrow by radiation, drugs, poisons, or leukemia can cause **thrombocytopenia,** a platelet count below $100,000/\mu L$. People with this disorder exhibit small hemorrhagic spots under the skin, and relatively slight blows cause large **hematomas**[23] (bruises or other masses of clotted blood in the tissues).

---

23. *oma* = mass or tumor

A deficiency of any clotting factor can shut down the coagulation cascade. This happens in **hemophilia,** a family of hereditary diseases characterized by deficiencies of one factor or another. Because of its sex-linked recessive mechanism of heredity, hemophilia occurs predominantly in males. They can inherit it only from their mothers, however, as happened with the descendants of Queen Victoria. The lack of factor VIII causes *classical hemophilia (hemophilia A),* which accounts for about 83% of cases and afflicts 1 in 5,000 males worldwide. Lack of factor IX causes *hemophilia B,* which accounts for 15% of cases and occurs in about 1 out of 30,000 males. Factors VIII and IX are therefore known as *antihemophilic factors A and B.*

Before purified factor VIII became available in the 1960s, more than half of those with hemophilia died before age 5 and only 10% lived to age 21. Physical exertion causes bleeding into the muscles and joints. Intramuscular and joint hematomas can be excruciatingly painful and can eventually immobilize the joints. Hemophilia varies in severity, however. Half of the normal level of clotting factor is enough to prevent the symptoms, and the symptoms are mild even in individuals with as little as 30% of the normal amount. Such cases may go undetected even into adulthood. Bleeding can be relieved for a few days by transfusion of plasma or purified clotting factors. Factor VIII is now produced by transgenic bacteria (see chapter 5 essay, p. 164).

Think About It

Why is it important for people with hemophilia not to use aspirin? (Hint: See chapter 17.)

Far more people die from unwanted blood clotting than from failure of the blood to clot. Most strokes and heart attacks are due to **thrombosis**—the abnormal clotting of blood in an unbroken vessel. A **thrombus** (clot) may grow large enough to obstruct a small vessel, or a piece of it may break loose and begin to travel in the bloodstream as an **embolus.**[24] An embolus may lodge in a small artery and block blood flow from that point on. If that vessel supplies a vital organ such as the heart, brain, lung, or kidney, *infarction* (tissue death) may result. About 650,000 Americans die annually of *thromboembolism* (traveling blood clots) in the cerebral, coronary, and pulmonary arteries.

Thrombosis is more likely to occur in veins than in arteries because blood flows more slowly in the veins and does not dilute thrombin and fibrin as rapidly. It is especially common in the leg veins of inactive people and patients immobilized in a wheelchair or bed. Most venous blood flows directly to the heart and then to the lungs. Therefore, blood clots arising in the legs or arms commonly lodge in the lungs, causing *pulmonary embolism.* When blood cannot circulate freely through the lungs, it cannot receive oxygen and a person may die of hypoxia.

Key Point Review

(25) What are the three basic mechanisms of hemostasis?

(26) How do the extrinsic and intrinsic mechanisms of coagulation differ? What do they have in common?

(27) In what respect does blood clotting represent a negative feedback loop? What part of it is a positive feedback loop?

(28) Describe some of the mechanisms that prevent clotting in undamaged vessels.

(29) Describe a common source and effect of pulmonary embolism.

CHAPTER ESSAY

Clinical Control of Coagulation

For many cardiovascular patients, the goal of treatment is to prevent clotting or to dissolve clots that have already formed. Several strategies employ inorganic salts and products of bacteria, plants, and animals with anticoagulant and clot-dissolving effects.

Preventing Clots from Forming

Since calcium is an essential requirement for blood clotting, blood samples can be kept from clotting by adding a few crystals of sodium oxalate, sodium citrate, or EDTA[25]—salts that bind calcium ions and prevent them from participating in the coagulation reactions. Blood-collection equipment such as hematocrit tubes may also be coated with heparin, a natural anticoagulant whose action was explained earlier.

Since vitamin K is required for the synthesis of clotting factors, anything that antagonizes vitamin K usage makes the blood clot less readily. One vitamin K antagonist is **coumarin**[26] (COO-muh-rin), a sweet-smelling extract of tonka beans, sweet clover, and other plants, used in perfume making. Taken orally by patients at risk for thrombosis, coumarin takes up to 2 days to act, but it has longer-lasting effects than heparin. A similar vitamin K antagonist is the pharmaceutical preparation **Warfarin**[27] **(Coumadin),** which was originally developed as a pesticide—it makes rats bleed to death. Obviously, such anticoagulants must be used in humans with great care.

As explained in the previous chapter, aspirin suppresses the formation of prostaglandins including thromboxane A_2, a factor in platelet aggregation. Low daily doses of aspirin can therefore suppress thrombosis and prevent heart attacks.

Many parasites feed on the blood of vertebrates and secrete anticoagulants to keep the blood flowing. Among these are segmented worms known as leeches. Leeches secrete a local anesthetic that makes their bites painless; therefore, as early as 1567 B.C.E., physicians used them for bloodletting. This method was less painful and repugnant to their patients than *phlebotomy*[28]—cutting a vein—and indeed, leeching became very popular. In seventeenth-century France it was quite the rage; tremendous numbers of leeches were used to treat headaches, insomnia, whooping cough, obesity, tumors, menstrual cramps, mental illness, and almost anything else doctors or their patients imagined to be caused by "bad blood."

The first known anticoagulant was discovered in the saliva of the medicinal leech, *Hirudo medicinalis,* in 1884. Named *hirudin,* it is a peptide of 65 amino acids that prevents clotting by inhibiting thrombin. It causes the blood to flow freely while the leech feeds and for as long as an hour thereafter. While the doctrine of bad blood is now discredited, leeches have lately reentered medical usage (fig. E.1). A major problem in reattaching a severed body part

—continued

24. *em* = in, within + *bolus* = ball, mass
25. Ethylenediaminetetraacetic acid
26. From the Spanish, *coumarú,* the tonka bean tree
27. Acronym for *Wisconsin Alumni Research Foundation*

28. *phlebo* = vein + *tomy* = cutting

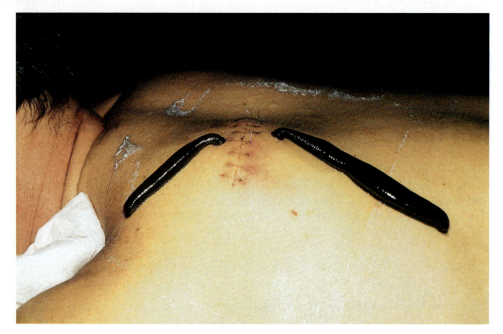

Figure E.1 A modern use of leeching. Two medicinal leeches are being used to remove clotted blood from a postsurgical hematoma. These leeches grow up to 20 cm long.

such as a finger or ear is that the tiny veins draining these organs are too small to reattach surgically. Since arterial blood flows into the reattached organ and cannot flow out, it pools and clots there. This inhibits the regrowth of veins and the flow of fresh blood through the organ and thus often leads to necrosis. Some vascular surgeons now place leeches on the reattached part. Their anticoagulant keeps the blood flowing freely and allows new veins to grow. After 5 to 7 days, venous drainage is restored and leeching can be stopped.

Anticoagulants also occur in the venom of some snakes. *Arvin,* for example, is obtained from the venom of the Malayan viper. It rapidly breaks down fibrinogen and may have potential as a clinical anticoagulant.

Dissolving Clots That Have Already Formed

When a clot has already formed, it can be treated with clot-dissolving drugs such as *streptokinase,* an enzyme made by certain bacteria (streptococci). Intravenous streptokinase is used to dissolve blood clots in coronary vessels, for example. It is nonspecific, however, and digests almost any protein. *Tissue plasminogen activator (TPA)* works faster, is more specific, and is now available from transgenic bacteria. Some anticoagulants of animal origin also work by dissolving fibrin. A giant Amazon leech, *Haementeria,* produces one such anticoagulant named *hementin.* This, too, has been successfully produced by genetically engineered bacteria and used to dissolve blood clots in cardiac patients.▲

Chapter Review   Study Outline

Functions and Properties of Blood (pp. 640–642)
1. Functions (table 18.1)
2. Composition
   a. Plasma
   b. Formed elements
3. Physical properties
   a. Viscosity
   b. Osmolarity

Plasma (pp. 642–645)
1. Plasma and serum
2. Plasma proteins (table 18.3)
   a. Albumins
   b. Globulins
   c. Fibrinogen

3. Nonprotein nitrogenous substances
4. Nutrients
5. Gases
6. Electrolytes

Blood Cell Formation (pp. 645–649)
1. Hemopoietic tissues at various ages
2. Types of hemopoiesis
   a. Myeloid
   b. Lymphoid
3. Erythropoiesis
   a. Role of erythropoietin
   b. Stages in RBC differentiation
   c. Response to hypoxemia
   d. Nutritional requirements
   e. Iron metabolism

4. Leukopoiesis
   a. Pathways of WBC differentiation
   b. Role of colony-stimulating factors
   c. Fate of WBCs after entering circulation
5. Thrombopoiesis

Erythrocytes (pp. 649–655)
1. Form and function
   a. Gas transport functions
   b. Size and shape
   c. Membrane proteins
   d. Lack of organelles
   e. Carbonic anhydrase
2. Hemoglobin
   a. α- and β-globin chains
   b. Heme group

c. Binding of O_2 and CO_2
d. Adult and fetal hemoglobins

3. Quantities of erythrocytes and hemoglobin
   a. Hematocrit
   b. Hemoglobin concentration
   c. RBC count
   d. Gender differences

4. Erythrocyte destruction
   a. Aging and hemolysis of RBCs
   b. Disposal of globin
   c. Iron recycling and storage
   d. Heme and bile pigments

5. Polycythemia
   a. Primary polycythemia
   b. Secondary polycythemia
   c. Effects of chronic polycythemia

6. Anemia
   a. Symptoms and effects
   b. Hemorrhagic anemia
   c. Anemia due to depressed erythropoiesis
      • Insufficient exercise
      • Aging
      • Nutritional anemia
      • Pernicious anemia
      • Hypoplastic and aplastic anemia
   d. Hemolytic anemia
      • Sickle-cell anemia
      • Thalassemia

Blood Types (pp. 655–660)

1. Blood type antigens and antibodies

2. ABO blood group
   a. A and B agglutinogens
   b. Anti-A and anti-B agglutinins
   c. Blood typing procedure
   d. Comparison of ABO types (table 18.7)
   e. Transfusion compatibility

3. Rh blood group
   a. Rh$^+$ and Rh$^-$ blood types
   b. Transfusions and Rh type
   c. Hemolytic disease of the newborn

4. Other blood groups

Leukocytes (pp. 660–663)

1. General appearance and function

2. Granulocytes
   a. Neutrophils
   b. Eosinophils
   c. Basophils

3. Agranulocytes
   a. Lymphocytes
   b. Monocytes

4. Abnormalities of leukocyte count
   a. Leukopenia
   b. Leukocytosis
   c. Leukemia
      • Myelocytic and lymphocytic
      • Chronic and acute
      • Effects of leukemia

Hemostasis (pp. 663–669)

1. Meaning and general mechanisms

2. Platelets
   a. Structure and number
   b. Functions

3. Mechanisms of hemostasis
   a. Vascular spasm
   b. Platelet plug formation
   c. Coagulation

4. Details of blood coagulation
   a. General goal—fibrin production
   b. Intrinsic and extrinsic mechanisms
   c. Clotting factors (procoagulants)
   d. Initiation of coagulation
      • Extrinsic pathway
      • Intrinsic pathway
   e. Completion of coagulation

5. Fate of blood clots
   a. Clot retraction
   b. Role of platelet-derived growth factor
   c. Clot digestion

6. Prevention of inappropriate coagulation
   a. Platelet repulsion
   b. Dilution of thrombin
   c. Anticoagulants

7. Coagulation disorders
   a. Thrombocytopenia
   b. Hemophilia
   c. Thrombosis and embolism

Selected Vocabulary

hematology 640
plasma 640
formed element 640
viscosity 641
osmolarity 642
colloid osmotic pressure (COP) 642
serum 642
albumin 643
globulin 643
fibrinogen 643
nitrogenous wastes 644
myeloid hemopoiesis 645
lymphoid hemopoiesis 645
hemocytoblast 645
erythropoiesis 645
erythropoietin 645
hypoxia 647
hypoxemia 647
gastroferritin 647
transferrin 647
ferritin 647
leukopoiesis 648
macrophage 648

thrombopoiesis 648
megakaryocyte 648
erythrocyte 649
hemoglobin (Hb) 650
globin 651
heme 651
hematocrit 652
hemoglobin concentration 652
RBC count 652
hemolysis 652
biliverdin 652
bilirubin 653
bile pigments 653
primary polycythemia 653
secondary polycythemia 653
hemorrhagic anemia 654
hemolytic anemia 654
sickle-cell anemia 654
thalassemia 655
agglutinogen 655
agglutinin 655
ABO blood group 656
agglutination 657
transfusion reaction 657

Rh blood group 658
hemolytic disease of the newborn (HDN) 659
leukocyte 660
granulocyte 660
agranulocyte 660
neutrophil 660
eosinophil 660
basophil 660
lymphocyte 660
monocyte 662
leukopenia 662
leukocytosis 662
myelocytic leukemia 662
lymphocytic leukemia 662
acute leukemia 662
chronic leukemia 662
hemostasis 663
platelet 663
vascular spasm 663
prostacyclin 663
platelet plug 663
degranulation 664
thromboxane A_2 664

coagulation 664
fibrin 664
extrinsic mechanism 664
intrinsic mechanism 665
procoagulant 665
anticoagulant 665
reaction cascade 665
tissue thromboplastin 665
prothrombin 666
thrombin 666
clot retraction 667
fibrinolysis 667
kallikrein 667
plasmin 667
antithrombin 668
heparin 668
thrombocytopenia 668
hematoma 668
hemophilia 668
thrombosis 668
thrombus 668
embolus 669

1. Antibodies belong to a class of plasma proteins called
   a. albumins.
   b. γ globulins.
   c. α globulins.
   d. procoagulants.
   e. agglutinins.

2. Serum is essentially blood plasma minus
   a. sodium ions.
   b. calcium ions.
   c. clotting proteins.
   d. globulins.
   e. albumins.

3. Which of the following conditions is most likely to cause hemolytic anemia?
   a. folic acid deficiency
   b. iron deficiency
   c. mushroom poisoning
   d. alcoholism
   e. hypoxemia

4. It would be impossible for a type O⁺ baby to have a/an _____ mother.
   a. AB⁻
   b. O⁻
   c. O⁺
   d. A⁺
   e. B⁺

5. Which of the following is *not* a component of hemostasis?
   a. platelet plug formation
   b. agglutination
   c. clot retraction
   d. a vascular spasm
   e. degranulation

6. _____ contribute(s) more to the viscosity of blood than any of these other factors.
   a. Albumin
   b. Sodium
   c. Globulins
   d. Erythrocytes
   e. Fibrin

7. Which of the following is a granulocyte?
   a. a monocyte
   b. a lymphocyte
   c. a macrophage
   d. an eosinophil
   e. an erythrocyte

8. Excess iron is stored in the liver in the form of a complex called
   a. gastroferritin.
   b. transferrin.
   c. ferritin.
   d. hepatoferritin.
   e. erythropoietin.

9. Pernicious anemia would be the result of
   a. hypoxemia.
   b. iron deficiency.
   c. malaria.
   d. lack of intrinsic factor.
   e. hemolytic disease of the newborn.

10. The first clotting factor that the intrinsic and extrinsic pathways have in common is
    a. thromboplastin.
    b. Hageman factor.
    c. factor X.
    d. prothrombin activator.
    e. factor VIII.

11. Production of all the formed elements of the blood is called _____.

12. The percentage of the blood volume composed of RBCs is called the _____.

13. The extrinsic pathway of blood clotting is activated by tissue _____ from damaged perivascular tissues.

14. The erythrocyte antigens that determine transfusion compatibility are called _____.

15. The hereditary lack of factor VIII causes a disease called _____.

16. The cessation of bleeding, due to a combination of mechanisms, is called _____.

17. _____ results from a mutation that changes one amino acid in adult hemoglobin.

18. An excessive RBC count is called _____.

19. Intrinsic factor enables the intestine to absorb _____.

20. The kidney hormone that stimulates red blood cell formation is called _____.

1. Why would erythropoiesis not correct hypoxemia resulting from lung cancer?

2. People with chronic kidney disease often have hematocrits of less than half the normal value. Explain why this may be the case.

3. An older white woman is hit by a city bus and severely injured. Accident investigators are informed that she lives in an abandoned warehouse, where her few personal effects include several empty wine bottles and an expired driver's license indicating she is 72 years old. She is found to be severely anemic. List all the factors you can think of that could contribute to her anemia.

4. How is coagulation different from agglutination?

5. Although fibrinogen and prothrombin are equally necessary for clotting, fibrinogen constitutes about 4% of all plasma protein while prothrombin is a very small fraction. In light of the roles of these clotting factors and your knowledge of enzymes, explain the difference in their abundance.

Web Site Link

For a listing of the most current web sites related to this chapter, please visit the Saladin homepage at:

http://www.mhhe.com/sciencemath/biology/saladin/

[The Circulatory System: The Heart

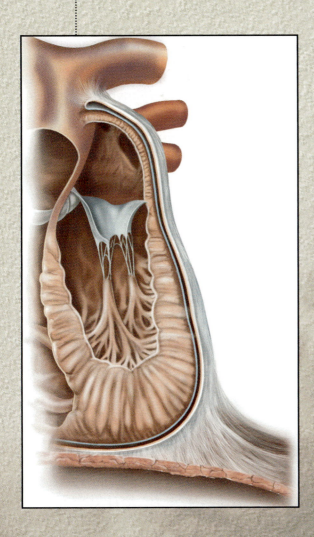

Brushing up

To understand this chapter, it is essential that you understand or brush up on the following concepts:

▶ Desmosomes and gap junctions (pp. 192–193)
▶ Ultrastructure of striated muscle (pp. 395–398)
▶ Excitation-contraction coupling in muscle (p. 404)
▶ Length-tension relationship in muscle fibers (pp. 409–410)
▶ Generator potentials and action potentials (pp. 440–443)

We are more conscious of our heart than we are of most organs and more wary of its possible failure. Speculation about the heart is at least as old as written history. Some ancient Chinese, Egyptian, Greek, and Roman scholars correctly surmised that the heart is a pump for filling the vessels with blood. Aristotle's interpretations, however, were a step backward. Perhaps because the heart quickens its pace when we are emotionally aroused, and because grief causes "heartache," he regarded it primarily as the seat of emotion, doubling as a source of heat to aid digestion. During the Middle Ages, Western medical schools clung dogmatically to the ideas of Aristotle and other ancient authorities. Perhaps the only significant advance came from Muslim medicine, when thirteenth-century physician Ibn an-Nafis described the role of the coronary blood vessels in nourishing the heart. The sixteenth-century dissections and anatomical charts of Vesalius, however, greatly improved knowledge of cardiovascular anatomy and set the stage for a more scientific study of the heart and treatment of its disorders—a science we now call **cardiology.**[1]

In the early decades of the twentieth century, little could be recommended for heart disease other than bed rest. Then nitroglycerin was found to restore coronary circulation and relieve the pain resulting from physical exertion; digitalis proved effective for treating abnormal heart rhythms; and diuretics were first used to promote excretion of water and to reduce hypertension. Coronary bypass surgery, replacement of diseased valves, clot-dissolving enzymes, heart transplants, artificial pacemakers, and artificial hearts have made cardiology one of the most dramatic and attention-getting field of medicine in the last quarter century.

Gross Anatomy

▼Objectives

When you have completed this section, you should be able to

- describe the relationship of the heart to the thoracic cavity and pericardium;
- identify the chambers and valves of the heart and the features of the heart wall;
- trace the flow of blood through the heart chambers and describe the blood supply to the myocardium itself; and
- explain how the structure and function of the adult heart relate to its evolutionary and embryonic developmental histories.

1. *cardio* = heart + *logy* = study

Overview of the Cardiovascular System

The heart and blood vessels constitute the **cardiovascular system,** which transports blood throughout the body. The term **circulatory system** refers to these organs plus the blood and is sometimes meant also to include the lymphatic system (see chapter 21). The cardiovascular system has two major divisions: a **pulmonary circuit,** which serves only the alveoli of the lungs, and a **systemic circuit,** which supplies blood to every organ of the body (fig. 19.1). The right heart serves the pulmonary circuit. It receives blood that has circulated throughout the body, unloaded its oxygen and nutrients, and picked up a load of carbon dioxide. It pumps this *deoxygenated* blood into a large artery, the *pulmonary trunk,* which immediately divides into *right* and *left pulmonary arteries.* These transport blood to the alveoli, where carbon dioxide is unloaded and oxygen is picked up. The freshly *oxygenated* blood then flows to the left heart by way of two *pulmonary veins* on each side.

The left heart serves the systemic circuit. Oxygenated blood leaves the left heart by way of another large artery, the *aorta.* The aorta takes a sharp U-turn, the *aortic arch,* and passes downward behind the heart. From the aortic arch arise arteries that supply the head, neck, and arms. The aorta then travels through the thoracic and abdominal cavities, issuing smaller arteries to all the other organs. After circulating through the body, the now-deoxygenated systemic blood returns to the right heart mainly by way of two large veins, the *superior vena cava* (draining the head, neck, arms, and thoracic organs) and *inferior vena cava* (draining the organs below the diaphragm). The major arteries and veins close to the heart are called the *great vessels* because of their relatively large diameters.

Size, Shape, and Position of the Heart

The heart lies at the center of the thoracic cavity in the mediastinum, the area between the lungs (fig. 19.2). Its broad superior portion, the **base,** is the point of attachment for the great vessels described previously. Its inferior end, the **apex,** tilts to the left and tapers to a blunt point (fig. 19.3). The adult heart is about 9 cm (3.5 in.) wide at the base, 13 cm (5 in.) from base to apex, and 6 cm (2.5 in.) from anterior to posterior at its thickest point—roughly the size of a clenched fist. It weighs about 300 g (10 oz). About two-thirds of the heart lies to the left of the midsagittal plane.

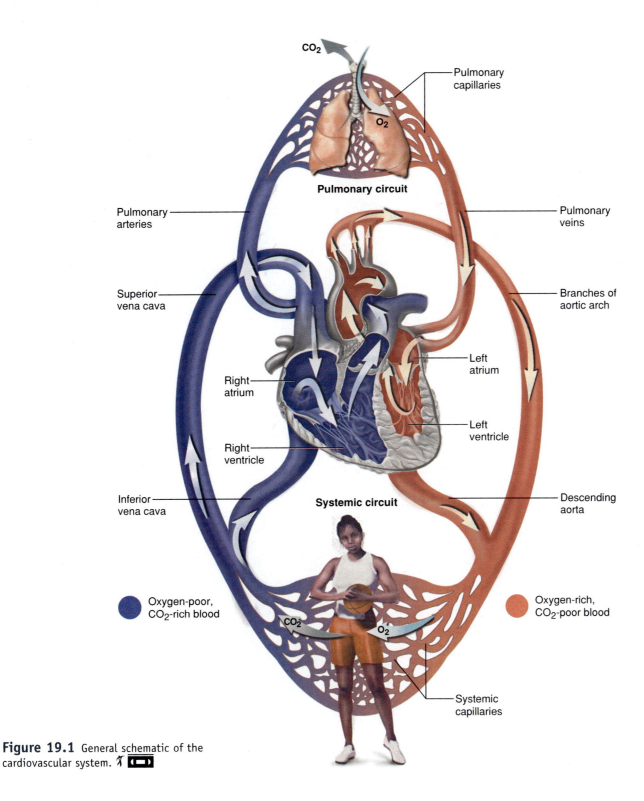

CO₂

Pulmonary
capillaries

O₂

Pulmonary circuit

Pulmonary
arteries

Pulmonary
veins

Superior
vena cava

Branches of
aortic arch

Right
atrium

Left
atrium

Right
ventricle

Left
ventricle

Inferior
vena cava

Systemic circuit

Descending
aorta

Oxygen-poor,
CO₂-rich blood

Oxygen-rich,
CO₂-poor blood

CO₂

O₂

Systemic
capillaries

Figure 19.1 General schematic of the
cardiovascular system.

The Pericardium

The heart is enclosed in a double-walled sac called the
pericardium,[2] which is anchored to the diaphragm
below and to the connective tissue of the great vessels
above the heart (fig. 19.4). The **parietal pericardium,**

---

2. *peri* = around

or *pericardial sac,* consists of a tough *fibrous layer* of
dense, irregular connective tissue and a thin, smooth,
moist *serous layer.* The serous layer turns inward at
the base of the heart and forms the **visceral pericardium** (*epicardium*) covering the heart surface. Between the parietal and visceral pericardia is a space
called the **pericardial cavity.** It contains 5 to 30 mL of
pericardial fluid, an exudate of the serous pericardium

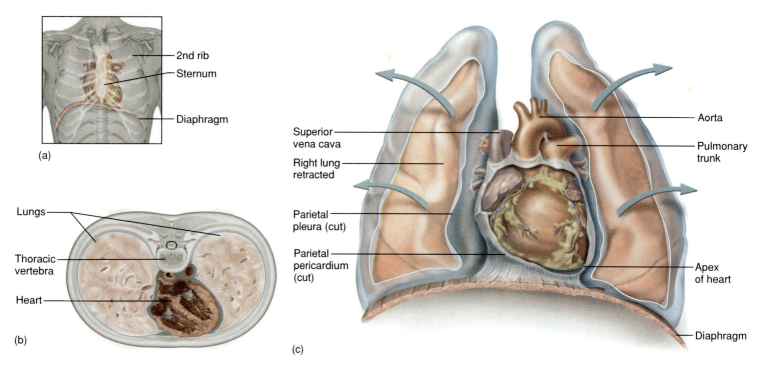

Figure 19.2 Position of the heart in the thoracic cavity. (*a*) Relationship to the thoracic cage. (*b*) Cross section of the thorax at the level of the heart. (*c*) Frontal section of the thoracic cavity with the lungs slightly retracted (*arrows*) and the pericardial sac opened. ⅄

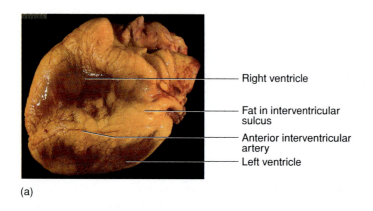

(a)

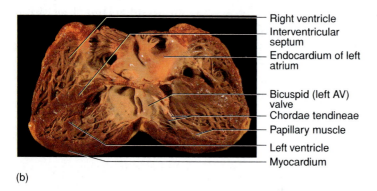

(b)

Figure 19.3 The human heart. (*a*) Anterior aspect; (*b*) internal anatomy.

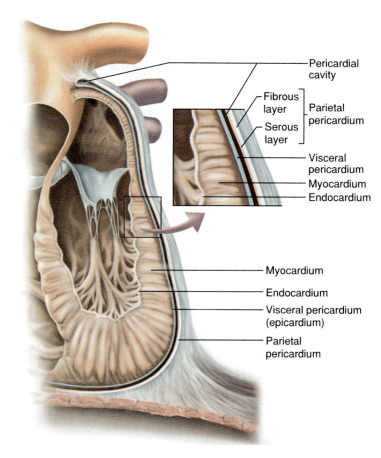

Figure 19.4 Anatomy of the pericardium and heart wall. Layers of the heart wall and its relationship to the pericardium. ⅄

that lubricates the membranes and allows the heart to beat almost without friction. In *pericarditis*—inflammation of the pericardium—the membranes may become dry and produce a painful *friction rub* with each heartbeat. In addition to reducing friction, the pericardium isolates the heart from other thoracic organs, allows it room to expand when it fills with blood, and resists excessive expansion.

The Heart Wall

The heart wall consists of three layers—the epicardium, myocardium, and endocardium (fig. 19.4). In limited regions, the **epicardium**[3] is a translucent serous membrane composed of a simple squamous epithelium overlying a thin layer of areolar tissue. Over much of the heart, however, it has thick deposits of adipose tissue that fill grooves in the heart surface and protect the coronary blood vessels. The **myocardium**,[4] by far the thickest layer, is composed of cardiac muscle and performs the work of the heart. Its muscle fibers spiral around the heart and are bound together by a meshwork of collagenous and elastic fibers that make up the **fibrous skeleton.** The fibrous skeleton has at least three functions: to provide structural support for the heart, especially around the valves and the openings of the great vessels; to give the muscle something to pull against; and, as a nonconductor of electricity, to limit the routes by which electrical excitation can travel through the heart. This is important in the timing and coordination of electrical and contractile activity. Elastic recoil of the fibrous skeleton may also aid the heart in refilling with blood after each beat, but physiologists are not in complete agreement about this. The **endocardium**[5] consists of a simple squamous endothelium overlying a thin areolar tissue layer. It forms the smooth inner lining of the chambers and valves and is continuous with the endothelium of the blood vessels.

The Chambers

The heart has four **chambers** (fig. 19.5*a, b*). The two superior chambers, the **right** and **left atria** (AY-tree-uh; singular *atrium*[6]), receive blood returning to the heart. The atria are mostly posterior in position; therefore, only a small portion of each is visible from the anterior aspect. Each atrium has a small earlike extension called an *auricle*[7] that slightly increases its volume. The two inferior chambers, the **right** and **left ventricles,**[8] are the pumps that eject blood into the arteries. The right ventricle constitutes most of the anterior portion of the heart, while the left ventricle forms the apex and inferoposterior portion.

The heart is crisscrossed by sulci (grooves) that indicate the boundaries of the four chambers. They are occupied largely by fat and coronary blood vessels. The **atrioventricular (coronary**[9]**) sulcus** encircles the heart near its base, separating the atria from the ventricles. The **anterior** and **posterior interventricular sulci** extend vertically from the coronary sulcus toward the apex; they separate the right and left ventricles on the anterior and posterior sides of the heart, respectively.

The four chambers are best seen in frontal section (fig. 19.5*c*). The atria then exhibit thin flaccid walls corresponding to their light workload—all they do is pump blood into the ventricles immediately below. They are separated from each other by the **interatrial septum.** The ventricles are separated by the **interventricular septum.** The right ventricle pumps blood only to the lungs and back, so its wall is only moderately thick and muscular. The left ventricle is two to four times as thick because it bears the greatest workload of all four chambers, pumping blood through the entire body. Both ventricles exhibit internal ridges and folds called **trabeculae carneae**[10] (trah-BEC-you-lee CAR-nee-ee).

The Valves

To function as an effective pump, the heart must have valves that ensure a one-way flow of blood. There is a valve between each atrium and ventricle and at the exit from each ventricle into its great artery (fig. 19.5*c*). Each valve consists of two or three **cusps**—fibrous flaps covered with endothelium.

The **pulmonary valve** guards the opening from the right ventricle into the pulmonary trunk, and the **aortic valve** guards the opening from the left ventricle into the aorta. These are also known as **semilunar**[11] **valves** because of the moonlike shape of their three cusps (fig. 19.6). The **atrioventricular (AV) valves** occupy the openings between the atria and ventricles. The **right AV valve,** also known as the **tricuspid valve,** has three cusps. The **left AV valve,** also known as the **bicuspid,** or **mitral**[12] (MY-trul), **valve,** has two. Stringlike **chordae tendineae** (COR-dee ten-DIN-ee-ee), reminiscent of the shroud lines of a parachute, connect these cusps to conical **papillary muscles** on the floor of the ventricle.

The opening and closing of heart valves is the result of pressure gradients from one side of the valve

---

3. *epi* = upon
4. *myo* = muscle
5. *endo* = internal
6. *atrium* = entryway
7. *auricle* = little ear
8. *ventr* = belly, lower part + *icle* = little

---

9. *coron* = crown + *ary* = pertaining to
10. *trabecula* = little beam + *carne* = flesh, meat
11. *semi* = half + *lunar* = like the moon
12. Named for a bishop's headdress (miter)

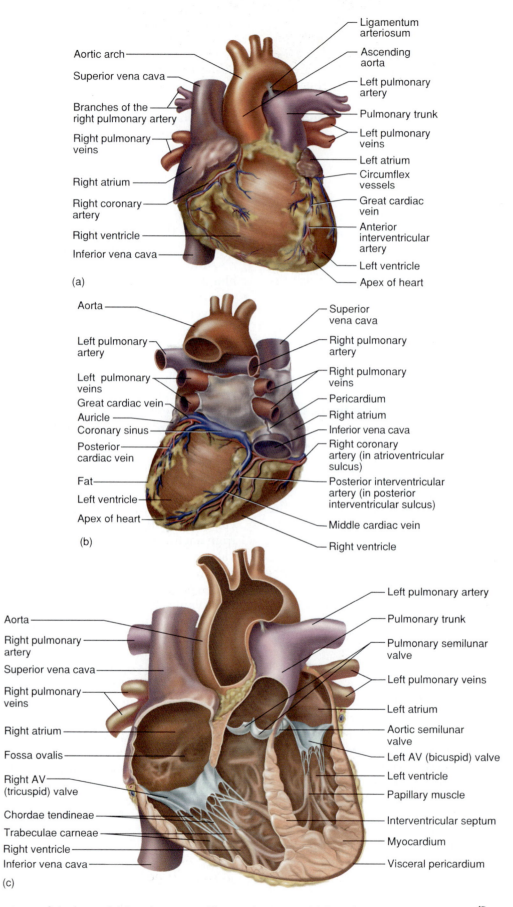

Figure 19.5 Gross anatomy of the heart. (*a*) Anterior aspect; (*b*) posterior aspect; (*c*) frontal section, anterior view. ꭓ

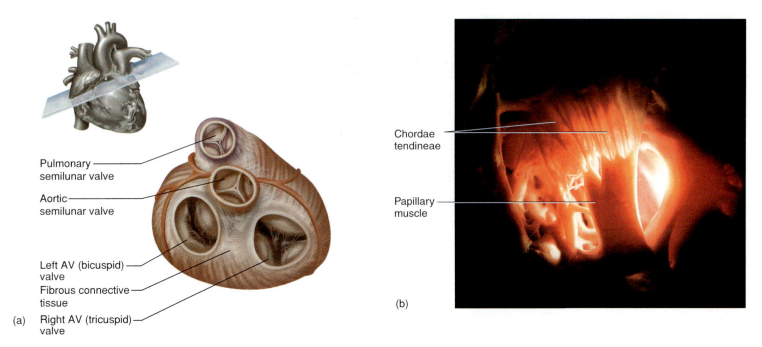

Pulmonary semilunar valve

Aortic semilunar valve

Left AV (bicuspid) valve

Fibrous connective tissue

(a) Right AV (tricuspid) valve

Chordae tendineae

Papillary muscle

(b)

Figure 19.6 The heart valves. (*a*) Superior view of the heart with the atria removed. (*b*) Papillary muscle and chordae tendineae seen from within the right ventricle. The upper ends of the chordae tendineae are attached to the cusps of the right AV valve.

Special Topic Valvular Insufficiency 19.1

Valvular insufficiency (incompetence) refers to any failure of a valve to prevent *reflux (regurgitation)*—the backward flow of blood. In one form of insufficiency, **valvular stenosis**,[13] scar tissue narrows the opening and stiffens the cusps. This increases the workload on the heart as it tries to force blood through the narrow opening, and it allows blood to regurgitate through the valve. Stenosis frequently results from *rheumatic*

fever. A streptococcus infection, often starting as a sore throat, stimulates the formation of antibodies that fail to distinguish bacterial antigens from similar antigens of the mitral and aortic valves. The valves become scarred and constricted, the reflux of blood through them creates turbulence that can be heard as a *heart murmur*.

Mitral valve prolapse (MVP) is an insufficiency in which one or more mitral valve

cusps are pushed back into the atrium during ventricular contraction. It is often hereditary and affects up to 10% of young people. In many cases, it causes no serious dysfunction, but in some people it causes chest pain, fatigue, and shortness of breath. An incompetent valve, most often the mitral valve, may be replaced with an artificial valve or a valve transplanted from a pig heart.

cusps to the other (fig. 19.7*a*). The semilunar valves are forced open when blood pressure in the contracting ventricles is higher than the opposing blood pressure in the great arteries. As the ventricles relax and their pressure falls below that in the arteries, arterial blood flows backward briefly and fills the pocketlike valve cusps. The cusps meet in the middle of the orifice and seal it (fig. 19.7*b*). The AV valves function slightly differently. When the ventricles are relaxed, the AV valve cusps hang down limply and both valves are open (fig. 19.7*c*). Thus, blood can flow freely from the atria into the ventricles. When the ventricles contract, they force blood against the valves. This pushes their cusps together, seals the openings, and prevents blood from flowing back into the atria. The papillary muscles contract with the rest of the ventricular myocardium and tug on the

chordae tendineae, which prevents the valves from being forced backward (prolapsing) into the atria like windblown umbrellas (see special topic 19.1).

Think About It

How would prolapse of the mitral valve affect the amount of blood pumped into the aorta? How might this affect a person's physical stamina? Explain your reasoning.

Blood Flow Through the Heart Chambers

Until the sixteenth century, it was believed that blood flowed directly from the right ventricle into the left

13. *steno* = narrow

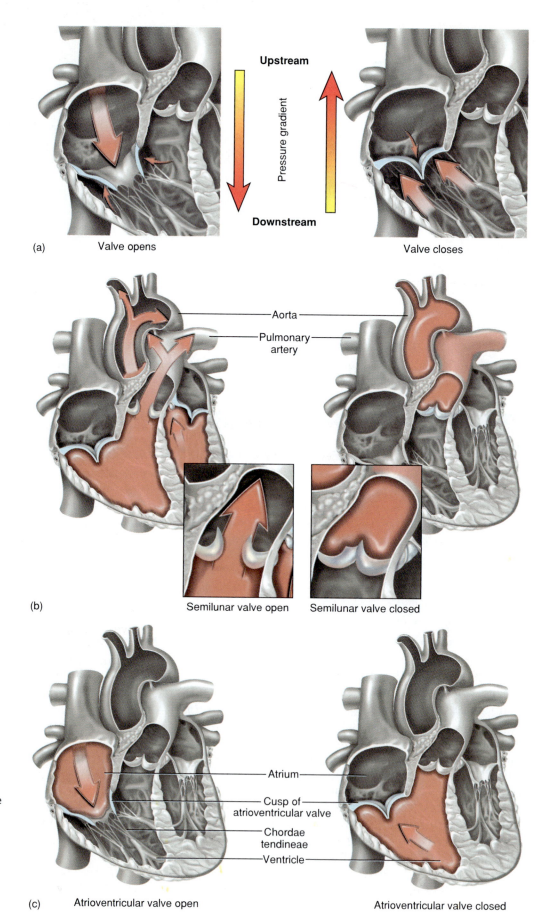

Upstream

Pressure gradient

Downstream

(a) Valve opens Valve closes

Aorta
Pulmonary artery

(b) Semilunar valve open Semilunar valve closed

Atrium
Cusp of atrioventricular valve
Chordae tendineae
Ventricle

(c) Atrioventricular valve open Atrioventricular valve closed

Figure 19.7 Operation of the heart valves. (*a*) A valve is forced open when the blood pressure upstream from it exceeds the pressure downstream from the valve, and it is then forced closed when the action of the heart reverses the pressure gradient. (*b*) Opening and closing of the semilunar valves. (*c*) Opening and closing of the AV valves.

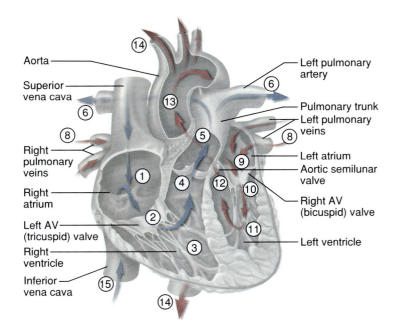

Figure 19.8 The pathway of a single red blood cell from the right atrium and back. (*1*) right atrium → (*2*) right AV valve → (*3*) right ventricle → (*4*) pulmonary valve → (*5*) pulmonary trunk → (*6*) pulmonary arteries → (*7*) lungs (not shown) → (*8*) pulmonary veins → (*9*) left atrium → (*10*) left AV valve → (*11*) left ventricle → (*12*) aortic valve → (*13*) aorta → (*14*) other systemic vessels → (*15*) inferior and superior venae cavae → (*16*) back to the right atrium. The pathway from *5* to *8* is the pulmonary circuit, and the pathway from *13* to *15* is the systemic circuit. ⟨cassette icon⟩

through invisible pores in the septum. This of course is not true. Blood on the right and left sides of the heart is kept entirely separate. The pathway of a single red blood cell as it travels from the right atrium through the body and back to this starting point is shown in figure 19.8.

Blood Flow Through the Myocardium

The heart is a remarkable and dependable little pump. If your heart beats an average of 75 times a minute for 80 years, it will beat more than 3 billion times and pump more than 200 million liters of blood. Understandably, it has a high metabolic rate and requires an abundant supply of oxygen and nutrients. Although it accounts for only 0.5% of the body's weight, the heart uses 5% of its output to meet its own needs. The myocardium thus has an extensive network of blood vessels, the **coronary circulation,** to ensure that fresh blood quickly reaches every muscle cell. At rest, these vessels *perfuse* (supply) the myocardium with about 250 mL of blood per minute.

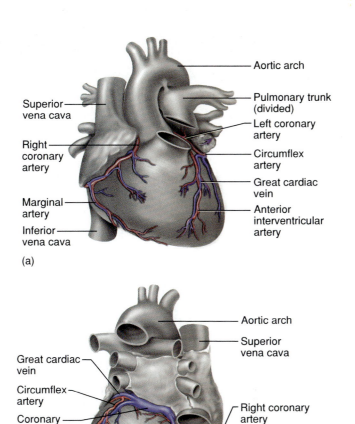

(a)

(b)

Figure 19.9 The coronary blood vessels. (*a*) Anterior aspect of the heart; (*b*) posterior aspect of the heart. ⟨symbol⟩

Arterial Supply

Immediately after the aorta leaves the left ventricle, it gives off right and left coronary arteries (figs. 19.5*a, b* and 19.9) whose openings are tucked behind the cusps of the aortic valve. The **left coronary artery** passes under the left auricle and divides into two branches: the anterior interventricular and circumflex arteries. The **anterior interventricular artery** travels down the anterior interventricular sulcus toward the apex. It issues smaller branches to the interventricular septum and anterior walls of both ventricle. The **circumflex artery** continues around the left side of the heart in the coronary sulcus. It supplies blood to the left atrium and posterior wall of the left ventricle.

The **right coronary artery** supplies the right atrium, continues along the coronary sulcus under the right auricle, and then gives off two branches: the marginal and posterior interventricular arteries. The **marginal artery** supplies the lateral aspect of the right atrium and ventricle. The **posterior interventricular**

Myocardial ischemia[15] (iss-KEE-me-uh) is responsible for about half of all deaths in the United States. It results from blockage of a coronary artery by thrombosis or atherosclerosis and is aggravated by coronary vasoconstriction. Temporary ischemia produces a sense of heaviness or pain in the chest called **angina pectoris**[16] (an-JY-na PEC-toe-riss). As the myocardium becomes hypoxic, it relies increasingly on anaerobic fermentation. This generates lactic acid, which stimulates pain receptors.

Untreated myocardial ischemia can lead to MI. MI produces a sense of heavy pressure or squeezing pain in the chest, often radiating to the shoulder and arm. Infarctions weaken the heart wall and disrupt electrical conduction pathways. This can cause fibrillation and cardiac arrest (discussed later in this chapter). Coronary thrombosis is commonly treated with clot-dissolving agents such as streptokinase or tissue plasminogen activator to restore circulation through the coronary arteries. These may be given in an ambulance on the way to a hospital, since it is critical to restore circulation quickly.

---

artery travels down the corresponding sulcus and supplies the posterior walls of both ventricles.

Think About It

Which ventricle receives the greatest coronary blood supply? Why should it receive a greater supply than the other? List the vessels that supply it.

The energy demand of the cardiac muscle is so critical that an interruption of the blood supply to any part of the myocardium can cause death within minutes. A lack of blood flow to this hardworking muscle, most often caused by a fatty deposit or a blood clot in a coronary artery, can cause a region of tissue death called a **myocardial infarction**[14] **(MI)** (see special topic 19.2 and chapter essay, p. 699). The coronary system has an anatomical feature that minimizes the chance of an MI—*anastomoses* (ah-NASS-tih-MO-seez), points where two arteries come together and combine their blood flow to points farther downstream. The most important anastomosis in the coronary circulation is the point at which the circumflex artery and right coronary artery meet on the posterior side of the heart and combine their blood flow into the posterior interventricular artery. Another is the meeting of the anterior and posterior interventricular arteries at the apex of the heart.

Venous Drainage

Venous drainage refers to the route by which blood leaves an organ. After flowing through the capillaries, coronary blood collects in small veins. These lead to two major *cardiac veins:* the great and middle cardiac veins. The **great cardiac vein** collects blood from the anterior aspect of the heart and travels alongside the anterior interventricular artery in its sulcus. The **middle cardiac vein,** found in the posterior sulcus, collects blood from the posterior aspect of the heart. These and smaller cardiac veins (fig. 19.9) drain into the **coronary sinus,** which passes across the posterior aspect of the heart in the coronary sulcus and empties into the right atrium.

Perfusion in Relation to the Cardiac Cycle

In most of the body, arterial blood flow is greater when the ventricles are contracting than it is when they relax. In the coronary arteries, however, flow is greater when the ventricles relax. There are two reasons for this. First, contraction of the myocardium compresses its own blood vessels and impedes flow. Second, when the ventricles relax, aortic blood flows back into the cusps of the semilunar valve; since the openings of the coronary arteries are right behind these cusps, some of this blood flows from the cusps into the coronary arteries.

Development of the Heart

The structure and function of the human heart are best understood from the standpoint of its developmental history—both evolutionary and embryonic. In fishes, the heart is little more than a pulsating tube, divided into a series of four chambers. They are aligned in a straight row in the embryo and become folded into an S by the time of birth. Blood is taken into one end of the tube and expelled from the other under slightly higher pressure. With the development of lungs in amphibians, there emerged a *double circulation*. In this system, blood is pumped to the lungs for oxygenation and returned to the heart to be repressurized before being pumped to the rest of the body. Amphibian hearts have two atria and a single ventricle, thus oxygenated and deoxygenated blood mix in the ventricle—sufficient for their purposes, but not sufficient to meet the high metabolic demands of a mammal such as ourselves. A

---

14. *infarct* = to stuff

15. *isch* = to hold back + *em* = blood
16. *angina* = to choke, strangle + *pectoris* = of the chest

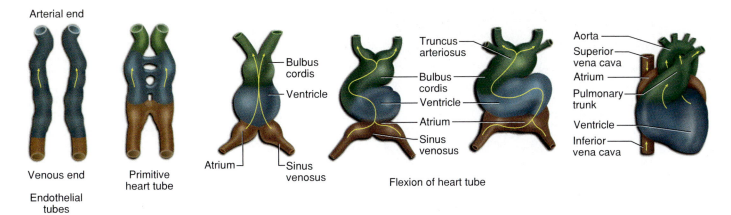

Figure 19.10 Embryonic development of the heart. Arrows indicate direction of blood flow, and colors trace the origin and development of major regions of the heart and great vessels.

four-chambered heart with complete separation of the pulmonary and systemic circuits is achieved only in birds, mammals, and a few reptiles. This ensures that the organs on the systemic circuit receive only "fresh" fully oxygenated blood that has not mixed with "stale" deoxygenated blood. The return of pulmonary blood to the heart to be repressurized by the left ventricle also ensures the speedy delivery of oxygenated blood to the remote reaches of the circulatory system.

Thus, the four chambers of the human heart are not essential to circulation per se; they are essential, however, for the support of our high metabolic rate. Without a four-chambered heart, it is doubtful that birds or mammals (including humans) would exist. This conjecture is supported by some congenital defects of the heart. Some children are born with defects of the interatrial and interventricular septa that allow mixing of oxygenated and deoxygenated blood. They are prone to hypoxia and cyanosis and cannot tolerate physical exertion. Without surgical repair of the heart, they die at a young age.

The evolutionary history of the human heart is somewhat replayed in its embryonic development. Around the third week, the mesoderm forms a pair of ventral *endothelial tubes* (fig. 19.10) that unite side by side to form a primitive *heart tube*. This divides into five regions, some of which correspond to later heart chambers. Two of them, the *ventricle* and *bulbus cordis*, grow especially rapidly, causing the heart to bend into a U, then an S shape reminiscent of the fish heart. Two other regions, the *atrium* and *sinus venosus*, come to lie superior to these. By day 22, the heart begins beating. Over the next 3 weeks it becomes increasingly contorted and divides into four chambers by the formation of interatrial and interventricular septa. The bulbus cordis and *truncus arteriosus* (the fifth primitive chamber) differentiate into the aorta and pulmonary trunk.

In fishes, the heart's one receiving chamber, the *sinus venosus,* contracts first and sets the rhythm for the rest of the heart. In mammals, the sinus venosus becomes greatly reduced and is incorporated into the atrium during embryonic development, forming the *sinoatrial node*—the pacemaker to be discussed later. Chapter 29 discusses changes in the human heart that occur at birth.

Chapter 29 discusses changes in the human heart that occur at birth.

> **Key Point Review**
>
> ① Make a two-color sketch of the pericardium; use one color for the fibrous pericardium and another for the serous pericardium and show their relationship to the heart wall.
>
> ② Trace the flow of blood through the heart, naming each chamber and valve in order.
>
> ③ Define *pulmonary* and *systemic circuit.*
>
> ④ Trace the flow of blood from the left coronary artery to the apex and then to the coronary sinus.

Cardiac Muscle and the Cardiac Conduction System

▼Objectives
When you have completed this section, you should be able to
- describe how the physiological properties of cardiac muscle differ from those of skeletal muscle and relate this to their structural differences;
- explain how the contraction of cardiac muscle is related to its electrophysiology; and
- describe the structure and function of the heart's conduction system.

Structure of Cardiac Muscle

Cardiac muscle behaves much differently from skeletal muscle; to understand why, it is necessary to understand its cellular structure. Cardiac muscle cells, or

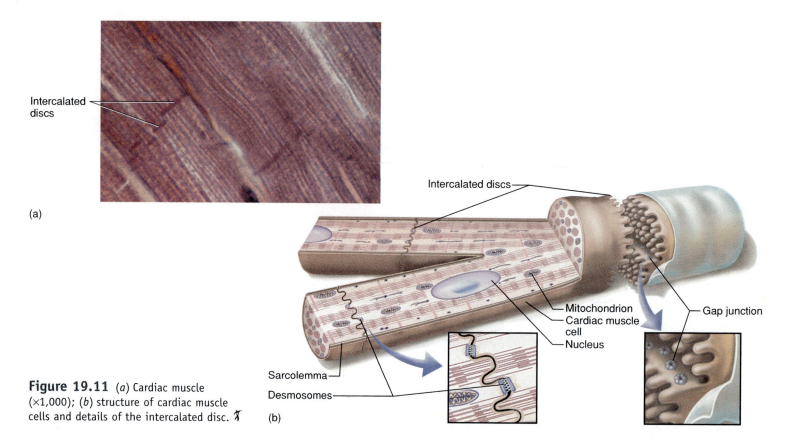

Figure 19.11 (a) Cardiac muscle (×1,000); (b) structure of cardiac muscle cells and details of the intercalated disc.

Labels in figure:
- Intercalated discs
- (a)
- Intercalated discs
- Sarcolemma
- Desmosomes
- Mitochondrion
- Cardiac muscle cell
- Nucleus
- Gap junction
- (b)

myocytes,[17] are different from skeletal muscle fibers in many ways. They are not long fibers like those of skeletal muscle but are relatively short, thick, branching cells (fig. 19.11). They usually have only one nucleus, located near the middle of the fiber. The sarcoplasmic reticulum (SR) is less developed than in skeletal muscle, lacks terminal cisternae, and releases less calcium to the myocytes. However, the T tubules are much larger than in skeletal muscle and admit calcium ions from the extracellular fluid (ECF) to the sarcoplasm during the excitation of the cell. About 25% of the cell volume is filled with especially large mitochondria, compared with 2%, and smaller mitochondria, in skeletal muscle.

Each myocyte is surrounded by a connective tissue endomysium, which allows access for blood capillaries. The endomysium is continuous with the fibrous skeleton. The myocytes are joined end to end by thick connections called **intercalated discs,** which have three distinctive features:

1. The plasma membranes of the adjacent cells display interdigitating folds, like two layers of corrugated cardboard. These folds increase the surface area of intercellular contact.
2. The cells are tightly joined by *desmosomes,* which keep them from pulling apart when they contract (fig. 19.11b).

3. There are *gap junctions* (electrical synapses) between the cells—ion channels from the cytosol of one cell to the next. In contrast to skeletal muscle, in which each cell must be separately stimulated by a nerve fiber, gap junctions allow myocytes to electrically stimulate each other. Thus the entire myocardium of the atria, and that of the ventricles, each acts almost as if it were a single cell. The muscle of a given heart chamber is sometimes described as a *functional syncytium*[18] (sin-SISH-ee-um). A true syncytium is a multinucleate mass formed by the fusion of two or more cells, such as a skeletal muscle fiber. Cardiac myocytes are not a true syncytium, but their electrical junctions enable them to behave as one. Their unified action is essential for the effective pumping of a heart chamber.

Metabolism of Cardiac Muscle

The metabolism of cardiac muscle is almost entirely aerobic and relatively adaptable with respect to the organic fuels used. At rest, the heart gets about 60% of its energy from fatty acids, 35% from glucose, and 5% from other fuels such as ketones, lactic acid, and amino acids. Cardiac muscle is more vulnerable to an oxygen

17. *myo* = muscle + *cyte* = cell

18. *syn* = together + *cyt* = cell

deficiency than it is to the lack of any specific fuel. Because it makes little use of anaerobic fermentation or the oxygen debt mechanism, it is not prone to fatigue. You can easily appreciate this fact by clenching and opening your fist once every second for a minute or two. You will soon feel weakness and fatigue in your skeletal muscles and perhaps feel all the more grateful that cardiac muscle can maintain its rhythm, without fatigue, for a lifetime.

The Cardiac Conduction System

Among invertebrates such as clams, crabs, and insects, each heartbeat is triggered by a pacemaker in the nervous system. Vertebrate hearts, however, are said to be **myogenic** because the pacemaker is in the heart itself and is derived from cardiac muscle. Autonomic nerve fibers to the heart can modify its rhythm, but they do not create it—the heart goes on beating even if all nerve connections to it are severed. Indeed, we can remove a vertebrate heart from the body, keep it in aerated physiological saline, and it will beat for hours to days. Cut the heart into little pieces, and each piece continues its own rhythmic pulsations.

Cardiac myocytes are said to be **autorhythmic**[19] because they depolarize spontaneously at regular time intervals without external stimulation. Some of them lose their ability to contract and become specialized, instead, for generating and transmitting action potentials. These constitute the **cardiac conduction system,** which consists of the following components (fig. 19.12):

1. The **sinoatrial (SA) node** is a patch of modified myocytes in the right atrium, near the superior vena cava, just deep to the epicardium. It is the **pacemaker** that initiates each heartbeat and normally determines the heart rate. Signals from the SA node spread throughout the atria, as shown by the yellow arrows in figure 19.12.
2. The **atrioventricular (AV) node** is located at the lower end of the interatrial septum, near the tricuspid (right AV) valve. It acts as an electrical gateway to the ventricles because the fibrous skeleton acts as an insulator to prevent passage of currents from the atria to the ventricles by any other route.
3. The **atrioventricular (AV) bundle,** or *bundle of His,*[20] is a route in the posterior part of the interatrial septum by which signals leave the AV node.
4. The AV bundle quickly divides into **right** and **left bundle branches** that enter the interventricular septum and descend toward the apex.

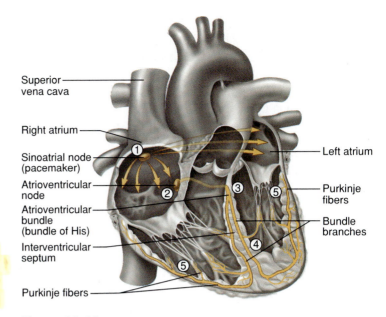

Figure 19.12 The cardiac conduction system. (*1*) SA node; (*2*) AV node; (*3*) AV bundle; (*4*) bundle branches; (*5*) Purkinje fibers. ⚲

5. As it approaches the apex, each bundle branch frays into numerous nervelike **Purkinje**[21] (pur-KIN-jee) **fibers.** At the apex, these turn upward and branch throughout the ventricular myocardium, forming a more elaborate network in the left ventricle than in the right.

In addition to setting the frequency of the heartbeat, the cardiac conduction system controls the route and timing of electrical conduction to ensure that the four chambers are coordinated with each other. The failure of any part of the cardiac conduction system to transmit signals is called **heart block;** usually it results from disease and degeneration of conduction system fibers. A *bundle branch block,* for example, is due to damage to one or both bundle branches. Damage to the AV node causes *total heart block,* in which signals from the atria fail to reach the ventricles and the ventricles beat at their own intrinsic rhythm of 20 to 40 beats per minute (bpm) (see the discussion of ectopic foci in the next section).

········· **Key Point Review** ·········

5 What organelle(s) are less developed in cardiac muscle than in skeletal muscle? What organelle(s) are more developed? What is the functional significance of these differences?

6 Name two types of cell junctions in the intercalated discs and explain their functional importance.

7 Why is the human heart described as myogenic? Where is its pacemaker and what is it called?

8 List the components of the cardiac conduction system in the order traveled by signals from the pacemaker.

---

19. *auto* = self
20. Wilhelm His, Jr. (1863–1934), German physiologist

21. Johannes E. Purkinje (1787–1869), Bohemian physiologist

Atrial flutter is a condition in which ectopic foci in the atria set off extra contractions and the atria beat 200 to 400 times/minute. Early firing of an ectopic focus can also cause **premature ventricular contractions (PVCs)**, which occur singly or in bursts. PVCs are often due to irritation of the heart by stimulants, emotional stress, or lack of sleep, but they sometimes indicate more serious pathology.

PVCs can lead to **ventricular fibrillation,** an arrhythmia caused by electrical signals arriving at different regions of the myocardium at widely different times. A fibrillating ventricle exhibits squirming, uncoordinated contractions; it has been described as looking like a "bag of worms." Since it does not pump blood, there is no coronary perfusion and the myocardium rapidly dies of ischemia. **Cardiac arrest** refers to a cessation of cardiac output, with the ventricles either motionless or in fibrillation. Death is imminent if fibrillation cannot be stopped.

Defibrillation is an emergency procedure in which the heart is given a strong electric shock with a pair of electrodes. Its objective is to depolarize the entire myocardium and stop the fibrillation in the hope that the SA node will resume its sinus rhythm. This does not correct the underlying cause of the arrhythmia, but it may sustain a patient's life long enough to allow for other corrective action.

Electrical and Contractile Activity of the Heart

▼Objectives

When you have completed this section, you should be able to
- explain why the SA node fires spontaneously and rhythmically;
- explain how the SA node excites the myocardium;
- describe the unusual action potentials of cardiac myocytes and relate them to the contractile behavior of this type of muscle; and
- interpret a normal electrocardiogram in relation to the electrical activities of the myocardium.

The contractions of the heart are initiated by depolarization of the SA node and governed by the electrophysiological properties of the cardiac conduction system and myocytes. In this section, we examine how the electrical events in the heart produce its cycle of contraction and relaxation.

The contraction of any chamber of the heart, called **systole** (SIS-toe-lee), results from the depolarization of its myocytes. Relaxation of any chamber, called **diastole** (dy-ASS-toe-lee), occurs when the myocytes repolarize. Used without reference to specific chambers, the words *systole* and *diastole* refer to the more conspicuous and important ventricular action, which ejects blood from the heart.

The Cardiac Rhythm

The normal heartbeat generated by the SA node is called the **sinus rhythm.** It typically shows a resting rate of 70 to 80 bpm. As discussed later, this rhythm would be faster if not for the inhibitory effect of the vagus nerves, which slow down the SA node. Other parts of the conduction system, however, fire spontaneously if not stimulated by the SA node first. Any region of spontaneous firing other than the SA node is called an **ectopic[22] focus.** Ectopic foci sometimes produce an extra heartbeat (*extrasystole*) as a result of caffeine, nicotine, other drugs, electrolyte imbalance, or hypoxia.

If the SA node is damaged, an ectopic focus may take over the governance of the heart rhythm. The most common ectopic focus is the AV node, which has a slower **nodal rhythm** of 40 to 50 bpm. This is sufficient to sustain life. If neither the SA nor AV node is functioning, however, other ectopic foci fire at rates of 20 to 40 bpm. This is too slow for survival, mainly because of inadequate perfusion of the brain. This condition calls for an artificial pacemaker. Any abnormal cardiac rhythm is called **arrhythmia[23]** (see special topic 19.3).

Physiology of the SA Node

Why does the SA node spontaneously fire every 0.8 seconds? Unlike neurons, skeletal muscle fibers, or even the contractile myocytes of the heart, the autorhythmic cells of the SA node do not have a stable resting membrane potential (RMP). Their membrane potential starts at about −60 mV and drifts spontaneously upward. This gradual depolarization is called the **pacemaker potential** (fig. 19.13). The reason for it is uncertain. One theory is that membrane K^+ channels close, and the slow influx of Na^+ without a compensating efflux of K^+ causes gradual depolarization.

When the pacemaker potential reaches a threshold of −40 mV, voltage-regulated **fast calcium channels** open and Ca^{2+} rushes in from the ECF. This produces the rising phase of the action potential, which peaks slightly above 0 mV. At this point, K^+ channels open and potassium ions rush out of the cell down their concentration gradient. This makes the cytosol increasingly

---

22. *ec* = out of + *top* = place
23. *a* = without + *rhythm* + *ia* = condition

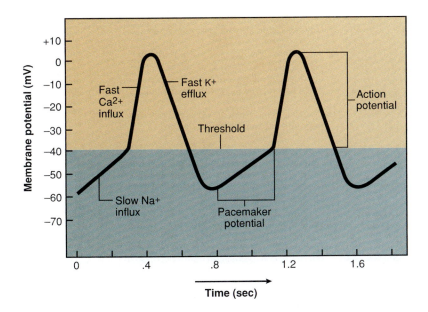

Figure 19.13 Pacemaker potentials and action potentials of the SA node.

negative and creates the falling phase of the action potential. When repolarization is complete, the K+ channels close again and the pacemaker potential starts over, on its way to producing the next heartbeat.

One depolarization of the SA node sets off one heartbeat. Although all cells of the cardiac conduction system are autorhythmic, the SA node is generally first to depolarize. When it does, it depolarizes all the other components in the series; the SA node thus serves as the system's pacemaker. At rest and with normal *vagal tone* (the steady background firing rate of the vagus nerves), the SA node fires every 0.8 second or so, creating a heart rate of about 75 bpm.

Impulse Conduction to the Myocardium

Firing of the SA node excites atrial myocytes and stimulates the two atria to contract almost simultaneously. Impulses from the SA node reach the AV node in about 50 msec. While impulses travel about 1 m/sec through the atria, the AV node has much thinner myocytes that slow the impulse down to about 0.05 m/sec. The signal is therefore delayed at the AV node for about 100 msec—like highway traffic slowing down at a small town. This delay is essential because it gives the ventricles time to fill with blood before they begin to contract.

The AV bundle and Purkinje fibers have a conduction speed of 4 m/sec, the fastest in the conduction system. Ordinary myocytes of the ventricles have conduction speeds of only 0.3 to 0.5 m/sec. If they were the only route of travel through the ventricular myocardium, some myocytes would be stimulated much sooner than others. Ventricular contraction would not be synchronized and the pumping effectiveness of the

ventricles would be severely compromised. Since the signals travel through the fast Purkinje fibers, however, the entire ventricular myocardium depolarizes within 200 msec after the SA node fires, causing the ventricles to contract in near unison.

Signals reach the papillary muscles before the rest of the myocardium. Thus, these muscles contract and begin taking up slack in the chordae tendineae an instant before ventricular contraction causes blood to surge against the AV valves. Ventricular contraction begins at the apex, which is first to be stimulated, and progresses upward—pushing the blood upward toward the semilunar valves. Because of the spiral arrangement of ventricular muscle fibers, the ventricles twist slightly as they contract, like someone wringing out a towel.

Electrical Behavior of the Myocardium

Contractile myocytes exhibit action potentials that are significantly different from those of neurons and skeletal muscle (fig. 19.14). Unlike the autorhythmic cells of the heart, they have a stable RMP of −90 mV and depolarize only when stimulated. A stimulus opens voltage-regulated sodium gates, causing a Na+ influx and depolarizing the cell to its threshold. This rapidly opens additional Na+ gates and triggers a positive feedback cycle like the one seen in the firing of a neuron (see p. 441). The action potential peaks at nearly +30 mV. The Na+ gates close quickly, and the rising phase of the action potential is very brief.

Depolarization of a myocyte triggers Ca2+ release from the SR. In contrast to the events in skeletal muscle, however, the relatively sparse SR of cardiac muscle does not release enough calcium to bring about contraction.

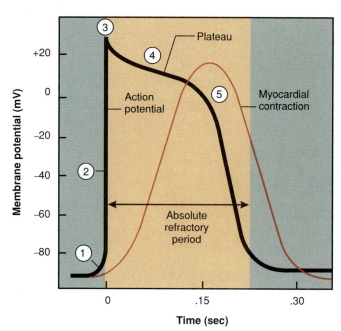

1. Voltage-gated sodium channels open.

2. A Na$^+$ influx depolarizes the membrane and triggers the opening of still more Na$^+$ channels, creating a positive feedback cycle and rapidly rising membrane voltage.

3. Sodium channels close when the cell depolarizes, and the voltage peaks at nearly +30 mV.

4. Calcium ions entering through slow calcium channels prolong the depolarization of the membrane, creating a plateau. Potassium channels are closed during this phase, preventing K$^+$ from leaving the cell and repolarizing the membrane.

5. Calcium channels close, Ca^{2+} is pumped out of the cell, potassium channels open, K$^+$ diffuses out of the cell, and the membrane returns to its resting membrane potential.

Figure 19.14 Action potential of a ventricular myocyte, and a curve of the tension generated by the myocardial contraction.

Supplemental Ca^{2+} must come from the ECF. Thus depolarization also opens **slow calcium channels** in the plasma membrane—so named because their opening is delayed compared to that of the faster Ca^{2+} channels of the SR and autorhythmic cells. Calcium binds to troponin and activates the sliding filament mechanism of contraction, just as it does in the excitation-contraction coupling of skeletal muscle.

In skeletal muscle and neurons, an action potential falls back to the RMP within 2 msec. In cardiac muscle, however, the slow Ca^{2+} channels remain open long after the Na$^+$ channels close, thus prolonging depolarization of the cell. At a resting heart rate of 70 to 80 bpm, this produces a plateau 200 to 250 msec long. Potassium channels also close during this period, which contributes to the plateau. If K$^+$ were allowed to diffuse rapidly out of the cell, it would repolarize the membrane.

As long as the action potential is in its plateau and Ca^{2+} is entering the myocytes, the myocytes contract. Thus, in figure 19.14, you can see the development of muscle tension (myocardial contraction) following closely behind the depolarization and plateau. Rather than showing a brief twitch like skeletal muscle, cardiac muscle has the more sustained contraction necessary for expulsion of blood from the heart chambers. Both atrial and ventricular myocytes exhibit these plateaus, but they are more pronounced in the ventricles.

After 200 to 250 msec, the Ca^{2+} channels close and K$^+$ channels open. K$^+$ diffuses out of the cell and Ca^{2+} is pumped back into the ECF and SR. Membrane voltage drops rapidly, and muscle tension declines soon afterward.

Cardiac muscle has an *absolute refractory period* of 250 msec, compared with 1 to 2 msec in skeletal muscle. This prevents wave summation and tetany, which would stop the pumping action of the heart.

Think About It

With regard to the ions involved, how does the falling (repolarization) phase of a myocardial action potential differ from that of a neuron's action potential? (see p. 441.)

The Electrocardiogram

Electrical currents generated in the heart can be detected by recording electrodes (leads) on the skin. An instrument called the *electrocardiograph* amplifies these signals and produces a record, usually on a moving paper chart, called an **electrocardiogram**[24] (**ECG** or **EKG**[25]). To record an ECG, electrodes are attached to the arms, legs, and six locations on the chest. Several simultaneous recordings can be made from electrodes at different distances from the heart; collectively, they provide a comprehensive image of the heart's electrical activity.

Figure 19.15 shows a typical ECG. An ECG is a composite recording of all the action potentials produced by the nodal and myocardial cells—it should not be misconstrued as a tracing of a single action potential. It shows three principal deflections above and below the baseline: the *P wave, QRS complex,* and *T wave.* Figure 19.16 shows how these correspond to

24. *graph* = recording instrument; *graphy* = recording procedure; *gram* = record of
25. EKG is from the German spelling, Elektrokardiogramm

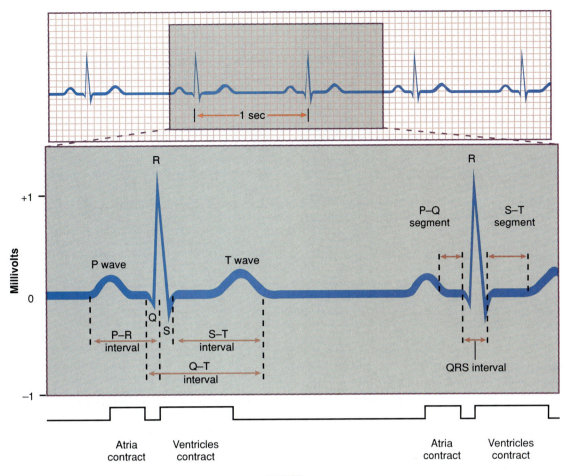

Figure 19.15 Components of the normal electrocardiogram (ECG). 🎞️

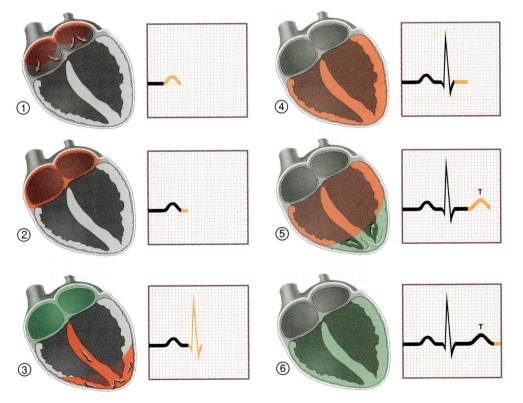

1. Atria begin depolarizing.
2. Atrial depolarization complete.
3. Ventricular depolarization begins at apex and progresses superiorly as atria repolarize.
4. Ventricular depolarization complete
5. Ventricular repolarization begins at apex and progresses superiorly.
6. Ventricular repolarization complete; heart is ready for the next cycle.

Figure 19.16 Relationship of the ECG to electrical activity in the myocardium. Each heart diagram indicates the events occurring at the time of the colored segment of the ECG. Red indicates depolarizing or depolarized myocardium, and green indicates repolarizing or repolarized myocardium. Arrows indicate the direction in which a wave of depolarization or repolarization is traveling. 🎞️

Table 19.1 Examples of the Diagnostic Interpretation of Abnormal Electrocardiograms

Appearance	Suggested Meaning
Enlarged P wave	Atrial hypertrophy, often a result of mitral valve stenosis
Missing or inverted P wave	SA node damage; AV node has taken over pacemaker role
Two or more P waves per cycle	Extrasystole; heart block
Enlarged Q wave	Myocardial infarction
Enlarged R wave	Ventricular hypertrophy
Abnormal T waves	Flattened in hypoxia; elevated in hyperkalemia (K^+ excess)
Abnormally long P–Q interval	Scarring of atrial myocardium, forcing impulses to bypass normal conduction pathways and take slower alternative routes to AV node
Abnormal S–T segment	Elevated above baseline in myocardial infarction; depressed below baseline in myocardial hypoxia

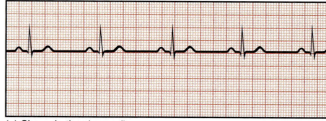

(a) Sinus rhythm (normal)

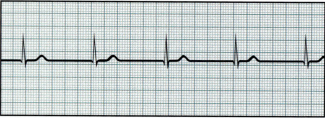

(b) Nodal rhythm – no SA node activity

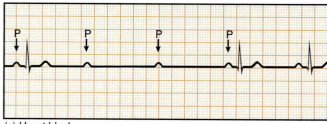

(c) Heart block

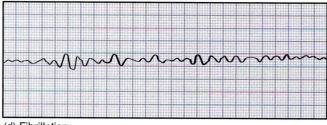

(d) Fibrillation

Figure 19.17 Normal and pathological electrocardiograms. (*a*) Normal sinus rhythm. (*b*) Nodal rhythm generated by the AV node in the absence of SA node activity; note the lack of P waves. (*c*) Heart block, in which some P waves are not transmitted through the AV node and do not generate QRS complexes. (*d*) Ventricular fibrillation, with grossly irregular waves of depolarization.

regions of the heart undergoing depolarization and repolarization.

The **P wave** is produced when a signal from the SA node spreads through the atria and depolarizes them. Atrial systole begins about 100 msec after the P wave begins, during the *P–Q segment*. This segment is about 160 msec long and represents the time required for impulses to travel from the SA node to the AV node.

The **QRS complex** consists of a small downward deflection (Q), a tall sharp peak (R), and a final downward deflection (S). It marks the firing of the AV node and the onset of ventricular depolarization. Its complex shape is due to the different sizes of the two ventricles and the different times required for them to depolarize. Ventricular systole begins shortly after the QRS complex in the *S–T segment*. Atrial repolarization and diastole also occur during the QRS interval, but atrial repolarization sends a relatively weak signal that is obscured by the activity of the more muscular ventricles. The S–T segment corresponds to the plateau in the myocardial action potential and thus represents the time during which the ventricles contract and eject blood.

The **T wave** is generated by ventricular repolarization immediately before diastole. The ventricles take longer to repolarize than to depolarize; the T wave is therefore smaller and more spread out, and it has a rounder peak than the QRS complex. Even in cases where the T wave is taller than the QRS complex, it can be recognized by its relatively rounded peak.

The ECG affords a wealth of information about the normal electrical activity of the heart. Deviations from normal are invaluable to the clinician in diagnosing abnormalities in the conduction pathways, myocardial infarction, enlargement of the heart, and electrolyte and hormone imbalances. A few examples are given in table 19.1 and figure 19.17.

Key Point Review

9 Define *systole* and *diastole*.

10 How does the pacemaker potential of the SA node differ from the RMP of skeletal muscle and neurons? Why is this important in creating the heart rhythm?

11. In what way is excitation-contraction coupling in cardiac muscle similar to that of skeletal muscle? In what way is it different?

12. What produces the plateau in the action potentials of myocytes? Why is this important with respect to the pumping ability of the heart?

13. Name the waves of the ECG and explain what myocardial events produce each wave.

Blood Flow, Heart Sounds, and the Cardiac Cycle

▼ Objectives

When you have completed this section, you should be able to
- explain how pressure and resistance determine the flow of a fluid such as blood;
- explain what causes the sounds of the beating heart;
- describe what is occurring in the heart in each phase of the cardiac cycle; and
- relate the events of the cardiac cycle to the volume of blood entering and leaving the heart.

A **cardiac cycle** consists of one complete cycle of contraction and relaxation. Its major events, in order of occurrence, are atrial systole, ventricular systole, ventricular diastole, and a *quiescent period* before the next atrial systole begins. In this section, we examine these events more closely to see how they relate to the entry and expulsion of blood. First, however, we consider two related issues: (1) some general principles of pressure changes and how they affect the flow of blood, and (2) the heart sounds produced during the cardiac cycle, which we can then relate to the stages of the cycle.

Principles of Pressure and Flow

A fluid is any liquid or gas—a state of matter that can flow in bulk from one place to another. In this and some forthcoming chapters, we are concerned with factors that govern the flow of fluids such as blood, lymph, air, and urine. Some basic principles of fluid movement (*fluid dynamics*) are therefore important to understand at this time. Fluid dynamics are essentially a matter of pressure versus resistance. Pressure can cause a fluid to flow, and resistance opposes flow.

Measurement of Pressure

Pressure is often measured by observing how high it can push a column of mercury (Hg) up an evacuated tube called a *manometer*. Mercury is used because it is very dense and enables us to measure pressure with shorter columns than we would need with a less dense liquid such as water. Because pressures are compared to the force generated by a column of mercury, they are expressed in terms of millimeters of mercury (mmHg).

Blood pressure is usually measured with a **sphygmomanometer**[26] (SFIG-mo-ma-NOM-eh-tur)—a calibrated tube filled with mercury and attached to an inflatable pressure cuff wrapped around the arm. Blood pressure and the method of measuring it are discussed in greater detail in the next chapter.

Pressure Gradients and Flow

Any change in the volume of a container creates a **pressure gradient,** or difference, between the inside and outside of the container. If there is an opening in the container, fluid flows in or out, "down the gradient," from point A, where pressure is higher, to point B, where pressure is lower. This raises the pressure at point B and lowers the pressure at point A until the two are equal. At that time, there is no more pressure gradient and flow stops. Flow also stops, of course, if it is obstructed by the closure of a passage between point A and B—a matter of obvious relevance where the heart valves are concerned.

Suppose you pull back the plunger of a syringe, for example. The volume in the syringe barrel increases and its pressure falls (fig. 19.18). Since air pressure outside the syringe is greater than the pressure inside, air flows into it until the pressures inside and outside are equal. If you then push the plunger in, pressure inside rises above the pressure outside, and air flows out—again following a gradient from high pressure to low.

The syringe barrel is analogous to a heart chamber such as the left ventricle. When the ventricle is expanding, its internal pressure falls. If the AV valve is open, blood flows into the ventricle from the atrium above. When the ventricle contracts, its internal pressure rises. When the aortic valve opens, blood is ejected from the ventricle into the aorta.

A pressure difference does not guarantee that a fluid will flow. There is always a positive blood pressure in the aorta, and if it is greater than the pressure in the ventricle, it holds the aortic valve closed and prevents the expulsion of blood. When continuing ventricular contraction causes its internal pressure to rise above aortic pressure, however, the valve is forced open and blood is ejected into the aorta.

Heart Sounds

As we follow events through the cardiac cycle, we will note the occurrence of the **heart sounds.** Listening to sounds made by the body is called **auscultation** (AWS-cul-TAY-shun). Each cardiac cycle generates two or three sounds that are audible with a stethoscope. The **first** and **second heart sounds,** symbolized S1 and S2,

---

26. *sphygmo* = pulse + *mano* = rare, sparse, roomy

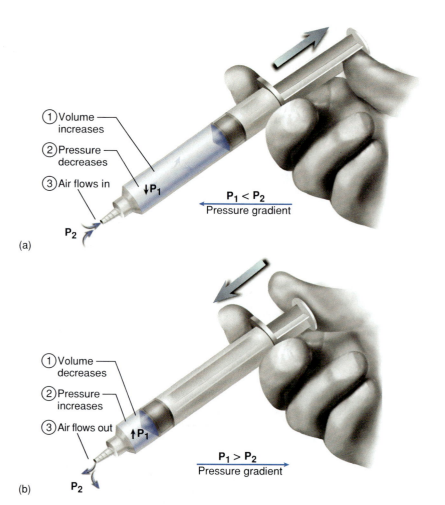

Figure 19.18 A syringe illustrates simple principles of volume, pressure, and flow that govern the functioning of heart valves and the flow of blood. (*a*) As the plunger is pulled back, the volume of the enclosed space increases, its pressure falls, and pressure inside the syringe (P_1) is lower than the pressure outside (P_2). The pressure gradient causes air to flow inward until the pressures are equal. This is analogous to the filling of a heart chamber. (*b*) As the plunger is depressed, the volume of the enclosed space decreases, P_1 rises, and air flows out until the pressures are equal. This is analogous to the ejection of blood from a heart chamber. In both cases, air flows down the pressure gradient.

are often described as a "lubb-dupp"—S1 is louder and a little longer and S2 a little softer and sharper. In children and adolescents, it is normal to hear a **third heart sound** (S3). This is rarely audible in people older than 30, but when it is, the heartbeat is said to show a *triple rhythm* or *gallop*. If the normal rhythm is roughly simulated by drumming two fingers on a table, a triple rhythm sounds a little like drumming with three fingers. The heart valves themselves operate silently, but S1 and S2 occur in conjunction with the closing of the valves as a result of turbulence in the bloodstream and movements of the heart wall. The cause of each sound is not known with certainty, but the probable factors are discussed in the respective phases of the cardiac cycle.

Phases of the Cardiac Cycle

We now examine the phases of the cardiac cycle, the pressure changes that occur, and how the pressure changes and valves govern the flow of blood. A substantial amount of information about these events is summarized in figure 19.19, which is divided into colored bars numbered to correspond to the phases described here. Closely follow the figure as you study the following

text. Where to begin when describing a circular chain of events is somewhat arbitrary. However, in this presentation we begin with the quiescent period, when the heart appears to be preparing for its next beat. Remember that all these events are completed in less than 1 second.

1. The **quiescent period.** This is a period in which none of the heart chambers are contracting. Blood is flowing into the atria, and since the AV valves are open, much of it flows directly through into the ventricles. As the ventricles fill, the flaccid cusps of the AV valves begin to drift up toward the closed position.

2. **Atrial systole.** The SA node fires and the atria depolarize, producing the P wave of the ECG and stimulating the onset of atrial systole. The right atrium contracts slightly before the left because it is the first to receive the signal from the SA node. Blood pressure in the atria rises, forcing additional blood into the ventricles. At the end of this phase, each ventricle contains an **end-diastolic volume (EDV)** of about 130 mL of blood. About 70% of this enters passively during the quiescent period; the last 30% is added by atrial systole.

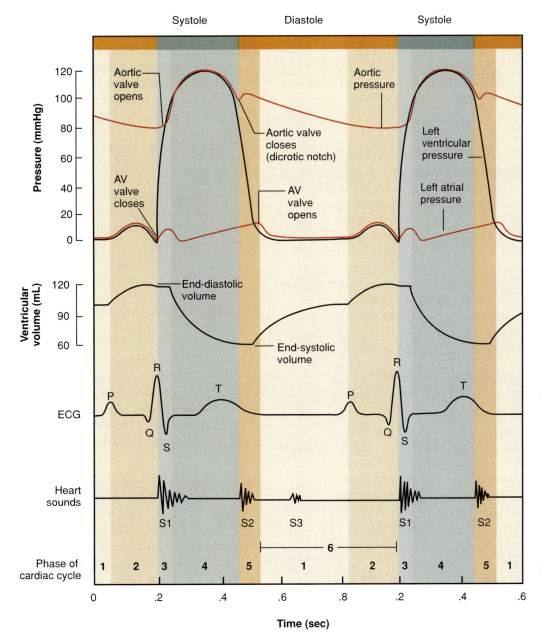

Figure 19.19 Major events of the cardiac cycle, showing two complete cycles. The six phases of the cycle are numbered along the bottom to correspond to the description in the text. 𝒳

3. **Isovolumetric contraction.** The atria relax and remain in diastole for the rest of the cardiac cycle. The ventricles depolarize, generate the QRS complex, and begin to contract. Pressure in the ventricles rises sharply, reversing the pressure gradient between atria and ventricles. The AV valves close as ventricular blood surges back against the cusps. Heart sound S1 occurs at the beginning of this phase and is produced mainly by the left ventricle; the right ventricle is thought to make little or no audible contribution. Causes of the sound are thought to include the tensing of ventricular tissues, acceleration of the ventricular

wall, turbulence in the blood as it surges against the closed AV valves, and impact of the heart against the chest wall.

This phase is called *isovolumetric*[27] because even though the ventricles contract they do not yet eject blood and there is no change in their volume. This is because pressures in the aorta (80 mmHg) and pulmonary trunk (10 mmHg) are still greater than the pressures in the respective ventricles and thus oppose the opening of the semilunar valves.

27. *iso* = same

The myocytes exert force, but they cannot shorten because liquids cannot be compressed and, with all four valves closed, the blood cannot go anywhere.

Think About It

Are the ventricular myocytes in isometric or isotonic contraction during the phase just described?

4. **Ventricular ejection.** The ejection of blood begins when ventricular pressure exceeds arterial pressure and forces the semilunar valves open. Blood spurts out of each ventricle rapidly at first (*rapid ejection*) and then flows out more slowly under less pressure (*reduced ejection*). By analogy, suppose you were to shake up a bottle of soda pop and remove the cap. The soda would spurt out rapidly at high pressure and then more would dribble out a lower pressure, much like the blood leaving the ventricles. Ventricular pressure peaks at 120 mmHg on the left and 25 mmHg on the right. Ventricular ejection lasts about 200 to 250 msec, which corresponds to the plateau of the myocardial action potentials but lags somewhat behind it (review the tension curve in fig. 19.14).

The ventricles do not expel all their blood. In an average resting heart, each ventricle contains an EDV of 130 mL and ejects about 70 mL of it, a quantity called the **stroke volume (SV).** SV is about 54% of the EDV—a percentage called the **ejection fraction.** The blood remaining behind, about 60 mL in this case, is called the **end-systolic volume (ESV).** Note that EDV − SV = ESV. In vigorous exercise, the ejection fraction may be as high as 90%. Ejection fraction is an important measure of cardiac health.

5. **Isovolumetric relaxation.** This is early ventricular diastole, when the T wave appears and the ventricles repolarize and begin to expand. There are competing theories as to how they expand. One is that the blood flowing into the ventricles "inflates" them. Another is that contraction of the ventricles deforms the fibrous skeleton, which subsequently springs back like a rubber ball that has been squeezed and released. This elastic recoil and expansion would cause pressure to drop rapidly and suck blood into the ventricles.

For a brief moment in ventricular diastole, blood from the aorta and pulmonary trunk flows backward through the semilunar valves. This reflux, however, quickly fills the valve cusps and closes them, creating a slight pressure rebound that appears as the *dicrotic notch* of the aortic pressure curve (fig. 19.19). Heart sound S2 occurs as blood rebounds from the closed semilunar valves and the ventricles expand. This phase is again called *isovolumetric* because the semilunar valves are

closed, the AV valves have not yet opened, and the ventricles are therefore not taking in blood.

6. **Ventricular filling.** Diastolic expansion of the ventricles causes their pressure to drop below that of the atria. As a result, the AV valves open and blood flows rapidly into the ventricles, causing ventricular pressure to rise and atrial pressure to fall. Heart sound S3, if it occurs, is thought to result from the transition from expansion of the empty ventricles to their sudden filling with atrial blood. Three subphases of ventricular filling occur: (1) The first one-third is *rapid ventricular filling,* when blood enters especially quickly by the same principle as the rapid ejection explained in phase 4 (the soda pop analogy). (2) The second one-third, called *diastasis* (di-ASS-tuh-sis), is marked by slower filling. The P wave occurs at the end of diastasis. (3) During the last one-third, atrial systole completes the filling process. Note, therefore, that phase 6 of the cardiac cycle, as outlined here, includes phases 1 and 2 described earlier. Diastasis occurs in the latter part of phase 1 and atrial systole in phase 2.

In a resting person, atrial systole lasts about 0.1 second; ventricular systole, 0.3 second; and the quiescent period, 0.4 second. Total duration of the cardiac cycle is therefore 0.8 second (800 msec) in a heart beating at 75 bpm.

Overview of Volume Changes

An additional perspective on the cardiac cycle can be gained if we review the volume changes that occur. This "balance sheet" is from the standpoint of the left ventricle, but for reasons explained shortly, these numbers also must be true of the right. The volumes vary from one individual to another and are rounded to the nearest 10 mL.

End-systolic volume (ESV, left from previous heartbeat)	60 mL
Passively added to the ventricle during atrial diastole	+ 30 mL
Added by atrial systole	+ 40 mL
Total: end-diastolic volume (EDV)	130 mL
Stroke volume (SV) ejected by ventricular systole	− 70 mL
Leaves: end-systolic volume (ESV)	60 mL

Notice that the ventricle pumps as much blood as it received during diastole—70 mL in this example.

The two ventricles eject the same amount of blood even though pressure in the right ventricle is only about one-fifth the pressure in the left. This is because the right ventricle has much less arterial blood pressure to overcome than the left. It is essential that both ventricles have the same output. If the right ventricle pumped more

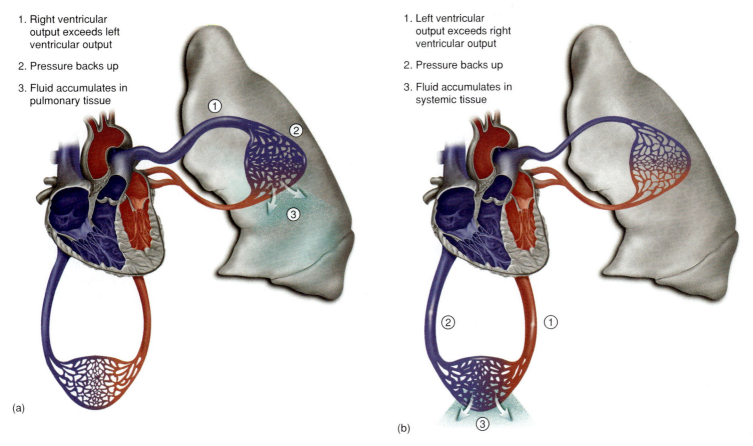

1. Right ventricular
   output exceeds left
   ventricular output

2. Pressure backs up

3. Fluid accumulates in
   pulmonary tissue

1. Left ventricular
   output exceeds right
   ventricular output

2. Pressure backs up

3. Fluid accumulates in
   systemic tissue

(a)

(b)

Figure 19.20 The necessity of balanced ventricular output. (*a*) If the left ventricle pumps less blood than the right, blood pressure backs up into the lungs and causes pulmonary edema. (*b*) If the left ventricle pumps more blood than the right, pressure backs up in the systemic circulation and causes systemic edema. To maintain homeostasis, both ventricles must pump the same average amount of blood.

Special Topic Congestive Heart Failure   19.4

Congestive heart failure (CHF) results from the failure of either ventricle to eject blood effectively. It is usually due to a heart weakened by myocardial infarction, chronic hypertension, valvular insufficiency, or congenital defects in cardiac structure. If the left ventricle fails first, blood backs up into the lungs and causes pulmonary edema (fluid in the lungs), shortness of breath, and a sense of suffocation. If the right ventricle is first to fail, blood backs up into the venae cavae and causes systemic, or generalized, edema (formerly called *dropsy*). Systemic edema is marked by enlargement of the liver, ascites, distension of the jugular veins, and swelling of the fingers, ankles, and feet. Failure of one ventricle eventually increases the workload on the other ventricle, which stresses it and leads to its eventual failure as well.

blood into the lungs than the left side of the heart could handle on return, it would cause hypertension and edema in the pulmonary circuit (fig. 19.20). This would put a person at risk for suffocation as the lungs swelled with fluid and gas exchange was impaired. Conversely, if the left ventricle pumped out more blood than the right heart could handle on return, edema and hypertension in the systemic circuit would result. Over the long term, this could lead to aneurysms (weakened, bulging arteries), stroke, kidney failure, or heart failure (see special topic 19.4). To maintain homeostasis, there must be a balance between the output of the two ventricles.

Key Point Review

14 Explain how a pressure gradient across a heart valve affects whether or not a ventricle will eject blood.

15 What factors are thought to cause the first and second heart sounds? When do these sounds occur?

16 What phases of the cardiac cycle are described as isovolumetric? Explain what this means.

17 If one of a child's ventricles has an EDV of 90 mL and an ESV of 60 mL, what is the stroke volume and what is the ejection fraction? Do you suspect normal health or a diseased heart?

▼Objectives

When you have completed this section, you should be able to

- define cardiac output and explain its importance;
- identify the factors that govern heart rate, contraction strength, and cardiac output;
- discuss some of the nervous and chemical factors that affect heart rate, stroke volume, and cardiac output; and
- explain how the right and left ventricles achieve balanced output.

All of the heart's activity we have considered to this point serves to pump blood into the arteries. **Cardiac output (CO)** is the volume pumped by each ventricle per minute. It is the product of heart rate (HR, beats/min) and stroke volume (SV, mL/beat): CO = HR × SV. At typical resting values, CO = 75 beats/min × 70 mL/beat = 5,250 mL/min = 5.25 L/min. Thus, the body's total volume of blood (4–6 L) passes through the heart every minute. From another perspective, an RBC leaving the left ventricle will, on average, arrive back at the left ventricle in about 1 minute.

Cardiac output is adjusted to the needs of the body. Vigorous exercise increases CO to as much as 21 L/min in a person in good condition and up to 35 L/min in world-class athletes. A person's **cardiac reserve** is the difference between maximum and resting cardiac output. The maximum CO is four or five times the resting CO in an average person in good health but up to seven or eight times in highly trained endurance athletes. In people with severe heart disease, there may be little or no cardiac reserve and little tolerance of physical exertion.

Given that CO = HR × SV, you can see that there are only two ways to change CO—change the heart rate or change the stroke volume. These are somewhat interdependent and usually change together. Agents that raise and lower the heart rate are called positive and negative **chronotropic**[28] agents, respectively. Agents that increase and decrease the amount of force generated by ventricular systole, and thus affect stroke volume most directly, are called positive and negative **inotropic**[29] agents, respectively. (Do not confuse *inotropic* with *ionotropic,* which refers to certain neurotransmitter actions discussed in chapter 13; note the slight difference in spelling.) In the following discussion, we consider a number of chronotropic and inotropic agents and then integrate this information into an overall view of cardiac output.

Heart Rate

The resting heart rate is commonly 120 bpm or greater in newborn infants. It declines steadily with age, averaging 72 to 80 bpm in young adult females and 64 to 72 bpm in young adult males. It tends to rise again in the elderly.

Tachycardia[30]—a persistent, resting adult heart rate above 100 bpm—can be caused by stress, anxiety, drugs, heart disease, or elevated body temperature (due to fever or exercise). As discussed in the next chapter, heart rate also rises to compensate to some extent for a drop in stroke volume—for example, when the body has lost a significant quantity of blood or when there is damage to the myocardium (myocardial infarction).

Bradycardia[31]—a persistent, resting adult heart rate below 60 bpm—is common during sleep and in endurance-trained athletes. Endurance training enlarges the heart and increases its stroke volume. Thus, it can maintain the same cardiac output with fewer beats. Hypothermia (low body temperature) also slows the heart rate and may be deliberately induced in preparation for cardiac surgery. Diving mammals such as whales and seals exhibit bradycardia during the dive, as do humans to some extent when the face is immersed in cool water.

Heart rate responds principally to nervous and chemical stimuli. We next consider some chronotropic effects of the autonomic nervous system, hormones, electrolytes, and blood gases.

Chronotropic Effects of the Autonomic Nervous System

Although the nervous system does not initiate the heartbeat, it does modulate its rhythm and force. The **cardiac center** of the medulla oblongata receives input from the cerebral cortex, limbic system, and hypothalamus; therefore, many sensory and emotional stimuli can alter the heart rate. It can climb even as you anticipate taking the first plunge on a roller coaster or competing in an athletic event and in response to emotions such as love and anger. The cardiac center also receives input from peripheral receptors in the muscles, joints, aortic arch, and carotid arteries.

- **Proprioceptors** in the muscles and joints (see chapter 16) quickly inform the cardiac center of changes in physical activity. Thus, the heart can increase its output even before the metabolic demands of the muscles rise.
- **Chemoreceptors** in the aortic arch and carotid arteries (the major arteries of the neck) provide

---

28. *chrono* = time + *trop* = turn, change, influence
29. *ino* = fiber

30. *tachy* = speed, fast + *card* = heart + *ia* = condition
31. *brady* = slow

information on blood pH and O_2 and CO_2 concentrations. *Hypercapnia* (an elevated CO_2 level) and *acidosis* (a low pH) suggest that the tissues are not being perfused well enough to remove CO_2 as fast as it is being produced. The cardiac center responds by increasing heart rate to increase perfusion. Hypoxemia has a similar but weaker effect.

- **Baroreceptors,** found in approximately the same locations as the chemoreceptors, monitor changes in arterial blood pressure. If it drops, they inform the cardiac center, which, in turn, raises heart rate and tends to restore pressure. Conversely, high blood pressure acts through the baroreceptors to lower the heart rate. The relationship of the baroreceptors to the CNS and heart is shown in figure 15.16 (p. 531). Baroreceptors and chemoreceptors are discussed more fully in the next chapter.

The cardiac center is subdivided into two neuronal pools, a cardioacceleratory center and cardioinhibitory center. Impulses from the **cardioacceleratory center** leave the thoracic spinal cord and travel by way of sympathetic **cardiac accelerator nerves** to the SA and AV nodes and myocardium. These nerves secrete norepinephrine, which binds to β_1 adrenergic receptors in the heart and has a positive chronotropic effect. Cardiac output peaks when the heart rate is 160 to 180 bpm. Maximum sympathetic stimulation can raise the rate to as high as 230 bpm, a limit set mainly by the refractory period of the AV node. At such a high rate, however, the ventricles beat so rapidly that they have little time to fill between beats; therefore, the stroke volume and cardiac output are less than they are at rest. Ventricular diastole lasts about 0.62 seconds at a heart rate of 65 bpm, but only 0.14 seconds at a rate of 200 bpm, so you can see that less time becomes available for adequate refilling between beats.

Output from the **cardioinhibitory center** travels by way of the vagus nerves to the SA and AV nodes. The right vagus nerve innervates mainly the SA node; the left vagus nerve the AV node. The vagus nerves have a background firing rate called **vagal tone** that inhibits the nodes. The nerves secrete acetylcholine, which binds to muscarinic receptors and opens K^+ channels in the nodal cells. Potassium ions leave the cells, which causes them to become hyperpolarized so they do not fire as frequently. The "natural," or intrinsic, firing frequency of the SA node is 100 times per minute, but vagal tone holds this down to the usual sinus rhythm of 70 to 80 bpm. Maximum vagal stimulation can reduce the heart rate to as low as 20 to 30 bpm.

Chronotropic Effects of Chemicals

The catecholamines epinephrine and norepinephrine are potent cardiac stimulants. Arousal, stress, and exer-

cise stimulate their secretion by the adrenal medulla, and they mimic the adrenergic effect of the cardiac accelerator nerves. Nicotine has a positive chronotropic effect because it stimulates catecholamine secretion. Catecholamines act through cAMP as a second messenger. Caffeine, theophylline, and theobromine—the stimulants in coffee, tea, and chocolate, respectively—inhibit cAMP breakdown, prolong the effect of the catecholamines, and thus have positive chronotropic effects. Thyroid hormone also increases the heart rate. One symptom of hyperthyroidism is tachycardia, which can weaken the heart in the long run and cause it to fail.

Three electrolytes with especially strong chronotropic effects are sodium, potassium, and calcium. *Hypernatremia*[32] (Na^+ excess) lowers the heart rate by blocking the influx of Ca^{2+} into nodal cells, which makes it harder for the SA node to fire. *Hyperkalemia*[33] (K^+ excess) inhibits firing of the SA node by hyperpolarizing it. It can kill quickly by causing the heart to arrest in diastole; it is used in some states for execution by lethal injection. *Hypercalcemia* (Ca^{2+} excess) increases the heart rate because it steepens the concentration gradient from ECF to cytosol and enhances the influx of Ca^{2+} through the fast calcium channels. Highly elevated blood calcium levels can cause the heart to arrest in systole.

Stroke Volume

Stroke volume is governed mainly by three factors: *preload, contractility,* and *afterload.* Increased preload or contractility increases stroke volume, while increased afterload opposes the emptying of the ventricles and reduces stroke volume.

Preload

The amount of tension in the ventricular myocardium immediately before it begins to contract is called the **preload.** To understand how it influences stroke volume, imagine yourself engaged in heavy exercise. As active muscles massage your veins, they drive more blood back to your heart—that is, they increase *venous return.* As more blood enters your heart, it stretches the myocardium. Due to the length-tension relationship of striated muscle explained in chapter 12, moderate stretch enables myocytes to generate more tension when they begin to contract—that is, it increases the myocardial preload. If the ventricles contract more forcefully, they expel more blood, thus adjusting your cardiac output to the increase in venous return.

---

32. *hyper* = excess, above normal + *natr* = sodium (Latin, *natrium*) + *emia* = blood condition
33. *kal* = potassium (Latin, *kalium*)

This theory is summarized by the **Frank–Starling law of the heart**.[34] In a concise, symbolic way, it states that SV ∝ EDV. In other words, the ventricles tend to pump out all the blood that entered them. Within limits, the more they are stretched, the harder they contract when stimulated.

While relaxed skeletal muscle is normally at an optimum length for the most forceful contraction, relaxed cardiac muscles is at less than optimum length. Any additional stretch therefore produces a significant increase in contraction force on the next beat. This helps balance the output of the two ventricles. For example, if the right ventricle begins to pump an increased amount of blood, this soon arrives at the left ventricle, stretches it more than before, and causes it to increase its stroke volume to match that of the right.

Contractility

The **contractility** of the myocardium refers to its strength of contraction *for a given preload.* It does not describe an increase in tension resulting from increased stretch but rather an increase caused by inotropic factors that make the myocytes more responsive to stimulation. Remember that Ca^{2+} is essential to the excitation-contraction coupling of muscle and prolongs the plateau of the myocardial action potential. Calcium therefore has a positive inotropic effect, as do agents that increase its availability to the myofilaments. Epinephrine and norepinephrine act through cAMP to open Ca^{2+} channels. By increasing the supply of Ca^{2+} to the myofilaments, they have a positive inotropic effect. Glucagon acts by stimulating the formation of cAMP; a solution of glucagon and calcium chloride is a standard emergency treatment for heart attacks. Digitalis, a cardiac stimulant from the foxglove plant, is used to treat congestive heart failure. It acts indirectly by inhibiting the Na^+-K^+ pumps of the myocardium, raising intracellular Na^+ concentration, and increasing the amount of Ca^{2+} in the sarcoplasm.

Some negative inotropic agents include myocardial hypoxia, hypercapnia, acidosis, and barbiturates. The vagus nerves have a negative inotropic effect on the atria, but they provide so little innervation to the ventricular myocytes that they have little effect on the ventricles. There are other chronotropic and inotropic agents too numerous to mention here. The ones we have discussed are summarized in table 19.2.

Table 19.2	Some Chronotropic and Inotropic Agents
Positive	**Negative**
Chronotropic	
Sympathetic stimulation	Parasympathetic stimulation
Epinephrine and norepinephrine	Acetylcholine
Thyroid hormone	Hypernatremia
Digitalis	Hyperkalemia
Hypercalcemia	Hypocalcemia
Hypoxia, hypercapnia, and acidosis	
Inotropic	
Sympathetic stimulation	(Parasympathetic effect negligible)
Epinephrine and norepinephrine	Hypernatremia
Glucagon	Hyperkalemia
Digitalis	Hypocalcemia
Hypercalcemia	Myocardial hypoxia, hypercapnia

Think About It

Suppose a person has a heart rate of 70 bpm and a stroke volume of 70 mL. A negative inotropic agent then reduces the stroke volume to 50 mL. What would the new heart rate have to be to maintain the same cardiac output?

Afterload

The blood pressure in the arteries just outside the semilunar valves, called the **afterload,** opposes the opening of these valves. An increased afterload therefore reduces stroke volume. Anything that impedes arterial circulation can increase the afterload. For example, in some lung diseases, scar tissue forms in the lungs and restricts pulmonary circulation. This increases the afterload in the pulmonary trunk and opposes emptying of the right ventricle. As the ventricle works harder to overcome this resistance, it gets larger like any other muscle. Stress and hypertrophy of a ventricle can eventually cause it to weaken and fail. Right ventricular failure due to obstructed pulmonary circulation is called *cor pulmonale*[35] (CORE PUL-mo-NAY-lee). It is a common complication of emphysema, chronic bronchitis, and black lung disease (see chapter 22).

Exercise and Cardiac Output

It is no secret that exercise makes the heart work harder, and it should come as no surprise that this increases

---

34. Otto Frank (1865–1944), German physiologist; Ernest Henry Starling (1866–1927), English physiologist

35. *cor* = heart + *pulmo* = lung

cardiac output. The main reason the heart rate increases at the beginning of exercise is that proprioceptors in the muscles and joints transmit signals to the cardiac center signifying that the muscles are active and will quickly need an increased rate of O_2 delivery and CO_2 removal. As the exercise progresses, muscular activity increases venous return. This increases the preload on the right ventricle and is soon reflected in the left ventricle as more blood flows through the pulmonary circuit and reaches the left heart. As the heart rate and stroke volume rise, cardiac output rises, which compensates for the increased venous return.

A sustained program of exercise causes hypertrophy of the ventricles, which increases their stroke volume. As explained earlier, this allows the heart to beat more slowly and still maintain a normal resting cardiac output. Endurance athletes commonly have low heart rates of 40 to 60 bpm, but because of the higher stroke volume, their resting cardiac output is about the same as that of an untrained person. They have greater cardiac reserve, so they can tolerate more exertion than a sedentary person can.

Key Point Review

18 Define *cardiac output* in words and with a simple formula.

19 Describe the cardiac center and innervation of the heart.

20 Explain what is meant by positive and negative chronotropic and inotropic agents. Give two examples of each.

21 How do preload, contractility, and afterload influence stroke volume and cardiac output?

22 Explain the principle behind the Frank–Starling law of the heart. How does this mechanism normally prevent pulmonary or systemic congestion?

CHAPTER ESSAY

Coronary Atherosclerosis

Atherosclerosis[36] is a disorder in which fatty deposits form in an artery, obstruct the lumen, and cause deterioration of the arterial wall. It is especially critical when it occurs in the coronary arteries and threatens to cut off the blood supply to the myocardium. Atherosclerosis is also a leading contributor to stroke and kidney failure.

Cause and Pathogenesis

According to one theory, the stage is set for atherosclerosis when the endothelium of a blood vessel is damaged by hypertension, viral infection, diabetes mellitus, or other causes. Monocytes adhere to the damaged endothelium, penetrate into the tunica intima, and transform into macrophages. Macrophages and smooth muscle cells absorb cholesterol and neutral fats from the blood and acquire a frothy appearance; they are then called *foam cells* and are visible as a *fatty streak* on the vessel wall.

Platelets also adhere to areas of endothelial damage, degranulate, and release platelet-derived growth factor (PDGF); some PDGF also comes from macrophages and endothelial cells. PDGF stimulates mitosis of smooth muscle, leading eventually to an **atheroma (atherosclerotic plaque)**—a mass of lipid, smooth muscle, and macrophages. The artery becomes increasingly necrotic as its muscle and elastic tissue are replaced with scar tissue. When atheromas become calcified, they are called **complicated plaques.** Advanced atherosclerosis is marked by a state of arterial degeneration and rigidity called **arteriosclerosis.**

The primary culprits in atherosclerosis are **low-density lipoproteins (LDLs)** in the blood plasma. LDLs are small protein-coated droplets of cholesterol, neutral fat, free fatty acids, and phospholipids (see chapter 26). Most cells have LDL receptors that take up these lipids from the blood when they are needed and stop when the cells have enough cholesterol. In atherosclerosis, arterial cells have dysfunctional receptors that continue taking up plasma lipids, causing the cells to accumulate excess cholesterol.

As an atheroma grows, more and more of the arterial lumen becomes obstructed (fig. E.1). Angina pectoris and other symptoms begin to occur when the lumen of a major coronary artery is reduced by at least 75%. When platelets adhere to lesions of the arterial wall, they also release clotting factors. An atheroma can therefore become a focus for thrombosis. A clot can block much of what remains of the lumen, or it can break free and become an embolus that travels downstream until it lodges in a smaller artery. Part of an atheroma itself can also break loose and travel as a *fatty embolus.*

Atheromas also contribute to coronary artery spasms. Healthy endothelial cells secrete nitric oxide (NO), which causes the arteries to dilate. Atheromas reduce NO secretion, and the coronary arteries then exhibit spasms. With much of the lumen already obstructed by the atheroma and perhaps a thrombus, an arterial spasm can temporarily shut off the remaining flow and precipitate an attack of angina.

36. *athero* = fat, fatty + *sclerosis* = hardening

—continued

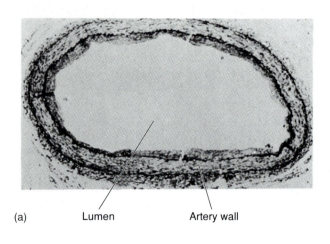

(a) Lumen   Artery wall

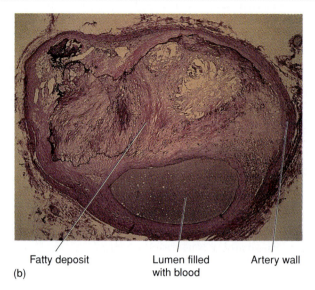

Fatty deposit   Lumen filled with blood   Artery wall
(b)

Figure E.1 Atherosclerosis. (*a*) Cross section of a healthy artery. (*b*) Cross section of an artery with advanced atherosclerosis. The greatly reduced lumen in the diseased artery can easily be blocked by thrombosis or embolism.

Risk and Prevention

Risk factors are elements of the environment or personal characteristics that predispose an individual to a particular disease. Some risk factors for atherosclerosis are preventable and some are not. The latter include aging, heredity, and being male. One form of hereditary atherosclerosis is *familial hypercholesterolemia* ("elevated blood cholesterol levels running in the family"). Most people have two recessive alleles (*hh*) of the gene for LDL receptors, which leads to the synthesis of normal receptors. One person in 500 is heterozygous (*Hh*) and makes only half the normal number of LDL receptors; one in a million is homozygous dominant (*HH*) and makes no LDL receptors. When the body's cells have few or no LDL receptors, they fail to absorb LDLs from the blood. Blood LDL levels therefore remain high—six times normal in *HH* individuals. (Note the resemblance to type II diabetes mellitus.) Foam cells, however, absorb LDLs even without these receptors, and with excess LDL in the blood plasma, atheromas grow rapidly. Heterozygous individuals usually suffer heart attacks by age 35, and homozygous dominant individuals usually have heart attacks in childhood, sometimes before age 2.

Many risk factors for atherosclerosis are preventable. A sedentary lifestyle promotes LDL formation, whereas exercise promotes the formation of *high-density lipoproteins (HDLs),* which do not contribute to coronary disease. Exercise also reduces obesity, which is itself a risk factor. Aggressiveness, anxiety, and emotional stress promote hypertension and atherosclerosis. Smoking is another avoidable risk factor. The incidence of coronary heart disease is proportional to the number of cigarettes smoked per day and the number of years a person has been a smoker. This is reversible; people who quit smoking drop to normal risk levels within 5 years.

Diet, of course, is an overwhelmingly important factor. Eating animal fat reduces the number of LDL receptors and raises plasma LDL levels. Foods high in soluble fiber (such as beans, apples, and oat bran) lower blood cholesterol by an interesting mechanism. The liver normally converts cholesterol to bile salts, which it secretes into the small intestine to aid fat digestion. The bile salts are resorbed farther down the intestine and recycled to the liver for reuse. Soluble fiber, however, binds bile salts and carries them out in the feces. To replace them, the liver must synthesize more, thus using more cholesterol and lowering the blood cholesterol.

In the 1970s, scientists found that the Eskimos of Greenland had unusually low rates of coronary atherosclerosis despite the fact that their diet consisted entirely of meat—averaging a pound of whale meat and a pound of fish per day. Japanese and other groups with large amounts of fish in their diets also show low blood cholesterol levels. It is suspected that this is due to *omega-3 polyunsaturated fatty acids (PUFAs)* in fish oil. PUFAs increase the fluidity of plasma membranes and enable cells to remove more lipid from the blood. However, a daily capsule of fish oil does not hold much promise for controlling cholesterol. Doses of PUFAs high enough to reduce blood cholesterol would be prohibitively expensive and have undesirable side effects, including suppression of the immune system. Studies on the effectiveness of PUFAs remain inconclusive.

Treatment Options

The first pioneering approach to treating athero-sclerosis, and still a common standby, is *coronary artery bypass surgery.* Sections of the great saphenous vein of the leg or small arteries from the thoracic cavity are used to connect the aorta directly to a point on a coronary artery below the obstruction, providing a detour for the blood.

Balloon angioplasty[37] is a technique in which a thin, flexible catheter is threaded into a coronary artery to the point of obstruction, and then a balloon at its tip is inflated to press the atheroma against the arterial

wall, opening up the lumen. Its usefulness is limited to well-localized atheromas. In another recent method, *laser angioplasty,* an illuminated catheter enables the surgeon to see inside a diseased artery on a monitor and to use a laser to vaporize atheromas and reopen the artery. These methods are cheaper and less risky than bypass surgery. Unfortunately, they are often followed by *restenosis*—atheromas grow back and reobstruct the artery months later. There is also some concern that these procedures may even cause new injuries to the arterial walls, which may be foci for the development of new atheromas.

Clearly, prevention is the least expensive, least risky, and most effective approach to the threat of the coronary artery disease.▲

---

37. *angio* = vessel + *plasty* = surgical repair

Chapter Review Study Outline

Gross Anatomy (pp. 674–683)

1. Cardiovascular system overview
   a. Pulmonary circuit
      - Served by right heart
      - Supplies blood to pulmonary alveoli only
      - Serves for O_2 loading and CO_2 unloading
   b. Systemic circuit
      - Served by left heart
      - Supplies blood to all organs
      - Delivers oxygen and nutrients; removes CO_2
2. Heart overview
   a. Location
   b. Base and apex
3. Pericardium
   a. Parietal pericardium
   b. Visceral pericardium
   c. Pericardial cavity and fluid
4. Heart wall
   a. Epicardium, myocardium, and endocardium
   b. Fibrous skeleton
5. Chambers
   a. Right and left atria
   b. Right and left ventricles
   c. Atrioventricular (coronary) sulcus
   d. Interventricular sulci
   e. Interatrial and interventricular septa
   f. Trabeculae carneae
6. Valves
   a. Semilunar valves
      - Pulmonary semilunar (right)
      - Aortic semilunar (left)
      - Three pocketlike cusps
   b. Atrioventricular (AV) valves
      - Tricuspid (right)
      - Bicuspid or mitral (left)
      - Chordae tendineae
      - Papillary muscles
   c. Operation of AV and semilunar valves
7. Blood flow through heart chambers
8. Blood flow through myocardium
   a. Left coronary artery
      - Anterior interventricular artery
      - Circumflex artery
   b. Right coronary artery
      - Posterior interventricular artery
      - Marginal artery
   c. Arterial anastomoses
   d. Venous drainage
      - Great cardiac vein (anterior)
      - Middle cardiac vein (posterior)
      - Coronary sinus
   e. Coronary perfusion
9. Development of the heart
   a. Evolution
      - Importance of double circulation
      - Importance of separate ventricles
   b. Embryology
      - Endothelial tube fusion; heart tube
      - Flexion into U and S shape
      - SA node development from sinus venosus

Cardiac Muscle and the Cardiac Conduction System (pp. 683–685)

1. Structure of cardiac muscle
   a. Short branching cells
   b. Scanty sarcoplasmic reticulum
   c. Numerous large mitochondria
   d. Intercalated discs
      - Interdigitating folds of membrane
      - Desmosomes
      - Gap junctions (electrical synapses)
   e. Behavior as a functional syncytium
2. Metabolism of cardiac muscle
   a. Variety of energy substrates
   b. Fatigue resistance
3. Myogenic control and autorhythmicity
4. Cardiac conduction system
   a. Sinoatrial (SA) node—pacemaker
   b. Atrioventricular (AV) node
   c. Atrioventricular bundle
   d. Right and left bundle branches
   e. Purkinje fibers

Electrical and Contractile Activity of the Heart (pp. 686–691)

1. Systole and diastole
2. The cardiac rhythm
   a. Normal sinus rhythm
   b. Ectopic foci
   c. Nodal rhythm
3. Physiology of SA node
   a. Production of the pacemaker potential

b. Production of the action potential

c. Influence of vagal tone

4. Impulse conduction to the myocardium
   a. Route of the conduction pathway
   b. Variations in conduction speed

5. Electrical behavior of the myocardium
   a. Stable RMP of −90 mV
   b. Opening of voltage-regulated Na⁺ gates
   c. Positive feedback cycle and depolarization
   d. Action potential peaks at +30 mV
   e. Ca^{2+} release from Sr
   f. Opening of slow Ca^{2+} channels; influx from ECF
   g. Plateau of 200 to 250 msec
   h. Sustained contraction
   i. Long absolute refractory period

6. Electrocardiogram
   a. P wave: atrial depolarization
   b. P–Q segment: atrial systole
   c. QRS complex
      • Ventricular depolarization
      • Atrial repolarization
   d. S–T segment: ventricular systole
   e. T wave: ventricular repolarization

Blood Flow, Heart Sounds, and the Cardiac Cycle (pp. 691–695)

1. The cardiac cycle—definition and major phases

2. Principles of pressure and flow
   a. Measurement of pressure
   b. Pressure gradients and flow
   c. Relationship to heart valve operation

3. Heart sounds S1 to S3

4. Phases of cardiac cycle
   a. Quiescent period (late diastole)
      • All chambers relaxed
      • AV valves open
      • Ventricles fill to 70% of EDV
   b. Atrial systole

• SA node fires and atria depolarize, producing P wave of ECG

• Atrial systole adds 30% of EDV

c. Isovolumetric contraction
   • Ventricles depolarize, producing QRS complex of ECG
   • Ventricular systole
   • Ventricular pressure < arterial pressure
   • AV valves close; semilunar valves remain closed
   • Heart sound S1
   • No blood ejected

d. Ventricular ejection
   • Ventricular pressure > arterial pressure
   • Semilunar valves open
   • Rapid and reduced ejection
   • Stroke volume (SV) about 70 mL
   • Ejection fraction = SV/EDV
   • End-systolic volume (ESV) remains

e. Isovolumetric relaxation
   • Ventricular repolarization, producing T wave of ECG
   • Ventricular diastole
   • Semilunar valves close
   • Heart sound S2
   • No ventricular filling

f. Ventricular filling
   • Atrial pressure > ventricular pressure
   • AV valves open
   • Rapid ventricular filling
   • Slower ventricular filling (diastasis)
   • Atrial systole

5. Representative volume changes
   a. End-systolic volume (ESV), 60 mL
   b. End-diastolic volume (EDV), 130 mL
   c. Stroke volume (SV), 70 mL
   d. Importance of balanced output
      • Left failure and pulmonary congestion

• Right failure and systemic congestion

Cardiac Output (pp. 696–699)

1. Cardiac output (CO)
   a. Definition: mL/min ejected by each ventricle
   b. CO = heart rate × stroke volume
   c. Cardiac reserve
   d. Chronotropic and inotropic agents

2. Heart rate
   a. Normal rates relative to age and sex
   b. Tachycardia and bradycardia
   c. Effects of autonomic nervous system
      • Cardiac center of medulla oblongata
      • Input from other parts of brain
      • Input from peripheral receptors
        •• Proprioceptors
        •• Chemoreceptors
        •• Baroreceptors
   d. Cardioacceleratory center and sympathetic afferents
   e. Cardioinhibitory center and parasympathetic (vagal) afferents
   f. Chronotropic effects of chemicals
      • Epinephrine and norepinephrine
      • Thyroid hormone
      • Electrolytes

3. Stroke volume
   a. Preload
      • Role of venous return
      • Frank-Starling law of the heart
   b. Contractility and inotropic agents
      • Calcium ions
      • Epinephrine and norepinephrine
      • Glucagon
      • Digitalis
      • Negative inotropic agents
   c. Afterload—resistance to ventricular ejection

4. Exercise and cardiac output

Testing Your Recall   Answers in Appendix C

1. The cardiac conduction system includes all of the following *except*
   a. the SA node.
   b. the AV node.
   c. the bundle branches.
   d. the trabeculae carneae.
   e. the Purkinje fibers.

2. To get from right atrium to right ventricle, blood must flow through
   a. the pulmonary semilunar valve.
   b. the tricuspid valve.
   c. the bicuspid valve.
   d. the aortic semilunar valve.
   e. the mitral valve.

3. In both evolution and embryology of the mammalian heart, the pacemaker arises from
   a. the bulbus cordis.
   b. the foramen ovale.
   c. the truncus arteriosus.
   d. the sinus venosus.
   e. the fossa ovalis.

4. A heart rate of 45 bpm and an absence of P waves from the ECG would suggest
   a. damage to the SA node.
   b. ventricular fibrillation.
   c. cor pulmonale.
   d. extrasystole.
   e. heart block.

5. The fast rising phase of the SA node action potential is due to
   a. opening of slow calcium channels.
   b. closing of potassium gates.
   c. potassium efflux.

   d. potassium influx.
   e. calcium influx.

6. Cardiac muscle does not exhibit tetany because it has
   a. fast calcium channels.
   b. scanty sarcoplasmic reticulum.
   c. a long absolute refractory period.
   d. electrical synapses.
   e. exclusively aerobic catabolism.

7. The atria contract during
   a. the first heart sound.
   b. the second heart sound.
   c. the QRS complex.
   d. the P–Q segment.
   e. the S–T segment.

8. Ventricular pressure peaks during
   a. the first heart sound.
   b. the second heart sound.
   c. the QRS complex.
   d. the P–Q segment.
   e. the S–T segment.

9. The blood contained in the ventricle during isovolumetric relaxation is
   a. the end-systolic volume.
   b. the end-diastolic volume.
   c. the stroke volume.
   d. the ejection fraction.
   e. none of the above; the ventricle is empty then.

10. Drugs that increase the heart rate are said to have a _____ effect.
    a. myogenic
    b. negative inotropic
    c. positive inotropic
    d. negative chronotropic
    e. positive chronotropic

11. The contraction phase of any heart chamber is called _____, and the relaxation phase is called _____.

12. The circulatory route from aorta to venae cavae is called the _____ circuit.

13. The circumflex artery travels in a groove called the _____.

14. The pacemaker potential of autorhythmic cells is due to the slow influx of _____.

15. Electrical signals spread quickly from one cardiac myocyte to another through _____ in the intercalated discs.

16. Repolarization of the ventricles produces the _____ of the electrocardiogram.

17. Closing of the _____ valves produces turbulence in the bloodstream, which contributes to the second heart sound.

18. The procedure for listening to the heart sounds is called cardiac _____.

19. The end-diastolic volume of blood stretches the ventricles and creates myocardial tension called the _____.

20. The Frank-Starling law of the heart explains why the _____ of the left ventricle is the same as that of the right.

1. Verapamil is a calcium channel blocker used to treat hypertension. It selectively blocks slow calcium channels. Would you expect it to have a positive or negative inotropic effect? Explain. (See p. 114 to review calcium channel blockers if necessary.)

2. To temporarily treat tachycardia and restore normal sinus rhythm, a physician may massage a patient's carotid artery near the angle of the mandible. Propose a mechanism by which this treatment would have the desired effect.

3. Explain how the meager sarcoplasmic reticulum of cardiac muscle relates to the existence of slow calcium channels in the plasma membrane. Why are slow calcium channels significant with respect to cardiac muscle but not skeletal muscle?

4. In each ventricular systole, the left ventricle is the first to begin contracting, but the right ventricle is the first to begin expelling blood. Aside from the obvious fact that the pulmonary valve opens before the aortic valve, how would you explain this?

5. The action potential of a cardiac myocyte looks very different from the action potential of a nerve fiber. Sketch the two and explain the ionic basis for this difference.

Web Site Link

For a listing of the most current web sites related to this chapter, please visit the Saladin homepage at:

http://www.mhhe.com/sciencemath/biology/saladin/

20

The Circulatory System: Blood Vessels and Circulation

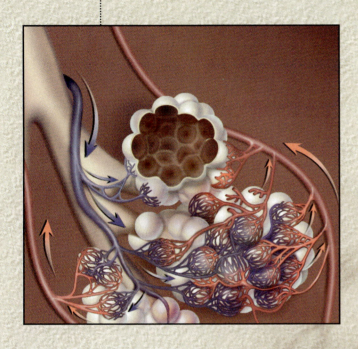

Special/Topics

Brushing up

To understand this chapter, it is essential that you understand or brush up on the following concepts:

- Set point and dynamic equilibrium in homeostasis (p. 21)
- Diffusion (pp. 125–126)
- Equilibrium between filtration and osmosis (pp. 126–127)
- Transcytosis (pp. 132–133)
- Viscosity and osmolarity of blood (pp. 641–642)
- Principles of pressure and flow (p. 691)
- Autonomic effects on the heart (p. 697)

The route taken by blood after it leaves the heart was a point of much confusion until the seventeenth century. Chinese emperor Huang Ti (2697–2597 B.C.E.) correctly believed that it flowed in a complete circle around the body and back to the heart. But in the second century, Roman physician Claudius Galen argued that it flowed back and forth in the veins, like air in the bronchial tubes. He believed that the liver received food from the small intestine and converted it to blood; the heart pumped the blood through the veins to all other organs; and those organs consumed the blood.

Huang Ti was right, but the first experimental demonstration of this did not come until the seventeenth century. English physician William Harvey (1578–1657) studied the filling and emptying of the heart in snakes, tied off the vessels above and below the heart to observe the effects on cardiac filling and output, and measured cardiac output in a variety of living animals. He concluded that (1) the heart pumps more blood in half an hour than there is in the entire body, (2) not enough food is consumed to account for the continual production and consumption of so much blood, and (3) since the planets orbit around the sun and (as he believed) the human body is modeled after the solar system, it follows that the blood orbits around the body. Thus, for a combination of experimental and superstitious reasons, Harvey argued that the blood must return to the heart rather than being consumed by the peripheral organs. He could not explain how it did so, however, since the microscope had yet to be invented and he did not know of capillaries—later discovered by Antony van Leeuwenhoek and Marcello Malpighi.

Harvey published his findings in 1628 in a short but elegant book entitled *Exercitio Anatomica de Motu Cordis et Sanguinis in Animalibus (Anatomical Studies on the Motion of the Heart and Blood in Animals)*. This landmark in the history of biology and medicine was the first experimental study of animal physiology. However, so entrenched were the ideas of Aristotle and Galen in the medical community, and so strange was the idea of doing experiments on living animals, that Harvey's ideas were rejected. Indeed, some of his colleagues regarded him as a crackpot because his conclusion flew in the face of common sense—if the blood was continually recirculated and not consumed by the tissues, they reasoned, then what purpose could it possibly serve?

Harvey lived to a ripe old age, served as physician to the kings of England, and later did important work in embryology. His case is one of the most interesting in biomedical history, for it shows how empirical science overthrows old theories and spawns better ones and how both common sense and blind allegiance to authority sometimes interfere with acceptance of the truth. But most importantly, Harvey's contributions represent the birth of experimental physiology—the method that generated most of the information in this book.

General Anatomy

▼**Objectives**

When you have completed this section, you should be able to
- trace the usual route taken by the blood from the heart and back again, and describe the common variations on this route;
- describe the structure of a blood vessel; and
- describe the different types of arteries, capillaries, and veins.

Circulatory Routes

The usual route of blood flow around the body is heart → arteries → arterioles → capillaries → venules → veins → heart. An **artery** is any vessel that carries blood away from the heart; arterioles are the smallest arteries. A **vein** is any vessel that carries blood toward the heart; the smallest of these are called venules. Microscopic capillaries connect the arterioles to the venules.

Blood usually passes through one network of capillaries only from the time it leaves the heart until the time it returns (fig. 20.1*a*). There are exceptions, however—notably portal systems and anastomoses. In a **portal system** (fig. 20.1*b*), blood flows through two consecutive capillary networks before returning to the heart. One portal system connects the hypothalamus and anterior pituitary (see chapter 17). Others are found in the kidneys and between the intestines and liver; the latter system is detailed in table 20.13.

Anastomoses are places where two veins or arteries merge with each other. In **arteriovenous anastomoses (shunts),** blood flows from an artery directly into a vein, bypassing the capillaries (fig. 20.1*c*). Shunts occur in the fingers, palms, toes, and ears, where they reduce heat loss in cold weather by allowing warm blood to bypass these exposed surfaces. Unfortunately, this makes these poorly perfused areas more susceptible to frostbite. In **arterial anastomoses,** two arteries merge, providing *collateral* (alternative) routes of blood supply to a tissue (fig. 20.1*d*). Those of the coronary circulation were mentioned in the previous chapter. They are also common near diarthrotic joints, where joint movement might temporarily obstruct one pathway. **Venous anastomoses** are more

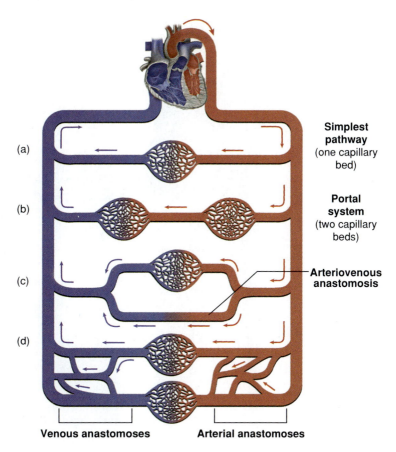

Figure 20.1 Variations in circulatory pathways.

Simplest pathway (one capillary bed)

Portal system (two capillary beds)

Arteriovenous anastomosis

Venous anastomoses   Arterial anastomoses

common. They provide several alternative routes of drainage from an organ, and so blockage of a vein is rarely as life-threatening as blockage of an artery. Several arterial and venous anastomoses are described later in this chapter.

The Structure of Blood Vessels

As shown in figures 20.2 and 20.3, three tissue layers called *tunics* occur in the walls of arteries and veins:

1. The **tunica externa (tunica adventitia[1])** is the outermost layer. It consists of loose connective tissue that often merges into the adventitia of neighboring blood vessels, nerves, or other organs. It anchors the vessel and provides passage for small nerves, lymphatic vessels, and smaller blood vessels. Small vessels called the **vasa vasorum**[2] (VAY-za vay-SO-rum) supply blood to at least the outer half of the wall of a larger vessel. Tissues of the inner half of the wall are thought to be nourished by diffusion from blood in the lumen.

2. The **tunica media,** the middle layer, is usually the thickest layer of a vessel. It consists of smooth muscle, elastic tissue, and collagen. The smooth muscle is responsible for the **vasomotion** (dilation and constriction) of blood vessels.

3. The **tunica interna (tunica intima)** faces the lumen of the vessel. It consists of a simple squamous **endothelium** overlying a basement membrane and a sparse layer of fibroconnective tissue. The endothelium provides a smooth inner lining to which blood cells and platelets usually will not adhere. It also acts as a selectively permeable barrier to blood solutes, and it secretes vasoconstrictors and vasodilators, to be considered later.

1. *advent* = added to
2. *vasa* = vessels + *vasorum* = of the vessels

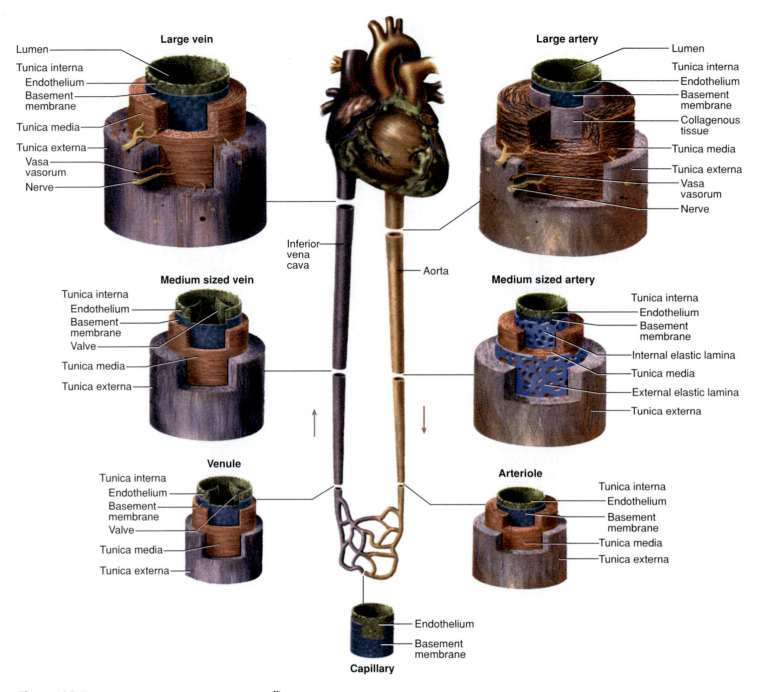

Figure 20.2 The structure of arteries and veins.

Arteries and Metarterioles

Arteries are constructed to withstand the surges of blood pressure generated by ventricular systole. They are more muscular than veins and appear relatively round in tissue sections. They are divided into three categories by size, but of course there is a smooth gradation from one category to the next.

1. **Conducting (elastic) arteries** are the largest. Examples include the pulmonary arteries, aorta, and common carotid arteries. Their tunica media consists of numerous sheets of elastic tissue, perforated like slices of Swiss cheese, alternating with thin layers of smooth muscle, collagen, and elastic fibers. Conducting arteries expand when the

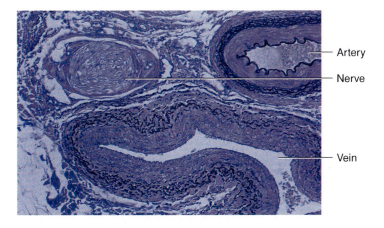

Figure 20.3 A small artery, small vein, and nerve bound together as a neurovascular bundle (×200).

ventricles eject blood during systole, and they recoil during diastole. This lessens the fluctuations in blood pressure exerted on smaller arteries downstream.

2. **Distributing (muscular) arteries** are smaller branches farther away from the heart that distribute blood to specific organs. You could compare a conducting artery to an interstate highway and a distributing artery to the exit ramps and state highways that serve specific cities. While smooth muscle constitutes about one-third of the wall of a conducting artery, it constitutes about three-quarters of the wall of a distributing artery, usually in 25 to 40 layers of muscle cells. Most arteries to which we give names are in these first two size classes. Some distributing arteries include the brachial, femoral, and radial arteries.

3. **Resistance (small) arteries** are usually too variable in number and location to be given names. They exhibit up to 25 layers of smooth muscle cells and relatively little elastic tissue. Their tunica media is thicker in proportion to the lumen than that of larger arteries. The smallest arteries, about 40 to 200 μm in diameter and with only one or two layers of smooth muscle, are the **arterioles.** For reasons discussed later, they are the primary means of controlling the routes of blood flow.

Metarterioles[3] (fig. 20.3) are short vessels that link arterioles and capillaries. Instead of a continuous tunica media, they have individual muscle cells spaced a short distance apart.

3. *meta* = beyond, next in a series

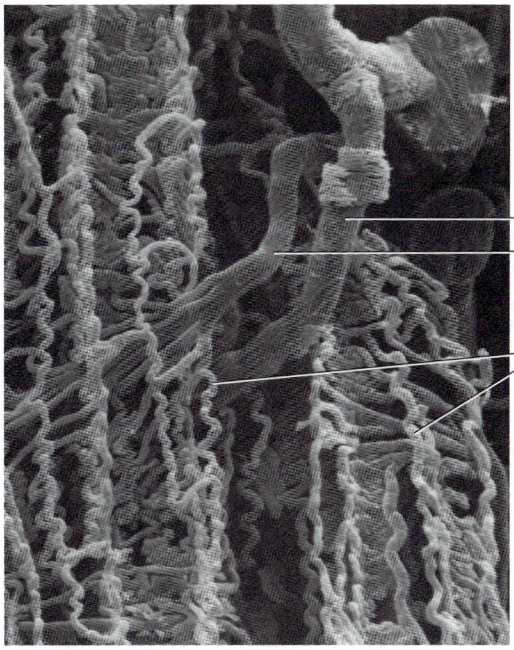

Figure 20.4 A vascular cast of blood vessels of human skeletal muscle, prepared by injecting the vessels with a polymer and then digesting away all tissue to leave a replica of the vessels; photographed by SEM.

From R. G. Kessel and R. H. Kardon, *Scanning Electron Microscopy of Tissues and Organs,* 1979.

Capillaries

Capillaries are the "business end" of the circulatory system (fig. 20.4; see also figs. 6.6 and 18.8). All the rest of the system exists to serve them, since it is across their walls that materials are exchanged between the blood and tissue fluid. Capillaries are ideally suited to their role. They consist of endothelium only and have walls as thin as 0.2 to 0.4 μm. They average about 5 μm in diameter at the proximal (arterial) end, widen to about 9 μm in diameter at the distal (venous) end, and often branch along the way. (Recall that an erythrocyte is about 7 μm in diameter.) The number of capillaries has been estimated at a billion and their total surface area at 6,300 m². But a more important point is that scarcely any cell in the body is more than 60 to 80 μm away from the nearest capillary. There are a few exceptions. Capillaries are scarce in tendons and ligaments and absent from cartilage, epithelia, and the cornea and lens of the eye.

Capillary Beds

Capillaries are organized in groups called **capillary beds**—usually 10 to 100 capillaries supplied by a single metarteriole (fig. 20.5). The metarteriole continues through the bed as a **thoroughfare channel** leading directly to a venule. Capillaries arise from the proximal end of the metarteriole and flow into the distal end of the thoroughfare channel or directly into the venule.

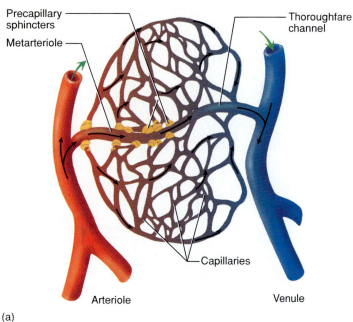

Sphincters open

Precapillary sphincters

Metarteriole

Thoroughfare channel

Capillaries

Arteriole

Venule

(a)

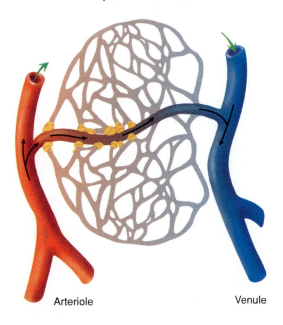

Sphincters closed

Arteriole

Venule

(b)

Figure 20.5 Control of perfusion of a capillary bed. (*a*) Precapillary sphincters dilated and capillaries well perfused. (*b*) Precapillary sphincters closed, with blood bypassing the capillaries.

At the origin of each capillary is a circular band of muscle called a **precapillary sphincter.** When the sphincters are open, the capillaries are well perfused with blood and they engage in exchanges with the tissue fluid. When the sphincters are closed, blood bypasses the capillaries, flows through the thoroughfare channel to a venule, and does not engage in fluid ex-

change. There is not enough blood in the body to fill the entire vascular system at once; consequently, about three-quarters of the body's capillaries are closed at any given time. The shifting of blood flow from one capillary bed to another is discussed later in the chapter.

Types of Capillaries

Two types of capillaries are distinguished by the sizes of gaps between or through the endothelial cells and the corresponding ease with which they allow substances to pass through their walls.

1. **Continuous capillaries** occur in most tissues, such as skeletal muscle. Their endothelial cells, held together by tight junctions, form an uninterrupted tube. The endothelial cells usually have narrow **intercellular clefts** about 4 nm wide between them. Small solutes, such as glucose, can pass through these clefts, but plasma proteins, other large molecules, and formed elements are held back. The continuous capillaries of the brain lack intercellular clefts and have more complete tight junctions that form the blood-brain barrier discussed in chapter 14.
2. **Fenestrated capillaries** differ from the continuous variety in that their endothelial cells are riddled with holes called **fenestrations**[4] **(filtration pores)** (fig. 20.6). Fenestrations are about 20 to 100 nm in diameter and are covered by a thin mucoprotein diaphragm. They allow for especially rapid passage of small molecules but still retain proteins and larger particles in the bloodstream. Fenestrated capillaries are important in organs that engage in rapid absorption or filtration—the kidneys, endocrine glands, small intestine, and choroid plexuses of the brain, for example.

Sinusoids are irregular blood-filled spaces in the liver, bone marrow, spleen, and some other organs. They are twisted, tortuous passageways that conform to the shape of the surrounding tissue. Some of them are continuous capillaries with very thin walls; others are fenestrated capillaries with extraordinarily large pores that allow the blood plasma to come into direct contact with the perivascular cells. Even proteins and blood cells can readily pass through these—this is how albumin, clotting factors, and other proteins synthesized by the liver enter the blood and how newly formed blood cells enter the circulation from the bone marrow and lymphatic organs.

4. *fenestra* = window

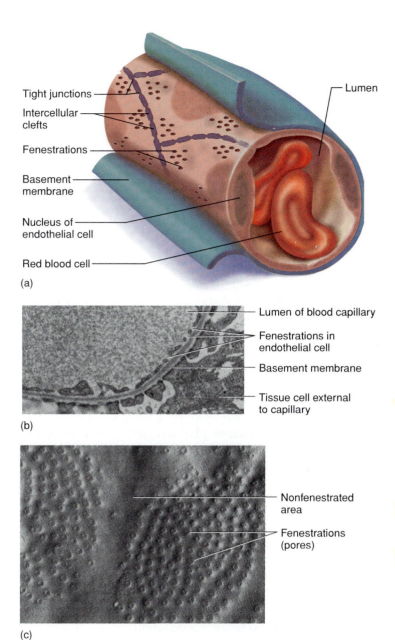

Tight junctions
Intercellular clefts
Fenestrations
Basement membrane
Nucleus of endothelial cell
Red blood cell
Lumen
(a)

Lumen of blood capillary
Fenestrations in endothelial cell
Basement membrane
Tissue cell external to capillary
(b)

Nonfenestrated area
Fenestrations (pores)
(c)

Figure 20.6 (*a*) The structure of a fenestrated capillary. (*b*) The wall of a fenestrated capillary of the kidney (TEM). All the fenestrations shown are in a single endothelial cell. (*c*) Surface view of a fenstrated endothelial cell (SEM). The cell has patches of fenestrations separated by nonfenestrated areas. 𝒯

Veins

After flowing through the capillaries, blood collects in the distal end of the thoroughfare channel and flows into a venule. In the venous circulation, blood flows from smaller vessels into progressively larger ones; hence, instead of giving off *branches* as arteries do, veins receive smaller *tributaries,* just as a river receives water from the many streams that form its tributaries.

Venules range from about 15 to 100 μm in diameter. The proximal part of a venule has only a few fibro-

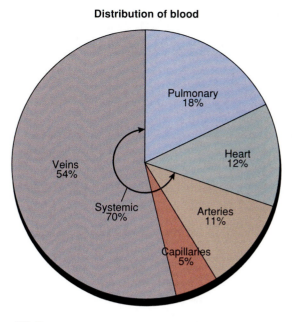

Figure 20.7 Average distribution of the blood in a resting adult.

blasts around it and is quite porous; therefore, fluid can escape the bloodstream here as well as at the capillaries. Farther along, a venule acquires a smooth muscle tunica media. Even the largest veins, however, have relatively sparse muscular and elastic tissue compared to arteries.

Venous sinuses are veins with especially thin walls, large lumens, and no smooth muscle or vasomotion. Examples include the coronary sinus of the heart and the dural sinuses associated with the brain.

Because they are farther away from the heart, veins have much lower blood pressure than arteries. In large arteries, the blood pressure averages 90 to 100 mmHg and surges to 120 mmHg during systole, whereas in veins it averages about 10 mmHg and does not fluctuate with the heartbeat. This has significant implications for the form and function of veins:

- Since they need not withstand high pressure, veins have thinner walls than arteries, with less muscular and elastic tissue. They collapse when empty and look relatively flattened or irregular in histological sections (see fig. 20.3).
- Since their walls are so thin, veins can expand more easily and accommodate more blood than arteries do. About 54% of the blood is found in the systemic veins at rest (fig. 20.7); veins are therefore called *capacitance vessels.*
- The pressure in the veins is not high enough to push blood upward against the pull of gravity to the heart. The upward flow of blood depends in part on the massaging action of skeletal muscles and the presence of one-way **venous valves** that keep the blood from dropping down again when the muscles relax (see fig. 20.2). These valves, similar to the

In people who stand much of the time, such as dentists and hairdressers, blood tends to pool in the lower extremities and stretch the veins. This especially affects veins close to the surface, which are not surrounded by supportive tissue. Stretching pulls the cusps of the venous valves farther apart until the valves become incompetent—unable to prevent the backflow of blood. As the veins become further distended, their walls grow weak and they develop into **varicose veins** with irregular dilations and twisted pathways. Obesity and pregnancy also promote development of varicose veins by putting pressure on large veins of the pelvic region and obstructing drainage from the legs. Varicose veins sometimes develop because of hereditary weakness of the valves. With less drainage of blood, tissues of the leg and foot may become edematous and painful. **Hemorrhoids** are varicose veins of the anal canal.

semilunar valves of the heart, occur especially in medium veins of the arms and legs; they are absent from very small and very large veins, veins of the ventral body cavity, and veins of the brain. Varicose veins result in part from the failure of these valves (see special topic 20.1).

Think About It

From figure 20.7 and your other knowledge, calculate the approximate volume of blood in the *systemic* capillaries. What percentage of the systemic blood is this?

Key Point Review

1. Explain how an anastomosis and a portal system differ from the simple artery → capillary → vein scheme of circulation.
2. Name the three tunics of a typical blood vessel and explain how they differ from each other.
3. Describe the route of blood flow through a capillary bed.
4. Contrast the two types of capillaries.
5. Explain why many veins have valves but arteries do not.

Blood Flow, Pressure, and Resistance

▼Objectives

When you have completed this section, you should be able to

- explain the relationship between blood pressure, resistance, and flow;
- describe how blood pressure is expressed and how pulse pressure and mean arterial blood pressure are calculated;
- describe three factors that determine resistance to blood flow; and
- explain how vasomotion influences blood flow and describe some local, neural, and hormonal influences on vasomotion.

Flow

Blood flow is the amount of blood flowing through an organ, tissue, or blood vessel in a given time (such as mL/min). **Perfusion** is the rate of blood flow per given volume or mass of tissue (such as mL/min/g). The principal importance of flow is that it governs the speed of oxygen and nutrient delivery to a tissue and the speed of waste removal. If flow does not keep pace with the metabolic rate of a tissue, the likely result is tissue necrosis and possibly death of the individual. In a resting individual, *total* flow is quite constant and is equal to cardiac output (typically 5.25 L/min). Flow through individual organs, however, varies from minute to minute as blood is redirected from one organ to another. Great variations in regional flow can occur with little or no change in total flow.

Hemodynamics, the physical principles of blood flow, are based mainly on pressure and resistance. The greater the pressure difference (ΔP) between two points, the greater the flow; the greater the resistance (R), the less the flow—in summary, $F \propto \Delta P/R$. Therefore, to understand the flow of blood, we must consider the factors that affect pressure and resistance.

Blood Pressure

Blood pressure (BP) can be measured within a blood vessel or heart chamber by inserting a catheter or needle connected to an external manometer. For routine clinical purposes, however, the measurement of greatest interest is the systemic arterial BP at a point close to the heart. As mentioned in the previous chapter, it is customarily measured with a sphygmomanometer at the brachial artery of the arm (fig. 20.8). This artery is easily encircled and compressed by a pressure cuff, and it is sufficiently close to the heart to reflect the maximum arterial BP to be found anywhere in the systemic circuit.

Two pressures are recorded: (1) **Systolic pressure** is the peak arterial BP attained during ventricular systole. (2) **Diastolic pressure** is the minimum arterial BP between heartbeats. For a healthy person in his or her 20s, these pressures are typically about 120 and 75 mmHg, respectively. Arterial BP is written as a ratio of systolic over diastolic pressure: 120/75.

The difference between systolic and diastolic pressure is called **pulse pressure** (not to be confused with pulse rate). For the preceding BP, pulse pressure would

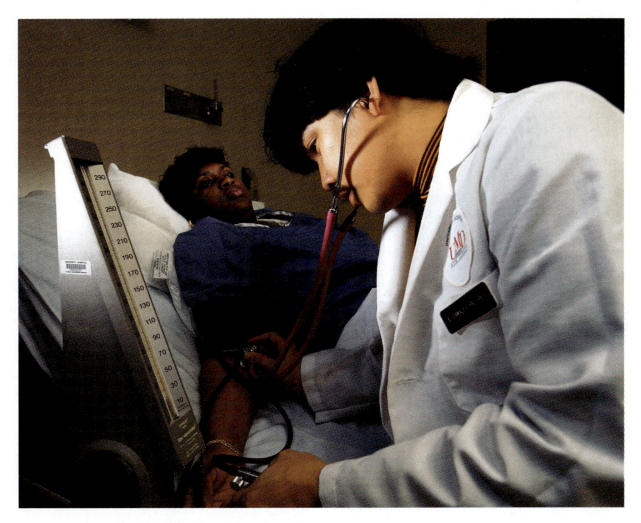

Figure 20.8 Blood pressure is recorded while listening for sounds in the brachial artery that result from compression by the cuff of a sphygmomanometer. Systolic and diastolic pressures are noted at the times that these sounds begin and end.

be 120 − 75 = 45 mmHg. This is an important measure of the stress exerted on small arteries by the pressure surges generated by the heart. Another measure of stress on the blood vessels is the **mean arterial blood pressure (MABP)**—the mean pressure you would obtain if you took measurements at several intervals (say every 0.1 sec) throughout the cardiac cycle. Since diastole lasts longer than systole, MABP is not simply the average of systolic and diastolic pressures. It is best estimated as the sum of diastolic pressure and one-third of the pulse pressure. For a blood pressure of 120/75, MABP ≈ 75 + 45/3 = 90 mmHg. This is typical for vessels at the level of the heart. MABP varies, however, with the influence of gravity. In a standing adult, it is about 62 mmHg in the major arteries of the head and 180 mmHg in major arteries of the ankle.

Hypertension (high BP) is commonly considered to be a chronic resting systolic pressure higher than 150 mmHg or diastolic pressure higher than 90 mmHg (see chapter essay, p. 749). (*Transient* high BP resulting from emotion or exercise is not hypertension.) Among other

effects, it can weaken the small arteries and cause **aneurysms**[5] (AN-you-rizm) (see special topic 20.2). **Hypotension** is chronic low resting BP. It may be a consequence of blood loss, dehydration, anemia, or other factors and is normal in people approaching the moment of death.

The ability of the arteries to distend and recoil during the cardiac cycle is important in the modulation of arterial BP. If the arteries were rigid tubes, pressure would rise much higher during systole and it would drop to nearly zero in diastole. Blood throughout the circulatory system would flow and stop, flow and stop, putting great stress on the small vessels. But when the conducting arteries are healthy, they expand with each systole, absorbing some of the force of the ejected blood. Then, when the heart is in diastole, their elastic recoil exerts pressure on the bloodstream and prevents it from dropping to zero. The combination of expansion and

5. *aneurysm* = widening

An aneurysm is a weak point in a blood vessel or in the heart wall. It forms a thin-walled, bulging sac that pulsates with each beat of the heart and may eventually rupture. In a *dissecting aneurysm,* blood pools between the tunics of a vessel and separates them, usually as a result of degeneration of the tunica media. The most common sites of aneurysms are the abdominal aorta, renal arteries, and the circle of Willis at the base of the brain. Even without hemorrhaging, aneurysms can cause pain or death by putting pressure on brain tissue, nerves, adjacent veins, pulmonary air passages, or the esophagus. Consequences include neurological disorders, difficulty in breathing or swallowing, chronic cough, or congestion of the tissues with blood. Aneurysms sometimes result from congenital weakness of the blood vessels and sometimes from trauma or bacterial infections such as syphilis. The most common cause, however, is the combination of atherosclerosis and hypertension.

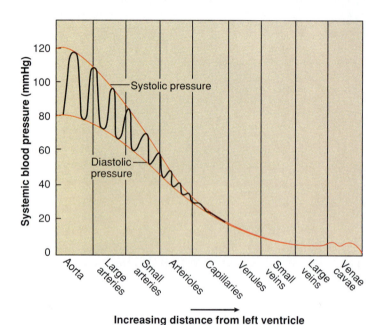

Table 20.1	Normal Arterial Blood Pressure at Various Ages	
Age (years)	Male	Female
1	96/66	95/65
5	92/62	92/62
10	103/69	103/70
15	112/75	112/76
20–24	123/76	116/72
25–29	125/78	117/74
30–34	126/79	120/75
40–44	129/81	127/80
50–54	135/83	137/84
60–64	142/85	144/85
70–74	145/82	159/85
80–84	145/82	157/83

Figure 20.9 All measures of blood pressure decline with increasing distance from the heart, including systolic pressure, diastolic pressure, pulse pressure, and mean arterial blood pressure. There is no pulse pressure beyond the arterioles, but there are slight pressure oscillations in the vena cava caused by the respiratory pump.

recoil maintains a steady flow of blood in the capillaries throughout the cardiac cycle. Thus, the elastic arteries "smooth out" the pressure fluctuations and reduce stress on the smaller arteries.

Nevertheless, blood flow in the arteries is *pulsatile.* Blood in the aorta rushes forward at a velocity of 120 cm/sec during systole and has an average speed of 40 cm/sec over the cardiac cycle. When measured at points farther away from the heart, systolic and diastolic pressures are lower and there is less difference between them (fig. 20.9). In capillaries and veins, the blood flows at a steady speed without pulsation because the pressure surges have been damped out by the distance traveled and the elasticity of the arteries. This is why an injured vein exhibits relatively slow, steady bleeding, whereas blood spurts intermittently from a severed artery. In the inferior vena cava near the heart, however, venous flow fluctuates with the respiratory cycle for reasons explained later.

Blood pressure rises with age (table 20.1) as the arteries become less distensible and absorb less systolic force. Atherosclerosis stiffens the arteries and also leads to a rise in BP.

Blood pressure is determined mainly by cardiac output, blood volume, and peripheral resistance. The regulation of cardiac output and blood volume are discussed in chapters 19 and 24, respectively. Here, we turn our attention to peripheral resistance.

Resistance

A moving fluid has no pressure unless it encounters at least some resistance. Thus, pressure and resistance are not independent factors in blood flow—rather, pressure is affected by resistance, and flow is affected by both.

Peripheral resistance is the resistance that the blood encounters in the vessels as it travels away from the heart. It results from the friction of blood against the walls of the vessels and is proportional to three variables: *blood viscosity, vessel length,* and *vessel radius.*

Blood Viscosity

Blood viscosity ("thickness" of the blood) is due mainly to erythrocytes and albumin (see chapter 18). A deficiency of erythrocytes (anemia) or albumin (hypoproteinemia) decreases peripheral resistance, leading to a drop in BP and reduced perfusion of the organs. If viscosity increases (as a result of polycythemia or dehydration, for example), resistance increases and flow declines.

Vessel Length

The farther a liquid travels through a tube, the more cumulative friction it encounters; thus, pressure and flow decline with distance. Partly for this reason, if you were to measure MABP in a recumbent person, you would obtain a higher value in the arm, for example, than at the ankle. (This would not be true in a standing person because of the influence of gravity explained earlier.) A strong pulse in the dorsal pedal artery of the foot is a good sign of adequate cardiac output. If perfusion is good at that distance from the heart, it is likely to be good elsewhere in the systemic circulation.

Vessel Radius

In a healthy individual, blood viscosity is quite stable, and of course vessel lengths do not change in the short term. Therefore, the only significant way of controlling peripheral resistance from moment to moment is by vasomotion—changing the radius of the vessels.

The blood normally exhibits smooth, silent **laminar[6] flow.** It flows faster near the center of a vessel than it does closer to the walls because it encounters less friction at the center. You can observe a similar effect from the vantage point of a riverbank. A current may be very swift in the middle of a river but quite sluggish near shore, where the water encounters more friction against the riverbank and bottom. When a blood vessel dilates, a greater portion of the blood is in the middle of the stream and the average flow may be quite swift. When the vessel constricts, more of the blood is close to the wall and the average flow is slower (fig. 20.10).

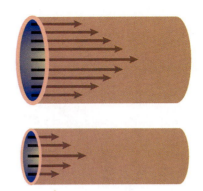

Figure 20.10 Laminar flow and the effect of vessel radius. Blood flows more slowly near the vessel wall, as indicated by shorter arrows, than it does near the center of the vessel. (*a*) When vessel radius is large, the average velocity of flow is high. (*b*) When the radius is less, the average velocity is lower because a larger portion of the blood is slowed down by friction against the vessel wall.

Thus the radius of a vessel markedly affects blood velocity. As you already know, smooth muscle of the tunica media makes vasomotion possible. The arterioles are the most significant point of control over peripheral resistance because (1) they are on the proximal sides of the capillary beds, so that they are best positioned to regulate flow into the capillaries; (2) they greatly outnumber any other class of arteries and thus provide the most numerous control points; and (3) they are more muscular in proportion to their diameters than any other class of blood vessels and are highly capable of vasomotion. Arterioles alone account for about half of the total peripheral resistance of the circulatory system. However, larger arteries and veins are also capable of considerable vasomotion and control of peripheral resistance. Vasomotion can result from the influence of sympathetic nervous stimulation, hormonal stimulation, or local conditions, as we will see shortly.

Blood flow is proportional not merely to vessel radius but to the *fourth power* of radius—that is, $F \propto r^4$. This makes vessel radius a very potent factor in the control of flow. The radius of an arteriole can show at least a threefold increase or decrease (fig. 20.11). For the sake of simplicity, consider a hypothetical blood vessel with a 3 mm radius when completely relaxed and a 1 mm radius when maximally constricted. At a 1 mm radius, suppose the bloodstream travels 1 mm/sec. By the formula $F \propto r^4$, consider how the velocity would change as radius changed:

$$r = 1 \text{ mm} \quad r^4 = 1^4 = 1 \quad F = 1 \text{ mm/sec (given)}$$
$$r = 2 \text{ mm} \quad r^4 = 2^4 = 16 \quad F = 16 \text{ mm/sec}$$
$$r = 3 \text{ mm} \quad r^4 = 3^4 = 81 \quad F = 81 \text{ mm/sec}$$

These actual numbers do not matter; what matters is that a mere 3-fold increase in radius has produced an 81-fold increase in velocity—a demonstration that

---

6. *lamina* = layer

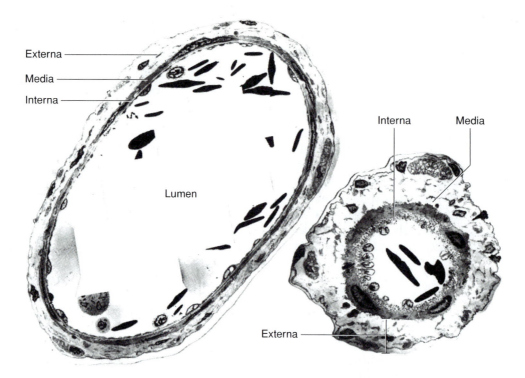

Externa
Media
Interna
Lumen

Interna   Media

Externa

Figure 20.11 Micrographs of a single arteriole at dilated and constricted points less than 1 mm apart. Just before the arteriole was fixed, a single drop of epinephrine was applied to one portion of it, causing vasoconstriction in that region. The diameter of the dilated region is about three times that of the constricted region.

vessel radius exerts a very powerful influence over flow; moreover, it is the most adjustable of all variables that govern peripheral resistance.

Poiseuille's Law

We have demonstrated that flow (F) is directly proportional to the pressure gradient (ΔP) and to the fourth power of the vessel radius (r^4), while it is inversely proportional to the vessel length (L) and blood viscosity (η). Two constants, pi (π) and 1/8, convert these relationships to an equation called **Poiseuille's**[7] (pwa-ZUH-yez) **law,** which sums up much of what you have just learned:

$$F = \frac{\Delta P \pi r^4}{8 \eta L}$$

To integrate this information, consider how the velocity of blood flow differs from one part of the systemic circuit to another (table 20.2). Flow is fastest in the aorta because it is a large vessel close to the pressure head, the left ventricle. From aorta to capillaries, velocity diminishes for three reasons: (1) The blood has traveled a greater distance, so friction has slowed it down. (2) The arterioles and capillaries have smaller radii and there-

Table 20.2	Blood Velocity in the Systemic Circuit	
Vessel	**Typical Lumen Diameter**	**Velocity***
Aorta	2.5 cm	1,200 mm/sec
Arterioles	20–50 μm	15 mm/sec
Capillaries	5–9 μm	0.4 mm/sec
Venules	20 μm	5 mm/sec
Inferior vena cava	3 cm	80 mm/sec

*Peak systolic velocity in the aorta; mean or steady velocity in other vessels.

fore put up more resistance. (3) Even though the radii of individual vessels become smaller as we progress farther from the heart, the number of vessels and their *total* cross-sectional area becomes greater and greater. The aorta has a cross-sectional area of 3 to 5 cm², while the total cross-sectional area of all the capillaries is about 4,500 to 6,000 cm². Thus, a given volume of aortic blood is distributed over a greater total area in the capillaries, which *collectively* form a wider path in the bloodstream. Just as water slows down when a narrow mountain stream flows into a lake, blood slows down as it enters the wider paths in the circulatory system.

From capillaries to vena cava, velocity rises again. One reason for this is that the veins have larger diameters than the capillaries, and so the blood in an individual vein is less impeded by friction. Furthermore, since

7. Jean Marie Poiseuille (1799–1869), French physiologist

many capillaries converge on one venule, and many venules on a larger vein, a large amount of blood is being forced into a progressively smaller channel—like water flowing from a lake into an outlet stream and thus flowing faster again. Note, however, that blood in the veins never regains the velocity it had in the large arteries. This is because the veins are farther from the heart and the pressure is much lower here.

Regulation of Peripheral Resistance

As we mentioned earlier, vasomotion is subject to local, neural, and hormonal controls. We now consider each of these three influences in turn.

Local Control

Autoregulation is the ability of tissues to regulate their own blood supply. According to the *metabolic theory of autoregulation,* if a tissue is inadequately perfused, it becomes hypoxic and its metabolites (waste products) accumulate—CO_2, H^+, K^+, lactic acid, and adenosine, for example. These factors stimulate vasodilation, which increases perfusion. As the bloodstream delivers oxygen and carries away the metabolites, the vessels constrict. Thus, a homeostatic dynamic equilibrium is established that adjusts perfusion to the tissue's metabolic needs. Under conditions such as trauma, inflammation, and exercise, vasodilation may also be triggered by chemicals such as prostaglandins, kinins, and histamine. The rising temperature of active muscles is also thought to promote vasodilation.

If a tissue's blood supply is cut off for a time and then restored, it often exhibits **reactive hyperemia**—an increase above the normal level of flow. This may be due to the accumulation of metabolites during the period of ischemia. Reactive hyperemia can be seen when the skin flushes after a person comes in from the cold; it occurs in the forearm if a blood pressure cuff is inflated for too long and then loosened.

Endothelial cells and platelets secrete *vasoactive chemicals* that promote vasomotion. Two vasodilators of endothelial origin are prostacyclin and nitric oxide (NO). NO is sometimes called *endothelium-derived relaxing factor (EDRF),* a name assigned before its chemical identity was known. Some vasoconstrictors of endothelial origin are the *endothelins,* a family of four similar polypeptides. The platelets produce two vasoconstrictors, serotonin and thromboxane A_2, in response to vessel injury.

In the long run, a hypoxic tissue can increase its own perfusion by **angiogenesis**[8]—the growth of new blood vessels. (This term also refers to embryonic development of the vascular system.) This is a common mechanism for bypassing obstructed arteries of the coronary circulation; it produces a higher density of capillaries in the muscles of trained athletes, and it is involved in regrowth of the uterine lining after every menstrual period. Several growth factors and inhibitors control angiogenesis, but physiologists are not yet sure how it is regulated.

Neural Control

In addition to local control, the blood vessels are under remote control by hormones and the autonomic nervous system. The **vasomotor center** of the medulla oblongata exerts sympathetic control over blood vessels throughout the body. (Precapillary sphincters have no innervation, however, and respond only to local and hormonal stimuli.) Most sympathetic fibers to the blood vessels induce vasoconstriction, but some induce vasodilation, notably in the skeletal muscles. The role of sympathetic and vasomotor tone in controlling vessel diameter is explained in chapter 15. The vasomotor center is an integrating center for three autonomic reflexes—*baroreflexes, chemoreflexes,* and the *medullary ischemic reflex.*

A **baroreflex** is an autonomic response to changes in blood pressure. The changes are detected by **baroreceptors**[9]—stretch receptors in the wall of the heart and in the tunica externa of the great vessels above it. They are branched, knobby nerve fibers somewhat resembling Golgi tendon organs (see p. 530). Those in the aortic arch and carotid sinus (fig. 20.12) are the most important in monitoring arterial BP. A rise in pressure stretches these vessels and increases the firing rate of afferent fibers from the baroreceptors to the brainstem. The effects of this are (1) to *inhibit* the sympathetic cardiac and vasomotor neurons and reduce sympathetic tone and (2) to *excite* the vagal fibers to the heart. Heart rate and cardiac output drop, arteries and veins dilate, and BP drops—a classic negative feedback loop that produces homeostasis (fig. 20.13). When BP drops below normal, on the other hand, the opposite reactions occur and raise it (see p. 531).

The baroreceptors are important chiefly in short-term regulation of BP. A good example of their action is seen in cardiovascular adjustments to postural changes. Perhaps, on occasion, you have jumped quickly out of bed and felt a little dizzy for a moment. This is because gravity draws the blood into the large veins of the abdomen and lower extremities when you stand, which reduces venous return to the heart and cardiac output to

---

8. *angio* = vessel + *genesis* = production of

9. *baro* = pressure

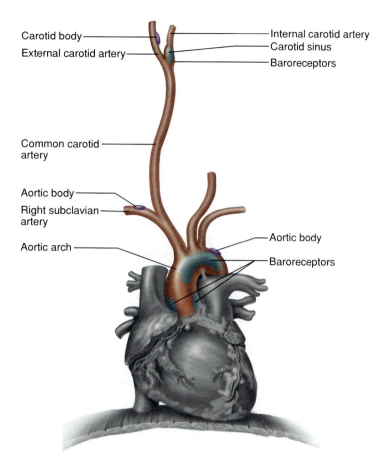

Figure 20.12 Locations of the baroreceptors and chemoreceptors in the arteries superior to the heart. Chemoreceptors are located in the carotid bodies and aortic bodies. Baroreceptors are located in the ascending aorta, aortic arch, and carotid sinus. The structures shown here in the right carotid arteries are repeated in the left carotids.

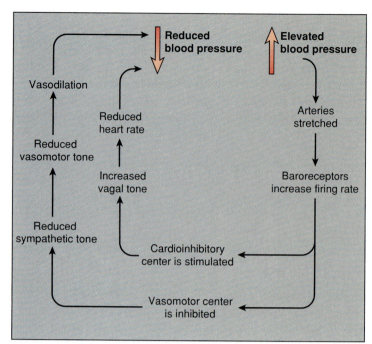

Figure 20.13 Negative feedback loops that correct for an increase in blood pressure. The opposite responses occur when blood pressure drops below normal.

the brain. Normally, the baroreceptors respond quickly to this drop in pressure and restore cerebral perfusion. Baroreflexes are not effective in correcting chronic hypertension, however. Apparently they adjust their set point to the higher BP and maintain dynamic equilibrium at this new level.

A **chemoreflex** is an autonomic response to changes in blood chemistry, especially its pH and concentrations of O_2 and CO_2. It is initiated by chemoreceptors within small organs called the **aortic** and **carotid bodies,** located in the aortic arch, subclavian arteries, and external carotid arteries. The primary role of chemoreflexes is to adjust respiration to changes in blood chemistry, but they have a secondary role in vasomotion. In response to hypoxemia, hypercapnia, or acidosis, they act through the vasomotor center to cause generalized vasoconstriction. This increases BP, thus increasing perfusion of the lungs and the rate of gas exchange. They also trigger an increase in the rate and depth of breathing, which allows ventilation of the lungs to keep pace with perfusion. Increasing one without the other would be of little use.

The **medullary ischemic** (iss-KEE-mic) **reflex** is an autonomic response to insufficient perfusion (ischemia) of the brainstem. Hypoxia and hypercapnia act on the vasomotor center to stimulate vasoconstriction in the lower parts of the body. This redirects more blood to the upper body and improves perfusion of the brain.

The vasomotor center also receives input from other brain centers. Stress, anger, and sexual arousal, for example, cause vasoconstriction and raise BP. The hypothalamus acts through the vasomotor center to redirect blood flow in response to exercise or changes in body temperature.

Hormonal Control

Several hormones affect vasomotion. One of them, **angiotensin II,** is made from a prohormone called *angiotensinogen,* which is secreted by the liver and continually present in the blood. In response to hypotension, the kidneys secrete an enzyme, **renin,** which converts angiotensinogen to *angiotensin I.* As angiotensin I passes through the lungs, another enzyme, **angiotensin-converting enzyme (ACE),** converts it to angiotensin II. Angiotensin II is a very potent vasoconstrictor. Its action is further explored in chapter 24. Epinephrine and norepinephrine also constrict most vessels, but they dilate those of the coronary circulation and skeletal muscles, thus increasing the perfusion of the heart and muscles when their workload demands it. Atrial natriuretic factor (ANF) is known primarily for

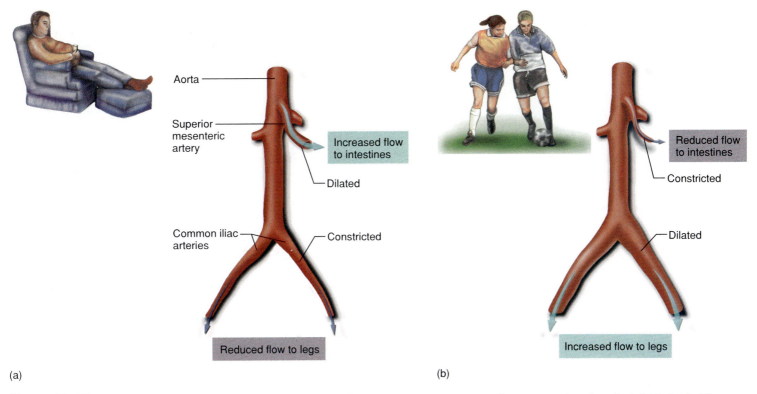

(a)   (b)

Figure 20.14 Vasoconstriction and vasodilation shift blood flow from one organ system to another to meet changing physiological priorities. (a) After a meal, the intestines receive priority and the skeletal muscles receive relatively little flow. (b) During exercise, the muscles receive higher priority.

increasing sodium excretion, but it also has a generalized vasodilator effect. Antidiuretic hormone (ADH) primarily functions to promote water retention, but at pathologically high concentrations it is also a vasoconstrictor—hence its alternate name, *vasopressin*.

Vasomotion and Redirection of Blood Flow

If a vasoconstrictor such as epinephrine causes widespread vasoconstriction, or if it causes vasoconstriction in a large system such as the integumentary or digestive system, it can produce an overall rise in blood pressure. Localized vasoconstriction, however, has a very different effect. If a particular artery constricts, pressure downstream from the constriction drops and pressure upstream from it rises. If blood can travel by either of two routes and one route puts up more resistance than the other, most blood follows the path of least resistance. This mechanism enables the body to redirect blood from one organ to another.

For example, if you are dozing in an armchair after a big meal, 90% or more of the capillaries in your leg muscles are shut down by vasoconstriction. This raises the aortic BP above the legs, where the aorta gives off a branch, the superior mesenteric artery, supplying the small intestine. High resistance in the circulation of the legs and low resistance in the superior mesenteric artery routes blood to the small intestine, where it is needed to absorb digested nutrients (fig. 20.14a).

On the other hand, during vigorous exercise, the arteries in your leg muscles dilate. To make blood available to the muscles, flow must be reduced elsewhere—notably in major areas such as the digestive tract (fig. 20.14b). Thus, changes in peripheral resistance can shift blood flow from one organ system to another to meet the changing metabolic priorities of the body. The state of rest or exercise affects not only total cardiac output but also the distribution of blood to different organ systems. Physical exertion increases perfusion of the lungs, myocardium, and skeletal muscles while reducing perfusion of the digestive tract, skin, and kidneys (fig. 20.15).

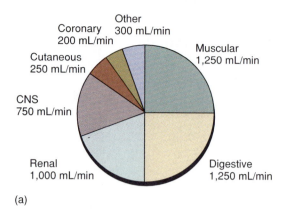

At rest
Total cardiac output 5 L/min

Other
300 mL/min

Coronary
200 mL/min

Cutaneous
250 mL/min

CNS
750 mL/min

Muscular
1,250 mL/min

Renal
1,000 mL/min

Digestive
1,250 mL/min

(a)

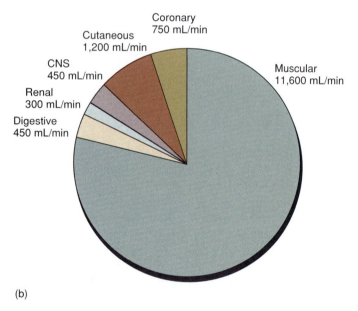

Heavy work
Total cardiac output 15 L/min

Coronary
750 mL/min

Cutaneous
1,200 mL/min

CNS
450 mL/min

Renal
300 mL/min

Digestive
450 mL/min

Muscular
11,600 mL/min

(b)

Figure 20.15 Relative distribution of systemic blood flow during rest and heavy exercise. Areas of the two charts are proportional to total cardiac output under the two conditions.

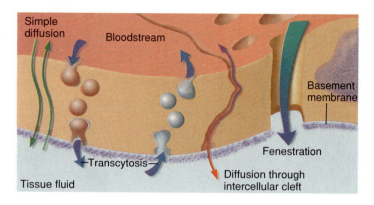

Figure 20.16 Modes and pathways of capillary fluid exchange.

Only 250 to 300 mL of blood is in the capillaries at any given time. This is the most important blood in the body, however, for it is only across capillary walls that any appreciable exchanges occur between the blood and surrounding tissues. **Capillary exchange** refers to this two-way movement of fluid.

The mechanisms of capillary exchange are difficult to study quantitatively because it is hard to measure pressure and flow in such small vessels. For this reason, theories of capillary exchange remain in dispute. Few capillaries of the human body are accessible to direct, noninvasive observation, but those of the fingernail bed and eponychium (cuticle) at the base of the nails can be observed with a stereomicroscope and have been the basis for a number of studies. Their BP has been measured at 32 mmHg at the arterial end and 15 mmHg at the venous end, 1 mm away. Capillary BP drops rapidly because of the substantial friction the blood encounters in such narrow vessels. Blood flows at about 0.7 mm/sec in nail bed capillaries; it takes 1 to 2 seconds for an RBC to pass through one.

There are four routes by which the capillary blood and tissue fluid can exchange matter: (1) through the intercellular clefts between endothelial cells, (2) through the pores of fenestrated capillaries, (3) by way of pinocytotic vesicles, and (4) through the endothelial cell plasma membranes (fig. 20.16). The mechanisms involved are *diffusion, transcytosis, filtration,* and *reabsorption,* which we examine in that order.

Diffusion

The most important mechanism of exchange is diffusion. Solutes that are more concentrated in the blood than in the tissue fluid, such as glucose and oxygen, diffuse out of the blood. Solutes that are more concentrated in the tissue fluid, such as carbon dioxide and nitrogenous wastes, diffuse into the blood. Such diffusion is only possible if the solute can either permeate the plasma membranes of the endothelial cells or find passages large enough to pass through—namely, the fenestrations and intercellular clefts. Such lipid-soluble substances as steroid hormones, O_2, and CO_2 diffuse easily through the plasma membranes. Substances insoluble in lipids, however, must pass through membrane channels, fenestrations, or intercellular clefts. These include water, glucose, and electrolytes. Large molecules such as proteins are usually held back by the small size of these passages.

Transcytosis

Transcytosis is a process in which endothelial cells pick up droplets of fluid on one side of the plasma membrane by pinocytosis, transport the vesicles across the cell, and discharge the fluid on the other side by exocytosis (see fig. 4.31, p. 133). This probably accounts for only a small fraction of solute exchange across the capillary wall, but fatty acids, albumin, and some hormones such as insulin move across the capillary barrier by this mechanism.

Filtration and Reabsorption

The equilibrium between filtration and osmosis discussed in chapter 4 becomes particularly relevant when we consider capillary fluid exchange. Fluid filters out of the arterial end of a capillary and usually reenters it at the venous end (fig. 20.17). This fluid delivers materials to the cells and removes their metabolic wastes.

It may seem odd that a capillary could give off fluid at one point and reabsorb it at another. This comes about as the result of a shifting balance between hydrostatic and osmotic forces. A typical capillary has a blood (hydrostatic) pressure of about 30 mmHg at the arterial end. The hydrostatic pressure of the interstitial space has been difficult to measure and remains a point of controversy, but an average value accepted by many authorities is about −3 mmHg—that is, a slight suction that helps draw fluid out of the capillary. In this case, the positive hydrostatic pressure within the capillary and the negative interstitial pressure work in the same direction, creating a total force of about 33 mmHg that causes fluid to leave the capillary.

These forces are opposed by **colloid osmotic pressure (COP)**. The blood has a COP of about 28 mmHg, due mainly to albumin. Tissue fluid has less than one-third the protein concentration of blood plasma, producing a COP of about 8 mmHg. The difference between the COP of blood and COP of tissue fluid is called **oncotic pressure**: $28_{in} − 8_{out} = 20_{in}$. Oncotic pressure tends to draw water into the capillary byosmosis, opposing **hydrostatic pressure**. These

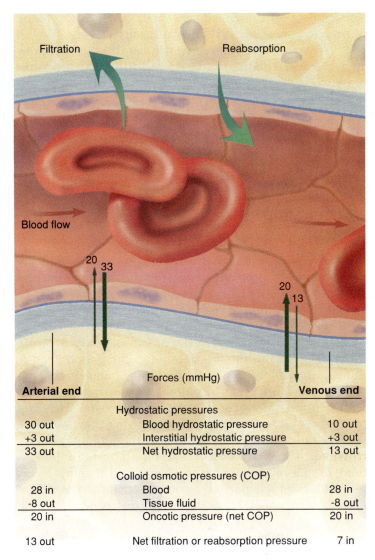

Forces (mmHg)		
Arterial end		**Venous end**
	Hydrostatic pressures	
30 out	Blood hydrostatic pressure	10 out
+3 out	Interstitial hydrostatic pressure	+3 out
33 out	Net hydrostatic pressure	13 out
	Colloid osmotic pressures (COP)	
28 in	Blood	28 in
-8 out	Tissue fluid	-8 out
20 in	Oncotic pressure (net COP)	20 in
13 out	Net filtration or reabsorption pressure	7 in

Figure 20.17 The forces of capillary filtration and reabsorption. Note the shift from net filtration at the arterial (*left*) end to net reabsorption at the venous (*right*) end.

opposing forces leave a **net filtration pressure (NFP)** as follows:

Hydrostatic pressure

Blood pressure	30_{out}
Interstitial pressure	$+ 3_{out}$
Net hydrostatic pressure	33_{out}

Colloid osmotic pressure

Blood	28_{in}
Tissue fluid	$− 8_{out}$
Oncotic pressure	20_{in}

Net filtration pressure

Net hydrostatic pressure	33_{out}
Oncotic pressure	$− 20_{in}$
Net filtration pressure	13_{out}

The NFP of 13 mmHg causes about 0.5% of the blood plasma to leave the capillaries at the arterial end.

At the venous end, however, capillary blood pressure is lower—about 10 mmHg. All the other pressures are unchanged. Thus, we get:

Hydrostatic pressure

Blood pressure	10_{out}
Interstitial pressure	$+ 3_{out}$
Net hydrostatic pressure	13_{out}

Net reabsorption pressure

Oncotic pressure	20_{in}
Net hydrostatic pressure	$- 13_{out}$
Net reabsorption pressure	7_{in}

The prevailing force is inward at the venous end because osmotic pressure overrides filtration pressure. The **net reabsorption pressure** of 7 mmHg inward causes the capillary to reabsorb fluid at this end.

Now you can see why a capillary gives off fluid at one end and reabsorbs it at the other. The only pressure that changes from the arterial end to the venous end is the capillary blood pressure, and this change is responsible for the shift from filtration to reabsorption. With a reabsorption pressure of 7 mmHg and a net filtration pressure of 13 mmHg, it might appear that far more fluid would leave the capillaries than reenter them. However, since capillaries branch along their length, there are more of them at the venous end than at the arterial end, which partially compensates for the difference between filtration and reabsorption pressures. Consequently, capillaries reabsorb about 85% of the fluid they filter. The other 15% is absorbed and returned to the blood by way of the lymphatic system, as described in the next chapter.

Of course, water is not the only substance that crosses the capillary wall by filtration and reabsorption. It carries along many of the solutes dissolved in it. This process is called **solvent drag.**

Variations in Capillary Filtration and Reabsorption

The figures used in the preceding discussion serve only as examples; circumstances differ from place to place in the body and from time to time in the same capillaries. Capillaries usually reabsorb most of the fluid they filter, but this is not always the case. The kidneys have capillaries called *glomeruli* in which there is little or no reabsorption; they are entirely devoted to filtration. Alveolar capillaries of the lungs, by contrast, are almost entirely dedicated to absorption.

Capillary activity also varies from moment to moment. In a resting tissue, most metarterioles and precapillary sphincters are constricted and the capillaries are collapsed. Capillary BP is very low (if there is any flow at all), and reabsorption predominates. When a tissue becomes metabolically active, its capillary flow increases. In active muscles, capillary pressure rises to the point that it overrides reabsorption along the entire length of the capillary. Fluid accumulates in the muscle, and exercising muscles increase in size by as much as 25%. Capillary permeability is also subject to chemical influences. Traumatized tissue releases such chemicals as substance P, bradykinin, and histamine, which increase permeability and filtration.

Key Point Review

10 List the three mechanisms of capillary exchange and relate each one to the structure of capillary walls.

11 What forces favor capillary filtration? What forces favor reabsorption?

12 How can a capillary be shifted from a predominantly filtering to a predominantly reabsorbing role?

Venous Return and Circulatory Shock

▼Objectives
When you have completed this section, you should be able to
- explain how blood in the veins is returned to the heart;
- discuss the importance of physical activity in venous return; and
- discuss several causes of circulatory shock and name and describe the stages of shock.

Hieronymus Fabricius (1537–1619) discovered the valves of the veins. He concluded from their structure that they would allow blood to flow in only one direction, not back and forth as Galen had thought. One of his medical students was William Harvey, who performed simple experiments on the valves that you can easily reproduce. Figure 20.18 is one of Harvey's original illustrations. In the top figure, the experimenter has pressed on a vein at point *H* to block flow from the wrist toward the elbow. With another finger, he has milked the blood out of it up to point *O,* the first valve proximal to *H.* When the vein is milked downward back to point *O,* blood stops at that valve and can go no farther, causing the vein to swell at that point. Blood can flow from right to left through the valve but not from left to right. Figure 20.18*b* is explained in Harvey's own words.

You can easily demonstrate the action of these valves in your own hand. Hold your hand still, below waist level, until veins stand up on the back of it. (Do not apply a tourniquet!) Press on a vein close to your knuckles, and while holding it down, use another finger to milk that vein toward the wrist. It collapses as you force the blood out of it, and if you remove the second finger, it will not refill. The valves prevent blood from flowing back into it from above. When you remove the first finger, however, the vein fills from below.

Injury to the dural sinuses or jugular veins presents less danger from loss of blood than from air sucked into the circulatory system. Air traveling in the bloodstream is called **air embolism.** This is an important concern to neurosurgeons, who sometimes must operate with the patient in a sitting position. If a dural sinus is punctured, air can be sucked into the sinus and accumulate in the heart chambers, blocking cardiac output and caus-ing sudden death. Smaller air bubbles in the systemic circulation can cut off blood flow to the brain, myocardium, and other vital tissues.

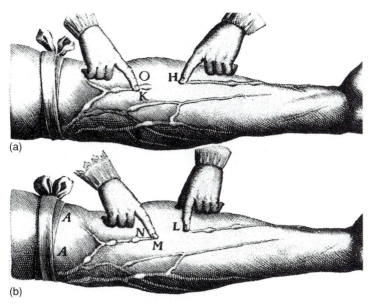

(a)

(b)

Figure 20.18 An illustration from William Harvey's *De Motu Cordis* (1628), demonstrating the existence of one-way valves in veins of the arms. (*a*) See text for explanation. (*b*) In Harvey's words, "With the arm bound as before and the veins swollen, if you will press on a vein a little below a swelling or valve and then squeeze blood upward beyond the valve with another finger, you will see that this part of the vein stays empty, and that no backflow can occur through the valve. But as soon as the finger is removed, the vein is filled from below."

Mechanisms of Venous Return

The flow of blood back to the heart, called **venous return,** is achieved by five mechanisms:

1. The **pressure gradient.** Pressure generated by the heart is the most important force in venous flow, even though it is substantially weaker than in the arteries. Pressure in the venules ranges from 12 to 18 mmHg, and pressure at the point where the venae cavae enter the heart, called **central venous pressure,** averages 4.6 mmHg. Thus, there is a venous pressure gradient (ΔP) of about 7 to 13 mmHg, favoring the flow of blood back to the heart.

2. The **thoracic (respiratory) pump.** This mechanism aids the flow of venous blood from the abdominal to the thoracic cavity. When you inhale, your thoracic cavity expands, creating a negative pressure, while downward movement of the diaphragm increases the pressure in your abdominal cavity. The *inferior vena cava (IVC)* is a flexible tube passing through both of these cavities. If abdominal pressure on the IVC rises while thoracic pressure on it drops, then blood is squeezed upward toward the heart. It is not forced back into the legs because the valves of the leg veins prevent this. Because of the thoracic pump, central venous pressure fluctuates from 2 mmHg during inhalation to 6 mmHg during exhalation, and blood flow is faster during inhalation.

3. **Cardiac suction.** During ventricular systole, the chordae tendineae pull the AV valve cusps downward, slightly expanding the atrial space. This creates a slight suction that draws blood into the atria from the venae cavae and pulmonary veins.

4. The **skeletal muscle pump.** In the extremities, the veins are surrounded and massaged by the muscles. This squeezes the blood out of the compressed part of a vein, and the valves ensure that this blood travels in one direction only—toward the heart (fig. 20.19).

5. **Gravity.** When you are sitting or standing, blood from your head and neck returns to the heart simply by "flowing downhill." Thus the large veins of the neck are normally collapsed or nearly so, and their venous pressure is close to zero. The dural sinuses, however, have more rigid walls and cannot collapse. Their pressure is as low as −10 mmHg, creating a risk of *air embolism* if they are punctured (see special topic 20.3).

Venous Return and Physical Activity

Exercise increases venous return for many reasons. The heart beats faster and harder, increasing cardiac output and blood pressure. Blood vessels of the skeletal muscles, lungs, and heart dilate, increasing flow. The increase in respiratory rate and depth enhances the action of the thoracic pump. Muscle contractions increase

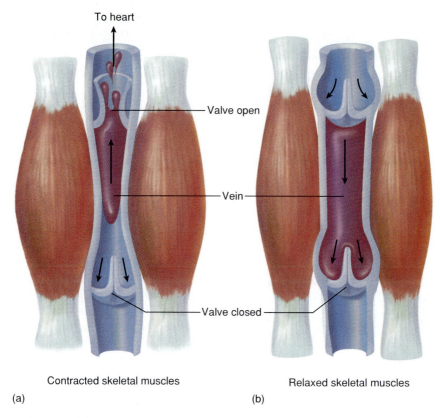

To heart

Valve open

Vein

Valve closed

Contracted skeletal muscles

(a)

Relaxed skeletal muscles

(b)

Figure 20.19 The skeletal muscle pump. (*a*) When the muscles contract and compress a vein, blood is squeezed out of it and flows upward toward the heart; valves below the point of compression prevent backflow of the blood. (*b*) When the muscles relax, blood flows back downward under the pull of gravity but can only flow as far as the nearest valve.

venous return by the skeletal muscle pump mechanism. Increased venous return increases cardiac output, which is important in perfusion of the muscles just when they need it most.

Conversely, when a person is still, blood accumulates in the extremities because venous pressure is not high enough to override the weight of the blood and drive it upward. Such accumulation of blood is called **venous pooling.** To demonstrate this effect, hold one hand above your head and the other below your waist for about a minute. Then, quickly bring your two hands together and compare the palms. The hand held above your head usually appears pale because its blood has drained down to the heart; the hand held below the waist appears redder than normal because of venous pooling in its veins and capillaries. Venous pooling is troublesome to people who must stand for prolonged periods. If enough blood accumulates in the extremities, cardiac output may become so low that the brain is inadequately perfused and a person may become dizzy or faint. This can usually be prevented by periodically tensing the calf and other muscles to keep the skeletal muscle pump active.

 Think About It

Why is venous pooling not a problem when you are sleeping and the skeletal muscle pump is inactive?

Circulatory Shock

Circulatory shock (not to be confused with electrical or spinal shock) is any state in which cardiac output is insufficient to meet the body's metabolic needs. **Cardiogenic shock** is a form of circulatory shock caused by inadequate pumping by the heart, usually as a result of myocardial infarction. All other forms fall under the broad heading of **low venous return (LVR) shock,** in which cardiac output is low because too little blood is entering the heart.

Types and Causes of LVR Shock

There are three principal forms of LVR shock:

1. **Hypovolemic shock,** the most common form, is produced by a loss of blood volume as a result of hemorrhage, trauma, bleeding ulcers, burns, or dehydration. Dehydration is a major cause of death

from heat exposure. In hot weather, the body produces as much as 1.5 L of sweat per hour. Water transfers from the bloodstream to the tissue fluid to replace this; consequently, blood volume may drop too low to maintain adequate circulation.

2. **Obstructed venous return shock** occurs when a growing tumor or aneurysm, for example, compresses a neighboring vein and impedes its blood flow.

3. **Venous pooling (vascular) shock** occurs when blood volume is normal in the body as a whole, but too much of it accumulates in the extremities. This can result from long periods of standing or sitting or from widespread vasodilation. **Neurogenic shock** occurs when there is a sudden loss of vasomotor tone, which allows the vessels to dilate. This can result from causes as severe as brainstem trauma or as slight as an emotional shock. **Syncope** (SIN-co-pee), or fainting, often results from the sudden drop in brain perfusion. **Septic shock** occurs when bacterial toxins trigger vasodilation and increased capillary permeability. **Anaphylactic shock,** discussed more fully in the next chapter, results from exposure to an antigen to which a person is allergic, such as bee venom. Antigen-antibody complexes trigger widespread release of histamine, which causes generalized vasodilation and increased capillary permeability. Septic and anaphylactic shock combine elements of venous pooling and hypovolemic shock in that they involve not only vasodilation but also a loss of fluid through the exceptionally permeable capillaries.

Responses to Circulatory Shock

Several homeostatic mechanisms can compensate for mild circulatory shock and bring about spontaneous recovery. If a person faints, for example, blood flow to the brain is restored because the body usually becomes horizontal. Elevating the feet can improve venous return and aid recovery. The baroreflex improves perfusion superior to the heart and helps to maintain or restore cerebral circulation. The hypotension resulting from low cardiac output also indirectly triggers the production of angiotensin II, which counteracts shock by causing vasoconstriction. When such mechanisms can still restore homeostasis, a person is said to be in **compensated shock.**

If they prove inadequate, however, **decompensated shock** ensues and several life-threatening positive feedback loops occur. Poor cardiac output results in myocardial ischemia and infarction, which further reduces output. Slow circulation of the blood allows clots to form. As the vessels become congested with clotted blood, venous return worsens still further.

Ischemia and acidosis in the brainstem depress the vasomotor and cardiac centers, causing loss of vasomotor tone, further vasodilation, and further drop in BP and cardiac output. Before long, damage to the cardiac and brain tissues may be too great to be undone.

Key Point Review

13 Explain how respiration aids venous return.

14 Explain how muscular activity and venous valves aid venous return.

15 Define *circulatory shock*. What are some of the causes of low venous return shock?

Special Circulatory Routes

▼**Objectives**
When you have completed this section, you should be able to
- explain how the brain maintains stable perfusion;
- discuss the causes and effects of strokes and transient ischemic attacks;
- explain the mechanisms that increase muscular perfusion during exercise; and
- contrast the blood pressure of the pulmonary circuit with that of the systemic circuit, and explain why the pressure difference is important in pulmonary function.

Certain circulatory pathways have special physiological properties adapted to the functions of their organs. Two of these are described in other chapters: the coronary circulation in chapter 19 and fetal and placental circulation in chapter 29. Here we take a closer look at the circulation to the brain, skeletal muscles, and lungs.

Brain

The flow of blood through the brain fluctuates less than the flow through any other organ. Such constancy is particularly important to the brain; even a few seconds of oxygen deprivation causes loss of consciousness, and 4 or 5 minutes of anoxia is time enough to cause irreversible brain damage. While total cerebral perfusion is fairly stable (about 750–900 mL/min), blood flow can be shifted from one part of the brain to another in a matter of seconds as different parts engage in motor, sensory, or cognitive functions.

Cerebral flow is governed by autoregulation in response to changes in BP and chemistry. The cerebral arteries dilate when the systemic BP drops and constrict when BP rises, thus minimizing fluctuations in cerebral BP. This mechanism maintains a stable cerebral blood flow over a range of MABPs from 60 to 140 mmHg. Syncope occurs at an MABP below

60 mmHg and cerebral edema occurs at an MABP above 160 mmHg.

The main chemical stimulus for cerebral autoregulation is pH. Hypercapnia lowers the pH and causes vasodilation up to a point—that is, poor cerebral perfusion leads to an accumulation of CO_2, which lowers the pH and triggers corrective vasodilation. Extreme hypercapnia, however, depresses neural activity. Hypocapnia raises the pH and causes vasoconstriction. Hyperventilation (exhaling CO_2 faster than the body produces it) induces cerebral vasoconstriction, ischemia, a feeling of dizziness, and sometimes syncope.

A stroke, or **cerebrovascular accident (CVA),** is death of brain tissue caused by ischemia. It may result from cerebral atherosclerosis, thrombosis, or a ruptured aneurysm. The consequences of CVAs range from the unnoticeable to sudden death, depending on the extent of tissue damage and the function of the affected tissue. Blindness, paralysis, loss of sensation, and loss of speech are common. Recovery depends on the ability of neighboring neurons to take over the lost functions and on the extent of collateral circulation to regions surrounding the cerebral infarction. A **transient ischemic attack (TIA)** is an episode of cerebral ischemia producing temporary dizziness, light-headedness, loss of vision or other senses, weakness, paralysis, headache, or aphasia. A TIA may result from spasms of diseased cerebral arteries. It lasts from just a moment to a few hours and is often an early warning of an impending stroke.

Skeletal Muscles

In contrast to the brain, the skeletal muscles receive a highly variable blood flow depending on their state of exertion. At rest, the arterioles are constricted, most of the capillary beds are shut down, and total perfusion of the muscular system is about 1 L/min. During exercise, adrenal epinephrine and sympathetic nerves stimulate vasodilation. Precapillary sphincters, which lack innervation, dilate in response to metabolites produced by the muscles. Blood flow increases as much as 100-fold during exercise, requiring that blood be diverted from other organs such as the digestive tract, skin, and kidneys to meet the needs of the working muscles.

Muscular contraction compresses the blood vessels and impedes flow. For this reason, isometric contraction causes fatigue more quickly than intermittent isotonic contraction. If you squeeze a rubber ball as hard as you can without relaxing your grip, you feel the muscles fatigue more quickly than if you intermittently squeeze and relax.

Lungs

After birth, the pulmonary circuit is the only route in which the arteries carry deoxygenated blood and the veins carry oxygenated blood. The pulmonary arteries have thin distensible walls with less elastic tissue than the systemic arteries. Thus, they have a BP of only 25/10. Capillary hydrostatic pressure is about 10 mmHg in the pulmonary circuit as compared with an average of 17 mmHg in systemic capillaries. This lower pressure has two implications for pulmonary circulation: (1) blood flows more slowly through the pulmonary capillaries, and therefore it has more time for gas exchange; and (2) oncotic pressure overrides hydrostatic pressure, so that these capillaries are engaged almost entirely in absorption. This prevents fluid accumulation in the alveolar walls and lumens, which would interfere with gas exchange. In a condition such as mitral valve stenosis, however, BP may back up into the pulmonary circuit, raising the capillary hydrostatic pressure and causing pulmonary edema, congestion, and hypoxemia.

 Think About It

What abnormal skin coloration would result from pulmonary edema?

Another unique characteristic of the pulmonary arteries is their response to hypoxia. Systemic arteries dilate in response to local hypoxia and improve tissue perfusion. By contrast, hypoxia stimulates pulmonary arteries to constrict. Pulmonary hypoxia indicates that part of the lung is not being ventilated well, perhaps because of mucous congestion of the airway or a degenerative lung disease. Vasoconstriction in poorly ventilated regions of the lung redirects blood flow to better ventilated regions.

Key Point Review

16 In what conspicuous way does blood flow to the brain differ from flow to the skeletal muscles?

17 How does a stroke differ from a transient ischemic attack? Which of these bears closer resemblance to a myocardial infarction?

18 How does the low hydrostatic blood pressure in the pulmonary circuit affect the fluid dynamics of the capillaries there?

19 Contrast the vasomotor responses to hypoxia in the lungs and skeletal muscles.

Anatomy of the Pulmonary Circuit

▼**Objective**

When you have completed this section, you should be able to
• trace the route taken by blood in the pulmonary circuit.

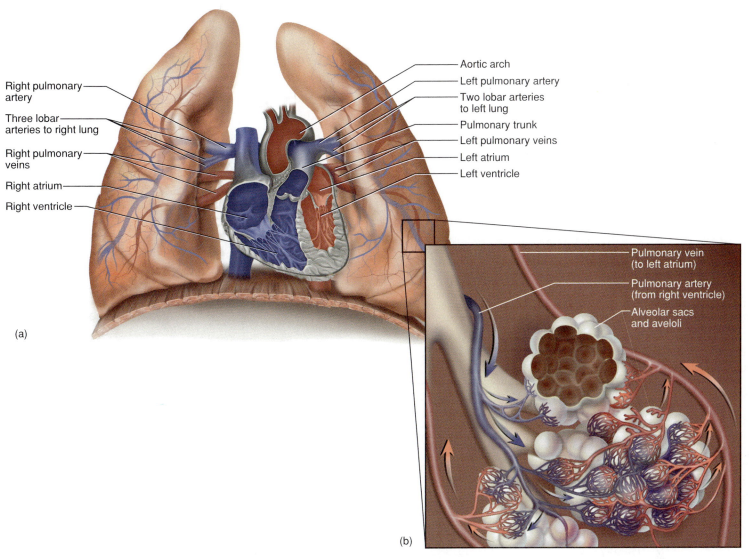

Figure 20.20 (*a*) Gross anatomy of the pulmonary circuit. (*b*) Microscopic anatomy of the blood vessels surrounding the pulmonary alveoli.

The remainder of this chapter centers on the names and pathways of the principal arteries and veins. The pulmonary circuit is described here; the systemic arteries and veins are treated in the two sections that follow.

The pulmonary circuit (fig. 20.20) begins with the **pulmonary trunk,** a large vessel that ascends diagonally from the right ventricle and branches into the **right** and **left pulmonary arteries.** Each pulmonary artery enters a medial indentation of the lung called the *hilus* and branches into **lobar arteries.** The right lung has three lobes each with one lobar artery, and the left lung has two lobes, each with one lobar artery. These arteries lead ultimately to small basketlike capillary beds that surround each pulmonary alveolus. After leaving the alveolar capillaries, the pulmonary blood flows into venules and veins, ultimately leading to the **pulmonary veins** that exit the lung at the hilus. Two pulmonary veins from each lung drain into the left atrium of the heart.

The purpose of the pulmonary circuit is to unload CO_2 and load O_2 at the alveolar capillaries. The pulmonary circuit does not serve the metabolic needs of the lung tissue itself; there is a separate systemic supply to the lungs for that purpose, the *bronchial arteries* discussed later.

- - - - - - - - - - - - - - **Key Point Review** - - - - - - - - - - - - - -

20 Trace the flow of an RBC from right ventricle to left atrium, naming the vessels along the way.

21 The lungs have two separate arterial supplies. Explain their functions.

▼**Objectives**

When you have completed this section, you should be able to
- identify the principal arteries of the systemic circuit; and
- trace the flow of blood from the heart to any major organ.

The systemic circuit supplies oxygen and nutrients to all the organs and removes their metabolic wastes. Part of it, the coronary circulation, was described in chapter 19. The other systemic arteries (fig. 20.21) are described in tables 20.3 through 20.8. The names of

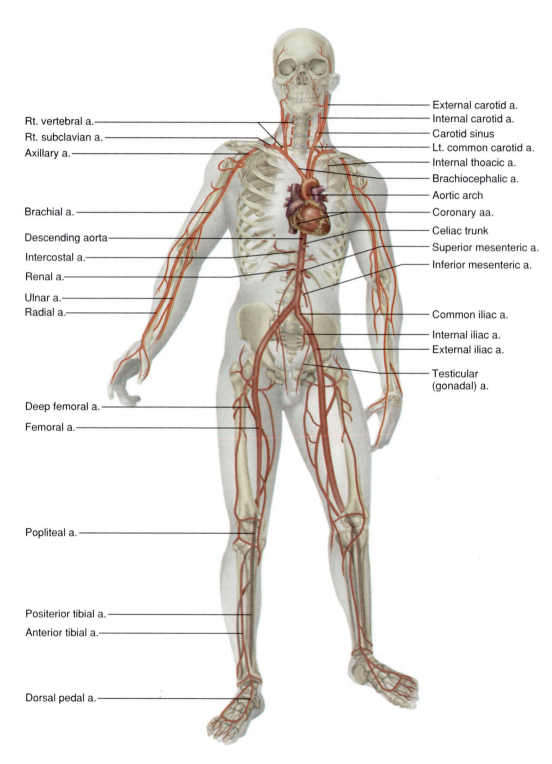

Rt. vertebral a.
Rt. subclavian a.
Axillary a.

External carotid a.
Internal carotid a.
Carotid sinus
Lt. common carotid a.
Internal thoacic a.
Brachiocephalic a.
Aortic arch

Brachial a.

Coronary aa.

Descending aorta
Intercostal a.
Renal a.

Celiac trunk
Superior mesenteric a.
Inferior mesenteric a.

Ulnar a.
Radial a.

Common iliac a.
Internal iliac a.
External iliac a.

Testicular
(gonadal) a.

Deep femoral a.
Femoral a.

Popliteal a.

Positerior tibial a.
Anterior tibial a.

Dorsal pedal a.

Figure 20.21 The major systemic arteries. (R. = right; L. = left; a. = artery.)

the blood vessels often describe their location by indicating the body region traversed (as in the *axillary* artery or *femoral* artery); an adjacent bone (as in *radial* artery or *temporal* artery); or the organ supplied or drained by the vessel (as in *hepatic* artery or *renal* artery).

Table 20.3 The Aorta and Its Major Branches

All systemic arteries arise from the aorta, which has three principal regions (see figure):

1. The **ascending aorta** rises about 5 cm above the left ventricle. Its only branches are the coronary arteries, which arise behind two cusps of the aortic semilunar valve. Opposite each semilunar valve cusp is an **aortic sinus** containing baroreceptors.

2. The **aortic arch** curves to the left like an inverted U superior to the heart. It gives off three major head-neck arteries in this order: the **brachiocephalic**[10] (BRAY-kee-oh-seh-FAL-ic), **left common carotid** (cah-ROT-id), and **left subclavian**[11] (sub-CLAY-vee-un) **arteries,** which are further traced in tables 20.4 and 20.5.

3. The **descending aorta** passes downward behind the heart, at first to the left of the vertebral column and then anterior to it, through the thoracic and abdominal cavities. It is called the **thoracic aorta** above the diaphragm and the **abdominal aorta** below. It ends in the lower abdominal cavity by forking into the *right* and *left common iliac arteries,* which are further traced in table 20.8.

10. *brachio* = arm + *cephal* = head
11. *sub* = below + *clavi* = clavicle, collarbone

Table 20.4 Arterial Supply to the Head and Neck

Origins of the Head-Neck Arteries

The head and neck receive blood from four pairs of arteries (see figure):

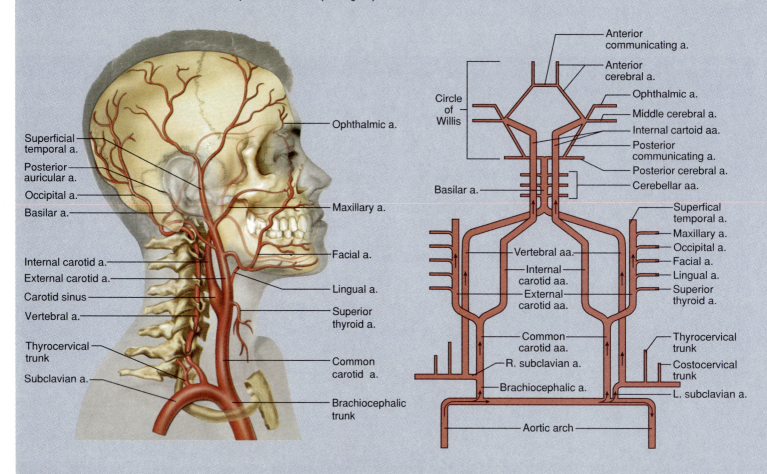

1. The **common carotid arteries.** The brachiocephalic artery divides shortly after leaving the aortic arch, giving rise to the *right subclavian* and *right common carotid arteries*. The *left common carotid* artery arises directly from the aortic arch. The common carotids pass up the anterolateral aspect of the neck, alongside the trachea.

2. The **vertebral arteries** arise from the right and left subclavian arteries. Each travels up the neck through the transverse foramina of the cervical vertebrae and enters the cranial cavity through the foramen magnum.

3. The **thyrocervical**[12] **trunks** are tiny arteries that arise from the subclavian arteries lateral to the origin of the vertebral arteries; they supply the thyroid gland and some scapular muscles.

4. The **costocervical**[13] **trunks** (also illustrated in table 20.6) arise from the subclavian arteries a little farther laterally. They perfuse the deep neck muscles and some of the intercostal muscles of the superior rib cage.

Continuation of the Common Carotid Arteries

The common carotid arteries have the most extensive distribution of all the head-neck arteries. Near the laryngeal prominence (Adam's apple), each common carotid branches into an *external carotid* and an *internal carotid*.

1. The **external carotid artery** ascends along the side of the head external to the cranium and supplies most external head structures except the orbits. Its major branches are listed here, in ascending order. The external carotid terminates at the point where it gives rise to the last two of these:

   a. the **superior thyroid artery** to the thyroid gland and larynx;

   b. the **lingual artery** to the tongue;

   c. the **facial artery** to the skin and muscles of the face;

---

12. *thyro* = thyroid gland + *cerv* = neck
13. *costo* = rib

d. the **occipital artery** to the posterior scalp;

e. the **maxillary artery** to the teeth, maxilla, buccal cavity, and external ear; and

f. the **superficial temporal artery** to the chewing muscles, nasal cavity, lateral aspect of the face, most of the scalp, and the dura mater surrounding the brain.

2. The **internal carotid artery** passes medial to the angle of the mandible and enters the cranial cavity through the carotid canal of the temporal bone. It supplies the orbits and about 80% of the cerebrum. Compressing the internal carotids near the mandible can therefore cause loss of consciousness.[14] The carotid sinus is located in the internal carotid just above the branch point; the carotid body is nearby. After entering the cranial cavity, each internal carotid artery gives rise to the following branches:

a. the **ophthalmic artery** to the orbits, nose, and forehead;

b. the **anterior cerebral artery** to the medial aspect of the cerebral hemisphere (see *circle of Willis*); and

c. the **middle cerebral artery,** which travels in the lateral fissure of the cerebrum and supplies the lateral aspect of the temporal and parietal lobes.

Continuation of the Vertebral Arteries

The vertebral arteries give rise to small branches in the neck that supply the spinal cord and other neck structures; they then enter the foramen magnum and merge to form a single **basilar artery** along the anterior aspect of the brainstem. Branches of the basilar artery supply the cerebellum, pons, and inner ear. At the pons-midbrain junction, the basilar artery divides and gives rise to the *circle of Willis*.

The Circle of Willis

Blood supply to the brain is so critical that it is furnished by several arterial anastomoses, especially an array of arteries called the **circle of Willis**[15] (see figure), which surrounds the pituitary gland and optic chiasma. The circle of Willis receives blood from the internal carotid and basilar arteries. It consists of

1. two **posterior cerebral arteries,**
2. two **posterior communicating arteries,**
3. two **anterior cerebral arteries,** and
4. a single **anterior communicating artery.**

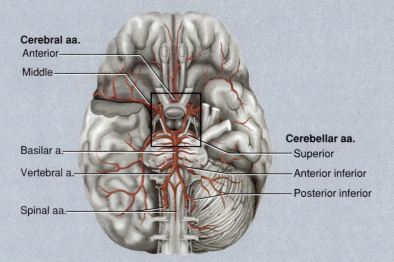

Cerebral aa.
Anterior
Middle

Basilar a.
Vertebral a.
Spinal aa.

Cerebellar aa.
Superior
Anterior inferior
Posterior inferior

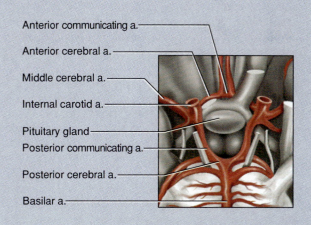

Anterior communicating a.
Anterior cerebral a.
Middle cerebral a.
Internal carotid a.
Pituitary gland
Posterior communicating a.
Posterior cerebral a.
Basilar a.

---

14. *carot* = stupor
15. Thomas Willis (1621–75), English anatomist

Table 20.5   Arterial Supply to the Upper Extremity

The Shoulder and Arm (Brachium)

The origins of the subclavian arteries were described and illustrated in table 20.3. We now trace these further to examine the blood supply to the upper extremity (see figure). This begins with a large artery that changes name from *subclavian* to *axillary* to *brachial* along its course.

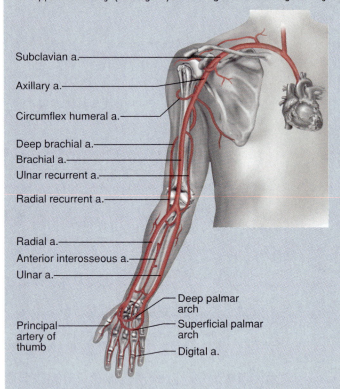

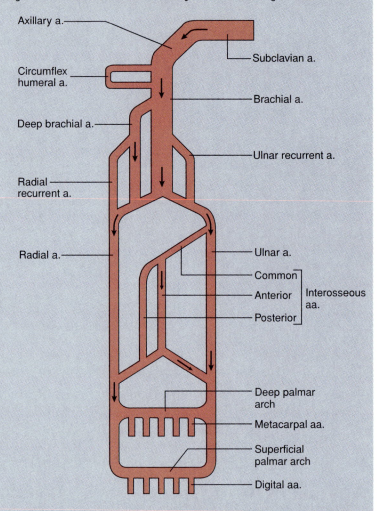

1. The **subclavian**[16] **artery** travels between the clavicle and first rib. It gives off several small branches to the thoracic wall and viscera, considered later.

2. The **axillary artery** is the continuation of the subclavian artery through the axillary region. It also gives off small thoracic branches, discussed later, and then ends at the neck of the humerus. Here, it gives off the **circumflex humeral artery,** which encircles the humerus. This loop supplies blood to the shoulder joint and deltoid muscle.

3. The **brachial** (BRAY-kee-ul) **artery** is the continuation of the axillary artery beyond the circumflex. It travels down the medial side of the humerus and ends just distal to the elbow, supplying the anterior flexor muscles of the brachium along the way. It exhibits several anastomoses near the elbow, two of which are noted next. This artery is the one most commonly used for routine BP measurements.

4. The **deep brachial artery** arises from the proximal end of the brachial artery and supplies the triceps brachii muscle.

5. The **ulnar recurrent artery** arises about midway along the brachial artery and anastomoses distally with the ulnar artery. It supplies the elbow joint and the triceps brachii.

6. The **radial recurrent artery** leads from the deep brachial artery to the radial artery and supplies the elbow joint and forearm muscles.

The Forearm (Antebrachium)

Just distal to the elbow, the brachial artery divides into the **radial artery** and **ulnar artery,** which travel alongside the radius and

ulna, respectively. The most common place to take a pulse is at the radial artery, just proximal to the thumb. Near its origin, the radial artery receives the deep brachial artery. The ulnar artery gives rise, near its origin, to the **anterior** and **posterior interosseous**[17] **arteries,** which travel between the radius and ulna. Structures supplied by these three arteries are as follows:

1. Radial artery: lateral forearm muscles, wrist, thumb, and index finger.

2. Ulnar artery: medial forearm muscles, digits 3 to 5, and medial aspect of index finger.

3. Interosseous arteries: deep flexors and extensors.

The Hand

At the wrist, the radial and ulnar arteries anastomose to form two *palmar arches:*

1. The **deep palmar arch** gives rise to the **metacarpal arteries** of the hand.

2. The **superficial palmar arch** gives rise to the **digital arteries** of the fingers.

---

16. *sub* = below + *clavi* = clavicle

17. *inter* = between + *osse* = bones

Table 20.6 Arterial Supply to the Thorax

The thoracic aorta begins distal to the aortic arch and ends at the **aortic hiatus** (hy-AY-tus), an opening through which it penetrates the diaphragm. Along the way, it sends off numerous small branches to viscera and structures of the body wall (see figure).

Visceral Branches

These supply the viscera of the thoracic cavity:

1. **Bronchial arteries.** Two of these on the left and one on the right supply the visceral pleura, esophagus, and bronchi of the lungs. They are the systemic blood supply to the lungs mentioned earlier.

2. **Esophageal arteries.** Four or five of these supply the esophagus.

3. **Mediastinal arteries.** Many small mediastinal arteries (not illustrated) supply structures of the posterior mediastinum.

Parietal Branches

The following branches supply chiefly the muscles, bones, and skin of the chest wall; only the first is illustrated.

1. **Posterior intercostal arteries.** Nine pairs of these course around the posterior aspect of the rib cage between the ribs and then anastomose with the anterior intercostal arteries (see following). They supply the skin and subcutaneous tissue, mammary glands, spinal cord and meninges, and the pectoralis, intercostal, and some abdominal muscles.

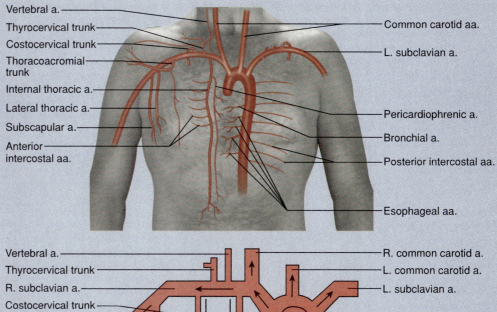

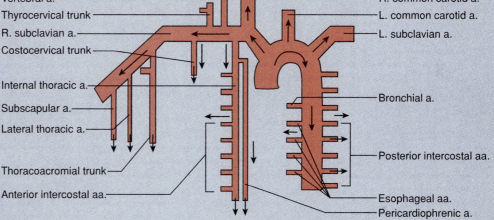

2. **Subcostal arteries.** A pair of these arise from the aorta, inferior to the twelfth rib, and supply the posterior intercostal tissues, vertebrae, spinal cord, and deep muscles of the back.

3. **Superior phrenic**[18] (FREN-ic) **arteries.** These supply the posterior and superior aspects of the diaphragm.

The thoracic wall is also supplied by several other arteries, listed next. The first of these arises from the subclavian artery and the other three from the axillary artery:

1. The **internal thoracic (mammary) artery** supplies the mammary gland and anterior thoracic wall and issues finer branches to the diaphragm and abdominal wall. Near its origin, it gives rise to the **pericardiophrenic artery**, which supplies the pericardium and diaphragm. As the internal thoracic descends alongside the sternum, it gives rise to **anterior intercostal arteries** that travel between the ribs and supply the ribs and intercostal muscles.

2. The **thoracoacromial**[19] (THOR-uh-co-uh-CRO-me-ul) **trunk** supplies the superior shoulder and pectoral regions.

3. The **lateral thoracic artery** supplies the lateral thoracic wall.

4. The **subscapular artery** supplies the scapula, latissimus dorsi, and posterior wall of the thorax.

---

18. *phren* = diaphragm
19. *thoraco* = chest + *acr* = tip + *om* = shoulder

Table 20.7   Arterial Supply to the Abdomen

Major Branches of Abdominal Aorta

After passing through the aortic hiatus, the aorta descends through the abdominal cavity and gives off arteries in the order listed here. Those indicated in the plural are paired right and left, and those indicated in the singular are single median arteries.

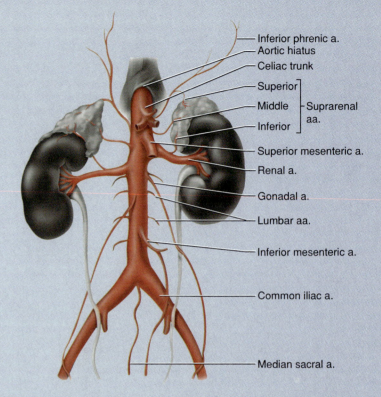

1. The **inferior phrenic arteries** supply the inferior surface of diaphragm and issue a small **superior suprarenal artery** to each adrenal (suprarenal) gland.

2. The **celiac**[20] (SEE-lee-ac) **trunk** issues several branches to the upper abdominal viscera, further traced later in this table.

3. The **superior mesenteric artery** supplies the intestines (see mesenteric circulation later in this table).

4. The **middle suprarenal arteries** arise on either side of the superior mesenteric artery and supply the adrenal glands.

5. The **renal arteries** supply the kidneys and issue a small **inferior suprarenal artery** to each adrenal gland.

6. The **gonadal arteries** are long, narrow, winding arteries that descend from the midabdominal region to the female pelvic cavity or male scrotum. They are called the **ovarian arteries** in females and **testicular arteries** in males. The gonads begin their embryonic development near the kidneys. These arteries acquire their peculiar length and course as the gonads descend to the pelvic cavity during fetal development.

7. The **inferior mesenteric artery** supplies the distal end of the large intestine (see mesenteric circulation).

8. The **lumbar arteries** arise from the lower aorta in four pairs and supply the posterior abdominal wall.

9. The **median sacral artery,** a tiny medial artery at the inferior end of the aorta, supplies the sacrum and coccyx.

10. The **common iliac arteries** arise as the aorta forks at its inferior end. They supply the lower abdominal wall, pelvic viscera (chiefly the urinary and reproductive organs), and lower extremities. They are further traced in table 20.8.

Branches of the Celiac Trunk

The celiac circulation to the upper abdominal viscera is perhaps the most complex route off the abdominal aorta. Because it has numerous anastomoses, the bloodstream does not follow a simple linear path but divides and rejoins itself at several points. As you study the following description, locate these branches in the figure and identify the points of anastomosis. The short, stubby celiac trunk is a median branch of the aorta. It immediately gives rise to three principal subdivisions—the *common hepatic, left gastric,* and *splenic arteries.*

1. The **common hepatic artery** issues two main branches:
   a. the **gastroduodenal artery,** which supplies the stomach, anastomoses with the right gastroepiploic artery (see following), and then continues as the **inferior pancreaticoduodenal** (PAN-cree-AT-ih-co-dew-ODD-eh-nul) **artery,** which supplies the duodenum and pancreas before anastomosing with the superior mesenteric artery; and
   b. the **hepatic artery,** which is the continuation of the common hepatic artery after it gives off the gastroduodenal artery. It enters the inferior surface of the liver and supplies the liver and gallbladder.

2. The **left gastric artery** supplies the stomach and lower esophagus, arcs around the *lesser curvature* of the stomach, becomes the **right gastric artery** (which supplies the stomach and duodenum), and then anastomoses with the hepatic artery.

3. The **splenic artery** gives off the following main branches:
   a. the **pancreatic arteries** (not illustrated), which supply the pancreas; and
   b. the **left gastroepiploic**[21] (GAS-tro-EP-ih-PLO-ic) **artery,** which arcs around the *greater curvature* of the stomach, becomes the **right gastroepiploic artery,** and then anastomoses with the gastroduodenal artery. Along the way, it supplies blood to the stomach and *greater omentum* (a fatty membrane suspended from the greater curvature).

....

20. *celi* = belly, abdomen
21. *gastro* = stomach + *epi* = upon, above + *ploic* = pertaining to the greater omentum

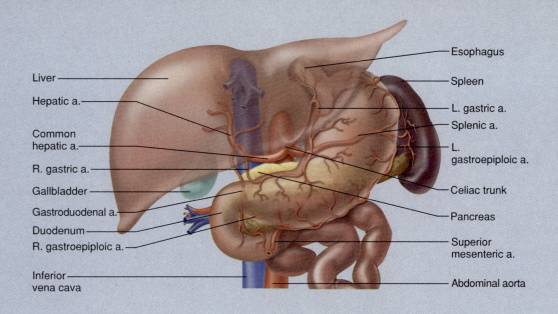

Liver

Hepatic a.

Common
hepatic a.

R. gastric a.

Gallbladder

Gastroduodenal a.

Duodenum

R. gastroepiploic a.

Inferior
vena cava

Esophagus

Spleen

L. gastric a.

Splenic a.

L.
gastroepiploic a.

Celiac trunk

Pancreas

Superior
mesenteric a.

Abdominal aorta

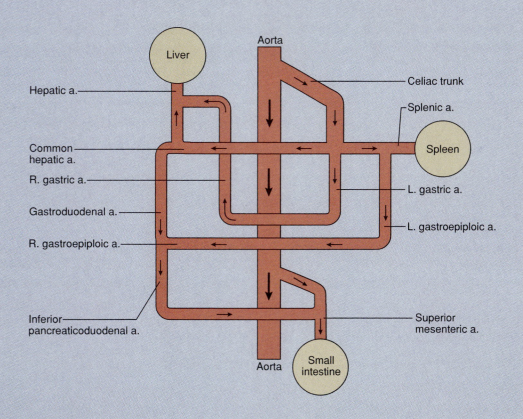

Liver

Aorta

Hepatic a.

Celiac trunk

Splenic a.

Spleen

Common
hepatic a.

R. gastric a.

L. gastric a.

Gastroduodenal a.

L. gastroepiploic a.

R. gastroepiploic a.

Inferior
pancreaticoduodenal a.

Superior
mesenteric a.

Aorta

Small
intestine

continued next page

Table 20.7   Continued

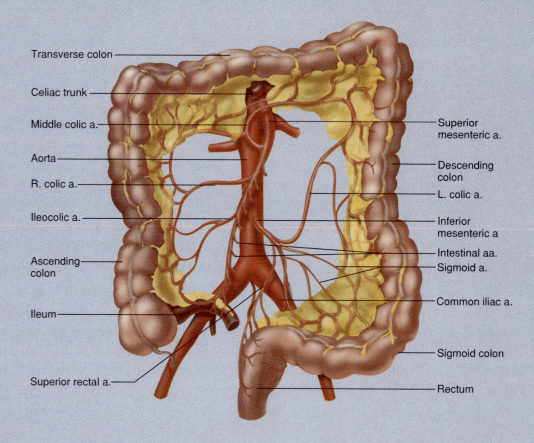

Mesenteric Circulation

The mesentery (see Atlas A, p. 39) contains numerous mesenteric arteries, veins, and lymphatic vessels that perfuse and drain the intestines. The arterial supply issues from the *superior* and *inferior mesenteric arteries* (see figure); numerous anastomoses between these ensure collateral circulation and adequate perfusion of the intestinal tract even if one route becomes obstructed. The following branches of the **superior mesenteric artery** serve the small intestine and most of the large intestine, among other organs.

1. The **inferior pancreaticoduodenal artery,** already mentioned, is an anatomosis from the gastroduodenal to the superior mesenteric artery; it supplies the pancreas and duodenum.
2. The **intestinal arteries** supply nearly all of the small intestine (jejunum and ileum).
3. The **ileocolic** (ILL-ee-oh-CO-lic) **artery** supplies the ileum of the small intestine and the appendix, cecum, and ascending colon.
4. The **right colic artery** supplies the ascending colon.
5. The **middle colic artery** supplies the transverse colon.

Branches of the *inferior mesenteric artery* serve the distal part of the large intestine:

1. The **left colic artery** supplies the transverse and descending colon.
2. The **sigmoid arteries** supply the descending and sigmoid colon.
3. The **superior rectal artery** supplies the rectum.

Table 20.8 Arterial Supply to the Pelvic Region and Lower Extremity

The common iliac arteries arise from the aorta at the level of vertebra L4 and continue for about 5 cm. At the level of the sacroiliac joint, each divides into an internal and an external iliac artery. The **internal iliac** supplies mainly the pelvic wall and viscera, and the **external iliac** supplies mainly the lower extremity (see figure).

Branches of the Internal Iliac Artery

1. The **iliolumbar** and **lateral sacral arteries** supply the wall of the pelvic region.
2. The **middle rectal artery** supplies the rectum.
3. The **superior** and **inferior vesical**[22] **arteries** supply the urinary bladder.
4. The **uterine** and **vaginal arteries** supply the uterus and vagina.
5. The **superior** and **inferior gluteal arteries** supply the gluteal muscles.
6. The **obturator artery** supplies the adductor muscles of the medial thigh.
7. The **internal pudendal**[23] (pyu-DEN-dul) **artery** serves the perineum and external genitals; it supplies the blood for vascular engorgement during sexual arousal.

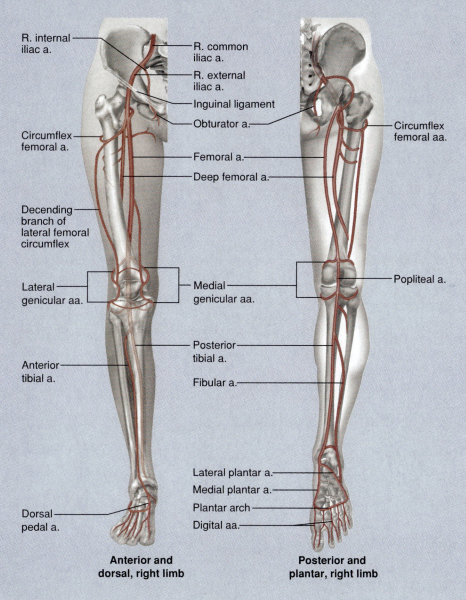

R. internal iliac a.
R. common iliac a.
R. external iliac a.
Inguinal ligament
Obturator a.
Circumflex femoral a.
Circumflex femoral aa.
Femoral a.
Deep femoral a.
Decending branch of lateral femoral circumflex
Lateral genicular aa.
Medial genicular aa.
Popliteal a.
Posterior tibial a.
Anterior tibial a.
Fibular a.
Lateral plantar a.
Medial plantar a.
Plantar arch
Digital aa.
Dorsal pedal a.

Anterior and dorsal, right limb

Posterior and plantar, right limb

22. *vesic* = bladder
23. *pudend* = literally, "shameful parts"; the external genitals

continued next page

Table 20.8 Continued

Brnaches of the External Iliac Artery

The external iliac artery sends branches to the skin and muscles of the abdominal wall and pelvic girdle. It then passes under the inguinal ligament and gives rise to branches that serve mainly the lower extremities.

1. The **femoral artery** passes through the femoral triangle of the upper medial thigh, where its pulse can be palpated. It gives off the following branches to supply the thigh region:

   a. The **deep femoral artery,** which supplies the hamstring muscles; and

   b. The **circumflex femoral arteries,** which encircle the neck of the femur and supply the femur and hamstring muscles.

2. The **popliteal artery** is a continuation of the femoral artery in the popliteal fossa at the rear of the knee. It produces anastomoses **(genicular arteries)** that supply the knee and then divides into the anterior and posterior tibial arteries.

3. The **anterior tibial artery** travels lateral to the tibia in the anterior compartment of the leg, where it supplies the extensor muscles. It gives rise to

   a. the **dorsal pedal artery,** which traverses the ankle and dorsum of the foot; and

   b. the **arcuate artery,** a continuation of the dorsal pedal artery that gives off the **metatarsal arteries** of the foot.

4. The **posterior tibial artery** travels through the posteromedial part of the leg and supplies the flexor muscles. It gives rise to

   a. the **fibular (peroneal) artery,** which arises from the proximal end of the posterior tibial artery and supplies the lateral peroneal muscles;

   b. the **lateral** and **medial plantar arteries,** which arise by bifurcation of the posterior tibial artery at the ankle and supply the plantar surface of the foot; and

   c. the **plantar arch,** an anastomosis from the lateral plantar artery to the dorsal pedal artery that gives rise to the **digital arteries** of the toes.

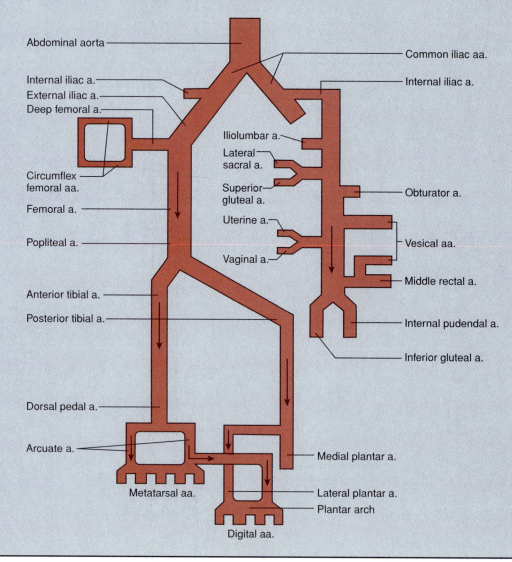

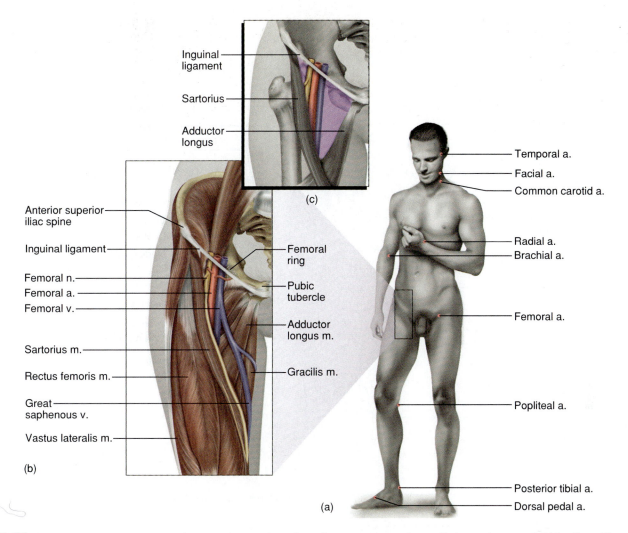

Inguinal ligament

Sartorius

Adductor longus

(c)

Anterior superior iliac spine

Inguinal ligament

Femoral n.

Femoral a.

Femoral v.

Sartorius m.

Rectus femoris m.

Great saphenous v.

Vastus lateralis m.

(b)

Femoral ring

Pubic tubercle

Adductor longus m.

Gracilis m.

Temporal a.

Facial a.

Common carotid a.

Radial a.
Brachial a.

Femoral a.

Popliteal a.

Posterior tibial a.

Dorsal pedal a.

(a)

Figure 20.22 (*a*) Arterial pressure points, where a pulse can be palpated or pressure can be applied to reduce arterial bleeding. (*b*) Structures in the femoral triangle. (*c*) Boundaries of the femoral triangle.

Think About It

There are certain similarities between the arteries of the hand and foot. What arteries of the wrist and hand are most comparable in arrangement and function to the arcuate artery and plantar arch of the foot?

In some places, major arteries come close enough to the body surface to be palpated. These places can be used to take a person's pulse. They may also serve as emergency **pressure points** (fig. 20.22*a*), where firm pressure can be applied to temporarily reduce arterial bleeding. One of these points is the **femoral triangle** of the upper medial thigh (fig. 20.22*b*). This is an important landmark for arterial supply, venous drainage, and innervation of the lower extremity. Its boundaries are

the sartorius muscle laterally, the inguinal ligament superiorly, and the adductor longus muscle medially (fig. 20.22*c*). The femoral artery, vein, and nerve run close to the surface at this point.

.... Key Point Review ....

22 Concisely contrast the destinations of the external and internal carotid arteries.

23 Briefly state the tissues that are supplied with blood by (a) the circle of Willis, (b) the celiac trunk, (c) the superior mesenteric artery, and (d) the external iliac artery.

24 Trace the path of an RBC from the left ventricle to the metatarsal arteries. State two places along this path where you can palpate the arterial pulse.

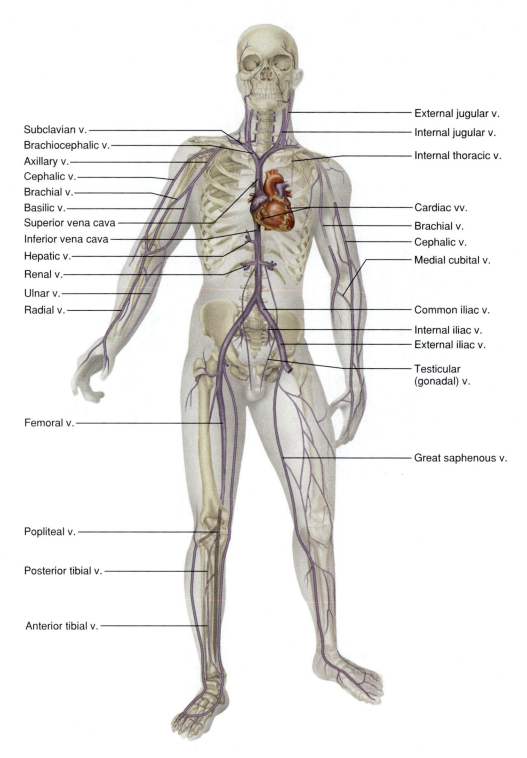

Subclavian v.
Brachiocephalic v.
Axillary v.
Cephalic v.
Brachial v.
Basilic v.
Superior vena cava
Inferior vena cava
Hepatic v.
Renal v.
Ulnar v.
Radial v.

External jugular v.
Internal jugular v.
Internal thoracic v.

Cardiac vv.
Brachial v.
Cephalic v.
Medial cubital v.

Common iliac v.
Internal iliac v.
External iliac v.

Testicular
(gonadal) v.

Femoral v.

Great saphenous v.

Popliteal v.

Posterior tibial v.

Anterior tibial v.

Figure 20.23 The major systemic veins.

Anatomy of the Systemic Veins

▼Objective
When you have completed this section, you should be able to
• identify the principle veins of the systemic circuit and trace the
  flow of blood from any major organ to the heart.

The principal veins of the systemic circuit (fig. 20.23) are detailed in tables 20.9 through 20.14. While arteries are usually deep and well protected, veins occur in both deep and superficial groups; you may be able to see quite a few of them in your arms and hands. Deep veins run parallel to the arteries and often have similar names (*femoral artery* and *femoral vein,* for example); this is not true of the superficial veins, however. The deep veins are not described in as much detail as the arteries were, since it can usually be assumed that they drain the same structures as the corresponding arteries supply.

In general, we began the study of arteries with those lying close to the heart and progressing away. In the venous system, by contrast, we begin with those that are remote from the heart and follow the flow of blood as they join each other and approach the heart. Venous pathways have more anastomoses than arterial pathways, so that the route of blood flow is often not as clear. Many anastomoses are omitted from the following figures for clarity.

Table 20.9 Venous Drainage of the Head and Neck

Most blood of the head and neck is drained by three pairs of veins—the *internal jugular, external jugular,* and *vertebral veins.* This table traces their origins and drainage and follows them to the formation of the *brachiocephalic veins* and *superior vena cava.*

Venous Sinuses

Large thin-walled veins called **dural sinuses** occur within the cranial cavity between layers of dura mater. They receive blood from the brain and face and empty into the internal jugular veins.

1. The **superior** and **inferior sagittal sinuses** are found in the falx cerebri between the cerebral hemispheres; they receive blood that has circulated through the brain.

2. The **cavernous sinuses** occur on each side of the body of the sphenoid bone; they receive blood from the **ophthalmic vein** draining the orbit and the **facial vein** draining the nose and upper lip.

3. The **transverse (lateral) sinuses** encircle the inside of the occipital bone and lead to the jugular foramen on each side. They receive blood from the previously mentioned sinuses and empty into the internal jugular veins.

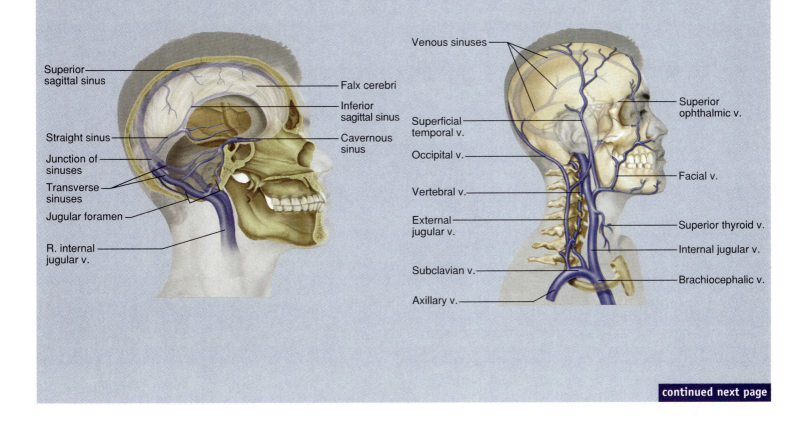

continued next page

Table 20.9   Continued

Major Veins of the Neck

Blood flows down the neck mainly through three veins on each side, all of which empty into the subclavian vein (see figure).

1. The **internal jugular**[24] (JUG-you-lur) **vein** courses down the neck, alongside the internal carotid artery, deep to the sternocleidomastoid muscle. It receives most of the blood from the brain, picks up blood from the **facial vein** and **superficial temporal vein** along the way, passes behind the clavicle, and joins the subclavian vein. (Note that the facial vein empties into both the cavernous sinus and the internal jugular vein.)

2. The **external jugular vein** drains tributaries from the parotid gland, facial muscles, scalp, and other superficial structures perfused by the external carotid artery. Some of this blood also follows venous anastomoses to the internal jugular vein. The external jugular vein courses down the side of the neck superficial to the sternocleidomastoid muscle and empties into the subclavian vein.

3. The **vertebral vein** travels with the vertebral artery in the transverse foramina of the cervical vertebrae. Although the companion artery leads to the brain, the vertebral vein does not come from there. It drains the cervical vertebrae, spinal cord, and some of the small deep muscles of the neck.

Drainage from Shoulder to Heart

From the shoulder region, blood takes the following path to the heart:

1. The **subclavian vein** drains the arm and travels beneath the clavicle; receives the external jugular, vertebral, and internal jugular veins in that order; and ends where it receives the internal jugular.

2. The **brachiocephalic vein** is formed by union of the subclavian and internal jugular veins. It continues medially and receives tributaries draining the upper thoracic wall and mammary gland. There is a brachiocephalic *vein* on both the right and left, even though there is a brachiocephalic *artery* only on the right.

3. The **superior vena cava** is formed by the union of the right and left brachiocephalic veins. It travels inferiorly for about 7.5 cm and empties into the right atrium. It drains all structures superior to the diaphragm except the pulmonary circuit and coronary circulation. It also receives considerable drainage from the abdominal cavity by way of the azygos system (see table 20.11).

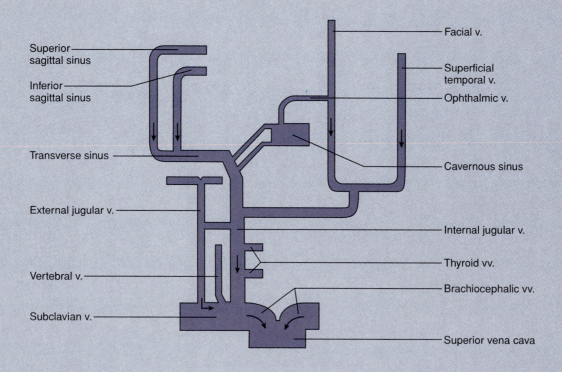

Superior sagittal sinus
Inferior sagittal sinus
Transverse sinus
External jugular v.
Vertebral v.
Subclavian v.

Facial v.
Superficial temporal v.
Ophthalmic v.
Cavernous sinus
Internal jugular v.
Thyroid vv.
Brachiocephalic vv.
Superior vena cava

---

24. *jugul* = neck, throat

Table 20.10 Venous Drainage of the Upper Extremity

Table 20.9 briefly noted the subclavian veins that drain each arm. This table begins distally in the arm and traces its venous drainage to the subclavian vein (see figure).

Deep Veins

1. The **digital veins** drain each finger into the **superficial palmar venous arch.**

2. The **metacarpal veins** parallel the metacarpal bones and drain blood from the hand into the **deep palmar venous arch.** Both the superficial and deep palmar venous arches are anastomoses between the next two veins, which are the major deep veins of the forearm.

3. The **radial vein** receives blood from the lateral side of both palmar arches and courses up the forearm alongside the radius.

4. The **ulnar vein** receives blood from the medial side of both palmar arches and courses up the forearm alongside the ulna.

5. The **brachial vein** is formed by the union of the radial and ulnar veins at the elbow; it courses up the brachium.

6. The **axillary vein** is formed at the axilla by the union of the brachial and basilic veins (see following page).

7. The **subclavian vein** is a continuation of the axillary vein into the shoulder beneath the clavicle. The further course of the subclavian is explained in the previous table.

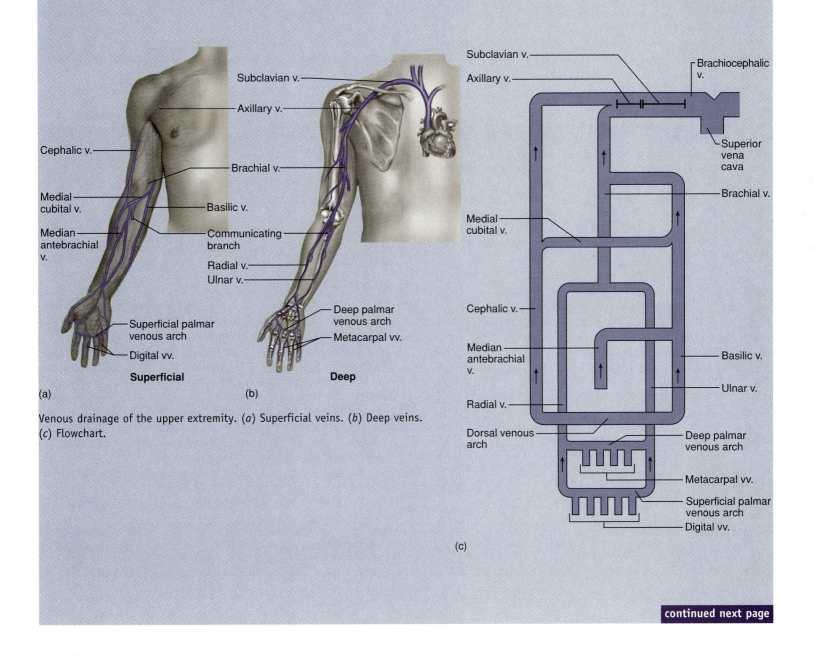

Venous drainage of the upper extremity. (*a*) Superficial veins. (*b*) Deep veins. (*c*) Flowchart.

continued next page

chapter 20   The Circulatory System: Blood Vessels and Circulation   743

Table 20.10   Continued

Superficial Veins

These are easily seen through the skin of most people and are larger in diameter than the deep veins.

1. The **dorsal venous arch** (not illustrated) is a plexus of veins visible on the back of the hand; it empties into the major superficial veins of the forearm, the cephalic and basilic.

2. The **cephalic vein** arises from the lateral side of the dorsal venous arch, winds around the radius as it travels up the forearm, continues up the lateral aspect of the brachium to the shoulder, and joins the axillary vein there.

3. The **basilic**[25] (bah-SIL-ic) **vein** arises from the medial side of the dorsal venous arch, travels up the posterior aspect of the forearm, and continues into the brachium. About midway up the brachium it turns deeper and runs beside the brachial artery. At the axilla it joins the brachial vein, and the union of these two gives rise to the axillary vein.

4. The **median cubital vein** is a short anastomosis between the cephalic and basilic veins that obliquely crosses the cubital fossa (anterior bend of the elbow). It is clearly visible through the skin and is the most common site for drawing blood and administering intravenous fluids and blood transfusions.

5. The **median antebrachial vein** originates near the base of the thumb, travels up the forearm between the radial and ulnar veins, and terminates at the elbow, emptying into the cephalic vein in some people and into the basilic vein in others.

25. *basilic* = royal, prominent, important

Table 20.11   The Azygos System

The superior vena cava receives extensive drainage from the thoracic and abdominal walls by way of the **azygos** (AZ-ih-goss) system (see figure).

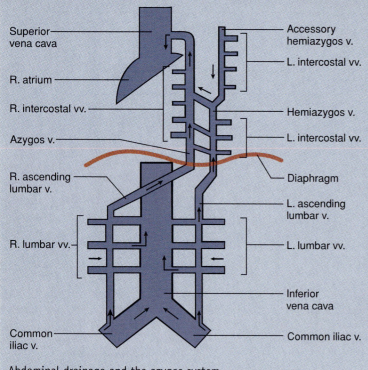

Superior vena cava
R. atrium
R. intercostal vv.
Azygos v.
R. ascending lumbar v.
R. lumbar vv.
Common iliac v.

Accessory hemiazygos v.
L. intercostal vv.
Hemiazygos v.
L. intercostal vv.
Diaphragm
L. ascending lumbar v.
L. lumbar vv.
Inferior vena cava
Common iliac v.

Abdominal drainage and the azygos system.

Drainage of the Abdominal Wall

A series of short horizontal **lumbar veins** drain the abdominal wall and lead into a pair of vertical **ascending lumbar veins.** The ascending lumbar veins also receive blood from the common iliac veins. They anatomose with the inferior vena cava beside them as well as ascend through the diaphragm into the thoracic cavity.

Drainage of the Thorax

Right side. After penetrating the diaphragm, the right ascending lumbar vein becomes the **azygos**[26] **vein** of the thorax. The azygos receives blood from the right **posterior intercostal veins,** which drain the chest muscles, and from the **esophageal, mediastinal, pericardial,** and **right bronchial veins.** It then empties into the superior vena cava at the level of vertebra T4.

Left side. The left ascending lumbar vein continues into the thorax as the **hemiazygos**[27] **vein.** The hemiazygos drains the ninth through eleventh posterior intercostal veins and some esophageal and mediastinal veins on the left. At midthorax, it crosses over to the right side and empties into the azygos vein.

The **accessory hemiazygos vein** is a superior extension of the hemiazygos. It drains the fourth through eighth posterior intercostal veins and the left bronchial vein. It also crosses to the right side and empties into the azygos vein.

26. Unpaired; from *a* = without + *zygo* = union, mate
27. *hemi* = half

Table 20.12 Major Tributaries of the Inferior Vena Cava

The **inferior vena cava (IVC)** is formed by the union of the right and left common iliac veins at the level of vertebra L5. It is retroperitoneal and lies immediately to the right of the aorta. Its diameter of 3.5 cm is the largest of any vessel in the body. As it ascends the abdominal cavity, the IVC picks up blood from numerous tributaries in the order listed here (see figure):

1. Some **lumbar veins** empty into the IVC as well as into the ascending lumbar veins described in table 20.11.

2. The **gonadal veins** (**ovarian veins** in the female and **testicular veins** in the male) drain the gonads. The right gonadal vein empties directly into the IVC, whereas the **left gonadal vein** empties into the left renal vein.

3. The **renal veins** drain the kidneys into the IVC. The left renal vein also receives blood from the left gonadal and left suprarenal veins.

4. The **suprarenal veins** drain the adrenal (suprarenal) glands. The right suprarenal empties directly into the IVC, and the left suprarenal empties into the renal vein.

5. The **hepatic veins** drain the liver, extending a short distance from its superior surface to the IVC.

6. The **inferior phrenic veins** drain the inferior aspect of the diaphragm.

After receiving these inputs, the IVC penetrates the diaphragm and enters the right atrium from below. It does not receive any thoracic drainage.

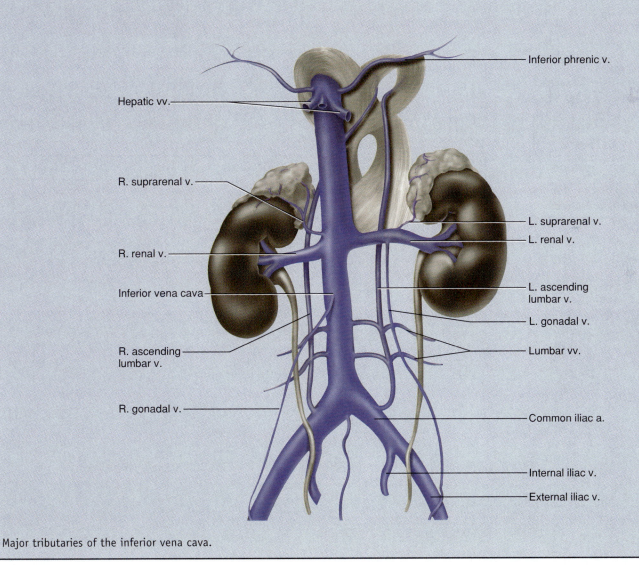

Major tributaries of the inferior vena cava.

Table 20.13   The Hepatic Portal System

The **hepatic portal system** connects capillaries of the intestines and other digestive organs to the **hepatic sinusoids** of the liver. The intestinal blood is richly laden with nutrients for a few hours following a meal. This circulatory arrangement gives the liver "first claim" to these nutrients before the blood is distributed to the rest of the body. It also allows the blood to be cleansed of bacteria and toxins picked up from the intestines, an important function of the liver. The route from the intestines to the inferior vena cava follows (see figure):

1. The **inferior mesenteric vein** receives blood from the rectum and distal part of the large intestine. It converges in a fanlike array in the mesentery and empties into the splenic vein.

2. The **superior mesenteric vein** receives blood from the entire small intestine, the ascending colon, the transverse colon, and the stomach. It, too, exhibits a fanlike arrangement in the mesentery and then joins the splenic vein.

3. The **splenic vein** drains the spleen and travels across the abdominal cavity toward the liver. Along the way, it picks up the **pancreatic veins** from the pancreas and the inferior mesenteric vein.

4. The **hepatic portal vein** is formed by convergence of the splenic and superior mesenteric veins. It travels about 8 cm up and to the right and then enters the inferior surface of the liver. Near this point it receives the **cystic vein** from the gallbladder. In the liver, the hepatic portal vein ultimately leads to the innumerable microscopic hepatic sinusoids. Blood from the sinusoids empties into the hepatic veins described earlier. Circulation within the liver is described in more detail in chapter 25.

5. The left and right **gastric veins** form an arch along the lesser curvature of the stomach and empty into the hepatic portal vein.

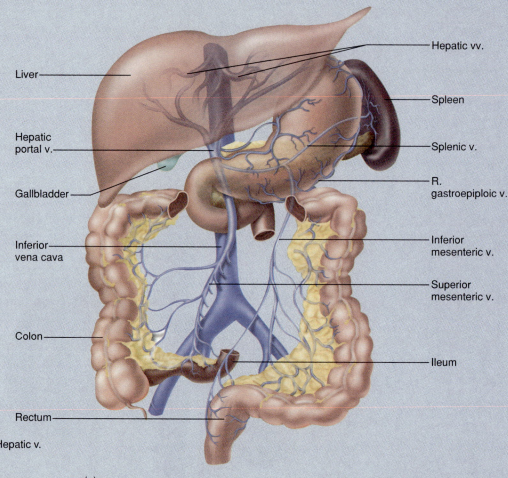

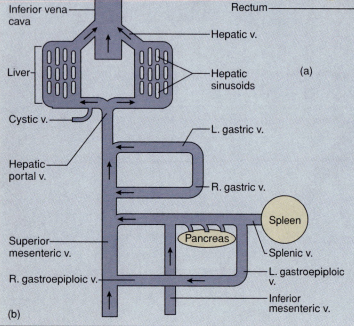

The hepatic portal system. (a) Anatomy. (b) Flowchart. ⚕

Table 20.14   Venous Drainage of the Lower Extremity and Pelvic Organs

Drainage of the lower extremity is described starting at the toes and following the flow of blood to the inferior vena cava (see figure). As in the upper extremity, there are deep and superficial veins with anastomoses between them.

Deep Veins

1. The **plantar arch** drains the plantar aspect of the foot, receives blood from the **digital veins** of the toes, and gives rise to the next vein.

2. The **posterior tibial vein** drains the plantar arch and passes up the leg embedded deep in the calf muscles, receiving drainage along the way from the **fibular (peroneal) vein.**

3. The **dorsal pedal vein** drains the dorsum of the foot.

4. The **anterior tibial vein** is a continuation of the dorsal pedal vein. It travels up the anterior compartment of the leg between the tibia and fibula.

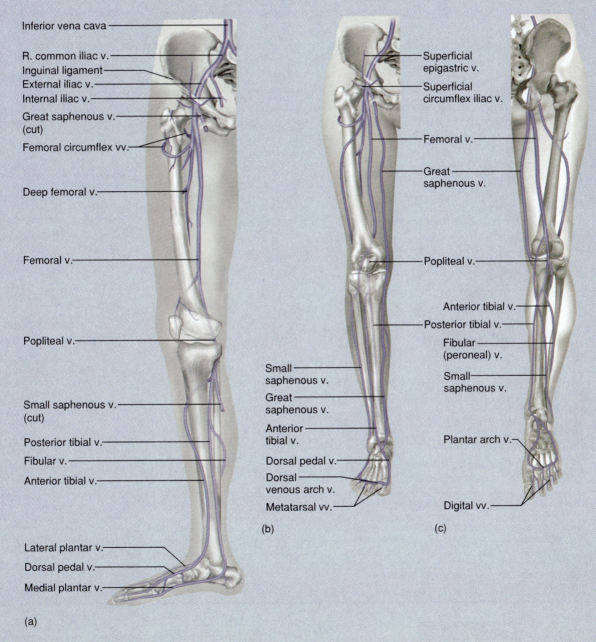

Venous drainage of the lower extremity. (*a*) Deep veins, anteromedial view of right limb. (*b*) Anterior aspect of right limb and dorsal aspect of foot. (*c*) Posterior aspect of right limb and plantar aspect of foot.

continued next page

Table 20.14 Continued

5. The **popliteal vein** is formed at the back of the knee by the union of the anterior and posterior tibial veins.

6. The **femoral vein** is a continuation of the popliteal vein into the thigh. It receives drainage from the deep thigh muscles and femur.

7. The **external iliac vein**, superior to the inguinal ligament, is formed by the union of the femoral vein and great saphenous vein (one of the superficial veins described next).

8. The **internal iliac vein** follows the course of the internal iliac artery and its distribution. Its tributaries drain the gluteal muscles; the medial aspect of the thigh; the urinary bladder, rectum, prostate, and ductus deferens in the male; and the uterus and vagina in the female.

9. The **common iliac vein** is formed by the union of the external and internal iliac veins; it also receives blood from the ascending lumbar vein. The right and left common iliacs then unite to form the inferior vena cava.

Superficial Veins

1. The **dorsal venous arch** is visible through the skin on the dorsum of the foot. It has numerous anastomoses similar to the dorsal venous arch of the hand.

2. The **great saphenous**[28] (sah-FEE-nus) **vein**, the longest vein in the body, arises from the medial side of the dorsal venous arch. It traverses the medial aspect of the leg and thigh and terminates by emptying into the femoral vein, just distal to the inguinal ligament. It is commonly used as a site for the long-term administration of intravenous fluids; it is a relatively accessible vein in infants and in patients in shock whose veins have collapsed. Portions of this vein are commonly excised and used as grafts in coronary bypass surgery.

3. The **small saphenous vein** arises from the lateral side of the dorsal venous arch, courses up the lateral aspect of the foot and through the calf muscles, and terminates at the knee by emptying into the popliteal vein. It has numerous anastomoses with the great saphenous vein. The great and small saphenous veins are among the most common sites of varicose veins.

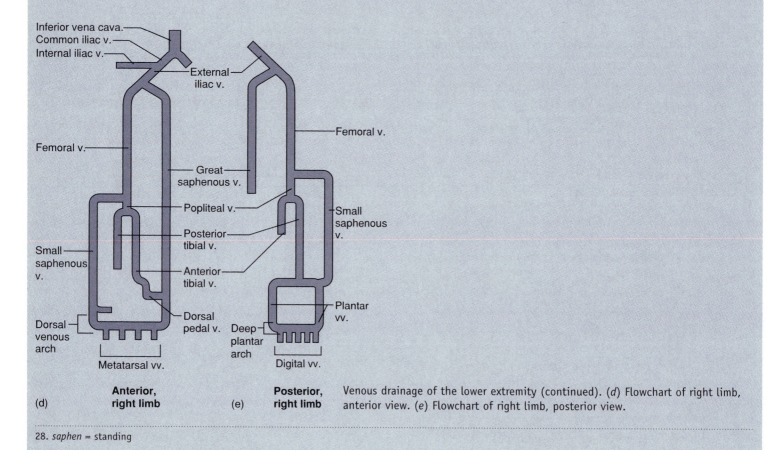

(d) **Anterior, right limb**   (e) **Posterior, right limb**   Venous drainage of the lower extremity (continued). (*d*) Flowchart of right limb, anterior view. (*e*) Flowchart of right limb, posterior view.

28. *saphen* = standing

Key Point Review

25 If you were dissecting a cadaver, where would you look for the internal and external jugular veins? What muscle would help you distinguish one from the other?

26 How do the vertebral veins differ from the vertebral arteries in their superior terminations?

27 By what route does blood from the abdominal wall reach the superior vena cava?

28 Trace one possible path of an RBC from the fingertips to the right atrium, naming the veins along the way.

29 State two ways in which the great saphenous vein has special clinical significance. Where is this vein located?

CHAPTER ESSAY

Hypertension—The "Silent Killer"

Hypertension, the most common cardiovascular disease, affects about 30% of Americans over age 50 and 50% by age 74. It is a "silent killer" that can wreak its destructive effects for 10 to 20 years before the first symptoms are noticed. Hypertension is the major cause of heart failure, stroke, and kidney failure. It damages the heart because it increases the afterload, which makes the ventricles work harder to expel blood. The myocardium enlarges up to a point (the *hypertrophic response*), but eventually it becomes excessively stretched and less efficient. Hypertension strains the blood vessels and tears the endothelium, creating lesions that become focal points of atherosclerosis. Atherosclerosis then worsens the hypertension and establishes an insidious positive feedback cycle.

Another positive feedback cycle is created in the effect of hypertension on the kidneys. Their arterioles thicken in response to the stress, their lumens become narrower, and renal perfusion declines. When the kidneys detect a drop in blood pressure, they release renin, which leads to the formation of the vasoconstrictor angiotensin II and *aldosterone,* a hormone that promotes salt retention (described in detail in chapter 24). These effects worsen the hypertension that already existed. If diastolic pressure exceeds 120 mmHg, blood vessels of the eye hemorrhage, blindness ensues, the kidneys and heart deteriorate rapidly, and death usually follows within 2 years.

Primary hypertension, which accounts for 90% of cases, results from such a complex web of behavioral, hereditary, and other factors that it is difficult to sort out any specific underlying cause. It was once considered such a normal part of the "essence" of aging that it continues to be called by another name, *essential hypertension.* That term suggests a fatalistic resignation to hypertension as a fact of life, but this need not be. Many risk factors have been identified, and most of them are controllable.

One of the chief culprits is obesity. Each pound of extra fat requires miles of additional blood vessels to serve it. As you can infer from Poiseuille's law, this increases peripheral resistance and pressure. Just carrying around extra weight, of course, also increases the workload on the heart. Even a small weight loss can significantly reduce blood pressure. Sedentary behavior is another risk factor. Aerobic exercise helps to reduce hypertension by controlling weight, reducing emotional tension, and stimulating vasodilation.

Dietary factors are also significant contributors to hypertension. Diets high in salt promote increased blood volume and pressure, and high cholesterol and saturated fat intake contributes to atherosclerosis. Potassium, magnesium, and calcium all reduce blood pressure; thus, diets deficient in these minerals promote hypertension. Nicotine makes a particularly devastating contribution because it stimulates the myocardium to beat faster and harder; it stimulates vasoconstriction and thus increases the afterload against which the myocardium must work. Just when the heart needs extra oxygen, nicotine causes coronary vasoconstriction and promotes myocardial ischemia.

Some risk factors cannot be changed at will—race, heredity, and sex. Hypertension runs in some families. A person whose parents or siblings have hypertension is more likely than average to develop it. The incidence of hypertension is about 30% higher, and the incidence of strokes about twice as high, among blacks as among whites. From ages 18 to 54, hypertension is more common in men, but above age 65, it is more common in women. Even people at risk from these factors, however, can minimize their chances of hypertension by changing risky behaviors.

Treatments for primary hypertension include weight loss; reduced intake of cholesterol, saturated fat, and salt; and various drugs. Diuretics lower blood volume and pressure by promoting urination. ACE inhibitors block the action of *angiotensin-converting enzyme,* which catalyzes a step in the formation of angiotensin II. Beta-blockers such as propanolol block the vasoconstrictive action of the sympathetic nervous system. Calcium channel blockers inhibit the influx of calcium into cardiac and smooth muscle, thus inhibiting their contraction and promoting vasodilation and reduced cardiac workload.

Secondary hypertension, accounting for about 10% of cases, is high blood pressure that results from (is secondary to) identifiable disorders. These include kidney disease (which may cause renin hypersecretion), atherosclerosis, hyperthyroidism, Cushing disease, and polycythemia. Secondary hypertension is corrected by treating the underlying disease. ▲

Interactions Between the CIRCULATORY SYSTEM and Other Organ Systems

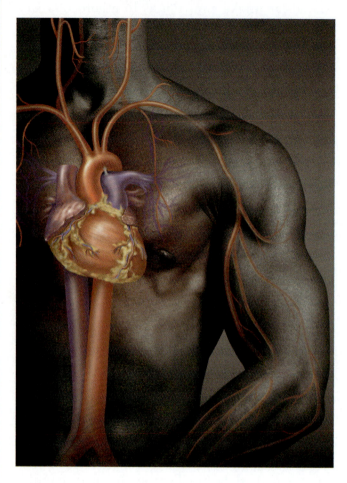

Muscular System
• Helps to regulate blood temperature; respiratory and limb muscles aid venous return; aerobic exercise enhances circulatory efficiency
• Circulatory system removes heat generated by exercise

Nervous System
• Regulates heart rate and strength of contraction; regulates diameters of blood vessels; governs routing of blood flow; monitors blood pressure and composition and activates other systems to regulate these
• Endothelial cells maintain blood-brain barrier and help to generate CSF

Endocrine System
• Helps to regulate blood volume and pressure; stimulates hemopoiesis
• Blood transports hormones to their target cells

Lymphatic/Immune Systems
• Lymphatic system works with circulatory system to regulate fluid balance; returns lymph to bloodstream; spleen serves as reservoir of RBCs; produces lymphocytes; immune cells protect circulatory system from specific pathogens
• Circulatory system produces tissue fluid, which in turn becomes lymph; provides the WBCs and plasma proteins involved in immunity

Respiratory System
• Serves as site of exchange for blood gases; helps to regulate blood pH; thoracic pump aids venous return
• Circulatory system delivers and carries away respiratory gases; low capillary blood pressure keeps alveoli dry

Urinary System
• Controls blood volume, pressure, and composition; initiates renin-angiotensin-aldosterone mechanism; kidneys secrete erythropoietin
• Blood pressure maintains kidney function

Digestive System
• Provides nutrients for hemopoiesis; affects blood composition
• Circulatory system carries away absorbed nutrients and transports most to liver for processing; involved in absorption and recycling of bile salts and minerals from intestines

Reproductive System
• Estrogens may slow development of atherosclerosis in women; testosterone stimulates erythropoiesis
• Circulatory system distributes sex hormones; local vasodilation causes erection of penis and clitoris

All Systems
The circulatory system delivers O_2 and nutrients to all other systems and carries away wastes.

Integumentary System
• Serves as blood reservoir; helps to regulate blood temperature
• Dermal blood flow affects sweat production

Skeletal System
• Provides site (bone marrow) for hemopoiesis; provides protective enclosure for heart and thoracic vessels; serves as reservoir of calcium needed for cardiac muscle contractions
• Provides calcium and phosphate ions for bone deposition; delivers erythropoietin to bone marrow and delivers hormones that regulate skeletal growth

General Anatomy (pp. 706–712)

1. Circulatory routes
   a. Most common route
   b. Portal systems
   c. Anastomoses
2. Structure of blood vessels
   a. Tunica externa
   b. Tunica media
   c. Tunica interna
3. Arteries and metarterioles
   a. Conducting (elastic) arteries
   b. Distributing (muscular) arteries
   c. Resistance (small) arteries
   d. Metarterioles
4. Capillaries
   a. Capillary beds
      • Thoroughfare channels
      • Precapillary sphincters
   b. Types of capillaries
      • Continuous
      • Fenestrated
   c. Sinusoids
5. Veins
   a. Size gradation
   b. Venous sinuses
   c. Low blood pressure
      • Thin walls
      • High capacitance
      • Venous valves

Blood Flow, Pressure, and Resistance (pp. 712–719)

1. Flow and perfusion
   a. Total flow = cardiac output
   b. Hemodynamics: principles of flow
2. Blood pressure (BP)
   a. Measurement
   b. Systolic and diastolic pressures
   c. Pulse pressure
   d. Mean arterial blood pressure (MABP)
   e. Hypertension and hypotension
   f. Importance of arterial elasticity
   g. Pulsatile flow and velocity
   h. Age and blood pressure
3. Resistance
   a. Meaning of peripheral resistance
   b. Factors affecting resistance
      • Blood viscosity
      • Vessel length
      • Vessel radius
        •• Relation to laminar flow
        •• Relation to vasomotion
        •• Special importance of arterioles
4. Poiseuille's law
5. Blood velocity and distance from heart

6. Regulation of peripheral resistance
   a. Local control
      • Autoregulation
        •• Metabolic theory
        •• Reactive hyperemia
        •• Local vasoactive secretions
      • Angiogenesis
   b. Neural control
      • Vasomotor center
      • Innervation of blood vessels
      • Baroreflexes
      • Chemoreflexes
      • Medullary ischemic reflex
   c. Hormonal control of vasomotion
      • Angiotensin II
      • Epinephrine and norepinephrine
      • Atrial natriuretic factor
      • Antidiuretic hormone
7. Vasomotion and redirection of blood flow

Capillary Exchange (pp. 720–722)

1. Routes of exchange
   a. Intercellular clefts
   b. Fenestrations
   c. Pinocytotic vesicles
   d. Plasma membrane
2. Mechanisms of exchange
   a. Diffusion
   b. Transcytosis
   c. Filtration and resorption
      • Forces favoring filtration
      • Forces favoring reabsorption
      • Changes from arterial to venous end
      • Variations in filtration and reabsorption
        •• Range of interstitial hydrostatic pressure
        •• Capillaries dedicated solely to filtration or absorption
        •• Temporal changes in function

Venous Return and Circulatory Shock (pp. 722–725)

1. Mechanisms of venous return
   a. Pressure gradient
   b. Thoracic pump
   c. Cardiac suction
   d. Skeletal muscle pump
   e. Gravity
2. Venous return and physical activity
   a. Increase in cardiac output
   b. Increased action of thoracic pump
   c. Action of skeletal muscle pump
   d. Inactivity and venous pooling

3. Circulatory shock
   a. Cardiogenic shock
   b. Low venous return shock
      • Hypovolemic shock
      • Obstructed venous return shock
      • Venous pooling shock
        •• Neurogenic
        •• Septic
        •• Anaphylactic
   c. Responses to circulatory shock
      • Fainting
      • Baroreflex
      • Production of angiotensin II
   d. Positive feedback in advanced shock

Special Circulatory Routes (pp. 725–726)

1. Flow to the brain
   a. Constancy of total flow
   b. Primacy of autoregulation
   c. Chemical stimuli and vasomotion
   d. Shifting flow within the brain
   e. Cerebrovascular accident
   f. Transient ischemic attack
2. Flow to the skeletal muscles
   a. Variability of total flow
   b. Perfusion at rest
   c. Effects of exercise
   d. Autoregulation and metabolites
   e. Muscle contraction and blood flow
3. Flow to the lungs
   a. Low blood pressure and thin-walled arteries
   b. Low velocity
   c. Oncotic pressure overriding filtration
   d. Response to hypoxia

Anatomy of the Pulmonary Circuit (pp. 726–727)

1. Pulmonary trunk and arteries
2. Lobar arteries
3. Alveolar capillaries
4. Pulmonary veins

Anatomy of the Systemic Arteries (pp. 728–739)

1. The aorta and its major branches (table 20.3)
2. Arterial supply to the head and neck (table 20.4)
3. Arterial supply to the upper extremity (table 20.5)
4. Arterial supply to the thorax (table 20.6)
5. Arterial supply to the abdomen (table 20.7)
6. Arterial supply to the pelvic region and lower extremity (table 20.8)

7. Emergency pressure points
8. The femoral triangle

**Anatomy of the Systemic Veins
(pp. 740–748)**
1. General characteristics of venous system
2. Venous drainage of the head and neck (table 20.9)
3. Venous drainage of the upper extremity (table 20.10)
4. The azygous system (table 20.11)
5. Major tributaries of the inferior vena cava (table 20.12)
6. The hepatic portal system (table 20.13)
7. Venous drainage of the lower extremity and pelvic organs (table 20.14)

Selected Vocabulary

Also review the terms in tables 20.3 through 20.14, which are not included here.

artery 706
vein 706
portal system 706
arteriovenous anastomosis 706
arterial anastomosis 706
venous anastomosis 706
tunica externa 707
vasa vasorum 707
tunica media 707
vasomotion 707
tunica interna 707
endothelium 707
conducting artery 708
distributing artery 709
resistance artery 709
metarteriole 709
capillary bed 709
thoroughfare channel 709
precapillary sphincter 710

continuous capillary 710
intercellular cleft 710
fenestrated capillary 710
fenestration 710
sinusoid 710
venule 711
venous sinus 711
venous valve 711
blood flow 712
perfusion 712
systolic pressure 712
diastolic pressure 712
pulse pressure 712
mean arterial blood pressure (MABP) 713
hypertension 713
aneurysm 713
hypotension 713
peripheral resistance 715
laminar flow 715
Poiseuille's law 716
autoregulation 717
reactive hyperemia 717
angiogenesis 717

vasomotor center 717
baroreflex 717
baroreceptor 717
chemoreflex 718
aortic body 718
carotid body 718
medullary ischemic reflex 718
angiotensin II 718
renin 718
angiotensin-converting enzyme (ACE) 718
capillary exchange 720
colloid osmotic pressure (COP) 721
oncotic pressure 721
net filtration pressure (NFP) 721
net reabsorption pressure 722
solvent drag 722
venous return 723
central venous pressure 723
thoracic pump 723
skeletal muscle pump 723
venous pooling 724

circulatory shock 724
cardiogenic shock 724
low venous return (LVR) shock 724
hypovolemic shock 724
obstructed venous return shock 725
venous pooling shock 725
neurogenic shock 725
syncope 725
septic shock 725
anaphylactic shock 725
compensated shock 725
decompensated shock 725
cerebrovascular accident (CVA) 726
transient ischemic attack (TIA) 726
pulmonary trunk 727
pulmonary artery 727
lobar artery 727
pulmonary vein 727
pressure point 739
femoral triangle 739

Testing Your Recall   Answers in Appendix C

1. Blood normally flows into a capillary bed from
   a. the distributing arteries.
   b. the conducting arteries.
   c. the metarterioles.
   d. a thoroughfare channel.
   e. the venules.

2. Plasma solutes enter the tissue fluid most easily from
   a. continuous capillaries.
   b. fenestrated capillaries.
   c. arteriovenous anastomoses.
   d. collateral vessels.
   e. venous anastomoses.

3. A blood vessel adapted to withstand a high pulse pressure would be expected to have
   a. an elastic tunica media.
   b. a thick tunica adventitia.

   c. one-way valves.
   d. a flexible endothelium.
   e. a relatively thin wall.

4. The substance most likely to produce a rapid drop in blood pressure is
   a. epinephrine.
   b. norepinephrine.
   c. angiotensin II.
   d. serotonin.
   e. histamine.

5. A person with a systolic pressure of 130 mmHg and a diastolic pressure of 85 mmHg would have a mean arterial blood pressure of about
   a. 85 mmHg.
   b. 100 mmHg.
   c. 108 mmHg.

   d. 115 mmHg.
   e. 130 mmHg.

6. According to Poiseuille's law, the velocity of blood flow decreases if
   a. vessel radius increases.
   b. the value of P increases.
   c. viscosity increases.
   d. pressure increases.
   e. afterload increases.

7. Blood flows faster in a venule than in a capillary because venules
   a. have one-way valves.
   b. exhibit vasomotion.
   c. are closer to the heart.
   d. have higher blood pressure.
   e. have larger diameters.

8. In a case where interstitial hydrostatic pressure is negative, the only force causing capillaries to reabsorb fluid is
   a. colloid osmotic pressure of the blood.
   b. colloid osmotic pressure of the tissue fluid.
   c. capillary hydrostatic pressure.
   d. interstitial hydrostatic pressure.
   e. net filtration pressure.

9. Intestinal blood flows to the liver by way of
   a. the superior mesenteric artery.
   b. the celiac trunk.
   c. the inferior vena cava.
   d. the azygos system.
   e. the hepatic portal system.

10. The brain receives blood from all of the following vessels *except* the _____ artery or vein.
    a. basilar
    b. vertebral
    c. internal carotid
    d. internal jugular
    e. anterior communicating

11. The highest arterial blood pressure attained during ventricular contraction is called _____ pressure, and the lowest attained during ventricular relaxation is called _____ pressure.

12. The capillaries of skeletal muscle are of the structural type called _____.

13. _____ shock occurs as a result of exposure to an antigen to which one is hypersensitive.

14. The role of breathing in venous return is called the _____.

15. The difference between the colloid osmotic pressure of blood and that of the tissue fluid is called _____.

16. Movement across the capillary endothelium by uptake and release of fluid droplets is called _____.

17. The pressure sensors in the major arteries near the heart are called _____.

18. All efferent fibers of the vasomotor center belong to the _____ division of the autonomic nervous system.

19. Most blood supply to the brain is from a ring of arterial anastomoses called the _____.

20. The major superficial veins of the arm are the _____ on the medial side and _____ on the lateral side.

1. It is a common lay perception that systolic blood pressure should be 100 plus a person's age. Evaluate the validity of this statement.

2. Calculate the filtration pressure at a point in a hypothetical capillary assuming a hydrostatic blood pressure of 28 mmHg, an interstitial hydrostatic pressure of –2 mmHg, a blood COP of 25 mmHg, and an interstitial COP of 4 mmHg. Give the magnitude (mmHg) and direction (in or out) of the net filtration pressure.

3. Aldosterone secreted by the adrenal gland must be delivered to the kidney immediately below. Trace the route that an aldosterone molecule must take from adrenal to kidney, naming all major blood vessels in the order traveled.

4. People in shock commonly exhibit paleness, cool skin, tachycardia, and a weak pulse. Explain the physiological basis for each of these symptoms.

5. Discuss why it is advantageous to have baroreceptors in the aortic arch and carotid sinus rather than in some other location, such as the common iliac arteries.

Web Site Link

For a listing of the most current web sites related to this chapter, please visit the Saladin homepage at:

http://www.mhhe.com/sciencemath/biology/saladin/

The Lymphatic and Immune Systems

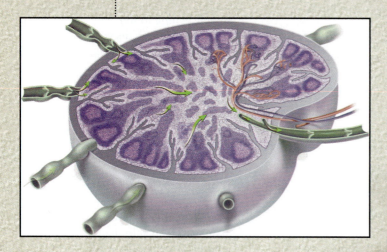

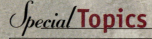

Brushing up

O f all the body systems, the lymphatic system is perhaps the least familiar—yet, without it, the circulatory system would quickly shut down from hypovolemic shock and the immune system would be seriously compromised. This chapter discusses the role of the lymphatic system in maintaining fluid balance in the body and protecting it against infection and disease.

The lymphatic system is closely allied with the immune system in providing us with a defense against foreign invaders. As the lymphatic system collects tissue fluid, it picks up pathogens that may be present in it. On its way back to the bloodstream, the fluid flows through numerous lymph nodes, where dense populations of lymphocytes and macrophages are strategically positioned to detect such pathogens and mount a quick response. Without this defense, the human body would be an ideal place for microorganisms to live. Our homeostatic mechanisms would ensure a constant warm temperature, ample water, and a continual supply of nutrients. Because the lymphatic system enables the immune system to function efficiently, it is appropriate that we consider these two systems together.

The Lymphatic System

▼**Objectives**
When you have completed this section, you should be able to
• list the functions of the lymphatic system;
• explain how lymph is formed and returned to the bloodstream; and
• describe the form and function of the lymph nodes, tonsils, thymus, and spleen.

Blood capillaries continually exude and reabsorb fluid, but they do not reabsorb it all. An excess of 2 to 4 L of water and one-quarter to one-half of the plasma protein enters the interstitial spaces each day. The **lymphatic system** (fig. 21.1) absorbs this excess interstitial fluid and returns it to the bloodstream by way of a network of lymphatic vessels. If not for fluid recovery by the lym-

phatic system, the circulatory system would not have sufficient blood volume to operate properly. Moreover, even partial interference with lymphatic drainage can lead to severe edema (fig. 21.2). Excess protein in the tissue fluid increases its colloid osmotic pressure and contributes to edema by shifting the balance of capillary fluid exchange farther toward filtration. In the small intestine, the lymphatic system has the additional role of absorbing dietary fat, which is not absorbed by blood capillaries (see chapter 25).

The principal components of the lymphatic system are (1) *lymphatic tissue,* consisting of aggregates of lymphocytes and macrophages that populate many organs of the body; (2) *lymphatic organs,* in which these cells are especially concentrated; (3) *lymphatic vessels,* which transport recovered fluid back to the bloodstream; and (4) *lymph,* the fluid contained in these vessels.

Lymph and the Lymphatic Vessels

Once tissue fluid has entered the vessels of the lymphatic system it is called **lymph.** Lymph is usually a clear, colorless fluid, similar to blood plasma but with much less protein. Its composition varies substantially from place to place. After a meal, for example, lymph draining from the small intestine has a milky appearance because of its high lipid content. Lymph leaving the lymph nodes contains a large number of lymphocytes—indeed, this is the main supply of lymphocytes to the bloodstream. Lymph may also contain bacteria, viruses, cellular debris, or even traveling cancer cells.

Origin of Lymph

Lymph is produced by the absorption of tissue fluid into microscopic vessels called **lymphatic capillaries.** These vessels penetrate nearly every tissue of the body, although they are absent from the central nervous system, cartilage, bone, and bone marrow. They are closely associated with blood capillaries (fig. 21.3*a*), but unlike them, they are closed at one end. A lymphatic capillary consists of a sac of thin endothelial cells that loosely overlap each other like the shingles of a roof. The cells are tethered to surrounding tissue cells by protein filaments that prevent the sac from collapsing (fig. 21.3*b*). The overlap between endothelial cells creates valvelike flaps that allow fluid and other matter to enter the lymphatic capillary. Unlike the endothelial cells of blood capillaries, lymphatic endothelial cells are not joined by tight junctions. The gaps between them are so large that bacteria and other cells can enter along with the fluid. When interstitial fluid pressure is high, it pushes the

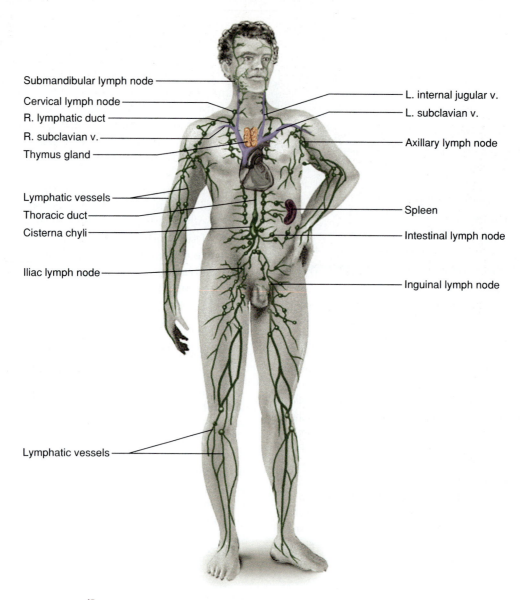

Figure 21.1 The lymphatic system. ✗

flaps inward (open) and fluid flows into the lymphatic capillary. When pressure is higher in the lymphatic capillary than in the tissue fluid, the flaps are pressed outward (closed).

Think About It

Contrast the structure of a lymphatic capillary with that of a continuous blood capillary. Explain why their structural difference is related to their functional difference.

Lymphatic Vessels

Lymphatic vessels form in the embryo by budding from the developing veins, so it is not surprising that the larger ones have a similar histology. They have three

tunics—a *tunica interna,* with an endothelium and valves (fig. 21.4); a *tunica media,* with elastic fibers and smooth muscle; and a thin outer *tunica externa.* Their walls are thinner and their valves are more numerous than those of the veins.

Lymph takes the following route from the tissues back to the bloodstream: lymphatic capillaries → collecting vessels → lymph nodes → lymphatic trunks → collecting ducts → subclavian veins. Thus, there is a continual recycling of fluid from blood to tissue fluid to lymph and back to the blood (fig. 21.5).

The lymphatic capillaries converge to form larger **collecting vessels.** These often travel alongside veins and arteries and share a common connective tissue sheath with them. Numerous lymph nodes occur along

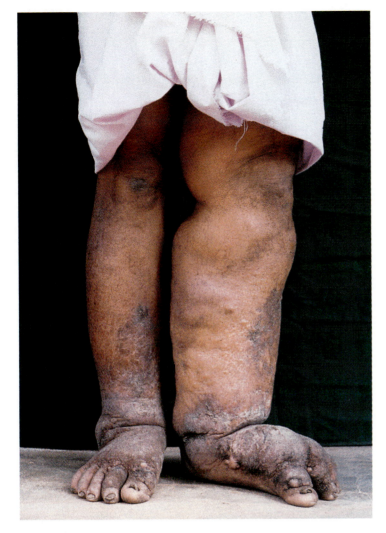

Figure 21.2 Elephantiasis is a tropical disease caused by mosquito-borne roundworms that infect and block the lymph nodes. By interfering with the flow of lymph and the recovery of tissue fluid and protein, the infection causes severe edema. Chronic edema leads to fibrosis and elephant-like thickening of the skin. The extremities are typically affected as shown here; the scrotum of men and breasts of women are often similarly affected.

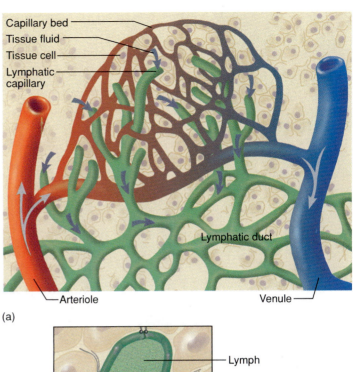

(a)

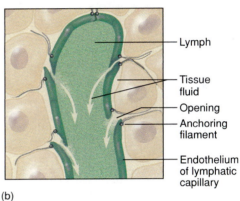

(b)

Figure 21.3 (a) Relationship of the lymphatic capillaries to a bed of blood capillaries. (b) Uptake of tissue fluid by a lymphatic capillary. 🕱

the course of the collecting vessels, receiving and filtering the lymph, as we will discuss shortly. The collecting vessels converge to form larger **lymphatic trunks,** each of which drains a major portion of the body. The principal lymphatic trunks are the *lumbar, intestinal, intercostal, bronchomediastinal, subclavian,* and *jugular trunks.* Their names indicate their locations and parts of the body they drain; the lumbar trunk also drains the lower extremities.

The lymphatic trunks converge to form two **collecting ducts,** the largest of the lymphatic vessels. The **right lymphatic duct** begins in the right thoracic cavity with the union of the right jugular, subclavian, and bronchomediastinal trunks; it receives lymphatic

drainage from the right arm and right side of the thorax and head (fig. 21.6a). The duct on the left, called the **thoracic duct,** is larger and longer. It begins as a prominent sac in the abdominal cavity called the **cisterna chyli** (sis-TUR-nuh KY-lye), and then passes through the diaphragm and up the mediastinum. It receives lymph from all parts of the body below the diaphragm and from the left arm and left side of the head, neck, and thorax (fig. 21.6b). Each collecting duct drains into the subclavian vein on its respective side.

Flow of Lymph

Lymph flows under forces similar to those that govern venous return, except that the lymphatic system has no pump analogous to the heart. Like venous blood, lymph flows at relatively low pressure and speed and is prevented from flowing backward by semilunar

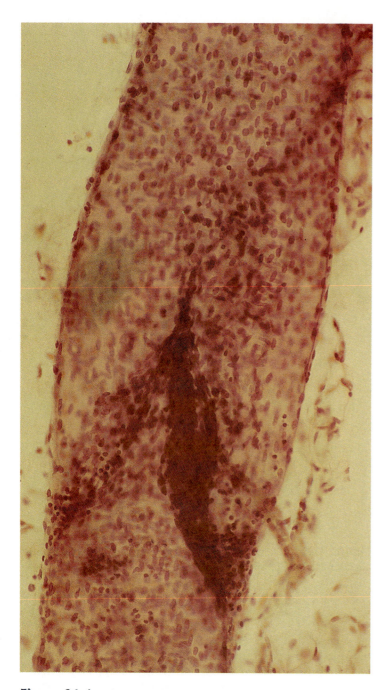

Figure 21.4 Valve in a lymphatic vessel (×200).

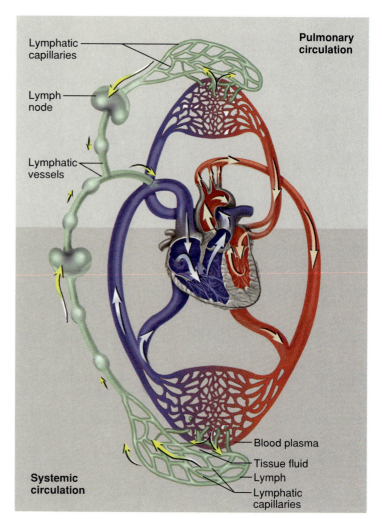

Figure 21.5 The fluid cycle from bloodstream to tissue fluid to lymph and back to the bloodstream. ✗

valves. Lymph is moved primarily by rhythmic contractions of the lymphatic vessels themselves. Stretching of the vessels with fluid stimulates them to contract. But the lymphatic vessels, like the veins, are also aided by a skeletal muscle pump that squeezes them and moves the lymph along. Since lymphatic vessels are often wrapped with an artery in a common sheath, arterial pulsation may also contribute to lymph flow. A thoracic (respiratory) pump aids the flow of lymph from the abdominal to the thoracic cavity, just as it does in venous return. At the point where the col-

lecting ducts join the subclavian veins, the rapidly flowing bloodstream draws the lymph into it. Considering these mechanisms of lymph flow, it should be apparent that physical exercise significantly increases the rate of lymphatic return.

Lymphatic Tissue

The simplest form of lymphatic tissue is **diffuse lymphatic tissue**—a sprinkling of lymphocytes throughout the mucous membranes and connective tissues of many organs. It is particularly prevalent in the mucous membranes of the respiratory, digestive, urinary, and reproductive tracts, where it is called **mucosa-associated lymphatic tissue (MALT)**. In some places, the lymphocytes congregate in dense oval masses called **lymphatic nodules (follicles)**. These nodules come and go as pathogens attempt to invade the tissues and the immune system answers the challenge. They are, however,

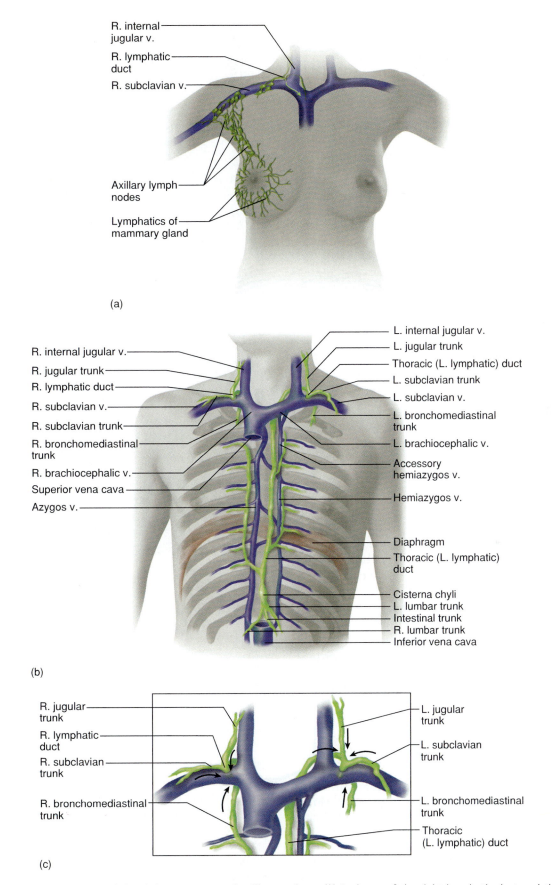

Figure 21.6 (*a*) Lymphatic drainage of the right mammary and axillary regions. (*b*) Drainage of the right lymphatic duct and thoracic duct into the subclavian veins. (*c*) Detail of the points of lymphatic drainage into the subclavian veins.

a relatively constant feature of the small intestine at its junction with the large intestine, where numerous lymphatic nodules form clusters called **Peyer**[1] **patches.**

Lymph Nodes

In contrast to the diffuse lymphatic tissue, the lymphatic organs have well-defined anatomical sites. There are hundreds of **lymph nodes** (fig. 21.7) in the body, mostly embedded in the connective tissues. They are especially concentrated in the following groups (see fig. 21.1):

- **cervical lymph nodes** behind and below the ear, along the inferior margin of the mandible, and down the side of the neck;
- **axillary lymph nodes** from the brachium, through the armpit, to the thoracic wall, breast, and superior abdominal wall;
- **thoracic lymph nodes** along the trachea and bronchi;
- **abdominal lymph nodes** along the abdominal aorta and mesenteric arteries;
- **pelvic lymph nodes** along the iliac arteries and veins; and
- **inguinal lymph nodes** in the groin.

A lymph node is an elongated or bean-shaped structure, usually less than 3 cm long, often with an indentation called the *hilum* on one side. It is enclosed in a fibrous capsule, extensions of which form trabeculae that subdivide the interior of the node into compartments (fig. 21.8). The interior consists of a *stroma* (connective tissue framework) of reticular fibers and phagocytic reticular cells and a *parenchyma* of lymphocytes and mobile macrophages. Between the capsule and the parenchyma is a narrow space called the *cortical sinus,* which is relatively clear of lymphocytes. The parenchyma is divided into an outer *cortex* and an inner *medulla.* The cortex consists mainly of ovoid lymphatic nodules. When the node is fighting off a pathogen, these nodules acquire light-staining **germinal centers,** where they rapidly produce B lymphocytes. The other cells of the cortex are mainly T lymphocytes. (The distinctions between B lymphocytes and T lymphocytes are discussed later.) The medulla consists of a loose system of irregular spaces called *medullary sinuses* that converge on the hilum. Macrophages are concentrated in the medulla.

Several **afferent lymphatic vessels** lead into the node along its convex surface. Lymph flows from these vessels into the cortical sinus, filters slowly through the nodules and medullary sinuses, and leaves the

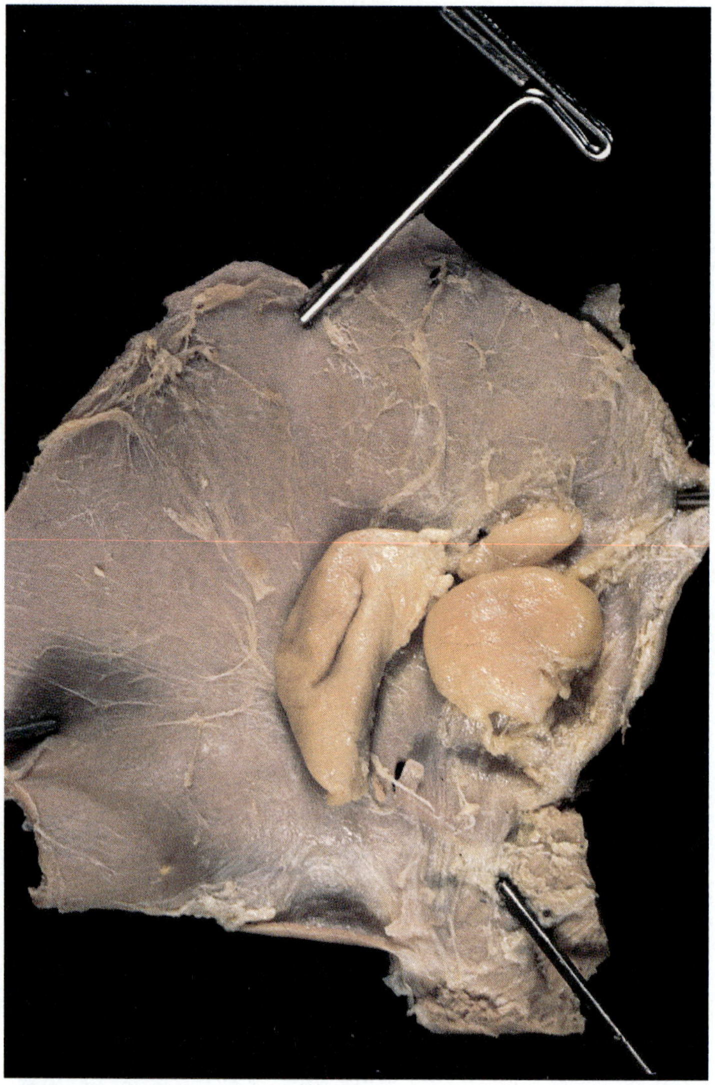

Figure 21.7 Three lymph nodes.

node through one to three **efferent lymphatic vessels** that emerge from the hilum. No other lymphatic organs have afferent lymphatic vessels; lymph nodes are the only organs that filter lymph as it flows along its course.

The lymph that flows into a node may contain bacteria, antigen molecules, or other undesirable matter. As it filters through the node, reticular cells and other macrophages phagocytize foreign matter, and lymphocytes may respond to antigens by activating an immune attack against them. Lymph flows through several nodes in series before returning to the bloodstream, with each node along the way removing more impurities. An infection or other pathogenic challenge can cause the nodes to become swollen and painful (see special topic 21.1).

[1]. Johann Conrad Peyer (1653–1712), Swiss anatomist

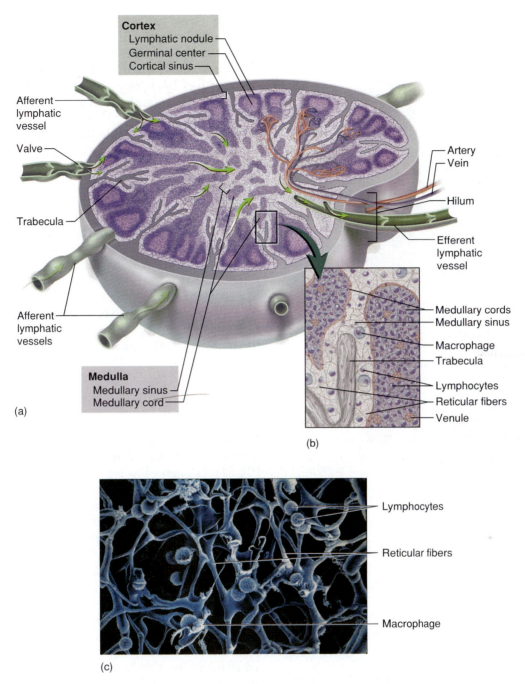

Cortex
Lymphatic nodule
Germinal center
Cortical sinus

Afferent lymphatic vessel

Valve

Trabecula

Afferent lymphatic vessels

Medulla
Medullary sinus
Medullary cord

(a)

Artery
Vein

Hilum

Efferent lymphatic vessel

Medullary cords
Medullary sinus
Macrophage
Trabecula
Lymphocytes
Reticular fibers
Venule

(b)

Lymphocytes

Reticular fibers

Macrophage

(c)

Figure 21.8 (*a*) Anatomy of a lymph node (bisected). (*b*) Detail of the boxed region in *a*. (*c*) Stroma and immune cells in a medullary sinus (SEM). ⅄

Special Topic (Lymph Node Diseases)   21.1

Lymphadenitis is inflammation of a lymph node caused by infection; inflamed lymph nodes are swollen and usually painful when palpated. **Lymphadenopathy** is a collective term for all diseases of the lymph nodes.

Lymph nodes are common sites of metastatic cancer because cancer cells from almost any organ can break loose, enter the lymphatic capillaries, and lodge in the nodes. Cancerous lymph nodes are swollen but relatively firm and usually painless. With a knowledge of lym-

phatic system anatomy and the direction of lymph flow, the sites of metastatic tumors can be predicted. When breast cancer is detected, for example, axillary lymph nodes are routinely examined (biopsied) to determine whether cells of the tumor have metastasized.

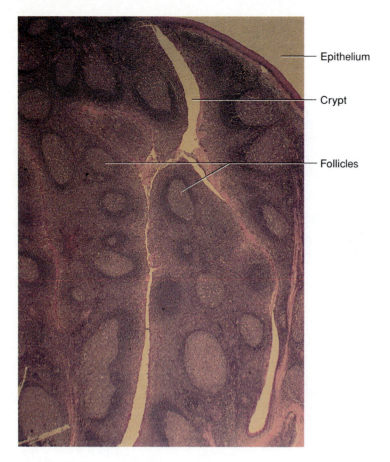

Epithelium

Crypt

Follicles

Figure 21.9 Palatine tonsil; a vertical section showing a branched tonsillar crypt and lymphatic follicles (×20).

Tonsils

The **tonsils** are patches of lymphatic tissue located at the entrance to the pharynx. Each is covered by an epithelium penetrated by **tonsillar crypts**—deep pits lined by lymphatic nodules (fig. 21.9). Ingested and inhaled pathogens get into the crypts, penetrate the epithelium, and encounter the lymphocytes. Many lymphocytes also get through the epithelium and enter the crypts and the saliva.

There are three main sets of tonsils: (1) a pair of **palatine tonsils** at the posterior margin of the oral cavity, (2) a pair of **lingual tonsils** in the root of the tongue, and (3) a single medial **pharyngeal tonsil** on the posterior wall of the pharynx. The palatine tonsils are the largest and the most often infected. Their surgical removal, called *tonsillectomy,* used to be one of the most common surgical procedures performed on children, but it is done less often today. A hypertrophied pharyngeal tonsil is called an *adenoid.*

Thymus

The **thymus** is a bilobed organ positioned between the sternum and aortic arch in the superior mediastinum. It

is very large in the fetus and grows slightly during childhood, when it is most active. After age 14, however, it begins to undergo **involution** (shrinkage) so that it is quite small in adults (fig. 21.10). In the elderly, the thymus has been replaced almost entirely by fibrous and fatty tissue and is barely distinguishable from the surrounding tissues.

The fibrous capsule of the thymus has extensions (trabeculae) that divide the parenchyma into several angular lobules. Each lobule has a cortex and medulla populated by lymphocytes (fig. 21.11). *Hassall[2] corpuscles,* small groups of epithelial cells, are seen in the lighter-staining medullary regions; their function is unknown. The thymus is both a lymphatic and an endocrine organ. It secretes **thymopoietin** and **thymosins,** hormones that stimulate the development and later activity of T lymphocytes and render them capable of fighting pathogens. If the thymus is removed from newborn mammals, they waste away and never develop immunity. Other lymphatic organs also seem to depend on thymosin and develop poorly in such animals.

Spleen

The **spleen** is located in the left hypochondriac region, just inferior to the diaphragm and behind the stomach (fig. 21.12; see also fig. A.31, p. 45). It has a medial hilum through which the splenic artery and vein enter and leave. Its parenchyma exhibits two types of tissue named for their appearance in fresh specimens (not in stained sections): **red pulp,** consisting of sinuses gorged with concentrated erythrocytes, and **white pulp,** consisting of lymphocytes and macrophages aggregated like sleeves along small branches of the splenic artery.

The two tissue types of the spleen reflect the multiple functions of this organ. It produces blood cells in the fetus and sometimes resumes this role in adulthood in cases of extreme anemia. The adult spleen is also a blood reservoir. Its capillaries are very permeable; they allow RBCs to leave the bloodstream, accumulate in the sinuses of the red pulp, and reenter the bloodstream later. If hemorrhagic anemia, for example, should create a need for RBCs, the spleen contracts and forces a few hundred milliliters of them into circulation. The adult spleen is also an "erythrocyte graveyard." Old, fragile RBCs tend to rupture as they squeeze through the capillary walls

2. Arthur H. Hassall (1817–94), British chemist and physician

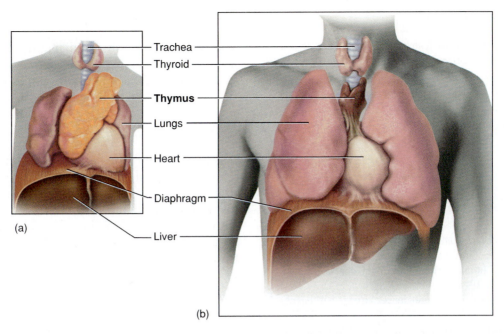

(a)

(b)

Figure 21.10 (a) The thymus is extremely large in a newborn infant. (b) The adult thymus is atrophied and often barely noticeable. ✗

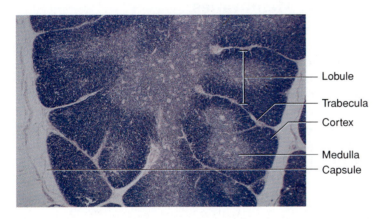

- Lobule
- Trabecula
- Cortex
- Medulla
- Capsule

Figure 21.11 Histological section of the thymus (×40). ✗

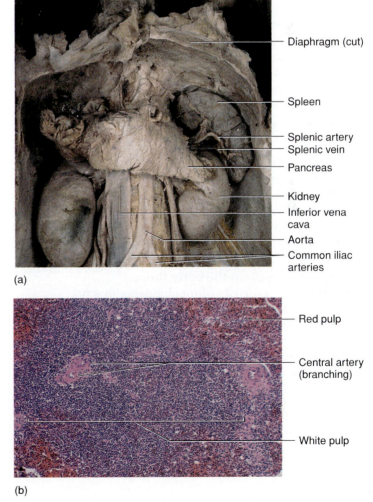

- Diaphragm (cut)
- Spleen
- Splenic artery
- Splenic vein
- Pancreas
- Kidney
- Inferior vena cava
- Aorta
- Common iliac arteries

(a)

- Red pulp
- Central artery (branching)
- White pulp

(b)

Figure 21.12 The spleen. (a) Position of the spleen in the upper left quadrant of the abdominal cavity. (b) Histological section (×20). ✗

into the sinuses. Splenic macrophages phagocytize their remains, just as they dispose of blood-borne bacteria and other cellular debris. Thus, the spleen filters the blood somewhat like the lymph nodes do the lymph. Lymphocytes and macrophages of the white pulp are quick to detect foreign antigens in the blood and lymph and help to activate defensive immune reactions.

The spleen is highly vascular and vulnerable to trauma and infection. Hemorrhage of the spleen can require immediate *splenectomy* (surgical removal), but the spleen is a relatively dispensable organ whose functions can be taken over by the liver and bone marrow.

1. List the primary functions of the lymphatic system. What do you think would be the most noticeable effect of clamping the right lymphatic duct closed?

2. How does fluid get into the lymphatic system? What prevents it from draining back out?

3. Predict the relative seriousness of removing the following damaged or diseased organs from a 2-year-old child: (a) a lymph node, (b) the spleen, (c) the thymus, (d) the palatine tonsils.

Nonspecific Resistance

▼**Objectives**

When you have completed this section, you should be able to
- contrast nonspecific resistance with immunity;
- describe the process of inflammation and explain what accounts for its cardinal signs; and
- Describe the body's other nonspecific defenses against pathogens.

The human body harbors a huge number of microorganisms. Indeed, the bacteria in your body outnumber your own cells by about 10,000 to 1. We are constantly exposed to countless organisms in the food and water we consume, in the air we breathe, and on the objects we touch. Upon death, the body is quickly overrun by microbial life, but as long as we are alive and maintaining homeostasis, we resist these hordes of would-be invaders. The reason is that we have several barriers to invasion and a standing army of cells engaged in a continual search-and-destroy mission against pathogens.

Our defenses fall into two categories:

1. **Nonspecific resistance** consists of mechanisms that guard against a wide variety of pathogens and often keep them from entering the tissues. These mechanisms are present at birth and are effective even against pathogens to which the body has never before been exposed. Examples range from the physical barrier of the skin to the macrophages that engulf bacteria in the body fluids.

2. **Immunity**[3] consists of mechanisms that are activated by exposure to a pathogen in the tissues. It provides specific opposition to future encounters with that pathogen or closely related ones. It is based on the actions of antibodies, lymphocytes, and other cells of the immune system.

In this section we study mechanisms of nonspecific resistance—physical and chemical barriers, phagocytic cells, inflammation, antimicrobial proteins, and fever. These mechanisms are summarized in table 21.3, at the end of the section.

3. *immuno* = free

Physical Barriers

Three important physical barriers—the *skin, mucous membranes,* and *connective tissue gel*—make it mechanically difficult for pathogenic organisms to enter the body or travel in the tissues.

Skin

If you suffer a scrape or an animal bite, prevention of infection is a primary concern; it is also one of the most urgent issues in caring for burn patients. Such facts attest to the importance of intact skin as a barrier to infection. The surface of the skin is composed mainly of keratin, a tough protein that few pathogens can penetrate (see "The Skin as a Barrier," p. 209). With exceptions such as the axillary and pubic areas, most of the skin lacks sufficient moisture and nutrients to support appreciable microbial growth. Some toxins, however, are absorbed through unbroken skin.

Mucous Membranes

The digestive, respiratory, urinary, and reproductive tracts are open to the exterior, but their mucous membranes protect them from invasion. Mucus traps microbes and contains protective chemicals, described shortly. Bacteria trapped in the respiratory mucus are moved by cilia to the pharynx, swallowed, and destroyed by stomach acid. Bacteria are flushed from the upper digestive tract by saliva and from the lower urinary tract by urine.

Connective Tissue Gel

The matrix of the fibroconnective tissues ranges from slightly viscous to rather stiff in consistency. Hyaluronic acid, a glycosaminoglycan (see p. 178), forms an especially viscous gel that inhibits the spread of pathogens through the connective tissues. Some organisms overcome this, however, by producing an enzyme called *hyaluronidase,* which breaks down hyaluronic acid to a thinner consistency that is more easily penetrated (see special topic 6.1, p. 179). Hyaluronidase occurs in some snake venoms and bacterial toxins, and it is produced by *Entamoeba histolytica,* an ameba of the large intestine that sometimes causes severe diarrhea and bleeding (amebic dysentery).

Chemical Barriers

Certain body secretions inhibit the survival, multiplication, or spread of microorganisms by their chemical effects, not their physical ones. Stomach acid, for example, destroys most ingested pathogens. Lactic acid in the perspiration forms a thin antibacterial film on the skin, the *acid mantle.* Acid also inhibits microbial growth in the urethra and vagina. Tears, saliva, and mucus contain **lysozyme,** a polysaccharide surfactant that dissolves bacterial cell walls.

Leukocytes and Macrophages

Leukocytes and macrophages have voracious appetites for foreign matter and play important roles in both non-specific defense and specific immunity. Those that are specialized for phagocytosis—especially neutrophils and macrophages—are collectively called **phagocytes.**

Leukocytes

The five types of leukocytes are described in chapter 18, and their functions are summarized in figure 18.17, p. 661. It is appropriate to revisit them now and examine their contributions to resistance and immunity.

1. **Neutrophils** are highly mobile cells that spend most of their lives wandering in the connective tissues phagocytizing bacteria. *Neutrophilia,* an elevated neutrophil count, is a common sign of bacterial infection. In addition to their phagocytic activity, neutrophils destroy nearby bacteria by creating a chemical **killing zone** around themselves. This process begins with a response called **degranulation,** in which lysosomes migrate to the cell surface and discharge their contents into the tissue fluid. By activating an enzyme in the plasma membrane, this triggers a reaction called the **respiratory burst**—the cell rapidly takes up oxygen and adds one electron to it, forming superoxide, a highly toxic free radical. Superoxide radicals ($O_2 \bullet^-$) and hydrogen ions (H^+) react to form hydrogen peroxide, H_2O_2. Neutrophils also release an enzyme that uses chloride ion in the tissue fluid to produce hypochlorite, $HClO$ (the active ingredient in chlorine bleach). Superoxide, hydrogen peroxide, and hypochlorite are highly toxic to bacteria, but they are also deadly to the neutrophils themselves, which die in the course of their attack on the bacteria. These chemicals damage connective tissues as well and contribute to rheumatoid arthritis.

2. **Eosinophils** are less avidly phagocytic than neutrophils, but they do phagocytize antigen-antibody complexes, allergens (allergy-causing antigens), and inflammatory chemicals, as we will discuss shortly. They are especially abundant in the mucosae of the respiratory, digestive, and lower urinary tracts. In infections with hookworms, tapeworms, and other parasites too large to phagocytize, eosinophils aggregate near the parasites and release enzymes that weaken or destroy them. Allergies and many parasitic infections are marked by *eosinophilia,* a rise in eosinophil count.

3. **Basophils** are not primarily phagocytic in function, but they aid the mobility and action of other leukocytes by secreting the vasodilator histamine and the anticoagulant heparin. Thus, they increase the blood flow to an infected tissue and they inhibit clotting, which could otherwise immobilize the phagocytes.

4. **Lymphocytes** all look more or less alike in blood films, but there are several functional types. **Natural killer (NK) cells** are large lymphocytes that attack and lyse *host cells* (cells of your own body) that have either turned cancerous or become infected with viruses. They do not attack foreign pathogens or depend on specific recognition of enemy cells, unlike the T and B lymphocytes discussed later.

5. **Monocytes** are the circulating precursors of macrophages, discussed next.

Macrophages

The **macrophage system** comprises all of the body's phagocytic cells except leukocytes. It is also known as the *lymphoid-macrophage (L-M),* or *reticuloendothelial (R-E), system.* Some of the macrophages are wandering cells that actively seek pathogens, while others are fixed in place and can phagocytize only those pathogens that come to them—although they are strategically positioned for this to occur. Macrophages include the following cell types, the first two of which are introduced in this chapter:

- **histiocytes,** wandering macrophages of the loose connective tissues;
- **reticular cells** in the stroma of the lymphatic organs;
- **Langerhans cells** of the epidermis (see chapter 7);
- **microglia** in the central nervous system (see chapter 14);
- **endothelial cells** of the blood vessels (see chapter 20);
- **alveolar macrophages** in the pulmonary alveoli (see chapter 22); and
- **Kupffer cells** lining the liver sinusoids (see chapter 25).

Note that the only way a pathogen can invade the body is through a breach in an epithelium, and nearly all epithelia rest on a layer of areolar tissue. Because of its loose organization and fluid-filled spaces, areolar tissue provides an arena in which granulocytes and histiocytes can move quickly to locate and destroy pathogens.

Inflammation

Inflammation is a response to tissue injury of any kind, including trauma and infection. Words ending in the suffix *-itis* denote inflammation of specific organs and

Table 21.1	Cellular Agents of Inflammation
Agents	**Action**
Basophils	Secrete histamine, heparin, bradykinin, serotonin, and leukotrienes
Endothelial cells	Produce cell adhesion molecules to recruit leukocytes; synthesize platelet-derived growth factor to stimulate tissue repair
Fibroblasts	Promote tissue repair by secreting collagen, ground substance, and other tissue components; produce scar tissue
Histiocytes	Phagocytize bacteria, tissue debris, dead and dying leukocytes and pathogens; act as antigen-presenting cells, which activate specific immunity
Mast cells	Same actions as basophils
Neutrophils	Phagocytize bacteria; secrete bactericidal oxidizing agents into tissue fluid
Platelets	Secrete clotting factors, which initiate tissue fluid coagulation, and platelet-derived growth factor, which promotes fibroblast activity and tissue repair

Table 21.2	Inflammatory Chemicals	
Substance	**Sources**	**Effects***
Bradykinin	Plasma and basophils	Pain; vasodilation; increased capillary permeability
Clotting factors	Plasma, platelets, and damaged cells	Coagulation in tissue fluid; isolation of pathogens; formation of temporary framework for tissue repair
Complement	Plasma	Enhanced histamine release; neutrophil chemotaxis; promotion of phagocytosis
Histamine	Basophils and mast cells	Vasodilation; increased capillary permeability
Leukocytosis-promoting factor	Damaged cells	Release of leukocytes from reservoir in red bone marrow
Leukotrienes	Damaged cells, mast cells, and basophils	Vasodilation; increased capillary permeability; neutrophil chemotaxis
Lymphokines	Lymphocytes	Enhanced monocyte and histiocyte activity; leukocyte chemotaxis; intercellular signaling among leukocyte types
Platelet-derived growth factor	Platelets and endothelial cells	Stimulation of fibroblast activity; cell division; and replacement of damaged tissue
Prostaglandins	Damaged cells	Pain; enhanced action of histamine and bradykinin; neutrophil diapedesis

*Some inflammatory chemicals have additional roles in specific immunity, as described in table 21.7.

tissues: *arthritis, encephalitis, peritonitis, gingivitis,* and *dermatitis,* for example. Inflammation can occur anywhere in the body, but it is most common and observable in the skin, which is subject to more trauma than any other organ. Examples of cutaneous inflammation include an itchy mosquito bite, a bee sting, sunburn, poison ivy rash, and the blisters raised by manual labor or tight skates.

Inflammation is characterized by five **cardinal signs**—redness, swelling, heat, pain, and impaired function. The general purposes of inflammation are (1) to limit the spread of pathogens and ultimately to destroy them, (2) to remove the debris of damaged tissue, and (3) to initiate tissue repair. The following discussion explains how the cardinal signs are produced and how the purposes of inflammation are achieved. These processes are mediated by the cells listed in table 21.1 and by the **inflammatory chemicals** listed in table 21.2. Inflammatory chemicals are secreted by damaged cells, mast cells, basophils, lymphocytes, macrophages, and platelets; some of them are also components of the blood plasma.

Pain and Loss of Function

Pain arises from direct injury to the nerve endings (nociceptors), their stimulation by inflammatory chemicals, and the pressure caused by tissue swelling. One of the most potent pain stimuli is **bradykinin,** which is secreted by basophils and mast cells and produced from *kininogen,* a protein in the blood plasma.

Prostaglandins and some bacterial toxins are also important pain stimuli. Pain is not merely an undesirable side effect of inflammation; it is an important alarm signal that calls attention to tissue injury, causes us to favor an injured body part, or makes it so uncomfortable to use that we keep it still. This gives it the rest it needs for recovery.

 Think About It

Review eicosanoid synthesis in chapter 17 and explain why aspirin eases the pain of inflammation.

Hyperemia, Swelling, Redness, and Heat

Bradykinin, histamine, and leukotrienes stimulate vasodilation, which in turn leads to *hyperemia* (increased blood flow) and accounts for the heat of inflamed tissue.

Heat increases the local metabolic rate, thus promoting cell multiplication and healing. The increased blood flow dilutes toxins and provides the oxygen, nutrients, and waste removal needed to sustain the elevated metabolic rate.

Histamine and leukotrienes also increase the permeability of blood capillaries, so that they exude more fluid into the tissue. This produces swelling, but more importantly it allows helpful plasma chemicals and blood cells to get into the tissue. The chemicals include antibodies, complement proteins (described shortly), and fibrinogen, which causes tissue fluid to clot. Clotting sequesters (isolates) bacteria and retards their spread to adjacent tissues, while the fibrin of the clot forms a scaffold for the organization of tissue repair. Swelling reduces venous drainage and increases lymphatic drainage, which favors the removal of bacteria and cellular debris. The redness of the tissue results from hyperemia and from the accumulation of RBCs that escape from the unusually permeable capillaries.

Leukocyte Deployment

Hyperemia and capillary permeability favor the deployment of leukocytes to the inflamed tissue in a three-step process (fig. 21.13):

1. **Margination.** In response to chemicals from damaged tissues, endothelial cells produce *cell adhesion molecules* that make their membranes "sticky." This leads to margination: rather than rushing by in the bloodstream, leukocytes adhere to the endothelium and slowly tumble along it, sometimes coating the endothelium so thickly that they obstruct blood flow.
2. **Diapedesis**[4] (DY-uh-peh-DEE-sis). Adherent leukocytes squeeze between the endothelial cells to enter the interstitial space.
3. **Chemotaxis.** Leukocytes are attracted to bradykinin, leukotrienes, and other inflammatory chemicals that guide them to the site of infection or injury.

Neutrophils are the quickest leukocytes to respond. Hordes of them arrive at the damage site within an hour, avidly phagocytize bacteria, and secrete the oxidizing agents described earlier. Since this is suicidal to the neutrophils themselves, they require replacements. Damaged tissues release **leukocytosis-promoting factor,** which stimulates a quick release of neutrophils from storage in the red bone marrow. Within a few hours, the neutrophil count in the circulating blood may rise to four or five times normal.

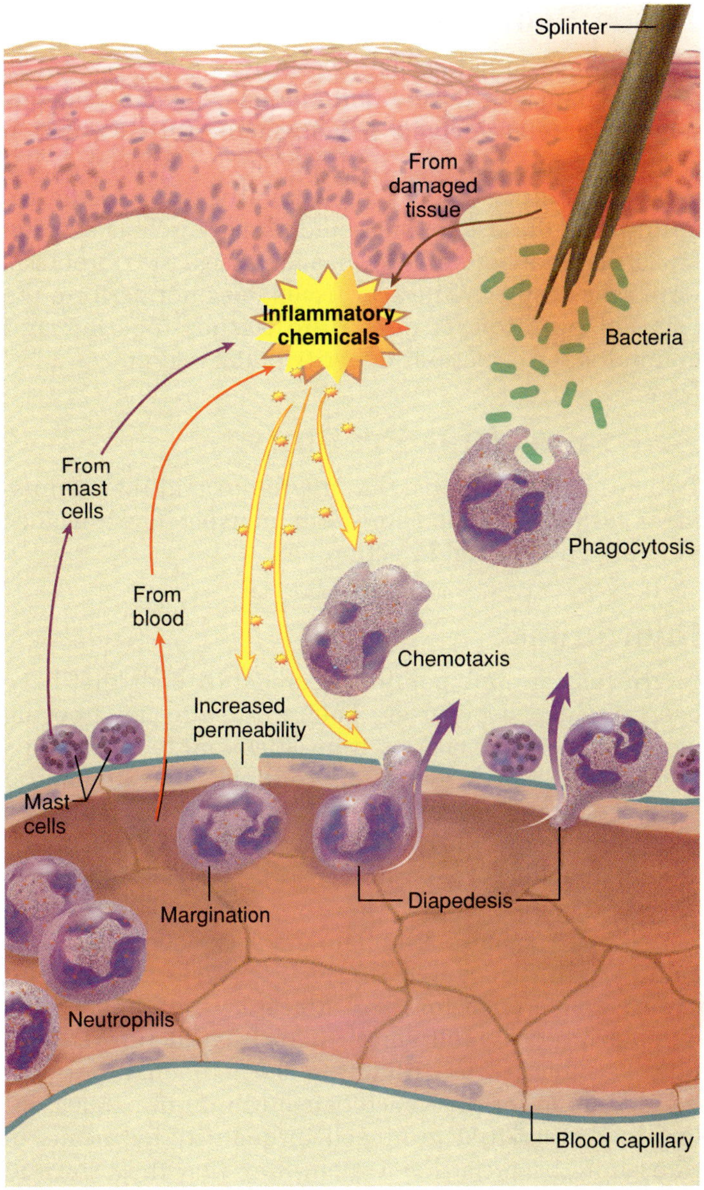

---

4. *dia* = through + *pedesis* = stepping

Figure 21.13 At a site of inflammation, chemical messengers are released by basophils, mast cells, the blood plasma, and damaged tissue. These inflammatory chemicals stimulate leukocyte margination, diapedesis, chemotaxis, and phagocytosis.

Basophils provide a potent mix of inflammatory chemicals (table 21.1). Eosinophils are attracted in cases of parasitism and allergy. Monocytes arrive within 8 to 12 hours, differentiate into histiocytes, and eventually become the primary agents of cleanup. They phagocytize bacteria, damaged host cells, and dead and dying neutrophils and play a vital role in active immunity as *antigen-presenting cells,* discussed later.

The mixture of tissue fluid, cellular debris, and dead and dying neutrophils and microbes is called **pus.** Pus is usually absorbed as an inflammation subsides, but it sometimes accumulates in a blister and may be released by its rupture.

Tissue Repair

Aside from combating pathogens, it is necessary to repair the damaged tissue. Endothelial cells and platelets secrete **platelet-derived growth factor (PDGF),** which stimulates fibroblasts to multiply and synthesize collagen fibers and matrix. This generally leads to the rebuilding of the tissue, but some tissues such as skeletal muscle cannot be replaced. In such cases, the fibroblasts form scar tissue. While scar tissue does not perform the original function, it does bind tissues together and maintains the structural integrity of the region.

Antimicrobial Proteins

Two groups of proteins, the *interferons* and the *complement system,* provide short-term, nonspecific resistance to viral and bacterial infection.

Interferons

Interferons are polypeptides secreted by cells that have been invaded by viruses. They diffuse to neighboring cells and prevent viruses from attaching to them, thus blocking viral invasion. Interferons also activate natural killer cells and macrophages, which destroy infected host cells before they release more viruses. Interferons are not specific for a particular virus but provide generalized protection.

Interferons also stimulate the destruction of cancer cells. When discovered in 1957, they seemed to hold promise as a cure for viral diseases and cancer. The body produces such small quantities of interferons, however, that it was difficult to isolate enough for testing, let alone for pharmacological use. In the 1980s, interferons were first produced in quantity by means of transgenic bacteria. Even then, however, they showed little effect against cancer except for one rare form of leukemia and Kaposi's sarcoma, a common cancer of AIDS patients. Interferons have been used with limited success in controlling hepatitis C and some other viral infections, but control of HIV is still in an experimental stage (see chapter essay, p. 787).

Complement

The **complement system** is a group of 20 or more β globulins of blood plasma that aid (complement) nonspecific resistance and specific immunity. These proteins are always present in the plasma but must be activated by pathogens to exert their effects. There are two pathways of complement activation (fig. 21.14). In the **classical pathway,** antibodies bind to pathogenic organisms and then bind a complex of three complement proteins called C1, C2, and C4. This step is called **complement fixation.** The **alternate pathway** begins with

three complement proteins—factors B, D, and P—binding to the surface polysaccharides of microbes. Both pathways converge on a step where complement C3 is split into two fragments, C3a and C3b. From this point on, the pathway to completion, whether initiated by the classical or alternate pathway, is the same, as shown in the figure.

Complement helps destroy pathogens in three ways:

1. **Enhanced inflammation.** Complement C3a stimulates mast cells and basophils to secrete inflammatory chemicals.
2. **Opsonization.** Complement C3b makes bacteria easier to phagocytize by coating their surfaces and serving as binding sites for macrophages and neutrophils.
3. **Cytolysis.** Complement C3b leads to the rupture of target cells. It binds to a cell and triggers the insertion of a group of proteins called the *membrane attack complex (MAC).* The MAC forms a doughnutlike ring in the microbial plasma membrane, allowing the cytoplasm to escape (fig. 21.15).

Fever

Fever (*pyrexia*) can result from trauma, drug reactions, infection, or several other causes. Because of variations in human body temperature, there is no exact criterion for what constitutes a fever—a fever for one person may be a normal temperature for another. Fever was long regarded as an undesirable side effect of illness, and efforts were made to reduce it for the sake of comfort (see special topic 21.2). It is now recognized, however, as an adaptive defense mechanism that, in moderation, does more good than harm. People with colds, for example, recover more quickly and are less infective to others when they allow a fever to run its course rather than using antipyretic (antifever) medications such as aspirin. Fever is beneficial in that (1) it promotes interferon activity, (2) it elevates metabolic rate and accelerates tissue repair, and (3) it inhibits reproduction of bacteria and viruses.

In bacterial infections, macrophages secrete *interleukin-1 (Il-1).* According to one theory, Il-1 is a **pyrogen**[5] (fever-producing agent) that stimulates the anterior hypothalamus to secrete prostaglandin E (PGE). PGE, in turn, stimulates the hypothalamic thermostat to raise the body temperature to a higher set point—say 39°C (102°F) instead of the usual 37°C. Aspirin and ibuprofen reduce fever by inhibiting PGE synthesis.

5. *pyro* = fire, heat + *gen* = producing

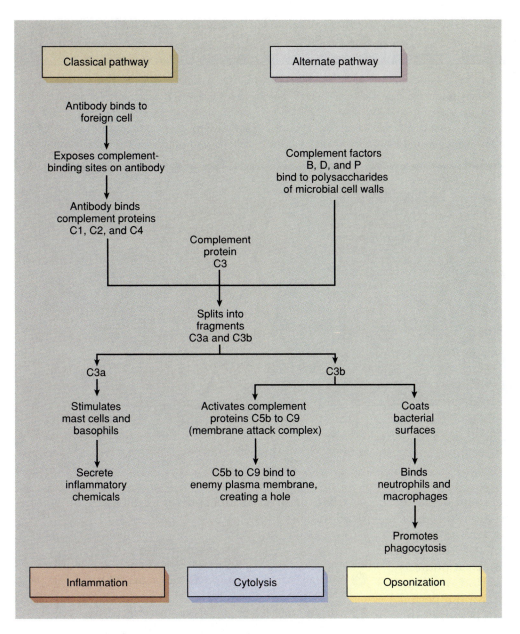

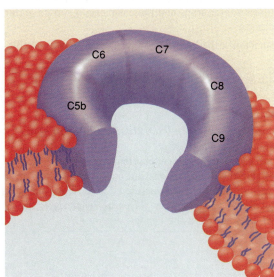

Classical pathway

Antibody binds to
foreign cell

↓

Exposes complement-
binding sites on antibody

↓

Antibody binds
complement proteins
C1, C2, and C4

Alternate pathway

Complement factors
B, D, and P
bind to polysaccharides
of microbial cell walls

Complement
protein
C3

↓

Splits into
fragments
C3a and C3b

C3a

↓

Stimulates
mast cells and
basophils

↓

Secrete
inflammatory
chemicals

C3b

Activates complement
proteins C5b to C9
(membrane attack complex)

↓

C5b to C9 bind to
enemy plasma membrane,
creating a hole

Coats
bacterial
surfaces

↓

Binds
neutrophils and
macrophages

↓

Promotes
phagocytosis

Inflammation

Cytolysis

Opsonization

Figure 21.14 The two pathways of complement activation, and the roles of complement in inflammation, cytolosis, and opsonization.

Figure 21.15 The membrane attack complex of complement proteins C5b to C9 forms a doughnutlike ring in the plasma membrane of an enemy cell, causing cytolysis.

In children under 15, an acute viral infection such as chickenpox or influenza is sometimes followed by a serious disorder called **Reye[6] syndrome.** First recognized in 1963, this disease is characterized by swelling of brain neu-rons and by fatty infiltration of the liver and other viscera. Neurons die from hypoxia and the pressure of the swelling brain, resulting in nausea, vomiting, disorientation, seizures, and coma. About 30% of victims die, and the survivors sometimes suffer mental retardation. Reye syndrome is sometimes triggered by the use of aspirin to control fever; parents are now strictly advised never to give aspirin to children with chickenpox or flulike symptoms.

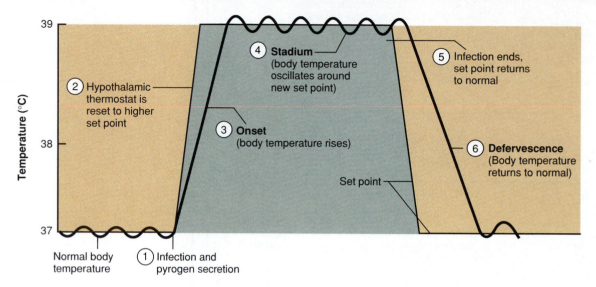

Figure 21.16 The course of a fever.

Resetting of the thermostat triggers shivering to produce heat and cutaneous vasoconstriction to reduce heat loss. The person has chills, the skin feels cold and clammy, and the temperature rises (fig. 21.16). The body temperature then oscillates around the new set point as long as the pathogen is present. The period of rising temperature is called the *onset* of fever, and the period of relatively stable elevated temperature is called the *stadium.*

The elevated temperature stimulates the liver and spleen to hoard zinc and iron, depriving bacteria of minerals needed for their multiplication. When the infection is defeated, pyrogen secretion ceases and the hypothalamic thermostat is set back to normal. This activates heat-losing mechanisms, especially cutaneous vasodilation and sweating. The phase of falling temperature is called *defervescence* in general, *crisis* if the temperature drops abruptly, or *lysis* if it falls slowly. The skin is warm and flushed during this phase.

Even though most fevers are beneficial, excessively high temperature can be dangerous because it causes protein denaturation and cellular dysfunction. Fevers above 40.5°C (105°F) can make a person delirious. Convulsions and coma ensue at higher temperatures, and death or irreversible brain damage commonly results from fevers that range from 44° to 46°C (111° to 115°F).

Mechanisms of nonspecific resistance are summarized in table 21.3.

Key Point Review

4. What are macrophages? Give four examples and state where they are found.

5. List the cardinal signs of inflammation and state the cause of each.

6. How do interferons and the complement system protect against disease?

7. Summarize the benefits of fever and the limits of this benefit.

General Aspects of Specific Immunity

▼**Objectives**
When you have completed this section, you should be able to
• define *specific immunity;*
• contrast humoral and cellular immunity;
• describe the chemical constitution of antigens and antibodies and sketch or describe the structure of an antibody molecule; and
• describe the general roles played by lymphocytes, macrophages, and interleukins in the immune response.

6. R. Douglas Reye (1912–77), Australian pathologist

Table 21.3 Mechanisms of Nonspecific Resistance

Category/Components	Protective Mechanisms
Physical Barriers	
Skin	Keratin is impenetrable to most pathogens; insufficient moisture and nutrients in most areas of the skin surface preclude pathogen growth
Mucous Membranes	Mucus traps most pathogens before they can invade underlying tissues
Connective Tissue Gel	Gelatinous consistency of glycosaminoglycans (GAGs) in connective tissue limits spread of pathogens
Chemical Barriers	
Acidity	Acidic film (acid mantle) on skin inhibits bacterial growth; acidity of vagina, urinary tract, and stomach inhibits or destroys pathogens
Lysozyme	Polysaccharide in tears, saliva, and mucus; dissolves bacterial cell walls
Leukocytes	
Neutrophils	Phagocytize bacteria and secrete strong oxidizing agents that destroy extracellular bacteria
Eosinophils	Phagocytize antigen-antibody complexes, allergens, and inflammatory chemicals; secrete enzymes that destroy pathogens too large to phagocytize
Basophils	Secrete inflammatory chemicals; see table 21.1
Monocytes	Become histiocytes and other types of macrophages; see histiocytes in table 21.1
Natural Killer Cells	Destroy virus-infected or cancerous host cells; other lymphocyte types are involved in specific immunity
Macrophage System	Widespread phagocytic cells of several types; actively phagocytize and destroy pathogens; secrete pyrogens and promote fever; also essential in specific immune responses as antigen-presenting cells
Inflammation	Sequesters pathogens at site of invasion; recruits leukocytes and macrophages to destroy them; removes debris of damaged tissue; initiates tissue repair
Antimicrobial Proteins	
Complement	Lyses pathogenic microbes; enhances action of phagocytes; stimulates release of inflammatory chemicals
Interferons	Produced by virus-infected cells; prevent viruses from adhering to neighboring cells; activate macrophages and natural killer cells
Fever	Inhibits multiplication of bacteria and viruses; promotes interferon secretion; elevates metabolic rate and accelerates tissue repair

The remainder of this chapter is concerned with the immune system and specific immunity. The **immune system** is not an organ system but an array of widely distributed cells that recognize foreign substances and act to neutralize or destroy them. Two characteristics distinguish immunity from nonspecific resistance:

1. **Specificity.** Immunity is directed against a particular pathogen, and immunity to one pathogen usually does not confer immunity to others.
2. **Memory.** When reexposed to the same pathogen, the body reacts so quickly that there is no noticeable illness. The reaction time for inflammation and other nonspecific defenses, by contrast, is just as long for later exposures as it was for the initial one.

These properties of immunity were recognized even in the fifth century B.C.E., when the Greek historian Thucydides remarked that people who recover from a disease often become immune to that one but remain susceptible to others. A person might be immune to measles but still susceptible to chickenpox, for example. In the late 1800s, it was discovered that immunity can be transferred from one animal to another by way of the blood serum. In the mid-1900s, however, it was found that serum does not always confer immunity; sometimes only donor lymphocytes do so. Thus, we now recognize two types of immunity, called humoral and cellular immunity, although these two mechanisms interact extensively and often respond to the same pathogen.

Humoral (antibody-mediated) immunity, named for the fluids, or "humors," of the body, is based on the action of soluble proteins called *antibodies.* Antibodies occur in the body fluids and on the plasma membranes of lymphocytes. Circulating antibodies bind to bacteria, toxins, and extracellular viruses, "tagging" them for destruction by mechanisms described later.

Cellular (cell-mediated) immunity is based on the action of lymphocytes that directly attack diseased or "suspicious" cells, including those of transplanted tissues, cells infected with viruses or parasites, and cancer cells. Lymphocytes can lyse these cells or release chemicals that enhance other defenses such as inflammation.

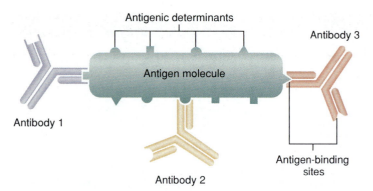

Figure 21.17 A single antigen molecule has multiple antigenic determinants that can stimulate formation of a variety of antibodies. 🎥

Specific immunity has the elements of a drama with a large cast of characters—antigens, antibodies, lymphocytes, macrophages, and chemical messengers. These agents are described before we examine the details of humoral and cellular immunity.

Antigens

An **antigen**[7] **(Ag)** is any molecule that triggers an immune response. Some antigens are free molecules such as venoms and toxins; others are components of plasma membranes and bacterial cell walls. Antigens are generally large, complex molecules such as proteins, polysaccharides, glycoproteins, and glycolipids with structures that are unique to each individual. Their uniqueness enables the body to distinguish its own "self" molecules from those of any other individual or organism ("nonself"). The immune system develops this ability prior to birth; thereafter, it normally attacks only nonself antigens. Small universal biomolecules such as glucose and amino acids are not antigenic; if they were, our immune systems would attack the nutrients and other molecules essential to our very survival. Most antigens have molecular weights over 10,000 amu.

A macromolecule typically has several regions called **antigenic determinants** that can trigger separate immune responses. One region of a glycoprotein molecule may stimulate the formation of one antibody, and a different region of the same molecule may stimulate the formation of a different antibody (fig. 21.17).

Some molecules called **haptens**[8] are too small to be antigenic in themselves, but they can stimulate an immune response by binding to a host macromolecule and creating a unique complex that the body recognizes as foreign. After the first exposure, a hapten may stimulate an immune response without needing to bind to another molecule. Cosmetics, detergents, industrial

7. acronym from *anti*body *gen*erating
8. from *haptein* = to fasten

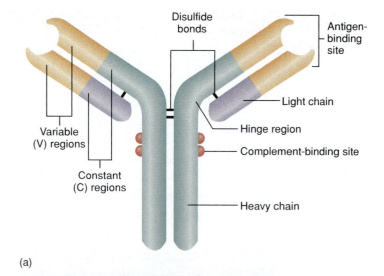

(a)

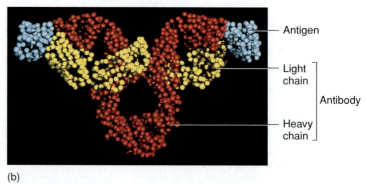

(b)

Figure 21.18 Antibody structure. (*a*) A molecule of IgG, a monomer. Only IgG and IgM have complement-binding sites. (*b*) Computer-generated image of IgG bound to an antigen (lysozyme). 🎥

chemicals, poison ivy, and animal dander contain haptens that cause allergies in many people. Penicillin is a hapten that sometimes binds to erythrocytes and triggers an allergic reaction.

Antibodies

An **antibody (Ab),** or *immunoglobulin,* is a γ globulin found in the blood plasma, body secretions, and membranes of some leukocytes. The basic structural unit of an antibody, an **antibody monomer,** is composed of four polypeptide chains linked by disulfide (—S—S—) bonds (fig. 21.18). Two of them are **heavy chains** of about 400 amino acids, and two are **light chains** about half that long. Each heavy chain has a hinge region where the antibody is bent, giving the monomer a T or Y shape.

All four chains have a **variable region** (V region), which gives an antibody its uniqueness. The V regions of a heavy chain and light chain combine to form an **antigen-binding site** on each arm, which attaches to the antigenic determinant of an antigen molecule. The rest

Class	Structure	Location and Function
Table 21.4	**The Five Classes of Antibodies**	
IgA	Monomer and dimer forms	Plasma IgA is a monomer found in blood plasma; secretory IgA is a dimer found in mucus, saliva, tears, milk, and intestinal secretions. IgA prevents pathogens from adhering to epithelia and penetrating the underlying tissues.
IgD	Monomer	An integral protein of the B cell membrane; acts as an antigen receptor.
IgE	Monomer	Found mainly in tonsils, skin, and mucous membranes. Stimulates mast cells and basophils to release histamine and other chemical mediators of inflammation and allergy; attracts eosinophils to sites of parasitic infection.
IgG	Monomer	Constitutes 75% to 85% of circulating antibodies in plasma. Crosses placenta and confers temporary immunity on the fetus. Includes the anti-D antibodies of the Rh blood group. The predominant antibody secreted in the secondary immune response. IgG and IgM are the only antibodies able to bind complement.
IgM	Monomer and pentamer forms	Monomer is an antigen receptor of the B cell membrane; pentamer occurs in blood plasma. The predominant antibody secreted in the primary immune response. Very strong agglutinating ability. Includes the anti-A and anti-B agglutinins of the ABO blood group.

of each chain is a **constant region** (C region), which has the same amino acid sequence in all of a person's antibodies of a given class. The C region determines the mechanism of an antibody's action—for example, whether it can bind complement proteins.

There are five classes of antibodies named **IgA, IgD, IgE, IgG,** and **IgM** (table 21.4). These names refer to the structures of their C regions (*alpha, delta, epsilon, gamma,* and *mu*). IgD, IgE, and IgG are monomers. IgA has a monomeric form as well as a dimer composed of two cojoined monomers. IgM is a pentamer composed of five monomers. IgG is particularly important in the immunity of the newborn because it is the only immunoglobulin that crosses the placenta. Thus, it transfers some of the mother's immunity to her fetus. In addition, an infant acquires some maternal IgA through breast milk and colostrum (the secretion an infant feeds on for the first 2 or 3 days, until true milk is secreted).

Passive and Active Immunity

Any immunity that an individual acquires from the antibodies or lymphocytes of another person is called **passive immunity.** It is conferred naturally on a fetus when the mother's antibodies cross the placenta, and artificially in the use of antiserum for the treatment of snakebite, botulism, tetanus, rabies, and other emergency conditions. It does not apply to preventive vaccinations against such diseases as tetanus or rabies. Passive immunity typically lasts only 2 or 3 weeks; it does not confer lasting protection against a pathogen. **Active immunity** refers to the production of one's own antibodies or lymphocytes against an antigen. It includes immunity that results from natural exposure to antigens, as well as the immunity artificially induced by vaccination.

Lymphocytes

The major cells of the immune system are lymphocytes and macrophages, which are especially concentrated at strategic places such as the lymphatic organs and mucous membranes. Macrophages are active in nonspecific resistance and were discussed earlier, but they are also crucial in specific immunity. Most lymphocytes can be classified as either **T lymphocytes (T cells)** or **B lymphocytes (B cells).**

T Lymphocytes

During fetal development, the bone marrow releases undifferentiated stem cells into the blood. Some of these colonize the thymus for 2 or 3 days; the *T* in T lymphocytes stands for thymus-dependent. Hormones of the thymus, called thymosins, stimulate maturing T cells to synthesize about 10,000 to 100,000 identical plasma membrane proteins that serve as antigen receptors. When these receptors are in place, a T cell is said to be **immunocompetent;** that is, it can now bind to antigens "presented" to it by other cells.

An immunocompetent T cell divides rapidly, forming a **clone** of T cells with identical receptors, "programmed" to respond to one particular antigen. All clones that have yet to encounter an antigen are collectively called the **virgin lymphocyte pool.**

Some of the clones in a fetus are capable of responding to self-antigens. In **clonal deletion,** these self-reactive clones are destroyed, leaving the body in a state of **self-tolerance** in which the immune system is unresponsive to the person's own tissues. Clonal deletion leaves only the T cells programmed to respond to foreign antigens. They account for 70% to 80% of the lymphocytes in the blood and colonize lymphatic tissues and organs everywhere.

Think About It

Is clonal deletion a case of apoptosis or necrosis? (Review these concepts in chapters 4 and 6 if necessary.)

B Lymphocytes

Fetal stem cells that do not colonize the thymus settle temporarily in the liver, the bone marrow, the junction of the small intestine and large intestine, the appendix, and other organs, where they differentiate into B lymphocytes (B cells). B cells are named for a lymphatic organ of chickens, the *bursa of Fabricius,*[9] where they were first discovered. The sites of B cell establishment in humans are sometimes called *bursa equivalents.* Here, the B cells synthesize their antigen receptors (IgD and IgM molecules in the plasma membrane), divide rapidly, and produce immunocompetent clones.

Each clone can react to only one antigen—tetanus toxin, for example. If you became infected with the tetanus bacterium, this clone would be called into service. Many clones are never used because the individual is never exposed to the corresponding antigens. For example, you are not likely to contract smallpox, but you do have clones of B cells ready to respond to it. Likewise, you have other clones programmed for a mul-

titude of other antigens you may or may not encounter. These clones are evolutionary adaptations to antigens that have afflicted humans throughout their existence. All an antigen does is determine which of the preexisting clones is activated.

Immunocompetent B cells disperse through the body to colonize the same organs as T cells. They form 20% to 30% of the lymphocytes of the blood, and they are abundant in the lymph nodes, spleen, bone marrow, glands, and MALT.

Antigen-Presenting Cells

In addition to their other roles, B cells and macrophages function as **antigen-presenting cells (APCs).** As you will see later, T cells cannot recognize free antigen molecules. The role of an APC is to phagocytize an antigen, digest it into molecular fragments, and "display" some of these fragments on its surface. Wandering T cells regularly inspect APCs for displayed antigens. If a T cell finds fragments for which it is genetically programmed, it sets in motion other immune processes.

Interleukins

With so many cell types involved in immunity, it is not surprising that they require a special system of chemical messengers to coordinate their activities. **Interleukins**[10] are hormonelike messengers between leukocytes or leukocyte derivatives. Those produced by lymphocytes are called **lymphokines,**[11] and those produced by macrophages are called **monokines** (after monocytes, the macrophage precursors).

Since the terminology of immune cells and chemicals is quite complex, you may find it helpful to refer often to tables 21.6 and 21.7 (pp. 783–84) as you read the following discussions of humoral and cellular immunity.

Key Point Review

8. How does specific immunity differ from nonspecific defense?

9. How does humoral immunity differ from cellular immunity?

10. What structural properties distinguish antigenic molecules from those that are not antigenic?

11. Explain the distinction between the variable region and the antigen-binding site of an immunoglobulin.

12. What is an immunocompetent lymphocyte? What does a lymphocyte have to produce in order to be immunocompetent?

13. Define *T cell* and *B cell.*

14. Define *interleukin, lymphokine,* and *monokine.*

---

9. Hieronymus Fabricius (Girolamo Fabrizzi) (1537–1619), Italian anatomist

10. *inter* = between + *leuk* = leukocytes
11. *kine* = motion, action

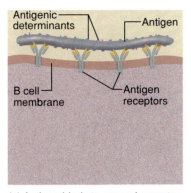 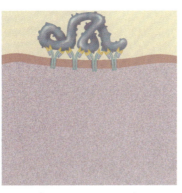 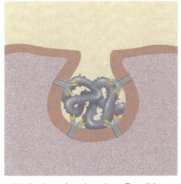 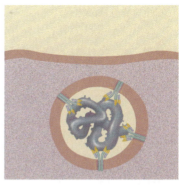

(a) Antigen binds to several receptor sites on membrane of B cell

(b) Receptor sites are drawn together into a cluster

(c) Antigen is taken into B cell by receptor-mediated endocytosis

(d) Antigen is fully internalized

Figure 21.19 Capping and endocytosis in the activation of a B cell by an antigen.

Humoral Immunity

▼Objectives

When you have completed this section, you should be able to
- explain how B cells recognize and respond to an antigen;
- describe the actions of antibodies; and
- explain how immune memory prevents disease upon reexposure to an antigen.

With the foregoing introduction to the "actors" of the immune system, we can now discuss the "plot"—the humoral and cellular mechanisms of immunity. Each mechanism is like a play in three acts—recognition, attack, and memory. In humoral immunity, the essential features of each stage are as follows:

- **Recognition.** B cells recognize an antigen and divide repeatedly; most of the daughter cells differentiate into *plasma cells,* which synthesize antibodies specific to that antigen.
- **Attack.** Plasma cells release their antibodies, which bind to the antigen, render it harmless, and "tag" it for destruction by other agents.
- **Memory.** Some B cells differentiate into *memory cells,* which provide lasting protection against future exposures to the same pathogen.

You might also think of these as "the three Rs of immunity"—*recognize, react,* and *remember.* We now consider some details of these three stages.

Recognition

The process of antigen recognition involves five steps: *capping, endocytosis, display, clonal selection,* and *plasma cell differentiation.*

1. **Capping.** An immunocompetent B cell has thousands of surface receptors (IgD and IgM molecules) for one antigenic determinant. Since an

antigen is a large molecule with numerous copies of each antigenic determinant, it binds to several receptors on a B cell and links them together (fig. 21.19a). This *capping* process activates the B cell. It helps to explain the action of haptens, which become antigenic only if they bind to host molecules that create a complex large enough to produce a capping reaction. Small molecules cannot reach from receptor to receptor and link them together.

2. **Endocytosis.** The capped receptors of the B cell become drawn into a cluster (fig. 21.19b). A depression forms in the membrane at this point, culminating in receptor-mediated endocytosis of the antigen-receptor complex (fig. 21.19c, d).

3. **Display.** The internalized antigen is digested into molecular fragments. Some of these are voided from the cell as waste, whereas others are linked to certain **MHC proteins,** transported to the plasma membrane, and *displayed* there (fig. 21.20). MHC proteins are named for a family of genes on chromosome 6, the *major histocompatibility complex (MHC),* that codes for them. Except for identical twins, no two people have identical MHC proteins; they are unique "identification tags" that label every cell of your body as belonging to you. An MHC protein somewhat resembles a hotdog bun—it has a distinct central groove in which it carries antigen fragments and displays them to other immune cells. If it carries a "self" antigen, the immune system disregards it. If it carries a "nonself" (foreign) antigen, however, it stimulates an attack.

For the response to go any further, another type of lymphocyte called a *helper T cell* must become involved. Helper T cells roam the tissues and crawl about on antigen-presenting cells as if "looking for trouble" (fig. 21.21). This state of continual inspection and "alertness" is called **immunological**

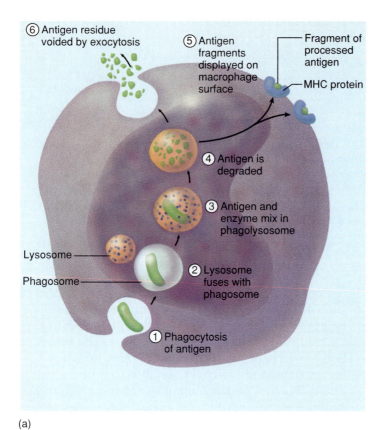

(a)

(b)

Figure 21.20 (a) Stages in the processing and presentation of an antigen by an antigen-presenting cell (APC), such as a B cell or macrophage. (b) Macrophages phagocytizing bacteria (SEM). Filamentous extensions of the macrophage snare the rod-shaped bacteria and draw them to the cell surface, where they are engulfed. ✗ ▭

surveillance. T cells cannot recognize free antigen molecules. However, when a nonself antigen is displayed (presented) by an MHC protein, the two form a powerful stimulatory complex that induces helper T cells to secrete interleukin-2 and **helper factors.** These lymphokines stimulate the B cell to go on to the next step.

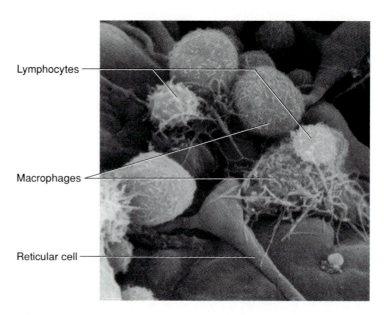

Figure 21.21 Scanning electron micrograph of T cells "inspecting" macrophages in a lymph node for antigen presentation. ✗

From *Tissues and Organs: A Text Atlas of Scanning Electron Microscopy,* © Richard G. Kessel and Randy H. Kardon. Published by W. H. Freeman and Co. 1977.

4. **Clonal selection.** A helper factor stimulates a B cell to divide repeatedly into a battalion of identical cells (fig. 21.22). This process is called *clonal selection* because the antigen "selects" only immunocompetent B cells with receptors for it and stimulates them to divide into a clone of B cells with the same receptors.

5. **Plasma cell differentiation.** Most cells of a clone differentiate into **plasma cells.** These are larger than B cells and contain a high density of rough endoplasmic reticulum (fig. 21.23). A plasma cell produces antibodies at the remarkable rate of 2,000 molecules per second over a brief lifespan of 4 to 5 days. Antibodies are soluble proteins that become distributed throughout the body fluids. The first time you are exposed to a particular antigen, your plasma cells produce mainly IgM. In later exposures to the same antigen, they produce mainly IgG.

Antibody Diversity

The immune system is thought to be capable of producing as many as 2 million different antibodies. This is an astonishing number considering that normally each protein in the body is encoded by a different gene and that we have a total of only 100,000 genes, most of which have functions unrelated to immunity. Obviously there cannot be a different gene for each antibody.

Instead, the genome contains several hundred DNA segments that can be combined in various ways to produce antibody genes. While a B cell is becoming

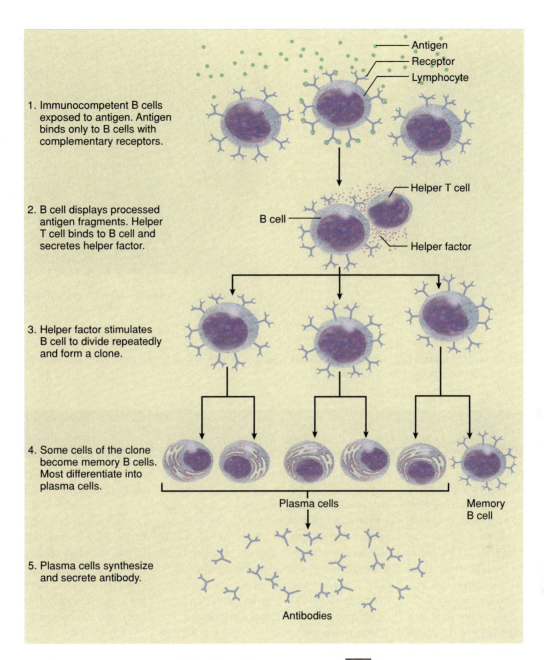

1. Immunocompetent B cells exposed to antigen. Antigen binds only to B cells with complementary receptors.

2. B cell displays processed antigen fragments. Helper T cell binds to B cell and secretes helper factor.

3. Helper factor stimulates B cell to divide repeatedly and form a clone.

4. Some cells of the clone become memory B cells. Most differentiate into plasma cells.

5. Plasma cells synthesize and secrete antibody.

Antigen
Receptor
Lymphocyte

Helper T cell
B cell
Helper factor

Plasma cells
Memory B cell

Antibodies

Figure 21.22 Clonal selection and ensuing events of the humoral immune response.

immunocompetent, it shuffles and recombines these segments to form genes unique to that cell. These genes code for the receptors and antibodies unique to the clone that will arise from that B cell. This process is called **somatic recombination** because it forms new combinations of DNA base sequences in somatic (nonreproductive) cells. This explains how B cells can produce a tremendous variety of antibodies with a relatively limited number of genes.

Antibodies have long been produced commercially for clinical and research purposes. Special topic 21.3 describes a relatively new method for producing purer antibodies against specific antigens.

Attack

Once released by a plasma cell, antibodies use four different mechanisms to render antigens harmless:

1. **Neutralization** is relatively simple. Only certain regions of an antigen are pathogenic—for example, the parts of a toxin molecule and attachment sites of a virus that enable these agents to bind to human cells. Antibodies mask those parts of the antigen to make it harmless.

2. **Complement fixation** is the binding of complement protein to IgM or IgG. When an antigen binds to one of these antibodies, the antibody changes shape

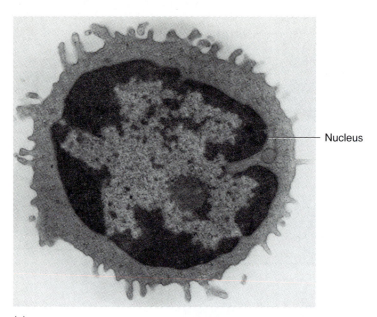

(a)

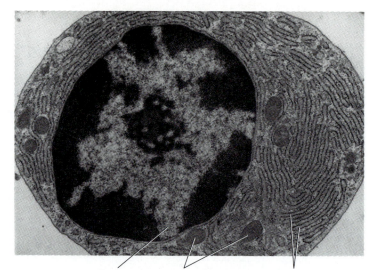

Nucleus

Nucleus   Mitochondria   Rough endoplasmic reticulum

(b)

Figure 21.23 (*a*) A B cell and (*b*) a plasma cell, which differentiates from a B cell (TEM). Note the abundance of rough endoplasmic reticulum in the plasma cell. The rough ER is the site of antibody synthesis.

Special Topic — Monoclonal Antibodies 21.3

Commercial antibody production was originally accomplished by injecting animals with purified antigens and later taking blood serum from them. Since each antigen, however, has multiple antigenic determinants capable of stimulating production of different antibodies, this procedure yielded an antibody mixture. The method for making pure antibodies is relatively new.

In the current technology, mice are injected with an antigen and sensitized B cells are later isolated from their spleens. To enable these B cells to survive and multiply, they are fused with certain cancer cells and then grown in thousands of tissue cultures. The hybrid cells multiply rapidly, producing a clone called a **hybridoma** that secretes only one antibody against the antigen to which the original B cell was sensitized. Out of thousands of clones, the one producing the desired antibody is selected and the others are discarded. The selected cell line goes on producing large amounts of pure **monoclonal antibodies.**

Monoclonal antibodies are more specific and more sensitive than the antibody mixtures produced by earlier techniques. They are now used to diagnose pregnancy, rabies, hepatitis, some sexually transmitted diseases, and cancer. By combining cancer drugs with monoclonal antibodies against cancer antigens, the drugs can be targeted more specifically to tumors and can destroy cancer cells with fewer effects on other cells. Some success has been achieved in treating leukemia and lymphoma (lymph node cancers) with monoclonal antibodies.

and exposes its complement binding site (see fig. 21.18*a*). This initiates the binding of complement proteins to an enemy cell surface (see the *classical pathway,* fig. 21.14) and leads to cytolysis, opsonization of bacteria, and enhanced inflammation as described earlier. Complement fixation is the primary mechanism of defense against such foreign cells as bacteria and mismatched erythrocytes.

3. **Agglutination** was described in chapter 18 in the discussion of ABO and Rh blood types. It is effective not only in mismatched blood transfusions but more importantly as a defense against bacteria and other pathogenic matter. An antibody molecule has 2 to 10 binding sites; thus, it can bind to antigen molecules on two or more cells at once and stick the cells together (fig. 21.24*a*).

4. **Precipitation** begins with a similar process in which antibodies link antigen molecules (not whole cells) together. This creates an antigen-antibody complex (fig. 21.24*b*) that is too large to remain in solution. The complex precipitates and an eosinophil may then phagocytize it.

Think About It

Explain why IgM has a stronger power of agglutination than antibodies of any other class.

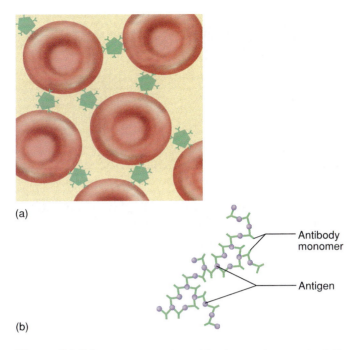

(a)

(b)

Figure 21.24 (a) Agglutination of foreign erythrocytes by IgM. (b) An antigen-antibody complex.

You will note that antibodies do not directly destroy an antigen in any of these mechanisms. They render it harmless by covering its pathogenic sites or agglutinating it, and they mark it for destruction by other agents such as complement, macrophages, eosinophils, or the cell-mediated processes discussed in the next section.

Memory

When a person is exposed to a particular antigen for the first time, the foregoing reactions are called the **primary immune response.** It takes 3 to 6 days for the virgin B cells to proliferate and produce plasma cells. The **antibody titer** (level of antibody in the blood plasma) then rises sharply, peaks in about 10 days, and gradually declines (fig. 21.25). In the primary response, the main antibody is IgM.

During clonal selection, some members of the clone become **memory cells** rather than plasma cells. Memory cells are long-lived and much more numerous than the B cells of a virgin lymphocyte pool. Consequently, if a person is reexposed to the same antigen, memory cells are likely to detect it more quickly than the smaller population of virgin lymphocytes that existed at the time of first exposure. They produce the much quicker **secondary immune response.** Plasma cells form within hours, the antibody titer peaks in about 2 days, and it remains elevated for weeks to years. The antibody is mainly IgG this time. The secondary response is more intense than the primary response,

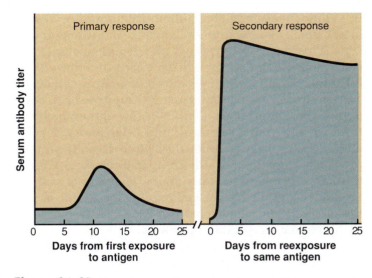

Figure 21.25 The primary and secondary (anamnestic) responses in humoral immunity. The individual is exposed to antigen on day 0 in both cases. Note the differences in the speed of response, the height of the antibody titer, and the rate of decline in antibody titer.

owing to a quicker reaction time and a much higher antibody titer. It is so rapid that the antigen has little chance to exert a noticeable effect on the body, and no illness results. It is also called the **anamnestic**[12] (an-am-NESS-tic) **response**—a reference to immune memory.

Key Point Review

15 What are the three phases of an immune response?

16 What is an MHC protein? How does it relate to B cell and helper T cell function?

17 What is the difference between a B cell and a plasma cell?

18 Describe four ways in which an antibody reacts against an antigen.

19 Why does the secondary immune response prevent a pathogen from causing disease, while the primary immune response does not?

Cellular Immunity

▼Objectives
When you have completed this section, you should be able to
- list the types of lymphocytes involved in cellular immunity and describe the roles they play;
- describe the role of MHC proteins in the activation of helper and cytotoxic T cells;
- explain how interleukins coordinate the actions of immune system cells;
- contrast the roles played by helper, cytotoxic, and suppressor T cells; and
- compare and contrast the action of memory cells in cellular and humoral immunity.

12. *ana* = back + *mnes* = remember

In cellular (cell-mediated) immunity, lymphocytes directly attack and destroy foreign cells and diseased host cells. This involves four types of T lymphocytes: (1) **cytotoxic (killer) T cells,** which carry out the attack; (2) **helper T cells,** which promote cytotoxic T cell action and other defense mechanisms; (3) **suppressor T cells,** which limit the attack and keep the immune system from running out of control; and (4) **memory T cells,** which are descended from the cytotoxic cells and are responsible for memory in cellular immunity. We can think of the cytotoxic cells as the *effectors* of cellular immunity and the helper and suppressor T cells as *regulatory* lymphocytes that act somewhat like an accelerator and brake on cellular immunity.

Helper T cells are also known as T_H, T_4, or CD4 cells; they have a glycoprotein on their surface called CD4 (CD stands for *cluster designation*—a classification system for many surface molecules). Cytotoxic (T_c) and suppressor (T_s) cells are collectively known as CD8 or T_8 cells after their surface glycoprotein, CD8. The roles of these glycoproteins are discussed later.

Like humoral immunity, cellular immunity occurs in stages of recognition, attack, and memory.

Recognition

The recognition phase has two aspects: (1) antigen presentation, as in humoral immunity, and (2) activation of T cells, so that they can mount the attack.

Antigen Presentation

As in humoral immunity, T cells respond only to antigen fragments displayed by APCs, not to free antigens. Cellular immunity, however, presents some new aspects of antigen recognition: (1) Both helper and cytotoxic T cells are involved in surveillance. (2) T cells examine not only B cells but all cells with MHC proteins. (3) Two kinds of MHC proteins are involved—MHC-I and MHC-II.

MHC-I (class I) proteins occur on nearly every cell of the body; they are not limited to APCs. They display fragments of antigens made within the host cell itself, which may be normal self-antigens, abnormal antigens made by cancer cells, or viral proteins. (When a virus invades a host cell, it causes the cell to manufacture proteins that become part of the structure of new viruses.) When an MHC-I protein is bound to a nonself antigen, it stimulates cytotoxic T cells to attack the displaying cell by a mechanism explained later. The antigen-MHC protein complex is like a tag on the host cell that says, "I'm diseased; kill me." Infected or malignant cells are then destroyed before they can do further harm to the body.

MHC-II (class II) proteins occur only on the surfaces of APCs, including B cells, macrophages, and

Table 21.5	Comparison of the Responses of Cytotoxic and Helper T Cells	
Characteristic	Cytotoxic (Killer) T Cells	Helper T Cells
Cells capable of stimulating a response	Nearly any cell	Antigen-presenting cells (macrophages, B cells, some T cells)
MHC protein	MHC-I	MHC-II

some T cells. An MHC-II protein displaying an antigen fragment stimulates helper T cells that come into contact with it.

Cytotoxic T cells respond only to MHC-I proteins, and helper T cells respond only to MHC-II—and in both cases, only if the MHC proteins are bound to antigens. While T cells are maturing in the thymus, they undergo a process called **MHC restriction** in which they develop receptors that can bind only to MHC-I or MHC-II proteins. Antigen presentation is similar for cytotoxic and helper T cells, but the presenting cells and MHC proteins differ (table 21.5).

It is still unclear how suppressor T cells are activated. Some exhibit MHC restriction and require the aid of APCs, while others bind to free antigens without need of presentation by APCs.

T Cell Activation

CD4 and CD8 proteins are cell adhesion molecules that bind a T cell to a target cell during antigen presentation. In addition to their role in cell adhesion, they are linked to a second-messenger system on the intracellular side of the T cell membrane. This involves a protein kinase that hydrolyzes ATP and phosphorylates cytoplasmic enzymes. This process triggers clonal selection: an activated T cell enlarges, multiplies, and forms a clone of identical T cells in much the same manner as the clonal selection of B cells.

Clonal selection requires **costimulation,** a process in which an APC and a T cell communicate with each other by means of interleukins. The binding of a helper T cell to a macrophage stimulates the macrophage to release interleukin-1. Il-1 stimulates the T cell to secrete Il-2 and synthesize Il-2 receptors. Il-2 stimulates more T cells to multiply and secrete still more Il-2. Thus, a large population of activated T cells builds up rapidly. These T cells then enlarge and divide to form clones of effector and memory cells.

Attack

Helper, cytotoxic, and suppressor T cells play different roles in the attack phase.

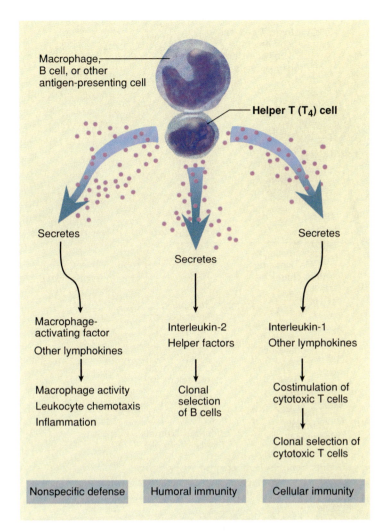

Figure 21.26 The central role of helper T cells in nonspecific defense and specific immunity. Note how extensively the body's defenses would be impaired if the helper T cells were destroyed, as happens in AIDS. ✗ ⬛

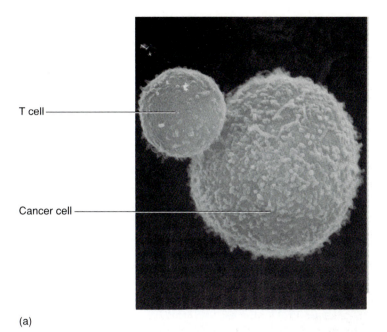

(a)

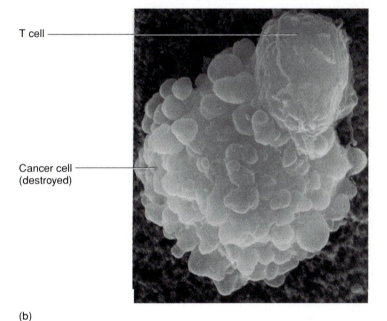

(b)

Figure 21.27 (a) A cytotoxic T cell binding to a cancer cell. (b) Death of the cancer cell due to the lethal hit by the T cell. ✗ ⬛

Helper T Cells

Most immune responses require the action of helper T cells, which play a central coordinating role in both humoral and cellular immunity (fig. 21.26). When a helper T cell recognizes an MHC-antigen complex, it secretes a variety of lymphokines that attract neutrophils and promote inflammation and other nonspecific responses. One of these lymphokines, **macrophage-activating factor (MAF),** enhances the phagocytic activity of macrophages.

Cytotoxic T Cells

Cytotoxic (killer) T cells are the only T lymphocytes that directly attack and kill other cells. They are particularly responsive to host cells that are infected with viruses, intracellular parasites, and bacteria; cells of transplanted tissues and organs; and cancer cells (fig. 21.27). (They are not to be confused with the *natural killer,* or *NK, lymphocytes* of nonspecific defense.)

When a cytotoxic T cell recognizes a complex of antigen and MHC-I protein on a diseased or foreign cell, it "docks" on that cell and delivers a **lethal hit** that ultimately kills it. To do this, it must be in actual contact with the enemy cell, in contrast to B cells, which attack indirectly. Although the mechanism of the lethal hit is not entirely clear, there is evidence that some cytotoxic T cells degranulate and release a protein, **perforin,** that

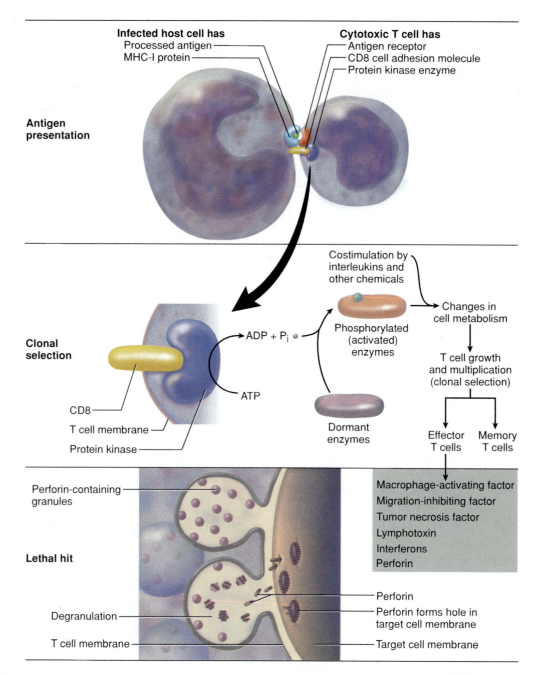

Infected host cell has
Processed antigen
MHC-I protein

Cytotoxic T cell has
Antigen receptor
CD8 cell adhesion molecule
Protein kinase enzyme

Antigen presentation

Costimulation by interleukins and other chemicals

Clonal selection

ADP + P$_i$

Phosphorylated (activated) enzymes

Changes in cell metabolism

T cell growth and multiplication (clonal selection)

CD8

T cell membrane

Protein kinase

ATP

Dormant enzymes

Effector T cells

Memory T cells

Macrophage-activating factor
Migration-inhibiting factor
Tumor necrosis factor
Lymphotoxin
Interferons
Perforin

Lethal hit

Perforin-containing granules

Degranulation

T cell membrane

Perforin

Perforin forms hole in target cell membrane

Target cell membrane

Figure 21.28 Stages in the activation and attack of a cytotoxic T cell on an infected host (target) cell.

is inserted into the enemy plasma membrane. The T cell detaches and leaves, in search of other prey, while the perforin gradually polymerizes and creates holes in the enemy cell, similar to the way complement proteins act (fig. 21.28).

Other cytotoxic T cells lack perforin but kill target cells by different means. Some secrete **lymphotoxin,** which destroys the target cell's DNA; others secrete **tumor necrosis factor (TNF),** which kills cancer cells in 2 to 3 days by unknown mechanisms. Cytotoxic T cells also secrete interferon, macrophage-activating factor, and **migration-inhibiting factor (MIF).** MIF is a lymphokine that suppresses the migration of macrophages, thus keeping them in the area where their service is most needed.

Think About It

How is a killer T cell like a natural killer (NK) cell? How are they different?

Table 21.6 Cellular Agents of Immunity

Cell Type	Action
Macrophages	Phagocytize bacteria, viruses, and expended or damaged host cells; act as antigen-presenting cells (APCs), which display antigen fragments to helper T cells; secrete interleukin-1 (see table 21.7)
Lymphocytes	
Natural Killer (NK) Cells	Attack host cells that have become malignant or infected with viruses; do not depend on specific immune recognition
Virgin Lymphocyte Pool	A reserve of immunocomponent lymphocytes, capable of responding to an antigen but not yet having encountered one
B Cells	Develop in bone marrow; as APCs, recognize and internalize antigens and display antigen fragments on cell surface; give rise to plasma cells
Plasma Cells	Develop from B cells that have been activated by helper T cells; synthesize antibodies
Memory B Cells	Activated B cells that do not immediately differentiate into plasma cells; act as a pool of lymphocytes that can execute a quick secondary, humoral immune response upon reexposure to the same antigen that activated them initially
Cytotoxic (Killer) T Cells	Effector cells of cellular immunity; the only lymphocytes that directly attack other cells; bind to host cells displaying antigen fragments on MHC-I proteins; produce perforin, lymphotoxin, interferon, macrophage-activating factors, migration-inhibiting factor, and tumor necrosis factor; produce the lethal hit that destroys target cells
Helper T Cells	Play a central regulatory role in nonspecific defense and humoral and cellular immunity; cannot recognize free antigens but recognize antigen fragments displayed by antigen-presenting cells with MHC-II proteins; secrete interleukin-2 and other lymphokines that stimulate other immune cells (see actions of lymphokines in table 21.7)
Suppressor T Cells	Release lymphokines that inhibit B cell and T cell activity; help "wind down" the immune response as pathogen is defeated
Memory T Cells	Activated T lymphocytes that do not immediately differentiate into effector T cells; act as a pool of lymphocytes that can execute a quick T cell recall response upon reexposure to the same antigen that activated them initially
CD4 (T_4) Cells	T lymphocytes with CD4 surface glycoproteins; primarily helper T cells
CD8 (T_8) Cells	T lymphocytes with CD8 surface glycoproteins, including cytotoxic and suppressor T cells

Suppressor T Cells

As the pathogen is defeated and disappears from the tissues, suppressor T cells release lymphokines that inhibit T cell and B cell activity. This slows down the immune reaction and keeps it from running out of control. Suppressor T cells may also help prevent autoimmune diseases, as we will see shortly.

Memory

As more and more cells are recruited by helper T cells, the immune response exerts an overwhelming force against the pathogen. On first exposure to a particular pathogen, this constitutes the primary response, which peaks in about a week and then gradually declines. The primary response is followed by immune memory, which works in much the same way as in humoral immunity although it tends to last longer. Following clonal selection, some T cells become memory cells. Upon re-exposure to the same pathogen later in life, memory cells can mount a quick attack called the **T cell recall response,** analogous to the secondary response of humoral immunity.

Tables 21.6 and 21.7 summarize many of the cellular and chemical agents involved in humoral and cellular immunity.

Key Point Review

20 Name four types of lymphocytes that are involved in cellular immunity. Which of these is also essential to humoral immunity?

21 Explain why cytotoxic T cells are activated by a broader range of the body's cells than are helper T cells.

22 What positive feedback cycle can you identify in T cell clonal selection?

23 Why would a viral attack on helper T cells be especially devastating to a person's health?

24 Describe one way in which cytotoxic T cells destroy target cells.

Table 21.7 Chemical Agents of Immunity

Substance	Source and Action
Pathogenic Agents	
Antigen (Ag)	Molecule capable of triggering a response by the immune system; usually a protein, polysaccharide, or glycolipid
Hapten	Small molecule unable to trigger an immune reaction by itself but able to bind to host molecules and produce a complex that is antigenic
Protective Agents	
Antibody (Ab)	Immunoglobulin, members of the γ globulin class of plasma proteins; produced by plasma cells in response to an antigen and able to interfere with the Ag's pathogenicity by means of complement fixation, neutralization of toxins, agglutination, and precipitation
Complement	Plasma proteins that help to destroy pathogens when activated by certain antibodies
Interleukin	Hormonelike messenger produced by macrophages and leukocytes to stimulate other leukocytes
Lymphokine	Interleukin produced by lymphocytes
Monokine	Interleukin produced by macrophages
Interleukin-1 (I1-1)	A monokine that stimulates helper T cells
Interleukin-2 (I1-2)	A lymphokine secreted by helper T cells; stimulates macrophages, making them more avidly phagocytic and inducing them to secrete bactericidal compounds; also stimulates T cells to divide and produce more I1-2
Helper Factor	Lymphokine produced by helper T cells; stimulates B cells to differentiate into plasma cells and synthesize antibodies
Perforin	A protein produced by cytotoxic T cells that binds to target cell walls, produces a hole, and causes cytolysis
Macrophage-Activating Factor (MAF)	A lymphokine produced by cytotoxic and helper T cells; enhances phagocytic activity of macrophages
Migration-Inhibiting Factor (MIF)	A lymphokine produced by cytotoxic T cells; suppresses macrophage migration, causing macrophages to remain in area of immune challenge
Tumor Necrosis Factor	Secreted by cytotoxic T cells; kills cancer cells by unknown mechanism
Lymphotoxin	Secreted by cytotoxic T cells; destroys DNA of target cells

Note: Some chemical agents of immunity have additional roles in inflammation, as described in table 21.2.

Immune System Disorders

▼**Objectives**

When you have completed this section, you should be able to

- distinguish between the four types of hypersensitivity and give an example of each;
- explain the cause of anaphylaxis and distinguish local anaphylaxis from anaphylactic shock;
- explain the cause of asthma and describe how its symptoms are produced;
- state some reasons immune self-tolerance may fail, and give examples of the resulting diseases; and
- contrast severe combined immunodeficiency disease (SCID) with AIDS.

Because the immune system involves a complex cellular interplay controlled by numerous chemical messengers, there are many points at which things can go wrong. The immune response may be too vigorous, too weak, or misdirected against the wrong targets. A few disorders are summarized here to illustrate the consequences.

Hypersensitivity

Hypersensitivity (allergy[13]**)** is an excessive immune reaction against antigens that most people tolerate. Such antigens, called **allergens,** occur in bee and wasp venoms; pollen; toxins from poison ivy and other plants; foods such as milk, nuts, eggs, and shellfish; mold; dust; and vaccines. Drugs such as penicillin, tetracycline, and insulin are allergenic to some people.

Of the four types of hypersensitivity that are recognized, type I is described as *acute,* or *immediate,* because the inflammatory response is extremely rapid. Types II and III are described as *subacute* because they exhibit a slower onset (1–3 hours after exposure) and last longer (10–15 hours) than acute hypersensitivity. Type IV is a cellular immune response whereas the other three are humoral.

- **Type I (acute) hypersensitivity** is the most common form. It begins within seconds of exposure and usually subsides within 30 minutes.

13. *allo* = altered + *erg* = action, reaction

Systemic lupus erythematosus (SLE) is an autoimmune disease that occurs most often in young women, but it can affect anyone. Its cause is unclear but seems to entail a combination of heredity and other risk factors such as injury, infection, emotional stress, drug reactions, or excessive exposure to sunlight. It involves the formation of autoantibodies against DNA and other nuclear components. Antigen-antibody complexes deposit in blood vessels and other organs, where they trigger the release of inflammatory chemicals that lead to widespread inflammation of the connective tissues. The disease was named for skin lesions once likened to a wolf bite.[16] Symptoms include fever, fatigue, joint pain, weight loss, hair loss, intolerance of bright light, and sometimes a "butterfly rash" across the nose and cheeks. The skin, liver, spleen, kidneys, lungs, heart, and central nervous system often become inflamed, and death may ensue from renal failure.

Anaphylaxis[14] (AN-uh-fih-LAC-sis) is a type I reaction that occurs in sensitized people when an allergen caps IgE molecules on the membranes of mast cells and basophils. The cells degranulate and release inflammatory chemicals that cause local edema, mucus hypersecretion, and congestion. Red itchy skin (hives), watery eyes, and a runny nose are typical symptoms. Food and drug allergens may also cause cramps, vomiting, and diarrhea. Local anaphylaxis can be relieved with antihistamines.

Asthma[15] is a local anaphylactic reaction to inhaled allergens. It is the most common chronic illness of children, especially boys. Allergen exposure and even emotional distress can trigger a massive release of histamine and other chemicals, causing spasmodic constriction of the bronchioles. The consequences include severe coughing and wheezing, difficult breathing, and sometimes fatal suffocation. Epinephrine and other β-adrenergic stimulants are used to dilate the airway.

Anaphylactic shock is a severe, widespread response that occurs when an allergen such as bee venom or penicillin is introduced to the bloodstream of a hypersensitive individual. Its effects include bronchiolar constriction, difficult breathing, widespread vasodilation, circulatory shock, and sometimes sudden death. Epinephrine relieves the symptoms by dilating the bronchioles and increasing cardiac output.

- **Type II (antibody-dependent cytotoxic) hypersensitivity** occurs when IgG or IgM binds to antigens on certain cells and lyses them. Blood transfusion reactions are an example. IgM agglutinates mismatched erythrocytes, and complement fixation leads to their lysis.
- **Type III (immune complex) hypersensitivity,** also mediated by IgG or IgM, stems from the formation of large amounts of antigen-antibody complex throughout the body. Some complexes become trapped in the basement membrane under the endothelium of blood vessels and trigger intense inflammation. Type III reactions are involved in systemic lupus erythematosus (see special topic 21.4) and acute glomerulonephritis, both of which can lead to kidney failure.
- **Type IV (delayed) hypersensitivity** is a reaction that appears from 12 to 72 hours after exposure. It occurs when APCs display antigens to helper T cells in the lymph nodes. These T cells then secrete γ interferon and other lymphokines that activate cytotoxic T cells and macrophages, thus triggering a spectrum of nonspecific and immune responses. Haptens in cosmetics and poison ivy can stimulate type IV reactions. The skin test for tuberculosis also involves a delayed hypersensitivity reaction.

Autoimmune Diseases

Autoimmune diseases are failures of self-tolerance—the immune system fails to distinguish self-antigens from foreign antigens and produces **autoantibodies** that attack the body's own tissues. There are at least three reasons why self-tolerance may fail:

1. *Cross-reactivity.* Some antibodies against foreign antigens react to similar self-antigens. In rheumatic fever, for example, a streptococcus infection stimulates production of antibodies that react not only against the bacteria but also against antigens of the heart tissue. It often results in scarring and stenosis (narrowing) of the mitral valve.
2. *Abnormal exposure of self-antigens to the blood.* Some of our native antigens are normally not exposed to the blood. For example, a blood-testis barrier (BTB) normally isolates sperm cells from the blood. Breakdown of the BTB can cause sterility when, in adolescence, sperm are first formed and trigger the production of autoantibodies.

---

14. *ana* = against + *phylax* = protection
15. *asthma* = panting

16. *lupus* = wolf + *erythema* = redness

3. *Change in the structure of self-antigens.* Viruses and drugs may change the structure of self-antigens and cause the immune system to perceive them as foreign. Type I diabetes occurs when viruses alter the antigens of β cells in the pancreatic islets and the β cells are attacked by autoantibodies.

Immunodeficiency Diseases

In the foregoing diseases, the immune system reacts too vigorously or directs its attack against the wrong targets. In immunodeficiency diseases, by contrast, the immune system fails to respond vigorously enough.

Severe combined immunodeficiency disease (SCID) is a congenital scarcity of both T and B cells.

Figure 21.29 Children with severe combined immunodeficiency disease (SCID), born without an immune system, must spend their lives in sterile enclosures to protect against infection. David, nicknamed the "bubble boy," lived with SCID from 1971 to 1984. He is seen here at the age of six, when he first received a portable life support system that enabled him to leave the hospital.

Children with SCID are highly vulnerable to opportunistic infections and must live in protective enclosures. Perhaps the most publicized case was David, the "bubble boy" (fig. 21.29). He spent his life in sterile plastic chambers, finally succumbing at age 12 to cancer triggered by a viral infection. Children with SCID are sometimes helped by transplants of bone marrow or fetal thymus, but in some cases the transplanted cells fail to survive and multiply, or transplanted T cells attack the patient's tissues (the *graft-versus-host response*). In David's case, the fatal virus was contracted from his sister through a bone marrow transplant.

Acquired immunodeficiency diseases are those that are contracted after birth. The most well-known example is **acquired immunodeficiency syndrome (AIDS),** a group of conditions that indicate a severely depressed immune response resulting from infection with the **human immunodeficiency virus (HIV).** HIV destroys helper T cells and thus strikes at the central coordinating element of both humoral and cellular immunity. AIDS is further discussed in the chapter essay.

Key Point Review

25 Contrast local anaphylaxis and anaphylactic shock. Include an explanation of why the latter is more serious.

26 Explain why anaphylactic shock occurs immediately upon antigen exposure while the response to poison ivy takes much longer to develop.

27 Give three reasons why the immune system may become activated by the body's own antigens. Name or briefly describe a disease that exemplifies each mechanism.

28 How are SCID and AIDS similar? How are they different?

CHAPTER ESSAY

Acquired Immunodeficiency Syndrome (AIDS)

The first warnings of the AIDS epidemic came as early as 1976, when a rare form of pneumonia, caused by a fungus called *Pneumocystis,* began appearing in homosexual men and intravenous drug users of both sexes in the Los Angeles area. At the same time, there was a mystifying increase in the incidence of a type of cancer called *Kaposi[17] sarcoma (KS)* in Los Angeles and New York. Although KS was formerly known only as a rare and mild form of cancer, in AIDS victims it was common and severe. AIDS had been known years earlier in Africa, although it wasn't named; HIV has been found in blood samples preserved since 1959 from several African countries.

The Human Immunodeficiency Virus (HIV)

HIV was first isolated and identified in 1983. The virus contains two molecules of RNA and two molecules of an enzyme called *reverse transcriptase.* Surrounding these molecules is a protein coat called the *capsid,* which in turn is surrounded by another type of protein called P18 and finally by an outer *envelope* of glycoproteins (fig. E.1a).

Like other viruses, HIV cannot replicate outside a living cell. HIV invades helper T (CD4) cells, macrophages, neutrophils, and brain cells. Its envelope glycoproteins bind to CD4 proteins and the virus enters the host cell by receptor-mediated endocytosis. Within the host cell, its reverse transcriptase synthesizes DNA using instructions encoded in the viral RNA. This is the reverse of the usual process of genetic transcription, which uses the DNA code to make RNA; viruses that go from RNA to DNA are called **retroviruses.**[18] The new DNA is inserted into the DNA of the host cell, where it may remain dormant for months to years. When activated, however, it causes the host cell to produce new viruses. These can shut down the cell's normal metabolism and cause it to be attacked and destroyed by immune cells. HIV also lyses host cells, releasing a multitude of viruses to invade new cells and repeat the process (fig. E.1b).

Pathogenesis—Development of Disease

HIV especially targets and destroys CD4 cells and thus suppresses the nonspecific defenses, humoral immunity, and cellular immunity normally coordinated by them. The incubation period—from the time of infection to the time of the first symptoms—can range from a few months to 12 years. Flulike episodes of chills and fever occur as HIV attacks CD4 cells. At first, antibodies against HIV are produced and the CD4 count returns nearly to normal. As the virus destroys more and more cells, however, some people experience a syndrome called **AIDS-related complex (ARC),** characterized by night sweats, fatigue, headache, extreme weight loss, and lymphadenitis.

Normally there are 600 to 1,200 CD4 cells per microliter of blood. In AIDS, by contrast, the CD4

—continued

---

17. Moritz Kaposi (1837–1902), Austrian physician

18. *retro* = backward

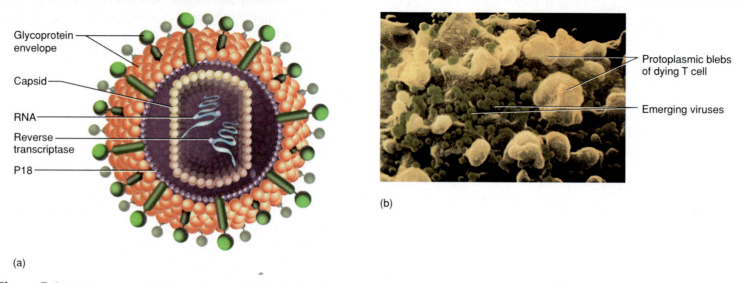

(a)

(b)

Figure E.1 (*a*) Structure of the human immunodeficiency virus (HIV). (*b*) Human immunodeficiency viruses emerging from an infected, dying helper T cell. Each virus can now invade a new helper T cell and produce a similar number of descendents.

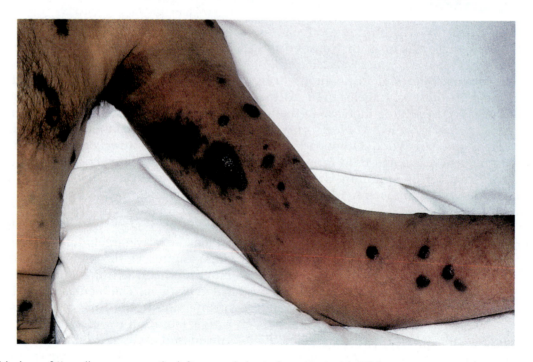

Figure E.2 Typical lesions of Kaposi's sarcoma on the left arm and chest of a patient with AIDS.

count is less than 200 cells/μL. With such severe depletion of CD4 cells, a person succumbs to opportunistic infections with *Pneumocystis, Toxoplasma* (a protozoan previously known mainly for causing birth defects), herpes simplex, cytomegalovirus (which can cause blindness), or tuberculosis bacteria. White patches may appear on the mucous membranes, a condition called *leukoplakia.* Kaposi sarcoma originates in the endothelial cells of the blood vessels and causes bruiselike purple lesions,

visible in the skin (fig. E.2). Patients with full-blown AIDS show no response to standard skin tests for delayed hypersensitivity. Slurred speech, loss of motor and cognitive functions, and dementia may occur as HIV invades the brain. Death from cancer or infection is inevitable, usually within a few months but sometimes as long as 8 years after diagnosis. Some people, however, have been diagnosed as HIV-positive and yet have survived for 10 years or longer without developing AIDS.

Transmission and Prevention

HIV is transmitted through blood, semen, vaginal secretions, and breast milk. It also occurs in saliva and tears, but it is not believed to be transmitted by those body fluids. It may be transmitted from mother to fetus through the placenta or from mother to infant during childbirth or nursing. Sexual intercourse, whether vaginal, anal, or oral; contaminated blood products; and drug injections with contaminated needles are other known means of transmission. Worldwide, about 75% of HIV infections are contracted through heterosexual, predominantly vaginal, intercourse. In the United States, homosexual men still account for the majority of cases, but drug use remains the chief means of transmission in urban ghettos. Adolescents are the fastest rising group of AIDS patients because of the increasing prevalence of unprotected sexual intercourse and exchange of sex for drugs.

AIDS is *not* known to be transmitted through casual contact—for example, to family members, friends, coworkers, classmates, or medical personnel in charge of AIDS patients. It is not transmitted by kissing. Despite some speculation and fear, it has not been found to be transmitted by mosquitoes or other blood-sucking arthropods. Many hemophiliacs became infected with HIV through blood transfusions before precautions with blood products were implemented in 1984, but all donated blood is now tested for HIV and risk of infection is less than 1%. HIV cannot be contracted by donating blood, but irrational fear has resulted in an alarming drop in blood donors.

HIV survives poorly outside the human body. It is destroyed by laundering, dishwashing, exposure to heat (50°C [135°F] for at least 10 minutes), chlorination of swimming pools and hot tubs, and disinfectants such as bleach, Lysol, hydrogen peroxide, rubbing alcohol, and germicidal skin cleansers (Betadine and Hibiclens, for example).

A properly used, undamaged latex condom is an effective barrier to HIV, but animal skin condoms are not. The spermicide nonoxynol-9 enhances protection.

Treatment

The AIDS epidemic has triggered an effort of unprecedented intensity to find a vaccine or cure. The strategies against HIV include efforts to prevent its binding to CD4 proteins, disrupting the action of reverse transcriptase, or inhibiting the assembly of new viruses or their release from host cells. HIV is a difficult pathogen to attack. Since it "hides" within host cells, it usually escapes recognition by the immune system. HIV in the brain is protected by the blood-brain barrier. Even when immune cells do become sensitized to it, HIV soon mutates and produces new surface antigens that escape recognition.

Until recently, the only drug approved by the Food and Drug Administration (FDA) was AZT (azidothymidine, or Retrovir), which inhibits reverse transcriptase and prolongs the lives of some HIV-positive individuals. AZT is now recommended for any patient with a CD4 count below 500 cells/μL. AZT damages the bone marrow, however, and causes anemia. The FDA has now approved other drugs, including ddI (dideoxyinosine) and ddC (dideoxycytidine) for patients who do not respond to AZT, but these drugs also may have severe side effects.

Another class of drugs—protease blockers—inhibit other enzymes (proteases) that HIV needs in order to replicate. Recently, a combination of two reverse transcriptase inhibitors with a protease inhibitor has been demonstrated to be highly effective at inhibiting viral replication. In addition, α interferon has been used with some success to inhibit viral replication and slow the progression of Kaposi sarcoma.

Efforts to develop a vaccine against subunits of the HIV envelope are currently underway, but there are two obstacles to this approach. One is the high mutation rate of HIV, which would quickly make today's vaccine ineffective against tomorrow's strain of virus. The other is the lack of animal models for AIDS research; most animals are not susceptible to HIV. The chimpanzee is an exception, but chimpanzees are difficult to maintain, and there are economic obstacles and ethical issues surrounding their use.

There remain not only these vexing clinical problems but also a number of unanswered questions about the basic biology of HIV. It remains unknown, for example, why there are such strikingly different patterns of heterosexual versus homosexual transmission in different countries and why some people succumb so rapidly to infection, while others are HIV-positive for years without symptoms.▲

Interactions Between the LYMPHATIC/IMMUNE SYSTEMS and Other Organ Systems

Nearly All Systems
The lymphatic system drains excess interstitial fluid and removes cellular debris and pathogens. The immune system provides defenses against pathogens and immune surveillance against cancer.

Integumentary System
- Skin provides mechanical barrier to pathogens; has antigen-presenting cells in epidermis and dermis; a common site of inflammation

Skeletal System
- Source of lymphocytes and macrophages; protects thymus and spleen

Muscular System
- Skeletal muscle pump moves lymph through lymphatic vessels; protects some lymph nodes and lymphatic vessels

Nervous System
- Neuropeptides and emotional states affect immune function

Endocrine System
- Hormones from thymus stimulate development of lymphatic organs and T lymphocytes; stress hormones depress immunity and increase susceptibility to infection
- Lymph transports some hormones

Circulatory System
- Blood plasma gives rise to lymph; lymphatic vessels develop from embryonic veins; arteries may aid flow of lymph in adjacent lymphatic vessels; WBCs serve as immune cells; bloodstream transports immune cells, antibodies, complement, interferon, and other immune chemicals; capillary endothelial cells signal areas of tissue injury and trigger margination and diapedesis; blood clotting restricts spread of pathogens
- Without return of fluid and protein by lymphatic system, cardiovascular system would quickly stop functioning; spleen disposes of old RBCs and recycles iron; lymphatic organs prevent accumulation of debris and pathogens in the blood

Respiratory System
- Provides immune cells with O_2 and removes CO_2; thoracic pump aids lymph flow; pharynx houses lymphoid organs (tonsils)
- Alveolar macrophages remove inhaled dust from lungs

Urinary System
- Eliminates wastes and maintains electrolyte balance needed for lymphoid/immune cell function; some pathogens flushed out of body in urine; acid pH of urine protects against urinary tract infection
- Lymphatic absorption of fluid and protein in kidneys essential for kidneys to produce concentrated urine

Digestive System
- Nourishes lymphatic system and affects lymph composition, especially that draining the intestines; stomach acid serves as barrier to pathogens
- Lymph absorbs and transports dietary lipids to bloodstream

Reproductive System
- Vaginal acidity inhibits spread of pathogens
- Immune system requires that testes have a blood-testis barrier to prevent autoimmune destruction of sperm

The Lymphatic System (pp. 755–764)
1. Functions of the lymphatic system
   a. Absorption of excess interstitial fluid
   b. Protection against pathogens
   c. Transport of dietary lipids
2. Lymph and the lymphatic vessels
   a. Composition of lymph
   b. Origin of lymph
      • Structure of lymphatic capillaries
      • Mechanism of fluid uptake
   c. Lymphatic vessels
      • Collecting vessels
      • Lymphatic trunks
      • Collecting ducts
   d. Flow of lymph
      • Contractions of lymphatic vessels
      • Skeletal muscle pump
      • Pulsation of adjacent arteries
      • Thoracic pump
3. Lymphatic tissue
   a. Diffuse lymphatic tissue
   b. Lymphatic nodules (follicles)
4. Lymph nodes
   a. Locations
   b. Structure
      • Size and shape
      • Capsule and trabeculae
      • Stroma and parenchyma
      • Cortex and medulla
      • Lymphatic nodules
      • Medullary sinuses
      • Afferent and efferent vessels
   c. Function
5. Tonsils
   a. Histology
   b. Three major sets
6. Thymus
   a. Location
   b. Age-related changes
   c. Histology
   d. Relationship to T cell development
7. Spleen
   a. Location
   b. Red and white pulp
   c. Role in erythrocyte production, storage, and disposal
   d. Roles in defense and immunity

Nonspecific Resistance (pp. 764–770)
1. Physical barriers
   a. Skin
   b. Mucous membranes
   c. Connective tissue gel
2. Chemical barriers
   a. Acids
   b. Lysozyme
3. Leukocytes and macrophages
   a. Leukocyte functions
   b. Macrophages
4. Inflammation
   a. Cardinal signs
   b. Causes of pain and functional impairment
   c. Causes of hyperemia, swelling, redness, and heat
   d. Leukocyte deployment
      • Margination, diapedesis, and chemotaxis
      • Leukocytosis-promoting factor
      • Involvement of basophils, eosinophils, and monocytes
   e. Tissue repair
5. Antimicrobial proteins
   a. Interferons
   b. Complement system
      • Classical pathway
      • Alternate pathway
      • Actions of complement proteins
        •• Enhanced inflammation
        •• Opsonization
        •• Cytolysis
6. Fever
   a. Beneficial effects
   b. Role of pyrogens and hypothalamus
   c. Stages
   d. Dangers

General Aspects of Specific Immunity (pp. 770–774)
1. General characteristics
   a. Specificity and memory
   b. Humoral and cellular immunity
2. Antigens
   a. Types of antigenic molecules
   b. Recognition of self and nonself
   c. Antigenic determinants
   d. Haptens
3. Antibodies
   a. Chemical nature and location
   b. Structure of antibody monomer
      • Heavy and light chains
      • Variable and constant regions
      • Antigen-binding sites
   c. Five antibody classes
4. Passive and active immunity
5. Lymphocytes
   a. T lymphocytes (T cells)
      • Development in thymus
      • Receptors and immunocompetence
      • Virgin lymphocyte pool
      • Clonal deletion and self-tolerance
   b. B lymphocytes (B cells)
      • Development in bursa equivalents
      • Receptors and immunocompetence
      • B cell diversity
      • Distribution in body
6. Antigen-presenting cells (APCs)
7. Interleukins
   a. Lymphokines
   b. Monokines

Humoral Immunity (pp. 775–779)
1. Recognition phase
   a. Capping of receptors
   b. Endocytosis of antigen
   c. Display of processed antigen
      • Role of MHC protein
      • Binding of helper T cell
      • Secretion of helper factors
   d. Clonal selection
   e. Plasma cell differentiation
      • Antibody synthesis
   f. Somatic recombination and antibody diversity
2. Attack phase
   a. Neutralization
   b. Complement fixation
   c. Agglutination
   d. Precipitation
3. Memory phase
   a. Primary and secondary responses
   b. Memory B cells

Cellular Immunity (pp. 779–784)
1. Lymphocytes involved
   a. Helper (CD4 or T_4) cells
   b. CD8 (T_8) cells
      • Cytotoxic (killer) T cells
      • Suppressor T cells
   c. Memory T cells
2. Recognition phase
   a. Antigen presentation
      • Antigen-presenting cells
      • MHC-I proteins and cytotoxic T cells
      • MHC-II proteins and helper T cells
      • MHC restriction
   b. T cell activation
      • CD4 and CD8 cell adhesion molecules
      • Role of protein kinase

- Costimulation by Il-1 and Il-2
- Clonal selection
3. Attack phase
  a. Role of helper T cells
     - Recognition of presented antigen
     - Secretion of lymphokines
     - Role of macrophage-activating factor
  b. Role of cytotoxic T cells
     - Docking and lethal hit
     - Other actions
       •• Lymphotoxin
       •• Tumor necrosis factor
       •• Interferon
       •• Macrophage-activating factor
       •• Migration-inhibiting factor
  c. Role of suppressor T cells
4. Memory phase
   a. Memory T cells
   b. Primary response and T cell recall response

Immune System Disorders (pp. 784–787)
1. Hypersensitivity (allergy)
   a. Type I (acute) reaction
      - Anaphylaxis
      - Asthma
      - Anaphylactic shock
   b. Type II (antibody-dependent cytotoxic) hypersensitivity
   c. Type III (immune complex) hypersensitivity
   d. Type IV (delayed) hypersensitivity
2. Autoimmune diseases
   a. Causes
      - Antigen cross-reactivity
      - Abnormal exposure of self-antigens
      - Changes in structure of self-antigens
   b. Examples
      - Rheumatic fever
      - Sterility from autoimmunity
      - Type I diabetes
3. Immunodeficiency diseases
   a. Severe combined immunodeficiency disease (SCID)
   b. Acquired immunodeficiency syndrome (AIDS)

Selected Vocabulary

lymphatic system 755
lymph 755
lymphatic capillary 755
collecting vessel 756
lymphatic trunk 757
collecting duct 757
right lymphatic duct 757
thoracic duct 757
cisterna chyli 757
diffuse lymphatic tissue 758
mucosa-associated lymphatic tissue (MALT) 758
lymphatic nodule 758
Peyer patches 760
lymph node 760
germinal center 760
afferent lymphatic vessel 760
efferent lymphatic vessel 760
tonsils 762
tonsillar crypts 762
palatine tonsil 762
lingual tonsil 762
pharyngeal tonsil 762
thymus 762
involution 762
thymopoietin 762
thymosin 762
spleen 762
red pulp 762
white pulp 762
nonspecific resistance 764
immunity 764
lysozyme 765
phagocyte 765
killing zone 765
degranulation 765

respiratory burst 765
natural killer (NK) cell 765
macrophage system 765
histiocyte 765
inflammation 765
cardinal signs 766
inflammatory chemicals 766
bradykinin 766
margination 767
diapedesis 767
chemotaxis 767
leukocytosis-promoting factor 767
pus 767
platelet-derived growth factor (PDGF) 768
interferon 768
complement system 768
classical pathway 768
complement fixation 768
alternate pathway 768
opsonization 768
cytolysis 768
fever 768
pyrogen 768
immune system 771
humoral immunity 771
cellular immunity 771
antigen (Ag) 772
antigenic determinant 772
hapten 772
antibody (Ab) 772
antibody monomer 772
heavy chain 772
light chain 772
variable region 772

antigen-binding site 772
constant region 773
IgA, IgD, IgE, IgG, IgM 773
passive immunity 773
active immunity 773
T lymphocyte 773
B lymphocyte 773
immunocompetent 774
clone 774
virgin lymphocyte pool 774
clonal deletion 774
self-tolerance 774
antigen-presenting cell (APC) 774
interleukin 774
lymphokine 774
monokine 774
capping 775
antigen display 775
MHC protein 775
immunological surveillance 775
helper factor 776
clonal selection 776
plasma cell 776
somatic recombination 777
neutralization 777
agglutination 778
precipitation 778
primary immune response 779
antibody titer 779
memory cell 779
secondary (anamnestic) immune response 779
cytotoxic T cell 780

helper T cell 780
suppressor T cell 780
memory T cell 780
MHC-I and MHC-II proteins 780
MHC restriction 780
costimulation 780
macrophage-activating factor (MAF) 781
lethal hit 781
perforin 781
lymphotoxin 782
tumor necrosis factor (TNF) 782
migration-inhibiting factor (MIF) 782
T cell recall response 783
hypersensitivity 784
allergen 784
type I hypersensitivity 784
anaphylaxis 785
asthma 785
anaphylactic shock 785
type II hypersensitivity 785
type III hypersensitivity 785
type IV hypersensitivity 785
autoimmune disease 785
autoantibody 785
severe combined immunodeficiency disease (SCID) 786
acquired immunodeficiency syndrome (AIDS) 787
human immunodeficiency virus (HIV) 787

1. The only lymphatic organ to have both afferent and efferent lymphatic vessels is
   a. the spleen.
   b. a lymph node.
   c. a tonsil.
   d. a Peyer patch.
   e. the thymus.

2. Which of the following cells are involved in nonspecific defense but not in specific immunity?
   a. helper T cells
   b. cytotoxic T cells
   c. natural killer cells
   d. B cells
   e. plasma cells

3. The respiratory burst is a mechanism used by _____ to kill bacteria.
   a. neutrophils
   b. basophils
   c. mast cells
   d. NK cells
   e. cytotoxic T cells

4. All the following are macrophages except
   a. microglia.
   b. Langerhans cells.
   c. reticular cells.
   d. histiocytes.
   e. mast cells.

5. The cytolytic action of the complement system most resembles the action of
   a. interleukin-1.
   b. platelet-derived growth factor.
   c. lymphotoxin.
   d. perforin.
   e. IgE.

6. _____ can become antigenic by binding to larger host molecules.
   a. Antigenic determinants
   b. Haptens
   c. Lymphokines
   d. Pyrogens
   e. Cell adhesion molecules

7. Which of the following correctly states the order of events in humoral immunity?
   a. capping–clonal deletion–antigen display–endocytosis–antibody secretion
   b. endocytosis–capping–antigen display–antibody secretion–clonal deletion
   c. capping–endocytosis–antigen display–clonal deletion–antibody secretion
   d. endocytosis–capping–antigen display–clonal deletion–antibody secretion
   e. antigen display–antibody secretion–clonal deletion–capping–endocytosis

8. Loss of self-tolerance would most likely result from a deficiency of
   a. immunocompetent B cells.
   b. MHC-II proteins.
   c. plasma cells.
   d. self-antigens.
   e. suppressor T cells.

9. A helper T cell can only bind to another cell that has
   a. MHC-II proteins.
   b. an antigenic determinant.
   c. an antigen-binding site.
   d. a complement-binding site.
   e. a CD4 protein.

10. Which of the following diseases results from a lack of self-tolerance?
    a. severe combined immunodeficiency disease
    b. acquired immunodeficiency disease
    c. systemic lupus erythematosus
    d. anaphylaxis
    e. asthma

11. Any organism or substance capable of causing disease is called a/an _____.

12. Mucous membranes contain an antibacterial surfactant called _____.

13. The pain of inflammation is due partly to a polypeptide called _____.

14. The migration of leukocytes through the capillary endothelium is called _____.

15. _____ is a process in which complement proteins coat bacteria and then serve as binding sites for phagocytes.

16. Any substance that triggers a fever is called a/an _____.

17. The hormonelike chemicals produced by lymphocytes to stimulate other leukocytes are called _____. Those produced by macrophages are called _____.

18. Part of an antibody molecule called its _____ binds to part of an antigen molecule called its _____.

19. Self-tolerance results from a process called _____, in which lymphocytes programmed to react against self-antigens are destroyed.

20. Wandering macrophages of the connective tissues are called _____.

1. Anti-D antibodies of an Rh⁻ woman sometimes cross the placenta and hemolyze the RBCs of an Rh⁺ fetus (see p. 659). Yet the anti-B antibodies of a type A mother do not affect the RBCs of a type B fetus. Explain this difference based on a knowledge of the five immunoglobulin classes.

2. In treating a woman for malignancy in the right breast, the surgeon removes some of her axillary lymph nodes. Following surgery, she experiences edema of the right upper extremity. Explain why.

3. A girl with a defective heart receives a new heart transplanted from another child who was killed in an accident. She is given an antilymphocyte serum containing antibodies against her lymphocytes. The transplanted heart is not rejected, but the girl dies of an overwhelming bacterial infection. Explain the intended effect of the antilymphocyte serum and the underlying reason for the girl's death.

4. A burn research center uses mice for studies of skin grafting. To prevent graft rejection, the mice are thymectomized at birth. Even though B cells do not develop in the thymus, these mice show no B cell (humoral) immune response and are very susceptible to bacterial infection. Explain why removal of the thymus would improve the success of skin grafts and why it would affect humoral immunity.

5. Contrast the structure of a B cell with that of a plasma cell and give a functional explanation of the difference.

Web Site Link

For a listing of the most current web sites related to this chapter, please visit the Saladin homepage at:

http://www.mhhe.com/sciencemath/biology/saladin/

The Respiratory System

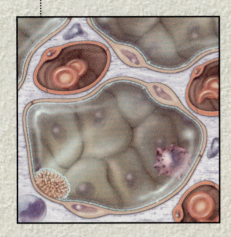

Special Topics

the unity of form and function

Brushing up

To understand this chapter, it is essential that you understand or brush up on the following concepts:

▶ Factors that affect simple diffusion (pp. 125–126)
▶ The muscles of respiration (p. 342)
▶ The structure of hemoglobin (p. 651)
▶ Principles of fluid pressure and flow (p. 691)
▶ Capillary filtration and reabsorption (pp. 720–722)
▶ Pulmonary blood circulation (pp. 726–727)

M ost metabolic processes of the body depend on ATP, and most ATP production requires oxygen and generates carbon dioxide as a waste product. Oxygen is supplied to the tissues and their carbon dioxide is removed by the combined action of the cardiovascular and respiratory systems. Not only do these two systems have a close spatial relationship in the thoracic cavity, they also have such a close functional relationship that they are often considered jointly under the heading *cardiopulmonary*. A disorder that affects the lungs has direct and pronounced effects on the heart, and vice versa. Furthermore, as discussed in the next two chapters, the respiratory system works closely with the urinary system to regulate the body's acid-base balance. Changes in the blood pH, in turn, trigger autonomic adjustments of the heart rate and blood pressure. Thus, the cardiovascular, respiratory, and urinary systems have an especially close physiological relationship. It is important that we now address the respiratory and urinary systems and their roles in the homeostatic control of blood gases, pH, blood pressure, and other variables related to the body fluids.

The term **respiration** has three meanings: (1) ventilation of the lungs (breathing), (2) the exchange of gases between air and blood and between blood and tissue fluid, and (3) the use of oxygen in cellular metabolism. In this chapter, we are concerned with the first two processes. Cellular respiration was introduced in chapter 3 and is considered more fully in chapter 26.

Anatomy of the Respiratory System

▼**Objectives**
When you have completed this section, you should be able to
• trace the flow of air from the nose to the pulmonary alveoli; and
• relate the function of any portion of the respiratory tract to its gross and microscopic anatomy.

The principal organs of the **respiratory system** are the nose, pharynx, larynx, trachea, bronchi, and lungs (fig. 22.1). These organs serve to draw in air, exchange gases with the blood, and expel the modified air. Within the lungs, air flows along a dead-end pathway consisting essentially of bronchi→bronchioles→alveoli (with some refinements to be introduced later). Incoming air stops in the alveoli, exchanges gases with the blood-

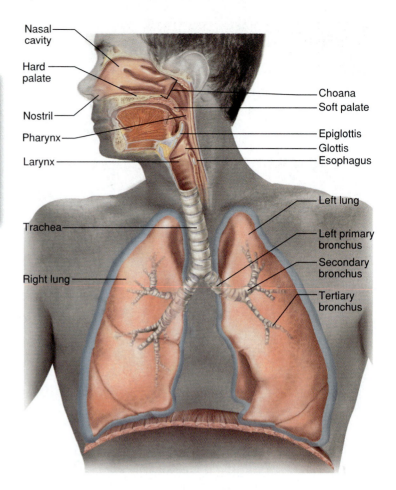

Figure 22.1 The respiratory system. ꭙ

stream across the alveolar wall, and then flows back out. The **conducting division** of the respiratory system consists of those passages that serve only for airflow, essentially from the nostrils through the bronchioles. The **respiratory division** consists of the alveoli and other distal gas-exchange regions. The airway from the nose through the pharynx is often called the **upper respiratory tract,** and the regions from the larynx through the lungs compose the **lower respiratory tract.**

The Nose

The **nose** has several functions: it warms, cleanses, and humidifies inhaled air; it detects odors in the airstream; and it serves as a resonating chamber that modifies the voice. The external, protruding part of the nose is supported and shaped by a framework of bone and cartilage. Its superior half is supported by the nasal bones medially and the maxillae laterally. The inferior half is supported by the **lateral** and **alar cartilages** (fig. 22.2). Dense fibroconnective tissue shapes the flared portion called the **ala nasi,** which forms the lateral wall of each nostril.

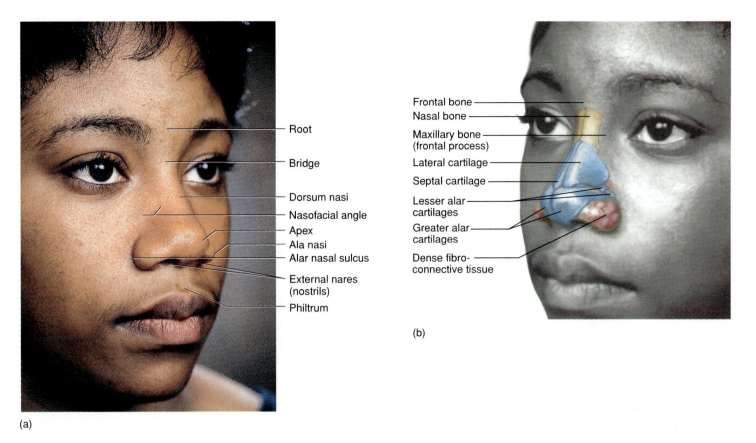

Labels for image (a):
- Root
- Bridge
- Dorsum nasi
- Nasofacial angle
- Apex
- Ala nasi
- Alar nasal sulcus
- External nares (nostrils)
- Philtrum

(a)

Labels for image (b):
- Frontal bone
- Nasal bone
- Maxillary bone (frontal process)
- Lateral cartilage
- Septal cartilage
- Lesser alar cartilages
- Greater alar cartilages
- Dense fibro-connective tissue

(b)

Figure 22.2 Anatomy of the nasal region. (*a*) External anatomy. (*b*) Connective tissues that shape the nose.

The **nasal cavity** (fig. 22.3) extends from the **anterior (external) nares** (NERR-eez) (singular, *naris*), or **nostrils,** to the **posterior (interior) nares,** or **choanae**[1] (co-AH-nee). The dilated chamber inside the ala nasi is called the **vestibule.** It is lined with stratified squamous epithelium and has stiff **vibrissae** (vy-BRISS-ee), or **guard hairs,** that block the inhalation of large particles.

The **nasal septum** divides the nasal cavity into right and left chambers called **nasal fossae** (FOSS-ee). The vomer forms the inferior part of the septum, the perpendicular plate of the ethmoid bone forms its superior part, and the *septal cartilage* forms its anterior part. The ethmoid and sphenoid bones compose the roof of the nasal cavity and the palate forms its floor. The palate separates the nasal cavity from the oral cavity, allowing you to breathe while there is food in your mouth.

There is little open space in a nasal fossa because its lateral wall gives rise to three folds of tissue—the **superior, middle,** and **inferior conchae**[2] (CON-kee)—that project toward the septum and occupy most of the fossa. They consist of mucous membranes supported by thin scroll-like **turbinate bones.** Beneath each concha is a narrow air passage called a **meatus** (me-AY-tus). The narrowness of these passages and the turbulence caused by the conchae ensure that most air contacts the mucous membrane on its way through, enabling the nose to cleanse, warm, and humidify it.

The *olfactory epithelium,* concerned with the sense of smell, lines the roof of the nasal fossa and extends over part of the septum and superior concha. The rest of the cavity is lined by ciliated pseudostratified *respiratory epithelium.* The nasal mucosa has an important defensive role. Goblet cells in the epithelium and glands in the lamina propria secrete a layer of mucus that traps inhaled particles. Bacteria are destroyed by lysozyme in the nasal mucus. Additional protection against bacteria is contributed by lymphocytes, which populate the lamina propria in large numbers, and by antibodies (IgE) secreted by plasma cells.

Cilia of the respiratory epithelium continually beat toward the posterior nares and drive debris-laden mucus into the pharynx to be swallowed and digested. Cold air inhibits ciliary motion; thus, in cold weather, some mucus drains from the anterior nares and produces a "runny nose." The four pairs of paranasal sinuses (see chapter 9) and the nasolacrimal ducts of the orbits drain into the nasal cavity and may also contribute to this cold-weather annoyance.

1. *choana* = funnel
2. *concha* = seashell

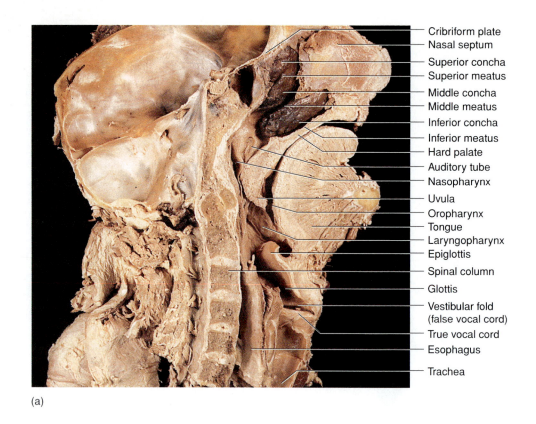

Cribriform plate
Nasal septum
Superior concha
Superior meatus
Middle concha
Middle meatus
Inferior concha
Inferior meatus
Hard palate
Auditory tube
Nasopharynx
Uvula
Oropharynx
Tongue
Laryngopharynx
Epiglottis
Spinal column
Glottis
Vestibular fold
(false vocal cord)
True vocal cord
Esophagus
Trachea

(a)

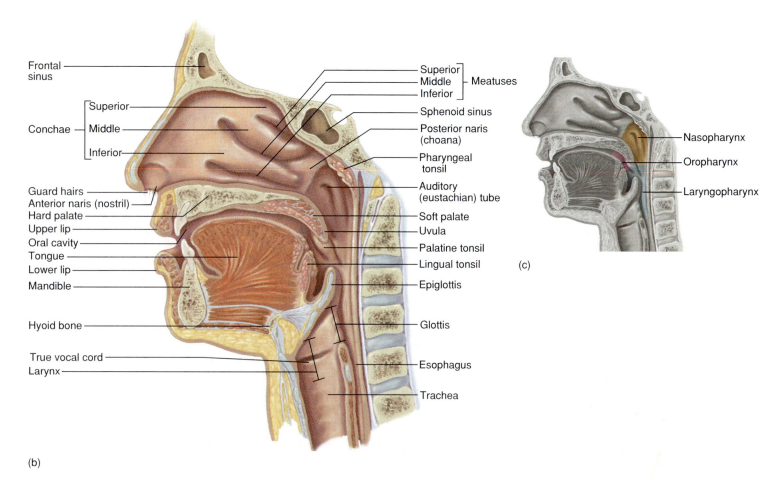

Frontal sinus

Conchae
— Superior
— Middle
— Inferior

Guard hairs
Anterior naris (nostril)
Hard palate
Upper lip
Oral cavity
Tongue
Lower lip
Mandible

Hyoid bone

True vocal cord
Larynx

Superior
Middle — Meatuses
Inferior

Sphenoid sinus
Posterior naris (choana)
Pharyngeal tonsil
Auditory (eustachian) tube
Soft palate
Uvula
Palatine tonsil
Lingual tonsil
Epiglottis

Glottis

Esophagus

Trachea

(b)

Nasopharynx
Oropharynx
Laryngopharynx

(c)

Figure 22.3 Anatomy of the upper respiratory tract. (*a*) Midsagittal section of the head. (*b*) Internal anatomy. (*c*) Regions of the pharynx.

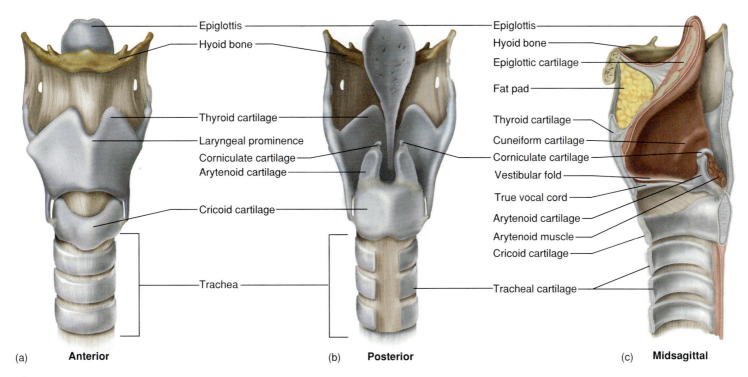

Figure 22.4 Anatomy of the larynx. (*a*) Anterior view. (*b*) Posterior view. (*c*) Midsagittal section, viewed from the left.

The lamina propria contains large blood vessels that help to warm the air. The inferior concha has an especially extensive venous plexus called the **erectile tissue** *(swell body)*. Every 30 to 60 minutes, the erectile tissue on one side becomes engorged with blood and restricts airflow through that fossa. Most air is then directed through the other naris and fossa, allowing the engorged side time to recover from drying. Thus the preponderant flow of air shifts between the right and left nares once or twice each hour. The inferior concha is the most common site of *epistaxis* (nosebleed). Spontaneous epistaxis is sometimes a sign of hypertension.

The Pharynx

The **pharynx** (FAIR-inks) is a muscular funnel extending about 13 cm (5 in.) from the choanae to the larynx. It has three regions: the *nasopharynx, oropharynx,* and *laryngopharynx* (fig. 22.3*c*).

The **nasopharynx,** which lies posterior to the soft palate, receives the auditory (eustachian) tubes from the middle ears and houses the pharyngeal tonsil. Inhaled air turns 90° downward as it passes through the nasopharynx. Dust particles larger than 10 *μ*m generally cannot make the turn because of their inertia. They col-

lide with the posterior wall of the nasopharynx and stick to the mucosa near the tonsil, which is well positioned to respond to airborne pathogens.

The **oropharynx** lies between the soft palate and hyoid bone and contains the palatine and lingual tonsils. Its anterior border is formed by the base of the tongue and the *fauces* (FAW-seez), the opening of the oral cavity into the pharynx. The nasopharynx and oropharynx meet at the level of the hyoid bone, where the soft palate ends, and form the **laryngopharynx** (la-RING-go-FAIR-inks), which extends to the larynx. The laryngopharynx marks the end of the upper respiratory tract. The nasopharynx passes only air and is lined by pseudostratified epithelium, whereas the oropharynx and laryngopharynx pass air, food, and drink and are lined by stratified squamous epithelium.

The Larynx

The **larynx** (LAIR-inks), or "voicebox" (figs. 22.4 and 22.5), is a cartilaginous chamber about 4 cm (1.5 in.) long. Its primary function is to keep food and drink out of the airway, but it has evolved the additional role of producing sound. In infants, the larynx is positioned above the pharynx, enabling them to swallow and

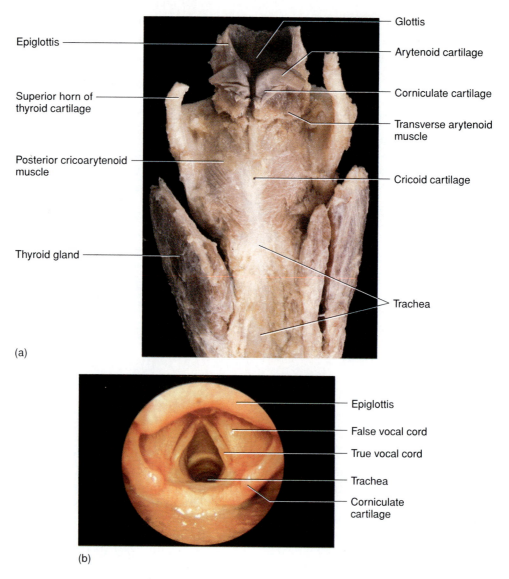

Figure 22.5 Photographs of the larynx. (a) Posterior view. (b) Superior view in a living person, as seen with a laryngoscope.

breathe at the same time. By age two, it descends to a position below the pharynx.

The superior opening of the larynx, the **glottis**,[3] marks the beginning of the lower respiratory tract. It is guarded by a flap of tissue called the **epiglottis.** During swallowing, *extrinsic muscles* of the larynx pull it upward toward the epiglottis, the tongue pushes the epiglottis downward to meet it, and the epiglottis directs food and drink into the esophagus behind the airway. The *vestibular folds* of the larynx, discussed shortly, play a greater role in keeping food and drink out of the airway, however. People who have had their epiglottis removed because of cancer do not choke any more than when it was present.

The framework of the larynx consists of nine cartilages. The first three are relatively large and unpaired. The most superior one, the **epiglottic cartilage,** is a spoon-shaped plate that supports the epiglottis. The largest, the **thyroid cartilage,** is named for its shieldlike shape. It has an anterior peak, the *laryngeal prominence,* commonly known as the Adam's apple. Testosterone stimulates the growth of this prominence, which is therefore significantly larger in males than in females. Below the thyroid cartilage is a ringlike **cricoid**[4] (CRY-coyd) **cartilage,** which connects the larynx to the trachea.

The remaining cartilages are smaller and occur in three pairs. Posterior to the thyroid cartilage are the two

---

3. *glottis* = back of the tongue

4. *crico* = ring + *oid* = resembling

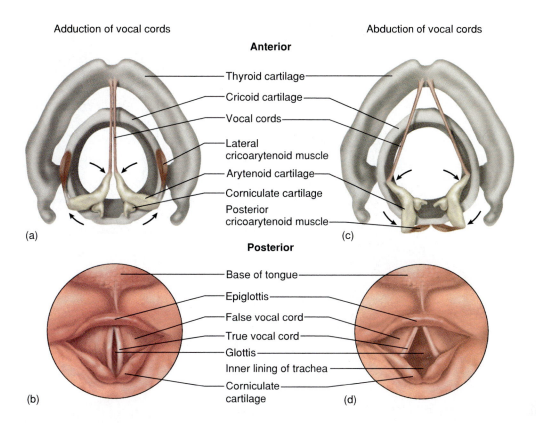

Adduction of vocal cords   Abduction of vocal cords

Anterior

- Thyroid cartilage
- Cricoid cartilage
- Vocal cords
- Lateral cricoarytenoid muscle
- Arytenoid cartilage
- Corniculate cartilage
- Posterior cricoarytenoid muscle

(a)   (c)

Posterior

- Base of tongue
- Epiglottis
- False vocal cord
- True vocal cord
- Glottis
- Inner lining of trachea
- Corniculate cartilage

(b)   (d)

Figure 22.6 Action of some of the intrinsic laryngeal muscles on the vocal cords; superior view. (*a*) Adduction of the vocal cords by the lateral cricoarytenoid muscles. (*b*) Laryngoscopic view of adducted vocal cords. (*c*) Abduction of the vocal cords by the posterior cricoarytenoid muscles. (*d*) Laryngoscopic view of the abducted vocal cords. ✗

arytenoid[5] (AR-ih-TEE-noyd) **cartilages,** and attached to their upper ends are a pair of little horns, the **corniculate**[6] (cor-NICK-you-late) **cartilages.** The arytenoid and corniculate cartilages function in speech, as explained shortly. A pair of **cuneiform**[7] (cue-NEE-ih-form) **cartilages** support the soft tissues between the arytenoids and the epiglottis. The epiglottic cartilage is elastic cartilage; all the others are hyaline.

The walls of the larynx are also quite muscular. The deep **intrinsic muscles** operate the vocal cords, and the superficial **extrinsic muscles** connect the larynx to the hyoid bone and elevate the larynx during swallowing. The extrinsic muscles, also called the *infrahyoid group,* are named and described in chapter 11.

The interior wall of the larynx has two folds on each side that stretch from the thyroid cartilage in front to the arytenoid cartilages in back. The superior pair, called the **vestibular folds** (*false vocal cords*) (fig. 22.5*b*), play no role in speech but close the glottis during swallowing. The inferior pair, the **true vocal cords,** produce sound when air passes between them. They are covered with stratified squamous epithelium, best suited to resist the stresses of vibration and contact between the cords.

The intrinsic muscles control the vocal cords by pulling on the corniculate and arytenoid cartilages, causing the cartilages to pivot. Depending on their direction of rotation, the arytenoid cartilages abduct or adduct the vocal cords (fig. 22.6). Air forced between the adducted vocal cords vibrates them, producing a high-pitched sound when the cords are relatively taut and a lower-pitched sound when they are more relaxed. In males, the vocal cords are longer and thicker, vibrate more slowly, and produce lower-pitched sounds than in females. Loudness is determined by the force of the air passing between the vocal cords. The crude sounds of the vocal cords are formed into words by actions of the pharynx, oral cavity, tongue, and lips.

The Trachea

The **trachea** (TRAY-kee-uh), or "windpipe," is a rigid tube about 12 cm (4.5 in.) long and 2.5 cm (1 in.) in diameter, lying anterior to the esophagus (fig. 22.7*a*). It is supported by 16 to 20 C-shaped rings of hyaline

5. *aryten* = ladle
6. *corni* = horn + *cul* = little + *ate* = possessing
7. *cune* = wedge + *form* = shape

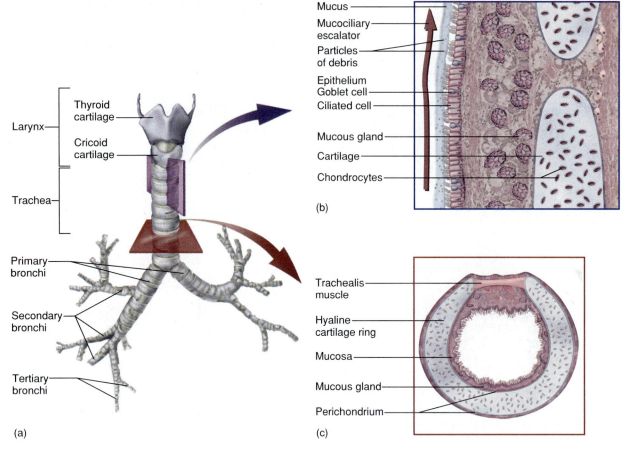

Figure 22.7 Anatomy of the lower respiratory tract. (*a*) Anterior view. (*b*) Longitudinal section of the trachea, showing the action of the mucociliary escalator. (*c*) Cross section of the trachea, showing the C-shaped tracheal cartilage. ✗

Labels for image:
- Larynx
  - Thyroid cartilage
  - Cricoid cartilage
- Trachea
- Primary bronchi
- Secondary bronchi
- Tertiary bronchi
- (a)
- Mucus
- Mucociliary escalator
- Particles of debris
- Epithelium
- Goblet cell
- Ciliated cell
- Mucous gland
- Cartilage
- Chondrocytes
- (b)
- Trachealis muscle
- Hyaline cartilage ring
- Mucosa
- Mucous gland
- Perichondrium
- (c)

Special Topic — Tracheostomy   22.1

If the airway is obstructed with secretions or foreign matter, it may be necessary to create a temporary opening in the trachea below the larynx and insert a tube to allow airflow—a procedure called **tracheostomy.** This prevents asphyxiation, but the inhaled air bypasses the nasal cavity and thus is not humidified. If the opening is left for long, the mucous membranes of the respiratory tract can dry out and become encrusted, interfering with the clearance of mucus from the tract and leading to severe infection. The functional importance of the nasal cavity becomes especially apparent in such cases.

cartilage, some of which you can palpate between your larynx and sternum. Like the wire spiral in a vacuum cleaner hose, the cartilage rings reinforce the trachea and keep it from collapsing when you inhale (see special topic 22.1). The open part of the C faces posteriorly, where it is spanned by a smooth muscle, the **trachealis** (fig. 22.7*c*). The gap in the C allows room for the esophagus to expand as swallowed food passes by. The trachealis muscles can contract or relax to adjust tracheal airflow.

The larynx and trachea are lined mostly by ciliated pseudostratified epithelium (figs. 22.7*b* and 22.8), which provides a **mucociliary escalator** for removal of debris that gets trapped in the mucus. The cilia beat up-ward and drive the debris-laden mucus to the pharynx, where it is swallowed.

The Lungs, Bronchial Tree, and Alveoli

Each **lung** (fig. 22.9) is a somewhat conical organ with a broad, concave **base** resting on the diaphragm and a blunt peak called the **apex (cupola)** projecting slightly above the clavicle. The broad **costal surface** is pressed against the rib cage, and the smaller concave **mediastinal surface** faces medially. The lung receives the bronchus, blood vessels, lymphatic vessels, and nerves through its **hilum**, a slit in the mediastinal surface (see fig. 22.27, p. 824). These structures entering the hilum constitute the **root** of the lung. Because the heart tilts to the left, the left lung is a little smaller than the right and has an indentation called the **cardiac notch** to accommodate it. The left lung has a **superior lobe** and an **inferior lobe** with a deep fissure between them; the right lung, by contrast, has three lobes (**superior, middle,** and **inferior**) separated by two fissures.

The Bronchial Tree

The lung has a spongy parenchyma containing the **bronchial tree** (fig 22.10), a highly branched system of air tubes extending from the primary bronchus to about 65,000 *terminal bronchioles.* Two **primary bronchi** (BRONK-eye) arise from the trachea at the level of the angle of the sternum. Each continues for 2 to 3 cm and enters the hilum of its respective lung. The right bronchus is slightly more vertical than the left; consequently, *aspirated* (inhaled) foreign objects lodge in the right bronchus more often than in the left. Like the trachea, the primary bronchi are supported by C-shaped hyaline cartilages. All divisions of the bronchial tree also have a substantial amount of elastic connective tissue, which is important in expelling air from the lungs.

After entering the hilum, the primary bronchus branches into one **secondary (lobar) bronchus** for each pulmonary lobe. Thus, there are two secondary bronchi in the left lung and three in the right.

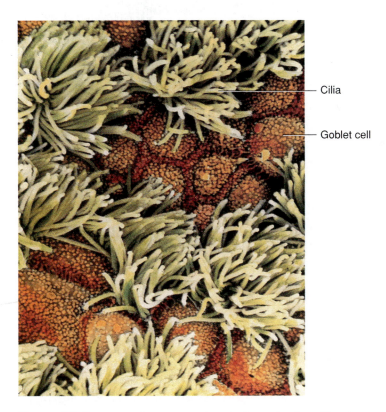

Figure 22.8 Surface of the tracheal epithelium, showing ciliated cells and nonciliated goblet cells. The small bumps on the goblet cells are microvilli. (Colorized SEM micrograph.)

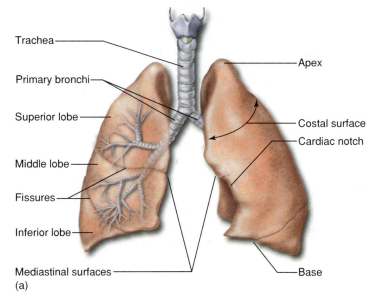

(a)

(b)

Figure 22.9 Anatomy of the lungs. (*a*) Anterior view. (*b*) Cross section through the thorax of a cadaver, showing the heart, lungs, and pleurae. ⚕

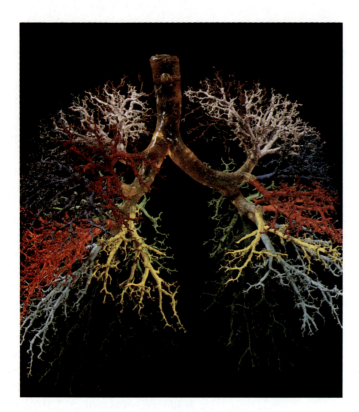

Figure 22.10 The bronchial trees. Each color identifies a bronchopulmonary segment supplied by a tertiary bronchus.

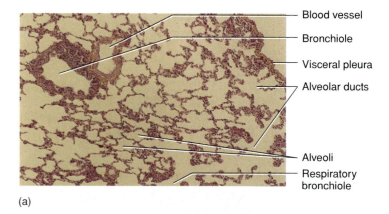

- Blood vessel
- Bronchiole
- Visceral pleura
- Alveolar ducts
- Alveoli
- Respiratory bronchiole

(a)

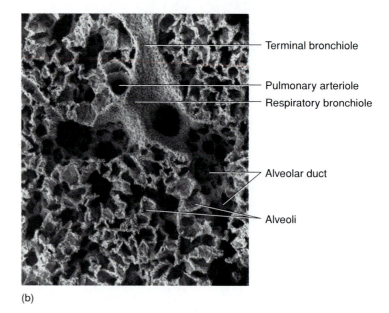

- Terminal bronchiole
- Pulmonary arteriole
- Respiratory bronchiole
- Alveolar duct
- Alveoli

(b)

Figure 22.11 Tissue of the lung. (*a*) Light micrograph (×100); (*b*) SEM micrograph. Note the spongy texture of the lung. ✗

Each secondary bronchus divides into **tertiary (segmental) bronchi**—10 in the right lung and 8 in the left. The portion of the lung supplied by each tertiary bronchus is called a **bronchopulmonary segment.** Secondary and tertiary bronchi are supported by overlapping plates of cartilage, not rings. Branches of the *pulmonary artery* closely follow the bronchial tree on their way to the alveoli. The bronchial tree itself is nourished by the *bronchial artery,* which arises from the aorta and carries systemic blood.

Bronchioles are continuations of the airway that are 1 mm or less in diameter and lack cartilage. A well-developed layer of smooth muscle in their walls enables them to dilate or constrict, as discussed later. Spasmodic contractions of this muscle at death cause the bronchioles to exhibit a wavy lumen in most histological sections. The portion of the lung ventilated by one bronchiole is called a **pulmonary lobule.**

Each bronchiole divides into 50 to 80 **terminal bronchioles,** the final branches of the conducting division. They measure 0.5 mm or less in diameter and have no mucous glands or goblet cells. They do have cilia, however, so that mucus draining into them from the higher passages can be driven back by the mucociliary escalator, preventing congestion of the terminal bronchioles and alveoli.

Each terminal bronchiole gives off two or more smaller **respiratory bronchioles,** which mark the beginning of the respiratory division. They have scanty smooth muscle and the smallest of them are nonciliated. Each respiratory bronchiole divides into 2 to 10 elongated, thin-walled passages called **alveolar ducts** that end in irregularly shaped spaces called **alveolar sacs** (fig. 22.11). Alveoli bud from the walls of the respiratory bronchioles, alveolar ducts, and alveolar sacs. The presence of alveoli defines the respiratory division.

The epithelium of the bronchial tree is pseudostratified in the bronchi, simple cuboidal in the bronchioles, and simple squamous in the alveolar ducts, sacs, and alveoli. It is ciliated except in the distal reaches of the respiratory bronchioles and beyond.

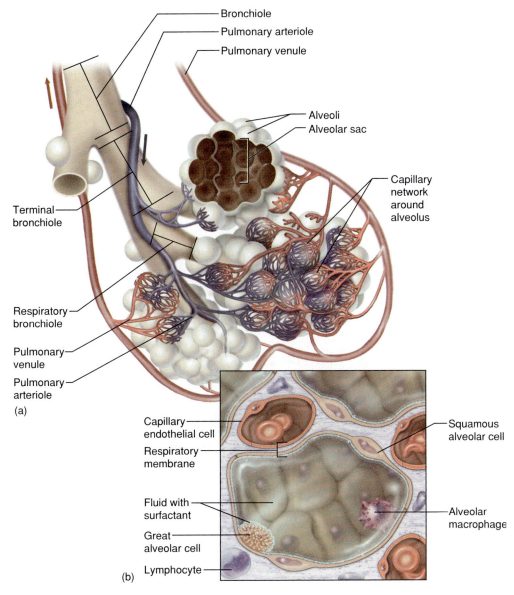

An **alveolus** (AL-vee-OH-lus) (fig. 22.12) is a pouch about 0.2 to 0.5 mm in diameter. Its wall consists predominantly of **squamous (type I) alveolar cells**—thin cells that allow for rapid gas diffusion between the alveolus and bloodstream. About 5% of the alveolar cells are round to cuboidal **great (type II) alveolar cells.** They secrete a detergent-like lipoprotein called **pulmonary surfactant,** which forms a thin film on the insides of the alveoli and bronchioles. Its function is discussed later.

Alveolar macrophages (dust cells) wander the lumens of the alveoli and the connective tissue between them. They are the last line of defense against inhaled matter. Particles measuring 10 μm or larger in diameter are usually strained out by the nasal vibrissae or trapped in the mucus of the upper respiratory tract. Most particles 2 to 10 μm in diameter are trapped in the mucus of the bronchi and bronchioles, where the airflow is relatively slow, and then removed by the mucociliary escalator. Many particles smaller than 2 μm, however, make their way into the alveoli, where they are phagocytized by the macrophages. In lungs that are infected or bleeding, the macrophages also phagocytize bacteria and loose blood cells. Alveolar macrophages greatly outnumber all other cell types in the lung; as many as 50 million of them perish each day as they ride up the mucociliary escalator to be swallowed.

Each alveolus is surrounded by a basket of blood capillaries supplied by the pulmonary artery. The barrier between the alveolar air and blood, called the **respiratory membrane,** consists only of the squamous type I alveolar cell, the squamous endothelial cell of the capillary, and their fused basement membranes.

The pulmonary circulation has very low blood pressure. In alveolar capillaries, the mean blood pressure is 10 mmHg and the oncotic pressure is 25 mmHg. The osmotic uptake of water thus overrides filtration and keeps

Figure 22.12 (*a*) The distal end of the airway, with bronchioles, alveoli, and associated blood vessels. (*b*) Structure of an alveolus. ⚡

Alveoli

The functional importance of human lung structure is best appreciated by comparison to the lungs of a few other animals. In frogs and other amphibians, the lung is a simple hollow sac lined with blood vessels. This is sufficient to meet the oxygen needs of animals with relatively low metabolic rates. Mammals, with their high metabolic rates, could never have evolved with such a simple lung. Rather than consisting of one large sac, each human lung is a spongy mass composed of 150 million little sacs, the alveoli, providing about 70 square meters of surface for gas exchange.

the alveoli free of fluid. The lungs also have a more extensive lymphatic drainage than any other organ in the body.

The Pleurae

The surface of the lung is covered by a moist serous membrane, the **visceral pleura** (PLOOR-uh), which extends into the fissures. At the hilum, the pleura turns back on itself and forms the **parietal pleura,** which adheres to the mediastinum, superior surface of the diaphragm, and inner surface of the rib cage (see fig. 22.9*b*). An extension of the parietal pleura, the *pulmonary ligament,* extends from the base of each lung to the diaphragm. The space between the parietal and visceral pleurae is called the **pleural cavity.** The two membranes are normally separated only by a film of slippery **pleural fluid;** thus, the pleural cavity is only a *potential space.*

The pleurae and pleural fluid have three functions:

1. **Reduction of friction.** Pleural fluid acts as a lubricant that enables the lungs to expand and contract with a minimum of friction. In some forms of *pleurisy,* the pleurae are dry and inflamed and each breath gives painful testimony to the function that the fluid should be serving.
2. **Creation of pressure gradient.** Pressure in the pleural cavity is lower than atmospheric pressure; as explained later, this assists in inflation of the lungs.
3. **Compartmentalization.** The pleurae, mediastinum, and pericardium compartmentalize the thoracic organs and prevent infections of one organ from spreading easily to neighboring organs.

Think About It

In what ways do the structure and function of the pleurae resemble the structure and function of the pericardium?

Key Point Review

1. A dust particle is inhaled and gets into an alveolus without being trapped along the way. Describe the path it takes, naming all air passages from external naris to alveolus. What would happen to it after arrival in the alveolus?
2. Describe the histology of the epithelium and lamina propria of the nasal cavity and the functions of the cell types present.
3. Describe the roles of the intrinsic muscles, corniculate cartilages, and arytenoid cartilages in speech.
4. Contrast the epithelium of the bronchioles with that of the alveoli and explain how the structural difference is related to functional differences.

Mechanics of Ventilation

▼Objectives

When you have completed this section, you should be able to
- explain how pressure differences between the atmosphere, pulmonary alveoli, and pleural cavity account for the flow of air in and out of the lungs;
- explain how the respiratory muscles produce these pressure differences;
- explain how pulmonary compliance and elasticity contribute to ventilation;
- explain the role of pulmonary surfactant; and
- define terms used to quantify and describe pulmonary ventilation.

A resting adult breathes 10 to 15 times per minute, inhaling about 500 mL of air during **inspiration** and exhaling it again during **expiration**. In this section, we examine the muscular actions and pressure gradients that produce this airflow.

Pressure and Flow

Airflow is governed by the same principles of flow, pressure, and resistance that govern blood flow (see chapter 20). The pressure that drives respiration is **atmospheric (barometric) pressure**—the weight of the air above us. At sea level, a column of air as thick as the atmosphere (60 mi) and 1 in. square weighs 14.7 lb; it is thus said to exert a force of 14.7 pounds per square inch (psi). In Standard International units, this is a column of air 100 km high exerting a force of 1.013×10^6 dynes/cm^2. This pressure, called *1 atmosphere* (1 atm), is enough to force a column of mercury 760 mm up an evacuated tube; therefore, 1 atm = 760 mmHg. This is the average atmospheric pressure at sea level; it fluctuates from day to day and is lower at higher altitudes.

One way to change the pressure of a gas, and thus to change flow, is to change the volume of its container. **Boyle's**[8] **law** states that *the pressure of a given quantity of gas is inversely proportional to its volume* (assuming a constant temperature). If the lungs contain a quantity of gas and their volume is increased, it reduces their **intrapulmonary pressure**—the pressure within the alveoli. If their volume is decreased, intrapulmonary pressure rises. (Compare this to the syringe analogy on p. 692). To make air flow into the lungs, it is necessary only to lower the intrapulmonary pressure below the atmospheric pressure. Raising the intrapulmonary pressure above the atmospheric pressure makes air flow out again. These changes are created as skeletal muscles of the thoracic and abdominal walls change the volume of the thoracic cavity.

---

8. Robert Boyle (1627–91), English physicist

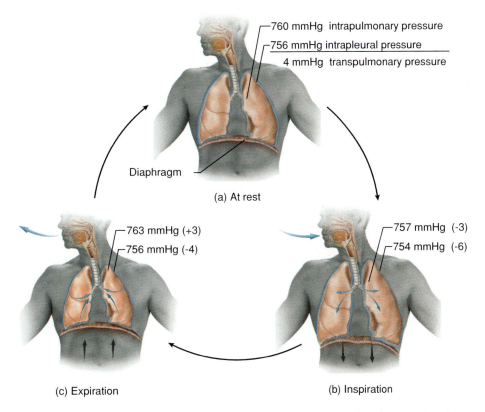

760 mmHg intrapulmonary pressure
756 mmHg intrapleural pressure
4 mmHg transpulmonary pressure

Diaphragm

(a) At rest

763 mmHg (+3)
756 mmHg (-4)

(c) Expiration

757 mmHg (-3)
754 mmHg (-6)

(b) Inspiration

Figure 22.13 The cycle of pressure changes causing ventilation of the lungs. The pressures given here are based on an assumed atmospheric pressure of 760 mmHg (1 atm). Values in parentheses are relative to atmospheric pressure. ⚕

What matters to flow is the *difference* between atmospheric pressure and intrapulmonary pressure. Since atmospheric pressures vary from one place and time to another, it is more useful for our discussion to refer to relative pressures. A relative pressure of −3 mmHg, for example, means 3 mmHg below atmospheric pressure; a relative pressure of +3 mmHg is 3 mmHg above atmospheric pressure. If the atmospheric pressure were 760 mmHg, these would represent absolute pressures of 757 and 763 mmHg, respectively.

Inspiration

The mechanical process of pulmonary ventilation is achieved by rhythmically changing the pressure in the thoracic cavity so that it drops slightly below atmospheric pressure, drawing air in, and then rises slightly above atmospheric pressure, expelling this air. This is achieved mainly by the dome-shaped diaphragm. When stimulated by the phrenic nerves, the diaphragm flattens somewhat and drops about 1.5 cm in quiet respiration (as much as 7 cm in deep breathing). This enlarges the thoracic cavity and reduces its pressure. In addition, the intercostal nerves stimulate the external intercostal muscles to contract. When the scalenes fix the first pair of ribs, the external intercostals elevate the second through twelfth pairs like bucket handles (see p. 342),

making them swing slightly up and out. Forced (deep) inspiration is aided by the pectoralis minor, sternocleidomastoid, and erector spinae muscles.

As the rib cage expands and the diaphragm drops, the parietal pleura clings to them. The **intrapleural pressure** between the parietal and visceral pleurae is about 4 mmHg less than atmospheric pressure at rest (fig. 22.13a) and drops to −6 mmHg during inspiration. Because of the surface tension of the pleural fluid, the visceral pleura clings to the parietal pleura, like two sheets of wet paper. Thus, when the parietal pleura is pulled outward, the visceral pleura and lung are pulled along with it, causing the lung to expand. Not all the pressure change in the pleural cavity is transferred to the interior of the lungs, but the intrapulmonary pressure drops to about −3 mmHg. The difference between the intrapleural and intrapulmonary pressures, called **transpulmonary pressure,** is a critical force in inflating the lung. In the case under discussion, this would be −6 − (−3) = −3 mmHg. At an atmospheric pressure of 760 mmHg, intrapleural pressure would be 754 mmHg and intrapulmonary pressure would be 757 mmHg. With a gradient of 760 → 757 mmHg from atmospheric to intrapulmonary pressure, air moves into the lungs, and the transpulmonary gradient of 757 → 754 mmHg helps the lungs expand in the enlarging thoracic cavity.

Another force that expands the lungs is warming of the inhaled air. **Charles'[9] law** states that *the volume of a given quantity of gas is directly proportional to its absolute temperature*. On a day when the ambient temperature is 21°C (70°F), inhaled air is 16°C warmer (37°C) by the time it reaches the alveoli. As the inhaled air expands, it helps to inflate the lungs.

When the respiratory muscles stop contracting, the inflowing air quickly achieves an intrapulmonary pressure equal to atmospheric pressure, and flow stops. The dimensions of the thoracic cage increase by only a few millimeters in each direction, but this is enough to increase its total volume by 500 mL. Thus, 500 mL of air flows into the respiratory tract during quiet breathing.

Inspiration does more than fill the lungs. Descent of the diaphragm increases pressure in the abdominal cavity. Taking a deep breath, holding it by closing the glottis, and contracting the abdominal muscles—the **Valsalva[10] maneuver**—aids in childbirth and the expulsion of urine and feces. Inspiration also aids the flow of blood and lymph from abdominal to thoracic vessels, as described in earlier chapters.

Expiration

It takes a muscular effort to inhale, and therefore an expenditure of ATP and calories. By contrast, normal expiration during quiet breathing is an energy-saving passive process that requires no muscular contraction. It is achieved by the elasticity of the lungs and thoracic cage—the tendency to return to their original dimensions when released from tension. The bronchial tree has a substantial amount of elastic connective tissue in its walls. The attachments of the ribs to the spine and sternum, and the tendons of the diaphragm and other respiratory muscles, also have a degree of elasticity that causes them to spring back when muscular contraction ceases. As these structures recoil, the thoracic cage diminishes in size. In accordance with Boyle's law, this raises the intrapulmonary pressure; it peaks at about +3 mmHg and expels air from the lungs (fig. 22.13c).

When inspiration ceases, the phrenic nerves continue to stimulate the diaphragm for a little while longer. This produces a slight braking action that prevents the lungs from recoiling too suddenly and makes the transition from inspiration to expiration smoother. In relaxed breathing, inspiration usually lasts about 2 seconds and expiration about 3 seconds.

To force a deeper expiration, the internal intercostal muscles contract and depress the ribs. The abdominal muscles (internal and external obliques, transversus abdominis, and rectus abdominis) contract and raise the intra-abdominal pressure, forcing the viscera and diaphragm upward and putting pressure on the thoracic cavity. Intrapulmonary pressure rises as high as 20 to 30 mmHg, causing faster and deeper evacuation of the lungs. Abdominal control of expiration is important in singing and public speaking.

The effect of pulmonary elasticity is evident in a pathological state of pneumothorax and atelectasis. **Pneumothorax** is the presence of air in the pleural cavity. If the thoracic wall is punctured, for example, air can be sucked through the wound into the pleural cavity during inspiration, separating the visceral and parietal pleurae. Without the negative intrapleural pressure to keep the lungs inflated, elastic recoil of the lungs causes **atelectasis[11]** (AT-eh-LEC-ta-sis)—the collapse of a lung or part of it. Atelectasis can also result from airway obstruction—for example, by a lung tumor, aneurysm, swollen lymph node, or aspirated object. Blood absorbs gases from the alveoli distal to the obstruction, and that part of the lung collapses because it cannot be reventilated. Diseases that reduce pulmonary elasticity interfere with expiration, as we will see in the discussion of emphysema.

Resistance to Airflow

In discussing blood circulation, we noted that flow = pressure/resistance ($F = P/R$). We now turn to the role of resistance in airflow. One factor that affects resistance is **pulmonary compliance**—the distensibility of the lungs, or ease with which they expand. More exactly, compliance means the change in lung volume relative to a given change in transpulmonary pressure. The lungs normally inflate with ease, but compliance can be reduced by airway obstructions or degenerative lung diseases that cause pulmonary fibrosis. In such conditions, the thoracic cage expands normally and transpulmonary pressure falls, but the lungs expand relatively little.

Another factor that governs resistance to airflow is the diameter of the bronchioles. Bronchioles are analogous to arterioles in that their large number, small diameters, and ability to change diameter make them the primary means of controlling resistance. Their smooth muscle allows for considerable **bronchoconstriction** and **bronchodilation**—changes in diameter that reduce or increase airflow, respectively. Bronchoconstriction can be triggered by airborne irritants, cold air, parasympathetic stimulation, or histamine. Many people have died of extreme bronchoconstriction in episodes of asthma and anaphylactic shock. Sympathetic stimulation and epinephrine produce bronchodilation. Epinephrine

---

9. Jacques A. C. Charles (1746–1823), French physicist
10. Antonio Maria Valsalva (1666–1723), Italian anatomist

11. *atel* = imperfect, incomplete + *ectasis* = extension

inhalants were widely used in the past to halt asthma attacks, but they have been replaced by more selective drugs.

Alveolar Surface Tension

Yet another factor that resists inspiration and promotes expiration is the surface tension of the water in the alveoli and distal bronchioles. Although the alveoli are relatively dry, they have a thin film of water over the epithelium that creates a potential problem for both inspiration and expiration. Water molecules are mutually attracted by hydrogen bonds, creating a surface tension that draws the walls of the alveoli inward toward the lumen. If this went unchecked, the alveoli would collapse with each expiration and would strongly resist reinflation.

The smaller the diameter of an alveolus, the more crowded the water molecules become, increasing their surface tension and the likelihood of alveolar collapse. This is summarized by the **law of Laplace**[12]—*the force drawing the alveolus in on itself is directly proportional to surface tension and inversely proportional to the radius of the alveolus.* If F = force, T = surface tension, and r = radius, this law can be summarized $F = 2T/r$. If surface tension remained the same and r decreased as an alveolus deflated, the force acting on the alveolar wall would increase; the smaller the alveolus became, the more it would tend to collapse.

The solution to this problem takes us back to the great alveolar cells and their surfactant. A *surfactant* is an agent that disrupts the hydrogen bonds of water and reduces surface tension; soaps and detergents are everyday examples. Pulmonary surfactant spreads over the alveolar epithelium and up the alveolar ducts and smallest bronchioles. As these passages contract during expiration, the surfactant molecules are pushed closer together; as the local concentration of surfactant increases, it exerts a stronger effect. Therefore, as alveoli shrink during expiration, surface tension decreases to nearly zero; the numerator ($2T$) of the fraction decreases when the denominator (r) does, and thus there is no increase in force (F) favoring alveolar collapse. The importance of this surfactant is especially apparent when it is lacking. Premature infants often have a deficiency of pulmonary surfactant and experience great difficulty breathing (see chapter 29).

Alveolar Ventilation

Air that actually enters the alveoli becomes available for gas exchange, but not all inhaled air gets that far. About 150 mL of it fills the conducting division of the airway. Since this air cannot exchange gases with the blood, it is called **dead air,** and the conducting division is called the **anatomic dead space.** In pulmonary diseases, some alveoli may be unable to exchange gases with the blood because of lack of blood flow to those alveoli or a thickening of the pulmonary membrane by edema. **Physiologic (total) dead space** is the sum of anatomic dead space and any pathological alveolar dead space that may exist. In healthy people, few alveoli are nonfunctional, and the anatomic and physiologic dead spaces are identical.

In a state of relaxation, the parasympathetic nervous system stimulates bronchoconstriction. This minimizes the dead space so that more of the inhaled air ventilates the alveoli. In a state of arousal, the sympathetic nervous system stimulates bronchodilation, which increases airflow. The rate of airflow then takes priority over the amount of air that is "wasted" by filling the increased dead space.

If a person inhales 500 mL of air and 150 mL of it is dead air, then 350 mL of air ventilates the alveoli. Multiplying this by the respiratory rate gives the **alveolar ventilation rate (AVR)**—for example, 350 mL/breath × 12 breaths/min = 4,200 mL/min. Of all measures of pulmonary ventilation, this one is most directly relevant to the body's ability to get oxygen to the tissues and dispose of carbon dioxide.

Nonrespiratory Air Movements

Some air movements serve purposes other than ventilating the alveoli. These include speaking, expressing emotion (laughing, crying), yawning, hiccuping, expelling noxious fumes, coughing, and sneezing. Coughing is triggered by irritants in the lower respiratory tract. To cough, the glottis is closed and the muscles of expiration contract, producing high pressure in the lower respiratory tract. The glottis is then suddenly opened, releasing an explosive burst of air at speeds over 900 km/hr (600 mi/hr). This drives mucus and foreign matter toward the pharynx and mouth. Sneezing is triggered by irritants in the nasal cavity. Its mechanism is similar to coughing except that the glottis is continually open, the soft palate and tongue block the flow of air while thoracic pressure builds, and then the uvula is depressed to direct part of the airstream through the nose. These actions are coordinated by coughing and sneezing centers in the medulla oblongata.

Measurements of Ventilation

Clinical measurements of ventilation are often made by having a subject breathe from a device called a **spirometer**[13] (fig. 22.14), which recaptures the expired breath and records such variables as the rate and depth of

---

12. Pierre Simon, Marquis de Laplace (1749–1827), French astronomer and mathematician

13. *spiro* = breath + *meter* = measuring device

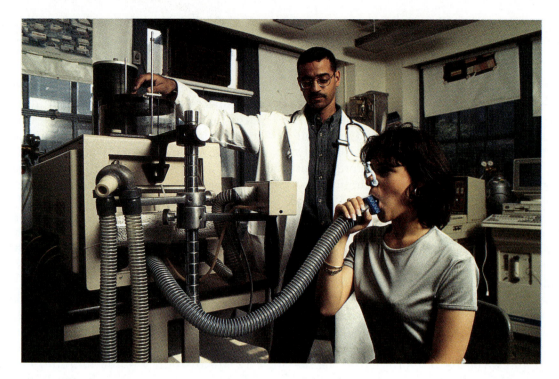

Figure 22.14 Measuring a person's pulmonary ventilation with the use of a spirometer. ⊀

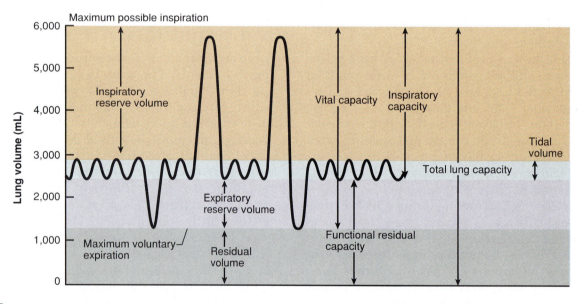

Figure 22.15 Lung volumes and capacities. The wavy line, like a spirogram, indicates inspiration when it rises and expiration when it falls. Compare table 22.1. ⊀

breathing, speed of expiration, and rate of oxygen consumption. Four spirometric measures are called **respiratory volumes: tidal volume, inspiratory reserve volume, expiratory reserve volume,** and **residual volume.** Four others, called **respiratory capacities,** are obtained by adding two or more of the respiratory volumes: **vital capacity, inspiratory capacity, functional residual capacity,** and **total lung capacity.** Defini-

tions and representative values for these are given in table 22.1 and figure 22.15.

In general, respiratory volumes and capacities are proportional to body size; consequently, they are generally lower for women than for men. Some values change with age because the lungs become less compliant and the respiratory muscles weaken. Vital capacity then declines, residual volume rises, and the

Table 22.1   Respiratory Volumes and Capacities for an Average Young Adult Male

Respiratory Volumes

Measurement	Value	Definition
Tidal volume (TV)	500 mL	Amount of air inhaled or exhaled in one breath during relaxed, quiet breathing
Inspiratory reserve volume (IRV)	3,000 mL	Amount of air in excess of tidal inspiration that can be inhaled with maximum effort
Expiratory reserve volume (ERV)	1,200 mL	Amount of air in excess of tidal expiration that can be exhaled with maximum effort
Residual volume (RV)	1,200 mL	Amount of air remaining in the lungs after maximum expiration; keeps alveoli inflated between breaths and mixes with fresh air on next inspiration

Respiratory Capacities

Measurement	Value	Definition
Vital capacity (VC)	4,700 mL	Amount of air that can be exhaled with maximum effort after maximum inspiration (ERV + TV + IRV); used to assess strength of thoracic muscles as well as pulmonary function
Inspiratory capacity (IC)	3,500 mL	Maximum amount of air that can be inhaled after a normal tidal expiration (TV + IRV)
Functional residual capacity (FRC)	2,400 mL	Amount of air remaining in the lungs after a normal tidal expiration (RV + ERV)
Total lung capacity (TLC)	5,900 mL	Maximum amount of air the lungs can contain (RV + VC)

lungs are ventilated with less fresh air in a given time. Regular aerobic exercise slows this degeneration by maintaining the strength of the respiratory muscles.

The measurement of respiratory volumes and capacities is important in assessing the severity of a respiratory disease and monitoring improvement or deterioration in a patient's pulmonary functioning. **Restrictive disorders** of the respiratory system, such as pulmonary fibrosis, reduce compliance and vital capacity. **Obstructive disorders** do not reduce respiratory volumes but reduce the speed of airflow. This is measured by having the subject exhale as rapidly as possible into a spirometer and measuring **forced expiratory volume (FEV)**—the percentage of the vital capacity that can be exhaled in a given time interval. A healthy adult should be able to expel 75% to 85% of the vital capacity in 1.0 second (called the $FEV_{1.0}$). Significantly lower values may indicate obstruction of the airway by mucus, a tumor, or bronchoconstriction (as in asthma).

The amount of air inhaled per minute is called the **minute respiratory volume (MRV)**. It can be measured directly with a spirometer or obtained by multiplying tidal volume by respiratory rate. For example, if a person has a tidal volume of 500 mL per breath and a rate of 12 breaths per minute, his or her MRV would be $500 \times 12 = 6,000$ mL/min. During heavy exercise, MRV may be as high as 125 to 170 L/min—this is called **maximum voluntary ventilation (MVV)**, formerly called *maximum breathing capacity.*

Table 22.2   Clinical Terminology of Ventilation

Eupnea[14] (yoop-NEE-uh)	Normal, relaxed, quiet breathing; typically 500 mL/breath, 12 to 15 breaths/min
Dyspnea[15] (DISP-nee-uh)	Labored, gasping breathing; shortness of breath
Apnea (AP-nee-uh)	Temporary cessation of breathing (one or more skipped breaths)
Respiratory Arrest	Permanent cessation of breathing (unless there is medical intervention)
Hyperpnea (HY-purp-NEE-uh)	Increased rate and depth of breathing in response to exercise, pain, or other conditions
Hyperventilation	Increased pulmonary ventilation in excess of metabolic demand, frequently associated with anxiety; expels CO_2 faster than it is produced, thus lowering the blood CO_2 concentration
Hypoventilation	Reduced pulmonary ventilation; leads to an increase in blood CO_2 concentration if ventilation is insufficient to expel CO_2 as fast as it is produced

Patterns of Breathing

Some variations in the rhythm of breathing are defined in table 22.2. You should familiarize yourself with these terms before proceeding further in this chapter.

---

14. *eu* = easy, normal + *pnea*
15. *dys* = difficult, abnormal, painful

5 Name the major muscles and nerves involved in inspiration.

6 Relate the action of the respiratory muscles to Boyle's law.

7 Explain the relevance of compliance and elasticity to pulmonary ventilation and describe some conditions that reduce compliance and elasticity.

8 Explain how pulmonary surfactant and the law of Laplace relate to compliance.

9 Define *vital capacity*. Express it in terms of a formula and define each of the variables.

Neural Control of Ventilation

▼Objectives

When you have completed this section, you should be able to

- name the brainstem centers that regulate respiration and describe their locations and functions;
- contrast the neural pathways for voluntary and automatic control of the major respiratory muscles; and
- describe the stimuli and afferent pathways to the brainstem nuclei that modify the respiratory rhythm.

The heartbeat and breathing are the two most conspicuously rhythmic processes in the body. The heart has an internal pacemaker and goes on beating even if all nerves to it are severed. Breathing, by contrast, depends on repetitive stimulation from the brain. There are two reasons for this: (1) Skeletal muscles do not contract without nervous stimulation. (2) Breathing involves the coordinated action of multiple muscles and thus requires a central coordinating mechanism to ensure that they all work together.

This section describes the central nervous system mechanisms that regulate pulmonary ventilation. Neurons in the medulla oblongata and pons provide automatic control of unconscious breathing, whereas neurons in the motor cortex of the cerebrum provide voluntary control.

Control Centers in the Medulla Oblongata

The medulla oblongata contains **inspiratory (I) neurons,** which fire during inspiration, and **expiratory (E) neurons,** which fire during forced expiration (but not during eupnea). Fibers from these neurons travel down the spinal cord and synapse with lower motor neurons in the cervical to thoracic regions. The phrenic and intercostal nerves carry signals from the lower motor neurons to the diaphragm and intercostal muscles, respectively. No pacemaker neurons have been found that are analogous to the autorhythmic cells of the heart, and the exact mechanism for setting the rhythm of respiration remains unknown despite intensive research.

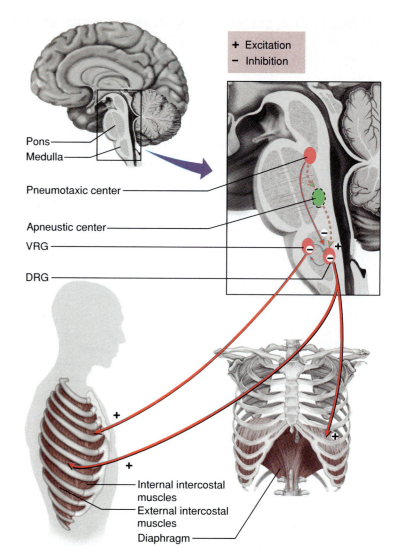

+ Excitation
− Inhibition

Pons
Medulla
Pneumotaxic center
Apneustic center
VRG
DRG

Internal intercostal muscles
External intercostal muscles
Diaphragm

Figure 22.16 The respiratory control centers of the brainstem and some of their efferent connections to the respiratory muscles. The apneustic center and its connections, shown in green, are hypothetical.

The medulla has two respiratory nuclei (fig. 22.16). One of them, called the **inspiratory center,** or **dorsal respiratory group (DRG),** is composed primarily of I neurons. These neurons stimulate the muscles of inspiration. The more frequently they fire, the more motor units are recruited and the more deeply you inhale. If they fire longer than usual, each breath is prolonged and the respiratory rate is slower. When they stop firing, elastic recoil of the lungs and thoracic cage produces passive expiration.

The other medullary nucleus is called the **expiratory center,** or **ventral respiratory group (VRG).** It has I neurons in its midregion and E neurons at its anterior and posterior ends. It is not involved in eupnea, but its I neurons inhibit the inspiratory center when deeper expiration is needed. Conversely, the inspiratory center inhibits the expiratory center when an unusually deep inspiration is needed.

Control Centers in the Pons

The pons regulates ventilation by means of a *pneumotaxic center* in the upper pons and possibly an *apneustic center* in the lower pons. The **pneumotaxic center** sends a continual stream of impulses to the inspiratory center of the medulla. When impulse frequency rises, inspiration lasts as little as 0.5 second and the breathing thus becomes faster and shallower. Conversely, when impulse frequency declines, breathing is slower and deeper, with inspiration lasting as long as 5 seconds.

The role of the **apneustic** (ap-NEW-stic) **center** is still hypothetical. It produces no noticeable effect except when experimenters sever the vagus nerves and the connection from the pneumotaxic center to the medulla. Then, signals from the apneustic center to the inspiratory center greatly prolong inspiration, causing exaggerated filling of the lungs interrupted only by occasional, brief expirations. The normal function of the apneustic center remains unclear.

Think About It

Do you think the fibers from the pneumotaxic center produce EPSPs or IPSPs at their synapses in the inspiratory center? Explain.

Afferent Connections to the Brainstem Nuclei

The brainstem respiratory centers receive input from the limbic system, hypothalamus, chemoreceptors, and the lungs themselves. Input from the limbic system and hypothalamus allow pain and emotions to affect respiration—for example, in gasping, crying and laughing. Anxiety often triggers an uncontrollable bout of hyperventilation. This expels CO_2 from the body faster than it is produced. As blood CO_2 levels drop, the cerebral arteries constrict, perfusion of the brain drops, and dizziness and fainting may result. Hyperventilation can be brought under control by having a person rebreathe the expired CO_2 from a paper bag.

Chemoreceptors monitor the O_2 level, CO_2 level, and pH of the blood. They transmit signals to the brainstem that enable pulmonary ventilation to be adjusted to keep these variables within homeostatic limits. Chemoreceptors are discussed more extensively later in the chapter.

The vagus nerves transmit sensory signals from the respiratory system to the inspiratory center of the medulla. Irritants in the airway, such as smoke, dust, noxious fumes, or mucus, stimulate these vagal afferent fibers. The medulla then returns signals that result in bronchoconstriction or coughing. Stretch receptors in the bronchial tree and visceral pleura monitor inflation of the lungs. Excessive inflation triggers the **inflation (Hering–Breuer**[16]**) reflex,** a protective somatic reflex that strongly inhibits the I neurons and stops inspiration. In infants, this may be a normal mechanism of transition from inspiration to expiration, but after infancy it is activated only by extreme stretching of the lungs.

Voluntary Control

Although breathing usually occurs automatically, without our conscious attention, we obviously can hold our breath, take a deep breath, and control ventilation while speaking or singing. This control originates in the motor cortex of the frontal lobe of the cerebrum, which sends impulses down the corticospinal tracts to the respiratory neurons in the spinal cord, bypassing the brainstem respiratory centers.

There are limits to voluntary control. Temperamental children may threaten to hold their breath until they die, but it is impossible to do so. Holding the breath causes the level of carbon dioxide in the blood to rise and the level of oxygen to fall. A *breaking point* is finally reached where automatic controls override volition, forcing breathing to resume even if a person has lost consciousness (see special topic 22.2).

16. Heinrich Ewald Hering (1866–1948), German physiologist; Josef Breuer (1842–1925), Austrian physician

10 Which of the brainstem respiratory nuclei is (are) indispensable to respiration? What do the other nuclei do?

11 Where do voluntary respiratory commands originate? What pathways do they take to the respiratory muscles?

Gas Exchange and Transport

▼Objectives

When you have completed this section, you should be able to

- explain what is meant by partial pressure and state Dalton's law;
- contrast the composition of inspired air and alveolar air;
- discuss how partial pressure affects the amount of gas that dissolves in the blood;
- explain how the blood transports O_2 and CO_2;
- describe the factors that govern gas exchange in the lungs and systemic capillaries; and
- explain how gas exchange is adjusted to the metabolic needs of different tissues.

We now consider the stages in which oxygen is obtained from inspired air and delivered to the tissues, while carbon dioxide is removed from the tissues and released into the air to be expired. First, however, it is necessary to understand the composition of air and the behavior of gases in contact with water.

Composition of Air

Air is a mixture of gases, each of which contributes a share of the total atmospheric pressure called its **partial pressure** (table 23.3). Partial pressure is abbreviated P with the formula of the gas as a subscript. The partial pressure of nitrogen is P_{N_2}, for example. Nitrogen constitutes about 78.6% of the atmosphere; thus at 1 atm of pressure, $P_{N_2} = 78.6\% \times 760$ mmHg = 597 mmHg. **Dalton's[17] law** states that *the total pressure of a gas mixture is the sum of the partial pressures of the individual gases.* That is, $P_{N_2} + P_{O_2} + P_{CO_2} + P_{H_2O} = 597.0 + 159.0 + 3.7 + 0.3 = 760.0$ mmHg. These partial pressures are important because they determine the rate of diffusion of a gas and therefore strongly affect the rate of gas exchange between the blood and alveolar air.

Alveolar air can be sampled with an apparatus that collects the last 10 mL of expired tidal air. Its gaseous makeup differs from that of the atmosphere because of three influences: (1) humidification by the airway, (2) exchanges of oxygen and carbon dioxide with the blood, and (3) the mixing of freshly inspired air with residual air from the previous respiratory cycle. These factors produce the composition shown in table 22.3.

Table 22.3 Composition of Inspired (Atmospheric) and Alveolar Air

Gas	Inspired Air*		Alveolar Air	
N_2	78.62%	597.0 mmHg	74.9%	569.0 mmHg
O_2	20.84%	159.0 mmHg	13.6%	104.0 mmHg
H_2O	0.50%	3.7 mmHg	6.2%	47.0 mmHg
CO_2	0.04%	0.3 mmHg	5.3%	40.0 mmHg
Total	100.00%	760.0 mmHg	100.0%	760.0 mmHg

*Typical values for a cool clear day; will vary with temperature and humidity. Other gases, present in small amounts, are disregarded.

Think About It

Expired air, considered as a whole (not just the last 10 mL), contains about 116 mmHg O_2 and 32 mmHg CO_2. Why do you think these values differ from the values for alveolar air?

The Air-Water Interface

When air and water are in contact with each other, as in the pulmonary alveolus, gases diffuse down their concentration gradients until the partial pressure of each gas in the air is equal to its partial pressure in the water. If a gas is more abundant in the water than in the air, it diffuses into the air; the smell of chlorine near a swimming pool is evidence of this. If it is more abundant in the air, it diffuses into the water.

Henry's[18] law states that *at the air-water interface, for a given temperature, the amount of gas that dissolves in the water is determined by its solubility in water and its partial pressure in the air* (fig. 22.17). Thus, the greater the P_{O_2} in the alveolar air, the more O_2 dissolves in the blood. And, since the blood arriving at an alveolus has a higher P_{CO_2} than alveolar air, the blood releases CO_2 into the air. At the alveolus, the blood is said to *unload* CO_2 and *load* O_2. Each gas in a mixture behaves independently; the diffusion of one gas does not influence the diffusion of another.

Alveolar Gas Exchange

Alveolar gas exchange is a process of O_2 loading and CO_2 unloading in the lungs. Since both processes depend on erythrocytes (RBCs), their efficiency depends on how long an erythrocyte spends in an alveolar capillary compared to how long it takes for O_2 and CO_2 to reach equilibrium concentrations in the capillary blood. It takes only 0.25 second for the gases to equilibrate, whereas, at rest, it takes an RBC about 0.75 second to pass through an

---

17. John Dalton (1766–1844), British chemist

18. William Henry (1774–1836), British chemist

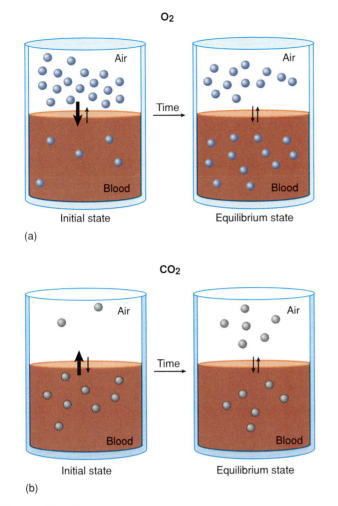

O₂

Air

Blood

Initial state

Air

Blood

Equilibrium state

Time

(a)

CO₂

Air

Blood

Initial state

Air

Blood

Equilibrium state

Time

(b)

Figure 22.17 The relationship of Henry's law to gas exchange in the pulmonary alveoli. (*a*) As blood arrives at the alveolus, alveolar P_{O_2} is initially higher than blood P_{O_2}. Oxygen diffuses into the blood until the two are in equilibrium. (*b*) Blood P_{CO_2} is initially higher than alveolar P_{CO_2}. Carbon dioxide diffuses into the alveolus until the two are in equilibrium. It takes about 0.25 second for both gases to reach equilibrium.

alveolar capillary. Even in vigorous exercise when the blood is flowing faster, the transit time for an RBC is at least 0.3 second. Therefore, even at the fastest blood flow, an RBC spends enough time in a capillary to load as much O_2 and unload as much CO_2 as it possibly can.

The following factors especially affect the efficiency of alveolar gas exchange:

- **Concentration gradients of the gases.** The P_{O_2} is about 104 mmHg in the alveolar air and 40 mmHg in the blood arriving at an alveolus. Oxygen therefore diffuses from the air into the blood, reaching a P_{O_2} of 104 mmHg in the bloodstream. Before the blood leaves the lung, however, this drops to about 95 mmHg because blood in the pulmonary veins receives some oxygen-poor blood from the bronchial veins by way of anastomoses. The P_{CO_2} is about 46 mmHg in the arriving venous blood and 40 mmHg

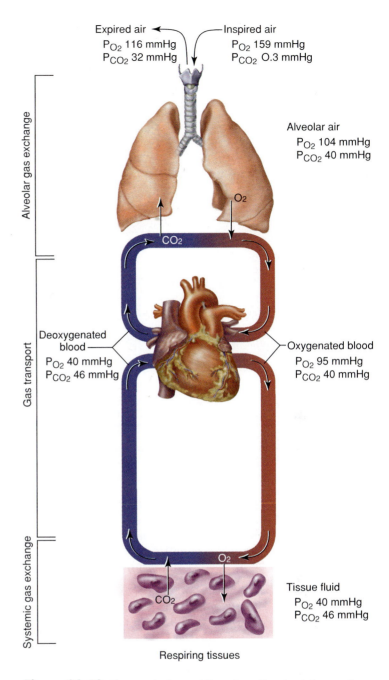

Expired air
P_{O_2} 116 mmHg
P_{CO_2} 32 mmHg

Inspired air
P_{O_2} 159 mmHg
P_{CO_2} 0.3 mmHg

Alveolar air
P_{O_2} 104 mmHg
P_{CO_2} 40 mmHg

Alveolar gas exchange

CO_2

O_2

Gas transport

Deoxygenated blood
P_{O_2} 40 mmHg
P_{CO_2} 46 mmHg

Oxygenated blood
P_{O_2} 95 mmHg
P_{CO_2} 40 mmHg

Systemic gas exchange

CO_2

O_2

Tissue fluid
P_{O_2} 40 mmHg
P_{CO_2} 46 mmHg

Respiring tissues

Figure 22.18 Changes in P_{O_2} and P_{CO_2} along the circulatory route. Trace the partial pressure of oxygen from inspired air to expired air and explain each change in P_{O_2} along the way. Do the same for P_{CO_2}.

in the alveolar air. Carbon dioxide therefore diffuses from the blood to the alveoli. These changes are summarized here and at the top of figure 22.18.

Blood Entering Lungs		**Blood Leaving Lungs**	
P_{O_2}	40.0 mmHg	P_{O_2}	95.0 mmHg
P_{CO_2}	46.0 mmHg	P_{CO_2}	40.0 mmHg

These gradients differ under special circumstances such as high altitudes and in *hyperbaric oxygen therapy* (treatment with oxygen at greater than 1 atm of pressure) (fig. 22.19). At high altitudes, the partial

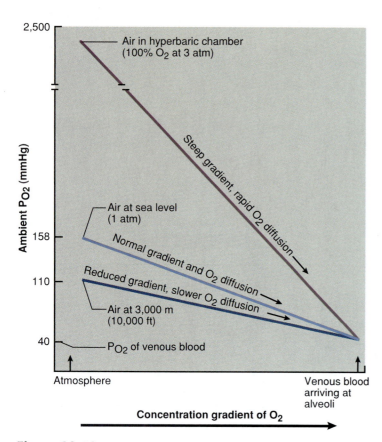

Figure 22.19 The rate of oxygen loading depends on the steepness of the gradient from alveolar air to the venous blood arriving at the alveolar capillaries. Compared to the oxygen gradient at sea level, the gradient is less at high altitude because the P_{O_2} of the atmosphere is lower. Thus oxygen loading of the pulmonary blood is slower. In a hyperbaric chamber with 100% oxygen, the gradient from air to blood is very steep and oxygen loading is correspondingly rapid. This is an illustration of Henry's law and has important effects in diving, aviation, mountain climbing, and oxygen therapy.

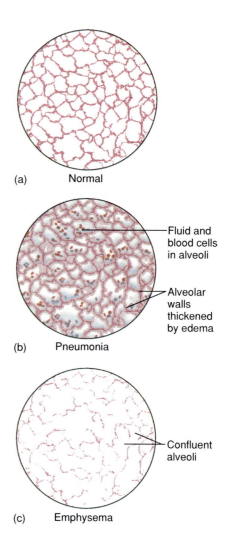

Figure 22.20 Comparison of the pulmonary alveoli in (a) a healthy lung, with small alveoli and thin respiratory membranes; (b) pneumonia, with thickened, edematous respiratory membranes and blood cells and fluid in the alveoli; and (c) emphysema, in which alveolar membranes break down and neighboring alveoli join to form larger, fewer alveoli with less total surface area.

pressures of all atmospheric gases are lower. Atmospheric P_{O_2}, for example, is 159 mmHg at sea level and 110 mmHg at 3,000 m (10,000 ft). The O_2 gradient from air to blood is proportionately less, and as we can predict from Henry's law, less O_2 diffuses into the blood. In a hyperbaric oxygen chamber, by contrast, a patient is exposed to 3 to 4 atmospheres of oxygen to treat such conditions as gangrene (to kill the anaerobic bacteria that cause it) and carbon monoxide poisoning (to displace the carbon monoxide from hemoglobin). The P_{O_2} ranges between 2,300 and 3,000 mmHg. Thus, there is a very steep gradient of P_{O_2} from alveolus to blood and diffusion into the blood is accelerated.

- **Solubility of the gases.** Gases differ in their ability to dissolve in water. Carbon dioxide is about 20 times as soluble as oxygen, and oxygen is about twice as soluble as nitrogen. Even though the concentration gradient of O_2 is much greater than that of CO_2 across the respiratory membrane, equal amounts of the two gases are exchanged because CO_2 is so much more soluble and diffuses more rapidly.

- **Membrane thickness.** The respiratory membrane between the blood and alveolar air is only 0.5 μm thick in most places—much thinner than the 7 to 8 μm diameter of a single RBC. Thus, it presents little obstacle to diffusion (fig. 22.20a). In such heart conditions as left ventricular failure, however, blood pressure backs up into the lungs and promotes capillary filtration into the connective tissues, causing the respiratory membranes to become edematous and thickened (fig. 22.20b). The gases have farther to travel between blood and air and cannot equilibrate fast enough to keep pace with blood flow. Blood leaving the lungs therefore has a relatively high P_{CO_2} and low P_{O_2}.

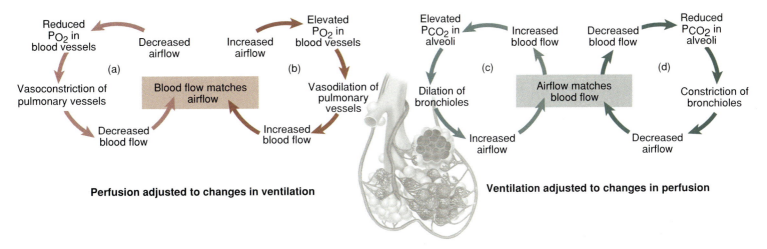

Perfusion adjusted to changes in ventilation

Ventilation adjusted to changes in perfusion

Figure 22.21 Negative feedback loops in the adjustment of ventilation-perfusion ratio. (*a*) Adjustment of perfusion to reduced ventilation. (*b*) Adjustment of perfusion to increased ventilation. (*c*) Adjustment of ventilation to increased perfusion. (*d*) Adjustment of ventilation to reduced perfusion.

- **Membrane area.** In good health, each lung has about 70 m² of respiratory membrane available for gas exchange. Since the alveolar capillaries of the lungs contain a total of only 100 mL of blood at any one time, this blood is spread very thinly. (Imagine trying to paint an area the size of a tennis court evenly with only 100 mL of paint!) Several pulmonary diseases, however, decrease the alveolar surface area and thus lead to hypoxemia—for example, lung cancer, tuberculosis, and emphysema (fig. 22.20*c*).
- **Ventilation-perfusion coupling.** Gas exchange not only requires good ventilation of the alveolus but also good perfusion of its capillaries. As a whole, the lungs have a *ventilation-perfusion ratio* of about 0.8—a flow of 4.2 L of air and 5.5 L of blood per minute (at rest). The ratio is somewhat higher in the apex of the lung and lower in the base because more blood is drawn toward the base by gravity. *Ventilation-perfusion coupling* is the ability to adjust, or match, ventilation and perfusion to each other (fig. 22.21). If part of a lung is poorly ventilated because of tissue destruction or airway obstruction, there is little point in directing the usual amount of blood there. This blood would leave the lung carrying less oxygen than it should. But poor ventilation causes local constriction of the pulmonary arteries, reducing blood flow to that area and redirecting this blood to better ventilated alveoli. Good ventilation, by contrast, dilates the arteries and increases perfusion. This is opposite to the reactions of systemic arteries, where hypoxia causes vasodilation. Ventilation is also adjustable. Poor ventilation causes local CO_2 accumulation, which stimulates local bronchodilation and improves airflow. Low P_{CO_2} causes local bronchoconstriction.

Gas Transport

Gas transport is the process of carrying gases from the alveoli to the systemic tissues and vice versa. Here we examine how blood loads and transports oxygen and carbon dioxide.

Oxygen

Systemic arterial blood normally carries about 20 mL of O_2 per dL of blood. About 98.5% of this is bound to hemoglobin and 1.5% is dissolved in the blood plasma. Hemoglobin consists of four protein (globin) chains, each with one heme group (see fig. 18.10). Each heme group can bind 1 O_2 to the ferrous ion at its center; thus, one hemoglobin molecule can carry up to 4 O_2. If even one molecule of O_2 is bound to hemoglobin, the compound is called **oxyhemoglobin (HbO$_2$).** When hemoglobin is 100% saturated, every molecule of it carries 4 O_2; if it is 75% saturated, there is an average of 3 O_2 per hemoglobin molecule; if it is 50% saturated, there is an average of 2 O_2 per hemoglobin; and so forth. The poisonous effect of carbon monoxide stems from its competition for the same binding site as O_2 (see special topic 22.3).

The relationship between percent hemoglobin saturation and P_{O_2} is shown by an **oxyhemoglobin dissociation curve** (fig. 22.22). As you can see, it is not a simple linear relationship. At low P_{O_2}, the curve rises slowly; then there is a rapid increase in oxygen loading as P_{O_2} rises further; finally, at high P_{O_2}, the curve levels off as the hemoglobin approaches 100% saturation. This reflects the way hemoglobin loads oxygen. When the first heme group binds a molecule of O_2, hemoglobin

The lethal effect of carbon monoxide (CO) is well known. This colorless, odorless gas occurs in cigarette smoke, engine exhaust, and fumes from furnaces and space heaters. It binds to the ferrous ion of hemoglobin to form **carboxyhemoglobin, HbCO.** Not only does it occupy the oxygen-binding site and make it unavailable to transport oxygen, but it binds 210 times as tightly as oxygen. Thus, CO tends to tie up hemoglobin for a long time. Nonsmokers usually have less than 1.5% HbCO. Residents of heavily polluted cities, however, may have up to 3% and heavy smokers up to 10%. An atmospheric concentration of 0.1% CO, as in a closed garage, is enough to bind 50% of a person's hemoglobin, and an atmospheric concentration of 0.2% is quickly lethal.

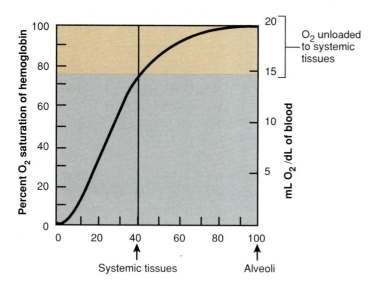

Figure 22.22 The oxyhemoglobin dissociation curve, relating the oxygen saturation of hemoglobin to the ambient partial pressure of oxygen. Hemoglobin becomes saturated as it passes through the alveolar capillaries and typically unloads about 22% of its oxygen as it passes through the systemic capillaries, where the P_{O_2} of the tissue fluids is lower. ✗

changes shape in a way that facilitates uptake of the second O_2 molecule by another heme group. This, in turn, promotes the uptake of the third and then the fourth O_2—hence the rapidly rising midportion of the curve.

Carbon Dioxide

Carbon dioxide is transported in three forms:

1. About 70% of it is hydrated (reacts with water) to form **carbonic acid,** which then dissociates into bicarbonate and hydrogen ions:

$$CO_2 + H_2O \rightarrow H_2CO_3 \rightarrow HCO_3^- + H^+$$

More will be said about this reaction shortly.

2. About 23% binds to the amino groups of plasma proteins and hemoglobin, forming **carbamino compounds**—chiefly, **carbaminohemoglobin (HbCO$_2$).** The reaction with hemoglobin can be summarized Hb + CO$_2$ → HbCO$_2$. Carbon dioxide does not compete with oxygen because CO_2 and O_2 bind to different sites on the hemoglobin molecule—oxygen to the heme moiety and CO_2 to the polypeptide chains. Hemoglobin can therefore transport both O_2 and CO_2 simultaneously. As we will see, however, each somewhat inhibits transport of the other.

3. The remaining 7% of the CO_2 is carried in the blood as dissolved gas, like the CO_2 in soda pop.

Systemic Gas Exchange

Systemic gas exchange is the unloading of O_2 and loading of CO_2 at the systemic capillaries (see fig. 22.18 [*bottom*] and fig. 22.23).

Carbon Dioxide Loading

Aerobic respiration produces a molecule of CO_2 for every molecule of O_2 it consumes. The tissue fluid therefore contains a relatively high P_{CO_2} and there is typically a CO_2 gradient of 46 → 40 mmHg from tissue fluid to blood. Consequently, CO_2 diffuses into the bloodstream, where it is carried in the three forms noted (fig. 22.23). Most of it reacts with water, producing bicarbonate (HCO$_3^-$) and hydrogen (H$^+$) ions. This reaction occurs slowly in the blood plasma but much faster in the RBCs, where it is catalyzed by the enzyme *carbonic anhydrase.* In an exchange called the **chloride shift,** most of the HCO$_3^-$ diffuses out of the RBCs in exchange for chloride ions (Cl$^-$) diffusing in. Most of the hydrogen ions bind to hemoglobin or oxyhemoglobin, which thus buffers the intracellular pH.

Oxygen Unloading

When H$^+$ binds to oxyhemoglobin (HbO$_2$), it reduces the affinity of hemoglobin for O_2 and tends to make hemoglobin release it. Oxygen consumption by respiring tissues keeps the P_{O_2} of tissue fluid relatively low, and so

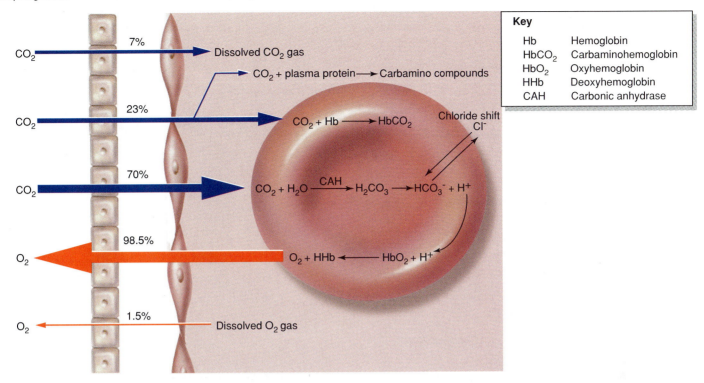

Respiring tissue

Capillary blood

7% CO_2 → Dissolved CO_2 gas

CO_2 + plasma protein → Carbamino compounds

23% CO_2 + Hb → $HbCO_2$

Chloride shift Cl^-

70% CO_2 + H_2O →CAH→ H_2CO_3 → HCO_3^- + H^+

98.5% O_2 + HHb ← HbO_2 + H^+

1.5% Dissolved O_2 gas

Key

Hb	Hemoglobin
$HbCO_2$	Carbaminohemoglobin
HbO_2	Oxyhemoglobin
HHb	Deoxyhemoglobin
CAH	Carbonic anhydrase

Figure 22.23 Systemic gas exchange. Blue arrows show the three mechanisms of CO_2 loading and transport; their thickness represents the relative amounts of CO_2 transported in each of the three forms. Red arrows show the two mechanisms of O_2 unloading; their thickness indicates the relative amounts unloaded by each mechanism. Note that CO_2 loading releases hydrogen ions in the erythrocyte, and hydrogen ions promote O_2 unloading. 𝒳

there is typically a concentration gradient of 95 → 40 mmHg of oxygen from the arterial blood to the tissue fluid. Thus, the liberated oxygen—along with some that was carried as dissolved gas in the plasma—diffuses from the blood into the tissue fluid. Hemoglobin that has unloaded oxygen is called **deoxyhemoglobin (HHb).**

As blood arrives at the systemic capillaries, its oxygen concentration is about 20 mL/dL and the hemoglobin is about 95% saturated. As it leaves the capillaries of a typical resting tissue, its oxygen concentration is about 15.6 mL/dL and the hemoglobin is about 75% saturated. Thus, it has given up 4.4 mL/dL—about 22% of its oxygen load. This fraction is called the **utilization coefficient.** The oxygen remaining in the blood after it passes through the capillary bed provides a **venous reserve** of oxygen sufficient to sustain life for 4 to 5 minutes even in the event of respiratory arrest. At rest, the circulatory system releases oxygen to the tissues at an overall rate of about 250 mL/min.

Alveolar Gas Exchange Revisited

The processes illustrated in figure 22.23 make it easier to understand alveolar exchange more fully. As shown in figure 22.24, the reactions that occur in the lungs are essentially the reverse of systemic gas exchange. As hemoglobin loads oxygen, its affinity for H^+ declines. Hydrogen ions dissociate from the hemoglobin and bind

with bicarbonate (HCO_3^-) ions diffusing from the plasma into the RBCs. Chloride ions diffuse back out of the RBC (the **reverse chloride shift**). The reaction of H^+ and HCO_3^- reverses the hydration reaction and generates free CO_2. This diffuses into the alveolus to be exhaled—as does CO_2 released from carbaminohemoglobin and CO_2 gas that was dissolved in the plasma.

Adjustment to the Metabolic Needs of Individual Tissues

Hemoglobin does not unload the same amount of oxygen to all tissues. Some tissues need more and some less, depending on their state of activity. Hemoglobin responds to such variations and unloads more oxygen to the tissues that need it most. In exercising skeletal muscles, for example, the utilization coefficient may be as high as 80%. Four factors adjust the rate of oxygen unloading to the metabolic rates of different tissues:

1. **Ambient P_{O_2}.** Since an active tissue consumes oxygen rapidly, the P_{O_2} of its tissue fluid remains low. From the oxyhemoglobin dissociation curve (see fig. 22.22), you can see that at a low P_{O_2}, HbO_2 releases more oxygen.

2. **Temperature.** When temperature rises, the oxyhemoglobin dissociation curve shifts to the right

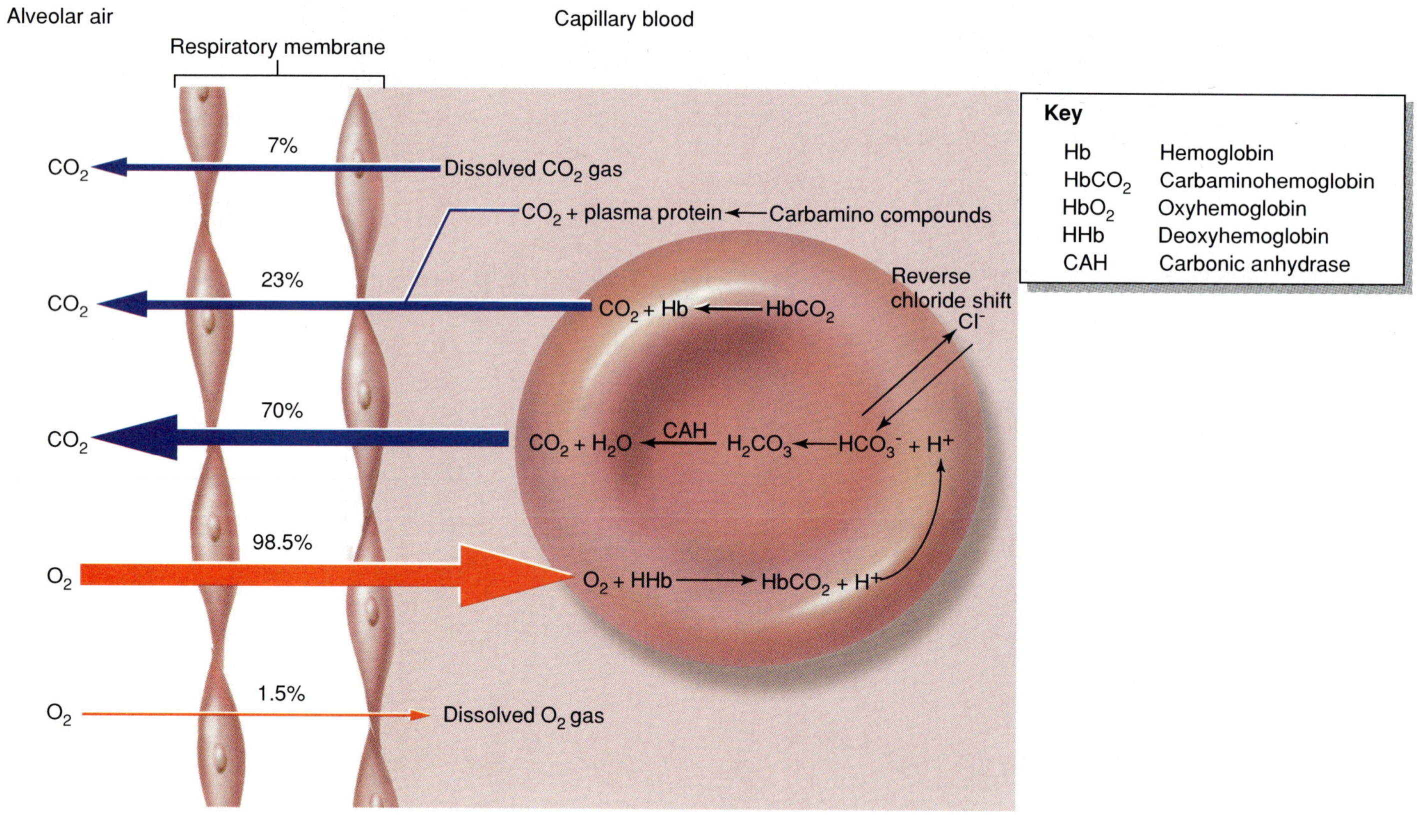

Figure 22.24 Alveolar gas exchange. In what fundamental way does this differ from the preceding figure? Following alveolar gas exchange, will the blood contain a higher or lower concentration of bicarbonate ions than it did before? ✗

(fig. 22.25a); in other words, elevated temperature promotes oxygen unloading. Active tissues are warmer than less active ones and thus extract more oxygen from the blood passing through them.

3. **The Bohr effect.** Active tissues also generate extra CO_2, which results in an elevated H^+ concentration and lower pH. Like elevated temperatures, a drop in pH shifts the oxygen-hemoglobin dissociation curve to the right (fig. 22.25b) and promotes oxygen unloading. Increased HbO_2 dissociation in response to low pH is called the **Bohr**[19] **effect.** This effect is less pronounced at the high P_{O_2} present in the lungs, so pH has relatively little effect on pulmonary oxygen loading. In the systemic capillaries, however, P_{O_2} is lower and the Bohr effect is more pronounced.

4. **DPG.** Erythrocytes have no mitochondria and meet their energy needs solely by anaerobic metabolism. One of their metabolic intermediates is **DPG** (2,3-diphosphoglycerate), which binds to hemoglobin and promotes oxygen unloading. An elevated body temperature (as in fever) stimulates DPG synthesis, as do thyroxine, growth hormone, testosterone, and epinephrine. All of these hormones thus promote oxygen unloading to the tissues.

The rate of carbon dioxide loading is also adjusted to varying needs of the tissues. A low level of oxyhemoglobin (HbO_2) enables the blood to transport more CO_2, a phenomenon known as the **Haldane effect.**[20] It occurs for two reasons: (1) HbO_2 does not bind CO_2 as well as deoxyhemoglobin (HHb) does. (2) HHb binds more hydrogen ions than HbO_2 does, and by removing H^+ from solution, HHb shifts the $H_2O + CO_2 \rightarrow HCO_3^- + H^+$ reaction to the right. A high metabolic rate keeps oxyhemoglobin levels relatively low and thus allows more CO_2 to be transported by these two mechanisms.

Key Point Review

12. Why is the gaseous makeup of alveolar air different from that of the atmosphere?

13. What four factors affect the efficiency of alveolar gas exchange?

14. Explain how perfusion of a pulmonary lobule changes if it is inadequately ventilated. How is ventilation of a lobule affected by high P_{CO_2}?

15. Describe how oxygen is transported in the blood, and explain why carbon monoxide interferes with this.

16. What are the three ways in which blood transports CO_2?

17. Describe the role of the chloride shift in CO_2 loading.

18. Give two reasons why highly active tissues can extract more oxygen from the blood than less active tissues.

19. Christian Bohr (1855–1911), Danish physiologist

20. John Scott Haldane (1860–1936), British physiologist

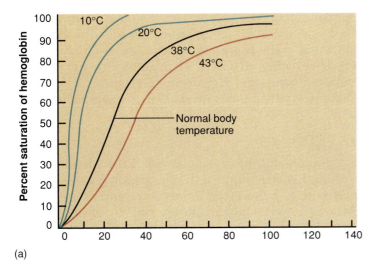

(a)

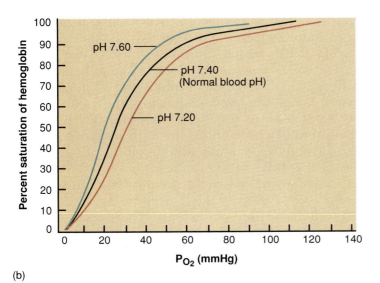

(b)

Figure 22.25 Effects of temperature and pH on the oxyhemoglobin dissociation curve. (a) For a given P_{O_2}, hemoglobin unloads more oxygen at higher temperatures. (b) For a given P_{O_2}, hemoglobin unloads more oxygen at lower pH (the Bohr effect). Both mechanisms cause hemoglobin to release more oxygen to tissues with higher metabolic rates. ⟨

Blood Gases and the Respiratory Rhythm

▼Objectives

When you have completed this section, you should be able to
- explain how variations in blood P_{O_2} and P_{CO_2} affect the respiratory centers of the brainstem; and
- explain how homeostatic control of blood gases and pH is achieved by the reactions of these respiratory centers.

The purpose of respiration is to maintain the pH, P_{CO_2}, and P_{O_2} of the body fluids within homeostatic limits. Consequently, the brainstem respiratory centers are sensitive to these conditions in the blood, which they monitor by means of peripheral and central chemoreceptors. The **peripheral chemoreceptors** are the **aortic bodies**

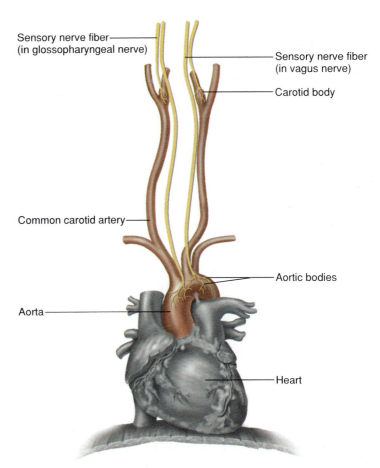

Figure 22.26 Nervous pathways from the peripheral chemoreceptors to the respiratory centers of the medulla oblongata.

and **carotid bodies** located in the aortic arch and near the branch of the carotid arteries (fig. 22.26). (These chemoreceptors are not to be confused with the aortic and carotid sinuses, which monitor blood pressure.) Although very small, the aortic and carotid bodies are richly supplied with capillaries and receive almost 40 times as much blood per gram of tissue as the brain does. The aortic bodies send signals to the medulla by way of the vagus nerves, and the carotid bodies do so by way of the glossopharyngeal nerves. The **central chemoreceptors** are paired areas very close to the surface of the medulla oblongata, ventral to the inspiratory center. They primarily monitor the pH of the cerebrospinal fluid (CSF) that surrounds the medulla and the extracellular fluid of the brain tissue.

Let's now consider how hydrogen ions, carbon dioxide, and oxygen individually affect respiration.

Hydrogen Ions

The pH of the CSF is the strongest stimulus to respiration. Hydrogen ions in the bloodstream cannot cross the blood-CSF barrier very easily, but blood CO_2 crosses easily. Once it is in the CSF, CO_2 reacts with water and releases H[+], just as we have seen that it does in the blood plasma and RBCs. Unlike blood, the CSF contains relatively little protein and cannot buffer this H[+]. H[+] thus remains free in the CSF and strongly stimulates the central chemoreceptors, which in turn stimulate the inspiratory center.

It is important to ask how we know it is the H[+] that does this and not the CO_2 that diffuses into the CSF. Experimentally, it is possible to vary the pH or the P_{CO_2} of the CSF while holding the other variable steady. When pH alone changes, there is a strong effect on respiration; when P_{CO_2} alone changes, there is only a weak effect. Therefore, even though these two variables usually change together, we can see that the chemoreceptors react primarily to the hydrogen ions.

The pH of the blood is normally 7.40 ± 0.05. If the blood pH falls below 7.35, a state of **acidosis** exists, and if it rises above 7.45, we have a state of **alkalosis**. The P_{CO_2} of the blood is normally 40 ± 3 mmHg. The most common cause of acidosis is **hypercapnia**,[21] a $P_{CO_2} > 43$ mmHg; the most common cause of alkalosis is **hypocapnia**, a $P_{CO_2} < 37$ mmHg. When there is hyper- or hypocapnia of the blood, CO_2 diffusion across the blood-CSF barrier causes parallel shifts in the pH of the CSF.

Acidosis triggers hyperventilation, "blowing off" CO_2 faster than the body produces it. This shifts the carbonic acid reaction to the left, lowers the H[+] concentration, and raises the pH back to normal:

$$CO_2 + H_2O \leftarrow H_2CO_3 \leftarrow HCO_3^- + H^+$$

Alkalosis inhibits respiration, allowing metabolic CO_2 to accumulate faster than the body exhales it. Hypoventilation shifts the reaction to the right, raises the H[+] concentration, and lowers the pH to normal:

$$CO_2 + H_2O \rightarrow H_2CO_3 \rightarrow HCO_3^- + H^+$$

Although pH usually is a reflection of P_{CO_2}, other factors can change it. In diabetes mellitus (DM), for example, fat oxidation releases acidic ketone bodies, causing an abnormally low pH called *ketoacidosis* (see chapter 17). Hyperpnea cannot reduce the level of ketone bodies in the blood, but by blowing off CO_2, it reduces the concentration of CO_2-generated H[+] and compensates to some degree for the H[+] released by the ketone bodies. Hyperpnea is therefore characteristic of untreated DM.

Carbon Dioxide

Although the arterial P_{CO_2} has a strong influence on respiration, we have seen that it is mostly an indirect one, mediated through its effects on the pH of the CSF. Yet the experimental evidence described earlier shows that CO_2 has some effect even when pH remains stable. It seems that at the beginning of exercise, the rising blood CO_2 level may directly stimulate the peripheral chemoreceptors and trigger an increase in ventilation more quickly than the central chemoreceptors do.

Oxygen

Oxygen concentration usually has little effect on respiration. Even in eupnea, the hemoglobin is 97% saturated with O_2; therefore, increased ventilation cannot add very much. Only if the arterial P_{O_2} drops below 60 mmHg does it significantly affect ventilation, and such a low P_{O_2} seldom occurs even in prolonged holding of the breath. A moderate drop in P_{O_2} does stimulate the peripheral chemoreceptors, but another effect overrides this: as the level of HbO_2 falls, hemoglobin binds more hydrogen ions (see fig. 22.23). This raises the blood pH, which indirectly inhibits respiration. Only at a $P_{O_2} < 60$ mmHg does the stimulatory effect of hypoxemia override the inhibitory effect of pH increase. Long-term hypoxemia can significantly stimulate ventilation in situations such as emphysema and pneumonia, which interfere with alveolar gas exchange, and in mountain climbing if the climb lasts at least 2 or 3 days.

In summary, the main chemical stimulus to pulmonary ventilation is the concentration of H[+] in the CSF and tissue fluid of the brain. These hydrogen ions arise mainly from CO_2 diffusing into the brain and generating H[+] through the carbonic acid reaction. Therefore the P_{CO_2} of the arterial blood is an important driving force in respiration, even though its action on the chemoreceptors is indirect. Ventilation is adjusted to maintain arterial pH at about 7.40 and arterial P_{CO_2} at about 40 mmHg. This automatically ensures that the blood is at least 97% saturated with O_2 as well. Under ordinary circumstances, arterial P_{O_2} has relatively little effect on respiration. When it drops below 60 mmHg, however, it excites the peripheral chemoreceptors and stimulates an increase in ventilation. This can be significant at high altitudes and in certain lung diseases.

| Key Point Review |

19 Describe the locations of the chemoreceptors that monitor blood pH and gas concentrations.

20 Define *hypocapnia* and *hypercapnia*. Name the pH imbalances that result from these conditions and explain the relationship between P_{CO_2} and pH.

21 Explain how variations in pulmonary ventilation can either cause or correct pH imbalances.

21. *capn* = smoke

▼Objectives
When you have completed this section, you should be able to
- name and describe the various forms of hypoxia and discuss the dangers of oxygen excess;
- describe the chronic obstructive pulmonary diseases and their consequences;
- explain how lung cancer begins, progresses, and exerts its lethal effects; and
- contrast the common cold, pneumonia, and tuberculosis as to cause and effect.

The delicate lungs are exposed to a wide variety of inhaled pathogens and debris; thus, it is not surprising that they are prone to a host of diseases. Several already have been mentioned in this chapter.

Oxygen Imbalances

Hypoxia, discussed in previous chapters, is a deficiency of oxygen in a tissue or the inability to use oxygen. It is not a respiratory disease in itself but is often a consequence of respiratory diseases. Hypoxia is classified according to cause:

- **Hypoxemic hypoxia,** a state of low arterial P_{O_2}, is usually due to inadequate pulmonary gas exchange. Some of its root causes include atmospheric deficiencies of oxygen at high altitudes; impaired pulmonary ventilation, as in drowning; aspiration of foreign matter; respiratory arrest; and the degenerative lung diseases discussed shortly. It also occurs in carbon monoxide (CO) poisoning because CO prevents hemoglobin from transporting oxygen.
- **Ischemic hypoxia** results from inadequate circulation of the blood, as in congestive heart failure.
- **Anemic hypoxia** is due to anemia and the resulting inability of the blood to carry adequate oxygen.
- **Histotoxic hypoxia** occurs when the tissues are unable to use the oxygen delivered to them because of a metabolic poison, such as cyanide.

An oxygen excess is also dangerous. You can safely breathe 100% oxygen at 1 atm for a few hours, but **oxygen toxicity** rapidly develops when pure oxygen is breathed at 2.5 atm or greater. Excess oxygen generates hydrogen peroxide and free radicals that destroy enzymes and damage nervous tissue; thus it can lead to seizures, coma, and death. This is why scuba divers breathe a mixture of oxygen and nitrogen rather than pure compressed oxygen (see chapter essay, p. 825). Hyperbaric oxygen was formerly used to treat premature infants for respiratory distress, but it caused retinal deterioration and blinded many infants before the practice was discontinued.

Chronic Obstructive Pulmonary Diseases

Chronic obstructive pulmonary disease (COPD) refers to any disorder in which there is a long-term obstruction of airflow and a substantial reduction in pulmonary ventilation. The major COPDs are *asthma, chronic bronchitis,* and *emphysema.* In asthma, an allergen triggers the release of histamine and other inflammatory chemicals that cause intense bronchoconstriction and sometimes suffocation (see chapter 21). The other COPDs are almost always caused by cigarette smoking but occasionally result from air pollution or occupational exposure to airborne irritants.

Beginning smokers exhibit inflammation and hyperplasia of the bronchial mucosa. In **chronic bronchitis,** the cilia are immobilized and reduced in number, while goblet cells enlarge and produce excess mucus. With extra mucus and fewer cilia to dislodge it, smokers develop a chronic cough that brings up **sputum** (SPEW-tum), a mixture of mucus and cellular debris. Thick, stagnant mucus in the respiratory tract provides a growth medium for bacteria, while cigarette smoke incapacitates the alveolar macrophages and reduces defense mechanisms against respiratory infections. Smokers therefore develop chronic infection and bronchial inflammation, with symptoms that include dyspnea, hypoxia, cyanosis, and attacks of coughing.

In **emphysema**[22] (EM-fih-SEE-muh), alveolar walls break down and the lung exhibits larger but fewer alveoli (see fig. 22.20*c*). Thus, there is much less respiratory membrane for gas exchange. The lungs become fibrotic and less elastic. The air passages open adequately during inspiration, but they tend to collapse and obstruct the outflow of air. Air therefore becomes trapped in the lungs, and over a period of time a person becomes barrel-chested. The overly stretched thoracic muscles contract weakly, which further contributes to the difficulty of expiration. People with emphysema become exhausted because they expend three to four times the normal amount of energy just to breathe. Even slight physical exertion, such as walking across a room, can cause severe shortness of breath.

Think About It

Explain how the length-tension relationship of skeletal muscle (see chapter 12) accounts for the weakness of the respiratory muscles in emphysema.

All of the COPDs tend to reduce pulmonary compliance and vital capacity and cause hypoxemia, hypercapnia, and respiratory acidosis. Hypoxemia stimulates the kidneys to secrete erythropoietin, which leads to

---

22. *emphys* = inflamed

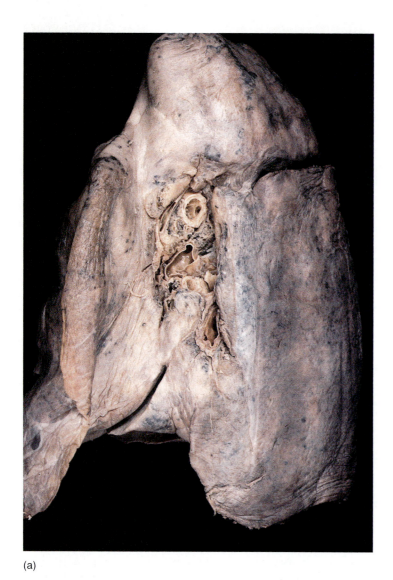

(a)

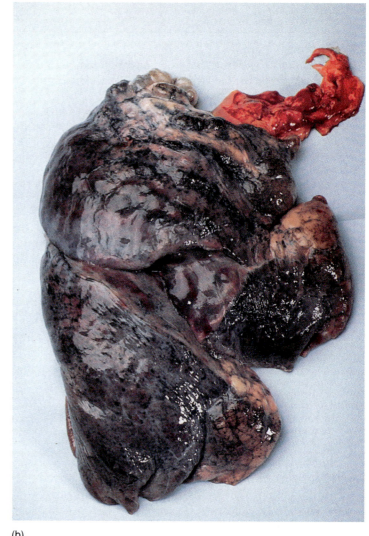

(b)

Figure 22.27 (*a*) A healthy adult lung. (*b*) A smoker's lung with emphysema.

accelerated erythrocyte production and polycythemia, as discussed in chapter 18. COPD also leads to *cor pulmonale*—hypertrophy and potential failure of the right heart due to obstruction of the pulmonary circulation (see chapter 19).

Smoking and Lung Cancer

Lung cancer (fig. 22.27) is the most common type of cancer and accounts for more deaths than any other form. The most important cause of lung cancer is cigarette smoking, distantly followed by air pollution. Cigarette smoke contains at least 15 carcinogenic compounds. Lung cancer commonly follows or accompanies COPD.

There are three forms of lung cancer, the most common of which is **squamous-cell carcinoma.** This begins with the multiplication of basal cells of the bronchial epithelium and transformation of the ciliated

pseudostratified epithelium into the stratified squamous type. As the dividing epithelial cells invade the underlying tissues of the bronchial wall, the bronchus develops bleeding lesions. Dense swirled masses of keratin appear in the lung parenchyma, replacing functional respiratory tissue. A second form of lung cancer, nearly as common, is **adenocarcinoma,**[23] which originates in mucous glands of the lamina propria. The least common (10%–20% of malignancies) but most dangerous form is **small-cell (oat-cell) carcinoma,** named for clusters of small cells that resemble oat grains. This originates in the primary bronchi but invades the mediastinum and metastasizes quickly to other organs.

Over 90% of lung tumors originate in the mucous membranes of the large bronchi. As a tumor invades the bronchial wall and grows around it, it compresses the air-

23. *adeno* = gland

way and may cause atelectasis (collapse) of more distal parts of the lung. Growth of the tumor produces a cough, but coughing is such an everyday occurrence among smokers it seldom causes much alarm. Often, the first sign of serious trouble is the coughing up of blood. Lung cancer metastasizes so rapidly that it has usually spread to other organs by the time it is diagnosed. Common sites of metastasis are the pericardium, heart, bones, liver, lymph nodes, and brain. The chance of recovery is poor, with only 7% of patients surviving for 5 years after diagnosis.

Infectious Diseases

Infectious disease is largely beyond the scope of this book, but we will briefly consider the common cold, pneumonia, and tuberculosis. **Acute coryza** (co-RY-zuh), the common cold, is caused by many types of viruses that infect the upper respiratory tract, especially the nasal cavity. The symptoms include congestion, nasal secretion, sneezing, and a dry cough. The 10 most common causes of colds are the fingers and thumbs—contaminated hands contacting the mucous membranes. The best prevention is to wash the hands frequently and avoid touching the nose and eyes. Cold viruses are not transmitted orally.

Pneumonia is a lower respiratory infection caused by any of several bacteria, viruses, fungi, or protozoans—most often the bacterium *Streptococcus pneumoniae.* The alveoli fill with fluid and dead leukocytes, and the respiratory membranes become thickened with edema (see fig. 22. 20*b*), interfering with gas exchange and causing hypoxemia. Pneumonia is especially dangerous to infants, the elderly, and people with compromised immune systems, such as AIDS and leukemia patients.

Tuberculosis (TB) is caused by the bacterium *Mycobacterium tuberculosis,* which enters the lungs by way of the air, blood, or lymph. The lung tissue forms fibrous nodules called *tubercles* around the bacteria. As fibrosis progresses, it compromises the elastic recoil and ventilation of the lungs. TB is especially common among the homeless and impoverished and is showing increasing incidence among people with AIDS.

Key Point Review

22 Describe the four classes of hypoxia.

23 Name and compare two COPDs and describe some pathological effects they have in common.

24 In what lung tissue does lung cancer originate? How does it kill?

25 How is pneumonia different from a common cold?

CHAPTER ESSAY

Diving Physiology and Decompression Sickness

Because of the rise in popularity of scuba diving in recent years, many people now know something about the scientific aspects of breathing under high pressure. But diving is by no means a new fascination. As early as the fifth century B.C.E., Aristotle described divers using snorkels and taking containers of air underwater in order to stay down longer. Some Renaissance artists depicted divers many meters deep, breathing from tubes to the water surface. In reality, this would be physically impossible. For one thing, such tubes would have so much dead space that fresh air from the surface would not reach the diver. The short snorkels used today are about the maximum length that will work for surface breathing. Another reason snorkels cannot be used at greater depths is that water pressure doubles for every 11 meters of depth, and even at 1 meter the pressure is so great that a diver cannot expand the chest muscles without help. This is why scuba divers use pressurized air tanks. The tanks create a positive intrapulmonary pressure and enable the diver to inhale with only slight assistance from the thoracic muscles. Scuba tanks also have regulators that adjust the outflow pressure to the diver's depth and opposing pressure of the surrounding water.

But breathing pressurized (hyperbaric) gas presents its own problems. Divers cannot use pure oxygen because of the problem of oxygen toxicity. Instead, they use compressed air—a mixture of 21% oxygen and 79% nitrogen. On land, nitrogen presents no physiological problems; it dissolves poorly in blood and it is physiologically inert. But under hyperbaric conditions, larger amounts of nitrogen dissolve in the blood. (Which of the gas laws applies here?) Even more dissolves in adipose tissue and the myelin of the brain, since nitrogen is more soluble in lipids. In the brain, it causes **nitrogen narcosis,** or what Jacques Cousteau termed "rapture of the deep." A diver can become dizzy, euphoric, and dangerously disoriented; for every 15 to 20 meters of depth, the effect is said to be equivalent to one martini on an empty stomach.

Strong currents, equipment failure, and other hazards sometimes cause scuba divers to panic, hold their breath, and quickly swim to the surface (a *breath-hold ascent*). Ambient (surrounding) pressure falls rapidly as a diver ascends, and the air in the lungs expands just as rapidly. (Which gas law is

—continued

demonstrated here?) It is imperative that an ascending diver keep the airway open to exhale the expanding gas; otherwise it is likely to cause **pulmonary barotrauma**—ruptured alveoli. Then, when the diver takes a breath of air at the surface, alveolar air goes directly into the bloodstream and causes air embolism. After passing through the heart, the emboli tend to enter the cerebral circulation because the diver is head-up and air bubbles rise in liquid. The resulting cerebral embolism can cause motor and sensory dysfunction, seizures, unconsciousness, and drowning.

Barotrauma can be fatal even at the depths of a backyard swimming pool. In one case, children trapped air in a bucket 1 meter underwater and then swam under the bucket to breathe from the air space. Because the bucket was under water, the air in it was compressed. One child filled his lungs under the bucket, did a "mere" 1-meter breath-hold ascent, and his alveoli ruptured. He died in the hospital, partly because the case was mistaken for drowning and not treated for what it really was. This would not have happened to a person who inhaled at the surface, did a breath-hold dive, and then resurfaced—nor is barotrauma a problem for those who do breath-hold dives to several meters. (Why? What is the difference?)

Even when not holding the breath, but letting the expanding air escape from the mouth, a diver must ascend slowly and carefully to allow for decompression of the nitrogen that has dissolved in the tissues. *Decompression tables* prescribe safe rates of ascent based on the depth and the length of time a diver has been down. When pressure drops, nitrogen dissolved in the tissues can go either of two places—it can diffuse into the alveoli and be exhaled, or it can form bubbles like the CO_2 in a bottle of soda when the cap is removed. The diver's objective is to ascend slowly, allowing for the former and preventing the latter. If a diver ascends too rapidly, nitrogen "boils" from the tissues—especially in the 3 meters just below the surface, where the relative pressure change is greatest. A diver may double over in pain from bubbles in the joints, bones, and muscles—a disease called the "bends," or **decompression sickness (DCS).** Nitrogen bubbles in the pulmonary capillaries cause "chokes"—substernal pain, coughing, and dyspnea. DCS is sometimes accompanied by mood changes, seizures, numbness, and itching. These symptoms usually occur within an hour of surfacing, but they are sometimes delayed for up to 36 hours. DCS is treated by putting the individual in a hyperbaric chamber to be recompressed, and then *slowly* decompressed.

DCS is also called **caisson disease.** A caisson is a watertight underwater chamber filled with pressurized air. Caissons are used in underwater construction work on bridges, tunnels, ships' hulls, and so forth. Caisson disease was first reported in the late 1800s among workmen building the foundations of the Brooklyn Bridge.▲

Interactions Between the RESPIRATORY SYSTEM and Other Organ Systems

All Systems
The respiratory system serves all other systems by supplying O_2, removing CO_2, and maintaining pH balance.

Integumentary System
• Nasal guard hairs prevent entry of dust and other foreign material

Skeletal System
• Thoracic cage provides protective enclosure for lungs; movements of ribs essential in pulmonary ventilation

Muscular System
• Muscles ventilate lungs and control position of larynx during swallowing; control tension on vocal cords during speech; exercise strongly stimulates pulmonary ventilation by generating CO_2

Nervous System
• Monitors changes in blood gases and pH; monitors distension of lungs; brainstem centers establish ventilation rhythm; hypothalamus and cerebrum modify ventilation; phrenic, intercostal, and other nerves innervate respiratory muscles

Endocrine System
• Epinephrine and norepinephrine dilate bronchioles and stimulate ventilation
• Lungs contain angiotensin-converting enzyme (ACE), which converts angiotensin I to angiotensin II

Circulatory System
• Blood transports respiratory gases; mitral stenosis or left heart failure causes pulmonary hypertension and edema; emboli from legs and other peripheral sites often lodge in lungs
• Activation of angiotensin II by ACE in lungs important in regulating blood volume and pressure; buffers blood pH; thoracic pump aids venous return of blood; lungs produce platelets; obstruction of pulmonary circulation leads to right heart failure

Lymphatic/Immune Systems
• Extensive lymphatic drainage from lungs essential for avoiding fluid accumulation; immune cells provide protection against infection
• Thoracic pump promotes lymph flow into thoracic vessels and subclavian veins

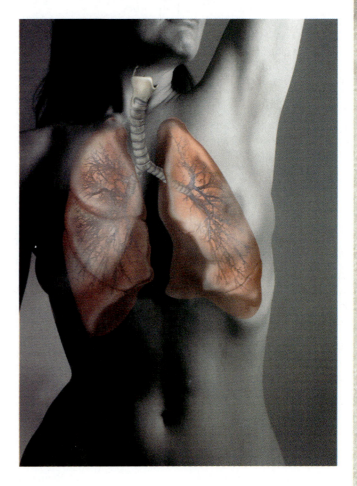

Urinary System
• Disposes of metabolic wastes from respiratory organs; participates with lungs in regulation of blood pH
• Expiratory effort against closed glottis (Valsalva maneuver) helps to empty bladder

Digestive System
• Provides nutrients for growth and maintenance of respiratory system
• Valsalva maneuver aids defecation

Reproductive System
• Sexual arousal stimulates respiratory rate and depth
• Valsalva maneuver aids childbirth

Anatomy of the Respiratory System (pp. 796–806)

1. Conducting and respiratory divisions

2. Upper and lower respiratory tract

3. The nose
   a. Structure
      • Bone and cartilage framework
      • Anterior and posterior nares
      • Vestibule and vibrissae
      • Nasal septum and fossae
      • Conchae and meatuses
   b. Functional aspects
      • Olfaction
      • Entrapment of inhaled debris
      • Ciliary movement of mucus
      • Erectile tissues of inferior concha

4. The pharynx
   a. Naso-, oro-, and laryngopharynx
   b. Passage of air, food, and drink

5. The larynx
   a. Epiglottis and glottis
   b. Unpaired cartilages
      • Epiglottic cartilage
      • Thyroid cartilage
      • Cricoid cartilage
   c. Paired cartilages
      • Arytenoid cartilages
      • Corniculate cartilages
      • Cuneiform cartilages
   d. Muscles
      • Extrinsic—elevate larynx
      • Intrinsic—speech
   e. Vocal cords and speech mechanism

6. The trachea
   a. Cartilage rings and trachealis muscle
   b. Mucociliary escalator

7. The lungs
   a. Four surfaces of lung
   b. Left lung: two lobes and cardiac notch
   c. Right lung: three lobes

8. The bronchial tree
   a. Primary bronchi
   b. Secondary bronchi
   c. Tertiary bronchi
   d. Bronchioles
      • Diameter 1 mm or less
      • Lack of cartilage
      • Well-developed smooth muscle
      • Major point of air resistance
   e. Terminal bronchioles
      • Diameter of 0.5 mm or less
      • Lack of mucous glands and goblet cells
   f. Respiratory bronchioles
      • Beginning of respiratory division
      • Smallest ones unciliated
      • Give off some alveoli

   g. Alveolar ducts and sacs

9. Alveoli
   a. Pouches 0.2 to 0.5 mm diameter
   b. Squamous alveolar cells
      • Gas diffusion role
   c. Great alveolar cells
      • Secrete pulmonary surfactant
   d. Alveolar macrophages
   e. Respiratory membrane
      • Squamous alveolar cells
      • Capillary endothelium
      • Fused basement membranes

10. The pleurae
    a. Visceral and parietal pleurae
    b. Pleural cavity and fluid
    c. Functions
       • Reduce friction
       • Create pressure gradient
       • Compartmentalize thoracic organs

Mechanics of Ventilation (pp. 806–812)

1. Pressure and flow
   a. Atmospheric pressure
   b. Boyle's law and intrapulmonary pressure

2. Inspiration
   a. Action of thoracic muscles
   b. Role of intrapleural pressure
   c. Transpulmonary pressure and airflow
   d. Charles' law and thermal expansion
   e. Effects on abdominal pressure

3. Expiration
   a. Usually passive
   b. Importance of elasticity
   c. Forced expiration
   d. Pneumothorax and atelectasis

4. Resistance to airflow
   a. Importance of compliance
   b. Bronchoconstriction
   c. Bronchodilation

5. Alveolar surface tension
   a. Law of Laplace
   b. Role of pulmonary surfactant

6. Alveolar ventilation
   a. Dead space
   b. Alveolar ventilation rate

7. Nonrespiratory air movements

8. Measurements of ventilation
   a. Respiratory volumes and capacities (table 22.1)
   b. Restrictive and obstructive disorders
   c. Forced expiratory volume
   d. Minute respiratory volume
   e. Maximum voluntary ventilation

9. Patterns of breathing (table 22.2)

Neural Control of Ventilation (pp. 812–814)

1. Control centers in the medulla oblongata
   a. I neurons and E neurons
   b. Inspiratory center
   c. Expiratory center

2. Control centers in the pons
   a. Pneumotaxic center
   b. Apneustic center

3. Afferent connections to brainstem nuclei
   a. Limbic system and hypothalamus
   b. Chemoreceptors
   c. Inflation reflex

4. Voluntary control of ventilation

Gas Exchange and Transport (pp. 814–820)

1. Composition of air
   a. Partial pressures and Dalton's law
   b. Inspired and alveolar air
   c. Henry's law and the gas-liquid interface

2. Factors in alveolar gas exchange
   a. Speed of RBC travel
   b. Concentration gradients
   c. Gas solubility
   d. Membrane thickness
   e. Membrane area
   f. Ventilation-perfusion coupling

3. Gas transport
   a. Oxygen
      • Binding to hemoglobin
      • Oxyhemoglobin dissociation curve
   b. Carbon dioxide
      • Carbonic acid reaction
      • Carbamino compounds
      • Dissolved gas

4. Systemic gas exchange
   a. Carbon dioxide loading
      • Hydration of CO_2
      • Chloride shift
   b. Oxygen unloading
      • Oxyhemoglobin dissociation
      • Utilization coefficient
   c. Reverse chloride shift
   d. Adjustment to metabolic needs of tissues
      • Ambient P_{O_2}
      • Effects of temperature
      • The Bohr effect of pH
      • Effects of DPG
      • Haldane effect

Blood Gases and the Respiratory Rhythm (pp. 821–822)

1. Chemoreceptors

2. Hydrogen ions
   a. Acidosis and alkalosis
   b. Hypercapnia and hypocapnia
3. Carbon dioxide
4. Oxygen

Respiratory Disorders (pp. 823–825)
1. Oxygen deficiency and excess
   a. Forms of hypoxia
   b. Oxygen toxicity

2. Chronic obstructive pulmonary diseases
   a. Chronic bronchitis
   b. Emphysema
3. Smoking and lung cancer
4. Infectious diseases

Selected Vocabulary

Also review the terms in tables 22.1 and 22.2, which are not repeated here.

respiration 796
respiratory system 796
conducting division 796
respiratory division 796
upper respiratory tract 796
lower respiratory tract 796
nose 796
lateral cartilages 796
alar cartilages 796
ala nasi 796
nasal cavity 797
anterior nares 797
posterior nares 797
vestibule 797
vibrissae 797
nasal septum 797
nasal fossae 797
nasal conchae 797
turbinate bones 797
meatus 797
erectile tissues 799
pharynx 799
nasopharynx 799
oropharynx 799
laryngopharynx 799
larynx 799
glottis 800
epiglottis 800
epiglottic cartilage 800
thyroid cartilage 800
cricoid cartilage 800
arytenoid cartilages 801
corniculate cartilages 801
cuneiform cartilages 801
intrinsic laryngeal muscles 801

extrinsic laryngeal
   muscles 801
vestibular folds 801
true vocal cords 801
trachea 801
trachealis muscle 802
mucociliary escalator 802
base and apex of lung 802
costal surface 802
mediastinal surface 802
hilum 803
root 803
cardiac notch 803
pulmonary lobes 803
bronchial tree 803
primary bronchus 803
secondary bronchus 803
tertiary bronchus 804
bronchopulmonary
   segment 804
bronchiole 804
pulmonary lobule 804
terminal bronchiole 804
respiratory bronchiole 804
alveolar duct 804
alveolar sac 804
alveolus 805
squamous alveolar cells 805
great alveolar cells 805
pulmonary surfactant 805
alveolar macrophages 805
respiratory membrane 805
visceral pleura 806
parietal pleura 806
pleural cavity 806
pleural fluid 806
inspiration 806
expiration 806
Boyle's law 806

intrapulmonary pressure 806
intrapleural pressure 807
transpulmonary pressure 807
Charles' law 808
Valsalva maneuver 808
pneumothorax 808
atelectasis 808
pulmonary compliance 808
bronchoconstriction 808
bronchodilation 808
law of Laplace 809
dead air 809
anatomic dead space 809
physiologic dead space 809
alveolar ventilation rate
   (AVR) 809
spirometer 809
respiratory volumes 810
respiratory capacities 810
restrictive disorder 811
obstructive disorder 811
forced expiratory volume
   (FEV) 811
minute respiratory volume
   (MRV) 811
maximum voluntary
   ventilation (MVV) 811
inspiratory (I) neurons 812
expiratory (E) neurons 812
inspiratory center 812
expiratory center 812
pneumotaxic center 813
apneustic center 813
inflation reflex 813
partial pressure 814
Dalton's law 814
Henry's law 814
ventilation-perfusion
   coupling 817

gas transport 817
oxyhemoglobin (HbO_2) 817
oxyhemoglobin dissociation
   curve 817
carbaminohemoglobin
   ($HbCO_2$) 818
chloride shift 818
deoxyhemoglobin (HHb) 819
utilization coefficient 819
venous reserve 819
reverse chloride shift 819
Bohr effect 820
DPG 820
Haldane effect 820
peripheral
   chemoreceptors 821
aortic body 821
carotid body 821
central chemoreceptors 821
acidosis 822
alkalosis 822
hypercapnia 822
hypocapnia 822
hypoxemic hypoxia 823
ischemic hypoxia 823
anemic hypoxia 823
histotoxic hypoxia 823
oxygen toxicity 823
chronic obstructive pulmonary
   disease (COPD) 823
chronic bronchitis 823
sputum 823
emphysema 823
squamous-cell carcinoma 824
adenocarcinoma 824
small-cell carcinoma 824
acute coryza 825
pneumonia 825
tuberculosis (TB) 825

Testing Your Recall   Answers in Appendix C

1. The nasal cavity is divided by the nasal septum into right and left
   a. nares.
   b. vestibules.
   c. fossae.
   d. choanae.
   e. conchae.

2. The intrinsic laryngeal muscles regulate speech by rotating
   a. the extrinsic laryngeal muscles.
   b. the corniculate cartilages.
   c. the arytenoid cartilages.
   d. the hyoid bone.
   e. the vocal cords.

3. The largest air passages that can engage in gas exchange with the blood are
   a. the respiratory bronchioles.
   b. the terminal bronchioles.
   c. the primary bronchi.
   d. the alveolar ducts.
   e. the alveoli.

4. The law of Laplace describes
   a. the tendency of alveoli to collapse.
   b. the dissolving of gases in liquid.
   c. the effect of pH on O_2 unloading.
   d. the binding of CO_2 to hemoglobin.
   e. the compliance of the lungs.

5. Which of these values would normally be highest?
   a. tidal volume
   b. inspiratory reserve volume
   c. expiratory reserve volume
   d. residual volume
   e. vital capacity

6. The _____ protects the lungs from injury by excessive inspiration.
   a. pleura
   b. rib cage
   c. inflation reflex
   d. Haldane effect
   e. Bohr effect

7. According to _____, the amount of gas that dissolves in a liquid is proportional to the partial pressure of that gas in the air when the temperature remains constant.
   a. Dalton's law
   b. the law of Laplace

   c. the Hering–Breuer law
   d. Haldane's law
   e. Henry's law

8. The blood transports most carbon dioxide in the form of
   a. dissolved gas.
   b. carbaminohemoglobin.
   c. carboxyhemoglobin.
   d. bicarbonate ion.
   e. deoxyhemoglobin.

9. The Bohr effect relates
   a. partial pressures to total pressure.
   b. oxygen unloading to pH.
   c. oxygen unloading to temperature.
   d. oxyhemoglobin dissociation to P_{O_2}.
   e. CO_2 loading to temperature.

10. Carbon dioxide has a greater effect on the pH of CSF than on the pH of blood because
    a. CSF contains less protein than blood.
    b. CO_2 cannot cross the blood-brain barrier.
    c. all CO_2 crosses the blood-brain barrier, so none remains in the blood.
    d. only the blood contains carbonic anhydrase.
    e. the chloride shift occurs in the red blood cells.

11. The superior opening into the larynx is called the _____.

12. Within each lung, the airway forms a branching complex called the _____.

13. The great alveolar cells secrete a lipoprotein called _____.

14. Intrapulmonary pressure must be greater than _____ pressure for inspiration to occur but greater than _____ pressure for expiration to occur.

15. _____ disorders reduce the flow of air through the airway.

16. Some inhaled air does not participate in gas exchange because it fills the _____ of the respiratory tract.

17. Inspiration depends on the ease of pulmonary inflation, called _____, whereas expiration depends on _____, which causes pulmonary recoil.

18. Inspiration is caused by the firing of I neurons in the _____ of the medulla.

19. The matching of airflow to blood flow in any region of the lung is called _____.

20. A blood pH > 7.45 is called _____; it can be caused by a CO_2 deficiency called _____.

Testing Your Comprehension   Answers in *Study Guide*

1. Discuss how different functions of the conducting division and respiratory division relate to differences in their histology.

2. State whether the blood P_{O_2}, P_{CO_2}, and pH would be raised, lowered, or unaffected by hyperventilation and explain why. Do the same for whether these variables would be affected by emphysema.

3. Some competitive swimmers hyperventilate before an underwater race, thinking they can "load up on extra oxygen" and hold their breaths

longer underwater. While they can indeed hold their breaths longer, it is not for the reason they think. Furthermore, many have fainted and drowned because of this practice. What is wrong with the thinking behind this strategy, and what accounts for the loss of consciousness and drowning?

4. Consider a man in good health with a 650 mL tidal volume and a respiratory rate of 11 breaths/minute. Report his minute respiratory volume in liters per minute. Assuming an anatomic dead

space of 185 mL, calculate his alveolar ventilation rate in liters per minute.

5. An 83-year-old woman is admitted to the hospital, where a critical care nurse attempts to insert a "stomach tube" (nasoenteric tube) for feeding. The patient begins to exhibit dyspnea, and a chest X ray reveals air in the pleural space and a collapsed right lung. The patient dies 5 days later from respiratory complications. Name the conditions revealed by the X ray and explain how they could have resulted from the nurse's procedure.

Web Site Link

For a listing of the most current web sites related to this chapter, please visit the Saladin homepage at:

http://www.mhhe.com/sciencemath/biology/saladin/

[The Urinary System

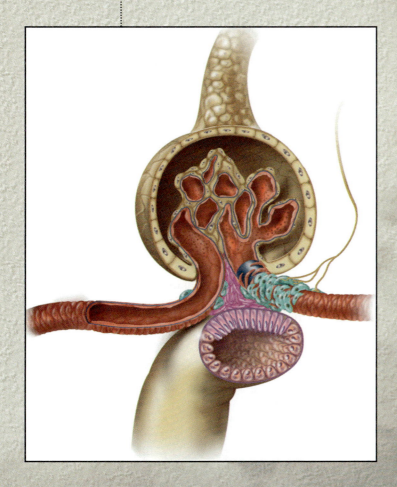

Brushing up

To understand this chapter, it is essential that you understand or brush up on the following concepts:

▶ Osmosis and tonicity (pp. 126–127)
▶ Carrier-mediated transport mechanisms (pp. 128–131)
▶ Osmotic diuresis (p. 616)
▶ Blood pressure, resistance, and flow (pp. 712–717)
▶ Capillary filtration and reabsorption (pp. 720–722)
▶ Solvent drag (p. 722)

The urinary system is well known for eliminating wastes from the body, but its role in maintaining homeostasis goes far beyond that. As you will see shortly, the kidneys also detoxify poisons, synthesize glucose, and play indispensable roles in controlling electrolyte and acid-base balance, blood pressure, erythrocyte count, and the P_{O_2} and P_{CO_2} of the blood. The urinary system thus has a very close physiological relationship with the endocrine, circulatory, and respiratory systems, covered in the preceding chapters.

Anatomically, the urinary system is closely associated with the reproductive system. In many animals the eggs and sperm are emitted through the urinary tract, and the two systems have a shared embryonic development and adult anatomical relationship. This is reflected in humans, where the systems develop together in the embryo and, in the male, the urethra continues to serve for elimination of both urine and sperm. Thus the urinary and reproductive systems are often collectively called the *urogenital (U-G) system,* and *urologists* treat both urinary and male reproductive disorders. We examine the anatomical relationship

between the urinary and reproductive systems in chapter 27, but the physiological link to the circulatory and respiratory systems is more important to consider at this time.

Functions of the Urinary System

▼**Objectives**
When you have completed this section, you should be able to
• name and locate the organs of the urinary system;
• list several functions of the kidneys in addition to urine formation;
• name the major nitrogenous wastes and identify their sources;
• define *excretion* and identify the systems that excrete wastes; and
• define *osmolarity* and state its relevance to urinary function.

The **urinary system** consists of six organs: two **kidneys,** two **ureters**, the **urinary bladder,** and the **urethra** (fig. 23.1). Most of our focus in this chapter is on renal[1] form and function.

Renal Functions

The most fundamental role of the kidneys is homeostatic regulation of the volume and composition of the body fluids. All of the following processes contribute to this basic purpose:

• The kidneys filter blood plasma, separate wastes from the useful chemicals, and eliminate the wastes while returning the rest to the bloodstream.

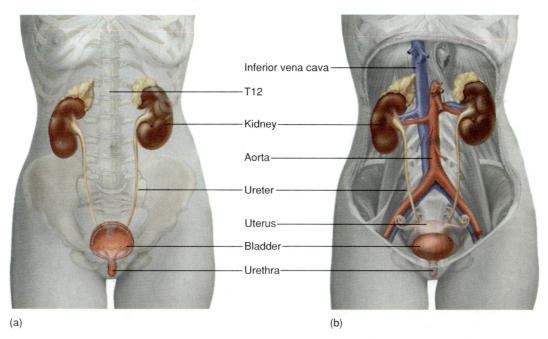

(a)   (b)

Figure 23.1 The urinary system. (*a*) Position in the abdominopelvic cavity. (*b*) Organs of the urinary system in relation to the ribs and vertebrae. 𝒶

1. *ren* = kidney + *al* = pertaining to

- They regulate blood volume and pressure by eliminating or conserving water as necessary.
- They regulate the osmolarity of the body fluids by controlling the relative amounts of water and solutes eliminated.
- They secrete the enzyme *renin,* which activates hormonal mechanisms (angiotensin, aldosterone) that control blood pressure and electrolyte balance.
- They secrete the hormone *erythropoietin,* which controls the red blood cell count and oxygen-carrying capacity of the blood.
- They function with the lungs to regulate the P_{CO_2} and acid-base balance of the body fluids.
- They control calcium homeostasis through their role in synthesizing calcitriol (vitamin D) (see chapter 8).
- They detoxify superoxides, free radicals, and drugs with the use of peroxisomes.
- In times of starvation, they *deaminate* amino acids— remove the —NH_2 group, excrete it as ammonia (NH_3), and make the rest available for glucose synthesis (*gluconeogenesis*).

Thus the functions of the kidneys are more diverse than you might expect.

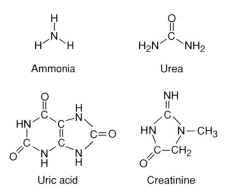

Figure 23.2 The major nitrogenous wastes.

Nitrogenous Wastes

A **waste** is any substance of no use to the body or present in excess of the body's needs. A **metabolic waste,** more specifically, is a waste substance produced by the metabolism. Thus the food residue in feces, for example, is a waste but not a metabolic waste, since it was not produced by the body and, indeed, never entered the body's tissues.

Metabolism produces a great quantity of wastes that are lethal to cells if allowed to accumulate. Two of the most toxic metabolic wastes are carbon dioxide and small nitrogen-containing organic compounds called **nitrogenous wastes** (fig. 23.2). The respiratory system excretes the former and the urinary system excretes both. About 50% of the nitrogenous waste is **urea,** a product of protein catabolism. After proteins are hydrolyzed, the free amino acids are deaminated. The ammonia produced by this reaction is exceedingly toxic and is quickly converted to a urea, a less harmful waste, by a reaction in the liver:

$$2\ NH_3 + CO_2 \rightarrow H_2N\overset{\overset{\displaystyle O}{\|}}{—C—}NH_2 + H_2O$$

Other nitrogenous wastes in the urine include **uric acid** and **creatinine** (cree-AT-ih-neen), produced by the catabolism of nucleic acids and creatine phosphate, respectively. Although less toxic than ammonia and less abundant than urea, these wastes are far from harmless.

Renal failure can lead to **azotemia**[2] (AZ-oh-TEE-me-uh), the accumulation of nitrogenous wastes in the blood. Azotemia may progress to **uremia** (you-REE-me-uh), a syndrome of diarrhea, vomiting, dyspnea, and cardiac arrhythmia stemming from the toxic effects of these wastes. Convulsions, coma, and death can follow within a few days. Unless a kidney transplant is available, renal failure requires *hemodialysis* to remove nitrogenous wastes from the blood (see chapter essay, p. 858).

Excretion

Excretion is the process of separating wastes from the body fluids and eliminating them. It is carried out by four organ systems:

1. The respiratory system excretes carbon dioxide, small amounts of other gases, and water.
2. The integumentary system excretes water, inorganic salts, lactic acid, and urea in the sweat.
3. The digestive system not only *eliminates* food residue (which is not a process of excretion) but also actively *excretes* water, salts, carbon dioxide, lipids, bile pigments, cholesterol, and other metabolic wastes.
4. The urinary system excretes a broad variety of metabolic wastes, toxins, drugs, hormones, salts, hydrogen ions, and water.

Lesser amounts of waste are eliminated by vomiting, spitting, and the shedding of hairs and dead skin cells.

Osmolarity

Much of renal physiology depends on osmosis, and one of the major functions of the kidneys is to regulate the osmotic concentration of the body fluids. Thus it is important to understand the units in which osmotic concentration is expressed.

2. *azot* = nitrogen + *emia* = blood condition

One **osmole** is 1 mole of dissolved solute particles. If a solute does not ionize in water, then 1 mole of the solute yields 1 osmole (osm) of dissolved particles. A solution of 1 molar (1 M) glucose, for example, is also 1 osm/L. If a solute does ionize, it yields two or more dissolved particles in solution. A 1 M solution of NaCl, for example, yields 1 mole of sodium ions and 1 mole of chloride ions per liter. Both ions affect osmosis and must be separately counted in a measure of osmotic concentration. Thus 1 M NaCl = 2 osm/L. Calcium chloride, $CaCl_2$, would yield three ions if it completely ionized in solution (one Ca^{2+} and two Cl^-), so 1 M $CaCl_2$ = 3 osm/L.

Osmolality is the number of osmoles of solute *per kilogram of water,* whereas **osmolarity** is the number of osmoles *per liter of solution.* Most clinical calculations are based on osmolarity since it is easier to measure the volume of a solution than the weight of water it contains. At the concentrations of human body fluids, there is less than 1% concentration difference between osmolality and osmolarity, and the two terms are nearly interchangeable. All body fluids and many clinical solutions are mixtures of many chemicals. The osmolarity of such a solution is the total osmotic concentration of all of its dissolved particles.

A concentration of 1 osm/L is substantially higher than we find in most body fluids. Physiological concentrations are therefore expressed in terms of 1/1,000 of this, or milliosmoles per liter (mOsm/L). Blood plasma, tissue fluid, and intracellular fluid have osmotic concentrations of approximately 300 mOsm/L. Urine, however, may have a concentration as high as 1,200 mOsm/L. Two important aspects of renal function are the ability to (1) produce such a concentrated (hypertonic) urine, thus eliminating a large quantity of waste from the body without excessive loss of water, and (2) maintain the body's extracellular fluids at a concentration of 300 mOsm/L, which is isotonic to the cells and prevents cells from osmotically gaining or losing excess water.

Key Point Review
---

1. State four functions of the kidneys other than forming urine.
2. List four nitrogenous wastes and their metabolic sources.
3. Name some wastes eliminated by three systems other than the urinary system.
4. Calculate the osmolarity of a solution that contains 0.1 M urea, 0.2 M NaCl, and 0.5 M $CaCl_2$. Assume urea does not ionize and NaCl and $CaCl_2$ ionize completely.

Anatomy of the Kidney

▼**Objectives**
When you have completed this section, you should be able to
- identify the major external and internal features of the kidney;
- trace the flow of blood through the kidney; and
- trace the flow of fluid through the renal tubules.

Gross Anatomy

The kidneys lie against the posterior abdominal wall at the level of vertebrae T12 to L3. The right kidney is slightly lower than the left because of the space occupied by the liver above it. Each kidney weighs about 160 g and measures about 12 cm long, 6 cm wide, and 3 cm thick—about the size of a bar of bath soap. The lateral surface is convex while the medial surface is concave and has a slit, the **hilum,** where it receives the renal nerves and blood vessels. The adrenal glands rest on the superior poles of the kidneys. The kidneys, adrenal glands, and ureters are retroperitoneal (fig. 23.3).

The kidney is protected by three layers of connective tissue: (1) a fibrous **renal fascia,** immediately deep to the parietal peritoneum, which binds the kidney and associated organs to the abdominal wall; (2) the **adipose capsule,** a layer of fat that cushions the kidney and holds it in place (see special topic 23.1); and (3) the **renal capsule,** a fibrous sac that is anchored at the hilum and encloses the rest of the kidney like a cellophane wrapper. It protects the kidney from trauma and infection.

The renal parenchyma—the glandular tissue that forms the urine—appears C-shaped in frontal section (fig. 23.4). It encircles a medial space, the **renal sinus,** occupied by blood and lymphatic vessels, nerves, and urine-collecting structures. Adipose tissue fills the remaining space and holds these structures in place.

The parenchyma is divided into an outer **cortex** about 1 cm thick and an inner **medulla** facing the sinus. Extensions of the cortex called **renal columns** project toward the sinus, dividing the medulla into 6 to 10 **renal (medullary) pyramids.** Each pyramid is conical, with a broad base facing the cortex and a blunt point called the **papilla** facing the sinus. One pyramid and the overlying cortex constitute one *lobe* of the kidney.

The papilla of a renal pyramid is nestled in a cup called a **minor calyx**[3] (CAY-lix), which collects its urine. Two or three minor calices (CAY-lih-seez) converge to form a **major calyx,** and two or three major calices converge in the sinus to form the funnel-like **renal pelvis.**[4] The ureter is a tubular continuation of the renal pelvis that drains the urine down to the urinary bladder.

The Nephron

Each kidney contains about 1.2 million functional units called **nephrons** (NEF-rons). Nearly all that is necessary to understand the function of the kidney is to understand one nephron. A nephron consists of three principal parts: (1) blood vessels, (2) a *renal corpuscle,* and

---

3. *calyx* = cup
4. *pelvis* = basin

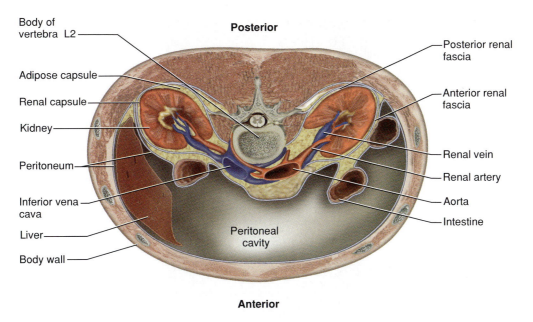

Body of
vertebra L2

Adipose capsule

Renal capsule

Kidney

Peritoneum

Inferior vena
cava

Liver

Body wall

Posterior

Posterior renal
fascia

Anterior renal
fascia

Renal vein

Renal artery

Aorta

Intestine

Peritoneal
cavity

Anterior

Figure 23.3 Cross section of the abdominal cavity showing the retroperitoneal position of the kidneys. The kidneys are posterior to the peritoneum and covered by the renal fascia, adipose capsule, and renal capsule.

Special Topic (Floating Kidney)   23.1

The kidneys may slip downward when subjected to continual vibration—as in truck drivers, motorcyclists, and equestrians—and in people who are very thin and have too little fat in the adipose capsule to hold the kidney in place. The slippage of the kidney is called **nephroptosis**[5] (NEF-rop-TOE-sis), or *floating kidney*. It may twist or kink the ureter, causing pain and interfering with urine drainage. This can lead in turn to **hydronephrosis**[6]—fluid pressure backing up into the kidney and potentially causing renal failure.

(3) a long *renal tubule* leading to a minor calyx (fig. 23.5). Most components of a nephron are found in the cortex. Nephrons close to the medulla, called **juxtamedullary**[7] **nephrons,** differ slightly in structure and function from those closer to the kidney surface, called **cortical nephrons.** About 70% to 80% of the nephrons are cortical.

Renal Circulation

Although the kidneys account for only 0.4% of the body weight, their share of the cardiac output—the *renal fraction*—is about 21% (1.2 L of blood per minute). This attests to their importance in controlling blood volume and composition.

Each kidney is supplied by a **renal artery** (occasionally two or more) arising from the aorta. Just before

or after entering the hilum, the renal artery divides and eventually gives rise to a few **interlobar arteries** (fig. 23.5). One interlobar artery penetrates each renal column and travels between the pyramids to the *corticomedullary junction (CMJ),* the boundary between the cortex and medulla. Here it branches again to form the **arcuate arteries,** which make a sharp 90° bend and travel along the base of the pyramid. Several **interlobular arteries** arise from the arcuate artery and pass upward into the cortex.

As an interlobular artery ascends through the cortex, a series of **afferent arterioles** arise from it like the limbs of a pine tree. Each afferent arteriole supplies blood to a single nephron. It ends in a rounded cluster of capillaries called a **glomerulus**[8] (glo-MERR-you-lus), where urine production begins with capillary filtration. The glomerulus is drained by an **efferent arteriole,**

---

5. *nephro* = kidney + *ptosis* = sagging, falling
6. *hydro* = water + *nephr* = kidney + *osis* = medical condition
7. *juxta* = next to

8. *glomer* = ball + *ulus* = little

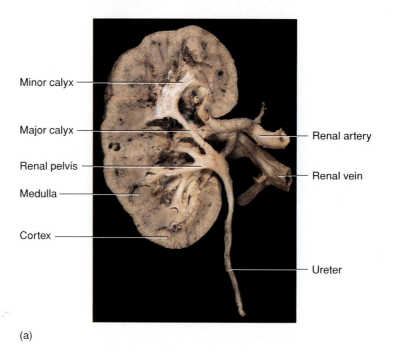

Minor calyx

Major calyx —————————————————— Renal artery

Renal pelvis —————————————————— Renal vein

Medulla

Cortex

—————————————————— Ureter

(a)

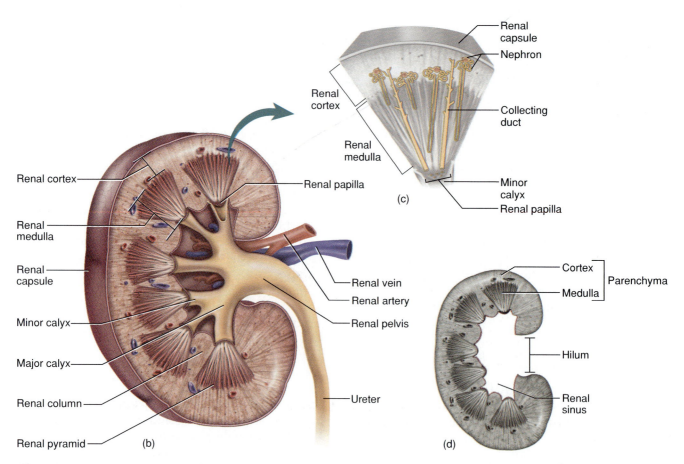

Renal capsule

Nephron

Renal cortex

Renal medulla

Collecting duct

Minor calyx

Renal papilla

(c)

Renal cortex —————————————————— Renal papilla

Renal medulla

Renal capsule

Minor calyx —————————————————— Renal vein

Major calyx —————————————————— Renal artery

Renal column —————————————————— Renal pelvis

—————————————————— Ureter

Renal pyramid ————— (b)

Cortex —————— Parenchyma

Medulla

Hilum

Renal sinus

(d)

Figure 23.4 Gross anatomy of the kidney. (*a*) Photograph of frontal section. (*b*) Major anatomical features. (*c*) Organization of one lobe of the kidney. (*d*) Ureter, renal pelvis and calices, and blood vessels removed to show the renal sinus. ↗

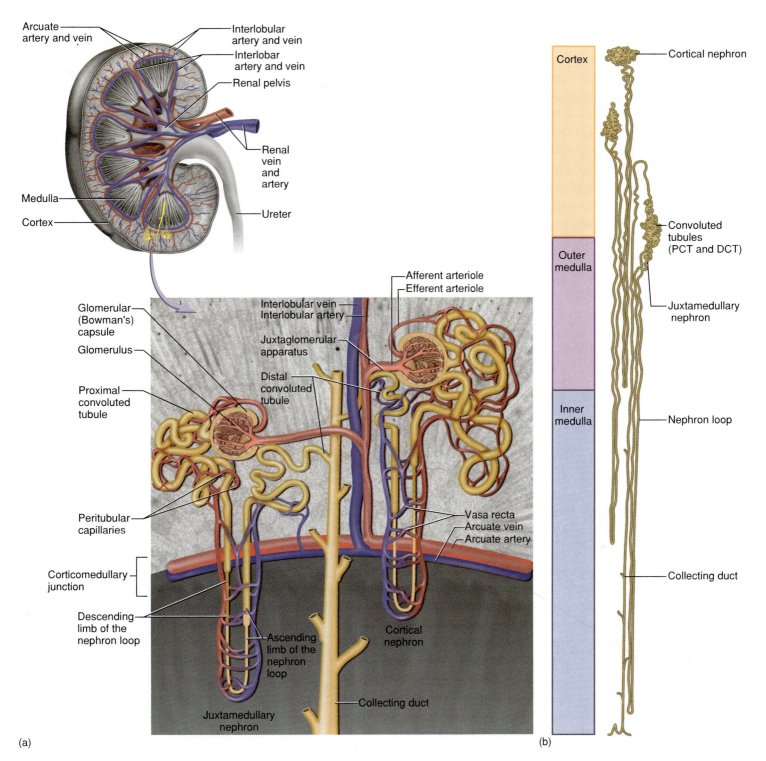

Figure 23.5 (*a*) Structure of the nephron, with the associated blood vessels and collecting duct. Cortical nephrons are farther away from the corticomedullary junction and have shorter nephron loops than juxtamedullary nephrons. (*b*) The true proportions of the nephron loops relative to the convoluted tubules.

which leads next to a plexus of **peritubular capillaries.** These are named for the fact that they form a network amid the renal tubules described shortly (fig. 23.5). Blood flows from the peritubular capillaries into the **interlobular veins, arcuate veins, interlobar veins,** and the **renal vein,** which travel parallel to the arteries of the same names. The renal vein leaves the hilum and drains into the inferior vena cava.

The route of blood flow can be traced in figure 23.5 as follows: renal artery → interlobar artery → arcuate artery → interlobular artery → afferent arteriole → glomerulus → efferent arteriole → peritubular capillaries

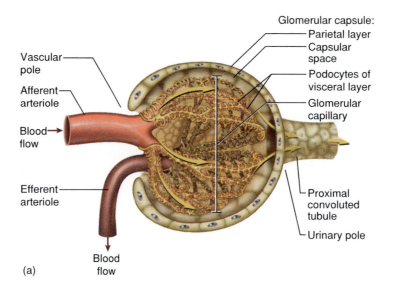

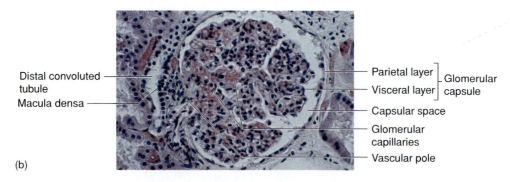

Figure 23.6 (a) Structure of the renal corpuscle; (b) light micrograph (×400).

→ interlobular vein → arcuate vein → interlobar vein → renal vein → inferior vena cava. The renal medulla is supplied by the **vasa recta**[9]—long straight vessels that arise from the efferent arterioles of the juxtamedullary nephrons. The medulla receives only 1% to 2% of the total renal blood flow.

Think About It

Can you identify a portal system in the renal circulation?

The Renal Corpuscle

The **renal corpuscle** (fig. 23.6) consists of a glomerulus enclosed in a two-layered **glomerular (Bowman's**[10]**) capsule.** The parietal layer of the capsule is a simple squamous epithelium, while the visceral layer consists of elaborate cells called **podocytes**[11] wrapped around the capillaries. The podocytes are described in detail later. The fluid that filters from the glomerular capillar-

ies collects in the **capsular space** between the parietal and visceral layers and then flows into the renal tubule at one side of the capsule called the **urinary pole.** The afferent and efferent arterioles join the glomerulus at the opposite side, called its **vascular pole.**

The Renal Tubule

The **renal (uriniferous**[12]**) tubule** is a duct that leads away from the glomerular capsule and ends at the tip of a medullary pyramid. It is about 3 cm long and divided into four major regions: the *proximal convoluted tubule, nephron loop, distal convoluted tubule,* and *collecting duct* (see fig. 23.5). The first three of these belong to an individual nephron, whereas many nephrons drain into each collecting duct. Each region has unique physiological properties and roles in the conversion of filtrate to urine.

The Proximal Convoluted Tubule The **proximal convoluted tubule (PCT)** arises from the glomerular capsule. It is the longest and most coiled of the four regions and

---

9. *vasa* = vessels + *recta* = straight
10. Sir William Bowman (1816–92), British physician
11. *podo* = foot + *cyte* = cell

12. *urin* = urine + *fer* = to carry

thus dominates histological sections of renal cortex. The PCT has a simple cuboidal epithelium with prominent microvilli that attest to the great deal of absorption that occurs here. The microvilli give the epithelium a distinctively shaggy look in tissue sections.

The Nephron Loop After coiling extensively near the renal corpuscle, the PCT straightens out and forms a long U-shaped **nephron loop (loop of Henle[13]).** The first portion of the loop, the **descending limb,** passes from the cortex into the medulla. At its lower end it turns 180° and forms an **ascending limb** that returns to the cortex. Cortical nephrons have short nephron loops that barely enter the outer medulla, while juxtamedullary nephrons have loops that dip deeply into the medulla and play a major role in concentrating the urine.

The descending limb has a simple cuboidal epithelium at its beginning, but this subsequently becomes simple squamous, forming the **thin segment** of the loop. The cells here have minimal metabolic activity but are very permeable to water. The thin segment rounds the bend and sometimes continues into the ascending limb. The epithelium then turns cuboidal again and forms the **thick segment** of the loop. The cells here have very high metabolic activity; they are heavily engaged in active transport of salts.

The Distal Convoluted Tubule When the nephron loop returns to the cortex, it coils again and forms the **distal convoluted tubule (DCT).** This is shorter and less convoluted than the PCT, so fewer sections of it are seen in histological sections. It has a cuboidal epithelium with smooth-surfaced cells nearly devoid of microvilli. The DCT is the end of the nephron.

The Collecting Duct The DCTs of several nephrons drain into a straight tubule called the **collecting duct,** which passes down into the medulla. Near the papilla, several collecting ducts merge to form a larger **papillary duct;** about 30 of these drain from each papilla into its minor calyx. The collecting and papillary ducts are lined with simple cuboidal epithelium.

To the naked eye, the renal cortex has a finely granular appearance because it consists mainly of the round renal corpuscles and a spaghetti-like tangle of convoluted tubules and capillaries. The medulla has a striated appearance because it is composed mainly of parallel nephron loops, collecting ducts, and vasa recta.

The flow of fluid from the point where the plasma filtrate is formed to the point where urine leaves the body is: glomerular capsule → proximal convoluted tubule →

nephron loop → distal convoluted tubule → collecting duct → papillary duct → minor calyx → major calyx → renal pelvis → ureter → urinary bladder → urethra.

The Juxtaglomerular Apparatus

The **juxtaglomerular** (JUX-tuh-glo-MER-you-lur) **apparatus (JGA)** is a structure near the vascular pole of a renal corpuscle, where the initial part of the DCT meets the afferent and efferent arterioles (see fig. 23.5a). It enables a nephron to monitor and stabilize its own performance and to compensate for fluctuations in blood pressure. It will be described in detail when we consider renal autoregulation.

.... **Key Point Review** ....

5 Arrange the following in order from the most numerous to the least numerous structures in a kidney: glomeruli, major calices, minor calices, interlobular arteries, interlobar arteries.

6 Trace the path taken by one red blood cell from the renal artery to the renal vein.

7 Consider one molecule of urea in the urine. Trace the route that it took from the point where it left the bloodstream to the point where it left the body.

Urine Formation I: Glomerular Filtration

▼**Objectives**

When you have completed this section, you should be able to

- describe the glomerular filtration membrane and how it excludes blood cells and proteins from the filtrate;
- explain the forces that promote and oppose glomerular filtration, and calculate net filtration pressure if given the magnitude of these forces; and
- describe how glomerular filtration rate is modified by renal autoregulation, the sympathetic nervous system, and hormones.

The kidney converts blood plasma to urine by essentially four processes: glomerular filtration, tubular reabsorption, tubular secretion, and concentration (fig. 23.7). **Glomerular filtration,** discussed in this section, is a special case of the capillary fluid exchange process described in chapter 20.

As we trace fluid through the nephron, we will refer to it by different names that reflect its changing composition: (1) The fluid in the capsular space, called **glomerular filtrate,** is similar to blood plasma except that it has almost no protein. (2) The fluid from the proximal tubule through the distal tubule will be called **tubular fluid.** It differs from the glomerular filtrate because of substances removed and added by the tubule cells. (3) The fluid will be called **urine** once it enters the collecting duct.

---

13. Friedrich G. J. Henle (1809–85), German anatomist

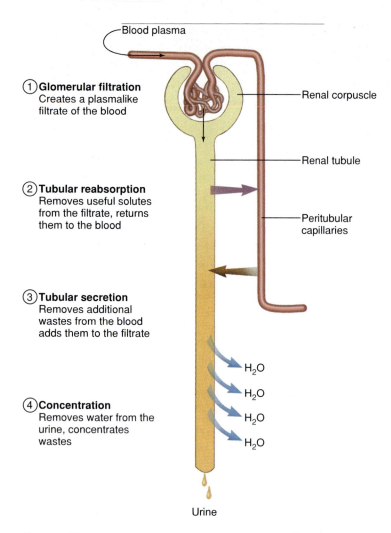

① **Glomerular filtration**
Creates a plasmalike filtrate of the blood

Blood plasma

Renal corpuscle

Renal tubule

② **Tubular reabsorption**
Removes useful solutes from the filtrate, returns them to the blood

Peritubular capillaries

③ **Tubular secretion**
Removes additional wastes from the blood adds them to the filtrate

H_2O
H_2O
H_2O
H_2O

④ **Concentration**
Removes water from the urine, concentrates wastes

Urine

Figure 23.7 Basic steps in the formation of urine. ⅄

The Filtration Membrane

To get from the bloodstream to the capsular space, fluid passes through three barriers that constitute the **filtration membrane** (fig. 23.8):

1. **The fenestrated endothelium of the capillary.** Endothelial cells of the glomerular capillaries look like honeycombs with large pores about 70 to 90 nm in diameter (see fig. 20.6, p. 710). They are much more permeable than endothelial cells elsewhere, although their pores are small enough to exclude blood cells from the filtrate.

2. **The basement membrane.** This is a proteoglycan gel. For large molecules to pass through it is like trying to pass sand through a kitchen sponge. A few particles may get through, but most are held back. On the basis of size alone, the basement membrane would exclude any molecules larger than 8 nm. Some smaller molecules, however, are also prevented from passing by a negative electrical

charge on the proteoglycans. Blood albumin is slightly less than 7 nm in diameter, but it is also negatively charged and thus repelled by the basement membrane. While the blood plasma is 7% protein, the glomerular filtrate is only 0.03% protein. It has traces of albumin and smaller polypeptides, including some hormones.

3. **Filtration slits.** The podocytes of the glomerular capsule are shaped somewhat like octopi, having bulbous cell bodies with several thick arms. Each arm has numerous little extensions called **foot processes (pedicels[14])** that wrap around the capillaries and interdigitate with each other, like wrapping your hands around a pipe and lacing your fingers together. The foot processes have negatively charged **filtration slits** about 30 nm wide between them, which are an additional obstacle to large anions.

Almost any molecule smaller than 3 nm can pass freely through the filtration membrane into the capsular space. This includes water, electrolytes, glucose, amino acids, nitrogenous wastes, and vitamins. Such substances have about the same concentration in the glomerular filtrate as in the blood plasma. Some substances of low molecular weight are retained in the bloodstream, however, because they are bound to plasma proteins that cannot get through the membrane—calcium, iron, fatty acids, and thyroid hormones, for example.

Kidney infections and trauma commonly damage the filtration membrane and allow albumin or blood cells to filter through. Albumin in the urine is called **albuminuria**, and blood in the urine is **hematuria**.

Filtration Pressure

Glomerular filtration follows the same principles that govern filtration in other blood capillaries, but there are significant differences in the magnitude of the forces involved:

- The blood pressure (BP) is much higher here than elsewhere—about 60 mmHg compared with 10 to 15 mmHg in most other capillaries. This results from the fact that the afferent arteriole is substantially larger than the efferent arteriole, giving the glomerulus has a large inlet and small outlet (fig. 23.9).
- The hydrostatic pressure in the capsular space is about 18 mmHg, compared with the slightly negative interstitial pressures elsewhere. This results from the high rate of filtration occurring here and the continual accumulation of fluid in the capsule.

14. *pedi* = foot + *cel* = little

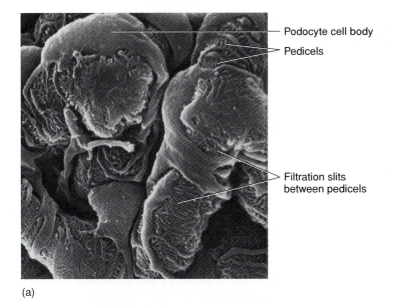

Podocyte cell body

Pedicels

Filtration slits
between pedicels

(a)

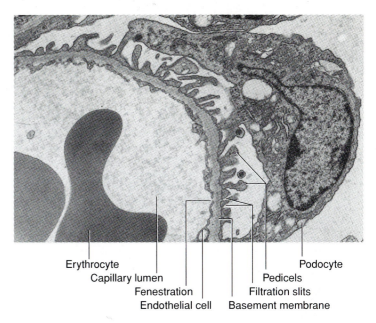

Erythrocyte

Capillary lumen

Fenestration

Endothelial cell

Podocyte

Pedicels

Filtration slits

Basement membrane

(b)

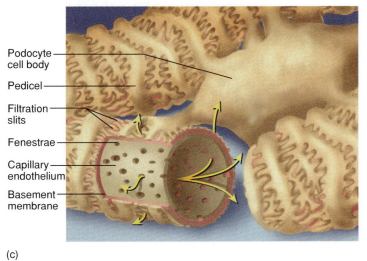

Podocyte
cell body

Pedicel

Filtration
slits

Fenestrae

Capillary
endothelium

Basement
membrane

(c)

Figure 23.8 Structure and function of the glomerulus. (*a*) Blood capillaries of the glomerulus closely wrapped in the spidery podocytes that form the visceral layer of the glomerular capsule (SEM). (*b*) A blood capillary and podocyte, showing fenestrations and filtration slits (TEM). (*c*) The production of glomerular filtrate by the passage of fluid through the fenestrations and filtration slits.

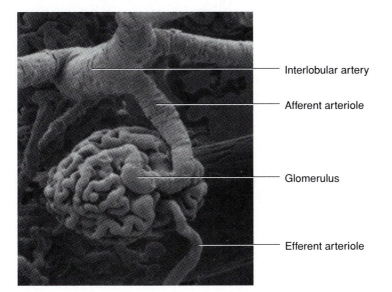

Interlobular artery

Afferent arteriole

Glomerulus

Efferent arteriole

Figure 23.9 Vascular cast of a glomerulus and its associated arteries. All the tissues have been removed, leaving only a resin cast of the blood vessels. Note that the afferent arteriole is substantially larger than the efferent arteriole, causing blood pressure in the glomerular capillaries to be unusually high (SEM). ✗

From R. G. Kessel and R. H. Kardon, *Scanning Electron Microscopy of Tissues and Organs,* 1979.

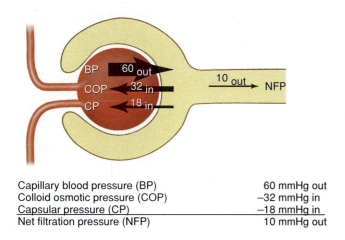

Capillary blood pressure (BP)	60 mmHg out
Colloid osmotic pressure (COP)	−32 mmHg in
Capsular pressure (CP)	−18 mmHg in
Net filtration pressure (NFP)	10 mmHg out

Figure 23.10 The forces involved in glomerular filtration.

- The colloid osmotic pressure (COP) of the blood is about the same here as anywhere else, 32 mmHg.
- The glomerular filtrate is almost protein-free and has no significant COP. (This can change markedly in kidney diseases that allow substantial amounts of protein to leave the glomerular capillaries.)

On balance, then, we have a high outward pressure of 60 mmHg, opposed by two inward pressures of 18 and 32 mmHg (fig. 23.10), giving a net filtration pressure (NFP) of

$$60_{out} - (18_{in} + 32_{in}) = 10 \text{ mmHg}_{out}$$

In most blood capillaries, the BP drops low enough at the venous end that osmosis overrides filtration and the capillaries reabsorb fluid. Although BP also drops along the course of the glomerular capillaries, it remains high enough that these capillaries are engaged solely in filtration. They reabsorb little or no fluid.

Think About It

How low would capillary BP have to be for glomerular filtration and urine output to cease?

Glomerular Filtration Rate

Glomerular filtration rate (GFR) is the amount of filtrate formed per minute by the two kidneys combined. For every 1 mmHg of net filtration pressure, the kidneys produce about 12.5 mL of filtrate per minute. This value, called the *filtration coefficient* (K_f), depends on the permeability and surface area of the filtration barrier. K_f is about 10% lower in women than in men. For an average adult male,

$$GFR = NFP \times K_f = 10 \times 12.5 = 125 \text{ mL/min}$$

This is a rate of 180 L/day—an impressive number considering that it is about 60 times the amount of blood plasma in the body and 60 times the amount of filtrate produced by all other capillaries combined. Obviously only a small portion of this is eliminated as urine. An average adult reabsorbs 99% of the filtrate and excretes 1 to 2 L of urine per day.

Regulation of Glomerular Filtration

GFR must be precisely controlled. If it is too high, fluid flows through the renal tubules too rapidly for them to reabsorb the usual amount of water and solutes. Urine output rises and creates a threat of dehydration and electrolyte depletion. If GFR is too low, fluid flows sluggishly through the tubules, they reabsorb wastes that should be eliminated in the urine, and azotemia may occur. The only way to adjust GFR from moment to moment is to change glomerular blood pressure. This is achieved by three homeostatic mechanisms: renal autoregulation, sympathetic control, and a hormonal mechanism involving renin and angiotensin.

Renal Autoregulation

Renal autoregulation is the ability of the kidneys to maintain a relatively stable GFR in spite of changes in arterial blood pressure. If the mean arterial blood pressure (MABP) rose from 100 to 125 mmHg and there were no renal autoregulation, urine output would increase from the normal 1 to 2 L/day to more than 45 L/day. Because of renal autoregulation, however, urine output increases only a few percent even if MABP rises as high as 160 mmHg.

The nephron has two ways to prevent drastic changes in GFR when blood pressure rises: it can constrict the afferent arteriole and reduce blood flow into the glomerulus, or it can dilate the efferent arteriole and allow the blood to flow out more easily. Conversely, if blood pressure falls, the nephron can compensate by either dilating the afferent arteriole or constricting the efferent arteriole. In actuality, the two arterioles work in unison to fine-tune glomerular blood pressure.

The juxtaglomerular apparatus (JGA) enables each nephron to respond to blood pressure fluctuations. It consists of three types of cells (fig. 23.11):

1. The **juxtaglomerular (JG) cells** are enlarged smooth muscle cells found in the afferent arteriole and to some extent in the efferent arteriole. When stimulated by changes in blood pressure or by the macula densa (discussed next), they dilate or constrict the arterioles accordingly. They also contain granules of renin, which they secrete in response to a drop in blood pressure. This initiates negative feedback mechanisms, described later, that raise blood pressure.

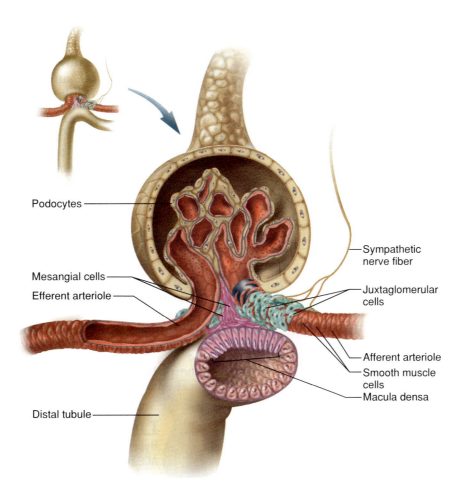

Figure 23.11 The juxtaglomerular apparatus.

Podocytes

Mesangial cells

Efferent arteriole

Distal tubule

Sympathetic nerve fiber

Juxtaglomerular cells

Afferent arteriole

Smooth muscle cells

Macula densa

2. The **macula densa**[15] is a patch of slender, closely spaced epithelial cells in the distal tubule, directly across from the juxtaglomerular cells. It is thought to monitor the salinity of tubular fluid in the DCT.

3. **Mesangial[16] cells** are found in the cleft between the afferent and efferent arterioles and among capillaries of the glomerulus. Their role is not yet clearly understood, but they are connected to the macula densa and juxtaglomerular cells by gap junctions and perhaps mediate communication between those cells.

The mode of action of the juxtaglomerular apparatus has not been fully determined yet. One hypothesis is that a high GFR causes fluid to flow too rapidly through the renal tubule to allow for normal reabsorption, resulting in an abnormally high NaCl concentration detected by the macula densa. The macula densa may then release a paracrine secretion that stimulates the af-

ferent arteriole to constrict, stimulates the efferent arteriole to dilate, or both. This would reduce the GFR to a normal level and complete a negative feedback loop (fig. 23.12).

Think About It

Describe or diagram a similar negative feedback loop to show how the macula densa could compensate for a drop in systemic blood pressure.

Two important points must be noted about renal autoregulation. First, it does not completely prevent changes in the GFR. Like any other homeostatic mechanism, it maintains a *dynamic equilibrium;* the GFR fluctuates within narrow limits. Changes in blood pressure do affect the GFR and urine output. When you drink a large amount of water, for example, it increases your blood pressure and GFR, causing *water diuresis*[17] (DY-you-REE-sis)—an increased urine output stemming from increased water intake.

---

15. *macula* = spot, patch + *densa* = dense
16. *mes* = in the middle + *angi* = vessel

17. *diuresis* = passing urine

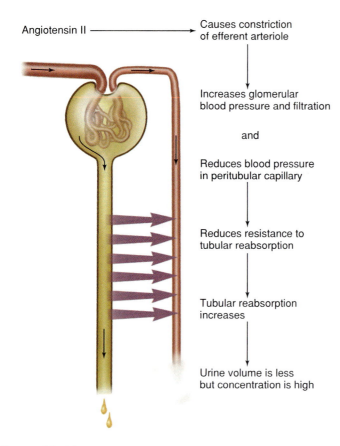

Angiotensin II ⟶ Causes constriction of efferent arteriole

⟶ Increases glomerular blood pressure and filtration

and

Reduces blood pressure in peritubular capillary

Reduces resistance to tubular reabsorption

Tubular reabsorption increases

Urine volume is less but concentration is high

Figure 23.14 The effect of angiotensin II on urine volume and concentration.

from them are engaged mainly in **tubular reabsorption.** One reason for this shift in function is that approximately 20% of the water in the blood is filtered out by the glomerulus while the protein stays behind. Therefore, the blood has an unusually high colloid osmotic pressure (COP) downstream from this point. Furthermore, the narrowness of the efferent arteriole reduces the blood pressure (BP) from about 60 mmHg in the glomerular capillaries to only 8 mmHg in the peritubular capillaries—even lower when this arteriole is constricted by antiogensin II (fig. 23.14). With an elevated COP and reduced BP, the balance of forces in the peritubular capillaries favors reabsorption. Water is reabsorbed by osmosis and carries along other solutes by solvent drag.

The PCT reabsorbs a greater variety of chemicals than any other part of the nephron. Some substances are taken up by the tubule epithelial cells, pass through their cytoplasm, and are then pumped out, or diffuse out, the base of the cell. This is called the **transcellular[18] route** of reabsorption. Other substances pass between the epithelial cells, a path called the **paracellular[19] route** of reabsorption. Even though the

18. *trans* = across
19. *para* = next to

epithelial cells are held together by tight junctions, these junctions are quite leaky and allow significant amounts of water, minerals, urea, and other matter to pass between the cells. Either way, such materials enter the ECF at the base of the epithelium, and from there they can be taken up by the peritubular capillaries. In the following discussion and figure 23.15, we examine mechanisms for the reabsorption of water and individual solutes.

Sodium Sodium, the most abundant cation in the glomerular filtrate, is reabsorbed by both transcellular and paracellular routes. It has a concentration of 140 mEq/L in the fluid entering the PCT and only 12 mEq/L in the cytoplasm of the PCT epithelial cells. Thus there is a steep concentration gradient favoring the diffusion of Na^+ into the epithelial cell. Some enters the cell by simple diffusion through membrane channels, and some is reabsorbed by facilitated diffusion using transport proteins in the apical plasma membrane.

This explains how Na^+ enters the epithelial cell, but we also must explain how the cell maintains the low intracellular Na^+ concentration that makes this possible. This is achieved by Na^+–K^+ pumps in the basal and lateral plasma membrane, which continually remove Na^+ from the cytoplasm and pump it into the ECF around the peritubular capillaries. As water enters the peritubular capillaries, it takes sodium with it by solvent drag. Sodium reabsorption is especially important to renal function because it creates a concentration gradient that makes it possible to reabsorb other substances, as we shall see next.

Glucose Some of the apical Na^+ carriers bind simultaneously to a sodium ion and a glucose molecule and transport both of them into the cell. Such a carrier protein is called the **sodium-dependent glucose transporter (SGLT).** Glucose leaves the basal surface of the cell by facilitated diffusion and enters the peritubular capillaries by solvent drag. Normally all glucose in the tubular fluid is reabsorbed and there is none in the urine.

Glucose reabsorption exemplifies certain principles of membrane transport described in chapter 4. It is an example of a *symport* system because the SGLT moves both solutes in the same direction. It is also a case of *secondary active transport,* a system that does not directly consume ATP but depends on other transporters that do. Glucose is more concentrated within the cell than in the tubular fluid, so when it enters the cell it is moving up its concentration gradient. The SGLT itself does not use ATP in doing this, but it depends on the low intracellular Na^+ concentration maintained by the Na^+–K^+ pumps at the base of the cell. Thus the absorption of glucose indirectly depends on ATP.

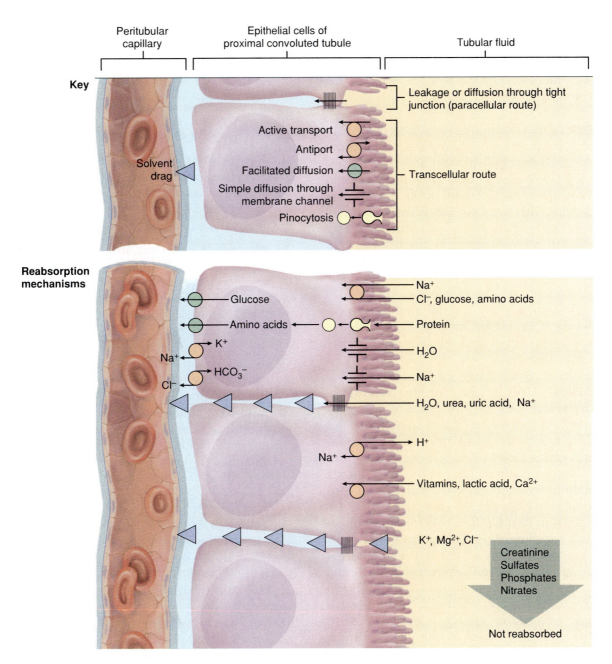

Figure 23.15 Mechanisms of reabsorption in the proximal convoluted tubule.

Amino Acids Amino acids are reabsorbed in the same way as glucose. Some apical Na$^+$ carriers cotransport amino acids into the cell. The amino acids leave the base of the cell by facilitated diffusion and enter the capillaries by solvent drag.

Water Sodium reabsorption makes the ICF and ECF hypertonic to the tubular fluid. Water follows Na$^+$ by diffusion through the paracellular route and by osmosis through the transcellular route. Blood is hypertonic to the ECF, so after passing through the epithelial cells, water enters the peritubular capillaries by osmosis. Because the PCT reabsorbs proportionate amounts of salt

and water, the osmolarity of the tubular fluid remains unchanged in the PCT. Elsewhere in the nephron, the amount of water reabsorption is continually modulated by hormones according to the body's state of hydration. In the PCT, however, water is reabsorbed at a constant rate called **obligatory water reabsorption.**

Chloride Chloride is reabsorbed along both the paracellular and transcellular routes. Its reabsorption is favored by two factors: (1) negative chloride ions tend to follow the positive sodium ions by electrostatic attraction, and (2) water reabsorption raises the Cl$^-$ concentration in the tubular fluid, creating a gradient favorable to Cl^- reabsorption. In the transcellular route, Cl^- is apically absorbed by Na^+-linked cotransport. At the base of the cell, chloride pumps remove Cl^- in exchange for HCO_3^-; this is a *chloride shift* mechanism like the one we saw in red blood cells (see fig. 22.24, p. 818).

Other Electrolytes   Potassium, magnesium, and variable amounts of calcium pass through the paracellular route by solvent drag. Sulfates, phosphates, and nitrates are not reabsorbed, however; thus they pass in the urine. Substantial amounts of bicarbonate enter the tubular fluid by glomerular filtration, yet the urine is usually bicarbonate-free. No bicarbonate ions are reabsorbed by the renal tubule, but all bicarbonate in the fluid is usually neutralized by acid in the tubular fluid (see chapter 24 for details).

Protein   The glomerulus filters a small amount of protein from the blood. The PCT reclaims it by pinocytosis, hydrolyzes it to amino acids, and releases these to the ECF by facilitated diffusion.

Nitrogenous Wastes   Urea diffuses through the tubule epithelium with water. The nephron as a whole reabsorbs 40% to 60% of the urea in the tubular fluid, but since it reabsorbs 99% of the water, urine has a substantially higher urea concentration than blood or glomerular filtrate. When blood enters the kidney, it has about 20 mg of urea per deciliter; when it leaves the kidney, the concentration is down to 10.4 mg/dL. Thus the kidney removes about half of it, keeping urea concentration down to a safe level but not completely clearing the blood of it.

The PCT reabsorbs nearly all the uric acid entering it, but later parts of the nephron secrete it back into the tubular fluid. Creatinine is not reabsorbed at all. It is too large to diffuse through water channels in the plasma membrane, and there are no transport proteins for it. Therefore, all creatinine filtered by the glomerulus is excreted in the urine.

The Transport Maximum

There is a limit to the amount of solute that the renal tubule can reabsorb because there are a limited number of transport proteins in the plasma membranes. If all the transporters are occupied as solute molecules pass through, some solute will remain in the tubular fluid and appear in the urine. The maximum rate of reabsorption is the *transport maximum* (T_m), which is reached when the transporters are saturated (see chapter 4). Each organic solute reabsorbed by the renal tubule has its own T_m. For glucose, for example, $T_m = 320$ mg/min. Glucose normally enters the renal tubule at a rate of 125 mg/min, well within the T_m; thus all of it is reabsorbed. When the plasma concentration of glucose reaches a

threshold of about 220 mg/dL, however, more glucose is filtered than the tubule reabsorbs and we begin to see **glycosuria**[20] (GLY-co-soo-ree-uh) (glucose in the urine). In untreated diabetes mellitus, the plasma glucose concentration may exceed 400 mg/dL.

Tubular Secretion

Tubular secretion is a process in which the renal tubule extracts chemicals from the capillary blood and secretes them into the tubular fluid (see fig. 23.7). Tubular secretion in the distal convoluted tubule is discussed shortly. In the proximal convoluted tubule, it serves two purposes:

1. **Waste removal.** Urea, uric acid, bile salts, ammonia, catecholamines, and a little creatinine are secreted into the tubule. Tubular secretion of uric acid compensates for its reabsorption earlier in the PCT and accounts for all of the uric acid in the urine. Tubular secretion also clears the blood of penicillin, aspirin, and other drugs. It is necessary to take a drug such as penicillin four times a day to compensate for this and maintain an effective concentration in the blood.
2. **Acid-base balance.** Tubular secretion of hydrogen and bicarbonate ions serves to regulate the pH of the body fluids. The details are discussed in chapter 24.

The Nephron Loop

The primary purpose of the nephron loop is to enable the collecting duct to concentrate the urine and conserve water, as discussed later. But in addition, it reabsorbs about 25% of the Na^+, K^+, and Cl^- and 20% of the water in the glomerular filtrate. Cells in the thick segment of the loop have proteins in the apical membranes that simultaneously bind 1 Na^+, 1 K^+, and 2 Cl^- from the tubular fluid and cotransport them into the cytoplasm. These ions leave the basolateral cell surfaces by active transport of Na^+ and passive diffusion of K^+ and Cl^-. The thick segment is impermeable to water; thus water cannot follow the reabsorbed electrolytes, and tubular fluid becomes very dilute by the time it passes from the nephron loop into the distal convoluted tubule.

The Distal Convoluted Tubule and Collecting Duct

Fluid arriving in the DCT still contains about 20% of the water and 10% of the salts from the glomerular filtrate. If this were all passed as urine, it would amount to 36 L/day, so obviously there is still a great deal of

---

20. *glycos* = sugar + *uria* = urine condition

fluid reabsorption to occur in the DCT and collecting duct. A distinguishing feature of these parts of the renal tubule is that they are subject to hormonal control.

Aldosterone

Aldosterone, the "salt-retaining hormone," is a steroid secreted by the adrenal cortex. A drop in blood Na^+ concentration or a rise in K^+ concentration directly stimulates aldosterone secretion. A drop in blood pressure does so indirectly—it stimulates the kidney to secrete renin, this leads to the production of angiotensin II, and angiotensin II stimulates aldosterone secretion (see fig. 23.14). The mechanism of aldosterone action on the kidney tubule is detailed in the following chapter, but its general effect is to cause the DCT and cortical portion of the collecting duct to reabsorb more Na^+ (which is followed by Cl^- and water) and to secrete more K^+. Thus the urine volume is reduced, and it contains more K^+ but less NaCl. Salt and water reabsorption helps to maintain blood volume and pressure.

Atrial Natriuretic Factor

Atrial natriuretic factor (ANF) is secreted by the atrial myocardium of the heart in response to high blood pressure. To some degree its effects are opposite those of aldosterone: it inhibits Na^+ and water reabsorption, thus increasing urine and Na^+ output and reducing blood volume and pressure.

In summary, the PCT reabsorbs about 65% of the glomerular filtrate and returns it to the blood of the peritubular capillaries. Much of this reabsorption occurs by osmotic and cotransport mechanisms linked to the active transport of sodium ions. The nephron loop reabsorbs another 25% of the solutes, although its primary purpose, detailed later, is to aid the function of the collecting duct. The DCT reabsorbs more sodium, chloride, and water, but its rates of reabsorption are subject to control by hormones, especially aldosterone and ANF. These tubules also extract drugs, wastes, and some other solutes from the blood and secrete them into the tubular fluid. The DCT essentially completes the process of determining the chemical composition of the urine. The principal function left to the collecting duct is to concentrate it and prevent excessive water loss from the body

Key Point Review

12 The reabsorption of water, Cl^-, and glucose by the PCT are all linked to the reabsorption of Na^+, but in three very different ways. Contrast these three mechanisms.

13 Explain why a substance appears in the urine if its concentration in the glomerular filtrate exceeds the T_m of the renal tubule.

14 Contrast the effects of aldosterone and ANF on the renal tubule.

Urine Formation III: Concentrating the Urine

▼Objectives

When you have completed this section, you should be able to
- explain how the collecting duct and antidiuretic hormone regulate the volume and concentration of urine; and
- explain how the countercurrent multiplier and countercurrent exchange systems maintain the osmotic gradient of the renal medulla.

The Collecting Duct

The collecting duct (CD) begins in the cortex, where it receives tubular fluid from numerous nephrons. As it passes through the medulla, it reabsorbs water and concentrates the urine. Urine is isotonic with blood plasma (300 mOsm/L) when it enters the upper end of the CD, and it can be up to four times as concentrated by the time it leaves the lower end. This ability to concentrate wastes and control water loss was crucial to the evolution of terrestrial animals such as ourselves (see special topic 23.2).

Special Topic — The Kidney and Life on Dry Land — 23.2

Physiologists first suspected that the nephron loop plays a role in water conservation because of their studies of a variety of animal species. Animals that must conserve water have longer, more numerous nephron loops than animals with little need to conserve it. Fish and amphibians lack nephron loops and produce urine that is isotonic to their blood plasma. Aquatic mammals such as beavers have short nephron loops and only slightly hypertonic urine.

But the kangaroo rat, a desert rodent, provides an instructive contrast. It lives on seeds and other dry foods and never drinks water. The water produced by its aerobic respiration is enough to meet its needs because its kidneys are extraordinarily efficient at conserving water. They have extremely long nephron loops and produce urine that is 10 to 14 times as concentrated as their blood plasma (compared with 4 times, at most, in humans).

Comparative studies thus suggested a hypothesis for the function of the nephron loop and prompted many years of difficult research that led to the discovery of the countercurrent exchange mechanism. This shows how comparative anatomy provides suggestions and insights into function and why physiologists do not study human function in isolation from other species.

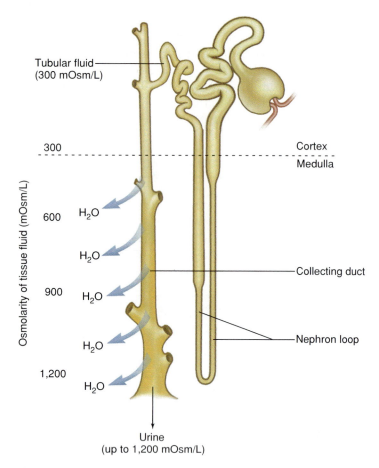

Tubular fluid (300 mOsm/L)

300

Cortex
Medulla

Osmolarity of tissue fluid (mOsm/L)

600   H_2O

H_2O

900   H_2O

Collecting duct

H_2O

Nephron loop

1,200

H_2O

Urine
(up to 1,200 mOsm/L)

Figure 23.16 Reabsorption of water by the collecting duct. Note that the osmolarity of the tissue fluid increases fourfold from 300 mOsm/L in the cortex to 1,200 mOsm/L deep in the medulla. When the collecting duct has open water channels, water leaves the duct by osmosis and urine concentration increases. ✗

Two facts enable the collecting duct to produce such hypertonic urine: (1) the osmolarity of the extracellular fluid is four times as high near the renal papilla as it is in the cortex, and (2) the medullary portion of the CD is permeable to water but not to NaCl. Therefore, as urine passes down the CD through the increasingly salty medulla, water leaves the tubule by osmosis, NaCl and other wastes remain in the tubule, and the urine becomes more and more concentrated (fig. 23.16).

Control of Concentration

Just *how* concentrated the urine becomes depends on the body's state of hydration. For example, if you drink a large volume of water, you will soon produce a large volume of hypotonic urine (water diuresis). Under such conditions, the cortical portion of the CD reabsorbs NaCl but is impermeable to water. Thus salt is removed from the urine, water stays in the CD, and urine concentration may be as low as 50 mOsm/L.

If you are dehydrated, your urine will be scanty and more concentrated. One reason for this effect is that when your body has lost a lot of fluid, your blood pressure is lower and your glomerular filtration rate (GFR) drops. When the GFR is low, fluid flows more slowly through the renal tubules and there is more time for tubular reabsorption. Less salt remains in the urine as it enters the collecting duct, so there is less opposition to the osmosis of water out of the duct and into the ECF. More water is reabsorbed and less urine is produced.

In addition, the high blood osmolarity of a dehydrated person stimulates the release of antidiuretic hormone (ADH) from the posterior lobe of the pituitary gland. The mechanism of ADH action on the collecting duct is explained in the next chapter, but the general effect is that water permeability of the CD increases. Its cortical portion, especially, reabsorbs more water, which is carried away by the peritubular capillaries. Urine output is consequently reduced. By contrast, ADH secretion is inhibited when you are well hydrated. The duct is then less permeable to water, so more of it remains in the duct and you produce abundant, dilute urine.

The Countercurrent Multiplier

The ability of the CD to concentrate urine depends on the salinity gradient of the renal medulla. It may seem surprising that the ECF is four times as salty deep in the medulla as it is at the corticomedullary junction. We would expect the salt to diffuse toward the cortex until it was evenly distributed through the kidney. However, there is a mechanism that overrides this (fig. 23.17)— the nephron loop acts as a **countercurrent multiplier,** which continually returns salt to the deep medullary tissue. It is called a *multiplier* because it multiplies the salinity deep in the medulla and a *countercurrent* mechanism because it is based on fluid flowing in opposite directions in two adjacent tubules—downward in the descending limb and upward in the ascending limb.

The descending limb is very permeable to water but not to NaCl. Therefore, more and more water leaves as the tubular fluid descends into the increasingly salty medulla, while NaCl remains behind in the tubule. As the fluid reaches the lower end of the loop, it has a concentration of about 1,200 mOsm/L. The ascending limb, by contrast, is impermeable to water, but the thick segment actively transports Na^+ into the ECF and cotransports K^+ and Cl^- with it. The more salt (NaCl and KCl) there is in the tubular fluid, the more is pumped into the ECF. This generates the high osmolarity of the renal medulla. Since water cannot follow NaCl out of the ascending limb, the tubular fluid becomes more and more dilute as it approaches the cortex. It is about 100 mOsm/L at the top of the loop. The essence of the *countercurrent* mechanism is that the two limbs of the nephron loop are close enough

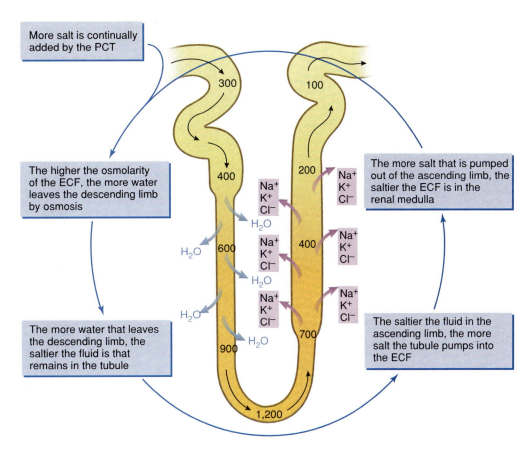

Figure 23.17 The countercurrent multiplier mechanism of the nephron loop.

to influence each other through the positive feedback relationship shown in fig. 23.17.

The collecting duct also helps to maintain the osmotic gradient (fig. 23.18). Its lower end is somewhat permeable to urea, which diffuses down its concentration gradient, out of the duct and into the tissue fluid. Some of this urea enters the descending thin segment of the nephron loop and travels to the distal convoluted tubule. Neither the thick segment of the loop nor the distal tubule is permeable to urea, so urea remains in the tubules and returns to the collecting duct. Combined with new urea being added continually by the glomerular filtrate, urea becomes more and more concentrated in the fluid of the collecting duct, and still more diffuses out into the medulla. Thus there is a continual recycling of urea from the collecting duct to the medulla and back. Urea accounts for about 40% of the high osmolarity deep in the medulla.

The Countercurrent Exchanger

The renal medulla must have a blood supply to meet its metabolic needs, and this creates a potential problem—capillaries of the medulla could carry away the urea and salt that produce the high medullary osmolarity. The vasa recta that supply the medulla, however, form a countercurrent system of their own that prevents this from happening. Blood flows in opposite directions in adjacent parallel capillaries. As the blood descends into the medulla, water diffuses out and salt diffuses into it, making the blood nearly isotonic with the ECF. As the blood flows back up toward the cortex, salt diffuses out and water diffuses in—each of them simply following its concentration gradient. Thus the vasa recta do not carry salt away from the medulla, and they reabsorb as much water on the way out as they unload on the way in. The vasa recta system is called a **countercurrent exchanger** because it does not increase the osmolarity of the medulla, but the capillaries are arranged in a way that prevents them from subtracting from it.

To summarize what we have studied in this section, the collecting duct is able to adjust water reabsorption to produce urine as hypotonic as 50 mOsm/L or as hypertonic as 1,200 mOsm/L, depending on the body's need for water conservation or removal. In a state of dehydration, ADH is not secreted and the cortical part of the CD reabsorbs salt without reabsorbing water; it leaves the latter to be excreted in the dilute urine. In a state of dehydration, ADH is secreted, the medullary part of the CD reabsorbs water, and the urine is more

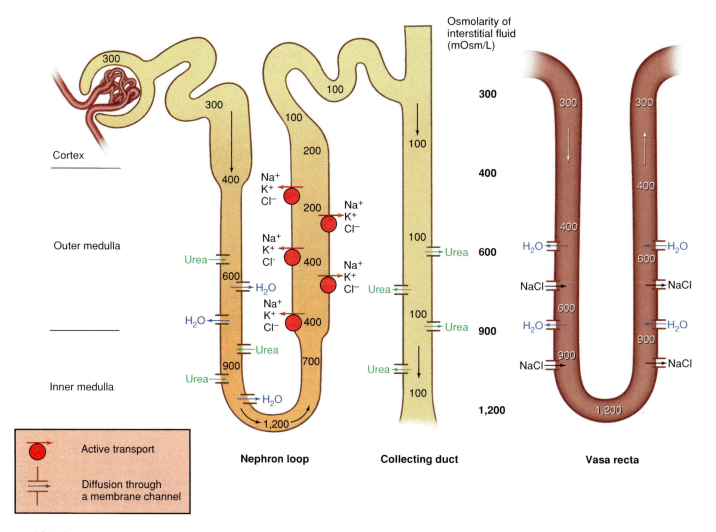

Osmolarity of
interstitial fluid
(mOsm/L)

Nephron loop   **Collecting duct**   **Vasa recta**

Active transport

Diffusion through
a membrane channel

Figure 23.18 The physiological relationship of the nephron loop, vasa recta, and collecting duct in maintaining a gradient of osmolarity in the renal medulla.

concentrated. The CD is able to do this because it passes through a salinity gradient in the medulla from 300 mOsm/L near the cortex to 1,200 mOsm/L near the papilla. This gradient is produced by a countercurrent multiplier of the nephron loop, which concentrates NaCl in the lower medulla, and by the diffusion of urea from the collecting duct into the medulla. The vasa recta are arranged as a countercurrent exchange system that enables them to supply blood to the medulla without subtracting from its salinity gradient.

Figure 23.19 summarizes the major solutes reabsorbed and secreted in each part of the renal tubule.

Key Point Review

15 Predict and explain the effect of ADH hypersecretion on the sodium concentration of the urine.

16 Concisely contrast the role of the countercurrent multiplier with that of the countercurrent exchanger.

17 How would the function of the collecting duct change if the nephron loop did not exist?

Urine and Renal Function Tests

▼**Objectives**

When you have completed this section, you should be able to
• describe the composition and properties of urine; and
• carry out some calculations to evaluate renal function.

Medical diagnosis often rests on determining the current and recent physiological state of the tissues. No two fluids are as valuable for this purpose as blood and urine. **Urinalysis,** the observation of the physical and chemical properties of urine, is therefore one of the most routine procedures in medical examinations. The principal characteristics of urine and certain tests used to evaluate renal function are described here.

Composition and Properties of Urine

Appearance. The yellow color of urine is due to **urochrome,** a pigment from the breakdown of hemoglobin. Urine varies from almost colorless to deep amber, depending on the body's state of hydration. Pink, green, brown, black, and other colors result from certain foods, vitamins, drugs, and metabolic diseases. Urine is normally clear but turns cloudy upon standing due to bacterial growth. Pus in the urine (**pyuria**) makes it cloudy and suggests kidney infection. Blood in the urine (hematuria) may be due to a urinary tract infection, trauma, or kidney stones. Cloudiness or blood in a urine specimen sometimes, however, simply indicates contamination with semen or menstrual fluid.

Odor. Fresh urine has a distinctive but not repellent odor. As it stands, however, bacteria multiply, degrade urea to ammonia, and produce the pungent odor typical of stale wet diapers. Asparagus and other foods can impart distinctive aromas to the urine. Diabetes mellitus gives it a sweet, "fruity" odor of acetone. A "mousy" odor suggests phenylketonuria (PKU), and a foul odor may indicate urinary tract infection.

Specific gravity. This is a ratio of the density (g/mL) of a substance to the density of distilled water. Distilled water has a specific gravity of 1.000, and urine ranges from 1.001 when it is very dilute to 1.035 when it is very concentrated. Multiplying the last two digits of the specific gravity by a proportionality constant of 2.6 gives an estimate of the grams of solid matter per liter of urine. For example, a specific gravity of 1.025 indicates a solute concentration of $25 \times 2.6 = 65$ g/L.

Osmolarity. Urine can have an osmolarity as low as 50 mOsm/L in a very hydrated person or as high as 1,200 mOsm/L in a dehydrated person. Compared to the osmolarity of blood (300 mOsm/L), then, urine can be either hypotonic or hypertonic under different conditions.

pH. The pH of urine ranges from 4.5 to 8.2 but is usually about 6.0 (mildly acidic). The regulation of urine pH is discussed extensively in chapter 24.

Chemical composition. Urine averages 95% water and 5% solutes by volume (table 23.1). The most abundant solute is urea, followed by sodium and potassium chlorides and lesser amounts of creatinine, uric acid, phosphates, sulfates, and traces of calcium, magnesium, and sometimes bicarbonate. Glucose, free hemoglobin, albumin, ketones, and bile pigments in the urine are important indicators of disease.

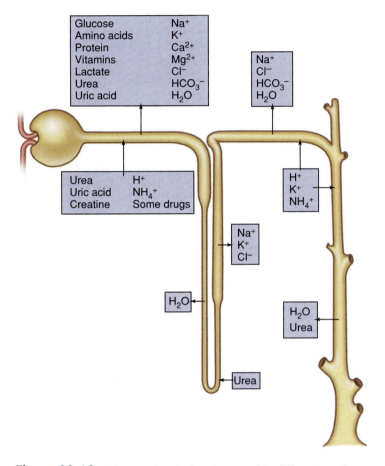

Figure 23.19 Solutes reabsorbed and secreted in different portions of the renal tubule.

Think About It

In table 23.1, what process accounts for most of the difference between values in the middle column and those on the right?

Table 23.1	Average Composition of Blood Plasma, Glomerular Filtrate, and Urine		
Substance	**Blood Plasma (total amount)**	**Glomerular Filtrate (amount per day)**	**Urine (amount per day)**
Water	3 L	180 L	1–2 L
Urea	4.8 g	53 g	25 g
Chloride	10.7 g	639 g	6.3 g
Sodium	9.7 g	580 g	4.6 g
Potassium	0.5 g	30 g	2.0 g
Creatinine	0.03 g	1.6 g	1.6 g
Uric acid	0.15 g	8.5 g	0.8 g
Protein	200 g	2 g	0.1 g
Bicarbonate	4.6 g	275 g	0 g
Glucose	3 g	180 g	0 g

Urine Volume

An average adult produces 1 to 2 liters of urine per day. Any output in excess of this is called **polyuria**[21] (POL-ee-YOU-ree-uh), or **diuresis** (DIE-you-REE-sis). Fluid intake and some drugs can temporarily increase output to as much as 20 L/day. Chronic diseases such as diabetes (see next) can do so over a long term. **Oliguria**[22] (oll-ih-GUE-ree-uh) is an output of less than 500 mL/day, and **anuria**[23] is an output of 0 to 100 mL/day. Low output can result from kidney disease, dehydration, circulatory shock, prostate enlargement, and other causes. An output of less than 400 mL/day is insufficient to maintain a safe, homeostatic concentration of wastes in the blood plasma and can lead to azotemia (see *obligatory water loss* in the next chapter).

Diabetes

Diabetes[24] is chronic polyuria resulting from various metabolic disorders. In most cases, the polyuria results from a high concentration of glucose in the renal tubule. Glucose opposes the osmotic reabsorption of water, so more water is passed in the urine (*osmotic diuresis*) and a person may become severely dehydrated. In three forms of diabetes, the high glucose concentration in the tubule is a result of hyperglycemia, a high concentration of glucose in the blood: (1) *diabetes mellitus,* caused by insulin hyposecretion or inactivity (see chapter 17); (2) *pituitary diabetes,* caused by growth hormone hypersecretion; and (3) *adrenal diabetes,* caused by cortisol hypersecretion. In *renal diabetes,* blood glucose level is not elevated, but there is a hereditary deficiency of glucose transporters in the PCT and glucose remains in the tubular fluid.

All of the preceding forms result in glycosuria. Before chemical tests for urine glucose were developed, physicians diagnosed diabetes mellitus by tasting the patient's urine for sweetness.[25] Tests for glycosuria are now as simple as dipping a chemical test strip into the urine specimen—an advance in medical technology for which urologists are no doubt grateful. In yet another type of diabetes, *diabetes insipidus,*[26] the urine contains no glucose and, by the old diagnostic method, does not taste sweet. This disease results from ADH hyposecretion; the collecting duct does not reabsorb as much water as usual, so more water is passed in the urine.

---

21. *poly* = many, much
22. *oligo* = few, a little
23. *an* = without
24. *diabetes* = passing through
25. *melli* = honey, sweet
26. *insipid* = tasteless

Diuretics

Diuretics are chemicals that increase urine volume. They are used for treating hypertension and congestive heart failure because they reduce the body's fluid volume and blood pressure. Diuretics work by one of two mechanisms—increasing glomerular filtration or reducing tubular reabsorption. For example, caffeine, in the former category, dilates the afferent arteriole and increases GFR. Alcohol, in the latter category, inhibits ADH secretion. Also in the latter category are many osmotic diuretics, which reduce water reabsorption by increasing the osmolarity of the tubular fluid. Many diuretic drugs, such as furosemide (Lasix), produce osmotic diuresis by inhibiting sodium reabsorption.

Renal Function Tests

There are several tests for diagnosing kidney diseases, evaluating their severity, and monitoring their progress. Here we examine two methods used to determine renal clearance and glomerular filtration rate.

Renal Clearance

Renal clearance is the volume of blood plasma from which a particular waste is completely removed in 1 minute. It represents the net effect of three processes:

Glomerular filtration of the waste
+ Amount added by tubular secretion
− Amount removed by tubular reabsorption

= Renal clearance

In principle, we could determine renal clearance by sampling blood entering and leaving the kidney and comparing their waste concentrations. In practice, it is not practical to draw blood samples from the renal vessels, but clearance can be assessed indirectly by collecting samples of blood and urine, measuring the waste concentration in each, and measuring the rate of urine output.

Suppose the following values were obtained for urea:

U (urea concentration in urine)   = 6.0 mg/mL
P (urea concentration in plasma) = 0.2 mg/mL
V (rate of urine output)   = 2 mL/min

Renal clearance (C) is

$$C = \frac{UV}{P} = \frac{(6.0 \text{ mg/mL}) \times (2 \text{ mL/min})}{0.2 \text{ mg/mL}} = 60 \text{ mL/min}$$

This means the equivalent of 60 mL of blood plasma is completely cleared of urea per minute. If this person has a normal GFR of 125 mL/min, then the kidneys have cleared urea from only 60/125 = 48% of the glomerular filtrate. This is a normal rate of urea clearance, however, and is sufficient to maintain safe levels of urea in the blood.

Think About It

What would you expect the value of renal clearance of glucose to be in a healthy individual? Why?

Glomerular Filtration Rate

Assessment of kidney disease often calls for a measurement of glomerular filtration rate (GFR). We cannot determine GFR from urea excretion for two reasons: (1) some of the urea in the urine is secreted by the renal tubule, not filtered by the glomerulus, and (2) much of the urea filtered by the glomerulus is reabsorbed by the tubule. To measure GFR requires a substance that is not secreted or reabsorbed at all, so that all of it in the urine gets there by glomerular filtration.

There doesn't appear to be a single urine solute produced by the body that is not secreted or reabsorbed to some degree. However, several plants, including garlic and artichoke, produce a chemical that is useful for GFR measurement: inulin (IN-you-lin), a polymer of fructose. All inulin filtered by the glomerulus remains in the renal tubule and appears in the urine; none is reabsorbed, nor does the tubule secrete it. To measure GFR, inulin is injected into the bloodstream, the rate of urine output is measured, and the concentrations of inulin in blood and urine samples are determined. For this solute, GFR is equal to the renal clearance.

Suppose, for example, that a patient's plasma concentration of inulin is P = 0.5 mg/mL, the urine concentration is U = 30 mg/mL, and urine output is V = 2 mL/min. This person has a normal GFR:

$$C = \frac{UV}{P} = \frac{(6.0 \text{ mg/mL}) \times (2 \text{ mL/min})}{0.2 \text{ mg/mL}} = 60 \text{ mL/min}$$

A solute that is reabsorbed by the renal tubules will have a renal clearance *less* than the GFR (provided its tubular secretion is less than its rate of reabsorption). This is why renal clearance of urea is about 70 mL/min. A solute that is secreted by the renal tubules will have a renal clearance *greater* than the GFR (provided its reabsorption does not exceed its secretion). Creatinine, for example, has a renal clearance of 140 mL/min.

.... **Key Point Review** ....

18 Define *oliguria* and *polyuria*. Which of these is characteristic of diabetes?

19 Identify two causes of glycosuria other than diabetes mellitus.

20 How is the diuresis produced by furosemide like the diuresis produced by diabetes mellitus? How are they different?

21 Explain why GFR could not be determined from measurement of the amount of NaCl in the urine.

Urine Storage and Elimination

▼Objectives

When you have completed this section, you should be able to
- describe the functional anatomy of the ureters, urinary bladder, and male and female urethra; and
- explain how the nervous system and dual urethral sphincters control the voiding of urine.

Urine is produced continually, but fortunately it does not drain continually from the body. Urination is episodic—occurring when we allow it. This is made possible by an apparatus for storing urine and neural controls for its timely release.

The Ureters

The renal pelvis funnels urine into the ureter, a muscular tube that extends to the urinary bladder. The ureter is about 25 cm long and reaches a maximum diameter of about 1.7 cm near the bladder. It enters the bladder from below, passes obliquely through its muscular wall, and opens onto its floor. As pressure builds in the bladder, it compresses the ureters and prevents urine from being forced back to the kidneys.

The ureter has three layers: an adventitia, muscularis, and mucosa. The adventitia is a connective tissue layer that binds it to the surrounding tissues. The muscularis consists of two layers of smooth muscle. When urine enters the ureter and stretches it, the muscularis contracts and initiates a peristaltic wave that "milks" the urine down to the bladder. These contractions occur every few seconds to few minutes, proportional to the rate at which urine enters the ureter. The mucosa has a transitional epithelium continuous with that of the renal pelvis above and urinary bladder below. The lumen of the ureter is very narrow and is easily obstructed or injured by kidney stones (see special topic 23.3).

The Urinary Bladder

The urinary bladder (fig. 23.20) is a muscular sac on the floor of the pelvic cavity, inferior to the peritoneum and posterior to the pubic symphysis. It is covered by parietal peritoneum on its flattened superior surface and by a fibrous adventitia elsewhere. Its muscularis, called the **detrusor**[27] (deh-TROO-zur) **muscle,** consists of three layers of smooth muscle. The mucosa has a transitional epithelium, and in the relaxed bladder it has conspicuous wrinkles called **rugae**[28] (ROO-gee). The openings of the two ureters and the urethra mark a smooth-surfaced triangular area called the **trigone**[29] on the bladder floor.

---

27. *de* = down + *trus* = push
28. *ruga* = fold, wrinkle
29. *tri* = three + *gon* = angle

A **renal calculus**[30] (kidney stone) is a hard granule of calcium, phosphate, uric acid, and protein. Renal calculi form in the renal pelvis and are usually small enough to pass unnoticed in the urine flow. Some, however, grow to several centimeters in diameter and block the renal pelvis or ureter, which can lead to destruction of nephrons as pressure builds in the kidney. A large, jagged calculus passing down the ureter stimulates strong contractions that can be excruciatingly painful. Causes of renal calculi include hypercalcemia, dehydration, pH imbalances, frequent urinary tract infections, or an enlarged prostate gland causing urine retention. Calculi are sometimes treated with stone-dissolving drugs, but often require surgical removal. A nonsurgical technique called *lithotripsy*[31] uses ultrasonic vibration to pulverize the calculi into fine granules easily passed in the urine.

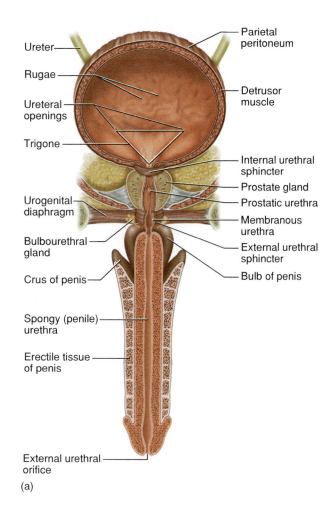

(a)

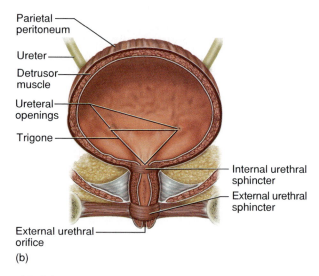

(b)

Figure 23.20 Anatomy of the urinary bladder and urethra. (*a*) Male. (*b*) Female. ✗

This is a common site of bladder infection (see special topic 23.4). For photographs of the relationship of the bladder and urethra to other pelvic organs in both sexes, see atlas A, figures A.38 and A.39.

The bladder is highly distensible. As it fills, it expands superiorly (fig. 23.21), the rugae flatten, and the wall becomes quite thin. A moderately full bladder contains about 500 mL of urine and extends about 12.5 cm from top to bottom. The maximum capacity is 700 to 800 mL.

The Urethra

The urethra conveys urine out of the body. In the female, it is a tube 3 to 4 cm long bound to the anterior wall of the vagina by fibroconnective tissue. Its opening, the **external urethral orifice,** lies between the vaginal orifice and clitoris. The male urethra is about 18 cm long and has three regions: (1) The **prostatic urethra** begins at the urinary bladder and passes for about 2.5 cm through the prostate gland. During orgasm, it receives semen from the reproductive glands. (2) The **membranous urethra** is a short (0.5 cm), thin-walled portion where the urethra passes through the muscular floor of the pelvic cavity. (3) The **penile urethra** is about 15 cm long and passes through the penis to the external urethral orifice. The male urethra assumes an S-shape: it passes downward from the bladder, turns anteriorly as it enters the root of the penis, and then turns about

30. *calc* = calcium, stone + *ul* = little
31. *litho* = stone + *tripsy* = crushing

Infection of the urinary bladder is called **cystitis.**[32] It is especially common in females because bacteria such as *Escherichia coli* can travel easily from the perineum up the short urethra. Because of this risk, young girls should be taught never to wipe the anus in a forward direction. If cystitis is untreated, bacteria can spread up the ureters and cause **pyelitis,**[33] infection of the renal pelvis. If it reaches the renal cortex and nephrons, it is called **pyelonephritis.** Kidney infections can also result from invasion by blood-borne bacteria. Urine stagnation due to renal calculi or prostate enlargement increases the risk of infection.

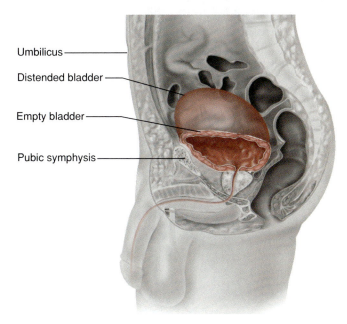

Umbilicus

Distended bladder

Empty bladder

Pubic symphysis

Figure 23.21 Extent of the bladder when empty and full.

90° downward again as it enters the external, pendant part of the penis. The mucosa has a transitional epithelium near the bladder, a pseudostratified epithelium for most of its length, and finally a stratified squamous epithelium near the external urethral orifice. There are mucous **urethral glands** in its wall.

In both sexes, the detrusor muscle is thickened near the urethra to form an **internal urethral sphincter,** which compresses the urethra and retains urine in the bladder. Since this sphincter is composed of smooth muscle, it is under involuntary control. Where the urethra passes through the pelvic floor, it is encircled by an **external urethral sphincter** of skeletal muscle, which provides voluntary control over the voiding of urine.

Voiding Urine

The neural pathways that control urination are shown in figure 23.22, which is numbered to correspond to the following description:

(1) When the bladder contains about 200 mL of urine, stretch receptors in the wall send afferent nerve impulses to the spinal cord by way of the pelvic nerves.

(2) By way of a parasympathetic reflex arc through sacral segments S2 to S3 of the cord, signals return to the bladder and stimulate contraction of the detrusor muscle **(3)** and relaxation of the internal urinary sphincter **(4).** This **micturition**[34] (MIC-too-RISH-un) **reflex** is the predominant mechanism that voids the bladder in infants and young children.

As the cerebral cortex and spinal cord mature, however, we acquire voluntary control over the external urethral sphincter, and emptying of the bladder is controlled predominantly by a **micturition center** in the pons. This center receives signals from the stretch receptors **(5)** and integrates this information with cortical input concerning the appropriateness of urinating at the moment. It sends back impulses **(6)** that excite the detrusor and relax the internal urinary sphincter. **(7)** At times when it is inappropriate to urinate, a steady train of nerve impulses travel from the brainstem through the pudendal nerve to the external urinary sphincter, keeping it contracted. When you wish to urinate, these impulses are inhibited, the external sphincter relaxes **(8),** and contractions of the detrusor muscle expel the urine. The Valsalva maneuver (p. 808) also aids in expulsion of urine by increasing pressure on the bladder. In males, the bulbocavernosus muscle encircling the base of the penis is voluntarily contracted to expel the last few milliliters of urine.

When it is desirable to urinate (for example, before a long trip) but the urge does not yet exist, the Valsalva maneuver can activate the micturition reflex. Contraction of the abdominal muscles compresses the bladder and may excite the stretch receptors even if there is less than 200 mL of urine in the bladder.

Key Point Review

22 Describe the location and function of the detrusor muscle.

23 Compare and contrast the functions of the internal and external urethral sphincters.

24 How would micturition be affected by a spinal cord lesion that prevented voluntary nerve impulses from reaching the sacral part of the cord?

32. *cyst* = bladder + *itis* = inflammation
33. *pyel* = pelvis
34. *mictur* = to urinate

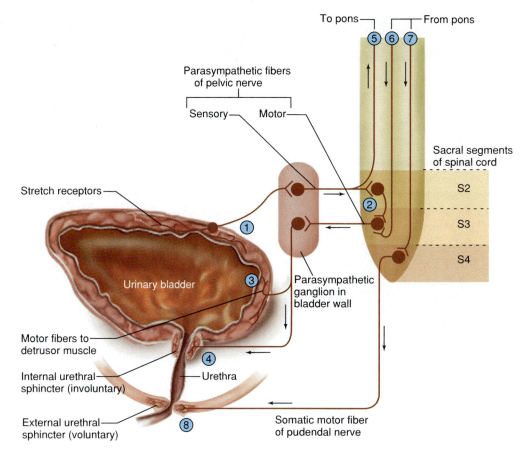

Figure 23.22 Neural control of micturition. Circled numbers correspond to text description.

Renal Insufficiency and Hemodialysis

Renal insufficiency is a state in which the kidneys cannot maintain homeostasis due to extensive destruction of their nephrons. Some causes of nephron destruction include the following:

- Chronic or repetitive kidney infections.
- Trauma from such causes as blows to the lower back or continual vibration from machinery.
- Poisoning by heavy metals, such as mercury and lead, and solvents, such as carbon tetrachloride, acetone, and paint thinners. These are absorbed into the blood from inhaled fumes or by skin contact and then filtered by the glomeruli. They kill renal tubule cells.
- Blockage of renal tubules with proteins small enough to be filtered by the glomerulus—for example, myoglobin released by skeletal muscle damage and hemoglobin released by a transfusion reaction.
- Atherosclerosis, which reduces blood flow to the kidney.
- Glomerulonephritis, an autoimmune disease of the glomerular capillaries.

Nephrons can regenerate and kidney function can recover after short-term injuries. Even when some of the nephrons are irreversibly destroyed, others hypertrophy and compensate for their lost function. Indeed, a person can survive on as little as one-third of one kidney. When 75% of the nephrons are lost, however, urine output may be as low as 30 mL/hr compared with the normal rate of 50 to 60 mL/hr. This is insufficient to maintain homeostasis and is accompanied by azotemia and acidosis. Uremia develops when there is 90% loss of renal function.

Acute renal failure is an abrupt decline in renal function, often due to traumatic damage to the nephrons or a loss of blood flow stemming from hemorrhage or thrombosis. **Chronic renal failure** is the long-term, progressive, irreversible loss of nephrons. Nephrons can regenerate and restore normal function after acute failure, but chronic failure requires a kidney transplant or hemodialysis.

Hemodialysis is a procedure for artificially clearing wastes from the blood when renal clearance is inadequate (fig. E.1). Blood is pumped from the radial artery to a *dialysis machine* (artificial kidney) and returned to the patient by way of a vein. In the dialysis machine, the blood flows through a semipermeable cellophane tube surrounded by isotonic dialysis fluid. Urea, potassium, and other solutes that are more concentrated in the blood than in the dialysis fluid diffuse through the membrane into the fluid, which is discarded. Glucose, electrolytes, and drugs can be administered by adding them to the dialysis fluid so they will diffuse through the membrane into the blood. Hemodialysis is not limited to treatment of renal insufficiency. It is also used in liver disease to remove wastes normally detoxified by the liver.

Hemodialysis sessions typically require 4 to 8 hours three times per week. In addition to inconvenience, it carries risks of infection and thrombosis. Blood tends to clot when it is exposed to foreign surfaces, so an anticoagulant such as heparin is added during hemodialysis. Unfortunately, this inhibits clotting in the patient's body as well, and hemodialysis patients sometimes suffer internal bleeding.

A procedure called *continuous ambulatory peritoneal dialysis (CAPD)* is more convenient. It can be carried out at home by the patient, who is provided with plastic bags of dialysis fluid. Fluid is introduced into the abdominal cavity through an indwelling catheter. Here, the peritoneum provides over 2 square meters of blood-rich semipermeable membrane. The fluid is left in the body cavity for 15 to 60 minutes to allow the blood to equilibrate with it; then it is drained, discarded, and replaced with fresh dialysis fluid. The patient is not limited by a stationary dialysis machine and can go about most normal activities. CAPD is less expensive and promotes better morale than conventional hemodialysis, but it is less efficient in removing wastes and it is more often complicated by infection.▲

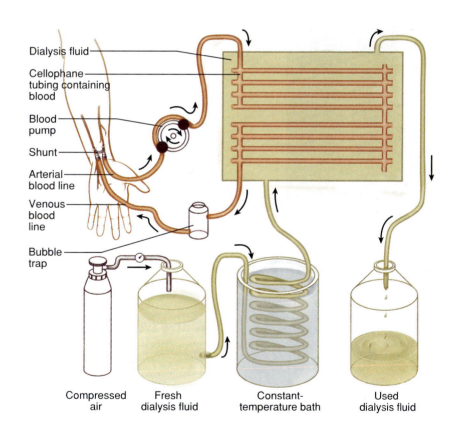

Dialysis fluid
Cellophane tubing containing blood
Blood pump
Shunt
Arterial blood line
Venous blood line
Bubble trap

Compressed air Fresh dialysis fluid Constant-temperature bath Used dialysis fluid

Figure E.1 Hemodialysis.

Interactions Between the URINARY SYSTEM and Other Organ Systems

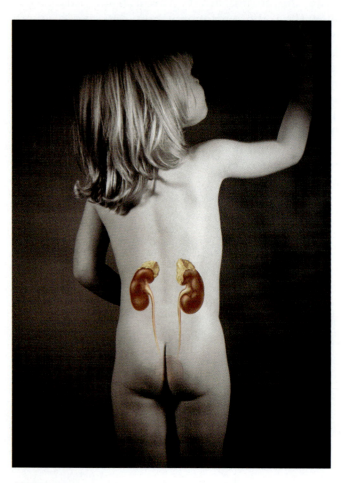

Nervous System
- Regulates glomerular filtration and micturition
- Nervous system especially sensitive to electrolyte and acid-base imbalances that may result from renal dysfunction

Endocrine System
- Regulates renal function through angiotensin II, aldosterone, ANF, and ADH
- Renin produced by kidneys regulates angiotensin and aldosterone production; kidneys produce erythropoietin

Circulatory System
- Perfuses kidneys so that wastes can be filtered from blood; blood pressure vital for glomerular filtration; blood reabsorbs water and solutes from renal tubules
- Kidneys modulate blood pressure more than any other organ does; kidneys control blood composition; regulate hematocrit by secreting erythropoietin; cardiac rhythm is especially sensitive to electrolyte imbalances that may result from renal dysfunction

Lymphatic/Immune Systems
- Return of fluid and protein to bloodstream maintains blood pressure and fluid balance essential for renal function; immune cells protect kidneys from infection
- Acidity of urine provides nonspecific defense against urinary tract infection

Respiratory System
- High metabolic rate of kidneys strongly depends on O_2 supply; inadequate pulmonary ventilation may require compensation by kidneys to maintain acid-base balance; kidneys may be damaged by inhaled toxic fumes
- Rate of acid excretion by kidneys affects pH and may therefore affect pulmonary ventilation

Digestive System
- Liver synthesizes urea, the major nitrogenous waste eliminated by kidneys; urea contributes to osmotic gradient of renal medulla; liver metabolizes blood-borne hormones to forms excreted in urine
- Kidneys excrete toxins absorbed by GI tract; calcitrol synthesized by kidneys regulates Ca^{2+} absorption by small intestine

Reproductive System
- Enlarged prostate can cause urine retention and kidney damage in males; pregnant uterus compresses bladder and increases micturition frequency in females
- Urethra serves as common passageway for urine and sperm cells in males; urine flushes residual semen from male urethra; mother's urinary system eliminates metabolic wastes of fetus

All Systems
The urinary system serves all other systems by maintaining water, electrolyte, and acid-base balance and eliminating metabolic wastes.

Integumentary System
- Epidermis is a barrier to fluid loss, although substantial fluid loss may occur through sweating, leading to oligura; skin eliminates wastes; skin and kidneys have complementary roles in vitamin D synthesis
- Renal control of fluid balance essential for sweat production

Skeletal System
- Lower ribs and pelvis protect some organs of urinary system
- Renal control of calcium and phosphate balance and role of kidneys in vitamin D synthesis are essential for bone deposition

Muscular System
- External urinary sphincter and bulbocavernosus muscle (in males) regulate micturition; abdominal muscles aid in emptying bladder; muscles of pelvic floor support bladder
- Muscle contraction depends on control of Na^+, K^+, and Ca^{2+} balance by kidneys

**Functions of the Urinary System
(pp. 832–834)**
1. Organs of the urinary system
2. Renal functions
3. Nitrogenous wastes
4. Excretion
5. Osmolarity

Anatomy of the Kidney (pp. 834–839)
1. Gross anatomy
   a. Location, size, weight, shape
   b. Protective coverings
      • Renal fascia
      • Adipose capsule
      • Renal capsule
   c. Renal parenchyma and sinus
   d. Renal cortex and medulla
   e. Renal calices and pelvis
2. The nephron
   a. Juxtamedullary and cortical
      nephrons
   b. Renal circulation
      • Renal artery
      • Interlobar arteries
      • Arcuate arteries
      • Interlobular arteries
      • Afferent arteriole
      • Glomerulus
      • Efferent arteriole
      • Peritubular capillaries
      • Interlobular veins
      • Arcuate veins
      • Interlobar veins
      • Renal vein
      • Vasa recta
   c. Renal corpuscle
      • Glomerulus
      • Glomerular capsule
      • Parietal and visceral layers
      • Urinary and vascular poles
   d. Renal tubule
      • Proximal convoluted tubule
      • Nephron loop
         •• Descending and ascending
            limbs
         •• Thin and thick segments
      • Distal convoluted tubule
      • Collecting duct
3. Juxtaglomerular apparatus

**Urine Formation I: Glomerular Filtration
(pp. 839–845)**
1. Filtration membrane
   a. Fenestrated endothelium

   b. Basement membrane
      • Particle exclusion by size
      • Particle exclusion by charge
   c. Filtration slits of podocytes
2. Filtration pressure
   a. Blood pressure
   b. Capsular hydrostatic pressure
   c. Colloid osmotic pressure
   d. Net filtration pressure
3. Glomerular filtration rate (GFR)
4. Regulation of glomerular
   filtration
   a. Renal autoregulation
      • Juxtaglomerular apparatus
         •• Juxtaglomerular cells
         •• Macula densa
         •• Mesangial cells
         •• Mechanism of action
   b. Sympathetic control
   c. Renin-angiotensin
      mechanism

**Urine Formation II: Tubular
Reabsorption and Secretion
(pp. 845–849)**
1. Proximal convoluted tubule
   a. Tubular reabsorption
   b. Role of solvent drag
   c. Reabsorption of specific
      substances
      • Sodium
         •• Creation of concentration
            gradient
         •• Transcellular and
            paracellular reabsorption
         •• Role in reabsorption of other
            substances
      • Glucose
      • Amino acids
      • Water
      • Chloride
      • Other electrolytes
      • Protein
      • Nitrogenous wastes
   d. Transport maximum
   e. Tubular secretion
      • Waste removal
      • Acid-base balance
2. Nephron loop
3. Distal convoluted tubule and
   collecting duct
   a. Effects of aldosterone
   b. Effect of atrial natriuretic
      factor

**Urine Formation III: Concentrating the
Urine (pp. 849–852)**
1. Collecting duct
2. Control of concentration
   a. Producing hypotonic urine
   b. Producing hypertonic urine
      • Role of glomerular filtration rate
      • Role of antidiuretic hormone
3. Countercurrent multiplier
   a. Water loss by descending limb
   b. NaCl reabsorption by ascending
      limb
   c. Contribution of urea to medullary
      osmolarity
4. Countercurrent exchanger of the vasa
   recta

**Urine and Renal Function Tests
(pp. 852–855)**
1. Composition and properties of urine
   a. Appearance
   b. Odor
   c. Specific gravity
   d. Osmolarity
   e. pH
   f. Chemical composition
2. Urine volume
   a. Forms of diabetes
   b. Diuretics
3. Renal function tests
   a. Calculating renal clearance
   b. Calculating glomerular filtration
      rate

**Urine Storage and Elimination
(pp. 855–857)**
1. Ureters
2. Urinary bladder
   a. Detrusor muscle
   b. Trigone
   c. Capacity of the bladder
3. Urethra
   a. Female
   b. Male
      • Prostatic portion
      • Membranous portion
      • Penile portion
   c. Urethral sphincters
      • Involuntary internal sphincter
      • Voluntary external sphincter
4. Voiding urine
   a. Stretch receptors of bladder
   b. Micturition reflex
   c. Voluntary control

Selected Vocabulary

urinary system 832
kidney 832
ureter 832
urinary bladder 832
urethra 832
waste 833
metabolic waste 833
nitrogenous waste 833
urea 833
uric acid 833
creatinine 833
azotemia 833
uremia 833
excretion 833
osmole 834
osmolality 834
osmolarity 834
hilum 834
renal fascia 834
adipose capsule 834
renal capsule 834
renal sinus 834
renal cortex 834
renal medulla 834
renal columns 834
renal pyramids 834
papilla 834
minor calyx 834
major calyx 834
renal pelvis 834

nephron 834
juxtamedullary nephron 835
cortical nephron 835
renal artery and vein 835
interlobar artery and vein 835
arcuate artery and vein 835
interlobular artery and
    vein 835
afferent arteriole 835
glomerulus 835
efferent arteriole 835
peritubular capillaries 837
vasa recta 838
renal corpuscle 838
glomerular capsule 838
podocyte 838
capsular space 838
urinary pole 838
vascular pole 838
renal tubule 838
proximal convoluted tubule
    (PCT) 838
nephron loop 839
descending limb 839
ascending limb 839
thin segment 839
thick segment 839
distal convoluted tubule
    (DCT) 839
collecting duct 839

papillary duct 839
juxtaglomerular apparatus
    (JGA) 839
glomerular filtration 839
glomerular filtrate 839
tubular fluid 839
urine 839
filtration membrane 840
foot process 840
filtration slit 840
albuminuria 840
hematuria 840
glomerular filtration rate
    (GFR) 842
renal autoregulation 842
juxtaglomerular (JG)
    cell 842
macula densa 843
mesangial cell 843
renin 844
angiotensin-converting
    enzyme (ACE) 844
angiotensin II 844
tubular reabsorption 846
transcellular route 846
paracellular route 846
sodium-dependent glucose
    transporter (SGLT) 846
obligatory water
    reabsorption 847

glycosuria 848
tubular secretion 848
countercurrent multiplier 850
countercurrent exchanger 851
urinalysis 852
urochrome 852
pyuria 853
specific gravity 853
polyuria 854
diuresis 854
oliguria 854
anuria 854
diabetes 854
diuretic 854
renal clearance 854
detrusor muscle 855
rugae 855
trigone 855
external urethral
    orifice 856
prostatic urethra 856
membranous urethra 856
penile urethra 856
urethral glands 857
internal urethral
    sphincter 857
external urethral
    sphincter 857
micturition reflex 857
micturition center 857

Testing Your Recall   Answers in Appendix C

1. Urination occurs when the _____ contracts.
   a. detrusor muscle
   b. internal urethral sphincter
   c. external urethral sphincter
   d. muscularis of the ureters
   e. all of the above

2. The compact ball of capillaries in a nephron is called
   a. the glomerular capsule.
   b. the peritubular capillary plexus.
   c. the renal corpuscle.
   d. the glomerulus.
   e. the vasa recta.

3. Which of these is the most abundant nitrogenous waste in the blood?
   a. uric acid
   b. urea
   c. ammonia
   d. creatinine
   e. albumin

4. Which of these lies closest to the renal cortex?
   a. the parietal peritoneum
   b. the renal fascia
   c. the renal capsule
   d. the adipose capsule
   e. the renal pelvis

5. Most sodium is reabsorbed from the glomerular filtrate by
   a. the vasa recta.
   b. the proximal convoluted tubule.
   c. the nephron loop.
   d. the distal convoluted tubule.
   e. the collecting duct.

6. A glomerulus and glomerular capsule make up one
   a. renal capsule.
   b. renal corpuscle.
   c. kidney lobule.
   d. kidney lobe.
   e. nephron.

7. The kidney has more _____ than any of the other structures listed.
   a. arcuate arteries
   b. minor calices
   c. medullary pyramids
   d. afferent arterioles
   e. collecting ducts

8. The renal clearance of _____ is normally zero.
   a. sodium
   b. potassium
   c. uric acid
   d. urea
   e. amino acids

9. Beavers have relatively little need to conserve body water and could therefore be expected to have _____ than humans.
   a. fewer nephrons
   b. longer nephron loops
   c. shorter nephron loops
   d. longer collecting ducts
   e. longer proximal convoluted tubules

10. Increased ADH secretion should cause the urine to have
    a. a higher specific gravity.
    b. a lighter color.
    c. a higher pH.
    d. less urea per milliliter.
    e. less potassium per milliliter.

11. The _____ reflex is an autonomic reflex activated by pressure in the urinary bladder.

12. _____ is the ability of a nephron to regulate GFR independently of external nervous or hormonal influences.

13. The two ureters and the urethra form the boundaries of a smooth area called the _____ on the floor of the urinary bladder.

14. The _____ is a group of epithelial cells of the distal convoluted tubule that monitors the flow or composition of tubular fluid.

15. To enter the capsular space, filtrate must pass between foot processes of the _____, cells that form the visceral layer of the glomerular capsule.

16. Glycosuria occurs if the concentration of glucose in the glomerular filtrate exceeds the _____ of the proximal tubule.

17. _____ is a hormone that regulates the amount of water reabsorbed by the collecting duct.

18. The _____ sphincter is under involuntary control and relaxes during the micturition reflex.

19. Very little _____ is found in the glomerular filtrate because it is negatively charged and repelled by the basement membrane of the glomerulus.

20. Blood flows through the _____ arteries just before entering the interlobular arteries.

Testing Your Comprehension

Answers in *Study Guide*

1. How would glomerular filtration rate be affected by a severe deficiency of protein in the diet and blood plasma?

2. A patient produces 55 mL of urine per hour. Urea concentration is 0.25 mg/mL in her blood plasma and 8.6 mg/mL in her urine. (a) What is her rate of renal clearance for urea? (b) About 95% of adults excrete 12.6 to 28.6 g of urea per day. Is this patient above, within, or below this range? Show how you calculated your answers.

3. A patient with poor renal perfusion is treated with an ACE (angiotensin converting enzyme) inhibitor and goes into renal failure. Explain why.

4. Suppose the ureters entered the bladder at its superior surface instead of from below. What problems might occur as the bladder filled with urine?

5. In what ways do the proximal and distal convoluted tubules differ in structure? How is this related to the difference in their function?

Web Site Link

For a listing of the most current web sites related to this chapter, please visit the Saladin homepage at:

http://www.mhhe.com/sciencemath/biology/saladin/

Water, Electrolyte, and Acid-Base Balance

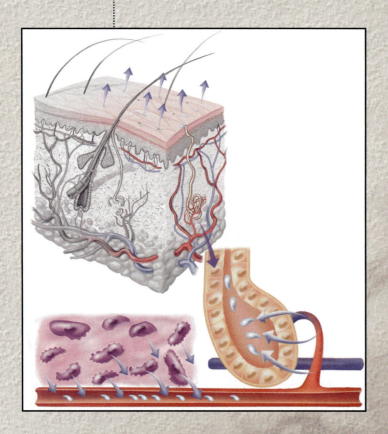

To understand this chapter, it is essential that you understand or brush up on the following concepts:

▸ Electrolytes and milliequivalents/liter (p. 64)
▸ Acids, bases, and the pH scale (pp. 64–65)
▸ Role of electrolytes in plasma membrane potentials (pp. 400–403)
▸ Depolarization and hyperpolarization of plasma membranes (p. 453)
▸ The hypothalamus and posterior pituitary (pp. 602–604)
▸ Influence of CO_2 and pH on pulmonary ventilation (pp. 821–822)
▸ Osmolarity (pp. 833–834)
▸ Structure and physiology of the nephron (pp. 834–852)

C ellular function requires a fluid medium with a carefully controlled composition. If the quantity, osmolarity, electrolyte concentrations, or pH of this medium is altered, life-threatening disorders of cellular function may result. Consequently, the body has several homeostatic mechanisms for maintaining these variables within narrow limits. These involve the urinary, respiratory, digestive, integumentary, endocrine, nervous, cardiovascular, and lymphatic systems. This chapter describes the homeostatic regulation of water, electrolyte, and acid-base balance and shows their close relationship.

Water Balance

▼Objectives

When you have completed this section, you should be able to
• name the major fluid compartments and explain how water moves from one to another;
• list the body's sources of water and routes of water loss;
• describe the mechanisms of regulating water intake and output; and
• describe some conditions in which the body has a deficiency or excess of water or an improper distribution of water among the fluid compartments.

We enter the world in a rather soggy condition, having swallowed, excreted, and floated in amniotic fluid for months. At birth, a baby is as much as 75% water by weight; infants normally lose a little weight in the first day or two as they excrete the excess. Young adult men average 55% to 60% water; women average slightly less because they have more fat, and adipose tissue is nearly free of water. Obese and elderly people are as little as 45% water by weight (see special topic 24.1).

Fluid Compartments

A 70 kg (150 lb) young male has about 40 L of **total body water (TBW).** This is distributed among the following **fluid compartments,** which are separated by selectively permeable membranes and differ from each other in chemical composition:

65% *intracellular fluid (ICF),* and

35% *extracellular fluid (ECF),* subdivided into

25% *tissue (interstitial) fluid,*

8% *blood plasma* and *lymph,* and

2% *transcellular fluid,* a catch-all category for cerebrospinal, synovial, peritoneal, pleural, and pericardial fluids; vitreous and aqueous humors of the eye; and fluid in the digestive, urinary, and respiratory tracts.

Fluid is continually exchanged between compartments by way of capillary walls and plasma membranes (fig. 24.1). Water moves by osmosis from the digestive tract to the bloodstream and by capillary filtration from the blood to the tissue fluid. From the tissue fluid, it may be reabsorbed by the capillaries, osmotically absorbed into cells, or taken up by the lymphatic system, which returns it to the bloodstream.

Because water moves so easily through plasma membranes, osmotic gradients between the ICF and ECF never last for very long. If a local imbalance arises, osmosis usually restores the balance within seconds so that intracellular and extracellular osmolarity are equal. If the osmolarity of the tissue fluid rises, water moves out of the cells; if it falls, water moves into the cells.

Osmosis from one fluid compartment to another is determined by the relative concentration of solutes in each compartment. The most abundant solute particles

Special Topic — Bias in Physiological Values — 24.1

Many physiological averages refer to the "young adult male." Decades of military recruitment provided a large sample of young, healthy men subject to almost every conceivable anatomical and physiological measurement. This generated a larger pool of data on them than on women or people of other age groups. Clinical problems can occur, however, when female and older patients are treated with therapies based on values typical for young men. Others may respond very differently to the same treatments, and assumptions based on young males can lead a health care provider to overlook or misinterpret important clinical signs in other patients.

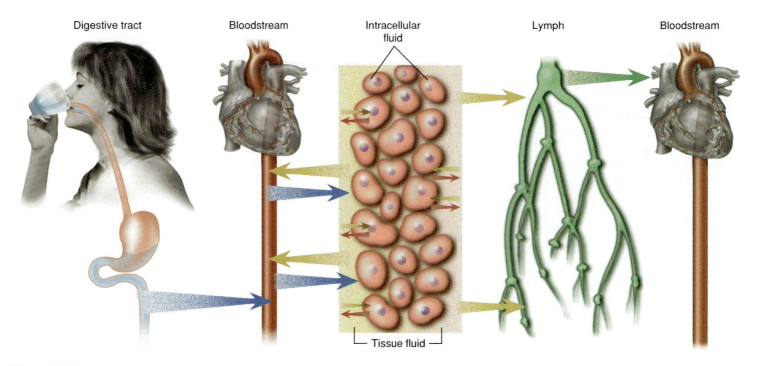

Figure 24.1 The movement of water between the major fluid compartments. Ingested water is absorbed by the bloodstream. There is a two-way exchange of water between the blood and tissue fluid and between the tissue and intracellular fluids. Excess tissue fluid is picked up by the lymphatic system, which returns it to the bloodstream.

by far are the electrolytes—especially sodium salts in the ECF and potassium salts in the ICF. Electrolytes therefore play the principal role in governing the body's water distribution and total water content; the subjects of water and electrolyte balance are inseparable.

Water Gain and Loss

A person is in a state of **fluid balance** when daily gains and losses are equal—about 2,500 mL on average (fig. 24.2). Water is gained from two sources: **metabolic water** (about 200 mL/day), which is produced as a by-product of aerobic respiration, and **preformed water,** which is ingested in food (700 mL/day) and drink (1,600 mL/day).

The routes of water loss are more varied:

- 1,500 mL/day is excreted as urine.
- 200 mL/day is eliminated in the feces.
- 300 mL/day is lost in the expired breath. You can easily visualize this by breathing onto a cool surface such as a mirror.
- 100 mL/day of sweat is secreted by a resting adult at a temperature of 20 °C (68 °F).
- 400 mL/day is lost as **cutaneous transpiration,**[1] water that diffuses through the epidermis and evaporates. This is not the same as sweat; it is not a

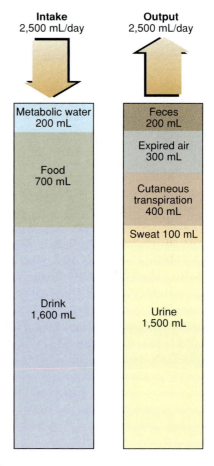

Figure 24.2 Typical water intake and output in a state of fluid balance.

....

1. *trans* = across, through + *spir* = to breathe

glandular secretion. A simple way to observe it is to cup the palm of your hand for a minute against a cool nonporous surface such as a laboratory benchtop or mirror. When you take your hand away, you will notice the water that transpired through the skin and condensed on that surface.

Such losses vary greatly with physical activity and environmental conditions. Respiratory loss increases in cold weather, for example, because cold air is drier and absorbs more body water from the respiratory tract. Hot, humid weather slightly reduces the respiratory loss but increases perspiration to as much as 1,200 mL/day. Prolonged, heavy work can raise the respiratory loss to 650 mL/day and perspiration to as much as 5 L/day, though it reduces urine output by nearly two-thirds.

Output through the breath and cutaneous transpiration is called **insensible water loss** because we are not usually conscious of it. **Obligatory water loss** is output that is relatively unavoidable: expired air, cutaneous transpiration, sweat, fecal moisture, and the minimum urine output, about 400 mL/day, needed to prevent azotemia. Even dehydrated individuals cannot prevent such losses; thus they become further dehydrated.

Regulation of Intake

Fluid intake is governed mainly by thirst, which is controlled by the mechanisms shown in figure 24.3. Dehydration reduces the blood volume and pressure and raises its osmolarity. For several reasons, this results in a smaller volume and greater viscosity of saliva. One reason is that most of the saliva is produced by capillary filtration, which is opposed by the lower hydrostatic pressure and higher osmolarity of the blood. Another reason is that salivation is inhibited by sympathetic output from the **thirst center** of the hypothalamus. The thirst center responds to stimuli of several kinds: angiotensin II, which is produced in response to falling blood pressure; antidiuretic hormone, which is released when blood osmolarity rises; and signals from the hypothalamic osmoreceptors, neurons that continually monitor the osmolarity of the ECF.

Reduced salivation gives us a dry, sticky-feeling mouth, but it is by no means certain that this is our primary motivation to drink. People who do not secrete saliva and experimental animals that have the salivary ducts tied off do not drink any more than normal individuals except when eating, when they need water to moisten the food.

Long-term satiation of thirst depends on absorbing water from the small intestine and lowering the osmolarity of the blood. This makes the saliva more abundant and watery by promoting capillary filtration and stopping the osmoreceptor response. However, these changes require 30 minutes or longer to take effect, and it would

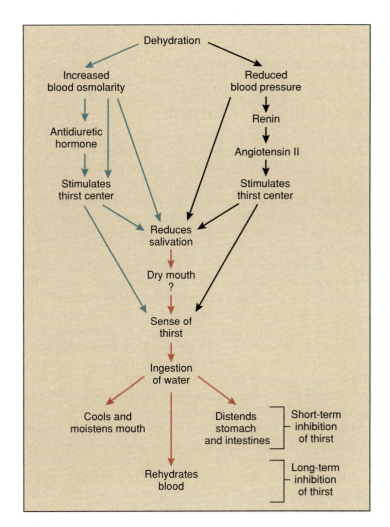

Figure 24.3 Mechanisms that lead from dehydration to thirst and rehydration.

be rather impractical if we had to drink that long while waiting to feel satisfied. Water intake would be grossly excessive. Fortunately, there are mechanisms that act more quickly to temporarily quench the thirst, allowing time for the change in blood osmolarity to occur.

Experiments with rats and dogs have independently tested the stimuli that quench the thirst. One of these is cooling and moistening the mouth; rats drink less if their water is cool than if it is warm, and simply moistening the mouth temporarily satisfies an animal even if the water is drained from its esophagus before it reaches the stomach. Distension of the stomach and small intestine is another inhibitor of thirst. If a dog is allowed to drink while the water is drained from its esophagus but its stomach is inflated with a balloon, its thirst is satisfied for a time. If the water is drained away but the stomach is not inflated, satiation does not last as long. Such fast-acting stimuli as coolness, moisture, and filling of the stomach stop an animal (and presumably a human) from drinking an excessive amount of liquid, but they are effective for only 30 to 45 minutes. If they are not soon followed by absorption of water into the bloodstream, thirst soon returns. Only a drop in blood osmolarity produces a lasting effect.

Regulation of Output

The only way to control water output significantly is through variations in urine volume. It must be realized, however, that the kidneys cannot completely prevent loss, nor can they replace lost water or electrolytes. Therefore, they never restore fluid volume or osmolarity, but in dehydration they can support existing fluid levels and slow down the rate of loss until water and electrolytes are ingested.

To understand the effect of the kidneys on water and electrolyte balance, it is also important to bear in mind that if a substance is reabsorbed by the kidneys, it is kept in the body and returned to the ECF, where it will be reflected in fluid volume and composition. If a substance is filtered by the glomerulus or secreted by the renal tubules and not reabsorbed, then it is excreted in the urine and lost from the body fluids.

Changes in urine volume are usually linked to adjustments in sodium reabsorption. As sodium is reabsorbed or excreted, proportionate amounts of water accompany it. The total volume of fluid remaining in the body may change, but its osmolarity remains stable. Controlling water balance by controlling sodium excretion is best understood in the context of electrolyte balance, discussed later in the chapter.

Antidiuretic hormone (ADH), however, provides a means of controlling water output independently of sodium. In true dehydration (discussed shortly), blood volume declines and sodium concentration rises. The increased osmolarity of the blood stimulates the hypothalamic osmoreceptors, which stimulate the posterior pituitary to release ADH. In response to ADH, cells of the collecting duct synthesize membrane proteins called *aquaporins*. When installed in the plasma membrane, these serve as channels that allow water to diffuse out of the duct into the hypertonic tissue fluid of the renal medulla. Thus the kidneys reabsorb more water and produce less urine. Sodium continues to be excreted, so the *ratio* of sodium to water in the urine increases (the urine becomes more concentrated). By helping the kidneys retain water, ADH slows down the decline in blood volume and the rise in its osmolarity. Thus the ADH mechanism forms a negative feedback loop (fig. 24.4).

Conversely, if blood volume and pressure are excessive or blood osmolarity is too low, ADH release is inhibited. The renal tubules reabsorb less water, urine output increases, and total body water declines. This is an effective way of compensating for hypertension. Since the lack of ADH increases the ratio of water to

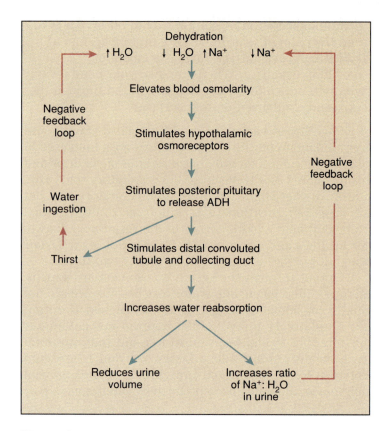

Figure 24.4 The release of antidiuretic hormone, and its effects, in response to dehydration. Pathways shown in red represent negative feedback.

sodium in the urine, it raises the sodium concentration and osmolarity of the blood.

Disorders of Water Balance

The body is in a state of fluid imbalance if there is an abnormality of total fluid *volume*, fluid *concentration*, or fluid *distribution* among the compartments.

Fluid Deficiency

Fluid deficiency arises when output exceeds intake over a prolonged period. There are two kinds of deficiency, called volume depletion and dehydration, which differ with respect to the relative loss of water and electrolytes and the resulting osmolarity of the ECF. This is an important distinction that calls for different strategies of fluid replacement therapy (see chapter essay, p. 882).

Volume depletion (hypovolemia[2]) occurs when proportionate amounts of water *and* sodium are lost without replacement. Total fluid volume declines but osmolarity remains normal. Volume depletion occurs in

---

2. *hypo* = below normal + *vol* = volume + *emia* = blood condition

Hot weather and profuse sweating are obvious threats to fluid balance, but so is cold weather. The body conserves heat by constricting the blood vessels of the skin and subcutaneous tissue, forcing blood into the deeper circulation. This raises blood pressure, which inhibits the secretion of antidiuretic hormone and increases the secretion of atrial natriuretic factor. These hormones increase urine output and reduce blood volume. In ad-

dition, cold air is relatively dry and increases respiratory water loss. This is why exercise causes the respiratory tract to "burn" more in cold weather than in warm.

These cold-weather respiratory and urinary losses can cause significant hypovolemia. Furthermore, the onset of exercise stimulates vasodilation in the skeletal muscles. In a hypovolemic state, there may not be enough blood to supply them and a person may expe-

rience weakness, fatigue, or fainting (hypovolemic shock). In winter sports and other activities such as snow shoveling, it is important to maintain fluid balance. Even if you do not feel thirsty, it is beneficial to take ample amounts of warm liquids such as soup or cider. Coffee, tea, and alcohol, however, have diuretic effects that defeat the purpose of fluid intake.

cases of hemorrhage, severe burns, and chronic vomiting or diarrhea. A less common cause is aldosterone hyposecretion (Addison disease), which results in inadequate sodium and water reabsorption.

Dehydration (negative water balance) occurs when the body eliminates significantly more water than sodium and the ECF osmolarity rises. The simplest cause of dehydration is a lack of drinking water; for example, when stranded in a desert or at sea. It can be a serious problem for elderly and bedridden people who depend on others to provide them with water—especially for those who cannot express their need or whose caretakers are insensitive to it. Diabetes mellitus, hyposecretion of ADH (diabetes insipidus), and overuse of diuretics are additional causes of dehydration. Prolonged exposure to cold weather can dehydrate a person just as much as exposure to hot weather (see special topic 24.2).

Dehydration from diarrhea is a major cause of infant mortality, especially under unsanitary conditions that lead to intestinal infections. For three reasons, infants are more vulnerable to dehydration than adults: (1) Their rapid growth and high metabolic rate produces toxic metabolites faster, and they must excrete more water to eliminate them. (2) Their kidneys are not fully mature and cannot concentrate urine as effectively. (3) They have a greater ratio of body surface to volume; consequently, compared to adults, they lose twice as much water per kilogram of body weight by evaporation.

Dehydration affects all fluid compartments. Suppose, for example, that you play a strenuous tennis match on a hot summer day and lose a liter of sweat per hour. Where does this fluid come from? Most of it filters out of the bloodstream through the capillaries of the sweat glands. In principle, 1 L of sweat would amount to about one-third of the blood plasma. However, as the blood loses water its osmolarity rises and water from the tissue fluid enters the bloodstream to balance the loss. This raises the osmolarity of the tissue fluid, and water moves out of the cells to balance that (fig. 24.5).

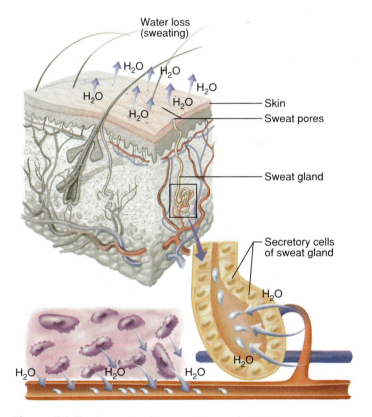

Figure 24.5 Effects of profuse sweating on the fluid compartments. Sweat is produced by filtration from the blood capillaries. Profuse sweating results in a drop in blood volume and pressure. The blood absorbs tissue fluid to replace this, and tissue fluid, in turn, is replaced by water from the ICF. This can result in cell shrinkage and malfunction.

So ultimately, all three fluid compartments lose water. To excrete 1 L of sweat, about 300 mL of water would come from the ECF and 700 mL from the ICF. Immoderate exercise without fluid replacement can lead to even greater loss than 1 L per hour. The most serious effects of volume depletion are circulatory shock due to loss of blood volume, and neurological dysfunction due to dehydration of brain cells.

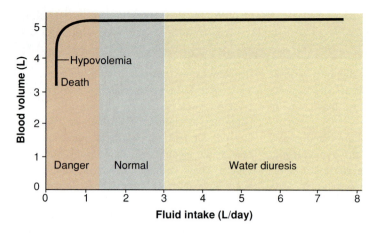

Figure 24.6 The relationship of blood volume to daily fluid intake. Note that the kidneys are able to compensate for a fluid excess much better than they can compensate for a fluid deficiency.

Fluid Excess

Fluid excess is less common than fluid deficiency because the kidneys are highly effective at compensating for excessive intake by excreting more urine (fig. 24.6). Renal failure and other causes, however, can lead to excess fluid retention.

Fluid excesses are of two types, called volume excess and hypotonic hydration. In **volume excess,** both sodium and water are retained and the ECF remains isotonic. Volume excess can result from aldosterone hypersecretion or from renal failure. In **hypotonic hydration** (also called **water intoxication** or **positive water balance**), more water than sodium is retained or ingested and the ECF becomes hypotonic. This can occur if you lose a large amount of water *and* salt through urine and sweat and you replace it by drinking plain water. Without a proportionate intake of electrolytes, water dilutes the ECF, makes it hypotonic, and causes cellular swelling. ADH hypersecretion can cause hypotonic hydration by stimulating excessive water retention as sodium continues to be excreted. Among the most serious effects of either type of fluid excess are pulmonary and cerebral edema (discussed shortly).

Fluid Sequestration

Fluid sequestration[3] (seh-ques-TRAY-shun) is a condition in which excess fluid accumulates in a particular location. Total body water may be normal, but the volume of circulating blood may drop to the point of causing circulatory shock. In the case of a hemorrhage, for example, blood that pools and clots in the tissues is lost to circulation. Pulmonary infections can cause *pleural effusion,* in which several liters of fluid accumulate in

the pleural cavity. But the most common form of sequestration is *edema,* the accumulation of serous fluid in the interstitial spaces. Edema is typically marked by swelling of the face, fingers, abdomen, or ankles. It has three fundamental causes:

1. **Increased capillary filtration.** This results from increases in capillary blood pressure or permeability. Poor venous return, for example, causes pressure to back up into the capillaries. Congestive heart failure and incompetent heart valves can impede venous return from the lungs and cause pulmonary edema. Systemic edema is a common problem when a person is confined to a bed or wheelchair, with insufficient muscular activity to promote venous and lymphatic return. Kidney failure can lead to edema by causing water retention and hypertension. Histamine causes edema by dilating the arterioles that supply the capillaries and by making the capillaries more permeable. Capillary permeability also increases with age, which puts older people at risk of edema.

2. **Reduced capillary reabsorption.** Capillary reabsorption depends on oncotic pressure—the difference in osmotic pressure between the blood and tissue fluid. Oncotic pressure is proportional to the concentration of blood albumin; therefore, an albumin deficiency (hypoproteinemia) produces edema. Since blood albumin is produced by the liver, liver diseases such as cirrhosis tend to lead to hypoproteinemia and edema. Edema is commonly seen in regions of famine due to dietary protein deficiency. Hypoproteinemia also commonly results from severe burns, radiation sickness (p. 71), and kidney diseases that allow protein to filter through the glomerulus and escape in the urine.

3. **Obstructed lymphatic drainage.** The lymphatic system returns tissue fluid to the bloodstream and recovers the small amounts of protein that filter from the capillaries. Obstruction of the lymphatic vessels or the surgical removal of lymph nodes can interfere with lymphatic drainage and lead to the accumulation of protein in the tissue fluid distal to the obstruction. Protein raises the osmolarity of the tissue fluid and causes water to accumulate (see elephantiasis, fig. 21.2).

Whatever the cause of edema, it can result in a loss of blood volume and pressure, which creates the potential for circulatory shock. And, as the tissues become swollen with fluid, oxygen delivery and waste removal are impaired and tissue necrosis may occur. Pulmonary edema presents a threat of suffocation. Cerebral edema produce headaches, nausea, and sometimes seizures and coma.

---

3. *sequestr* = to isolate

Key Point Review

1. Name the two major fluid compartments and the subcategories of one of these. Approximately what fraction of total body water does each compartment contain?

2. List five routes of water loss. Which one accounts for the greatest loss? Which one is most controllable?

3. Explain why even a severely dehydrated person inevitably experiences further fluid loss.

4. Suppose there were no mechanisms to stop the sense of thirst until the blood became sufficiently hydrated. Explain why we would routinely suffer hypotonic hydration.

Electrolyte Balance

▼**Objectives**

When you have completed this section, you should be able to

- state the physiological roles of sodium, potassium, calcium, chloride, and phosphate;
- describe the hormonal and renal mechanisms that regulate the concentrations of these electrolytes; and
- state the term for an excess or deficiency of each electrolyte and describe the consequences of these imbalances.

Electrolytes are physiologically important for two principal reasons: they are chemically reactive and participate in all metabolism, and they strongly affect the osmolarity of the body fluids and the body's water content and distribution. The major cations are sodium (Na^+), potassium (K^+), calcium (Ca^{2+}), and hydrogen (H^+), and the major anions are chloride (Cl^-), bicarbonate (HCO_3^-), and phosphate (PO_4^{3-}). Hydrogen and bicarbonate regulation are discussed later under acid-base balance. Here we focus on the other five.

Their typical concentrations and the terms for their imbalances are listed in table 24.1. Blood plasma is the most accessible fluid for measurements of electrolyte concentration, so excesses and deficiencies are defined with reference to normal plasma concentrations. Concentrations in the interstitial fluid differ only slightly from those in the plasma. The prefix *normo-* denotes a normal electrolyte concentration (for example, *normokalemia*), and *hyper-* and *hypo-* denote concentrations that are sufficiently above or below normal to cause physiological disorders.

Sodium

Functions. Sodium is one of the principal ions responsible for the resting membrane potentials of cells, and the influx of sodium through gated membrane channels is an essential event in the depolarization that underlies nerve and muscle function. Sodium is the principal cation of the ECF; sodium salts account for 90% to 95% of its osmolarity. Sodium is therefore the most significant solute in determining total body water and the distribution of water among fluid compartments. Sodium gradients across the plasma membrane provide the potential energy that is tapped to cotransport other solutes such as glucose, potassium, and calcium. The Na^+-K^+ pump is an important mechanism for generating body heat. Sodium bicarbonate ($NaHCO_3$) plays a major role in buffering the pH of the ECF.

Homeostasis. An adult needs about 0.5 g of sodium per day, whereas the typical American diet contains 10 to 15 g/day. Thus a dietary sodium deficiency is rare, and the primary concern is adequate renal excretion of the excess. This is one of the most important roles of the kidneys. There are multiple mechanisms for controlling sodium concentration, tied to its effects on blood pressure and osmolarity and coordinated by three hormones: aldosterone, antidiuretic hormone, and atrial natriuretic factor.

Aldosterone, the "salt-retaining hormone," plays the primary role in adjustment of sodium excretion. Hyponatremia and hyperkalemia directly stimulate the

Table 24.1	Electrolyte Concentrations and the Terminology of Electrolyte Imbalances			
	Mean Concentration (mEq/L)			
Electrolyte	**Plasma**	**ICF**	**Deficiency**	**Excess**
Sodium (Na^+)	142	10	Hyponatremia	Hypernatremia[4]
Potassium (K^+)	5	141	Hypokalemia	Hyperkalemia[5]
Calcium (Ca^+)	5	<1	Hypocalcemia	Hypercalcemia
Chloride (Cl^-)	103	4	Hypochloremia	Hyperchloremia
Phosphate (PO_4^{3-})	4	75	Hypophosphatemia	Hyperphosphatemia

4. *natr* = sodium + *emia* = blood condition
5. *kal* = potassium

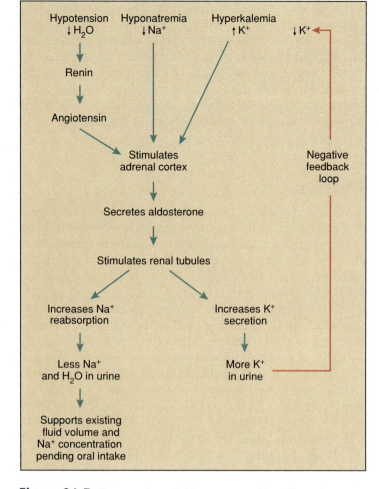

Figure 24.7 The secretion of aldosterone, and its effects, in response to fluid and electrolyte imbalances. The pathway shown in red represents negative feedback.

adrenal cortex to secrete aldosterone, and hypotension stimulates its secretion by way of the renin-angiotensin mechanism (fig. 24.7).

Only cells of the distal convoluted tubule (DCT) and cortical part of the collecting duct (CD) have aldosterone receptors. Aldosterone, a steroid, diffuses into these cells, binds to nuclear receptors, and activates transcription of a gene for the Na$^+$-K$^+$ pump. In 10 to 30 minutes, enough Na$^+$-K$^+$ pumps have been synthesized and installed in the plasma membrane to produce a noticeable effect—sodium concentration in the urine begins to fall and potassium concentration rises as the tubules reabsorb more Na$^+$ and secrete more H$^+$ and K$^+$. Water and Cl$^-$ passively follow Na$^+$. Thus the primary effects of aldosterone are that the urine contains less NaCl and more K$^+$ and has a lower pH. An average adult male excretes 5 g of sodium per day, but the urine can be virtually sodium-free when aldosterone level is high. Although aldosterone strongly influences sodium reabsorption, it has little effect on plasma sodium *concen-tration* because reabsorbed sodium is accompanied by a proportionate amount of water.

Hypertension inhibits the renin-angiotensin-aldosterone mechanism. The kidneys then reabsorb almost no sodium beyond the proximal convoluted tubule (PCT), and the urine contains up to 30 g of sodium per day.

Aldosterone has only slight effects on urine volume, blood volume, and blood pressure in spite of the tendency of water to follow sodium osmotically. Even in aldosterone hypersecretion (Cushing disease), blood volume is rarely more than 5% to 10% above normal. This is because an increase in blood volume increases blood pressure and glomerular filtration rate (GFR). Even though aldosterone increases the tubular reabsorption of sodium and water, this is offset by the resulting rise in GFR and there is only a small drop in urine output.

Antidiuretic hormone modifies water excretion independently of sodium excretion. Thus, unlike aldosterone, it can change sodium *concentration*. A high concentration of sodium in the blood stimulates ADH release by the posterior pituitary. Thus the kidneys reabsorb more water, which helps to slow down any further increase in blood sodium concentration. ADH alone cannot lower the blood sodium concentration; this requires water ingestion, but remember that ADH also stimulates thirst. A drop in sodium concentration, by contrast, inhibits ADH release. More water is excreted and this raises the concentration of the sodium that remains in the blood.

Atrial natriuretic factor (ANF) is secreted by endocrine cells of the atrial myocardium in response to hypertension. It inhibits sodium and water reabsorption and the secretion of renin and ADH. The kidneys thus eliminate more sodium and water and lower the blood pressure.

Several other hormones also affect sodium homeostasis. Estrogens mimic the effect of aldosterone and cause women to retain water during pregnancy and certain phases of the menstrual cycle. Progesterone reduces sodium reabsorption and has a diuretic effect. High levels of glucocorticoids promote sodium reabsorption and edema.

In some cases, sodium homeostasis is achieved by regulation of salt intake. A craving for salt occurs in people who are depleted of sodium; for example, by blood loss or Addison disease. Pregnant women sometimes develop a craving for salty foods. Salt craving is not limited to humans; many animals ranging from elephants to butterflies seek out salty soil where they can obtain this vital mineral.

Imbalances. True imbalances in sodium concentration are relatively rare because sodium excess or depletion is almost always accompanied by proportionate changes in water volume. **Hypernatremia** is a plasma sodium concentration in excess of 145 mEq/L. It can result from the administration of intravenous saline (see chapter essay, p. 882). Its major consequences are water retention, hypertension, and edema. **Hyponatremia** (less than 130 mEq/L) is usually the result of excess body water rather than excess sodium excretion, as in the case mentioned earlier of a person who loses considerable volumes of sweat or urine and replaces it by drinking plain water. Usually, hyponatremia is quickly corrected by excretion of the excess water, but if uncorrected it produces the symptoms of hypotonic hydration described earlier.

Potassium

Functions. Potassium is the most abundant cation of the ICF and is the greatest contributor to intracellular osmolarity and cell volume. Along with sodium, it produces the resting membrane potentials and action potentials of nerve and muscle cells (fig. 24.8a). Potassium is as important as sodium to the Na⁺-K⁺ pump and its functions of cotransport and thermogenesis (heat production). It is an essential cofactor for protein synthesis and some other metabolic processes.

Homeostasis. Potassium homeostasis is closely linked to that of sodium. Regardless of the body's state of potassium balance, about 90% of the K⁺ filtered by the glomerulus is reabsorbed by the PCT and the rest is excreted in the urine. Variations in potassium excretion are controlled later in the nephron by changing the amount of potassium secreted into the tubular fluid by the DCT and cortical portion of the CD. When K⁺ concentration is high, they secrete more K⁺ into the filtrate and the urine may contain more K⁺ than the glomerulus filtered from the blood. When blood K⁺ level is low, the CD secretes less. Certain *intercalated cells* of the CD can reabsorb potassium when necessary.

Aldosterone regulates potassium excretion and reabsorption along with sodium (see fig. 24.7). A rise in K⁺ concentration stimulates the adrenal cortex to secrete aldosterone. Aldosterone stimulates renal secretion of K⁺ at the same time that it stimulates reabsorption of sodium. The more sodium there is in the urine, the less potassium, and vice versa.

Imbalances. Hyperkalemia (> 5.5 mEq/L) can result from aldosterone deficiency, renal failure, or acidosis. In crush injuries and hemolytic anemia, large amounts of K⁺ may be released from ruptured cells and cause hyperkalemia. Hyperkalemia sometimes follows transfusions with stored blood because K⁺ tends to leak from erythrocytes into the plasma during storage. When the ECF concentration of K⁺ is abnormally high, K⁺ diffuses into cells and partially depolarizes them (fig. 24.8b). Hyperkalemia is a very dangerous condition that can quickly produce cardiac arrest.

Most diets contain ample amounts of potassium, so **hypokalemia** (< 3.5 mEq/L) seldom results from dietary deficiency except in people with depressed appetites. It more often results from heavy sweating, chronic vomiting or diarrhea, excessive use of laxatives, aldosterone hypersecretion, or alkalosis. (The relationship of acid-base imbalances to potassium imbalances is explained later.) As ECF potassium concentration falls, K⁺ moves from the ICF to the ECF. With the loss of these cations from the cytoplasm, cells become hyperpolarized and nerve and muscle cells are less excitable (fig. 24.8c). This is reflected in muscle weakness, loss of muscle tone, depressed reflexes, and irregular electrical activity of the heart.

Think About It

Some tumors of the adrenal cortex secrete excess aldosterone and may cause paralysis. Explain this effect and identify the electrolyte and fluid imbalances you would expect to observe in such a case.

Chloride

Functions. Chloride ions are the most abundant anions of the ECF and thus make a major contribution to its osmolarity. Chloride ions are required for the formation of stomach acid (HCl), and they are involved in the chloride shift mechanism that accompanies carbon dioxide loading and unloading by the erythrocytes (see chapter 22). By a similar mechanism explained later, Cl⁻ plays a major role in the regulation of body pH.

Homeostasis. Cl⁻ is strongly attracted to Na⁺, K⁺, and Ca²⁺. It would require great expenditure of energy to keep it separate from these cations, so Cl⁻ homeostasis is achieved primarily as an effect of Na⁺ homeostasis—as sodium is retained or excreted, Cl⁻ passively follows.

Imbalances. Hyperchloremia (> 105 mEq/L) is usually the result of dietary excess or administration of intravenous saline. **Hypochloremia** (< 95 mEq/L) is usually a side effect of hyponatremia but sometimes results from hypokalemia. The kidneys retain K⁺ by excreting more Na⁺, and Na⁺ takes Cl⁻ with it. The primary effects of chloride imbalances are disturbances in acid-base balance, but this works both ways—a pH imbalance arising from some other cause can produce a chloride imbalance. Chloride balance is therefore discussed further in connection with acid-base balance.

Calcium and phosphate homeostasis were extensively discussed in chapter 8 and are summarized only briefly here.

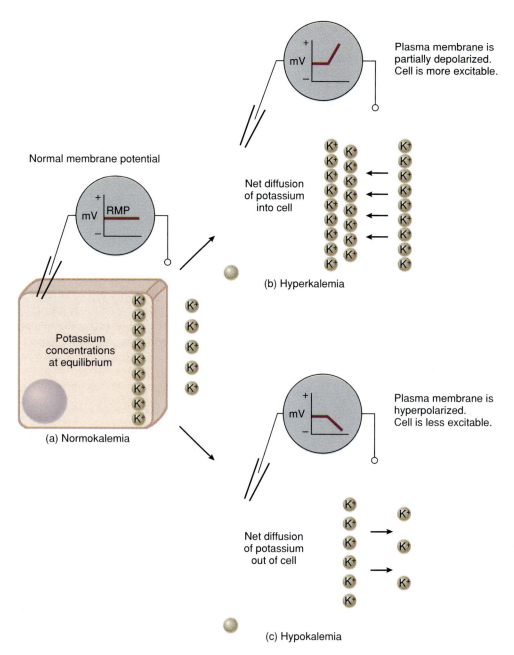

Normal membrane potential

Plasma membrane is partially depolarized. Cell is more excitable.

Net diffusion of potassium into cell

(b) Hyperkalemia

Potassium concentrations at equilibrium

(a) Normokalemia

Plasma membrane is hyperpolarized. Cell is less excitable.

Net diffusion of potassium out of cell

(c) Hypokalemia

Figure 24.8 Effects of potassium imbalances on cell membrane potentials. The diagram at the top of each cell represents the voltage measured across the plasma membrane. (*a*) At normal ECF potassium concentrations (normokalemia), cells have a normal resting membrane potential (RMP). (*b*) In hyperkalemia, K+ diffuses into the cell and reduces the membrane voltage. The cell is partially depolarized and more excitable than normal. (*c*) In hypokalemia, K+ diffuses out of the cell and increases the membrane voltage. The cell is hyperpolarized and less excitable than normal.

Calcium

Functions. Calcium lends strength to the skeleton, activates the sliding filament mechanism of muscle contraction, serves as a second messenger for some hormones and neurotransmitters, activates exocytosis of neurotransmitters and other cellular secretions, and is an essential factor in blood clotting. Cells maintain a very low intracellular calcium concentration because they require a high concentration of phosphate ions (for rea-

sons discussed shortly). If calcium and phosphate were both very concentrated in a cell, the product of their concentrations, $[Ca^{2+}] \bullet [PO_4{}^{3-}]$, could reach the critical *solubility product* where calcium phosphate crystals precipitate in the cytoplasm (as described in chapter 8). To maintain a high phosphate concentration but avoid crystallization of calcium phosphate, cells must pump out Ca^{2+} and keep it at a low intracellular concentration or else sequester Ca^{2+} in the smooth ER and release it only when needed. Cells that store Ca^{2+} often have a protein called *calsequestrin,* which binds the stored Ca^{2+} and keeps it chemically unreactive.

Homeostasis. The homeostatic control of Ca^{2+} concentration was discussed extensively in chapter 8. It is regulated chiefly by parathyroid hormone and, in children, by calcitonin. These hormones regulate blood calcium concentration through their effects on bone deposition and resorption, intestinal absorption of calcium, and urinary excretion.

Imbalances. Hypocalcemia (< 4.5 mEq/L) can result from vitamin D deficiency, diarrhea, pregnancy, lactation, acidosis, hypoparathyroidism, or hyperthyroidism. It increases the Na^+ permeability of plasma membranes, causing the nervous and muscular systems to be overly excitable. Tetany occurs when calcium concentration drops to 6 mg/dL and may be lethal at 4 mg/dL due to laryngospasm and suffocation.

Hypercalcemia (> 5.8 mEq/L) can result from alkalosis, hyperparathyroidism, or hypothyroidism. It reduces the Na^+ permeability of plasma membranes and inhibits the depolarization of nerve and muscle cells. At concentrations of $\geq$ 12 mg/dL, hypercalcemia causes muscular weakness, depressed reflexes, and cardiac arrhythmia.

Phosphates

Functions. The phosphates of the body fluids are an equilibrium mixture of phosphate (PO_4^{3-}), monohydrogen phosphate (HPO_4^{2-}), and dihydrogen phosphate ($H_2PO_4^-$) ions. Phosphates are relatively concentrated in the ICF, where they are generated by the hydrolysis of ATP and other phosphate compounds. They are needed for the synthesis of ATP, other nucleotide phosphates, nucleic acids, and phospholipids. Every process that depends on ATP depends on phosphate ions. Phosphates activate many metabolic pathways by phosphorylating enzymes or substrates such as glucose. They are also important as buffers that help stabilize the pH of body fluids.

Homeostasis. The average diet provides ample amounts of phosphate ions, which are readily absorbed by the small intestine. Plasma phosphate concentration is usually maintained at about 4 mEq/L, with continual loss of excess phosphate by glomerular filtration. If plasma phosphate concentration drops much below this level, however, the renal tubules reabsorb all filtered phosphate.

Parathyroid hormone increases the excretion of phosphate as part of the mechanism for increasing the concentration of free calcium ions in the ECF. Lowering the ECF phosphate concentration minimizes the formation of calcium phosphate and thus helps support plasma calcium concentration. Rates of phosphate excretion are also strongly affected by the pH of the urine, as discussed shortly.

Imbalances. Phosphate homeostasis is not as critical as that of other electrolytes. The body can tolerate broad variations several times above or below normal concentration with little immediate effect on physiology.

5 Which of these do you think would have the most serious effect, and why—a 5 mEq/L increase in the plasma concentration of sodium, potassium, chloride, or calcium?

6 Answer the same question for a 5 mEq/L *decrease*.

7 Explain why ADH is more likely than aldosterone to change the osmolarity of the blood plasma.

8 Explain why aldosterone hyposecretion could cause hypochloremia.

9 Why are more phosphate ions required in the ICF than in the ECF? How does this affect the distribution of calcium ions between these fluid compartments?

Acid-Base Balance

▼Objectives

When you have completed this section, you should be able to

- define *buffer* and write chemical equations for the bicarbonate, phosphate, and protein buffer systems;
- discuss the relationship between pulmonary ventilation, pH of the extracellular fluids, and the bicarbonate buffer system;
- explain how the kidneys secrete hydrogen ions and how these ions are buffered in the tubular fluid;
- identify some causes of respiratory and metabolic acidosis and alkalosis, and describe the effects of these pH imbalances; and
- explain how the respiratory and urinary systems correct acidosis and alkalosis, and compare the effectiveness and limitations of the two systems.

As we saw in chapter 3, metabolism depends on the function of enzymes, and enzymes are very sensitive to pH. Slight deviations from the normal pH can denature them and shut down metabolic pathways, as well as alter the structure and function of other macromolecules. Consequently, acid-base balance is one of the most important aspects of homeostasis.

The blood and tissue fluid normally have a pH of 7.35 to 7.45. Such a narrow range of variation is remarkable considering that our metabolism constantly produces acid: lactic acid from anaerobic fermentation, phosphoric acids from nucleic acid catabolism, fatty acids and ketones from fat catabolism, and carbonic acid from carbon dioxide. Here we examine mechanisms by which the body resists these challenges and maintains acid-base balance.

Think About It

In the systemic circulation, arterial blood has a pH of 7.40 and venous blood has a pH of 7.35. What do you think causes this difference?

Acids, Bases, and Buffers

Only free hydrogen ions (H^+) determine the pH of a solution. An acid is any chemical that releases H^+ in solution. A **strong acid** such as hydrochloric acid (HCl) ionizes freely, gives up most of its hydrogen ions, and can markedly lower the pH of a solution. A **weak acid** such as carbonic acid (H_2CO_3) ionizes only slightly, keeping most hydrogen in a chemically bound form that does not affect pH. A base is any chemical that accepts H^+. A **strong base** such as the hydroxyl ion (OH^-) has a strong tendency to bind H^+ and raise the pH, whereas a **weak base** such as the bicarbonate ion (HCO_3^-) binds only a small portion of the available H^+ and has less effect on pH.

A **buffer,** broadly speaking, is any mechanism that resists changes in pH by converting a strong acid or base to a weak one. The body has both physiological and chemical buffers. A **physiological buffer** is a system—namely the respiratory or urinary system—that stabilizes pH by controlling the body's output of acids, bases, or CO_2. Of all buffer systems, the urinary system buffers the greatest quantity of acid or base, but it requires several hours to days to exert an effect. The respiratory system exerts an effect within a few minutes but cannot alter the pH as much as the urinary system can.

A **chemical buffer** is a substance that binds H^+ and removes it from solution as its concentration begins to rise, or releases H^+ into solution as its concentration falls. Chemical buffers can restore normal pH within a fraction of a second. They function as mixtures called **buffer systems** composed of a weak acid and a weak base. The three major chemical buffer systems of the body are the bicarbonate, phosphate, and protein systems.

The amount of acid or base that can be neutralized by a chemical buffer system depends on two factors: the concentration of the buffers and the pH of their working environment. Each system has an optimum pH at which it can function; its effectiveness is greatly reduced if the pH of its environment deviates too far from this. The relevance of these factors will become apparent as you study the following buffer systems.

The Bicarbonate Buffer System

The **bicarbonate buffer system** is a solution of carbonic acid and bicarbonate ions. Carbonic acid (H_2CO_3) forms by the hydration of carbon dioxide and then dissociates into bicarbonate (HCO_3^-) and H^+:

$$CO_2 + H_2O \leftrightarrow H_2CO_3 \leftrightarrow HCO_3^- + H^+$$

This is a reversible reaction. When it proceeds to the right, carbonic acid acts as a weak acid by releasing H^+ and lowering pH. When the reaction proceeds to the left, bicarbonate acts as a weak base by binding H^+, removing the ions from solution, and raising pH.

The bicarbonate system does not have a particularly strong buffering capacity because its optimum pH is 6.1, relatively far from the pH 7.4 of the ECF. If a strong acid were added to a beaker of carbonic acid–bicarbonate solution at pH 7.4, the preceding reaction would shift only slightly to the left. Much surplus H^+ would remain and the pH would be substantially lower. The bicarbonate system works well in the body, however, because (1) these buffers are more concentrated than any other extracellular buffers and (2) more importantly, the lungs and kidneys constantly remove CO_2 and prevent an equilibrium from being reached. This keeps the reaction moving to the left, and more H^+ is neutralized. Conversely, if there is a need to lower the pH, the kidneys excrete HCO_3^-, keep this reaction moving to the right, and elevate the H^+ concentration of the ECF. Thus you can see that the physiological and chemical buffers of the body function together in maintaining acid-base balance.

The Phosphate Buffer System

The **phosphate buffer system** is a solution of HPO_4^{2-} and $H_2PO_4^-$. It works in much the same way as the bicarbonate system. The following reaction can proceed to the right to liberate H^+ and lower pH, or it can proceed to the left to bind H^+ and raise pH:

$$H_2PO_4^- \leftrightarrow HPO_4^{2-} + H^+$$

The optimal pH for this system is 6.8, closer to the actual pH of the ECF. Thus the phosphate buffer system has a stronger buffering effect than an equal amount of bicarbonate buffer. However, phosphates are much less concentrated in the ECF than bicarbonate, so they are less important in buffering the ECF. They are more important in the renal tubules and ICF, where not only are they more concentrated, but the pH is lower and closer to the optimum for them to function. In the ICF, the constant production of metabolic acids creates pH values ranging from 4.5 to 7.4, probably averaging 7.0. The reason for the low pH in the renal tubules is discussed later.

The Protein Buffer System

Proteins are more concentrated than either bicarbonate or phosphate buffers, especially in the ICF and blood plasma. The **protein buffer system** accounts for about three-quarters of all chemical buffering ability of the body fluids. The buffering ability of proteins is due to certain side groups of their amino acid residues. Some have carboxyl (—COOH) side groups, which release H^+ when pH begins to rise and thus lower pH:

$$-COOH \rightarrow -COO^- + H^+$$

Others have amino (NH_2) side groups, which bind hydrogen ions when pH falls too low, thus raising pH toward normal:

$$—NH_2 + H^+ \rightarrow —NH_3^+$$

Think About It

What protein do you think is the most important buffer in blood plasma? In erythrocytes?

Respiratory Control of pH

The equation for the bicarbonate buffer system shows that the addition of CO_2 to the body fluids raises H^+ concentration and lowers pH, while the removal of CO_2 has the opposite effects. This is the basis for the strong buffering capacity of the respiratory system. Indeed, this system can neutralize two or three times as much acid as the chemical buffers can.

Carbon dioxide is constantly produced by aerobic metabolism and is normally eliminated by the lungs at an equivalent rate. As explained in chapter 22, rising CO_2 concentration and falling pH stimulate peripheral and central chemoreceptors, which stimulate an increase in pulmonary ventilation. This expels excess CO_2 and thus reduces H^+ concentration. Essentially, the free H^+ becomes part of the water molecules produced by this reaction:

$$HCO_3^- + H^+ \rightarrow H_2CO_3 \rightarrow H_2O + CO_2 \text{ (expired)}$$

Conversely, a drop in H^+ concentration raises pH and reduces pulmonary ventilation. This allows metabolic CO_2 to accumulate in the ECF faster than it is expelled, thus lowering pH to normal.

These are classic negative feedback mechanisms that result in acid-base homeostasis. Respiratory control of pH has some limitations, however, which are discussed later under acid-base imbalances.

Renal Control of pH

The kidneys can neutralize more acid or base than either the respiratory system or the chemical buffers. The essence of this mechanism is that the renal tubules secrete H^+ into the tubular fluid, where most of it binds to bicarbonate, ammonia, and phosphate buffers. Bound and free H^+ are then excreted in the urine. Thus the kidneys are the only organs that actually expel H^+ from the body. The other buffer systems only reduce its concentration by binding it to another chemical.

Figure 24.9 shows the process of H^+ secretion and neutralization. It is numbered to correspond to the following description, and the hydrogen ions are shown in

color so you can trace them through the system from blood to urine.

1. Hydrogen ions in the blood are neutralized in two ways: by reacting with bicarbonate ions to produce carbonic acid and with hydroxyl ions to produce water.
2. Carbonic acid dissociates into water and carbon dioxide, which diffuse into the tubule cells.
3. The tubule cells obtain CO_2 from three sources: the blood, the tubular fluid, and their own aerobic respiration.
4. Within the tubule cell, carbonic anhydrase (CAH) catalyzes the reaction of CO_2 and H_2O to produce carbonic acid.
5. Carbonic acid dissociates into bicarbonate and hydrogen ions.
6. The bicarbonate ions diffuse back into the bloodstream and may reenter the reaction cycle.
7. The hydrogen ions are pumped into the tubule lumen by an antiport that exchanges them for sodium ions.
8. Sodium bicarbonate ($NaHCO_3$) in the glomerular filtrate reacts with these hydrogen ions, producing free sodium ions and carbonic acid.
9. The sodium ions are pumped into the tubule cells by the antiport at step 7 and then transferred to the blood by a Na^+-K^+ pump in the basal plasma membrane.
10. The carbonic acid in the tubular fluid dissociates into carbon dioxide and water. (The role of CAH is discussed shortly.) The CO_2 is recycled into the tubule cell and the water may be passed in the urine. Thus the hydrogen ions removed from the blood at step 1 are now part of the water molecules excreted in the urine at step 10.

Tubular secretion of H^+ (step 7) continues only as long as there is a sufficient concentration gradient between a high H^+ concentration in the tubule cells and a lower H^+ concentration in the tubular fluid. If the pH of the tubular fluid drops any lower than 4.5, tubular secretion of H^+ (step 7) ceases for lack of a sufficient gradient. Thus, pH 4.5 is the **limiting pH** for tubular secretion of H^+. This has added significance later in our discussion.

In a person with normal acid-base balance, the tubules secrete enough H^+ to neutralize all HCO_3^- in the tubular fluid; thus there is no HCO_3^- in the urine. Bicarbonate ions are filtered by the glomerulus, gradually disappear from the tubular fluid, and appear in the peritubular capillary blood. It *appears* as if HCO_3^- were reabsorbed by the renal tubules, but this is not the case; indeed, the renal tubules are incapable of HCO_3^- reabsorption. The cells of the proximal convoluted tubule, however, have carbonic anhydrase (CAH) on their brush

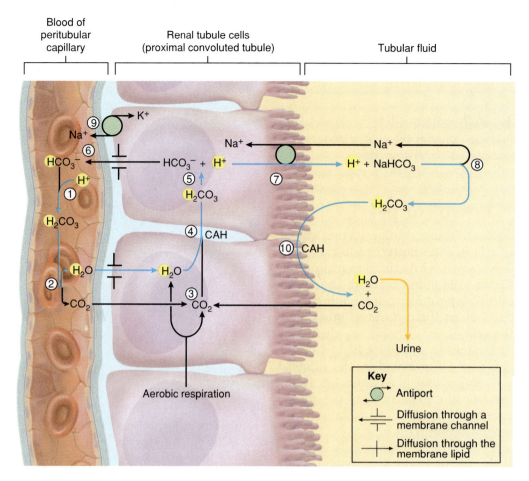

Figure 24.9 Secretion and neutralization of hydrogen ions in the kidneys. Circled numbers correspond to the explanation in the text. The colored arrows and hydrogen symbols allow you to trace hydrogen from H+ in the blood to H_2O in the urine.

borders facing the lumen. This breaks down the H_2CO_3 in the tubular fluid to CO_2 + H_2O (step 10). It is the CO_2 that is reabsorbed, not the bicarbonate. For every CO_2 reabsorbed, however, a *new* bicarbonate ion is formed in the tubule cell and released into the blood (steps 5–6). The effect is the same as if the tubule cells had reabsorbed bicarbonate itself.

Note that for every bicarbonate ion that enters the peritubular capillaries, a sodium ion does too. Thus the reabsorption of Na+ by the renal tubules is part of the process of neutralizing acid. The more acid the kidneys excrete, the less sodium the urine contains.

The tubules secrete somewhat more H+ than the available bicarbonate can neutralize. The urine therefore contains a slight excess of free H+, which gives it a pH of about 5 to 6. Yet if all of the excess H+ secreted by the tubules remained in this free ionic form, the pH of the tubular fluid would drop far below the limiting pH of 4.5, and H+ secretion would stop. This must be prevented, and there are additional buffers in the tubular fluid to do so.

The glomerular filtrate contains Na_2HPO_4 (dibasic sodium phosphate), which reacts with some of the H+

(fig. 24.10). A hydrogen ion replaces one of the sodium ions in the buffer, forming NaH_2PO_4 (monobasic sodium phosphate). This is passed in the urine and the displaced Na+ is transported into the tubule cell and from there to the bloodstream.

In addition, tubular cells catabolize certain amino acids and release ammonia, NH_3, as a product (fig. 24.10). Ammonia diffuses into the tubular fluid, where it acts as yet another buffer. It reacts with H+ and Cl− (the most abundant anion in the glomerular filtrate) to form ammonium chloride, NH_4Cl, which is passed in the urine.

Since there is so much chloride in the tubular fluid, you might ask why H+ is not simply excreted as hydrochloric acid, HCl. Why involve ammonia? The reason is that HCl is a strong acid—it dissociates almost completely, so most of its hydrogen would be in the form of free H+. The pH of the tubular fluid would drop below the limiting pH and prevent excretion of more acid. Ammonium chloride, by contrast, is a weak acid—most of its hydrogen remains bound to it and does not lower the pH of the tubular fluid.

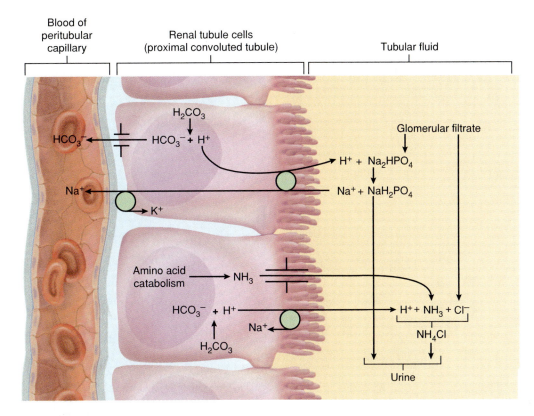

Figure 24.10 Mechanisms of buffering excess hydrogen ions in the urine. Reactions in the tubule cells are the same as in figure 24.9 but are simplified in this diagram. The essential differences are the buffering mechanisms shown in the tubular fluid.

Disorders of Acid-Base Balance

At pH 7.4, the ECF has a 20:1 ratio of HCO_3^- to H_2CO_3 (fig. 24.11). An excess of H_2CO_3 tips the balance to a lower pH, and if the pH falls below 7.35 a state of **acidosis** exists. An excess of HCO_3^- tips the balance to a higher pH. A pH above 7.45 is a state of **alkalosis.** Either of these imbalances has potentially fatal effects. A person cannot live more than a few hours if the blood pH is below 7.0 or above 7.7; a pH below 6.8 or above 8.0 is quickly fatal.

In acidosis, H^+ diffuses down its concentration gradient from the ECF to the ICF, and to maintain electrical balance, K^+ diffuses out of the cells (fig. 24.12a). The H^+ is buffered by intracellular proteins, so there is a net loss of cations from the cell. This makes the resting membrane potential more negative than usual (hyperpolarized) and makes nerve and muscle cells more difficult to stimulate. Thus acidosis depresses the central nervous system and causes such symptoms as confusion, disorientation, and coma.

In alkalosis, the extracellular H^+ concentration is low. Hydrogen ions diffuse out of the cells and K^+ diffuses in to replace them (fig. 24.12b). The net gain in positive intracellular charges shifts the membrane potential

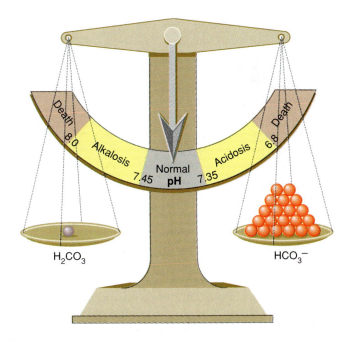

Figure 24.11 At a normal pH of 7.40, there is a 20:1 ratio of bicarbonate ions (HCO_3^-) to carbonic acid (H_2CO_3) in the blood plasma. An excess of HCO_3^- tips the balance toward alkalosis, whereas an excess of H_2CO_3 tips it toward acidosis.

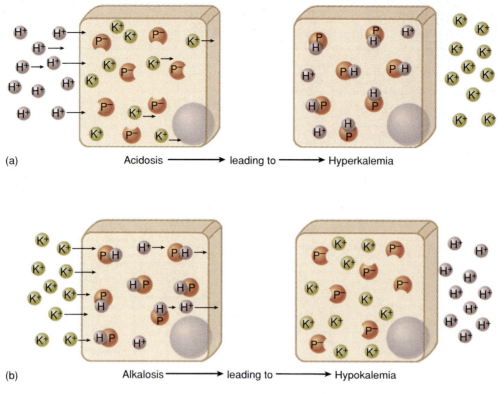

(a) Acidosis ⟶ leading to ⟶ Hyperkalemia

(b) Alkalosis ⟶ leading to ⟶ Hypokalemia

Figure 24.12 The relationship between acid-base imbalances and potassium imbalances. (*a*) In acidosis, H^+ diffuses into the cells and drives out K^+, elevating the K^+ concentration of the ECF. (*b*) In alkalosis, H^+ diffuses out of the cells and K^+ diffuses in to replace it, lowering the K^+ concentration of the ECF.

Table 24.2	Some Causes of Acidosis and Alkalosis	
	Acidosis	**Alkalosis**
Respiratory	Hypoventilation, apnea, or respiratory arrest; asthma, emphysema, chronic bronchitis	Hyperventilation due to emotions or oxygen deficiency (as at high altitudes)
Metabolic	Excess production of organic acids, as in diabetes mellitus and long-term anaerobic fermentation; drugs such as aspirin and laxatives; chronic diarrhea	Rare but can result from chronic vomiting or overuse of bicarbonates (antacids)

closer to firing level and makes the nervous system hyperexcitable. Neurons fire spontaneously and overstimulate skeletal muscles, causing muscle spasms, tetany, convulsions, or respiratory paralysis.

Acid-base imbalances are classified as respiratory or metabolic (table 24.2). **Respiratory acidosis** occurs when the rate of alveolar ventilation fails to keep pace with the body's rate of CO_2 production. Carbon dioxide accumulates in the ECF and lowers its pH. **Respiratory alkalosis** results from excessive ventilation (hyperventilation), in which CO_2 is eliminated faster than it is produced.

Metabolic acidosis can result from elevated production of organic acids, such as lactic acid in anaerobic fermentation and ketone bodies in alcoholism and diabetes mellitus. It can also result from the ingestion of acidic drugs such as aspirin or from the loss of base due to chronic diarrhea or overuse of laxatives. Dying persons also typically exhibit acidosis. **Metabolic alkalosis** is rare but can result from overuse of bicarbonates (such as oral antacids and intravenous bicarbonate solutions) or from the loss of stomach acid in chronic vomiting.

Compensation for Acid-Base Imbalances

Uncompensated acidosis or alkalosis is a pH imbalance that the body fails to correct. **Compensated** acidosis or alkalosis is a more temporary condition that the body corrects by means of its physiological and chemical buffers.

The respiratory system compensates for some states of acid-base imbalance by adjusting the partial pressure of carbon dioxide (P_{CO_2}) in the ECF. Hypercapnia (a CO_2 excess) stimulates increased pulmonary ventilation, and hypocapnia (a CO_2 deficiency) reduces it, by the mechanisms discussed in chapter 22. This is very effective in correcting pH imbalances due to abnormal P_{CO2} but not

Table 24.3 Some Relationships Among Fluid, Electrolyte, and Acid-Base Imbalances

Cause	Possible Effect	Reason
Acidosis →	Hyperkalemia	H^+ diffuses into cells and displaces K^+ (see fig. 24.12a, b). As K^+ leaves the ICF, concentration in the ECF rises.
Hyperkalemia →	Acidosis	Opposite of the above; a high K^+ concentration in the ECF causes K^+ to diffuse into cells and H^+ to diffuse out, lowering extracellular pH (see fig. 24.12c, d).
Alkalosis →	Hypokalemia	H^+ diffuses down its concentration gradient from ICF to ECF. K^+ diffuses into the cells to replace H^+, causing K^+ concentration in the ECF to drop.
Hypokalemia →	Alkalosis	Opposite of the above; low K^+ concentration in the ECF causes K^+ to diffuse out of cells. H^+ diffuses inward to replace them, lowering the H^+ concentration of the ECF and raising its pH.
Acidosis →	Hypochloremia	More Cl^- is excreted as NH_4Cl in order to buffer the excess acid in the renal tubules. This leaves less Cl^- in the ECF.
Alkalosis →	Hyperchloremia	More Cl^- is reabsorbed from the renal tubules in alkalosis, so ingested Cl^- accumulates in the ECF rather than being excreted.
Hyperchloremia →	Acidosis	More H^+ is retained in the blood plasma to balance the negative charges of the excess Cl^-, causing hyperchloremic acidosis.
Hypovolemia →	Alkalosis	More Na^+ is reabsorbed. Na^+ reabsorption is indirectly coupled to H^+ secretion (see fig. 24.9), so more H^+ is secreted and pH of the ECF rises.
Hypervolemia →	Acidosis	Less Na^+ is reabsorbed, so less H^+ is secreted into the renal tubules. H^+ retained in the ECF causes acidosis.
Acidosis →	Hypocalcemia	Acidosis causes more Ca^{2+} to bind to plasma protein and citrate ions, lowering the concentration of free, ionized calcium and causing symptoms of hypocalcemia.
Alkalosis →	Hypercalcemia	Alkalosis causes Ca^{2+} to dissociate from plasma protein and citrate, raising the concentration of free Ca^{2+}.

very effective in correcting other causes of acidosis and alkalosis. In diabetic acidosis, for example, the lungs cannot reduce the concentration of ketone bodies in the blood, although it can somewhat compensate for the H^+ released by ketone bodies by increasing pulmonary ventilation and exhausting extra CO_2. The respiratory system can adjust a blood pH of 7.0 back to 7.2 or 7.3 but not all the way back to the normal 7.4. Although the respiratory system has a very powerful buffering effect, its ability to stabilize pH is therefore limited.

The kidneys are slower to respond to pH imbalances but better at restoring a fully normal pH. Urine usually has a pH of 5 to 6, but in acidosis it may fall as low as 4.5 due to the excess of unbound H^+, whereas in alkalosis it may rise as high as 8.2 due to an excess of HCO_3^-. The kidneys cannot act quickly enough to compensate for short-term pH imbalances, such as the acidosis that might result from an asthmatic attack lasting an hour or two or the alkalosis resulting from a brief episode of emotional hyperventilation. They are effective, however, at compensating for pH imbalances that last for a few days or longer.

In acidosis, the renal tubules increase the rate of H^+ secretion. The extra H^+ in the tubular fluid must be buffered; otherwise, the fluid pH could exceed the limiting pH and H^+ secretion would stop. Therefore, in acidosis, the renal tubules secrete more ammonia to buffer

the added H^+, and the amount of ammonium chloride in the urine may rise to 7 to 10 times normal.

 Think About It

Suppose you measured the pH and ammonium chloride concentration of urine from a person with emphysema and urine from a healthy individual. How would you expect the two to differ, and why?

In alkalosis, the bicarbonate concentration and pH of the urine are elevated. This is partly because there is more HCO_3^- in the blood and glomerular filtrate and partly because there is not enough H^+ in the tubular fluid to neutralize all the HCO_3^- in the filtrate.

pH Imbalances in Relation to Electrolyte and Water Imbalances

The foregoing discussion once again stresses a point made early in this chapter—we cannot understand or treat imbalances of water, electrolyte, or acid-base balance in isolation from each other, because each of these frequently affects the other two. Table 24.3 itemizes and explains a few of these interactions. This is by no means a complete list of how fluid, electrolytes, and pH affect each other, but it does demonstrate their interdependence. Note that many of these relationships are reciprocal—for example, acidosis can cause hyperkalemia, and conversely, hyperkalemia can cause acidosis.

Key Point Review

10 Write two chemical equations that show how the bicarbonate buffer system compensates for acidosis and alkalosis and two equations that show how the phosphate buffer system compensates for these imbalances.

11 Why are phosphate buffers more effective in the cytoplasm than in the blood plasma?

12 Renal tubules cannot reabsorb HCO_3^-, and yet HCO_3^- concentration in the tubular fluid falls while in the blood plasma it rises. Explain this apparent contradiction.

13 In acidosis, the renal tubules secrete more ammonia. Why?

CHAPTER ESSAY

Fluid Replacement Therapy

One of the most significant problems in the treatment of seriously ill patients is the restoration and maintenance of proper fluid volume, composition, and distribution among the fluid compartments. Fluids may be administered to replenish total body water, restore blood volume and pressure, shift water from one fluid compartment to another, or restore and maintain electrolyte and acid-base balance.

Drinking water is the simplest method of fluid replacement, but it does not replace electrolytes. Heat exhaustion can occur when you lose water and salt in the sweat and replace the fluid by drinking plain water. Broths, juices, and isotonic sports drinks such as Gatorade replace both water and electrolytes. Sports drinks contain a mixture of salts, lactate, and glucose comparable to ECF.

If a patient cannot take fluids by mouth, they must be administered by alternative routes. Some can be given by enema and absorbed through the colon. **Parenteral**[6] routes are all those other than the digestive tract. The most common of these is the intravenous (I.V.) route, but for various reasons, including inability to find a suitable vein, fluids are sometimes given by other parenteral routes including subcutaneous (sub-Q) and intramuscular (I.M.). Many kinds of sterile solutions are available to meet the fluid replacement needs of different patients.

In cases of extensive blood loss, there may not be time to type and cross-match blood for a transfusion. The more urgent need is to replenish blood volume and pressure. **Normal saline** (isotonic, 0.9% NaCl) is a relatively quick and simple way to raise blood volume while maintaining normal osmolarity, but it has significant shortcomings. It takes three to five times as much saline as whole blood to rebuild normal volume because much of the saline escapes the circulation into the interstitial fluid compartment or is excreted by the kid-

neys. In addition, normal saline can induce hypernatremia and hyperchloremia, because the body excretes the water but retains much of the NaCl. Hyperchloremia can, in turn, produce acidosis. Normal saline also lacks potassium, magnesium, and calcium. Indeed, it dilutes those electrolytes that are already present and creates a risk of cardiac arrest from hypocalcemia. Saline also dilutes plasma albumin and RBCs, creating still greater risks for patients who have suffered extensive blood loss. Nevertheless, the emergency maintenance of blood volume sometimes takes temporary precedence over these other considerations.

Fluid therapy is also used to correct pH imbalances. Acidosis may be treated with **Ringer's lactate solution,** which includes sodium to rebuild ECF volume, potassium to rebuild ICF volume, lactate to balance the cations, and enough glucose to make the solution isotonic. Alkalosis can be treated with potassium chloride. This must be administered very carefully, because potassium ions can cause painful venous spasms, and even a small potassium excess can cause cardiac arrest. High-potassium solutions should never be given to patients in renal failure or whose renal status is unknown, because in the absence of renal excretion of potassium they can bring on lethal hyperkalemia. Ringer's lactate or potassium chloride also must be administered very cautiously, with close monitoring of blood pH, to avoid causing a pH imbalance opposite the one that was meant to be corrected. Too much Ringer's lactate causes alkalosis and too much KCl causes acidosis.

Plasma volume expanders are hypertonic solutions of chemicals that are retained in the bloodstream and draw interstitial water into it by osmosis. They include albumin, sucrose, mannitol, and dextran. These are also used to combat hypotonic hydration by drawing water out of swollen cells and averting such problems as seizures and coma. A plasma expander can draw several liters of water out of the intracellular compartment within a few minutes.

---

6. *para* = beside + *enter* = intestine

Patients who cannot eat are often given isotonic 5% dextrose (glucose). A fasting patient loses 70 to 85 g of protein per day from the tissues. Giving 100 to 150 g of I.V. glucose per day reduces this by half and is said to have a *protein-sparing effect*. More than glucose is needed in some cases—for example, if a patient has not eaten for several days and cannot be fed by nasogastric tube (due to lesions of the digestive tract, for example) or if large amounts of nutrients are needed for tissue repair following severe trauma, burns, or infections. In **total parenteral nutrition (TPN),** or **hyperalimentation,**[7] a patient is provided with complete I.V. nutritional support, including a protein hydrolysate (amino acid mixture), vitamins, electrolytes, 20% to 25% glucose, and on alternate days, a fat emulsion.

The water from parenteral solutions is normally excreted by the kidneys. If the patient has renal insufficiency, however, excretion may not keep pace with intake, and there is a risk of hypotonic hydration. Intravenous fluids are usually given slowly, by **I.V. drip,** to avoid abrupt changes or overcompensation for the patient's condition. In addition to pH, the patient's pulse rate, blood pressure, hematocrit, and plasma electrolyte concentrations are monitored, and the patient is examined periodically for any respiratory sounds indicating pulmonary edema.

The delicacy of fluid replacement therapy underscores the close relationships among fluids, electrolytes, and pH. It is dangerous to manipulate any one of these variables without close attention to the others. Parenteral fluid therapy is usually used for persons who are seriously ill. Their homeostatic mechanisms are already compromised and leave less room for error than in a healthy person.▲

---

7. *hyper* = above normal + *aliment* = nourishment

Chapter Review   Study Outline

Water Balance (pp. 865–871)
1. Fluid compartments
   a. Total body water
   b. Intracellular fluid
   c. Extracellular fluid
      • Tissue fluid
      • Blood plasma and lymph
      • Transcellular fluid
   d. Movement among fluid compartments
2. Water gain and loss
3. Regulation of intake
   a. Thirst
      • Hyposalivation
      • Response of the thirst center
   b. Satiation of thirst
      • Moistening and cooling of mouth
      • Distension of stomach and intestine
      • Drop in blood osmolarity
4. Regulation of output
   a. Relationship to Na⁺ reabsorption
   b. Control by antidiuretic hormone
5. Disorders of water balance
   a. Fluid deficiency
      • Volume depletion
      • Dehydration
   b. Fluid excess
      • Volume excess
      • Hypotonic hydration

   c. Fluid sequestration
      • Hemorrhage
      • Pleural effusion
      • Edema

Electrolyte Balance (pp. 871–875)
1. Major electrolytes
2. Terminology of imbalances
3. Sodium
   a. Functions
   b. Homeostasis
      • Aldosterone
      • Antidiuretic hormone
      • Atrial natriuretic factor
      • Other hormones
      • Salt craving
   c. Imbalances
      • Hypernatremia
      • Hyponatremia
4. Potassium
   a. Functions
   b. Homeostasis
      • Tubular secretion
      • Aldosterone
   c. Imbalances
      • Hyperkalemia
      • Hypokalemia
5. Chloride
   a. Functions
   b. Homeostasis

   c. Imbalances
      • Hyperchloremia
      • Hypochloremia
6. Calcium
   a. Functions
   b. Homeostasis
   c. Imbalances
      • Hypercalcemia
      • Hypocalcemia
7. Phosphates
   a. Functions
   b. Homeostasis
   c. Imbalances

Acid-Base Balance (pp. 875–882)
1. Importance of stability
2. Acids, bases, and buffers
   a. Strong and weak acids and bases
   b. Physiological buffers
      • Urinary system
      • Respiratory system
   c. Chemical buffer systems
      • Bicarbonate buffer system
      • Phosphate buffer system
      • Protein buffer system
3. Respiratory control of pH
4. Renal control of pH
   a. H⁺ secretion
      • Mechanism
      • Limiting pH
      • Bicarbonate exchanges

b. Buffers in the tubular fluid
  • Sodium phosphate
  • Ammonia
5. Disorders of acid-base balance
  a. Effects of acidosis and alkalosis

b. Respiratory acidosis and alkalosis
c. Metabolic acidosis and alkalosis
d. Compensation for acid-base imbalances
  • Respiratory compensation

  • Urinary compensation
e. pH imbalances in relation to electrolyte and water imbalances

Selected Vocabulary

total body water (TBW) 865
fluid compartment 865
fluid balance 866
metabolic water 866
preformed water 866
cutaneous transpiration 866
insensible water loss 867
obligatory water loss 867
thirst center 867
volume depletion 868
dehydration 869

volume excess 870
hypotonic hydration 870
fluid sequestration 870
hypernatremia 872
hyponatremia 873
hyperkalemia 873
hypokalemia 873
hyperchloremia 873
hypochloremia 873
hypocalcemia 875
hypercalcemia 875

strong acid 876
weak acid 876
strong base 876
weak base 876
buffer 876
physiological buffer 876
chemical buffer 876
buffer system 876
bicarbonate buffer system 876
phosphate buffer system 876

protein buffer system 876
limiting pH 877
respiratory acidosis 880
respiratory alkalosis 880
metabolic acidosis 880
metabolic alkalosis 880
uncompensated 880
compensated 880

Testing Your Recall   Answers in Appendix C

1. The greatest percentage of the body's water is in the
   a. blood plasma.
   b. lymph.
   c. intracellular fluid.
   d. interstitial fluid.
   e. extracellular fluid.

2. High blood pressure is likely to increase the secretion of
   a. atrial natriuretic factor.
   b. antidiuretic hormone.
   c. bicarbonate.
   d. aldosterone.
   e. ammonia.

3. _____ increases water reabsorption without increasing sodium reabsorption.
   a. Antidiuretic hormone
   b. Aldosterone
   c. Atrial natriuretic factor
   d. Parathyroid hormone
   e. Calcitonin

4. Hypotonic hydration can result from
   a. ADH hypersecretion.
   b. ADH hyposecretion.
   c. aldosterone hypersecretion.
   d. aldosterone hyposecretion.
   e. *a* and *d* only.

5. Tetany is most likely to result from
   a. hypernatremia.
   b. hypokalemia.
   c. hyperkalemia.
   d. hypocalcemia.
   e. *c* and *d* only.

6. The principal determinant of intracellular osmolarity and cellular volume is
   a. protein.
   b. phosphate.
   c. potassium.
   d. sodium.
   e. chloride.

7. Increased excretion of ammonium chloride in the urine most likely indicates
   a. hypercalcemia.
   b. hyponatremia.
   c. hypochloremia.
   d. alkalosis.
   e. acidosis.

8. The most effective buffer in the intracellular fluid is
   a. phosphate.
   b. protein.
   c. bicarbonate.
   d. carbonic acid.
   e. ammonia.

9. Tubular secretion of hydrogen is directly linked to
   a. tubular secretion of potassium.
   b. tubular secretion of sodium.
   c. tubular reabsorption of potassium.
   d. tubular reabsorption of sodium.
   e. tubular secretion of chloride.

10. Hyperchloremia is most likely to result in
    a. alkalosis.
    b. acidosis.

c. hypernatremia.
d. hyperkalemia.
e. hypovolemia.

11. The most abundant cation in the extracellular fluid is _____.

12. The most abundant cation in the intracellular fluid is _____.

13. Water produced by the body's chemical reactions is called _____.

14. The skin loses water by two processes, sweating and _____.

15. Any abnormal accumulation of fluid in a particular place in the body is called _____.

16. An excessive concentration of potassium ions in the blood is called _____.

17. A deficiency of sodium ions in the blood is called _____.

18. A blood pH of 7.2 caused by inadequate pulmonary ventilation would be classified as _____.

19. Tubular secretion of hydrogen ions would cease if the acidity of the tubular fluid fell below a value called the _____.

20. Long-term satiation of thirst depends on a reduction of the _____ of the blood.

1. A duck hunter is admitted to the hospital with a shotgun injury to the abdomen. He has suffered extensive blood loss but is conscious. He complains of being intensely thirsty. Explain the physiological mechanism leading from his injury to his thirst.

2. A woman living at poverty level finds bottled water at the grocery store next to the infant formula. The label on the water states that it is made especially for infants, and she construes this to mean it can be used as a nutritional supplement. Because the water is much cheaper, she gives her baby several ounces of bottled water a day as a partial substitute for formula. After several days the baby has seizures and is taken to the hospital. Here, the baby is found to have edema, acidosis, and a plasma sodium concentration of 116 mEq/L. The baby is treated with anticonvulsants followed by normal saline and recovers. Explain each of the symptoms.

3. Explain why the respiratory and urinary systems are necessary for the bicarbonate buffer system to work very well in the blood plasma.

4. The left column indicates some changes (increases or decreases) in blood plasma values. In the right column, replace the question mark with an up or down arrow to indicate the expected effect. Explain each effect.

Cause	Effect
a. $\uparrow H_2O$	? Na^+
b. $\uparrow Na^+$	? Cl^-
c. $\downarrow K^+$	? H^+
d. $\uparrow H^+$	? K^+
e. $\downarrow Ca^{2+}$	? PO_4^{3-}

5. A 4-year-old child is caught up in tribal warfare in Africa. In a refugee camp, there is only sewage-contaminated water to drink. He soon contracts severe diarrhea and dies 10 days later of cardiac arrest. Explain the possible physiological cause(s) of his death.

Web Site Link

For a listing of the most current web sites related to this chapter, please visit the Saladin homepage at:

http://www.mhhe.com/sciencemath/biology/saladin/

[The Digestive System

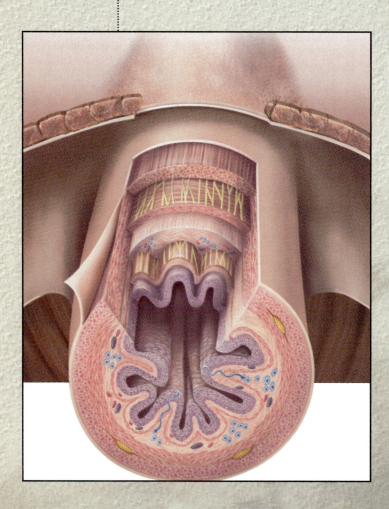

$\mathcal{Special}$Topics

t h e   u n i t y   o f   f o r m   a n d   f u n c t i o

Brushing up

Most of the nutrients we eat cannot be used in their existing form. They must be broken down into smaller components, such as amino acids and monosaccharides, that are universal to all species. Consider what happens if you eat a piece of beef, for example. The myosin of the beef differs very little from that of your own muscles. Yet the two are not identical, and even if they were, beef myosin could not be absorbed, transported in the blood, and incorporated into your muscles. Dietary protein must be broken down into amino acids before it can be used. Since all proteins are made of a limited "alphabet" of the same 20 amino acids, those of beef proteins might indeed become part of your own myosin but could equally well wind up in your insulin, fibrinogen, collagen, or any other protein. The primary purpose of the digestive system is to break nutrients down into forms that can be used by the body and absorb them so they can be distributed to the tissues.

General Anatomy and Digestive Processes

▼ Objectives

When you have completed this section, you should be able to
- list the regions of the digestive tract and the accessory organs of the digestive system;
- list the functions and major physiological processes of the digestive system;
- distinguish between mechanical and chemical digestion; and
- describe the basic chemical process underlying all chemical digestion, and name the major substrates and products of this process.

Subdivisions of the Digestive System

The digestive system has two anatomical subdivisions, the accessory organs and the digestive tract (fig. 25.1). The **accessory organs** are the teeth, tongue, salivary glands, liver, gallbladder, and pancreas. The **digestive tract** is a tube extending from mouth to anus, measuring about 9 m (30 ft) long in the cadaver. It is also known as the *gastrointestinal (GI) tract,* or *alimentary*[1]

---

1. *aliment* = food

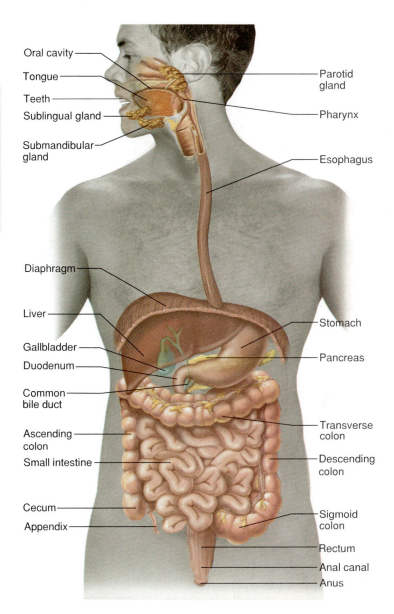

Figure 25.1 The digestive system. ✗

canal. It includes the oral cavity, pharynx, esophagus, stomach, small intestine, and large intestine. The small intestine consists of the *duodenum,* a short segment leading from the stomach; a much longer middle segment, the *jejunum;* and the longest of all, the *ileum,* terminating at the large intestine. The large intestine consists of the *ascending colon* rising up the right side of the abdominal cavity; the *transverse colon* crossing from right to left; the *descending colon* passing down the left side of the abdominal cavity; a short S-shaped *sigmoid colon;* a straight *rectum;* and an *anal canal* terminating at the anus.

The digestive tract is open to the environment at both ends. Most of the material in it has not entered any body tissues and is considered to be external to the

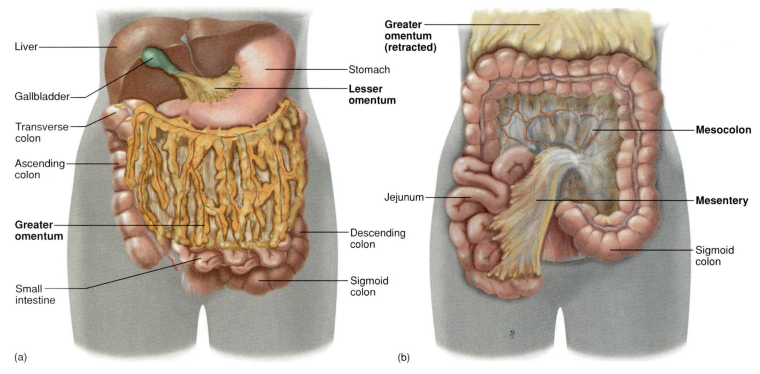

Figure 25.2 The serous membranes associated with the abdominal portion of the digestive tract. (*a*) The greater and lesser omenta. The greater omentum is shown translucent so the intestines can be seen behind it. (*b*) Greater omentum and small intestine retracted to show the mesocolon and mesentery. These membranes contain the mesenteric arteries and veins.

body until it is absorbed by epithelial cells of the alimentary canal. In the strict sense, food residue that is defecated never entered the body.

Relationship of the Digestive Tract to the Peritoneum

Most of the digestive tract is within the peritoneal cavity; the relationship of the stomach and intestines to the peritoneum is shown in figure A.15 (p. 39). Along the dorsal wall of the abdominal cavity, the parietal peritoneum turns inward and forms a sheet of tissue, the **dorsal mesentery,** extending to the digestive tract. The membrane then forms the outer covering, or **serosa,** of the stomach and most parts of the intestines. In some places it continues beyond the digestive organ as a sheet of tissue called the **ventral mesentery,** which may hang freely in the abdominal cavity or attach to the ventral abdominal wall or to other organs.

Along the right superior margin (*lesser curvature*) of the stomach, the serosae of the anterior and posterior stomach surfaces meet and continue as a ventral mesentery, the **lesser omentum,** extending from the stomach to the liver (fig. 25.2). Along the left inferior margin (*greater curvature*) of the stomach, the serosae form the **greater omentum,** which hangs loosely over the small intestine like an apron. At its inferior margin, the

greater omentum turns back on itself, passes upward, and forms serous membranes around the spleen and transverse colon. Beyond the transverse colon, it continues as a mesentery called the **mesocolon,** which anchors the colon to the posterior abdominal wall.

The omenta have a loosely organized, lacy appearance due partly to many holes or gaps in the membranes and partly to an irregular distribution of fatty tissue. They also contain many lymph nodes, lymphatic vessels, blood vessels, and nerves. The omenta adhere to perforations or inflamed areas of the stomach or intestines, contribute immune cells to the site, and isolate infections that might otherwise give rise to peritonitis.

Some portions of the digestive system are retroperitoneal—notably the duodenum, most of the pancreas, and parts of the large intestine.

Digestive Functions and Processes

The digestive system has four functions:

1. **ingestion,** the selective intake of nutrients;
2. **digestion,** the breakdown of large molecules into smaller ones;
3. **absorption,** the uptake of nutrient molecules, first into the epithelial cells of the digestive tract and then into the blood or lymph; and finally
4. **defecation,** the elimination of undigested residue.

These functions are carried out through three principal processes:

1. **motility,** muscular contractions that break up food and propel it through the canal;
2. **secretion** of enzymes, hormones, and other products that carry out or regulate digestion; and
3. **membrane transport,** mechanisms of absorbing nutrients and transferring them to the blood and lymph.

Stages of Digestion

There are two stages of digestion: mechanical and chemical. **Mechanical digestion** is achieved by the cutting and grinding action of the teeth and the churning contractions of the stomach and small intestine. These actions break food items into smaller particles to expose more of their surface to the action of digestive enzymes.

Chemical digestion consists solely of hydrolysis reactions that break the dietary macromolecules into their monomers (residues): polysaccharides into monosaccharides, proteins into amino acids, fats into glycerol and fatty acids, and nucleic acids into nucleotides. It is carried out by digestive enzymes produced by the salivary glands, stomach, pancreas, and small intestine. Some nutrients are already present in usable form in the ingested food and are absorbed without being digested: vitamins, free amino acids, minerals, cholesterol, and water.

Key Point Review

1. What must nutrients such as proteins be digested into before they can be used by the human body?
2. Which physiological process of the digestive system truly moves a nutrient from the outside to the inside of the body?
3. What one type of reaction is responsible for all chemical digestion?
4. Name some nutrients that are absorbed without being digested.

The Mouth Through Esophagus

▼**Objectives**
When you have completed this section, you should be able to
- describe the gross anatomy of the digestive tract from the mouth through the esophagus;
- describe the composition and functions of saliva;
- describe the neural control of salivation and swallowing; and
- describe the histological layers of the esophagus and explain the functional significance of each.

The Mouth

The mouth is also known as the **oral,** or **buccal** (BUCK-ul), **cavity.** Its functions include:

- **ingestion,** or food intake;
- **sensory responses** to taste, texture, temperature, and other qualities of food;

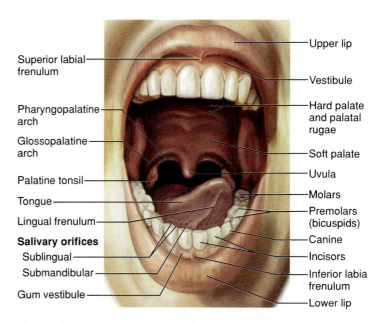

Figure 25.3 The oral (buccal) cavity. For a photographic midsagittal view see figure A.33 (atlas A). ✗

Labels on figure:
Superior labial frenulum — Pharyngopalatine arch — Glossopalatine arch — Palatine tonsil — Tongue — Lingual frenulum — **Salivary orifices** Sublingual — Submandibular — Gum vestibule — Upper lip — Vestibule — Hard palate and palatal rugae — Soft palate — Uvula — Molars — Premolars (bicuspids) — Canine — Incisors — Inferior labia frenulum — Lower lip

- **mastication,** the mechanical breakup of food;
- **chemical digestion,** the partial hydrolysis of starch while food is being chewed;
- **swallowing,** which is initiated by the tongue;
- **speech,** which involves the tongue, cheeks, and lips; and
- **respiration,** in which the mouth supplements the nose as a passage for airflow.

The mouth is enclosed by the cheeks, lips, palate, and tongue (fig. 25.3). Its anterior opening is the **oral orifice** and its posterior opening is the **fauces**[2] (FAW-seez). The oral mucosa is lined with stratified squamous epithelium.

The Cheeks and Lips

The cheeks and lips retain food and push it between the teeth for mastication. They are essential for articulate speech and for sucking and blowing actions, including suckling by infants. Their fleshiness is due mainly to subcutaneous fat, the buccinator muscles of the cheeks, and the orbicularis oris muscle of the lips. Externally the lips are divided into a *cutaneous area* and *red area (vermilion)*. The cutaneous area is colored like the rest of the face and has hair follicles and sebaceous glands; it is the mustache region of the upper lip and the corresponding region of the lower lip. The red area, to which a woman might apply lipstick, lacks hair and sebaceous

---

2. *fauces* = throat

glands but has unusually tall dermal papillae. Blood capillaries and nerve fibers come close to the surface in the papillae, making this area redder and more sensitive than the cutaneous area.

Each lip is attached to the gum behind it by a midsagittal fold called the **labial frenulum.**[3] The **vestibule** is the space between the teeth and the cheeks and lips (where you insert a toothbrush to clean the outer surfaces of the teeth).

The Tongue

The tongue (see fig. 16.4, p. 559) is a muscular and bulky but remarkably agile and sensitive organ. It manipulates food between the teeth while it avoids being bitten, it can extract food particles from between the teeth after a meal, and it is sensitive enough to feel a single stray hair in a bite of food. Its surface is covered with stratified squamous epithelium and exhibits innumerable bumps and projections, the **lingual papillae,** most of which have taste buds. The types of papillae, sense of taste, and other aspects of the flavor of food were discussed in chapter 16.

Think About It

Why is proprioception important in protecting the tongue from being bitten?

The anterior two-thirds of the tongue, called the **body,** occupies the oral cavity and the posterior one-third, the **root,** occupies the oropharynx. The boundary between them is marked by a V-shaped row of circumvallate papillae and, behind them, a groove called the **terminal sulcus.** The body is attached to the floor of the mouth by a midsagittal fold called the **lingual frenulum.**

The mass of the tongue is composed mainly of two groups of **lingual muscles** composed of skeletal muscle tissue. The *intrinsic muscles* lie entirely within the body and are oriented in dorsal-to-ventral, right-to-left, and anterior-to-posterior directions, enabling the tongue to flex in all directions. The *extrinsic muscles* of the root connect the tongue to the hyoid bone and skull; they are described and illustrated on page 337. Amid the muscles are serous and mucous **lingual glands,** which secrete a portion of the saliva. The lingual tonsils are contained in the root.

The Palate

The palate, separating the oral cavity from the nasal cavity, makes it possible to breathe while chewing food. Its anterior portion, the **hard (bony) palate,** is supported by the palatine processes of the maxillae and the smaller palatine bones. It has transverse *friction ridges (palatal rugae)* that aid the tongue in holding and manipulating food. Posterior to this is the **soft palate,** which has a more spongy texture and is composed mainly of skeletal muscle and glandular tissue, but no bone. It has a conical medial projection, the **uvula,**[4] visible at the rear of the oral cavity.

A pair of muscular folds on each side of the oral cavity begin dorsally near the uvula and arch along the wall of the oral cavity to its floor. The anterior one is the **glossopalatine arch,** and the posterior one is the **pharyngopalatine arch.** The latter arch marks the beginning of the pharynx. The palatine tonsils are located on the wall between the arches.

The Teeth

An adult normally has 16 teeth in the mandible and 16 in the maxilla. Collectively they are called the **dentition.** Both the upper and lower jaws exhibit two incisors, a canine, two premolars, and three molars on each side (fig. 25.4*a*). The **incisors** are chisel-like cutting teeth used to bite off a piece of food. The **canines** are more pointed and act to puncture and shred it. They serve as weapons in many mammals but became reduced in the course of human evolution until they now project barely above the other teeth. The **premolars** and **molars** have relatively broad surfaces adapted to crushing and grinding.

The meeting of the teeth when the mouth closes is called **occlusion** (uh-CLUE-zhun), and the surfaces where they meet are called the **occlusal** (uh-CLUE-zul) **surfaces.** The occlusal surface of a premolar has two rounded bumps called **cusps;** thus the premolars are also known as **bicuspids.** The molars have four to five cusps. Cusps of the upper and lower premolars and molars mesh when the jaws are closed and slide over each other as the jaw makes lateral chewing motions. This grinds and tears food more effectively than if the occlusal surfaces were flat.

Teeth develop beneath the gums and **erupt** (emerge) in predictable order. Twenty **deciduous teeth** (*milk teeth,* or *baby teeth*) erupt from the ages of 6 to 30 months, beginning with the incisors (fig. 25.4*b*). From 6 to 25 years of age, these are replaced by the 32 **permanent teeth.** As a permanent tooth grows below a deciduous tooth (fig. 25.5), the root of the deciduous tooth dissolves, leaving little more than the crown by the time it falls out. The third molars (wisdom teeth) erupt around ages 17 to 25, if at all. Over the course of human evolution, the face became flatter and the jaws shorter.

---

3. *labi* = lip + *frenulum* = little bridle

4. *uvula* = little grape

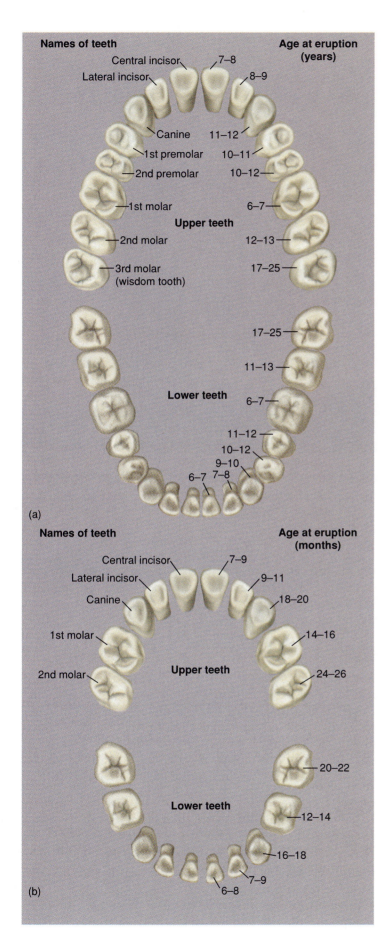

Names of teeth

Central incisor
Lateral incisor
Canine
1st premolar
2nd premolar
1st molar
2nd molar
3rd molar
(wisdom tooth)

Upper teeth

Lower teeth

Age at eruption (years)

7–8
8–9
11–12
10–11
10–12
6–7
12–13
17–25

17–25
11–13
6–7
11–12
10–12
9–10
6–7 7–8

(a)

Names of teeth

Central incisor
Lateral incisor
Canine
1st molar
2nd molar

Upper teeth

Lower teeth

Age at eruption (months)

7–9
9–11
18–20
14–16
24–26

20–22
12–14
16–18
7–9
6–8

(b)

Figure 25.4 The dentition and ages at which the teeth erupt. (a) Permanent teeth; (b) deciduous (baby) teeth.

This left little room for the third molars to erupt, so they often remain below the gum and become *impacted*—so crowded against neighboring teeth and bone that they cannot erupt.

Each tooth is embedded in a socket called an **alveolus,** forming a joint called a *gomphosis* between the tooth and bone. The alveolus is lined by a **periodontal** (PERR-ee-oh-DON-tul) **membrane,** a modified periosteum whose collagen fibers extend into the jaw bone on one side and into the tooth on the other. This anchors the tooth very firmly in the alveolus. The gum, or **gingiva** (JIN-jih-vuh), covers the alveolar bone. Regions of a tooth are defined by their relationship to the gingiva: the **crown** is the portion visible above the gumline, the **neck** is the portion from the margin of the gum to the beginning of the alveolar bone, and the **root** is the portion inserted into the alveolus (fig. 25.6). The space between the neck and gum is the **gingival sulcus.** The hygiene of this sulcus is especially important to dental health (see special topic 25.1).

Most of a tooth consists of hard yellowish tissue called **dentin,** covered with **enamel** in the crown and **cementum** in the root. Dentin and cementum are living connective tissues with cells or cell processes embedded in a calcified matrix. Cells of the cementum (*cementocytes*) are scattered more or less at random and occupy tiny cavities similar to the lacunae of bone. Cells of the dentin (*odontoblasts*) line the pulp cavity and have slender processes that travel through tiny parallel tunnels in the dentin.

Enamel is not a tissue but a noncellular secretion produced before the tooth erupts. Damaged dentin and cementum can regenerate, but damaged enamel cannot—it must be artificially repaired.

Internally, a tooth has a dilated **pulp cavity** in the crown and a narrow **root canal** in the root. These spaces are occupied by **pulp,** composed of loose connective tissue, blood and lymphatic vessels, and nerves. These nerves and vessels enter the tooth through the **apical foramen** at the inferior end of each root canal.

Mastication

Mastication (chewing) breaks food into pieces small enough to be swallowed and exposes more surface to the action of digestive enzymes. It is the first step in mechanical digestion. The tongue, buccinator, and orbicularis oris muscles manipulate food and push it between the teeth. The masseter and temporalis muscles produce the up-and-down crushing action of the teeth, and the

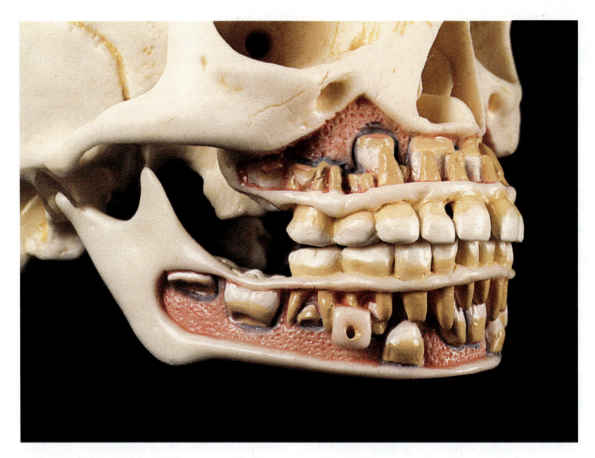

Figure 25.5 Model of a child's skull, showing permanent and deciduous teeth. See if you can estimate the child's age from the teeth that have erupted and those that have not, using information in the previous figure.

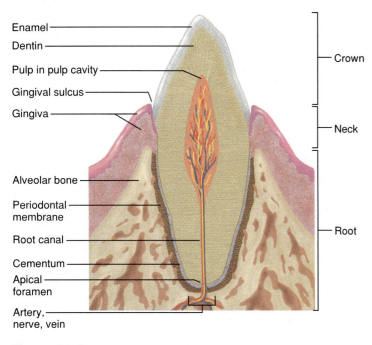

Enamel
Dentin
Pulp in pulp cavity
Gingival sulcus
Gingiva
Alveolar bone
Periodontal membrane
Root canal
Cementum
Apical foramen
Artery, nerve, vein

Crown
Neck
Root

Figure 25.6 Medial section of a canine tooth and its alveolus, showing typical anatomy of a tooth and periodontal tissues. ✗

lateral and medial pterygoid muscles and masseters produce side-to-side grinding action. Mastication requires little thought because food stimulates receptors that trigger an involuntary chewing reflex.

Saliva and the Salivary Glands

Saliva moistens the mouth, digests a little starch and fat, cleanses the teeth, inhibits bacterial growth, dissolves molecules so they can stimulate the taste buds, and moistens food and binds particles together to aid in swallowing. It is secreted by salivary glands, some located in the vicinity of the mouth and connected to it by ducts, and others located within the oral tissues themselves.

Saliva

Saliva is a hypotonic solution composed of 97.0% to 99.5% water and the following solutes:

- **salivary amylase,** an enzyme that begins starch digestion in the mouth;
- **lingual lipase,** an enzyme that is activated by stomach acid and digests fat after the food is swallowed;

Food leaves a sticky residue on the teeth called **plaque,** composed mainly of bacteria and sugars. If plaque is not thoroughly removed by brushing and flossing, bacteria accumulate, metabolize the sugars, and release lactic acid and other acids. These acids dissolve the minerals of enamel and dentin, and the bacteria enzymatically digest the collagen and other organic components. The eroded "cavities" of the tooth are known as dental **caries.**[5] If not repaired, caries may fully penetrate the dentin and spread to the pulp cavity. They then require either extraction of the tooth or **root canal therapy,** in which all pulp is removed and replaced with inert material.

When plaque calcifies on the tooth surface, it is called **calculus,** or **tartar.** Calculus in the gingival sulcus wedges the tooth and gum apart and allows bacterial invasion of the sulcus. This leads to **gingivitis,** or gum inflammation. Gingivitis is reversible if the calculus is removed, but otherwise, bacteria spread from the sulcus into the alveolar bone and begin dissolving it. This produces **periodontal disease,** which accounts for 80% to 90% of adult tooth loss.

- **mucus,** which binds and lubricates the food mass and aids in swallowing;
- **lysozyme,** which kills bacteria;
- **immunoglobulin A** (IgA), an antibody that inhibits bacterial growth; and
- **electrolytes,** including sodium, potassium, chloride, phosphate, and bicarbonate ions.

Saliva has a pH of 6.8 to 7.0. There are striking differences in pH from one region of the digestive tract to another, with a powerful influence on the activity and deactivation of digestive enzymes. For example, salivary amylase works well at a neutral pH and is deactivated by the low pH of the stomach, whereas lingual lipase does not act in the mouth at all but is activated by the acidity of the stomach. Thus saliva begins to digest starch before the food is swallowed and fat after it is swallowed.

The Salivary Glands

The **intrinsic salivary glands,** located within the oral tissues, include the *lingual glands* embedded in the tongue, *labial glands* on the inner aspect of the lips, and *buccal glands* on the inner aspect of the cheeks. They secrete relatively small amounts of saliva at a fairly constant rate, which keeps the mouth moist and inhibits bacterial growth. They produce lingual lipase and lysozyme.

The **extrinsic salivary glands** are situated outside the oral cavity but convey saliva to it by way of their ducts. They secrete salivary amylase, electrolytes, and more mucus than the intrinsic glands. They increase their secretion substantially in response to food. There are three of these on each side (fig. 25.7):

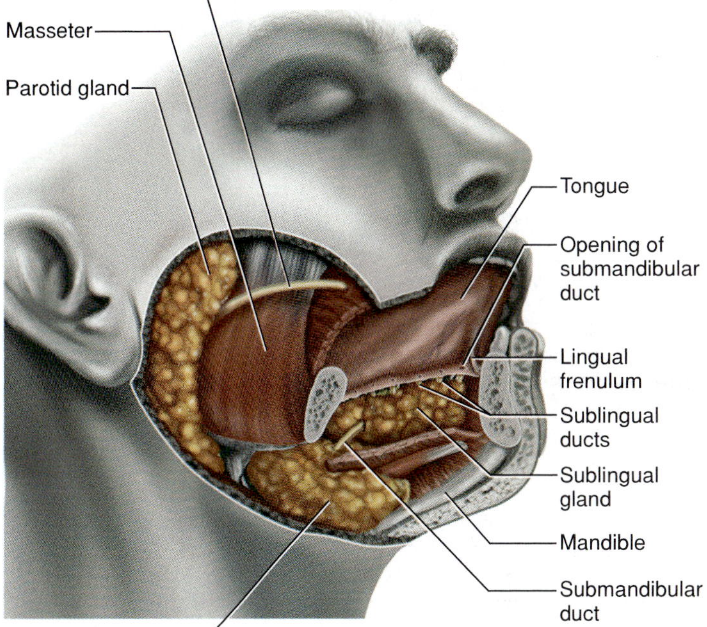

1. The **parotid**[6] **gland** lies between the skin and masseter muscles anterior to the earlobes, where it can be palpated as a slightly spongy mass. Its duct passes superficially over the masseter muscle,

Figure 25.7 The extrinsic salivary glands. ⸙

Parotid duct — Masseter — Parotid gland — Tongue — Opening of submandibular duct — Lingual frenulum — Sublingual ducts — Sublingual gland — Mandible — Submandibular duct — Submandibular gland

pierces the buccinator, and opens into the mouth opposite the second upper molar on each side. Mumps is a viral inflammation and swelling of the parotid glands.

2. The **submandibular gland** is located halfway along the body of the mandible, medial to its margin, just deep to the mylohyoid muscle. Its duct empties into the mouth at a papilla on the side of the lingual frenulum, near the lower central incisors.

3. The **sublingual gland** is located in the floor of the mouth. It has multiple *sublingual ducts* emptying onto the floor posterior to the papilla of the submandibular duct.

These are all compound tubuloacinar glands with a treelike arrangement of branching ducts ending in acini

5. *caries* = rottenness
6. *par* = next to + *ot* = ear

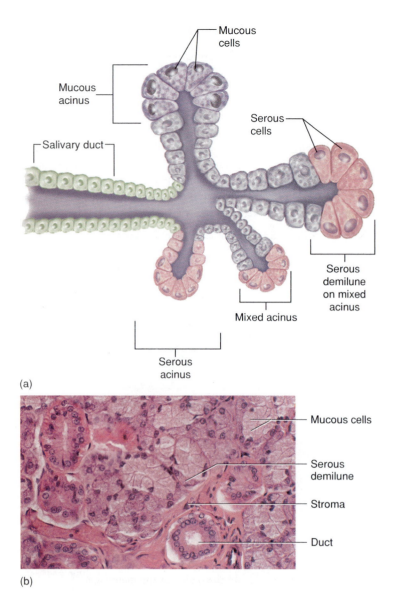

(a)

(b)

Figure 25.8 Microscopic anatomy of the salivary glands. (*a*) Duct and acini of a generalized salivary gland with a mixture of mucous and serous cells. Serous cells often form crescent-shaped caps called serous demilunes over the ends of mucous acini. (*b*) Histology of the sublingual salivary gland. ✗

(see chapter 6). Some acini have only mucous cells, some have only serous cells, and some have a mixture of both (fig. 25.8). Mucous cells secrete salivary mucus, and serous cells secrete a thinner fluid rich in amylase.

Salivation

The extrinsic salivary glands secrete about 1.0 to 1.5 L of saliva per day. Cells of the acini filter water and electrolytes from the blood capillaries and add amylase, mucin, and lysozyme to it. The ducts slightly modify its electrolyte composition.

Food stimulates tactile, pressure, and taste receptors in the mouth, which transmit signals to a group of salivatory nuclei in the medulla oblongata and pons. These nuclei also receive input from higher brain centers, so even the odor, sight, or thought of food stimulates salivation. Irritation of the stomach and esophagus by spicy foods, stomach acid, or toxins also stimulates salivation, perhaps to dilute and rinse away the irritants.

The salivatory nuclei send autonomic signals to the glands by way of the facial and glossopharyngeal nerves. Sympathetic stimulation constricts the blood vessels, reduces capillary filtration, and thus reduces the amount of saliva produced. The saliva is then relatively thick, with more mucous than serous content. This is why the mouth may feel sticky or dry under conditions of stress. Dehydration also reduces capillary filtration and salivation. Parasympathetic stimulation produces thinner saliva with more amylase.

Salivary amylase begins to digest starch as the food is chewed, while the mucus of saliva binds food particles into a soft, slippery, easily swallowed mass called a **bolus.** Without mucus, a person must drink a much larger volume of fluid to swallow food.

The Pharynx

The pharynx, described in chapter 22, has a deep layer of longitudinally oriented skeletal muscle and a superficial layer of circular skeletal muscle. The circular muscle is divided into **superior, middle,** and **inferior pharyngeal constrictors,** which force food downward during swallowing. When food is not being swallowed, the inferior constrictor remains contracted to exclude air from the esophagus.

The Esophagus

The **esophagus** is a straight muscular tube 25 to 30 cm long. It begins posterior to the larynx at the level of the cricoid cartilage and travels downward through the mediastinum of the thorax. It penetrates the diaphragm at an opening called the *esophageal hiatus,* continues another 3 to 4 cm, and meets the stomach at an opening called the **cardiac orifice** (named for its proximity to the heart).

The wall of the esophagus is organized into the following tissue layers (fig. 25.9):

Mucosa

    Epithelium

    Lamina propria

    Muscularis mucosae

Submucosa

Muscularis externa

    Inner circular layer

    Outer longitudinal layer

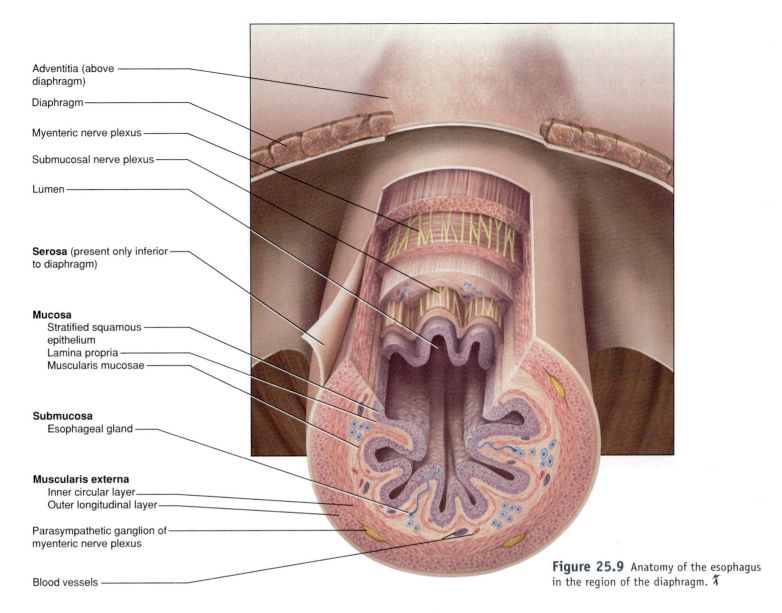

Adventitia (above diaphragm)

Diaphragm

Myenteric nerve plexus

Submucosal nerve plexus

Lumen

Serosa (present only inferior to diaphragm)

Mucosa
  Stratified squamous epithelium
  Lamina propria
  Muscularis mucosae

Submucosa
  Esophageal gland

Muscularis externa
  Inner circular layer
  Outer longitudinal layer

Parasympathetic ganglion of myenteric nerve plexus

Blood vessels

Figure 25.9 Anatomy of the esophagus in the region of the diaphragm. ⨏

Adventitia (above the diaphragm), or

Serosa (below the diaphragm)

　Areolar tissue

　Mesothelium

With some variations to be noted later, these layers are found throughout the rest of the digestive tract. You should familiarize yourself with these terms because they are also used in later discussion of the stomach and intestines.

The **mucosa,** lining the lumen, consists of a stratified squamous epithelium, a loose connective tissue layer called the **lamina propria,** and a thin layer of smooth muscle called the **muscularis mucosae** (MUSS-cue-LERR-is mew-CO-see).

The **submucosa** is a thicker layer of loose connective tissue containing blood vessels, lymphatic vessels, a nerve plexus, and **esophageal glands** that secrete lubricating mucus into the lumen. When the esophagus is empty, the mucosa and submucosa are deeply folded into longitudinal ridges, which gives the lumen a star-like shape in cross section.

The **muscularis externa** is a double layer of muscle near the outer surface. The inner fibers encircle the esophagus while the outer layer consists of longitudinal fibers. In the upper one-third of the esophagus, the muscularis externa consists of skeletal muscle, in the middle one-third it is a mixture of skeletal and smooth muscle, and in the lower one-third it is smooth muscle only.

Most of the esophagus is in the mediastinum, where it is covered with a connective tissue **adventitia.** This merges into the adventitias of the trachea and thoracic aorta. The short segment below the diaphragm is covered by a **serosa,** which consists of a simple squamous mesothelium overlying a thin layer of areolar tissue. Most of the digestive tract from here through the large intestine is covered with such a serosa, but some exceptions will be noted.

The esophagus, stomach, and intestines have a nervous network called the **enteric[7] nervous system,** which regulates the system's motility, secretion, and blood flow. Its major subdivisions are the **submucosal (Meissner[8]) plexus** in the submucosa and the **myenteric (Auerbach[9]) plexus** between the two layers of the muscularis externa. These plexuses mediate **short reflexes** in which a stimulus to the mucosa triggers changes in the motility or secretion of nearby regions of the tract. **Long reflexes** act through autonomic nerve fibers between the digestive tract and central nervous system.

The inferior end of the esophagus is more constricted than the rest, forming the **gastroesophageal sphincter.** This is not an anatomical feature—the muscularis externa is no thicker here than it is higher up—but is a physiological constriction that helps close the cardiac orifice. Reflux (backflow) of stomach contents into the esophagus is prevented in part by the tonus of this sphincter but more importantly by constriction of the diaphragm around the lower esophagus.

Swallowing

Swallowing, or **deglutition** (DEE-glu-TISH-un), is a complex action involving over 22 muscles in the mouth, pharynx, and esophagus. They are coordinated by the **swallowing center,** a nucleus in the medulla oblongata and pons. It communicates with muscles of the pharynx and esophagus by way of the trigeminal, facial, and hypoglossal nerves (cranial nerves V, VII, and XII).

Swallowing occurs in stages called the *buccal* and *pharyngeal-esophageal phases* (fig. 25.10). In the buccal phase, the tongue collects food, presses it against the palate, forms a bolus, and pushes it back into the oropharynx. Here the bolus stimulates tactile receptors and activates the pharyngeal-esophageal phase. In that phase, three actions prevent food or drink from reentering the mouth or entering the nasal cavity or larynx: (1) the root of the tongue blocks the oral cavity, (2) the soft palate rises and blocks the nasopharynx, and (3) the infrahyoid muscles pull the larynx up, the epiglottis covers its opening, and the vestibular folds are adducted to close the airway.

The food bolus is driven downward by constriction of the upper, then the middle, and finally the lower pharyngeal constrictors. As the bolus slides off the epiglottis into the esophagus, it stretches the esophagus and triggers **peristalsis,** a wave of contraction in the muscularis externa (fig. 25.10d).

Peristalsis is moderated partly by a short reflex through the myenteric nerve plexus. The bolus stimulates stretch receptors that feed into the nerve plexus, which transmits signals to the muscularis externa behind and ahead of the bolus. The circular muscle behind the bolus constricts and pushes it downward. Ahead of the bolus, the circular muscle relaxes while the longitudinal muscle contracts. The latter action pulls the wall of the esophagus slightly upward, which makes the esophagus a little shorter and wider and able to receive the descending food.

In an erect position, most food and liquid drop through the esophagus by gravity faster than the peri-

Table 25.1	Anatomical Checklist of the Digestive System from the Mouth through the Esophagus
Oral (Buccal) Cavity	Dental tissues
Oral orifice	Dentin
Fauces	Enamel
Cheeks	Cementum
Lips	Pulp
Cutaneous area	Anatomical features
Red area (vermilion)	Crown
Labial frenulum	Occlusal surface
Vestibule	Cusps
Tongue	Neck
Body	Root
Root	Pulp cavity
Terminal sulcus	Root canal
Lingual frenulum	Apical foramen
Lingual papillae	Periodontal tissues
Taste buds	Alveolus
Lingual muscles	Periodontal membrane
Lingual tonsils	Gingiva (gum)
Lingual glands	Gingival sulcus
Palate	**Salivary Glands**
Hard palate	Intrinsic (in palate and tongue)
Soft palate	Extrinsic
Uvula	Parotid
Glossopalatine arch	Submandibular
Pharyngopalatine arch	Sublingual
Dentition (teeth)	**Pharynx**
Developmental types	Pharyngeal constrictors
Deciduous teeth	**Esophagus**
Permanent teeth	Esophageal glands
Functional types	Gastroesophageal sphincter
Incisors	Cardiac orifice
Canines	
Premolars (bicuspids)	
Molars	

7. *enter* = intestine
8. Georg Meissner (1829–1905), German histologist
9. Leopold Auerbach (1828–97), German anatomist

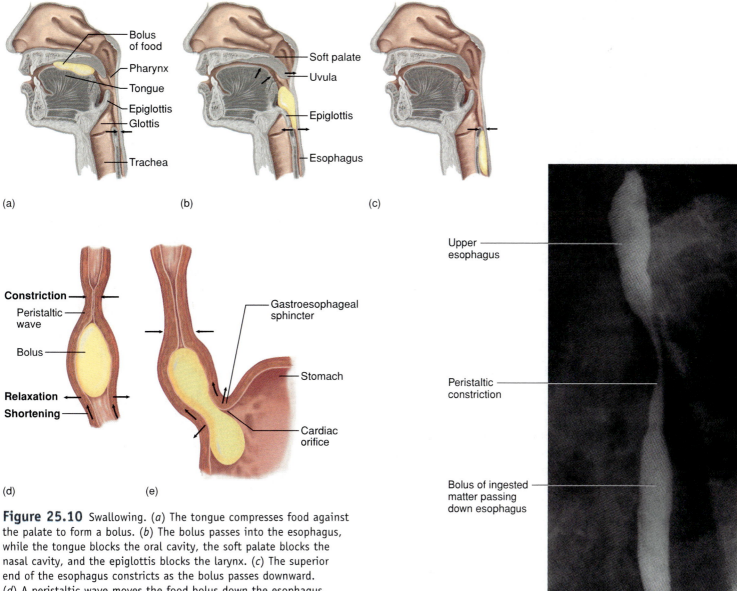

(a)   (b)   (c)

Bolus of food
Pharynx
Tongue
Epiglottis
Glottis
Trachea

Soft palate
Uvula
Epiglottis
Esophagus

Constriction
Peristaltic wave
Bolus
Relaxation
Shortening

Gastroesophageal sphincter
Stomach
Cardiac orifice

(d)   (e)

Upper esophagus
Peristaltic constriction
Bolus of ingested matter passing down esophagus

(f)

Figure 25.10 Swallowing. (*a*) The tongue compresses food against the palate to form a bolus. (*b*) The bolus passes into the esophagus, while the tongue blocks the oral cavity, the soft palate blocks the nasal cavity, and the epiglottis blocks the larynx. (*c*) The superior end of the esophagus constricts as the bolus passes downward. (*d*) A peristaltic wave moves the food bolus down the esophagus. The esophagus constricts behind the bolus, while it dilates and shortens in front of the bolus. (*e*) The gastroesophageal sphincter relaxes to admit the bolus to the stomach. (*f*) X ray of esophagus showing a peristaltic constriction above a bolus of ingested material. ✂ ▭

staltic wave travels. Peristalsis, however, propels more solid food pieces and ensures that you can swallow regardless of the body's position—even standing on your head! Liquid normally reaches the stomach in 1 to 2 seconds and a food bolus in 4 to 8 seconds. As a bolus reaches the lower end of the esophagus, the gastroesophageal sphincter relaxes to let it pass into the stomach.

The anatomical features covered up to this point are summarized in table 25.1.

Key Point Review

5 List as many functions of the tongue as you can.

6 Imagine a line from the mandibular bone to the root canal of a tooth. Name the tissues, in order, through which this line would pass.

7 What is the difference in function and location between intrinsic and extrinsic salivary glands? Name the extrinsic salivary glands and describe their locations.

8 Describe the muscularis externa of the esophagus and its action in peristalsis.

9 Describe the mechanisms for preventing food from entering the nasal cavity and larynx during swallowing.

▼Objectives
When you have completed this section, you should be able to
- describe the gross and microscopic anatomy of the stomach;
- state the function of each type of epithelial cell in the gastric mucosa;
- describe the components of gastric juice and state their functions;
- explain how the stomach produces hydrochloric acid and pepsin;
- describe the contractile responses of the stomach to food; and
- describe the three phases of gastric function and how gastric activity is activated and subsequently inhibited.

The stomach is a muscular sac in the upper left abdominal cavity immediately inferior to the diaphragm. It functions primarily as a food storage organ, with a volume of about 50 mL when empty and 1.0 to 1.5 L after a typical meal. When extremely full, it may hold up to 4 L and extend nearly as far as the pelvis.

Prior to the work of William Beaumont in the nineteenth century (see chapter essay, p. 925), authorities regarded the stomach as essentially a grinding chamber, fermentation vat, or cooking pot. Some even attributed digestion to a supernatural spirit in the stomach. We now know that it mechanically breaks up food particles, liquefies the food, and begins the chemical digestion of proteins and a small amount of fat. This produces a soupy or pasty mixture of semidigested food called **chyme**[10] (kime). Most digestion occurs after the chyme passes on to the small intestine.

Gross Anatomy

The stomach is J-shaped (fig. 25.11), relatively vertical in tall people, and more horizontal in short people. Its medial margin is called the **lesser curvature** and its longer lateral margin is the **greater curvature.**

The stomach is divided into four regions: (1) The **cardiac region (cardia)** is a small area immediately inside the cardiac orifice. (2) The **fundic region (fundus)** is the dome-shaped portion superior to the esophageal attachment. (3) The **body (corpus)** makes up the greatest part of the stomach inferior to the cardiac orifice. (4) The **pyloric region** is a slightly narrower pouch at the inferior end; it is subdivided into a funnel-like **antrum**[11] and a narrower **pyloric canal.** The latter terminates at the **pylorus,**[12] a narrow passage into the duodenum. The pylorus is surrounded by a thick ring of smooth muscle, the **pyloric (gastroduodenal) sphincter,** which regulates the admission of chyme into the duodenum.

---

10. *chyme* = juice
11. *antrum* = cavity
12. *pylorus* = gatekeeper

Innervation and Circulation

The stomach receives parasympathetic nerve fibers from the vagus nerves and sympathetic fibers from the celiac plexus (see p. 538). It is supplied with blood by branches of the celiac trunk (see p. 734). All blood drained from the stomach and intestines enters the hepatic portal circulation and filters through the liver before returning to the heart.

The Stomach Wall

The stomach wall has tissue layers similar to those of the esophagus, with some variations. The mucosa is covered with a simple columnar glandular epithelium (fig. 25.12). The apical regions of the surface cells are filled with mucin. The mucosa and submucosa are flat and smooth when the stomach is full, but as it empties, these layers form conspicuous longitudinal wrinkles called **rugae** (ROO-gee). The lamina propria is almost entirely occupied by tubular glands to be described shortly. The muscularis externa has three layers, rather than two—an outer longitudinal, middle circular, and inner oblique layer (see fig. 25.11).

Think About It

Contrast the epithelium of the esophagus with that of the stomach. Why is each epithelial type best suited to the function of its respective organ?

The gastric mucosa is pocked with depressions called **gastric pits** (fig. 25.12) lined with the same columnar epithelium as the surface. Cells near the bottom of the gastric pits divide repeatedly, producing new epithelial cells that continually migrate upward toward the lumen and replace old epithelial cells that are sloughed off into the chyme.

Two or three tubular glands open into the bottom of each gastric pit and span the rest of the lamina propria (fig. 25.12*a, b*). In the cardiac and pyloric regions they are called **cardiac and pyloric glands,** respectively, and secrete only mucus. In the rest of the stomach, they are called **gastric glands** and have a greater variety of cell types and secretions:

- **Mucous neck cells,** which secrete mucus, are found in the narrow *neck* of each gland where it opens into the gastric pit.
- **Chief cells** are so-named because they are the most numerous. They secrete rennin and lipase in infancy and pepsinogen throughout life.
- **Parietal cells** are scattered among the chief cells and are much less numerous; they secrete hydrochloric acid and intrinsic factor.
- **Enteroendocrine cells,** located deep in the gastric glands, secrete hormones and paracrines that regulate digestion.

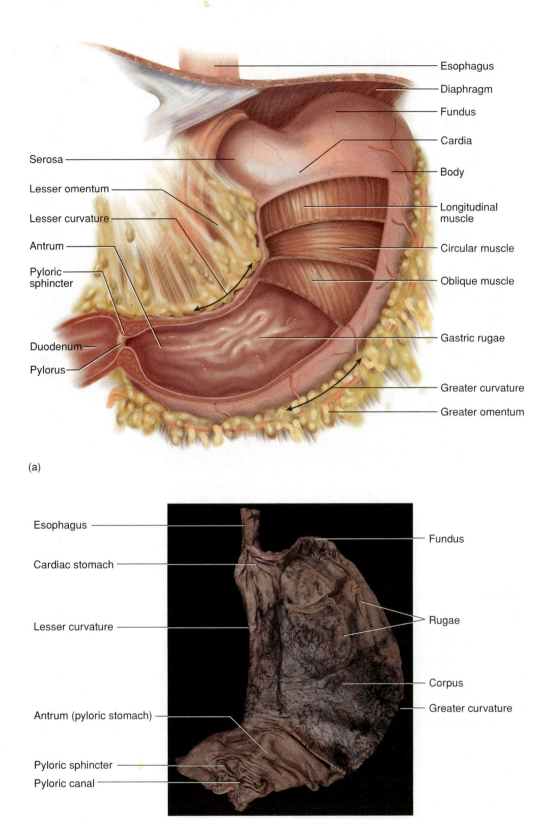

(a)

Esophagus
Diaphragm
Fundus
Cardia
Serosa
Body
Lesser omentum
Longitudinal muscle
Lesser curvature
Circular muscle
Antrum
Oblique muscle
Pyloric sphincter
Duodenum
Gastric rugae
Pylorus
Greater curvature
Greater omentum

(b)

Esophagus
Fundus
Cardiac stomach
Lesser curvature
Rugae
Antrum (pyloric stomach)
Corpus
Pyloric sphincter
Greater curvature
Pyloric canal

Figure 25.11 (*a*) Gross anatomy of the stomach. (*b*) Photograph of the interior of the stomach. ✗

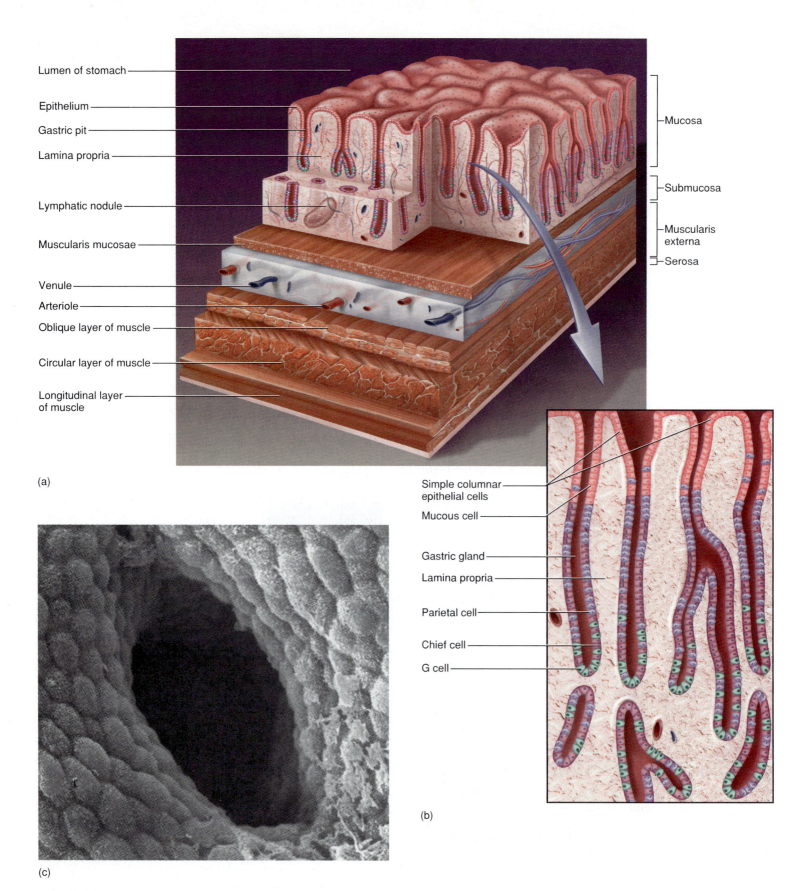

Lumen of stomach

Epithelium

Gastric pit

Lamina propria

Lymphatic nodule

Muscularis mucosae

Venule

Arteriole

Oblique layer of muscle

Circular layer of muscle

Longitudinal layer of muscle

Mucosa

Submucosa

Muscularis externa

Serosa

(a)

Simple columnar epithelial cells

Mucous cell

Gastric gland

Lamina propria

Parietal cell

Chief cell

G cell

(b)

(c)

Figure 25.12 Microscopic anatomy of the stomach wall. (a) A block of stomach tissue showing all layers from the mucosa (top) to the serosa (bottom). (b) Detail of the gastric pits and gastric glands. (c) The opening of a gastric pit into the stomach, surrounded by the rounded apical surfaces of the columnar epithelial cells of the mucosa (SEM). ⚡

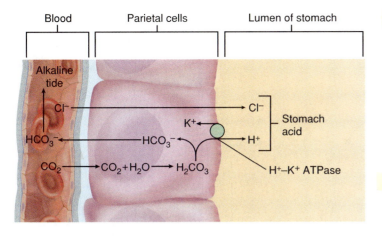

Figure 25.13 Mechanism of hydrochloric acid secretion by gastric parietal cells.

Gastric Secretions

The gastric glands produce 2 to 3 L of **gastric juice** per day, composed mainly of water, hydrochloric acid, and pepsin.

Hydrochloric Acid

Gastric juice has a high concentration of hydrochloric acid (HCl) and a pH as low as 0.8. Such concentrated acid could cause a serious chemical burn if splashed on the skin. How then, does the stomach produce and tolerate such acidity?

The reactions that produce HCl (fig. 25.13) may seem familiar by now because they have been discussed in previous chapters—most recently in connection with renal excretion of H^+ in chapter 24. Parietal cells contain carbonic anhydrase (CAH), which catalyzes the first step in the following reaction:

$$CO_2 + H_2O \xrightarrow{CAH} H_2CO_3 \rightarrow HCO_3^- + H^+$$

The H^+ produced by this reaction is pumped into the lumen of a gastric gland by an active transport protein similar to the Na^+-K^+ pump, called **H^+-K^+ ATPase.** This is an antiport that uses the energy of ATP to pump H^+ out and K^+ into the cell. HCl secretion does not affect the pH within the parietal cell because H^+ is pumped out as fast as it is generated. The bicarbonate ions (HCO_3^-) are exchanged for chloride ions (Cl^-) from the blood plasma—the same process that occurs in renal tubules and red blood cells—and the Cl^- is pumped into the lumen of the gastric gland.

Thus HCl accumulates in the stomach while bicarbonate ions accumulate in the blood. This significantly raises the pH of the blood leaving the stomach; thus the hepatic-portal blood is said to exhibit an **alkaline tide** when digestion is underway and the stomach is secreting acid.

Stomach acid has several functions: (1) It activates the enzymes pepsin and lingual lipase, as discussed shortly. (2) It breaks up connective tissues and plant cell walls, helping to liquefy food and form chyme. (3) It converts ingested ferric (Fe^{3+}) ions to ferrous ions (Fe^{2+}), a form of iron that can be absorbed and used for hemoglobin synthesis. (4) It contributes to nonspecific disease resistance by destroying ingested bacteria and other pathogens.

Intrinsic Factor

Parietal cells also secrete a glycoprotein called **intrinsic factor** that is essential to the absorption of vitamin B_{12} by the small intestine. Without vitamin B_{12}, hemoglobin cannot be synthesized and pernicious anemia develops (see chapter 18). The secretion of intrinsic factor is the only indispensable function of the stomach. Digestion can continue following removal of the stomach (*gastrectomy*), but a person must then take vitamin B_{12} and intrinsic factor either orally or by injection. As we age, the gastric mucosa atrophies, less intrinsic factor is secreted, and the risk of pernicious anemia rises.

Pepsin

Several enzymes are secreted as inactive proteins called **zymogens.** After they are secreted, some of their amino acids are removed and they become active enzymes. In the stomach, chief cells secrete a zymogen called **pepsinogen.** Hydrochloric acid removes some of its amino acids and converts it to **pepsin.** Since pepsin digests protein, and pepsinogen itself is a protein, pepsin has an *autocatalytic* effect—as some pepsin is formed, it catalyzes the conversion of pepsinogen to more pepsin (fig. 25.14). The ultimate function of pepsin, however, is to digest dietary proteins to shorter peptide chains, which then pass to the small intestine, where their digestion is completed.

Other Enzymes

In infants, the chief cells also secrete **gastric lipase** and **rennin** (not to be confused with renin, the hormone secreted by the kidneys). Gastric lipase digests some of the *butterfat* of milk, and rennin coagulates *casein* (milk protein) to form a soft curd.

Chemical Messengers

Gastric glands have various kinds of enteroendocrine cells that collectively produce as many as 20 secretions.

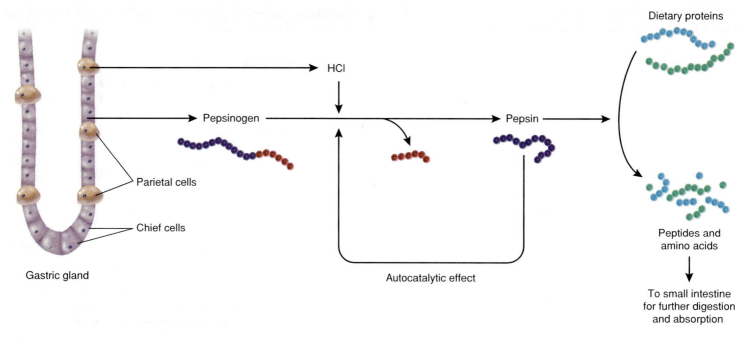

Figure 25.14 The production and digestive action of pepsin. The chief cells secrete pepsinogen and the parietal cells secrete HCl. HCl removes some of the amino acids from pepsinogen, coverting it to pepsin. Pepsin catalyzes the production of more pepsin (autocatalytic effect), as well as partially digesting dietary proteins.

Most of these secretions behave as hormones—they travel in the bloodstream and stimulate distant target cells. Some also behave as paracrine secretions, diffusing a short distance away and stimulating other cells in the gastric mucosa. *Gastrin,* for example, shows both effects. It travels in the blood and stimulates motility of the large intestine, and it diffuses to the nearby parietal and chief cells and stimulates secretion of hydrochloric acid and enzymes. Some of these secretions are also produced by neurons of the enteric nervous system and behave as neurotransmitters.

Several of the gastric secretions are summarized in table 25.2. Some of the functions listed there are explained later in the chapter.

Gastric Motility

As you begin to swallow, signals from the swallowing center of the medulla oblongata stimulate the stomach to relax, preparing it to receive the food. When food enters the stomach, it stretches it and activates the *stress-relaxation response* of smooth muscle, described on page 421. This further relaxes the stomach and enables it to contain more food.

Soon, however, the stomach begins to show a rhythm of peristaltic contractions. These are governed by pacemaker cells in the longitudinal layer of the muscularis externa of the greater curvature. About every 20 seconds, a gentle ripple begins in the fundus and becomes stronger as it progresses toward the pyloric region, where the muscularis externa is thicker. After 30 minutes or so, these contractions become quite strong. They churn the food, mix it with gastric juice, and contribute to its physical breakup.

The antrum holds about 30 mL of chyme. As a peristaltic wave passes down the antrum, it squirts about 3 mL of chyme into the duodenum. When the wave reaches the pyloric sphincter, the pylorus is squeezed shut and chyme that did not get through is turned back into the antrum and body of the stomach for further digestion. Allowing only small amounts into the duodenum at a time enables the duodenum to neutralize the stomach acid and digest nutrients little by little. If the duodenum is overfilled, it inhibits gastric motility and postpones receiving more chyme; the mechanism for this is discussed shortly. A typical meal is emptied from the stomach in about 4 hours, but it takes less time if the meal is more liquid and as long as 6 hours if the meal is high in fat.

Vomiting

Vomiting is induced by excessive stretching of the stomach, psychological stimuli, and chemical irritants such

Table 25.2	Major Secretions of the Gastric Glands	
Cells	Secretions	Functions
Mucous neck cells	Mucus	Protects gastric mucosa from action of HCl and enzymes
Parietal cells	Hydrochloric acid	Activates pepsin and lingual lipase; helps break up plant cells and connective tissues and liquefy food; converts dietary iron to usable form (F^{2+}); destroys ingested pathogens
	Intrinsic factor	Required for absorption of vitamin B_{12}
Chief cells	Pepsinogen	Converted by HCl to pepsin, which digests protein
	Rennin	Milk-curdling secretion of the infant stomach; not secreted in adults
	Gastric lipase	Fat-digesting enzyme of the infant stomach; not secreted in adults
Enteroendocrine cells	Gastrin	Stimulates gastric glands to secrete HCl and enzymes; stimulates intestinal motility; relaxes ileocecal valve between small and large intestines
	Serotonin	Stimulates gastric motility
	Histamine	Stimulates HCl secretion
	Somatostatin	Inhibits gastric secretion and motility; delays emptying of stomach; inhibits secretion by pancreas; inhibits gallbladder contraction and secretion of bile; reduces blood circulation and nutrient absorption in small intestine

as alcohol and bacterial toxins. These factors stimulate the **emetic**[13] **center** in the medulla oblongata, which in turn stimulates the gastroesophageal sphincter to relax and the diaphragm and abdominal muscles to contract. These muscles squeeze the stomach like a bagpipe and force its contents up the esophagus. Vomiting is not caused by contraction of the gastric muscle itself. Severe vomiting may expel even the contents of the small intestine.

Digestion and Absorption

Protein, starch, and fat are partially digested in the stomach by salivary and gastric enzymes and then passed to the small intestine, where most digestion and nearly all nutrient absorption occur. The stomach does not absorb any significant amount of nutrients but does absorb aspirin and some lipid-soluble drugs. Alcohol is absorbed mainly by the small intestine, so its intoxicating effect depends partly on how rapidly the stomach is emptied.

Protection of the Stomach

You may wonder why the stomach does not digest itself. Animal stomachs (tripe) can be digested as readily as any other meat. The living stomach, however, is protected in three ways from the harsh acidic and enzymatic environment it creates:

1. **Mucous coat.** The thick, highly alkaline mucus resists the action of acid and enzymes.
2. **Epithelial cell replacement.** The stomach's epithelial cells live only 3 to 6 days and are then sloughed off into the chyme and digested with the food. They are replaced just as rapidly, however, by cell division in the gastric pits.
3. **Tight junctions.** The epithelial cells have tight junctions between them that prevent gastric juice from seeping through to the connective tissue of the lamina propria or beyond.

The breakdown of these protective mechanisms can result in inflammation and peptic ulcer (see special topic 25.2).

Regulation of Gastric Function

Because of the coordinated actions of the endocrine and nervous systems, gastric secretion and motility rise when food is eaten and decline as the stomach empties. Gastric activity is divided into three stages called the cephalic, gastric, and intestinal phases (fig. 25.15), but these phases overlap and all three can occur simultaneously.

The Cephalic Phase

The **cephalic phase** (fig. 25.15a) is stimulated by the sight, smell, taste, or mere thought of food. These

13. *emet* = vomiting

Inflammation of the stomach, called **gastritis,** can lead to a **peptic ulcer** (fig. 1) as pepsin and hydrochloric acid erode the stomach wall. Less commonly, peptic ulcers occur in the duodenum or esophagus. If untreated, they can perforate the organ and cause potentially fatal hemorrhaging or peritonitis. Most such fatalities are in people over age 65.

Although peptic ulcers are commonly assumed to result from psychological stress, there is no evidence to support this. Hypersecretion of acid and pepsin is sometimes involved, but normal secretion can cause ulceration if mucosal defense is compromised by other causes. Many or most cases involve an acid-resistant bacterium, *Helicobacter pylori,* that invades the mucosa of the stomach and duodenum and opens the way to chemical damage to the tissue. Other risk factors include smoking and the use of aspirin and other nonsteroidal anti-inflammatory drugs (NSAIDs). NSAIDs suppress the synthesis of prostaglandins, which normally stimulate the secretion of protective mucus and acid-neutralizing bicarbonate. Aspirin itself is an acid that directly irritates the gastric mucosa.

Until recently, the most widely prescribed drug in the United States was Cimetidine (Tagamet), which was designed to treat peptic ulcers by reducing acid secretion. Histamine stimu-

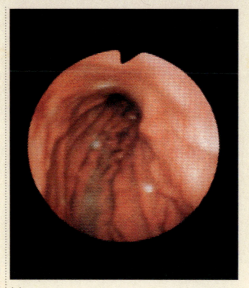
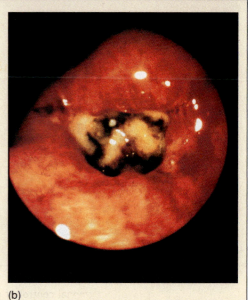

(a)   (b)

Figure 1 Endoscopic views of the gastroesophageal junction. The esophagus can be seen opening into the cardiac stomach. (*a*) A healthy gastric mucosa; the small white spots are reflections of light from the endoscope. (*b*) A bleeding peptic ulcer. A peptic ulcer typically has an oval shape and yellow-white color. Here the yellowish floor of the ulcer is partially obscured by black blood clots, and fresh blood is visible around the margin of the ulcer.

lates acid secretion by binding to sites on the parietal cells called **H₂ receptors;** Cimetidine, an *H₂ blocker,* prevents this binding. Lately, however, ulcers have been treated more successfully with antibiotics against *Helicobacter* combined with bismuth suspensions such as Pepto-Bismol. This is a much shorter and less expensive course of treatment and permanently cures about 90% of peptic ulcers, as compared with a cure rate of only 20% to 30% for H₂ blockers.

sensory and mental inputs converge on the hypothalamus, which transmits signals by way of the medulla oblongata and vagus nerves to the stomach. Gastric secretion thus begins even before food is swallowed.

The Gastric Phase

The **gastric phase** (fig. 25.15*b*) is stimulated by food in the stomach itself and accounts for about two-thirds of all gastric secretion. Two reflexes stimulate secretion: a short **myenteric reflex** mediated through the myenteric nerve plexus and a long **vagovagal reflex** mediated through the medulla oblongata and sensory and motor fibers of the vagus nerves.

Gastric secretion is also under hormonal and paracrine control. Ingested food raises the pH in the stomach, which stimulates enteroendocrine **G cells** to secrete the hormone **gastrin.** Gastrin reaches the nearby chief and parietal cells by diffusion through the tissue fluid and by way of the bloodstream. It stimulates pepsinogen and especially HCl secretion. As dietary protein is digested, it produces smaller peptides, which directly stimulate the G cells to secrete more gastrin—a *positive* feedback mechanism that accelerates protein digestion (fig. 25.16). Peptides also buffer stomach acid so the pH does not fall excessively low.

As digestion proceeds and more and more chyme is expelled into the small intestine, however, the stomach contains less of these buffering peptides and the pH drops lower and lower. When it falls below 2, gastrin secretion is inhibited—a *negative* feedback mechanism

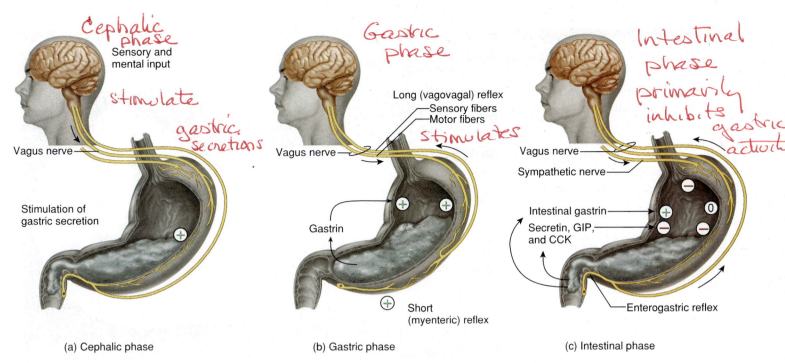

Cephalic phase

Sensory and mental input

stimulate

gastric secretions

Vagus nerve

Stimulation of gastric secretion

(a) Cephalic phase

Gastric phase

Long (vagovagal) reflex
—Sensory fibers
—Motor fibers

stimulates

Vagus nerve

Gastrin

⊕ Short (myenteric) reflex

(b) Gastric phase

Intestinal phase primarily inhibits gastric activity

Vagus nerve

Sympathetic nerve

Intestinal gastrin
Secretin, GIP, and CCK

Enterogastric reflex

(c) Intestinal phase

Figure 25.15 Neural and hormonal control of gastric secretion. Plus signs indicate stimulation and minus signs indicate inhibition.

that winds down gastric activity as the need for pepsin and HCl declines.

The Intestinal Phase

The **intestinal phase** (fig. 25.15*c*) is stimulated by chyme entering the duodenum. For a short time, enteroendocrine cells here secrete *intestinal gastrin,* a synergist of the stomach's own gastrin that further enhances gastric secretion and motility.

Soon, however, signals from the small intestine to the stomach become inhibitory. Hydrochloric acid, fats, and peptides in the duodenum trigger the **enterogastric reflex.** Inhibitory signals pass from the duodenum to the stomach by way of the submucosal and myenteric plexuses. In addition, collateral nerve fibers from these plexuses inhibit the vagal nuclei of the medulla, reduce vagal stimulation to the stomach, and stimulate sympathetic neurons. Sympathetic signals inhibit gastric motility and secretion. The effects of all this are that gastrin secretion declines and the pyloric sphincter contracts tightly to limit the admission of more chyme into the duodenum. This gives the duodenum time to work on the chyme it has already received before being loaded with more.

Chyme in the duodenum also stimulates its enterendocrine cells to release **secretin, cholecystokinin** (CO-lee-SIS-toe-KY-nin) **(CCK),** and **gastric inhibitory peptide (GIP).** Secretin and CCK primarily stimulate the

pancreas and gallbladder, as discussed later, but all three of these hormones suppress gastric secretion and motility.

The Liver, Gallbladder, and Pancreas

▼**Objectives**

When you have completed this section, you should be able to

• describe the gross and microscopic anatomy of the liver, gallbladder, bile duct system, and pancreas;

• describe the digestive secretions and functions of the liver, gallbladder, and pancreas; and

• describe the effects of CCK and secretin on the secretion of bile and pancreatic juice.

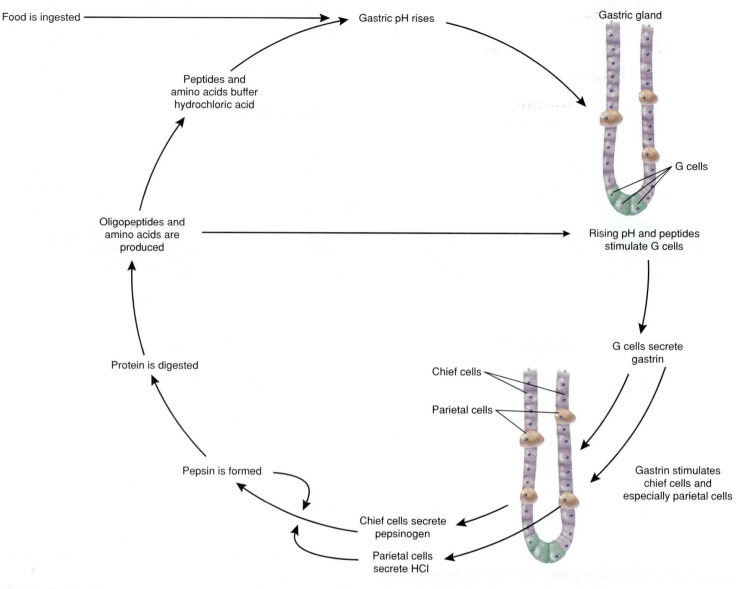

Figure 25.16 Positive feedback control of gastric secretion. This feedback cycle declines and stops as the stomach is emptied and the pH drops.

The small intestine receives not only chyme from the stomach but also secretions from the liver and pancreas. These secretions are so important to the digestive processes of the small intestine that it is necessary to understand them before continuing with intestinal physiology.

The Liver

The liver (fig. 25.17) is a reddish brown gland located immediately inferior to the diaphragm, filling most of the right hypochondriac and epigastric regions. It is the body's largest gland, weighing about 1.4 kg (3 lb). The liver has a tremendous variety of functions discussed in the next chapter. Only one of these, the secretion of bile, contributes to digestion.

Gross Anatomy

The liver has four lobes called the right, left, quadrate, and caudate lobes. The large **right lobe** and smaller **left lobe** are visible anteriorly. They are separated from each other by the **falciform**[14] **ligament,** a sheet of mesentery that suspends the liver from the diaphragm and anterior abdominal wall. The **round ligament** *(ligamentum teres),* also visible anteriorly, is a fibrous remnant of the umbilical vein, which carries blood from the umbilical cord to the liver of a fetus.

In the inferior aspect, we see a squarish **quadrate lobe** next to the gallbladder and a tail-like **caudate**[15]

14. *falci* = sickle + *form* = shape
15. *caud* = tail

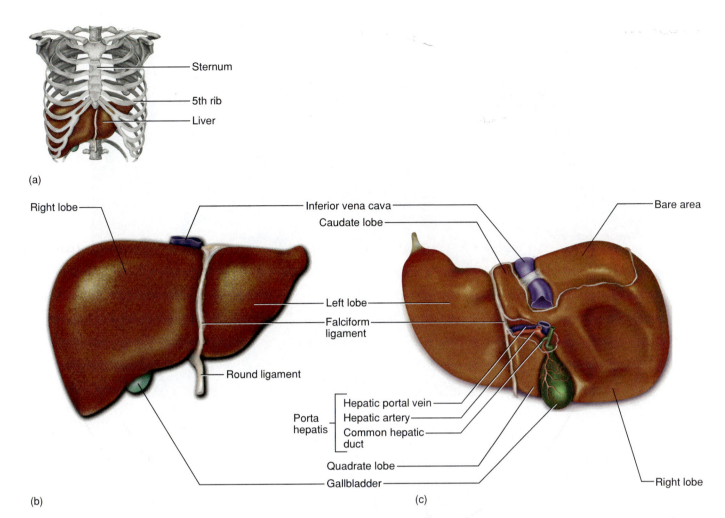

Figure 25.17 Gross anatomy of the liver. (*a*) Relationship of the liver to the thoracic cage. (*b*) Anterior aspect. (*c*) Inferior aspect. ⚕

lobe posterior to that. An irregular opening between these lobes, the **porta hepatis** ("gateway to the liver"), is a point of entry for the hepatic portal vein and hepatic artery and a point of exit for the common bile duct, all of which travel in the lesser omentum. The gallbladder adheres to a depression on the inferior surface of the liver between the right and quadrate lobes.

The posterior aspect of the liver has a deep groove (sulcus) that accommodates the inferior vena cava. Most of the liver is covered by a serosa, but this is lacking from the *bare area* of its superior surface, where the liver is attached to the diaphragm.

Microscopic Anatomy

The liver parenchyma consists mainly of cuboidal cells called **hepatocytes,** arranged in innumerable cylinders called **hepatic lobules** (fig. 25.18). Each lobule is about 1 mm in diameter and 2 mm long and has a **central vein** passing through its core. The hepatocytes form thin epithelial plates that fan out from the central vein. These plates are separated from each other by blood-filled

channels (discontinuous capillaries) called **hepatic sinusoids.** The hepatocytes are in direct contact with the blood; they continually extract some substances from it while adding other substances to it. The sinusoids also have phagocytic cells called **hepatic macrophages (Kupffer[16] cells),** which remove bacteria and debris from the blood.

The hepatic lobules are separated by a sparse connective tissue stroma. In cross sections, the stroma is especially visible in the triangular areas at the intersections of three or more lobules, where it frequently contains a **hepatic triad** of two blood vessels and a bile ductule. The blood vessels are small branches of the hepatic artery and hepatic portal vein. Both of them supply blood to the sinusoids, which therefore receive a mixture of nutrient-laden venous blood from the intestines and freshly oxygenated arterial blood from the celiac trunk. As the blood filters through the sinusoids, it is modified by the hepatocytes and

16. Karl W. von Kupffer (1829–1902), German anatomist

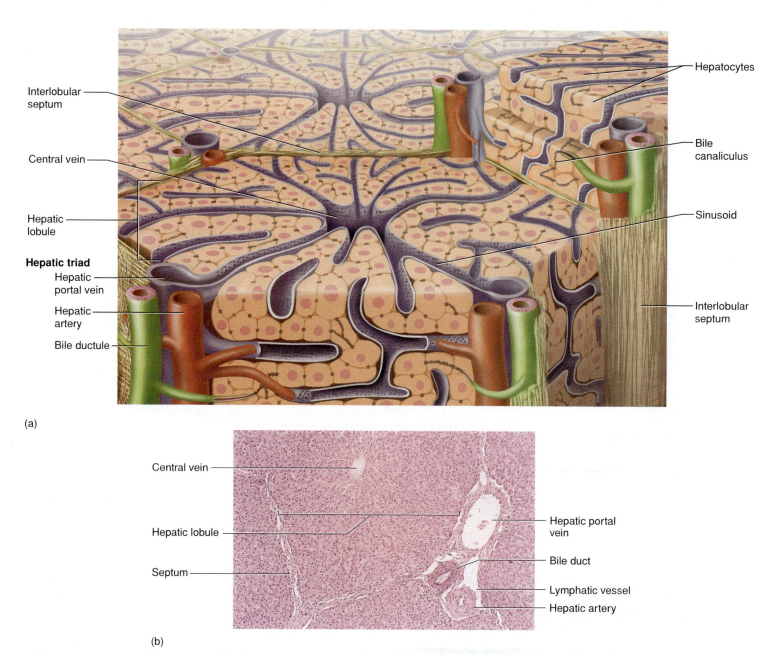

Interlobular septum

Central vein

Hepatic lobule

Hepatic triad
Hepatic portal vein

Hepatic artery

Bile ductule

Hepatocytes

Bile canaliculus

Sinusoid

Interlobular septum

(a)

Central vein

Hepatic lobule

Septum

Hepatic portal vein

Bile duct

Lymphatic vessel

Hepatic artery

(b)

Figure 25.18 Microscopic anatomy of the liver. (*a*) The hepatic lobules and their relationship to the blood vessels and bile tributaries. (*b*) Histological section of the liver (×400). ✗

cleaned up by the phagocytic action of the macrophages. This blood collects in the central vein of the lobule, from which it ultimately flows into the right and left hepatic veins. These leave the liver at its superior surface and drain immediately into the inferior vena cava.

The liver secretes bile into narrow channels, the **bile canaliculi,** between sheets of hepatocytes. Bile passes from there into the small **bile ductules** of the triads and ultimately into the **right** and **left hepatic ducts.** The two hepatic ducts converge to form the **common hepatic duct,** which then joins the **cystic duct** coming from the gallbladder (fig. 25.19). The **common**

bile duct formed by their union descends through the lesser omentum and joins the duct of the pancreas, forming an expanded chamber called the **hepatopancreatic ampulla.** The ampulla terminates at a fold of tissue, the **major duodenal papilla,** in the duodenal wall. This papilla contains a muscular **hepatopancreatic sphincter (sphincter of Oddi[17]),** which regulates the passage of bile and pancreatic secretion into the duodenum.

17. Ruggero Oddi (1864–1913), Italian physician

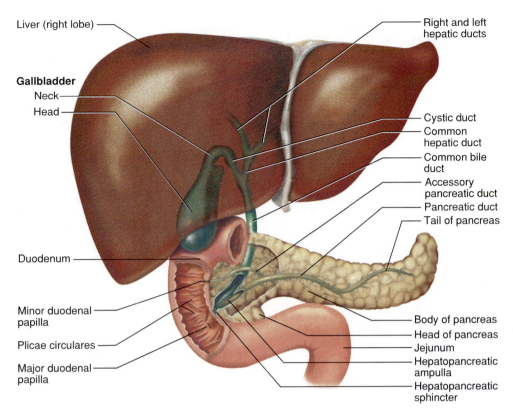

Figure 25.19 Gross anatomy of the gallbladder and pancreas and the route of bile and pancreatic secretion. ⚵

The Gallbladder and Bile

The gallbladder is a greenish sac about 10 cm long, internally lined by a highly folded mucosa with a simple columnar epithelium. Its head (*fundus*) usually projects slightly beyond the inferior margin of the liver. Its neck (*cervix*) is continuous with the cystic duct, which leads into the common bile duct. When bile is not needed for digestion, the hepatopancreatic sphincter is closed. Bile then fills up the common bile duct and spills over into the gallbladder. The gallbladder absorbs water and ions and stores the concentrated bile for later use.

The liver produces 500 to 1,000 mL of bile per day. It is a yellow-green fluid containing minerals, bile pigments, bile salts, cholesterol, neutral fats, and phospholipids. The principal bile pigment is **bilirubin,** derived from the decomposition of hemoglobin. Bacteria of the small intestine metabolize bilirubin to **urobilinogen,** which is responsible for the brown color of feces. In the absence of bile secretion, the feces are grayish white and marked with streaks of undigested fat.

Bile salts are steroids synthesized from cholesterol. Bile salts and phospholipids aid in fat digestion and absorption, as discussed later; their secretion is the only role the liver plays in digestion. All other components of the bile are waste, destined for excretion by way of the intestines. When these waste products become excessively concentrated, they may form gallstones (see special topic 25.3).

Bile salts are not excreted in the feces but are reabsorbed in the ileum, returned to the liver by way of the

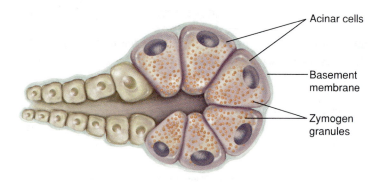

Acinar cells

Basement membrane

Zymogen granules

(a)

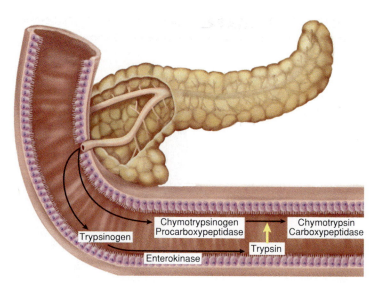

Figure 25.21 The activation of pancreatic enzymes in the small intestine. 𝓧

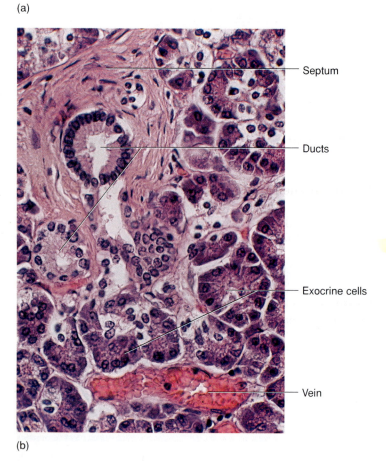

Septum

Ducts

Exocrine cells

Vein

(b)

Figure 25.20 (a) Microscopic anatomy of a pancreatic acinus. (b) Histology of the exocrine tissue of the pancreas (x400).

hepatic portal blood, and resecreted. This pathway from the liver to the ileum and back is called the **enterohepatic circulation.** Bile secretion increases in response to fats in the duodenum and to increased enterohepatic return of the bile salts.

The Pancreas

The pancreas (fig. 25.19) is a soft, spongy, pink gland posterior to the greater curvature of the stomach and outside the peritoneal cavity. It has a globose *head* encircled by the duodenum, a midportion called the *body,*

and a blunt, tapered *tail* on the left. The pancreas is both an endocrine and exocrine gland. Its endocrine part is composed of the pancreatic islets, which secrete insulin and glucagon. Most of the pancreas is exocrine tissue, which secretes 1,200 to 1,500 mL of **pancreatic juice** per day. The cells of the secretory acini exhibit a high density of rough ER and zymogen granules, which are vesicles filled with secretion (fig. 25.20). These acini open into a system of larger and larger ducts that eventually converge on the main **pancreatic duct.** This duct runs lengthwise through the middle of the gland and joins the common bile duct at the hepatopancreatic ampulla. The hepatopancreatic sphincter thus controls the release of both bile and pancreatic juice into the duodenum. In many people, however, pancreatic juice can bypass this sphincter by way of a smaller **accessory pancreatic duct,** which branches from the pancreatic duct and opens independently into the duodenum at the **minor duodenal papilla,** proximal to the major papilla.

Pancreatic juice is an alkaline mixture (pH 8) of water, electrolytes (especially sodium bicarbonate), enzymes, and zymogens. The acinar cells secrete the enzymes and zymogens, whereas the gland ducts secrete the sodium bicarbonate. Bicarbonate buffers hydrochloric acid from the stomach.

The pancreatic zymogens are **trypsinogen** (trip-SIN-oh-jen), **chymotrypsinogen** (KY-mo-trip-SIN-o-jen), and **procarboxypeptidase** (PRO-car-BOC-see-PEP-tih-dase). After trypsinogen is secreted into the intestinal lumen, it is converted to trypsin by **enterokinase,** an enzyme on the surface of the intestinal epithelial cells (fig. 25.21). Trypsin then converts the other two zymogens into chymotrypsin and carboxypeptidase, in addition to its primary role of digesting dietary protein.

Table 25.3	Exocrine Secretions of the Pancreas
Secretion	**Function**
Sodium bicarbonate	Neutralizes HCl
Zymogens	Converted to active digestive enzymes after secretion
Trypsinogen	Becomes trypsin, which digests protein
Chymotrypsinogen	Becomes chymotrypsin, which digests protein
Procarboxypeptidase	Becomes carboxypeptidase, which hydrolyzes the terminal amino acid from the carboxyl (—COOH) end of small peptides
Enzymes	
Pancreatic amylase	Digests starch
Pancreatic lipase	Digests fat
Ribonuclease	Digests RNA
Deoxyribonuclease	Digests DNA

Other pancreatic enzymes include **pancreatic amylase,** which digests starch; **pancreatic lipase,** which digests fat; and **ribonuclease** and **deoxyribonuclease,** which digest RNA and DNA, respectively. Unlike the zymogens, these enzymes are not altered after secretion. They become active, however, only upon exposure to bile and ions in the intestinal lumen.

The exocrine secretions of the pancreas are summarized in table 25.3. Their specific digestive functions are explained later in more detail.

Regulation of Secretion

Bile and pancreatic juice are secreted in response to similar stimuli. During the cephalic and gastric phases of gastric secretion, the vagus nerves also stimulate pancreatic secretion. As chyme enters the duodenum laden with acid and fat, it stimulates the duodenal mucosa to secrete cholecystokinin (CCK). CCK stimulates three processes: (1) contraction of the gallbladder, forcing bile into the common bile duct; (2) secretion of pancreatic enzymes; and (3) relaxation of the hepatopancreatic sphincter, allowing bile and pancreatic juice to be released into the duodenum.

Acidic chyme also stimulates the duodenal mucosa to release **secretin,** the first hormone ever discovered. Secretin stimulates the production of bicarbonate ions by the hepatic bile ducts and pancreatic ducts, so both the bile and pancreatic juice contain more bicarbonate and neutralize stomach acid more effectively.

Think About It
Make a diagram showing how secretin exerts negative feedback control over duodenal pH.

Key Point Review

14 What contribution does the liver make to digestion?

15 Trace the pathway taken by a molecule of bile salt from the liver and back. What is this pathway called?

16 Name two hormones, four enzymes, and one buffer secreted by the pancreas, and state the function of each.

17 What causes the duodenum to secrete cholecystokinin (CCK), and what effects does this have on other parts of the digestive system?

The Small Intestine

▼Objectives
When you have completed this section, you should be able to
• describe the gross and microscopic anatomy of the small intestine;
• state how the mucosa of the small intestine differs from that of the stomach, and explain the functional significance of the differences;
• define *contact digestion* and state where it occurs; and
• describe the types of motility seen in the small intestine.

Nearly all chemical digestion and nutrient absorption occur in the small intestine. To perform these roles well, the small intestine must have a large surface area exposed to the chyme. This requirement is fulfilled in two ways. One is its length; the small intestine is a coiled mass filling most of the abdominal cavity inferior to the stomach and liver. It is about 2 m long in a living person and 6 to 7 m long in a cadaver. The name *small* intestine refers not to its length but to its diameter— about 2.5 cm (1 in.). The other factor in its large surface area is the folding of the mucosa into *circular folds, villi,* and *microvilli,* described shortly.

Think About It
Why do you think the small intestine is so much longer in a cadaver than in a living person?

Gross Anatomy

The small intestine is divided into three regions (fig. 25.22); the measurements given here are based on the cadaver. The **duodenum** (DEW-oh-DEE-num; dew-ODD-eh-num) constitutes the first 25 cm (10 in.). It begins at the pyloric valve, arcs around the head of the pancreas and passes to the left, and ends at a sharp bend called the **duodenojejunal flexure.** Its name refers to its length, about equal to the width of 12 fingers.[18] The duodenum receives the stomach contents, pancreatic juice, and bile. Stomach acid is neutralized here, pepsin is inactivated by the elevated pH, and pancreatic enzymes take over the job of chemical digestion.

18. *duoden* = 12

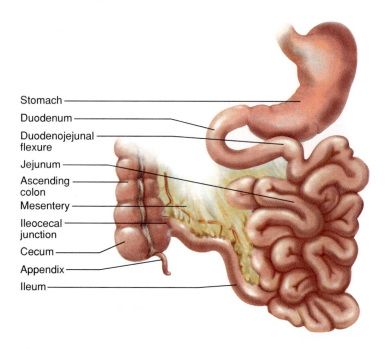

Stomach
Duodenum
Duodenojejunal flexure
Jejunum
Ascending colon
Mesentery
Ileocecal junction
Cecum
Appendix
Ileum

Figure 25.22 Gross anatomy of the small intestine.

The **jejunum** (jeh-JOO-num) comprises the next 2.5 m (8 ft). Its name refers to the fact that early anatomists typically found it to be empty.[19] The **ileum**[20] forms the last 3.6 m (12 ft). The jejunum and ileum are not separated by any conspicuous anatomical landmark, but the jejunum is located largely toward the upper left of the intestinal coils and the ileum to the lower right. The ileum ends at the **ileocecal** (ILL-ee-oh-SEE-cul) **junction,** where it joins the cecum, the first part of the large intestine.

Most of the duodenum is retroperitoneal and is surrounded by a fibrous adventitia. The jejunum and ileum, however, are within the peritoneal cavity and covered externally with serosa, which is continuous with the complex, folded mesentery that suspends the small intestine from the dorsal abdominal wall.

Microscopic Anatomy

The largest folds of the intestinal wall are transverse to spiral ridges, up to 10 mm high, called **circular folds** (*plicae circulares*) (see fig. 25.19). These involve the mucosa and submucosa but are not visible on the external surface, which is smooth. They occur from the duodenum to the middle of the ileum, where they cause the chyme to flow on a spiral path along the intestine. This

slows its progress, causes more contact with the mucosa, and promotes more thorough mixing and nutrient absorption. Circular folds are not found in the distal half of the ileum, but most nutrient absorption is completed by that point.

If the mucosa is examined closely it appears fuzzy, like a terrycloth towel. This is due to projections called **villi** (VIL-eye; singular, *villus*), about 0.5 to 1.0 mm high, with tongue- to fingerlike shapes (fig. 25.23). The villi are largest in the duodenum and become progressively smaller in more distal regions of the small intestine. A villus is covered with two kinds of epithelial cells—columnar **absorptive cells** and mucus-secreting **goblet cells.** Like epithelial cells of the stomach, those of the small intestine are joined by tight junctions that prevent digestive enzymes from seeping between them.

The core of a villus is filled with areolar tissue of the lamina propria. Embedded in this tissue are an arteriole, a capillary network, a venule, and a lymphatic capillary called a **lacteal** (LAC-tee-ul). Most nutrients are absorbed by the blood capillaries, but fats are absorbed by the lacteal and give its contents a milky appearance for which the lacteal is named.[21] The core of the villus also has a few fibers of smooth muscle that contract periodically. This enhances mixing of the chyme in the intestinal lumen and milks lymph down the lacteal to the larger lymphatic vessels of the submucosa.

Each epithelial cell of a villus has a fuzzy brush border of microvilli about 1 μm high. As in the kidney tubules and elsewhere, these greatly increase surface area and the efficiency of absorption. In the small intestine, they have an additional role: some of the proteins in their plasma membranes are **brush border enzymes.** One of these, enterokinase, activates pancreatic enzymes as explained earlier. Others carry out some of the final stages of enzymatic digestion. They are not released into the lumen; instead, the chyme must contact the brush border for digestion to occur. This process, called **contact digestion,** is one reason that thorough mixing of the chyme is so important.

On the floor of the small intestine, between the bases of the villi, there are numerous pores that open into tubular glands called **intestinal crypts (crypts of Lieberkühn;**[22] LEE-ber-koohn). These crypts, similar to the gastric glands, extend as far as the muscularis mucosae. In the upper half they consist of absorptive and goblet cells like those of the villi. The lower half is dominated by dividing epithelial cells. In its life span of 3 to 6 days, an epithelial cell migrates up the crypt to the tip of the villus, where it is sloughed off and digested. A few **Paneth**[23] **cells** are clustered at the base of

---

19. *jejun* = empty, dry
20. from *eilos* = twisted

21. *lact* = milk
22. Johann N. Lieberkühn (1711–56), German anatomist
23. Josef Paneth (1857–90), Austrian physician

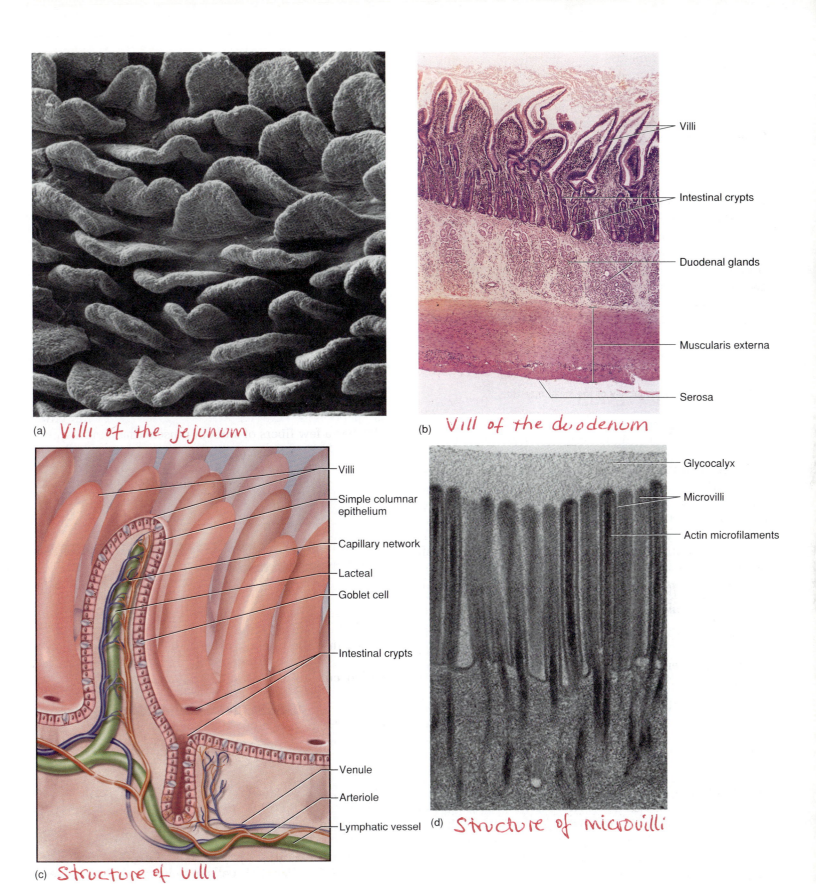

(a) Villi of the jejunum

(b) Villi of the duodenum

- Villi
- Intestinal crypts
- Duodenal glands
- Muscularis externa
- Serosa

(c) Structure of villi

- Villi
- Simple columnar epithelium
- Capillary network
- Lacteal
- Goblet cell
- Intestinal crypts
- Venule
- Arteriole
- Lymphatic vessel

(d) Structure of microvilli

- Glycocalyx
- Microvilli
- Actin microfilaments

Figure 25.23 Intestinal villi. (*a*) Tongue-shaped villi of the jejunum (SEM). The specimen has been flexed to spread the villi apart and reveal the intestinal floor between them, where openings into the intestinal crypts can be seen as tiny pores. (*b*) Histology of the duodenum, showing villi, intestinal crypts, and duodenal glands (LM, ×400). (*c*) Structure of a villus. (*d*) Microvilli on the apical surface of an absorptive cell. Actin microfilaments can be seen forming a dark supportive core in each microvillus and extending into the apical cytoplasm of the cell (TEM).

(a) From R. G. Kessel and R. H. Kardon, *Scanning Electron Microscopy of Tissues and Organs,* 1979.

each crypt. Their function is uncertain, but they apparently secrete lysozyme as a defense against bacteria.

The duodenum has prominent **duodenal (Brunner[24]) glands** in the submucosa. They secrete an abundance of bicarbonate-rich mucus, which neutralizes stomach acid while shielding the mucosa from its corrosive effects. Throughout the small intestine, the lamina propria and submucosa have a large population of lymphocytes that intercept pathogens before they can invade the bloodstream. In some places these are aggregated into conspicuous lymphatic follicles. These become more and more numerous closer to the large intestine, where the bacterial population is greatest. In the distal portion of the ileum, the follicles form aggregates called **Peyer[25] patches** on one side of the intestinal wall.

The muscularis externa consists of a relatively thick inner circular layer and a thinner outer longitudinal layer. Ganglia of the myenteric nerve plexus occur between these layers.

Intestinal Secretion

The intestinal crypts secrete 1 to 2 L of **intestinal juice** per day, especially in response to acid, hypertonic chyme, and distension of the intestine. This fluid has a pH of 7.4 to 7.8 and contains water and mucus but relatively little enzyme. Most enzymes that function in the small intestine are found in the brush border and pancreatic juice.

Intestinal Motility

Contractions of the small intestine serve three functions: (1) to mix chyme with intestinal juice, bile, and pancreatic juice, allowing these fluids to neutralize acid and digest nutrients more effectively; (2) to churn chyme and bring it into contact with the mucosa for contact digestion and nutrient absorption; and (3) to move residue toward the large intestine.

Segmentation is the most common type of movement of the small intestine. Ringlike constrictions appear at several places along the intestine and then relax as new constrictions form elsewhere (fig. 25.24a). The effect is to knead or churn the contents, which promotes the mixing of food and digestive secretions and enhances contact digestion. The rhythm of segmentation is determined by pacemaker cells of the muscularis externa. Contractions occur about 12 times per minute in the duodenum and 8 to 9 times per minute

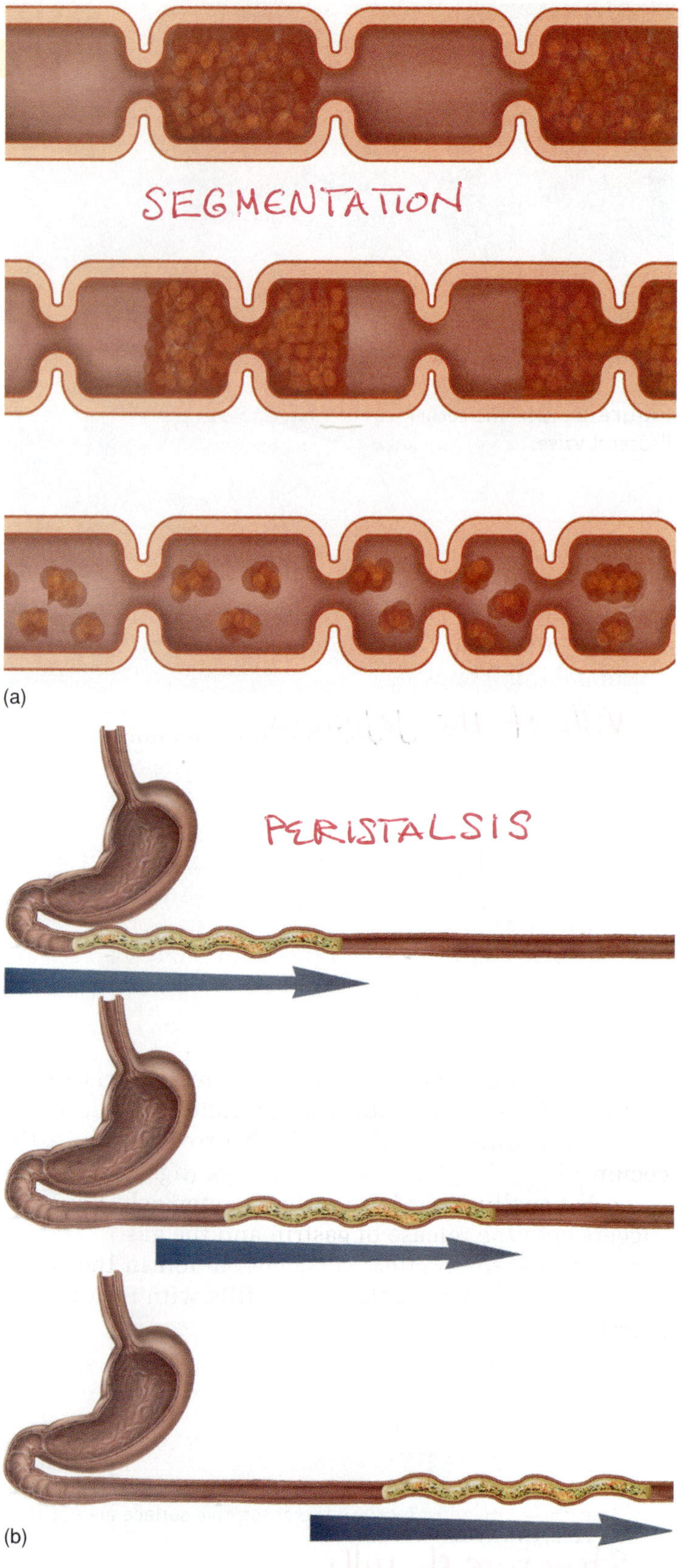

(a)

(b)

Figure 25.24 Contractions of the small intestine. (*a*) Segmentation, in which circular constrictions of the intestine cut into the contents, churning and mixing them. (*b*) The migrating motor complex of peristalsis, in which successive waves of peristalsis overlap each other. Each wave travels partway down the intestine, milking the contents toward the colon.

---

24. Johann C. Brunner (1653–1727), Swiss anatomist
25. Johann K. Peyer (1653–1712), Swiss anatomist

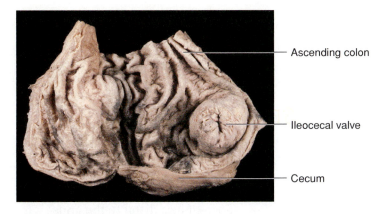

Figure 25.25 The cecum, cut and folded open to show the ileocecal valve.

in the ileum. Since the contractions are less frequent distally, segmentation causes slow progression of the chyme toward the colon. The intensity (but not frequency) of contractions is modified by nervous and hormonal influences.

When most nutrients have been absorbed and little remains but undigested residue, segmentation declines and peristalsis begins. A peristaltic wave begins in the duodenum, travels 10 to 70 cm, and dies out. Successive, overlapping waves of contraction, called a **migrating motor complex,** milk the chyme toward the colon over a period of about 2 hours (fig. 25.24*b*). A second complex then expels residue and bacteria from the small intestine, helping to limit bacterial colonization. Refilling of the stomach at the next meal suppresses peristalsis and reactivates segmentation.

At the ileocecal junction, the muscularis of the ileum is thickened to form a sphincter, the **ileocecal** (ILL-ee-oh-SEE-cul) **valve,** which protrudes into the cecum like a doughnut or pair of lips (fig. 25.25). This valve is usually closed. Food in the stomach, however, triggers both the release of gastrin and the **gastroileal reflex,** both of which enhance segmentation in the ileum and relax the valve. As the cecum fills with residue, the pressure pinches the valve shut, preventing the reflux of cecal contents into the ileum.

<hr>

Key Point Review

18 What three structures increase the absorptive surface area of the small intestine?

19 Sketch a villus and label its epithelium, brush border, lamina propria, blood capillaries, and lacteal.

20 Distinguish between segmentation and the migrating motor complex of the small intestine. How do these differ in function?

The Large Intestine

▼Objectives

When you have completed this section, you should be able to
- describe the gross anatomy of the large intestine;
- contrast the mucosa of the colon with that of the small intestine;
- state the physiological significance of intestinal bacteria;
- discuss the types of contractions that occur in the colon; and
- explain the neurological control of defecation.

The large intestine receives food residue, absorbs water and salts, and eliminates the remainder by defecation. In the cadaver it measures about 1.5 m (5 ft) long and 6.5 cm (2.5 in.) in diameter.

Gross Anatomy

The large intestine (fig. 25.26) begins with the **cecum,**[26] a blind pouch in the lower right abdominal quadrant inferior to the ileocecal valve. Attached to its lower end is the **vermiform**[27] **appendix,** a blind tube 2 to 7 cm long. The appendix is densely populated with lymphocytes and is an important source of immune cells.

The **ascending colon** begins at the ileocecal valve and passes up the right side of the abdominal cavity. It makes a 90° turn at the **hepatic (right colic) flexure,** near the right lobe of the liver, and becomes the **transverse colon.** This passes horizontally across the upper abdominal cavity and turns 90° downward at the **splenic (left colic) flexure** near the spleen. Here it becomes the **descending colon,** which passes down the left side of the abdominal cavity. Ascending, transverse, and descending colons thus form a squarish, three-sided frame around the small intestine.

The pelvic cavity is narrower than the abdominal cavity, so at the pelvic inlet the colon turns medially and downward, forming a roughly S-shaped portion called the **sigmoid**[28] **colon.** (Visual examination of this region is performed with an instrument called a *sigmoidoscope.*) In the pelvic cavity, the large intestine straightens and forms the **rectum.**[29] The rectum has three internal folds called **rectal valves** that enable it to retain feces while passing gas.

The final 3 cm of the large intestine is the **anal canal** (fig. 25.26*b*), which passes through the levator ani muscle of the pelvic floor and terminates at the anus. Here, the mucosa forms longitudinal ridges called **anal columns** with depressions between them called **anal sinuses.** As feces pass

<hr>

26. *cec* = blind
27. *vermi* = worm + *form* = shaped
28. *sigm* = sigma or S + *oid* = resembling
29. *rect* = straight

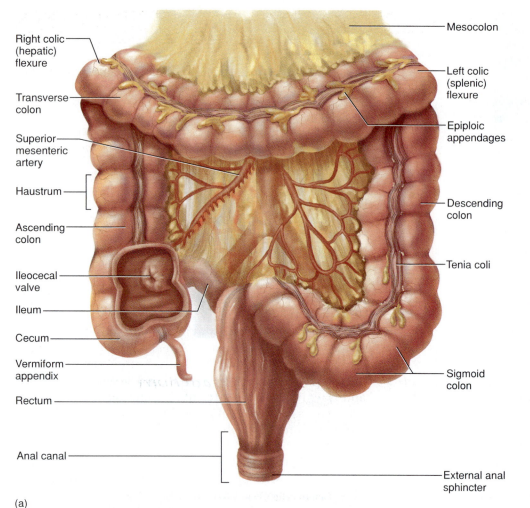

Right colic (hepatic) flexure

Transverse colon

Superior mesenteric artery

Haustrum

Ascending colon

Ileocecal valve

Ileum

Cecum

Vermiform appendix

Rectum

Anal canal

(a)

Mesocolon

Left colic (splenic) flexure

Epiploic appendages

Descending colon

Tenia coli

Sigmoid colon

External anal sphincter

Rectum

Rectal valve

Anal canal

Hemorrhoidal veins

Levator ani muscle

Internal anal sphincter

External anal sphincter

Anus   Anal sinuses   Anal columns

(b)

Figure 25.26 (a) Gross anatomy of the large intestine. (b) Detail of the anal canal. ⅄

through the canal, they press against the sinuses and cause them to exude extra mucus, which lubricates the canal during defecation. Large **hemorrhoidal veins** form superficial plexuses in the anal columns and around the orifice. Unlike veins in the extremities, they lack valves and are particularly subject to distension and venous pooling. **Hemorrhoids** are permanently distended veins that protrude into the anal canal or form bulges distal to the anus.

The muscularis externa of the colon is unusual in that its longitudinal fibers do not encircle the colon but are divided into three ribbonlike strips called the **teniae coli** (TEE-nee-ee CO-lye). The tonus of the teniae coli contracts the colon lengthwise and causes its wall to form pouches called **haustra**[30] (HAW-stra; singular, *haustrum*). In the rectum and anal canal, however, the longitudinal muscle forms a continuous sheet and haustra are absent. The anus, like the urethra, is regulated by two sphincters—an **internal anal sphincter** composed of smooth muscle of the muscularis externa and an **external anal sphincter** composed of skeletal muscle.

The ascending and descending colon are retroperitoneal, whereas the transverse and sigmoid colon are covered with serosa and anchored to the abdominal wall by a mesentery called the **mesocolon.** The serosa of these regions often has **epiploic**[31] **appendages,** clublike fatty pouches of peritoneum of unknown function.

Microscopic Anatomy

The mucosa of the large intestine has a simple columnar epithelium in all regions except the anal canal, where it is stratified squamous. The latter provides optimal resistance to the abrasion caused by the passage of feces. There are no circular folds or villi in the large intestine, but there are intestinal crypts. They are deeper than in the small intestine and have a greater density of goblet cells; mucus is their only significant secretion.

The anatomical features of the stomach, intestines, and accessory digestive glands of the abdomen are summarized in table 25.4.

---

30. *haustr* = to draw
31. *epiploic* = pertaining to an omentum

Table 25.4 Anatomical Checklist of the Stomach, Intestines, and Accessory Organs

Stomach	Quadrate	Exocrine acini	**Large Intestine**
General features	Caudate	Pancreatic islets	Regions
Lesser curvature	Ligaments	Pancreatic duct	Cecum
Greater curvature	Falciform ligament	Accessory pancreatic duct	Vermiform appendix
Cardiac region	Round ligament	**Small Intestine**	Ascending colon
Fundic region	Porta hepatis	Regions	Hepatic flexure
Body (corpus)	Microscopic anatomy	Duodenum	Transverse colon
Pyloric region	Hepatocytes	Major duodenal papilla	Splenic flexure
Antrum	Hepatic lobules	Minor duodenal papilla	Descending colon
Pyloric canal	Central vein	Duodenal glands	Sigmoid colon
Pylorus	Hepatic sinusoids	Duodenojejunal flexure	Rectum
Pyloric sphincter	Hepatic macrophages	Jejunum	Rectal valves
Mucosa	Hepatic triads	Ileum	Anal canal
Rugae	Bile tributaries	Peyer patches	Internal anal sphincter
Gastric pits	Bile canaliculi	Ileocecal valve	External anal sphincter
Cardiac glands	Bile ductules	Ileocecal junction	Anal columns
Pyloric glands	Right and left hepatic ducts	Mucosa	Anal sinuses
Gastric glands	Common hepatic duct	Circular folds	Hemorrhoidal veins
Mucous neck cells	**Gallbladder and Bile Duct**	Villi	External anatomy
Chief cells	Head (fundus)	Goblet cells	Teniae coli
Parietal cells	Neck (cervix)	Absorptive cells	Haustra
Enteroendocrine cells	Cystic duct	Microvilli	Epiploic appendages
Submucosa	Common bile duct	Lacteal	Mesocolon
Muscularis externa	Hepatopancreatic ampulla	Intestinal crypts	
Serosa	Hepatopancreatic sphincter	Paneth cells	
Liver	**Pancreas**		
Lobes	Head		
Right	Body		
Left	Tail		

Bacterial Flora and Intestinal Gas

The large intestine is densely populated with several species of bacteria collectively called the **bacterial flora**.[32] They ferment cellulose and other undigested carbohydrates and synthesize B vitamins and vitamin K, which are absorbed by the colon. This vitamin K is especially important because the diet alone usually does not provide enough to ensure adequate blood clotting.

The average person expels about 500 mL of **flatus** (gas) per day. Most of this is swallowed air that has worked its way through the digestive tract, but the bacterial flora add to it. Painful cramping can result when undigested nutrients pass into the colon and furnish an abnormal substrate for bacterial action—for example, in *lactose intolerance* (see special topic 25.4). Flatus is composed of nitrogen (N_2), carbon dioxide (CO_2), hydrogen (H_2), methane (CH_4), hydrogen sulfide (H_2S), and two amines: indole and skatole. The last three of these produce the odor of flatus and feces, whereas the others are odorless.

Absorption and Motility

Each day, about 500 mL of food residue enters the large intestine. It undergoes no further chemical digestion, but its volume is reduced over the next 12 to 24 hours as the colon absorbs water and electrolytes (especially NaCl) from it. The average adult voids about 150 mL of feces per day, consisting of 75% water and 25% solid matter. The latter is about 30% bacteria, 30% undigested dietary fiber, 10% to 20% fat, and smaller amounts of protein, sloughed epithelial cells, salts, mucus, and other digestive secretions.

The most common type of colonic motility is a type of segmentation called **haustral contractions,** which

---

32. *flora* = flowers, plants

Humans are a strange species. Unique among mammals, they go on drinking milk in adulthood, and, moreover, they drink the milk of other species! This peculiar habit is largely limited, however, to Europeans, a few pastoral tribes of Africa, and their descendants. Only they produce lactase as adults—a trait that seems to have evolved in the 10,000 years that people have milked domestic animals.

People without lactase have **lactose intolerance.** If they consume milk, lactose passes undigested into the large intestine, increases the osmolarity of the intestinal contents, and causes colonic water retention and diarrhea. In addition, lactose fermentation by the bacterial flora produces gas, resulting in painful cramps and flatulence.

Lactose intolerance occurs in about 15% of American whites, nearly all people of Asian descent, and about 90% of American blacks, who are predominantly descended from non-pastoral African tribes. People with lactose intolerance can consume products such as yogurt and cheese, in which bacteria have broken down the lactose, and they can digest milk with the aid of lactase drops or tablets.

occur about every 30 minutes. Distension of a haustrum with feces stimulates it to contract. This churns and mixes the residue, promotes water and salt absorption, and passes the residue distally to another haustrum.

Stronger contractions called **mass movements** occur one to three times a day, last about 15 minutes, and occur especially within an hour after breakfast. They are often triggered by the **gastrocolic** and **duodenocolic reflexes,** in which filling of the stomach and duodenum stimulates motility of the colon. Mass movements occur especially in the transverse to sigmoid colon, moving residue for several centimeters. Stretching of the rectum stimulates the defecation reflexes, which account for the urge to defecate that is often felt soon after a meal. The predictability of this reflex is useful in house-training pets and toilet-training children.

Defecation

In the **intrinsic defecation reflex** (fig. 25.27), stretch signals travel by the myenteric nerve plexus to the muscularis of the descending and sigmoid colons and the rectum. This triggers a peristaltic wave that drives feces downward, and it relaxes the internal anal sphincter. Defecation occurs only if the external anal sphincter is voluntarily relaxed at the same time.

The intrinsic reflex is relatively weak and usually requires the cooperative action of a stronger **parasympathetic defecation reflex** involving the spinal cord. Stretch signals to the sacral segments of the cord are returned by way of the pelvic nerves, leading to the same regions that are stimulated by the intrinsic reflex. This intensifies the peristaltic response and helps relax the internal anal sphincter.

There are several forms of conscious control over defecation, including the decision whether to relax the external anal sphincter and allow feces to pass. As involuntary contractions of the rectum push the feces downward, voluntary contractions of the levator ani pull the anal canal upward and allow the feces to fall away. Defecation is also aided by the Valsalva maneuver, which increases abdominal pressure, compresses

the rectum, and squeezes the feces from it. This maneuver can also initiate the defecation reflex by forcing feces from the descending colon into the rectum.

If the defecation urge is suppressed, contractions cease in a few minutes and the rectum relaxes. The defecation reflexes reoccur a few hours later or when another mass movement propels more feces into the rectum.

Key Point Review

21 How does the mucosa of the large intestine differ from that of the small intestine? How does the muscularis externa differ?

22 Name and briefly describe two types of contractions that occur in the colon and nowhere else in the alimentary canal.

23 Describe the reflexes that would cause defecation in an infant. Describe the additional neural controls that would function following toilet-training.

Chemical Digestion and Absorption

▼Objectives

When you have completed this section, you should be able to

- describe how polysaccharides, proteins, and fats are chemically digested, name the enzymes involved, and discuss the functional differences among these enzymes;
- describe how monosaccharides, amino acids, and lipids are absorbed by the small intestine; and
- describe the processing of nucleic acids, vitamins, minerals, and water by the digestive system.

Nearly all calories in the diet come from carbohydrates, proteins, and fats. Here we trace the digestion and absorption of these nutrients throughout the alimentary canal and more briefly consider the other classes of nutrients.

Carbohydrates

Most digestible dietary carbohydrate is starch. Cellulose is indigestible and is not considered here, although its importance as dietary fiber is discussed in the next chapter. The amount of glycogen in the diet is negligible, but it is digested in the same manner as starch.

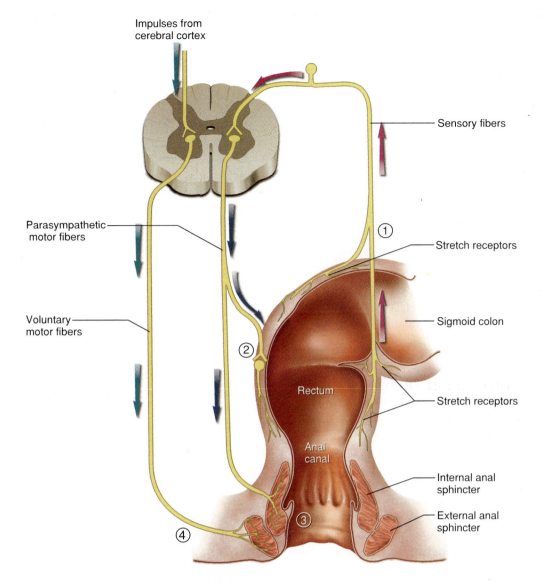

1. Stretching of the sigmoid colon and rectum triggers afferent signals to the spinal cord.

2. Efferent signals return to the rectum and stimulate its contraction.

3. Efferent signals stimulate relaxation of the internal anal sphincter.
   Steps 1-3 constitute the parasympathetic defecation reflex.

4. If it is appropriate to defecate, voluntary signals stimulate relaxation
   of the external anal sphincter, allowing the large intestine to expel the feces.

Figure 25.27 Neural control of defecation. (*1*) Filling of the rectum with feces stimulates stretch receptors, which transmit impulses to the spinal cord. (*2*) A spinal reflex stimulates contractions of the rectum and relaxation of the internal anal sphincter. (*3*) Defecation normally does not occur unless voluntary impulses relax the external anal sphincter.

Carbohydrate Digestion

Starch is digested first to oligosaccharides two to eight glucose residues long, then into the disaccharide maltose, and finally to glucose, which is absorbed by the small intestine. The process begins in the mouth, where salivary amylase hydrolyzes starch into oligosaccharides. Salivary amylase functions best at pH 6.8 to 7.0, typical of the oral cavity. It is quickly denatured upon contact with stomach acid, but it can digest starch for as long as 1 to 2 hours in the stomach as long as it is in the middle of a food mass and escapes contact with the acid. Amylase therefore works longer when the meal is larger, especially in the fundus, where gastric motility is weakest and a food bolus takes longer to break up. As acid, pepsin, and the churning contractions of the stomach break up the bolus, amylase is denatured; it does not function at a pH of 4.5 or lower. Being a protein, amylase is then digested by pepsin along with the dietary proteins.

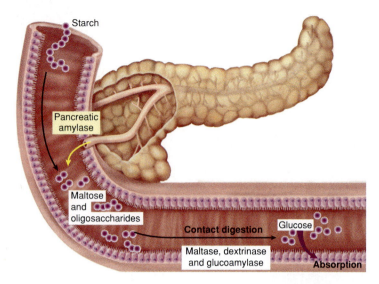

Figure 25.28 Starch digestion in the small intestine. Pancreatic amylase digests starch into maltose and small oligosaccharides. Brush border enzymes digest these to glucose, which is absorbed by the epithelial cells. ✗ 🔲

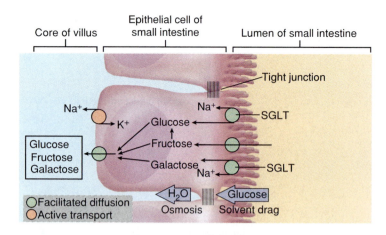

Figure 25.29 Mechanisms of monosaccharide absorption by the small intestine. ✗

Starch digestion resumes in the small intestine when the chyme mixes with pancreatic amylase (fig. 25.28). It is entirely converted to oligosaccharides and maltose within 10 minutes. Its digestion is completed by brush border enzymes. Two of these, **dextrinase** and **glucoamylase,** hydrolyze oligosaccharides that are three or more residues long. The third, **maltase,** hydrolyzes maltose to glucose.

Maltose is also present in some foods, but the major dietary disaccharides are sucrose (cane sugar) and lactose (milk sugar). They are digested by the brush border enzymes **sucrase** and **lactase,** respectively, and the resulting monosaccharides (glucose and fructose from the former; glucose and galactose from the latter) are immediately absorbed. In most humans, however, lactase is no longer produced after the age of four and lactose is indigestible past that age (see special topic 25.4).

Carbohydrate Absorption

In the plasma membrane adjacent to the brush border enzymes, there are transport proteins that absorb monosaccharides as soon as they are produced (fig. 25.29). About 80% of the absorbed sugar is glucose, which is taken up by a sodium-dependent glucose transporter (SGLT) like that of the kidney tubules (see p. 130). After a high-carbohydrate meal, however, two to three times as much glucose is absorbed by solvent drag as by the SGLT. When sugars enter the extracellular fluid (ECF) at the base of the intestinal epithelium, they increase its osmolarity. Water then passes osmotically from the lumen, through the tight junctions between the epithelial cells, into the ECF, carrying glucose and other nutrients with it.

The SGLT also absorbs galactose, whereas fructose is absorbed by facilitated diffusion using a separate carrier that does not depend on an Na⁺ gradient. Inside the epithelial cell, most fructose is converted to glucose. Glucose, galactose, and the small amount of remaining fructose are then transported out the base of the cell by facilitated diffusion and are absorbed by the blood capillaries of the villus. The hepatic portal system delivers them to the liver; chapter 26 follows the fate of these sugars from there.

Proteins

The amino acids absorbed by the small intestine come from three sources: (1) dietary proteins, (2) digestive enzymes digested by each other, and (3) sloughed epithelial cells digested by these enzymes. The endogenous amino acids from the last two sources total about 30 g/day.

Enzymes that digest proteins are called **proteases,** or **peptidases.** They are absent from the saliva but begin their job in the stomach. Here, pepsin hydrolyzes any peptide bond between tyrosine and phenylalanine, thus digesting 10% to 15% of the dietary protein into shorter polypeptides and a small amount of free amino acids (fig. 25.30). Pepsin has an optimal pH of 1.5 to 3.5; thus it is inactivated when it passes into the duodenum and mixes with the alkaline pancreatic juice.

In the small intestine, the pancreatic enzymes trypsin and chymotrypsin take over protein digestion by hydrolyzing polypeptides into even shorter oligopeptides. Finally, these are taken apart one amino acid at a time by three more enzymes: (1) **carboxypeptidase** removes amino acids from the—COOH end of the chain; (2) **aminopeptidase** removes them from the —NH₂ end; and (3) **dipeptidase** splits dipeptides in the middle, releasing the last two free amino acids. All three of these enzymes are found on the brush border, while carboxypeptidase also occurs in the pancreatic juice.

Amino acid absorption is similar to that of monosaccharides. There are several sodium-dependent amino

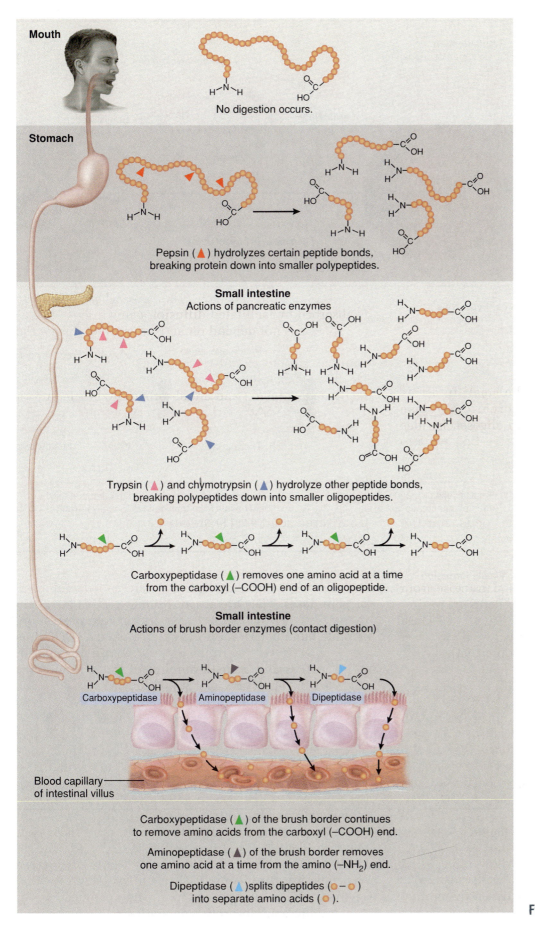

Mouth

No digestion occurs.

Stomach

Pepsin (▲) hydrolyzes certain peptide bonds,
breaking protein down into smaller polypeptides.

Small intestine
Actions of pancreatic enzymes

Trypsin (▲) and chymotrypsin (▲) hydrolyze other peptide bonds,
breaking polypeptides down into smaller oligopeptides.

Carboxypeptidase (▲) removes one amino acid at a time
from the carboxyl (−COOH) end of an oligopeptide.

Small intestine
Actions of brush border enzymes (contact digestion)

Carboxypeptidase   Aminopeptidase   Dipeptidase

Blood capillary
of intestinal villus

Carboxypeptidase (▲) of the brush border continues
to remove amino acids from the carboxyl (−COOH) end.

Aminopeptidase (▲) of the brush border removes
one amino acid at a time from the amino (−NH$_2$) end.

Dipeptidase (▲) splits dipeptides (●−●)
into separate amino acids (●).

Figure 25.30 Protein digestion.

acid cotransporters for different classes of amino acids. Dipeptides and tripeptides can also be absorbed, but they are hydrolyzed within the cytoplasm of the epithelial cells before their amino acids are released to the bloodstream. At the basal surfaces of the cells, amino acids behave like the monosaccharides discussed previously—they leave the cell by facilitated diffusion, enter the capillaries of the villus, and are thus carried away in the hepatic portal circulation.

The absorptive cells of infants are able to take up intact proteins by pinocytosis and release them to the blood by exocytosis. This allows IgA from breast milk to pass into an infant's bloodstream and confer passive immunity from mother to infant. It has the disadvantage, however, that intact proteins entering the infant's blood are detected as foreign antigens and sometimes trigger food allergies. As the intestine matures, its ability to pinocytose proteins declines but never completely ceases.

Lipids

Fats are digested by enzymes called **lipases.** Lingual lipase, secreted by the intrinsic salivary glands of the tongue, is activated by acid in the stomach, where it digests as much as 10% of the ingested fat. In infants, the stomach also secretes gastric lipase. Most fat digestion, however, occurs in the small intestine through the action of pancreatic lipase.

The hydrophobic quality of lipids makes their digestion and absorption more complicated than that of carbohydrates and proteins (fig. 25.31). As chyme enters the duodenum, lipids are in large globules exposed to lipase only at their surfaces. Fat digestion would be rather slow and inefficient if it remained this

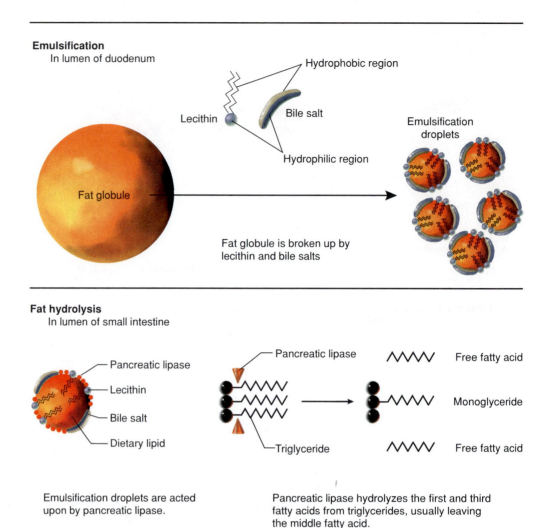

Emulsification
In lumen of duodenum

Hydrophobic region

Lecithin

Bile salt

Hydrophilic region

Fat globule

Emulsification droplets

Fat globule is broken up by lecithin and bile salts

Fat hydrolysis
In lumen of small intestine

Pancreatic lipase

Lecithin

Bile salt

Dietary lipid

Emulsification droplets are acted upon by pancreatic lipase.

Pancreatic lipase

Free fatty acid

Triglyceride

Monoglyceride

Free fatty acid

Pancreatic lipase hydrolyzes the first and third fatty acids from triglycerides, usually leaving the middle fatty acid.

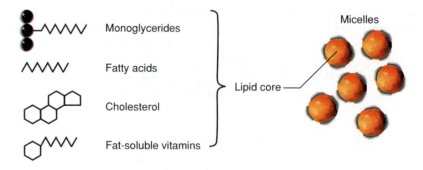

Micelle formation
In lumen of small intestine

Monoglycerides

Fatty acids

Cholesterol

Fat-soluble vitamins

Micelles

Lipid core

Several types of lipids form micelles coated with bile salts.

Figure 25.31 Fat digestion and absorption—*Cont'd.*

Chylomicron formation
   By intestinal epithelial cells

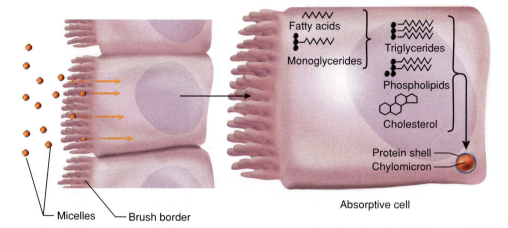

Chylomicron exocytosis and lymphatic uptake

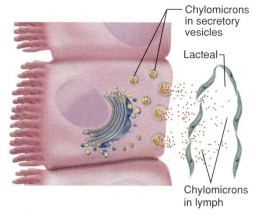

Micelles pass between microvilli to reach surfaces of absorptive cells.

Lipids of the micelle core enter absorptive cell; bile salts do not.

Fatty acids and monoglycerides are re-synthesized into neutral fats. Neutral fats combine with cholesterol and phospholipid and aquire a hydrophillic protein shell to become a chylomicron.

Golgi apparatus packages chylomicrons into secretory vesicles; chylomicrons are released from basal cell membrane by exocytosis and enter the lacteal (lymphatic capillary) of the villus.

Figure 25.31—Cont'd

way, but instead, it is broken up into smaller **emulsification droplets** by certain components of the bile— lecithin (a phospholipid) and bile salts (steroids). These agents have hydrophobic regions that dissolve in the surface of a fat globule and hydrophilic regions that are attracted to the surrounding water. Combined with the agitation produced by intestinal segmentation contractions, this breaks the fat up into droplets as small as 1 μm in diameter and exposes far more surface to enzymatic action.

When lipase acts on a triglyceride, it removes the first and third fatty acid from the glycerol backbone and usually leaves the middle fatty acid. The products of lipase action are therefore two free fatty acids (FFAs) and a monoglyceride. Bile salts coat these and other lipids and form droplets called **micelles**[33] (my-SELLS) as small as 3 to 6 nm in diameter. Within the shell of bile salts, a micell contains FFAs, monoglycerides, cholesterol, and fat-soluble vitamins. Micelles pass between the microvilli of the brush border, and upon reaching the surface of the epithelial cell they release their lipids. The lipids diffuse freely through the phospholipid plasma membrane.

Within the cell, the FFAs and monoglycerides are resynthesized into triglycerides. These are combined with small amounts of cholesterol and phospholipid and coated with a film of protein, forming droplets about 1 μm in diameter called **chylomicrons**[34] (KY-lo-MY-crons). The Golgi complex packages chylomicrons into secretory vesicles that migrate to the basal surface of the cell and release their contents into the core of the villus. Although some free fatty acids enter the blood capillaries, chylomicrons are too large to do so. They are taken up by the more porous lacteal (lymphatic vessel) into the lymph. The white, fatty intestinal lymph (*chyle*) flows through larger and larger lymphatic vessels of the mesenteries, eventually reaching the thoracic duct and entering the bloodstream at the left subclavian vein. The further fate of dietary fat is described in chapter 26.

Think About It
Explain why the right lymphatic duct does not empty dietary fat into the bloodstream.

Nucleic Acids

The nucleic acids, DNA and RNA, are present in much smaller quantities than the polymers discussed previously. The **nucleases** (ribonuclease and deoxyribonuclease) of

33. *mic* = grain, crumb + *elle* = little

34. *chyl* = juice + *micr* = small

pancreatic juice hydrolyze these to their constituent nucleotides. **Nucleosidases** and **phosphatases** of the brush border then decompose the nucleotides into phosphate ions, ribose (from RNA) or deoxyribose (from DNA), and nitrogenous bases. These products are transported across the intestinal epithelium by membrane carriers and enter the capillary blood of the villus.

Vitamins

Vitamins are absorbed unchanged. The fat-soluble vitamins A, D, E, and K are absorbed with other lipids as just described. Therefore, if they are ingested without fat-containing food, they are not absorbed at all but are passed in the feces and wasted. Water-soluble vitamins (the B complex and vitamin C) are absorbed by simple diffusion, with the exception of vitamin B_{12}. This is an unusually large molecule that can only be absorbed if it binds to intrinsic factor from the stomach. The B_{12}-intrinsic factor complex then binds to receptors on absorptive cells of the distal ileum, where it is taken up by receptor-mediated endocytosis.

Minerals

Minerals (electrolytes) are absorbed along the entire length of the small intestine. Sodium ions are cotransported with sugars and amino acids. Chloride ions are actively transported in the distal ileum by a pump that exchanges them for bicarbonate ions, reversing the chloride-bicarbonate exchange that occurs in the stomach. Potassium ions are absorbed by simple diffusion. As water is absorbed from the chyme, its K^+ rises and creates a gradient favorable to its absorption. In diarrhea, when water absorption is hindered, potassium ions are also lost in the feces; thus chronic diarrhea can lead to hypokalemia.

Iron and calcium are unusual in that they are absorbed in proportion to the body's need, whereas other minerals are absorbed at fairly constant rates regardless of need, leaving it to the kidneys to excrete any excess. Furthermore, iron and calcium absorption is largely limited to the duodenum.

Iron enters the absorptive cells by active transport and binds to the cytoplasmic protein ferritin. The iron-ferritin complex forms an intracellular reservoir of iron that is lost when the epithelial cells are sloughed off, unless the iron is needed sooner. In that case intracellular iron is released to the blood and carried to the bone marrow by the plasma protein transferrin (see chapter 18).

Think About It

Young adult women have four times as many iron transport proteins in the intestinal mucosa as men have. Can you explain this?

The rate of calcium absorption is determined by the availability of the cofactor calcitriol (vitamin D). The synthesis of calcitriol is under the control of parathyroid hormone (PTH) and thus adjusted to the body's state of calcium balance (see chapter 8).

Water

The digestive system is one of several systems involved in water balance. The digestive tract receives about 9 L of water per day—0.7 L in food, 1.6 L in drink, 6.7 L in the gastrointestinal secretions: saliva, gastric juice, bile, pancreatic juice, and intestinal juice. About 8 L of this is absorbed by the small intestine and 0.8 L by the large intestine, leaving 0.2 L voided in the daily fecal output. Water is absorbed by osmosis, following the absorption of salts and organic nutrients that create an osmotic gradient from the intestinal lumen to the ECF.

Diarrhea occurs when the large intestine absorbs too little water from the feces. This occurs when the intestine is irritated by bacteria and feces pass through too quickly for adequate reabsorption, or when the feces contain abnormally high concentrations of a solute such as lactose that opposes osmotic absorption of water. *Constipation* occurs when fecal movement is slow, too much water is reabsorbed, and the feces become hardened. This can result from lack of dietary fiber, lack of exercise, emotional upset, or long-term laxative abuse.

Think About It

Magnesium sulfate (epsom salt) is poorly absorbed by the intestines. In light of this, explain why it has a laxative effect.

········· Key Point Review ·········

24 What three polymers account for most of the dietary calories? What are the end products of enzymatic digestion of each?

25 What two nutrients are digested by saliva? Why is only one of them digested in the mouth?

26 Name as many enzymes of the intestinal brush border as you can, and identify the substrate or function of each.

27 Explain the distinctions between an emulsification droplet, a micelle, and a chylomicron.

28 What happens to digestive enzymes after they have done their job? What happens to dead epithelial cells that slough off the gastrointestinal mucosa? Explain.

The Man with a Hole in His Stomach

Perhaps the most famous episode in the history of digestive physiology began in 1822 on Mackinac Island in the strait between Lake Michigan and Lake Huron. Alexis St. Martin, a 19-year-old fur trapper, was standing outside a trading post when he was accidentally hit by a shotgun blast from 3 feet away. An Army doctor stationed at Fort Mackinac, William Beaumont (1785–1853), was summoned to examine St. Martin. As Beaumont later wrote, "a portion of the lung as large as a turkey's egg" protruded through St. Martin's lacerated and burnt flesh. Below that was a portion of the stomach with a puncture in it "large enough to receive my forefinger." Beaumont did his best to pick out bone fragments and dress the wound, though he did not expect St. Martin to survive.

Surprisingly, St. Martin lived. Over a period of months the wound extruded pieces of bone, cartilage, gunshot, and gun wadding. As the wound healed, a fistula (hole) remained in the stomach, so large that Beaumont had to cover it with a compress to prevent food from coming out. A fold of tissue later grew over the fistula, but it was easily opened. A year later, St. Martin was still feeble. Town authorities decided they could no longer support him on public funds and wanted to ship him 2,000 miles to his home in Canada.

Beaumont, however, was imbued with a passionate sense of destiny. Very little was known about digestion, and he saw the accident as a unique opportunity to learn. He took St. Martin in at his personal expense and performed 238 experiments on him over several years. Beaumont had never attended medical school and had little idea how scientists work, yet he proved to be an astute experimenter. Under crude frontier conditions and with almost no equipment, he discovered many of the basic facts of gastric physiology discussed in this chapter.

"I can look directly into the cavity of the stomach, observe its motion, and almost see the process of digestion," Beaumont wrote. "I can pour in water with a funnel and put in food with a spoon, and draw them out again with a siphon." He put pieces of meat on a string into the stomach and removed them hourly for examination. He sent vials of gastric juice to the leading chemists of America and Europe, who could do little but report that it contained hydrochloric acid. He proved that digestion required HCl and could even occur outside the stomach, but he found that HCl alone did not digest meat; gastric juice must contain some other digestive ingredient. Theodor Schwann, one of the founders of the cell theory (see chapter 4), identified that ingredient as pepsin. Beaumont also demonstrated that gastric juice is secreted only in response to food; it did not accumulate between meals as previously thought. He disproved the idea that hunger is caused by the walls of the empty stomach rubbing against each other.

For his part, St. Martin felt helpless and humiliated by Beaumont's experiments. His fellow trappers taunted him as "the man with a hole in his stomach," and he longed to return to hunting and trapping in the wilderness. He had a wife and daughter in Canada whom he rarely got to see, and he ran away repeatedly to join them. He was once gone for 4 years before his poverty and physical disability made him yield to Beaumont's financial enticement to come back. Beaumont despised St. Martin for his drunkenness and profanity and was quite insensitive to St. Martin's embarrassment and discomfort over the experiments. Yet St. Martin's temper enabled Beaumont to make the first direct observations of the relationship between emotion and digestion. When St. Martin was particularly distressed, Beaumont noted little digestion occurring—as we now know, the sympathetic nervous system inhibits digestive activity.

Beaumont published a book in 1833 that laid the foundation for modern gastric physiology and dietetics. It was enthusiastically received by the medical community and had no equal until Russian physiologist Ivan Pavlov (1849–1936) performed his celebrated experiments on digestion in animals. Building on the methods pioneered by Beaumont, Pavlov received the 1904 Nobel Prize for Physiology or Medicine.

In 1853, Beaumont slipped on some ice, suffered a blow to the base of his skull, and died a few weeks later. St. Martin continued to tour medical schools and submit to experiments by other physiologists, whose conclusions were often less correct than Beaumont's. Some, for example, attributed chemical digestion to lactic acid instead of hydrochloric acid. St. Martin lived in wretched poverty in a tiny shack with his wife and several children, and died 28 years after Beaumont. By then he was senile, believing he had been to see Paris, where Beaumont had often promised to take him.▲

Interactions Between the DIGESTIVE SYSTEM and Other Organ Systems

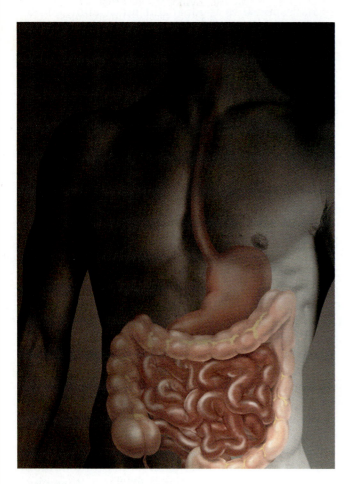

Nervous System

- Enteric and autonomic nervous systems regulate GI motility and secretion; somatic nervous system controls chewing, swallowing, and defecation; sense organs involved in food selection and ingestion; hypothalamus contains centers for hunger, thirst, and satiation

Endocrine System

- Hormones regulate GI motility, secretion, and processing of absorbed nutrients
- Liver deactivates hormones; enteroendocrine cells and pancreatic islets produce many hormones

Circulatory System

- Circulates hormones that regulate GI function; absorbs and distributes nutrients; adjusts salivation by vasomotion of salivary gland blood vessels
- Digestive system absorbs fluid to maintain normal blood volume and cardiac function; liver excretes heme from dead erythrocytes, synthesizes clotting proteins, plasma albumin, and all other plasma proteins except antibodies, and regulates blood glucose levels; intestinal epithelium stores and releases iron as needed for erythropoiesis

Lymphatic/Immune Systems

- Lymphatic system absorbs dietary lipids; immune cells protect GI tract from infection
- GI mucosa serves as site of lymphocyte production; acid, lysozyme, and enzymes secreted by GI tract provide nonspecific defense against microbes

Respiratory System

- Provides O_2 and removes CO_2; Valsalva maneuver aids defecation
- Pressure of digestive organs against inferior surface of diaphragm aids expiration when abdominal muscles are contracted.

Urinary System

- Excretes bile pigments and other products of liver metabolism; completes synthesis of calcitriol, needed for intestinal calcium absorption
- Intestines complement the kidneys in water and electrolyte reabsorption; liver carries out the last step before the kidneys in calcitriol synthesis; liver synthesizes urea, the main nitrogenous waste of the urine

Reproductive System

- Developing fetus may crowd digestive organs; heartburn and constipation common during pregnancy
- Digestive system provides nutrients for fetal growth

All Systems

The digestive system provides all other systems with nutrients in usable form for cellular metabolism and building of tissues.

Integumentary System

- Skin plays role in synthesis of vitamin D needed for calcium and phosphorus absorption in small intestine

Skeletal System

- Provides protective enclosure for some digestive organs, physical support for the teeth, and movements of mastication
- Small intestine adjusts calcium absorption in proportion to demands of skeletal system

Muscular System

- Essential for chewing, swallowing, and voluntary control of defecation; abdominal muscles protect lower GI organs
- Liver disposes of lactic acid generated by anaerobic fermentation of muscle, thus promotes recovery from muscle fatigue

Study Outline

General Anatomy and Digestive Processes (pp. 887–889)

1. Subdivisions of the digestive system
   a. Accessory organs
   b. Digestive tract
2. Relationship of the digestive tract to the peritoneum
   a. Mesentery
   b. Serosa
   c. Omenta
   d. Mesocolon
3. Digestive functions and processes
4. Stages of digestion
   a. Mechanical digestion
   b. Chemical digestion

The Mouth Through Esophagus (pp. 889–897)

1. Oral (buccal) cavity
   a. Functions
   b. Cheeks and lips
   c. Tongue
      • Lingual papillae
      • Body and root
      • Lingual muscles
      • Lingual glands
   d. Palate
      • Hard and soft palates
      • Glossopalatine and pharyngopalatine arches
   e. Teeth (dentition)
      • Types
      • Development and eruption
      • Periodontal tissues
      • Tooth structure
2. Mastication
3. Saliva and salivary glands
   a. Composition of saliva
   b. Salivary glands
      • Intrinsic
      • Extrinsic
   c. Salivation
4. Pharynx
5. Esophagus
   a. General structure
   b. Tissue layers
   c. Enteric nervous system
      • Submucosal (Meissner) plexus
      • Myenteric (Auerbach) plexus
      • Short and long reflexes
6. Swallowing (deglutition)
   a. Swallowing center
   b. Buccal phase
   c. Pharyngeal-esophageal phase
   d. Mechanism of peristalsis

The Stomach (pp. 898–905)

1. Functions

2. Gross anatomy
   a. Curvatures
   b. Four regions of stomach
   c. Pyloric sphincter
3. Innervation and circulation
4. Stomach wall
   a. Rugae
   b. Gastric pits
   c. Glands
   d. Cell types of gastric glands
5. Gastric secretions
   a. Hydrochloric acid
   b. Intrinsic factor
   c. Pepsinogen and pepsin
   d. Gastric lipase and rennin
   e. Hormones and paracrines
6. Gastric motility
   a. Stress-relaxation response
   b. Gastric peristalsis
   c. Entry of chyme into duodenum
7. Vomiting
8. Digestion and absorption
9. Protective mechanisms
   a. Mucous coat
   b. Epithelial replacement
   c. Tight junctions
10. Regulation of gastric function
    a. Cephalic phase
    b. Gastric phase
       • Myenteric reflex
       • Vagovagal reflex
       • G cells and gastrin
       • Negative feedback control
    c. Intestinal phase
       • Intestinal gastrin
       • Enterogastric reflex
       • Secretin
       • Cholecystokinin (CCK)
       • Gastric inhibitory peptide (GIP)

The Liver, Gallbladder, and Pancreas (pp. 905–911)

1. Liver
   a. Gross anatomy
      • Lobes and ligaments
      • Porta hepatis
   b. Microscopic anatomy
      • Hepatic lobules
      • Hepatic sinusoids
      • Hepatic triads
      • Bile passageways
2. Gallbladder and bile
   a. Gallbladder structure
   b. Composition of bile
   c. Digestive role of bile salts
   d. Enterohepatic circulation

3. Pancreas
   a. Gross anatomy
   b. Endocrine and exocrine components
   c. Ducts and duodenal papillae
   d. Pancreatic juice
   e. Zymogens and their activation
   f. Other pancreatic enzymes
4. Regulation of secretion
   a. Cholecystokinin
   b. Secretin

The Small Intestine (pp. 911–915)

1. Importance of surface area
2. Gross anatomy
   a. Duodenum
   b. Jejunum
   c. Ileum
   d. Ileocecal junction
3. Microscopic anatomy
   a. Circular folds
   b. Villi
      • Absorptive and goblet cells
      • Blood capillaries and lacteal
      • Brush border
   c. Intestinal crypts
      • Absorptive and goblet cells
      • Paneth cells
   d. Duodenal glands
   e. Peyer patches
4. Secretion (intestinal juice)
5. Motility
   a. Functions of intestinal motility
   b. Segmentation
   c. Peristalsis: the migrating motor complex
   d. Ileocecal valve
   e. Gastroileal reflex

The Large Intestine (pp. 915–918)

1. Functions
2. Gross anatomy
   a. Cecum and appendix
   b. Ascending, transverse, and descending colon
   c. Sigmoid colon, rectum, and anal canal
   d. Teniae coli and haustra
   e. Anal sphincters
   f. Mesocolon
   g. Epiploic appendages
3. Microscopic anatomy
4. Bacterial flora and intestinal gas
5. Absorption and motility
   a. Absorption
   b. Quantity and composition of feces

c. Motility
- Haustral contractions
- Mass movements

d. Defecation reflexes
- Intrinsic reflex
- Parasympathetic reflex
- Voluntary controls

Chemical Digestion and Absorption (pp. 918–924)

1. Carbohydrates
   a. Digestion
      - Action of salivary amylase
      - Action of pancreatic amylase
      - Action of brush border enzymes
        - •• Dextrinase
        - •• Glucoamylase
        - •• Maltase
        - •• Sucrase
        - •• Lactase
   b. Absorption
      - Glucose, the SGLT, and solvent drag
      - Galactose
      - Fructose

2. Proteins
   a. Sources of amino acids
   b. Action of pepsin
   c. Action of trypsin and chymotrypsin
   d. Contact digestion
   e. Amino acid absorption
   f. Protein pinocytosis

3. Lipids
   a. Action of lingual and gastric lipase
   b. Emulsification
   c. Action of pancreatic lipase
   d. Formation of micelles
   e. Absorption by epithelial cells
   f. Formation of chylomicrons
   g. Absorption by lacteals

4. Nucleic acids

5. Vitamins
   a. Fat-soluble vitamins
   b. Water-soluble vitamins
   c. Vitamin B_{12} and intrinsic factor

6. Minerals
   a. Na^+, K^+, and Cl^- absorption
   b. Iron absorption and storage
   c. Calcium absorption

7. Water

Selected Vocabulary

Also review the terms listed in tables 25.1 and 25.4, which are not repeated here

accessory organs 887
digestive tract 887
dorsal mesentery 888
serosa 888
ventral mesentery 888
lesser omentum 888
greater omentum 888
mesocolon 888
mechanical digestion 889
chemical digestion 889
occlusion 890
eruption of a tooth 890
mastication 891
salivary amylase 892
lingual lipase 892
salivatory nuclei 894
bolus 894
mucosa 895
lamina propria 895
muscularis mucosae 895
submucosa 895
muscularis externa 895
adventitia 895

serosa 895
enteric nervous system 896
submucosal plexus 896
myenteric plexus 896
short reflex 896
long reflex 896
gastroesophageal sphincter 896
deglutition 896
swallowing center 896
peristalsis 896
chyme 898
gastric juice 901
H^+-K^+ ATPase 901
alkaline tide 901
intrinsic factor 901
zymogen 901
pepsinogen 901
pepsin 901
gastric lipase 901
rennin 901
emetic center 903
cephalic phase 903
gastric phase 904
myenteric reflex 904
vagovagal reflex 904
G cells 904

gastrin 904
intestinal phase 905
enterogastric reflex 905
secretin 905
cholecystokinin (CCK) 905
gastric inhibitory peptide (GIP) 905
bilirubin 909
urobilinogen 909
bile salts 909
enterohepatic circulation 910
pancreatic juice 910
trypsinogen 910
chymotrypsinogen 910
procarboxypeptidase 910
enterokinase 910
pancreatic amylase 911
pancreatic lipase 911
ribonuclease 911
deoxyribonuclease 911
secretin 911
brush border enzymes 912
contact digestion 912
intestinal juice 914
segmentation 914
migrating motor complex 915
gastroileal reflex 915

hemorrhoids 916
bacterial flora 917
flatus 917
haustral contraction 917
mass movement 918
gastrocolic reflex 918
duodenocolic reflex 918
intrinsic defecation reflex 918
parasympathetic defecation reflex 918
dextrinase 920
glucoamylase 920
maltase 920
sucrase 920
lactase 920
protease 920
carboxypeptidase 920
aminopeptidase 920
dipeptidase 920
lipase 922
emulsification droplet 923
micelle 923
chylomicron 923
nuclease 923
nucleosidase 924
phosphatase 924

Testing Your Recall   Answers in Appendix C

1. Which of the following enzymes hydrolyzes nutrients in the stomach?
   a. chymotrypsin
   b. lingual lipase
   c. carboxypeptidase
   d. enterokinase
   e. dextrinase

2. Which of the following enzymes does *not* hydrolyze any nutrients?
   a. chymotrypsin
   b. lingual lipase
   c. carboxypeptidase
   d. enterokinase
   e. dextrinase

3. Which of the following is *not* an enzyme?
   a. chymotrypsin
   b. enterokinase
   c. secretin
   d. pepsin
   e. nucleosidase

4. The substance in question 3 that is not an enzyme is
   a. a zymogen.
   b. a nutrient.
   c. an emulsifier.
   d. a neurotransmitter.
   e. a hormone.

5. The lacteals absorb
   a. chylomicrons.
   b. micelles.
   c. emulsification droplets.
   d. amino acids.
   e. monosaccharides.

6. All of the following *except* ___ contribute to the absorptive surface area of the small intestine.
   a. its length
   b. the brush border
   c. haustra
   d. circular folds
   e. villi

7. Which of the following is a periodontal tissue?
   a. the gingiva
   b. the enamel
   c. the cementum
   d. the pulp
   e. the dentin

8. The ___ of the stomach most closely resemble the ___ of the small intestine.
   a. gastric pits, intestinal crypts
   b. pyloric glands, intestinal crypts
   c. rugae, Peyer patches
   d. parietal cells, goblet cells
   e. gastric glands, duodenal glands

9. Which of the following cells secrete digestive enzymes?
   a. chief cells
   b. mucous neck cells
   c. parietal cells
   d. goblet cells
   e. enteroendocrine cells

10. What phase of gastric regulation includes inhibition by the enterogastric reflex?
   a. the intestinal phase
   b. the gastric phase
   c. the buccal phase
   d. the cephalic phase
   e. the pharyngo-esophageal phase

11. Cusps are a feature of the ___ surfaces of the premolars and molars.

12. The acidity of the stomach deactivates ___ but activates ___ of the saliva.

13. The ___ salivary gland is named for its proximity to the ear.

14. The submucosal and myenteric nerve plexuses collectively make up the ___ nervous system.

15. Nervous stimulation of gastrointestinal activity is mediated mainly through parasympathetic fibers of the ___ nerves.

16. Food in the stomach causes G cells to secrete ___, which in turn stimulates secretion of HCl and pepsinogen.

17. Hepatic macrophages are found in blood-filled spaces of the liver called ___.

18. The brush border enzyme that finishes the job of starch digestion, producing glucose, is called ___. Its substrate is ___.

19. Fats are transported in the lymph and blood in the form of droplets called ___.

20. Within the absorptive cells of the small intestine, ferritin binds the nutrient ___.

Testing Your Comprehension

Answers in *Study Guide*

1. A physician plans to attend a cocktail party, but he is also on call that night and must avoid intoxication. Therefore he drinks a glass of cream just before leaving for the party. Explain.

2. Which of these do you think would have the most severe effect on digestion: surgical removal of the stomach, gallbladder, or pancreas. Why?

3. What do carboxypeptidase and aminopeptidase have in common? Identify as many differences between them as you can.

4. What do micelles and chylomicrons have in common? Identify as many differences between them as you can.

5. Explain why dietary lipids cannot be absorbed by the blood capillaries of a villus as sugars and amino acids can.

Web Site Link

For a listing of the most current web sites related to this chapter, please visit the Saladin homepage at:

http://www.mhhe.com/sciencemath/biology/saladin/

chapter twenty-six

26

[Nutrition and Metabolism

the   unity   of   form   and   functio

Nutrition is the starting point and basis for all human form and function. From the time a single-celled, fertilized egg divides in two, nutrition provides the matter needed for cell division, growth, and development. It is the source of fuel that provides the energy for all biological work and of the raw materials for replacement of worn-out biomolecules and cells. The fact that it provides only the *raw* materials means, further, that chemical change—metabolism—lies at the foundation of form and function. In the previous chapter, we saw how the digestive system breaks nutrients down into usable form and absorbs them into the blood and lymph. We now consider these nutrients in more depth, follow their fate after absorption, and explore related issues of metabolism and body heat.

Nutrition

▼Objectives

When you have completed this section, you should be able to

- describe the mechanisms of hunger and satiety;
- define *nutrient* and list the six major categories of nutrients;
- state the function of each class of macronutrients, the approximate amounts required in the diet, and some major dietary sources of each;
- name the serum lipoproteins, state their functions, and describe how they differ from each other; and
- name the major vitamins and minerals required by the body and the general functions they serve.

Body Weight and Energy Balance

The subject of nutrition quickly brings to mind the subject of body weight and the popular desire to control it. While this is only one aspect of the subject, it is a significant consideration. Weight is determined by the body's energy balance—if our energy intake and output are equal, our weight is stable. We gain weight if intake exceeds output and lose weight if output exceeds intake.

Although weight is determined by this ratio, it remains quite stable over many years' time and seems to have a homeostatic set point. This has been experimentally demonstrated in animals. If an animal is force-fed until it becomes quite obese and then allowed to feed at will, it voluntarily reduces its intake and quickly returns to its former weight. Similarly, if an animal is undernourished until it loses much of its weight and then allowed to feed at will, it increases its intake and again returns quickly to its former weight.

In humans the set point varies greatly from person to person and body weight results from a combination of hereditary and environmental influences. From studies of identical twins and other subjects, it appears that about 30% of the variation in weight among people is due to variation in environmental influences, and 70% is due to hereditary variation. That is, the difference in weight between you and another person is likely to be largely hereditary, but within the range set by heredity, your weight will vary according to your eating and exercise habits.

Appetite

The struggle for weight control often seems to be a struggle against the appetite; but despite decades of research, we are still far from a complete understanding of how appetite is regulated. In the 1940s, it was discovered that a region in the lateral area of the hypothalamus seems to trigger the desire for food. When this **feeding center** is destroyed in animals, they exhibit drastic **anorexia**[1] (loss of appetite) and starve to death if not force-fed. The ventromedial hypothalamus has a **satiety center;** damage here causes **hyperphagia**[2] (overeating) and extreme obesity (fig. 26.1).

The satiety center has neurons called **glucostats** that rapidly absorb blood glucose after a meal. One hypothesis on hunger is that glucose uptake causes the satiety center to send inhibitory signals to the hunger center and thus suppresses the appetite. Some weight loss drugs work by inhibiting the hunger center. According to the glucostat hypothesis, as blood glucose concentration drops hours after a meal, inhibitory signals from the satiety center decline or cease, the feeding center becomes active, and the sense of hunger returns. But this is certainly far from the whole story. Hunger and satiety are regulated by a complex interaction of multiple brain centers, hormones, and sensory and motor pathways. Experimental lesions disrupt multiple pathways and may produce abnormal feeding behavior for a variety of reasons.

Other factors influence appetite in ways similar to the control of thirst (see p. 867). Merely chewing and swallowing food briefly satisfies the appetite, even if the food is removed through an esophageal fistula (opening) before reaching the stomach. Inflating the

---

1. *an* = without + *orexia* = appetite
2. *hyper* = excessive + *phagia* = eating

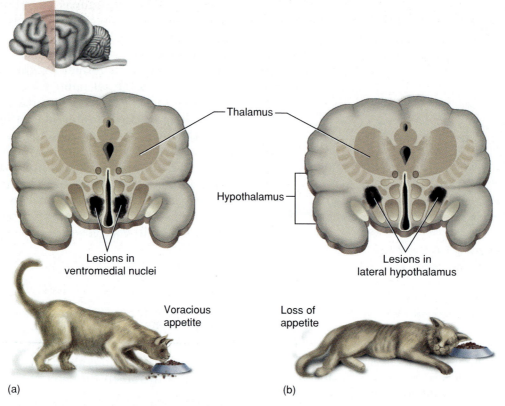

Figure 26.1 Regions of the hypothalamus associated with appetite. (*a*) Destruction of the ventromedial nuclei destroys the sense of satiety and results in extreme overeating. (*b*) Destruction of the lateral hypothalamus destroys the sense of hunger.

stomach with a balloon inhibits hunger even in an animal that has not actually swallowed any food.

Hunger is stimulated partly by gastric peristalsis. Mild **hunger contractions** begin soon after the stomach is emptied and increase in intensity over a period of hours. They can become quite a painful and powerful incentive to eat, yet they do not affect the amount of food consumed—this remains much the same even when nervous connections to the stomach and intestines are severed to cut off all conscious perception of hunger contractions.

Hormones also play a role in appetite regulation. Amino acids and fatty acids stimulate the duodenum to secrete cholecystokinin (CCK). CCK, a polypeptide, is a well-known appetite suppressant that exerts its effect by stimulating the brain and digestive organs. There is also some evidence that adipocytes secrete appetite-stimulating hormones when their lipid stores decline.

Think About It
Suppose CCK could be economically produced and packaged as tablets to be taken orally. Would this be an effective diet drug? Why or why not?

Appetite is not merely a question of *how much* but also *what kind* of food is consumed. Even animals shift their diets from one kind of food to another, apparently because some foods provide nutrients that others do not. Hu-

mans are known to seek foods that satisfy certain nutritional deficiencies even when the missing nutrients have no flavor. Different neurotransmitters also seem to affect the appetite for different classes of nutrients. For example, norepinephrine stimulates the appetite for carbohydrates, galanin for fatty foods, and endorphins for protein.

Calories

One calorie is the amount of heat that will raise the temperature of 1 g of water 1°C. One thousand calories is called a Calorie (capital C) in dietetics and a **kilocalorie** (kcal) in biochemistry. The relevance of calories to physiology is that they are a measure of the capacity to do biological work.

Nearly all dietary calories come from carbohydrates, proteins, and fats. Carbohydrates and proteins yield about 4 kcal/g when they are completely oxidized, and fats yield about 9 kcal/g. Some foods such as sugar and alcohol are said to provide "empty calories" because they provide nothing useful except for calories (see chapter essay, p. 957). By lessening the appetite but failing to provide other nutrients the body requires, they can contribute to malnutrition. In sound nutrition, the body's energy needs are met by more complex foods that simultaneously meet the need for proteins, lipids, vitamins, and other nutrients.

Table 26.1	Nutrient Classes and Their Principal Functions	
Nutrient	**Daily Requirement**	**Representative Functions**
Water	2.5 L	Solvent; coolant; reactant or product in many metabolic reactions (especially hydrolysis and condensation); dilutes and eliminates metabolic wastes; supports blood volume and pressure
Carbohydrates	125–175 g	Fuel; a component of nucleic acids, ATP and other nucleotides, glycoproteins, and glycolipids
Lipids	80–100 g	Fuel; plasma membrane structure; myelin sheaths of nerve fibers; hormones; eicosanoids; bile salts; insulation; protective padding around organs; absorption of fat-soluble vitamins; vitamin D synthesis; some blood-clotting factors
Proteins	44–60 g	Muscle contraction; ciliary and flagellar motility; structure of cellular membranes and extracellular material; enzymes; major component of connective tissues; transport of plasma lipids; some hormones; oxygen binding and transport; blood-clotting factors; blood viscosity and osmolarity; antibodies; immune recognition; neuromodulators; buffers; emergency fuel
Minerals	0.05–3,300 mg	Structure of bones and teeth; component of some structural proteins, hormones, ATP, phospholipids, and other chemicals; cofactors for many enzymes; electrolytes; oxygen transport by hemoglobin and myoglobin; buffers; stomach acid; osmolarity of body fluids
Vitamins	0.002–60 mg	Coenzymes for many metabolic pathways; antioxidants; component of visual pigment; one hormone (vitamin D)

When a chemical is described as **fuel** in this chapter, we mean it is oxidized solely or primarily to extract energy from it. The extracted energy is usually used to make adenosine triphosphate (ATP), which then transfers the energy to other physiological processes (see fig. 3.25).

Nutrients

A **nutrient** is any ingested chemical that is used for growth, repair, or maintenance of the body. Nutrients fall into six major classes: water, carbohydrates, lipids, proteins, minerals, and vitamins (table 26.1). Water, carbohydrates, lipids, and proteins are considered **macronutrients** because they must be consumed in relatively large quantities. Minerals and vitamins are called **micronutrients** because they are required in small quantities.

Recommended daily allowances (RDAs) of nutrients were first developed in 1943 by the National Research Council and National Academy of Sciences; they have been revised several times since. An RDA is a liberal but safe estimate of the daily intake that would meet the nutritional needs of most healthy people. Consuming less than the RDA of a nutrient does not necessarily mean you will be malnourished, but the probability of malnutrition increases in proportion to the amount of the deficit and how long it lasts.

Many nutrients can be synthesized by the body when they are unavailable from the diet. The body, however, is incapable of synthesizing minerals, most vitamins, eight or nine of the amino acids, and one or two of the fatty acids. These are called **essential nutrients** because it is essential that they be included in the diet.

Carbohydrates

A well-nourished adult has about 375 to 475 g of carbohydrate in the body, most of it in three places: about 325 g of muscle glycogen, 90 to 100 g of liver glycogen, and 15 to 20 g of blood glucose.

Sugars function as a structural component of other molecules including nucleic acids, glycoproteins, glycolipids, ATP, and related nucleotides (GTP, cAMP, etc.), and they can be converted to amino acids and fats. Most of the body's carbohydrate, however, serves as fuel—an easily oxidized source of chemical energy. Most cells meet their energy needs from a combination of carbohydrates and fats, but some cells, such as neurons and erythrocytes, depend almost exclusively on carbohydrates. Even a brief period of **hypoglycemia**[3] (deficiency of blood glucose) causes nervous system disturbances felt as weakness or dizziness.

Blood glucose concentration is therefore carefully regulated, mainly through the interplay of insulin and glucagon (see chapter 17 and later in this chapter). Among other effects, these hormones regulate the balance between glycogen and free blood glucose. If blood glucose concentration drops too low, the body draws on its stores of glycogen to meet its energy needs. If glycogen stores are depleted, physical endurance is greatly reduced. Thus it is important to consume enough carbohydrate to ensure that the body maintains adequate stores of glycogen for periods of exercise and fasting (including sleep).

3. *hypo* = below normal + *glyc* = sugar + *emia* = blood condition

Carbohydrate intake also influences the metabolism of other nutrients. Excess carbohydrate is converted to fat and conversely, fat is oxidized as fuel when glucose and glycogen levels are too low to meet our energy needs. This is why the consumption of starchy and sugary foods has a pronounced effect on body weight. It is unwise, however, to try to "burn off fat" by excessively reducing carbohydrate intake. As shown later in this chapter, the complete and efficient oxidation of fats depends on adequate carbohydrate intake and the presence of certain intermediates of carbohydrate metabolism. If these are lacking, fats are incompletely oxidized to ketone bodies, which may cause metabolic acidosis.

Requirements

Because carbohydrates are rapidly oxidized, they are required in greater amounts than any other nutrient. The RDA is 125 to 175 g. The brain alone consumes about 120 g of glucose per day. Most Americans get about 40% to 50% of their calories from carbohydrates, but highly active people should get up to 60% from this source.

Carbohydrate consumption in the United States has become quite excessive over the past century due to a combination of fondness for sweets, increased use of sugar in processed foods, and reduced levels of physical activity (see special topic 26.1). A century ago, Americans consumed an average of 1.8 kg (4 lb) of sugar per year. Now, with sucrose and high-fructose corn syrup so widely used in foods and beverages, the average American ingests 200 to 300 g of carbohydrate per day, and the equivalent of 27 kg (60 lb) of table sugar and 21 kg (46 lb) of corn syrup per year. A single nondiet soft drink contains 38 to 43 g (about 8 teaspoons) of sugar per 355 mL (12 oz) serving.

Dietary carbohydrates come in three principal forms discussed in earlier chapters: monosaccharides, disaccharides, and polysaccharides (complex carbohydrates). The only nutritionally significant polysaccharide is starch. Although glycogen is in this category, only trivial amounts of it are present in cooked meats. Cellulose, another polysaccharide, is not considered a nutrient because it is not digested and never enters the human tissues. Its importance as dietary fiber, however, is discussed shortly.

The three disaccharides are sucrose, lactose, and maltose. The monosaccharides—glucose, galactose, and fructose—arise mainly from the digestion of starch and disaccharides. The small intestine and liver convert fructose and galactose to glucose, so ultimately all carbohydrate digestion generates glucose. Outside of the hepatic portal system, glucose is the only monosaccharide present in the blood in significant quantity; thus it is known as *blood sugar*. Its concentration is normally maintained at 70 to 110 mg/dL in peripheral venous blood.

Think About It
Glucose concentration is about 15 to 30 mg/dL higher in arterial blood than in venous blood. Explain why.

Ideally, most carbohydrate intake should be in the form of complex carbohydrates, primarily starch. This is partly because foods that provide these also usually provide other nutrients. Simple sugars not only provide empty calories but also promote tooth decay. A typical American, however, now obtains only 50% of his or her carbohydrates from starch and the other 50% from sucrose and corn syrup.

Dietary Sources

Nearly all dietary carbohydrates come from plants—particularly grains, legumes, fruits, and root vegetables. Sucrose is refined from sugarcane and sugar beets. Fructose is present in fruits, maltose is present in some foods such as germinating cereal grains, and lactose is the most abundant solute in cow's milk (milk is about 4.6% lactose by weight).

Fiber

Dietary fiber refers to all fibrous materials of plant and animal origin that resist digestion. Most is plant matter—the carbohydrates cellulose and pectin and such noncarbohydrates as gums and lignin. Although it is not a nutrient, fiber is an essential component of the diet. The recommended daily allowance is about 30 g, but average intake varies greatly from country to country—from 40 to 150 g/day in India and Africa to only 12 g/day in the United States.

Fiber in the intestines absorbs water, swells, softens the stool, and increases its bulk by 40% to 100%. The last effect stretches the colon and stimulates peristalsis, thereby quickening the passage of feces and removal of carcinogenic chemicals from the colon. Fiber itself binds and carries away toxins and other undesirable chemicals, thus reducing the risk of colon cancer.

Pectin is a **water-soluble fiber** found in oats, beans, peas, carrots, brown rice, and fruits. Soluble fiber reduces blood cholesterol and low-density lipoprotein (LDL) levels. Cellulose, hemicellulose, and lignin, called **water-insoluble fibers,** apparently have no effect on cholesterol or LDLs. Supplementing the diet with oat bran, high in soluble fiber, significantly reduces cholesterol and LDL levels, whereas a comparable supplement with wheat bran, high in cellulose, has no such effect.

While a certain amount of fiber is beneficial to the health, you should not ingest too much. Excess fiber interferes with the absorption of iron, calcium, magnesium, phosphorus, and some trace elements.

Lipids

By weight, the adult body averages about 15% fat in males and 25% fat in females. Fat accounts for most of the body's stored energy. Lesser amounts of phospholipid, cholesterol, and other lipids also play vital structural and physiological roles.

A well-nourished adult meets 80% to 90% of his or her resting energy needs from fat. Fat is superior to carbohydrates for energy storage for two reasons: (1) carbohydrates are hydrophilic, absorb water, and thus expand and occupy more space in the tissues. Fat, however, is hydrophobic, contains almost no water, and is a more compact energy storage substance. (2) Fat is less oxidized than carbohydrate and contains over twice as much energy (9 kcal/g of fat compared with 4 kcal/g of carbohydrate). A man's typical fat reserves contain enough energy for 119 hours of running, whereas his carbohydrate stores would suffice for only 1.6 hours.

Fat has **glucose-sparing** and **protein-sparing effects**—as long as enough fat is available to meet the energy needs of the tissues, protein is not catabolized for fuel and glucose is spared for consumption by cells that cannot use fat, such as neurons.

Vitamins A, D, E, and K are fat-soluble vitamins, which depend on dietary fat for their absorption by the intestine. People who ingest less than 20 g of fat per day are at risk of vitamin deficiency because there is not enough fat in the intestine to transport these vitamins into the body tissues.

Phospholipids and cholesterol are major structural components of plasma membranes and myelin. Cholesterol is also important as a precursor of steroid hormones, bile salts, and vitamin D. Thromboplastin, an essential blood-clotting factor, is a lipoprotein. Prostaglandins and other eicosanoids are derived from a fatty acid, linoleic acid.

In addition to these metabolic effects, fat has important protective and insulating roles described for adipose tissue in chapter 6.

Requirements

The average adult needs 80 to 100 g of dietary fat per day. Fat should account for no more than 30% of your daily caloric intake; no more than 10% of your fat intake should be saturated fat; and average cholesterol intake should not exceed 300 mg/day (the amount in one egg yolk). A typical American consumes 30 to 150 g of fat per day, obtains 40% to 50% of his or her calories from fat, and ingests twice as much cholesterol as the recommended limit.

Most fatty acids can be synthesized by the body, but linoleic and linolenic acids cannot. They must be present in the diet and are therefore called **essential fatty acids.**

Sources

Saturated fats are predominantly of animal origin. They occur in meat, egg yolks, and dairy products but also in some plant products such as coconut and palm oil (common in nondairy coffee creamers and other products). Processed foods such as hydrogenated oils and vegetable shortening are also high in saturated fat, which is therefore abundant in many baked goods. Unsaturated fats predominate in nuts, seeds, and most vegetable oils. The essential fatty acids are amply provided by the vegetable oils present in mayonnaise, salad dressings, and margarine and by whole grains and vegetables.

The richest source of cholesterol is egg yolks, but it is also prevalent in milk products, shellfish (especially shrimp), organ meats such as kidneys, liver, and brains, and other mammalian meat. Cholesterol does not occur in foods of plant origin. It must be noted, however, that saturated fat stimulates cholesterol synthesis. A food may be truthfully advertised as cholesterol-free, but if it is high in saturated fat it can nevertheless raise your blood cholesterol level. Excessive consumption of saturated and unsaturated fats is a risk factor for diabetes mellitus, cardiovascular disease, and breast and colon cancer.

Table 26.2

Table 26.2		The Major Classes of Serum Lipoproteins, Listed in Order of Increasing Density				
		Percent Composition				
Type	Size (nm)	Protein	Total Lipid	Cholesterol	Triglyceride	Phospholipid
Chylomicrons	75–1,000	2	98	5	90	3
VLDLs	30–80	8	92	20	55	17
LDLs	20	20	80	53	6	21
HDLs	7.5–10.0	50	50	20	5	25

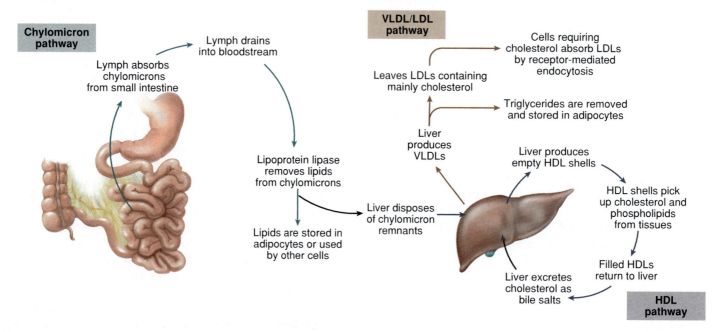

Figure 26.2 Pathways of lipoprotein processing.

Cholesterol and Serum Lipoproteins

Lipids are an important part of the diet and must be transported to all cells of the body, yet they are hydrophobic and will not dissolve in the aqueous blood plasma. This problem is overcome by complexes called **lipoproteins**—tiny droplets with a core of cholesterol and triglycerides and a coating of proteins and phospholipids. The coating not only enables the lipids to remain suspended in the blood but also serves as a recognition marker for cells that absorb them.

Lipoproteins are classified into four major categories (and some lesser ones) by their density: **chylomicrons, high-density lipoproteins (HDLs), low-density lipoproteins (LDLs),** and **very low-density lipoproteins (VLDLs)**. In table 26.2, notice that the higher the proportion of protein to lipid, the higher the density. (What other trends do you see correlated with their density?) Lipoproteins vary not only in size and density but more importantly in composition and function. Figure 26.2

shows the three primary pathways by which they are made and processed.

Chylomicrons form in the absorptive cells of the small intestine and then pass into the lymphatic system and ultimately the bloodstream (see chapter 25). Endothelial cells of the blood capillaries have a surface enzyme called **lipoprotein lipase** that hydrolyzes triglycerides into glycerol and free fatty acids (FFAs). These products can then pass through the capillary walls into adipocytes, where they are recondensed into storage triglycerides. Some FFAs, however, remain in the blood plasma bound to albumin. The remainder of a chylomicron after the triglycerides have been extracted, called a *chylomicron remnant,* is removed and degraded by the liver.

VLDLs, produced by the liver, transport lipids to the adipose tissue for storage. Lipoprotein lipase removes the triglycerides, converting VLDLs to LDLs, which contain mostly cholesterol. Cells that need cholesterol (usually for membrane structure or steroid hormone synthesis) have membrane receptors for LDLs.

Some bodybuilders and other "power athletes" take powdered or liquid amino acid mixtures ("predigested protein") in the belief that simple amino acids are absorbed more easily and rapidly or that they somehow contribute more to muscle building. Such beliefs are un- founded. Dietary proteins are rapidly digested and absorbed, and there is no added benefit to taking predigested protein. Amino acid supplements have not been shown to increase muscle mass, strength, or endurance. More- over, they may be harmful to the health. Con- centrated amino acid solutions osmotically re- tain water in the intestines and cause cramps and diarrhea. The catabolism of excess amino acids produces extra nitrogenous waste, which places undesirable stress on the liver and kidneys.

They absorb LDLs by receptor-mediated endocytosis, di- gest them with their lysosomes, and release the choles- terol for intracellular use.

HDL production begins in the liver, which pro- duces and releases an empty, collapsed, protein shell. This shell travels in the blood and picks up cholesterol and phospholipids from other organs. The next time it circulates through the liver, the liver removes the cho- lesterol and converts it to bile salts. HDLs are therefore a vehicle for removing excess cholesterol from the body.

It is essential to maintain serum cholesterol within homeostatic limits, but it is not yet universally agreed what a desirable cholesterol level should be. Some au- thorities feel it should not exceed 200 mg/dL, while oth- ers feel levels of 230 to 240 mg/dL are normal and acceptable. Excess cholesterol is absorbed by macrophages, which then become the *foam cells* seen in atherosclerotic blood vessels (see p. 699). A deficiency of cholesterol is as dangerous as an excess, however. At concentrations less than 190 mg/dL for men and 178 mg/dL for women, there is increased incidence of hemorrhages and strokes.

Most of the body's cholesterol is endogenous (inter- nally synthesized) rather than dietary, and to some extent the body compensates for variations in dietary intake. High intake somewhat inhibits hepatic cholesterol syn- thesis. However, the liver synthesizes a certain amount of cholesterol regardless of intake, and severe restriction of dietary cholesterol does not necessarily result in a propor- tionate drop in blood cholesterol. Dietary fatty acids also strongly influence cholesterol levels. A moderate decrease in saturated fatty acids can lower blood cholesterol by 15% to 20%, while unsaturated fatty acids reduce blood cholesterol by enhancing its catabolism and excretion.

Blood cholesterol is not the only important mea- sure of healthy lipid concentrations, however. A high LDL concentration is a warning sign because, as you can see from the function of LDLs described previously, it signifies a high rate of cholesterol deposition in the ar- teries. Factors that contribute to high LDL levels include not only saturated fats but also cigarette smoking, cof- fee, and stress. A high proportion of HDL, on the other hand, is beneficial because it indicates that cholesterol is being removed from the arteries and transported to the liver for disposal. Thus it is desirable to increase your ratio of HDL to LDL. This is best done with a diet low in calories and saturated fats and is promoted by regular aerobic exercise.

Proteins

Protein constitutes about 12% to 15% of the body's mass; 65% of it is in the skeletal muscles. Proteins are responsi- ble for muscle contraction and the motility of cilia and fla- gella. They are a major structural component of all cellular membranes, with multiple important roles such as mem- brane receptors, pumps, ion channels, and cellular recog- nition markers. Fibrous proteins such as collagen, elastin, and keratin make up much of the structure of bone, carti- lage, tendons, ligaments, skin, hair, and nails. Globular proteins include antibodies, hormones, neuromodulators, hemoglobin, myoglobin, and about 2,000 enzymes that control nearly every aspect of cellular metabolism. They also include the albumin and other plasma proteins that maintain blood viscosity and osmolarity and transport lipids and some other plasma solutes. Proteins buffer the intra- and extracellular pH and contribute to the resting membrane potentials of all cells. No other class of biomol- ecules has such a broad variety of functions.

Requirements

For persons of average weight, the RDA of protein is 44 to 60 g, depending on age and sex. Multiplying your weight in pounds by 0.37 or your weight in kilograms by 0.8 gives an estimate of your RDA of protein in grams. A higher intake is recommended, however, under conditions of stress, infection, injury, and preg- nancy. Infants and children require more protein than adults relative to body weight.

Some athletes ingest large protein supplements in the belief that they will promote the building of mus- cles, but there is no evidence that this has any benefit and indeed more evidence that it can be harmful (see special topic 26.2). The excess protein is either burned for energy or converted to fat.

Protein Quality   Total protein intake is not the only significant measure of dietary adequacy. The nutritional value of a protein depends on whether it supplies the right amino acids in the proportions needed to construct human proteins. Adults can synthesize 12 of the 20 amino acids from other organic compounds when they are not available from the diet, but there are **8 essential amino acids** that we cannot synthesize: isoleucine, leucine, lysine, methionine, phenylalanine, threonine, tryptophan, and valine. (Infants also require histidine.) In addition, there are 2 amino acids that can only be synthesized from essential amino acids—cystine from methionine and tyrosine from phenylalanine. The other 10 (9 in infants) are called **inessential amino acids**—not because the body does not require them but because it can synthesize its own when the diet does not supply them.

Cells do not store surplus amino acids for later use. When a protein is to be synthesized, all of the amino acids necessary must be present at once, and if even one is missing, the protein cannot be made. High-quality **complete proteins** are those that provide all of the essential amino acids in the necessary proportions for human tissue growth, maintenance, and nitrogen balance. Lower-quality **incomplete proteins** lack one or more essential amino acids. For example, cereals are low in lysine, and legumes are low in methionine.

Protein quality is also determined by **net protein utilization**—the percent of the amino acids in a protein that the human body uses. We typically use 70% to 90% of animal protein but only 40% to 70% of plant protein. It therefore takes a larger serving of plant protein than animal protein to meet our needs—for example, we need 400 g (about 14 oz) of rice and beans to provide as much usable protein as 115 g (about 4 oz) of hamburger. However, reducing meat intake and increasing plant intake has advantages. Among other considerations, plant foods provide more vitamins, minerals, and fiber; less saturated fat; no cholesterol; and fewer pesticides. In an increasingly crowded world, it must also be borne in mind that it requires far more land to produce meat than to produce plant crops.

Dietary Sources

The animal proteins of meat, eggs, and dairy products closely match human proteins in amino acid composition. Thus animal products provide high-quality complete protein, whereas plant proteins are incomplete. Nevertheless, this does not mean that your dietary protein *must* come from meat; indeed, about two-thirds of the world's population receives adequate protein nutrition from diets containing very little meat. We can combine plant foods so that one provides what another lacks—beans and rice, for example, are a complementary combination of legume and cereal. Beans provide the isoleucine and lysine lacking in grains, while rice provides the tryptophan and cysteine lacking in beans.

An exclusively vegetarian diet cannot meet all of your nutritional needs unless supplemented with vitamin B_{12} and essential amino acids. Supplementing a vegetarian diet with milk and one egg a day (an *ovolactovegetarian* diet), however, can provide an adequate intake of complete protein.

Nitrogen Balance   Proteins are our chief dietary source of nitrogen. **Nitrogen balance** is a state in which the rate of nitrogen ingestion equals the rate of excretion (chiefly as nitrogenous wastes). Growing children exhibit a state of **positive nitrogen balance** because they ingest more than they excrete, thus retaining protein for tissue growth. Pregnant women and athletes in resistance training also show positive nitrogen balance. When excretion exceeds ingestion, a person is in a state of **negative nitrogen balance.** This indicates that body proteins are being broken down and used as fuel. Proteins of the muscles and liver are more easily broken down than others; thus negative nitrogen balance tends to be associated with muscle atrophy. Negative nitrogen balance may occur if carbohydrate and fat intake are insufficient to meet the need for energy. Carbohydrates and fats are said to have a protein-sparing effect because they prevent protein catabolism when present in sufficient amounts to meet energy needs.

Nitrogen balance is affected by some hormones. Growth hormones and sex steroids promote protein synthesis and positive nitrogen balance during childhood, adolescence, and pregnancy. Glucocorticoids, on the other hand, promote protein catabolism and negative nitrogen balance in states of stress.

Think About It
Would you expect a person recovering from a long infectious disease to be in a state of positive or negative nitrogen balance? Why?

Minerals and Vitamins

Minerals are inorganic elements and vitamins are small organic compounds. Neither is used as fuel, but both are essential to our ability to use other nutrients. With the exception of a few vitamins, these nutrients cannot be synthesized by the body and must be included in the diet. They are, however, required in relatively small quantities. Mineral RDAs range from 0.05 mg of chromium to 3,300 mg of sodium. Vitamin RDAs range from about 0.002 mg of vitamin B_{12} to 60 mg of vitamin C. Despite the small quantities involved, minerals and vitamins have very potent effects on physiology. Indeed, excessive amounts are toxic and potentially lethal.

Table 26.3 Mineral Requirements and Some Dietary Sources*

Mineral	RDA (mg)	Some Dietary Sources
Major Minerals		
Sodium	3,300	Table salt, processed foods; usually present in excess
Calcium	800	Milk, fish, shellfish, greens, tofu, orange juice
Phosphorus	800	Red meat, poultry, fish, eggs, milk, legumes, whole grains, nuts
Chloride	700	Table salt, some vegetables; usually present in excess
Magnesium	Men: 350	Milk, greens, whole grains, nuts, legumes, chocolate
	Women: 280	
Potassium	Unknown	Red meat, poultry, fish, cereals, spinach, squash, bananas, apricots
Sulfur	Unknown	Meats, milk, eggs, legumes; almost any proteins
Trace Minerals		
Zinc	12–15	Red meat, seafood, cereals, wheat germ, legumes, nuts, yeast
Iron	Men: 10	Red meat, liver, shellfish, eggs, dried fruits, legumes, nuts, molasses
	Women: 15	
Manganese	2.5–5.0	Greens, fruits, legumes, whole grains, nuts
Copper	1.5–3.0	Red meat, liver, shellfish, legumes, whole grains, nuts, cocoa
Fluoride	1.5–4.0	Fluoridated water and toothpaste, tea, seafood, seaweed
Iodine	0.15	Marine fish, fish oils, shellfish, iodized salt
Molybdenum	0.07–0.25	Beans, whole grains, nuts
Chromium	0.05–2.0	Meats, liver, cheese, eggs, whole grains, yeast, wine
Selenium	0.05–0.07	Red meats, organ meats, fish, shellfish, eggs, cereals
Cobalt	Unknown	Red meat, poultry, fish, liver, milk

*The RDAs given here are for adult males unless otherwise stated; RDAs for females are usually the same or slightly lower. A range is given where a precise RDA has not been established. *Red meat* refers to mammalian muscle such as beef and pork. *Organ meat* refers to brain, pancreas, heart, kidney, etc. Liver is specified separately and refers to beef, pork, and chicken livers, which are similar for most nutrients.

Minerals

Minerals constitute about 4% of the body mass, with three-quarters of this being the calcium and phosphorus in the bones and teeth. Phosphorus is also a key structural component of phospholipids, ATP, cAMP, GTP, and creatine phosphate and is the basis of the phosphate buffer system (see chapter 24). Calcium, iron, magnesium, and manganese function as cofactors for enzymes. Iron is essential to the oxygen-carrying capacity of hemoglobin and myoglobin. Chlorine is a component of stomach acid (HCl). Many mineral salts function as electrolytes and thus govern the function of nerve and muscle cells, osmotically regulate the content and distribution of water in the body, and maintain blood volume.

Table 26.3 summarizes adult mineral requirements and dietary sources. Broadly speaking, the best sources of minerals are vegetables, legumes, milk, eggs, fish, shellfish, and some meats. Cereals are a relatively poor source.

Sodium chloride has been both a prized commodity and a curse. Animal tissues contain relatively large amounts of salt, and carnivores rarely lack ample salt in their diets. Plants, however, are relatively poor in salt and herbivores often must supplement their diet by ingesting salt from the soil. As humans developed agriculture and became more dependent on plants, they also became increasingly dependent on supplemental salt. Salt has often been used as a form of payment for goods and services—the word *salary* comes from *sal* (salt). Our fondness for salt and high sensitivity to it undoubtedly stem from its physiological importance and its scarcity in a largely vegetarian diet.

Now, however, this fondness has become a bane. The recommended sodium intake is 1.1 to 3.3 g/day, but a typical American diet contains about 4.5 g/day. This is due not just to the use of table salt but more significantly to the large amounts of salt in processed foods, much of it "disguised" as soy sauce, MSG (monosodium glutamate), baking soda, and baking powder. In some areas of Japan, salt intake averages 27 g/day and the great majority of people die before age 70 of stroke and other complications of hypertension. Hypertension is a leading cause of death among

The notorious link between sodium and hypertension is an especially serious medical problem for American blacks. They have twice the risk of hypertension and 10 times the risk of dying from it than American whites do. While they constitute about 10% of the U.S. population, blacks account for two-thirds of cases of hypertensive kidney failure. The reason is not excessive salt consumption; there is no significant difference in average salt consumption between the races. Rather, the kidneys of American blacks retain more salt than do the kidneys of whites.

One hypothesis traces this to a combination of evolution and slavery. Most American blacks are descended from slaves from West Africa—a hot climate where salt loss through profuse sweating would favor the evolution of highly effective salt-retaining mechanisms.

Added to this is the fact that about 70% of African captives died during their captivity, transport, and first 2 or 3 years in the Americas. Much of this mortality was related to salt loss through sweating, vomiting from

seasickness, and diarrhea stemming from the crowded, unsanitary conditions in the cargo holds of slave ships. People with highly efficient, salt-retaining kidneys would have been the most likely to survive and to have passed that trait to their descendents. Combine this with the high salt content of today's average American diet, and we can see plausible historical reason why hypertension so disproportionately affects American blacks today.

Table 26.4 — Vitamin Requirements and Some Dietary Sources

Vitamin	RDA (mg)	Some Dietary Sources
Water-Soluble Vitamins		
Ascorbic acid (C)	60	Citrus fruits, strawberries, tomatoes, greens, cabbage, cauliflower, broccoli, Brussels sprouts
B complex		
Thiamine (B_1)	1.5	Red meat, organ meats, liver, eggs, greens, asparagus, legumes, whole grains, seeds, yeast
Riboflavin (B_2)	1.7	Widely distributed, and deficiencies are rare; all types of meat, milk, eggs, greens, whole grains, apricots, legumes, mushrooms, yeast
Pyridoxine (B_6)	2.0	Red meat, organ meats, fish, liver, greens, apricots, legumes, whole grains, seeds
Cobalamin (B_{12})	0.002	Red meat, organ meats, liver, shellfish, eggs, milk; absent from food plants
Niacin (nicotinic acid)	19	Readily synthesized from tryptophan, which is present in any diet with adequate protein; red meat, organ meats, liver, poultry, fish, apricots, legumes, whole grains, mushrooms
Pantothenic acid	4–7	Widely distributed, and deficiencies are rare; red meat, organ meats, liver, eggs, green and yellow vegetables, legumes, whole grains, mushrooms, yeast
Folic acid (folacin)	0.2	Eggs, liver, greens, citrus fruits, legumes, whole grains, seeds
Biotin	0.03–0.10	Red meat, organ meats, liver, eggs, cheese, cabbage, cauliflower, bananas, legumes, nuts
Fat-Soluble Vitamins		
Vitamin A (retinol)	1.0	Fish oils, eggs, cheese, milk, greens, other green and yellow vegetables and fruits, margarine
Vitamin D (calcitriol)	0.01	Formed by exposure of skin to sunlight; fish, fish oils, milk
Vitamin E (α-tocopherol)	10	Fish oils, greens, seeds, wheat germ, vegetable oils, margarine, nuts
Vitamin K (phylloquinone)	0.08	Most of RDA is met by synthesis by intestinal flora; liver, greens, cabbage, cauliflower

American blacks, perhaps an unfortunate legacy of the institution of slavery (see special topic 26.3).

Vitamins

Vitamins were originally named with letters in the order of their discovery, but they also have chemically descriptive names such as ascorbic acid (vitamin C) and riboflavin (vitamin B_2). Most vitamins must be obtained from the diet (table 26.4), but the body synthesizes some

of them from precursors called *provitamins*—niacin from the amino acid tryptophan, vitamin D from cholesterol, and vitamin A from carotene, which is abundantly present in carrots, squash, and other yellow vegetables and fruits. Vitamin K, pantothenic acid, biotin, and folic acid are produced by the bacterial flora of the large intestine. The feces contain more biotin than food does.

Vitamins are classified as water-soluble or fat-soluble. *Water-soluble vitamins* are absorbed with water from the small intestine, dissolve freely in the body fluids, and are readily excreted by the kidneys. They cannot be stored in the body and therefore seldom accumulate to excess. The water-soluble vitamins are ascorbic acid (vitamin C) and the B vitamins. Ascorbic acid promotes collagen synthesis and sound connective tissue structure, and it is an antioxidant that scavenges free radicals and possibly reduces the risk of cancer. The B vitamins function as coenzymes or parts of coenzyme molecules; they assist enzymes by transferring electrons from one metabolic reaction to another, making it possible for enzymes to catalyze these reactions. Some of their functions arise later in this chapter as we consider carbohydrate metabolism.

Fat-soluble vitamins are incorporated into lipid micelles in the small intestine and absorbed with dietary lipids. They are more varied in function than water-soluble vitamins. Vitamin A is a component of the visual pigments and promotes mucopolysaccharide synthesis and epithelial maintenance. Vitamin D promotes calcium absorption and bone mineralization. Vitamin K is essential to prothrombin synthesis and blood clotting. Vitamins A and E are antioxidants, like ascorbic acid.

It is common knowledge that various diseases result from vitamin deficiencies, but it is less commonly known that **hypervitaminosis** (vitamin excesses) also causes disease. A *deficiency* of vitamin A, for example, can result in night blindness, dry skin and hair, a dry conjunctiva and cloudy cornea, and increased incidence of urinary, digestive, and respiratory infections. This is the world's most common vitamin deficiency. An *excess* of vitamin A, however, may cause anorexia, nausea and vomiting, headache, pain and fragility of the bones, hair loss, and an enlarged liver and spleen. Toxic hypervitaminosis is also known for vitamins B_6, C, D, E, and niacin.

Some athletes take *megavitamins*—doses 10 to 1,000 times the RDA—thinking that they will improve performance. Since vitamins are not burned as fuel, and small amounts fully meet the body's metabolic needs, there is no evidence that vitamin supplements improve performance except when used to correct a dietary deficiency. Indeed, megadoses of fat-soluble vitamins can be extremely harmful.

Key Point Review

1. What regions of the hypothalamus regulate hunger and satiety? Why would it be wrong to say these are the sole controls over appetite?

2. Explain the following statement: Cellulose is an important part of a healthy diet but it is not a nutrient.

3. What class of nutrients provides most of the calories in the diet? What class of nutrients provides the body's major reserves of stored energy?

4. Contrast the functions of VLDLs, LDLs, and HDLs. Explain how this is related to the fact that a high blood HDL level is desirable, but a high VLDL-LDL level is undesirable?

5. Why do some proteins have more nutritional value than others?

Carbohydrate Metabolism

▼**Objectives**

When you have completed this section, you should be able to
- describe the principal reactants and products at each major step of glucose oxidation;
- contrast the purposes and products of anaerobic fermentation and aerobic respiration;
- explain where and how cells produce ATP; and
- describe the production, function, and use of glycogen.

Most dietary carbohydrate is burned as fuel within a few hours of absorption. Although three monosaccharides are absorbed from digested food—glucose, galactose, and fructose—the last two are quickly converted to glucose, and all oxidative carbohydrate consumption is essentially a matter of glucose catabolism. The overall reaction for this is

$$C_6H_{12}O_6 + 6\ O_2 \rightarrow 6\ CO_2 + 6\ H_2O$$

The purpose of this reaction is not to produce carbon dioxide and water but to transfer energy from glucose to ATP.

Along the pathway of glucose oxidation are several links through which other nutrients—especially fats and amino acids—can also be oxidized as fuel. Carbohydrate catabolism therefore provides a central vantage point from which we can view the catabolism of all fuels and the generation of ATP.

Glucose Catabolism

If the preceding reaction were carried out in a single step, it would generate a short, intense burst of heat—like the burning of paper, which has the same chemical equation. Not only would this be useless to the body's metabolism, it would kill the cells. In the body, however, the process is carried out in a series of small steps, each controlled by a separate enzyme. Energy is released in small manageable amounts, and as much as possible is transferred to ATP. The rest is released as heat.

There are three major pathways of glucose catabolism:

1. **glycolysis,** which splits a glucose molecule into two molecules of pyruvic acid;
2. **anaerobic fermentation,** which occurs in the absence of oxygen and reduces pyruvic acid to lactic acid; and
3. **aerobic respiration,** which occurs in the presence of oxygen and oxidizes pyruvic acid to carbon dioxide and water.

You may find it helpful to review figure 3.26 (see p. 101) for a broad overview of these processes and their relationship to ATP production. Figures 26.3 to

26.6 examine these processes in closer detail. The first two figures are labeled with numbers that correspond to reaction steps described shortly.

Coenzymes are vitally important to these reactions. Enzymes remove electrons (as hydrogen atoms) from the intermediate compounds of these pathways, but they do not bind them. Instead, they transfer the hydrogen atoms to coenzymes, and the coenzymes donate them to other compounds later in one of the reaction pathways. Thus the enzymes of glucose catabolism cannot function without their coenzymes.

The two coenzymes of special importance to glucose catabolism are **NAD+** (nicotinamide adenine dinucleotide) and **FAD** (flavin adenine dinucleotide). Both are derived from B vitamins: NAD+ from niacin and FAD from riboflavin. Hydrogen atoms are removed from metabolic intermediates in pairs—that is, two protons and two electrons (2 H+ and 2 e−) at a time—and transferred to a coenzyme. This produces a reduced coenzyme with a higher free energy content than it had before the reaction. Coenzymes thus become the temporary carriers of the energy extracted from glucose metabolites. The reactions for this are

$$FAD + 2\ H \rightarrow FADH_2$$
and
$$NAD^+ + 2\ H \rightarrow NADH + H^+$$

FAD binds two protons and two electrons to become $FADH_2$. NAD+, however, binds the two electrons but only one of the protons to become NADH. The other proton remains a free hydrogen ion, H+ (or H_3O^+, but it will be represented in this chapter as H+).

Glycolysis

Upon entering a cell, glucose begins a series of conversions called **glycolysis**[4] (fig. 26.3):

Step 1: phosphorylation. The enzyme *hexokinase* transfers an inorganic phosphate group (P_i) from ATP to glucose, producing glucose 6-phosphate (G6P). This has two effects:

- It keeps the intracellular concentration of glucose very low, thus maintaining a concentration gradient that favors the continued diffusion of more glucose into the cell.
- Phosphorylated compounds cannot pass through the plasma membrane, so this prevents glucose from leaving the cell. In most cells, step 1 is irreversible because they lack the enzyme to convert G6P back to glucose. The few exceptions are cells that must be able to release free glucose to the blood: absorptive cells of the small intestine, proximal convoluted tubule cells in the kidney, and liver cells.

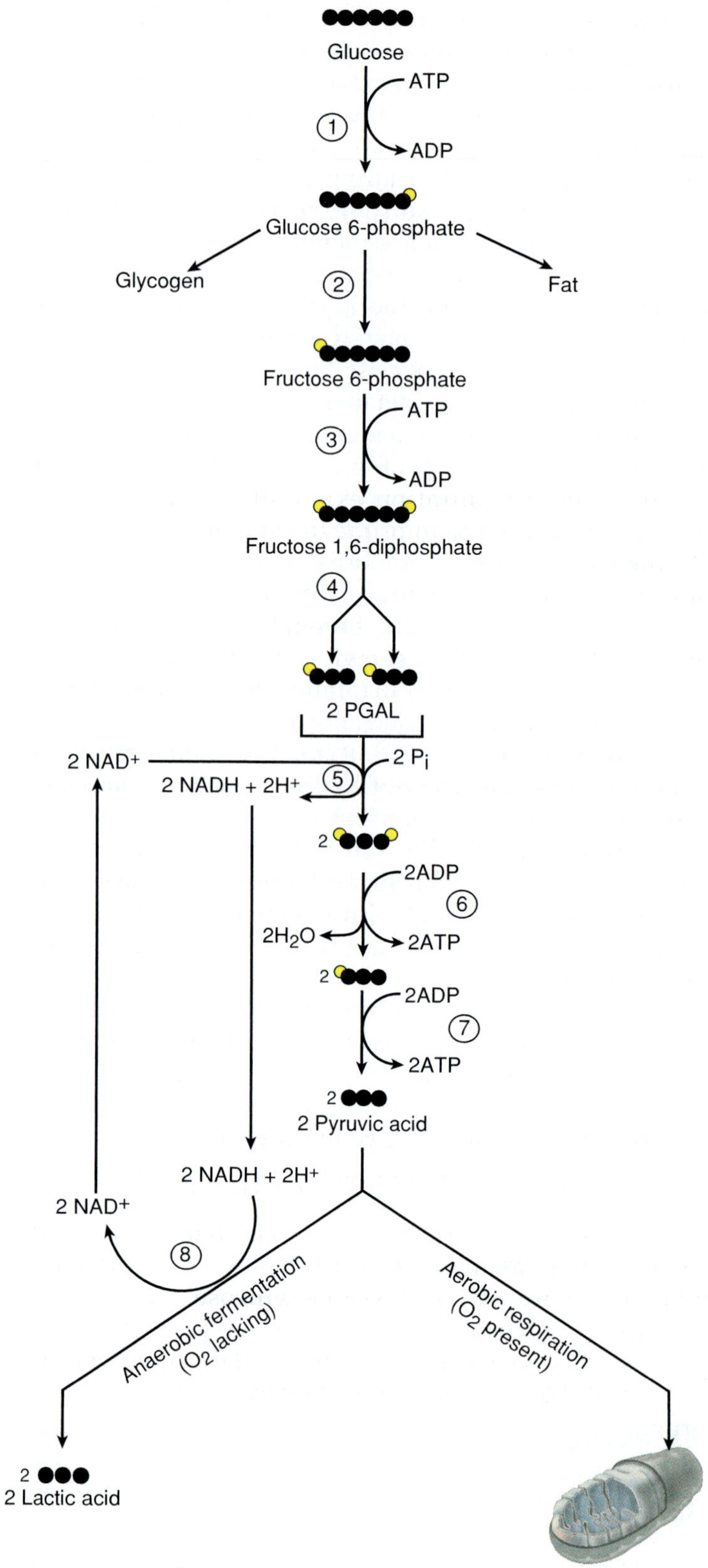

Figure 26.3 Glycolysis and anaerobic fermentation. The black circles represent carbon atoms in the carbon skeleton of each molecule; the yellow dots indicate phosphate groups. Numbered reaction steps are explained in the text.

4. *glyco* = sugar + *lysis* = splitting

G6P is a versatile molecule that can be converted to fat or amino acids, polymerized to form glycogen for storage, or further oxidized to extract its energy. For now, we are mainly concerned with its further oxidation (glycolysis), the general effect of which is to split G6P (a six-carbon sugar, C_6) into two three-carbon (C_3) molecules of **pyruvic acid (pyruvate).** Continue tracing these steps in figure 26.3 as you read.

Steps 2 and 3: priming. G6P is rearranged (isomerized) to form fructose 6-phosphate, which is phosphorylated again to form fructose 1,6-diphosphate. This "primes" glucose by providing activation energy, somewhat like the heat of a match used to light a fireplace. Two molecules of ATP have already been consumed, but just as a fire gives back more heat than it takes to start it, aerobic respiration eventually gives back far more ATP than it takes to prime the glucose.

Step 4: cleavage. The "lysis" part of glycolysis occurs when fructose 1,6-diphosphate is split into two three-carbon (C_3) molecules. Through a slight rearrangement of one of them (not shown in the figure), this generates two molecules of **PGAL (phosphoglyceraldehyde, or glyceraldehyde 3-phosphate).**

Step 5: oxidation. Each PGAL molecule is then oxidized by removing a pair of hydrogen atoms. The electrons and one proton are picked up by NAD^+ and the other proton is released into the cytosol, yielding NADH + H^+. At this step, a phosphate (P_i) group is also added to each of the C_3 fragments. Unlike the earlier steps, this P_i is not supplied by ATP but comes from the cell's pool of free phosphate ions.

Steps 6 and 7: dephosphorylation. In the next two steps, phosphate groups are taken from the glycolysis intermediates and transferred to ADP, phosphorylating it to ATP. This converts the C_3 compound to pyruvic acid. The end products of glycolysis are therefore

$$2 \text{ pyruvic acid} + 2 \text{ NADH} + 2 \text{ H}^+ + 2 \text{ ATP}$$

Note that 4 ATP are actually produced (steps 6–7), but 2 ATP were consumed to initiate glycolysis (steps 1 and 3), so the net gain is 2 ATP per glucose.

Some of the energy originally in the glucose is contained in this ATP, some is in the NADH, and some is lost as heat. Most of the energy, however, remains in the pyruvic acid.

Anaerobic Fermentation

The fate of pyruvic acid depends on whether or not oxygen is available. In an exercising muscle, the demand for ATP may exceed the supply of oxygen, and the only ATP the cells can make under these circumstances is the 2 ATP produced by glycolysis. Cells that lack mitochondria, such as erythrocytes, are also restricted to making ATP by this method.

But glycolysis would quickly come to a halt if the reaction stopped at pyruvic acid. Why? Because it would use up the supply of NAD^+, which is needed to accept electrons at step 5 and keep glycolysis going. NAD^+ must be replenished.

In the absence of oxygen, a cell resorts to a one-step reaction called anaerobic fermentation. (This is often inaccurately called *anaerobic respiration,* but strictly speaking, human cells do not carry out anaerobic respiration; that is a process found only in certain bacteria.) In this pathway (fig. 26.3, **step 8**), NADH denotes a pair of electrons to pyruvic acid, reducing it to **lactic acid** and regenerating NAD^+.

Think About It
Does lactic acid have more free energy than pyruvic acid or less? Why?

Lactic acid leaves the cells that generate it and travels by way of the bloodstream to the liver. When oxygen becomes available again, the liver oxidizes lactic acid back to pyruvic acid, which can then enter the aerobic pathway described shortly. The oxygen required to do this is part of the *oxygen debt* created by exercising skeletal muscles (see p. 415). The liver can also convert lactic acid back to G6P and can do either of two things with it: (1) polymerize it to form glycogen for storage or (2) remove the phosphate group and release free glucose into the blood.

Although anaerobic fermentation keeps glycolysis running a little longer, it has some drawbacks. One is that it is wasteful, because most of the energy of glucose is still in the lactic acid and has contributed no useful work. The other is that lactic acid is toxic and contributes to muscle fatigue.

Skeletal muscle is relatively tolerant of anaerobic fermentation, and cardiac muscle is less so. The brain employs almost no anaerobic fermentation. During birth, when the infant's blood supply is cut off, almost every organ of the body switches to anaerobic fermentation; thus they do not compete with the brain for the limited supply of oxygen.

Aerobic Respiration

Most ATP is generated in the mitochondria, which require oxygen as the final electron acceptor. In the presence of oxygen, pyruvic acid enters the mitochondria and is oxidized by aerobic respiration. This occurs in two principal steps:

- a group of reactions we will call the **matrix reactions,** because their controlling enzymes are in the fluid of the mitochondrial matrix; and
- reactions we will call the **membrane reactions,** because their controlling enzymes are bound to the membranes of the mitochondrial cristae.

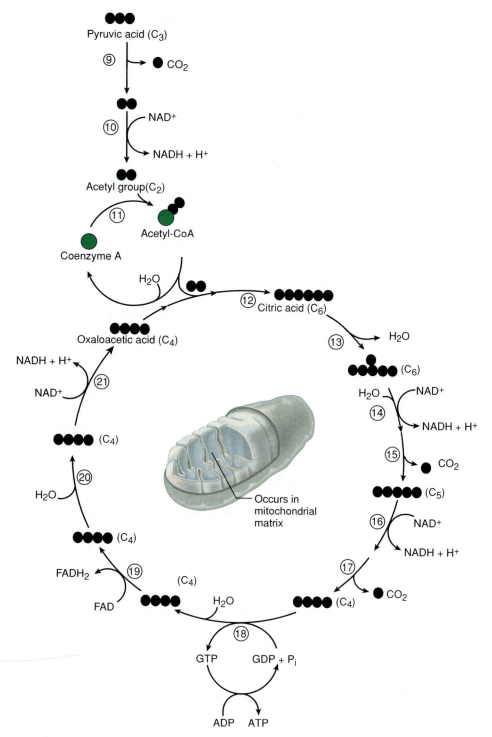

Pyruvic acid (C₃)

⑨ → CO₂

⑩ NAD⁺ → NADH + H⁺

Acetyl group(C₂)

⑪ Coenzyme A → Acetyl-CoA

H₂O

Oxaloacetic acid (C₄)

⑫ Citric acid (C₆)

⑬ → H₂O

(C₆)

H₂O → NAD⁺

⑭ → NADH + H⁺

⑮ → CO₂

(C₅)

⑯ NAD⁺ → NADH + H⁺

⑰ → CO₂

(C₄)

NADH + H⁺

NAD⁺

㉑

(C₄)

㉒ Occurs in mitochondrial matrix

H₂O

㉑

(C₄)

(C₄)

FADH₂

FAD

⑲

(C₄)

H₂O

⑱

GTP   GDP + Pᵢ

ADP   ATP

Figure 26.4 The mitochondrial matrix reactions. Black circles represent carbon atoms in the carbon skeleton of each molecule. Numbered reaction steps are explained in the text. [▭]

The Matrix Reactions

The matrix reactions are shown in figure 26.4, where the reaction steps are numbered to resume where figure 26.3 ended. Most of the matrix reactions constitute a se-

ries called the **citric acid (Krebs[5]) cycle.** Preceding this, however, are three steps that prepare pyruvic acid to enter the cycle and thus link glycolysis to it.

Step 9. Pyruvic acid is *decarboxylated*—CO_2 is removed and pyruvic acid (C_3) becomes a C_2 compound.

Step 10. Hydrogen atoms are removed (an oxidation reaction) and accepted by NAD^+. This converts the C_2 compound to an **acetyl group (acetic acid).**

Step 11. The acetyl group binds to coenzyme A, a derivative of pantothenic acid (a B vitamin). The result is **acetyl-coenzyme A (acetyl-CoA).** At this stage the C_2 remnant of the original glucose molecule is ready to enter the citric acid cycle.

Step 12. At the beginning of the citric acid cycle, CoA hands off the acetyl (C_2) group to a C_4 compound, **oxaloacetic acid.** This produces the C_6 compound **citric acid,** for which the cycle is named.

Step 13. Water is removed and the citric acid molecule is reorganized but still retains its six carbon atoms.

Step 14. Hydrogen atoms are removed and accepted by NAD^+.

Step 15. Another CO_2 is removed and the substrate becomes a five-carbon chain.

Steps 16 and 17. Processes 14 and 15 are essentially repeated, generating another free CO_2 molecule and leaving a four-carbon chain. No more carbon atoms are removed beyond this point; the substrate remains a series of C_4 compounds from here back to the start of the cycle. The three carbon atoms of pyruvic acid have all been removed as CO_2 at steps 9, 15, and 17. These *decarboxylation reactions* are the source of most of the CO_2 in your breath.

Step 18. Some of the energy in the C_4 substrate goes to phosphorylate guanosine diphosphate (GDP), converting it to guanosine triphosphate (GTP), a molecule similar to ATP. GTP quickly transfers the P_i group to ADP to make ATP. Coenzyme A participates again in this step but is not shown in the figure.

---

5. Sir Hans Krebs (1900–1981), German biochemist

Step 19. Two hydrogen atoms are removed and accepted by the coenzyme FAD.

Step 20. Water is added.

Step 21. A final two hydrogen atoms are removed and transferred to NAD^+. This reaction generates oxaloacetic acid, which is available to start the cycle all over again.

It is important to remember that for every glucose molecule that entered glycolysis, all of these matrix reactions occur twice (once for each pyruvic acid). The matrix reactions can be summarized:

$$
\begin{array}{rcl}
2 \text{ pyruvate} + 6 \text{ } H_2O & \rightarrow & 6 \text{ } CO_2 \\
+ 2 \text{ ADP} + 2 \text{ } P_i & \rightarrow & 2 \text{ ATP} \\
+ 8 \text{ NAD} + 8 \text{ } H_2 & \rightarrow & 8 \text{ NADH} + 8 \text{ } H^+ \\
+ 2 \text{ FAD} + 2 \text{ } H_2 & \rightarrow & 2 \text{ } FADH_2
\end{array}
$$

There is nothing left of the organic matter of the glucose anymore; its carbon atoms have all been carried away as CO_2 and exhaled. Although still more of its energy is lost as heat along the way, some is stored in the additional 2 ATP, and most of it, by far, is in the reduced coenzymes—8 NADH and 2 $FADH_2$ molecules generated by the matrix reactions, and 2 NADH generated by glycolysis. These must be oxidized to extract the energy from them.

The citric acid cycle not only oxidizes glucose metabolites but is also a pathway and a source of intermediates for the synthesis of fats and nonessential amino acids. The connections between the citric acid cycle and the metabolism of other nutrients will be discussed later.

The Membrane Reactions

The membrane reactions have two purposes: (1) to further oxidize NADH and $FADH_2$ and transfer their energy to ATP and (2) to regenerate NAD^+ and FADH and make them available again to earlier reaction steps. The membrane reactions are carried out by a series of compounds called the **electron-transport chain.** Most members of the chain are bound to the inner mitochondrial membrane. They are arranged in a precise order that enables each one to receive a pair of electrons from the member on one side of it (or, in two cases, from NADH and $FADH_2$) and pass these electrons along to the member on the other side—like a row of people passing along a hot potato. By the time the "potato" reaches the last member in the chain it is relatively "cool"—its energy has been used to make ATP.

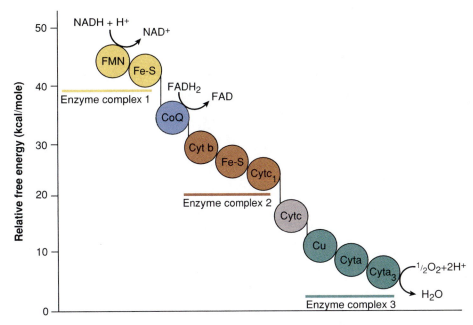

Figure 26.5 The mitochondrial electron-transport system. Transport molecules are grouped into three enzyme complexes, each of which acts as a proton pump. Molecules at the upper left of the figure have a relatively high free energy content, and molecules at the lower left are relatively low in energy. ◼️

The members of this transport chain are as follows:

- **Flavin mononucleotide (FMN),** a derivative of riboflavin similar to FAD, bound to a membrane protein. FMN accepts electrons from NADH.
- **Iron-sulfur (Fe-S) centers,** complexes of iron and sulfur atoms bound to membrane proteins.
- **Coenzyme Q (CoQ),** which accepts electrons from $FADH_2$. Unlike the other members, this is a relatively small, mobile molecule that moves about in the membrane.
- **Copper (Cu) ions,** bound to two membrane proteins.
- **Cytochromes,**[6] five enzymes with iron cofactors, so-named because they are brightly colored in pure form. In order of participation in the chain, they are cytochromes b, c_1, c, a, and a_3.

Electron Transport   Figure 26.5 shows the order in which electrons are passed along the chain. Hydrogen atoms are split apart as they are transferred from coenzymes to the chain. The protons are released to the mitochondrial matrix and the electrons travel in pairs (2 e^-) along the transport chain. Each electron carrier in the chain becomes reduced when it receives an electron pair and oxidized again when it passes the electrons along to the next carrier. Energy is liberated at each transfer.

---

6. *cyto* = cell + *chrom* = color

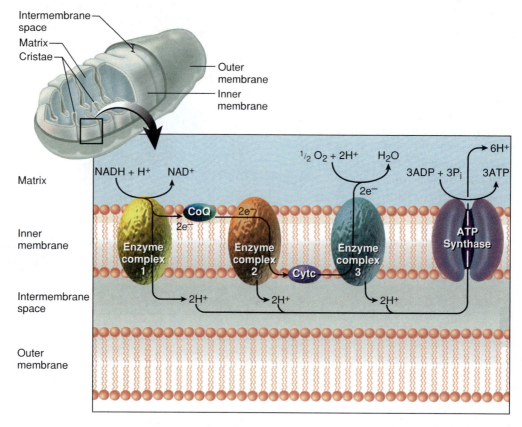

Figure 26.6 The chemiosmotic mechanism of mitochondrial ATP synthesis. Each enzyme complex pumps hydrogen ions into the space between the mitochondrial membranes. These hydrogen ions diffuse back into the matrix by way of ATP synthase, which taps their energy to synthesize ATP. In what ways would you change this figure to show the role of $FADH_2$ in place of NADH?

The final electron acceptor in the chain is oxygen. Each oxygen atom (half of an O_2 molecule) accepts two electrons (2 e^-) from cytochrome a_3 and two protons (2 H^+) from the mitochondrial matrix. The result is a molecule of water:

$$1/2 \; O_2 + 2 \; e^- + 2 \; H^+ \rightarrow H_2O$$

This is the body's primary source of *metabolic water*—water synthesized in the body rather than ingested in food and drink. This reaction also explains why the body requires oxygen. Without it, this reaction stops and creates a bottleneck, which, like a traffic jam, stops all the other processes leading to it. As a result, a cell produces too little ATP to sustain life, and death can ensue within a few minutes.

The Chemiosmotic Mechanism Of primary importance is what happens to the energy liberated by the electrons as they pass along the chain. Some of it is unavoidably lost as heat, but some of it drives **proton pumps.** The carriers of the electron-transport chain are clustered in three groups called **respiratory enzyme complexes.** The first complex includes FMN and five or more Fe-S centers; the second complex includes cytochromes b and c_1

and an Fe-S center; and the third complex includes two copper centers and cytochromes a and a_3. Each complex collectively acts as a proton pump that removes H^+ from the mitochondrial matrix and pumps it into the space between the inner and outer mitochondrial membranes (fig. 26.6). Coenzyme Q is a shuttle that transfers electrons from the first pump to the second, and cytochrome c shuttles electrons from the second pump to the third.

These pumps create a very high H^+ concentration (low pH) and positive charge between the membranes compared to a low H^+ concentration and negative charge in the mitochondrial matrix. That is, they create a steep electrochemical gradient across the inner mitochondrial membrane. If the inner membrane were freely permeable to H^+, these ions would have a strong tendency to diffuse down this gradient and back into the matrix.

The inner membrane, however, is permeable to H^+ only through specific channel proteins called **ATP synthase** (separate from the electron-transport system). As H^+ flows through these channels, it creates an electrical current (which, you may recall, is simply moving charged particles). ATP synthase harnesses the energy of this current to drive ATP synthesis. This process is called the **chemiosmotic**[7] **mechanism,** suggesting the "push" created by the electrochemical H^+ gradient.

Overview of ATP Production

NADH releases its electron pairs (as hydrogen atoms) to FMN in the first proton pump of the electron-transport system. From there to the end of the chain, this generates enough energy to synthesize 3 ATP molecules per electron pair. $FADH_2$ releases its electron pairs to coenzyme Q, the shuttle between the first and second proton pumps. Therefore, it enters the chain at a point beyond the first pump and does not contribute energy to that pump. Each $FADH_2$ contributes enough energy to synthesize 2 ATP.

With that in mind, we will draw up an energy balance sheet to see how much ATP is produced by the complete aerobic oxidation of glucose to CO_2 and H_2O and where the ATP comes from; see also figure 26.7. This summary refers back to the reaction steps 1 to 21 in figures 26.3 and 26.4. For each glucose molecule, there are

      10 NADH produced
         at steps 5, 10, 14, 16, and 21
$\times$ 3 ATP per NADH produced
         by the electron-transport chain

      30 ATP generated by NADH

Plus: 2 $FADH_2$ produced at step 19
     $\times$ 2 ATP per $FADH_2$ produced
          by the electron-transport chain

=   **4 ATP** generated by $FADH_2$

Plus: **2 ATP** net amount generated by
          glycolysis (steps 6–7 offset by step 3)
       2 ATP generated by the
          matrix reactions (step 18)

Total: **38 ATP** per glucose

This should be viewed as a theoretical maximum. There is some uncertainty about how much H^+ must be pumped between the mitochondrial membranes to generate 1 ATP, and some of the energy from the H^+ current is consumed by pumping ATP from the mitochondrial matrix into the cytosol and exchanging it for more raw materials—ADP and P_i pumped from the cytosol into the mitochondria.

Furthermore, the NADH generated by glycolysis cannot enter the mitochondria and donate its electrons directly to the electron-transport chain. In liver, kidney, and myocardial cells, NADH passes its electrons to *malate,* a "shuttle" molecule that delivers the electrons

7. *chemi* = chemical + *osmo* = push

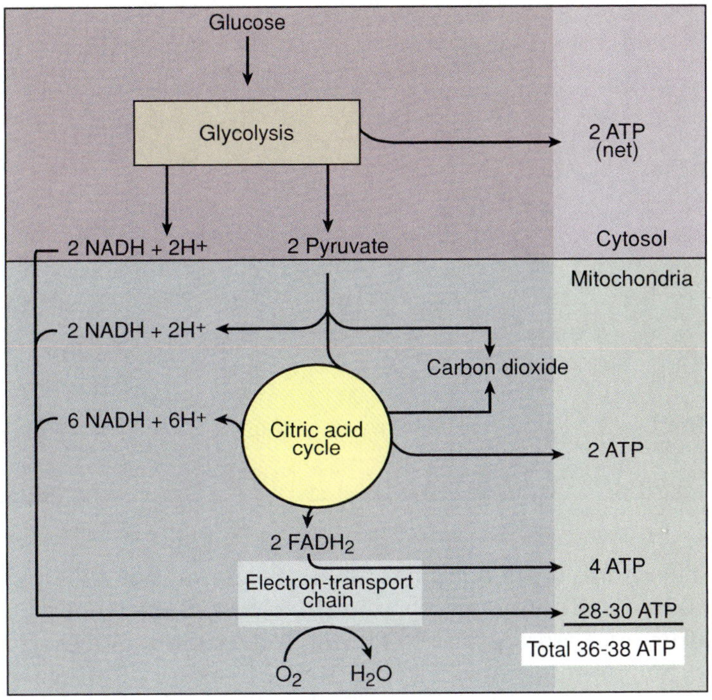

Figure 26.7 Summary of the sources of ATP generated by the complete oxidation of glucose.

to the beginning of the electron-transport chain. In this case, each NADH yields enough energy to generate 3 ATP. In skeletal muscle and brain cells, however, the glycolytic NADH transfers its electrons to *glycerol phosphate,* a different shuttle that donates the electrons farther down the electron-transport chain and results in the production of only 2 ATP. Therefore the amount of ATP produced per NADH differs from one cell type to another and is still unknown for others.

But if we assume the maximum ATP yield, every mole (180 g) of glucose releases enough energy to synthesize 38 moles of ATP. Glucose has an energy content of 686 kcal/mole and ATP has 7.3 kcal/mole (277.4 kcal in 38 moles). This means that aerobic respiration has an **efficiency** (a ratio of energy output to input) of up to 277.4 kcal / 686 kcal = 40%. The other 60% (408.6 kcal) is body heat.

The pathways of glucose catabolism are summarized in table 26.5. The aerobic respiration of glucose can be represented in the summary equation:

$$C_6H_{12}O_6 + 6\ O_2 \longrightarrow 6\ CO_2 + 6\ H_2O$$

$$36\text{–}38\ ADP + P_i \qquad 36\text{–}38\ ATP$$

Glycogen Metabolism

ATP is quickly used after it is synthesized—it is an *energy transfer* molecule, not an *energy storage* molecule.

Table 26.5 Pathways of Glucose Catabolism

Stage	Principal Reactants	Principal Products	Purpose
Glycolysis	Glucose, 2 ADP, 2 P_i, 2 NAD^+	2 pyruvic acid, 2 ATP, 2 NADH, 2 H_2O	To reorganize glucose and split it in two in preparation for further oxidation by the mitochondria; sole source of ATP under anaerobic conditions
Lactic acid fermentation	2 pyruvic acid, 2 NADH	2 lactic acid, 2 NAD^+	To regenerate NAD^+ so glycolysis can continue to function (and produce a little ATP) in the absence of oxygen
Matrix reactions	2 pyruvic acid, 8 NAD^+, 2 FAD, 2 ADP, 2 P_i, 8 H_2O	6 CO_2, 8 NADH, 2 $FADH_2$, 2 ATP, 2 H_2O	To remove electrons from pyruvate and transfer them to coenzymes NAD^+ and FAD; produces some ATP
Membrane reactions	10 NADH, 2 $FADH_2$, 6 O_2	32–34 ATP, 12 H_2O	To complete oxidation and produce most of the ATP of cellular respiration

Therefore, if the body has an ample amount of ATP and there is still more glucose in the blood, it does not produce and store excess ATP but converts the glucose to other compounds better suited for energy storage—namely glycogen and fat. Fat synthesis is considered later. Here we consider the synthesis and use of glycogen. The average adult body contains about 400 to 450 g of glycogen: nearly one-quarter of it in the liver, three-quarters of it in the skeletal muscles, and small amounts in cardiac muscle and other tissues.

Glycogenesis, the synthesis of glycogen, is stimulated by insulin. Glucose 6-phosphate (G6P) is isomerized to glucose 1-phosphate (G1P). The enzyme *glycogen synthase* then cleaves off the phosphate group and attaches the glucose to the growing polysaccharide chain.

Glycogenolysis, the hydrolysis of glycogen, releases glucose between meals when new glucose is not being ingested. It is stimulated by glucagon and epinephrine. The enzyme *glycogen phosphorylase* begins by phosphorylating a glucose residue and splitting it off the glycogen molecule as G1P. This is isomerized to G6P, which can then enter the pathway of glycolysis.

G6P usually cannot leave the cells that produce it. Hepatocytes, however, have an enzyme called *glucose 6-phosphatase,* which removes the phosphate group and produces free glucose. This can diffuse out of the cell into the blood, where it is available to any cells in the body. Although muscle cells cannot directly release glucose into the blood, they contribute indirectly to blood glucose concentration because they release pyruvic and lactic acids, which are converted to glucose by the liver.

Gluconeogenesis[8] is the synthesis of glucose from noncarbohydrates such as fats and amino acids. It occurs chiefly in the liver, but after several

8. *gluco* = sugar, glucose + *neo* = new + *genesis* = production of

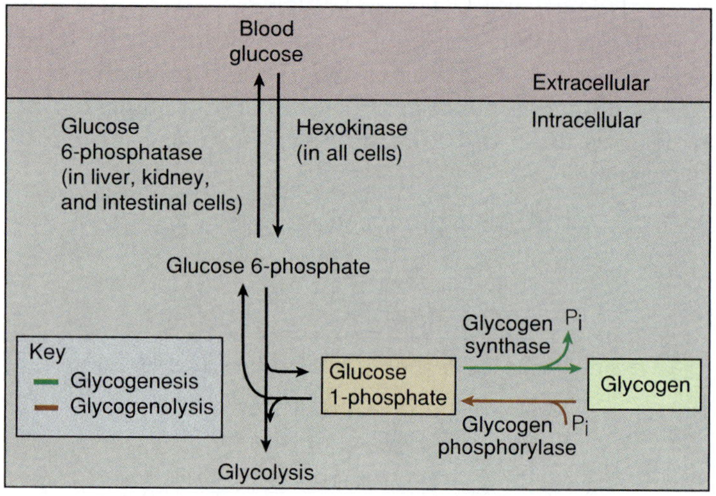

Figure 26.8 Major pathways of glucose storage and use. In most cells, the glucose 1-phosphate generated by glycogenolysis can only undergo glycolysis. In liver, kidney, and intestinal cells, it can be converted back to free glucose and released into circulation.

weeks of fasting the kidneys also undertake this process and eventually produce just as much glucose as the liver does.

The processes described here are summarized in figure 26.8, and the distinctions among those similar terms are summarized in table 26.6.

Functions of the Liver

We have seen that the liver plays a central role in carbohydrate metabolism. Additional liver functions (table 26.7) were described in previous chapters and will be described later in this chapter. Except for phagocytosis, all of these are performed by the cuboidal hepatocytes described in chapter 25. Such functional diversity is remarkable in light of the monotonously uniform structure of these cells.

Table 26.6   Some Terminology Related to Glucose and Glycogen Metabolism.

Anabolic (Synthesis) Reactions

Glycogenesis	The synthesis of glycogen by polymerizing glucose
Gluconeogenesis	The synthesis of glucose from noncarbohydrates such as fats and amino acids

Catabolic (Breakdown) Reactions

Glycolysis	The splitting of glucose into two molecules of pyruvic acid in preparation for anaerobic fermentation or aerobic respiration
Glycogenolysis	The hydrolysis of glycogen to release free glucose or glucose 1-phosphate

We can dispense with most of these functions and still live, although with some difficulty. The liver's protein metabolism, however, is indispensible; we cannot survive for more than a few days if those functions are lost. Degenerative liver diseases such as hepatitis, cirrhosis, and liver cancer are therefore especially life-threatening.

----- **Key Point Review** -----

6 Identify the reaction steps in figures 26.3 and 26.5 at which vitamins are essential to glucose catabolism.

7 In the laboratory, glucose can be oxidized in a single step to CO_2 and H_2O. Why is it done in so many little steps in cells?

8 Explain the origin of the word *glycolysis* and why this is an appropriate name for the purpose of that reaction pathway.

9 What are two advantages of aerobic respiration over anaerobic fermentation?

10 What important enzyme is found in the inner mitochondrial membrane other than those of the electron-transport chain? Explain how its function depends on the electron-transport chain.

11 Describe how the liver responds to (a) an excess and (b) a deficiency of blood glucose.

Lipid and Protein Metabolism

▼Objectives
When you have completed this section, you should be able to
• discuss some advantages of lipids over carbohydrates for energy storage and fuel;
• describe the processes of lipid catabolism and anabolism;
• describe the processes of protein catabolism and anabolism; and
• explain the metabolic source of ammonia and how the body disposes of it.

In the foregoing discussion, glycolysis and the mitochondrial reactions were treated from the standpoint of carbohydrate oxidation. These pathways also serve for

Table 26.7   Functions of the Liver

Carbohydrate Metabolism

Converts dietary fructose and galactose to glucose. Stabilizes blood glucose concentration by storing excess glucose as glycogen (glycogenesis), releasing glucose from glycogen when needed (glycogenolysis), and synthesizing glucose from fats and amino acids (gluconeogenesis) when glucose demand exceeds glycogen reserves. Receives lactic acid generated by anaerobic fermentation in skeletal muscle and other tissues and converts it back to pyruvic acid or glucose 6-phosphate.

Lipid Metabolism

Degrades chylomicron remnants. Carries out most of the body's lipogenesis (fat synthesis) and synthesizes cholesterol and phospholipids; produces VLDLs to transport lipids to adipose tissue and other tissues for storage or use; and stores fat in its own cells. Carries out most β-oxidation of fatty acids; produces ketone bodies from excess acetyl CoA. Produces HDL shells, which pick up excess cholesterol from other tissues and return it to the liver; excretes the excess cholesterol in bile.

Protein and Amino Acid Metabolism

Carries out most deamination and transamination of amino acids. Removes $-NH_2$ from glutamic acid and converts the resulting ammonia to urea by means of the ornithine cycle. Synthesizes nonessential amino acids by transamination reactions.

Synthesis of Plasma Proteins

Synthesizes nearly all the proteins of blood plasma, including albumin, α and β globulins, fibrinogen, prothrombin, and several other clotting factors. (Does not synthesize plasma enzymes, peptide hormones, or γ globulins.)

Vitamin and Mineral Metabolism

Converts vitamin D_3 to calcidiol, a step in the synthesis of calcitriol; stores a 3- to 4-month supply of vitamin D. Stores a 10-month supply of vitamin A and enough vitamin B_{12} to last one to several years. Stores iron in ferritin and releases it as needed. Excretes excess calcium by way of the bile.

Digestion

Synthesizes bile salts, which emulsify fat and promote its digestion.

Disposal of Drugs, Toxins, and Hormones

Detoxifies alcohol, antibiotics, and many other drugs. Metabolizes bilirubin from RBC breakdown and excretes it as bile pigments. Deactivates thyroxine and steroid hormones and excretes them or converts them to a form more easily excreted by the kidneys.

Phagocytosis

Macrophages cleanse blood of bacteria and other foreign matter; blood entering liver is laden with intestinal bacteria, and blood leaving liver is virtually sterile.

the oxidation of proteins and lipids as fuel and as a source of metabolic intermediates that can be used for protein and lipid synthesis. Here we examine these related metabolic pathways.

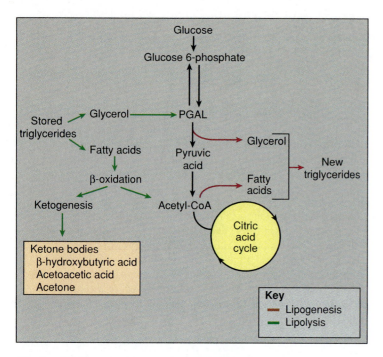

Figure 26.9 Pathways of lipolysis (*left*) and lipogenesis (*right*) in relation to glycolysis and the citric acid cycle.

Lipids

Triglycerides are stored primarily in the body's adipocytes, where they remain for about 2 to 3 weeks. Although the total amount of stored triglyceride remains quite constant, there is a continual turnover as lipids are released, transported in the blood, and either oxidized for energy or redeposited in other adipocytes. Synthesizing triglycerides from other types of molecules is called **lipogenesis,** and breaking down fat to be used as fuel is called **lipolysis** (lih-POL-ih-sis).

Lipogenesis

It is common knowledge that a diet high in sugars causes us to put on fat and gain weight. Lipogenesis employs compounds such as sugars and amino acids to synthesize the triglyceride precursors, glycerol and fatty acids. PGAL, one of the intermediates of glucose oxidation, can be converted to glycerol. As glucose and amino acids enter the citric acid cycle by way of acetyl-CoA, the acetyl-CoA can also be diverted to make fatty acids. The glycerol and fatty acids can then be condensed to form a triglyceride, which can be stored in the adipose tissue or converted to other lipids. These pathways are summarized in figure 26.9.

Lipolysis

Lipolysis, also shown in figure 26.9, begins with the hydrolysis of a triglyceride into glycerol and fatty acids—a process stimulated by epinephrine, norepinephrine,

glucocorticoids, thyroid hormone, and growth hormone. The glycerol and fatty acids are further oxidized by separate pathways. Glycerol is easily converted to PGAL and thus enters the pathway of glycolysis. It generates only half as much ATP as glucose, however, because it is a C_3 compound compared to glucose, C_6; thus it leads to the production of only half as much pyruvic acid.

The fatty acid component is catabolized in the mitochondrial matrix by a process called **β-oxidation,** which removes two carbon atoms at a time from the fatty acid. The resulting acetyl (C_2) groups are bonded to coenzyme A to make acetyl-CoA—the entry point into the citric acid cycle. A fatty acid of 16 carbon atoms can yield 129 molecules of ATP—obviously a much richer source of energy than a glucose molecule.

Excess acetyl groups can be metabolized by the liver in a process called **ketogenesis.** Two acetyl groups are condensed to form acetoacetic acid, and some of this is further converted to β-hydroxybutyric acid and acetone. These three products are the *ketone bodies.* Some cells convert acetoacetic acid back to acetyl-CoA and thus feed the C_2 fragments into the citric acid cycle to extract their energy. Cardiac muscle and the renal cortex, in fact, use acetoacetic acid as their principal fuel. When the body is rapidly oxidizing fats, however, excess ketone bodies accumulate and may cause the ketoacidosis typical of insulin-dependent diabetes mellitus.

Acetyl-CoA cannot go backward up the glycolytic pathway and produce glucose, because this pathway is irreversible past the point of pyruvic acid. While glycerol can be used for gluconeogenesis, fatty acids cannot.

It was mentioned earlier that inadequate carbohydrate intake interferes with the complete fat oxidation. This is because the mitochondrial reactions cannot proceed without oxaloacetic acid as a "pickup molecule" in the citric acid cycle. When carbohydrate intake is deficient, oxaloacetic acid is converted to glucose and becomes unavailable to the citric acid cycle. Under such circumstances, fat oxidation leads to ketogenesis and potentially to ketoacidosis. This is a serious risk of poorly planned diets with excessively restricted carbohydrate ingestion.

Proteins

About 100 g of tissue protein breaks down each day into free amino acids. These combine with the amino acids from the diet to form an **amino acid pool** that cells can draw upon to make new proteins. The fastest rate of tissue protein turnover is in the intestinal mucosa, where epithelial cells are replaced at a very high rate. Dead cells are digested along with the food and thus contribute to the amino acid pool. Of all the amino acid absorbed by the small intestine, about 50% is from the diet, 25% from dead epithelial cells, and 25% from enzymes that have digested each other.

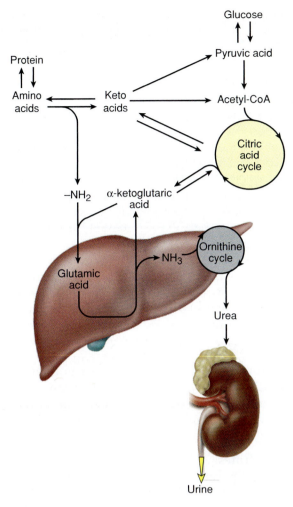

Figure 26.10 Pathways of amino acid metabolism in relation to glycolysis and the citric acid cycle.

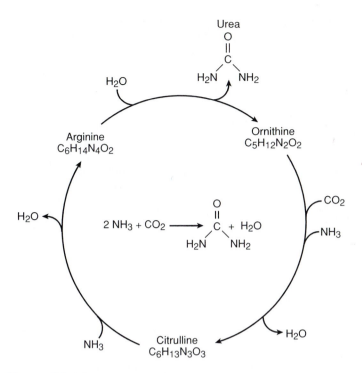

Figure 26.11 The ornithine cycle of the liver, which converts ammonia and carbon dioxide to urea.

Some amino acids in the pool can be converted to others. Free amino acids also can be converted to glucose and fat or directly used as fuel. Such conversions involve three processes: (1) **deamination,** the removal of an amino group (—NH₂); (2) **amination,** the addition of an amino group; or (3) **transamination,** the transfer of an amino group from one molecule to another. The following discussion shows how these processes are involved in amino acid metabolism.

Use as Fuel

The first step in using amino acids as fuel is to deaminate them. After removal of the —NH₂ group, the remainder of the molecule is called a *keto acid.* Depending on which amino acid is involved, the resulting keto acid may be converted to pyruvic acid, acetyl-CoA, or one of the acids of the citric acid cycle (fig. 26.10). It is important to note that some of these reactions are reversible. When there is a deficiency of amino acids in the body, citric acid cycle intermediates can be ami-

nated and converted to amino acids, which are then available for protein synthesis. In gluconeogenesis, keto acids are used to synthesize glucose, essentially through a reversal of the glycolysis reactions.

Transamination, Ammonia, and Urea

When an amino acid is deaminated, its amino group is transferred to a citric acid cycle intermediate, α-ketoglutaric acid, converting it to glutamic acid. Such transamination reactions are the route by which several amino acids enter the citric acid cycle.

Glutamic acid can travel from any of the body's cells to the liver. Here its —NH₂ group is removed, converting it back to α-ketoglutaric acid. The —NH₂ becomes ammonia, NH₃, which is extremely toxic to cells and cannot be allowed to accumulate. The liver quickly converts ammonia to a less toxic form, urea, by a pathway called the **ornithine cycle** (fig. 26.11). Urea is then excreted in the urine as one of the body's nitrogenous wastes (see chapter 23). Other nitrogenous wastes and their sources are described in chapter 23 (see p. 000). When a diseased liver cannot carry out the ornithine cycle, NH₃ accumulates in the blood and death from *hepatic coma* may ensue within a few days.

Protein Synthesis

Protein synthesis is a complex process involving DNA, mRNA, tRNA, ribosomes, and often the rough ER—all

Table 26.8	Major Aspects of the Absorptive and Postabsorptive States	
	Absorptive	**Postabsorptive**
Regulatory hormones	Principally insulin Also gastrin, secretin, CCK	Principally glucagon Also epinephrine, growth hormone
Carbohydrate metabolism	Blood glucose concentration rising Glucose stored by glycogenesis Gluconeogenesis suppressed	Blood glucose concentration falling Glucose released by glycogenolysis Gluconeogenesis stimulated
Lipid metabolism	Lipogenesis occurring Lipid uptake from chylomicrons Lipid storage in fat and muscle	Lipolysis occurring Fatty acids oxidized for fuel Glycerol used for gluconeogenesis
Protein metabolism	Amino acid uptake and protein synthesis Excess amino acids burned as fuel	Amino acids oxidized if glycogen and fat stores are inadequate for energy needs

of which were described in chapter 5. It is stimulated by growth hormone, thyroid hormones, and insulin, and it requires a supply of all the amino acids necessary for a particular protein. The liver can make many of these amino acids from other amino acids or from citric acid cycle intermediates by transamination reactions. The essential amino acids, however, must be obtained from the diet.

.... **Key Point Review** ....

12 Which of the processes in table 26.6 is most comparable to lipogenesis? Which is most comparable to lipolysis? Explain.

13 When fats are converted to glucose, only the glycerol component is used in this way, not the fatty acid. Explain why and what happens to the fatty acids.

14 What metabolic process produces ammonia? How does the body dispose of ammonia?

Metabolic States and Metabolic Rate

▼**Objectives**

When you have completed this section, you should be able to
• define the absorptive and postabsorptive states;
• explain what happens to carbohydrates, fats, and proteins in each of these states;
• describe the hormonal and nervous regulation of each state;
• define *metabolic rate* and *basal metabolic rate;* and
• describe some factors that alter the metabolic rate.

Your nutrient metabolism changes from hour to hour depending on how long it has been since your last meal. The **absorptive (fed) state** lasts about 4 hours during and after a meal. This is a time in which nutrients are being absorbed and may be used immediately to meet energy and other needs. The **postabsorptive (fasting) state** prevails in the late morning, late afternoon, and overnight. During this time the stomach and small intestine are empty and the body's energy needs are met

from stored fuels. The two states are compared in table 26.8 and explained in the following discussion.

Absorptive State

In the absorptive state, blood glucose is readily available for ATP synthesis. It serves as the primary fuel and spares the body from having to draw on stored fuels. The status of major nutrient classes during this phase is as follows:

• **Carbohydrates.** Absorbed sugars are transported by the hepatic portal system to the liver. Most glucose passes through the liver and becomes available to cells everywhere in the body. Glucose in excess of immediate need, however, is absorbed by the liver and may be converted to glycogen or fat. Most fat synthesized in the liver is released into the circulation; its further fate is comparable to that of dietary fats, discussed next.
• **Fats.** Fats enter the lymph as chylomicrons and initially bypass the liver. As described earlier, lipoprotein lipase removes fats from the chylomicrons for uptake by the tissues, especially adipose and muscular tissue. The liver disposes of the chylomicron remnants. Fats are the primary energy substrate for hepatocytes, adipocytes, and muscle cells.
• **Amino acids.** Amino acids, like sugars, circulate first to the liver. Most pass through and become available to other cells for protein synthesis. Some, however, are removed by the liver and have one of the following fates: (1) to be used for protein synthesis; (2) to be deaminated and used as fuel for ATP synthesis; or (3) to be deaminated and used for fatty acid synthesis.

Regulation of the Absorptive State

The absorptive state is regulated largely by insulin, which is secreted in response to elevated blood glucose and amino acid levels and to the intestinal hormones gastrin, secretin, and cholecystokinin. Insulin stimulates nearly all cells of the body except neurons, which absorb glucose at a rate independent of it. On other target cells, insulin has the following effects:

- Within minutes, it increases the cellular uptake of glucose by as much as 20-fold. As cells absorb glucose, the blood glucose concentration falls.
- It stimulates glucose oxidation, glycogenesis, and lipogenesis.
- It inhibits gluconeogenesis, which makes sense since blood glucose concentration is already high and there is no immediate need for more.
- It stimulates the active transport of amino acids into the cells and promotes protein synthesis.

Following a high-protein, low-carbohydrate meal, it may seem that the amino acids would stimulate insulin secretion, insulin would accelerate both amino acid and glucose uptake, and since there was relatively little glucose in the ingested food, this would create a risk of hypoglycemia. In actuality, this is prevented by the fact that a high amino acid level stimulates the secretion of *both* insulin and glucagon. Glucagon, you may recall, is an insulin antagonist (see chapter 17). Thus cells do not absorb too much glucose but leave enough for the brain.

Postabsorptive State

The essence of the postabsorptive state is to homeostatically regulate blood glucose concentration within about 90 to 100 mg/dL. This is especially critical to the brain, which cannot use alternative energy substrates except in cases of prolonged fasting. The postabsorptive status of major nutrients is as follows:

- **Carbohydrates.** Glucose is drawn from the body's glycogen reserves (glycogenolysis) or synthesized from other compounds (gluconeogenesis). There is usually enough glycogen stored in the liver to support 4 hours of postabsorptive metabolism before significant gluconeogenesis occurs.
- **Fats.** Adipocytes and hepatocytes hydrolyze fats and convert the glycerol to glucose. The free fatty acids (FFAs) cannot be converted to glucose, but they can favorably effect blood glucose concentration. As the liver oxidizes them to ketone bodies, other cells absorb and use these, or use FFAs directly, as their source of energy. By switching from glucose to fatty acid catabolism, they conserve glucose for use by the brain (the glucose-sparing effect). After 4 to 5 days of fasting, the brain begins to use ketone bodies as supplemental fuel.
- **Proteins.** If glycogen and fat reserves are depleted, the body begins to use proteins as fuel. Some proteins are more resistant to catabolism than others. The first to go are skeletal muscle proteins, and starvation therefore results in wasting away of the muscles. Collagen, by contrast, is almost never broken down for fuel.

Regulation of the Postabsorptive State

Postabsorptive metabolism is regulated mainly by the sympathetic nervous system and glucagon, but several other hormones are also involved and the regulation of this state is more complex than that of the absorptive state. As blood glucose level drops, insulin secretion declines and the pancreatic α cells secrete glucagon. Glucagon promotes glycogenolysis and gluconeogenesis, raising the blood glucose level, and it promotes lipolysis and a rise in FFA levels. Thus it makes both glucose and lipids available for fuel.

The sympathoadrenal system also promotes glycogenolysis and lipolysis. Adipose tissue is richly innervated by the sympathetic nervous system, while adipocytes, hepatocytes, and muscle cells also respond to epinephrine from the adrenal medulla. The sympathoadrenal effects on these tissues are activated by such conditions as stress, injury, fear, or anger. In circumstances where there is likely to be tissue injury and a need for repair, the sympathoadrenal system therefore mobilizes stored energy reserves and makes them available to meet the demands of tissue repair.

Growth hormone is secreted in response to rapid drops in blood glucose level and in conditions of prolonged fasting. It opposes the effects of insulin and raises blood glucose concentration.

Metabolic Rate

Metabolic rate (MR) means the amount of energy liberated in the body per unit of time, expressed in such terms as kcal/hr or kcal/day. Metabolic rate can be measured directly by putting a person in a closed chamber called a **calorimeter** with water-filled walls that absorb the heat given off by the body. The rate of energy release is measured from the temperature change of the water. Metabolic rate can also be measured indirectly with a spirometer, an apparatus described in chapter 22 that measures the amount of air (or other gases) inhaled and exhaled by a person. If a subject breathes pure oxygen from the spirometer, the spirometer can measure the rate of oxygen consumption. For every liter of oxygen consumed, approximately 4.82 kcal of energy is released from organic nutrients. This is only an estimate, however, because the number of kilocalories per liter of oxygen varies slightly with the type of nutrients the person is predominantly oxidizing at the time of measurement.

Metabolic rate depends on your physical activity, mental state, absorptive or postabsorptive status, and other factors. The **basal metabolic rate (BMR)** is a baseline or standard of comparison that minimizes the effects of such variables. It is your metabolic rate when you are awake but relaxed, in a room at comfortable temperature, in a postabsorptive state 12 to 14 hours since your last meal. It is not the minimum metabolic rate needed to sustain life, however. When you are asleep, your metabolic rate is slightly lower than your BMR. **Total metabolic rate** is the sum of BMR and energy expenditure for voluntary activities, especially muscular contractions.

The BMR of an average adult is about 2,000 kcal/day for a male and slightly less for a female. Roughly speaking, an average male must therefore consume at least 2,000 kcal/day to fuel his essential metabolic tasks—active transport, muscle tone, brain activity, cardiac and respiratory rhythms, renal function, and other essential processes. Even a relatively sedentary lifestyle requires another 500 kcal/day to support a low level of physical activity, and someone who does hard physical labor may require as much as 5,000 kcal/day.

Aside from physical activity, some factors that raise the metabolic rate and caloric requirements include pregnancy, anxiety (which stimulates epinephrine release and muscle tension), fever (MR rises about 14% for each 1°C of body temperature), eating (MR rises after a meal), and the catecholamine and thyroid hormones. MR is relatively high in children and declines with age. Therefore, as we reach middle age we often find ourselves gaining weight with no apparent change in food intake.

Some factors that lower metabolic rate include apathy, depression, and prolonged starvation. In weight loss diets, loss is often rapid at first and then goes more slowly. This is partly because the initial loss is largely water and partly because the metabolic rate drops over time, fewer dietary calories are "burned off," and there is more lipogenesis even with the same caloric intake.

Body Heat and Thermoregulation

▼**Objectives**

When you have completed this section, you should be able to
- identify the principal sources of body heat;
- describe variations in body temperature with age, sex, method and time of measurement, and state of activity;
- define and contrast radiation, conduction, and evaporative cooling and their relative importance to total heat loss;
- describe how the hypothalamus monitors and controls body temperature; and
- state the terms for excessively high and low body temperatures, and describe the consequences of these conditions.

Heat generation must be matched by heat loss in order to maintain a stable internal body temperature. **Hypothermia,** an excessively low body temperature, can cause metabolism to slow down to the point that it cannot sustain life, whereas **hyperthermia,** an excessively high temperature, can denature enzymes, shut down metabolic pathways, and also lead to death. The balance between heat production and loss, called **thermoregulation,** is a critically important aspect of homeostasis. Thermoregulation was introduced in chapter 1 as an example of homeostasis.

Body Temperature

"Normal" body temperature depends on when, where, and in whom it is measured. Body temperature fluctuates about 1°C (1.8°F) in a 24-hour cycle. It tends to be lowest in the early morning and highest in the late afternoon. Temperature also varies from one place in the body to another.

The most important body temperature is the **core temperature**—the temperature of organs in the cranial, thoracic, and abdominal cavities. Rectal temperature is relatively easy to measure and gives an estimate of core temperature: usually 37.2° to 37.6°C (99.0°–99.7°F). It may be as high as 38.5°C (101°F) in active children and some adults.

Shell temperature is the temperature closer to the surface, especially skin and oral temperature. Here, heat is lost from the body and temperatures are slightly lower than rectal temperature. Adult oral temperature is typically 36.6° to 37.0°C (97.9° to 98.6°F) but may be as high as 40°C (104°F) during hard exercise.

Heat Production and Loss

Most body heat comes from exergonic (energy-releasing) chemical reactions such as nutrient oxidation and ATP use. A little heat is generated by joint friction, blood flow, and other movements. At rest, most heat is generated by the brain, heart, liver, and endocrine glands; the skeletal muscles contribute about 20% to 30% of the total resting heat. Increased muscle tone or exercise greatly increases heat generation in the muscles, however; in vigorous exercise, they produce 30 to 40 times as much heat as the rest of the body.

The body loses heat in three ways: radiation, conduction, and evaporation.

1. **Radiation.** In essence, heat means molecular motion. All molecular motion produces radiation in the infrared (IR) region of the electromagnetic spectrum. When IR radiation is absorbed by any object, it increases its molecular motion and raises its temperature. Therefore IR radiation removes heat from its source and adds heat to anything that absorbs it. The heat lamps used in bathrooms and restaurants work on this principle. Our bodies continually receive IR from the objects around us and give off IR to our surroundings. Since we are usually warmer than the objects around us, we lose more heat this way than we gain.

2. **Conduction.** As the molecules of our tissues vibrate with heat energy, they collide with other molecules and transfer kinetic energy to them. The warmth of your body therefore adds to the molecular motion and temperature of the clothes you wear, the chair you sit in, and the air around you. Conductive heat loss is aided by **convection,** the motion of a fluid due to uneven heating. Air is a fluid that becomes less dense and therefore rises as it is heated. Thus warm air rises from the body and is replaced by cooler air from below. The same is true of water; for example, when you swim in a lake or take a cool bath. You can easily see convection when you heat water in a clear container. Convection of the air around the human body can be seen by a technique called schlieren photography (fig. 26.12). Convection is not a separate category of heat loss in itself, but it increases the efficiency of conduction.

3. **Evaporation.** The cohesion of water molecules hampers their vibratory movement in response to heat input. If the temperature of water is raised sufficiently, however, its molecular motion becomes great enough for molecules to break free and evaporate. Evaporation of water thus carries a substantial amount of heat with it (0.5 kcal/g). This is the significance of perspiration. Sweat wets the skin surface and its evaporation carries heat away. In extreme conditions, the body can lose up to 4 L of sweat per hour, representing a heat loss of 2,000 kcal/hr. When you are sweaty and stand in front of a fan, you may feel refreshingly cooled or even uncomfortably cold. The rapid movement of heat-laden air away from the body (**forced convection**) accelerates heat removal, so a breeze or a fan enhances heat loss by conduction or evaporation. It has no effect on radiation.

Figure 26.12 Heat loss from the body demonstrated by schlieren photography. The body loses heat to the surrounding air by conduction, and the rising column of warm air then carries the heat away by convection, replacing it with cooler air from below.

The relative amounts of heat lost by different methods depend on prevailing conditions. A nude body at an air temperature of 21°C (70°F) loses about 60% of its heat by radiation, 18% by conduction, and 22% by evaporation. If air temperature is higher than skin temperature, evaporation becomes the only means of heat loss because radiation and conduction add more heat to the body than they remove. Hot, humid weather hinders even evaporative cooling because there is less of a humidity gradient from skin to air. Such conditions increase the risk of heatstroke (discussed shortly).

Thermoregulation

Thermoregulation is achieved through several negative feedback loops. The preoptic area of the hypothalamus (anterior to the optic chiasma) has a nucleus of neurons called the **hypothalamic thermostat.** It monitors the temperature of the blood and receives signals also from **peripheral thermoreceptors** in the skin. In turn, it sends appropriate signals either to the **heat-losing center,** a nucleus still farther anterior in the hypothalamus, or to the **heat-promoting center,** a more posterior nucleus.

When the heat-losing center senses that the blood temperature is too high, it activates heat-losing mechanisms. The first and simplest of these is dilation of the dermal arterioles, which increases blood flow close to the body surface and thus promotes heat loss. If this fails to restore normal temperature, the heat-losing center triggers sweating. It also inhibits the heat-promoting center.

When the heat-promoting center senses that the blood temperature is too low, it activates mechanisms to conserve body heat or to generate more. By way of the sympathetic nervous system, it causes dermal vasoconstriction. Warm blood is then retained deeper in the body and less heat is lost through the skin. The sympathetic nervous system in other mammals also stimulates the arrector pili muscles, which make the hair stand on end. This traps an insulating blanket of still air near the skin. The human sympathetic nervous system attempts to do this as well, but since our body hair is so scanty, the only noticeable effect of this is goosebumps.

If dermal vasoconstriction cannot restore or maintain normal core temperature, the body resorts to **shivering thermogenesis.** If you leave a warm house on a cold day, you may notice that your muscles become tense, sometimes even painfully taut, and you begin to shiver. Shivering involves a spinal reflex that causes tiny alternating contractions in antagonistic muscle pairs. It produces heat because every muscle contraction releases energy from ATP. Shivering increases the body's heat production as much as fourfold.

Nonshivering thermogenesis is a more long-term mechanism for generating heat, used especially in the colder seasons of the year. At first the sympathetic nervous system, and then thyroid hormone, stimulates an increase in metabolic rate. After several weeks of cold weather, the metabolic rate rises by 20% to 30%. More nutrients are burned as fuel, we consume more calories to "stoke the furnace," and consequently, we have greater appetites in the winter than in the summer.

In summary, you can see that thermoregulation is a function of multiple organs: the brain (hypothalamus), autonomic nerves, thyroid gland, skin, blood vessels, and skeletal muscles.

Disturbances of Thermoregulation

Chapter 21 described the mechanism of fever and its importance in combating infection. Fever is a normal protective mechanism that should be allowed to run its course if it is not excessively high. If the body temperature rises above 42° to 43°C (108°–110°F), however, a dangerous positive feedback loop can occur. The high temperature elevates the metabolic rate, and the elevated metabolic rate generates heat faster than the body's heat-losing mechanisms can disperse it (see fig. 1.18). Thus the metabolic rate increases the fever and the fever increases the metabolic rate. A core body temperature of 44° to 45°C (111°–113°F) can be fatal due to enzyme denaturation and neurological damage.

Exposure to excessive heat causes heat cramps, heat exhaustion, and heatstroke (sunstroke). **Heat cramps** are painful muscle spasms that result from excessive electrolyte loss in the sweat. They occur especially when a person begins to relax after strenuous exertion and heavy sweating. **Heat exhaustion** results from more severe electrolyte loss and is characterized by hypotension, dizziness, vomiting, and sometimes fainting. Prolonged heat waves, especially if accompanied by high humidity, bring on many deaths from **heatstroke.** The body gains heat by radiation and conduction, while the humidity retards evaporative cooling. The body temperature can rise as high as 43°C (110°F). Brain cells malfunction, hypothalamic mechanisms break down, and convulsions, coma, and death may suddenly occur.

Hypothermia can result from exposure to cold weather or immersion in icy water. It, too, entails life-threatening positive feedback loops. If the core temperature falls below 33°C (91°F), the metabolic rate drops so low that heat production cannot keep pace with heat loss, and the temperature falls even more. Death from cardiac fibrillation may occur below 32°C (90°F), but some people survive body temperatures as low as 29°C (84°F) in a state of suspended animation. A body temperature below 24°C (75°F) is usually fatal. It is dangerous to give alcohol to someone in a state of hypothermia because it produces an illusion of warmth but actually accelerates heat loss by dilating cutaneous blood vessels.

Key Point Review

19 What is the primary source of body heat? What are some lesser sources?

20 What mechanisms of heat loss are aided by convection?

21 Describe the major heat-promoting and heat-losing mechanisms of the body.

CHAPTER ESSAY

Alcohol and Alcoholism

Alcohol is not only a popular mind-altering drug but is also regarded in many cultures as a food staple. As a source of empty calories, an addictive drug, and a toxin, it can have a broad spectrum of adverse effects on the body.

Absorption and Metabolism

Alcohol is rapidly absorbed from the digestive tract—about 20% of it in the stomach and 80% in the proximal small intestine. Carbonation, as in beer and sparkling wines, increases its rate of absorption, whereas food reduces its absorption by delaying gastric emptying so the alcohol takes longer to reach its place of maximum absorption. Alcohol is soluble in both water and fat, so it is rapidly distributed to all body tissues and it easily crosses the blood-brain barrier to exert its intoxicating effects on the brain.

Alcohol is detoxified by the hepatic enzyme *alcohol dehydrogenase,* which oxidizes it to acetaldehyde. This enters the citric acid cycle and is oxidized to CO_2 and H_2O. The average adult male can clear the blood of about 10 mL of 100% alcohol per hour—about the amount in 30 mL (1 oz) of whiskey or 355 mL (12 oz) of beer. Women have less alcohol dehydrogenase and clear alcohol from the bloodstream less quickly. They are also more vulnerable to alcohol-related illnesses such as cirrhosis of the liver (discussed shortly).

Tolerance to alcohol, the ability to "hold your liquor," results from two factors: behavioral modification, such as giving in less readily to lowered inhibitions, and increased levels of alcohol dehydrogenase in response to routine alcohol consumption. Alcohol dehydrogenase also deactivates other drugs, and drug dosages must be adjusted to compensate for this when treating alcoholics for other diseases.

Physiological Effects

Nervous System

Alcohol is a depressant that inhibits the release of norepinephrine and disrupts the function of GABA receptors. In low doses, it depresses inhibitory synapses and creates sensations of confidence, euphoria, and giddiness; but it subsequently depresses excitatory synapses and then produces feelings of relaxation and drowsiness. At a blood alcohol level of 100 mg/dL—the legal criterion of intoxication in most states—alcohol impairs judgment, motor coordination, and reaction time. Above 400 mg/dL, it disrupts the plasma membranes and electrophysiology of neurons and may induce coma and death.

Liver

The liver's role in metabolizing alcohol makes it especially susceptible to long-term toxic effects. Heavy drinking stresses the liver with a high load of acetaldehyde and acetate; this depletes its oxidizing agents and reduces its ability to catabolize these intermediates as well as fatty acids. Alcoholism often produces a greatly enlarged and fatty liver for multiple reasons: the calories provided by alcohol make it unnecessary to burn fat as fuel, fatty acids are poorly oxidized, and acetaldehyde is converted to new fatty acids. Acetaldehyde also causes inflammation of the liver and pancreas (**hepatitis** and **pancreatitis**), leading to disruption of digestive function. Acetaldehyde and other toxic intermediates destroy hepatocytes faster than they can be regenerated, and the liver exhibits extensively fibrous scarring and a lumpy or nodular surface—a state called **cirrhosis.** Many symptoms of alcoholism stem from deterioration of liver functions. Hepatic coma may occur as the liver becomes unable to produce urea, which allows ammonia to accumulate in the blood. Jaundice results from the liver's inability to excrete bilirubin.

Circulatory System

Deteriorating liver functions exert several effects on the blood and cardiovascular system. Blood clotting is impaired because the liver cannot synthesize clotting factors adequately. Edema results from inadequate synthesis of blood albumin. Cirrhosis obstructs the hepatic portal blood circulation. Portal hypertension results, and combined with hypoproteinemia, this causes the liver and other organs to "sweat" serous fluid into the peritoneal cavity. The abdomen may become swollen with several liters of pooled fluid (ascites). The combination of hypertension and impaired clotting often leads to hemorrhaging. The subject may vomit blood as enlarged veins of the esophagus hemorrhage into its lumen. Alcohol abuse also destroys myocardial tissue, reduces contractility of the heart, and causes cardiac arrhythmia.

Digestive System and Nutrition

Alcohol breaks down the protective mucus barrier of the stomach and the tight junctions between its epithelial cells. Thus it may cause gastritis and bleeding. Alcohol is commonly believed to be a factor in peptic ulcers, but there is little concrete evidence of this. Heavy drinking, especially in combination with

—continued

smoking, increases the incidence of esophageal cancer. Malnutrition is a typical complication of alcoholism, partly because the empty calories of alcohol suppress the appetite for more nutritious foods. The average American gets about 4.5% of his or her calories from alcohol (more when nondrinkers are excluded), but heavy drinkers may obtain half or more of their calories from alcohol and have less appetite for foods that would meet their other nutritional requirements. In addition, acetaldehyde interferes with vitamin absorption and use. Thiamine deficiency is common in alcoholism, and thiamine is routinely given to alcoholics in treatment.

Addiction

Alcohol is the most widely available addictive drug in America. In many respects it is almost identical to barbiturates in its toxic effects, its potential for tolerance and dependence, and the risk of overdose. The difference is that obtaining barbiturates usually requires a prescription, while alcohol requires only proof of age.

Alcoholism is defined by a combination of criteria, including the pathological changes just described; physiological tolerance of high concentrations; impaired physiological, psychological, and social functionality; and withdrawal symptoms occurring when intake is reduced or stopped. Heavy drinking followed by a period of abstinence—for example, when a patient is admitted to the hospital and cannot get access to alcohol—may trigger **delirium tremens (DT),** characterized by restlessness, insomnia, confusion, irritability, tremors, incoherent speech, hallucinations, convulsions, and coma. DT has a 5% to 15% mortality rate.

Most alcoholism (type I) sets in after age 25 and is usually associated with stress or peer pressure. These influences lead to increased drinking, which can start a vicious cycle of illness, reduced job performance, family and social problems, arrest, and other stresses leading to still more drinking. A minority of alcoholics have type II alcoholism, which is at least partially hereditary. This appears predominantly in men who become addicted before age 25, especially the sons of men who were also addicted at an early age. There is little evidence for hereditary alcoholism in women. Type II alcoholics show abnormally rapid increases in blood acetaldehyde levels when they drink, and they have unusual brain waves (EEGs) even when not drinking. Male children of alcoholics have a higher than average incidence of becoming alcoholic even when raised by nonalcoholic foster parents. It is by no means inevitable that such people will become alcoholics, but stress or peer pressure can trigger alcoholism more easily in those who are genetically predisposed to it.

Alcoholism is treated primarily through behavior modification—abstinence, peer support, avoidance or correction of the stresses that encourage drinking, and sometimes psychotherapy. The drug disulfiram (Antabuse) is used to support behavior modification programs. It inhibits the breakdown of acetaldehyde and thus heightens its short-term toxic effects. A person who takes Antabuse and drinks alcohol experiences headache, vomiting, tachycardia, chest pain, and hyperventilation and is less likely to look upon alcohol as a means of pleasure or escape.▲

Chapter Review Study Outline

Nutrition (pp. 931–941)
1. Body weight and energy balance
   a. Homeostatic set point
   b. Heredity and environment
2. Appetite
   a. Feeding and satiety centers
   b. Glucostats
   c. Other factors in appetite
3. Calories
4. Nutrients
   a. Macro- and micronutrients
   b. Recommended daily allowances
5. Carbohydrates
   a. Functions
   b. Requirements
   c. Types in the diet
   d. Dietary sources
   e. Fiber

6. Lipids
   a. Functions
   b. Requirements
   c. Dietary sources
   d. Cholesterol and serum lipoproteins
      • Lipoprotein classes
      • Processing of chylomicrons
      • Processing of VLDLs and LDLs
      • Processing of HDLs
      • Cholesterol and lipoprotein levels
7. Proteins
   a. Functions
   b. Requirements
   c. Protein quality
   d. Dietary sources
   e. Nitrogen balance

8. Minerals
   a. Functions
   b. Requirements
   c. Dietary sources
   d. Sodium
9. Vitamins
   a. Requirements
   b. Water-soluble vitamins
   c. Fat-soluble vitamins
   d. Vitamin excesses

Carbohydrate Metabolism (pp. 941–949)
1. Overview of glucose catabolism
   a. Summary equation
   b. Phases of catabolism
   c. Role of coenzymes
2. Glycolysis

Selected Vocabulary

1. _____ are not used as fuel and are required in relatively small quantities.
   a. Micronutrients
   b. Macronutrients
   c. Essential nutrients
   d. Proteins
   e. Lipids

2. The only significant digestible polysaccharide in the diet is
   a. glycogen.
   b. cellulose.
   c. starch.
   d. maltose.
   e. fiber.

3. The greatest amount of energy can be stored in the smallest tissue space in the form of
   a. glucose.
   b. triglycerides.
   c. glycogen.
   d. proteins.
   e. vitamins.

4. The lipoproteins that remove cholesterol from the tissues are
   a. chylomicrons.
   b. lipoprotein lipases.
   c. VLDLs.
   d. LDLs.
   e. HDLs.

5. Proteins serve all of the following functions *except* that they do not act as
   a. enzymes.
   b. coenzymes.
   c. hormones.
   d. antibodies.
   e. structural support for tissues.

6. The primary function of B-complex vitamins is to act as
   a. structural components of cells.
   b. sources of energy.
   c. components of pigment molecules.
   d. antioxidants.
   e. coenzymes.

7. FAD is reduced to $FADH_2$ in
   a. glycolysis.
   b. anaerobic fermentation.
   c. the citric acid cycle.
   d. the electron-transport chain.
   e. β-oxidation of lipids.

8. The primary, direct benefit of anaerobic fermentation is to
   a. regenerate NAD^+.
   b. produce $FADH_2$.
   c. produce lactic acid.
   d. dispose of pyruvic acid.
   e. produce more ATP than glycolysis does.

9. Which of the following occurs in the matrix of a mitochondrion?
   a. glycolysis
   b. chemiosmosis
   c. the cytochrome reactions
   d. the citric acid cycle
   e. anaerobic fermentation

10. When the body emits more infrared (IR) energy than it absorbs, it is losing heat by
    a. convection.
    b. forced convection.
    c. conduction.
    d. radiation.
    e. evaporation.

11. A/an _____ protein lacks one or more essential amino acids.

12. In the postabsorptive state, glycogen is hydrolyzed to liberate glucose. This process is called _____.

13. Synthesis of glucose from amino acids or triglycerides is called _____.

14. The major nitrogenous waste resulting from protein catabolism is _____.

15. The organ that synthesizes the nitrogenous waste in the preceding question is the _____.

16. The absorptive state is regulated mainly by the hormone _____.

17. The temperature of organs in the body cavities is called _____.

18. The feeding center, satiety center, heat-losing center, and heat-promoting center are nuclei located in part of the brain called the _____.

19. The brightly colored, iron-containing, electron-transfer molecules of the mitochondrial inner membrane are called _____.

20. The flow of H^+ from the intermembrane space to the mitochondrial matrix creates an electrical current used by the enzyme _____ to make ATP.

1. Cyanide blocks the transfer of electrons from cytochrome a_3 to oxygen and causes sudden death. Explain why it would do so. Also explain whether or not cyanide poisoning could be treated by giving a patient supplemental oxygen, and justify your answer.

2. Chapter 17 defines and describes some hormone actions that are synergistic and antagonistic. Identify some synergistic and antagonistic interactions in the postabsorptive state of metabolism.

3. Mrs. Jones, a 42-year-old, complains that, "Everything I eat goes to fat," whereas, "My husband and my son eat twice as much as I do, and they're both as skinny as can be." How would you explain this to her?

4. A television advertisement proclaims, "Feeling tired? Need more energy? Order your supply of Zippy Megavitamins and feel better fast!" Your friend Cathy is about to send in her order, and you try to talk her out of wasting her money. Summarize the argument you would use.

5. Explain why a patient whose liver has been extensively damaged by hepatitis could show elevated concentrations of thyroid hormone and bilirubin in the blood.

Web Site Link

For a listing of the most current web sites related to this chapter, please visit the Saladin homepage at:

http://www.mhhe.com/sciencemath/biology/saladin/

[The Male Reproductive System

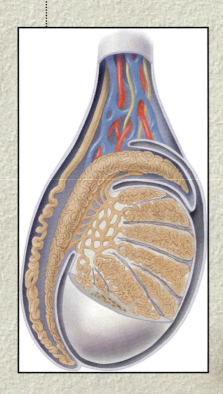

*Special***Topics**

Brushing up

To understand this chapter, it is essential that you understand or brush up on the following concepts:

- Stages of mitosis (pp. 154–156)
- Chromosome structure (p. 156)
- The karyotype (p. 159)
- Muscles of the pelvic floor (pp. 347–349)
- Hypothalamic releasing factors and pituitary gonadotropins (pp. 604–606)
- Negative feedback inhibition of the pituitary (p. 607)
- Androgens (pp. 613, 617)

From all we have learned of the structure and function of the human body, it seems a wonder that it works at all! The fact is, however, that even with modern medicine we cannot keep it working forever. The body suffers various degenerative changes as we age, and eventually we expire. Yet our genes live on in new containers—our offspring. Their production is the subject of these final chapters.

Sexual Reproduction

▼**Objectives**
When you have completed this section, you should be able to
- define *sexual reproduction;*
- state the most fundamental biological distinction between male and female; and
- define *primary sex organs, secondary sex organs,* and *secondary sex characteristics.*

The Essence of Sex

Reproduction is one of the fundamental properties of all living things. Many organisms reproduce simply by fission (splitting in two) or by budding off miniature replicas of themselves. These are asexual processes that produce exact genetic copies of the parent. Most organisms, however, reproduce sexually. **Sexual reproduction** does not necessarily entail copulation or even physical contact between the parents—many species merely shed sex cells into the sea and the parents never meet. The essence of sexual reproduction is that each offspring has two parents and a combination of genes from both. Thus the offspring are not genetically identical to their parents and usually not even to each other. Genetic diversity provides the foundation for the survival and evolution of a species and has been such a substantial advantage that it is employed by the great majority of organisms.

The Two Sexes

To reproduce sexually means that the parents must produce **gametes**[1] (sex cells) that can meet and combine their genes in a **zygote**[2] (fertilized egg). The gametes must provide two things for reproduction to be successful: (1) motility, so they can achieve contact, and (2) nutrients for the developing embryo. A single cell cannot perform both of these roles very well, because to contain ample nutrients means to be relatively large and heavy, and this is inconsistent with the need for motility. Therefore, these tasks are apportioned to two kinds of gametes. The small motile one—little more than DNA with a propeller—is the **spermatozoon (sperm cell),** while the large nutrient-laden one is the **egg (ovum).**

It is usually easy to distinguish a human male and female from each other, but this is not so obvious in many other species. By definition, an individual that produces eggs is female and one that produces sperm is male. Even in humans, this criterion is not always that simple—for example, when considering some abnormalities in sexual development to be discussed later. Genetically, any individual with a Y sex chromosome is classified as male and any individual lacking a Y chromosome (usually but not always having two X chromosomes) is classified as female.

In mammals, the female is also the parent that provides a sheltered internal environment for the development and prenatal nutrition of the embryo. For fertilization and development to occur in the female, the male must have a copulatory organ, the penis, for introducing his gametes into the female reproductive tract and the female must have a copulatory organ, the vagina, for receiving it. This is the most obvious difference between the sexes, but appearances can be deceiving (see fig. 17.14, p. 614 and special topic 27.1).

Overview of the Reproductive System

The reproductive system consists of primary and secondary sex organs. The **primary sex organs** are those that produce the gametes. Also known collectively as the **gonads,** they are the **testes (testicles)** of the male and **ovaries** of the female. The **secondary sex organs** are organs other than the gonads that are essential to reproduction. In the male, they constitute a system of ducts, glands, and the penis, concerned with the storage, survival, and conveyance of sperm. In the female, they

---

1. *gam* = marriage, union
2. *zygo* = yoke, union

include the uterine tubes, uterus, and vagina, concerned with uniting the sperm and egg and harboring the developing fetus.

Secondary sex characteristics are features that are not essential for reproduction but that attract the sexes to each other. From the call of a bullfrog to the tail of a peacock, these are well known in the animal kingdom. In humans there is room for opinion on what makes one sex attractive to the other, but the differences in physique, voice, and body hair; the apocrine scent glands (see chapter 7); the beards of males; and the breasts of females are considered to be secondary sex characteristics.

.... **Key Point Review** ....

1. Define *gonad* and *gamete*. Explain the relationship between the terms.
2. Define *male, female, sperm,* and *egg.*

Sex Determination and Development

▼**Objectives**
When you have completed this section, you should be able to
- explain the role of the sex chromosomes in determining sex;
- explain how the Y chromosome determines the response of the fetal gonad to prenatal hormones;
- describe which of the male and female external genitals are homologous to each other; and
- describe the descent of the testes and explain why it is important.

Role of the Sex Chromosomes

Most of our cells have 23 pairs of chromosomes: 22 pairs of *autosomes* and 1 pair of *sex chromosomes* (see fig. 5.17). A sex chromosome can either be a large X chromosome or a small Y chromosome. Every egg contains an X chromosome, but half of the sperm carry an X while the other half carry a Y. If an egg is fertilized with an X-bearing sperm, it produces an XX zygote that is destined to become a female. If it is fertilized with a Y-bearing sperm, it produces an XY zygote destined to become a male. Thus the sex of a child is determined at conception (fertilization), and not by the mother's egg but by the sperm that fertilizes it (fig. 27.1).

Hormones and Sex Differentiation

Sex determination does not end with fertilization, however. It requires an interaction between genetics and the hormones produced by the mother and fetus. Just as we have seen with other hormones (see chapter 17), those involved here require specific receptors on their target cells to exert an effect.

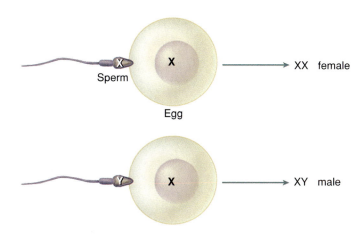

Figure 27.1 Chromosomal sex determination. All eggs carry the X chromosome. The sex of a child is determined by whether the egg is fertilized by an X-bearing sperm or a Y-bearing sperm.

Up to a point, a fetus is sexually undifferentiated, or "noncommittal" as to which sex it will become. Its gonads begin to develop at 5 to 6 weeks as **gonadal ridges** near the kidneys. Adjacent to them are two ducts, the **mesonephric**[3] (MEZ-oh-NEF-ric) **ducts (wolffian**[4] **ducts)** and the **paramesonephric**[5] **ducts (müllerian**[6] **ducts).** In males, the mesonephric ducts develop into the male reproductive tract and the paramesonephric ducts degenerate. In females, the opposite occurs (fig. 27.2).

But why? First, the Y chromosome has a gene for **testis-determining factor (TDF),** which codes for hormone receptors in the embryonic gonad. Second, the placenta (a disc of tissue that attaches the fetus to the mother's uterus) secretes a hormone called **human chorionic** (COR-ee-ON-ic) **gonadotropin (HCG).** If the fetus is XX, its gonad has no HCG receptors and it does not respond to the hormone. In such a fetus the mesonephric ducts degenerate and the paramesonephric ducts develop into the female reproductive tract (fig. 27.3).

If the fetus is XY, however, the gonad has HCG receptors and responds to the hormone. By 8 to 10 weeks, it begins to secrete **müllerian-inhibiting factor (MIF),** and androgens (primarily testosterone and a lesser amount of dihydrotestosterone). Androgens stimulate the mesonephric duct to develop into the male reproductive tract, while MIF causes the paramesonephric ducts to regress. Even an adult male, however, retains a tiny Y-shaped vestige of the paramesonephric ducts,

....

3. *meso* = middle + *nephr* = kidney; named for a temporary embryonic kidney, the mesonephros
4. Kaspar F. Wolff (1733–94), German anatomist
5. *para* = next to
6. Johannes P. Müller (1801–58), German physician

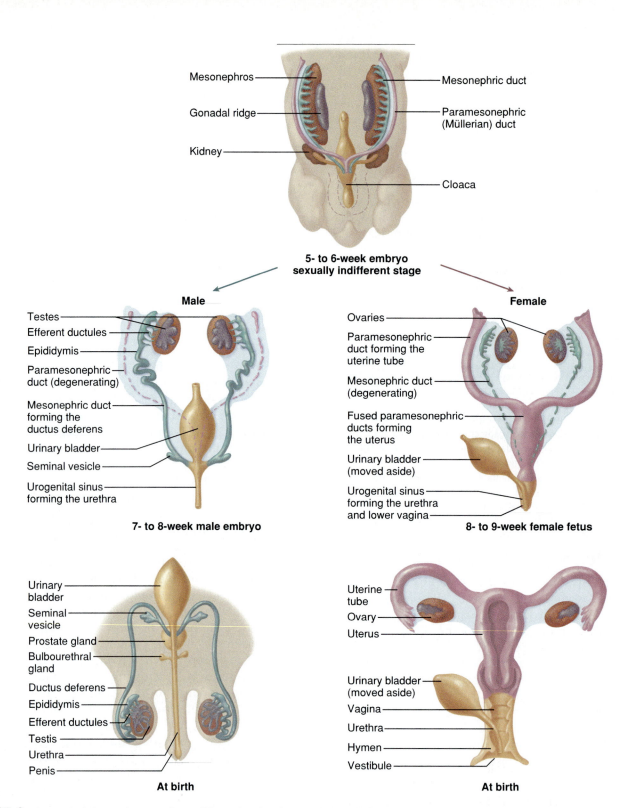

Mesonephros

Gonadal ridge

Kidney

Mesonephric duct

Paramesonephric (Müllerian) duct

Cloaca

5- to 6-week embryo sexually indifferent stage

Male

Testes

Efferent ductules

Epididymis

Paramesonephric duct (degenerating)

Mesonephric duct forming the ductus deferens

Urinary bladder

Seminal vesicle

Urogenital sinus forming the urethra

7- to 8-week male embryo

Female

Ovaries

Paramesonephric duct forming the uterine tube

Mesonephric duct (degenerating)

Fused paramesonephric ducts forming the uterus

Urinary bladder (moved aside)

Urogenital sinus forming the urethra and lower vagina

8- to 9-week female fetus

Urinary bladder

Seminal vesicle

Prostate gland

Bulbourethral gland

Ductus deferens

Epididymis

Efferent ductules

Testis

Urethra

Penis

At birth

Uterine tube

Ovary

Uterus

Urinary bladder (moved aside)

Vagina

Urethra

Hymen

Vestibule

At birth

Figure 27.2 The genital ducts of a sexually undifferentiated embryo and their differentiation into male and female reproductive organs. The gonad begins as a gonadal ridge adjacent to the mesonephros, an embryonic kidney that later degenerates. The mesonephric duct that originally served this kidney becomes the ductus deferens in the male.

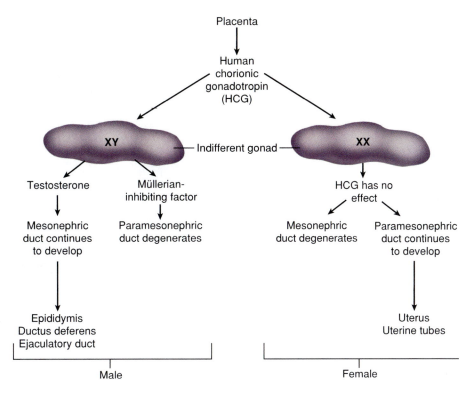

Figure 27.3 The XX or XY genetic constitution determines whether the indifferent gonad responds to human chorionic gonadotropin, which in turn determines whether the embryo takes a male or female course of development.

like a vestigial uterus and uterine tubes, in the area of the prostatic urethra. It is named the *uterus masculinus.*

It may seem as if androgens should induce the formation of a male reproductive tract and estrogens induce a female reproductive tract. However, estrogen levels are always high during pregnancy; if this mechanism were the case, they would have a feminizing effect on all fetuses. Thus the development of a female results from the absence of androgens, not the presence of estrogens (see special topic 27.1).

Development of the External Genitals

Most people regard the external genitals as the most definitive characteristics of a male or female. Yet there is more similarity between the male and female genitals than many people realize. In the fetus, they begin developing from identical structures in both sexes. By 8 weeks, the fetus has the following (fig. 27.4):

- a **phallus,** a small shaft of tissue with a swollen *glans* (head);
- **urogenital folds,** a pair of medial tissue folds slightly posterior to the phallus; and
- **labioscrotal folds,** a larger pair of tissue folds lateral to the urogenital folds.

By the end of week 9, the fetus begins to show sexual differentiation, and either male or female genitalia are fully formed by the end of week 12. In the female, the three structures listed become the clitoris, labia minora, and labia majora, respectively; all of these are more fully described in the next chapter. In the male, the phallus elongates to form the penis, the urogenital folds fuse to enclose the urethra within the penis, and the labioscrotal folds fuse to form the scrotum, a sac that will later contain the testes. Occasionally, the urogenital folds incompletely fuse, leaving a urinary orifice on the ventral side of the penis. This condition, called **hypospadias,**[7] is usually surgically corrected around 12 months of age.

Male and female organs that develop from the same embryonic structure are said to be **homologous.** Thus the penis is homologous to the clitoris and the scrotum is homologous to the labia majora. This becomes strikingly evident in some abnormalities of sexual development. In the presence of excess androgen, the clitoris may become greatly enlarged and resemble a small penis (see fig. 17.14, p. 614). In other cases, the ovaries descend into the labia majora as if they were testes descending into a scrotum. Such abnormalities sometimes result in mistaken identification of the sex of an infant at birth.

7. *hypo* = below + *spad* = to draw off (the urine)

Occasionally, a girl shows all the usual changes of puberty except that she fails to menstruate; physical examination shows the presence of testes in the abdomen; and a karyotype reveals that she has the XY chromosomes of a male. The testes produce normal male levels of testosterone, but the target cells lack receptors for it. This is called **androgen-insensitivity syndrome (AIS),** or **testicular feminization.**

The external genitals develop female anatomy as if no testosterone were present. At puberty, breasts and other feminine secondary sex characteristics develop (fig. 1) because the testes secrete small amounts of estrogen and there is no overriding influence of testosterone in AIS individuals. However, there is no uterus or menstruation. If the abdominal testes are not removed, the person has a high risk of testicular cancer.

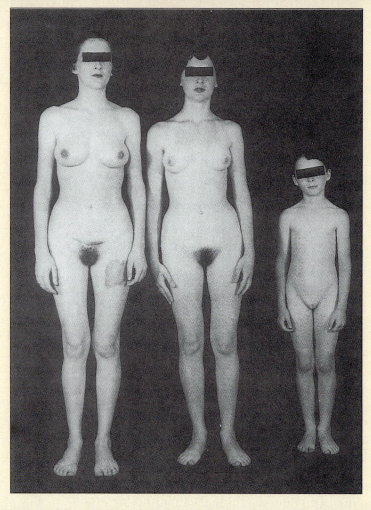

Figure 1 Androgen insensitivity syndrome. These three siblings are genetically male (XY). Testes are present and secrete testosterone, but the target cells lack receptors for it and the testosterone cannot exert its masculinizing effects. The external genitalia and secondary sex characteristics are feminine, but there are no ovaries, uterus, or vagina.

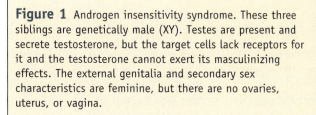

Think About It

In what way is androgen-insensitivity syndrome (see special topic 27.1) similar to noninsulin-dependent diabetes mellitus and Laron dwarfism (discussed in chapter 17)?

Descent of the Testes

The testes begin development near the kidney. How do they wind up in the scrotum, and why? In the embryo, a cord of muscle and connective tissue called the **gubernaculum**[8] (GYU-bur-NACK-you-lum) extends from the floor of the scrotum to the gonad. It shortens as the fetus grows and guides the testis through a passageway in the groin called the **inguinal canal.** This **descent** of the testes (fig. 27.5) begins in weeks 6 to 10; by 28 weeks, the testes enter the scrotum. As they descend, they are accompanied by ever-elongating testicular arteries and veins and by lymphatic vessels, nerves, sperm ducts, and extensions of the internal abdominal oblique muscle.

In the scrotum, the testes are kept at a temperature of about 35°C, about 2°C cooler than in the pelvic cavity. This is essential for sperm production. About 3% of boys are born with one or both testes still in the pelvic cavity—a condition called **cryptorchidism**[9] (crip-TOR-kid-izm). It is too warm in the pelvic cavity

---

8 . *gubern* = rudder, to steer, guide

9. *crypt* = hidden + *orchid* = testes

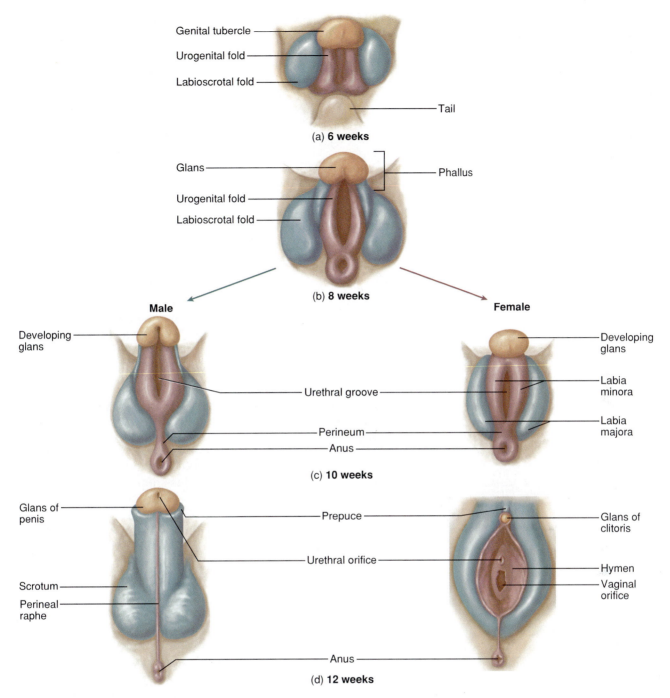

Figure 27.4 Fetal development of the external genitalia. (*a*) By 6 weeks, the embryo has three primordial structures—the phallus, urogenital folds, and labioscrotal folds—that will become the male or female genitalia. (*b*) At 8 weeks these structures have grown, but the sexes are still indistinguishable. (*c*) Slight sexual differentiation is noticeable at 10 weeks. (*d*) The sexes are fully distinguishable by 12 weeks. Matching colors identify homologous structures of the male and female.

for sperm to be produced; consequently, a male will be sterile if this is not corrected. In about 80% of cases, the testes descend spontaneously in the first year of infancy, and in another 10% of cases they descend by puberty. If spontaneous descent does not occur, hormone treatment or surgery is used to move the testes into the scrotum.

Key Point Review

3 What are mesonephric and paramesonephric ducts? What factors determine which one develops and which one regresses in the fetus?

4 What male structures develop from the phallus and labioscrotal folds?

5 Define the *gubernaculum* and describe its function.

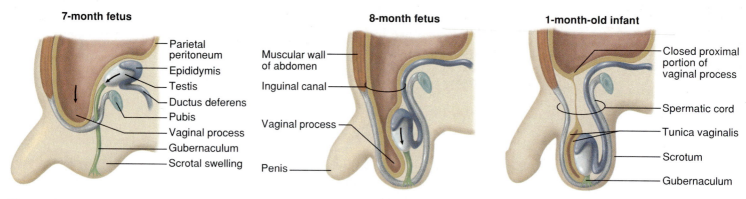

7-month fetus
- Parietal peritoneum
- Epididymis
- Testis
- Ductus deferens
- Pubis
- Vaginal process
- Gubernaculum
- Scrotal swelling

8-month fetus
- Muscular wall of abdomen
- Inguinal canal
- Vaginal process
- Penis

1-month-old infant
- Closed proximal portion of vaginal process
- Spermatic cord
- Tunica vaginalis
- Scrotum
- Gubernaculum

Figure 27.5 Descent of the testis. Note that the testis and spermatic ducts are retroperitoneal. An extension of the peritoneum called the vaginal process follows the testis through the inguinal canal and becomes the tunica vaginalis.

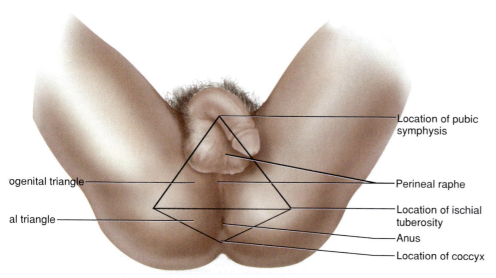

- Location of pubic symphysis
- ogenital triangle
- Perineal raphe
- Location of ischial tuberosity
- al triangle
- Anus
- Location of coccyx

Figure 27.6 The male perineum, inferior aspect.

Reproductive Anatomy

▼Objectives

When you have completed this section, you should be able to
- describe the anatomy of the testes, scrotum, and penis;
- describe the pathway taken by a sperm cell from its formation to its ejaculation; and
- describe the accessory glands and other secondary sex organs of the male, and state their functions.

The external genitalia occupy the **perineum** (PERR-ih-NEE-um), a diamond-shaped region marked by the pubic symphysis, ischial tuberosities, and coccyx (fig. 27.6). The internal reproductive anatomy of the male is shown in figure 27.7.

Testes

Each testis is oval, about 4 cm long and 2.5 cm in diameter, and weighs 10 to 14 g (fig. 27.8). Its anterior and lateral surfaces are covered by the **tunica vaginalis,**[10] a saclike extension of the peritoneum that follows the testis as it descends into the scrotum. The testis itself has a white fibrous capsule called the **tunica albuginea**[11] (TOO-nih-ca AL-byu-JIN-ee-uh). Connective tissue septa divide the organ into 250 to 300 wedge-shaped lobules.

Each lobule contains one to three **seminiferous**[12] (SEM-ih-NIF-er-us) **tubules**—slender ducts up to 70 cm long in which the sperm are produced. A seminiferous tubule has a narrow lumen lined by a thick **germinal epithelium** (fig. 27.9). The epithelium consists of several layers of **germ cells** in the process of becoming sperm and a much smaller number of tall **sustentacular**[13] (**Sertoli**[14]) **cells,** which protect the germ cells and promote their development. The germ cells depend on the sustentacular cells for nutrients, waste removal, growth factors, and other needs.

A sustentacular cell is shaped a little like a tree trunk whose roots spread out over the basement membrane, forming the boundary of the tubule, and whose thick trunk reaches to the tubule lumen. Tight junctions between adjacent sustentacular cells form a **blood-testis barrier (BTB),** which prevents proteins and other large molecules in the blood and intercellular fluid from getting to the germ cells. This is important because the

10. *tunica* = coat + *vagina* = sheath
11. *alb* = white
12. *semin* = seed, sperm + *fer* = to carry
13. *sustentacul* = support
14. Enrico Sertoli (1842–1910), Italian histologist

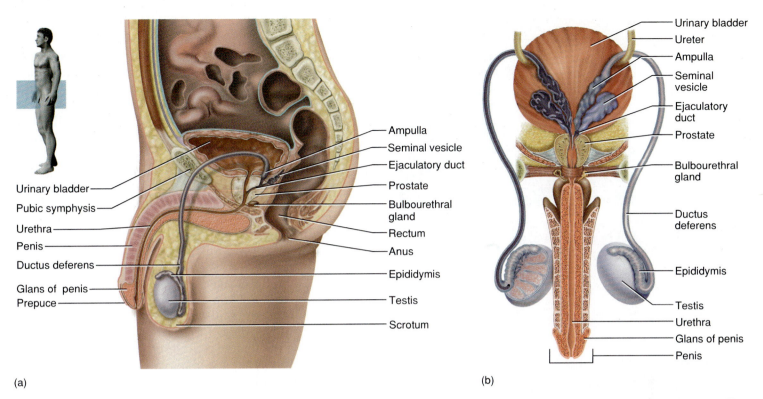

Figure 27.7 (*a*) The male reproductive system; midsagittal section viewed from the left. (*b*) Posterior view of the male reproductive organs. ✗

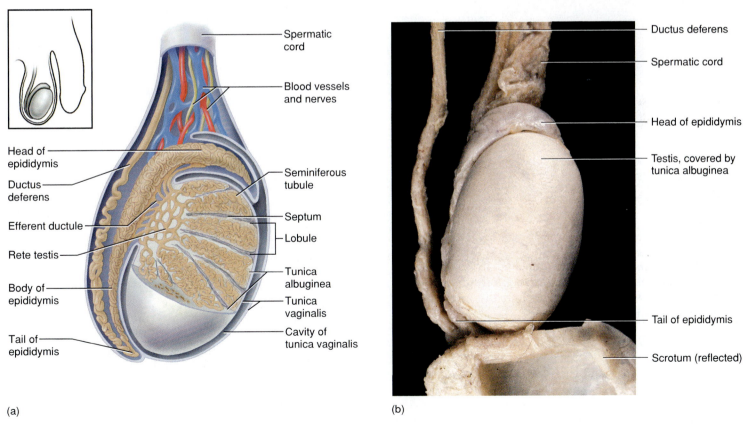

Figure 27.8 (*a*) Anatomy of the testis and epididymis. (*b*) The testis and associated structures dissected free of the scrotum. ✗

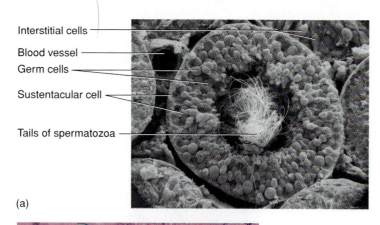

Interstitial cells

Blood vessel

Germ cells

Sustentacular cell

Tails of spermatozoa

(a)

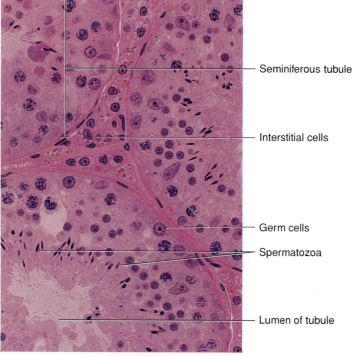

Seminiferous tubule

Interstitial cells

Germ cells

Spermatozoa

Lumen of tubule

(b)

Figure 27.9 Histology of the testis, showing the seminiferous tubules and interstitial cells. (*a*) SEM; (*b*) LM (×400). ⚡

(*a*) From R. G. Kessel and R. H. Kardon, *Scanning Electron Microscopy of Tissues and Organs*, 1979.

germ cells, being genetically different from other cells of the body, would otherwise be attacked by the immune system. Some cases of sterility occur when the BTB fails to form adequately in adolescence and the immune system produces autoantibodies against the germ cells.

 Think About It

Would you expect to find blood capillaries in the walls of the seminiferous tubules? Why or why not?

Between the seminiferous tubules are clusters of **interstitial (Leydig[15]) cells,** the source of testosterone.

15. Franz von Leydig (1821–1908), German anatomist

The testes also secrete small amounts of estrogen, the source and purpose of which are unknown.

The seminiferous tubules lead into a network called the **rete**[16] (REE-tee) **testis,** embedded in the capsule on the posterior side. Sperm partially mature in the rete. They are moved along by the flow of fluid secreted by the sustentacular cells and possibly by the cilia seen on some rete cells. Sperm do not swim while they are in the male reproductive tract.

Each testis is supplied by a **testicular artery** that arises from the abdominal aorta just below the renal artery. This is a very long, slender artery that winds its way down the posterior abdominal wall before passing through the inguinal canal into the scrotum. Its blood pressure is very low, and indeed this is one of the few arteries to have no pulse. Consequently, blood flow to the testes is quite meager and the testes receive a poor oxygen supply. Perhaps to compensate for this, the sperm develop unusually large mitochondria, which may precondition them for survival in the hypoxic environment of the female reproductive tract.

Blood leaves the testis by way of a **testicular vein.** The right testicular vein drains into the inferior vena cava and the left one drains into the left renal vein. Lymphatic vessels also drain each testis and lead to the inguinal lymph nodes. **Testicular nerves** lead to the gonads from spinal cord segment T10. They are mixed sensory and motor nerves containing predominantly sympathetic but also some parasympathetic fibers.

Scrotum

The testes are contained in a pendulous sac, the **scrotum**[17] (fig. 27.10). The left testis is usually suspended lower than the right so the two are not compressed against each other between the thighs. The skin of the scrotum has sparse hair, sebaceous glands, rich sensory innervation, and somewhat darker pigmentation than skin elsewhere. The scrotum is divided into right and left compartments by an internal **median septum,** which protects each testis from infections of the other one. The location of the septum is externally marked by a seam called the **perineal raphe**[18] (RAY-fee), which also extends anteriorly along the ventral side of the penis and posteriorly as far as the margin of the anus (see fig. 27.6).

A structure called the **spermatic cord** passes up the back of the scrotum, then anterior to the pubis, and through the *inguinal ring* into the inguinal canal. The canal is a passage about 4 cm long through the abdominal muscle to the pelvic cavity. The spermatic cord can be palpated through the skin on the back of the scrotum

16. *rete* = network
17. *scrotum* = bag
18. *raphe* = seam

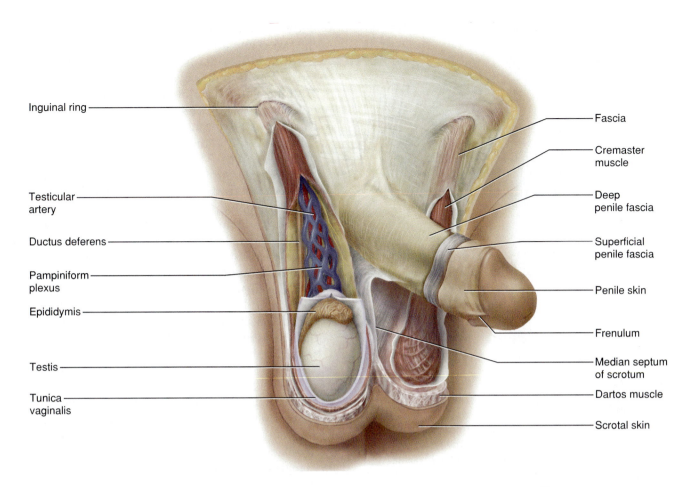

Inguinal ring

Testicular artery

Ductus deferens

Pampiniform plexus

Epididymis

Testis

Tunica vaginalis

Fascia

Cremaster muscle

Deep penile fascia

Superficial penile fascia

Penile skin

Frenulum

Median septum of scrotum

Dartos muscle

Scrotal skin

Figure 27.10 Dissection of the perineum to show anatomy of the scrotum, testis, spermatic cord, and penis. ⟋

and anterior to the pubic bone. It contains a sperm duct called the *ductus deferens* and other structures that descended with the testis as it passed through the inguinal canal—the aforementioned blood and lymphatic vessels, testicular nerves, and so forth.

The original reason that a scrotum evolved has long been a subject of debate among reproductive biologists and still has no universally accepted answer. Now that human testes reside in the scrotum, however, they have adapted to this cooler environment and cannot produce sperm at the core body temperature. The scrotum has three mechanisms for regulating the temperature of the testes:

1. The **dartos[19] muscle (tunica dartos)** is a subcutaneous layer of smooth muscle. When it is cold, the dartos contracts and draws the testes closer to the body to keep them warm; the scrotum then becomes taut and wrinkled. When it is warm, the dartos relaxes and the testes are suspended farther from the body.
2. The **cremaster[20] muscle** consists of strips of the internal abdominal oblique muscle that enmesh the spermatic cord. This, too, reflexively contracts when it is cold and relaxes when it is warm, elevating or lowering the testes as needed.
3. The **pampiniform[21] plexus** is an extensive network of veins from the testis that surround the testicular artery in the spermatic cord. As they pass through the inguinal canal, these veins converge to form the testicular vein, which emerges from the inguinal canal into the pelvic cavity. Without the pampiniform plexus, warm arterial blood would heat the testis and inhibit spermatogenesis. The pampiniform plexus, however, prevents this by acting as a *countercurrent heat exchanger* (fig. 27.11). Imagine that a house had uninsulated hot and cold water pipes running close to each other. Much of the heat from the hot water pipe would be absorbed by the cold water pipe next to it—especially if the water flowed in opposite directions so the cold water carried the heat away instead of running alongside the hot water. In the spermatic cord, such a mechanism removes heat from the descending arterial blood, so by the time it reaches the testis this blood is 1.5° to 2.5°C cooler than the core body temperature.

19. *dartos* = skinned
20. *cremaster* = suspender

21. *pampin* = tendril + *form* = shape

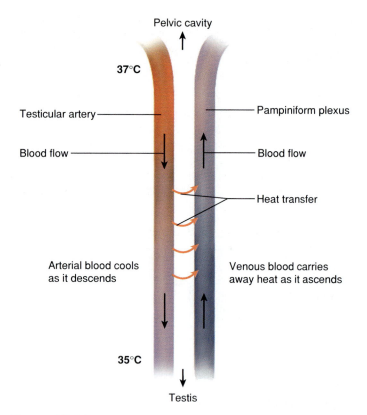

Figure 27.11 The mechanism of countercurrent heat exchange in the pampiniform plexus. Warm blood flowing down the testicular artery loses some of its heat to the cooler blood flowing in the opposite direction through the venous pampiniform plexus (represented as a single vessel for simplicity). Arterial blood is about 2°C cooler by the time it reaches the testis.

Spermatic Ducts

After leaving the testis, the sperm travel through a series of **spermatic ducts** to reach the urethra. These include the following:

- **Efferent ductules.** About 12 small efferent ductules arise from the posterior side of each testis and carry sperm to the epididymis. They have clusters of ciliated cells that help drive the sperm along.
- **Epididymis.** The epididymis[22] (EP-ih-DID-ih-miss; plural, *epididymides*) adheres to the posterior side of the testis (see fig. 27.8). It is about 7.5 cm long and consists of a clublike *head* at the superior pole of the testis, a long middle *body,* and a slender *tail* at its inferior end. It contains a single coiled duct embedded in fibroconnective tissue. If straightened out, the duct would be over 6 m (18 ft) long. The epididymis reabsorbs about 90% of the fluid secreted by the testis. Sperm are physiologically immature (incapable of fertilizing an egg) when they leave the testis but mature as they travel through the head and

22. *epi* = upon + *didym* = twins, testes

body of the epididymis. In 20 days or so, they reach the tail. They are stored here and in the adjacent portion of the ductus deferens. Stored sperm remain fertile for 40 to 60 days, but if they become too old without being ejaculated, they disintegrate and the epididymis reabsorbs them.

- **Ductus (vas) deferens.** The duct of the epididymis straightens out at the tail, turns 180°, and becomes the ductus deferens. This is a muscular tube about 45 cm long and 2.5 mm in diameter. It passes upward from the scrotum, travels through the inguinal canal, and enters the pelvic cavity (see fig. 27.7). There, it turns medially and approaches the urinary bladder. After passing between the bladder and ureter, the duct turns downward behind the bladder and widens into a terminal **ampulla.** The ductus deferens ends by uniting with the duct of the seminal vesicle, a gland considered later. Surgical sterilization of the male, called **vasectomy,** is done by making a small incision in the posterior side of the scrotum, clipping out a segment of the ductus deferens, and tying the cut ends.

  The ductus deferens has a very narrow lumen and a thick wall of smooth muscle well innervated by sympathetic nerve fibers. During orgasm, the sympathetic nervous system stimulates peristaltic contractions that drive sperm from the epididymis to the ejaculatory duct.
- **Ejaculatory duct.** Where the ductus deferens and duct of the seminal vesicle meet, they form a short (2 cm) ejaculatory duct, which passes through the prostate gland and empties into the urethra. The ejaculatory duct is the last of the spermatic ducts.

  The male urethra is shared by the reproductive and urinary systems. It is about 20 cm long and consists of three regions: the *prostatic, membranous,* and *penile urethra* (see fig. 23.21). Although it serves both urinary and reproductive roles, it cannot pass urine and semen simultaneously for reasons discussed later.

Accessory Glands

There are three sets of **accessory glands** in the male reproductive system: the seminal vesicles, prostate gland, and bulbourethral glands.

- **Seminal vesicles.** The seminal vesicles are a pair of glands posterior to the urinary bladder; one is associated with each ductus deferens. A seminal vesicle is about 5 cm long, or approximately the dimensions of your little finger. It has a connective tissue capsule and underlying layer of smooth muscle. The secretory portion is a very convoluted duct with numerous branches that form a complex labyrinth. The duct empties into the ampulla of the ductus deferens. The secretion of the seminal

The prostate gland weighs about 20 g by age 20, remains at that weight until age 45 or so, and then begins to grow slowly again. By age 70, over 90% of men show some degree of **benign prostatic hyperplasia**—noncancerous enlargement of the gland. The major complication of this is that it compresses the urethra, obstructs the flow of urine, and may promote bladder and kidney infections.

Prostate cancer is the second most common cancer in men (after lung cancer), affecting about 9% of men over the age of 50. Prostate tumors tend to form near the periphery of the gland, where they do not obstruct urine flow; therefore, they often go unnoticed until they cause pain. Prostate cancer often metastasizes to nearby lymph nodes and then to the lungs and other organs. It is more common among American blacks than whites and very uncommon among Japanese. It is diagnosed by digital rectal examination and by detecting *prostate specific antigen (PSA)* and *acid phosphatase* (a prostatic enzyme) in the blood. Up to 80% of men with prostate cancer survive when it is detected and treated early, but only 10% to 50% survive if it spreads beyond the prostatic capsule.

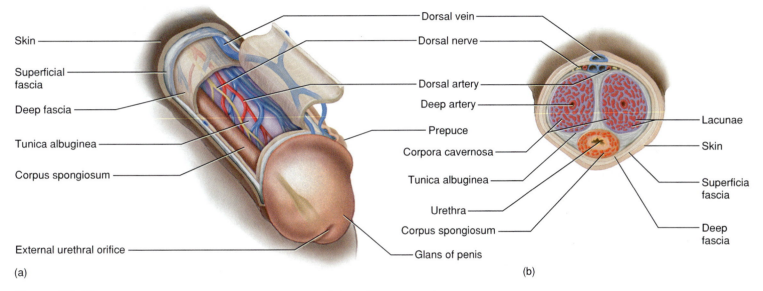

Figure 27.12 Anatomy of the penis. (*a*) Dissection in lateral view. (*b*) Cross section.

vesicles constitutes about 60% of the semen; its composition and functions are discussed later.

- **Prostate.** The prostate[23] (PROSS-tate) surrounds the urethra immediately inferior to the urinary bladder. It measures about 2 by 4 by 3 cm and is an aggregate of 30 to 50 compound tubuloacinar glands enclosed in a single fibrous capsule. These glands empty through about 20 pores in the urethral wall. The stroma of the prostate consists of connective tissue and smooth muscle, like that of the seminal vesicles. The thin, milky secretion of the prostate contributes about 30% of the semen. Its functions, too, are considered later. The position of the prostate immediately anterior to the rectum allows it to be palpated through the rectal wall to check for lumps suggestive of prostate cancer. This procedure is known as *digital rectal examination (DRE)* (see special topic 27.2).

- **Bulbourethral (Cowper[24]) glands.** The name of these glands refers to their position near a dilated bulb at the inner end of the penis and their short (2.5 cm) ducts leading into the penile urethra. They are brownish, spherical glands about 1 cm in diameter. During sexual arousal, they produce a clear slippery fluid that lubricates the head of the penis in preparation for intercourse. Perhaps more importantly, though, it neutralizes the acidity of residual urine in the urethra, which would be harmful to the sperm.

Penis

The **penis**[25] serves to deposit semen in the vagina. Half of it is an internal **root** and half is the externally visible **shaft** (see figs. 27.7 and 27.12). The external portion is

---

23. *pro* = before + *stat* = to stand; commonly misspelled and mispronounced "prostrate"

24. William Cowper (1666–1709), British anatomist
25. *penis* = tail

about 8 to 10 cm (3–4 in.) long and 3 cm in diameter when flaccid (nonerect); the typical dimensions of an erect penis are 13 to 18 cm (5–7 in.) long and 4 cm in diameter. Distally, the penis ends in a swollen head called the **glans**,[26] which has the external urinary meatus at its tip. The proximal margin of the glans is the **corona.**

The skin is very loosely attached to the shaft, allowing for expansion during erection. It continues over the glans as the **prepuce**, or foreskin, which is often removed by circumcision. A ventral fold of tissue called the **frenulum** attaches the skin to the glans. The skin of the glans itself is thinner and firmly attached to the underlying erectile tissue. The glans and facing surface of the prepuce have sebaceous glands that produce a waxy secretion called **smegma.**

The penis consists mainly of three cylindrical bodies called **erectile tissues,** which fill with blood during sexual arousal and account for its enlargement and erection. A single erectile body, the **corpus spongiosum,** passes along the ventral side of the penis and encloses the penile urethra. It dilates at the distal end to fill the entire glans. On the dorsal side of the penis there are a pair of **corpora cavernosa** (singular, *corpus cavernosum*). Each is ensheathed in a fibrous **tunica albuginea,** and they are separated from each other by a **median septum.** (Note that the testes also have a tunica albuginea and the scrotum also has a median septum.)

All three erectile tissues are spongy in appearance and contain numerous tiny blood sinuses called **lacunae.** The partitions between lacunae, called **trabeculae,** are composed of fibroconnective tissue and smooth **trabecular muscle.** In the flaccid penis, tonus of the trabecular muscle collapses the lacunae, which appear as tiny slits in the tissue.

At the body surface, the penis turns 90° and continues inward as the root. The corpus spongiosum terminates internally as a dilated **bulb,** which is ensheathed in the bulbospongiosus muscle and attached to the lower surface of the perineal membrane within the urogenital triangle (see pp. 347–348). The corpora cavernosa diverge like the arms of a Y. Each arm, called a **crus** (cruss; plural, *crura*), attaches the penis to the pubic arch (ischiopubic ramus) and perineal membrane on its respective side. The crura are enveloped by the ischiocavernosus muscle. The innervation and blood supply to the penis are discussed later in connection with the mechanism of erection.

-------- **Key Point Review** --------

6 State the names and locations of two muscles that help regulate the temperature of the testes.

7 Name three types of cells in the testis, and describe their locations and functions.

26. *glans* = acorn

8 Name all the ducts that the sperm follow, in order, from the time they leave the testis to the time of ejaculation.

9 Describe the locations and functions of the seminal vesicles, prostate, and bulbourethral glands.

10 Name the erectile tissues of the penis, and describe their locations relative to each other.

Puberty and Climacteric

▼**Objectives**
When you have completed this section, you should be able to
• describe the hormonal control of puberty;
• describe the resulting changes in the male body; and
• define and describe male climacteric and the effect of aging on male reproductive function.

Unlike any other organ system, the reproductive system remains dormant for several years after birth. Around age 10 to 12 in boys and 8 to 10 in girls, however, a surge of pituitary gonadotropins "awakens" the reproductive system and begins preparing it for adult reproductive function. This is **puberty,**[27] a period of transition in which the reproductive organs grow to mature size and begin producing gametes. In boys, puberty lasts approximately until age 13.

Endocrine Control of Puberty

The testes secrete substantial amounts of testosterone in the first trimester (3 months) of fetal development. Even in the first few months of infancy, testosterone levels are about as high as they are in midpuberty, but then the testes become dormant for the rest of infancy and childhood. From puberty through adulthood, reproductive function is regulated by hormonal links between the hypothalamus, pituitary, and gonads—the **brain-testicular axis** (fig. 27.13).

As the hypothalamus matures, it begins producing **gonadotropin-releasing hormone (GnRH),** which travels by way of the hypothalamo-hypophyseal portal system to the anterior lobe of the pituitary. Here it stimulates cells called *gonadotropes* to secrete **follicle-stimulating hormone (FSH)** and **interstitial cell–stimulating hormone (ICSH),** two *gonadotropins* that stimulate different cells in the testis (ICSH is the same as luteinizing hormone).

ICSH stimulates the interstitial cells of the testis to secrete androgens (mainly testosterone). FSH is needed for testosterone to have an effect on the testis. It stimulates the sustentacular cells to secrete **androgen-binding protein (ABP).** ABP is thought to raise androgen levels in the seminiferous tubules and epididymis, but this

27. *puber* = grown up

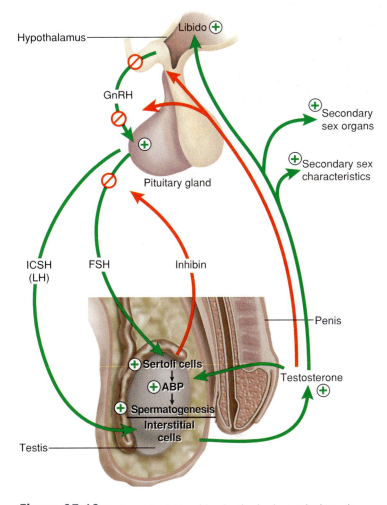

Figure 27.13 Hormonal relationships in the brain-testicular axis.

stimulates development of the pubic, the axillary, and later the facial hair. The skin becomes darker and thicker and secretes more sebum, which often leads to acne. The skin of acne patients contains 2 to 20 times as much DHT as normal.

Testosterone stimulates increased secretion of growth hormone. This produces a spurt of growth, an increase in muscle mass, a higher basal metabolic rate, and a larger larynx. The last of these effects deepens the voice and makes the thyroid cartilage more prominent at the front of the neck. Testosterone also stimulates erythropoiesis, which gives men a higher hematocrit and RBC count than those of women. It causes the ducts and accessory glands of the reproductive system to enlarge. Erections occur frequently and ejaculation may occur during sleep (nocturnal emissions, or "wet dreams").

Testosterone also stimulates the brain and awakens the **libido** (sex drive)—although, perhaps surprisingly, the neurons convert it to estrogen (a supposedly "female" sex hormone), which is what affects the behavior. Throughout adulthood, testosterone sustains the reproductive tract, spermatogenesis, and libido.

It is necessary at times to reduce FSH secretion without reducing ICSH secretion, particularly when sperm are produced faster than they are used. **Inhibin,** a hormone from the sustentacular cells, selectively suppresses FSH output from the pituitary. This slows down sperm production without inhibiting testosterone secretion. If sperm count drops below 20 million sperm/mL, however, inhibin secretion drops and FSH secretion rises.

Think About It

If a male animal is castrated, would you expect FSH and ICSH levels to rise, fall, or be unaffected. Why?

Aging and Sexual Function

Testosterone secretion peaks at about 7 mg/day at age 20 and then declines steadily to as little as one-fifth of this level by age 80. Correspondingly, there is a gradual decline in the number and secretory activity of the interstitial cells (the source of testosterone) and sustentacular cells (the source of inhibin). As testosterone and inhibin levels decline, so does feedback inhibition of the pituitary. Consequently, FSH and ICSH levels rise significantly in the late 40s to early 50s, producing changes called **male climacteric.** Most men pass through climacteric with little or no effect, but in a few cases there are mood changes, hot flashes, and illusions of suffocation—symptoms similar to those of menopause in women. Despite references to "male menopause," the term *menopause* refers to the cessation of menstruation and therefore makes little sense in the context of male physiology.

remains unproven. Germ cells have no androgen receptors and do not respond to it. Nevertheless, the androgens have three principal effects:

1. They stimulate spermatogenesis in the presence of ABP. If testosterone secretion ceases, the sperm count and semen volume decline rapidly and a male becomes sterile.
2. They suppress the secretion of GnRH by the hypothalamus and reduce the sensitivity of the pituitary to GnRH. Consequently, FSH and LH secretion are held in check. During puberty, however, the pituitary becomes less sensitive to this negative feedback, which is why these gonadotropin levels rise.
3. They stimulate development of the secondary sex characteristics and other somatic changes characteristic of puberty. Many of the physiological processes of puberty go unnoticed at first. The first visible sign of puberty is usually enlargement of the testes and scrotum around age 13. The penis continues to grow for about two more years. One of the androgens, dihydrotestosterone (DHT),

About 20% of men in their 60s and 50% of men in their 80s experience **impotence,** the inability to maintain an erection rigid enough to enter the vagina in half or more of their sexual attempts. Over 90% of impotent men, however, are able to ejaculate. Impotence and declining sexual activity can have a major impact on older people's perception of the quality of life. Impotence is due in part to the declining level of testosterone but also stems from a complex array of other causes: cultural beliefs, fear of failure, death of a spouse, depression, fewer available partners, physical frailty, cardiovascular and neurological diseases, diabetes mellitus, and many medications.

.... **Key Point Review** ....

11 State the source, target organ, and effect of GnRH.

12 Identify the target cells and effects of FSH and ICSH.

13 Explain how testicular hormones affect the secretion of FSH and ICSH.

14 Describe the major effects of androgens on the body.

15 Define *climacteric* and *impotence*.

Spermatogenesis, Spermatozoa, and Semen

▼Objectives

When you have completed this section, you should be able to
- describe the stages of meiosis and contrast meiosis with mitosis;
- describe the sequence of cell types in spermatogenesis, and relate these to the stages of meiosis;
- describe the role of the sustentacular cell in spermatogenesis;
- draw or describe a sperm cell; and
- describe the composition of semen and functions of its components.

The most significant event of puberty is the onset of **spermatogenesis.** This is a process in which germ cells undergo two divisions called **meiosis** and the four daughter cells differentiate into spermatozoa. It occurs in the seminiferous tubules.

Meiosis

In the eukaryotes (all living organisms except bacteria), there are two forms of cell division—mitosis and meiosis. Mitosis, described in chapter 5, is the basis for division of the single-celled fertilized egg, growth of an embryo, and all postnatal growth and tissue repair. It is essentially the splitting of a cell with a distribution of chromatids that results in two genetically identical daughter cells. It consists of four stages—prophase, metaphase, anaphase, and telophase.

You may find it beneficial to review figure 5.15 (p. 155) because of the important similarities and differences between mitosis and meiosis. There are three important differences:

1. In mitosis, each chromosome changes from a double-stranded to a single-stranded form, but each daughter cell still has 46 of them (23 pairs). Meiosis, by contrast, reduces the chromosome number by half. The parent cell is **diploid (2n),** meaning it has 46 chromosomes in 23 homologous pairs, whereas the daughter cells are **haploid (n),** having 23 unpaired chromosomes.

2. In mitosis, the chromosomes do not change their genetic makeup. In an early stage of meiosis, however, the chromosomes of each homologous pair join and exchange portions of their DNA. This creates new combinations of genes, so the chromosomes passed on to the gametes and offspring are not the same chromosomes that were inherited from the parents.

3. In mitosis, each parent cell produces only two daughter cells. In meiosis, it produces four. In the male, four sperm therefore develop from each original germ cell. The situation is somewhat different in the female, as we will see in the next chapter.

Why use such a relatively complicated process for gametogenesis? Why not use mitosis, as we do for all other cell replication in the body? The answer is basically that sexual reproduction is, by definition, biparental. If we are going to combine gametes from two parents to make an offspring, there must be a mechanism for keeping the chromosome number constant from generation to generation. Mitosis would produce eggs and sperm with 46 chromosomes each. If these gametes combined, the zygote and the next generation would have 92 chromosomes per cell, the generation after that would have 184, and so forth. To prevent the chromosome number from doubling in every generation, the number is reduced by half during gametogenesis. Meiosis is sometimes called *reduction division* for this reason.[28]

The stages of meiosis are fundamentally the same in both sexes. Briefly, it consists of two cell divisions in succession and occurs in the following phases: prophase I, metaphase I, anaphase I, telophase I, interkinesis, prophase II, metaphase II, anaphase II, and telophase II. These events are shown in detail in figure 27.14, but let us note some of its most unique and important aspects.

In prophase I, each pair of homologous chromosomes line up side by side, forming a **tetrad** (*tetra* denoting the four chromatids). One chromosome of each

28. *meio* = less, fewer

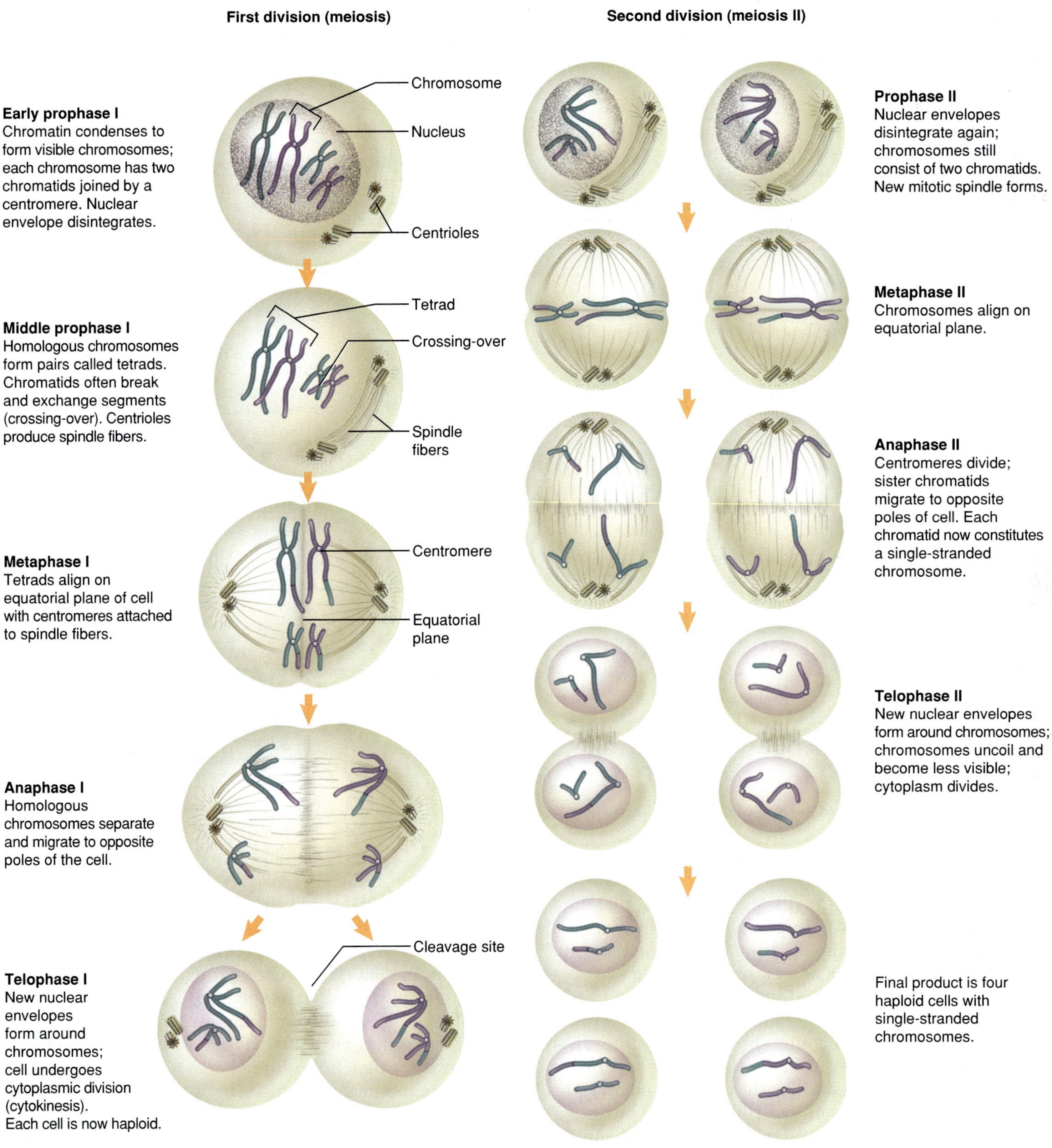

First division (meiosis)

Second division (meiosis II)

Early prophase I
Chromatin condenses to form visible chromosomes; each chromosome has two chromatids joined by a centromere. Nuclear envelope disintegrates.

Chromosome

Nucleus

Centrioles

Middle prophase I
Homologous chromosomes form pairs called tetrads. Chromatids often break and exchange segments (crossing-over). Centrioles produce spindle fibers.

Tetrad

Crossing-over

Spindle fibers

Metaphase I
Tetrads align on equatorial plane of cell with centromeres attached to spindle fibers.

Centromere

Equatorial plane

Anaphase I
Homologous chromosomes separate and migrate to opposite poles of the cell.

Telophase I
New nuclear envelopes form around chromosomes; cell undergoes cytoplasmic division (cytokinesis). Each cell is now haploid.

Cleavage site

Prophase II
Nuclear envelopes disintegrate again; chromosomes still consist of two chromatids. New mitotic spindle forms.

Metaphase II
Chromosomes align on equatorial plane.

Anaphase II
Centromeres divide; sister chromatids migrate to opposite poles of cell. Each chromatid now constitutes a single-stranded chromosome.

Telophase II
New nuclear envelopes form around chromosomes; chromosomes uncoil and become less visible; cytoplasm divides.

Final product is four haploid cells with single-stranded chromosomes.

Figure 27.14 Meiosis. For simplicity, the cell is shown with only two pairs of homologous chromosomes. Human cells begin meiosis with 23 pairs. ▭

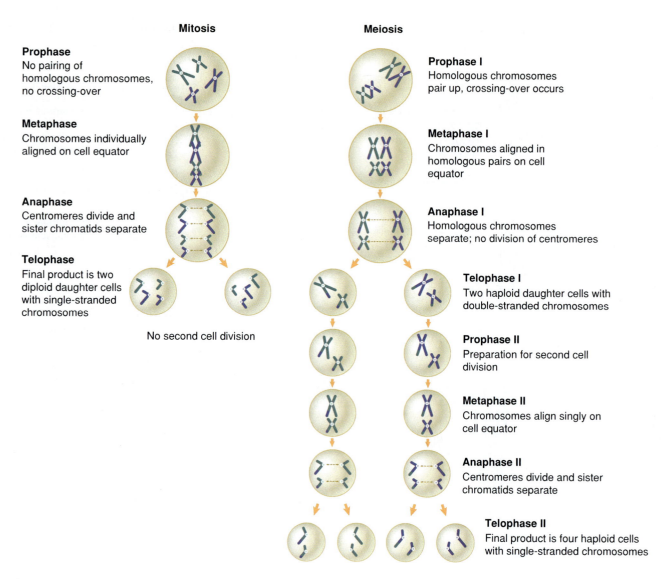

Mitosis

Prophase
No pairing of homologous chromosomes, no crossing-over

Metaphase
Chromosomes individually aligned on cell equator

Anaphase
Centromeres divide and sister chromatids separate

Telophase
Final product is two diploid daughter cells with single-stranded chromosomes

No second cell division

Meiosis

Prophase I
Homologous chromosomes pair up, crossing-over occurs

Metaphase I
Chromosomes aligned in homologous pairs on cell equator

Anaphase I
Homologous chromosomes separate; no division of centromeres

Telophase I
Two haploid daughter cells with double-stranded chromosomes

Prophase II
Preparation for second cell division

Metaphase II
Chromosomes align singly on cell equator

Anaphase II
Centromeres divide and sister chromatids separate

Telophase II
Final product is four haploid cells with single-stranded chromosomes

Figure 27.15 Comparison of the major features of mitosis and meiosis.

tetrad is from the individual's father (the paternal chromosome) and the other is from the mother (the maternal chromosome). The paternal and maternal chromosomes exchange segments of DNA in a process called **crossing-over.** This creates new combinations of genes and thus contributes to genetic variety in the offspring.

After crossing-over, the chromosomes line up at the midline of the cell in metaphase, they separate at anaphase, and the cell divides in two at telophase. This looks superficially like mitosis, but there is an important difference: the centromeres do not divide and the chromatids do not separate from each other at anaphase I; rather, each homologous chromosome parts company with its twin. Therefore, at the conclusion of meiosis I, each chromosome is still double-stranded, but each daughter cell has only 23 chromosomes—it has become haploid.

Meiosis II is more like mitosis—the chromosomes line up on the cell equator at metaphase, the centromeres divide, and each chromosome separates into two chromatids. These chromatids are drawn to opposite poles of the cell at anaphase II. At the end of meiosis II, there are four haploid cells, each containing 23 single-stranded chromosomes. Fertilization combines 23 chromosomes from the father with 23 chromosomes from the mother, reestablishing the diploid number of 46 in the zygote. These contrasts between mitosis and meiosis are summarized in figure 27.15.

Spermatogenesis

Now let's consider how meiosis relates to sperm production (fig. 27.16). The first cells destined to become sperm are **primordial germ cells.** Like the first blood cells, these form in the yolk sac, a membrane associated with the developing embryo. In the fifth to sixth week of development, they migrate into the embryo itself and colonize the gonadal ridges. Here they differentiate into **spermatogonia**, which lie along the periphery of the seminiferous tubule, outside the blood-testis barrier (BTB).

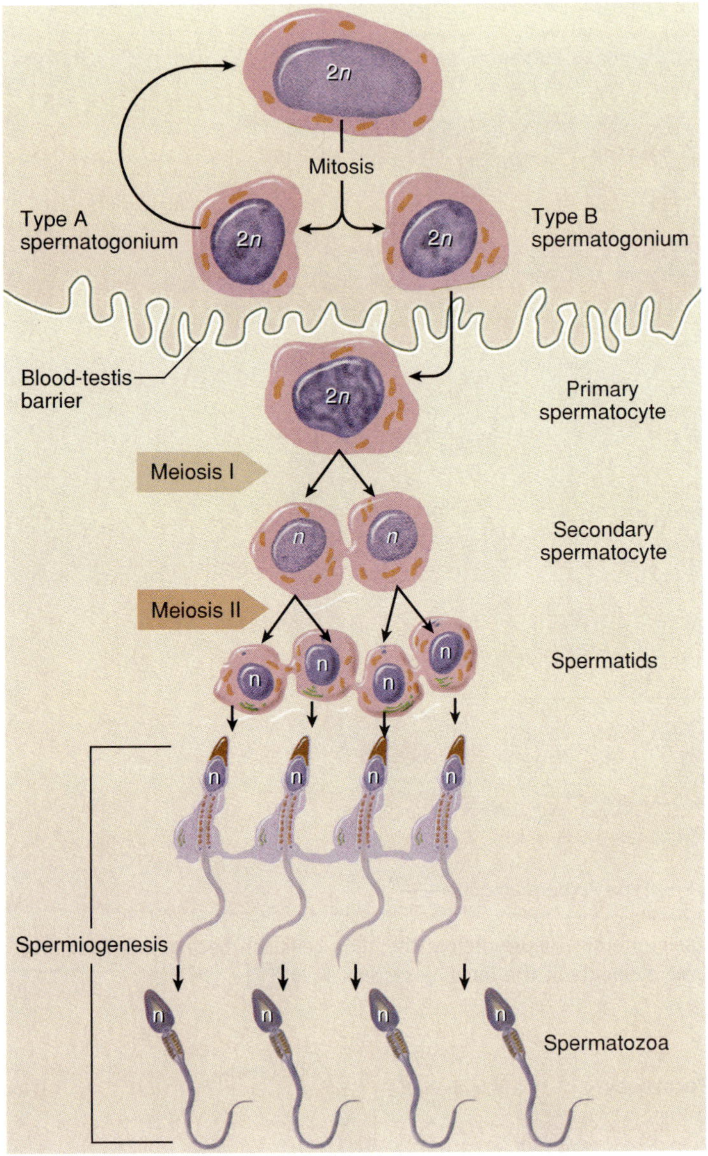

Spermatogonia multiply by mitosis, producing two types of daughter cells called type A and type B spermatogonia. Type A cells remain outside the BTB and continue to multiply from puberty until death. Thus men never exhaust their supply of gametes and normally remain fertile throughout old age.

Figure 27.16 Spermatogenesis. $2n$ indicates diploid cells and n indicates haploid cells. Note that the daughter cells from secondary spermatocytes through spermatids remain connected by slender cytoplasmic processes until spermatogenesis is complete and individual spermatozoa are released. 🎞 ♂

Type B spermatogonia migrate closer to the tubule lumen and differentiate into slightly larger cells called **primary spermatocytes**. Sustentacular cells transfer these cells through the BTB toward the inside of the tubule (fig. 27.17). To do so, they dismantle the tight junction ahead of a spermatocyte and assemble a new tight junction behind it. The spermatocyte moves through the BTB a bit like an astronaut moving through the double doors of an airlock.

After the BTB closes behind it, the primary spermatocyte undergoes meiosis I, giving rise to two equal-sized, haploid **secondary spermatocytes**. Each of these undergoes meiosis II, dividing into two **spermatids**—or a total of four for each spermatogonium. Each stage is a little closer to the lumen of the tubule than the earlier stages. All stages on the lumenal side of the BTB are bound to the sustentacular cells by tight junctions and gap junctions and are closely enveloped in tendrils of the sustentacular cells. Throughout these meiotic divisions, the daughter cells also remain connected to each other by narrow cytoplasmic bridges.

The rest of spermatogenesis is called **spermiogenesis** (fig. 27.18). It involves no further cell division, but a gradual transformation of each spermatid into a spermatozoon. A spermatid sprouts a flagellum (tail) and discards most of its cytoplasm, becoming as small and lightweight as possible. Eventually, the sperm cell is released and is washed down the tubule by fluid from the sustentacular cells. It takes about 74 days for a spermatogonium to become a mature spermatozoon. A young man produces about 300,000 sperm per minute, or 400 million per day.

The Spermatozoon

The spermatozoon has two parts: a pear-shaped head and a long tail (fig. 27.19). The **head**, about 4 to 5 μm long and 3 μm wide at its broadest part, contains three structures: a nucleus, acrosome, and flagellar basal body. The most important of these is the nucleus, which fills most of the head and contains a haploid set of condensed, genetically inactive chromosomes. The acrosome[29] is a lysosome in the form of a thin cap covering the apical half of the nucleus. It contains enzymes that are later used to penetrate the egg if the sperm is successful. The basal body of the tail flagellum is nested in an indentation at the basal end of the nucleus.

The **tail** is divided into three regions called the midpiece, principal piece, and endpiece. The **midpiece**, a cylinder about 5 to 9 μm long and half as wide as the head, is the thickest part. It contains numerous large mitochondria that spiral tightly around the axoneme of the

29. *acro* = tip, peak + *some* = body

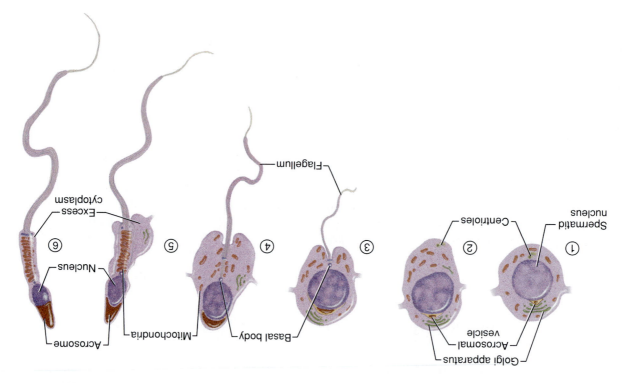

Figure 27.18 Spermiogenesis, in which the spermatids discard excess cytoplasm, grow tails, and become spermatozoa.

Figure 27.17 Spermatogenesis in relation to the sustentacular cells of the seminiferous tubule. Before the germ cells can become haploid, they must be passed through the blood-testis barrier and be isolated from blood-borne elements of the immune system. ⚡

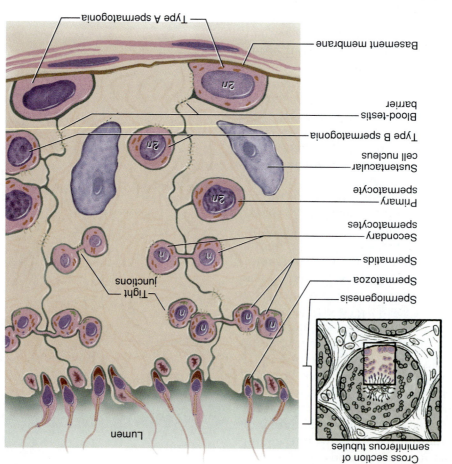

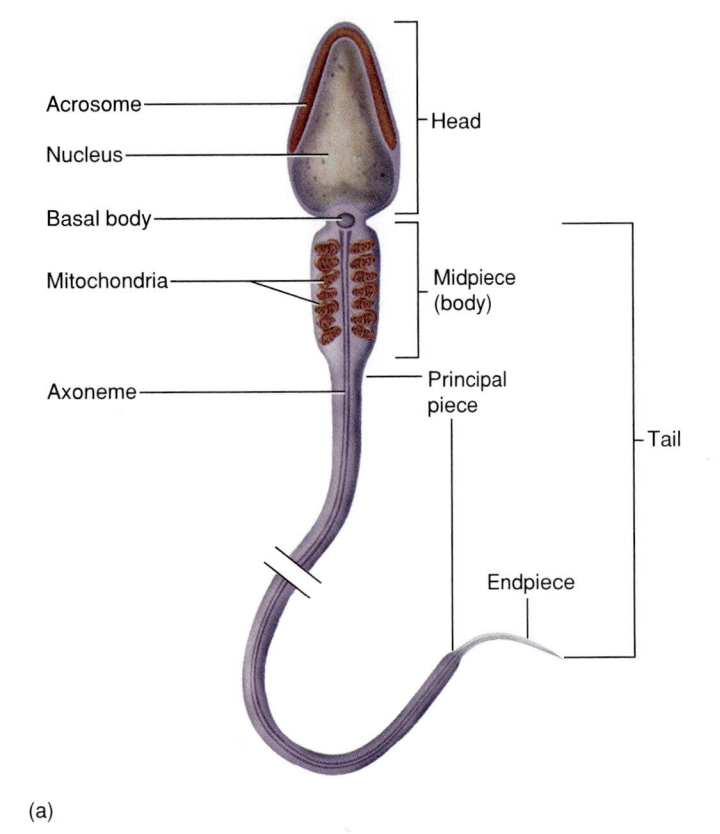

(a)

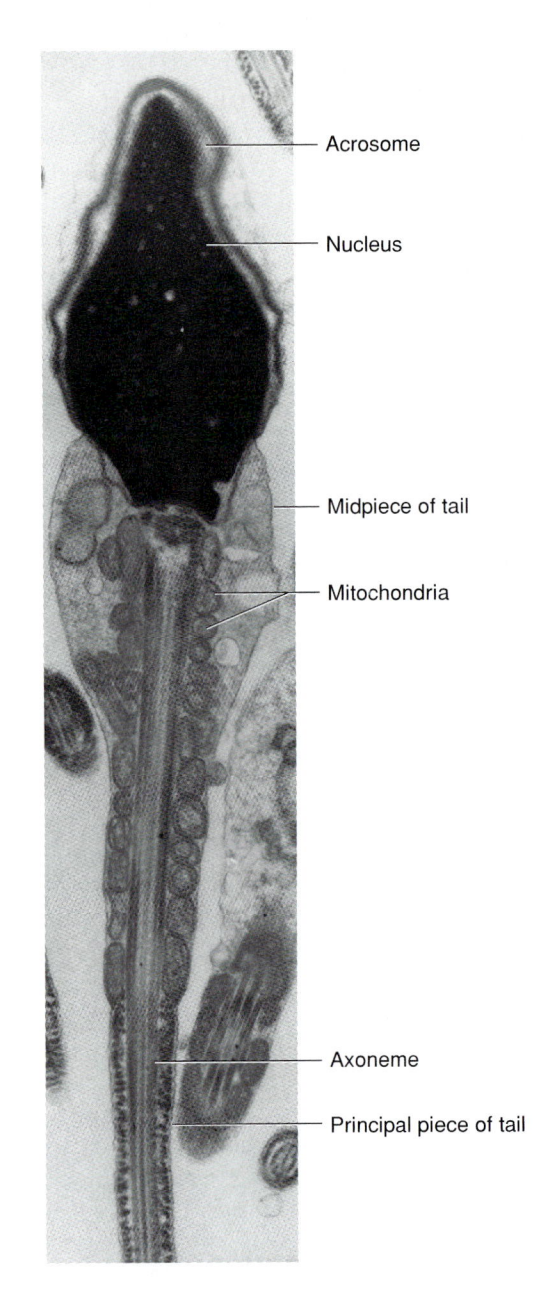

(b)

Figure 27.19 (*a*) Structure of a mature spermatozoon. (*b*) Head and part of the tail of a spermatozoon (TEM).

flagellum. They produce the ATP needed for the beating of the tail when the sperm migrates up the female reproductive tract. The **principal piece,** 40 to 45 μm long, constitutes most of the tail and consists of the axoneme surrounded by a sheath of fibers. The **endpiece,** 4 to 5 μm long, consists of flagellum only and is the narrowest part of the sperm.

Semen

The fluid expelled during orgasm is called **semen,**[30] or **seminal fluid.** A typical ejaculation discharges 2 to 5 mL of semen, composed of about 60% seminal

vesicle fluid, 30% prostatic fluid, 10% sperm and spermatic duct secretions, and a trace of bulbourethral gland secretion.

The major constituents of semen are as follows:

- **Spermatozoa.** The semen usually has a **sperm count** of 50 to 120 million sperm/mL. A sperm count any lower than 20 to 25 million sperm/mL is usually associated with **sterility,** the inability to fertilize an egg. Sterility also occurs, however, when the sperm count is normal but more than 20% of the sperm show deformity (two heads or tails, defective tails, etc.) or poor motility. Sterility can result from poor nutrition, gonorrhea and other infections (see chapter essay, p. 986), testosterone deficiency, and perhaps environmental pollutants (see special topic 27.3).

---

30. *semen* = seed

Special Topic 27.3 Feminizing Effects of Pollution

In recent decades, wildlife biologists have noticed increasing numbers of male birds, fish, and alligators with "feminizing" abnormalities of reproductive development. Evidence is mounting that humans, too, are showing declining fertility and increasing anatomical abnormalities due to feminizing pollutants in water, meat, vegetables, and even breast milk and the uterine environment. Cryptorchidism and hypospadias have increased at alarming rates in the last 50 years, and the rate of testicular cancer has more than tripled. Data on 15,000 men from several countries show a sharp drop in average sperm count—from 113 million/mL in 1940 to only 66 million/mL in 1990. Total sperm production has decreased even more, because the average volume of semen per ejaculate has dropped 19% over this period.

The pollutants implicated in this trend include a wide array of common herbicides, insecticides, industrial chemicals, and breakdown products of materials ranging from plastics to dishwashing detergents. Some of these act by mimicking estrogens, while others block the action of testosterone by binding to its receptors.

- **Fructose.** This sugar, produced by the seminal vesicles, provides a source of energy for sperm motility.
- **Clotting and anticoagulant factors.** The seminal vesicles secrete fibrinogen and the prostate produces a clotting enzyme. When these secretions mix during ejaculation, the clotting enzyme converts fibrinogen to fibrin, causing the semen to clot like blood. It becomes very sticky and adheres to the deep recesses of the vagina rather than draining back out. About 15 to 30 minutes later, fibrinolysin in the prostatic fluid dissolves the clot. Sperm are then liberated and able to begin their migration up the female reproductive tract.
- **Prostaglandins.** Prostaglandins, produced by the prostate and seminal vesicles, stimulate peristaltic contractions of the female reproductive tract that may help draw semen into the uterus or spread it through the uterus. Prostaglandins also reduce the viscosity of the mucus in the cervical canal of the female, making it easier for sperm to travel up this passageway.
- **Spermine.** Spermine and other bases give the semen a pH of 7.2 to 7.6. This is important because the vagina has a pH of about 3.5 to 4.0. Sperm motility is poor at a pH less than 6 and thus requires neutralization of vaginal acidity by the semen.

Key Point Review

16. State how many chromosomes a cell has, and how many chromatids each chromosome has, at the conclusion of meiosis I and meiosis II.
17. Name the stages of spermatogenesis from spermatogonium to spermatozoon.
18. Describe the three major parts of a spermatozoon and state what organelles or cytoskeletal components are contained in each.
19. List the major contributions of the seminal vesicles and prostate gland to the semen, and state the functions of these components.

Sexual Intercourse

▲ **Objectives**

When you have completed this section, you should be able to
- describe the blood and nerve supply to the penis; and
- explain how these govern erection and ejaculation.

The physiology of sexual intercourse was unexplored territory before the 1950s because of repressive attitudes toward the subject. British psychologist Havelock Ellis (1859–1939) suffered severe professional sanctions merely for surveying people on their sexual behavior. In the 1950s, William Masters and Virginia Johnson daringly launched the first physiological studies of sexual response in the laboratory. In 1966, they published *Human Sexual Response*, detailing measurements and observations on more than 10,000 sexual acts by nearly 700 volunteer men and women. Masters and Johnson then turned their attention to disorders of sexual function and pioneered modern therapy for sexual dysfunctions.

Sexual intercourse is also known as **coitus, coition,**[31] or **copulation.**[32] Masters and Johnson divided intercourse into four recognizable phases, which they called excitement, plateau, orgasm, and resolution. The following discussion is organized around this model, although other authorities have modified or proposed alternatives to it.

Anatomical Considerations

To understand male sexual function, we must give closer attention to the blood circulation and nerve supply to the penis.

31. *coit* = to come together
32. *copul* = link, bond

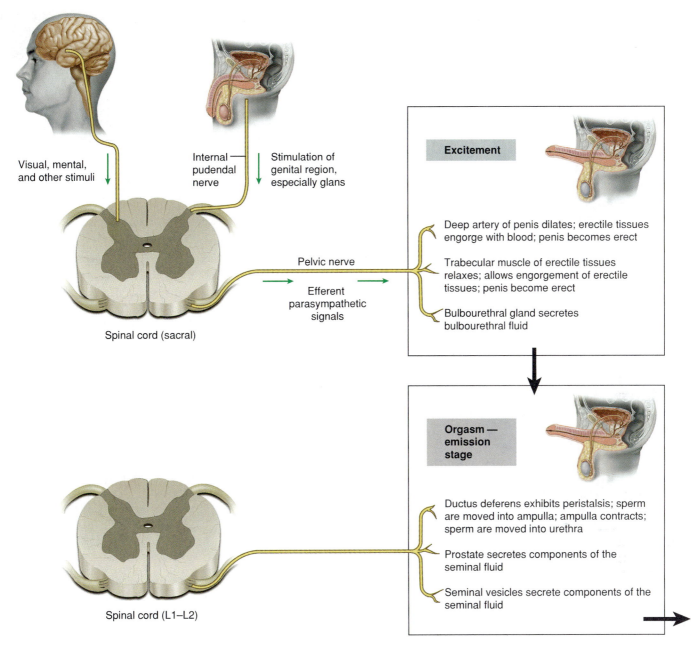

Visual, mental, and other stimuli

Internal pudendal nerve

Stimulation of genital region, especially glans

Spinal cord (sacral)

Pelvic nerve

Efferent parasympathetic signals

Excitement

Deep artery of penis dilates; erectile tissues engorge with blood; penis becomes erect

Trabecular muscle of erectile tissues relaxes; allows engorgement of erectile tissues; penis become erect

Bulbourethral gland secretes bulbourethral fluid

Orgasm— emission stage

Ductus deferens exhibits peristalsis; sperm are moved into ampulla; ampulla contracts; sperm are moved into urethra

Prostate secretes components of the seminal fluid

Seminal vesicles secrete components of the seminal fluid

Spinal cord (L1–L2)

Figure 27.20 Neural control of male sexual response—*Cont'd on next page.*

Each internal iliac artery gives rise to an **internal pudendal (penile) artery,** which enters the root of the penis and divides in two. One branch, the **dorsal artery,** travels dorsally along the penis not far beneath the skin (see fig. 27.12). The other branch, the **deep artery,** travels through the core of the corpus cavernosum, giving off smaller **helicine**[33] **arteries,** which penetrate the trabeculae and empty into the lacunae. When the deep artery dilates, the lacunae fill with blood and the penis

becomes erect. When the penis is flaccid, most of its blood supply comes from the dorsal arteries.

The penis is richly innervated by sensory and motor nerve fibers. The glans has an abundance of tactile, pressure, and temperature receptors, especially on the corona and frenulum. Sensory fibers of the shaft, scrotum, perineum, and elsewhere are also highly important to erotic stimulation. They lead by way of a prominent **dorsal nerve** of the penis to the **internal pudendal nerve,** then to the sacral plexus, and finally to the sacral region of the spinal cord (fig. 27.20).

33. *helic* = coil, helix

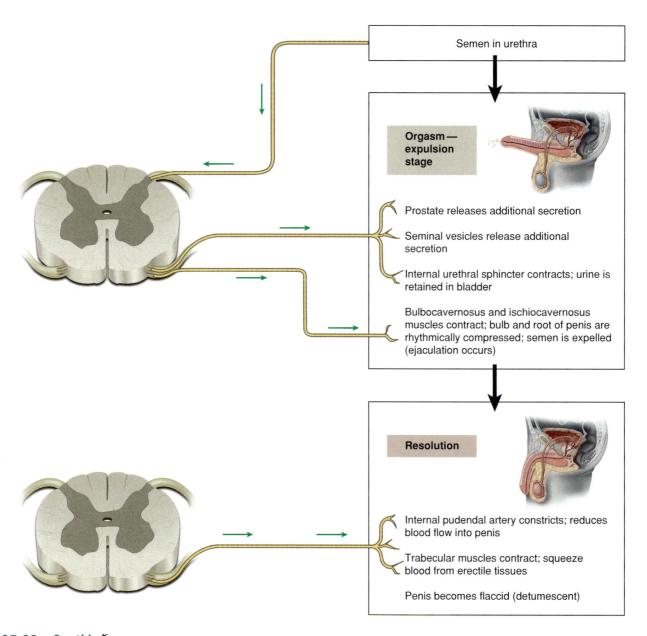

Figure 27.20—Cont'd. ✗

Both autonomic and somatic motor fibers carry impulses from integrating centers in the spinal cord to the penis and other pelvic organs. Sympathetic fibers arise from levels T12 to L2, pass through the hypogastric and pelvic nerve plexuses, and innervate the penile arteries, trabecular muscle, spermatic ducts, and accessory glands. They have a vasodilating effect on the penile arteries and can bring about erection even when the sacral region of the spinal cord is damaged. They are the initial mediators of erection induced by thoughts and by input to the special senses.

Parasympathetic fibers extend from segments S2 to S4 of the spinal cord through the pudendal nerves to the arteries of the penis. They are involved in an autonomic reflex arc that causes erection in response to direct stimulation of the penis and other perineal organs.

Excitement and Plateau

The **excitement** phase is characterized by **vasocongestion** (swelling of the genitals with blood), **myotonia** (muscle tension), and increases in heart rate, blood pressure, and pulmonary ventilation. The bulbourethral glands secrete their fluid during this phase. The excitement phase can be initiated by a broad spectrum of erotic stimuli—sights, sounds, aromas, touch—and even by dreams or the mere thought of sex. Conversely, emotions can inhibit sexual response and make it difficult to function when a person is anxious, stressed, or preoccupied with other thoughts.

The most obvious manifestation of male sexual excitement is **erection** of the penis, which makes entry of the vagina possible. Erection is an autonomic reflex mediated predominantly by parasympathetic nerve fibers that travel alongside the deep and helicine arteries of the penis. These fibers apparently produce no organic neurotransmitter but secrete nitric oxide (NO), which relaxes and dilates the deep arteries and lacunae. Whether this is enough to cause erection, or whether it is also necessary to block the outflow of blood from the penis, is still being debated. According to one hypothesis, as lacunae close to the deep arteries fill with blood, they expand and compress lacunae closer to the periphery of the erectile tissue. This is where blood leaves the erectile tissues, so the compression of the peripheral lacunae helps retain blood in the penis. Their compression is aided by the fact that each corpus cavernosum is wrapped in a tunica albuginea, which fits over the erectile tissue like a tight fibrous sleeve and contributes to its tension and firmness.

As the corpora cavernosa expand, the penis becomes enlarged, rigid, and elevated to an angle conducive to entry of the vagina. Once **intromission** (entry) is achieved, the tactile and pressure sensations produced by vaginal massaging of the penis further accentuate the erection reflex.

The corpus spongiosum has neither a central artery nor a tunica albuginea. It swells and becomes more visible as a cordlike ridge along the ventral surface of the penis, but it does not become nearly as engorged and hardened as the corpora cavernosa. Vasocongestion is not limited to the penis; the testes also become as much as 50% larger during excitement.

In the **plateau** phase, variables such as respiratory rate, heart rate, and blood pressure are sustained at a high level, or rise slightly, for a few seconds to a few minutes before orgasm. It may be marked by increased vasocongestion and myotonia.

Think About It

Why is it important that the corpus spongiosum not become as engorged and rigid as the corpora cavernosa?

Orgasm and Ejaculation

The **orgasm,**[34] or **climax,** is a short but intense reaction, lasting 3 to 15 seconds and usually marked by the discharge of semen. The heart rate increases to as high as 180 beats per minute, blood pressure rises proportionately, and the respiratory rate becomes as high as 40

breaths per minute. From the standpoint of producing offspring, the most significant aspect of male orgasm is the **ejaculation**[35] of semen into the vagina.

Ejaculation occurs in two stages called emission and expulsion. In **emission,** the sympathetic nervous system causes peristalsis of the ductus deferens, which propels sperm from the tail of the epididymis, along the ductus, and into the ampulla. Contractions of the ampulla propel the sperm into the prostatic urethra, and contractions of smooth muscle in the prostate gland express prostatic fluid into the urethra. Secretions of the seminal vesicles join the semen soon after the prostatic secretion. The contractions and seminal flow of this phase create an urgent sensation that ejaculation is inevitable.

Semen in the urethra activates somatic and sympathetic reflexes that produce **expulsion** of the semen. Sensory impulses travel to the spinal cord via the pudendal nerve. Efferent signals come from a reflex center in the upper lumbar region of the spinal cord by way of sympathetic fibers to the prostate and seminal vesicles, causing further glandular secretion. The sympathetic reflex also constricts the internal urinary sphincter so urine cannot enter the urethra.

Somatic motor impulses leave the third and fourth sacral segments of the cord and travel to the bulbospongiosus, ischiocavernosus, and levator ani muscles. The bulbospongiosus, which envelops the root of the corpus spongiosum, undergoes five or six strong, spasmodic contractions that compress the urethra and forcibly expel the semen. Most sperm are ejected in the first one or two spurts of semen, mixed primarily with prostatic fluid. The seminal vesicle secretion follows and flushes most remaining sperm from the ejaculatory ducts and urethra. Some sperm may seep from the penis prior to ejaculation. For this and other reasons, the withdrawal method of birth control is notoriously unreliable, and pregnancy can result from sexual foreplay even without ejaculation.

Ejaculation is accompanied by an intense feeling of release from pent-up tension. Ejaculation and orgasm are not the same. Although they usually occur together, it is possible to have all of the sensations of orgasm without ejaculating, and ejaculation occasionally occurs with little or no sensation of orgasm.

Resolution

Immediately following orgasm comes the **resolution** phase. Discharge of the sympathetic nervous system constricts the internal pudendal artery and reduces the flow of blood into the penis. It also causes contraction of the

---

34. *orgasm* = swelling

35. *e = ex* = out + *jacul* = to throw

trabecular muscles, which squeeze blood from the lacunae of the erectile tissues. The penis may remain semi-erect long enough to continue intercourse, which may be important to the female's attainment of climax, but gradually the penis undergoes **detumescence**—it becomes soft and flaccid again. The resolution phase is also a time in which cardiovascular and respiratory functions return to normal. Many people break out in a "cold sweat" during the resolution phase. In men, resolution is followed by a **refractory period,** lasting anywhere from 10 minutes to a few hours, in which it is usually impossible to attain another erection and orgasm.

Men and women have many similarities and a few significant differences in sexual response. The response cycle of women is described in the next chapter.

Key Point Review

20 Explain how penile blood circulation changes during sexual arousal and why the penis becomes enlarged and stiffened.

21 State the roles of the sympathetic and parasympathetic nervous systems in male sexual response.

CHAPTER ESSAY

Sexually Transmitted Diseases

Sexually transmitted diseases (STDs) have been well known since the writings of Hippocrates and Galen and have been called by a number of euphemisms aimed at downplaying their mode of transmission—for example, "social diseases" and "venereal diseases" (after Venus, the goddess of love). Despite the main topic of this chapter, the STDs discussed here concern men and women equally, as does the essay on contraception at the end of the next chapter.

Here we discuss three bacterial STDs—chlamydia, gonorrhea, and syphilis—and two viral STDs—genital herpes and genital warts. AIDS, another important viral STD, is discussed at the end of chapter 21.

All of these STDs have certain points in common. They have an **incubation period** in which the pathogen multiplies in the body and begins to produce disease but symptoms have not appeared yet, and they have a **communicable period** in which an infected person can transmit the disease to others. The communicable period sometimes begins before symptoms are noticed, after symptoms have disappeared, or in **symptomless carriers,** who show no evidence of the disease. STDs usually do not stimulate an immune response, and there are no vaccines against them. STDs often cause fetal deformity, stillbirth, and neonatal death. The focus here is on adults, whereas STDs of the newborn are discussed in chapter 29.

Chlamydia (cla-MID-ee-uh) is the most common bacterial STD in America today, with 3 to 5 million cases per year. It is caused by *Chlamydia trachomatis.* After an incubation period of 1 to 3 weeks, there appears a scanty, watery discharge from the urethra and pain in the testes or in the rectal or abdominal region. **Nongonococcal urethritis (NGU)** is an STD caused by agents other than the gonorrhea bacterium (see next)—for example, *Chlamydia, Mycoplasma hominis,* and *Ureaplasma urealyticum.* NGU is characterized by inflammation of the urethra, causing pain or discomfort on urination.

Gonorrhea (GON-oh-REE-uh), nicknamed the "clap" or the "drip," is caused by the bacterium *Neisseria gonorrhoeae.* It acquired the name *gonorrhea* ("flow of seed") because Galen thought the pus discharged from the penis was semen. In addition to penile or vaginal discharge, gonorrhea causes abdominal discomfort, genital pain, painful urination (dysuria), and abnormal uterine bleeding. It may cause the oviducts to become scarred and obstructed, resulting in infertility. About 20% of infected women, however, are asymptomatic. Gonorrhea is treated with antibiotics. Gonorrhea and chlamydia frequently occur together and require treatment with one antibiotic for chlamydia and a different one for gonorrhea.

Pelvic inflammatory disease (PID) is a bacterial infection of the female pelvic organs, usually with *Chlamydia* or *Neisseria.* It often results in sterility and may require surgical removal of infected oviducts or other organs. The incidence of PID in America has increased from 17,800 cases in 1970 to about a million cases per year currently. PID has rendered tens of thousands of American women sterile.

Syphilis (SIFF-ih-liss), one of the most potentially devastating STDs, is named for a shepherd boy in a sixteenth-century poem by the physician Fracastoro. It is caused by a corkscrew-shaped bacterium named *Treponema pallidum.* After an incubation period of 2 to 6 weeks, an ulcer called a **chancre** (SHAN-kur) appears at the site of infection—usually on the penis of a male but sometimes out of sight in the vagina of a female. A chancre is a small, hard lesion with no discharge. It disappears in 4 to 6 weeks, ending the first stage of syphilis and often creating an illusion of recovery. In the second stage, however, the disease reappears, with a pink rash over the body, other skin eruptions, fever, joint pain, and hair loss. These symptoms disappear in 3 to 12 weeks. Symptoms then come and go for up to 5 years, but the infection is detectable by a blood test and the infected person is contagious even when symptoms are not occurring. The disease may progress to a third stage, *tertiary syphilis,* or *neurosyphilis,* in which there is damage to the blood vessels and heart valves, thickening of the meninges, and lesions of the brain that can cause paralysis and dementia. Syphilis is treated with antibiotics.

Genital herpes is the most common STD in America, with 20 to 40 million infected people and 500,000 new cases per year. It is usually caused by the herpes simplex virus type 2 (HSV-2). A close relative, HSV-1, causes cold sores (fever blisters) of the mouth and occasionally causes genital herpes, probably transmitted through oral-genital sex. After an incubation period of 4 to 10 days, the virus causes red blisters on the penis of the male; on the labia, vagina, or cervix of the female; and sometimes on the thighs and buttocks of either sex. Over a period of 2 to 10 days, these blisters rupture, seep fluid, and begin to form scabs. The initial infection may be painless or may cause intense pain, urethritis, and watery discharge from the penis or vagina. The lesions heal in 2 to 3 weeks and leave no scars.

During this time, however, the viruses travel by way of sensory nerve fibers to the dorsal root ganglia, where they become dormant. Later, they can migrate along the nerves and cause small epithelial lesions at various places on the body. The movement from place to place is the basis of the name *herpes.*[36] These secondary lesions are generally smaller and heal more quickly than the primary lesions. Most patients have five to seven recurrences, ranging from several times a year to several years apart. An infected person is contagious to a sexual partner when the lesions are present and sometimes even when they are not. Although mostly a painful nuisance, HSV has been implicated as a risk factor in cervical cancer. The drug Acyclovir reduces the shedding of viruses and the spread of lesions but does not necessarily prevent recurrences.

Genital warts (condylomas) are one of the most rapidly increasing STDs in America today. There are about a million new cases per year, especially among promiscuous young adults. Genital warts are caused by 60 or more viruses collectively called *human papillomaviruses (HPVs).* In the male, lesions usually appear on the penis, perineum, or anus, and in the female they are usually on the cervix, vaginal wall, perineum, or anus. Lesions are sometimes small and virtually invisible. A few of the 60 varieties of HPV have been implicated in cervical cancer and cancer of the penis, vagina, and anus. HPV is found in about 90% of all cervical cancers tested. About 90% of cases of genital warts, however, involve forms of HPV that have not been linked to cancer. There is still considerable difference of opinion on how to treat genital warts; they are sometimes treated with cryosurgery (freezing and excision), laser surgery, or interferon.▲

36. *herp* = to creep

Chapter Review Study Outline

Sexual Reproduction (pp. 962–963)
1. The essence of sex
2. The two sexes
   a. Difference in gametes
   b. Difference in sex chromosomes
3. Overview of the reproductive system
   a. Primary sex organs (gonads)
   b. Secondary sex organs
   c. Secondary sex characteristics

Sex Determination and Development (pp. 963–968)
1. Role of the sex chromosomes
2. Hormones and sex differentiation
   a. Embryonic genital ducts
   b. Testis-determining factor

   c. Human chorionic gonadotropin
   d. Androgens
   e. Müllerian-inhibiting factor
3. Development of the external genitals
   a. Phallus
   b. Urogenital folds
   c. Labioscrotal folds
4. Descent of the testes
   a. Role of the gubernaculum
   b. Cryptorchidism

Reproductive Anatomy (pp. 968–974)
1. Perineum
2. Testes
   a. Tunics and stroma

   - Tunica vaginalis
   - Tunica albuginea
   - Mediastinum testis
   - Septa and lobules
   b. Seminiferous tubules
   - Germ cells
   - Sustentacular cells
   - The blood-testis barrier
   c. Interstitial (Leydig) cells
   d. Rete testis
   e. Testicular artery, vein, and nerves
3. Scrotum
   a. Median septum and perineal raphe
   b. Spermatic cord
   c. Thermoregulatory mechanisms
   - Dartos muscle

- Cremaster muscle
- Pampiniform plexus

4. Spermatic ducts
   a. Efferent ductules
   b. Epididymis and its duct
   c. Ductus (vas) deferens
   d. Ejaculatory duct
5. The urethra
6. The accessory glands
   a. Seminal vesicles
   b. Prostate
   c. Bulbourethral glands
7. Penis
   a. Root, shaft, and glans
   b. Prepuce and frenulum
   c. Erectile tissues
      - Corpus spongiosum
      - Corpora cavernosa
      - Lacunae and trabeculae
   d. Bulb and bulbocavernosus muscle
   e. Crura and ischiocavernosus muscle

Puberty and Climacteric (pp. 974–976)

1. Endocrine control of puberty
   a. Gonadotropin-releasing hormone (GnRH)
      - Secreted by hypothalamus
      - Stimulates FSH and ICSH secretion
   b. Interstitial cell–stimulating hormone (ICSH)
      - Secreted by anterior pituitary
      - Stimulates interstitial cells to secrete androgens
   c. Follicle-stimulating hormone (FSH)
      - Secreted by anterior pituitary
      - Stimulates sustentacular cells to secrete ABP
   d. Androgen-binding protein (ABP)
      - Uncertain function
   e. Effects of androgens
      - Spermatogenesis

- Inhibition of GnRH, FSH, and LH secretion
- Development of secondary sex characteristics
- Stimulation of growth
- Stimulation of libido
   f. Inhibin
      - Secreted by sustentacular cells
      - Selectively inhibits FSH secretion

2. Aging and sexual function
   a. Climacteric
   b. Impotence

Spermatogenesis, Spermatozoa, and Semen (pp. 976–982)

1. Meiosis
   a. Comparisons to mitosis
      - Reduces chromosome number
      - Produces four daughter cells
      - Crossing-over between homologous chromosomes
   b. Meiosis I
      - Prophase I
        - •• Formation of tetrads
        - •• Crossing-over
      - Metaphase I
      - Anaphase I
      - Telophase I
   c. Interphase
   d. Meiosis II
      - Prophase II
      - Metaphase II
      - Anaphase II
      - Telophase II
2. Spermatogenesis
   a. Spermatogonia
      - Arise from primordial germ cells
      - Multiply by mitosis
      - Type A and type B spermatogonia
   b. Primary spermatocytes
      - Differentiate from type B spermatogonia

- Relationship to blood-testis barrier
- Still diploid
- Undergo meiosis I
   c. Secondary spermatocytes
      - Haploid
      - Undergo meiosis II
   d. Spermatids
      - Undergo spermiogenesis
      - Become spermatozoa
3. The spermatozoon
   a. Head
      - Acrosome
      - Nucleus
      - Basal body
   b. Tail
      - Midpiece
      - Principal piece
      - Endpiece
4. Semen
   a. Spermatozoa
   b. Fructose
   c. Clotting and anticoagulant factors
   d. Prostaglandins
   e. Spermine and other bases

Sexual Intercourse (pp. 982–986)

1. Anatomical considerations
   a. Circulation
   b. Innervation
2. Excitement and plateau
   a. Neurological mechanisms
   b. Respiratory and cardiovascular effects
   c. Bulbourethral secretion
   d. Mechanisms of erection
3. Orgasm and Ejaculation
   a. Emission
   b. Expulsion
4. Resolution
   a. Detumescence
   b. Refractory period

Selected Vocabulary

sexual reproduction 962
gamete 962
zygote 962
spermatozoon 962
egg 962
primary sex organ 962
gonad 962
testis 962
ovary 962
secondary sex organ 962
secondary sex characteristic 963
gonadal ridge 963
mesonephric duct 963

paramesonephric duct 963
testis-determining factor (TDF) 963
human chorionic gonadotropin (HCG) 963
müllerian-inhibiting factor 963
phallus 965
urogenital folds 965
labioscrotal folds 965
hypospadias 965
homologous structures 965
gubernaculum 966
inguinal canal 966
descent of testes 966

cryptorchidism 966
perineum 968
tunica vaginalis 968
tunica albuginea of testis 968
seminiferous tubule 968
germinal epithelium 968
germ cell 968
sustentacular cell 968
blood-testis barrier (BTB) 968
interstitial cell 970
rete testis 970
testicular artery 970
testicular vein 970
testicular nerve 970

scrotum 970
median septum of scrotum 970
perineal raphe 970
spermatic cord 970
dartos muscle 971
cremaster muscle 971
pampiniform plexus 971
spermatic ducts 972
efferent ductule 972
epididymis 972
ductus (vas) deferens 972
ampulla 972
vasectomy 972
ejaculatory duct 972

Testing Your Recall   Answers in Appendix C

1. The scrotum develops from the
   _____ of the embryo.
   a. mesonephric duct
   b. paramesonephric duct
   c. phallus
   d. labioscrotal folds
   e. urogenital folds

2. Descent of the testes is achieved by
   contraction of a fibromuscular cord
   called
   a. the gubernaculum.
   b. the spermatic cord.
   c. the vas deferens.
   d. the pampiniform plexus.
   e. the mediastinum testis.

3. The male target cells for luteinizing
   hormone are
   a. pituitary gonadotropes.
   b. sustentacular cells.
   c. spermatozoa.
   d. spermatogonia.
   e. interstitial cells.

4. Prior to ejaculation, sperm are stored
   primarily in
   a. the seminiferous tubules.
   b. the rete testis.
   c. the epididymis.
   d. the seminal vesicles.
   e. the ejaculatory ducts.

5. The penis is attached to the pubic
   arch by crura of
   a. the corpora cavernosa.
   b. the corpus spongiosum.
   c. the perineal membrane.
   d. the bulbocavernosus muscle.
   e. the ischiocavernosus muscle.

6. The earliest hormone to be secreted
   at the onset of puberty is
   a. follicle-stimulating hormone.
   b. interstitial cell–stimulating
      hormone.
   c. human chorionic gonadotropin.
   d. gonadotropin-releasing hormone.
   e. testosterone.

7. When it is necessary to reduce the rate
   of sperm production without reducing
   the rate of testosterone secretion, the
   sustentacular cells secrete
   a. dihydrotestosterone.
   b. androgen-binding protein.
   c. ICSH.
   d. FSH.
   e. inhibin.

8. Four spermatozoa arise from each
   a. primordial germ cell.
   b. type A spermatogonium.
   c. type B spermatogonium.
   d. secondary spermatocyte.
   e. spermatid.

9. The point in meiosis at which
   centromeres divide and sister
   chromatids separate from each other is
   a. prophase I.
   b. metaphase I.
   c. anaphase I.
   d. anaphase II.
   e. telophase II.

10. Blood is forced out of the penile
    lacunae by contraction of the _____
    muscle.
    a. bulbocavernosus
    b. ischiocavernosus

    c. cremaster
    d. trabecular
    e. dartos

11. Under the influence of androgens, the
    embryonic _____ duct develops
    into the male reproductive tract.

12. Spermatozoa obtain energy for
    locomotion from _____ in the
    semen.

13. The _____ , a network of veins in
    the spermatic cord, helps keep the
    testes cooler than core body
    temperature.

14. All germ cells beginning with the
    _____ are genetically different from
    the rest of the body and therefore
    must be protected by the blood-testis
    barrier.

15. The corpora cavernosa as well as the
    testes have a fibrous capsule called
    the _____ .

16. Over half the semen consists of
    secretions from a pair of glands
    called the _____ .

17. The blood-testis barrier is formed by
    tight junctions between the _____
    cells.

18. The earliest haploid stage of
    spermatogenesis is the _____ .

19. Erection of the penis results when
    nitric oxide causes dilation of the
    _____ arteries of the penis.

20. The _____ of a sperm contains
    enzymes used to penetrate the egg.

1. Explain why testosterone may be considered both an endocrine and paracrine secretion of the testes. Review paracrines in chapter 17 if necessary.

2. Suppose you were a physician who had a midadolescent patient who looked and felt completely feminine but who was concerned that she had not started her menstrual periods and had grown no pubic hair. Tests revealed that she was an XY individual with androgen-insensitivity syndrome. Would you consider the patient to be a boy or a girl? What would you tell your patient? Discuss your rationale.

3. Considering the temperature in the scrotum, would you expect hemoglobin to unload more oxygen to the testes, or less, than it unloads to the warmer internal organs? (Hint: see fig. 22.25). Why? How would you expect this to influence sperm development?

4. Why is it possible for spermatogonia to be outside the blood-testis barrier but essential for primary spermatocytes and later stages to be within the barrier (isolated from the blood)? Explain this contrast.

5. A 68-year-old man taking medication for hypertension complains to his physician that it has made him impotent. Explain why this could be an effect of antihypertensive drugs.

Web Site Link

For a listing of the most current web sites related to this chapter, please visit the Saladin homepage at:

http://www.mhhe.com/sciencemath/biology/saladin/

The Female Reproductive System

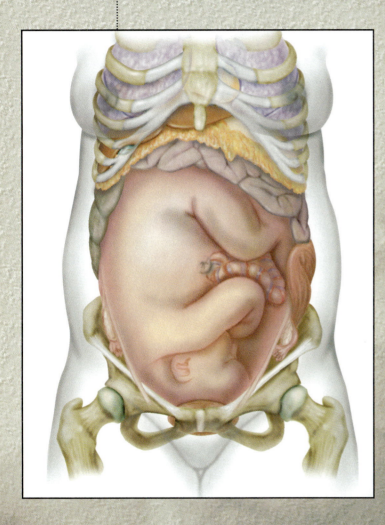

Special **Topics**

Brushing up

To understand this chapter, it is essential that you understand or brush up on the following concepts:

► Negative feedback inhibition of the pituitary (p. 607)
► Synergistic, permissive, and antagonistic hormone interactions (pp. 628–629)
► Fetal development of the reproductive system (pp. 963–966)
► Meiosis (pp. 976–978)
► The physiological phases of intercourse (pp. 985–986)

Reproductive Anatomy

▼Objectives

When you have completed this section, you should be able to
- describe the structure of the ovary and compare it to the testis;
- trace the female reproductive tract from the uterine tube to the vagina, describing the gross anatomy and histology of each organ;
- identify the ligaments that support the female reproductive organs;
- describe the blood supply to the female reproductive tract;
- identify the external genitals of the female; and
- describe the structure of the nonlactating breast.

T he female reproductive system is more complex than the male's because it serves more purposes. Whereas the male needs only to produce and deliver gametes, the female must do this as well as provide nutrition and safe harbor for fetal development and then give birth and nourish the infant. Furthermore, female reproductive physiology is cyclic and female hormones are secreted in a more complex sequence compared to the relatively steady, simultaneous secretion of regulatory hormones in the male.

This chapter discusses the anatomy of the female reproductive system; the production of gametes and how it relates to the ovarian and menstrual cycles; the female sexual response; and the physiology of pregnancy, birth, and lactation. Embryonic and fetal development are treated in the next chapter.

Sex Differentiation

The female reproductive system (fig. 28.1) is conspicuously different from that of the male, but as we saw earlier the two sexes are indistinguishable for the first 8 to 10 weeks of development (see fig. 27.4, p. 967). Having no Y chromosome, a female fetus has no receptors for human chorionic gonadotropin (HCG) and her ovaries produce no androgens. In the absence of that influence the phallus becomes a clitoris, the urogenital folds develop into labia minora, and the labioscrotal folds develop into labia majora. The indifferent gonad becomes an ovary, the mesonephric duct degenerates, and the paramesonephric duct becomes the uterus and uterine

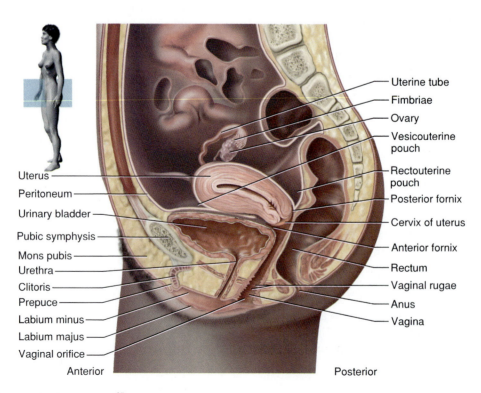

Figure 28.1 The female reproductive system.

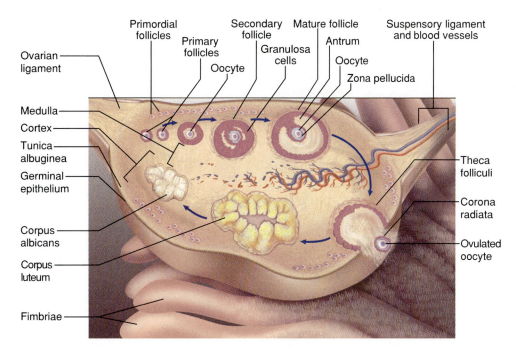

Figure 28.2 Structure of the ovary, showing the sequence of development of the ovarian follicles. ✗

tubes. This developmental pattern can be disrupted, however, by abnormal hormonal exposure before birth, as happens in adrenogenital syndrome (see p. 614).

The Ovary

The female gonads are the **ovaries,**[1] which produce egg cells (ova) and sex steroids. The ovary is an almond-shaped organ nestled in the **ovarian fossa,** a depression of the posterior pelvic wall. It measures about 3 cm long, 1.5 cm wide, and 1 cm thick. Its capsule, like that of the testis, is called the **tunica albuginea.** Overlying this is a simple cuboidal **germinal epithelium,** so-named because of a mistaken impression that it produces the germ cells—on the contrary, it is simply the parietal peritoneum. The interior of the ovary is indistinctly divided into an outer **cortex,** where the germ cells develop, and a central **medulla** occupied by the major arteries and veins (fig. 28.2).

The ovary lacks ducts comparable to the seminiferous tubules. Instead, each egg develops in its own fluid-filled **follicle** and is released by **ovulation,** the bursting of the follicle. Figure 28.2 shows several types of follicles that coincide with different stages of egg maturation, as discussed later.

The ovary is held in place by several connective tissue ligaments (fig. 28.3). Its medial pole is attached to the uterus by the **ovarian ligament,** and its lateral pole is attached to the pelvic wall by the **suspensory liga-**ment. A sheet of peritoneum called the **broad ligament** flanks the uterus on each side and encloses the oviduct in its superior margin. The anterior margin of the ovary is anchored to the broad ligament by a peritoneal fold called the **mesovarium.**[2]

The ovary is supplied with an **ovarian artery, ovarian veins,** and **ovarian nerves,** which travel through the suspensory ligament. It receives an additional blood supply from the ovarian branches of the uterine arteries.

Secondary Sex Organs (Genitalia)

The **internal genitalia** are the uterine tubes, uterus, and vagina, which constitute a duct system from the vicinity of the ovary to the outside of the body. The **external genitalia** include principally the clitoris, labia minora, and labia majora. These occupy the perineum, which is defined by the same skeletal landmarks as in the male. Beneath the skin of the perineum are several accessory glands that provide most of the lubrication for intercourse.

The Uterine Tubes

The **uterine tube,** also called the **oviduct** or **fallopian tube,**[3] is a canal about 10 cm long from the ovary to the uterus. At the distal (ovarian) end it flares into a

---

1. *ov* = egg + *ary* = place for

2. *mes* = middle + *ovari* = ovary
3. Gabriele Fallopio (1523–62), Italian anatomist and physician

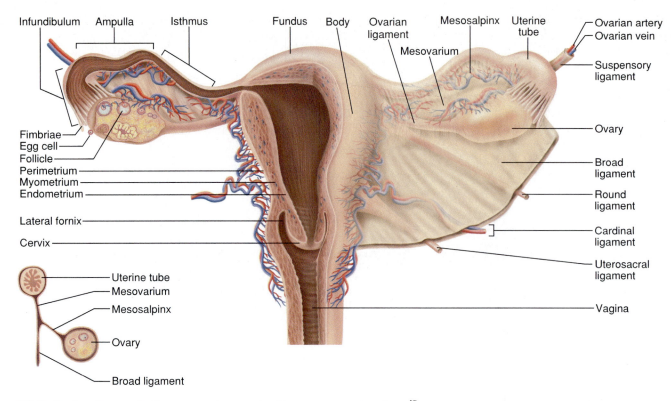

Infundibulum   Ampulla   Isthmus   Fundus   Body   Ovarian ligament   Mesosalpinx   Uterine tube   Ovarian artery   Ovarian vein   Mesovarium   Suspensory ligament

Fimbriae
Egg cell
Follicle
Perimetrium
Myometrium
Endometrium
Lateral fornix
Cervix

Ovary
Broad ligament
Round ligament
Cardinal ligament
Uterosacral ligament
Vagina

Uterine tube
Mesovarium
Mesosalpinx
Ovary
Broad ligament

Figure 28.3 The female reproductive tract and supportive ligaments, posterior view. 𝄁

trumpet-shaped **infundibulum**[4] with feathery projections called **fimbriae**[5] (FIM-bree-ee); the long middle part of the tube is the **ampulla;** and near the uterus it forms a narrower **isthmus.** The uterine tube is enclosed in the **mesosalpinx**[6] (MEZ-oh-SAL-pinks), which is the superior margin of the broad ligament.

The wall of the uterine tube is well endowed with smooth muscle. Its mucosa is extremely folded and convoluted and has an epithelium of ciliated cells and a smaller number of secretory cells (fig. 28.4). The cilia beat toward the uterus and, with the help of muscular contractions of the tube, convey the egg in that direction. It takes about 3 days for an egg to travel the length of the uterine tube, but an unfertilized egg lives only 24 hours. If unfertilized, it will die before it arrives in the uterus.

The Uterus

The **uterus**[7] (fig. 28.5) is a thick muscular chamber that opens onto the roof of the vagina and tilts forward over the urinary bladder. Its function is to harbor the embryo, provide a source of nutrition, and expel the fetus at the end of its development. It is somewhat pear-shaped,

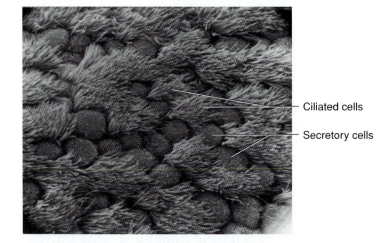

Ciliated cells
Secretory cells

Figure 28.4 Epithelium of the uterine tube. Note the mixture of domelike secretory cells with microvilli and ciliated cells, which propel the egg to the uterus (SEM). 𝄁

From R. G. Kessel and R. H. Kardon, *Scanning Electron Microscopy of Tissues and Organs,* 1979.

with a broad superior curvature called the **fundus,** a midportion called the **body (corpus),** a short constricted **isthmus,** and a cylindrical inferior end called the **cervix** (see fig. 28.3). The uterus measures about 7 cm from cervix to fundus, 4 cm wide at its broadest point, and 2.5 cm thick, but it is somewhat larger in women who have been pregnant. A Pap smear is a specimen of cells from the cervix and vagina (see special topic 28.1).

---

4. *infundibulum* = funnel
5. *fimbria* = fringe
6. *meso* = mesentery + *salpin* = trumpet
7. *uterus* = womb

Cervical cancer is common among women from ages 30 to 50, especially those who smoke, who began sexual activity at an early age, and who have histories of frequent sexually transmitted diseases or cervical inflammation. It begins in the epithelial cells of the lower cervix, develops slowly, and remains a local, easily removed lesion for several years. If the cancerous cells spread to the subepithelial connective tissue, however, the cancer is said to be *invasive* and is much more dangerous, potentially requiring **hysterectomy**[8] (removal of the uterus).

The best protection against cervical cancer is a routine **Pap**[9] **smear**—a procedure in which loose cells are scraped from the cervix and vagina and microscopically examined. The findings are rated on a five-point scale:

Class I—no abnormal cells seen
Class II—atypical cells suggestive of inflammation, infection, or irritation
Class III—nonmalignant but mildly abnormal cell growth (*dysplasia*)
Class IV—cells typical of localized cancer
Class V—cells typical of invasive cancer

An average woman is typically advised to have three annual Pap smears and may then have them less often at the discretion of her physician. Women with any of the risk factors listed may be advised to have more frequent examinations.

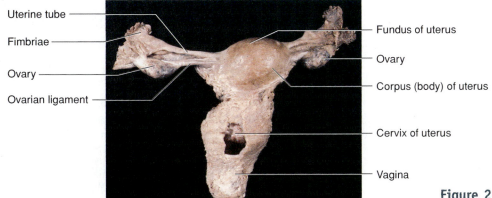

Uterine tube
Fimbriae
Ovary
Ovarian ligament

Fundus of uterus
Ovary
Corpus (body) of uterus
Cervix of uterus
Vagina

Figure 28.5 The female reproductive tract. A window has been cut in the upper vagina to show the cervix.

The lumen is roughly triangular, with its two upper corners opening into the uterine tubes. It communicates with the vagina by way of a narrow passage through the cervix called the **cervical canal.** The superior opening of this canal into the body of the uterus is the **internal os**[10] (oss) and its opening into the vagina is the **external os.** The canal contains **cervical glands** that secrete a fairly thick mucus thought to prevent the spread of microorganisms from the vagina into the uterus. Near the time of ovulation, however, the mucus becomes thinner and allows easier passage for sperm.

Uterine Wall   The uterine wall consists of three layers called the perimetrium, myometrium, and endometrium.[11] The **perimetrium** is the serosa. The **my-**ometrium is a layer of smooth muscle constituting most of the wall; it is about 1.25 cm thick in the nonpregnant uterus. The myometrium has bundles of smooth muscle running in all directions, but it is less muscular and more fibrous near the cervix, and the cervix itself is almost entirely collagenous. The smooth muscle fibers of the myometrium are about 40 μm long immediately after menstruation, but they are twice this long at the middle of the menstrual cycle and 10 times as long in pregnancy. The function of the myometrium is to produce the labor contractions that expel the fetus.

The **endometrium** is the mucosa. It has a simple columnar epithelium and compound tubular glands. The superficial half to two-thirds of it, called the **stratum functionalis,** is shed in each menstrual period. The deeper layer, called the **stratum basalis,** stays behind and regenerates a new functionalis in the next cycle. When pregnancy occurs, the endometrium is the site of attachment of the embryo and forms the maternal part of the *placenta* from which the fetus is nourished.

Blood Supply   The uterine blood supply is particularly important to the menstrual cycle and pregnancy. A **uterine**

8. *hyster* = uterus + *ectomy* = cutting out
9. George N. Papanicolaou (1883–1962), Greek-American physician and cytologist
10. *os* = mouth
11. *peri* = around; *myo* = muscle; *endo* = within; *metr* = uterus

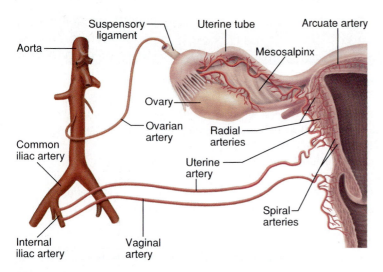

Figure 28.6 Blood supply to the female reproductive tract.

artery arises from each internal iliac artery, travels through the broad ligament, and approaches the uterus (fig. 28.6). It gives off several branches that penetrate into the myometrium and lead to **arcuate arteries.** Each arcuate artery travels in a circle around the uterus and anastomoses with the arcuate artery on the other side. Along its course, it gives rise to smaller arteries that penetrate the rest of the way through the myometrium, into the endometrium, and produce the **spiral arteries.** The spiral arteries wind tortuously between the endometrial glands, toward the surface of the mucosa. They rhythmically constrict and dilate, making the mucosa alternately blanch and flush with blood.

Ligaments   The uterus is supported mainly by the muscular floor of the pelvic outlet but also by folds of peritoneum that form supportive ligaments around the organ, as they do for the ovary and uterine tube (see fig. 28.3). The broad ligament has two parts: the mesosalpinx mentioned earlier and the **mesometrium** on each side of the uterus. The cervix and superior part of the vagina are supported by the **cardinal (lateral cervical) ligaments** extending to the pelvic wall. A pair of **uterosacral ligaments** attach the posterior side of the uterus to the sacrum, and a pair of **round ligaments** attach the anterior surface of the uterus to the abdominal wall. The round ligaments continue through the inguinal canals and terminate in the labia majora, much like the gubernaculum of the male terminating in the scrotum.

As the peritoneum folds around the various pelvic organs, it creates several dead-end recesses and pouches. Two major ones are the **vesicouterine**[12] **pouch,** which forms the space between the uterus and urinary bladder, and **rectouterine pouch** between the uterus and rectum (see fig. 28.1).

Vagina

The **vagina,**[13] or birth canal, is a tube about 8 to 10 cm (3–4 in.) long that allows for the discharge of menstrual fluid, receipt of the penis and semen, and birth of a baby. The vaginal wall is thin but very distensible. It consists of an outer adventitia, a middle muscularis, and an inner mucosa. The vagina tilts posteriorly between the urethra and rectum; the urethra is embedded in its anterior wall. The vagina has no glands, but it is lubricated by the *transudation* ("vaginal sweating") of serous fluid through its walls and by mucus from the cervical glands above it.

The vagina extends slightly beyond the cervix. The recess beyond the cervix is called the **posterior fornix,**[14] the recesses that pass around the cervix on either side are the **lateral fornices** (FOR-nih-sees), and the recess below the cervix is the **anterior fornix** (see figs. 28.1 and 28.3).

At its lower end, the mucosa folds inward and forms a membrane, the **hymen,** which stretches across the orifice. It has one or more openings to allow menstrual fluid to pass through, but it usually must be ruptured to allow for intercourse. A little bleeding often accompanies the first act of intercourse; however, the hymen is commonly ruptured before then by tampons, medical examinations, or strenuous exercise. The lower end of the vagina also has transverse friction ridges, or **vaginal rugae,** which stimulate the penis and help induce ejaculation.

The vaginal epithelium is simple cuboidal in childhood, but the estrogens of puberty stimulate it to transform into a stratified squamous epithelium. This is an example of *metaplasia,* the transformation of one tissue type to another. The epithelial cells are rich in glycogen. Bacteria ferment this to lactic acid, resulting in a low vaginal pH (about 3.5–4.0) that inhibits the growth of pathogens. Recall from the previous chapter that this acidity is neutralized by the semen so it does not harm the sperm. The mucosa also has antigen-presenting cells (APCs) called **dendritic cells,** which are thought to be a route of HIV invasion of the female body from infected semen.

 Think About It

Why do you think the vaginal epithelium changes type at puberty? Of all types of epithelium it might become, why stratified squamous?

---

12. *vesico* = bladder

13. *vagina* = sheath
14. *fornix* = arch, vault

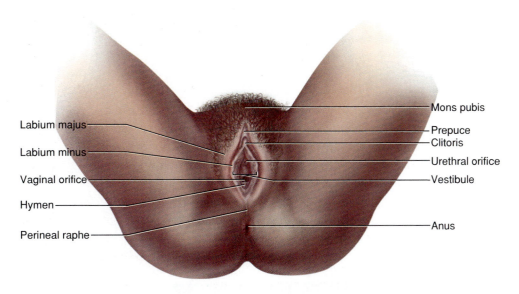

Figure 28.7 The female perineum.

Labium majus
Labium minus
Vaginal orifice
Hymen
Perineal raphe

Mons pubis
Prepuce
Clitoris
Urethral orifice
Vestibule
Anus

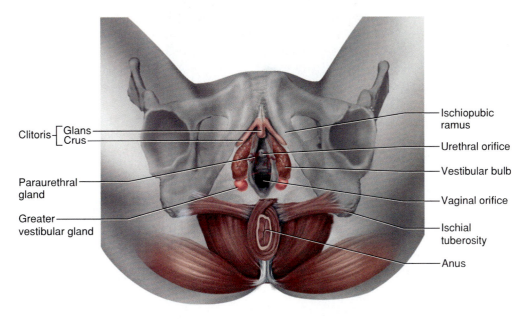

Figure 28.8 Subcutaneous structures of the female perineum.

Clitoris — Glans, Crus
Paraurethral gland
Greater vestibular gland

Ischiopubic ramus
Urethral orifice
Vestibular bulb
Vaginal orifice
Ischial tuberosity
Anus

The Vulva (Pudendum)

The **vulva**[15] (**pudendum**[16]) includes the mons pubis, labia majora and minora, and clitoris (fig. 28.7). The **mons**[17] **pubis** consists mainly of a mound of adipose tissue overlying the pubic symphysis. The **labia majora**[18] (singular, *labium majus*) are a pair of thick folds of skin and adipose tissue inferior to the mons; the slit between them is the **pudendal cleft.** Pubic hair grows on the mons

pubis and lateral surfaces of the labia majora at puberty, but the medial surfaces of the labia remain hairless. Medial to the labia majora are the much thinner, entirely hairless **labia minora**[19] (singular, *labium minus*). The area enclosed by them, called the **vestibule,** contains the urinary and vaginal orifices. At the anterior margin of the vestibule, the labia minora meet and form a hoodlike **prepuce.**

The **clitoris** is structured like a miniature penis except that it is almost entirely internal, it has no corpus spongiosum, and it does not enclose the urethra. Essentially, it is a pair of corpora cavernosa enclosed in connective tissue. Its **glans** protrudes slightly from the prepuce. The **body (corpus)** passes internally, inferior to the pubic symphysis (see fig. 28.1). At its internal end, the corpora cavernosa diverge like a Y as a pair of **crura,** which, like those of the penis, attach the clitoris to the ischiopubic ramus on each side of the pubic arch. Like the penis, the clitoris is supplied by the internal pudendal arteries, also called the **clitoral arteries** in the female. The clitoris is the primary center of erotic stimulation.

Accessory Glands and Erectile Tissues

On each side of the vagina is a pea-sized **greater vestibular (Bartholin**[20]**) gland** with a short duct opening into the vestibule or lower vagina (fig. 28.8). These glands are homologous to the bulbourethral glands of the male. They keep the vulva moist, and during sexual excitement they provide most of the lubrication for intercourse. The vestibule is also lubricated by a number of **lesser vestibular glands.** A pair of mucous **paraurethral (Skene**[21]**) glands,** homologous to the male prostate, open into the vestibule near the external urethral orifice.

15. *vulva* = covering
16. *pudend* = shameful
17. *mons* = mountain
18. *labi* = lip + *major* = larger, greater

19. *minor* = smaller, lesser
20. Caspar Bartholin (1655–1738), Danish anatomist
21. Alexander J. C. Skene (1838–1900), American gynecologist

Just deep to the labia majora, a pair of subcutaneous erectile tissues called the **vestibular bulbs** bracket the vagina like a pair of parentheses. They become vasocongested during sexual excitement and cause the vagina to tighten somewhat around the penis, enhancing sexual stimulation.

Secondary Sex Characteristics

In the female as in the male, there is room for interpretation as to what constitutes the secondary sex characteristics—that is, what features arouse the sexual interest of potential mates. This is partly a matter of cultural conditioning. In Victorian England a glimpse of ankle was considered highly arousing, but we would hesitate to classify the ankles as secondary sex characteristics. More widely recognized as such are the feminine physique, determined largely by the distribution of body fat and flare of the pelvis; the relatively fine body hair; a voice pitched higher than the male's; and the breasts.

The Breasts

The breast is a mound of tissue overlying the pectoralis major. It has two principle regions: the conical to pendulous **body,** with the nipple at its apex, and an extension toward the armpit called the **axillary tail** (fig. 28.9a). Lymphatics of the axillary tail are especially important as a route of breast cancer metastasis.

The nipple is surrounded by a circular colored zone, the **areola.** Dermal blood capillaries and nerves come closer to the surface here than in the surrounding skin, making the areola more sensitive and more reddish in color than the surrounding skin. In pregnancy, the areola and nipple tend to become melanized, making the nipple more visible to the indistinct vision of a nursing infant. Sensory nerve fibers of the areola are important in triggering a *milk ejection reflex* when an infant nurses. The areola has sparse hairs and **areolar glands,** visible as small bumps on the surface. These are intermediate between sweat glands and mammary glands in development. When a woman is nursing, their secretion minimizes chapping and cracking of the areola. The dermis of this region has smooth muscle fibers that contract in response to cold, touch, and sexual arousal, wrinkling the areola and erecting the nipple.

Internally, the nonlactating breast consists mostly of adipose and collagenous tissue (fig. 28.9b). Breast size is determined by the amount of adipose tissue and has no relationship to the amount of milk the mammary gland can produce. **Suspensory ligaments** attach the breast to the dermis of the overlying skin and to the fascia of the pectoralis major. The nonlactating breast contains very little mammary gland, but it does have a

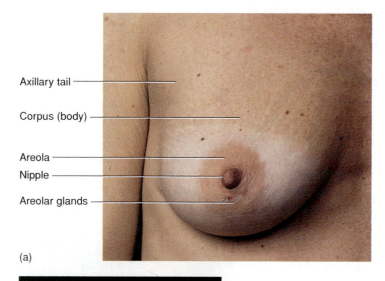

(a)

Axillary tail
Corpus (body)
Areola
Nipple
Areolar glands

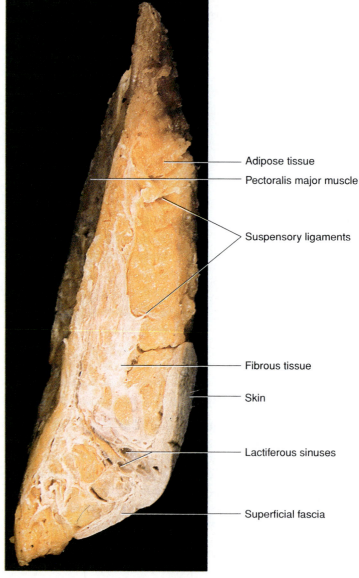

Adipose tissue
Pectoralis major muscle
Suspensory ligaments
Fibrous tissue
Skin
Lactiferous sinuses
Superficial fascia

(b)

Figure 28.9 Structure of the breast. (a) Surface anatomy, living subject. (b) Sagittal section of the breast of a cadaver.—*Cont'd*

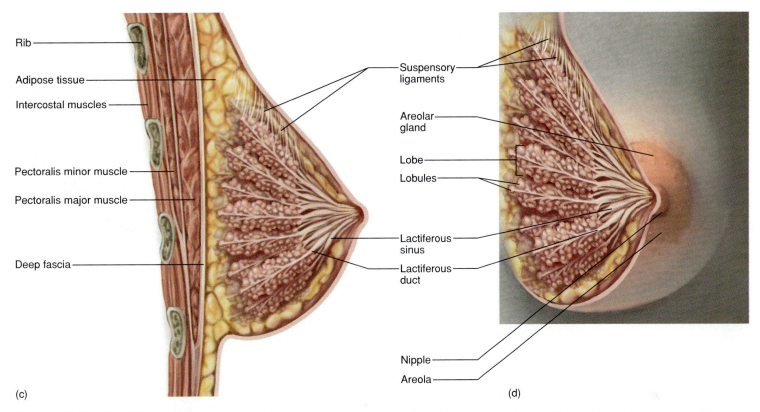

Labels (lateral view, c):
Rib
Adipose tissue
Intercostal muscles
Pectoralis minor muscle
Pectoralis major muscle
Deep fascia

Labels (anterior view, d):
Suspensory ligaments
Areolar gland
Lobe
Lobules
Lactiferous sinus
Lactiferous duct
Nipple
Areola

(c)   (d)

Figure 28.9—Cont'd (*c*) Internal anatomy of the lactating breast, lateral view. (*d*) Internal anatomy of the lactating breast, anterior view.

system of ducts branching through its connective tissue stroma and converging on the nipple. When the mammary gland develops during pregnancy, it exhibits 15 to 20 lobes arranged radially around the nipple, separated from each other by stroma (fig. 28.9*c, d*). Each lobe is drained by a **lactiferous[22] duct,** which dilates to form a **lactiferous sinus** opening onto the nipple. Lactation and the associated changes in breast structure are described at the end of this chapter.

Breast Cancer

Breast cancer (fig. 28.10) occurs in one out of every eight or nine American women and is one of the leading causes of female mortality. Breast tumors begin with cells of the mammary ducts and may metastasize to other organs by way of the mammary and axillary lymphatics. Symptoms of breast cancer include a palpable lump (the tumor), puckering of the skin, changes in skin texture, and drainage from the nipple.

Two breast cancer genes, named BRCA1 and BRCA2, were discovered in the 1990s, but not all breast cancer is hereditary. Some breast tumors are stimulated by estrogen. Consequently, breast cancer is more common among women with early menarche and late menopause—that is, a longer reproductive life with a longer period of estrogen exposure. Other risk factors include aging, exposure to ionizing radiation and carcinogenic chemicals, excessive alcohol and fat intake, and smoking. Over 70% of cases, however, lack any identifiable risk factors.

The majority of tumors are discovered during breast self-examination (BSE), which should be a monthly routine for all women. *Mammograms* (breast X rays), however, can detect tumors too small to be noticed by BSE. Although opinions vary, a mammogram schedule commonly recommended is to have a baseline mammogram in the late 30s and then have one every 2 years from ages 40 to 49 and every year beginning at age 50.

Treatment of breast cancer is usually by *lumpectomy* (removal of the tumor only) or *simple mastectomy* (removal of the breast tissue only or breast tissue and some axillary lymph nodes). *Radical mastectomy,* rarely done since the 1970s, involves the removal of not only the breast but also the underlying muscle, fascia, and lymph nodes. Although very disfiguring, it proved to be no more effective than simple mastectomy or lumpectomy. Surgery is generally followed by radiation or chemotherapy, and estrogen-sensitive tumors may also

22. *lact* = milk + *fer* = to carry

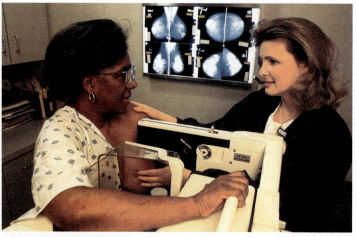
(a)

(b)

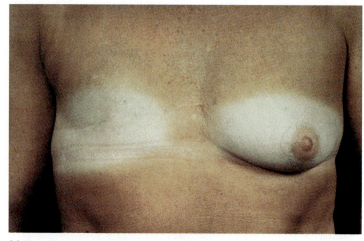
(c)

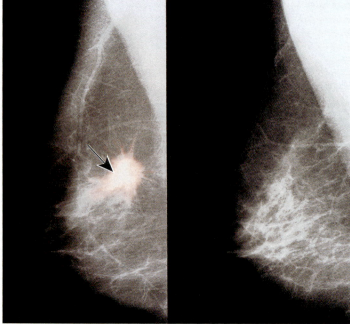
(d)

Figure 28.10 Breast cancer screening and treatment. (*a*) Nurse assisting a patient in mammography. (*b*) Mammogram of a breast with a tumor visible at the arrow (*left*), compared to the appearance of normal fibroconnective tissue of the breast (*right*). (*c*) Patient following mastectomy of the right breast. (*d*) The same patient following surgical breast reconstruction.

be treated with an estrogen blocker such as tamoxifen. A natural-looking breast can often be reconstructed from skin, fat, and muscle from other parts of the body (fig. 28.10*c, d*).

Key Point Review

1 How do the site of female gamete production and mode of release from the gonad differ from those in the male?

2 How is the structure of the uterine tube mucosa related to its function?

3 Contrast the function of the endometrium with that of the myometrium.

4 Describe the similarities and differences between the clitoris and penis.

Puberty and Menopause

▼Objectives

When you have completed this section, you should be able to
• name the hormones that regulate female reproductive function, and state their roles;
• define *thelarche, pubarche,* and *menarche,* the milestones of puberty;
• describe the hormonal changes of female climacteric and their effects; and
• define and describe menopause, and distinguish menopause from climacteric.

Puberty and menopause are physiological transitions at the beginning and end of a female's reproductive life.

Puberty

Puberty begins at ages 8 to 13 for most girls in the United States and Europe but significantly later in many countries. It is triggered by the same hypothalamic and pituitary hormones in girls as it is in boys. Rising levels of gonadotropin-releasing hormone (GnRH) stimulate the anterior pituitary to secrete follicle-stimulating hormone (FSH) and luteinizing hormone (LH). FSH, especially, stimulates development of the ovarian follicles, which, in turn, secrete estrogens, progesterone, inhibin, and a small amount of androgen. These hormone levels rise gradually from ages 8 to 12 and then more sharply in the early teens. The **estrogens**[23] are feminizing hormones with widespread effects on the body. They include **estradiol** (the most abundant), **estriol**, and **estrone.** Most of the visible changes at puberty result from estradiol and androgens.

The earliest noticeable change is **thelarche**[24] (thee-LAR-kee), the development of the breasts. The areolae and nipples become elevated **breast buds** and the areolae increase in diameter. The breasts develop rapidly in adolescence and assume a hemispherical shape as fat is deposited and the ducts develop. The areola typically forms an elevated secondary mound lasting until late adolescence or early adulthood, but then flattens so only the nipple protrudes on the mature breast.

Soon after thelarche comes **pubarche** (pyu-BAR-kee)—development of the pubic and axillary hair, sebaceous glands, and apocrine glands. Pubarche is induced by androgens from the ovaries and adrenal cortex. Androgens are also responsible for the libido of both sexes. Women secrete about 0.5 mg of androgens per day, compared with 6 to 8 mg/day in men.

Next comes **menarche**[25] (men-AR-kee), the first menstrual period. In Europe and America, the average age at menarche declined from age 16.5 in 1860 to age 13 in 1960, probably due in part to improved nutrition. Menarche cannot occur until a girl has reached at least 17% body fat, so it is delayed until the mid- to late teens in some athletes and dancers. Adult menstruation generally ceases if a woman drops below 22% body fat. This is about the minimum needed to sustain pregnancy and lactation; thus the body reacts as if to prevent a futile pregnancy when it is too lean. Menarche does not necessarily signify fertility. For the first year, a girl's menstrual periods are typically *anovulatory* (no egg is ovulated). Most girls begin ovulating about a year after they begin menstruating.

Estradiol stimulates many other changes of puberty. It causes the vaginal metaplasia described earlier.

It stimulates growth of the ovaries and secondary sex organs. It stimulates osteoblasts, causing a girl to grow rapidly in height and her pelvis to widen. Estradiol is largely responsible for the feminine physique because it stimulates fat deposition in the mons pubis, labia majora, hips, thighs, buttocks, and breasts. It makes a girl's skin thicker, but it remains thinner, smoother, softer, and warmer than in males of corresponding age.

Progesterone[26] acts primarily on the uterus, preparing it for possible pregnancy in the second half of each menstrual cycle and playing roles in pregnancy discussed later. Estrogens and progesterone also suppress FSH and LH secretion through negative feedback inhibition of the anterior pituitary. **Inhibin** selectively suppresses FSH secretion.

Thus we see many hormonal similarities in males and females from puberty onward. The sexes differ less in the hormones that are present than in the relative amounts of those hormones—high levels of androgens and low levels of estrogens in males and the opposite in females. Another difference is that these hormones are secreted more or less continually and simultaneously in males, whereas in females secretion is distinctly cyclic and the hormones are secreted in sequence rather than simultaneously. This will be very apparent as you read about the ovarian and menstrual cycles.

Climacteric and Menopause

Women, like men, go through a midlife change in hormone secretion called the **climacteric.** In women, it is accompanied by **menopause,** the cessation of menstruation (see special topic 28.2).

With age, the ovaries have fewer remaining follicles and those that remain are less responsive to gonadotropins. Consequently, they secrete less estrogen and progesterone. Without these steroids, the uterus, vagina, and breasts atrophy. Intercourse may become uncomfortable, and vaginal infections are more common, as the vagina becomes thinner, less distensible, and drier. The skin becomes thinner, blood cholesterol levels rise (increasing the risk of cardiovascular disease), and bone mass declines (increasing the risk of osteoporosis). Blood vessels constrict and dilate in response to shifting hormone balances, and the sudden dilation of cutaneous arteries often causes **hot flashes**—a spreading sense of heat from the abdomen to the thorax, neck, and face. Hot flashes may occur several times a day, sometimes accompanied by headaches resulting from the sudden vasodilation of arteries in the head. In some people, the changing hormonal profile also causes depression, irritability, or other mood changes.

---

23. *estro* = desire, frenzy + *gen* = to produce
24. *thel* = breast, nipple + *arche* = beginning
25. *men* = monthly

26. *pro* = favoring + *gest* = pregnancy + *sterone* = steroid hormone

There has been considerable speculation about why women do not remain fertile to the end of their lives, as men do. Some theorists argue that menopause served a biological purpose for our prehistoric foremothers. Human offspring take a long time to rear. Beyond a certain point, the frailties of age make it unlikely that a woman could rear another infant to maturity or even survive the stress of pregnancy.

She might do better in the long run to become infertile and finish rearing her last child instead of having another one. In this view, menopause was biologically advantageous for our ancestors—in other words, an evolutionary adaptation.

Others argue against this hypothesis on the grounds that Pleistocene (Ice Age) skeletons indicate that early hominids rarely lived past age 40. If this is true, menopause setting in at 45 to 55 years of age could have served little purpose. In this view, Ice Age women may indeed have been fertile to the end of their lives; menopause now may be just an artifact of modern nutrition and medicine, which have made it possible for us to live much longer than our ancestors did.

Think About It

FSH and LH secretion *rise* at climacteric and these hormones attain high concentrations in the blood. Explain this using the preceding information and what you know about the pituitary-gonadal axis.

Menopause is the cessation of menstrual cycles, usually occurring between the ages of 45 and 55. The average age has increased steadily in the last century and is now about 52. It is difficult to precisely establish the time of menopause because the periods can stop for several months and then begin again. It is generally considered to have occurred when there has been no menstruation for a year or more.

To ease the transitions of climacteric, many physicians prescribe hormone replacement therapy (HRT)—low doses of estrogen and progesterone taken orally or by a skin patch. The risks and benefits of HRT are still being debated.

Key Point Review

5 Describe the similarities and differences between male and female puberty.

6 Describe the major changes that occur in female climacteric and the principal cause of these changes.

7 What is the difference between climacteric and menopause?

Oogenesis and the Sexual Cycle

▼**Objectives**

When you have completed this section, you should be able to
- describe the sequence of cell types in oogenesis and relate these to meiosis;
- describe how the ovarian follicles change in relation to oogenesis;
- describe the hormonal events that regulate the ovarian cycle;
- describe how the uterus changes during the ovarian cycle; and
- construct a chart of the phases of the monthly reproductive cycle showing the hormonal, ovarian, and uterine events of each phase.

Egg production is called **oogenesis**[27] (OH-oh-JEN-eh-sis) (fig. 28.11). Like spermatogenesis, it produces a haploid gamete by means of meiosis. There are, however, numerous differences between oogenesis and spermatogenesis. The most obvious, perhaps, is that spermatogenesis goes on continually, while oogenesis is a distinctly cyclic event that normally produces only one egg per month. Oogenesis is accompanied by cyclic changes in hormone secretion and histological structure of the ovaries and uterus; the uterine changes result in the monthly menstrual flow.

Oogenesis

The female germ cells arise, like those of the male, from the yolk sac of the embryo. They colonize the gonadal ridges in the first 5 to 6 weeks of development and then differentiate into **oogonia** (OH-oh-GO-nee-uh). Oogonia multiply until the fifth month of fetal development, reach 6 to 7 million in number, and then go into a state of arrested development until shortly before birth. At that time, some of them transform into **primary oocytes** and go as far as early meiosis I. Any stage from the primary oocyte to the time of fertilization can be called an egg, or ovum.

Most primary oocytes undergo a process of degeneration called **atresia** (ah-TREE-zhee-uh), before a girl is born. Only 2 million remain at the time of birth, and most of those undergo atresia during childhood. By puberty, only 400,000 oocytes remain. This is the female's lifetime supply of gametes, but it is more than ample; even if she ovulated every 28 days from the ages of 14 to 50, she would ovulate only 480 times.

Beginning in adolescence, FSH stimulates the primary oocytes to complete meiosis I, which yields two haploid daughter cells of unequal size and different destinies. In oogenesis it is important to produce an egg

---

27. *oo* = egg + *genesis* = production

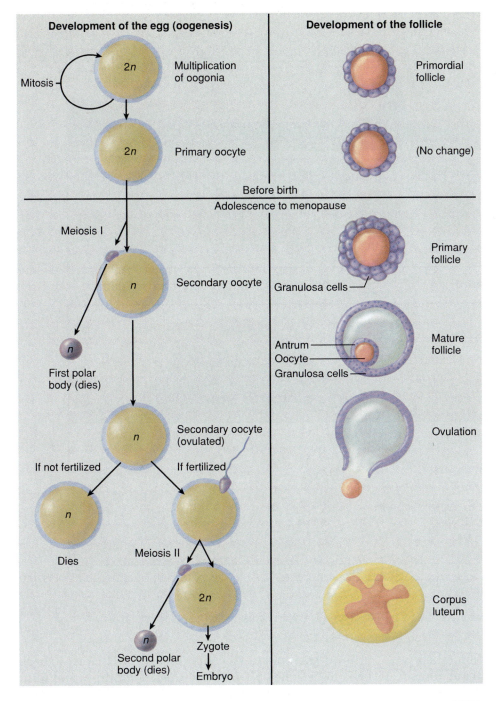

Development of the egg (oogenesis)

Mitosis

2n — Multiplication of oogonia

2n — Primary oocyte

Development of the follicle

Primordial follicle

(No change)

Before birth
Adolescence to menopause

Meiosis I

n — Secondary oocyte

n — First polar body (dies)

Primary follicle

Granulosa cells

Mature follicle

Antrum
Oocyte
Granulosa cells

n — Secondary oocyte (ovulated)

If not fertilized   If fertilized

n — Dies

Ovulation

Meiosis II

2n

n — Second polar body (dies)

Zygote

Embryo

Corpus luteum

Figure 28.11 Oogenesis (*left*) and corresponding development of the follicle (*right*).

with as much cytoplasm as possible, because if it is fertilized it must divide repeatedly and produce numerous daughter cells. Splitting each oocyte into four equal parts would run counter to this purpose. Rather, meiosis produces a large daughter cell called the **secondary oocyte** and a much smaller one called the **first polar body.** The polar body sometimes undergoes meiosis II, but whether it does or not, it or its daughter cells disintegrate. It is nothing more than a means of discarding the extra haploid set of chromosomes.

The secondary oocyte proceeds as far as metaphase II and then arrests until ovulation. If it is not fertilized, it dies after ovulation and never finishes meiosis. If it is fertilized, it completes meiosis II and produces a **second polar body,** which disposes of one chromatid from each chromosome. The large remaining egg unites its chromosomes with those of the sperm and ultimately produces a new person. Its fate from fertilization to birth is discussed in the next chapter.

Each developing germ cell is contained in a follicle, surrounded by one or more layers of follicular cells. The development and types of follicles are described shortly.

The Sexual Cycle

The female **sexual cycle** is a monthly sequence of changes caused by shifting patterns of hormone secretion. It ranges from 20 to 45 days long, but the timetable in the following discussion is based on the average 28-day cycle. We focus here on changes in the ovaries called the **ovarian cycle** and parallel changes in the uterus called the **menstrual cycle.**

In brief, the average cycle begins with a 2-week *follicular phase.* Menstruation occurs during the first 5 days of this phase, and then the stratum functionalis of the uterus regenerates by mitosis. The ovarian follicles grow during this phase, and typically one of them ovulates around day 14. Over the next 2 weeks, called the *postovulatory phase,* the stratum functionalis accumulates secretions and increases in thickness. If pregnancy does not occur, it breaks down again in the last 2 days. As loose tissue and blood accumulate, menstruation begins and the cycle starts over. The following discussion examines these events in more detail. They are summarized in figure 28.12 and table 28.1.

The Follicular Phase

The **follicular phase** extends from the beginning of menstruation until ovulation. It averages 14 days long, but this is the most variable part of the cycle and it is

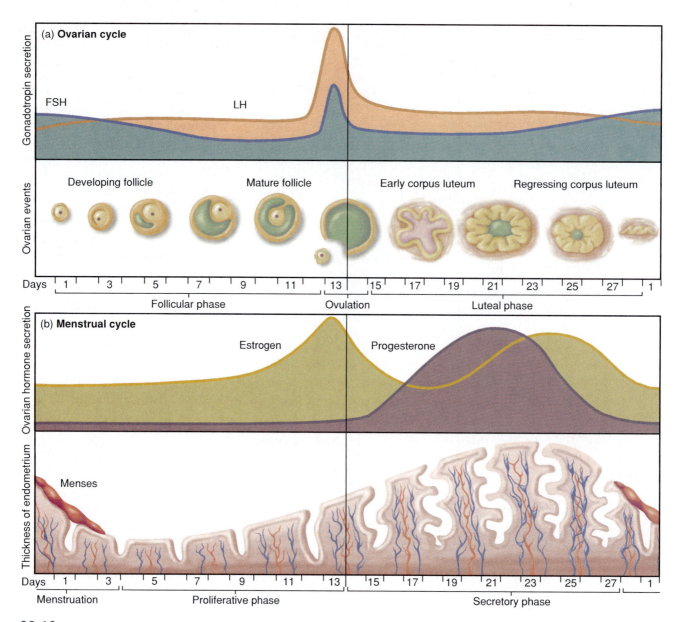

Figure 28.12 The female sexual cycle. (*a*) The ovarian cycle (events in the ovary); (*b*) the menstrual cycle (events in the uterus). ⚶

seldom possible to predict the date of ovulation reliably. This phase is subdivided into the **menstrual phase** (days 1–5) and **preovulatory phase** (days 6–14).

We will consider the mechanism of menstruation later, because it is best understood after you become acquainted with the buildup of endometrial tissue over the earlier parts of the cycle. The menstrual phase, however, is also a time of important developments in the ovaries, as follows.

The Menstrual Phase Around day 25 of the previous cycle, FSH secretion rises and stimulates 20 to 25 primary oocytes to go through meiosis I. The follicles containing these oocytes also undergo changes. The primary oocyte is enclosed in a **primordial follicle,** which has a single layer of squamous follicular cells

around the oocyte (see fig. 28.11). FSH causes these cells to become cuboidal and then to multiply and form a stratified epithelium. They are now called **granulosa cells** and the follicle is called a **primary follicle.** The follicle as a whole becomes larger, and connective tissue begins to condense around it to form the **theca[28] folliculi** (THEE-ca fol-IC-you-lye). Its outer layer, the **theca externa,** becomes a fibrous capsule. Its inner layer, the **theca interna,** secretes androgen, which the granulosa cells convert to estrogen.

In the first few days of the follicular phase, the granulosa cells begin to secrete an estrogen-rich **follicular fluid,** which accumulates in little pools amid the

---

28. *theca* = box, case

Table 28.1 Phases of the Female Sexual Cycle

Days	Phase	Major Events
1–14	**Follicular Phase**	
1–5	Menstrual phase	Menstruation occurring; FSH level high; primordial follicles developing into primary and then secondary follicles
6–13	Preovulatory (proliferative) phase	Rapid growth of one follicle and atresia of the lagging follicles; drop in FSH level; endometrial regeneration and growth by cell proliferation; development of mature follicle; completion of meiosis I, producing secondary oocyte, which arrests at metaphase II; sharp rise in LH level
14	**Ovulation**	Rupture of follicle and release of oocyte
15–28	**Postovulatory Phase**	
15–26	Luteal (secretory) phase	Formation of corpus luteum; secretion of progesterone; mucus and glycogen secretion by endometrium, causing endometrial thickening; involution of corpus luteum, producing corpus albicans; falling progesterone level
26–28	Premenstrual (ischemic) phase	Endometrial involution; vascular spasms causing endometrial ischemia and necrosis; sloughing of necrotic tissue from uterine wall, mixing with blood and forming menstrual fluid

cells. These pools soon merge and become a fluid-filled cavity, the **antrum.** The follicle is now called a **secondary (antral) follicle** (fig. 28.13). A mound of follicular cells called the **cumulus oophorus**[29] (CUE-mew-lus oh-OFF-ur-us) covers the oocyte and secures it to the follicle wall. Between the ovum and the cumulus oophorus is a clear layer of gel called the **zona pellucida.**[30] The innermost layer of cumulus cells, in contact with the zona pellucida, is the **corona radiata.**[31] This is the state of follicular development when menstruation ceases around day 5.

The estrogen produced by a developing follicle has two seemingly contrary effects on the ovaries. It stimulates granulosa cells in its own follicle to develop increasing numbers of FSH receptors, making them more sensitive to FSH. FSH in turn causes this follicle to produce still more estrogen, completing a positive feedback loop. At the same time, however,

29. *cumulus* = mound, heap + *oo* = egg + *phor* = to bear, carry
30. *zona* = zone + *pellucid* = clear, transparent
31. *corona* = crown + *radiata* = radiating

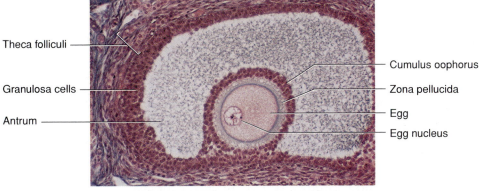

Figure 28.13 Ovarian follicles. (*a*) Primary follicles consist of an egg cell surrounded by a simple squamous epithelium of follicular cells. The secondary follicle has a stratified cuboidal epithelium of follicular cells and has begun to develop pools of follicular fluid that soon merge to form an antrum. (*b*) A mature (graafian) follicle.

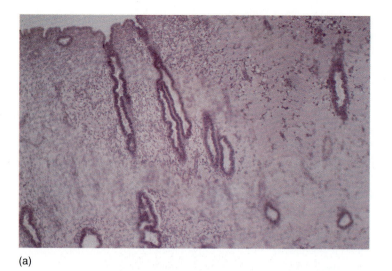

(a)

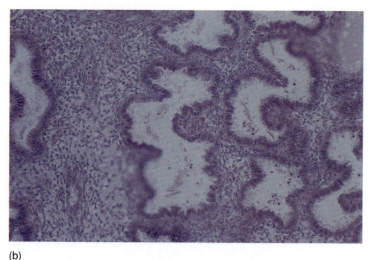

(b)

Figure 28.14 Histology of the endometrium. (*a*) In the proliferative phase, around day 10, the endometrium is about 4 mm thick and the glands are straight and narrow. (*b*) In the late secretory phase, around day 25, the endometrium is about 6 mm thick and the glands are wide and coiled.

Special Topic — Signs of Ovulation — 28.3

If a couple is attempting to conceive a child or attempting to avoid pregnancy, it is important to be able to tell when ovulation occurs. The signs are subtle but detectable. For one, the cervical mucus becomes thinner and more stretchy. Also, the resting body temperature (*basal temperature*) rises 0.2° to 0.3°C (0.4°–0.6°F). This is best measured first thing in the morning, before rising from bed; the change can be detected if basal temperatures are recorded for several days before ovulation in order to see the difference. The LH surge that occurs about 24 hours before ovulation can be detected with a home testing kit. Finally, some women experience twinges of ovarian pain known by the German name, *mittelschmerz*,[32] lasting from a few hours to a day or so at the time of ovulation. The best time to become pregnant is within 24 hours after the cervical mucus changes consistency and the basal temperature rises.

circulating estrogen inhibits FSH secretion by the pituitary. Therefore, circulating FSH levels go down; the less developed, less sensitive follicles undergo atresia; and the most developed, highly sensitive follicle continues to grow. In addition, FSH and estrogens stimulate the granulosa cells to produce LH receptors, which are important to the next phase of the cycle.

Think About It

Review the phenomenon of up-regulation in chapter 17 and explain its relevance to the processes just described in the ovaries.

The Preovulatory Phase   This is the period between menstruation and ovulation—days 6 to 14. It is also called the **proliferative phase** because it includes the regrowth of endometrial tissue following menstruation. At the start of the preovulatory phase, the endometrium is 2 to 3 mm thick and has only a stratum basalis. Estrogen stimulates mitosis, the prolific growth of blood vessels, and the formation of a new stratum functionalis (fig. 28.14*a*). Immediately after menstruation, the endometrium is about 2 mm thick; by day 14, it is twice that thickness. Estrogen also stimulates the en-

dometrium to develop progesterone receptors, preparing it for the progesterone-dominated postovulatory phase (the *permissive* effect of one hormone on another; see chapter 17).

One follicle rapidly outpaces the others and attains a diameter of up to 2.5 cm. This follicle, called a **mature, vesicular,**[33] or **graafian**[34] **follicle,** protrudes from the surface of the ovary like a blister. As it develops, the primary oocyte completes meiosis I, producing a secondary oocyte. This cell begins meiosis II but stops at metaphase II. It is now ready for ovulation.

Ovulation

Ovulation typically occurs on day 14, takes only 2 to 3 minutes, and separates the preovulatory from the postovulatory phase. Some ways to detect its occurrence are described in special topic 28.3. Ovulation is triggered by shifting patterns of hormone secretion. In the

32. *mittel* = in the middle + *schmerz* = pain
33. *vesic* = bladder, blister
34. Reijnier de Graaf (1641–73), Dutch physiologist and histologist

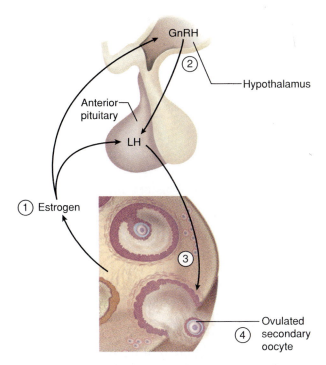

Figure 28.15 Control of ovulation by hormones of the pituitary-ovarian axis. (*1*) The maturing follicle secretes high levels of estrogen, which stimulates the hypothalamus and anterior pituitary. (*2*) The hypothalamus secretes GnRH. (*3*) In response to estrogen and GnRH, the anterior pituitary secretes LH. (*4*) LH triggers ovulation.

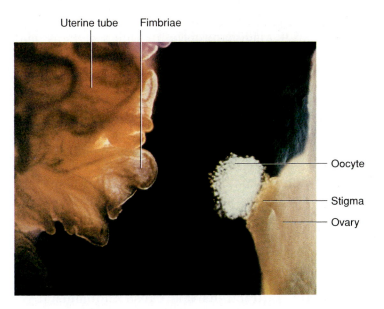

Figure 28.16 Ovulation of a human follicle, viewed by endoscopy.

last day or two of the preovulatory phase, the estrogen level is high; it stimulates increased output of GnRH by the hypothalamus and FSH and LH by the anterior pituitary. Estrogen also stimulates LH secretion directly (fig. 28.15). Therefore, the FSH level rises in the last day or two before ovulation, but the LH level rises even more sharply.

LH induces ovulation. It causes hyperemia in the follicle—increased blood flow, more serous fluid exuding from the capillaries into the antrum, and rapid swelling of the follicle in its last day or two. Meanwhile, LH causes the theca interna to secrete collagenase, an enzyme that weakens the ovarian wall over the swelling follicle. A nipplelike **stigma** develops over the follicle, follicular fluid seeps from the stigma for 1 to 2 minutes, and then the stigma ruptures. The remaining follicular fluid oozes out, carrying the secondary oocyte covered by the corona radiata (fig. 28.16).

As ovulation approaches, the uterine tube becomes edematous, its fimbriae envelop and caress the ovary, and its cilia create a gentle current in the nearby peritoneal fluid. The ovulated egg is usually caught up in this current and swept into the tube, but many oocytes fall into the pelvic cavity and die.

An oocyte has only 24 hours to be fertilized, or else it dies in the uterine tube. The chance of fertilization is enhanced by changes in the cervical mucus at the time of ovulation. It becomes thinner and more stringy, with strands oriented lengthwise in the cervical canal. If there are any sperm in the vagina, these strands guide them into the uterus.

Postovulatory Phase

The **postovulatory phase** spans days 15 to 28, from ovulation to the beginning of menstruation. It is the most predictable phase of the cycle. The first 12 days are called the **luteal phase** and the last 2 days are the **premenstrual phase.**

Luteal Phase   When the follicle expels the oocyte and follicular fluid, it collapses and bleeds into the antrum. As the clotted blood is slowly absorbed, granulosa and theca interna cells multiply and fill the antrum and a dense bed of blood capillaries grows amid them. The theca interna cells accumulate a yellow lipid and are then called **lutein cells.** The structure that has developed from the ovulated follicle is called the **corpus luteum.**[35] Its activity is controlled by LH, which continues to be secreted by the anterior pituitary after ovulation. Because of this effect, LH is also called *luteotropic hormone.*

The lutein cells produce mainly androgen, and the granulosa cells now convert most of this to progesterone and a smaller amount of estrogen. Progesterone prepares the uterus for possible pregnancy. Remember that in the preovulatory phase, estrogens primed the uterus by stimulating it to develop progesterone receptors. Progesterone now causes the endometrium to thicken further by secreting and accumulating mucus and glycogen. The

---

35. *corpus* = body + *lute* = yellow

endometrial glands become wider, longer, and coiled; the cells of the lamina propria fill with secretions; and the endometrium becomes suffused with increased tissue fluid (see fig. 28.14b). The endometrium attains a thickness of up to 6 mm—a soft, wet, nutrient-rich bed for embryonic development. Lutein cells also produce inhibin, which suppresses FSH secretion and prevents new follicles from developing. If pregnancy occurs, the corpus luteum remains active for about 3 months.

Premenstrual Phase In the absence of pregnancy, the corpus luteum begins to degenerate in about 10 days because its own rising output of progesterone inhibits GnRH and LH secretion. Declining levels of LH result in involution (shrinkage) of the corpus luteum, which results in rapidly falling progesterone levels. In the absence of progesterone, the spiral arteries undergo spasmodic contractions that cause endometrial ischemia (interrupted blood flow)—hence this phase is also known as the **ischemic** (iss-KEE-mic) **phase.** Ischemia leads to tissue necrosis. As the endometrial stroma, glands, and blood vessels degenerate, pools of blood accumulate in the functionalis. Necrotic tissue falls away from the uterine wall, mixes with blood in the lumen, and forms the **menses** (menstrual fluid). Involution of the corpus luteum is complete by day 26, leaving an inactive scar called the **corpus albicans.**

Menstrual Phase The first day of vaginal discharge is considered to be day 1 of a new cycle. The average woman discharges about 40 mL of blood and 35 mL of serous fluid over a 5-day period of menstruation. Menstrual fluid contains fibrinolysin, therefore it normally does not clot. Vaginal discharge of clotted blood may indicate uterine pathology rather than normal menstruation.

Even as the endometrium is undergoing necrosis, the stage is set for renewal. Involution of the corpus luteum ends the negative feedback effect of progesterone on the hypothalamus. Consequently, GnRH secretion rises, followed by FSH and the development of a new crop of ovarian follicles.

.... **Key Point Review** ....

8 Name the sequence of cell types in oogenesis and identify the ways oogenesis differs from spermatogenesis.

9 Distinguish between a primordial, primary, and secondary follicle. Describe the major structures of a mature follicle.

10 Describe what happens in the ovary during the follicular and postovulatory phases.

11 Describe what happens in the uterus during the menstrual, proliferative, secretory, and premenstrual phases.

12 Describe the effects of FSH and LH on the ovary.

13 Describe the effects of estrogen and progesterone on the uterus, hypothalamus, and anterior pituitary.

Female Sexual Response

▼**Objectives**
When you have completed this section, you should be able to
• describe the female sexual response at each phase; and
• compare and contrast the female and male responses.

Female sexual response, the physiological changes that occur during intercourse, may be viewed in terms of the four phases identified by Masters and Johnson and discussed in the previous chapter—excitement, plateau, orgasm, and resolution. The neurological and vascular controls of female sexual response are essentially the same as in the male and need not be repeated here. The emphasis here is on ways the female response differs from that of the male.

Excitement and Plateau

Excitement is marked by myotonia, vasocongestion, and increased heart rate, blood pressure, and respiratory rate. Although vasocongestion works by the same mechanism in both sexes, its effects are quite different. In females, the labia minora become congested and often protrude beyond the labia majora. The labia majora become reddened and enlarged and then flatten and spread away from the vaginal orifice.

The vaginal wall becomes purple due to hyperemia, and serous fluid called the **vaginal transudate** seeps through the wall into the canal. Along with secretions of the greater vestibular glands, this moistens the vestibule and provides lubrication. The inner end of the vagina dilates and becomes cavernous, while the lower one-third of it constricts to form a narrow passage called the **orgasmic platform.** Thus it grips the penis more tightly, and combined with the vaginal rugae (friction ridges), this enhances stimulation and helps induce orgasm in both partners (fig. 28.17).

The uterus, which normally tilts forward over the urinary bladder, stands more erect during excitement and the cervix withdraws from the vagina. In plateau, the uterus is nearly vertical and extends into the false pelvis. This is called the **tenting** effect.

Although the vagina is the female copulatory organ, the clitoris is most comparable to the penis in structure, physiology, and importance as the primary focus of erotic stimulation. It has a high concentration of sensory nerve endings, which, by contrast, are almost absent from the vagina. Recall that the penis and clitoris are homologous structures, both arising from the phallus of the fetus. Both have a pair of corpora cavernosa with central (deep) arteries, and both become engorged by the same mechanism. The glans and shaft of the clitoris swell to two or three times their unstimulated size,

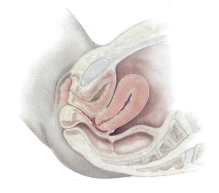

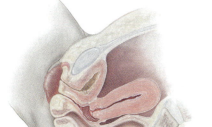

Unstimulated

Uterus tilts forward over urinary bladder; vagina relatively narrow; labia minora retracted

Excitement

Uterus stands more vertically; inner end of vagina dilates; labia minora vasocongested, may extend beyond labia majora; labia minora and vaginal mucosa reddened or violet due to hyperemia; vaginal transudate moistens vagina and vestibule

Plateau

Uterus is tented (vertical) and cervix is withdrawn from vagina; orgasmic platform (lower one-third) of vagina constricts penis; clitoris is engorged and its glans is withdrawn beneath prepuce; labia bright red or violet

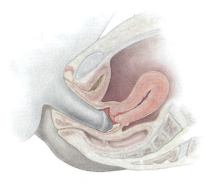

Orgasm

Orgasmic platform contracts rhythmically; uterus exhibits peristaltic contractions; anal and urinary sphincters constrict

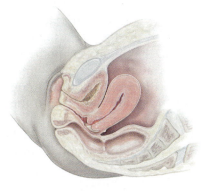

Resolution

Uterus returns to original position; cervix may dip into pool of semen; orgasmic platform relaxes; inner end of vagina constricts and returns to original dimensions

Figure 28.17 Stages of female sexual response.

but since the clitoris cannot swing upward away from the body like the penis, it tends to withdraw beneath the prepuce. Thrusting of the penis in the vagina tugs on the labia minora and, by extension, pulls on the prepuce and stimulates the clitoris. The clitoris may also be stimulated by pressure between the pubic symphyses of the partners.

The breasts also become congested and swollen during the excitement phase, and the nipples become erect. Stimulation of the breasts also enhances sexual arousal.

Orgasm

Late in plateau, many women experience involuntary pelvic thrusting, followed by 1 to 2 seconds of "suspension" or "stillness" preceding orgasm. Orgasm is commonly described as an intense sensation spreading from the clitoris through the pelvis, sometimes with pelvic throbbing and a spreading sense of warmth. The orgasmic platform gives three to five strong contractions about 0.8 seconds apart. The anal and urinary sphincters contract, and the uterus undergoes peristaltic waves from fundus to cervix, similar to labor contractions. The paraurethral glands, which are homologous to the prostate, sometimes expel fluid similar to prostatic fluid. Tachycardia and hyperventilation occur; the breasts enlarge still more and the areolae often become engorged; and in many women a reddish, rashlike *sex flush* appears on the lower abdomen, chest, neck, and face.

Resolution

During resolution the uterus drops forward to its resting position and the cervix protrudes into the vagina once more—into the pool of semen if present. The orgasmic platform quickly relaxes, while the inner end of the vagina returns more slowly to its normal dimensions. The sex flush disappears quickly and the areolae and nipples undergo rapid detumescence, but it may take 5 to 10 minutes for the breasts to return to their normal size. In many women (and men) there is a postorgasmic outbreak of perspiration. Unlike men, women do not have a refractory period and may quickly experience additional orgasms.

.... Key Point Review ....

14 What are the female sources of lubrication in coitus?

15 What female tissues and organs become vasocongested?

16 Describe the changes in uterine position through the sexual response cycle.

This section treats pregnancy from the maternal standpoint—that is, adjustments of the woman's body to pregnancy and the mechanism of childbirth. Development of the human embryo and fetus are described in the next chapter.

Gestation (pregnancy) usually lasts for 270 days from conception to childbirth, but the gestational calendar is usually measured from the first day of the woman's last menstrual period (LMP). Thus the birth is predicted to occur 284 days (about 40 weeks) from LMP. The duration of pregnancy, called its *term,* is commonly described with reference to 3-month intervals called **trimesters.**

Prenatal Development

A few fundamental facts of fetal development are introduced here as a foundation for understanding maternal physiology. Fertilization must occur in the distal half of the uterine tube if it is to occur at all, since an unfertilized egg does not live long enough to reach the uterus alive. If fertilized, the egg divides five or six times before it reaches the uterus. All the products of conception—the embryo or fetus as well as the placenta and membranes associated with it—are collectively called the *conceptus.* The developing individual is called a *blastocyst* for most of the first 2 weeks, an *embryo* from 3 to 8 weeks, and a *fetus* from week 9 to birth. The exact meanings and distinctions among these terms are detailed in the next chapter. For the first 6 weeks after birth, the infant is called a *neonate.*[36]

Soon after it arrives in the uterus, the conceptus becomes a *blastocyst,* which consists of an inner cell mass destined to develop into the embryo and an outer cell mass, the *trophoblast,* which plays various supporting roles. About 4 days after arrival, the conceptus attaches to the endometrium by means of rootlike growths

....

36. *neo* = new + *nat* = born, birth

of the trophoblast cells. This process is called *implantation.* It is soon followed by development of the *placenta,* a disclike organ composed of a combination of maternal (uterine) and trophoblastic tissues. This has several functions, including fetal nutrition, waste disposal, and secretion of hormones that regulate pregnancy, mammary development, and fetal development. It is connected to the fetus by way of the *umbilical cord.* The fetus is enclosed in a transparent sac called the *amnion* and floats in the *amniotic fluid* filling the sac. External to the amnion is a membrane called the *chorion* (CO-ree-on), which is the source of the placental hormones.

Hormones of Pregnancy

The hormones with the strongest influences on pregnancy are estrogens, progesterone, human chorionic gonadotropin, and human chorionic somatomammotropin. These are secreted primarily by the placenta, but the corpus luteum is an important source of hormones in the first 7 to 12 weeks. If the corpus luteum is removed before the seventh week, abortion almost always occurs. From weeks 7 to 17, the corpus luteum degenerates and the placenta takes over its endocrine functions.

Human Chorionic Gonadotropin

Human chorionic gonadotropin (HCG) is secreted by the trophoblast cells. Its presence in the urine is the basis of pregnancy tests and can be detected with home testing kits as early as 8 to 9 days after conception. HCG secretion peaks around 10 to 12 weeks and then falls to a relatively low level for the rest of gestation (fig. 28.18). Like LH, it stimulates growth of the corpus luteum, which doubles in size and secretes increasing amounts of progesterone and estrogen. HCG also stimulates the male fetal gonad to secrete testosterone; thus it indirectly induces formation of the male reproductive system and descent of the testes, as we saw in the previous chapter.

Estrogens

Estrogen secretion increases to about 30 times normal by the end of gestation. The corpus luteum is an important source of estrogen for the first 12 weeks; after that, it comes mainly from the placenta. The adrenal glands of the mother and fetus secrete androgens, which the placenta converts to estrogens. The most abundant estrogen of pregnancy is estriol, but its effects are relatively weak; estradiol is less abundant but accounts for most of the estrogenic effects in pregnancy.

Estrogens stimulate tissue growth in the fetus and mother. They cause the mother's uterus and external

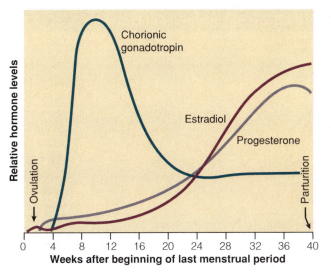

Figure 28.18 Hormone levels over the course of pregnancy.

genitalia to enlarge, the mammary ducts to grow, and the breasts to increase to nearly twice their former size. They make the pubic symphysis more elastic and the sacroiliac joints more limber, so the pelvis widens during pregnancy and the pelvic outlet expands during childbirth.

Progesterone

The placenta secretes a great deal of progesterone. Progesterone and estrogens suppress pituitary secretion of FSH and LH, preventing follicles from developing during pregnancy. (This is the basis for contraceptive pills and implants; see chapter essay, p. 1019.) Progesterone also suppresses uterine contractions so the conceptus is not prematurely expelled. It prevents menstruation and promotes the proliferation of *decidual cells* of the endometrium, an essential source of nutrition for the blastocyst. Once estrogen has stimulated growth of the mammary ducts, progesterone stimulates development of the secretory acini—another step toward lactation.

Human Chorionic Somatomammotropin

The amount of **human chorionic somatomammotropin (HCS)** secreted in pregnancy is several times that of all the other hormones combined, yet its function is the least understood. The placenta begins secreting HCS around the fifth week and HCS output increases steadily from then until term, in direct proportion to the size of the placenta.

HCS is sometimes called *human placental lactogen* because, in other mammals, it causes mammary

Table 28.2 The Hormones of Pregnancy

Hormone	Effects
Human chorionic gonadotropin (HCG)	Prevents involution of corpus luteum and stimulates its growth; stimulates male fetal gonad to secrete testosterone and indirectly induces development of male reproductive system; basis of pregnancy tests
Estrogens	Stimulate maternal and fetal tissue growth, including enlargement of uterus and maternal genitalia; stimulate development of mammary ducts; soften pubic symphysis and sacroiliac joints, facilitating pelvic expansion in pregnancy and childbirth; suppress FSH and LH secretion
Progesterone	Suppresses premature uterine contractions; prevents menstruation; stimulates proliferation of decidual cells, which nourish embryo; stimulates development of mammary acini; suppresses FSH and LH secretion
Human chorionic somatomammotropin (HCS)	Weak growth-stimulating effects similar to growth hormone; glucose-sparing effect on mother, making glucose more available to fetus; mobilization of fatty acids as maternal fuel
Pituitary thyrotropin	Stimulates thyroid activity and metabolic rate
Human chorionic thyrotropin	Same effect as pituitary thyrotropin
Parathyroid hormone	Stimulates osteoclasts and mobilizes maternal calcium
Adrenocorticotropic hormone	Stimulates glucocorticoid secretion; thought to mobilize amino acids for fetal protein synthesis
Aldosterone	Causes fluid retention, contributing to increased maternal blood volume
Relaxin	Uncertain function; may soften the cervix to facilitate childbirth

development and lactation; however, it does not induce lactation in humans. Its effects seem similar to those of growth hormone, but weaker. It also seems to reduce the mother's insulin sensitivity and glucose usage such that the mother consumes less glucose and leaves more of it for use by the fetus. HCS promotes the release of free fatty acids from the mother's adipose tissue, providing an alternative energy substrate for her cells to use in lieu of glucose.

Other Hormones

Pregnancy affects many other aspects of endocrine function. A woman's pituitary gland grows about 50% larger during pregnancy and produces markedly elevated levels of thyrotropin, prolactin, and ACTH. The thyroid gland also becomes about 50% larger under the influence of HCG, pituitary thyrotropin, and *human chorionic thyrotropin* from the placenta. Elevated thyroid secretion increases the metabolic rate of the mother and fetus. The parathyroid glands enlarge and stimulate osteoclast activity, liberating calcium from the mother's bones for fetal use. ACTH stimulates glucocorticoid secretion, which may serve primarily to mobilize amino acids for fetal protein synthesis. Aldosterone secretion rises and promotes fluid retention. The corpus luteum and placenta secrete *relaxin*, which relaxes the pubic symphysis in other animals but does not seem to have a

significant effect in humans. It is thought by some to soften the cervix in preparation for childbirth.

The hormones of pregnancy are summarized in table 28.2.

Adjustments to Pregnancy

Pregnancy places a considerable stress on a woman's body and requires adjustments in nearly all the organ systems. A few of the major adjustments and effects of pregnancy are described here.

Digestive System, Nutrition, and Metabolism

Many women experience nausea (morning sickness) in the first few months of pregnancy—one of the first signs that a women may be pregnant. The cause of morning sickness is unknown. One hypothesis is that it stems from the reduced intestinal motility caused by the steroids of pregnancy. Another is that it is an evolutionary adaptation to protect the fetus from toxins. The fetus is most vulnerable to toxins at the same time that morning sickness peaks. Women with morning sickness tend to prefer bland foods and to avoid spicy and pungent foods, which are highest in toxic compounds. In some women, the nausea progresses to vomiting. In **hyperemesis gravidarum**[37] (HI-per-EM-eh-sis GRAV-ih-DAH-rum), vomiting is so severe it may require hospitalization to stabilize the electrolyte and acid-base balance.

Constipation and heartburn are common in pregnancy. The former is another result of reduced intestinal motility. The latter is due to the enlarging uterus pressing upward on the stomach, causing the reflux of gastric contents into the esophagus.

The basal metabolic rate rises about 15% in the second half of gestation. Pregnant women often feel overheated because of this and the effort of carrying the extra weight. The appetite may be strongly stimulated, but a pregnant woman only needs an extra 300 kcal/day even in the last trimester. With poor prenatal care and little self-control, however, some women greatly overeat and gain as much as 34 kg (75 lb) of weight compared with a healthy average of 11 kg (24 lb). Maternal nutrition should emphasize the quality of food eaten, not quantity.

During the last trimester, the fetus needs more nutrients than the mother's digestive tract can absorb. In preparation for this, the placenta stores nutrients early in gestation and releases them in the final trimester. The demand is especially high for protein, iron, calcium, and phosphates. A pregnant woman needs an extra 600 mg of iron for her own hemopoiesis and 375 mg for the fetus. She is likely to become anemic if she does not ingest enough iron during late pregnancy. Supplemental vitamin K is often given late in pregnancy to promote prothrombin synthesis in the fetus. This reduces the risk of neonatal hemorrhage, especially in the brain, caused by the stresses of birth. Supplemental folic acid reduces the risk of neurological disorders in the fetus, such as spina bifida and anencephalia. A vitamin D supplement helps to ensure adequate calcium absorption to meet fetal demands.

Circulatory System

By term, the placenta requires a maternal blood supply of about 625 mL/min. The mother's blood volume rises about 30% during pregnancy due to fluid retention and hemopoiesis; she eventually has about 1 to 2 L of extra blood. Cardiac output rises about 30% to 40% above normal by 27 weeks, but for unknown reasons, it falls almost to normal in the last 8 weeks. As the pregnant uterus puts pressure on the large pelvic blood vessels, it interferes with venous return from the legs and pelvic region. This can result in hemorrhoids and varicose veins.

Respiratory System

Minute ventilation increases about 50% during pregnancy for two reasons: (1) Oxygen demands are about 20% higher by late pregnancy in order to supply the fetus and support the mother's increased metabolic rate. (2) Progesterone increases the sensitivity of the respiratory chemoreceptors to carbon dioxide, and ventilation is adjusted to keep the arterial P_{CO_2} lower than normal. While there is a demand for increased ventilation, the expanding uterus pushes the abdominal viscera up against the diaphragm and interferes with inspiration. Consequently, the respiratory rate increases to compensate for the lack of depth. Pressure on the diaphragm may be great enough to cause breathing difficulty (dyspnea) by late pregnancy. In the last month, however, the pelvis expands enough for the fetus to drop lower in the abdominopelvic cavity, taking some pressure off the diaphragm and allowing the woman to breathe more easily.

Urinary System

Aldosterone and the other steroids of pregnancy promote water and salt retention by the kidneys. Nevertheless, urine output increases slightly because the glomerular filtration rate increases by about 50%. Increased urine output is necessary because the mother disposes of both her own and the fetus's wastes. As the pregnant uterus compresses the bladder and reduces its capacity, urination becomes more frequent and sometimes uncontrollable (stress incontinence).

Integumentary System

The skin must grow to accommodate expansion of the abdomen and breasts and the added fat deposition in the hips and thighs. Stretching of the dermis often tears the connective tissue and causes *stretch marks*, (see fig. 7.5, p. 206). These appear reddish at first but fade to a silvery white color after pregnancy. Melanocyte activity increases in some areas and causes darkening of the areolae and linea alba. The latter often becomes a dark line, the **linea nigra**[38] (LIN-ee-uh NY-gruh), from the umbilical to the pubic region. Some women also acquire a temporary blotchy darkening of the skin over the nose and cheeks called the "mask of pregnancy," or **chloasma**[39] (clo-AZ-muh). This disappears when the pregnancy is over.

Uterine Growth and Weight Gain

The uterus weighs about 50 g when a woman is not pregnant and 1,100 g by the end of pregnancy. Its

---

37. *hyper* = excessive + *emesis* = vomiting + *gravida* = pregnant woman

38. *linea* = line + *nigra* = black
39. *chloasma* = to be green

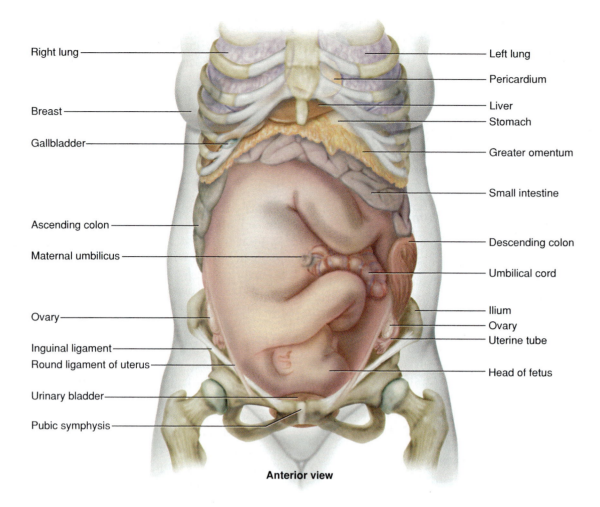

Right lung

Breast

Gallbladder

Ascending colon

Maternal umbilicus

Ovary

Inguinal ligament

Round ligament of uterus

Urinary bladder

Pubic symphysis

Left lung

Pericardium

Liver

Stomach

Greater omentum

Small intestine

Descending colon

Umbilical cord

Ilium

Ovary

Uterine tube

Head of fetus

Anterior view

Figure 28.19 The full-term fetus in vertex position.

Table 28.3	Average Weight Gain in Pregnancy
Fetus	3 kg (7 lb)
Placenta, fetal membranes, and amniotic fluid	1.8 kg (4 lb)
Blood and tissue fluid	2.7 kg (6 lb)
Fat	1.4 kg (3 lb)
Uterus	0.9 kg (2 lb)
Breasts	0.9 kg (2 lb)
Total	**11 kg (24 lb)**

growth is monitored by palpating the fundus, which eventually reaches almost to the xiphoid process (fig. 28.19). Table 28.3 shows a breakdown of weight gain in pregnancy.

Childbirth

The process of giving birth is called **parturition** (PAR-too-RISH-un). The ancients thought that the fetus kicks against the uterus and pushes itself out head first. The fetus, however, is a rather passive player in its own birth; its expulsion is achieved only by the contractions of the mother's uterine and abdominal muscles. Yet there is evidence that the fetus may play some role in its birth by chemically stimulating labor contractions and perhaps even sending chemical messages that signify when it is ready to be born.

Uterine Contractility

Over the course of gestation, the uterus exhibits relatively weak peristaltic **Braxton Hicks**[40] **contractions.** These become stronger in late pregnancy and often send women rushing to the hospital with "false labor." At term, however, these contractions transform suddenly into the more powerful **labor contractions.**

Progesterone and estrogen balance may be one factor in this pattern of increasing contractility. Both hormone levels increase over the course of gestation, and

40. John Braxton Hicks (1823–97), British gynecologist

progesterone inhibits uterine contractions. Beginning in the seventh month, however, progesterone secretion levels off or declines slightly, while estrogen secretion continues to rise (see fig. 28.18). Estrogen stimulates uterine contractions and may be one reason the uterus becomes more irritable in late pregnancy.

Also, as the pregnancy nears full term, the posterior pituitary releases more oxytocin and the uterus produces more oxytocin receptors. Oxytocin promotes labor in two ways: (1) it directly stimulates muscle of the myometrium, and (2) it stimulates the fetal membranes to secrete prostaglandins, which are synergists of oxytocin in producing labor contractions. Labor is prolonged if oxytocin or prostaglandins are lacking, and it may be induced or accelerated by giving a vaginal prostaglandin suppository or an intravenous "drip" of oxytocin. The conceptus itself may produce chemical stimuli promoting its own birth. Fetal cortisol secretion rises in late pregnancy and may enhance estrogen secretion by the placenta. The fetal pituitary gland also produces oxytocin, which does not enter the maternal circulation but may stimulate the fetal membranes to secrete prostaglandins.

Uterine stretching is also thought to play a role in initiating labor. Stretching any smooth muscle increases its contractility, and movements of the fetus produce the sort of intermittent stretch that is especially stimulatory to the myometrium. Twins are born an average of 19 days earlier than single infants, probably due to the greater stretching of the uterus. When the fetus is in the *vertex position* (head-down), its head pushes against the cervix, which is especially sensitive to stretch.

Labor Contractions

Labor contractions begin about 30 minutes apart. As labor progresses, they become more intense and eventually occur every 1 to 3 minutes. It is important that they be intermittent rather than one long, continual contraction. Each contraction sharply reduces maternal blood flow to the placenta, so the uterus must periodically relax to restore flow and oxygen delivery to the fetus. Contractions are strongest in the fundus and body of the uterus and weaker near the cervix, thus pushing the fetus downward.

According to the **positive feedback theory of labor,** true labor contractions are induced by stretching the cervix. This triggers a reflex contraction of the uterine body that pushes the fetus downward and stretches the cervix still more. Thus there is a self-amplifying cycle of stretch and contraction. In addition, cervical stretching induces a neuroendocrine reflex through the spinal cord, hypothalamus, and posterior pituitary. The posterior pituitary releases oxytocin, which is carried in the blood and stimulates the uterine muscle both directly and through the action of prostaglandins. This, too, is a positive feedback cycle: cervical stretching → oxytocin secretion → uterine contraction → cervical stretching (see fig. 1.17, p. 22).

As labor progresses, a woman feels a growing urge to "bear down." A reflex arc extends from the uterus to the spinal cord and back to the skeletal muscles of the abdomen. Contraction of these muscles—partly reflexive and partly voluntary—aids in expelling the fetus.

The pain of labor is mainly due at first to hypoxia of the myometrium—muscle hurts when deprived of blood, and each labor contraction temporarily restricts uterine circulation. As the fetus enters the vaginal canal, the pain becomes stronger due to increased stretching of the cervix, vagina, and perineum and sometimes the tearing of vaginal tissue. At this stage the obstetrician may perform an **episiotomy**—an incision to widen the vaginal canal. It is easier to repair a neat incision than random tears in the vagina. The pain of human childbirth, compared to the relative ease with which other mammals give birth, is an evolutionary product of the narrowing of the pelvic outlet as hominids adopted bipedal locomotion (see p. 289).

Stages of Labor

Labor occurs in three stages that are punctuated by distinct events. The durations of these stages tend to be longer in a **primipara** (a woman giving birth for the first time) than in a **multipara** (a woman who has previously given birth).

Dilation (First) Stage This is the longest stage, lasting 8 to 24 hours in a primipara but as little as a few minutes in a multipara. It is marked by the **dilation** (widening) of the cervical canal and by **effacement** (thinning) of the cervix (fig. 28.20*a, b*). The cervix reaches a maximum diameter of about 10 cm (the diameter of the baby's head). Usually at this time, the fetal membranes rupture and the amniotic fluid is discharged (the "breaking of the waters").

Expulsion (Second) Stage This stage typically lasts about 30 minutes in a primipara and as little as 1 minute in a multipara. It begins when the baby's head enters the vagina and lasts until the baby is entirely expelled (fig. 28.20*c*). The head acts as a wedge to widen the cervix, vagina, and vulva. The baby is said to be **crowning** when the top of its head is visible, stretching the vulva (fig. 28.21*a*). Delivery of the head is the most difficult part, with the rest of the body following much more easily. An episiotomy may be performed during this stage, and mucus is removed from the baby's mouth and nose by suction. When the baby is fully expelled, the umbilical cord is clamped in two places and cut between the clamps.

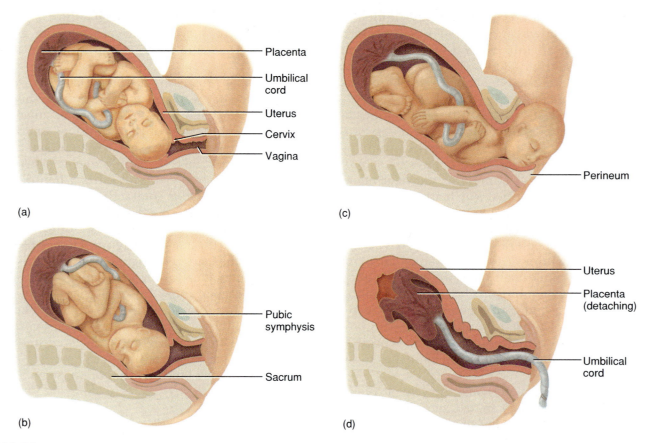

Figure 28.20 The stages of childbirth. (*a*) Early dilation. (*b*) Late dilation. (*c*) Expulsion. (*d*) The placental stage.

Placental (Third) Stage The uterus continues to contract after expulsion of the baby. The placenta, however, is a nonmuscular organ that cannot contract, so it buckles and shears away from the uterine wall (see fig. 28.20*d*). About 350 mL of blood is typically lost at this stage, but contractions of the myometrium compress the blood vessels and prevent more extensive bleeding. The placenta, amnion, and other fetal membranes are expelled by uterine contractions, which may be aided by a gentle pull on the umbilical cord. The membranes (*afterbirth*) must be carefully inspected to be sure everything has been expelled (fig. 28.21*c*). If any of these structures remain in the uterus, they can cause postpartum hemorrhaging. Also, the umbilical blood vessels are counted because an abnormal number in the cord may indicate cardiovascular abnormalities in the infant.

Puerperium

The first 6 weeks **postpartum** (after birth) are called the **puerperium**[41] (PYU-er-PEER-ee-um), a period in which the mother's anatomy and physiology stabilize and the reproductive organs return nearly to the pregravid state. The shrinkage of the uterus during this period is called **involution.** In a lactating woman, it loses about 50% of its weight in the first week and is nearly at its pregravid weight in 4 weeks. Involution is achieved through the **autolysis** (self-digestion) of uterine cells by their own lysosomal enzymes. For about 10 days, this produces a vaginal discharge called **lochia,** which is bloody at first and then turns clear and serous. Lactation promotes involution by (1) suppressing the secretion of estrogen, which would otherwise cause the uterus to remain more flaccid; and (2) stimulating the secretion of oxytocin, which causes the myometrium to contract and firm up the uterus sooner. It is important for the puerperium to be undisturbed, as emotional upset can suppress lactation in some women.

.... Key Point Review ....

17 List the roles of HCG, estrogen, progesterone, and HCS in pregnancy.

18 What is the role of the corpus luteum in pregnancy? What eventually takes over this role?

19 List and briefly explain the special nutritional requirements of pregnancy.

20 How much weight does the average woman gain in pregnancy? What contributes to this weight gain other than the fetus?

21 Describe the positive feedback theory of labor.

22 What major events define the three stages of labor?

....
41. *puer* = child + *per,* from *par* = birth

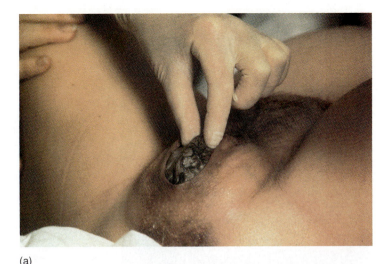

(a)

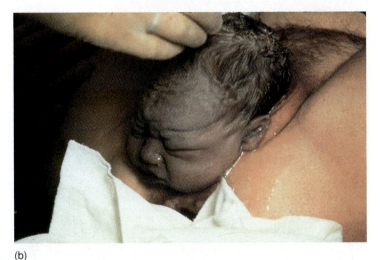

(b)

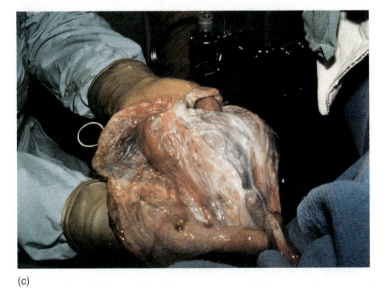

(c)

Figure 28.21 Childbirth. (*a*) Crowning. The baby's head, between the attendant's fingers, has begun to dilate the vulva. (*b*) Emergence of the head. (*c*) The placental and fetal membranes (afterbirth).

▼**Objectives**

When you have completed this section, you should be able to
- describe development of the breasts in pregnancy;
- describe the shifting hormonal balance that regulates the onset and continuation of lactation;
- describe the mechanism of milk ejection;
- contrast colostrum with breast milk; and
- discuss the benefits of breast-feeding.

La Leche League, an organization that advocates breast-feeding, reports that the average worldwide age at weaning is 4.2 years. **Lactation,** the synthesis and ejection of milk from the mammary glands, can continue indefinitely as long as the breast is stimulated by a nursing infant or a mechanical device (breast pump).

Development of the Mammary Glands in Pregnancy

The high estrogen level in pregnancy causes the ducts of the mammary gland to grow and branch extensively. Growth hormone, insulin, glucocorticoids, and prolactin also contribute to this development. Once the ducts are complete, progesterone stimulates the budding and development of acini at the ends of the ducts. The fully developed mammary gland is of the compound tubuloacinar type. The acini are organized into grapelike clusters (lobules) within each lobe of the breast (see fig. 28.9*c, d*).

Colostrum and Milk Synthesis

In late pregnancy, the mammary acini and ducts become distended with a secretion called **colostrum.** This is similar to breast milk in protein and lactose content but contains about one-third less fat. It is the infant's only natural source of nutrition for the first 2 to 3 days postpartum. Colostrum has a thin watery consistency and a cloudy yellowish color. The amount of colostrum secreted per day is at most 1% of the amount of milk secreted later, but since infants are born with excess body water and ample fat, high calorie and fluid intake are not required at first. A major benefit of colostrum is that it contains immunoglobulins, especially IgA. IgA resists digestion and may protect the infant from gastroenteritis. It is also thought to be pinocytosed by the small intestine and to confer wider, systemic immunity to the neonate.

Milk synthesis is promoted by prolactin, a hormone of the anterior pituitary. In the nonpregnant state, prolactin secretion is suppressed by **prolactin-inhibiting factor (PIF),** or dopamine, from the hypothalamus. Prolactin

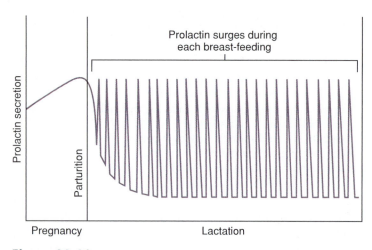

Figure 28.22 Hormone secretion in the lactating female.

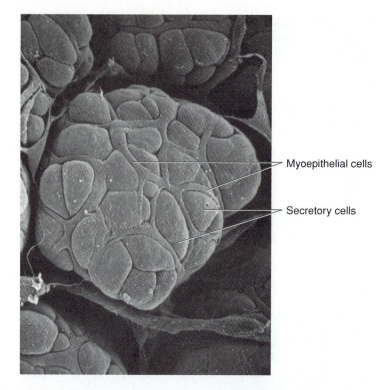

Figure 28.23 Myoepithelial cells forming a mesh around an acinus of the mammary gland. These are the cells that respond to oxytocin and squeeze milk from the acinus into the duct.

secretion begins 5 weeks into the pregnancy, and by full term it is 10 to 20 times its initial level. Even so, prolactin has little effect on the mammary glands until after birth. While the steroids of pregnancy prepare the mammary glands for lactation, they antagonize the effect of prolactin and suppress milk synthesis. When the placenta is discharged at birth, the steroid levels abruptly drop and allow prolactin to have a stronger effect. Milk is synthesized in increasing quantity over the following week. Around 3 days postpartum the milk "comes in" and is available to the infant. Milk synthesis also requires the action of growth hormone, cortisol, parathyroid hormone, and insulin to mobilize the necessary amino acids, fatty acids, glucose, and calcium.

At the time of birth, baseline prolactin secretion drops to the nonpregnant level. Every time the infant nurses, however, it jumps to 10 to 20 times this level for the next hour and stimulates the synthesis of milk for the next feeding (fig. 28.22). These prolactin surges are accompanied by smaller increases in estrogen and progesterone secretion. If the mother does not nurse or these hormone surges are absent (due to pituitary damage, for example), the mammary glands stop producing milk in about a week. Even if she does nurse, the rate of milk production declines after 7 to 9 months, although a woman can continue to lactate for years.

Only 5% to 10% of women become pregnant again while nursing an infant. Apparently, either prolactin or nerve signals from the breast inhibit GnRH secretion, which, in turn, results in reduced gonadotropin secretion and ovarian cycling. This mechanism may have evolved as a natural means of spacing births, but breast-feeding is not a reliable means of contraception. Even in women who breast-feed, the ovarian cycle sometimes resumes several months postpartum. In those who do not breast-feed, the cycles resume in a few weeks, but for the first 6 months they are anovulatory.

Milk Ejection

Milk is continually secreted into the mammary acini, but it does not easily flow into the ducts. **Milk ejection** ("let-down") is mediated by a neuroendocrine reflex. The infant's suckling stimulates nerve endings of the nipple and areola, which in turn signal the hypothalamus and posterior pituitary to release oxytocin. Oxytocin stimulates the contraction of **myoepithelial cells** in the mammary gland. These cells are of epithelial origin but are actin-packed contractile cells that behave like smooth muscle. They form a basketlike mesh around the acinus (fig. 28.23) and essentially squeeze milk from the acinus into the duct. The infant does not get any milk for the first 30 to 60 seconds of suckling, but milk soon fills the ducts and lactiferous sinuses and is then easily sucked out.

Think About It

When a woman is nursing her baby at one breast, would you expect only that breast, or both breasts, to eject milk? Explain why.

Breast Milk

Table 28.4 compares the composition of human milk, colostrum, and cow's milk. Breast milk changes composition over the first 2 weeks, varies from one time of day to another, and changes even during the course of a

Table 28.4 A Comparison of Colostrum, Human Milk, and Cow's Milk*

	Human Colostrum	Human Milk	Cow's Milk
Total protein (g/L)	22.9	10.6	30.9
Lactalbumin (g/L)	—	3.7	25.0
Casein (g/L)	—	3.6	2.3
Immunoglobulins (g/L)	19.4	0.09	0.8
Fat (g/L)	29.5	45.4	38.0
Lactose (g/L)	57	71	47
Calcium (mg/L)	481	344	1370
Phosphorus (mg/L)	157	141	910

*Colostrum data are for the first day postpartum, and human milk data are for "mature milk" at ≥15 days postpartum.

single feeding. For example, at the end of a feeding there is less lactose and protein in the milk but six times as much fat as there was at the beginning.

Cow's milk is not a good substitute for human milk. It has one-third less lactose but three to five times as much protein and minerals. The excess protein forms a harder curd in the infant's stomach, and cow's milk is not digested and absorbed as efficiently as mother's milk. The excess protein also increases the infant's nitrogenous waste excretion above normal, and this increases the incidence and severity of diaper rash, particularly as bacteria in the diaper break urea down to ammonia.

Colostrum and milk have a laxative effect that helps to clear the neonatal intestine of *meconium,* a greenish black, sticky fecal matter composed of bile, epithelial cells, and other wastes that accumulated during fetal development. By clearing bile and bilirubin from the system, breast-feeding also reduces the incidence and degree of jaundice in neonates. Breast milk promotes colonization of the neonatal intestine with beneficial bacteria and continues to supply antibodies that lend protection against infection by pathogenic bacteria. Breast-feeding also tends to promote a closer bond between mother and infant.

A woman nursing one baby eventually produces about 1.5 L of milk per day; women with twins produce even more. Lactation places a great metabolic demand on the mother. It is equivalent to losing 50 g of fat, 100 g of lactose (made from her blood glucose), and 2 to 3 g of calcium phosphate per day. A woman is at greater risk of bone loss when breast-feeding than when she is pregnant, because much of the infant's skeleton is still cartilage at birth and becomes mineralized at her expense in the first year postpartum. If a nursing mother does not have enough calcium and vitamin D in her own diet, lactation stimulates parathyroid hormone secretion and osteoclast activity, taking calcium from her bones to supply her baby.

Key Point Review

23 Why is little or no milk secreted while a woman is pregnant?

24 How does a lactating breast differ from a nonlactating breast in structure? What stimulates these differences to develop during pregnancy?

25 What is colostrum and what is its significance?

26 How does suckling stimulate milk ejection?

27 Why is breast milk superior to cow's milk for an infant?

CHAPTER ESSAY

Methods of Contraception

The term **contraception** is used here to mean any procedure or device intended to prevent pregnancy (the presence of an implanted conceptus in the uterus). This essay describes the most common methods of contraception, some issues involved in choosing among them, and the relative reliability of the various methods. Several of those contraceptive options are shown in figure E.1.

Behavioral Methods

Abstinence (refraining from intercourse) is, obviously, a completely reliable method if used consistently. Other behavioral methods are notoriously unreliable.

The **rhythm method** is based on avoiding intercourse near the time of expected ovulation. It has a high failure rate, partly due to lack of restraint and partly because it is difficult to predict the exact date of ovulation. Intercourse must be avoided for at least 48 hours before ovulation so there will be no surviving sperm in the reproductive tract when the egg is ovulated, and for at least 24 hours after ovulation so there will be no fertile egg present when sperm are introduced. The rhythm method is valuable, however, for couples who are trying to conceive a child by having intercourse at the time of apparent ovulation.

Withdrawal (coitus interruptus) requires the male to withdraw the penis before ejaculation. This often fails because of lack of willpower and because some sperm are present in the preejaculatory fluid.

—continued

Contraceptive foam with vaginal applicator

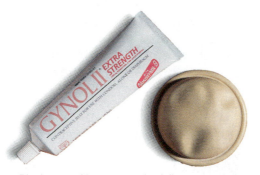
Diaphragm with contraceptive jelly

Male condom

Female condom

Birth control pills

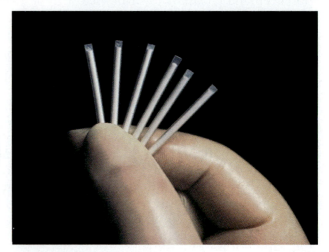
Norplant

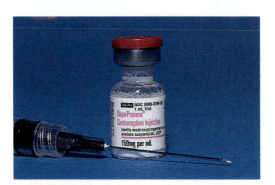
Depo-Provera

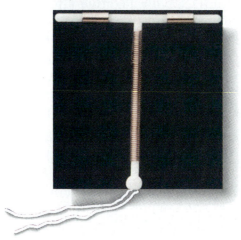
Intrauterine device (IUD)

Figure E.1 Contraceptive devices.

Barrier and Spermicidal Methods

Barrier methods are designed to prevent sperm from getting into or beyond the vagina; they are most effective when used with chemical **spermicides,** available as over-the-counter foams, creams, and jellies.

The **male condom** is a sheath of latex, rubber, or animal membrane (lamb intestine) that is unrolled over the erect penis and collects the semen. It is inexpensive, convenient, and very reliable. About 25% of American couples who use contraceptives use only condoms, which rank second to birth-control pills in popularity.

The **female condom** is less used. It is a polyurethane sheath with a flexible ring at each end. The inner ring fits over the cervix and the outer ring covers the external genitalia. Male and female condoms are the only contraceptives that also protect against disease transmission. Animal membrane condoms, however, are porous to HIV and hepatitis B viruses and do not afford dependable protection from disease.

The **diaphragm** is a latex or rubber dome that is placed over the cervix to block sperm transmission. It requires a physical examination and prescription to ensure proper fit but is otherwise comparable to the condom in convenience and reliability, provided it is used with a spermicide. Without a spermicide, it is not very effective.

Hormonal Methods

Birth-control pills are composed of estrogen and progesterone. They mimic the negative feedback effect of the ovarian hormones on FSH secretion, thus preventing follicle development and ovulation. They are effective for most women up to 35 years old with minimal complications, but they can increase the risk of heart attack or stroke in smokers and in women with a history of diabetes, hypertension, or clotting disorders. Birth-control pills are the most widely used contraceptives in the United States.

A **male birth-control pill** has been the goal of considerable research, but it is more difficult to stop the production of billions of sperm per day than to stop the development of one egg per month. Some drugs that inhibit spermatogenesis also suppress the sex drive, and some cause irreversible loss of fertility and other undesirable side effects.

Norplant is a porous silicone tube that is inserted under a woman's skin, usually on the upper arm. Six tubes the size of wooden matchsticks are implanted at one time. They effectively prevent pregnancy for up to 5 years by releasing a synthetic progesterone that suppresses ovulation. Even if ovulation should occur, Norplant interferes with egg transport by altering the cilia in the uterine tube, and it makes the endometrium unreceptive to implantation.

Depo-Provera is a synthetic progesterone administered by injection two to four times per year. It provides highly reliable, long-term contraception without the need of daily pills or implants, although in some women it causes headaches, nausea, or weight gain, and fertility may not return immediately when its use is discontinued.

Surgical Sterilization

People who are confident they do not want more children (or any) often elect to be surgically sterilized. This entails the cutting and tying or clamping of the genital ducts, thus blocking the passage of sperm or eggs. Surgical sterilization has the advantage of convenience, since it requires no further attention. Its initial cost is higher, however, and for people who later change their minds, reversibility is uncertain and the attempt is much more expensive than the original procedure. **Vasectomy** is the severing of the ductus (vas) deferens, done through a small incision in the back of the scrotum. In **tubal ligation,** the uterine tubes are cut. This can be done through small abdominal incisions to admit the cutting tool and a laparoscope (viewing device).

Preventing Implantation

The **intrauterine device (IUD)** is a springy plastic device inserted through the cervical canal and left in place for 1 to 4 years. It prevents the implantation of a blastocyst in the uterine wall. Some IUDs were removed from the market because they caused serious complications such as uterine perforation, and many other models were removed for fear of legal liability. Only two T-shaped models are currently available in the United States.

RU486, the "morning after pill," can be taken after intercourse has occurred around the expected time of ovulation and any time within 10 days of when the next menstrual period would ordinarily occur. By blocking the action of progesterone, it induces menstruation and the discharge of any conceptus that might have formed. RU486 has been widely used in France without complications, but its availability in the United States remains uncertain because of ethical controversy.

Issues in Choosing a Contraceptive

Many issues enter into the appropriate choice of a contraceptive, including personal preference, pattern of sexual activity, medical history, religious views, convenience, initial and ongoing costs, and disease prevention. For most people, however, the two primary issues are safety and reliability.

—continued

The following table shows the expected rates of failure for several types of contraception. Each column shows the number of sexually active women who typically become pregnant within 1 year while they or their partners are using the indicated contraceptives. The lowest rate is for those who use the method correctly and consistently, while the higher rate is based on random surveys of users and takes human error (lapses and incorrect usage) into account.

We have not considered all the currently available methods of contraception or all the issues important to the choice of a contraceptive. No one contraceptive method can be recommended as best for all people. Further information necessary to a sound choice and proper use of contraceptives should be sought from a health department, college health service, physician, or other such sources.▲

Failure Rates of Contraceptive Methods

Method	Rate of Failure (pregnancies per 100 users)	
	Lowest	Typical
No protection	85	85
Rhythm method	2–9	20
Withdrawal	4	18
Spermicide alone	3	21
Condom (male or female)	2	12
Diaphragm with spermicide	6	18
Birth-control pill	0.1	3
Norplant	0.04	0.2–0.4
Depo-Provera	0.3	0.3–0.4
Vasectomy	0.1	0.15
Tubal ligation	0.2	0.4
Intrauterine device	0.8–2.0	3

Interactions Between the REPRODUCTIVE SYSTEM and Other Organ Systems

Integumentary System
- Sensory impulses from skin important in erotic stimulation; mammary glands nourish infant
- Androgens cause skin to become thicker and darker and stimulate growth of body hair and sebaceous secretion; estrogens cause skin to become thinner and smoother and promote feminine distribution of subcutaneous fat; pregnancy may cause stretch marks and pigmentation changes

Skeletal System
- Encloses and protects pelvic organs; pelvis may hinder vaginal delivery of infant; provides minerals for fetal growth and for lactation
- Gonadal steroids stimulate adolescent growth; estrogens maintain bone mass in females

Muscular System
- Muscles of pelvic floor support reproductive organs, aid erection of penis and clitoris, and function in orgasm and ejaculation; abdominal muscles important in childbirth; dartos and cremaster muscles help to maintain proper temperature of testes
- Gonadal steroids stimulate muscular growth in adolescence

Nervous System
- Hypothalamus initiates gonadotropin function and lactation; sense organs, somatic nervous system, and autonomic nervous system involved in sexual arousal and orgasm
- Androgens stimulate libido; gonads and placenta secrete hormones that exert negative feedback effects on hypothalamus

Endocrine System
- Hormones regulate puberty, gametogenesis, libido, pregnancy, lactation, and climacteric
- Gonads and placenta constitute part of endocrine system

Circulatory System
- Changes in blood flow cause erection and other vasocongestive processes during sexual arousal; blood distributes sex hormones, supplies nutrients to fetus, and removes its wastes
- Androgens stimulate erythropoiesis; estrogens may slow development of atherosclerosis in females; pregnancy stimulates increase in blood volume and cardiac output and may contribute to varicose veins

Lymphatic/Immune System
- Testes, scrotum, and breasts have extensive lymphatic drainage; immune cells protect reproductive organs from infection; IgA in colostrum and breast milk protects neonate
- Blood-testis barrier isolates sperm cells and protects them from immune system

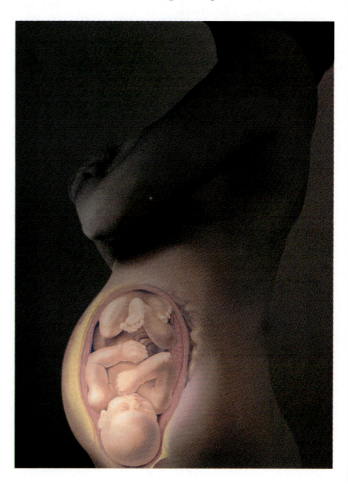

Respiratory System
- Provides O_2 and removes CO_2; Valsalva maneuver aids childbirth
- Sexual arousal increases pulmonary ventilation; pregnancy reduces depth of inspiration but increases respiratory rate

Urinary System
- Disposes of wastes and maintains electrolyte and pH balance of blood of mother and fetus; urethra is the common passageway for urine and semen in males
- Sexual arousal constricts internal urinary sphincter; prostatic hyperplasia may impede urine flow; pregnancy crowds urinary bladder

Digestive System
- Provides nutrients for gametogenesis and fetal development
- Developing fetus crowds digestive organs, which may cause heartburn and constipation

Reproductive Anatomy (pp. 992–1000)

1. Sex differentiation
2. Ovary
   a. Superficial coverings
   b. Cortex and medulla
   c. Follicles
   d. Supportive ligaments
   e. Blood vessels and innervation
3. Secondary sex organs
   a. Uterine tubes (fallopian tubes)
      • Major anatomical regions
      • Mesosalpinx
      • Structure of mucosa
   b. Uterus
      • Major anatomical regions
      • Layers of the wall
      • Endometrial strata
      • Blood supply
      • Supportive ligaments
      • Peritoneal pouches
   c. Vagina
      • Functions
      • Gross anatomy
      • Fornices
      • Elaborations of mucosa
         •• Hymen
         •• Rugae
   d. Vulva (pudendum)
      • Mons pubis
      • Labia
      • Vestibule
      • Prepuce
      • Clitoris
   e. Accessory glands
      • Vestibular glands
      • Paraurethral glands
   f. Vestibular bulbs
4. Secondary sex characteristics
   a. Physique and voice
   b. Breasts
      • Body and axillary tail
      • Areola and nipple
      • Stroma and ligaments
      • Lactiferous ducts
      • Breast cancer
         •• Causes
         •• Detection
         •• Treatment

Puberty and Menopause (pp. 1000–1002)

1. Hormones
   a. GnRH and gonadotropins
   b. Estrogens and androgens
      • Thelarche
      • Pubarche
      • Menarche
      • Other effects

   c. Progesterone
   d. Inhibin
   e. Comparison of female and male puberty
2. Climacteric and menopause
   a. Effects of climacteric
   b. Cessation of menses

Oogenesis and the Sexual Cycle (pp. 1002–1008)

1. Oogenesis
   a. Oogonia
   b. Primary oocytes
   c. Atresia
   d. Secondary oocyte
   e. First polar body
   f. Ovulation and fertilization
   g. Second polar body
2. The sexual cycle
   a. Follicular phase
      • Menstrual phase
         •• FSH secretion
         •• Development of primary and secondary follicles
      • Preovulatory (proliferative) phase
         •• Endometrial regeneration
         •• Growth of one follicle
         •• Meiosis I completed
         •• Meiosis II begins, arrests
   b. Ovulation
      • Effects of LH
      • Rupture of follicle
   c. Postovulatory phase
      • Luteal (secretory) phase
         •• Formation of corpus luteum
         •• Progesterone secretion
         •• Endometrial secretion
      • Premenstrual (ischemic) phase
         •• Involution of corpus luteum
         •• Progesterone level falls
         •• Endometrial ischemia
         •• Tissue necrosis and menses
   d. Menstruation

Female Sexual Response (pp. 1008–1010)

1. Excitement and plateau
   a. Vasocongestion
   b. Vaginal transudation
   c. Orgasmic platform
   d. Tenting of uterus
   e. Reactions of clitoris
2. Orgasm
   a. Muscular spasms
   b. Uterine peristalsis
   c. Paraurethral secretion
3. Resolution

Pregnancy and Childbirth (pp. 1010–1016)

1. Timetable of gestation
2. Fundamentals of prenatal development
3. Hormones of pregnancy
   a. Human chorionic gonadotropin (HCG)
   b. Estrogens
   c. Progesterone
   d. Human chorionic somatomammotropin (HCS)
   e. Other hormones
4. Adjustments to pregnancy
   a. Digestive system, nutrition, and metabolism
      • Morning sickness
      • Constipation and heartburn
      • Metabolic rate and caloric demands
      • Storage role of placenta
      • Nutrient needs
         •• Iron
         •• Vitamin K
         •• Folic acid
         •• Vitamin D
   b. Circulatory system
      • Increased blood volume
      • Changes in cardiac output
      • Varicose veins, hemorrhoids
   c. Respiratory system
      • Increased O_2 consumption
      • Increased CO_2 sensitivity
      • Pressure on diaphragm
   d. Urinary system
      • Increased urine output
      • Reduced bladder capacity
   e. Integumentary system
      • Striae (stretch marks)
      • Pigmentation changes
   f. Uterine growth and weight gain
5. Childbirth (parturition)
   a. Uterine contractility
      • Braxton Hicks contractions
      • Labor contractions
      • Progesterone and estrogen balance
      • Role of oxytocin
      • Role of prostaglandins
      • Role of mechanical stimulation
   b. Labor contractions
      • Nature and frequency
      • Positive feedback theory
      • Role of voluntary abdominal muscles
      • Labor pains

c. Stages of labor
- Dilation (first) stage
  - •• Dilation
  - •• Effacement
- Expulsion (second) stage
  - •• Crowning
  - •• Delivery of infant
- Placental (third) stage

d. Puerperium
- Involution of uterus
- Discharge of lochia
- Effect of lactation on uterus

Lactation (pp. 1017–1019)

1. Development of mammary gland
   a. Growth of duct system
   b. Development of acini

2. Colostrum and Milk synthesis
   a. Colostrum
   b. Role of prolactin
   c. Effect on fertility

3. Milk ejection
   a. Neuroendocrine reflex
   b. Myoepithelial cells

4. Breast milk
   a. Composition
   - Compared to colostrum
   - Compared to cow's milk
   b. Benefits of breast-feeding
   c. Metabolic demands on mother

Selected Vocabulary

ovary 993
ovarian fossa 993
tunica albuginea 993
germinal epithelium 993
cortex 993
medulla 993
follicle 993
ovulation 993
ovarian ligament 993
suspensory ligament
   of ovary 993
broad ligament 993
mesovarium 993
ovarian artery 993
ovarian veins 993
ovarian nerves 993
internal genitalia 993
external genitalia 993
uterine tube 993
infundibulum 994
fimbriae 994
ampulla 994
isthmus of uterine tube 994
mesosalpinx 994
uterus 994
fundus 994
body of uterus 994
isthmus of uterus 994
cervix 994
cervical canal 995
internal os 995
external os 995
cervical gland 995
perimetrium 995
myometrium 995
endometrium 995
stratum functionalis 995
stratum basalis 995
uterine artery 995–996
arcuate artery 996
spiral artery 996
mesometrium 996

cardinal ligament 996
uterosacral ligament 996
round ligament 996
vesicouterine pouch 996
rectouterine pouch 996
vagina 996
posterior fornix 996
lateral fornix 996
anterior fornix 996
hymen 996
vaginal rugae 996
dendritic cells 996
vulva 997
mons pubis 997
labia majora 997
pudendal cleft 997
labia minora 997
vestibule 997
prepuce 997
clitoris 997
glans 997
body of clitoris 997
crura 997
clitoral artery 997
greater vestibular glands 997
lesser vestibular glands 997
paraurethral glands 997
vestibular bulbs 998
body of breast 998
axillary tail 998
areola 998
areolar glands 998
suspensory ligament of
   breast 998
lactiferous duct 999
lactiferous sinus 999
estrogens 1001
estradiol 1001
estriol 1001
estrone 1001
thelarche 1001
breast bud 1001

pubarche 1001
menarche 1001
progesterone 1001
inhibin 1001
climacteric 1001
menopause 1001
hot flashes 1001
oogenesis 1002
oogonium 1002
primary oocyte 1002
atresia 1002
secondary oocyte 1003
first polar body 1003
second polar body 1003
sexual cycle 1003
ovarian cycle 1003
menstrual cycle 1003
follicular phase 1003
menstrual phase 1004
preovulatory phase 1004
primordial follicle 1004
granulosa cells 1004
primary follicle 1004
theca folliculi 1004
theca externa 1004
theca interna 1004
follicular fluid 1004
antrum 1005
secondary follicle 1005
cumulus oophorus 1005
zona pellucida 1005
corona radiata 1005
proliferative phase 1006
mature (graafian) follicle 1006
stigma 1007
postovulatory phase 1007
luteal phase 1007
premenstrual phase 1007
lutein cells 1007
corpus luteum 1007
ischemic phase 1008
menses 1008

corpus albicans 1008
vaginal transudate 1008
orgasmic platform 1008
tenting 1008
gestation 1010
trimester 1010
human chorionic
   gonadotropin (HCG) 1011
human chorionic
   somatomammotropin
   (HCS) 1011
hyperemesis gravidarum 1013
striae 1013
linea nigra 1013
chloasma 1013
parturition 1014
Braxton Hicks
   contractions 1014
labor contractions 1014
positive feedback theory of
   labor 1015
episiotomy 1015
primipara 1015
multipara 1015
dilation stage 1015
dilation 1015
effacement 1015
expulsion stage 1015
crowning 1015
placental stage 1016
postpartum 1016
puerperium 1016
involution 1016
autolysis 1016
lochia 1016
lactation 1017
colostrum 1017
prolactin-inhibiting factor
   (PIF) 1017
milk ejection 1018
myoepithelial cells 1018

1. Of the following organs, the one most comparable to the penis is the
   a. clitoris.
   b. vagina.
   c. vestibular bulbs.
   d. labia majora.
   e. prepuce.

2. The ovaries secrete all of the following except
   a. estradiol.
   b. progesterone.
   c. androgens.
   d. follicle-stimulating hormone.
   e. inhibin.

3. The earliest haploid stage of oogenesis is
   a. the oogonium.
   b. the primary oocyte.
   c. the secondary oocyte.
   d. the second polar body.
   e. the zygote.

4. FSH secretion begins to rise and new follicles begin to grow during
   a. the premenstrual phase.
   b. the menstrual phase.
   c. the preovulatory phase.
   d. the luteal phase.
   e. the plateau phase.

5. The secretory phase of the menstrual cycle is under the most direct influence of
   a. HCG.
   b. FSH.
   c. LH.
   d. estrone.
   e. progesterone.

6. Excessive vomiting in pregnancy is called
   a. chloasma.
   b. episiotomy.
   c. parturition.
   d. hyperemesis gravidarum.
   e. puerperium.

7. Before secreting milk, the mammary glands secrete
   a. prolactin.
   b. colostrum.
   c. lochia.
   d. meconium.
   e. chloasma.

8. Few women become pregnant while nursing because _____ inhibits GnRH secretion.
   a. FSH
   b. prolactin
   c. prostaglandins
   d. oxytocin
   e. HCG

9. Smooth muscle cells of the myometrium and myoepithelial cells of the mammary glands are the target cells for
   a. prostaglandins.
   b. LH.
   c. oxytocin.
   d. progesterone.
   e. FSH.

10. Which of these is not true of the luteal phase of the sexual cycle?
    a. Progesterone level is high.
    b. The endometrium synthesizes glycogen.

    c. Ovulation occurs.
    d. Fertilization may occur.
    e. The endometrial glands become wider and more spiraled.

11. Each egg cell develops in its own fluid-filled vesicle called a _____.

12. The mucosa of the uterus is called the _____.

13. A girl's first menstrual period is called _____.

14. The source of progesterone in the secretory phase is a yellowish structure called the _____.

15. The layer of cells immediately around a mature secondary oocyte is called the _____.

16. A secondary follicle differs from a primary follicle in that it has a cavity called the _____.

17. Menopause occurs during a midlife period of changing hormonal secretion called _____.

18. The fornices are recesses where the _____ of the uterus protrudes into the vagina.

19. The funnel-like distal end of the uterine tube is called the _____ and has feathery processes called _____.

20. Postpartum uterine involution results in a vaginal discharge called _____.

1. Would you expect puberty to create a state of positive or negative nitrogen balance? Explain.

2. Aspirin and ibuprofen can prevent the onset of labor and are sometimes used to prevent premature birth. Review your knowledge of these drugs and the mechanisms of labor, and explain this effect.

3. At 6 months postpartum, a nursing mother is in an automobile accident

that fractures her skull and severs the hypothalamo-hypophyseal portal vessels. How would you expect this to affect her milk production? How would you expect it to affect her future ovarian cycles? Explain the contrast.

4. If the ovaries are removed in the first 6 weeks of pregnancy (for example, in treating ovarian cancer), the embryo will abort. If the ovaries are removed later in pregnancy, it has no

effect and the pregnancy can go to full term. Explain the difference.

5. A breast-feeding mother leaves her infant at home with her husband and goes to the grocery store. There, she hears another woman's baby crying in the next aisle and notices her blouse becoming wet with a little exuded milk. Explain the physiological links between hearing this sound and milk ejection.

Web Site Link

For a listing of the most current web sites related to this chapter, please visit the Saladin homepage at:

http://www.mhhe.com/sciencemath/biology/saladin/

[Human Development

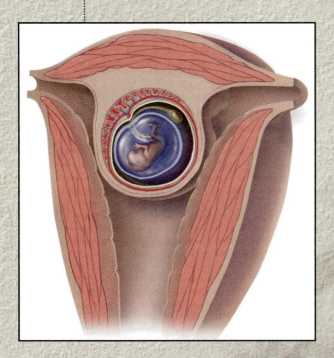

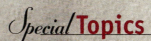

P erhaps the most dramatic, seemingly miraculous aspect of human life is the transformation of a one-celled fertilized egg into an independent, fully developed individual. From the beginning of recorded thought, people have pondered how a baby forms in the mother's body and how two parents can produce another human being who, although unique, possesses characteristics of each of them. Aristotle dissected the embryos of various birds, established their sequence of organ development, and speculated that the hereditary traits of a child resulted from the mixing of the male's semen with the female's menstrual blood. The human egg was first observed and the modern science of **developmental biology** was born in the nineteenth century. In its broad sense, developmental biology embraces the scientific study of changes in form and function from fertilized egg through old age—which is the scope of this chapter.

Fertilization and Preembryonic Development

▼**Objectives**

When you have completed this section, you should be able to
• describe sperm migration, capacitation, fertilization, and the egg's barriers to additional sperm;
• describe the major events that transform a fertilized egg into an embryo; and
• describe the implantation of the preembryo in the uterine wall.

Sperm Migration

An egg must be fertilized within 12 to 24 hours of ovulation if it is to survive. Because the egg takes about 72 hours to reach the uterus, spermatozoa must migrate up the female reproductive tract and meet it somewhere in the distal one-third of the uterine tube. The vast majority of spermatozoa never make it. Many are destroyed by vaginal acid or drain out of the vagina. Others fail to penetrate the mucus of the cervical canal, and those that do are often destroyed by leukocytes in the uterus. Half of the remainder are likely to go up the wrong uterine tube. Finally, 2,000 to 3,000 spermatozoa reach the vicinity of the egg—not many of the 300 million that were ejaculated.

Spermatozoa migrate mainly by means of the snakelike lashing of their tails as they crawl along the female mucosa, but they are assisted by certain aspects of female physiology. Strands of mucus guide them through the cervical canal. Although female orgasm is not required for fertilization, the uterine contractions of orgasm may suck semen from the vagina and spread it throughout the uterus, like hand lotion pressed between your palms. Sperm may be chemically attracted to the egg within close range of it; this has been demonstrated for some animals but remains unproven for humans.

Capacitation

Spermatozoa can reach the distal uterine tube within 5 to 10 minutes of ejaculation, but they cannot fertilize an egg for about 10 hours. During their migration, they undergo a process of **capacitation** that makes it possible to penetrate an egg. Prior to ejaculation, the membrane of the spermatozoon head contains a substantial amount of cholesterol, which toughens it and prevents premature release of the acrosomal enzymes. This avoids wastage of sperm and enzymatic damage to the spermatic ducts. After ejaculation, however, fluids of the female reproductive tract wash away cholesterol and other inhibitory factors in the semen. The membrane of the spermatozoon head becomes more fragile and more permeable to calcium ions, which diffuse into the spermatozoon and cause more powerful lashing of the tail.

Most spermatozoa are fertile for a maximum of 48 hours after ejaculation, so there is little chance of fertilizing an egg if intercourse occurs more than 48 hours before ovulation. In fact, fertilization is unlikely if intercourse takes place more than 14 hours after ovulation because the egg would no longer be viable by the time the sperm became capacitated. For those wishing to conceive a child, the "window of opportunity" is therefore 48 hours before ovulation to 14 hours after. Those wishing to avoid pregnancy, however, should allow a wider margin of safety for variations in sperm and egg longevity, capacitation time, and time of ovulation—variations that make the rhythm method of contraception so unreliable.

Modern reproductive technology has created several options for people unable to conceive a child in "the old-fashioned way." Several of these are described in the chapter essay (p. 1055).

Fertilization

When a spermatozoon and egg meet (fig. 29.1), the spermatozoon undergoes an **acrosomal reaction** that releases the enzymes necessary for penetration of the egg. To fertilize the egg proper, a spermatozoon must first penetrate the granulosa cells and zona pellucida that surround it (fig. 29.2). Indeed, the first spermatozoon to reach an egg is *not* the one that fertilizes it—it may require hundreds of spermatozoa to clear a path for the one that penetrates the egg.

Two of the acrosomal enzymes are **hyaluronidase,** which digests the hyaluronic acid that binds granulosa cells together, and **acrosin,** a protease similar to the trypsin of pancreatic juice. When a path has been cleared, a spermatozoon binds to the zona pellucida and releases its enzymes by exocytosis. It digests a pathway through the zona pellucida until it contacts the egg itself. The membranes of the two gametes then fuse, and the sperm nucleus enters the egg cytoplasm.

Fertilization combines the haploid (n) set of sperm chromosomes with the haploid set of egg chromosomes, producing a diploid ($2n$) set. Fertilization by two or more spermatozoa, called **polyspermy,** would produce a triploid ($3n$) or larger set of chromosomes and the egg would die. To prevent this, the egg has two mechanisms called a fast block and slow block. In the **fast block,** binding of the spermatozoon to the egg opens Na^+ channels in the egg membrane. The rapid influx of Na^+ depolarizes the egg membrane and inhibits the binding of any more sperm to it. The **slow block** involves secretory vesicles called **cortical granules** just beneath the membrane. Sperm penetration triggers an influx of Ca^{2+}; this, in turn, triggers a **cortical reaction** in which the cortical granules release their secretion beneath the zona pellucida. The secretion swells with water, pushes any remaining sperm away from the egg, and creates an impenetrable **fertilization membrane.**

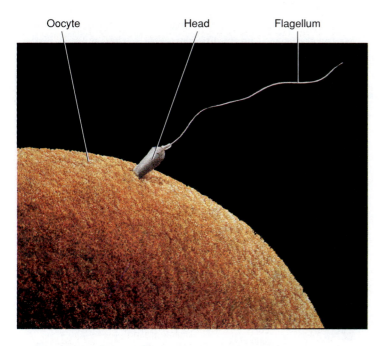

Oocyte   Head   Flagellum

Figure 29.1 Boy meets girl—a human spermatozoon bound to the surface of the oocyte.

Think About It

What similarity can you see between the slow block to polyspermy and the release of acetylcholine from synaptic vesicles of a neuron? (Compare p. 450.)

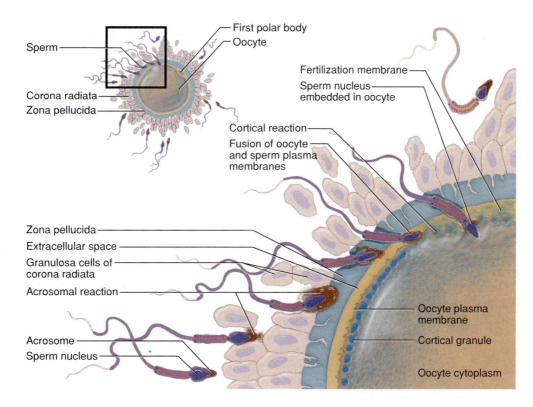

Sperm — First polar body — Oocyte

Corona radiata —
Zona pellucida —

Fertilization membrane —
Sperm nucleus embedded in oocyte —

Cortical reaction —
Fusion of oocyte and sperm plasma membranes —

Zona pellucida —
Extracellular space —
Granulosa cells of corona radiata —
Acrosomal reaction —

Acrosome —
Sperm nucleus —

Oocyte plasma membrane
Cortical granule
Oocyte cytoplasm

Figure 29.2 Stages in fertilization and the oocyte cortical reaction (slow block to polyspermy).

The first two blastomeres produced during cleavage each retain the potential to develop into a complete individual. Occasionally, they become separated and develop into **monozygotic (MZ) twins.** Because MZ twins are genetically identical, they are of the same sex and nearly identical appearance. Reproductive biologists now question whether MZ twins are truly genetically identical, however. They have suggested that one blastomere may undergo a mutation in the process of DNA replication, and the separation of the blastomeres may represent an attempt of each cell to reject the other one as "foreign."

Dizygotic (DZ) twins result when two eggs are ovulated and fertilized by separate spermatozoa. They are no more (or less) genetically similar than any other siblings and may be of different sexes. Other multiple births such as triplets and quadruplets also result from multiple ovulation.

Meiosis II

A secondary oocyte begins meiosis II before ovulation and completes it only if fertilized. Through the formation of a second polar body, the egg discards one chromatid from each chromosome. The sperm and egg nuclei then swell and become **pronuclei.** A mitotic spindle forms between them, each pronucleus ruptures, and the chromosomes of the two gametes mix into a single diploid set (fig. 29.3). The fertilized egg is now called a **zygote.**

The Preembryonic Stage

The **preembryonic stage** consists of three processes that occupy the first 2 weeks of development: cleavage, implantation, and embryogenesis.

Cleavage

Cleavage refers to the mitotic divisions that occur in the first 3 days after fertilization and produce daughter cells called **blastomeres.**[1] The first cleavage occurs about 30 hours after fertilization and produces the first two blastomeres (see special topic 29.1). These divide simultaneously at shorter and shorter time intervals and double the number of blastomeres each time. By the time the conceptus arrives in the uterus, about 72 hours after ovulation, it consists of 16 or more cells and somewhat resembles a mulberry—hence it is called a **morula** ("little mulberry") (fig. 29.4). The morula is no larger than the zygote; cleavage merely produces smaller and smaller blastomeres. This increases the ratio of cell surface area to volume, which favors rapid nutrient uptake and waste removal, and it produces a larger number of cells from which to form different types of embryonic tissue.

The morula lies free in the uterine cavity for 4 to 5 days and divides into 100 cells or so. During this time,

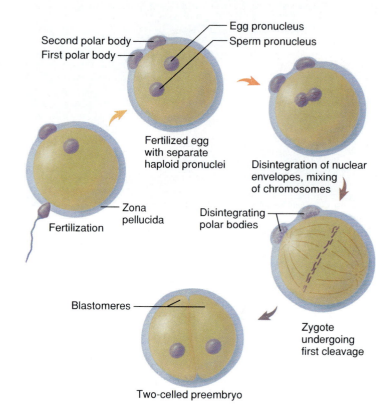

Figure 29.3 Major events from fertilization to first cleavage.

it is nourished by an endometrial secretion called **uterine milk,** which accumulates in a cavity in the morula. Meanwhile, the zona pellucida disintegrates and releases the conceptus, which is now at a stage called the **blastocyst**—a hollow sphere with an outer layer of squamous cells called the **trophoblast**[2] and an inner cell mass called the **embryoblast** at one side of the cavity. The trophoblast is destined to form part of the placenta and play an important role in nourishment of the embryo, whereas the embryoblast is destined to become the embryo itself.

---

1. *blast* = bud, precursor + *mer* = segment, part

2. *troph* = food, nourishment

In about 1 out of 300 pregnancies, the blastocyst implants somewhere other than the uterus, producing an **ectopic**[3] **pregnancy.** Most cases begin as **tubal pregnancies,** in which the blastocyst implants in the uterine tube. This usually occurs because it has encountered an obstruction such as a constriction resulting from earlier pelvic inflammatory disease, tubal surgery, previous ectopic pregnancies, or repeated abortions. The uterine tube cannot expand enough to accommodate the growing conceptus for long; if the situation is not detected and treated early, the tube usually ruptures within 12 weeks. This can kill the mother, or the conceptus can reimplant in the abdominopelvic cavity, producing an **abdominal pregnancy.** About 1 pregnancy in 7,000 is abdominal. The conceptus can grow anywhere it finds an adequate blood supply—for example, on the broad ligament or the outside of the uterus, colon, or bladder. This is a serious threat to the mother's life and usually requires abortion, but about 9% of abdominal pregnancies result in live birth by cesarian section.

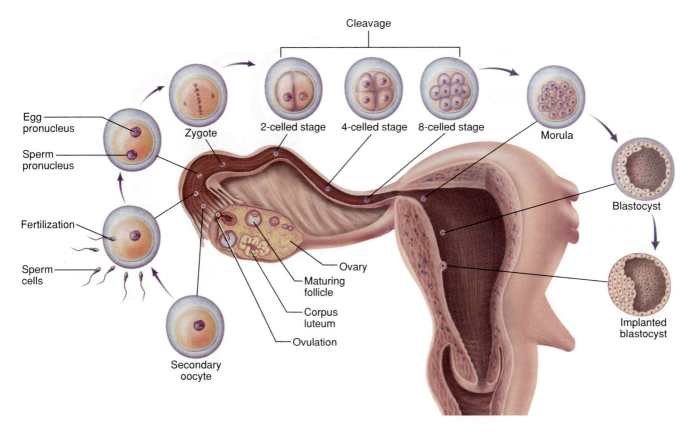

Figure 29.4 The events from ovulation to implantation of a blastocyst and where they occur in the reproductive tract. These processes occur during the first week after ovulation.

Implantation

About 6 days after ovulation, the blastocyst attaches to the endometrium, usually on the posterior wall of the fundus or body of the uterus (see special topic 29.2 for exceptions). This is the beginning of **implantation,** a process in which the conceptus becomes buried in the endometrium (fig. 29.5). The trophoblast cells overlying the embryoblast secrete enzymes that stimulate thickening of the adjacent endometrium and they separate into two layers. The deep layer is called the **cytotrophoblast** because it retains individual cells divided by membranes. In the superficial layer, however, the plasma membranes break down and the cells fuse into a multinucleate mass called the **syncytiotrophoblast**[4] (sin-SISH-ee-oh-TRO-foe-blast). (A syncytium is any body of protoplasm containing multiple nuclei.)

The syncytiotrophoblast grows into the endometrium like little roots, digesting endometrial cells along the way. The endometrium reacts to this injury by growing over the trophoblast and eventually enclosing

3. *ec* = outside + *top* = place
4. *syn* = together + *cyt* = cell

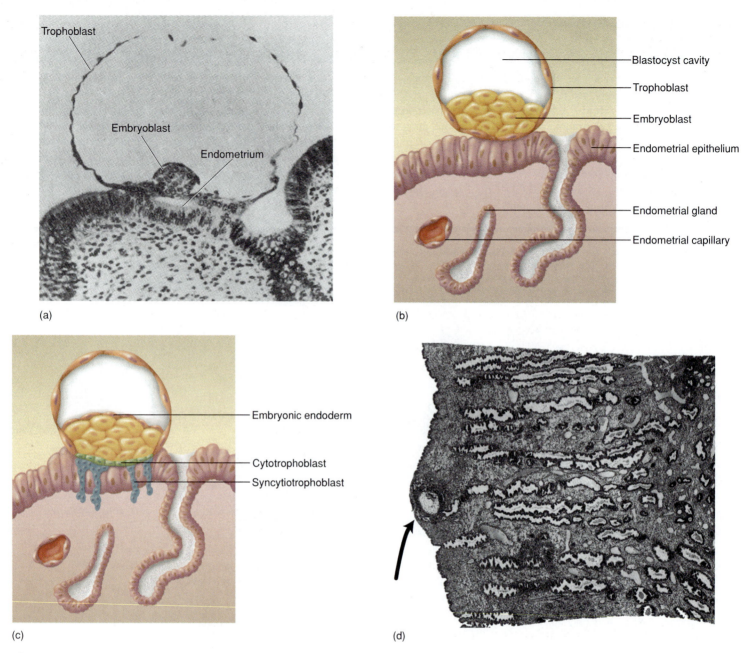

(a)

(b)

(c)

(d)

Figure 29.5 Implantation. (*a*) Attachment of the blastocyst to the uterine wall (LM). (*b*) Structure of the blastocyst about 6 to 7 days after ovulation. (*c*) The progress of implantation about 1 day later; the syncytiotrophoblast has begun growing rootlets, which penetrate the endometrium. (*d*) A blastocyst (*at arrow*) fully embedded in the endometrium (LM).

it (fig. 29.5*d*). Implantation takes about a week and is completed about the time the next menstrual period would have occurred if the woman had not become pregnant.

The trophoblast also secretes human chorionic gonadotropin (HCG). HCG stimulates the corpus luteum to secrete estrogen and progesterone, and progesterone suppresses menstruation. The level of HCG in the mother's blood rises until the end of the second month. During this time, the trophoblast develops into a membrane called the *chorion,* which takes over the role of

the corpus luteum and makes HCG unnecessary. The ovaries then become inactive for the rest of the pregnancy, but estrogen and progesterone levels rise dramatically as they are secreted by the ever-growing chorion (see fig. 28.18, p. 1011).

Embryogenesis

During implantation, the embryoblast undergoes **embryogenesis,** in which the blastomeres become organized into the three primary germ layers—ectoderm, mesoderm, and endoderm. This begins with the formation of a narrow space called the **amniotic cavity** between the embryoblast and cytotrophoblast (fig. 29.6). The embryoblast flattens into an **embryonic disc** composed of ectoderm and endoderm. As the disc elongates, a raised groove called the **primitive streak** forms along the midline of the ectoderm. Cells on the surface migrate medially toward this groove, down into it, and then laterally between the ectoderm and endoderm. This forms the middle germ layer, the mesoderm (fig. 29.7). The ectoderm and endoderm consist of epithelia composed of tightly joined cells, but the mesoderm is a more loosely organized, gelatinous connective tissue called **mesenchyme.** Its cells are more mobile and wispy than those of the ectoderm and endoderm (see fig. 6.5, p. 175). At the conclusion of embryogenesis, the individual is an **embryo,** about 2 mm long and 2 weeks old.

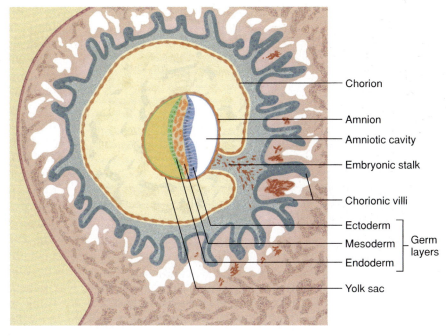

Figure 29.6 The implanted conceptus at 2 weeks. The primary germ layers and extraembryonic membranes have formed, and the conceptus is now an embryo.

Key Point Review

---

1. How soon can a sperm reach an egg after ejaculation? How soon can it fertilize an egg? What accounts for the difference?
2. Describe two ways a fertilized egg prevents the entry of excess sperm.
3. In the blastocyst, what are the cells called that eventually give rise to the embryo? What are the cells that are responsible for implantation?
4. What major characteristic distinguishes an embryo from a preembryo?

Embryonic and Fetal Development

▼Objectives

When you have completed this section, you should be able to
- describe the formation and functions of the placenta;
- explain how the conceptus is nourished before the placenta takes over this function;
- describe the four embryonic membranes and their functions;
- itemize the major tissues derived from the primary germ layers;
- describe the major events of fetal development; and
- explain how blood circulation differs before and after birth.

Two weeks after conception, the germ layers are present and the embryonic stage of development begins. Over the next 6 weeks, the conceptus forms a set of membranes external to the embryo, the embryo begins receiving its nutrition primarily from the placenta, and the germ layers differentiate into organs and organ systems. Although these organs are still far from

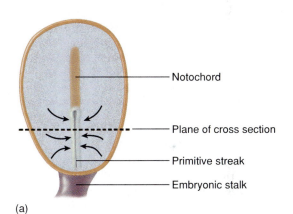

(a)

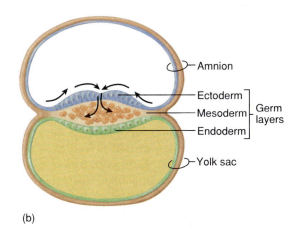

(b)

Figure 29.7 Formation of the primary germ layers. (*a*) Dorsal view of the embryonic disc at 16 days, showing the migration of ectodermal cells into the primitive streak. (*b*) Cross section of the embryonic disc at the level indicated in *a*, showing ectodermal cells entering the primitive streak and forming the mesodermal layer.

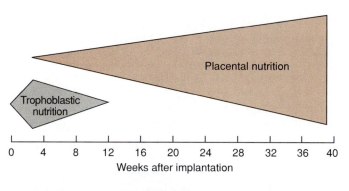

Figure 29.8 The timetable of trophoblastic and placental nutrition. Trophoblastic nutrition peaks at 2 weeks and ends by 12 weeks. Placental nutrition begins at 2 weeks and becomes increasingly important until birth, 39 weeks after implantation. The transition from the trophoblastic phase to the placental phase of pregnancy, named for the mode that supplies most nutrition to the conceptus, is at 8 weeks.

functional, it is their presence at 8 weeks that marks the transition from the embryonic stage to the fetal stage.

Prenatal Nutrition

One of the important changes to occur in the embryonic phase is the mode of nutrition. In **trophoblastic (deciduous) nutrition,** the conceptus is nourished by digesting cells of the mother's endometrium. Progesterone stimulates the development of **decidual[5] cells,** which are rich in glycogen, proteins, and lipids. The trophoblast digests these cells and the embryoblast imbibes the resulting nutritious fluid. This is gradually followed by **placental nutrition,** in which the conceptus is nourished by diffusion of nutrients from the mother's bloodstream through the **placenta,**[6] a vascular organ that develops on the uterine wall. Figure 29.8 shows the time course of the transition from trophoblastic to placental nutrition. Trophoblastic nutrition is the only means of nutrition for the first week after implantation. The placenta begins to develop about 11 days after conception, but trophoblastic nutrition remains dominant for 8 weeks, called the **trophoblastic phase** of the pregnancy. After that, the placenta provides most nutrition; the trophoblastic mode declines and, by 12 weeks, it ends. The **placental phase** of the pregnancy lasts from the beginning of week 9 until birth.

---

5. *decid* = falling off
6. *placenta* = flat cake

Placentation, the formation of the placenta, extends from 11 days through 12 weeks after conception. Most development occurs in the embryonic stage. It begins when extensions of the syncytiotrophoblast, called **chorionic villi,** penetrate more and more deeply into the endometrium, like the roots of a tree penetrating into the nourishing "soil" of the uterus (fig. 29.9). As they digest their way through uterine blood vessels, the villi become surrounded by pools of free blood that eventually merge to form the **placental sinus.** This maternal blood stimulates increasingly rapid growth of the chorionic villi. Mesenchyme grows into their cores and develops into blood vessels. The smooth surface of the placenta facing the fetus gives rise to the **umbilical cord** (fig. 29.10), which contains three blood vessels that connect the fetus to the placenta. The surface attached to the uterine wall is rough due to its chorionic villi. At birth, the placenta is about 20 cm in diameter, 3 cm thick, and weighs about one-sixth as much as the baby.

The fetal heart pumps blood to the placenta by way of two umbilical arteries. This blood flows into the capillaries of the villi and then back to the fetus by way of a single umbilical vein. The villi are *filled with* fetal blood and *surrounded by* maternal blood; the two bloodstreams do not mix unless there is damage to the placental barrier. The barrier, however, is only 3.5 μm thick—half the diameter of a single red blood cell. Early in development, the chorionic villi have relatively thick membranes, they are not very permeable to nutrients and wastes, and their total surface area is relatively small. As the villi grow and branch, their surface area increases and the membranes become thinner and more permeable. Thus there is a dramatic increase in **placental conductivity,** the rate at which substances diffuse through the membrane. Oxygen and nutrients pass from the maternal blood to the fetal blood, while fetal wastes pass the other way to be eliminated by the mother.

Gases, electrolytes, fatty acids, and steroids pass through the membrane by simple diffusion; glucose passes by facilitated diffusion; amino acids pass by active transport; and insulin is transported to the fetus by receptor-mediated endocytosis. Unfortunately, the placenta is also permeable to nicotine, alcohol, and most other drugs in the maternal bloodstream. Nutrition, excretion, and other functions of the placenta are summarized in table 29.1.

Embryonic Membranes

The placenta is not the only accessory organ of the conceptus. There are also four membranes—the *amnion, yolk sac, allantois,* and *chorion* (fig. 29.11). To understand these, it helps to realize that all mammals evolved from egg-laying reptiles. Within the shelled, self-contained egg of a reptile, the embryo rests atop a yolk, which is

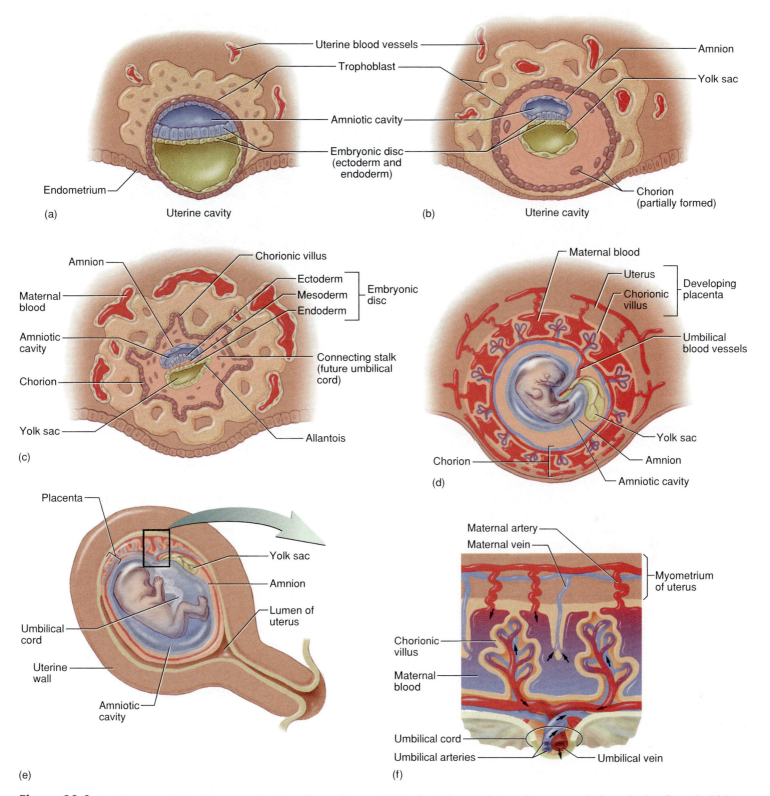

Figure 29.9 Development of the placenta and embryonic membranes. (*a*) Implantation nearly complete; an amniotic cavity has formed within the embryonic disc, lined by cells that will become the amnion. (*b*) Conceptus about 9 days after fertilization; embryonic disc cells have now formed the yolk sac. (*c*) Conceptus at 16 days; the allantois is beginning to form, and the chorion is forming from the trophoblast and embryonic mesoderm. (*d*) Embryo at 4.5 weeks, enclosed in the amnion and chorion. (*e*) Embryo at 13.5 weeks; placentation is complete. (*f*) A portion of the mature placenta and umbilical cord.

Figure 29.10 The placenta and umbilical cord viewed from the fetal side.

Table 29.1	Functions of the Placenta
Nutritional Roles	Transports nutrients such as glucose, amino acids, fatty acids, minerals, and vitamins from the maternal blood to the fetal blood; stores nutrients such as carbohydrates, protein, iron, and calcium in early pregnancy and releases them to the fetus later, when fetal demand is greater than the mother can absorb from the diet
Excretory Roles	Transports nitrogenous wastes such as ammonia, urea, uric acid, and creatinine from the fetal blood to the maternal blood
Respiratory Roles	Transports O_2 from mother to fetus and CO_2 from fetus to mother
Endocrine Roles	Secretes estrogens, progesterone, relaxin, human chorionic gonadotropin, and human chorionic somatomammotropin; allows other hormones synthesized by the conceptus to pass into the mother's blood and maternal hormones to pass into the fetal blood

enclosed in the yolk sac; it floats in a little sea of liquid contained in the amnion; it stores its toxic wastes in the allantois; and to breathe, it has a chorion permeable to gases (this is the "skin" found just beneath the shell of a boiled egg). All of these membranes persist in mammals, including humans, but are modified in their functions.

The **amnion** is a transparent sac that develops from cells of the embryonic disc. It grows to completely enclose the embryo and is penetrated only by the umbilical cord. The amnion becomes filled with **amniotic fluid,** which protects the embryo from trauma and temperature fluctuations, allows the freedom of movement important to muscle development, enables the embryo to develop symmetrically, and keeps its surface tissues from adhering to each other. At first, the amniotic fluid forms by filtration of the mother's blood plasma, but beginning at 8 to 9 weeks, the fetus urinates into the amniotic cavity about once an hour and contributes substantially to the fluid volume. The fetus swallows amniotic fluid at an equivalent rate, thus maintaining a stable volume. At term, the amnion contains 700 to 1,000 mL of fluid.

The **yolk sac** arises partly from cells of the embryonic disc, opposite the amnion. It is a small sac suspended from the ventral side of the embryo. It contributes to the formation of the digestive tract and produces the first blood cells and germ cells.

The **allantois** (ah-LON-toe-iss) is an outpocketing of the posterior end of the yolk sac. It forms the foundation for the umbilical cord and becomes part of the urinary bladder. It can be seen in proximal cross sections of the cord.

The **chorion** is the outermost membrane, enclosing all the rest of the membranes and the embryo. Initially it has villi around its entire surface, but as the pregnancy

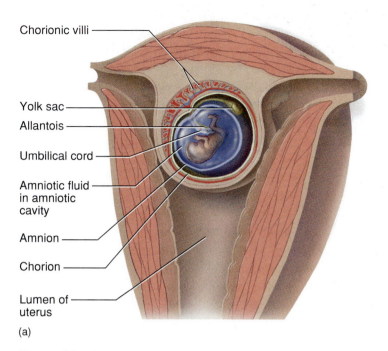

Chorionic villi

Yolk sac

Allantois

Umbilical cord

Amniotic fluid
in amniotic
cavity

Amnion

Chorion

Lumen of
uterus

(a)

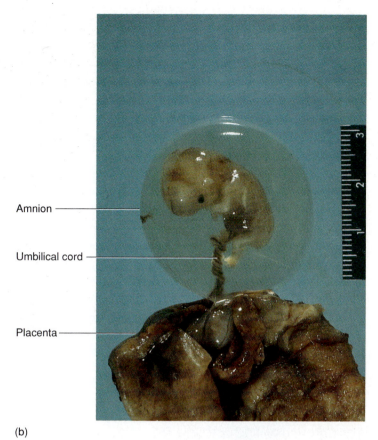

Amnion

Umbilical cord

Placenta

(b)

Figure 29.11 The embryonic membranes. (*a*) Diagram of a frontal section of the uterus. (*b*) Photograph of a human fetus at 8 weeks of gestation. The chorion has been stripped back to reveal the transparent amnion and the fetus. Major divisions of the scale bar are centimeters.

advances, the villi of the placenta grow and branch while those of the rest of the chorion degenerate. At the placental attachment the chorion is then called the *villous chorion,* and over the rest of the surface it is called the *smooth chorion.* The chorion forms the fetal portion of the placenta; its functions are essentially those listed in table 29.1.

Organogenesis

Organogenesis is the formation of organs and organ systems from the primary germ layers. This is the primary developmental process to occur in the embryo itself. The major structures that arise from the primary germ layers are listed in table 29.2.

 Think About It

List the four primary tissue types of the adult body (see chapter 6) and identify which of the three primary germ layers of the embryo predominantly gives rise to each.

At the end of 8 weeks, all of the organ systems are present, the individual is about 3 cm long, and it is now considered a **fetus** (figs. 29.11 and 29.12). Its bones have just begun to ossify and the skeletal muscles exhibit

Table 29.2	Derivatives of the Three Primary Germ Layers
Layer	**Major Derivatives**
Ectoderm	Epidermis; hair follicles and arrector pili muscles; cutaneous glands; nervous system; adrenal medulla; pineal and pituitary glands; lens, cornea, and intrinsic muscles of the eye; internal and external ear; salivary glands; epithelia of the nasal cavity, oral cavity, and anal canal
Mesoderm	Skeleton; skeletal, cardiac, and most smooth muscle; cartilage; adrenal cortex; middle ear; dermis; blood; blood and lymphatic vessels; bone marrow; lymphoid tissue; epithelium of kidneys, ureters, gonads, and genital ducts; mesothelium of ventral body cavity
Endoderm	Most of the mucosal epithelium of the digestive and respiratory tracts; epithelial components of accessory reproductive and digestive glands (except salivary glands); thyroid and parathyroid glands; thymus

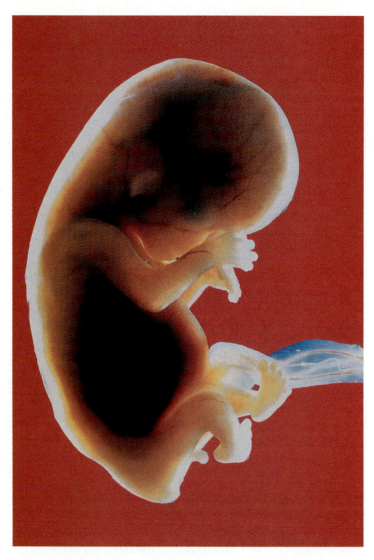

Figure 29.12 The human fetus at 11 weeks of gestation. The face has begun to look human and the fingers and toes have begun to develop. Total length is about 7 cm.

Table 29.3		The Stages of Prenatal Development
Stage	**Age***	**Major Developments and Defining Characteristics**
Zygote	24–30 hours	A single diploid cell formed by the union of egg and sperm
Cleavage	30–72 hours	Mitotic division of the zygote into smaller, identical blastomeres
Morula	3–4 days	A hollow, spherical stage consisting of 16 or more blastomeres
Blastocyst	5–14 days	A fluid-filled, spherical stage with an outer mass of trophoblast cells and inner mass of embryoblast cells; becomes implanted in the endometrium; inner cell mass forms an embryonic disc and differentiates into the three primary germ layers
Embryo	2–8 weeks	A stage in which the primary germ layers differentiate into organs and organ systems; ends when all organ systems are present
Fetus	9–40 weeks	A stage in which organs grow and mature at a cellular level to the point of being capable of supporting life independently of the mother

* From the time of ovulation.

spontaneous contractions, although these are too weak to be felt by the mother. The heart, which has been beating since the fourth week, now circulates blood. The heart and liver are very large and form a prominent ventral bulge; the head is nearly half the total body length.

The stages of prenatal development are summarized in table 29.3.

Fetal Development

The fetus is the final stage of prenatal development, from 9 weeks to birth. The organs that formed during the embryonic stage now undergo growth and cellular differentiation, acquiring the functional capability to support life outside the mother.

The circulatory system shows the most conspicuous anatomical changes from a prenatal state, dependent on the placenta, to the independent neonatal (newborn) state. The first trace of its development is the appearance of small spaces in the mesoderm before the third week. These become lined with endothelium and merge with each other to form the future blood vessels, lymphatic vessels, and heart. Two side-by-side endothelial tubes fuse to form a heart tube, which folds into the four-chambered heart as described on page 683.

The unique aspects of fetal circulation (fig. 29.13) are the umbilical-placental route and the presence of three circulatory shortcuts called *shunts.* The internal iliac arteries give rise to a pair of **umbilical arteries,** which pass on either side of the bladder into the umbilical cord. The blood in these arteries is low in oxygen and high in carbon dioxide and other fetal wastes. It discharges these wastes in the placenta, loads oxygen and nutrients, and returns to the fetus by way of a single **umbilical vein,** which leads toward the liver. Some of this blood filters through the liver to nourish it, but most of it bypasses the liver by way of a shunt called the **ductus venosus,** which leads directly to the inferior vena cava. The immature liver is not capable of performing many of its postpartum functions; many of its functions are performed by the placenta.

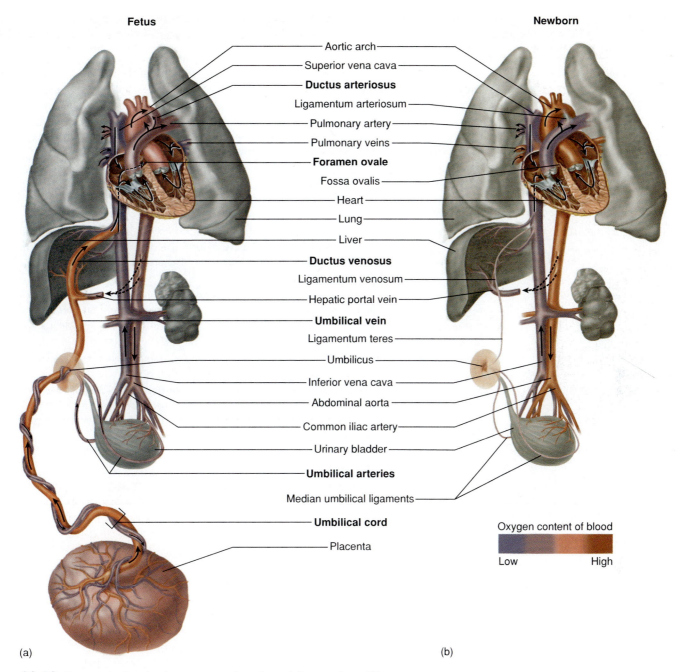

Fetus

Aortic arch
Superior vena cava
Ductus arteriosus
Ligamentum arteriosum
Pulmonary artery
Pulmonary veins
Foramen ovale
Fossa ovalis
Heart
Lung
Liver
Ductus venosus
Ligamentum venosum
Hepatic portal vein
Umbilical vein
Ligamentum teres
Umbilicus
Inferior vena cava
Abdominal aorta
Common iliac artery
Urinary bladder
Umbilical arteries
Median umbilical ligaments
Umbilical cord
Placenta

Newborn

Oxygen content of blood

Low   High

(a)   (b)

Figure 29.13 Changes in the circulatory system from fetus (*a*) to newborn (*b*).

In the inferior vena cava, placental blood mixes with venous blood from the fetus's body and flows to the right atrium of the heart. While the right heart normally pumps all of its blood into the lungs after birth, there is little need for this in the fetus because the lungs are not yet functional. Most of the fetal blood therefore bypasses the pulmonary circuit. Some of it passes from the right atrium through the **foramen ovale** (a hole in the interatrial septum) directly into the left atrium. Some goes into the right ventricle and is pumped into the pulmonary trunk, but most of this is shunted directly into the aorta by way of a short passage called the **ductus arteriosus.** This occurs because blood flows from a point of high pressure to a point of lower pressure, and the collapsed state of the fetal lungs causes resistance and pressure in the pulmonary circuit to be higher than in the aorta. The lungs receive only a trickle of blood, sufficient to meet their metabolic needs during development. Blood leaving the left ventricle enters the general systemic circulation, and some of this returns to the placenta.

Table 29.4   Major Events of Prenatal Development, with Emphasis on the Fetal Stage

Month	Length and Weight	Developmental Events
1	0.6 cm	Spinal column and central nervous system begin to form; appendages represented by small **limb buds;** heart begins beating; no visible eyes, nose, or ears
2	3 cm; 1 g	Eyes form, eyelids fused shut; nose flat, nostrils evident but plugged with mucus; head nearly as large as the rest of the body; brain waves detectable; ossification begins; limb buds form paddlelike hands and feet with ridges called **digital rays,** which then separate into distinct fingers and toes; blood cells and major blood vessels form; genitals present but sexes not yet distinguishable
3	9 cm; 30 g	Eyes well developed, eyelids still fused; nose develops bridge; external ears present; limbs well formed, digits exhibit nails; fetus swallows amniotic fluid and produces urine; fetus moves, but too weakly for mother to feel it; liver is prominent and produces bile; palate is fusing; sexes can be distinguished
4	14 cm; 100 g	Face looks more distinctly human; body larger in proportion to head; skin is bright pink, scalp has hair; joints forming; lips exhibit sucking movements; kidneys well formed; digestive glands forming and **meconium**[7] (fetal feces) accumulating in intestinal lumen; heartbeat can be heard with a stethoscope
5	19 cm; 200–450 g	Body covered with fine hair called **lanugo**[8] and cheeselike sebaceous secretion called **vernix caseosa;**[9] skin bright pink; brown fat forms and will be used for postpartum heat production; fetus is now bent forward into "fetal position" due to crowding; **quickening** occurs—mother can feel fetal movements
6	27–35 cm; 550–800 g	Eyes open, eyelashes form; skin wrinkled, pink, and translucent; lungs begin producing surfactant; rapid weight gain
7	32–42 cm; 1,100–1,350 g	Skin wrinkled and red; fetus turns into upside-down **vertex position;** can survive if born prematurely, but thermoregulation is poor and respiratory distress is common due to insufficient pulmonary surfactant; bone marrow is now the sole site of hemopoiesis; testes descend into scrotum
8	41–45 cm; 2,000–2,300 g	Subcutaneous fat deposition gives fetus a more plump, babyish appearance, with lighter, less wrinkled skin; twins usually born at this stage
9	50 cm; 3,200–3,400 g	More subcutaneous fat deposited; lanugo is shed; nails extend to or beyond fingertips

7. *mecon* = poppy juice, opium
8. *lan* = down, wool
9. *vernix* = varnish + *caseo* = cheese

Other major aspects of embryonic and fetal development are listed in table 29.4 and depicted in figures 29.14 and 29.15.

Key Point Review

5 Distinguish between trophoblastic and placental nutrition.

6 Identify the two sources of blood to the placenta. Where do these two bloodstreams come closest to each other? What keeps them separated?

7 State the functions of the placenta, amnion, chorion, yolk sac, and allantois.

8 What developmental characteristic distinguishes a fetus from an embryo? At what gestational age is this attained?

9 Identify the three circulatory shunts of the fetus. Why does the blood take these "shortcuts" before birth?

The Neonate

▼Objectives

When you have completed this section, you should be able to
- describe how and why the circulatory system changes at birth;
- explain why the first breaths of air are relatively difficult for a neonate;
- describe the major physiological problems of a premature infant; and
- discuss some common causes of birth defects.

Development is by no means complete at birth. The liver and kidneys still are not fully functional, most joints are not yet ossified, and myelination of the nervous system is not completed until adolescence, for example. Indeed, humans are born in a very immature state compared to other newborn mammals.

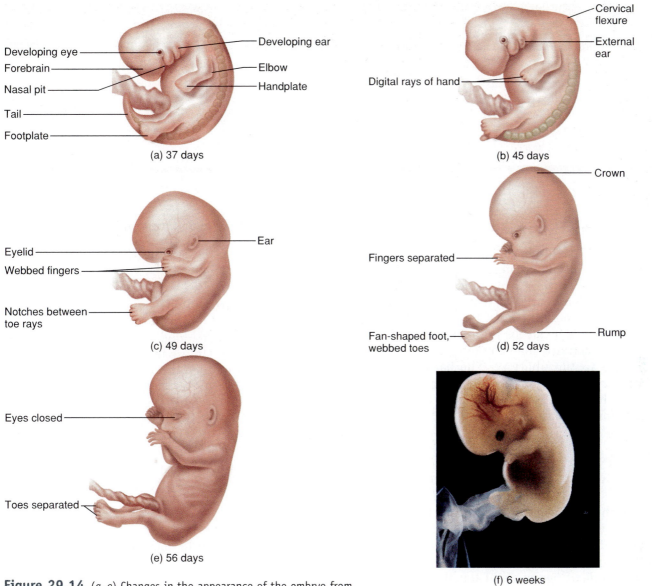

(a) 37 days

Developing eye
Forebrain
Nasal pit
Tail
Footplate
Developing ear
Elbow
Handplate

(b) 45 days

Cervical flexure
External ear
Digital rays of hand

(c) 49 days

Eyelid
Webbed fingers
Notches between toe rays
Ear

(d) 52 days

Crown
Fingers separated
Fan-shaped foot, webbed toes
Rump

(e) 56 days

Eyes closed
Toes separated

(f) 6 weeks

Figure 29.14 (*a–e*) Changes in the appearance of the embryo from 37 to 56 days (5–8 weeks). (*f*) Photograph of a 6-week-old embryo.

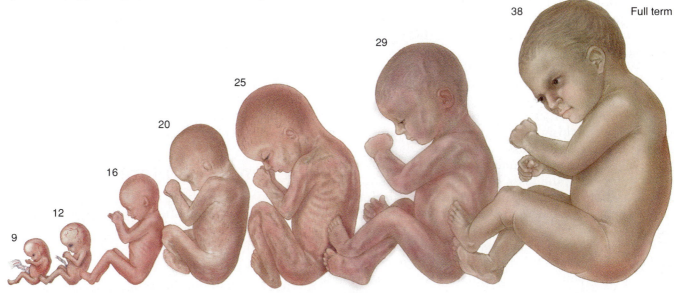

9
12
16
20
25
29
38
Full term

Figure 29.15 Appearance of the fetus from 9 weeks to 38 weeks (full term).

A newborn infant is immediately evaluated for general appearance, vital signs (temperature, pulse, and respiratory rate), weight, length, and head circumference and other dimensions and is screened for congenital disorders such as phenylketonuria (PKU). At 1 minute and 5 minutes after birth, the heart rate, respiratory effort, muscle tone, reflexes, and skin color are noted and given a score of 0 (poor), 1, or 2 (excellent). The total (0–10), called the **Apgar**[10] **score,** is a good predictor of infant survival. Infants with low Apgar scores may have neurological damage and need immediate attention if they are to survive. A low score at 1 minute suggests asphyxiation and may demand assisted ventilation. A low score at 5 minutes indicates a high probability of death.

The Transitional Period

The period immediately following birth is a crisis in which the neonate suddenly must adapt to life outside the mother's body. The first 6 to 8 hours are a **transitional period** in which the heart and respiratory rates increase and the body temperature falls. Physical activity then declines and the baby sleeps for about 3 hours. In its second period of activity, the baby often gags due to mucus and debris in the pharynx. The baby then sleeps again, becomes more stable, and begins a cycle of waking every 3 to 4 hours to feed. The first 6 weeks of life constitute the **neonatal period.**

Circulatory Adaptations

After the umbilical cord is clamped and cut, the umbilical arteries and veins collapse and become fibrotic. The proximal part of each umbilical artery becomes the *superior vesical artery,* which remains to supply the bladder. Other obliterated vessels become fibrous cords or ligaments: the distal parts of the umbilical arteries become the *medial umbilical ligaments* of the abdominal wall; the umbilical vein becomes the *round ligament (ligamentum teres)* of the liver; and the ductus venosus (the former shunt around the liver) becomes the *ligamentum venosum* on the inferior surface of the liver (see fig. 29.13*b*).

When the lungs expand with air, blood pressure in the pulmonary circuit drops rapidly and pressure in the right heart falls below that in the left. Blood flows briefly from the left atrium to the right through the foramen ovale, pushing two flaps of tissue into place to close this shunt. In most people these flaps fuse and permanently seal the foramen during the first year, leaving a depression, the **fossa ovalis,** in the interatrial septum. In about 25% of people, however, the foramen ovale remains unsealed and the flaps are held in place only by the relatively high blood pressure in the left atrium. The pressure changes in the pulmonary trunk and aorta also cause the ductus arteriosus to collapse. It closes permanently around 3 months of age, leaving a permanent cord, the *ligamentum arteriosum,* between the two vessels.

Respiratory Adaptations

It is an old and untrue cliché that a neonate must be spanked to stimulate it to breathe. During birth, CO_2 accumulates in the baby's blood and strongly stimulates the respiratory center. Unless the infant is depressed by oversedation of the mother, it normally begins breathing spontaneously. It requires a great effort, however, to take the first few breaths and inflate the collapsed alveoli. For the first 2 weeks, a baby takes about 45 breaths per minute, but subsequently stabilizes at about 12 breaths per minute. Respiratory rate is one of the *vital signs* used to assess the health of the neonate (see special topic 29.3).

Other Adaptations

Thermoregulation and fluid balance are also critical aspects of neonatal physiology. An infant has a larger ratio of surface area to volume than an adult does, so it loses heat more easily. As a baby grows, its metabolic rate increases and it accumulates more subcutaneous fat, thus producing and retaining more heat. Nevertheless, body temperature is more variable in infants and children than in adults.

The kidneys are not fully developed at birth and cannot concentrate the urine as much as a mature kidney can. Consequently, infants have a relatively high rate of water loss and require more fluid intake, relative to body weight, than adults do.

Premature Infants

Neonates weighing under 2.5 kg (5.5 lb) are generally considered **premature.** They have multiple difficulties involving respiration, thermoregulation, excretion, digestion, and liver function.

---

10. Virginia Apgar (1909–74), American anesthesiologist

The respiratory system is adequately developed by 7 months' gestation to support independent life. Infants born before this have a deficiency of pulmonary surfactant, causing **respiratory distress syndrome** (**RDS;** also called *hyaline membrane disease*). The alveoli collapse each time the infant exhales and a great effort is needed to reinflate them. The infant becomes very fatigued by the high energy demand of breathing. RDS may be treated by ventilating the lungs with oxygen-enriched air at a positive pressure to keep the lungs inflated between breaths and by administering surfactant as an inhalant. Nevertheless, RDS remains the most common cause of neonatal death.

Due to incomplete development of the hypothalamus, a premature infant cannot thermoregulate effectively. Body temperature must be controlled by placing the infant in a warmer. If the infant is more than 8 weeks premature, its digestive tract is too poorly developed for a normal diet of breast milk. The baby requires a low-fat formula because it cannot absorb fats very well, and it is given calcium and vitamin D supplements to promote ossification.

The liver is also poorly developed, and bearing in mind its very diverse functions (see table 26.7, p. 949), you can probably understand why this would have several serious consequences. Because of its low rate of albumin synthesis, the premature baby suffers hypoproteinemia. This upsets the balance between capillary filtration and reabsorption, leading to edema. The infant bleeds easily due to a deficiency of the clotting factors synthesized by the liver. This is true to some degree even in full-term infants, however, because the baby's intestines are not yet colonized by the bacteria that synthesize vitamin K, which is essential for the synthesis of clotting factors. Jaundice is common in neonates, especially premature babies, because the liver cannot dispose of bile pigments efficiently.

Congenital Anomalies

The most anxious moment for many new parents is awaiting reassurance that the child has no visible birth defects. Any abnormality present at birth is called a **congenital anomaly.** Here we consider some anomalies that result from infectious diseases, teratogens, mutagens, and genetic defects.

Infectious Diseases

Infectious diseases are largely beyond the scope of this book, but it must be noted at least briefly that several microorganisms can cross the placenta and cause serious congenital anomalies, stillbirth, or neonatal death. Common viral infections of the fetus and newborn include herpes simplex, rubella, cytomegalovirus, and

Figure 29.16 Infant with undeveloped arms as a result of thalidomide, a sedative taken by the mother in early pregnancy.

human immunodeficiency virus (HIV). Congenital bacterial infections include gonorrhea and syphilis. *Toxoplasma,* a protozoan contracted from meat, unpasteurized milk, and housecats, is another common cause of fetal deformity. Some of these pathogens have relatively mild effects on adults, but because of its immature immune system, the fetus is vulnerable to devastating effects such as blindness, hydrocephalus, cerebral palsy, seizures, and profound physical and mental retardation. These diseases are treated in greater detail in microbiology textbooks.

Teratogens

Teratogens[11] are viruses, chemicals, and other agents that cause anatomical deformities in the fetus. Perhaps the most notorious teratogenic drug is thalidomide, a sleeping pill first marketed in 1957. Thalidomide was taken by women in early pregnancy, often before they knew they were pregnant, and caused over 5,000 babies to be born with unformed arms or legs (fig. 29.16) and often with defects of the ears, heart, and intestines. It was taken off the market in 1961. Many teratogens produce less obvious effects, including physical or mental retardation, hyperirritability, inattention, strokes, seizures, respiratory arrest, crib death, and cancer. A general lesson to be learned from the thalidomide tragedy and other cases is that pregnant women should avoid all sedatives, barbiturates, and opiates. Even the acne medicine Acutane, however, has caused severe birth defects.

Alcohol causes more birth defects than any other drug. Even one drink a day has noticeable effects on fetal and postpartum development, some of which are not noticed until a child begins school. Alcohol abuse during pregnancy can cause **fetal alcohol syndrome (FAS),** characterized by a small head, malformed facial features, cardiac and central nervous system defects,

---

11. *terato* = monster + *gen* = producing

stunted growth, and behavioral symptoms such as hyperactivity, nervousness, and a poor attention span. Cigarette smoking also contributes to fetal and infant mortality, ectopic pregnancy, anencephaly (failure of the cerebrum to develop), cleft lip and palate, and cardiac abnormalities. Diagnostic X rays should be avoided during pregnancy because radiation can have teratogenic effects.

Mutagens and Genetic Anomalies

A **mutagen** is any agent that alters DNA or chromosome structure. Ionizing radiation and some chemicals have mutagenic, teratogenic, and carcinogenic effects, with extremely diverse results. Prenatal exposure to mutagens may result, for example, in stillbirths or in increased risk of childhood cancer.

Some of the most common genetic disorders result not from mutagens, however, but from the failure of homologous chromosomes to separate during meiosis. Recall that homologous chromosomes pair up during prophase I and normally separate from each other at anaphase I (see p. 977). This separation, called **disjunction**, produces daughter cells with 23 chromosomes each. In **nondisjunction**, a pair of chromosomes fails to separate. Both chromosomes then go to the same daughter cell, which receives 24 chromosomes while the other daughter cell receives 22. **Aneuploidy**[12] (AN-you-PLOY-dee), the presence of an extra chromosome or lack of one, accounts for about 50% of spontaneous abortions. It can be detected prior to birth by *amniocentesis,* the examination of cells in a sample of amniotic fluid, or by *chorionic villus sampling,* the removal and examination of cells from the chorion.

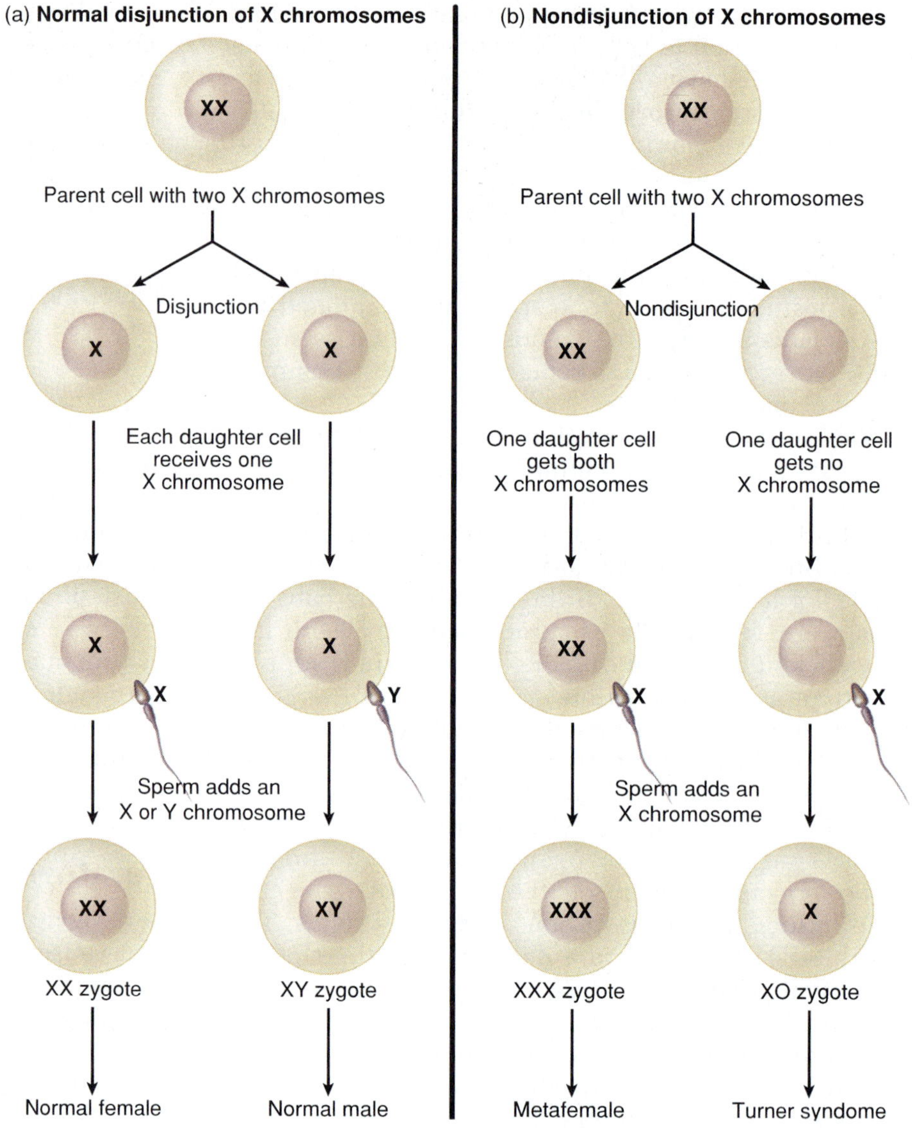

Figure 29.17 (*a*) The chromosomal results of normal disjunction and fertilization by X- or Y-bearing spermatozoa. (*b*) Two of the possible chromosomal results of nondisjunction followed by fertilization with an X-bearing spermatozoon. If the spermatozoon had carried a Y chromosome, it would have produced either an XXY individual with Klinefelter syndrome or a YO zygote that would die with little further development.

Figure 29.17 compares normal disjunction of the X chromosomes with some effects of nondisjunction. In nondisjunction, an egg cell may receive both X chromosomes. If it is fertilized by an X-bearing sperm, the result is an XXX zygote and a set of anomalies called the **triplo-X syndrome.** Such females are sometimes infertile and may have mild intellectual impairments. If an XX egg is fertilized by a Y-bearing sperm, the result is an XXY combination and **Klinefelter**[13]

12. *an* = not, without + *eu* = true, normal + *ploid,* from *diplo* = double, paired

13. Harry F. Klinefelter, Jr. (1912–), American physician

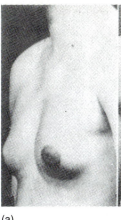

(a)

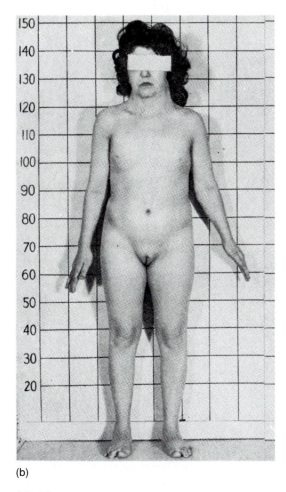

(b)

Figure 29.18 (a) Gynecomastia in a male with Klinefelter syndrome. (b) A 22-year-old woman with Turner syndrome. Note her short stature (about 145 cm, or 4 ft 9 in.), lack of sexual development, webbed neck, and widely spaced nipples.

syndrome. People with Klinefelter syndrome are sterile males, usually of average intelligence, but with undeveloped testes, sparse body hair, unusually long arms and legs, and enlarged breasts (gynecomastia) (fig. 29.18a). This syndrome commonly goes undetected until puberty, when failure to develop the secondary sex characteristics may prompt genetic testing.

The other possible outcome of X chromosome nondisjunction is that an egg cell may receive no X chromosome (both X chromosomes are discarded in the first polar body). If fertilized by a Y-bearing sperm, it dies for lack of the indispensable genes on the X chromosome. If such an egg is fertilized by an X-bearing sperm, however, the result is **Turner**[14] **syndrome,** with an XO combination (O represents the absence of one sex chromosome). Only 3% of those with Turner syndrome survive to birth. Girls with Turner syndrome show no serious impairments as children but tend to have a webbed neck and widely spaced nipples. At puberty, the secondary sex characteristics fail to develop (fig. 29.18b). The ovaries are nearly absent, the girl remains sterile, and she usually has a short stature.

The other 22 pairs of chromosomes (the autosomes) are also subject to nondisjunction. The most common autosomal anomaly is **Down**[15] **syndrome (trisomy-21),** resulting from nondisjunction of chromosome 21. The symptoms include retarded physical development; short stature; a relatively flat face with a flat nasal bridge; low-set ears; *epicanthal folds* that give the eyes an Asian appearance; an enlarged, protruding tongue; stubby fingers; and a short broad hand with only one palmar crease (fig. 29.19). People with Down syndrome tend to have outgoing, affectionate personalities. Mental retardation is common and sometimes severe, but is not inevitable. Nearly 50% of victims die within the first year of life due to immune deficiency or abnormalities of the heart or kidneys. Down syndrome occurs in about 1 out of 700 to 800 live births in the United States and increases in proportion to the age of the mother. The chance of having a child with Down syndrome is about 1 in 3,000 for a woman under 30, 1 in 365 by age 35, and 1 in 9 by age 48.

Key Point Review

10 How does inflation of the lungs at birth affect the route of blood flow through the heart?

11 Why is respiratory distress syndrome common in premature infants?

12 Define nondisjunction and explain how it causes aneuploidy. Name two syndromes resulting from aneuploidy.

14. Henry H. Turner (1892–1970), American endocrinologist
15. John Langdon H. Down (1828–96), British physician

(a)

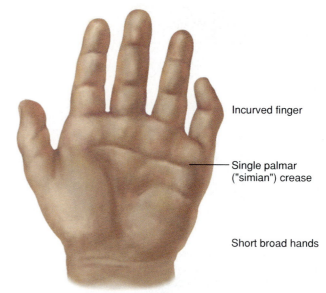

Incurved finger

Single palmar ("simian") crease

Short broad hands

(b)

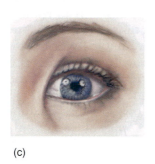

(c)

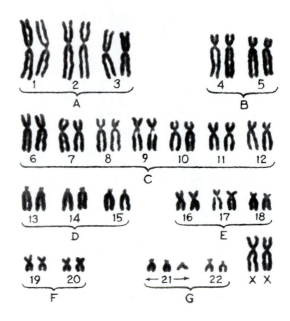

(d)

Figure 29.19 Down syndrome. (*a*) A child with Down syndrome (*right*) plays with her big sister. (*b*) Characteristics of the hand seen in Down syndrome. (*c*) The epicanthal fold over the medial commissure (canthus) of the left eye. (*d*) The karyotype of Down syndrome, showing the trisomy of chromosome 21.

Aging and Senescence

▼**Objectives**
When you have completed this section, you should be able to
- define senescence and distinguish it from aging;
- describe the major senescent changes that occur in each organ system;
- summarize some current theories of senescence; and
- be able to explain how exercise and other factors can slow the rate of senescence.

Like Ponce de León searching for the legendary fountain of youth in the Florida wilderness, people yearn for a way to preserve their youthful appearance and function. Our deepest concern, however, is not aging but senescence. The term **aging** is used in various ways but is taken here to mean all changes that occur in the body with the passage of time—including the growth, development, and increasing functional efficiency that occur from childhood to adulthood, as well as the degenerative changes that occur later in life. **Senescence** refers to the degenerative changes that occur in an organ system after the age of peak functional efficiency. It includes a gradual loss of reserve capacities, reduced ability to repair damage and compensate for stress, and increased susceptibility to disease.

Senescence is not just a personal concern but an important issue for health-care providers. One in nine Americans is 65 or older. As the average age of the population rises, health-care professionals will find

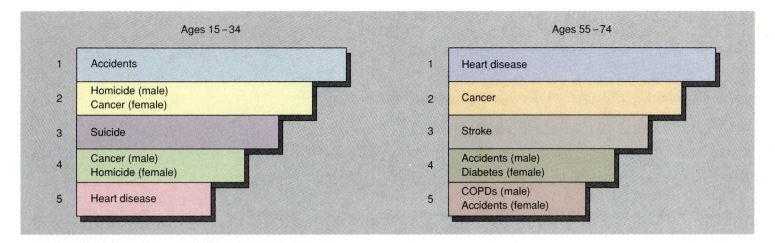

Figure 29.20 The five major causes of death in the United States, in rank order, for individuals 15 to 34 years old and those 55 to 74 years old. Notice the relative contribution of senescence in each group. COPD is chronic obstructure pulmonary disease (chronic bronchitis and emphysema). Bar lengths indicate only rank and not relative numbers of deaths.

themselves increasingly occupied by the prevention and treatment of the diseases of age. The leading causes of death change markedly with age (fig. 29.20) and are clearly related, in old age, to senescence of the organ systems. The causes of senescence, however, remain as much a scientific mystery today as cancer was 50 years ago and heredity was 100 years ago.

As we survey the senescence of the organ systems, you should notice many points relevant not only to caring for an aging population but also to personal health and fitness practices that can lessen the rate and effects of senescence. In addition, the study of senescence calls renewed attention to the multiple interactions among organ systems. As you will see, the senescence of one organ system typically leads to the senescence of others. Your study of this topic will bring together many concepts introduced in earlier chapters of the book. You may find the glossary helpful in refreshing your memory of concepts revisited in the following discussion.

Such terms as *middle age, old age,* and *elderly* have no precise universal definitions, but for present purposes we will consider ages 40 to 59 to be middle age and ages 60 and over to be old age. Unless otherwise stated, when describing the functional decline in the organ systems, we are comparing average, healthy people in their 60s to people in their 20s.

Senescence of the Organ Systems

Organ systems do not all degenerate at the same rate. For example, from ages 30 to 80, the speed of nerve conduction declines only 10% to 15%, but the number of functional glomeruli in the kidneys declines about 60%. Some physiological functions show only moderate changes at rest but more pronounced differences when tested under exercise conditions. The organ systems

also vary widely in the age at which senescence becomes noticeable. There are traces of atherosclerosis, for example, even in infants. Sensory functions such as visual acuity and auditory sensitivity begin to decline soon after puberty. By contrast, the female reproductive system does not show significant senescence until menopause and then its decline is relatively abrupt. Aside from these unusual examples, most physiological measures of performance peak between the late teens and age 30 and then decline at a rate influenced by the level of use of the organs.

Integumentary System

Two-thirds of people aged 50 and over, and nearly all people over age 70, have medical concerns or complaints about their skin. Senescence of the integumentary system often becomes most noticeable in the late 40s. The hair turns grayer and thinner as melanocytes die out, mitosis slows down, and dead hairs are not replaced. The atrophy of sebaceous glands leaves the skin and hair drier. As epidermal mitosis declines and collagen is lost from the dermis, the skin becomes almost paper-thin and translucent. It becomes looser because of a loss of elastic fibers and flattening of the dermal papillae, which normally form a stress-resistant corrugated boundary between the dermis and epidermis. If you pinch a fold of skin on the back of a child's hand, it quickly springs back when you let go; do the same on an older person, and the skinfold remains longer. Because of its loss of elasticity, aged skin sags to various degrees and may hang loosely from the arm.

Aged skin has fewer blood vessels than younger skin, and those that remain are more fragile. The skin can become reddened as broken vessels leak into the connective tissue. Many older people exhibit **rosacea**—patchy networks of tiny, dilated blood vessels visible especially on the nose and cheeks. Because of the fragility of the dermal blood vessels, aged skin bruises more easily. Injuries to the skin are more common and severe in old age, partly because the cutaneous nerve endings decline by two-thirds from age 20 to age 80, leaving a person less aware of touch, pressure, and injurious stimuli. Injured skin heals slowly in old age because of poorer circulation and a relative scarcity of immune cells and fibroblasts. Antigen-presenting Langerhans cells decline by as much as 40% in the aged epidermis, leaving the skin more susceptible to recurring infections.

Thermoregulation is a serious problem in old age because of the atrophy of cutaneous blood vessels, sweat glands, and subcutaneous fat. Older people are more vulnerable to hypothermia in cold weather and to heatstroke in hot weather. Heat waves and cold spells take an especially heavy toll among the elderly poor, who suffer from a combination of reduced homeostasis and inadequate housing.

These are all "normal" changes in the skin, or **intrinsic aging**—changes that occur more or less inevitably with the passage of time. In addition to these, there is **photoaging**—degenerative changes in proportion to a person's lifetime exposure to ultraviolet radiation. UV radiation accounts for more than 90% of the integumentary changes that people find medically troubling or cosmetically disagreeable: skin cancer; yellowing and mottling of the skin; age spots, which resemble enlarged freckles on the back of the hand and other sun-exposed areas; and wrinkling, which affects the face, hands, and arms more than areas of the body that receive less exposure. A lifetime of outdoor activity can give the skin a leathery, deeply wrinkled, "outdoorsy" appearance (fig. 29.21), but beneath this rugged exterior is a less happy histological appearance. The sun-damaged skin shows many malignant and premalignant cells, extensive damage to the dermal blood vessels, and dense masses of coarse, frayed elastic fibers underlying the surface wrinkles and creases.

Senescence of the skin has far-reaching effects on other organ systems. Cutaneous vitamin D production declines as much as 75% in old age. This is all the more significant because the elderly spend less time outdoors and, due to increasing lactose intolerance, tend to avoid dairy products, the only dietary source of vitamin D. Consequently, the elderly are at high risk of insufficient calcium absorption, which, in turn, contributes to bone loss, muscle weakness, and impaired glandular secretion and synaptic transmission.

Skeletal System

After age 30, osteoblasts become less active than osteoclasts. This imbalance results in **osteopenia**, the loss of

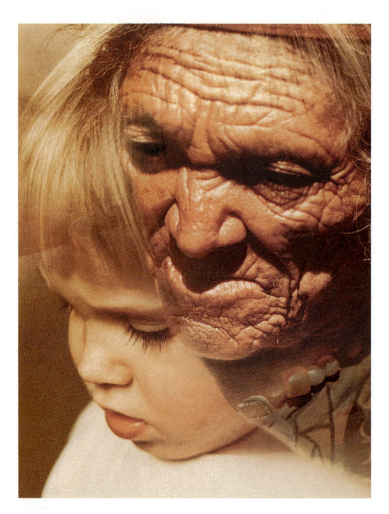

Figure 29.21 With age, the skin exhibits both intrinsic aging and photoaging. The deep creases seen here result mainly from photoaging.

Photograph of child from *The 1974 Science Year*. 1973 Field Enterprises Educational Corporation. By permission of World Book, Inc.

bone; when the loss is severe enough to compromise a person's physical activity and health, it is called **osteoporosis** (see p. 248). After age 40, women lose about 8% of their bone mass per decade and men about 3%. Bone loss from the jaws is a contributing factor in tooth loss.

Not only does bone density decline with age, but the bones become more brittle as the cells synthesize less protein. Fractures occur more easily and heal more slowly. A fracture may impose a long period of immobility, which makes a person more vulnerable to pneumonia and other infectious diseases.

People notice more stiffness and pain in the synovial joints as they age, and degenerative joint diseases affect the lifestyle of 85% of people over age 75. Synovial fluid is less abundant and the articular cartilage is thinner or absent. Exposed bone surfaces abrade each other, causing friction, pain, and reduced mobility. **Osteoarthritis** is the most common joint disease of older people and one of the most common causes of physical disability (p. 321). Even breathing becomes more difficult and tiring in old age because expansion of the thorax is restricted by calcification of the sternocostal joints. Degeneration of the intervertebral discs causes back pain and stiffness, but herniated discs are less common in old age because the discs become more fibrous and stronger, with less nucleus pulposus.

Muscular System

One of the most noticeable changes we experience with age is the replacement of lean body mass (muscle) with fat. The change is dramatically exemplified by CT scans of the thigh. In a young well-conditioned male, muscle accounts for 90% of the cross-sectional area of the midthigh, whereas in a frail 90-year-old woman, it is only 30%. Muscular strength and mass peak in the 20s; by the age of 80, most people have only half as much strength and fatigue resistance. A large percentage of people over age 75 cannot lift a 4.5 kg (10 lb) weight with their arms; such simple tasks as carrying a sack of groceries into the house may become impossible. The loss of strength is a major contributor to falls, fractures, and dependence on others for the routine activities of daily living. Fast twitch fibers exhibit the earliest and most severe atrophy, limiting reaction time and coordination.

There are multiple reasons for the loss of strength. Aged muscle fibers have fewer myofibrils, so they are smaller and weaker. The sarcomeres are increasingly disorganized, and muscle mitochondria are smaller and have reduced quantities of oxidative enzymes. Aged muscle has less ATP, creatine phosphate, glycogen, and myoglobin; consequently, it fatigues quickly. Aging muscles also exhibit more fat and collagen deposition (fibrosis) with age, which limits their movement and blood circulation. With reduced circulation, muscle injuries heal more slowly and with more scar tissue.

But the weakness and easy fatigue of aged muscle also stems from the senescence of other organ systems. There are fewer motor neurons in the spinal cord, and some muscle shrinkage may represent denervation atrophy. The remaining neurons produce less acetylcholine and show less efficient synaptic transmission, which makes the muscles slower to respond to stimulation. As muscle atrophies, motor units have fewer muscle fibers per motor neuron, and more motor units must be recruited to perform a given task. Tasks that used to be easy, such as buttoning the clothes or eating a meal, take more time and effort. The sympathetic nervous system is also less efficient in old age; consequently, blood flow to the muscles does not respond efficiently to exercise and this contributes to their rapid fatigue.

Nervous System

The nervous system reaches its peak development around age 30. From age 35 on, it is estimated that 100,000 brain cells die each day, and since neurons are amitotic, they cannot be replaced. The average brain weighs 56% less at age 75 than at age 30. The cerebral gyri are narrower, the sulci are wider, the cortex is thinner, and there is more space between the brain and meninges. The remaining cortical neurons have fewer synapses and for multiple reasons, synaptic transmission is less efficient: the neurons produce less neurotransmitter, they have fewer receptors, and the neuroglia around the synapses is more leaky and allows neurotransmitter to diffuse away. The degeneration of myelin sheaths with age also slows down axonal transmission.

The somas exhibit less rough ER and Golgi complex, indicating that their metabolism is slowing down. Old neurons accumulate lipofuscin pigment and show more neurofibrillary tangles—dense mats of cytoskeletal elements in their cytoplasm. In the extracellular material, plaques of fibrillar protein (amyloid) appear, especially in people with Down syndrome and Alzheimer disease (AD). AD is the most common nervous disability of old age (p. 458).

Not all functions of the central nervous system are equally affected by senescence. Motor coordination, intellectual function, and short-term memory decline more than language skills and long-term memory. Elderly people are often better at remembering things in the distant past than remembering recent events.

The sympathetic nervous system loses adrenergic receptors with age and becomes less sensitive to norepinephrine. This contributes to the decline of homeostatic control of such variables as body temperature and blood pressure. Many elderly people experience *orthostatic hypotension*—a drop in blood pressure when they stand, which sometimes results in dizziness, loss of balance, or fainting.

Sense Organs

Some sensory functions decline shortly after adolescence. Presbyopia (loss of flexibility in the lenses) makes it more difficult for the eyes to focus on nearby objects. Visual acuity declines and often requires corrective lenses by middle age. Cataracts (cloudiness of the lenses) are more common in old age. Night vision is impaired as more and more light is needed to stimulate the retina. This has several causes: there are fewer receptor cells in the retina, the vitreous body becomes less transparent, and the pupil becomes narrower as the pupillary dilators atrophy. Dark adaptation takes longer as the enzymatic reactions of the photoreceptor cells become slower. Changes in the structure of the iris and ciliary body may interfere with the reabsorption of aqueous humor, increasing the risk of glaucoma. Having to give up reading and driving can be among the most difficult changes of lifestyle in old age.

Auditory sensitivity peaks in adolescence and declines afterward. The tympanic membrane and the joints between the auditory ossicles become stiffer, so vibrations are transferred less effectively to the inner ear, creating a degree of conduction deafness. Nerve deafness occurs as the number of cochlear hair cells and auditory nerve fibers declines. Sensitivity to high-frequency sounds is most diminished. The death of receptor cells in the semicircular ducts, utricle, and saccule, and of nerve fibers in the vestibular nerve and neurons in the cerebellum, results in poor balance and dizziness—another factor in falls and fractures.

The senses of taste and smell are blunted as the number of taste buds, olfactory cells, and second-order neurons in the olfactory bulbs decline. Food may lose its appeal, and thus declining sensory function can be a factor in malnutrition.

Endocrine System

The endocrine system degenerates less than any other organ system. The reproductive hormones drop sharply and growth hormone and thyroid hormone secretion decline steadily after adolescence, but other hormones continue to be secreted at fairly stable levels even into old age. Target cell sensitivity declines, however, so some hormones have less effect. For example, the pituitary gland is less sensitive to negative feedback inhibition by adrenal glucocorticoids; consequently, the response to stress is more prolonged than usual. Diabetes mellitus is more common in old age, largely because the target cells have fewer insulin receptors. In part, this is an effect of the greater percentage of body fat in the elderly. The more fat at any age, the less sensitive other cells are to insulin. Body fat increases as the muscles atrophy, and muscle is one of the body's most significant glucose-buffering tissues. Because of the blunted insulin response, glucose levels remain elevated longer than normal after a meal.

Circulatory System

Cardiovascular disease is a leading cause of death in old age. Senescence has multiple effects on the blood, heart, arteries, and veins. Anemia is common and may result from nutritional deficiencies, inadequate exercise, disease, and other causes. The factors that can cause anemia in older people are so complicated it is almost impossible to control them enough to determine whether aging alone causes anemia. Evidence suggests that there is no change in the baseline rate of erythropoiesis in old age. Hemoglobin concentration, cell counts, and other variables are about the same among healthy people in their 70s as in the 30s. However, older people do not adapt well to stress on the hemo-poietic system, perhaps because of the senescence of other organ systems. As the gastric mucosa atrophies, for example, it produces less of the intrinsic factor needed for vitamin B_{12} absorption. This increases the incidence of pernicious anemia. As the kidneys age and the number of nephrons declines, less erythropoietin is secreted. There may also be a limit to how many times the hemopoietic stem cells can divide and continue giving rise to new blood cells. Whatever its cause, anemia limits the amount of oxygen that can be transported and thus contributes to the atrophy of tissues everywhere in the body.

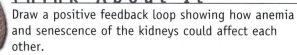

Think About It
Draw a positive feedback loop showing how anemia and senescence of the kidneys could affect each other.

Everyone exhibits coronary atherosclerosis with age. Consequently, myocardial cells die, angina pectoris and myocardial infarction become more common, the heart wall becomes thinner and weaker, and stroke volume, cardiac output, and cardiac reserve decline. Like other connective tissues, the fibrous cardiac skeleton becomes less elastic. This limits its distension and reduces the force of cardiac systole. Degenerative changes in the nodes and conduction pathways of the heart lead to a higher incidence of cardiac arrhythmia and heart block. Physical endurance is compromised by the drop in cardiac output.

As atherosclerosis stiffens the arteries, they cannot expand as effectively to accommodate the pressure surges of cardiac systole. Blood pressure therefore rises steadily with age (see table 20.1, p. 714). Atherosclerosis also narrows the arteries and reduces the perfusion of most organs. The effects of reduced circulation on the skin, skeletal muscles, and brain have already been noted. The combination of atherosclerosis and hypertension also weakens the arteries and increases the risk of aneurysm and stroke.

Atherosclerotic plaques trigger thrombosis, especially in the lower extremities where flow is relatively slow and the blood clots more easily. About 25% of people over age 50 experience venous blockage by thrombosis—especially people who do not exercise regularly.

Degenerative changes in the veins are most evident in the extremities. The valves become weaker and less able to stop the backflow of blood. Blood pools in the legs and feet, raises capillary blood pressure, and causes edema. Chronic stretching of the vessels often produces varicose veins and hemorrhoids. Support hose can reduce edema by compressing the tissues and forcing tissue fluid to return to the bloodstream, but physical activity is even more important in promoting venous return.

Immune System

The amounts of lymphatic tissue and red bone marrow decline with age; consequently there are fewer hemopoietic stem cells, disease-fighting leukocytes, and antigen-presenting cells (APCs). Also, the lymphocytes produced in these tissues often fail to mature and become immunocompetent. Both humoral and cellular immunity depend on APCs and helper T cells, and therefore both types of immune response are blunted. As a result, an older person has less protection against cancer and infectious diseases. It becomes especially important in old age to be vaccinated against influenza and other acute seasonal infections.

Respiratory System

Pulmonary ventilation declines steadily after the 20s and is one of several factors in the gradual loss of stamina. The costal cartilages and joints of the thoracic cage become less flexible, the lungs have less elastic tissue, and the lungs have fewer alveoli. Vital capacity, minute respiratory volume, and forced expiratory volume fall. The elderly are also less capable of clearing the lungs of irritants and pathogens and are therefore increasingly vulnerable to respiratory infections. Pneumonia causes more deaths than any other infectious disease and is often contracted in hospitals and nursing homes.

The chronic obstructive pulmonary diseases (COPDs)—emphysema and chronic bronchitis—are more common in old age since they represent the cumulative effects of a lifetime of degenerative change. They are among the five leading causes of death in old age (see fig. 29.20). Pulmonary obstruction also contributes to cardiovascular disease, hypoxemia, and hypoxic degeneration in all the organ systems. Respiratory health is therefore a major concern in aging.

Urinary System

The kidneys exhibit a striking degree of atrophy with age. From ages 25 to 85, the number of nephrons declines 30% to 40% and up to a third of the remaining glomeruli become atherosclerotic, bloodless, and nonfunctional. The kidneys of a 90-year-old are 20% to 40% smaller than those of a 30-year-old and receive only half as much blood. The glomerular filtration rate is proportionately lower and the kidneys are less efficient at clearing wastes from the blood. Although baseline renal function is adequate even in old age, there is little reserve capacity; thus other diseases can lead to surprisingly rapid renal failure. Drug doses often need to be reduced in old age because the kidneys cannot clear drugs from the blood as rapidly; this is a contributing factor in overmedication among the aged.

Water balance becomes more precarious in old age because the kidneys are less responsive to antidiuretic hormone and because the sense of thirst is sharply reduced. Even when given free access to water, elderly people may not drink enough to maintain normal blood osmolarity. Dehydration is therefore common. It is often said that aged kidneys are deficient in maintaining electrolyte balance, but the evidence for this remains questionable.

Voiding and bladder control become a problem for both men and women. About 80% of men over the age of 80 are affected by benign prostatic hyperplasia. The enlarged prostate compresses the urethra and interferes with emptying of the bladder. Urine retention causes pressure to back up in the kidneys, aggravating the failure of the nephrons. Older women are subject to incontinence (leakage of urine), especially if their history of pregnancy and childbearing has weakened the pelvic muscles and urethral sphincters. Senescence of the sympathetic nervous system and nervous disorders such as stroke and Alzheimer disease can also cause incontinence.

Digestive System and Nutrition

Less saliva is secreted in old age, making food less flavorful, swallowing more difficult, and the teeth more prone to caries. Nearly half of people over age 65 wear dentures because they have lost their teeth to caries and periodontitis. The stratified squamous epithelium of the oral cavity and esophagus is thinner and more vulnerable to abrasion. The gastric mucosa atrophies and secretes less acid and intrinsic factor. Acid deficiency reduces the absorption of calcium, iron, zinc, and folic acid. Heartburn becomes more common as the weakening gastroesophageal sphincter fails to prevent reflux into the esophagus. The most common digestive complaint of older people is constipation, which results from the reduced muscle tone and weaker peristalsis of the colon. This seems to stem from a combination of factors: atrophy of the muscularis externa, reduced sensitivity to neurotransmitters, less fiber and water in the diet, and less exercise. The liver, gallbladder, and pancreas show only slightly reduced function. Any drop in liver function, however, makes it harder to detoxify drugs and can contribute to overmedication.

Older people tend to reduce their food intake because of lower energy demand and appetite, because declining sensory functions make food less appealing, and because reduced mobility makes it more troublesome to shop and prepare meals. However, they need fewer calories than younger people because they have lower basal metabolic rates and tend to be less physically active. Protein, vitamin, and mineral requirements remain essentially unchanged. Vitamin and mineral supplements may be needed to compensate for reduced food intake and intestinal absorption. Malnutrition is common among older people and is an important factor in anemia and reduced immunity.

Reproductive System

In men, the senescent changes in the reproductive system are relatively gradual; they include declining sperm count, testosterone secretion, and libido. By age 65, sperm count is about one-third of what it was in a man's 20s. Men remain fertile (capable of fathering a child) well into old age, but impotence (inability to maintain an erection) can occur because of atherosclerosis, hypertension, or medication.

In women, the changes are more pronounced and develop more rapidly, over the course of menopause. Gametogenesis ceases and the ovaries stop producing sex steroids. This may result in vaginal dryness, genital atrophy, and reduced libido, making sex less enjoyable. With the loss of ovarian steroids, a postmenopausal woman has an elevated risk of osteoporosis and atherosclerosis.

Exercise and Senescence

Other than the mere passage of time, the greatest contributing factors in senescence are obesity and insufficient exercise. Conversely, good nutrition and exercise are the best ways to slow its progress.

There is no clear evidence that exercise will prolong your life, but there is little doubt that it improves the quality of life in old age. It maintains endurance, strength, and joint mobility while it reduces the incidence and severity of hypertension, osteoporosis, obesity, and diabetes mellitus. This is especially true if you begin a program of regular physical exercise early in life and continue it throughout old age. If you stop exercising regularly after middle age, the body rapidly becomes deconditioned, although appreciable reconditioning can be achieved even when an exercise program is begun late in life. A person in his or her 90s can increase muscle strength two- or threefold in 6 months with as little as 40 minutes of isometric exercise a week. The improvement results from a combination of muscle hypertrophy and neural efficiency.

Resistance training may be the most effective way of reducing accidental injuries such as bone fractures in old age, whereas endurance training reduces body fat and increases muscle mass, cardiac output, and maximum rate of oxygen uptake. A general guideline for ideal endurance training is to have three to five periods of aerobic exercise per week, each 20 to 60 minutes long and vigorous enough to reach 60% to 90% of your maximum heart rate. The maximum is best determined by a stress test but averages about 220 beats per minute minus your age in years.

An exercise program should ideally be preceded by a complete physical examination and stress test. Warm-up and cool-down periods are especially important in avoiding soft tissue injuries. Because of their lower capacity for thermoregulation, older people must be careful not to overdo exercise, especially in hot weather. At the outset of a new exercise program, it is best to "start low and go slow."

Hypotheses on Senescence

Life expectancy, the average length of life in a given population, has steadily increased in industrialized countries. People born in the United States at the beginning of this century had a life expectancy of only 45 to 50 years; nearly half of them died of infectious disease. The average boy born today can expect to live 72 years and the average girl 79 years. This is due mostly to victories over infant and child mortality, not to advances at the other end of the life span. **Life span,** the maximum age attainable by humans, has not increased for many centuries and there seems to be little prospect that it ever will. There is no credible record of anyone living past the age of 120 years.

Why do we inevitably die? There still is no general theory on this. The question actually comes down to two issues: What are the mechanisms that cause the organs to deteriorate with age, and why hasn't natural selection eliminated these and produced bodies capable of longer life?

Mechanisms of Senescence

Numerous hypotheses have been proposed and discarded to explain why organ function degenerates with age. Some authorities maintain that senescence is an intrinsic process governed by inevitable or even preprogrammed changes in cellular function. Others attribute senescence to extrinsic (environmental) factors that progressively damage our cells over the course of a lifetime. Quite likely, no one hypothesis explains all forms of senescence, but let's briefly examine some of them.

Programmed Cell Death Hypothesis Some developmental biologists believe that atrophy of the organs is an effect of programmed cell death (apoptosis). After a certain number of cell divisions or at a certain age, cells may activate a "suicide program" that destroys their own DNA. There is good evidence that the cause of senescence is at least partially genetic. Unusually long and short lives tend to run in families. Monozygotic twins are more likely than dizygotic twins to die at a

Figure 29.22 Progeria is a genetic disorder in which senescence appears to be greatly accelerated. The individuals here, from left to right, are 15, 12, and 26 years old. Few people with progeria live as long as the woman on the right.

similar age. One striking genetic defect called *progeria*[16] is characterized by greatly accelerated senescence (fig. 29.22). Symptoms begin to appear by age 2. The child's growth rate declines, the muscles and skin become flaccid, most lose their hair, and most die in early adolescence from advanced atherosclerosis. There is some controversy over the relevance or similarity of progeria to normal senescence, but it does demonstrate that many of the changes associated with old age can be brought on by a genetic anomaly.

The Hayflick Phenomenon   Normal organ function usually depends on a rate of cell renewal that keeps pace with cell death. There may be a limit, however, to how many times cells can divide. In 1950, Leonard Hayflick discovered that cultured fibroblasts from human fetuses divided about 50 times and then degenerated and died. If cells in the body have similar limitations on the number of divisions, it could mean an end of cell renewal and the onset of organ atrophy and functional decline.

Telomere Hypothesis   All chromosomes have a "cap" at each end called a **telomere**,[17] like the plastic tip of a shoelace. In humans, it consists of the nucleotide sequence CCCTAA repeated 1,000 times or more. It has no genes but apparently stabilizes the chromosome and prevents it from unraveling or sticking to other chromosomes. Every time DNA is replicated, however, 50 to 100 bases are lost from the telomere. Consequently, old human cells have shorter telomeres than young cells. Old chromosomes may therefore be more vulnerable to damage, thus causing old cells to be increasingly dysfunctional. This might be a factor in the Hayflick limit on cell divisions.

Cross-Linking Hypothesis   About one-fourth of the body's protein is collagen. With age, collagen molecules become cross-linked by more and more disulfide bonds, making the fibers less soluble and more stiff. This is thought to be a factor in several of the most noticeable changes of the aging body, including stiffening of the joints, lenses, and arteries. Similar cross-linking of DNA and enzyme molecules could progressively impair their functions as well.

Other Protein Abnormalities   Not only collagen but also many other proteins exhibit increasingly abnormal structure in older tissues and cells. The changes are not in amino acid sequence—therefore not attributable to DNA mutations—but lie in the way the proteins are folded and in other moieties such as carbohydrates attached to them. This is another reason that cells accumulate more dysfunctional proteins as they age.

Free Radical Hypothesis   Free radicals have very destructive effects on macromolecules (see chapter 2). We have a number of antioxidant chemicals to protect us from their effects, but it is hypothesized that some of these antioxidants become less abundant with age and are eventually overwhelmed by the free radicals, or that some of the molecules damaged by free radicals are long-lived and accumulate in cells. Free radical damage may therefore be a contributing factor in some of the other mechanisms of senescence discussed here.

16. *pro* = before + *ger* = old age

17. *telo* = end + *mer* = piece

Autoimmune Hypothesis   Some of the altered macromolecules described previously may be recognized as foreign antigens, stimulating lymphocytes to mount an immune response against the body's own tissues. Autoimmune diseases such as rheumatoid arthritis do, in fact, become more common in old age.

Evolution and Senescence

If certain genes contribute to senescence, it raises an evolutionary question—why doesn't natural selection eliminate them? In an attempt to answer this, biologists once postulated that senescence and death are for the good of the species—a way for older, worn-out individuals to make way for younger, healthier ones. We can see the importance of death for the human population by imagining that science had put an end to senescence and people died at a rate of 1 per 1,000 per year regardless of age (the rate at which American 18-year-olds now die). If so, half the human population would live to age 163 and 13% would live to be 2,000 years old. The implications for world population and competition for resources would be staggering. Thus it is easy to understand why death was once interpreted as a self-sacrificing phenomenon for the good of the species.

But this hypothesis has several weaknesses. One of them is the fact that natural selection works exclusively through the effects of genes on the reproductive rates of individuals. A species evolves only because some members reproduce more than others. A gene that does not affect reproductive rate can be neither eliminated nor favored by natural selection. Genes for disorders such as Alzheimer disease have little or no effect until a person is past reproductive age. Our prehistoric and even fairly recent ancestors usually died of accidents, predators, starvation, weather, and infectious diseases at an early age. Few people lived long enough to be affected by atherosclerosis, colon cancer, or Alzheimer disease. Natural selection would have been "blind" to such death-dealing genes, which would escape the selection process and remain with us today.

Death

There is no definable instant of biological death. Some organs function for an hour or more after the heart stops beating. During this time, even if a person is declared legally dead, living organs may be removed for transplant. For legal purposes, death was once defined as the loss of a spontaneous heartbeat and respiration. Now that cardiopulmonary functions can be artificially maintained, this criterion is less distinct. Clinical death is now widely defined in terms of **brain death**—a lack of cerebral activity indicated by a flat electroencephalogram for 30 minutes to 24 hours (depending on state laws), accompanied by a lack of reflexes or lack of spontaneous respiration and heartbeat.

Death usually results from the failure of a particular organ, which then has a cascading effect on other organs. Kidney failure, for example, leads to the accumulation of toxic wastes in the blood, which in turn leads to loss of consciousness, brain function, respiration, and heartbeat.

Ninety-nine percent of us will die before age 100, and there is little chance that this outlook will change within our lifetimes. We cannot presently foresee any "cure for old age" or significant extension of the human life span. The real issue is to maintain the best possible quality of life, and when the time comes to die, to do so in comfort and dignity.

Key Point Review

13 Define *aging* and *senescence*.

14 List some tissues or organs in which changes in collagenous and elastic connective tissues lead to senescence.

15 Many older people have difficulty with mobility and simple self-maintenance tasks such as dressing and cooking. Name some organ systems whose senescence is most relevant to these limitations.

16 Explain why both endurance and resistance exercises are important in old age.

17 Summarize any five cellular mechanisms that may be responsible for senescence.

Reproductive Technology—Making Babies in the Laboratory

Fertile heterosexual couples who have frequent intercourse and use no contraception have an 85% chance of conceiving within 1 year. About one in six American couples, however, are **infertile**—unable to conceive. Infertility can sometimes be corrected by hormone therapy or surgery, but when this fails, parenthood may still be possible through other reproductive technologies discussed here.

Artificial Insemination

If only the male is infertile, the oldest and simplest solution is **artificial insemination (AI),** in which a physician introduces donor semen into or near the cervix. This was first done in the 1890s, when the donor was often the physician himself or a medical student who donated semen for payment. In 1953, a technique was developed for storing semen in glass ampules frozen in liquid nitrogen; the first commercial sperm banks opened in 1970. Most women use sperm from anonymous donors but are able to select from a catalog that specifies the donors' physical and intellectual traits. A man with a low sperm count, however, can donate semen at intervals over a course of several weeks and have it pooled, concentrated, and used to artificially inseminate his partner. Men planning vasectomies sometimes donate sperm for storage as insurance against the death of a child, divorce and remarriage, or a change in family planning. Some cases of infertility are due to sperm destruction by the woman's immune system. This can sometimes be resolved by *sperm washing*—a technique in which the sperm are collected, washed to remove antigenic proteins from their surfaces, and then introduced by AI.

Oocyte Donation

The counterpart to sperm donation is **oocyte donation,** in which fresh oocytes are obtained from one woman, fertilized, and transplanted to the uterus of another woman. A woman who has reached menopause, or who has a known hereditary disorder she does not want to pass to her children, can become pregnant with a donor oocyte fertilized by her partner's sperm. The donated oocytes are sometimes provided by a relative or may be left over from another woman's in vitro fertilization (see next). The first baby conceived by oocyte donation was born in 1984. This procedure has a success rate of 20% to 50%.

In Vitro Fertilization

In some women the uterus is normal but the uterine tubes are scarred by pelvic inflammatory disease or other causes. **In vitro fertilization (IVF)** is an option in some of these cases. The woman is given gonadotropins to induce the "superovulation" of multiple eggs. The physician views the ovary with a laparoscope and removes eggs by suction. These are placed in a solution that mimics the chemical environment of her reproductive tract, and sperm are added to the dish. The term *in vitro fertilization* refers to the fact that fertilization occurs in laboratory glassware; children conceived by IVF are often misleadingly called "test-tube babies." In some cases, fertilization is assisted by piercing the zona pellucida before the sperm are added (*zona drilling*) or by injecting sperm directly into the egg through a micropipet. By the day after fertilization, some of the preembryos will have reached the 8- to 16-celled stage. Several of these are transferred to the mother's uterus through the cervix and her blood HCG level is monitored to determine whether implantation has occurred. Excess IVF preembryos may be donated to other infertile couples or frozen and used in later attempts. In cases where a woman has lost her ovaries to disease, the oocytes may be provided by another donor, often a relative.

IVF costs up to $10,000 per attempt and succeeds only 14% of the time. A couple can easily spend up to $100,000 before IVF is successful, and then some attempts are "too successful." Multiple preembryos are usually introduced to the uterus as insurance against the low probability that any one of them will implant and survive. Sometimes, however, this results in multiple births—quintuplets in one famous case. One advantage of IVF is that when the preembryo reaches the eight-celled stage, one or two cells can be removed and tested for genetic defects before the preembryo is introduced to the uterus.

IVF has been used in animal breeding since the 1950s, but the first child conceived this way was Joy Louise Brown (fig. E.1), born in England in 1978. It is now estimated that worldwide, about one IVF child is born every day.

Surrogate Mothers

IVF is an option only for women who have a functional uterus. A woman who has had a hysterectomy or is otherwise unable to become pregnant or maintain a pregnancy may opt to contract with a **surrogate mother** who provides a "uterus for hire." Some surrogates are both genetic and gestational mothers, and others gestational only. In the former case, the surrogate is artificially inseminated by a man's sperm and

—continued

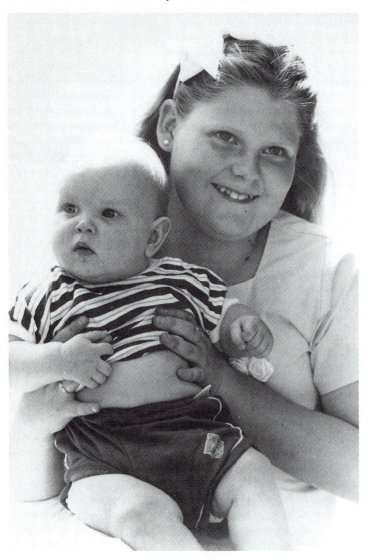

Figure E.1 New beginnings through reproductive technology. Joy Louise Brown, shown here at age 10, was the first child ever conceived by in vitro fertilization (IVF). She is holding Andrew Macheta, another IVF baby, at a 10-year anniversary celebration at the clinic near London where both were conceived.

agrees to give the baby to the man and his partner at birth. In the latter case, oocytes are collected from one woman's ovaries, fertilized in vitro, and the preembryos are placed in the surrogate's uterus. This is typical of cases in which a woman has functional ovaries but no functional uterus. A surrogate typically receives a fee of about $10,000 plus medical and legal costs. Several hundred babies have been produced this way in the United States. In at least one case, a woman carried the child of her infertile daughter, thus giving birth to her own granddaughter.

Gamete Intrafallopian Transfer

The low success rate of IVF led to a search for more reliable and cost-effective techniques. **Gamete intrafallopian transfer (GIFT)** was developed in the mid-1980s on the conjecture that pregnancy would be more successful if the oocyte were fertilized and began cleavage in a more natural environment. Eggs are obtained from a woman after a weeklong course of superovulatory drug treatment. The most active sperm cells are isolated from the semen, and the eggs and sperm are introduced into her uterine tube proximal to any existing obstruction. GIFT is about half as expensive as IVF and succeeds about 40% of the time. In a modification called **zygote intrafallopian transfer (ZIFT),** fertilization occurs in vitro and the preembryo is introduced into the uterine tube. Traveling down the uterine tube seems to improve the chance of implantation when the conceptus reaches the uterus.

Embryo Adoption

Embryo adoption is used when a woman has malfunctioning ovaries but a normal uterus. A man's sperm are used to artificially inseminate another woman. A few days later, the preembryo is flushed from the donor's uterus before it implants and is transferred to the uterus of the woman who wishes to have a child.

Ethical and Legal Issues

Like many other advances in medicine, reproductive technology has created its own ethical and legal dilemmas, some of which are especially confounding. Perhaps the most common problem is the surrogate mother who changes her mind. She has entered into a contract to surrender the baby to a couple at birth, but after carrying a baby for 9 months and giving birth, she may feel differently. This raises questions about the definition of motherhood, especially if she is the gestational but not the genetic mother.

The converse problem is illustrated by a case in which the child had hydrocephalus and neither the contracting couple nor the surrogate mother wanted it. In this case, genetic testing showed that the child actually had been fathered by the surrogate's husband, not the man who had contracted for her service. The surrogate then accepted the baby as her own. Nevertheless, the case raised the question of whether the birth of a genetically defective child constituted fulfillment of the contract and obligated the contracting couple to accept the child, or whether such a contract implies that the surrogate mother must produce a healthy child.

In another case, a wealthy couple was killed in an accident and left some frozen embryos in an IVF clinic. A lawsuit was filed on behalf of the embryos on the grounds that they were heirs to the couple's estate and should be carried to birth by a surrogate mother so they could inherit it. The court ruled against the suit and the embryos were allowed to die. In still another widely publicized case, a man sued his wife for custody of their frozen embryos as part of a divorce settlement.

IVF also creates a question of what to do with the excess preembryos. Some people view their disposal as a form of abortion, even if the preembryo is only a mass of 8 to 16 undifferentiated cells. On the other hand, there are those who see such excess preembryos as a research opportunity to obtain information that could not be obtained in any other way. In 1996, an IVF clinic in England was allowed to destroy 3,300 unclaimed preembryos, but only after heated public controversy.

It is common for scientific advances to require new advances in law and ethics. The parallel development of these disciplines is necessary if we are to benefit from the developments of science and ensure that knowledge is applied in an ethical and humane manner.▲

Chapter Review Study Outline

Fertilization and Preembryonic Development (pp. 1028–1033)
1. Sperm migration
2. Capacitation
3. Fertilization
   a. Acrosomal reaction
   b. Blocks to polyspermy
4. Meiosis II
5. Preembryonic stage
   a. Cleavage
   b. Loss of zona pellucida
   c. Blastocyst
      • Trophoblast
      • Embryoblast
   d. Implantation
      • Cytotrophoblast
      • Syncytiotrophoblast
      • Formation of chorion
   e. Embryogenesis

Embryonic and Fetal Development (pp. 1033–1040)
1. Prenatal nutrition
   a. Trophoblastic nutrition
   b. Placental nutrition
   c. Placentation
      • Chorionic villi
      • Placental sinus
      • Placental conductivity
      • Functions of the placenta
2. Embryonic membranes
   a. Amnion
   b. Yolk sac

   c. Allantois
   d. Chorion
3. Organogenesis
4. Fetal development
   a. Fetal circulation
      • Umbilical arteries
      • Umbilical vein
      • Shunts
        •• Ductus venosus
        •• Foramen ovale
        •• Ductus arteriosus
   b. Developmental calendar (table 29.4)

The Neonate (pp. 1040–1045)
1. Transitional period
2. Circulatory adaptations
3. Respiratory adaptations
4. Thermoregulation
5. Fluid balance
6. Premature infants
   a. Respiratory distress syndrome
   b. Thermoregulation
   c. Digestive system deficiencies
7. Congenital anomalies
   a. Infectious diseases
   b. Teratogens
   c. Mutagens and genetic anomalies
      • Nondisjunction and aneuploidy
      • Metafemale syndrome
      • Klinefelter syndrome

      • Turner syndrome
      • Down syndrome

Aging and Senescence (pp. 1046–1054)
1. Definitions
2. Senescence of the organ systems
   a. Integumentary system
   b. Skeletal system
   c. Muscular system
   d. Nervous system
   e. Sense organs
   f. Endocrine system
   g. Circulatory system
   h. Immune system
   i. Respiratory system
   j. Urinary system
   k. Digestive system and nutrition
   l. Reproductive system
3. Exercise and senescence
4. Hypotheses on senescence
   a. Life expectancy and life span
   b. Mechanisms of senescence
      • Programmed cell death hypothesis
      • The Hayflick phenomenon
      • Telomere hypothesis
      • Cross-linking hypothesis
      • Other protein abnormalities
      • Free radical hypothesis
      • Autoimmune hypothesis
5. Evolution and senescence
6. Death

Selected Vocabulary

developmental biology 1028
capacitation 1028
acrosomal reaction 1028

hyaluronidase 1029
acrosin 1029
polyspermy 1029

fast block 1029
slow block 1029
cortical granules 1029

cortical reaction 1029
fertilization membrane 1029
pronucleus 1030

Testing Your Recall   Answers in Appendix C

1. When a conceptus arrives in the uterus it is at what stage of development?
   a. zygote
   b. morula
   c. blastomere
   d. blastocyst
   e. embryo

2. The entry of a sperm nucleus into an egg must be preceded by
   a. the cortical reaction.
   b. the acrosomal reaction.
   c. the fast block.
   d. cleavage.
   e. implantation.

3. The stage of a conceptus that implants in the uterine wall is
   a. a blastomere.
   b. a morula.
   c. a blastocyst.
   d. an embryo.
   e. a zygote.

4. Chorionic villi develop from
   a. the zona pellucida.
   b. the endometrium.
   c. the syncytiotrophoblast.
   d. the embryoblast.
   e. the corona radiata.

5. Which of these is a result of aneuploidy?
   a. Turner syndrome
   b. fetal alcohol syndrome
   c. nondisjunction
   d. progeria
   e. rubella

6. Fetal urine accumulates in the _____ and contributes to the fluid there.
   a. placental sinus
   b. yolk sac
   c. allantois
   d. chorionic villi
   e. amnion

7. One hypothesis of senescence is that it results from a lifetime of cellular damage by
   a. teratogens.
   b. aneuploidy.
   c. free radicals.
   d. cytomegalovirus.
   e. nondisjunction.

8. Photoaging is a major factor in senescence of
   a. the integumentary system.
   b. the sense organs.
   c. the nervous system.
   d. the skeletal system.
   e. the cardiovascular system.

9. Senescence of the skin most directly contributes to senescence of
   a. the urinary system.
   b. the sense organs.
   c. the nervous system.
   d. the skeletal system.
   e. the cardiovascular system.

10. The top five causes of death in old age include all of the following *except*
    a. cancer.
    b. diabetes mellitus.

    c. pulmonary diseases.
    d. suicide.
    e. stroke.

11. Viruses and chemicals that cause congenital anatomical deformities are called _____.

12. Aneuploidy is caused by _____, the failure of two homologous chromosomes to separate in meiosis.

13. The maximum age attained by humans is called the human _____.

14. For the first 8 weeks after implantation, the conceptus gets most of its nutrition by digesting _____ cells of the endometrium.

15. Fetal blood flows through growths called _____, which project into the placental sinus.

16. The enzymes with which a sperm penetrates an egg are contained in an organelle called the _____.

17. Stiffening of the arteries, joints, and lenses in old age may be due to cross-linking between _____ molecules.

18. An enlarged tongue, epicanthal folds of the eyes, and mental retardation are characteristic of a genetic anomaly called _____.

19. The fossa ovalis is a remnant of a fetal shunt called the _____.

20. A developing individual is first classified as a/an _____ when the three primary germ layers have formed.

1. Suppose a woman had a mutation resulting in a tough zona pellucida that did not disintegrate after her oocyte was fertilized. How would this affect her fertility and why?

2. Suppose a drug were developed that could slow down the rate of collagen cross-linking with age. What diseases of old age could potentially be eliminated or made less severe with such a drug?

3. Some health-food stores market the enzyme superoxide dismutase (SOD) as an oral antioxidant to retard senescence. Explain why it would be a waste of your money to take it.

4. In some children the ductus arteriosus fails to close after birth—a condition that eventually requires surgical correction. Predict how this condition would affect (a) pulmonary blood pressure, (b) systemic diastolic pressure, and (c) the right ventricle of the heart.

5. Only one spermatozoon fertilizes an egg, and yet men who ejaculate fewer than 10 million spermatozoa are usually infertile. Explain this seeming contradiction. For a man who does ejaculate 10 million spermatozoa, predict about how many are likely to come within close range of the oocyte. How likely is it that one of those sperm would fertilize the oocyte?

Web Site Link

For a listing of the most current web sites related to this chapter, please visit the Saladin homepage at:

http://www.mhhe.com/sciencemath/biology/saladin/

Periodic Table of the Elements

Nineteenth-century chemists discovered that when they arranged the known elements by atomic weight, certain properties reappeared periodically. In 1869, Russian chemist Dmitri Mendeleev published the first modern periodic table of the elements, leaving gaps for those that had not yet been discovered. He accurately predicted properties of the missing elements, which helped other chemists discover and isolate them.

Each row in the table is a *period* and each column is a *group (family)*. Each period has one electron shell more than the period above it, and as we progress from left to right within a period, each element has one more proton and electron than the one before. The dark steplike line from boron (5) to astatine (85) separates the metals to the left of it (except hydrogen) from the nonmetals to the right. Each period begins with a soft, light, highly reactive *alkali metal*, with one valence electron, in family IA. Progressing from left to right, the

metallic properties of the elements become less and less pronounced. Elements in family VIIA are highly reactive gases called *halogens*, with seven valence electrons. Elements in family VIIIA, called *noble (inert) gases*, have a full valence shell of eight electrons, making them chemically unreactive. Elements 1–96 occur naturally, whereas 97–109 are short-lived elements known only in the laboratory. The names of elements 104–109 are still in dispute. The names shown in this table are the ones recommended by their discoverers, but the International Union of Pure and Applied Chemistry has recommended different names for these elements; the disagreement is under review.

The 24 elements with normal roles in human physiology are color-coded according to their relative abundance in the body (see chapter 2). Others, however, may be present as contaminants with very destructive effects (such as arsenic, lead, and radiation poisoning).

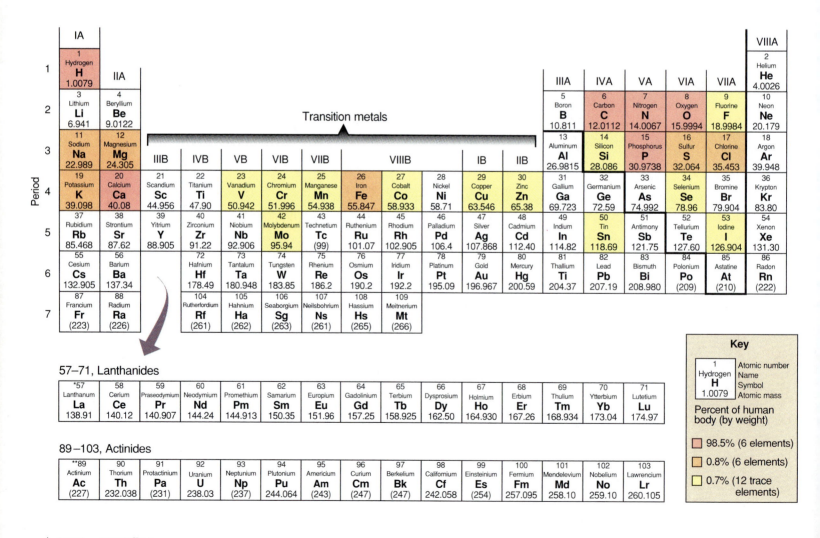

57–71, Lanthanides

89–103, Actinides

Answers to Testing Your Recall Questions

Answers to the objective chapter review questions are given here. Answers to the Think About It questions are in the *Instructor's Manual* and the answers to the Test-ing Your Comprehension questions are printed in the Instructor's Manual and the *Study Guide*.

Chapter 1
1. a
2. b
3. d
4. a
5. c
6. c
7. a
8. c
9. e
10. b
11. dissection
12. Robert Hooke
13. deduction
14. psychosomatic
15. homeostasis
16. set point
17. negative feedback
18. organ
19. stereoscopic
20. opposable

Atlas A
1. d
2. c
3. e
4. d
5. d
6. a
7. a
8. d
9. b
10. e
11. supine
12. parietal
13. mediastinum
14. occipital, cervical
15. sagittal
16. meninges
17. retroperitoneal
18. peripheral
19. inferior
20. cubital, popliteal

Chapter 2
1. b
2. c
3. a
4. c
5. a
6. a
7. e
8. d
9. a
10. e
11. valence
12. cation
13. polar covalent
14. isomers
15. conservation of mass
16. catalyst
17. emulsion
18. mole
19. 10^{-9} M
20. anabolism

Chapter 3
1. a
2. e
3. e
4. a
5. c
6. b
7. c
8. e
9. b
10. d
11. hydration spheres
12. condensation
13. functional groups
14. -ose, -ase
15. anaerobic fermentation
16. hydrogen
17. peptide
18. phospholipids
19. allosteric inhibition
20. monosaccharides, amino acids

Chapter 4
1. e
2. b
3. d
4. b
5. e
6. e
7. a
8. a
9. c
10. b
11. micrometers (µm)
12. second messenger
13. voltage-gated
14. hydrostatic pressure
15. hypertonic
16. exocytosis
17. mitochondria, nucleus
18. peroxisomes, smooth ER
19. ligand-gated channel
20. cisterna

Chapter 5
1. a
2. e
3. c
4. c
5. e
6. b
7. b
8. d
9. d
10. a
11. cytokinesis
12. alleles
13. genetic code
14. polysome
15. carcinogens
16. chaperones
17. 46, 92, 92
18. malignant (cancerous)
19. cyclins
20. autosomes

Chapter 6
1. a
2. b
3. c
4. e
5. c
6. a
7. a
8. e
9. b
10. b
11. necrosis
12. mesothelium
13. lacunae
14. fibers
15. collagen
16. skeletal
17. basement membrane
18. matrix (extracellular material)
19. proteoglycan
20. simple

Chapter 7
1. d
2. c
3. c
4. b
5. a
6. e
7. c
8. a
9. a
10. d
11. insensible
12. arrector pili
13. debridement
14. testosterone
15. cyanosis
16. dermal papillae
17. earwax
18. sebaceous
19. hyponychium
20. hair papilla

Chapter 8
1. e
2. b
3. a
4. d
5. c
6. c
7. d
8. a
9. a
10. d
11. hydroxyapatite
12. canaliculi
13. appositional
14. collagen fibers
15. hypocalcemia
16. osteoblasts
17. calcitriol
18. osteoporosis
19. metaphysis
20. osteomalacia

Chapter 9
1. b
2. e
3. a
4. d
5. a
6. e
7. c
8. b
9. e
10. b
11. fontanelles
12. temporal
13. incus
14. sphenoid
15. annulus fibrosus
16. dens (odontoid process)
17. auricular
18. styloid
19. pollex, hallux
20. longitudinal

Chapter 10
1. c
2. b
3. a
4. e
5. c
6. c
7. a
8. d
9. b
10. d
11. synovial fluid
12. bursa
13. pivot
14. kinesiology
15. gomphosis
16. serrate
17. extension
18. range of motion
19. labrum
20. menisci

Chapter 11
1. b
2. e
3. a
4. c
5. e
6. e
7. b
8. a
9. d
10. c
11. origin
12. fascicle
13. prime mover (agonist)
14. hamstring
15. retinacula
16. urogenital triangle
17. linea alba
18. synergist
19. bipennate
20. sphincter

Atlas B
1. f
2. b

3. k
4. p
5. h
6. z
7. o
8. x
9. c
10. a
11. y
12. m
13. n
14. e
15. g
16. v
17. f
18. c
19. y
20. x
21. k
22. d
23. s
24. b
25. a
26. u
27. j
28. i
29. g
30. q

Chapter 12
1. a
2. d
3. b
4. a
5. c
6. e
7. c
8. c
9. e
10. d
11. threshold
12. tetanus
13. terminal cisternae
14. myosin
15. acetylcholine
16. myoglobin
17. Z discs
18. varicosities
19. tonus
20. lactic acid

Chapter 13
1. e
2. d
3. d
4. a
5. c
6. e
7. d
8. a
9. d
10. b
11. afferent
12. conductivity
13. absolute refractory period
14. dendrites
15. Schwann
16. nodes of Ranvier
17. axon hillock, initial segment

18. metabotropic
19. facilitated zone
20. neuromodulators

Chapter 14
1. d
2. d
3. a
4. b
5. c
6. e
7. c
8. a
9. c
10. e
11. corpus callosum
12. ventricles
13. arbor vitae
14. decussation
15. choroid plexus
16. precentral
17. limbic system
18. association areas
19. categorical
20. Broca's area

Chapter 15
1. b
2. a
3. e
4. c
5. d
6. d
7. b
8. d
9. a
10. c
11. ganglia
12. ramus
13. phrenic
14. vagus
15. intrafusal fibers
16. tibial, common peroneal
17. sympathetic
18. acetylcholine
19. cAMP
20. sympathetic tone, vasomotor tone

Chapter 16
1. a
2. c
3. a
4. a
5. e
6. e
7. d
8. c
9. c
10. b
11. fovea centralis
12. ganglion
13. Na$^+$
14. otoliths
15. hair cells
16. stapes
17. colliculi
18. taste hairs

19. olfactory bulbs
20. referred pain

Chapter 17
1. b
2. d
3. a
4. c
5. e
6. c
7. d
8. c
9. e
10. e
11. adenohypophysis
12. tyrosine
13. acromegaly
14. parathyroid
15. glucocorticoids
16. granulosa, interstitial
17. negative feedback inhibition
18. hypothalamo-hypophyseal portal system
19. permissive
20. eicosanoids

Chapter 18
1. b
2. c
3. c
4. a
5. b
6. d
7. d
8. c
9. d
10. c
11. hemopoiesis
12. hematocrit
13. thromboplastin
14. agglutinogens
15. hemophilia
16. hemostasis
17. sickle-cell anemia
18. polycythemia
19. vitamin B$_{12}$
20. erythropoietin

Chapter 19
1. d
2. b
3. d
4. a
5. e
6. c
7. d
8. e
9. a
10. e
11. systole, diastole
12. systemic
13. coronary sulcus
14. Na$^+$
15. gap junctions
16. T wave
17. semilunar
18. auscultation
19. preload
20. cardiac output

Chapter 20

1. c
2. b
3. a
4. e
5. b
6. c
7. e
8. a
9. e
10. d
11. systolic, diastolic
12. continuous capillaries
13. anaphylactic
14. thoracic pump
15. oncotic pressure
16. transcytosis
17. baroreceptors
18. sympathetic
19. circle of Willis
20. basilic, cephalic

Chapter 21

1. b
2. c
3. a
4. e
5. d
6. b
7. c
8. e
9. a
10. c
11. pathogen
12. lysozyme
13. bradykinin
14. diapedesis
15. opsonization
16. pyrogen
17. lymphokines, monokines
18. antigen-binding site, antigenic determinant
19. clonal deletion
20. histiocytes

Chapter 22

1. c
2. c
3. a
4. a
5. e
6. c
7. e
8. d
9. b
10. a

11. glottis
12. bronchial tree
13. pulmonary surfactant
14. intrapleural, atmospheric
15. obstructive
16. anatomic dead space
17. compliance
18. inspiratory center
19. ventilation-perfusion coupling
20. hypocapnia

Chapter 23

1. a
2. d
3. b
4. c
5. b
6. b
7. d
8. e
9. c
10. a
11. micturition
12. renal autoregulation
13. trigone
14. macula densa
15. podocytes
16. transport maximum
17. antidiuretic hormone
18. internal urinary
19. protein
20. arcuate

Chapter 24

1. c
2. a
3. a
4. a
5. d
6. c
7. e
8. b
9. d
10. b
11. Na^+
12. K^+
13. metabolic water
14. cutaneous transpiration
15. fluid sequestration
16. hyperkalemia
17. hyponatremia
18. respiratory acidosis
19. limiting pH
20. osmolarity

Chapter 25

1. b
2. d
3. c
4. e
5. a
6. c
7. a
8. a
9. a
10. a
11. occlusal
12. amylase, lipase
13. parotid
14. enteric
15. vagus
16. gastrin
17. sinusoids
18. maltase, maltose
19. chylomicrons
20. iron

Chapter 26

1. a
2. c
3. b
4. e
5. b
6. e
7. c
8. a
9. d
10. d
11. incomplete
12. glycogenolysis
13. gluconeogenesis
14. urea
15. liver
16. insulin
17. core temperature
18. hypothalamus
19. cytochromes
20. ATP synthase

Chapter 27

1. d
2. a
3. e
4. c
5. a
6. d
7. e
8. c

9. d
10. d
11. mesonephric
12. fructose
13. pampiniform plexus
14. secondary spermatocytes
15. tunica albuginea
16. seminal vesicles
17. sustentacular
18. secondary spermatocytes
19. deep
20. acrosome

Chapter 28

1. a
2. d
3. c
4. a
5. e
6. d
7. b
8. b
9. c
10. c
11. follicle
12. endometrium
13. menarche
14. corpus luteum
15. corona radiata
16. antrum
17. climacteric
18. cervix
19. infundibulum, fimbriae
20. lochia

Chapter 29

1. b
2. b
3. c
4. c
5. a
6. e
7. c
8. a
9. d
10. d
11. mutagens
12. nondisjunction
13. life span
14. decidual
15. chorionic villi
16. acrosome
17. collagen
18. Down syndrome
19. foramen ovale
20. embryo

Understanding Biomedical Vocabulary

One of the greatest challenges confronted by students of anatomy and physiology is the scientific vocabulary. Biomedical science has long had an especially large vocabulary of technical terms, which continues to grow to match the fast pace of scientific discovery. *Stedman's Medical Dictionary,* for example, added 12,000 new terms from 1990 to 1995.

The purpose of this appendix is to share with you some basic skills that can help you feel more comfortable with the technical language of medicine. People who find scientific terms confusing and difficult to pronounce, spell, and remember usually feel more confident once they realize the logic of how terms are composed. The editors of *Stedman's* point out that about 90% of scientific terms are composed of various combinations of just 1,200 Greek and Latin word elements. A term such as *hyponatremia* is less forbidding once we recognize that it is composed of three common word elements: *hypo-* (below normal), *natr-* (sodium), and *-emia* (blood condition). Thus, hyponatremia is a deficiency of sodium in the blood. Those three word elements appear over and over in many other terms in this book: *hypothermia, natriuretic, anemia,* and so on. Once you learn the meanings of *hypo-, natri-,* and *-emia,* you already have the tools at least to partially understand hundreds of other biomedical terms.

This appendix explains how scientific words are composed. Inside the back cover, you will find a lexicon of the 400 word elements most commonly footnoted in this book. The more familiar you are with these principles of etymology (the study of word origins) and this list of elements, the more easily you can pronounce, spell, define, and remember the terms you encounter in this and other biomedical textbooks. This knowledge can also enhance your ability to communicate with your peers and supervisors in the health professions.

Basic Word Elements

Scientific terms are typically composed of one or more of the following elements:

- At least one **root,** or **stem,** that bears the core meaning of the word. In *cardiology,* for example, the root is *card-* (heart). Many words have two or more roots. In *cytochrome,* the roots are *cyt-* (cell) and *chrom-* (color).

- **Combining vowels** that are often inserted between roots to join them and make the word easier to pronounce. In *cytochrome,* for example, the letter *o* is a combining vowel. Although *o* is the most common combining vowel, all vowels of the alphabet are used in this way, such as *a* in *ligament, e* in *vitreous,* the first *i* in *spermicidal, u* in *ovulation,* and *y* in *tachycardia.* Some words, however, have no combining vowel. *Periosteum,* for example, does not need one because the vowels that belong to the roots already join them together in an easy-to-pronounce form. A combination of a root and combining vowel is called a **combining form:** for example, *ost* (bone) + *e* (a combining vowel) make the combining form *oste-* (also interpreted "bone"), as in *osteology.*

- A **prefix** may be present to modify the core meaning of the word. For example, *gastric* (pertaining to the stomach or to the belly of a muscle) takes on a wide variety of new meanings when prefixes are added to it: *epigastric* (above the stomach), *hypogastric* (below the stomach), *endogastric* (within the stomach), and *digastric* (a muscle with two bellies).

- A **suffix** may be added to the end of a word to modify its core meaning. For example, *microscope, microscopy, microscopic,* and *microscopist* have different meanings because of their suffixes alone. Often two or more suffixes, or a root and suffix, occur so commonly together that they are treated jointly as a **compound suffix;** for example, *log* (study) + *y* (process) form the compound suffix *-logy* (the study of).

To summarize these basic principles, consider the word *gastroenterology,* a branch of medicine dealing with the stomach and small intestine. It breaks down into:

gastr/o/enter/o/log/y

gastro = a combining form meaning "stomach"
entero = a combining form meaning "small intestine"
logy = a compound suffix meaning "the study of"

Obscure Word Origins

Before you get *too* excited about your new insight, you should be warned that the literal translation of a word doesn't always provide great insight into its modern meaning. The history of language is full of twists and turns that are fascinating in their own right and say much about the history of the whole of human culture, but they can create confusion for students. For example, the *amnion* is a transparent sac that forms around the developing fetus. The word is derived from *amnos*, from the Greek for "lamb." From this origin, *amnos* came to mean a bowl for catching the blood of sacrificial lambs, and from there the word found its way into biomedical usage for the membrane that emerges (quite bloody) as part of the afterbirth. The *acetabulum*, the socket of the hip joint, literally means "vinegar cup." Apparently the hip socket reminded early anatomists of the little cups used to serve vinegar as a condiment on dining tables in ancient Rome. The word *testicles* literally means "little witnesses"; the history of medical vocabulary has several amusing conjectures as to why this word was chosen to name the male gonads.

Many anatomical terms and medical conditions are named in honor of authorities in the field, such as *Alzheimer disease* and the *fallopian tubes*. Terms named after people are called **eponyms** (from *epo* = after, related to + *nym* = name). These, too, afford little clue as to what a structure or condition is. There is an increasing trend in medical education to avoid eponyms in favor of more descriptive terms (such as *uterine tubes* instead of *fallopian tubes*). This book avoids eponyms when possible.

Another case where an attempt at literal translation may lead you to a dead end is in **acronyms**—words composed of the first letter, or first few letters, of a series of words. For example, the word *calmodulin* is cobbled together from a few letters of each of three words: *calcium modulating protein*.

Words from Other Languages

Although Latin and Greek words form the basis for most biomedical terms, it is important not to overlook the contributions of other cultures. Many of the most important medical advances in the Middle Ages were made in the Muslim world, and such modern terms as *alcohol, alkaline, chemistry,* and *sugar* are Arabic in origin. Some medical terms are from modern foreign lan-

guages, such as *treppe* from German and *tic douloureux* from French. Some are derived from the languages of the aborigines of various continents, such as *curare* from the language of the Carib Indians and *kwashiorkor* from the Bantu language of East Africa.

Singular and Plural Noun Terminals

A point of confusion for many beginning students is to recognize the plural forms of medical terms. Few people would fail to recognize that *ovaries* is the plural of *ovary*, but the connection is harder to make in some other cases: for example, the plural of *cortex* is *cortices* (COR-ti-sees), the plural of *corpus* is *corpora*, and the plural of *epididymis* is *epididymides* (EP-ih-DID-ih-MID-ees). The following reference table will help you make the connection between common singular and plural noun terminals.

Singular and Plural Forms of Some Noun Terminals		
Singular Ending	**Plural Ending**	**Examples**
-a	-ae	axilla, axillae
-ax	-aces	thorax, thoraces
-en	-ina	lumen, lumina
-ex	-ices	cortex, cortices
-is	-es	diagnosis, diagnoses
-is	-ides	epididymis, epididymides
-ix	-ices	appendix, appendices
-ma	-mata	carcinoma, carcinomata
-on	-a	ganglion, ganglia
-um	-a	septum, septa
-us	-era	viscus, viscera
-us	-i	villus, villi
-us	-ora	corpus, corpora
-x	-ges	phalanx, phalanges
-y	-ies	ovary, ovaries
-yx	-ices	calyx, calices

glossary

The 1,000 terms defined here are not necessarily the most important ones in the book, but are terms that are reintroduced most often and, for lack of space, are not redefined each time they arise. The index indicates where you can find definitions or explanations of additional terms. Terms are defined only in the sense that they are used in this book. Some have broader meanings, even within biology and medicine, that are beyond its scope. Terms that are commonly abbreviated, such as *ATP* and *PET scan,* are defined under the full spelling. See the list of abbreviations inside the front cover for complete spellings. The glossary gives pronunciation guides for many terms. Syllables in capital letters should be accented, and common letter sequences should be pronounced as shown in the following list.

ah	as in father	err	as in merry	na	as in corona	ruh	as in rugby
al	as in pal	fal	as in fallacy	nerr	as in nary	serr	as in serration
ay	as in day	few	as in fuse	new	as in news	sterr	as in stereo
bry	as in bribe	ih	as in fit	nuh	as in nothing	sy	as in siren
byu	as in bureau	iss	as in sister	odj	as in dodger	terr	as in terrain
c	as in calculus	lerr	as in lair	oe	as in go	tirr	as in tyranny
cue	as in ridiculous	lur	as in learn	oh	as in home	thee	as in theme
cuh	as in cousin	ma	as in man	ol	as in alcohol	uh	as in mother
cul	as in bicycle	mah	as in mama	oll	as in doll	ul	as in bicycle
cus	as in custard	me	as in meat	ose	as in gross	verr	as in very
dew	as in dual	merr	as in merry	oss	as in floss	y	as in why
eez	as in ease	mew	as in music	perr	as in pair	zh	as in measure
eh	as in feather	muh	as in mother	pew	as in pewter	zy	as in enzyme

A

abdominal cavity　The body cavity between the diaphragm and pelvic brim. fig. A.12

abduction (ab-DUC-shun)　Movement of a body part away from the midsagittal plane, as in raising an arm away from the side of the body. fig. 10.10

absorption　**1.** Process in which a chemical passes through a membrane or tissue surface and becomes incorporated into a body fluid or tissue. **2.** Any process in which one substance passes into another and becomes a part of it. *Compare* adsorption.

acetate　A two-carbon carboxylic acid; the ionized form of acetic acid (CH_3COO^-). The monomer of fatty acids and the intermediate of aerobic metabolism that enters the citric acid cycle.

acetylcholine (ASS-eh-till-CO-leen) **(ACh)**　A neurotransmitter released by somatic motor fibers, parasympathetic fibers, and some other neurons, composed of choline and an acetyl group. fig. 13.17

acetylcholinesterase (ASS-eh-till-CO-lin-ESS-ter-ase) **(AChE)**　An enzyme that hydrolyzes acetylcholine, thus halting signal transmission at a cholinergic synapse.

acid　A proton (H^+) donor; a chemical that releases protons into solution.

acidosis　An acid-base imbalance in which the blood pH is lower than 7.35.

acinus (ASS-ih-nus)　A sac of secretory cells at the inner end of a gland duct. fig. 6.20

acquired immunodeficiency syndrome (AIDS)　A group of conditions that indicate severe immunosuppression related to infection with the human immunodeficiency virus (HIV); typically characterized by a very low T_4 lymphocyte count and high susceptibility to certain forms of cancer and opportunistic infections.

actin　A filamentous intracellular protein that provides cytoskeletal support and interacts with other proteins, especially myosin, to cause cellular movement; important in muscle contraction, ciliary and flagellar beating, and membrane actions such as phagocytosis, ameboid movement, and cytokinesis.

action　The movement produced by the contraction of a particular muscle.

action potential　A rapid voltage change in which a plasma membrane briefly reverses electrical polarity; has a self-propagating effect that produces a traveling wave of excitation in nerve and muscle cells.

active site　The region of a protein that binds to a ligand, such as the substrate-binding site of an enzyme or the hormone-binding site of a receptor.

active transport　The movement of a solute through a cellular membrane, against its concentration gradient, involving a carrier protein that expends ATP.

acute　Pertaining to a disease with abrupt onset, intense symptoms, and short duration. *Compare* chronic.

adaptation　**1.** An evolutionary process leading to the establishment of species characteristics that favor survival and reproduction. **2.** Any characteristic of anatomy, physiology, or behavior that promotes survival and reproduction. **3.** A sensory process in which a receptor adjusts its sensitivity or response to the prevailing level of stimulation, such as dark adaptation of the eye.

adduction (ah-DUC-shun)　Movement of a body part toward the midsagittal plane, such as bringing the feet together from a spread-legged position. fig. 10.10

adenine (AD-eh-neen)　A double-ringed nitrogenous base (purine) found in such molecules as DNA, RNA, and ATP; one of the four bases of the genetic code; complementary to thymine in the double helix of DNA. fig. 5.3

adenosine (ah-DEN-oh-seen) **triphosphate (ATP)**　A molecule composed of adenine, ribose, and three phosphate groups that functions as a universal energy-transfer molecule; yields adenosine diphosphate (ADP) and an inorganic phosphate group (P_i) upon hydrolysis.

adenylate cyclase (ah-DEN-ih-late SY-clase)　An enzyme of the plasma membrane that removes two phosphate molecules from ATP and makes cyclic adenosine monophosphate (cAMP); important in the activation of the cAMP second messenger system.

adipocyte (AD-ih-po-site)   A fat cell.

adipose tissue   A connective tissue composed predominantly of adipocytes; fat.

adrenal (ah-DREE-nul) **gland**   An endocrine gland on the superior pole of each kidney. fig. 17.13

adrenergic (AD-reh-NUR-jic)   Pertaining to epinephrine (adrenalin) or norepinephrine (noradrenalin), as in adrenergic neurons that secrete one of these chemicals or adrenergic effects on a target organ.

adrenocorticotropic (ah-DREE-no-COR-tih-co-TRO-pic) **hormone (ACTH)**   A hormone secreted by the anterior pituitary gland that stimulates the adrenal cortex.

adsorption   The binding of one substance to the surface of another without becoming a part of the latter. *Compare* absorption.

adventitia (AD-ven-TISH-uh)   Loose fibroconnective tissue forming the outermost sheath around organs such as a blood vessel or the esophagus. fig. 20.2

aerobic (air-OH-bic) **exercise**   Exercise in which oxygen is used to produce ATP; endurance exercise.

aerobic respiration   Oxidation of organic compounds in a reaction series that requires oxygen and produces ATP.

afferent (AFF-uh-rent)   Carrying toward, as in *afferent neurons,* which carry signals toward the central nervous system, and *afferent arterioles,* which carry blood toward a tissue.

afterload   The force exerted by arterial blood pressure that opposes the openings of the aortic and pulmonary valves of the heart.

agglutination (ah-GLUE-tih-NAY-shun)   Clumping of cells or molecules by antibodies. fig. 18.13

aging   Any changes in the body that occur with the passage of time, including growth, development, and senescence.

agonist   *See* prime mover.

agranulocyte   Either of the two leukocyte types (lymphocytes and monocytes) that lack prominent cytoplasmic granules.

albumin (al-BYU-min)   A class of small proteins constituting about 60% of the protein fraction of the blood plasma; plays roles in blood viscosity, colloid osmotic pressure, and solute transport.

aldosterone (AL-doe-steh-RONE, al-DOSS-teh-rone)   A steroid hormone secreted by the adrenal cortex that acts on the kidneys to promote sodium retention and potassium excretion.

alkalosis   An acid-base imbalance in which the blood pH is higher than 7.45.

allele (ah-LEEL)   Any of the alternative forms that one gene can take, such as dominant and recessive alleles.

all-or-none law   A statement pertaining to action potentials and the contraction of

muscle fibers that a cell either produces its maximum response or no response at all, depending on whether the stimulus is above or below threshold.

alveolus (AL-vee-OH-lus)   **1.** A microscopic air sac of the lung. **2.** A gland acinus. **3.** A tooth socket. **4.** Any small anatomical space.

Alzheimer (ALTS-hy-mur) **disease (AD)**   A degenerative disease of the senescent brain, typically beginning with memory lapses, progressing to severe losses of mental and motor functions and ultimately death.

ameboid (ah-ME-boyd) **movement**   Movement of a cell by means of pseudopods, in a manner similar to that of an ameba; seen in leukocytes and some macrophages.

amino acids   Small organic molecules with an amino group and a carboxyl group; the monomers of which proteins are composed.

amino group   A functional group with the formula —NH_2, found in amino acids and some other organic molecules.

amitotic (AY-my-TOT-ic)   Incapable of mitosis, as in neurons.

ampulla (am-PULL-uh)   A wide or saclike portion of a tubular organ such as a semicircular duct or uterine tube.

anabolism (ah-NAB-oh-lizm)   Any metabolic reactions that consume energy and construct more complex molecules with higher free energy from less complex molecules with lower free energy; for example, the synthesis of proteins from amino acids. *Compare* catabolism.

anaerobic (AN-err-OH-bic) **fermentation**   A reduction reaction independent of oxygen that converts pyruvic acid to lactic acid and enables glycolysis to continue under anaerobic conditions.

anaphylactic shock   A severe systemic form of anaphylaxis involving bronchiolar constriction, impaired breathing, vasodilation, and a rapid drop in blood pressure with a threat of circulatory failure.

anaphylaxis (AN-uh-fih-LAC-sis)   A form of immediate hypersensitivity in which an antigen triggers the release of inflammatory chemicals, causing edema, congestion, hives, and other, usually local, symptoms.

anastomosis (ah-NASS-tih-MO-sis)   An anatomical convergence, the opposite of a branch; a point where two blood vessels merge and combine their bloodstreams or where two nerves or ducts converge. fig. 20.1

anatomical position   A reference posture that allows for standardized anatomical terminology. A subject in anatomical position is standing with the feet flat on the floor and slightly apart, arms down to the sides, and the palms and eyes directed forward. fig. A.1

anatomy   **1.** Structure of the body. **2.** The study of structure.

androgen (AN-dro-jen)   Testosterone or a related steroid hormone. Stimulates somatic changes at puberty in both sexes, adult libido in both sexes, development of male anatomy in the fetus and adolescent, and spermatogenesis.

anemia (ah-NEE-me-uh)   A deficiency of erythrocytes or hemoglobin.

aneurysm (AN-you-rizm)   A weak, bulging point in the wall of a heart chamber or blood vessel, presenting a threat of hemorrhage.

angiogenesis (AN-jee-oh-GEN-eh-sis)   The growth of new blood vessels.

angiotensin (AN-jee-oh-TEN-sin) **II**   A hormone produced from angiotensinogen (a plasma protein) by the kidneys and lungs; raises blood pressure by stimulating vasoconstriction and stimulating the adrenal cortex to secrete aldosterone.

anion (AN-eye-on)   An ion with more electrons than protons and consequently a net negative charge.

antagonist   **1.** A muscle that opposes the agonist at a joint. **2.** Any agent, such as a hormone or drug, that opposes another.

antebrachium (AN-teh-BRAY-kee-um)   The region from elbow to wrist; the forearm.

anterior   Pertaining to the front (facial-abdominal aspect) of the body; ventral.

antibody   A protein of the gamma globulin class that reacts with an antigen; found in the blood plasma, in other body fluids, and on the surfaces of certain leukocytes and their derivatives.

anticoagulant (AN-tee-co-AG-you-lent)   A chemical agent that opposes blood clotting.

antidiuretic (AN-tee-DYE-you-RET-ic) **hormone (ADH)**   A hormone released by the posterior pituitary gland in response to low blood pressure; promotes water retention by the kidneys. Also known as *vasopressin.*

antigen (AN-tih-jen)   Any large molecule capable of binding to an antibody and triggering an immune response.

antigen-presenting cell (APC)   A cell that phagocytizes an antigen and displays fragments of it on its surface for recognition by other cells of the immune system; chiefly macrophages and B lymphocytes.

antioxidant   A chemical that binds and neutralizes free radicals, minimizing their oxidative damage to a cell; for example, selenium and vitamin E.

antiport   An active transport carrier that moves two or more solutes in opposite directions through a cellular membrane; for example, the Na^+–K^+ pump.

aorta   A large artery that extends from the left ventricle to the lower abdominal cavity and gives rise to all arteries of the systemic circulation. fig. 20.21

apical surface   The uppermost surface of an epithelial cell, usually exposed to the lumen of an organ. fig. 4.4

apocrine　Pertaining to certain sweat glands with large lumens and relatively thick, aromatic secretions and to similar glands such as the mammary gland; formerly thought to form secretions by pinching off bits of apical cytoplasm.

apoptosis (AP-oh-TOE-sis)　Programmed cell death; the normal death of cells that have completed their function. *Compare* necrosis.

appendicular (AP-en-DIC-you-lur)　Pertaining to the extremities and their supporting skeletal girdles. fig. 8.1

aqueous humor　A serous fluid between the cornea and lens of the eye.

arcuate (AR-cue-et)　Making a sharp L- or U-shaped bend (arc), as in the *arcuate arteries* of the kidneys and uterus.

areolar (AIR-ee-OH-lur) **tissue**　A fibro-connective tissue with loosely organized, widely spaced fibers and cells and an abundance of fluid-filled space; found under nearly every epithelium, among other places. fig. 6.6

arrector pili (ah-REC-tur PY-lye) (plural, *arrectores pilorum;* ah-rec-TORE-eez py-LOR-um)　A bundle of smooth muscle cells associated with a hair follicle, responsible for erection of the hair. fig. 7.9

arrhythmia (ah-RITH-me-uh)　An irregularity in the cardiac rhythm.

arteriole (ar-TEER-ee-ole)　A small artery that empties into a metarteriole or capillary.

artery　Any blood vessel that conducts blood away from the heart.

articular cartilage　A thin layer of hyaline cartilage covering the articular surface of a bone at a synovial joint, serving to reduce friction and ease joint movement. fig. 8.3

articulation　A skeletal joint; any point at which two bones meet. May or may not be movable.

ascorbic acid　Vitamin C; a dietary antioxidant.

aspect　A particular view of the body or one of its structures, or a part that faces in a particular direction, such as the anterior aspect.

atheroma (ATH-ur-OH-muh)　A fatty deposit (plaque) in a blood vessel, consisting of lipid, smooth muscle, and macrophages; characteristic of atherosclerosis. fig. E.1, p. 700

atherosclerosis (ATH-ur-oh-skleh-ROE-sis)　A degenerative disease of the blood vessels characterized by the presence of atheromas and often leading to calcification of the vessel wall.

atrial natriuretic (AY-tree-ul NAY-tree-you-RET-ic) **factor (ANF)**　A hormone secreted by the heart that lowers blood pressure by promoting sodium excretion and antagonizing aldosterone.

atrioventricular (AY-tree-oh-ven-TRIC-you-lur) **(AV) node**　A group of autorhythmic cells in the interatrial septum of the heart that relays excitation from the atria to the ventricles.

atrioventricular (AV) valves　The bicuspid (right) and tricuspid (left) valves between the atria and ventricles of the heart.

atrophy (AT-ruh-fee)　Shrinkage of a tissue due to age, disuse, or disease.

auditory ossicles　Three small middle-ear bones that transfer vibrations from the tympanic membrane to the inner ear; the malleus, incus, and stapes.

autoimmune disease　Any disease in which antibodies fail to distinguish between foreign and self-antigens and attack the body's own tissues; for example, systemic lupus erythematosus and rheumatic fever.

autolysis (aw-TOLL-ih-sis)　Digestion of cells by their own internal enzymes.

autonomic (AW-toe-NOM-ic) **nervous system (ANS)**　A division of the nervous system that innervates glands, smooth muscle, and cardiac muscle; consists of sympathetic and parasympathetic divisions and functions largely without voluntary control. *Compare* somatic nervous system.

autoregulation　The ability of a tissue to control its own blood supply through vasomotion or angiogenesis.

autorhythmic (AW-toe-RITH-mic)　Pertaining to cells that spontaneously produce action potentials at regular time intervals, chiefly cardiac and smooth muscle cells.

autosome (AW-toe-some)　Any chromosome except the sex chromosomes. Genes on the autosomes are inherited without regard to the sex of the individual.

axial (AC-see-ul)　Pertaining to the head, neck, and trunk; the part of the body excluding the appendicular portion.

axillary (ACK-sih-LERR-ee)　Pertaining to the armpit.

axolemma　The plasma membrane of a nerve fiber.

axon　A process of a neuron that transmits action potentials; also called a *nerve fiber*. There is only one axon to a neuron, and it is usually much longer and much less branched than the dendrites. fig. 13.4

axoneme (AC-so-neem)　The core of microtubules, usually in a "9 + 2" array, at the center of a cilium or flagellum. fig. 4.9

B

baroreceptors (BAR-oh-re-SEP-turz)　Pressure sensors located in the heart, aortic arch, and carotid sinuses that trigger autonomic reflexes in response to fluctuations in blood pressure.

basal metabolic rate (BMR)　The rate of energy consumption of a person who is awake, relaxed, at a comfortable temperature, and has not eaten for 12 to 14 hours; usually expressed as kilocalories per square meter of body surface area per hour.

basal nuclei　Masses of deep cerebral gray matter that play a role in the coordination of posture and movement. fig. 14.22

base　**1.** A chemical that binds protons from solution; a proton acceptor. **2.** Any of the purines or pyrimidines of a nucleic acid (adenine, thymine, guanine, cytosine, or uracil), serving in part to code for protein structure. **3.** The broadest part of a tapered organ such as the uterus or the inferior aspect of an organ such as the brain.

basement membrane　A thin layer of glycoproteins, collagen, and glycosaminoglycans beneath the deepest cells of an epithelium, serving to bind the epithelium to the underlying tissue. fig. 6.23

base triplet　A sequence of three DNA nucleotides that codes indirectly (through mRNA) for one amino acid of a protein.

basophil (BAY-so-fill)　A granulocyte with coarse cytoplasmic granules that produces heparin, histamine, and other chemicals involved in inflammation. fig. 18.17

belly　The thick part of a skeletal muscle between its origin and insertion. fig. 11.3

bicarbonate buffer system　An equilibrium mixture of carbonic acid, bicarbonate ions, and hydrogen ions ($H_2CO_3 \leftrightarrow HCO_3^- + H^+$) that stabilizes the pH of the body fluids.

bicarbonate ion　An anion, HCO_3^-, that functions as a base in the buffering of body fluids.

bile　A secretion produced by the liver, concentrated and stored in the gallbladder, and released into the small intestine; consists mainly of wastes such as excess cholesterol, salts, and bile pigments but also contains lecithin and bile salts, which aid in fat digestion.

bile pigments　Strongly colored organic compounds produced by the breakdown of hemoglobin, including biliverdin and bilirubin.

bilirubin (BIL-ih-ROO-bin)　A yellow to orange bile pigment produced by the breakdown of hemoglobin and excreted in the bile; causes jaundice and neurotoxic effects if present in excessive concentration.

biogenic amines　A class of chemical messengers with neurotransmitter and hormonal functions, synthesized from amino acids and retaining an amino group; also called *monoamines*. Examples include epinephrine and thyroxine.

bipedalism　The habit of walking on two legs; a defining characteristic of the family Hominidae that underlies many skeletal and other characteristics of humans.

blood-brain barrier (BBB)　A barrier between the bloodstream and nervous tissue of the brain that is impermeable to many blood solutes and thus prevents them from affecting the brain tissue; formed by the tight junctions between capillary endothelial cells, the basement membrane of the endothelium, and the perivascular feet of astrocytes.

B lymphocyte A lymphocyte that functions as an antigen-presenting cell and, in humoral immunity, differentiates into an antibody-producing plasma cell; also called a *B cell.*

body 1. The entire organism. 2. Part of a cell, such as a neuron, containing the nucleus and most other organelles. 3. The largest or principal part of an organ such as the stomach or uterus; also called the *corpus.*

bolus A mass of matter, especially food or feces traveling through the digestive tract.

bone 1. A calcified connective tissue; also called *osseous tissue.* 2. An organ of the skeleton composed of osseous tissue, fibroconnective tissue, marrow, cartilage, and other tissues.

Bowman's capsule *See* glomerular capsule.

brachial (BRAY-kee-ul) Pertaining to the arm proper, the region from shoulder to elbow.

bradykinin (BRAD-ee-KY-nin) An oligopeptide produced in inflammation that stimulates vasodilation, increases capillary permeability, and stimulates pain receptors.

brainstem The stalklike lower portion of the brain, composed of all of the brain except the cerebrum and cerebellum. (Many authorities also exclude the diencephalon and regard only the medulla oblongata, pons, and midbrain as the brainstem.) fig. 14.13

bronchiole (BRONK-ee-ole) A pulmonary air passage that is usually 1 mm or less in diameter and lacks cartilage, but has relatively abundant smooth muscle, elastic tissue, and a simple cuboidal, usually ciliated epithelium.

bronchus (BRONK-us) A relatively large pulmonary air passage with supportive cartilage in the wall; any passage beginning with the primary bronchus at the fork in the trachea and ending with tertiary bronchi, from which air continues into the bronchioles.

brush border A fringe of microvilli on the apical surface of an epithelial cell, serving to enhance surface area and promote absorption. fig. 25.23

buffer A mixture of chemicals that resists changes in pH when acid or base is added to the solution.

bulk transport The movement of particles or fluid droplets through the plasma membrane by the process of endocytosis or exocytosis.

bursa A sac filled with synovial fluid at a diarthrosis, serving to facilitate muscle or joint action. fig. 10.21

C

calcaneal (cal-CAY-nee-ul) **tendon** A thick tendon at the heel that attaches the triceps surae muscles to the calcaneus; also called the *Achilles tendon.* fig. 11.38

calcification The hardening of a tissue due to the deposition of calcium salts.

calcitonin (CAL-sih-TOE-nin) A hormone secreted by C cells of the thyroid gland that promotes calcium deposition in the skeleton and lowers blood calcium concentration.

calmodulin An intracellular protein that binds calcium ions and mediates many of the second messenger effects of calcium.

calorie The amount of thermal energy that will raise the temperature of 1 g of water by 1°C. Also called a *small calorie.*

Calorie *See* kilocalorie.

calorigenic (ca-LOR-ih-JEN-ic) Heat-producing, as in the calorigenic effect of thyroid hormone.

calsequestrin A protein found in smooth endoplasmic reticulum that reversibly binds and stores calcium ions, rendering calcium chemically unreactive until needed for such processes as muscle contraction.

calvaria (cal-VERR-ee-uh) The rounded bony dome that forms the roof of the cranium; skullcap.

calyx (CAY-lix) (plural, *calices*) A cuplike structure, as in the kidneys. fig. 23.4

canaliculus (CAN-uh-LIC-you-lus) A microscopic canal, as in osseous tissue. fig. 8.7

capillary (CAP-ih-LERR-ee) The narrowest type of vessel in the cardiovascular and lymphatic systems; engages in fluid exchanges with surrounding tissues.

capillary exchange The process of fluid transfer between the bloodstream and tissue fluid.

capsule The fibrous covering of a structure such as the spleen or a diarthrosis.

carbohydrate A hydrophilic organic compound composed of carbon and a 2:1 ratio of hydrogen to oxygen; includes sugars, starches, glycogen, and cellulose.

carbonic anhydrase An enzyme found in erythrocytes and kidney tubule cells that catalyzes the decomposition of carbonic acid into carbon dioxide and water or the reverse reaction ($H_2CO_3 \leftrightarrow CO_2 + H_2O$).

carboxyl (car-BOC-sil) **group** An organic functional group with the formula —COOH, found in many organic acids such as amino acids and fatty acids.

carcinogen (car-SIN-oh-jen) An agent capable of causing cancer, including certain chemicals, viruses, and ionizing radiation.

cardiac center A neuronal pool in the medulla oblongata that regulates autonomic reflexes for controlling the rate and strength of the heartbeat.

cardiac cycle One complete cycle of cardiac systole and diastole.

cardiac muscle Striated involuntary muscle of the heart.

cardiac output (CO) The amount of blood pumped by each ventricle of the heart in 1 minute.

cardiac reserve The difference between maximum and resting cardiac output, which determines a person's tolerance for exercise.

cardiovascular system An organ system consisting of the heart and blood vessels, serving for the transport of blood. *Compare* circulatory system.

carotid (ca-ROT-id) **body** A small cellular mass immediately superior to the branch in the common carotid artery, containing sensory cells that detect changes in blood pH and carbon dioxide and oxygen content. fig. 20.12

carotid sinus A dilation of the common carotid artery at the point where it branches into the internal and external carotids; contains baroreceptors, which monitor changes in blood pressure.

carpal Pertaining to the wrist (carpus).

carrier 1. A protein in a cellular membrane that performs carrier-mediated transport. 2. A person who is heterozygous for a recessive allele and does not exhibit the associated phenotype but may transmit this allele to his or her children; for example, a carrier for sickle-cell anemia.

carrier-mediated transport A process of transporting materials through a cellular membrane that involves reversible binding to a membrane protein.

cartilage A connective tissue with a rubbery matrix, cells (chondrocytes) contained in lacunae, and no blood vessels; covers the articular surfaces of many bones and supports organs such as the ear and larynx.

catabolism (ca-TAB-oh-lizm) Any metabolic reactions that release energy and break relatively complex molecules with high free energy into less complex molecules with lower free energy; for example, digestion and glycolysis. *Compare* anabolism.

catalyst (CAT-uh-list) Any chemical that lowers the activation energy of a chemical reaction and thus makes the reaction proceed more rapidly; a role served in cells by enzymes.

catecholamine (CAT-eh-COAL-uh-meen) A subclass of biogenic amines that includes epinephrine, norepinephrine, and dopamine. fig. 13.17

cation (CAT-eye-on) An ion with more protons than electrons and consequently a net positive charge.

caudal (CAW-dul) 1. Pertaining to a tail or narrow tail-like part of an organ. 2. Pertaining to the inferior part of the trunk of the body, where the tail of other animals arises. *Compare* cranial.

celiac (SEE-lee-ac) Pertaining to the abdomen.

cell The smallest subdivision of a tissue considered to be alive; consists of a plasma membrane enclosing cytoplasm and, in most cases, a nucleus.

cellular membrane Any unit membrane enclosing a cell or organelle. *See also* unit membrane.

central Located relatively close to the medial axis of the body, as in the central nervous system; opposite of peripheral.

central nervous system (CNS) The brain and spinal cord.

centriole (SEN-tree-ole) An organelle composed of two short perpendicular cylinders of microtubules; origin of the mitotic spindle. fig. 4.17

cephalic (seh-FAL-ic) Pertaining to the head.

cerebellum (SERR-eh-BEL-um) A large portion of the brain posterior to the brainstem and inferior to the cerebrum, responsible for equilibrium and motor coordination. fig. 14.14

cerebrospinal (SERR-eh-bro-SPY-nul) **fluid (CSF)** A liquid that fills the ventricles of the brain, the central canal of the spinal cord, and the space between the CNS and dura mater.

cerebrovascular (SERR-eh-bro-VASS-cue-lur) **accident (CVA)** The loss of blood flow to any part of the brain due to obstruction or hemorrhage of an artery, leading to necrosis of nervous tissue; also called *stroke* or *apoplexy.*

cerebrum (SERR-eh-brum, seh-REE-brum) The largest and most superior part of the brain, divided into two convoluted cerebral hemispheres separated by a deep longitudinal fissure.

cervical (SUR-vih-cul) Pertaining to the neck or any cervix.

cervix (SUR-vix) **1.** The neck. **2.** A narrow or necklike part of an organ such as the uterus and gallbladder. fig. 28.3

channel protein A protein in the plasma membrane that has a pore through it for the passage of materials between the cytoplasm and extracellular fluid. fig. 4.6

chemical bond A force that attracts one atom to another, such as their opposite charges or the sharing of electrons.

chemical digestion Hydrolysis reactions that occur in the digestive tract and covert dietary polymers into monomers that can be absorbed by the small intestine.

chemical synapse A meeting of a nerve fiber and another cell with which the neuron communicates by releasing neurotransmitters. fig. 13.16

chemoreceptor An organ or cell specialized to detect chemicals, as in the carotid bodies and taste buds.

chemotaxis (KEM-oh-TAC-sis) The movement of a cell along a chemical concentration gradient, especially the attraction of neutrophils to chemicals released by pathogens or inflamed tissues.

chief cells The majority type of cell in an organ or tissue such as the parathyroid glands or gastric glands.

choanae (co-AH-nee) Openings of the nasal cavity into the pharynx; also called *posterior nares.* fig. 22.3

cholecystokinin (CO-lee-SIS-toe-KY-nin) **(CCK)** A polypeptide employed as a hormone and neurotransmitter, secreted by some brain neurons and cells of the digestive tract.

cholesterol (co-LESS-tur-ol) A steroid that functions as part of the plasma membrane and as a precursor for all other steroids in the body.

cholinergic (CO-lin-UR-jic) Pertaining to acetylcholine (ACh), as in cholinergic nerve fibers that secrete ACh, cholinergic receptors that bind it, or cholinergic effects on a target organ.

chondrocyte (CON-dro-site) A cartilage cell; a former chondroblast that has become enclosed in a lacuna in the cartilage matrix. fig. 8.11

chorion (CO-ree-on) A fetal membrane external to the amnion; forms part of the placenta and has diverse functions including fetal nutrition, waste removal, and hormone secretion. fig. 29.11

chromatid (CRO-muh-tid) One of two genetically identical rodlike bodies of a metaphase chromosome, joined to its sister chromatid at the centromere. fig. 5.16

chromatin (CRO-muh-tin) Filamentous nuclear material composed of DNA and associated proteins.

chromosome A condensed form of chromatin visible in a cell undergoing mitosis or meiosis. fig. 5.16

chronic **1.** Long-lasting. **2.** Pertaining to a disease that progresses slowly and has a long duration. *Compare* acute.

chronic bronchitis A chronic obstructive pulmonary disease characterized by damaged and immobilized respiratory cilia, excessive mucus secretion, infection of the lower respiratory tract, and bronchial inflammation; caused especially by cigarette smoking. *See also* chronic obstructive pulmonary disease.

chronic obstructive pulmonary disease (COPD) A group of lung diseases (asthma, chronic bronchitis, and emphysema) that result in long-term obstruction of airflow and substantially reduced pulmonary ventilation; one of the leading causes of death in old age.

chylomicron (KY-lo-MY-cron) A protein-coated lipid droplet formed in the small intestine and found in the lymph and blood after a meal; a means of lipid transport in the bloodstream and lymph.

chyme (kime) A slurry of partially digested food in the stomach and small intestine.

cilium (SIL-ee-um) A hairlike process, with an axoneme, projecting from the apical surface of an epithelial cell; usually motile and serving to propel matter across the surface of an epithelium but sometimes serving sensory roles. fig. 4.10

circulatory shock A state of cardiac output inadequate to meet the metabolic needs of the body.

circulatory system An organ system consisting of the heart, blood vessels, and blood. *Compare* cardiovascular system.

circumduction A joint movement in which one end of an appendage remains relatively stationary and the other end is moved in a circle. fig. 10.12

cirrhosis (sih-RO-sis) A degenerative liver disease characterized by replacement of functional parenchyma with fibrous and adipose tissue; causes include alcohol, other poisons, and viral and bacterial inflammation.

cisterna (sis-TUR-nuh) A fluid-filled space or sac, such as the cisterna chyli of the lymphatic system and a cisterna of the endoplasmic reticulum or Golgi complex. fig. 4.12

citric acid cycle A cyclic reaction series involving several carboxylic acids in the mitochondrial matrix; oxidizes acetyl groups to carbon dioxide while reducing NAD^+ to NADH and FADH to $FADH_2$, making these reduced coenzymes available for ATP synthesis. Also called the *Krebs cycle* or *tricarboxylic acid (TCA) cycle.* fig. 26.4

climacteric A period in the lives of men and women, usually in the late 40s to early 50s, marked by pronounced hormonal changes, a variety of somatic and psychological effects, and in women, cessation of ovulation and menstruation (menopause).

clone A population of cells that are mitotically descended from the same parent cell and are identical to each other genetically or in other respects.

coagulation (co-AG-you-LAY-shun) The clotting of blood, lymph, tissue fluid, or semen.

codominant (co-DOM-ih-nent) A condition in which neither of two alleles is dominant over the other, and both are phenotypically expressed when both are present in an individual. For example, blood type alleles I^A and I^B produce blood type AB when inherited together.

codon A series of three nucleotides in mRNA that codes for one amino acid in a protein or signals the end of a gene.

coenzyme (co-EN-zime)   A small organic molecule, usually derived from a vitamin, that is needed to make an enzyme catalytically active; acts by accepting electrons from an enzymatic reaction and transferring them to a different reaction chain.

cofactor   A metal ion that binds to an enzyme and activates its catalytic function.

cohesion   The clinging of identical molecules such as water to each other.

collagen (COLL-uh-jen)   The most abundant protein in the body, forming the fibers of many connective tissues in places such as the dermis, tendons, and bones.

colloid   An aqueous mixture of particles that are too large to pass through most selectively permeable membranes but small enough to remain evenly dispersed through the solvent by the thermal motion of solvent particles; for example, the proteins in blood plasma.

colloid osmotic pressure (COP)   A portion of the osmotic pressure of a body fluid that is due to its protein. *Compare* oncotic pressure.

colostrum (co-LOS-trum)   A watery, low-fat secretion of the mammary gland that nourishes and immunizes an infant for the first 2 to 3 days postpartum, until true milk is secreted.

commissure (COM-ih-shur)   **1.** A bundle of nerve fibers that crosses from one side of the brain or spinal cord to the other. **2.** A corner or angle at which the eyelids, lips, or genital labia meet; in the eye, also called the *canthus.* fig. 16.22

complement   **1.** To complete or enhance the structure or function of something else, as in the coordinated action of two different hormones. **2.** A system of plasma proteins involved in nonspecific defense against pathogens.

computerized tomography (CT)   A method of medical imaging that uses X rays and a computer to create an image of a thin section of the body; also called a *CT scan.*

concentration gradient   A gradual change in chemical concentration from one point to another.

conception   The fertilization of an egg, producing a zygote.

conceptus   All products of conception, ranging from a fertilized egg to the full-term fetus with its extraembryonic membranes, placenta, and umbilical cord.

condyle (CON-dile)   A rounded knob on a bone serving to produce smooth motion at a joint. fig. 9.4

conformation   The three-dimensional structure of a protein that results from interaction among its amino acid side groups, its interactions with water, and the formation of disulfide bonds.

congenital   Present at birth; for example, an anatomical defect, a syphilis infection, or a hereditary disease.

conjugated   A state in which one organic compound is bound to another compound of a different class, such as a protein conjugated with a carbohydrate to form a glycoprotein.

connective tissue   A tissue usually composed of more extracellular than cellular volume and usually with a substantial amount of extracellular fiber; forms supportive frameworks and capsules for organs, binds structures together, holds them in place, stores energy (as in adipose tissue), or transports materials (as in blood).

contractility   The amount of force that a contracting muscle fiber generates for a given stimulus; may be increased by epinephrine, for example, while stimulus strength remains constant.

contralateral   On opposite sides of the body, as in reflex arcs where the stimulus comes from one side of the body and a response is given by muscles on the other side.

convergent   Coming together, as in a convergent muscle and a converging neuronal circuit.

cooperative effects   Effects in which two hormones, or both divisions of the autonomic nervous system, work together to produce a single overall result.

cornified   Having a heavy deposit of keratin, as in the stratum corneum of the epidermis.

corona   A halo- or crownlike structure, as in the corona radiata or the coronal suture of the skull.

coronal plane   *See* frontal plane.

corona radiata   **1.** An array of nerve tracts in the brain that arise mainly from the thalamus and fan out to different regions of the cerebral cortex. **2.** The first layer of cuboidal cells immediately external to the zona pellucida around an egg cell.

coronary circulation   A system of blood vessels that serve the wall of the heart. fig. 19.19

corpus   Body or mass; the main part of an organ, as opposed to such regions as a head, tail, or cervix.

corpus callosum (COR-pus ca-LO-sum)   A prominent C-shaped band of nerve tracts that connect the right and left cerebral hemispheres to each other, visible in a midsagittal section of the brain superior to the third ventricle. fig. 14.2

corpus luteum (LOO-tee-um)   A yellowish cellular mass that forms in the ovary from a follicle that has ovulated; secretes progesterone, hormonally regulates the second half of the menstrual cycle, and is essential to sustaining the first 7 weeks of pregnancy.

cortex (plural, *cortices*)   The outer layer of some organs such as the adrenal gland, cerebrum, lymph node, and ovary; usually covers or encloses tissue called the medulla.

corticosteroid (COR-tih-co-STERR-oyd)   Any steroid hormone secreted by the adrenal cortex, such as aldosterone, cortisol, and sex steroids.

costal (COSS-tul)   Pertaining to the ribs.

costal cartilage   A bladelike plate of hyaline cartilage that attaches the distal end of a rib to the sternum.

cotransport   *See* secondary active transport.

countercurrent   A situation in which two fluids flow side by side in opposite directions, as in the countercurrent multiplier of the kidney and the countercurrent heat exchanger of the scrotum.

cranial (CRAY-nee-ul)   **1.** Pertaining to the cranium. **2.** In a position relatively close to the head or a direction toward the head. *Compare* caudal.

cranial nerve   Any of 12 pairs of nerves connected to the base of the brain and passing through foramina of the cranium.

creatine phosphate (CREE-uh-tin FOSS-fate) **(CP)**   An energy storage molecule in muscle tissue that donates a phosphate group to ADP and thus regenerates ATP in periods of hypoxia.

crista   A crestlike structure, such as the crista galli of the ethmoid bone or the crista of a mitochondrion.

cross section   A cut perpendicular to the long axis of the body or an organ.

crural (CROO-rul)   Pertaining to the leg proper or to the crus of a organ. *See* crus.

crus (cruss) (plural, *crura*)   **1.** The leg proper; the region from the knee to the ankle. **2.** A leglike extension of an organ such as the penis and clitoris. fig. 28.8

cuboidal (cue-BOY-dul)   A cellular shape that is roughly like a cube or in which the height and width are about equal.

cumulus oophorus (CUE-mew-lus oh-OFF-ur-rus)   A multilayered aggregation of cells that surround an egg before and after ovulation. fig. 28.13

cuneiform (cue-NEE-ih-form)   Wedge-shaped, as in the cuneiform cartilages of the larynx and cuneiform bone of the wrist.

current   A moving stream of charged particles such as ions or electrons.

current sink   An open channel in a plasma membrane through which ions flow.

cusp   **1.** One of the flaps of a valve of the heart, veins, and lymphatic vessels. **2.** A conical projection on the occlusal surface of a premolar or molar tooth.

cutaneous (cue-TAY-nee-us)   Pertaining to the skin.

cyanosis (SY-uh-NO-sis)   A bluish color of the skin due to ischemia or hypoxemia.

cyclic adenosine monophosphate (cAMP)   A cyclic molecule produced from ATP by the action of adenylate cyclase; serves as a second messenger in many hormonal and neurotransmitter actions.

cyclooxygenase   An enzyme that converts arachidonic acid to prostacyclin, prostaglandins, and thromboxanes.

cytochromes   Enzymes on the mitochondrial cristae that transfer electrons in the final reaction chain of aerobic respiration.

cytokinesis (SY-toe-kih-NEE-sis)   Division of the cytoplasm of a cell into two cells following nuclear division.

cytology   The study of cell structure and function.

cytolysis (sy-TOL-ih-sis)   The rupture and destruction of a cell by such agents as complement proteins and hypotonic solutions.

cytoplasm   The contents of a cell between its plasma membrane and its nuclear envelope, consisting of cytosol, organelles, inclusions, and the cytoskeleton.

cytosine   A single-ringed nitrogenous base (pyrimidine) found in DNA; one of the four bases of the genetic code; complementary to guanine in the double helix of DNA. fig. 5.3

cytoskeleton   A system of protein microfilaments, intermediate filaments, and microtubules in a cell, serving in physical support, cellular movement, and the routing of molecules and organelles to their destinations within the cell. fig. 4.18

cytosol   A clear, featureless, gelatinous colloid in which the organelles and other internal structures of a cell are embedded.

cytotoxic T cell   A T lymphocyte that directly attacks and destroys infected body cells, cancerous cells, and the cells of transplanted tissues.

D

daughter cells   Cells that arise from a parent cell by mitosis or meiosis.

deamination (dee-AM-ih-NAY-shun)   Removal of an amino group from an organic molecule; a step in the catabolism of amino acids.

decomposition reaction   A chemical reaction in which a larger molecule is broken down into smaller ones. *Compare* synthesis reaction.

decussation (DEE-cuh-SAY-shun)   The crossing of nerve fibers from the right side of the central nervous system to the left or vice versa, especially in the spinal cord, medulla oblongata, and optic chiasma.

deep   Relatively far from the body surface; opposite of *superficial*. For example, the bones are deep to the skeletal muscles.

degranulation   Exocytosis and disappearance of cytoplasmic granules, especially in platelets and granulocytes.

dehydration synthesis   A reaction in which two chemical monomers are joined together with water produced as a by-product; also called a *condensation reaction. Compare* hydrolysis.

denaturation   A change in the three-dimensional conformation of a protein that destroys its enzymatic or other functional properties, usually caused by extremes of temperature or pH.

dendrites   Processes of a neuron that receive information from other cells or from environmental stimuli and conduct signals to the soma. Dendrites are usually shorter, more branched, and more numerous than the axon and are incapable of producing action potentials. fig. 13.4

denervation atrophy   The shrinkage of skeletal muscle that occurs when the motor neuron dies or is severed from the muscle.

dense connective tissue   A fibroconnective tissue with a high density of fiber, relatively little ground substance, and scanty cells; seen in tendons and the dermis, for example.

deoxyribonucleic (dee-OCK-see-RY-bo-new-CLAY-ic) **acid (DNA)**   A very large nucleotide polymer that carries the genes of a cell; composed of a double helix of intertwined chains of deoxyribose and phosphate, with complementary pairs of nitrogenous bases facing each other between the helices. fig. 5.4

depolarization   A shift in the electrical potential across a plasma membrane toward 0 mV, associated with excitation of a nerve or muscle cell. *Compare* hyperpolarization.

dermal papillae   Bumps or ridges of dermis that extend upward to interdigitate with the epidermis, creating a wavy boundary that resists stress and slippage of the epidermis.

dermis   The deeper of the two layers of the skin, underlying the epidermis and composed of fibroconnective tissue.

desmosome (DEZ-mo-some)   A patchlike intercellular junction that mechanically links two cells together. fig. 6.19

dextrose   The D-isomer of glucose; the only form of glucose with a normal role in physiology.

diabetes (DY-uh-BEE-teez)   Any disease characterized by chronic polyuria of metabolic origin; diabetes mellitus unless otherwise specified.

diabetes insipidus (in-SIP-ih-dus)   A form of diabetes that results from hyposecretion of antidiuretic hormone; unlike other forms, it is not characterized by hyperglycemia or glycosuria.

diabetes mellitus (mel-EYE-tus) **(DM)**   A form of diabetes that results from hyposecretion of insulin or from a deficient target cell response to it; symptoms include hyperglycemia and glycosuria.

dialysis (dy-AL-ih-sis)   **1.** The separation of some solute particles from others by diffusion through a selectively permeable membrane. **2.** Hemodialysis, the process of separating wastes from the bloodstream and sometimes adding other substances to it (such as drugs and nutrients) by circulating the blood through a machine with a selectively permeable membrane, used to treat cases of renal or hepatic insufficiency.

diapedesis (DY-uh-peh-DEE-sis)   Migration of formed elements of the blood through a capillary wall into the interstitial space. fig. 21.13

diaphysis (dy-AFF-ih-sis)   The shaft of a long bone. fig. 8.3

diarthrosis (DY-ar-THRO-sis)   A freely movable synovial joint such as the knuckles, elbows, shoulders, or knees.

diastole (dy-ASS-tuh-lee)   A period in which a heart chamber relaxes and fills with blood; especially ventricular relaxation.

diencephalon (DY-en-SEFF-uh-lon)   A portion of the brain between the midbrain and corpus callosum; composed of the thalamus, epithalamus, and hypothalamus. fig. 14.18

differentiation   Development of a relatively unspecialized cell into one with a more specific structure and function.

diffusion   The movement of molecules resulting from their spontaneous thermal vibration; *net diffusion* is the movement of more molecules in one direction than in another.

dilation (dy-LAY-shun)   Widening of an organ or passageway such as a blood vessel or the pupil of the eye.

diploid (2n)   Pertaining to a cell or organism with chromosomes in homologous pairs.

disaccharide (dy-SAC-uh-ride)   A carbohydrate composed of two simple sugars (monosaccharides) joined by a glycosidic bond; for example, lactose, sucrose, and maltose. fig. 3.5

distal   Relatively distant from a point of origin or attachment; for example, the wrist is distal to the elbow. *Compare* proximal.

disulfide bond   A covalent bond between the sulfur atoms of two cysteine residues, serving to link one polypeptide chain to another or to hold a single chain in its three-dimensional conformation.

diuretic (DY-you-RET-ic)   A chemical that increases urine output.

dizygotic (DZ) twins   Two individuals who developed simultaneously in one uterus but originated from separate fertilized eggs and therefore are not genetically identical.

dominant   **1.** Pertaining to a genetic allele that is phenotypically expressed in the presence of any other allele. **2.** Pertaining to a trait that results from a dominant allele.

dopamine (DOE-puh-meen)   An inhibitory catecholamine neurotransmitter of the central nervous system, especially of the basal nuclei, where it acts to suppress unwanted motor activity.

dorsal   Toward the back (spinal) side of the body.

dorsal root   A branch of a spinal nerve that enters the spinal cord on its dorsal side, composed of sensory fibers. fig. 15.2

dorsiflexion (DOR-sih-FLEC-shun)   A movement of the ankle that reduces the joint angle and raises the toes. fig. 10.14

Down syndrome   *See* trisomy-21.

duodenum (DEW-oh-DEE-num, dew-ODD-eh-num)   The first portion of the small intestine, extending for about 25 cm from the pyloric valve of the stomach to a sharp bend called the duodenojejunal flexure; receives chyme from the stomach and secretions from the liver and pancreas. fig. 25.22

dynamic equilibrium   **1.** A state of continual change that is controlled within narrow limits, as in homeostasis and chemical equilibrium. **2.** The sense of motion or acceleration of the body.

dynein (DINE-een) A motor protein involved in the beating of cilia and flagella and in the movement of molecules and organelles within cells, as in retrograde transport in a nerve fiber.

E

ectoderm   The outermost of the three primary germ layers of an embryo; gives rise to the nervous system and epidermis.

ectopic (ec-TOP-ic)   In an abnormal location; for example, ectopic pregnancy and ectopic pacemakers of the heart.

edema (eh-DEE-muh)   Abnormal accumulation of tissue fluid resulting in swelling of the tissue.

effector   A molecule, cell, or organ that carries out a response to a stimulus.

efferent (EFF-ur-unt)   Carrying away or out, such as a blood vessel that carries blood away from a tissue or a nerve fiber that conducts signals away from the central nervous system.

eicosanoids (eye-CO-sah-noyds)   Twenty-carbon derivatives of arachidonic acid that function as intercellular messengers; includes prostaglandins, prostacyclin, leukotrienes, and thromboxanes.

elastic fiber   A connective tissue fiber, composed of the protein elastin, that stretches under tension and returns to its original length when released; responsible for the resilience of organs such as the skin and lungs.

elasticity   The tendency of a stretched structure to return to its original dimensions when tension is released.

electrical synapse   A gap junction that enables one cell to stimulate another directly, without the intermediary action

of a neurotransmitter; such synapses connect the cells of cardiac muscle and single-unit smooth muscle.

electrochemical gradient   A difference in ion concentration from one point to another (especially across a plasma membrane), resulting in a gradient of both chemical concentration and electrical charge.

electrolyte   A salt that ionizes in water and produces a solution that conducts electricity; loosely speaking, any ion that results from the dissociation of such salts, such as sodium, potassium, calcium, chloride, and bicarbonate ions.

elevation   A joint movement that raises a body part, as in hunching the shoulders or closing the mouth.

embolism (EM-bo-lizm)   The obstruction of a blood vessel by an embolus.

embolus (EM-bo-lus)   Any abnormal traveling object in the bloodstream, such as agglutinated bacteria or blood cells, a blood clot, or an air bubble.

embryo   A developing individual from the end of the second week of gestation when the three primary germ layers have formed, through the end of the eighth week when all of the organ systems are present. *Compare* conceptus, fetus.

emphysema (EM-fih-SEE-muh)   A degenerative lung disease characterized by a breakdown of alveoli and diminishing surface area available for gas exchange; occurs with aging of the lungs but is greatly accelerated by smoking or air pollution.

emulsion   A suspension of one liquid in another, such as oil in water or fat in the lymph.

endocrine (EN-doe-crin) **gland**   A ductless gland that secretes hormones into the bloodstream; for example, the thyroid and adrenal glands. Compare *exocrine gland*.

endocytosis (EN-doe-sy-TOE-sis)   Any process in which a cell forms vesicles from its plasma membrane and takes in large particles, molecules, or droplets of extracellular fluid; for example, phagocytosis and pinocytosis.

endoderm   The innermost of the three primary germ layers of an embryo; gives rise to the mucosae of the digestive and respiratory tracts and to their associated glands.

endogenous (en-DODJ-eh-nus)   Originating internally, such as the endogenous cholesterol synthesized in the body in contrast to the exogenous cholesterol coming from the diet.

endometrium (EN-doe-MEE-tree-um)   The mucosa of the uterus; the site of implantation and source of menstrual discharge.

endoplasmic reticulum (EN-doe-PLAZ-mic reh-TIC-you-lum) **(ER)**   An extensive system of interconnected cytoplasmic tubules or

channels; classified as rough ER or smooth ER depending on the presence or absence of ribosomes on its membrane. figs. 4.12, 4.16

endothelium (EN-doe-THEEL-ee-um)   A simple squamous epithelium that lines the lumens of the blood vessels, heart, and lymphatic vessels.

endurance exercise   A form of physical exercise, such as running or swimming, that promotes cardiopulmonary efficiency and fatigue resistance more than muscular strength. *Compare* resistance exercise.

enteric (en-TERR-ic)   Pertaining to the small intestine, as in enteric hormones.

enzyme   A protein that functions as a catalyst.

enzyme amplification   A series of chemical reactions in which the product of one step is an enzyme that produces an even greater number of product molecules at the next step, resulting in a rapidly increasing amount of reaction product. Seen in hormone action and blood clotting, for example.

eosinophil (EE-oh-SIN-oh-fill)   A granulocyte with a large, often bilobed nucleus and coarse cytoplasmic granules that stain with eosin; phagocytizes antigen-antibody complexes, allergens, and inflammatory chemicals and secretes enzymes that combat parasitic infections. fig. 18.17

epidermis   A stratified squamous epithelium that constitutes the superficial layer of the skin, overlying the dermis. fig. 7.2

epinephrine (EP-ih-NEFF-rin)   A catecholamine that functions as a neurotransmitter in the sympathetic nervous system and as a hormone secreted by the adrenal medulla; also called *adrenalin.*

epiphyseal (EP-ih-FIZZ-ee-ul) **plate**   A plate of hyaline cartilage between the epiphysis and diaphysis of a long bone in a child or adolescent, serving as a growth zone for bone elongation. fig. 8.13

epiphysis (eh-PIF-ih-sis)   **1.** The head of a long bone. **2.** The pineal gland (epiphysis cerebri).

epithelium   A type of tissue consisting of one or more layers of closely adhering cells with little intercellular material and no blood vessels; forms the coverings and linings of many organs and the parenchyma of the glands.

erectile tissue   A tissue that functions by swelling with blood, as in the penis and clitoris and inferior concha of the nasal cavity.

erythema (ERR-ih-THEE-muh)   Abnormal redness of the skin due to such causes as burns, inflammation, and vasodilation.

erythrocyte (eh-RITH-ro-site)   A red blood cell.

erythropoiesis (eh-RITH-ro-poy-EE-sis)   The production of erythrocytes.

erythropoietin (eh-RITH-ro-POY-eh-tin)   A hormone that is secreted by the kidneys in response to hypoxemia and stimulates erythropoiesis.

estrogens (ESS-tro-jenz)   A family of steroid hormones known especially for producing female secondary sex characteristics and regulating various aspects of the menstrual cycle and pregnancy; major forms are estradiol, estriol, and estrone.

evolution   A change in the relative frequencies of alleles in a population over a period of time; the mechanism that produces adaptations in human form and function. *See also* adaptation.

excitability   The ability of a cell to respond to a stimulus, especially the ability of nerve and muscle cells to produce membrane voltage changes in response to stimuli; irritability.

excitation-contraction coupling   Events that link the synaptic stimulation of a muscle cell to the onset of contraction.

excitatory postsynaptic potential (EPSP)   A partial depolarization of a cell in response to a neurotransmitter, making it more likely to reach threshold and produce an action potential.

excretion   The process of eliminating metabolic waste products from a cell or from the body. *Compare* secretion.

exocrine (EC-so-crin) **gland**   A gland that secretes its products into another organ or onto the body surface by way of a duct; for example, salivary and gastric glands.

exocytosis (EC-so-sy-TOE-sis)   A process in which a vesicle in the cytoplasm of a cell fuses with the plasma membrane and releases its contents from the cell; used in the elimination of cellular wastes and in the release of gland products and neurotransmitters.

exogenous (ec-SODJ-eh-nus)   Originating externally, such as exogenous (dietary) cholesterol; extrinsic. *Compare* endogenous.

expiration   1. Exhaling. 2. Dying.

extension   Movement of a joint that increases the angle between articulating bones (straightens the joint). *Compare* flexion. fig. 10.9

extracellular fluid (ECF)   Any body fluid that is not contained in the cells; for example, blood, lymph, and tissue fluid.

extrinsic (ec-STRIN-sic)   1. Originating externally, such as extrinsic blood-clotting factors; exogenous. 2. Not fully contained within an organ but acting on it, such as the extrinsic muscles of the hand and eye. *Compare* intrinsic.

exude (ec-SUDE)   To seep out, such as fluid filtering from blood capillaries.

F

facilitated diffusion   The process of transporting a chemical through a cellular membrane, down its concentration gradient, with the aid of a membrane carrier that does not consume ATP; enables substances to diffuse through the membrane that would do so poorly, or not at all, without a carrier.

facilitation   Making a process more likely to occur, such as the firing of a neuron, or making it occur more easily or rapidly, as in facilitated diffusion.

fallopian (fah-LO-pee-un) **tube**   *See* uterine tube.

fascia (FASH-ee-uh)   A layer of fibroconnective tissue between the muscles (deep fascia) or separating the muscles from the skin (superficial fascia). fig. 11.2

fascicle (FASS-ih-cul)   A bundle of muscle or nerve fibers ensheathed in connective tissue; multiple fascicles bound together constitute a muscle or nerve as a whole. fig. 11.2

fat   1. A triglyceride (triacylglycerol) molecule. 2. Adipose tissue.

fatty acid   An organic molecule composed of a chain of an even number of carbon atoms with a carboxyl group at one end and a methyl group at the other; one of the structural subunits of triglycerides and phospholipids.

fenestrated (FEN-eh-stray-ted)   Perforated with holes or slits, as in fenestrated blood capillaries and the elastic sheets of large arteries. fig. 20.6

fetus   In human development, an individual from the beginning of the ninth week when all of the organ systems are present, through the time of birth. *Compare* conceptus, embryo.

fibrin (FY-brin)   A sticky fibrous protein formed from fibrinogen in blood, tissue fluid, lymph, and semen; forms the matrix of a blood clot.

fibroblast   A connective tissue cell that produces collagen fibers and ground substance; the only type of cell in tendons and ligaments.

fibroconnective tissue   Any connective tissue with a preponderance of fiber; also called *fibrous connective tissue.*

fibrosis   A degenerative tissue change in which an excessive amount of fiber is produced, for example, in tuberculosis.

filtrate   A fluid formed by filtration.

filtration   A process in which hydrostatic pressure forces a fluid through a selectively permeable membrane (especially a capillary wall).

fire   To produce an action potential, as in nerve and muscle cells.

fix   1. To hold a structure in place, for example, by fixator muscles that prevent unwanted joint movements. 2. To preserve a tissue by means of a fixative.

fixative   A chemical that preserves tissues from decay, such as formalin.

flagellum (fla-JEL-um)   A long, motile, usually single hairlike extension of a cell; the tail of a sperm cell is the only functional flagellum in humans. fig. 27.18

flexion   A joint movement that, in most cases, decreases the angle between two bones. *Compare* extension. fig. 10.9

fluid balance   An equilibrium between fluid intake and output or between the amounts of fluid contained in the body's different fluid compartments.

fluid compartments   Any of the major categories of fluid in the body, separated by selectively permeable membranes and differing from each other in chemical composition. Primary examples are the intracellular fluid, the tissue fluid, and the blood and lymph.

fluid-mosaic model   The current theory of the structure of a plasma membrane, depicting it as a bilayer of phospholipids with embedded proteins, many of which are able to move about in the lipid film. fig. 4.6

fluid-phase pinocytosis   A form of endocytosis in which the cell imbibes droplets of extracellular fluid without modifying its composition or concentration.

follicle (FOLL-ih-cul)   1. A small space, such as a hair follicle, thyroid follicle, or ovarian follicle. 2. An aggregation of lymphocytes in a lymphatic organ or mucous membrane.

follicle-stimulating hormone (FSH)   A hormone secreted by the anterior pituitary gland that stimulates development of the ovarian follicles and egg cells.

foramen (fo-RAY-men)   A hole through a bone or other organ, in many cases providing passage for blood vessels and nerves.

formed element   An erythrocyte, leukocyte, or platelet; any cellular component of blood or lymph as opposed to the extracellular fluid component.

fossa (FOSS-uh)   A depression in an organ or tissue, such as the fossa ovalis of the heart or a cranial fossa of the skull.

fovea (FOE-vee-uh)   A small pit, such as the fovea capitis of the femur or fovea centralis of the retina.

free energy   The potential energy in a chemical that is available to do work.

free radical   An uncharged particle derived from an atom or molecule, having an unpaired electron that makes it highly reactive and destructive to cells; produced by intrinsic processes such as aerobic respiration and by extrinsic agents such as chemicals and ionizing radiation.

frontal plane   An anatomical plane that passes through the body or an organ from right to left and superior to inferior; also called a *coronal plane.* fig. A.3

functional group   A group of atoms, such as a carboxyl or amino group, that determines the functional characteristics of an organic molecule.

fundus The base, the broadest part, or the part farthest from the opening of certain viscera such as the stomach and uterus.

fusiform (FEW-zih-form) Spindle-shaped; elongated, thick in the middle, and tapered at both ends, such as the shape of a smooth muscle cell or a muscle spindle.

G

gamete (GAM-eet) An egg or sperm cell.

gametogenesis (GAM-eh-toe-JEN-eh-sis) The production of eggs or sperm.

gamma-(γ-)aminobutyric (ah-MEE-no-byu-TIRR-ic) **acid (GABA)** An inhibitory neurotransmitter of the central nervous system in the biogenic amine class. fig. 13.17

gamma (γ) globulins (GLOB-you-lins) A class of relatively large proteins found in the blood plasma and on the surfaces of immune cells, functioning as antibodies. *See also* globulin.

ganglion (GANG-glee-un) A cluster of nerve cell bodies in the peripheral nervous system, often resembling a knot in a string.

gap junction A junction between two cells consisting of a pore surrounded by a ring of proteins in the plasma membrane of each cell, allowing solutes to diffuse from the cytoplasm of one cell to the next; functions include cell-to-cell nutrient transfer in the developing embryo and electrical communication between cells of cardiac and smooth muscle. *See also* electrical synapse. fig. 6.19

gastric Pertaining to the stomach.

gate A protein channel in a unit membrane that can open or close in response to chemical or electrical stimuli, thus controlling when substances are allowed to pass through the membrane.

gene A segment of DNA that codes for the synthesis of one protein.

gene locus The site on a chromosome where a given gene is located.

generator potential A graded, reversible rise in the local voltage across the plasma membrane of a nerve or muscle cell in response to a stimulus; triggers an action potential if it reaches threshold.

genetic engineering Any of several techniques that alter the genetic constitution of a cell or organism, including recombinant DNA technology and gene substitution therapy.

genome (JEE-nome) All the genes of one individual, estimated at 100,000 genes in humans.

genotype (JEE-no-type) The pair of alleles possessed by an individual at one gene locus on a pair of homologous chromosomes; strongly influences the individual's phenotype for a given trait.

germ cell A gamete or any precursor cell destined to become a gamete.

germ layer Any of the three original tissue layers of an embryo; ectoderm, mesoderm, or endoderm.

gestation (jess-TAY-shun) Pregnancy.

gland Any organ specialized to produce a secretion; in some cases a single cell, such as a goblet cell.

glaucoma (glaw-CO-muh) A visual disease in which an excessive amount of aqueous humor accumulates, creating pressure that is transmitted through the lens and vitreous body to the retina; pressure on the blood vessels of the choroid causes ischemia, retinal necrosis, and blindness.

globulin (GLOB-you-lin) A globular protein such as an enzyme, antibody, or albumin; especially a family of proteins in the blood plasma that includes albumin, antibodies, fibrinogen, and prothrombin.

glomerular (glo-MERR-you-lur) **capsule** A double-walled capsule around each glomerulus of the kidney; receives glomerular filtrate and empties into the proximal convoluted tubule. Also called *Bowman's capsule*. fig. 23.6

glomerulus A spheroid mass of blood capillaries in the kidney that filters plasma and produces glomerular filtrate, which is further processed to form the urine. fig. 23.6

glucagon (GLUE-ca-gon) A hormone secreted by α cells of the pancreatic islets in response to hypoglycemia; promotes glycogenolysis and other effects that raise blood glucose concentration.

glucocorticoid (GLUE-co-COR-tih-coyd) Any hormone of the adrenal cortex that affects carbohydrate, fat, and protein metabolism; chiefly cortisol and corticosterone.

gluconeogenesis (GLUE-co-NEE-oh-JEN-eh-sis) The synthesis of glucose from noncarbohydrates such as fats and amino acids.

glucose A monosaccharide ($C_6H_{12}O_6$) also known as blood sugar; glycogen, starch, cellulose, and maltose are made entirely of glucose, and glucose constitutes half of a sucrose or lactose molecule. The isomer involved in human physiology is also called *dextrose*.

glucose-sparing effect An effect of fats or other energy substrates in which they are used as fuel by most cells, so that those cells do not consume glucose; this makes more glucose available to cells such as neurons that cannot use alternative energy substrates.

glycerol (GLISS-er-ol) A viscous three-carbon alcohol that forms the structural backbone of triglyceride and phospholipid molecules; also called *glycerin*.

glycocalyx (GLY-co-CAY-licks) A layer of carbohydrate molecules covalently bonded to the phospholipid and protein molecules of a plasma membrane; forms a surface coat on all human cells.

glycogen (GLY-co-jen) A glucose polymer synthesized by liver, muscle, uterine, and vaginal cells that serves as an energy-storage polysaccharide.

glycogenesis (GLY-co-JEN-eh-sis) The synthesis of glycogen.

glycogenolysis (GLY-co-jeh-NOLL-ih-sis) The hydrolysis of glycogen, releasing glucose.

glycolipid (GLY-co-LIP-id) A phospholipid molecule with a carbohydrate covalently bonded to it, found in the plasma membranes of cells.

glycolysis (gly-COLL-ih-sis) A series of anaerobic oxidation reactions that break a glucose molecule into two molecules of pyruvic acid and produce a small amount of ATP.

glycoprotein (GLY-co-PRO-teen) A protein molecule with a smaller carbohydrate covalently bonded to it; found in mucus and the glycocalyx of cells, for example.

glycosaminoglycan (GLY-cose-am-ih-no-GLY-can) **(GAG)** A polysaccharide composed of modified sugars with amino groups; the major component of a proteoglycan. GAGs are largely responsible for the viscous consistency of tissue gel and the stiffness of cartilage.

glycosuria (GLY-co-SOOR-ee-uh) The presence of glucose in the urine, typically indicative of a kidney disease, diabetes mellitus, or other endocrine disorders.

goblet cell A mucus-secreting gland cell, shaped somewhat like a wineglass, found in the epithelia of many mucous membranes. fig. 6.23

Golgi (GOAL-jee) **complex** An organelle composed of several parallel cisternae, somewhat like a stack of saucers, that modifies and packages newly synthesized proteins and synthesizes carbohydrates. fig. 4.13

Golgi vesicle A membrane-bounded vesicle pinched from the Golgi complex, containing its chemical product; may be retained in the cell as a lysosome or become a secretory vesicle that releases the product by exocytosis.

gonad The ovary or testis.

gonadotropin (go-NAD-oh-TRO-pin) A pituitary hormone that stimulates the gonads; specifically FSH and LH.

G protein A protein of the plasma membrane that is activated by a membrane receptor and, in turn, opens an ion channel or activates an intracellular physiological response; important in linking ligand-receptor binding to second messenger systems.

graded potential A variable change in voltage across a plasma membrane, as opposed to the all-or-none quality of an action potential.

gradient A difference or change in any variable, such as pressure or chemical concentration, from one point in space to another; provides a basis for molecular movements such as gas exchange, osmosis, and facilitated diffusion.

granulocyte (GRAN-you-lo-site)   Any of three types of leukocytes (neutrophils, eosinophils, or basophils) with prominent cytoplasmic granules. fig. 18.17

granulosa cells   Cells that form a stratified cuboidal epithelium lining an ovarian follicle; source of steroid sex hormones. fig. 28.13

gray matter   A zone or layer of tissue in the central nervous system where the neuron cell bodies, dendrites, and synapses are found; forms the core of the spinal cord, nuclei of the brainstem, basal nuclei of the cerebrum, and the cerebral and cerebellar cortices. fig. 14.6

gross anatomy   Bodily structure that can be observed without magnification.

growth factor   A chemical messenger that stimulates mitosis and differentiation of target cells that have receptors for it; important in such processes as fetal development, tissue maintenance and repair, and hemopoiesis; sometimes a contributing factor in cancer.

growth hormone (GH)   A hormone of the anterior pituitary gland with multiple effects on many tissues, generally promoting tissue growth.

guanine   A double-ringed nitrogenous base (purine) found in DNA and RNA; one of the four bases of the genetic code; complementary to cytosine in the double helix of DNA. fig. 5.3

gyrus (JY-rus)   A wrinkle or fold in the cortex of the cerebrum or cerebellum.

H

hair cells   Sensory cells of the cochlea, semicircular ducts, utricle, and saccule, with a fringe of surface microvilli that respond to the relative motion of a gelatinous membrane at their tips; responsible for the senses of hearing and equilibrium.

hair follicle   An oblique epidermal pit that contains a hair and extends into the dermis or hypodermis.

half-life (T$_{1/2}$)   1. The time required for one-half of a quantity of a radioactive element to decay to a stable isotope (*physical half-life)* or to be cleared from the body through a combination of radioactive decay and physiological excretion (*biological half-life*). 2. The time required for one-half of a quantity of hormone to be cleared from the bloodstream.

haploid (*n***)**   In humans, having 23 unpaired chromosomes instead of the usual 46 chromosomes in homologous pairs; in any organism or cell, having half

the normal diploid number of chromosomes for that species.

helper T cell   A type of lymphocyte that performs a central coordinating role in humoral and cellular immunity; target of the human immunodeficiency virus (HIV).

hematocrit (he-MAT-oh-crit)   The percentage of blood volume that is composed of erythrocytes.

hematoma (HE-muh-TOE-muh)   A mass of clotted blood in the tissues; forms a bruise when visible through the skin.

heme (heem)   The nonprotein, iron-containing prosthetic group of hemoglobin or myoglobin; oxygen binds to its ferrous ion. fig. 18.10

hemocytoblast (HE-mo-SY-toe-blast)   An undifferentiated stem cell of the bone marrow that can give rise to any of the formed elements of the blood. fig. 18.3

hemoglobin (HE-mo-GLO-bin)   The red gas transport pigment of an erythrocyte.

hemolysis (he-MOLL-ih-sis)   The rupturing of erythrocytes from such causes as a hypotonic medium, parasitic infection, or a complement reaction.

hemopoiesis (HE-mo-poy-EE-sis)   Production of any of the formed elements of blood.

heparin (HEP-uh-rin)   A polysaccharide secreted by basophils and mast cells that inhibits blood clotting.

hepatic (heh-PAT-ic)   Pertaining to the liver.

hepatic portal system   A network of blood vessels that connect capillaries of the intestines to capillaries (sinusoids) of the liver, thus delivering newly absorbed nutrients directly to the liver.

hepatitis (HEP-uh-TY-tiss)   Inflammation of the liver.

heterozygous (HET-er-oh-ZY-gus)   Having nonidentical alleles at the same gene locus of two homologous chromosomes.

hiatus (hy-AY-tus)   An opening or gap, such as the esophageal hiatus through the diaphragm.

high-density lipoprotein (HDL)   A lipoprotein of the blood plasma that is about 50% lipid and 50% protein; functions to transport phospholipids and cholesterol from other organs to the liver for disposal. A high proportion of HDLs to low-density lipoproteins is desirable for cardiovascular health.

hilum (HY-lum)   A point on the surface of an organ where blood vessels, lymphatic vessels, or nerves enter and leave, usually marked by a depression and slit; the midpoint of the concave surface of any organ that is roughly bean-shaped, such as the lymph nodes, kidneys, and lungs. Also called the *hilus.* fig. 23.4

histamine (HISS-ta-meen)   An amino acid derivative secreted by basophils, mast cells, and some neurons; functions as a paracrine secretion and neurotransmitter to stimulate

effects such as gastric secretion, bronchoconstriction, and vasodilation.

histiocyte (HISS-tee-oh-site)   A mobile connective tissue macrophage derived from a monocyte.

histological section   A thin slice of tissue, usually mounted on a slide and artificially stained to make its microscopic structure more visible.

histology   1. The microscopic structure of tissues and organs. 2. The study of such structure.

homeostasis (HO-me-oh-STAY-sis)   The tendency of a living body to maintain relatively stable internal conditions in spite of greater changes in its external environment.

homologous (ho-MOLL-uh-gus)   1. Having the same embryonic or evolutionary origin but not necessarily the same function, such as the scrotum and labia majora. 2. Pertaining to two chromosomes with identical structures and gene loci but not necessarily identical alleles; each member of the pair is inherited from a different parent.

homozygous (HO-mo-ZY-gus)   Having identical alleles at the same gene locus of two homologous chromosomes.

hormone   A chemical messenger that is secreted into the blood by an endocrine gland or isolated gland cell and triggers a physiological response in distant cells with receptors for it.

host cell   Any cell belonging to the human body, as opposed to foreign cells introduced to it by such causes as infections and tissue transplants.

human   Any species of primate classified in the family Hominidae, characterized by bipedal locomotion, relatively large brains, and usually articulate speech; currently represented only by *Homo sapiens* but including extinct species of *Homo* and *Australopithecus.*

human chorionic (COR-ee-ON-ic) **gonadotropin (HCG)**   A hormone of pregnancy secreted by the chorion that stimulates continued growth of the corpus luteum and secretion of its hormones. HCG in urine is the basis for pregnancy testing.

human immunodeficiency virus (HIV)   A virus that infects human helper T cells and causes AIDS.

hyaline (HY-uh-lin) **cartilage**   A form of cartilage with a relatively clear matrix and fine collagen fibers but no conspicuous elastic fibers or coarse collagen bundles as in other types of cartilage.

hyaluronic (HY-uh-loo-RON-ic) **acid**   A glycosaminoglycan that is particularly abundant in connective tissues, where it becomes hydrated and forms the tissue gel.

hydrogen bond   A weak attraction between a slightly positive hydrogen atom on one molecule and a slightly negative oxygen or nitrogen atom on another molecule, or between such atoms on different parts of the same molecule; responsible for the cohesion of water and the coiling of protein and DNA molecules, for example.

hydrolysis (hy-DROL-ih-sis)   A chemical reaction that breaks a covalent bond in a molecule by adding an —OH group to one side of the bond and —H to the other side, thus consuming a water molecule. *Compare* dehydration synthesis.

hydrophilic (HY-dro-FILL-ic)   Pertaining to molecules that attract water or dissolve in it because of their polar nature.

hydrophobic (HY-dro-FOE-bic)   Pertaining to molecules that do not attract water or dissolve in it because of their nonpolar nature; such molecules tend to dissolve in lipids and other nonpolar solvents.

hydrostatic pressure   The physical force generated by a liquid such as blood or tissue fluid, as opposed to osmotic and atmospheric pressures.

hydroxyl (hy-DROCK-sil) **group**   A functional group with the formula —OH found on many organic molecules such as carbohydrates and alcohols.

hypercalcemia (HY-pur-cal-SEE-me-uh)   An excess of calcium ions in the blood.

hypercapnia (HY-pur-CAP-nee-uh)   An excess of carbon dioxide in the blood.

hyperextension   A joint movement that increases the angle between two bones beyond 180°. fig. 10.9

hyperglycemia (HY-pur-gly-SEE-me-uh)   An excess of glucose in the blood.

hyperkalemia (HY-pur-ka-LEE-me-uh)   An excess of potassium ions in the blood.

hypernatremia (HY-pur-na-TREE-me-uh)   An excess of sodium ions in the blood.

hyperplasia (HY-pur-PLAY-zhuh)   The growth of a tissue through cellular multiplication, not cellular enlargement. *Compare* hypertrophy.

hyperpolarization   A shift in the electrical potential across a plasma membrane to a value more negative than the resting membrane potential, tending to inhibit a nerve or muscle cell. *Compare* depolarization.

hypersecretion   Excessive secretion of a hormone or other gland product; can lead to endocrine disorders such as Addison disease or gigantism, for example.

hypertension   Excessively high blood pressure; criteria vary but it is often considered to be a condition in which systolic pressure exceeds 150 mmHg or diastolic pressure exceeds 90 mmHg.

hyperthermia   Excessively high core body temperature, as in heatstroke or fever.

hypertonic   Having a higher osmotic pressure than human cells or some other reference solution and tending to cause osmotic shrinkage of cells.

hypertrophy (hy-PUR-truh-fee)   The growth of a tissue through cellular enlargement, not cellular multiplication; for example,

the growth of muscle under the influence of exercise. Compare *hyperplasia.*

hypocalcemia (HY-po-cal-SEE-me-uh)   A deficiency of calcium ions in the blood.

hypocapnia (HY-po-CAP-nee-uh)   A deficiency of carbon dioxide in the blood.

hypodermis (HY-po-DUR-miss)   A layer of fibroconnective tissue deep to the skin; also called *superficial fascia, subcutaneous tissue,* or when it is predominantly adipose, *subcutaneous fat.*

hypoglycemia (HY-po-gly-SEE-me-uh)   A deficiency of glucose in the blood.

hypokalemia (HY-po-ka-LEE-me-uh)   A deficiency of potassium ions in the blood.

hyponatremia (HY-po-na-TREE-me-uh)   A deficiency of sodium ions in the blood.

hyposecretion   Inadequate secretion of a hormone or other gland product; can lead to endocrine disorders such as diabetes mellitus or pituitary dwarfism, for example.

hypothalamic (HY-po-thuh-LAM-ic) **thermostat**   A neuronal pool in the hypothalamus that monitors body temperature and sends afferent signals to hypothalamic heat-promoting or heat-losing centers to maintain thermal homeostasis.

hypothalamus (HY-po-THAL-uh-mus)   The inferior portion of the diencephalon of the brain, forming the walls and floor of the third ventricle and giving rise to the posterior pituitary gland; controls many fundamental physiological functions such as appetite, thirst, and body temperature, exerting many of its effects through the endocrine and autonomic nervous systems. fig. 14.18

hypothermia (HY-po-THUR-me-uh)   A state of abnormally low core body temperature.

hypothesis   An informed conjecture about a phenomenon that is capable of being tested and potentially falsified by experimentation or data collection.

hypotonic   Having a lower osmotic pressure than human cells or some other reference solution and tending to cause osmotic swelling and lysis of cells.

hypovolemic (HY-po-vo-LEE-mic) **shock**   Insufficient cardiac output resulting from a drop in blood volume. *See also* shock.

hypoxemia (HY-pock-SEE-me-uh)   A deficiency of oxygen in the bloodstream.

hypoxia (hy-POCK-see-uh)   A deficiency of oxygen in any tissue.

I

immune system   A population of cells, including leukocytes and macrophages, that occur in most organs of the body and protect against foreign organisms, some foreign chemicals, and cancerous or other aberrant host cells.

immunity   The ability to ward off a specific infection or disease, usually as a

result of prior exposure and the body's production of antibodies or lymphocytes against a pathogen. *Compare* resistance.

immunoglobulin (IM-you-no-GLOB-you-lin)   *See* antibody.

implantation   The attachment of a conceptus to the endometrium of the uterus.

inclusion   Any visible object in the cytoplasm of a cell other than an organelle or cytoskeletal element; usually a foreign body or a stored cell product, such as a virus, dust particle, lipid droplet, glycogen granule, or pigment.

infarction (in-FARK-shun)   A patch of tissue in an organ such as the heart or brain that has died from lack of blood perfusion; also called an *infarct.*

inferior   Lower than another structure or point of reference from the perspective of anatomical position; for example, the stomach is inferior to the diaphragm.

inflammation (IN-fluh-MAY-shun)   A complex of tissue responses to trauma or infection serving to ward off a pathogen and initiate tissue repair; recognized by the cardinal signs of redness, heat, swelling, pain, and compromised function.

infundibulum (IN-fun-DIB-you-lum)   Any funnel-shaped passage or structure, such as the distal portion of the uterine tube and the stalk that attaches the pituitary gland to the hypothalamus.

inguinal (IN-gwih-nul)   Pertaining to the groin.

inhibin   A hormone produced by the testes and ovaries that inhibits the secretion of FSH.

inhibitory postsynaptic potential (IPSP)   Hyperpolarization of a postsynaptic cell in response to a neurotransmitter, making it less likely to reach threshold and fire.

innervation (IN-ur-VAY-shun)   The nerve supply to an organ.

insertion   The point at which a muscle attaches to another tissue (usually a bone) and produces movement, opposite from its stationary origin; the origin and insertion of a given muscle sometimes depend on what muscle action is being considered.

in situ (in SY-too)   In the normal anatomical location (Latin, *in place*).

inspiration   Inhaling.

insulin (IN-suh-lin)   A hormone produced by β cells of the pancreatic islets in response to a rise in blood glucose concentration; accelerates glucose uptake and metabolism by most cells of the body, thus lowering blood glucose concentration.

integral (transmembrane) protein   A protein that extends through a plasma membrane and contacts both the extracellular and intracellular fluid. fig. 4.6

integration   A process in which a neuron receives input from multiple sources and their combined effects determine its output; the cellular basis of information processing by the nervous system.

integumentary (in-TEG-you-MEN-tuh-ree) **system** An organ system consisting of the skin, cutaneous glands, hair, and nails.

interatrial (IN-tur-AY-tree-ul) **septum** The wall between the atria of the heart.

intercalated (in-TUR-kuh-LAY-tid) **disc** A complex of gap junctions and desmosomes that join two cardiac muscle cells end to end, microscopically visible as a dark line, which helps to histologically distinguish this muscle type; functions as a mechanical and electrical link between cells. fig. 19.11

intercellular Between cells.

intercostal (IN-tur-COSS-tul) Between the ribs, as in the intercostal muscles, arteries, veins, and nerves.

interdigitate (IN-tur-DIDJ-ih-tate) To fit together like the fingers of two folded hands; for example, at the dermis-epidermis boundary, intercalated discs of the heart, and pedicels of the podocytes in the kidney. fig. 23.8

interleukin (IN-tur-LOO-kin) A hormonelike chemical messenger from one leukocyte to another, serving as a means of communication and coordination during immune responses.

interneuron (IN-tur-NEW-ron) A neuron that is contained entirely in the central nervous system and, in the path of signal conduction, lies anywhere between an afferent pathway and an efferent pathway.

interosseous (IN-tur-OSS-ee-us) **membrane** A fibrous membrane that connects the radius to the ulna and the tibia to the fibula along most of the shaft of each bone. fig. 9.31

interphase That part of the cell cycle between one mitotic phase and the next, from the end of cytokinesis to the beginning of the next prophase.

interstitial (IN-tur-STISH-ul) **1.** Pertaining to the extracellular spaces in a tissue. **2.** Located between other structures, as in the interstitial cells of the testis.

interstitial cell–stimulating hormone (ICSH) Another name for luteinizing hormone, emphasizing its effect on the interstitial cells of the testis.

interstitial fluid Fluid in the interstitial spaces of a tissue, also called *tissue fluid.*

intervertebral disc A cartilaginous disc between two adjacent vertebrae.

intracellular Within a cell.

intracellular fluid (ICF) The fluid contained in the cells; one of the major fluid compartments.

intravenous (I.V.) **1.** Present or occurring within a vein, such as an intravenous blood clot. **2.** Introduced directly into a vein, such as an intravenous injection or I.V. drip.

intrinsic (in-TRIN-sic) **1.** Arising from within, such as intrinsic blood-clotting factors; endogenous. **2.** Fully contained within an organ, such as the intrinsic

muscles of the hand and eye. *Compare* extrinsic.

intrinsic factor A secretion of the gastric glands required for the intestinal absorption of vitamin B_{12}. Hyposecretion of intrinsic factor leads to pernicious anemia.

in vitro (in VEE-tro) In a laboratory container; removed from the body and observed in isolation (Latin, *in glass*).

in vivo (in VEE-vo) In the living state; in the body (Latin, *in life*).

involuntary Not under conscious control, including tissues such as smooth and cardiac muscle and events such as reflexes.

involution (IN-vo-LOO-shun) Shrinkage of a tissue or organ by autolysis, such as involution of the thymus after childhood and of the uterus after pregnancy.

ion A chemical particle with unequal numbers of electrons or protons and consequently a net negative or positive charge; it may have a single atomic nucleus as in a sodium ion or a few atoms as in a bicarbonate ion, or it may be a large molecule such as a protein.

ionic bond The force that binds a cation to an anion.

ionizing radiation High-energy electromagnetic rays that eject electrons from atoms or molecules and convert them to ions, frequently causing cellular damage; for example, X rays and gamma rays.

ionotropic (eye-ON-oh-TRO-pic) **effect** Effect in which a neurotransmitter or hormone directly causes a gated channel to open in a plasma membrane, thus triggering ion flow and voltage changes across the target cell membrane.

ipsilateral (IP-sih-LAT-ur-ul) On the same side of the body, as in reflex arcs in which a muscular response occurs on the same side of the body as the stimulus. *Compare* contralateral.

ischemia (iss-KEE-me-uh) Insufficient blood flow to a tissue, typically resulting in metabolite accumulation and possibly tissue death.

isometric contraction A muscle contraction in which internal tension rises but the muscle does not shorten.

isotonic Having the same osmotic pressure as human cells or some other reference solution.

isotonic contraction A muscle contraction in which the muscle shortens and moves a load while its internal tension remains constant.

J

jaundice (JAWN-diss) A yellowish color of the skin, corneas, mucous membranes, and body fluids due to an excessive

concentration of bilirubin; usually indicative of a liver disease, obstructed bile secretion, or hemolytic disease.

K

ketone (KEE-tone) Any organic compound with a carbonyl (C=O) group covalently bonded to two other carbons.

ketone bodies Certain ketones (acetone, acetoacetic acid, and β-hydroxybutyric acid) produced by the incomplete oxidation of fats, especially when fats are being rapidly catabolized. *See also* ketosis.

ketonuria (KEE-toe-NEW-ree-uh) The abnormal presence of ketones in the urine as an effect of ketosis.

ketosis (kee-TOE-sis) An abnormally high concentration of ketone bodies in the blood, occurring in pregnancy, starvation, diabetes mellitus, and other conditions; tends to cause acidosis and to depress the nervous system.

kilocalorie The amount of heat energy needed to raise the temperature of 1 kg of water by 1°C; 1,000 calories. Also called a *Calorie* or *large calorie.*

Kupffer (COOP-fur) **cell** A stationary macrophage in the sinusoids of the liver.

L

labium (LAY-bee-um) A lip, such as those of the mouth and the labia majora and minora of the vulva.

lactation The secretion of milk.

lactic acid A small organic acid produced as an end product of the anaerobic fermentation of pyruvic acid; a contributing factor in muscle fatigue.

lacuna (la-CUE-nuh) A small cavity or depression in a tissue such as bone, cartilage, and the erectile tissues.

lamella (la-MELL-uh) A little plate, such as the lamellae of bone. fig. 8.7

lamina (LAM-ih-nuh) A thin layer, such as the lamina of a vertebra or the lamina propria of a mucous membrane. fig. 9.23

lamina propria (PRO-pree-uh) A thin layer of areolar tissue immediately deep to the epithelium of a mucous membrane. fig. 6.23

Langerhans cell An antigen-presenting cell of the epidermis. fig. 7.3

larynx (LAIR-inks) A cartilaginous chamber in the neck containing the vocal cords; the voicebox.

latent period The interval between a stimulus and response, especially in the action of nerve and muscle cells.

lateral Away from the midline of an organ or midsagittal plane of the body; toward the side. *Compare* medial.

law A verbal or mathematical description of a predictable natural phenomenon or of the relationships between variables; for example, Poiseuille's law and the second law of thermodynamics.

law of mass action A law that states that the speed and direction of a reversible chemical reaction is determined by the relative quantities of the reactants. A reversible reaction A + B ↔ C + D proceeds left to right if the quantity of A + B is greater than the quantity of C + D and right to left if the latter is greater. This principle governs such reactions as the binding and dissociation of oxygen and hemoglobin.

leader sequence A chain of hydrophobic amino acids that guides a newly synthesized protein into the cisterna of the rough endoplasmic reticulum and is then cleaved off. In the synthesis of peptide hormones, removal of the leader sequence converts the preprohormone into the prohormone.

length-tension relationship A law that relates the tension generated by muscle contraction to the length of the muscle fiber prior to stimulation; it shows that the greatest tension is generated when the fiber exhibits an intermediate degree of stretch before stimulation.

lesion A circumscribed zone of tissue injury, such as a skin abrasion or myocardial infarction.

leukocyte (LOO-co-site) A white blood cell.

leukotrienes (LOO-co-TRY-eens) Eicosanoids that promote allergic and inflammatory responses such as vasodilation and neutrophil chemotaxis; secreted by basophils, mast cells, and damaged tissues.

libido (lih-BEE-do) Sex drive.

ligament A band of tough collagenous tissue joining one bone to another.

ligand (LIG-and, LY-gand) A chemical that binds to a receptor site on a protein, such as a neurotransmitter that binds to a membrane receptor or a substrate that binds to an enzyme.

ligand-gated channel A channel protein in a plasma membrane that opens or closes when a ligand binds to it, enabling the ligand to determine when substances can enter or leave the cell.

light microscope (LM) A microscope that produces images with visible light.

linea (LIN-ee-uh) An anatomical line, such as the linea albicans.

lingual (LING-gwul) Pertaining to the tongue, as in lingual papillae.

lipase (LY-pace) An enzyme that hydrolyzes a triglyceride into fatty acids and glycerol.

lipid A hydrophobic organic compound composed mainly of carbon and a high ratio of hydrogen to oxygen; includes fatty acids, fats, phospholipids, steroids, and prostaglandins.

lipoprotein (LIP-oh-PRO-teen) A protein-coated lipid droplet in the blood plasma or lymph, serving as a means of lipid transport; for example, chylomicrons and the high- and low-density lipoproteins.

load **1.** To pick up a gas for transport in the bloodstream. **2.** The resistance acted upon by a muscle.

lobe **1.** A structural subdivision of an organ such as a gland, a lung, or the brain, bounded by a visible landmark such as a fissure or septum. **2.** The inferior, noncartilaginous, often pendant part of the ear pinna.

lobule (LOB-yool) A small subdivision of an organ or of a lobe of an organ, especially of a gland.

locus *See* gene locus.

long bone A bone such as the femur or humerus that is markedly longer than it is wide and that generally serves as a lever.

longitudinal Oriented along the longest dimension of the body or of an organ.

loose connective tissue *See* areolar tissue.

low-density lipoprotein (LDL) A blood-borne droplet of about 20% protein and 80% lipid (mainly cholesterol) that transports cholesterol from the liver to other tissues.

lower extremity The appendage that arises from the hip, consisting of the thigh from hip to knee; the crural region from knee to ankle; the ankle; and the foot; loosely called the leg, although that term properly refers only to the crural region.

lumbar Pertaining to the lower back and sides, between the thoracic cage and pelvis.

lumen (LOO-men) The internal space of a hollow organ such as a blood vessel or the esophagus, or a space surrounded by cells as in a gland acinus.

luteinizing (LOO-tee-in-eye-zing) **hormone (LH)** A hormone of the anterior pituitary gland that stimulates ovulation in females and testosterone secretion in males. *See also* interstitial cell–stimulating hormone.

lymph The fluid contained in lymphatic vessels and lymph nodes, produced by the absorption of tissue fluid.

lymphatic (lim-FAT-ic) **system** An organ system consisting of lymphatic vessels, lymph nodes, the tonsils, spleen, and thymus; functions include tissue fluid recovery and immunity.

lymph node A small organ found along the course of a lymphatic vessel that filters the lymph and contains lymphocytes and macrophages, which respond to antigens in the lymph. fig. 21.8

lymphocytes (LIM-foe-sites) A class of relatively small agranulocytes with numerous types and roles in nonspecific defense, humoral immunity, and cellular immunity. fig. 18.17

lymphokine Any interleukin secreted by a lymphocyte.

lysosome (LY-so-some) A membrane-bounded organelle containing a mixture of enzymes with a variety of intracellular and extracellular roles in digesting foreign matter, pathogens, and expired organelles.

lysozyme (LY-so-zime) A secretion found in tears, milk, saliva, mucus, and other body fluids that destroys bacteria by disrupting their cell walls; variously regarded as a hydrolytic enzyme or as a polysaccharide surfactant. Also called *muramidase.*

M

macromolecule Any molecule of large size and high molecular weight, such as a protein, nucleic acid, polysaccharide, or triglyceride.

macrophage (MAC-ro-faje) Any cell of the body, other than a leukocyte, that is specialized for phagocytosis; usually derived from blood monocytes and often functioning as antigen-presenting cells.

macula (MAC-you-luh) A patch or spot, such as the *macula lutea* of the retina.

malignant (muh-LIG-nent) Pertaining to a cell or tumor that is cancerous; capable of metastasis.

maltose A disaccharide composed of two glucose monomers.

mammary gland The milk-secreting gland that develops within the breast in pregnancy and lactation; absent or only minimally developed in the breast of a nonpregnant or nonlactating woman.

mast cell A connective tissue cell, similar to a basophil, that secretes histamine, heparin, and other chemicals involved in inflammation; often concentrated along the course of blood capillaries. fig. 6.6

matrix **1.** The extracellular material of a tissue. **2.** The fluid within a mitochondrion containing enzymes of the citric acid cycle. **3.** The substance or framework within which other structures are embedded, such as the fibrous matrix of a blood clot. **4.** A patch of thickened epidermis from which a nail root develops.

mechanoreceptor A sensory nerve ending or organ specialized to detect mechanical stimuli such as touch, pressure, and vibration.

medial Toward the midline of an organ or midsagittal plane of the body. *Compare* lateral.

mediastinum (ME-dee-ass-TY-num) The thick medial partition of the thoracic cavity that separates one pleural cavity from the other and contains the heart, great blood vessels, and thymus. fig. A.13

medulla (meh-DULL-uh) Tissue deep to the cortex of certain two-layered organs such as the adrenal glands, lymph nodes, hairs, and kidneys.

medulla oblongata (OB-long-GAH-ta) The most inferior part of the brainstem, immediately superior to the foramen magnum of the skull, connecting the spinal cord to the rest of the brain. fig. 14.1

meiosis (my-OH-sis) A form of cell division in which a diploid cell divides twice and produces four haploid daughter cells; occurs only in gametogenesis.

melanocyte A cell of the stratum basale of the epidermis that synthesizes melanin and transfers it to the keratinocytes.

meninges (meh-NIN-jeez) (singular, *meninx*) The three fibrous membranes between the central nervous system and surrounding bone; the dura mater, arachnoid, and pia mater. fig. 14.5

menopause Cessation of the menstrual cycles, occurring during female climacteric.

merocrine (MERR-oh-crin) Pertaining to gland cells that release their product by exocytosis; also called *eccrine*.

mesenchyme (MEZ-en-kime) A gelatinous embryonic connective tissue derived from the mesoderm; differentiates into all permanent connective tissues and most muscle.

mesentery (MESS-en-tare-ee) A serous membrane that binds the intestines together and suspends them from the abdominal wall; the visceral continuation of the peritoneum. fig. 25.2

mesoderm (MEZ-oh-durm) The middle layer of the three primary germ layers of an embryo; gives rise to muscle and connective tissue.

mesothelium (MEZ-oh-THEE-lee-um) A simple squamous epithelium that covers the serous membranes.

metabolic pathway A series of linked chemical reactions, most of which are catalyzed by a separate enzyme; glycolysis, for example.

metabolic rate The overall rate of the body's metabolic reactions at any given time, determining the rates of nutrient and oxygen consumption; often measured from the rate of oxygen consumption or heat production. *Compare* basal metabolic rate.

metabolic waste A product of metabolism that is not useful to the body but is potentially toxic and must be excreted.

metabolism (meh-TAB-oh-lizm) The sum of all chemical reactions in the body.

metabolite (meh-TAB-oh-lite) Any chemical produced by metabolism.

metabotropic (meh-TAB-oh-TRO-pic) **effect** An effect in which a neurotransmitter or hormone changes the metabolism of a target cell but does not directly stimulate the opening of ion gates in the plasma membrane.

metastasis (meh-TASS-tuh-sis) The spread of cancer cells from the original tumor to a new location, where they seed the development of a new tumor.

microtubule An intracellular cylinder composed of the protein tubulin, forming the axonemes of cilia and flagella and part of the cytoskeleton.

microvillus An outgrowth of the plasma membrane that increases the surface area of a cell and functions in absorption and some sensory processes; distinguished from cilia and flagella by its smaller size and lack of an axoneme.

midsagittal (MID-SADJ-ih-tul) **plane** The plane that divides the body into equal right and left halves. *Compare* sagittal plane. fig. A.3

milliequivalent One one-thousandth of an equivalent, which is the amount of an electrolyte that would neutralize 1 mole of H^+ or OH^-. Electrolyte concentrations are commonly expressed in milliequivalents per liter (mEq/L).

mineralocorticoid (MIN-ur-uh-lo-COR-tih-coyd) A steroid hormone, chiefly aldosterone, that is secreted by the adrenal cortex and acts to regulate electrolyte balance.

mitochondrion (MY-toe-CON-dree-un) An organelle specialized to synthesize ATP, enclosed in a double unit membrane with infoldings of the inner membrane called cristae.

mitosis (my-TOE-sis) A form of cell division in which a cell divides once and produces two genetically identical daughter cells; sometimes used to refer only to the division of the genetic material or nucleus and not to include cytokinesis, the subsequent division of the cytoplasm.

moiety (MOY-eh-tee) A chemically distinct subunit of a macromolecule, such as the heme and globin moieties of hemoglobin or the lipid and carbohydrate moieties of a glycolipid.

molarity A measure of chemical concentration expressed as moles of solute per liter of solution.

mole The mass of a chemical equal to its molecular weight in grams, containing 6.023×10^{23} molecules.

monocyte An agranulocyte specialized to migrate into the tissues and transform into a macrophage. fig. 18.17

monokine Any interleukin secreted by a monocyte or macrophage.

monomer (MON-oh-mur) **1.** One of the identical or similar subunits of a larger molecule in the dimer to polymer range; for example, the glucose monomers of starch, the amino acids of a protein, or the nucleotides of DNA. **2.** One subunit of an antibody molecule, composed of four polypeptides.

monosaccharide (MON-oh-SAC-uh-ride) A simple sugar, or sugar monomer; chiefly glucose, fructose, and galactose.

monozygotic (MZ) twins Two individuals who developed from the same fertilized egg and are therefore genetically identical.

motor end plate A depression in a muscle fiber where it has synaptic contact with a nerve fiber and has a high density of neurotransmitter receptors. fig. 12.8

motor neuron A neuron that transmits signals from the central nervous system to any effector (muscle or gland cell); its axon is an efferent nerve fiber.

motor protein Any protein that produces movements of a cell or its components, owing to its ability to undergo quick repetitive changes in conformation and to bind reversibly to other molecules; for example, myosin, dynein, and kinesin.

motor unit One motor neuron and all the skeletal muscle fibers innervated by it.

mucosa (mew-CO-suh) A tissue layer that forms the inner lining of an anatomical tract that is open to the exterior, such as the respiratory, digestive, urinary, and reproductive tracts. Composed of epithelium, connective tissue (lamina propria), and often smooth muscle (muscularis mucosae). fig. 6.23

mucous membrane A mucosa.

mucus A viscous, slimy or sticky secretion produced by mucous cells and mucous membranes, consisting of a hydrated glycoprotein, mucin; serves to bind particles together, such as bits of masticated food, and to protect the mucous membranes from infection and abrasion.

muscle fiber One muscle cell, especially of skeletal muscle.

muscle tone A state of continual, partial contraction of resting skeletal or smooth muscle.

muscularis externa The external muscular wall of certain viscera such as the esophagus and small intestine. fig. 25.9

muscularis mucosae (MUSS-cue-LERR-iss mew-CO-see) A layer of smooth muscle immediately deep to the lamina propria of a mucosa. fig. 6.23

muscular system An organ system composed of the skeletal muscles, specialized mainly for maintaining postural support and producing movements of the bones.

muscular tissue A tissue composed of elongated, electrically excitable cells specialized for contraction; the three types are skeletal, cardiac, and smooth muscle.

mutagen (MEW-tuh-jen) Any agent that causes a mutation, including viruses, chemicals, and ionizing radiation.

mutation Any change in the structure of a chromosome or a DNA molecule, often resulting in a change of organismal structure or function.

myelin (MY-eh-lin) A lipid sheath around a nerve fiber, formed from closely spaced spiral layers of the plasma membrane of a Schwann cell or oligodendrocyte. fig. 13.6

myocardium (MY-oh-CAR-dee-um) The middle, muscular layer of the heart.

myocyte A muscle cell, especially a cell of cardiac or smooth muscle.

myoepithelial cell An epithelial cell that has become specialized to contract like a muscle cell; important in dilation of the pupil and ejection of secretions from gland acini.

myofibril (MY-oh-FY-bril) A bundle of myofilaments forming an internal subdivision of a cardiac or skeletal muscle cell. fig. 12.2

myofilament A protein microfilament responsible for the contraction of a muscle cell, composed mainly of myosin or actin. fig. 12.3

myoglobin (MY-oh-GLO-bin) A red oxygen-storage pigment of muscle; supplements hemoglobin in providing oxygen for aerobic muscle metabolism.

myosin A motor protein that constitutes the thick myofilaments of muscle and has globular, mobile heads of ATPase that bind to actin molecules.

N

necrosis (neh-CRO-sis) Pathological tissue death due to such causes as infection, trauma, or hypoxia. *Compare* apoptosis.

negative feedback A self-corrective mechanism that underlies most homeostasis, in which a bodily change is detected and responses are activated that reverse the change.

negative feedback inhibition A mechanism for limiting the secretion of a pituitary tropic hormone. The tropic hormone stimulates another endocrine gland to secrete its own hormone, and that hormone inhibits further release of the tropic hormone.

neonate (NEE-oh-nate) An infant up to 6 weeks old.

neoplasia (NEE-oh-PLAY-zee-uh) Abnormal growth of new tissue, such as a tumor, with no useful function.

nephron One of approximately 1 million blood-filtering, urine-producing units in each kidney; consists of a glomerulus, glomerular capsule, proximal convoluted tubule, nephron loop, distal convoluted tubule, and the blood vessels supplying them. fig. 23.5

nerve A cordlike organ of the peripheral nervous system composed of multiple nerve fibers ensheathed in connective tissue.

nerve fiber The axon of a single neuron.

nerve impulse A wave of self-propagating action potentials traveling along a nerve fiber.

nervous system An organ system composed of the brain, spinal cord, nerves, and ganglia, specialized for rapid communication of information.

nervous tissue A tissue composed of neurons and neuroglia.

net filtration pressure A net force favoring filtration of fluid from a capillary when all the hydrostatic and osmotic pressures of the blood and tissue fluids are taken into account.

neural tube A dorsal hollow tube in the embryo that develops into the central nervous system. fig. 14.3

neuroglia (noo-ROG-lee-uh) All cells of nervous tissue except neurons; cells that perform various supportive and protective roles for the neurons.

neuromuscular junction A synapse between a nerve fiber and a muscle fiber. fig. 12.8

neuron (NOOR-on) A nerve cell; an electrically excitable cell specialized for producing and transmitting action potentials and secreting chemicals that stimulate adjacent cells.

neuronal (noor-OH-nul) **pool** A group of interconnected neurons of the central nervous system that perform a single collective function; for example, the vasomotor center of the brainstem and speech centers of the cerebral cortex.

neurotransmitter A chemical released at the distal end of an axon that stimulates an adjacent cell; for example, acetylcholine, norepinephrine, or serotonin.

neutral fat A triglyceride.

neutrophil (NOO-tro-fill) A granulocyte, usually with a multilobed nucleus, that serves especially to destroy bacteria by means of phagocytosis, intracellular digestion, and secretion of bactericidal chemicals. fig. 18.17

nitrogenous (ny-TRODJ-eh-nus) **base** An organic molecule with a single or double carbon-nitrogen ring that forms one of the building blocks of ATP, other nucleotides, and nucleic acids; the basis of the genetic code. fig. 5.3

nitrogenous waste Any nitrogen-containing substance produced as a metabolic waste and excreted in the urine; chiefly ammonia, urea, uric acid, and creatinine.

nociceptor (NO-sih-SEP-tur) A nerve ending specialized to detect tissue damage and produce a sensation of pain; pain receptor.

norepinephrine (nor-EP-ih-NEF-rin) A catecholamine that functions as a neurotransmitter and adrenal hormone, especially in the sympathetic nervous system. fig. 13.17

nuclear (NEW-clee-ur) **envelope** A pair of unit membranes enclosing the nucleus of a cell, with prominent pores allowing traffic of molecules between the nucleoplasm and cytoplasm. fig. 5.1

nucleic (new-CLAY-ic) **acid** An acidic polymer of nucleotides found or produced in the nucleus, functioning in heredity and protein synthesis; of two types, DNA and RNA.

nucleotide (NEW-clee-oh-tide) An organic molecule composed of a nitrogenous base, a monosaccharide, and a phosphate group; the monomer of a nucleic acid.

nucleus (NEW-clee-us) **1.** A cellular organelle containing DNA and surrounded by a double unit membrane. **2.** A mass of neurons (gray matter) surrounded by white matter of the brain, including the basal nuclei and brainstem nuclei. **3.** The positively charged core of an atom, consisting of protons and neutrons. **4.** A central structure, such as the nucleus pulposus of an intervertebral disc.

nucleus pulposus The gelatinous center of an intervertebral disc.

O

olfaction (ole-FAC-shun) The sense of smell.

oncotic (ong-COT-ic) **pressure** The difference between the colloid osmotic pressure of the blood and that of the tissue fluid, usually favoring fluid absorption by the blood capillaries. *Compare* colloid osmotic pressure.

oocyte (OH-oh-site) In the development of an egg cell, any haploid stage between meiosis I and fertilization.

oogenesis (OH-oh-JEN-eh-sis) The production of a fertilizable egg cell through a series of mitotic and meiotic cell divisions; female gametogenesis.

ophthalmic (off-THAL-mic) Pertaining to the eye or vision; optic.

opposition A movement of the thumb in which it touches any fingertip of the same hand.

optic Pertaining to the eye or vision.

orbit The eye socket of the skull.

organ Any anatomical structure that is composed of at least two different tissue types, has recognizable structural boundaries, and has a discrete function different from the structures around it. Many organs are microscopic and many organs contain smaller organs, such as the skin containing numerous microscopic sense organs.

organelle Any structure within a cell that carries out one of its metabolic roles, such as mitochondria, centrioles, endoplasmic reticulum, and the nucleus; an intracellular structure other than the cytoskeleton and inclusions.

organic Pertaining to compounds of carbon.

origin The relatively stationary attachment of a skeletal muscle. *Compare* insertion.

osmolality (OZ-mo-LAL-ih-tee)   The molar concentration of dissolved particles in 1 kg of water.

osmolarity (OZ-mo-LERR-ih-tee)   The molar concentration of dissolved particles in 1 L of solution.

osmoreceptor (OZ-mo-re-SEP-tur)   A neuron of the hypothalamus that responds to changes in the osmolarity of the extracellular fluid.

osmosis (oz-MO-sis)   The net diffusion of water through a selectively permeable membrane.

osmotic diuresis (oz-MOT-ic DY-you-REE-sis)   Increased urine output due to an increase in the concentration of osmotically active particles in the tubular fluid.

osmotic pressure   The amount of pressure that would have to be applied to one side of a selectively permeable membrane to stop osmosis; proportional to the concentration of nonpermeating solutes on that side and therefore serving as an indicator of solute concentration.

osseous (OSS-ee-us)   Pertaining to bone.

ossification (OSS-ih-fih-CAY-shun)   Bone formation.

osteoarthritis (OA)   A chronic degenerative joint disease characterized by loss of articular cartilage, growth of bone spurs, and impaired movement; occurs to various degrees in almost all people with age.

osteoblasts   Bone-forming cells that arise from osteogenic cells, deposit bone matrix, and eventually become osteocytes.

osteoclasts   Macrophages of the bone surface that dissolve the matrix and return minerals to the extracellular fluid.

osteocyte   A mature bone cell formed when an osteoblast becomes surrounded by its own matrix and entrapped in a lacuna.

osteon   A structural unit of compact bone consisting of a central canal surrounded by concentric cylindrical lamellae of matrix. fig. 8.7

osteoporosis (OSS-tee-oh-pore-OH-sis)   A degenerative bone disease characterized by a loss of bone mass, increasing susceptibility to spontaneous fractures, and sometimes deformity of the vertebral column; causes include aging, estrogen hyposecretion, and insufficient weight-bearing exercise.

ovary   The female gonad; produces eggs, estrogen, and progesterone.

ovulation (OV-you-LAY-shun)   The release of a mature oocyte by the bursting of an ovarian follicle.

ovum   Any stage of the female gamete from the conclusion of meiosis I until fertilization; a primary or secondary oocyte; an egg.

oxidation   A chemical reaction in which one or more electrons are removed from a molecule, lowering its free energy content; opposite of reduction and always linked to a reduction reaction.

oxytocin (OCK-see-TOE-sin) **(OT)**   A hormone released by the posterior pituitary gland that stimulates labor contractions and milk release.

P

pancreas (PAN-cree-us)   A gland of the upper abdominal cavity, near the stomach, that secretes digestive enzymes and sodium bicarbonate into the duodenum and secretes hormones into the blood.

pancreatic islets (PAN-cree-AT-ic EYE-lets)   Small clusters of endocrine cells in the pancreas that secrete insulin, glucagon, somatostatin, and other intercellular messengers; also called *islets of Langerhans.* fig. 17.16

papilla (pa-PILL-uh)   A conical or nipplelike structure, such as a lingual papilla of the tongue or the papilla of a hair bulb.

papillary (PAP-ih-lerr-ee)   Pertaining to or shaped like a nipple, such as the papillary muscles of the heart; or having papillae, such as the papillary layer of the dermis.

paracrine (PERR-uh-crin)   **1.** A chemical messenger similar to a hormone whose effects are restricted to the immediate vicinity of the cells that secrete it; sometimes called a local hormone. **2.** Pertaining to such a secretion, as opposed to *endocrine.*

parasympathetic (PERR-uh-SIM-pa-THET-ic) **nervous system**   A division of the autonomic nervous system that issues efferent fibers through the cranial and sacral nerves and exerts cholinergic effects on its target organs.

parathyroid (PERR-uh-THY-royd) **glands**   Small endocrine glands, usually four in number, adhering to the posterior side of the thyroid gland. fig. 17.12

parathyroid hormone (PTH)   A hormone secreted by the parathyroid glands that raises blood calcium concentration by stimulating bone resorption by osteoclasts, promoting intestinal absorption of calcium, and inhibiting urinary excretion of calcium.

parenchyma (pa-REN-kih-muh)   The tissue that performs the main physiological functions of an organ, especially a gland, as opposed to the tissues (stroma) that mainly provide structural support.

parietal (pa-RY-eh-tul)   **1.** Pertaining to a wall, as in the parietal cells of the gastric glands and parietal bone of the skull. **2.** The outer or more superficial layer of a two-layered membrane such as the pleura, pericardium, or glomerular capsule. *Compare* visceral. fig. A.13

pathogen   Any disease-causing organism or chemical.

pedicle (PED-ih-cul)   A small footlike process, as in the vertebrae and the renal podocytes; also called a *pedicel.*

pelvis   A basinlike structure such as the pelvic girdle of the skeleton or the urine-collecting space near the hilum of the kidney. fig. 23.4

peptide   Any chain of two or more amino acids. *See also* polypeptide, protein.

peptide bond   A group of four covalently bonded atoms (a —C═O group bonded to an —NH group) that links two amino acids in a peptide. fig. 3.13

perfusion   The amount of blood supplied to a tissue in a given period of time.

perichondrium (PERR-ih-CON-dree-um)   A layer of fibroconnective tissue covering the surface of hyaline or elastic cartilage.

perineum (PERR-ih-NEE-um)   The region between the thighs bordered by the coccyx, pubic symphysis, and ischial tuberosities; contains the orifices of the urinary, reproductive, and digestive systems. fig. 11.19

periosteum (PERR-ee-OSS-tee-um)   A layer of fibroconnective tissue covering the surface of a bone. fig. 8.3

peripheral (peh-RIF-eh-rul)   Away from the center of the body or of an organ, as in peripheral vision and peripheral blood vessels.

peripheral nervous system (PNS)   A subdivision of the nervous system composed of all nerves and ganglia; all of the nervous system except the central nervous system.

peristalsis (PERR-ih-STAL-sis)   A wave of constriction traveling along a tubular organ such as the esophagus or ureter, serving to propel its contents.

peritoneum (PERR-ih-toe-NEE-um)   A serous membrane that lines the peritoneal cavity of the abdomen and covers the mesenteries and viscera.

perivascular (PERR-ih-VASS-cue-lur)   Pertaining to the region surrounding a blood vessel.

pernicious anemia   A deficiency of hemoglobin synthesis resulting from a deficiency of vitamin B_{12} ingestion or absorption.

phagocytosis (FAG-oh-sy-TOE-sis)   A form of endocytosis in which a cell surrounds a foreign particle with pseudopods and engulfs it, enclosing it in a cytoplasmic vesicle called a phagosome.

pharynx (FAIR-inks)   A muscular passage in the throat at which the respiratory and digestive tracts cross.

physiology   **1.** The functional processes of the body. **2.** The study of such function.

pineal (PIN-ee-ul) **gland**   A small conical endocrine gland arising from the roof of the third ventricle of the brain; produces melatonin and serotonin and may be involved in timing the onset of puberty. fig. 14.18

pinocytosis (PIN-oh-sy-TOE-sis)   A form of endocytosis in which the plasma membrane sinks inward and imbibes droplets of extracellular fluid or specific molecules concentrated from that fluid. *See also* fluid-phase pinocytosis, receptor-mediated pinocytosis.

pituitary (pih-TOO-ih-terr-ee) **gland**   An endocrine gland suspended from the hypothalamus and housed in the sella turcica of the sphenoid bone; secretes numerous hormones, most of which regulate the activities of other glands. fig. 17.4

placenta (pla-SEN-tuh)   A thick discoid organ on the wall of the pregnant uterus, composed of a combination of maternal and fetal tissues, serving multiple functions in pregnancy including nutrient and waste transfer between mother and fetus. fig. 29.10

plantar (PLAN-tur)   Pertaining to the sole of the foot.

plaque   A small scale or plate of matter, such as dental plaque, the fatty plaques of atherosclerosis, and the amyloid plaques of Alzheimer disease.

plasma   The noncellular portion of the blood.

plasma membrane   The unit membrane that encloses a cell and controls the traffic of molecules in and out of the cell. fig. 4.6

platelet   A formed element of the blood derived from the peripheral cytoplasm of a megakaryocyte, known especially for its role in stopping bleeding but also serves in dissolving blood clots, stimulating inflammation, promoting tissue growth, and destroying bacteria.

pleura (PLOOR-uh)   A double-walled serous membrane that encloses each lung.

plexus   A network of blood vessels, lymphatic vessels, or nerves, such as a choroid plexus of the brain or brachial plexus of spinal nerves.

polymer   A molecule that consists of a long chain of identical or similar subunits, such as protein, DNA, or starch.

polypeptide   Any chain of more than 10 or 15 amino acids.

polysaccharide (POL-ee-SAC-uh-ride)   A polymer of simple sugars; for example, glycogen, starch, and cellulose.

polyuria (POL-ee-YOU-ree-uh)   Excessive output of urine.

popliteal (po-LIT-ee-ul)   Pertaining to the pit on the posterior aspect of the knee.

position emission tomography (PET)   A method of producing a computerized image of the physiological state of a tissue using injected radioisotopes that emit positrons.

posterior   Near or pertaining to the back or spinal side of the body; dorsal.

postganglionic (POST-gang-glee-ON-ic)   Pertaining to a neuron that carries signals away from a ganglion to a more distal target organ.

postsynaptic (POST-sih-NAP-tic)   Pertaining to a neuron or other cell that receives signals from the presynaptic neuron at a synapse. fig. 13.16

potential   A difference in electrical charge from one point to another, especially on opposite sides of a plasma membrane; usually measured in millivolts.

potential space   An anatomical space that is usually obliterated by contact between two membranes but that opens up if air, fluid, or other matter comes between the membranes. Examples include the pleural cavity and the lumen of the uterus.

preganglionic (PRE-gang-glee-ON-ic)   Pertaining to a neuron through which signals travel from the central nervous system to a ganglion.

presynaptic (PRE-sih-NAP-tic)   Pertaining to a neuron by which signals arrive at a synapse. fig. 13.16

prime mover   The muscle primarily responsible for a given joint action; agonist.

prolactin (PRL)   A pituitary hormone that promotes milk synthesis.

pronation (pro-NAY-shun)   A rotational movement of the forearm that turns the palm downward or posteriorly. fig. 10.13

proprioceptor (PRO-pree-oh-SEP-tur)   A sensory receptor that detects muscle contractions, joint movements, or changes in the orientation or velocity of the head.

prostaglandins (PROSS-ta-GLAN-dinz)   A family of eicosanoids with a five-sided carbon ring in the middle of a hydrocarbon chain, playing a variety of roles in inflammation, neurotransmission, vasomotion, reproduction, and metabolism. fig. 3.10

prostate (PROSS-tate) **gland**   A male reproductive gland that encircles the urethra immediately inferior to the bladder and contributes to the semen. fig. 27.7

protein   A large polypeptide; while criteria for a protein are somewhat subjective and variable, polypeptides over 100 amino acids long are generally classified as proteins.

proximal   Relatively near a point of origin or attachment; for example, the shoulder is proximal to the elbow. *Compare* distal.

pseudopod (SOO-doe-pod)   A temporary cytoplasmic extension of a cell used for locomotion (ameboid movement) and phagocytosis.

pulmonary   Pertaining to the lungs.

pulmonary circuit   A route of blood flow that supplies blood to the pulmonary alveoli for gas exchange and then returns it to the heart; all blood vessels from the right ventricle to the left atrium.

pyrogen (PY-ro-jen)   A fever-producing agent.

pyruvic acid   The three-carbon end product of glycolysis; occurs at the branch point between glycolysis, anaerobic fermentation, and aerobic respiration and is thus an important metabolic intermediate linking these pathways to each other.

R

ramus (RAY-mus)   An anatomical branch, as in a nerve or in the pubis.

receptor   1. A cell or organ specialized to detect a stimulus, such as a taste cell or the eye. 2. A protein molecule that binds and responds to a chemical such as a hormone, neurotransmitter, or odor molecule.

receptor-mediated pinocytosis   A form of endocytosis in which certain molecules in the extracellular fluid bind to receptors in the plasma membrane, these receptors become gathered together, the membrane sinks inward at that point, and the molecules become incorporated into vesicles in the cytoplasm.

receptor potential   A variable change in membrane voltage produced by a stimulus acting on a receptor cell; generates an action potential if it reaches threshold.

recombinant DNA (rDNA)   A molecule composed of the DNA of two different species spliced together, such as a combination of bacterial and human DNA used to produce transgenic bacteria that synthesize human proteins.

reduction   1. A chemical reaction in which one or more electrons are added to a molecule, raising its free energy content; opposite of oxidation and always linked to an oxidation reaction. 2. Treatment of a fracture by restoring the broken parts of a bone to their proper alignment.

reflex   A stereotyped, automatic, involuntary response to a stimulus; includes somatic reflexes, in which the effectors are skeletal muscles, and visceral (autonomic) reflexes, in which the effectors are usually visceral muscle, cardiac muscle, or glands.

reflex arc   A simple neural pathway that mediates a reflex; involves a receptor, an afferent nerve fiber, sometimes one or more interneurons, an efferent nerve fiber, and an effector.

reflux   A backward flow, such as the movement of stomach contents back into the esophagus.

refractory period   1. A period of time after a nerve or muscle cell has responded to a stimulus in which it cannot be reexcited by a threshold stimulus. 2. A period of time after male orgasm when it is not possible to reattain erection or ejaculation.

renal (REE-nul)   Pertaining to the kidney.

renin (REE-nin)   An enzyme secreted by the kidneys in response to hypotension; converts the plasma protein angiotensinogen to angiotensin I, leading indirectly to a rise in blood pressure.

repolarization   Reattainment of the resting membrane potential after a nerve or muscle cell has depolarized.

reproductive system   An organ system specialized for the production of offspring.

resistance   **1.** A nonspecific ability to ward off infection or disease regardless of whether the body has been previously exposed to it. **2.** A force that opposes the flow of a fluid such as air or blood. **3.** A force, or load, that opposes the action of a muscle or lever.

resistance exercise   A physical exercise such as weight lifting that promotes muscle strength more than it promotes cardiopulmonary efficiency, endurance, or fatigue resistance.

respiratory system   An organ system specialized for the intake of air and exchange of gases with the blood, consisting of the lungs and the air passages from the nose to the bronchi.

resting membrane potential (RMP)   A stable voltage across the plasma membrane of an unstimulated cell.

reticular (reh-TIC-you-lur) **cell**   A delicate, branching macrophage found in the reticular connective tissue of the lymphatic organs.

reticular fiber   A fine, branching collagen fiber coated with glycoprotein, found in the stroma of lymphatic organs and some other tissues and organs.

reticular tissue   A connective tissue composed of reticular cells and reticular fibers, found in bone marrow, lymphatic organs, and in lesser amounts elsewhere.

ribonucleic (RY-bo-new-CLAY-ic) **acid**   Any of three types of nucleotide polymers smaller than DNA that play various roles in protein synthesis. Composed of ribose, phosphate, adenine, uracil, cytosine, and guanine forming a single nucleotide chain.

ribosome   A granule found free in the cytoplasm or attached to the rough endoplasmic reticulum, composed of ribosomal RNA and enzymes; specialized to read the nucleotide sequence of messenger RNA and assemble a corresponding sequence of amino acids to make a protein.

risk factor   Any environmental factor or characteristic of the individual that increases a person's chance of developing a particular disease; includes such intrinsic factors as age, sex, and race and such extrinsic factors as diet, smoking, and occupation.

ruga (ROO-ga)   **1.** An internal fold or wrinkle in the mucosa of a hollow organ such as the stomach and urinary bladder;

typically present when the organ is empty and relaxed but not when the organ is full and stretched. **2.** Tissue ridges in such locations as the hard palate and vagina. fig. 25.11

S

saccule (SAC-yule)   A saclike receptor in the inner ear with a vertical patch of hair cells, the macula sacculi; senses the orientation of the head and responds to vertical acceleration, as when riding in an elevator or standing up. fig. 16.13

sagittal (SADJ-ih-tul) **plane**   Any plane that extends from ventral to dorsal and cephalic to caudal, dividing the body into right and left portions. *Compare* midsagittal plane. fig. A.3

sarcomere (SAR-co-meer)   In skeletal and cardiac muscle, the portion of a myofibril from one Z disc to the next, constituting one contractile unit. fig. 12.5

sarcoplasmic reticulum (SR)   The smooth endoplasmic reticulum of a muscle cell, serving as a calcium reservoir. fig. 12.2

scanning electron microscope (SEM)   A microscope that uses an electron beam in place of light to form high-resolution, three-dimensional images of the surfaces of objects; capable of much higher magnifications than a light microscope. fig. E.1, p. 136

sclerosis (scleh-RO-sis)   Hardening or stiffening of a tissue, as in multiple sclerosis of the central nervous system or atherosclerosis of the blood vessels.

sebum (SEE-bum)   An oily secretion of the sebaceous glands that keeps the skin and hair pliable.

secondary active transport   A mechanism in which solutes are moved through a plasma membrane by a carrier that does not itself use ATP but depends on a concentration gradient established by an active transport pump elsewhere in the cell. Also called *cotransport*.

secondary sex characteristic   Any feature that is inessential to reproduction but otherwise distinguishes male from female and promotes attraction between the sexes; examples include the distribution of subcutaneous fat, pitch of the voice, female breasts, and male facial hair.

secondary sex organ   An organ other than the ovaries and testes that is essential to reproduction, such as the external genitalia, internal genital ducts, and accessory reproductive glands.

second messenger   A chemical that is produced within a cell (such as cAMP) or that enters a cell (such as calcium ions) in response to the binding of a messenger to a membrane receptor and that triggers a metabolic reaction in the cell.

secretion   **1.** A chemical released by a cell to serve a physiological function, such

as a hormone or digestive enzyme. **2.** The process of releasing such a chemical, often by exocytosis. *Compare* excretion.

section   *See* histological section.

selectively permeable membrane   A membrane that allows some substances to pass through while excluding others; for example, the plasma membrane and dialysis membranes.

semen (SEE-men)   The fluid ejaculated by a male, including spermatozoa and the secretions of the prostate and seminal vesicles.

semicircular ducts   Three ring-shaped, fluid-filled tubes of the inner ear that detect angular accelerations of the head; each is enclosed in a bony passage called the semicircular canal. fig. 16.21

semilunar valve   A valve that consists of crescent-shaped cusps, including the aortic and pulmonary valves of the heart and valves of the veins and lymphatic vessels. fig. 19.6

semipermeable membrane   *See* selectively permeable membrane.

senescence (seh-NESS-ense)   Degenerative changes that occur with age.

sensation   Conscious perception of a stimulus; pain, taste, and color, for example, are not stimuli but sensations resulting from stimuli.

sensory nerve fiber   An axon that conducts information from a receptor to the central nervous system; an afferent nerve fiber.

serosa (seer-OH-sa)   *See* serous membrane.

serous (SEER-us) **fluid**   A watery, low-protein fluid similar to blood serum, formed as a filtrate of the blood or tissue fluid or as a secretion of serous gland cells; moistens the serous membranes.

serous membrane   A membrane such as the peritoneum, pleura, or pericardium that lines a body cavity or covers the external surfaces of the viscera; composed of a simple squamous mesothelium and a thin layer of areolar connective tissue.

serum   **1.** The fluid that remains after blood has clotted and the solids have been removed; essentially the same as blood plasma except for a lack of fibrinogen. Used as a vehicle for vaccines. **2.** Serous fluid.

sex chromosomes   The X and Y chromosomes, which determine the sex of an individual.

shock   **1.** Circulatory shock, a state of cardiac output that is insufficient to meet the body's physiological needs, with consequences ranging from fainting to death. **2.** Insulin shock, a state of severe hypoglycemia caused by administration of insulin. **3.** Spinal shock, a state of depressed or lost reflex activity inferior to a point of spinal cord injury. **4.** Electrical shock, the effect of a current of electricity passing through the body, often causing muscular spasm and cardiac arrhythmia or arrest.

sinus **1.** An air-filled space in the cranium. **2.** A modified, relatively dilated vein that lacks smooth muscle and is incapable of vasomotion, such as the dural sinuses of the cerebral circulation and coronary sinus of the heart. **3.** A small fluid-filled space in an organ such as the spleen and lymph nodes. **4.** Pertaining to the sinoatrial node of the heart, as in *sinus rhythm.*

skeletal muscle Striated voluntary muscle, almost all of which is attached to the bones.

skeletal system An organ system consisting of the bones, ligaments, bone marrow, periosteum, articular cartilages, and other tissues associated with the bones.

smooth muscle Nonstriated involuntary muscle found in the walls of the blood vessels, many of the viscera, and other places.

somatic **1.** Pertaining to the body as a whole. **2.** Pertaining to the skin, bones, and skeletal muscles as opposed to the viscera. **3.** Pertaining to cells other than germ cells.

somatic nervous system A division of the nervous system that includes efferent fibers mainly from the skin, muscles, and skeleton and afferent fibers to the skeletal muscles. *Compare* autonomic nervous system.

somesthetic **1.** Pertaining to widely distributed senses in the skin, muscles, tendons, joint capsules, and viscera, as opposed to the special senses found in the head only; also called *somatosensory, general senses,* or *somatic senses.* **2.** Pertaining to the cerebral cortex of the postcentral gyrus, which receives input from such receptors.

sperm **1.** The fluid ejaculated by the male; semen. Contains spermatozoa and glandular secretions. **2.** A spermatozoon.

spermatogenesis (SPUR-ma-toe-JEN-eh-sis) The production of sperm cells through a series of mitotic and meiotic cell divisions; male gametogenesis.

spermatozoon (SPUR-ma-toe-ZOE-on) A sperm cell.

sphincter (SFINK-tur) A ring of muscle that opens or closes an opening or passageway; found, for example, in the eyelids, around the urinary orifice, and at the beginning of a blood capillary.

spinal column A dorsal series of usually 33 vertebrae; encloses the spinal cord, supports the skull and thoracic cage, and provides attachment for the limbs and postural muscles. Also called *spine, vertebral column.*

spinal cord The nerve cord that passes through the spinal column and constitutes all of the central nervous system except the brain.

spinal nerve Any of the 31 pairs of nerves that arise from the spinal cord and pass through the intervertebral foramina.

spindle **1.** An elongated structure that is thick in the middle and tapered at the ends (fusiform). **2.** A football-shaped complex of microtubules that guide the movement of chromosomes in mitosis and meiosis. fig. 5.15 **3.** A stretch receptor in the skeletal muscles. fig. 15.11

spine **1.** The spinal (vertebral) column. **2.** A pointed process or sharp ridge on a bone, such as the styloid process of the cranium and spine of the scapula.

splanchnic (SPLANK-nic) Pertaining to the digestive tract.

stem cell Any undifferentiated cell that can divide and differentiate into more functionally specific cell types such as blood cells and germ cells.

stenosis (steh-NO-sis) The narrowing of a passageway such as a heart valve or uterine tube; a permanent, pathological constriction as opposed to physiological constriction of a passageway.

stereocilium An unusually long, sometimes branched microvillus, lacking the axoneme and motility of a true cilium; serves such roles as absorption in the epididymis and sensory transduction in the inner ear.

steroid (STERR-oyd, STEER-oyd) A lipid molecule that consists of four interconnected carbon rings; cholesterol and several of its derivatives.

stimulus A chemical or physical agent in a cell's surroundings that is capable of creating a physiological response in the cell; especially agents detected by sensory cells, such as chemicals, light, and pressure.

stress **1.** A mechanical force applied to any part of the body; important in stimulating bone growth, for example. **2.** A condition in which any environmental influence disturbs the homeostatic equilibrium of the body and stimulates a physiological response, especially involving the increased secretion of hormones of the pituitary-adrenal axis.

stroke *See* cerebrovascular accident.

stroke volume The volume of blood ejected by one ventricle of the heart in one contraction.

stroma The connective tissue framework of a gland, lymphatic organ, or certain other viscera, as opposed to the tissue (parenchyma) that performs the physiological functions of the organ.

subcutaneous (SUB-cue-TAY-nee-us) Beneath the skin.

substrate **1.** A chemical that is acted upon and changed by an enzyme. **2.** A chemical used as a source of energy, such as glucose and fatty acids.

substrate specificity The ability of an enzyme to bind only one substrate or a limited range of related substrates.

sulcus (SUL-cuss) A groove in the surface of an organ, as in the cerebrum or heart.

summation **1.** A phenomenon in which multiple stimuli combine their effects on a cell to produce a response; seen especially in nerve and muscle cells. **2.** A phenomenon in which multiple muscle twitches occur so closely together a muscle fiber cannot fully relax between twitches but develops more tension than a single twitch produces. fig. 12.18

superficial Relatively close to the surface; opposite of deep. For example, the ribs are superficial to the lungs.

superior Higher than another structure or point of reference from the perspective of anatomical position; for example, the lungs are superior to the diaphragm.

supination (SOO-pih-NAY-shun) A rotational movement of the forearm that turns the palm so that it faces upward or forward. fig. 10.13

surfactant (sur-FAC-tent) A chemical that reduces the surface tension of water and enables it to penetrate other substances more effectively. Examples include pulmonary surfactant and bile salts.

sympathetic nervous system A division of the autonomic nervous system that issues efferent fibers through the thoracic and lumbar nerves and usually exerts adrenergic effects on its target organs; includes a chain of paravertebral ganglia adjacent to the spinal column, and the adrenal medulla.

symphysis (SIM-fih-sis) **1.** A joint in which two bones are held together by fibrocartilage; for example, between bodies of the vertebrae and between the right and left pubic bones. **2.** The medial line of fusion between the two halves of the mandible, which is a true symphysis in early infancy but becomes a synostosis by the age of 1 year.

symport A carrier protein that moves two solutes simultaneously through a plasma membrane in the same direction, such as the sodium-dependent glucose transporter of the small intestine.

synapse (SIN-aps) **1.** A junction at the end of an axon where it stimulates another cell. **2.** A gap junction between two cardiac or smooth muscle cells at which one cell electrically stimulates the other; called an *electrical synapse.*

synaptic (sih-NAP-tic) **cleft** A narrow space between the synaptic knob of an axon and the adjacent cell, across which a neurotransmitter diffuses. fig. 13.16

synaptic knobs The swollen tips of the distal branches of an axon; the site of synaptic vesicles and neurotransmitter release. fig. 13.4

synaptic vesicle A spheroid organelle in a synaptic knob containing neurotransmitter.

synergist (SIN-ur-jist)   A muscle that works with the agonist to contribute to the same overall action at a joint.

synergistic   An effect in which two agents working together (such as two hormones) exert an effect that is greater than the sum of their separate effects. For example, neither follicle-stimulating hormone nor testosterone alone stimulates significant sperm production, but the two of them together stimulate production of vast numbers of sperm cells.

synovial (sih-NO-vee-ul) **fluid**   A lubricating fluid similar to eggwhite in consistency, found in the synovial joint cavities and bursae.

synovial joint   A point where two bones are separated by a narrow, encapsulated space filled with lubricating synovial fluid; most such joints are relatively mobile.

synthesis reaction   A chemical reaction in which smaller molecules are combined to form a larger one. *Compare* decomposition reaction.

systemic (sis-TEM-ic)   Widespread or pertaining to the body as a whole, as in the systemic circulation.

systemic circuit   All blood vessels that convey blood from the left ventricle to all organs of the body and back to the right atrium of the heart; all of the cardiovascular system except the heart and pulmonary circuit.

systole (SIS-toe-lee)   The contraction of any heart chamber; ventricular contraction unless otherwise specified.

systolic (sis-TOLL-ic) **pressure**   The peak arterial blood pressure measured during ventricular systole.

T

target cell   A cell acted upon by a nerve fiber, hormone, or other chemical messenger.

tarsal   Pertaining to the ankle (tarsus).

T cell   A type of lymphocyte involved in nonspecific defense, humoral immunity, and cellular immunity; occurs in several forms including helper, cytotoxic, and suppressor T cells and natural killer cells.

tendon   A collagenous band or cord that connects a muscle to a bone and transfers muscular tension to it.

testis   The male gonad; produces spermatozoa and testosterone.

tetanus   1. A state of sustained muscle contraction produced by temporal summation as a normal part of contraction; also called *tetany*. 2. Spastic muscle paralysis produced by the toxin of the bacterium *Clostridium tetani*.

tetraiodothyronine (TET-ra-EYE-oh-doe-THY-ro-neen)   *See* thyroxine.

thalamus (THAL-uh-muss)   The largest part of the diencephalon, located immediately inferior to the corpus callosum and bulging into each lateral ventricle; a point of synaptic relay of nearly all signals passing from lower levels of the CNS to the cerebrum. fig. 14.18

theory   An explanatory statement, or set of statements, that concisely summarizes the state of knowledge on a phenomenon and provides direction for further study; for example, the fluid mosaic theory of the plasma membrane and the sliding filament theory of muscle contraction.

thermogenesis   The production of heat, for example, by shivering or by the action of thyroid hormones.

thermoreceptor   A neuron specialized to respond to heat or cold, found in the skin and hypothalamus, for example.

thermoregulation   Homeostatic regulation of the body temperature within a narrow range by adjustments of heat-promoting and heat-losing mechanisms.

thorax   A region of the trunk between the neck and the diaphragm; the chest.

threshold   1. The minimum voltage to which the plasma membrane of a nerve or muscle cell must be depolarized before it produces an action potential. 2. The minimum combination of stimulus intensity and duration needed to generate an afferent signal from a sensory receptor.

thrombosis (throm-BO-sis)   The formation or presence of a thrombus.

thrombus   A clot that forms in a blood vessel or heart chamber; may break free and travel in the bloodstream as a thromboembolus.

thymine   A single-ringed nitrogenous base (pyrimidine) found in DNA, complementary to adenine in the double helix of DNA. fig. 5.3

thymus   A lymphatic organ in the mediastinum superior to the heart; the site where T lymphocytes differentiate and become immunocompetent. fig. 21.10

thyroid gland   An endocrine gland in the neck, partially encircling the trachea immediately inferior to the larynx. fig. 17.10

thyroid hormone   Either of two similar hormones, thyroxine and triiodothyronine, synthesized from iodine and tyrosine.

thyroid-stimulating hormone (TSH)   A hormone of the anterior pituitary gland that stimulates the thyroid gland.

thyroxine (thy-ROCK-seen) **(T₄)**   The thyroid hormone secreted in greatest quantity, with four iodine atoms; also called *tetraiodothyronine*. fig. 17.20

tight junction   A zipperlike junction between epithelial cells that limits the passage of substances between them. fig. 6.19

tissue   An aggregation of cells and extracellular materials, usually forming part of an organ and performing some discrete function for it; the four primary classes are muscular, nervous, connective, and epithelial tissue.

tissue gel   The viscous colloid that forms the ground substance of many tissues; gets its consistency from hyaluronic acid or other glycosaminoglycans.

trabecula (tra-BEC-you-la)   A thin plate or layer of tissue, such as the calcified trabeculae of spongy bone or the fibrous trabeculae that subdivide a gland. fig. 8.3

trachea (TRAY-kee-uh)   A cartilage-supported tube from the inferior end of the larynx to the origin of the primary bronchi; conveys air to and from the lungs; the "windpipe."

transcription   The process of enzymatically reading the nucleotide sequence of a gene and synthesizing a pre-mRNA molecule with a complementary sequence.

transducer   Any device that converts one form of energy to another, such as a sense organ, which converts a stimulus into an encoded pattern of action potentials.

transgenic bacteria   Genetically engineered bacteria that contain genes from humans or other species and produce proteins of that species; used commercially to produce clotting factors, interferon, insulin, and other products.

translation   The process of enzymatically reading an mRNA molecule and synthesizing the protein encoded in its nucleotide sequence.

transmission electron microscope (TEM)   A microscope that uses an electron beam in place of light to form high-resolution, two-dimensional images of ultrathin slices of cells or tissues; capable of extremely high magnification. fig. E.1, p. 136

triglyceride (try-GLISS-ur-ide)   A lipid composed of three fatty acids joined to a glycerol; also called a *triacylglycerol* or *neutral fat*. fig. 3.8

triiodothyronine (try-EYE-oh-doe-THY-ro-neen) **(T₃)**   A thyroid hormone with three iodine atoms, secreted in much lesser quantities than thyroxine. fig. 17.20

trisomy-21   The presence of three copies of chromosome 21 instead of the usual two; causes variable degrees of mental retardation, a shortened life expectancy, and structural anomalies of the face and hands.

tropic (TROPE-ic) **hormone**   A hormone of the anterior pituitary gland that stimulates secretion by another endocrine gland. The four tropic hormones are FSH, LH, TSH, and ACTH.

trunk   1. That part of the body excluding the head, neck, and appendages. 2. A major blood vessel, lymphatic vessel, or nerve that gives rise to smaller branches; for example, the pulmonary trunk and spinal nerve trunks.

T tubule   A tubular extension of the plasma membrane of a muscle cell that conducts action potentials into the sarcoplasm and excites the sarcoplasmic reticulum. fig. 12.2

tunic (TOO-nic)   A layer that encircles or encloses an organ, such as the tunics of a blood vessel or eyeball.

tympanic membrane   The eardrum.

U

ultraviolet radiation   Invisible, ionizing, electromagnetic radiation with shorter wavelength and higher energy than violet light; causes skin cancer and photoaging of the skin but is required in moderate amounts for the synthesis of vitamin D.

umbilical (um-BIL-ih-cul)   1. Pertaining to the cord that connects a fetus to the placenta. 2. Pertaining to the navel (umbilicus).

unit membrane   Any membrane composed of a bilayer of phospholipids and embedded proteins. A single unit membrane comprises the plasma membrane and encloses many organelles of a cell, whereas double unit membranes enclose the nucleus and mitochondria.

unmyelinated (un-MY-eh-lih-nay-ted) Lacking a myelin sheath. fig. 13.6

upper extremity   The appendage that arises from the shoulder, consisting of the brachium from shoulder to elbow, the antebrachium from elbow to wrist, the wrist, and the hand; loosely called the *arm,* but that term properly refers only to the brachium.

uracil   A single-ringed nitrogenous base (pyrimidine) found in RNA; one of the four bases of the genetic code; in RNA, occupies the place that thymine does in DNA. fig. 5.3

urea (you-REE-uh)   A nitrogenous waste produced from two ammonia molecules and carbon dioxide; the most abundant nitrogenous waste in the blood and urine. fig. 23.2

urinary system   An organ system specialized to filter the blood plasma, excrete waste products from it, and regulate the body's water, acid-base, and electrolyte balance.

uterine tube   A duct that extends from the ovary to the uterus and conveys an egg or conceptus to the uterus; also called *fallopian tube* or *oviduct.*

utricle (YOU-trih-cul)   A saclike receptor in the inner ear with a horizontal patch of hair cells, the macula utriculi; senses the orientation of the head and responds to horizontal acceleration, as when riding in a car that starts and stops. fig. 16.13

V

varicose vein   A vein that has become permanently distended and convoluted due to a loss of competence of the venous valves; especially common in the lower extremity, esophagus, and anal canal (where they are called hemorrhoids).

vas (vass) (plural, *vasa*)   A vessel or duct.

vascular   Pertaining to blood vessels.

vasoconstriction (VAY-zo-con-STRIC-shun)   The narrowing of a blood vessel due to muscular constriction of its tunica media.

vasodilation (VAY-zo-dy-LAY-shun)   The widening of a blood vessel due to relaxation of the muscle of its tunica media and the outward pressure of the blood exerted against the wall.

vasomotion (VAY-zo-MO-shun)   Collective term for vasoconstriction and vasodilation.

vasomotor center   A neuronal pool in the medulla oblongata that transmits efferent signals to the blood vessels and regulates vasomotion.

vein   Any blood vessel that carries blood toward either atrium of the heart.

ventral   Pertaining to the front of the body, the regions of the chest and abdomen; anterior.

ventral (anterior) root   The branch of a spinal nerve that emerges from the anterior side of the spinal cord and carries efferent (motor) nerve fibers.

ventricle (VEN-trih-cul)   A fluid-filled chamber of the brain or heart.

venule (VEN-yool)   The smallest type of vein, receiving drainage from capillaries.

vertebra (VUR-teh-bra)   One of the bones of the spinal column.

vertebral (VUR-teh-brul) **column**   *See* spinal column.

vesicle (VESS-ih-cul)   A fluid-filled tissue sac or an organelle such as a synaptic or secretory vesicle.

viscera (VISS-er-uh) (singular, *viscus*)   The organs contained in the dorsal and ventral body cavities, such as the brain, heart, lungs, stomach, intestines, and kidneys.

visceral (VISS-er-ul)   1. Pertaining to the viscera. 2. The inner or deeper layer of a two-layered membrane such as the pleura, pericardium, or glomerular capsule. *Compare* parietal. fig. A.13

visceral muscle   Single-unit smooth muscle found in the walls of blood vessels and the digestive, respiratory, urinary, and reproductive tracts.

viscosity   The resistance of a fluid to flow; the thickness or stickiness of a fluid.

vitamins   A small organic nutrient that is absorbed undigested and serves a purpose other than being oxidized for energy; serve as coenzymes. Most vitamins cannot be synthesized by the body and are therefore a dietary necessity.

vitreous (VIT-ree-us) **body**   A transparent, gelatinous mass that fills the space between the lens and retina of the eye.

voluntary muscle   Muscle that is usually under conscious control; skeletal muscle.

vulva   The female external genitalia; the labia majora and all external structures between them.

W

white matter   White myelinated nervous tissue deep to the cortex of the cerebrum and cerebellum and superficial to the gray matter of the spinal cord.

X

X chromosome   The larger of the two sex chromosomes; males have one X chromosome and females have two in each somatic cell.

xiphoid (ZIF-oyd, ZYE-foyd) **process**   A small pointed cartilaginous or bony process at the inferior end of the sternum.

X ray   1. A high-energy, penetrating electromagnetic ray with wavelengths in the range of 0.1 to 10 nm; used in diagnosis and therapy. 2. A photograph made with X rays; radiograph.

Y

Y chromosome   Smaller of the two sex chromosomes, found only in males and having little if any genetic function except development of the testis.

yolk sac   An embryonic membrane that encloses the yolk in vertebrates that lay eggs and serves in humans as the origin of the first blood and germ cells.

Z

zygomatic arch   An arch of bone anterior to the ear, formed by the zygomatic processes of the temporal, frontal, and zygomatic bones; origin of the masseter muscle.

zygote   A single-celled, fertilized egg.

credits

Photographs

Chapter 1

1.1: © The McGraw-Hill Companies, Inc./Dennis Strete, photographer **1.8:** © Corbis/Bettmann **1.10:** © Tim Davis/Photo Researchers, Inc. **1.12:** © Stock Montage **1.13a:** © Corbis/Bettmann **1.13b-c:** © Kathy Talaro/Visuals Unlimited **1.14a** Courtesy Armed Forces Institute of Pathology **1.14b:** © Stock Montage **E1.1a:** © U.H.B. Trust/Tony Stone Images **E1.1b:** Richard Anderson **E1.2a:** Alexander Tsiaras/Photo Research, 5W 4857 **E1.2b:** © Joan Creager **E1.3:** © CNRI/Phototake **E1.4:** © Tony Stone Images **E1.5:** © Monte S. Buchsbaum, M.D. **E3.1:** © Pete Saloutos/Tony Stone Images

Atlas A

A.1, A.3, A.8a-b, A.9a-b: © The McGraw-Hill Companies, Inc./Joe DeGrandis, Photographer **A.33, A.34, A.35a, A.36, A.37a:** © The McGraw-Hill Companies, Inc. **A.38, A.39:** © The McGraw-Hill Companies, Inc./Dennis Strete, photographer

Chapter 2

2.7b: © The McGraw-Hill Companies, Inc./Dennis Strete, photographer **2.11:** © Ken Saladin **E2.1:** © K.G. Murti/Visuals Unlimited **E2.2:** © Science VU/Visuals Unlimited **E2.3b:** © SIU/Peter Arnold, Inc.

Chapter 4

4.3: © K.G. Murti/Visuals Unlimited **4.5a-b:** From *Cell Ultrastructure* by William A. Jenson and Roderick B. Park. © 1967 By Wadsworth Publishing Co., Inc. Reprinted by Permission of the Publisher. **4.6a:** © Don Fawcett/Photo Researchers, Inc. **4.8:** Courtesy of S. Ito **4.9a,c:** © Biophoto Associates/Photo Researchers, Inc. **4.10:** © Don Fawcett/Photo Researchers, Inc. **4.12a:** © Don Fawcett/Photo Researchers, Inc. **4.13a:** © Visuals Unlimited **4.14a:** © D.W.

Fawcett/Visuals Unlimited **4.14b:** © D. Friend/D. Fawcett/Visuals Unlimited **4.15a:** Courtesy of Dr. Keith Porter **4.16a:** © Don Fawcett/Photo Researchers, Inc. **4.17a:** © 1998 Warren Rosenberg/Biological Photo Service **4.18b:** © K.G. Murti/Visuals Unlimited **4.18a1:** © K.G. Murti/Visuals Unlimited **4.18a2:** © Biology Media/Photo Researchers, Inc. **4.23a,b,c:** © David M. Phillips/Visuals Unlimited. **4.30a-d2:** M.M. Perry and A.B. Gilbert, *Journal of Cell Science,* 39:257-272, 1979 **4.31a:** © Don W. Fawcett/Visuals Unlimited. **4.32c:** Courtesy of Dr. Birgit H. Satir **E4.1a:** © The McGraw-Hill Companies, Inc./Dennis Strete, photographer **E4.1b-c:** © David M. Phillips/Photo Researchers, Inc.

Chapter 5

5.1a: © Richard Chao **5.1b:** © E.G. Pollack **5.2a:** © Don Fawcett/H. Ris & A. Olins/Photo Researchers, Inc. **5.10:** © E.V. Kiseleva, February 5 Letter, 257:251-253, 1989 **5.15a:** © David M. Phillips/Photo Researchers, Inc. **5.15b-c:** © The McGraw-Hill Companies, Inc./Dennis Strete, photographer **5.15d:** © David M. Phillips/Photo Researchers, Inc. **5.16b-c:** © The McGraw-Hill Companies, Inc./Dennis Strete, photographer **5.17:** © Bio Photo Associates/Photo Researchers, Inc. **ST5.2-1a:** Courtesy of Cold Spring Harbor Laboratory **ST5.2-1b:** Courtesy Kings College, London **ST5.2-1c:** © Corbis/Bettmann

Chapter 6

6.3a-c, 6.4, 6.5, 6.6a, 6.7: © The McGraw-Hill Companies, Inc./Dennis Strete, photographer **6.8:** From Buckwalter, J.A. Rosenberg, L., Coll. Rel. Res. (1983) 3:489-504 **6.9a-b, 6.10a-b, 6.11a-c, 6.12, 6.13, 6.15a-b, 6.16a-c, 6.17a-c:** © The McGraw-Hill Companies, Inc./Dennis Strete, photographer **6.18a-b:** From R.G. Kessel and R.H. Kardon, *Scanning Electron Microscopy of Tissues and Organs,* 1979

Chapter 7

7.2b, 7.4a: © The McGraw-Hill Companies, Inc./Dennis Strete, photographer **7.4b:** From R.G. Kessel and R.H. Kardon, *Scanning Electron Microscopy of Tissues and Organs,* 1979 **7.4c:** © SPL/Custom Medical Stock Photos **7.5:** © Dr. Sheril D. Burton **7.7a-b:** © The McGraw-Hill Companies, Inc./Dennis Strete, photographer **7.8-1:** © Tom McHugh/Photo Researcher, Inc. **7.8-2:** © Manoj Shah/Tony Stone Images **7.8-3:** © Art Wolfe/Tony Stone Images **7.8-4-6:** © The McGraw-Hill Companies, Inc./Joe DeGrandis, Photographer **7.9b:** © The McGraw-Hill Companies, Inc./Dennis Strete, photographer **7.9c:** © P.M. Motta/SPL/Custom Medical Stock Photos **7.10a-d:** © The McGraw-Hill Companies, Inc./Joe DeGrandis, Photographer **7.12a-c:** © The McGraw-Hill Companies, Inc./Dennis Strete, photographer **7.13a:** © NMSB/Custom Medical Stock Photos **7.13b:** © Biophoto Associates/Photo Researchers, Inc. **7.13c:** © James Stevenson/SPL/Photo Researchers, Inc. **7.14a:** © SPL/Custom Medical Stock Photos **7.14b-c:** © John Radcliffe/Photo Researchers, Inc. **ST7.1-1:** © Catherine Ellis/Photo Researchers, Inc.

Chapter 8

8.6a-c: Courtesy Trent Stephens **8.7a:** © Don W. Fawcett/Visuals Unlimited. **8.7c:** © The McGraw-Hill Companies, Inc./Dennis Strete, photographer **8.7d:** From: *A Text-Atlas of Scanning Electron Microscopy,* by R.G. Kessel and R. Kardon, 1979. **8.11a:** © The McGraw-Hill Companies, Inc./Dennis Strete, photographer **8.12:** © Biophoto Associates/Photo Researchers, Inc. **8.13:** Courtesy of Utah Valley Regional Medical Center, Dept. of Radiology **8.15:** © The McGraw-Hill Companies, Inc./Joe DeGrandis, Photographer **8.20a:** © Doug Sizemore/Visuals Unlimited **8.20b:** © SIU/Visuals Unlimited **E8.1a:** © Michael

Klein/Peter Arnold, Inc. **E8.1b:** © Biophoto Associates/Photo Researchers, Inc. **E8.2:** Reproduced with permission from the National Osteoporosis Foundation, Washington, DC

Chapter 9

9.18: © The McGraw-Hill Companies, Inc./Bob Coyle, Photographer **9.32b:** © NHS Trust/Tony Stone Images **9.35a:** © The McGraw-Hill Companies, Inc./Joe DeGrandis, photographer **9.35b:** © The McGraw-Hill Companies, Inc./Joe DeGrandis, Photographer **9.40b:** © Walter Reiter/Phototake

Chapter 10

10.1: © Bruce Curtis/Peter Arnold, Inc. **10.15:** © The McGraw-Hill Companies, Inc./Joe DeGrandis, Photographer **10.18:** © Greg Hodson/Visuals Unlimited **10.21d, 10.23d:** © The McGraw-Hill Companies, Inc./Dennis Strete, photographer **10.24:** © Richard Anderson **10.25b:** © The McGraw-Hill Companies, Inc./Dennis Strete, photographer **10.26:** © Mark D. Phillips/Photo Researchers, Inc. **10.27:** Vidic/Saurez: Photographic Atlas of the Human Body, Mosby, St. Louis, 1984 **10.32a:** © SIU/Visuals Unlimited **10.32b:** © Ron Mensching/Phototake **10.32c:** © SIU/Peter Arnold, Inc. **ST10.4-1a:** © SIU/Photo Researchers, Inc. **ST10.4-1b:** © Alexander Tsiaras/Photo Researchers, Inc. **E10.1a-b:** © Richard Anderson **E10.2a:** © SIU/Visuals Unlimited **E10.2b:** © Ron Mensching/Phototake **E10.2c:** © SIU/Peter Arnold, Inc. **E10.2d:** Courtesy Dr. Jerome P. Zechmann, Olympia Orthopedic Associates

Chapter 11

11.2a: © The McGraw-Hill Companies, Inc./Dennis Strete, photographer **11.5a-d:** © The McGraw-Hill Companies, Inc./Joe DeGrandis, photographer **11.7:** © From *A Stereoscopic Atlas of Anatomy* by David L. Bassett/

Courtesy Dr. Robert A. Chase, M.D. **11.11b:** © The McGraw-Hill Companies, Inc./Joe DeGrandis, photographer **11.13:** © From *A Stereoscopic Atlas of Anatomy* by David L. Bassett/Courtesy Dr. Robert A. Chase, M.D. **11.16:** © The McGraw-Hill Companies, Inc. **11.18:** © The McGraw-Hill Companies, Inc./ Dennis Strete, photographer **11.23a:** © From *A Stereoscopic Atlas of Anatomy* by David L. Bassett/Courtesy Dr. Robert A. Chase, M.D. **11.23b:** © The McGraw-Hill Companies, Inc./ Dennis Strete, photographer **11.24b, 11.25b:** © Bruce Curtis/ Peter Arnold, Inc. **11.34:** © The McGraw-Hill Companies, Inc./ Dennis Strete, photographer **11.39:** © The McGraw-Hill Companies, Inc.

Atlas B
B.1a-b, B.2a-b, B.3, B.4a-b, B.5, B.6a-b, B.7a-b, B.8a-b, B.9a-c, B.10a-d, B.11a-b: © The McGraw-Hill Companies, Inc./Joe DeGrandis, Photographer

Chapter 12
12.4: © The McGraw-Hill Companies, Inc./Dennis Strete, photographer **12.5a:** © Don W. Fawcett/Visuals Unlimited. **12.6:** © The McGraw-Hill Companies, Inc./Dennis Strete, photographer **E12.1:** © Yoav Levy/ Phototake **E12.2:** © NMSB/Custom Medical Stock Photos

Chapter 13
13.6a: © The McGraw-Hill Companies, Inc./Dr. Dennis Emery, Dept. of Zoology and Genetics, Iowa State **13.15:** © E.R. Lewis, Y.Y. Zeevi, T.E. Everhart, Univ. Of California/BPS **E13.1:** © Martin M. Rotker/Photo Researchers, Inc.

Chapter 14
14.2a-b: © The McGraw-Hill Companies, Inc./Dennis Strete, photographer **14.6c:** © The McGraw-Hill Companies, Inc./ Dennis Strete, photographer **14.24a:** © The McGraw-Hill Companies, Inc./Bob Coyle, photographer **ST14.1-1:** Alfred I. duPont Institute. **ST14.3-1:** Corbis-Bettmann **E14.1:** Courtesy Marcus E. Raichle, M.D., Washington School of Medicine

Chapter 15
15.1: © The McGraw-Hill Companies, Inc./Dennis Strete,

photographer **15.2b:** From Richard G. Kessel and Randy H. Kardon, *Tissues and Organs: A Text-Atlas of Scanning Electron Microscopy,* 1979, W. H. Freeman and Co. **15.4b:** © The McGraw-Hill Companies, Inc./Dennis Strete, photographer **15.7:** © From *A Stereoscopic Atlas of Anatomy* by David L. Bassett/Courtesy Dr. Robert A. Chase, M.D. **15.9:** Vidic/Saurez: Photographic Atlas of the Human Body, Mosby, St. Louis, 1984 **15.13a-d:** © The McGraw-Hill Companies, Inc./ Joe DeGrandis, Photographer **15.18:** © From *A Stereoscopic Atlas of Anatomy* by David L. Bassett/Courtesy Dr. Robert A. Chase, M.D. **15.24:** © Lisa Klancher **TA15.5:** © NHS Trust/ Tony Stone Images **TA15.7:** © From *A Stereoscopic Atlas of Anatomy* by David L. Bassett/ Courtesy Dr. Robert A. Chase, M.D. **TA15.7c:** © The McGraw-Hill Companies, Inc./Joe DeGrandis, Photographer

Chapter 16
16.4c: © The McGraw-Hill Companies, Inc./Dennis Strete, photographer **16.10:** © The McGraw-Hill Companies, Inc./ Joe DeGrandis, Photographer **16.15:** From R.G. Kessel and R.H. Kardon, Scanning Electron Microscopy of Tissues and Organs, 1979 **16.22:** © The McGraw-Hill Companies, Inc./ Joe DeGrandis, Photographer **16.27c:** © Ralph C. Eagle/MD/ Photo Researchers, Inc. **16.29a:** © The McGraw-Hill Companies, Inc./Dennis Strete, photographer **16.30a:** Courtesy of Beckman Vision Center at UCSF School of Medicine/ D. Copeahagen, S. Miltman, and M. Maglio **16.31a:** © Lisa Klancher

Chapter 17
17.7a: © Daniel Margulies **17.7b:** © Bettina/Cirone/Photo Researchers, Inc. **17.8a-d:** From Clinical Pathological Conference on Acromegaly, Diabetes, Hypermetabolism, Protein Use and Heart Failure, *American Journal of Medicine* 20:133, 1956 **17.10b:** © The McGraw-Hill Companies, Inc./Dennis Strete, photographer **17.11a:** © Lester Bergman & Associates **17.11b:** © CNRI/Phototake **17.11c:** © NMSB/Custom Medical Stock Photos **17.14:** © John Money, Ph.D., John Hopkins School Of

Medicine **17.15:** Addison's/ Custom Medical Stock **17.16b:** © The McGraw-Hill Companies, Inc./Dennis Strete, photographer

Chapter 18
18.1b: © Kessel & Kardon/Peter Arnold, Inc. **18.2:** © The McGraw-Hill Companies, Inc./Dennis Strete, photographer **18.7b:** 1991, Polaroid International Instant Photomicrography Competition. **18.8:** © Ed Reschke/Peter Arnold, Inc. **18.9a-b:** © Don W. Fawcett/Visuals Unlimited. **18.12:** © Walter Reinhart/Phototake **18.14:** © Claude Revy, Phototake, NYC **18.17a-b:** © The McGraw-Hill Companies, Inc./Dennis Strete, photographer **18.17c:** © U.S. Public Health Service **18.17d-e, 18.18a-b:** © The McGraw-Hill Companies, Inc./Dennis Strete, photographer **18.20:** © David M. Phillips/Photo Researchers, Inc. **ST18.1-1:** © John Paul Kay/Peter Arnold, Inc. **ST18.3-1:** Schomburg Center for Research in Black Culture, The New York Public Library **E18.1:** © SPL/Photo Researchers, Inc.

Chapter 19
19.3a: © The McGraw-Hill Companies, Inc./Dennis Strete, photographer **19.6b:** © The McGraw-Hill Companies, Inc. **19.6c:** © Lennart Nilsson/ Beholdman, Bonnierforlagen **19.11a:** © The McGraw-Hill Companies, Inc./Dennis Strete, photographer **19.18:** © The McGraw-Hill Companies, Inc./Bob Coyle, photographer **E19.1a-b:** © Carroll Weiss/Camera M.D. Studios

Chapter 20
20.4: From R.G. Kessel and R.H. Kardon, Scanning Electron Microscopy of Tissues and Organs, 1979 **20.6b:** © Don W. Fawcett/Visuals Unlimited. **20.6c:** Courtesy of S. McNutt **20.8:** © Yoav Levy/Phototake **20.11:** P.C. Phelps and J.H. Luft, 1969, Am. Anat. 125:3999 **20.18a: 20.3:** © The McGraw-Hill Companies, Inc./Dennis Strete, photographer

Chapter 21
21.11, 21.12a-b: © The McGraw-Hill Companies, Inc./Dennis Strete, photographer **21.18b:** © Len Lessin/Peter Arnold, Inc. **21.2:** © SPL/Custom Medical

Stock Photos **21.4:** © The McGraw-Hill Companies, Inc./ Joe DeGrandis, Photographer **21.7:** © The McGraw-Hill Companies, Inc./Dennis Strete, photographer **21.8c:** © Francis Leroy, SPL/Photo Researchers, Inc. **21.9:** © The McGraw-Hill Companies, Inc./Dennis Strete, photographer **21.20b:** © Manfred Kage/Peter Arnold, Inc. **21.21:** From Tissues and Organs: A Text Atlas of Scanning Electron Microscopy, © Richard G. Kessel and Randy H. Kardon, Published by W.H. Freeman and Company 1971 **21.23a-b:** © Don W. Fawcett/ Visuals Unlimited. **21.27a-b:** Dr. Andrejs Liepins **21.29:** © Corbis/ Bettmann **E21.1b:** © SPL/Custom Medical Stock Photos **E21.2:** © SPL/Photo Researchers, Inc.

Chapter 22
22.3a, 22.5a: © The McGraw-Hill Companies, Inc./Dennis Strete, photographer **22.5b:** © CNRI/ Phototake **22.8:** © SPL/Custom Medical Stock Photos **22.9b:** © The McGraw-Hill Companies, Inc. **22.10:** John Watney Photo Library **22.11a:** © The McGraw-Hill Companies, Inc./Dennis Strete, photographer **22.11b:** © C.Y. Shih-R. Kessel/Visuals Unlimited **22.14:** © David Grossmann/Phototake **22.2a:** © The McGraw-Hill Companies, Inc./Joe DeGrandis, Photographer **22.27a:** © The McGraw-Hill Companies, Inc./ Dennis Strete, photographer **22.27b:** © Matt Meadows/Peter Arnold, Inc.

Chapter 23
23.4a, 23.6b: © The McGraw-Hill Companies, Inc./Dennis Strete, photographer **23.8a:** © Don Fawcett/Photo Researchers, Inc. **23.8b:** © B.F. King, U. Of California School Of Medicine/ Biological Photo Service **23.9:** From R.G. Kessel and R.H. Kardon, *Scanning Electron Microscopy of Tissues and Organs,* 1979

Chapter 25
25.5, 25.8b: © The McGraw-Hill Companies, Inc./Dennis Strete, photographer **25.10f:** © The McGraw-Hill Companies, Inc./Jim Shaffer **25.11b:** © The McGraw-Hill Companies, Inc./Dennis Strete, photographer **25.12c:** © Roberts Caughem/Visuals Unlimited **25.18b, 25.20b:** © The McGraw-Hill Companies,

Inc./Dennis Strete, photographer **25.23a:** From R.G. Kessel and R.H. Kardon, *Scanning Electron Microscopy of Tissues and Organs,* 1979 **25.23b:** © The McGraw-Hill Companies, Inc./ Dennis Strete, photographer **25.23d:** © S. Ito-D. Fawcett/ Visuals Unlimited **25.25b:** © The McGraw-Hill Companies, Inc./ Dennis Strete, photographer **ST25.2-1a-b:** © CNRI/SPL/Photo Researchers, Inc.

Chapter 26
26.12: © R.P. Clark & M. Goff/ Photo Researchers, Inc.

Chapter 27
27.8b: © The McGraw-Hill Companies, Inc./Dennis Strete, photographer **27.9a:** From R.G. Kessel and R.H. Kardon, *Scanning Electron Microscopy of Tissues and Organs,* 1979 **27.9b:** © The

McGraw-Hill Companies, Inc./ Dennis Strete, photographer **27.19b:** © Visuals Unlimited **ST27.1-1:** Courtesy Mihaly Bartalos

Chapter 28
28.1: © The McGraw-Hill Companies, Inc./Joe DeGrandis, Photographer **28.4:** From R.G. Kessel and R.H. Kardon, *Scanning Electron Microscopy of Tissues and Organs,* 1979 **28.5:** © The McGraw-Hill Companies, Inc./ Dennis Strete, photographer **28.10a:** © Charles Gupton/Tony Stone Images, Inc. **28.10b:** © SBHS/ Tony Stone Images **28.10c-d:** © Biophoto Associates/Photo Researchers, Inc. **28.13:** © The McGraw-Hill Companies, Inc./Dennis Strete, photographer **28.13a:** © David M. Phillips/ Visuals Unlimited. **28.14a-b:** © The McGraw-Hill Companies, Inc./

Dennis Strete, photographer **28.16:** © Dr. Landrum Shettles. **28.21a-b:** © D. Van Rossum/Photo Researchers, Inc. **28.21c:** © Bruce Curtis/Visuals Unlimited **28.23:** T. Nagato, 1980, *Cell and Tissue Research* 209:1, Springer-Verlag **28.9b:** © The McGraw-Hill Companies, Inc./Dennis Strete, photographer **E28.1a-f:** © The McGraw-Hill Companies, Inc./ Bob Coyle, photographer **E28.1g:** © Hauk Morgan/Photo Researchers, Inc. **E28.1h:** Courtesy GynoPharma, Inc.

Chapter 29
29.1: © Francis Leroy, Biocosmos/ SPL/Photo Researchers, Inc. **29.5a-b:** © Carnegie Institution of Washington, Department of Embryology, Davis Division/ Ronan O'Rahilly **29.10:** © SIU/ Visuals Unlimited **29.11b:** © Martin Rotker/Phototake

29.12: © Garry Watson/SPL/Photo Researchers, Inc. **29.14:** © Neil Harding/Tony Stone Images **29.16:** Courtesy Dr. W. Lenz, University of Munster, Germany **29.18a:** From M. Bartalos and T. A. Baranski, *Medical Cytogenetics,* 1976, The Williams & Wilkins Co. **29.19a:** © Steve King/Photo Researchers, Inc. **29.19d:** © Biophoto Associates/Photo Researchers, Inc. **29.21:** © (Child): From Science Year, The World Book Science Annual **29.22:** © Corbis/Bettmann **29.23:** © AP/Wide World Photos

Connective Issues
CI7, CI8, CI11, CI13, CI17, CI19, CI21, CI22, CI23, CI25, CI27, Images copyright 1997 PhotoDisc, Inc.

alveoli, 802–3
alveolus, 262, 805–6, 891
alveolus glands, 194
Alzheimer disease (AD), 458
amacrine cells, 583
ambient P_{O2}, 819
amination, 951
amino acid pool, 950
amino acids, 90, 448, 847, 952
aminopeptidase, 920
ammonia, 951
amnion, 1036
amniotic cavity, 1033
amniotic fluid, 1036
amphetamines, 545
amphiarthrosis, 295
amphiphilic molecules, 86
ampulla, 573, 972, 994
amygdala, 487
amygdaloid nucleus, 497
amylase, 84
β-amyloid protein, 458
amylopectin, 84
amylose, 84
amyotrophic lateral sclerosis (ALS), 473
anabolic steroids, 102
anabolism, 71, 72
anaerobic fermentation, 101, 941, 943
anal canal, 915
anal columns, 915
analgesic, 557
anal sinuses, 915
anal triangle, 348
anamnestic response, 779
anaphase, 156
anaphylactic shock, 725, 785
anaphylaxis, 785
anaplasia, 158
anastomoses, 706
anatomical neck, 277
anatomical planes, 32–33
anatomical plates, 42–50
anatomical position, 31
anatomic dead space, 809
anatomy, 2–3
anatomy and physiology, 1–23
    biomedical science origins, 15–20
    homeostasis and feedback, 20–23
    human evolution, 10–13
    major themes, 23
    nature of human life, 3–10
    scientific method, 13–15
anconeus, 356
andrenal cortex, 613–14
androgen-binding protein (ABP), 974
androgen-insensitivity syndrome (AIS), 625, 966
androgens, 613
anemia, 653, 654
anemic hypoxia, 823
anesthesia, 595–96
aneuploidy, 1044
aneurysms, 713, 714
angina pectoris, 682
angiogenesis, 717
angiotensin-converting enzyme (ACE), 718, 844
angiotensin II, 614, 718, 844
angle, 263, 275

angle of incidence, 586
angular gyrus, 495
animal characteristics in human life, 4–5
animal electricity, 411
Animalia (kingdom), 4–5
anion, 56
ankle, 35, 286–88
ankle joint, 310, 317–20
ankylosis, 321
annular ligament, 313
annulus fibrosus, 269
anomic aphasia, 496
anorexia, 931
antagonist, 332
antagonistic effects, 541
    hormones, 629
antagonist pair, 332
antebrachium, 35, 277
anterior, 265, 317, 573
anterior arch, 271
anterior cerebral arteries, 731
anterior chamber, 581
anterior clinoid processes, 260
anterior commissures, 484
anterior communicating artery, 731
anterior cranial fossa, 258
anterior crest, 285
anterior direction, 33
anterior (external) nares, 797
anterior fornix, 996
anterior inferior spines, 282
anterior intercostal arteries, 733
anterior interosseous ateries, 732
anterior interventricular artery, 681
anterior interventricular sulci, 677
anterior lobe, 602
    hypothalamic control, 605–6
anterior lobe hormones, 604–5
anterior median fissure, 472, 478
anterior scalenes, 340
anterior superior spine, 282
anterior tibial artery, 738
anterior tibial vein, 747
anterior tibiofibular ligaments, 318
anterior triangles, 341
anterior tubercle, 271
anterior vagal trunks, 539
anterograde transport, 436
anterolateral system, 475
antibodies (Ab), 771, 772–73
    classes, 773
    diversity, 776–77
antibody monomer, 772
antibody titer, 779
anticoagulants, 669–70
anticodon, 148
antidiuretic hormone (ADH), 605
antigen-binding site, 772
antigenic determinants, 772
antigen presentation, 780
antigen-presenting cells (APCs), 774
antigens (Ag), 772
anti-inflammatory drugs, 631
antimicrobial proteins, 768
antioxidants, 57
antiport systems, 130–31
antithrombin, 668

antrum, 898, 1005
anuria, 854
aorta, 729
aortic arch, 729
aortic bodies, 718, 821
aortic hiatus, 733
aortic sinus, 729
aortic valve, 677
apex, 285, 286, 674, 802
Apgar score, 1042
aphasia, 496
apical foramen, 891
apical surface, 108
aplastic anemia, 654
apneustic center, 813
apocrine glands, 195
apocrine sweat glands, 216
apolipoprotein-E, 458
aponeurosis, 330
apoptosis, 118–19
appendages, 211
appendicular region, 35
appendicular skeleton, 227
appetite, 931–32
applied science, 15
appositional growth, 239
aprosodia, 495
aqueous humor, 581
arabinosylcytosine, 634
arachidonic acid, 630
arachnoid membrane, 468
arachnoid villi, 470
arboreal habitat, 11–12
arbor vitae, 480
archicortex, 486
archondroplastic dwarfism, 240
arcuate arteries, 835, 996
arcuate artery, 738
arcuate popliteal ligament, 317
arcuate veins, 837
areola, 998
areolar connective tissue, 179
areolar glands, 998
Aristotle, 16
arm, 35
arousal, reticular formation, 482
arrector pili, 11, 214
arrhythmia, 686
arterial anastomoses, 706
arterial supply
    abdomen, 734–36
    head and neck, 730–31
    myocardium, 681–82
    pelvic region and lower extremity, 737–38
    thorax, 733
    upper extremity, 732
arteries, 706, 708–9, 729
arterioles, 709
arteriosclerosis, 699
arteriovenous anastomoses (shunts), 706
arthritis, 321–23
arthrology, 295
arthroplasty, 322
articular cartilage, 182, 230
    exercise and, 299
articular disc, 311
articular facets, 285

G proteins, 113
graafian follicle, 1006
gracilis, 367
graded local potential, 439
graded strength of muscle twitch, 411
granulation tissue, 245
granulocytes, 660
granulosa cells, 617, 1004
gravity, 723
gray commissure, 473
gray communicating ramus, 533
gray matter, 463
    spinal cord, 473–74
Gray's Anatomy, 17
great cardiac vein, 682
greater curvature, 898
greater (false) pelvis, 281
greater omentum, 38, 888
greater palatine foramina, 263
greater sciatic notch, 282
greater vestibular (Bartholin) gland, 997
great saphenous vein, 748
great (type II) alveolar cells, 805
Greco-Roman contributions to biomedical
        science, 16–17
gross anatomy, 2, 834, 898, 906–7, 911–12,
        915–16
ground substance, 178
growth factors, cancer, 158
growth hormone (GH), 605, 608
growth in human life, 3
guanine, 141
guanosine triphosphate (GTP), 101
guard hairs, 215, 797
gubernaculum, 966
gustation, 558
gustatory nucleus, 560
gut-brain peptides, 452
gyri, 463

H

habenula, 484
habituation, 482
Hack, Gary, 339
hair, 211–15
    analysis, 213
    functions, 214–15
    growth, 214
hair cells, 568
hair receptors (peritrichial endings), 214, 554
Haldane effect, 820
half-life, 621
hallux, 287
hamate, 279
hamulus, 279
hand, 35
    anatomy, 385
    intrinsic muscles of, 360–63
    muscles acting on, 356–60
haploid cells, 159
haploid (*n*), 976
haptens, 772
hard (bony) callus, 245
hard (bony) palate, 890
hard keratin, 212

hard palate, 262
Harvey, William, 17–18, 706
haustra, 916
haustral contractions, 917
haversian canal, 184, 233
Hawking, Stephen, 473
Hayflick phenomenon, 1053
H band, 397
head, 35, 47, 275, 277, 279, 331, 979
    anatomy, 380
    muscles acting on, 340–41
    muscular system, 333–42
healing fractures, 245–47
hearing, 562–76
    auditory projection pathway, 570–72
    cochlear tuning, 570
    middle ear, 568–69
    sensory coding, 569–70
    stimulation of cochlear hair cells, 569
heart, 673–704
    blood flow, heart sounds, cardiac cycle,
        691–95
    development, 682–83
    size, shape, position, 674
heart block, 685
heart chambers, 677
heart rate, 696
heart sounds, 691–92
heart valves, 677–79
heart wall, 677
heat, 68
heat capacity, 79
heat cramps, 956
heat exhaustion, 956
heat-losing center, 956
heat production, 328, 954–56
    connective tissue, 175
    sodium potassium (Na+-K+) pump, 131
heat-promoting center, 956
heat-shock proteins, 150
heatstroke, 956
heavy chains, antibodies, 772
heavy lifting, back injuries and, 347
helicine arteries, 983
α helix, 92
helper factors, 776
helper T cells, 780, 781
hemagiomas, 208
hematocrit, 652
hematology, 640
hematomas, 208, 668
    formation, 245
hematuria, 840
heme, 651
hementin, 670
hemiazygos vein, 744
hemidecussation, 593
hemocytoblasts, 645
hemodialysis, 858–59
hemoglobin concentration, 652
hemoglobin (Hb), 92, 93, 206, 650–51
    packaging, 651
hemolysis, 652
hemolytic anemia, 654
hemolytic disease of newborn (HDN), 659
hemophilia, 668
hemopoiesis, 645

hemopoietic, 234
hemopoietic syndrome, 71
hemopoietic tissues, 645
hemorrhagic anemia, 654
hemorrhoidal veins, 916
hemorrhoids, 712, 916
hemostasis, 663–69
    blood clots, 667
    coagulation, 664–69
    platelet plug formation, 663–64
    platelets, 663
    vascular spasm, 663
Henry's law, 814
heparin, 177, 668
hepatic artery, 734
hepatic lobules, 907
hepatic macrophages (Kupffer cells), 907
hepatic portal system, 746
hepatic portal vein, 746
hepatic (right colic) flexure, 915
hepatic sinusoids, 746, 907
hepatic triad, 907
hepatic veins, 745
hepatitis, 957
hepatocytes, 907
hepatopancreatic ampulla, 908
hepatopancreatic sphincter (sphincter of
        Oddi), 908
heredity, 159
    terminology, 163
hernias, 350
herpes zoster virus and shingles, 524
heterodonty, 5
heterograft, 221
heterotrophic beings, 4
heterozygous chromosomes, 160
hiatal hernia, 350
hierarchy of structure, 23
high-density lipoproteins (HDLs), 700, 936
higher brain center influence, 607
hilum, 803, 834
hindbrain, 465, 477–82
hinge joints, 301
hippocampus, 497
Hippocrates, 16, 20
hirsutism, 214
hirudin, 669
histamine, 177
histiocytes, 176, 765
histological sections, 170
histology, 9, 169–200
    compact bone, 233–34
    embryonic and fibrous connective tissues,
        175–82
    epithelial tissue, 185–91
    excitable tissues, 172–75
    intercellular junctions, glands, membranes,
        191–96
    osseous tissue, 232–36
    spongy bone, 234
    supportive and fluid connective tissues,
        182–85
    tissue study, 170–72
    tissue transformations, 196–97
histones, 141
histotoxic hypoxia, 823
H+-K+ ATPase, 901

ligaments, 226, 300, 996
ligamentum teres, 314
ligand, 93
ligand-gated channels, 113
light, vision and, 577
light adaptation, 591
light chains, antibodies, 772
light microscope (LM), 108, 111, 135
limb-girdle dystrophy, 423
limbic system, 488
limiting pH, 877
linea alba, 343
linea albicantes, 206
linea aspera, 284
linea nigra, 1013
lingual artery, 730
lingual frenulum, 890
lingual glands, 890
lingual lipase, 892
lingual muscles, 890
lingual papillae, 558, 890
lingual tonsils, 762
lipases, 86, 922
lipid and protein metabolism
    nutrition and metabolism, 949–52
lipid metabolism, 608
lipids, 85, 89, 922–23, 935–37, 950
    molecules of life, 85–89
lipofuscin, 432
lipogenesis, 950
lipolysis, 950
lipoprotein lipase, 936
lipoproteins, 89, 936
lipoxygenase, 630
lips, 889–90
lithotripsy, 909
liver, 906, 957
    digestive system, 905–11
liver disease, 668
lobar arteries, 727
lobes, 193
lobotomy, 494
lobules, 194
local anesthetics, 596
local control, peripheral resistance, 717
local hormones, 629
local potentials, neurons, 439–40
location, sensory receptors, 552
lochia, 1016
locomotion, spinal cord, 471
locus, 159
Loewi, Otto, 446
Long, Crawford W., 595
long bones, 229
longissimus thoracis, 344
longitudinal fissure, 463
longitudinal section, 170
long reflexes, 896
long-term energy transfer, 414–15
long-term memory (LTM), 497
long-term potentiation (LTP), 455
loose connective tissue, 179
lordosis, 269
loudness, sound, 563–64
Lou Gehrig disease, 473
low-density lipoproteins (LDLs), 699, 936

lower extremities, 35, 284–88
    muscular system, 363–73
lower motor neurons, 480
lower respiratory tract, 796
low venous return (LVR) shock, 724–25
lumbar arteries, 734
lumbar curvatures, 267
lumbar enlargement, 472
lumbar plexus, 518, 522
lumbar puncture, 468
lumbar region, 471
lumbar veins, 744, 745
lumbar vertebrae, 272
lumbodorsal fascia, 344
lumborum, 344
lumbrical, 372
lumbrical muscles, 363
lumpectomy, 999
lunate, 279
lunate facet, 278
lung cancer, smoking and, 824–25
lungs, 802–3, 815
    circulatory routes, 726
luteal phase, 1007–8
lutein cells, 1007
luteinizing hormone (LH), 604
lymph, 185–87, 755
lymphadenitis, 761
lymphadenopathy, 761
lymph and lymphatic vessels, 755–58
lymphatic and immune systems, 754–94
lymphatic capillaries, 755
lymphatic nodules (follicles), 758
lymphatic system, 41, 755–64
lymphatic tissue, 758–60
lymphatic trunks, 757
lymphatic vessels, 756–57
lymph flow, 757–58
lymph node diseases, 761
lymph nodes, 760–61
lymphocytes, 660–62, 765, 773
lymphocytic leukemia, 662
lymphoid hemopoiesis, 645
lymphokines, 774
lymph origin, 755–56
lymphotoxin, 782
lysis, 770
lysosomes, 118–19
lysozyme, 765, 893

M

M, mitotic phase of cells, 154
macronutrients, 933
macrophage-activating factor (MAF), 781
macrophages, 176, 648, 765
macrophage system, 765
macula, 573
macula densa, 843
macula lutea, 584
macula sacculi, 573
macula utriculi, 573
magnetic resonance imaging (MRI),
    25–26, 500
major calyx, 834

major duodenal papilla, 908
male birth-control pills, 1021
male body, 34, 43–44
male climacteric, 975
male condom, 1021
male pelvic cavity, 50
male reproductive system, 961–90
    puberty and climacteric, 974–76
    reproductive anatomy, 968–74
    sex determination and development,
        963–68
    sexual intercourse, 982–86
    sexual reproduction, 962–63
    spermatogenesis, spermatozoa, semen,
        976–82
malignant melanoma, 219
malignant tumors, 157
malleus, 264, 565
maltase, 920
maltose, 83, 84
Mammalia class, 5
mammary glands, 218
mammillary bodies, 483–84
mammograms, 999
mandible, 263–64
mandibular condyle, 263
mandibular foramen, 264
mandibular fossa, 259
mandibular notch, 263
manometer, 691
manubrium, 274
manus, 35, 277
marcomolecules, 81
Marfan syndrome, 197
marginal artery, 681
margination, 767
markings of skin, 208–9
mass activation, 535
masseter muscle, 339
mass movements, 918
mast cells, 177
mastectomy, 999
mastication, 889, 891–92
mastoid fontanels, 265
mastoid foramen, 259
mastoid notch, 259
mastoid part, 259
mastoid process, 259
matrix, 172
    osseous tissue, 233
matrix reactions, 943, 944
matter, 53
matter and energy, 52–76
    acids, bases, pH, 64–66
    chemical elements and atomic structure,
        53–57
    chemical reactions, 66–68
    energy, 68–70
    mixtures, 61–64
    molecules, compounds, chemical bonds,
        58–61
    thermodynamics and metabolism, 70–72
mature follicle, 1006
mature-onset DM, 617
maxillae, 262
maxillary artery, 731

Lexicon of Biomedical Word Elements

a- no, not, without (atom, agranulocyte)
ab- away (abducens, abduction)
acetabulo- small cup (acetabulum)
acro- tip, extremity, peak (acromion, acromegaly, acrosome)
ad- to, toward, near (adsorption, adrenal)
adeno- gland (lymphadenitis, adenohypophysis, adenoids)
aero- air, oxygen (aerobic, anaerobe, aerophagy)
af- toward (afferent)
ag- together (agglutination)
-al pertaining to (parietal, pharyngeal, temporal)
ala- wing (ala nasi)
albi- white (albicans, linea alba, albino)
algi- pain (analgesic, myalgia)
aliment- nourishment (alimentary, hyperalimentation)
allo- other, different (allele, allosteric)
amphi- both, either (amphiphilic, amphiarthrosis)
an- without (anaerobic, anemic)
ana- 1. up, build up (anabolic, anaphylaxis). 2. apart (anaphase, anatomy). 3. back (anaplasia)
andro- male (androgen)
angi- vessel (angiogram, angioplasty, hemangioma)
ante- before, in front (antebrachium)
antero- forward (anterior, anterograde)
anti- against (antidiuretic, antibody, antagonist)
apo- from, off, away, above (apocrine, aponeurosis)
arbor- tree (arboreal, arborization)
artic- 1. joint (articulation). 2. speech (articulate)
-ary pertaining to (axillary, coronary)
-ase enzyme (polymerase, kinase, amylase)
ast-, astro- star (aster, astrocyte)
-ata, -ate 1. possessing (Chordata, corniculate). 2. plural of -a (stomata, carcinomata)
athero- fat (atheroma, atherosclerosis)
atrio- entryway (atrium, atrioventricular)
auri- ear (auricle, auricular)
auto- self (autolysis, autoimmune)
axi- axis, straight line (axial, axoneme, axon)
baro- pressure (baroreceptor, hyperbaric)
bene- good, well (benign, beneficial)
bi- two (bipedal, biceps, bifid)
bili- bile (biliary, bilirubin)
bio- life, living (biology, biopsy, microbial)
blasto- precursor, bud, producer (fibroblast, osteoblast, blastomere)
brachi- arm (brachium, brachialis, antebrachium)
brady- slow (bradycardia, bradypnea)
bucco- cheek (buccal, buccinator)
burso- purse (bursa, bursitis)
calc- calcium, stone (calcified, calcaneus, hypocalcemia)
callo- thick (callus, callosum)
calori- heat (calorie, calorimetry, calorigenic)
calv-, calvari- bald, skull (calvaria)
calyx- cup, vessel, chalice (glycocalyx, renal calyx)
capito- head (capitis, capitate, capitulum)
capni- smoke, carbon dioxide (hypocapnia)
carcino- cancer (carcinogen, carcinoma)
cardi- heart (cardiac, cardiology, pericardium)
carot- 1. carrot (carotene). 2. stupor (carotid)
carpo- wrist, seize (carpus, metacarpal)
case- cheese (caseosa, casein)
cata- down, break down (catabolism)
cauda- tail (cauda equina, caudate nucleus)
-cel little (pedicel)
celi- belly, abdomen (celiac)
centri- center, middle (centromere, centriole)
cephalo- head (cephalic, encephalitis)
cervi- neck, narrow part (cervix, cervical)
chiasmo- cross, X (optic chiasma)
choano- funnel (choana)
chole- bile (cholecystokinin, cholelithotripsy)
chondro- 1. grain (mitochondria). 2. cartilage, gristle (chondrocyte, perichondrium)

chromo- color (dichromat, chromatin, cytochrome)
chrono- time (chronotropic, chronic)
cili- eyelash (cilium, supraciliary)
circa- about, around (circadian, circumduction)
cis- cut (incision, incisor)
cisterna- reservoir (cisterna chyli)
clast- break down, destroy (osteoclast)
clavi- hammer, club (clavicle, supraclavicular)
-cle little (tubercle, corpuscle)
cleido- clavicle (sternocleidomastoid)
cnemo- lower leg (gastrocnemius)
co- together (coenzyme, cotransport)
collo- 1. hill (colliculus). 2. glue (colloid, collagen)
contra- opposite (contralateral)
corni- horn (cornified, corniculate, cornu)
corono- crown (coronary, corona, coronal)
corpo- body (corpus luteum, corpora quadrigemina)
corti- bark, rind (cortex, cortical)
costa- rib (intercostal, subcostal)
coxa- hip (os coxae, coxal)
crani- helmet (cranium, epicranius)
cribri- sieve, strainer (cribriform, area cribrosa)
crino- separate, secrete (holocrine, endocrinology)
crista- crest (crista ampullaris, mitochondrial crista)
crito- to separate (hematocrit)
cruci- cross (cruciate ligament)
-cule, -culus small (canaliculus, trabecula, auricular)
cune- wedge (cuneiform, cuneatus)
cutane-, cuti- skin (subcutaneous, cuticle)
cysto- bladder (cystitis, cholecystectomy)
cyto- cell (cytology, cytokinesis, monocyte)
de- down (defecate, deglutition)
demi- half (demifacet, demilune)
den-, denti- tooth (dentition, dens, dental)
dendro- tree, branch (dendrite, oligodendrocyte)
derma-, dermato- skin (ectoderm, dermatology, hypodermic)
desmo- band, bond, ligament (desmosome, syndesmosis)
dia- 1. across, through, separate (diaphragm, dialysis). 2. day (circadian)
dis- 1. apart (dissect, dissociate). 2. opposite, absence (disinfect, disability)
diure- pass through, urinate (diuretic, diuresis)
dorsi- back (dorsal, dorsum, latissimus dorsi)
duc- to carry (duct, adduction, abducens)
dys- bad, abnormal, painful (dyspnea, dystrophy)
e- out (ejaculate, eversion)
-eal pertaining to (hypophyseal, arboreal)
ec-, ecto- outside, out of, external (ectopic, ectoderm, splenectomy)
ef- out of (efferent, effusion)
-el, -elle small (fontanel, organelle, micelle)
electro- electricity (electrocardiogram, electrolyte)
em- in, within (embolism, embedded)
emesi-, emeti- vomiting (emetic, hyperemesis)
-emia blood condition (anemia, hypoxemia, hypovolemia)
en- in, into (enzyme, parenchyma)
encephalo- brain (encephalitis, telencephalon)
enchymo- poured in (mesenchyme, parenchyma)
endo- within, into, internal (endocrine, endocytosis)
entero- gut, intestine (mesentery, myenteric)
epi- upon, above (epidermis, epiphysis, epididymis)
ergo- work, energy, action (allergy, adrenergic)
eryth-, erythro- red (erythema, erythrocyte)
esthesio- sensation, feeling (anesthesia, somesthetic)
eu- good, true, normal, easy (eukaryote, eupnea, aneuploidy)
exo- out (exopeptidase, exocytosis, exocrine)
facili- easy (facilitated)
fasci- band, bundle (fascia, fascicle)
fenestr- window (fenestrated, fenestra vestibuli)
fer- to carry (efferent, uriniferous)

ferri- iron (ferritin, transferrin)
fibro- fiber (fibroblast, fibrosis)
fili- thread (myofilament, filiform)
flagello- whip (flagellum)
foli- leaf (folic acid, folia)
-form shape (cuneiform, fusiform)
fove- pit, depression (fovea)
funiculo- little rope, cord (funiculus)
fusi- 1. spindle (fusiform). 2. pour out (perfusion)
gamo- marriage, union (monogamy, gamete)
gastro- belly, stomach (digastric, gastrointestinal)
-gen, -genic, -genesis producing, giving rise to (pathogen, carcinogenic, glycogenesis)
genio- chin (geniohyoid, genioglossus)
germi- 1. sprout, bud (germinal, germinativum). 2. microbe (germicide)
gero- old age (progeria, geriatrics)
gesto- pregnancy (gestation, progesterone)
glia- glue (neuroglia, microglia)
globu- ball, sphere (globulin, hemoglobin)
glom- ball (glomerulus)
glosso- tongue (hypoglossal, glossopharyngeal)
glyco- sugar (glycogen, glycolysis, hypoglycemia)
gono- 1. angle, corner (trigone). 2. seed, sex cell, generation (gonad, oogonium, gonorrhea)
gradi- walk, step (retrograde, gradient)
-gram recording of (sonogram, electrocardiogram)
-graph recording instrument (sonograph, electrocardiograph)
-graphy recording process (sonography, radiography)
gravi- severe, heavy (gravid, myasthenia gravis)
gyro- turn, twist (gyrus)
hallu- great toe (hallux, hallucis)
hemi- half (hemidesmosome, hemisphere, hemiazygos)
-hemia blood condition (polycythemia)
hemo- blood (hemophilia, hemoglobin, hematology)
hetero- different, other, various (heterotrophic, heterozygous)
histo- tissue, web (histology, histone)
holo- whole, entire (holistic, holocrine)
homeo- constant, unchanging, uniform (homeostasis, homeothermic)
homo- same, alike (homologous, homozygous)
hyalo- clear, glassy (hyaline, hyaluronic acid)
hydro- water (dehydration, hydrolysis, hydrophobic)
hyper- above, above normal, excessive (hyperkalemia, hypertonic)
hypo- below, below normal, deficient (hypogastric, hyponatremia, hypophysis)
-ia condition (anemia, hypocalcemia, osteomalacia)
-ic pertaining to (isotonic, hemolytic, antigenic)
-icle, -icul small (ossicle, canaliculus, reticular)
ilia- flank, loin (ilium, iliac)
-illa, -illus little (bacillus)
-in protein (trypsin, fibrin, globulin)
infra- below (infraspinous, infrared)
ino- fiber (inotropic, inositol)
insulo- island (insula, insulin)
inter- between (intercellular, intercalated, intervertebral)
intra- within (intracellular, intraocular)
iono- ion (ionotropic, cationic)
ischi- to hold back (ischium, ischemia)
-ism 1. process, state, condition (metabolism, rheumatism). 2. doctrine, belief, theory (holism, reductionism, naturalism)
iso- same, equal (isometric, isotonic, isomer)
-issimus most, greatest (latissimus, longissimus)
-ite little (dendrite, somite)
-itis inflammation (dermatitis, gingivitis)
jug- to join (conjugated, jugular)
juxta- next to (juxtamedullary, juxtaglomerular)
kali- potassium (hypokalemia)
karyo- seed, nucleus (megakaryocyte, karyotype, eukaryote)